new ed. jp 55/00   10⁰⁰

FLORA OF THE BRITISH ISLES

FLORA
OF THE
BRITISH ISLES

BY

A. R. CLAPHAM
University of Sheffield

T. G. TUTIN
University of Leicester

AND

E. F. WARBURG
University of Oxford

SECOND EDITION

CAMBRIDGE
AT THE UNIVERSITY PRESS
1962

PUBLISHED BY
THE SYNDICS OF THE CAMBRIDGE UNIVERSITY PRESS
Bentley House, 200 Euston Road, London, N.W. 1
American Branch: 32 East 57th Street, New York 22, N.Y.
West African Office: P.O. Box 33, Ibadan, Nigeria

CAMBRIDGE UNIVERSITY PRESS
1962

First Edition   1952
Reprinted 1957, 1958
Second Edition 1962

Printed in Great Britain at the University Press, Cambridge
(Brooke Crutchley, University Printer)

To

HUMPHREY GILBERT-CARTER

*To whose stimulating teaching
and wide knowledge of plants we, his pupils,
owe so much*

CONTENTS

FOREWORD

By Professor A. G. Tansley

A new British flora has been a desideratum for the past half century and urgently needed during the last thirty years. What has been particularly required is a flora not primarily for specialists but a book of limited size easily usable by students and by everyone interested in our wild plants who is willing to learn the comparatively few technical terms necessary for the accurate description of species. The absence of such a flora has seriously hampered the teaching and learning of field botany. Time and again I have been asked by visiting foreign botanists to recommend a good modern British flora and have been ashamed to confess that no such thing existed. In this whole sphere the lack of an adequate handbook has indeed been something of a national scandal. Several attempts have been made to fill the gap but none has been carried through to success, largely because they were all too ambitious, aiming at a completeness and exhaustiveness unattainable except through many years of laborious effort and the collaboration of a large body of specialists.

It is often taken for granted by those who are unacquainted with the subject that the comparatively small British plant population is more or less completely known and has been fully and accurately described in the existing floras. It is not realized that modern work during the past half century, and increasingly since the end of the first world war, has revealed the existence of many distinct forms—species, subspecies and varieties—that had not previously been clearly recognized, or had not been recognized at all. Some of these were formerly described under the names of continental types which they resembled, but deeper knowledge and closer comparison have established that the British forms are in reality quite distinct. At the same time much new knowledge has been gained about many well-known species, especially about their genetics and ecology. This has been the result of the great revival of interest in field observation and work in the experimental garden among professional botanists and academic students of the subject. During the latter part of last century and the beginning of this the study of British plants was very largely in the hands of enthusiastic amateurs to whom the subject owes a great deal, several of them having become the leading specialists in particular groups. With the rise of ecology and genetics, interest in British plants spread to the universities, and thus aroused the renewed attention to taxonomy among academically trained botanists which has been a marked feature of recent years.

At last it has been possible to stimulate three men, scarcely more than entering upon middle life, all with the modern training, all keenly interested in plants as they grow in the field, in ecology and genetics, to undertake the production of the much needed flora as a matter of urgency. Though closely occupied with teaching they have carried through the task in little more than three years and in a manner that seems to me excellently adapted to meet the need. A comparison of their book with any of the previous floras will make plain the distance that has been traversed since those were written.

Readers will find a good many unfamiliar specific names, but these changes were necessary if the rule of priority in nomenclature was to be followed. Personally, I should like to see the principle of *nomina conservanda* applied to specific epithets as well as to the names of genera, so that well-known names that have been in use for many years might be retained and cease to be subject to the risk of perpetual displacement as the result of literary research, often in obscure historical works. But botanists are not yet in agreement on this point, and the discovery of 'prior' names must, one supposes, come to an end some day. Meanwhile, the present generation of students has still to suffer in this respect, though the suffering may, we hope, be transient.

A. G. T.

GRANTCHESTER

January 1951

PREFACE TO THE FIRST EDITION

The reason for the addition of yet another flora to the long series which began in the seventeenth century is perhaps best explained by a brief historical survey.

Although many records of British plants are to be found in herbals, the first attempt at a true flora of these islands was John Ray's *Catalogus Plantarum Angliae et Insularum Adjacentium*, published in 1670. William Hudson's *Flora Anglica* (1762), and thus nearly a century later than Ray's *Catalogus*, was a worthy successor to that pioneer work and notable for the introduction of binomial nomenclature and the Linnean system of classification into British floras. This was followed by Withering's *Botanical Arrangement of all the Vegetables naturally growing in Great Britain* (1776–92), the first of many floras written primarily for the amateur, and one which enjoyed considerable popularity.

Sowerby's *English Botany*, the first edition of which was published between 1790 and 1820, occupies a unique place. It presented for the first time a complete set of coloured illustrations of our plants, illustrations which are still unsurpassed in their delicacy of line and colouring. The third edition (1863–72), in which the text was completely rearranged, has inferior illustrations but is still a valuable work of reference more than 150 years after the first edition appeared.

The nineteenth century saw the production of three floras, all still in regular use, and a number of others which are now seldom seen. Bentham's famous *Handbook of the British Flora* (1858; revised by J. D. Hooker, 1886) was written as 'a before-breakfast relaxation' and was deliberately intended 'for the use of beginners and amateurs'. Its keys, a new feature in British floras, make it of value to anyone who desires to identify plants easily and with the minimum of previous knowledge, but its treatment of species in many groups makes it of limited use to the ecologist, cytologist or serious taxonomist.

J. D. Hooker's *Student's Flora of the British Islands* (1870), familiar to many generations of botanists, has beautifully clear and concise descriptions but has not been revised since 1884. Babington's *Manual of British Botany* (1843) treats certain groups in greater detail than any other easily accessible work and was last revised as recently as 1922, but its scanty and frequently not very clear descriptions make it unsuitable for the average student of botany and particularly for the beginner.

During the past fifty years such advances have been made in all branches of botany that these floras are no longer adequate. The rise of ecology to a position of recognized importance has led to a demand not only for clear descriptions of species but for information of a kind not essential to identification, though of value to everyone interested in plants as living organisms. There has also been among botanists a change of outlook so marked as to affect very seriously the usefulness of the existing books. When Babington and Hooker wrote their floras 'systematic botany' was almost or (to Babington) quite synonymous with 'botany' and consequently these works are not primarily intended to permit the correct identification of plants but to teach the principles of classification and the technical characters of families and genera. Taxonomy is now only one branch, though an important and indeed a fundamental branch of botany, and many people who are not primarily taxonomists have need to identify plants correctly. Further, within the province of taxonomy itself there have been great changes. We now believe that the best way of learning general principles is by the recognition and study of individual species, so that from the point of view of the taxonomist also, a flora should provide a ready means of identifying plants. In the technique of description the value of measurements has been recognized and the general acceptance of the metric system has facilitated their use. To a systematic botanist a millimetre scale is now as essential a piece of equipment as a hand-lens. The experimental approach to taxonomic problems combined with the application of cytology and genetics provides a new method of attack. Though there are as yet only a few problems to which this method has been applied it has yielded valuable results and has greatly increased our understanding of certain species and their relationships.

There have also been changes in the flora itself, as well as in our knowledge of it, many of which will be apparent to every field botanist. A considerable number of introduced plants have become well established and some of them are now widespread. All those which persistently occur in natural or semi-natural communities must be regarded as integral parts of the flora of the country and so should be included in any account of it. Others, which only maintain themselves by repeated reintroduction, are of frequent occurrence on rubbish tips, near ports and in railway sidings. These, though in a different category from the naturalized plants and of less importance to the ecologist, are of interest to the systematist and should also be included in a British flora.

It is thus clear that there is a great need for a new flora of the British Isles and this need, at the suggestion of Professor A. G. Tansley, we have attempted to meet. Our aim has been to make accessible to students and amateurs a portion of the increased knowledge of our flora which has been gained since Hooker and Babington wrote. We have also included a considerably larger number of introduced plants, either because they are naturalized or because they are of frequent occurrence. Some information is also given about the time of flowering, fruiting and germination, the pollination and seed-dispersal mechanism as well as the life form and chromosome number.

It is necessary here to say something of the limitations of this book. In the first place it is intended primarily for students and amateur botanists who desire to gain an introduction to British plants and for botanists who are not taxonomic specialists but need to identify species without going into great detail in the so-called critical genera. It does not attempt to describe all named varieties or to give other details which a specialist might reasonably desire. Since it seemed desirable to complete the book as soon as possible, it has been written in the course of three and a half years in the intervals of teaching and other duties. Consequently, there has not been time to elucidate more than a few of the problems which have arisen and the work is, up to a point, a compilation of existing knowledge. In some groups we have had the benefit of expert help, but there remain a considerable number of families and genera where specialist knowledge was lacking; there is therefore some unevenness in the treatment of the different groups, and in a few (e.g. *Salicornia* and *Rhinanthus*) where existing knowledge is manifestly inadequate, the account given is necessarily unsatisfactory though, we hope, the best at present available.

The descriptions have, with few exceptions, been drawn up afresh from living material or herbarium specimens and the keys, wherever possible, make use of characters at least as easy to observe in the field as in the herbarium. For some of the larger families synopses of classification have been given, while in others descriptions of Tribes, etc., will be found in the text. It is hoped that the text-figures, drawn by Miss S. J. Roles, will prove of use in aiding identification.

The arrangement of families is in general similar to that adopted by Bentham and Hooker though we have made a number of alterations to try to bring it more into line with modern ideas, and have always kept the doctrine of evolution in mind. Thus, instead of placing the Pteridophyta at the end, we start with them, as they are clearly the most primitive plants included in the book. It must be borne in mind, however, that no linear sequence of organisms, a sequence which must be used in a book, can be natural; often, particularly among the families of flowering plants, an arbitrary order has to be adopted. Within families and genera we have also followed the principle of starting with what appear to be the most primitive representatives in the British flora, though in some groups of which we have no intimate personal knowledge we have adopted the arrangement of a standard monograph.

In matters of nomenclature we have in general followed the *Check List of British Vascular Plants* issued by the British Ecological Society in 1946 and have also given synonyms in current use in British floras and in Druce's *British Plant List* (2nd ed. 1928).

In the spelling of certain specific epithets it has been customary to use an initial capital letter when the epithet concerned is derived from a personal name or is a noun, e.g. the name of another genus, or the pre-Linnean name for the plant. This custom is not made obligatory by the International Rules of Nomenclature but is mentioned in a recommendation attached to these Rules. The use of the initial capital has certain advantages; for instance it conveys some information about the origin of the name and explains the apparent lack of grammatical agreement between a generic name and a specific epithet which appears when written with a small initial letter to be adjectival (e.g. *Selinum Carvifolia*). We found upon inquiry, however, that many botanists in this country prefer, as a matter of convenience, to drop the initial capital. We have therefore adopted small initial letters for all specific epithets in the body of the book, but have indicated those which are commonly spelled with capitals.

English names have been given wherever possible for the benefit of agriculturists and others who prefer them for their special purposes. English names are frequently only of local use, and they give no reliable information of the relationship or otherwise of the plants, while frequently one name includes a number of distinct entities or is applied to different plants in various parts of the country. In addition only a small number of plants have English names which are in common use, though many others have names, often translations of the scientific names, which have been given to them mainly by the writers of nineteenth-century floras. We have tried to distinguish between the genuine English names and the invented ones by putting the latter in quotation marks. It cannot be too strongly emphasized that the scientific system of nomenclature has so many advantages over English names that it should be taught to University student and schoolchild alike.

Up to a point the limits of families and genera are a matter of personal opinion. For instance Oxalidaceae and Balsaminaceae can be included in Geraniaceae or can be regarded as separate families. In such instances we have preferred to take the narrower view of family or generic limits when by doing so the groups obtained are more natural and are consequently easier to recognize. The genus *Antirrhinum* as established by Linnaeus was a large one from which the majority of species were soon removed and placed in the genus *Linaria*. This left *Antirrhinum* as a small homogeneous group in no way comparable with the vast assemblage of plants included in *Linaria*. There are only two reasonable courses open in such a case, either to keep the one large heterogeneous genus or to divide it up into a number of comparable and reasonably homogeneous groups. Wherever it seemed possible and convenient to do so we have adopted the latter course.

As has already been pointed out no attempt has been made to describe all the numerous named varieties of British plants, but when plants which are morphologically similar have been shown to differ cytologically or in geographical distribution or ecological preferences we have not hesitated to recognize them as subspecies.

In some genera we have placed two or more superficially similar species in an aggregate (agg.). This is simply a device for the convenience of those who do not wish to go into minute detail, and is of no taxonomic significance.

Hybrids between species have as far as possible been mentioned; descriptions have been given where the hybrid is common, usually owing to abundant vegetative reproduction (e.g. in *Mentha* and *Potamogeton*), where it is a highly distinct plant which has in the past been regarded as a species (e.g. × *Agropogon littoralis*), or where it is liable to lead to confusion between species (e.g. *Alopecurus* × *hybridus*). We have discarded as far as possible all names which appear to us to be ambiguous, either because there is doubt about what plant was originally intended by that name (e.g. *Orchis latifolia* L.), or because the name is currently applied to two or more distinct species (e.g. *Carex leporina* L.).

A volume of illustrations is in course of preparation but as it cannot be ready for some time yet, references to illustrations in easily accessible floras have been included wherever these drawings were sufficiently satisfactory to be a real aid to identification.

While we hope that this flora will prove useful, we are fully aware that it has many deficiencies and will doubtless be found to contain errors. As Bentham wrote nearly a hundred years ago 'the aptness of a botanical description, like the beauty of a work of imagination, will always vary with the style and genius of the author'. We should be most grateful if users of the book who detect any errors would inform us.

ACKNOWLEDGEMENTS

We should like to express our thanks to Professor Tansley for his constant encouragement, and to the many botanists who have given us the benefit of their expert advice. Among these we should specially like to name W. T. Stearn for much help with nomenclature and with the petaloid monocotyledons, and for reading the proofs; S. M. Walters for great assistance with *Alchemilla, Aphanes, Montia, Eleocharis*, etc.; and E. K. Horwood whose continuous help has enabled the work to be completed much more rapidly than would otherwise have been possible.

We are also greatly indebted to the following for help with special problems: A. H. G. Alston (Pteridophyta), Miss K. B. Blackburn, J. P. M. Brenan (*Chenopodium*), B. L. Burtt, Miss M. S. Campbell (*Salicornia*), J. L. Crosby (*Anagallis*), J. E. Dandy (Hydrocharitaceae, Najadaceae, Potamogetonaceae), J. S. L. Gilmour, C. E. Hubbard (Gramineae), Miss I. Manton, R. Melville (*Ulmus*, especially the key and the originals of Figs. 48–9), E. Nelmes (*Carex*), C. D. Pigott, H. W. Pugsley (*Hieracium*), N. Y. Sandwith, H. K. Airy Shaw, W. A. Sledge, T. A. Sprague (classification), V. S. Summerhayes, G. Taylor, D. H. Valentine (*Primula, Viola*), A. E. Wade (Boraginaceae), W. C. R. Watson (*Rubus*), D. A. Webb, F. H. Whitehead, A. J. Wilmott (*Salicornia*, etc.), and many others who have assisted to a lesser extent in various ways. It should be added that these specialists cannot be held responsible for all the views expressed or implied in the accounts of those genera about which they have so freely given us their advice.

We should also like to express our indebtedness to the Director of the Royal Botanic Gardens, Kew, the Regius Keeper of the Royal Botanic Garden, Edinburgh, the Keeper of the Department of Botany at the British Museum, the Professors of Botany at Oxford and Cambridge, and the Director of the Leicester City Museum for the loan of specimens and other help.

We are greatly indebted to P. W. Richards for the account of the Juncaceae.

A. R. C.
T. G. T.
E. F. W.

November 1948

Two of us wish at this point to acknowledge the special contribution of T. G. Tutin, who besides writing a substantial part of this flora undertook in addition the arduous task of acting as general editor. It was he who collected and collated the various sections as they were completed, who strove to secure uniformity of treatment, who wrestled with text-figures, glossary and index, and who urged us on when we flagged. The work owes much to his patient and devoted labour.

A. R. C.
E. F. W.

November 1948

PREFACE TO THE SECOND EDITION

Since the writing of the first edition of this flora was completed considerable progress has been made with the study of British plants. The most spectacular, though in some respects not the most important, addition to our knowledge has been the discovery of some plants not hitherto known to grow in these islands but occurring, to all appearances as native species, in a few localities. Among these *Artemisia norvegica* and *Diapensia lapponica* deserve special mention not only for their phytogeographical interest but for the aesthetic pleasure that the field botanist will obtain from them.

Perhaps the greatest advance in the last few years has been brought about by the combination of intensive experimental, field and herbarium studies of variable species and 'critical groups'. This has resulted in a considerable clarification of the taxonomy of certain genera, for example *Dactylorchis* and *Polypodium* but, though much has been achieved, still more remains to be done.

Knowledge of the distribution of vascular plants in Britain has also been materially improved, largely as a result of the field work undertaken by the numerous helpers with the Botanical Society's Distribution Maps Scheme, which was made possible by the generosity of the Nature Conservancy and the Nuffield Foundation. The detailed results of this scheme are not yet fully available, but the major discoveries are incorporated here.

There have also been changes of various kinds in the flora itself. It is sad to record that *Schoenus ferrugineus* has apparently become extinct in Scotland, but to be set against this is the discovery of *Spiranthes romanzoffiana* in two new localities and the almost annual appearance of *Epipogium aphyllum*.

Changes among the alien plants have, of course, been even greater; some have become newly established while others have failed to maintain their footing. These changes are reflected, as accurately as our information permits, in this edition.

All this new work has resulted in the complete rewriting of many pages and innumerable minor alterations and additions. Several of the keys have also been considerably modified and, it is hoped, made easier to use and more certain to lead to correct identification.

The publication of Mr J. E. Dandy's *List of British Vascular Plants* marks another step towards nomenclatural stability, though changes of name for taxonomic reasons will inevitably continue to occur from time to time. We have in general followed this list, though differences of taxonomic opinion will be found here and there and, very occasionally, divergences of purely nomenclatural origin. We are, once again, greatly indebted to Mr Dandy for his help.

To those whose help was acknowledged in the first edition we should like to add the following: H. G. Baker (*Limonium*), P. W. Ball (*Cakile, Salicornia*), C. D. K. Cook (subgen. *Batrachium* and *Sparganium*), E. W. Davies (*Asparagus*), K. M. Goodway (*Galium*), R. A. Graham (*Mentha*), G. Halliday (*Minuartia*), J. Heslop-Harrison (*Dactylorchis*), I. H. McNaughton (*Papaver*), R. Melville (*Epilobium*), P. A. Padmore (*Ranunculus*), A. Pettet (*Viola*), C. D. Pigott (*Polemonium* and *Cirsium*), M. C. F. Proctor (*Helianthemum*),

N. M. Pritchard (*Gentianella*), Peter Raven (Onagraceae), B. T. Styles (*Polygonum*), S. Walker (*Dryopteris*), D. P. Young (*Oxalis*).

In addition we should like to thank those correspondents, too numerous to mention individually, who pointed out errors, provided additional information or made suggestions for improvement.

A. R. C.
T. G. T.
E. F. W.

November 1958

SYNOPSIS OF CLASSIFICATION

For signs and abbreviations see p. xlvii

PTERIDOPHYTA

Plants with an alternation of free-living generations. Sporophyte with vascular tissue, reproducing by spores which give rise to the small filamentous or thalloid gametophyte (prothallus) bearing archegonia and antheridia either on the same or different prothalli.

LYCOPODINEAE

Stems simple or dichotomously branched. Lvs normally small, spirally arranged; no lf-gap in stele. Sporangia borne singly in the axil of a lf (sporophyll) or on its upper-surface near the base.

LYCOPODIALES

Stems long with numerous small lvs; secondary thickening 0. Ligule 0. Homosporous. Spermatozoids biciliate.
1. Lycopodiaceae.

SELAGINELLALES

Stems long, with numerous small lvs; secondary thickening 0. Ligule present. Heterosporous. Spermatozoids biciliate.
2. Selaginellaceae.

ISOETALES

Stem short, tuberous, with secondary thickening. Lvs subulate. Ligule present. Heterosporous. Spermatozoids multiciliate.
3. Isoetaceae.

EQUISETINEAE

Stem simple or with whorls of branches. Lvs small, in whorls. No lf-gap. Sporangia several on peltate sporangiophores borne in cones. Spermatozoids multiciliate.

EQUISETALES

Herbs. Isosporous.
4. Equisetaceae.

FILICINEAE

Lvs usually large, often compound, spirally arranged; lf-gap present. Sporangia often grouped in sori, borne on the underside of the lvs or on special lf-segments. Spermatozoids multiciliate.

FILICALES

Lvs large, flat, circinate in bud. Sporangia with wall of 1 layer of cells borne on the lf-surface (sometimes ± modified). Homosporous.
5. Osmundaceae.   7. Polypodiaceae.
6. Hymenophyllaceae.

MARSILEALES

Plants rooted. Lvs circinate in bud, not fern-like. Sporangia with wall of 1 layer of cells, borne in thick-walled sporocarps containing several sori. Heterosporous.
8. Marsileaceae.

SALVINIALES

Free floating. Lvs small, not circinate. Sporangia with wall of 1 layer of cells borne in thin-walled sporocarps containing 1 sorus. Heterosporous.
  9. Azollaceae.

OPHIOGLOSSALES

Lvs flat, not circinate. Sporangia with wall of several layers of cells, borne on special spike- or panicle-like lf-segments. Homosporous.
  10. Ophioglossaceae.

GYMNOSPERMAE

Gametophyte not free-living. Woody plants with secondary thickening. Ovules not enclosed in an ovary. Female prothallus well developed, of numerous cells forming the food reserve of the seed. Xylem without vessels (except in Gnetales).

CONIFERAE

Stem usually freely branched. Lvs simple, usually small. Pollen sacs borne on the under-surface of microsporophylls arranged in cones. Fertilization effected by means of a pollen-tube; male gametes not motile. Ovules usually on the surface of a scale.
  11. Pinaceae.   13. Taxaceae.
  12. Cupressaceae.

ANGIOSPERMAE

Ovules completely enclosed in an ovary which is usually crowned by a style and stigma. Microspores (pollen grains) adhering to the stigma and fertilization effected by means of a pollen-tube. Xylem containing vessels (except in some Winteraceae and Trochodendraceae).

DICOTYLEDONES

Embryo with 2 cotyledons (rarely one by reduction). Vascular bundles of the stem usually arranged in a single ring, cambium usually present. Lvs rarely parallel-veined. Fls typically 5–4-merous.

ARCHICHLAMYDEAE

Petals free from each other or 0, rarely united into a tube.

RANALES

Herbs, often with numerous vascular bundles and little or no cambium in the stems, less frequently woody. Lvs alternate (very rarely opposite), nearly always exstipulate. Fls hermaphrodite (rarely unisexual) hypogynous or rarely perigynous, actinomorphic (rarely zygomorphic). Perianth present (very rarely 0). Stamens numerous or less frequently definite in number, often spirally arranged. Ovary apocarpous; carpels often numerous and spirally arranged. Fr. various, but rarely fleshy. Seeds with copious endosperm and small embryo.
  14. Ranunculaceae.   17. Nymphaeaceae.
  15. Paeoniaceae.   18. Ceratophyllaceae.
  16. Berberidaceae.

RHOEADALES

Herbs or rarely ± woody. Lvs alternate (rarely opposite), exstipulate or with small stipules. Fls hermaphrodite, rarely unisexual, hypogynous, actinomorphic to zygomorphic. Petals and sepals usually in whorls of 2 or 4. Stamens numerous or 6 or 4, rarely 3 or 2. Ovary syncarpous (rarely apocarpous) 1–several-celled, often

2-celled and divided by a false septum; placentation parietal. Fruit dry. Seeds either with little or no endosperm and a large embryo or with abundant endosperm and a minute embryo.

19. Papaveraceae.   21. Cruciferae.
20. Fumariaceae.   22. Resedaceae.

VIOLALES

Herbs, shrubs or small trees. Lvs alternate (rarely opposite), stipulate. Fls hermaphrodite (rarely polygamous), hypogynous, actinomorphic to zygomorphic. Sepals 5, lowermost petal often larger and spurred. Stamens 5, ±connivent in a ring round the ovary. Ovary syncarpous, 1-celled, with 3–5 parietal placentae. Fr. a capsule or sometimes fleshy. Seeds with endosperm and a straight embryo.

23. Violaceae.

POLYGALALES

Herbs, shrubs and trees. Lvs alternate or rarely opposite; stipules 0 or small. Fls hermaphrodite, hypogynous to subperigynous, zygomorphic or rarely actinomorphic. Sepals 5, often unequal. Petals 1–5, free or sometimes some joined. Stamens up to 12, sometimes only 1 fertile, sometimes monadelphous. Anthers often opening by pores. Ovary syncarpous, 1–3-celled; placentation axile or apical. Fr. a capsule, drupe or samara. Seeds with a straight embryo; endosperm present or 0.

24. Polygalaceae.

CISTIFLORAE

Herbs or more often woody, juice often coloured. Lvs opposite or sometimes alternate; stipules usually present. Fls hermaphrodite, hypogynous, actinomorphic, usually large and showy. Sepals 3–5. Petals 5 (rarely fewer or 0). Stamens numerous, sometimes with their filaments joined in bundles. Ovary syncarpous, 1-celled or sometimes 3–5-celled; placentation parietal or sometimes axile or apical. Fr. a capsule, rarely fleshy. Seeds with a straight or, more often, curved or bent embryo; endosperm usually present and often abundant.

25. Hypericaceae.   26. Cistaceae.

TAMARICALES

Trees or shrubs, rarely herbs. Lvs opposite or alternate, small and scale-like or ericoid; stipules present or 0. Fls usually hermaphrodite, hypogynous, usually small, actinomorphic. Sepals 4–6. Petals 4–6. Stamens 5–10. Ovary syncarpous, 1-celled; placentation parietal or basal. Fr. a capsule. Seeds with a straight embryo; endosperm present or 0.

27. Tamaricaceae.   28. Frankeniaceae.

CENTROSPERMAE

Herbs, rarely soft-wooded shrubs or trees. Lvs opposite or verticillate, or sometimes alternate; stipules present or 0. Fls hermaphrodite or occasionally unisexual, hypogynous to perigynous, actinomorphic or rarely slightly zygomorphic. Perianth in 2 whorls or 1 (rarely several) the outer sepaloid, the inner petaloid or sepaloid. Stamens usually definite. Ovary usually syncarpous, 1(–3–several)-celled; placentation axile to free-central or basal; ovules usually campylotropous. Seeds usually with endosperm and a curved embryo.

29. Elatinaceae.   33. Amaranthaceae.
30. Caryophyllaceae.   34. Chenopodiaceae.
31. Portulacaceae.   35. Phytolaccaceae.
32. Aizoaceae.

MALVALES

Trees, shrubs or herbs usually with stellate hairs. Lvs usually alternate, stipulate, often mucilaginous. Fls hermaphrodite or unisexual, hypogynous, actinomorphic. Calyx-lobes usually valvate. Petals present or, less frequently, 0. Stamens numerous, free or monadelphous, often some sterile or anthers 1-celled. Ovary syncarpous, 2 (or more)-celled; placentation axile. Fr. various. Seeds usually with endosperm; embryo straight or curved.

36. Tiliaceae.   37. Malvaceae.

GERANIALES

Herbs or small shrubs, rarely trees. Lvs usually alternate or radical, generally stipulate. Fls hermaphrodite, very rarely unisexual, hypogynous, actinomorphic to zygomorphic. Sepals imbricate or rarely valvate. Petals contorted or sometimes imbricate, often clawed, rarely some ±joined, rarely 0. Stamens as many to three times as many as petals, commonly twice as many. Ovary syncarpous, 3–5-celled; placentation axile. Fr. various, but very rarely fleshy. Seeds usually without endosperm; embryo straight.

38. Linaceae.   40. Oxalidaceae.
39. Geraniaceae.   41. Balsaminaceae.

RUTALES

Trees, shrubs or climbers, rarely herbs. Lvs usually alternate, often compound, and frequently gland-dotted. Fls hermaphrodite, rarely unisexual, hypogynous or weakly perigynous, usually actinomorphic. Sepals usually imbricate. Petals contorted or sometimes valvate, free or joined at base. Disk usually conspicuous. Stamens as many or twice as many as the petals. Ovary syncarpous, 1–5-celled; ovules usually 1–2 in each cell. Fr. various. Seeds with or without endosperm; embryo straight or curved.

42. Simaroubaceae.

SAPINDALES

Trees or shrubs. Lvs usually pinnate and exstipulate. Fls often polygamous or unisexual, hypogynous to slightly perigynous, sometimes zygomorphic, usually small. Sepals usually 4–5, imbricate. Petals usually 4–5, rarely 0. Stamens often twice as many as petals. Disk present. Ovary syncarpous (rarely apocarpous) with 1–2 ovules in each cell; placentation axile. Fr. various. Seeds usually without endosperm; embryo curved or variously bent.

43. Aceraceae.   45. Hippocastanaceae.
44. Staphyleaceae.

CELASTRALES

Trees or shrubs. Lvs simple, often entire; stipules small or 0. Fls hermaphrodite, rarely unisexual, hypogynous to perigynous, actinomorphic, usually small. Sepals usually imbricate. Petals often 4–5, rarely 0. Disk present or 0. Stamens often 4–5, opposite the sepals. Ovary syncarpous usually with 1–2 ovules in each cell; placentation axile or apical. Fr. various. Seeds usually with abundant endosperm; embryo straight.

46. Aquifoliaceae.   48. Buxaceae.
47. Celastraceae.

RHAMNALES

Trees, shrubs or woody climbers. Lvs usually stipulate. Fls similar to Celastrales but stamens 4–5, opposite the petals or alternating with the sepals in apetalous spp. Fr. usually a drupe or berry. Seeds with endosperm; embryo usually straight.

49. Rhamnaceae.   50. Vitaceae.

Leguminosae

Trees, shrubs or herbs. Lvs often pinnate or bipinnate, sometimes trifoliate or simple; stipules present or 0. Fls hermaphrodite, hypogynous to perigynous, actinomorphic to zygomorphic, often large and showy. Sepals often 5, often ± united into a tube. Petals usually 5, rarely 0, occasionally united. Stamens often 10, sometimes numerous, often monadelphous or diadelphous. Ovary of one carpel. Fr. a legume, often dehiscent. Seeds usually with little or no endosperm, rarely with abundant endosperm; embryo large.

51. Papilionaceae.

Rosales

Trees, shrubs or herbs. Lvs simple or compound. Fls hermaphrodite or rarely unisexual, perigynous to epigynous, actinomorphic (rarely ± zygomorphic). Sepals usually 4–5, free or united. Petals usually 4–5 (rarely 0), occasionally united. Stamens numerous to definite. Ovary apocarpous to syncarpous, with one or more ovules in each cell; placentation often axile. Fr. various. Seeds with or without endosperm.

52. Rosaceae.
53. Platanaceae.
54. Crassulaceae.
55. Saxifragaceae.
56. Parnassiaceae.

57. Hydrangeaceae.
58. Escalloniaceae.
59. Grossulariaceae.
60. Pittosporaceae.

(marginalia: p 440, p. 447 beside 54 and 55)

Sarraceniales

Herbs. Lvs tubular or covered with viscid glands, adapted for trapping insects. Fls hermaphrodite, hypogynous to perigynous, actinomorphic. Sepals 4–5, ± united at base. Petals 5, rarely 0. Stamens 4–numerous. Ovary syncarpous; ovules usually numerous; placentation axile to parietal. Fr. a capsule. Seeds with endosperm; embryo straight.

61. Droseraceae.

62. Sarraceniaceae.

Myrtales

Trees, shrubs or herbs often with bicollateral vascular bundles. Lvs often opposite, usually exstipulate. Fls hermaphrodite or rarely unisexual, perigynous to epigynous often with a long receptacle, actinomorphic. Calyx-tube usually ± adnate to the ovary, lobes mostly 4–5, often valvate. Petals commonly 4–6 (rarely 0), sometimes united. Stamens 1–many, often 4 or 8. Ovary syncarpous, 1–many-celled; ovules numerous to 1; placentation axile or rarely parietal, apical or basal. Fr. various. Seeds with or without endosperm.

63. Lythraceae.
64. Thymelaeaceae.
65. Elaeagnaceae.
66. Onagraceae.

67. Haloragaceae.
68. Hippuridaceae.
69. Callitrichaceae.

Santalales

Trees, shrubs or herbs often parasitic on other angiosperms or rarely on gymnosperms. Lvs usually opposite, sometimes scale-like, exstipulate. Fls hermaphrodite or unisexual, epigynous, actinomorphic. Calyx with valvate lobes or often reduced, sometimes to a ring. Petals present or 0, sometimes united into a tube. Stamens the same number as the calyx-lobes and opposite them or opposite the petals when present. Ovary 1-celled; ovules few, often imperfectly differentiated; placentation axile. Fr. a drupe or berry, less frequently a nut. Seeds with endosperm and a straight embryo.

70. Loranthaceae.

71. Santalaceae.

UMBELLALES

Trees, shrubs or herbs. Lvs usually alternate, often much-divided; stipules present or 0. Fls hermaphrodite or unisexual, epigynous, actinomorphic to weakly zygomorphic, usually small and arranged in umbels or heads. Calyx small, truncate or 4–10-toothed. Petals usually 4–5, rarely 0. Stamens usually the same number as the petals and alternate with them. Ovary usually 1–2-celled, sometimes many-celled; ovules solitary in each cell, pendulous from the apex. Seeds usually with copious endosperm.

72. Cornaceae.
73. Araliaceae.
74. Hydrocotylaceae.
75. Umbelliferae.

CUCURBITALES

Herbs, or sometimes small trees, often with bicollateral vascular bundles, and frequently climbing by tendrils. Lvs usually alternate, often large and deeply lobed or compound. Fls unisexual, epigynous, actinomorphic or rarely zygomorphic, often showy. Calyx variously lobed. Petals free or united into a tube. Stamens 1–numerous, free or variously united, sometimes epipetalous. Ovary 1–4-celled; ovules numerous, very rarely few; placentation parietal or axile. Fr. a capsule or berry. Seeds with little or no endosperm.

76. Cucurbitaceae.

ARISTOLOCHIALES

Woody climbers with broad medullary rays, or parasites, or epiphytes, rarely erect herbs. Lvs alternate, simple, exstipulate, sometimes 0. Fls hermaphrodite or unisexual, hypogynous to epigynous, actinomorphic or zygomorphic. Per. segs in one whorl, usually petaloid. Stamens numerous to few. Ovary 1–6-celled; ovules numerous in each cell; placentation parietal or axile. Fr. a capsule or sometimes fleshy. Seeds with or without endosperm.

77. Aristolochiaceae.

EUPHORBIALES

Trees, shrubs or herbs. Lvs usually alternate and stipulate, simple or compound, sometimes reduced. Fls unisexual, hypogynous, actinomorphic. Sepals usually present. Petals usually 0. Stamens numerous to solitary, free or united. Ovary usually 3-celled; ovules 1–2 in each cell; placentation axile. Fr. a capsule or drupe. Seeds with copious endosperm.

78. Euphorbiaceae.

POLYGONALES

Herbs, shrubs or climbers, rarely trees. Lvs usually alternate, often with sheathing stipules. Fls hermaphrodite or unisexual, hypogynous, actinomorphic. Per. segs 3–6, sepaloid or petaloid, free or united. Stamens usually 6–9. Ovary 1-celled with a solitary basal ovule. Fr. a trigonous or lenticular nut. Seeds with copious endosperm.

79. Polygonaceae.

URTICALES

Trees, shrubs or herbs. Lvs usually alternate and stipulate. Fls hermaphrodite or unisexual, hypogynous, actinomorphic. Per. segs usually 4–5 ± united, sepaloid. Stamens usually the same number as and opposite to the per. segs, erect or inflexed in bud. Ovary 1–2-celled; ovule solitary, erect or pendulous. Fr. various. Seeds with or without endosperm.

80. Urticaceae.
81. Cannabiaceae.
82. Ulmaceae.
83. Moraceae.

JUGLANDALES

Trees, often resinous and aromatic. Lvs alternate, pinnate, exstipulate. Fls unisexual, epigynous, actinomorphic. Perianth small and sepaloid or 0. Stamens 3–40. Ovary 1-celled; ovule solitary, erect. Fr. a drupe or rarely a nut. Seeds without endosperm.

84. Juglandaceae.

MYRICALES

Aromatic trees or shrubs. Lvs alternate, simple, exstipulate. Fls unisexual, arranged in dense bracteate spikes. Perianth 0. Stamens 2–many, free or connate. Ovary 1-celled; ovule solitary basal. Fr. a drupe. Seeds without endosperm.

85. Myricaceae.

FAGALES

Trees or shrubs. Lvs alternate, simple, stipulate. Fls monoecious, epigynous or the female devoid of perianth, in catkins (rarely heads) or the female in cone-like spikes or few, often appearing before the lvs. Perianth very small or 0 in one sex; female fls often surrounded by an involucre of bracts. Stamens 2–many. Ovary 2–6-celled; ovules 1–2 in each cell, pendulous. Fr. a nut, sometimes winged. Seeds without endosperm.

86. Betulaceae.   88. Fagaceae.
87. Corylaceae.

SALICALES

Trees or shrubs. Lvs alternate, simple, usually stipulate. Fls dioecious, in catkins, often appearing before the lvs. Perianth 0, or very small. Stamens 2 or more. Ovary 1-celled; ovules numerous; placentation parietal. Fr. a capsule. Seeds without endosperm.

89. Salicaceae.

METACHLAMYDEAE

Petals united into a longer or shorter tube, very rarely free or 0.

ERICALES

Shrubs, rarely trees or herbs. Lvs simple, exstipulate, usually alternate. Fls hermaphrodite, rarely unisexual, hypogynous or epigynous, actinomorphic to zygomorphic. Calyx usually 4–6-lobed, sometimes of free sepals. Petals united, rarely free or 0. Stamens usually twice as many as the corolla-lobes, free, anthers often opening by pores. Ovary 3–several-celled; ovules 1–many in each cell; placentation usually axile. Fr. a capsule, berry or drupe. Seeds with abundant endosperm and small straight embryo.

90. Ericaceae.   93. Empetraceae.
91. Pyrolaceae.   94. Diapensiaceae.
92. Monotropaceae.

PLUMBAGINALES

Herbs, small shrubs, or sometimes climbers. Lvs alternate or radical, exstipulate. Fls hermaphrodite, hypogynous, actinomorphic. Calyx commonly 5-lobed, often strongly ribbed and membranous between the lobes. Petals 5, united into a longer or shorter tube, rarely free. Stamens 5, opposite corolla-lobes and ±adnate to the tube. Ovary 1-celled; ovule solitary; placentation basal; styles 5. Fr. dry, usually indehiscent. Seeds with or without endosperm.

95. Plumbaginaceae.

PRIMULALES

Herbs, shrubs or trees. Lvs variously arranged, but often radical, exstipulate. Fls hermaphrodite, hypogynous or very rarely perigynous, actinomorphic or very rarely zygomorphic. Calyx 4–9-lobed, persistent. Corolla 4–9-lobed, very rarely two-lipped or 0. Stamens as many as and opposite the corolla-lobes, adnate to the tube. Ovary 1-celled, very rarely adnate to the calyx; ovules 2–many on a free-central placenta; style 1. Fr. a capsule, variously dehiscent. Seeds with copious endosperm.

96. Primulaceae.

CONTORTAE

Trees, shrubs or herbs. Lvs simple, often opposite, usually exstipulate. Fls hermaphrodite (rarely unisexual), hypogynous, actinomorphic. Calyx tubular or rarely composed of separate sepals or 0. Corolla usually 4–5-lobed (rarely of free petals or 0), lobes contorted or valvate in bud. Stamens usually epipetalous, the same number as and alternate with the corolla-lobes, rarely fewer. Ovary mostly 1–2-celled, sometimes of 2 separate carpels; ovules numerous to 1 in each cell; placentation usually parietal or axile. Fr. various. Seeds with endosperm; embryo straight, often small.

97. Buddlejaceae.   100. Gentianaceae.
98. Oleaceae.   101. Menyanthaceae.
99. Apocynaceae.

TUBIFLORAE

Herbs, less frequently trees, shrubs or woody climbers. Lvs often opposite, usually exstipulate. Fls usually hermaphrodite, hypogynous (rarely epigynous), actino-morphic to zygomorphic. Calyx usually 4–5-lobed, sometimes 2-lipped. Corolla 4–5-lobed, often 2-lipped. Stamens epipetalous, as many as or fewer than the corolla-lobes and alternate with them. Ovary usually 1–2-celled; ovules numerous to 1 in each cell; placentation usually axile, parietal or basal. Fr. various. Seeds with or without endosperm.

102. Polemoniaceae.   107. Orobanchaceae.
103. Boraginaceae.   108. Lentibulariaceae.
104. Convolvulaceae.   109. Acanthaceae.
105. Solanaceae.   110. Verbenaceae.
106. Scrophulariaceae.   111. Labiatae.

PLANTAGINALES

Herbs. Lvs simple, often sheathing at base. Fls usually hermaphrodite, hypo-gynous, actinomorphic. Calyx 4-lobed. Corolla 3–4-lobed, scarious. Stamens epipetalous, usually 4. Ovary 1–4-celled; ovules 1–several in each cell; placentation axile or basal. Fr. a capsule or nut. Seeds with endosperm.

112. Plantaginaceae.

CAMPANALES

Usually herbs. Lvs mostly alternate, simple, exstipulate. Fls hermaphrodite, rarely unisexual, epigynous (rarely hypogynous), actinomorphic to zygomorphic. Calyx usually 5-lobed. Corolla often 5-lobed, sometimes 2-lipped. Stamens as many as the corolla-lobes and alternate with them, free or inserted on the corolla-tube near its base; anthers often connivent and sometimes adhering in a tube. Ovary (1–)2–10-celled; ovules usually numerous; placentation axile. Fr. various. Seeds with endosperm.

113. Campanulaceae.   114. Lobeliaceae.

RUBIALES

Trees, shrubs or herbs. Lvs usually opposite, stipulate or not. Fls usually hermaphrodite, epigynous, actinomorphic to zygomorphic. Calyx often 4–5-lobed or reduced to a rim. Corolla sometimes 2-lipped. Stamens epipetalous, the same number as and alternate with the corolla-lobes, rarely fewer; anthers not connivent or cohering. Ovary (1–)2 (or more)-celled; ovules numerous to 1 in each cell; placentation axile or apical, rarely basal. Fr. various. Seeds with or without endosperm.

115. Rubiaceae.
116. Caprifoliaceae.
117. Adoxaceae.
118. Valerianaceae.
119. Dipsacaceae.

ASTERALES

Herbs, shrubs, or rarely trees or woody climbers. Lvs exstipulate. Fls hermaphrodite or unisexual, epigynous, actinomorphic to zygomorphic, crowded in heads (rarely solitary) surrounded by 1 or more series of free or connate bracts. Calyx small, often with thread-like lobes (pappus). Corolla usually 4–5-lobed. Stamens epipetalous, 5(–4); anthers connate (rarely imperfectly so). Ovary 1-celled; ovule solitary; placentation basal. Fr. an achene. Seeds without endosperm.

120. Compositae.

MONOCOTYLEDONES

Embryo with one cotyledon. Vascular bundles of the stem usually in several series or ±irregularly arranged, cambium usually 0. Lvs usually parallel-veined. Fls typically 3-merous.

ALISMATALES

Herbs living in water or wet places, sometimes marine. Fls actinomorphic, hermaphrodite or unisexual. Perianth in two whorls, the outer usually sepaloid, the inner petaloid. Stamens 3, 6 or numerous. Ovary apocarpous and superior or syncarpous and inferior. Ovules 1–numerous, basal, parietal or scattered. Seeds without endosperm.

121. Alismataceae.
122. Butomaceae.
123. Hydrocharitaceae.

NAJADALES

Herbs living in water or wet places, sometimes marine. Lvs linear, with scales (*squamulae intravaginales*) in their axils. Fls hypogynous, hermaphrodite or unisexual. Perianth 0 or of one whorl, less often of 2 similar whorls. Stamens 1–6, rarely more. Ovary of few (often only one) free or ±connate carpels; ovules 1, rarely more; placentation usually basal or apical. Fr. usually dry. Seeds with little or no endosperm.

124. Scheuchzeriaceae.
125. Juncaginaceae.
126. Aponogetonaceae.
127. Zosteraceae.
128. Potamogetonaceae.
129. Ruppiaceae.
130. Zannichelliaceae.
131. Najadaceae.

ERIOCAULALES

Herbs with narrow lvs. Fls small, unisexual, arranged in heads. Perianth scarious or membranous, segments in 2 whorls, inner often united. Ovary superior, 3–2-celled. Ovules solitary, pendulous. Seeds with endosperm.

132. Eriocaulaceae.

LILIIFLORAE

Herbs, often with corms, bulbs or rhizomes, rarely shrubs or small trees. Lvs mostly linear. Fls hermaphrodite or sometimes unisexual, hypogynous to epigynous, actinomorphic to zygomorphic. Perianth of two whorls, usually both petaloid, rarely both sepaloid, very rarely unlike. Stamens in one or 2 whorls, commonly 3 or 6. Ovary syncarpous, usually 3-celled; ovules 1–many in each cell; placentation axile or parietal. Seeds with endosperm.

133. Liliaceae.
134. Trilliaceae.
135. Pontederiaceae.
136. Juncaceae.

137. Amaryllidaceae.
138. Iridaceae.
139. Dioscoreaceae.

ORCHIDALES

Herbs without bulbs but often with tubers, often epiphytes or saprophytes. Lvs simple, often rather thick. Fls mostly hermaphrodite, epigynous, zygomorphic. Perianth of two whorls, usually both petaloid, but sometimes the outer sepaloid. Stamens 2 or 1; pollen usually agglutinated into masses (*pollinia*). Ovary usually 1-celled often twisted through 180°; ovules numerous; placentation parietal. Fr. usually a capsule. Seeds minute, without endosperm and with undifferentiated embryo.

140. Orchidaceae.

ARALES

Herbs or occasionally woody climbers, rarely floating aquatics. Fls very small, hermaphrodite or unisexual, hypogynous, densely crowded on a spadix or rarely few together, infl. usually ±enclosed in a large bract (spathe). Perianth present and small, or 0. Ovary 1–many-celled; placentation various. Fr. usually a berry. Seeds with endosperm.

141. Araceae.

142. Lemnaceae.

TYPHALES

Rhizomatous marsh or aquatic herbs. Lvs linear, sheathing at base. Fls unisexual, hypogynous, small, densely crowded in spikes or heads. Perianth small, sepaloid, often of scales or threads. Stamens 2 or more. Ovary 1-celled; ovule solitary, pendulous. Fr. dry. Seeds with endosperm.

143. Sparganiaceae.

144. Typhaceae.

CYPERALES

Mostly rhizomatous perennial herbs with solid stems. Lvs usually linear and sheathing at base, sometimes reduced to sheaths. Fls hermaphrodite or unisexual, hypogynous, small, crowded in heads or spikes and each subtended by a bract. Perianth of scales, bristles or 0. Stamens usually 3; anthers basifixed. Ovary 1-celled; ovule solitary, erect. Fr. dry, indehiscent. Seeds with endosperm.

145. Cyperaceae.

GLUMIFLORAE

Annual or more often perennial herbs, rarely woody; stems often hollow. Lvs usually linear and sheathing at base. Fls hermaphrodite or unisexual, hypogynous, small, distichously arranged, usually enclosed between 2 bracts. Perianth 0 or perhaps represented by minute scales. Stamens often 3; anthers versatile. Ovary 1-celled; ovule solitary, often adnate to the side of the carpel. Fr. a caryopsis, rarely a nut or berry. Seeds with endosperm.

146. Gramineae.

ARTIFICIAL KEY TO FAMILIES

1 Plant free-floating on or below surface of water. 2
     Land plants, or if aquatic, rooted at bottom of water. 3

2 Plant floating on surface of water, consisting of a discoid thallus (1–
     15 mm. diam.) with or without roots from lower surface; or floating
     below the surface with roots from its lower surface; propagation
     mainly by vegetative budding, so that several plants are often found
     joined together; fls minute, rarely produced.
     142. LEMNACEAE (p. 1052)
     Plant floating on surface of water, much-branched and with numerous
     small imbricate lvs. **9. AZOLLACEAE** (p. 41)
     Plant submerged, bearing small bladders either on the lvs or on colourless
     apparently lfless shoots; lvs divided into filiform segments.
     108. LENTIBULARIACEAE (p. 725)
     Lvs in rosettes, either linear and distinctly spinulose-serrate (*Stratiotes*),
     or orbicular, cordate, entire (*Hydrocharis*); fls, when present, with
     3 sepals and 3 petals and an inferior ovary.
     123. HYDROCHARITACEAE (p. 936)
     Not as above. 3

3 Plant reproducing by spores; fls 0; always herbs. 4
     Plant reproducing by seeds; fls with stamens or carpels or both; often
     woody. 9

4 Stems jointed, unbranched or with whorls of branches at the nodes;
     lvs not green, united into a scarious sheath; sporangia in terminal
     cones. **4. EQUISETACEAE** (p. 5)
     Lvs small, numerous, ± triangular, green, densely clothing the stems;
     sporangia solitary on the upper surface of lvs (sporophylls) which
     either resemble the foliage lvs or are smaller and grouped in terminal
     cones; stems dichotomously branched. 5
     Lvs linear, ± rush-like. 6
     Lvs relatively large and flat, often several times pinnate. 7

5 Sporangia of one kind only, containing numerous small spores; lvs with-
     out ligule; plant relatively robust. **1. LYCOPODIACEAE** (p. 1)
     Sporangia of two kinds, microsporangia containing numerous small
     spores, and megasporangia with 4 large ones, plants either with both
     sorts or frequently with megasporangia only; lvs with minute ligule on
     upper surface; plant slender. **2. SELAGINELLACEAE** (p. 3)

6 Plant with creeping rhizome; sporangia 4 together in hard pill-like
     sporocarps at the lf bases. **8. MARSILEACEAE** (p. 41)
     Stock short, erect; sporangia solitary and immersed in the upper surface
     of the lvs at their bases, not enclosed in hard sporocarps.
     3. ISOETACEAE (p. 4)

7 Fertile lvs consisting of two pinnae (which are themselves either simple
     or pinnate), the upper fertile, stalked, not green, the lower sterile, green
     (the lf appears to be a stem with a terminal fertile portion and a single
     lateral lf). **10. OPHIOGLOSSACEAE** (p. 42)

Fertile lvs bipinnate with several pinnae, the pinnules of the upper part of the lf not green, ±spike-like, and densely covered with sporangia.
5. OSMUNDACEAE (p. 11)
Fertile lvs either not differing from sterile or sometimes with narrower pinnae which are green throughout, though the green lower surface may be ±concealed by brown sporangia. *8*

8 Sporangia borne in flask-like indusia projecting from the margins of the lf; lvs thin and translucent. **6. HYMENOPHYLLACEAE** (p. 12)
Sporangia not projecting from the margins of the lf; lvs not at all translucent. **7. POLYPODIACEAE** (p. 14)

9 Ovules naked, either on the upper surface of scales arranged in cones or solitary and terminal on a short scaly axillary shoot; pollen sacs two or more on the lower surface of a flat sporophyll, or several pendant from the apex of a peltate sporophyll, the male sporophylls always in cones; monoecious or dioecious trees or shrubs with small needle-like or scale-like (but green) lvs; perianth 0. **(CONIFERAE)** *10*
Ovules completely enclosed in a carpel; pollen sacs 4 (or occasionally fewer) surrounding and adnate to a connective at the apex of a usually slender filament. **(ANGIOSPERMAE)** *12*

10 Lvs opposite or whorled; short shoots 0.
12. CUPRESSACEAE (p. 52)
Lvs alternate or in clusters on short lateral shoots. *11*

11 Ovules on the surface of scales arranged in cones; pollen sacs two on the lower surface of a flat sporophyll; trunk usually single.
11. PINACEAE (p. 45)
Ovules solitary and terminal on short axillary shoots; pollen sacs several on a peltate sporophyll; trunks usually several.
13. TAXACEAE (p. 55)

12 Herbs without chlorophyll, the lvs reduced to scales. *242* (J)
Green plants (if lfless at flowering time either trees or shrubs, or else herbs with only the fls showing above ground). *13*

13 Perianth of two (rarely more) distinct whorls, differing markedly from each other in shape, size or colour. *14*
Perianth 0, or of 1 whorl, or of 2 or more similar whorls, or segments numerous and spirally arranged. *18*
Small herb with lvs linear and all radical; fls solitary, monoecious, axillary; male on long stalks with perianth of two whorls and 4 long-exserted stamens; female sessile, ±hidden, with perianth of one whorl and long, projecting style (*Littorella*).
112. PLANTAGINACEAE (p. 764)

14 Petals free (very rarely cohering at apex, free at base). *15*
Petals united at least at the base. *17*

15 Ovary superior. *16*
Ovary inferior or partly so. *81* (C)

16 Carpels and styles free, or carpels slightly united at the extreme base. *20* (A)
Carpels or styles or both obviously united, or ovary of one carpel. *28* (B)

17 Ovary superior. *100* (D)
Ovary inferior. *129* (E)

18 Perianth corolla-like, at least the inner segments usually brightly-coloured
    or white.   *141* (F)
    Perianth green and calyx-like, or scarious, or 0.   *19*

19 Trees or shrubs.   *168* (G)
    Herbs.   *190* (H)

GROUP A

Petals free, ovary superior, carpels and styles free or nearly so.

20 Tree with palmately lobed lvs; fls in globose heads; perianth incon-
    spicuous.   **53. PLATANACEAE** (p. 440)
    Shrubs or herbs; fls not in globose heads.   *21*

21 Sepals and petals each 3.   *22*
    Sepals and petals more than 3.   *23*

22 Aquatic plants; fls conspicuous; at least the upper lvs broad, flat, stalked;
    carpels ± numerous.   **121. ALISMATACEAE** (p. 932)
    Small land plants of mossy appearance; fls axillary, inconspicuous; lvs
    small, oblong, rather fleshy, sessile; carpels 3 (*Crassula*).
    54. CRASSULACEAE (p. 440)

23 Stamens numerous.   *24*
    Stamens twice as many as petals or fewer.   *26*

24 Herbs; stipules 0; fls hypogynous.   *25*
    Herbs with stipules, or else shrubs; fls perigynous (sometimes only
    slightly so).   **52. ROSACEAE** (p. 363)

25 Fls c. 10 cm. diam.   **15. PAEONIACEAE** (p. 88)
    Fls much smaller.   **14. RANUNCULACEAE** (p. 56)

26 Lvs ternate, not fleshy; alpine plant (*Sibbaldia*).
    52. ROSACEAE (p. 363)
    Lvs simple.   *27*

27 Lvs linear, not fleshy; carpels numerous on an elongated receptacle
    (*Myosurus*).   **14. RANUNCULACEAE** (p. 56)
    Lvs fleshy (except in 1 very rare aquatic sp.); carpels in one whorl,
    equalling petals in number.   **54. CRASSULACEAE** (p. 440)

GROUP B

Petals free, ovary superior, carpels or styles or both united,
or ovary of one carpel.

28 Fls regular.   *29*
    Fls zygomorphic.   *73*

29 Stamens more than twice as many as the petals (always more than 6),
    or stamens and petals both numerous.   *30*
    Stamens twice as many as petals, or fewer (never more than 12), or
    petals 2, stamens 6.   *38*

30 Aquatic plants with large cordate, floating lvs and floating fls; petals
    more than 10.   **17. NYMPHAEACEAE** (p. 90)
    Plant not aquatic.   *31*

31 Stamens all united below into a tube; fls pink or purple; lvs usually
    palmately lobed.   **37. MALVACEAE** (p. 292)
    Stamens free or in bundles; lvs never palmately lobed.   *32*

32   Lvs very succulent, trigonous; petals numerous.
                                       32. AIZOACEAE (p. 270)
      Lvs not succulent; petals 5 or fewer.   *33*

33   Fls markedly perigynous; carpel 1, ovule 1; trees or shrubs with white
        fls (*Prunus*).   **52. ROSACEAE** (p. 363)
      Fls hypogynous; ovules numerous, carpels 2 or more, if 1 then plant
        not woody.   *34*

34   Petals ± conspicuous, sepals not petaloid; lvs not ternate.   *35*
      Petals very small, sepals petaloid; lvs ternate (*Actaea*).
                           14. RANUNCULACEAE (p. 56)

35   Trees; infl. with a conspicuous bract partly adnate to peduncle.
                             36. TILIAGEAE (p. 290)
      Herbs or shrubs; bracts, if present, not adnate to peduncle.   *36*

36   Styles free; stamens united into bundles below.
                         25. HYPERICACEAE (p. 200)
      Styles 1 or 0; stigma simple; stamens free.   *37*

37   Sepals 2; petals 4; lvs dentate to pinnate.
                       19. PAPAVERACEAE (p. 94)
      Sepals 5; petals 5; lvs entire.   **26. CISTACEAE** (p. 206)

38   Trees, shrubs or woody climbers.   *39*
      Herbs.   *50*

39   Trees or shrubs, not climbing.   *40*
      Woody climbers with tendrils; petals cohering at apex and thrown off
        as fl. opens.   **50. VITACEAE** (p. 327)

40   Fls on the middle of lf-like cladodes; true lvs scale-like, colourless
        (*Ruscus*).   **133. LILIACEAE** (p. 962)
      Fls not on cladodes; lvs green.   *41*

41   Fls 3-merous.   *42*
      Fls not 3-merous.   *43*

42   Per. segs yellow or orange; stamens 6; lvs broad.
                      16. BERBERIDACEAE (p. 89)
      Per. segs not yellow; stamens 3; lvs small, linear.
                      93. EMPETRACEAE (p. 619)

43   Lvs small and scale-like; fls numerous in dense spikes.
                     27. TAMARICACEAE (p. 209)
      Lvs not scale-like, not particularly small.   *44*

44   Lvs opposite.   *45*
      Lvs alternate.   *47*

45   Lvs undivided.   **47. CELASTRACEAE** (p. 324)
      Lvs deeply lobed or pinnate.   *46*

46   Lvs palmately lobed.   **43. ACERACEAE** (p. 319)
      Lvs pinnate.   **44. STAPHYLEACEAE** (p. 322)

47   Lvs pinnate.   **42. SIMAROUBACEAE** (p. 319)
      Lvs simple.   *48*

48   Lvs evergreen, coriaceous, persistently densely tomentose beneath; fls
        not greenish.   *49*
      Lvs deciduous, not tomentose, at least when mature; fls greenish.
                      49. RHAMNACEAE (p. 325)

49 Fls cream; stamens more than 5; ovary 5-celled with axile placentas
(*Ledum*). **90. ERICACEAE** (p. 602)
Fls purplish-red; stamens 5; ovary with 3 parietal placentas.
60. PITTOSPORACEAE (p. 463)

50 Sepals 2, petals 5. **31. PORTULACACEAE** (p. 266)
Sepals more than 2, sepals and petals equal in number. 51

51 Lvs modified into pitchers; stigma very large, umbrella-like.
62. SARRACENIACEAE (p. 465)
Lvs not modified into pitchers. 52

52 Fls perigynous with a long tubular or campanulate receptacle.
63. LYTHRACEAE (p. 465)
Fls hypogynous or, if perigynous, receptacle flat to cup-shaped. 53

53 Lvs opposite or whorled. 54
Lvs alternate or all radical. 61

54 Lvs compound or lobed. **39. GERANIACEAE** (p. 300)
Lvs entire. 55

55 Lvs in a single whorl of usually 4 on the stem; fl. solitary, terminal.
134. TRILLIACEAE (p. 982)
Lvs opposite or in numerous whorls. 56

56 Stipules present. 57
Stipules 0. 58

57 Stipules not scarious; submerged aquatics.
29. ELATINACEAE (p. 210)
Stipules scarious; land plants.
30. CARYOPHYLLACEAE (p. 211)

58 Sepals free or united at the base; petals always white. 59
Sepals united to above the middle; petals white, pink or purple. 60

59 Ovary 1-celled with free-central placentation; stamens usually twice as
many as petals, if as many or fewer then lvs narrowly linear or plant
± hairy or sepals scarious margined.
30. CARYOPHYLLACEAE (p. 211)
Ovary 4–5-celled with axile placentation; fertile stamens as many as
petals; lvs obovate to oval; plant glabrous; sepals not scarious.
38. LINACEAE (p. 297)

60 Style long, simple (but stigmas free); placentation parietal.
28. FRANKENIACEAE (p. 210)
Styles free; placentation free-central.
30. CARYOPHYLLACEAE (p. 211)

61 One whorl of per. segs spurred; radical lvs biternate (*Epimedium*).
16. BERBERIDACEAE (p. 89)
Per. segs not spurred; lvs not biternate. 62

62 Lvs ternate (rarely 4-nate) with obcordate or cuneiform and emarginate
lflets. **40. OXALIDACEAE** (p. 314)
Lvs not ternate. 63

63 Sepals and petals 2–3; fls greenish or reddish, in many-fld terminal
panicles. **79. POLYGONACEAE** (p. 542)
Sepals and petals 4–5. 64

2

64   Both floral whorls green and sepal-like (calyx and epicalyx); fls small, with conspicuous hollow receptacle; lvs palmate or palmately lobed (*Alchemilla* and *Aphanes*).   **52. ROSACEAE** (p. 363)
     Petals ± brightly-coloured, never sepal-like.   *65*

65   Sepals and petals 4; stamens 6, rarely 4.
         21. CRUCIFERAE (p. 110)
     Sepals and petals 5; stamens 5 or 10.   *66*

66   Lvs covered with conspicuous red insectivorous glandular hairs.
         61. DROSERACEAE (p. 463)
     Lvs not conspicuously glandular.   *67*

67   Style 1, stigma simple or shallowly lobed; anthers opening by pores.
         91. PYROLACEAE (p. 615)
     Styles, or at least the stigmas, more than 1, free; anthers opening by slits.   *68*

68   Stigmas 5; petals blue, pink or purple, rarely white.   *69*
     Stigmas 2–4, petals white or yellow.   *71*

69   Lvs lobed or pinnate.   **39. GERANIACEAE** (p. 300)
     Lvs entire.   *70*

70   Calyx funnel-shaped or obconic, scarious; lvs all ± radical; fls in heads or panicles.   **95. PLUMBAGINACEAE** (p. 621)
     Sepals free, not scarious or scarious only at the margins; stem lfy; fls in loose corymbs.   **38. LINACEAE** (p. 297)

71   Fls with conspicuous glandular-fimbriate staminodes; lvs ovate, cordate, entire.   **56. PARNASSIACEAE** (p. 457)
     Staminodes 0; lvs not as above.   *72*

72   Stamens 5; procumbent plant; lvs entire, linear-lanceolate; stipules scarious; fls very small (*Corrigiola*).
         30. CARYOPHYLLACEAE (p. 211)
     Stamens 10; fls conspicuous; other characters not as above.
         55. SAXIFRAGACEAE (p. 447)

73   Fl. saccate or spurred at base.   *74*
     Fl. not saccate or spurred.   *76*

74   Lvs much divided; corolla (apparently) laterally compressed; stamens 2, each with 3 branches bearing anthers, not connivent.
         20. FUMARIACEAE (p. 102)
     Lvs simple; corolla not compressed; stamens 5, connivent round the style.   *75*

75   Sepals 5, ± equal, not spurred; petals 5, one spurred; stipules present; fls solitary, axillary; stem not translucent.
         23. VIOLACEAE (p. 188)
     Sepals 3, very unequal, one spurred; petals 3, not spurred; stipules 0; fls in few-fld cymes; stem ± translucent.
         41. BALSAMINACEAE (p. 317)

76   Stamens 8 or more, all, or all but 1, united into a long tube; fls very zygomorphic, the petals ± erect.   *77*
     Stamens free; fls less zygomorphic, petals spreading.   *78*

77   Fl. with upper sepal; anthers opening by pores; stigma tufted.
         24. POLYGALACEAE (p. 198)
     Fl. with upper petal; anthers opening by slits; stigma not tufted.
         51. PAPILIONACEAE (p. 327)

78 Trees; lvs palmate.   **45. HIPPOCASTANACEAE** (p. 322)
    Herbs; lvs not palmate.   *79*

79 Fls in cymes (often umbel-like); ovary 5-lobed with long beak.
                               39. GERANIACEAE (p. 300)
    Fls in racemes; ovary not lobed or 2-lobed, rarely beaked.   *80*

80 Petals fimbriate or lobed; stamens more than 6.
                               22. RESEDACEAE (p. 186)
    Petals entire or emarginate; stamens 6.
                               21. CRUCIFERAE (p. 110)

GROUP C
Petals free, ovary inferior or partly so.

81 Petals numerous.   *82*
    Petals 5 or fewer.   *83*

82 Aquatic plants with floating lvs and fls.
                               17. NYMPHAEACEAE (p. 90)
    Land plants with very succulent lvs.
                               32. AIZOACEAE (p. 270)

83 Petals and sepals 3.   *84*
    Petals and sepals 2, 4 or 5.   *87*

84 Fls zygomorphic.   **140. ORCHIDACEAE** (p. 1009)
    Fls regular.   *85*

85 Both whorls of per. segs petaloid.   *86*
    Outer or both whorls of per. segs sepaloid.
                               123. HYDROCHARITACEAE (p. 936)

86 Stamens 6.   **137. AMARYLLIDACEAE** (p. 998)
    Stamens 3.   **138. IRIDACEAE** (p. 1002)

87 Stamens numerous.   *88*
    Stamens 10 or fewer.   *89*

88 Lvs alternate.   **52. ROSACEAE** (p. 363)
    Lvs opposite.   **57. HYDRANGEACEAE** (p. 458)

89 Submerged aquatic with lvs pinnately divided into filiform segments;
    fls monoecious or polygamous, in terminal spikes projecting above
    surface.   **67. HALORAGACEAE** (p. 483)
    Land plants or, if aquatic, fls hermaphrodite and in umbels.   *90*

90 Trees or shrubs.   *91*
    Herbs.   *96*

91 Woody climber; fls in subglobose umbels, green.
                               73. ARALIACEAE (p. 494)
    Not climbing; fls not in umbels.   *92*

92 Lvs palmately lobed; petals shorter than sepals.
                               59. GROSSULARIACEAE (p. 459)
    Lvs simple, not lobed.   *93*

93 Both perianth whorls petaloid; receptacle long and tubular (*Fuchsia*).
                               66. ONAGRACEAE (p. 469)
    Outer perianth whorl sepaloid.   *94*

94   Calyx-teeth very small; fls in corymbs; carpels 2, each with one ovule.
                              72. CORNACEAE (p. 492)
      Calyx-teeth large; fls not in corymbs; ovules numerous in each carpel.   *95*

95   Lvs opposite, not glandular.   **57. HYDRANGEACEAE** (p. 458)
      Lvs alternate, gland-dotted.   **58. ESCALLONIACEAE** (p. 459)

96   Both perianth whorls green and sepaloid (calyx and epicalyx), or with
      an epicalyx as well as sepals and petals, or with a crown of long
      spines on the receptacle below the calyx; carpels 1 or 2, free from the
      receptacle and thus not truly inferior.   **52. ROSACEAE** (p. 363)
      Inner perianth whorl always petaloid, no epicalyx or crown of spines;
      ovary truly inferior.   *97*

97   Petals 5; styles normally 2, rarely 3.   *98*
      Petals 4 or 2; style simple.   *99*

98   Fls in heads or umbels; stamens 5; ovules one in each carpel.
                              75. UMBELLIFERAE (p. 495)
      Fls not in heads or umbels; stamens 10; ovules numerous.
                              55. SAXIFRAGACEAE (p. 447)

99   Fls deep purple, in umbels subtended by 4 conspicuous white petaloid
      involucral bracts.   **72. CORNACEAE** (p. 492)
      Fls not in umbels; petaloid involucral bracts 0.
                              66. ONAGRACEAE (p. 469)

GROUP D

Petals united, ovary superior.

100   Stamens numerous; sepals petaloid, longer than corolla (*Delphinium*).
                              14. RANUNCULACEAE (p. 56)
      Stamens 8–10, united into a tube; fls strongly zygomorphic.   *101*
      Stamens twice as many as corolla-lobes (8 or 10), not united; fls regular
      or slightly zygomorphic.   *102*
      Stamens as many as or fewer than corolla-lobes.   *103*

101   Lvs entire; fl. with upper sepal; stamens 8.
                              24. POLYGALACEAE (p. 198)
      Lvs ternate; fl. with upper petal; stamens 10.
                              51. PAPILIONACEAE (p. 327)

102   Shrubs or trees; stamens free; lvs not peltate; carpels united.
                              90. ERICACEAE (p. 602)
      Succulent herb; stamens inserted on corolla-tube; lvs peltate; carpels
      free (*Umbilicus*).   **54. CRASSULACEAE** (p. 440)

103   Sepals 2, petals 5; fls regular, not arranged in heads; lvs flat.
                              31. PORTULACACEAE (p. 266)
      Sepals and petals each 2; fls regular, arranged in heads; lvs linear,
      terete.   **132. ERIOCAULACEAE** (p. 961)
      Sepals more than 2, or if 2 then fls zygomorphic (sometimes 2 con-
      spicuous sepal-like bracts occur outside the calyx).   *104*

104   Ovary deeply 4-lobed with one ovule in each lobe.   *105*
      Ovary not 4-lobed.   *106*

105   Lvs alternate.   **103. BORAGINACEAE** (p. 650)
      Lvs opposite.   **111. LABIATAE** (p. 730)

106 Fls zygomorphic.   107
    Fls regular.   112

107 Shrub; stamens free; anthers opening by pores (*Rhododendron*).
        90. ERICACEAE (p. 602)
    Herbs; stamens inserted on corolla; anthers opening by slits.   108

108 Ovary 1-celled, with numerous ovules on a free-central placenta; corolla
    spurred; insectivorous bog or aquatic plants with lvs either all radical
    or else deeply divided into filiform segments.
        108. LENTIBULARIACEAE (p. 725)
    Ovary 2-celled; plants not insectivorous; if corolla spurred then lvs
    neither all radical nor divided into filiform segments.   109

109 Ovules 4; fls spicate; bracts small, c. $\frac{1}{2}$ as long as calyx.
        110. VERBENACEAE (p. 729)
    Ovules numerous; if fls spicate then bracts longer than calyx or else
    lf-like.   110

110 Bracts spine-toothed; corolla 1-lipped; plant very robust.
        109. ACANTHACEAE (p. 728)
    Bracts not spine-toothed; corolla 2-lipped or several-lobed.   111

111 Stamens 4 or 2.   **106. SCROPHULARIACEAE** (p. 672)
    Stamens 5; fls only slightly zygomorphic.   128

112 Stamens fewer than corolla-lobes.   113
    Stamens equalling corolla-lobes in number.   114

113 Shrubs; ovules 4 or fewer.   **98. OLEACEAE** (p. 637)
    Herbs; ovules numerous (*Hebe* is shrubby but the fls are slightly zygo-
    morphic).   **106. SCROPHULARIACEAE** (p. 672)

114 Stamens opposite corolla-lobes.   115
    Stamens alternate with corolla-lobes.   116

115 Ovules numerous; style simple, stigma capitate.
        96. PRIMULACEAE (p. 627)
    Ovule 1; styles or at least stigmas more than 1.
        95. PLUMBAGINACEAE (p. 621)

116 Lvs opposite.   117
    Lvs alternate or all radical.   122

117 Carpels 2, free; style 1, expanded into a ring below the stigma; trailing
    undershrubs with conspicuous, usually violet fls.
        99. APOCYNACEAE (p. 638)
    Carpels united; style without ring below stigma.   118

118 Shrubs or undershrubs.   119
    Herbs.   121

119 Shrub with large deciduous lvs; fls violet, in many-fld panicles.
        97. BUDDLEJACEAE (p. 636)
    Undershrubs with small, evergreen lvs (mountains).   120

120 Creeping; lvs oval or oblong; fls 4–5 mm., pink (*Loiseleuria*).
        90. ERICACEAE (p. 602)
    Cushion-like; lvs spathulate; fls 8–15 mm., white.
        94. DIAPENSIACEAE (p. 620)

121 Aquatic plants with floating lvs on long stalks and fls at surface
    (*Nymphoides*).   **101. MENYANTHACEAE** (p. 649)
    Land plants; lvs sessile.   **100. GENTIANACEAE** (p. 640)

122   Sepals, petals and stamens 4.   *123*
        Sepals, petals and stamens 5, rarely sepals fewer.   *124*

123   Shrub with evergreen spiny lvs.   **46. AQUIFOLIACEAE** (p. 323)
        Herbs; fls in spikes or solitary; lvs all radical.
                112. PLANTAGINACEAE (p. 764)

124   Ovary 3-celled, stigmas 3, or 1 but 3-lobed.   *125*
        Ovary 2-celled, stigmas 2 or 1, not 3-lobed.   *126*

125   Erect herb; lvs pinnate.   **102. POLEMONIACEAE** (p. 650)
        Cushion-like undershrub; lvs simple.
                94. DIAPENSIACEAE (p. 620)

126   Ovules 4 or fewer; twining or prostrate herbs; lvs cordate or hastate at
        base; corolla shallowly lobed.
                104. CONVOLVULACEAE (p. 665)
        Ovules numerous; ± erect herbs or woody climbers; corolla-lobes con-
        spicuous.   *127*

127   Aquatic or bog plants; lvs orbicular or ternate; corolla fringed.
                101. MENYANTHACEAE (p. 649)
        Land plants; lvs not orbicular nor all ternate (if some ternate, woody
        climber); corolla not fringed.   *128*

128   Fls numerous, in terminal spikes or racemes (sometimes aggregated into
        panicles); corolla-tube very short; stamens spreading (*Verbascum*).
             106. SCROPHULARIACEAE (p. 672)
        Fls solitary or in cymes (sometimes scorpioid); corolla-tube long, or if
        short anthers connivent.   **105. SOLANACEAE** (p. 668)

GROUP E

Petals united, ovary inferior.

129   Stamens 8–10 or 4–5 with filaments divided to the base.   *130*
        Stamens 5 or fewer, filaments not divided.   *131*

130   Herb; fls in heads, green; lvs ternate.   **117. ADOXACEAE** (p. 792)
        Low shrubs or prostrate creeping undershrubs; fls pink or white, not
        in heads; lvs simple (*Vaccinium*).   **90. ERICACEAE** (p. 602)

131   Fls in heads surrounded by an involucre; herbs (rarely slightly woody).   *132*
        Fls not in heads, or if in heads then with 2 bracts only and plant a woody
        climber.   *135*

132   Anthers coherent into a tube round the style.   *133*
        Anthers free.   *134*

133   Ovules numerous; calyx-lobes conspicuous, green; fls blue (*Jasione*).
                113. CAMPANULACEAE (p. 767)
        Ovule 1; calyx represented by hairs or scales; fls rarely blue.
                120. COMPOSITAE (p. 800)

134   Ovules numerous; corolla-lobes long and narrow, longer than tube.
                113. CAMPANULACEAE (p. 767)
        Ovule 1; corolla-lobes much shorter than tube.
                119. DIPSACACEAE (p. 796)

135   Lvs in whorls of 4 or more; fls regular; petals 4.
                115. RUBIACEAE (p. 775)
        Lvs not in whorls; fls zygomorphic or, if regular, petals 5.   *136*

136 Anthers coherent into a tube round the style; fls zygomorphic, in
     terminal racemes.   **114. LOBELIACEAE** (p. 774)
     Anthers 2, free, the pollen cohering in pollinia; fls zygomorphic, in
     terminal racemes or spikes.   **140. ORCHIDACEAE** (p. 1009)
     Anthers free; fls regular or, if zygomorphic, not in racemes.   137
137 Herb; climbing by tendrils.   **76. CUCURBITACEAE** (p. 531)
     Herbs, shrubs or woody climbers; tendrils 0.   138
138 Lvs opposite.   139
     Lvs alternate.   140
139 Stamens 4 or 5; usually shrubs or woody climbers; if herbs either
     prostrate and creeping or with lf-like stipules.
                                116. CAPRIFOLIACEAE (p. 786)
     Stamens 1–3; herbs, ± erect and without lf-like stipules.
                                118. VALERIANACEAE (p. 792)
140 Stamens opposite corolla-lobes; stigmas capitate; fls white (*Samolus*).
                                96. PRIMULACEAE (p. 627)
     Stamens alternating with corolla-lobes; stigmas 2–5; fls normally blue
     or purple.   **113. CAMPANULACEAE** (p. 767)

GROUP F

Perianth entirely petaloid or in several series, the inner petaloid.

141 Stamens numerous.   142
     Stamens 12 or fewer, or fls female.   144
142 Aquatic plants with floating lvs and fls.
                                17. NYMPHAEACEAE (p. 90)
     Succulent prostrate plant with trigonous lvs.
                                32. AIZOACEAE (p. 270)
     Lvs neither floating nor trigonous.   143
143 Carpels free, rarely united and then per. segs numerous.
                                14. RANUNCULACEAE (p. 56)
     Carpels united; petals c. 4; sepals 2, falling as fl. opens.
                                19. PAPAVERACEAE (p. 94)
144 Fls crimson, in ovoid heads without an involucre; lvs pinnate (*San-
     guisorba*).   **52. ROSACEAE** (p. 363)
     Fls not in heads or, if so, with an involucre.   145
145 Ovary superior.   146
     Ovary inferior or fls male.   158
146 Per. seg. 1, lateral, bract-like; aquatic plant.
                                126. APONOGETONACEAE (p. 940)
     Per. segs more than 1, but often joined below.   147
147 Perianth strongly zygomorphic, spurred or saccate at base; stamens 2,
     each with 3 anther-bearing branches; lvs much divided (sepals
     present, 2, but bract-like and soon falling).
                                20. FUMARIACEAE (p. 102)
     Perianth regular or somewhat zygomorphic, and then not spurred nor
     saccate.   148
148 Shrubs.   149
     Herbs.   152
149 Fls borne on the surface of lf-like cladodes; true lvs small and scale-like
     (*Ruscus*).   **133. LILIACEAE** (p. 962)
     Fls not on cladodes.   150

150   Per. segs 4, continued below into a coloured or green receptacle.
                    64. THYMELAEACEAE (p. 467)
     Per. segs 6 or more, free.   *151*

151   Low heath-like shrub with inconspicuous axillary fls (if per. segs 8, purple, in 2 differing whorls, see *Calluna* in Ericaceae).
                    93. EMPETRACEAE (p. 619)
     Tall shrub with yellow fls in racemes or panicles.
                    16. BERBERIDACEAE (p. 89)

152   Lvs ternate; one whorl of per. segs spurred (*Epimedium*).
                    16. BERBERIDACEAE (p. 89)
     Lvs simple; per. segs not spurred.   *153*

153   Per. segs 5.   *154*
     Per. segs 6, rarely 4.   *156*

154   Stigma 1; small herb; stipules 0 (*Glaux*).
                    96. PRIMULACEAE (p. 627)
     Stigmas more than 1; stipules present, scarious or, if 0, a large herb.   *155*

155   Stigmas 2–3; stipules sheathing, scarious.
                    79. POLYGONACEAE (p. 542)
     Stigmas c. 10, carpels free or nearly so; stipules 0.
                    35. PHYTOLACCACEAE (p. 289)

156   Stamens 9; ovules scattered over the whole inner surface of carpels; aquatic.   **122. BUTOMACEAE** (p. 935)
     Stamens 6, rarely 4; ovules not scattered.   *157*

157   Lvs cordate at base; infl. subtended by a bract resembling a lf-sheath; aquatic.   **135. PONTEDERIACEAE** (p. 983)
     Lvs not cordate; bracts 0 or not as above; land or marsh plants.
                    133. LILIACEAE (p. 962)

158   Trees or shrubs; calyx present but very small and rim-like or with minute teeth. See 129 (Group E).
     Herbs.   *159*

159   Lvs in whorls of 4 or more.   **115. RUBIACEAE** (p. 775)
     Lvs not in whorls.   *160*

160   Fls in heads surrounded by a common involucre.   *161*
     Fls not in heads, though sometimes shortly stalked in compact umbels.   *162*

161   Stamens free; fls hermaphrodite.   **119. DIPSACACEAE** (p. 796)
     Anthers cohering in a tube round the style, or fls unisexual.
                    120. COMPOSITAE (p. 800)

162   Per. segs 3, or perianth with a long tube swollen below and a unilateral entire limb; lvs ± orbicular, cordate, entire.
                    77. ARISTOLOCHIACEAE (p. 533)
     Per. segs 5 or 6; lvs not as above.   *163*

163   Per. segs 5; fls small; ovules 1 or 2.   *164*
     Per. segs 6; fls large, few, solitary or in simple infl.; ovules numerous.   *167*

164   Fls in simple cymes; lvs alternate, narrow-linear, small.
                    71. SANTALACEAE (p. 491)
     Fls in umbels or superposed whorls or, if in cymes, lvs opposite.   *165*

165   Stamens 5; per. segs free; fls in umbels or superposed whorls; lvs alternate.   *166*
     Stamens 1–3; per. segs united; fls in cymes or panicles; lvs opposite.
                    118. VALERIANACEAE (p. 792)

166 Stems erect or if creeping and rooting at nodes then lvs not peltate; fls in umbels; inner layer of fr.-wall not woody.
75. UMBELLIFERAE (p. 495)
Stems creeping and rooting at nodes; lvs peltate; fls in whorls, often superposed; inner layer of fr.-wall woody.
74. HYDROCOTYLACEAE (p. 495)

167 Stamens 6. **137. AMARYLLIDACEAE** (p. 998)
Stamens 3. **138. IRIDACEAE** (p. 1002)

GROUP G

Trees or shrubs; perianth sepaloid or 0.

168 Parasitic on the branches of trees; lvs opposite, obovate or oblong, thick, evergreen, leathery; stems green.
70. LORANTHACEAE (p. 490)
Not as above. *169*

169 Climbers. *170*
Not climbing. *171*

170 Climbing by tendrils; deciduous. **50. VITACEAE** (p. 327)
Climbing by roots; evergreen. **73. ARALIACEAE** (p. 494)

171 Fls borne on the surface of flattened evergreen lf-like cladodes; true lvs colourless, scale-like (*Ruscus*). **133. LILIACEAE** (p. 962)
Fls not on cladodes; lvs green. *172*

172 Lvs opposite or subopposite. *173*
Lvs alternate. *177*

173 Lvs evergreen, thick, coriaceous, entire; styles 3.
48. BUXACEAE (p. 324)
Lvs deciduous; styles 4, 2 or 1. *174*

174 Fls in catkins. **89. SALICACEAE** (p. 578)
Fls not in catkins. *175*

175 Lvs pinnate; perianth 0; stamens 2 (*Fraxinus*).
98. OLEACEAE (p. 637)
Lvs simple; perianth present; stamens 4 or more. *176*

176 Lvs palmately lobed; unarmed. **43. ACERACEAE** (p. 319)
Lvs simple; thorny. **49. RHAMNACEAE** (p. 325)

177 Lvs evergreen, small (less than 10×2 mm.), dense, oblong or linear, entire; shrubs to 1 m. or less. *178*
Lvs relatively large (longer or broader), not particularly dense, usually deciduous and if evergreen 3 cm. or more. *179*

178 Procumbent; stamens 3; stigmas 6–9; lvs coriaceous; moors, etc.
93. EMPETRACEAE (p. 619)
Erect; stamens 5; stigmas 2; lvs fleshy; maritime (*Suaeda*).
34. CHENOPODIACEAE (p. 271)

179 Lvs pinnate (present at flowering time).
84. JUGLANDACEAE (p. 567)
Lvs simple (sometimes 0 at flowering time). *180*

180 Lvs palmately lobed, present at fl.-time. *181*
Lvs not lobed, sometimes 0 at fl.-time. *182*

181   Bark coming off in flakes; lf-lobes triangular; fls in globose heads.
                                      53. PLATANACEAE (p. 440)
      Bark not flaking; lf-lobes obovate; fls inside a pear-shaped axis.
                                       83. MORACEAE (p. 566)

182   Fls, at least the male, in catkins or in tassel-like heads on long pendulous
      peduncles.   *183*
      Fls not in catkins nor in peduncled heads.   *188*

183   Fls dioecious; perianth 0; fls always solitary in each bract.   *184*
      Fls monoecious, though usually in separate infl.; perianth present at
      least in fls of one sex.   *186*

184   Scales of catkins fimbriate or lobed at apex; fls of both sexes with a
      cup-like disk; ovules numerous (*Populus*).
                                      89. SALICACEAE (p. 578)
      Scales of catkins entire; disk 0.   *185*

185   Ovules numerous; lvs without resin glands, not aromatic when crushed;
      fls of both sexes without bracteoles but with nectaries at the base,
      placed above or below the fl.; stamens with long filaments (*Salix*).
                                       89. SALICACEAE (p. 578)
      Ovule 1; lvs dotted with resin glands, strongly aromatic when crushed;
      male fl. without nectaries or bracteoles, female fl. with 2 lateral
      bracteoles; filaments short.   **85. MYRICACEAE** (p. 568)

186   Fls of both sexes with perianth; styles 3 or more; fr. large and nut-like
      partly or completely enclosed in a hard cup or shell.
                                       88. FAGACEAE (p. 573)
      Perianth present in one sex only; styles 2; fr. small or large and nut-like;
      cup, if present, papery or lf-like.   *187*

187   Male fls 3 to each bract; perianth present; fr. small, in the axils of the
      accrescent bracts which persist till maturity and form cone-like
      structures.   **86. BETULACEAE** (p. 569)
      Male fls solitary in each bract; perianth 0; fr. not borne in cones, sur-
      rounded by a papery or lf-like cup formed from the bracts.
                                       87. CORYLACEAE (p. 572)

188   Lvs and twigs densely covered with silvery or brown peltate scales;
      fls very small, dioecious; male with 2 free per. segs; female with
      tubular perianth with 2 small lobes at apex.
                                       65. ELAEAGNACEAE (p. 468)
      Plant without peltate scales; fls hermaphrodite; per. segs 4 or more.   *189*

189   Deciduous trees; fls in sessile clusters, appearing before the lvs; perianth
      ± campanulate, the stamens inserted at its base; styles 2.
                                       82. ULMACEAE (p. 562)
      Evergreen shrub; fls in short stalked racemes; perianth with long
      cylindric tube, the stamens inserted high on the tube; style 1.
                                   64. THYMELAEACEAE (p. 467)

GROUP H

Herbs; perianth sepaloid or 0.

190   Perianth 0 or represented by scales or bristles, minute in fl. but some-
      times elongating in fr.; the fls in the axils of specialized chaffy bracts
      which are usually arranged along the rhachis of spikelets, sometimes

themselves aggregated into compound infl.; lvs always ±linear and
grass-like, sheathing below.   *191*
Perianth present or, if minute or 0, fls not arranged in spikelets nor the
bracts chaffy; lvs various.   *192*

191   Fls with a bract above and below; lvs ±jointed at the junction with
the sheath, usually with a prominent projecting ligule; sheaths usually
open; stems terete or flattened, usually with hollow internodes.
                   146. GRAMINEAE (p. 1115)
Fls with a bract below only; lvs not jointed at the junction with the
sheath; ligule, if present, not projecting; sheaths usually closed; stems
often trigonous; internodes nearly always solid.
                   145. CYPERACEAE (p. 1057)

192   Aquatic plants; lvs submerged or floating; infl. sometimes rising above
the surface of the water.   *193*
Land plants or, if aquatic, with stiffly erect stems and with lvs as well
as fls rising above the surface of the water.   *206*

193   Lvs divided into numerous filiform segments.   *194*
Lvs entire or toothed.   *195*

194   Lvs pinnately divided; fls in a terminal spike (bracts sometimes lf-like).
               67. HALORAGACEAE (p. 483)
Lvs dichotomously divided; fls solitary, axillary.
           18. CERATOPHYLLACEAE (p. 94)

195   Fls solitary in the axils of 'petaloid bracts'.
           126. APONOGETONACEAE (p. 940)
Fls in a spike surrounded by a petaloid spathe.
             141. ARACEAE (p. 1050)
Without petaloid bracts or spathe.   *196*

196   Fls in spikes, sometimes short and subumbellate and then carpels long-
stalked in fr.   *197*
Fls in heads; heads on a long scape or forming a compound infl.   *199*
Fls axillary, solitary or in few-fld clusters.   *200*

197   Fls monoecious, arranged on one side of a flattened spadix; perianth 0;
marine.   **127. ZOSTERACEAE** (p. 941)
Fls hermaphrodite, arranged all round or on 2 sides of a terete rhachis;
fresh or brackish water but not truly marine.   *198*

198   Per. segs 4; carpels remaining sessile; usually fresh water.
          128. POTAMOGETONACEAE (p. 942)
Perianth 0; fruiting carpels on long stalks; brackish pools and ditches.
            129. RUPPIACEAE (p. 959)

199   Fls hermaphrodite; heads few-fld, arranged in cymes.
             136. JUNCACEAE (p. 983)
Fls monoecious, the 2 sexes mixed in the same many-fld head; head
solitary on a long scape.   **132. ERIOCAULACEAE** (p. 961)
Fls monoecious, the 2 sexes in separate many-fld heads, the heads
arranged in a raceme; stems lfy.
           143. SPARGANIACEAE (p. 1054)

200   Female fls with very long filiform stalk raising them to the surface of
the water.   **123. HYDROCHARITACEAE** (p. 936)
Tube and pedicel short or 0.   *201*

201   Carpels 2–6, free; lvs narrowly linear, quite entire, not whorled.
                    130. ZANNICHELLIACEAE (p. 960)
     Carpels united or 1 only; lvs broader or, if narrowly linear, denticulate
       or whorled.   202

202   Perianth with 4–6 segments; stamens 4 or more.   203
     Perianth 0, or entire, or with 2 segments; stamen 1.   204

203   Per. segs 4; ovary inferior; lvs ovate (*Ludwigia*).
                         66. ONAGRACEAE (p. 469)
     Per. segs 6; ovary superior; lvs obovate (*Peplis*).
                         63. LYTHRACEAE (p. 465)

204   Lvs in whorls of 8 or more; fls hermaphrodite; style 1.
                         68. HIPPURIDACEAE (p. 486)
     Lvs opposite or in whorls of 3; fls unisexual; styles 2–3.   205

205   Lvs narrow-linear with sheathing base, spinulose-serrulate (sometimes
       minutely so), apex acute; ovary terete, not lobed.
                         131. NAJADACEAE (p. 961)
     Lvs (at least the upper) usually spathulate; if all linear, entire and with
       an emarginate apex; base not sheathing; ovary flattened, 4-lobed.
                      69. CALLITRICHACEAE (p. 487)

206   Twining plants; fls unisexual.   207
     Not climbing or, if climbing, fls hermaphrodite.   208

207   Lvs opposite, palmately lobed; per. segs 5.
                         81. CANNABIACEAE (p. 561)
     Lvs alternate, cordate, entire; per. segs 6.
                       139. DIOSCOREACEAE (p. 1008)

208   Lvs linear, ±grass-, rush-, or iris-like; plants of wet places.   209
     Lvs not linear or, if so, small and not at all grass-like.   214

209   Fls monoecious, the male and female in separate infl., or parts of the
       same infl.   210
     Fls hermaphrodite.   211

210   Fls in globose heads, the male and female in separate heads.
                    143. SPARGANIACEAE (p. 1054)
     Fls in dense cylindrical spikes, male above, female below.
                         144. TYPHACEAE (p. 1056)

211   Fls in dense spikes borne laterally on a flattened lf-like stem (*Acorus*).
                         141. ARACEAE (p. 1050)
     Infl. not as above.   212

212   Carpels united only at extreme base; fls in racemes.
                   124. SCHEUCHZERIACEAE (p. 939)
     Carpels ±completely united.   213

213   Fls in spikes; perianth herbaceous.
                     125. JUNCAGINACEAE (p. 939)
     Fls not in spikes or racemes; perianth scarious.
                         136. JUNCACEAE (p. 983)

214   Lvs compound.   215
     Lvs simple or 0.   219

215   Fls in compound umbels (*Angelica*).
                         75. UMBELLIFERAE (p. 495)
     Fls not in compound umbels.   216

216   Fls in heads.   217
      Fls in racemes, panicles or cymes.   218

217   Lvs simply pinnate; style 1 (rarely 2); stamens 4 or numerous.
                   52. ROSACEAE (p. 363)
      Lvs ternate (sometimes 2 or 3 times); styles 3–5; stamens apparently
        8–10 (4 or 5 with filaments divided to base).
                   117. ADOXACEAE (p. 792)

218   Stamens numerous; no epicalyx.
                   14. RANUNCULACEAE (p. 56)
      Stamens 4 or 5 (rarely 10); epicalyx present.
                   52. ROSACEAE (p. 363)

219   Juice milky; infl. compound, the branches dichotomously branched,
      ending in fl.-like cyathia (*Euphorbia*).
                   78. EUPHORBIACEAE (p. 534)
      Juice not milky; infl. not as above.   220

220   Infl. a dense spike surrounded by a pale green or yellow spathe.
                   141. ARACEAE (p. 1050)
      Infl. not as above; spathe 0.   221

221   Lvs 0; stems green and succulent, jointed; perianth flush with the stem;
      salt-marsh plants (*Salicornia*).
                   34. CHENOPODIACEAE (p. 271)
      Lvs obvious, green; stems not succulent.   222

222   Lvs alternate or all radical (rarely the lower opposite).   223
      Lvs all opposite or whorled.   233

223   Stamens numerous; carpels free; perianth of 5 large segments with a
      whorl of honey-lvs within (*Helleborus*).
                   14. RANUNCULACEAE (p. 56)
      Stamens 12 or fewer; carpels united or 1 only; fls small or per. segs 3;
      honey-lvs 0.   224

224   Stamens 8 or fewer; fls small.   225
      Stamens 12; per. segs 3, rather large, brown within (*Asarum*).
                   77. ARISTOLOCHIACEAE (p. 533)

225   Stipules ± scarious, united into a sheath.
                   79. POLYGONACEAE (p. 542)
      Stipules free or 0.   226

226   Lvs large and rhubarb-like, all radical; fls in dense many-fld spikes from
      the base, much shorter than the lvs (*Gunnera*).
                   67. HALORAGACEAE (p. 483)
      Lvs not rhubarb-like; fls not in basal spikes.   227

227   Stamens twice as many as per. segs; lvs reniform, cordate (*Chryso-
      splenium*).   **55. SAXIFRAGACEAE** (p. 447)
      Stamens as many as per. segs or fewer; lvs not reniform nor cordate.   228

228   Stipules lf-like; perianth of 4 segments with an epicalyx of 4 segments
      outside; lvs palmately lobed (*Aphanes* and *Alchemilla*).
                   52. ROSACEAE (p. 363)
      Stipules very small or 0; perianth without epicalyx.   229

229   Ovary inferior; fls in simple bracteate cymes.
                   71. SANTALACEAE (p. 491)
      Ovary superior; fls in simple racemes without bracts or in panicles.   230

230 Infl. a simple raceme (*Lepidium*). **21. CRUCIFERAE** (p. 110)
Infl. branched. *231*

231 Styles 2 or more, free or united below; stigmas simple; fls mostly
5-merous. *232*
Style 1; stigma feathery, tufted; fls 4-merous (*Parietaria*).
80. URTICACEAE (p. 559)

232 Perianth herbaceous. **34. CHENOPODIACEAE** (p. 271)
Perianth scarious. **33. AMARANTHACEAE** (p. 270)

233 Lvs toothed or lobed. *234*
Lvs entire. *237*

234 Fls hermaphrodite; stems creeping or decumbent. *235*
Fls unisexual; aerial stems erect. *236*

235 Ovary inferior, not lobed; styles 2; fls in dichotomous cymes (*Chrysosplenium*). **55. SAXIFRAGACEAE** (p. 447)
Ovary superior, 5-lobed, prolonged into a long beak bearing 5 stigmas;
fls solitary or very few on long axillary peduncles (*Erodium*).
39. GERANIACEAE (p. 300)

236 Plant with stinging hairs; per. segs 4 or 2; stamens 4; style 1; stigmas
feathery (*Urtica*). **80. URTICACEAE** (p. 559)
Plant without stinging hairs; per. segs 3; stamens 9 or more; styles 2,
simple (*Mercurialis*). **78. EUPHORBIACEAE** (p. 534)

237 Perianth of 2 segments or obscurely 2-lobed or 0; stamen 1 (plants
± aquatic). *238*
Perianth of 3 or more segments; stamens 3 or more. *239*

238 Lvs whorled; fls hermaphrodite; style 1.
68. HIPPURIDACEAE (p. 486)
Lvs opposite; fls monoecious; styles 2.
69. CALLITRICHACEAE (p. 487)

239 Perianth of 4 or more segments. *240*
Perianth of 3 segments (*Koenigia*). **79. POLYGONACEAE** (p. 542)

240 Ovary inferior; style 1; per. segs 4 (*Ludwigia*).
66. ONAGRACEAE (p. 469)
Ovary superior. *241*

241 Per. segs 6 or 12 inserted on a campanulate receptacle; style 1; plant
± aquatic; lvs obovate (*Peplis*). **63. LYTHRACEAE** (p. 465)
Per. segs 4 or 5, usually free (if on a campanulate receptacle lvs linear);
styles 2 or more, free; land plants.
30. CARYOPHYLLACEAE (p. 211)

GROUP J

Herbs without chlorophyll, lvs scale-like.

242 Fls zygomorphic. *243*
Fls regular. *244*

243 Per. segs free. **140. ORCHIDACEAE** (p. 1009)
Per. segs united into a tubular corolla.
107. OROBANCHACEAE (p. 718)

244 Erect saprophytes. **92. MONOTROPACEAE** (p. 618)
Twining parasites (*Cuscuta*). **104. CONVOLVULACEAE** (p. 665)

SIGNS AND ABBREVIATIONS

agg.	aggregate, incl. 2 or more spp. which resemble each other closely.
C.	central.
c.	about (circa).
Ch.	Chamaephyte; see p. 1196.
f.	forma, *filius*.
ff.	fragments (of chromosomes).
fl.	flower, flowering time; plural fls.
-fld	-flowered.
fr.	fruit, fruiting.
G.	Geophyte; see p. 1196.
Germ.	time of germination.
H.	Hemicryptophyte; see p. 1196.
Hel.	Helophyte; see p. 1196.
Hyd.	Hydrophyte; see p. 1196.
incl.	including.
infl.	inflorescence, inflorescences.
lf	leaf; plural lvs.
lfless	leafless.
lflet	leaflet.
lfy	leafy.
M.	Microphanerophyte; see p. 1196.
MM.	Mega- or Mesophanerophyte; see p. 1196.
N.	Nanophanerophyte; see p. 1196.
0	absent.
per. seg.	perianth segment.
p.p.	*pro parte.*
Rep. B.E.C.	See Bibliography under *The Botanical Society and Exchange Club of the British Isles.*
sp.	species; plural spp.
ssp.	subspecies; plural sspp.
Th.	Therophyte; see p. 1196.
var.	variety.
×	Preceding the name of a genus or sp. indicates a hybrid.
±	more or less.
*	Preceding the name of a sp. or genus indicates that it is certainly introduced.
$2n$	The diploid chromosome number; when the number is followed by an asterisk it indicates that it refers to British material.
μ	1/1000 mm. (micron).

Measurements without qualification (e.g. lvs 4–7 cm.) refer to lengths; lvs 4–7 × 1–2 cm. means lvs 4–7 cm. long and 1–2 cm. wide. Measurements or numbers enclosed in brackets (e.g. lvs 4–7(–10) cm.) are exceptional ones outside the normal range. Numbers and letters in the accounts of the distributions of spp. (e.g. 102, H30, S.) refer to the total number of vice-counties from which the plant has been recorded in Great Britain (112 vice-counties), Ireland (Hibernia, 40 vice-counties) and the Channel Is. (Sarnia). Thus 102, H30, S means that the plant has been recorded from 102 out of 112 vice-counties in Great Britain, 30 out of 40 in Ireland and from the Channel Is. For further information see Druce, G. C. (1932). In the body of the book families are printed in bold capitals, genera and subgenera with a large initial capital followed by small capitals, and species in bold upper and lower case. Passages in italics indicate distinguishing features of the plant being described.

PTERIDOPHYTA

1. LYCOPODIACEAE

Herbs. Lvs small. Sporangia all alike, borne near the base of the upper surface of sporophylls, which vary from being like the foliage lvs and arranged among them to being strongly differentiated from the foliage lvs and arranged in terminal cones. Sporangia unilocular, compressed, dehiscing by a split; spores numerous. Prothallus very different in different spp., always with endotrophic mycorrhiza, either (i) without chlorophyll, massive and subterranean; or (ii) with a massive conical subterranean base and an apical aboveground green portion, lobed or not; or (iii) in epiphytic spp. much branched, without chlorophyll, growing in dead bark; antheridia in the centre of the apical part of the prothallus, containing numerous biciliate spermatozoids; archegonia in a ring round the antheridia.

Two genera, the following and the monotypic *Phylloglossum* from Australia and New Zealand.

1. Lycopodium L.

Stems long, dichotomously branched, ± densely clothed with small lvs. About 180 spp., cosmopolitan.

1	Stems suberect, rooting at the base only; sporophylls not differing from the lvs and not arranged in cones.	**1. selago**
	Stems creeping and rooting; sporangia borne in terminal cones.	*2*
2	Cones 1–2(–3) on long peduncles which bear distant appressed scale-like lvs; lvs with long filiform hair-points.	**4. clavatum**
	Cones sessile at the apices of ordinary shoots; lvs without hair-points.	*3*
3	Lvs 4-ranked, appressed, glaucous.	**5. alpinum**
	Lvs not 4-ranked, spreading, green.	*4*
4	Sporophylls like the foliage lvs, not bordered; stem 5–20 cm., little branched; bogs.	**2. inundatum**
	Sporophylls with broad scarious denticulate margins, differing markedly from the foliage lvs; stem 30–60 cm., moderately branched; moors on mountains.	**3. annotinum**

Subgenus 1. UROSTACHYA Pritzel.

Stems regularly dichotomously branched, rooting at the base only.

1. L. selago L.   'Fir Clubmoss.'

Stems 5–25 cm., *erect* from a decumbent base. Lvs 4–8 mm., suberect to spreading, lanceolate or linear-lanceolate, acute, entire or very minutely serrulate, dull green, often bearing in their axils bud-like gemmae. *Sporangia* borne in the axils of many of the lvs, *not forming a terminal cone* but usually in fertile zones alternating with sterile ones on the stem. Spores ripe 6–8, but not functional, reproduction only by gemmae. $2n=$c. 260*. Chh.

Native. Heaths, moors, mountain grassland, rock ledges and mountain tops, ascending to nearly 4300 ft., usually in open habitats; common in the mountains, very rare and decreasing in lowland areas; throughout the British Is., but absent from a number of counties in S., E. and C. England. 89, H 40.

Arctic (to 81° 43′ N. in Greenland) and north temperate zone (mainly on mountains), south to N.W. Spain, C. Italy, Montenegro, Bulgaria, Himalaya, N. Carolina and Oregon; Madeira, Azores (a ssp.); the sp. or allied forms (sspp.?) also in scattered places in the southern hemisphere.

Subgenus 2. LYCOPODIUM.

Stems with most of the branches apparently monopodial, owing to unequal development, creeping and rooting. Sporophylls forming cones borne terminally on erect branches.

2. L. inundatum L.   'Marsh Clubmoss.'

Stems 5–20 *cm.*, closely prostrate, *sparingly branched. Lvs* 4–6 mm., *spirally inserted* but secund towards the upper side of the stem, linear-subulate, acute, entire, green. Fertile branches 3–10 cm., their lvs spirally arranged, suberect. *Cones* 1–3 cm., *sessile*, solitary; *sporophylls similar to the foliage lvs* but more spreading, toothed, and somewhat broadened at the base. Spores ripe 6–9. $2n = 156^*$. Chh.

Native. Wet heaths in the lowlands, local; from Orkney southward (not Outer Hebrides), very rare in S. Scotland, N.E. England, Wales; Kerry, Cork, Galway, Mayo, Offaly, Wicklow. 61, H5. Europe from Scandinavia and N. Russia (Ladoga-Ilmen and Volga-Kama regions) to Portugal (rare) and N.W. Spain, N. Italy, Roumania, Black Sea region and W. Caucasus (not in Balkans); N. America from Newfoundland and Alaska to Pennsylvania, Idaho and Oregon.

3. L. annotinum L.   'Interrupted Clubmoss.'

Stems 30–60 *cm.*, *moderately branched* with many of the branches ascending. *Lvs* 4–6 mm., *spirally arranged*, denser on the branches than on the main stem, ± spreading, linear-lanceolate, acuminate, *ending in a short stiff point*, minutely serrulate or entire, dull green. Fertile branches 10–25 cm. *Cones* 1·5–3 cm., *sessile*, solitary; *sporophylls ovate, acuminate, with broad scarious denticulate margins*. Spores ripe 6–8. $2n = $ c. 68^*. Chh.

Native. Moors on mountains from 150 to 2700 ft., local; Yorks, Westmorland, Cumberland, Dumfries, Peebles, Arran; Mull, Perth and Angus to Orkney; Caernarvon (extinct). 24. Arctic and north temperate zone (confined to the mountains except in the colder parts), extending south to C. Spain, N. Apennines, Bosnia, Roumania, Caucasus, Himalaya, Oregon, Colorado and Maryland.

4. L. clavatum L.   Stag's-Horn Moss, 'Common Clubmoss'.

Stems 30–100 cm., much-branched, the branches all (except the fertile ones) procumbent. *Lvs* 3–5 mm., spirally arranged, dense, somewhat appressed or incurved, linear, acuminate, *ending in a long white flexuous hair-point*, minutely serrulate, rather bright green. Fertile branches 10–25 cm. *Cones* 1–2(–3), 2–5 cm., *on the ends of long peduncles*; peduncles with distant, pale, appressed, linear-subulate, scale-like lvs; sporophylls ovate with long filiform points, with broad scarious denticulate margins. Spores ripe 6–9. $2n = 68^*$. Chh.

Native. Heaths, moors and mountain grassland ascending to 2760 ft., common in mountain districts, rare and decreasing in lowland areas. Throughout the British Is., but absent from a number of scattered counties. 101,

H 27. Arctic and north temperate zone extending south in Europe to C. Spain, N. Italy, Serbia, Roumania, and the Caucasus; also in many places in the mountains of the tropics and south temperate zone.

5. L. alpinum L.   'Alpine Clubmoss.'

Stems 15–50 cm., much-branched; branches mostly suberect, densely tufted. *Lvs* 2–4 mm., distant on the main stems, appressed and *strongly 4-ranked on the branches*, becoming ± equal, the ventral row well-developed, oblong-lanceolate, acute or acuminate, often with a short broad hyaline point, concave, entire, *glaucous*. Fertile branches 4–7 cm. Cones 1–2 cm., sessile, solitary; sporophylls lanceolate or ovate-lanceolate, acute or acuminate, gradually tapering, with narrow scarious denticulate margins. Spores ripe 6–8. $2n = $ c. 48*. Chh.

Native. Moors, mountain grassland and mountain tops, ascending to 4000 ft., rather common; Devon, Somerset; Wales; Cheshire, Derby and Yorks northwards; Kerry; Dublin, Wicklow; Galway, Sligo, Tyrone and Down northwards. 67, H 13. Arctic and high mountains of north temperate zone south to the Pyrenees, N. Apennines, Bosnia, Bulgaria, Asia Minor, Caucasus, Altai, Japan, British Columbia and Quebec.

L. issleri (Rouy) Lawalrée

L. complanatum auct. brit., *L. alpinum* var. *decipiens* Syme

Like *L. alpinum* but more robust; branches prostrate, flattened not densely tufted. Lvs often green, the ventral row on the branches somewhat smaller than the lateral rows. Sporophylls ovate, suddenly acuminate.

Native. Heaths and moors, very rare; extinct in Hants?; Gloucester and Worcester; N. Devon; N. Wales; Scottish Highlands; Belgium and France to Austria and Poland; perhaps elsewhere.

Further investigation is needed to see if this plant is distinct from *L. alpinum*. Intermediates appear to occur, as do intermediates with *L. complanatum* L. on the Continent.

2. SELAGINELLACEAE

Herbs with long, usually creeping stems producing lfless branches (rhizophores) which bear the roots. Lvs small with a minute ligule at the base, spirally arranged or 4-ranked and of two kinds. Sporangia of two kinds borne near the base of the upper surface of sporophylls forming terminal cones, usually with the megasporangia in the lower, the microsporangia in the upper part. Megaspores (1–)4(–42). Microspores numerous. Male prothallus contained in the microspore until maturity with a vegetative cell and an antheridium containing numerous biciliate spermatozoids. Female prothallus many-celled filling the megaspore and protruding from its split top; archegonia several, at the top of the prothallus. Fertilization occasionally taking place before the shedding of the megaspore.

One genus.

1. SELAGINELLA Beauv.

The only genus. About 700 spp., cosmopolitan.

Subgenus 1. SELAGINELLA.

Lvs all alike. Sporophylls all alike.

1. S. selaginoides (L.) Link   'Lesser Clubmoss.'

S. spinosa Beauv.; *S. spinulosa* A.Br.

Stems 3–15 cm. decumbent, slender, with short sterile and long ascending more robust fertile branches. Lvs 2–4 mm., spirally arranged, spreading or somewhat appressed; lanceolate, acute, spinulose-ciliate. Fertile branches 2–6 cm., suberect, their lvs larger than those of the stems and sterile branches. Cones sessile, solitary, 1–1·5 cm.; sporophylls similar to the lvs but larger; megasporangia occupying the greater part of the cone; microsporangia few, in the upper part of cone, often 0. Spores ripe 6–8. $2n=18*$. Chh.

Native. Damp grassy or mossy ground, mainly on mountains, ascending to 3500 ft., rather common; Merioneth, Cheshire, Derby and Lincoln northwards; Cork; Clare and Wexford northwards. 61, H33. Arctic and north temperate zones of Europe, Asia and America (in the latter almost always in the mountains), south to N. Spain, Italian Alps, Montenegro, Roumania, Caucasus, Angara-Sayan region, Kamchatka, Colorado and New Hampshire.

Subgenus 2. STACHYGYNANDRUM.

Lvs 4-ranked; those of the two ranks on the upper side of the dorsiventral stem appressed and directed towards the stem-apex; those of the two lower ranks larger and spreading laterally. Sporophylls all alike.

S. kraussiana (Kunze) A.Br.

Stems creeping, jointed at the nodes. Lateral lvs c. 2 mm., ovate-lanceolate, acute, somewhat unequal, rounded at base; lvs on upper side of stem scarcely more than 1 mm., very unequal at the base, with a rounded auricle on the outer margin. Cones short, 4-sided; sporophylls ovate, cuspidate, keeled.

Commonly grown in greenhouses, escaped and naturalized in Cornwall, Ireland and perhaps elsewhere. Native of tropical and S. Africa and the Azores.

3. ISOETACEAE

Aquatic or terrestrial perennial heterosporous plants with short stout stems. Roots arising from the 2- or 3-lobed stem-base, slender, dichotomously branched. Stems with 1, rarely 2, rings of meristematic cells producing secondary tissue. Lvs crowded in a dense rosette, subulate or filiform, usually terete or subterete, often tubular and septate, sheathing at base. Ligule present. The first-produced lvs in any season bearing megasporangia, the next microsporangia, and the last sterile. Sporangia sessile, ±embedded in the lf-base below the ligule, usually covered by an indusium formed from the lf-base. Megasporangia traversed by strands of tissue. Outer layers of spore wall impregnated with silica. Spores on germination giving rise to prothalli. Male prothallus (from microspore) of 1 vegetative cell and an antheridium with a 4-celled wall surrounding 2 cells which give rise to 4 spermatozoids. The multi-ciliate spermatozoids liberated by the dehiscence of the spore and the breaking down of the antheridium wall. Female prothallus (from megaspore) many-celled, filling megaspore and bearing archegonia the necks of which protrude from the split top of the megaspore. The young plant developing without a resting stage from the fertilized archegonium.

Two genera and about 70 spp., distributed throughout the world.

1. Isoetes L.

The only genus, except for the monotypic *Stylites* from the Andes of Peru.

1 Plant aquatic, never completely dormant and lfless; stem without persistent
   lf-bases; lvs 4–20 cm., 2–3 mm. wide.   2

   Plant terrestrial, dormant and lfless in summer; stem ±covered by persistent lf-bases; lvs up to c. 3 cm., 1 mm. wide.   **3 histrix**

2 Lvs very stiff; megaspores with short blunt tubercles.   **1. lacustris**
   Lvs rather flaccid; megaspores with long sharp spines.   **2. echinospora**

1. I. lacustris L.   Quill-wort.

A submerged aquatic. Stem without persistent lf-bases. *Lvs* 8–20(–45) cm.,
2–3 mm. wide, subulate, subterete, with 4 longitudinal septate tubes, *stiff*, dark
green. Stomata 0. *Megaspores* yellowish, rarely white, *covered with short
blunt tubercles*. Microspores granulate. Spores ripe 5–7. $2n = $ c. 100*. Hyd.

Native. In lakes and tarns with water poor in dissolved salts, on substrata
of stones with little silt, boulder clay, sand, or rarely thin peat, locally abundant.
42, H18. Mountain districts of Wales; Shropshire, S.E. Yorks, Lake District;
scattered throughout Scotland and Ireland north to Shetland. Iceland to
C. Europe and the eastern Pyrenees.

2. I. echinospora Durieu   Quill-wort.

Similar to *I. lacustris* but usually smaller. *Lvs* 4–12 cm. × 2 mm., *rather flaccid*,
pale green. *Megaspores* white or yellowish, *covered with long sharp fragile
spines*. Spores ripe 5–7. $2n = $ c. 100*. Hyd.

Native. In lakes and tarns, usually on peaty substrata, local. 21, H5.
E. Cornwall, S. Devon, Dorset, Glamorgan, Merioneth, Caernarvon; scattered throughout Scotland from Perth northward to Shetland; Ireland, mainly
in the west. Europe from N. Italy and France northwards; Iceland; Greenland; ?N. America.

3. I. histrix Bory

Terrestrial. *Stem covered with persistent* short blackish *lf-bases* each with two
long points. Lvs 1–3 cm. × 1 mm., ½-terete, dark green, shiny. Stomata
present. *Megaspores ornamented with a regular net-like pattern.* Spores ripe
4–5. Period of vegetative growth 10–4. $2n = 20$*. Hr.

Native. In peaty and sandy places, damp in winter but dry in summer, very
local. Lizard district, W. Cornwall; Channel Is. W. France, Spain, Portugal
and the Mediterranean region to Asia Minor.

4. EQUISETACEAE

Perennial herbs with creeping rhizome, giving rise to aerial stems at intervals.
Stems either all alike, green and assimilating or of two kinds, green and
assimilating sterile stems and fertile stems without chlorophyll. Stem grooved,
simple or branched from near the base, the branches resembling the stem or
with whorls of slender green branches from the nodes; traversed by a central
cavity and with a ring of smaller cavities ('vallecular canals') in the cortex;
a third ring of much smaller 'carinal canals' associated with the protoxylem

of the vascular bundles alternating with the vallecular canals. Lvs very small, usually not green, combined into sheaths above the nodes, the sheaths ending in free teeth, usually of the same number as the grooves on the stem; sheaths of the branches much smaller with fewer teeth. Spores all alike, overlaid by two spiral bands ('elaters') which show hygroscopic movement, numerous, in sporangia borne several together round the under surface of a peltate sporangiophore. Sporangiophores in whorls, closely aggregated together to form a cone terminal on the main stem and occasionally on the branches also. Archegonia and antheridia borne on separate prothalli, the female being larger, or successively on the same prothallus. Prothalli with a cushion-like base with lobed green flat structures arising from the upper surface. Sex organs borne on the upper surface of the cushion. Spermatozoids multiciliate.

A single living genus.

1. Equisetum L.

The only living genus. About 23 spp., almost cosmopolitan, but absent from Australia, New Zealand, etc.

1   Fertile stems present, not green; green sterile stems 0.   *2*
     Sterile stems present (fertile stems present or not, like or unlike the sterile stems).   *3*

2   Sheaths numerous with numerous (20–30) teeth; cone 4–8 cm.
                                             11. telmateia
     Sheaths few (4–6) with 6–12 teeth; cone 1–4 cm.   **10. arvense**

3   Sterile stems with axillary whorls of branches.   *4*
     Sterile stems simple, or with few branches resembling the stems.   *10*

4   Sterile stem whitish, robust (c. 1 cm. diam.); grooves 20–40, fine; branches
       very numerous, regular.   **11. telmateia**
     Sterile stem green, more slender (7 mm. diam. or less, rarely to 12 mm.
       in *E. fluviatile*); grooves less than 20 or if more (*E. fluviatile*) the
       branches relatively few and ceasing several nodes below the stem-apex.   *5*

5   Branches usually again branched; stem sheaths with 3–6 broad subacute
       lobes, fewer than the grooves.   **8. sylvaticum**
     Branches usually simple; stem sheaths with subulate teeth, as many as the
       grooves.   *6*

6   Stem with 10–30 very fine grooves, the ridges between them not pro-
       minent; teeth not ribbed; fertile and sterile stems alike; central hollow
       $\frac{4}{5}$ diam. of stem or more.   **6. fluviatile**
     Stem with 4–20 deep grooves with prominent ridges between; teeth
       ribbed; central hollow less than $\frac{3}{4}$ diam. of stem.   *7*

7   Lowest internode of branches much shorter than stem sheath; fertile and
       sterile stems alike; branches hollow.   *8*
     Lowest internode of branches as long as or longer than stem sheath
       (except sometimes in lower part of stem); fertile and sterile stems
       normally differing; branches solid.   *9*

8   Stem bright green with 4–8 grooves; central hollow less than $\frac{1}{2}$ diam. of
       stem; cone obtuse; common.   **7. palustre**
     Stem dull green with 8–20 grooves; central hollow more than $\frac{1}{2}$ diam.
       of stem; cone apiculate; Lincolnshire.   **3. ramosissimum**

9   Branches mostly 4-angled, teeth of branch sheaths triangular-lanceolate, acuminate; fertile stems dying after fr.   **10. arvense**
        Branches mostly 3-angled, teeth of branch sheaths deltoid, acute; fertile stems continuing to grow and branching after fr.   **9. pratense**

10   Sheaths finally with a blackish band at the top and bottom; teeth caducous, usually quickly so; cone apiculate.   *11*
       Sheaths without blackish band at bottom (though sometimes wholly black); teeth persistent.   *12*

11   Sheaths about as broad as long; internodes somewhat swollen; stems persisting through the winter; widespread.   **1. hyemale**
       Sheaths longer than broad; internodes not swollen; teeth less quickly caducous; stems dying down in autumn; Wicklow and Wexford, Surrey.   **2. × moorei**

12   Stems smooth or slightly rough, dying down in autumn; cone obtuse.   *13*
       Stems very rough, persisting through the winter; cone apiculate.   *14*

13   Stems with 6–8 deep grooves; teeth 1-ribbed; central hollow scarcely larger than the outer ones.   **7. palustre**
       Stems with 10–30 fine grooves; teeth not ribbed; central hollow $\frac{4}{5}$ diam. of stem or more.   **6. fluviatile**

14   Sheaths green with black band at top, loose; teeth finally obtuse, triangular-ovate or triangular-lanceolate, scarious, with broad white margin.   **5. variegatum**
       Sheaths finally black, appressed; teeth acute, subulate, black.   **4. × trachyodon**

Subgenus 1. SCLEROCAULON.

Stomata sunk below the other epidermal cells. Spikes apiculate. Stems all alike (as in *Aestivalia*), usually persisting through the winter.

1. E. hyemale L.   Dutch Rush.

Stems 30–100 cm., erect, 4–6 mm. diam., glaucous-green, *simple, persisting through the winter*; internodes somewhat inflated; *ridges rough with two regular rows of tubercles; grooves* 10–30, *moderate; sheaths* 3–9 mm., *about as long as broad*, soon whitish *with a black band at top and bottom, appressed; teeth as many as the grooves, very quickly caducous*, leaving a crenulate upper edge to the sheath; central hollow c. $\frac{2}{3}$ or more diam. of stem. Cone 8–15 mm. Spores ripe 7–8. $2n = $ c. 216*. Grh.

    Native. Shady stream-banks, etc., ascending to 1750 ft.; from Somerset and Kent to Sutherland, widespread but local in the northern part of its range, very rare in Wales and S. England; scattered over Ireland. 50, H20. Europe from Iceland and N. (not arctic) Russia to C. Spain, Corsica, N. Italy, Crete and the Caucasus; N. and C. Asia; western N. America south to California and New Mexico.

2. E. × moorei Newm.

E. occidentale (Hy) Coste

Differs from *E. hyemale* as follows: *Stems* more slender, *dying down in autumn* at least to near the base; *internodes not inflated*; ridges rough with irregular rows of tubercles or with cross-bands; *sheaths somewhat loose, longer than broad*, remaining green for longer; teeth more persistent, dark brown; central

hollow c. $\frac{2}{3}$ diam. of stem. Spores abortive. Grh. Probably a hybrid of
E. hyemale and some other sp.

Native. Dunes and banks by the sea in Wicklow and Wexford; Surrey
(1 locality, possibly introduced). W. and C. Europe (exact distribution
uncertain owing to confusion with *E. hyemale*).

3. E. ramosissimum Desf.

Stems 50–75 cm., *greyish-green,* ± persistent, *with numerous branches in the
lower half,* rough with scattered tubercles; *grooves* 8–20, moderate; *sheaths*
c. 8 mm., at first green, then brown with a black band at the bottom; teeth
black with narrow white margins and a ± persistent hair-like apex; *central
hollow* $\frac{1}{2}$–$\frac{2}{3}$ *diam. of stem. Branches hollow; lowest internode c.* $\frac{1}{3}$ *length of stem
sheath.* Cone 6–12 mm. Spores ripe 5–8. $2n=$ c. 216*. Grh.

Probably native. Long grass by river near Boston, S. Lincoln; first found
1947. Europe from the Loire, S. Bavaria and C. Russia (Volga-Don region)
southwards and in isolated localities in Brittany, the Netherlands and N.
Germany; Asia; Africa; America.

4. E. × trachyodon A.Br.   'Mackay's Horsetail.'

Stems 30–100 cm., 2–4 mm. diam., erect to decumbent, greyish-green, *simple*
or with a few branches resembling the stem, *without axillary whorls, very
rough,* persisting through the winter; *grooves* 8–12, moderate; *sheaths* 2–4 mm.,
at first green with a black band at the top, *becoming wholly black, appressed;
teeth subulate,* persistent, *usually black,* 4-ribbed; central hollow c. $\frac{1}{3}$ diam. of
stem. Cone c. 5 mm., often sterile or partly so. Spores abortive. $2n=$ c. 216*.
Grh. Probably a hybrid, *E. hyemale* × *variegatum.*

Native. Shady banks of lakes and streams. Aberdeen and Kincardine by
the R. Dee; Perth; widespread but very local in Ireland and absent from the
south-east and most of the centre. 3, H16. Iceland, Norway, Upper Rhineland (and perhaps elsewhere in Europe); S.W. Greenland; northern temperate
N. America.

5. E. variegatum Schleich. ex Web. & Mohr   'Variegated Horsetail.'

Stems 15–60 cm. decumbent or less often erect, to 3 mm. diam., green, *simple
or branched at the base, without axillary whorls,* persisting through the winter;
grooves 4–10, moderate; ridges finely 1-grooved, rough with 2 regular rows
of minute tubercles; *sheaths* c. 2–4 mm., *green with a black band round the
top, rather loose; teeth* as many as the grooves, *scarious, whitish* except for
a blackish centre, *triangular-ovate or triangular-lanceolate,* at first subulate,
the tip falling and leaving an *obtuse* apex, 4-ribbed; central hollow c. $\frac{1}{3}$ diam.
of stem. Cone short, 5–7 mm. Spores ripe 7–8. $2n=$ c. 216*. Grh.

A variable sp., of which several varieties have been described: var. *wilsonii*
(Newm.) Milde, sometimes considered a distinct sp., differs in its tall erect
smoother stems (60–100 cm.) with less angular ridges.

Native. Dunes, river-banks, wet ground on mountains, etc., ascending to
1600 ft. in Kerry. Devon, Somerset, Berks, Wales; Cheshire and Yorks to
Sutherland and Outer Hebrides, very local; widespread in C. Ireland, rare
elsewhere; var. *wilsonii* in Kerry only, usually in shallow water. 34, H21.
Arctic and north temperate zones of Europe, Asia and N. America extending

from 82° 17′ N. in Greenland to the Pyrenees, N. Italy, Thrace, Caucasus, Mongolia, Connecticut and Oregon.

Subgenus 2. EQUISETUM.

Stomata not sunk below the other epidermal cells. Spikes obtuse. Stems dying down in autumn.

Section 1. *Aestivalia* A.Br. Fertile and sterile stems alike, green, produced at the same time. Branches hollow.

6. E. fluviatile L.   'Water Horsetail.'
E. limosum L.; *E. heleocharis* Ehrh.

Rhizome glabrous. *Stems* 50–140 cm., ± erect, 2–12 mm. diam., green, *simple or with irregular whorls of branches in the middle, smooth; grooves* 10–30, *very fine*; sheaths 5–10 mm., green, scarcely loose; *teeth* as many as the grooves, subulate, small (c. 1 mm.), black at least at the tip, *not ribbed*, margins not or scarcely scarious; *central hollow* ⅘ *diam. of stem or more. Branches* ascending, slender, simple, usually 5-angled, lowest internode about as long as stem sheath or shorter; sheaths with 4–5 moderate, subulate, green or blackish teeth. Cone 1–2 cm. Spores ripe 6–7. 2*n*=c. 216*. Hel.

Native. In shallow water at the edges of lakes, ponds and ditches, frequently dominant in these swamp communities; less often in marshes and fens, ascending to 3000 ft. 110, H40, S. Common throughout the British Is. Europe from Iceland and arctic Russia to C. Spain, Italy, Macedonia and the Caucasus; temperate Asia; N. America from Labrador and Alaska to Virginia and Oregon.

E. arvense × fluviatile = E. × litorale Kühlew. ex Rupr.

Stems all alike, resembling those of *E. fluviatile* but more deeply grooved, and with twice the number of green bands, more branches and with loose sheaths. Teeth appressed with minute black apex. Central hollow ½–⅔ diam. of stem. Cone short; spores abortive. Occurs occasionally with the parents, but often in considerable quantity.

7. E. palustre L.   'Marsh Horsetail.'

Rhizome glabrous. Stems 10–60 cm., erect or decumbent, 1–3 mm. diam., green, usually branched, often rather irregularly, but occasionally simple, slightly rough; *grooves* 4–8, *deep*; sheaths 4–12 mm., green, loose; *teeth* 4–8, triangular-subulate, blackish with narrow whitish scarious margins, 1-*ribbed; central hollow small, scarcely larger than the outer ones.* Branches spreading to suberect, often short, simple, 4–5-grooved; *lowest internode much shorter than the stem sheath*; sheaths with 4 short appressed black-tipped teeth. Cone 1–3 cm. Spores ripe 5–7. 2*n*=c. 216*. Grh.

Native. Bogs, fens, marshes and wet heaths, woods and meadows, ascending to 3000 ft. 111, H40, S. Common throughout the British Is. Europe from Iceland and N. (not arctic) Russia to C. Spain and N. Portugal, Sicily, Greece and the Caucasus; temperate Asia; N. America from Newfoundland and Alaska to Connecticut and Oregon.

Section 2. *Subvernalia* A.Br. Fertile and sterile stems produced at the same time. Fertile stems at first unbranched and without chlorophyll, becoming green and branching after spore dispersal and then resembling the sterile stems. Branches solid.

8. E. sylvaticum L.   'Wood Horsetail.'

Sterile stems 10–80 cm., erect, c. 1–4 mm. diam., green, smooth or nearly so; grooves 10–18; sheaths 5–10 mm., green below; *teeth united into 3–6, broad, subacute,* brown *lobes,* each with several ribs; central hollow c. $\frac{1}{2}$ diam. of stem. *Branches drooping at ends,* numerous, *branched* except in small plants, regular, 3–4-grooved; lowest internode longer than stem sheath; sheaths with 3–4 long subulate teeth. Fertile stem 10–40 cm., usually with short branches when the spores are ripe; sheaths loose, numerous, greenish below, brown above with 3–6 broad brown teeth. Cone 1·5–2·5 cm. Spores ripe 4–5. $2n=$ c. 216*. Grh.

Native. Damp woods on acid soils, moors, etc., ascending to 3000 ft., throughout the British Is., common in Scotland, N. England and N. Ireland, becoming very local southwards and absent from several midland and southern counties and from the Channel Is. 103, H37. Europe from Iceland and Scandinavia to N. Spain, N. Italy, Crete and the Caucasus; temperate Asia; N. America from Newfoundland and Alaska to Virginia and Iowa; S. Greenland.

9. E. pratense Ehrh.   'Shady Horsetail.'

Sterile stems 20–60 cm., erect, c. 1–2 mm. diam., green, rough; *grooves 8–20, deep*; sheaths 3–8 mm.; teeth as many as the grooves, brown with a blackish rib, subulate, acute; central hollow c. $\frac{1}{2}$ diam. of stem or rather more. *Branches* spreading or sometimes somewhat drooping, numerous, simple, regular, 3(–4)-*grooved; lowest internode longer than the stem sheath* (sometimes shorter in the lower part of stem); sheaths pale, 3(–4)-toothed; *teeth deltoid, acute.* Fertile stem 10–25 cm., simple or with short branches when the spores are ripe; sheaths loose, numerous, yellowish-white with 10–20 pale teeth, the ribs dark. Cone 1·5–4 cm. Spores ripe 4. $2n=$ c. 216*. Grh.

Native. Grassy stream banks, etc., ascending to over 3000 ft. In scattered localities from Yorks and Westmorland to Orkney, very local and mainly in the east; Fermanagh, Donegal, Antrim. 23, H4. Europe from Iceland and arctic Russia to E. France (Savoy), Tirol, Roumania and the Caucasus; N. and C. Asia; N. America from Nova Scotia and Alaska to New Jersey and Colorado.

Section 3. *Vernalia* A.Br. Fertile stems without chlorophyll, produced earlier than the sterile ones and dying after spore-dispersal. Branches solid.

10. E. arvense L.   'Common Horsetail.'

Rhizome pubescent, with ovoid tubers. Sterile stems 20–80 cm., erect or decumbent, c. 3–5 mm. diam., green, slightly rough; *grooves 6–19, deep*; sheaths 3–8 mm., green; teeth as many as grooves, subulate, acute, green below with blackish tips, 1-ribbed; central hollow less than $\frac{1}{2}$ diam. of stem. *Branches* spreading, numerous, usually simple, regular, (3–)4-*grooved; lowest internode longer than stem sheath*; sheaths pale, 4-toothed, *teeth triangular-lanceolate, acuminate,* somewhat spreading, pale. Fertile stem 10–25 cm., simple, brown; *sheaths* loose, pale brown *with 6–12 darker teeth, few* (4–6), distant. Cone 1–4 cm. Spores ripe 4. $2n=$ c. 216*. Grh.

Native. Fields, hedgebanks, waste places, dune-slacks, etc., ascending to over 3000 ft. 112, H40, S. Common throughout the British Is. Arctic and north temperate zones from 82° 29′ N. in Greenland to S. Spain, Crete, C. China, Virginia, Alabama and California.

11. E. telmateia Ehrh.   'Great Horsetail.'

E. maximum auct.

Rhizome pubescent, often with pyriform tubers. *Sterile stems* 1–2 m., erect, c. 10–12 mm. diam., *dirty white*, smooth; *grooves* 20–40, *fine*; sheaths 1·5–4 cm., ± appressed, pale, blackish above; teeth c. 5 mm. or more, as many as the grooves, blackish, subulate, 2-ribbed; central hollow $\frac{2}{3}$ diam. of stem or more. *Branches* spreading, *numerous*, simple, *regular*, 4-grooved; lowest internode shorter than stem sheath; sheaths short, pale, 4-toothed. Fertile stem 20–40 cm., simple, pale brown; *sheaths* loose, *numerous*, close together, pale brown *with* 20–30 dark *teeth*. Cone 4–8 cm. Spores ripe 4. $2n = $ c. 216*. Grh.

Native. Damp shady banks, etc., ascending to 1200 ft. 86, H40, S. Throughout England, Wales and Ireland, rather local; more local in Scotland and absent from the north-east, extending north to Fife and W. Sutherland; Alderney. Europe from Sweden, Denmark and the Black Sea region southwards; Asia Minor and Caucasus; N. Africa; Azores, Madeira; N.W. America from British Columbia to California.

5. OSMUNDACEAE

Rhizome large, erect, not scaly. Lvs pinnately divided, not jointed to the rhizome, expanded at the base, the expansions covered with glandular hairs; veins free. Sporangia marginal or superficial, all alike and developing simultaneously; indusium 0; annulus consisting of a group of thick-walled cells near the apex, sporangia dehiscing by a slit running from the annulus across the apex. Spores rather numerous (up to 500), green. Prothallus green, cordate, fleshy.

Three genera and about 20 spp., cosmopolitan.

1. OSMUNDA L.

Sporangia marginal on reduced pinnules without chlorophyll or flat blade, the fertile pinnules either occupying the top or middle portion of the lf or the whole lf. Outer lvs sterile.

12 spp., absent from Australasia.

1. O. regalis L.   Royal Fern.

Rhizome short, erect, massive. Lvs 30–300 cm., tufted, 2-pinnate, the outer sterile, the inner with the lower pinnae sterile, the upper fertile (often with a transition region with pinnae with some sterile and some fertile pinnules); blade glabrous, ± lanceolate in outline, in the fertile lvs with c. 2–3 pairs of sterile pinnae and 5–14 pairs of fertile ones markedly decreasing in size upwards; petiole hairy when very young, soon glabrous; sterile pinnae with 5–13 pairs of pinnules, rhachis narrowly winged; pinnules 2–6·5 cm., ± oblong, subobtuse, ± truncate at the base, often with a rounded lobe on the lower side at the base and occasionally shallowly crenately lobed on both sides, minutely and irregularly crenulate-serrulate, veins prominent on both surfaces branching by repeated dichotomy and reaching the margin; fertile pinnules up to 3 cm., 2–4 mm. broad, without blade, densely covered with clusters of brown sporangia. Spores ripe 6–8. $2n=44*$. H.

Native. Fens, bogs, wet heaths and woods, on peaty soil but in well-drained places, ascending to 1200 ft. 86, H38, S. Throughout the British Is., local, especially in the east and now almost extinct in most heavily populated areas owing to the depredations of collectors. Europe from Norway (extreme south, very rare) southwards, eastwards not reaching Russia and rare in Balkans; Asia Minor, Transcaucasia; N. Africa; India, etc.; S. Africa, Madagascar, etc.; eastern N. America from Newfoundland and Saskatchewan southwards; C. and S. America to Uruguay.

6. HYMENOPHYLLACEAE

Rhizome usually creeping. Lvs nearly always thin and translucent, of 1 layer of cells without stomata, entire or divided, with the ultimate segments 1-nerved. Sori marginal on the vein endings, often projecting from the lf; indusium ± cup-like, entire or 2-lipped or 2-valved, surrounding the base or whole of the sorus; sporangia all alike, shortly stalked, developed successively from apex to base; annulus oblique, without definite stomium, the sporangia opening laterally by a long slit. Spores 32–420. Prothallus green, either filamentous or strap-like.

Four genera and about 630 spp., mostly tropical, a few in warm moist parts of the temperate zones, absent from dry areas.

Indusium narrowly campanulate, not valved, receptacle projecting from it as a
  long bristle; pinnae 1–2-pinnatisect; ultimate lobes short; lvs mostly over
  10 cm., rhizome over 1 mm. diam.   1. TRICHOMANES
Indusium 2-valved, receptacle not projecting; pinnae irregularly dichotomously divided, ultimate lobes oblong; lvs rarely reaching 10 cm., usually
  much less; rhizome filiform.   2. HYMENOPHYLLUM

1. TRICHOMANES L.

Lvs thin and translucent. Indusium cylindric to campanulate, sometimes 2-lipped but not 2-valved. Receptacle often exserted. Prothallus filamentous.

About 330 spp., distribution of the family.

1. T. speciosum Willd.   'Bristle Fern.'
T. radicans auct.

Rhizome creeping, c. 1–3 mm. diam., clothed with blackish hairs. Lvs 7–35 cm., ± irregularly 2–3-pinnatisect, persistent for some years; petiole occupying ⅓–½ length, naked, winged above; blade ovate-triangular in outline, dark green, rhachis winged; pinnae ± lanceolate in outline, rhachis winged; pinnules or segments rather irregularly pinnatifid or pinnately lobed; lobes c. 1 mm. or less, 1-veined, entire. Sori projecting from the margins of the upper pinnae; indusium 1–2 mm., narrowly campanulate; receptacle bristle-like exserted. Spores ripe 7–9. $2n=144*$. Chh.

Native. Among shady rocks in places with a very humid atmosphere, ascending to 1500 ft. in Kerry; very rare in Great Britain; N. Wales, Yorks (extinct), ?Westmorland, ?Cumberland, Arran, Argyll (extinct?); more widespread in Ireland and formerly abundant in some places but now rare owing to the depredations of collectors; Kerry to Clare and Waterford; Wicklow (extinct?), Sligo, Fermanagh, Donegal, Tyrone, Antrim. 5, H14. W. French Pyrenees, W. Spain, Portugal (very rare in all); Macaronesia.

2. HYMENOPHYLLUM Sm.

Lvs thin and translucent. *Indusium of 2 valves*. Receptacle included. *Prothallus flat.*

About 300 spp., distribution of the family.

Valves of indusium orbicular, toothed; lf ±flat. **1. tunbrigense**
Valves of indusium ovate, entire; pinnae bent back from the rhachis.
　　　　　　　　　　　　　　　　　　　　　　　　　　　2. wilsonii

1. H. tunbrigense (L.) Sm.　　　　　　　'Tunbridge Filmy Fern.'

Habit moss-like. Rhizome creeping, filiform. *Lvs* 2·5–8(–12) cm., pinnate, the pinnae divided ±dichotomously but very irregularly into oblong segments, ±*flat*, persistent for some years; petiole occupying $\frac{1}{3}$–$\frac{1}{2}$ length, wiry, naked or with a few hairs; blade oblong or ovate-oblong in outline, 12–20 mm. broad, rhachis winged; segments up to 3 mm., oblong, sharply and remotely serrulate, 1-veined, *the vein ceasing slightly below the apex*. Sori marginal on the tips of the segments mostly near the rhachis of the lf; *indusium* c. 1 mm., flattened, *the valves* ±orbicular, *with a wide irregularly and sharply toothed mouth*. Spores ripe 6–7. $2n=26^*$. Chh.

Native. Rocks, tree trunks, etc., in a moist atmosphere, ascending to 1000 ft. in Wales, local but often abundant where it occurs. Cornwall to Somerset; Sussex, Kent (extinct); Hereford; Wales; Lancashire to W. Inverness and Mull, east to Yorks, Peebles and Stirling; throughout Ireland but absent from many areas. 33, H24. N.W. France, Luxembourg (1 loc.), Belgium, Saxony (2 locs), Corsica, Italy (Tuscany), Spain (Navarre); Macaronesia. The same or related spp. in N. America, Australia, etc.

2. H. wilsonii Hook.　　　　　　　'Wilson's Filmy Fern.'

H. peltatum auct., vix Desv.; *H. unilaterale* auct.

Differs from *H. tunbrigense* as follows: Lvs usually narrower and appearing considerably so because the *pinnae* are *bent back from the rhachis*; pinnae usually with fewer and more unilateral segments; *vein reaching apex of segments*. Sorus somewhat projecting; *indusium ovoid*, not flattened, *the valves entire*. Spores ripe 6–7. $2n=36^*$. Chh.

Native. In similar places to *H. tunbrigense*, requiring, in general, less sheltered conditions, and thus commoner, though in some places extending less far east, ascending to 3300 ft. in Kerry; Cornwall, Devon, Stafford, Shropshire (?extinct), Wales, Isle of Man; Lancashire and Yorks to Fair Is. (absent from N.E. Scotland and from several other eastern counties); throughout Ireland but absent from many areas. 52, H27. Faeroes, W. Norway (from c. 63° N. southwards), Normandy, Brittany.

7. POLYPODIACEAE

Of varied habit and lf-shape. Lvs not very thin nor translucent. Sporangia all alike, ripening over a period but not in regular succession (very rarely developing from apex to base), long-stalked, usually grouped into definite sori; annulus vertical, incomplete, with a stomium of thin-walled cells on one side at the base across which the sporangium opens by a transverse slit. Spores few, usually 32 or 64. Prothallus green, flat, cordate.

About 170 genera and 7000 spp., cosmopolitan.

In the following descriptions the degree of division always applies to the best developed (lower) part of the lf or pinna, and 'pinnate' is used throughout although 'pinnatisect' would be more strictly correct. The upper part is almost always less divided. Lengths of pinnae, pinnules and segments apply also to the best developed ones of a lf and lf-lengths (unless otherwise stated) to fertile lvs only. Young plants often differ much in appearance from mature ones and their determination (and that of sterile ferns in general) should not be attempted by the beginner. For a more detailed account (with illustrations) of the British spp. see Hyde & Wade, *Welsh Ferns*.

Many of the British ferns vary considerably and numerous varieties of ± abnormal appearance occur occasionally as isolated individuals. These were formerly much sought after by fern-growers. No attempt is made to cover these in the following descriptions.

1 Lvs densely covered beneath with brown overlapping scales completely
   concealing the lf-surface; small plant with deeply pinnatifid lvs.
                                                          8. CETERACH
   Lvs not or sparsely scaly beneath, the lf-surface clearly visible.   2

2 Lvs entire.   6. PHYLLITIS
   Lvs deeply pinnatifid to several times pinnate.   3

3 Sori on or near the margin of the lower surface of the lf, covered by the
   inrolled indusium-like margin or apparent margin of the lf; lvs not
   simply pinnate.   4
   Sori on the lower surface of the lf, not covered by the margin (the
   indusium may appear to be the margin in *Blechnum* with simply pinnate lvs).   6

4 Segments fan-shaped, recurved margin interrupted.   4. ADIANTUM
   Segments ± oblong, recurved margin continuous along the segment.   5

5 Lvs all alike, solitary from a creeping underground rhizome, usually
   over 30 cm. (–400 cm.).   1. PTERIDIUM
   Plant tufted, fertile lvs with narrower segments than the sterile, 10–30 cm.
                                                          2. CRYPTOGRAMMA

6 Sori a continuous line along either side of the midrib of the linear pinna;
   fertile lvs with much narrower pinnae than the sterile. 5. BLECHNUM
   Sori not as above; fertile lvs not differing markedly from sterile.   7

7 Sori oblong to linear.   8
   Sori ± orbicular.   10

8 Indusium 0; slender plant; Jersey and Guernsey.   3. ANOGRAMMA
   Indusium present.   9

9   Sori oblong, lower margin of the indusium bent in the middle; lvs
    2-pinnate, pinnules oblong.   9. ATHYRIUM
    Sori linear, lower margin of the indusium straight; (if lvs 2-pinnate,
    pinnules not oblong).   7. ASPLENIUM

10 Lvs deeply pinnatifid, the segments toothed but not lobed or pinnatifid;
    sori large, naked, yellow when ripe.   15. POLYPODIUM
    Lvs pinnate, the pinnae lobed, pinnatifid, or 2–3-pinnate; sori brown or
    blackish.   11

11 Lvs solitary from a creeping underground rhizome.   12
    Lvs in a crown at the apex of the rhizome.   14

12 Lvs pinnate with deeply pinnatifid pinnae; indusium reniform or 0.
                                                            14. THELYPTERIS
    Lvs 2-pinnate with deeply pinnatifid pinnae or 3-pinnate.   13

13 Indusium 0.   14. THELYPTERIS
    Indusium flap-like, persistent; Scottish mountains.   10. CYSTOPTERIS

14 Indusium peltate; teeth of lf-segments spinulose.   13. POLYSTICHUM
    Indusium not peltate or 0; teeth not spinulose.   15

15 Indusium consisting of numerous hairs; lf-segments with few or many
    scales beneath; small, very rare mountain plants with pinnate lvs with
    lobed to deeply pinnatifid pinnae.   11. WOODSIA
    Indusium ± continuous or 0; scales confined to rhachis and petiole. 16

16 Indusium flap-like; slender fragile plants; scales on petiole few.
                                                            10. CYSTOPTERIS
    Indusium reniform or 0; robust plants, not fragile; scales on petiole
    many and conspicuous.   17

17 Lvs pinnate with deeply pinnatifid pinnae or 2-pinnate, the pinnules
    oblong, toothed to nearly entire.   18
    Lvs more deeply divided than above.   19

18 Sori large, about midway between the midrib and margin of the segment;
    not smelling of lemon.   12. DRYOPTERIS
    Sori small, near the margin of the segment; lvs smelling of lemon when
    crushed.   14. THELYPTERIS

19 Indusium 0; Scottish mountains.   9. ATHYRIUM
    Indusium present, reniform.   12. DRYOPTERIS

Subfamily 1. PTERIDOIDEAE. Terrestrial. Rhizome with hairs or scales.
Lvs not jointed to the rhizome, nearly always pinnately divided; veins free or
joined without free vein-endings inside the network. Sori on or just within
the lf-margin, usually connecting the ends of the other veins, covered by the
reflexed lf-margin, continuous on a marginal vein, sometimes with an inner
indusium also. Spores without perispore.

1. PTERIDIUM Scop.

Rhizome creeping, *with hairs but without scales. Petiole with numerous vascular
bundles*; veins connected only by the marginal receptacle. Sori continuous
round the margin of the lf-segment. *Inner indusium present.* Sporangia at
first developing from apex to base of the sorus, later irregularly.
    One sp.

1. P. aquilinum (L.) Kuhn   Bracken.

Pteris aquilina L.; *Eupteris aquilina* (L.) Newm.

Rhizome underground, creeping for long distances, stout, tomentose. Lvs (15–)30–180(–400) cm., (2–)3-pinnate, solitary, erect below with the blade bent towards the horizontal, pubescent and with numerous brown scales when young, dying in autumn; petiole to 2 m. and c. 1 cm. diam., about as long as blade, dark and tomentose at base, ½-cylindrical and soon glabrous above; blade ± deltoid in outline; pinnae lanceolate or oblong; segments 5–15 mm., oblong, pectinately arranged, sessile, broad-based, usually subacute, entire or the larger ones lobed at the base, rather thick, subglabrous above, pubescent below, at least on the main veins at maturity. Sori running all round the margins of the segments, both the recurved margin and the inner indusium membranous and ciliate. Spores ripe 7–8. $2n=104$*. Grh.

Native. Woods, heaths, etc., mainly on light acid soils, rare on limestone and on wet peat; the commonest dominant in the field-layer of woods on acid soils, dominant also over considerable areas formerly occupied by acid grass-land or heather, spreading long distances vegetatively and favoured by the grazing of sheep or rabbits, neither of which animals eat it, and by fire; not tolerant of exposure, though ascending to 2000 ft. in sheltered valleys in Scotland, nor of deep shade; common throughout the British Is. 112, H40, S. Cosmopolitan except for temperate S. America and the Arctic, though just passing the Arctic Circle in Norway; divisible into a number of geographical sspp.

Subfamily 2. GYMNOGRAMMOIDEAE. Rhizome with hairs or scales. Lvs not jointed to rhizome. Sori on the lower surface, without indusia but often protected by the reflexed lf-margin, not borne on a vein connecting the other vein-endings (except in *Onychium* Kaulf.). Spores without perispore.

2. CRYPTOGRAMMA R.Br.

Rhizome scaly. Fertile and sterile lvs differing, 2- or more pinnate; veins free. *Sori borne on the apical part of the veins,* ± oblong, *protected by the reflexed, continuous lf-margin.*

Seven spp. North temperate zone, temperate S. America, S. Africa.

1. C. crispa (L.) R.Br. ex Hook.   Parsley Fern.

Allosorus crispus (L.) Röhl.

Rhizome short, creeping or ascending. Lvs densely tufted, outer sterile, inner fertile. Sterile lvs 5–15 cm., 3-pinnatisect, naked except for a few brownish scales at the base of the petiole which is from 1–3 times as long as blade; blade triangular-ovate, pinnate; pinnae 3–7 on each side, usually 2-pin-natisect; segments c. 5–10 mm., obovate, obtuse, cuneate at base, pinnately lobed or toothed. Fertile lvs 10–30 cm. (the blade about the same size as in the sterile lvs, the petiole much longer); blade ovate, 3–4-pinnate; segments oblong-linear, stalked, appearing entire, margins recurved, shallowly sinuately lobed, at first almost meeting and hiding the sori. Sori oblong, at first distinct but appearing to form a continuous band when mature. Spores ripe 6–8. $2n=120$*. H.

Native. Screes, etc., on acid soils on mountains from 300–4000 ft.; Somerset (artificial habitats, very rare); Wales; Cheshire, Derby and Yorks to Caithness and the Outer Hebrides; locally abundant but absent from some areas; very rare in Ireland, in Wicklow and from Louth and Fermanagh northwards. 55, H8. Mountains of Europe from Iceland, Scandinavia and the Urals to the Sierra Nevada (Spain), Corsica, Apennines, Macedonia and the Caucasus; W. Siberia (Ob region).

3. ANOGRAMMA Link

Small annuals but with perennial prothallus. *Rhizome very short with few scales. Fertile and sterile lvs somewhat different*, thin, 2–3 pinnate; *veins free. Sori linear, running along the length of the veins, lf-margin flat.*

Seven spp. scattered through the tropics and south temperate zone, only the following reaching the north temperate zone.

1. A. leptophylla (L.) Link

Gymnogramma leptophylla (L.) Desv.; *Grammitis leptophylla* (L.) Sw.

Rhizome very short, with a few narrow scales when young. Lvs few, slightly hairy when young, soon glabrous; outer sterile, 7 cm. or less, pinnate, pinnae c. 1 cm. long and almost as broad, deeply pinnatifid or almost palmatifid, segments lobed; inner lvs fertile, but not clearly marked off from the sterile ones, 3–20 cm., ovate-oblong, 2–3-pinnate; pinnules or segments c. 5–10 mm., obovate, obtuse, cuneate at base, pinnately lobed or pinnatifid. Sori linear, along the ultimate veins, appearing confluent at maturity. Spores ripe 3–5. $2n=58*$. Th.

Native. Hedgebanks in the Channel Is., rare. Mediterranean region, north to Caucasus, Crimea, Switzerland (S. Tessin) and W. France; Macaronesia; Abyssinia, S. Africa, Madagascar; India; Australia, New Zealand; America from Mexico to Argentina.

4. ADIANTUM L.

Rhizome scaly. Lvs all alike, usually with black glossy petiole and broad ± fan-shaped segments. *Sori* close to the ends of the veins *borne on the reflexed lf-margins*.

About 200 spp., tropical and warm temperate, only the following in Europe.

1. A. capillus-veneris L.   Maidenhair-fern.

Rhizome creeping, densely covered with narrow brown scales. Lvs close together, 6–30 cm., 2–3-pinnate; petiole (and rhachis) black and shining, as long as or shorter than blade; blade ± ovate in outline, pinnae and pinnules wide-spaced with slender black stalks; pinnules 5–15 mm., fan-shaped, often broader than long, crenately lobed in the upper part with broad rounded or truncate lobes which are recurved on the fertile pinnules, veins dichotomously branched, free, midrib 0. Sori borne close together along the veins of the recurved part of the lobes, 2–10 on each lobe. Spores ripe 5–9. $2n=60*$. H.

Native. Damp crevices of sea-cliffs and basic rocks, almost always near the sea, very local and rare. Cornwall, Devon, Dorset, Glamorgan, Westmorland, Isle of Man; W. Ireland from Clare to Donegal; Channel Is.; also found rarely as an escape on walls, etc., elsewhere. 8, H 7, S. Tropical and warm temperate zones of nearly the whole world; north in Europe to Caucasus, Crimea, S. Switzerland and W. France.

Subfamily ONOCLEOIDEAE. Rhizome scaly. Lvs not jointed to rhizome, markedly dimorphic, the fertile brown and much smaller and more densely pinnate than the sterile. Sori globose covered by the firm incurved lf-margin and by an inner indusium.
The two following are grown in gardens and rarely naturalized.

*Matteuccia struthiopteris (L.) Tod. Stock erect with crown of lvs at apex, but with long underground rhizomes. Sterile lvs up to 1 m., 2-pinnatifid, resembling those of *Dryopteris filix-mas*; vein endings free; fertile somewhat feather-like. N. temperate zone.

*Onoclea sensibilis L. Rhizome creeping. Sterile lvs up to 1 m., ovate-triangular, pinnate, with lobed pinnae, veins anastomosing, petiole longer than blade, fertile much smaller. Eastern N. America, N. Asia.

Subfamily 3. BLECHNOIDEAE. Terrestrial. Rhizome scaly. Lvs not jointed to rhizome. Sori short, in a single (rarely –3) row or joined into a continuous line, on a vein running parallel to the midrib; indusium present, outside the sorus, on the surface or margin of the lf.

5. BLECHNUM L.

Veins (of sterile pinnae) usually *free. Sori joined into a continuous* (rarely interrupted) *line.* Spores without perispore.

About 180 spp., cosmopolitan, but only the following in the north temperate zone.

1. B. spicant (L.) Roth   Hard-fern.

Lomaria Spicant (L.) Desv.

Rhizome short, robust, ± erect, densely clothed with brown lanceolate scales. Lvs numerous, tufted, outer sterile and spreading, inner fertile and suberect. Sterile lvs 10–50 cm., pinnate, glabrous; petiole not more than ⅓ as long as blade and often much shorter, scaly at base, dark brown; blade narrowly lanceolate, rhachis green; pinnae numerous, close together, pectinately arranged, somewhat coriaceous, entire or nearly so; longest pinnae about the middle of the blade, 1–2 cm. × 3–5 mm., linear-oblong and somewhat curved, acute or apiculate, widening at the base and there contiguous; lowest pinnae short, often broader than long, more distant. Fertile lvs 15–75 cm.; rhachis blackish (except near apex); pinnae linear, the longest to 2·5 cm. × 1–2 mm., distant, very suddenly widened near the base but not contiguous (except near the apex of the lf). Sori forming a continuous line down the whole length of the pinna on either side of the midrib. Sporangia appearing to cover the entire under-surface of the pinna after the opening of the indusium which is at first whitish, then brownish; margin of the lf outside indusium very narrow. Spores ripe 6–8. $2n = 68*$. H.

Native. Woods, heaths, moors, mountain grassland and rocks ascending to 3900 ft., calcifuge. 112, H 40, S. Throughout the British Is., common in hilly districts, less so in lowland ones. Europe from Iceland and Scandinavia southwards, east to S.W. Russia (Middle Dnieper region); Caucasus; Madeira, Azores, N. Morocco; Japan; western N. America from Alaska to California.

Subfamily 4. ASPLENIOIDEAE. Usually terrestrial. Rhizome scaly. Lvs not jointed to rhizome. Sori usually oblong or linear, borne along one or both sides of a lateral vein on the surface of the lf; indusium usually present, single or double; spores with perispore (except *Cystopteris*).

6. PHYLLITIS Hill

Rhizome scales firm, cell walls dark. *Lvs entire* or slightly lobed. *Sori borne direct on the vein, paired*, each pair appearing like a single sorus, one member of each pair being borne on the uppermost fork of one main vein, the other from the lowermost fork of an adjoining one; indusia of the paired sori opening towards each other.

About 14 spp., north temperate zone, E. Asia, C. and S. America.

1. P. scolopendrium (L.) Newm.   Hart's-tongue Fern.

Scolopendrium vulgare Sm.

Rhizome short, stout, ±erect, densely clothed with narrow brown scales. Lvs 10–60 cm., tufted, persistent, entire; petiole up to half as long as blade, usually much less, scaly; blade strap-shaped, cordate at base, tapered towards the usually obtuse apex, with scattered subulate scales at least when young; veins dichotomous, parallel, free. Pairs of sori linear, usually occupying more than half the width of the lf, with a conspicuous membranous indusium on either side. Spores ripe 7–8. $2n=72^*$. H.

Native. Rocky woods and hedgebanks, shady rocks and walls. 107, H40, S. Throughout the British Is.; common in the wetter districts, less so in the drier ones but rare only in N. Scotland. Europe from Scandinavia (rare), C. Germany and S.W. Russia (Middle Dnieper region) southwards; Morocco (mountains); Macaronesia; Asia Minor to Caucasus and Persia; Japan.

× ASPLENOPHYLLITIS Alston

Hybrids between *Asplenium* and *Phyllitis*. Sori of various types on the same plant, paired (as in *Phyllitis*), simple (as in *Asplenium*) or hooked (as in *Athyrium*). Lvs (at least in the British hybrids) simply pinnate with overlapping pinnules and with a long apical lobed portion. The following have occurred (all very rarely and not recently) in Britain: *A. trichomanes* × *P. scolopendrium* = × *A. confluens* (Lowe) Alston, *A. obovatum* × *P. scolopendrium* = × *A. microdon* (T. Moore) Alston, *A. adiantum-nigrum* × *P. scolopendrium* = × *A. jacksonii* Alston.

7. ASPLENIUM L.

Rhizome scales firm, cell walls dark. Lvs various (always divided in the British spp.) not densely scaly below; veins usually (always in the British spp.) free. *Sori oval to linear borne direct on the vein, simple, solitary* (rarely in some foreign spp. double or paired); indusium well developed, of the same shape as the sorus, usually opening towards the midrib.

About 650 spp., cosmopolitan.

1   Lvs irregularly dichotomously forked into very narrow cuneiform segments.   **7. septentrionale**

Lvs 1–3-pinnate, segments broader.   *2*

2 Lvs simply pinnate, pinnae toothed or (rarely) lobed. 3
  Lvs 2–3-pinnate. 6

3 Pinnae cuneiform without distinct midrib. × alternifolium
  Pinnae orbicular, ovate or oblong, with midrib. 4

4 Rhachis with a conspicuous green wing; pinnae 1 cm. or more.
  3. marinum
  Rhachis not winged or with a narrow brownish wing; pinnae less than
  1 cm. 5

5 Rhachis black. 4. trichomanes
  Rhachis green. 5. viride

6 Pinnae decreasing in length towards the base of the lf. 2. obovatum
  Basal pair of pinnae the longest. 7

7 Blade and petiole dark dull green (if only the basal pinnae compound,
  see also *A.* × *alternifolium*). 6. ruta-muraria
  Blade bright green, petiole blackish. 1. adiantum-nigrum

1. A. adiantum-nigrum L. 'Black Spleenwort.'

Rhizome short, creeping or decumbent, clothed when young with subulate, dark brown scales. *Lvs* 10–50 cm., tufted, persistent, somewhat coriaceous, 2–3-*pinnate; petiole* scaly at the extreme base, *blackish*, about as long as blade; *blade* triangular-ovate to triangular-lanceolate, *naked, bright green*; rhachis winged, blackish on the underside in the lower part; *pinnae* up to 15 on each side *decreasing in size upwards*, the lowest 2–6 cm., triangular-ovate or triangular-lanceolate, stalked; pinnules varying from lobed to pinnate; segments c. 4–12 mm., varying much in shape on different plants, ovate, obovate, elliptic or lanceolate, nearly always with a tendency to rhombic, acute or obtuse; cuneate at the narrow base, serrate. *Sori* 1–2 mm., linear-oblong or linear *occupying the greater part of the lateral veins, nearer the indistinct midrib than the margin of the segment*, finally confluent and occupying the greater part of the segment; indusium whitish, entire or sinuate. Spores ripe 6–10. H.

Native. Rocky woods and hedgebanks and shady walls and rocks. 112, H 40, S. Ascending to nearly 2000 ft. Europe from the Faeroes, Scandinavia, C. Germany and S. Russia (Middle Dnieper region and Crimea) southwards; S.W. Asia to the Himalaya and Pamir-Alai; Macaronesia; N. Africa; mountains of tropical Africa, S. Africa, Réunion; Formosa; Hawaii; Colorado.

Ssp. adiantum-nigrum

Blades of lvs and pinnae almost exactly triangular in outline, straight-sided, thick; petiole not longer than the blade; pinnae straight; segments ovate to lanceolate with acute teeth. $2n=144*$.

Throughout the British Is.; common in general but rather local in eastern and C. England. C. Europe, becoming rare and confined to the mountains in the Mediterranean region (extra-European distribution?).

Ssp. onopteris (L.) Luerss.

A. adiantum-nigrum var. *acutum* Poll.

Blades of lvs (which are always 3-pinnate) and pinnae somewhat concave-sided above with a long tapering point, less thick; petiole often longer than blade; pinnae somewhat curved towards apex of lf; segments narrowly lanceolate with long acuminate teeth. $2n = 72^*$.

S. and W. Ireland from Cork to Mayo, local; Kilkenny, Wicklow, Down, very local; S.W. England (very local). Mediterranean region, where it is the common form, north to Switzerland (S. Ticino) and S. France; Macaronesia (elsewhere?).

2. A. obovatum Viv.   'Lanceolate Spleenwort.'

A. lanceolatum Huds., p.p.

Rhizome short, erect to decumbent, densely tufted, with narrow, subulate, dark brown scales. Lvs 10–30 cm., tufted, persistent, 2-pinnate; *petiole* scaly at the extreme base, *blackish*, from about ½ as long to nearly as long as the blade; *blade* lanceolate, *with few scattered dark hair-like scales mainly on the rhachis*, bright green; rhachis not or very narrowly winged, blackish on the underside in the lower part; *pinnae* up to 20 on each side, *longest* (c. 1–5 cm.) *about the middle of the blade but the lower ones little shorter*, subsessile; pinnules 4–10 mm., ovate or obovate, obtuse, cuneate at the base, dentate, the teeth mucronate, the lowest pinnule on the upper side of the pinna often larger than the others and lobed. *Sori* 1–2 mm., oblong, *nearer the margin than the midrib of the pinnule* (measured along the lateral vein), finally sometimes confluent; indusium whitish, entire. Spores ripe 6–9. $2n = 144^*$. H.

Native. Rocks, walls and hedgebanks, usually near the sea, very local; S. and W. coastal counties from W. Kent to Caernarvon (but absent from several); Yorks, Cumberland; Kintyre; Kerry, Cork, Wexford, Carlow, Wicklow; Channel Is. 17, H 6, S. Mediterranean region, mainly in the littoral, extending north only in the west to C. and W. France and to the Vosges (1 loc.); Macaronesia.

*A. fontanum (L.) Bernh.

Similar to *A. obovatum* but smaller and slenderer; petiole and rhachis green; lowest pinnae much shorter than the middle ones. Several times found as an escape on walls, but not recently. $2n = 72$.

Native of Europe.

3. A. marinum L.   'Sea Spleenwort.'

Rhizome short, ± erect, densely clothed with blackish subulate scales. *Lvs* 6–30(–100) cm., tufted, persistent, *coriaceous, simply pinnate*; petiole scaly at extreme base, brown, ⅓–½ as long as blade; blade lanceolate, naked, dark green; *rhachis* brown in the lower part, *winged, the wings green; pinnae* up to 20 on each side, *longest* (1–)1·5–4 *cm.* about the middle of the blade, oblong or ovate-oblong, obtuse, markedly asymmetric (more developed on the upper side where there is sometimes a small lobe) at the truncate or broad-cuneate narrow base, crenate-serrate (sometimes doubly), the apex of the frond often with a long lobed or pinnatifid point. Sori 3–5 mm., linear, on the upper fork of the secondary veins, about midway between the midrib and margin; indusium brownish, entire. Spores ripe 6–9. $2n = 72^*$. H.

Native. Crevices of sea-cliffs from Sussex (probably extinct) and Isle of Wight to Cornwall, thence to Shetland and southwards to N. Yorks, local, commonest along the west coast; all round the coast of Ireland and inland near Killarney (Kerry). 55, H21, S. Atlantic coast from W. Norway (very rare) south to N. Morocco and W. Mediterranean coast to Corsica, Sardinia, Pantellaria and Algeria; Macaronesia.

4. A. trichomanes L.　　　　　　　　　　　　　　　　　'Maidenhair Spleenwort.'

Rhizome short, creeping to ± erect, with dark narrow scales. *Lvs* 4–20 (–40) cm., tufted, persistent, *not coriaceous*, simply pinnate; *petiole*, not scaly, ¼ as long as blade or less, *like the rhachis blackish and with a narrow brownish wing*; blade linear, deep green; *pinnae* 3–7(–10) *mm.*, 15–40 on each side, ± equally long in the middle of the blade for some distance, *oval or oblong*, obtuse, somewhat asymmetric at the truncate or cuneate base, crenate or crenate-dentate round the apex and on the upper margin, usually with a few short hairs below, finally falling from the rhachis. *Sori* 1–2 mm., oblong-linear, *situated mainly on the upper branch of the veins*, though sometimes continuing below the fork, about midway between the midrib and margin; indusium whitish, entire or nearly so. Spores ripe 5–10. $2n = 72^*, 144^*$. H.

　　Native. Walls and crevices of mainly basic rocks, ascending to 2850 ft.; common throughout the British Is. (absent only from Orkney). 111, H40, S. N. and S. temperate zones and mountains of the tropics; in Europe extending north to the Ladoga-Ilmen and Volga-Kama regions of Russia, W. Norway (Möre) and the Faeroes.

5. A. viride Huds.　　　　　　　　　　　　　　　　　　'Green Spleenwort.'

Differs from *A. trichomanes* as follows: Lvs sometimes not persistent; petiole usually relatively longer, brownish or blackish near the base, green above; *rhachis green, not winged; pinnae orbicular or ovate-orbicular*, less unequal at the base, more deeply toothed all round, paler green, not falling from the rhachis, glabrous. *Sori* nearer the midrib than the margin of the pinna, *situated mainly below the fork of the veins* though sometimes extending along the upper fork. Spores ripe 6–9. $2n = 72^*$. H.

　　Native. Crevices of basic rocks in hilly districts, local, ascending to 3150 ft.; Monmouth, Hereford, Stafford, Wales; Lancashire, Derby and Yorks northwards; S. and W. Ireland from Kerry to Donegal, Waterford and Tipperary. 49, H12. Mountains of Europe from Iceland, Finland and N. (not arctic) Russia to Sierra Nevada (Spain), Corsica, Italian Alps, Greece, Crimea and Caucasus; Morocco (Atlas); W. Asia east to Yenisei region of Siberia, south to the Himalaya; N. America south to Vermont and Washington.

6. A. ruta-muraria L.　　　　　　　　　　　　　　　　　　Wall-Rue.

Rhizome short, creeping, clothed when young with dark subulate scales. *Lvs* 3–12(–15) cm., tufted, persistent, coriaceous, *dark dull green throughout*, except for the blackish base of the petiole, 2(–3)-pinnate; petiole 1–2 times as long as blade, glandular and with a few hair-like scales when young; blade triangular-ovate or triangular-lanceolate, naked; rhachis narrowly winged; *pinnae* 1–3 cm., 3–5 on each side, *decreasing in size from the base upwards*, stalked; pinnules rarely more than 5 and often only 3, even on the lowest pinnae, usually undivided but sometimes the lowest trisect; segments 2–8 mm., varying considerably in shape on different plants, obovate-cuneiform to rhombic-lanceolate, obtuse, cuneate at the narrow base, crenate or dentate above the middle, veins dichotomous, without midrib. Sori c. 2 mm., linear, nearer the base than the apex of the segment, finally confluent; *indusium* whitish, *finely crenulate*. Spores ripe 6–10. $2n = 144*$. H.

Native. Walls and mainly basic rocks, ascending to 2000 ft. 112, H40, S. Common throughout the British Is. Europe from E. Norway and the Ladoga-Ilmen and Volga-Kama regions of Russia southwards; Mediterranean region; N. and S. Asia to the Himalaya and Pacific; eastern N. America from S. Ontario to Alabama and Missouri.

7. A. septentrionale (L.) Hoffm.   'Forked Spleenwort.'

Rhizome short, creeping, clothed when young with dark subulate scales. *Lvs* 4–15 cm., tufted, persistent, dark dull green throughout, except for the blackish base of the petiole, *dichotomously divided*, the forks often unequal; petiole usually several times as long as the blade (rarely only as long), with a few small hairs when young; blade 1–2(–3) times forked, the primary divisions stalked, glabrous; *segments linear-cuneiform*, 0·5–3 cm., very long-attenuate at the base, with a few long narrow teeth at the apex, or subentire without distinct midrib, veins dichotomous. Sori 5–20 mm., narrowly linear, covering almost the whole surface; indusium whitish, entire. Spores ripe 6–10. $2n = 144*$. H.

Native. Crevices of hard rocks, seldom on walls, rare; ascending to 3000 ft.; Devon, Somerset, N. and C. Wales, Northumberland, Cumberland, Peebles, Roxburgh, Midlothian, E. Perth, S. Aberdeen, Argyll, W. Ross, N. Inner Hebrides. 19. Europe from Scandinavia and N. Russia to the Sierra Nevada (Spain), Corsica, Mt Etna (Sicily), Macedonia, Crimea and Caucasus; W. Asia from W. Siberia to the Himalaya and N. Syria; Morocco (Atlas and Rif); Japan; Rocky Mountains.

A. septentrionale × trichomanes = A. × alternifolium Wulf.

A. germanicum auct., *A. breynii* auct.

Habit and colour of *A. septentrionale. Lvs simply pinnate* (rarely the lowest pair of pinnae divided into 3 pinnules though frequently trifid); blade narrowly lanceolate or triangular-lanceolate; *pinnae* 5–15 mm., 3–5 pairs, *oblanceolate-cuneiform*, often 2–3-lobed, cuneate at the narrow base, sharply and deeply toothed at the apex, veins dichotomous, without distinct midrib. Sori linear; *indusium entire*. $2n = $ c. $100*$ (?108).

Often occurring where the parents occur together.

A. ruta-muraria × septentrionale = A. × murbeckii Dörfl.

A. ruta-muraria var. *cuneatum* T. Moore

Similar to *A. × alternifolium* but the blade triangular-ovate, the lowest pinnae often with 3 pinnules and with usually only 2 pairs of pinnae, teeth blunt and shallow. Indusium crenulate. Very rare.

A. ruta-muraria × trichomanes = A. × clermontiae Syme. Once found on a wall in Louth. **A. obovatum × trichomanes = A. × refractum** T. Moore. Very rare.

8. Ceterach DC.

Differs from *Asplenium* as follows: Lvs densely scaly beneath with overlapping scales, veins anastomosing; indusium 0 or rudimentary.

Three or four spp. Europe, W. Asia, Africa.

1. C. officinarum DC.   Rusty-back Fern.
Asplenium Ceterach L.

Rhizome short, ± erect, clothed with dark narrow scales. Lvs 3–20 cm., tufted, persistent, coriaceous, simply pinnate; petiole ¼ as long as blade or less, scaly; blade dull green above, below entirely covered by light brown, ovate, overlapping scales; pinnae up to 2 cm., ovate or oblong, rounded at apex, somewhat widened at the broad base, entire or crenate. Sori c. 2 mm., linear; indusium 0. Spores ripe 4–10. $2n=144*$. H.

Native. Crevices of limestone rocks and mortared walls, ascending to 1450 ft. in Wales. 75, H 40, S. Rather common in S. and W. England, Wales and Ireland; very local in E. England and absent from several counties; in Scotland very local and mainly in the south-west extending north to Perth, Argyll and Iona, in the south-east only recorded from Berwick. Mediterranean region extending east and north to the Himalaya, Tien-Shan, Caucasus, Crimea, Gotland, C. Germany and Belgium; Madeira.

9. Athyrium Roth

Rhizome scales large and soft, cell walls thin. Lvs 1–3-pinnate usually, rather limp, not hairy; veins free. *Sori borne on receptacles with a vascular strand branched off from the vein. Indusium in typical spp. hooked* (i.e. consisting of two unequal arms placed back to back, the longer arm opening towards the midrib of the pinnule, crossing the vein at the outer end and continuous with the shorter arm which opens outwards), but in other spp. of various other shapes or 0. *Perispore present.* Stele of rhachis U-shaped.

About 180 spp., cosmopolitan, mainly temperate E. Asia.

Sorus oblong; indusium conspicuous.   **1. filix-femina**
Sorus orbicular; indusium absent at maturity.   **2. alpestre**

1. A. filix-femina (L.) Roth   Lady-fern.
Asplenium Filix-femina (L.) Bernh.

Rhizome short, ± erect, stout, densely clothed with brown lanceolate scales. Lvs forming a crown at the apex of the rhizome, dying in autumn, usually spreading and often drooping at the ends, but sometimes suberect, 20–100 (–150) cm., 2-pinnate, rarely 3-pinnate; petiole ¼–⅓ as long as blade, scaly, at least in the lower part, with brown lanceolate scales; blade thin and rather flaccid, light green, lanceolate; rhachis green or purplish-red, naked or with scattered hairs or scales, very rarely the midribs also with short whitish hairs; pinnae up to 30 on each side, the longest 3–25 cm., about the middle of the lf, the lowest considerably shorter, linear-lanceolate, tapered at the apex to an acute point, rhachis winged; pinnules regularly arranged, oblong or oblong-lanceolate, 3–20 mm., subobtuse to acute, sessile, truncate at the narrow base, pinnately lobed or pinnatifid, the lobes often toothed. *Sori* c. 1 mm., forming a row down either side of the pinnule, nearer the midrib than the margin, ± *oblong*, the lower ones hooked, the upper nearly straight; *indusium persistent*, covering the sorus till the spores are ripe, whitish, toothed. Very variable. Spores ripe 7–8. $2n=80^*$. H.

Native. Damp woods and hedgebanks, shady rocks, screes, marshes, etc.; calcifuge, ascending to 3300 ft. in Caernarvon; throughout the British Is., common in most districts but rather local in E. England. 111, H40, S. Throughout the north temperate zone and south to the mountains of India and Java and in tropical America to Peru and Argentina.

2. A. alpestre Clairv.   'Alpine Lady-fern.'
Polypodium alpestre (Hoppe) Spenn.

Differs from *A. filix-femina*, from which it is only certainly distinguishable by the sorus, as follows: Scales of petiole shorter. Lobes of segments usually broader and blunter. *Sori* nearer the margin of the pinnule, *orbicular*, very small; *indusium rudimentary, falling long before the spores are ripe*. Spores ripe 7–8. $2n=80^*$. H.

Var. *flexile* (Newm.) Milde (*A. flexile* (Newm.) Druce)
Differs in the very short (1–2 cm.) petioles, lvs smaller and relatively narrower, suddenly bent near the base, pinnae short, sori borne mainly in lower part of lf. Often regarded as a distinct sp. and needing further study. $2n=80^*$.

Native. Screes and rocks, usually acid, on mountains from 1200–3600 ft. 14. From Argyll, Stirling and Angus to Sutherland on the mainland and mainly in the east, local; var. *flexile* rare. N. Europe and Asia from Iceland to Kamchatka and in the high mountains of Europe south to the Pyrenees, Piedmont, Bosnia, Roumania and the Caucasus.

10. CYSTOPTERIS Bernh.

Rhizome scales soft, cell walls thin. Veins free. *Sori* orbicular *borne on receptacles with a vascular strand branched off from the vein. Indusium attached at the base of the sorus, flap-like*, vaulted and at first covering the sorus like a hood, later becoming reflexed and exposing the sporangia. *Perispore* 0.

About 18 spp., cosmopolitan.

1 Rhizome far-creeping, lvs solitary; lowest pair of pinnae the longest.
                                                                         3. montana
  Rhizome short, lvs tufted; pinnae decreasing in length towards the base.   **2**
2 Spores rugose; pinnules overlapping, crenate or shallowly crenately lobed.
  Kincardine.   **2. dickieana**
  Spores with numerous acute tubercles; pinnules not (or only a few of
    them) overlapping, dentate to deeply pinnatifid.   **1. fragilis**

1. C. fragilis (L.) Bernh.   'Brittle Bladder-fern.'
Rhizome short, rather stout, ± decumbent, clothed with thin brown lanceolate scales. *Lvs* 5–35(–45) cm., *tufted*, dying in autumn, suberect, 2(–3)-pinnate; petiole with a few scales at the base and usually a very few hair-like scales above, from ⅓ as long to as long as the blade, dark brown at the base, paler

above, slender and brittle; blade thin, naked, lanceolate; pinnae up to 15 on each side, at least the lowest distant, the *longest* 1–4 cm., *about the middle of the lf* or rather below, ovate or lanceolate; *pinnules* 4–10 mm., *not* (or only a few of them) *overlapping*, ovate, lanceolate, or oblong, acute or obtuse, cuneate at the narrow base, ± decurrent, very variable in toothing from *dentate to deeply pinnatifid*; teeth obtuse or acute, rarely retuse or bidentate at the apex, but frequently toothed on the margins, the veins usually ending in the apices of the teeth. Sori in two rows on either side of the midrib of the pinnule, small; indusium whitish, ovate, acuminate, exceeding the sorus. *Spores with numerous narrow acute tubercles.* Very variable. Spores ripe 7–8. H. $2n=168^*, 252^*$.

Native. Rocky woods, shady rocks and walls, especially on basic rocks, ascending to 4000 ft.; rather common in Scotland, N. England, Wales and N. and W. Ireland, becoming more local southwards and eastwards and absent from several counties in S. and E. England and S. and E. Ireland. 90, H 34, S. Cosmopolitan, only on mountains in the tropics; the most widespread of all ferns, extending from 81° 47′ N. in Greenland to Kerguelen. The plants with different chromosome numbers are possibly morphologically separable but the details have not yet been worked out.

C. regia (L.) Desv. Like *C. fragilis* but lvs 3–4-pinnate with truncate or emarginate apices of the segments; veins ending in the emarginations. Formerly in Upper Teesdale but long extinct. Mountains of Europe, Caucasus.

2. C. dickieana Sim

Differs from *C. fragilis* as follows: Pinnae often all overlapping, rhachis broadly winged; *pinnules overlapping, obtuse, crenate or shallowly crenately lobed*; lobes sometimes indistinctly emarginate, the veins then ending in the notch. *Spores rugose.* Spores ripe 7–8. H. $2n=168^*$.

Native. Sea-caves in Kincardine, very rare. 1. In a few isolated localities in arctic Europe and Asia and Siberia.

3. C. montana (Lam.) Desv.   'Mountain Bladder-fern.'

Rhizome long and creeping, blackish, rather slender (c. 2 mm. diam.) with a few scattered ovate scales when young. *Lvs* 10–30(–45) cm., *solitary, distant*, dying in autumn, 3-pinnate; petiole with a few ovate-lanceolate scales mainly near the base, longer than the blade, dark brown at the base, pale above; blade deltoid, sparsely glandular below; pinnae up to 13 on each side, *the lowest much the longest*, 2·5–7 cm., *triangular-ovate*; segments c. 7 mm., ovate or oblong, obtuse, ± cuneate at the narrow base, pinnately lobed or pinnatifid, the lobes usually bidentate at the apex and often with 1 or 2 teeth on the margin. Sori small and widely separated; indusium whitish ovate-orbicular, acute or subobtuse, irregularly toothed. Spores ripe 7–8. $2n=168$. Grh.?

Native. Damp, usually basic, rocks on mountains from 2300 to 3600 ft.; Westmorland; Argyll, Stirling and Angus to Inverness and Skye; rare and very local. 10. N. Europe from Norway to N. (not arctic) Russia; Jura, Alps (to Bosnia), Pyrenees, Roumania, Caucasus; Siberia (east to the Yenisei region); N. America, south to Ontario and British Columbia.

Subfamily 5. WOODSIOIDEAE. Small, terrestrial. Rhizome short, scaly.

Lvs not jointed to the rhizome but sometimes (as in the British spp.) with a joint in the petiole, 1–2-pinnate, veins free. Sorus orbicular, borne on the lower surface of the lvs near the vein-endings; indusium surrounding the base of the sorus either (as in the British spp.) split up from the first into narrow segments or at first cup-shaped, later splitting. Spores 2-sided, without perispore.

11. WOODSIA R.Br.

The only genus of the subfamily. 38 spp., arctic and mountains of north temperate zone (to Himalaya) extending down the Andes to temperate S. America, 1 in S. Africa.

Largest pinnae oblong or ovate-oblong, 1½–2 times as long as broad; rhachis and underside of pinnae densely scaly with long (2–3 mm.) scales.

1. ilvensis

Largest pinnae triangular-ovate, 1–1½ times as long as broad; rhachis and underside of pinnae with sparse, short (c. 1 mm.) scales, or the latter without scales.

2. alpina

1. **W. ilvensis** (L.) R.Br.

Rhizome short, ± erect, sparsely scaly above. Lvs tufted, dying in autumn, suberect, 5–15 cm., dull green, simply pinnate or almost 2-pinnate; petiole clothed with brown lanceolate scales below and subulate scales above and with flexuous hairs throughout, pale reddish-brown, jointed near the middle, from ½ as long to as long as blade; blade oblong-lanceolate, *rhachis and under surface* (at least the veins) ± *densely clothed with long* (2–3 mm.) *subulate pale brown scales* and flexuous hairs; pinnae c. 7–15 on each side, the longest about the middle of the lf, 7–17 mm. (the lowest sometimes scarcely shorter, sometimes considerably so), *oblong or ovate-oblong*, 1½–2 times as long as broad, obtuse or subobtuse, ± truncate at the narrow base, deeply pinnatifid into 7–13 oblong, obtuse, ± crenate lobes. Sori near the margin of the lobes; indusium surrounding the base of the sorus, divided nearly to the base into numerous irregular lobes which end in long jointed hair points which are longer than the rest of the indusium and arch over the sporangia. Spores ripe 7–8. $2n = 82^* (84^*)$. H.

Native. Rock crevices on mountains from 1200 to 2700 ft., very rare; Merioneth, Caernarvon, Durham, Westmorland, Cumberland, Dumfries, Perth, Angus, mid Inner Hebrides. 9. Arctic and high mountains of north temperate zone, south to the S. Alps, S.W. Russia (Middle Dnieper region), Altai, Iowa and North Carolina.

2. **W. alpina** (Bolton) S. F. Gray   'Alpine Woodsia.'

W. hyperborea (Lilj.) R.Br.

Differs from *W. ilvensis* as follows: Usually smaller. Lvs 3–15 cm., petiole with few scales, glabrescent, ¼–⅔ as long as blade; blade oblong-linear, *rhachis sparsely clothed with short* (c. 1 mm.) *subulate* (or a few of them lanceolate) *scales, the undersurface of the pinnae without or with very few* (up to c. 5) *scales*, hairs also few; *pinnae* 5–12 mm., *triangular-ovate*, 1–1½ *times as long as broad*, obtuse, pinnately lobed or pinnatifid into 3–7 obovate or oblong-obtuse, ± crenate lobes. Spores ripe 7–8. $2n = 164^* (168^*)$. H.

Native. Rock crevices on mountains from 1900 to 3000 ft., very rare. Caernarvon, Perth, Angus, Argyll, N. Inner Hebrides. 6. Arctic and high mountains of north temperate zone, south to the Pyrenees, S. Alps, S. Russia (Black Sea region), Altai, W. Ontario and New York.

Subfamily 6. DRYOPTERIDOIDEAE. Usually terrestrial. Rhizome scaly. Lvs not jointed to rhizome. Sori usually orbicular, on or sometimes terminating a vein on the under surface of the lf; indusium at first below the sorus but soon becoming central, and either reniform or peltate, sometimes 0; spores 2-sided, with perispore.

12. DRYOPTERIS Adans.

Rhizome short, stout, *densely scaly*, with broad soft scales. Lvs forming a crown, deeply 2-pinnatifid to 3-pinnate (rarely 1-pinnate) *with ± numerous scales and without hairs*, veins free; *petiole with* 4 *or more vascular bundles. Sori large; indusium reniform,* nearer the midrib or about midway between the midrib and margin of the segment. Stele of rhachis of several strands.

About 150 spp., mainly north temperate, also tropical and S. Africa, tropical Asia, tropical America.

1 Lvs deeply 2-pinnatifid or 2-pinnate with toothed pinnules; pinnae decreasing markedly in length towards the base; pinnules (except sometimes the lowest pair of each pinna) not at all narrowed at the base.   *2*

Lvs 3-pinnate or 2-pinnate with the pinnules lobed to nearly half-way; basal pinnae not or little shorter than the longest pair; pinnules markedly contracted at the base or stalked.   *5*

2 Pinnules and lobes of the longest pinnae more than 15 on each side, not more than 5 mm. broad, teeth not mucronate.   (**filix-mas** agg.)   *3*

Pinnules and lobes of longest pinnae less than 15 on each side, 7 mm. broad or more, teeth usually mucronate.   **5. cristata**

3 Pinnae dark brown or blackish above at the junction with the rhachis; petiole and rhachis densely covered with orange-brown scales; pinnules toothed mainly at the apex, the inner (or both) margins entire or nearly so (occasionally with a few irregular lobes).   **2. borreri**

Pinnae not discoloured at the junction with the rhachis; petiole and rhachis less densely scaly with pale brown scales; pinnules toothed all round or less toothed at the apex.   *4*

4 Plant large, with 1 or few crowns; pinnules with acute teeth and ± flat margins; sori of the large pinnules mostly 4 or 5 on each side of the midrib.   **1. filix-mas**

Plant small, usually with several crowns; pinnules slightly concave with blunt teeth and ± undulate margins giving the lf a somewhat crisped appearance; sori rarely more than 3 on any pinnule and often only 1. Mountains.   **3. abbreviata**

5 Lowest pinnule on the basal side of the lowest pinna much longer than the lowest one on the upper side of the same pinna and also longer than the adjoining one (or occasionally the second pinnule also long but then the third shorter); lvs 3-pinnate or nearly so.   *6*

Lowest 3 or 4 pinnules on both sides of the lowest pinna all about equal in length; lvs 2-pinnate with pinnatifid pinnules, glandular on both sides.   **4. villarii**

6   Scales on the petioles darker in the middle than near the margin (rarely
    concolorous in alpine plants and then indusium glandular and petiole
    pale above); pinnules flat or convex.   (dilatata agg.)   7
    Scales on the petioles uniformly coloured (either indusium eglandular or
    petiole dark brown throughout); pinnules flat or concave.   8
7   Spores dark brown with dense blunt spinules; blade relatively firm; basal
    pinnule on lower side of lowest pinna nearly always less than half as
    long as pinna. Common.   **7. dilatata**
    Spores pale brown with more widely spaced smaller acute spinules; blade
    thin; basal pinnule on lower side of lowest pinna mostly half as long
    as pinna or more. Scottish mountains, very rare.   **8. assimilis**
8   Lf triangular-ovate or triangular-lanceolate in outline, minutely glandular;
    segments concave; indusium toothed, margin glandular; petiole dark
    brown throughout, with narrow-lanceolate scales.   **9. aemula**
    Lf lanceolate or ovate-lanceolate in outline, eglandular; segments flat,
    indusium entire, eglandular; petiole dark below, pale above, with ovate
    or ovate-lanceolate scales.   **6. carthusiana**

(1–3). D. filix-mas agg.   Male Fern.

Rhizome ± erect. *Lvs* 15–150(–180) cm., dying in autumn or subpersistent,
suberect or ± spreading, *pinnate with deeply pinnatifid pinnae or 2-pinnate*;
petiole $\frac{1}{6}$–$\frac{1}{2}$ (or rather more) as long as blade, ± scaly with pale brown or
orange-brown uniformly coloured scales; blade rather firm, bright deep green,
lanceolate, naked except for scales on the rhachis and midribs and sometimes
some minute sessile glands on the under surface; *pinnae* 20–35 on each side,
the longest 5–15 cm., linear-lanceolate, about the middle of the lf, *decreasing
considerably in length downwards; pinnules* or segments c. 15–25 on each side
of the longest pinnae, regularly arranged, ± equal for some distance above the
base (except sometimes for a larger lowest pinna on the upper side), oblong,
4–22 mm. × 2–5 mm., all (except the lowest pair of each pinna) *attached at
the base by their whole width and often somewhat widened* and decurrent, obtuse
or subtruncate, subentire or toothed (or rarely) lobed, the *lobes not reaching
half-way to the midrib and entire or obscurely crenate or serrate*. Sori forming
a row down each side of the pinnule, c. $\frac{1}{2}$–2 mm. diam.; indusium entire,
sometimes with minute marginal glands. Spores ripe 7–8. H.

Woods, hedgebanks, rocks and screes. 112, H 40, S. Common throughout
the British Is., ascending to 3150 ft. in Kerry. Europe; temperate Asia and
in the mountains to Java; Morocco; Madeira; Mascarene Is., Madagascar;
America, south to Peru and Brazil; Hawaii.

The following three plants are distinct both morphologically and cytologically.
Their distribution both in this country and abroad is still imperfectly known.
Hybrids between them appear to occur.

1. D. filix-mas (L.) Schott

Lastrea Filix-mas (L.) C. Presl; *Aspidium Filix-mas* (L.) Sw.; *Nephrodium Filix-
mas* (L.) Strempel

Crowns solitary or few. Lvs large (usually over 50 cm. on well-developed
plants), usually dying in winter; *petiole* usually $\frac{1}{4}$–$\frac{1}{3}$ as long as blade, *sparsely
or moderately covered with pale brown scales*, the larger ones ovate, long-
acuminate; *blade not glandular*, rhachis and midribs sparsely (or not at all)
scaly, *pinnae not dark above at the junction with the rhachis; pinnules ±flat,*

obtuse, ± *equally crenate-serrate* (to lobed) *all round* (the upper teeth sometimes smaller and sharper); *the teeth somewhat curved towards the apex of the pinnule or appressed, acute*; lowest pinnule on the upper side of each pinna markedly (4 mm.) longer than its neighbour. *Sori 5–6 on each side of the midrib of the largest pinnules and at least 3 on the majority of the pinnules*, large (c. 1 mm.); indusium not glandular, not embracing the sporangia. $2n = 164*$. Reproducing sexually.

Native. Throughout the British Is., in general the commonest of the 3 spp. but less common on screes, etc. Apparently throughout Europe and temperate Asia (distribution elsewhere uncertain).

2. D. borreri Newm.

Lastrea pseudo-mas Wollaston; *D. Filix-mas* var. *paleacea* auct.

Crowns solitary or few. Lvs large (usually over 50 cm.) on well-developed plants, mostly remaining green through the winter, thicker and usually more yellowish than in *D. filix-mas; petiole and rhachis* (at least in the lower part) *densely covered with orange-brown scales*, the larger ones lanceolate; petiole usually less than ¼ as long as blade; blade not glandular; *pinnae dark brown or blackish above at the point of junction with the rhachis; pinnules ± flat, subtruncate, with a few small sharp triangular teeth at the apex, the rest of the margins all subentire* or the largest pinnules with a few irregular lobes on the outer margin and occasionally 1 or 2 on the inner; lowest pinnule on the upper side of each pinna not or scarcely larger than its neighbour. *Sori (3–)4–5 on each side of the midrib of the largest pinnules* (and at least 3 on the majority of the pinnules), usually smaller than in *D. filix-mas*; indusium not glandular, embracing the sporangia when young. $2n = 82*, 123*$. Reproducing apogamously.

Native. Throughout the British Is., common in many places, in woods and on screes, usually on acid soils and frequently in damper habitats than *D. filix-mas*. S. and W. Europe and S.W. Asia, north to the Caucasus. S. Tirol and Norway; Madeira; (elsewhere?).

D. borreri × *filix-mas* occurs occasionally as isolated individuals. $2n = 164*, 205*$.

3. D. abbreviata (DC.) Newm.

Lastrea propinqua Wollaston; *D. Filix-mas* var. *pumila* (T. Moore) Druce

Crowns several, tufted. Lvs relatively small (usually 30–50 cm., rarely up to 120 cm.), dying in winter; *petiole* usually ¼–½ as long as the blade, *sparsely or moderately* (densely at the base) *covered with pale brown scales*, the larger ones ovate-lanceolate; *blade covered with minute glands beneath; pinnae not dark at the junction with the rhachis; pinnules somewhat concave giving the lf a crinkly appearance, obtuse, sinuate-crenate or crenately lobed, the teeth broad, obtuse or truncate, near the apex usually less deeply toothed or subentire*; lowest pinnule on the upperside of at least some of the pinnae markedly longer than its neighbour. *Sori 2–4 on each side of the midrib of the longest pinnules, not more than 2 and often only 1 on the majority of the pinnules*, rather large; indusium with minute glands on the margin. $2n = 82*$. Reproducing sexually.

Native. Screes etc. on mountains, apparently widespread but rather local. Iceland, France (Auvergne), Spain (Pyrenees, Sierra de Guadarrama), doubtless also elsewhere.

4. D. villarii (Bell.) Woynar 'Rigid Buckler-fern.'

Lastrea rigida (Sw.) C. Presl; *Aspidium rigidum* Sw.; *D. rigida* (Sw.) A. Gray; *Nephrodium rigidum* (Sw.) Desv.

Rhizome decumbent or ascending. *Lvs* 20–60 cm., dying in autumn, erect or ± spreading, 2-*pinnate*, stiff and rather wiry; petiole from ⅓ to nearly as long as blade, densely scaly at base, sparsely so above, with pale-brown uniformly coloured, lanceolate scales; blade dull green, lanceolate or triangular-lanceolate, ± densely glandular on both surfaces with short-stalked glands, and with narrow scales on the rhachis; *pinnae* c. 15–25 on each side, *usually several basal pairs about equally long*, the lowest 4–7 cm., from somewhat longer to slightly shorter than the next, ± lanceolate; *pinnules* 8–12 × 3–5 mm., ± *equal in length for some distance above the base* (sometimes the lowest one on the upper side longer), oblong, several pairs on the lower pinnae rounded to broad-cuneate and *narrowed at the base* (only the upper segments wide and decurrent), obtuse, *regularly crenately lobed half-way to the midrib or nearly so, each lobe with 2–4 conspicuous acute teeth at the apex*. Sori 4–6 in a row down each side of the larger pinnules, crowded, c. 1 mm. diam.; indusium glandular on the surface and margin. Spores ripe 7–8. $2n = 82, 164^*$. H.

Native. Clefts of limestone pavement, ascending to 1550 ft., very local; Caernarvon, Denbigh, Lancashire, Yorks, Westmorland. 6. Mountains of Europe and the Mediterranean region from the Jura, Alps and Caucasus to N. Africa, Palestine and Afghanistan; the more southern plants differ sub-specifically.

5. D. cristata (L.) A. Gray 'Crested Buckler-fern.'

Lastrea cristata (L.) C. Presl; *Aspidium cristatum* (L.) Sw.; *Nephrodium cristatum* Michx.

Rhizome decumbent or shortly creeping. Lvs 30–100 cm., the outer sterile, ± spreading, shorter than the suberect fertile inner ones, dying in autumn, *pinnate with deeply pinnatifid pinnae or 2-pinnate*; petiole ⅓ as long as blade, with pale-brown uniformly coloured ovate scales, numerous at the base, few above; blade dull green, lanceolate or oblanceolate, naked except for a few scales on the rhachis; *pinnae* c. 10–20 on each side, the longest 5–10 cm., triangular-ovate to lanceolate, about the middle of the frond, *decreasing in length downwards; pinnules* or segments c. 5–10 *on each side of the longest pinna*, ± regularly decreasing in size above the base, the lowest pair 10–25 × 7–10 mm., ± equal in length or that of the lower side longer, oblong or ovate-oblong, *attached at the base by their whole width* (rarely the basal pair somewhat narrowed), decurrent, obtuse, the larger *shallowly pinnately lobed*, all *serrate with acute mucronate teeth* which are often curved inwards. Sori forming a row down each side of the pinnule or more irregular on the larger pinnules, large; indusium entire or sinuate, without glands. Spores ripe 7–8. $2n = 164^*$. H.

Native. Wet heaths and marshes, very local and rare, decreasing; Kent, Surrey, Suffolk, Norfolk, Huntingdon, Cheshire; extinct in Notts and Yorks; Renfrew. 13. Europe from S.E. Norway and N. (not arctic) Russia to N. France, N. Italy, Roumania and S. Russia (not Crimea nor Caucasus), W. Siberia; eastern N. America from Newfoundland to Saskatchewan, south to Virginia and Idaho.

6. D. carthusiana (Villar) H. P. Fuchs　　　　　'Narrow Buckler-fern.'

Lastrea spinulosa C. Presl; *Nephrodium spinulosum* Strempel; *Aspidium spinulosum* Sw.; *D. spinulosa* Watt.

Rhizome decumbent or shortly creeping. *Lvs* 30–120(–150) cm., dying in autumn, the base of the petiole decaying first, 2-*pinnate with deeply pinnatifid pinnules or 3-pinnate*; petiole from ⅓ as long to rather longer than the blade, *dark brown at the base, pale or green above, with pale brown, uniformly coloured, ovate or ovate-lanceolate, entire scales*, which are few above, more numerous (but not dense) at the base; *blade light or yellowish green, lanceolate or ovate-lanceolate, naked* except for a few scales on the rhachis; *pinnae* c. 15–25 on each side, 5–10 cm., the 3 *or 4 basal pairs about equal in length* (the lowest slightly longer or slightly shorter than the next), triangular-ovate or triangular-lanceolate; *pinnules of the lowest pinna* c. 7–12, *the basal one* (or sometimes 2) *on the lower side markedly longer than that on the upper and than its neighbour*, c. 2–3 cm., ± lanceolate, *pinnatifid nearly to the midrib or pinnate, narrowed at the attachment, ± flat; segments* ± oblong, serrate *with incurved* acuminate, mucronate or aristate *teeth*. Sori c. ½–1 mm., forming a row down either side of the segment; *indusium* entire or sinuate, *without glands*. Spores ripe 7–9. 2*n* = 164*. H.

　　Native. Damp and wet woods, marshes and wet heaths, from Sutherland and the Hebrides southwards and throughout Ireland, rather common but unrecorded for several counties and from the Channel Is. 99, H38. Europe from Scandinavia and N. (not arctic) Russia to Portugal and N.W. Spain, Corsica, Italy, N. Greece and the Caucasus; Siberia (east to the Yenisei region); N. America from Quebec to British Columbia, south to Virginia and Mississippi.

(7–8). D. dilatata agg.　　　　　　　　　　　'Broad Buckler-fern.'

Rhizome erect or ascending. *Lvs* (7–)30–150(–180) cm., dying in autumn or subpersistent, the base of the petiole decaying first, 3-*pinnate; petiole with ovate-lanceolate or lanceolate entire scales which are dark brown in the centre, pale brown at the edge*, and usually dense at the base, less so above, from rather less than ½ as long as blade to about as long, dark brown at base, pale or green above; *blade* usually *dark green*, from triangular-ovate to lanceolate, sparsely or moderately scaly on the rhachis and sometimes glandular on the rhachis and lower surface; pinnae 6–20 cm., c. 15–25 on each side, 3 *or more of the basal pairs about equal in length* (the lowest rather longer to slightly shorter than the next), triangular-ovate or triangular-lanceolate; pinnules of the lowest pinna c. 10–20, *the basal one on the lower side markedly longer than that on the upper* and usually longer than its neighbour, c. 2–5 cm., ± lanceolate, *pinnate, narrowed at the attachment, convex or flat; segments* ± oblong, toothed or pinnately lobed, *with incurved* mucronate or aristate *teeth*. Sori c. ½–1 mm., in a row down either side of the segment; *indusium fringed with stalked glands* (sometimes indistinctly so) and usually irregularly toothed. Spores ripe 7–9. H.

Variable. Mountain forms are usually lighter green, with flat pinnules and with the scales of the petiole less markedly dark-centred, and so more resemble *D. carthusiana*

7. D. dilatata (Hoffm.) A. Gray

D. austriaca auct., *Lastrea dilatata* (Hoffm.) C. Presl, *D. aristata* Druce

Very variable. As the agg. but: Blade relatively firm. Basal pinnule on lower side of lowest pinna very rarely as much as half as long as the pinna and usually less than twice as long as that on the upper (though in some plants more). *Spores with thick dark brown perispore with dense rather irregular obtuse spinules*, up to 2μ. $2n=164*$.

Native. Woods, hedgebanks, wet heaths, shady rock ledges and crevices, etc., ascending to 3700 ft. 112, H 40, S. Common throughout the British Is. Europe and temperate Asia; in Europe from Iceland and N. Russia south to Portugal and N. Spain, Corsica, Italy, Macedonia and Caucasus.

8. D. assimilis S. Walker

D. dilatata var. *alpina* T. Moore

Much less variable. Small. Lvs 7–60 cm.; scales of petiole usually dark brown in centre but sometimes uniformly coloured; *blade* light green, *thin and membranous*; pinnules of lowest pinna c. 10–12, *basal one on lower side very long, mostly twice as long or more than twice as long as that on upper side, and at least half as long as the pinna*. Sori usually relatively large, c. $\frac{1}{2}$–$1\frac{1}{2}$ mm. diam., regularly arranged in 2 lines along the segments; indusium small, fugacious. *Spores with thin pale brown perispore with smaller*, $\pm$ *widely spaced, acute spinules, up to* 1μ. $2n=82*$.

Native. At present only certainly known from Ben Lawers but probably occurring elsewhere. Norway, Sweden, Switzerland (incomplete).

9. D. aemula (Ait.) O. Kuntze   'Hay-scented Buckler-fern.'

Lastrea aemula (Ait.) Brack.; *Nephrodium aemulum* (Ait.) Baker; *Aspidium aemulum* (Ait.) Sw.

Rhizome erect or ascending. *Lvs* 15–60 cm., persistent, finally decaying from the apex downwards, 3-*pinnate* or almost 4-pinnate; *petiole with* few or numerous *narrow-lanceolate, lacerate, uniformly coloured, reddish-brown scales*, about as long as blade, *dark brown throughout; blade bright green, triangular-ovate or triangular-lanceolate*, scaly below on the rhachis and on the midribs of the segments below and *sprinkled beneath or on both surfaces with minute sessile glands*; pinnae c. 15–20 on each side, *the lowest pair usually the longest*, 5–15 cm., triangular-ovate or triangular-lanceolate; *pinnules of the lowest pinna* c. 10–20 on each side, *the basal one on the lower side markedly longer than that on the upper* and usually longer than its neighbour (or about equal and the third shorter), c. 2–6 cm., triangular-ovate to lanceolate, *pinnate, narrow at the attachment* and often shortly stalked, *somewhat concave and thus giving the frond a somewhat crinkly appearance*; segments lobed or almost pinnate, and toothed with narrow acuminate, scarcely or shortly mucronate *teeth of which some are* $\pm$ *straight, some somewhat incurved, and some somewhat curved outwards*. Sori c. $\frac{1}{2}$–1 mm. forming a row down each side of the midrib of the segment; *indusium irregularly toothed and fringed with sessile glands*. Spores ripe 7–9. $2n=82*$. H.

Native. Woods, hedgebanks, shady rocks, etc., ascending to 3300 ft., local and mainly in the west; Cornwall to Somerset and Dorset; Sussex, Kent,

Surrey; Wales and the border counties; N. Yorks, Westmorland, Cumberland, Northumberland; Isle of Man; Ayr; Kintyre and E. Inverness to Orkney; throughout Ireland but rare in the centre. 40, H37. N.W. France, N. Spain, Madeira, Azores.

The following hybrids occur: *D. carthusiana* × *cristata* = *D.* × *uliginosa* (Newm.) O. Kuntze ex Druce, usually to be found where *D. cristata* occurs; *D. dilatata* × *filix-mas*, very rare; *D. carthusiana* × *dilatata*, probably fairly frequent; *D. carthusiana* × *filix-mas* very rare.

13. POLYSTICHUM Roth

Differs from *Dryopteris* as follows: Lvs usually coriaceous, pinnules unequal-sided, teeth aristate. *Indusium peltate*, the top orbicular.

About 225 spp., cosmopolitan.

1 Lvs 2-pinnate or at least with one free lobe at the base of some of the lower
         pinnae.   2
      Lvs simply pinnate, pinnae never with a free lobe.   **3. lonchitis**

2 Lvs flaccid; margins at the base of the upper pinnules meeting each other
         at about a right angle, those of the lower pinnules at an obtuse angle;
         pinnules c. 12–20 on each side of the longest pinna.   **1. setiferum**
      Lvs rigid, somewhat coriaceous; margins at the base of the pinnules
         meeting each other at an acute angle (or of the lower pinnules at about
         a right angle); pinnules up to c. 15 on each side of the longest pinnae
         and often only 1.   **2. aculeatum**

1. P. setiferum (Forsk.) Woynar   'Soft Shield-fern.'

P. angulare (Willd.) C. Presl; *Aspidium angulare* Kit. ex Willd.

Rhizome stout, erect or ascending. *Lvs* 30–150 cm., persistent or subpersistent, 2-*pinnate*; petiole c. $\frac{1}{4}$–$\frac{1}{2}$ as long as blade, clothed with ovate-lanceolate brown scales; *blade* deep green above, paler below, *rather flaccid and usually arching or drooping*, lanceolate, scaly on the rhachis, the lower surface with or without a few hair-like scales; pinnae c. 30–40 on each side, the longest c. 6–8 cm., about the middle of the lf, the lowest much shorter, linear-lanceolate, fully pinnate, straight or the upper curved towards the apex; *pinnules c.* 12–20 *on each side*, the basal one on the upper side usually the longest, 4–10 mm., ± ovate, curved, unequally cuneate at the shortly stalked base with the *margins meeting each other at about a right angle on most of the pinnules and at much more on one or more of the lowest ones of each pinna*, acute and spine-pointed at the apex with a rounded or deltoid lobe on the outer margin at the base reaching nearly to the midrib on the lowest pinnules, serrate with the teeth rounded on the outer edge but ending in a straight spine. Sori c. $\frac{1}{2}$–1 mm., in a row down each side of the midrib of the pinnule and of its basal lobe; *vein on which the sorus is borne not or only shortly continued beyond it*. Spores ripe 7–8. $2n = 82^*$. H.

Native. Woods, hedgebanks, etc. 77, H40, S. From Argyll, Perth and Midlothian southwards, common in S.W. England becoming more local northwards and eastwards; common throughout Ireland. Temperate and tropical regions of the whole world (range perhaps including allied spp.); extending north in Europe to Caucasus, Crimea, Bohemia, S.W. Germany and N. France.

2. P. aculeatum (L.) Roth   'Hard Shield-fern.'

P. lobatum (Huds.) Chevall.; *Aspidium aculeatum* (L.) Sw.

Rhizome stout, erect or ascending. *Lvs* 20–100 cm., persistent, *2-pinnate*; petiole usually c. $\frac{1}{3}$ as long as blade but sometimes c. $\frac{1}{2}$ as long, clothed with ± ovate brown scales; *blade* dark green above and somewhat glossy, paler below, *rigid, somewhat coriaceous,* lanceolate or linear-lanceolate, scaly on the rhachis and with hair-like scales on the surface below; *pinnae* c. 25–50 on each side, the lowest much shorter, ovate-lanceolate to linear-lanceolate, *varying on different plants from fully pinnate to pinnately lobed with only a single free pinnule* and that sometimes only on a few of the lower pinnae, straight or curved towards the apex; *pinnules up to c. 15 on each side,* the basal one on the upper side of the pinnule the longest, 5–15 mm., rhombic-ovate to rhombic-lanceolate, somewhat curved, unequally cuneate at the narrow but scarcely stalked base *with the margins meeting each other at less than a right angle* (or sometimes at a right angle in the lowest pinnule of each pinna), acute and spine-pointed at the apex, often with a small deltoid lobe on the outer margin at the base, serrate with spine-pointed straight or somewhat incurved teeth. Sori c. $\frac{1}{2}$–1 mm., in a row down each side of the midrib of the pinnule and sometimes also on its basal lobe; *vein on which the sorus is borne continuing well beyond the sorus.* Spores ripe 7–8. $2n=164*$. H.

Native. Woods, hedgebanks, etc., ascending to nearly 2500 ft.; from Orkney southwards, rather common in the wetter districts, local in the drier ones; throughout Ireland but rather local. 106, H39. Europe from S. Scandinavia, Finland and S. Russia (Middle Dnieper region) to Portugal and N. Spain, Italy, Greece and the Caucasus; N. Africa (mountains); S.W. Asia to India and the Tien Shan; China, Japan.

P. aculeatum × setiferum = P. × bicknellii (Christ) Hahne occurs occasionally.

3. P. lonchitis (L.) Roth   Holly Fern.

Aspidium Lonchitis (L.) Sw.

Rhizome stout, ascending. *Lvs* 10–60 cm., persistent, *simply pinnate*; petiole $\frac{1}{6}$ as long as blade or less, clothed with ± ovate, reddish-brown scales; *blade deep green* above, paler below, *coriaceous,* linear-lanceolate or linear, with few or numerous scales on the rhachis and narrow hair-like ones on the lower surface; pinnae c. 20–40 on each side, the longest c. 1–3 cm., about the middle of the lf, decreasing in size downwards, the lowest 5 mm. or less, close together and often overlapping, ovate or ovate-lanceolate and somewhat curved towards the apex, acute, very unequal at the narrow short-stalked base where the upper margin runs ± parallel with the rhachis and the lower diverges from the rhachis at an angle of c. 45°, with a single ± deltoid lobe on the upper side at the base, serrate with straight spine-pointed teeth, the teeth often crenulate or serrulate. Sori c. $\frac{1}{2}$–1 mm., in a row down each side of the midrib of the pinna (nearer the midrib than the margin) and a double row down the basal lobe, usually confined to the upper part of the lf; indusium irregularly toothed. Spores ripe 6–8. $2n=82*$. H.

Native. Crevices of basic rocks on mountains, from 200 ft. in Sutherland to 3500 ft. in Perth, local; Merioneth, Caernarvon; Yorks and Westmorland to Northumberland and Dumfries; Dunbarton and Stirling to Orkney;

Kerry, Galway, Sligo, Leitrim, Donegal. 30, H6. Arctic and mountains of
north temperate zone, south to Sierra Nevada (Spain), Corsica, Italy, Crete,
Caucasus, Himalaya, N. California and Nova Scotia.

*Cyrtomium falcatum (L. f.) C. Presl. Sori like *Polystichum* but lvs with anastomosing
veins. Lvs pinnate, pinnae 6–10 cm., leathery, often with pointed lobe on one side
at base. Much grown as a pot plant and reported naturalized. E. Asia, S. Africa,
Madagascar, Hawaii.

14. THELYPTERIS Schmidel

Rhizome and lvs with relatively few, usually hairy, *scales*. Lvs usually deeply
2-pinnatifid, *nearly always ± pubescent*, veins free; *petiole with 1–2 vascular
bundles*. *Sori small*, near the margin of segment, *indusium reniform or* 0.

About 500 spp., cosmopolitan. The last two of the spp. below are unusual
in their 3-pinnate lvs without hairs and are sometimes separated under
Gymnocarpium Newm.

1 Lvs pinnate with deeply pinnatifid pinnae.   *2*
    Lvs 2-pinnate with deeply pinnatifid pinnae or 3-pinnate.   *4*

2 Lvs not bent at junction of rhachis and petiole; pinnae parallel and all in
       one plane, attached to the rhachis by a narrow base, with a few hairs
       beneath only or subglabrous; indusium usually present.   *3*
    Lvs bent at junction of rhachis and petiole; lowest pair of pinnae bent
       backwards from the others; pinnae (except lowest pair) attached to the
       rhachis by a broad base, hairy on both sides; indusium 0.
                                                            3. phegopteris

3 Rhizome stout, erect or ascending, with a crown of lvs at the apex; lvs with
       numerous sessile glands beneath, smelling of lemon when crushed.
                                                            1. limbosperma
    Rhizome creeping underground; lvs solitary, not glandular nor smelling
       of lemon.   **2. palustris**

4 Lvs glabrous, bright or clear green; lowest pinnule on lower side of lowest
       pinna about equalling each of 3rd pair of pinnae from base.
                                                            4. dryopteris
    Lvs glandular, dull green; lowest pinnule on lower side of lowest pinna
       considerably smaller than each of the 3rd pair of pinnae from base.
                                                            5. robertiana

1. T. limbosperma (All.) H. P. Fuchs   'Mountain Fern.'

Dryopteris Oreopteris (Ehrh.) Maxon; *Lastrea Oreopteris* (Ehrh.) Bory; *Aspidium Oreopteris* (Ehrh.) Sw.; *Nephrodium Oreopteris* (Ehrh.) Desv.; *Lastrea
montana* Newm.; *Thelypteris Oreopteris* (Ehrh.) Slosson

Rhizome short, stout, ascending, clothed when young with brown lanceolate
scales. *Lvs 30–100 cm., forming a crown at apex of rhizome*, suberect not bent,
pinnate with the pinnae pinnatifid nearly to their rhachises, dying in autumn;
petiole rather stout, up to c. ¼ as long as blade, sparsely clothed with pale
brown ovate or ovate-lanceolate scales which are most numerous below;
blade lanceolate or oblanceolate, firm, yellowish-green, *smelling of lemon when
crushed*, sparsely scaly on the rhachis below, the *lower surface covered with
numerous brownish-yellow glands*, the rhachis of the pinnae with short white

hairs at least when young; *pinnae* c. 20–30 on each side, all in one plane and parallel, longest c. 5–12 cm. linear-lanceolate, about the middle of the lf, *narrow at the base* but scarcely stalked, decreasing in length downwards, the *lowest distant, deltoid, short* (c. 1 cm.); segments c. 15–25 on each side of the longest pinnae, ±equal for some distance above the base or often the basal pair or the basal one on the upper side longer, c. 7–12 mm., oblong, wide at the base, obtuse or subacute, sinuate-crenate or subentire, the margins often recurved. Sori ½ mm. or less, close to the margin of the segment; indusium small, thin, falling early, irregularly toothed, sometimes 0. Spores ripe 7–8. $2n = 68^*$. H.

Native. Woods, mountain pastures, screes, etc., especially characteristic of steep banks above streams, absent from limestone; ascending to 3000 ft.; throughout Great Britain, common in the wetter districts, very local in the drier ones and absent from several eastern counties; widespread in Ireland but rather local and absent from several counties. 105, H 31. Europe from Scandinavia and Denmark to N. Spain, Corsica, Sicily and Serbia, east to S.W. Russia (Middle Dnieper region); Madeira; E. Siberia (Angara-Sayan region). A ssp. in Kamchatka, Japan and western N. America from Alaska to N. Washington.

2. T. palustris Schott   'Marsh Fern.'

Dryopteris Thelypteris (L.) A. Gray; *Lastrea Thelypteris* (L.) Bory; *Aspidium Thelypteris* (L.) Sw.; *Nephrodium Thelypteris* (L.) Strempel

Rhizome long, slender, creeping below the ground, with a few small scales which soon disappear. *Lvs solitary* (rarely a few in a tuft), erect not bent, 15–120 cm., the sterile often smaller with shorter petioles and often broader segments than the fertile, *pinnate with the pinnae pinnatifid nearly to the rhachis*, dying in autumn; petiole (of fertile lf) about as long as blade or longer, blackish at base, slender, brittle, without or with very few scales; *blade lanceolate*, thin, light green, *not smelling of lemon when crushed, eglandular*, very sparingly or not scaly, with a few white hairs at least when young on rhachis, midribs and margins and sometimes lower surface; *pinnae* c. 15–25 on each side, *all in one plane and parallel*, the longest c. 5–9 cm., linear-lanceolate, about the middle of the lf, *shortly stalked*, decreasing in length downwards but never short and deltoid, or sometimes nearly uniform in length from the middle to the base of the lf; segments c. 10–20 on each side of the longest pinnae, ±equal for some distance above the base or often the basal pair or the basal one on the lower side longer, c. 6–12 mm., oblong to triangular-lanceolate, wide at base, obtuse or acute, margins sinuate and occasionally with 1 or 2 small lobes at the base, usually flat in sterile lvs and sharply reflexed in the fertile ones. Sori forming a row on each side of the segment, about midway between the midrib and margin (but usually appearing close to the latter because of the reflexing), very small; indusium small, thin, irregularly toothed. Spores ripe 7–8. $2n = 70^*$. Grh.

Native. Marshes and fens, often abundant in carr or alderwood; from Devon and Kent to Isle of Man, Cumberland and Yorks, local; Perth, Angus; N. Kerry, Clare and Wicklow northwards, local. 49, H 16. Europe from S.E. Norway, Sweden and N. (not arctic) Russia to C. Portugal and N. Spain,

Corsica, Italy, Greece and Caucasus; temperate Asia to Syria, Himalaya, S. China and Sakhalin; S. India; Algeria (very rare); tropical Africa; eastern N. America from New Brunswick and Manitoba to Florida and Texas. A ssp. in S. Africa and New Zealand.

3. T. phegopteris (L.) Slosson   Beech Fern.

Dryopteris Phegopteris (L.) C.Chr.; *Polypodium Phegopteris* L.; *Phegopteris polypodioides* Fée

Rhizome long, *slender, creeping below the ground*, with brown lanceolate scales when young. *Lvs solitary*, 10–50 cm., *pinnate with the pinnae pinnatifid nearly to the rhachis*, dying in autumn; petiole erect, as long to twice as long as blade or more, slender, brittle, with a few scales at the base and sometimes a few small ones near the apex, the upper part usually clothed with reflexed white hairs; *blade triangular-ovate*, bent at nearly a right angle from the petiole, rather thin, light dull green, rhachis scaly and ± hairy, *the pinnae ± hairy on both surfaces*, especially the midrib below, eglandular; pinnae c. 10–20 on each side, *the lowest pair bent backwards away from the others*, 4–11 cm., lanceolate, *longer or slightly shorter than the one above it, the remainder decreasing rapidly in length*, all (except the lowest pair) *attached to the rhachis by a broad decurrent base*; segments c. 10–20 on each side of the lowest pinnae, longest near the middle of the pinna c. 5–17 mm., oblong, wide at base, obtuse or subacute, usually entire or subentire but occasionally crenate or dentate. Sori close to the sometimes somewhat reflexed margin of the segment, small (c. $\frac{1}{2}$ mm. or less); indusium 0. Spores ripe 6–8. $2n=90^{*}$. Apogamous. Grh.

Native. Damp woods and shady rocks, ascending to 3680 ft., absent from limestone; Cornwall to Somerset, Sussex and Berks; Wales and the border counties; Cheshire, Derby and Yorks northwards; rather common in the north, rare in S. England; Clare, Leix and Wicklow northwards; Kerry, W. Cork. 84, H21. Europe from Iceland, Scandinavia and N. (not arctic) Russia to the mountains of S. France, Corsica, N. Italy, Macedonia and the Caucasus; temperate Asia to Asia Minor, Himalaya and Japan; N. America, south to Virginia and Oregon.

4. T. dryopteris (L.) Slosson   Oak Fern.

Gymnocarpium Dryopteris (L.) Newm.; *Phegopteris Dryopteris* (L.) Fée; *Polypodium Dryopteris* L.; *Dryopteris Linnaeana* C.Chr.

Rhizome long, slender, creeping below the ground with a few brown ovate scales when young. *Lvs solitary*, 10–40 cm., *3-pinnate or 2-pinnate with deeply pinnatifid pinnae*, when young rolled up so as to resemble 3 small balls, dying in autumn; petiole erect, $1\frac{1}{2}$–3 times as long as blade, slender, brittle, with a few scales near the base; *blade deltoid*, bent at nearly a right angle to petiole, *brilliant yellowish-green when young*, becoming less bright *but remaining a clear green, thin, naked*; pinnae c. 5–10 on each side, lowest pair much longer than the others, not deflexed, 3–15 cm., long-stalked, triangular-ovate; *lowest pinnule on lower side of lowest pinna much the longest*, 1–6 cm., *about equalling in size the 3rd pinna from the base on either side*; upper pinnae not more than pinnate with lobed pinnules, sessile or shortly stalked; segments ± oblong, but not markedly parallel-sided, c. 5–15 mm., the lower ones rounded at the

narrow base, entire, toothed or lobed, margins flat; sori rather small (some-
imes nearly 1 mm.), near the margin of the segments; indusium 0. Spores
ripe 7–8. $2n = 160^*$. Grh.

Native. Damp woods and shady rocks, ascending to 3000 ft.; absent from
limestone; E. Cornwall to Somerset and S. Hants; Berks, Bucks, Oxford;
W. Suffolk; Wales and the border counties; Cheshire, Derby and N. Lincoln
northwards; rather common in the north, rare in S. England; very rare in
Ireland, Clare, Wicklow, Sligo, Leitrim, Antrim. 84, H5. N. Europe and
Asia from Iceland to E. Siberia south in the mountains to N.W. Spain, Corsica,
Italy, Macedonia; Caucasus, W. Himalaya, China and Japan; N. America
south to Virginia and Oregon.

5. T. robertiana (Hoffm.) Slosson 'Limestone Fern.'

Gymnocarpium Robertianum (Hoffm.) Newm.; *Phegopteris Robertiana* (Hoffm.)
A.Br.; *Polypodium Robertianum* Hoffm.; *P. calcareum* Sm.; *Dryopteris
Robertiana* (Hoffm.) C.Chr.

Differs from *T. dryopteris* as follows: Lvs rather larger on an average (15–
55 cm.), when young each pinnule and pinna rolled up separately, the lf then
rolled up as a whole; scales often ascending to nearly middle of petiole; *blade
little bent relative to the petiole, dull green, rather firm, glandular on the rhachis
and rhachises of the pinnae and on the surface beneath*, the glands short-stalked
and appearing mealy; pinnae up to c. 15; *lowest pinnule on lower side of the
lowest pinna the longest but less markedly so than in T. dryopteris* and *considerably smaller than the 3rd pinna from the base*; 2nd pair of pinnae often
long-stalked; segments parallel-sided, margins usually somewhat recurved.
Spores ripe 7–8. $2n = c.\ 160^*$. Grh.

Native. Limestone screes and rocks, ascending to 2000 ft., local; Somerset,
Sussex and Norfolk to Cumberland and Durham; Perth, Sutherland; E. Mayo.
32, H1. Europe from Scandinavia and the Ladoga-Ilmen region of Russia
to N. Spain, Corsica, Italy, Thessaly, Crimea and Caucasus; mountains of
Asia from Afghanistan to the Far East of Russia; N. America from Labrador
to Alaska south to New Brunswick and Iowa.

Subfamily 7. POLYPODIOIDEAE. Often epiphytic. Lvs nearly always jointed
to the rhizome. Sori orbicular or oblong on the under surface of the lf, near
the ends of the veins if the latter are free, occasionally sporangia scattered or
sori fused; indusium 0.

15. POLYPODIUM L.

Rhizome creeping, fleshy, *scaly*. Lvs usually pinnatifid or 1-pinnate; *veins
usually regularly anastomosing with free endings inside the loops, sometimes
free*. Sori terminal on the veins in 1(–3) rows on each side of the midrib.

About 50 spp., cosmopolitan, mainly tropical America, Asia and Polynesia.

1 Sori orbicular when young; lf-blade oblong; annulus of sporangium with
  11–15 thick-walled cells. **1. vulgare**
  Sori oval when young; lf-blade ovate to ovate-lanceolate; annulus of
  sporangium with 4–10 thick-walled cells. 2

2 Paraphyses 0; annulus of sporangium with 7–10 thick-walled cells.

2. interjectum

Paraphyses present; annulus of sporangium with 4–6 thick-walled cells.

3. australe

P. vulgare agg. (spp. 1–3).   'Polypody.'

Rhizome creeping on or below the surface, rather stout, densely clothed when young with reddish-brown lanceolate scales. Lvs solitary, pinnatifid nearly to the rhachis or pinnate, persistent; petiole from about ⅓ to nearly as long as blade, naked, jointed to the rhizome, ± erect; blade 5–45 cm., from ovate-lanceolate to oblong or linear-lanceolate, suberect to drooping, somewhat coriaceous, dull green; pinnae or segments c. 5–25 on each side, usually equally long for a considerable distance in the middle of the lf, the lowest somewhat shorter, oblong to linear-lanceolate, acute or obtuse, wide-based, subentire to deeply crenate or serrate (sometimes doubly so); veins free. Sori on the ends of the lowest fork on the upper side of the main veins, about midway between the midrib and the margin, large (1·5–3 mm.), orbicular or shortly oval, often bright yellow. Grh. or Ch.

Native. Woods, often on trees but also on the ground, rocks and walls, ascending to 2800 ft. in Kerry; throughout the British Is., common in the wetter districts, less so in the drier ones. 112, H40, S. All Europe and Mediterranean region; Macaronesia; W. Siberia, Tibet, China, Japan; eastern N. America from Newfoundland and Keewatin to Georgia and Missouri; a ssp. or allied sp. in western N. America; S. Africa; Kerguelen.

1. P. vulgare L.

Lf-blade (5–)10–25 cm., oblong (i.e. with the pinnae ± equal in length from the base to near the apex). Longest pinnae (10–)15–35(–45) mm., oblong, rounded at apex, subentire. Sori orbicular when young (sometimes becoming oval); paraphyses 0. Annulus of sporangium with 11–15 thick-walled cells. Becoming fertile in spring but long remaining so. $2n = 148^*$.

Native. Throughout the British Is.

2. P. interjectum Shivas

Usually more robust than *P. vulgare*. Lf-blade (5–)15–30(–40) cm., ovate-lanceolate or lanceolate (i.e. with the longest pinnae ⅓ or ½ distance from the base, the pinnae becoming shorter upwards and downwards). Longest pinnae (20–)40–70(–90) mm., oblong or oblong-lanceolate, rounded or tapered to apex, subentire to deeply toothed. Sori oval when young; paraphyses 0. Annulus of sporangium with 7–10 thick-walled cells. Becoming fertile in spring but long remaining so. $2n = 222^*$.

Native. Throughout the British Is.

3. P. australe Fée

Lf-blade (5–)10–15 cm., ovate, ovate-lanceolate or triangular-lanceolate (lowest pinna usually hardly shorter than those above), often with the pinnae bent upwards from the rhachis. Longest pinnae (35–)45–60(–75) mm., oblong or oblong-lanceolate, rounded or tapered to apex, always toothed. Sori oval when young; paraphyses present. Annulus of sporangium with 4–6 thick-walled cells. Becoming fertile in late summer. $2n = 74*$.

Native. Very local on limestone in S.W. England, Wales and S.W. Ireland.

8. MARSILEACEAE

Aquatic or subaquatic perennial herbs with long, creeping, hairy rhizomes. Lvs alternate, 2-ranked, circinate in bud, subulate and entire or with 2 or 4, palmately arranged, broad, cuneiform lflts on a long petiole. Sporangia borne in globose or ovoid-oblong, hard, hairy 'sporocarps' at the base of the petiole. Each sporocarp containing 2 or more sori, each of which is surrounded by an indusium. Sporangia of two kinds, both occurring in the same sorus, the megasporangia below the microsporangia, developing in regular sequence from apex to base of the sorus. Megasporangia with 1 megaspore; microsporangia with numerous microspores; annulus 0 or rudimentary. Male prothallus consisting of 2 vegetative cells and 2 antheridia each producing 16 multi-ciliate spermatozoids. Female prothallus with many vegetative cells and a single archegonium which is situated on a projection from the megaspore.

Three genera and about 70 spp., cosmopolitan (except Arctic).

1. Pilularia L.

Lvs subulate, entire. Sporocarp solitary, divided by the indusia into 2 or 4 compartments, each containing 1 sorus and splitting separately longitudinally at maturity and releasing the sporangia in a mass of mucilage.

Six spp., Europe and Mediterranean region, America (temperate, and tropical mountains), Australia and New Zealand.

1. P. globulifera L.   Pillwort.

Rhizome to 50 cm., slender, creeping, often with short axillary branches. Lvs 3–15 cm., subulate. Sporocarps c. 3 mm. diam., globose, shortly (up to 1 mm.) stalked, brown at maturity, hairy. Sori 4. Spores ripe 6–9. $2n = 26$. Hel. or Hyd.

Native. Edges of ponds and lakes, often submerged, on acid soils. 69, H 6, S. From Moray and the Outer Hebrides southwards, local, absent from many counties; Kerry, Galway, Tyrone, Donegal, Antrim and Londonderry; Jersey. Europe from S. Scandinavia to France (not south), N. Italy and Serbia, east to the Urals; Portugal; local everywhere and becoming more so eastwards.

9. AZOLLACEAE

Plant small, free-floating, with branched stems bearing roots and lvs. Lvs alternate, 2-ranked, imbricate, 2-lobed, the upper lobe floating, green and assimilating and with a hollow filled with mucilage and containing the threads of the blue-green alga *Anabaena*; lower lobe thin, submerged and bearing the sori in pairs on cylindrical receptacles, covered by a flange of the upper lobe of the lf. Sori surrounded by an indusium from the base and consisting of either numerous microsporangia developed successively from the apex to base of sorus or a single megasporangium; annulus 0; megaspore 1 in each sporangium, with 3 floats; microspores 64, grouped together into 'massulae', each massula being in some spp. furnished with projecting barbed hairs ('glochidia'). Massulae becoming fixed to the megaspore by the glochidia (when these are present), the spores germinating within them; male prothallus consisting of a few vegetative cells and 8 spermatozoid mother cells. Female prothallus small, green.

One genus.

1. Azolla Lam.

The only genus. Six spp., tropical and warm temperate (not Europe).

*1. A. filiculoides Lam.

Plant 1–5 cm. diam., often growing in large masses, bluish-green, usually becoming red in autumn. Upper lf-lobes c. 1 mm., ovate, obtuse, covered above with unicellular hairs which make the surface non-wettable, margins hyaline. Sori in each pair either both containing megasporangia or one of each kind of sporangium. Glochidia present, not divided into cells by transverse walls. Spores ripe 6–9. Hyd.

Introduced. Ditches, etc. Naturalized in many places, mainly in S. England, suffering severely in hard winters. Native of western N. America from Washington southwards, C. and S. America; naturalized in C. and S. Europe.

A. caroliniana Willd. has been recorded as naturalized on insufficient evidence. It is usually a smaller plant with shorter dichotomous branches, but can only be certainly distinguished microscopically. The hairs on the lf-surface are 2-celled and the glochidia divided by transverse walls. Native of America from E. United States to Brazil.

10. OPHIOGLOSSACEAE

Terrestrial (rarely epiphytic) herbs with short, usually erect, rhizome without scales; roots fleshy. Lvs one or more, stalked, not circinate in bud. Fertile lvs consisting of a sterile blade and one or more fertile spikes or a fertile panicle; (in our species the fertile spike or panicle appears as if terminal on a stem on which the sterile blade is borne laterally). Sporangia all alike, borne in two rows on the margins of the fertile spike or panicle-branches, sessile or nearly so, each derived from a group of cells; wall of several layers of cells; annulus 0; spores very numerous (1500–15,000). Prothallus usually subterranean, massive, without chlorophyll but with endotrophic mycorrhiza, bearing organs of both sexes, the antheridia sunk in the tissues.

Three genera, the two following and the monotypic *Helminthostachys* Kaulf. from tropical Asia and Australasia.

Sterile blade pinnate; fertile portion a panicle.   1. Botrychium
Sterile blade simple, entire; fertile portion a spike.   2. Ophioglossum

1. BOTRYCHIUM Sw.

Sterile blade pinnately lobed or *pinnate* (sometimes several times); veins free. *Sporangia subsessile, arranged in a panicle*, opening by a transverse slit.

About 30 spp., cosmopolitan, mainly north temperate regions. Three spp. besides the following have been reported from Scotland, each as found on one occasion only (none recently); all have bipinnate leaves.

1. B. lunaria (L.) Sw.   Moonwort.

Rhizome underground, erect, very short, usually unbranched. Lvs (2–)5–15(–30) cm., solitary, rarely 2, erect, sheathed at the base by the brown remains of the previous year's lvs; sterile blade (1–)2–5(–12) cm., usually inserted about the middle of the lf, oblong in outline, pinnate; pinnae (2–)4–7(–9) pairs, fan-shaped, entire or shallowly and irregularly (rarely deeply) crenate, without midrib; fertile panicle (0·5–)1–5 cm. (excluding stalk), overtopping sterile blade, 1–3 times branched. Spores ripe 6–8. $2n=90^*$. Grh.

Native. Dry grassland and rock ledges throughout the British Is., ascending to 3350 ft. in Perth, rather local, though unrecorded for only a few counties. 107, H 38, S. Arctic and north temperate zones, south to the Azores, Morocco (Atlas), Sicily (Etna), Roumania, Asia Minor, Himalaya, California and New York; Australia, Tasmania, New Zealand.

2. OPHIOGLOSSUM L.

Sterile blade simple (rarely palmately lobed); veins reticulate. *Sporangia sunken, arranged in a simple spike* (or spikes), opening by a transverse slit.

About 50 spp., cosmopolitan.

Sterile blade with free vein endings inside the network, ovate to ovate-
    lanceolate or oblong; spores ripe 5–8; widespread.   **1. vulgatum**
Sterile blade without free vein endings inside the network, oblanceolate or
    linear-lanceolate; spores ripe 1–2; Channel and Scilly Is.   **2. lusitanicum**

1. O. vulgatum L.   Adder's Tongue.

Rhizome underground, erect, very short. Roots producing new plants from adventitious buds. *Lvs* 1(–3), (4–)8–20(–45) *cm.*, erect; *sterile blade* (1·5–)4–12(–15) *cm.*, *ovate to ovate-lanceolate or oblong*, entire, obtuse or acute, sheathing the stalk of the fertile spike at the base, *with free vein endings inside the meshes of the network*, epidermal cells with sinuate walls; fertile spike 2–5(–7) cm. (excluding stalk), overtopping sterile blade at maturity; apex sterile, acute. Spores tubercled. Spores ripe 5–8. $2n=500$–520^*. Grh.

Europe from Iceland, S.E. Norway and N. (not arctic) Russia to C. Spain and Portugal, Corsica, Sicily, Macedonia, the Caucasus; Madeira, Azores; N. and W. Asia east to Kamchatka, south to Lebanon; N. Africa; N. America from Prince Edward Is. to Alaska, south to Florida and Washington.

Ssp. vulgatum

Lf solitary (exceptionally 2), rarely less than 8 cm.; sterile blade 3–15 cm., obtuse or subacute, usually broad at the base. Sporangia (12–)16–40 on each side of the spike.

Native. Damp grassland, fens and scrub throughout the British Is.; rather common in England, Wales and Ireland, less so in Scotland. 104, H 40, S. Nearly the whole range of the species but absent from Iceland, Azores (etc. ?).

Ssp. **ambiguum** (Coss. & Germ.) E. F. Warb.

Ssp. *polyphyllum* E. F. Warb., p.p.; var. *ambiguum* Coss. & Germ.

Lvs (1–)2–3 together, 4–10 cm.; sterile blade 1·5–3·5 cm., usually acute, tapered at the base or contracted so as to appear stalked. Sporangia 6–14 on each side of the spike.

Native. Sandy ground and short turf near the sea, very local; Scilly Is., Lundy Is., Dorset, Glamorgan, Merioneth, Anglesey, Orkney, Fair Is., Kerry, Donegal, Guernsey. Iceland, W. France, Azores (etc. ?).

2. O. lusitanicum L.

Differs from *O. vulgatum* as follows: *Lvs 1–3, 2–6(–10) cm.; sterile blade* 1–2·5 cm., narrowly *oblanceolate or linear-lanceolate*, acute, tapered at base, shortly sheathing, *without free vein endings inside the network*, epidermal cells with straight walls; spike 7–15 mm. Sporangia 5–8(–12) on each side of the spike. Spores smooth. Spores ripe 1–2. $2n = 250$–260^*. Grh.

Native. Grassy cliff-tops in Scilly Is. and Guernsey. Mediterranean region and W. Europe north to W. France, Canaries, Azores.

GYMNOSPERMAE

11. PINACEAE

Trees, rarely shrubs; branches usually in regular whorls at the end of each year's growth. Buds scaly. Lvs spirally arranged, linear, entire or minutely serrulate, usually evergreen. Fls monoecious, both the male and female in cones, formed of numerous scales, spirally arranged. Scales of male cone (microsporophylls) bearing 2 pollen sacs on the under-surface. Scales of female cones double, consisting of an ovuliferous scale bearing 2 inverted ovules on its upper surface, borne in the axil of a bract, the bract and scale free except at the extreme base. Pollen usually with 2 conspicuous bladder-like wings. Cones usually large and ± woody in fr.; seeds winged.

Nine genera and about 180 spp., northern hemisphere, mainly temperate.

1 Lvs all solitary on the main stems; short shoots 0.   2
   Assimilating lvs all or mainly on short lateral shoots.   5

2 Lvs without persistent woody decurrent prominent bases, the scars not or
      scarcely projecting after the lvs have fallen; bract large, often exserted
      in the fr. cone.   3
   Lvs with persistent woody decurrent bases, projecting as pegs after the
      lvs have fallen; bract small, never exserted in the fr. cone.   4

3 Cones erect; scales falling from the axis with the seeds; lf scars quite flat;
      lvs rigid; buds ovoid, obtuse.   1. ABIES
   Cones pendulous; scales and bracts persistent after the fall of the seeds;
      lf scars slightly prominent; lvs rather soft; buds fusiform, acute.
                                                                     2. PSEUDOTSUGA

4 Branches on the trunk all in whorls; lvs without a stalk above the per-
      sistent base, not 2-ranked on the upper side of the shoot; cones usually
      large.   3. PICEA
   Branches on the trunk not all whorled; lvs with a distinct stalk above the
      persistent base, spreading laterally and so appearing 2-ranked; cones
      small.   4. TSUGA

5 Short shoots prominent, with numerous lvs, continuing to grow for several
      years and producing fresh lvs each year; lvs deciduous (if evergreen,
      see *Cedrus*); lvs on the long shoots green and assimilating.   5. LARIX
   Short shoots much reduced, appearing as a cluster of 2(–5) lvs with a few
      scales at the base and finally falling as a whole; lvs evergreen; lvs on the
      long shoots brownish and scale-like.   6. PINUS

1. ABIES Mill.

Evergreen trees of pyramidal habit, the branches in annual whorls. Bark of young trees smooth, often fissured on older ones. Twigs smooth (rarely grooved). Buds usually ovoid, obtuse or acute (fusiform only in *A. venusta* (Dougl.) K. Koch). *Lvs spirally inserted on the stems* (short shoots 0) but often spreading into 2 lateral sets, ± flat, with 2 whitish waxy bands containing the stomata below and occasionally also above. contracted above the base, inserted directly on the stem and leaving a flat, ±circular scar on falling. *Ripe cones erect; scales* thin, densely imbricate, narrowed into a stalk below, *finally falling from the persistent axis*; *bracts long*, included or exserted. Seeds winged, ripening the first year. Cotyledons 4–10.

About 40 spp., north temperate zone. A number of species besides the following are sometimes planted in parks, etc.

*A. alba Mill. (*A. pectinata* DC.)   Silver Fir.

Bark grey, smooth, becoming scaly on old trees. *Buds* small, ovoid, obtuse, *not resinous.* Twigs grey, pubescent. *Lvs* 1·2–3 *cm.*, *arranged in two lateral sets, the lower spreading horizontally, the upper* much shorter, *pointing upwards* and outwards, notched at apex, *dark shining green above.* Cones 10–14 cm., cylindric, greenish when young, brown when ripe; bracts exserted, reflexed.

Commonly planted for ornament and formerly for forestry but now little used, though it thrives well in E. Anglia. Native of the mountains of C. and S. Europe from S. Germany to the Pyrenees, Corsica, Apennines, Albania and Macedonia.

*A. grandis Lindl.   'Giant Fir.'

Bark becoming dark brown and scaly. *Buds* small, ovoid, obtuse, *resinous.* Twigs olive-green or brown, pubescent. *Lvs* 2–5 *cm.*, *arranged in two lateral sets, all spreading horizontally*, the upper shorter than the lower, notched at apex, *dark shining green above.* Cones 5–10 cm., cylindric, green; bracts hidden.

Planted for ornament and recently for forestry on an experimental scale. Native of western N. America from British Columbia to California and Montana.

*A. procera Rehd. (*A. nobilis* (Dougl. ex D. Don) Lindl., non A. Dietr.)   'Noble Fir.'

Bark reddish-brown and ridged on old trees. Buds small, ovoid, obtuse, resinous above. Twigs rusty-pubescent. Lvs 1·5–3·5 cm., the lower ones spreading laterally, *the upper ones appressed to the shoot for a short distance, then curving upwards* (so that the upper side of the shoot is hidden), entire or slightly notched at apex, *bluish-green and with stomata on both surfaces.* Cones 14–25 cm., oblong-cylindric, purplish-brown at maturity. Bracts long-exserted, reflexed.

Commonly planted for ornament in Scotland and occasionally met with as small plantations. Native of western N. America from Washington to N. California.

2. PSEUDOTSUGA Carr.

Similar to *Abies*. Bark of old trees thick, corky, furrowed. *Buds fusiform*, acute. Lf-scars slightly prominent after the fall of the lf. *Ripe cones pendulous; scales* rigid, concave, *persistent; bracts* 3-lobed, *exserted.*

About 5 spp., western N. America and E. Asia.

*1. P. menziesii (Mirb.) Franco   Douglas-fir.

P. Douglasii (Lindl.) Carr.; *P. taxifolia* Britton

Twigs yellowish, becoming brown, usually pubescent. Lvs 2–4 cm., ± spreading and arranged in two lateral sets, entire at apex, acute or obtuse, dark or bluish-green above. Cones 5–10 cm., ovoid. $2n = 26$. MM.

Introduced. Fairly frequently planted for forestry and commonly for ornament and sometimes self-sown. Native of western N. America from British Columbia to California, W. Texas and N. Mexico.

3. PICEA A. Dietr.

Evergreen trees of pyramidal habit, the branches in annual whorls. Bark scaly. *Twigs covered with lf-cushions which are separated by grooves and end in prominent peg-like projections on which the lvs are inserted and which remain after the lvs fall.* Buds ovoid or conic. *Lvs* spirally inserted on the stems (short shoots 0), *sessile* on the peg-like bases, 4-angled with stomata on all sides, or flat with stomata only below (but on the morphologically upper surface), *not divided into 2 lateral sets when seen from the upper side of the shoot*; resin canals 2. *Ripe cones pendulous; scales persistent*, rather thin; *bracts minute.* Seeds winged, ripening the first year. Cotyledons 4–15.

About 30 spp., north temperate zone. Several are planted as ornamental trees.

Lvs 4-sided with inconspicuous stomatal lines on each side.   **1. abies**
Lvs flat, dark green above, with conspicuous glaucous lines beneath.
                                                                          2. sitchensis

*1. P. abies (L.) Karst.   Norway Spruce.
P. excelsa Link

Tree to 40 m. Bark brown. Buds ovoid, acute, not resinous. Twigs yellowish- or reddish-brown, glabrous or slightly pubescent. *Lvs* 1–2 cm., lower ranks spreading laterally, upper ranks pointing forward, 4-*angled and rhombic in section, with 2 or 3 rather inconspicuous glaucous stomatal lines on each side*, acute and with a short horny point. Cones 10–15 cm., cylindric; scales thin, ±rhombic, irregularly toothed at apex. Fl. 5–6. Fr. 10. $2n=24, 48$. MM.

Introduced. Commonly planted for forestry but apparently rarely re-generating. Native of N. and C. Europe from Scandinavia and N. Russia to the Pyrenees, Italian Alps, Albania and Macedonia.

*2. P. sitchensis (Bong.) Carr.   Sitka Spruce.
P. Menziesii (Dougl. ex D. Don) Carr.

Tree to 40 m. Bark brown. Buds ovoid, acute, resinous. Twigs light brown, glabrous. *Lvs* 1·5–2·5 cm., *flat*, arranged as in *P. abies, dark green above, with 2 conspicuous glaucous stomatal bands beneath*, acute and with a pungent horny point. Cones 5–10 cm., oblong-cylindric; scales thin, rhombic-oblong, irregularly toothed at apex. $2n=24$. MM.

Introduced. Frequently planted for forestry especially in the north and west, not reported as regenerating. Native of western N. America from Alaska to California.

4. Tsuga (Endl.) Carr.

Evergreen trees of pyramidal habit, branching irregular, leading shoot and branches often pendulous at tips. Bark furrowed, red-brown. *Twigs and lf-insertion as in Picea* but grooves often less conspicuous. Buds globose or ovoid, not resinous. *Lvs* spirally inserted on the stems (short shoots 0), *shortly stalked*, ±flat and usually with stomata only beneath, usually *spreading laterally into 2 sets*; resin canal 1. *Ripe cones pendulous; scales persistent; bracts inconspicuous.* Seeds winged, ripening the first year. Cotyledons 3–6.

About 10 spp., temperate N. America and E. Asia. Several spp. are sometimes planted for ornament.

*T. heterophylla (Raf.) Sarg.   Western Hemlock.
T. Albertiana (A. Murr.) Sénécl.; *T. Mertensiana* auct.

Tree to 70 m. Buds globose-ovoid, obtuse. Twigs light brown, densely pubescent for several years. Lvs 6–18 mm., spreading laterally into 2 sets, dark green, grooved above, with 2 broad, ill-defined glaucous bands beneath, obtuse, minutely serrulate. Cones 1·5–2·5 cm., scales obovate, concave, entire.

Sometimes planted for forestry. Native of western N. America from S. Alaska to California and Idaho.

5. LARIX Mill.

Deciduous trees, ± pyramidal when young, branches horizontal. Bark scaly. *Twigs of two kinds,* long shoots with spirally arranged lvs and *short lateral spurs* increasing very slowly and producing a fresh tuft of numerous lvs at the apex each year. Lvs flat. Cones erect; scales persistent, suborbicular to oblong; bracts large, conspicuous in fl. and often brightly coloured. Seeds winged, ripening the first year. Cotyledons c. 6.

About 10 spp., cool north temperate regions.

Twigs yellowish, not pruinose; lvs bright green, without white bands beneath; ripe cone scales erect.   **1. decidua**

Twigs reddish, pruinose; lvs glaucous, with 2 conspicuous white bands beneath; ripe cone scales curved back near the apex.   **2. leptolepis**

*1. L. decidua Mill.   European Larch.

L. europaea DC.

Tree to 50 m. Bark greyish-brown, shed in small plates. *Young long shoots yellowish,* glabrous. *Lvs* of short shoots 12–30 mm., *bright green,* 30–40 on each shoot. Bracts bright pink (rarely cream) in fl. Cones 2–3·5 cm., ovoid; *scales erect,* 40–50, *suborbicular,* entire, pubescent outside; bracts almost concealed by the scale. Fl. 3–4. Fr. 9. $2n = 24, 48$. MM.

Introduced. Commonly planted for forestry and ornament, frequently self-sown and naturalized in a number of places. Native of the mountains of C. Europe from the Alps to S. Poland, Carpathians and Croatia.

*2. L. leptolepis (Sieb. & Zucc.) Gord.   Japanese Larch.

L. Kaempferi Sarg., non Carr.

Tree to 30 m., of stiffer habit than *L. decidua.* Bark reddish-brown, shed in flakes or strips usually rather larger than those of *L. decidua. Young long shoots reddish with a glaucous bloom,* slightly pubescent or glabrous. *Lvs* of short shoots 15–35 mm., *bluish-green with 2 whitish bands beneath,* broader than those of *L. decidua,* c. 40 on each shoot. Bracts cream tinged with pink in fl. Cones 1·5–3·5 cm., ovoid; *scales* numerous, *somewhat spreading and with the margin rolled back at the apex* so that the cone has a rather rosette-like appearance, truncate or slightly emarginate at apex, slightly pubescent outside; bracts mostly concealed. Fl. 3–4. $2n = 24$. MM.

Introduced. Rather frequently planted for forestry, not known to be naturalized. Native of Japan.

CEDRUS Trew

Evergreen trees, pyramidal when young, becoming spreading and massive. Twigs and lf-arrangement as in *Larix.* Cones erect, taking 2 or 3 years to ripen, large; scales densely imbricate, falling from the persistent axis at maturity; bracts minute. Cotyledons 9–10.

Four closely allied spp., Mediterranean region and Himalaya; three of them are frequently planted and are occasionally met with in woods, etc., the following being the most common.

*C. libani A. Rich. (*C. libanensis* Mirb.)   Cedar of Lebanon.

Leading shoot of young trees usually somewhat curved, branches spreading, not drooping. Lvs mostly 2·5–3 cm. Cones 8–10 × 4–6 cm. $2n = 24$.

    Native of Asia Minor and Lebanon.

6. PINUS L.

Evergreen trees with regularly whorled branches, of pyramidal habit when young but often with spreading crowns at maturity. Bark rough, furrowed or scaly. Buds usually large. *Shoots of 2 kinds, ordinary long shoots bearing spirally arranged scale-lvs without chlorophyll and with woody decurrent bases, and short shoots borne in the axils of the scale-lvs and apparently consisting only of a definite number (usually 2, 3 or 5) of green needle-like lvs surrounded by sheathing scale-lvs at the base, the short shoot not growing further and finally falling off as a whole.* Male cones replacing short shoots at the base of a year's growth. Female cones replacing long shoots, taking 2(–3) years to ripen. Ripe cones with thin or thick and woody scales; in thick-scaled spp. the exposed part of the scale much thickened and provided with a prominent protuberance or scar ('umbo'); in other spp. the scale is flat with a small terminal umbo; bracts minute. Seeds usually winged. Cotyledons 4–15. Seedlings with green spirally arranged lvs.

    About 80 spp., northern hemisphere, in the tropics mainly in the mountains. A number of spp. besides the following are planted for ornament.

1 Lvs glaucous, less than 10 cm.; bark of upper part of trunk bright reddish-
    brown.   **1. sylvestris**
    Lvs dark green, mostly more than 10 cm.; bark not bright reddish-brown.   *2*

2 Buds resinous, with appressed scales; lvs not 2 mm. broad; bark of 3-year
    old twigs conspicuously divided into plates; cone 5–8 cm.   **2. nigra**
    Buds not resinous, the tips of the scales strongly recurved; lvs very stout,
    more than 2 mm. broad; bark of 3-year old twigs not conspicuously
    divided into plates; cone 9–18 cm.   **3. pinaster**

1. P. sylvestris L.   Scots Pine.

Tree to 30(–50) m.; old trees with a flat crown. *Bark on the upper part of the trunk bright reddish-brown or orange,* shed in thin scales; on the lower part thick, dark brown, and fissured into irregular longitudinally elongated plates. Twigs greenish-brown when young, becoming greyish-brown in the 2nd year. *Buds* 6–12 mm., oblong-ovoid, reddish-brown, *resinous, with the upper scales free (but not reflexed) at the tips.* Lvs 2 on each short shoot, 3–8(–10) *cm.* × 1–2 *mm.*, stiff, usually twisted, *blue-green* from the continuous glaucous stomatal lines on the inner surface and the interrupted ones on the outer, finely serrulate; resin-canals marginal; sheath at first c. 8 mm., whitish, becoming grey and shortening. Cones 3–7 cm., ovoid-conic, symmetrical or somewhat asymmetrical, dull brown; scales oblong, the exposed portion flat or somewhat projecting, umbo small, with a small prickle or its remains. Seed c. 3–5 mm., with a wing about 3 times its length. Fl. 5–6. $2n = 24$. MM.

    Native. The dominant tree of considerable areas in the Highlands from Perthshire to Sutherland, mainly in the east, ascending to 2200 ft.; usually supposed to be introduced in England and Ireland but possibly persisting in small quantity since the Boreal period; now often forming woods on sandy soils in S.E. England, less frequent elsewhere but planted and naturalized in many places. Europe from Scandinavia to the mountains of Portugal, S. Spain, N. Italy, Albania and Thrace; temperate Asia. Several geographical sspp. exist.

Ssp. **scotica** (Schott) E. F. Warb. Remaining pyramidal for a long period, then with a round crown. Lvs c. 4 cm. Cones c. 4 cm., symmetrical.

The native Scottish plant. 14. English plants usually more quickly become flat-topped and have longer needles and cones, but are of too mixed origin to be referred to any ssp.

***P. mugo** Turra   Mountain Pine.
P. montana Mill.

Usually a *low, spreading shrub* with irregular branches. Bark dark grey, scaly. *Buds* 6–12 mm., *very resinous*, scales appressed. *Lvs* 2 on each short shoot, 3–8 *cm.*, rigid, *curved but not twisted, dark green*, finely serrulate; resin canals marginal; *sheath* at first 12–15 *mm*. Cones 2·5–5 cm., symmetrical or asymmetrical, yellowish or dark brown; scales with exposed portion flat or pyramidal, umbo small with a small prickle. $2n=24$.

Sometimes planted for shelter in exposed situations. Native of the mountains of C. and S. Europe.

*2. P. nigra Arnold

Tree to 50 m. with short, spreading branches, usually remaining ± pyramidal. Bark of old trees thick, dark grey, deeply fissured. Twigs light brown, *the lf-bases persisting for several years and conspicuous as scale-like plates. Buds* 12–25 mm., oblong-ovoid or cylindric, acuminate, light brown, *resinous; scales appressed. Lvs* 2 on each short shoot, 8–16 *cm.* × 1–2 *mm.*, stiff, twisted or not, *dark green*, the stomatal lines interrupted and inconspicuous, finely serrulate, resin canals deep in the tissues; sheath at first c. 12 mm., becoming shorter. *Cones* 5–8 *cm.*, ovoid-conic, subsymmetrical, shining yellowish-brown; scales oblong, the exposed portion ± flat, transversely keeled, umbo rather small, usually with a small prickle. Fl. 5–6. $2n=24$. MM.

Introduced. Native of S. Europe, Asia Minor and Morocco. A variable sp., divisible into several geographical sspp. of which the two following are commonly grown in this country.

Ssp. **nigra**   Austrian Pine.
P. laricio var. *nigricans* (Host) Parl.

Habit rather gaunt and irregular. Lvs 8–12 cm., very rigid, nearly straight, dark dull green.

Commonly planted for making quick shelter but not suitable for forestry. S.E. Europe from Austria to Crete.

Ssp. **laricio** (Poir.) Palibin   Corsican Pine.
P. nigra var. *Poiretiana* (Antoine) Aschers. & Graebn.; *P. laricio* Poir.

Habit more regular. Lvs 10–14 cm., somewhat twisted, rather lighter in colour.

Fairly frequently planted for forestry and also for ornament. S.W. Europe.

*3. P. pinaster Ait   Maritime Pine.

Tree to 30 m., when mature usually bare of branches for the greater part of its height. Bark thick, dark reddish-brown, deeply fissured. Twigs light brown. *Buds* 20–25 mm., cylindric, acute, brown, *not resinous; the scales strongly reflexed at the tips. Lvs* 2 on each short shoot, 10–20 × 2–3 *mm.*, very stout and rigid, curved, *dark green* and somewhat shining, the stomatal lines interrupted and rather inconspicuous, finely serrulate; resin canals deep in the tissues; sheath at first c. 2 cm. *Cones* often clustered, 8–18 *cm.*, ovoid-conic, subsymmetrical, bright shining brown, often remaining closed for several years; scales oblong, the exposed portion ± pyramidal, transversely keeled, umbo prominent, with a prickle at first. Fl. 5–6. $2n=24$. MM.

Introduced. Frequently planted on poor sandy soils, especially near the sea in S. England and completely naturalized near Bournemouth. W. and S. France, Corsica, Spain, Portugal, C. and S. Italy; Algeria, Morocco.

*P. contorta Dougl. ex Loud.

Tree to 25(–50) m. Bark dark red-brown, deeply fissured into plates. *Buds* c. 12 mm., *very resinous*, scales appressed. *Lvs* 2 on each short shoot, 3–8 *cm.*, rigid, *twisted, dark green*, very obscurely serrulate; resin canals deep in the tissues; *sheath* at first 3–6 *mm.* Cones 2–5 cm., asymmetrical, light yellowish-brown; scales with exposed portion prominent on upperside of cone, umbo with a slender, often deciduous prickle. $2n=24$.

Sometimes planted for forestry on an experimental scale. Native of western N. America from Alaska to California and Colorado.

Of spp. with 3 needles on each short shoot **P. radiata** D. Don (*P. insignis* Dougl.) with bright green soft lvs 10–15 cm., from S. California, is frequently grown in the south-west.

The two following spp. have 5 slender soft bluish-green lvs on each short shoot:

*P. strobus L. Twigs with minute tufts of hairs below the lf-insertions, not pruinose. Lvs 6–14 cm. Cones 8–20 cm., cylindric, often curved; scales thin. $2n=24$.

Formerly planted for forestry, but unsuccessful in this country; old specimens are, however, met with in woods, etc. Native of eastern N. America from Newfoundland and Manitoba to Georgia and Iowa.

*P. chylla Swarz (*P. excelsa* Wall. ex D. Don, non Lam.). Twigs glabrous, pruinose. Lvs 10–20 cm. Cones 12–25 cm., cylindric; scales thickened at apex. $2n=24$.

Rather commonly planted for ornament in parks, etc. Native of the Himalaya.

TAXODIACEAE

Trees, rarely shrubs. Buds usually naked. Lvs spirally inserted, linear and needle-like or small and scale-like, usually evergreen. Fls monoecious, both sexes in cones formed of spirally arranged scales. Microsporophylls with 2–8 pollen sacs on the under-surface. Scales of female cone ± woody when ripe, bearing 2–12 erect or inverted ovules on their upper surface, the bract wholly or partially united with the scale. Pollen not winged.

Nine genera and about 15 spp., N. America, E. Asia, Tasmania. Spp. of several of the genera are planted in gardens.

SEQUOIA Endl.

Evergreen trees of pyramidal habit with horizontal or slightly drooping branches. Bark very thick, spongy, fibrous. Twigs green. Buds small. Lvs spirally arranged. Cones small (to 8 cm.), short, pendulous; scales with flattened woody ±rhombic tops persistent; ovules 3–12 on each side. Seeds winged.

Two spp., California.

*S. wellingtonia Seem. (*S. gigantea* (Lindl.) Decne, non Endl.)

Wellingtonia, Big Tree.

Tree to 100 m. in nature, with very wide tapering trunk. Buds naked. Lvs 3–6 mm., lanceolate, arranged all round the stem, densely imbricate, with flat bases adherent to the twig and completely covering it, tips free. Cone 5–8 cm., ovoid. $2n=22$.

Commonly planted in parks and gardens and occasionally among native vegetation.

*S. sempervirens (Lamb.) Endl.   Redwood.

Tree to 110 m. in nature. Trunk not narrowly tapering. Buds scaly. Lvs of lateral twigs 6–18 mm., linear, spreading into 2 lateral rows, free, dark green above, with 2 whitish bands beneath. Lvs of leading and cone-bearing twigs c. 6 mm., spirally arranged, scale-like and appressed. Cone 2–2·5 cm. $2n=66$.

Less frequently planted for ornament than *S. gigantea* but has been planted experimentally for forestry.

12. CUPRESSACEAE

Evergreen trees or shrubs. Buds naked. Lvs opposite or in whorls of 3 or 4, usually small and scale-like, appressed to and completely hiding the stem, on young plants (and in some spp. throughout their life) needle-like, rarely the lvs of the sterile twigs spirally arranged. Fls monoecious or sometimes dioecious, in small cones formed of opposite or whorled scales. Microsporophylls in 4–8 whorls in each cone, each with 3–5 pollen sacs on the under surface. Scales of female cone woody, rarely fleshy when ripe, bearing 2–many erect ovules on their upper surface, the bract united with the scale. Pollen not winged.

About 15 genera and over 100 spp.; N. and S. temperate regions and mountains of the tropics.

THUJA L.

Trees of narrow pyramidal habit. Bark scaly. Branchlets much divided near their tips into fine, ultimately deciduous, twigs arranged in one plane. Lvs scale-like, opposite, 4-ranked, those of the lateral ranks keeled or rounded at their back and partly concealing the flattened or grooved facial ones. Fls monoecious. Microsporophylls 3–6 pairs. Cones ± ovoid, composed of 3–6 pairs of imbricate scales of which only the 2 or 3 middle pairs are fertile; scales oblong, ±flat, thickened at apex; seeds 2–3 to each scale, usually winged.

Six spp., N. America and E. Asia. Several are commonly grown in gardens.

*T. plicata D. Don.   Western Red Cedar, Giant Arbor-Vitae.

Tree to 60 m., the trunk buttressed at the base. Bark cinnamon-red, fissured into scaly ridges. Main axes of branchlet systems ±terete, their lvs up to 6 mm., ovate, long-acuminate, with an inconspicuous sunken resin gland. Ultimate twigs c. 2 mm. across, their lvs up to c. 3 mm., scarcely acuminate, mostly eglandular, deep green on the upper side of the twig, with whitish marks on the lower, aromatic when bruised. Cones c. 12 mm.; scales 5–6 pairs, 3 of them usually fertile, thin. Seeds winged. $2n=22$.

Planted for forestry. Native of western N. America from Alaska to N. California and Montana.

CHAMAECYPARIS Spach

Habit, branchlets and lvs as in *Thuja*. Fls monoecious. Cones globose, ripening the 1st year, up to 12 mm.; scales 3–6 pairs, peltate, fitting together by their margins until maturity, with a central boss; seeds 1–5, winged. Cotyledons 2.

Six spp., N. America, Japan and Formosa, most of which are ± frequently grown in gardens.

*C. lawsoniana (A. Murr.) Parl.   Lawson's Cypress.
Cupressus Lawsoniana A. Murray

Tree to 60 m. Bark red-brown, very thick and spongy, fissured into scaly ridges. Lvs of main axes of branchlet systems up to 6 mm., the lateral pair acuminate. Ultimate twigs 1–1·5 mm. across; lvs up to 2 mm., with a sunken gland, deep green on the upper side of the twig, with indistinct whitish marks on the lower. Cones c. 8 mm., reddish-brown, pruinose; scales 8; seeds 2–5 on each scale. $2n=22$.

Very commonly planted in parks, gardens, etc., and sometimes on a small scale for forestry. Numerous forms differing in habit, colour of foliage, etc., are grown. Native of S.W. Oregon and N.W. California.

CUPRESSUS L.

Trees, rarely shrubs. Twigs in most spp. not arranged in one plane. Lvs scale-like, opposite, 4-ranked. Fls monoecious. Cones as in *Chamaecyparis* but larger and ripening the 2nd year; seeds usually 6–many, winged. Cotyledons 2–5.

About 12 spp., warm north temperate regions.

*C. macrocarpa Hartw. ex Gord.   Monterey Cypress.

Tree to 25 m., either pyramidal or becoming flat-topped. Branchlet systems not in 1 plane. Lvs all alike, 1–2 mm., deltoid, obtuse, deep green, not glandular. Cones 2·5–3·5 cm.; seeds c. 20 on each side.

Commonly planted near the sea in the south and west as an ornamental tree, windbreak or hedge-plant. Native of S. California.

1. JUNIPERUS L.

Trees or shrubs. Bark thin, usually shedding in longitudinal strips. Lvs either needle-like, subulate and spreading (and always so on young plants), in whorls of 3, or scale-like, appressed, opposite and resembling *Cupressus*. Fls dioecious or monoecious. Female fls of 3–8 *scales* which are opposite or in whorls of 3, and *become fleshy and coalescent forming a berry-like fr.*

40–50 spp., northern hemisphere extending south to Mexico, W. Indies, mountains of E. Africa, Himalaya and Formosa. Several spp. (including some with scale-like lvs) are sometimes planted in gardens.

1. J. communis L.   Juniper.

Shrub with the habit varying from procumbent to erect and narrow, rarely a small tree to 10 m. Bark reddish-brown, shedding. Lvs in whorls of 3, 5–19 mm., linear, sessile, jointed at the base, with a spiny point at the apex, spreading or ascending, entire, concave and with a broad white band above, green and keeled beneath. Dioecious. Male cones c. 8 mm., solitary, cylindric, with 5–6 whorls of scales. Female cones c. 2 mm. in fl., solitary. Fr. ripening the 2nd or 3rd year, c. 5–6 mm., globose or rather longer than broad, blue-black, pruinose, with 1–6 seeds. Fl. 5–6. Fr. 9–10. Germ. 3–4. $2n = 22$.

Native. Chalk downs, heaths, moors, pine- and birchwoods, ascending to 3200 ft., often dominant in scrub on chalk, limestone and slate. Throughout Great Britain but rather local and absent from a number of counties; in Ireland in the west and north from Kerry to Down, absent from the south-east. 82, H18. Arctic and north temperate zones, south to the mountains of N. Africa, Himalaya, N. California and Pennsylvania.

A variable sp. probably divisible into several ecological and geographical sspp. The two following extremes are often regarded as distinct spp., but many intermediate forms occur.

Ssp. communis

Erect or spreading. Prickly to the touch. Lvs spreading almost at right angles to the stem, 8–19 × c. 1 mm., gradually tapering to a long point. Fr. globose. N. The lowland form.

Ssp. nana Syme

J. nana Willd.; *J. sibirica* Burgsdorf

Procumbent. Scarcely prickly to the touch. Lvs ascending or loosely appressed, 4–10 × c. 1·5 mm., more suddenly contracted to a shorter point. Fr. longer than broad. $2n = 22$. Chw.

Rocks and moors on mountains and lowland bogs in W. Ireland; N. Wales, N. England, Scotland, Ireland.

ARAUCARIACEAE

Evergreen trees with whorled branches. Lvs spirally arranged, needle-like or broad and flat. Fls usually dioecious, both sexes in cones of spirally arranged scales. Microsporophylls with 5–15 pendulous pollen-sacs. Scales of female cone numerous, the bract large and woody, the scale ('ligule') fused to it; ovule 1, centrally placed, inverted.

Two genera and about 25 spp., Australasia, Malaya and S. America.

ARAUCARIA Juss.

Lvs needle-like or flat and lanceolate. Seeds adnate to the bract and falling with it.

About 10 spp., Australasia and S. America.

*A. araucana (Molina) C. Koch   Monkey Puzzle.

A. imbricata Pav.

Tree to 50 m. Lvs 2·5–5 cm., lanceolate, spiny pointed, spreading all round the stem and persisting many years. Cones globose, 14–20 cm.

Commonly planted in parks and gardens and occasionally among native vegetation. Native of Chile and western Argentina.

13. TAXACEAE

Evergreen trees or shrubs. Lvs linear, spirally inserted. Fls usually dioecious. Male fls in small cones (spikes or heads). Female fls solitary or in pairs in the lf-axils, not forming cones; ovules erect, not borne on scales but wholly or partly surrounded when ripe by a fleshy aril.

Three genera and about 15 spp., north temperate zone and New Caledonia.

1. Taxus L.

Trees or shrubs, the branches numerous and not regularly whorled. Buds small, with imbricate scales. Lvs ±spreading in 2 lateral ranks, linear, without resin canals. Normally dioecious. Male cones axillary, head-like stalked, surrounded by scales at the base; *microsporophylls* 6–14, peltate, each *with 5–9 pollen sacs. Ovule solitary*, axillary, with scales at the base. *Seed surrounded by a red, fleshy, cup-like aril*, ripening the 1st year.

About 7 very closely related spp., north temperate zone.

1. T. baccata L. Yew.

Tree to 20 m. with massive trunk and usually rounded outline, capable of producing lfy branches from stools or old trunks (unlike most other gymnosperms). Bark reddish-brown, thin, scaly. Twigs green. Lvs 1–3 cm., shortly stalked, mucronate, dark green above, paler and yellowish beneath, midrib prominent on both sides, margins recurved. Seeds c. 6 mm., olive-brown, ellipsoid. Fl. 3–4. Fr. 8–9. $2n=24$. M.

Native. Woods and scrub, mainly on limestone, sometimes forming pure woods in sheltered places on the chalk in S.E., and on limestone in N.W. England, tolerant of considerable shade; from Perth and Argyll southwards, rather local; in Ireland, mainly in the west, extending east to Cork, Offaly and Antrim; also Wicklow but possibly introduced. 56, H20. Europe from Scandinavia (62° 45′ N.) and Estonia to the mountains of Spain and Portugal, Sicily and Greece; Crimea, Caucasus; N. Persia, W. Himalaya; mountains of N. Africa.

ANGIOSPERMAE: DICOTYLEDONES: ARCHICHLAMYDEAE

14. RANUNCULACEAE

Herbs or woody plants and some woody climbers. Lvs usually spirally arranged and exstipulate, but opposite in *Clematis* and stipulate in *Thalictrum* and *Ranunculus* subgenus *Batrachium*; often palmately lobed or dissected. Fls solitary or in terminal infl., usually hermaphrodite, actinomorphic and hypogynous; usually nectar-secreting and entomophilous. Perianth undifferentiated or of ± distinct calyx and corolla, the latter of variously shaped, free, nectar-secreting organs ('nectaries', 'honey-leaves', 'petals'). Stamens numerous, usually arranged spirally on the conical receptacle, their anthers dehiscing outwards. Carpels 1–numerous, usually free and usually arranged spirally; ovules 1–numerous. Fr. various, but usually 1 or more follicles or a cluster of achenes. Seeds with a small embryo in oily endosperm. Germination usually epigeal.

48 genera and c. 1300 spp., cosmopolitan but chiefly in northern temperate and arctic zones.

Mostly acrid and often very poisonous plants owing to the presence of alkaloids some of which are of medicinal importance. Many have showy fls and are grown in gardens.

In floral structure the Ranunculaceae show many features which are regarded as primitive amongst Angiosperms (e.g. spiral arrangement of numerous free parts, undifferentiated perianth). Vegetatively and ecologically, however, they are amongst the most specialized Angiosperms, including advanced geophytic, hydrophytic, alpine and arctic types. They include also many species which show striking morphological and anatomical resemblances to Monocotyledons.

1	Lvs opposite; woody climber.	10. CLEMATIS
	Lvs spirally arranged or distichous (rarely ± opposite on creeping stems); not climbers.	2
2	1 carpel.	3
	2 or more carpels, free or united below.	4
3	Fr. a berry; fls small, whitish, not spurred.	8. ACTAEA
	Fr. a follicle; fls blue, spurred.	7. DELPHINIUM
4	Each carpel ripening to a 1-seeded achene.	5
	Each carpel ripening to a many-seeded follicle, or fr. a capsule.	10
5	Fl.-stems with a whorl of 3 lvs below the fl.; petals 0; per. segs petaloid, conspicuous.	9. ANEMONE
	Fl.-stems not with a whorl of 3 lvs below the fl.; petals present or 0.	6
6	Petals 5 or more, equalling or exceeding the sepals.	7
	Petals fewer than 5, or shorter than the sepals, or 0.	8
7	Petals red (or yellow), with no basal nectary; lvs bi- or tri-pinnate with linear segments.	12. ADONIS
	Petals yellow or white, each with a nectary near its base; lvs simple or palmately lobed or divided.	11. RANUNCULUS
8	Fl. solitary, terminal on a lfless stem; all lvs linear; sepals greenish, spurred; achenes in a long slender spike.	13. MYOSURUS

     Fls not solitary, on lfy fl.-stems; lvs not linear; sepals not spurred;
     achenes in a ±globular cluster.   **9**

9   All or most lvs palmately lobed or divided.   11. RANUNCULUS
    Lvs 2–4 times pinnate or biternate.   15. THALICTRUM

10   Fls strongly zygomorphic.   6. ACONITUM
    Fls actinomorphic.   **11**

11   Each of the 5 petals spurred.   14. AQUILEGIA
    Petals, if present, not spurred.   **12**

12   Fls with a single series of yellow per. segs.   1. CALTHA
    Fls with petals or 'honey-lvs' as well as sepals.   **13**

13   Petals represented by tubular nectaries.   **14**
    Petals not tubular.   **15**

14   Fl. solitary with a green involucre of deeply cut lvs; sepals and nectaries
    yellow.   4. ERANTHIS
    Fls not solitary and without an involucre; nectaries greenish.
                        3. HELLEBORUS

15   Sepals usually blue; fr. of 2–5 carpels united below to varying extents, or
    a capsule.   5. NIGELLA
    Sepals yellow (forming a globular fl.); fr. a spherical cluster of many
    small follicles.   2. TROLLIUS

Tribe 1. HELLEBOREAE. Fr. of one or more follicles, rarely a berry; chromosomes large, basic number 8.

1. CALTHA L.

Perennial herbs with stout creeping rhizomes and ±cordate lvs which in some spp. of the S. hemisphere have curiously prolonged basal lobes. Fls in a few-fld terminal cymose panicle, hermaphrodite, actinomorphic, hypogynous, with all parts spirally arranged. *Per. segs 5 or more, petaloid, yellow* or white; stamens numerous; carpels few, nectar-secreting, with numerous ovules in 2 rows. Fr. a group of follicles. N. and S. temperate regions.

1. C. palustris L.   Kingcup, Marsh Marigold, May Blobs.

A perennial glabrous herb with stout creeping rhizomes and erect ascending or prostrate aerial stems. Lvs chiefly basal, rounded, reniform or deltoid, ±cordate at the base, long-stalked; upper lvs reniform-deltoid, subsessile; all lvs ±crenate or toothed. Fls 16–50 mm. diam. Per. segs 5–8, 10–25 mm., bright golden-yellow above, often greenish beneath. Stamens 50–100. Carpels 5–13, erect or spreading; ovules numerous. Follicles 9–18 mm., dehiscing before they are dry. Seeds up to 2·5 mm. Fl. 3–7. Homogamous. Visited by a great variety of insects for pollen and for the nectar secreted from small depressions, one on each side of each carpel. $2n = 28, 32, 48, 56$. Hel.

    A highly polymorphic species, showing continuous and independent variation in many features. British forms may be grouped into two sspp.:

Ssp. **palustris**: *stems ascending or erect; fls at least* 30 *mm. diam.* This is the common lowland form in Britain. Forms with non-contiguous per. segs and spreading carpels which when ripe are narrowed into a beak (1·5–2·0 mm.) are placed in var. *cornuta* (Schott, Nyman & Kotschy) Rouy & Fouc. (incl. *C. guerangerii* Bor.).

Ssp. **minor** (Mill.) Clapham: includes smaller plants with *decumbent or pro-cumbent stems*, often rooting at the nodes, and 1 or a few *fls* 16–30 *mm. diam.*, with narrow non-contiguous per. segs. These are common northern and mountain forms of the British Is. In var. *radicans* (T. F. Forst.) Huth (incl. *C. radicans* T. F. Forst., with small, deltoid, hardly cordate, acutely crenate lvs, and *C. palustris* var. *zetlandica* Beeby) the procumbent stem roots at the nodes.

Native. In marshes, fens, ditches and wet woods, becoming most luxuriant in partial shade; rare on very base-poor peat. Reaches 3600 ft. in Scotland. 112, H40. Common throughout the British Is. except Channel Is. Ssp. *minor* is chiefly on mountains and in the north. Var. *radicans* occurs in Scotland and is common on lake margins throughout Ireland. Europe, incl. Iceland and arctic Russia; temperate and arctic Asia; N. America.

The flower of *Caltha palustris* is perhaps the most 'primitive' in the British flora.

2. TROLLIUS L.

Perennial herbs with ± erect woody stocks and spirally arranged palmately-lobed lvs. Fls large, in 1–3-fld cymes, hermaphrodite, actinomorphic, hypo-gynous, with all parts spirally arranged. *Sepals 5–15, petaloid,* imbricate in bud, *yellow; nectaries 5–15, small, narrow, yellow, with a basal nectar-secreting depression*; stamens numerous; carpels numerous, sessile, free, each with numerous ovules in 2 rows. *Fr. a group of many-seeded follicles.*

About 12 spp. in temperate and arctic Europe, Asia and N. America.

All species are acrid and poisonous.

1. T. europaeus L.   'Globe Flower.'

A perennial herb with a short erect woody stock, fibrous above, and an erect glabrous lfy usually simple shoot 10–60 cm. Basal lvs stalked, pentagonal in outline, palmately 3–5-lobed with the cuneate lobes ± deeply cut and toothed; stem-lvs ± sessile and usually 3-lobed; all lvs glabrous, dark green above, paler beneath. *Fls* 2·5–3 cm. diam., terminal, solitary or rarely 2–3, ± *globose*. *Sepals* c. 10 (5–15), pale or greenish-yellow, ± orbicular, very concave, *incurved and imbricate. Nectaries 5–15, yellow, equalling the stamens and hidden by the sepals,* ligulate, clawed, with a nectar-secreting cavity at the junction of blade and claw. Carpels numerous. Follicles c. 12 mm., keeled, transversely wrinkled, the persistent style forming a subulate beak c. 3 mm. Seeds blackish, shining, 1·5 mm. Fl. 6–8. Homogamous. Visited by various small insects. $2n = 16$. Hp.

Native. Locally common in wet pastures, scrub and woods in mountain districts northwards from S. Wales, Monmouth and Derby, and in N.W. Ireland; reaching 3700 ft. in Scotland. 56, H3. Throughout Europe to 71° N. in Norway, Caucasus and arctic America.

T. caucasicus Stev., with *spreading yellow sepals* and *nectaries equalling the stamens,* and *T. asiaticus* L. (*T. giganteus* hort.), with *spreading orange sepals* and *nectaries longer than the stamens,* are also grown in gardens.

3. HELLEBORUS L.

Perennial herbs with stout erect or ascending stocks and *digitately or pedately divided lvs with serrate margins.* Fls in cymes, hermaphrodite, actinomorphic, hypogynous, with all parts spirally arranged. *Sepals 5, green* or petaloid, imbricate, *persistent; nectaries 5–20, stalked, tubular, 2-lipped, green*; stamens numerous; carpels 3–10, sessile, free or slightly joined below; ovules numerous. Fr. a group of many-seeded follicles; seeds at first with a fleshy white ridge ('elaiosome') along the raphe.

About 20 spp. in Europe and W. Asia, all calcicolous.

All the hellebores have a burning taste and are highly poisonous owing to the presence of the glycosides helleborin and helleborein. Both *H. viridis* and *H. foetidus* were formerly officinal as violent cathartics and emetics, but their use has long been discontinued.

No radical lvs; uppermost stem lvs (bracts) simple, entire; fls numerous; perianth almost globular. **1. foetidus**
Radical lvs usually 2; uppermost stem lvs digitately lobed; serrate; fls 2–4; perianth spreading, almost flat. **2. viridis**

1. H. foetidus L. Bear's-foot, 'Stinking Hellebore'.

A perennial *foetid* herb with a stout blackish ascending stock and a robust *overwintering* branched lfy stem 20–80 cm., glabrous below, glandular above. *No radical lvs. Lower stem-lvs evergreen*, pedate, long-stalked, with sheathing base; lf segments 3–9, narrowly lanceolate, acute, serrate; middle stem-lvs with enlarged sheaths and reduced blades, transitional to the *uppermost* (bracts) which are *broadly ovate* and *entire* or with a small vestigial blade. *Fls 1–3 cm. diam., numerous*, drooping, in a corymb-like branched cyme. *Perianth campanulate or globular*, the erect concave broadly ovate sepals yellowish-green, usually bordered with reddish-purple. Nectaries 5–10, green, c. ½ as long as the stamens, short-stalked, curved, the outer lip slightly longer than the inner, both irregularly toothed. Stamens 30–55, equalling the perianth. Carpels 2–5, usually 3, slightly joined below. Follicles wrinkled, glandular, the persistent style forming a subulate beak ⅓ of the total length. Seeds black, smooth, with a fleshy white ridge ('elaiosome'). Fl. 3–4. Protogynous. Visited by early bees and other insects. Seeds said to be dispersed by ants. Germ. spring. $2n = 32$. Chh.

Native. A local plant of woods and scrub on shallow calcareous soils and scree in S. and W. England and Wales, probably native northwards to Lancashire and Northants but naturalized as a garden escape in N. England and E. Scotland to Aberdeen. 32. W. and S. Europe from Belgium to Spain, S. Italy and Styria.

2. H. viridis L. Bear's-foot, 'Green Hellebore'.

A perennial herb with a short stout simple or branched ascending blackish stock and an erect stem 20–40 cm., glabrous or sparsely hairy above, *not overwintering*, slightly branched distally and lfless below the lowest branch at the time of flowering. *Radical lvs usually 2, arising with the flowers and dying before winter*, long-stalked, digitate-pedate, with 7–11 sessile segments, the central free, the lateral connected at their bases, all narrowly elliptical, acute or shortly acuminate, serrate, with veins prominent on the paler lower surface. *Stem-lvs* (bracts) *smaller, sessile, ± digitate with narrow serrate segments. Fls usually* 2–4, *half-drooping*, 3–5 cm. diam. *Perianth spreading, almost flat, the sepals broadly elliptical, or ovate-acuminate, ± imbricate, yellowish-green. Nectaries 9–12, green, c. ⅔ as long as the stamens, shortly stalked, curved, the outer lip slightly longer than the inner, both irregularly toothed. Carpels 3, slightly joined at the base. Follicles more than ½ as broad as long, the persistent style forming a subulate beak ⅓ of the total length. Fl. 3–4. Protogynous. Visited by early bees. 2n = 32. Hp.

The British plant described above is ssp. **occidentalis** (Reut.) Schiffn., with glabrous lvs, not hairy on the veins beneath, and with fls smaller than in the C. European forms.

Native. A rather local plant of moist calcareous woods and scrub chiefly in S. and W. England and Wales, and probably native as far north as Westmorland and N. Yorks, but more widely naturalized as a relic of cultivation and reaching Aberdeen. 45. Ssp. *occidentalis* is native also in W. Germany, Belgium, France and Spain, where it replaces ssp. *viridis* of C. Europe from N.W. France to Switzerland and Hungary.

Forms of *H. niger* L. (C. and S. Europe, W. Asia), with pedate radical lvs, small entire bract-like stem-lvs and spreading white or pink-tinged sepals, bloom in winter and are grown in gardens as 'Christmas Roses'. The mostly later-flowering 'Lent Roses' consist of various spp. native in S.E. Europe and W. Asia and in particular of their numerous garden hybrids. They differ from *H. niger* in having toothed ± pedate bracts and are usually taller plants with fls of a wide range of colours from white or cream to deep purple. *H. argutifolius* Viv., like *H. foetidus* but with spinescent lvs and spreading yellowish-green sepals, is also seen in gardens.

4. ERANTHIS Salisb.

Perennial herbs with short tuberous rhizomes and stalked palmately divided radical lvs. *Stem-lvs* 3, *sessile, deeply palmate-lobed, forming an involucre-like whorl just beneath the solitary terminal fl.* Fls hermaphrodite, actinomorphic, hypogynous, with all parts spirally arranged. *Perianth of* 5–8, *usually* 6, *yellow sepals; nectaries* stalked, *tubular, 2-lipped, yellow*; stamens numerous; carpels c. 6, free, stalked, with many ovules in 1 row. *Fr. a group of many-seeded stalked follicles surrounded by the persistent stem lvs.*

Eight spp. in S.E. Europe and Asia.

All spp. have a burning taste and are poisonous owing to the presence of an alkaloid.

*1. E. hyemalis (L.) Salisb. 'Winter Aconite.'
Helleborus hyemalis L.

A glabrous perennial herb with irregular tuberous rhizome and erect flowering stems 5–15 cm. Radical lvs arising after the fls, borne singly (or rarely 2) on dwarf vegetative shoots, long-stalked, orbicular in outline, palmately 3–5-lobed, the lobes further cut into contiguous segments; stem-lvs spreading horizontally. Fl. 20–30 mm. diam. Sepals 10–15 mm., usually 6, yellow, narrowly ovate, enlarging during flowering. Nectaries 6 (5–9), shorter than the stamens, their outer lip considerably longer than the inner. Stamens c. 30.

Carpels 3–11, usually 6, stalked (2 mm.). Follicles brown, to 15 mm., each with numerous yellowish-brown seeds, 2·5 mm. Fl. 1–3. Homogamous. Fls very temperature-sensitive, opening above 10° C. Visited by hive-bees and flies. Follicles dehisce in May, the lvs dying soon after. Germ. spring. $2n=16$. Grh.

Introduced. A native of S. Europe, from S. France and C. Italy to Serbia, which has been naturalized in parks, plantations and woods in many parts of Great Britain.

The closely allied *E. cilicica* Schott & Kotschy, of western Asia, is often grown in gardens for its larger fls.

5. NIGELLA L.

Annual herbs with spirally arranged bi- or tri-pinnate lvs with linear or filiform segments. Fls usually solitary, terminal; hermaphrodite, actinomorphic, hypogynous. *Sepals 5, petaloid; nectaries 5, clawed,* opposite to and usually much smaller than the sepals; stamens numerous. *Fr. a group of partly joined follicles or a capsule.*

About 20 spp. in Europe and W. Asia, chiefly Mediterranean. None native in the British Is. Three blue-fld spp. which are sometimes found as casuals may be distinguished as follows:

1   Fls with an involucre of much dissected lvs; carpels 5, united almost
      throughout their length; fr. a globular capsule.
                            ('Love-in-a-Mist.')   *N. damascena L.
      Fls not surrounded by an involucre; fr. an angular capsule.   2

2   Carpels united only below the middle.   *N. arvensis L.
      Carpels united almost to their summits.   *N. gallica Jord.
                            (*N. hispanica* auct., non L.)

N. damascena, and less frequently *N. gallica,* are grown in gardens.

6. ACONITUM L.

Perennial herbs with erect tuberous stocks and spirally arranged *palmately lobed or divided lvs. Fls in terminal racemes,* hermaphrodite, *zygomorphic,* hypogynous, with all parts spirally arranged. *Sepals 5, petaloid, the posterior sepal forming a large erect helmet-shaped hood; nectaries 2–8, the posterior pair included within the hood, long-clawed and with limbs prolonged into nectar-secreting spurs, the remainder very small or* 0; stamens numerous; carpels 3–5, sessile, free or slightly joined at the base, with numerous ovules. *Fr. a group of many-seeded follicles.*

Perhaps c. 100 spp. throughout the north temperate zone.

All species are highly poisonous and have often proved fatal owing to the presence of the powerful and deadly alkaloid aconitin and of other associated alkaloids. *A. napellus* L. has long been officinal as a narcotic and analgesic.

1. A. anglicum Stapf   'Monkshood.'

A. Napellus, auct. angl., non L.

A perennial herb often with paired blackish tuberous taproot-like stocks, up to 9 cm. long by 3 cm. diam. at the top, and erect lfy minutely downy flowering stems 50–100 cm., usually simple. *Lower stem-lvs* up to 15 cm. across, short-stalked, pentagonal in outline, *palmately 3–5-partite, the middle segment wedge-shaped at the base then deeply laciniate with ± linear lobes,* the lateral segments similar but less divided (Fig. 1); upper lvs smaller, ± sessile; all *light green, soft,* and minutely hairy to glabrous. Infl. erect, moderately dense. *Fls mauve to blue-mauve,* minutely downy. Lower sepals strongly deflexed; *helmet 18–20 mm. high, produced into a small point. Nectaries almost horizontal on the forward-curving tips of their erect stalks, their spurs forwardly directed with upturned capitate ends, their 2-lobed lips recurved.* Stamen filaments hairy above. Car-

Fig. 1. Leaf of *Aconitum anglicum.* ×⅓.

pels 3, almost parallel, glabrous. Ripe follicles c. 2 cm. Seeds 4–5 mm., with a wide dorsal and narrower frontal wings. Fl. 5–6. Protandrous, the stamens erecting and dehiscing successively. Visited by long-tongued bumble-bees. $2n=32*$. Grh.

Native. Local in S.W. England and Wales, on shady stream banks. 15.

A. napellus L. is a polymorphic aggregate of slightly differing geographical and ecological units which are often treated as separate spp. *A. anglicum* is one of the segregates and appears to be endemic in Great Britain. The aggregate has a wide range in Europe and N.W. Asia to the Himalaya. Of the cultivated forms and hybrids of *A. napellus* agg. which are grown in gardens and occasionally escape, **A. compactum* (Rchb.) Gáyer, with ultimate lf-lobes linear to lanceolate, fl.-stalks very short and appressed to the infl.-axis, and infl.-axis, fl.-stalks and helmet shortly appressed-hairy, is reported as established locally. Most cultivated forms have darker green and less narrowly divided lvs than *A. anglicum* and have darker blue fls opening later in the season.

Besides *A. napellus* agg., *A. variegatum* L., with larger, often whitish fls in looser racemes, straight-clawed nectaries and seeds with prominent undulate ridges; and *A. vulparia* Rchb. (*A. lycoctonum* auct., non L.), with pale yellow fls and a very tall conical-cylindrical helmet, are also seen in gardens.

7. DELPHINIUM L.

Annual or perennial herbs with slender tap-roots or erect stocks and spirally arranged palmately lobed or divided lvs. *Fls* in racemes or panicles, hermaphrodite, *zygomorphic,* hypogynous, all the parts spirally arranged. Sepals 5, petaloid, the *posterior alone prolonged into a conical spur*; nectaries 4, either all united into 1 spur or all free with only the 2 posterior spurred and the remaining 2 unspurred; the spurs nectar-secreting and contained within the sepal spur; stamens numerous; carpels 1 or 3–9, each with numerous ovules. Fr. of 1 or a few many-seeded follicles.

About 200 spp., all in the northern hemisphere except for a few African spp.

All *Delphinium* spp. are poisonous owing to the presence of alkaloids of which the most commonly occurring is delphinin.

1 Follicle glabrous; plant diffusely branched; racemes few-fld. **3. consolida**
Follicle pubescent; plant single-stemmed; racemes many-fld. **2**

2 Branches ±erect; bracteoles inserted above middle of pedicel, reaching base of fl.; follicle abruptly beaked. **2. orientale**
Branches almost horizontal; bracteoles inserted below middle of pedicel, not reaching base of fl.; follicle gradually tapering. **1. ambiguum**

*1. D. ambiguum L. Larkspur.

D. Ajacis L., sec. Gay; *D. Consolida* L., sec. Sm.; *D. Gayanum* Wilmott

An annual *pubescent* herb with slender tap-root and erect stem 25–60 cm., usually with a few ascending branches. Lower stem-lvs long-stalked, deeply palmate-cut into many narrow-linear acute segments; upper stem-lvs similar but ±sessile; all lf segments finely pubescent. Fls c. 2–5 mm. diam., bright blue, rarely white or pink, in terminal 4–16-fld racemes; *lowest bracts at least equalling the pubescent fl.-stalks*; upper bracts entire. Sepals c. 16 mm., bright blue, ovate, clawed, the posterior sepal with a slender spur c. 15 mm. Nectaries purplish-blue united into a posterior 3-lobed limb with a single spur fitting into the sepal spur; central lobe of limb narrow, deeply notched and marked with a few dark lines (resembling the letters AIA, whence the name *ajacis* for this and related spp.); lateral lobes broader. Stamens with broad filaments. *Follicle pubescent*, 1·5–2·5 cm., ±cylindrical, *gradually narrowed into a beak c. 3 mm. Seeds* c. 2·5 mm., roundish, nearly black, *with numerous ±continuous transverse ridges.* Fl. 6–7. Protandrous. Fls visited by bumble-bees: short-tongued spp. often perforate the spur. 2*n*=16. Th.

Introduced. Formerly naturalized in Cambridgeshire and still found as a rare casual in cornfields on light soils in many parts of the country. Native of the Mediterranean region but extensively naturalized.

*2. D. orientale Gay 'Eastern Larkspur.'

D. Ajacis L. sec. Wilmott

An annual pubescent herb much resembling *D. ambiguum* and often confused with it but differing in its nearly erect branches, with glandular hairs above and the spur usually short and about as long as the sepals, and the *follicles narrowed abruptly into the beak.* Fl. 6–7. 2*n*=16. Th.

Introduced. Rarely found as a casual or garden-escape in cornfields and waste places, especially near ports. S. Europe, N. Africa and W. Asia. A common 'Larkspur' of gardens.

*3. D. consolida L. 'Forking Larkspur.'

An annual pubescent herb differing from *D. ambiguum* in having *spreading branches*, fewer-fld racemes with *simple or at most 3-lobed bracts, all much shorter than their fl.-stalks*, and quite *glabrous follicles* whose *seeds*, c. 2 mm., have *broken transverse ridges* like rows of scales. Fl. 6–7. 2*n*=16. Th.

Introduced. A rare casual of cornfields and waste places, especially near ports. Europe and W. Asia.

Besides the annual 'Larkspurs' of section *Consolida*, with one follicle and with all nectaries united into a single petal spur, there are perennial garden forms grown as 'Delphiniums'. These have 4 free nectaries, the 2 posterior spurred, and 3 or more follicles. They are chiefly derived by hybridization, probably from the European *D. elatum* L. and the Asiatic *D. grandiflorum* L. and *D. cheilanthum* Fisch. *D. staphisagria* L. ('Stavesacre') of the Mediterranean region, very short-spurred, was formerly officinal and was much grown for the delphinin in its seeds.

8. ACTAEA L.

Perennial herbs with blackish rhizomes and bi- or tri-pinnate lvs with adnate stipules. *Fls small, in short racemes*; hermaphrodite, *actinomorphic*, hypogynous. Sepals 3–5, rather unequal, petaloid, deciduous; nectaries 4–10, small, or 0. Stamens numerous; *carpel* 1; ovules numerous. *Fr. a single berry with several flattened seeds.*

About 8 spp. in colder parts of the north temperate zone. Very poisonous. The rhizome of *A. spicata* was formerly officinal as *Radix Christophorianae* and was used against skin diseases and asthma.

1. A. spicata L.   Baneberry, Herb Christopher.

A perennial foetid herb with stout blackish obliquely ascending rhizome and an erect glabrous flowering stem 30–65 cm. Radical lvs large, long-stalked, biternate or bipinnate, the secondary lflets ovate, acute, often 3-lobed, inciseserrate; stem-lvs 1–4, much smaller than the radical; all dark green above, paler beneath and ± hairy on the veins beneath. Raceme 25–50 cm., terminal or occasionally also axillary, dense-fld, elongating in fr., its axis and the fl.-stalks pubescent. Sepals 3–6, usually 4, whitish, blunt, concave, soon falling. *Petals* 4–6 or 0, *white, spathulate, clawed, shorter than the stamens,* not nectarsecreting. *Stamens white, clavate,* the filaments dilating upwards into the broad connective; anther-lobes small, distant, ± spherical; carpel pyriform; stigma broad, sessile. *Berry c.* 1 *cm., ovoid, at first green, then blackish, shining.* Seeds c. 4 mm. wide, semicircular, flattened. Fl. 5–6. Protogynous. Visited by small pollen-eating insects. $2n = 16$, c. 32. Grh.

Native. A local plant of ashwoods on limestone and of limestone pavements in Yorks, Lancashire and Westmorland; reaches 1650 ft. in N. Yorks. 7. Europe, northwards to Norway; temperate and arctic Asia to China.

Tribe 2. ANEMONEAE. Fr. a group of achenes, usually 1-seeded; chromosomes large, basic number 8.

9. ANEMONE L.

Perennial herbs or rarely small shrubs with rhizomes or woody stocks and spirally arranged usually palmately lobed or divided radical lvs; *stem-lvs* 3, *whorled,* resembling the radical lvs or not, sometimes close beneath the fl. and then calyx-like. Fl. solitary terminal (rarely axillary) or in a few-fld cymose umbel, usually hermaphrodite, actinomorphic, hypogynous. *Per. segs* 4–20, *petaloid*; stamens numerous, the outermost sometimes staminodal and nectarsecreting; carpels numerous, each with 1 functional ovule. Fr. a cluster of achenes, sometimes with persistent elongated hairy style (*Pulsatilla*).

About 120 spp., cosmopolitan but chiefly in northern extra-tropical regions, and with many arctic, alpine and steppe types.

Sharp-tasting plants, poisonous owing to the presence of the narcotic anemonin and dangerous to cattle.

1	Fl. bell-shaped, violet; achenes plumed.	**4. pulsatilla**
	Fls not bell-shaped; achenes not plumed.	2
2	Per. segs 10–15, blue.	**3. apennina**
	Per. segs 5–9, rarely blue.	3
3	Per. segs usually 6–7, whitish or pinkish.	**1. nemorosa**
	Per. segs usually 5, yellow.	**2. ranunculoides**

Subgenus 1. ANEMONE.

Stem-lvs at some distance below the fl.; all stamens fertile; achenes not plumed.

1. A. nemorosa L. 'Wood Anemone.'

A perennial herb with slender brown rhizome and erect simple flowering stems 6–30 cm., glabrous or sparsely hairy. Radical lvs 1–2, borne directly on the rhizome and appearing after flowering, long-stalked, palmately 3-lobed, the lobes further cut or divided into cuneate coarsely toothed acute segments; stem-lvs 3, borne at about ⅔ up from the base of the flowering stem, stalked, palmately 3-lobed, resembling the radical lvs. Fl. 2–4 cm. diam., solitary, terminal. *Per segs 5–9, usually* 6 *or* 7, glabrous, oblong-elliptical, *white or pink-tinged*, or occasionally reddish-purple, rarely almost blue. Stamens 50–70, all fertile. Carpels 10–30. Achenes 4–4·5 mm., in a globular cluster, downy; beak curved, rather shorter than achene. Fl. 3–5. Homogamous. No nectar. Visited for pollen by various bees and flies. Germ. spring; hypogeal. $2n = 32$ (39); 28–32; c. 45. Grh.

Native. An abundant gregarious herb especially of deciduous woodland on all but the most base-deficient or water-logged soils. 109, H40, S. Throughout Great Britain except the Outer Hebrides, Orkney and Shetland; Ireland; Channel Is. Throughout the northern temperate zone of C. Europe and W. Asia.

The somewhat similar white-fld *A. sylvestris* L. of C. Europe and W. Asia, often cultivated, may be distinguished by the radical lvs at the base of the flowering stem, the larger fls (4–7 cm. diam.) with usually 5 per. segs, downy below, and the elongated woolly head of achenes.

*2. A. ranunculoides L. 'Yellow Wood Anemone.'

Resembling *A. nemorosa* in habit but with *very shortly stalked* deeply 3-lobed *stem-lvs* and usually 5 broadly ovate *yellow per. segs*. Fls occasionally more than 1. Achenes downy with a short glabrous beak. Fl. 4 $2n = 32$. Grh.

Introduced. Naturalized in a few scattered localities in England. Native in most of Europe except the north-west, and in W. Asia.

*3. A. apennina L. 'Blue Anemone.'

Resembling *A. nemorosa* in habit, with blackish tuberous rhizome and stem-lvs all stalked and biternate, triangular in outline, almost glabrous. Fls solitary with 10–15 *very narrow blue, rarely white, per. segs* which are somewhat pubescent near the base of the underside. Achenes broader than long, shortly pubescent with a short abruptly curved beak. Fl. 4. $2n = 16$. Grh.

Introduced. Much grown in gardens; occasionally escaping and perhaps naturalized in a few localities. S. Europe.

The closely allied *A. blanda* Schott & Kotschy (E. Mediterranean region), often grown in gardens, differs in having the lvs ±circular in outline and the blue, pink or white per. segs usually glabrous beneath; achenes c. as broad as long. Fl. 1–3.

Other members of subgen. *Anemone* often seen in gardens are *A.* × *hybrida* Paxton (*A. Japonica* auct.), a tall (up to 1 m.) autumn-flowering plant with large white or reddish fls whose per. segs are silky beneath; and the brightly coloured florists' anemones, all with sessile stem-lvs, which are derived from *A. coronaria* L., *A. pavonina* Lam., *A.* × *fulgens* Gay, and *A. hortensis* L.

Subgenus 2. PULSATILLA (Mill).

Stem-lvs close below the fl.; outer stamens sterile; achenes with elongated plumed styles.

4. A. pulsatilla L.   'Pasque Flower.'

Pulsatilla vulgaris Mill.

A perennial herb with an erect blackish tap-root-like stock and flowering stem 10–30 cm. *Basal lvs* in a rosette, *long-stalked, bipinnate, the lflets pinnately cut with linear segments*; stem-lvs sessile, ±erect, deeply divided into long linear segments; all lvs hairy, the radical with woolly stalks. *Fl.* solitary, terminal, erect at first, later drooping, *bell-shaped. Per. segs* 6, elliptical, *violet-purple*, paler and silky on the outside. *Stamens* numerous, *about ½ as long as the per. segs, the sterile outer ones nectar-secreting.* The internode between stem-lvs and fl. elongates after flowering. *Achenes silky, with a plume* 3·5–5 cm. Fl. 4–5. Protogynous. Visited by many bees for pollen and nectar. Andro- and gyno-monoecious and dioecious forms have been reported from the Continent. $2n = 32$. H.

Native. A local plant of dry calcareous grassy slopes chiefly of E. England from Gloucester and Essex to N. Lincs. 13. N. and C. Europe and W. Asia.

Various forms of *A. pulsatilla* some with reddish or white fls, and *A. vernalis* L., with fls whitish within and silky brown outside, are grown in gardens.

The subgenus *Hepatica*, in which the stem-lvs are small and entire and arise close beneath the fl., simulating a calyx, is represented by the commonly cultivated *A. hepatica* L. (*Hepatica nobilis* Mill.), with pretty evergreen 3-lobed cordate lvs and usually blue fls with 6–7 per. segs.

10. CLEMATIS L.

Perennial woody climbers or shrubs, occasionally perennial herbs, with usually *compound* exstipulate *lvs in opposite pairs*; lvs often ending in tendrils or with twining petiole and rhachis. Fls solitary or in panicled infl., usually hermaphrodite; actinomorphic, hypogynous. *Per. segs valvate, usually* 4, ±petaloid; staminodes usually present; stamens and carpels numerous. Achenes with *persistent long plumose styles*.

About 230 spp., widely distributed in temperate regions especially of the N. hemisphere.

1. C. vitalba L.   Traveller's Joy, Old Man's Beard.

A perennial woody climber with stems up to 30 m. Lvs pinnate usually with (3–)5 rather distant lflets; lflets 3–10 cm., narrowly ovate, acute or acuminate, rounded or subcordate at the base, coarsely toothed or entire, glabrous or slightly pubescent. Fls c. 2 cm. diam., in terminal and axillary panicles, fragrant. Per. segs greenish-white, densely pubescent outside. Achenes in large heads, pubescent, with long whitish plumose styles. Fl. 7–8. Slightly protogynous. Visited by pollen-collecting bees and pollen-eating flies, especially Syrphids. $2n=16^*$. M.

Native. In hedgerows, thickets and wood-margins chiefly on calcareous rocks or soils from Denbigh, Stafford and S. Yorks southwards; not native in Ireland but widely naturalized. 53. Europe from the Netherlands southwards; N. Africa; Caucasus.

The following, sometimes escaping from gardens, are established locally:

*C. flammula L. (Mediterranean region to Persia), resembling *C. vitalba* but with bipinnate lvs and pure white (not greenish) fls, 12–30 mm. diam.

*C. montana Buch.-Ham. ex DC. (Himalaya), with ternate lvs, often grown for its profusion of white or pink fls, 4–5 cm. diam.

*C. viticella L. (S. Europe and S.W. Asia), with ± bipinnate lvs and bluish to crimson fls, 3–5 cm. diam., solitary or in threes.

The climbing Clematises most commonly grown in gardens, with large blue-purple to reddish fls, are chiefly hybrids of the Chinese *C. lanuginosa* Lindl. with *C. viticella* (*C. × jackmanii* Moore) or with the Chinese *C. patens* Morr. & Decne (*C. × lawsoniana* Moore & Jackman). The *viticella* hybrids are later-flowering than the *patens* hybrids and have usually only 4 per. segs instead of 6–8. Non-climbing spp. often grown include *C. integrifolia* L., *C. heracleifolia* DC., and *C. recta* L.

The genus *Clematopsis* (Africa and Madagascar), with imbricate per. segs, connects *Clematis* with *Anemone*.

11. RANUNCULUS L.

Annual or perennial herbs with spirally arranged or distichous lvs, stipulate or exstipulate, often palmately lobed or divided, sometimes simple and ± entire. Fls solitary and terminal or in cymose panicles, hermaphrodite, actinomorphic, hypogynous, with all the parts spirally arranged. *Sepals* 3–5; *petals* 5 or more, rarely 0, usually *yellow or white, each with a nectar-secreting depression near the base*; stamens numerous; carpels numerous, each with 1 basally attached ascending ovule. *Fr. a head of achenes.*

About 300 spp., cosmopolitan, but chiefly in northern extra-tropical regions.

All species are acrid and poisonous and are dangerous to cattle, but are ordinarily avoided by all grazing animals. The poisonous constituent is probably anemonin.

1	Aquatic plants with white fls; achenes strongly wrinkled transversely.	*15*
	Terrestrial or marsh plants with yellow fls or petals 0; achenes not strongly wrinkled transversely.	*2*
2	Sepals 3; petals 7–12; lvs simple, cordate.	**23. ficaria**
	Sepals 5; petals 5 or fewer, rarely 0; lvs various.	*3*
3	Lvs palmately lobed or divided.	*4*
	Lvs simple, lanceolate or spathulate, entire or toothed.	*12*
4	Achenes smooth or downy or with small tubercles.	*5*
	Achenes with spines or hooked bristles.	*11*
5	Sepals strongly reflexed during flowering.	*6*
	Sepals not strongly reflexed.	*7*

6 Stem tuberous at base, ± appressed-hairy above; achenes minutely pitted, with hooked beak.   **3. bulbosus**
   Stem not tuberous, with spreading hairs throughout; achenes usually with a few tubercles just within the conspicuous border and with almost straight beak.   **6. sardous**

7 Lvs glabrous or nearly so.   8
   Lvs hairy.   9

8 Fls 0·5–1 cm. diam.; achenes in an oblong head; in wet places, and on bare mud.   **13. sceleratus**
   Fls 1·5–2·5 cm. diam. but less when, as often, petals are reduced or 0; achenes in a spherical cluster; woods and shady hedgebanks.
                                                                                 8. auricomus

9 Plant with fleshy root-tubers at the base of a short erect stock; fls 2·5–3 cm. diam. with very glossy petals; stem-lvs 1(–2), small.   **4. paludosus**
   Plant without root-tubers; stem-lvs usually more than 2, the lower ones large.   10

10 Plant with long runners; fl.-stalk furrowed.   **2. repens**
   Plant without runners; fl.-stalk not furrowed.   **1. acris**

11 Plant decumbent or ascending, diffusely branched; fls 3–6 mm. diam.; achenes with shortly-hooked tubercles on the faces.   **7. parviflorus**
   Plant erect with erect branches; fls 4–12 mm. diam.; achenes spiny with the largest spines on the margin.   **5. arvensis**

12 Plant 60–90 cm. high; fls 2–3 cm. diam.   **9. lingua**
   Plant not exceeding 60 cm. high; fls less than 2 cm. diam.   13

13 Plant erect with broadly ovate ± cordate basal lvs and fls 6(–9) mm. diam.; achenes tubercled.   **12. ophioglossifolius**
   Plant ascending, decumbent or creeping; basal lvs not cordate; fls 5–18 mm. diam; achenes not tubercled.   14

14 Plant with a filiform stem rooting at every node and arching in the internodes; fls solitary, 5–10 mm. diam.; achenes 1 mm.   **11. reptans**
   Plant, if creeping, not rooting at every node; fls 1–several; achenes 1·5–1·8 mm.   **10. flammula**

15 Plant with no finely dissected submerged lvs.   16
   Finely dissected submerged lvs present.   18

16 Lvs deeply 3(–5)-lobed, the lobes cuneate, distant; receptacle pubescent.
                                                                                 16. tripartitus
   Lvs shallowly lobed; receptacle glabrous.   17

17 Lf-lobes broadest at their base; fls 3–6 mm. diam.   **14. hederaceus**
   Lf-lobes narrowest at their base; fls 8–12 mm. diam.   **15. omiophyllus**

18 Petals usually less than 6 mm.; stamens 5–15.   19
   Petals usually more than 6 mm.; stamens 12–numerous.   20

19 Floating lvs never present; achenes pubescent to hispid (at least while immature).   **19. trichophyllus**
   Floating lvs always present, usually less than 1·5 cm. across, deeply 3(–5)-lobed; submerged lvs few; achenes glabrous.   **16. tripartitus**

20 Submerged lvs very long (8–30 cm.) and parallel; floating lvs 0; receptacle glabrous.   **17. fluitans**
   Submerged lvs very rarely exceeding 8 cm.; floating lvs present or 0; receptacle pubescent.   21

21   Submerged lvs usually 0·5–2 cm. diam., sessile, circular in outline, their short rigid segments all lying in one plane; floating lvs 0.

                                                        18. circinatus

     Submerged lvs with segments not lying in one plane.   *22*

22   Achenes 1–1·4 mm., numerous (40–100), glabrous, membranous-winged; usually in brackish water.   **22. baudotii**

     Achenes exceeding 1·5 mm., rarely more than 40, ± pubescent or hispid (at least while immature), not winged.   *23*

23   Petals less than 10 mm.; nectaries circular (Fig. 7A); fr.-stalks usually 2–5 cm.; floating lvs commonly subcircular with cuneate dentate lobes, or 0.   **20. aquatilis**

     Petals usually 10 mm. or longer; nectaries elongated, ± pyriform (Fig. 7c); fr.-stalks usually exceeding 5 cm.; floating lvs usually reniform or truncate with rounded crenate lobes, or 0.   **21. peltatus**

Subgenus RANUNCULUS.

Annual or perennial. Sepals usually 5, falling during or soon after flowering; petals yellow, rarely red or white; achenes with a firm thick wall, neither transversely ridged nor striate nor nerved.

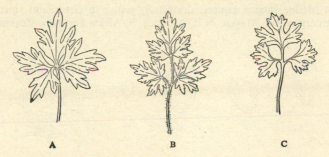

Fig. 2. Leaves of *Ranunculus*. A, *R. acris*; B, *R. repens*; C, *R. bulbosus.* × ½.

Section 1. *Ranunculus.* Nectary covered by a flap ± free laterally; achenes distinctly beaked, strongly compressed; fruiting receptacle up to 3 times its length in the fl.

1. R. acris L.   'Meadow Buttercup.'

R. acer auct. plur.

A perennial herb with erect or creeping stock and a much branched hairy stem 15–100 cm., hollow below. *Not stoloniferous*: the stock creeps for a short distance, then turns up; and the creeping part may or may not persist until flowering. *Basal lvs* (Fig. 2A) long-stalked, *pentagonal or roundish in outline*, palmately 2–7-lobed, the *terminal lobe sessile*, all lobes further cut into 3-toothed segments; lower stem-lvs similar but shorter stalked; uppermost sessile, deeply cut into linear segments; all ± hairy. Fls 18–25 mm. diam., terminating the branches of the irregular cymose infl., their *stalks hairy, not furrowed. Sepals* hairy, appressed to the petals, *not reflexed.* Petals

6–11 mm., roundish, glossy, bright yellow, sometimes paler or even white. *Receptacle glabrous.* Achenes 2·5–3 mm., rounded, bordered, glabrous, with a short hooked beak (Fig. 3 A). Fl. 5–7(–10). Protogynous and often self-sterile. Plants may show the whole range from completely hermaphrodite to female (no viable pollen) or male (no functional ovules), and occasional plants produce neither viable pollen nor functional ovules. Visited by various insects, especially flies and small bees. $2n = 14^{*}$, 28, 29–32. Hp.

A polymorphic species with many named varieties and subspecies differing chiefly in the length and direction of growth of the rhizome, the angle of branching and hairiness of the stem and the degree of dissection and hairiness of the lvs. Mountain plants are commonly of small stature, little branched and with only a few large fls. British forms of the species have not yet been studied experimentally and their taxonomic status is unknown.

Native. Found commonly and often abundantly in damp meadows and pastures on calcareous or circumneutral substrata throughout the British Is.; frequent on damp rock-ledges and in gullies and occasional on mountain-top detritus. Reaches 4000 ft. in Scotland. *R. acris* sens. lat. has a wide distribution throughout temperate and arctic Eurasia from Greenland and Spain to China and Japan but is absent from Portugal, N. Africa, S. Italy and the eastern Mediterranean region; doubtfully native in Greenland; introduced in Spitsbergen, Jan Mayen, N.E. America, S. Africa and New Zealand.

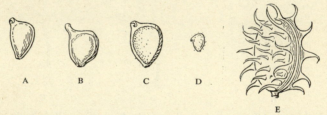

Fig. 3. Achenes of *Ranunculus.* A, *R. acris*; B. *R. repens*; C, *R. bulbosus*; D, *R. paludosus*; E, *R. arvensis.* × 3.

2. R. repens L.   'Creeping Buttercup.'

A perennial herb with long stout roots and *strong lfy* epigeal *stolons* (runners) which root at the nodes; flowering stems 15–60 cm., erect, lfy, hairy. *Basal* and lower stem *lvs* (Fig. 2 B) stalked, *triangular-ovate* in outline, divided into 3 lobes of which the *middle lobe is long-stalked* and projects beyond the others, the lobes further cut or divided into 3 toothed segments; upper lvs sessile with narrow ± entire segments; all lvs usually hairy. Infl. as in *R. acris.* Fls 2–3 cm. diam., their *stalks hairy, furrowed. Sepals* hairy, *not reflexed. Petals* 5, 6–12 mm., ovate, *suberect,* golden yellow, glossy. *Receptacle hairy.* Achenes 3 mm., roundish, bordered, glabrous, with a short curved beak (Fig. 3 B). Fl. 5–8. Pollination as in *R. acris.* $2n = 16^{*}$, 24, 32. Hs.

Native. Common in wet meadows, pastures and woods, in dune-slacks and on gravel-heaps, etc., and as a weed especially on heavy soils. Reaches 3400 ft. on Snowdon. 112, H40, S. Throughout the British Is. and the whole

of Europe, except Iceland, Sardinia, southern Balkans and northern Russia;
extends in an interrupted broad band across Siberia to Kamchatka and Japan;
introduced in eastern N., C. and S. America, and New Zealand.

3. R. bulbosus L.   'Bulbous Buttercup.'

A perennial herb, 15–40 cm. Stems erect or ascending, ± hairy, with spreading
hairs below and *appressed hairs above*, arising from a *corm-like stem-tuber*,
0·6–1·5(–4·5) cm., rounded or flattened and with enlarged lf-bases above and
bearing fleshy contractile roots beneath. Basal and lower stem-*lvs* (Fig. 2c)
stalked, *ovate* in outline, 3-lobed, the *middle lobe long-stalked* (sessile in
ssp. **bulbifer** (Jord.) Neves); upper stem-lvs ± sessile, deeply cut into narrow
often linear segments; all lvs usually hairy. Fls 1·5–3 cm. diam., in an irregular
and often corymbose cyme, rarely solitary, their *stalks* ± *hairy, furrowed.*
Sepals 5, pale yellowish, hairy beneath, *strongly reflexed.* Petals 5, 6–20 mm.,
broadly obovate-cuneate, bright yellow (rarely paler or white), glossy.
Receptacle hairy. Achenes c. 3 mm., obovate, their faces finely pitted, dark
brown with a paler border; beak short, somewhat hooked (Fig. 3c). Fl.
(3–)5–6. Pollination as in *R. acris*; often ± gynomonoecious. $2n=16^*$. Hs.

A very variable species, especially in size of corm and number of stems arising from
it, direction of growth of aerial stems (forms with very large corms, var. *macrorhizus*
Godr., may have several widely divaricating or almost prostrate stems), and density
and length of hairs on the stems, lf-stalks and fl.-stalks. British forms have not yet
been studied experimentally. The above description is of ssp. **bulbosus.**

Native. Throughout most of the British Is. (but not in Shetland) and
abundant in dry pastures, and on grassy slopes and fixed dunes, especially
on calcareous substrata in the south; becoming less common in the north.
110, H40, S. Throughout Europe to c. 60° N. and 24–28° E., but absent from
the extreme south of Italy and most of the Balkan Peninsula; local in Morocco,
Tunis, Georgia and Azerbaijan; introduced in America and New Zealand.

The earliest flowering of our three common buttercups, and growing in
drier places than *R. acris* and *R. repens.* Easily distinguished by the stem-
tubers and the reflexed sepals.

Section 2. *Ranunculastrum* DC. Perennial herbs with both fleshy and fibrous
roots, often with filiform stolons; fls axillary; achenes compressed, strongly
keeled, beaked; receptacle elongating in fr.

4. R. paludosus Poir.   'Fan-leaved Buttercup.'

R. chaerophyllos auct., non L.; *R. flabellatus* Desf.

A perennial stoloniferous herb with a short erect stock bearing both fibrous
roots and a *cluster of fleshy root-tubers* 4–8 mm. Stolons hypogeal, very
slender, with tiny scale-lvs. Stem 10–30 cm., erect, simple or little branched,
silky-hairy. Lvs chiefly basal, stalked, the lowest often simply 3-lobed, the
rest 3-lobed with the central lobe long-stalked and all lobes further divided
into narrow toothed segments; stem-lvs 1–2, small, sessile. Fls 1–4, 2·5–3 cm.
diam., their *stalks hairy, not furrowed. Sepals* 5, hairy, *not reflexed.* Petals 5,
bright yellow, very glossy. *Receptacle glabrous, much elongating in fr.* Achenes

c. 2 mm., roundish, glabrous, with an acute straight beak flattened below and about half as long as the rest of the achene (Fig. 3D). Fl. 5. 2*n* = 32. G.–H.

Native. Found only in dry places near St Aubyns, Jersey. W. and S. Europe, N. Africa, and W. Asia to India.

Section 3. *Echinella* DC. Annual or perennial. Nectary covered by an entire truncate scale, attached laterally or free to the base; achenes distinctly beaked, moderately compressed, often conspicuously bordered, their faces with spines or hooked hairs or rarely only papillate; receptacle usually elongating little in fr.

5. R. arvensis L.   'Corn Crowfoot.'

An annual herb with fibrous roots and an erect branching lfy stem 15–60 cm., ± hairy. Lvs stalked, the lowest simple, broadly spathulate to obovate, toothed near the tip; the rest ± deeply 3-lobed, ternate or biternate, with narrow segments; upper stem-lvs short-stalked with few linear segments. Fls 4–12 mm. diam., axillary and in a terminal cyme, their *stalks ± hairy, not furrowed*. Sepals 5, pale yellowish-green, *spreading*. Petals 5, ovate, *bright lemon-yellow*, the nectary covered by a free flap broader than that part of the petals. Stamens 10–13 or fewer. *Receptacle hairy*. Achenes 4–8, 6–8 mm., ovate, *conspicuously bordered, spiny*, the longest spines on the border; beak 3–4 mm., straight (Fig. 3E). Fl. 6–7. Protandrous or homogamous, sometimes gynomonoecious. Visited by small flies. 2*n* = 32. Th.

Native or introduced. Long established as a cornfield weed, especially on calcareous soils, and perhaps native. 71, S. Common in the south but less so towards the north and perhaps a recent introduction north of Perth and in Ireland, where it is rare; Channel Is. Throughout Europe except in the further north, N. Africa, W. Asia to India.

*****R. muricatus** L., an annual almost glabrous herb with a stout much-branched spreading or ascending stem 10–40 cm.; roundish-cordate long-stalked basal lvs with 3(–5) shallow toothed lobes, upper stem-lvs cuneate at base with 3 narrow, cut or toothed divisions; fls 8–10 mm. diam., petals longer than the reflexed sepals; and the 8–16 *achenes* large, flattened, strongly keeled, *with short spines confined to the faces* (not on the strongly grooved border) *and tapering abruptly into a broad curved beak* almost half as long as the rest of the achene, is sometimes found as a casual and is locally established, as in the Scilly Islands. Mediterranean region, but widely naturalized.

*****R. marginatus** D'Urv. ssp. **trachycarpus** (Fisch. & Mey.) Hayek, resembling *R. muricatus* in habit but ± sparsely hairy, lower lvs pinnately or palmately 3-lobed, the lobes further cut or toothed; fls small, petals slightly exceeding the spreading sepals; *achenes* 30–40, small, flattened, *with blunt wrinkled tubercles on the faces and a short beak*, is also a casual and occasionally established. Western Asia, but widely naturalized.

6. R. sardous Crantz   'Hairy Buttercup.'

R. Philonotis Ehrh.; incl. *R. hirsutus* Curt. and *R. parvulus* L.

An annual herb with erect branched lfy ± hairy stem 10–45 cm., like *R. bulbosus* but with *no stem-tuber* and with spreading hairs throughout. *Basal lvs* stalked, *shining*, ± *deeply* 3-*lobed, or ternate* with stalked 3-lobed irregularly cut or toothed lflets, the middle lflet longest-stalked; stem-lvs shorter-stalked, the uppermost sessile and with fewer and narrower segments; all ±hairy. Fls many, 1·2–2·5 cm. diam., their *stalks hairy, furrowed. Sepals 5,* hairy, *reflexed.* Petals 5 or more, 8–12 mm., ovate, pale yellow. *Receptacle hairy. Achenes 3–4 mm.,* roundish, their brownish faces usually with *a ring of tubercles close to the conspicuous green border*; beak short, slightly curved (Fig. 4A). Fl. 6–10. Slightly protogynous. Visited by flies and small bees. 2*n* = 16, 48. Th.

Variable in size and hairiness and in the size and distribution of tubercles on the achenes. *R. parvulus* L. is merely a dwarf form, f. *parvulus* (L.) Hegi.

Native. A local and often casual weed of damp arable and waste land. 80, S. From Argyll and Angus southwards; Inner Hebrides; Channel Is. C. and S. Europe, N. Africa, W. Asia.

The very closely related *R. trilobus Desf., ±glabrous,* with *narrow toothed lf segments,* smaller fls, and *tubercles all over the faces of the achenes,* is sometimes found as a casual.

Fig. 4. Achenes of *Ranunculus.* A, *R. sardous*; B, *R. parviflorus*;
C, D, *R. auricomus.* × 3.

7. R. parviflorus L.　　　　　　　　　　　　　'Small-flowered Buttercup.'

An annual herb with fibrous roots and numerous *spreading, ascending or decumbent,* hairy, little branched *lfy stems* 10–40 cm. Basal lvs roundish-cordate, ± 3–5-lobed with cuneate toothed lobes; upper lvs with fewer and narrower lobes, uppermost oblong; all softly hairy, yellowish-green. Fls 3–6 mm. diam., their *stalks opposite the lvs or in the forks of the branches,* hairy, *furrowed. Sepals* 5, hairy, *reflexed.* Petals 5 or fewer, 1–2 mm., oblong, pale yellow. *Receptacle glabrous. Achenes* few, 2·5–3 mm., roundish, *narrowly bordered,* with *shortly-hooked tubercles* all over the reddish-brown face; beak short, curved (Fig. 4B). Fl. 5–6. 2*n* = 28. Th.

Native. A local lowland plant of dry grassy banks and path-sides. 63, H7, S. Throughout England and Wales; rare in Ireland; Channel Is. S. and S.W. Europe, N. Africa, W. Asia.

Section 4. *Epirotes* (Prantl) L. Benson. Perennials. Sepals always purple-tinged; nectary scale laterally attached forming a pocket, often very small; achenes beaked, little compressed, smooth, usually glabrous; fruiting receptacle 3–15 times its length in the fl.

8. R. auricomus L.　　　　　　　　　　　　　　　　　　Goldilocks.

A perennial herb with a short premorse stock and numerous fibrous roots. Stems 10–40 cm., numerous, ± erect, slightly branched, sparsely hairy. *Lvs variable* in form, chiefly basal; *the lowest long-stalked, roundish or reniform, crenate or coarsely toothed but hardly lobed*; the rest ± deeply 3-lobed, the lobes crenate or further cut; stem-lvs few, ± sessile, deeply divided into narrow segments. Fls few, 1·5–2·5 cm. diam. when petals are present, their *stalks hairy, not furrowed.* Sepals hairy, appressed to the petals. *Petals 5 or fewer or 0,* 5–10 mm., obovate, golden yellow, the *nectary scale small or abortive. Receptacle with elongated tubercles* to which the achenes are attached. *Achenes* 3·5–4 mm., roundish, narrowly bordered, *downy; beak fairly long, hooked* (Fig. 4c, d). Fl. 4–5. Homogamous or protogynous. Visited by various flies and small bees but apomictic. $2n = 16, 32*, 48$. Hs. or G.

Native. A common woodland herb, occasionally on rocks. 87, H 36, S. Throughout Great Britain but not in the Outer Hebrides, Orkney or Shetland; Ireland; Channel Is.; reaches 1590 ft. in Scotland. Europe, N. Asia.

Section 5. *Flammula.* Webb & Berth. Terrestrial or aquatic annuals or perennials; basal and stem-lvs simple, entire or toothed, not palmately lobed or divided; nectary scale forming a pocket; achenes distinctly beaked, not greatly compressed, usually glabrous; fruiting receptacle much elongating.

9. R. lingua L.   Great Spearwort.

A perennial stoloniferous herb with stout stems 50–120 cm., creeping in mud below, then erect, branching above, hollow, glabrous or with a few appressed hairs. Basal lvs up to 20×8 cm., produced in autumn and often submerged, long-stalked, ovate or ovate-oblong cordate, blunt, disappearing before flowering; *stem-lvs* up to $25 \times 2·5$ cm., *distichous, short-stalked or sessile, half-clasping, oblong-lanceolate,* acute or acuminate, entire or remotely denticulate. Infl. a loose few-fld cyme. *Fls 2–5 cm. diam.,* their *stalks* ± hairy, *not furrowed.* Sepals 5, subglabrous. Petals 5, roundish-ovate, bright yellow, glossy. Receptacle glabrous. Achenes c. 2·5 mm., obovate, glabrous, minutely pitted, bordered; beak short, broad, somewhat curved (Fig. 5A). Fl. 6–9. Protogynous. Visited by various flies. $2n = 128$.

Native. A local plant of marshes and fens. 86, H 33, S. From Caithness southwards; Ireland; Channel Is. Europe, Siberia.

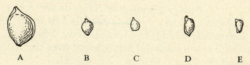

Fig. 5. Achenes of *Ranunculus.* A, *R. lingua*; B, *R. flammula*; C. *R. reptans*; D, *R. ophioglossifolius*; E. *R. sceleratus.* × 3.

10. R. flammula L.   Lesser Spearwort.

A perennial herb with stems 8–50(–80) cm., erect, ascending or creeping, commonly rooting at least at the lower nodes, sometimes at irregular intervals throughout (f. *tenuifolius*), hollow, slightly branched, ± glabrous. *Basal lvs* up to 4(–5) × 2·5(–3) cm., stalked, *subulate or lanceolate to broadly ovate,* rounded to cordate at base; *stem-lvs* short-stalked, the uppermost sessile, *broadly lanceolate to linear-lanceolate,* ± acute, subentire to distinctly toothed, parallel-veined. *Fls 8–20(–25) mm. diam.,* solitary or in a few-fld cyme, their *stalks* slightly hairy, *furrowed.* Sepals 5, greenish-yellow. Petals 5, obovate, pale yellow, glossy. Receptacle glabrous. Achenes 20–50(–60) in a globose head, each 1–2 mm. long (excluding the short blunt beak), ovate, minutely pitted, obscurely bordered (Fig. 5B). Fl. 5–9. Protandrous. Visited by various flies and small bees. $2n = 32^*$. Three sspp. may be distinguished:

Ssp. **flammula**: stem 8–50(–80) cm., erect to creeping; *basal lvs* usually 1–4 × 0·8–2·5 cm., *lanceolate to broadly ovate* (but narrowly elliptical in sterile submerged forms); achene about ⅓ longer than broad).

Ssp. **minimus** (A. Benn.) Padmore: stem 3–8(–14) cm., semi-prostrate but not rooting, with very short internodes; *basal lvs* short-stalked, *at least as broad as long, cordate at base*, thick and fleshy; fls 15 mm. diam. or more; achenes about as broad as long.

Ssp. **scoticus** (E. S. Marshall) Clapham: stem erect, 20–60 cm.; *lowest lvs subulate* (*with no expanded blade*), caducous; later lvs sub-persistent but readily detachable, with short blunt linear-oblong blade; upper stem-lvs lanceolate, ± sessile; fls 1(–4), 10–15 mm. diam.; achenes as in ssp. *flammula*.

Native. Common in wet places throughout the British Is. 112, H40, S. Europe, temperate Asia, Azores. Ssp. *minimus* in exposed places by the sea, often forming dense mats, in Caithness, Outer Hebrides, Orkney, Shetland and Co. Clare. 4, H1. Ssp. *scoticus* on lake-shores in Argyll and Inner Hebrides and perhaps in Co. Mayo. 2, H1.

11. R. reptans L.   'Slender Creeping Spearwort.'

A perennial stoloniferous herb resembling the slender creeping forms of *R. flammula* but with the *filiform stem* 5–20(–50) cm., *arching in the internodes and rooting at every node*. *Lvs* 0·5–2 cm., in small tufts at each rooting point, long-stalked, *spathulate or narrowly elliptical*. *Fl.* 5–10 mm. diam., *solitary* on the ascending tip of a stolon. Petals 5, narrowly obovate, pale yellow, glossy. *Achenes* 1–1·5 mm., ovate, glabrous, *with a curved ± terminal slender beak about ¼ as long as the rest of the achene* (Fig. 5c) (c. ⅛ in *R. flammula*). Fl. 6–8. $2n = 32$. H.

Plants referred to this sp. on the shores of Ullswater (Lake District) and Loch Leven (Kinross) do not show uniform agreement with the above description, based on continental material, but vary in size and in the extent of nodal rooting; some are indistinguishable from *R. flammula* f. *tenuifolius* Wallr. and these become erect in cultivation, unlike the more extreme forms. All plants show some pollen sterility and it is thought that the populations are of hybrid origin (*R. flammula* × *reptans*) with no remaining pure *reptans*, though some plants approach it closely.

Native or formerly native. A rare plant of lake margins in the Lake District and Scotland. 5. N. and C. Europe.

12. R. ophioglossifolius Vill.   'Snaketongue Crowfoot.'

An *annual* herb with a fibrous root system and an erect or ascending branching stem 10–40 cm., hollow, furrowed, glabrous or somewhat hairy above. Basal lvs long-stalked, roundish-ovate-cordate, to 20 × 12 mm. or larger if floating; upper lvs becoming narrower and more shortly stalked upwards, the uppermost ± sessile, narrowly elliptical; all lvs obscurely and distantly toothed or ± entire. *Fls* 6–9 *mm. diam.*, many, in axillary cymes, or few and opposite the lvs, their stalks glabrous, somewhat furrowed. Petals 5, pale yellow, slightly exceeding the glabrous sepals. Receptacle glabrous. Achenes c.

1·5 mm., ovate, compressed, obscurely bordered, their *faces covered with small tubercles*, very shortly beaked (Fig. 5D). Fl. 6–7. $2n=16$. Th.

Native. A rare plant of marshes, found only in Gloucester; formerly also in Dorset and Jersey but now extinct. 3. Gotland, France, S. Europe, N. Africa, W. Asia.

Section 6. *Hecatonia* (Lour.) DC. Annual or perennial marsh or water plants. Nectary scale often forked or completely surrounding the nectary. Achenes not beaked, moderately compressed, often with a corky thickening on the keel, their faces smooth or somewhat wrinkled.

13. R. sceleratus L.   'Celery-leaved Crowfoot.'

Annual or overwintering herb with a fibrous root system and a stout erect stem 20–60 cm., hollow, branched above, ± glabrous, furrowed. Lower lvs long-stalked, reniform or pentagonal in outline, ± deeply palmately 3-lobed, the lateral lobes often again 2–3-lobed, all crenate; stem-lvs short-stalked, more deeply divided into narrower segments, the uppermost sessile, with 3 or fewer entire segments; lower *lvs* ± glabrous, *shining*, uppermost slightly hairy below. Fls 5–10 mm. diam., numerous, in branching cymes, their stalks glabrous, furrowed. Sepals reflexed, hairy below. Petals narrowly ovate, hardly exceeding the sepals, pale yellow, the open nectary surrounded by its scale. Receptacle slightly hairy. *Achenes very numerous* (70–100), c. 1 mm., roundish, compressed, glabrous, each face with a faintly wrinkled central area (Fig. 5E); the *head of ripe achenes oblong-ovoid*, 6–10 mm. Fl. 5–9. Protogynous. Visited by flies. $2n=32$. Th.–Hel.

Native. Common in and by slow streams, ditches and shallow ponds of mineral-rich water with a muddy bottom. 104, H 39, S. Throughout most of Great Britain but rare in N. Scotland and not in Shetland; Ireland; Channel Is. Europe, especially north and central.

Subgenus BATRACHIUM (DC.) A. Gray.

Aquatic or subaquatic annuals and perennials. Lvs stipulate, all with a distinct blade ('floating lvs'), or all finely dissected ('submerged lvs'), or of both kinds. Fls solitary, lf-opposed. Petals usually 5, white, but commonly with a yellow claw (as in all British spp.); nectary-scale minute or 0. Achenes usually not strongly compressed, transversely ridged.

A difficult group, not yet adequately understood. Many of the forms with finely dissected submerged lvs may sometimes be found in very shallow water or in places where there is no standing water for at least part of the year. The segments of their dissected lvs are then commonly shorter, stouter and more rigid and diagnostic characters based on their behaviour on removal from water (collapsing or not) cease to be of value: identification must depend on features of the fl. and fr.

14. R. hederaceus L.   'Ivy-leaved Water Crowfoot.'

An annual or perennial herb with branching stem 10–40 cm., creeping in mud or the upper part floating. Lvs 1–3 cm. wide, usually opposite, stalked, reniform or roundish-cordate, often with dark markings near the base, shallowly 3–5-lobed, the *lobes broadest at their base* (Fig. 6A, B) roundish or triangular, blunt, entire; stipules membranous, shorter than wide, almost entirely adnate. *No dissected lvs.* Fls 3–6 mm. diam. Petals scarcely exceeding

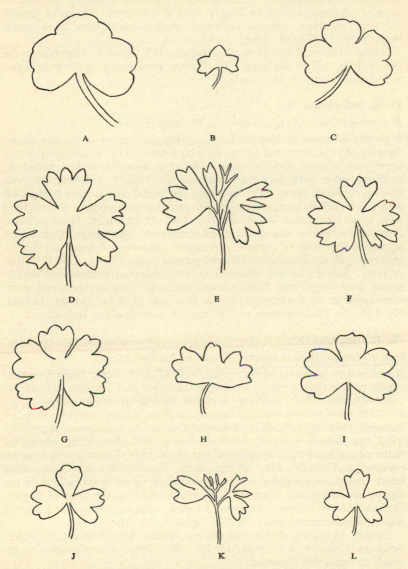

Fig. 6. Leaves of *Ranunculus* subgenus *Batrachium*. *R. hederaceus*: A, robust floating form; B, small form on mud. *R. omiophyllus*: C. *R. aquatilis*: D, F, floating lvs; E, transitional lf. *R. peltatus*: G, I, floating lvs; H, var. *truncatus*. *R. baudotii*: J, floating lf; K, transitional lf. *R. tripartitus*: L, floating lf. All ×c. 1·5.

the sepals, not contiguous. Stamens 6–10, somewhat exceeding the head of carpels. *Receptacle glabrous*, rarely with a few hairs. Fl.-stalks usually 1–2·5 cm., recurved. *Achenes* 1–1·5 mm., *glabrous*, with short blunt lateral beak. Fl. 6–9. $2n=16*$. Hyd.

Native. On mud and in shallow water. 112, H40, S. Throughout the British Isles, but rather local. W. Europe northwards to Denmark and S. Sweden; doubtfully in C. Italy.

15. R. omiophyllus Ten.

R. coenosus Guss.; *R. Lenormandii* F. W. Schultz

A prostrate annual or biennial herb growing on mud or in shallow water. Stem usually 5–25 cm. but longer in robust floating forms, creeping or the upper part floating. Lvs 8–30 mm. wide, often opposite, stalked, roundish-reniform, never with dark markings near the base, 3(–5)-lobed, the *lobes shallow*, rounded, *narrowest at their base*, the lateral ones slightly 2-lobed, all with a few shallow crenations (Fig. 6c); stipules membranous, shorter than wide, about half adnate. *No dissected lvs*. Fls 8–12 mm. diam. *Sepals reflexed*. *Petals* about *twice as long as sepals*, not contiguous. Stamens 8–10, somewhat exceeding the head of carpels. *Receptacle glabrous*. *Achenes* 1–1·5 mm., *glabrous*, with an *almost central* curved pointed *beak*. Fl. 6–8. $2n=32*$. Hel. or Hyd. Native. Locally common in non-calcareous streams and muddy places from Argyll and Stirling southwards, but less common and more interrupted in its distribution on the east side of Great Britain; Ireland. 63, H13. W. Mediterranean and W. Europe northwards to Ireland.

16. R. tripartitus DC.   'Three-lobed Water-Crowfoot.'

Incl. *R. lutarius* (Revel) Bouvet; *R. intermedius* auct. angl.

An annual or perennial herb growing on mud or in water. Stem 5–60 cm., simple or branched, creeping in mud or the upper part floating. *Floating lvs* 5–20(–40) mm. across, *reniform or almost circular* in outline, *deeply* 3(–5)-*lobed* (to more than half-way), the *lobes cuneate*, *distant*, entire or with 2–4 rounded crenations (Fig. 6L); *submerged dissected lvs* few, distant, repeatedly trifid, the *ultimate segments extremely slender*, collapsing, rarely somewhat flattened and more rigid, sometimes absent in terrestrial plants; stipules large, rounded, half-adnate. Fls 3–10 mm. diam. Sepals with a blue subterminal band. *Petals not exceeding* 6 *mm.*, equalling or up to twice as long as the sepals, *not contiguous*. Stamens 5–10, about equalling the head of carpels. *Receptacle hairy, globose*. *Achenes* c. 1·5 mm., inflated, *glabrous*, with a small lateral or subterminal beak. Fl. 4–6. $2n=48*$; 32. Hyd.

Native. In muddy ditches and shallow ponds. S.W. England, S. Wales, Anglesey, Cheshire; Co. Cork. 18, H1. W. Europe from Cheshire southwards to Spain and Portugal.

R. lutarius (Revel) Bouvet, described as differing from *R. tripartitus* in its larger fls and in the usual absence of dissected lvs, these when present having flattened non-collapsing segments, appears to consist partly of forms inseparable from *R. tripartitus* and partly of hybrids of *R. tripartitus* with *R. omiophyllus* together with products of selfing and back-crossing. The populations in the New Forest, upon which the earliest British records were based, are almost certainly of this hybrid origin.

17. R. fluitans Lam.

A large robust perennial herb with creeping rhizome and submerged lfy stems which may reach 6 m. in length. *Submerged lvs* usually 8–30 cm., *longer than the stem-internodes*, greenish-black, about twice trifid, with rather few *very long* firm slender *sub-parallel* segments which are twice or three times forked; *floating lvs* 0. *Fls* 20–30 *mm. diam.* Petals 5–10, overlapping persistent. Stamens numerous, *equalling or falling short of the head of carpels. Receptacle nearly or quite glabrous.* Fr.-stalks stout, usually not exceeding the lvs, often projecting from the water. *Achenes* large, inflated, usually *glabrous.* Fl. 6–8. $2n = 16^*, 31^*$. Hyd.

Var. *bachii* (Wirtg.) Wirtg. (probably of hybrid origin) is a more slender plant with shorter subsessile lvs, smaller fls with only 5 petals and fr.-stalks larger than the lvs.

Native. In rapidly flowing streams and rivers throughout Great Britain from the Clyde southwards; in Ireland only in Antrim (replaced by *R. peltatus* ssp. *pseudofluitans*). Var. *bachii* is very local in England and S. Scotland. Europe northwards to Denmark; absent from Iceland and Fennoscandinavia.

18. R. circinatus Sibth.

R. divaricatus sensu Coste; *R. foeniculaceus* auct.

A slender perennial with usually *erect* little-branched *submerged stems. Submerged lvs* 0·5–2(–3) *cm. diam., sessile, circular in outline,* trifid and then repeatedly trifid, *the short rigid divergent segments all lying in one plane; floating lvs* 0. Fls 8–18 mm. diam. *Petals up to 10 mm.,* barely contiguous, persistent. Stamens 15–20, exceeding the head of carpels. Receptacle pubescent. Fr.-stalks 3–8 cm., much exceeding the lvs, tapering upwards. Achenes 1·2–1·5 mm., pubescent (at least while immature), somewhat compressed, with a distinct beak. Fl. 6–8. $2n = 16^*$. Hyd.

Native. Locally common in ditches, canals, slow streams, ponds and lakes with a high mineral content. From Aberdeen southwards; Ireland from Limerick and Carlow northwards. Europe northwards to c. 61° N. in Sweden and Finland; absent from Norway and southern Balkans.

19. R. trichophyllus Chaix

R. divaricatus Schrank; *R. flaccidus* Pers.; *R. paucistamineus* Tausch

An annual or perennial herb with *short* (2–4 cm.) *sessile or rather short-stalked submerged lvs* which are repeatedly trifid, the *ultimate segments* commonly dark green, forked, *not lying in one plane,* rigid or collapsing; *no floating lvs.* Fls small (5–)8–10(–15) mm. diam. *Petals rarely exceeding* 5 mm., *not contiguous,* fugacious; *nectaries lunate* (Fig. 7B). Stamens 5–15, equalling or exceeding the head of carpels. Receptacle pubescent. *Fr.-stalk usually less than 4 cm.,* strongly recurved close to the base. Achenes compressed, pubescent or bristly above (at least while immature) and with a very short lateral beak. Fl. 5–6. $2n = 16^*, 32^*$. Hyd.

The British form is placed in **ssp. trichophyllus**, with achenes c. 2 mm., ovate.

Variable in the fineness and rigidity of the lf-segments and in the extent to which hairs persist on the maturing achenes. British forms which have been referred to *R. drouetii* F. W. Schultz, said to have collapsing lf-segments and glabrous achenes, seem to be assignable either to *R. trichophyllus* or to *R. aquatilis*.

Native. In ponds, ditches and slow streams throughout the British Is. Europe; Asia; N. America.

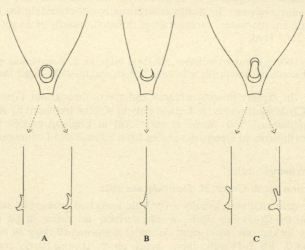

Fig. 7. Surface views and vertical sections of nectaries of *Ranunculus* subgenus *Batrachium*. A, 'circular' nectary of *R. aquatilis*; B, 'lunate' nectary of *R. trichophyllus*; C, '± pyriform' nectary of *R. peltatus*.

20. R. aquatilis L.

R. heterophyllus Weber; *R. diversifolius* Gilib.; incl. *R. radians* Revel

Annual or perennial herbs with branched submerged stems and finely dissected submerged lvs, 3–6(–8) cm., shorter than the stem-internodes, their segments not lying in one plane, rigid or collapsing; *floating lvs subcircular,* ± deeply lobed or cut into *cuneate usually straight-sided and often dentate segments* (Fig. 6D, F); transitional lvs (Fig. 6E) between submerged and floating types often present; floating lvs sometimes 0 (var. *submersus* Bab.). Fls 12–18 mm. diam. *Petals usually 5–10 mm.,* persistent; *nectaries circular* (Fig. 7A). Stamens 13 or more, exceeding the head of carpels. Receptacle pubescent. *Fr-stalk* 2–5 cm., rarely longer, and usually *shorter than the stalk of the opposed lf,* not or slightly tapering, recurved below. Achenes 1·5–2 mm., ± distinctly pubescent (at least while immature), shortly beaked. Fl. 5–6. $2n = 16^*$; 48. Hyd.

Variable in the length, fineness and rigidity of the segments of the submerged lvs, in the number of transitional and floating lvs and in the density, coarseness and persistence of hairs on the achenes. *R. radians* Revel comprises forms having submerged lvs with short rigid segments as well as transitional and floating lvs and having achenes bristly at maturity, but no clear separation can be made. Forms which have been named *R. drouetii* F. W. Schultz are commonly referable to *R. aquatilis* var. *submersus* Godr., distinguishable from *R. trichophyllus* by the longer petals with circular (not lunate) nectaries and the usually longer lvs (4–8 cm.) with collapsing segments.

Native. In ponds, ditches and streams throughout lowland Great Britain and Ireland. Europe.

21. R. peltatus Schrank

Annual or perennial herbs with branched submerged stems and finely dissected submerged lvs whose segments do not lie in one plane; *floating lvs*, when present, usually *reniform to semicircular* in outline, 3–5-lobed, the *lobes crenate and with rounded sides* (Fig. 6G, H, I). Fls 15–20(–30) mm. diam. *Petals 8–15(–18) mm.*, often more than 5, usually broad and contiguous, persistent; *nectaries elongated, ±pyriform* (Fig. 7c). Stamens c. 30, exceeding the head of carpels. Receptacle pubescent. *Fr.-stalk usually 5–15 cm., often longer than the stalk of the opposed lf,* usually tapering above, curving downwards through most of its length. Achenes 2–2·5 mm., sparsely puberulous to hispid (at least while immature), shortly beaked. Fl. 5–8. $2n=48*$ (?32). Hyd.

Very variable. The two following sspp. are commonly recognized, but they are connected by intermediate forms:

Ssp. **peltatus**: *Submerged lvs* usually shorter than the stem-internodes, their segments usually divergent, *not collapsing; floating lvs usually present,* reniform in outline (or almost semicircular in var. *truncatus* Koch, Fig. 6H). Fr.-stalk usually tapering gradually upwards (but not tapering and often quite short in var. *floribundus* (Bab.) Druce, whose large fls have non-contiguous petals).

Ssp. **pseudofluitans** (Syme) C. Cook: *Submerged lvs* often longer than the stem-internodes, with long parallel *collapsing* segments; *floating lvs* with wide basal sinus, but *commonly* 0. Fr.-stalk very long (to 15 cm.), stout, tapering only near the top. *R. peltatus* var. *penicillatus* Hiern, less robust, with finer lf-segments and often with floating lvs, may be placed here.

Native. Ssp. *peltatus* common in lakes, ponds and slow streams from Ross southwards; Ireland. Ssp. *pseudofluitans* in fast-flowing streams from Perth southwards; common in Ireland where it largely replaces *R. fluitans*. Europe, N. Africa.

R. aquatilis and *R. peltatus* are highly variable species which have caused a great deal of taxonomic confusion. Further research is needed, especially in relation to the occurrence of natural hybrids. In general the form of the floating lvs, the size of the flowers, the shape of the nectaries and the length of the fr.-stalks afford satisfactory distinguishing characters, but some forms present difficulties and may be of hybrid ancestry. *R. peltatus* ssp. *pseudofluitans* has sometimes been confused with *R. fluitans*, but even when floating lvs are absent its shorter submerged lvs (rarely exceeding 8 cm.), its longer stamens and pubescent receptacle should prevent misidentification; just as its larger fls and longer lvs and fr.-stalks distinguish it from *R. aquatilis* var. *submersus*.

22. R. baudotii Godr.

Batrachium marinum Fr.

An annual or perennial herb usually with a thick succulent submerged stem. *Submerged lvs with spreading non-collapsing segments; floating lvs* small, ± reniform in outline, *deeply 3-lobed* (to ⅔ or more), the lobes distant, cuneate, entire or crenate (Fig. 6 J); transitional lvs sometimes present (Fig. 6 K); floating lvs sometimes 0. Fls 12–18 mm. diam. Petals 6–10 mm., contiguous, persistent. Stamens 15–20, usually not exceeding the head of carpels. *Receptacle* pubescent, *ovoid.* Fr.-stalks 5–8(–12) cm., stout, tapering, usually much longer than the stalks of the opposed lvs, strongly recurved (often with a double curve). *Achenes small*, 1–1·4 mm., *very numerous* (40–100) on the ovoid receptacle, *glabrous*, inflated, with a short lateral beak and usually *with a thin membranous wing* on the keel below the beak. Fl. 5–9. $2n = 32^*$. Hyd.

Variable in robustness, in the presence or absence of floating lvs, etc. Slender forms with very deeply lobed floating lvs, almost semicircular in outline, stamens exceeding the head of carpels and less strongly recurved fr.-stalks have been named var. *confusus* (Godr.) Syme, but they grade into the type. 'Var. *marinus* Fr.' of British authors comprises very robust forms with succulent stems, no floating lvs and very long fr.-stalks. Again there is no clear distinction from the type.

Native. In brackish streams, ditches and ponds in coastal districts throughout the British Is.; rarely inland. Coasts of most of Europe northwards to c. 64° N. on the Baltic coast of Sweden and Finland; absent from Norway and Iceland.

Hybrids of *R. baudotii* with *R. trichophyllus* and *R. aquatilis* have been reported.

Subgenus 3. FICARIA (Huds.) L. Benson

Perennial terrestrial herbs with fusiform root tubers and simple ± entire lvs. Sepals 3, greenish, falling. Petals 7–12 or more, yellow. Nectary scale a shallow pocket. Achenes not compressed, ± beakless. Cotyledon apparently 1.

23. R. ficaria L.   Lesser Celandine, Pilewort.

Ficaria verna Huds.; *F. ranunculoides* Moench

A perennial herb with fibrous roots and numerous fusiform or clavate root-tubers 10–25 mm. Stems 5–25 cm., branched, ascending, often rooting at the decumbent base. Lvs all stalked, with sheathing bases, those of the lowest very broad; *basal lvs* 1–4 cm. long and wide, in a rosette, long-stalked, *cordate,* bluntly angled or crenate, rarely toothed; stem-lvs similar but smaller and shorter-stalked; all fleshy, dark green, glabrous. Fls 2–3 cm. diam., solitary, terminal on each stem. *Sepals* 3, rarely more, ovate, concave, green. *Petals* 8–12, rarely 0, narrowly ovate, c. twice as long as the sepals, bright golden yellow, fading white. Receptacle hairy. Achenes often aborting, numerous, to 2·5 mm., ± spherical, keeled, ± downy, minutely beaked. Fl. 3–5. Protandrous. Some fls without stamens. Visited by various flies and bees, etc., but often setting little fr. (see below). Grt. Our two cytologically different forms seem often, but not always, morphologically distinct:

Ssp. **ficaria:** lvs with no axillary bulbils; fls 20–30 mm. diam., their petals broad, overlapping; pollen largely viable; carpels mostly fertile and yielding well-developed achenes. $2n=16^*$. The commoner form, especially in sunny places.

Ssp. **bulbifer** (Marsden-Jones) Lawalrée: lvs with small axillary bulbils which reproduce the plant vegetatively; fls up to 20 mm. diam., their petals usually narrow, not overlapping; pollen largely non-viable; carpels rarely fertile, yielding up to c. 6 well-developed achenes per fl. $2n=32^*$ (hence a tetraploid). A more local plant, chiefly in shade.

Native. A common plant of woods, meadows, grassy banks and stream-sides. 112, H40, S. Throughout Great Britain and Ireland; Channel Is.; reaches 2400 ft. in Wales. Europe, W. Asia. Introduced in N. America.

12. ADONIS L.

Perennial or annual herbs with lfy stems and bi- or tri-pinnate lvs with linear segments. Fls usually solitary, terminal; hermaphrodite, actinomorphic, hypogynous. Sepals 5(–8), ± petaloid; *petals 3–20, yellow or red, not nectar-secreting*; stamens numerous; carpels numerous, each with 1 pendulous ovule. Fr. an elongated head of *wrinkled short-beaked achenes*.

About 20 spp. in Europe and temperate Asia.

*1. A. annua L.   Pheasant's Eye.

A. autumnalis L.

An annual herb with slender tap-root and erect usually branched stem 10–40 cm., usually glabrous. Lvs tripinnate with linear acute segments. Fls 15–25 mm. diam., terminal on stem and branches, erect. Sepals 5, ovate, spreading, green or purplish. *Petals 5–8, suberect*, somewhat longer than the sepals, ovate, *bright scarlet with a dark basal spot*. Filaments dark violet. *Achenes* arranged in a ± lax head c. 18 mm., each *with a short straight beak and with no tooth at the base of the upper face* (Fig. 8 B). Fl. 7. Homogamous. Visited by bees. $2n=32$. Th.

Fig. 8.

Achenes of A, *Adonis aestivalis* and B, *A. annua*. × 2.

Introduced. Naturalized as a cornfield weed in a few southern counties of England from Dorset to Oxford and Kent, and sometimes found as a casual elsewhere in England and in the Channel Is.; a rare casual in S. Ireland. S. Europe and S.W. Asia.

The related *A. aestivalis* L. sometimes occurs as a casual and may be distinguished by its sepals being closely appressed to the ± spreading petals (in *A. annua* they do not touch the suberect petals), and by the sharp tooth at the base of the upper face of the achene (Fig. 8A). *A. vernalis* L., a perennial herb with 10–15 narrow yellow petals and rounded pubescent achenes with a strongly curved beak, is often grown in gardens.

13. MYOSURUS L.

Small annual herbs with linear lvs confined to a basal rosette. Fls terminal, solitary, small, hermaphrodite, actinomorphic, hypogynous. Sepals 5 or more, with a small basal spur; *petals 5–7 or 0, tubular*, nectar-secreting; stamens few; carpels numerous, each with 1 pendulous ovule. *Fr. of numerous 1-seeded achenes in an elongated spike*. 7 spp., in the north temperate zone, Chile and New Zealand.

1. M. minimus L.   Mouse-tail.

A glabrous annual herb with a basal rosette of linear, entire, somewhat fleshy lvs and numerous erect lfless flowering stems 5–12·5 cm. Fls very small, pale greenish-yellow. Sepals 5, rarely 6–7, 3–4 mm., narrowly oblong, with their basal spurs appressed to the stem. Petals 5, greenish, tubular with a short strap-shaped limb about equalling the sepals. Stamens 5–10. Achenes 1–1·5 mm., numerous, brownish, keeled, shortly beaked, on a filiform receptacle 2·5–7 cm. Fl. 6–7. Homogamous or slightly protandrous. Visited by small flies for the nectar secreted in the petals, but usually self-pollinated. $2n=28$. Th.

Probably native. Very locally in damp arable fields. 46, S. Throughout lowland England from Northumberland to Devon and Kent, and in N. Wales; Channel Is. C. and S. Europe, N. Africa, S.W. Asia. Naturalized in N. America and Australia.

Tribe 3. THALICTREAE. Fr. of follicles or achenes; chromosomes small; basic number 7.

14. AQUILEGIA L.

Perennial herbs with erect woody stocks and spirally arranged *bi- or triternately compound lvs. Fls* terminal, solitary, or in panicles; hermaphrodite, actinomorphic, hypogynous, *with all parts arranged in 5-merous whorls. Sepals 5, petaloid; petals 5, each with a long hollow backwardly directed nectar-secreting spur*; stamens numerous, the innermost staminodal; carpels 5, rarely 10, sessile, free, each with many ovules. *Fr. a group of many-seeded follicles.*

About 55 spp., chiefly in the north temperate zone. Poisonous.

Secondary lflets stalked; spur of petal strongly curved into a knobbed hook;
   ripe follicles 15–20 mm.   **1. vulgaris**
Secondary lflets sessile; spur of petal almost straight; ripe follicles 12–15 mm.
                                      2. pyrenaica

1. A. vulgaris L.   Columbine.

A perennial herb with a short stout erect often branched blackish stock and erect lfy flowering stems 40–100 cm., glabrous or softly hairy, branched above. *Radical lvs long-stalked, biternate*, with the *secondary lflets usually stalked*, irregularly 3-lobed, the lobes crenate; stem lvs smaller, short-stalked; the uppermost (bracts) ± sessile and narrowly 3-lobed; all lvs ± glabrous, somewhat glaucous above, paler and greener beneath, their stalks ± hairy, broadening into a sheathing base. Infl. irregularly cymose. Fls 3–5 cm. diam., drooping, both sepals and petals usually blue but sometimes white or reddish. Sepals 15–30 mm., ovate, acute. Limb of petal oblong-truncate almost equalling the *spur*, which is *sharply curved and knobbed at its tip*; overall length of petals c. 30 mm. Fertile stamens c. 50, exceeding the petal-limbs; staminodes c. 10, white, blunt, crimped. Carpels 5(–10). shortly hairy. *Follicles 15–25 mm.*, erect, beaked, dehiscing while green. Seeds 2–2·5 mm., black, shining. Fl. 5–6. Protandrous, the stamens rising and dehiscing from outside inwards. Visited by long-tongued humble-bees for pollen and nectar. $2n=14$. Hp.

Native. A local plant of woods and wet places on calcareous soil or fen peat, and probably native throughout England, Wales and S. Scotland but naturalized further north to Caithness; Ireland; Channel Is.; reaches 3000 ft. in Angus. 66, H31, S. S. and C. Europe and reaching S. Sweden; N. Africa; temperate Asia to China. Naturalized in N. America.

*2. A. pyrenaica DC.   'Pyrenean Columbine.'

A. alpina auct. angl.

Stem 20–25 cm., slender, with small ternate or biternate radical lvs, the *secondary lflets sessile*, crenate or entire; stem-lvs much smaller. Fls 4·5–5 cm. diam., 1–3, darker blue than in *A. vulgaris*. Petal-limb obovate-cuneate, rounded; *spur slender, very slightly incurved*. Stamens exceeding the petal-limb. *Follicles* 12–15 *mm*. Fl. 8–9. 2n=14. Hp.

Introduced. Naturalized in Caenlochan Glen, Angus. Native in the Pyrenees.

Several columbines besides *A. vulgaris* are grown in gardens. The 'long-spurred' types, with straight spurs much exceeding the petal limbs, are apparently derived by hybridization chiefly from the American *A. coerulea* James (pale blue), *A. chrysantha* A. Gray (yellow, often with red spurs) and *A. formosa* Fisch. (sepals and spurs red, petal-limbs yellow).

15. THALICTRUM L.

Perennial herbs usually with repeatedly pinnate stipulate lvs. Fls numerous, small, in terminal or axillary racemes or panicles; hermaphrodite, actino-morphic, hypogynous. No nectaries. Per. segs 4–5, ± petaloid, spreading, readily falling; petals 0; stamens numerous with long erect or drooping fila-ments; carpels free, usually few, often stalked, with a sessile stigma. Fr. a group of sessile or stalked 1-seeded achenes.

About 85 spp., chiefly in the north temperate zone.

The genus is interesting in including some insect-pollinated fls, fragrant, with erect and conspicuous filaments (e.g. *T. aquilegifolium*); some wind-pollinated, non-fragrant, with drooping filaments (e.g. *T. alpinum*), and some which are intermediate and apparently both insect- and wind-pollinated (e.g. *T. flavum, T. minus*).

1 Fls in a simple raceme; plant usually not exceeding 15 cm.   **2. alpinum**
   Fls panicled; plant usually exceeding 15 cm.   2

2 Fls in dense clusters; stamens erect; achenes 1·5–2·5 mm.; ultimate lflets
     of middle and upper lvs much longer than broad.   **1. flavum**
   Fls not densely clustered; stamens drooping; achenes 3–6 mm.; ultimate
     lflets usually roundish, about as long as broad.   **3. minus**

1. T. flavum L.   'Common Meadow Rue.'

A perennial herb with creeping stoloniferous rhizome and erect robust fur-rowed usually simple stem 50–100 cm. Lower stem-lvs stalked, upper sessile, all bi- or tripinnate, with stipule-like structures at each branching; ultimate *lflets longer than broad, obovate-cuneate*, or oblong-lanceolate in the upper lvs, 3–4-lobed distally, dark green above, paler and almost glabrous beneath. *Panicle compact. Fls in dense clusters, erect*, fragrant, with 4 narrow whitish per. segs. *Stamens ± erect, yellow; anthers not apiculate*. Achenes 1·5–2·5 mm., ovoid to elliptical, 6-ribbed, glabrous. Fl. 7–8. Homogamous or slightly protogynous. Pollen smooth. Visited by various bees and syrphids but probably wind- as well as insect-pollinated. $2n = 28, 84$. H.

There is variation in the shape of the panicle and the size and shape of the achenes, and several varieties have been described.

Native. Common in meadows and fens by streams; to 1000 ft. in Derby. Great Britain northwards to Inverness. 73, H 29. Europe and temperate Asia.

2. T. alpinum L.   'Alpine Meadow Rue.'

A perennial shortly stoloniferous herb with short slender rhizome and an erect slender wiry stem 8–15 cm., rarely taller. *Lvs* chiefly radical, stalked, *biternate*, the *ultimate lflets roundish*, shallowly and bluntly lobed or crenate, dark green above, whitish below. *Fls in a simple raceme*, at first drooping, then erect. Per. segs 4, c. 3 mm., pale purplish, spreading. *Stamens 8–20, long and pendulous, with slender pale violet filaments and yellow anthers.* Carpels shortly stalked. *Fl. stalks recurved in fr.* Achenes 2–3, 3–3·5 mm., stalked, narrowly oblong, curved, ribbed, with a short hooked beak. Fl. 6–7. Protogynous. Anemophilous. $2n = 14$. H.

Native. Rocky slopes and ledges especially on mountains but descending almost to sea-level in the north-west; to 3980 ft. in Scotland. Great Britain northwards from Westmorland, Yorks, and Durham; N. Wales; Inner and Outer Hebrides; Orkney; Shetland. 32, H 6. Arctic and alpine Europe; Asia; N. America.

3. T. minus L.   'Lesser Meadow Rue.'

Perennial caespitose to stoloniferous herb with horizontal or ascending rhizome and erect rigid often flexuous lfy stems 15–150 cm., ± terete or furrowed, green or pruinose, often glandular above. Lvs stipulate, 3 or 4 times pinnate, the ultimate *lflets* variable in size and shape but usually *about as broad as long*, 3–7-lobed or toothed, green or glaucous, glabrous or with stalked glands especially beneath. Fls in a loose spreading panicle, drooping at first but sooner or later erecting. Per. segs 4, yellowish- or purplish-green. *Stamens* numerous, long and ± *pendulous*, with *apiculate anthers*. *Achenes* 3–6 mm., 3–15 per fl., *sessile*, ± *erect*, variable in shape from roundly and symmetrically ovoid to ± narrowly and asymmetrically ovoid-oblong with the ventral side gibbous and the dorsal side ± straight or gibbous only below, not or somewhat compressed, 8–10-ribbed, glabrous or with stalked glands. Fl. 6–8. Protogynous. Chiefly anemophilous but sometimes visited by insects; some races probably apomictic. $2n = 28, 42$. H.

A satisfactory treatment of this highly polymorphic group cannot be effected until it has been much more closely studied by modern taxonomic methods. It is not yet known to what extent the observed variation arises from genetic or from environmental differences, nor whether the group is cytologically uniform. Populations fall roughly into groups corresponding with the three main habitats; limestone rocks and grassland, dunes, and streamside or lakeside gravel and shingle. Within these groups plants show considerable inter- and intra-population variability in stature, mode of growth, leafiness of the stem-base, shape, colour and glandulosity of the lflets, size and shape of stigmas and achenes, etc. A statistical comparison of populations on the East Anglian chalk with those on the western limestones or the Pennine limestones would doubtless show different averages in respect of some of these characters but would show so much overlapping as to make specific or even subspecific distinction difficult to achieve or to justify. Unless better criteria of discrimination can be discovered it seems best merely to treat as subspecies the three habitat groups referred to, while recognizing that the distinctions are not clear-cut and are partly or even perhaps largely determined by the habitat differences.

Ssp. **minus** (including *T. babingtonii* Butcher, *T. montanum*, *T. collinum*, *T. saxatile*, *T. calcareum*, etc., auct. angl.). Commonly subcaespitose but often stoloniferous on soft or loose substrata. *Stem* (12–)25–50(–100) *cm.*, branched, lfy to the base or with brown lfless sheaths below, terete or somewhat furrowed, pruinose or not, ± glabrous or with some stalked glands at least above. Lowest lvs c. 10–25 cm.; lflets 4–15 mm. across, rounded-cuneate to cordate at the base, green or glaucous, pruinose or not, ± glabrous or with stalked glands especially on the underside. *Panicle usually branching from above the middle of the stem.* Achenes 3–6 mm., varying in shape even in the same population from roundly and symmetrically ovoid to ± narrowly and asymmetrically ovoid-oblong, the longer achenes being usually the narrower, more asymmetrical and more compressed.

Native. Dry limestone slopes, limestone rocks, cliff-ledges, scree or shingle, chalk quarries and banks, etc.; to 2800 ft. in Wales. Locally in suitable habitats throughout the British Is. except Shetland and Channel Is. Europe.

Ssp. **arenarium** (Butcher) Clapham (including *T. arenarium* Butcher = *T. dunense* auct., non Dum.). Far-creeping by underground stolons. *Stem commonly* 15–40 *cm.*, usually lfy to the base, green or glaucous. Lowest lvs commonly 5–15 cm.; lflets often only 4–8 mm. across, rounded-cuneate to cordate at the base, green or glaucous, usually densely covered especially beneath with stalked glands. *Panicle usually branching from near or below the middle of the stem*, almost as broad as long. Achenes 3·5–5 mm., ± narrowly and asymmetrically ovoid-oblong.

Native. Open or closed dunes, especially in the north and west. Coasts of Great Britain, northwards from Devon and Cornwall in the west and from Suffolk in the east; Inner and Outer Hebrides; Orkney. N. and E. coasts of Ireland. N.W. Europe from France to Scandinavia.

Ssp. **majus** (Crantz) Clapham (including *T. umbrosum* Butcher, *T. expansum* Jord. *T. flexuosum* p.p., *T. kochii*, *T. capillare*, etc., auct. angl.). Caespitose to stoloniferous. *Stems commonly* 50–120 *cm.*, lfy to the base often markedly striate or furrowed, ± glandular at least above. Lowest lvs commonly 20–40 cm.; lflets commonly 10–30 mm. across, truncate or cordate at the base, acutely or bluntly lobed, ± glabrous to densely covered with stalked glands, especially beneath. *Panicle often branching from about the middle of the stem*, about as broad as long. Stigma lanceolate to suborbicular. Achenes 3–5·5 mm., roundly and symmetrically ovoid to ± narrowly and asymmetrically ovoid-oblong.

This group is perhaps the most heterogeneous of the three, with the strongest tendency to differentiation between local populations.

Native. Damp places chiefly by streams and lakes and commonly in shade, local; to 2000 ft. in Wales. Wales and W. and N. England northwards to Perth. Europe.

15. PAEONIACEAE

Large herbs or shrubs with spirally arranged exstipulate lvs. Fls large, terminal, usually solitary; hermaphrodite, actinomorphic, hypogynous. Calyx of 5 free sepals; corolla of 5–10(–13) large free petals; stamens numerous, often united below into a ring; carpels 2–5, free, with fleshy walls and mounted on a fleshy disk; ovules several, each with 2 integuments of which the outer projects beyond the inner. Fr. a group of 2–5 large follicles each with several seeds. Chromosomes large; basic number = 5.

One genus, *Paeonia*.

Formerly included in Ranunculaceae but differing markedly in anatomy, in the structure of the gynaecium and the peculiar outer integument of the ovules, as well as in the morphology and basic number of the chromosomes.

1. PAEONIA L.

Perennial herbs or shrubs with erect tuberous stocks and fleshy roots. Lvs large, ternately or pinnately divided, commonly biternate. Outer sepals often with a rudimentary lamina and inner sepals often grading into petals; petals red, purplish or white, rarely yellow; carpels with a short thick style and broad stigma. Fls visited by various insects chiefly for pollen, but some lick the fleshy disk at the base of the carpels. Follicles often hairy; seeds in 2 rows; at first red, then dark blue and shining.

About 50 spp. especially in C. and E. Asia, the Mediterranean region and S. Europe; 2 in western N. America.

*1. P. mascula (L.) Mill.   Peony.

P. corallina Retz.

A perennial herb with tuberous stock and fleshy roots, and an erect simple lfy glabrous shoot to 50 cm. Lvs biternate, the lflets ovate or elliptical, entire, dark-green, glabrous and shining above, but glaucous and finely hairy below. Fl. solitary, terminal, c. 10 cm. diam. Sepals 5, green, broadly ovate, imbricate, persistent, the outermost often transitional to the uppermost lvs. Petals 5–8, 4–5(–6) cm., broadly ovate, ± irregularly entire, deep purple-red, rarely whitish or yellowish. Stamens with crimson filaments and yellow anthers. Carpels usually 5, covered with white downy hairs; stigmas red, hooked or coiled. Follicles divergent, usually recurved. Seeds round, smooth, shining, at first red, then dark blue, finally black. Fl. 4–5. Protogynous. $2n = 20$. H.–G.

Introduced. Naturalized on Steep Holm (Severn Estuary). Native in S. Europe from Spain to Greece, Crete and the Near East.

The cultivated 'tree peonies' are derived from the woody *P. suffruticosa* Andr. (*P. moutan* Sims), and the herbaceous types chiefly from *P. lactiflora* Pallas, both from N.E. Asia. *P. lactiflora* can be distinguished from *P. mascula* by its several-fld stem and usually ± glabrous carpels.

16. BERBERIDACEAE

Herbs or shrubs. Lvs simple or compound, alternate, usually exstipulate. Fls hermaphrodite, regular, hypogynous, solitary or in cymes, racemes or panicles, usually 3-, rarely 2-merous. Per. segs free, in 2–7 whorls, the whorls often differing. Stamens usually in 2 (3 in *Achlys*) whorls opposite the inner pcr. segs; anthers usually opening by valves, rarely by slits (*Podophyllum, Nandina*). Ovary of 1 carpel with 1–many anatropous ovules on a basal or lateral placenta; style short or 0. Fr. a berry or capsule; endosperm copious.

10 genera and about 300 spp., confined to the northern hemisphere (mainly temperate), except for *Berberis*.

A family with the incompletely differentiated perianth characteristic of many *Ranales*, distinguished from the other families by the single carpel.

Herbs; lvs 2-ternate; fr. a capsule.	1. EPIMEDIUM
Shrubs; fr. a berry.	2
Lvs simple; stem spiny.	2. BERBERIS
Lvs pinnate; stem unarmed.	3. MAHONIA

1. EPIMEDIUM L.

Perennial *herbs* with creeping rhizome. Lvs usually biternate, less often pinnate. *Fls 2-merous.* Perianth lvs in 6 whorls, the 2 outer whorls small, sepaloid; the two next petaloid; the two innermost petaloid and usually spurred, bearing nectar in the spur. *Ovules numerous. Fr. a capsule.*

Over 20 spp. in the Mediterranean region and temperate E. Asia. Several spp. and many hybrids are cultivated.

*1. E. alpinum L.   Barren-wort.

Rhizome long. Lvs radical and cauline, usually biternate, rarely ternate; lflets to 13 cm., ovate, acute or acuminate, cordate at base, stalked, glabrous above, pubescent beneath at first, becoming subglabrous, remotely spinulose-serrate. Fl. stems 6–30 cm., with a single lf. Infl. a panicle, glandular, 8–26-fld; fls 9–13 mm. across; pedicels 5–15 mm. Outer per. segs 2·5–4 mm., greyish; middle 5–7 mm., dull red, ovate; inner c. 4 mm., yellow, slipper-shaped. Fr. c. 15 mm. Fl. 6–7. Pollinated by insects, protogynous. $2n=12*$. Hs.

Introduced. Cultivated and sometimes escaping but apparently not thoroughly naturalized. Native of N.E. Italy, S. Austria (Carinthia) and N.W. Balkans (to Serbia and Albania).

2. BERBERIS L.

Evergreen or deciduous *shrubs.* Wood yellow. *Shoots of 2 sorts, long shoots with the lvs represented by (usually tripartite) spines and short axillary shoots bearing clusters of simple lvs.* Fls in panicles, racemes, fascicles or solitary, 3-merous. Perianth lvs yellow or orange, usually in 5 whorls; the outermost small (sometimes with 1 member absent); the 2 inner usually smaller than the intermediate, each bearing 2 nectaries near the base. Stamens springing inwards when touched at the base. *Ovules 1 or few, basal. Fr. a berry.*

Over 200 spp., mainly in C. and E. Asia and S. America, a few in Europe, the Mediterranean region, mts of tropical E. Africa and temperate N. America. Many other spp. and hybrids are ± commonly cultivated.

The following are rarely naturalized:

*B. buxifolia Pers. Lvs 1–2·5 cm., obovate, evergreen, entire. Fls 1–2, axillary, orange. Fr. subglobose, dark purple, pruinose. Chile.

*B. glaucocarpa Stapf. Lvs 3–5 cm., obovate-elliptic, somewhat shiny, subevergreen, with a few spiny teeth. Infl. a 15–25-fld. erect raceme, 2·5–4(–5·5) cm. Fr. 9 mm., oblong, dark blue, very pruinose. Naturalized in Somerset. Himalaya.

1. B. vulgaris L.   Barberry.
Shrub 1–2·5 m.; twigs grooved, yellowish. Spines (1–)3(–7)-partite, 1–2 cm. Lvs 2–4 cm., obovate or oblong-obovate, usually obtuse, spinulose-serrate, reticulate; petiole to 1 cm., but usually much less. Infl. a pendulous raceme 4–6 cm.; pedicels 5–12 mm. Fls 6–8 mm. diam., yellow. Fr. 8–12 mm., oblong, red. Fl. 5–6. Pollinated by various insects, possibly sometimes selfed, homogamous. Fr. 9–10. $2n = 28$. N.

Introduced? Hedges, etc., throughout Great Britain but everywhere very local and in small quantity; formerly much planted for its edible fr. and still often grown for ornament; possibly native in a few places in England; introduced in Ireland. 104. Europe from Scandinavia to C. Spain, Italy, Greece and the Caucasus.

3. Mahonia Nutt.

Evergreen *shrubs. Lvs all alike, pinnate.* Infl. a many-fld raceme or panicle. Fls and fr. as in *Berberis.*

About 50 spp. in E. Asia and N. and C. America. Several others are sometimes grown in gardens.

*1. M. aquifolium (Pursh) Nutt.   Oregon Grape.
Berberis Aquifolium Pursh
Shrub 1–2 m. Lflets 5–9, ovate, 3·5–8 cm., stiff, coriaceous, dark green and glossy above, sinuately spiny dentate. Fls in clustered terminal racemes, yellow. Fr. c. 8 mm., globose, blue-black, pruinose. Fl. 1–5. Pollinated by various insects, possibly sometimes selfed, homogamous. Fr. 9. $2n = 28$. N.

Introduced. Commonly planted for pheasant cover and naturalized in a number of places. Native of western N. America from British Columbia to Oregon.

17. NYMPHAEACEAE

Water-lilies. Perennial water or marsh plants usually with stout creeping rhizomes and floating lvs; often also with submerged or with aerial lvs. Submerged lvs simple, thin, translucent; floating lvs peltate or cordate, usually long-stalked. Fls solitary, terminal, generally floating, hermaphrodite, actinomorphic, hypogynous to epigynous, the parts variously arranged. Perianth usually differentiated into 3–6 green sepals and 3 to many petals which may pass gradually into the usually numerous stamens with introrse anthers, and are either hypogynous or inserted at various heights on the wall of the ± inferior ovary. Carpels 8 or more, either free (but then sunk in the receptacle) or united and superior, or united into a many-celled ovary and inferior. Ovules 1–many, inserted all over the inner walls of the carpels. Fr. a group of achenes sunk in the receptacle or a spongy capsule dehiscing regularly or irregularly by the swelling of internal mucilage. Seeds often arillate, with or without endosperm, sometimes with perisperm as well as endosperm.

About 6 genera with c. 60 spp., cosmopolitan.

Sepals 4, green; petals white, the outermost longer than the sepals; lateral veins of lvs anastomosing at the margin. 1. NYMPHAEA
Sepals 5–6, yellowish; petals yellow, much shorter than the sepals; lateral veins of lvs forking near the margin. 2. NUPHAR

1. NYMPHAEA L.

Rhizome stout. *Lvs stipulate*, ± circular or oval, rarely eccentrically peltate. Fls usually floating, often large and showy. *Sepals* 4, hypogynous, green; *petals numerous*, inserted at successively higher levels on the side of the ½-inferior ovary, *the outermost much longer than the sepals*, the innermost shorter and narrower and grading into the stamens; stamens numerous, the outer with broad petaloid filaments, all inserted towards the top of the ovary; ovary syncarpous, many-celled with numerous ovules in each cell; summit of the ovary concave but with a central boss from which the stigmatic surfaces radiate, and with marginal stylar processes. Fr. a spongy berry-like capsule ripening under water and splitting by the swelling of internal mucilage to release the seeds, which float because of the air-containing aril and seed-wall.

About 40 spp., cosmopolitan.

The cultivated water-lilies are chiefly varieties and hybrids of *N. alba* (white and red fls), *N. lotus* (white fls and toothed lvs), *N. rubra* (red fls and toothed lvs), *N. caerulea* (blue fls, carpels free laterally, entire lvs), *N. capensis* (blue fls, carpels free laterally, sinuate lvs) and *N. mexicana* (yellow fls) etc.

Fls 5–12 cm. diam.; fr. globose, its upper part devoid of stamen-scars.
ssp. **occidentalis**
Fls 9–20 cm. diam.; fr. usually obovoid with stamen-scars extending to the top. ssp. **alba**

1. N. alba L. ssp. alba   White Water-lily.

Castalia alba (L.) Wood; *C. speciosa* Salisb.

Lvs all floating, almost circular, 10–30 cm., entire, with a deep basal sinus; basal lobes ⅕–⅓ the length of the lf, the primary veins diverging or parallel; lf dark green above, paler and often reddish beneath. *Fls 10–20 cm. diam.*, floating, their stalks reaching 3 m. Sepals lanceolate, white inside; petals usually 20–25, white; filaments of inner stamens ± linear; pollen grains with spines and warts of different lengths; *stigmatic rays usually* 15–20, yellow. Fr. 18–40 *mm. diam.*, obovoid, sometimes ± globose, *with stamen scars almost to the top*. Seeds c. 3 mm. Fl. 7–8. Homogamous. Visited by few insects and probably self-pollinated. $2n$=c. 84*, c. 112*. Hyd.

Native. In lakes and ponds throughout the British Is., reaching 1100 ft. in W. Ireland. 98, H12. Europe.

Ssp. occidentalis Ostenf.   'Lesser White Water-lily.'

? *N. alba* var. *minor* DC.

Like ssp. *alba* but smaller in all its parts. Lvs 10–13 cm., their basal lobes with primary veins converging. *Fls 5–12 cm. diam.*; pollen grains usually with uniformly short projections; *stigmatic rays usually 10–15. Fr. 16–28 mm. diam., ± globose, its upper part without stamen-scars.* Seeds c. 3 mm. Fl. 7–8. Hyd.

Native. In lakes in N. Scotland, Inner and Outer Hebrides, Shetland, and near Roundstone and Recess in W. Ireland. 8, H1. W. Norway, Denmark.

Ssp. *occidentalis* was based on populations in W. Ireland and in Perthshire composed of small individuals, but it grades into typical *N. alba*.

2. NUPHAR Sm.

Rhizome stout. *Lvs not stipulate*, all floating or some submerged, broadly elliptical or oblong, the lateral veins three times forked. Fls yellow, globose, rising above the water. *Sepals 5–6, yellowish-green, hypogynous; petals numerous, yellow, usually shorter than the sepals, spathulate,* all hypogynous, with a nectary on the lower side; stamens numerous, all hypogynous, with broad filaments; ovary superior, syncarpous, many-celled, with many ovules in each cell; summit of ovary ± convex with a central depression and many radiating sessile stigmas. Insect-pollinated. Fr. flask-shaped, berry-like, ripening above water, splitting irregularly into parts each comprising the internal tissue and seeds of a single ovary-cell. Seeds without aril, with little endosperm and much perisperm.

About 12 spp. in temperate regions of the northern hemisphere.

1   Fls 4–6 cm. diam.; stigma-rays 10–25; stigmatic disk circular, with entire
        margin.   **1. lutea**
    Fls 1·5–4 cm. diam.; stigma-rays 7–14; margin of stigmatic disk wavy to
        deeply lobed.   2

2   Stigma-rays 7–12; stigmatic disk 6–8·5 mm. diam., ± stellate, its margin
        deeply lobed.   **2. pumila**
    Stigma-rays 9–14; stigmatic disk 7·5–11 mm. diam., circular with a wavy
        margin.   **3. × spennerana**

1. N. lutea (L.) Sm.   Yellow Water-lily, Brandy-bottle.

Nymphaea lutea L.

Rhizome 3–8 cm. diam., branched. Floating lvs 12–40 × 9–30 cm., ovate-oblong, with a deep basal sinus, the basal lobes being about ¼ the length of the lf, thick and leathery; submerged lvs shorter-stalked, broadly ovate or round, cordate, thin and translucent. *Fls 4–6 cm. diam.*, rising out of the water, their stalks up to 2 m., rarely to 3 m. Sepals 2–3 cm., broadly obovate, persistent, bright yellow within. Petals broadly spathulate, one-third as long as the sepals. Stamens shorter than the ovary. Stigmatic disk wider than the top of the ovary (10–15 mm. wide), with usually 15–20 *stigmatic rays which* *do not reach the entire margin* (Fig. 9A). Fr. 3·5–6 cm., flask-shaped. Seeds c. 5 mm. Fl. 6–8. Homogamous or protogynous. Visited by small flies. Smelling of alcohol ('Brandy-bottle'). $2n = 34^*$. Hyd.

Native. In lakes, ponds and streams throughout the British Is., but less common in N. Scotland than elsewhere; reaching 1672 ft. in Wales. 96, H38. Europe, N. Asia. Very rare in N. Africa.

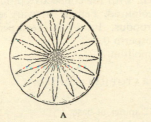

A   B

Fig. 9. Stigmatic disks of A, *Nuphar lutea* and B, *N.* × *spennerana.* × 2·5.

2. N. pumila (Timm) DC.   'Least Yellow Water-lily.'

Rhizome 1–3 cm. diam. Floating lvs 4–14 × 3·5–13 cm., broadly oval, with a deep basal sinus and lobes which are usually more divergent than in *N. lutea*; submerged lvs as in *N. lutea* but much smaller. *Fls 1·5–3·5 cm. diam.* Sepals 4–5, roundish, yellow, persistent. Petals narrowly spathulate, one-third as long as the sepals. Stamens shorter than the ovary. Stigmatic disk 6–8·5 mm. wide, little wider than the top of the ovary, with usually 8–10 *stigma-rays which reach the ends of the deep ± acute lobes of the margin.* Fr. pear-shaped, 2–4·5 cm. Seeds 2·5–4 mm. Fl. 7–8. Less fragrant than *N. lutea.* $2n = 34^*$. Hyd.

Native. A local plant of lakes chiefly in N. Scotland, rare in Shropshire and Merioneth. 10. N. and C. Europe and N. Asia.

3. N. × **spennerana** Gaud.   'Hybrid Yellow Water-lily.'

N. lutea var. *minor* Syme; *N. intermedia* Ledeb.

Intermediate between *N. lutea* and *N. pumila* and probably a hybrid between them, since the pollen is partly non-viable and fr. and seeds are produced sparingly. Lvs intermediate in size. *Fls 3–4 cm. diam.* Petals broadly obovate. *Stigmatic disk 7·5–11 mm. wide, circular with a wavy margin* and with usually 10–14 *stigma-rays* not quite reaching the margin (Fig. 9B). Fl. 6–8. $2n = 34^*$. Hyd.

Native. A local plant of lakes and ponds in N. England and Scotland. 8. Usually with the parents but in some N. English localities *N. pumila* is not present. N. and C. Europe.

18. CERATOPHYLLACEAE

Submerged aquatic herbs. Lvs in whorls, divided, segments linear or filiform. Fls unisexual, sessile, solitary in the whorls of lvs, the male and female fls at different nodes. Perianth herbaceous, divided into numerous narrow segments, often dentate or lacerate at apex. Male fls: stamens 10–20 on a flat torus; anthers subsessile, 2-celled, with the connective produced beyond the cells, opening lengthwise, often coloured. Female fls: ovary sessile, unilocular; ovule solitary, pendulous, anatropous. Fr. a nut. Seed pendulous; embryo straight, cotyledons equal; endosperm 0.

One genus and 3 spp., cosmopolitan.

1. CERATOPHYLLUM L.   Horn-wort.

The only genus.

Lvs once to twice forked; fr. with 2 spines at base.   **1. demersum**
Lvs thrice forked; fr. without spines at base.   **2. submersum**

1. C. demersum L.

A dark green rather stiff densely lfy perennial. Stems 20–100 cm., slender. *Lvs* 1–2 cm., *once to twice forked*, segments linear, rather closely but irregularly denticulate. *Fr.* c. 4 mm., ovoid, ± warty, somewhat shorter than or equalling the persistent style, *with two spines at base when ripe*. Fl. 7–9. $2n=$ c. 24. Hyd.

Native. In ponds and ditches, local. 66, H12. Scattered throughout England and Ireland, rare in Wales and Scotland. Europe, except the arctic.

2. C. submersum L.

Similar to *C. demersum* but softer and brighter green. *Lvs thrice forked*, segments sparingly denticulate. *Fr.* warty, longer than the persistent style, *devoid of spines*. Fl. 7–9. $2n=40$; 72. Hyd.

There has been much confusion between the two spp., but this plant appears to be the rarer.

Native. In ponds and ditches. Distribution imperfectly known, but apparently mainly southern and eastern in England; recorded from Somerset and coastal counties from Sussex to Lincoln. 21. Europe north to Denmark; Asia, N. Africa.

19. PAPAVERACEAE

Annual or perennial herbs, rarely shrubs, with spirally arranged exstipulate often deeply lobed or divided lvs. All parts of the plant exude a milky or ± coloured latex when cut. Fls hermaphrodite, actinomorphic, hypogynous or rarely perigynous, their parts whorled. Sepals 2(–3), free, falling before the fl. is fully open; petals 2+2 (or 3+3, rarely 0), overlapping, often crumpled at first; stamens usually numerous, free; ovary superior, syncarpous, of 2 or more united carpels, 1-celled, the numerous ovules on ± long-projecting parietal placentae, the carpels rarely separating in fr. Fr. usually a capsule opening by valves or pores, 1-celled or rarely (*Glaucium*) 2-celled by meeting of placentae. Seeds small with a minute embryo in oily endosperm.

About 25 genera and 115 spp., cosmopolitan but chiefly in temperate and subtropical regions of the northern hemisphere.

Many genera are cultivated for their showy fls. Apart from native or adventive British genera these include the shrubby *Romneya*, with large solitary white fls; *Macleaya*, with numerous small apetalous fls in large panicles (*M.* (*Bocconia*) *cordata* is the Plume Poppy of gardens); and *Argemone*, the annual Prickly Poppies, the yellow-fld, prickly and glaucous *A. mexicana* being a weed of cultivation in many parts of the world.

1	Stigmas sessile on a disk at the top of the ovary; fr. a capsule opening by pores just beneath the stigmatic disk.	1. PAPAVER
	Stigmas not sessile on an expanded disk; fr. a long pod-like capsule opening by valves.	2
2	Valves of fr. short and tooth-like, not extending the whole length of the capsule.	2. MECONOPSIS
	Valves of fr. extending to the base.	3
3	Fr. opening by 3–4 valves; fls violet.	3. ROEMERIA
	Fr. opening by 2 valves; fls yellow or red.	4
4	Sepals joined to form a hood which is pushed off as the petals open; receptacle forming a distinct ledge round the base of the ovary; latex watery.	6. ESCHSCHOLZIA
	Sepals free; no receptacle-ledge; latex orange.	5
5	Fls 35–90 mm. diam.; capsule 2-celled.	4. GLAUCIUM
	Fls 18–25 mm. diam.; capsule 1-celled.	5. CHELIDONIUM

1. PAPAVER L.

Annual to perennial herbs usually with *white latex* and toothed or pinnately lobed or divided lvs, often hispid. Fls solitary, terminal or axillary, showy. Sepals 2, free, falling as the fl. opens; petals 2+2, fugacious, crumpled in bud; stamens very numerous, with anthers dehiscing outwards; *stigmas 4–20, sessile and over the placentae*; ovules very numerous. *Fr. a capsule opening by pore-like valves just beneath the persistent stigmatic disk*. Seeds small, without appendage.

About 120 spp. in the northern hemisphere with a few tropical spp. and one in Australia. Many are arctic or subarctic, and many are common weeds of cultivation.

1	Perennials; rosette-lvs lanceolate, ±coarsely toothed or at most shallowly pinnatifid, stiffly hairy, not glaucous; petals orange to brick-coloured; rare aliens.	2
	Annuals; rosette-lvs usually deeply pinnatifid to one or more times pinnatisect or, if less deeply lobed, then ovate-oblong and glaucous; petals scarlet, pale lilac or white.	3
2	Petals dull orange or brownish; capsules to 25 mm., clavate, tapering from just beneath the disk.	7. atlanticum
	Petals apricot- to brick-coloured; capsules to 20 mm., obovate-clavate, tapering from a widest point well below the disk.	8. lateritium
3	Lvs toothed or slightly lobed, clasping the stem at their base; plant glaucous.	6. somniferum
	Lvs once or twice pinnately lobed or divided; plant green.	4

4 Capsule glabrous.   *5*
     Capsule with stiff hairs or bristles.   *7*

5 Capsule almost globose, about as long as wide.   **1. rhoeas**
     Capsule obovoid-oblong, at least twice as long as wide.   *6*

6 Latex white; petal-bases overlapping.   **2. dubium**
     Latex turning yellow in the air; petal-bases not overlapping. **3. lecoqii**

7 Capsule almost globose; bristles spreading, numerous.   **4. hybridum**
     Capsule narrowly obovoid-oblong, bristles few, erect.   **5. argemone**

1. P. rhoeas L.   'Field Poppy.'

An annual or rarely biennial herb with a slender tap-root and erect or
ascending usually branched stems 20–60 cm., with stiff spreading hairs or
glabrous. Basal lvs stalked, once or twice pinnately cut or divided, with
narrow acute ± toothed bristle-pointed segments (Fig. 10 A, B); upper lvs sessile,

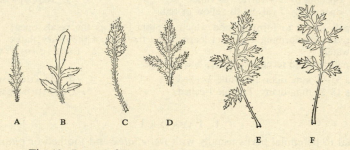

Fig. 10. Leaves of *Papaver*. A, B, *P. rhoeas*; C, D, *P. dubium*;
E, *P. hybridum*; F, *P. argemone*. ×⅓.

usually 3-lobed, the *central lobe elongate lanceolate*, all further pinnately lobed,
cut or toothed; all lvs green, stiffly hairy. Fls 7–10 cm. diam., axillary, their
stalks with spreading hairs. Sepals bristly. Petals 2–4 cm. wide, roundish,
scarlet, rarely pink or white, often with blotch at the extreme base. Filaments
slender; anthers bluish. *Capsule* 1–2 cm., *glabrous, sub-globose to broadly
obovoid, at most twice as long as wide*; stigma-rays 8–12 on a disk with as many
overlapping marginal lobes. Seed dark brown. Fl. 6–8. Homogamous.
Self-incompatible. No nectar. Visited by various pollen-collecting insects,
especially bees. $2n = 14$. Th.

Native or introduced. A weed of arable fields and waste places. 104,
H 39, S. Throughout the British Is., common in the south but rare and local
in N. Scotland and not in Outer Hebrides, Orkney or Shetland; Channel Is.
Europe, except the far north; N. Africa; temperate Asia. Introduced in
N. America, Australia and New Zealand.

P. rhoeas is very variable in shape and hairiness of the lvs, colour and blotching of
the petals, shape of capsule, etc. The Shirley Poppies of gardens comprise a wide
range of colour forms found occasionally in wild populations of the species. The
fertile and otherwise typical plants which have appressed hairs on the fl.-stalks,

variously named *P. strigosum* (Boenn.) Schur, *P.* × *strigosum* (Boenn.) Schur and *P. rhoeas* f. *strigosum* Boenn., seem similarly to fall within the natural range of variation of *P. rhoeas*, differing by only a single gene from plants with the more usual spreading hairs. Hybrids between *P. rhoeas* and *P. dubium* have been produced artificially and have been recognized in mixed wild populations. They are very variable, with either appressed or spreading hairs on the fl.-stalks, and are quite sterile: it seems very unlikely that *P. rhoeas* var. *strigosum* can be of this hybrid origin.

P. commutatum Fisch. & Mey. (E. Europe and W. Asia), found occasionally as a casual or garden escape, with appressed hairs on the fl.-stalks and narrow petals, at least as long as broad, which have a rounded or quadrangular blotch in the middle (not at the extreme base), does not seem to merit full specific distinction from *P. rhoeas*.

2. P. dubium L.   'Long-headed Poppy.'

P. Lamottei Bor.

An annual herb with a slender tap-root and erect usually branched stems 20–60 cm., with stiff hairs spreading below, appressed above. *Latex white*, probably always so, plants with yellowish latex being assignable to *P. lecoqii*. Lvs as in *P. rhoeas* but with the segments shorter, broader and more abruptly acute and with a smaller terminal segment (Fig. 10c, D); *pubescent*, glaucous. Fls 3–7 cm. diam., their *stalks with appressed hairs*. Sepals bristly. *Petals* 1·5–4 cm. wide, *always overlapping at base*, *orange-pink* (paler than in *P. rhoeas*). Filaments slender; anthers *usually falling short of the stigmatic disk*; pollen lemon-yellow. *Capsule* up to 2·5 cm., *glabrous*, *obovoid-oblong*, narrowing gradually from near the top and *more than twice as long as wide*; stigma-rays (4–)7–9(–12), each falling a little short of the margin of a shallow lobe of the stigmatic disk. *Seeds bluish-black*. Fl. 6–7. Homogamous and self-compatible but often outcrossed. $2n=42$*. Th.

  Native or introduced. A weed of arable land (especially in cereal crops) and waste places throughout the British Is., including Orkney and Shetland, and more common in the north than *P. rhoeas*, perhaps more tolerant of poor soils. 106, H19, S. C. and S. Europe.

3. P. lecoqii Lamotte   'Babington's Poppy.'

An annual herb closely resembling *P. dubium* but with *latex turning yellowish after exposure to the air*; *lvs sparsely hairy*, hardly glaucous, with deeper, narrower and more acute segments; *petals not overlapping at base*, deeper red than in *P. dubium*; *anthers equalling the stigmatic disk*; *seeds chocolate brown*. Fl. 6–7. Homogamous and self-compatible; more readily self-pollinated than *P. dubium* because of position of anthers. $2n=28$*. Th.

  Native or introduced. An infrequent weed of road-verges, quarries, headlands of arable fields, etc., but not found amongst cereals; often on calcareous soils. Probably confined to the southern and eastern counties of England, but detailed distribution unknown owing to confusion with *P. dubium*. Europe.

4. P. hybridum L.   'Round Prickly-headed Poppy.'

P. hispidum Lam.

An annual herb with slender tap-root and erect branching stems 15–50 cm., stiffly hairy with spreading or appressed hairs. Latex white. Lvs 2–3 times pinnately lobed, the ultimate segments acuminate, bristle-pointed (Fig. 10E);

basal lvs stalked, upper sessile. Fls 2–5 cm. diam., their stalks with appressed hairs. Sepals bristly. Petals roundish-obovate, crimson with a blackish spot at the base. Filaments violet, thickened above; anthers bluish. *Capsule* 1–1·5 cm., ± *globose, densely covered with stiff yellow bristles*; stigma-disk convex with 4–8 rays. Seeds grey-brown. Fl. 6–7. Homogamous and self-compatible. $2n = 14$. Th.

Native or introduced. A rare weed of arable fields and waste places. 45, H11, S. England and S. Scotland; Ireland; Channel Is. C. and S. Europe; C. Asia.

5. P. argemone L.   'Long Prickly-headed Poppy.'

An annual herb with a slender tap-root and erect or ascending ± branched stems 15–45 cm., stiffly hairy with appressed hairs. Lvs twice pinnately lobed, the ultimate segments suddenly acuminate, bristle-pointed (Fig. 10 F); basal lvs stalked, upper sessile with longer and narrower segments, all ± stiffly hairy. Fls 2–6 cm. diam., their stalks with appressed hairs. Sepals bristly. Petals obovate, not contiguous, scarlet with a dark base. Filaments violet, thickened above; anthers bluish. *Capsule* 1·5–2·5 cm., *narrowly obovoid-oblong, strongly ribbed, with a few erect bristles*, especially above; stigma-disk convex with 4–6 rays curving downwards at their ends. Fl. 6–7. Self-compatible and often selfed in bud. $2n = 12^*, 42$. Th.

Native or introduced. A common weed on light soils in the south, becoming rare in the north. 96, H25, S. Throughout Great Britain and Ireland (not in Orkney and Shetland), Channel Is. C. and S. Europe.

*6. P. somniferum L.   Opium Poppy.

An annual herb with a tap-root and a single simple or branched erect stem 30–100 cm., glabrous or with a few spreading bristles. *Lvs undulate, ovate-oblong*, ± *pinnately lobed*, the lobes shallow, irregular, coarsely toothed; lowest lvs narrowed into a short stalk, upper sessile clasping the stem; all *glaucous*, glabrous or with a few stiff hairs (ssp. *setigerum*). Fls to 18 cm. diam., on glabrous or somewhat hairy stalks. *Petals white or pale lilac*, with or without a basal blotch. Filaments thickened above; anthers bluish. Capsule globular or ovoid, large but very variable in size; stigma-disk with deep non-contiguous marginal lobes; stigma-rays not quite reaching the ends of the lobes. Capsule dehiscent or not; seeds black or white. Fl. 7–8. $2n = 20, 22$. Th.

A very variable species, variously subdivided by taxonomists, but all combinations of characters may be found. The cultivated races are commonly separated into two sspp.

Ssp. **hortense** Hussenot. Glabrous, very glaucous. Petals pale lilac, spotted at the base. Capsule globular, dehiscent. Seeds dark grey to black. This is the complex of races whose seeds yield poppy-seed oil.

Ssp. **somniferum**. Sparingly hairy, less glaucous. Petals white unspotted. Capsule very large, ovoid or sub-globular, indehiscent. Seeds white. This is the group from whose immature capsules an alkaloid-containing latex is obtained by making incisions immediately after the petals have fallen. The crude drug is called opium, and contains about 20 different alkaloids of which the most important are the narcotics morphine and codeine. *P. somniferum* is cultivated for opium on a large scale both in Europe and E. Asia.

Besides these cultivated forms there is a wild type which is often treated as a distinct species:

Ssp. **setigerum** (DC.) Thell. Lvs scarcely glaucous ('dull smoky green'). Lvs, sepals, and fl.-stalks with stiff hairs. Capsule pear-shaped or oblong, dehiscent.

Introduced. Ssp. *hortense* still occurs as a relic of cultivation in various parts of England, especially in the Fens. It is also grown in gardens and occasionally escapes. Sspp. *somniferum* and *setigerum* are rare casuals. Europe, Asia.

*7. P. atlanticum (Ball) Coss.

A perennial herb with a stout stock crowned by the remains of lvs of former years. Rosette-lvs 12–25 cm., oblong-lanceolate, gradually narrowed into a stalk-like base, coarsely and irregularly crenate-serrate or shallowly pinnatifid, covered on both sides with whitish bristle-like hairs; stem-lvs only on lower half of stem or 0, subsessile. Sepals hispid. *Petals* 3–4 cm., obovate, *dull orange* to reddish. Stigma-rays usually 6. *Capsule* to 2·5 cm., glabrous, *long-clavate, tapering from close beneath the stigmatic disk.* Fl. 6–9. Hr.

Introduced. Established in gardens and waste places in S. England, especially in Surrey. Morocco. Said to be introduced in bird-seed.

*8. P. lateritium C. Koch

Perennial herb resembling *P. atlanticum* but with larger and more brightly coloured *apricot- to brick-coloured petals* and with the *capsule obovate-clavate*, the widest point being well beneath the stigmatic disk.

Introduced and reported as established locally. Mountains of Armenia.

P. orientale L. (Mediterranean region and Near East), the Oriental Poppy of gardens, is a robust hispid perennial with large pinnately-cut lvs and fls c. 15 cm. diam., the petals scarlet with a basal black spot. In *P. nudicaule* L. (Siberia, C. Asia, and N. Pacific coasts), the Iceland Poppy of gardens, the lvs are confined to a basal rosette and the fls 3–8 cm. diam., have yellow, orange or reddish petals. The closely related *P. radicatum* Rottb. (circumpolar and one of the 4 plant spp. found at 83° 24′ N. on the N. coast of Greenland) has yellow, not white latex, smaller fls and a more rounded capsule. It is polymorphic with numerous sspp. in Scandinavia which seem to have arisen through isolation during the Ice Age.

2. MECONOPSIS Vig.

Annual or perennial herbs with yellow latex. Lvs entire, toothed or pinnately lobed. Fls variously arranged. Sepals usually 2, soon falling; petals usually 4, rarely to 10; stamens numerous; *capsule* subglobose to narrow-cylindrical with a distinct style and 4–6 *stigmas* opposite the placentae, *dehiscing by valves which usually reach only a short distance below the top.* Seeds numerous, crested or not. 41 spp., one confined to W. Europe, the rest to south-central temperate Asia, and especially the Himalaya.

Several of the Asiatic species are high alpines, and *M. horridula* was found at 19,000 ft. on Everest. This species, known to horticulturists as *M. racemosa*

Maxim., is grown in gardens for its racemes of pale blue fls, but the more spectacular 'Blue Poppy' is *M. betonicifolia* Franch. (*M. baileyi* Prain) with sky-blue fls up to 7·5 cm. diam., native at 10,000–13,000 ft. in S.E. Tibet, N.W. Yunnan and N. Burma.

1. M. cambrica (L.) Vig.   Welsh Poppy.
Papaver cambricum L.

A perennial herb with a branched tufted stock covered with the persistent lf-bases and erect branched leafy ± glabrous stems 30–60 cm. Basal lvs long-stalked, pinnately divided with pinnately lobed ovate acute segments, ± glabrous; upper lvs similar but shortly stalked. Fls 5–7·5 cm. diam., arising singly in the axils of the upper lvs. Sepals hairy. Petals 4, yellow. Stamens with filiform filaments and yellow anthers. Capsule 2·5–3 cm., ovoid to ellipsoid, 4–6 ribbed, with 4–6 stigma-lobes, and splitting into 4–6 valves for about ¼ its length. Seeds pitted. Fl. 6–8. $2n=22^*$, 28. H.

Native. In damp, shady, rocky places in S.W. England and Wales, up to 2000 ft. W. Ireland. 21, H 13. Extensively naturalized. W. France (to Massif Central), Portugal and N. Spain.

3. ROEMERIA Medic.

Annual herbs with yellow latex and bi- or tri-pinnately cut lvs. Fls solitary, axillary. Sepals 2, free, petals 4; stamens numerous; stigmas 2–4, sessile over the placentae; ovules many. *Capsule linear, 1-celled, opening to the base by 2–4 va ves.* Seeds without an appendage. 2–3 spp., resembling annual poppies, in S. and C. Europe.

*1. R. hybrida (L.) DC.   'Violet Horned-poppy.'
Chelidonium hybridum L.

Stem 20–40 cm., erect, branched, ± hairy. Lvs tripinnately cut, the ultimate segments ± linear, bristle-pointed; basal lvs stalked, upper sessile. Fls 5–7·5 cm. diam. Sepals hairy. *Petals roundish, deep violet-b ue* with a basal dark blotch. Capsule 5–7·5 cm., usually opening by 3 valves. Fl. 5–6. $2n=22$. Th.

Introduced. Formerly naturalized in E. Anglia as a cornfield weed, now very rare but found as a casual near ports. C. and S. Europe.

4. GLAUCIUM Mill.

Annual to perennial glaucous herbs with yellow latex. Lvs pinnately lobed or cut. Fls axillary, large. Sepals 2, caducous; petals 4, yellow to red; stamens numerous; stigma ± sessile, 2-lobed, over the placentae which meet in the centre of the ovary; ovules many. *Capsule linear, 2-celled, opening almost to the base by 2 valves, and leaving the seeds embedded in the septum.* Seeds pitted, without an appendage.

About 21 spp., chiefly Mediterranean.

Stem and capsule ± glabrous; fls yellow, 6–9 cm. diam.   **1. flavum**
Stem and capsule stiffly hairy; fls orange to scarlet, 3–5 cm. diam.
                                                              2. corniculatum

1. G. flavum Crantz   'Yellow Horned-poppy.'

G. luteum Crantz; *Chelidonium Glaucium* L.

A perennial or biennial herb with a deep stout tap-root and an erect branched stem 30–90 cm., glaucous, ± glabrous. Basal lvs roughly hairy, stalked, pinnately lobed or divided, sublyrate, the lobes pointing various ways and further lobed or coarsely toothed; upper *lvs* sessile, half-clasping, less deeply lobed or sinuate, rough; all *glaucous*. Fls 6–9 cm. diam., their stalks short, glabrous. Sepals hairy. *Petals* roundish, *yellow*. Capsule 15–30 cm., glabrous but rough. Fl. 6–9. Protogynous. Visited by various flies and some small bees. $2n = 12$. Hs.

Native. A maritime plant, chiefly of shingle banks, all round the coast southwards from Argyll and Kincardine; formerly in Shetland; Ireland; Channel Is. 51, H16, S. Mediterranean region, W. Europe northwards to Norway and Sweden; W. Asia. Not exclusively maritime in continental Europe.

*2. G. corniculatum (L.) Rudolph   'Red Horned-poppy.'

G. phoeniceum Crantz; *Chelidonium corniculatum* L.

An annual herb with an erect branched stem 25–30 cm., with stiff spreading hairs. Basal lvs stalked, sublyrate and pinnately and deeply lobed, the lobes distant, oblong, toothed; upper lvs sessile half-clasping, less deeply lobed, the lobes remotely toothed; all ± hairy. Fls 3–5 cm. diam., their stalks very short. Sepals softly hairy. *Petals* roundish, *bright scarlet or orange-red* with a black spot at the base. Capsule 10–22 cm. slightly curved, hairy. Fl. 6–7. $2n = 12$. Th.

Introduced. Perhaps once established in Norfolk and at Portland but now a frequently occurring casual of waste places near ports. S. Europe and Mediterranean region.

5. CHELIDONIUM L.

Perennial herbs with short branching stocks and erect branched lfy stems. *Latex bright orange.* Lvs pinnately cut or lobed. Sepals 2, free, caducous; petals 4; stamens numerous; style very short with two spreading adnate stigma-lobes over the placentae. *Capsule linear, 1-celled, with no septum, opening from below by 2 valves which separate from the placentae*; seeds with a fleshy crested appendage on the raphe.

Two spp. in Europe and N. Asia.

1. C. majus L.   Greater Celandine.

A perennial herb with its branched woody stock covered with persistent lf bases and with erect branched lfy *brittle stems* 30–90 cm., slightly glaucous and sparsely hairy. Lvs almost pinnate with 5–7 ovate to oblong lflets; terminal lflet often 3-lobed, laterals usually with a stipule-like lobe on the lower side; all crenately-toothed and ± glabrous, somewhat glaucous beneath. Fls 2–2·5 cm. diam., terminal. Sepals greenish-yellow, ± hairy. Petals bright yellow, broadly obovate. Stamens yellow; filaments thickened above. Capsule 3–5 cm. Seed black with a white appendage. Fl. 5–8. Homogamous. Visited by pollen-collecting flies and bees. Germ. spring. $2n = 12$*. Hs.

Native or introduced. A frequent plant of banks, hedgerows and walls chiefly near habitations. 97, H40, S. From Inverness southwards; Ireland; Channel Is. Europe and N. Asia.

Formerly a herbalist's remedy for warts and eye troubles. The orange latex contains several alkaloids, including chelidonin and chelerythrin, and is poisonous.

Var. *laciniatum* (Mill.) Koch (*C. laciniatum* Mill.) is sometimes found as a casual. It has the lobes of the lvs deeply and narrowly pinnately cut, and petals with laciniate margins. Arose in a garden in Heidelberg in 1590.

6. Eschscholzia Cham.

Annual or perennial glabrous and glaucous herbs with watery latex and very much divided lvs whose ultimate segments are ± linear. Fls solitary, terminal. *Receptacle forming a distinct ledge round the ovary. Sepals 2, coherent into a hood* which is pushed off when the petals open; petals 4; stamens numerous; style very short; stigma 4–6-lobed, spreading; ovules many. *Capsule linear, 1-celled, opening from below by two valves which separate from the placentae.*
About 15 spp., in western N. America.

*1. E. californica Cham.   Californian Poppy.
An annual herb (in this country) with a deep tap-root and erect or diffuse stems 20–60 cm. Lvs ternately dissected, the ultimate segments linear. Fls 5–7·5 cm. diam., yellow to orange (ivory to scarlet in garden races), deeper-coloured at the base of the petals. Capsule 7–10 cm., ribbed. Fl. 7–9. $2n = 12$. Th.
Introduced. A frequent garden-escape. Native in California and Oregon.

20. FUMARIACEAE

Herbs with usually brittle stems, sometimes climbing; juice watery. Lvs alternate, usually much divided. Fls in racemes, or spikes, rarely solitary, usually zygomorphic, hermaphrodite, hypogynous. Sepals 2, small caducous. Petals 4 in two dissimilar whorls, one or both of the outer whorl spurred or saccate, ± connivent, the inner narrower and often cohering. Stamens 2, tripartite, the central branch bearing a 2-celled anther, the lateral branches each bearing a 1-celled anther. Ovary 1-celled with 2 parietal placentae, each bearing 1–many anatropous ovules. Fr. a capsule or nutlet. Seeds with small embryo and copious endosperm.

Sixteen genera and over 400 spp. in the north temperate zone and S.E. Africa. Closely allied to Papaveraceae, differing mainly in the corolla, stamens and watery juice. The fls when zygomorphic are transversely so, but the axis is twisted so that one of the outer petals appears to be at the top of the fl.; it is called the 'upper petal' in the following account.

Fr. a many-seeded capsule; petals without dark tips.   1. Corydalis
Fr. a 1-seeded nutlet; inner petals with dark purple tips.   2. Fumaria

Dicentra Bernh.

Erect herbs with much-divided lvs. Fls racemose, actinomorphic; both the outer petals spurred or saccate. Fr. a capsule. Several spp. are grown in gardens and some have been found as escapes. The most common is* **D. spectabilis** (L.) Lemaire (Bleeding Heart). Roots fibrous. Glabrous, branched. Lf-lobes ovate or oblong. Fls pendulous; petals free, outer pink, reflexed at apex, saccate; inner white, exserted. Rarely naturalized. Native of China, etc.

1. CORYDALIS Vent.

(*Capnoides* Mill.)

Glabrous ± glaucous herbs. Lvs variously divided. Fls zygomorphic, in bracteolate racemes; only the upper petal spurred. *Ovules ± numerous. Fr. a 2-valved capsule.*

About 300 spp., north temperate, except for 1 sp. in the mountains of E. Africa; a few spp. are sometimes grown in gardens.

```
1  Stems simple or nearly so; stock tuberous; racemes terminal; fls purple;
      spur long (nearly as long as rest of corolla).            1. solida
   Stems usually much branched; tuberous stock 0; racemes lf-opposed;
      fl. yellow or cream; spur short.                                 2
2  Lvs ending in a branched tendril; fl. 5–6 mm., cream.   2. claviculata
   Lvs without tendrils; fl. 12–18 mm., yellow.               3. lutea
```

*1. C. solida (L.) Sw.

C. bulbosa auct.

Perennial herb 10–20 cm. *Tuber* globose, *solid*. Stem erect, usually simple with fleshy ovate scales below the lvs. Lvs 2-ternate, segments cuneate, lobed. *Infl.* solitary, *terminal*; bracts cuneate, incised. *Corolla 15–22 mm., dull purple, with a long nearly straight spur.* Fl. 4–5. Pollinated by long-tongued bees, homogamous, self-sterile. $2n=16, 24*$. Grt.

   Introduced. Sometimes grown in gardens, escaped and ± naturalized in a few places. Europe from Sweden and Finland to the Pyrenees, Italy, Serbia and Thrace; N. and W. Asia.

*C. cava (L.) Schweigg. & Koerte

C. bulbosa (L.) DC., nom. ambig.

Resembling *C. solida*. Tuber hollow. Stem without scales at base. Bracts ovate, entire. Corolla 22–30 mm., spur curved at apex. $2n=16$. A rare escape from gardens. Native of C. and S. Europe.

2. C. claviculata (L.) DC.   'White Climbing Fumitory.'

Annual climbing much-branched delicate herb 20–80 cm., roots fibrous. Lvs pinnate; *the rhachis ending in a branched tendril*; lflets distant, long-stalked, digitately divided into 3–5 segments; segments 5–12 mm., elliptic, mucronate, entire. *Infl.* c. 6-fld, *lf-opposed. Fls 5–6 mm.; petals cream, the upper with a very short obtuse spur*; pedicels c. 1 mm. Fr. c. 6 mm., with 2–3 seeds. Fl. 6–9. Pollinated by bees, perhaps more often selfed. $2n=32$. Th.

   Native. Woods and shady rocks on acid soils over most of Great Britain from Caithness southwards, ascending to 1830 ft. in Aberdeen, rather local especially in the east; S.E. Ireland to Waterford and Dublin; E. Donegal. 98, H6, S. S.W. Norway, Holland, Belgium, N.W. Germany, Denmark, W., C. and S. France, N. and C. Spain and Portugal.

*3. C. lutea (L.) DC.   'Yellow Fumitory.'

Perennial herb 15–30 cm., much branched; root fibrous. *Lvs* pinnate, *ending in a lflet*; lflets distant, long-stalked, ternately or pinnately divided into 3–5 segments; segments 8–20 mm., 2–3(–rarely more)-lobed, the smaller entire; lobes mucronate. *Infl.* 6–10-fld, *lf-opposed. Fls* 12–18 *mm.; petals yellow, the upper with a short obtuse spur directed downwards*; pedicels to 10 mm. in fr. Fl. 5–8. Pollinated by bees or selfed, self-fertile. $2n = 28$, 56? Hp.

Introduced. Commonly cultivated and naturalized on old walls in a number of places scattered over the British Is. Europe from the S. Alps to C. Italy and Croatia; doubtfully native in S. France and C. and E. Spain; naturalized in C. Europe.

2. Fumaria L.

Glabrous, ± glaucous annual herbs, often climbing by the petioles. Lvs irregularly 2–4-pinnatisect, all cauline. Fls zygomorphic, in lf-opposed bracteolate racemes; only the upper petal spurred. *Ovules 1, or 1 on each placenta. Fr. a nutlet*, with two apical pits when dry.

About 60 spp. mainly in Europe and the Mediterranean region but extending to Mongolia, N.W. India, the mountains of E. Africa, and the Cape Verde Is.

Although the spp. are well-defined, their determination is sometimes difficult. Ill-grown or shaded specimens often produce ± colourless, cleistogamous fls much smaller than normal and in shaded plants the pedicels may be straight in spp. where they are normally recurved. These states are ignored in the following account. The corolla provides important characters based on the colour and shape, best observed when fresh. The upper petal has a greenish keel and lateral, usually dark, wings which may be reflexed upwards so as to hide the keel or be ± spreading; towards the base of the petal the wings leave the margin and form a lateral ridge varying in position with the species and making the tube apparently dorsally or laterally compressed. The presence of dark tips on the lateral petals appears to be constant in the British spp., the dark colour often suffusing over other parts of the corolla after pollination. The shape of the lower petal is often very important; it is always provided with a keel and ± definite 'margins' but the outline may be ± spathulate or parallel-sided for most of its length. The direction of the margins in the latter may be ± erect, when the whole petal appears boat-shaped, or they may spread out laterally to form a rim. The fr. is sometimes joined to the expanded top of the pedicel by a distinct narrow fleshy neck which must be observed when fresh. The rugosity of the fr. in those spp. where it occurs can, however, only be seen when dry; the same is true of the apical pits.

Most spp. vary considerably. For a full account of the British spp. and varieties, see Pugsley, *Journ. Bot.* 50, Suppl. 1 (1912).

1 Fls 9 mm. or more; lower petal not spathulate (except *F.* × *painteri*); lf-lobes oblong, lanceolate or cuneiform.   2

   Fls 5–8(–9) mm.; lower petal distinctly spathulate; lf-lobes lanceolate, linear-oblong or linear.   8

2 Lower petal with broad spreading margins; fr. c. 3 mm. (Cornwall).
                                                      1. occidentalis

   Lower petal with narrow margins; fr. 2–2½(–2¾) mm.   3

3 Pedicels rigidly recurved in fr.; peduncles equalling or longer than
    raceme; fls numerous (c. 20 or more); fr. with a distinct fleshy neck
    when fresh.   4
    Pedicels rarely recurved in fr. and then usually flexuous; either peduncles
    shorter than raceme or fls few (c. 12); fr. with an indistinct fleshy neck.   5

4 Fls white; infl. rather dense; upper petal laterally compressed, wings not
    concealing keel.   **2. capreolata**
    Fls purple; infl. rather lax; upper petal not laterally compressed, wings
    concealing keel.   **3. purpurea**

5 Lower petal not spathulate.   6
    Lower petal subspathulate; very rare (see also *muralis* var. *cornubiensis*).
                                                                      × **painteri**

6 Lower petal with spreading margins; fls numerous (c. 20).   7
    Lower petal with erect margins; fls usually few (c. 12); fr. smooth when
    dry.   **6. muralis**

7 Fr. rugose when dry; fls 9–11(–12) mm.; upper petal laterally com-
    pressed, often without dark tip; sepals serrate.   **4. bastardii**
    Fr. smooth even when dry; fls 11–13 mm.; upper petal not laterally com-
    pressed (always dark-tipped); sepals subentire.   **5. martinii**

8 Sepals at least 2 × 1 mm.; fls at least 6 mm.   9
    Sepals not more than 1½ × 1 mm.; fls 5–6 mm.   10

9 Bracts longer than pedicels; fr. rounded at apex; lf-segments channelled.
                                                                     7. micrantha
    Bracts shorter than pedicels; fr. truncate or emarginate at apex;
    lf-segments flat.   **8. officinalis**

10 Lf-segments flat; bracts shorter than pedicels in fr.; fls pink; racemes
     shortly peduncled.   **9. vaillantii**
     Lf-segments channelled; bracts equalling pedicels in fr.; fls white or
     tinged with pink; racemes subsessile.   **10. parviflora**

Section 1. *Grandiflora* Pugsl. Lf-lobes flat, relatively broad (from oval to
lanceolate or cuneiform). Fls 9 mm., or more. Wings of upper petal ± reflexed
upwards. Lower petal not spathulate.

1. F. occidentalis Pugsl.

Very robust, often climbing. Lf-segments with oblong-lanceolate lobes. Infl.
rather lax, 12–20-fld, about equalling its peduncle. Bracts lanceolate, acuminate, usually nearly equalling pedicels (rarely much shorter). Pedicels stout,
straight and suberect to arcuate and slightly decurved, apex much dilated.
Sepals 4–5½ × 2–3½ mm., ovate, acute or shortly acuminate, dentate towards
the base. *Corolla* 12–14 *mm.*, white at first, becoming bright pink, wings and
tips blackish-red, the *wings at first white-edged; upper petal dorsally compressed,*
subacute, wings concealing keel; *lower petal with broad spreading margins,*
which are sometimes slightly dilated above. *Fr. fully* 3 *mm.*, suborbicular;
when fresh *subacute* with an indistinct fleshy neck; *when dry* with distinct
lateral keel and short apical beak, *coarsely rugose.* Fl. 5–10. Th.
    Native. Arable land and waste places in several localities in Cornwall. 2.
Endemic.

2. F. capreolata L.   'Ramping Fumitory.'

F. pallidiflora Jord.

Robust, climbing, often to 1 m. Lf-segments with oblong or cuneiform lobes. *Infl. rather dense, many (c. 20)-fld, mostly shorter than its peduncle.* Bracts linear-lanceolate, acuminate, equalling or rather shorter than the fr. pedicels. *Pedicels stout, rigidly arcuate-recurved in fr.*, apex much dilated. *Sepals* 4–6 × 2½–3 mm., ± oval, *acute or acuminate*, toothed at base, entire above. *Corolla* 10–12(–14) *mm.*, creamy *white*, wings and tips blackish-red, sometimes becoming pink after pollination; *upper petal acute, strongly laterally compressed, wings not concealing keel; lower petal with erect narrow margins. Fr.* (2–)2½ × 2 *mm.*, subrectangular or suborbicular, *truncate* (rarely very obtuse but not truncate), *with a distinct fleshy neck, smooth or faintly rugulose when dry.* Fl. 5–9. 2n = 56. Th.

Native. Cultivated and waste ground and hedgebanks, scattered over the British Is. but local and absent from many districts, especially inland, commoner in the west than in the east. 53, H28, S. Mediterranean region and C. and W. Europe extending north to the Tirol, Denmark and Scandinavia (where it is only casual). Introduced in Florida and S. America.

3. F. purpurea Pugsl.

Differs from *F. capreolata* as follows: Usually shorter and more branched. Lf-lobes slightly narrower. *Infl. c. 15–25-fld, rather lax, about as long as its peduncle.* Lowest bracts sometimes lf-like. *Pedicels less recurved* (rarely divaricate). *Sepals* (4½–)5–6½ × 2–3 mm., oblong (rarely broadly oval), *obtuse or shortly acute. Corolla* 10–13 mm., pinkish-*purple*, wings and tips dark purple; *upper petal not laterally compressed, wings broader concealing keel. Fr. c. 2½ mm. long, as broad or rather broader, subquadrate, truncate, faintly rugulose when dry; apical pits broader and shallower. Fl. 7–10. Th.

Native. Cultivated and waste ground and hedgebanks, scattered over the British Is. (not Channel Is.), but very local, especially so in E. England and in Ireland. 42, H11. Endemic. Of interest as an almost certainly endemic plant, entirely confined to artificial habitats.

4. F. bastardii Bor.

F. confusa Jord.

Rather robust, suberect or diffuse, not or scarcely climbing. Lf-segments with oblong lobes. *Infl. rather lax, usually 15–25-fld, longer than its peduncle.* Bracts linear-oblong, cuspidate, shorter than fr. pedicels. *Pedicels straight and suberect or ascending in fr.*, apex somewhat dilated. *Sepals* c. 3 × 1½ mm., oval, acute, ± *serrate* nearly all round (the teeth directed forwards). *Corolla* 9–11(–12) *mm., pink*, tips of lateral petals blackish-red, *wings* blackish-red or more *frequently pink like the rest of the corolla; upper petal laterally compressed*, obtuse or acute; *lower petal with narrow spreading margins. Fr.* c. (2–)2½ mm., *suborbicular*, subacute or obtuse, often scarcely narrowed below with the indistinct neck as broad as the apex of the pedicel, but sometimes narrowed; *rugose* with broad shallow apical pits *when dry.* Fl. 4–10. Th.

Native. Cultivated ground and waste places. Widespread in W. England, Wales and Ireland and common in many places; scattered over Scotland;

very rare in E. England and absent from the Midlands; Channel Is. 53, H 32, S. Mediterranean region (rare in the east) extending north in W. Europe to W. and C. France; Madeira, Canaries; introduced in Australia.

5. F. martinii Clav.

F. paradoxa Pugsl.

Robust, sometimes climbing. Lf-segments small with oblong or cuneiform lobes. *Infl.* lax, 15–20-*fld*, *longer than its peduncle*. Bracts linear-oblong, cuspidate, much shorter than fr. pedicels. *Pedicels usually arcuate-recurved in fl.*, *ascending or spreading*, sometimes flexuous, *in fr.*, the apex somewhat dilated. *Sepals* 3–5 × 1½–2½ mm., oval, acute, *subentire* or with a few small teeth at the base. *Corolla* 11–13 *mm.*, *pink*, tips and wings blackish-red; *upper petal broad but not compressed*, usually subacute; *lower petal with narrow spreading margins*. *Fr.* 2½–2¾ × 2–2½ mm., *oval*, acute (sometimes obtuse when dry), with a very indistinct neck; *smooth*, or rarely rugulose, *when dry*, with large shallow apical pits. Fl. 5–10. Th.

Native. Cultivated ground, very rare. 4, S. W. Cornwall, Guernsey; has also occurred in Devon, Somerset and Surrey, but is perhaps impermanent in these. W. France, Spain (uncommon in both).

6. F. muralis Sond. ex Koch

Slender to robust, suberect or diffuse or climbing. Lf-segments with oblong, lanceolate or broadly cuneiform lobes. Infl. rather lax. *Pedicels slender*, *usually straight and ascending or spreading*, rarely flexuous and recurved, the apex somewhat dilated. *Sepals* 3–5 × 1½–3 mm., *ovate*, rarely oval, *usually dentate towards the base*, entire above, rarely subentire (the teeth directed outwards). *Corolla* 9–12 *mm.*, *pink*, tips and wings blackish-red; *upper petals dorsally compressed*, spathulately dilated, wings concealing keel; *lower petals with narrow erect margins*. *Fr.* 2–2½ mm. with an indistinct fleshy neck; *smooth or finely rugulose when dry*, with small apical pits. Fl. 5–10. Th.

A very variable species divisible into the three following sspp. of which the first two are themselves variable.

Ssp. muralis

Slender. Infl. few (c. 12)-fld, about equalling peduncles. Bracts linear-lanceolate, acuminate, c. ⅔ as long as fr. pedicels. *Sepals* 3–4 × 1½–2½ mm., *ovate*, ± acuminate, always toothed. Corolla 9–11 mm.; upper petal not particularly broad, apiculate. *Fr.* 2–2½ mm. long, rather less broad, ovate-orbicular, *subacute or apiculate*; quite smooth when dry, the apical pits faint. $2n = 28$.

Native. Cultivated ground, etc. 19, S. Rare and local in S. and W. England and Wales. W. Europe from Norway to Spain and Portugal, N. Germany (?native); Morocco; Algeria; Macaronesia where it is common and very variable. Introduced in St Helena, Ascension, S. Africa, Mauritius, S. India, Java, New Zealand, Bermuda, S. America.

Var. *cornubiensis* Pugsl. Fls small (c. 9 mm.), lilac-pink with broad wings; lower petal subspathulate with spreading margins (thus the plant may be confused with other spp.). Mevagissey, Cornwall.

Ssp. **boraei** (Jord.) Pugsl.

F. Boraei Jord.

Robust to rather slender but always more robust than ssp. muralis. Infl.
c. 12–15-fld, from shorter than to nearly equalling peduncles. Bracts linear-
lanceolate, acuminate, usually a little shorter than the fr. pedicels but varying
from much shorter to longer. *Sepals* 3–5 × 2–3 mm., *ovate*, acute or acuminate.
Corolla 10–20 mm., upper petal broad, acute, acuminate or rarely obtuse.
Fr. 2–2½ mm. long, less than to nearly as broad, *obovate* to orbicular-obovate,
occasionally subquadrate, *obtuse* at least at maturity; smooth or finely
rugulose when dry, apical pits usually distinct.

Native. Cultivated and waste ground, hedgebanks and old walls. 87,
H17, S. Spread over the whole of Great Britain and the commonest member
of this section, common in many places in the west, less so in the east; in
Ireland less common and absent from the centre and west coast. W. Europe
from Norway and W. Germany to C. Spain and Portugal and Sardinia.

Ssp. **neglecta** Pugsl.

Robust, suberect and ascending. *Infl. many* (c. 20)*-fld*, longer than peduncles.
Bracts linear-oblong, cuspidate, about half as long as fr. pedicels, which are
never recurved. *Sepals c.* 3 × 1½–2 mm., *broadly oval, subentire* or with a few
shallow teeth. Corolla c. 10 mm.; upper petal not particularly broad, obtuse.
Fr. c. 2 × 2 mm., orbicular-obovate, *almost truncate*, smooth or finely rugulose
when dry with distinct apical pits.

Native. Only known in a cultivated field near Penryn (W. Cornwall).
1. Endemic.

F. × painteri Pugsl. (*F. muralis* ssp. *Boraei × F. officinalis*?)

Robust, climbing. Lf-lobes of *F. muralis* ssp. *boraei* but rather narrower.
Infl. rather lax, many (c. 20)-fld, longer than peduncles. Bracts linear-
lanceolate, acuminate, nearly equalling fr. pedicels. Fr. pedicels ascending,
somewhat thickened at apex. *Sepals* 3–3½ × 1½ mm., *ovate-lanceolate*, acuminate, irregularly toothed below, entire above. Corolla 10–11 mm., pale
pink, tips and wings blackish-red; upper petal dorsally compressed, wings
spathulately dilated, concealing keel; *lower petal subspathulate with narrow
spreading margins*. Fr. c. 2½ mm., subquadrate, truncate but shortly apiculate
with an obscure fleshy neck; faintly rugulose when dry, with broad shallow
apical pits.

Only known from two places in Shropshire. This plant needs further study.
It is the only fertile hybrid known in the genus and a sterile hybrid of the
same parentage is known from Guernsey and S. England. It is possible that
it will prove to be a distinct sp.

Section 2. *Fumaria* Pugsl. Lf-lobes relatively narrow (from lanceolate to
linear). Fls 5–8(–9) mm. Wings of upper petal less reflexed. Lower petal
± spathulate, margins spreading. Fr. with an obscure fleshy neck when fresh,
± rugose when dry, with shallow apical pits.

7. F. micrantha Lag.

F. densiflora auct.

A ± robust, usually suberect, rarely diffuse or climbing herb. *Lf-segments with channelled* linear or linear-oblong *lobes*. Infl. very dense at first, becoming lax in fr., many (c. 20–25)-fld, much longer than peduncles. *Bracts* linear-oblong, cuspidate, *longer than fr. pedicels* which are ascending and dilated above. *Sepals* (2–)2½–3½ × (1–)2–3 *mm.*, *orbicular or broadly ovate* (rarely ovate), subentire or toothed at base, acute or mucronate. *Corolla* 6–7 *mm.*, *pink*, tips and wings blackish-red; upper petal somewhat laterally compressed, obtuse or subacute, wings ascending. *Fr.* 2–2½ mm., *subglobose*, very little compressed but somewhat keeled, *obtuse.* Fl. 6–10. $2n = 28$. Th.

Native. Arable land on dry soils in E. England and E. Scotland; very rare in the west and absent from Devon and Cornwall, S. and C. Wales and W. Scotland (except the south-west); only in the northern half of Ireland, very local. 57, H8. Mediterranean region extending eastwards to Persia and north (in the west only) to Germany and the Netherlands.

8. F. officinalis L.   'Common Fumitory.'

A ± robust, suberect, diffuse or climbing herb. *Lf-segments with flat*, lanceolate or linear-oblong *lobes*. Infl. at first dense, becoming lax, 10–40 (usually more than 20)-fld, longer than peduncles. *Bracts* linear-lanceolate, acuminate, *shorter than the ascending fr. pedicels. Sepals* 2–3½ × 1–1½ mm., *ovate or ovate-lanceolate*, irregularly toothed at base, acuminate or cuspidate. *Corolla* 7–8(–9) *mm.*, *pink*, tips and wings blackish-red; upper petal dorsally compressed, obtuse (rarely apiculate), wings concealing keel. *Fr.* 2–2½ mm., *usually considerably broader* but rarely scarcely so, little compressed and obscurely keeled, *truncate or retuse* sometimes apiculate. Variable. Fl. 5–10. Pollinated by bees or probably more frequently selfed, homogamous, self-fertile. $2n = 28$. Th.

Native. Cultivated ground on the lighter soils. 110, H33, S. Throughout the British Is., common in the east, rather less so in the west and in Ireland. Europe (except Iceland, etc.); Mediterranean region, east to Persia; Canary Is.; introduced in America.

9. F. vaillantii Lois.

Slender, dwarf, much-branched, often very glaucous. *Lf-segments* distant *with flat* linear-oblong or lanceolate *lobes*. Infl. rather lax, c. 6–16-fld, longer than the *short peduncle. Bracts* linear-lanceolate, *acuminate, shorter than* the suberect or ascending *fr. pedicels. Sepals not more than* 1 × 0·3–0·5 mm., lanceolate, acuminate, ± laciniate-serrate. *Corolla* 5–6 mm., pale *pink* with tips of lateral petals blackish-red, the wings less markedly dark; upper petal dorsally compressed, emarginate but apiculate. *Fr.* c. 2 mm., suborbicular, compressed but *obscurely keeled*, obtuse. Fl. 6–9. Th.

Native. Arable land, usually on chalk. From Kent to Wilts, E. Gloucester and Norfolk, very local; ?Yorks; ?West Lothian. 21. Europe (except Iceland, Norway, Portugal, etc.); Mediterranean region (rare in N. Africa); temperate Asia to the Altai and Kashmir.

10. F. parviflora Lam.

Robust, suberect, diffuse or occasionally climbing, markedly glaucous. *Lf-segments with channelled* linear or subulate *lobes. Infl.* dense in fl., lax in fr., often c. 20-fld, *subsessile. Bracts* linear-oblong, *cuspidate, about equalling* the suberect or ascending *fr. pedicels. Sepals* $1–1\frac{1}{2} \times 0·6–0·8$ *mm.*, broadly ovate, acute, $\pm$ laciniate-dentate. *Corolla* 5–6 *mm.*, *white or flushed with pink* with a blotch at the base of the wings and with the tips of the lateral petals blackish-red; upper petal dorsally compressed, truncate. *Fr.* c. 2 mm., suborbicular or ovate-orbicular, little compressed but *distinctly keeled*, obtuse or subacute and apiculate or beaked. Fl. 6–9. $2n=28$. Th.

Native. Arable land, usually on chalk. From Kent to Dorset, E. Gloucester, Worcester and Norfolk, local; W. Cornwall; Yorks; S.E. Scotland. 32. Mediterranean region extending to Baluchistan, Arabia and the Sahara and north to the Caucasus, the Crimea, Hungary, Germany and Belgium; Sweden (casual); Madeira; Canaries; introduced in Mexico and S. America.

The following hybrids have occurred. but very rarely; with the exception of *F. × painteri* (see above) all are nearly or quite sterile: *F. bastardii × muralis* ssp. *boraei*, *F. micrantha × officinalis*, *F. muralis* ssp. *boraei × officinalis* (see also *F. × painteri*), *F. officinalis × parviflora*, *F. officinalis × vaillantii*.

21. CRUCIFERAE

Annual to perennial herbs, rarely woody plants, with spirally arranged exstipulate lvs and racemose usually ebracteate infl. Fls usually hermaphrodite, actinomorphic and hypogynous. Sepals 4, in 2 decussate pairs, the inner pair often with saccate bases in which nectar collects; petals 4, free, clawed, commonly white or yellow, placed diagonally and so alternating with the sepals; stamens usually 6, an outer transverse pair with short filaments and 2 inner pairs with long filaments, one pair on the anterior and one on the posterior side; stamens sometimes 4, by suppression of the outer pair, or fewer; filaments sometimes with tooth-like or wing-like appendages; ovary syncarpous with 1–many ovules on each of 2 parietal placentae, usually 2-celled, a false septum being formed by the meeting of outgrowths from the placentae; style single, rudimentary to long; stigma capitate, discoid, or $\pm$ 2-lobed, the lobes standing over the placentae or sometimes with long decurrent lobes alternating with the placentae. Fr. usually a specialized capsule (called a *siliqua*, or, if less than 3 times as long as wide, a *silicula*) opening from below by 2 valves which leave the seeds attached to a framework consisting of the placentae and adjacent wall tissue (*replum*) and the false septum; sometimes 1 or more seeds develop also in an indehiscent beak at the base of the style, and the relative size of the dehiscent and indehiscent parts varies greatly, some Crucifers having their seeds confined to the beak which may then break transversely at maturity into 1-seeded joints (*lomentum*); sometimes again the valves are indehiscent and the fr. may break into 1-seeded halves or remain intact, when it may be 1–many-celled and often has spines or wings which facilitate dispersal. Seeds usually in 1 or 2 rows in each cell, non-endospermic, the relative positions of the radicle and cotyledons affording characters which have been used for taxonomic purposes. When the radicle is bent round so as to lie along the edges of the cotyledons, $\underset{||}{\circ}$, it is said to be *accumbent*; when it lies on the face of one cotyledon, $\underline{\circ}$, *incumbent*; in the *Brassiceae* and a few other genera the radicle is incumbent but the cotyledons are folded longitudinally, $\circ$». The fls usually secrete nectar from glands, variable in size, shape and position, which lie round the bases of the stamens and ovary. Visited by insects, but spp. with small fls are commonly self-pollinated.

About 1900 spp. in 220 genera, cosmopolitan but chiefly in north temperate regions. Many are annual or ephemeral herbs of dry open habitats, or weeds of cultivation.

The Cruciferae are closely related to the Papaveraceae and Capparidaceae and appear to be a recently evolved specialized offshoot from the Rhoeadalian stock.

Synopsis of Classification of Native and Introduced Genera*

(After O. E. SCHULZ)

See also artificial key (p. 115)

Tribe 1. BRASSICEAE. Hairs simple or 0; stamen filaments very rarely appendaged (*Crambe*); median nectaries present, opposite the pair of long stamens; stigma capitate to 2-lobed, rarely with long decurrent lobes (*Eruca, Moricandia*); fr. very variable but commonly a siliqua with a distinct closed beak usually containing 1 or more seeds, rarely a silicula (*Succowia*); in many genera the fr. is divided into a small often sterile basal segment and an indehiscent seed-containing beak which may break transversely at maturity; rarely the basal segment is alone fertile but indehiscent, or the fr. is non-segmented and indehiscent (*Calepina*); cotyledons longitudinally folded or bent round the incumbent radicle.

I. Fr. opening from below by 2 valves.
  A. Fr. a $\pm$ beaked siliqua more than 1 cm. and at least three times as long as wide.
    1. Plants glaucous and quite glabrous with fully amplexicaul entire stem-lvs.
      Stigma deeply 2-lobed; fls violet. (Fig. 12D)   MORICANDIA
      Stigma shortly 2-lobed; fls cream or pale yellow.   12. CONRINGIA
    2. Plants rarely both glaucous and quite glabrous, but if so with the stem-lvs not more than $\frac{1}{4}$ clasping.
    a. Stigma $\pm$ capitate or shortly 2-lobed; beak often seed-bearing.
      a. Seeds in 1 row in each cell.
       i. Valves each with a conspicuous midrib and much weaker lateral veins, and so $\pm$ 1-veined.
         Siliquae usually with convex rounded valves; seeds spherical.
              1. BRASSICA
        Siliquae with strongly keeled valves; seeds ovoid or ellipsoidal.
             2. ERUCASTRUM
       ii. Valves each with a midrib and only slightly weaker lateral veins and so 3–7-veined at least when young.
        Blade of petal contracted abruptly into a somewhat longer filiform claw; ovary with 14–54 ovules.   3. RHYNCHOSINAPIS
        Blade of petal narrowed gradually into a short claw; ovary with 4–17 ovules.
          Sepals spreading; seeds spherical.   4. SINAPIS
          Sepals suberect; seeds ovoid or oblong.   5. HIRSCHFELDIA

* Several genera with species occurring rarely as casuals are included here, although they are not described in the text. Fruits of some are figured on pp. 119–21.

      b. Seeds in 2 rows in each cell.   6. DIPLOTAXIS
      b. Stigma deeply 2-lobed; beak flat, seedless.   7. ERUCA
  B. Fr. an almost spherical silicula opening by 2 spiny valves; style long,
        persistent. (Fig. 12c).   SUCCOWIA
  C. Fr. less than 1 cm. with 2 distinct ±equal segments of which only the
        lower opens by 2 valves.
        Upper segment usually 2-seeded. (Fig. 12A).   ERUCARIA
        Upper segment seedless, broad and flat. (Fig. 12B). CARRICHTERA

II. Fr. remaining intact or breaking transversely into joints, but not opening
    by valves.

  A. Fr. of 2 or more distinct segments separated by transverse constrictions,
      the lowest segment often slender and seedless and then resembling a
      short stalk.
    1. Fr. elongated, usually more than 3 times as long as wide.
        Lowest segment of fr. seedless, stalk-like; upper part either ±cylin-
          drical and indehiscent or constricted between the seeds and then
          often breaking into 1-seeded joints.   8. RAPHANUS
        Fr. of 2 indehiscent segments, the lower with 1–3, the upper with 3 or
          more seeds. (Fig. 11D).   ENARTHROCARPUS
    2. Fr. of 2 indehiscent segments, the upper less than 3 times as long as wide.
        Lower segment seedless, stalk-like; upper ±spherical, smooth-walled,
          non-septate, 1-seeded; large cabbage-like plants.   9. CRAMBE
        Lower segment 0–2-seeded, variable in diam.; upper segment ovoid
          or spherical, usually ribbed and rugose, septate but 1-seeded; plants
          not cabbage-like.   10. RAPISTRUM
        Lowest segment small, top-shaped, ±flattened, 1-seeded; upper
          larger, 4-angled, narrowed upwards, 1-seeded, separating from the
          lower at maturity; succulent maritime plants.   11. CAKILE
  B. Fr. not jointed, 1-seeded, indehiscent, ovoid with a short broadly rounded
      conical apex. (Fig. 11E).   CALEPINA

Tribe 2. LEPIDIEAE. Hairs usually simple or 0; stamen filaments often appendaged;
median nectaries 0 or, if present, sometimes fused with the lateral; stigma capitate
or shortly 2-lobed; fr. usually an angustiseptate silicula (i.e. laterally compressed,
with narrow septum), but sometimes not or little compressed (*Subularia,Cochlearia*
spp.) and sometimes breaking into 1-seeded halves or indehiscent; radicle incumbent
or accumbent.

I. Fr. a dehiscent silicula.

  A. Filaments with tooth- or wing-like appendages.
        Lvs all in a basal rosette, toothed to pinnate.   19. TEESDALIA
        Lvs not confined to a basal rosette, small, narrow, entire. AETHIONEMA
  B. Filaments not appendaged.
    1. Cells of fr. each with only 1 seed.
        Fr. of 2 ±circular halves, resembling a pair of spectacles; petals
          equal, yellow. (Fig. 12K).   BISCUTELLA
        Fr. with winged valves; petals unequal, not yellow.   17. IBERIS
        Fr. with winged or strongly keeled valves; petals equal, whitish, or 0.
                         13. LEPIDIUM
    2. Cells of fr. each with 2 or more seeds.
      a. Valves of fr. turgid, not strongly compressed.
        Small submerged aquatic plants with subulate lvs. 23. SUBULARIA
        Maritime or alpine plants; lvs not subulate.   22. COCHLEARIA

   b. Valves of fr. strongly compressed.
     i. Valves winged.   18. THLASPI
     ii. Valves not winged.
        Fr. elliptical to broadly obovate; stem lvs not sagittate.
                                        21. HORNUNGIA
        Fr. triangular-obcordate, rarely elliptical; stem lvs sagittate at the
          base.   20. CAPSELLA

II. Fr. breaking into 1-seeded halves or indehiscent.
  A. Infl. borne opposite the lvs.   14. CORONOPUS
  B. Infl. terminal or axillary.
      Fr. pendulous, flattened, winged; fls yellow.   16. ISATIS
      Fr. ± turgid, broadly cordate or deltoid, not winged; fls white.
                                       15. CARDARIA

Tribe 3. EUCLIDIEAE. Hairs simple, branched, glandular or 0; stamen filaments without appendages; median nectaries present or 0; stigma capitate or shortly 2-lobed, rarely with erect or spreading lobes; fr. indehiscent, neither beaked nor segmented, 1–4-seeded, hard-walled, often appendaged; radicle incumbent or accumbent.

Fr. small, spherical.   24. NESLIA
Fr. ovoid, densely covered with branched hairs. (Fig. 13 c). EUCLIDIUM
Fr. 1-celled and 1-seeded, ovoid-quadrangular with broad wavy wings along
  the angles. (Fig. 13 A).   BOREAVA
Fr. 1–4-celled with 1–4 seeds, irregularly ovoid, warty or with 4 irregularly
  toothed and crested wings.   25. BUNIAS
Fr. shortly and broadly compressed-clavate, 3-celled, the 2 upper side by side,
  empty. (Fig. 13 B).   MYAGRUM

Tribe 4. LUNARIEAE. Hairs simple or 0; stamen filaments without appendages, sometimes broadened; median nectaries present or 0; stigma capitate or 2-lobed; fr. a large latiseptate silicula (i.e. with broad septum); radicle accumbent.

Fr. with flat thin-walled translucent valves and a thin shining white septum;
  seeds flat.   26. LUNARIA

Tribe 5. ALYSSEAE. Hairs usually branched; stamen filaments sometimes appendaged; median nectaries 0; stigma usually shortly 2-lobed; fr. a small latiseptate silicula; epidermal cells of fr. septum with parallel walls; radicle accumbent.

Hairs stellate; petals yellow; filaments of at least some of the stamens appendaged; valves faintly net-veined.   27. ALYSSUM
Hairs bifid; petals white, entire; filaments without appendages; valves with
  a slender midrib.   28. LOBULARIA
Hairs stellate; petals white, deeply bifid; filaments toothed or winged; valves
  indistinctly veined.   29. BERTEROA

Tribe 6. DRABEAE. Hairs simple, or branched, or 0; stamen filaments sometimes appendaged; median nectaries present or 0; stigma capitate or shortly 2-lobed; fr. a latiseptate silicula; epidermal cells of fr. septum not with parallel walls; radicle accumbent.

I. Valves of fr. flat with the midrib vanishing above the middle.
    Petals entire or slightly notched.   30. DRABA
    Petals deeply bifid.   31. EROPHILA
II. Valves of the ± spherical fr. strongly convex with an indistinct network
    of veins.   32. ARMORACIA

Tribe 7. ARABIDEAE. Hairs simple, or branched, or 0; sepals somewhat spreading; stamen filaments without appendages; median nectaries present or 0; stigma capitate, ±2-lobed; fr. a siliqua; radicle accumbent.

I. Siliqua flat, opening suddenly, the valves coiling spirally from the base and flinging out the seeds; lvs often pinnately or palmately compound.

  Perennial herbs with whitish rhizomes bearing fleshy triangular scales; seeds with dilated ± triangular funicle.   34. DENTARIA
  Annual to perennial herbs, not with scaly rhizomes; seeds not with dilated funicle.   33. CARDAMINE

II. Siliqua neither opening suddenly nor the valves coiling spirally from below.

 A. Siliquae with strongly compressed valves.

   Septum of fr. thick and rigid; valves nodular (i.e. seeds forming protuberances on the faces of the valves); lvs commonly lyrate.   36. CARDAMINOPSIS
   Septum thin; valves flat, not nodular; lvs not lyrate.   37. ARABIS
 B. Siliquae with convex or keeled valves.
  1. Valves with distinct midrib.
   Siliqua bluntly 4-angled; plant grass-green.   35. BARBAREA
   Siliqua ± cylindrical; plant glaucous.   38. TURRITIS
  2. Valves indistinctly veined.   39. RORIPPA

Tribe 8. MATTHIOLEAE. Simple and branched hairs usually present; sepals erect; stamen filaments usually without appendages; median nectaries present or 0; stigma usually with long decurrent lobes; fr. a siliqua, rarely breaking transversely; radicle accumbent.

Fr. a short ellipsoidal siliqua with convex stellate-hairy valves; stigma capitate.   AUBRIETA
Fr. a short (8–10 mm.) curved 4-angled siliqua ending in 4 spreading horn-like points.   TETRACME
Fr. a long (1–12 cm.) compressed siliqua; stigma of 2 erect lobes each with a dorsal swelling or horn-like process which often enlarges in fr.   40. MATTHIOLA
Fr. long (3–4 cm.), cylindrical, constricted between 2-seeded joints which separate and split vertically into 1-seeded nutlets at maturity. (Fig. 13G).   CHORISPORA

Tribe 9. HESPERIDEAE. Hairs simple, or branched, or 0; sepals erect; stamen filaments sometimes broadened but not appendaged; median nectaries present or 0; stigma ± 2-lobed or with long decurrent lobes; fr. a siliqua or breaking transversely; radicle incumbent (or sometimes accumbent in *Erysimum* and *Cheiranthus*).

I. Fr. a siliqua.
 A. Median nectaries present, usually confluent with the lateral nectaries.
   Stigma capitate or shortly 2-lobed.   42. ERYSIMUM
 B. Median nectaries 0.
   Stigma deeply 2-lobed, the lobes erect, joined below and forming a conical pointed beak; valves of fr. 3-veined.   MALCOLMIA
   Stigma deeply 2-lobed, the lobes ± erect and facing each other; valves of fr. 1-veined; plant ± glandular; radicle incumbent. 41. HESPERIS
   Stigma often with 2 deep spreading lobes; valves of fr. 1-veined; plant with branched appressed hairs, not glandular; radicle ± accumbent.   43. CHEIRANTHUS

II. Fr. 4-angled, indehiscent or constricted and breaking transversely into
    1-seeded joints. (Fig. 13J).   GOLDBACHIA

Tribe 10. SISYMBRIEAE. Hairs simple, or branched, or 0; sepals spreading; stamen filaments without appendages; median nectaries present, confluent with the lateral; stigma capitate, often shortly 2-lobed; fr. a siliqua, rarely a silicula; radicle incumbent.

I. Fr. a siliqua.
  A. Plants glabrous or with simple hairs; valves of fr. ± distinctly 3-veined.
      Valves keeled; fls white.   44. ALLIARIA
      Valves convex; fls usually yellow.   45. SISYMBRIUM
  B. Lvs with branched hairs; valves of fr. 1-veined with faint lateral veins
      or an indistinct lateral network.
      Lvs entire or distantly toothed; fls white.   46. ARABIDOPSIS
      Lvs finely pinnatisect; fls yellow.   48. DESCURAINIA

II. Fr. a latiseptate silicula.
    Silicula obovoid or pear-shaped.   47. CAMELINA

Key to Genera

1   A woodland plant with a white scaly rhizome, pinnate basal lvs, ternate
    or simple stem-lvs bearing brownish-violet axillary bulbils, and purple,
    pink or rarely white fls; fr. a flat veinless siliqua which rarely ripens
    in this country.   34. DENTARIA
    Not as above.   *2*

2   A robust plant of gardens, waysides and waste places, with a very stout
    tap-root, large oblong waved lvs up to 60 cm., white fls in a large
    panicle, and obovoid fr. which does not ripen in this country.
                    32. ARMORACIA
    Not as above.   *3*

3   A small submerged aquatic plant with subulate lvs, minute white fls and
    ± ovoid scarcely compressed siliculae.   23. SUBULARIA
    Not as above.   *4*

4   Infls borne opposite the pinnately cut lvs (as well as terminal and at
    forking of the stem); small ± prostrate plants with whitish fls.
                    14. CORONOPUS
    Infls not lf-opposed.   *5*

5   Fr. 8–20 mm., pendulous, flattened, winged; tall plants with yellow fls.
                    16. ISATIS
    Fr. not pendulous, flattened and winged.   *6*

6   Fr. at least 3 times as long as wide.   *7*
    Fr. less than 3 times as long as wide.   *33*

7   Stigma deeply 2-lobed or with long decurrent lobes, not ± capitate or
    discoid.   *8*
    Stigma ± capitate or discoid, at most shallowly 2-lobed.   *11*

8   Stigma-lobes spreading; fls yellow or red, without deep violet veins.
                    43. CHEIRANTHUS

Stigma-lobes erect, facing one another, not spreading, or spreading only
at their tips; fls purplish or white, or, if pale yellow, then with deep
violet veins.　　　　　　　　　　　　　　　　　　　　　　　　　　　9

9 Stigma-lobes thickened or horned at the back; maritime plants.
　　　　　　　　　　　　　　　　　　　　　　40. MATTHIOLA
Stigma-lobes not thickened at the back.　　　　　　　　　　　　　10

10 Biennial to perennial, with simple toothed lvs; petals violet or white,
not veined with deep violet; fr. not beaked.　　　41. HESPERIS
Annual or overwintering, with lyrate-pinnatifid lvs; petals pale yellow
or white, with deep violet veins; fr. with a broad flat beak.　7. ERUCA

11 Fr. strongly flattened.　　　　　　　　　　　　　　　　　　　12
Fr. quadrangular, cylindrical or slightly flattened.　　　　　　　14

12 Valves of fr. veinless, rolling suddenly into spirals on dehiscence; lvs
pinnate; fls white, greenish or mauve.　　　33. CARDAMINE
Valves of fr. with a ±conspicuous central vein, not rolling into spirals
on dehiscence; lvs simple, sometimes pinnatifid but not pinnate; fls
white or very pale yellow.　　　　　　　　　　　　　　　　　13

13 Valves of fr. with a strong central vein and prominent seeds; basal lvs
distinctly long-stalked, often lyrate-pinnatifid; a rare alpine plant.
　　　　　　　　　　　　　　　　　36. CARDAMINOPSIS
Valves of fr. with a weak central vein; seeds not prominent; basal lvs not
pinnatifid, narrowed gradually into a short stalk.　　37. ARABIS

14 Fls white or lilac.　　　　　　　　　　　　　　　　　　　　15
Fls yellow, apricot or cream.　　　　　　　　　　　　　　　　19

15 Lvs simple.　　　　　　　　　　　　　　　　　　　　　　16
At least the basal lvs pinnate or pinnatifid.　　　　　　　　　　17

16 Plant with both simple and branched hairs; basal lvs small, elliptical or
spathulate; plant not smelling of garlic when crushed.
　　　　　　　　　　　　　　　　　46. ARABIDOPSIS
Plant glabrous or with simple hairs; basal lvs reniform or cordate, long-
stalked; plant smelling of garlic when crushed.　　44. ALLIARIA

17 Fls white or lilac and dark-veined, or white at first but later turning lilac;
weeds of arable or waste land, with pinnatifid lvs.　　　　　　　18
Fls white; plants of wet places, with pinnate lvs.　　39. RORIPPA

18 Fr. either distinctly constricted between the seeds or inflated and with
a spongy wall, not flattened, indehiscent.　　8. RAPHANUS
Fr. a siliqua with somewhat flattened valves and seeds in 2 rows in each
cell, dehiscent.　　　　　　　　　　　　6. DIPLOTAXIS

19 Plant with branched or stellate hairs.　　　　　　　　　　　　20
Plant glabrous or with simple hairs only.　　　　　　　　　　　21

20 Lvs deeply and finely pinnatisect.　　　　　48. DESCURAINIA
Lvs simple, entire or toothed.　　　　　　　　42. ERYSIMUM

21 Fr. evidently with 2 valves (by which it will eventually open upwards
from the bottom).　　　　　　　　　　　　　　　　　　　22
Fr. not valved (and not opening from below upwards), cylindrical, constricted between the seeds and ultimately breaking into 1-seeded joints.
　　　　　　　　　　　　　　　　　　　8. RAPHANUS

22 Valves of fr. veinless or with an indistinct central vein which vanishes
above.　　　　　　　　　　　　　　　　39. RORIPPA
Valves of fr. with a distinct central vein.　　　　　　　　　　23

23 Valves of fr. 1-veined, or with lateral veins very much weaker than the
      central so that the valves appear 1-veined.   24
    Valves of fr. with lateral veins only slightly weaker than the central vein
      and so appearing 3–7-veined, at least when young.   30

24 Seeds in 2 rows in each cell of the fr.   25
    Seeds in 1 row in each cell of the fr.   26

25 Upper stem-lvs sagittate at base, clasping the stem; fls cream or very
      pale yellow; fr. erect, appressed to the stem.   38. TURRITIS
    Upper stem-lvs not clasping the stem; fls bright yellow; fr. not appressed
      to the stem.   6. DIPLOTAXIS

26 Fr. with convex rounded valves.   1. BRASSICA
    Central veins of valves prominent so that the valves are keeled and the
      fr. ± quadrangular in section.   27

27 Stem and lvs ± glabrous.   28
    Stem hairy, at least below.   29

28 Lvs all glaucous, entire; fls cream-coloured.   12. CONRINGIA
    Lvs green, the basal lyrate-pinnatifid; fls yellow.   35. BARBAREA

29 Stem-lvs deeply pinnatifid; fr. curving upwards but not stiffly erect nor
      appressed to the stem.   2. ERUCASTRUM
    Stem-lvs shallowly lobed or ± entire; fr. held stiffly erect, appressed to
      the stem when ripe.   1. *Brassica nigra*

30 Fr. with a distinct beak, often containing 1 or more seeds, between the
      top end of the valves and the stigma.   31
    Fr. not beaked and never with seeds beyond the ends of the valves, which
      open almost to the stigma.   45. SISYMBRIUM

31 Fr. short-stalked, erect, appressed to the stem; beak short, swollen;
      seeds ovoid.   5. HIRSCHFELDIA
    Fr. not appressed to the stem; beak flat or conical, not swollen; seeds
      spherical.   32

32 Sepals spreading horizontally; weeds of arable land and waste places.
                                                             4. SINAPIS
    Sepals erect; maritime plants, rarely casuals in waste places.
                                                          3. RHYNCHOSINAPIS

33 Fls yellow.   34
    Fls not yellow, or petals 0.   41

34 Fr. of 2 distinct segments, the lower resembling a short stalk, the upper
      ovoid or almost spherical; weeds of arable land or waste places.
                                                          10. RAPISTRUM
    Fr. not of 2 distinct segments.   35

35 Fr. almost spherical, tiny (1·5–3 mm. diam.); casuals.   36
    Fr. not spherical and tiny.   37

36 Lvs grey-green with stellate hairs, ± entire; fr. 1·5–2 mm. diam.
                                                             24. NESLIA
    Lvs glabrous or with simple hairs, irregularly toothed; fr. 1·5–3 mm.
      diam.   39. *Rorippa austriaca*

37 Fr. distinctly flattened.   38
    Fr. not or slightly flattened.   39

38 Tufted plant with lvs confined to basal rosettes; fl.-stems lfless; fr.
      elliptical.   30. *Draba aizoides*
    Fl-stems lfy; fr. almost circular emarginate.   27. ALYSSUM

39   Lvs entire or nearly so; fr. pear-shaped.   47. CAMELINA
      At least the basal lvs pinnatifid or distinctly toothed.   *40*

40   Fr. irregularly ovoid, either with crested wings or warty prominences.
                                                 25. BUNIAS
      Fr. neither winged nor warty.   39. RORIPPA

41   Large glaucous cabbage-like maritime plant with much-branched infl. of
      greenish-white fls and fr. having 2 segments, the lower slender, stalk-
      like, the upper almost spherical (and indehiscent).   9. CRAMBE
      Not as above.   *42*

42   Annual herb of sandy or shingly sea-shores with succulent lvs, at least
      the lower usually pinnatifid; fr. of 2 segments, the lower small, top-
      shaped, the upper larger, mitre-shaped.   11. CAKILE
      Not as above.   *43*

43   Petals bifid almost to half-way, white.   *44*
      Petals entire, or notched but not deeply bifid, or 0.   *45*

44   Stem lfless, 2–20 cm.   31. EROPHILA
      Stem lfy, 20–60 cm.   29. BERTEROA

45   Two adjacent petals of each fl. conspicuously larger than the other two.   *46*
      Petals of equal size, or 0.   *47*

46   Lvs scattered up the stem; smaller petals twice as long as sepals; fr.
      4–5 mm., distinctly winged.   17. IBERIS
      Lvs almost or quite confined to a basal rosette; smaller petals scarcely
      longer than sepals; fr. 3–4 mm., very narrowly winged.
                                                 19. TEESDALIA

47   Fr. rounded, not or slightly flattened.   22. COCHLEARIA
      Fr. distinctly flattened.   *48*

48   Cells of fr. 1-seeded.   *49*
      Cells of fr. each with 2 or more seeds.   *51*

49   Lvs all simple, narrow, entire, not clasping the stem; fr. latiseptate,
      i.e. with the septum across the widest diam.   28. LOBULARIA
      At least some lvs pinnatifid or toothed or clasping the stem; fr. angusti-
      septate, i.e. with the septum across the narrowest diam.   *50*

50   Infl. a corymbose panicle of small white fls; fr. broadly cordate, deltoid
      or ovate, tapering above into the persistent style, not winged, in-
      dehiscent.   15. CARDARIA
      Infl. not corymbose; fr. not or hardly cordate below, often winged,
      dehiscent.   13. LEPIDIUM

51   Fr. 2–9 cm., with thin-walled translucent valves and silvery septum;
      petals purple (rarely white); lvs cordate.   26. LUNARIA
      Fr. usually less than 2 cm., its valves not translucent.   *52*

52   Fr. winged.   18. THLASPI
      Fr. not winged.   *53*

53   Fr. triangular-obcordate.   20. CAPSELLA
      Fr. oval or elliptical.   *54*

54   Lvs deeply pinnatifid to pinnate; fr. angustiseptate, i.e. with the septum
      across the narrowest diam.   21. HORNUNGIA
      Lvs simple, entire or toothed; fr. latiseptate, i.e. with the septum across
      the widest diam.   30. DRABA

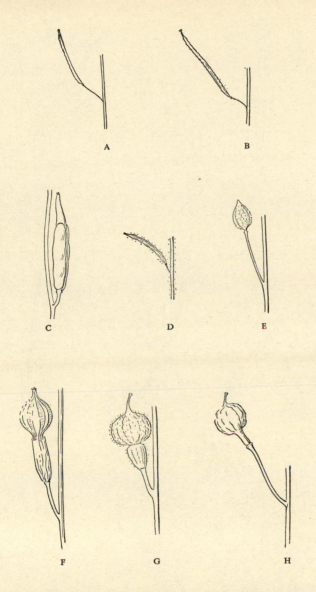

Fig. 11. Fruits of Cruciferae. A, *Erucastrum nasturtiifolium*; B, *E. gallicum*; C, *Eruca sativa*; D, *Enarthrocarpus lyratus*; E, *Calepina irregularis*; F, *Rapistrum perenne*; G, *R. rugosum*; H, *R. hispanicum*. A–D ×1; E ×2·5; F–H ×2.

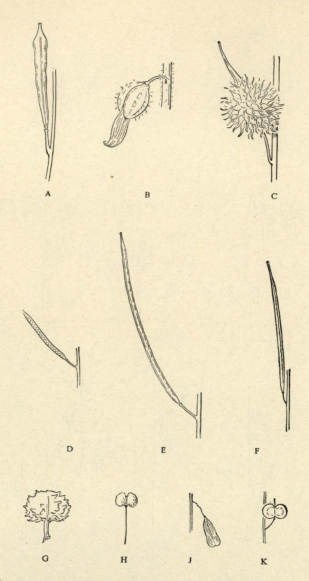

Fig. 12. Fruits of Cruciferae. A, *Erucaria myagroides*; B, *Carrichtera annua*; C, *Succowia balearica*; D, *Moricandia arvensis*; E, *Conringia orientalis*; F, *C. austriaca*; G, *Coronopus squamatus*; H, *C. didymus*; J, *Isatis tinctoria*; K, *Biscutella laevigata*. A ×2·5; B, C, G, H ×2; K ×1; D–F, J ×½.

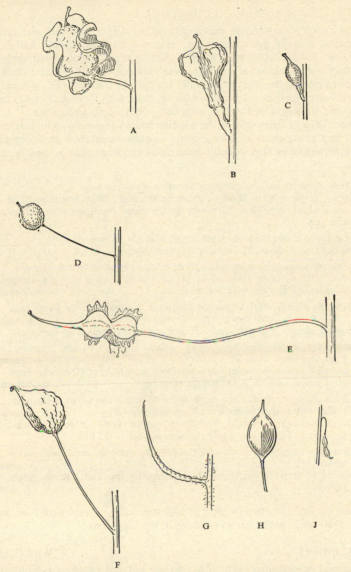

Fig. 13. Fruits of Cruciferae. A, *Boreava orientalis*; B, *Myagrum perfoliatum*; C, *Euclidium syriacum*; D, *Neslia paniculata*; E, *Bunias erucago*; F, *B. orientalis*; G, *Chorispora tenella*; H, *Camelina sativa*; J, *Goldbachia laevigata*. A, C, H ×2·5; B, D–F ×2; G, J ×½.

1. Brassica L.

Annual or biennial, rarely perennial, herbs with a tap-root and erect or ascending, usually branched stems, glabrous or with simple hairs, often pruinose. Infl. an ebracteate raceme. Sepals erect or somewhat spreading, the inner pair ±saccate at the base; petals clawed, usually yellow; stamens 6, without appendages; stigma capitate, or slightly 2-lobed. *Fr.* ±linear *with convex valves each with* 1 *prominent vein*; beak with 0–3 seeds; seeds almost *spherical*, unwinged, *in* 1 *row* in each cell. Cotyledons obcordate.

About 40 spp. especially in the Mediterranean region, usually calcicolous.

An important genus including many valuable vegetables and crop plants, and taxonomically very difficult because of the multitude of closely related races.

1 Upper stem-lvs stalked or narrowed into a stalk-like base.   *2*
  Upper lvs rounded or deeply cordate at the base, often broadened and
  ±clasping; sometimes narrowed at the base, but then with convex
  margins.   *6*
2 Fr. stalked above the sepal scars; plant biennial or perennial.   *3*
  Fr. not stalked above the sepal scars; plant annual.   *4*
3 Lower lvs densely covered with curved bristles; beak of fr. 0·5–2 mm.
                                                                                                      4. elongata
  Lower lvs ±glabrous; beak 3–6 mm.   **5. fruticulosa**
4 Fl.-stalks usually shorter than the sepals; fr. appressed to the stem; beak
  up to 4 mm., usually shorter.   **8. nigra**
  Fl.-stalks longer than the sepals; fr. not appressed; beak 6–16 mm.   *5*
5 Lower lvs lyrate-pinnatifid, sparsely bristly; beak 4–10 mm., narrower
  than the stigma at its tip.   **6. juncea**
  Lower lvs runcinate-pinnatifid, ciliate and densely bristly beneath; beak
  10–16 mm., as wide as the stigma at its tip.   **7. tournefortii**
6 All lvs glabrous; middle and upper stem-lvs never more than ⅓-clasping;
  buds overtopping the open fls; all stamens erect.   **1. oleracea**
  Lowest lvs always ±bristly; middle and upper stem-lvs cordate, at least
  ½-clasping; filaments of outer stamens curved at the base.   *7*
7 All lvs glaucous; buds slightly overtopping the open fls; petals pale yellow
  or buff.   **2. napus**
  Lowest lvs grass green; open fls overtopping the buds; petals bright
  yellow.   **3. rapa**

Section 1. *Brassica.* Sepals erect or half-spreading. Ovary with 9–45 ovules Fr. 1·5–10 cm., usually with 1–2 seeds in the long beak.

1. B. oleracea L.   Wild Cabbage.

A biennial or perennial herb with a strong but not tuberous tap-root and a *thick* ±*decumbent* and irregularly curved *stem* which becomes *woody* and *covered with conspicuous lf scars below.* Basal lvs stalked, broad and rounded, with sinuate margins, occasionally lyrate-pinnatifid with a few small basal lobes; upper lvs oblong, entire, sessile or up to ⅓-clasping, *not broadened at the base; all lvs glabrous and glaucous.* Infl. lengthening so that *buds overtop the opened fls* (Fig. 14A). Sepals erect. Petals 12–25 mm., c. twice as long as the sepals, lemon yellow. *All stamens erect. Fr.* 5–10 cm., ± cylindrical, *with a short* tapering usually seedless *beak* 5–10 mm. Seeds 8–16 in each cell, dark grey-brown, 2–4 mm. diam. Fl. 5–8. Homogamous. Visited by hive bees and other insects, $2n = 18$ (18, 36, 72 in cultivated races). Ch.

Probably native. The wild cabbage appears to be indigenous as a maritime cliff plant in S. and S.W. England and Wales, but occurs further north as an escape from cultivation. 15, S. Mediterranean region and Atlantic coast of Europe to France and Great Britain.

There are very many cultivated races of *B. oleracea* which agree with the wild species in fls. fr. and seeds but vary widely in vegetative morphology and in the form of the immature infl. They include the various types of cabbage, kale, cauliflower, broccoli, brussels sprouts and kohlrabi.

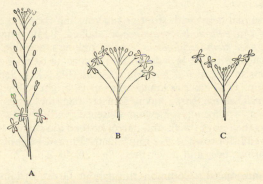

Fig. 14. Inflorescences of *Brassica*. A, *B. oleracea*; B, *B. napus*; C, *B. rapa*. × ½.

*2. B. napus L.   Rape, Cole; and Swedish Turnip or Swede.
Incl. *B. Napobrassica* (L.) Mill. and *B. chinensis* L.

Annual or biennial herbs with a strong and often tuberous tap-root and erect ± branched stems up to 1 m. with a few *not very conspicuous lf scars* below. *Lvs all glaucous*, not very fleshy: lowest lvs stalked, *sparsely bristly*; middle and upper stem-lvs sessile, oblong-lanceolate with a broadened base which is more than half-clasping. Infl. lengthening so that *buds overtop the opened fls* (Fig. 14 B). Sepals half-spreading; petals 11–14 mm., almost twice as long as the sepals, pale yellow or buff; *outer stamens curved outwards at the base*, shorter than the inner stamens. Fr. obliquely erect, 4·5–11 cm., distinctly flattened, with a beak about ¼ as long as the rest of the fr. and narrower than the stigma at its tip. Seeds blue-black with a whitish bloom, 1·5–2·5 mm. diam. Fl. 5–8. Slightly protogynous. Visited by various bees. $2n = 38$ (38, 57, 76 in other cultivated races): an allopolyploid from *B. oleracea × rapa*.

?Introduced. Of the various cultivated varieties and races the one most often found escaped or naturalized in this country is the biennial race of var. *arvensis* (Lam.) Thellung (*B. oleifera* Moench) known as Rape, Cole or Coleseed, which is grown as a fodder crop and is often found as an arable or wayside weed and on the banks of streams and ditches. It has a non-tuberous root, lyrate-pinnatifid lowest lvs which usually fall before flowering, and almost entire middle and upper stem-lvs. The seeds are very rich in oil and this and an annual race are much cultivated on the Continent for the rape or colza oil expressed from them. Rape-seed cake is made from the still oil-rich residue after extraction. Var. *napobrassica* (L.) Rchb. is the Swedish turnip or swede, with a tuberized stem-base and tap-root which is yellow-fleshed and either violet, white or yellow (var. *rutabaga* Léveillé) on the outside, and differs from the common turnip in having a small 'neck' formed by the tuberized base of the epicotyl. It is sometimes found as an escape or relic from cultivation. Var. *chinensis* (L.) O. E. Schulz (*B. chinensis* L.), Chinese Cabbage or 'Pak-choi', is a distinct cultivated form with obovate ±entire basal lvs narrowing gradually into a broad stalk, and with rather smaller fls than in the other varieties.

3. B. rapa L.   Turnip, Navew.

Incl. *B. campestris* L.

Annual or biennial herbs with stout or tuberous tap-root and erect branched stems up to 1 m. *Basal lvs grass-green*, stalked, lyrate-pinnatifid, *bristly*; middle and *upper stem-lvs glaucous*, sessile, oblong-lanceolate, ±completely clasping the stem with the broadened deeply cordate base, ±glabrous. Infl. not lengthening so that *opened fls overtop the buds* (Fig. 14c). Sepals spreading; petals 6–10 mm., c. 1½ times as long as the sepals, bright yellow; *outer stamens curved outwards at the base*, much shorter than the inner stamens. Fr. obliquely erect, 4–6·5 cm., somewhat flattened, with a long tapering beak ⅓–½ as long as the rest of the fr. and narrower than the stigma at its tip. Seeds blackish or reddish-brown, 1·5–2 mm. diam. Fl. 5–8, slightly protogynous. Visited by various bees and hover-flies. $2n = 20, 40$. Th.–Ch.

Probably introduced. Common throughout Great Britain and Ireland. 111, H40, S.

Wild plants mostly belong to ssp. **campestris** (L.) Clapham, with a non-tuberous tap-root. They closely resemble the races grown on the Continent for their oil-containing seeds (colza, turnip-like rape), but have rather smaller grey or blackish, not red-brown seeds. There are two chief habitats, stream-banks and arable or waste land; plants of the former being usually biennial with a well-defined basal rosette (var. *sylvestris* H. C. Wats. ex Briggs), while those growing as weeds are often annual and lack the basal rosette (var. *briggsii* H. C. Wats. ex Briggs). It is doubtful whether the two forms are genetically distinct. Similar plants are apparently wild throughout Europe, N. Africa, W. Asia and China, and have been introduced elsewhere. It is possible that their original home lies in N. and C. Europe.

The cultivated turnip is ssp. **rapa**, with a tuberous white-fleshed 'tap-root' lacking the epicotyledonary neck of *B. napus* var. *napobrassica*. It is occasionally found as an escape from cultivation.

The following four casuals, *all with ±stalked non-clasping upper lvs*, also belong to section *Brassica*:

*4. B. elongata Ehrh.

Biennial to perennial, to 1 m. high. *Lower lvs grass-green*, stalked, up to 20 cm., *elliptical in outline, sinuate or shallowly pinnatifid, densely covered with curved bristles*; *middle and upper lvs diminishing rapidly in size*, glaucous, narrowly oblong or lanceolate, narrowed into a long stalk-like base. Opened fls overtopping the buds. Sepals ±erect; petals pale yellow, 6–10 mm., twice as long as the sepals. Fr. obliquely erect, 1–2 cm., with a *stalk 1–3 mm.*, *above the sepal scars* and a *tapering seedless beak* 0·5–2 mm. Seeds c. 6–8 in each cell, brown, 1–1·5 mm. diam. Fl. 6–8. 2*n*=22. Ch.

Introduced. A casual in cornfields and near ports. S.E. Europe and the Near East. *B. elongata* resembles a *Diplotaxis* but has seeds in 1 row in each cell of its fr.

*5. B. fruticulosa Cyr.

Biennial or perennial, to 30 cm. high, ±*woody at the base*. *Lvs stalked, lyrate-pinnatifid* with a large rounded terminal segment, *very sparsely bristly*; *upper lvs diminishing rapidly in size, the uppermost linear*, ±*entire*. Opened fls overtopping the buds. *Sepals erect*; petals pale yellow, later whitish, twice as long as the sepals. *Fr.* almost erect, with *a stalk 1–3 mm.*, *above the sepal scars*, and with 0–2 seeds in the *ovate-lanceolate beak, 3–6 mm.* Seeds brown. Fl. 5–8. Ch.

Introduced. A rare casual, chiefly near ports. Mediterranean region.

*6. B. juncea (L.) Czern. *Sinapis juncea* L.

Annual, to 1 m. high, ±*glabrous*, with *numerous almost erect branches*. *Lower and middle lvs* stalked, up to 20 cm., usually lyrate-pinnatifid with a very *large ovate terminal segment* and only 1–2 small lateral segments on each side; upper lvs smaller, narrower, ±entire; all lvs ±glaucous. Open fls at about the same level as the buds. Sepals yellowish, half-spreading; petals pale yellow, 7–9 mm., almost twice as long as the sepals. *Fr.* ascending, 3–5 cm., *not stalked above the sepal scars, with a tapering seedless beak 5–10 mm., narrower than the stigma at its tip*. Seeds 8–12 in each cell, yellowish- or reddish-brown, 1·5 mm. diam. Fl. 6–8. 2*n*=36. Th.

Introduced. A casual in cornfields and near ports. Asia. Cultivated, as Indian or Chinese Mustard, for the oil from its seeds.

*B. integrifolia (West) O. E. Schulz, very like *B. juncea* but basal lvs often obovate entire and fr. with a shorter (4–6 mm.) broad flat beak, has been found as a casual in London.

*7. B. tournefortii Gouan

Annual. *Lower lvs runcinate-pinnatifid, densely bristly beneath, ciliate; upper lvs diminishing rapidly in size, the uppermost bract-like.* Sepals erect; *petals* pale yellow, later whitish, 5–8 mm., *only* 1·5 *mm. wide. Fr.* 3–7 cm., *not stalked above the sepal scars*, with a *beak 10–16 mm., as wide as the stigma at its tip*. Fl. 6–8. 2*n*=20. Th.

Introduced. A rare casual, chiefly near ports. Mediterranean region and Asia.

Section 2. *Melanosinapis* (DC.) Boiss. Sepals half-spreading. Ovary usually with 5–11 ovules. Siliquae short, 0·5–3 cm., with a slender seedless beak.

8. B. nigra (L.) Koch   Black Mustard.

Sinapis nigra L.

An annual herb with a slender tap-root and an erect shoot up to 1 m. high, bristly below, glabrous and glaucous above, with numerous ascending branches. *Lvs all stalked*; the lowest lyrate-pinnatifid with a large terminal lobe; grass-green, bristly, up to 16 cm.; middle lvs sinuate, and uppermost lanceolate or narrowly elliptical, entire, glabrous, ±glaucous. Infl. sub-corymbose. *Fl.-stalks shorter than the calyx.* Sepals half-spreading. Petals c. 8 mm., twice as long as the sepals, bright yellow. *Fr.* 12–20 mm., held *erect and appressed* to the stem on *stalks 2–3 mm.*, ±quadrangular, with *strongly keeled valves* and with a slender *beak, 1·5–3 mm., always seedless*. Seeds 2–5 in each cell, dark red-brown. Chiefly visited by Diptera. 2*n*=16. Th.

Var. *bracteolata* (Fisch. & Mey.) Spach has the 1–6 lowest fls of each raceme subtended by linear bracts, the stem-lvs hastate or 5-lobed, the fl.-stalk equalling the calyx, fr. up to 2·5 cm., and seeds 1·5 mm. diam.

Probably native. Apparently wild on cliffs by the sea, especially in S.W. England, and on stream-banks throughout England and Wales; but common, probably as an escape, in waysides and waste places. Only in S. Scotland and S. and E. Ireland. 70, H10, S. Widespread in C. and S. Europe and most regions with a temperate climate. Long cultivated for its seeds which yield the black mustard of commerce and also an oil used in medicine and soap-making. The original home of the species is unknown.

2. ERUCASTRUM Presl

Annual to perennial herbs with a tap-root and erect ascending stems, usually with numerous simple hairs. *Lower lvs lyrate-pinnatifid*, uppermost often linear. Infl. often with bracts, at least below. Sepals erect or spreading, the inner pair somewhat saccate; petals clawed, yellow; stamens 6, without appendages; stigma capitate or slightly 2-lobed. *Fr.* linear, ± *quadrangular*, *with keeled valves each with* 1 *prominent dorsal vein* and a lateral network; *beak* ± *conical*, with 0–3 seeds. *Seeds ovoid or ellipsoidal*, in 1 row in each cell. Sixteen spp. in C. and S. Europe, N. Africa and the Canary Is.

Basal pinnae of the upper stem-lvs downwardly directed and clasping the stem; infl. ebracteate; sepals spreading; fls bright yellow; fr. distinctly stalked above the sepal scars.   **1. nasturtiifolium**

Basal pinnae of the upper stem-lvs not clasping; infl. bracteate below; sepals erect; fls yellowish-white; fr. not stalked above the sepal scars.   **2. gallicum**

*1. E. nasturtiifolium (Poir.) O. E. Schulz

E. obtusangulum (Willd.) Rchb.; *Brassica Erucastrum* sensu Vill. et auct. plur., vix L. A biennial or perennial herb with an erect stem, 30–80 cm., densely covered at least below, with short white downwardly directed simple hairs. Basal lvs in a rosette, often falling before flowering, oblong-lanceolate in outline, pinnatifid with oblong lobes; *stem-lvs deeply pinnatifid with two basal lobes downwardly directed and clasping the stem like auricles*; uppermost smaller and less divided; all ± hairy. *Infl. ebracteate*, or the lowest fl. bracteate. *Sepals spreading. Petals bright yellow*, 8–12 mm., twice as long as the sepals. *Fr.* (Fig. 11A) 2–4 cm., *ascending, making an angle with their spreading stalks*, and *with a distinct stalk* (0·5–1 mm.) above the sepal scars; *beak* 3–4 mm., flattened, tapering, *usually with* 1 *seed*. Seeds red-brown, 1·3 × 0·7 mm. Fl. 5–9. Homogamous. Visited by bees and flies. $2n = 32$. H. (biennial)–Ch.

Introduced. A casual, found rarely near ports; C. and S.W. Europe.

*2. E. gallicum (Willd.) O. E. Schulz

Sisymbrium Erucastrum sensu Poll.; *Erucastrum Pollichii* Schimper & Spenner; *Diplotaxis bracteata* Godr.

An annual or biennial herb with an erect stem 20–60 cm., densely covered with short white downwardly directed hairs. Basal lvs lyrate with small lateral lobes; *stem-lvs deeply pinnatifid with rather distant oblong lobes, the basal pair not clasping the stem. Infl. usually bracteate below. Sepals almost erect; petals pale or whitish yellow*, 7–8 mm., twice as long as the sepals.

Fr. (Fig. 11 B) 2–4 cm., *curving upwards and continuing the direction of their stalks, not stalked above the sepal scars*; *beak* 3–4 mm., slender, *seedless*. Seeds red-brown, 1·3 × 0·7 mm. Fl. 5–9. $2n = 30$. Th.–Ch.

Introduced. A frequent casual in waste places and near ports, and occasionally establishing itself for a period. W. and C. Europe.

3. RHYNCHOSINAPIS Hayek

Herbs with a long slender tap-root and stems becoming woody below, glabrous or with simple hairs. Infl. ebracteate. *Sepals erect*; the inner pair distinctly saccate at the base; petals yellow; stamens 6, without appendages; ovary with 16–54 ovules; stigma somewhat 2-lobed. Fr. a linear siliqua with *convex valves each distinctly 3-veined* and a *flattened sword-shaped beak with 1–6 seeds. Seeds spherical, in 1 row in each cell.*

Eight spp., chiefly in W. Europe.

1 Plant glabrous, except for a few hairs on the sepals; stem ascending, with spreading branches from the basal rosette; few stem lvs.   **1. monensis**
   Stem hairy, at least below; lvs hairy, at least beneath; stem erect, branched above; several stem-lvs.   2

2 Stem with scattered hairs; lvs hairy beneath, rarely also above; sepals equalling or exceeding the fl.-stalks; ovary glabrous.   **3. cheiranthos**
   Stem densely hairy; lvs hairy on both sides; sepals shorter than the fl.-stalks; ovary hairy.   **2. wrightii**

1. R. monensis (L.) Dandy   'Isle of Man Cabbage.'

Sisymbrium monense L.; *Brassicella monensis* (L.) O. E. Schulz

A biennial herb with slender tap-root and ascending ± *glabrous stems*, 15–30 cm., branched at the base, the branches spreading-ascending. *Lvs glabrous* and glaucous, almost confined to a basal rosette, very deeply pinnatifid, the short oblong segments again lobed or coarsely toothed, the lobes bristle-pointed; *stem-lvs* 0–2, small, deeply pinnatifid. Infl. very lax. Fls about 18 mm. diam. Sepals hairy at the tip, equalling or exceeding the fl.-stalks. Petals pale yellow, twice as long as the sepals. Ovary glabrous. Siliquae 4–7 cm. × 2 mm., spreading, on spreading stalks 6–10 mm.; beak about ⅓ of the overall length, with up to 5 seeds. Seeds 1·3–2 mm., dark brown. Fl. 6–8. $2n = 24^*$. Hs. (biennial).

Native. Locally common on the W. coast of Great Britain from N. Devon and Glamorgan to Kintyre, and in the Isle of Man and Clyde Is. 18. Confined to the British Is.

2. R. wrightii (O. E. Schultz) Dandy   'Lundy Cabbage.'

Brassicella wrightii O. E. Schulz

A short-lived perennial herb with slender tap-root and erect stout *stems*, 20–90 cm., branched above, woody below, *densely covered with simple deflexed hairs*, often purple early in the year. Lvs all stalked and hairy; basal lvs c. 12 cm., lyrate-pinnatifid with the lowest segments often runcinate, the segments irregularly toothed or lobed; *stem-lvs several*, the lower pinnatifid, the uppermost linear, entire, glaucous. Infl. lax. Fls up to 25 mm. diam. *Sepals hairy, equalling or shorter than the fl.-stalks.* Petals yellow, twice as long as the sepals. *Ovary hairy.* Siliquae 6·5–8 mm., spreading or somewhat recurved on spreading stalks 10–12 mm.; beak about ⅓ of the overall length, with 1–3 seeds. Seeds 1·5 mm. diam., purple-black. Fl. 6–8. Much visited by beetles (*Meligethes* spp.), but self-fertile. Siliquae dehisce very late in the season. $2n = 24*$. Ch.–N.

Native. On cliffs and slopes of the E. side of Lundy Island; known nowhere else.

*3. R. cheiranthos (Vill.) Dandy   'Tall Wallflower Cabbage.'

R. Erucastrum Dandy, p.p.; *Brassica Erucastrum* L., sec. O. E. Schulz; *B. Cheiranthos* Vill.; *Brassicella Erucastrum* auct.

A usually biennial herb with slender tap-root and 1 or more erect *stems* 30–90 cm., branched above, ± *hispid below* with scattered spreading or deflexed hairs, glabrous and somewhat glaucous above. Basal and lower stem-lvs about 10 cm., stalked, lyrate-pinnatifid, with 3–5 long, narrow, coarsely toothed or lobed segments on each side, hispid on the margins and on the veins beneath, rarely also above, grass-green or slightly glaucous; middle stem-lvs with fewer and narrower segments, uppermost lvs lanceolate, ± entire. Fls c. 18 mm. diam. *Sepals equalling or exceeding the fl.-stalks,* usually hispid near the top. Petals pale yellow, almost twice as long as the sepals. *Ovary usually glabrous.* Siliquae 4–7 cm. × 2 mm. spreading or somewhat recurved on spreading stalks 6–10 mm.; beak about ⅓ of the overall length, 1–3-seeded. Seeds 1·3–2 mm., dark brown. Fl. 6–8. Homogamous. Visited by butterflies. $2n = 48$. Ch.

Introduced. A casual of waste places in S. and E. England and S. Wales, and established in Channel Is. 13, S. S. and C. Europe from Portugal to N.W. Germany and Italy.

4. SINAPIS L.

Usually annual herbs with slender tap-root and stems with simple hairs. Infl. ebracteate. *Sepals spreading,* non-saccate; petals bright yellow, *clawed, the claw shorter than the limb*; stamens 6, without appendages; *ovary with 4–17 ovules*; stigma somewhat 2-lobed. *Siliquae* linear *with convex valves, each distinctly 3–7 veined, and a long beak* with 0–9 seeds. Seeds spherical, in 1 row in each cell.

Ten spp., chiefly in the Mediterranean region.

1 Upper lvs lanceolate, toothed; beak of fr. conical, straight, rather more
        than half as long as the valves.   **1. arvensis**
    All lvs pinnately lobed or cut; beak strongly compressed, sabre-shaped,
        equalling or exceeding the valves.   *2*

2 Lvs lyrate-pinnatifid, the terminal lobe much larger than the laterals, all
        lobes coarsely toothed or sinuate; plant hairy.   **2. alba**
    Lvs twice pinnatifid, the primary lobes narrow, distant, the terminal
        equalling the laterals; plant glabrous or sparsely hairy. **3. dissecta**

1. S. arvensis L. Charlock, 'Wild Mustard.'

Brassica arvensis (L.) Rabenh., non L.; *B. Sinapistrum* Boiss.; *B. Sinapis* Vis.; *B. kaber* (DC.) L. C. Wheeler

An annual herb with slender tap-root and erect simple or branched stem 30–80 cm., usually stiffly hairy at least at the base, but sometimes glabrous. Lvs up to 20 cm., all roughly hairy; *lower lvs stalked, lyrate*, with a large very coarsely toothed terminal lobe and usually with a few smaller lateral lobes; *upper lvs sessile*, usually simple, lanceolate, coarsely toothed. Petals 9–12 mm. Fr. 25–40 × 2·5–3 mm., spreading; *the conical straight beak rather more than* ½ *as long as the valves*, often with 1 seed; valves glabrous or stiffly hairy but with glabrous beak in var. *orientalis* (L.) Koch & Ziz (*S. orientalis* L.), strongly 3(–5)-veined with 6–12 dark red-brown seeds in each cell. Fl. 5–7. Homogamous or protogynous. Freely visited by various flies and bees. $2n = 18$. Th.

Probably native. A weed of arable land, especially on calcareous and heavy soils. 112, H40, S. Throughout Great Britain and Ireland. Throughout Europe, N. Africa, S.W. Asia, Siberia. Introduced in N. and S. America, S. Africa, Australia and New Zealand. One of our most serious weeds, more troublesome in spring-sown than in autumn-sown crops. It may be controlled by spraying with acids, copper salts, or certain selective weed-killers.

*2. S. alba L. White Mustard.

Brassica hirta Moench; *B. alba* (L.) Rabenh.

An annual herb with a pale slender tap-root and erect simple or branched stem 30–80 cm., glabrous or, more commonly, with stiff downwardly directed simple hairs. *Lvs* up to 15 cm., usually *stiffly hairy*, all stalked, *all lyrate-pinnatifid or pinnate with the terminal lobe larger than the laterals*. Petals 10–15 mm. Fr. 25–40 × 3–4 mm., spreading; the *strongly compressed sabre-like often curved beak equalling or exceeding the valves*, narrowing upwards from a broad base, often 1-seeded; *valves* usually *stiffly hairy, strongly* 3-*veined*, with 1–4 yellowish or pale brown seeds in each cell. Fl. 6–8. Homogamous. The vanilla-scented fls are visited by flies and bees. $2n = 24$. Th.

Introduced. Grown as a green fodder crop or green manure as well as for the 'mustard' derived from its ground seeds; and naturalized as a weed of arable and waste land, especially on calcareous soils. 101, H37, S. Throughout Great Britain and Ireland. Probably native in the Mediterranean region and Near East, but introduced in N.W. Europe (to 67° N. in Norway), Japan, N. and S. America and New Zealand.

*3. S. dissecta Lag.

An annual herb closely related to *S. alba* from which it differs in its *twice pinnatifid lvs, not lyrate*, the terminal lobe being little or no larger than the long narrow pinnatifid lateral lobes; and in the *broader fr.* (4–7 *mm. wide*) with a gradually tapering beak. The *valves are less strongly hairy or* ±*glabrous*, and less strongly 3-veined; and the seeds more coarsely punctate.

Introduced. Occasional as a casual in arable or waste land. Mediterranean region.

*S. flexuosa Poir., like *S. alba* but with scabrid lvs and the stem covered with reflexed bristles, also occurs as a casual. It is native in Spain.

5. Hirschfeldia Moench

Annual or overwintering herbs, very hairy below, with terminal ebracteate racemes. Sepals almost erect, the inner pair slightly saccate; petals yellow or white; stamens 6, without appendages; ovary with 8–13 ovules; stigma capitate. *Fr* a siliqua *with a short swollen beak* containing 0–2 seeds; *valves distinctly 3-veined when young*, with a strong midrib and two weaker laterals, *but obscurely veined when ripe. Seeds ovoid, in 1 row in each cell.*

Two spp., one in the Mediterranean region and one in Socotra.

*1. H. incana (L.) Lagrèze-Fossat   'Hoary Mustard.'

Brassica incana (L.) Meigen., non, Ten.; *B. adpressa* Boiss.

An annual or overwintering herb with a slender tap-root and a simple or branching *stem* 30–100 cm., *densely covered below with short downwardly-directed stiff white hairs. Rosette- and lower stem-lvs grey with dense hairs*, stalked, deeply pinnately lobed or divided, with a large blunt broadly ovate terminal lobe, all lobes coarsely toothed; upper stem-lvs smaller, sessile, simple, narrowly lanceolate, glabrous or nearly so. Open fls overtopping the buds, their stalks shorter than the sepals; petals pale yellow, often with dark veins, 6–8 mm., twice as long as the sepals. *Fr.* 8–12 × 1–1·5 mm., *erect and appressed to the stem on short* (2–3 mm.) *club-shaped stalks*, with a *short basally swollen 0–2-seeded beak* about half as long as the valves; valves thick-walled, obscurely 3-veined, hairy or glabrous. Seeds 3–6 in each cell, reddish-brown, about 1 × 0·7 mm. Fl. 6–9. Visited by bees. $2n = 14$. Th. Resembles *Brassica nigra* in habit but is greyer owing to the dense white hairs.

Introduced. Native in the Mediterranean region and Near East, where it is a troublesome weed, and naturalized in France and in sandy places in Jersey and Alderney. A casual in parts of S. England, the Netherlands, S. Germany, California, Australia, New Zealand, etc.

6. Diplotaxis DC.

Annual to perennial herbs usually with pinnatifid lvs and terminal ebracteate racemes. Sepals somewhat spreading, the inner pair not or slightly saccate; petals clawed, yellow, white or lilac; stamens 6, without appendages; *stigma shortly 2-lobed. Fr. a long slender short-beaked siliqua with flattened 1-veined valves. Seeds numerous, ovoid, in two rows in each cell.*

Twenty-three spp., chiefly in C. Europe and the Mediterranean region but reaching India.

1   Fls white or pale lilac; beak of fr. conical.   **3. erucoides**
    Fls yellow; beak of fr. ±cylindrical.   2

2   Stem glabrous at base; lvs glaucous, the lower pinnatifid with long narrow
      lobes; fr. stipitate, i.e. distinctly stalked above the insertion-scars of the
      sepals.   **2. tenuifolia**
    Stem sparsely hairy towards base; lvs yellowish-green with short triangular
      teeth; fr. not stipitate.   **1. muralis**

*1. D. muralis (L.) DC.   Wall Rocket, Wall Mustard, Stinkweed.

Sisymbrium murale L.; *Brassica muralis* (L.) Huds.

An annual or biennial herb, occasionally perennial, with slender taproot and erect or ascending *stems*, 15–60 cm., branched from the base, *usually with sparse stiff hairs below. Lvs* up to 10 cm., elliptical-spathulate, *toothed or with triangular lobes up to twice as long as broad, yellowish-green*, foetid when crushed, narrowed gradually into a long stalk; lobes variable in depth, the terminal longest, all entire or with a few distant teeth. Lvs at first confined to a basal rosette, but these dying and replaced by stem-lvs in perennial plants (f. *caulescens* Kit.). *Fls lemon-yellow*, c. 10 mm. diam.; sepals half-spreading; *petals 5–8 mm., twice as long as the sepals. Fr.* (Fig. 15 B) 30–40 × 24 mm., narrowed at both ends, *not stalked above the insertion-scars of the sepals, ascending to make an angle with the much shorter spreading stalk*; beak slender, 2 mm., seedless; *stigma hardly broader than style. Fr.-stalk much shorter than fr.* Seeds yellow-brown, c. 1·2 × 0·6 mm. Fl. 6–9. Homogamous, fragrant. Visited by various flies and bees. $2n=42$. Th.

Introduced. Naturalized especially in S. England on limestone rocks and walls and as a weed of arable and waste land. By railways in Ireland. 73, H 30, S. S. and C. Europe.

*D. viminea (L.) DC., a small annual 10–30 cm. high, closely resembles *D. muralis* but has *petals only 3–4 mm., little longer than the sepals*, and fr.-stalks only ¼ as long as the ripe fruits (⅓–½ in *D. muralis*). It is sometimes found as a casual in S. England, but is native in the Mediterranean region.

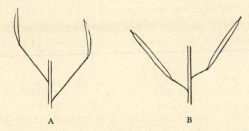

Fig. 15. Siliquae of A, *Diplotaxis tenuifolia* and B, *D. muralis.* × ½.

2. D. tenuifolia (L.) DC.   'Perennial Wall Rocket.'

Sisymbrium tenuifolium L.; *Brassica tenuifolia* (L.) Fr.

A perennial herb with long stout tap-root and erect branching stem, 30–80 cm., entirely glabrous, glaucous. Rosette-lvs 0; *lower stem-lvs* narrowed into a stalk-like base, *deeply and narrowly pinnatifid with ± linear lobes never less than 3 times as long as broad*, entire or with a few coarse teeth, the terminal lobe longer but little wider than the laterals; upper lvs less deeply divided or almost entire, linear-lanceolate; all glabrous, *glaucous* and foetid when crushed. Fls lemon-yellow; outer sepals half-spreading, inner erect; *petals 8–15 mm.,* twice as long as the sepals. *Fr.* (Fig. 15 A) 25–35 × 2 mm., *conspicuously stalked* (1–3 mm.) *above the insertion-scars of the sepals; stigma much broader than style; fr.-stalk almost as long as fr*; beak 2–2·5 mm., slender, seedless. Seeds

as in *D. muralis*. Fl. 5–9. Homogamous, fragrant. Visited by various insects.
$2n = 22*$. H.–Ch.

Doubtfully native. On old walls and in waste places in S. England. 53, S.
A casual further north and in Scotland and Ireland. Native in S. and C.
Europe.

*3. D. erucoides (L.) DC.   'White Wall Rocket.'

Sinapis erucoides L.; *Brassica erucoides* (L.) Boiss., non Hornem.

An annual or overwintering herb with slender tap-root and erect stem
30–50 cm., branching from near the base, hairy or glabrous. *Lvs almost confined to a basal rosette* in the first season but lost and replaced by stem-lvs in
overwintering plants; all usually pinnately lobed or divided with a large
terminal lobe, the *lobes with coarse and irregular whitish horny-tipped teeth*;
basal lvs 5–15 × 1·5–3 cm.; stem-lvs smaller, often ± hastate and somewhat
amplexicaul; uppermost bract-like. *Fls white, becoming lilac*, 15 mm. diam.;
sepals almost erect; petals 7–11 mm., twice as long as the sepals. Fr. 25–
40 × 2 mm., ascending to make an angle with the much shorter spreading
stalk; *beak conical*, 2–4 mm., usually seedless. Seeds as in *D. muraiis*. Fl. 5–9.
$2n = 14$. Th.

Introduced. Naturalized as a weed of arable and waste land in S. England
and frequent as a casual. Mediterranean region; a serious weed in S. Europe.

*D. catholica (L.) DC. is found occasionally as a casual. It resembles *D. erucoides*
in the *conical beak* of its fr. but has *yellow fls* and its *stem lvs are narrowed at the
base*, not hastate or amplexicaul. $2n = 18$.
Native in Spain, Portugal and Morocco.

*D. tenuisiliqua Del. (*D. auriculata* Durieu), also a rare casual, has yellow fls and
ovate to oblong-lanceolate deeply cordate and amplexicaul stem lvs. Native in N.
Africa.

7. ERUCA Mill.

Annual to perennial herbs usually with pinnatifid lvs and terminal ebracteate
racemes. Sepals ± erect, the inner pair somewhat saccate at the base; *petals
long-clawed, usually yellowish with violet veins*; stamens 6, without appendages;
stigma strongly 2-lobed. Fr. a short *siliqua with strongly 1-veined valves* and a
broad flat seedless beak. Seeds spherical or ovoid, *in 2 rows in each cell*.

Five spp. in the Mediterranean region.

*1. E. sativa Mill.

Brassica Eruca L.

An annual or overwintering *foetid* herb with slender tap-root and an erect
usually stiff hairy stem, 10–60 cm., ± branched above. Lower lvs stalked,
upper ± sessile; all lyrate-pinnatifid or rarely pinnate, with a large oblong or
obovate terminal lobe and 2–5 narrow laterals on each side, all coarsely
toothed or lobed, rarely entire. *Petals* 12–20 mm., *pale yellow or whitish, with
deep violet veins*, twice as long as the sepals. *Fr.* (Fig. 11 c) 12–25 × 3–5 mm.,
erect and ± appressed to the stem on short erect stalks, with a *sabre-shaped
beak about half as long as the valves*. Seeds 1·5–2 mm., yellow-brown or
reddish. Fl. 5–8. Homogamous. $2n = 22$. Th.

A very variable species. In var. *cappadocica* (Reut.) Thell. (*E. cappadocica* Reut.) the fr. is 4·5–6 mm. wide with beak about one-third as long as the valves, and the seeds are 2·5–3 mm.

Introduced. A frequent casual, sometimes establishing itself for a period. Mediterranean region and E. Asia. Long cultivated as a salad plant and for the medicinal oil expressed from its seeds.

*E. vesicaria (L.) Cav., differing in the calyx ± persistent in fr. and the narrower beak about as long as the valves, is a rare casual. Spain and N. Africa.

8. RAPHANUS L.

Annual to perennial herbs somewhat glaucous above, stiffly hairy below, with lyrate-pinnatifid lvs and ebracteate terminal racemes. Sepals usually erect; petals long-clawed; stamens 6, without appendages; stigma capitate, somewhat 2-lobed. *Fr. elongated, jointed, the lowest joint* (corresponding with the

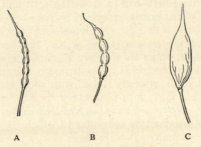

Fig. 16. Fruits of *Raphanus*. A, *R. raphanistrum*;
B, *R. maritimus*; C, *R. sativus*. × ½.

valves of a typical siliqua) *short, slender, seedless, resembling a short stalk; the upper part of the fr. either ± cylindrical and indehiscent, or constricted between the seeds and then often breaking into 1-seeded joints at maturity;* the apex narrowing into a seedless beak. Seeds ± spherical or ovoid, reddish-brown, pitted.

Eight spp. chiefly in the Mediterranean region.

1   Tuberous 'tap-root'; fr. not markedly constricted between the seeds.
                                                                    3. sativus
    Tap-root not tuberous; fr. markedly constricted between the seeds.   2
2   Fr. breaking readily into 1-seeded joints; beak up to 5 times as long as the
        top joint.   **1. raphanistrum**
    Fr. not readily breaking into 1-seeded joints; beak not more than twice
        as long as the top joint.   **2. maritimus**

1. R. raphanistrum L.   Wild Radish, White Charlock, Runch.

An annual herb with a slender whitish tap-root and an erect simple or branched stem, 20–60 cm., rough with spreading or reflexed bristles, especially below, and somewhat glaucous above. *Lower lvs lyrate-pinnatifid with a large rounded terminal lobe and usually 1–4 pairs of much smaller distant laterals*; upper lvs smaller, oblong, pinnately lobed or toothed; all grass green and ± bristly. Petals 12–20 mm., twice as long as the sepals, yellow, lilac or white, usually dark-veined, but golden yellow and unveined in var. *aureus* Wilmott. *Fr. 3–6 mm. diam. with long but not very deep constrictions between the 3–8 seeds, firm-walled, breaking readily into weakly ribbed 1-seeded joints; beak slender,* reaching 4–5 times the length of the uppermost joints (Fig. 16 A). Seeds round-ovoid, 1·5–3 mm. diam. Fl. 5–9. Homogamous. Visited especially by bees and flies. $2n=18$. Th.

Doubtfully native. A common and troublesome weed, especially of non-calcareous soils. 111, H 37, S. Throughout Great Britain and Ireland. Var. *aureus* in N. Scotland and Hebrides; Isle of Man. Found throughout Europe, N. Africa, Australia, N. and S. America and Japan. The yellow-fld form is commoner in the northern and the white-fld in the southern part of the range.

2. R. maritimus Sm.   'Sea Radish.'

A biennial or perennial herb with a stout tap-root and an erect simple or branched stem, 20–80 cm., bristly, especially above. *Lower lvs lyrate-pinnatifid or interruptedly pinnate with a large terminal lobe and usually 4–8 pairs of contiguous lateral lobes, often alternating in size, the basal lobes downwardly directed and very small*; upper lvs smaller; all *dark green*. Petals c. 20 mm., usually yellow (white in Channel Is.), less distinctly veined than in *R. raphanistrum*. *Fr. 5–8 mm. diam.*, strongly ribbed with short but deep constrictions between the 1–5 *seeds*, fairly firm-walled, *not readily breaking into 1-seeded joints; beak* slender, *rarely more than twice as long as the top joint* (Fig. 16 B). Fl. 6–8. The fr. are dispersed by sea-water in which they will float for 7–10 days without loss of viability. $2n=18$. Hs.

Native. A plant of the drift-line and cliffs on sandy and rocky shores. 37, H 15, S. From Argyll and Durham southward; Hebrides; all round the Irish coast, and in the Channel Is. Native along the Atlantic coast of Europe from the Netherlands to N. Spain, and on the coasts of the Mediterranean and Black Seas.

*R. landra Moretti ex DC., which resembles *R. maritimus* but has non-contiguous and irregularly spaced lateral lobes to the basal lvs, petals not more than 15 mm., and a beak at least twice as long as the top joint, occurs as a casual near ports. It is close to *R. maritimus* and *R. sativus*, and the latter is perhaps derived from hybrids between the other two species.

*3. R. sativus L.   Radish.

An annual or biennial bristly herb with a tuberous white or brightly coloured 'tap-root' and an erect simple or branched stem, 20–100 cm. *Petals white, lilac or violet*, rarely mixed purple and yellow, never pure yellow; usually with dark veins. *Fr. inflated, up to 15 mm. diam., hardly or irregularly constricted between the 6–12 seeds and not breaking into 1-seeded joints; wall of fr. ± spongy*; beak long-conical (Fig. 16 C). Seeds c. 3 mm. diam. Fl. 6–8. Homogamous. Visited by bees, flies, etc. $2n=18, 36$. Th.–Hs.

Introduced. A cultivated plant of doubtful origin and nowhere found wild but not infrequent as an escape from gardens in the British Is. and elsewhere.

9. CRAMBE L.

Annual to perennial cabbage-like herbs, often woody below, with swollen tap-roots and large entire or pinnately lobed lvs, and with white fls in large much-branched racemose infls. Sepals spreading, the inner hardly saccate; petals short-clawed; *stamens 6, the inner with toothed appendages*; stigma capitate, sessile. *Fr. indehiscent, 2-jointed, the lower joint slender, stalk-like, seedless; the upper ± spherical or ovoid, 1-seeded.*

Twenty spp. in C. Europe, Mediterranean region, N. Tropical Africa, and W. Asia. Often very large herbs.

1. C. maritima L.   Seakale.

A perennial herb with branched fleshy root-stock sprouting readily after burial under shingle. Stem erect, 40–60 × 2–3 cm., branching below, glabrous. *Lower lvs up to 30 cm., ovate, long-stalked, glaucous, glabrous,* ± pinnately lobed, with irregular toothed and wavy margins; upper lvs narrow, the upper-most very narrow, entire and bract-like. Partial infl. corymbose. Fls 10–16 mm. diam.; *petals 6–9 mm., white with green claws.* Fr. 12–14 × 8 mm., ascending to make an angle with the spreading stalk 20–25 mm. Fl. 6–8. Fr. dispersed by sea-water in which it will float for many days without loss of viability. $2n=30$; $60*$. Grh. or Hs.

Native. On coastal sands, shingle, rocks and cliffs. 41, H8, S. From Fife and Islay southwards and in Ireland. Often a plant of the drift-line. Generally distributed along the Atlantic coast of Europe from Oslo Fjord to N. Spain, along the Baltic coast and round the Black Sea, but not round the Mediter-ranean Sea. Cultivated for its asparagus-like shoots, blanched by darkening the sprouting stocks.

10. RAPISTRUM Crantz

Annual to perennial herbs with stiff bristly hairs, pinnately lobed or cut lvs, and fls in terminal ebracteate branching racemes. Sepals half-spreading, the inner pair slightly saccate; petals short-clawed; stamens 6, without appendages; stigma capitate, weakly 2-lobed. *Fr. of 2 joints*; the lower ± slender, with 0–2 seeds and therefore variable in diam. and length, separated by a con-striction from the *larger upper joint which is usually 1-seeded and falls at maturity.* Seeds ovoid, that in the upper joint larger than that in the lower joint.

Three spp. in C. Europe, Mediterranean region and W. Asia, widely intro-duced as weeds of arable and waste land.

1   Fr. with upper joint narrowing gradually into a broadly conical beak
     0·5–1 mm.; perennial.   **1. perenne**
     Upper joint of fr. surmounted by the slender persistent style 1–5 mm.;
     annual.   *2*

2   Fr.-stalk 1–3 times as long as the lower joint; upper joint strongly rugose
     and ribbed, narrowing suddenly into the slender style.   **2. rugosum**
     Fr.-stalk 2–4 times as long as the lower joint; upper joint hardly rugose,
     slightly ribbed, narrowing gradually into the slender style.
                                            3. hispanicum

*1. R. perenne (L.) All.

A *biennial or perennial* herb with a deep stout tap-root and 1 or several branching stems, 30–80 cm., covered below with dense downwardly directed stiffish white hairs, but glabrous above. *Lower lvs* 10–15 × 3–6 cm., hairy, stalked, *with about 6 pairs of lateral lobes* smaller towards the base, and a terminal lobe somewhat larger than the adjacent laterals, all irregularly and coarsely toothed, the teeth horny-tipped; *upper lvs* glabrous, *short-stalked to sessile*, less deeply lobed or merely toothed. Infl. branched, much lengthening in fr. *Petals* twice as long as the sepals, *bright yellow* with darker veins. Fr.-stalk up to twice as long as the lower joint of the fr. Fr. (Fig. 11 F) 7–10 mm.; *upper joint* ovoid, strongly ribbed, *narrowing gradually into the short broadly conical beak*, 0·5–1 mm.; lower joint narrower, usually 1-seeded. Fl. 6–8. Homogamous. G.

Introduced. Naturalized as a weed of arable land and waste places in some localities; more frequent as a casual. Native of C. and S.E. Europe and perhaps of Siberia, but introduced in France, Belgium, Holland, Germany, etc., as well as Great Britain. A steppe species.

*2. R. rugosum (L.) All.

An *annual* or rarely biennial herb with slender tap-root and simple or branched stem 15–60 cm., ± glaucous, stiffly hairy at least below. Basal and *lower lvs* stalked, pinnately lobed *with* a large terminal lobe and *about 3 pairs of much smaller laterals*, irregularly and coarsely toothed, the teeth inconspicuously horny-tipped; *upper lvs* narrower less deeply lobed or merely toothed, *narrowed into a stalk-like base*; all dark green, stiffly hairy at least beneath. Infl. branched, much lengthening in fr. *Petals* twice as long as the sepals, *lemon-yellow* with darker veins. Fr.-stalk ± erect, 1–3 times as long as the lower joint of the fr. Fr. (Fig. 11 G) 3–10 mm., hairy or becoming glabrous; lower joint cylindrical, seedless or with 1(–2) seed; *upper joint* 1-*seeded, ovoid, strongly rugose and ribbed, not more than twice as long as the slender style into which it narrows suddenly*. Fl. 5–9. Homogamous. ?Self-sterile. $2n = 16$. Th.

The plants most commonly found in Great Britain are referable to ssp. **rugosum**, with fr.-stalk 1–1½ times as long as the usually 1-seeded lower joint and with the upper joint ovoid.

Sometimes found as a casual is ssp. **orientale** (L.) Rouy & Fouc. (*R. orientale* (L.) Crantz), with the fr. stalk 1½–3 times as long as the slender seedless lower joint and with the upper joint spherical and very strongly rugose.

Introduced. Naturalized as a weed of arable and waste land in some localities, and frequent as a casual. Native in the Mediterranean region but widely naturalized in C. Europe, N. and S. America, S. Africa, Australia and New Zealand.

*3. R. hispanicum (L.) Crantz (*R. rugosum* ssp. *linnaeanum* (Boiss. & Reut.) Rouy & Fouc.) differs from *R. rugosum* in the longer fr.-stalk, 2–4 times as long as the slender seedless lower joint, and the more nearly spherical and smooth or slightly ribbed upper joint which narrows more gradually into a style of variable length; fr. (Fig. 11 H) glabrous or very hairy (var. *hirsutum* (Cariot) O. E. Schulz).

Introduced. A rare casual, native in the Mediterranean region, from Portugal to Greece. Var. *microcarpum* (Loret) O. E. Schulz, with fr. only 1·5–2 mm. diam., also occurs as a casual.

11. CAKILE Mill.

Annual herbs with *glabrous succulent* simple or pinnately divided *lvs* and fls in terminal and axillary ebracteate racemes lengthening in fr. Sepals erect, the inner pair somewhat saccate; *petals* clawed, *violet, pink or white*; stamens 6, without appendages; stigma small, capitate. *Fr. indehiscent, of 2 unequal 1-seeded joints; the upper larger, ovoid, narrowing upwards; the lower smaller, top-shaped, ±flattened. At maturity the upper joint breaks off*, the lower remaining attached to the plant.

Four spp.; three on sandy shores throughout Europe, N. Africa, W. Asia, N. America and Australia and one in C. Arabia.

1. C. maritima Scop.   Sea Rocket.

Bunias Cakile L.

An annual herb with a very long slender tap-root and a prostrate or ascending branched stem, 15–45 cm. Lower lvs 3–6 cm., narrowed into a stalk-like base, ±entire, obovate or oblanceolate to deeply pinnate-lobed, the lobes oblong, distant, entire or distantly toothed; upper lvs less lobed or entire, sessile. Infl. dense, many-fld, terminating the main stem and branches. Petals 6–10 mm. purple, lilac or white, twice as long as the sepals. *Fr. 10–25 mm. overall* on short thick stalks 2–5 mm.; upper joint up to twice as long as the lower, usually mitre-shaped with two broadly triangular basal teeth fitting over the convex top of the lower joint, which usually has two lateral projections just below the join; lower joint sometimes seedless, and then small and stalk-like. Seeds yellow-brown, smooth. Fl. 6–8. Homogamous. Visited by various flies, bees and other insects. Fr. dispersed by floating in sea-water without loss of viability. $2n=18^*$. Th.

The above description is of ssp. **integrifolia** (Hornem.) Hyland. Plants from N. Scotland, Shetland and the Outer Hebrides commonly have ±entire lvs and the lower joint of the fr. without lateral teeth, but intermediates occur. These extreme forms have been referred to *C. edentula* (Bigel.) Hook., but it is doubtful if this is correct.

Native. A drift-line plant in sandy and shingly places all round the British Is. 69, H19, S.

Ssp. *integrifolia* on Atlantic coasts of Europe; other sspp. in the Baltic, Mediterranean and Black Seas.

C. edentula in N. America, Iceland and Arctic Russia.

12. CONRINGIA Adans.

Annual or overwintering herbs, *glabrous and glaucous*, with rounded lvs, entire and with transparent hoary margins. Basal lvs shortly stalked, middle and upper lvs sessile, clasping the stem with rounded basal lobes. Infl. a terminal ebracteate raceme. Sepals erect, the inner pair somewhat saccate; style short, stigma slightly 2-lobed. *Fr. a long slender 4- or 8-angled siliqua*; valves with strong midrib. Seeds dark brown in 1 row in each cell.

Six spp. in the Mediterranean region, C. Europe and C. Asia.

Petals pale yellowish-white; fr. 4-angled.	**1. orientalis**
Petals lemon yellow; fr. 8-angled.	**2. austriaca**

*1. C. orientalis (L.) Dum.   'Hare's-ear Cabbage.'

Brassica orientalis L.; *Erysimum orientale* (L.) Cr., non Mill.; *E. perfoliatum* Cr.
An annual or overwintering herb with stout tap-root and usually simple erect stem 10–50 cm. Basal lvs long-obovate, blunt; *stem-lvs* 2·5–4 cm., obovate-oblong, blunt, with a *broadly clasping base*. Fls 10–12 mm. diam. *Petals yellowish- or greenish-white*. Siliquae (Fig. 12E) 6–10 cm. × 2–2·5 mm., curving upwards on spreading stalks 6–12 mm.; valves with prominent midrib so that the siliqua is 4-*angled*. Seeds ovoid, dark brown, 2–2·5 mm. Fl. 5–7. Homogamous. Visited by hover-flies and Lepidoptera. $2n=14$. Th.
   Introduced. A frequent casual of arable and waste land and of cliffs by the sea, especially on calcareous and clayey soils, and occasionally established. Probably native in the E. Mediterranean region, but widely introduced in Europe and N. Africa.

*2. C. austriaca (Jacq.) Sweet

Erysimum austriacum (Jacq.) Roth
An annual or overwintering herb with slender tap-root and simple erect stem 10–100 cm. high. Basal lvs long-obovate, blunt; stem-lvs ovate, blunt, with a narrow clasping base. Fls 6–8 mm. diam. *Petals lemon yellow*. Siliquae (Fig. 12F) 5–8 cm. × 2·5–3 mm., held almost erect on stiffly erect stalks 4–5 mm.; valves 3-veined, with midrib and two lateral veins prominent, so that the siliqua is 8-*angled*. Seeds dark-brown, c. 3 mm. Fl. 5–7. Th.
   Introduced. A casual of waste and arable land and of cliffs by the sea. Native in S.E. Europe, Asia Minor and the Caucasus.

13. LEPIDIUM L.

Annual to perennial herbs, sometimes woody below, often with simple hairs. Fls small, whitish, in dense terminal ebracteate racemes. Sepals non-saccate; petals sometimes shorter than the sepals or 0; stamens 2, 4 or 6, without appendages; ovary with 2 ovules, style short or 0, stigma capitate, sometimes slightly 2-lobed. Fr. an angustiseptate silicula, the valves strongly keeled or winged. Seeds usually 1 in each cell, hanging from the apex.
   About 130 spp., cosmopolitan.

1   Fr. equalling or longer than its stalk, broadly winged above, the wing
         involving the lower part of the style.   2
      Fr. shorter than its stalk, not winged or with the style free from the
         narrow wing.   6

2   Upper stem-lvs not clasping the stem.   **1. sativum**
      Upper stem-lvs clasping the stem.   3

3   Fr. densely covered with small scale-like vesicles; style usually not projecting beyond the apical notch of the wing; annual or biennial.
                                                                     2. campestre
      Vesicles on fr. few or 0; style projecting beyond the apical notch of the
         wing; perennial.   4

4   Fr. hairy, at least when young.   **5. hirtum**
      Fr. glabrous, even when young.   5

5	Fr.-stalks glabrous.	**4. pratense**
	Fr.-stalks hairy.	**3. heterophyllum**
6	Upper stem-lvs broad, clasping the stem, strikingly different from the lower stem-lvs.	**11. perfoliatum**
	Upper stem-lvs not clasping the stem, not strikingly different from the lower stem-lvs.	7
7	Fr. with style not projecting beyond the top of the apical notch.	8
	Fr. not deeply notched, with style projecting beyond the top of the notch.	11
8	Fr. not more than 2 mm. wide.	**6. ruderale**
	Fr. 2–3 mm. wide.	9
9	Petals longer than the sepals.	**7. virginicum**
	Petals shorter than the sepals or 0.	10
10	Upper stem-lvs entire.	**9. neglectum**
	Upper stem-lvs distantly toothed.	**8. densiflorum**
11	Sepals broadly white-margined from the base; fr. orbicular, 2 × 2 mm.	**10. latifolium**
	Sepals narrowly white-margined above the middle; fr. ovoid, 2·5–4 × 1·5–3 mm.	**12. graminifolium**

Section 1. *Cardamon* DC. Cotyledons deeply 3-lobed, the central lobe longest; fr. winged, longer than the ascending fr.-stalk; plant glaucous above.

*1. L. sativum L.   Garden Cress.

An annual herb with pale slender tap-root and a single erect stem 20–40 cm. Basal lvs soon falling, long-stalked, lyrate with toothed obovate lobes; stem-lvs once or twice pinnate; *uppermost leaves* sessile, *not clasping stem*, linear, acute, entire. Petals white or reddish, twice as long as the sepals. Stamens 6. Fr.-stalk little more than half as long as the fr., ascending; ripe siliculae 5–6 × 3–4 mm., broadly elliptical or nearly orbicular, narrowly winged above, with deep apical notch; style not projecting beyond the notch. Fl. 6–7. Protogynous. Fragrant and visited by small insects. $2n = 16$. Th.

Introduced. Occasionally established. Cultivated as a salad plant all over the world. Wild forms seem native in Egypt and W. Asia.

Section 2. *Lepia* DC. Fr. broadly winged above, the wing involving the lower part of the style; fr.-stalk horizontal, shorter than or equalling the fr.; middle and upper stem lvs clasping the stem.

2. L. campestre (L.) R.Br.   Pepperwort.

Annual or biennial herb with pale slender tap-root and a single erect stem 20–60 cm., usually grey-green with dense short spreading hairs, branched above the middle, the branches curving upwards. Basal lvs entire or lyrate, falling before the fls open; lower stem-lvs narrowed into a short stalk; middle and upper stem-lvs narrowly triangular, sessile, clasping the stem with long narrow pointed basal lobes; all lvs softly hairy with small distant marginal teeth. Fls inconspicuous, 2–2·5 mm. diam. Petals white, little longer than the sepals. Stamens 6, anthers yellow. Fr.-stalks hairy; *ripe siliculae* 5 × 4 mm., *densely covered with small white vesicles* becoming scale-like when dry; *style* c. 0·5 mm., *not or slightly projecting* beyond the apical notch of the fr. Seeds 2·5 mm. (Fig. 17A). Fl. 5–8, slightly protogynous. $2n = 16$. Th.

Native. In dry pastures, on walls and banks, by waysides and in arable and waste land. 92, H11, S. Throughout Great Britain from Moray and Lanark southwards, rare in Scotland and Ireland. Throughout Europe and in Asia Minor and the Caucasus. Introduced in N. America.

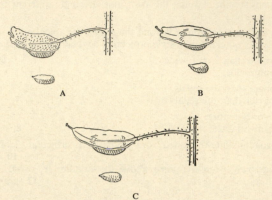

Fig. 17. Siliquae and seeds of *Lepidium*. A, *L. campestre*; B, *L. heterophyllum*; C, *L. heterophyllum* var. *alatostylum*. × 2·5.

3. L. heterophyllum Benth.   'Smith's Cress.'

L. Smithii Hook.; *L. hirtum* Sm., p.p.

A *perennial* herb with a stout woody rootstock and numerous ascending stems 15–45 cm., grey-green with short spreading hairs. Stems often branched below as well as above the middle, the branches curving upwards. Basal lvs oblanceolate or elliptical, falling before the fls open; lower stem-lvs narrowed into a short stalk; middle and upper *stem-lvs* narrowly triangular, sessile, *clasping the stem* with long narrow pointed basal lobes. Fls 3–3·5 mm. diam. Petals white, half as long again as the sepals. Stamens 6, with violet anthers. *Fr.-stalks hairy; siliculae* 5 × 4 mm. *with few or no vesicles*; the fr. wing accounts for about one-third of the total length of the fr.; *style* at least 1 mm., *projecting* beyond the apical notch of the fr. Seeds 2 mm. (Fig. 17 B). In var. *alatostylum* (Towns.) Thell. the fr. has no apical notch and the fr. wing in consequence involves a greater length of the style (Fig. 17 C). The lvs vary greatly in hairiness. The Pyrenean plant is almost glabrous and the British plant is often referred to *L. heterophyllum* var. *canescens* Godr. Fl. 5–8. 2n = 16*. H.

Native. In arable fields, on dry banks and waysides. 99, H27, S. Throughout Great Britain from Moray and Dunbarton southwards; Ireland, except the north-west. Native only in W. Europe (Spain, Portugal, France and the British Is.), but introduced in Belgium and Germany and in N. America and New Zealand.

*4. L. pratense Serres (*L. heterophyllum* var. *pratense* (Serres) F. W. Schultz).

Distinguishable from *L. heterophyllum* by the long-stalked broadly ovate entire basal lvs and glabrous fr.-stalk. The fr.-wing accounts for about one-quarter of the total length of the fr.

Introduced. A casual, native in the French Alps.

*5. L. hirtum (L.) Sm.

A W. Mediterranean sp. resembling *L. heterophyllum* but with hairy not glabrous fr., whose wing accounts for about one-half the total length of the fr.

Introduced. A casual.

Section 3. *Dileptium* (Raf.) DC. Fr. either not winged, or, if narrowly winged above, with the style quite free; style not exceeding the apical notch; fr.-stalk ascending, equalling or longer than the fr.; stamens 2–4; middle and upper stem-lvs narrowed at the base, not clasping the stem.

6. L. ruderale L.   'Narrow-leaved Pepperwort.'

An annual or biennial ± *foetid* herb with pale slender tap-root and a single erect or ascending stem 10–30 cm., *almost glabrous or with sparse short spreading hairs.* Stem branched above, the branches curving upwards. Basal lvs 5–7 cm., long-stalked, deeply pinnately divided, the narrow segments often again pinnately divided or lobed; lower stem-lvs pinnate with narrow entire segments; middle and upper *stem-lvs* sessile, simple, narrowly oblong up to 20 × 2 mm., blunt, *with entire margins.* Fls inconspicuous, greenish. *Petals usually* 0. Stamens 2, occasionally 4. Fr.-stalks hairy, about half as long again as the fr.; *siliculae* 2–2·5 × 1·5–2 mm., ovate or broadly elliptical, *deeply notched* above; valves sharply keeled and narrowly winged above; style very short, at the base of the notch. Fl. 5–7. Automatically self-pollinated. $2n = 32^*$. Th.–H.

Native. In waste places and by waysides, generally near the sea. 66. Throughout England but especially in E. Anglia. Rare and doubtfully native in Scotland; apparently absent from Ireland and Channel Is. Native throughout Europe except the far north, and in S.W. Asia. Introduced in N. Africa, N. America and Australia.

*7. L. virginicum L.

Annual or biennial herb with single erect stem 30–50 cm., ± downy with long downwardly curved, *appressed, simple hairs* on raised bases. Basal lvs to 8 cm., rough with short curved bristles, lyrate or pinnate with large rounded terminal lobe and numerous much smaller lateral lobes; middle and upper *stem-lvs* simple, with *sharply toothed* and hairy margins, the uppermost only 15 × 2 mm. Fls inconspicuous, white. *Petals up to twice as long as the sepals* in the first opening fls, becoming shorter. Stamens 2–4. Fr.-stalks up to half as long again as the fr.; ripe *siliculae* 3–4 × 2·5–3 mm., orbicular *with a broad but shallow apical notch*; valves sharply keeled below, narrowly winged above, shining when ripe; style very short, at the base of the notch. Fl. 5–7. Th.–H.

Introduced. A casual, native in N. America, but introduced in many parts.

*8. L. densiflorum Schrad.

An annual or biennial herb with single erect stem, branched above, grey with dense short *upwardly curved or spreading simple hairs* 3–5 times as long as wide. Basal lvs long-stalked, elliptic in outline and usually deeply toothed but sometimes pinnately lobed; *upper stem-lvs* sessile, linear-lanceolate, with conspicuous lateral veins, *distantly toothed* and hairy at the margin; all lvs grey-green with short stout hairs. *Petals* filiform, *shorter than sepals, or* 0. Stamens 2–4. Fr.-stalks ascending, about as long as the fr.; *siliculae* 3–4 × 2·5–3 mm., ovate or orbicular, *with a narrow and shallow apical notch*; valves sharply keeled below, narrowly winged in the upper third; style very short, at the base of the notch. Fl. 5–7. $2n=32$. Th.–H.

Introduced. A casual, native in N. America, but introduced in many parts of the world.

*9. L. neglectum Thell.

Annual or biennial herb with a general resemblance to *L. densiflorum* and *L. virginicum*. Stem ± downy with *short straight spreading ± papillose hairs* 2–3 times as long as wide. Lower stem-lvs hairy; *upper stem-lvs* linear, 1-veined, *entire*, ciliate with short stout hairs. *Petals rudimentary.* Stamens 2–4. Fr.-stalk equalling or slightly longer than the fr.; *siliculae* 3×3 mm., orbicular, narrowly winged above, *with a deep* but narrow apical *notch and very short style*. Fl. 5–6.

Introduced. A casual, occasionally established. N. America, introduced in many parts of the world.

Section 4. *Lepidium.* Fr. not or very slightly winged, not or slightly emarginate; style short but usually longer than the notch; fr.-stalks slender, equalling or exceeding the fr.; stamens 6.

10. L. latifolium L.   Dittander, 'Broad-leaved Pepperwort.'

Perennial herb with thick branched rootstock from which subterranean stolons arise; from each branch of the rootstock a single erect stem 50–130 cm., glabrous, much branched above. Basal lvs to 30 cm., long-stalked, simple and ovate with toothed margin or pinnately lobed with large terminal and 2 or more smaller lateral lobes, the lobes all rounded; lower stem-lvs like the basal but shorter stalked; middle and upper stem-lvs $5–10 \times 1–2$ cm., sessile, ovate or ovate-lanceolate, acute, entire or with distant teeth; uppermost lvs bract-like, white-margined near the apex. Infl. a large dense pyramidal panicle. Fls 2·5 mm. diam. *Sepals broadly white-margined.* Petals white, up to twice as long as the sepals. Fr.-stalks very slender, twice as long as the fr.; *siliculae* 2×2 mm., broadly elliptical or orbicular, *with no or very slight apical notch*; valves somewhat keeled but not winged; style very short, with large rounded stigma. Fl. 6–7. $2n=24$. H.

Native. In salt-marshes and wet sand. 26, H 5, S. N.E. England and from Norfolk and Wales southwards; S. Ireland. Throughout Europe except the far north, N. Africa, S.W. Asia. Formerly cultivated as a condiment.

*11. L. perfoliatum L.

An annual or biennial herb with single erect stem 20–40 cm., sparsely hairy, usually branched above. *Basal lvs* to 10 cm., long-stalked, *bipinnate*, the ultimate segments less than 1 mm. wide; middle and *upper stem-lvs* $1–1·5 \times 1–1·5$ cm., *broadly ovate*, acute, *entire, clasping the stem* with large basal lobes. Petals pale yellow, half as long again as the sepals. Fr.-stalks ascending, glabrous, equalling the fr.; siliculae variable in shape but usually about as broad as long, 3–4 mm.; valves keeled below, very narrowly winged above; style usually projecting beyond the apical notch of the fr. Fl. 5–6. $2n=16$. Th.–H.

Introduced. A casual from E. Europe and W. Asia, introduced throughout the rest of Europe and in N. Africa and N. America.

*12. L. graminifolium L.

A perennial herb with several stiffly erect stems densely to sparsely hairy with short thick hairs, branched above. Basal lvs to 10 cm., long-stalked, often densely hairy, simple, lanceolate-spathulate, toothed or pinnately lobed; upper stem-lvs linear, entire. *Sepals narrowly white-margined* above. Petals white, half as long again as the sepals. Fr.-stalks ascending, equalling or slightly longer than the fr.; *siliculae ovate,* 2·5–4 × 1·5–3 mm., *with no apical notch;* valves keeled but not or scarcely winged; style short but projecting beyond the valves. Fl. 6–7. $2n=16$. H.

Introduced. A casual, occasionally established. Mediterranean region and the Near East, naturalized in C. Europe.

Of the numerous other *Lepidium* spp. which have reported as casuals the most frequently encountered are perhaps *L. bonariense* L. (S. America) with orbicular siliculae like those of *L. virginicum* but with all lvs pinnatifid; and *L. ramosissimum* A. Nelson with ovate siliculae shorter than their stalks like those of *L. ruderale* but with petals present and the lower stem-lvs simple with a few distant long teeth.

14. CORONOPUS Zinn

Annual to perennial herbs often with deeply pinnatifid lvs. *Infl. often opposite the lvs* (i.e. terminating the segments of a sympodium). Fls inconspicuous; sepals half-spreading, non-saccate; petals whitish, often small or 0; stamens 6, 4 or 2, usually without appendages; ovary with 2 ovules, stigma somewhat 2-lobed. *Fr. indehiscent or splitting into* 1-*seeded halves; valves hemispherical, reticulate-pitted; septum narrow.*

Ten spp., almost cosmopolitan.

Fr. not notched above, longer than its stalk.	**1. squamatus**
Fr. with an apical notch, shorter than its stalk.	**2. didymus**

1. C. squamatus (Forsk.) Aschers.   Swine-cress, Wart-cress.

Cochlearia Coronopus L.; *Senebiera Coronopus* (L.) Poir.; *Coronopus procumbens* Gilib.; *C. Ruellii* All.

An annual or biennial herb with slender tap-root and prostrate branched lfy stems 5–30 cm. Lvs stalked, deeply pinnatisect, the segments of the lower lvs obovate or oblanceolate, ± pinnatifid, with short lobes especially on the upper side, those of the upper lvs narrower, ± entire. Infl. sessile, one terminating the main stem, the others opposite the lvs. Fls c. 2·5 mm. diam. Petals white, longer than the sepals. *Fertile stamens usually* 6. *Fr.* 2·5–3 × 4 mm., *longer than its stalk, emarginate below but not above where it narrows abruptly into the short, pointed style;* valves rounded, reticulate-pitted or strongly and irregularly ridged, somewhat constricted at the septum, indehiscent (Fig. 12 G). Seeds 2–2·5 mm. Fl. 6–9. Protogynous. Rarely visited by small flies, and automatically self-pollinated. $2n=32^*$. Th.

Native. Waste ground, and especially trampled places such as gateways. 86, H 29, S. Common in S. England but becoming infrequent in the north though reaching Inverness and Ross; Ireland. Throughout Europe; Mediterranean region; Canary Is. Introduced in N. America, S. Africa and Australia.

*2. C. didymus (L.) Sm.   'Lesser Swine-cress.'

Lepidium didymum L.; *Senebiera didyma* (L.) Pers.

An annual or biennial foetid ± glabrous herb with slender tap-root and prostrate or ascending branched lfy stems 15–30 cm. Basal and lower stem-lvs stalked, very deeply pinnatisect, the 3–5 pairs of segments oblanceolate pinnatifid with slender acute lobes especially on the upper side; upper stem-lvs sessile, the segments narrower and ± entire. Main infl. terminal on the main stem, the others opposite the lvs or in the forks of branches. Fls 1–1·5 mm. diam. Petals white, shorter than the sepals, or more usually 0. *Usually 2 fertile stamens,* rarely 4. *Fr.* c. 1·5 × 2·5 mm., *shorter than its stalk, emarginate above and below and constricted at the septum*; valves rounded, reticulate-pitted; style 0; the valves separating at maturity into 1-seeded achene-like nutlets (Fig. 12 H). Seeds 1–2 mm. Fl. 7–9. Automatically self-pollinated. $2n = 32^*$. Th.

Introduced. A weed of cultivated and waste ground. 66, H 23, S. Widespread, especially in S. England, and extending northwards to Ross. Ireland. Probably native only in S. America but widely introduced.

15. CARDARIA Desv.

A perennial stoloniferous herb with deep tap-root and branched woody stock and several erect densely lfy stems with simple hairs. Infl. an ebracteate *corymbose panicle.* Sepals non-saccate; petals white, clawed; stamens 6, without appendages; ovary with 2(–4) ovules; style distinct; stigma capitate, broader than the style. *Fr.* broadly ovate to *cordate, indehiscent,* 2(–1)-seeded.

One species. Differs from *Lepidium* in the indehiscent fr.

*1. C. draba (L.) Desv.   'Hoary Cress', 'Hoary Pepperwort'.

Lepidium Draba L.

Flowering stems 30–90 cm., flexuous, branched above, ± glabrous or with short appressed simple hairs. Basal lvs withering before the fls open, stalked, obovate, ± sinuate, toothed; middle and upper stem-lvs ovate-oblong somewhat narrowed towards the base then enlarging into auricles which clasp the stem; uppermost ovate amplexicaul; all ± sinuate-toothed, ± hairy. Fls 5–6 mm. diam. Petals white, twice as long as the sepals. Fr. c. 4 mm. long and wide, broadly cordate, deltoid, or ovate, tapering above into the persistent style; valves somewhat turgid, often keeled above, reticulate when dry, not dehiscing but often separating into 1-seeded portions; when only 1 seed matures the fr. is asymmetrical. Seeds 1·5–2 mm. Fl. 5–6. Slightly protogynous. Visited by small bees and other insects, and automatically self-pollinated. $2n = 64^*$. H.–G.

Ssp. **draba**: fr. cordate or deltoid, usually emarginate below.

Ssp. **chalepensis** (L.) Clapham: fr. ovate, somewhat tapering at base as well as apex.

Introduced. Ssp. *draba* is spreading rapidly as a weed of arable land and is now found throughout England and Wales to Yorkshire and Cumberland; Kintyre; ssp. *chalepensis* is much more locally established northwards to

Cumberland. 52, H11. A very troublesome weed spreading by root-buds as well as seeds. Mediterranean region and W. Asia, but widely introduced as a weed. The seeds were formerly ground as a substitute for pepper.

16. ISATIS L.

Annual to perennial herbs with tall often glabrous and glaucous lfy shoots. Stem lvs sessile, sagittate, amplexicaul. Infl. a corymbose panicle. Sepals non-saccate; petals yellow, short-clawed; stamens 6, without appendages; ovary with 2 ovules, one aborting; style 0; stigma slightly 2-lobed. *Fr. indehiscent, flattened, broadly winged, 1- (rarely 2-)seeded.* Seeds large, unwinged.

About 30 spp. in C. Europe, the Mediterranean region, W. and C. Asia.

1. I. tinctoria L.   Woad.

A biennial or perennial herb with a stout tap-root and branched stock bearing several rosettes and flowering shoots; the latter lfy, branched above, 50–120 cm., softly hairy below, glabrous and glaucous above. Basal lvs lanceolate narrowed into a long stalk, ± sinuate-toothed, softly hairy; stem-lvs sessile, sagittate, amplexicaul, oblong-lanceolate, ± entire, the uppermost bract-like, with smaller basal lobes, all ± *glabrous, glaucous.* Infl. a much-branched corymbose panicle. Fls c. 4 mm. diam. Petals yellow, up to twice as long as the sepals. *Fr.* (Fig. 12J) glabrous, purple-brown, 8–20 × 3–6 mm., *pendulous on deflexed stalks* 4–7 mm., slender below, clavate above; fr. oblong, usually broadest beyond the middle, truncate or rounded at the end, the 1-seeded cell surrounded by a broad thick wing. Seeds ellipsoidal, c. 3 mm. Fl. 7–8. Homogamous. Visited by various small insects. $2n = 28$. H.

Probably introduced. Formerly cultivated for the blue pigments (woad) formed when the partially dried leaves are crushed into a paste and exposed to air. Naturalized on cliffs of the Severn valley and in cornfields in various localities in S. and C. England. 9. C. and S. Europe, and introduced elsewhere as the result of its cultivation since prehistoric times.

*I. aleppica Scop., with very long and narrow cuneate-based fr., has occurred as a casual.

17. IBERIS L.

Annual to perennial herbs with linear or spathulate lvs and simple hairs. Infl. ± corymbose. Sepals not saccate; petals white or pinkish, the 2 *towards the outside of the infl. much larger than the other* 2; stamens 6, without appendages; ovary with 2 ovules; stigma capitate, sometimes 2-lobed. Fr. an *angustiseptate silicula, the valves keeled and usually winged above.* Seeds large, flat, often winged, 1 in each cell.

About 30 spp., chiefly Mediterranean.

1. I. amara L.   Wild Candytuft.

An annual herb with slender tap-root and ± erect leafy stem, 10–30 cm., corymbosely branched, especially above, ± hairy below. Lvs scattered, the lower spathulate, narrowing into a stalk-like base, the upper oblanceolate-cuneate, sessile; all distantly pinnatifid or toothed, sometimes entire, ± ciliate but otherwise nearly glabrous. Fls 6–8 mm. diam., white or mauve, in corymbs which elongate in fr. Outer petals 4 times and inner twice as long as the sepals. Siliculae 4–5 mm., suborbicular, with wings broadening upwards and ending in triangular lobes which leave a deep triangular apical notch; style about equalling the notch; valves convex, net-veined towards the margins. Seeds 2·5–3 mm., semiorbicular, slightly winged below, reddish-brown. Fl. 7–8. Homogamous. Visited by small bees. $2n = 14$; 16. Th.

Native. Locally common on dry calcareous soils of hillsides and cornfields. Sometimes abundant on the bare soil of rabbit-infested chalk slopes, with *Sedum acre, Arenaria serpyllifolia*, etc. S. England and northwards to Gloucester and S. Lincs; naturalized further north to Fife. W. and S. Europe; N. Africa. Introduced in S. Russia, Balkans, Japan, S. America and New Zealand. Large-fld races are cultivated as *I. coronaria* hort. 'Rocket Candytuft' includes hybrids with *I. umbellata*.

Several spp. of *Iberis* occur as casuals or garden escapes. These include the annual or biennial *I. umbellata* L. with ± entire narrowly elliptical lvs, infl. not lengthening in fr., mauve fls, and fr. with the wings prolonged upwards into long triangular-acuminate lobes which equal or exceed the valves; and the perennial *I. sempervirens* L., Perennial Candytuft, woody below and evergreen, with entire linear to oblong-lanceolate lvs, infl. lengthening in fr., white fls and style exceeding the ± blunt apical lobes of the wing.

18. Thlaspi L.

Annual to perennial, usually glabrous herbs with lfy stems and simple lvs, the stem lvs usually ± amplexicaul. Infl. ebracteate. Sepals non-saccate; petals short-clawed; *stamens 6, without appendages*, ovary with 2–16 ovules, stigma capitate, somewhat 2-lobed. Fr. an angustiseptate silicula, the *valves* keeled and *usually winged*, obscurely veined. *Seeds 1–5 in each cell*.

About 60 spp. in temperate Europe and Asia, N. America and one in S. America.

1	Fr. almost circular, 12–22 mm.	**1. arvense**
	Fr. obovate or obcordate, less than 8 mm.	2
2	Annuals; style less than 0·5 mm.; included within the apical notch of the fr.	3
	Biennial to perennial, with short branched woody stock; style usually more than 0·5 mm. and equalling or exceeding the apical notch.	**3. alpestre**
3	Stem 5–25 cm., glabrous, terete; stem-lvs glaucous, with blunt auricles; fr. broadly winged.	**2. perfoliatum**
	Stem 20–60 cm., hairy below, grooved; stem-lvs with acute auricles; fr. narrowly winged.	**4. alliaceum**

1. T. arvense L.   'Field Penny-cress.'

An annual or overwintering glabrous herb with slender tap-root and erect lfy stem, 10–60 cm., simple or branched above, foetid when crushed. Basal lvs not in a compact rosette, oblanceolate or obovate, narrowed to a stalk-like base; stem-lvs oblong or lanceolate, sessile with sagittate amplexicaul base; all lvs glabrous, entire or distantly sinuate-toothed. Infl. greatly lengthening in fr. Fls 4–6 mm. diam. Petals white, twice as long as the sepals. *Siliculae*

12–22 *mm. diam.*, *almost circular*, on upwardly curving stalks 5–15 mm.; valves strongly compressed, with wings broadening upwards and leaving a *deep narrow apical notch which includes the very short style.* Seeds 1·5–2 mm., brownish-black, 5–8 in each cell. Fl. 5–7. Homogamous. Visited by flies and small bees and automatically self-pollinated. $2n=14^*$. Th.

Doubtfully native. A weed of arable land and waste places. 101, H19, S. Throughout the British Is., except the Outer Hebrides and Shetland, and sometimes a serious pest. Throughout Europe to 79° N., N. Africa, W. Asia. Siberia and Japan. Introduced in N. America.

2. T. perfoliatum L.   'Perfoliate Penny-cress.'

An overwintering or annual herb with slender tap-root and one or more erect lfy glaucous usually simple stems, 5–25 cm. Basal lvs in a loose rosette, obovate, stalked; stem-lvs sessile, ovate-cordate, amplexicaul with contiguous rounded basal lobes; all lvs glabrous, entire or with very small distant teeth. Infl. greatly elongating in fr. Fls 2–2·5 mm. diam. Petals white, narrow, c. twice as long as the sepals. *Siliculae* 4–6 × 3–5 mm., *broadly obcordate*, on slender spreading stalks 3–6 mm.; wings of valves very narrow at the base,

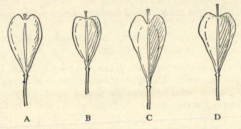

Fig. 18. Siliquae of *Thlaspi alpestre*: A, from Teesdale; B, from Matlock; C, from Mendips; D, from N. Wales. × 2·5.

broadening above and leaving a *wide notch* between the broadly rounded apical shoulder; valves and wings conspicuously net-veined; *style very short at the base of the apical notch.* Seeds c. 1·5 mm., yellow-brown, $2n=c. 70$. Th.

Native. Confined to limestone spoil in Oxford, Gloucester, Wilts and Worcester: casual elsewhere. 5. Throughout Europe except for the far north, N. Africa, Near East. Introduced in N. America.

3. T. alpestre L.   'Alpine Penny-cress.'

Biennial to perennial herbs with a short branched woody stock and erect lfy flowering stems to 40 cm. Basal lvs in a rosette, spathulate, contracted abruptly into the long stalk, ± entire; stem-lvs narrowly ovate-cordate, amplexicaul with usually subacute auricles, entire or sinuate-toothed. Petals white or lilac. Anthers often violet. *Siliculae* (Fig. 18) *obovate-obcordate*, 5–8 mm., with valves winged from just above the base, the wings broadening upwards to apical lobes very variable in size and shape and causing variability in the depth, breadth and shape of the apical notch; *style usually equalling or exceeding the notch.* Seeds c. 1·5 mm., yellowish-brown, 4–6 in each cell. Fl. 4–8. Homogamous. Visited by various insects. $2n = 14$. Ch.

Native. A local subalpine or alpine plant of limestone and other basic rocks in various parts of Great Britain and in the Inner Hebrides. Reaches 3000 ft. on Helvellyn. 17. Mountains of C. and S. Europe, and S. Sweden. Related types in America.

A highly polymorphic sp. or aggregate of spp. The British representatives fall into 4 types often identified with named Continental types which, however, they do not match satisfactorily:

(*a*) A type ('*T. virens* Jord.', '*T. calaminare* Lej. & Court.') with a dense fruiting raceme equalling or falling short of the rest of the stem, siliculae (Fig. 18 B) narrowly obovate, *shallowly notched or even truncate,* and with a *long style* (*to 2 mm.*) *usually much exceeding the notch.* Fl. 4–7. Matlock and other localities in Derby, and Inner Hebrides.

(*b*) A type ('*T. sylvestre* Jord.') with a much elongating fruiting raceme which exceeds the rest of the stem, and the siliculae (Fig. 18 A) obcordate, the *apical lobes of the wing rounded* and the *style about equalling the notch.* Teesdale (Yorks), Northumberland, Scotland (?).

(*c*) A type ('*T. occitanicum* Jord.') with the fruiting raceme equalling or falling short of the rest of the stem, and the siliculae (Fig. 18 D) obcordate, the *apical lobes of the wing divaricate,* rounded on the outside but ± straight on the inner side, and thus *forming a wide straight-sided shallow notch exceeded by the style.* Lake District, Yorks, and N. Wales.

(*d*) A type with ± *densely tufted* rosettes, occasionally stoloniferous, and *lilac fls,* otherwise resembling (*c*). (Fig. 18 C) Mendips.

*4. T. alliaceum L., an annual herb, 20–60 cm., *smelling of garlic,* has the stem hairy below, grooved, and its *stem-lvs have acute auricles.* The fls are white and the *narrowly obovate fr.* has a narrow wing and shallow apical notch. Introduced and locally established as a weed of arable land in S. England. C. and S.E. Europe.

19. TEESDALIA R.Br.

Annual herbs, glabrous or with short simple hairs and with usually pinnatifid *lvs confined to a basal rosette* or with 1–3 stem-lvs on lateral shoots. Infl. ebracteate. Sepals half-spreading, non-saccate; petals white, the pair towards the outside sometimes larger than the inner pair; stamens 4 or 6, their *filaments with a white basal scale;* ovary with 4 ovules, style very short, stigma capitate. *Fr. an angustiseptate roundish-cordate silicula;* the valves thin-walled, *narrowly winged above,* concave on the upper and convex on the lower face; style very short or 0. Seeds 2 in each cell.

Two spp. in C. Europe and the Mediterranean region.

Lvs lyrate-pinnatifid with blunt lobes; petals unequal; stamens 6.

1. nudicaulis

Lvs entire, or pinnatifid with acute lobes; petals subequal; stamens 4.

2. coronopifolia

1. T. nudicaulis (L.) R.Br.   'Shepherd's Cress.'
Iberis nudicaulis L.

An annual ± glabrous or shortly pubescent herb with short tap-root and erect stem 8–45 cm., often with several ascending basal branches. *Lvs 2–5 cm.,* mostly in a basal rosette, stalked, *narrowly lyrate-pinnatifid* with a few short *rounded lateral lobes* and a broader often 3-lobed terminal segment; stem-lvs 1–3 on the lateral stems (0 on the erect central stem), less lobed than the rosette lvs, the uppermost ±entire. Fls c. 2 mm. diam. *Inner petals slightly longer than the sepals, outer twice as long.* Ripe siliculae 3–4 mm., broadly elliptical-obovate, somewhat broader above the middle, narrowly winged, emarginate at the apex, style very short; fr.-stalks ±spreading, about as long as the fr. Seeds 1–1·2 mm., ovoid, pale brown. Fl. 4–6. Homogamous. $2n=36*$. Th.

Native. Locally common on sand and gravel. 85, H4, S. Throughout Great Britain northwards to Ross, but rare in Scotland; rare and local in Ireland; Europe to Norway and Sweden; N. Africa. Often associated with *Hypochoeris glabra, Ornithopus perpusillus, Rumex acetosella, Aira* spp., *Trifolium arvense*, etc.

2. T. coronopifolia (Bergeret) Thell.   'Lesser Shepherd's Cress.'
T. Lepidium DC.

An annual herb, slender, glabrous, shining, differing from *T. nudicaulis* in the *narrowly lanceolate rosette lvs*, pinnatifid with *acute lobes*, sometimes merely toothed or entire; the shorter flowering stems (3–12 cm.), lfless or the laterals with 1–2 very small lvs; the *subequal petals* not exceeding the sepals; the 4 stamens, and the siliculae only 3 mm., broadest at the middle; style 0. Fl. 4–6. Th.

Reported from sandy ground on Eigg in the Inner Hebrides; Mediterranean region.

20. CAPSELLA Medic.

Annual or biennial herbs with slender tap-root, and entire or pinnatifid basal lvs, and amplexicaul stem-lvs. Sepals non-saccate; petals usually white; stamens 6, without appendages; ovary with 12–24 ovules, style short, stigma capitate. *Fr. an angustiseptate silicula, obcordate* (rarely ovoid), the *valves keeled, net-veined*. Seeds unwinged, several in each cell.

Five (–10) spp., of which *C. bursa-pastoris* is a cosmopolitan weed and the others are restricted to the Mediterranean region, E. Europe and W. Asia.

1. C. bursa-pastoris (L.) Medic.   Shepherd's Purse.
Thlaspi Bursa-pastoris L.

An annual to biennial herb, 3–40 cm., glabrous or with simple and branched hairs. Basal lvs in a rosette, oblanceolate in outline, narrowed into a stalk, varying from very deeply pinnatifid to quite entire; stem-lvs also very variable in shape and lobing but always clasping the stem with basal ±acute auricles. Fls c. 2·5 mm. diam., white. Petals up to twice as long as the sepals. Siliculae 6–9 mm., triangular-obcordate, emarginate above, on spreading stalks 5–20 mm.; style c. 0·5 mm. Seeds 0·8–1 mm., pale brown, up to 12 in each cell. Fl. 1–12. Homogamous. Stamens often reduced or abortive in cold weather. Visited by small insects and automatically self-pollinated. $2n=32$. Th.

Native. Common everywhere on cultivated land, waysides and waste places. 112, H40, S. Throughout the British Is. Cosmopolitan.

Very variable with a strong tendency for distinctive local populations to arise because of self-pollination. Many of these have been named.

The Mediterranean *C. rubella* Reut. has been recorded from Surrey and Bucks, presumably as a casual. It may be recognized by its smaller fls with petals hardly exceeding the sepals and its concave-sided siliculae with shallow apical notch. The fl. buds, and sometimes also the lf-lobes, are bright red.

21. HORNUNGIA Rchb.

Annual herb with *pinnatisect lvs*. Infl. ebracteate. Sepals non-saccate; stamens 4 or 6, without appendages; ovary with 4 ovules, stigma small, sessile. *Fr. an oval unwinged angustiseptate silicula with 2 seeds in each cell.*
     One species.

1. H. petraea (L.) Rchb.   'Rock Hutchinsia.'
Lepidium petraeum L.; *Hutchinsia petraea* (L.) R.Br.

An overwintering annual herb with slender tap-root and simple or basally branching erect or ascending slender stems, 5–15 cm., glabrous or with small stellate hairs. Basal *lvs* in a rosette, stalked, *deeply pinnatisect with small elliptical segments*; stem-lvs numerous, similar but sessile. Fls c. 1·3 mm. diam. Petals greenish-white, little longer than the sepals. Siliculae 2–4 mm., varying in shape from narrowly elliptical to oblong-obovate, not or hardly emarginate, strongly compressed, on horizontally spreading stalks 3–6 mm. Seeds ovoid, 0·6 mm., pale brown. Fl. 3–5. Homogamous. $2n = 12^*$. Th.
     Native. A rare plant of limestone rocks and calcareous dunes. 12, S. W. England and Wales from Somerset to N.W. Yorks and ?Dumfries; Jersey. Europe northwards to S. Norway and S. Sweden, Asia Minor, N. Africa.

22. COCHLEARIA L.

Annual to perennial, often maritime, plants with usually simple lvs, glabrous or with simple hairs. Sepals non-saccate; petals usually white or mauve, short-clawed; stamens 6, without appendages; ovary with 2–32 ovules; stigma capitate. *Fr. a swollen or angustiseptate silicula, the valves strongly convex* with a strong midrib and usually with a conspicuous lateral network. Seeds in 2 rows in each cell.
     About 25 spp. in Europe, Asia and N. America.

1   Basal lvs cuneate at base; petals c. 5–7 mm.; silicula much compressed
         laterally, its septum at least 3 times as long as wide.   **4. anglica**
      Basal lvs not cuneate; petals usually not more than 5 mm.; silicula little
         compressed laterally, its septum usually not more than twice as long
         as wide.   *2*

2   Fls 3–5 mm. diam.; seeds 1–1·5 mm.; upper lvs usually stalked.   **3. danica**
      Fls usually exceeding 5 mm. diam.; seeds exceeding 1·5 mm.; upper lvs
         usually sessile.   *3*

3   Basal lvs reniform to triangular-ovate, truncate at base, not cordate;
         petals pale mauve; dwarf compact maritime plants of N. and W. Ireland,
         Isle of Man, N. Scotland and the western and northern islands.
                                                                              2. scotica
      Basal lvs cordate at base; petals usually white; widely distributed maritime
         and inland plants of very variable size and habit.   **1. officinalis**

1. C. officinalis L. Scurvy-grass.

Biennial to perennial herbs with a long tap-root, becoming stout in old plants, and one or more procumbent to ascending glabrous shoots, 5–50 cm. *Basal lvs* in a loose rosette, long-stalked, *orbicular to reniform*, cordate at base, ±entire; *stem-lvs* oblong to triangular-ovate, entire to coarsely toothed or sinuate-lobed, only the lowest stalked, the remainder *sessile and clasping the stem*; all lvs glabrous. Fls usually 8–10 mm. diam. Petals white, rarely lilac, usually 2–3 times as long as the sepals. *Siliculae* 3–7 × 2·5–6 mm., on spreading stalks 4–7 mm., *subglobose and rounded below* but usually narrowed above into the short persistent style, *or ovoid-elliptical, broadest near the middle and tapering at both ends; valves turgid* with prominent midrib and ±conspicuous lateral network when fully ripe; septum up to twice as long as wide. Seeds 1·5–2 mm., reddish-brown, 2–4(–6) per cell. Fl. 5–8. Homogamous. The fragrant fls are visited chiefly by flies and beetles. 2n=28*. Hs.

Native. A maritime and subalpine plant of W., N. and C. Europe, Iceland, Spitzbergen and Novaya Zemlya.

This very variable sp. may conveniently be divided into 2 sspp., though some inland populations are difficult to assign:

Ssp. **officinalis**: *lvs markedly thick and fleshy*: the basal commonly ±orbicular, rarely less than 1·5 cm. across, entire or with shallow distant teeth; stem-lvs oblong or ovate, coarsely and distantly toothed or sinuate, amplexicaul with broadly cordate base; *fr. usually subglobose, rounded below* and ±narrowed into the style, less commonly ovoid-elliptical, narrowed below as well as above. Widely distributed on drier salt- and brackish-marshes and on cliffs and banks by the sea throughout the British Is.; some inland populations also seem to belong here. 87, H25, S. Formerly eaten, especially by sailors for its pleasantly sharp-tasting lvs, which are valuable as a source of ascorbic acid (vitamin C).

Ssp. **alpina** (Bab.) Hook.: *lvs firm but not markedly fleshy*; the basal commonly ±reniform, very variable in size (c. 4–40 mm. across), entire or sinuate-lobed; stem-lvs ±sessile, the upper ±triangular or ovate and with 2–8 teeth or angles, amplexicaul with ±conspicuous auricles; *fr. usually ovoid-elliptical, tapering at both ends*, less commonly subglobose and rounded below. Small-leaved forms with upper stem-lvs less toothed and lacking distinct auricles and with fr. said not to be net-veined when ripe, have been named *C. micacea* E. S. Marshall but do not seem clearly distinct. Local in mountain streamlets and flushes and on wet rock-ledges of upland and mountain areas in Wales, N. England, Scotland and Iceland, and in the Inner and Outer Hebrides, Orkney and Shetland. 35, H9. Reaches 3170 ft. in Wales but descends to 50 ft. in Shetland. Resembles forms in the mountains of C. Europe.

2. C. scotica Druce 'Scottish Scurvy-grass.'
C. groenlandica auct. angl., non L.

A biennial to perennial herb forming a small compact tuft usually 5–10 cm. diam. with ±prostrate flowering stems. *Basal lvs* in a rosette, long-stalked, varying in shape from reniform to triangular-ovate *with a truncate base*, rarely with a truly cordate base, c. 1 cm. wide, *thick and fleshy; stem-lvs* sessile or short-stalked, elliptic to rhomboidal, *entire or 2-toothed*, not clasping the stem. Fls 5–6 mm. diam. Petals pale mauve, with the blade almost square, abruptly contracted into the short claw. *Silicula* c. 3 × 2 mm., *broadly ovoid to broadly ellipsoidal*, narrowed at both ends; valves turgid, *net-veined when perfectly ripe*. Seeds reddish-brown, 1–2 mm. Fl. 6–8. $2n=14*$. H.–Ch.

Native. A local maritime plant of the Isle of Man, N. Scotland, the Hebrides, Orkney and Shetland; N. and W. Ireland. 18, H12. Apparently confined to the British Is., but close to other arctic segregates from *C. officinalis* sensu lato. The British material identified as *C. scotica* is not homogeneous but shows variation especially in the size of the seeds. It seems probable that there is a large-seeded species (*C. scotica*) distinct from and forming sterile hybrids with *C. officinalis*; and also small-seeded types of unknown taxonomic status which extend southwards to Berwick. The arctic *C. officinalis* 'var. *groenlandica* (L.) Gelert apud Andersson & Hesselman' is small-seeded.

3. C. danica L.   'Danish Scurvy-grass.'

An overwintering annual herb with slender tap-root and ascending flowering stems 10–20 cm. *Basal lvs* long-stalked, roundish or ± triangular, c. 1 cm. wide, *cordate at the base*, fleshy; *stem-lvs* mostly stalked, not amplexicaul, distantly toothed or entire; the lowest *3–7-lobed* (resembling ivy lvs), the middle triangular-hastate, and the uppermost sometimes oblong-lanceolate. *Fls 4–5 mm. diam.* Petals mauve or whitish, with elliptical blade and short claw, hardly twice as long as the sepals. Silicula variable in shape, ovoid to ellipsoidal, ± narrowed at both ends, 3–5·5 × 2·5–4 mm.; valves turgid, finely net-veined. *Seeds* reddish-brown, c. 1 mm. Fl. 1–6. Homogamous. The fls are non-fragrant and yield no nectar. Visited by a few small insects and automatically self-pollinated. $2n=28, 42$. Th.

Native. Locally common on sandy and rocky shores and on walls and banks near the sea, also on railway ballast inland. 61, H25, S. All round the British Is. N. and W. Europe from S. Norway, Sweden and Finland to Atlantic Spain and Portugal.

Striking dwarf forms have been reported with basal lvs only 3 mm. wide and slender stems 3–5 cm. high; or 'like a moss', with stems less than 1 cm. high and vivid lilac fls.

4. C. anglica L.   'Long-leaved Scurvy-grass.'

A biennial or perennial herb with slender tap-root and stiffish ± erect stems 8–35 cm. *Basal lvs* in a rosette, ovate, oblong or obovate, ± *cuneate at the base*, never cordate, entire or with a few distant teeth, ± fleshy; stem-lvs ovate or elliptical, toothed or entire, mostly sessile, the *upper lvs clasping the stem*. *Fls 10–14 mm. diam.* Petals white or pale mauve, broadly ovate, narrowed abruptly into the short claw. Siliculae ovoid-oblong, 8–15 mm., much compressed laterally and strongly furrowed at the narrow septum which is 3–5 times as long as wide; valves turgid, conspicuously reticulate. *Seeds* reddish-brown, 2–2·5 mm., 5–6 in each cell. Fl. 4–7. $2n=36–50$. Hs.

Native. Locally common on muddy shores and in estuaries all round the British Is., but not in Orkney or Shetland. 53, H26. Atlantic and North Sea coasts of N.W. Europe.

Hybrids between *C. officinalis* and *C. danica*, intermediate between the parents, and between *C. officinalis* and *C. anglica* have been reported. The latter is probably *C. anglica* var. *hortii* Syme, a large plant with basal lvs rounded, truncate or even somewhat cordate at the base, and fruits intermediate in size and shape between those of *anglica* and *officinalis*.

23. SUBULARIA L.

Small annual or biennial *aquatic herbs* with very short stock and fibrous root system; *lvs* confined to a basal rosette, *subulate*. Infl. a few-fld ebracteate raceme, often submerged. Receptacle concave; sepals non-saccate; petals white, sometimes 0; stamens 6, without appendages; *ovary slightly sunk in the receptacle, its base surrounded by a fleshy ring*; 8–14 ovules; stigma sessile. *Fr. a latiseptate silicula with convex 1-veined valves.* Seeds 2–6 in each cell.

Two spp., one in Europe, N. Asia and N. America, the other in the mountains of E. Africa.

1. S. aquatica L. Awlwort.

A dwarf aquatic annual herb with flowering stems 2–8 cm. *Lvs* 2–7 cm., numerous, *terete, subulate, entire, glabrous*. Fls 2–8, c. 2·5 mm. diam., often submerged. Petals twice as long as the sepals. Siliculae 2–5 × 1·5–2·5 mm., oblong-elliptical, on ascending stalks of about the same length; valves very convex, with a strong midrib. Seeds 2–6, c. 0·7 mm., in 2 rows in each cell. Fl. 6–8. Homogamous. Nectarless and rarely visited by insects; automatically self-pollinated and sometimes cleistogamous. Th.

Native. A local plant of base-poor lakes and pools in Wales, N. England and Scotland; W. Ireland. 32, H6. Reaches 2000 ft. in W. Ross. Often with *Isoetes lacustris, Littorella uniflora, Lobelia dortmanna*, etc. Mountains of W. Europe and in N. Europe from Ireland and Scandinavia to N. Russia; Siberia, Greenland, N. America.

24. NESLIA Desv.

Two spp. in Mediterranean region and W. Asia.

*1. N. paniculata (L.) Desv. (*Myagrum paniculatum* L.; *Vogelia paniculata* (L.) Hornem.)

An annual herb with slender tap-root and erect simple or branched stem, 15–80 cm. high, with stellate hairs especially below. Basal lvs oblong, narrowed below into a stalk; middle and upper lvs lanceolate, acute, sessile, with pointed basal lobes clasping the stem; all entire or distantly toothed, and grey-green with stellate hairs. Infl. ebracteate, much elongating in fr. Fls 4–5 mm. diam. Sepals erect, yellowish-green, non-saccate; petals golden yellow; ovary with 4 ovules, style filiform, stigma entire. *Fr.* (Fig. 13 D) *indehiscent*, usually 1-seeded, *almost spherical*, 1·5–2 mm. diam., with a fine network of wrinkles, readily falling when ripe; stalks spreading, 6–12 mm. Fl. 6–9. Homogamous. Automatically self-pollinated. $2n = 14$. Th. Introduced. A casual of waste and arable land. Naturalized throughout Europe except N. Scandinavia, and in N. Africa, but probably native in the Near East.

25. BUNIAS L.

Annual to perennial herbs with branched and simple hairs and some stout glandular hairs. Lvs ± pinnatifid, hairy. Infl. much elongating in fr. Sepals half-spreading, the inner pair not or slightly saccate; petals white or yellow; stamens 6, without appendages; ovary with 2 or 4 ovules; stigma capitate, slightly 2-lobed. *Fr. indehiscent,* irregularly ovoid, *warty or with wing-like crests,* 1–4-celled with 1–4 seeds.

Six spp. in the Mediterranean region and W. Asia, widely introduced.

Fr. with 4 irregularly crested wings and a long slender persistent style.
1. erucago
Fr. irregularly ovoid, warty, with a short, broad, asymmetrically-placed style.
2. orientalis

*1. B. erucago L.

An annual to biennial herb, 30–60 cm., roughly glandular-hairy, with the basal lvs runcinate-pinnatifid and the upper lvs oblong, entire or toothed, sessile, not amplexicaul. Fls yellow. *Fr.* (Fig. 13E) c. 11 mm., *quadrangular with irregularly toothed and crested wings on the angles,* and 4-celled with 1 seed in each cell; persistent style c. 5 mm.; stalks ± horizontal. Fl. 5–8. Homogamous. Visited by various bees and automatically self-pollinated. $2n = 14$. Th.–H.

Introduced. A casual, scarcely established. Native in S. Europe but widespread as an adventive. The roots and shoots are eaten in Greece.

*2. B. orientalis L.

A biennial to perennial herb, 25–100 cm., sparsely glandular, with ± pinnatifid lvs, the lower stalked, upper ± sessile. Fls yellow. *Fr.* (Fig. 13F) 6–10 mm., *asymmetrically ovoid, unwinged, covered with warty prominences,* 1–2-celled with 1–2 seeds; style short, ± laterally placed. Fl. 5–8. Homogamous. Visited by various flies and small bees and automatically self-pollinated. $2n = 14, 42$. H.–G.

Introduced. Established in a number of places. Native in E. Europe and W. Asia, adventive in C. and W. Europe. Used as a salad and fodder plant.

26. LUNARIA L.

Annual to perennial herbs with simple hairs and large cordate lvs. Inner sepals deeply saccate; petals usually purple; stamens 6, the 4 inner with broad filaments; ovary with 4–6 ovules, style short, stigma ± deeply 2-lobed. *Fr. a large latiseptate silicula, with quite flat thin-walled translucent net-veined valves and a thin shining white septum.* Seeds large, strongly compressed, in 2 rows in each cell.

Three spp. in C. and S.E. Europe.

Fr. narrowly elliptical; all lvs distinctly stalked.   **2. rediviva**
Fr. broadly elliptical or almost circular; upper lvs subsessile.   **1. annua**

*1. L. annua L.   Honesty.

L. biennis Moench

A usually biennial herb (1–3 years) with a stiffly hairy stem, 30–100 cm. and irregularly and coarsely toothed broadly cordate-acuminate lvs, rough with appressed hairs, the lower long-stalked, the *upper subsessile.* Fls c. 3 cm. diam., in a loose raceme, scentless. Petals reddish-purple, rarely white. *Fr. broadly elliptical,* oblong, *or almost circular,* rounded at the ends, 30–45 × 20–25 mm. Seeds uniform, winged, 5–8 mm. diam. Fl. 4–6. Homogamous. Visited for nectar by butterflies and long-tongued bees (working distance = 10 mm.) and for pollen by smaller insects; automatically self-pollinated also. $2n = 28 + 2$ ff. H.

Introduced. Much grown in gardens and often found as an escape. S.E. Europe. The infructescences with persistent silvery septa are used for winter decoration.

*2. L. rediviva L.

A perennial herb differing from *L. annua* in the *distinctly stalked* and finely toothed *upper lvs,* the fragrant fls, the *narrowly elliptical fr.* 35–90 × 15–35 mm., and the larger seeds, 7–10 mm. wide. Fl. 5–8. $2n = 28 + 2$ ff. H.

Introduced. A rare garden-escape. Native in moist shady woods throughout Europe from Portugal, C. France, Belgium, Denmark and S. Sweden to the Balkans, S. Russia and W. Siberia.

27. Alyssum L.

Annual to perennial, often *mat-forming herbs with stems and simple lvs densely covered with usually stellate hairs.* Infl. ebracteate. Sepals non-saccate; *petals yellow,* clawed; *stamens* 6, some or all *with filament-appendages*; ovary with 1–16 ovules, style short, stigma small, somewhat 2-lobed. *Fr. a circular or oval latiseptate silicula;* the *valves flat, faintly net-veined.* Seeds compressed, winged.

About 100 spp., chiefly Mediterranean but some in C. Europe and C. Asia.

*1. A. alyssoides (L.) L.   'Small Alison.'

A. calycinum L.

A usually annual herb with slender tap-root and an erect or ascending stem, 7–25 cm., with several ascending branches near the base, the whole shoot grey-pubescent with stellate hairs. Lvs 6–18 mm., few, oblanceolate, narrowed below into a short stalk, entire, grey with stellate down. Fl. c. 3 mm. diam. *Sepals erect, stellate-hairy, persistent in fr.* Petals pale yellow but becoming whitish, up to twice as long as the sepals, stellate-hairy on the outside. Outer stamens each with a pair of slender appendages from near the base of their filaments. *Siliculae* 3–4 mm., *almost circular,* emarginate above, on spreading stalks 2–5 mm., the valves bordered, stellate-hairy. Seeds 1–2 in each cell, reddish-brown, obovate, 1·2–1·5 mm., narrowly winged. Fl. 5–6. Slightly protogynous. Said to be devoid of nectar and little visited by insects. $2n = 32$. Th.–H.

Introduced. A rare plant of grassy fields and arable land in scattered localities throughout S. and E. England and E. Scotland northwards to Angus.

Europe, except the north-west (where it is introduced), and W. Asia. Introduced in New Zealand.

*A. saxatile L. (Golden Alyssum), a spring-flowering perennial with greyish oblanceolate lvs, 7–10 cm., and corymbose panicles of golden-yellow or lemon-yellow fls, is a native of C. Europe much cultivated in gardens and has been found as a casual.

28. LOBULARIA Desv.

Annual to perennial herbs with *narrow entire lvs, the whole shoot covered with bipartite hairs*. Infl. ebracteate. Sepals non-saccate; *petals white, entire*, short-clawed; stamens 6, without appendages; ovary with 2–10 ovules, style short, stigma capitate. Fr. a circular or oval latiseptate silicula, the *valves ± flat with slender midrib*. Seeds compressed, narrowly winged.

Five spp. in the Mediterranean region and Macaronesia.

*1. L. maritima (L.) Desv.   'Sweet Alison.'

Clypeola maritima L.; *Alyssum maritimum* (L.) Lam.

An annual to perennial herb with slender tap-root and ascending stem 10–30 cm., branching freely near the base, the whole shoot greyish with dense bipartite appressed hairs. *Lvs* 2–4 cm., *scattered, linear-lanceolate*, subacute, narrowed below, subsessile. Fls c. 6 mm. diam. *Sepals not persisting in fr.* Petals white, nearly twice as long as the sepals, with ± circular spreading limb. *Siliculae obovate*, c. 2·5 mm. excluding the persistent style, on spreading stalks, the *valves slightly convex*, 1-*veined, pubescent. Seeds* pale reddish-brown, narrowly winged, 1 *in each cell*. Fl. 6–9. Homogamous. The fragrant nectar-bearing fls are much visited by small insects and are said to be self-sterile. $2n = 24$. Th.–H.

Introduced. An escape from cultivation naturalized in many scattered coastal localities especially in S.W. England, but reaching northwards to Angus. Native of the Mediterranean and Macaronesian regions, but widely naturalized through cultivation. Much used as an edging plant.

29. BERTEROA DC.

Seven spp. from N. and C. Europe eastwards to C. Asia.

*1. B. incana (L.) DC.

Alyssum incanum L.; *Farsetia incana* (L.) R.Br.

An annual or overwintering herb with pale slender tap-root and erect branching stem, 20–60 cm., stellate-hairy, often reddish below. *Lvs lanceolate, narrowed below, subsessile, entire or distantly toothed, grey with stellate hairs*. Infl. ebracteate. Sepals non-saccate, hairy, not persistent. *Petals white, deeply bifid*, more than twice as long as the sepals. Stamens 6, the filaments of the outer pair toothed at the base, those of the 4 inner winged below. Ovary with 4–12 ovules; style long, stigma small, capitate. Fr. a broadly elliptical latiseptate silicula 7–10 mm., on a ± erect stalk 5–10 mm., the valves somewhat convex, stellate-hairy, indistinctly veined; persistent style 1·5–3 mm. Seeds brown, roundish, 1·5–2 mm., very narrowly winged. Fl. 6–9. Homogamous. Visited by hover-flies, etc. $2n = 16$. Th.

Introduced. Naturalized in waste places and a casual in cultivated ground in many scattered localities. Native in C., N. and E. Europe and W. Asia, and naturalized in W. Europe.

30. DRABA L.

Annual to perennial, often densely tufted herbs or dwarf shrubs with stellate-hairy, rarely glabrous, shoots. Basal lvs in a rosette; stem-lvs sessile or 0; all lvs simple, entire, toothed or lobed. Infl. an ebracteate corymbose raceme lengthening in fr. Fls small, white or yellow; outer sepals ±saccate; *petals entire or slightly notched*; stamens 6 or 4, without appendages or rarely the filaments of the outer stamens toothed; ovary with 4–80 ovules; style distinct, stigma capitate or ±2-lobed. Fr. a latiseptate silicula, the *valves ±flat with a midrib conspicuous only in the lower half*. Seeds numerous, in 2 rows in each cell.

About 270 spp. of arctic and alpine plants of the northern hemisphere and in C. and S. America.

1	Lvs glabrous apart from marginal bristles; fls yellow.	**1. aizoides**	
	Lvs stellate-hairy; fls white.		2
2	Flowering stem lfless or occasionally with 1–2 small lvs.	**2. norvegica**	
	Flowering stem lfy.		3
3	Stem-lvs lanceolate or narrowly ovate, not or hardly amplexicaul; fr. twisted.	**3. incana**	
	Stem-lvs broadly ovate, amplexicaul; fr. straight.	**4. muralis**	

1. D. aizoides L. (z)   'Yellow Whitlow Grass.'

A perennial tufted herb with slender tap-root and much-branched *glabrous* stems, 5–15 cm. Lvs confined to compact rosettes, rigid, linear, narrowed at each end, the midrib forming a keel below, terminating in a long white bristle and the margins fringed with similar bristles; dead lvs persisting. *Flowering stem* rigid, *lfless*. Infl. few-fld. Fls 8–9 mm. diam. Sepals yellowish, usually non-saccate. *Petals bright yellow*, exceeding the sepals. Siliculae 6–12 × 2·5–4 mm., elliptical, compressed, narrowing above into the persistent style 2·5–4 mm., and held almost erect on ascending stalks 5–15 mm.; valves usually glabrous, almost flat, with inconspicuous midrib and a faint lateral network. Seeds yellowish-brown, 1·5 mm., 6–12 in each cell. Fl. 3–5. Protogynous. Visited by flies and Lepidoptera. Ch.

Doubtfully native. Found only on limestone rocks and walls west of Port Eynon, Glamorgan. A calcicolous alpine plant of C. and S.E. Europe from the Pyrenees to the Balkans.

2. D. norvegica Gunn.   'Rock Whitlow Grass.'

D. hirta auct.; *D. rupestris* R.Br.

A perennial tufted herb with slender tap-root and short much-branched stellate-hairy stems bearing close basal rosettes of oblong-lanceolate lvs, hairy with mostly simple hairs, ciliate, usually entire. *Flowering stems* 2–5 cm., *usually lfless* but sometimes with 1–2 lvs, sessile, ovate-lanceolate, hairy and ciliate. Infl. few-fld, ±corymbose, lengthening in fr. Fls 4–5 mm. diam. Petals white, slightly notched, nearly twice as long as the sepals. Stigma

2-lobed. *Siliculae* 5–6 × 2–2·5 mm., *straight*, elliptical, compressed, sparsely stellate-hairy, held obliquely on ascending stellate-hairy stalks c. 3 mm. long; valves ± flat with inconspicuous midrib and lateral network; style very short. Seeds pale reddish-brown, 4–6 in each cell. Fl. 7–8. $2n=48$. Ch.

Native. Found near the tops of a few Scottish mountains including Ben Lawers, Cairngorm and Ben Hope, and reaching c. 3980 ft. 6. The British plant belongs to a difficult arctic complex but is probably conspecific with plants in Norway and Greenland.

3. D. incana L.   'Hoary Whitlow Grass.'

A biennial to perennial herb with slender tap-root and short prostrate occasionally branched stock bearing the remains of dead rosettes. Basal lvs in a loose rosette, oblong-lanceolate, narrowing to a stalk, entire or distantly toothed, densely stellate-hairy, ciliate, dying before the fr. ripen. Flowering stems erect, 7–50 cm., simple or branched, stellate-hairy with numerous, ± erect, sessile, *narrowly elliptical or ovate stem-lvs, rounded at the base* or slightly amplexicaul, densely stellate-hairy, their margins ciliate and usually coarsely toothed, sometimes entire. Infl. dense, much elongating later. Fls 3–5 mm. diam. Petals white, very slightly notched, twice as long as the sepals. Stigma ± entire. *Siliculae* 7–9 × 2–2·5 mm., variable in shape, elliptical to lanceolate, *twisted*, glabrous or stellate-hairy, held erect on ascending stalks 2–9 mm. long; valves with conspicuous midrib; style not exceeding 1 mm. Seeds numerous, yellow-brown, 0·8–1 mm. Fl. 6–7. Homogamous. Self-pollinated. $2n=32$. Ch.

Plants with stellate-hairy fr. have been named *D. confusa* Ehrh., but appear to be merely extreme types of a very variable species. Very small plants ('var. *nana* Lindbl.') are sometimes mistaken for *D. norvegica*.

Native. On rocky screes and cliffs and shelly dunes northwards from Caernarvon, Stafford and Derby to Orkney and Shetland. Ireland. 32, H9. From sea-level to 3550 ft. on Ben Lawers. Mountains of C. Europe and C. Asia, arctic and subarctic Europe, Iceland, Greenland.

4. D. muralis L.   'Wall Whitlow Grass.'

An annual or biennial herb with slender tap-root and an ascending simple or branched lfy stem 8–30 cm., densely stellate-hairy below, sparsely above. Basal lvs in a rosette, obovate or oblanceolate, narrowed into a stalk, entire or toothed; *stem-lvs sessile, broadly ovate with broad rounded semi-amplexicaul base* and sharply toothed margins; all lvs rough with stellate and a few simple hairs. Infl. many-fld, much elongating in fr. Fls 2·5–3 mm. diam. Sepals hairy. Petals white, narrow, entire, almost twice as long as the sepals. *Siliculae* 3–6 × 1·5–2 mm., elliptical-oblong, *straight*, spreading on almost horizontal slender glabrous stalks 5–9 mm.; valves glabrous, flat with distinct midrib; style almost 0. Seeds pale brown, c. 0·8 mm., 6–8 in each cell. Fl. 4–5. $2n=$ c. 32. Th.–H.

Native. A rare plant of limestone rocks and walls from Somerset to Yorks and Westmorland, and a weed in nursery gardens in Cornwall. Probably introduced in Scotland and Ireland. 30. Widely distributed in Europe from C. Scandinavia and Finland to S. Spain and Italy; N.W. Africa, W. Asia.

31. EROPHILA DC.

Small annual or overwintering herbs with lvs confined to a basal rosette. Infl. ebracteate. Sepals non-saccate; *petals* white, *deeply bifid*; stamens 6, without appendages; ovary with 10–60 ovules; style very short, stigma flat, entire. *Fr. a latiseptate silicula,* the valves with a thin midrib vanishing above the middle. Seeds very small, in 2 rows in each cell.

Perhaps 8 spp., some highly polymorphic, in Europe, W. Asia and N. Africa and 1 doubtful sp. in Peru.

Distinguished from *Draba* by the deeply bifid petals. A taxonomically difficult genus some of whose spp. have been much subdivided. The three following seem satisfactorily distinct.

1 Siliculae oblanceolate with 40–60 ovules.   **1. verna**
   Siliculae obovate or suborbicular, with 18–40 ovules.   2

2 Lvs minutely hairy above with slender forked and stellate hairs; seeds
   0·3–0·4 mm.   **2. spathulata**
   Lvs coarsely hairy above with mostly simple and some forked hairs; seeds
   0·5 mm.   **3. praecox**

1. E. verna (L.) Chevall.   Whitlow Grass.
Draba verna L.

Lvs 1–1·5 cm., broadly lanceolate or elliptical, ± acute, entire or rarely with 1–2 teeth, narrowed below into a broad stalk, minutely hairy with slender forked and stellate hairs. Flowering stems 2–20 cm., shortly hairy below.

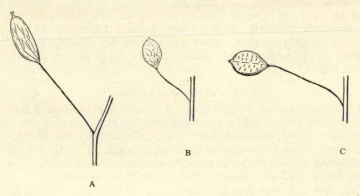

Fig. 19. Siliquae of *Erophila*. A, *E. verna*; B, *E. spathulata*; C, *E. praecox.* ×2·5.

Fls 3–6 mm. diam. Petals 2·5 mm. *Ovary with 40–60 ovules. Siliculae* (Fig. 19 A) 3–9 × 1·5–3 mm., *oblanceolate or elliptical,* compressed. Seeds 0·3–0·4 mm. Fl. 3–6. Rarely visited by flies and small bees; automatically self-pollinated. 2*n* = 14, 24, 30, 40, 64, 94. Th.

Extremely variable in the size and shape of the siliculae, the various types coming ± true from seed owing to prevalent self-pollination. Var. *oedocarpa*

(Drabble) O. E. Schulz, widely distributed in Great Britain, has fruits 3–4 × 2–2·5 mm., with only about 30 seeds, and is often separated as a species.

Native. Common and widely distributed on rocks, walls and dry places. 68, H40, S. Great Britain and Ireland, but not in the Hebrides, Orkney or Shetland. Europe, Asia, N. Africa. Introduced in N. America.

2. E. spathulata Láng

E. Boerhaavii (Van Hall) Dum.

Lvs short (4–10 mm. including stalk), obovate-spathulate, narrowing below into a stalk, ± acute, entire or with 1–3 teeth on each side, *covered with very short fine forked and stellate hairs* and with longer simple hairs on the margins. Flowering stems up to 10 cm., hairy below. Petals 2 mm. *Ovary with* 18–32 *ovules. Siliculae* (Fig. 19 b) 4–5 × 2·5–3 mm., *obovoid*, rounded at the top with short spathulate style. *Seeds* 0·4 *mm.* Fl. 3–6. Th.

Var. *inflata* (Bab.) O. E. Schulz differs in its more inflated fr., 3–3·5 × 1·5 mm., ovoid or ellipsoidal, almost circular in section, and its linear-lanceolate lvs.

Native. Scattered on rocks walls, etc., throughout the British Is. northwards to Orkney. 54, S. Var. *inflata* more local but reaching 3990 ft. on Ben Lawers. Europe, Asia, N. America. The hybrid *E. spathulata × verna* has been reported from Britain.

3. E. praecox (Stev.) DC.

Lvs short, obovate-spathulate, narrowing below into a stalk, ± acute, entire or with 1–3 teeth on each side, *densely hairy with coarse mostly simple and a few forked hairs.* Petals 2–2·5 mm. *Ovary with* 24–40 *ovules. Siliculae* (Fig. 19 c) 4–6 × 2–3 mm., *shortly obovoid*, compressed, rounded at the top with very short apiculate style. *Seeds* 0·5 *mm.*, distinctly tuberclate. Fl. 3–6. Th.

Native. Locally on rocks, walls and exposed dry places in England, apparently rare in Scotland and Ireland. 24, H1, S. C. Europe and Mediterranean region.

32. ARMORACIA Gilib.

Tall glabrous perennial herbs with stout cylindrical sharp-tasting roots and lfy branching stems. Infl. a racemose panicle. Sepals non-saccate; *petals white* short clawed; stamens 6, without appendages; ovary with 18–50 ovules; style short, *stigma large, discoid*, shallowly 2-lobed. *Fr. an almost spherical or ellipsoidal silicula, the valves strongly convex with an indistinct network of veins.* Seeds ovoid-compressed, unwinged, in 2 rows in each cell.

Three spp. in Europe and Siberia.

*1. A. rusticana Gaertn., Mey. & Scherb.   Horse-radish.

Cochlearia Armoracia L.; *A. lapathifolia* Gilib.

A stout glabrous perennial herb with thick tap-root-like stocks, branched above and continuous downwards with the very long fleshy cylindrical roots. From the stocks arise subterranean stolons and erect lfy stems up to 125 cm., with numerous ± erect slender branches above. *Radical lvs* 30–50 *cm.*, *ovate or ovate-oblong, long-stalked (to 30 cm.), crenate-serrate*; stem-lvs short-stalked or sessile, elliptical or oblong-lanceolate, the lower often ±pinnatifid, the upper coarsely toothed or entire. *Infl. a much-branched corymbose panicle.* Fls 8–9 mm. diam. Petals white, twice as long as the sepals. Siliculae 4–6 mm., spherical to obovoid, on slender ascending stalks 8–20 mm., their convex valves ±unveined. Seeds 8–12 in 2 rows in each cell. Fl. 5–6. Homogamous. Visited by various small insects. $2n=32*$. H.

Introduced. Often cultivated for the condiment prepared from the roots and widely naturalized throughout Great Britain, northwards to Moray, in fields and waste places and on stream banks. Rare in Ireland. 89, H6. Probably native in S.E. Europe and W. Asia, but naturalized as an escape from cultivation throughout Europe and in N. America, New Zealand, etc. Rarely ripens its fr. in the British Is.

33. CARDAMINE L.

Annual or perennial often glabrous herbs with *trifoliolate or pinnate lvs.* Infl. an ebracteate raceme. Inner sepals hardly saccate; petals white or purple; stamens 4–6, without appendages; stigma entire or slightly 2-lobed. *Fr.* a siliqua with compressed inconspicuously veined *valves* which *open suddenly and coil spirally from the base,* thus flinging the seeds to some distance. Seeds in 1 row in each cell.

About 100 spp., cosmopolitan, chiefly in damp habitats, the tropical spp. confined to mountains.

1	Petals broad, spreading, c. three times as long as the sepals.	2
	Petals narrow, ±erect, little longer than the sepals.	4
2	Lvs trifoliolate.	**6. trifolia**
	Lvs pinnate.	3
3	Petals usually white; anthers purple.	**2. amara**
	Petals lilac, rarely white; anthers yellow.	**1. pratensis**
4	Lvs with slender sagittate basal lobes which clasp the stem.	**3. impatiens**
	Lvs not with basal lobes.	5
5	Annual; stamens 4.	**5. hirsuta**
	Biennial (often perennating); stamens 6.	**4. flexuosa**

1. C. pratensis L.   Cuckoo Flower; Lady's Smock.

A perennial herb with a short prostrate or ascending stock bearing numerous roots and occasional stolons, and erect or ascending usually simple terete and ±glabrous stems, 15–60 cm. Basal lvs in a rosette, long-stalked, pinnate, the lflets ovate or roundish, the terminal ±reniform and much larger than the lateral, all with distant gland-tipped teeth; stem lvs short-stalked, lflets narrowly lanceolate, ±entire, or terminal lflet 3-toothed; all sparsely hairy. In damp situations vegetative reproduction occurs by small plantlets from buds in the axils of basal lvs. Infl. of 7–20 fls, at first corymbose then lengthening. Fls 12–18 mm. diam., on stalks 8–15 mm. Sepals scarious-margined with violet tip; *petals* clawed, c. 3 *times as long as the sepals, lilac* or rarely white; *anthers yellow.* Siliquae 25–40(−55) × 1–1·5 mm., obliquely erect on stalks 12–25 mm. Fl. 4–6. Protogynous. Probably self-sterile. Seeds thrown 1·5–2 m. $2n=16, 24, 28, 30*, 32*, 33–37, 38*, 40–46, 48*, 52–55, 56*, 57*, 58*, 59–63, 64*, 67–71, 72*, 73–96$. Hs.

A very variable complex in which forms with higher chromosome numbers ($2n = 56$–72) tend to be taller but weaker plants having stem-lvs with stalked and sometimes markedly toothed lflets (sessile and ± entire in forms with low chromosome numbers) and large whitish fls (commonly smaller and a deeper lilac in forms with low chromosome number). These correlations do not seem sufficiently clear and constant to warrant a subdivision of the sp.

Native. Common throughout the British Is. in damp meadows and pastures and by streams, etc. Channel Is. Reaches 3300 ft. in Scotland. 112, H40, S. Forms with low chromosome numbers appear to favour drier habitats in S. England. Europe, N. Asia, N. America.

2. C. amara L.   'Large Bitter-cress.'

A perennial herb with a horizontal creeping stock from which roots and stolons arise. Stems prostrate at the base then erect or ascending 10–60 cm., angled, usually glabrous. Basal lvs not in a rosette, stalked, pinnate, the 5–9 lflets ovate or orbicular, ± cordate, short-stalked, the terminal largest; upper lvs very shortly stalked, the 5–11 lflets ovate to lanceolate; all lflets angular in outline or ± crenate. Racemes 10–20-fld; fls c. 12 mm. diam.; *petals white*, rarely purple, spreading, clawed, *twice as long as the sepals; anthers violet*; style long, slender. Siliquae 20–40 × 1–2 mm., narrowing upwards, straight, held obliquely on slender stalks 10–20 mm.; seeds pale brown. Fl. 4–6. Homogamous. Visited by various small insects. $2n = 16$. Hs.

Native. A locally abundant plant of springs, flushes, fens and stream-sides, chiefly on peat, throughout Great Britain from Aberdeen southwards, and in N.E. Ireland. Reaches 1500 ft. in Scotland. 79, H6. Throughout Europe from 64° 30′ N. in Sweden to the Pyrenees, C. Apennines and N. Balkans; Asia Minor; Altai. A ground-flora dominant in alder woods with moving ground-water, and occurring in other types of damp woodland.

3. C. impatiens L.   'Narrow-leaved Bitter-cress.'

A biennial or sometimes annual herb with a stout tap-root and an erect very lfy glabrous shoot, 25–60 cm., branched only above. Lvs pinnate with a winged rhachis; basal lvs in a rosette, dying in the second season, long-stalked, with 2–4 pairs of ± ovate, deeply toothed lflets and a larger ± lobed terminal lflet; stem-lvs ± sessile with 6–9 pairs of narrower toothed or entire lflets; all with *basal stipule-like auricles clasping the stem*, ciliate but otherwise glabrous. Infl. many-fld, inconspicuous. *Fls 6 mm. diam.; petals white, narrow, scarcely longer than the sepals, often 0*; stamens 6; *anthers greenish-yellow*. Siliquae 18–30 × c. 1 mm., at first ± erect, later wide-spreading, on stalks 3–5 mm.; seeds unwinged, yellow-brown. Fl. 5–8. Little visited by insects. $2n = 16^*$. Th.–H.

Native. Very local in shady woods, especially of ash, and on moist limestone rocks and scree, chiefly in the west of Great Britain. Angus and from Cumberland and Yorks to Wales, Somerset and Devon. 28. Europe from the Faeroes and C. Sweden to C. Spain, Italy and the N. Balkans; Asia to Japan. In C. Europe it occurs especially in montane woods and shows no strong preference for calcareous rocks.

4. C. flexuosa With. 'Wood Bitter-cress.'

C. sylvatica Link

An annual, biennial or usually *perennial herb* with no persistent tap-root but with a short ± branched ascending stock and an erect or ascending very lfy flexuose stem, 10–50 cm., hairy especially below. *Basal lvs few, in a loose rosette*, stalked, pinnate with c. 5 pairs of ovate, rounded or reniform lflets and a larger terminal lflet; stem-lvs short-stalked or sessile, with 5 or more pairs of lflets becoming narrower up the stem; all lflets gland-toothed or ± lobed and sparsely ciliate. Infl. at first corymbose. Sepals greenish-violet, white-margined. Petals white, narrow, twice as long as the sepals. *Stamens* 6. Style c. 1 mm. long. Young siliquae usually not overtopping the unopened fls. Ripe siliquae 12–25 × 1 mm., ± erect on slender upwardly curved stalks 6–13 mm. Seeds 1–1·2 mm., very narrowly winged. Fl. 4–9. $2n=32^*$. Th.–H.

Native. Common in moist shady places, by streams, etc., throughout the British Is. except the Isle of Man and Shetland. 110, H40, S. Reaches 3900 ft. on Ben Lawers. Europe, Asia Minor, China, Japan. Probably introduced in N. America.

5. C. hirsuta L. 'Hairy Bitter-cress.'

An annual herb with slender tap-root and erect simple or basally branched usually glabrous stem, 7–30 cm. *Basal lvs numerous in a compact rosette*, stalked, pinnate with 3–7 pairs of obovate or orbicular lflets and a larger ± reniform terminal lflet; stem-lvs few, almost sessile, with smaller and narrower lflets; all lflets ± lobed or angled and (especially the basal lvs) sparsely hairy above and on the margins. Infl. at first corymbose. Sepals greenish-violet with narrow white margins. Petals white, narrow, twice as long as the sepals, sometimes 0. *Stamens* 4. Style c. 0·5 mm. Young siliquae usually overtopping the unopened fls. Ripe siliquae 18–25 × c. 1 mm., slightly beaded, ± erect on slender upwardly curved stalks 5–10 mm. Seeds 1 mm., very narrowly winged. Fl. 4–8. Homogamous. Rarely visited by small insects and automatically self-pollinated. $2n=16^*$. Th.

Native. Common throughout the British Is. on bare ground, rocks, screes, walls, etc. 111, H40, S. Reaches 3800 ft. in Breadalbane. Through most of the northern hemisphere.

*6. C. trifolia L.

A perennial herb with branched creeping scaly stock and ascending simple ± glabrous stem, 20–30 cm. *Basal lvs* very long-stalked with half-sheathing bases, *pinnately trifoliolate* with roundish angled or lobed lflets, sparsely hairy above, violet-tinged below. Stem-lvs 1–2, trifoliolate or simple. Infl. few-fld. Sepals green, white margined, the inner pair strongly saccate. Petals white or pale pink, four times as long as the sepals, spreading. Stamens 6; anthers yellow. Siliquae 20–25 × c. 2 mm., ascending on long stalks. Seeds 2·5–3 mm., not winged. Fl. 4–6. H.

Introduced. An occasional garden-escape. Moist shady woods in the mountains of C. Europe, especially on calcareous substrata.

*C. raphanifolia Pourr., a robust glabrous perennial, 30–60 cm., resembles *C. pratensis* but has all its lvs lyrate with large orbicular terminal lflet; fls large, lilac; fr. broader than in *C. pratensis*, winged. Occasionally established. France, Spain, Italy.

34. DENTARIA L.

Perennial herbs with *creeping ±fleshy scaly rhizomes*. Lvs not in a basal rosette, pinnate or palmate. Fls large, purple or white. Stamens 6. Fr. a siliqua as in *Cardamine*. Seeds with dilated ± triangular funicle, not winged.

Sixteen spp. in C. and E. Europe, Caucasus, Siberia, E. Asia, and eastern N. America.

1. D. bulbifera L.   Coral-wort.

Cardamine bulbifera (L.) Crantz

A perennial herb with a creeping branched *whitish rhizome bearing triangular fleshy scales* with laciniate margins and minute rudimentary pinnae at their tips. Stem simple, erect or ascending, 30–70 cm., glabrous, lfless below. Lower stem-lvs short-stalked, pinnate with a terminal lflet and 2–3 pairs of smaller lateral lflets; middle stem-lvs ternate, uppermost simple; all lflets lanceolate or narrowly oblong, acute, entire or ± irregularly gland-toothed, finely ciliate and often with sparse appressed hairs on the upper surface. *Upper lvs with brownish-violet axillary bulbils*. Infl. a few-fld corymb. Fls 12–18 mm. diam. Sepals violet-tipped, the inner pair saccate. Petals purple, pale pink or rarely white, 3 times as long as the sepals. Siliquae 20–35 × 2·5 mm., but very rarely ripening in the British Is. Seeds 2·5 mm. Fl. 4–6. The fls are rarely visited by insects and do not ordinarily set seed, the plant being reproduced by bulbils. $2n = $ c. 96*. G.

Native. Very local in woods, usually on calcareous soil. Devon, S.E. England, the Chiltern Hills, Stafford and Ayr. 13. C. Europe from C. France eastwards to Italy and the Balkans in the south, and to S. Sweden and S. Finland in the north; S.W. Russia, Asia Minor; the Caucasus. A characteristic species of base-rich beechwoods in C. Europe.

35. BARBAREA R.Br.

Biennial or perennial herbs with erect angular stems, glabrous or with sparse simple hairs. Basal lvs in a rosette, lyrate-pinnatifid; stem-lvs ± sessile, amplexicaul. Infl. dense. Inner sepals somewhat saccate; petals clawed, yellow; stamens 6, without appendages; ovary with 24–32 ovules; style longish, stigma slightly 2-lobed. *Fr. a bluntly 4-angled siliqua, the valves with strong midrib* and a lateral network. Seeds ellipsoidal, unwinged, in 1 row in each cell.

About 12 spp. in Europe, the Mediterranean region, N. Asia and N. America.

1	Upper stem-lvs simple, toothed or shallowly lobed.	2
	Upper stem-lvs pinnately divided or cut.	3
2	Widest part of basal lvs usually falling below the rounded ±cordate terminal lobe; fl.-buds glabrous; siliquae with persistent style 2–3 mm.	**1. vulgaris**
	Widest part of basal lvs falling within the very large ovate-oblong non-cordate terminal lobe; fl.-buds hairy; siliquae with persistent style 0·5–1·5 mm.	**2. stricta**
3	Basal lvs with large terminal and 3–5 pairs of lateral lobes; siliquae 1–3 cm.	**3. intermedia**
	Basal lvs with small terminal and 6–10 pairs of lateral lobes; siliquae 3–6 cm.	**4. verna**

1. B. vulgaris R.Br. 'Winter Cress', 'Yellow Rocket'.

Erysimum Barbarea L.

A biennial or perennial herb with stout yellowish tap-root and erect, branching, glabrous stem, 30–90 cm. *Rosette lvs stalked, lyrate-pinnatisect with a rounded often cordate terminal lobe shorter than the rest of the lf*, and *5–9 oblong lateral lobes*, the *uppermost pair at least equalling the width of the terminal lobe*; lower stem-lvs lyrate-pinnatisect with a few small lateral lobes; *uppermost simple, ovate*; all *deep-green*, shining, glabrous, with coarsely toothed or sinuate margins. Infl. at first very dense, the width of the flowering part exceeding its length. Fl. 7–9 mm. diam. Buds glabrous. Petals bright yellow, about twice as long as the sepals. *Siliquae* (Fig. 20 A) 15–25 × 1·5–2 mm., standing ± *erect*

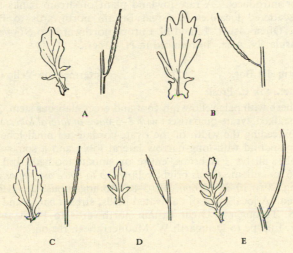

Fig. 20. Stem-leaves and siliquae of *Barbarea*. A, *B. vulgaris*; B, *B. vulgaris* var. *arcuata*; C, *B. stricta*; D, *B. intermedia*; E, *B. verna*. × ½.

on their 4–5 mm. long stalks; *persistent style 2–3 mm*. Seeds yellowish-brown, ovoid, 1·3–1·5 mm. Fl. 5–8. Homogamous. Visited by hive-bees, beetles and flies. $2n = 16^*$. H.

Var. *arcuata* (Opiz) Fr. (*B. arcuata* (Opiz) Rchb.) differs from var. *vulgaris* in its *yellow-green* colour, the *cuneate base of the terminal* lobe of the rosette lvs, the laxer raceme whose length exceeds its breadth during flowering, the somewhat longer petals, and the *upwardly curving siliquae* (Fig. 20 B), *5–8 times the length of their spreading stalks* and the *persistent style only c. 2 mm*. (2·5–3 mm. in var. *vulgaris*).

Native. A plant of hedges, stream banks, wayside and other damp places. 107, H40, S. Throughout Great Britain and Ireland, common in the south but less so further north and absent from the Outer Hebrides and Shetland. Europe to 65° N. in Finland, N. Africa, Asia. Introduced in N. America, Australia and New Zealand.

2. B. stricta Andrz.   'Small-flowered Yellow Rocket.'

A biennial herb with stout yellow tap-root and erect simple or branched glabrous stem, 60–100 cm., the *branches stiffly ascending*. Rosette *lvs stalked, lyrate-pinnatifid with a large ovate-oblong non-cordate terminal lobe longer than the rest of the lf* and 1–3 *small lateral lobes, the uppermost pair falling short of the width of the terminal lobe*; lower stem-lvs pinnately lobed; *uppermost simple*, obovate, coarsely sinuate-toothed; all lvs yellowish-green, shining, glabrous. Infl. at first dense. Fls 5–6 mm. diam. Buds hairy. Petals bright yellow, about half as long again as the sepals. *Siliquae* (Fig. 20 c) 20–30 × 1·5–2 mm., *held stiffly erect and appressed to the stem* on thick stalks 3–5 mm. long; *persistent style* 0·5–1·5 *mm*. Seeds reddish-brown, ovoid, 1·5 mm. Fl. 5–8. Homogamous. $2n = 16^*$, 18^*. Hs.

Native or introduced. A rare lowland plant of stream banks and waste places in scattered localities in Great Britain northwards to Perth. 14. Channel Is. Often casual. E. and N. Europe northwards to Norway and the Kola Peninsula; N. Asia. Introduced in N. America.

*3. B. intermedia Bor.   'Intermediate Yellow Rocket.'

B. sicula auct., vix C. Presl

A biennial herb with pale yellow tap-root and erect glabrous stem, 30–60 cm. *Rosette lvs* stalked, lyrate-pinnatisect *with 3–5 pairs of lateral lobes*, the uppermost pair exceeding the width of the ovate-cordate terminal lobe; stem-lvs all deeply pinnatifid with long narrow lateral lobes and a somewhat larger terminal lobe; all lvs ± glabrous, entire or sinuate-toothed. Infl. not very dense. Fl. 6 mm. diam. *Petals* bright yellow, *up to twice as long as the sepals*. *Siliquae* (Fig. 20 d) 10–30 × 2 *mm*. Fl. 5–8. Homogamous. $2n = 16^*$. Hs.

Introduced. A local plant of cultivated fields, stream banks and waysides. 35, H14, S. Throughout Great Britain northwards to Perth and Angus; Ireland. W. Europe to Denmark; W. Mediterranean region.

*4. B. verna (Mill.) Aschers.   'Early-flowering Yellow Rocket', Land-cress.

B. praecox (Sm.) R.Br.

A biennial herb with yellow tap-root and erect usually branching ± glabrous stem 20–70 cm. *Rosette lvs stalked, oblong in outline*, pinnatisect, *with 6–10 pairs of lateral lobes*, the uppermost pair at least equalling the width of the oblong-cordate terminal lobe; *stem-lvs all deeply pinnatifid* with 5–8 pairs of long narrow lobes. Infl. not dense. Fls 7–10 mm. diam. *Petals* bright yellow, *about three times as long as the sepals*. *Siliquae* (Fig. 20 e) 30–60 × 1·5–2 *mm*., *curving upwards on thick stalks* 4–8 mm. long; *persistent style* 1–2 *mm*. Seeds dark red-brown, ovoid, to 2·5 mm. Fl. 5–7. Homogamous. $2n = 16^*$. Hs.

Introduced. Widespread but local in waste and cultivated ground throughout England and Wales, rare in S. Scotland to Perth; Ireland. 54, H11, S. Probably native in the W. Mediterranean region, Canary Is., Madeira, etc., widely naturalized in W. and C. Europe and in N. America, S. Africa, Japan and New Zealand.

36. CARDAMINOPSIS (C. A. Mey.) Hayek

Biennial or perennial herbs with *simple or forked hairs* below and a *basal rosette of usually lyrate lvs.* Inner sepals somewhat saccate; petals clawed, white or pinkish; stamens 6, without appendages; ovary with 12–14 ovules; style short, stigma capitate. Fr. a *compressed unbeaked siliqua,* the *valves with a conspicuous midrib and prominent seeds.* Seeds compressed, not or slightly winged, in 1 row in each cell.

Ten spp. of arctic and alpine plants of the northern hemisphere. Differs from the closely related *Arabis* in the usually lyrate lvs and in the fr. whose septum is thicker and more rigid than in *Arabis,* so that the seeds cannot sink into it but instead bulge the valve-walls outwards.

1. C. petraea (L.) Hiit.   'Northern Rock-cress.'

Cardamine petraea L.; *Arabis hispida* L.; *A. petraea* (L.) Lam.

A perennial herb with a stout tap-root and a branching stock from which arise shoots with rosettes of basal lvs. Flowering stems erect or ascending, 10–26 cm., simple or branched, hispid below with spreading simple hairs, ± glabrous and pruinose above. *Basal lvs lyrate-pinnatifid* or oblong-ovate, *long-stalked,* glabrous or ± hispid with simple and stellate hairs; *stem-lvs spathulate or narrowly oblong,* entire or with a few distant teeth, *narrowed into a short stalk,* ± glabrous. Infl. few-fld, corymbose, elongating in fr. Fls c. 6 mm. diam. Petals white or purplish, twice as long as the sepals. Siliquae 12–30 × 1·5 mm., curving upwards to continue the line of the slender stalks 6–10 mm. long; *valves* 3-*veined* with conspicuous midrib and less distinct branching lateral veins. Seeds yellow, 1–1·5 mm., winged at the apex. Fl. 6–8. Homogamous. Probably self-pollinated. $2n=16$. Ch.

Native. A local plant of alpine rocks in N. Wales, Scotland, the Inner and Outer Hebrides, and Shetland; Ireland: only on the Galtee Mountains (Tipperary) and the Glenade Mountains (Leitrim). 20, H2. Extending from near sea-level to over 4000 ft. in Scotland. An arctic-alpine, preferring calcareous rocks, on the mountains of C. Europe and in the Faeroes, Iceland, Scandinavia, Finland, Siberia and N. America.

37. ARABIS L.

Annual to perennial *calcicolous herbs,* usually hairy *with simple and branched hairs. Stem lvs numerous, sessile.* Infl. with basal 1–2 fls often bracteate. Inner sepals often saccate; petals clawed, variously coloured; stamens 6, without appendages; ovary with 20–60 ovules, style very short, stigma ± entire. *Fr. a strongly compressed unbeaked siliqua,* the *flat valves with a slender indistinct midrib* and a lateral network. *Seeds* ovoid or compressed, often winged, *in 1 row in each cell.*

About 100 spp. throughout the north temperate zone and in Africa.

1	Fls yellowish; siliquae downwardly curved.	**1. turrita**
	Fls white; siliquae ± erect.	2
2	Petals 7·5–15 mm.	3
	Petals c. 5 mm.	4

3 Lvs with 3–6 marginal teeth on each side; petals not more than 10 mm.
                                                                  2. alpina
    Lvs with 2–3 marginal teeth on each side; petals 10–15 mm.   **3. caucasica**
4 Lvs hairy only on their margins; stem glabrous; seeds unwinged.   **5. brownii**
    Lvs hairy on their surfaces; stem hispid, at least below; seeds winged.   5
5 Fls 3–4 mm. diam. in a many-fld raceme; seeds roundish, 1·2–1·5 mm.
                                                                  4. hirsuta
    Fls 5–6 mm. diam. in a 3–6-fld raceme; seeds ovoid, 1·6 mm.   **6. stricta**

Section 1. *Turrita* (Wallr.) Rchb. Stem-lvs amplexicaul; sepals hardly saccate; petals yellowish; siliquae very long, downwardly curved; midrib of valves indistinct; seeds narrowly winged.

*1. A. turrita L.   'Tower Rock-cress', 'Tower-cress'.
A biennial to perennial herb with slender tap-root and a horizontal or ascending stock from which arise non-flowering rosettes and 1 or more erect flowering stems, 20–70 cm., pubescent with simple and stellate hairs, often purple below. *Basal lvs* in a rosette, broadly *elliptical, sinuate-toothed, narrowing below into a long stalk; stem-lvs* 3–5 cm., *numerous, oblong, irregularly toothed, sessile, clasping the stem* with ± rounded basal lobes; *all lvs grey with short dense stellate hairs. Petals very pale yellow,* twice as long as the sepals. *Siliquae* 8–12 cm. × 2–2·5 mm., *all twisted to one side and curving downwards* on erect stalks 4–7 mm. long; valves thick-walled, with many prominent veins but no distinct midrib. Seeds 2·5–3 mm., brown, with a membranous wing. Fl. 5–8. Homogamous. Probably self-pollinated. 2*n* = 16. Hs.
    Introduced. Naturalized on old walls at Cambridge and formerly at Oxford and Cleish Castle, Kinross. Native in C. and S. Europe, Asia Minor and Algeria.

Section 2. *Arabis.* Stoloniferous; stem-lvs amplexicaul; inner sepals distinctly saccate; midrib of valves vanishing unwards; seeds narrowly winged.

2. A. alpina L.   'Alpine Rock-cress.'
A perennial *stoloniferous mat-forming herb* with long slender tap-root and erect flowering stems, 6–40 cm., from whose basal rosettes arise prostrate stolons ending in daughter-rosettes. Basal lvs (Fig. 21 A) obovate or oblong, narrowed into a short stalk-like base; *stem-lvs* oblong acute, sessile, *clasping the stem with rounded basal lobes*; stem and lvs with simple and stellate hairs, and all *lvs* coarsely and irregularly toothed *with 3–6 teeth on each margin.* Infl. dense. Fls 6–10 mm. diam. Sepals with scarious margin. Petals white, twice as long as the sepals. Siliquae 20–60 × 1·5–2 mm., held obliquely on almost horizontal stalks 8–14 mm. long; valves flattened, with indistinct midrib and a faint lateral network. Seeds 1–1·5 mm., brown, winged. Fl. 6–8. Homogamous. Probably self-pollinated, but visited by some insects. 2*n* = 16. Ch.

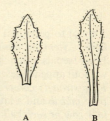

Fig. 21. Leaves of A, *Arabis alpina* and B, *A. caucasica.* × ⅔.

Native. Found first in 1887 and known only from the Cuillin Mountains in Skye at altitudes of 2700–2800 ft. A widespread arctic-alpine plant of the northern hemisphere. In the mountains of S. and C. Europe, Faeroes, Iceland, Scandinavia, Finland, N. Russia, Novaya Zemlya, Alaska, Siberia and the Himalaya.

*3. A. caucasica Willd.   'Garden Arabis.'
A. albida Stev. ex Jacq. f.
Closely allied to *A. alpina* and resembling it in mode of growth, but with only 2–3 *teeth on each lf-margin* (Fig. 21 B); the *stem-lvs with pointed*, not rounded, *basal lobes*; and all the *lvs grey-green or whitish*, more densely hairy than in *A. alpina*. The fragrant fls are larger (c. 15 mm. diam.) and the siliquae have a more prominent midrib. The seeds are at most very narrowly winged. Fl. 3–5. Homogamous. Visited by small bees and other insects. Ch.
Introduced. A native of mountains of the Mediterranean region and Near East to Persia. Much cultivated on rock gardens and walls and occasionally naturalized.

Section 3. *Turritella* C. A. Mey. Midrib of valves weak. No stolons; stem-lvs usually amplexicaul; sepals non-saccate; petals white.

4. A. hirsuta (L.) Scop.   'Hairy Rock-cress.'
Turritis hirsuta L.; incl. *A. sagittata* (Bert.) DC. and *A. ciliata* R.Br.
A biennial or perennial herb with long slender tap-root and 1 or more erect usually unbranched stems, 10–60 cm., rough with many simple and some

Fig. 22. Leaves of *Arabis*. A, *A. hirsuta*; B, *A. brownii*; C, *A. stricta*. Leaves × ⅔.

stellate hairs. Basal lvs (Fig. 22 A) in a rosette, obovate, narrowing gradually into a stalk-like base; *stem-lvs* numerous, erect, ovate to narrowly oblong, blunt, *sessile, truncate at the base or half-clasping* the stem with cordate or sagittate base; all lvs hispid with simple and stellate hairs, and entire or with distant marginal teeth. Infl. dense. Fls 3–4 mm. diam. Sepals often violet with a white margin. *Petals white*, twice as long as the sepals. Siliquae 15–50 × 1·2–1·5 mm., straight, held erect on erect stalks 3–8 mm. long; valves flat, with ± distinct midrib, glabrous. Seeds 1·2–1·5 mm., reddish-brown, with a wing all round but broadest at the apex, 12–18 in each cell. Fl. 6–8. Homogamous. Visited by small bees and other insects. $2n = 32$. H.
Variable, especially in hairiness and in the size and shape of the lvs. Dwarf coastal plants from Wales and the Scottish Islands often have non-amplexicaul

stem-lvs, rounded at the base, and were formerly named *A. ciliata* by British botanists.

Native. Chalk and limestone slopes, limestone rocks and walls, dunes and dry banks throughout Great Britain and Ireland. 103, H28, S. Throughout Europe to 61° N., N. Africa, N. Asia, Japan, N. America.

5. A. brownii Jord.   'Fringed Rock-cress.'

A. ciliata R.Br., p.p.; *A. hibernica* Wilmott

A biennial herb with long slender tap-root and 1 or more erect usually unbranched ± glabrous stems, 7–25 cm. Basal lvs (Fig. 22B) in a rosette, obovate or broadly elliptical, narrowed into a stalk-like base; *stem-lvs ovate, subsessile, rounded at the base, not amplexicaul*; all lvs ciliate but elsewhere ± glabrous, and entire or with distant marginal teeth. Infl. and fls much as in small forms of *A. hirsuta*, but siliquae fewer and broader and seeds unwinged, rounded both at base and apex. Fl. 7–8. H.

Native. Endemic in Ireland where it is confined to dunes and other sandy places along the W. coast from Donegal to Cork. H6. Plants from near Tenby (Pembroke) formerly regarded as conspecific appear to be local variants of *A. hirsuta*. Putative hybrids with *A. hirsuta* (*A. ciliata* var. *hybrida* Syme and *A. sagittata* var. *glabrata* Syme) have been recorded from W. Ireland.

6. A. stricta Huds.   'Bristol Rock-cress.'

A. scabra All.

A perennial herb with slender tap-root and 1 or more erect usually unbranched stems, 8–25 cm., rough below with many simple and a few stellate hairs, glabrous above. *Basal lvs* (Fig. 22C) in a compact rosette, *oblanceolate, sinuate-lobed*, narrowing into a short stalk-like base, dark green; *stem-lvs few, oblong, distantly sinuate-toothed, sessile, with cordate half-clasping base*; all lvs hispid and ciliate with simple and stellate hairs. Infl. of 3–6 fls, each 5–6 mm. diam. Sepals white and scarious at the margin. *Petals cream-coloured*, twice as long as the sepals. Siliquae 25–40 mm., held erect on slender stalks 5–7 mm. long; valves somewhat convex, with conspicuous midrib and a faint lateral network. Seeds dark brown, narrowly winged at the apex. Fl. 3–5. Ch.

Native. Confined to the carboniferous limestone at St Vincent's Rocks and a few other places near Bristol. 2. Mountains of Spain, S. France and the Jura. Introduced in N. America.

38. Turritis L.

Tall glaucous biennial herbs with stiffly erect very lfy stems and with branched and simple hairs only near the base. *Stem-lvs amplexicaul, glabrous.* Infl. ebracteate. Inner sepals barely saccate; petals white or yellowish; stamens 6, without appendages; ovary with 130–200 ovules; style short; stigma capitate, somewhat 2-lobed. *Fr. a long slender ± cylindrical unbeaked siliqua, the convex 1-veined valves with prominent midribs. Seeds ovoid in 1–2 rows* in each cell.

Three spp. in Europe, Asia and Africa.

1. T. glabra L.   Tower Mustard.

Arabis glabra (L.) Bernh.; *A. perfoliata* Lam.

An overwintering or biennial *glaucous* herb with slender tap-root and a usually simple stem, 30–100 cm., sparsely hairy at the base with soft mostly simple hairs. Basal lvs in a rosette, dying before fls appear, obovate or oblong narrowing into a stalk, entire or sinuate-toothed, downy with mostly forked hairs; stem-lvs 2–8 cm., ovate-lanceolate acute, entire, ± glabrous, *sessile, sagittate at the base, amplexicaul.* Fls c. 6 mm. diam. *Petals yellowish- or greenish-white*, less than twice as long as the sepals. *Siliquae* 30–60 × 1–1·5 mm., *held erect on their slender erect appressed stalks* 7–10 mm. long. *Seeds in 2 rows in each cell*, brown, unwinged but with a dark brown edge. Fl. 5–7. Homogamous. Automatically self-pollinated and rarely visited by insects. $2n = 16$, 32. H.

Native. A local plant of dry banks, cliffs and rocks, roadsides and waste places, especially in E. England; rare in S.W. England and E. Scotland to Perth, and not in Wales. 41. Throughout Europe to 67° N., through temperate Asia to Japan, and in Africa. Probably introduced in N. America and certainly so in Australia.

39. RORIPPA Scop.

Annual to perennial herbs, glabrous or with simple hairs. Infl. usually ebracteate. Inner sepals saccate; petals white or yellow; stamens 6, without appendages; ovary with many ovules in each cell; style short, stigma somewhat 2-lobed. *Fr.* variable in shape but always *with convex valves whose midrib is very slender or indistinct or vanishes above the middle*; seeds in 1 or 2 rows in each cell.

About 90 spp., widespread in the northern hemisphere.

1	Fls white.	*2*
	Fls yellow.	*4*
2	Fr. a well-formed siliqua with many good seeds.	*3*
	Fr. dwarfed and misshapen with few or no good seeds.	
	3. microphylla × nasturtium-aquaticum	
3	Fr. 13–18 mm.; seeds distinctly 2-rowed with c. 25 depressions in each face of the seed-coat. **1. nasturtium-aquaticum**	
	Fr. 16–22 mm.; seeds ± in 1 row with c. 100 depressions in each face. **2. microphylla**	
4	Fr. 9–18 mm. **4. sylvestris**	
	Fr. usually less than 9 mm.	*5*
5	Fr. 1·5–3 mm., spherical, with persistent style almost equalling rest of fr. **7. austriaca**	
	Fr. 3–9 mm., ovoid or oblong, with style not more than half as long as rest of fr.	*6*
6	Fr. 3–6 mm., ovoid, on stalks at least twice as long; petals about twice as long as sepals. **6. amphibia**	
	Fr. 4–9 mm., oblong, about equalling their stalks; petals not or slightly longer than sepals. **5. islandica**	

1. R. nasturtium-aquaticum (L.) Hayek   Watercress.

Sisymbrium Nasturtium-aquaticum L.; *Nasturtium officinale* R.Br.

A perennial herb, rarely annual, with non-persistent tap-root and hollow angular glabrous shoots usually 10–60 cm., procumbent and rooting below, then ascending or floating. Lvs lyrate-pinnate, the lowest stalked with 1–3, the upper sessile, auricled, with 5–9 or more lflets; terminal lflets roundish or broadly cordate, lateral lflets broadly elliptical or ovate, all entire or sinuate-toothed. *Lvs and stems remain green in autumn.* Fls 4–6 mm. diam. Petals white, nearly twice as long as the sepals. Siliquae usually 13–18 mm., somewhat ascending on horizontal or slightly deflexed stalks 8–12 mm., beaded, the valves with a distinct slender midrib and a faint lateral network. *Seeds in 2 distinct rows* in each cell, ovoid, 1 mm., with c. 25 polygonal depressions on each face. Fl. 5–10. Homogamous. Visited by various flies and small bees. $2n=32*$. Hel.

Native. A common lowland plant throughout Great Britain in streams, ditches, flushes, etc., with moving water; Ireland; Channel Is. 71, H13, S. S. and C. Europe northwards to Scotland, S. Sweden and Denmark, and eastwards to Posen, Lower Silesia and the Carpathians; N. Africa, W. Asia. Introduced in many parts of the world, and a serious river-weed in New Zealand.

2. R. microphylla (Boenn.) Hyland.   'One-rowed Watercress.'

Nasturtium uniseriatum Howard & Manton; *N. microphyllum* (Boenn.) Rchb.

A perennial herb scarcely distinguishable vegetatively from *R. nasturtium-aquaticum* except by observations of the stomatal index of the lvs (ratio of number of stomata to total number of epidermal initials, expressed as a percentage), which for the lower epidermis is c. 11 % for *R. microphylla* and c. 18 % for *R. nasturtium-aquaticum.* *Lvs and stems turn purple-brown in autumn.* Fls as in the preceding but slightly larger. Siliquae usually 16–22 mm., on stalks 11–15 mm., both siliquae and stalks more slender than in *R. nasturtium-aquaticum.* *Seeds ± in 1 row* in each cell, slightly smaller than in *R. nasturtium-aquaticum*, with c. 100 polygonal depressions on each face. Fl. 5–10, about 2 weeks later than *R. nasturtium-aquaticum.* $2n=64*$. Hel. Apparently an allotetraploid hybrid of *R. nasturtium-aquaticum*, perhaps with a *Cardamine* sp.

Native. Common throughout Great Britain, and perhaps more abundant than *R. nasturtium-aquaticum* in Scotland; Ireland. 61, H10, S. Europe, W. Asia.

3. R. microphylla × nasturtium-aquaticum (*R. × sterilis* Airy-Shaw). The triploid hybrid ($2n=48$) between *R. microphylla* and *R. nasturtium-aquaticum* has been found wild and has also been raised by crossing the parent species. It resembles *R. microphylla* in that the stem and lvs turn purple-brown in autumn. Its stomatal index (lower epidermis) is c. 15 % and its fr. are dwarfed and deformed, with an average of less than 1 good seed per fr. The good seeds are intermediate in size and marking between those of the parents.

Native. A common plant, particularly in N. England, Scotland and Ireland. 37, H6. France and C. Europe (?introduced).

'*N. officinale* var. *siifolium* Rchb.', very robust with ±erect stems to 2 m. and lvs with narrowly elliptical terminal and lateral lflets, may be merely a luxuriant growth form, usually of *R. nasturtium-aquaticum* but sometimes of *R. microphylla*.

'*N. officinale* var. *microphyllum* Rchb.', dwarfed with small lflets, consists of starved terrestrial growth forms of *R. nasturtium-aquaticum*, *R. microphylla* and their hybrids.

Two types of water-cress are in cultivation, green or summer cress, which is *R. nasturtium-aquaticum*, and brown or winter cress, which is *R. microphylla*× *R. nasturtium-aquaticum*. Both types are propagated by cuttings.

4. R. sylvestris (L.) Besser   'Creeping Yellow-cress.'

Sisymbrium sylvestre L.; *Nasturtium sylvestre* (L.) R.Br.

A perennial *stoloniferous* herb with *horizontal or ascending stock* and erect or ascending branched shoots, 20–50 cm., angled, ± glabrous. *Lower lvs* stalked, *pinnate or deeply pinnatifid* with oblong or lanceolate ± toothed or lobed

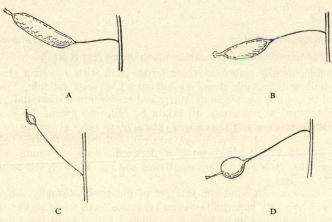

Fig. 23. Fruits of *Rorippa*. A, *R. sylvestris*; B, *R. islandica*; C, *R. amphibia*; D, *R. austriaca*. × 2.

segments, upper lvs sessile, usually pinnatifid with narrower segments, sometimes almost entire. *Infl.* a corymbose panicle *with flexuous axis*. Fls c. 5 mm. diam. *Petals yellow, twice as long as the sepals. Siliquae* (Fig. 23 A) 9–18 *mm.*, linear, ± ascending on horizontal or somewhat deflexed stalks 5–12 mm. Seeds indistinctly 2-rowed, reddish-brown, c. 0·7 mm. Fl. 6–8. Homogamous. Visited by various small bees and flies: apparently self-incompatible. $2n = 48$*. Hel.–H.

Native. Frequent on moist ground by streams and brooks and areas with standing water only in winter, occasionally a persistent garden weed. 81, H15, S. Northwards to Argyll and Angus. Europe northwards to Denmark, S. Sweden and Esthonia; N. Africa. In C. Europe, as in the British Is., often with *Chenopodium rubrum*, *Bidens* spp. *Rorippa islandica* and *Ranunculus sceleratus*.

5. R. islandica (Oeder) Borbás   'Marsh Yellow-cress.'

Sisymbrium amphibium L., p.p.; *S. islandicum* Oeder; *Nasturtium palustre* (L.) DC., non Crantz

An annual or biennial herb with pale slender tap-root and ± erect hollow angled stem, 8–60 cm., simple or branching, glabrous or sparsely hairy below. Lower lvs stalked, deeply lyrate-pinnatifid with narrow irregularly sinuate-toothed lateral segments and an ovate lobed terminal segment; upper lvs similar but short-stalked or sessile, auricled. Fls c. 3 mm. diam., in lax corymbs. *Petals pale yellow, equalling or slightly shorter than the sepals. Fr.* (Fig. 23 B) 4–9 × 1·5–2 *mm.*, *oblong, turgid, curved*, contracted abruptly into the short style; their *stalks* 4–10 *mm.*, horizontal or slightly deflexed. Seeds c. 0·7 mm. Fl. 6–9. Homogamous. Visited by occasional flies and small bees; automatically self-pollinated. $2n = 16$, 32^*. Th.–Hel.

Native. In moist places, especially where water stands only in winter, throughout the British Is., but not in the Hebrides. 93, H 38, S. Cosmopolitan. In the British Is., and Europe often with *Chenopodium rubrum*, etc. (see *R. sylvestris.*)

6. R. amphibia (L.) Besser   'Great Yellow-cress.'

Sisymbrium amphibium L.; *Nasturtium amphibium* (L.) R.Br.

A perennial, usually glabrous, stoloniferous herb with a ± erect stout furrowed hollow branching stem, 40–120 cm. Lvs very variable; lower lvs elliptical or broadly oblanceolate, narrowed below into a short stalk, entire, sinuate, toothed or pinnatifid; all bright or yellowish green. Fls c. 6 mm. diam. Sepals spreading. *Petals bright yellow, about twice as long as the sepals. Fr.* (Fig. 23 C) 3–6 × 1–3 *mm.*, *ovoid, straight*, with persistent style 1–2 mm.; their *stalks* 16–17 *mm.*, horizontal or deflexed. Seeds c. 1 mm. Fl. 6–8. Homogamous. Visited by various small bees and automatically self-pollinated. $2n = 16^*$, 32. Hel.

Var. *variifolium* DC. has the first lvs of the season (usually submerged) deeply and narrowly pinnatifid or even pectinate, with linear segments.

Native. Locally common by ponds, ditches and streams from Somerset and Kent northwards to Westmorland and Berwick, but occasional and probably adventive further north; Ireland. 54, H 30. Europe, Siberia, N. Africa. Often with *Carex riparia, C. acutiformis, C. acuta, Galium palustre, Alisma plantago-aquatica, Glyceria maxima, Phalaris arundinacea, Mentha aquatica*, etc.

*7. R. austriaca (Crantz) Besser   'Austrian Yellow-cress.'

A perennial *stoloniferous* herb with non-persistent tap-root and creeping stock. Stems erect, usually branched, 30–90 cm., ± glabrous. *Lvs elliptical, irregularly toothed*, ± glabrous; the lower lvs broader, stalked, auricled; upper narrower, sessile, *clasping the stem with cordate base.* Fls 3–4 mm. diam. Sepals yellow. Petals yellow, little longer than the sepals. *Fr.* (Fig. 23 D) 1·5–3 *mm.*, *spherical*, with persistent style almost equalling the rest of the fr.; their *stalks* 7–15 *mm.*, slender, horizontally spreading. Seeds c. 1 mm., 6–12 in each cell. Fl. 6–8. Self-incompatible. $2n = 16$. H.

Introduced. A not infrequent casual naturalized in a few localities near ports. Europe (eastwards from the Elbe and Austria) and W. Asia.

The hybrids *R. amphibia* × *R. islandica* (2*n*=24), and *R. amphibia* × *R. sylvestris* have been recorded for the British Is.

***Aubrieta deltoidea** (L.) DC., 'Aubretia', is a *mat-forming* perennial with procumbent shoots and short-stalked *spathulate or rhomboidal lvs* with 1–2 teeth on each side, *the stems and lvs covered with stellate down*. Infl. ebracteate, few-fld. Fls c. 15 mm. diam. Inner sepals deeply saccate. *Petals* usually *deep mauve or red-violet*, long-clawed. Stamens 6, the filaments of the outer pair with a toothed appendage just beneath the anther. Ovary with c. 30 ovules, style long, stigma capitate. Fr. a *short ellipsoidal siliqua* c. 1 cm., the *valves convex, stellate-hairy*, 1-*veined*, with a long persistent style. Seeds ovoid, somewhat compressed, unwinged, in 2 rows in each cell. Fl. 4–5. 2*n*=16. Ch.

Introduced. Much cultivated on rock-gardens and walls for its profusion of purple, red or pink fls, and occasionally escaping. Native in Balkans and Asia Minor.

40. MATTHIOLA R.Br.

Annual to perennial herbs or dwarf shrubs with entire, sinuate or pinnatifid lvs, the *stems and lvs grey with branched hairs*. Infl. an ebracteate raceme. Fls large; sepals erect, the inner pair *saccate*; petals long-clawed; stamens 6, without appendages but the filaments of the inner stamens branched; ovary with numerous ovules, *stigma of 2 erect lobes each with a dorsal swelling or horn-like process*. *Fr. a siliqua with hairy* 1-*veined valves*. Seeds compressed, in 1 row in each cell.

About 50 spp., chiefly in the Mediterranean region, some in C. Asia and S. Africa.

Stem woody below; all lvs entire; siliquae eglandular.   **1. incana**
Stem not woody; lower lvs sinuate or pinnatifid; siliquae glandular.

                                                                          2. sinuata

1. M. incana (L.) R.Br.   Stock, Gilliflower.

Cheiranthus incanus L.

Annual to perennial with stout tap-root and *erect shoot* to 80 cm., stout, glandular-hairy, *woody and lfless* below, with several ascending branches from near the base. *Lvs* in rosettes on old branches and scattered on shoots of current season, *narrowly lanceolate*, lower stalked, upper ± sessile, *all entire or nearly so*, hoary. Fls 2·5–5·0 cm. diam., in a loose raceme. Sepals hairy, with scarious margins. Petals c. 25 mm., purple, red or white. Siliquae 4·5–13 cm. × 3–4 mm., held ± erect on ascending stalks 1–2·5 cm., the *valves* ± compressed, 1-veined, *downy but not glandular*; the persistent stigma-lobes with dorsal processes up to 3 mm. Seeds brown, 3 mm., compressed, broadly winged. Fl. 5–7. Homogamous. The fragrant fls are visited by butterflies. 2*n*=14. Th.–Ch.

Doubtfully native. A rare plant of sea-cliffs in a few localities in S. England (including Isle of Wight) and S. Wales, and recently reported in Durham. 9, S. Apparently native in S. Europe, Canary Isles, N. Africa, Asia Minor, but widely naturalized through cultivation. The 'Ten Weeks Stock' of gardens is var. *annua* Voss. Double-fld forms are also in cultivation.

2. M. sinuata (L.) R.Br.   'Sea Stock.'

Cheiranthus sinuatus L.

A biennial herb with stout tap-root and diffuse shoots, 20–60 cm., very lfy below, glandular-hairy. *Basal lvs* ± oblanceolate, narrowing below into a stalk, *sinuate, or pinnatifid with narrow oblong lobes*; stem-lvs narrowly elliptical, the uppermost linear-lanceolate and entire; all tomentose and *glandular*. Fls 2–2·5 cm. diam., in a loose raceme. Sepals tomentose. Petals 20–25 mm., pale purple. Siliquae 7–12 cm. × 3–5 mm. held ± erect on ascending stalks c. 1 cm., the *valves* compressed, 1-veined, tomentose and roughly *glandular*; the persistent stigma-lobes with prominent tooth-like dorsal processes. Seeds 3–4 mm., brown, oval, much compressed, broadly winged. Fl. 6–8. Homogamous. Fragrant, especially in the evening. 2*n*=14. H.

Native. A rare plant of sea-cliffs and dunes in N. Devon, Kent, Glamorgan, Pembroke; Ireland: Clare, Wexford. 8, H2, S. Formerly in N. Wales and Anglesey but apparently extinct. W. Europe and W. Mediterranean region.

Also found as casuals are *M. bicornis* (Sibth. & Sm.) DC. (Night-scented Stock), from Greece and Asia Minor, whose siliquae, 8–30 cm. long, look like branches and have two long slender horn-like processes from the backs of the stigma-lobes; *M. oxyceras* DC., from the Mediterranean, similar but with shorter processes; *M. tricuspidata* (L.) R.Br., a Mediterranean annual whose cylindrical siliquae appear 3-horned; and *M. fruticulosa* (L.) Maire with narrow entire or sinuate lvs, rusty-red fls, and slender cylindrical non-glandular siliquae, from the Mediterranean region.

Malcolmia maritima (L.) R.Br., often grown in gardens as Virginia Stock, is a diffuse pubescent annual, 10–40 cm., with narrow entire lvs, small violet, pink or white fls and siliquae c. 5 cm., which are slender, flexuous, patent or downwardly curved and constricted between the seeds.

A frequent casual near gardens. Mediterranean region.

41. Hesperis L.

Tall biennial to perennial herbs with toothed or pinnatifid lvs; stem and lvs with simple, branched and glandular hairs. Infl. sometimes bracteate below. Inner sepals saccate; petals large, long-clawed; stamens 6, the filaments of the inner 4 broadened, ± winged; ovary with 4–32 ovules; style short; *stigma deeply 2-lobed, the lobes ± erect, facing each other, not appendaged at the back.* Fr. a linear, cylindrical or somewhat 4-angled siliqua, the valves with distinct midrib and ± distinct lateral veins, beaded. Seeds many in 1 row in each cell.

About 24 spp. chiefly in the Mediterranean region, some in C. Europe and C. Asia.

Lvs toothed, all stalked; fl.-stalks equalling or exceeding the sepals.
                                                        1. matronalis

Lower lvs pinnatifid near the base; upper lvs sessile; fl.-stalks about ⅓ as long as the sepals.   **2. laciniata**

*1. H. matronalis L.   Dame's Violet.

A biennial or perennial herb with tap-root and branching ± woody stock and 1 or more erect lfy stems 40–90 cm., usually branched, ± hairy with short simple and stellate hairs. *Lvs* oblong-ovate to lanceolate, narrowing up the stem, *all short-stalked, finely toothed* and roughly hairy, the upper lvs usually with two glands at the base. Fls c. 18 mm. diam., violet or white; their *stalks equalling or exceeding the sepals.* Siliquae 9 cm., curving upwards on spreading stalks 10–30 mm.; valves ± glabrous, constricted between the seeds. Seeds brown, 3 mm. Fl. 5–7. Homogamous. The fls are fragrant, especially in the evening, and are visited by many kinds of insects including Lepidoptera. $2n = 24, 26, 28$. H.

Introduced. A garden-escape occasionally naturalized in meadows, hedgerows, plantations, grass verges, etc. Europe; W. and C. Asia. Much cultivated.

*2. **H. laciniata** All. differs from *H. matronalis* in having the lower lvs somewhat pinnatifid near the base, and the upper lvs broad and sessile. The fl.-stalks are only about ⅓ as long as the sepals, and the petals are reddish-violet, rarely pale yellow. The siliquae are spreading and glandular-hairy.

Introduced. A casual from the W. Mediterranean region.

42. ERYSIMUM L.

Annual to perennial herbs, sometimes woody below, with very lfy shoots covered with *branched appressed hairs. Lvs usually narrow.* Infl. ebracteate, often ± corymbose. Outer sepals often with a horny projection just beneath the tip, the inner saccate; petals long-clawed, variously coloured; stamens 6, without appendages; *nectaries ± encircling the bases of the outer stamens and also outside the inner stamens;* ovary with 32–72 ovules; style short; stigma capitate, somewhat 2-lobed. *Fr. a ± 4-angled siliqua,* the *valves strongly 1-veined,* hairy. *Seeds in 1 row* in each cell, sometimes winged.

About 80 spp., chiefly European and Mediterranean but some in C. Asia, N. America and Mexico.

Differs from *Cheiranthus* in the presence of median nectaries, the slightly 2-lobed stigma, and the 1-rowed seeds.

1. E. cheiranthoides L. Treacle Mustard.

An annual or overwintering herb with short tap-root and one or more erect ± branched lfy stems, 15–90 cm., with scattered, *short, appressed, 2–3-branched hairs.* Basal lvs in a rosette, dying before the fls open, oblong-lanceolate, acute, narrowed below into a short stalk, hairy, usually irregularly sinuate-toothed; upper lvs narrower, sessile, ± toothed. Fls c. 6 mm. diam. Sepals with branched hairs. *Petals yellow,* twice as long as the sepals. Median nectaries ± abortive. Siliquae 12–25 × c. 1 mm., 4-angled, slightly curved, on slender ascending stalks 5–11 mm.; their valves conspicuously 1-veined, stellate-hairy. Seeds pale brown, 1–1·2 mm., shortly winged at the apex. Fl. 6–8. Homogamous. Visited by insects and automatically self-pollinated. $2n = 16$. Th.

Probably introduced. Locally common as a weed of cultivated ground and waste places at low altitudes especially in S. England, but rarer in the north; reaching Caithness but absent from the Scottish Highlands, Hebrides, Orkney and Shetland. 76, H18, S. Throughout Europe to 68° 30′ N., N. Africa, N. Asia, N. America.

The following key will assist in recognizing spp. of *Erysimum* which have been recorded as casuals in Britain:

1 Fls large, orange.   **E. perofskianum** Fisch. & Mey.
   Fls yellow.   *2*

2 Lower lvs runcinate; fls very small.   **E. aurigeranum** Jeanb. & Timb.-Lagr.
   Lower lvs not runcinate.   *3*

3 Lvs with mostly 3-branched hairs.   *4*
   Lvs with mostly 2-branched hairs.   *5*

4 Resembling *Cheiranthus cheiri* (which, however, has 2-branched hairs);
        stigma ± deeply lobed.   **E. suffruticosum** Spreng.
   Stigma shallowly lobed; fl.-stalks almost equalling the sepals (exceeding
        them in *E. cheiranthoides*).   **E. hieracifolium** L.

5 Annual; lvs narrowly lanceolate; fr.-stalks short, stout, spreading ± hori-
        zontally.   **E. repandum** L.
   Perennial; lvs ± linear, entire; fr.-stalks slender, ascending.
                                                            E. canescens Roth

E. linifolium (Pers.) Gay, an Iberian species with linear lvs and purple fls, is some-times grown in gardens.

The Siberian Wallflower, called by horticulturists *Cheiranthus allionii*, appears in reality to be an *Erysimum*, probably of hybrid origin, but its status and provenance are obscure.

43. CHEIRANTHUS L.

Perennial herbaceous or suffruticose plants with the stems and *narrow entire lvs* covered with *appressed branched hairs*. Infl. an ebracteate raceme. Sepals erect, the inner pair saccate; petals long-clawed; stamens 6, without appendages; *nectaries only round the base of each of the 2 outer stamens*; ovary with 16–60 ovules, style short, *stigma often with 2 deep spreading lobes*. Fr. a ± flattened siliqua, the valves 1-veined, with conspicuous midrib and faint lateral network. Seeds compressed, in 1–2 rows in each cell.

About 10 spp. from Madeira and the Canary Is. to the Himalaya and in N. America.

Very close to *Erysimum*, and best distinguished by the absence of median nectaries.

*1. C. cheiri L.   Wallflower.

A perennial herb with slender tap-root and an erect or ascending branched lfy stem, 20–60 cm., woody below, angled, and covered with forked appressed hairs. Basal lvs in a rosette, 5–10 cm., short-stalked; stem-lvs crowded, subsessile; all oblong-lanceolate, ± entire, with forked hairs especially beneath. Fls c. 2·5 cm. diam. Petals usually bright orange-yellow, at least twice as long as the sepals. Siliquae 2·5–7 cm. × 2–4 mm., ± erect on ascending stalks 5–15 mm., the valves conspicuously 1-veined, flattened, hairy. Seeds pale-brown, ± spherical, 3 mm., winged at the apex, 1-rowed or irregularly 2-rowed. Fls 4–6. Homogamous. The fragrant fls are visited freely by various bees and hover-flies. $2n = 14$. Ch.

Introduced. Well established on walls throughout lowland Great Britain, but not in Isle of Man, the Scottish Highlands, Hebrides, Orkney or Shetland.

Widespread in Ireland. Channel Is. Probably native in the E. Mediterranean region, but widely naturalized through cultivation.

Besides *C. cheiri*, grown in many colour-varieties, *C. mutabilis* L'Hér. (Madeira, Tenerife), with petals at first whitish then passing through orange-brown to deep violet and fading to whitish again, is sometimes seen in gardens. (For Siberian Wallflower (*C. allionii* hort.) see *Erysimum*.)

44. ALLIARIA Scop.

Annual to perennial herbs with slender tap-root and *stems with simple hairs*. Infl. ± ebracteate. Sepals non-saccate; *petals white*, short-clawed; stamens 6, without appendages; ovary with 3–18 ovules, style short, stigma capitate. *Fr. a linear ± 4-angled unbeaked siliqua with 3-veined valves.* Seeds in 1 row in each cell.

Two spp., one widespread in Europe, N. Africa and Asia, the other confined to the Caucasus.

1. A. petiolata (Bieb.) Cavara & Grande   Garlic Mustard, Jack-by-the-Hedge, Hedge Garlic.

Erysimum Alliaria L.; *Sisymbrium Alliaria* (L.) Scop.; *A. officinalis* Bieb.

A biennial herb, often persisting through the formation of adventitious buds on the roots, with a *tap-root smelling strongly of garlic* and an erect, usually unbranched stem, 20–120 cm., sparsely hairy below but glabrous and pruinose above. *Basal lvs* in a rosette, *long-stalked, reniform, cordate*, the margin sinuate or distantly toothed; stem-lvs short-stalked, triangular-ovate, ± cordate, deeply and irregularly sinuate-toothed; *all lvs* thin, pale green, and *smelling of garlic when crushed.* Racemes terminating the main stem and branches, bractless except for the basal 1–2 fls. Fls 6 mm. diam. Petals white, about twice as long as the sepals. Siliquae 35–60 × 2 mm., ± 4-angled, curving at the base so as to stand almost erect on their spreading short thick stalks 4–6 mm.; valves glabrous, 3-veined, with prominent midrib and faint lateral veins. Seeds almost black. Fl. 4–6. Homogamous. Visited by various small insects but automatically self-pollinated. $2n=36^*$, c. 42. Hp.

Native. A common herb of hedgerows and wood margins, shady gardens, wall bases, etc., and locally frequent in beechwoods on chalk. Great Britain northwards to Ross; Ireland; Channel Is. 101, H37, S. Throughout Europe to 68° 30′ N., N. Africa, Caucasus, Asia Minor to Himalaya, in hedgerows, shady waysides, etc., and as a ground-flora herb in deciduous woods on base-rich soils.

45. SISYMBRIUM L.

Annual to perennial herbs with simple hairs and entire or pinnatifid lvs. Sepals not or slightly saccate; petals yellow or white, clawed; stamens 6, without appendages; stigma not or shallowly 2-lobed. Fr. a long slender siliqua whose convex valves have a strong midrib and usually 2 weaker lateral veins; beak very short or 0. Seeds numerous, not winged.

Eighty spp. in the temperate zones of both hemispheres.

1 Infl. bracteate, later with 2–3 siliquae in the axil of each lf-like bract.
                                                                7. **polyceratium**
  Infl. ebracteate.   2

2 Lvs simple, toothed or entire.   6. **strictissimum**
  Lvs ± deeply lobed or divided.   3

3 Siliquae straight, stiffly erect and appressed to the infl. axis.   1. **officinale**
  Siliquae not appressed to the infl. axis.   4

4 Uppermost lvs ± sessile, pinnately divided into linear or filiform segments.
                                                                5. **altissimum**
  Uppermost lvs stalked or narrowed into a stalk-like base, lanceolate or
    hastate, or if pinnatifid not with linear or filiform segments.   5

5 Siliquae 2–5 cm., usually glabrous.   6
  Siliquae 5–10 cm., hairy when young.   4. **orientale**

6 Siliquae with persistent style 1–2 mm.   3*b*. **austriacum**
  Siliquae with persistent style not more than 0·5 mm.   7

7 Perennial; tips of outer sepals horny; valves of siliquae 1-veined (the lateral
    veins very weak).   3*a*. **volgense**
  Annual or overwintering; tips of outer sepals not horny; valves of siliquae
    distinctly 3-veined.   8

8 Young fr. overtopping the open fls.   2. **irio**
  Young fr. not overtopping the open fls.   3. **loeselii**

1. S. officinale (L.) Scop.   Hedge Mustard.
Erysimum officinale L.

An annual or overwintering herb with slender tap-root and stiffly erect stem,
30–90 cm., branched above, usually bristly with downwardly directed hairs.
Basal lvs in a rosette, 5–8 cm., *deeply pinnatifid* with a round terminal lobe
and 4–5 smaller lateral lobes on each side, all ± toothed; stem-lvs with a long
hastate terminal lobe and 1–3 small oblong lateral lobes. *Infl. bractless*, at
first corymbose but lengthening in fr. Fls 3 mm. diam., short stalked. Petals
pale yellow, half as long again as the sepals. *Siliquae* 10–15 × 1 mm., held
stiffly erect on short (2 mm.) stalks and *± appressed to the axis*; their valves
hairy (var. *officinale*) or glabrous (var. *leiocarpum* DC.), 3-veined. Seeds
c. 1 mm., ovoid, orange-brown, c. 6 in 1 row in each cell. Fl. 6–7. Homogamous. Rarely visited by flies and small bees, and automatically self-
pollinated. $2n = 14$, $14 + 4$ ff. Th.
  Native. In hedgebanks, by roadsides and in waste places and as a weed of
arable land throughout Great Britain and Ireland except for Shetland. 111,
H40, S. Native throughout Europe to c. 63° N., and in N. Africa and the
Near East. Naturalized in N. and S. America, S. Africa, Australia and
N. Zealand, Siberia and Greenland.

2. S. irio L.   London Rocket.

An annual or overwintering herb with slender tap-root and erect usually
branched stem 10–60 cm., glabrous or with short appressed hairs. Lower *lvs*
not in a distinct rosette, stalked, deeply *pinnately lobed*, the terminal lobe
larger than the 2–6 distant lateral lobes, all usually somewhat toothed; *upper
lvs stalked* or narrowed into a stalk-like base, with hastate terminal lobe and

a few smaller lateral lobes (Fig. 24A); uppermost lvs sometimes simply hastate. *Infl. ebracteate. Young fr. overtopping the open fls.* Fls 3–4 mm. diam. Petals pale yellow, little longer than the sepals. *Siliquae 3–5 cm. × 1 mm., glabrous,* narrowed at both ends, usually curved and ascending on slender stalks 7–10 mm. (Fig. 24A); *valves 3-veined,* thin-walled and made nodular by the seeds which are visible through the translucent wall; *style not more than 0·5 mm.* Seeds ovoid, hardly 1 mm., yellowish-brown, c. 40, in a single row in each cell. Fl. 6–8. $2n=14+\text{ff}$. Th.

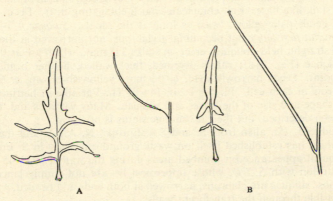

Fig. 24. Fruits and leaves of A, *Sisymbrium irio* and B. *S. orientale.* × ½.

Doubtfully native. On roadsides, walls and waste places in various parts of Great Britain and in the suburbs of Dublin. 21, H1. Probably native in the Mediterranean region, but now widespread in Europe (apart from the far north), N. Africa, the Near East and N. America.

Called London Rocket from its abundance after the Great Fire of 1666; *S. orientale,* not *S. irio,* occurred on bombed sites in 1940–5.

*3. S. loeselii L. differs from *S. irio* in having the bright yellow petals about twice as long as the sepals, the young siliquae not overtopping the open fls, the ripe siliquae only 2–4 cm., and the seeds only c. 0·7 mm. $2n=14$. A casual, occasionally established. Native in S.E. Europe, eastwards from Lombardy and Austria, and W. Asia.

*3a. S. volgense Bieb., a perennial glaucous herb, hairy only below; lower stem-lvs pinnatifid with a large triangular sub-hastate terminal lobe and 2–3 pairs of long narrow lateral lobes; uppermost lvs linear-lanceolate, ±entire (not pinnatifid or hastate as in *S. irio* and *S. loeselii*). Petals bright yellow, c. twice as long as the sepals. Siliquae 2·5–4 cm., 1-veined, the lateral veins being much less distinct than in *S. irio* and *S. loeselii.* Seeds c. 1 mm. A casual, native in S.E. Russia.

*3b. S. austriacum Jacq. (*S. pyrenaicum* (L.) Vill., non L.) which has also occurred as a casual, is a very polymorphic species distinguishable from *S. irio. loeselii* and *volgense* by the longer style (1–2(–3) mm. in fr.). Uppermost lvs pinnatifid or hastate; petals golden-yellow, twice as long as the sepals; siliquae and their clavate stalks so curved and twisted as to give an impression of irregular crowding. S., C. and S.E. Europe. The siliquae are shorter in the Pyrenean than in the C. European subspecies.

*4. S. orientale L.　　　　　　　　　　　　　　　'Eastern Rocket.'

S. Columnae Jacq.

An annual or overwintering herb with slender tap-root and erect branched stem 25–90 cm., ± hairy with short downwardly directed hairs. Basal lvs in a rosette but usually dying before flowering, long-stalked with large terminal lobe and c. 4 pairs of broadly triangular lateral lobes; *stem-lvs* with hastate terminal lobe and fewer and smaller lateral lobes (Fig. 24B); *uppermost* lvs hastate or simply *lanceolate*, entire; all hairy, grey-green, *stalked* or narrowing into a stalk-like base. Infl. ebracteate, much elongating in fr. Fls c. 7 mm. diam. Petals pale yellow, twice as long as the sepals. *Siliquae* 4–10 cm. × 1–2 mm., *at first hairy* but becoming ± glabrous, not narrowed at the lower end, ± straight, held obliquely erect on stalks 3–5 mm., which are as thick as the siliquae (Fig. 24B); valves 3-veined, thick-walled, neither beaded nor translucent. Seeds narrowly ovid, c. 0·9 mm., yellowish-brown, c. 60 in a single row in each cell. Fl. 6–8. $2n=14+\text{ff}$. Th. Variable in hairiness and in the shape and size of the leaves and siliquae. Many varieties and 'forms' have been described, but their taxonomic status is not clear.

Introduced. An alien from S. and S.E. Europe, N. Africa and the Near East which has established itself on waste ground especially in S. England. Its frequent appearance on bombed areas during the war of 1939–45 led to its confusion with *S. irio*, whose uppermost lvs are not simply lanceolate, and whose siliquae are glabrous, narrowed at both ends and beaded, with the seeds visible through the translucent walls.

*5. S. altissimum L.　　　　　　　　　　　　　　　'Tall Rocket.'

S. pannonicum Jacq.; *S. Sinapistrum* Crantz

An annual or overwintering herb with slender tap-root and erect branched stem, 20–100 cm., ± hairy below but usually glabrous and pruinose above. Basal lvs dying before flowering, stalked, roughly hairy, runcinate-pinnatifid with 6–8 pairs of narrowly triangular distantly toothed lobes; middle stem-lvs deeply pinnatisect with narrow distantly and irregularly toothed or hastate lobes; *uppermost sessile*, glabrous, *deeply pinnatisect with linear or filiform entire segments*. Infl. ebracteate, much elongating in fr. Fls c. 11 mm. diam. Petals pale yellow, about twice as long as the spreading sepals. *Siliquae* 5–10 cm. × 1–1·5 mm., ± glabrous, straight, and *held obliquely* erect on stalks 6–10 mm.; valves thick-walled, with prominent midrib and much weaker lateral veins. Seeds narrowly ovoid, 0·8–1 mm., yellowish-brown, c. 60 in 1 row in each cell. Fl. 6–8. Homogamous. $2n=14$. Th.

Introduced. A native of E. Europe and the Near East which has established itself in waste places in several parts of Great Britain and elsewhere in N.W. Europe and in N. America. 96, H6, S.

*6. S. strictissimum L.

A *perennial* herb with a stout root-stock tasting like horse-radish and a stiffly erect, branched, very lfy stem reaching 1 m. or more, ± pubescent below with short downwardly directed hairs. *Lvs* 3–8 × 1–3 cm., *simple, elliptical or ovate-lanceolate*, acuminate, entire or glandular-toothed, hairy beneath, shortly stalked; uppermost lvs lanceolate, sessile. Infl. of numerous racemose panicles terminating the main stem and branches. Fls 4–6 mm. diam. Petals bright yellow, less than twice as long as the sepals. Siliquae 3–5 cm. × 1–1·5 mm., reddish-brown, glabrous, curving upwards on stalks 4–6 mm.; valves thick-walled, strongly 3(–5)-veined. Seeds narrowly ovoid, c. 2 mm., reddish-brown, 15–20 in 1 row in each cell. Fl. 6–8. Homogamous. H.

Introduced. A native of C. and E. Europe which has established itself in waste places in various parts of Great Britain.

*7. S. polyceratium L.

An annual or overwintering foetid herb with slender tap-root and ascending ± glabrous stem, 20–40 cm., branched at the base, the branches decumbent. Basal lvs in a rosette, pinnatifid with broadly triangular lobes; stem-lvs with irregularly hastate terminal lobe and 1–3 pairs of triangular lateral lobes; all lvs ± glabrous, dull green. *Fls* 2 mm. diam., 1–3 *together* on short stalks *in the axils of the upper lvs.* Petals pale yellow, about half as long again as the sepals. Siliquae c. 2 cm. × 1 mm., sparsely hairy, somewhat beaded, short-stalked, narrowing gradually from near the base and curving outwards; valves 3-veined. Seeds c. 10 in 1 row in each cell. Fl. 6–8. Homogamous. H.

Introduced. A native of the Mediterranean region and the Near East, said to have been sown by Dr Goodenough at Bury St Edmunds (Suffolk) and found near docks as a casual.

Microsisymbrium lasiophyllum (Hook. & Arn.) O. E. Schulz, with cream-coloured fls and slender deflexed siliquae; and *Lycocarpus fugax* (Lag.) O. E. Schulz (*Sisymbrium fugax* Lag.) with the siliquae clavate and hooked distally, also occur as casuals.

46. ARABIDOPSIS (DC.) Heynh.

Slender annual to perennial herbs. Stems with both *simple and branched hairs.* Infl. often bracteate. Inner sepals not or hardly saccate; petals white, lilac or yellow; stamens 6, without appendages; ovary with 20–72 ovules, style short, stigma entire or shortly 2-lobed. Fr. a *slender siliqua, the convex* 1-*veined valves having ± prominent midribs. Seeds ovoid, usually in* 1 *row in each cell.*

Thirteen spp. chiefly in Europe, Asia and N. Africa, some in N. America.

1. A. thaliana (L.) Heynh.   Thale Cress, 'Common Wall Cress.'

Arabis Thaliana L.; *Sisymbrium Thalianum* (L.) Gay

An annual or occasionally biennial herb with slender tap-root and 1 or more erect stems, 5–50 cm., branched above in large plants, roughly hairy below with mostly simple hairs, ± glabrous above. Basal lvs in a rosette, elliptical or spathulate, stalked, grey-green with simple and branched hairs; stem-lvs sessile, narrowly oblong or lanceolate, narrowed to a non-clasping base, ± glabrous or with branched hairs beneath and on the margins; lower lvs distantly toothed, upper stem lvs entire. Fls 3 mm. diam. Petals white, about twice as long as the sepals. Siliquae 10–18 × 0·8 mm., somewhat curved, held obliquely erect on slender spreading stalks 5–10 mm.; valves glabrous, with conspicuous midribs and very faint lateral veins. Seeds yellowish-brown, 0·5 mm., 25–40 ± in 1 row in each cell. Fl. 4–5 and sometimes 9–10. Homogamous. Sometimes gynomonoecious. Visited by various small insects and automatically self-pollinated. $2n = 10$. Th.–H.

Native. Fairly common on walls and banks, in hedgerows and waste places, and on dry soils throughout the British Is. 109, H33, S. Throughout Europe to 68° 30′ N. in Sweden, Mediterranean region, E. Africa, N. and C. Asia to Japan. Introduced in N. America, S. Africa and Australia.

47. CAMELINA Crantz

Annual or overwintering herbs with erect stems and simple hastate sessile amplexicaul stem-lvs. Infl. ebracteate. Sepals non-saccate; petals yellow or whitish, spathulate; stamens 6, without appendages; ovary with 8–24 ovules, style distinct, stigma capitate. *Fr.* an *obovoid or pear-shaped latiseptate silicula* with a long style, the *valves convex, strongly keeled and ± winged, with a strong midrib vanishing above.* Seeds numerous, ovoid-angular, in 2 rows in each cell.

Ten spp. in C. Europe, the Mediterranean region and C. Asia.

1   Silicula truncate or slightly emarginate above, the valves very turgid,
      compressible even when ripe.   **3. alyssum**
    Silicula rounded above, the valves convex but not turgid, hard and
      imcompressible when ripe.   *2*

2   Silicula (excl. style) about 1½ times as long as wide, yellowish, with
      prominent midrib; seeds 1–2 mm.   **1. sativa**
    Silicula (excl. style) slightly longer than wide, grey-green, with indistinct
      midrib; seeds not more than 1 mm.   **2. microcarpa**

*1. C. sativa (L.) Crantz   Gold of Pleasure.

Myagrum sativum L.

An annual or overwintering herb with slender yellow tap-root and erect lfy stems 30–80 cm., branched above, ± glabrous or with simple and branched hairs. Basal lvs oblong-lanceolate, narrowed to the sessile base, entire or sinuate-toothed, rarely pinnate-lobed; middle and upper lvs lanceolate or linear-oblong, sessile, semi-amplexicaul, with short acute auricles, entire or sinuate-toothed; all ± glabrous or hairy. Fls 3 mm. diam. Petals yellow, 1½ times as long as the sepals. *Siliculae* (Fig. 13H) pale yellowish, obovoid, *rounded above*, 6–9 mm. excluding the style, about 1½ *times as long as wide*, on ascending stalks 10–25 mm.; valves convex, reticulate, with prominent midrib and with the margins flattened and appressed to form a narrow wing; style 1·5–2 mm. *Seeds* brown, 1–2 *mm.*, several in each cell. Fl. 6–7. Homogamous. Visited by bees and automatically self-pollinated. $2n=28, 40, 42.$ Th.

Var. *pilosa* DC. has stem and lvs with long simple as well as stellate hairs, and siliculae only 6–7 mm. excluding the style. The type has short simple and stellate hairs and siliculae 7–9 mm.

Introduced. Occasionally found as a weed in corn, flax and lucerne fields, in corn siftings, etc. Throughout Great Britain northwards to Stirling and Fife, and in Ireland, and occasionally naturalized. 62, H9, S. Var. *pilosa* has also been recorded. Probably native in E. Europe and W. Asia, but widespread as a weed especially of flax and through cultivation for its oil-yielding seeds.

*2. C. microcarpa Andrz. ex DC.

C. sylvestris Wallr., ?p.p.

A usually overwintering herb whose stems and lvs are grey-green with dense long simple and short stellate hairs. *Siliculae* obovoid, c. 5 mm. excluding the style, *rounded above, almost as wide as long*, grey-green, with broad wings and indistinct midribs. *Seeds at most* 1 *mm.* $2n = 40$.

Introduced. A casual, occasionally established. C. and S. Europe, N. Africa, W. Asia.

*3. C. alyssum (Mill.) Thell.

Myagrum Alyssum Mill.; *M. foetidum* Bergeret

An annual herb with the lvs usually sinuate-toothed and sometimes pinnatifid; stem and lvs with short simple and stellate hairs. *Siliculae* broadly obovoid or top-shaped, dark olive when ripe, *truncate or slightly emarginate above,* valves very turgid, thinner and softer-walled than in *C. sativa* so as to be *compressible even when ripe.* Seeds 2–2·5 mm. $2n = 40$.

Introduced. A not uncommon casual in flax fields, on ballast heaps and in waste places. S. and C. Europe, but widely introduced.

48. DESCURAINIA Webb & Berth.

Usually annual herbs with *finely pinnatisect lvs* and *branched or stellate* and often glandular as well as simple *hairs.* Infl. ebracteate. Fls small; sepals non-saccate; petals yellowish, usually not exceeding the sepals; stamens 6, without appendages, often exceeding the petals; ovary with 6–85 ovules; style very short, stigma capitate. *Fr. a short siliqua,* the somewhat *convex valves with a strong midrib* and a lateral network of smaller veins. Seeds in 1–2 rows in each cell.

Forty-six spp. chiefly in N. and S. America with a few in Europe, Asia and Macaronesia.

Differs from *Sisymbrium* in the finely pinnatisect lvs and branched hairs.

1. D. sophia (L.) Webb ex Prantl   Flixweed.

Sisymbrium Sophia L.

An annual or overwintering herb with slender tap-root and erect terete stem 30–80 cm., branched above, usually grey with stellate hairs below but ± glabrous above. Lvs greyish-green, ± stellate-hairy, bi- or tri-pinnatisect, the segments ± linear; uppermost lvs sometimes almost simply pinnate with linear segments. Fls 3 mm. diam. Petals pale yellow, about as long as the sepals. Siliquae 15–25 × c. 1 mm., ± cylindrical, curving upwards and held almost erect, making an angle with their spreading very slender stalks c. 1 cm. Seeds narrowly ovoid, c. 1 mm., orange-brown, 10–15 in 1 row in each cell. Fl. 6–8. Said to be self-pollinated. $2n = 28, (56)$. Th.

Doubtfully native. On roadsides and waste places throughout the British Is. except the extreme north, and in the Channel Is., but nowhere common and rare in the north. 80, H9, S. Apparently native throughout Europe to c. 65° N., in N. Africa and across Asia to China and Japan. Introduced in N. and S. America and N. Zealand.

Several American spp. of *Descurainia* have been reported as casuals, amongst them
D. multifida (Pursh) O. E. Schulz, with short ±clavate siliquae; and *D. richardsonii*
(Sweet) O. E. Schulz, densely stellate-hairy, with erect linear siliquae only 6–9 mm.

22. RESEDACEAE

Annual to perennial herbs, rarely woody plants, with spirally arranged simple
or pinnately divided often glandular-stipulate lvs and racemose or spicate infl.
Fls 4–7-merous, zygomorphic, hypogynous or perigynous, hermaphrodite or
rarely unisexual. Calyx usually zygomorphic, of 4–7 persistent sepals; corolla
of usually 4–7, free, entire or laciniate petals, those at the posterior (top) side
being largest; stamens 3–40, inserted on the zygomorphic disk, those at the
anterior side largest and most crowded; carpels 2–6, superior, free or united
below into a 1-celled ovary which often remains open at the top; ovules
numerous, campylotropous, on 2–6 parietal placentae. Entomophilous;
nectar secreted by a hypogynous disk. Fr. a capsule open at the top, rarely
a berry or group of follicles; seeds numerous, with a curved embryo and no
endosperm.

Six genera and about 70 spp., chiefly in the Mediterranean region but
extending to Somaliland and India and with a few spp. in the Canary Is., the
Cape, and California and New Mexico.

1. RESEDA L.

Annual to perennial herbs with simple, pinnatifid or pinnate lvs with glandular
stipules. Fls in spike-like racemes, hypogynous, usually hermaphrodite.
Sepals 4–7. Petals 4–7, those at the back with larger and more deeply and
repeatedly lobed limbs than those at the front. Stamens 7–40, ± free, crowded
to the front, inserted on the fleshy nectar-secreting disk which is broadest at
the back. Carpels 3–6, united below, each with an apical stigma-bearing lobe,
the ovary being open between these lobes. Fr. a 1-celled capsule opening
more widely by the spreading of the apical lobes, but not splitting.

About 60 spp. chiefly in the Mediterranean region and E. Africa.

1	Ovary and capsule usually with 4 apical lobes; petals white.	**3. alba**	
	Ovary and capsule with 3 apical lobes.		2
2	Petals whitish; capsules drooping.	**4. phyteuma**	
	Petals yellow or green; capsules erect.		3
3	Lvs simple, entire; sepals and petals usually 4; capsule 5–6 mm., sub-globular.	**1. luteola**	
	Lvs pinnately lobed or divided; sepals and petals usually 6; capsule 12–18 mm., oblong.	**2. lutea**	

1. R. luteola L.   Dyer's Rocket, Weld.

A biennial glabrous herb with a deep tap-root, producing a rosette of lvs in
the first season and a flowering stem in the second. *Stem* 50–150 cm., stiffly
erect, ribbed, hollow, *simple* or with a few erect branches. Rosette lvs commonly 5–8 cm., narrowly oblanceolate, sessile; stem *lvs* narrowly oblong, the
lower narrowed into a stalk-like base, the upper sessile; all *with entire*

± *undulate margins*. Fls 4–5 mm. diam., in long slender spike-like terminal racemes with or without shorter lateral racemes, yellowish-green, on ascending stalks hardly equalling the sepals. Sepals 4, not accrescent. Petals 4 (3–5) those at the back and sides with the limb divided into 3 or more lobes, the front petal usually entire, linear; all with a small scale-like claw. Stamens 20–25, ± downwardly curved. Capsule 5–6 mm., ovoid to subglobular, divided nearly halfway into 3 acuminate lobes. Seeds 0·8–1 mm., brownish-black, smooth, shining. Fl. 6–8. Homogamous; visited by small bees and self-pollinated. $2n = 24$–28. Hs. (biennial).

Native. A common plant of disturbed ground, arable land, walls, etc., especially on calcareous substrata; to 1000 ft. in Ireland. Throughout Great Britain but rare and local in the north. Throughout Ireland; Channel Is. 98, H40, S. C. and S. Europe northwards to Denmark and S. Sweden; W. Asia; N. Africa; Canary Is. Introduced in America. Formerly much cultivated for the dye, a flavone, which colours textiles yellow.

2. R. lutea L.   Wild Mignonette.

A biennial to perennial glabrous herb with a deep tap-root, a woody stock and erect or ascending *diffusely branched stems*, 30–75 cm., ribbed, solid, ± rough with whitish tubercles, with ascending branches. Basal lvs commonly 2·5–8 cm., in a rosette withering early; stem *lvs* numerous; all ± *pinnatifid* with 1–3(–5) pairs of narrowly oblong, blunt, entire or pinnatifid lobes diverging rather widely from the central part of the lf which is narrowly elliptical to broadly oblanceolate in outline but constricted above the insertions of the lobes; the whole with ± undulate margins. Fls c. 6 mm. diam., greenish-yellow, in short compact conical racemes; stalks erect, tubercled, exceeding the sepals. Sepals usually 6, linear, unequal. Petals usually 6, with ± rounded claws, the 2 upper with 3-fid, the 2 lateral with 2–3-fid, the 2 lower with entire linear limbs. Stamens 12–20, ± downwardly curved, inserted on the nectar-secreting disk which is broadest at the back of the fl. Capsule 12–18 mm., oblong, short-stalked, tubercled, opening by the further divergence of the 3 stigma-bearing lobes. Seeds 1·6–2 mm., black, smooth, shining. Fl. 6–8. ± Homogamous; visited by small bees and other insects, and self-pollinated. $2n = 48$. H.

Variable in lf-shape and in roughness of the stem and capsules.

Native. Waste places, disturbed ground and arable land, especially on calcareous substrata. Great Britain northwards to Lancs and Durham, and probably adventive further north. Ireland. Jersey. 58, H24, S. C. and S. Europe northwards to Denmark and S. Sweden; Asia Minor; N. Africa. Introduced in N. America.

*3. R. alba L.   'Upright Mignonette.'

Incl. *R. suffruticulosa* L.

An annual to perennial glabrous and somewhat glaucous herb with a deep tap-root and a branched ± woody stock from which arise several erect stems, 30–75 cm., ribbed, hollow. *Lvs deeply pinnatifid* with 5–8 pairs of narrow unequal decurrent lobes whose margins are entire and ± undulate. *Fls* c. 9 mm. diam., *whitish*, in dense conical racemes, their ascending stalks about equalling the sepals. Sepals usually 5, linear, acute. Petals usually 5 with short roundish claws and 3-fid subequal limbs. Stamens 11–14 inserted on the funnel-shaped disk which is broadest at the back. Capsule c. 12 mm., oblong, contracted above, opening by the further divergence of the usually 4 short stigma-bearing lobes. Seeds c. 1 mm., reniform, brownish, dull and tubercled. Fl. 6–8. $2n = 20$. (Th.–)H., sometimes biennial.

Introduced. Waste places. A frequent casual near ports and occasionally establishing itself, especially in S.W. England. Mediterranean region; W. Asia to Persia.

*4. R. phyteuma L.

An annual to biennial herb with ascending branched stems, 20–30 cm., somewhat pubescent, ribbed, the ribs rough with tubercles. Lvs obovate-oblong in outline, entire or with 1 pair of narrowly oblong blunt lobes in the upper half, pale green, ± pubescent. *Fls whitish*, in elongating racemes, their stalks about equalling the sepals. Sepals 6, oblong, blunt, accrescent. Petals 6, not exceeding the sepals, their limbs deeply divided into numerous linear lobes. Stamens 16–20. *Capsule* ovoid-oblong, *drooping*, opening by the further divergence of the 3 acuminate stigma-bearing lobes. Seeds greyish, rugose. Fl. 6–8. $2n = 12$. Th.–H. (biennial).

Introduced. A not infrequent casual, well established in Surrey. C. and S. Europe northwards to Austria; W. Asia; N. Africa.

**R. odorata* L., the fragrant Mignonette of gardens, occurs occasionally as an escape. *R. stricta* Pers. was recorded in error.

**Astrocarpus sesamoides* (L.) DC., with 5–6 free carpels spreading widely in fr., each with a single seed, is recorded as a casual.

23. VIOLACEAE

Trees, shrubs or herbs. Lvs usually stipulate, simple. Infl. of racemose type or fls solitary. Pedicels with 2 bracteoles. Fls 5-merous (except gynaecium) regular or zygomorphic, hermaphrodite, hypogynous, heterochlamydeous, corolla spurred when zygomorphic. Stamens 5, alternate with the petals, the 2 lower spurred in zygomorphic fls; anthers connivent round the ovary, introrse; filaments short; connective often elongated. Ovary 1-celled with (2–)3(–5) parietal placentae with 1–many anatropous ovules; style simple, often curved or thickened above; stigma very variously developed. Fr. a capsule or berry; endosperm fleshy, copious; embryo straight.

About 16 genera and 800 spp., almost cosmopolitan. A natural family not clearly related to any other.

1. Viola L.

Herbs, rarely undershrubs. Lvs alternate, stipulate (stalked in the British spp.). *Fls solitary*, rarely 2. *Sepals prolonged into appendages below their insertion. Corolla zygomorphic, the lower petal spurred.* Two lower stamens spurred; connective broad. Ovary with 3 placentae; ovules numerous; style thickened above, straight or curved. Fr. a 3-valved capsule; valves elastic.

About 400 spp. mainly temperate.

1 Style hooked or obliquely truncate; stipules entire to fimbriate, not lobed;
     lateral petals spreading ±horizontal.   *2*
     Style expanded above into a globose head; stipules pinnatifid or pal-
     matifid; lateral petals directed towards the top of the fl.   *11*

2 Style hooked at apex.   *3*
     Style straight, obliquely truncate at apex; lvs orbicular-reniform; plant
     with long slender creeping rhizome.   **9. palustris**

3 Sepals obtuse; plant acaulous (i.e. lvs and fls all radical); petioles and
     capsules pubescent.   *4*
     Sepals acute; plant normally caulescent though in small forms often
     appearing acaulous; petioles and capsules glabrous (except *rupestris*).   *5*

4 Plant with long stolons; fls sweet-scented, dark violet or white; hairs on
     petioles deflexed.   **1. odorata**
     Plant without stolons; fls odourless, usually blue-violet; hairs on petioles
     spreading.   **2. hirta**

5 Petioles and capsules pubescent; lvs all or mostly obtuse; (Teesdale).
                                                                   3. rupestris
     Petioles and capsules glabrous; lvs all or mostly acute or acuminate.   *6*

6 Main axis ending in a rosette of lvs, not growing out into a fl. stem; lvs
     ovate-orbicular; teeth of stipules usually filiform, flexuous, spreading.   *7*
     Main axis without basal rosette, growing out into a fl. stem; lvs ovate to
     lanceolate; teeth of stipules triangular-subulate, ±straight, ascending.   *8*

7 Appendages of sepals large, accrescent in fr.; corolla blue-violet, spur
     paler, stout, furrowed or notched.   **4. riviniana**
     Appendages of sepals small, obsolete in fr.; corolla lilac, spur darker,
     slender, not furrowed or notched.   **5. reichenbachiana**

8 Corolla deep or bright blue; stipules rarely more than ⅓ as long as petiole;
     lvs usually ovate.   **6. canina**
     Corolla pale or nearly white; middle and upper stipules usually half as
     long as petiole or more; lvs lanceolate.   *9*

9 Lvs rounded or broad-cuneate at base, widest at about ⅓ of the distance
     to the apex, rather thick. Heaths.   **7. lactea**
     Lvs truncate or cordate at base, widest very near base. Fens.   *10*

10 Spur short, ±conical, not or scarcely longer than the appendages of the
     calyx; plant creeping below the ground; lvs thin.   **8. stagnina**
     Spur about twice as long as the appendages of calyx; plant not creeping;
     lvs thick.   **6. canina**

11 Perennials; stylar flap (see p. 195) large and conspicuous; spur usually
     twice as long as appendages of sepals or more; petals much longer than
     sepals.   *12*
     Annuals; stylar flap 0 to distinct; spur usually less than twice as long as
     appendages; petals shorter or longer than sepals.   *13*

12 Plant with long creeping rhizome; corolla large (2–3·5 cm. vertically) on
     long (5–9 cm.) pedicels.   **10. lutea**
     Rhizome 0 or short and scarcely creeping, plants ±tufted; corolla
     moderate (1·5–2·5 cm. vertically) on often shorter pedicels.
                                                                  11. tricolor

13 Corolla blue-violet or mainly so, moderate (1·5–2·5 cm. vertically); petals
     longer than sepals; stylar flap distinct; mid-lobe of stipules usually
     lanceolate and entire, not lf-like.   **11. tricolor**

Corolla cream or occasionally tinged blue in upper part (rarely pale blue-
violet but then very small), usually smaller (0·5–2 cm. vertically); petals
usually shorter than or as long as sepals but sometimes longer; stylar
flap 0 or indistinct; mid-lobe of stipules (at least the lower) usually
ovate-lanceolate, crenate-serrate, lf-like.   *14*

14 Corolla 8–20 mm. vertically, ±flat; plant usually large and much
    branched.   **12. arvensis**
    Corolla c. 5 mm. vertically, concave; plant small, not or sparingly
    branched. (Scilly and Channel Is.)   **13. kitaibeliana**

Subgenus 1. VIOLA   Violet.

Stipules not lf-like. Lateral petals spreading ±horizontally. Style hooked or
obliquely truncate at apex. Cleistogamous apetalous fertile fls produced in
summer after the normal ones have ripened seed (fl. times refer to chasmogamous fls). The normal fls of all spp. are pollinated by bees or, less often,
by other insects (apparently never selfed in nature though self-fertile).

Section 1. *Viola*. Style hooked at apex, tapered, the hook about as
long as the diam. of the style. Plants acaulous, i.e. with lvs and fls all
radical. Sepals obtuse. Capsules lying on the ground and spilling out their
seeds.

1. V. odorata L.   Sweet Violet.

Perennial herb with short thick rhizome and *long procumbent stolons rooting
at the ends.* Lf-blades 1·5–6 cm. ovate-orbicular or the summer ones broadly
ovate, deeply cordate at base, obtuse or the summer ones acute, crenate-
serrate, dark green, sparingly hairy or subglabrous; *petioles* long, *with short
deflexed hairs* or those of the spring lvs sub-glabrous; stipules ovate or ovate-
lanceolate, glandular-fimbriate-toothed. Pedicels about equalling lvs, bracteoles about or above the middle. *Fls sweet-scented.* Sepals oblong,
appendages spreading. *Corolla c.* 1·5 *cm., deep violet or* as commonly *white*
with pale lilac or violet spur, rarely purple or pink, very rarely apricot; the
base whitish; four upper petals obovate-oblong. Capsule globose, obscurely
trigonous, pubescent. Fl. (9–)2–4. $2n=20^*$. Hr.
    Native. Hedgebanks, scrub and plantations (rarely natural woods), usually
on calcareous soils. 87, H39, S. From Dunbarton and Angus southwards
and all over Ireland, rather common; Jersey; sometimes escaped but often
native. Europe (except Iceland); Asia Minor, Caucasus, Syria, Palestine;
N. Africa; Macaronesia.

V. hirta × odorata (*V. × permixta* Jord.)

A rather frequent hybrid combining the characters of the parents in various
ways. Fertility variable in different plants, usually partially sterile. Often
very vigorous and floriferous.

2. V. hirta L.   'Hairy Violet.'

Differs from *V. odorata* as follows: *No stolons. Lvs narrower*, triangular-ovate
or oblong-ovate, hairier, lighter green, small at fl., increasing greatly in size
later; *petioles with* longer, more numerous, *spreading hairs*; stipules lanceolate.

Pedicels often exceeding lvs; bracteoles usually below the middle. *Fls without scent.* Appendages of sepals appressed. *Corolla blue-violet* (paler and bluer) rarely white or white streaked with violet or pink. Capsule pubescent, rarely glabrous. Europe from S. Scandinavia and C. Russia to Spain, Corsica, Italy and Macedonia; Caucasus, Turkestan, Siberia (to 95° E.).

Ssp. **hirta**

Lvs usually much longer than broad, flat at maturity. Corolla 1–1·5 cm., not cross-like; spur much longer than the appendages, usually somewhat curved upwards. Fl. 4. $2n=20*$. Hr.

Native. Calcareous pastures, scrub and open woods from Kirkcudbright and Kincardine southwards, widespread and rather common on suitable soils, ascending to 1950 ft.; rare in Ireland (Limerick, Clare, Leix, Wexford, Kildare, Dublin). 77, H6. N. and C. (not S.) Europe.

Ssp. **calcarea** (Bab.) E. F. Warb.

V. hirta var. *calcarea* Bab.; *V. calcarea* (Bab.) Gregory

Lvs often scarcely longer than broad, folded upwards about the midrib at maturity. Corolla less than 1 cm., petals narrow; the four upper arranged like a St Andrew's cross; spur very short; shorter or scarcely longer than the appendages, straight, conical. Fl. 4–5 (1–2 weeks later than the ssp. *hirta*). Hr.

Native. Calcareous pastures from Cornwall and Kent to Glamorgan, N. Lancs and Yorks, local; Down; usually occurring with ssp. *hirta* and apparently connected with it by intermediate forms. 24, H1. France and probably elsewhere.

Section 2. *Rostratae* (Kupffer) E. F. Warb. Dog Violet. Style hooked at apex, the hook shorter than the diam. of the style. Plants caulescent, the fls solitary, axillary (dwarf specimens sometimes appear acaulous). Sepals acute. Capsule erect, shooting out its seeds.

3. V. rupestris Schmidt   'Teesdale Violet.'

V. arenaria DC.

Small perennial tufted herb, 2–4 cm. in fl., to 10 cm. in fr., *pubescent all over*; stock thick, clothed with the remains of dead lvs, not creeping; *with central non-flowering rosette*, fls on axillary branches. *Lf-blades* 5–10 mm., *reniform or ovate-orbicular, obtuse* (rarely a few shortly acuminate), *truncate or shallowly cordate at base,* crenulate; stipules ovate-lanceolate, fimbriate, much shorter than *pubescent petiole.* Sepals ovate-lanceolate, appendages subquadrate in fr. Corolla pale blue-violet, spur paler, lower petal without dark zone and with the veins fainter and shorter than in *V. riviniana,* petals obovate-oblong; spur thick, ± furrowed, at least twice as long as appendages. *Capsule* c. 6 mm., ovoid-oblong, trigonous, *pubescent*. Fl. 5. $2n=20*$. Hs.

Native. Open mossy sheep-grazed turf or bare ground on sugar limestone in Upper Teesdale (Durham). 1. N. Europe and the mountains of C. Europe from 70° 20′ N. in Norway to N. Spain, Corsica, Italian Alps and Macedonia; N. and C. Asia to c. 145° E., south to the Himalaya and Caucasus; N. America from E. Quebec to Alaska, south to Maine and Oregon.

Glabrous forms (rather rarely) occur on the Continent but all the records from this country are errors.

4. V. riviniana Rchb.   'Common Violet.'

Perennial herb 2–20 cm. in fl., to 40 cm. in fr., *glabrous or somewhat pubescent;*
stock rather slender, short, erect, not creeping, but shoots arise in some plants
from adventitious buds on the roots; *with central, non-flowering rosette,* fls on
axillary branches. *Lf-blades* 0·5–8 cm., *ovate-orbicular* (slightly longer than
broad), shortly and often bluntly acuminate (or those of the rosette obtuse),
deeply cordate at base, crenate; *stipules* much shorter than *glabrous petiole,*
lanceolate, *fimbriate,* or the upper almost entire, the fimbriae long, spreading,
filiform, often flexuous. Peduncle glabrous or hairy. Sepals 7–12 mm.,
lanceolate; *appendages large,* about ¼ to ⅓ as long as sepals, often emarginate,
the lowest subquadrate, *accrescent in fr. Corolla* (Fig. 25 A) 14–22 mm., usually
blue-violet but variable in colour, *spur paler,* whitish or pale violet, or more
rarely yellowish, lower petal usually with dark zone outside the whitish base
which has numerous long dark veins; petals broadly obovate, usually overlapping (the lowest to 7 mm. broad); *spur* c. 5 mm., *thick,* scarcely tapering,
furrowed or notched at apex. Capsule 6–13 mm., trigonous, acute, *glabrous.*
Very variable. Fl. 4–6 (rarely again 8–10). Hs.

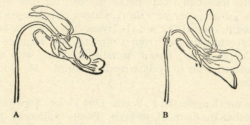

Fig. 25. Flowers of A, *Viola riviniana* and B, *V. reichenbachiana.* × 1·5.

Native. Woods, hedgebanks, heaths, pastures and mountain rocks, ascending
to 3350 ft., on all types of soil if not very wet. 112, H40, S. Common and
often abundant throughout the British Is. Europe from Iceland and Scandinavia (70° N.) to Portugal and N. Spain, Corsica, C. Italy and Greece;
Morocco (Atlas); Madeira.

Ssp. **riviniana**

Lvs large (c. 3 cm.). Fl. branches usually more than 15 cm. Fls 14–22 mm.,
often more than 20 mm. Capsule 9–13 mm. $2n=40$–46^* (usually 40). In the
more sheltered habitats (woods, etc.).

Ssp. **minor** (Gregory) Valentine

Lvs mostly 1·5–2 cm. Fl. branches c. 10 cm. Fls c. 15 mm., petals relatively
narrow. Capsule 6–8·5 mm. $2n=40^*$. In the more exposed habitats.

The distribution of the sspp. is incompletely known. Both are widespread
but ssp. *riviniana* is probably the commoner both in the British Is. and abroad.

5. V. reichenbachiana Jord. ex Bor.   'Pale Wood Violet.'

V. sylvestris Lam., p.p.

Scarcely distinguishable from *V. riviniana* except by the fls or fr. but often
shorter (8–15 cm. in fl., to 25 cm. in fr.), more slender and less hairy, the

peduncles always glabrous, stipules narrower and the upper more fimbriate. *Appendages of sepals small, very short, entire, obsolete in fr. Corolla* (Fig. 25 b) 12–15 mm., *lilac*, rarely pink or white, *spur dark purple*, lower petal with dark zone and with few short dark veins; petals narrower, not overlapping; *spur slender, laterally compressed*, tapering, *not furrowed or notched*. Fl. 3–5 (earlier than *V. riviniana*). $2n = 20^*$. Hs.

Native. Woods, hedgebanks, etc., usually on calcareous soils. 62, H 26. Rather common in S., C. and E. England, more local in N. England, Wales and Ireland, very rare in Scotland. Europe from S. Sweden to Spain and Portugal, Sicily and Greece; Caucasus, Kashmir; Morocco (Atlas); Madeira, Canaries (a ssp.).

6. V. canina L.   'Heath Violet.'

Perennial herb 2–30 cm. (–40 cm. in fr.), glabrous or sparingly and shortly pubescent; stems decumbent to erect, solitary to many together from a short creeping rhizome; *no central non-flowering rosette. Lf-blades* 0·5–8 cm., *ovate to ovate-lanceolate* often ± triangular in outline, the upper sometimes markedly narrower than the lower, obtuse or subacute or a few bluntly acuminate, *truncate or widely and shallowly, rarely deeply, cordate at base*, often somewhat decurrent, shallowly crenate or crenate-serrate, bright or dark green, rather thick; *stipules* ± lanceolate, *subentire, distantly serrate-dentate or fimbriate-dentate, the teeth usually ascending, straighter, stouter, shorter and fewer than those of V. riviniana*, ⅓ *the length of the glabrous petiole or less* (to ½ as long at the middle of the stem, longer in the upper part, in some fen forms). Sepals ± lanceolate, appendages rather large. *Corolla* 7–18(–22) mm., *blue* with very little violet tint, *spur usually yellowish* (occasionally greenish-white), lower petal without a dark zone outside the whitish base, petals obovate (usually about twice as long as broad); spur thick, usually straight, obtuse, sometimes furrowed, usually about twice as long as the appendages of the calyx but sometimes little longer, ± cylindric. Capsule c. 8–9 mm., glabrous, trigonous, blunt, often apiculate. Fl. 4–6 (later than *V. riviniana*). Hp.

A very variable sp., probably divisible into several sspp. The forms of the drier habitats are often ± constant and characteristic but the fen plants are extremely variable. Experimental work is much needed. The sp. is usually recognizable by its flower colour when once known (apart from other characters). The two following sspp. are generally recognized.

Ssp. canina

Upper margin of lf concave; stipules rarely long. Corolla 7–18 mm., deep or bright blue. Very variable. $2n = 40^*$, 40–47. Heaths, dry grassland, dunes and fens, ascending to 1400 ft. 106, H 38, S. Throughout the British Is. but rather local. Europe from Iceland and Scandinavia to C. Spain and Portugal, N. Italy and Macedonia; Greenland; N.W. Asia Minor, C. Asia (rare).

Ssp. montana (L.) Hartm.

Stems erect. Lvs ovate-lanceolate-triangular, often asymmetric, margins straight or convex; middle stipules half as long as petiole, upper often as long. Corolla large, 15–22 mm., pale blue, petals oblong; spur greenish, curved upwards. Only known from fens in Cambridge and Huntingdon. 2. Europe from Iceland and Scandinavia to C. Spain (not in W. France or Portugal), N. Italy, Montenegro and Bulgaria; N. Asia to Kamchatka and Manchuria.

7. V. lactea Sm.   'Pale Heath Violet.'

Perennial herb 4–20 cm., subglabrous; stems ascending, solitary or few; *no central non-flowering rosette*; foliage sometimes purplish tinged. *Lf-blades* (except the small ovate lowest ones which usually soon disappear) 1–4·5 cm., *ovate-lanceolate or lanceolate*, subacute, *rounded or broadly cuneate*, or occasionally subcordate, *at base, broadest at about* ⅓ *of the distance from base to apex*, shallowly crenate-serrate, dark green, *rather thick*; stipules (except the lower) large, the middle ± lanceolate, c. ½ as long as the rather short petioles, *the upper ovate-lanceolate, conspicuous, coarsely and irregularly fimbriate-serrate* or dentate, *as long or somewhat longer than the petioles.* Sepals lanceolate (rarely ovate-lanceolate); appendages rather large. *Corolla* 12–20 mm., *pale greyish-violet*, spur yellowish or greenish; *petals oblong, subacute, c. 3–4 times as long as broad; spur* thick, obtuse, *about twice as long as appendages of calyx*, cylindric. Capsule glabrous, acuminate. Not a variable sp. Fl. 5–6 (later than *V. canina*). $2n = 60*$. Hp.

Native. On heaths; in scattered localities from Yorks and the Isle of Man southwards, very local, commonest in S.W. England; Kerry, W. Cork, Clare. 31, H4. W. and C. France, N.W. Spain, N. Portugal.

8. V. stagnina Kit.   'Fen Violet.'

V. persicifolia auct. ?Schreb.

Perennial herb 10–25 cm., subglabrous, *creeping underground* and sending up stems at intervals from adventitious buds on the roots; *no central non-flowering rosette*. *Lf-blades* 2–4 cm., *triangular-lanceolate*, subacute, *truncate or subcordate at base*, often somewhat decurrent, shallowly crenate-serrate, light green, *rather thin*; stipules lanceolate, subentire or distantly serrate-dentate or fimbriate, the *middle ones usually c.* ½ *as long as the petiole*, but variable, the upper often as long. Sepals lanceolate; appendages large. *Corolla* 10–15 mm., appearing almost circular in front view, *bluish-white or white*, spur greenish; *petals obovate-orbicular, scarcely longer than broad; spur* obtuse, *not or scarcely longer than the appendages of the calyx*, ± conical. Capsule c. 7 mm., glabrous, ovoid, acute. Not a variable sp. Fl. 5–6. $2n = 20*$. Hp.

Native. Fens in Oxford, Norfolk, Suffolk, Cambridge, Huntingdon, Lincoln, Nottingham and Yorks; damp grassy hollows on the limestone in Clare, E. Galway, Longford, Roscommon, E. Mayo and Fermanagh; very local. 8, H6. Europe from S. Scandinavia to E. and C. France, Switzerland, Austria, Hungary and W. Russia; C. Siberia to the Altai.

The following hybrids are often to be found where the parents occur together. They are often very vigorous and in general sterile, though some may produce a small amount of seed: *V. canina × lactea, V. canina × riviniana, V. canina × stagnina, V. lactea × riviniana, V. reichenbachiana × riviniana, V. riviniana × rupestris*.

Section 3. *Stolonosae* (Kuppfer). Style straight, obliquely truncate at apex. Plants acaulous with very long creeping rhizome. Capsule erect, shooting out its seeds.

9. V. palustris L.   'Marsh Violet.'

Perennial herb with long slender creeping rhizome, emitting lvs at the nodes; *aerial stems* 0. *Lvs* usually 3–4, blades 1–4 cm., *orbicular-reniform*, usually very obtuse, sometimes some subacute or bluntly acuminate, cordate at base, obscurely crenate; stipules ovate, glandular-denticulate. *Sepals* oval, *obtuse.* *Corolla* 10–15 mm., *lilac*, rarely white, with darker veins; petals obovate; spur obtuse, longer than appendages. Capsule glabrous, subtrigonous. Fl. 4–7. Hr.

Native. Bogs, fens, marshes and wet heaths ascending to 4000 ft.; common almost throughout the British Is., but absent from several eastern and Midland counties and the Channel Is. Europe from Iceland and Scandinavia to Spain and Portugal, N. Italy and Bosnia; Morocco (mountains); Azores; Greenland; N. America, south to the mountains of New England and Washington.

Ssp. **palustris.** Lvs all obtuse, glabrous, bracteoles below middle of pedicel. Spur slightly longer than appendages. $2n = 48$. The common form. 104, H40.

Ssp. **juressii** (Neves) P. Coutinho

V. epipsila auct. angl.; *V. Juressii* Link ex Neves

Summer lvs subacute or shortly and bluntly acuminate, usually with spreading hairs on the petioles but sometimes glabrous; veins and teeth more prominent. Bracteoles often about the middle of the pedicel. Fls rather larger with longer spur. In scattered localities mostly in S. England and Wales but recorded as far north as Perth and Angus (distribution incompletely known); Ireland; usually growing with ssp. *palustris* and connected with it by intermediate forms. 22, H10. Portugal (the only ssp.), Spain, S. France and perhaps elsewhere.

Subgenus 2. MELANIUM (DC.) Hegi

Stipules lf-like, pinnatifid or palmatifid or at least deeply lobed. Lateral petals directed upwards; lowest very broad, cuneiform. Style geniculate near base, expanded at the apex into a subglobose head with a hollow at one side, with or without a flap below the hollow.

A difficult group which cyto-genetical investigations have helped to elucidate to a considerable extent.

*V. cornuta L.

Perennial herb. Stipules lobed. Fls pale violet, very large, white at base of lower petal; spur c. 6 times as long as appendages. Often grown in gardens, sometimes naturalized. Native of the Pyrenees.

10. V. lutea Huds.   'Mountain Pansy.'

Perennial herb 7–20 cm. with slender creeping rhizome sending up usually solitary, simple, slender, ± flexuous stems. Lowest lvs ovate, obtuse; *lvs* rapidly becoming narrower upwards; upper 1–2 cm., oblong-lanceolate, subacute, cuneate at base, crenate or crenate-serrate, *sparsely clothed with short stiff hairs* at least on the margins and veins below; stipules ± palmatipartite, *the middle lobe oblanceolate-linear, entire,* c. 2 mm. broad. *Pedicels long* (5–)6–9 *cm.,* 1–2(–4) on each stem. Sepals triangular-lanceolate, acute; appendages toothed. *Corolla large, 2–3·5 cm. vertically,* flat, *bright yellow, blue-violet or red-violet, or of these colours variously combined,* always yellow at the base of the lower petal; *spur 2–3 times as long as appendages.* 80–90 % *of pollen-grains with 4 pores,* remainder with 3 or 5. Stylar flap large and conspicuous. Fl. (4–)5–8. Insect pollinated. $2n=48*$. Hp.

Native. Grassland and rock ledges in hilly districts, especially on base-rich but not strongly basic soils; ascending to c. 3500 ft. From Monmouth, Shropshire, Nottingham and Yorks northwards and westwards (not Orkney or Shetland), widespread and locally common; W. Cork, Clare, Carlow, Wicklow, Dublin, Kildare. 66, H7. Belgium, France, W. Germany (Rhine Province), Switzerland. A ssp. further east (to the Tatra).

11. V. tricolor L.   Wild Pansy.

Annual or perennial herb, usually much branched, glabrous or pubescent, *rhizome 0 or short and scarcely creeping.* Lvs very variable, lowest oval or ovate, obtuse, becoming slowly or rapidly narrower upwards; upper ovate, oblong-lanceolate, lanceolate or elliptic, obtuse to subacute, ± cuneate at base, crenate, crenate-serrate or crenate-dentate, glabrous or ± pubescent; stipules *variable but often palmately lobed, mid-lobe usually lanceolate and entire, not lf-like. Pedicels* 2–8 *cm.,* several on each stem. Sepals triangular-lanceolate to linear-lanceolate, acute; appendages variable. *Corolla moderate,* 1·5–2·5 *cm. vertically, longer than sepals, flat, blue-violet* (rarely pink) *or yellow or of combinations of these; spur longer than appendages* (up to twice as long). 80–90 % of *pollen-grains with 4 pores,* remainder with 3 or 5. *Stylar flap distinct.* Fl. 4–9. Pollinated mainly by long-tongued bees.

Ssp. tricolor

Incl. ssp. *saxatilis* (Schmidt) E. F. Warb., *V. Lloydii* auct., *V. orcadensis* Drabble, *V. Lejeunei* auct. p.p., *V. variata* auct. p.p., *V. cantiana* Drabble, *V. lepida* auct.

Annual, or more rarely perennial with several long (15–45 cm.) stems, but scarcely tufted; stems usually much branched. Fls blue-violet or mainly so, not or very rarely all yellow. Spur rather longer than appendages. $2n=26*$. Th. or Hp.

Native. Cultivated and waste ground and short grassland, mainly on acid and neutral soils, ascending to 1460 ft. Throughout the British Is., common in some districts, local in others. 112, H36, S. Europe from Iceland and Scandinavia (70° 16′ N.) to Corsica (mountains), N. Italy and Macedonia; Asia Minor, Caucasus, Siberia (to c. 103° E.), Himalaya.

Ssp. curtisii (E. Forst.) Syme

Ssp. *maritima* (Schweigg.) Hyl.; *V. Curtisii* E. Forst., incl. *V. Pesneaui* Lloyd

Perennial 3–15 cm., usually branched from the base and ± tufted, the stock vertical, not or scarcely creeping, producing many slender lateral branches. Corolla less than 2 cm. vertically, blue-violet, yellow or parti-coloured. Spur variable in length, often twice as long as appendages or rather more. $2n = 26^*$. Hp.

Native. Dunes and grassy places near the sea; S. Devon, Cornwall, whole west coast of Great Britain, east coast south to Northumberland; nearly the whole coast of Ireland; inland in the breckland of Norfolk and Suffolk and by a few lakes in N. Ireland. 44, H20. Shores of the Baltic, North Sea and English Channel.

*V. × wittrockiana Gams   Garden Pansy.

V. tricolor var. *hortensis* auct., ?DC.

A group of hybrids originating about 1830 probably from *V. tricolor* and *V. lutea* ssp. *sudetica* (Willd.) W. Becker. Sometimes escaping and hybridizing with *V. tricolor*. They can usually be distinguished by their large fls with overlapping petals.

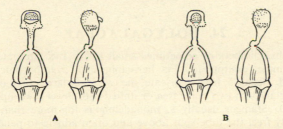

A   B

Fig. 26. Front and side views of the gynoecium of *Viola tricolor* (A) and *V. arvensis* (B), both × 6.

12. V. arvensis Murr.   'Field Pansy.'

Incl. *V. agrestis* auct., *V. segetalis* auct., *V. obtusifolia* auct., *V. latifolia* Drabble, *V. ruralis* auct., *V. Deseglisii* auct., *V. subtilis* auct., *V. anglica* Drabble, *V. arvatica* auct., *V. derelicta* auct., *V. contempta* auct., *V. monticola* auct., *V. vectensis* Williams, *V. variata* var. *sulphurea* Drabble, *V. Lejeunei* auct. p.p.

Similar vegetatively to annual forms of *V. tricolor* ssp. *tricolor* but with the *mid-lobe of stipules usually ovate-lanceolate or ovate, crenate-serrate, lf-like*. Corolla 8–20 mm. vertically, usually small and not longer than calyx but sometimes longer, *flat, cream* or with some blue-violet on the upper and lateral petals but never mainly blue, the spur sometimes violet; spur about equalling appendages. 85–95 % *of the pollen-grains with 5 pores*, remainder with 4 or 6. *Stylar flap 0 or indistinct*. Very variable. Fl. 4–10. Pollinated by various insects, often selfed. $2n = 34^*$. Th.

Native. Cultivated and waste ground, mainly on basic and neutral soils. 112, H40, S. Common throughout the British Is. Europe (except Iceland, Greece, Crete, etc.); Siberia (to c. 93° E.), Turkestan, Caucasus, Asia Minor, Persia, Iraq; N. Africa; Madeira.

13. V. kitaibeliana Schult.

Incl. *V. nana* (DC.) Godr.

An erect annual, 3–10 cm., simple or little-branched, ±grey pubescent all over. Lowest lvs suborbicular, obtuse; upper to 2 cm., oblong-lanceolate, subacute, crenate-serrate; stipules, at least the lower, with large ovate-lanceolate, crenate-serrate lf-like mid-lobe and few lateral lobes. *Corolla very small, c. 5 mm. vertically, about half the length of the sepals, somewhat concave, cream or pale blue-violet*; spur slightly longer than appendages. *70–85 % of pollen-grains with 5 pores*, remainder with 4 or 6. *Stylar flap 0*. Fl. 4–7. $2n = 48*$. Th.

Native. Dunes in the Scilly Is. and Channel Is. 1, S. W. France, etc. Sometimes regarded as a separate sp., but this complex of forms, mainly from S. and E. Europe, requires further study.

The hybrids *V. arvensis × tricolor* and *V. lutea × tricolor* are recorded but appear to be rare.

24. POLYGALACEAE

Herbs, shrubs, climbers or sometimes small trees. Lvs alternate or rarely opposite, simple, exstipulate. Fls hermaphrodite, zygomorphic. Pedicels often jointed. Sepals 5, free, imbricate, the two inner larger than outer, often petaloid. Petals 3–5; two outer free or united with lower; two upper free or minute or 0. Stamens usually 8, monadelphous for more than half their length, rarely free, the tube split above and often adnate to petals; anthers usually opening by an apical pore. Ovary superior, usually 2-celled; style simple. Fr. a capsule, or sometimes fleshy. Seeds often hairy, with a conspicuous strophiole; endosperm usually present; embryo straight.

About 10 genera and 700 spp., throughout the world, except New Zealand, Polynesia and the Arctic.

1. POLYGALA L.   Milkwort

Herbs or small shrubs. Fls bracteate and bibracteolate, in terminal or lateral racemes or spikes, rarely axillary. Inner sepals petaloid, much larger than outer. Petals 3, outer united with lower and adnate to staminal tube. Stamens 8; anthers 1–2-celled, opening by pores. Capsule 2-celled, compressed, obcordate and narrowly winged in our spp., splitting loculicidally at the edges. Seeds one in each cell, pendulous, ±hairy, strophiole 3-lobed.

About 500 spp. Distribution of the family.

The strophiole is sometimes used in distinguishing the spp.; its shape and size, however, seem to vary independently of other specific characters. Experimental work is desirable to elucidate the status of the British spp.

1 Lower lvs smaller than upper, not forming a rosette; veins of inner sepals much-branched, anastomosing.   *2*

    Lower lvs larger than upper, forming a rosette; veins of inner sepals little-branched, not anastomosing.   *3*

2   At least the lower lvs opposite.   **2. serpyllifolia**
    All lvs alternate.   **1. vulgaris**

3   Fls 6–7 mm.; rosette usually not at base of stem, the portion below the
    rosette being lfless; stem lvs usually less than 1 cm.   **3. calcarea**
    Fls not more than 5 mm.; rosette at base of stem; stem lvs usually more
    than 1 cm.   **4. amara**

1. P. vulgaris L.   'Common Milkwort.'

Incl. *P. oxyptera* Rchb.

A perennial herb, 10–30 cm. Stems woody at base, erect or ascending, much-branched. *Lvs scattered, alternate*, lower 5–10 mm., narrowly obovate or

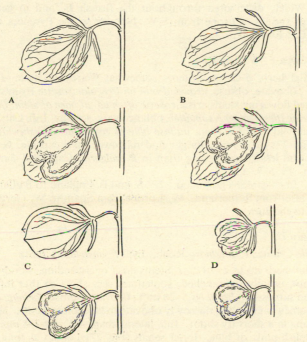

Fig. 27. Fruits of *Polygala*. A, *P. vulgaris*; B, *P. serpyllifolia*; C, *P. calcarea*;
D, *P. amara*. × 3.

spathulate, *upper longer* (up to c. 35 mm.), lanceolate or linear-lanceolate, all
± acute. Infl. usually many-fld, lax. Fls 6–8 mm., blue, pink, or white. Outer
sepals c. 3 mm., green with coloured borders; inner sepals (Fig. 27 A) c. 6 mm.,
ovate, apiculate, lateral veins much-branched, anastomosing. Capsule c.
5 mm. Seeds 2·5–3 mm. Fl. 5–9. $2n=48–56$; c. 70. Chh.

A variable plant.

Native. In grassland, on heaths, dunes, etc. 108, H40, S. Widely distributed throughout the British Is. Europe to c. 68° N. in Norway, W. Asia,
N. Africa.

2. P. serpyllifolia Hose   'Common Milkwort.'

P. serpyllacea Weihe; *P. depressa* Wender.

Similar to *P. vulgaris* but usually smaller and more slender. Stems filiform, scarcely woody at base. *Lower lvs opposite*, ovate, upper larger, often opposite, elliptic-lanceolate. Infl. 3–8-fld, usually short and rather dense. Fls 5–6 mm., commonly gentian blue or slate blue. Inner sepals (Fig. 27 B) 4·5–5·5 mm., longer and slightly broader than capsule, veins much-branched and anastomosing. Capsule 4–5 mm. Seed c. 2·5 mm. Fl. 5–8. Chh.

Some Scottish forms, referred to this species, differ considerably from southern plants.

Native. On heaths and in grassy places, usually on lime-free soils. 110, H 40, S. Widely distributed throughout the British Is. and in general the commonest species. Europe from S.W. Norway to the Pyrenees and eastwards to Czechoslovakia.

3. P. calcarea F. W. Schultz

A perennial herb, 5–20 cm. Stems somewhat woody at base. *Lower lvs* 5–20 mm., obovate, obtuse, *crowded into an irregular rosette* from which the unbranched flowering stems arise, *rosette often not at base of stem, the portion below it being lfless; upper lvs smaller*, oblanceolate, ± acute. Infl. usually rather dense, many-fld. *Fls 6–7 mm., intense blue or sometimes bluish-white*. Inner sepals (Fig. 27 C) c. 5 mm., longer and narrower than capsule, veins little-branched and less anastomosing than in *P. vulgaris*. Capsule 4–5 mm. Seed c. 2·5 mm. Fl. 5–7. Ch.

Native. In calcareous grassland. 21. S. and E. England to Rutland, local; recorded also from Sutherland. W. Europe from Spain to W. Germany and eastward to the Jura.

4. P. amara L.

P. amarella Crantz; *P. uliginosa* Rchb. Incl. *P. austriaca* Crantz

A perennial herb up to c. 10 cm. Stems erect or ascending, woody at base, rather stout, usually unbranched, sometimes very short so that infl. is condensed and subsessile. *Basal lvs* 5–20 mm., obovate, obtuse, *forming a rosette*; stem lvs somewhat smaller, narrowly obovate to lanceolate, acute. Main stem often ending in a short lfy shoot. Infl. lateral, many-fld. *Fls 2–5 mm.*, usually pink, purplish-pink, or blue. Inner sepals (Fig. 27 D) about as long and half as wide as capsule; veins nearly or quite unbranched. Capsule c. 4 mm. Seed 2 mm., oblong. Fl. 6–8. Ch.

Native. In chalk grassland and damp mountain pastures on limestone, rare. 6. Kent, Surrey (extinct); N.W. and Mid-W. Yorks, Durham. Europe from Norway southwards.

25. HYPERICACEAE

Herbs, shrubs or trees with resinous juice. Lvs opposite, simple, usually entire, often gland-dotted, exstipulate. Fls showy, usually terminal, solitary or in branched cymes, actinomorphic. Sepals usually 5, imbricate in bud. Petals usually 5, imbricate or contorted in bud. Stamens numerous, often

± connate in bundles; anthers versatile. Ovary superior, 1-, 3- or 5-celled; ovules usually numerous, anatropous, on axile or parietal placentae. Fr. a septicidal capsule, rarely a berry or drupe.

About 8 genera and 270 spp. in temperate regions; mainly but not exclusively on mountains in the tropics.

1. HYPERICUM L.   St John's Wort

Herbs, shrubs or small trees. Lvs sessile or nearly so. Fls yellow, without nectar. Sepals 5. Petals 5, generally very oblique. Ovary 1-celled with 3 or 5 parietal placentae, or 3- or 5-celled with axile placentae; styles 3 or 5, free or connate. Fr. a capsule, rarely a berry. Pollinated by a considerable variety of insects; eventually self-pollinated if crossing has not previously occurred.

About 220 spp. in temperate regions.

1	Plant ± shrubby; petals deciduous; stamens in 5 bundles united at base only.	*2*
	Stems not or slightly woody and then at base only; petals persistent; stamens in 3 bundles (very rarely 5).	*5*
2	Plant not rhizomatous; stems freely branched; stamens longer than or nearly equalling petals; styles 3.	*3*
	Plant rhizomatous; stems simple or nearly so; stamens distinctly shorter than petals; styles 5.	**4. calycinum**
3	Stems quadrangular; lvs with strong goat-like smell when crushed; sepals deciduous; stamens exceeding petals.	**3. hircinum**
	Stems 2-edged or nearly terete; lvs ± aromatic; sepals persistent; stamens about equalling petals.	*4*
4	Stems with 2 raised lines; sepals about equalling petals; infl. 1–9-fld; styles shorter than stamens; fr. fleshy, indehiscent.	**1. androsaemum**
	Stems slightly 2-edged; sepals distinctly shorter than petals; infl. several- to many-fld; styles longer than stamens; fr. dry, dehiscent.	**2. elatum**
5	Plant pubescent, at least on the lvs beneath.	*6*
	Plant glabrous.	*8*
6	Lvs glabrous above, puberulous beneath; infl. dense, often subcapitate.	**13. montanum**
	Lvs pubescent on both sides; infl. lax.	*7*
7	Stems stiff, erect, ± branched; infl. many-fld; stamens free except at base (dry places).	**12. hirsutum**
	Stems soft, creeping and rooting below; infl. few-fld; stamens united in 3 bundles ¼ way up (wet places).	**14. elodes**
8	Stems quadrangular.	*9*
	Stems terete or with 2 raised lines.	*12*
9	Petals shorter or little longer than sepals.	*10*
	Petals 2–3 times as long as sepals.	*11*
10	Stems up to 20 cm., very slender; petals deep golden yellow.	**15. canadense**
	Stems usually 30–70 cm., stout; petals pale yellow.	**8. tetrapterum**
11	Lvs abruptly narrowed at base, not amplexicaul, glandular dots few or 0; sepals obtuse.	**6. maculatum**
	Lvs ½-amplexicaul, abundantly gland-dotted; sepals acute.	**7. undulatum**

12   Petals little longer than sepals; stems slender, prostrate or ascending.
                                                                9. humifusum
     Petals at least twice as long as sepals; stems usually stout and erect.   *13*

13   Sepals broadly ovate, rounded at apex; lvs of main stems truncate to
         cordate at base.   **11. pulchrum**
     Sepals lanceolate, acute; lvs of main stems narrowed at base.   *14*

14   Sepals entire, not fringed with black stalked glands (common).
                                                                5. perforatum
     Sepals fringed with black stalked glands (rare).   **10. linarifolium**

Section 1. *Androsaemum* (Durham.) Endl. Sepals free, unequal. Petals deciduous. Stamens connate in 5 bundles of 10–25, united at base only, without
scales alternating with the bundles. Ovary incompletely 3-celled.

1. H. androsaemum L.   Tutsan (from French *Toute-saine*).

A glabrous, *slightly aromatic*, half-evergreen, shrubby, branched perennial,
40–100 cm. *Stems with 2 raised lines.* Lvs 5–10 cm., sessile, ovate, obtuse,
with minute translucent glands. *Infl. few-fld.* Fls c. 2 cm. diam.; *sepals* ovate
obtuse, *very unequal, the larger about equalling the petals.* Stamens about
equalling petals. *Styles shorter than stamens.* *Fr. fleshy*, red, turning purple-
black when ripe, *indehiscent.* Fl. 6–8. $2n = 40$. N.
    Native. In damp woods and hedges, local. 82, H 40, S. Scattered through-
out the British Is., northwards to Ross, rarer in the north and east. Western
and southern Europe; Caucasus; Asia Minor; Algeria.

*2. H. elatum Ait.

Similar to *H. androsaemum* in general appearance but taller, up to 150 cm.
Stems slightly 2-edged. *Lvs* 3–8 cm., ovate to ovate-oblong, with a *distinct
aromatic smell when crushed.* Fls 2–3 cm. diam.; *sepals* ovate, *distinctly shorter
than the petals*, persistent and reflexed in fr. Stamens about equalling petals.
Styles longer than stamens. Fr. dehiscent. Fl. 7–8. $2n = 16$. N.
    Introduced. Naturalized in hedges and thickets, particularly in S.W.
England and W. Scotland. Madeira and Canaries.

*3. H. hircinum L.

Similar to *H. androsaemum* in general appearance. *Stems quadrangular.*
Lvs 2·5–6 cm., ovate to ovate-lanceolate, acute, *with a strong goat-like smell
when crushed.* Fls c. 3 cm. diam.; *sepals* lanceolate, acute, shorter than petals,
deciduous in fr. Stamens longer than petals. Styles longer than stamens. Fr.
dehiscent. Fl. 5–8. $2n = 40$. N.
    Introduced. Planted in woods in several localities. Europe from Corsica
to Greece and western Asia, naturalized elsewhere.

Section 2. *Eremanthe* Spach. Similar to *Androsaemum* but with 60–100
stamens in each bundle and ovary 5-celled.

*4. H. calycinum L.   Rose of Sharon, Aaron's Beard.

An extensively creeping rhizomatous evergreen shrub up to 60 cm. Stems
bluntly quadrangular. Lvs 5–10 cm., ovate-oblong to elliptic, subsessile. *Fls*

usually solitary, 7–8 *cm. diam.*; sepals obovate. Styles shorter than stamens. Fl. 7–9. $2n = 20*$. Chw. or N.

Introduced. Naturalized in shrubberies, parks, and by roads in many places. S.E. Europe.

Section 3. *Hypericum.* Sepals connate at base, rarely free, generally subequal. Petals persistent. Stamens connate in 3 bundles at base only, without scales alternating with the bundles. Ovary completely 3-celled.

5. H. perforatum L.   'Common St John's Wort.'

An erect glabrous rhizomatous perennial, 30–90 cm. *Stems* woody at base, *with 2 raised lines. Lvs* 1–2 cm., sessile, elliptic, oblong or linear, obtuse, sometimes mucronulate, *abundantly furnished with translucent glandular dots.* Infl. many-fld. Fls c. 2 cm. diam.; sepals entire, lanceolate, glandular, much shorter than petals. Fl. 6–9. $2n = 32, 36$ (apomictic). Hp. Variable.

Native. In open woods, hedgebanks and grassland; common, especially on calcareous soils. 105, H 38, S. Throughout most of the British Is., absent from a number of northern counties. Europe (except arctic), W. Asia, N. Africa; Madeira, Azores.

6. H. maculatum Crantz   'Imperforate St John's Wort.'
H. quadrangulum auct.; *H. dubium* Leers

An erect glabrous rhizomatous perennial, 20–60 cm. *Stems quadrangular, not winged. Lvs* 1–2 cm., sessile, elliptic, obtuse, *abruptly narrowed at base*, not clasping stem, *glandular dots* 0 *or few.* Infl. several-fld. Fls c. 2 cm. diam.; petals golden yellow; *sepals* entire, *ovate, obtuse*, glandular, $\frac{1}{8}$–$\frac{1}{4}$ length of petals. Fl. 6–8. Hp.

Native. In damp places at margins of woods and in hedgebanks, local. 92, H 37. Scattered throughout the British Is. except the north of Scotland. C. and N. Europe, western Siberia.

Ssp. **maculatum**: Plant slender and rather sparsely branched. Lvs usually without pellucid glands, venation densely reticulate. Infl. branches making an angle of c. 30° with stem. Sepals ovate, entire. Petals entire; dark glands on margin 0, on surface in form of dots. $2n = 16$.
Apparently very local and mainly in W. Scotland.

Ssp. **obtusiusculum** (Tourlet) Hayek: Plant stouter, branches numerous. Lvs often with pellucid glands, venation less dense. Infl. branches making an angle of c. 50° with stem. Sepals often lanceolate, margins eroded. Petals sometimes crenate; dark glands on margin sometimes present, on surface mainly in form of lines. $2n = 32$.
Widely distributed.

H. × **desetangsii** Lamotte

Probably *H. maculatum* × *perforatum*. Stem with 2 very distinct and 2 faint raised lines. Lvs ovate-oblong, sessile, with few large pellucid glandular dots, or sometimes without them.

Native. In hedgebanks and open woods, rare or perhaps overlooked. Recorded from Germany, Belgium, France, Spain and Italy.

7. H. undulatum Schousb. ex Willd.

H. boeticum Boiss.

An erect glabrous stoloniferous perennial, 30–60 cm. *Stems quadrangular, narrowly winged.* Lvs 1–2 cm., sessile, oblong or ovate-oblong, obtuse, ½-amplexicaul, *margins strongly undulate, profusely furnished with translucent glandular dots.* Infl. lax, several-fld. Fls c. 2 cm. diam.; *petals* rather narrow, one half *red-tinged beneath; sepals* entire, *ovate, acute,* glandular, c. ⅓ length of petals. Fl. 8–9. 2*n*=32. Hp. A beautiful plant.

Native. In small boggy patches or marshy ground beside streams on non-calcareous soils, very local. 7. Cornwall, Devon, Pembroke and Cardigan, Merioneth. Western Spain, Portugal, Azores.

8. H. tetrapterum Fr.   'Square-stemmed St John's Wort.'

H. acutum Moench; *H. quadrangulum* L., nom. ambig.; *H. quadratum* Stokes

An erect glabrous perennial, 30–70 cm. *Stems* with slender stolons at base, *quadrangular, angles winged.* Lvs 1–2 cm., sessile, ovate, obtuse, ½-amplexi-caul, furnished with small translucent glands. Infl. many-fld. *Fls c. 1 cm. diam.; petals pale yellow; sepals* entire, *lanceolate, acute, ⅔ length of petals.* Fl. 6–9. 2*n*=16. Hp.

Native. In damp meadows, grassy places beside rivers and ponds and in marshes. 104, H40, S. Throughout the British Is. except the north of Scotland. Europe, N. Africa, Madeira.

9. H. humifusum L.   'Trailing St John's Wort.'

A glabrous *procumbent* or ascending perennial (?sometimes biennial), 5–20 cm. *Stems very slender,* rather woody at base, with 2 raised lines. Lvs 0·5–1(–1·5) cm., elliptic, oblong or obovate, *with translucent glands. Infl. few-fld.* Fls c. 1 cm. diam.; *sepals unequal,* oblong to lanceolate, obtuse or acute, glandular, entire or toothed, teeth sometimes glandular at tip. Fl. 6–9. 2*n*=16. ?Chh.

Native. On heaths, dry moors and in open woods on non-calcareous soils. 105, H38, S. In suitable places throughout the British Is. Western and C. Europe; Madeira, Azores.

10. H. linarifolium Vahl   'Flax-leaved St John's Wort.'

An ascending glabrous perennial, 20–50 cm. Stems subterete, often reddish. *Lvs 1–2·5 cm., linear or linear-oblong,* obtuse, *scarcely gland-dotted, margins revolute.* Infl. few-fld. Fls c. 1 cm. diam.; petals red-tinged, bordered with black glands; *sepals* lanceolate, acute or subobtuse, ⅓–½ length of petals, dotted with black glands and *fringed with slender black, stalked glands* Fl. 6–7. Hp or Chh.

Native. On dry rocky slopes on acid soils, rare. 6, S. W. Cornwall (?extinct), S. Devon, Radnor, Merioneth, Anglesey and Caernarvon; Guernsey, Alderney. W. and S.W. France, Spain, Portugal and Madeira.

11. H. pulchrum L.   'Slender St John's Wort.'

A glabrous erect or ascending perennial, 30–60 cm. Stems terete, often reddish. *Lvs 0·5–1 cm., sessile, broadly ovate-cordate,* obtuse, ½-amplexicaul, *dotted with pellucid glands.* Infl. several-fld. Fls c. 1·5 cm. diam.; petals red-tinged, with a row of black glands near margin; *sepals ovate*, obtuse, ⅓–¼ length of petals, *margins with shortly stalked black glands.* Fl. 6–8. 2n=18. Hp. A very elegant plant.

Native. In dry woods and rough grassy places on non-calcareous soils, local. 111, H40, S. Generally distributed throughout the British Is., except for Huntingdon. Europe, mainly western and central, north to S.W. Norway and Sweden.

12. H. hirsutum L.   'Hairy St John's Wort.'

An erect *pubescent* perennial, 40–100 cm. *Stems* terete, *little branched. Lvs* 2–5 cm., subsessile, ovate, obtuse, glandular-punctate but *without black marginal glands. Infl. lax, many-fld.* Bracteoles with short-stalked black glands. Fls c. 1·5 cm. diam.; petals pale yellow, sparsely glandular at tips; sepals oblong-lanceolate, subacute, ½ length of petals, margins with short stalked black glands. Fl. 7–8. 2n=18. Hp.

Native. In woods and damp grassland, mostly on basic soils. 92, H4. Scattered throughout the British Is. Europe, mainly north and central to c. 68° N. in Norway; W. Asia; Siberia.

13. H. montanum L.   'Mountain St John's Wort.'

An erect slightly pubescent perennial, 40–80 cm. *Stems* terete, rigid, *simple or nearly so, upper internodes longer than lvs. Lvs* 3–5 cm., sessile, ovate to elliptic, obtuse, ½-amplexicaul, *puberulous and with a marginal row of black glands beneath. Infl. dense, often subcapitate,* several-fld. Bracts and bracteoles with glandular teeth. Fls 1–1·5 cm. diam., fragrant; petals pale yellow, nearly eglandular; sepals lanceolate, acute, ½–⅓ length of petals, strongly glandular-toothed. Fl. 6–8. 2n=16. Hp.

Native. In woods, scrub and hedgebanks on calcareous or gravelly soils, local. 53. Scattered throughout England and Wales; Scotland, Ayr and Ross only. Europe to S.W. Norway, W. Asia, Algeria.

Section 4. *Elodes* (Adans.) Koch. Sepals connate at base, nearly equal. Petals persistent. Stamens connate one-third way up in three bundles alternating with bifid scales. Ovary incompletely 3-celled.

14. H. elodes L.   'Marsh St John's Wort.'

A decumbent stoloniferous *tomentose* perennial, 10–30 cm. *Stems soft*, terete, *rooting at nodes below. Lvs* sessile, *suborbicular or oblong, cordate*, half-amplexicaul. Infl. few-fld. Fls c. 1·5 cm. diam.; *sepals* elliptic, obtuse or acute, *with fine, red or purplish glandular teeth.* Fl. 6–9. Hel. The hairs prevent the plant from being wetted when submerged.

Native. In bogs and wet places beside ponds and streams on acid soils. 64, H25, S. Scattered throughout the British Is. in suitable habitats, commoner in the west. Lost through drainage in some localities. Western Europe from Britain and Portugal to Germany, and Italy; Azores.

Section 5. *Brathys* (Mutis) Choisy. Stamens arranged in 3–5 bundles or not definitely grouped, without scales alternating with the bundles. Ovary 1-celled.

15. H. canadense L.

A *slender erect* glabrous annual (sometimes perennial), ± tinged with purplish-red, (3–)12–20 cm. *Stems quadrangular,* ± winged. *Lvs* up to 8 mm., elliptical to linear-oblanceolate, *mostly 3-veined,* semi-amplexicaul, with numerous translucent glands. Infl. 3–9(–30)-fld. Sepals 4 mm., ovate-lanceolate, with pale or reddish streaks, but without black glands. *Petals* 3–4 mm., deep golden yellow, usually with a crimson line on the back, *widely separated so that the fl. has a star-like appearance.*

Native. In wet peaty places near Lough Mask. H2. France, N. America.

26. CISTACEAE

Shrubs or herbs. Lvs usually opposite, less frequently alternate, stipulate or not, simple, mostly with stellate indumentum. Fls hermaphrodite, actino-morphic, hypogynous, solitary or in cymes. Sepals 5 or 3. Petals 5, rarely 3 or 0, usually quickly caducous (often lasting one day only). Stamens numerous. Ovary 1-celled or septate at base with 3 or 5 (rarely 10), often intrusive, parietal placentae; ovules few to numerous, usually orthotropous, rarely anatropous; style simple or 0, stigmas 3 or 5, free or united. Fr. a loculicidal capsule. Seeds with ± curved embryo; endosperm present.

Eight genera and c. 170 spp., mainly Mediterranean region, a few in N. and C. Europe, C. Asia and N. and S. America.

Style 0; annual herb.   1. TUBERARIA
Style filiform; perennial undershrubs.   2. HELIANTHEMUM

CISTUS L.

Differs from *Helianthemum* in being taller shrubs (to 2 m.) and in the 5 (rarely 10)-valved capsule. About 19 spp. in Mediterranean region and Canary Is. Most species and numerous hybrids are grown in gardens.

*C. incanus L. Shrub c. 1 m. Lvs 1–7 cm., stalked, ovate, elliptic or obovate, ± undulate. Sepals 5, ± equal. Petals large, purplish-pink. Style long, straight. Mediterranean region, rarely naturalized.

1. TUBERARIA (Dunal) Spach

Herbs with basal rosette. Infl. cymose. Sepals 5 unequal. Petals 5. *Style* 0 *or very short.* Placentae 3; ovules orthotropous; funicle stout, swollen in the middle; embryo curved but not plicate. About 10 spp., Mediterranean region, only the following more widespread.

1. T. guttata (L.) Fourr.   'Annual Rockrose.'

Helianthemum guttatum (L.) Mill.

Erect or diffuse *annual* 6–30 cm., simple or branched, stellate-pubescent and pilose. Basal lvs forming a rosette, usually dead at fl.; lower cauline lvs 1·5–5 cm., elliptic-lanceolate, 3-veined, subacute, sessile, entire, opposite, exstipulate; upper shorter and narrower, alternate, with conspicuous linear stipules. Fls 5–12 in lax unilateral raceme-like cymes; pedicels slender, bracts conspicuous, oblong or ovate, or 0. Inner sepals ovate, pilose, accrescent;

outer oblong, c. ½ as long as inner. Corolla 8–12 mm. diam., pale yellow, often with a red spot at the base of the petals. Capsule ovoid. Fl. 5–8. $2n = 36*$, 48. Th.

Ssp. **guttata**

Erect, simple or branched. Upper lvs oblong-linear, stipulate, much narrower than lower. Bracts 0.

Native. Dry cliffs, etc. Jersey, Alderney; very local and rare. S. Mediterranean region extending north (in W. Europe only) to N.W. Germany; Canaries.

Ssp. **breweri** (Planch.) E. F. Warb.

Helianthemum Breweri Planch.

Often diffuse and branched from the base, 10 cm. high or less. Upper lvs usually exstipulate, oblong, scarcely narrower than the lower. Bracts usually present and conspicuous.

Native. Exposed rocky moorland near the sea; Anglesey, Caernarvon; W. Cork, W. Galway, W. Mayo; very local and rare. 2, H3. ?Endemic.

2. HELIANTHEMUM Mill.

Shrubs or annual herbs. Infl. cymose. Sepals 5, the 2 outer smaller than the 3 inner. Petals 5. Placentae 3; ovules orthotropous; funicle stout, thickened at the apex; style usually filiform; stigmas large capitate; embryo plicate. Capsule 3-valved. About 80 spp., mainly Mediterranean region, a few in N. and C. Europe, C. Asia and the Cape Verde Is.

The garden 'rockroses', with fls of varied colours, are hybrids derived from *H. chamaecistus*, *H. apenninum* and allied spp.

1   Style nearly straight; stipules present; corolla normally large (2 cm. diam. or more).   2
    Style markedly bent in the middle; stipules 0; corolla small (1–1·5 cm. diam.).   **3. canum**

2   Lvs green above; corolla yellow.   **1. chamaecistus**
    Lvs grey-tomentose above; corolla white.   **2. apenninum**

Subgenus 1. HELIANTHEMUM.

Herbs or shrubs. Style filiform, straight or curved at base, then ascending, shorter than or equalling stamens. Embryo once plicate, cotyledons straight.

1. **H. chamaecistus** Mill.   Common Rockrose.

H. vulgare Gaertn.; *H. nummularium* auct.

Undershrub 5–30 cm., with ± numerous, procumbent or ascending, often rooting branches from a thick woody stock with vertical tap-root. *Lvs* 0·5–2 cm., oblong or the lower oval, *green* and glabrescent or somewhat pubescent *above*, densely white-tomentose beneath, obtuse or subacute, shortly stalked, margins entire, not or slightly revolute, *all stipulate*; stipules lanceolate or linear-lanceolate, c. twice as long as petiole. Fls 1–12 in lax unilateral raceme-like cymes, rhachis and pedicels tomentose; bracts lanceolate, shorter than pedicels. Inner sepals ovate, c. 6 mm., prominently veined, ± pubescent, outer subulate c. ⅓ as long. *Corolla* 2–2·5(–3) *cm. diam.*, rarely considerably smaller, *bright yellow*, sometimes with a small orange spot at base, rarely cream, very rarely white or copper-coloured. Capsule ovoid, tomentose, about equalling the spreading calyx. Fl. 6–9. Pollinated by various insects or selfed, stamens irritable. $2n = 20^*$. Chw.

Native. Basic grassland and scrub; ascending to 2100 ft. Common over most of Great Britain, but absent from Cornwall, Isle of Man, N.W. Scotland, Orkney, Shetland and one or two other counties; in Ireland known only from one locality in Donegal. 93, H1. Europe from Sweden and Finland to C. Spain and Portugal, Italy and the Aegean Is.; Asia Minor, Caucasus, N. Persia.

H. apenninum × *chamaecistus* (*H.* × *sulphureum* Willd.) with pale yellow fls frequently occurs with the parents.

2. H. apenninum (L.) Mill.   White Rockrose.

H. polifolium Mill.

Differs from *H. chamaecistus* as follows: *Lvs* oblong-linear, *grey-tomentose above*, more densely so beneath, margins often more revolute; stipules linear-subulate, the lower scarcely longer than the petiole. Sepals densely grey-tomentose. *Corolla white.* Fl. 5–7. Pollinated by various insects or selfed, stamens irritable. $2n = 20^*$. Chw.

Native. Limestone rocks, Brean Down (Somerset) and near Torbay (Devon). Belgium, W. Germany, France, Spain, Portugal, Switzerland (very rare), N. and C. Italy, Albania, Greece, Crete; Asia Minor; Morocco.

Subgenus 2. PLECTOLOBUM Willk.

Shrubs. Style filiform with a strong sigmoid curve in the middle, longer than the stamens. Embryo biplicate, the cotyledons sharply reflexed in the middle.

3. H. canum (L.) Baumg.   'Hoary Rockrose.'

Undershrub 4–20 cm. with ± numerous procumbent or ascending branches from a central stock. Lvs 2–12 mm., elliptic, obovate-elliptic or ovate, green or occasionally greyish above, densely grey or white-tomentose beneath, obtuse to acute, shortly stalked, margins entire, not or slightly revolute; stipules 0. Fls 1–6(–10) in rather lax, ± unilateral, raceme-like cymes; bracts ovate-lanceolate, small. Inner sepals oval, obtuse, prominently veined, ± pubescent; outer oblong-linear, less than half as long. *Corolla* 1–1·5 *cm. diam.*, bright yellow. Capsule ovoid, about equalling the appressed sepals, pilose. Fl. 5–7. Visited by various bees, stamens irritable. $2n = 22^*$. Chw.

Native. Rocky limestone pastures, very local; ascending to 1750 ft. Glamorgan, N. Wales; Yorks, Westmorland; Clare, W. Galway. 8, H2. Europe from France, S. Sweden (Öland) and C. and S. Germany to Spain, Sicily and Macedonia; Asia Minor, Caucasus; Morocco, Algeria.

Ssp. **canum.** ± prostrate. Lvs with few or many long bristles above, sometimes stellate-pubescent as well (British plants are, in general, hairier than Irish); the longest 6–12 mm. Infl. c. 6–10 cm. usually 2–6-fld. Sepals stellate-pubescent and pilose.

Distribution of the species, but absent from Upper Teesdale. 7, H2. Foreign distribution uncertain but widespread on the Continent.

Ssp. **levigatum** M.C.F. Proctor. Closely prostrate (more so than in ssp. *canum*). Lvs small, the longest c. 6 mm., dark green and glabrous or nearly so above. Infl. c. 3–6 cm., mostly 1–3-fld. Sepals stellate-pubescent and slightly pilose. Cronkley Fell (Upper Teesdale), Yorks only. 1. Endemic.

27. TAMARICACEAE

Trees or shrubs. Lvs small, scale-like or needle-like, exstipulate, alternate. Fls regular, hermaphrodite, hypogynous, solitary or in slender spikes or racemes, 4- or 5-merous. Disk present. Petals imbricate, free. Stamens as many as petals and alternate with them or twice as many and diplostemonous, sometimes united below, anthers usually extrorse. Ovary unilocular with 2–5 basal or parietal placentae each with 2–many anatropous ovules; styles free or united below or stigmas sessile. Fr. a capsule. Seeds with long hairs, with or without endosperm; embryo straight.

Four genera and over 100 spp., mainly Mediterranean region and C. Asia, extending to S.E. Asia, S. Africa and W. Europe.

1. TAMARIX L.

Deciduous, the smaller twigs falling with lvs. Fls in long catkin-like racemes. *Petals without ligules. Stamens free or nearly so; anthers extrorse. Styles short and thick. Seeds with a sessile tuft of hairs at apex; endosperm 0.*

About 75 spp., mainly E. Mediterranean extending to C. and S.E. Asia, S. Africa and W. Europe. A few others are sometimes grown.

Green; lvs of larger twigs acute; buds ovoid; lobes of disk gradually tapered
    to insertion of stamens. **1. anglica**

Glaucous; lvs of larger twigs acuminate; buds globose; lobes of disk rounded
    on either side of insertion of stamens. **2. gallica**

*1. T. anglica Webb   Tamarisk.
T. gallica auct. brit. p.p.

Feathery shrub 1–3 m. *Lvs* of larger twigs c. 2 mm., triangular-lanceolate *acute*; those of ultimate twigs much smaller and densely imbricate, ovate-lanceolate, sessile, acute, *green*, margins scarious. Infl. terminal on the current year's growth or the rhachis continuing as a lfy shoot, consisting of cylindric dense-fld spike-like racemes, 1–3 cm., arranged in a panicle. *Buds ovoid.* Fls c. 3 mm., pink or white; sepals, petals and stamens 5; styles 3; disk 5-lobed, the stamens inserted on the *lobes which gradually taper to the insertion.* Capsule ovoid, trigonous, acuminate. Fl. 7–9. M. or N.

Introduced. Often planted near the sea and naturalized in a number of places along the S. and E. coasts of England from Cornwall to Suffolk; Channel Is. W. France (probably not native north of Finisterre), N.W. Spain, Portugal.

*2. T. gallica L.   Tamarisk.

Differs from *T. anglica* as follows: *Lvs* of larger twigs relatively shorter, mostly *acuminate*; those of the smaller twigs *glaucous*. Infl. more slender. *Buds globose. Lobes of the disk rounded on either side of the insertion of the stamens* so that the disk is sometimes described as 10-lobed. Fl. 7–9. $2n=24$. M.

   Introduced. In similar places to *T. anglica*; from Cornwall to Kent, but less common. S. Europe from Portugal eastwards; Morocco, Algeria; Canaries.

28. FRANKENIACEAE

Herbs or small shrubs. Lvs opposite, exstipulate, often ericoid. Fls actinomorphic, hermaphrodite or rarely unisexual, solitary or cymose. Sepals 4–6, persistent, connate at base. Petals 4–6, clawed, often with a scale-like appendage on the claw. Stamens usually 6; anthers 2-celled, opening lengthwise. Ovary superior, 1-celled, with 2–4 parietal placentae; ovules numerous. Capsule enclosed in the calyx, opening by valves. Seeds endospermous, embryo straight, axile.

   Four genera with about 60 spp., mostly in salt-marshes and salty deserts in temperate and subtropical regions.

1. FRANKENIA L.

Fls nearly always hermaphrodite. Petals and sepals 5 (rarely 4), petals usually with a scale-like appendage on the claw. Stamens in two whorls, the outer shorter; anthers emarginate at both ends. Ovary usually of 3 carpels with all the placentae bearing ovules.

   About 50 spp. with the distribution of the family.

1. F. laevis L.   'Sea Heath.'

A *procumbent* slightly pubescent rather wiry perennial, dark green tinged with reddish-brown. Stems up to 15 cm., rarely more, puberulous, woody at base. *Lvs* 2–4 mm., *heath-like*, linear with revolute margins, glabrous above, puberulous beneath, ciliate at base, *mostly densely crowded on short lateral shoots*. Fls c. 5 mm. diam., solitary, sessile, usually in the forks of the branches or terminating the short lateral shoots. Calyx somewhat fleshy; teeth erect, narrow, acute. *Petals* cuneate-obovate, obtuse and faintly crenate, *pink*. Stamens usually 6. Capsule trigonous-conical, concealed in calyx-tube and surrounded by the persistent flattened filaments. Fl. 7–8. Chh.

   Native. At the landward margins of salt-marshes, usually on rather sandy or gravelly soils. 10, S. Coast of S. and E. England, from Hants to W. Norfolk, Channel Is., local; occasionally naturalized elsewhere. W. Europe from the English Channel southwards, Mediterranean region, Madeira, W. Asia.

29. ELATINACEAE

Small herbs of wet places, or under-shrubs. Lvs opposite or verticillate, simple, stipulate. Fls small, axillary, solitary or cymose. Sepals 3–5, free, imbricate. Petals 3–5, imbricate. Stamens as many or twice as many as petals; anthers versatile. Ovary superior, 3–5-celled; styles 3–5, free; ovules numerous; placentation axile. Fr. a septicidal capsule.

Two genera and about 40 spp., cosmopolitan.

1. ELATINE L.

Small, submerged, glabrous herbs. Sepals 3–4, membranous. Petals 3–4. Capsule ± globose, membranous.

About 15 spp., cosmopolitan.

Fl. pedicelled, 3-merous; seeds 8–12 in each cell.   **1. hexandra**
Fl. sessile or subsessile, 4-merous; seeds 4 in each cell.   **2. hydropiper**

1. E. hexandra (Lapierre) DC.

A slender prostrate or decumbent annual, 2·5–10 cm. Stems rooting at nodes. Lvs opposite, spathulate, entire, *petiole shorter than blade*. *Fls pedicelled, 3-merous*. Petals pinkish-white. *Stamens* 6. Pedicels in fr. as long as or somewhat longer than fr. *Seeds* 8–12 *in each cell, slightly curved*. Fl. 7–9. Hel. or Hyd.

Native. In ponds and on wet mud, very local. 33, H12, S. Scattered throughout the British Is. north to Shetland, but absent from East Anglia. Europe from Norway to Spain and N. Italy; Azores.

2. E. hydropiper L.

Similar in general appearance to *E. hexandra*. Petiole usually as long as or longer than the blade. Fls sessile or subsessile, 4-merous. Stamens 8. Seeds 4 in each cell, unequally horseshoe-shaped. Fl. 7–8. $2n = $ c. 40. Hyd.

Native. In ponds and small lakes, rare and local. 8, H4. Hants, Sussex, Surrey, Berks, Worcester, Staffs, Denbigh, Anglesey, Armagh, Down, Antrim, Derry. Europe; W. Asia; N. Africa; N. America.

30. CARYOPHYLLACEAE

Annual to perennial herbs, sometimes suffruticose, rarely shrubs. *Lvs usually in opposite and decussate pairs,* rarely whorled, sometimes spirally arranged; *simple*, *entire*, usually narrow, commonly exstipulate but sometimes with scarious stipules. *Fls commonly in terminal bracteate dichasia*, sometimes in raceme-like cymes or dense cymose clusters, rarely solitary, terminal; usually hermaphrodite, hypogynous or sometimes perigynous, actinomorphic. Sepals 4–5, free or gamosepalous below; petals sometimes 0, usually 4–5, free, often bifid or emarginate, sometimes with coronal scales; stamens commonly twice as many or as many as the sepals, sometimes an intermediate number, occasionally 1–3; *ovary superior, syncarpous,* 1-*celled at least above*, with 1–many *campylotropous ovules on a basal or free-central placenta*; styles 2–5, free to the base or joined below (*Polycarpon*). Nectar is usually secreted by hypogynous glands and the fls are cross-pollinated by insects or self-pollinated. Fr. a capsule opening by as many or twice as many teeth or valves as styles, the valves rarely remaining joined above (*Paronychia* spp., *Illecebrum*); sometimes a berry (*Cucubalus*), or an indehiscent 1-seeded nutlet (*Corrigiola*,

Scleranthus). Seeds usually with a curved embryo surrounding perisperm; embryo rarely straight (*Dianthus, Kohlrauschia, Illecebrum*, etc.).

About 1450 spp. in 70 genera, chiefly in temperate regions of the northern hemisphere.

A well-defined family with relationships to Portulacaceae, Phytolaccaceae, Amaranthaceae, Chenopodiaceae etc. and perhaps to Polygonaceae, Primulaceae and Plumbaginaceae.

Synopsis of Classification

Calyx gamosepalous: *Subfamily* SILENOIDEAE

  Styles 3–5   *Tribe* LYCHNIDEAE

    Fr. a capsule, rarely indehiscent; plants not scrambling.

      Capsule dehiscing by as many teeth as styles.

        Styles alternating with calyx-segments; no carpophore; petals entire; coronal scales 0; calyx with long herbaceous teeth.   3. AGROSTEMMA

        Styles opposite the calyx-segments; carpophore usually well developed; petals usually bifid and with coronal scales; calyx-teeth short.

                                              2. LYCHNIS

      Capsule dehiscing by twice as many teeth as styles.   1. SILENE

    Fr. berry-like; plants scrambling.   4. CUCUBALUS

  Styles 2   *Tribe* DIANTHEAE

    Calyx-tube with white scarious seams between the teeth.

      Individual fls or heads of fls with an involucre of 2–5 or more membranous bracts; seeds peltate, convex above, concave below; embryo straight.   8. KOHLRAUSCHIA

      Involucre 0; seeds reniform, convex on both sides; embryo curved.

                                         9. GYPSOPHILA

    Calyx-tube lacking scarious seams between the teeth.

      Petals with coronal scales.   7. SAPONARIA

      Coronal scales 0.

      Epicalyx present; embryo straight.   5. DIANTHUS

      Epicalyx 0; embryo curved.   6. VACCARIA

Calyx of free sepals.   *Subfamily* ALSINOIDEAE

  Fr. a capsule opening by teeth or valves.

    Styles free to the base.

      Stipules 0.   *Tribe* ALSINEAE

        Petals bifid or deeply emarginate; sometimes 0.

          Styles 5, opposite the sepals, rarely 3, 4 or 6; capsule cylindrical.

                                       10. CERASTIUM

          Styles 5, opposite the petals; capsule ovoid, opening by 5 bifid teeth.

                                       11. MYOSOTON

          Styles 3; capsule ovoid, opening by 6 teeth.   12. STELLARIA

          Petals irregularly toothed or jagged.   13. HOLOSTEUM

        Petals entire or weakly emarginate; sometimes minute or 0.

          Styles as many as sepals.

            Styles opposite the sepals; capsule opening by 8 short teeth.

                                       14. MOENCHIA

            Styles alternating with the sepals; capsule opening by 4–5 valves.

                                       15. SAGINA

        Styles fewer than sepals, usually 3.

          Capsule opening by as many teeth as styles.

            Capsule ovoid or cylindrical; lvs linear or lanceolate.

            Nectaries not conspicuous.   16. MINUARTIA

Nectaries 10, emarginate, conspicuous; an alpine cushion-plant.

17. CHERLERIA

Capsule ±spherical; lvs ovate, fleshy; a maritime plant.

18. HONKENYA

Capsule opening by twice as many teeth as styles.
Seeds with an oily appendage (strophiole or elaiosome).

19. MOEHRINGIA

Seeds without an oily appendage. 20. ARENARIA

Lvs with small scarious stipules. *Tribe* SPERGULEAE
Styles 5. 21. SPERGULA
Styles 3.
Ovary 1-celled throughout. 22. SPERGULARIA
Ovary 3-celled below. 23. TELEPHIUM
Styles joined below. *Tribe* POLYCARPEAE
Styles 3; sepals entire. keeled; petals 5. 24. POLYCARPON
Fr. indehiscent or opening by valves which remain joined above.
Lvs with stipules. *Tribe* PARONYCHIEAE
Embryo curved.
Styles 3. 25. CORRIGIOLA
Styles 2.
Sepals ±hooded, usually with dorsal points; fls in terminal or lateral
clusters with scarious and often very conspicuous bracts.

26. PARONYCHIA

Sepals not hooded, blunt; fls in lateral clusters with herbaceous bracts.

27. HERNIARIA

Embryo straight.
Sepals white, hooded, with dorsal points, thick and becoming hard in fr.;
fls in axillary whorls. 28. ILLECEBRUM
Lvs without stipules. *Tribe* SCLERANTHEAE
Perigynous; fr. 1-seeded. 29. SCLERANTHUS

Key to Genera

1	Lvs alternate or spirally arranged, not in opposite pairs or whorled.	*2*
	At least the lower lvs in opposite pairs or rarely whorled.	*3*
2	Annual; lvs oblong-lanceolate or strap-shaped; fr. trigonous, 1-seeded, indehiscent.	25. CORRIGIOLA
	Perennial; lvs ovate; fr. a capsule with many seeds.	23. TELEPHIUM
3	Stipules 0.	*4*
	Stipules present.	*28*
4	Fls perigynous; petals 0; fr. 1-seeded. dry and indehiscent, enclosed in the perigynous tube; ±prostrate herbs with subulate connate lvs.	29. SCLERANTHUS
	Fls hypogynous; fr. a capsule or berry with few to many seeds.	*5*
5	Calyx of joined sepals.	*6*
	Calyx of free sepals.	*14*
6	Styles 2.	*7*
	Styles 3–5 (or fls with stamens only).	*11*
7	Calyx-tube with whitish scarious seams alternating with the free teeth.	*8*
	Calyx-tube lacking scarious seams.	*9*
8	Individual fls or heads of fls with an epicalyx or involucre respectively of 2 to many membranous scales.	8. KOHLRAUSCHIA
	No epicalyx or involucre.	9. GYPSOPHILA

9   Base of calyx tightly enclosed by an epicalyx of 1–3 pairs of bracteoles;
      petals with no scale-like appendages at the top of the claw (coronal
      scales).   5. DIANTHUS
      Epicalyx 0.   10

10 Calyx-tube winged; coronal scales 0.   6. VACCARIA
      Calyx-tube not winged: coronal scales present.   7. SAPONARIA

11 Fr. black, berry-like; plant scrambling.   4. CUCUBALUS
      Fr. a capsule; plant not scrambling.   12

12 Calyx with long narrow herbaceous teeth much longer than the petals.
                                                        3. AGROSTEMMA
      Calyx-teeth not exceeding the petals.   13

13 Styles 3, or (2–)5 in fls lacking stamens, or fls with stamens only; capsule
      opening by twice as many teeth as styles.   1. SILENE
      Styles 5; fls always with both stamens and styles; capsule opening by
      5 teeth.   2. LYCHNIS

14 Petals present.   15
      Petals 0.   26

15 Petals ± deeply bifid.   16
      Petals entire, emarginate or irregularly toothed.   19

16 Styles 5.   17
      Styles 3, or varying from 3 to 6   18

17 Lvs 2–5 cm., ovate-cordate; petals bifid almost to the base.
                                                        11. MYOSOTON
      Lvs rarely exceeding 2·5 cm., not cordate; petals bifid to less than
      half-way.   10. CERASTIUM

18 Styles 3–6; a decumbent alpine plant with small elliptical lvs; rarely
      descending below 1500 ft.; capsule oblong.   10. *Cerastium cerastoides*
      Styles 3; not exclusively alpine plants; capsule narrowly to broadly
      ovoid.   12. STELLARIA

19 Petals irregularly toothed or jagged.   13. HOLOSTEUM
      Petals entire or slightly emarginate.   20

20 As many styles as sepals (4–5).   21
      Fewer styles (2–3) than sepals.   22

21 Styles opposite the sepals; capsule opening by 8 short teeth; a small
      glaucous herb with strap-shaped or narrowly lanceolate lvs.
                                                        14. MOENCHIA
      Styles alternating with the sepals; capsule splitting to the base by 4–5
      valves; small non-glaucous herbs with linear lvs.   15. SAGINA

22 Succulent maritime plants with broad lvs, greenish fls and ± spherical
      capsules.   18. HONKENYA
      Not as above.   23

23 Lvs linear.   24
      Lvs not linear.   25

24 Fls greenish; petals minute; nectaries 10, conspicuous; a densely
      caespitose alpine plant.   17. CHERLERIA
      Fls with white petals; nectaries not conspicuous.   16. MINUARTIA

25 Lvs ovate, 1–2·5 cm., 3-veined; seeds with a fleshy oily appendage
      (elaiosome); a slender woodland plant.   19. MOEHRINGIA
      Lvs not exceeding 1 cm.; seeds without elaiosome.   20. ARENARIA

26 Lvs linear.   27
      Lvs not linear.   12. STELLARIA

27 Styles 3; nectaries 10, conspicuous; a densely caespitose alpine plant.
                                                          17. CHERLERIA
      Styles 4–5; nectaries not conspicuous.   15. SAGINA

28 Lower lvs usually in closely approximated pairs resembling whorls of 4,
      obovate; a small decumbent herb with tiny fls c. 3 mm. diam.; styles
      joined below.   24. POLYCARPON
      At least the lower lvs in opposite pairs; styles free to the base.   *29*

29 Fls in terminal dichasia or monochasia; petals ± equalling the sepals or
      exceeding them.   *30*
      Fls aggregated in dense terminal, lateral or axillary clusters; petals
      minute or 0.   *31*

30 Styles 5.   21. SPERGULA
      Styles 3.   22. SPERGULARIA

31 Fls in terminal clusters made conspicuous by white scarious bracts.
                                                          26. PARONYCHIA
      Fls not in terminal clusters; bracts not conspicuous.   *32*

32 Fls white, in axillary clusters or false whorls; sepals hooded with dorsal
      points and resembling follicles of *Sedum*; upper lvs in equal pairs.
                                                          28. ILLECEBRUM
      Fls greenish, in lateral clusters; sepals not hooded; upper lvs commonly
      in unequal pairs.   27. HERINARIA

1. SILENE L.

Annual to perennial herbs of very various habit, or rarely dwarf shrubs.
Lvs opposite, exstipulate. Fls in cymose infls or solitary, hermaphrodite or
unisexual, with the bracteoles not forming an epicalyx. Sepals joined below,
commissures not scarious, with 5 free teeth and either with 10 veins (i.e.
alternating with the teeth as well as opposite them) or with 20, 30 or 60;
petals long-clawed, usually bifid or emarginate and with or without coronal
scales at the throat; stamens 10; *ovary 1-celled, at least above, usually 3–5-
celled at base; styles 3, or 5 alternating with the sepals.* Fr. a *capsule opening
by twice as many teeth as there are styles,* usually with a stalk (carpophore)
between its base and the base of the calyx. Seeds numerous, reniform.
      About 300 spp. in Europe, extra-tropical Asia and Africa, and N. America.

1   Styles 5; dioecious.   *2*
      Styles usually 3; fls usually hermaphrodite (but styles varying 2–5 and
      plants ± dioecious in *S. otites*).   *3*

2   Fls red; capsule-teeth revolute.   **1. dioica**
      Fls white; capsule-teeth erect.   **2. alba**
      (Fls pink.   **alba × dioica**)

3   Fruiting calyx with 20–30 veins, strongly inflated at least below.   *4*
      Fruiting calyx with 10 veins, not strongly inflated.   *8*

4   Perennial herbs; fruiting calyx usually glabrous, bladdery, subglobular
      or broadly ellipsoidal, conspicuously net-veined; teeth broadly
      triangular.   *5*
      Annuals; fruiting calyx pubescent, swollen below but not bladdery,
      conical, strongly ribbed but not conspicuously net-veined; teeth nar-
      rowly triangular-acuminate or subulate.   *7*

5   Upper lvs linear-lanceolate; fls dull red, rarely white.   **5. linearis**
    Upper lvs elliptic-lanceolate to ovate; fls white.   6

6   Plants with prostrate non-flowering shoots; fls 1–4; calyx hardly con-
    tracted; petals with distinct coronal scales; capsule with recurved teeth.
                                                                      6. maritima
    Plants lacking non-flowering shoots; fls many; calyx contracted at its
    mouth; petals with very small coronal scales or bosses; capsule with
    erect teeth.   **4. vulgaris**

7   Lower lvs linear-oblanceolate; petals bifid; capsule shorter than the calyx.
                                                                      7. conica
    Lower lvs broadly lanceolate; petals entire or slightly emarginate;
    capsule equalling the calyx.   **8. conoidea**

8   Calyx glabrous.   9
    Calyx hairy, downy or viscid.   12

9   Perennial plants with non-flowering shoots.   10
    Annual plants lacking non-flowering shoots.   11

10 Lvs linear; fls solitary, pink or rarely white; a cushion-plant of mountains
    and maritime cliffs in the north.   **12 acaulis**
    Basal lvs spathulate; fls in a long narrow panicle, yellowish-green; an
    erect herb of dry sandy soils in E. Anglia.   **14. otites**

11 Upper lvs ovate-amplexicaul, glaucous; fls in a crowded sub-corymbose
    cyme.   **13. armeria**
    Upper lvs linear-lanceolate; fls in a loose dichasial cyme.   **17. muscipula**

12 Perennial plants with non-flowering shoots; infl. a lax panicle of opposite
    stalked dichasia with a terminal fl. or dichasium.   13
    Annual plants lacking non-flowering shoots; infl. various.   14

13 Fls ± erect; carpophore as long as the capsule.   **16. italica**
    Fls horizontal or drooping; carpophore ⅓ as long as the capsule.
                                                                      15. nutans

14 Infl. a few-fld dichasium; calyx 20–25 mm. long; petals rolled inwards
    during the day.   **3. noctiflora**
    Fls in long raceme-like cymes, at least above; calyx less than 15 mm.
    long; petals expanded during the day.   15

15 Fls ± erect.   **10. gallica**
    Fls horizontal or drooping.   16

16 Upper lvs ovate-lanceolate; infl. usually simple; fls pink.   **11. pendula**
    Upper lvs lanceolate; infl. dichasial below; fls whitish.   **9. dichotoma**

Section 1. *Melandriformes* Boiss. Annual to perennial. Fls in dichasial
cymes, unisexual (plants dioecious) or hermaphrodite. Calyx 10–20-veined,
not bladdery in fr. Styles 3 or 5. Ovary usually 1-celled to the base. Carpophore short.

1. S. dioica (L.) Clairv.   Red Campion.

Lychnis dioica L., p.p.; *L. diurna* Sibth.; *Melandrium rubrum* (Weigel) Garcke;
M. dioicum (L.) Coss. & Germ.

A biennial to perennial herb with a slender creeping stock producing numerous
decumbent non-flowering shoots up to 20 cm. and erect flowering stems

30–90 cm. covered with soft spreading hairs and sometimes slightly viscid above. Basal lvs obovate, narrowed into a long winged stalk; upper lvs obovate-oblong short-stalked or ± sessile; all with blade 4–10 cm., acute or acuminate, hairy. *Fls* 18–25 mm. diam., unisexual, *scentless*, in many-fld terminal dichasia, their stalks 0·5–1·5 cm. Calyx-tube 10–15 mm., hairy and slightly viscid; of male fl. cylindrical, faintly 10-veined; of female fl. ovoid and 20-veined, becoming rounded in fr.; *teeth triangular-acute*, 2–2·5 mm., broader in the female fl. *Petals* bright *rose-coloured*, rarely white, the limb broadly obovate, deeply bifid into narrow segments with 2 narrow acute scales at the base; claw auricled above. Styles 5. *Capsule* (Fig. 28 A) *broadly ovoid, opening widely by* 10 *revolute teeth*; carpophore

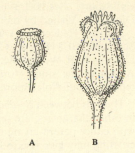

Fig. 28. Dehisced capsules of A, *Silene dioica* and B, *S. alba*. × 1.

extremely short. Seeds black, densely and acutely tubercled. Fl. 5–6. Dioecious. Visited by long-tongued humble-bees and hover-flies. $2n=24^*$, 48. Hp.–Chh.

The above description is of ssp. **dioica**, the widespread form.

Ssp. **zetlandica** (Compton) Clapham, found in Orkney and Shetland, has a very stout densely hairy stem, basal lvs narrowing abruptly into a slender stalk, stem-lvs narrower than in the type and very downy on both sides, fls in dense subsessile terminal and axillary clusters, larger and more conical capsules, and heavier seeds (average weight 1·4 mg.; of type 0·5 mg.). It may be identical with Turesson's coastal ecotype.

Native. In woods on well-drained and at least moderately base-rich soils and rocky slopes, and especially marginally or in clearings and on limestone; also in hedgerows and on cliff-ledges and stabilized scree-slopes; preferring substrata rich in nitrates and common on bird-cliffs. Reaches 3500 ft. in Scotland. 112, H 34, S. Locally abundant throughout the British Is., though very rare in some districts. Europe, N. Africa, W. Asia.

2. S. alba (Mill.) E. H. L. Krause   White Campion.

Lychnis alba Mill.; *L. vespertina* Sibth.; *Melandrium album* (Mill.) Garcke

A short-lived perennial or sometimes annual or biennial herb with a thick almost woody stock from which arise a few short non-flowering shoots and erect flowering shoots 30–100 cm., covered with soft spreading hairs and slightly glandular-viscid above. Lower lvs and those of barren shoots oblanceolate or elliptical, narrowed into long unwinged stalks; upper lvs lanceolate or elliptical, acuminate, sessile; all 3–10 cm., hairy. *Fls* 25–30 mm. diam., unisexual, slightly *evening-scented* in a few-fld terminal dichasium. Calyx-tube 18–25 mm., downy and viscid; of the male fl. cylindrical, 10-veined; of the female narrowly ovoid, 20-veined; *teeth linear-lanceolate blunt* c. 6·5 (5–8) mm. in the female fls, c. 4·8 mm. in male. *Petals white*, the limb obovate, deeply bifid into broad segments with 2-lobed coronal scales, the claw auricled above. Styles 5. *Capsule* (Fig. 28 B) *ovoid-conical*, often breaking the calyx-tube, opening narrowly by 10 *suberect teeth*; carpophore very short. Seeds 1·3–1·5 mm. diam., grey, bluntly tubercled. Fl. 5–9. Dioecious. Visited by moths. $2n=24*$. Hp.

Native. Hedgerows, waste places and cultivated land, reaching 1400 ft. in Scotland. 107, H 26, S. Common throughout the British Is., except Shetland and much of W. Ireland. Europe, N. Africa, W. Asia. Introduced in N. America. The smut *Ustilago violacea* causes the formation of stamens in female fls but violet spores of the fungus replace pollen-grains.

Hybrids between *S. alba* and *S. dioica*, often with pink fls and intermediate in other characters, are frequently seen. They are fertile and may back-cross with either parent giving types more closely resembling that parent. Further crossing may give rise to intermediates of all kinds.

*S. macrocarpa (Boiss. & Reut.) E. H. L. Krause, resembling *S. alba* in its white fls, differs in having acute calyx-teeth and a large capsule opening by teeth which are ultimately revolute. A not infrequent casual, native in the Mediterranean region.

3. S. noctiflora L.   'Night-flowering Campion.'
Melandrium noctiflorum (L.) Fr.

An annual herb 15–60 cm., with an erect simple or branched flowering stem, with soft spreading hairs, glandular and viscid especially above. Lower lvs obovate or ovate-lanceolate narrowed to a stalk-like base, uppermost narrowly oblong-lanceolate, sessile, very acute; all 5–10 cm., with scattered hairs *Fls* c. 18 mm. diam., *hermaphrodite*, in a terminal few-fld dichasium. Calyx-tube c. 20 mm., woolly and viscid, cylindrical-ovoid, becoming swollen in fr., membranous and whitish between the 10 broad green veins, and with 5 slender ciliate teeth. *Petals yellowish beneath and rosy above, rolled inwards during the day*, spreading and scented at night, deeply bifid, with 2 coronal scales at the base of the limb and auricles at the top of the claw. *Styles* 3. Capsule ovoid-conical often bursting the calyx, opening by 6 ± recurved teeth; carpophore about ⅙ as long as the capsule. Seeds 1–1·2 mm. across, tubercled. Fl. 7–9. Strongly protandrous. Probably visited by moths. $2n=24*$. Th.

Native. A rather local weed of lowland arable fields especially on sandy soils. Throughout Great Britain northwards to Aberdeen; introduced in S. and E. Ireland. 70, H 10. Europe northwards to Norway and Finland; W. and C. Asia. Introduced in N. America.

Section 2. *Inflatae* Boiss. Perennial. Fls in dichasia, often few-fld, hermaphrodite or unisexual. Petals not convolute-contorted in bud. Calyx bladdery in fr., faintly 20-veined. Styles 3. Ovary septate at base. Carpophore well-developed.

4. S. vulgaris (Moench) Garcke   Bladder Campion.
Cucubalus Behen L.; *S. cucubalus* Wibel; *S. inflata* Sm.

A perennial herb 25–90 cm., with a branching woody stock and several erect or ascending shoots, all flowering, simple or branched above, usually glabrous but sometimes hairy. Lvs elliptic-lanceolate to ovate, the middle lvs averaging 4·5 × 1·2 cm.; lowest short-stalked, upper sessile, all acute, glabrous or ciliate, often somewhat glaucous. *Infl. a many-fld subcorymbose cyme with scarious bracts.* Fls c. 18 mm. diam., drooping, polygamous. Calyx (Fig. 29 B) pale green or reddish, bladdery, ovoid to subglobular with the mouth narrower than the base, 20-veined with a strong connecting network of veins, glabrous or downy, with 5 broadly triangular teeth. *Petals* white, 14 mm. wide, deeply bifid, usually *with small and inconspicuous coronal scales or bosses.* Styles 3. *Capsule* enclosed by the calyx, broadly ovoid-ellipsoid *with 6 erect teeth*; carpophore about ⅓ as long as the capsule. Seeds white when young, 1·2–1·5 mm. across, usually with dense acute tubercles. Fl. 6–8. Plants of 3, or elsewhere 5, types have been observed bearing hermaphrodite, male and female fls only, andromonoecious and gynomonoecious respectively. Hermaphrodite fls protandrous. Fragrant and visited by long-tongued bees and night-flying moths. 2*n* = 24*. Hp.

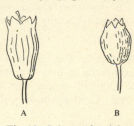

Fig. 29. Calyces of A, *Silene maritima* and B, *S. vulgaris.* × 1.

Native. Common on grassy slopes, arable land, roadsides and broken ground through most of Great Britain and Ireland but thinning out in the north and absent from the Outer Hebrides and Shetland. 106, H40, S. Reaches 1000 ft. in Yorks. Europe, temperate Asia, N. Africa.

*5. S. linearis Sweet   'Narrow-leaved Bladder Campion.'

S. angustifolia Guss., non Poir.

Like *S. vulgaris* but stoloniferous and with the lower lvs linear-oblong or narrowly oblanceolate, upper lvs linear-lanceolate, all acute. The calyx is less globular, more herbaceous and less clearly net-veined than in *S. vulgaris*, the fls are dull red (rarely white) and rather large, and the seeds are twice as large as in *S. vulgaris*.

Introduced and established on Plymouth Hoe, Devon. Mediterranean region.

6. S. maritima With.   Sea Campion.

A perennial, usually maritime, herb 8–25 cm., with a branching woody stock from which arise ascending flowering shoots, and *prostrate non-flowering shoots forming a loose cushion.* Lvs much as in *S. vulgaris* but smaller (middle lvs averaging 2 × 0·8 cm.), rather narrower, somewhat stiffer and fleshier and distinctly more glaucous; usually glabrous but often ciliate. *Infl. 1–4-fld, with herbaceous bracts.* Fls 20–25 mm. diam., erect. Calyx (Fig. 29 A) broadly ellipsoidal with its mouth broader than its base, otherwise as in *S. vulgaris.* *Petals* white, 14 mm. wide, deeply bifid, each *with 2 distinct coronal scales* at the base of the blade. Styles 3. *Capsule* broadly ovoid *with 6 recurved teeth*; carpophore about half as long as the capsule. Seeds pink when young, 1·0–1·2 mm. across, usually with blunt tubercles. Fl. 6–8. Protandrous. 2*n* = 24. Hp.

Native. Locally abundant on maritime shingle, cliffs and stony ground all round the British Is.; rarely on cliff-ledges, gravelly lake shores or by alpine streams inland. 82, H25, S. Reaches 3180 ft. in Scotland. Atlantic Europe from Norway to Spain.

S. maritima and *S. vulgaris* are very closely related spp. which can be crossed artificially with little loss of fertility. Natural hybrids are nevertheless rare.

Section 3. *Conoimorpha* Otth. Annuals. Fls in dichasia. Petals convolute-contorted in bud. Fruiting calyx conical, swollen below but not bladdery, 20–30(–60)-veined, with long-acuminate teeth. Styles 3. Ovary septate at base. Carpophore very short.

7. S. conica L.   'Striated Catchfly.'

An annual herb 10–35 cm., with a slender tap-root and an erect simple or branched densely glandular and viscid stem. Lower lvs linear-oblanceolate subacute, upper linear-lanceolate acuminate; all downy. Infl. 1- to few-fld. Fls c. 4–5 mm. diam. and 12–15 mm. long, short-stalked, erect. *Calyx* at first ± cylindrical, later *ovoid-conical*, swollen below, 30-veined, *densely glandular-hairy*, with *long subulate teeth*. Petals rose-coloured, long-clawed, the blade bifid with coronal scales at the base, the claw auricled above. Styles 3. Capsule somewhat shorter than the calyx, ovoid-conical, opening by 6 ± erect teeth; carpophore very short. Seeds pale grey, reniform, 0·7–0·8 mm. across, tubercled. Fl. 5–6. Protandrous. $2n=24*$. Th.

Native. A local plant of stabilized dunes, sandy pastures and waste places chiefly by the coast of S. and E. England and E. Scotland northwards to Angus; also in S. Wales and inland in East Anglia and the S. Midlands. 23, S. W., C. and S. Europe, N. Africa and W. Asia. Often associated with other small annuals such as *Aira praecox*, *Teesdalia nudicaulis*, *Cerastium semidecandrum* and *Hypochaeris glabra*.

*8. S. conoidea L.

An annual herb closely resembling *S. conica* but with the lvs broadly lanceolate and ± glabrous, the petals entire or slightly emarginate, and the capsule equalling the calyx, globular below and long-attenuate above. Fl. 6–7. $2n=24$. Th.

Introduced. A not infrequent casual near ports and established in Jersey. Mediterranean region and eastwards to India.

Section 4. *Dichotomae* (Rohrb.) Chowdhuri. Annuals. Fls almost sessile in paired raceme-like cymes. Petals convolute-contorted in bud. Fruiting calyx 10-veined. Styles 3. Ovary septate at base. Carpophore well developed.

*9. S. dichotoma Ehrh.   'Forked Catchfly.'

An annual herb with a stiffly erect sparsely hairy branching stem, 20–60 cm. Lower lvs spathulate, stalked, middle 3–5 cm., ovate-lanceolate, upper lanceolate acute; all 3–5-veined, hairy. Infl. of *long terminal cymes dichasial below (with fls in the forks) then raceme-like and one-sided*. Fls 15–18 mm. long and wide, ± horizontal, short-stalked, erecting in fr. Calyx 11–15 mm., cylindrical, 10-veined, hairy, the veins stiffly ciliate. Petals whitish, deeply bifid, usually with short coronal scales. Styles 3. Capsule ovoid, opening

by 6 acute teeth; carpophore $\frac{1}{5}$ to $\frac{1}{4}$ as long as the capsule. Seeds 1·3 mm., reniform, furrowed along the back. Fl. 7–8. Protandrous. $2n=24$. Th.
Introduced. A frequent casual. E. and S.E. Europe and W. Asia.

Section 5. *Silene*. Annuals. Fls almost sessile in simple raceme-like cymes. Petals convolute-contorted in bud. Fruiting calyx 10-veined. Styles 3. Ovary septate at base.

10. S. gallica L.   'Small-flowered Catchfly.'

Incl. *S. anglica* L. and *S. quinquevulnera* L.

An annual herb with slender tap-root and an erect or ascending stem, 15–45 cm., simple or branched, hairy, viscid above. Lower lvs spathulate, stalked; middle 2·5–5 cm., lanceolate, sessile; upper narrowly lanceolate; all hairy. *Infl. of simple terminal raceme-like subsecund cymes with the fls short-stalked and alternating.* Fls 10–12 mm., ± erect. Calyx cylindrical-ovoid in fr., viscid, with 10 hairy veins and 5 long acuminate teeth. Petals white or pale pink, sometimes with a basal dark-red blotch, entire or emarginate, with 2 acute coronal scales. Stamen filaments hairy. Styles 3. Capsule ovoid, about as long as the calyx; carpophore very short. Seeds 0·8 mm. across, blackish, reniform with depressed faces. Fl. 6–10. $2n=24$. Th. Native, and introduced (see below). C. and S. Europe, but widely introduced. A very variable sp. The following varieties occur in Britain, chiefly on waste ground (from Lousley, *B.E.C. Report*, 1936, pp. 395–6).

Var. *anglica* (L.) Clapham. Stem branched, spreading. Petals small, dingy white, yellowish or pale pink. Lower fruiting stalks exceeding the calyx, often spreading or reflexed. Widespread but local as a native or casual in sandy or gravelly fields and waste places throughout most of Great Britain northwards to Inverness, and in Ireland. Channel Is.

Var. *gallica* (*S. eu-gallica* Syme; *S. anglica* var. *stricta* Bromf.). Stem simple or with a few erect branches. Petals with laminae at least 5 × 5 mm., pale pink or rose, almost entire. Fruiting stalks shorter than the calyx, erect. Native in the Channel Is. and established in the Scilly Is., casual elsewhere; rare.

Var. *quinquevulnera* (L.) Mert. & Koch (*S. quinquevulnera* L.). Like var. *anglica* but petals with a deep crimson spot near the base. Intermediates between this and *anglica* occur. Native in the Channel Is. and casual in S. England and near Edinburgh. Formerly cultivated.

Section 6. *Erectorefractae* Chowdhuri. Annuals. Fls in lax dichasia or simple raceme-like cymes, their longish stalks spreading or recurved in fr. Petals convolute-contorted in bud. Styles 3. Ovary septate at base.

*11. S. pendula L.   'Drooping Catchfly.'

An annual herb with ascending branched pubescent stems to 45 cm., viscid above. Lower lvs spathulate, blunt; upper ovate-lanceolate, sessile, acute. Fls in usually simple raceme-like cymes, ± drooping, short-stalked. Calyx cylindrical, becoming obovoid, viscid and hairy, 10-veined, somewhat inflated in fr., with 5 blunt teeth. Petals pale pink, notched, with 2 acute coronal scales. Fl. 6–7. Th.
Introduced. Frequently grown in gardens and occasionally escaping. Mediterranean region.

Section 7. *Nanosilene* Otth. Caespitose perennials. Fls solitary. Petals convolute-contorted in bud. Fruiting calyx faintly 10-veined. Styles 3. Ovary septate at base.

12. S. acaulis (L.) Jacq.   Moss Campion.

A perennial, densely tufted and much branching, ± glabrous, somewhat woody below, forming *bright green moss-like cushions* 2–10 cm. high. *Lvs* 6–12 mm., *in dense rosettes, linear-acute*, channelled above, stiffly ciliate especially towards the base. Hermaphrodite, male and female fls on different plants. Fls 9–12 mm. diam., solitary, erect, their stalks very short at first then lengthening. Calyx often reddish, campanulate, faintly 10-veined, glabrous, its 5 blunt teeth with scarious margins. *Petals deep rose* or whitish, the blade bifid with a bifid scale at its base, the claw without auricles. Styles 3. Capsule up to twice as long as the calyx, subcylindrical, opening by 6 erect teeth; carpophore downy, shorter than the capsule. Seeds pale yellowish, reniform. Fl. 7–8. Protandrous. Visited by various lepidoptera and other insects. $2n = 24$. Chc.

Native. On mountain cliffs, ledges and scree in N. Wales, the Lake District and Scotland, and on rocks at low levels in the Hebrides, Orkney and Shetland; W. and N. Ireland. Reaches 4200 ft. in Scotland. 25, H5. An arctic-alpine plant of the mountains of W. and C. Europe and N. America, and of the arctic regions of America, Europe and E. Asia, to 83° 6′ N. in N. Greenland.

Section 8. *Compactae* Boiss. Annuals. Fls in compact subcorymbose cymes. Petals convolute-contorted in bud. Fruiting calyx membranous, 10-veined. Styles 3 or 5. Ovary septate at base.

*13. S. armeria L.   'Sweet-William Catchfly.'

An annual herb with slender tap-root and erect hollow stems, 10–60 cm., simple or branched above, slightly viscid above and glabrous, glaucous. *Lvs* 2–5 cm., sessile, the lowest rather crowded, lanceolate, narrowed below, the upper *ovate-amplexicaul acute*; all *glabrous and glaucous*. Fls short-stalked, erect, c. 15 mm. diam. and 18 mm. long, in rather crowded subcorymbose dichasial cymes. Calyx reddish, 10-veined, cylindrical-clavate with short blunt teeth. Petals deep rose, emarginate, with 2 narrow erect scales at the base of the lamina, the claw not auricled. Styles 3. Capsule cylindric-ovoid just enclosed by the calyx, opening by 6 revolute teeth; carpophore almost equalling the capsule. Seeds small, blackish, furrowed down the back. Fl. 6–7. Protandrous. Nectar accessible only to Lepidoptera. $2n = 24$. Th.

Introduced. A garden escape in a few places in S. England and occasionally establishing itself. S. and C. Europe, but widely introduced by cultivation in gardens.

Section 9. *Otites* Otth. Perennials. Fls in opposite pairs of cymes simulating whorls. Petals convolute-contorted in bud. Fruiting calyx faintly 10-veined. Styles 3 (2–5). Ovary septate at base.

14. S. otites (L.) Wibel

Spanish Catchfly.

Cucubalus Otites L.

A perennial herb with a thick branching woody stock from which arise non-flowering rosettes and erect simple flowering stems, 20–90 cm., shortly hairy below and viscid up to about the middle. Basal lvs 1·5–8 cm., narrowly spathulate, narrowing to a stalk; lower stem lvs similar but smaller; upper stem lvs in distant pairs, linear-lanceolate, sessile; all with dense short hairs. *Infl. of numerous fls in a long narrow panicle, interrupted below, the opposite pairs of short cymes simulating false whorls.* ± Dioecious, but some male plants with a few hermaphrodite fls. *Fls 3–4 mm. diam.* and 4–5 mm. long, erect, their glabrous stalks not much exceeding the calyx. Calyx glabrous, narrowly campanulate, faintly 10-veined, with 5 short blunt teeth. *Petals pale yellowish-green*, narrow, entire, with no coronal scales, the claw without auricles. Female fls have no stamens and male fls a usually vestigial ovary. Stamens and the 3 (2–5) styles exserted. Capsule ovoid, longer than and rupturing the calyx, opening by 6 (4–10) short teeth; carpophore almost 0. Seeds reniform, furrowed along the back, finely rugose, 0·7 mm. across. Fl. 6. Hermaphrodite fls protandrous. Evening-scented and nectar-secreting and visited in some districts by Lepidoptera, but perhaps also wind-pollinated (male fls said not to secrete nectar in C. Germany). $2n = 24$. Hs.

Native. Confined to the Breckland heaths of Norfolk, W. Suffolk and Cambridgeshire, but occasionally casual elsewhere. 4. S. and C. Europe and W. Asia. A 'steppe' species.

Section 10. *Siphonomorpha* Otth. Perennials. Fls drooping, usually in a panicle with paired dichasia. Petals convolute-contorted in bud. Fruiting calyx 10-veined, ± cylindrical. Styles 3. Ovary septate at base.

15. S. nutans L.

Nottingham Catchfly.

A perennial herb with a slender branched woody stock from which arise short non-flowering shoots and erect flowering shoots, 25–80 cm., downy below and viscid above. Basal lvs spathulate narrowing to a long stalk, c. 7·5 cm.; upper lvs narrowly lanceolate, subsessile, acute; all softly hairy and ciliate. Infl. a lax subsecund panicle with opposite branches ending in 3–7-fld dichasial cymes. Fls hermaphrodite or unisexual, 18 mm. diam., c. 12 mm. long, drooping, their stalks short, viscid. Calyx cylindrical-clavate, glandular-pubescent, with 10 purplish veins and 5 acute white-margined teeth. *Petals white or pink-tinged (rarely red or yellow), bifid, with narrow ± inrolled lobes* and 2 *acute basal scales, the claws not auricled.* Styles 3. Capsule ovoid, somewhat exceeding the calyx and opening by 6 spreading or reflexed teeth; *carpophore downy, about ⅓ as long as the capsule.* Seeds 1–2·2 mm. across, reniform, furrowed along the back, tubercled. Fl. 5–7. Fls heterostylous, nectarless, protandrous, 'opening and fragrant for 3 nights, 5 stamens ripening on each of the two first nights, the styles protruding on the third'. Visited by Lepidoptera and humble-bees. $2n = 24*$. Hs.

British plants fall into two distinct varieties: var. *salmoniana* Hepper, with narrow sparsely hairy lvs and ripe capsules 11 14 mm. (carpophore 3–4 mm.); and var.

smithiana Moss, with usually broader, more pubescent, less acute lvs and ripe capsules 8–10 mm. (carpophore 2–2·5 mm.). Individual populations, especially of var. *smithiana*, show further intra-varietal differences.

Native. A local plant of dry slopes, rocks, cliff-ledges, shingle, walls and field-borders. 18, S. Var. *salmoniana* in Kent, Sussex, S. Hants, Isle of Wight and Surrey; var. *smithiana* in Kent, Dorset, Devon, Caernarvon, Denbigh, Staffs, Derby, Notts, W. Yorks, Fife, Angus, Kincardine. The form in Jersey is close to var. *salmoniana*. *S. nutans* is also introduced in several localities. Europe, Canary Is., N. Africa, Caucasus, N. Asia to Japan.

Section 11. *Paniculatae* Boiss. Perennials. Like *Siphonomorpha* but fls erect.

*16. S. italica (L.) Pers.   'Italian Catchfly.'
Cucubalus italicus L.

A perennial herb with slender branched woody stock from which arise long non-flowering shoots and erect flowering shoots 25–70 cm., softly hairy and viscid. Lower lvs 3–5 cm., lanceolate-spathulate narrowed to a long stalk; uppermost linear-lanceolate sessile; all pubescent. Infl. a pyramidal panicle, the opposite, long, ascending, viscid branches ending in c. 3-fld dichasia. Fls c. 18 mm. diam. and 20 mm. long, ± erect, their stalks viscid. Calyx narrowly clavate, glandular-pubescent, 10-veined with 5 blunt ovate teeth. *Petals* yellowish-white above, often reddish-green beneath, deeply bifid *with two small bosses at the base* of the blade, the *claw auricled* above. Styles 3. Capsule ovoid, about equalling the calyx, opening by 6 spreading teeth; *carpophore at least equalling the capsule.* Fl. 6–7. Fls heterostylous, opening only in the evening, and then fragrant. $2n = 24$. H.

Probably introduced. A rare and local plant of quarry-sides in a very few scattered localities northwards to Kincardine. 7. Mediterranean region and Near East.

Section 12. *Behenantha* Otth. Annuals. Fls solitary or in lax dichasia. Petals convolute-contorted in bud. Fruiting calyx 10-veined, bladdery above but contracted at the mouth. Styles 3. Ovary septate at base.

*17. S. muscipula L.

A glabrous annual herb with slender tap-root and stiffly erect branching stems, 20–40 cm., very viscid above. Lower lvs obovate blunt, upper linear-lanceolate acute. Fls erect, very short-stalked, in a loose dichasial cyme. Calyx glabrous, 10-veined, with acute teeth, in fr. broadening above but contracted at the mouth. Petals red, bifid, with long blunt scales at the base of the blade and with the claw auricled above. Styles 3. Capsule oblong-ovoid opening by 6 short teeth; carpophore downy, less than half as long as the capsule. Fl. 6–8. Th.

Introduced. A frequent casual especially near ports. Mediterranean region.

Other spp. which commonly occur as casuals are *S. cserei* Baumg., resembling *S. vulgaris* but the calyx with only 10 veins and the petals deeply and narrowly bifid (Balkans, Asia Minor); and *S. cretica* L. (incl. *S. annulata* Thore), a tall rigid annual with linear-lanceolate stem-lvs, a few-fld cymose panicle, a glabrous calyx with broadly scarious-margined teeth, small rose-coloured shortly bifid petals with coronal scales, an ovoid capsule, and seeds 1·5–2 mm. across.

Others grown in gardens and sometimes escaping are *S. coeli-rosa* (L.) A.Br. (Mediterranean), a glabrous annual with linear-lanceolate lvs, rose-pink petals paler beneath, bifid and with long linear coronal scales, 5 styles, and a carpophore equalling the glabrous oblong-ovoid capsule; and *S. schafta* S. G. Gmel., a tufted rock-garden perennial to 15 cm. high with rosettes of small spathulate lvs and rose-coloured fls borne singly or in pairs, the notched petals with coronal scales (Caucasus).

2. LYCHNIS L.

Annual to perennial herbs with opposite lvs and cymose infls which are sometimes spike-like or head-like. Fls hermaphrodite, 5-merous. Epicalyx 0. *Sepals joined below* into a calyx-tube and with 5 free teeth above; *petals* red or white, long-clawed, *with coronal scales* at the base of the limb; stamens 10; ovary 1-celled throughout or 5-celled only at the base; *styles usually* 5. *Capsule opening* loculicidally, usually *by 5 teeth*; carpophore present but sometimes rudimentary; seeds tubercled or rugose, often very small.

About 15 spp. in the northern hemisphere.

1   Stems usually less than 15 cm.; fls in a compact head-like panicle.
                                                                    2. alpina
    Stems usually more than 30 cm.; fls in a spike-like panicle or long-stalked
        dichasia.   2

2   Stems very viscid beneath each node; fls in an interrupted spike-like
        panicle; petals slightly notched.   **3. viscaria**
    Stems not viscid beneath the nodes; fls in long-stalked dichasia; petals
        deeply 4-cleft with narrow segments.   **1. flos-cuculi**

1. L. flos-cuculi L.   Ragged Robin.

A perennial herb with slender branching stock from which arise decumbent non-flowering shoots to 15 cm. and erect flowering shoots 30–75 cm., rough above and with a few downwardly directed hairs. Lvs 2–10 cm., glabrous, somewhat rough; lower lvs and those of barren shoots oblanceolate, acute, narrowing to a stalk-like base; upper stem lvs narrower, oblong-lanceolate subsessile. Fls 3–4 cm. diam. on stalks 1–1·5 cm., in terminal and lateral long-stalked dichasia. Calyx-tube c. 6 mm., reddish, sub-membranous, strongly 10-veined, with 5 ovate-acuminate teeth 3 mm. *Petals* rose-red, rarely white, *deeply* 4-*cleft* with narrow spreading segments and 2 bifid subulate coronal scales at the base of the limb. Capsule broadly ovoid, enclosed by the calyx, 1-celled throughout, opening by 5 short acute revolute teeth; *carpophore extremely short*. Seeds 0·5–0·7 mm. across, brown, tubercled. Fl. 5–6. Protandrous. Visited by butterflies and long-tongued bees and flies. $2n=24$. Hp.

Native. A common plant of damp meadows, marshes, fens and wet woods throughout the British Is. 112, H 40, S. Reaches 2000 ft. in Scotland. Europe to Norway and Iceland, Siberia, Caucasus. Introduced in N. America.

2. L. alpina L.   'Red Alpine Catchfly.'

Viscaria alpina (L.) G. Don

A perennial tufted herb with a tap-root and a woody stock whose short branches end in non-flowering rosettes or erect lfy *glabrous flowering stems* 5–15(–20) *cm*. Lvs 1·5–5 cm., sessile, oblong-lanceolate or linear, acute, in dense basal rosettes with 1–6 pairs on the flowering stems; all glabrous or slightly ciliate at the base. *Fls* 6–12 mm. diam., crowded into *almost head-like corymbose panicles*, their stalks very short. Bracts ovate-acuminate, rose-coloured. Calyx narrowly campanulate, its tube faintly veined, glabrous, the 5 teeth short broad and rounded with purplish scarious margins. *Petals* rose-coloured, long-clawed, the limb obovate *deeply bifid* with 2 short *tubercle-like coronal scales* at its base. Capsule ovoid, 5-celled at base, opening by 5 recurved teeth, on a *carpophore less than half its length*. Seeds 0·3–0·4 mm., dark brown, bluntly tubercled. Fl. 6–7. Protandrous. Larger fls hermaphrodite, smaller usually with abortive stamens. Visited by butterflies and automatically self-pollinated. 2*n* = 24. Chh.

Native. A very rare alpine plant of the Lake District and Angus, reaching 2853 ft. in Angus. 3. A subarctic-alpine species found in the Alps and Pyrenees and in subarctic Europe, W. Asia and N. America, reaching 73° 10′ N. in E. Greenland. Frequently on serpentine and other substrata with unusually high content of copper, zinc, nickel or other heavy metals.

3. L. viscaria L.   'Red German Catchfly.'
Viscaria vulgaris Bernh.

A perennial tufted herb with a woody stock whose short erect branches end in very short densely lfy non-flowering shoots or erect lfy *flowering shoots*, 30–60 *cm*. *Stems* dark green or purplish, glabrous, *very viscid beneath each node*. Lvs of barren shoots 5–12·5 cm., narrowly elliptic-lanceolate, acute or acuminate narrowing below into a long stalk-like base; upper lvs oblong-lanceolate; all with woolly margins at the base, otherwise glabrous. *Infl. an interrupted spike-like panicle* of axillary cymes which simulate whorls. Bracts broadly lanceolate, acuminate. Fls 18–20 mm., diam., their stalks extremely short. Calyx c. 12 mm., cylindrical-clavate, membranous, purplish, 10-ribbed, with short ovate teeth. *Petals* purple-red, claw long, auricled above, limb obovate, *slightly notched, with 2 conspicuous coronal scales* 3 *mm. long* at its base. Capsule ovoid, 5-celled at base, bursting the calyx, opening by 5 ± spreading teeth, on a *carpophore almost equalling it in length*. Seeds 0·5 mm. across, acutely tubercled. Fl. 6–8. Protandrous. Visited by butterflies and long-tongued bumble-bees. 2*n* = 24. Chh.

Native. A very rare and local plant of cliffs, dry rocks and rock débris, chiefly on igneous rocks, in a few places in Wales and in Scotland from Roxburgh and Edinburgh northwards to Perth and Angus. 11. Reaches 1400 ft. in Scotland. Europe northwards to S. Scandinavia and Finland; W. Asia. Said to be calcifuge on the Continent, and characteristic of open sandy habitats.

3. AGROSTEMMA L.

Annual herbs with tall erect stems and narrow opposite lvs. Fls conspicuous, solitary or in few-fld dichasia. *Epicalyx* 0; sepals joined below into a 10-ribbed tube and *with five narrow spreading teeth much longer than the petals*; *coronal scales* 0; stamens 10; styles 5, alternating with the sepals; ovary 1-celled to the base. Capsule opening by 5 teeth, carpophore 0; seeds numerous, black.

Two spp., *A. gracile* Boiss., confined to Asia Minor, and *A. githago* L.

*1. A. githago L. Corn Cockle.

Lychnis Githago (L.) Scop.; *Githago segetum* Link

An annual herb with a strong tap-root and an erect simple or sparingly branched flowering stem 30–100 cm., covered with appressed white hairs. Lvs 5–12·5 cm., linear-lanceolate, acute, with appressed hairs. Fls 3–5 cm. diam., usually solitary at the ends of main stem and branches, their stalks long, hairy. *Calyx-tube* cylindrical-ovoid, coriaceous, woolly, 10-ribbed, *with long spreading linear acute lf-like teeth* 3–5 cm. long. *Petals pale reddish-purple*, long-clawed, the limb shorter than the sepals, obovate, *slightly notched, with no coronal scales.* Capsule ovoid, exceeding the calyx-tube, opening by 5 ± erect teeth. Seeds 3–3·5 mm. across, black, tubercled. Fl. 6–8. Protandrous to homogamous. Visited by butterflies and automatically self-pollinated. $2n = 24, 48$. Th.

Introduced. A decreasing cornfield weed. 104, H 22, S. Throughout Great Britain; absent from a few districts in N. Wales and Scotland, the Outer Hebrides and Shetland. Probably native in the Mediterranean region and perhaps derived from *A. gracile* Boiss., but now naturalized as a weed in most temperate regions of the world. The seeds are said to be poisonous and also to affect deleteriously the physical properties of wheat flour.

4. CUCUBALUS L.

Lvs opposite. Fls hermaphrodite, in lax dichasia terminating the main stem and branches. *Epicalyx* 0; *sepals joined* below, with 5 teeth; *petals* 5, long-clawed, *with coronal scales*; stamens 10; ovary 3-celled below; styles 3. *Fr. berry-like*, indehiscent, becoming dry when mature; seeds reniform, black.

One sp.

*1. C. baccifer L. 'Berry Catchfly.'

A large perennial herb 60–100 cm., with a branched creeping stock and brittle diffusely branched, scrambling flowering shoots, pubescent with short curved hairs. Lvs ovate, acuminate, narrowed into a short stalk, sparsely hairy, entire or somewhat sinuate. Fls c. 18 mm. diam., shortly stalked, ± drooping. Calyx-tube 8–15 mm., widely campanulate, indistinctly veined and rough with reflexed points, unequally 5-toothed above, the teeth almost twice as long as the tube, blunt. Petals greenish-white, their spreading distant limbs narrowly spathulate and deeply bifid, with 2 boss-like scales at the base, their claws cylindrical, broadening at the junction with the limb. *Fr.* 6–8 mm. diam., *black, globular*, on a short carpophore, not enclosed by the widely open ± rotate calyx with revolute teeth. Seeds 1·5 mm. across, white at first, turning yellow then black. Fl. 7–9. Protandrous. Fr. taken by birds. $2n = 24$. Hp.

Introduced. Naturalized in a few places; e.g. on the Isle of Dogs, Kent. C. and S. Europe and across Asia to Japan.

5. DIANTHUS L.

Perennial, rarely annual, herbs usually with ± linear and often glaucous opposite lvs. Fls solitary or in cymose infl., hermaphrodite or unisexual. Base of calyx tightly enclosed by an *epicalyx* of 1–3 pairs of usually mucronate or awned scales (bracteoles); sepals joined below into a *cylindrical calyx-tube* which is neither strongly ribbed nor has scarious seams; calyx-teeth 5, usually short; petals pink or red, rarely white, long-clawed; *coronal scales* 0; stamens 10; *styles* 2. *Capsule 1-celled, opening by* 4 *teeth*; seeds compressed, concave on one side.

About 270 spp. in Europe, Asia, Africa and N. America. Many spp. are grown in gardens as 'carnations' and 'pinks'.

1 Fls in ± head-like cymose clusters surrounded by an involucre of bracts.   2
   Fls solitary or 2–5 in a lax cyme.   4

2 Involucral bracts and epicalyx scales hairy.   **1. armeria**
   Bracts and epicalyx glabrous.   3

3 Lvs broadly lanceolate or elliptical, their sheaths as long as broad
   (5–8 mm.); epicalyx herbaceous.   **2. barbatus**
   Lvs linear, their sheaths c. four times as long as broad (15 mm.); epicalyx
   toughly membranous.   **3. carthusianorum**

4 Petals deeply cut into long narrow segments ⅓ to ½ of the length of the
   limb.   5
   Petals entire, crenate or toothed, not deeply cut.   6

5 Stem minutely downy below; epicalyx scales oblong; calyx-teeth not
   ciliate; petals cut to c. ⅓ their length.   **6. gallicus**
   Stem glabrous below; epicalyx scales ovate; calyx-teeth ciliate; petals cut
   to the middle.   **4. plumarius**

6 Stem downy below; epicalyx scales long-awned; fls not fragrant; petals
   with paler dots and dark transverse bars.   **8. deltoides**
   Stem glabrous; epicalyx scales abruptly and shortly mucronate; petals not
   with pale dots.   7

7 Stems 10–20 cm.; lvs with rough edges; calyx-tube c. 20 mm.; petals
   bearded.   **7. gratianopolitanus**
   Stems 20–50 cm.; lvs with smooth edges; calyx-tube 25–30 mm.; petals
   smooth.   **5. caryophyllus**

1. D. armeria L.   Deptford Pink.

An *annual* or biennial herb 30–60 cm., with a slender tap-root and a rigidly erect simple or branched stem, branched above and sometimes also below, dark green, not glaucous, shortly hairy at least above. Basal lvs linear-oblanceolate, in a rosette; stem lvs linear-lanceolate acute, keeled, 3–5 cm. × 1–3 mm., obliquely ascending; all with short hairs. Fls 8–13 mm. diam., ± sessile in terminal and lateral *2–10-fld short-stalked cymose clusters*. Involucral bracts erect, lfy, hairy, equalling the fl.-clusters. *Epicalyx* scales 2, lanceolate-subulate, *hairy*, ribbed, equalling the *calyx-tube* which is 13–20 mm., cylindrical, narrowing upwards, *woolly*, strongly ribbed, with 5 lanceolate-acute teeth. Petals bright rose-red with pale dots; limb 4–5 mm., narrowly ovate, shallowly and irregularly toothed, not contiguous; claw white. Capsule ± cylindrical, equalling the calyx. Seeds 1·5 mm. across. Fl. 7–8. Protandrous. Stamens of one or both whorls abort in some fls. Infrequently visited by butterflies and automatically self-pollinated. 2*n* = 30. Th.

Native. A rare and local lowland plant of hedgerows, waysides and dry pastures especially on light sandy soil. 54, S. Throughout England and Wales and in a few localities in Scotland northwards to Perth and Angus. Europe northwards to Denmark and S. Sweden; Transcaucasia. Introduced in N. America.

*2. D. barbatus L.   Sweet William.

A perennial herb 30–70 cm., with an oblique stock and one or more erect stems, simple or branched above. Lvs broadly elliptical, 7–9 × 1·5–2 cm., sessile; base sheathing, 5–8 mm., margins rough. *Fls almost sessile in dense compact cymose heads* of 3–30 enclosed by linear-lanceolate or linear acute rough-edged bracts. *Epicalyx* equalling or exceeding the calyx, *of 4 herbaceous glabrous ovate long-awned scales.* Calyx 15–18 mm., cylindrical, green below and often violet above, glabrous, with narrow awned teeth. Petals dark red or pink, variously spotted and transversely barred in cultivated forms; limb triangular-obovate, toothed. Capsule oblong. Fl. 7–8. Protandrous. Visited by butterflies and day-flying hawkmoths. $2n = 30$. Hp.

Introduced. A favourite garden plant, sometimes escaping. S. Europe from Spain and S. France to the Balkans and S. Russia.

*3. D. carthusianorum L., a closely related perennial herb, with stiffly erect stems 20–50 cm., differs in its linear-acute lvs with sheaths c. 15 mm., and toughly membranous epicalyx scales about half as long as the dark purplish calyx. It is sometimes cultivated and is found occasionally as a garden escape or casual. Native in C. and S. Europe southwards from Denmark and the Netherlands.

*4. D. plumarius L.   Common Pink.

A perennial tufted herb 15–30 cm., with a woody stock, procumbent rooting non-flowering shoots and erect or ascending 4-angled flowering shoots, glabrous and glaucous, usually branched above. *Lvs* to 5 cm., ascending, linear-subulate, very acute, *with rough margins. Fls* 1–5 in a lax cyme, 25–35 mm. diam., *strongly fragrant.* Epicalyx ¼ to ⅓ as long as the calyx, of 4–6 broadly ovate, shortly cuspidate, herbaceous, scarious-margined scales. Calyx-tube 25–30 × 3–4 mm., narrowing upwards, glabrous, violet-coloured, with narrow ciliate teeth. *Petals* pale pink or white, the *limb digitately cut almost to the middle* into slender lobes, the entire part obovate, *hairy at the base*; claw up to 25 mm. Capsule cylindrical, slightly longer than the calyx. Seeds flat, ± orbicular. Fl. 6–8. Protandrous. $2n = 30, 60, 90$. Ch.

Introduced. Naturalized on old walls in many parts of the country. S.E. Europe southwards from N. Italy, Styria and Bohemia. It and its hybrids are much cultivated for their fragrant fls.

*5. D. caryophyllus L.   Clove Pink, Carnation.

A perennial tufted herb with the habit of *D. plumarius* but with shoots 20–50 cm. and linear-lanceolate *lvs with their edges smooth* or rough only near the base. *Fls* 1–5 in a lax cyme, 35–40 mm. diam., *strongly fragrant.* Epicalyx ¼ as long as the calyx, of 4–6 broadly ovate, abruptly mucronate, membranous scales with herbaceous tips. Calyx-tube 25–30 mm., its teeth not ciliate. *Petals* rose-pink, the limb *obovate, crenate or dentate*, the teeth not exceeding ⅓ of its length. Capsule cylindrical-ovoid, longer than the calyx. Seeds pear-shaped, nearly flat. Fl. 7–8. Protandrous. Visited by butterflies and hawk-moths, and said to be self-sterile. $2n=30, 90$. Ch.

Introduced. Occasionally naturalized on old walls. S. Europe and N. Africa. The cultivated carnations are derived from *D. caryophyllus* and its hybrids with *D. plumarius*, etc.

*6. D. gallicus Pers.   'Western Pink.'

A perennial tufted herb with erect flowering stems 15–25 cm., glaucous, *downy below with minute hairs*. Lvs linear, short, stiff, ± blunt, 3-veined. Fls 1–3, fragrant. Epicalyx ¼ as long as the calyx, of 4 ovate-oblong abruptly mucronate scales. Calyx cylindrical, striate. Petals pink, the *limb cut to* ⅓ *of its length into narrow lobes*, the entire part ± orbicular. Capsule cylindrical. Fl. 6–8. $2n=60, 90$. Ch.

Probably introduced. Known only on dry grassy dunes in Jersey. Atlantic coast of Europe from Brittany to Spain and Portugal.

7. D. gratianopolitanus Vill.   Cheddar Pink.

D. caesius Sm.; *D. glaucus* sensu Huds., non L.; *D. caespitosus* Poir.

A perennial *densely tufted* herb with a woody stock, *long procumbent non-flowering shoots* and ascending flowering shoots, 10–20 cm., *glabrous and glaucous*. Lvs 2–6 cm., linear, bluntish, rough at the edges. Fls usually solitary, c. 25 mm. diam., fragrant. Epicalyx hardly ¼ as long as the calyx, its 4–6 scales herbaceous, roundish, abruptly mucronate or blunt. Calyx-tube 16–20 mm., striate, glabrous, usually violet in colour, teeth broadly triangular. Petals pale to deep rose, the *limb* ± obovate *with irregular teeth reaching not more than* ⅙ *of its length*, ± hairy at the base; claw c. 15 mm. Capsule cylindrical. Seeds ovate. Fl. 6–7. Protandrous. Strongly clove-scented and visited by butterflies and day-flying hawkmoths. $2n=90$. Ch.

Native. A rare and local plant confined to Carboniferous Limestone cliffs at Cheddar Gorge (N. Somerset); occasionally naturalized elsewhere. W. and C. Europe from France to Moravia and Hungary.

8. D. deltoides L.   Maiden Pink.

A perennial *loosely tufted* green or glaucous herb with a creeping branched slender stock, *short procumbent non-flowering shoots* and decumbent then ± erect flowering shoots, 15–45 cm., *rough with short hairs*. Lower lvs and those of barren shoots 10–16 mm., narrowly oblanceolate, blunt; upper *lvs* 10–25 mm., linear-lanceolate, acute; all *roughly hairy on the margins* and on the underside of the midrib. Fls c. 18 mm. diam., scentless, solitary or rarely 2–3 terminating the main stem and branches. *Epicalyx about* ½ *as long as the calyx*, of 2–4 broadly ovate, long-cuspidate scales, herbaceous with scarious margins. Calyx 12–17 mm., cylindrical, glabrous, green or reddish; teeth lanceolate, acute. Petals rose or white with pale spots and a dark basal band, the limb shortly and irregularly toothed. Capsule equalling or somewhat exceeding the calyx. Seeds 2–2·5 mm. obovate, black. Fl. 6–9. Protandrous. Visited by butterflies and moths. $2n=30, 60$. Ch.

Varies in the number of epicalyx scales and in glaucousness.

Native. A local lowland plant of dry grassy fields and banks and hilly pastures throughout Great Britain northwards to Inverness. 58, S. Throughout Europe northwards to Norway, Finland and N. Russia; temperate Asia. Introduced in N. America.

D. chinensis L., and especially the annual var. *heddewigii* Regel, is much grown in gardens for its large strikingly patterned fls. The epicalyx bracts are lfy and as long as the calyx, and the toothed petals are hairy at the base. *D. superbus* L., with petals cut more than half-way and hairy at the base, is also grown in gardens.

6. VACCARIA Medic.

Differs from *Saponaria* in the winged calyx-tube and absence of coronal scales.
    Three spp. in the Mediterranean region.
*1. V. pyramidata Medic. (*Saponaria Vaccaria* L.; *S. segetalis* Neck.). An annual herb with a slender tap-root and glabrous branched flowering stems 30–60 cm. Basal lvs oblong-lanceolate, upper ovate-lanceolate cordate, all sessile, acute, glabrous and glaucous. Fls in loose dichasia. Epicalyx 0. *Calyx-tube* inflated, glabrous, *with 5 sharp angles or wings* and 5 triangular teeth. Petals pale rose-coloured, the limb cuneate, rounded or somewhat emarginate, toothed; *coronal scales* 0. Stamens 10. Styles 2. Capsule globular, 4-celled below, opening by 4 teeth. Seeds 2 mm. across, black. Fl. 6. ±Homogamous. Visited by butterflies and automatically self-pollinated. $2n=30$. Th.
    Introduced. A not infrequent casual, occasionally becoming established. Europe northwards to Denmark and Sweden; Asia. Introduced in N. America, Australia and New Zealand.

7. SAPONARIA L.

Annual or perennial herbs with opposite lvs. Fls in loose or condensed dichasia. *Epicalyx* 0; *sepals joined* below into a green *tube without scarious seams* and with 5 teeth above; petals 5, with the limb narrowing abruptly into the long claw; coronal scales present; stamens 10; *styles* 2(–3); ovary 1-celled. Capsule opening by 4(–6) teeth; carpophore short; seeds reniform.
    About 20 spp., especially in the Mediterranean region.

1. S. officinalis L.   Soapwort, Bouncing Bett.

A perennial herb with a stout branched creeping rhizome from which arise long stolons and erect or ascending ±glabrous, flowering shoots 30–90 cm., simple or branched above. Lvs 5–10 cm., broadly ovate to elliptical, acute, 3(–5)-veined, ±glabrous. Fls c. 2·5 cm. diam., in compact terminal corymbs on the main stem and branches. Calyx-tube 18–20 mm., cylindrical, often reddish, with 5 short triangular teeth. Petals pink or flesh-coloured, the claw exceeding the calyx-tube, the limb obovate, entire or slightly emarginate, not contiguous, with 2 small blunt coronal scales at the base. Capsule oblong-ovoid, equalling the calyx-tube, opening by 4(–5) ±unequal teeth, but often failing to ripen; carpophore short. Seeds 1·8 mm. across, blackish. Fl. 7–9. Protandrous. Visited chiefly by day- and night-flying hawkmoths. $2n=28$. Hp.
    Native or introduced. A fairly common plant of hedgebanks and waysides near villages, probably as an escape from cultivation, northwards to Aberdeen; and perhaps native in Devon and Cornwall and in N. Wales, where it grows by streams. Europe northwards to Scandinavia; Asia. Introduced in N. America. Naturally a plant of streamsides and damp alluvial woods.

S. ocymoides L., a perennial mat-forming plant with ± prostrate pubescent shoots bearing obovate or spathulate lvs and small bright pink fls, is much grown on rock-gardens and walls and occasionally escapes. Native in the mountains of C. and S.W. Europe.

8. KOHLRAUSCHIA Kunth

Annual to perennial herbs with opposite linear lvs and hermaphrodite or unisexual fls. Individual fls or heads of fls with an *epicalyx or involucre* respectively of 2 or more membranous bracts. *Sepals joined below into a 15-veined* but not strongly ribbed *calyx-tube, with whitish scarious seams* alternating with the free teeth; petals ± clawed; coronal scales 0; stamens 10; styles 2. Capsule 1-celled above but 4-celled below, opening by 4 teeth; seeds flat, rugose.

Three spp. in C. and S. Europe.

Differs from *Dianthus* in the whitish scarious commissures of the calyx-tube.

Plant annual; fls in ovoid heads almost completely enclosed in a brown membranous involucre.   **1. prolifera**

Plant perennial; fls not in heads, each with an epicalyx of 2–5 whitish scales.
                                                                  2. saxifraga

1. K. prolifera (L.) Kunth   'Proliferous Pink.'

Dianthus prolifer L.; *Tunica prolifera* (L.) Scop.

An annual herb with a slender tap-root and one or more erect wiry stems, 10–50 cm., simple or branched above, glabrous or puberulous, slightly glaucous. Lvs 10–20 mm., linear-lanceolate, ± acute, scabrid on the margins. *Fls* opening 1 at a time, 6–9 mm. diam., *in ovoid heads of* 1–11 *enclosed in a loose involucre* of several pairs *of broad shining membranous brown bracts* (each fl. except the terminal subtended by 1 bract), the lowest pair mucronate, the remainder blunt; all fls ± concealed by the involucre except when open. Calyx-tube 10–13 mm., faintly 15-veined, glabrous, reddish, with 5 blunt membranous teeth. Petals pale purplish-red, abruptly clawed; limb 2–3 mm., obovate, emarginate; claw long, slender. *Capsule ovoid-ellipsoidal*, shorter than the calyx. Seeds 1·5 mm. across, rugulose, narrowly winged. Fl. 7. Homogamous. Visited sparingly by butterflies, etc. $2n = 30$; 60*. Th.

Native. A rare lowland plant of sandy and gravelly places in S. England, southwards from Gloucester, Berks and Norfolk; Glamorgan; Jersey; also as a casual. 9, S. S. and C. Europe southwards from Denmark and S. Sweden; Caucasus; N. Africa.

*2. K. saxifraga (L.) Dandy

Tunica saxifraga (L.) Scop.

A perennial mat-forming herb with numerous decumbent or ascending branching shoots, 10–35 cm., glabrous or pubescent, ± woody at base. Lvs up to 1 cm. × 0·5 mm., linear, acute, rough at the margin. *Fls solitary* at the ends of branches; each with an *epicalyx of 2–5 whitish mucronate bracts*. Calyx 4–5 mm., narrowly campanulate, whitish scarious seams alternating with the 5 blunt scarious-margined teeth. Petals lilac to deep pink, the emarginate limb, 3–4 mm., narrowing gradually into the claw. Capsule ovoid, scarcely equalling the calyx; seeds c. 1 mm., peltate, concave on one side, convex on the other. Fl. 6–9. $2n = 60$. Chh.

Introduced and established near Tenby, Pembroke. C. and S. Europe, Asia Minor, Caucasus, Persia.

9. GYPSOPHILA L.

Annual to perennial herbs or dwarf shrubs, with opposite lvs. Fls hermaphrodite in cymose infls. Epicalyx 0; *sepals joined* below into a campanulate *tube with pale scarious seams between the main veins* and with 5 teeth above; petals with the *limb narrowing gradually into the claw*; coronal scales 0; stamens 10; styles 2, rarely 3. Capsule opening by 4 (6) teeth; seeds reniform.

About 10 spp. chiefly in S.E. Europe and W. Asia.

No sp. of *Gypsophila* is native in the British Is. The annual *G. muralis* L. (Europe) with somewhat glaucous shoots 4–18 cm. high, shortly hairy below, ± linear lvs to 2 cm., and pink fls 5 mm. diam. on long stalks; and the annual *G. porrigens* (L.) Boiss. (Near East) with diffusely branched procumbent stems covered with long dense spreading hairs, tiny fls and very large seeds, occur as casuals. The perennial *G. paniculata* L. (Maiden's Breath; C. Europe and W. Asia), 60–90 cm. high with lanceolate lvs and numerous white fls 4–5 mm. diam. in crowded corymbose panicles, is much grown in gardens and sometimes escapes, as does the annual *G. elegans* Bieb. (Caucasus), 30–45 cm. high, with lanceolate, acute lvs and white or pink fls.

10. CERASTIUM L.   CHICKWEED

Annual to perennial herbs or dwarf shrubs, usually hairy, with opposite, entire, sessile lvs. Fls usually in cymose infls, sometimes solitary; 5- or 4-merous. Sepals free, with membranous margins; petals white, bifid up to half-way or emarginate, sometimes 0; stamens usually 10 (8), sometimes 5 or fewer; nectaries present; ovary 1-celled; *styles usually 5, opposite the sepals,* sometimes 4, 3 or 6. *Fr.* an *oblong capsule exceeding the sepals,* usually ± curved, *opening by twice as many short teeth as styles.* Seeds numerous, spherical or reniform, rough. Perhaps c. 100 spp., cosmopolitan, but principally in the north temperate regions of the Old World.

1	Styles usually 3 (varying 4–6); capsule-teeth usually 6; a rare plant of high mountains.	**1. cerastoides**
	Styles 4–5; capsule-teeth 8 or 10.	2
2	Petals about twice as long as sepals; perennials with prostrate non-flowering shoots and fls at least 12 mm. diam.	3
	Petals not more than 1½ times as long as sepals; chiefly annuals; fls usually less than 12 mm. diam.	8
3	Lvs linear-lanceolate to narrowly oblong; lower lvs commonly with axillary lf-clusters; chiefly lowland plants.	4
	Lvs narrowly elliptical-oblong to nearly circular; lower lvs usually lacking axillary lf-clusters; chiefly alpine plants, at sea-level only in the extreme north.	5
4	Stem and lvs densely white-tomentose.	**3. tomentosum**
	Stem and lvs shortly hairy or almost glabrous.	**2. arvense**
5	Stem and young lvs ± densely covered with long soft white hairs; sepals oblong-lanceolate; capsule narrow and curved in its upper half; seeds 1–1·4 mm., tubercled.	**4. alpinum**

Stem and lvs with few or no long soft white hairs, though commonly
with short whitish hairs and sometimes also with short glandular
hairs; sepals broadly lanceolate to ovate-lanceolate.   6

6 Bracts and bracteoles wholly herbaceous; fls 18–30 mm. diam.; capsule
broad and nearly straight in its upper half; seeds more than 1·4 mm.,
rugose.   7
At least the upper bracteoles with narrow scarious margins and tips;
fls 12–25 mm. diam.; capsule narrow and distinctly curved in its upper
half; seeds not exceeding 1·4 mm., tubercled.
                                          7. holosteoides forms and hybrids

7 Lvs dark and purple-tinged, almost circular, covered with short stout
glands (Shetland).   **6. nigrescens**
Lvs not dark and purplish nor densely glandular, usually elliptical;
mountains (not in Shetland).   **5. arcticum**

8 Perennial usually eglandular herb with prostrate non-flowering shoots;
sepals with glabrous tips, eglandular; capsule 9–12 mm.   **7. holosteoides**
Annual or biennial herbs lacking non-flowering shoots, usually glandular
(except *brachypetalum*); sepals glabrous or hairy at the tip; capsule not
exceeding 10 mm., commonly much shorter.   9

9 Fls in compact clusters which remain compact in fr., so that the capsule-
stalk does not exceed the sepals.   **8. glomeratum**
Fls in spreading cymes; capsule-stalk usually much exceeding the sepals.   10

10 Plant shaggy with spreading hairs, eglandular or nearly so; petals about
half as long as sepals; stamens 10, their filaments hairy at the base.
                                    9. brachypetalum var. *eglandulosum*
Plant viscid with dense glandular hairs, not shaggy; petals more than
half as long as sepals; stamens 4–5, their filaments glabrous.   11

11 Bracts entirely herbaceous; fr.-stalks usually erect throughout; fls usually
4-merous, sometimes 5-merous.   **10. atrovirens**
At least the upper bracts with scarious tips or margins; fr.-stalks at first
recurved or sharply deflexed; fls 5-merous.   12

12 Bracts with the upper half scarious; petals about two-thirds as long as
sepals, slightly notched; fr.-stalk at first sharply deflexed from the base,
but not curved; plant often decumbent.   **12. semidecandrum**
Upper bracts with narrow scarious margins; petals about equalling sepals,
about ¼ bifid; fr.-stalk at first curved downwards, not sharply deflexed;
plant erect, reddish.   **11. pumilum**

1. C. cerastoides (L.) Britton   'Starwort Mouse-ear Chickweed.'

Stellaria cerastoides L.; *Cerastium trigynum* Vill.

A perennial herb with much branched creeping ± woody stems from which
arise prostrate non-flowering shoots, 5–15 cm., and decumbent or ascending
flowering shoots 5–10 cm.; all *shoots* rooting, *glabrous except for a line of
small hairs down each internode*. Lvs 6–12 mm., pale green, elliptic-oblong or
linear-lanceolate, blunt, usually curving to one side. Fls 9–12 mm. diam.,
1–3, their slender glandular stalks up to 8 cm. Bracts lf-like. Sepals 4–5 mm.,
narrowly lanceolate, 1-veined, with a narrow scarious margin. Petals white,
deeply bifid, almost twice as long as the sepals. Stamens 10. *Styles usually* 3,
sometimes 4–6. Capsule oblong, straight, up to twice as long as the calyx,

opening by 6(–10) spreading valves. Seeds brown, 0·5 mm. across. Fl. 7–8. ± Homogamous. Visited by flies and automatically self-pollinated. $2n = 38*$. Chh.

Native. A local alpine plant of high mountains in Cumberland and in Scotland northwards to Sutherland. 8. Arctic Europe, eastern N. America and W. Asia, and mountains of C. Europe and Asia. Reaches 73° 15′ N. in E. Greenland.

2. C. arvense L.   'Field Mouse-ear Chickweed.'

A perennial herb with a branched creeping stock from which arise long prostrate rooting non-flowering shoots up to 30 cm., and ascending flowering shoots 4–30 cm. high, prostrate below; all ± hairy and glandular. *Lower lvs commonly with axillary lf-clusters* (i.e. non-elongated axillary shoots). Lvs 5–20 mm., linear-lanceolate or narrowly oblong, hardly narrowed to the base, usually soft and downy but not woolly, ± glandular. Fls 12–20 mm. diam. in lax dichasia, their stalks glandular-hairy. Sepals 5–8 mm., oblong-lanceolate subacute, glandular-hairy, with membranous margins and tips. Petals white, obovate, bifid, about twice as long as the sepals. Stamens 10.

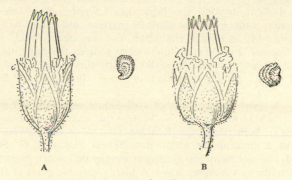

Fig. 30. Capsule and seeds of A, *Cerastium alpinum* and B, *C. arcticum*. Capsules × 2·5; seeds × 5.

Styles 5. Capsule cylindrical, slightly exceeding the calyx. Fr. stalks erect but curved just beneath the calyx. Seeds 0·8–1 mm. across, tubercled. Fl. 4–8. Protandrous. Visited by various insects, chiefly flies and small bees. $2n = 36, 72*$. Chh.

Variable. Var. *latifolium* Fenzl, with lvs up to 25 × 6 mm., has been collected in W. Norfolk.

Native. On dry banks and waysides and in grassland especially on calcareous or slightly acid sandy soils. 81, H 12. Through most of Great Britain to Sutherland, but with a distinct eastern tendency; Orkneys; very local in Ireland. Europe, N. Africa, W. and temperate Asia, N. America.

*3. C. tomentosum L.   Dusty Miller, Snow-in-Summer.

A perennial mat-forming herb resembling luxuriant *C. arvense* but with *densely white-tomentose* stems and lvs. Stems prostrate below, readily rooting;

flowering stems ascending, 15–30 cm. Lower lvs commonly with axillary lf-clusters. Lvs linear-lanceolate. Fls 12–18 mm. diam., in lax dichasia, their stalks and sepals white-tomentose. Petals white, bifid, twice as long as the sepals. Stamens 10. Styles 5. Fl. 5–8. Protandrous or homogamous. $2n=72*$. Chh.

Introduced. A commonly grown rock-garden and wall plant, often escaping. Native of S.E. Europe and the Caucasus.

4. C. alpinum L.   'Alpine Mouse-ear Chickweed.'

Incl. *C. lanatum* Lam.

A perennial mat-forming herb with prostrate woody stems from which arise short ± prostrate non-flowering shoots up to 6 cm., and decumbent then ascending flowering shoots 5–15 cm. Lvs commonly 10 × 5 mm. but reaching 18 × 7 mm.; those on barren shoots and at the base of flowering shoots the smallest, obovate to elliptic-oblanceolate, the others oval or elliptical. *Stems and lvs densely covered with long soft white hairs* and usually also with some short glandular hairs. *Bracts and especially bracteoles with narrow membranous margins.* Fls 1–4, 18–25 mm. diam., on hairy ± glandular stalks extending 1–4 cm. beyond the bracteoles. *Sepals oblong-lanceolate, 7·5– 10 mm.*, hairy, with ± narrow scarious margins and often with violet tips. Petals white, twice as long as the sepals, not very deeply bifid. Stamens 10. Styles 5. *Capsule* (Fig. 30 A) about twice as long as the calyx, the upper half *narrow and distinctly curved.* Fr. stalks slightly curved just below the capsule. Seeds 1–1·4 mm., acutely tubercled. Fl. 6–8. Protandrous. $2n=54, 72*$, 108. Ch.

Native. A local alpine plant of rock-ledges especially on granites and schists on high mountains in N. Wales, the Lake District and Scotland northwards to Sutherland. 20. Reaches 3980 ft. on Ben Lawers. An arctic-alpine with a circumpolar arctic distribution, reaching 83° 24′ N. in N. Greenland, the 'northernmost botanical locality on earth'; Norway, Sweden, Iceland; Pyrenees, Auvergne, Alps. A highly variable plant in the Arctic.

British forms seem all to be referable to ssp. **lanatum** (Lam.) Aschers. & Graebn. Putative hybrids with *C. arcticum* and *C. holosteoides* also occur.

5. C. arcticum Lange   'Arctic Mouse-ear Chickweed.,

C. Edmondstonii auct.; *C. latifolium* Sm., non L.

A perennial mat-forming herb with prostrate woody stems from which arise numerous ascending non-flowering shoots up to 7 cm. and ascending flowering shoots up to 15 cm. Lvs very variable in shape, from narrowly elliptic-lanceolate to broadly elliptical, commonly 10–15 × 4–5 mm., and narrower than in *C. alpinum*. *Stem and lvs covered with short whitish hairs,* shorter, stiffer and less dense than in *C. alpinum. Bracts and bracteoles wholly lf-like,* with no scarious margins. Fls 18–30 mm. diam., 1–3, their *stalks* extending 1–3 cm. beyond the bracteoles, usually *with spreading yellowish hairs* and often glandular. *Sepals 5·5–9 mm., broadly lanceolate to ovate-lanceolate,* hairy and sometimes glandular, with broad scarious margins. Petals white, twice or more as long as the sepals, shallowly bifid. Stamens 10. Styles 5.

Capsule (Fig. 30 B) up to twice as long as the calyx, the upper half *broad and nearly straight.* Fr. stalks often slightly curved just beneath the capsule. *Seeds* 1·5–1·8 mm., rugose. Fl. 6–8. Protandrous. $2n=108*$. Chh.

A very variable plant, especially in stature, lf-shape, and glandulosity.

Native. A local plant of rock ledges on high mountains in N. Wales (Snowdon), Scotland, Inner Hebrides (Skye); reaching c. 3500 ft. 13. Norway, Sweden, Finland, Faeroes, Iceland, and perhaps elsewhere in the Arctic.

6. C. nigrescens Edmondst. ex H. C. Wats. 'Shetland Mouse-ear Chickweed.'

C. edmondstonii (Edmondst.) Murb. & Ostenf.

A dwarf tufted perennial with purplish shoots hardly exceeding 5 cm. in height. *Lvs* 4–7 mm., *broadly ovate-elliptical to almost circular, dark and purple-tinged, densely* clothed, like the stems, with short stout spreading *glandular* hairs. Fls semiglobular, resembling those of *C. arcticum* but with the sepals blunt and somewhat broader. Fruits very freely, with capsule and seeds like those of *C. arcticum.* Fl. 6–8. Chh.

Native. Only on serpentine débris at 50 ft. on Unst, Shetland.

7. C. holosteoides Fr.   'Common Mouse-ear Chickweed.'

C. caespitosum Gilib.; *C. triviale* Link; *C. viscosum* L., nom. ambig.

A perennial (perhaps rarely annual) herb with a slender creeping stock from which arise decumbent non-flowering shoots up to 15 cm. and decumbent then erect flowering shoots up to 45 cm.; all *shoots* ± *hairy, very rarely glandular* above. *Lvs* 1–2·5(–4) cm., those on barren shoots oblanceolate, blunt, narrowed to a stalk-like base; on flowering shoots elliptical to ovate-oblong, subacute, sessile, in distant pairs; all *dark greyish-green,* covered densely with white hairs. Fls in dichasia which become lax in fr., their stalks hairy. Lower bracts wholly lf-like; upper bracts often with narrow scarious margins. Sepals 5–7 mm., ovate-lanceolate, hairy, with scarious margins and glabrous tips. *Petals* white, *equalling or somewhat exceeding the sepals,* rarely nearly twice as long, rather deeply bifid. Stamens 10, rarely 5. Styles 5. Capsule 9–12 mm., narrowly cylindrical, curved. Fr. stalks longer than the calyx and somewhat curved at their distal end. Seeds 0·8–0·9 mm., reddish-brown, bluntly tubercled. Fl. 4–9. Protandrous. Visited chiefly by flies and automatically self-pollinated. $2n=136*–152*$. Chh.(–Th.).

Alpine forms are usually dwarfer and more hairy and have larger fls than the type. They may constitute a distinct ecotype. Plants from Ben Lawers named '*C. vulgatum* var. *alpestre* (Lindbl.) Hartm.', with longish white hairs and large petals, do not fruit and are probably *C. alpinum* × *holosteoides* (*C. alpinum* var. *pubescens* Syme).

Var. *glabrescens* (G. W. F. Mey.) Hylander, a plant of wet places, has ± glabrous lvs and sepals and often two opposite hairy lines down each internode.

Native. A very common plant of grassland, shingle, dunes, waysides, waste places and cultivated ground throughout the British Is.; reaching nearly 4000 ft. in Scotland. 112, H 40, S. Cosmopolitan.

8. C. glomeratum Thuill.   'Sticky Mouse-ear Chickweed.'

C. viscosum auct.

An annual or overwintering herb with *pale yellowish-green lvs* and erect or ascending flowering shoots, 5–45 cm., hairy, *glandular* at least above, rarely eglandular. Lvs 0·5–2·5 cm.; basal lvs oblanceolate or obovate narrowed below; stem lvs broadly ovate or elliptic-ovate; all ± sessile, apiculate, covered with long white hairs. *Fls aggregated into compact cymose clusters*; stalks hairy, very short. Bracts herbaceous, hairy. *Sepals* 4–5 mm., lanceolate, very acute, with a narrow scarious margin, usually *with* glandular hairs as well as *long white hairs which extend to and project beyond the tip*. Petals white, about equalling sepals, c. ¾ bifid; rarely 0. Stamens 10. Styles 5. Capsule 7–10 mm., narrowly cylindrical, curving upwards out of the line of its stalk which it exceeds in length. Seeds 0·5–0·6 mm., pale brown, finely tubercled. Fl. 4–9. Homogamous. Little visited by insects; automatically self-pollinated and sometimes cleistogamous. $2n=72^*$. Th.

Native. A common weed of arable land and waste places, on walls and banks and on open sand-dunes. 112, H 40, S. Throughout the British Is. Cosmopolitan.

9. C. brachypetalum Pers. var. *eglandulosum* Fenzl

An annual herb with simple or basally branched stems, 5·5–30 cm., erect, *shaggy* with spreading-ascending hairs, greenish above, purple-tinged below. Lvs 4–14 mm., those at the base ± crowded, lanceolate-acute, narrowed into a rather broad stalk; stem lvs lanceolate or elliptic, acute, sessile; all thinly covered with long hairs. Infl. a lax dichasium of up to 30 fls on erect or ascending shaggy stalks, 6–15 mm. *Bracts entirely herbaceous*, thinly hairy. Fls 5-merous. *Sepals* 4–4·5 mm. lanceolate, concave, herbaceous and *thinly hairy to the tip*, with or without scarious margins. *Petals about half as long as the sepals*, bifid for ⅓ their length, sparsely ciliate below. Stamens 10, with a few long hairs near the base of the filament. Styles 5. Capsule 5–7 mm., slightly exceeding the sepals, broad, cylindrical, slightly curved near the apex, its stalk bent near the upper end. Seeds 0·5 mm. across, pale brown, acutely tubercled. Fl. 5. Probably self-pollinated. $2n=90^*$. Th.

Probably introduced. Found in 1947 on the bank of a railway-cutting between Sharnbrook and Irchester, Beds. C. Europe, from Spain, N. Italy and Roumania to Denmark. The range of the species is wider, from the Mediterranean to S. Scandinavia, and in N. Africa and the Caucasus.

The wholly herbaceous bracts distinguish it from *C. semidecandrum* and *C. pumilum*, the ciliate bases of petals and filaments from *C. atrovirens* and the lax infl. from *C. glomeratum*. From all it differs in the long spreading silvery hairs which give it a greyish appearance.

10. C. atrovirens Bab.   'Dark-green Mouse-ear Chickweed.'

C. tetrandrum Curt.

An annual herb with a slender tap-root and diffusely branched decumbent or ascending flowering shoots 7·5–30 cm., dull dark green in colour, densely covered and viscid with short glandular hairs. Lvs 5–10 mm.; basal oblanceolate; stem lvs ovate or oblong-ovate, acute; all covered with short glandular and non-glandular hairs. Fls 3–6 mm. diam., in few-fld dichasia, their stalks short, slender, glandular. *Bracts and bracteoles wholly lf-like. Fls usually 4-merous*, sometimes 5-merous. Sepals 4–7 mm., lanceolate, acute or acuminate, with narrowly scarious white margins and tips, glandular-hairy, but with hairs usually not projecting beyond the *glabrous tip*. Petals white, c. ¾ length of the sepals, *about ⅕ bifid*, with several branching veins. Stamens 4(–5). Styles 4(–5). Capsule 5–7·5 mm., narrowly cylindrical, often not much ex-

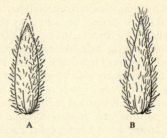

Fig. 31. Sepals of A, *Cerastium atrovirens* and B, *C. glomeratum* × 5.

ceeding the calyx, nearly straight. Fr. stalks much longer than the capsule, straight, ultimately erect. Seeds 0·5–0·7 mm., reddish-brown, bluntly tubercled. Fl. 5–7. Homogamous. Rarely visited by insects and automatically self-pollinated. $2n=72^*$. Th. Very variable in habit and robustness. Tall slender plants from Sutherland and the Shetlands have been identified as *C. subtetrandrum* (Lange) Murb., which they resemble in habit, in the long sepals and in the long capsules, but differ in that the seeds are large (0·7 mm.) and dark (0·5 mm. and pale yellowish-brown in *C. subtetrandrum*). British *C. atrovirens* shows independent variability in all these characters and it does not yet seem possible to recognize two distinct species.

Native. A locally common plant of sandy and stony places near the sea, rarely inland on railway ballast. All round the British Is. 88, H 37, S. Reaches nearly 950 ft. in Wilts. W. Europe from Spain to Norway and Sweden; Sardinia, Corsica, Capraia. Commonly on dunes.

11. C. pumilum Curt.   'Curtis's Mouse-ear Chickweed.'

Incl. *C. glutinosum* Fr.

An annual or overwintering herb with slender tap-root and erect or ascending stems, 2–12 cm., branching near the base, covered with glandular and non-glandular hairs, *often reddish below*. Lvs 0·5–1·5 cm.; the basal oblanceolate with a long stalk-like base; stem lvs elliptic or ovate-oblong; all hairy. Fls 6–7 mm. diam., in few-fld dichasia, their stalks short, glandular; lower bracts usually lf-like; *upper bracts small with narrow scarious margins and tip*. Fls 5-merous. Sepals 4–5 mm., oblong-lanceolate, acute, glandular-hairy with fairly broad scarious margins; hairs usually not projecting beyond the glabrous tip. *Petals* white or purple-tinged, about equalling the sepals, *about ¼ bifid*, with branching veins. Stamens 5. Styles 5. Capsule 6–7 mm., narrowly cylindrical, slightly curved upwards. *Fr. stalks* longer than the capsule, *at first curved downwards* then nearly erect but with a slight curve just beneath the capsule. Seeds 0·6–0·7 mm., dark brown, finely tubercled. Fl. 4–5. Homogamous. Automatically self-pollinated. $2n=90^*, 95^*$. Th.

Native. A rare and local plant of open calcareous grassland and chalk or limestone quarries, etc., in S. and C. England and Wales southwards from Caernarvon, Leicester and Suffolk. 20. Europe southwards from S. Scandinavia; N. Africa, W. Asia.

12. C. semidecandrum L.   'Little Mouse-ear Chickweed.'

An annual or overwintering herb with slender tap-root and erect or ascending stems, 1–20 cm., branching at the base, covered with short dense glandular and non-glandular hairs. Lvs 0·5–1·8 cm.; the basal narrowly oblanceolate with a long stalk-like base; stem lvs ovate to elliptic-oblong; all with short white hairs. Fls 5–7 mm. diam., in ± spreading dichasia, their stalks glandular. Lowest bracts with a broad scarious margin; *upper bracts largely scarious with a small green central portion.* Sepals 3–5 mm., narrowly lanceolate, acute, with broad scarious margins, glandular-hairy. *Petals white, c. ⅔ length of the sepals*, narrow, *slightly notched*, with *unbranched veins*. Stamens 5. Styles 5. Capsule 4·5–6·5 mm., cylindrical, very slightly curved. *Fr. stalks* longer than the capsule, *at first sharply deflexed from the base but not curved*, later almost erect. Seeds 0·4–0·6 mm., pale yellowish-brown, finely tubercled. Fl. 4–5. Homogamous. Sparingly visited by small insects, and automatically self-pollinated. $2n = 36^*$. Th. Variable in habit and stature.

Native. Common; dry open habitats especially on calcareous or sandy soils (including dunes) through most of Great Britain; mainly coastal in Ireland; not in Shetland. 103, H 17, S. Europe southwards from S. Scandinavia; N. Africa; W. Asia.

11. Myosoton Moench

An annual to perennial herb with opposite ovate-cordate lvs and a lfy cymose infl. Fls 5-merous. Sepals free; *petals* white, *bifid*; stamens 10; ovary 1-celled; *styles 5, alternating with the sepals. Fr.* an *ovoid capsule opening by 5 bifid teeth.* Seeds numerous, small.

One sp.

Differs from *Stellaria* in the 5 styles and from *Cerastium* in their alternation with the sepals and in the ovoid capsule; from both in the 5 bifid teeth by which the capsule opens.

1. M. aquaticum (L.) Moench   Water Chickweed.

Cerastium aquaticum L.; *Stellaria aquatica* (L.) Scop.; *Malachia aquatica* (L.) Fr.

A usually perennial herb with prostrate overwintering stem-bases from which lfy flowering and non-flowering shoots arise. Flowering shoots 20–100 cm., decumbent or ascending, weak and fragile and often trailing over other plants, ± glabrous below, glandular-hairy above. Lvs 2–5(–8) cm., thin, ovate-acuminate with a cordate base and often a wavy margin; lower lvs with a short ciliate stalk, upper ± sessile. Infl. a spreading dichasial cyme with lf-like bracts so that the fls appear solitary at forks of the stem. Fls 12–15 mm. diam., their stalks glandular. Sepals narrowly ovate, blunt, with broad membranous margins. Petals up to half as long again as the sepals, bifid almost to the base with diverging lobes. Capsule longer than the sepals, drooping, its stalk at first reflexed then ± horizontal. Seeds brown, 0·4 mm. across, densely covered with barbed papillae. Fl. 7–8. Protandrous. Visited by flies and small bees, etc. $2n = 28$. H.–Ch.

Native. In marshes and fens, streamsides, ditches and damp woods at low altitudes through England, Wales and S. Scotland northwards to Stirling. 63. Europe (except arctic) and temperate Asia.

12. STELLARIA L.

Annual to perennial herbs, usually slender and fragile and often glabrous. Lvs opposite, simple, entire. Fls in dichasial cymes, rarely solitary, usually 5-merous. Sepals free; *petals* white, usually *very deeply bifid*, sometimes 0; stamens 10 (8) or fewer; nectaries present; ovary 1-celled; *styles* 3. *Fr. a ± rounded capsule opening by* 6 *valves*. Seeds numerous, roundish-reniform.
About 85 spp., cosmopolitan.

1	Lvs ovate or cordate, at least the lower stalked; stem terete.	2
	Lvs narrow, linear to oblong, sessile; stem quadrangular.	5
2	Petals about twice as long as the sepals; capsule narrowly ovoid, up to twice as long as the sepals.	**1. nemorum**
	Petals not or little exceeding the sepals, or 0; capsule ovoid, longer than the sepals.	3
3	Sepals 5–6·5 mm.; stamens 10; seeds not less than 1·3 mm. diam.	**4. neglecta**
	Sepals 2–5 mm.; stamens 1–7; seeds not more than 1·3 mm. diam.	4
4	Sepals 4·5–5 mm.; petals usually present; stamens 3–7 with red-violet anthers; seeds 0·9–1·3 mm. diam.	**2. media**
	Sepals 2–3·5 mm.; petals 0; stamens 1–3(–5) with grey-violet anthers; seeds 0·6–0·8 mm. diam.	**3. pallida**
5	Petals deeply bifid with divergent lobes, distinctly shorter than the sepals; calyx with a funnel-shaped base.	**8. alsine**
	Petals equalling or exceeding the sepals; calyx with a rounded base.	6
6	Bracts wholly herbaceous; fls 20–30 mm. diam.; petals notched to less than half-way; capsule globular.	**5. holostea**
	Bracts with scarious margins; fls 5–18 mm. diam.; petals bifid at least half-way and usually to near the base; capsule oblong-ovoid.	7
7	Bracts ciliate; plant not glaucous; fls 5–12 mm. diam.	**7. graminea**
	Bracts not ciliate; plant usually glaucous; fls 12–18 mm. diam.	**6. palustris**

1. S. nemorum L. Wood Stitchwort, Wood Chickweed.

A *perennial* stoloniferous herb with weak decumbent or ascending pale green lfy shoots 15–60 cm., and subterranean stolons up to 15 cm. long. *Lfy shoots* terete, sparsely *hairy all round* or ± glabrous, slightly glandular above. *Lower lvs* and those of non-flowering shoots *long-stalked, ovate, ± cordate*; upper lvs 2·5–7·5 cm., ± sessile, ovate-acuminate; all thin, glabrous or sparsely hairy and often somewhat ciliate. Fls 13–18 mm. diam., in lax dichasia, their stalks slender, glandular-pubescent. Sepals c. 6 mm., broadly lanceolate, blunt, with narrow membranous margins. *Petals* white, *twice as long as the sepals*, deeply bifid with narrow divergent lobes. Stamens 10, rarely 0. Styles 3. Capsule narrowly ovoid, somewhat exceeding the calyx, opening almost to the base by 6 valves. Fruiting stalks spreading or reflexed. Seeds orange-brown, 1–1·3 mm. across. Fl. 5–6. Protandrous. Visited by flies and beetles. $2n=26$. H.

There are 2 sspp. in the British Is.: ssp. **nemorum**, with bracts decreasing gradually in size at each branching of the infl. and seeds with marginal rows of ± hemispherical tubercles; and ssp. **glochidisperma** Murb., with bracts decreasing abruptly in size after the first pair and seeds with long cylindrical papillae (c. 0·15 mm.) on the margin.

Native. In damp woods and by streams in the west and north of Great Britain. Ssp. *nemorum* from N. Devon, Derby and N. Lincs to Caithness, reaching 3000 ft. in Scotland; ssp. *glochidisperma* only in Wales and Monmouth. 57. Through most of Europe eastwards to the Caucasus, but chiefly a plant of montane and subalpine woods on base-rich substrata.

2. S. media (L.) Vill.   Chickweed.

Alsine media L.

An annual or overwintering herb with a slender tap-root and diffusely branched decumbent or ascending lfy *stems* 5–40 cm., *with a single line of hairs down each internode.* Lower lvs 3–20 mm., ovate-acute, long-stalked; upper ovate or broadly elliptical, usually larger (up to 25 mm.), acute or shortly acuminate, ± sessile; all ± glabrous or ciliate at the base. Fls numerous, in terminal dichasia, their stalks usually with a line of hairs. *Sepals* 4·5–5 *mm.*, ovate-lanceolate, with a narrow membranous margin, usually glandular-hairy. Petals white, not exceeding the sepals, deeply bifid, sometimes 0. *Stamens* 3–8 with red-violet anthers. Styles 3. Capsule ovoid-oblong, somewhat longer than the calyx. Fr. stalks downwardly curved, often wavy. *Seeds reddish-brown*, 0·9–1·3 *mm. across, usually with rounded or flat-topped tubercles.* Fl. 1–12. ± Homogamous. Visited by numerous flies and small bees, etc., and automatically self-pollinated. $2n = 42$. Th.

A very polymorphic species, varying in size, habit, hairiness, length of petals, number of stamens, size and surface-detail of seeds, etc.

Native. A weed of cultivated ground and waste places, abundant throughout the British Is. 112, H 40, S. Cosmopolitan.

3. S. pallida (Dum.) Piré   'Lesser Chickweed.'

S. apetala auct.; incl. *S. Boraeana* Jord.

An annual much-branched prostrate herb resembling *S. media* but with more slender often filiform stems and smaller pale green ovate *lvs* which are usually less than 7 mm. and *all or most* of which are *short-stalked.* Stem up to 40 cm., usually with a single line of hairs along each internode. Fls smaller than in *S. media*, never opening widely, their stalks hairy or almost glabrous. *Sepals* 2–3·5 *mm.*, lanceolate, hairy or glabrous. *Petals* 0 *or minute. Stamens* 1–3 with grey-violet anthers. Stigmas 3, sessile. Capsule exceeding the calyx. Fr. stalks short, usually not reflexed. *Seeds pale yellowish-brown*, rarely dark-brown, 0·6–0·8 *mm. across, with small blunt tubercles.* Fl. 3–5. Automatically self-pollinated and often cleistogamous. $2n = 22$. Th. Readily distinguished from apetalous *S. media* by the small sepals and small pale seeds.

Native. A locally common plant of dunes and an inland weed of waste places and cultivated ground on light sandy soils. Chiefly in S. and E. England but reaching Sutherland. Ireland. Channel Is. C. and E. Europe. A characteristic species of woods of *Pinus sylvestris* on light glacial sands.

4. S. neglecta Weihe   'Greater Chickweed.'

S. umbrosa Opiz & Rufr.; incl. *S. Elisabethae* F. W. Schultz

An overwintering annual or perennial herb resembling *S. media* but larger in all its parts, with weak branching stems 25–90 cm., procumbent below then ascending, each internode with a single line of hairs. Lower lvs ovate, subcordate, acute or shortly acuminate, long-stalked, with the blade 1–2·5 cm. and the stalk up to twice as long; upper lvs to 5 cm., ovate-oblong, acuminate, with short flattened stalks or subsessile; all glabrous. Fls c. 10 mm. diam., in lax dichasia, their long slender stalks glabrous or pubescent. *Sepals 5–6·5 mm.*, lanceolate, glabrous or pubescent. Petals white, deeply bifid, equalling or slightly exceeding the sepals, rarely 0. *Stamens usually* 10. Styles 3. Capsule exceeding the calyx. Fr. stalks 25–30 mm., at first spreading or deflexed, finally erect. *Seeds* 1·3–1·7 *mm. diam., dark reddish-brown, usually with slender acute tubercles.* Fl. 4–7. Protandrous. $2n = 22$. H.

Native. A local plant of hedgerows, wood-margins, streamsides and shady places. 62, H4. Found throughout Great Britain but commonest in the west. S.W. Ireland. Europe.

5. S. holostea L.   Satin Flower, Adders' Meat, 'Greater Stitchwort.'

A perennial herb with a slender creeping stock and weak brittle ascending stems, some short and non-flowering, others 15–60 cm., flowering. Stems with 4 rough angles, more slender below, glabrous or hairy above. *Lvs* 4–8 cm., slightly glaucous, rigid, *lanceolate-acuminate, tapering from a wide base to a long fine point, very rough at the margins* and on the underside of the midrib. Fls 20–30 mm. diam., long-stalked, in loose terminal dichasia, their *bracts lf-like.* Sepals 6–8 mm., ovate-lanceolate, inconspicuously 3-veined, with a narrow membranous margin. Petals 8–12 mm., white, bifid to about halfway, occasionally laciniate, rarely 0. Stamens 10, sometimes fewer by degeneration. Styles 3. Capsule subglobose, about equalling the calyx. Seeds 1·5–2 mm. across, reddish-brown, papillose. Fl. 4–6. Protandrous to homogamous. Visited by many flies, small bees and beetles. $2n = 26$. Chh.

Native. A common plant, especially of woods and hedgerows throughout the British Is. except Shetland. 111, H40, S. Reaches 2100 ft. in Scotland. Europe, N. Africa, the Near East. A woodland herb on a wide range of mull soils.

6. S. palustris Retz.   'Marsh Stitchwort.'

Incl. *S. Dilleniana* Moench, non Leers, and *S. glauca* With.

A perennial herb with slender creeping stock and glabrous *smoothly* 4-*angled* ± erect flowering and non-flowering *shoots*. Flowering shoots 20–60 cm., weak and brittle. *Lvs* 1·5–5 cm., usually glaucous, sessile, *linear-lanceolate*, ascending, glabrous and *with smooth edges*. Fls 12–18 *mm. diam.* in a terminal few-fld cyme, or solitary, axillary, their stalks 3–10 cm. *Bracts* narrowly lanceolate *with broad membranous non-ciliate margins and a narrow central green strip.* Sepals 6–8 mm., lanceolate-acute, distinctly 3-veined, with broad white membranous margins. Petals white, equalling or up to twice as long as the sepals, bifid almost to the base. Stamens 10. Styles 3. Capsule ovoid-oblong, about equalling the calyx. Seeds 1·2–1·4 mm. across, dark reddish-brown, with rounded tubercles. Fl. 5–7. Protandrous. Visited by flies. $2n = c.$ 130. Hel.

Native. A local plant of marshes and base-rich fens southwards from Perth. Ireland, chiefly in central districts and absent from the south-west. 67, H21. C. and N. Europe and temperate Asia. Associated with *Carex acuta, acutiformis, elata* and *riparia, Phalaris arundinacea, Galium palustre, Lythrum salicaria,* etc., in the sedge-dominated communities which follow the *Scirpus-Phragmites* reed-swamp.

7. S. graminea L.   'Lesser Stitchwort.'

A perennial herb with slender creeping stock from which flowering shoots and numerous non-flowering shoots arise. Flowering *shoots* slender and brittle, diffusely branched, decumbent or ascending, *smoothly 4-angled,* glabrous, 20–90 cm. Lvs 1·5–4 cm., not glaucous, *linear-lanceolate* or narrowly elliptical, acute, *smooth at the margins but often ciliate near the base. Fls* 5–12 *mm. diam.,* rarely larger, in a loose spreading terminal cyme, their stalks 1–3 cm., 4-angled, glabrous. *Bracts wholly scarious, margins ciliate.* Sepals 3–7 mm., lanceolate, acute, distinctly 3-veined, with membranous margins. Petals white, bifid more than half-way, equalling or exceeding the sepals, being largest in hermaphrodite and smallest in male-sterile fls. Stamens 10, all, some or 0 of which may be fully developed and fertile. Styles 3. Capsule ovoid-oblong, longer than the calyx. Seeds c. 1 mm. across, reddish-brown, rugulose. Fl. 5–8. Protandrous. Hermaphrodite and partially or completely male-sterile plants are found. Visited chiefly by flies. $2n = 26.$ H.

Native. A common plant of woods, heaths and grassland especially on light soils. 110, H40, S. Throughout most of the British Is. Europe and Asia.

8. S. alsine Grimm   'Bog Stitchwort.'

S. uliginosa Murr.

A perennial herb with a slender creeping stock and numerous decumbent and ascending smoothly 4-angled glabrous shoots 10–40 cm. Lvs 5–10(–20) mm., sessile, elliptical or oblanceolate, acute, slightly ciliate at the base, otherwise glabrous. Fls c. 6 mm. diam. in terminal few-fld cymes. Bracts scarious with a green central stripe. *Calyx funnel-shaped at the base*; sepals 2·5–3·5 mm., lanceolate, acute, 3-veined. *Petals* white, *shorter than the sepals, bifid almost to the base, with widely divergent lobes.* Stamens 10. *Capsule* equalling the calyx, ovoid, narrowed below and *with a short carpophore.* Fr. stalks 0·5–3 cm., at first reflexed, then erect. Seeds 0·3–0·4 mm. across, pale reddish-brown, with small tubercles. Fl. 5–6. Protandrous or homogamous. Sparingly visited by flies. $2n = 24.$ Hel.

Native. A frequent plant of stream-sides, flushes, wet tracks and woodland-rides, etc., throughout the British Is., reaching 3300 ft. in Scotland. 112, H40, S. Often with *Juncus bufonius, J. articulatus, Montia, Callitriche stagnalis.* Europe, temperate Asia, N. America.

13. HOLOSTEUM L.

Small annual herbs, glaucous and usually glabrous below but viscid with glandular hairs above. Lvs opposite, exstipulate. *Fls in umbel-like terminal cymes*, 5-merous. Sepals free; *petals* white, *irregularly toothed or jagged*; stamens 10 or fewer; nectaries present; ovary 1-celled; styles 3, rarely 4 or 5. *Fr. a cylindrical capsule opening by usually 6 revolute teeth. Seeds peltate*, papillose.

Six spp., in Europe, N. Africa and temperate Asia to the Himalaya.

1. H. umbellatum L.   'Jagged Chickweed.'

An annual herb with a simple or basally branched erect or ascending stem 3–20 cm., glaucous, glabrous below, viscid above. Basal lvs 10–25 mm., oblanceolate, narrowed into a short broad stalk; stem-lvs (2–3 pairs) oblong or elliptical, sessile; all acute, glabrous or glandular-ciliate. Petals white or pale pink, longer than the sepals. Stamens 3(–5). Capsule cylindrical, narrowing above, twice as long as the sepals, its stalk at first deflexed then re-erecting. Seeds reddish-brown, peltate, furrowed on one side and broadly keeled on the other, 0·5 mm. across. Fl. 4–5. ± Homogamous. Visited by small bees and flies and automatically self-pollinated. $2n = 20$. Th.

Doubtfully native. A very rare plant of walls, roofs and sandy soils. Formerly in Surrey, Norfolk and Suffolk, where it may still persist. Europe northwards to S. Sweden, N. Africa, W. Asia. Commonly with other small annuals such as *Erophila verna*, *Cerastium semidecandrum*, *Veronica* spp., and in C. Europe usually in man-made habitats.

14. MOENCHIA Ehrh.

Annual herbs, *glabrous and glaucous*, with opposite narrow entire lvs. Fls solitary or in few-fld spreading cymes, 4- or 5-merous. Sepals with broad membranous margins; *petals* white, *entire*; stamens 8 (10) or 4(–5); nectaries present; ovary 1-celled; *styles 4 or 5, opposite the sepals*. Fr. an oblong straight *capsule opening by* as many or twice as many *short* blunt *teeth* as styles. Seeds numerous, reniform, papillose.

About 5 spp. in Europe, N. Africa and the Near East.

1. M. erecta (L.) Gaertn., Mey. & Scherb.   'Upright Chickweed.'

Sagina erecta L.; *Cerastium quaternellum* Fenzl

An annual herb with erect main stem 3–12 cm., and usually with a few ascending or decumbent basal branches. Basal lvs 6–20 mm., strap-shaped, short-stalked; upper lvs shorter, narrowly lanceolate, sessile, ascending; all rigid, acute, glabrous and glaucous. Fls about 8 mm. diam., usually 5-merous, 1–3 on long stalks. *Sepals* lanceolate, acute, *with broad white margins*. Petals shorter than the sepals. Stamens usually 4. *Styles* 4, short and recurved. *Capsule* a little longer than the sepals, *opening by* 8 teeth. Seeds reddish-brown, papillose, c. 0·5 mm. across. Fl. 5–6. Protogynous. Visited by flies. Automatic self-pollination takes place in dull weather in fls which do not open. $2n = 36^*$. Th.

Native. A local plant of gravelly pastures, maritime cliffs and dunes throughout England and Wales to Northumberland and Cumberland 60, S. C. and S. Europe, northwards to the Netherlands and Brandenburg.

15. SAGINA L.   Pearlwort

Small annual or perennial herbs, often tufted, with slender prostrate or ascending flowering shoots and subulate exstipulate opposite and connate lvs. Fls ± spherical in bud, 4–5-merous, in dichasial cymes or solitary and terminal. Sepals free; petals white, entire, often minute, sometimes 0; stamens as many or twice as many as the sepals; ovary 1-celled; *styles 4–5, alternating with the sepals*; ovules many. Fr. a *capsule splitting to the base* into 4–5 valves. Seeds very small.

About 20 spp., chiefly in the north temperate zone.

1	Fls usually 4-merous; stamens 4; petals minute or 0.	*2*
	Fls usually 5-merous, sometimes 4-merous; stamens 10, rarely 8; petals ± obvious.	*6*
2	Plants densely tufted with crowded glabrous rigid strongly recurved lvs.	**5. boydii**
	Lvs not strongly recurved.	*3*
3	Perennial; branches long, procumbent, rooting, from a central rosette of lvs; main stem not flowering.	**4. procumbens**
	Annual; branches ascending or erect, not rooting; main stem flowering.	*4*
4	Lvs blunt or minutely mucronate but not awned.	**3. maritima**
	Lvs awned.	*5*
5	Sepals all blunt, spreading horizontally in ripe fr.	**1. apetala**
	Outer sepals usually pointed, inner blunt, all appressed to the ripe fr.	**2. ciliata**
6	Petals twice as long as the sepals.	**10. nodosa**
	Petals not exceeding the sepals.	*7*
7	Calyx glandular-hairy.	**9. subulata**
	Calyx glabrous.	*8*
8	Plant forming small tufts 1–3 cm. high; no basal lf-rosette after the first season; lvs 3–6 mm.; dehisced capsule 2·5(–3) mm., greenish-yellow, dull.	**8. intermedia**
	Plant forming tufts or mats usually more than 3 cm. high; basal lvs 1·5–3 cm.; dehisced capsule 3–4 mm., straw-coloured, shining.	*9*
9	Lvs of basal rosette up to 2 cm.; stems decumbent then ascending, 2–7 cm. high; stamens 10; capsule 3·5–4 mm., setting seeds freely.	**6. saginoides**
	Lvs of basal rosette up to 3 cm.; stems prostrate, rooting, 2·5–10(–15) cm.; stamens often fewer than 10; capsule 3–3·5 mm., but rarely enlarging fully and with few or no good seeds.	**7. normaniana**

1. S. apetala Ard.   'Common Pearlwort.'

An *annual* herb with a slender tap-root and a loose central rosette-like cluster of lvs which soon wither, *the main stem ± erect, flowering*, with several decumbent or ascending rarely prostrate lateral stems 3–18 cm. *Lvs* linear-subulate, flattened above, *tapering into an awn*, glabrous or ciliate towards the base, usually not longer than the internodes. Fl.-stalks ±erect, glandular-hairy or glabrous. Fls usually 4-merous. *Sepals* ovate, hooded and *blunt* at the apex, *spreading horizontally in ripe fr*. Petals minute. Capsule (Fig. 32A) about as long as the sepals. Seeds ±reniform, 0·3–0·4 mm. wide, tuberclate. Fl. 5–8. Slightly protandrous. Visited by small crawling insects and automatically self-pollinated. 2n=12*. Th.

Native. Common on bare soil, paths, walls, etc. 93, H40, S. Through most of Great Britain and Ireland, but infrequent in N. Scotland and the Hebrides and not in Shetland. Europe northwards to Denmark and S. Sweden; N. Africa; W. Asia; S. America.

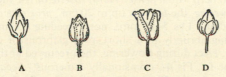

Fig. 32. Unripe capsules of *Sagina*. A, *S. apetala*; B, *S. ciliata*; C, *S. maritima*; D, *S. procumbens*. ×2·5.

2. S. ciliata Fr.   'Ciliate Pearlwort.'
?*S. patula* Jord.

An *annual* herb much like *S. apetala* but with the lvs and peduncles more usually glabrous though the lvs are sometimes ciliate towards the base. *Sepals* ovate-oblong, often glandular, the outer pair *pointed*, the inner blunt, *appressed to the ripe capsule* (Fig. 32B) which they almost equal. Petals minute or 0. Seeds larger than in *S. apetala*. Fl. 5–7. 2n=12*. Th.

Native. A frequent plant of dry grassland, heaths, bare ground, etc. 82, H15, S. Scattered throughout the British Is. northwards to Inverness. Europe northwards to S. Sweden; N. Africa.

Forms from S. and E. England with smaller fls and very slender stems which, with the lvs, fl.-stalks and sepals, are glandular-hairy, have been named *S. filicaulis* Jord. They do not appear to merit specific distinction and may be placed in var. *filicaulis* (Jord.) Corbière.

'*S. Reuteri* Boiss.' of British authors appears to be merely a dwarf growth-form of *S. ciliata* growing in dry places in S.W. England.

3. S. maritima Don   'Sea Pearlwort.'

An *annual* herb with slender tap-root with or without a central rosette of lvs, the *main stem flowering* and it and the numerous laterals varying from prostrate to ±erect and sometimes densely tufted; usually glabrous. *Lvs* linear-lanceolate, fleshy, blunt or apiculate but *not awned*, usually glabrous, rarely ciliate, shorter than the internodes. Fl.-stalks remaining erect, glabrous. Fls 4-merous. *Sepals hooded, blunt*, glabrous often with a purplish margin, *half-spreading in fr*. Petals white, minute or 0. Capsule (Fig. 32c) about equalling the sepals. Seeds 0·3–0·5 mm. wide, papillose. Fl. 5–9. Automatically self-pollinated. $2n = 28^*$; 22, 24. Th.

Native. A local maritime plant of dune-slacks, rocks and cliffs, and occasionally Scottish mountains to 4000 ft. (var. *alpina* Syme), all round the coasts of the British Is. 71, H 23, S. Coasts of Europe and N. Africa.

A very variable species. *S. debilis* Jord. and *S. densa* Jord. are names given to types with few weak slender decumbent branches and numerous densely tufted branches respectively, both described as having lvs more acute than in *S. maritima*.

4. S. procumbens L.   'Procumbent Pearlwort.'

A *perennial* tufted herb *with a dense central rosette* of lvs, the *main stem never* elongating or *flowering*, the *laterals* up to 20 cm., *prostrate and rooting at the base*, then ascending. Lvs 5–12 mm., not exceeding the internodes, linear-subulate narrowing abruptly to a short awn, glabrous or ciliate. Fl.-stalks glabrous, slender, erect during flowering, then recurved at the tip, and finally re-erecting in ripe fr. Fls 4-, occasionally 5-merous. *Sepals* ovate, *hooded, blunt, spreading in ripe fr.* Petals white, minute or 0. Stamens 4. Capsule (Fig. 32 D) longer than the sepals, opening with 4 blunt valves. Seeds 0·3–0·5 mm., brown, strongly papillose. Fl. 5–9. Homogamous. Automatically self-pollinated. $2n = 22$. H.

Native. A common plant of paths, lawns, grass verges, banks and stream-sides throughout the British Is. 112, H 40, S. Reaches 3800 ft. in Scotland. Europe, Asia, Greenland, N. America. Widely introduced in the southern hemisphere.

5. S. boydii F. B. White   'Boyd's Pearlwort.'

A dwarf *perennial densely tufted* glabrous herb with erect stems and *crowded* imbricating *rigid* short curved *lvs* which are somewhat fleshy and *strongly recurved*. Fl.-stalks short, slightly curved. Fls 4–5-merous. Sepals ovate, blunt, never opening widely. Petals 0 or very minute. Stamens 5–10. Stigmas often rudimentary. Capsule enclosed by the *sepals* which are *closely appressed* to it, never ripening seeds. Fl. 5–7. $2n = 22^*$. Ch.

Probably native. Presumed to have been collected near Braemar, Aberdeen. Not found since 1878 but retained in cultivation with its distinctive characters unchanged.

6. S. saginoides (L.) Karst.   'Alpine Pearlwort.'
S. Linnei Presl

A *perennial* tufted herb with a slender woody branching stock and numerous decumbent then ascending non-rooting glabrous lfy *shoots* 2–7 cm. high arising from *basal* rosettes of linear *lvs up to* 2 cm. Stem-lvs 0·5–1 cm., linear; all lvs glabrous, mucronate or awned. *Fls* usually 5-merous, c. 4 mm. diam., usually solitary *on slender stalks* 1–2·5 cm. Sepals ovate, rounded above, glabrous, with narrow scarious margins. Petals white, broad, rounded, sometimes slightly emarginate, contracted below into a short claw, equalling or somewhat shorter than the sepals. Stamens 10. Styles 5. *Capsule* 3·5–4 mm., ovoid, exceeding the sepals, and almost twice as long when the valves have straightened after dehiscence; sepals appressed to the ripe capsule. Fr. stalk at first recurved just beneath the capsule and ± prostrate, later becoming straight and erect. Seeds 0·3 mm. across, with rows of fine tubercles. Fl. 6–8. Homogamous. Visited by small flies, etc., and automatically self-pollinated; remaining closed in dull weather. $2n=22*$. Ch.

Native. A very rare arctic-alpine plant of rock-ledges on mountains in Scotland from Perth and Angus to Sutherland. Reaches 3980 ft. on Ben Lawers. 11. A subarctic circumboreal plant of Europe, Asia and N. America, reaching 71° N. in N. Norway; and an alpine plant of the mountains of C. Europe, Asia and Mexico.

7. S. normaniana Lagerh.

S. scotica (Druce) Druce; *S. saginoides* ssp. *scotica* (Druce) Clapham

A perennial loosely tufted or mat-forming herb differing from *S. saginoides* only in the *longer more slender ± prostrate rooting stems* 2·5–10(–15) cm., the longer lvs of the basal rosettes (0·5–3 cm.), the shorter sepals (2–2·5 mm.), shorter capsules (3·3–5 mm.), and longer fr.-stalks, 1·5–3(–4) cm., and the sepals somewhat spreading in ripe fr. Rarely sets seed and the great majority of the capsules remain undeveloped. Fl. 7–10. Chh.

Native. Occurs in the same mountain regions as *S. saginoides*, growing with it in wet subalpine mat-pasture or on water-splashed rocks. 9. Does not reach as high as *S. saginoides* (to 3500 ft. on Ben Lawers) and flowers later. The small differences in habit, lf-length and time of flowering are retained in cultivation. May be the hybrid *S. procumbens* × *saginoides* as suggested by its morphological features and many abortive capsules, but a higher proportion of well-developed capsules and good seed is set in cultivation.

8. S. intermedia Fenzl   'Lesser Alpine Pearlwort.'

S. nivalis auct., ?Fr.; *S. caespitosa* auct. angl., non Lange

A dwarf perennial herb with a much-branched woody stock and numerous densely tufted erect or ascending glabrous lfy shoots, forming a *small cushion* 1·5–5 *cm. across and* 1–3 *cm. high*. Basal lf-rosette only in first season; stem-lvs 3–6 mm., linear, acute, usually shortly mucronate, glabrous. *Fls* 3–4 mm. diam., 4–5-merous, borne singly *on short stalks* 2–5 *mm*. Sepals 1·5–2 mm., *glabrous*, ovate rounded at the apex, with narrow scarious often violet-coloured margins, strongly concave, appressed to the ripe capsule. *Petals* white, somewhat shorter than the sepals, *narrowly elliptical*. Stamens 8–10. Styles 4–5. Capsule 2·5–3 mm., about half as long again as the sepals, greenish- or whitish-yellow, dull. Fr. stalks usually erect. Seeds 0·5–0·6 mm. across, yellowish-brown, tubercled. Fl. 6–8. Homogamous. Rarely visited and automatically self-pollinated; remaining closed on dull days. Chh.

Native. A very rare alpine plant of Ben Lawers and neighbouring Scottish mountains, reaching 3700 ft. 2. Circumpolar, reaching 82° 29′ N. on the N. coast of Greenland; Norway, Sweden, Faeroes, Finland, Iceland, Alps (very rare).

9. S. subulata (Sw.) C. Presl   'Awl-leaved Pearlwort.'

A perennial mat-forming herb with a branched creeping stock and rosettes of linear lvs 0·5–1·5 cm., from which arise numerous decumbent then erect or ascending ±*glandular-hairy shoots* 2–7·5(–12·5) cm. high; the main stem short and non-flowering. Stem-*lvs* diminishing upwards, 0·3–1·2 cm., linear *narrowing to an awned tip* somewhat channelled above and keeled below, ± *ciliate* and with scattered hairs elsewhere. Fls 5-merous, usually solitary on glandular-hairy stalks 2–4 cm. *Sepals* 2–2·5 mm., oblong-oblanceolate, hooded and blunt at the apex, with narrow scarious margins, *glandular-hairy*, appressed to the ripe capsule. *Petals* white, ovate, blunt, short-clawed, *equalling or somewhat shorter than the sepals*. Stamens 10. Styles 5. Capsule c. 3 mm., somewhat longer than the sepals. Fr. stalks at first curved just beneath the capsule, then erect. Seeds 0·3–0·4 mm., yellowish-brown, tubercled. Fl. 6–8. Homogamous. Rarely visited and automatically self-pollinated. $2n = 22^*$; 18. Chh.

Native. A local plant of dry sandy, rocky or gravelly places. Throughout Great Britain, but more frequent in the north; Ireland: in the north and west. 70, H 10, S. Reaching 2700 ft. in Scotland. Europe southwards from Iceland, Faeroes and Norway.

10. S. nodosa (L.) Fenzl   'Knotted Pearl-wort.'
Spergula nodosa L.

A perennial tufted herb with an occasionally branched stock and numerous procumbent or ascending stems 5–15(–25) cm. high, glabrous or glandular-hairy above arising from basal lf-rosettes. Main stem short and non-flowering. Rosette-lvs 5–20 mm.; stem-lvs diminishing upwards from 10–15 mm. to 1–2 mm., with lateral lf-clusters in the axils of the upper lvs giving the characteristic 'knotted' appearance; all lvs linear-subulate, abruptly and shortly mucronate; glabrous or somewhat glandular at the base. *Fls* 0·5–1 *cm. diam.*, 1–3 at the ends of the stems and upper branches, 5-merous; their stalks glabrous or glandular, 3–10 mm. Sepals 2–4 mm., ovate, concave, blunt, glabrous or glandular, appressed to the ripe capsule. *Petals* white, narrowly ovate, *nearly twice as long as the sepals*. Stamens 10. Styles 5. Capsule 4 mm., ovoid, exceeding the sepals. Fr. stalks erect, straight. Seeds 0·4 mm., dark brown, tubercled. Fl. 7–9. Protandrous. Large fls are hermaphrodite, but smaller fls may lack fertile stamens. Sparingly visited by insects and automatically self-pollinated. $2n = 56^*$; 20–24. H.–Ch.

Var. *moniliformis* (G. F. W. Mey.) Lange has the stems ±procumbent and the upper axillary buds with their lf-fascicles become detached as bulbils which propagate the plant vegetatively.

Native. Frequent in damp sandy and peaty places throughout the British Is. 106, H 40, S. Reaching 2100 ft. on Cross Fell. Europe, including Iceland; W. Asia, E. and Central N. America, Greenland. Reaches 71° N. in N. Norway.

16. MINUARTIA L. SANDWORT

Annual to perennial small tufted or mat-forming herbs with usually *linear or subulate*, rarely broader, opposite exstipulate *lvs*. Fls usually 5-merous in dichasial cymes or solitary terminal. Sepals free; *petals* usually white, *entire or slightly emarginate*; stamens usually with nectaries at their base; *styles* 3, rarely more. *Capsule* ovoid or oblong, *opening by* 3(–5) *teeth*. Seeds reniform, papillose or spiny.

About 70 spp., in temperate and arctic regions of the northern hemisphere.

1 Annual herb of walls and dry places; all stems erect or ascending; non-flowering shoots 0; petals little more than half as long as sepals.
        4. hybrida
Perennial herbs with decumbent or ascending non-flowering shoots forming tufts or cushions; petals at least ⅔ as long as sepals.   2

2 Lvs 4–8 mm.; petals ⅔ as long as sepals (Scotland).   **2. rubella**
Lvs 6–15 mm.; petals equalling or exceeding sepals.   3

3 Lvs strongly 3-veined; pedicels ± glandular-hairy; widespread.   **1. verna**
Lvs indistinctly 1-veined; pedicels glabrous; only in Upper Teesdale.
        3. stricta

1. M. verna (L.) Hiern   'Vernal Sandwort.'

Arenaria verna L.; *Alsine verna* (L.) Wahlenb.

A *perennial* cushion-forming herb with fairly stout tap-root and a branching almost woody stock from which tufts of ascending flowering and non-flowering shoots arise. Flowering shoots 5–15 cm., glabrous or glandular-hairy. *Lvs* 6–15 *mm.*, linear-subulate, ± acute or apiculate, *strongly 3-veined*, rather rigid, often curved. Fls 8–12 *mm.* diam., in a few-fld cyme; *pedicels slender*, ± *glandular-hairy*. Sepals ovate-lanceolate, acute, strongly 3-veined, scarious-margined. *Petals* white, obovate, short-clawed, a little *longer than the sepals*. *Anthers red.* Seeds 0·6–0·9 mm., reniform, papillose. Fl. 5–9. Protandrous. $2n = 24*$, 48 (78). Chh.

Very variable, especially in ciliation and flower size. Plants from the Lizard (Cornwall) are smaller, more glaucous and ciliate, with the lower lvs closely appressed to the stem. They have been referred to ssp. *gerardii* (Willd.) Graebn., but many of the distinctive features disappear in cultivation.

Native. A local plant of base-rich rocks, screes and pastures, and often abundant on the rocky debris of old lead workings; scattered through the west and north of Great Britain from Cornwall and Somerset through Mid and North Wales to the Pennines and the Lake District, and in Scotland northwards to Banff; Clare, Antrim and Derry. 26, H3. Through most of Europe except Scandinavia; N. Africa, Caucasus, Siberia.

2. M. rubella (Wahlenb.) Hiern   'Alpine Sandwort.'

Alsine rubella Wahlenb.; *Arenaria rubella* (Wahlenb.) Sm.; *A. hirta* Wormsk.

A *perennial* herb forming tufts 2–8 cm. across, with a low leathery tap-root and woody branched stock from which arise decumbent non-flowering and ascending flowering stems 2–6(–9) cm. high. Lvs 4–8 mm., linear, glabrous or sparsely glandular, strongly 3-veined when dry, crowded below but distant on the glandular flowering stems. Bracts glandular, ovate. Fls 5–9 mm. diam., in 1–2(–3)-fld cymes; pedicels 4–15 mm., glandular. Sepals 3·5–5 mm., broadly lanceolate, acute, strongly 3-veined when dry, glandular at least below. *Petals* white, ovate-oblong, short-clawed, ⅔ *as long as sepals*. Anthers red. Styles 3–4(–5). Capsule ovoid, 3–4-valved, equalling or just exceeding sepals. Seeds 0·55–0·7 mm., brown, reniform, with low rounded papillae. Fl. 6–8. Homogamous. $2n = 24^*, 26$. Chh.

Native. A rare plant of base-rich rock ledges and detritus near the tops of a few Scottish mountains in E. and Mid Perth; probably extinct in E. and W. Sutherland and Shetland. 5. Circumpolar, in arctic and subarctic Europe, Siberia, Greenland and N. America south to Colorado.

3. M. stricta (Sw.) Hiern   'Bog Sandwort.'

Spergula stricta Sw.; *Alsine stricta* (Sw.) Wahlenb.; *Arenaria uliginosa* Schleich. ex DC.

A perennial glabrous herb forming small loose tufts, with slender tap-root and short somewhat woody stock from which arise short ascending non-flowering and slender erect flowering stems 5–10 cm. *Lvs 6–12 mm., filiform,* often curved, *obscurely 1-veined or apparently veinless*, rather crowded below but few and distant on the flowering stems. Fls 5–8 mm. diam. in 1–3(4)-fld cymes; *pedicels 15–50 mm., filiform, glabrous*. Sepals 2·5–4 mm., ovate, acute, veinless when fresh, 3-veined when dry. *Petals* white, elliptical-oblong, narrowed below, *about equalling sepals*. Anthers white. Styles 3. *Capsule* ovoid, equalling or slightly exceeding sepals, *divided nearly to the base by the 3 spreading blunt valves*. Seeds 0·75–0·85 mm., reddish-brown, finely reticulate, almost smooth. Fl. 6–7. Homogamous. $2n = 22^*, 26$. Chh.

Native. Confined to calcareous flushes on Widdybank Fell, Upper Teesdale (Durham), at 1500–1650 ft. 1. French and Swiss Jura, S. Germany; arctic Europe, Spitsbergen, Greenland and probably throughout arctic N. America to Alaska. An interesting 'arctic-prealpine', not found in the Alps.

4. M. hybrida (Vill.) Schischk.   'Fine-leaved Sandwort.'

Arenaria tenuifolia L.; *M. tenuifolia* (L.) Hiern, non Nees ex Mart.

An *annual* herb with a slender tap-root and slender erect or ascending stems 5–20 cm., branched especially below, usually glabrous, sometimes glandular above. Lvs 5–15 mm., crowded below, distant above, linear-subulate acute, 3(–5)-veined near the enlarged base, often recurved. Fls about 6 mm. diam., in much branched subcorymbose dichasial cymes, their stalks 5–15 mm., glabrous or glandular (var. *hybrida* (Vill.) Willk.). Sepals lanceolate-acuminate, 3-veined, with scarious margins, glabrous or glandular. *Petals* white, oblong, *little more than half as long as the sepals*. Stamens 10 or fewer (5 in var. *laxa* Jord.). Styles 3. Capsule oblong, about equalling the sepals. Seeds 0·4–0·5 mm., narrowly reniform, finely papillose. Fl. 5–6. Th.

Native. In dry rocky or stony places, on walls, railway-tracks, sandy arable fields, etc., southwards from York, Derby, Flint and Caernarvon, especially in E. England; S. and C. Ireland. 42, H20. Europe to 54° N.; N. Africa; W. Asia; Siberia.

17. CHERLERIA L.

Cushion-forming alpines with densely imbricating linear lvs. Fls 5-merous, solitary, terminal, ± sessile, polygamous to sub-dioecious. Sepals free; petals 0 or minute; stamens 10 or fewer; styles and stigmas 3. Capsule opening by 3 valves. Seeds reniform.

About 12 spp. in the mountains of Europe, Asia, and N. and S. America. Close to *Minuartia*, in which it is often included, but differing in the ± apetalous and usually unisexual fls.

1. C. sedoides L. Mossy Cyphel.

Alsine Cherleria Peterm.; *Arenaria sedoides* (L.) F. J. Hanb.

A perennial herb with very long tap-root and a branching woody stock from which arise *densely lfy shoots*, some ± prostrate and non-flowering, others short, erect, flowering, the whole *forming a yellow-green mossy cushion* 8–25 cm. diam. and 4–8 cm. high. Lvs 5–15 mm., somewhat fleshy, crowded, imbricating, linear-subulate, blunt, keeled beneath and channelled above, the margins horny and ciliate, otherwise glabrous. Fls 4–5 mm. diam. Sepals 5, ovate, blunt, 3-veined, with narrow membranous margins. *Petals usually 0,* but sometimes present in male fls and then minute and subulate. Stamens 10, shorter than the sepals, sometimes fewer, and 0 in female fls. Styles very short. Capsule up to twice as long as the sepals, but often abortive in fls with fertile stamens, opening half-way by 3 valves. Seeds few, c. 0·5 mm., nearly smooth. Fl. 6–8. Protandrous to homogamous. Visited for nectar by small flies. $2n=48*$; 26. Chh.

Native. Locally common on rocky slopes of the Scottish mountains from Stirling to Sutherland and in Skye and Rhum in the Inner Hebrides. Reaches 3900 ft. on Ben Lawers but is on shingle at sea-level near Montrose, Angus. 10. Pyrenees, Alps, Carpathians. One of our few alpine spp. not found also in the Arctic.

18. HONKENYA Ehrh.

A perennial *maritime* ± dioecious herb with opposite fleshy ovate glabrous lvs and greenish 5-merous fls. Sepals free; *petals entire*; stamens 10, abortive in female fls; nectaries present; ovary 1-celled, abortive in male plants; *styles usually 3. Fr. a globular capsule opening by 3 teeth.* Seeds few, large, pyriform.

One sp.

Differs from *Minuartia* in the shape of the capsule and the large seeds.

1. H. peploides (L.) Ehrh. 'Sea Sandwort.'

Arenaria peploides L.

A succulent stoloniferous herb with lfy flowering and non-flowering shoots, decumbent below, then erect, arising terminally from long pale slender *stolons creeping in sand or shingle.* Flowering shoots 5–25 cm., with *sessile, very fleshy, ovate-acute lvs,* 6–18 mm., having translucent wavy margins and downwardly pointing tips. Fls 6–10 mm. diam., solitary in the forks of the stem and in the axils of the upper lvs. Sepals ovate, blunt, fleshy, with membranous margins. *Petals greenish-white,* obovate, equalling the sepals in male fls but shorter in female. Stamens alternately longer and shorter, each with a yellow, oblong nectary-gland at its base. Styles 3, sometimes 4 or 5. Capsule up to

8 mm. diam., longer than the sepals, usually with 6 or fewer chestnut-coloured pear-shaped seeds 3–4 mm. across. Fl. 5–8. Protandrous. Rarely visited by flies; hermaphrodite fls are automatically self-pollinated when they close in dull weather. $2n=68$ (?48, 64, 66). H.

Native. Common all round the British Is. on mobile sand and sandy shingle. Forms miniature dunes on its own or more commonly is associated with *Agropyron junceiforme* on fore-dunes and persists to the 'yellow dune' stage. Tolerant of short periods of immersion in salt water. 83, H24, S. Coasts of the temperate and arctic regions of Eurasia and N. America to 78° 20′ N. in W. Greenland and Spitsbergen.

19. MOEHRINGIA L.

Annual to perennial usually slender herbs with opposite exstipulate lvs and 4–5-merous fls either solitary axillary or in terminal cymes. Sepals free; *petals* white, ± *entire*; stamens 8 or 10; nectaries present; ovary 1-celled; styles usually 2 or 3, rarely 4 or 5. *Fr. a rounded capsule opening by* 4 *or* 6 *recurved or revolute teeth. Seeds* dark, reniform, *with a variously-shaped oily appendage* (*elaiosome*).

About 20 spp. in temperate and arctic Eurasia and N. America.

Differs from *Arenaria* chiefly in the appendaged seeds.

1. M. trinervia (L.) Clairv.   'Three-nerved Sandwort.'

Arenaria trinervia L.

A usually annual herb with weak slender diffusely branching shoots 10–40 cm., prostrate or ascending, pubescent. *Lvs 6–25 mm., ovate, acute,* 3(–5)*-veined,* ciliate; lower stalked, upper subsessile. Fls c. 6 mm. diam., solitary in the forks of the stem and axils of the upper lvs, usually 5-merous; their stalks long, slender, pubescent. Sepals lanceolate-acuminate, 3-veined with the central vein hairy and the margins membranous and ciliate. Petals white, entire, shorter than the sepals. Stamens 10. Styles 3(–4). *Capsule almost globular,* shorter than the sepals, opening by 6(–8) revolute valves. *Seeds* c. 1 mm. across, blackish, shining, almost smooth with a *small laciniate appendage. Fl. 5–6. ±* Homogamous. Visited by small flies and beetles and automatically self-pollinated. $2n=24$. Th.

Native. A woodland herb of well-drained nitrate-rich mull soils. 106, H39, S. Throughout Great Britain and Ireland but not in the Inner and Outer Hebrides, Orkney or Shetland. Europe, W. Asia, Siberia.

20. ARENARIA L.

Annual to perennial herbs or dwarf shrubs with slender branched ascending shoots and small usually *ovate or lanceolate,* rarely linear, opposite exstipulate *lvs.* Fls solitary, terminal or in dichasial, often few-fld cymes, usually 5-merous. Sepals free; *petals* white or pink, *entire* or slightly emarginate; stamens 10; ovary 1-celled; *styles* 3, rarely 4 or 5. *Capsule opening by* 6 (*rarely* 8 *or* 10) *teeth,* or by 3 bifid teeth. Seeds numerous, reniform.

About 160 spp., cosmopolitan but chiefly in temperate and arctic regions of the northern hemisphere.

1 Sepals exceeding the petals.   2
   Sepals shorter than the petals.   3
2 Fls 5–8 mm. diam.; sepals ovate-lanceolate; capsule flask-shaped with
   curving sides; seeds 0·5–0·7 mm.   **1. serpyllifolia**
   Fls 3–5 mm. diam.; sepals lanceolate; capsule conical, straight-sided;
   seeds 0·3–0·5 mm.   **2. leptoclados**
3 Stems filiform; lvs suborbicular, all stalked.   **6. balearica**
   Stems not filiform; lvs ovate or spathulate, middle and upper sessile.   4
4 Lvs 12–30 mm.; fls c. 20 mm. diam.   **5. montana**
   Lvs usually less than 10 mm.; fls rarely more than 12 mm. diam.   5
5 Lvs ciliate for most of their length, their lateral veins usually visible in dry
   material; sepals ciliate on margins and back; fls 12–16 mm. diam.
               3. ciliata ssp. **hibernica**
   Lvs glabrous or ciliate only in the lower half, their lateral veins rarely
   visible even in dry material; sepals glabrous or nearly so; fls 9–12 mm.
   diam.   **(4. norvegica)**   6
6 Perennial; most shoots non-flowering; outer sepals usually quite glabrous;
   fls 9–10 mm. diam.   **4.** ssp. **norvegica**
   Winter annual or biennial; most of summer shoots flowering; outer sepals
   usually ciliate at base; fls 11–12 mm. diam.   **4.** ssp. **anglica**

1. A. serpyllifolia L.   'Thyme-leaved Sandwort.'

An annual to biennial herb 2·5–25 cm., with slender decumbent or ascending
grey-green shoots, often bushy in habit. Lvs up to 6 mm., ovate-acuminate,
subsessile, ± roughly hairy, ciliate. *Fls 5–8 mm. diam.*, numerous in dichasial
cymes, 5-merous. Sepals ovate-lanceolate, acute, hairy, 3–5-veined, the inner
scarious-margined. Petals white, narrowly obovate, shorter than the sepals.
Stamens 10 or fewer, sometimes 0. Styles 3. *Capsule* (Fig. 33 A) *flask-shaped,
with firm curving sides*, opening by 6 short teeth, its stalk ± erect. *Seeds
0·5–0·7 mm. wide*, black, papillose. Fl. 6–8. Homogamous. Nectar-secreting
and occasionally visited by small insects, but automatically self-pollinated. $2n=40^*$. Th.

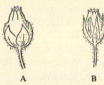

Native. Common on bare ground, arable fields,
walls, bare soil on chalk downs, cliffs, etc. 112,
H40, S. Throughout the British Is. Europe to
Scandinavia, but not in the Arctic nor in Iceland;
temperate Asia; N. America. Apparently disliked
by rabbits and common round their burrows.

Fig. 33. Capsules of A,
Arenaria serpyllifolia and
B, *A. leptoclados*. × 2·5.

2. A. leptoclados (Rchb.) Guss. 'Lesser Thyme-
leaved Sandwort.'

An annual herb resembling *A. serpyllifolia* but more slender and diffuse with
longer infls. *Fls 3–5 mm. diam.* Sepals lanceolate. *Capsule* (Fig. 33 B) *conical
straight-sided, thin-walled, not firm*; its stalk often curved at the top, spreading.
Seeds 0·3–0·5 mm. wide, dark red, papillose. Fl. 6–8. Homogamous. Automatically self-pollinated. $2n=20^*$. Th.

Native. Common in similar habitats to those of *A. serpyllifolia*, with which
it is often associated. 78, H35, S. Through most of the British Is., but rare
in the north and perhaps absent from the Outer Hebrides, Orkney and
Shetland. Europe, W. Asia.

3. A. ciliata L. ssp. **hibernica** Ostenf. & Dahl   'Irish Sandwort.'

A low *perennial* herb with *prostrate hairy non-flowering shoots*, somewhat woody at the base, and ascending flowering shoots to 5 cm. high. *Lvs* 4–6 mm., *spathulate*, slightly broader above, *ciliate usually to the tip*, prominently veined, the *lateral veins usually visible in dry material*. *Fls* 12–16 *mm. diam.*; cymes 1(–2)-fld. *Sepals* ovate-lanceolate with 3 prominent veins, *strongly ciliate on the margin and back.* Petals 7–7·5 mm., white, oblong, clawed, half as long again as sepals. Conspicuous *orange nectaries* at the base of the 5 outer stamens. Capsule ovoid, equalling or somewhat exceeding sepals, opening by 6 teeth. Seeds 0·95 mm., black, with low tubercles. Fl. 6–7; fr. 7–8. $2n = $ c. 40*. Chh.

Native. An endemic ssp. confined to the limestone cliffs of the N. side of the Ben Bulben range, Co. Sligo, Ireland, between 1200 and 1950 ft. H1. *A. ciliata* is very polymorphic, both morphologically and cytologically ($2n = 40, 80,$ c. 120, ?200). Ssp. *hibernica* and the high Arctic ssp. *pseudofrigida* appear to be related to ssp. *tenella* of the E. Alps and Carpathians, one of three central European sspp.

4. A. norvegica Gunn.

Annual to perennial herbs 3–6 cm. *Lvs obovate, glabrous or ciliate only in the lower half, obscurely veined. Shoots and pedicels almost glabrous.* Cymes 1–2(–4)-fld. *Sepals* ovate-lanceolate, acute, 3-veined but the laterals often obscure, *glabrous or with a few cilia at the basal margin.* Petals 4–5·5 mm., white, ovate-oblong, clawed, exceeding sepals. *Nectaries green.* Seeds 0·8–1·0 mm., black, with low broad tubercles. $2n = 80$*.

Ssp. **norvegica**   'Norwegian Sandwort.'
Perennial, with slender tap-root and short basal internodes; stem branching to the base forming a compact tuft; flowering shoots to 6 cm. high. Lvs 3–4·5(–6) mm., somewhat succulent, dark green. *Fls* 9–10 *mm. diam. Sepals glabrous* or rarely with a few cilia at the basal margin. Petals 4–4·5 mm., broadly ovate, slightly exceeding sepals. Styles 3–5. Capsule-teeth 6–10. Fl. 6–7; fr. 7–8. Chh.

Native. A rare and local plant of base-rich screes and river shingle in Argyll, Westerness, Rhum, W. Sutherland and Shetland. 5. Norway, Sweden, Iceland.

Ssp. **anglica** Halliday   'English Sandwort.'
Winter annual or biennial with few non-flowering but many short erect flowering shoots to 5 cm. high; laxer in habit and lighter green than ssp. *norvegica.* Lvs 4·5–5(–6) mm., narrowly obovate. *Fls* 11–12 *mm. diam. Outer sepals slightly ciliate at base*, rarely quite glabrous. Petals 4·5–5·5 mm., oblong. Styles 3. Capsule-teeth 6. Fl. 5–10. Th.–Chh.

Native. An endemic ssp. confined to limestone depressions and tracks in Upper Ribblesdale, Yorks. 1. Quite distinct from *A. gothica* Fr. ($2n = 100$) in which it was formerly included.

*5. A. montana L.   Mountain Sandwort.
A perennial grey-green herb with long prostrate non-flowering shoots and decumbent flowering shoots, 10–30 cm. all shortly pubescent and eglandular.

Lvs linear-lanceolate to narrowly ovate, 12–20 mm., acuminate, 1-veined, with ciliate margins. Fls to 2 cm. diam. or more. Sepals ovate-lanceolate, 1-veined. *Petals* white, obovate, *at least twice as long as the sepals*. Capsule broadly ovoid, about equalling the calyx. Fl. 5–7. Chh.

Introduced. Grown in gardens and occasionally establishing itself. France, Spain, Portugal.

*6. A. balearica L.   'Balearic Pearlwort.'

A perennial herb with slender stock and diffusely branching prostrate filiform ± hairy shoots, ascending only at the tips of flowering branches. *Lvs small, ovate or suborbicular, blunt, stalked, the stalks almost equalling the blade.* Fls 5-merous, solitary, terminal, on filiform stalks 6–10 times as long as the sepals. Sepals ovate, subacute. Petals white, obovate, longer than the sepals, 5 or sometimes fewer. Capsule ovoid, drooping, hardly exceeding the sepals. Fl. 6–8. Chh.

Introduced. Often grown on rock-gardens, paved garden-paths, etc., and naturalized in many scattered localities. Balearic Is., Corsica, Sardinia, Tyrrhenian Is.

21. SPERGULA L.

Annual herbs whose linear blunt opposite lvs have *small deciduous stipules* and conspicuous axillary clusters of lvs. Fls in loose terminal dichasia, 5-merous, hermaphrodite, their stalks after flowering at first deflexed then re-erected. Sepals 5, free; *petals* 5, white, *entire*; stamens 5–10; ovary 1-celled with 5 short styles. Fr. an ovoid capsule splitting deeply into 5 valves; seeds black, sharply keeled or winged.

About 5 spp. in the north temperate zone.

Stems weak, ascending; lvs furrowed beneath; seeds sharply keeled or with
     a very narrow wing.   **1. arvensis**
Stems usually stiffly erect from a decumbent base; lvs not furrowed beneath;
     seeds with a broad wing somewhat narrower than the diam. of the seed.
     2. morisonii

1. S. arvensis L.   Corn Spurrey.

An annual herb with ascending geniculate stems 7·5–40 cm., branching close to the base; internodes weak, about equally long throughout. *Lvs* 1–3 cm., grass-green to grey-green, linear, fleshy, blunt, convex above and *channelled beneath*. Stem and lvs slightly to strongly viscid with glandular hairs. Fls 4–7 mm. diam., in forked or umbel-like cymose panicles; stalks 1–2·5 cm.; bracts small, ± scarious. Sepals 3–5 mm., ovate, obtuse, faintly veined, ± glandular, with a narrow scarious margin. Petals white, obovate, slightly longer than the sepals. Stamens 10 or fewer. Capsule ovoid-conical, up to twice as long as the sepals. Fr. stalks at first strongly deflexed from the base, then re-erecting. *Seeds* 1·2–1·5 mm. across, blackish, ± *papillose; wing very narrow or* 0. Fl. 6–8. Homogamous. Visited by various flies and bees. $2n = 18$. Th. Very variable. The British forms belong to Zinger's *Spontaneae*, with seed-weight 0·3–0·4 mg. Two varieties may be recognized:

Var. *arvensis*. Seeds brownish-black, covered with pale club-shaped deciduous papillae; wing narrow or 0.

Var. *sativa* (Boenn.) Mert. & Koch. Seeds black, dull, minutely tubercled, narrowly winged.

These varieties are often raised to specific rank, other differential characters being added. But although plants of var. *arvensis* may tend to be grass-green and only slightly viscid, and those of var. *sativa* to be grey-green and highly viscid, the correlations are far from constant in British populations: all combinations of the characters may be found in a single field.

Native. A locally abundant and often troublesome calcifuge weed of arable land; reaching 1470 ft. in England. Throughout the British Is. northwards to Aberdeen; Channel Is. The percentages of densely hairy plants and of plants with non-papillate seeds increase towards the north and west. Almost cosmopolitan.

Certain forms grown as fodder plants have a higher seed-weight (0·7–1 mg.: Zinger's *Cultae*), and these may be encountered as casuals in Britain. Others occur as weeds of flax and have still larger seeds (seed-weight 1·5–2 mg.: Zinger's *Linicolae*).

2. S. morisonii Bor.

S. vernalis auct.; *S. pentandra* auct., non L.

An annual herb with a much-branched stem 10–20 cm., erect from a decumbent base, ± glandular above. *Lvs* to 1 cm., linear-subulate, dark green, apparently in dense whorls, slightly fleshy, *not furrowed beneath*, ± glandular, often ± deflexed; stipules short. Infl. usually of many fls on long slender stalks, spreading in fr. Sepals ovate, shortly acuminate, ± glabrous, with scarious margins. Petals white, elliptical, subacute, contiguous, ± equalling the sepals. Stamens 5 or 10. Styles 5. Capsule slightly exceeding the calyx, opening by 5 valves. *Seeds* 1–1·6 mm. across (including the wing), nearly circular, blackish-brown, flattened, *smooth* except for minute whitish marginal papillae, *with a scarious striate whitish or brownish wing*, whose breadth falls short of the diam. of the seed. Fl. 4–6. Probably self-pollinated. $2n=18$. Th.

Probably introduced. Found in 1943 in Sussex, on sandy cultivated ground on the heathland between Crowborough and Tunbridge Wells. C. Europe from Spain, N. Italy and Hungary northwards to S. Scandinavia, Germany and Poland.

The closely related *S. pentandra* L. has laxer whorls of longer and less deflexed lvs, narrower non-contiguous petals which exceed the sepals, and black non-papillose seeds with a broader wing which equals the diameter of the actual seed. *S. arvensis* is larger and coarser with longer, furrowed lvs and subglobose seeds with the wing narrow or lacking.

22. SPERGULARIA (Pers.) J. & C. Presl

Annual to perennial often decumbent herbs whose opposite linear lvs have *pale scarious stipules*. Fls hermaphrodite or unisexual, 5-merous, in cymose infls. Sepals free; *petals* white or pink, *entire*, sometimes 0; stamens 5–10; ovary 1-celled; *styles* 3. Capsule opening by 3 valves; seeds spherical or pyriform.

About 20 spp., cosmopolitan, mostly halophytic.

1 Seeds all broadly winged; capsule 7–11 mm.　　　　　　　　　**4. media**
　Seeds not winged, or a few winged at the base of the capsule; capsule not exceeding 7 mm.　　　　　　　　　　　　　　　　　　　　　　*2*

2　Perennial; densely glandular-pubescent throughout; fls 8–10 mm. diam.
　　　　　　　　　　　　　　　　　　　　　　　　　　　　3. rupicola
　　Annual; glabrous, or glandular only above; fls not exceeding 8 mm.
　　diam., often much smaller.　　　　　　　　　　　　　　　　　　　*3*

3　Lvs bright or yellowish-green, shortly mucronate or blunt, fleshy; seeds
　　winged or not, at least 0·6 mm.　　　　　　　　　　　　　**5. marina**
　　Lvs grey-green, awned, not fleshy; seeds unwinged, 0·4–0·5 mm.　　　*4*

4　Stipules lanceolate-acuminate, silvery; fls 3–5 mm. diam.; fl.-stalks longer
　　than the sepals.　　　　　　　　　　　　　　　　　　　　**1. rubra**
　　Stipules broadly triangular dull; fls 2 mm. diam.; fl.-stalks usually shorter
　　than the sepals; infl. branches lengthening to resemble many-fld 1-sided
　　racemes.　　　　　　　　　　　　　　　　　　　　　　**2. bocconii**

1. S. rubra (L.) J. & C. Presl　　　　　　　　　　　　Sand-spurrey.

Arenaria rubra L.; *A. campestris* auct.; *Buda rubra* (L.) Dum.

An annual or biennial herb with a slender tap-root and several decumbent
branched ± hairy stems 5–25 cm., usually glandular above. *Lvs* 4–25 mm.,
narrowly linear *tapering to an awned tip*, not fleshy; *stipules lanceolate
acuminate*, usually torn at the tip, *silvery* and conspicuous. Fls 3–5 mm. diam.,
in terminal rather few-fld cymes, their *stalks longer than the sepals*. Sepals
3–4 mm., ovate-lanceolate, usually glandular, with a broad scarious margin.
Petals rose-coloured, paler at the base, ovate, shorter than the sepals. Stamens
usually fewer than 10. Capsule about equalling the sepals. Fr. stalks spreading
or reflexed then becoming ± erect. Seeds 0·4–0·5 mm. across, brownish,
tubercled, unwinged, with a raised rim. Fl. 5–9. ± Homogamous. Visited by
flies and automatically self-pollinated; the fls sometimes fail to open.
$2n = 36$. Th.

　　Native. A common calcifuge plant of open sandy or gravelly habitats.
106, H 10, S. Throughout most of Great Britain; very local in Ireland, and
absent from the Outer Hebrides, Orkney and Shetland. Europe, N. Africa,
Asia, N. America. Introduced in Australia.

2. S. bocconii (Scheele) Aschers. & Graebn.　　'Boccone's Sand-spurrey.'

S. campestris auct., non Aschers.; *S. atheniensis* Aschers.

An annual or biennial herb resembling *S. rubra* but with stems and lvs very
glandular; lvs more shortly awned; *stipules broadly triangular, not silvery;
fls c. 2 mm. diam.*, ± *numerous, the infl. branches lengthening to resemble
1-sided racemes; fl.-stalks usually shorter than the sepals*. Sepals 2·5–3·5 mm.
Petals pale rose or whitish, shorter than the sepals. Seeds 0·4–0·5 mm. across,
pale greyish-brown, tubercled, unwinged. Fl. 5–9. Th.

　　Probably native. A rare and local plant of dry sandy and rocky coastal
areas in Cornwall, Devon, Essex, and Glamorgan. 6, S. S. and S.W. Europe,
N. Africa, Near East.

3. S. rupicola Lebel ex Le Jolis　　　　　　　　　　'Cliff Sand-spurrey.'

S. rupestris Lebel, non Cambess.

A *perennial* herb with a stoutish branched ± woody stock and numerous
decumbent shoots 5–15 cm. *Stems* ± *terete*, often dark purple, *densely
glandular-hairy*. Lvs 5–15 × 1·5–2 mm., linear-acute, fleshy, ± flattened, with

a horny tip prolonged into a short mucro, sparsely glandular-hairy especially near the base; *stipules* ovate-triangular, acuminate, somewhat *silvery*; the lvs of lateral shoots form axillary fascicles at each node. Infl. a terminal cyme becoming monochasial above, with up to 20 fls, each 8–10 mm. diam. on a glandular-hairy stalk 7–8 mm. Sepals 5 mm., lanceolate, blunt, glandular-hairy, with white scarious margins. Petals equalling or slightly exceeding the sepals, deep pink. Stamens 10, with yellow anthers and filaments broadened below. Capsule 5–7 mm., slightly exceeding the erected sepals but shorter than its stalk which reflexes strongly after flowering then gradually erects during ripening. Seeds c. 0·6 mm., pyriform-triangular, black, minutely tubercled, with thickened borders along two sides. Fl. 6–9. Chh.

Native. Maritime cliffs, rocks and walls, local, and chiefly in the south and west. 31, H 20, S. Coasts of Great Britain from Hants to Cornwall and up the W. coast to Ross; on the E. coast only in Norfolk, Mid and East Lothian and Aberdeen; Isle of Man, Inner and Outer Hebrides; coasts of Ireland. S.W. Europe from S. Italy and Spain to the N. coast of France.

4. S. media (L.) C. Presl

S. marginata Kittel; *S. Dillenii* Lebel

A *perennial* herb with a ± stout branching root-stock and many decumbent or geniculate-ascending, stout, flattened *shoots* to 30 cm. or more, usually *glabrous* except in the inflorescence. Lvs 1–2·5 cm. × 1–2 mm., linear, fleshy, horny-tipped, blunt to acute, often shortly mucronate, flat above and rounded beneath, usually glabrous; stipules broadly triangular, not silvery. Infl. lax, dichasial, its bracts becoming very small. Fls 7·5–12 mm. diam. Sepals 4–5 mm., blunt, glabrous or hairy. *Petals* 4·5–5·5 mm., *somewhat exceeding the sepals*, ovate, blunt, whitish, or pink above. Stamens 10 (7–11). Capsule 7–8·5 mm., exceeding the persistent erect sepals (5–6 mm.) and about equalling or exceeding its stalk. *Seeds* 1·5 mm. in overall diam., pale yellowish-brown, smooth, almost *all with a broad scarious border*. Fl. 6–9. Protandrous and visited by small flies; sometimes gynodioecious. $2n = 18$. G.–Chh.

Native. Muddy and sandy salt-marshes. 68, H 26, S. In suitable localities all round the coasts of the British Is. Muddy coasts of the temperate zones of both hemispheres.

5. S. marina (L.) Griseb.

S. salina J. & C. Presl

An annual herb with a slender tap-root and many prostrate or decumbent slender flattened shoots to 20 cm. or more, subglabrous or ± glandular-hairy above. Lvs 1–2·3 cm. × 1–2 mm., linear, very variable in stoutness and fleshiness, horny-tipped, ± acute, not or very shortly mucronate, flat above and rounded beneath, glabrous to glandular-hairy; stipules broadly triangular-acuminate, not silvery. Inflorescence dichasial at first but becoming ± monochasial, its bracts resembling the lvs or much smaller. Fls 6–8 mm. diam. Sepals 3·5–4 mm., lanceolate, blunt, often pink-tinged, glabrous or glandular-hairy. *Petals* 2·5–3 mm., *shorter than sepals*, ovate, blunt, commonly pink or deep rose with a white base. Stamens usually 4–8 or fewer. Capsule 4–5(–6) mm., somewhat exceeding the persistent sepals which are usually somewhat spreading at maturity, but distinctly shorter than its glabrous or glandular stalk. *Seeds* 0·8 mm. (unbordered) to 1·5 mm. (bordered) across, brownish, smooth, slightly rugose or strongly tubercled, *all unbordered or a small proportion of basal seeds with a broad scarious border* as in *S. media*. Fl. 6–8. Occasionally visited by small insects but usually self-pollinated. $2n = 36$. Th.

Native. In the drier zones of muddy and sandy coastal salt-marshes and of brackish marshes; also in inland salt areas and rarely as an inland adventive. 80, H 25, S. In suitable habitats all round the coasts of the British Is., in inland salty areas of Worcester and Cheshire, and as a casual in waste places in Herts, Sussex, etc. Coasts and inland salt areas of the temperate zone of the northern hemisphere.

S. marina, though very variable. is typically a much less robust plant than *S. media*, with more slender lvs and often with some glandular hairs, especially above. Its fls are conspicuously smaller and deeper pink than those of *S. media* its capsules considerably shorter, its sepals less closely appressed to the ripe capsule, apart from the differences in the seeds. It extends further from the coast into brackish marshes than does *S. media*, and is sometimes found inland. Intermediates are sometimes encountered where the two spp. occur together, and these may be hybrids, but further investigation is required.

23. TELEPHIUM L.

About 6 spp. in the Mediterranean region and northwards to Switzerland and Austria.

*1. T. imperati L.

A perennial dwarf shrub with a stout stock and numerous decumbent glabrous and glaucous shoots 15–40 cm. Lvs up to 13 mm., alternate, obovate, blunt, glaucous, somewhat fleshy, with small scarious stipules. Fls in a compact, terminal, cymose cluster. Sepals 5, oblong, blunt. Petals 5, white, a little longer than the sepals. Stamens 5. Styles 3(–4). Fr. a ± trigonous capsule 3(–4)-celled below, with numerous seeds, opening by 3(–4) valves. Fl. 6–8. Ch.

Introduced. A casual. Native in the W. Mediterranean region.

24. POLYCARPON L.

Small herbs with forking stems and obovate or oblong *opposite or whorled lvs with scarious stipules*. Fls small, in terminal dichasia, with scarious bracts; hermaphrodite, hypogynous, 5-merous. *Sepals keeled and hooded; petals narrow, shorter than the sepals*; stamens 3–5; ovary 1-celled; *styles trifid with 3 stigmas. Fr. a capsule with numerous seeds*, opening by 3 valves.

About 36 spp. in warm and temperate regions throughout the world.

1. P. tetraphyllum (L.) L.   'Four-leaved All-seed.'

Mollugo tetraphylla L.

An annual herb with a slender tap-root and a much-branched slender erect or ascending shoot 5–15 cm., glabrous but with rough angles. *Lvs* 8–13 mm., *obovate*, narrowed into a stalk-like base, blunt, in opposite pairs, but the *lower pairs approximated so as to simulate whorls of* 4; stipules very small, narrowly triangular-acuminate, scarious. Infl. a much-branched dichasium, the fls 2–3 mm. diam., short-stalked. Sepals with broad white scarious margins. Petals white, narrowly oblong, emarginate, soon falling. Stamens 3–5. Capsule ovoid, about equalling the calyx. Seeds brownish, finely papillose. Fl. 6–7. The homogamous fls are automatically self-pollinated and are often cleistogamous. Th.

Native. A rare and local plant of sandy and waste places in Cornwall, S. Devon and Dorset. 3, S. Mediterranean region and C. Europe, but widely introduced elsewhere in Europe and in Asia, Africa, Australia, and S. America.

25. CORRIGIOLA L.

Annual to perennial glabrous herbs with decumbent shoots and *alternate oblong or linear glaucous stipulate lvs.* Fls in axillary clusters often further aggregated at the ends of the stems and branches; hermaphrodite, slightly perigynous, 5-merous. Sepals with white margins; petals equalling or exceeding the sepals; stamens 5; ovary 1-celled with 1 basal long-stalked ovule; stigmas 3, subsessile. Fr. an indehiscent ± trigonous 1-seeded nutlet enclosed in the calyx.

About 10 spp. in Europe, Africa, W. Asia and America.

1. C. litoralis L.   Strapwort.

An annual, rarely biennial, *glaucous* herb with slender tap-root and several slender decumbent branching shoots 5–25 cm., often reddish. Lvs 0·5–3 cm., linear-oblanceolate, blunt, entire, slightly fleshy, narrowed gradually into a stalk-like base; stipules small, scarious, whitish, half-sagittate, acuminate, denticulate. Fls very small, in crowded terminal and axillary head-like cymes. Sepals c. 1 mm., blunt, green or red in the centre with broad white margins. Petals white or red-tipped, almost equalling the sepals. Anthers violet. Ovary surrounded at its base by the perigynous zone; styles small, sessile. Fr. c. 1 mm., obscurely trigonous. Seed papillose. Fl. 7–8. Homogamous. Automatically self-pollinated and often cleistogamous. $2n=18*$; 16. Th.

Native. Very local. On sandy and gravelly banks of pools at Slapton Ley (S. Devon) and near Helston (Cornwall); Channel Is. 2, S. Also adventive on railway tracks. S.W. Europe northwards to Denmark; N. and E. Africa; W. Asia.

26. PARONYCHIA (Mill.)

About 45 spp. in temperate and subtropical regions throughout the world.

*1. P. polygonifolia (Vill.) DC.

A perennial mat-forming plant with a stout stock and spreading prostrate hairy shoots. *Lvs opposite,* lanceolate, glabrous, with *large silvery scarious stipules.* Fls in dense cymose clusters with *conspicuous silvery scarious lanceolate bracts.* Sepals slightly hooded, awned. Petals 0. Stamens 5. Stigmas 2. Fr. dry, with 1 seed.

Introduced. A casual, native in S.E. Europe and W. Asia.

Other spp. of *Paronychia* sometimes occur as casuals, and can be recognized by the conspicuous silvery stipules and bracts and readily distinguished from *Illecebrum* by the non-corky sepals.

27. HERNIARIA L.

Annual or perennial mat-forming herbs with opposite stipulate lvs, the upper lvs often alternate through abortion of one member of a pair. *Fls in dense axillary clusters*, hermaphrodite or unisexual, *perigynous*, 5-merous; bracteoles scarious. Sepals 5, on the edge of the bowl-shaped perigynous zone; petals 5, subulate, shorter than the sepals; ovary free at the base of the perigynous zone which closely surrounds its lower half, 1-celled with 1 basal long-stalked ovule; styles 2-branched with 2 stigmas. Fr. an indehiscent nutlet with 1 shining black seed.

About 15 spp. in Europe, N. and S. Africa and Asia.

1 Calyx ±glabrous; plants green.	2
Calyx with spreading hairs; plant grey or whitish with hairs.	3

2 Stipules ovate, often greenish; fl.-clusters ±confluent on short lateral branches which resemble lfy spikes; fr. acute, considerably exceeding the calyx.   **1. glabra**
  Stipules broadly ovate-acuminate, white; fl.-clusters roundish, distinct; fr. blunt, little exceeding the calyx.   **2. ciliolata**

3 Plant prostrate; lvs lanceolate; lower lvs opposite; sepals hair-pointed.   **3. hirsuta**
  Branches ascending; lvs ovate-oblong, mostly alternate; sepals not hair-pointed.   **4. cinerea**

1. H. glabra L.   'Glabrous Rupture-wort.'

An annual or biennial, rarely perennial herb with tap-root and numerous prostrate *shoots* 6–15(–30) cm., rarely slightly woody at the base, *glabrous or slightly hairy all round*, green, with short alternate branchlets. Lvs 3–7(–10) mm., ovate-lanceolate, ±acute, narrowing to the base, glabrous or

Fig. 34. Fruits of *Herniaria*. A, *H. ciliolata*; B, *H. glabra*; C, *H. cinerea*. × 5.

ciliate; stipules small, ovate, shortly fringed, often greenish; upper lvs alternate. *Fls* c. 2 mm. diam. almost sessile, up to 10 in axillary clusters *on short lateral branches, the clusters ± confluent into oblong lfy spikes*. *Sepals* c. 6 mm., *ovate, blunt*, ±*glabrous*. Petals minute, white. Stigmas slightly divergent. *Fr.* (Fig. 34 B) *acute, considerably exceeding the sepals*. Seeds c. 6 mm., lenticular, shining, red then black. Fl. 7. Homogamous. Visited by various tiny insects and automatically self-pollinated. $2n = 18*$. Th.–H.

Native. A rare and local plant of dry sandy places in S. Devon, Cambridge, Suffolk, Norfolk, S. Lincoln and Cumberland, and perhaps elsewhere. 7, S. Europe northwards to S. Scandinavia; N. Africa; Asia.

2. H. ciliolata Meld.   'Ciliate Rupture-wort.'

H. glabra var. *scabrescens* Roem.; var. *setulosa* Beck; *H. ciliata* Bab., non Clairv.

A perennial evergreen dwarf shrub with stout erect branched woody stock and numerous spreading prostrate *shoots* 5–20(–40) cm., woody and rooting at the base, usually *hairy only on the upper side*, each with short alternate ascending branches. *Lvs* 2–6(–10) mm., broadly ovate to ovate-elliptical, usually *ciliate*, the upper lvs often alternate though less generally than in *H. glabra*; stipules ovate-acuminate, silvery-white, fringed, conspicuous. *Fls* c. 2 mm. diam., almost sessile in *distinct roundish axillary clusters* on the lateral branchlets. *Sepals* ± ciliate, *bristle-tipped*, but otherwise glabrous. Petals minute. Stigmas strongly divergent. Fr. (Fig. 34 A) obtuse, hardly exceeding the sepals. Seeds c. 0·7 mm., lenticular, black, shining. Fl. 7–8. Pollination as in *H. glabra*. $2n = 72^*$. Ch.

Var. *angustifolia* (Pugsl.) Meld. has narrowly elliptical lvs and is hairy all round the stem.

Native. A very rare plant of maritime sands and rocks; var. *ciliolata* known only from Lizard Point (Cornwall) and the Channel Is., var. *angustifolia* in Jersey. Coasts of W. Europe from Spain and Portugal to N. Germany.

*3. H. hirsuta L.   'Hairy Rupture-wort.'

An annual to perennial herb resembling *H. glabra* but with the wholly prostrate *shoots covered with dense straight spreading hairs*. Lvs lanceolate, the lower opposite. Fls in distinct roundish axillary clusters. *Calyx burr-like* with straight spreading hairs, and each *sepal ending in a long bristle*. Fl. 7–8. Pollination as in *H. glabra*. $2n = 36^*$. Th.–Ch.

Introduced. Naturalized on sandy ground at Christchurch, Hants. C. and S. Europe, Asia, Africa.

*4. H. cinerea DC., very like *H. hirsuta* but *ashy-grey* with most of the ovate-oblong lvs alternate and the hairy *sepals not ending in a conspicuously long bristle*, is naturalized in waste places at Burton-on-Trent, Stafford. S. Europe, N. Africa, W. Asia.

28. ILLECEBRUM L.

A small herb with opposite stipulate lvs and *white fls in axillary cymose clusters*. Fls hermaphrodite or unisexual, 5-merous, with 2 ± sessile stigmas and a 1-celled ovary with 1 ovule. *Fr.* 1-*seeded, dehiscent*.

One species.

1. I. verticillatum L.   'Illecebrum.'

An annual glabrous herb with a slender tap-root and many slender spreading decumbent branches 5–20 cm., rooting at the basal nodes, often reddish. Lvs 2–6 mm., obovate, blunt, entire; stipules ovate, scarious. Fls 4–5 mm. diam., 4–6 in each axillary cluster, the 2 clusters at a node forming a shining white whorl. Bracteoles scarious, silvery. Sepals 5, 2–2·5 mm., shining white thick and spongy, hooded, with a fine awn on the dorsal side just behind the incurved tip. Petals 5, white, filamentous, shorter than the sepals. Stamens 5, opposite the sepals, short, with roundish anthers. Styles very short, with 2 stigmas. Fr. a 1-seeded capsule enclosed by the persistent erect sepals and opening below by 5(–10) valves which remain cohering above. Seeds 0·8–1 mm., brown. Fl. 7–9. Homogamous. Automatically self-pollinated and sometimes cleistogamous. Th.

Native. A very local plant of moist sandy places in Cornwall, Hants, Kent and Berks. W. and C. Europe from Spain to Denmark, C. Germany, Bohemia and N. Italy; Canary Is.

The fruiting sepals resemble the carpels of *Sedum*.

29. Scleranthus L. KNAWEL

Annual to perennial herbs with diffusely branched stems and *opposite subulate slightly connate exstipulate lvs*. Fls very small, green or whitish, in dense terminal and axillary cymes; hermaphrodite, perigynous, (4–)5-merous. Sepals usually 5, inserted on the rim of the urceolate perigynous zone; *petals* 0; stamens 1–10; ovary with its apex hardly reaching the insertion-level of the sepals, 1-celled with 1(–2) basal long-stalked ovules; styles and stigmas 2. *Fr. an indehiscent* 1-*seeded nutlet* enclosed by the hardened wall of the perigynous zone and the persistent sepals, which are shed with it; seed lenticular, smooth.

Perhaps as many as 150 spp. in Europe, Asia, Africa and Australia.

Annual or biennial; sepals acute with narrow scarious margins, suberect
in fr.   **1. annuus**
Perennial, woody below; sepals blunt with broad white margins, incurved
in fr.   **2. perennis**

1. S. annuus L.   'Annual Knawel.'

An *annual or biennial* herb with slender tap-root and one or more branched decumbent or ascending glabrous or shortly hairy stems 2·5–25 cm. Lvs 5–15(–20) mm., subulate, acute, usually ciliate, the members of a pair slightly connate by their narrow scarious margins. Fls c. 4 mm., subsessile, solitary in the forks of the stem and in terminal and axillary clusters. Bracts usually longer than the fls. *Sepals triangular, ± acute, glabrous, narrowly scarious-margined.* Stamens 10 or fewer, much shorter than the sepals. *Sepals suberect* or slightly incurved *in fr.*, the ± glabrous perigynous *tube below them becoming ± deeply* 10-*furrowed* (Fig. 35 A). Fl. 6–8. Homogamous. Secretes a little nectar and is visited by a few insects: automatically self-pollinated. $2n = 22, 44$. Th.

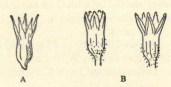

Fig. 35. Fruits of A, *Scleranthus annuus* and B, *S. perennis*. ×2·5.

Very variable in height, mode and extent of branching, size and orientation of leaves, width of scarious margin of sepals, depth of furrowing of perigynous tube, etc. It seems very doubtful whether distinct overwintering (var. *hibernus* Rchb.) or biennial (var. *biennis* Fr.) plants merit taxonomic recognition.

S. polycarpos L., a slender erect or decumbent plant, annual or biennial, with the short sepals (2 mm.) not or scarcely white-bordered and ± connivent in fr., has been reported, but it has not yet been convincingly demonstrated that the British plants are not variants of *S. annuus*.

Native. In dry sandy and gravelly places and in cultivated and waste ground on sandy soil. 106, H 19, S. Throughout Great Britain, but not in the Outer Hebrides, Orkney or Shetland; local in Ireland. Europe, N. Africa, Asia. Introduced in N. America. Calcifuge.

2. S. perennis L.   'Perennial Knawel.'

A *perennial* herb closely resembling *S. annuus* but somewhat *woody below* and usually more robust and more glaucous, and becoming reddish. *Lvs glaucous,* ± ciliate at the base, *often curved to one side of the stem* and with axillary lf-clusters. Bracts shorter than the fls. *Sepals oblong obtuse with broad white margins, incurved over the ripe fr.,* their tips in contact; *perigynous tube hairy* with 10 *shallow furrows* in fr. (Fig. 35 B). Stamens 10. Fl. 6–8. Homogamous. The more conspicuous fls which yield more nectar than those of *S. annuus* are visited by many small flies and automatically self-pollinated. 2*n*=44*. Ch.

Native. A rare plant of dry sandy fields in Norfolk and Suffolk and on rocks in Radnor. Europe, W. Asia.

31. PORTULACACEAE

Annual to perennial herbs, sometimes suffruticose, usually glabrous, often ± fleshy. Lvs spirally arranged or opposite and decussate and then sometimes connate, simple, entire; stipules scarious, bristle-like or 0. Fls hermaphrodite, actinomorphic. Sepals 2, free or united below, anterior-posterior; petals 2–6, free or united below, usually fugacious, sometimes minute or 0; stamens (1–)3–20 or rarely more, opposite the petals when equalling them in number, hypogynous or half-epigynous; ovary superior or half-inferior (*Portulaca*), usually 1-celled, with 1–many campylotropous ovules on a basal placenta; style simple or with 2 or more branches. Fr. a capsule opening by valves or transversely (*Portulaca*); seeds 1–many, with a curved embryo surrounding mealy perisperm.

About 200 spp. in 17 genera, chiefly in temperate and subtropical regions and especially in W. America.

Distinguished from Caryophyllaceae by the calyx of only 2 sepals.

1   Fls yellow, sessile; sepals deciduous; stamens 6–15; capsule opening
         transversely; seeds numerous.   3. PORTULACA
     Fls white or pink, stalked; sepals persistent; stamens 3–5; capsule opening
         by valves; seeds usually 3.   2

2   Small aquatic or subaquatic herbs with several pairs of stem-lvs.
                                                                   1. MONTIA —
     Herbs with 1 pair of stem-lvs, the members of a pair free or broadly
         connate.   3

3   At least the later-formed basal lvs with broadly elliptical or ovate blades.
                                                                   1. MONTIA —
     All basal lvs with linear or linear-oblanceolate blades.   2. CLAYTONIA

1. MONTIA L.

Annual to perennial glabrous herbs, *fibrous-rooted, rhizomatous or stoloniferous, never with corms*; stem-lvs in 1 or more opposite and decussate pairs; stipules 0. Fls in terminal infls which may appear lateral through overtopping by a branch. Sepals persistent. Petals (3–)5, sometimes unequal, free or united into a short tube. Stamens 3–5, adhering to the base of the petals. Ovary superior; styles 3-cleft. *Fr. a 3-valved capsule* with 1–3(–6) seeds.

About 40 spp., chiefly in America.

1 Stem-lvs many, narrowed into a short stalk-like base; fls 2–3 mm. diam.;
    stamens 3.   **1. fontana**
  Basal lvs long-stalked; stem-lvs 2, sessile or broadly connate; fls more
    than 4 mm. diam.; stamens 5.   2

2 Stem-lvs broadly connate; fls 5–8 mm. diam.; petals white, entire or
    slightly notched.   **2. perfoliata**
  Stem-lvs sessile but not broadly connate; fls 15–20 mm. diam.; petals
    pink or white, bifid.   **3. sibirica**

1. M. fontana L.   Blinks.

Annual to perennial herb with branching shoots 2–50 cm., short and erect in land forms, weaker, decumbent and rooting below in aquatic forms (which may have non-flowering shoots and behave as biennials or perennials); sometimes floating. *Stem-lvs in few to many pairs*, 2–20 × 1·5–6 mm., narrowly spathulate to obovate, *narrowed below into a stalk-like base*, the members of a pair free or somewhat connate for a short distance. Fls 2–3 mm. diam., inconspicuous, in terminal cymes which may be displaced to an apparently lateral position in vigorously growing shoots. *Petals 5, white, unequal, joined into a short tube which is cleft to the base in front. Stamens usually* 3, opposite the 3 smaller petals and adhering to their bases. Capsule 1·5–2 mm., exceeding the calyx, globose. Seeds 3, dark brown or black. Fl. 5–10. Homogamous; cleistogamous in dull weather; little visited by insects. $2n = 18$. Th.–Hel.–Hyd.

Native. Streamsides, springs, flushes, wet places among rocks, moist pastures, etc., especially on non-calcareous substrata; also in arable fields in S.W. England; to 3400 ft. in Scotland. Common in one or more of its forms throughout the British Is. Probably in temperate regions throughout the world.

M. fontana has often been divided into two or more spp. on the basis of differences in vegetative and seed-coat characters. The following treatment (S. M. Walters, *Watsonia*, 3, 1953) seems the most satisfactory. The Irish distribution of the sspp. is not yet known, but all may occur.

Ssp. **fontana** (*M. lamprosperma* Cham.; *M. rivularis* auct., ?C. C. Gmel.)
Plant variable in habit but usually loosely tufted and often submerged, green or yellow-green. Ripe *seed* 1·1–1·35 mm., *smooth and shining* as seen under a lens; under higher magnification the individual cells of the coat easily seen as a reticulate pattern, irregular in size and shape, those near the keel elongated and in rows; seed-coat thin and brittle. In trickles of water or very wet places

on acid soil or rock. Common in Scotland and N. England southwards to Cheshire and E. Yorks; N. Wales. Circumpolar Arctic and N. Temperate; also in the S. hemisphere into the Antarctic.

Ssp. **chondrosperma** (Fenzl) S. M. Walters (*M. verna* auct.; *M. minor* auct.)

Plant usually annual, tufted, yellowish-green, with short erect branches terminating in cymes. Rarely submerged and then looser and longer-lived. Ripe *seed* 1·0–1·2 mm., *dull and entirely covered with rather coarse tubercles*; individual cells more or less hexagonal. On light acid soils, usually sandy or gravelly, with a high water-table at least in spring. The common, and in some areas the only, ssp. in southern England; rarer in N. England and Scotland but reaching Moray and Ross. C. and S. Europe; Australia.

Ssp. **intermedia** (Beeby) S. M. Walters (*M. lusitanica* Sampaio; *M. limosa* Decker)

Plant usually loose in habit, often more or less aquatic with long bright-green trailing branches bearing apparently axillary cymes only. Ripe *seed* 0·85–1·1 mm., *finely tuberculate at the edge and rather shiny* under the lens; under higher magnification (2–)3–4(–5) rows of cells on each side of the keel are seen to bear small but relatively narrow and high tubercles; cells rather elongated. In trickles of water or very wet places on acid soil or rock. Common in W. England and Wales but extending to the south and east coast and northwards to Bute and Perth. W., C. and S. Europe; Australia.

Ssp. **variabilis** S. M. Walters (*M. rivularis* auct. mult., ?C. C. Gmel.)

Plant usually loose in habit, like ssp. *intermedia*. Ripe *seed* 0·9–1·1 mm., *more or less smooth* under the lens *but not as shining as in ssp. fontana*; under higher magnification showing variable development of small usually very low tubercles along the keel. In similar habitats to ssp. *intermedia*. Locally common in N. England and Wales and occurring throughout the area where both ssp. *fontana* and ssp. *intermedia* are found; in certain districts, e.g. Isle of Man, the common and perhaps the only ssp. W. and C. Europe.

*2. M. perfoliata (Willd.) Howell (*Claytonia perfoliata* Donn ex Willd.)

Annual glabrous herb with erect or ascending flowering stems 10–30 cm. Basal lvs very long-stalked, their blades 1–2·5 cm., elliptical to ovate-rhomboidal, entire, rather fleshy, faintly veined; *stem-lvs* 2, *opposite, broadly connate* to form a concave suborbicular involucre beneath the terminal raceme-like infl. which often has 1 or more separate fls at its base. Fls 5–8 mm. diam., on stalks about twice as long as the broadly ovate sepals. *Petals* 5, 2–3 *mm.*, somewhat exceeding the sepals, entire or slightly notched, white, ± equal, joined into a short complete basal tube. Stamens 5, opposite the petals. Capsule shorter than the sepals, subglobose. Seeds c. 2 mm., 1(–3) per capsule, black, shining. Fl. 5–7. Visited by small insects. $2n = 18$*. Th.

　　Introduced. Cultivated, disturbed and waste ground, especially on light sandy soils. 57, S. Scattered throughout Great Britain, and locally abundant, northwards to Aberdeen and Inverness; Channel Is. Pacific N. America from Alaska to Mexico; Cuba.

*3. M. sibirica (L.) Howell (*Claytonia sibirica* L.; *C. alsinoides* Sims)

Annual to perennial glabrous herb with erect or ascending flowering stems 15–40 cm. Basal lvs very long-stalked, their blades 1–3 cm., ovate-acuminate, entire, rather fleshy, distinctly veined; *stem-lvs* 2, *opposite, sessile but not connate*. Infl. a terminal bracteate raceme-like cyme. Fls 15–20 mm. diam., on stalks 2–3 cm. Sepals 4 mm., broadly ovate, blunt. *Petals* 5, 8–10 *mm.*, clawed, white or pink *with dark veins, deeply notched or bifid,* ± equal, joined into a very short complete basal tube. Stamens 5, opposite the petals. Capsule about as long as the sepals, ovoid. Seeds c. 2·5 mm. Fl. 4–7. Pro-tandrous and nectar-secreting; visited by flies and other insects. Th.–H.

Introduced. Damp woods, shaded streamsides, etc., especially on sandy soil. Scattered throughout western and northern Great Britain from Devon and Hants northwards to Aberdeen and Inverness; Inner Hebrides; very rare and casual in Ireland. 51. E. Siberia, Pacific N. America from Alaska to S. California.

2. CLAYTONIA L.

Perennial glabrous herbs with basal lvs and flowering stems from *deep-seated globose tubers* (*corms*). Flowering stems with 1 pair of opposite lvs and a terminal raceme or umbel of showy fls. Sepals ovate, persistent. Petals 5, free, equal. Stamens 5, opposite the petals and adherent to their clawed bases. Ovary superior, 1-celled; style 3-cleft. Fr. a capsule opening by 3 valves; seeds 3–6. About 160 spp. in America, Asia, Australia and New Zealand. N.E. USA – N. Jersey

*C. virginica L., with 2–40 flowering stems from a corm 1–5 cm. diam., linear to linear-oblanceolate basal lvs 4–15 cm., and racemes of fls with rose-coloured dark-veined petals which are usually longer than the blunt sepals, is grown in gardens and is established locally. Eastern N. America from Quebec to Texas.

3. PORTULACA L.

Fleshy low-growing herbs with spirally arranged or ± opposite lvs, the uppermost forming a kind of involucre below the fls; stipules scarious, or reduced to small bristles. Fls yellow or red, solitary terminal or in cymose infl. Sepals 2, the anterior larger and overlapping the posterior; petals 4–6, free or united basally, deliquescent after flowering; *stamens* 4–*many*, their filaments often hairy below. Style ± deeply divided into 3–8 branches. Ovary ½-inferior. *Capsule* 1-celled with a thin membranous wall, *opening by a transverse lid; seeds numerous*, on a free-central placenta which often has 3–8 branches.

About 20 spp. chiefly in tropical and subtropical America but with some in the Old World and some cosmopolitan.

*1. P. oleracea L.

An annual glabrous fleshy herb with prostrate or ascending branching stems 10–30 cm. Lvs 1–2 cm., spirally arranged or subopposite, the uppermost crowded beneath the fls, obovate-oblong with a cuneate sessile base, blunt, fleshy, shining; stipules often reduced to bristles. Fls 8–12 mm. diam., 1(–3), terminal or at the forkings of the stem. Sepals with blunt hooded tips, falling in fr. *Petals* 4–6, ± free, *yellow*, soon falling. Stamens 6–15. Style with 3–6 branches. Ovary half-inferior. Capsule 3–7 mm. Seeds 0·7 mm., brownish-black, bluntly tubercled, shining. Fl. 6–9. Homogamous; no nectar; visited by some small insects but opening only on sunny mornings and probably often self-pollinated. $2n = 54$. Th.

Introduced. A not infrequent casual, occasionally establishing itself for a time. A cosmopolitan weed of warm temperate and sub-tropical climates. Var. *sativa*, a more robust and erect plant, has long been cultivated as a pot-herb (Purslane).

*P. grandiflora Hook., with brightly coloured fls 2·5 cm. diam., is grown in gardens and sometimes escapes. Brazil.

32. AIZOACEAE

Herbs or ± woody, often fleshy. Lvs alternate or opposite. Fls usually hermaphrodite, actinomorphic. Calyx-tube free or adnate to ovary; lobes (1–)5–8, herbaceous and often fleshy. Petals numerous, inserted in the calyx-tube in 1 or more series, sometimes 0.

1. Carpobrotus N.E.Br.

Perennials. Stems trailing, branched. Lvs distinct, opposite, equal, 3-angled. Fls showy; stigmas 10–16; ovules borne on *placentae on the outer wall or floor of the cell. Fr. indehiscent, fleshy, edible.*

At least 30 spp., in S. Africa, Australia and temperate S. America.

*1. C. edulis (L.) N.E.Br.   Hottentot Fig.

Mesembryanthemum edule L.

A perennial herb. *Stems trailing*, woody, angled. *Lvs* 7–10 cm., narrow, upwardly curved, *triangular in section* and serrulate on the keel, *fleshy*, opposite and connate at base. Peduncle c. 3 cm., swollen upwards. *Fls* c. 5 cm. across, solitary, *magenta* or (less frequently) yellow. Calyx-tube adnate to ovary; lobes 5, unequal, ± lf-like. Petals numerous, linear. Stamens numerous. Fr. fleshy and edible. Fl. 5–7. $2n=18$. Ch.

Introduced. Naturalized and locally abundant on cliffs and banks by the sea. Cornwall, Devon, Dublin. Native of S. Africa but naturalized in many of the warmer temperate regions.

At least two other members of this family are ± naturalized near the sea in S.W. England. They are smaller than the above and have 4–6 stigmas. *Drosanthemum* has the lvs covered with glistening papillae which are absent in *Lampranthus*. **D. candens* (Haw.) Schwantes ($2n=36$) and **L. multiradiatus* (Jacq.) N.E.Br. seem to be the two common spp.

33. AMARANTHACEAE

Annual or rarely perennial herbs. Lvs opposite or alternate, entire, exstipulate. Fls hermaphrodite or unisexual. Perianth of 3–5 segments, free or connate at base, dry and scarious, often brightly coloured. Stamens 3–5, opposite the per. segs. Ovary unilocular; ovules 1–several, basal. Fr. dry, membranous, indehiscent or dehiscing irregularly.

About 40 genera and 500 spp., mainly in tropical and warm temperate regions.

1. Amaranthus L.

Infl. cymose, the small cymes forming short dense axillary spikes, or the upper spikes making a large, dense, terminal, lfless infl. Fls mostly unisexual. Per.

segs 3 or 5. Stamens free, usually as many as the per. segs. Stigmas 2–3. Fr. 1-seeded, splitting transversely or indehiscent.

About 50 spp. in tropical and temperate regions.

Stems pubescent; per. segs 5; terminal part of infl. nearly lfless.   **1. retroflexus**
Stems nearly or quite glabrous; per. segs 3; infl. of small axillary spikes,
   lfy to top.   **2. albus**

*1. A. retroflexus L.

A rather stout somewhat pubescent grey-green annual 15–90 cm. *Stem* erect or with spreading branches, rough and *shortly pubescent*. Lvs up to 15 cm., ovate or ovate-oblong, obtuse, mucronulate; petiole long, rough. *Infl.* much-branched, forming dense, stout spikes, *the upper part nearly or quite lfless*. Bracteoles equalling or exceeding the per. segs, stiff, acuminate or aristate. *Fls usually 5-merous.* Per. segs of female fls c. 2 mm., spathulate. Fr. c. 1·5 mm., suborbicular, compressed, crowned by 2–3 persistent stigmas, dehiscent. Fl. 7–9. $2n = 32, 34$. Th.

Introduced. A casual of cultivated land and waste places, rare and impermanent. N. America.

*2. A. albus L.

Differs from *A. retroflexus* as follows: Nearly or quite glabrous. Lvs obovate, oblong or spathulate, mucronate. Infl. of small axillary spikes, lfy to top. Fls 3-merous. Per. segs of female fls c. 1 mm.

Introduced. A rare and impermanent casual. N. America.

34. CHENOPODIACEAE

Annual or perennial herbs or shrubs, rarely arborescent, frequently ± succulent or with bladder-like hairs which give the plant a 'mealy' appearance. Lvs usually alternate, simple, exstipulate. Fls often bracteate, small and greenish, hermaphrodite or unisexual, usually actinomorphic. Perianth 3–5-lobed, rarely 0 in female fls, persistent, often accrescent in fr. Stamens usually the same number as the per. segs, usually free; anthers 2-celled, opening lengthwise. Ovary superior or half-inferior, 1-celled; stigmas 2–3, rarely 1; ovule solitary, basal. Fr. an achene or occasionally with circumscissile dehiscence. Perisperm present or 0. Embryo curved round the outside of the perisperm.

About 75 genera and at least 500 spp. Cosmopolitan but mainly in arid regions.

1	Lvs flattened, neither subterete (or ½-terete) and succulent, nor plant apparently lfless but with succulent stems.	2
	Lvs subterete (or ½-terete) and succulent or plant apparently lfless but stems succulent.	5
2	Fls mostly hermaphrodite; fr. ± surrounded by 2–5 persistent per. segs.	3
	Fls all unisexual; fr. enclosed between 2 ± vertical appressed bracteoles.	4

3   Lower lvs usually toothed or lobed, if entire either triangular and ± hastate
      or cordate at base or else not more than 5 cm.; per. segs neither con-
      spicuously thickened at base in fr. nor adhering in groups of 2–4.
                                                          1. CHENOPODIUM
      Lower lvs entire, ovate, often ± cuneate at base, some at least more than
      5 cm.; per. segs conspicuously thickened at base in fr. and adhering in
      groups of 2–4.   2. BETA

4   Lvs toothed or if entire not elliptic; bracteoles not united above the
      middle; annual herbs.   3. ATRIPLEX
      Lvs entire, elliptic or nearly so; bracteoles united above the middle; small
      shrub or annual herb with long-pedicelled fr. (the latter very rare).
                                                            4. HALIMIONE
5   Lvs alternate, spreading.   6
      Lvs opposite, appressed to stem and fused along the margins.
                                                          7. SALICORNIA

6   Lvs acute or obtuse, not spinescent at tip.   5. SUAEDA
      Lvs spinescent at tip.   6. SALSOLA

1. CHENOPODIUM L.

Herbs or small shrubs, very rarely arborescent, usually ± mealy and very
variable. Stem usually grooved or angled and often striped with white, red
or green. Lvs lobed or toothed, less frequently entire. Fls hermaphrodite or
female, in small cymes (glomerules) arranged in a ± branched infl. Perianth
herbaceous, of 2–5 segments joined at the base, or sometimes to half way up
or nearly to apex. Stamens (0–)2–5. Pericarp thin and membranous. Stigmas
2(–5). Seeds usually horizontal, often vertical in terminal fls, rarely all vertical;
testa variously sculptured. Generally in open communities on disturbed
ground, rubbish tips, or by the sea.
   About 110 spp., mainly in temperate regions.
   The markings on the testa of the seeds provide valuable specific characters.
These can be seen with the low power of a microscope when the pericarp has
been removed. The removal of the pericarp can often be effected by rubbing
the seed between the finger and thumb ('pericarp easily removable'), but in
some spp. it is necessary to boil the seed before the pericarp can be got off
by this means, or else to scrape it off with a needle ('pericarp persistent').

1   Perennial; lvs triangular-hastate; stigmas long, exserted; testa: Fig. 36 A.
                                                      1. bonus-henricus
      Annual; lvs rarely hastate; stigmas short.   *2*

2   Larger stem-lvs cordate or truncate at base, coarsely sinuate-dentate;
      testa with large, deep, nearly circular pits (Fig. 36 P).   **13. hybridum**
      Lvs never cordate, ± cuneate at base.   *3*

3   Infl. axis and perianth glabrous (rarely ± mealy in *C. urbicum* and then
      not completely enclosing fr.).   *4*
      Infl. axis and perianth mealy at least when young; fr. usually entirely or
      almost entirely enclosed by perianth.   *9*

4   Lvs entire or at most with a single obscure tooth on each side, green on
      both sides (or purple); stems 4-angled; seeds black; testa: Fig. 36 B.
                                                          2. polyspermum

Lvs, except the uppermost, not entire (very rarely entire or nearly so in *C. rubrum* and *C. botryodes* but then seeds red-brown); stems ridged but not 4-angled.   5

5   All fls with 5 per. segs and 5 stamens; seeds black, 1·2–1·5 mm. diam.; testa: Fig. 36o.   **12. urbicum**
All fls except the terminal ones with 2–4 per. segs and 2–3 stamens; seeds red-brown, 0·75–1·1 mm. diam.   6

6   Lvs mealy-glaucous beneath, green above; testa: Fig. 36R.   **16. glaucum**
Lvs green on both sides.   7

7   Fruiting perianth fleshy, turning scarlet; fls in sessile heads forming a spike, lfless at top; testa: Fig. 36s   **17. capitatum**
Fruiting perianth not fleshy, not turning scarlet; infl. branched; testa: Fig. 36Q.   8

8   Per. segs of lateral fls usually free to middle or below, not or only feebly ridged on back; lvs usually much toothed (not uncommon in various habitats).   **14. rubrum**
Per. segs of lateral fls connate almost to apex, forming a sack closely investing the fr., ±distinctly ridged or keeled on back (at least when young); lvs entire or ±deltoid and only slightly toothed; (rare, in salt-marshes).   **15. botryodes**

9   Lvs entire or almost so.   10
Lvs toothed or lobed.   12

10 Lvs ovate or rhomboid; per. segs rounded on back; plant stinking of bad fish, strongly grey-mealy; testa: Fig. 36c.   **3. vulvaria**
Lvs linear to linear-lanceolate or narrowly oval; per. segs strongly keeled on back; plant not stinking.   11

11 Lvs ±densely white-mealy beneath, linear to linear-oblong; petioles, even of larger cauline lvs, all short (up to c. 1 cm.); testa: Fig. 36M.
                                                                              10. pratericola
Lvs normally green and not strongly white-mealy beneath; petioles of larger cauline lvs long (usually much exceeding 1 cm.).   15

12 Lvs toothed, but not distinctly 3-lobed.*   13
At least some of the lvs distinctly 3-lobed.   16

13 Infl. lfy almost to top, its branches short, numerous, divaricate; pericarp strongly adherent; seeds dull, with sharp, rather prominent keel; testa densely covered with small pits (Fig. 36N).   **11. murale**
Infl. usually lfless in upper part, branches usually long and themselves little branched; seeds shining, obtuse or subacute at margin, but keel not prominent; testa never densely pitted.   14

14 Lvs usually markedly longer than broad, often longer than 3 cm., if nearly entire then scarcely mealy.   **(album** agg.)   15
Larger ste n lvs often as long as or broader than long, up to c. 3 cm., much toothed to nearly entire, but some at least ±3-lobed, usually grey-green and very glaucous-mealy when young (as is the infl.); stems never red; testa: Fig. 36J.   **7. opulifolium**

* From this point onwards in the key the student is recommended, at least until he is familiar with the appearance of the spp. (which is commonly plastic), to check all his determinations by examining the testa of the ripe seed, whose markings appear to be very constant.

15   Plant usually deep green (though often ±masked by grey meal); stems
     often reddish; lvs variable, usually ovate-lanceolate, toothed or entire;
     testa with shallow, spaced radial furrows (Fig. 36 D) or a ±quadrate
     reticulum (Fig. 36 E).   **4. album**
     Plant usually rather bright glaucescent green; stems not red; larger stem
     lvs always ovate-rhomboid with sharp ascending teeth; testa with
     more numerous, closer and deeper furrows than in *C. album* (Fig. 36 F).
                                                                  5. suecicum

16   Plant usually smelling strongly of bad fish; young shoots densely greyish-
     white and mealy; lvs deeply 3-lobed, lobes divergent; testa: Fig. 36 K.
                                                                  8. hircinum
     Plant not stinking (or rarely so in *C. berlandieri*), but then lvs weakly
     3-lobed with lateral lobes short.   17

17   At least the larger stem lvs as broad as or broader than long, lateral
     lobes short; plant usually grey-green; stems never red; testa: Fig. 36 J.
                                                                7. opulifolium
     Lvs usually markedly longer than broad.   18

18   Seeds c. 1·15 mm. diam.; testa with narrow, radially elongate pits (Fig.
     36 L); mid-lobe of lvs elongate, ±parallel-sided, often obtuse;
     glomerules small.   **9. ficifolium**
     Seeds 1·2–1·85 mm. diam.; mid-lobe of lvs very rarely (in var. of *C.
     album*) ±parallel-sided but then testa not pitted.   19

19   Testa without radial lines but with deep honeycomb-like pitting (Fig.
     36 G).   **6. berlandieri**
     Testa not at all pitted.   15
     Testa with mixture of radial lines and irregularly shaped pits (Fig. 36 H).
                                                                  × variabile

Section 1. *Agathophytum* (Moq.) Aschers. Perennial. Per. segs and stamens
4–5. Per. segs not or scarcely keeled on back. Stigmas 2–3, long. Seeds
vertical except in terminal fls.

*1. C. bonus-henricus L.   All-good, Mercury, Good King Henry.

An erect perennial 30–50 cm. *Lvs* up to 10 cm., mealy when young, *broadly
hastate*, obtuse or acute, margins sinuous, entire. Infl. mainly terminal, narrowly pyramidal and tapering, lfless except at base. Seeds 1·8–2·2 mm. diam.,
red-brown, not enclosed by the perianth. Pericarp persistent. Testa irregularly
roughened (Fig. 36 A). Fl. 5–7. $2n = 36$. Hs.

Introduced. In nitrogen-rich habitats, rich pastures, farm-yards, roadsides, etc., long-established and well naturalized, though usually near buildings.
101, H 39, S. Throughout England and Wales, northwards to southern and
eastern Scotland, rather local; rare in N. and W. Scotland and local in Ireland.
Europe north to C. Scandinavia; W. Asia; N. America.

Section 2. *Chenopodium.* Annual. Per. segs and stamens 5. Per. segs often
prominently keeled on the back. Stigmas short. Seeds all horizontal, black
(at least in our spp.).

2. C. polyspermum L.   All-seed.

An erect or decumbent very sparsely mealy to quite glabrous annual up to
c. 1 m. *Stems usually 4-angled. Lvs* up to c. 5 cm., *ovate or elliptic, entire* or

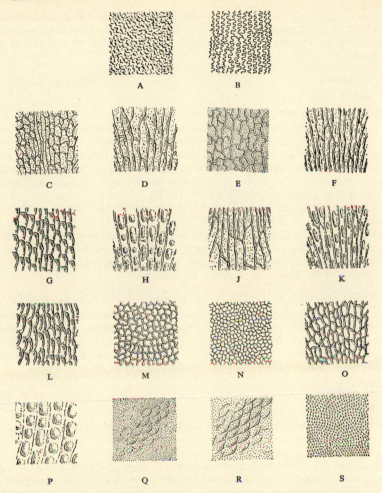

Fig. 36. Sculpturing of the testa of *Chenopodium* seeds. A, *C. bonus-henricus*; B, *C. polyspermum*; C, *C. vulvaria*; D, E, *C. album*; F, *C. suecicum*; G, *C. berlandieri*; H, *C.* × *variabile*; J, *C. opulifolium*; K, *C. hircinum*; L, *C. ficifolium*; M, *C. pratericola*; N, *C. murale*; O, *C. urbicum*; P, *C. hybridum*; Q, *C. rubrum*; R, *C. glaucum*; S, *C. capitatum*. All drawn with illumination from the left. × c. 40.

occasionally with a single tooth-like angle on one or both sides towards the base, obtuse or subacute, thin. Infl. elongate, lax, mostly of axillary, divaricately-branched cymes shorter than their subtending lvs; glomerules small and indistinct. Per. segs rounded on back. Seeds 1·1–1·25 mm. diam., not enclosed by the perianth. Pericarp easily detached. *Testa with close radially elongate pits with sinuous margins* (Fig. 36B). Fl. 7–10. 2*n* = 18. Th.

Native. In waste places and cultivated ground. 59, S. Cornwall and Kent to Cheshire and Lincoln, N. Yorks and Berwick, locally abundant; probably introduced in the north and in Ireland. Europe; Asia; introduced in N. America.

3. C. vulvaria L. 'Stinking Goosefoot.'

C. olidum Curt.

An ascending mealy annual 5–35 cm., *smelling strongly of bad fish. Lvs 1–2·5 cm., rhomboid or ovate, entire,* or with a single tooth-like angle on one or both sides towards the base, *acute, very mealy beneath.* Infl. 0·5–2·5 cm., terminal and axillary, lfless, usually subtended by a full-sized lf. Per. segs rounded on back. Seeds 1·4–1·5 mm. diam., enclosed by perianth. Pericarp rather persistent. *Testa with faint radial furrows and irregular thickenings between them* (Fig. 36c). Fl. 7–9. $2n = 18$. Th.

This plant, like *C. hircinum,* contains trimethylamine, which gives it the nauseous odour of stale salt fish.

Native. At the landward margins of salt-marshes and shingle beaches, inland in waste places. 57, S. Cornwall and Kent to Durham; introduced in southern Scotland; rather rare. Europe to southern Scandinavia and Russia; N. Africa; S.W. Asia; introduced in N. America.

C. album agg. (spp. 4–5).

An erect ± mealy annual up to c. 1 m. Lamina rhomboid to ovate, ovate-lanceolate or lanceolate, ± toothed or occasionally entire. Infl. ± branched, usually dense. Per. segs with a raised ridge or keel running down the back. Seeds 1·2–1·85 mm. diam., enclosed by perianth. Pericarp easily detached. Testa never closely pitted. $2n = 18; 36; 54$. 112, H 40, S.

4. C. album L. Fat Hen.

Incl. *C. reticulatum* Aell.

Plant usually ± deep green, ± mealy, usually with rather short strict branches. *Stem often reddish.* Lvs most often ovate-lanceolate and toothed but varying from rhomboid to lanceolate. Glomerules usually densely crowded, usually spicate but not infrequently cymose. Seeds (1·25–)1·5–1·85 mm. diam. *Testa very faintly radially striate, otherwise nearly smooth* (Fig. 36D), sometimes with raised lines forming a quadrate reticulum (Fig. 36E). Fl. 7–10. $2n = 54*$.

Very common and variable.

Native. In waste places and cultivated land. By far the commonest sp. throughout the British Is. Europe north to c. 71°; N. and S. Africa; Asia; America; Australia.

*5. C. suecicum J. Murr

C. viride auct.

Plant usually a rather bright glaucescent green, nearly or quite glabrous when mature, somewhat mealy when young and on the infl. *Stem without red coloration;* branches usually rather long, slender and ± spreading. *Larger stem lvs always ovate-rhomboid with few to several sharp ascending teeth,* sometimes somewhat 3-lobed; upper lvs oval to linear, ± toothed to entire.

Glomerules usually rather small, in rather lax cymes. Seeds 1·5–1·7 mm. diam. *Testa marked with more numerous and deeper furrows than in C. album* (Fig. 36 F). $2n = 18*$.

Introduced. Rather rare and most frequently on rubbish tips and in waste places. From N. Somerset and Essex to Selkirk; Down; very thinly scattered. Northern Europe and Asia.

*6. C. berlandieri Moq.

A mealy erect branched annual up to 1·5 m. *Lower lvs 3-lobed*, ovate or elliptic; upper entire or slightly toothed, ovate to oblong or linear-lanceolate, usually mucronate. Infl. much-branched; glomerules in lax spikes. *Per. segs prominently keeled on the back*. Seed c. 1·4 mm. diam. Pericarp easily detached. *Testa with honeycomb-like pitting* (Fig. 36 G). Our plant is ssp. **zschackei** (J. Murr) Zobel.

Introduced. A casual near docks and on rubbish-tips. Native of N. America.

C. × variabile Aell. (*C. album × berlandieri*) occurs not infrequently in waste places. Up to 2·6 m. Nodes of main stem usually marked with vivid amaranth-purple blotches. Lvs ± toothed, usually ± 3-lobed, some at least mucronate. Glomerules usually large, in ± compact spikes or cymes. Per. segs prominently keeled on back. Seed 1·25–1·6 mm. diam. Pericarp easily detachable. Testa with a variable combination of the furrows of *C. album* and the pits of *C. berlandieri* (Fig. 36 H). Our plant is var. *murrii* Aell. $2n = 36*$.

*7. C. opulifolium Schrad. ex Koch & Ziz

A variable, erect or decumbent mealy annual 30–80 cm. *Lvs (0·7–)2–3 cm., often broader than long*, ± *strongly glaucous-mealy beneath*, especially when young, rhomboid, cuneate and subentire below, *often 3-lobed; lateral lobes short*; middle lobe triangular to shortly half oval, obtuse or subacute, subentire to coarsely and irregularly toothed, teeth often shortly mucronate. Infl. usually very mealy, branched; glomerules in dense interrupted spikes. *Per. segs keeled on back*. Seed 1·25–1·5 mm. diam. Pericarp somewhat persistent *Testa with radial furrows and finely and irregularly papillose* (Fig. 36 J). Fl. 8–10. $2n = 54*; 36$. Th.

Introduced. A not infrequent alien in waste places, chiefly in the south of England. 41, S. Europe north to Belgium and Germany; N. Africa; E. tropical and S. Africa; Asia Minor; C. Asia; introduced in N. America.

*8. C. hircinum Schrad.

A ± mealy annual up to c. 1 m., *smelling of trimethylamine*. Young shoots and infl. conspicuously and densely white-mealy. Lvs up to 6 cm., 3-lobed; lobes large, divergent, entire or with rounded teeth. *Glomerules in short dense axillary spikes*, much shorter than the main stem, forming a lfy infl., or in dense terminal infl. Per. segs prominently keeled on back. Seed 1·4–1·8 mm. diam. Pericarp easily removable or persistent. *Testa with many radial furrows and irregular pits between them* (Fig. 36 K). Fl. 8–10.

Introduced. A casual on rubbish-tips and near docks. Native of S. America.

9. C. ficifolium Sm.   'Fig-leaved Goosefoot.'

C. serotinum auct., non L.

An erect or decumbent mealy annual 30–90 cm. *Blades of lower lvs up to c. 8 cm.*, *3-lobed; lateral lobes short, oblong or triangular, usually with 1 tooth on the lower margin; middle lobe oblong,* coarsely toothed or subentire; *lobes and teeth obtuse* or subacute; upper lvs slightly lobed or subentire. Infl. slender, much-branched; axillary branches longer than subtending lvs; *glomerules rather small* and distant. *Per. segs with* a ridge-like keel running down the back. Seed c. 1·15 mm. diam. Pericarp easily removable. *Testa with narrow radially elongate pits* (Fig. 36L). Fl. 7–9. $2n=18^*$. Th.

Native. On waste ground and arable land, particularly round manure heaps. 44, S. From Dorset and Kent to Somerset, Leicester and Lincoln; rare and probably casual in Wales, Ireland and northern England. Europe from Denmark southward; Asia east to the Altai; N. Africa.

*10. C. pratericola Rydb.

C. leptophyllum auct., non Nutt.

A mealy annual up to 1 m. Stem erect, not red; branches long, reaching nearly to top of main stem. *Lvs 15–50 mm., linear to linear-oblong, entire,* or the lowest with a few small teeth, acute or mucronate, *conspicuously glaucous-mealy beneath, the larger distinctly 3-nerved.* Infl. lfless except at base; glomerules in dense or ± interrupted spikes. *Per. segs prominently keeled on back.* Seed 1–1·2 mm. diam. Pericarp easily removable. *Testa marked with shallow grooves forming a close, rather irregular reticulum* (Fig. 36M). Fl. 6–10.

Introduced. A casual on rubbish-tips and near buildings and docks. Native of N. America.

11. C. murale L.   'Nettle-leaved Goosefoot.'

An erect slightly mealy annual up to 70 cm. *Lvs 1·5–6(–9) cm., broadly triangular or rhomboid, acute,* entire below, *coarsely and irregularly toothed above,* rarely nearly entire; *teeth acute and ± incurved.* Infl. of axillary and terminal, *cymosely and divaricately branched panicles* up to c. 5 cm.; glomerules rather densely crowded. *Per. segs bluntly keeled on back.* Seed 1·25–1·5 mm. diam. *Pericarp very persistent. Testa with dense, minute, not radially elongate pits* (Fig. 36N). Fl. 7–10. $2n=18^*$. Th.

Native. On dunes and in waste places, chiefly on light soils. 58, H4, S. Lowlands of England and Wales, local; a casual in southern Scotland. Europe north to southern Sweden; N. and S. Africa; S.W. and S. Asia; introduced in America and Australia.

12. C. urbicum L.   'Upright Goosefoot.'

An erect annual 30–70 cm., *glabrous or very nearly so.* Lvs up to c. 9 cm., lower triangular, truncate to cuneate at base, usually toothed, teeth often long and hooked, acute or subobtuse. Infl. branched, branches crowded, short, erect; axillary branches mostly shorter than the subtending lvs. Glomerules small, distant. *Per. segs not or scarcely keeled on back. Seed 1·2–1·5 mm. diam., black, not completely enclosed by the perianth.* Pericarp persistent. *Testa marked with shallow grooves forming a slightly elongate reticulum enclosing convex areas* (Fig. 36 O). Fl. 8–9. Th.

?Native. On waste ground and arable land especially round manure heaps, rare. 48. In lowland districts from Cornwall and Kent to Lancashire and Yorks; rare and casual in Wales, Scotland and Ireland. Europe to southern Scandinavia; S.W. and C. Asia; introduced in N. (and probably S.) America.

13. C. hybridum L. Sowbane.

An erect, scarcely mealy annual up to 1 m. *Lvs* up to 18 cm. (usually smaller), *broadly triangular to cordate-ovate*, acute or acuminate, *with few very large teeth. Infl.* lax, *cymose with divaricate branches*, nearly lfless. *Per. segs not or scarcely keeled on back.* Seed 2 mm. diam., black, not enclosed by the perianth. Pericarp usually easily removable, sometimes very persistent. *Testa with very large deep thick-walled pits* (Fig. 36 P). Fl. 8–10. $2n=18, 36$.

?Native. In waste places and cultivated ground, rare. 37. Dorset and Kent to Shropshire and Norfolk; casual elsewhere. Europe to southern Scandinavia; Asia Minor; C. Asia; N. Africa.

Section 3. *Pseudoblitum* (Gren.) Aschers. Annual. Perianth of terminal fls with 5, of lateral fls with 2–4 segments, not or scarcely ridged on back, or rarely ± distinctly keeled. Stamens equalling the per. segs in number. Stigmas short. Seeds all vertical or those of the terminal fls horizontal, red-brown (at least in our spp.).

14. C. rubrum L. 'Red Goosefoot.'

A prostrate, ascending or erect *nearly or quite glabrous, usually reddish* annual up to 70 cm. Lvs very variable in size and shape, commonly 2–5 cm., ovate to rhomboid or broadly triangular, coarsely and irregularly toothed, sinuate or subentire, teeth often blunt, apex acute to acuminate or sometimes blunt. Infl. variable, often dense, simple to much-branched, lfy or lfless; glomerules usually crowded. *Per. segs of lateral fls usually free to middle or almost to base, not or only weakly ridged on back.* Seed 0·7–1·1 mm. diam., red-brown. Pericarp easily removable. *Testa with shallow non-radial grooves forming a reticulum especially near the hilum, sometimes nearly smooth, but always with numerous minute pits* (Fig. 36 Q). Fl. (5–)7–9. $2n=36$. Th.

Native. In waste places, cultivated ground, and often near the sea; frequently abundant on rubbish-tips and in farm-yards. 80, H 16, S. England, not uncommon; Wales, Scotland and Ireland, rare and often only as a casual. Europe, Asia Minor, C. Asia, N. America.

15. C. botryodes Sm.

Incl. *C. crassifolium* Hornem.; ?*C. chenopodioides* (L.) Aell.

A glabrous or nearly glabrous annual 5–30 cm. Stems erect or, more often, with spreading branches from the base. Lvs broadly triangular, entire or slightly toothed, usually obtuse, rather thick. Glomerules usually in distinct, rather distant groups. *Perianth of lateral fls saccate, closely investing the fr., the segments connate almost to the apex, each one marked by a distinct ridge or keel in the upper part* (at least when young). Otherwise very similar to forms of *C. rubrum* and not distinguishable by the seeds. Fl. 7–9. Th.

Native. On the muddy margins of salt-marsh ditches and creeks by the sea, very local. 6, S. South and east coasts of England from Hants to Norfolk. Coasts of Europe from Denmark southwards; S. Africa; N. America.

16. C. glaucum L.   'Glaucous Goosefoot.'

A prostrate, ascending or erect annual 5–50 cm. *Lvs* 1–5 cm., lanceolate to narrowly rhomboid, obtuse or subacute, *sinuate to repand-dentate, mealy-glaucous beneath, green above*. Infl. of copious, axillary and terminal, little-branched, spiciform partial infl. up to c. 3 cm. Seed very similar to that of *C. rubrum* but differing slightly in the sculpturing of the testa (Fig. 36R). Fl. 6–9. $2n = 18$. Th.

?Native. On rich waste ground, rarely on sea-shores. 35, S. From Dorset and Kent to Northumberland, local and rare; a casual in S. Wales and southern Scotland. Europe, Asia, America; a different ssp. in S. Africa and Australia.

Section 4. *Morocarpus* Aschers. Similar to *Pseudoblitum* but the perianth becoming succulent and bacciform in fr.

*17. C. capitatum (L.) Aschers.

An erect nearly or quite glabrous annual 10–60 cm. *Lvs* long-petioled, *narrowly triangular, acuminate*, toothed or entire, *usually with 2 narrow spreading lateral lobes near the base. Infl. of dense sessile rather large subglobose heads of fls, becoming scarlet at maturity*, lfless in the upper part. Seed 1–1·1 mm. diam., red-brown, oval not circular. Pericarp very persistent. *Testa densely punctate-pitted* (Fig. 36s). Fl. 7–8. $2n = 18$. Th.

Introduced. Naturalized in fields in Caernarvon and Fermanagh, and as a rare casual on rubbish tips elsewhere. Origin unknown but naturalized in Europe from C. Scandinavia to France and Switzerland and, more rarely, in the south; also in N. America.

2. BETA L.   Beet

Herbs. Lvs almost entire. Fls hermaphrodite, in small cymes arranged in branched spike-like infl. *Perianth* of 5 segments, *becoming thicker especially towards the base as the fr. ripens*. Stamens 5. *Ovary half-inferior*. Fr. 1-seeded, adhering in groups (glomerules) by the swollen perianth bases.

About 10–13 spp. Europe and Asia.

1. B. vulgaris L.

A glabrous or slightly hairy annual, biennial or perennial of very varied habit, 30–120 cm. Root stout, conspicuously swollen or not. Stems decumbent, ascending or erect, ± branched and lfy. Lvs very varied in size, shape and colour, often dark green or reddish and rather shiny, frequently forming a radicle rosette. Infl. varied, usually large and ± branched; partial infl. sessile, subtended by a small narrow lf, each of 1–several, frequently 2–4, fls. Per. segs green. Fl. 7–9. Wind pollinated. $2n = 18$. Th. (biennial) or Hs.

Ssp. **vulgaris**. Root conspicuously swollen at junction with stem; stems ascending or erect; lvs up to c. 20 cm., often ovate and cordate at base,

3. ATRIPLEX

margins commonly wavy (except in spina̶ ̶ ̶ ̶ ̶
the nerves; partial infl. usually of 3–4 fls. Pet̶ ̶ ̶ ̶
green. Sugar-beet, beetroot, spinach-beet, chard,̶ ̶ ̶
sugar-beet type occur sporadically with ssp. *maritim̶ ̶ ̶*
to be interfertile and to cross freely when opportunity ̶ ̶

Ssp. **maritima** (L.) Thell. Root usually not conspicuously sw̶ ̶
with stem; stems commonly decumbent; lvs usually up to c. 10 cm̶,̶
thick and leathery, glossy or matt; partial infl. usually of 2–3 fls.̶
lamina and infl. axis usually coloured red by anthocyanin. The common̶
except in cultivation.
　　Native. On sea shores. 52, H25, S. Throughout the British Is. Europe to
S.W. Sweden; Azores; N. Africa; Asia Minor to the East Indies.

*****B. trigyna** Waldst. & Kit., with yellow to whitish per. segs, persisted for a
few years in a field near Cambridge and has also occurred in Surrey. $2n = 54$.

3. ATRIPLEX L.　　　　　　　　　　　Orache

Herbs or small shrubs, often mealy. Stems frequently striped white and green
or red and green. *Lvs* toothed or lobed, sometimes entire and then linear to
triangular but *not elliptic*. Infl. like that of *Chenopodium*. *Fls unisexual*; male
fls with a perianth of (3–)5 segments; female fls without perianth but enclosed
by two persistent bracteoles. *Bracteoles not fused above the middle, not
obdeltoid, entire or else with the lateral lobes smaller than the middle one.*
Stigmas 2. Seeds vertical (except in *A. hortensis*).
　　About 100 spp., cosmopolitan.
　　Self-fertilization is frequent and numerous genetically distinct forms are
thus perpetuated. The polymorphism of the genus is said to be further
increased by the not infrequent occurrence of hybrids (see Tureson, G., 1922,
Hereditas, **3**, 238–60).

1	Bracteoles united only at base.	*2*
	Bracteoles united up to middle.	*4*
2	Lower lvs linear to linear-oblong.	**1. littoralis**
	Lower lvs triangular or rhomboid, often hastate.	*3*
3	Lower lvs long-cuneate at base, narrowing gradually into the petiole.	**4. patula**
	Lower lvs truncate or shortly cuneate at base, abruptly contracted into the petiole.	**2. hastata**
4	Plant greenish, mealy; bracteoles in fr. not hardened below.	**3. glabriuscula**
	Plant white or silvery; bracteoles in fr. hardened below.	**5. laciniata**

Section 1. *Teutliopsis* Dum. Stems striped white or red and green. Bracteoles
united less than half-way up and not hardened in fr., except in *A. glabriuscula*
where they are united half-way up and become slightly hardened.

1. A. littoralis L.　　　　　　　　　　　'Shore Orache.'

A ± mealy annual up to 100 cm. Root deep and spreading. Stems usually
stout, much-branched ± erect. *Lvs linear to linear-oblong*, entire or dentate,

20 cm., spike-like, lfless
trongly muricate, sometimes
18. Th.
ostrata. 52, H8, S. Around
d and Ireland. Europe from

'Hastate Orache.'

in habit and branching but often
mealy. Root deep and very tough.
iangular-hastate, truncate or broadly
some making an angle of 90° or more
o the petiole; the uppermost lvs often
tiole. Bracteoles in fr. 3–12 mm., ovate,
subcordate at base, tubercled or not on
bac. $2n=18*$. Th.

Native.   adsides, etc. inland and near the sea on
sand, shingle an.   high-tide mark, common. 108, H31, S.
Throughout most of the   Is. Europe from Scandinavia southwards;
N. Africa; Asia; probably introduced in N. America.

3. A. glabriuscula Edmondst.   'Babington's Orache.'

Incl. *A. Babingtonii* Woods

Only certainly distinguishable from maritime forms of the preceding by
having the bracteoles united up to about the middle. Probably best regarded
as a ssp. of it. Fl. 7–9. $2n=18*$. Th.

Native. Usually on sandy or gravelly shores at or somewhat above high-
tide mark. 80, H24, S. Around the coasts of the British Is. Coasts of N.W.
Europe to the Loire.

4. A. patula L.   Iron-root, 'Common Orache.'

Similar to *A. hastata*, with which it often grows, and almost as variable; best
distinguished by lf-shape and chromosome number. Plant more frequently
prostrate. Lvs usually linear to linear-oblong, tapering gradually into the
rather short petiole, rarely broader with a cuneate base, the margins making
an angle of distinctly less than 90°; uppermost lvs sessile or nearly so. Fl. 7–9.
Fr. 8–10. $2n=36*$. Th.

Native. In cultivated ground, waste places and in open habitats near the
sea, where it is generally less common than *A. hastata*. 109, H39, S. Throughout most of the British Is. Europe from c. 71° N. southwards; N. Africa;
W. Asia; naturalized in N. America.

The triploid hybrid between this and *A. hastata* has been made in cultivation
and is almost sterile; it has not been found wild, presumably owing to the
frequency of self-pollination in the genus.

Section 2. *Obionopsis* Lange. Stems whitish or pale brown, occasionally with
red patches. Bracteoles united to the middle, hardened in the lower half.

5. A. laciniata L.   'Frosted Orache.'

A. sabulosa Rouy; *A. arenaria* Woods, non Nutt.

A mealy *white or almost silvery* decumbent annual up to 30 cm. or rarely more. Stems much-branched, yellowish or reddish. Lvs usually 1·5–2 cm., rhomboid to ovate, sinuate-dentate, obtuse, rather thick and very mealy on both surfaces. Infl. axillary, much shorter than the lvs. *Bracteoles* in fr. 6–7 mm. long and usually rather broader, lateral angles rounded and sometimes toothed, mealy, *hardened in the lower half*. Fl. 8–9. Fr. 9–10. 2*n* = 18. Th.

Native. On sandy and gravelly shores at about high-tide mark. 59, H 8, S. Scattered round the shores of Great Britain, local; south and east coasts of Ireland. S. Norway, Germany (Baltic coast), Belgium, northern France.

*****A. hortensis** L., Orache, with dimorphic fls, about ¼ without bracteoles but with perianth, the remainder with free, ovate bracteoles but no perianth, is cultivated and sometimes escapes.

*****A. nitens** Schkuhr is also found as a casual. It differs from the preceding in its broad, hastate, deeply dentate lvs.

*****A. halimus** L., a mealy shrub with ovate-rhomboid lvs, is planted near the sea in southern England and occasionally escapes and becomes naturalized.

4. HALIMIONE Aell.

Similar to *Atriplex* but differs as follows: Lvs entire, elliptic or nearly so. Bracteoles obdeltoid in fr., united nearly to the top, 3-lobed with lateral lobes usually larger than middle one.

Woody perennial; fr. sessile.	**1. portulacoides**
Annual; fr. long-pedicelled (very rare).	**2. pedunculata**

1. H. portulacoides (L.) Aell.   'Sea Purslane.'

Atriplex portulacoides L.; *Obione portulacoides* (L.) Moq.

A very mealy small *shrub* up to 80 or rarely 150 cm. Rhizome short, creeping. Stems decumbent, branches ascending, terete below, angled above. *Lower lvs opposite*, shortly (5–10 mm.) petioled, elliptic; upper linear, entire, obtuse or apiculate. Infl. of terminal and axillary compound spikes; partial infl. dense. *Fr. sessile*. Bracteoles in fr. 3–5 mm. long and rather broader, obdeltoid, united ⅔ of the way from the base, usually 3-lobed. Fl. 7–9. Fr. 9–10. 2*n* = 36. Chw. or N.

Native. In salt-marshes, especially fringing channels and pools, ordinarily flooded at high tide. 40, H 12, S. England north to Northumberland and Westmorland, locally abundant; Scotland: Wigtown, Ayr and Outer Hebrides; Ireland: Galway Bay, round the S. and E. coasts to Down. Europe from Denmark southward; N. Africa; Asia Minor; S. Africa; introduced in N. America.

2. H. pedunculata (L.) Aell.

Atriplex pedunculata L.; *Obione pedunculata* (L.) Moq.

A silvery-mealy erect *annual* up to 30 cm. *Lvs all alternate*, sessile or shortly petioled, elliptic to oblong, entire; apex rounded, apiculate. Partial infl. lax. *Fr. pedicelled*, pedicels up to 12 mm. when mature. Bracteoles in fr. obdeltoid, united almost to the top, 3-lobed, middle lobe very small, lateral lobes long and spreading. Fl. 8–9. Fr. 9–10. $2n = 18$. Th.

Native. In the drier parts of salt-marshes with *Puccinellia maritima*, very rare. 7. Kent, Suffolk, Norfolk, Cambridge and Lincoln; probably extinct in some counties. Occasionally also in other places introduced with ballast. Western Europe from southern Sweden; Baltic coast north to Estonia; C. Germany; S.E. Europe; Turkestan; Siberia.

5. SUAEDA Forsk. ex Scop.

Herbs or small shrubs growing in saline places. Lvs fleshy, alternate, terete or ½-terete, small. Fls unisexual or hermaphrodite, small, axillary; bracteoles 2–3, minute. Per. segs 5, not keeled, small and ±succulent. Stamens 5. Stigmas 3–5; achenes with a thin membranous pericarp. Seed horizontal or vertical; embryo in a flat spiral; endosperm present or 0.

About 40 spp., cosmopolitan.

Annual herb; lvs acute or subacute and narrowed at base; stigmas 2; seed
 horizontal.   **1. maritima**
Perennial, shrubby; lvs rounded at tip and base; stigmas 3; seed vertical.
                                                                         2. fruticosa

1. S. maritima (L.) Dum.   'Herbaceous Seablite.'

A prostrate to erect ± glaucous or red-tinged glabrous *annual* 7–30(–60) cm. *Lvs* 3–25 mm. × 1–2(–4) mm., ½-terete, *acute or subacute*. Fls 1–3 together in small axillary cymes. *Stigmas* 2. *Seed* 1·1–2 mm. diam., *biconvex*, nearly circular in outline but with a small curved beak, black, shining, *with fine reticulate sculpturing, horizontal*. Fl. 7–10. Fr. 8–11. Germ. spring. $2n = 36$. Th.

The following three varieties, differing considerably in appearance and fr. are recognized. Their status is not certain:

Var. *maritime*. Plant large, branches spreading. Seed not more than 1·5 mm. diam. Fl. 8–10.

Var. *macrocarpa* (Desv.) Moq. Plant decumbent or prostrate, rarely erect and then small. Lvs up to 10 mm. Seed c. 2 mm. diam. Fl. 7–8.

Var. *flexilis* (Focke) Rouy. Plant usually erect, often unbranched and never branched from the base; branches short, erect. Lvs 10–25 mm. Seed 1·1–1·4 mm. diam. Fl. 8–10.

Native. In salt-marshes and on seashores, usually below high-water mark spring tides, common. 74, H24, S. Around the coasts of the British Is., except Berwick, Banff, W. Sutherland and Caithness; ascends the Severn estuary to E. and W. Gloucester. Coasts of Europe (except the Arctic); inland in saline areas of C. Europe, Russia, E. Asia and E. Indies; N. America; sspp. or perhaps other closely allied spp. in S. America and Australia.

2. S. fruticosa Forsk.   'Shrubby Seablite.'

A small much-branched glabrous shrub 40–120 cm. Stems suberect or ascending, up to 5 cm. diam., very lfy, subterranean ones rooting freely. *Lvs* 5–18 × 1 mm., almost terete, *rounded at tip*, evergreen. Fls 1–3 together in small axillary cymes. *Stigmas* 3. *Seed* 1·7–1·8 mm., *ovoid*, beaked near the hilum, *smooth*, black, shining, *vertical*. Fl. 7–10. Fr. 9–11. Germ. spring. $2n = 36$. N.

Native. On shingle banks and other well-drained substrata by the sea but above high-water mark spring tides, local. 10, S. Dorset, Isle of Wight, E. and W. Kent, N. and S. Essex, E. Suffolk, E. and W. Norfolk, Lincoln, Glamorgan; extinct in S. Hants and N.E. Yorks and Channel Is. Coasts of Europe from N. France southwards, inland in Spain; Madeira, Canaries, St Helena, Angola, Somaliland, S.W. Asia; inland in S. Russia, Trans-caucasia, Afghanistan and India.

6. SALSOLA L.

Herbs. *Lvs sessile, succulent.* Fls small, sessile; per. segs (4–)5, usually developing a transverse dorsal wing in fr. Stamens (3–)5. Ovary subglobose, style elongate, stigmas 2–3. Seed horizontal.

About 40 spp. in Europe, temperate Asia and N. and S. Africa, mostly in saline habitats.

1. S. kali L.   Saltwort.

A decumbent or prostrate, seldom erect, prickly annual up to 60 cm. Stems with pale green or reddish stripes, usually much-branched. Lvs 1–4 cm., subulate, sessile, succulent, subterete, narrowed into a little spine at the tip. Fls usually solitary in the axil of a lf, each with 2 lf-like bracteoles. Per. segs becoming tough in fr. and thickened transversely about the middle, the thickening forming either a ridge or a horizontal wing of variable size. Fr. c. 3·5 mm., turbinate, enclosed in the persistent perianth. The stems and lvs are either asperous (var. *hirsuta* Hornem.) or nearly glabrous (var. *glabra* Detharding). Fl. 7–9. Fr. 8–10. $2n = 36$. Th.

Native. On sandy shores. 72, H 20, S. Around the coasts of the British Is. Europe to c. 60° N., Azores, N. Africa, Asia, N. America.

*S. pestifer A. Nels. (*S. tragus* auct.), a slender usually erect plant with almost filiform lvs and the perianth nearly always without wings, occurs occasionally as a casual on waste ground. $2n = 36$.

7. SALICORNIA L.   Glasswort, Marsh Samphire.

Woody perennials or annuals inhabiting salt-marshes. Stems usually much-branched. *Lvs succulent*, translucent and glabrous, *opposite, the pairs fused along their margins and enveloping the stem* forming the 'segments', tips usually free. *Infl.* of terminal, ± branched spikes with *axillary cymes* of 3, rarely 1 or, in foreign spp., several fls. Perianth indistinctly 3-lobed and ± immersed in the bracts. Bracteoles 0. *Stamens* 1–2. Seed covered with short curved or hooked hairs; radicle incumbent; endosperm 0 (in our spp.).

Probably at least 50 spp., cosmopolitan in saline districts. Many of the spp. are critical and need much further work for their elucidation. Herbarium specimens dried in the ordinary way are almost valueless.

Phenotypic variation is great and may render the identification of single specimens difficult, if not impossible. For instance, *S. ramosissima*, though typically much-branched, bushy and erect, becomes prostrate on hard stony mud and very much dwarfed and often quite unbranched when growing in crowded pure stands or in competition with other plants. Crowding reduces the degree of branching and this is often accompanied by an increase in the length of the terminal spike. Damage to or removal of the terminal spike often increases the abundance and size of the lateral branches and a similar effect is produced by the plant falling over. The typical colour of most species is not fully developed until the seed is nearly ripe and then only in the absence of shade. Until some familiarity with the genus has been obtained attempts at identification should be confined to well-grown specimens with undamaged main stems collected in September or October. Dried specimens, however carefully prepared, lose much of their character and are difficult to interpret. Preservation in a fluid, such as equal parts by volume of alcohol, glycerin and water or sea water with 2 % chrome alum and 2 % formaldehyde, is more satisfactory, particularly if accompanied by colour photographs or detailed notes of colours.

Much work remains to be done on the taxonomy and distribution of the species.

1 Perennial, uprooted with difficulty, with ± prostrate woody stems and
    ascending fleshy shoots.   **1. perennis**
    Annual, easily uprooted, erect or prostrate.   *2*

2 Cymes regularly 1-fld; terminal spikes up to c. 6 mm.   **5. pusilla**
    Cymes normally 3-fld; terminal spikes usually more than 6 mm.   *3*

3 Central fl. of cyme usually much larger than the visible part of the lateral
    fls; terminal spikes usually with up to 10 fertile segments; stamens
    normally 1.   *4*
    Central fl. of cyme usually little larger than the visible part of the lateral
    fls; terminal spikes usually with (6–)12 or more fertile segments; stamens
    normally 2.   *6*

4 Plant dark shining green, often becoming dark purplish-red; free part of
    lf with a conspicuous scarious border c. 0·2 mm. wide; normally much
    branched and bushy, sometimes simple but then dark purplish-red.
                                                                    2. ramosissima
    Plant green, becoming yellowish or sometimes pink; free part of lf with
    a narrow scarious border not more than 0·1 mm. wide; moderately to
    freely branched.   *5*

5 Plant dark green, fertile spikes pink when mature; bushy with lower
    primary branches up to as long as the main stem, tertiary branches
    usually present.   **3. europaea**
    Plant dull glaucous green, fertile spikes dull yellow, occasionally purplish
    round the fls when mature; lower primary branches $\frac{1}{4}(-\frac{1}{2})$ as long as the
    main stem, tertiary branches 0, secondary branches present or 0.
                                                                    4. obscura

6 Plant pale green or yellow with, at most, a slight purplish tinge when
    mature; fertile segments usually more than 3 mm.   *7*
    Plant brownish-purple to brownish-orange when mature; fertile segments
    rarely exceeding 3 mm.   **6. nitens**

7 Plant bushy and much-branched with numerous primary and usually
    secondary and some tertiary branches; lowest primary branches
    generally half as long as to as long as main stem.   *8*

Plant not bushy; tertiary branches usually 0; lowest primary branches ¼(–½) as long as main stem.

9

8 Sterile segments becoming dull yellow and finally brown; terminal spike distinctly tapering; fertile segments usually more than 12.

8. dolichostachya

Sterile segments becoming bright yellow; terminal spike almost cylindrical; fertile segments usually less than 12. **9. lutescens**

9 Sterile segments becoming dull yellow and finally brown; terminal spike with up to 20 segments, usually ± tapering. **7. fragilis**

Sterile segments becoming bright yellow; terminal spike usually with less than 12 fertile segments, almost cylindrical. **9. lutescens**

1. S. perennis Mill.

S. radicans Sm.; incl. *S. lignosa* Woods

A somewhat *woody perennial* often forming tussocks up to c. 1 m. diam. Stems up to c. 30 cm., ascending or decumbent; segments dark green becoming yellowish, basal keeled. Terminal spikes 10–20 mm., cylindrical, blunt; segments c. 8. Cymes 3-fld. Middle fl. completely separating the lateral ones. Stamens 2. *Stigma bifid.* Seeds ovoid, covered with short, slightly curved hairs. Fl. 8–9. Fr. 10. $2n = 18*$. Ch.

Native. Gravelly foreshores and salt-marshes. S. England. France southwards to Algeria.

2. S. ramosissima Woods

S. gracillima (Towns.) Moss, *S. prostrata* auct., *S. appressa* Dum., *S. smithiana* Moss

An erect or prostrate annual 3–40 cm., typically abundantly branched and bushy with the lowest branches c. as long as the main stem, but when crowded reduced to a single stem. *Plant dark green,* usually *becoming deep purplish-red. Segments with a conspicuous broad scarious border* c. 0·2 mm. wide. Terminal spike (5–)10–30(–40) mm., ± tapering, with (1–)4–9(–12) fertile segments Cymes 3-fld. Central fl. rounded-rhomboid to almost circular, much larger than the visible part of the lateral fls. Stamen 1, exserted or not. Seeds 1·2–1·5(–1·7) mm. Fl. 8–9. $2n = 18*$. Th.

Native. In the upper part of salt-marshes on bare, rather firm mud and behind sea walls. Common and widespread in suitable habitats in S. and E. England and S. Wales. Distribution elsewhere uncertain. A somewhat similar sp. occurs in S. and E. Ireland.

3. S. europaea L.

S. herbacea (L.) L., ?*S. stricta* Dum.

An erect annual (10–)15–30(–35) cm., *fairly richly branched,* with the *lowest branches up to c. as long as the main stem. Plant* dark green *becoming yellow-green and ultimately flushed with pink or red.* Segments with an inconspicuous scarious border not more than 0·1 mm. wide. Terminal spike (10–)15–50 mm., slightly tapering, obtuse, with (3–)5–9(–12) fertile segments. Cymes 3-fld. Central fl. rounded above, cuneate below, much larger than the visible part

of the lateral fls. Stamen usually 1, exserted or not. Seeds 1·2–1·8(–2·0) mm.
Fl. 8. $2n=18*$. Th.

Native. On open sandy mud in salt-marshes, local. S. and W. coasts of
England; ?Ireland. W. Europe.

4. S. obscura P. W. Ball & Tutin

An erect annual 10–40(–45) cm., *typically with primary branches only, the
lowest branches up to c. ½ as long as the main stem. Plant dull slightly glaucous
green with a matt surface, becoming dull yellow*, rarely purple round the pores
of the fls. Segments with an inconspicuous scarious border not more than
0·1 mm. wide. Terminal spike 10–40(–45) mm., nearly cylindrical, obtuse, with
(3–)5–9(–14) fertile segments. Cymes and fls rather similar to those of
S. europaea. Stamen 1, rarely exserted. Seeds 1·1–1·7 mm. Fl. 8–9. $2n=18*$. Th.

Native. On bare damp mud, in pans and at the sides of channels in salt-
marshes. In suitable habitats on the E., S. and W. coasts from Lincs to
Cheshire. Distribution elsewhere unknown.

5. S. pusilla Woods

S. disarticulata Moss

An erect or rarely prostrate annual up to 25 cm., usually *much-branched and
bushy. Plant yellowish-green, becoming brownish- or pinkish-yellow*, often with
bright pink tips to the branches. *Terminal spike up to c. 6 mm.*, with 2–4 fertile
segments. *Cymes 1-fld.* Fls almost circular. *Fertile segments all disarticulating*
shortly before the seeds are ripe. Stamen 1. Fl. 8–9. $2n=18*$. Th.

Native. In the drier parts of salt-marshes, chiefly along the drift-line.
Dorset to Kent, Essex, Norfolk; Carmarthen; Waterford. N.W. France.

6. S. nitens P. W. Ball & Tutin

An erect annual 5–25 cm., *typically with primary branches only*, the lowest
branches usually less than ¼ as long as the main stem. *Plant smooth, shining
and somewhat translucent, green or yellowish-green, becoming clear light
brownish-purple to brownish-orange. Sterile segments conspicuously swollen near
the top.* Terminal spike 12–40 mm., cylindrical, obtuse, with 4–9 fertile segments. Cymes 3-fld. Central fl. semicircular above, cuneate below, little
larger than the triangular visible part of the lateral fls. Stamens usually 2,
exserted. Seeds 1·5–1·7 mm. Fl. 9. Th.

Native. On bare damp mud and in pans in the upper parts of salt-marshes.
S. and E. England, apparently common in suitable habitats from E. Suffolk
to Hants. Distribution elsewhere unknown.

7. S. fragilis P. W. Ball & Tutin

An erect annual (10–)15–30(–40) cm., *usually with primary branches only*, the
*lowest branches normally less than ¼ as long as the main stem. Plant dull green
becoming dull yellowish-green. Terminal spike (20–)30–80(–100) mm.*, distinctly tapering, with (6–)8–16(–20) fertile segments. Cymes 3-fld. Central fl.
semicircular to triangular above, cuneate below, the visible part of the lateral

fls triangular and almost as long as the central fl. Stamens usually 2, exserted. Seeds 1·5–2·0 mm. Fl. 8–9. $2n=36*$. Th.

Native. On soft mud in the lower levels of salt-marshes and, particularly, on the sides of channels below the fringe of *Halimione*; often forming pure stands. E. Suffolk to Kent, ?elsewhere.

8. S. dolichostachya Moss

An erect annual 10–40(–45) cm., *abundantly branched and bushy*, the *lowest branches usually about as long as the main stem*. *Plant dark green, becoming paler green or dull yellow and finally brownish*, the fertile spikes occasionally with a slight purplish flush. *Terminal spike* (25–)50–120(–200) *mm., distinctly tapering* but often obtuse, with (7–)12–25(–32) fertile segments. Cymes and fls similar to *S. fragilis*. Stamens usually 2, exserted. Seeds 1·5–2·3 mm. Fl. 7–8. $2n=36*$. Th.

Native. Usually on rather firm mud and on muddy sand at the lowest levels in salt-marshes, occasionally on the sides of narrow channels at the middle levels. Coasts of Great Britain from Lancs to Devon, Kent and E. Ross; S. and E. coasts of Ireland, W. Galway. Netherlands, Denmark. Characterized by the bushy growth, dull coloration and very long tapering spikes but varying in habit from straggling to strongly fastigiate.

9. S. lutescens P. W. Ball & Tutin

An erect annual (10–)15–30(–40) cm., typically with the habit of *S. dolichostachya* from which it differs in the *bright green to yellow-green* colour, *soon becoming bright yellow*, the shorter *terminal spikes* (*usually* 25–60 *mm.* with 8–12 fertile segments) and in having the *spikes cylindrical* and obtuse. Fl. 7–8. Th.

Native. On firm and comparatively dry mud or muddy sand. Coasts of England from S. Lincs to S. Hants; Glamorgan; ?elsewhere.

35. PHYTOLACCACEAE

Trees, shrubs, herbs or woody climbers. Lvs alternate, entire; stipules 0 or very small. Fls usually regular, usually hypogynous. Perianth usually in 1 whorl: segs 4–5 free, or united at base; stamens 4–∞; carpels 1–10, free or united; ovule 1 in each carpel, campylotropous. Fr. various; embryo large, curved, surrounding the perisperm.

About 15 genera and 100 spp., mainly tropical.

1. PHYTOLACCA L.

Usually herbs. Fls regular, in lf-opposed racemes. Per. segs 5. Stamens 5–30. Carpels 5–16, free or joined at base. Fr. berry-like.

*P. americana L.   Pokeweed.

Coarse glabrous perennial up to 3 m. Lvs 10–30 cm. ovate to oblong-lanceolate, stalked. Infl. 10–20 cm. Fls c. 6 mm., stamens and carpels usually 10. Fr. dark purple. Cultivated and occasionally naturalized. Native of eastern N. America.

36. TILIACEAE

Trees, shrubs or rarely herbs. Lvs spirally arranged or distichous and alternate, rarely opposite; stipules usually small and caducous, often functioning as bud scales; sometimes 0. Infl. usually cymose. Fls hermaphrodite, actinomorphic, hypogynous. *Sepals* 5(–3), free or united, usually *valvate in bud*; petals as many as the sepals, free, rarely 0; stamens 10 or usually more, their *filaments free or united only at the base; anthers with* 4 *pollen sacs*; ovary superior 2–10-celled, each cell with 1–many anatropous ovules on axile placentae; style 1, with as many radiating stigma-lobes as ovary-cells, or the stigmas sessile. Fr. a capsule, a drupe or a nut with 1–5 seeds, rarely a berry or separating into drupelets. Seeds endospermic.

About 300 spp. in 35 genera, widely distributed in temperate to tropical regions.

Jute is made from the pericycle fibres of *Corchorus capsularis* and other spp., native in tropical Asia. *Sparmannia africana* is often grown as an ornamental plant in conservatories, etc. Distinguishable from Malvaceae by the absence of a stamen tube or more certainly by the stamens with 4 pollen sacs.

1. Tilia L.   Linden.

Deciduous trees with sympodial growth owing to the abortion of the terminal bud. Winter buds large, blunt. Lvs distichous, alternate, usually cordate or truncate at the base and with a slender stalk; stipules (bud-scales) caducous. Fls yellowish or whitish, fragrant, in cymose infl. whose stalk is adnate for about half-way to a large oblong ± membranous bracteole. Sepals 5, free; petals 5, free; stamens many, free or in bundles opposite the petals, the filaments often forked distally; epipetalous staminodes sometimes present; ovary 5-celled, each cell with 2 ovules; style slender, with a 5-lobed stigma. Entomophilous, many insects visiting the fls for their copious nectar. *Fr. ovoid, indehiscent and nutlike*, 1-celled, with usually 1–3 seeds. *The infructescence is shed as a whole, with the adnate bracteole acting as a wing.* Cotyledons broad, lobed.

About 30 spp. in the temperate regions of the northern hemisphere.

Lime timber was formerly in great demand for wood-carving. (G. Gibbons)

1 Lvs pubescent beneath and often also above; fls 2–5, usually 3; fr. strongly
   3–5 ribbed.   **1. platyphyllos**
   Lvs glabrous beneath except for tufts in the axils of the veins; fls 4–10;
   fr. slightly 3–5-ribbed or ribs 0.   2

2 Lvs 6–10 cm., broadly ovate-acuminate, bright green beneath, with the
   tertiary veins prominent; petioles 3–5 cm.; cymes pendulous; fr. slightly
   ribbed.   **3. × europaea**
   Lvs 3–6 cm., suborbicular, abruptly acuminate, often broader than long,
   ± glaucous beneath with the tertiary veins not prominent; petioles
   1·5–3 cm.; cymes obliquely erect; fr. not or barely ribbed.   **2. cordata**

1. T. platyphyllos Scop.   'Large-leaved Lime.'

T. grandifolia Ehrh. ex Hoffm.

A large tree reaching 30 m., with spreading branches and a ± smooth dark bark. Young *twigs pubescent*, rarely glabrous. *Lvs* 6–12 cm. broadly ovate, abruptly acuminate, obliquely cordate at the base, dark dull green and glabrescent above, pale green and *pubescent beneath* with simple hairs *all over the surface* as well as in whitish tufts in the axils of the veins; *tertiary veins prominent beneath*; margin sharp-toothed; stalk 1·5–5 cm., pubescent. Cymes usually 3-fld (2–5) *pendulous*, the adnate bracteole 5–12 cm., pubescent on the midrib below. Fls yellowish-white. Stamens exceeding the petals. *Fr.* 8–10 mm., subglobose to pyriform, apiculate, densely pubescent, *strongly 3–5-ribbed*, woody when mature. Fl. late June, before *T. cordata* and *T.* × *europaea.* Freely visited by bees. $2n = 82$. MM.

Probably native. In woods on good calcareous or base-rich soils and on limestone cliffs, but perhaps often planted originally since its pollen-grains have rarely been found in peat from this country. 10. Naturalized (or possibly native) in the Wye Valley and its neighbourhood (Gloucester, Hereford, Monmouth, Worcester), in Brecon and Radnor and on the Magnesian Limestone in S. Yorks, and established in old plantations northwards to Perth. C. and S.E. Europe from N. Spain, S. Italy and Greece northwards to c. 51° N. in France, Belgium, Germany and Poland; Crimea; Asia Minor; Caucasus.

2. T. cordata Mill. 'Small-leaved Lime.'

T. parvifolia Ehrh. ex Hoffm.

A large tree reaching 25 m. with spreading branches and a smooth bark. Young *twigs* usually downy at first, then becoming *glabrous. Lvs* 3–6 cm., suborbicular, abruptly acuminate, cordate at the base, dark ± shining green and glabrous above, somewhat glaucous and *glabrous* beneath *except for tufts of rusty hairs* in the axils of the veins; *tertiary veins not prominent*; margin sharp-toothed; stalk 1·5–3 cm., glabrous. *Cymes* 4–10 fld, *obliquely erect or spreading*, the adnate bracteole 3·5–8 cm., glabrous below. Fls yellowish-white. Stamens about equalling the petals. *Fr.* c. 6 mm., globose, apiculate, *thin-shelled, ribs obscure or* 0. Fl. early July. Visited by bees. $2n = 82$. MM.

Native. In woods on a wide range of fertile soils but especially over limestone; commonly on wooded limestone cliffs. 39. Scattered throughout England and Wales northwards to the Lake District and Yorkshire, but planted northwards to Perth. C. and E. Europe from N. Spain, N. Italy, N. Balkans and S. Russia northwards to c. 63° N. in Norway, Sweden and Finland, eastwards across C. Russia to c. 75° E. in Siberia; Crimea; Caucasus.

Distinguishable without much difficulty from *T. platyphyllos* and their hybrid by the smaller, relatively broader, tougher lvs, ± glaucous beneath, with relatively longer stalks, as well as by characters of the infl. and fr.

3. T. × europaea L. Common Lime.

T. cordata × *T. platyphyllos*; *T. vulgaris* Hayne; *T. intermedia* DC.

A large tree reaching 25 m., with arching lower branches and the trunk often covered with irregular bosses. Young *twigs* usually *glabrous. Lvs* 6–10 cm. broadly ovate, shortly acuminate, obliquely cordate or ± truncate at the base, dark green and glabrous above, light green and *glabrous* or nearly so beneath *except for tufts of whitish hairs* in the axils of the veins; *tertiary veins ± prominent* beneath; margin sharp-toothed; stalk 3–5 cm., ± glabrous. *Cymes* 4–10-fld, *pendulous.* Fls yellowish-white. Stamens equalling or somewhat exceeding the petals. *Fr.* c. 8 mm., subglobose to broadly ovoid, apiculate, pubescent, *slightly ribbed*, woody when mature, producing some viable seed. Fl. early July. Freely visited by bees. $2n = 82$. MM.

Introduced or very doubtfully native. Widely planted over a long period, especially in copses, parks, gardens, roadsides, etc. Throughout the British Is. except the far north. Europe.

A variable tree, reproducing by seed despite its hybrid origin. Much planted despite the unsightly bosses on its trunk and the frequent infestation of its lvs by aphids which cause a copious rain of honey-dew.

37. MALVACEAE

Herbs, shrubs or trees, usually with mucilage canals and stellate pubescence. Lvs spirally arranged, usually palmately veined and commonly palmately lobed; stipules free, usually small, often caducous. Fls in racemes or racemose panicles, or solitary, axillary; usually hermaphrodite, actinomorphic, hypogynous. Sepals 5(–3), free or united, valvate in bud; often with an 'epicalyx' of 3 to several segs resembling an outer set of sepals; petals 5, free, convolute, commonly adherent to the base of the staminal tube; stamens numerous (–5), the filaments united below into a staminal tube which divides above into branchlets each bearing a single 1-celled anther-lobe; ovary superior, with 2–many cells each with 1–many amphitropous ovules on axile placentae; styles and stigmas as many or rarely twice as many as cells. Entomophilous. Fr. a capsule or schizocarpic by separation of 1-seeded nutlets, rarely fleshy. Seeds with very little endosperm and a curved embryo; cotyledons lf-like.

About 900 spp. in 40 genera, widely distributed in temperate to tropical regions.

Cotton is spun from the hairs on the seed coat of *Gossypium* spp. belonging to the subfamily Hibisceae, with a capsular fr. *Hibiscus syriacus* L., an Asiatic shrub of the same subfamily, is often cultivated in this country for its variously coloured fls. Many members of the subfamily Malveae, with schizocarpic fr., are also grown for their fls. Native British representatives of the family all belong to Malveae.

Malvaceae are distinguished from Tiliaceae by the united filaments and 1-celled anthers of their stamens.

1	Epicalyx of 3 segs.	2
	Epicalyx of 6–9 segs.	3. ALTHAEA
2	Epicalyx-segs free to the base.	1. MALVA
	Epicalyx-segs united below.	2. LAVATERA

1. MALVA L.

Annual to perennial subglabrous herbs with palmately lobed or divided very mucilaginous lvs. Epicalyx of 3 segs free to the base, arising close beneath the calyx. Calyx of 5 sepals, united below. Petals 5, free, cuneate or obovate, emarginate or deeply notched, purple, rose or white. Fr. of numerous unbeaked 1-seeded nutlets arranged in a flat whorl round the short conical apex of the receptacle.

Thirty spp. in north temperate regions.

1 Stem-lvs deeply palmatisect with slender pinnatifid segments.
                                      1. moschata
    Stem-lvs ± palmately lobed or roundish-crenate.   2

2 Perennial; petals 20–45 mm., 3–4 times as long as the sepals.   **2. sylvestris**
    Annual to biennial; petals usually 4–13 mm. (rarely to 20 mm.), up to
      twice as long as the sepals (rarely three times as long).   3

3 Epicalyx-segments ovate-lanceolate; lvs not or slightly cordate at base.
                                     3. nicaeensis
    Epicalyx-segments linear-lanceolate; lvs distinctly cordate at base.   4

4 Petals 8–13(–20) mm., 2–3 times as long as the sepals; nutlets pubescent
      but neither reticulate nor ridged on their dorsal face.   **4. neglecta**
    Petals 3–9 mm., usually about as long as the sepals and never more than
      twice as long; nutlets distinctly reticulate or at least transversely ridged
      on their dorsal face.   5

5 Fls 1–1·5 cm. diam., in dense axillary clusters, their stalks, even in fr., not
      more than twice as long as the sepals; nutlets transversely ridged or
      obscurely reticulate; robust erect plants, sometimes with crisped lvs.
                                      7. verticillata
    Fls c. 0·5 cm. diam., not in dense clusters, their stalks in fr. many times
      longer than the sepals; nutlets distinctly reticulate.   6

6 Calyx herbaceous, little enlarged in fr., nutlets with acute but neither
      winged nor toothed dorsal angles.   **5. pusilla**
    Calyx almost scarious, much enlarged in fr.; nutlets with distinctly winged
      and toothed dorsal angles.   **6. parviflora**

1. M. moschata L.   Musk Mallow.

A *perennial* herb with a branching stock producing several erect lfy shoots
30–80 cm., terete, often purple-spotted, with sparse simple hairs. Basal lvs
5–8 cm. diam., reniform in outline, long-stalked, with 3 contiguous crenate
lobes; *stem-lvs* successively shorter-stalked and more deeply divided, the 3–7
primary divisions *deeply pinnatifid into ± linear ultimate segments*; all subglabrous; stipules small, lanceolate. Fls 3–6 cm. diam., usually solitary in the
axils of the upper lvs and in an irregularly racemose terminal cluster. Epicalyx-
segments linear-lanceolate, narrowed at both ends, half as long as the calyx.
Calyx-lobes ovate-deltoid, erect and enlarging in fr. Petals rose-pink, rarely
white, obovate-cuneate, truncate and deeply emarginate distally, c. three times
as long as the calyx. Nutlets blackish when ripe, not rugose, hispid on the
back, the back and sides not separated by an angle. Fl. 7–8. Visited by bees
and other insects. $2n = 42$. H.

A very variable plant, especially in the degree of cutting of the lvs.

Native. Grassy places, pastures, hedgebanks, etc.; not uncommon on the
more fertile soils. 92, H 34, S. Throughout Great Britain northwards to Perth
and the Inner Hebrides. Europe. N. Africa.

**M. alcea* L., with stellate pubescence, ovate epicalyx-segments and glabrous nutlets,
resembles *M. moschata* otherwise and sometimes occurs as a casual. Europe,
especially central and southern.

2. M. sylvestris L.   Common Mallow.

A *perennial* herb with an erect ascending or decumbent stem 45–90 cm., with
sparse spreading hairs. Basal lvs 5–10 cm. diam., roundish with very shallow

crenate lobes, cordate at base, somewhat folded, long-stalked; stem-lvs with 5–7 rather deep crenate lobes, sparsely hairy, ciliate. *Fls* 2·5–4 *cm. diam.*, stalked, in axillary clusters, the whole forming an irregular raceme. Epicalyx-segments oblong-lanceolate, two-thirds as long as the calyx. Calyx-lobes ovate-deltoid, connivent and not enlarging in fr. *Petals* rose-purple with darker stripes, obovate-cuneate, deeply emarginate, 2–4 *times as long as calyx*. Nutlets brownish-green when ripe, the reticulate rugose usually glabrous back separated by a sharp angle from the transversely wrinkled sides. Fl. 6–9. Visited by various insects, especially bees. $2n = 42$. H.

Var. *lasiocarpa* Druce has the nutlets hairy.

Native. Roadsides, waste places, etc.; common in the south, less so in the north. 102, H 40, S. Throughout the British Is., though local in N. Scotland. Not in Outer Hebrides, Orkney or Shetland. Throughout Europe.

*3. M. nicaeensis All.

An *annual* herb with ± ascending *hispid* stems 20–50 cm., and long-stalked palmately 5–7 lobed *lvs* which are *not or scarcely cordate* at the base; lobes of upper lvs ± acute. Fls in axillary clusters of 2–6, rarely solitary. *Epicalyx-segments ovate-oblong to broadly lanceolate*, about equalling the triangular-ovate sepals which enlarge little in fr. Petals 10–12 mm., narrowly cuneate, up to twice as long as the sepals, pale bluish-lilac. *Nutlets* glabrous or puberulent, *strongly reticulate*, the dorsal angles sharp but not toothed. Fl. 6–9. Th.

Introduced. A not infrequent casual, occasionally establishing itself. Mediterranean region and eastwards to Baluchistan, but introduced in C. and W. Europe.

4. M. neglecta Wallr.   'Dwarf Mallow.'

M. rotundifolia auct., non L.; *M. vulgaris* Fr., non Gray

An *annual* or longer-lived herb with stems 15–60 cm., decumbent or with the central stem ascending, densely clothed in stellate down. Lvs 4–7 cm. diam., long-stalked, roundish-reniform, deeply cordate, with 5–7 shallow acutely crenate lobes, ± pubescent; stipules ovate. *Fls* 1·8–2·5 *cm.* diam., in irregular racemes. *Epicalyx-segments linear-lanceolate*, c. $\frac{1}{2}$–$\frac{2}{3}$ as long as the calyx. *Calyx-lobes* broadly ovate-deltoid with stellate pubescence, slightly connivent with reflexed tips, *not enlarging in fr*. *Petals* whitish with lilac veins or pale lilac, obovate-cuneate *with bearded claws*, deeply emarginate, *c. twice* (or up to three times) *as long as the calyx*. Nutlets brownish-green, in a disk 6–7 mm. diam., with the pubescent but non-reticulate back and sides of each nutlet separated by a blunt angle, and with a straight line of contact between neighbours. Fl. 6–9. Little visited by insects and automatically self-pollinated. $2n = 42$. Th.–H.

Native. Waste places, roadsides, drift-lines, etc.; frequent in the south less so in the north. 90, H 25, S. Throughout the British Is. except Outer Hebrides and Shetland. Europe, Asia, N. Africa; introduced in N. America.

The following annual spp., often mistaken for *M. neglecta*, occur as casuals or locally naturalized aliens:

*5. M. pusilla Sm.

M. rotundifolia L., nom. ambig.; *M. borealis* Wallm.

An annual herb resembling *M. neglecta* but the *fls only c.* 0·5 *cm. diam.*, epicalyx segments equalling the deltoid calyx-lobes; petals pale rose to whitish, barely exceeding the calyx, claws bearded; and nutlets in a disk 7–9 mm. diam., each *nutlet reticulate*, ± glabrous or more usually pubescent (var. *lasiocarpa* Salmon), with *sharp but not winged or toothed dorsal angles. Calyx hardly enlarged in fr., herbaceous*, slightly connivent but with reflexed tips. Fl. 6–9. Self-pollinated. $2n = 42, 76$. Th.

Introduced. Waste places, foreshores, etc.; local northwards to Aberdeen. N. and C. Europe.

*6. M. parviflora L.

An annual herb with pale mauve fls 4–5 mm. diam., closely resembling *M. pusilla* but with broadly ovate calyx-lobes, glabrous petal-claws, spreading *calyx, scarious and greatly enlarging in fr.*, and glabrous or pubescent reticulate *nutlets with toothed and winged dorsal angles*, so that there is a wavy line of contact between neighbours.

Introduced. A casual, occasionally becoming established. Mediterranean and W. Asia, widely introduced.

*7. M. verticillata L.

A biennial herb with erect stems to 80 cm. or more, glandular-hairy to almost glabrous. Lower lvs long-stalked, shallowly 5–7-lobed, cordate at base, green and sparsely hairy above, bluish-green and more densely hairy beneath. *Fls* 1–1·5 cm. diam., *short-stalked in dense axillary clusters.* Epicalyx-segments linear. Calyx-lobes ovate, acute. Petals pale rose, up to twice as long as the calyx. Calyx enlarging in fr. *Nutlets* glabrous, obscurely reticulate, *with transverse ridges extending all or part of the way to the median longitudinal ridge down the dorsal face*; dorsal angles squarish but neither winged nor toothed. Fl. 7–9. $2n = c. 84$.

Var. *crispa* L., with crisped lvs and whitish petals hardly exceeding the calyx, has been grown as a salad plant.

Introduced. Casual and occasionally becoming established. E. Asia, but very widely naturalized owing to its use as a vegetable and as a medicinal plant.

2. LAVATERA L.

Annual to perennial herbs or woody plants usually covered with stellate pubescence. Closely resembling spp. of *Malva* but commonly larger and with the 3 *epicalyx segments united at the base*, forming an involucre-like lobed cup below the calyx.

About 25 spp., chiefly in the Mediterranean region but also in the Canary Is., Australia and the islands off S. and Lower California.

Plant 60–300 cm., erect, woody below; epicalyx enlarging in fr.; nutlets transversely wrinkled with acute raised dorsal angles.   **1. arborea**

Plant 50–150 cm., erect or ascending, herbaceous, hispid; epicalyx not enlarging in fr.; nutlets not transversely wrinkled, with blunt dorsal angles.
                                                                               2. cretica

1. L. arborea L.　　　　　　　　　　　　　　　　Tree Mallow.

A suffruticose biennial, almost tree-like, with stout erect stems 60–300 cm., reaching 2·5 cm. diam., *woody* below, thinly and softly stellate-pubescent above. Lvs to 20 cm. diam., roundish-cordate, stalked, with 5–7 broadly triangular ± acutely crenate lobes, softly velvety with stellate pubescence, ± folded like a fan. Fls 3–4 cm. diam. in terminal simple or compound racemes. *Epicalyx* with broadly ovate blunt segments *exceeding the calyx*, joined almost half-way, *much enlarging in fr.* Calyx-lobes ovate acute. Petals broadly obovate-cuneate, overlapping, 2–3 times as long as the calyx, pale rose-purple with broad deep purple veins confluent below. *Nutlets* yellowish, *transversely wrinkled* and with acute *raised dorsal angles*, glabrous or pubescent. Fl. 7–9. Protandrous. $2n = 36, 42$. H. (biennial).

Native. Maritime rocks or waste ground near the sea, to 500 ft., S. and W. coasts of Great Britain from Sussex to Cornwall and northwards to Anglesey, Isle of Man and Ayr; probably introduced on the east coast northwards to Fife. 23, H 8, S. Mediterranean and Atlantic coasts of Europe from Greece to Spain and northwards to the Channel coasts of France as far east as Calvados, and to Great Britain and Ireland; N. Africa; Canary Is.

2. L. cretica L.

L. sylvestris Brot.

An annual or biennial herb with erect or ascending (rarely prostrate) stellate-pubescent *herbaceous* stems 50–150 cm. Lower lvs roundish-cordate; upper ± truncate below, with 5 triangular-acute toothed lobes. Fls in irregular racemose panicles. *Epicalyx*-segments broadly ovate, spreading, *somewhat shorter than the calyx, not enlarging in fr.* Calyx-lobes ovate, abruptly acuminate. Petals obovate-cuneate, not contiguous, deeply emarginate, 2–3 times as long as the calyx, lilac. *Nutlets* yellowish, *almost smooth*, with *blunt dorsal angles*, glabrous or pubescent. Fl. 6–7. $2n = 40$–44; c. 112. Th.–H. (biennial).

Native. Waysides and waste places near the sea. W. Cornwall, Scilly Is. and Jersey. 1, S. Casual elsewhere. Mediterranean region and Near East; W. Europe northwards to Brittany and W. Cornwall.

3. ALTHAEA L.

Annual to perennial herbs resembling *Malva* and *Lavatera* but with an *epicalyx of 6–9 segments joined below into a cup-like involucre.*

About 15 spp. in temperate Europe, Asia and N. Africa.

1 Annual; hispid with swollen-based bristles.	**2. hirsuta**	
Biennial or perennial; not hispid.		2
2 Softly pubescent; lvs with shallow acute lobes.	**1. officinalis**	
Sparsely hairy; lvs with blunt lobes.	**3. rosea**	

1. A. officinalis L.　　　　　　　　　　　　　　Marsh Mallow.

A *perennial* herb with a thick fleshy pale stock tapering downwards into a tap-root, and erect densely stellate-pubescent velvety stems 60–120 cm., simple or slightly branched. Lower lvs 3–8 cm. across, roundish, stalked; upper *lvs* narrower and ± ovate; all slightly 3–5-lobed, the upper more deeply and acutely so, irregularly toothed, *velvety*, folded like a fan. Fls 2·5–4(–5) cm. diam., 1–3 in each upper lf-axil to form an irregularly racemose infl.; their *stalks shorter than the lvs*. *Epicalyx-segments* 5–7 *mm.*, 8–9, narrowly triangular. *Sepals* 8–10 mm., *ovate-acuminate*, velvety like the lvs and epicalyx. Petals 2–3 times as long as the sepals, obovate-cuneate, truncate and somewhat emarginate, pale pink. *Nutlets* brownish-green, *pubescent*, with the rounded *smooth* back separated from each face by a marginal ring. Calyx curved over the fr. Fl. 8–9. Protandrous; visited by various bees and automatically self-pollinated. $2n=42$. H.

Native. Upper margins of salt and brackish marshes, and ditch-sides and banks near the sea; locally common. 33, H15, S. Coasts of Great Britain northwards to Northumberland and to Dumfries, Kirkcudbright and Arran; Jersey. C. and S. Europe northwards to Denmark; N. Africa; W. Asia. Introduced in N. America.

The roots yield abundant mucilage used in confectionery and medicinally as a demulcent and emollient.

2. A. hirsuta L.   'Hispid Mallow.'

An *annual or biennial* herb with many ascending slender *hispid* stems 8–60 cm. Lower lvs 2–4 cm. across, long-stalked, reniform, ± 5-lobed, the lobes blunt to subacute, crenate; upper lvs short-stalked, deeply 3–5-lobed or palmately divided, the crenate-toothed segments becoming narrower and more acute towards the top of the stem. Fls 2·5 cm. diam., solitary, axillary and in a terminal racemose cluster, their *stalks* long, slender, *exceeding the lvs*. Epicalyx-segments 7–10 mm., narrowly triangular, hispid, enlarging in fr. *Sepals* 12–15 mm., narrowly *lanceolate-acuminate*, hispid, like the stems and epicalyx, with swollen-based bristles. Petals exceeding the sepals, obovate-truncate, pale rosy-purple, becoming bluish. *Nutlets* dark brownish-green, *glabrous*, with the rounded slightly keeled back and the lateral faces transversely ridged; *fruiting calyx erect*. Fl. 6–7. Th.–H. (biennial).

Doubtfully native. Borders of fields, and wood margins, to 800 ft.; rare and local. N. Somerset and W. Kent.; Jersey. Casual elsewhere. Europe northwards to Belgium; W. Asia.

*3. A. rosea (L.) Cav.   Hollyhock.

A tall biennial to perennial herb with stout erect hairy stems to 3 m. Lvs to 30 cm. across, long-stalked, roundish-cordate, 5–7-lobed or angled, crenate, roughly hairy. Fls 6–7 cm. diam., ± sessile in axillary clusters forming a long irregular spike-like infl. Petals variable in colour, often white, yellow or red, sometimes very dark. Fl. 7–9. H. (sometimes biennial).

Introduced. Native in China. Forms of *A. rosea* and of *A. ficifolia* (L.) Cav. (W. Asia), with deeply 5–9 lobed lvs, and of hybrids between them (*A.* × *cultorum* Bergmans) are cultivated and may escape.

38. LINACEAE

Herbs, shrubs or trees. Lvs usually alternate, rarely opposite, simple, entire or weakly toothed, with or without stipules. Infl. a cyme or raceme, rarely fls solitary. Fls hermaphrodite, often heterostylous, heterochlamydeous, actinomorphic, hypogynous, 5- (rarely 4-)merous. Sepals usually imbricate, persistent. Petals usually contorted, often fugacious. Stamens equal in number to and alternating with petals, often with small staminodes in addition, less often twice as many; filaments ±connate at base; anthers introrse. Ovary (2–)3–5-celled, cells often nearly halved by a false septum; ovules (1–)2 in each cell, axile, collateral (when 2), anatropous, micropyle directed upwards, raphe ventral; styles free, less often ±united. Fr. a loculicidal capsule or dehiscing also down the false septum (and so with twice as many lines of dehiscence as carpels), rarely a drupe. Endosperm usually scanty, sometimes copious or 0; embryo straight or slightly curved.

About 14 genera and 275 spp., tropical and temperate.

Most like the *Oxalidaceae* but can be distinguished from them by the simple lvs and dehiscence of the capsule and in our spp. by the fewer stamens.

Fls 5-merous; petals longer than calyx; sepals entire at apex.   1. LINUM
Fls 4-merous; petals not longer than calyx; sepals toothed or lobed at
   apex; plant small and inconspicuous.   2. RADIOLA

1. LINUM L.

Herbs, rarely undershrubs. *Lvs sessile, usually narrow, parallel-veined or with only the midrib prominent. Fls cymose, 5-merous. Sepals imbricate, entire.* Petals clawed. *Stamens 5, united into a tube at the base, with tooth-like staminodes between. Ovary 5-celled, ovules 2 in each cell, separated by false septa.* Fr. dehiscing by 10 valves. *Seeds flat.*

About 200 spp., subtropical and temperate, mainly northern hemisphere.

A number of spp. including the annual *L. grandiflorum* Desf. with crimson fls and others with yellow fls are grown in gardens.

1 Petals blue, 8 mm. or more; lvs linear, alternate.   2
   Petals white, 6 mm. or less; lvs oblong or obovate, opposite.
                                                       3. catharticum
2 Sepals all acuminate, more than half as long as capsule in fr., inner
     glandular-ciliate.   1. bienne
   Inner sepals very obtuse, outer obtuse to acuminate, all entire, less than
     half as long as capsule in fr.   2. anglicum

1. L. bienne Mill.   'Pale Flax.'

L. angustifolium Huds.

Glabrous and subglaucous annual, biennial or perennial 30–60 cm. with several stems from the base. Stems ascending or suberect, rather rigid, often flexuous, ±branched. *Lvs alternate*, numerous, *linear*, acute, 1–2·5 cm. and mostly 3-nerved on the main stem, entire. Fls in a loose cyme; bracts like the lvs but smaller; pedicels slender, c. 1–2 cm. *Sepals all ovate, acuminate* or the inner apiculate; *inner with a scarious glandular-ciliate border*, outer entire; *more than ½ as long as the capsule in fr. Petals* 8–12 *mm., pale blue*, obovate, with a short claw. *Stigmas club-shaped.* Capsule c. 6 mm., globose-conic. Seeds c. 3 mm., ovate, shining, not beaked. Fl. 5–9. $2n=30$. Th. or Hp.

Native. Dry grassland especially near the sea. From Isle of Man, Lancashire and Yorks southwards, local and with a western tendency, commonest in the south-west; Galway and Meath southwards. 42, H17, S. Mediterranean region, extending further north only in W. Europe to W. and C. France; Madeira, Canaries.

*L. usitatissimum L.   'Cultivated Flax.'

Differs from *L. bienne* as follows: More robust, always annual. Stem usually solitary. Lvs and fls larger (petals c. 1·5 cm.). Inner sepals ciliate or not. Capsule larger, 1 cm. or more, seeds with a short obtuse beak. $2n=30, 32$. Th. Cultivated for the production of linen or linseed and sometimes persisting for a year or two or occurring as a casual but scarcely naturalized. Origin unknown, possibly derived from *L. bienne*.

2. L. anglicum Mill.   'Perennial Flax.'

L. alpinum L. ssp. *anglicum* (Mill.) F. W. Schultz; *L. perenne* auct. angl.

Glabrous glaucous perennial 30–60 cm. with ±numerous stems. Stems ascending, rather rigid, often curved, ± branched or nearly simple. *Lvs alternate*, very numerous, 1–2 cm. on the main stems, linear, acute, 1-nerved, entire. Fls in a loose cyme; bracts like the lvs but smaller; pedicels slender, c. 1–2 cm. Outer sepals oval or elliptic, obtuse, acute or apiculate; *inner broader oval or obovate, rounded at apex*, with an *entire* scarious *margin; all less than ½ as long as capsule in fr. Petals* 15–20 *mm., sky blue*, obovate with a short claw. *Stigmas capitate.* Capsule c. 7 mm., globose or oval. Seeds 4–5 mm., oblong-ovate, matt, obscurely beaked. Fl. 6–7. Hp.

Native. Calcareous grassland; E. England from N. Essex to Durham, extending west to Cambridge, Leicester and Westmorland, very local. 15. E. and C. France, Baden (1 loc.).

3. L. catharticum L.   'Purging Flax.'

Slender glabrous annual 5–25 cm. Stems usually solitary, sometimes several, simple (except in infl.), erect, wiry. *Lvs* 5–12 mm., *opposite*, distant, *oblong or obovate*, subobtuse, 1-nerved, entire. Fls many, in a loose often dichotomous cyme; bracts partly alternate, upper small, linear, lower passing into the lvs; pedicels slender, c. 5–10 mm. Sepals ovate-lanceolate, acuminate. *Petals* 4–6 *mm., white*, narrowly obovate. Stigmas capitate. Capsule c. 3 mm., globose. Fl. 6–9. Pollinated by various insects or selfed, homogamous. $2n=16$, c. 57. Th.

Native. Grassland, heaths, moors, rock-ledges and dunes, ascending to 2800 ft., especially common and characteristic of calcareous grassland but by no means confined to basic soils. 112, H40, S. Common throughout the British Is. Europe from Iceland and Finland to N. Portugal and C. Spain, Corsica, C. Italy, Macedonia and the Caucasus; Asia Minor, Persia.

2. RADIOLA Hill

Small annual herb, differing from *Linum* as follows: *fls* 4-*merous; sepals* (2–)3(–4)-*lobed or toothed at the apex; seeds ovoid.*

One sp.

1. R. linoides Roth   All-seed.

R. Millegrana Sm.

Very small delicate annual 1·5–8 cm. Stems filiform, sometimes simple but more frequently several times dichotomously branched so that the habit is bushy. Lvs opposite, 3 mm. or less, ovate-elliptic or elliptic, acute to subobtuse. Fls very numerous in a dichotomous cyme; pedicels short; bracts not differentiated. Sepals scarcely 1 mm. Petals about equalling sepals, white. Capsule globose, scarcely 1 mm. Fl. 7–8. Probably almost always self-pollinated. $2n=18$. Th.

Native. Damp bare sandy or peaty ground on grassland and heaths. 93, H20, S. Spread over the whole of Great Britain but local everywhere and absent from a number of counties particularly in the Midlands and in S. Scotland; in Ireland frequent near the west coast, but rare elsewhere. Europe (not Iceland, Finland, N. Russia); Mediterranean region; Madeira, Tenerife; temperate Asia; mountains of tropical Africa.

39. GERANIACEAE

Herbs or undershrubs (very rarely tree-like). Lvs usually alternate, lobed or compound, lflets not jointed to rhachis, usually stipulate. Infl. cymose or fls solitary. Fls hermaphrodite (very rarely dioecious), heterochlamydeous, actinomorphic or slightly zygomorphic, 5- (rarely 4- or 8-)merous, hypogynous. Sepals persistent, usually imbricate. Petals usually imbricate, occasionally 0. Stamens usually obdiplostemonous, rarely in 3 whorls, sometimes some sterile. Filaments usually ± connate at the base. Ovary 3–5- (rarely 2- or 8-)celled, usually lobed and with a long beak ending in the free lingulate stigmas; ovules axile, usually 2 in each cell, superposed, rarely solitary, anatropous, micropyle directed upwards, raphe ventral. Fr. a lobed capsule, the lobes usually 1-seeded, usually opening septicidally from base to apex, sometimes separating into 1-seeded portions. Endosperm scanty or 0, embryo usually curved, rarely straight.

Eleven genera and about 700 spp., mainly temperate and subtropical. The garden scarlet and ivy-leaved 'Geraniums' belong to the genus *Pelargonium* L'Hérit., differing from *Geranium* in the zygomorphic fls with a spur adnate to the pedicel.

Lvs palmate or palmately lobed; beak of carpel rolling upwards in dehiscence
and releasing the seeds.   1. GERANIUM
Lvs pinnate or pinnately lobed; beak of carpel twisting spirally in dehiscence
and remaining attached to seeds.   2. ERODIUM

1. GERANIUM L.

Herbs. Lvs palmate or palmately lobed. *Fls* solitary or in pairs, actinomorphic or nearly so, *without spur. Stamens* 10, *all fertile* (except *G. pusillum*). Seeds 1 in each carpel; *beak of carpel usually rolling upwards at dehiscence remaining attached by its apex, releasing the seeds*, beak sometimes curling away at the apex also but never coiling spirally.

About 300 spp. mainly temperate. Some are grown in gardens and about 15 have been found as escapes or casuals. The relative lengths of the lobes and entire portion of the lvs in the following account are measured down the main axis of the lobes.

1 Perennial; corolla large (petals usually 10 mm. or more).   2
  Annual; corolla small (petals less than 10 mm., sometimes to 12 mm. but then with a claw as long or longer than limb).   11

2 Petals with claw about as long as limb; stamens curving downwards; stock very stout, creeping; corolla pink.   **8. macrorrhizum**
  Petals without or with short claw; stamens spreading radially.   3

3 Lvs reniform or orbicular, the lobes ±cuneiform; petals 7–10 mm., deeply emarginate; root vertical, fusiform; rhizome obsolete.
                                                          9. pyrenaicum
  Lf-lobes not cuneiform; petals 10 mm. or more; rhizome present, oblique or horizontal.   4

4 Fls solitary; petals purplish-crimson, rarely pink or white; lobes of lvs narrow throughout their length.   **7. sanguineum**
  Fls mostly in pairs; petals not purplish-crimson; lobes of lvs broader about the middle.   5

5 Petals blackish-purple, apiculate at apex, spreading or slightly reflexed so that the fl. is ±flat. Sepals mucronate (the point less than ½ mm.).
                                                              6. phaeum
  Petals not blackish-purple, rounded or emarginate at apex, less spreading, fl. ±cup-shaped. Sepals aristate (the point c. 1 mm.).   6

6 Petals rounded at apex (sometimes emarginate in *G. endressii*), veins not darker than rest of petal; lvs 5–7-lobed.   7
  Petals emarginate, veins darker than rest of petal; lvs 3–5-lobed.   9

7 Fls pink; rhizome horizontal, creeping; plant almost eglandular.
                                                            3. endressii
  Fls violet or blue; rhizome oblique, not creeping; plant glandular above.   8

8 Pedicels recurved after fl.; lowest segments of the lf-lobes c. 4–5 times as long as entire portion.   **1. pratense**
  Pedicels erect after fl.; lowest segments of lf-lobes 1–2 times as long as entire portion.   **2. sylvaticum**

9 Fls pink with darker veins.   endressii × **versicolor**
  Fls white or pale lilac with lilac veins.   10

10 Stems with spreading hairs; lobes of lvs lobed or deeply dentate.
                                                            4. versicolor
   Stems nearly glabrous; lobes of lvs shallowly crenate-dentate.
                                                             5. nodosum

11 Sepals ±spreading; lvs dull- or grey-green.   12
   Sepals erect, somewhat connivent near the apex; lvs bright green.   16

12 Lvs divided nearly to base (at least ⅚) into linear lobes; lobes (except sometimes those of the lowest lvs) pinnatifid with linear segments; sepals aristate (the point 1–2 mm.).   13
   Lvs divided to ¾ or less, lobes widened above, trifid or 3-lobed; sepals mucronate (the point less than 0·5 mm.).   14

13 Pedicels 2 cm. or more; carpels usually glabrous.   **10. columbinum**
   Pedicels not more than 1·5 cm.; carpels pubescent.   **11. dissectum**

14   Petals entire; seeds pitted.   **12. rotundifolium**
       Petals emarginate; seeds smooth.   *15*

15   Perennial; petals 7–10 mm.; carpels pubescent; stamens all fertile.
                                                             9. pyrenaicum
       Annual; petals 3–7 mm.; carpels glabrous; stamens all fertile.   **13. molle**
       Annual; petals 2–4 mm.; carpels pubescent; 5 of the stamens without
          anthers.   **14. pusillum**

16   Lvs 5–7-lobed to halfway or less.   **15. lucidum**
       Lvs ternate or palmate.   *17*

17   Petals (8–)9–12 mm.; anthers orange to purple.   **16. robertianum**
       Petals 6–9 mm.; anthers yellow.   **17. purpureum**

Section 1. *Geranium.* Perennial. Rhizome present, oblique or horizontal.
Lvs deeply 3–7-lobed; the radical lvs on long petioles; cauline lvs smaller
on shorter petioles, decreasing in size upwards, often with fewer lobes.
Fls normally in pairs. Sepals spreading; petals ± spreading, claw short.
Stamens spreading.

1. G. pratense L.   'Meadow Cranesbill.'

Perennial herb 30–80 cm.; *rhizome stout, oblique. Stems* erect or ascending
with short deflexed hairs below, densely pilose and *glandular-hairy above* (like
the fl.-stalks and calyx). *Radical lvs* (Fig. 37A) polygonal in outline, blade
7–15 cm. diam., appressed hairy on both sides, long-petioled, *deeply 5–7-lobed*
(lobes c. 5 times as long as entire portion) the 2 basal lobes often contiguous;
lobes 7–10 mm. across at base, ovate-rhombic in outline, pinnately lobed;
secondary lobes with the lower margin usually concave, *the basal ones 4–5
times as long as the entire portion, themselves lobed*; ultimate lobes ± oblong,
acute, apiculate; cauline lvs smaller, decreasing upwards, on shorter petioles,
the uppermost 3-lobed, subsessile. Fls in pairs on axillary peduncles. *Pedicels
reflexed after fl.*, becoming erect in fr. Sepals ovate, aristate. *Petals* 15–18 *mm.*,
of a beautiful *violet-blue*, obovate, *apex rounded*; claw very short; corolla
± cup-shaped. Filaments of stamens much widened below into a broadly
ovate base. Carpels glandular-pilose, smooth; beak c. 2·5 cm. in fr. Seeds
minutely reticulate. Fl. 6–9. Pollinated mainly by Hymenoptera, protandrous.
$2n = 28^*$. Hs.

   Native. Meadows and roadsides, ascending to c. 1750 ft., widespread but
rather local in England, Wales and S. and C. Scotland; rare in N. Scotland
(to Orkney); Antrim (native); an escape elsewhere in Ireland. 104, H1.
Europe from Scandinavia (64° 12′ N.) and Finland to N. Spain, N. Apennines
and Thrace; N. and C. Asia to Ussuri region, Japan and the W. Himalaya;
naturalized in N. America.

2. G. sylvaticum L.   'Wood Cranesbill.'

Perennial herb 30–80 cm.; *rhizome stout, oblique. Stems* erect, with short
deflexed hairs below, pilose and *glandular-hairy above* (like the fl. stalks and
calyx). *Radical lvs* (Fig. 37B) polygonal in outline, blade 7–12 cm. diam.,
hairy on both sides, long-petioled, *deeply* (5–)7-*lobed* (lobes c. 5 times as long
as entire portion) the 2 basal lobes separated; lobes 8–15 mm. across at base,
obovate-rhombic in outline, pinnately lobed; secondary lobes ± triangular in

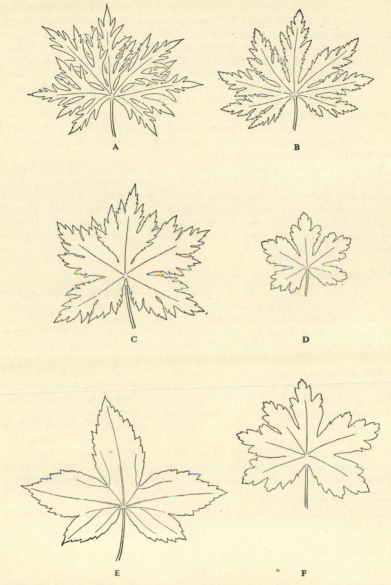

Fig. 37. Leaves of *Geranium*. A, *G. pratense*; B, *G. sylvaticum*; C, *G. endressii*;
D, *G. versicolor*; E, *G. nodosum*; F, *G. phaeum*. ×⅓.

outline (but with the lower edge much longer than the upper), the lower edge mostly straight, *the basal ones 1–2 times as long as the entire portion*, deeply and coarsely dentate but scarcely lobed; cauline lvs smaller, decreasing upwards, on shorter petioles, the uppermost subsessile. Fls in pairs on axillary peduncles, more numerous than in *G. pratense* and forming a loose cyme. *Pedicels erect after fl.* Sepals ovate-lanceolate, aristate. *Petals 12–18 mm., blue-violet* (usually smaller and more reddish than in *G. pratense*) obovate, *apex rounded*; claw very short; corolla ± cup-shaped. Filaments of stamens somewhat widened below into a lanceolate base. Carpels glandular-pilose, smooth; beak c. 2·5 cm. in fr. Seeds minutely reticulate. Fl. 6–7. Pollinated by various insects, protandrous; homogamous and unisexual fls also recorded. $2n = 28$. Hs.

Native. Meadows, hedgebanks, damp woods and mountain rock-ledges, ascending to c. 3300 ft.; from Caithness to Yorks, Derby, Gloucester, Glamorgan and Monmouth (but absent from most of Wales), widespread and often common but very rare in the south of its range; Antrim. 56, H1. Europe from Iceland and arctic Russia to the mountains of N. Spain, Italy, Albania and Thrace; Caucasus, Siberia to the Yenisei region; naturalized in N. America.

*3. G. endressii Gay

Perennial herb 30–80 cm.; *rhizome slender, horizontal,* long. Stems ± erect with ± spreading hairs, *not glandular* (but the pedicels and calyx somewhat glandular). Blades of radical lvs (Fig. 37 c) 5–8 cm. diam., polygonal in outline, hairy on both sides, long petioled, deeply 5-lobed (lobes c. 5 times as long as entire portion); lobes 10–15 mm. across at base, broadly ovate-rhombic in outline, pinnately lobed; secondary lobes with the lower margins straight or slightly concave, the basal ones 1–2 times as long as the entire portion, deeply dentate or lobed; cauline lvs smaller, decreasing upwards, the uppermost 3-lobed, subsessile. Fls in pairs on long peduncles. Pedicels erect after fl. Sepals ovate-lanceolate, aristate. *Petals c. 16 mm., deep pink, the veins not darker,* obovate, *apex rounded* or somewhat emarginate, tapered at base into a short claw; corolla ± cup-shaped. Carpels pubescent, somewhat glandular, smooth; beak c. 2 cm. in fr. Fl. 6–7. $2n = 28$. Hs.

Introduced. Grown in gardens, sometimes escaping and ± naturalized by roadsides, etc., in a number of places. 21. Native of the W. French Pyrenees.

G. endressii × versicolor.

Petals pale pink with darker veins, emarginate. Sometimes escaping independently of the parents.

*4. G. versicolor L.

G. striatum L.

Perennial herb 30–60 cm.; rhizome horizontal. *Stems* ± erect, *with spreading hairs,* not glandular. Blades of radical lvs (Fig. 37 D) 4–8 cm. diam., polygonal in outline, with scattered short hairs on both sides, deeply (3–)5-lobed (lobes 3–4 times as long as entire portion), basal sinus wide; *lobes* 10–15 mm. across at base, obovate in outline, *pinnately lobed*; secondary lobes with the lower margins ± straight, the basal ones less than half as long as the entire portion, coarsely dentate; cauline lvs often 5-lobed with the 2 basal lobes much shorter than the other 3. Fls in pairs on long peduncles. Pedicels erect after fl., with small sessile glands. Sepals lanceolate-aristate, sparingly hairy. *Petals 15–18 mm., white or pale lilac with violet veins*, obovate-cuneiform, *deeply emarginate*, tapered at the base into a short claw; corolla ± cup-shaped. Carpels sparingly pubescent, smooth; beak 2–3 cm. in fr. Fl. 5–9. Visited by bees. $2n = 28$. Hs.

Introduced. Grown in gardens and naturalized on hedgebanks, etc., in a number of places, mainly in S. England and Wales (common in Cornwall). 29, H6, S. Native of S. Italy, Sicily, Serbia and the Caucasus.

*5. G. nodosum L.

Perennial herb 20–50 cm. Differs from *G. versicolor* as follows: *Stems nearly glabrous, or with short reflexed and appressed hairs*, swollen at the nodes. Lvs (Fig. 37 E) more often 3-lobed, *the lobes crenate-dentate*, not lobed. Petals lilac with violet veins. Carpels with a transverse ridge at the apex. Fl. 5–9. $2n = 28$. Hs.

Introduced. Sometimes found as a garden escape and ± naturalized but apparently rare. It has been much confused with *G. endressii*. Native of the mountains of S. Europe from the S. Alps to the Pyrenees, C. Apennines and Serbia.

*6. G. phaeum L.   'Dusky Cranesbill.'

Perennial herb 30–60 cm.; rhizome stout, oblique. Stems erect, with spreading hairs, glandular above (like the fl.-stalks). Blades of radical lvs (Fig. 37 F) 7–12 cm., polygonal in outline, with numerous long hairs above, shortly pubescent beneath with a few long hairs mainly on the veins, deeply and often irregularly 5–7-lobed (deepest lobes usually c. twice as long as entire portion), often with a blackish blotch at the sinuses, basal sinus wide; lobes 10–20 mm. across at base, usually ± oblong in outline, irregularly lobed towards the apex; secondary lobes dentate; cauline lvs soon smaller, the uppermost sessile, very small. Fls in pairs on long axillary peduncles, ± numerous and forming a terminal infl. Pedicels erect or ascending after fl. *Sepals oblong-lanceolate, mucronate. Petals c. 1 cm., blackish-purple*, obovate-orbicular, *apiculate at apex*, claw very short or obsolete, *spreading at right angles so that the corolla is flat* or slightly reflexed. Carpels hairy, with several strong transverse ridges near the apex; beak c. 1·5 cm. in fr. Fl. 5–6. Pollinated by bees, protandrous. $2n = 28$. Hs.

Introduced. Grown in gardens and naturalized on hedgebanks, etc. in many places (very rarely in N. and W. Scotland and S. Ireland, more commonly elsewhere). 73, H11. Native of the mountains of C. and S. Europe from the Alps to the Pyrenees, Macedonia, Bessarabia and the Upper Dnieper region; naturalized in Belgium, Sweden, etc.

Section 2. *Sanguinea* R. Knuth. Perennial. Rhizome present, horizontal. Lvs deeply 5–7-lobed; radical 0; cauline scarcely decreasing in size upwards. Fls solitary. Sepals spreading. Petals ± spreading, with very short claw, forming a cup-shaped corolla. Stamens spreading.

7. G. sanguineum L.   'Bloody Cranesbill.'

Perennial herb 10–40 cm., often of rather bushy habit; rhizome stout, horizontal, creeping. Stems procumbent to erect, often branched from the base, and geniculate at the nodes, with numerous long, spreading or slightly deflexed hairs. Lf-blades (Fig. 38 A) 2–6 cm. diam., orbicular or somewhat polygonal in outline, with stiff appressed white hairs on both sides, petioled, deeply 5–7-lobed (the lobes c. 4–7 times as long as the entire portion); *lobes narrow*, not widened in the middle, 1·5–6 mm. across at the base, mostly trifid but some bifid; secondary lobes entire or themselves bifid or trifid; *ultimate lobes oblong or linear* or sometimes somewhat broadened towards the apex. Fls solitary or rarely a few in pairs, on long peduncles. Sepals oval or elliptic, aristate. *Petals* 12–18 *mm., bright purplish-crimson*, rarely pink or white, emarginate. Carpels sparingly hairy, smooth; beak c. 3 cm. in fr. Seeds minutely dotted. Fl. 7–8. Pollinated by various insects, protandrous but capable of self-pollination. $2n = 84*$. Hp.

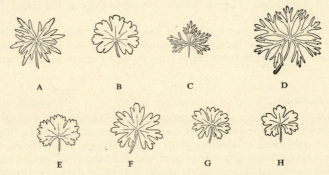

Fig. 38. Leaves of *Geranium*. A, *G. sanguineum*; B, *G. pyrenaicum*; C, *G. columbinum*; D, *G. dissectum*; E, *G. rotundifolium*; F, *G. molle*; G, *G. pusillum*; H, *G. lucidum*. $\times \frac{1}{3}$.

Native. Grassland, woods and among rocks on basic soils and on fixed dunes, etc., ascending to 1200 ft. 64, H11, S. Spread over the greater part of the British Is. from Caithness and the Outer Hebrides southwards, but local; absent from S.E. England, S. Ireland. Europe from S. Scandinavia to C. Spain and Portugal, Sicily and Greece, east to the Urals; Caucasus.

The form of maritime sands (var. *prostratum* (Cav.) Pers.) differs from the inland form in its procumbent (not ascending or erect) stems and its often broader lf-lobes; it occasionally occurs with pink fls (var. *lancastrense* (With.) Druce).

Section 3. *Unguiculata* Boiss. Perennial. Rhizome very thick, vertical, scaly. Radical lvs numerous, on long petioles, deeply 5–7-lobed; cauline smaller on shorter petioles, decreasing upwards. Fls in pairs. Sepals ± erect, petals with claw as long as limb. Stamens curved downwards, then upwards.

*8. G. macrorrhizum L.
Perennial 10–30 cm. Lf-blades 5–10 cm., orbicular in outline, shortly pubescent, deeply 5–7-lobed; lobes crenate-dentate or lobulate. Fls in pairs on rather short

peduncles, forming a corymbose cyme. Pedicels erect after fl. Sepals ovate or oval, aristate, reddish. *Petals c.* 15 *mm.* (including claw) *pink*; limb obovate-orbicular, rounded at apex. Carpels glabrous, with 2–3 ridges near the apex. Fl. 6–8. $2n=46$, 87–93. Hs.

Introduced. Escaped from gardens and established on village walls in a number of places in S. Devon. Native of S.E. France, Italy, Austria and the Balkans.

Section 4. *Batrachioides* Koch. Perennial. Root vertical, fusiform; rhizome obsolete. Radical lvs numerous, on long petioles, lobed to about half-way; cauline smaller, decreasing upwards on shorter petioles. Fls in pairs. Sepals spreading. Petals with short claw. Stamens spreading.

9. G. pyrenaicum Burm. f.   'Mountain Cranesbill.'

G. perenne Huds.

Perennial herb 25–60 cm. Stem erect, with numerous short glandular and numerous or few long spreading glandular hairs. Blades of radical lvs (Fig. 38 B) 5–8 cm., orbicular in outline, hairy on both sides, 5–9-lobed, the lobes 1–1½ times as long as the entire portion; lobes obovate-cuneiform in outline, irregularly 3-lobed at apex; margins of lower ⅔ quite entire; secondary lobes ¼–⅓ as long as the entire portion, themselves usually 3-lobed or -toothed at apex (additional lateral lobes sometimes occur); upper cauline lvs often 3-lobed. Fls in pairs, forming a loose cyme. Pedicels deflexed after fl., curving upwards near the apex so that the fr. is erect. Sepals ovate-oblong, mucronate. *Petals* 7–10 *mm.*, *purple*, less often purplish-white, obovate-cuneiform, *deeply emarginate. Carpels pubescent, smooth*; beak c. 1 cm. in fr. Seeds smooth. Fl. 6–8. Pollinated by various insects, protandrous but capable of self-pollination. $2n=26, 28*$. Hs.

Native? (first recorded 1762). Hedgebanks, waste places and field margins. 81, H 23, S. Common in S. and E. England, rarer westwards and northwards, extending north to E. Inverness and the Clyde Is.; scattered over Ireland (absent from the north-west). Europe from Sweden (naturalized in Norway) to the mountains of Spain and Portugal, Sicily and Greece; Asia Minor; N. Africa; naturalized in N. America.

Section 5. *Trygonium* Dum. Annual. Lvs palmately 5–9-lobed, dull or grey-green. Fls in pairs. Sepals spreading. Petals with short claw.

10. G. columbinum L.   'Long-stalked Cranesbill.'

Annual herb 10–60 cm., usually branched. Branches ascending or erect, with reflexed appressed hairs or subglabrous. Lower *lf-blades* (Fig. 38 c) 2–5 cm. diam., ± polygonal in outline, appressed hairy on both sides, on long petioles, *divided nearly or quite to the base* into 5–7 lobes; lobes 1–2 mm. across at base, pinnately lobed; secondary lobes c. 5 times as long as entire portion, linear or oblong-linear, usually entire but occasionally lobed; upper lvs smaller and sometimes less lobed, on shorter petioles. *Peduncles 2–12 cm.; pedicels 2–6 cm.*, spreading and often curving upwards after fl. *Sepals* ovate, *aristate* (awn c. 2 mm.), *with appressed hairs* mainly on the veins, *eglandular*. Petals 7–9 mm., purplish-pink, obovate-cuneiform, rounded or truncate and often apiculate or with a few small teeth at the apex. *Carpels glabrous or* *sparsely hairy* smooth; beak 1·5–2 cm. in fr. Seeds minutely pitted. Fl. 6–7. $2n=18^*$. Th.

Native. Open habitats in dry grassland and scrub, mainly on basic soils; throughout England but rather local; in Scotland local and mainly in the S.E., extending north to Moray and the S. Inner Hebrides; widespread in Ireland but local and mainly in the south. 81, H24, S. Europe (except the Arctic); Algeria, Tunisia; naturalized in N. America.

11. G. dissectum L.   'Cut-leaved Cranesbill.'

Annual herb 10–60 cm., usually branched and often of straggling habit. Branches ascending, densely clothed with reflexed but not appressed hairs. Lower lf-blades (Fig. 38 D) 2–7 cm. diam., orbicular or reniform in outline, ± hairy especially on the veins beneath, on long petioles, deeply divided into (5–)7 lobes; *lobes 5–10 times as long as entire portion*, 2–4 mm. across at the base, pinnately lobed (or those of the lowest lvs trifid); secondary lobes of the trifid lobes often oblong and only twice as long as entire portion but of the pinnate ones often linear and 7 times as long, often again lobed; upper lvs smaller, on shorter petioles, with narrower lobes so that they are similar to the corresponding lvs of *G. columbinum. Peduncles 0·5–2(–3) cm.; pedicels 0·5–1·5 cm.*, spreading or ascending after fl. *Sepals* ovate-lanceolate, *aristate* (awn 1–2 mm.), *pilose and glandular-hairy* (like the fl. stalks). Petals c. 5 mm., reddish-pink, obovate, emarginate at apex. *Carpels pubescent*, smooth; beak 7–12 mm. in fr. Seeds pitted. Fl. 5–8. Homogamous and protogynous, probably usually selfed, insect visitors few. $2n=22^*$. Th.

Native. Cultivated and waste ground, grassland, hedgebanks, etc., ascending to c. 1250 ft. 112, H40, S. Common throughout the British Is. Europe (except Iceland, only Crimea in European Russia); S.W. Asia to Persia and Pamir-Alai region; N. Africa, Macaronesia; naturalized in N. and S. America.

12. G. rotundifolium L.   'Round-leaved Cranesbill.'

Annual herb 10–40 cm., usually much-branched from the base. Branches erect or ascending, pubescent and glandular-hairy. Lower lf-blades (Fig. 38 E) 3–6 cm. diam., reniform in outline, pubescent on both sides, on long petioles, 5–7-*lobed to half-way or less*; lobes c. 1 cm. across at base, cuneiform, contiguous, 3-lobed at apex; secondary lobes short, obtuse, often 2- or 3-toothed or shortly lobed; upper lvs smaller, often rather more deeply and acutely lobed on shorter petioles. Fls numerous; peduncles 0·5–3 cm.; pedicels 0·5–1·5 cm., spreading or somewhat deflexed after fl., turning upwards at apex. *Sepals* ovate-oblong, mucronate (mucro not 0·5 mm.) pilose and glandular-hairy. *Petals* 5–7 mm., pink, obovate-cuneiform, *rounded or slightly retuse at apex.* Carpels pubescent, smooth; beak 10–15 mm. in fr. *Seeds pitted.* Fl. 6–7. Homogamous, probably usually selfed, insect visitors few. $2n=26$. Th.

Native. Hedgebanks, wall-tops, etc. 33, H6, S. From Cornwall and Kent to Carmarthen, Flint, Northampton and Lincoln, local; casual further north; Cork, Clare, Wexford, Kilkenny. Europe from Belgium, S.E. Netherlands, W. Germany and Upper Dnieper region southwards; S.W. and C. Asia to the W. Himalaya and the Balkhash region; N. Africa; Macaronesia; naturalized in N. and S. America.

13. G. molle L.   'Dove's-foot Cranesbill.'

Annual herb 10–40 cm., branched from the base, ± densely clothed with long soft white hairs, glandular above. Branches decumbent or ascending. Lower lf-blades (Fig. 38 F) 1–5 cm. diam., orbicular or reniform in outline, on long petioles, rather irregularly 5–9-lobed; *lobes from rather longer than to twice as long as the entire portion,* cuneiform, contiguous or with narrow sinuses, 2–5 mm. across at base, 3-lobed at apex; secondary lobes short, obtuse or subacute, often somewhat widened above, mostly entire; upper lvs smaller, on shorter petioles, more deeply and acutely lobed, with wider sinuses, the lobes often entire. Fls numerous; peduncles 0·5–3 cm.; pedicels 0·5–1·5 cm., ± ascending and often curved upwards after fl. *Sepals* ovate, *mucronulate* (or some without mucro), densely clothed with long white hairs and with some shorter glandular ones. *Petals* 3–6(–7) mm., bright rosy-purple, or sometimes whitish, *deeply emarginate. Carpels glabrous,* wrinkled or rarely (var. *aequale* Bab.) smooth, keeled; beak 5–8 mm. in fr. Seeds smooth. Fl. 4–9. Visited by various insects, ± homogamous, probably often selfed. $2n = 26^*$. Th.

Native. Dry grassland, dunes, waste places and cultivated ground, ascending to c. 1750 ft.; common throughout the British Is. 112, H 40, S. Var. *aequale* rare. Europe; S.W. Asia to the Himalaya; N. Africa; Macaronesia. Naturalized or escaped in N. and S. America New Zealand and Japan.

14. G. pusillum L.   'Small-flowered Cranesbill.'

Annual herb 10–40 cm. branched from the base, shortly and softly pubescent, glandular above. Branches decumbent or ascending. Lower lf-blades (Fig. 38 G) 1–4 cm. diam., ± orbicular in outline, on long petioles, 7–9-lobed; *lobes 2–3 times as long as entire portion,* somewhat widened above, not contiguous, 2–5 mm. across at base, trifid towards apex; secondary lobes oblong, mostly undivided, much longer than the entire portion; upper lvs smaller, on shorter petioles but divided like the lower. Fls rather numerous; peduncles 0·5–1(–3) cm.; pedicels 0·5–1·5 cm., spreading or ascending and often curved upwards after fl. *Sepals* ovate, *mucronulate,* clothed with long white hairs (but less densely than in *G. molle*) and shorter glands. Petals 2–4 mm., pale dingy lilac, deeply emarginate. Filaments of 5 of the stamens without anthers. *Carpels pubescent,* smooth; beak c. 5–6 mm. in fr. *Seeds smooth.* Fl. 6–9. Protogynous, probably usually selfed, insect visitors few. $2n = 26$.

Native. Cultivated and waste ground and open habitats in dry grassland. 87, H 10, S. Widespread and rather common in England and Wales; local and mostly only casual in Scotland (north to Orkney) and mainly in the east, rare in the north and west; in a few scattered places in Ireland, perhaps introduced. Europe from Scandinavia to Spain and Portugal, Corsica, C. Italy and Crete; Asia Minor to the Caucasus, Himalaya and Palestine. Naturalized in N. America.

Section 6. *Lucida* R. Knuth. Annual. Lvs palmately 5-lobed, bright green. Fls in pairs. Sepals erect. Petals with claw longer than limb.

15. G. lucidum L.   'Shining Cranesbill.'

Lfy annual 10–40 cm., usually branched from the base, bright green and

shining, often reddish tinged, fragile. Stems erect or ascending, nearly glabrous. Lower lf-blades (Fig. 38 H) 2–6 cm. diam., orbicular in outline, with
few short scattered hairs or nearly glabrous, on long petioles, *palmately
5-lobed to ½ or ¾*; lobes 5–12 mm., across at base, cuneiform, 3-lobed at apex;
secondary lobes shorter than entire portion, 2–3-lobed or toothed at apex,
teeth short, broad, obtuse, mucronate; upper lvs smaller on shorter petioles.
Pedicels spreading or ascending after fl., often curving upwards near apex.
Sepals oblong-ovate, aristate, subglabrous, erect and somewhat connivent
near apex, forming an angled calyx with transverse ridges towards the margins.
Petals 8–9 mm. (including claw), pink; limb 3–4 mm., obovate, rounded at
apex. Carpels 5-ridged at puberulent apex, the ridges continuing down the
back but less well-marked, the rest of the surface reticulate, glabrous; beak
c. 1 cm. in fr., the carpel not remaining attached at the apex. Seeds smooth.
Fl. 5–8. $2n=20^*$. Th.

Native. Shady rocks, walls and hedgebanks, calcicole, ascending to 2500 ft.
98, H 34, S. From Caithness southwards; locally common but uncommon
in many districts, rare in N. Scotland; widespread in Ireland but rare in the
north; Jersey. Europe (except the Arctic); S.W. Asia to the Himalaya;
N. Africa; Madeira.

Section 7. *Ruberta* Dum. Annual. Lvs palmate or ternate, bright green.
Fls in pairs. Sepals erect. Petals with claw longer than limb.

16. G. robertianum L.   Herb Robert.

Lfy annual or biennial 10–50 cm., usually branched from the base, bright or
dark green, usually ± reddish tinged, fragile, with a strong disagreeable smell.
Stems decumbent or ascending, ± clothed with dense hairs below, often nearly
glabrous above, but very variable in indumentum. *Lvs palmate*, lower mostly
with 5 lflets, polygonal in outline, with scattered appressed hairs on both
sides, on long petioles; lflets 1·5–6·5 cm., ± ovate in outline, deeply pinnatisect, the segments pinnately lobed; upper lvs mostly ternate, smaller, on
shorter petioles. Pedicels ascending after fl., mostly straight. Sepals oblong-
ovate, aristate, ± pilose and glandular, erect and somewhat connivent near
apex, without ridges; calyx not angled. *Petals* (8–)9–12 *mm.* (including claw),
bright pink or occasionally white; limb 4–6 mm., obovate-cuneiform, rounded
at apex. *Anthers orange or purple.* Carpels reticulately ridged (but often with
fewer longitudinal than transverse ridges), clothed with short stiff hairs, or
glabrous; beak 1–2 cm. in fr. Seeds smooth. Fl. 5–9. Visited by various
insects, protandrous, but self-pollination possible. $2n=64^*$ (all 3 sspp.).

Native. Woods, hedgebanks, among rocks and on shingle, etc. 112, H 40, S.
Common throughout the British Is., ascending to 2300 ft. Europe (except
Iceland) to 68° 12′ N. in Scandinavia; S. temperate Asia to Japan and the
Himalaya; N. Africa (rare and only in the mountains); Macaronesia;
naturalized in N. America; S. America.

Ssp. **robertianum**. Usually biennial with a rather dense mound-like rosette of
lvs in the first year. Fl. stems ascending. Stems and petioles dull deep red.
Lf-segments broad. Petals twice as long as sepals or more. Fr. ± hairy. The
common inland form.

Ssp. **maritimum** (Bab.) H. G. Baker. Usually biennial with an open procumbent rosette of lvs in the first year. Fl. stems procumbent or arcuately ascending. Lf-segments narrow, petioles shorter. Fls usually rather smaller. Fr. usually glabrous. Shingle beaches, less frequently cliffs or walls near the sea probably all round the British Is. W. Europe, Madeira.

Ssp. **celticum** Ostenf. Usually annual. Reddish only on the nodes and petiole-bases. Petioles rather short. Fls pale. Fr. hairy, large. Sunny limestone rocks. Glamorgan, Brecon, Carmarthen; Clare, ?Galway. Endemic?

17. G. purpureum Vill.

Differs from *G. robertianum* as follows: Often less reddish. Lf-segments often narrow. Pedicels and calyx more densely pilose. *Petals smaller, 6–9 mm. Anthers yellow. Carpels typically with much more pronounced and more numerous ridges.* Fl. 5–9. $2n = 32*$.

Ssp. **purpureum**. Erect or strongly ascending. Fr. as above. Native. Sunny, rocky places mainly near the sea. Cornwall to W. Sussex and Gloucester, very local; Carmarthen; Cork, Waterford; Channel Is. 8, H2, S.W. and S.W. Europe, Madeira.

Ssp. **forsteri** (Wilmott) H. G. Baker. Procumbent, stems ascending at tips. Fr. varying from that of ssp. *purpureum* to similar to that of *G. robertianum*. Native. Stabilized shingle beaches. S. Hants, W. Sussex; Guernsey. 2, S.

2. ERODIUM L'Hérit.

Herbs. Lvs usually pinnate or pinnately lobed. *Fls* solitary or in pairs, actinomorphic or slightly zygomorphic, *without spur. Stamens* 10, *those opposite the petals sterile and without anthers.* Seeds 1 in each carpel; *beak of carpel twisting spirally at maturity, seeds coming off enclosed in the carpel walls, the beaks remaining attached to them as awns.*

About 60 spp., mainly Mediterranean but extending over Europe and to C. Asia, a very few in S. Africa, N. America and Australia. Several have been found as casuals.

1 Lvs simple, lobed.   **1. maritimum**
   Lvs pinnate.   2

2 Stipules and bracts obtuse; filaments of fertile stamens with a tooth on
    each side at the base; plant densely glandular, smelling of musk; lflets
    incised-lobed, rarely to more than half-way to the midrib. **2. moschatum**
   Stipules and bracts acute to acuminate; filaments of fertile stamens enlarged at the base but not toothed; plant glandular or not, never smelling
    of musk; lflets pinnatifid, pinnatisect or 2-pinnatifid (*cicutarium* agg.).   3

3 Fls zygomorphic, mostly 12 mm. in diam. or more, bright rosy-purple
    often with a blackish spot at the base of the upper petals; fr. with a
    conspicuous pit, usually surrounded by a furrow, at the apex, beak
    22–40 mm.; peduncles mostly 3- or more-fld; indumentum variable.
                                                 3. cicutarium
   Fls scarcely zygomorphic, c. 7 mm. diam., pale pink, unspotted; fr. with
    a small pit at apex, furrow 0, beak 15–24 mm.; peduncles mostly 2–3-fld;
    plant always densely glandular.   **4. glutinosum**

1. E. maritimum (L.) L'Hérit.   'Sea Storksbill.'

Small annual with vertical tap-root. Stems decumbent, to 30 cm. but usually much less and often almost 0, usually several from the base, with rather stiff white hairs. *Lf-blades* 5–15 mm., ovate, *pinnately incised-lobed to about half-way*, the lobes coarsely toothed or lobulate, with white appressed hairs on both sides, on long petioles. Fls 1–2 on axillary peduncles about equalling the lf. Sepals c. 4 mm., oblong, mucronate, with white hairs. Petals pink, not longer than the calyx, often 0. Carpels hairy, the apical pit surrounded by a furrow, beak 1 cm. or less in fr. Fl. 5–9. Th.

Native. Fixed dunes and open habitats in short dry grassland, mainly near the sea. 37, H11, S. S. and W. coasts from Cornwall to Kent and Wigtown, local; Norfolk, Durham, Northumberland; inland in Worcester, Shropshire and Stafford; Surrey (perhaps extinct); coasts of Ireland from Down and Clare southwards, local; Channel Is. W. coast of France from the Somme to the Vendée. A ssp. or allied sp. in Corsica, Sardinia and Sicily.

2. E. moschatum (L.) L'Hérit.   'Musk Storksbill.'

Annual to 60 cm., usually robust, but sometimes dwarf and almost stemless, branching from base, covered ± all over with long white hairs and stalked glands, *smelling of musk*. Stems decumbent or ascending. *Lvs* 5–15 cm., *pinnate; lflets* ovate, *incised-dentate, the lower teeth rarely reaching half-way to the midrib; stipules* scarious, whitish, broad at the base, *obtuse*. Peduncles exceeding lvs, with 2–8 fls in a terminal umbel; bracts resembling the stipules. Sepals 4–7 mm., accrescent, ± oblong, mucronate. Petals rosy-purple, rather longer than the calyx. *Filaments of the fertile stamens with a tooth each side at the base*; pollen yellow. Carpels densely clothed with long hairs, the pit surrounded by a furrow, beak 2·5–4 cm. in fr. Fl. 5–7. $2n=20$. Th.

Native. Waste places, etc., mainly near the sea. Cornwall to Kent, S. Lancashire and the Isle of Man, local; also in several inland counties from Yorks southwards but scarcely native in them, perhaps only casual; round most of the Irish coast and in Offaly; Channel Is. 36, H20, S. Mediterranean region (south to Abyssinia) extending further north only in the west (to W. France); Macaronesia. Escaped in N. Europe, S. Africa, N. and S. America, Australia and New Zealand.

(3–4). E. cicutarium agg.   'Common Storksbill.'

Annual to 60 cm., very variable in habit, at first stemless, later usually branching from the base and often robust, with few to very many long white hairs and eglandular to densely glandular, *not smelling of musk*. Lvs 2–20 cm., *pinnate; lflets* ± ovate, *pinnatifid or pinnatisect or 2-pinnatifid; stipules* scarious, whitish, sometimes broad at base but less markedly so than in *E. moschatum, acute or acuminate*. Peduncles equalling or exceeding lvs with 1–9 fls in a terminal umbel; bracts resembling the stipules. Sepals 3–7 mm., accrescent. Petals rosy-purple sometimes with a blackish spot at the base of the two upper petals, or pink or white, longer than sepals. *Filaments of fertile stamens ± dilated below but not toothed*; pollen orange or more rarely yellow. Carpels ± hairy, pit surrounded by a furrow or not, beak 15–40 mm. in fr. Fl. 6–9.

Pollinated by various insects or selfed; protandrous to weakly protogynous, mechanism apparently varying with variety and locality. Th.

3. E. cicutarium (L.) L'Hérit.

Plant rather slender to very robust, eglandular or glandular. Lvs variable in size and division. *Peduncles* (2–)3–7(–9)-*fld*, 2–15 cm. in fr. *Fls* (8–)12–14(–17) *mm. diam., ± zygomorphic. Petals bright rosy-purple*, rarely white, *often with a blackish spot at the base of each of the two upper petals.* Fr. 5–6·5 mm., relatively dark brown, gradually tapered below; *pit at the apex conspicuous*, nearly equalling the diam. of the fr., *usually surrounded by a furrow* of varying distinctness; *beak 22–40 mm.* $2n=40*$.

Native. Dunes, dry grassland and arable fields and waste places, mainly on sandy soils, ascending to 1200 ft.; common near the sea all round the British Is.; inland, widespread in Great Britain but rather local, very rare in Ireland. 111, H25, S. Europe (except Iceland, etc.) to c. 70° N. in Norway; temperate Asia to N.W. Himalaya and Kamchatka; N. Africa to Abyssinia; Macaronesia; naturalized in N. and S. America.

The two following sspp. may be recognized. They are connected by intermediate forms:

Ssp. cicutarium

Ssp. *arvale* Andreas; incl. *E. pimpinellifolium* (With.) Sibth., *E. triviale* Jord., *E. Ballii* Jord., ?*E. dentatum* Dum. Usually sparsely pilose and eglandular or sparsely glandular, very robust at maturity. Peduncles mostly 4–7-fld. Fls 12–14(–17) mm. diam., often spotted. Fr. ± 6 mm., furrow conspicuous; beak 25–40 mm. Usually inland, occasionally near the sea.

Ssp. dunense Andreas

Incl. *E. Lebelii* Jord., ?*E. neglectum* Baker & Salmon. More densely glandular and less robust. Peduncles mostly 3–5-fld. Fls c. 12 mm. diam., rarely spotted. Fr. 5–6 mm., furrow inconspicuous or 0; beak 22–28 mm. Near the sea.

4. E. glutinosum Dum.

Plant more slender (to c. 25 cm.), greyish, *densely pilose and glandular, sticky and with sand grains adhering.* Lvs almost bipinnate, up to 3 cm. *Peduncles* 2–3(–5)-*fld*, up to 2·5(–3) cm. in fr. *Fls c. 7 mm. diam.*, scarcely zygomorphic. *Petals pale pink or white*, unspotted. Fr. 5 mm. or less, relatively light brown, more suddenly tapered below; *pit at apex small*, much less than diam. of fr., *furrow* 0; *beak 15–24 mm.* $2n=20*$.

Native. Dunes and sandy ground near the sea, local; British distribution incompletely known but occurring in a number of places, mainly on the W. coast. 8, H1, S. Coast of W. Europe from the Netherlands to Portugal.

*E. botrys (Cav.) Bertol.

Plant robust. Lvs pinnatifid. Fls 1–4, rather large (petals 10–15 mm.), dull bluish-purple. Fr. with 2 furrows round the pit; beak 70–110 mm. $2n=40$.

The most frequent of the casual spp. Native of the Mediterranean region.

40. OXALIDACEAE

Herbs often with fleshy stock, rarely undershrubs, very rarely trees. Lvs alternate, pinnate or palmate (sometimes 1-foliate), often showing sleep movements, with or without stipules; lflets entire, jointed to petiole or rhachis. Fls solitary or in cymes, hermaphrodite, often heterostylous, heterochlamydeous, actinomorphic, 5-merous, hypogynous. Sepals usually imbricate. Petals contorted. Stamens 10, weakly obdiplostemonous, rarely 15, connate at base. Ovary (3–)5-celled; ovules 1–many, axile, anatropous, micropyle directed upwards, raphe ventral; styles (3–)5; stigmas capitate. Fr. a loculicidal capsule, rarely a berry; seeds often with elastic testa; endosperm copious, fleshy; embryo straight.

Eight genera and nearly 900 spp., tropical and temperate.

Allied to Geraniaceae, differing in the lvs with separate lflets, the absence of a beak to the ovary, the separate styles, often numerous ovules, dehiscence of the capsule and presence of endosperm.

1. OXALIS L.

Lvs palmate; lflets 1–many (3 in our spp. except *O. tetraphylla*). *Stamens* 10. *Carpels 5, completely united. Fr. a capsule.*

About 800 spp., temperate and tropical, especially S. Africa, S. America and Mexico. Some are grown in gardens. Homogamous, some foreign spp. heterostylous.

1	Fls white, solitary in the axils of lvs borne on a slender creeping rhizome. Native plant of woods, etc.	**1. acetosella**
	Fls rarely white, in umbels or cymes or, if solitary, stems lfy, aerial and rhizome 0. Introduced plants of cultivated ground, walls, etc.	*2*
2	Fls yellow.	*3*
	Fls pink, lilac or red, rarely white.	*7*
3	Stem and leaves succulent; 3 sepals cordate.	**5. megalorrhiza**
	Not succulent; sepals not cordate.	*4*
4	Lvs all radical; bulbils present at base of plant.	**10. pes-caprae**
	Stems lfy; bulbils 0.	*5*
5	Stem creeping and rooting; lvs alternate; stipules oblong.	**2. corniculata**
	Stem ±erect, not rooting; lvs whorled or clustered; stipules 0 or inconspicuous.	*6*
6	Peduncles not reflexed after fl.; stipules 0.	**3. europaea**
	Peduncles reflexed after fl.; stipules narrow-oblong, adnate to petiole.	**4. dillenii**
7	Stem lfy, aerial; fls solitary.	**11. incarnata**
	Aerial stem 0 or a stout woody rhizome with lvs and fls from apex; fls cymose or umbellate.	*8*
8	Lflets 3.	*9*
	Lflets 4.	**9. tetraphylla**
9	Lvs and infl. from apex of stout woody rhizome.	**6. articulata**
	Stem 0; lvs and infls from bulbs or bulbils.	*10*

10 Lvs broadest near apex, lobes meeting each other at more than a right
   angle, nearly glabrous, dots 0.   **8. latifolia**
   Lflets broadest at about the middle, lobes meeting each other at a narrow
   angle, hairy, with dark dots near margin beneath.   **7. corymbosa**

1. O. acetosella L.   Wood-sorrel.

Perennial herb with *slender creeping rhizome* clothed with swollen, fleshy,
scale-like petiole bases, *bearing the lvs and fls on long stalks.* Lvs ternate;
petioles 5–15 cm.; lflets 1–2 cm., cuneiform-obcordate, broader than long,
entire except at apex, ciliate and with scattered appressed hairs, bright yellow-
green. *Fls solitary*; stalks about equalling petioles, with two small bracteoles
in the middle. Sepals oblong-lanceolate, subobtuse. *Petals* 10–16 mm., *white*
veined with lilac, rarely lilac or purple, obovate. Capsule 3–4 mm., ovoid-
globose, 5-angled, glabrous. Fl. 4–5(–8). Insect visitors apparently few;
cleistogamous fls occur abundantly in summer and produce most of the seed.
$2n = 22$. Hr. or Chh.
   Native. Woods, hedgebanks, shady rocks, etc., sometimes epiphytic, frequently growing in humus and very tolerant of shade, avoiding very heavy
and wet soils; ascending to 4000 ft.; a characteristic dominant of some phases
of plateau beechwoods; sometimes also dominant in sessile oakwoods. 112,
H40, S. Common and often abundant throughout the British Is. Europe
from Iceland and Scandinavia to the mountains of C. Spain, Italy and
Greece; N. and C. Asia to Sakhalin and Japan.

*2. O. corniculata L. var. *corniculata*   'Procumbent Yellow Sorrel.'

Annual or perennial herb 5–15 cm., rhizome and stolons 0. *Stems* many,
weak, *procumbent*, sometimes ascending at the ends, rooting at the nodes
below, ± densely clothed with long spreading hairs, lfy. *Lvs alternate*, oblong,
ternate, petioles 1–8 cm., *stipules c. 2 mm.*, oblong, adnate to petiole; *lflets*
8–15 mm., cuneiform-obcordate, broader than long, *bilobed at apex to* ¼ *or* ⅓
with a narrow sinus, otherwise entire, ± pilose on the margins and under-
surface, the lateral sometimes smaller than the terminal. *Fls 1–6 in umbels
on axillary peduncles*, equalling or exceeding lvs. *Bracteoles linear-lanceolate*,
mostly 2 mm. or more. *Pedicels* c. 1 cm., clothed with appressed hairs or subglabrous, ± *deflexed after fl*. Sepals ± lanceolate, acute. *Petals* 8–10 mm.,
yellow, narrowly cuneiform. Capsule 12–15 mm., cylindric, pubescent. Fl.
6–9. Probably usually self-pollinated. $2n = 24*$. Th. or Chh.
   Introduced. Waste places and as a garden weed; first recorded c. 1585;
rather common in S.W. England, recorded from a number of other English
and Welsh counties, a few Scottish and Irish ones. 64, H3, S. Tropical and
warmer temperate regions of the whole world (country of origin not definitely
known).

Var. *microphylla* Hook. f.

Much smaller in all its parts forming a low compact mat. Petioles 6–13 mm.,
lflets 2–5 mm. Peduncles 1-fld. Capsules short, 5–7 mm., abruptly narrowed
at top.
   A garden weed. 14, H2, S. New Zealand and Tasmania. Should probably
be regarded as a ssp.

*3. O. europaea Jord.   'Upright Yellow Sorrel.'

O. stricta auct.

Perennial herb 5–40 cm., with slender underground stolons, but probably often annual. *Stems* solitary, *erect*, simple or sparingly branched, glabrous or hairy, lfy. *Lvs whorled or clustered*, ternate; petioles 1–12 cm., *stipules* 0; *lflets* 8–20 mm., cuneiform-obcordate, broader than long, *bilobed at apex to c. ⅙*, with a broad or narrow sinus, otherwise entire, glabrous or appressed hairy. *Fls 2–5 in short cymes on axillary peduncles*, equalling or exceeding lvs. *Bracteoles subulate*, c. 1 mm. *Pedicels* c. 1 cm., glabrous or hairy, *erect to spreading after fl.* Sepals lanceolate, acute. *Petals* 4–10 mm., *yellow*, narrowly cuneiform. Capsule 8–12 mm., cylindric, glabrous or sparsely pubescent. Fl. 6–9. Probably usually self-pollinated. $2n=24$. Hp or Th.

Introduced. Waste places and as a garden, rarely arable, weed. Recorded from a number of counties, mainly in S. and C. England, 49, H4. Europe from Scandinavia and Finland to N. Spain, Italy, Hungary and Poland; N. China, Korea (but not native in the Old World); N. America from Nova Scotia to Colorado, Texas and Florida; escaped in Tropical Africa and New Zealand.

*4. O. dillenii Jacq.

O. stricta L., nom. ambig.

Similar to *O. europaea* but without stolons. *Stipules narrow-oblong*, adnate to petiole. Fls (1–)2–3(–4) *in umbels*. *Pedicels* often longer, *deflexed in fr.* Capsules 15–25 mm. Fl. 7–10. Th. or Chh.

Introduced. Arable fields near Pulborough, W. Sussex and in Sark. 1, S. Eastern N. America.

*O. valdiviensis Barnéoud, a glabrous annual with a short unbranched stem and yellow fls in long forked cymes, occurs as a casual or as a weed in gardens where it has been grown. $2n=18$. Chile.

*5. O. megalorrhiza Jacq. (*O. carnosa* auct.)

Succulent glabrous herb to 20 cm. Stems 1–2 cm. diam., fleshy, with lvs and fls at apices. Lflets up to 2 cm., broad, obcordate, fleshy, covered below with crystalline papillae. Fls 1–3(–5), in an umbel. *Three outer sepals cordate*. Petals c. 15 mm., bright *yellow*. Capsule 7 mm., oblong. Fl. 5–9. Chh.

Introduced. Walls and banks in Scilly Is. 1. Chile.

*6. O. articulata Savigny

O. floribunda Lehm.

Tufted perennial herb with *short thick fleshy rhizomes* up to 2 cm. diam. with lvs and fls at apices. Lflets 1–4 cm., obcordate, ±pubescent, with raised orange spots, especially near the margins beneath. Fls many, in an umbel. *Petals* 10–15 mm., *deep bright pink*, rarely pale or white. Capsule 10 mm. Fl. 5–10. Hr.

Introduced. Much grown in gardens and sometimes naturalized on waste ground, roadsides etc., occasionally among native vegetation, mainly in S.W. England. 10, H3, S. Native of eastern S. America from S. Brazil southwards.

*7. O. corymbosa DC.

Perennial herb with lvs and fls from a *bulb developing later into a mass of sessile bulbils*. Petioles 5–15 cm.; lflets 2·5–5 cm. wide, orbicular-obcordate, *lobes rounded with a deep narrow indentation, sparsely hairy, with small raised reddish spots near the margin beneath.* Fls in short cymes on peduncles up to 30 cm. *Petals* 15–20 mm., purplish-*pink*. Fr. unknown. Fl. 7–9. Gb.

Introduced. Naturalized in gardens in England, especially near London. 12, S. S. America.

*8. O. latifolia Kunth

Habit of *O. corymbosa* but *bulbils on stolons up to 2 cm.* Petioles 10–30 cm.; *lflets* 2–4·5 cm., wider, usually much broader than long, very broadly triangular *with a wide shallow indentation*, often with straight sides meeting at an obtuse angle, ciliate but otherwise *glabrous, raised spots* 0. Fls in umbels. *Petals* 8–13 mm., *pink*. Fr. unknown. Fl. 5–9. Gb.

Introduced. A horticultural weed, mainly in S.W. England and Jersey. 6, S. S. America and West Indies.

*9. O. tetraphylla Cav.

Similar to *O. latifolia* but stolons longer; *lflets* 4, thinly pubescent; *petals* c. 20 mm., *bright rose-red*. Fl. 5–9. Gb.

Introduced. Naturalized in fields in Jersey. S. Mexico.

*10. O. pes-caprae L.　　　　　　　　　　　　　　　　Bermuda Buttercup.

O. cernua Thunb.

Perennial herb with a *bulb producing* an underground stem bearing bulbils below and at ground level and *lvs and infls at ground level.* Petioles up to 20 cm.; lflets up to 2 cm. obcordate, sparsely hairy below. Fls in umbels on peduncles 10–30 cm. *Petals* 20–25 mm., *bright lemon-yellow*. Fr. not produced in Britain. Fl. 3–6. $2n=35$. Gb.

Introduced. A weed in bulb-fields in Scilly Is., occasionally in S. Devon and Channel Is. 2, S. S. Africa, naturalized in many of the warmer parts of the world.

*11. O. incarnata L.

Perennial, nearly glabrous herb 10–20 cm. with a bulb producing an *aerial branching stem bearing bulbils in the lf-axils.* Lvs opposite; petiole 2–6 cm., lflets 5–15 mm. obcordate. *Fls solitary*; peduncles 3–7 cm. *Petals* 12–20 mm., *pale lilac* with darker veins. Fr. not produced in Britain. Fl. 5–7. Gb.

Introduced. Naturalized on walls and hedgebanks in S.W. England and the Channel Is., occasionally escaped elsewhere. 8, S. S. Africa.

41. BALSAMINACEAE

Herbs with translucent stems. Lvs alternate, opposite or in whorls of 3, simple; stipules 0 or represented by glands. Fls solitary or in racemes, hermaphrodite, heterochlamydeous, strongly zygomorphic, hypogynous. Sepals 5 or 3, often petaloid; the lowest (the posterior sepal is at the bottom of the fl. because of the twisting of the pedicel) large and spurred; the lateral small. Petals 5, the upper large, the 4 lower usually united in pairs on each side of the fl. Stamens 5, alternate with the petals; filaments broad, short, connate above; anthers connate round the ovary. Ovary 5-celled; ovules axile, numerous (rarely 2–3) in 1 row in each cell, anatropous, the raphe in the reversed position as compared with allied families. Fr. a loculicidal capsule with the valves dehiscing elastically and coiling, very rarely a berry. Endosperm 0; embryo straight.

Two genera, one monotypic (S.E. Asia). Of different aspect from allied families and with very distinctive fls.

1. IMPATIENS L.

Sepals usually 3. Lower petals usually united in pairs, so petals apparently 3. *Ovules numerous. Fr. an elastically opening capsule.*

Over 220 spp., mainly tropical Asia and Africa, a very few in the north temperate zone and S. Africa.

1	Lvs alternate; fls yellow or orange.	2
	Lvs opposite or whorled; fls purplish-pink, rarely white.	**4. glandulifera**
2	Fls large, 2 cm. or more; lvs with 15 or fewer teeth on each side.	3
	Fls small, 1·5 cm. or less, pale yellow; lvs with 20 or more teeth on each side.	**3. parviflora**
3	Fls bright yellow, spur gradually contracted and curved; teeth of lvs c. 2–3 mm. deep, usually 10–15 on each side.	**1. noli-tangere**
	Fls orange, spur suddenly contracted and bent; teeth of lvs 1–2 mm. deep, rarely more than 10 on each side.	**2. capensis**

1. I. noli-tangere L.   Touch-me-not.

Erect glabrous annual 20–60 cm., stem swollen at nodes. *Lf-blades* 5–12 cm., alternate, ovate-oblong, subobtuse or subacute, cuneate at base, petiolate, coarsely crenate-serrate *with c.* 10–15 *teeth on each side* and a few ciliations at the base; *teeth* very obtuse, mucronate, *c. 2–3 mm. deep. Fls* in few-fld axillary racemes usually consisting of both normal and cleistogamous ones, *c.* 3·5 *cm.,* bright yellow with small brown spots. *Lower sepal* conical below, *tapered into a slender spur gradually curving* downwards or upwards, sometimes recurved, c. 2·5 cm. (including spur). Lateral lobes of corolla clawed, bilobed. Fl. 7–9. Pollinated by bees (normal fls). $2n = 20$. Th.

Native. By streams and in wet ground in woods, very local, probably native in the Lake District and N. Wales, and perhaps W. Yorks and Lancashire; recorded from many other counties but introduced and mostly only casual. c. 8. Europe from Scandinavia to C. France, the Apennines and Macedonia; temperate Asia to the Pacific.

*2. I. capensis Meerburgh   Orange Balsam.

I. biflora Walt.; *I. fulva* Nutt.

Differs from *I. noli-tangere* as follows: Lvs smaller, 3–8 cm., ovate-oblong or elliptic-oblong; *teeth fewer, usually* 10 *or less,* rather less obtuse, *shallower,* 1–2 *mm. deep;* margins often somewhat concave at base. *Fls* often all cleistogamous, the normal ones 2–3 *cm.* overall, *orange,* strongly blotched or spotted with reddish-brown inside. *Lower sepal* rather *suddenly contracted into a spur which is suddenly bent or curved* upwards or downwards at a right angle, or recurved, c. 2 cm. (including spur). Fl. 6–8. $2n = 20$. Th.

Introduced. Completely naturalized on river banks, etc., especially the Thames and its tributaries, extending to Leicester, Stafford, Glamorgan and Dorset; probably still spreading. 21. Native of eastern N. America from Newfoundland to Saskatchewan, Florida and Nebraska.

*3. I. parviflora DC.   'Small Balsam.'

Erect glabrous annual herb 30–100 cm.; stem simple or branched. *Lf-blades* 5–15 cm., alternate, ovate, acuminate, tapered at the base into a winged petiole, serrate or crenate-serrate *with numerous* (20 or more) *teeth* on each side and a few ciliations at the base; teeth directed forwards, mostly acute. *Fls* small, 4–10 in axillary racemes which often become elongated in fr. 5–15 *mm. overall, pale yellow*, not spotted. Lower sepal ± conical; spur straight or nearly so, very variable in length. Fl. 7–11. Visited mainly by hover-flies but probably also selfed. $2n = 20$, 24, 26 (?some errors). Th.

Introduced. Completely naturalized in woods and waste shady places, local; mainly in S. and E. England but found also in N. and W. England and Wales; rare in Scotland; probably still increasing. 36. Native of Siberia and Turkistan; naturalized in N. and C. Europe.

*4. I. glandulifera Royle   Policeman's Helmet.

I. Roylei Walp.

Erect glabrous robust annual herb 1–2 m.; stem reddish, stout. *Lvs opposite or in whorls of* 3, blades 6–15 cm., lanceolate or elliptic, acuminate, rounded or cuneate at base, stalked, sharply serrate with numerous teeth on each side. *Fls* large (2·5–4 cm.), *purplish-pink*, rarely white, 5–10 together in racemes on long peduncles in the axils of all the lvs of a few of the upper whorls, which are much smaller than the lower ones. Lower sepal very broad, obtuse, with a short tail-like spur. Fl. 7–10. Pollinated by humble-bees, cleistogamous fls not known to occur. $2n = 18$, 20, 26 (?some errors). Th.

Introduced. Completely naturalized on river banks and in waste places; locally common in N. and W. England and Wales, less common in S.E. England, Scotland and Ireland but still increasing. 48, H12. Native of the Himalaya.

42. SIMAROUBACEAE

Trees or shrubs, usually with bitter bark. Lvs usually alternate and pinnate. Fls regular, in panicles or spikes, 3–7-merous. Stamens usually obdiplostemonous. Ovary superior, surrounded by prominent disk. Carpels 2–5 often free below. Endosperm 0 or scanty.

About 28 genera and 150 spp., mainly tropical.

*Ailanthus altissima (Mill.) Swingle, Tree of Heaven. Tree to 20 m.; bark with pale stripes. Lvs 45–60 cm., pinnate; lflets 7–12 cm., ovate-lanceolate with 2–4 coarse teeth near base, ± glabrous. Infl. a large terminal panicle. Fls 7–8 mm. diam., greenish. Fr. a group of usually 3 reddish-brown samaras, 3–4 cm. Fl. 6–7.

Much planted in gardens and parks, rarely naturalized. China.

43. ACERACEAE

Trees or shrubs. Lvs opposite, simple (often palmately lobed) or pinnate, exstipulate. Infl. of racemose type. Fls regular, apparently andromonoecious (but functionally monoecious), androdioecious or dioecious, perigynous or hypogynous; disk usually present. Sepals 4 or 5. Petals equal in number to sepals or 0. Stamens 4–10, usually 8 in two whorls even in fls with a 5-merous perianth, inserted on or inside the disk. Ovary 2-celled and 2-lobed, compressed with the septum along the short axis; ovules orthotropous or

anatropous, 2 in each cell, axile; styles 2, free or united below. Fr. a schizocarp splitting into 2 samaras. Endosperm 0; embryo with flat, folded or
rolled cotyledons and long radicle.

Two genera, the other consisting of two Chinese spp.

1. ACER L.

Samaras winged on one side only.

About 115 spp. in the north temperate zone. Besides the following, a
number of others are ± frequently planted. Two of the commonest are
A. palmatum Thunb. (Japanese Maple), a shrub with 5–11-lobed lvs with the
lobes acuminate and doubly serrate, and *A. negundo* L. (Box Elder), from
N. America, with pinnate lvs. Both are much grown in small gardens, the
former often as a purple-lvd form, the latter as a variegated-lvd one.

The female fls in our spp. have apparently well-developed anthers but they
do not open. The four following spp. (and the two mentioned above) each
belong to a different section.

1 Twigs glabrous; lvs large (5–16 cm.) glabrous except in the axils of the veins
      beneath, lobes acute or acuminate; wings of fr. spreading at an acute
      angle.   2
   Twigs pubescent; lvs small (mostly 4–7 cm.), pubescent beneath, lobes
      obtuse; wings of fr. spreading horizontally.   **3. campestre**
2 Fls in pendulous panicles; lf-lobes irregularly crenate-serrate, juice not
      milky.   **1. pseudoplatanus**
   Fls in erect corymbs; lf-lobes sinuate-dentate with few large acuminate
      teeth; juice milky.   **2. platanoides**

*1. A. pseudoplatanus L.   Sycamore.

Large deciduous tree to 30 m., with broad spreading crown. Bark grey,
smooth for a long time, finally scaling. Buds rather large (to 12 × 8 mm.)
ovoid, acute; scales green with blackish margins, pubescent near the tip.
Twigs soon light brown, *glabrous*. *Lf-blades* 7–16 *cm.* and about as broad,
5-lobed to about half-way, cordate at base, dark green and glabrous above
(slightly pubescent on veins at first), *glaucescent* and soon *glabrous* except in
the axils of the veins *beneath; lobes* ± ovate, *acute, coarsely and irregularly
crenate-serrate*, often lobulate; *petiole* 10–20 cm., often red, *without milky
juice*. Fls 60–100 *in a narrow pendulous panicle* 5–20 cm., terminal on a short
lfy branch, appearing with or after the lvs. Fls c. 6 mm. diam., yellowish-
green, monoecious; sepals and petals 5; *stamens* 8, hypogynous, *inserted within
the disk*; ovary tomentose. *Fr.* glabrous, *wings spreading at an acute angle* or
incurved at apices; each samara 3·5–5 cm. Fl. 4–6, pollinated mainly by bees.
Germ. early spring. $2n = 52$. MM.

Introduced in the fifteenth or sixteenth century. Now common in woods,
hedges, plantations, etc., throughout the British Is., preferring deep, moist,
well-drained rich soils, but growing on all but very poor soils, very tolerant
of exposure and salt spray, ascending to 1600 ft.; often planted but completely
and commonly naturalized. Native of the mountains of C. and S. Europe
from N.E. France, S. Germany and Upper Dnieper region to N. Spain,
Corsica, Sicily, C. Greece; Asia Minor, Caucasus; often naturalized in the
lowlands of C. and N. Europe.

*2. A. platanoides L. Norway Maple.

Deciduous tree to 30 m. with broad spreading crown. Bark deep grey, with numerous short, shallow fissures, not scaling. Buds to c. 10 mm., ovoid; scales greenish at base, reddish-brown above. *Twigs* brownish, *glabrous. Lf-blades* 5–15 *cm.* and about as broad, 5(–7)-lobed to about ⅓, cordate at base, *bright green* and somewhat shining *on both surfaces, glabrous* except in the axils of the veins *beneath; lobes* triangular or parallel-sided below, *acuminate, sinuate-dentate with few large acuminate teeth; petiole* 5–20 cm., often red, *with milky juice. Fls many in a broad erect corymbose panicle,* terminal on a short lfy branch, opening before the lvs have expanded. Fls c. 8 mm. diam., bright greenish-yellow, monoecious; sepals and petals 5; *stamens* 8, perigynous, *inserted about the middle of the disk;* ovary glabrous. *Wings of fr. spreading at an acute angle,* divergent; each samara 3·5–5 cm. Fl. 4–5 (c. 3 weeks earlier than *A. pseudoplatanus),* pollinated mainly by bees. Fr. 9–10. 2n=26. MM.

Introduced. In similar situations to *A. pseudoplatanus* and liking similar conditions but much less commonly planted and so less often met with, though readily becoming naturalized. Mountains of Europe from S. Scandinavia and the Urals to N. Spain, N. Italy and Greece; Asia Minor, Caucasus, N. Persia.

3. A. campestre L. Common Maple.

Small deciduous tree 9–15(–25) m. with a ± ovoid crown, or a shrub. Bark light grey with shallow fissures, finally flaking off. Buds c. 5 mm., brown. *Twigs* brown, *pubescent,* often developing corky wings at 4 or 5 years. *Lf-blades* 4–7(–10) *cm.,* as broad or rather broader, (3–)5-lobed to about half-way, cordate at base, dull green above, pubescent at first, finally subglabrous, paler *beneath* and persistently *pubescent* especially on the veins; lobes ovate or obovate, *obtuse* (or on some lvs subacute), *entire or with a very few broad shallow rounded teeth or shallowly trilobed;* petiole 2–8 cm., pubescent, *juice milky. Fls* rather few (c. 10–20) *in an erect corymbose panicle,* terminal on a short lfy branch, appearing with the lvs. Fls c. 6 mm. diam., pale green, monoecious; sepals and petals 5; *stamens* 8, perigynous, *inserted on the disk near the inner edge. Fr.* pubescent or sometimes glabrous; *wings spreading horizontally,* collinear; each samara 2–4 cm. Fl. 5–6; visited by small insects. Fr. 9–10. Germ. late spring. 2n=26. MM.

Native. Woods, hedges and old scrub, mainly on basic soils and only abundant on these, frequently coppiced. 68, S. From Westmorland and Durham southwards; common in S., E. and C. England, becoming rarer westwards and northwards; rare in Scotland and probably introduced; rare and not native in Ireland. Europe from S. Sweden (rare), Denmark, Poland and Upper Volga region to C. Spain, Corsica, Sicily and N. Greece; Asia Minor and Caucasus to N. Persia and Turkistan; Algeria (very rare).

*A. saccharinum L. (*A. dasycarpum* Ehrh.) Silver Maple.

Tree. Lf-blades 8–14 cm., deeply 5-lobed, green above, silvery beneath; lobes deeply and doubly serrate. Fls before the lvs in lfless fascicles from axillary buds; petals 0.

Occasionally planted in parks, etc. as an isolated tree among native vegetation, but not naturalized. Native of eastern N. America.

44. STAPHYLEACEAE

Trees or shrubs. Lvs opposite or alternate, pinnate or ternate, with stipules and stipels. Fls in panicles, regular, usually hermaphrodite, disk present. Sepals and petals 5, imbricate. Stamens 5, alternating with petals, inserted on the edge of the disk, anthers introrse. Ovary 2–3-celled with 1–many axile anatropous ovules; styles free or united below; stigmas capitate. Fr. a capsule. Embryo large, straight; endosperm scanty.

Five genera and about 25 spp., north temperate zone and south to the E. Indies and Peru.

1. STAPHYLEA L.

Deciduous shrubs. *Lvs opposite.* Infl. terminal. *Ovules numerous. Fr. an inflated, membranous capsule. Seeds without aril.*

Eleven spp., north temperate.

*1. S. pinnata L.   Bladder-nut.

Shrub to 5 m. Lvs with 5–7 lflets; lflets 5–10 cm., ovate-oblong, acuminate, serrulate, glabrous. Panicles 5–12 cm., pendulous. Fls c. 1 cm., petals and sepals whitish, of about equal length, erect. Fr. 2·5–3 cm., subglobose, much inflated, 2–3-lobed. Fl. 5–6. Visited by Diptera. $2n=26$. N or M.

Introduced. Sometimes planted in shrubberies, etc., and perhaps ± naturalized in a few places. Native of C. and S. Europe.

45. HIPPOCASTANACEAE

Trees or shrubs. Lvs opposite, palmate, exstipulate. Fls in large terminal panicles, composed of scorpioid cymes, zygomorphic, andromonoecious; disk present. Sepals 5; petals 5 or 4, clawed. Stamens 5–8, hypogynous; anthers introrse. Ovary superior, 3-celled; ovules 2 in each cell, axile, anatropous; style simple, long; stigma simple. Fr. a large leathery, usually 1-seeded loculicidal capsule, opening by three valves. Seed large, nut-like, shining; hilum large; embryo curved; endosperm 0.

Two genera, the other consisting of 2 spp. from Mexico to northern S. America.

1. AESCULUS L.

Deciduous trees. Buds large. *Sepals united into a tubular or campanulate calyx.*

About 25 spp., E. Asia, S.E. Europe and N. America. Several spp. and hybrids are sometimes planted.

*1. A. hippocastanum L.   Horse-chestnut.

Large deciduous tree to 25 m. with broad crown. Bark dark greyish-brown, finally cast in scales. Buds to 3·5 cm., ovoid, deep red-brown, very sticky. Twigs pale grey or brown, glabrous. Lvs palmate with 5–7 lflets; lflets 8–20 cm., obovate-lanceolate, acuminate, long cuneate at base, irregularly crenate-serrate, dark green and glabrous above, somewhat woolly tomentose beneath when young, often glabrous at maturity; petiole long. Panicle 20–30 cm. Fls c. 2 cm. across, andromonoecious. Petals 4, white with basal spots which are at first yellow then pink. Stamens long, arched downwards; pollen red. Fr. large, c. 6 cm., subglobose, prickly; seeds 1 or 2. Fl. 5–6. Fr. 8–9. Pollinated by humble bees. Infl. protandrous, hermaphrodite fls protogynous. $2n=40$. MM.

Introduced. Commonly planted for ornament and often self-sown. Native of Albania and Greece.

*A. carnea Hayne

Usually smaller tree than *A. hippocastanum*. Lflets usually 5, darker and firmer. Panicles 12–20 cm. Fls pink or red. Fr. 3–4 cm. with few prickles. $2n=80$.

Often planted. Of garden origin, by hybridization between *A. hippocastanum* and *A. pavia* L. (from eastern N. America) and chromosome doubling, so that it now behaves as a species.

46. AQUIFOLIACEAE

Trees or shrubs. Lvs alternate, simple, exstipulate. Fls in few-fld axillary cymes (rarely solitary), regular, hermaphrodite or dioecious, 4–5-merous, hypogynous, disk 0. Sepals and petals imbricate; sepals small; petals free or united at base. Stamens equalling in number and alternate with corolla-lobes (rarely numerous), free or united with the extreme base of the petals. Ovary (2–)4(–many)-celled, with 1–2 apical ovules in each cell; style very short or 0. Fr. a drupe with 3 or more stones. Embryo small, straight; endosperm copious, fleshy.

Three to five genera and over 300 spp., cosmopolitan (except the Arctic).

1. Ilex L.

Fls dioecious, rarely polygamous. *Petals united below. Stamens the same number as the petals.* Carpels 4–6(–8).

About 300 spp., cosmopolitan (except the Arctic). Some spp. are occasionally cultivated.

1. I. aquifolium L.   Holly.

Evergreen small tree or shrub 3–15 m.; crown cylindric or conic. Bark grey, smooth for a long time, eventually finely fissured. Buds ovoid, c. 2–3 mm. Twigs green, glabrous or puberulous when young. Lvs 3–10 cm., thick and coriaceous, ovate, elliptic or oblong, margin undulate, sinuate-dentate with large triangular spine-pointed teeth or on old trees largely entire, dark green and glossy above, paler beneath, glabrous, with a narrow cartilaginous border; short-stalked. Cymes cluster-like, axillary on the old wood. Fls c. 6 mm. diam., white, 4-merous. Fr. 7–12 mm., scarlet, globose. Fl. 5–8. Visited by honey-bees. Fr. 9–3. $2n=40*$. MM. or M.

Native. Woods, scrub, hedges and among rocks, on all but wet soils, ascending to 1800 ft., sometimes dominant in the lower tree- or shrub-layer of woods, tolerating a considerable amount of shade; throughout the British Is. except Caithness, Orkney and Shetland, common. 107, H40, S. Often planted, as are numerous varieties with variegated lvs, etc. W. and C. Europe from Norway, Denmark and N.W. and S. Germany southwards; Mediterranean region (mainly in the mountains).

47. CELASTRACEAE

Shrubs, trees or woody climbers. Lvs alternate or opposite, simple, stipules small or 0. Fls usually cymose, small, regular, hermaphrodite or unisexual, 4–5-merous with a well-marked fleshy disk, ±perigynous. Sepals and petals imbricate (very rarely valvate), inserted on or below the margin of the disk (petals rarely 0). Stamens equalling the corolla-lobes in number and alternate with them (very rarely 10). Ovary 2–5-celled with (1–)2(–many) axile anatropous ovules in each cell; style very short; stigma ±capitate, often lobed. Fr. a loculicidal capsule, samara, drupe or berry; seeds often surrounded by an aril; embryo large, straight; endosperm fleshy.

About 45 genera and 450 spp., cosmopolitan (except the Arctic).

1. EUONYMUS L.

Trees or shrubs, rarely climbing by rootlets. Lvs opposite (rarely alternate). Fls 4–5-merous; *disk wide*, fleshy. Stamens short, inserted on disk. *Carpels isomerous with other parts. Ovules* 1–2 *in each cell. Fr.* a fleshy, often brightly coloured, loculicidal *capsule. Seed completely enclosed in a fleshy aril.*

About 170 spp., Europe, N. Africa (1 sp.), Madagascar (1 sp.), Asia, N. and C. America, Australia (1 sp.). Some are grown in gardens, especially *E. japonicus* L. f. from Japan, with glossy evergreen lvs, much planted near the sea, often as a variegated form.

1. E. europaeus L.   Spindle-tree.

Much-branched, deciduous, rather stiff, glabrous shrub or rarely small tree 2–6 m. Bark grey, smooth. Buds 2–4 mm., ovoid, greenish, ±4-angled. Twigs green, 4-angled. Lvs opposite, 3–13 cm., ovate-lanceolate to oblong-lanceolate or elliptic, acute or acuminate, crenate-serrulate, cuneate at base, often turning reddish in autumn; petiole 6–12 mm. Fls 8–10 mm. diam., 4-merous, hermaphrodite or polygamous, 3–10 together in axillary dichotomous peduncled cymes. Petals greenish, ±oblong, widely separated. Fr. 4-lobed, deep pink, 10–15 mm. across, exposing the bright orange aril after opening. Fl. 5–6. Pollinated by small insects. Fr. 9–10. $2n=64$. M.

Native. Woods and scrub, mostly on calcareous soil, ascending to 1200ft. 79, H40, S. Throughout England, Wales and Ireland, rather common; extending north in Scotland to the Clyde and Forth. Europe from Sweden and Latvia to C. Spain, Sicily, Greece and the Caucasus; W. Asia.

48. BUXACEAE

Evergreen shrubs or small trees, rarely herbs. Lvs simple, exstipulate. Fls regular, usually monoecious, small, without disk, in spikes, racemes or clusters. Calyx usually of 4 sepals, sometimes 0 or sepals more in female fl. Petals 0. Stamens 4, opposite the sepals, or more numerous. Ovary superior, (2–)3(–4)-celled with 1–2 pendulous anatropous ovules in each cell; raphe dorsal; styles free. Fr. a loculicidal capsule or drupe. Seeds with fleshy endosperm and straight embryo.

Six genera and about 40 spp., scattered over tropical and temperate regions. A family of obscure relationships, very variously placed by different authors.

1. BUXUS L.

Shrubs or small trees with *opposite entire* coriaceous *lvs*. Fls in axillary clusters consisting of a terminal female fl. and several male fls in the axils of bracteoles. Male fl. with 4 sepals, 4 *stamens*, and rudimentary ovary; filaments thick. Female fl. with several sepals and 3-celled ovary; styles short, thick, *ovules 2 in each cell*. Fr. a capsule, the valves 2-horned. Seeds black, shining, with small caruncle.

About 30 spp. in W. Europe, Mediterranean region, temperate E. Asia, Socotra, Madagascar, W. Indies and C. America.

1. B. sempervirens L.   Box.

Evergreen shrub or small tree 2–5(–10) m. Twigs pubescent, somewhat 4-angled. Lvs 1–2·5 cm., oblong or elliptic, shortly stalked, obtuse to emarginate. Fls whitish-green, stamens exserted. Fr. c. 8 mm., ovoid, 3-horned. Fl. 4–5. Pollinated by bees and flies. Fr. 9. $2n=28$. M.

Native. In beech woods and scrub on chalk and oolitic limestone in Kent, Surrey, Bucks and Gloucester, locally abundant; elsewhere commonly planted and sometimes naturalized. 5. S. Europe from C. Spain and Portugal, Corsica, C. Italy and Greece to France, S.W. Germany and Albania; N. Africa (in the mountains).

49. RHAMNACEAE

Trees or shrubs with simple, usually stipulate lvs. Infl. cymose. Fls small, green, yellow or blue, sometimes unisexual. Calyx tubular, 4–5-lobed, lobes valvate in bud. Petals 4–5, sometimes 0, small, inserted at mouth of calyx-tube and often hooded. Stamens 4–5, opposite the petals and often ± enclosed by them; anthers versatile. Ovary 2–4-celled, free or sunk in disk; ovules solitary, basal, erect, anatropous. Fr. often fleshy.

About 40 genera and 500 spp., cosmopolitan.

For an account of the ecology of the British spp. see H. Godwin, *J. Ecol.* **31** (1943), pp. 66–92.

Thorny; buds with scales; lvs serrate; blaze of bark orange.   1. RHAMNUS
Unarmed; buds naked; lvs entire; blaze of bark lemon-yellow. 2. FRANGULA

1. RHAMNUS L.

Trees or shrubs. Buds with scales. Lvs alternate or subopposite. Fls inconspicuous, greenish, usually 4-merous, polygamous or dioecious. Calyx-tube urceolate, adnate to base of ovary. Style cleft. Fr. 2–4-seeded. Germination epigeal.

1. R. catharticus L.   Buckthorn.

A rather *thorny* deciduous bush or small tree commonly 4–6 m., but sometimes up to nearly 10 m. *Branches* opposite, *spreading almost at right angles*

to main stem, many of the laterals forming short lfy spurs or ending in thorns.
Young twigs grey or brown, glabrous to densely pubescent; *bark of old
branches fissured and scaling, blaze orange; young wood whitish. Buds with
dark scales. Lvs* 3–6 cm., petioled, *ovate to nearly elliptic,* obtuse, sometimes
± cuspidate, *serrate,* somewhat pubescent to glabrous, *dull green turning
yellow or brownish in autumn; large lateral veins* 2–3 *pairs* curving upward and
running ± parallel nearly to tip of lf. *Fls* c. 4 mm. diam., on slender pedicels,
solitary or in axillary fascicles *on the previous year's wood of the short shoots.
Calyx* greenish, *lobes* 4, lanceolate. *Petals* 4, small. Fr. 6–10 mm. diam.,
changing from green to black on ripening, 3–4-seeded. Fl. 5–6. Pollinated
by various insects. Fr. 9–10. Germ. autumn and spring. $2n = 24$. N. or M.
Bark and berries purgative.

Native. On fen peat, in scrub, hedges, and ash and oak woods on calcareous
soils. 61, H23. England and Wales, except for Cornwall, Devon, Northumberland, Pembroke, Cardigan, Radnor, Merioneth and Montgomery; frequent in many midland, southern and eastern counties; Scotland: doubtfully
native; Ireland: frequent in the central plain, rare in the south and north.
Europe from southern Scandinavia to C. Spain and eastwards to western
Asia; N. Africa, on high mountains in Morocco and Algeria.

*R. alaternus** L., an evergreen shrub with elliptic to ovate-lanceolate lvs 2–5 cm.,
dark green and shiny above, paler but glabrous beneath, and very small clustered
fls, sometimes escapes from cultivation. Mediterranean region.

2. FRANGULA Mill.

Similar to *Rhamnus* but buds naked, fls usually 5-merous and hermaphrodite,
style entire, germination hypogeal.

1. F. alnus Mill.   Alder Buckthorn, Black Dogwood.

Rhamnus Frangula L.

An *unarmed* deciduous shrub or small tree commonly 4–5 m. *Branches* subopposite, *ascending at an acute angle to the main stem, without marked
distinction into long and short shoots.* Young twigs green becoming grey-brown,
appressed-puberulent; *bark of old branches smooth,* except in very old trees,
*blaze lemon-yellow; young wood dark brown. Buds without scales, densely
covered with brownish hairs. Lvs* 2–7 cm., petioled, *obovate,* bluntly apiculate,
entire, undulate, with a caducous brownish tomentum particularly beneath,
*shiny green turning clear yellow and red in autumn; large lateral veins about
7 pairs. Fls* c. 3 mm. diam., on rather stout pedicels, in axillary fascicles *on
the young wood. Calyx* greenish, *lobes* 5, ovate. *Petals* 5, small. Fr. 6–10 mm.
diam., changing from green to red and then violet-black on ripening, 2(–3)-
seeded. Fl. 5–6(–9). Pollinated by various insects, especially bees. Fr. 8–11.
Germ. spring. $2n = 20^*$, 26. N. or M. Purgative. Wood yields a high-grade
charcoal.

Native. In scrub on fen peat and around the margins of raised bogs and
valley bogs, on moist heaths and commons, in limestone scrub and as under-
growth in open woods; usually on damp and ± peaty soils. 73, H16.
Generally distributed throughout England and Wales except for N. Wilts.,
Durham, Northumberland, Isle of Man, Pembroke, Radnor, Montgomery

and Anglesey; local, though often abundant in suitable habitats; absent from Scotland; Ireland: rare, and absent from the south-eastern and north-western counties. Europe north to c. 67° in Sweden and 66° 50′ in Russian Lapland, eastwards to the Urals and Siberia; N. Africa, in Morocco and Algeria; rare in the Mediterranean region.

50. VITACEAE

Similar to Rhamnaceae but usually climbers with tendrils. Infl. lf-opposed. Fr. a berry.

About 11 genera and 600 spp., mostly in warm temperate or tropical regions.

***Vitis vinifera** L., the Grape Vine, a woody climber with palmately lobed lvs and tendrils which twine round any convenient support, is ±naturalized on the banks of the Thames at Kew.

Spp. of **Parthenocissus** (Virginia Creeper, 'Ampelopsis') are commonly cultivated. ***P. tricuspidata** (Sieb. & Zucc.) Planch., which has the tendrils ending in adhesive disks and the lvs varied in shape but usually simple, is the commonest. 'Hundreds of miles of suburban architecture are happily hidden in summer by its foliage, which turns red in autumn' (Gilbert-Carter, *British Trees and Shrubs*). This sp. is also ±naturalized by the Thames at Kew.

***P. quinquefolia** (L.) Planch. is also grown fairly frequently and sometimes escapes It has all the lvs 5-foliolate and the young branches terete.

51. PAPILIONACEAE

Herbs, less frequently shrubs or trees. Lvs simple or (usually) compound, often 3-foliolate or pinnate, sometimes ending in a tendril. Fls papilionate, with a large, often erect, adaxial petal (standard), 2 lateral petals (wings) and 2 lower petals usually ±connate by their lower margins (keel); standard outside and enclosing the other petals in bud. Sepals usually 5, ±connate into a tube. Stamens 10, either all fused (monadelphous) or 1 free and 9 fused (diadelphous), rarely all free. Fr. usually dehiscent. Seeds often large; endosperm 0 or very scanty.

A large family (more than 300 genera and 5500 spp.), mainly temperate in distribution. The fls are usually insect-pollinated. The related families, Mimosaceae and Caesalpiniaceae, are mainly tropical in distribution and contain few herbs.

1	Trees or shrubs, sometimes small and prostrate.	*2*
	Herbs, at most slightly woody at base.	*9*
2	Trees with pendulous racemes.	*3*
	Shrubs; fls in erect racemes or umbels or solitary in lf axils.	*4*
3	Fls white; lvs pinnate; bark of trunk deeply fissured.	14. ROBINIA
	Fls yellow; lvs palmate, bark of trunk ±smooth.	2. LABURNUM
4	Fls solitary in lf axils.	*5*
	Fls in erect racemes or umbels.	*7*
5	Spines branched.	4. ULEX
	Spines simple or 0.	*6*

6   Twigs terete; style curved; valves of pod not twisted after dehiscence.
                                    3. GENISTA (see also *Spartium*)
    Twigs angled; style spirally coiled; valves of pod twisted after dehiscence.
                                    5. SAROTHAMNUS

7   Fls in racemes.   *8*
    Fls in umbels.   19. CORONILLA

8   Lvs palmate; pod hard, not inflated.   1. LUPINUS
    Lvs pinnate; pod papery, strongly inflated.   15. COLUTEA

9   At least some lvs trifoliolate.   *10*
    No lvs trifoliolate.   *13*

10 Fls solitary in lf axils.   6. ONONIS
    Fls 2–many together in stalked heads or racemes.   *11*

11 Pod sickle-shaped or spirally coiled.   7. MEDICAGO
    Pod straight, usually shorter or little longer than calyx.   *12*

12 Infl. a lax raceme with many fls.   8. MELILOTUS
    Infl. a dense head or else few(–6)-fld.   9. TRIFOLIUM

13 Lvs palmate with 6 or more lflets.   1. LUPINUS
    Lvs pinnate, simple or 0.   *14*

14 Lvs paripinnate or simple, usually ending in tendrils.   *15*
    Lvs imparipinnate; tendrils 0.   *18*

15 Lflets 0 or 1 pair.   23. LATHYRUS
    Lflets 2 or more pairs.   *16*

16 Lflets (2–)3–7 cm., normally 2–4 pairs.   *17*
    Lflets rarely as much as 2 cm. and then more than 4 pairs on most lvs.
                                                    22. VICIA

17 Standard purple, wings white or fls less than 10 mm.   22. VICIA
    Not as above.   23. LATHYRUS

18 Fl. 25–30 mm., solitary, pale yellow.   12. TETRAGONOLOBUS
    Not as above.   *19*

19 Fls in umbels.   *20*
    Fls in racemes.   *24*

20 Umbels in pairs subtended by a lf-like involucre; calyx woolly, strongly
        inflated.   10. ANTHYLLIS
    Umbels solitary, if subtended by a lf-like bract then calyx neither woolly
        nor inflated.   *21*

21 Umbels 10–20-fld; fls usually pink, rarely white or purple.
                                                    19. CORONILLA
    Umbels up to 8(–12)-fld; fls yellow; if whitish closely subtended by a
        pinnate bract.   *22*

22 Lflets always 5, two lower at base of petiole; pod straight, not jointed.
                                                    11. LOTUS
    Lflets usually more than 5, the two lower not at base of petiole; pod
        curved or jointed.   *23*

23 Perennial; fls yellow; peduncles longer than subtending lvs; segments of
        pod horseshoe-shaped.   20. HIPPOCREPIS
    Annual; fls whitish or, if yellow, then peduncles shorter than subtending
        lvs; segments of pods not horseshoe-shaped.   18. ORNITHOPUS

24 Stipules scarious; fls bright pink or red.   21. ONOBRYCHIS
    Stipules green; fls not bright pink or red.   *25*

| 25 | Plant glabrous. | 26 |
| | Plant hairy. | 27 |

26 Plant stiffly erect; infl. at least as long as the subtending lf.
                                                                    13. Galega
   Plant prostrate or ascending; infl. much shorter than the subtending lf.
                                                                 16. Astragalus

27 Keel obtuse.   16. Astragalus
   Keel mucronate.   17. Oxytropis

Tribe 1. Genisteae. Shrubs or herbs, rarely small trees. Lvs simple, 3-foliolate, sometimes 0, rarely palmate; lflets entire. Stamens usually monadelphous, always so in our spp. Pod usually dehiscent.

1. Lupinus L.

Usually herbs. *Lvs palmate*; stipules adnate to base of petiole. *Fls in a terminal raceme, showy.* Calyx deeply 2-lipped; stamens monadelphous; style curved, stigma capitate. *Pod flattened* and often constricted between the seeds.

About 300 spp., chiefly American.

Herb; lflets 6–8; fls blue and white.   **1. nootkatensis**
Shrub; lflets 7–11; fls yellow or white, rarely bluish-tinged.   **2. arboreus**

*1. L. nootkatensis Donn ex Sims   Lupin.

A stout lfy pubescent *perennial. Lflets* 2·5–5 cm., 6–8, *cuneate-elliptic, mucronate. Petioles somewhat exceeding lflets*; stipules subulate to linear-acuminate. Infl. up to c. 10 cm. *Fls* 1·5–2 cm., ±whorled, *blue and white*; bracts caducous, exceeding the buds. *Pods c. 5 cm.*, brown and silky. Fl. 5–7. 2n=48. Hs.

   Introduced. Naturalized on river shingle in several parts of Scotland. N.E. Asia and N.W. America.

*2. L. arboreus Sims   Tree Lupin.

An upright *shrub* up to 3 m. *Lflets* c. 3 cm., 7–11, *oblanceolate*, mucronulate, glabrous above, silky beneath. *Petioles about equalling lflets.* Racemes up to 25 cm. *Fls yellow* or white, rarely bluish tinged. *Pod c. 8 cm.*, pubescent, 8–12-seeded. Fl. 6–9. 2n=40. N. or M.

   Introduced. Naturalized in waste places. Native of California.

L. polyphyllus Lindl. often persists for a time as a garden outcast. Perennial herb. Lflets 9–16, lanceolate to oblanceolate. Fls blue, pink or white, the keel often darker at tip. 2n=48. Native of N. America from California to Washington.

2. Laburnum Medic.

Small unarmed trees with smooth bark. Lvs 3-foliolate. Racemes simple, *lfless, pendulous.* Fls yellow. Calyx 2-lipped, shortly toothed. Stamens monadelphous, tube entire; anthers alternately long and basifixed and short and versatile. *Pod subterete, bulging over the seeds*; seeds several, poisonous, dispersed by explosive dehiscence.

   Two species in C. and S. Europe.

*1. L. anagyroides Medic.   Golden Rain, Laburnum.

Cytisus Laburnum L.; *L. vulgare* Bercht & Presl

A small unarmed tree up to 7 m. Bark dark brownish-green. Twigs appressed-pubescent. Petioles 5–8 cm.; lflets 4–8 cm., elliptic or elliptic-ovate, obtuse and mucronulate, appressed-pubescent beneath. Racemes 10–20 cm.; pedicels appressed-pubescent. Fls c. 20 mm. Pods 3–5 cm., appressed-pubescent, the upper suture thickened. Fl. 5–6. Pollination mainly by humble bees. The pods remain on the trees throughout the winter. Germ. spring. $2n=48$. M.

Introduced. Planted and ± naturalized in waste, bushy places. C. and S. Europe.

3. GENISTA L.

Small shrubs, *sometimes spinous, then with simple spines. Lvs* 1-*foliolate*, stipules minute or 0. Fls yellow, solitary in the axils of the lvs, bracteolate, bracteoles minute. Calyx shortly 2-lipped; *upper lip deeply bifid*, lower shortly 3-toothed. Wings oblong, deflexed after flowering; keel-petals separating and not resilient after deflection. Stamens monadelphous, tube entire; anthers alternately long and basifixed and short and versatile. Style curved; stigma oblique. Pod 2-valved and dehiscent; seeds several, dispersed by explosive dehiscence.

About 90 spp. mainly in the Mediterranean region.

Spiny (rarely unarmed); lvs ovate, glabrous; pod inflated.   **2. anglica**
Unarmed; lvs oblong-lanceolate, ciliate; fls glabrous or nearly so, on long
    branches.   **1. tinctoria**
Unarmed; lvs ovate, densely pubescent; fls pubescent, on dwarf lateral
    branches.   **3. pilosa**

1. G. tinctoria L.   Dyer's Greenweed.

An unarmed erect or ascending shrub 30–70 cm. Stems slender, branched, brown; *young twigs* green, sparsely hairy, *striate. Lvs* up to 30 mm., sub-sessile, *oblong-lanceolate, acute or mucronate, margins ciliate*, glabrescent when old. *Stipules* 1–2 mm., *thin, subulate.* Fls c. 15 mm., axillary, towards the ends of the main branches, without nectar; standard equalling keel. Pedicels 2–3 mm., almost glabrous; bracteoles minute, basal. Calyx and corolla glabrous, the former deciduous when the fr. is ripe, the standard persistent. *Pod* 25–30 mm., glabrous, *flat, tapering and obtuse at both ends.* Fl. 7–9. Pollination by diverse pollen-collecting insects. $2n=48$. N.

Native. In rough pastures. 78, S. Throughout England and Wales, rare in southern Scotland, and absent in the north and from Ireland. Mediterranean to southern Norway (absent from Portugal) and eastward to the Urals, Caucasus and Asia Minor.

Var. *prostrata* Bab. is procumbent, 15–25 cm., with ovate or oblong lvs and pods hairy on the back of each valve. Near the Lizard, Cornwall and St David's Head, Pembrokeshire. Probably a distinct ssp.

2. G. anglica L.   Needle Furze, Petty Whin.

A spiny, rarely unarmed (var. *subinermis* (Le Grand) Rouy) *erect or ascending shrub* 10–50(–100) cm. Stems slender, brown; *young twigs terete*, glabrous or pubescent; spines 1–2 cm., spreading or recurved, rarely branched, lfy when young. *Lvs* 2–8 mm., ovate, *acute or apiculate, glabrous and glaucous*; those on the spines linear-lanceolate. *Stipules* 0. Fls c. 8 mm., glabrous, axillary towards the ends of the main branches; standard shorter than keel. Pedicels c. 2 mm., sparsely hairy, bibracteolate. Calyx-lobes fringed; corolla glabrous. *Pod* 12–15(–20) × 5 mm., *glabrous, inflated, obliquely narrowed and acute at both ends.* Fl. 5–6. Pollinated by bees. $2n = 42^*$, 48. N.

Native. On dry heaths and moors. 91. Scattered throughout Great Britain; absent from Ireland. Western Europe from Spain and Portugal to Denmark and southern Sweden.

3. G. pilosa L.   'Hairy Greenweed.'

An unarmed prostrate shrub 10–40 cm. Stems rather stout, much-branched and tortuous, greyish when young, brown when old; *young twigs* pubescent, *grooved. Lvs* 3–5 mm., subsessile, ovate, *obtuse, with appressed hairs beneath, glabrous above. Stipules* 0·5 *mm., thick, ovate, obtuse.* Fls c. 10 mm., axillary, on dwarf lateral branches. Pedicels c. 5 mm., densely hairy, minutely bibracteolate. *Calyx and corolla pubescent. Pods* 14–18(–28) mm., *pubescent, not inflated but bulging over the seeds, rounded at the base, acute at the apex.* Fl. 5–6. Pollinated by honey-bees. $2n = 24$. Chw. or N.

Native. On cliffs and dry sandy and gravelly heaths on poor soils. 7. Cornwall, Sussex, Kent, Suffolk, Worcester, Pembroke and Merioneth; rare and very local. Atlantic Europe from Portugal to Jutland; S.W. Sweden and eastwards to S.W. Poland; northern Balkans to southern Italy and C. Spain.

4. ULEX L.

Densely spinous shrubs; spines green and branched. Lvs 3-foliolate on young plants, spinous or reduced to scales on mature plants. Stipules 0. Fls yellow, axillary, shortly pedicelled, bracteoles small. Calyx membranous, yellow, bipartite; lower lip minutely 3-toothed, *upper minutely 2-toothed.* Petals obtuse, keel pubescent. Stamens as in *Genista. Style somewhat curved*; stigma capitate. Pod 2-valved and dehiscent; seeds several, dispersed by explosive dehiscence. About 20 spp., in W. Europe and N.W. Africa. Pollination by insects.

1 Spines deeply furrowed; bracteoles 3–5 mm. (fl. winter and spring).
                                                                                    1. europaeus
   Spines faintly furrowed or striate; bracteoles 0·5 mm. (fl. late summer and
      autumn).   2

2 Spines rigid; calyx-teeth connivent; wings curved, longer than keel when
      straightened.   **2. gallii**
   Spines weak; calyx-teeth diverging; wings straight, shorter than keel.
                                                                                    3. minor

1. U. europaeus L.   Furze, Gorse, Whin.

A densely spinous rather glaucous shrub 60–200 cm. or sometimes more. Main branches erect or ascending, hairs rather sparse, blackish; *spines* 1·5–2·5 cm., rigid, *deeply furrowed.* Fls c. 15 mm. Pedicels 3–5 mm., densely velvety; *bracteoles* (2–)3–5 × 1·5–4 *mm., much wider than the pedicels. Calyx* ⅔ *length of corolla, with spreading hairs,* teeth connivent. Wings rather longer than keel. Pod c. 15 mm., bursting in summer, black, with grey or brown hairs. Fl. 3–6, and sporadically during a mild winter. Pollinated especially by humble-bees. Fr. 7. $2n = 64, 96$. N.

Our plant is ssp. **europaeus**.

Native. In rough grassy places and edges of heaths, usually on the lighter and less calcareous soils. 109, H40, S. Generally distributed throughout the British Is. though often planted in northern Scotland and rather infrequent in W. Ireland. Spain and Portugal, W. France, Belgium to S.W. Scandinavia; naturalized in much of C. Europe. Introduced in many parts of the world. Occasionally cut to the ground by severe frosts.

2. U. gallii Planch.   'Dwarf Furze.'

A densely spinous, dark green shrub 10–200 cm. Main branches usually ascending, hairs abundant, brown; spines 1–2·5 cm., faintly furrowed or striate. Fls 10–12 mm. Pedicels 3–5 mm., hairs appressed; bracteoles c. 0·6 mm., usually wider than the pedicels. *Calyx* ⅔ to ¾ length of corolla, hairs appressed, *teeth connivent. Wings* when straightened usually *exceeding keel.* Pod c. 10 mm., bursting in spring. Fl. 7–9. Fr. 4–5. $2n = 80$. N.

Native. On heaths and siliceous hill grasslands, strongly calcifuge. 67, H31, S. In suitable habitats throughout England and Wales, commoner in the west; S. Scotland north to Ayr; throughout Ireland but commonest in the south and east. N.W. France, Cantabria and Galicia.

3. U. minor Roth   'Dwarf Furze.'

U. nanus T. F. Forst.

A densely spinous shrub 5–100 cm. Main branches usually procumbent, hairs abundant, brown; spines c. 1 cm., faintly furrowed or striate. Fls 8–10 mm. Pedicels 3–5 mm., hairs appressed; bracteoles c. 0·7 mm., usually narrower than the pedicels. *Calyx* nearly as long as corolla, hairs appressed, rather sparse, *teeth divergent. Wings about as long as keel, straight.* Pod c. 7 mm., persistent for nearly a year. Fl. 7–9. $2n = 32, 64$. N.

Native. On heaths. 42, S. England and Wales, more frequent in the east; S. Scotland north to Ayr; Caithness. W. France, N. Spain, W. Portugal and S.W. Spain. Introduced in the Azores and Brazil.

5. SAROTHAMNUS Wimmer

Unarmed shrubs with small readily deciduous lvs and green stems. Lvs 1–3-foliolate. Fls axillary, yellow. Calyx herbaceous, 2-lipped, minutely toothed, lower lip 3-, upper 2-toothed. Stamens as in *Genista. Style long, spirally coiled;* stigma capitate. Pod 2-valved, *valves coiled after dehiscence;* seeds several, dispersed by explosive dehiscence.

About 10 spp., mostly in Spain and Portugal.

1. S. scoparius (L.) Wimmer ex Koch   Broom.

Cytisus scoparius (L.) Link

A much-branched erect shrub 60–200 cm. Twigs glabrous, green, strict, 5-angled. Lvs distinctly petioled; lflets narrowly elliptic to obovate, acute, hairs appressed. Stipules 0. Fls c. 20 mm., yellow. Pedicels up to 10 mm., slender, glabrous; bracteoles minute. Calyx ¼ length of corolla, glabrous. Pod 2·5–4 cm., black, with brown hairs on the margins. Fl. 5–6. Pollination by large bees. $2n=48*$. N.

Native. On heaths and waste ground and in woods, usually on sandy soils, strongly calcifuge. 110, H40, S. Generally distributed throughout the British Is. except for Orkney and Shetland. From Spain to S. Scandinavia, eastwards to Poland and Hungary; Madeira and Tenerife. Broom tops (*Scoparii Cacumina* of the Pharmacopoeia) are used as a diuretic.

Ssp. **maritimus** (Rouy) Ulbrich with prostrate stems and densely silky lvs and young twigs, occurs on cliffs in W. Cornwall, Pembroke, Lundy, W. Cork and the Channel Is. It maintains its characters in cultivation and breeds true. $2n=48* (24*)$.

*Spartium junceum L., a yellow-flowered shrub, somewhat resembling *Sarothamnus* but with terete stems and simple linear-lanceolate lvs, is ±naturalized locally in S. England.

Tribe 2. TRIFOLIEAE Herbs. Lvs digitately or pinnately 3-foliolate; lflets toothed, sometimes obscurely so. Stamens diadelphous, rarely monadelphous. Pod usually dehiscent.

6. ONONIS L.

Herbs or small shrubs. Lvs pinnately 3-foliolate, sometimes reduced to the terminal lflets, nerves ending in teeth. Stipules adnate to the petiole. *Fls* axillary, *pink* (in our spp.), without nectar. Standard broad, wings oblong, keel pointed. *Stamens monadelphous*; 5 or all the *filaments dilated above*. Style curved; stigma terminal. Pod 1-celled, 2-valved; seeds 1–many. Pollination by bees. About 70 spp.

1	Stout woody perennials; fls 10 mm. or more; pod erect.	2
	Slender annual; fls 7 mm. or less; pod deflexed.	**3. reclinata**
2	Stem hairy all round; wings equalling keel.	**1. repens**
	Stem with 2 lines of hairs; wings shorter than keel.	**2. spinosa**

1. O. repens L. ssp. **repens**   Restharrow.

O. arvensis auct.

A *procumbent* or rarely ascending perennial 30–60 cm. *Rhizomatous. Stems usually unarmed, rooting at the base, uniformly hairy*, woody and muchbranched, sometimes spiny above. Petioles 3–5 mm., pubescent; stipules 3–5 mm., clasping the stem, serrate and ±glandular-pubescent. Lvs up to 20 mm.; lflets obovate, serrate, pubescent and ±glandular. Fls 10–15 mm. Pedicels 1–5 mm. *Wings equalling keel. Pod* 1–4-seeded, *shorter than the enlarged calyx*. Fl. 6–9. $2n=32$. Hp. or Chh.

Native. In rough grassy places. 105, H33, S. Scattered throughout the British Is., common in calcareous districts. France and Belgium. Ssp. *procurrens* (Wallr.) Aschers. & Graebn. in N. Europe.

2. O. spinosa L.   Restharrow.

O. campestris Koch & Ziz

An *erect or ascending* perennial 30–60 cm. Similar to *O. repens* but *not rhizomatous. Stems usually spiny, not rooting at the base, with 2 lines of hairs.* Lflets generally narrower. *Wings shorter than keel. Pod exceeding calyx.* Fl. 6–9. $2n=32$. Hp. or Chh.

Native. In rough grassy places. 72. Scattered throughout England and Wales, rare in southern Scotland. Europe, except the extreme north and high mountain regions; Asia and N. Africa.

Intermediate forms (probably hybrids) between *O. repens* and *O. spinosa* occur.

3. O. reclinata L.   'Small Restharrow.'

An erect or ascending viscid *annual* 4–8 cm. *Stems slender*, unarmed, pubescent. Petioles slender, short; stipules 1–4 mm., pubescent, serrate, clasping the stem. Lvs 3–5 mm.; lflets narrowly obovate, cuneate, serrate at apex, densely pubescent and glandular. *Fls 5–7 mm. Pedicels strongly deflexed after flowering. Calyx as long as the fl. or longer. Pod* 14–18-*seeded*, glandular, pubescent, about equalling the calyx. Fl. 6–7. $2n=64, 60$. Th.

Native. By the sea in sandy soil. 2, S. Berry Head, Devon; Glamorgan; Alderney and Guernsey, Channel Is. Introduced, Wigtown. Mediterranean region eastwards to Persia, Syria and Palestine; south to Arabia and Abyssinia.

7. MEDICAGO L.

Herbs. Lvs pinnately trifoliolate, nerves ending in teeth. Stipules adnate to the petiole. Fls racemose, yellow or purple. Calyx-teeth 5, nearly equal. *Petals caducous*; keel obtuse, shorter than the wings. Upper stamen free; filaments not dilated; anthers all equal. Style glabrous; stigma subcapitate. *Pod spirally curved or coiled, rarely falcate, often spiny*, longer than the calyx, *usually indehiscent*, 1–many-seeded.

About 120 spp. in Europe, W. Asia, N. Africa.

1	Fls 7–9 mm., yellow or purple.	*2*
	Fls 2–5 mm., yellow.	*3*
2	Fl. yellow; pedicel longer than the calyx-tube; pod falcate or nearly straight.	**1. falcata**
	Fl. purple; pedicel shorter than calyx-tube; pod a spiral of 2–3 turns.	**2. sativa**
3	Racemes many-fld; pod unarmed, 1-seeded, black when ripe.	**3. lupulina**
	Racemes 1–5-fld; pod spiny or tubercled, several-seeded, brown when ripe.	*4*
4	Fr. densely pubescent.	**4. minima**
	Fr. glabrous.	*5*
5	Lflets not blotched; peduncle as long as lf.	**5. polymorpha**
	Lflets usually blotched; peduncle shorter than lf.	**6. arabica**

1. M. falcata L.   'Sickle Medick.'

A decumbent or sometimes erect perennial 30–60 cm. Lflets up to 15 mm., linear-lanceolate, emarginate, mucronate. Stipules lanceolate, acute to acuminate. Racemes up to 25 mm. *Fls* c. 8 mm., *yellow, pedicels longer than calyx-tube. Pod* pubescent or glabrescent, *falcate or nearly straight*, 2–5-seeded. Fl. 6–7. Pollinated mainly by bees. Fr. 8–9. Germ. autumn. $2n = 16, 32$. Hp. Seedlings flower in their 1st year.

Native. In grassy places on gravelly soil in the Breckland. 5. Norfolk, Suffolk and Cambridge. Introduced elsewhere. Europe (except the extreme north), N. Africa, W. Asia.

*2. M. sativa L.   Lucerne, Alfalfa.

A deep-rooted, erect or ascending perennial 30–90 cm. Lflets up to 30 mm., narrowly obovate, toothed in the upper third. Stipules linear-lanceolate, acuminate, ± toothed. Racemes up to 40 mm. *Fls* c. 8 mm., *purple; pedicels shorter than calyx-tube. Pod in a spiral of* 2–3 *turns*, pubescent or glabrous, 10–20-*seeded*. Fl. 6–7. Pollinated mainly by bees. Fr. 8–9. Germ. autumn. $2n = 16, 32, 64$. Hp.

Introduced. Planted and ± naturalized on waste ground. Probably native in the Mediterranean region and W. Asia.

M. × varia Martyn

M. falcata L. × *M. sativa* L.; *M. sylvestris* Fr.

This fertile hybrid frequently segregates and back-crosses with both parents giving a great variety of forms. Stems decumbent or erect. Lflets broad or narrow. Fls yellow, purple, or yellow changing to purple through a series of dark greens and black. Pods from almost straight to a spiral of 2–3 turns.

Spontaneous in East Anglia with the parents; elsewhere as an alien.

3. M. lupulina L.   Black Medick.

A procumbent or ascending, usually downy annual or short-lived perennial 5–50 cm. Lflets 3–20 mm., obovate, apiculate, finely serrate in the upper half. Stipules lanceolate, acuminate, half-cordate at base, shortly toothed. Racemes 3–8 mm., compact, peduncles exceeding the petioles. Fls 2–3 mm., bright yellow. *Pod* 2 mm. diam., *reniform*, reticulate, *coiled in almost* 1 *complete turn*, 1-*seeded, black when ripe*. Fl. 4–8. Mainly self-pollinated. Fr. 5–9. Germ. autumn or spring. $2n = 16, 32$. Th. or Hp.

Native. Generally distributed and common in grassy places and roadsides, up to 1200 ft. in Derbyshire. 111, H40, S. Europe, N. Africa, W. Asia, Atlantic Islands.

4. M. minima (L.) Bartal.   'Small Medick.'

An erect or procumbent pubescent annual 5–20 cm. Lflets 3–6 mm., narrowly or sometimes broadly obovate, often emarginate, apiculate, serrate at the apex. *Stipules* lanceolate to ovate, acuminate, *nearly entire*. Racemes 3–5 mm., peduncles equalling or somewhat exceeding the petioles. Fls 3–4 mm., bright yellow. *Pod* 4–5 mm. diam., *subglobose*, in a spiral of 4–5 turns, faintly reticulate, with a double row of hooked spines. Fl. 5–7. Fr. 6–8. Germ. autumn. $2n = 16$. Th.

Native. In sandy fields and on heaths. 9, S. Kent to Norfolk, Channel Islands. From the Mediterranean to S. Sweden, eastwards to western Asia; N. Africa, Abyssinia, Canaries.

5. M. polymorpha L.   'Hairy Medick.'

M. hispida Gaertn.; incl. *M. denticulata* Willd.

A procumbent nearly glabrous annual 5–60 cm. Lflets up to 25 mm., obovate or obcordate, serrate towards the top. *Stipules ovate-lanceolate*, acuminate, laciniate. Racemes 4–6 mm.; peduncles about equalling petioles. Fls 3–4 mm. *Pod* 4–6 mm. diam., *flat*, in a spiral of $1\frac{1}{2}$–4 turns, *strongly reticulate*, with a double row of longer or shorter hooked or curved spines, border thin. Fl. 5–8. Mainly self-pollinated. Fr. 6–9. Germ. spring or autumn. $2n = 14$, 16. Th. Very variable.

Native. In sandy or gravelly ground near the sea in eastern and southern England. c. 28, S. Introduced and naturalized elsewhere. C. and S. Europe, N. Africa, Asia.

6. M. arabica (L.) Huds.   'Spotted Medick.'

M. maculata Sibth.

A procumbent nearly glabrous annual 10–60 cm. *Lflets* up to 25 mm., obovate or obcordate, serrate towards the top, *usually blotched. Stipules half-cordate, toothed*, acuminate. Racemes 5–7 mm.; *peduncles shorter than petioles*. Fls 4–6 mm. *Pod* 4–6 mm. diam., *subglobose, faintly reticulate*, in a spiral of 3–5 turns, with a double row of curved or hooked spines. Fl. 4–8. Pollinated mainly by bees. Fr. 5–9. Germ. spring. $2n = 16$. Th.

Native. In grassy places and waste ground, especially on gravelly or sandy soils near the sea; Ireland, naturalized in a few localities. 55, H 6, S. Mediterranean region from Spain to W. Asia; introduced elsewhere.

*****M. praecox** DC. occurs as a casual. Similar to *M. polymorpha* but smaller. Peduncle much shorter than lf. Fr. not strongly reticulate; border thick.

*****M. laciniata** (L.) Mill. is another casual. Similar to *M. arabica* but lflets dentate to almost pinnatifid in the upper $\frac{1}{2}$, never blotched. Stipules deeply pectinate.

8. MELILOTUS Mill.

Annual, biennial or short-lived perennial herbs. Lvs pinnately 3-foliolate, nerves ending in teeth. Stipules adnate to the petiole. Fls racemose, small, yellow or white. Similar to *Medicago*, but *racemes usually longer and laxer and the pod short, straight, thick* and usually indehiscent, *never spiny*. Pollination mainly by bees. Many spp. smell strongly of coumarin (new-mown hay), especially when drying.

1 Wings and standard equal.   2
    Standard longer than wings.   3

2 Wings, standard and keel all equal; pod pubescent, acute, black when ripe.
                                             1. altissima
    Keel shorter than other petals; pod glabrous, obtuse, brown when ripe.
                                            2. officinalis

3 Fls 4–5 mm., white; pod 4–5 mm., brown when ripe.   **3. alba**
    Fls c. 2 mm., pale yellow; pod 2–3 mm., olive-green when ripe. **4. indica**

1. M. altissima Thuill.   'Tall Melilot.'

M. officinalis auct.

An erect branched biennial or short-lived perennial 60–120 cm. Lflets 15–20(–30) mm., oblong or obovate, those of the upper lvs nearly parallel-sided, serrate. Stipules subulate to setaceous. Racemes 20–50 mm., compact, lengthening in fr. Fls 5–6 mm., yellow; *wings, standard and keel all equal. Pod* 5–6 mm., *pubescent*, reticulate, compressed ovoid, *acute, black when ripe*; style long, persistent. Fl. 6–8. Fr. 8–10. $2n=16$. Hp.

   ?Introduced. Naturalized in waste places and woods. 90, H 10, S. Generally distributed but rarer in the north; Ireland, naturalized in the south and east. Throughout most of Europe north to Scandinavia; eastwards through Siberia to Japan.

*2. M. officinalis (L.) Pall.   'Common Melilot.'

M. arvensis Wallr.

A decumbent or erect branched biennial similar in appearance to *M. altissima*. Lflets of upper lvs oblong-elliptic narrowed at both ends. Racemes rather lax and slender. Fls 5–6 mm., yellowish; *wings and standard equal but longer than the keel. Pod* 3–5 mm., glabrous, *transversely rugose*, ovoid, slightly compressed, blunt, mucronate, brown when ripe; style usually deciduous. Fl. 7–9. $2n=16$. Hs.

   Introduced. Naturalized in S. England and in Ireland in fields and waste places. 66, H 7, S. Throughout most of Europe, though often only as a weed of cultivation, eastwards to W. China.

*3. M. alba Medic.   'White Melilot.'

An erect branched annual or biennial similar in appearance to *M. altissima*. Racemes rather lax and slender. *Fls 4–5 mm., white; wings and keel nearly equal, somewhat shorter than standard. Pod 4–5 mm.*, glabrous, reticulate, ovoid, compressed, mucronate, *brown when ripe*; style usually deciduous. Fl. 7–8. $2n=16$, 24. Th. or H.

   Introduced. Naturalized in fields and waste places, mainly in S. England and Wales. 79, S. Throughout most of Europe and Asia eastwards to Tibet, introduced in many places; introduced in America and Australia.

*4. M. indica (L.) All.   'Small-flowered Melilot.'

M. parviflora Desf.

An erect branched annual, similar in general appearance to, but smaller in all its parts than, *M. altissima*. Racemes slender, usually lax, sometimes dense in fr. *Fls c.* 2 *mm., pale yellow; wings and keel equal, shorter than standard. Pod* 2–3 *mm.*, glabrous, strongly reticulate, globular-ovoid, *olive green when ripe*; style usually persistent. Fl. 6–10. $2n=16$. Th.

   Introduced. Naturalized in fields and waste places, mainly in S. England and Wales. 57, S. Mediterranean region to India; introduced and naturalized in most of the rest of the world.

9. Trifolium L.   Clover, Trefoil.

Annual or perennial, usually low-growing herbs. Lvs digitately or sometimes pinnately trifoliolate; stipules adnate to petiole. *Fls* sessile or shortly pedicelled *in* usually *dense racemose heads.* Calyx-teeth 5, subequal or rarely the lower one much longer than the others. *Petals usually persistent,* wings longer than keel, the claws of both adnate to the staminal tube. Upper stamen free, *all or 5 of the filaments dilated at the top. Pod* small, 1–10-seeded, indehiscent, opening by 2 valves or by the top falling off, ± *enclosed in the calyx and often covered by the persistent standard.* The fls have nectar and are usually cross-pollinated by bees.

About 290 spp. in temperate and subtropical regions but mainly N. temperate.

1	Fls yellow, not exceeding 7 mm.	2
	Fls white, pink or purple, if yellowish then more than 10 mm.	5
2	Standard not folded; fls 5–7 mm.; heads c. 40-fld.	3
	Standard folded over the pod; fls 2–4 mm.; heads up to 26-fld.	4
3	Stipules half ovate.	**4. campestre**
	Stipules linear-oblong.	**5. aureum**
4	Pedicels rather stout, shorter than calyx-tube; standard entire; heads (4–)10–20-fld.	**3. dubium**
	Pedicels slender, equalling calyx-tube; standard notched; heads 2–6-fld.	**2. micranthum**
5	Fls not more than 6, usually fewer, in lax heads.	6
	Fls numerous, tightly crowded in dense heads.	7
6	Lvs glabrous; petals pinkish, persistent; pods several-seeded, longer than the non-accrescent calyces.	**1. ornithopodioides**
	Lvs pubescent; petals cream, caducous; pods 1-seeded, enclosed in the accrescent calyces which are ±concealed by the reflexed enlarged calyces of sterile fls.	**21. subterraneum**
7	Stems creeping and rooting at the nodes; heads all axillary.	8
	Stems not rooting at the nodes; some or all heads terminal.	9
8	Upper part of calyx becoming vesicular in fr.; lateral veins thickened towards margins of lflets.	**11. fragiferum**
	Calyx not inflated in fr.; lateral veins not thickened towards margins of lflets.	**7. repens**
9	Heads all terminal.	10
	Some heads axillary.	14
10	Calyx-tube glabrous or with scattered long hairs.	11
	Calyx-tube densely covered with appressed brownish hairs.	**17. incarnatum**
11	Corolla much longer than calyx; fls up to 15 mm.	12
	Corolla little longer than calyx; fls up to 7 mm. (pastures near the sea, very local).	**20. squamosum**
12	Fls whitish-yellow; lflets never spotted.	**19. ochroleucon**
	Fls pink or red, rarely white; lflets often spotted.	13
13	Free part of stipules triangular with a setaceous point; calyx-tube and teeth hairy.	**18. pratense**
	Free part of stipules subulate; calyx-tube glabrous or sparsely hairy, teeth hairy.	**12. medium**

14	Heads distinctly peduncled.	*15*
	Heads sessile.	*17*
15	Heads softly downy, cylindrical.	**13. arvense**
	Heads not downy, ovoid or globular.	*16*
16	Fls 8–10 mm.; stipules oblong, entire.	**6. hybridum**
	Fls 4–5 mm.; stipules ovate, toothed (rare).	**10. strictum**
17	Lateral veins thickened and backward-curved near margins of lflets.	**14. scabrum**
	Lateral veins thin and straight or slightly forward-curved near margins of lflets.	*18*
18	Lflets glabrous.	*19*
	Lflets pubescent, at least on the veins beneath.	*20*
19	Calyx-teeth broad, abruptly narrowed to a short fine point; corolla distinctly longer than calyx, purplish.	**8. glomeratum**
	Calyx-teeth narrow, gradually tapering; corolla shorter than calyx, whitish.	**9. suffocatum**
20	Lflets hairy above.	**16. striatum**
	Lflets glabrous above (Lizard cliffs, rare).	**15. bocconei**

Subgenus 1. ORNITHOPODA (Malladra) D. E. Coombe. Fls bracteolate. Calyx without a ring of hairs or thickening in the throat. Stamens with none of the filaments dilated. Pod exserted, curved, 5–10-seeded, not concealed by the petals. Plant glabrous or nearly so.

Section 1. *Ornithopoda.* Infl. few-fld. Fls subsessile. Calyx-teeth subequal. Petals white or pink.

1. T. ornithopodioides L.   'Birdsfoot Fenugreek.'

Trigonella ornithopodioides (L.) DC.; *T. purpurascens* Lam.; *Falcatula falso-trifolium* Brot.; *F. ornithopodioides* (L.) Brot. ex Bab.

A slender almost glabrous annual or short-lived perennial 2–20 cm. Petioles up to 25 mm.; stipules lanceolate, acuminate. Lflets 2–8 mm., obovate to obcordate, cuneate at base, sharply serrate. *Fls* 6–8 mm., *white and pink*. *Heads* 1–3(–5)-*fld*, axillary; peduncles shorter than petioles. Calyx-tube narrowly conical; teeth narrowly triangular, acuminate, about as long as the tube. *Pod* 5–7 *mm.*, oblong, *somewhat curved*, slightly pubescent, 5–8-*seeded*. Fl. 5–9. $2n=18*$. Th.

    Native. In sandy and gravelly places, chiefly near the coast. 42, H7, S. Local in suitable habitats north to Renfrew and Fife; south and east coasts of Ireland. S. and W. Europe from Denmark southwards; N.W. Africa.

Subgenus 2. TRIFOLIASTRUM (Ser.) Taub. Fls bracteolate. Calyx without a ring of hairs or thickening in the throat. All or 5 of the stamens with filaments flattened. Pod 1–6-seeded, generally concealed by the petals. Plants usually glabrous or glabrescent.

Section 2. *Chronosemium* Ser. Heads often few-fld. Infl. distinctly racemose. Fls shortly stalked. Calyx not enlarging in fr.; upper teeth shorter than lower. Petals yellow, standard becoming enlarged, scarious and folded over the pod. Pod stalked, exserted, 1–2-seeded. Lvs frequently pinnately trifoliolate.

2. T. micranthum Viv.   'Slender Trefoil.'

T. filiforme L., nom. ambig.

A slender procumbent or ascending sparsely hairy annual 2–10(–20) cm. Lflets c. 5 mm., obcordate to obovate. Lvs shortly petioled; stipules ovate. *Heads* axillary, 2–6-*fld*; peduncles filiform, equalling or exceeding lvs. Fls 2·5–3 mm., yellow; *standard deeply notched*, folded over the pod; *pedicels slender, about as long as calyx-tube.* Pod 2–2·5 mm., ovoid, usually 2-seeded; seeds dull brown, smooth. Fl. 6–7. Germ. spring. 2*n*=14. Th.

Native. In rather open grassy places on sandy and gravelly soils. 70, H 32, S. England, Wales and Ireland, rarer in the north; Scotland: Wigtown and Roxburgh. W. and S. Europe from S. Norway to the Caucasus.

3. T. dubium Sibth.   'Suckling Clover, Lesser Yellow Trefoil.'

T. minus Sm.

A slender procumbent or ascending ± hairy annual up to 25(–50) cm. Lflets up to 11 mm., obcordate or obovate, cuneate at base. Lvs pinnately trifoliate, shortly petioled; stipules broadly ovate, acuminate. *Heads* axillary, (4–)10–26-*fld*, usually 5–7 mm. diam.; peduncles exceeding petioles. Fls c. 3 mm., yellow, turning dark brown; *standard narrow, folded over pod; pedicels rather stout, shorter than calyx-tube.* Pod 2·5–3 mm., ovoid, usually 1-seeded; seeds light brown. Fl. 5–10. Germ. spring. 2*n*=14, 16, 28. Th.

Native. In grassy places and sometimes cultivated. 112, H 40, S. Throughout the British Is., rarer in N. and N.W. Scotland. Europe from Sweden southwards, east to the Caucasus.

4. T. campestre Schreb.   'Hop Trefoil.'

T. procumbens auct.

A rather stout erect or ascending ± hairy annual up to 35(–50) cm. Lflets commonly 8–10 mm., obovate, sometimes obcordate, cuneate at base. Lvs pinnately trifoliate, petioles up to 10 mm.; *stipules ½-ovate*, tips triangular, acute. Heads axillary, many-fld, (7–)10–15 mm.; peduncles exceeding petioles. Fls 4–5 mm., yellow, turning rather light brown; *standard broad, not folded, much exceeding pod.* Pedicels rather stout, c. half the length of the calyx-tube. *Pod* 2–2·5 mm., ovoid, usually 1-seeded, *several times as long as the style.* Seeds yellow, rather shiny. Fl. 6–9. Germ. spring and autumn. 2*n*=14. Th.

Native. In grassy places and roadsides. 110, H 40, S. Generally distributed throughout the British Is., except the extreme north. Europe, except the extreme north; W. Asia; N. Africa; Madeira and Canaries; introduced in N. America.

*5. T. aureum Poll.

T. agrarium auct.

Similar to *T. campestre* in general appearance, but usually larger. Stipules linear-oblong, acuminate. Fls c. 5 mm. Pod oblong, c. twice as long as style, usually 2-seeded. Fl. 7–8. Germ. spring and autumn. 2*n*=14. Th.

Introduced. Naturalized in fields and waste places, rather uncommon. 46, H 3, S. Scattered throughout the British Is., but casual in many places. Europe from N. Spain to Norway, eastward to C. Russia and Asia Minor.

Section 3. *Trifoliastrum*. Heads many-fld. Fls stalked or subsessile. Calyx not enlarging in fr.; upper teeth longer than lower. Petals white or pinkish, standard persistent and folded over the pod. Pod sessile, 2–6-seeded. Lvs palmately trifoliolate.

*6. T. hybridum L.   Alsike Clover.

An *erect or decumbent* nearly glabrous *perennial* up to 60 cm. Lflets 10–35 mm., obovate or elliptic. Petioles up to c. 10 cm.; *stipules* ovate to oblong *with long acuminate tips*. Heads all axillary, up to c. 25 mm., globular; peduncles up to 15 cm. Fls whitish or pink; pedicels up to 3 times as long as calyx-tube. *Calyx-tube* campanulate, *whitish, feebly ribbed; teeth subulate*, greenish, subequal, *nearly twice as long as tube*. Standard (5–)7–8 mm., c. twice as long as calyx, folded over the pod. Pod 3–4 mm., oblong, (1–)2-seeded. Fl. 6–9. $2n = 16$. Hs.

Introduced. Naturalized by roadsides etc. throughout the British Is. 96, H40, S. Apparently formerly rather local in Europe but cultivated since the 18th century and now widespread in temperate regions.

Ssp. **hybridum**: Usually erect, 30 cm. or more. Lflets mostly more than 20 mm. with c. 20 pairs of lateral veins. Fls numerous, whitish or pinkish, sometimes deep pink in stunted plants; standard 7–8 mm. Commonly cultivated and often naturalized.

Ssp. **elegans** (Savi) Aschers. & Graebn.; Usually decumbent, c. 15 cm. Lflets mostly less than 20 mm. with c. 40 pairs of lateral veins. Fls up to c. 30 in a head, bright pink; standard 5–6 mm. Rather uncommon.

7. T. repens L.   White Clover, Dutch Clover.

A *creeping* glabrous *perennial* up to 50 cm., *rooting at the nodes. Lflets* 10–30 mm., obovate or obcordate, *usually with a whitish angled band towards the base*, lateral veins thin and nearly straight (cf. *T. fragiferum*). Petioles erect, up to 14 cm. or sometimes more; *stipules* ovate to oblong *with short subulate points*. Heads 15–20(–35) mm., all axillary, globular. Peduncles up to 30 cm., usually shorter. Fls numerous, white or pink, rarely purple (var. *rubescens* Ser.); pedicels up to 3(–6) mm. *Calyx-tube* campanulate, *white with green veins; teeth narrowly triangular c. half as long as tube.* Standard 8–10 mm., folded over the pod. Pod 4–5 mm., oblong, (1–)3–6-seeded. Fl. 6–9. $2n = 32$, 48. H. or Chh.

Native. In grassy places, particularly common on clayey soils. 112, H40, S. Throughout the British Is. Europe to 71° N., ascending to 2750 m.; N. and W. Asia; N. Africa; introduced in S. Africa, Atlantic Islands, N. and S. America and E. Asia.

8. T. glomeratum L.   'Clustered Clover.'

A procumbent or ascending glabrous annual 5–25 cm. Lflets 5–7(–10) mm., obovate-cuneate, apiculate or emarginate, sharply serrate. Petioles up to c. 20 mm.; *stipules usually ovate with long points*. Heads 5–10 mm., terminal and axillary, globular, sessile. *Fls purplish*, subsessile. *Calyx-tube* glabrous, broadly campanulate, *strongly ribbed, whitish; teeth triangular, spinescent*, green, $\frac{1}{2}-\frac{1}{3}$ as long as tube. Standard c. 4 mm., c. 3 times as long as calyx-tube. Pod c. 2 mm., oblong, (1–)2-seeded. Fl. 6–8. $2n = 16$. Th.

Native. In grassy places on sandy and gravelly soils, rare and mainly near the sea. 20, H2, S. Cornwall and Kent to Carmarthen and Norfolk, but doubtfully native in Wales. Ireland: Wexford and Wicklow. W. France, Spain, Portugal, Mediterranean region.

9. T. suffocatum L. 'Suffocated Clover.'

A prostrate glabrous annual 2–10 cm. *Lflets* 3–5(–7) mm., *triangular*, sharply serrate and sometimes emarginate at the top. Petioles up to 20(–30) mm., slender; *stipules ovate, acuminate*, whitish when old. *Heads* terminal and axillary, *sessile, densely crowded on the short stems and often confluent.* Fls whitish, shortly pedicelled. *Calyx-tube* nearly cylindrical, *weakly ribbed; teeth subulate*, about as long as tube. *Standard* c. 2 mm., *shorter than calyx.* Pod 1·5–2 mm., ovoid, (1–)2-seeded. Fl. 4–8. Th.

Native. In grassy places on sandy and gravelly soils, rare. 17, S. Cornwall and Kent to Somerset, Norfolk and S.E. Yorks; Anglesey and Cheshire; mainly near the coast. W. Europe from the English Channel southwards, Mediterranean region.

Section 4. *Involucrarium* Hook. Heads many-fld. Fls subsessile; bracteoles short and broad, those of the lower fls forming a small involucre. Calyx enlarging somewhat in fr.; lower tooth somewhat longer than upper. Standard persistent but not enclosing fr. Pod sessile, asymmetrical, 1-seeded, with a corky thickening on one side.

10. T. strictum L.

T. laevigatum Poir.

An erect or ascending rather stiff glabrous annual 3–15(–50) cm. *Lflets* 5–15(–25) mm., *narrowly elliptic*, sharply toothed. Petioles 5–10(–40) mm., slender; *stipules broadly ovate, toothed, whitish.* Heads terminal and axillary, 5–10 mm., ovoid; peduncles stout, 10–20(–60) mm. Fls purplish. *Calyx-tube strongly ribbed and angled*, about half as long as the subulate spinescent teeth. Standard 4 mm., longer than calyx. Pod 2·5 mm., beaked, 1-seeded. Fl. 5–7. Th.

Native. In grassy places near the Lizard, Cornwall; Channel Is. 1, S. W. France to Greece, N. Africa.

Section 5. *Fragifera* Koch. Heads many-fld. Fls subsessile, bracteolate. Calyx 2-lipped, lower lip with 3 herbaceous teeth, the upper membranous, pubescent to tomentose, enlarging greatly after fl. and enclosing the fr. Standard persistent. Pod sessile, 1–2-seeded.

*T. resupinatum L. A slender annual with narrow obovate lvs, *pink fls twisted so that the standard is below*, and calyx becoming somewhat inflated in fr. occurs occasionally as a casual, chiefly near docks. $2n = 14$, 16. Probably native in Asia.

*T. tomentosum L., a slender annual with obovate lflets, pink fls c. 2·5 mm., peduncles shorter than petioles and densely tomentose fruiting calyces occurs as a casual. Mediterranean region, W. Asia.

11. T. fragiferum L. 'Strawberry Clover.'

A creeping nearly glabrous *perennial*, lfless in winter, *rooting at the nodes*, up to c. 30 cm. Lflets commonly 10–15 mm., obovate or obcordate with fine forward-directed teeth, lateral veins thickened markedly and curved backwards towards the lf margin (cf. *T. repens*). Petioles up to c. 10 cm., often shorter, erect; *stipules* ovate-oblong, *long-acuminate*. Heads c. 10–20 mm. diam. in fr., axillary; *peduncles longer than petioles*. Fls pinkish to purplish; *bracteoles about as long as calyx*. Upper lip of calyx strongly inflated, reticulate, downward-curved, pubescent. Standard 5–6 mm., longer than calyx. Pod 3 mm., oblong, (1–)2-seeded. Fl. 7–9. $2n=16$. H.

Native. In grassy places mainly on heavy clay and often rather saline soils. 73, H15, S. Scattered throughout Great Britain, except the north of Scotland, locally common and increasingly cultivated; in Ireland local, northwards to Louth and Sligo, chiefly on the coast. Europe, to about 60° N.; W. Asia; N. Africa; Madeira and Canaries.

Subgenus 3. TRIFOLIUM. Fls ebracteolate. Calyx often with a ring of hairs or thickening in the throat. All or 5 of the stamens with filaments flattened. Pod 1-seeded. Plants usually hairy.

Section 6. *Trifolium.* Heads many-fld, fls all similar and fertile.

A. Pod membranous, dehiscing longitudinally.

12. T. medium L. Zigzag Clover.

A straggling ascending ± appressed pubescent perennial with flexuous stems up to c. 50 cm. and extensive slender rhizomes. *Lflets commonly* 15–40 mm., *rather narrowly elliptic*, obtuse or acute, ciliate, often with a faint whitish spot. Petioles up to c. 8 cm., often less; *stipules oblong or linear-oblong* (the upper ovate), *free part subulate, spreading. Heads* terminal, 10–30 mm., *subglobose*, shortly peduncled and subtended by a pair of lvs. Fls reddish-purple. Calyx-tube campanulate, weakly ribbed, hairy but not thickened in the throat; teeth subulate, ciliate, spreading in fr., the lower longest. Corolla up to c. 18 mm., 2–3 times as long as calyx, tardily deciduous. Pod c. 2 mm., obovoid, truncate, asymmetrical. Seed c. 1·7 mm. Fl. 6–9. $2n=$ c. 84; c. 126. Hp.

Native. In grassy places. 109, H37, S. Widely distributed but rather local, commoner in the north but not recorded from the Isle of Man, W. Ross, Shetlands and three Irish counties. Europe, except the extreme north and south, eastwards to Siberia; naturalized in N. America.

B. Pod membranous, indehiscent.

13. T. arvense L. Hare's-foot.

A softly hairy erect or ascending annual (5–)10–20(–40) cm., usually with spreading branches. Lflets 10–15(–25) mm., narrowly obovate-oblong. Petioles 2–10 mm.; stipules ovate with long setaceous points. *Heads* terminal and axillary, up to 25 mm., *cylindrical and softly downy; peduncles equalling or exceeding the lvs*, elongating in fr. Fls white or pink. Calyx-tube campanulate, feebly ribbed, throat glabrous and unthickened; teeth setaceous, long-ciliate. Corolla c. 3·5 mm., c. $\frac{1}{2}$ as long as calyx, persistent. Pod c. 1·3 mm., ovoid. Seed c. 0·8 mm. Fl. 6–9. $2n=14$. Th.

Native. In sandy fields, pastures and on dunes. 102, H17, S. Scattered throughout the British Is. and locally common; Ireland, chiefly on the coast in the east. Throughout Europe, except the extreme north; north and west Asia; N. Africa; Atlantic islands; naturalized in N. America.

14. T. scabrum L.   'Rough Trefoil.'

An erect or decumbent ± pubescent annual up to c. 20 cm. *Lflets* 5–8 mm., obovate, apiculate, *pubescent on both surfaces; lateral nerves backward curving and thickened towards the margins.* Petioles up to c. 10 mm.; *stipules* rather rigid, *ovate, cuspidate, almost amplexicaul. Heads* terminal and axillary, up to c. 10 mm., sessile, *ovoid.* Fls white. Calyx-tube campanulate, coriaceous, ribbed, throat thickened and somewhat hairy; teeth triangular, erect in fl., rigid and recurved in fr. Corolla about as long as calyx, persistent. Pod c. 1·7 mm., obovate-oblong, asymmetrical; style slender, nearly central. Seed c. 1·5 mm., similar in shape to pod. Fl. 5–7. $2n=16$. Th.

Native. In dry places on shallow and sandy soils. 60, H4, S. Scattered throughout Great Britain, except N. Scotland; S.E. coast of Ireland. Western Europe from the Netherlands southwards, Mediterranean region; W. Asia; Atlantic islands.

15. T. bocconei Savi

A ± erect somewhat pubescent annual 5–10 cm. *Lflets* 5–15 mm., obovate, *glabrous above, sparsely appressed-pubescent beneath; lateral veins thin and straight.* Petioles c. 5 mm.; *stipules ovate-oblong with ciliate subulate points.* Heads terminal and axillary, up to 10 mm., sessile, ovoid. Fls white, turning pink and finally brown. *Calyx-tube* cylindrical, coriaceous, *strongly ribbed,* throat thickened, slightly hairy; teeth triangular, spinescent and erect in fr. Corolla rather longer than calyx, persistent. Pod c. 1·8 mm., obovate-oblong, asymmetrical; style slender, nearly central. Seed similar in size and shape to pod. Fl. 5–6. Th.

Native. In grassy places near the Lizard, Cornwall and in Jersey. 1, S. Mediterranean region and S.W. Europe north to the English Channel.

16. T. striatum L.   'Soft Trefoil.'

A softly hairy procumbent to erect annual 5–40 cm. Lflets 5–15 mm., obovate, emarginate to acute or apiculate, *lateral nerves nearly straight and thin.* Petioles of lower lvs up to 30 mm.; upper lvs often subsessile; stipules membranous, ovate or triangular with subulate points. *Heads* terminal and axillary, up to 15 mm., sessile, ovoid, ± *enfolded in the dilated stipules of the subtending lvs when young, becoming obtusely conical in fr.* Fls pink. Calyx-tube narrow-campanulate, becoming inflated in fr., ribbed, throat slightly thickened, hairy; teeth somewhat shorter than tube, subulate, spinescent and suberect in fr. Corolla somewhat longer than calyx, persistent. Pod 2–5 mm.; style short, slender, asymmetrical. Seed 2 mm., ovoid. Fl. 5–7. $2n=14$. Th.

Native. In rather open habitats on well-drained soils. 87, H10, S. Scattered throughout the British Is., but very local in Ireland and only on the coast. Mediterranean region and W. Europe to S. Scandinavia, widespread but local, eastwards to the lower Oder and S. Russia; N.W. Africa; Macaronesia.

*T. **stellatum** L., a pubescent annual with ovate subacute stipules, terminal peduncled heads, 2–4 cm. in fr., and densely villous calyx-tube with ciliate triangular-subulate teeth 3 times as long as the tube has persisted near Shoreham, Sussex for many years. S. Europe, W. Asia, N. Africa.

C. *Pod membranous in the lower part, thickened above and often rupturing along the junction between the two parts.*

17. T. incarnatum L.   'Crimson Clover.'

A pubescent *annual* up to 50 cm. Lflets 5–30 mm., broadly obovate or obcordate. Petioles up to c. 8 cm.; *stipules membranous, ovate, obtuse or acute. Heads* 15–40(–70) *mm.*, terminal, ovoid or cylindrical, subtended by a single lf; peduncles stout. Fls crimson, pink or pale cream, sessile. Calyx-tube nearly cylindrical, villous, ribbed; throat somewhat hairy but scarcely thickened; teeth setaceous 1½–2 times as long as tube, spreading in fr. Corolla 9–15 mm., exceeding calyx, deciduous. Pod 2·5 mm., style stout, ± central, thickened 'cap' small. Seed 2 mm. Fl. 5–9. $2n = 14$. Th.

*Ssp. **incarnatum**: Usually 30 cm. or more, erect. Lflets obovate; hairs on stems and petioles usually spreading. Fls crimson. Introduced. Cultivated and ± naturalized, especially in the south. 45, S. Mediterranean region.

Ssp. **molinerii** (Balb.) Hook. f.: Usually under 20 cm., decumbent or suberect. Lflets often obcordate; hairs on stems and petioles usually appressed. Fls pink or pale cream. Fl. 5–6. Native. Grassy places near the Lizard, Cornwall and in Jersey. 1, S. W. and S. Europe to the Balkan peninsula.

*T. **angustifolium** L., an annual, somewhat resembling *T. incarnatum*, and with similar pods, but differing in the narrow-oblong to linear acute lflets, shorter corolla and thickened throat to the calyx-tube, occurs as a casual. Native of S. and S.E. Europe, Asia Minor, N. Africa and Macaronesia.

18. T. pratense L.   Red Clover.

An erect or decumbent ± pubescent perennial up to 60 cm. *Lflets* 10–30(–50) mm., elliptic to obovate, obtuse, sometimes acute or emarginate, *often with a whitish crescentic spot towards the base.* Petioles up to c. 20 cm.; stipules ovate to oblong, *free portion triangular with a setaceous point, usually applied to the petiole. Heads* terminal, up to c. 30 mm., *globose* becoming ovoid, sessile and subtended by a pair of short-petioled lvs. Fls pink-purple, sometimes whitish, sessile. Calyx-tube narrow-campanulate, ribbed; throat hairy, with longitudinal folds; teeth unequal, the longest c. twice as long as tube, all ciliate. Corolla 15–18 mm., exceeding calyx, persistent. Pod 2·2 mm., obovoid with thickened ± central style-base and slender style. Seed 1·7 mm., ovoid with a blunt projection at one side. Very variable. The cultivated form, var. *sativum* (Crome) Schreb., has the lflets usually entire, the stems fistular, and is larger than var. *sylvestre* Syme, which has toothed lflets and usually solid stems. Fl. 5–9. $2n = 14^*$. Hp.

Native. Generally distributed in grassy places throughout the British Is.

Var. *sativum* cultivated for hay and often naturalized. Europe to 71° N.; W. Asia to the Altai; Baikal and Kashmir; Algeria. Introduced in N. and S. America and New Zealand.

19. T. ochroleucon Huds.   'Sulphur Clover.'

An erect or ascending ± pubescent perennial up to c. 50 cm. Lflets 10–30 mm., elliptic to obovate, sometimes emarginate. Petioles up to 10 cm.; *stipules oblong or ovate-oblong with subulate tips.* Heads terminal, 20–30 mm., globose becoming ovoid, subsessile and subtended by a pair of nearly sessile lvs. *Fls whitish-yellow,* sessile. Calyx-tube campanulate, strongly ribbed, throat hairy, becoming somewhat thickened in fr.; lowest calyx-tooth 2–3 times as long as the others, all spreading in fr. Corolla c. 15 mm., longer than calyx, eventually deciduous. Pod c. 2·5 mm., obovoid, with a ± central stout style-base and slender style. Seed c. 1·8 mm., obovoid. Fl. 6–7. 2*n* = 16. Hp.

Native. In grassy places mainly on boulder clay in eastern England from Essex to Lincolnshire and west to Northamptonshire, probably introduced elsewhere. 17. France and Germany, south and east to the Crimea and C. Poland; Asia Minor.

20. T. squamosum L.   'Sea Clover.'
T. maritimum Huds.

An erect or ascending ± pubescent annual up to c. 40 cm. Lflets 10–25 mm., usually narrowly obovate, often apiculate. Petioles up to c. 10 cm.; *stipules oblong, the free part linear-lanceolate,* spreading. Heads terminal, c. 10 mm., ovoid, shortly peduncled and subtended by a pair of lvs, enlarging after flowering. Fls pink. *Calyx-tube campanulate, strongly ribbed and coriaceous in fr.*; throat minutely hairy becoming strongly thickened in fr.; *teeth triangular,* subequal, *green, spreading in fr.* Corolla c. 7 mm., caducous. Pod 2·5 mm., obovoid, style ± central, slender. Seed c. 2 mm., obovoid. Fl. 6–7. 2*n* = 16. Th.

Native. In turf near the sea or tidal estuaries, local. 23, S. Coasts of southern England from Cornwall and Kent to Carmarthen and Lincolnshire; Caernarvon, Lancashire. W. and S. Europe; W. Asia; N. Africa.

Section 7. *Trichocephalum* Koch. Heads few-fld. Calyx glabrous and unthickened in the throat, accrescent, splitting down one side and covered by the enlarged calyces of sterile fls when in fr. Heads burying themselves after fl. Pod 1-seeded, strongly asymmetrical.

21. T. subterraneum L.   'Subterranean Trefoil.'

A hairy prostrate annual 3–20(–60) cm. Lflets 5–12(–20) mm., broadly obcordate. Petioles 1–5(–10) cm.; stipules broadly ovate, acute. *Heads* axillary, *few-fld*; peduncles usually shorter than petioles of subtending lvs. *Fertile fls cream-coloured,* subsessile, sometimes cleistogamous, reflexed after fl. Calyx-tube narrow-cylindrical, becoming ± spherical in fr.; teeth setaceous, ciliate, curved, subequal, c. ⅓ length of tube. Corolla 8–12 mm., exceeding calyx, caducous. *Sterile fls consisting of slender, rigid, palmately-lobed calyces* which enlarge after fl. Pod 2·5 mm., ± globose with an asymmetrical keel near the top, style ± central, small. Seed 2 mm., ovoid with a blunt projection on one side. Fl. 5–6. 2*n* = 12, 16. Th.

Native. In sandy and gravelly pastures and on cliff tops, local. 48, H1, S.

Cornwall and Kent to Cheshire and Lincolnshire; Wicklow. S. and W. Europe; widely cultivated in warm temperate regions.

Tribe 3. LOTEAE Herbs. Lvs (in our spp.) pinnate or pinnately 3-foliolate; lflets entire. Fls in heads. Stamens diadelphous, rarely monadelphous; 5 or all filaments dilated at apex. Pod 2-valved.

10. ANTHYLLIS L.

Herbs or shrubs with *imparipinnate lvs*, lower sometimes reduced to terminal lflet. *Fls in capitate cymes*, yellow or red and *surrounded by an involucre (in our sp.). Calyx inflated, mouth oblique*, shortly 5-toothed. Petals with long claws, the 4 lower adnate to the staminal tube; keel incurved, gibbous at the sides. Stamens united or the upper free; anthers uniform. *Pod enclosed by the calyx*, 1–3-*seeded*. Pollinated mainly by humble bees.

About 20 spp. in Europe, N. Africa and W. Asia.

1. A. vulneraria L.   Kidney-vetch, Ladies' Fingers.

An erect or decumbent pubescent perennial up to 60 cm. Lvs up to c. 14 cm., pinnate or the lower sometimes reduced to the terminal lflet. Lflets of lower lvs ovate or elliptic, acute or obtuse, alternate, the terminal one much the largest; of upper linear-oblong, acute or mucronate, opposite, all similar. Petioles short or 0. Infl. a capitate cyme; heads in pairs, rarely solitary, up to 4 cm. across, sessile or subsessile within a lfy involucre, each pair peduncled. Fls 12–15 mm., yellow or red. Calyx ± woolly, contracted at the mouth. Petals exceeding calyx. Pod c. 3 mm., semiorbicular, compressed, glabrous, reticulate, 1-seeded. Fl. 6–9. $2n=12^*$. Hs.

Native. In dry places on shallow soils. 112, H40, S. Generally distributed throughout the British Is., but more abundant on calcareous soils and near the sea. Europe to about 70° N., east to the Caucasus; N. Africa.

For an account of the varieties and genetics of this variable plant see *Journ. Bot.* **71** (1933), pp. 207–13 and *J. Genet.* **27** (1933), pp. 261–85.

11. LOTUS L.

Herbs or under-shrubs. *Lvs 5-foliolate*, margins entire; stipules brown, minute. *Fls in* axillary, peduncled, *cymose heads*, yellow or reddish (in our spp.), protandrous; bracts 3-foliolate. Calyx 5-toothed. Keel incurved, gibbous on each side. Upper stamen free; alternate filaments dilated at the top. Style attenuate at the top. *Pod elongate*, 2-valved, *many-seeded, septate between the seeds*. The fls have nectar and are usually cross-pollinated by bees.

About 70 spp. in temperate Europe, Asia, N. and S. Africa and Australia.

1	Perennial; fls 10 mm. or more.	2
	Annual; fls c. 8 mm.	4
2	Calyx-teeth erect in bud, 2 upper with an obtuse sinus.	3
	Calyx-teeth spreading in bud, 2 upper with an acute sinus.	3. pedunculatus
3	Lflets obovate, not acuminate; fls c. 15 mm.	1. corniculatus
	Lflets linear-lanceolate, acuminate (rarely narrowly obovate), fls c. 10 mm.	2. tenuis

4 Pods 6–12 mm.; peduncles longer than lvs.   **4. hispidus**
    Pods 20–30 mm.; peduncles usually shorter than lvs.   **5. angustissimus**

1. L. corniculatus L.   Birdsfoot-trefoil, Bacon and Eggs.

A decumbent almost glabrous or rarely hairy perennial 10–40 cm. Rootstock stout, scarcely stoloniferous, stem solid or nearly so. *Lflets* 3–10 mm., *obovate, obtuse or apiculate,* lower pair broadly ovate or lanceolate; petioles short. Heads (1–)2–6(–8)-fld; peduncles up to c. 8 cm., stout. *Fls c.* 15 *mm.,* yellow, often streaked or tipped with red, shortly pedicelled. Calyx-teeth triangular, erect in bud, two upper with an obtuse sinus. Petals 2–3 times as long as calyx. Pod up to 3 cm. Fl. 6–9. 2*n*=24*. Hp. Variable.

Native. In pastures and grassy places. 112, H40, S. Generally distributed throughout the British Is. Europe to about 71° N.; Asia; N. and E. Africa; in the tropics only on mountains.

2. L. tenuis Waldst. & Kit. ex Willd.   'Slender Birdsfoot-trefoil.'

Similar to *L. corniculatus* but stems slender, often taller (up to 90 cm.), and more branched. Lflets linear-lanceolate, acuminate, or rarely narrowly obovate. Heads rarely more than 4-fld; peduncles slender. Fls c. 10 mm.; calyxteeth narrower, often subulate. Stems much more wiry than in *L. corniculatus.* Fl. 6–8. 2*n*=12*. Hp.

Native. In dry grassy places. 71, H4, S. Scattered throughout the British Is., local in the south, rarer in the north, probably introduced in Ireland. Europe, from S.W. Sweden; W. Asia; N. Africa.

3. L. pedunculatus Cav.   'Large Birdsfoot-trefoil.'

L. uliginosus Schkuhr; *L. major* auct.

An erect or ascending glabrous or pubescent perennial 15–60(–100) cm. *Rootstock slender, producing numerous stolons. Stem hollow.* Lflets usually 15–20 mm., obovate, often obliquely so, obtuse or mucronate, lower pair ovate; petioles up to 10 mm. Heads (1–)5–12-fld; peduncles up to c. 15 cm., rather slender. Fls 10–12 mm. *Calyx-teeth spreading in bud, 2 upper with an acute sinus.* Otherwise much the same as *L. corniculatus.* Fl. 6–8. 2*n*=12*, 24. Hp.

Native. In damp grassy places. 107, H40, S. Throughout the British Is., except the extreme north, but less common than *L. corniculatus.* Europe from Spain and the northern Balkans to S. Scandinavia and C. Russia; Asia; N. Africa.

4. L. hispidus Desf. ex DC.   'Hairy Birdsfoot-trefoil.'

A villous much-branched perennial 3–30(–90) cm. Lflets up to 20 mm., narrowly obovate to lanceolate, often obliquely so; lower pair half-cordate. Heads (1–)3–4-fld; *peduncles equalling or exceeding the lvs.* Fls c. 8 mm., yellow. Calyx-teeth subulate, often longer than the tube. *Standard obovate. Pod* 6–12 × 1·5–2 mm., 1½–3 times length of calyx. Fl. 7–8. 2*n*=24*. Hp.

Native. In dry grassy places near the sea. 7, H1, S. Cornwall, Devon, Dorset, Hampshire, Pembroke and Channel Is. W. Europe from the English Channel southwards and east to Italy; Corsica, Sardinia, Sicily; N. Africa.

5. L. angustissimus L.   'Slender Birdsfoot-trefoil.'

L. diffusus Soland. ex Sm.

Similar to *L. hispidus* but annual, heads 1–2-fld, peduncles usually shorter than lvs, standard elliptic and pods $20–30 \times 1–1\cdot5$ mm., 4–7 times length of calyx. Fl. 7–8. $2n=12^{*}$. Th.

Native. In dry grassy places near the sea. 9, S. Cornwall, Devon, Hampshire, Sussex, E. Kent, Surrey and Channel Is. W. and S. Europe, W. Asia, N. Africa.

12. TETRAGONOLOBUS Scop.

Herbs similar to *Lotus* but with 3-foliolate lvs, large green stipules, pods 4-winged and tardily dehiscent and style thickened at the top. About 6 spp., mainly Mediterranean region.

*1. T. maritimus (L.) Roth

Lotus siliquosus L.; *T. siliquosus* Roth

Slightly pubescent perennial 10–25 cm. Lvs shortly petioled; middle lflet rhomboid-ovate, the lateral ones ovate, asymmetrical. Peduncles much exceeding the subtending lf, 1-fld with a sessile 3-foliolate bract. Pedicel c. 1 mm. Fls 25–30 mm., pale yellow with faint brown veins in the standard. Fr. 2·5–5 cm., oblong, acute, 4-winged, many-seeded, ultimately dehiscing.

Introduced. Very local in rough calcareous grassland from Somerset and Glamorgan to Sussex and Essex; Lincoln; well naturalized. Europe from S. Sweden to the Mediterranean region, east to Poland and the Crimea.

DORYCNIUM Mill.

Differs from *Lotus* in the small white fls with blackish keels crowded in dense heads, in the short rounded pod whose valves do not twist after dehiscence, and in the small number (2–4) of seeds in the pod. Three or four spp. in Europe, W. Asia and N. Africa.

*D. gracile Jord.

D. suffruticosum auct.

A greyish-pubescent perennial 30–60 cm., usually with appressed hairs. Stems herbaceous, ascending or erect. Lvs 3-foliolate, with lf-like stipules; lflets linear-lanceolate, entire. Heads with 5–15 fls. Pedicels about as long as calyx-tube. Calyx-teeth as long as tube, acuminate. Standard apiculate, keel bluish-black, curved, mucronate. Pod obtuse but mucronate. Fl. 6–7. $2n=14$. Hp.

Introduced. Naturalized in a few places in S.E. England. France, Spain.

Tribe 4. GALEGEAE. Herbs, shrubs or trees. Lvs (in our spp.) pinnate. Fls in racemes. Stamens diadelphous; filaments not dilated at apex. Pod 2-valved.

13. GALEGA L.

Erect perennial *herbs. Fls* blue, lilac or white, *in erect racemes*; petals clawed; calyx-teeth subequal. Pod linear, terete. The fls are without nectar.

Three spp. in S. Europe and W. Asia.

*1. G. officinalis L.   Goat's Rue, French Lilac.

A stout erect glabrous perennial 60–150 cm. Lflets oblong or ovate-oblong, mucronate or emarginate. Peduncles usually equalling or exceeding lvs, many-fld; infl. racemose. Fls 12–15 mm., white, lilac or sometimes pinkish; bracts subulate, persistent; calyx gibbous at base on upper side, teeth setaceous, shorter than tube. Pod 2–3 cm., linear, terete. Fl. 6–7. $2n=16$. Hp.

Introduced. Naturalized in waste places. Europe, from Italy and the Carpathians eastward; W. Asia.

14. Robinia L.

Deciduous trees or shrubs with scaleless buds hidden by the petiole base. *Stipules often spiny and persistent. Racemes pendulous. Pod compressed.* The fls have nectar and are pollinated by bees.

Twenty spp. in N. America.

*1. R. pseudoacacia L.   Acacia.

A deciduous tree with coarsely fissured bark, up to 27 m. Stipules often eventually spiny. Racemes pendulous. Fls fragrant, white or sometimes pink. Pods strongly compressed, persistent. Fl. 6. $2n=20$. M.

Introduced. Commonly cultivated and sometimes planted in thickets. Eastern N. America, from the Appalachians and Pennsylvania to N. Georgia; naturalized throughout much of N. America, Europe, N. Africa, Asia and New Zealand.

15. Colutea L.

Deciduous shrubs. Fls yellow or brownish red *in small erect racemes. Pod membranous, much inflated.* The fls have nectar and are cross-pollinated by humble bees.

About 10 spp. in Europe and Asia. Poisonous.

*1. C. arborescens L.   Bladder Senna.

A small shrub up to 4 m. Lflets silky beneath. Racemes 2–8-fld. Fls yellow, standard with red markings. Pods c. twice as long as broad, greenish, closed at apex. Fl. 5–7. $2n=16$. N. or M.

Introduced. Naturalized on railway banks and in waste places particularly in the neighbourhood of London. Mediterranean region.

16. Astragalus L.

Herbs or shrubs. Lvs imparipinnate, stipulate; lflets entire. Fls in axillary racemes, bracteoles small. Calyx tubular with 5 subequal teeth. *Keel obtuse.* Upper stamen free; anthers all similar. *Pod 2-valved, often longitudinally 2-celled, septum developed from suture next the keel*; seeds 2–many. Fls with nectar, usually cross-pollinated by bees.

About 1600 spp., cosmopolitan except for Australia.

1 Plant glabrous or nearly so; fls creamy with a greenish-grey tinge.

                                                                     3. glycyphyllos

   Plant pubescent; fls blue or purplish.   *2*

2 Stipules connate below; fls strict; peduncle usually much longer than
    subtending lf.   **1. danicus**
  Stipules free to base; fls spreading or deflexed; peduncles shorter to slightly
    longer than subtending lf.   **2. alpinus**

1. A. danicus Retz.   'Purple Milk-vetch.'

A. hypoglottis auct., non L.

A slender ascending perennial 5–35 cm. Rootstock slender, branched. Stems
and lvs clothed with soft white hairs. Lvs 3–7 cm.; lflets 5–12 mm.; *peduncles
usually much longer than the subtending lf*, pubescent with soft white hairs
below mixed with black above. *Fls 15 mm., blue-purple, strict*; bracteoles
oblong to triangular. Calyx with appressed black and white hairs. *Pod*
7–10 mm., *covered with crisped white hairs*, up to 7-seeded. Fl. 5–7. $2n=16$.
Hp.

    Native. In short turf on calcareous soils and dunes. 51, H1. Locally
abundant in suitable habitats from Wiltshire to Sutherland, but mainly in
the eastern half of the country; Ireland: only Aran Islands. Denmark and
S. Sweden eastwards to the Baikal region; S.W. Alps; N. America.

2. A. alpinus L.   'Alpine Milk-vetch.'

Similar in general appearance to *A. danicus* but more slender. Stipules free
to base. Fls c. 10 mm., pale blue tipped with purple, spreading or deflexed;
peduncles shorter to slightly longer than the subtending lf. Pod c. 10 mm.,
with short appressed brown or blackish hairs, few-seeded. Fl. 7. $2n=$c. 56.
Hp.

    Native. In grassy places on mountains from 2350 to 2600 ft., rare. 3.
E. Perth, Angus and S. Aberdeen. Arctic Europe and on mountains south
to the Pyrenees and Alps, Carpathians; Caucasus, Himalaya.

3. A. glycyphyllos L.   Milk-vetch.

A stout *glabrous* prostrate or ascending perennial 60–100 cm. Rootstock
short, stout. Lvs 10–20 cm.; lflets 15–40 mm., oblong-elliptic, obtuse or
mucronate; stipules c. 2 cm., free. Infl. 2–5 cm.; peduncles much shorter than
subtending lvs. *Fls* 10–15 mm., *creamy-white with a greenish-grey tinge*,
spreading; *bracteoles subulate. Pod* 25–35 *mm.*, slightly curved, acuminate,
many-seeded, 2-celled with a longitudinal septum. Fl. 7–8. $2n=16$. Hp.

    Native. In rough grassy and bushy places, local. 71. Scattered throughout
Great Britain. Europe to c. 65° N., rare in the Mediterranean region, eastwards to the Altai, Caucasus and Asia Minor.

17. OXYTROPIS DC.

Herbs or shrubs similar to *Astragalus* in habit, etc., but differing in *the
mucronate keel and in the septum in the pod being developed from the suture
next the standard*. The fls have nectar and are pollinated by bees. About
100 spp. in N. temperate regions.

Peduncles exceeding the lvs at flowering; fls pale purple.   **1. halleri**
Peduncles shorter than the lvs at flowering; fls yellow tinged with purple.
                                                     2. campestris

1. O. halleri Bunge   'Purple Oxytropis.'

O. sericea (DC.) Simonk., non Nutt.; *O. uralensis* auct.

A softly hairy perennial with a stout rootstock and very short branches. Lvs up to c. 10 cm.; lflets 5–8 mm., c. 10 pairs, elliptic, subacute; stipules lanceolate, persistent and clothing the rootstock. Infl. c. 3 cm., 6–10-fld; *peduncle* stout, erect, *exceeding the lvs.* Fls c. 2 cm., *pale purple*, keel tipped with dark purple. Pod c. 2·5 cm., pubescent, many-seeded. Fl. 6–7. $2n=16$. Hr.

Native. In dry rocky pastures up to 2000 ft., local. 9. Wigtown, Fife, Argyll to Angus, Ross, Sutherland and Caithness. Pyrenees, Alps from France to Austria, Carpathians and mountains of N. Balkans.

2. O. campestris (L.) DC.   'Yellow Oxytropis.'

Similar to *O. halleri* but larger. Lvs up to 15 cm.; lflets 10–20 mm., linear-oblong or oblong-lanceolate. Peduncles shorter than the lvs at flowering, lengthening afterwards. Fls yellow tinged with purple. Pod up to c. 1·6 cm. Fl. 6–7. $2n=36$. Hr.

Native. On rocky ledges up to 2000 ft., very rare. 2. E. Perth and Angus. S.E. Sweden, Finland and N. Russia, Alps, Pyrenees, C. Carpathians, Dinaric Alps; N. Asia to Sakhalin; N. America, Rocky Mountains, Labrador, Maine; absent from Greenland.

Tribe 5. HEDYSAREAE. Herbs or shrubs. Lvs (in our spp.) pinnate with a terminal lflet. Fls in umbels or short racemes. Stamens diadelphous. Pod indehiscent, several-seeded, jointed, breaking up into 1-seeded joints, rarely 1–2-seeded and not jointed.

18. ORNITHOPUS L.

Slender herbs. Lvs imparipinnate, minutely stipulate. Fls small, in axillary peduncled heads. Calyx 5-toothed, teeth equal or two upper connate. Keel obtuse, sometimes very short. Upper stamen free; alternate filaments dilated towards the top, anthers all similar. *Pod curved, ± constricted between the seeds* and breaking up into 1-seeded portions when ripe.

About 8 spp. in Europe, N. and tropical Africa, W. Asia and S. Brazil.

Plant pubescent; fls white; bracts pinnate; pods strongly jointed.
                                                                        1. perpusillus
Plant nearly glabrous; fls yellow; bracts 0; pods slightly jointed.   **2. pinnatus**

1. O. perpusillus L.   Birdsfoot.

A *finely pubescent* slender spreading prostrate annual 2–45 cm. Lvs 15–30 mm.; lflets up to 4 mm., 4–7(–13) pairs, elliptic to linear-oblong, *lowest pair often at the base of the petiole, distant from the others and recurved. Heads* 3–6-fld, *subtended by a sessile pinnate bract*; peduncles filiform, longer or shorter than lvs. *Fls 3–4 mm., white veined with red*; pedicels very short, stout. *Pod* 10–20 mm., curved and *strongly constricted between the seeds.* Fl. 5–8. Self-pollinated. $2n=14*$. Th.

Native. In dry sandy and gravelly places. 86, H5, S. Generally distributed in Great Britain, except the north of Scotland; Ireland: Cork, Wexford, Wicklow, Dublin and Down. Spain and W. Italy northwards to Denmark and S. Sweden, eastwards to Poland and European Russia.

2. O. pinnatus (Mill.) Druce

O. ebracteatus Brot.; *Artrolobium ebracteatum* (Brot.) Desv.

A *nearly glabrous* slender ascending annual 5–15 cm. Lvs 10–25 mm., lflets up to 5 mm., 2–4 pairs, linear-lanceolate to narrowly obovate, *the lowest pair always distant from the base of the petiole*. Heads 1–2-fld, rarely more, *ebracteate*; peduncles filiform, as long as the lvs. *Fls* 6–8 mm., yellow, veined with red; pedicels very short, slender. *Pod* up to c. 30 mm., slender, curved and *slightly constricted between the seeds*. Fl. 4–8. Th.

Native. In short open turf or disturbed ground on sandy soils. 1, S. Scilly Is. and Channel Is. Atlantic coast from France to Spain, Mediterranean region; Macaronesia.

19. CORONILLA L.

Herbs or shrubs. Fls yellow, purple or white, in axillary umbels; *calyx* short-campanulate, *teeth almost equal*. Pod terete or 4-angled, breaking up into 1-seeded joints. Pollinated by bees.

About 20 spp. in the Mediterranean region, W. Asia and the Canary Islands.

*1. C. varia L.   Crown Vetch.

A straggling glabrous perennial 30–60 cm. Lflets oblong-elliptic, mucronate. *Peduncles exceeding lvs, 10–20-fld; infl. capitate. Fls c.* 12 *mm., white, purple or pink*; calyx-tube broadly campanulate, teeth very short. Pod slender, breaking up into 1-seeded joints. Fl. 6. $2n=24$. Hp.

Introduced. Well naturalized in a number of localities scattered throughout Great Britain. 51, S. Native of C. and S. Europe.

C. glauca L., a small shrub with glaucous lvs, umbels of 5–8 yellow fls and pods with 2–3 joints, is naturalized locally in S. Devon. Mediterranean region.

20. HIPPOCREPIS L.

Nearly glabrous herbs. Lvs imparipinnate, lflets entire, stipules small or 0. Fls yellow, capitate; calyx 5-toothed, 2 *upper teeth connate below*; petals with long claws. Upper stamen free, alternate filaments dilated towards the top, anthers uniform. *Pod* several-seeded, breaking up into 3–6 *horseshoe-shaped segments*. Pollinated mainly by bees.

Twelve spp. in the Mediterranean region and Europe.

1. H. comosa L.   Horse-shoe Vetch.

An almost glabrous diffuse perennial 10–40 cm. Rootstock woody, branched. Lvs 3–5 cm.; lflets 5–8 mm., usually 4–5(–15) pairs, obovate to oblong; stipules lanceolate, spreading. Heads 5–8(–10)-fld; *peduncles exceeding lvs*, slender, curved. *Fls* c. 10 mm., yellow, shortly pedicelled and bracteate; pedicels and upper part of peduncles with a few appressed hairs. Pod c. 30 mm., compressed, minutely papillate, segments horseshoe-shaped; style curved, persistent. Fl. 5–7. $2n=28*$. Chh.

Native. In dry calcareous pasture and on cliffs, local. 50, S. In suitable habitats north to Kincardine. S. Europe and the Alps to the calcareous regions of Belgium and the Rhine.

21. ONOBRYCHIS Mill.

Herbs or shrubs. Lvs imparipinnate; lflets numerous, entire; *stipules scarious*. *Infl. racemose*, axillary. *Wings short*, keel obliquely truncate, equalling or exceeding standard. Upper stamen free, *filaments not dilated*, anthers all similar. *Pod indehiscent, not jointed*, often *spiny or tubercled*, 1–2-seeded. The fls have nectar and are usually cross-pollinated by bees.

About 130 spp. in temperate Europe and Asia.

1. O. viciifolia Scop.   Sainfoin.

O. sativa Lam.

An erect slightly pubescent perennial 30–60 cm. Lflets 1–3 cm., 6–12 pairs, obovate or linear-oblong, mucronate, shortly petiolate; stipules ovate-acuminate, scarious. Peduncle stout, exceeding lvs, up to c. 50-fld, racemes dense. Fls 10–12 mm., bright pink or red, wings half as long as calyx; calyx-teeth subulate, much longer than the short, often woolly tube. Pod 6–8 mm., strongly reticulate, pubescent, tubercled on the lower margin, 1-seeded. Fl. 6–8. $2n=28$. Hp.

Native. Often a relic of cultivation, but probably native in chalk and limestone grassland. 37, S. From Somerset and Kent to Shropshire and Norfolk; Glamorgan, Carmarthen and Flint; apparently not even naturalized elsewhere. Europe, north to France and Germany; Siberia east to the Baikal region, Caucasus, Persia and Asia Minor; cultivated and often naturalized elsewhere.

Tribe 6. VICIEAE. Herbs. Lvs pinnate without a terminal lflet, usually ending in a tendril or point. Fls in racemes, rarely solitary. Stamens diadelphous. Pod 2-valved, dehiscent.

22. VICIA L.   Vetch, Tare.

Climbing or scrambling annual or perennial herbs. Lvs pinnate without a terminal lflet; tendrils usually present, simple or branched. Lflets opposite or alternate in 2–many pairs. Fls in axillary racemes; wings adhering to keel. Calyx-teeth subequal or the lower longer. *Staminal tube obliquely truncate*, upper stamen ± free; filaments not dilated and anthers all similar. *Style* cylindrical or flattened, *glabrous or equally downy all round or bearded below the stigma*. Pod compressed, 2-valved, dehiscent, several-seeded. The fls have nectar and are usually cross-pollinated by bees.

About 150 spp. in north temperate regions and S. America.

1	Fls more than 10 mm., in long-peduncled racemes.	2
	Peduncles very short or 0, fls less than 10 mm.	5
2	Racemes 1–2-fld.	**11. bithynica**
	Racemes 6–40-fld.	3

| 3 | Tendrils 0; plant erect. | 5. orobus |
| | Tendrils present; plant trailing or climbing. | 4 |

| 4 | Fls blue-purple; upper calyx-teeth minute. | 4. cracca |
| | Fls white with blue or purple veins; upper calyx-teeth half length of lower. | 6. sylvatica |

| 5 | Fls long-peduncled or less than 8 mm. | 6 |
| | Peduncles very short or 0; fls not less than 10 mm. | 9 |

| 6 | Peduncles very short or 0. | 10. lathyroides |
| | Peduncles long. | 7 |

| 7 | Calyx-teeth subequal, somewhat exceeding tube; pod hairy. | 1. hirsuta |
| | Calyx-teeth unequal, upper two shorter than tube; pod glabrous. | 8 |

| 8 | Fls c. 4 mm.; pod 4-seeded; hilum oblong. | 2. tetrasperma |
| | Fls c. 8 mm.; pod 5–8-seeded; hilum almost circular. | 3. tenuissima |

| 9 | Fls pale dirty yellow or brownish-violet; pod densely hairy. | 8. lutea |
| | Fls bluish or purple; pod glabrous or sparsely hairy. | 10 |

| 10 | Standard purple, wings white; stipules c. 10 mm. | 11. bithynica |
| | Standard and wings purplish or bluish; stipules smaller. | 11 |

| 11 | Calyx-teeth unequal, the smaller ones shorter than tube. | 7. sepium |
| | Calyx-teeth subequal, as long as tube. | 9. sativa |

Section 1. *Ervum* (L.) S. F. Gray. Annuals. Lflets 3–8 pairs. Peduncles long, 1–8-fld. Fls not exceeding 8 mm. Calyx-tube not gibbous at base. Style pubescent all round or sometimes nearly glabrous.

1. V. hirsuta (L.) S. F. Gray   Hairy Tare.

A slender trailing nearly glabrous annual 20–30(–70) cm. Lflets 5–12 mm., 4–8(–10) pairs usually alternate, linear-oblong or rarely ovate-oblong, truncate to emarginate and often mucronulate; tendrils usually branched; stipules often 4-lobed. Racemes 1–9-fld; peduncles 1–3 cm., slender. Fls 4–5 mm., dirty white or purplish; *calyx-teeth subequal*, subulate, *somewhat exceeding tube*. *Pod c.* 10 *mm.*, oblong, *pubescent, usually* 2-*seeded*. Fl. 5–8. $2n = 14$. Th.

    Native. In grassy places. 111, H37, S. Throughout the British Is., except Shetland, local in Ireland. Europe to c. 70° N.; W. Asia; N. Africa. A weed of cultivation throughout much of the world.

2. V. tetrasperma (L.) Schreb.   'Smooth Tare.'

V. gemella Crantz

A slender ± glabrous annual (15–)30–60 cm. Lflets 10–20 mm., (3–)4–6 pairs usually alternate, linear-oblong, obtuse, mucronate, or rarely linear, acute (var. *tenuissima* Druce); tendrils usually simple; stipules half-arrow-shaped. Racemes 1–2-fld; fruiting peduncles 2–4 cm., about equalling the lvs. *Fls c.* 4 *mm.*, pale blue; *calyx-teeth unequal, triangular, upper 2 shorter than tube. Pod* (9–)12–15 *mm.*, *oblong, shortly stipitate, glabrous*, 4-*seeded; hilum oblong*. Fl. 5–8. $2n = 14$. Th.

    Native. In grassy places. 84, H6, S. Throughout England, Wales and southern Scotland, local in N. Scotland, naturalized in a few places in Ireland. Europe north to 61°; W. Asia; N. Africa.

3. V. tenuissima (M. Bieb.) Schinz & Thell. 'Slender Tare.'
V. gracilis Lois., non Soland.

Similar to *V. tetrasperma* but lflets up to 25 mm., 3–4 pairs, usually linear-acuminate; racemes 1–4-fld; fruiting peduncles up to 8 cm., longer than lvs; fls c. 8 mm.; pods 5–8-seeded; hilum almost circular. Fl. 6–8. $2n=14$. Th.

Native. In grassy places. 30. From Cornwall and Kent to Worcester and Cambridge, local. S. and W. Europe, introduced in C. Europe.

Section 2. *Cracca* S. F. Gray. Perennials. Lflets 6–12 pairs. Peduncles long, 6–40-fld. Fls more than 10 mm. Calyx-tube gibbous at base on upper side. Style equally pubescent all round.

4. V. cracca L. Tufted Vetch.

A somewhat pubescent scrambling perennial 60–200 cm. Lflets 10–25 mm., 6–12(–15) pairs, lowest almost at base of petiole, oblong-lanceolate to linear-lanceolate, acute or mucronate; tendrils branched; *stipules half-arrow-shaped, entire*. Racemes 2–10 cm., 10–40-fld, rather dense; peduncles 2–10 cm., stout. *Fls 10–12 mm.*, blue-purple, shortly pedicelled and drooping; calyx-teeth very unequal, upper minute; tube short. Pod 10–20 mm., ovate, obliquely truncate, with a 'stalk' shorter than the calyx-tube, glabrous, 2–6-seeded. Fl. 6–8. $2n=28$ (14). Hp.

Native. In grassy and bushy places. 112, H40, S. Generally distributed and common throughout the British Is. Northern Europe and Asia to Sakhalin and Japan; Iceland; Greenland; introduced in N. America.

The 4 following resemble *V. cracca* in general appearance:

*V. tenuifolia** Roth. Usually more hairy. Racemes rather lax. Fls larger (12–15 mm.), pale blue or violet with whitish wings. Pod more gradually narrowed at the base, with a 'stalk' equalling the calyx-tube. A casual on disturbed ground, and locally naturalized. Europe from Sweden southwards, W. Asia, N.W. Africa.

*V. cassubica** L. Calyx-tube scarcely gibbous at base, pod short, stipitate, with 2–3 seeds. A casual in waste places, naturalized locally. Europe, W. Asia.

*V. villosa** Roth (incl. *V. dasycarpa* Ten.). Racemes longer than the subtending lf, calyx strongly gibbous, claw of standard twice as long as limb and 'stalk' of pod longer than the calyx-tube. Variable in hairiness, size of lflets, number of fls in the raceme and pubescence of pod. Several spp. have been described, based on these characters and differing somewhat in distribution, but intermediates are of frequent occurrence. Not infrequent in cultivated land and waste places. C. and S. Europe, W. Asia, N. Africa.

*V. benghalensis** L. (*V. atropurpurea* Desf.) differs from the preceding in the shaggy stipules and calyx and in having racemes shorter than their subtending lvs. Field margins, waste places, etc. S. Europe, N.W. Africa.

5. V. orobus DC. 'Bitter Vetch.'
Orobus sylvaticus L.

An erect branched ± pubescent perennial 30–60 cm. Stem stout, base lfless or with reduced lvs. Lflets 10–20 mm., 6–9 pairs, elliptic-mucronate; *tendrils* 0; *stipules half-arrow-shaped, slightly toothed*. Racemes 1–3 cm., 6–20-fld, secund, rather dense; peduncles about as long as lvs. *Fls* 12–15 *mm., white tinged with purple*, distinctly pedicelled and drooping; calyx-teeth very unequal, upper minute. Pod 20–30 mm., oblong-lanceolate, acute at both ends, glabrous, few-seeded. Fl. 6–9. $2n = 12$. Hp.

Native. In rocky and wooded places. 41, H6. Scattered throughout the British Is. but very local; rare in Ireland. From N. Spain through S., W. and C. France to the Swiss Jura; Jutland; west coast of Norway.

6. V. sylvatica L.   'Wood Vetch.'

A trailing glabrous perennial 60–130 cm. Lflets 5–20 mm., 6–9(–12) pairs, oblong-elliptic, mucronate; tendrils much branched; *stipules* lanceolate, *semicircular with many setaceous teeth at base*. Racemes 1–7 cm., up to 18-fld, secund, rather lax; peduncles stout, up to c. 10 cm. *Fls* 15–20 mm., *white with blue or purple veins*, distinctly pedicelled and drooping; *calyx-teeth setaceous, upper about ½ length of lower*. *Pod* c. 30 mm., oblong-lanceolate, *acuminate at both ends*, glabrous, few-seeded, black when ripe. Fl. 6–8. $2n = 14$. Hp.

Var. *condensata* Druce is a small densely lfy few-fld form found on shingle banks.

Native. In rocky bushy places, woods and shingle and cliffs by the sea. 86, H24. Scattered throughout the British Is., local. S.E. France through Switzerland, Germany and Denmark to Scandinavia (67° 56′ N.), east to Siberia (Baikal region) and N. Balkans.

Section 3. *Vicia*. Annuals or rarely perennials. Lflets 3–9, rarely 2 pairs. Peduncles very short or almost 0, 1–6-fld. Fls 10–25 mm., rarely less. Calyx-tube usually gibbous at base on upper side. Style bearded below stigma.

7. V. sepium L.   'Bush Vetch.'

A climbing or trailing, nearly glabrous perennial 30–100 cm. Lflets 10–30 mm., 5–9 pairs, ovate to elliptic, acute, obtuse or truncate, mucronate; tendrils branched; stipules half-arrow-shaped, sometimes toothed. Racemes 1–2 cm., 2–6-fld, subsessile. Fls 12–15 mm., pale purple, shortly pedicelled; *calyx-teeth unequal, the lower much shorter than tube, connivent*. Pod 20–25 *mm.*, oblong-lanceolate, beaked, *glabrous*, 6–10-seeded, black when ripe. Fl. 5–8. $2n = 14$. Hp.

Native. In grassy places, hedges and thickets. 112, H40, S. Throughout the British Is., common. Europe, but rare and local in the Mediterranean region; W. Asia, Baikal, Caucasus and Kashmir.

8. V. lutea L.   'Yellow Vetch.'

A tufted prostrate nearly glabrous annual 10–45(–60) cm. Lflets up to c. 10 mm., 3–7 pairs, linear-oblong to elliptic, obtuse and mucronate, rarely acute; tendrils simple or branched; *stipules small, triangular, the lower with a basal lobe*. *Fls* solitary, rarely in pairs, 20–25 mm., *pale* dirty *yellow*, shortly pedicelled; *standard glabrous; calyx-teeth setaceous, very unequal, the lower exceeding the tube*. *Pod* 20–30 mm., oblong-lanceolate, beaked, *pubescent*, 4–8-seeded. Fl. 6–8. $2n = 14$. Th.

Native. On cliffs and shingle by the sea. 22, S. Scattered round the coasts of Great Britain north to Ayr and Kincardine, local. Mediterranean region; rare and perhaps not native in much of western and C. Europe.

The 3 following resemble *V. lutea* in general appearance:

V. hybrida L. has toothed stipules, emarginate lflets, the fls solitary, yellowish, the standard pubescent on the back and the pod with spreading hairs. Formerly on Glastonbury Tor but long extinct except as a casual. S. Europe.

*****V. pannonica** Crantz, with entire stipules, fls usually 2–4 together (often brownish-violet) and the pod with appressed hairs, is found occasionally. C. and S. Europe to Asia Minor.

V. laevigata Sm. with pale blue or whitish fls and glabrous pods was formerly found on Weymouth beach. Doubtfully distinct from *V. lutea*.

9. V. sativa L.   'Common Vetch.'

A tufted trailing or climbing sparsely hairy annual 15–120 cm. Lflets 10–20 mm., 4–8 pairs, linear to obovate, acute, obtuse or retuse and mucronate; tendrils simple or branched; *stipules* half-arrow-shaped, toothed or entire, *often with a dark blotch*. Fls solitary or in pairs (rarely 4), 10–30 mm., purple; *calyx-teeth subequal, as long as the tube*. Pod 25–80 mm., linear-oblong, beaked, sparsely hairy to glabrous, 4–12-seeded. Fl. 5–9. $2n=12$. Th.

*Ssp. **sativa** has stout stems to over 1 m., 4–8 pairs of oblong or obovate, retuse and mucronate lflets, fls 20–30 mm., and fr. 40–80 mm.

Ssp. **angustifolia** (L.) Gaud. has slender stems 10–40(–80) cm., 3–5(–7) pairs of narrow often obtuse or acute lflets, fls 10–16 mm., and fr. 35–50 mm.

Both sspp. are variable and intermediates occur.

Native. In hedges and grassy places. Ssp. *sativa* cultivated and naturalized in waste places. 112, H40, S. Europe, W. Asia, N. Africa.

10. V. lathyroides L.   'Spring Vetch.'

A slender ± pubescent spreading annual 5–20 cm. *Lflets* 4–10 mm., *2–3 pairs*, linear-oblong or obovate, obtuse or emarginate and mucronulate; tendrils small and unbranched or 0; stipules small, half-arrow-shaped. Fls solitary, 5–7 mm., lilac; *calyx-teeth equal, nearly as long as tube*. *Pod* 15–25 mm., tapering at both ends, *glabrous*, 8–12-seeded. Fl. 5–6. $2n=10, 12$. Th.

Native. In dry grassy places, particularly on sandy soils. 70, H6, S. Scattered throughout Great Britain; N. and E. coasts of Ireland. Widely distributed throughout Europe north to S. Scandinavia, eastwards to the Caucasus and Asia Minor, southwards to the Crimea, Balkans, Italy, Spain and the Atlas.

Section 4. *Arachus* (Medic.) Tutin. Perennials or annuals. Lflets 1–2 pairs. Peduncles long or medium, 1–3-fld. Fls c. 20 mm. Calyx equal at base. Style bearded below the stigma.

11. V. bithynica (L.) L.   'Bithynian Vetch.'

A trailing or climbing nearly glabrous tufted perennial 30–60 cm. Stem angled, ± winged. Lflets 20–50 mm., 1–2 pairs, elliptic or ovate, acute or obtuse and mucronate, or sometimes linear-acuminate; tendrils branched; *stipules large* (1 *cm. or more*) *ovate-acuminate, toothed.* Fls solitary or in pairs, c. 20 mm., standard purple, wings white; peduncle very short or up to c. 5 cm., stout; *pedicels as long as calyx-tube; calyx-teeth exceeding tube.* Pod 30–40 mm., abruptly beaked, 4–8-seeded. Fl. 5–6. $2n=14$. Hp.

Native. On bushy cliffs and in hedges. 20, S. From Cornwall and Brecon to Essex, mainly near the coast, Worcester, Denbigh, Flint and N.E. Yorks, local. Europe from France southwards; Asia Minor; Algeria.

*V. faba L. (*Faba bona* Medic.), Broad bean, Horse bean, an erect annual without tendrils, is cultivated and sometimes persists for a few years at the margins of fields. Stems square; fls white with a purplish-black blotch, in axillary clusters; pods 10–20 cm. with few large (2–3 cm.) seeds. Cultivated in the Mediterranean region since prehistoric times.

*V. narbonensis L. An erect annual with purplish fls, tendrils on the upper lvs only and a pod 5–6 cm. occurs as a casual and persists in some localities. It is rather similar to, but smaller in all its parts than, *V. faba.* S. Europe, N. Africa, W. Asia.

23. LATHYRUS L.

Similar to *Vicia* but usually with fewer lflets and winged or angled stems. *Staminal tube transversely truncate.* Style flattened and *bearded on its upper side.* The fls have nectar and are usually cross-pollinated by bees.

About 110 spp. in north temperate regions and the mountains of tropical Africa and S. America.

1	Lflets 0.	2
	Lflets present.	3
2	Stipules lf-like; fls yellow.	**1. aphaca**
	Stipules minute; phyllodes grass-like; fls crimson.	**2. nissolia**
3	Tendrils present on most lvs.	4
	Tendrils 0, lvs ending in short point.	9
4	Lflets 1 pair.	5
	Lflets 2 or more pairs.	8
5	Stem flattened and distinctly winged.	6
	Stem ± square and sharply angled.	7
6	Fls 15–17 mm., mostly 5–7 together; calyx-teeth shorter than tube; pod glabrous.	**6. sylvestris**
	Fls 10–12 mm., 1–2 together; calyx-teeth longer than tube; pod densely silky.	**3. hirsutus**
7	Lflets lanceolate; fls yellow; pod compressed.	**4. pratensis**
	Lflets obovate; fls crimson; pod nearly cylindrical.	**5. tuberosus**
8	Stem winged; stipules lanceolate; pod compressed.	**7. palustris**
	Stem angled; stipules broadly triangular; pod turgid.	**8. japonicus**
9	Stem winged; stipules lanceolate; lower calyx-teeth about as long as tube.	**9. montanus**
	Stem angled; stipules linear-lanceolate; lower calyx-teeth longer than tube.	**10. niger**

Section 1. *Aphaca* (Mill.) Dum. Annual. Lvs with tendrils but devoid of lflets except in young seedlings. Calyx not gibbous at base.

1. L. aphaca L.   'Yellow Vetchling.'

A glabrous scrambling annual up to 100 cm. Lflets 0 in mature plants; tendrils simple; *stipules 1–3 cm., ovate-hastate*, acute or obtuse, entire. Peduncles equalling or exceeding stipules, 1-fld. *Fls* 10–12 mm., *yellow*, usually erect; *calyx-teeth linear-lanceolate, nearly equalling corolla.* Pod 20–30 × 5–8 mm., *somewhat falcate, 6–8-seeded*; seeds smooth. Fl. 6–8. Germ. autumn. A large proportion of the young plants is killed by severe frost. $2n=14$. Th.

Native. In dry places on sand, gravel or chalk. 33. Devon and Kent to Worcester and Norfolk; Glamorgan and Denbigh; very local. Western Europe from Germany southward; Mediterranean region east to Persia, Afghanistan and Egypt.

Section 2. *Nissolia* (Mill.) Dum. Annual. Lvs reduced to grass-like phyllodes; tendrils 0. Calyx ± gibbous at base.

2. L. nissolia L.   'Grass Vetchling.'

An erect or ascending nearly glabrous annual 30–90 cm. Lflets 0; *tendrils 0; stipules minute; phyllodes up to c. 15 cm., grass-like.* Peduncles shorter than or equalling phyllodes, 1–2-fld. *Fls* c. 15 mm., *crimson*, usually erect; *calyx-teeth triangular, much shorter than corolla.* Pod 30–60 × 2–3 mm., *straight, 15–20-seeded*; seeds rough. Fl. 5–7. Germ. autumn. $2n=14$. Th.

Native. In grassy and bushy places. 46. From Cornwall and Kent to Cheshire and Norfolk; S.E. Yorkshire; very local. Mediterranean region, north to France and Germany, east to the Caucasus and Syria; introduced in Belgium, the Netherlands and N. America.

Section 3. *Lathyrus.* Perennial or rarely annual. Lvs always with lflets and tendrils. Calyx-tube gibbous at base.

3. L. hirsutus L.   'Hairy Vetchling.'

An almost glabrous scrambling annual 30–120 cm. Stem winged. Lflets c. 5 cm., linear-oblong, mucronate; tendrils branched; stipules 1–1·5 cm., subulate, half-arrow-shaped. Peduncles exceeding lvs, 1–2-fld. *Fls* 10–12 mm., *with crimson standard and pale blue wings; calyx-teeth* subulate, *longer than tube.* Pod 30–50 × 6–8 mm., *tubercled and densely silky*, 8–10-seeded; seeds rough. Fl. 5–7. $2n=14$.

?Native. In fields and waste places. Rare and usually casual; north to Edinburgh. Mediterranean region, east to Transcaucasia.

4. L. pratensis L.   'Meadow Vetchling.'

A scrambling finely pubescent perennial 30–120 cm. Stem angled. *Lflets* 1–3 cm., *lanceolate*, acute; veins parallel, scarcely anastomosing; tendrils simple or branched; stipules 1–2·5 cm., lf-like, sagittate. Peduncles exceeding lvs, stout, (2–)5–12-fld. *Fls* 15–18 mm., *yellow*; calyx-teeth triangular-subulate, equalling tube. Pod 25–35 mm., glabrous or finely pubescent, *compressed*, 5–10-seeded, seeds smooth. Fl. 5–8. $2n=14, 28^*$. Hp.

Native. In hedges and grassy places. 112, H 40, S. Throughout the British Is., common. Europe north to 70° 21′ in Scandinavia, rare in the Mediterranean region; Siberia to the arctic circle, south to the Himalaya; N. Africa, sinia; Abysintroduced in N. America.

*5. L. tuberosus L. 'Earth-nut Pea.'

A glabrous scrambling perennial 60–120 cm. Roots bearing small *tubers*. Stem angled. *Lflets* 1·5–3 cm., *obovate*, obtuse or subacute, mucronate, tendrils simple or branched; *stipules* up to 1·5 cm., *narrowly lanceolate*, half-arrow shaped. Peduncles exceeding lvs, 2–5-fld. *Fls* 15 mm., *crimson*; calyx-teeth triangular, equalling tube. *Pod* c. 25 mm., glabrous, *nearly cylindrical*. Fl. 7. $2n = 14$. Hp.

Introduced. Naturalized in cornfields and hedges since about 1800 at Fyfield, Essex. Rare and scattered, usually as a casual, north to Westmorland. Europe and W. Asia from Germany southwards, rare in the Mediterranean region and absent from the islands.

6. L. sylvestris L. 'Narrow-leaved Everlasting Pea.'

A glabrous often glaucous scrambling perennial 100–200 cm. *Stem broadly winged*. *Lflets* 7–15 cm., *ensiform*; tendrils usually large and branched; *stipules* up to 2 cm., less than $\frac{1}{2}$ as wide as stem, *narrowly ensiform with a spreading basal lobe*. Peduncles 10–20 cm., 3–8-fld. Fls 15–17 mm., rose-pink; *calyx-teeth* triangular, *shorter than tube*. Pod 5–7 cm., glabrous, compressed, narrowly winged along upper side, 8–14-seeded. Fl. 6–8. $2n = 14$. Hp.

Native. In thickets and woods; sometimes naturalized in hedges near dwellings. 70. Scattered throughout Great Britain, local. Europe from Scandinavia and N. France southwards, east to C. and S. Russia.

*L. latifolius L., Everlasting Pea, closely related to *L. sylvestris* but with ovate lflets, stipules more than $\frac{1}{2}$ as wide as stem, fls 20–30 mm., and calyx-teeth equalling or exceeding tube, is naturalized in some localities. $2n = 14$. S. Europe.

*L. odoratus L., Sweet Pea, is a widely cultivated plant with fls c. 30 mm. Probably native in S. Europe from Spain to Italy.

7. L. palustris L. 'Marsh Pea.'

A glabrous scrambling perennial 60–120 cm. *Stem winged*. *Lflets* 3·5–7 cm., 2–3 *pairs*, *narrowly lanceolate*, acute or obtuse and mucronate; tendrils usually branched; *stipules* 1·5–2 cm., *lanceolate*, half-arrow-shaped. *Peduncles usually longer than lvs*, 2–6-fld. Fls c. 18 mm., pale purplish-blue; lower calyx-teeth subulate, almost equalling tube. *Pod* 3–5 cm., *compressed*, glabrous, 3–8-seeded. Fl. 5–7. $2n = 42$. Hp.

Native. In fens and damp grassy and bushy places. 19, H 13. Scattered in England, Wales and Ireland, very local. Europe north to N. Norway, absent from the Mediterranean region; arctic Russia and Siberia east to Sakhalin and Japan; eastern N. America.

8. L. japonicus Willd. 'Sea Pea.'

L. maritimus Bigel.

A creeping and ascending glabrous glaucous perennial 30–90 cm. *Stem angled*. *Lflets* 2–4 cm., 3–4 *pairs*, *obovate*, obtuse; tendrils simple or branched, sometimes 0; *stipules* c. 2 cm., *broadly triangular*, half-hastate. *Peduncles usually shorter than lvs*, 5–15-fld. Fls c. 18 mm., purple to blue; lower calyx-teeth narrowly triangular, about equalling tube. *Pod* 3–5 cm., glabrous, *turgid*, 4–8-seeded. Fl. 6–8. $2n=14$. Hp.

Native. On shingle beaches. 11, H1. From Cornwall and Dorset to Lincoln; Glamorgan; Angus, W. Ross; Kerry; very local. Circumpolar; south to N. France, north Russian coast and south to Lakes Ladoga and Onega, Siberia, Kamchatka, Sakhalin, China and N. America.

Section 4. *Orobus* (L.) Godr. Perennial. Lf ending in a short point, without tendrils. Calyx gibbous at base.

9. L. montanus Bernh.   'Bitter Vetch.'

L. macrorrhizus Wimm.; *Orobus tuberosus* L.

An erect glabrous perennial 15–40 cm. *Rhizome creeping and tuberous. Stem winged.* Lflets 1–4 cm., 2–4 pairs, narrowly lanceolate to elliptic, acute or obtuse and mucronate; *stipules lanceolate*, half-arrow-shaped, *variable but usually somewhat toothed below. Peduncles* 2–6-fld, *glabrous. Fls* c. 12 mm., *lurid crimson becoming green or blue; calyx-teeth unequal, lower about equalling tube.* Pod 3–4 cm., subcylindric, glabrous, 4–6-seeded. Var. *tenuifolius* (Roth) Garcke (*Orobus tenuifolius* Roth) has linear acuminate lflets and narrow stipules Fl. 4–7. $2n=14$. Hp.

Native. In woods, thickets and hedgebanks in hilly country. 108, H38. Scattered throughout the British Is., but commoner in the west and north. Europe, except the extreme north, very rare in the south-east.

10. L. niger (L.) Bernh.

Orobus niger L.

An erect nearly glabrous perennial 30–60 cm. *Rhizome short. Stem angled.* Lflets 1–3 cm., 3–6 pairs, lanceolate or elliptic, acute or obtuse and mucronate; *stipules linear-lanceolate*, half-arrow-shaped, *entire. Peduncles* 2–8-fld, *pubescent with appressed crisped hairs. Fls* c. 12 mm., *livid purple becoming blue; calyx pubescent with appressed crisped hairs, teeth unequal, lower longer than tube.* Pod c. 5 cm., turgid, rugose, 6–8-seeded. Fl. 6–7. $2n=14$. Hp.

Native or introduced. Rocky woods in mountain valleys up to 1200 ft. Formerly perhaps native in a few localities in Scotland but now extinct or very rare; naturalized in E. Sussex. In most of Europe north to 63° 40′ in Norway, absent from N. Belgium, the Netherlands, N. Germany and Scandinavia; east to the Caucasus; Algeria, very rare.

24. Pisum L.

Similar to *Lathyrus* but stems terete, *wings adherent to keel, calyx-teeth* ± *lf-like* and style dilated at the top with reflexed margins. The fls have nectar but are very rarely visited by bees and are completely self-fertile.

About 6 spp. in the Mediterranean region and W. Asia.

*1. P. sativum L.   Garden Pea, Field Pea.

A climbing glabrous glaucous annual 30–200 cm. Lflets 1–3 pairs, often with whitish markings; tendrils branched; stipules large and lf-like. Peduncle

1–3-fld. Fls 1–2 cm., white or (var. *arvense* (L.) Poir.) coloured. Pod sub-terete, several-seeded. $2n=14$. Th.

Introduced. In waste places as an occasional escape from cultivation. Cultivated from ancient times, probably native in the Near East.

52. ROSACEAE

Trees, shrubs or herbs. Lvs nearly always alternate and stipulate. Fls regular, usually hermaphrodite, perigynous or epigynous. Epicalyx sometimes present outside the sepals. Sepals usually 5, usually imbricate. Petals equalling in number and alternate with sepals, imbricate, rarely convolute, sometimes 0. Stamens usually 2, 3 or 4 times as many as the sepals, rarely indefinite or 1–5 only. Carpels 1–many, free or sometimes united to each other and to the receptacle (very rarely united and free from the receptacle); ovules usually 2, sometimes 1 or more, anatropous; styles free, very rarely united. Fr. of one or more achenes, drupes or follicles or a pome (very rarely a capsule), the receptacle sometimes becoming coloured and fleshy; endosperm 0, rarely scanty.

About 90 genera and over 2000 spp., cosmopolitan, especially temperate.

A very diverse but natural family usually recognizable by its perigynous or epigynous fls with ±numerous stamens. Closest to the Saxifragaceae (q.v. for distinctions). Liable to be confused with the Ranunculaceae by beginners but apart from the perigynous fls, the presence of stipules will separate any member likely to be taken for a member of the Ranunculaceae.

1	Trees or upright shrubs.	*2*
	Herbs or sometimes slightly woody and prostrate.	*12*
2	Fls bright yellow; epicalyx present; lvs pinnate; low unarmed shrub.	
		4. POTENTILLA
	Fls white (occasionally tinged yellow), or pink; epicalyx 0.	*3*
3	Receptacle strongly convex; lvs compound.	3. RUBUS
	Receptacle concave.	*4*
4	Receptacle slightly concave; carpels 5; fls pink, very numerous in a dense panicle.	1. SPIRAEA
	Receptacle strongly concave (carpels 1 or numerous), or ovary truly inferior (carpels 1–5).	*5*
5	Carpels and styles numerous (the latter occasionally connate into a column); prickly shrubs with pinnate lvs.	16. ROSA
	Carpels and styles 1–5; unarmed or thorny shrubs or trees (if lvs pinnate, then unarmed).	*6*
6	Carpel 1, free from receptacle; lvs simple, not lobed.	17. PRUNUS
	Carpels 1–5 (if 1 then lvs lobed), united to the receptacle so that the ovary is inferior (at least partly).	*7*
7	Fls in compound corymbs; false septum 0.	*8*
	Fls solitary or 2–3 or in short simple umbel-like corymbs; false septum 0.	*9*
	Fls in racemes; carpels with false septum.	21. AMELANCHIER
8	Thorny; lvs lobed; carpel-wall hard in fr.	19. CRATAEGUS
	Unarmed; lvs entire; carpel-wall hard in fr.	18. COTONEASTER
	Unarmed; lvs toothed, lobed or pinnate; carpel-wall cartilaginous in fr.	22. SORBUS

9   Fls less than 1 cm. diam.; fr. c. 1 cm. diam., red.   18. COTONEASTER
    Fls c. 3 cm. diam. or more; fr. larger, brown or green (sometimes tinged
    red).   10

10 Fls solitary; carpel-wall hard in fr.; sepals long, often lf-like.
    20. MESPILUS
    Fls several; carpel-wall papery in fr.; sepals short, not lf-like.   11

11 Styles free; anthers purple; fr. gritty.   23. PYRUS
    Styles connate below; anthers yellow; fr. not gritty.   24. MALUS

12 Carpels (4–)5 or more on the surface of a convex to weakly concave
    receptacle.   13
    Carpels 1–2(–4) enclosed in the strongly concave receptacle.   19

13 Epicalyx 0.   14
    Epicalyx present.   16

14 Lvs pinnate; fls small in many-fld panicles.   2. FILIPENDULA
    Lvs not pinnate; fls not in many-fld panicles.   15

15 Petals c. 8; plant prostrate; fr. a group of achenes with long feathery
    awns.   8. DRYAS
    Petals (4–)5; plant not prostrate; fr. a group of druplets.   3. RUBUS

16 Style terminal, persistent on the fr. as a long, jointed awn; radical lvs
    pinnate with the terminal lflet much larger than any of the lateral ones.
    7. GEUM
    Style ± lateral, not persistent; if lvs pinnate, terminal lflet not or scarcely
    larger than lateral ones (though these are sometimes unequal).   17

17 Stamens and carpels few (10 or less); petals inconspicuous, lvs ternate,
    lflets tridentate; alpine plant.   5. SIBBALDIA
    Stamens and carpels usually numerous; petals usually conspicuous; if lvs
    ternate, lflets with more than 3 teeth.   18

18 Receptacle fleshy in fr.; fls white, lvs ternate, receptacle glabrous.
    6. FRAGARIA
    Receptacle dry in fr.; if fls white and lvs ternate, receptacle hairy.
    4. POTENTILLA

19 Petals present; fls in racemes or few-fld cymes.   20
    Petals 0; fls in heads or many-fld cymes, rarely spikes.   21

20 Epicalyx present; receptacle without spines.   10. AREMONIA
    Epicalyx 0; receptacle with a crown of spines.   9. AGRIMONIA

21 Fls in cymes (sometimes dense and head-like, then lf-opposed); epicalyx
    present.   24
    Fls in terminal heads or spikes; epicalyx 0.   22

22 Plant erect; receptacle not spiny in fr.   23
    Plant prostrate; fr. receptacle with conspicuous spines.   15. ACAENA

23 Fls hermaphrodite, stamens 4; infl. oblong, fls crimson (or cylindric,
    fls white).   13. SANGUISORBA
    Fls polygamous; stamens numerous; infl. globose, fls greenish.
    14. POTERIUM

24 Perennial; fls in terminal cymes; stamens 4.   11. ALCHEMILLA
    Annual; fls in dense lf-opposed clusters; stamen 1.   12. APHANES

Subfamily 1. Spiraeoideae. Unarmed shrubs, rarely herbs. Stipules usually 0. Fls usually small and numerous in compound infl. Receptacle flat or slightly concave. Carpels 1–12, usually 5, in a single whorl; ovules several (rarely 2) in each. Fr. a group (rarely 1) of follicles, rarely a group of achenes or a capsule.

Tribe 1. Spiraeeae. Fr. follicular; seeds not winged.

Species of *Neillia* D. Don, *Stephanandra* Sieb. & Zucc., *Sibiraea* Maxim., *Aruncus* Schaeff. and *Sorbaria* (Ser.) A.Br. are sometimes grown. The two other tribes of this sub-family are represented in gardens by *Exochorda* Lindl. (Tribe *Quillajeae*) with capsular fr. and winged seeds, and *Holodiscus* (C. Koch) Maxim. (Tribe *Holodisceae*) with the fr. of 5 achenes (stipules 0).

1. Spiraea L.

Deciduous shrubs. Lvs alternate, *simple, exstipulate.* Infl. a simple or compound corymb or a panicle. Receptacle somewhat concave. Sepals and petals 5. Stamens 15–many. *Carpels* 5, *free, alternate with the sepals. Fr. membranous, dehiscent along the inner edge*; seeds several, oblong, very small; endosperm 0. Nectar secreted by a ring inside the stamens.

About 80 spp., north temperate. Many spp. (and some hybrids) are grown in gardens and a few have been found as escapes.

*1. S. salicifolia L.   'Willow Spiraea.'

Shrub 1–2 m. with numerous erect suckering stems. Branches strict, yellowish-brown, somewhat angled, puberulous when young. Buds small, ovoid. Lvs 3–7 cm., oblanceolate-oblong or elliptic-oblong, cuneate at base, acute or subobtuse, sharply and sometimes doubly serrate, glabrous. Infl. a narrowly conical or cylindric many-fld terminal panicle, 2–10 cm., dense or lax below, the fls always dense on the branches; rhachis, etc., pubescent. Sepals triangular-ovate, puberulous. Corolla pink, c. 8 mm. diam. Stamens c. twice as long as petals. Carpels subglabrous, erect. Fl. 6–9. Visited by various insects. $2n = 36$. N.

Introduced. Frequently planted in hedges, etc., mainly in Wales, N. England and S. Scotland and ± naturalized in woods, etc., in many places. 43. Native of E. Europe and temperate Asia from Czechoslovakia and Hungary (?Austria) eastwards; naturalized in other parts of Europe.

*S. douglasii Hook., similar to *S. salicifolia* but more robust and with lvs tomentose beneath, is rarely naturalized. Western N. America.

Physocarpus (Cambess.) Maxim.

Deciduous shrubs. Lvs lobed; stipules caducous. Fls white or pinkish, in corymbs. Frs 1–5, dehiscing along both sutures.

*P. opulifolius (L.) Maxim. Lvs 2–7(–10) cm., 5-lobed. Carpels 4–5. Fr. glabrous. Grown in gardens and occasionally naturalized. Eastern N. America.

Subfamily 2. Rosoideae. Herbs or shrubs. Stipules present. Receptacle from very convex to very concave. Carpels usually numerous and spirally arranged, but sometimes few or 1, free from each other and from the receptacle but sometimes enclosed in it; ovules 1 or 2 in each carpel.

Fr. of 1 or more achenes or a group of drupes (druplets) adhering to each other; seeds always 1.

Tribe 2. ULMARIEAE. Receptacle flat or slightly concave. Filaments narrowed at base. Carpels 5–15, in one whorl; ovules 2, pendulous. Fr. a group of achenes.

2. FILIPENDULA Mill.

Perennial herbs with short rhizome. Lvs alternate, pinnate with small pinnae between the large ones. Infl. a cymose panicle. Sepals and petals 5–8. Stamens 20–40. Homogamous pollen-fls without nectar; both British spp. are visited by various insects but automatic self-pollination occurs if visits fail.

About 10 spp., north temperate zone.

Radical lvs with 8 or more pairs of the larger lflets; lflets less than 1·5 cm.;
  carpels straight, erect.   **1. vulgaris**
Radical lvs with 5 or fewer pairs of the larger lflets; lflets 2 cm. or more;
  carpels twisted together after fl.   **2. ulmaria**

1. F. vulgaris Moench   Dropwort.

F. hexapetala Gilib.; *Spiraea Filipendula* L.; *Ulmaria Filipendula* (L.) Hill.
Perennial herb (7–)15–80 cm., nearly glabrous. *Roots bearing ovoid tubers. Radical lvs* ± numerous, 2–25 cm. (including the short petiole) *with 8–20 pairs of main lflets; lflets* 5–15 *mm.,* ± oblong in outline, pinnately lobed with acute, sparingly dentate lobes, green on both sides; terminal lflets trifid, resembling 3 fused lflets; smaller lflets 1–3 mm.; *stem-lvs very few,* the upper simple, lobed, very small, the lower usually with 2 or 3 pairs of lflets, sometimes a few at the extreme base resembling the radical lvs. Fl.-stems usually simple. Fls ± numerous in an irregular, cymose panicle much broader than high. Sepals usually 6, triangular-ovate, spreading, then reflexed. Petals usually 6, 5–9 mm., obovate-spathulate, cream-white tinged reddish-purple outside. Stamens about as long. *Carpels* 6–12, *erect, pubescent,* c. 4 mm. in fr. Fl. 5–8. $2n=14, 15^*$. Hs.

Native. Calcareous grassland, ascending to 1200 ft. in Yorks, widespread in England but rather local though often abundant; N. Wales, Carmarthen; very local in Scotland but extending north to Caithness; Ireland: Clare and Galway only; Jersey. 67, H2, S. Europe from Scandinavia and N. (not arctic) Russia to Spain and Portugal, Italy, Greece and the Caucasus; N. Africa (mountains); Asia Minor, Siberia (to the Angara-Sayan region).

2. F. ulmaria (L.) Maxim.   Meadow-sweet.

Spiraea Ulmaria L.; *Ulmaria pentapetala* Gilib.
Perennial herb, 60–120 cm. *Roots not tuberous.* Radical lvs ± numerous, mostly 30–60 cm. (including the rather long petiole), *with 2–5 pairs of main lflets; lflets* 2–8 *cm.,* ovate, acute, sharply doubly serrate, dark green and glabrous above, usually white-tomentose beneath but frequently green and pubescent or glabrous; terminal lflet 3- (or obscurely 5-)lobed to about ⅔, resembling 3 fused lflets; smaller lflets 1–4 mm.; lower stem-lvs resembling radical but shorter and with fewer lflets, uppermost simple or with only small

lflets in addition to the terminal one. *Fl.-stems lfy*, simple or branched above, glabrous or nearly so. Fls very numerous in an irregular cymose panicle, usually rather higher than broad, the fls dense on the ultimate branches. Sepals usually 5, triangular-ovate, reflexed, pubescent (like the fl.-stalks). Petals usually 5 (sometimes 6), 2–5 mm., obovate, clawed, cream-white. Stamens c. twice as long. *Carpels* 6–10, erect in fl. but *soon becoming twisted together spirally*, c. 2 mm. in fr., *glabrous*. Fl. 6–9. $2n=14, 16$. Hs.

Native. Swamps, marshes, fens, wet woods and meadows, wet rock ledges and by rivers, etc. (absent from acid peat); ascending to nearly 3000 ft. 112, H40, S. Common throughout the British Is., sometimes locally dominant in fens and wet woods. Europe from Iceland and arctic Russia to N. Portugal, C. Spain, C. Italy, Macedonia and the Caucasus; temperate Asia to Asia Minor and Mongolia; an escape in eastern N. America.

Tribe **KERRIEAE**. Receptacle weakly concave. Stamens numerous; filaments with broad base. Carpels 4–6, in one whorl. Fr. a group of achenes, often with hard coat.

Kerria japonica (L.) DC. Lvs alternate. No epicalyx. Petals 5, yellow. C. and W. China.

Rhodotypos scandens (Thunb.) Makino. Lvs opposite. Epicalyx present. Petals 4, white. Japan, C. China.

Both are monotypic genera, commonly grown in gardens, *Kerria* most commonly as a double-fld form.

Tribe 3. **POTENTILLEAE**. Receptacle convex to slightly concave. Stamens and carpels numerous (except *Sibbaldia* where both are few). Filaments with broad base (filiform in some *Rubus*). Fr. a group of achenes or of druplets.

3. RUBUS L.

Herbs or shrubs; if the latter, usually with biennial stems dying after fl. Epicalyx 0. Sepals 5, imbricate. Petals 5(–8). Stamens numerous. Carpels ± numerous, spirally arranged; *ovules* 2 *in each carpel*. Fr. of ± *numerous* 1-*seeded druplets, aggregated together* into a compound fr. Cotyledons elliptic, ciliate.

Spp. variously estimated at from c. 400 upwards. Cosmopolitan, especially north temperate zone.

1   Lvs simple; herb with solitary terminal fls (if shrubby see Subgenus
      Anoplobatus).   **1. chamaemorus**
      Lvs compound; fls in infl.   2

2   Stems annual, herbaceous, fl. branches thus from ground level; stipules
      attached to the stem.   **2. saxatilis**
      Stems woody, usually biennial, bearing axillary fl. branches in their 2nd
      year; stipules attached to the petiole only.   3

3   Lvs pinnate with 3–7 lflets; fr. bright red, coming away from the receptacle
      when ripe.   **3. idaeus**
      Lvs palmate with 3–5 lflets (rarely with 7 and then the 4 lower ones from
      the same point), or all ternate; fr. black or blue-black (rarely deep red),
      coming away with the receptacle when ripe.   4

4 Fr. very pruinose, with 2–5 or 14–20 druplets; stems very pruinose, terete,
weak; lvs ternate; stipules lanceolate.   **4. caesius**
Fr. not or slightly pruinose, druplets usually more numerous; stipules
linear or linear-lanceolate (if the latter lvs not all ternate).
                                                               5. fruticosus agg.

Subgenus 1. CHAMAEMORUS (Focke) Focke

Stems annual, herbaceous. Lvs simple, lobed. Stipules attached to the stem,
slightly adherent to the petiole above. Fl. solitary, terminal, dioecious.
Filaments flattened, tapering. Stone smooth.

1. R. chamaemorus L.   Cloudberry.

Unarmed somewhat pubescent herb with creeping rhizome. Fl. stems
5–20 cm., annual, erect. *Lvs* few; blade 1·5–8 cm., *simple*, ± orbicular,
palmately 5–7-lobed, deeply cordate at base, somewhat rugose; lobes dentate,
those of the male plant reaching c. $\frac{1}{3}$ of the way to the base, those of the
female plant shallower; petiole 1–7 cm.; stipules ovate, scarious. Fl. solitary,
terminal, (4–)5-merous. Sepals ovate, acuminate. Petals 8–12 mm., obovate,
white, much longer than sepals. Fr. orange when ripe, red earlier; druplets
few, large. Fl. 6–8. Visited by humble-bees and flies. $2n = 56*$.

Native. Mountain moors and blanket bogs, locally abundant; ascending
to 3800 ft. N. Wales; Lancashire, Derby and Yorks to Caithness but absent
from the Isle of Man, S.W. Scotland, Outer and Inner Hebrides, etc.; Ireland:
only known from Tyrone, where it is very rare. 42, H1. N. Europe from
Scandinavia and arctic Russia to the mountains of S. Germany (not France)
and C. Russia (Volga-Don region); N. Asia (east to Kamchatka and Sakhalin);
N. America.

Subgenus 2. CYLACTIS (Raf.) Focke

Stems annual, herbaceous. Lvs ternate, rarely simple. Stipules attached to
the stem, slightly adhering to the petiole above. Fls cymose or solitary,
terminal and axillary, hermaphrodite. Filaments flattened below. Receptacle
flat. Stone smooth or reticulate.

2. R. saxatilis L.   'Stone Bramble.'

A herb with long aerial stolons. Fl. stems annual, ± erect, 8–40 cm.; sterile
stems procumbent, much longer, producing axillary branches which root at
the tips; both kinds pubescent, armed with weak prickles or unarmed. Lvs
ternate; *terminal lflet* 2·5–8 cm., *usually stalked*, ovate to obovate, acute or
subacute, broadly cuneate at base, irregularly and often doubly dentate,
green and glabrescent above, paler and pubescent beneath; lateral lflets subsessile, often with a shallow rounded lobe on the lower margin; petiole
2–7 cm., armed like the stem; stipules ovate, green. Fls 2–8 in a compact
terminal cyme, sometimes with axillary cymes in addition; peduncle from
very short to 2·5 cm. Sepals triangular-ovate, acute or acuminate, shortly
pubescent. *Petals* 3–5 mm., oblanceolate, *white*, shorter or scarcely longer
than the sepals. Fr. scarlet, translucent; druplets 2–6, large, glabrous, separate.
Fl. 6–8. Visited by bees and flies, self-pollination possible. $2n = 28*$.

Native. Stony woods and shady rocks, especially basic, in hilly districts, rather local; Cornwall to Somerset; Wales and the Border Counties; N. Lincoln, Derby and Lancashire northwards; widespread in Ireland from Kerry, Leix and Wicklow northwards. 70, H31. Europe from Iceland and arctic Russia to the Pyrenees, mountains of Italy, N. Greece and the Caucasus; N. Asia Minor, temperate Asia to the Himalaya and Japan; Greenland.

R. arcticus L.

Similar to *R. saxatilis*, but creeping underground and without aerial stolons. Lvs smaller; terminal lflet 1–5 cm., subsessile. Fls solitary. Petals 7–10 mm., bright pink. Fr. dark red.

Appears formerly to have occurred in the Scottish Highlands but has not been seen for many years. Arctic and subarctic Europe, Asia and America.

Subgenus ANOPLOBATUS Focke

Stems biennial, woody, unarmed; bark peeling. Lvs simple, palmately lobed. Stipules attached to the petiole. Fls hermaphrodite. Receptacle flat.

The two following spp. are grown for their fls and sometimes become naturalized:

*R. odoratus L.

Stems erect, up to 3 m. Lvs 10–30 cm. broad, 5-lobed. Infl. many-fld. Fls 3–5 cm. diam., purple. Native of eastern N. America.

*R. parviflorus Nutt. (*R. nutkanus* Moç. ex Ser.)

Stems erect, up to 2 m. Lvs 6–20 cm. broad, 3–5-lobed. Infl. 3–10-fld. Fls 3–6 cm. diam., white. Native of western N. America.

Subgenus 3. IDAEOBATUS Focke

Stems biennial, woody, normally prickly. Lvs pinnate, rarely palmate or simple. Stipules attached to the petiole. Fls cymose, normally hermaphrodite. Filaments subulate. Fr. red or yellow, coming away from the dry conical receptacle when ripe. Stone rugose.

3. R. idaeus L.   Raspberry.

Suckering by adventitious buds from the roots. Stems 100–160 cm., erect, terete, somewhat pruinose, armed with slender, straight, subulate prickles, rarely unarmed. Lvs pinnate with 3–5(–7) lflets; lflets 5–12 cm., the terminal one larger than the others, ovate or ovate-lanceolate, acuminate (rarely (var. *anomalus* Arrh.) orbicular, obtuse), rounded or subcordate at base, irregularly dentate with mucronate teeth, green and somewhat pubescent above, densely and closely white-tomentose beneath; petiole 2–7 cm. cylindric; stipules filiform. Fls 1–10 in dense axillary and terminal cymes, forming a compound infl. on a short lateral branch with ternate lvs. Sepals ovate-lanceolate-triangular with long acuminate tips. Petals oblanceolate or oblong, white, erect, about as long as sepals. Fr. red, rarely pale yellow, opaque; druplets numerous, pubescent. Fl. 6–8. Pollinated by various insects and selfed. $2n = 14^*$.

Native. Woods and heaths, especially in hilly districts; ascending to 2700 ft. Common almost throughout the British Is. but not recorded for W. Cornwall

and Pembroke. 109, H 40, S. Europe from Iceland and N. (not arctic) Russia
to C. Spain, Sicily, Greece and the Caucasus; N. Asia Minor, N. and C. Asia
east to the Yenisei region; sspp. or allied spp. in E. Asia and N. America.

*R. phoenicolasius Maxim.   Japanese Wineberry.
Stems densely covered with red gland-tipped bristles. Lflets 3(–5), ovate, green and
slightly pilose above, white-tomentose beneath. Petals pinkish, shorter than sepals,
appressed to stamens. Fr. orange-red, translucent, edible.
    Sometimes grown in gardens and occasionally naturalized. Native of Japan,
Korea and N. China.

*R. spectabilis Pursh
Stems with slender prickles below. Lflets 3, subglabrous. Fls solitary, bright rose-
purple, c. 2·5 cm. diam. Fr. large, orange, edible.
    Grown in gardens for its fls and sometimes naturalized. Native of western N.
America.

*R. loganobaccus L. H. Bailey   Loganberry.
Stems robust, long, arching. Lflets 5, pinnate (as in *R. idaeus*), large. Fr. purplish-
red, large, coming away with the receptacle (as in subgenus *Rubus*). $2n=42$.
    Commonly grown for its fr. and rarely naturalized. Originated in 1881 in a Cali-
fornian garden by hybridization between *R. idaeus* ssp. *strigosus* (Michx.) Focke
(an American ssp.) and *R. vitifolius* Cham. & Schlecht. (a Californian sp. of subgenus
Rubus); a hexaploid, now behaving as a sp.

Subgenus 4. Rubus   Blackberry, Bramble.

Stems biennial (sometimes lasting longer or flowering the first year), woody,
prickly. Lvs ternate or palmate (at least in their primary division). Stipules
attached to the petiole. Infl. usually compound. Fls hermaphrodite. Filaments filiform. Fr. black (occasionally deep red or blue with bloom), coming
away with the fleshy conical receptacle when ripe. Pollinated by various
insects. (Description applying to European spp. only.)

4. R. caesius L.   Dewberry.

Stem procumbent, rooting, *weak*, *terete*, glabrous, very *pruinose*; bark finally
exfoliating; stalked glands and acicles few or 0; *prickles very* weak, subulate,
straight to falcate, *scattered*. Lflets 3, sparsely pilose above, green and subglabrous or closely pubescent beneath, irregularly, coarsely and doubly
dentate or shallowly lobed; terminal variable in shape but ± ovate or rhombic,
acute or slightly acuminate, mostly emarginate at base, sometimes trilobed;
stipules lanceolate, those of the lower lvs broader, those of upper lvs narrower.
Infl. short, lax, few-fld, shortly pubescent, glandular or nearly eglandular and
with weak prickles; peduncles ascending, pedicels slender. Sepals grey-green,
white-margined, cuspidate with long points, clasping the fr. Petals elliptic to
suborbicular, white or pinkish. Stamens about equalling styles; filaments
white; pollen perfect. Styles greenish. Fr. densely pruinose and so appearing
bluish, partially separable from receptacle; lateral frs with few druplets,
terminal fr. with many more. Very variable. Fl. 6–9. $2n=28$.
    Native. Dry grassland and scrub mainly on basic soils, also common in
fen carr; widespread and common in England and Wales; local in Scotland,

especially in the north, extending to Caithness; rather local in Ireland. 86, H29, S. Europe from Scandinavia (62° N.) and N. Russia to Spain and Portugal, Sicily and Greece; W. Siberia and Russian C. Asia to Asia Minor.

5. R. fruticosus agg.

Bark not exfoliating. Lvs rarely all ternate. Stipules filiform, linear or linear-lanceolate. Fr. glabrous when ripe, black (rarely deep red), not or slightly pruinose adhering to the receptacle and coming away with it when ripe. Fl. 5–9. Pollinated by various insects.

Native. Woods, scrub, hedges and heaths. Common throughout the British Is. 112, H40, S. Europe, Mediterranean region, Macaronesia.

The spp. included in this aggregate are very difficult to determine and the number of British spp. has been very variously estimated. Bentham (*Handbook of the British Flora*) includes them all under one sp., *R. fruticosus* L., whereas the latest account by W. C. R. Watson (*Handbook of the Rubi of Great Britain and Ireland*, 1958) gives 386 spp. Though pollination is necessary for development, the vast majority of forms are apomictic (*R. ulmifolius* is the only known exception in the British Is.) and thus constant, but can on occasion reproduce sexually and give rise to hybrids (and presumably also at times to new apomicts). It does not seem possible at present to group the forms satisfactorily into a limited number of spp. In the following account the sections and series used by Watson are given, and a few of the common species mentioned. The number of spp. in each section and series is taken from his book. It should be noted that the 'sections' are not comparable with the sections in genera with completely or mainly sexual reproduction but are at a much lower taxonomic level.

It must also be remembered that the spp. appear to form a ±continuous network of forms from the *Suberecti* and *Discolores* on the one hand to the *Glandulosi* on the other and that it is often a matter of doubt to which section a particular sp. should be referred, although the central forms in each section are usually distinct enough. The following are some of the more important characters used in the group:

Habit. The majority of spp. have stems ±arching over and rooting at the tips but the height of the arch varies much, in some spp. reaching 3 or 4 ft., in others being negligible so that the stems lie along the ground unless supported. The *Suberecti* differ from the other sections in the suberect, not rooting, stems.

Angularity of stem varies from terete to sharply angled in different species and in proceeding from the base to the tip of the stem.

Armature and indumentum of first-year stem (descriptions of armature always refer to this and not to the branches). Apart from eglandular hairs and pruinosity, the organs clothing the stem may be grouped as follows:

Prickles. The largest type of arm. In the earlier sections they are confined to the angles of the stem, uniform in size and the only type present. In the later sections they are scattered round the stem, unequal, and pass gradually into the other types. Numerous intermediate stages occur. The size, shape, direction and abundance of the prickles are often most important characters.

Pricklets. Small prickles, often with stout bases, scattered round the stem.

Acicles. Very slender prickles or very stiff bristles, not stout-based, sometimes gland-tipped.

Stalked glands.

The last three types may be all ± equal in length to each other and among themselves and sharply marked off from the prickles (as in the *Radulae*) or very unequal and passing into each other and into the prickles (as e.g. in the *Hystrices*).

The quantity of stalked glands, acicles and pricklets present on the stem varies much from the lower part to the upper part of the same internode as well as from the lower part to the middle part of the same stem and also between a young plant and an older one of the same sp. Prickles always become more curved towards the tip of the stem. The quality of the armature is more constant.

The above characters of habit and armature are those mainly used for delimiting the sections.

Leaves. Number of lflets per lf. Nature and colour of the indumentum especially of lower surface. Shape of lflets especially of the terminal one. Depth and type of toothing. Length of stalks of lflets and especially whether the basal lflets are stalked or sessile. The lvs used are normally those about the middle of the first-year stems (not those of the branches).

Infl. The shape and branching of the infl., its armature and indumentum and leafiness.

Sepals. Armature and indumentum which varies considerably so that the sepals vary from green and sparingly hairy with grey-tomentose margins to uniformly grey- or white-tomentose. The direction of the sepals at the time the fr. swells is a character of considerable importance and it is this that is intended in the following descriptions unless otherwise mentioned.

Petals. Shape, colour (when first expanding) and clothing.

Stamens. Length relative to the styles. Colour (when first expanding) of filaments. Whether the anthers are glabrous or pilose (the latter occurs only in a few spp. and unless otherwise mentioned the anthers are glabrous).

Styles. Colour, especially at the base.

Specimens of brambles should consist of (1) pieces of stem from the middle of the first-year stem with lvs attached; (2) infl., preferably at two stages showing (*a*) buds and fls, and (*b*) fls and young fr.; they should be gathered so that there is at least one lf without an axillary peduncle below the infl.; (3) separate petals. Care should be taken to collect them from the same bush. Note should be made when collecting, of the habit of the plant, of the number of lflets per lf (many spp. have a variable number of lflets and this does not always show on the specimen), and of the colour of the various parts of the fl. in the opening bud and of the stem, etc. The direction of the sepals and the relative length of the stamens and styles are also not always easy to see when dry. The ideal procedure is to determine at least the section of the plant in the field and to prepare the specimen, bearing in mind the characters used for determination.

In view of what has been said above it will be recognized that great caution is necessary in using the key given here.

While many of the bramble spp. are widespread (some of the *Suberecti*, for example, extend from Britain to Russia), others are very restricted in range although they may be abundant where they occur. In any one locality the spp. are usually easily separable but in another neighbouring locality, an entirely different selection of spp. may occur.

For a full account of the plants of this genus, Watson's book should be consulted.

In the following key and descriptions 'stems' refers throughout to the first-year stem and 'lvs' to well-developed lvs from about the middle of such stems. If a character is qualified by 'usually' in the description of a section it means that the great majority of spp. in the section possess it.

Key to the Sections

1 Stalked glands absent from stem and infl.   2
    Stalked (or subsessile) glands present at least on infl.   4

2 Stem suberect or high-arching, not rooting, glabrous or nearly so; lvs green or somewhat greyish beneath; some lvs often with 7 lflets; infl. simple or only slightly branched; sepals green or greyish-green outside; fl. early.   **1. Suberecti**
    Stem arching and rooting; lflets never 7, white-felted beneath; infl. branched, fls large and usually brightly coloured; sepals grey or greyish-white, always reflexed; fl. late.   **4. Discolores**
    Not as either of above.   3

3 Stem usually low; prickles scattered; basal lflets usually sessile or sub-sessile; infl. usually corymbose with comparatively few large long-stalked fls; druplets often few, large, partly defective (common on calcareous soils).   **2. Triviales**
    Stems with the prickles confined to the angles; basal lflets usually stalked; fls usually numerous; druplets numerous (rare on calcareous soils).   **3. Silvatici**

4 Basal lflets usually sessile or subsessile, often overlapping the others; stipules usually linear-lanceolate; infl. usually corymbose with comparatively few large long-stalked fls; druplets often large, partly defective; stem usually low, terete or obscurely angled, prickles scattered.   **2. Triviales**
    Basal lflets nearly always stalked; stipules usually linear; other characters not combined as above.   5

5 Stem prickles equal or somewhat unequal, seated on or near angles of stem, much larger than and clearly marked off from the pricklets, acicles and stalked glands.   6
    Stem prickles very unequal, scattered all round the stem, usually connected with the pricklets, acicles and stalked glands by intermediates.   **7. Glandulosi**

6 Stalked glands, acicles and pricklets 0 or very few on the stem, not more numerous than the prickles.   7
    Stalked glands, acicles and pricklets (or some of them) ±numerous on the stem (at least about as many as prickles).   **6. Appendiculati**

7 Stems slender, obscurely angled, ±hairy, stalked glands, pricklets and
      acicles 0 or very few; fls small (less than 2 cm. diam.); sepals never
      reflexed; stamens shorter than styles.   **5. Sprengeliani**
   Stems mostly more robust, hairy or not, stalked glands, pricklets and
      acicles 0 or very few; fls more than 2 cm. diam.; stamens usually longer
      than styles.   **3. Silvatici**
   Stalked glands, acicles and pricklets few but some of them always present;
      stem and lvs usually very hairy; fls usually more than 2 cm. diam.
                                                    6. Appendiculati ser. *Vestiti*

Section 1. **Suberecti** P. J. Muell. Often spreading by suckers. *Stems suberect,*
arching only near the apex, *not rooting*, angled, glabrous or more rarely
slightly pilose at first or with sessile glands; *prickles confined to the angles*
(except in *R. scissus* W. Wats.), *all equal; pricklets, acicles and stalked glands* 0.
Lflets (3–)5(–7) but always 5 on most lvs, green on both sides, or occasionally
somewhat greyish beneath, the basal pair usually subsessile or shortly stalked.
Infl. usually nearly simple, subracemose; stalked glands 0. *Sepals green or
greyish-green* with conspicuous white-tomentose margins, spreading or reflexed
after fl. Petals usually white or pale pink. $2n = 21, 28, 42$. Fl. 5–7.
   Usually on heaths or in woods, especially high forest, usually on very acid
soils.
   A natural group of comparatively few (13) spp., clearly marked off from
all other groups except the *Silvatici*. They are believed to have originated by
hybridization with *R. idaeus*, but some spp. are very wide-ranging and they
are clearly of ancient origin.

The 3 following spp. are common over most of the British Is.

R. nessensis W. Hall

Prickles short (to 1·5 mm.), straight, purplish-black. Lflets large, 3–5(–7), green on
both sides. Petals glabrous, white. Stamens white, longer than greenish styles.
Fr. dark red.

R. scissus W. C. R. Wats.

Similar to *R. nessensis* but with scattered prickles, stamens shorter than styles and
pubescent petals.

R. plicatus Weihe & Nees

Prickles slender but stronger than in 2 preceding, falcate, yellow and red. Terminal
lflet broad, plicate. Petals hairy. Stamens about equalling styles. Fr. black.

Section 2. **Triviales** P. J. Muell. (*Corylifolii* Focke). Stems low-arching or
procumbent, rooting at end, terete or slightly angled, *prickles scattered round
the stem, usually straight and spreading*; stalked glands, acicles and pricklets
variable (from 0 to numerous). Lflets 3–5, *the basal ones sessile or subsessile*,
usually broad, green or grey beneath; *stipules usually lanceolate or linear-
lanceolate* (broader than in the other sections) but filiform in some spp. Infl.
various but mostly either short, or narrow with the branches few-fld, stalked
glands few or many, mostly unequal. Sepals grey, usually spreading or erect.
Petals usually large, often suborbicular. *Druplets usually large and few, partly
abortive, not, or rarely at length slightly, pruinose.* $2n = 28, 35, 42$. Fl. 5–9
(the fl. period of the individual spp. is also long).
   Mainly hedges and scrub, frequent on basic soils.

This section (20 spp.) is believed to have originated comparatively recently by hybridization between *R. caesius* and various members of the other sections.

Very local hybrids involving *R. caesius* or members of this section occur fairly frequently and only those with a fair distribution have been given names. Several spp. are widespread and common. The commonest may be distinguished as follows:

1 Prickles equal or nearly so.   2
   Prickles numerous, unequal; glands numerous.   4

2 Stalked glands 0 or very few.   3
   Stalked glands rather numerous; lvs large, soft; fls very large (to 5 cm.
   diam.), pink; stamens pink, anthers hairy; styles red. Fl. begins early
   May.   **R. balfourianus** Bloxam ex Bab.

3 Petals pink; prickles short; basal lflet stalked, terminal not lobed.
                                                           R. conjungens (Bab.) W. C. R. Wats.
   Petals white; prickles deep purple; terminal lflet often lobed.
                                                           R. sublustris Lees

4 Terminal lflet suborbicular; infl. broad.   **R. scabrosus** P. J. Muell.
   Terminal lflet ± oblong; infl. narrow.   **R. myriacanthus** Focke

Section 3. **Silvatici** P. J. Muell. *Stem arching*, often low or scrambling, *rooting at the end, angled* (occasionally only weakly so), *glabrescent or hairy; prickles confined to the angles, usually without pricklets, acicles or stalked glands*, rarely with a few. Lflets 5 on most lvs in the great majority of spp., green or greyish beneath, rarely greyish-white, the basal pair stalked except in a very few spp. Infl. varying in shape, from little- to much-branched and from eglandular to rather glandular with stalked glands. Sepals usually uniformly grey-tomentose, sometimes green or greyish-green with whiter margins. Stamens longer than the styles in the great majority of spp. $2n = 21$, 28, 42. Fl. 7–8.

Woods, scrub, heaths, hedges, etc.

A large and varied group (94 spp.). Variously subdivided by different authors. Watson divides it into 2 subsections and 7 series, the differences between which are ill-defined and difficult. It may be noted that all eglandular brambles which are clearly not *Suberecti*, *Triviales* or *Discolores* belong here.

The following are common spp.:

Subsection A. *Virescentes* Genev. Lvs green beneath, hairy, rarely the upper lvs of stem and panicle greyish felted (55 spp.).

R. carpinifolius Weihe & Nees
Eglandular. Stem tall. Prickles strong, numerous, yellow or reddish. Lflets 5(–7), unequally serrate, hairy beneath with longer hairs along the midrib and main veins. Panicle little-branched above, with numerous straight prickles. Fls rather large, faintly pink in bud. Sepals spreading.

R. nemoralis P. J. Muell.
Stem furrowed, becoming deep purple. Prickles strong, curved. Lflets 5(–7), serrate, pubescent to grey-felted beneath; terminal suborbicular. Panicle broad, lfy, interrupted, with scattered glands. Sepals loosely reflexed, then patent, with small spines and long lf-like tips. Petals and filaments pink. Carpels pilose.

*R. laciniatus Willd.

Lflets irregularly and deeply pinnatifid (and so unlike all other spp.). Origin unknown, not known as a wild plant but fairly frequent as a bird-sown escape.

R. lindleianus Lees

Very prickly. Stem very tall, shining, prickles large. Lflets greyish-felted beneath, irregularly toothed. Panicle very large, branches spreading at right angles, densely hairy with a few glands. Fls white, very late (7). Sepals reflexed.

R. macrophyllus Weihe & Nees

Very robust. Stem hairy, prickles short, subulate. Terminal lflet very large, ovate, acuminate with long point, unequally serrate. Infl. little branched, densely hairy with a few glands and weak prickles. Fls usually pink. Sepals reflexed or spreading. Receptacle hairy.

R. pyramidalis Kalt.

Stem low-arching, sometimes with a few glands and acicles. Prickles many, straight. Lflets softly hairy beneath, pectinately hairy on the veins, irregularly serrate. Infl. dense, pyramidal with straight narrow-based prickles. Sepals loosely reflexed or spreading. Petals pink; filaments white. Receptacle hairy.

Subsection B. *Discolorides* Genev. Lvs, at least the upper, greyish-white-felted beneath (39 spp.).

R. polyanthemus Lindeb.

Stem angled, somewhat pilose, yellow or red, slightly glandular. Lflets 3–6(–7) grey-felted beneath, simply serrate; terminal usually obovate, cuspidate. Panicle long, many-fld, with many prickles, glandular. Sepals reflexed, glandular and bristly. Petals pink. Carpels pilose.

R. cardiophyllus Muell. & Lefèv.

Eglandular. Stem red, glabrescent. Prickles rather few, straight. Lflets 5(–7) sharply and rather deeply serrate; terminal long-stalked, ovate-orbicular, cordate at base. Panicle rather short and broad. Sepals reflexed. Petals hairy, abruptly clawed, white or pinkish. Receptacle hairy.

Section 4. **Discolores** P. J. Muell. Eglandular and without acicles or pricklets. Stems arching, often scrambling, often red or purple, angled, rooting at ends. Prickles equal, confined to angles. *Lflets* 3–5, subcoriaceous, *white-felted beneath*, the basal pair stalked; stipules linear. Infl. compound and well-developed. Sepals grey-tomentose, reflexed. Petals large, often bright pink. Stamens longer than or equalling styles. Fl. 7–8. $2n = 14, 21, 28, 35$.

A small group in Britain (14 spp.), only likely to be confused with some of the *Silvatici*.

The following is very common and the only known diploid sexual sp. of bramble in Britain. It is the only common sp. in this section.

R. ulmifolius Schott

Stem arching, often climbing or forming dense tangled bushes, furrowed, very *pruinose*, with a very fine often scarcely visible stellate pubescence; prickles strong, spreading to recurved. *Lflets* 3–5, small, dark green, somewhat rugose and glabrescent above, *densely and closely white-tomentose beneath*. Infl. cylindrical, lfy only at base; branches spreading; prickles strong, falcate.

Petals crumpled, suborbicular, clawed, usually bright purplish-pink. *Stamens* pink, purplish or rarely white, *about equalling styles*. Styles usually pink or purple, sometimes greenish. Very variable. Fl. late. $2n=14^*$; sexually reproduced, largely (?entirely) self-sterile.

The commonest sp. of bramble in general (except in Scotland) and unlike most sp. growing on chalk and heavy clay soils. Frequently hybridizing with other spp. and forming fertile hybrids.

Netherlands and S.W. Germany to Spain and Portugal, Italy and Dalmatia; N.W. Africa; Macaronesia.

Section 5. **Sprengeliani** Focke. *Stems* arching, *slender*, procumbent or climbing, rooting at the ends, obscurely angled, glabrescent or hairy; prickles confined to the angles, often somewhat unequal, a few pricklets and stalked glands often present. Lflets 3 or 5, thin, green or more rarely greyish or greyish-white beneath, the basal pair stalked. Infl. compound, usually short and loose, sparsely glandular. *Sepals* uniformly grey or greyish-green with white margins, *spreading, or clasping the fr. Petals small*, or if larger, crumpled. *Stamens shorter than or equalling styles.* $2n=21, 28$. Fl. 7–8.

Woods, scrub, heaths, etc.

A small natural group (7 spp.), closest to some *Silvatici* and *Suberecti*, but differing in the combination of characters given above.

No sp. is common but *R. sprengelii* Weihe is widespread. It has slender prickles, 3–5 lflets, the terminal elliptic-obovate, a short broad infl., and narrow crumpled bright rose-pink petals. The other spp. are all very local or rare.

Section 6. **Appendiculati** Genev. Stems arching to procumbent, rooting at ends. Prickles varying from equal and confined to the angles to somewhat scattered. Pricklets, acicles and stalked glands always present but varying from few (comparable in number to prickles) to numerous but not passing into the prickles. Sepals grey or greenish-grey.

A very large and varied section (178 spp.), formerly grouped in several sections which are now reduced by Watson to series. As the range of variation is great, these series are considered separately below.

Series 1. *Vestiti* Focke. *Stems* low-arching or sometimes prostrate, rooting at the end, angled (occasionally nearly terete), *usually markedly hairy, prickles confined to the angles, pricklets, acicles and stalked glands present but few* ($\pm$ comparable in number to the prickles). Lflets 5 on most lvs in most spp., the basal pair stalked. *Infl.* $\pm$ glandular with stalked glands and aciculate, *mostly conspicuously pilose* with the hairs longer than the glands. Sepals usually grey-tomentose, usually reflexed but sometimes spreading. Stamens usually longer than styles. Fl. 7–8.

Woods, scrub, hedges, heaths, etc.

This series (30 spp.) links the *Silvatici* with the other series in this section, being less glandular than the latter but more so than the former. The plants are often, but not always, very hairy. The following is a very common and usually easily recognized sp.

R. vestitus Weihe & Nees
Stems densely hairy with short and long hairs. Lflets 3–5, feeling thick and soft, densely greenish-tomentose beneath and pectinately pilose on the veins; terminal

suborbicular, shallowly and evenly serrate-dentate with broad teeth, shortly cuspidate. Infl. long, with long slender declining prickles. Fls large. Sepals reflexed. Petals suborbicular, either like the filaments deep rose-pink or more commonly (var. *albiflorus* Boul.) pinkish and the filaments white. Common in England and Wales, rare or absent in Scotland, tolerant, unlike most spp., of some lime and of clay.

Series 2. *Mucronati* W. C. R. Wats. (Section *Rotundifolii* W. C. R. Wats.). Stem low- (rarely high-)arching, rooting at end, angled or subterete, glabrous or hairy, prickles confined to angles, *pricklets, acicles and stalked glands present*, unequal, but very variable in number, from very few (and usually few on the basal part of the stem) to rather numerous in the upper part of the stem. Lflets 3–5, grey or green beneath with *fine or rather fine, ± even, mucronate teeth; terminal orbicular or broadly ovate or obovate, cuspidate or abruptly acuminate.* Infl. ± glandular and aciculate. Sepals greenish-grey with white margins or wholly grey, spreading or reflexed. Stamens usually longer than styles. Fl. 7–8.

Mainly woods, scrub and hedges.

A small group (16 spp.) recently separated by Watson, partly characterized by variability in armature. Different members have been referred by various authors to *Sprengeliani*, *Vestiti* and *Apiculati*. The rounded terminal lflet is usually characteristic.

No sp. is more than locally common and several are rare. There are 3 Irish endemic spp.

Series 3. *Dispares* W. C. R. Wats. Stem arching, often angled, usually hairy, prickles usually numerous with *subsessile glands, short acicles and small pricklets present and irregularly distributed on stem and panicle* usually very few on lower part of stem, longer glands, acicles and pricklets present or 0. Lflets (3–)5, green or grey- or white-felted beneath, often coarsely serrate. Stamens longer than styles.

A small group (7 spp.), first separated by Watson in the *Handbook*. It appears to be chiefly characterized by the presence of subsessile glands. No sp. is common and most are rare.

Series 4. *Radulae* Focke. Stems arching, usually rather robust, rooting at end, angled, *prickles confined to angles; acicles and stalked glands numerous, short and ± equal*, sharply differentiated from the prickles, making the stem rough to the touch between the prickles; pricklets mostly few, equalling the acicles and glands. Lflets usually 5, grey or whitish beneath in most spp., but green in some, the basal pair stalked. Infl. usually well-developed, armed much like the stem but the arms more unequal, the glands mostly shorter than the hair. Sepals grey or greenish-grey. Stamens usually longer than styles. Fl. 7–8.

Mainly hedges, scrub and edges of woods.

A rather well-defined group (18 spp.); the clothing of the stem is characteristic in the commoner spp.

Several spp. are widespread and locally common. The 3 commonest can be distinguished as follows:

1 Stalked glands, acicles and pricklets all short and nearly equal; lvs finely
     but unequally serrate-dentate.   **R. radula** Weihe ex Boenn.

Stalked glands, acicles and pricklets more unequal; lvs coarsely and
    jaggedly compound-serrate.   2

2 Robust; stem hairy; infl. broad, petals broad, pink, entire; carpels
    glabrous.   **R. discerptus** P. J. Muell.
  Slender; stem glabrous; infl. narrow; petals narrow, emarginate, pinkish;
    carpels pilose.   **R. echinatoides** (Rogers) Sudre

Series 5. *Apiculati* Focke. *Stem usually low-arching or procumbent and not
very strong*, rooting at end, angled or subterete, *prickles subequal to somewhat
unequal and ± confined to the angles but not strictly so; pricklets, acicles and
stalked glands* (or some of them) *numerous and ± unequal but not passing into
the prickles*. Lflets 3–5, mostly green but often grey beneath, the basal pair
stalked (occasionally subsessile). Infl. often broad with a truncate top, clothed
with numerous stalked glands, and a varying number of prickles. Many spp.
have white petals and red styles. Sepals grey or greenish-grey, rarely with
long points. Fl. 7–8.

Mainly woods and scrub.

A large group (87 spp.) distinguished from series 1–4 by the much greater
development of glands. Few spp. extend to Scotland and Ireland but a
number are ± frequent in S.E. England. Two common, widespread and
usually easily recognized spp. are:

R. foliosus Weihe & Nees (*R. flexuosus* Muell. & Lefèv.)

Stem deep purple with numerous purple glands, acicles and pricklets. Lflets 3(–5),
thick, grey beneath, terminal ± elliptic, gradually acuminate. Infl. frequently narrow
with a zig-zag rhachis, but sometimes broad and pyramidal. Sepals long-pointed.
Petals usually pink, narrow, fimbriate; filaments usually white; styles usually red or
pinkish.

R. rufescens Muell. & Lefèv.

Lflets 3(–5), rather large, yellowish-green on both sides, densely and softly hairy
beneath with glands on the midrib. Panicle pyramidal with long pedicels. Sepals
erect and clasping. Petals clear pink, narrow; filaments white; styles red.

Series 6. *Grandifolii* Focke. Stem arching, usually more robust than in
Apiculati, rooting at end, ± angled, *prickles more unequal and often less strictly
confined to the angles than in the preceding sections; pricklets, acicles and
stalked glands numerous and very unequal but not passing into the prickles*.
Lflets 3–5, often rather sharply and irregularly serrate, the basal stalked.
Infl. well-developed, often pyramidal with racemose top with numerous unequal
stalked glands, and usually many prickles. *Sepals tapered at the apex into
long points*. Fl. 7–8.

A group (20 spp.) in armature ± intermediate between the *Radulae* and the
Hystrices and not very clearly marked off from them or from the *Apiculati*,
but generally with characteristic foliage and panicle.

No species is very common and some are rare.

Section 7. Glandulosi P. J. Muell. Stem low-arching to procumbent, rooting
at end. Stalked glands, acicles and pricklets numerous passing into each
other and into the prickles which are distributed all round the stem. Lflets
3–5, the basal pair stalked. Infl. densely glandular with unequal stalked

glands, the longer ones usually exceeding the diam. of the pedicel. Sepals grey or greenish-grey.

Mainly in woods and scrub.

A group (60 spp.) fairly easily distinguishable by the armature of the stem. Contains 2 series formerly regarded as distinct sections.

Series 1. *Hystrices* Focke. Relatively robust. Stem usually low-arching, angled; largest prickles strong and broad-based. Infl. well-developed, middle branches cymose. Fls usually large, often pink. Petals relatively broad. Stamens long.

40 spp. Several are widespread and frequent and the following is very common and widespread.

R. dasyphyllus (Rogers) Rogers

Stem red, hairy, with long glands, acicles and slender prickles. Lflets 3–5, thick, glabrous above, softly hairy beneath, unequally serrate. Infl. long, lax, narrow. Sepals spreading, then reflexed. Petals deep pink, hairy; filaments pink, styles salmon. Carpels hairy.

Series 2. *Euglandulosi* W. C. R. Wats. Slender. Stem usually procumbent, terete; prickles slender and weak. Infl. usually with middle branches racemose. Petals small, nearly always white. Stamens often short, styles often red.

Almost confined to woods.

Twenty spp., none of which is common. The series is not known north of Yorkshire in Great Britain and is rare in Ireland.

4. POTENTILLA L.

Perennial, rarely annual or biennial herbs, or small shrubs. Lvs palmate, pinnate or ternate. Fls solitary or in cymes, usually 5-, sometimes 4-merous, usually hermaphrodite. *Epicalyx present*; segments the same number as the sepals. *Stamens* 10–30 (often 20). *Carpels numerous*, indefinite (4–)10–80; *ovule pendulous; style* jointed at its insertion, ± *lateral* but varying in position from nearly basal to nearly apical, *withered or not persistent in fr. Fr. a group of achenes*, inserted on the hemispheric or conic, *dry or spongy*, *receptacle*. Usually homogamous, nectar secreted by a ring on the receptacle.

Over 300 spp., northern hemisphere (mainly temperate), a very few extending into the southern hemisphere, to Peru and New Guinea. Several spp. and hybrids are cultivated.

1	Lvs pinnate.	2
	Lvs palmate or ternate.	5
2	Shrub; lflets entire.	**1. fruticosa**
	Herbs, sometimes woody at base; lflets toothed.	3
3	Calyx and corolla purple; woody at base; bogs.	**2. palustris**
	Calyx green, corolla white or yellow; not woody; plants of relatively dry ground.	4
4	Fls white in terminal cymes.	**4. rupestris**
	Fls yellow, solitary.	**5. anserina**

5   Fls white; lvs ternate; carpels hairy. **3. sterilis**
    Fls yellow; carpels glabrous. 6

6   Lvs densely white-tomentose beneath. **6. argentea**
    Lvs green on both sides. 7

7   Fl.-stems stiffly erect, terminal; lflets often 7. **7. recta**
    Fl.-stems prostrate to ascending or erect and flexuous (if stiffly erect,
      lvs ternate); lflets rarely 7. 8

8   Fls axillary and terminal forming a terminal cyme, 5-merous. 9
    Fls solitary, or if forming a cyme all or mostly 4-merous. 12

9   Corolla shorter than or equalling calyx; fl.-stems terminal; fls usually
      numerous. 10
    Corolla longer than calyx, fl.-stems axillary from a basal rosette; fls few. 11

10 Lvs all ternate; calyx accrescent in fr. **8. norvegica**
    Lvs mostly palmate with 5 lflets; calyx not or scarcely accrescent.
                                9. intermedia

11 Stems prostrate, rooting, forming mats; free part of stipules of radical
      lvs long and linear. **10. tabernaemontani**
    Without long prostrate rooting stems; free part of stipules of radical lvs
      ovate. **11. crantzii**

12 Stems decumbent to erect, never rooting; lvs all ternate, the cauline
      sessile or with stalks less than 5 mm.; fls nearly all 4-merous, c. 10 mm.
      diam., carpels 4–20. **12. erecta**
    Stems decumbent except at the very first, rooting only late in the season;
      lvs partly palmate, partly ternate, the lower cauline with stalks 1–2 cm.;
      fls mostly 4-merous but some 5-merous, 15–18 mm. diam.; carpels 20–
      50 (see also hybrids). **13. anglica**
    Stems prostrate, rooting from the first; lvs mostly palmate with 5 lflets,
      the cauline with long stalks; fls 5-merous, 18–25 mm. diam.; carpels
      60–120. **14. reptans**

Section 1. *Comocarpa* Torr. & Gray. Shrubs. Lvs pinnate. Fls sometimes dioecious. Carpels densely pilose. Style clavate (gradually widening upwards and contracted below the stigma), about as long as mature carpels, sub-basal. Receptacle hairy, not spongy.

1. P. fruticosa L.   'Shrubby Cinquefoil.'

Deciduous *shrub* c. 1 m., ± pilose, much-branched, branches erect or ascending, bark peeling off in about the 3rd year. Lvs numerous, with (3–)5–(7) lflets; *lflets* 1–2 cm., oblong-lanceolate, acute, elliptic, *entire*, margins revolute; stipules scarious, sheathing, entire, persisting for 2 or 3 years; petiole 5–10 mm. Fls few in a terminal cyme, or solitary, 5-merous, dioecious but with the organs of both sexes conspicuous. Epicalyx-segments oblanceolate-linear, green, about equalling the triangular-ovate, yellowish sepals. Petals 8–12 mm., yellow, orbicular. Anthers oblong. Fl. 6–7. Visited by various insects. $2n = 14, 28$. N.

Native. Damp rocky ground nearly always on basic rocks, ascending to 2300 ft. on Helvellyn, very local; Upper Teesdale; Lake District (very rare); Clare, Galway, Mayo. 4, H4. S. Sweden, Esthonia, Latvia, Urals, Caucasus, Maritime Alps, Pyrenees; N. and C. Asia to Armenia, Himalaya and Japan; N. America from Labrador to Alaska south to New Jersey and (in the mountains) to California and New Mexico; Greenland.

Section 2. *Comarum* (L.) Ser. Plants with a persistent woody base and annual herbaceous stems. Fl.-stems terminal. Lvs pinnate. Carpels glabrous or hairy. Style filiform, longer than mature carpels, lateral. Receptacle hairy, spongy.

2. P. palustris (L.) Scop.   'Marsh Cinquefoil.'
Comarum palustre L.; *P. Comarum* Nestl.

Rhizome woody, long-creeping. Stems 15–45 cm., ascending, dying back nearly to the base in winter. Lower lvs with 5 or 7 lflets; lflets 3–6 cm., oblong, sharply and coarsely serrate, subglaucous beneath, from nearly glabrous to (rarely) densely villous beneath; petiole longer than blade; stipules scarious, adnate to petiole, long; upper lvs smaller and on shorter petioles passing into the ternate bracts, their stipules green, short. Fls in a loose terminal cyme, 5-merous; fl.-stalks glandular-pubescent; upper bracts small (c. 1 cm.). *Sepals* 1–1·5 cm., ovate, acuminate, *purplish*, accrescent; epicalyx-segments much smaller than sepals, linear. *Petals* ovate-lanceolate, acuminate, *deep purple*, shorter than sepals. Stamens, carpels and styles deep purple. Anthers ovate. Carpels glabrous. Fl. 5–7. Protandrous, visited by various insects, self-pollination apparently not possible. $2n = 28, 42, 64$. Hel.

Native. Fens, marshes, where it is occasionally locally dominant, bogs, wet heaths, and moors throughout the British Is., ascending to 3000 ft. in Perth; common except in S. and C. England. 107, H40, S. Europe from Iceland and arctic Russia to the mountains of C. Spain, N. Italy, Montenegro, Bulgaria and the Caucasus; temperate Asia to Armenia, Turkistan and Japan; N. America south to New Jersey and N. California; Greenland.

Section 3. *Fragariastrum* Ser. Perennial herbs. Fl.-stems lateral, axillary from a terminal rosette. Lvs palmate or ternate. Carpels ± hairy. Style filiform, longer than mature carpels, subterminal. Receptacle hairy, not spongy.

3. P. sterilis (L.) Garcke   'Barren Strawberry.'
P. fragariastrum Pers.

Perennial herb 5–15 cm., softly pilose. Stock thick, oblique, somewhat woody, ending in a rosette of lvs, frequently emitting prostrate stolons. Fl.-stems axillary, very slender, decumbent, 1–3-fld. Rosette *lvs ternate* with long petiole; lflets 0·5–2·5 cm., orbicular or broadly obovate, ± truncate at apex, crenate-dentate with 5–7 teeth on each side, the terminal tooth much smaller than its neighbours, bluish-green and pilose above, paler and more densely pilose beneath, somewhat silky at least when young, shortly stalked; stipules ovate-lanceolate, scarious, yellowish, partly adnate to petiole; cauline lvs 1–2, smaller, otherwise similar. Fls 10–15 mm. diam., 5-merous. Sepals triangular-ovate, acute; epicalyx-segments lanceolate, shorter than sepals. *Petals white*, obcordate, widely separated, about equalling or rather longer than sepals. Filaments glabrous. Carpels hairy only near the tip, style scarcely longer. Distinguishable in fl. from *Fragaria vesca* by the blue-green lvs with spreading hairs beneath, small terminal tooth, distant petals, hairy receptacle and the different stolons. Fl. 2–5. Visited by various insects, self-pollination finally possible. $2n = 28$. Hs.

Native. Scrub, wood margins, open woods, etc., usually on rather dry soils, ascending to over 2000 ft. 109, H40, S. Throughout the British Is. except Shetland, common except in N. Scotland. Europe from S. Sweden to N. Portugal, C. Spain, Sardinia, Sicily and Macedonia, becoming rare in E. Germany and Czechoslovakia and not reaching Russia; Asia Minor? Algeria?

Section 4. *Closterostyles* Torr. & Gray. Perennial herbs. Fl.-stems terminal. Lvs pinnate. Carpels glabrous. Style fusiform, about equalling mature carpels or longer, sub-basal. Receptacle hairy, not spongy.

4. P. rupestris L.   'Rock Cinquefoil.'

Pubescent perennial herb 20–50 cm. Stock thick, somewhat woody, branched. Fl.-stems terminal, erect, glandular-pubescent above. *Radical lvs* 7–15 cm., *pinnate* with 2–4 distant pairs of lflets, sometimes with smaller ones in addition, petiolate; lflets 2–6 cm., decreasing in size from apex to base of lf, ± ovate, obtuse, doubly dentate, the lateral asymmetric at the base, green and pubescent on both sides; stipules ovate, adnate, subscarious. Cauline lvs few, the upper ternate, passing into the small simple upper bracts, stipules herbaceous. Fls several in a loose dichotomous cyme, 5-merous. Sepals ovate-oblong, mucronate; epicalyx-segments linear-lanceolate, much smaller than sepals. *Petals white*, 5–7 mm., obovate, somewhat longer than calyx. Fl. 5–6. Insect visits few, probably usually self-pollinated. $2n = 14$. Hs.

Native. Basic rocks in Montgomery and Radnor, extremely rare. 2. Europe from S. Scandinavia to C. Spain, Sicily, Montenegro, Bulgaria and S.W. Russia; Morocco (Atlas, very rare); sspp. in Asia Minor, Caucasus, W. and C. Asia (to c. 110° E.) and in the Rocky Mountains from Canada to Nevada and California.

Section 5. *Anserina* Torr. & Gray. Perennial herbs. Fl.-stems axillary. Lvs pinnate. Carpels glabrous. Style filiform, shorter than or equalling mature carpel, lateral. Receptacle hairy.

5. P. anserina L.   Silverweed.

Perennial ± silky herb with short, thick, simple or branched stock, ending in a rosette of lvs and emitting long (to 80 cm.) creeping rooting and flowering stolons. Radical *lvs* 5–25 cm., *pinnate*, with 7–12 pairs of main lflets, alternating with smaller ones; *lflets* 1–6 cm., the lower smaller, oval or oblong, deeply and regularly serrate with narrow teeth, *silvery silky* on both sides or beneath only, rarely green and sparingly hairy or glabrous on both sides; smaller lflets 2–5 mm., 2–5-fid; stipules brownish, scarious, adnate, entire; lvs of stolons smaller, the upper much reduced or 0, stipules herbaceous, multifid, connate at the base and shortly sheathing. *Fls 5-merous, solitary, axillary*, on long stalks. Sepals ovate; epicalyx-segments triangular-lanceolate, often toothed, about as long as sepals. *Petals yellow*, c. 1 cm., obovate. Style shorter than carpel. Fl. 6–8. Visited by various insects, self-pollination possible. $2n = 28$, rarely 42. Hr.

Native. Waste places, roadsides, damp pastures, dunes, etc., ascending to 1400 ft., common throughout the British Is. 112, H40, S. Europe from Iceland and N. (not arctic) Russia to N. Portugal, C. Spain, N. Italy, Montenegro and the Dobruja; N. and C. Asia to the Caucasus, N. Persia,

W. Himalaya, Manchuria and Japan; Lebanon; Greenland; N. America south to New Jersey and N. California; S. America; Victoria, Tasmania, New Zealand.

Section 6. *Potentilla*. Perennial, rarely annual or biennial, herbs. Lvs usually palmate or ternate, sometimes pinnate. Carpels glabrous. Style gradually expanded either above or below (sometimes both), subterminal, usually about equalling the mature carpel. Receptacle glabrous or hairy, not spongy. All spp. studied homogamous, visited by various insects, but self-pollination possible if insect visits fail.

6. P. argentea L.   '*Hoary Cinquefoil.*'

Perennial herb 15–50 cm. with short thick branched stock. *Fl. stems terminal, decumbent or ascending*, ± tomentose. *Basal lvs palmate with 5 lflets; lflets* 1–3 cm., obovate-cuneiform, pinnately lobed with 2–5 lobes on each side, the lobes ± oblong, entire or more rarely with 1 or 2 small teeth, varying from green and glabrous to appressed pilose and greyish above, *densely white-tomentose beneath*, margins narrowly recurved; petiole longer than blade; stipules brownish, scarious, adnate; cauline lvs similar but the upper sometimes ternate, segments narrower, oblanceolate-cuneiform, petioles short; the upper lvs subsessile, stipules herbaceous, ± ovate, acuminate, entire or with a few small teeth. Fls 10–15 mm. across, 5-merous, ± numerous in a tomentose dichotomous cyme; upper bracts small, entire, passing into lvs below. Calyx tomentose; sepals ovate; epicalyx-segments oblong-lanceolate, about as long as sepals. Petals yellow, obovate, slightly longer than the calyx. Carpels minutely rugulose; style conic at base, usually papillose, rather shorter than mature carpel; stigma dilated. Fl. 6–9. $2n=14, 28, 35, 42, 56$. Some strains apomictic but pollination necessary for development, other strains amphimictic. Morphological differences exist between plants with different chromosome numbers and it is likely that several sp. or ssp. may ultimately be recognized. Hs.

Native. Dry sandy grassland from Moray and Cumberland southwards, local, especially so in the west; Jersey. 64, S. Europe from Scandinavia to C. Spain, C. Italy and Greece; W. and C. Asia to Asia Minor, Turkistan and L. Baikal; N. America from Nova Scotia to District of Columbia, Kansas and N. Dakota.

*P. canescens Bess. (*P. inclinata* auct.)

Resembling *P. intermedia* in habit but with the lvs grey-tomentose beneath (not white nor with the margins recurved as in *P. argentea*). Rather frequently reported as a casual. $2n=42$. Native of C. and E. Europe.

*7. P. recta L.

Perennial herb 30–70 cm. with short thick branched stock. *Fl.-stems terminal, stiffly erect*, simple except in the infl., pubescent with short stiff hairs and sparingly villous with long flexuous white hairs. Basal and lower cauline *lvs palmate with 5–7 lflets*; lflets large (the middle one 5–10 cm., the basal ones shorter), oblong or oblanceolate, regularly serrate-dentate with 7–17 teeth on each side, green and appressed hairy on both sides, the veins prominent beneath; petiole very long; stipules long-adnate; lvs becoming smaller and petioles shorter upwards, the uppermost ternate, sessile; stipules of upper lvs green, shortly adnate, ovate-lanceolate, entire or cut. Infl. a many-fld dichotomous cyme, glandular-pubescent and villous; bracts simple, rather small. Fls 5-merous, 20–25 mm. diam. Calyx glandular and ± densely villous; sepals ovate-lanceolate, acuminate; epicalyx-segments lanceolate, acute, about as long as sepals. Petals yellow, often sulphur-coloured, obovate-orbicular, deeply emarginate, from as long as to much longer than the calyx. Carpels rugose; style rather thick, more so at the base, shorter than mature carpel; stigma slightly dilated. A very variable sp. on the Continent. Fl. 6–7. $2n = 28, 42$. Hs.

Introduced. A garden escape or casual sometimes becoming ± naturalized in waste or grassy places. C. and S. Europe from C. France and Germany to C. Spain, Sardinia, Sicily and Macedonia; N. Africa (mountains); W. and C. Asia to c. 100° E., south to Asia Minor and N. Persia; casual in eastern N. America.

*8. P. norvegica L.

Annual, biennial or short-lived perennial herb 20–50 cm., with simple vertical root which is often rather thick in the biennial and perennial forms, ± hirsute and occasionally with scattered sessile glands. Stems 1 or several, terminal, erect or ascending, robust, often branched. Lvs all ternate; lflets 1–7 cm., obovate, elliptic or oblong, coarsely serrate or serrate-dentate, green on both sides; lower on long petioles, upper subsessile; stipules of lower lvs long-adnate, of upper large, ovate, entire or dentate. Infl. a dichotomous cyme, often much branched; bracts lf-like, the lower ternate, the upper simple; pedicels short or some of them longer (–2 cm.). Fls 5-merous. Calyx 7–8 mm. diam. in fl., accrescent in fr. to 15–20 mm., covered with long hairs; sepals ovate, acute; epicalyx-segments ± oblong, subobtuse, about as long as sepals in fl., becoming much longer in fr. Petals yellow, obovate, small, not longer than the calyx. Carpels rugulose; style with a very thick conical base, as long or shorter than mature carpels; stigma dilated. Fl. 6–9. $2n = 70$. Th. or Hs.

Introduced. Naturalized on waste ground in a number of places from Inverness southwards, mostly in S.E. England. 47. N. and C. Europe from Scandinavia to Switzerland, Hungary and C. Russia (but doubtfully native west of Finland and Hungary); N. Asia to Kamchatka; naturalized in W. and S. Europe; a ssp. in N. America from Newfoundland to S. Carolina, California and Alaska.

*9. P. intermedia L.

Biennial or perennial herb 20–50 cm. finally developing a thick root. Stems terminal, robust, ascending from an arcuate base, dichotomously branched from low down, softly pubescent. Lower lvs mostly palmate with 5 lflets, a few ternate; lflets 1–4 cm., obovate or obovate-oblong, serrate-dentate or incised-serrate, green and softly pilose on both sides or more densely pilose and greyish beneath; petiole long; stipules small, shortly adnate; upper cauline lvs ternate, subsessile, lflets oblong-lanceolate, stipules ovate, mostly incised-dentate on the outer margin, rarely entire. Infl. a lax many-fld *corymbose dichotomous cyme*, softly pubescent; upper bracts simple, often trifid, lower ternate; pedicels 0·5–2 cm. Fls 5-merous, c. 10 mm. across. Calyx pubescent and usually villous, not or scarcely accrescent; sepals ovate, acute; epicalyx-segments about as long as sepals, subobtuse or subacute. Petals yellow, about as long as calyx or rather shorter, obovate, slightly emarginate. Receptacle pilose. Carpels rugulose; style thickened and papillose at base, about as long as mature carpel. Fl. 6–9. 2*n*=28. Hs.

Introduced. A rather frequent casual, sometimes becoming ± naturalized. Native of N. and C. Russia; naturalized elsewhere in N. and C. Europe and in eastern N. America.

*P. thuringiaca Bernh. ex Link, with a thick stock, 7–9 lflets and short style, is sometimes naturalized. C. Europe.

10. P. tabernaemontani Aschers.   'Spring Cinquefoil.'

P. verna L., p.p.

Perennial herb 5–20 cm. *Stock* thick, much branched *emitting prostrate usually rooting branches and forming mats. Fl.-stems axillary* from the lvs of a terminal rosette, decumbent, branched from below the middle, pubescent and villous with ascending hairs. *Basal lvs palmate with 5 lflets*; lflets 0·5–2 cm., obovate-cuneiform, dentate or deeply serrate with 2–9 teeth on each side, the terminal tooth markedly smaller than its neighbours, green and ± appressed hairy on both sides or subglabrous above; petiole long; stipules long-adnate, *the free part long and linear-lanceolate*; cauline lvs few, the lower like the radical, the upper ternate, subsessile; stipules ovate, ovate-lanceolate or oblong, acute to subobtuse, usually entire. Infl. a lax few-fld cyme, not or little raised above the lvs, pubescent and villous; pedicels slender, 1–2 cm.; bracts simple. *Fls* 10–15 *mm. diam.*, 5-merous. Calyx pilose; sepals ovate, acute; epicalyx-segments oblong, obtuse, shorter than sepals. Petals yellow, obovate, deeply emarginate, longer than the calyx. Receptacle pilose. Carpels rugose, not keeled; style clavate, shorter than mature carpel; stigma dilated. Fl. 4–6. 2*n*=28, 42, 49, 63, 84. Apomictic but pollination necessary for seed development. Hs.

Native. Dry basic grassland and rocky outcrops on sunny slopes from Angus and Cumberland to Suffolk, Hants and Somerset, very local. 25. Europe from S. Scandinavia to C. Spain, Corsica, N. Italy, N.W. Hungary, Roumania, Poland and the Baltic States.

11. P. crantzii (Crantz) G. Beck ex Fritsch   'Alpine Cinquefoil.'

P. alpestris auct.; *P. maculata* auct.; *P. salisburgensis* auct.; *P. verna* L., p.p.

Perennial herb 5–25 cm. *Stock* thick, branched, *emitting short branches not or scarcely rooting and never forming mats. Fl.-stems axillary* from the lvs of a terminal rosette, arching and ascending, often flexuous, branched only in the infl. above the middle of the stem, pubescent, the hairs crisped or spreading. *Basal lvs palmate with 5 lflets*; lflets 1–2 cm., obovate-cuneiform, dentate with 2–5 teeth on each side, the terminal tooth scarcely smaller than its neighbours, green on both sides, subglabrous or hairy above, hairy with ± spreading hairs especially on the veins beneath; petiole long; stipules long-adnate, *the free part ovate often obtuse*; cauline lvs few, the upper ternate, subsessile, stipules ovate, acute to subobtuse, usually entire. Infl. a lax few-fld cyme (or fl.1), raised well above the lvs, pubescent like the stem; pedicels long (1·5–3·5 cm.) and slender; bracts simple. *Fls* 15–25 *mm. diam.*, 5-merous. Calyx pilose; sepals triangular-ovate, acute; epicalyx-segments oblong or oblong-lanceolate, acute or obtuse, nearly as long as sepals. Petals yellow, often with an orange spot at base, obovate, deeply emarginate, longer than calyx. Receptacle pilose. Carpels rugulose or nearly smooth; style clavate, shorter than mature carpel; stigma dilated. Fl. 6–7. $2n=42$ (49). Apomictic but pollination necessary for seed development. Hs.

Native. Mountain rock ledges and crevices and occasionally grassland, usually on basic soils from 400–3350 ft., very local. N. Wales; Yorks and Westmorland to Northumberland and Kirkcudbright; Argyll, Stirling and Angus to Skye and Sutherland. 20. Arctic and subarctic Europe, Asia and America but absent between c. 150° E. and 130° W.; high mountains of Europe to Pyrenees (also N.W. Portugal, very rare), C. Apennines, Albania, Bulgaria and Caucasus; Asia Minor, Persia.

12. P. erecta (L.) Räusch.   Common Tormentil.

P. Tormentilla Stokes

Perennial herb; stock very thick (1–3 cm.), woody, vertical to nearly horizontal, with reddish flesh, bearing a terminal *rosette of lvs which often withers and disappears before fl. time*. *Fl.-stems* (5–)10–30(–50) cm., several, axillary, slender, flexuous, decumbent to suberect, *never rooting, dichotomously branched above*, rather silkily loosely appressed-pilose. Lvs all *ternate* or rarely a few of the radical lvs palmate with 4 or 5 lflets; radical lvs on long petioles; lflets 5–10 mm., broadly obovate-cuneiform, coarsely dentate near the truncate apex, with 3–4 teeth on each side; stipules long-adnate, the free part lanceolate, entire; *cauline lvs sessile or subsessile*, lflets 1–2 cm., narrowly obovate to oblanceolate-oblong, incised-serrate above the middle, with (2–)3–5(–6) teeth on each side, narrowly cuneate at base, green and glabrous or sparingly pilose above, appressed silky pilose on the margins and veins beneath, stipules large, palmately lobed, appearing like extra lflets. *Fls many in loose terminal cymes; pedicels long and slender*, appressed pilose; bracts lf-like, the upper simple, the lower passing into the cauline lvs. *Fls all 4-merous* (sometimes a few 3-, 5- or 6-merous), 7–11(–15) *mm. diam.* Calyx loosely appressed-pilose; sepals ovate-lanceolate, acute to subobtuse; epicalyx-segments linear-oblong, shorter to longer than sepals. Petals yellow, cuneiform, obovate or orbicular, emarginate, usually rather longer than calyx. Stamens 14–20, usually 16. Receptacle pilose. Carpels 4–8(–20), ovoid, rugose, obscurely keeled; style somewhat dilated at the base or at the apex or nearly equally slender throughout, about equalling mature carpel. Fl. 6–9. $2n=28^{*}$. Hs.

Var. *sciaphila* (Zimm.) Druce is a small, slender decumbent form with the *cauline lvs shortly* (1–4 mm.) *petioled* and their *stipules entire* or bifid. It is liable to be confused with *P. anglica* but differs in the lvs all ternate and smaller fls. It deserves further study.

Native. Grassland, heaths, bogs, fens, mountain tops and sometimes open woods; absent or rare on heavy and strongly calcareous soils, very common on light acid ones; ascending to nearly 3500 ft. Throughout the British Is. 112, H40, S. Europe from the Faeroes, Scandinavia (70° 20′ N.) and arctic Russia to Spain and Portugal, Corsica, Italy and Macedonia; W. Asia from W. Siberia to N. Asia Minor; Algeria, Morocco; Azores.

13. P. anglica Laicharding   'Trailing Tormentil.'
P. procumbens Sibth.

Perennial herb; stock rather thick (3–10 mm.), branched, bearing a *persistent terminal rosette* of lvs. *Fl.-stems* 15–80 cm., axillary, slender, at first ascending, *soon decumbent, usually rooting at the nodes in late summer and autumn and producing new plants, usually dichotomously branched*, rarely simple, ± pilose. *Radical lvs* on long petioles, *partly* (c. 50%) *palmate with 5 lflets, the remainder ternate or with 4 lflets*; lflets 0·5–2 cm., obovate-cuneiform, rounded or truncate at apex, coarsely dentate with 4–6 teeth on each side; stipules shortly adnate, the free part ovate or lanceolate, entire. *Cauline lvs shortly* (1–2 cm. on the lower lvs, less on the upper) *petioled, mostly ternate* but some of the lower ones palmate with 4–5 lflets; lflets obovate or oblong-obovate (broader than in *P. erecta*) incised-serrate or serrate-dentate in the upper part, with 3–5(–6) teeth on each side, entire towards the cuneate base, subglabrous on both sides or ± pilose beneath; *stipules entire* or a few bifid or trifid. *Fls solitary on long slender pedicels, the upper often forming a few-fld cyme*, the upper bracts smaller than the lvs but rarely simple. *Fls 4-merous* (c. 75%) *and 5-merous mixed*, (10–)14–18 mm. diam. Sepals ovate-lanceolate, acute; epicalyx-segments linear-oblong or lanceolate, about as long as sepals. Petals yellow, obovate, emarginate, usually nearly twice as long as calyx. Stamens 15–20. Receptacle pilose. Carpels 20–50, oblong-ovoid, rugose; style slightly dilated at apex, about equalling mature carpel. Fl. 6–9. $2n$=c. 56*. Hr. or Hs.

Native. Woods, especially their edges, heaths, hedgebanks, etc., on similar soils to *P. erecta* but less tolerant of extreme acid conditions and not recorded above 1350 ft., widespread but rather local, becoming very rare in N. Scotland though extending to Orkney. 97, H40, S. S. Sweden, Finland, Denmark, Holland, Belgium, W. and C. France, ?N.W. Spain, N. and C. Germany, Bohemia and Moravia, Switzerland (rare), Corsica; Madeira, Azores.

14. P. reptans L.   'Creeping Cinquefoil.'

Perennial herb; stock rather thick (2–6 mm.), branched, bearing a persistent terminal rosette of lvs. *Fl.-stems* 30–100 cm. or more, axillary, stoloniform, *prostrate, quickly rooting at the nodes* and producing new plants, *simple*, pubescent or subglabrous. *Radical lvs* on long petioles, *mostly palmate with* 5(–7) *lflets*, only a few ternate or with 4 lflets; lflets 0·5–3 cm., obovate or oblong-obovate, dentate in the upper part or all round, with 6–10 teeth on each side, subglabrous or sparingly pilose on both sides; stipules long-adnate, the free part lanceolate, entire. *Cauline lvs scarcely differing* but with rather shorter petioles, mostly with 5 lflets, a few of the uppermost often with 3–4 lflets; *stipules* herbaceous, usually *entire*, rarely dentate. *Fls solitary* on long slender pedicels, never forming a cyme. *Fls 5-merous*, 17–25 mm. diam. Calyx often accrescent; sepals variable in shape, usually acute; epicalyx-segments usually obtuse, varying in length. Petals yellow, obovate, emarginate, to twice as long as calyx. Stamens c. 20. Receptacle pilose. Carpels 60–120 oblong-ovoid, rugose; style slightly dilated at apex, about as long as or rather shorter than mature carpel. Fl. 6–9. $2n=28$. Hr.

Native. Hedgebanks, waste places and sometimes grassland mainly on basic and neutral soils, ascending to 1400 ft. Common throughout the British Is. 109, H40, S. Europe (except Iceland); Mediterranean region; W. Siberia, Turkistan, Persia, Himalaya; introduced in N. and S. America.

P. anglica × *erecta*, *P. erecta* × *reptans* = *P.* × *italica* Lehm. and *P. anglica* × *reptans* all appear to occur rather frequently where the parents occur together. They are all highly (but apparently not completely) sterile. This sterility often affords the best means of distinguishing them, forms of *P.* × *italica* in particular being very difficult to distinguish from *P. anglica*.

5. Sibbaldia L.

Small perennial herbs. *Lvs palmate or ternate*. Fls in cymes, 5-merous, hermaphrodite or polygamous. *Epicalyx present. Stamens* 5, rarely 4 or 10. *Carpels* 5–12, otherwise as in *Potentilla*. Nectar secreted by the disc.

About 8 spp., all except the following in the higher mountains of Asia.

1. S. procumbens L.

Potentilla Sibbaldii Haller. f.; *P. procumbens* (L.) Clairv., non Sibth.

Compact perennial tufted herb 1–2 cm., clothed with stiff appressed hairs. Stock woody, branched, ending in a rosette of lvs. Fl.-stems axillary, often shorter than lvs. Lvs ternate, stalked; lflets 0·5–2 cm., obovate-cuneiform, tridentate at the truncate apex, the central tooth markedly smaller than the lateral, bluish-green, often purple-tinged beneath; stipules adnate, the free part ovate-lanceolate, entire, often purplish. Fls few, in a rather dense cyme, c. 5 mm. diam. Sepals oblong-lanceolate, acute, often purple-tinged; epicalyx-segments linear-lanceolate, shorter than sepals. Petals yellow, small, narrow, inconspicuous, or 0. Carpels rugulose, style basal. Fl. 7–8. Visited by flies, ants, etc., homogamous, self-pollination apparently difficult. $2n=14$. Hr.

Native. Mountain tops and grassland, rock crevices, etc., from 1500 to over 4000 ft. Westmorland; Kirkcudbright, Peebles; Dunbarton and Stirling to Sutherland on the mainland; Shetland, Skye. 22. Arctic Europe from Iceland to the Kola peninsula; high mountains of Europe, south to the Sierra Nevada, S. Alps and Macedonia; Greenland; sspp. or allied spp. in arctic Asia and N. America (in the mountains to New Hampshire and California).

6. Fragaria L.

Perennial stoloniferous herbs. Lvs ternate in a basal rosette. Fls in few-fld cymes on axillary scapes, 5-merous. Epicalyx present. Stamens numerous (c. 20). Receptacle glabrous. Carpels as in *Potentilla*, glabrous. *Fr. a group of achenes on the surface of a much enlarged, fleshy, juicy, brightly coloured receptacle*. Protogynous, nectar secreted by a ring on the receptacle.

About 35 spp., north temperate and subtropical and in S. America.

1 Lflets hairy above, the terminal one cuneate at base; fls small or moderate, 12–20(–25) mm. diam.; fr. small (2 cm. or less), with projecting achenes.   2

Lflets usually glabrous above, the terminal one rounded at base; fls large, 20–35 mm. diam.; fr. large (c. 3 cm.), with the achenes sunk in the flesh.

3. × ananassa

2   Pedicels (at least the upper) with appressed hairs; lateral lflets sessile or nearly so; fr. with achenes all over.   1. vesca

Pedicels with spreading hairs; lateral lflets stalked; fr. without achenes at base.   2. moschata

1. F. vesca L.   Wild Strawberry.

Perennial herb 5–30 cm., stock rather thick and woody, producing very long arching runners, rooting at the nodes and forming fresh plants. *Lflets* 1–6 cm., ovate, obovate or oblong, coarsely serrate-dentate, the terminal tooth not or scarcely shorter (though sometimes narrower) than its neighbours, *bright green above and somewhat pilose*, pale and glaucous beneath and clothed with silky appressed hairs, the lateral sessile or subsessile, the terminal one sessile or shortly stalked, cuneate at base; petiole long, clothed with spreading white hairs; stipules scarious, often purplish. Fl.-stems ± erect, not much exceeding lvs, clothed with spreading hairs; cauline lvs 0 but the lowest bracts usually lf-like; upper bracts ovate-lanceolate, entire; *pedicels, at least the upper, clothed with ± appressed hairs*. Fls 12–18 *mm. diam.*, hermaphrodite. *Calyx spreading or reflexed in fr.*; sepals ovate, acuminate; epicalyx-segments ± lanceolate, acute, about as long as sepals. Petals white, obovate, obtuse, nearly contiguous to overlapping. Fr. receptacle 1–2 cm., ovoid or sub-globose, red (rarely white), *covered all over with achenes which project from the surface*. Fl. 4–7. Visited by various insects. $2n=14$. Hr.

Native. Woods and scrub on base-rich soils and on basic grassland, sometimes becoming locally dominant in woods on calcareous soils; ascending to 2400 ft. Common throughout the British Is. 112, H40, S. Europe (except Crete, etc.); Asia to L. Baikal; Madeira, Azores; apparently native also in eastern N. America; introduced in other parts of the world. The Alpine Strawberry, still sometimes grown for its fr., is a form of this sp.

*2. F. moschata Duchesne   Hautbois Strawberry.

F. elatior Ehrh.

Differs from *F. vesca* as follows: More robust, 10–40 cm. high. Stolons few or 0. Fl.-stems much exceeding lvs. *Lflets all shortly stalked. Pedicels clothed with spreading or somewhat deflexed hairs*. Fls 15–25 *mm. diam.*, often dioecious or poly-gamous. *Fr. receptacle without achenes at the base* and more contracted below, purplish-red or partly greenish, of a musky flavour. Fl. 4–7. Visited by various insects. $2n=42$. Hr.

Introduced. Formerly grown for its fruit but now seldom seen. Reported as naturalized in many places in England and a few in Wales and Scotland but many of the records probably refer to *F.× ananassa*. 41? Native of C. Europe from N. France, C. Germany and N. Russia to N. Italy, Montenegro, Bulgaria and C. Russia (Volga-Don region).

*3. F. × ananassa Duchesne (*F. chiloensis × virginiana*).   Garden Strawberry.

F. chiloensis auct. angl.; *F. grandiflora* Ehrh., non Crantz.

Habit of *F. vesca* but much larger. *Lflets* 5–8 cm., orbicular or ovate, coarsely serrate-dentate, somewhat *bluish-green and glabrous above*, pale beneath and somewhat appressed hairy, *all stalked, the terminal one rounded at the base*; petiole long, sparsely hairy. Fl.-stems ascending in fl., usually decumbent in fr., not or scarcely longer than the lvs, sparingly hairy, the hairs or some of them appressed; bracts not lf-like. *Fls* relatively numerous, 20–35 *mm. diam.*, hermaphrodite, often more than 5-merous. *Calyx appressed to the fr. Fr. receptacle very large* (c. 3 cm.) *covered all over with achenes which do not project from the surface.* Fl. 5–7. 2*n*= 56. Hr.

Introduced. The commonly cultivated strawberry originated in France as a hybrid between the two octoploid spp., *F. virginiana* from eastern N. America and *F. chiloensis* (L.) Duchesne from Chile. It frequently escapes and often occurs ± naturalized on railway banks, etc.

F. virginiana Duchesne. Forms of this (e.g. 'Little Scarlet') are still sometimes grown for preserves. The fr. is much smaller than in *F.* × *ananassa* and with pink flesh and the fr. calyx spreading. It has not been reported as escaped.

*Duchesnea indica (Andr.) Focke

Resembling *Fragaria*. Spreading by runners. Lvs ternate. Lflets rhombic or elliptic, ± hairy. Fls solitary, axillary, on long stalks, bright yellow. Epicalyx segs. in fr. c. 5 × 5 mm., toothed at apex. Fr. receptacle red, but dry and insipid.

Grown in gardens and rarely naturalized. Native of mountains of Asia from Afghanistan to Indonesia and Japan.

7. GEUM L.

Perennial herbs. Lvs unequally pinnate. *Fls* solitary or cymose, 5-*merous*. *Epicalyx present*. Stamens numerous (20 or more). Receptacle flat. Carpels numerous (20 or more); *ovule* 1, *erect; style terminal, enlarged and* (all or the lower part) *persistent as an awn on the fr.* Fr. a group of achenes on a dry receptacle. Protogynous fls, nectar secreted by receptacle.

About 40 spp., temperate regions of both hemispheres (a few arctic). Cultivated forms or hybrids of *G. coccineum* Sibth. & Sm., often with double red fls, are much grown in gardens. The awn on our spp. is hooked and the fr. is dispersed by animals.

Fls erect; petals 5–9 mm., yellow, spreading, not emarginate nor clawed.

1. urbanum

Fls nodding; petals 10–15 mm., reddish, erect, emarginate, clawed.   **2. rivale**

1. G. urbanum L.   Herb Bennet, Wood Avens.

Perennial herb 20–60 cm., ± pubescent; rhizome short, thick. Stems ± erect. Radical lvs pinnate with 2–3 pairs of unequal lateral lflets, 5–10 mm. long, and large suborbicular lobed terminal lflet 5–8 cm., or with the upper pair of laterals also large; lflets all crenate or dentate. Cauline lvs large, the lower like the radical or ternate or deeply trilobed, the upper usually simple; *stipules large* (1–3 cm.), lf-like with large irregular triangular teeth or lobes. *Fls erect*, few, on long stalks in very open cymes. *Calyx green*; sepals triangular-lanceolate; epicalyx-segments oblong-linear, c. one-third as long as sepals. *Petals 5–9 mm., yellow, spreading, obovate or oblong, not clawed nor emarginate*, about as long as sepals. *Carpels* hirsute, *remaining as a sessile head in fr.*;

awn purplish 5–10 mm., jointed near the apex, the lower part hooked and persistent, glabrous throughout. Fl. 6–8. Visited by various insects but visitors few, self-pollination probably usual. $2n=42$. Hs.

Native. Woods, scrub, hedgebanks and shady places on good damp soils; ascending to 1700 ft. Common throughout the British Is. 112, H40, S. Europe (except Iceland and Crete, etc.), extending north to 68° 15′ in Norway; W. Asia from W. Siberia to the Himalaya and Syria; N. Africa.

2. G. rivale L.   Water Avens.

Perennial herb 20–60 cm., pubescent; rhizome short, thick. Radical lvs pinnate with 3–6 pairs of unequal lateral lflets 2–20 mm. long, and large suborbicular lobed terminal lflet 2–5 cm.; lflets all ± dentate. Cauline lvs few, small, simple or with a few very small lateral lflets, often trilobed; *stipules small* (c. 5 mm.) green but scarcely lf-like, dentate or entire. *Fls nodding*, few, in a narrow cyme. *Calyx purple*; sepals triangular-lanceolate; epicalyx-segments linear, c. one-third as long as sepals. *Petals* 1–1·5 cm., *dull orange-pink, erect, ± spathulate with a long claw, retuse or emarginate*, about as long as sepals. *Carpels* hirsute, *head becoming stalked in fr.*, awn jointed rather above the middle, the lower part glabrous, hooked and persistent, the upper joint hairy. Fl. 5–9. Visited by various insects, especially humble-bees; self-pollination possible if insect visits fail. $2n=42$. Hs.

Native. Marshes, streamsides, wet rock-ledges, and damp woods on base-rich soil, most frequently in shade; ascending to 3200 ft. Widespread and rather common in Scotland, N. England and Wales (absent only from the Outer Hebrides and the Isle of Man), local in S. England and very rare in the south-east; widespread in Ireland but rather local. 104, H35. Europe from Iceland and N. (not arctic) Russia to S. Spain (absent from W. France and Spain), N. Apennines, Macedonia and the Caucasus; Asia Minor, Siberia (to the Yenisei region); N. America from Newfoundland to British Columbia, Colorado and New Jersey.

G. rivale × urbanum (*G. × intermedium* Ehrh.) is usually to be found when the parents grow together (a rather infrequent occurrence). It is fertile and forms hybrid swarms showing a large range of variation. 76, H14.

*G. macrophyllum Willd.

Resembling *G. urbanum* but more robust. Radical lvs with 5–7 pairs of lateral lflets, terminal lflet often more rounded. Stipules of stem-lvs smaller and less toothed or entire. Fls usually rather larger. Carpels more numerous, bristly at apex; lower part of style minutely glandular.

Rarely naturalized. Native of N. America and N.E. Asia.

8. DRYAS L.

Evergreen prostrate undershrubs. Lvs simple. *Fls solitary, axillary, herma-phrodite or polygamous, 7–10- (often 8-)merous. Epicalyx 0. Stamens c. 20. Receptacle slightly concave. Carpels numerous; *ovule 1, erect; style terminal, persistent on the fr.* and covered with long white feathery hairs. Fr. a group of achenes on a dry receptacle. Nectar secreted by a ring below the stamens.

Three or more spp., Arctic and mountains of north temperate region.

1. D. octopetala L. 'Mountain Avens.'

Much branched tortuous, creeping undershrub. Lvs numerous, blade 0·5–2 cm., oblong or ovate-oblong, obtuse, rounded or truncate at base, stalked, deeply crenate or crenate-dentate, dark green and glabrous above with impressed veins, densely white tomentose beneath; stipules scarious, brownish, adnate. Fl.-stalk 2–8 cm., erect, tomentose, with blackish glandular hairs above. Fls 2·5–4 cm. diam. Sepals oblong, tomentose and with blackish glandular hairs. Petals oblong, white. Fl. 6–7. Homogamous or nearly so, visited by various insects, self-pollination possible. $2n = 18^*$. Chw.

Native. Ledges and crevices on mountains of basic rocks, local; ascending to 3200 ft., descending to sea level in Sutherland and Clare. ?Stafford, Caernarvon (very rare), W. Yorks, Westmorland; Argyll and Perth to Orkney; N. and W. Ireland from Antrim to Clare. 19, H10. Arctic and subarctic Europe, Asia, America; high mountains of Europe south to the Pyrenees, Apennines, and Macedonia; Rocky Mountains, south to Colorado.

Tribe 4. SANGUISORBEAE. Receptacle deeply concave, enclosing the carpels and (usually) becoming dry and hard in fr. Carpels 1–4, often 2. Filaments narrowed at base. Fr. of 1 or more achenes enclosed in the persistent receptacle and shed with it.

9. AGRIMONIA L.

Perennial herbs. Lvs unequally pinnate. Fls in terminal spike-like racemes, 5-merous, hermaphrodite. *Epicalyx* 0. *Petals* 5. Stamens 10–20. *Receptacle covered above with small spines*, becoming hard in fr. Carpels 2; *styles terminal*; ovule 1, pendulous. Homogamous pollen-fls without nectar.

About 15 spp., in north temperate regions and S. America.

Lvs not (rarely slightly) glandular beneath; fr. receptacle obconic, deeply
grooved throughout, basal spines spreading laterally. **1. eupatoria**
Lvs with numerous sessile glands beneath; fr. receptacle campanulate, without
grooves at the base, basal spines deflexed. **2. odorata**

1. A. eupatoria L. Common Agrimony.

Erect perennial herb 30–60 cm., ± villous, not or sparingly glandular. Stems usually simple, often reddish, densely lfy below, sparsely so above. Lower *lvs* pinnate with 3–6 pairs of main lflets which become larger upwards, and 2–3 pairs of smaller lflets between each pair of main ones; largest lflets 2–6 cm., ± elliptic, deeply and coarsely serrate or serrate-dentate, usually densely villous and often greyish beneath, *not glandular*; upper lvs smaller with fewer lflets; stipules lf-like. Fls numerous, c. 5–8 mm. diam., yellow. *Fr. receptacle obconic, deeply grooved almost throughout its length*, covered above with hooked spines; *lowest spines ascending or spreading horizontally*.

Var. *sepium* Bréb. is a more robust plant resembling *A. odorata* in habit and somewhat glandular. The glands on the lower surface of the lf are, however, absent or very sparse and the spines of the receptacle those of *A. eupatoria*.

Fl. 6–8. Visited by Diptera and Hymenoptera, self-pollination frequent. $2n = 28^*$. Hs.

Native. Hedgebanks, roadsides, edges of fields, etc.; ascending to 1600 ft. Common throughout the British Is. except N. Scotland where it is rare (absent from Orkney and Shetland). 108, H40, S. Europe (except Iceland, Crete, etc.) extending to 63° 35′ N. in Norway; Asia Minor, Persia; N. Africa; Macaronesia.

2. A. odorata (Gouan) Mill.   'Fragrant Agrimony.'

Differs from *A. eupatoria* as follows: More robust than most forms of that sp. (often 1 m. high), sweet-smelling. Stems often more lfy. *Lvs* larger; lflets larger and relatively narrower, with narrower more acute teeth, green and usually less hairy beneath, and always *with numerous small sessile shining glands* (which also occur on other parts of the plant). Fls larger (to 1 cm. diam.). *Fr. receptacle campanulate, more shallowly grooved, the grooves ceasing well above the base,* or almost without grooves; *lowest spines deflexed* (the others erect or ascending). Fl. 6–8. $2n=56^*$. Hs.

Native. In similar places to *A. eupatoria* but usually absent from calcareous soils. From Argyll and Inverness southwards; widespread but local in Ireland and rare in the Central Plain; Channel Is. 75, H31, S. Europe from S. Scandinavia and Finland to N. Portugal, Spain, N. Italy, Macedonia and S.W. Russia (but absent from many areas).

10. AREMONIA DC.

Differs from *Agrimonia* as follows: Fls in few-fld cymes, each fl. surrounded by an involucre. *Epicalyx present. Receptacle without spines.* Stamens 5–10.
    One sp.

*1. A. agrimonoides (L.) DC.

Agrimonia Agrimonoides L.

Perennial herb 20–40 cm.; stem ascending. Lvs pinnate, with 2–4 pairs of main lflets, the lower smaller than the upper, with smaller ones between; lflets serrate-dentate. Fls 7–10 mm. diam., yellow. Involucre with 6–10 entire sepal-like lobes, concealing the receptacle. Fl. 6–7. Hs.

Introduced. Naturalized in woods in C. Scotland. 7. Native of S.E. Europe, extending west to Italy and Austria; N. Asia Minor.

11. ALCHEMILLA L.

Perennial herbs. Lvs palmate or palmately lobed. Fls ± numerous, in cymes, 4-merous, small, green. *Epicalyx present. Petals* 0. *Stamens* 4(–5), *alternate with sepals, inserted on the outer margin of the disk; anthers introrse.* Receptacle not spiny, small and dry in fr. Carpel 1; *style basal,* stigma capitate; ovule 1, erect.

    About 200 spp., north temperate zone and the mountains of tropical Africa.

    Nectar secreted by a ring on the receptacle, and fls sometimes visited by various insects, but most of the European spp. are apomictic.

1 Lvs divided to at least $\frac{2}{3}$ and often quite to the base, green above, silvery-
        silky beneath, the hairs hiding the surface.   *2*
    Lvs divided to less than $\frac{1}{2}$, green on both sides (see p. 397 for notes on
        this aggregate).   **(vulgaris** agg.)   *3*

2 Lvs divided to the base or nearly so into oblong-oblanceolate segments
3–6 mm. broad.   **1. alpina**
Lvs divided to $\frac{2}{3}$ or $\frac{4}{5}$ into obovate or oblong segments 6–15 mm. broad.
                                                                      2. conjuncta

3 Stem (at least the lower part) and petioles ± densely clothed with spreading
hairs.   *4*
Stem and petioles with appressed hairs or subglabrous.   *12*

4 Lvs glabrous above.   *5*
Lvs hairy above, at least in the folds.   *6*

5 Lf-lobes rounded, with nearly equal curved teeth; lvs all glabrous above.
Widespread.   **10. xanthochlora**
Lf-lobes straight-sided, teeth ±straight, those in the middle of the
margins of the lobes much larger than those above and below; spring
lvs hairy above. Durham.   **9. acutiloba**

6 Whole plant, including pedicels, hairy.   *7*
Plant less hairy, the pedicels always glabrous (if plant very small see also
A. minima).   *9*

7 Basal sinus of lvs closed or nearly so; teeth subobtuse, nearly straight;
plant silvery and silky.   **3. glaucescens**
Basal sinus of lvs wide (60° or more); teeth acute, somewhat curved;
plant not or scarcely silky or silvery.   *8*

8 Sinuses between the lf-lobes ending in a toothless incision; stipules
brownish; plant very small (lvs not more than 2·5 cm.) and rather
sparingly hairy. Yorks.   **7. minima**
Lf-lobes toothed all round; stipules usually purplish tinged; plant usually
larger and more hairy than above.   **4. vestita**

9 Basal sinus of lvs wide; lvs, at least the summer ones, rather thinly hairy,
the hairs often restricted to the folds above and the veins beneath.   *10*
Basal sinus of lvs closed or nearly so; lvs usually 9-lobed, thickly hairy
on both surfaces. Teesdale and Weardale.   **8. monticola**

10 Stipules usually tinged purplish-red; receptacle often slightly hairy.
                                                                      5. filicaulis
Stipules colourless or brownish; fls glabrous. Teesdale and Weardale.   *11*

11 Lf-lobes ±straight-sided, triangular, with rather narrow straight-sided
teeth.   **9. acutiloba**
Lf-lobes short, semicircular, with broad blunt teeth.   **6. subcrenata**

12 Stem and petioles densely hairy with silky hairs; lvs hairy above; fl.-
clusters dense; fls glabrous.   **11. glomerulans**
Stem hairy only below or nearly glabrous; lvs glabrous above; fl.-clusters
usually lax.   *13*

13 Basal sinus of lvs wide; sinuses between the lobes toothed to the base,
without a narrow prolongation; teeth in the middle of the lobe larger
than those above and below, broad or rather broad.   **12. glabra**
Basal sinus of lvs closed or nearly so; sinuses between the lobes ending
in a narrow entire prolongation; teeth ±equal, narrow.   **13. wichurae**

1. A. alpina L.   'Alpine Lady's-Mantle.'

A. alpina ssp. *glomerata* (Tausch) Camus

Perennial herb 10–20 cm.; stock somewhat woody, rather thick, branched,
shortly creeping; lvs mostly radical; stems ascending, with few lvs. *Radical
lf-blades* (Fig. 39 A) 2·5–3·5 cm. diam., orbicular or reniform in outline,

palmately divided almost or quite to the base into 5–7 *oblanceolate-oblong segments,* green and glabrous above, *densely silvery-silky beneath; segments* 1–2 cm. × 3–6 *mm.,* sharply serrate at the extreme apex; petiole long; stipules brown, scarious. Cauline lvs few, small, sometimes with only 3 segments, on short petioles. Stems, fl.-stalks, receptacle and calyx appressed silky. Fls c. 3 mm. long and 3 mm. diam. in rather dense clusters forming a terminal cyme. Fl. 6–8. $2n=$ c. 120. Hs.

Fig. 39. Leaves of *Alchemilla.* A, *A. alpina*; B, *A. conjuncta*; C, *A. glaucescens.* × ½.

Native. Mountain grassland, where it is sometimes locally dominant, rock crevices, screes and mountain tops; ascending to over 4000 ft. in the Cairngorms, descending nearly to sea-level in Skye. Yorks, Westmorland, Cumberland; Stirling and Arran northwards, widespread and locally abundant; Kerry, Wicklow, very rare. 28, H2. N. Europe from Iceland and Spitzbergen to the Kola Peninsula and S. Scandinavia; Pyrenees, Alps; Greenland.

2. A. conjuncta Bab.

A. argentea (Trevelyan) D. Don, non Lam.

Similar to *A. alpina* but often more robust, reaching 30 cm. *Radical lf-blades* (Fig. 39B) 4–8 cm. diam., palmately divided to c. $\frac{2}{3}$–$\frac{4}{5}$ (the sinuses varying in depth on the same lf) *into* 7–9 *obovate or oblong segments; segments* 6–15 *mm. broad,* the serrations less sharp and extending further down the lf than in *A. alpina.* Fls larger (c. 4 mm. diam.), the clusters often larger. Fl. 6–7. Hs.

Possibly native among rocks by streams in Glen Clova (Angus) and Glen Sannox (Arran), very rare; sometimes occurring as an escape elsewhere. Alps of France (Savoy and Dauphiné) and of W. Switzerland; Jura.

(3–13). A. vulgaris agg.   Lady's Mantle.

Perennial herb 5–45 cm., variously hairy to nearly glabrous; stock very thick and woody; lvs mostly radical; stems ascending or decumbent with few lvs. *Radical lf-blades* 1–15 cm. diam., orbicular or reniform in outline, *palmately lobed to ½ or less, green on both sides*; lobes (5–)7–11, broad, serrate, teeth with a tuft of hairs even on otherwise glabrous lvs; petiole long; stipules scarious, brownish, sometimes purplish tinged. Cauline lvs few, smaller, on short petioles. Fls 3–4 mm. diam. in a compound terminal cyme, made up of dense or lax small cymes. Fl. 6–9. Hs.

Native. Damp grassland, open woods, rock-ledges, etc., especially on basic and neutral soils; ascending to nearly 4000 ft. Almost throughout the British Is., rather common in the north and west, becoming rare in S.E. England and absent from the Channel Is. 109, H40. Europe, N. and W. Asia, Greenland, eastern N. America.

The following spp. are in general quite clear-cut. The most useful characters are derived from the nature and distribution of the hairs, the shape of the lvs and their lobing and toothing, and to a lesser extent the infl. and fls, and the habit. In the following descriptions 'lvs', if not qualified, refers to well-developed summer rosette lvs. It should be noted that when hairs are said to be 'spreading' they all stick out from the stem at 45° or more, if 'appressed' most of the hairs are ± closely appressed to the stem though occasional individual hairs may stick out. The following account is based on the work of S. M. Walters (*Watsonia* I; 1948). He has done much to clear up the confusion existing in the group and has shown that records of various other continental species from this country are based on errors of identification. The distributions of the various spp. are, however, still in several instances incompletely known.

3. A. glaucescens Wallr.

A. hybrida auct.; *A. minor* Huds., p.p., sec. Wilmott; *A. anglica* Rothm.; *A. pubescens* auct.

Plant small. *Stems, pedicels, receptacle and petioles densely clothed with spreading silky hairs. Lvs* (Fig. 39 c) *hairy on both surfaces*, orbicular in outline, *basal sinus closed or nearly so*; lobes (5–)7–9, broad (c. twice as broad as long), rounded, often overlapping; *teeth 9–11 on each of the middle lobes, almost straight, subobtuse.* Fls c. 3 mm., in dense clusters, almost silvery outside. $2n =$ c. 103–110.

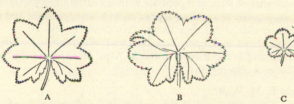

Fig. 40. Leaves of *Alchemilla*. A, *A. vestita*; B, *A. subcrenata*; C, *A. minima.* × ½.

Native. Limestone grassland in Yorks; W. Ross, W. Sutherland; Leitrim; an escape in one or two other places. 4, H1. Europe from Scandinavia (c. 68° N.) and S. Finland to the Cevennes, Corsica, Apennines, Bosnia, Bulgaria and the Upper Dnieper region, east to the western part of the Volga-Kama region. Quebec (introduced).

4. A. vestita (Buser) Raunk.

A. pseudominor Wilmott; *A. minor* auct. plur.

Plant small to medium-sized. *Stems, pedicels, receptacle and petioles rather densely covered with spreading hairs*, which are less dense and silky than those of *A. glaucescens*. *Lvs* (Fig. 40 A) *hairy all over both surfaces*, ± reniform in outline, *basal sinus wide* (60° or more, usually c. 90°); lobes usually 7 (sometimes with small basal ones in addition), not much broader than long, less rounded than in *A. glaucescens* but not straight-sided (those of the spring lvs more like *A. glaucescens*) not overlapping, toothed all round; *teeth* (9–)11(–13) on each of the middle lobes, *acute, somewhat curved towards apex of lobe*, all ± equal; *stipules purplish tinged*. Fls 3–4 mm., clusters not very dense, not silvery outside. $2n=$c. 102–10*.

Native. Grassland, etc.; ascending to 2750 ft. Throughout the British range of the agg., common in general and the most frequent sp. in S. England. 70, H34. Europe from Iceland and N. Scandinavia to N. Spain east to Finland (rare), Lithuania (1 loc.), Austria (Vorarlberg); S. Greenland, Newfoundland, Nova Scotia.

5. A. filicaulis Buser

A. Salmoniana Jaquet

Differs from A. vestita only in the hair covering. Lower part of stem and petioles with spreading hairs; upper part of stem and pedicels glabrous; receptacle glabrous or hairy. Lvs less densely hairy, often only on the folds above and the veins below. $2n=$c. 101–10*; c. 150*.

Native. Grassland in mountain districts; ascending to over 3000 ft., local. From Brecon and Derby northwards; Ireland: Ben Bulben. 28, H1. Europe from Iceland and Scandinavia to Belgium and the Alps of Savoy and Switzerland, east to N.W. Russia (Karelian Lapland and Ladoga-Ilmen regions), C. Russia (one isolated locality in Volga-Don region); Greenland (to c. 70° N.); Newfoundland, Labrador.

6. A. subcrenata Buser

Differs from *A. filicaulis*: *Stipules* brownish, *not purple-tinged*. Lvs (Fig. 40 B) undulate with rather broad lobes and *coarse broad teeth*, rather sparsely hairy above. Lower part of stem and petioles with spreading hairs with *some hairs usually directed slightly downward*. Upper part of stem, pedicels and receptacle glabrous. Fls smaller, 2·5–3 mm. diam. $2n=$c. 96; c. 104–10.

Native. Grassland in Upper Teesdale and Weardale. 2. Europe from N. Scandinavia and arctic Russia to the S. Alps, Bosnia and the Volga-Don region; W. Siberia (Ob region).

7. A. minima S. M. Walters

Plant very small. *Stems, pedicels, receptacle and petioles rather sparsely covered with spreading hairs*. Lf-blades (Fig. 40 C) 1–2·5 × 1·5–3 *cm*., reniform in outline, *hairy on the folds and edges above and veins beneath*, and sometimes sparsely on the rest of the surfaces, *basal sinus wide*; lobes usually 5 (sometimes with small basal ones in addition), broader than long, rounded above, not overlapping, *the sinuses between them ending in a deep toothless incision; teeth (7–)9–11 on each of the middle lobes*, subacute, *somewhat unequal-sided but scarcely curved towards the apex of the lobe; stipules brownish. Pedicels usually with very few hairs*, occasionally glabrous, receptacles more hairy, not at all silvery. Fls c. 2 mm. in small, rather dense clusters. $2n=$c. 103–9*.

Native. Limestone grassland on Ingleborough (Yorks) but only recently described and may occur elsewhere. ?Endemic.

8. A. monticola Opiz

A. pastoralis Buser

Plant medium-sized. *Stem, petioles, and usually peduncles, with spreading hairs; pedicels glabrous*; receptacle usually hairy. Lvs (Fig. 41 A) orbicular in

outline, *densely hairy on both surfaces, basal sinus closed*; lobes 9–11, the two
basal often small, broader than long, ± rounded, teeth 15–19 on each of the
middle lobes, acute, somewhat curved towards the apex. Fls 2–3 mm. diam.,
clusters dense. $2n=101$; c. 103–9.

Native. Grassland in Upper Teesdale and Weardale; Surrey and Bucks.
(probably escaped). 4. Europe from N. Scandinavia to the S. Alps, Albania,
Bulgaria and the Volga-Don region, east to W. Siberia.

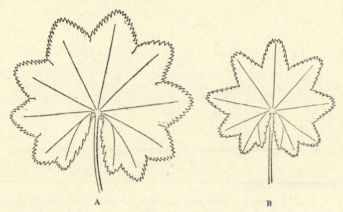

Fig. 41. Leaves of *Alchemilla*. A, *A. monticola*; B, *A. acutiloba*. ×½.

9. A. acutiloba Opiz

A. acutangula Buser

Plant rather large. *Lower part of stem and petioles with spreading hairs*, upper
part of stem, *peduncles, pedicels and receptacle glabrous*. Lvs (Fig. 41 B)
± reniform in outline, very variable in hairiness, early lvs ± hairy on both
sides, late summer lvs often glabrous, *basal sinus open; lobes* 7–11, somewhat
broader than long, *straight-sided*, ± truncate at apex, *teeth* (13–)15–19(–21)
on each of the larger lobes, acute, almost triangular, *straight* or nearly so,
*those at the middle of each margin of the lobe much larger than the top and
bottom ones*, the apical tooth markedly smaller than its neighbours. Fls.
c. 3–4 mm., clusters loose. $2n=$ c. 100, c. 105–9.

Native. Grassland in Upper Teesdale, Weardale and mid-Durham. 1. From
Scandinavia and N.W. Germany (very rare) east to W. Siberia (Ob region)
south to the Alps and the Volga-Don region.

10. A. xanthochlora Rothm.

A. pratensis auct., vix Opiz

Plant usually robust. *Stems and petioles densely clothed with spreading hairs*;
peduncles, pedicels and receptacle glabrous or very sparingly hairy. Lvs
(Fig. 42 A) reniform in outline, *glabrous above* (rarely with a very few hairs
on the folds), hairy on the veins and sometimes thinly on the surface beneath,
basal sinus wide (45° or more); *lobes* usually 9 (the two lowest small) *rounded*,
rather broad, not overlapping; *teeth* 13–15 on each of the middle lobes, acute,

somewhat curved towards apex of lobe, all ± equal; stipules brown. Fls
2·5–3 mm. diam., numerous, in dense clusters. $2n =$ c. 105.

Native. Grassland, etc., at low altitudes. Throughout the British range of
the agg. (except in Ireland, where it is not recorded from the south-east),
common in general. 66, H19. Europe from S.W. Norway (c. 62° N.),
southernmost Sweden and S. Latvia to C. Spain, Apennines and Croatia, east
to Poland and the Ladoga-Ilmen region; naturalized in eastern N. America.

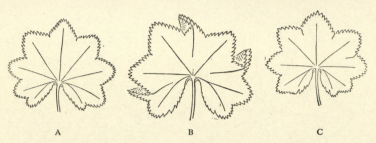

Fig. 42. Leaves of *Alchemilla*. A, *A. xanthochlora*; B, *A. glomerulans*;
C, *A. glabra*. ×½.

11. A. glomerulans Buser

Robust or medium-sized. *Stem and petioles densely clothed with appressed
rather silky hairs*; pedicels, receptacle and often peduncles glabrous. Lvs
(Fig. 42B) reniform in outline, usually hairy all over above, *less hairy beneath*
and with the hairs sometimes confined to the veins, spring lvs often less hairy
especially above, basal sinus wide; lobes usually 9, usually broad and rounded
(often c. twice as broad as long), often overlapping and folding when dried;
teeth 13–15 on each of the middle lobes, broad but subacute, somewhat
curved, those on either side of the apical tooth the largest, the lowest considerably smaller. Fls c. 4 mm. diam., in dense clusters, infl. short. $2n =$ c. 64,
96, c. 101–9.

Native. Damp rock-ledges, usually acid, in the mountains, ascending to
3000 ft. From Perth to Sutherland; Teesdale. 9. Iceland, Scandinavia (except
southernmost Sweden), Finland, Baltic states, Arctic and N. Russia (east to
the northern Urals); rarely in the high mountains of Europe in the Jura, Alps
(Savoy and Engadine) and Pyrenees; Greenland (to 70° 15′ N.), Labrador.

12. A. glabra Neygenfind

A. alpestris auct. plur.; *A. obtusa* auct. (incl. ed. I).

Robust to small. *Stems sparsely* and closely *appressed hairy on the lowest
1–2 internodes, or almost glabrous; upper part of plant glabrous*; petioles
appressed hairy or glabrous. Lvs (Fig. 42C) reniform in outline, glabrous
except for the apical portion of the veins beneath, *basal sinus wide* (usually
60° or more and often nearly 180°); lobes 7–9 rather broad (but rarely twice
as broad as long), rounded or straight-sided, not or rarely overlapping, the
sinus between them wide and *not suddenly narrowed at the base, toothed
throughout*; teeth 11–17 on each of the middle lobes, broad or rather broad,
somewhat curved, *those in the middle of the lobe larger and broader than the*

upper and lower ones, the apical tooth conspicuously narrower and usually shorter than its neighbours. Fls 3–4 mm., in loose clusters or scarcely clustered (except when the first fls expand). $2n=$ c. 96, c. 100, c. 102–110.

Native. Grassland, open woods and rock ledges, ascending to nearly 4000 ft. Almost throughout the British range of the agg., the commonest sp. on mountains but the rarest of the three widespread spp. in S. England and not known in S.E. Ireland. 70, H22. Europe from Norway (c. 70° N.) and S. Finland to N. Spain and N. Italy, east to the Ladoga-Ilmen region and Transylvania; N. Urals; N.E. America (?native).

British plants referred to *A. obtusa* Buser have been proved on cultivation to be forms of this sp.

13. A. wichurae (Buser) Stefánsson

A. acutidens auct.; *A. acutidens* var. *alpestriformis* C. E. Salmon

Rather small. *Stems appressed hairy on the two lowest internodes* (hairier than *A. glabra*), *glabrous or with a few hairs above; infl. glabrous*; petioles appressed hairy. Lvs (Fig. 43) orbicular in outline, glabrous except on the veins beneath, which are usually hairy throughout their length but sometimes only in the apical portion, *basal sinus closed or nearly so*; lobes 7–9, usually broad and rounded (often twice as long as broad), *sinus between them with a narrow* V-shaped (*or almost closed*) *entire prolongation at the base; teeth* 17–19 on each of the middle lobes, rather narrow, *strongly curved towards the apex, all ± uniform in size, terminal tooth about equalling its neighbours* (sometimes somewhat shorter). Fls 3–4 mm., in loose clusters or scarcely clustered.

Fig. 43. Leaf of *Alchemilla wichurae.* × ½.

Native. Grassland and rock-ledges on mountains, apparently confined to basic soils, ascending to 3000 ft., very local. Yorks, Westmorland, Perth to Sutherland. 7. Scandinavia (from 58° N. northwards), Finland (very rare), Faeroes, Iceland, ?Riesengebirge; E. Greenland (very rare).

Plants of the *acutidens* agg. not referable to *A. wichurae* occur in Scotland, particularly on the Ben Lawers range, and require further study.

*A. mollis (Buser) Rothm., a large very hairy plant with shallowly lobed lvs, differing from the British spp. in the epicalyx about equalling the calyx, is commonly grown in gardens and sometimes naturalized. Roumania, Asia Minor.

*A. tytthantha Juz. (A. multiflora Buser ex Rothm.) has recently been found naturalized in Selkirk and Berwick; plant medium to large; lvs hairy on both sides; petioles and stems with ± downward-directed hairs; fls very small, glabrous. Native in the Crimea.

12. APHANES L.

Differs from *Alchemilla* as follows: *Annual.* Fls in dense lf-opposed cluster-like cymes. *Stamen* 1(–2), *opposite a sepal, inserted on inner margin of disc; anthers extrorse.*

About 12 spp., cosmopolitan.

Lobes of stipules surrounding infl. triangular-ovate; fr. 2·2–2·6 mm., bottle-
shaped.   **1. arvensis**

Lobes of stipules surrounding infl. oblong; fr. 1·4–1·8 mm., calyx-teeth
converging.   **2. microcarpa**

(1–2). A. arvensis agg.   Parsley Piert.

Small, rather inconspicuous annual 2–20 cm., pilose, usually much-branched
from the base, pale green; branches decumbent or ascending. Lvs 2–10 mm.,
fan-shaped, shortly stalked, trisect, the segments divided at the apex into
3–5 oblong lobes; stipules fused into a lf-like cup, with 5–7 obtuse lobes.
Fls in sessile clusters, which are half-enclosed by the stipular cup, very small
(not 2 mm.); segments of epicalyx much smaller than the sepals, difficult to
see. Fl. 4–10. Th.

Native. Arable land and bare places in grassland, mainly on dry soils;
ascending to 1700 ft. 112, H40, S. Common throughout the British Is.

The distribution of the two following is incompletely known in this country
but both are widespread. *A. arvensis* is the commoner and occurs on both
acid and basic soils, whereas *A. microcarpa* appears to be confined to acid
sandy soils.

1. A. arvensis L.

Alchemilla arvensis (L.) Scop.

Usually relatively robust, greyish-green. *Lobes of stipules surrounding infl.
triangular-ovate, little longer than broad, c. half as long as the entire portion.
Fr. (including sepals) 2·2–2·6 mm., the sepals ascending so that the whole appears
bottle-shaped.* 2n=48. Apomictic.

Throughout the British Is. on both acid and basic soils. Europe from
S. Sweden and the Baltic States southwards (?absent from W. Spain and
Portugal, in Russia only in the Crimea and Caucasus); Asia Minor, Persia;
introduced in N. America.

2. A. microcarpa (Boiss. & Reut.) Rothm.

Usually more slender, not greyish. *Lobes of stipules surrounding infl. oblong,
c. twice as long as broad, nearly as long as the entire portion. Fr. (including
sepals) 1·4–1·8 mm., the sepals convergent so that the whole appears ovoid.*
2n=16; amphimictic.

Throughout the British Is. on acid sandy soils, though less common than
A. arvensis. Europe from S. Sweden and Denmark to Spain and Portugal,
Italy and Bohemia; Bulgaria, Thrace; Morocco, Algeria; Macaronesia;
N. America.

13. SANGUISORBA L.

Erect perennial herbs. Lvs pinnate. *Fls* in dense terminal spikes or heads,
hermaphrodite, with 2 or 3 bracteoles below each fl. *Epicalyx* 0. Sepals 4,
coloured. Petals 0. *Stamens* 4. *Carpel* 1; ovule 1, pendulous; style terminal,
stigma simple. *Receptacle* becoming corky in fr., enclosing the achene,
without spines. Nectar secreted by a ring round the style.

Two or three spp., in north temperate regions.

1. S. officinalis L. 'Great Burnet.'

Poterium officinale (L.) A. Gray

Glabrous perennial 30–100 cm.; stock thick. Stems erect, branched above. Radical and lower cauline lvs pinnate with 3–7 pairs of lflets; lflets increasing in size upwards, the larger 2–4 cm., ovate or oblong-ovate, stalked, mostly cordate at base, obtuse, crenate-dentate or serrate-dentate; stipules of radical lvs scarious, adnate, of cauline lf-like, dentate; upper cauline lvs small and few. Fls dull crimson, in oblong heads 1–2 cm. Stamens equalling or somewhat longer than calyx. Receptacle in fr. 4-winged, smooth between the wings. Fl. 6–9. Visited by Diptera and Lepidoptera. $2n=28$. Hs.

Native. Damp grassland, ascending to 1500 ft. From Hants, Surrey and Suffolk westwards and northwards to Ayr and Berwick, locally common; W. Mayo, Down, Antrim. 70, H4. Europe from Iceland and Finland to C. Spain, Apennines, Montenegro and Bulgaria; temperate Asia to N. Persia, China and Japan; escaped in N. America.

*S. *canadensis* L. with the white fls in cylindric spikes and stamens exserted is sometimes grown in gardens and has been found as an escape. Native of eastern N. America.

14. POTERIUM L.

Differs from *Sanguisorba* as follows: *Fls monoecious or polygamous*. Sepals green. *Stamens numerous,* exserted. *Carpels* 2(–3); stigmas feathery. No nectar, wind-pollinated.

About 25 spp., in Europe, W. Asia and N. Africa.

Fr. receptacle with 4 longitudinal entire ridges, the surface between them
    with fine raised reticulations. **1. sanguisorba**
Fr. receptacle with 4 longitudinal sinuate wings, the surface between them
    strongly and irregularly pitted and ridged, the ridges toothed.
                                                        2. polygamum

1. P. sanguisorba L. Salad Burnet.

P. dictyocarpum Spach; *Sanguisorba minor* Scop.

Subglaucous perennial herb 15–40(–60) cm., smelling of cucumber when crushed, glabrous except for some long flexuous hairs in the lower part; stock somewhat woody. Stems erect, simple or branched above. Radical lvs pinnate with 4–12 pairs of lflets; lflets increasing in size upwards, the larger 0·5–2 cm., orbicular or shortly oval, shortly stalked, rounded at base, ± truncate at apex, deeply serrate or serrate-dentate with the terminal tooth smaller than the lateral; stipules scarious, adnate. Cauline lvs few (sometimes 0), the lower like the radical, a few of the upper often with narrower (oblong) lflets, passing into the small bracts; stipules lf-like, small. Fls green, often purple-tinged, in globose heads 7–12 mm.

A   B

Fig. 44. Fruits of *Poterium*. A, *P. sanguisorba*; B, *P. polygamum*. × 2·5.

diam., the lower fls male, the middle hermaphrodite, the upper female. Styles purple-red. *Fr. receptacle c. 4 mm.,* ovoid, 4-angled, *with entire ridges down* *the angles, the surface between them with fine raised entire reticulations, not pitted* (Fig. 44 A). Fl. 5–8. $2n = 28^*$. Hs.

Native. Calcareous grassland, occasionally becoming locally dominant, less frequently neutral grassland; ascending to 1650 ft. in Yorks. Widespread and common in England and Wales (not Isle of Man); very local in Scotland, extending north to Dunbarton and Angus (a rare introduction farther north); widespread in S.E. and C. Ireland to mid Cork, E. Galway and Louth; Donegal. 76, H21, S. C. and S. Europe (rare in the S.W.) extending north to S. Sweden and the Volga-Don region; Armenia, N. Persia; Morocco (Atlas, sspp.); naturalized in N. America.

*2. P. polygamum Waldst. & Kit.

P. muricatum Spach

Differs from *P. sanguisorba* as follows: Taller (30–80 cm.), more robust and larger in most of its parts. Several cauline lvs with narrow deeply toothed lflets nearly always present. Fl.-heads larger and often somewhat longer than broad. *Fr. receptacle c. 6 mm.,* ovoid, 4-angled *with sinuate wings down the angles, the surface between them strongly rugose with coarse raised toothed ridges or reticulations* (Fig. 44 B). Fl. 6–8. Hs.

Introduced. Formerly grown for fodder and completely naturalized on field borders, etc., in many places in S. England and Wales and a few in N. England and S. Scotland, north to Lanark and Angus; Jersey. 51, S. Mediterranean region extending to Russian C. Asia; naturalized in C. Europe.

15. Acaena Mutis ex L.

Herbs or undershrubs. Lvs usually pinnate. Fls in terminal spikes or heads, hermaphrodite. *Epicalyx* 0. Sepals 3–4. *Petals* 0. Carpels 2; stigmas feathery. *Receptacle* dry in fr., *with 4 spines, 1 below each sepal, or more numerous scattered spines.*

About 40 spp., south temperate, extending north in America to Mexico. A few spp. are occasionally grown in gardens.

*1. A. anserinifolia (J. R. & G. Forst.) Druce

A. Sanguisorbae Vahl

Prostrate creeping much-branched undershrub, emitting short, ascending or erect lfy pilose stems 2–15 cm. high. Lvs pinnate with 3–4 pairs of lflets; lflets increasing in size upwards, the larger 5–10 mm., oblong, sessile, cuneate or rounded at base, obtuse, deeply crenate-serrate, glabrous above, appressed pilose beneath; stipules adnate and scarious with a free lf-like apex. Fls in globose heads which are solitary on erect peduncles 4–7 cm. long; heads 5–10 mm. diam., greenish, the spines of the receptacle not evident at fl. time. Heads larger in fr., the receptacle of each fl. with 4 long (c. 1 cm.) spines barbed at the tip. Fl. 6–7. Chw.

Introduced. Probably originally imported with wool, now completely naturalized in a few places. Native of E. Australia, New Zealand, etc.

Tribe 5. Roseae. Receptacle deeply concave, often almost closed at the mouth, becoming coloured and fleshy in fr. Carpels numerous. Fr. of numerous achenes enclosed in the fleshy receptacle ('hip').

16. ROSA L.

Shrubs, sometimes trailing or scrambling, usually deciduous. Lvs pinnate (very rarely simple), stipules (usually) adnate. Stems usually prickly. Fls terminal, solitary or in corymbs, hermaphrodite, (4–)5-merous. Stamens numerous. Ovule 1, pendulous. Fls homogamous, mostly without nectar, visited by various insects for pollen, self-pollination possible if insect visits fail.

Over 100 spp. (according to some authors many more), north temperate and subtropical regions. Our garden roses are hybrids of complex origin, derived from a number of spp., and have been developed over a very long period. A number of foreign spp. are also fairly often grown.

For a full account of the British forms see A. H. Wolley-Dod, *Journal. Bot.* 68–69 (1930–1), Suppl.

1.   Styles united into a column, which equals at least the shorter stamens; trailing shrub.   **1. arvensis**
    Styles free or united into a short column; not trailing.   *2*

2.   Fls always solitary, without bracts; lflets small, 3–5 pairs; stems densely prickly and bristly, not tomentose; sepals entire.   **2. pimpinellifolia**
    Fls 1 or more, bracts resembling the stipules; lflets 2–3(–4) pairs; stems not bristly or with a few bristles or, if with numerous bristles, tomentose.   *3*

3.   Stems prickly and bristly, tomentose; sepals entire; fls 6–8 cm. diam.; fr. 2 cm. or more.   **3. rugosa**
    Stems without or with few scattered bristles, not tomentose; outer sepals usually with pinnately arranged projecting lobes; fls c. 5 cm. diam. or less; fr. usually less than 2 cm.   *4*

4.   Styles united into a column at fl., becoming free in fr.; disk conical, prominent.   **4. stylosa**
    Styles free throughout; disk flat or nearly so.   *5*

5.   Lvs ± densely covered over the whole surface beneath with conspicuous brownish viscid fruity-scented glands, glabrous or somewhat pubescent but not tomentose; prickles hooked.   (**rubiginosa** agg.)   *11*
    Lvs not glandular beneath or with glands confined to the main veins or if with glands scattered over the whole surface, then usually ± densely tomentose; the glands smaller, not fruity-scented; prickles hooked to straight (if lvs densely glandular then prickles rarely hooked).   *6*

6.   Prickles hooked or strongly curved; lvs glabrous or pubescent, simply or doubly serrate.   (**canina** agg.)   *7*
    Prickles straight or slightly curved; lvs always pubescent, usually very tomentose, always doubly serrate.   (**villosa** agg.   *9*

7.   Sepals reflexed or spreading after fl., falling before the fr. ripens and usually before it reddens; stigmas in a ± conical head not concealing the disk; styles glabrous, pilose or villous; pedicels relatively long.   *8*
    Sepals usually erect, sometimes spreading or reflexed after fl., persistent at least till the fr. reddens; stigmas in a flat head concealing the disk; styles villous; pedicels short.   **6. dumalis**

8.   Outer sepals with narrow, usually entire lobes; lvs glabrous or pubescent, eglandular or rarely with a few glands on the main veins.   **5. canina**
    Outer sepals with broad, usually toothed or lobed lobes; lvs always pubescent at least on the veins beneath, usually glandular at least on the main veins but often eglandular; prickles more strongly hooked with stouter bases.   **7. obtusifolia**

9   Stems arching or flexuous; prickles somewhat curved, rarely straight,
     relatively stout; auricles of upper stipules straight or diverging; sepals
     long-persistent but falling before the fr. decays.   *10*
     Stems straight; prickles straight, relatively slender; auricles of upper
     stipules falcate and incurved; sepals erect, persistent till the fr. decays;
     styles always villous.   **10. villosa**

10   Stems flexuous (often zigzag) but scarcely arching; lvs bluish-green;
     pedicels relatively short; sepals erect or ascending, persistent till the
     fr. ripens; styles villous, rarely pilose; disk not more than 3½ times
     diameter of orifice.   **9. sherardii**
     Stems arching; lvs pale green, not bluish; pedicels relatively long; sepals
     erect to reflexed, falling before the fr. is ripe, though usually persistent
     till it reddens; styles pilose or glabrous, rarely villous; disk 4–6 times
     diameter of orifice.   **8. tomentosa**

11   Lflets rounded at base; pedicels glandular-hispid.   *12*
     Lflets cuneate at base; pedicels glabrous.   *13*

12   Stems erect; prickles usually unequal; styles pilose; sepals persistent at
     least till the fr. reddens, erect or spreading.   **11. rubiginosa**
     Stems arching; prickles ±equal; styles glabrous; sepals soon falling,
     usually reflexed.   **12. micrantha**

13   Stems erect; styles pilose or villous; sepals persistent till the fr. is ripe,
     spreading or suberect. Very rare.   **13. elliptica**
     Stems arching; styles glabrous or thinly pilose; sepals soon falling, usually
     reflexed.   **14. agrestis**

Section 1. *Synstylae* DC. Stems trailing, climbing or creeping; prickles
hooked. Fls in bracteate corymbs; two outer sepals usually with pinnately
arranged projecting lobes, all caducous. *Styles united into a column, reaching
at least the lower stamens.*

1. R. arvensis Huds.   'Field Rose.'

Deciduous shrub, glabrous or nearly so, with weak, trailing, subglaucous,
often purple-tinted stems, either decumbent and forming low bushes 50–
100 cm. high, or climbing over other shrubs, rarely more erect and reaching
2 m. Prickles hooked, all ±equal. Lflets 2–3 pairs, 1–3·5 cm., ovate or ovate-
elliptic, simply, rarely doubly, serrate, glabrous on both sides, or pubescent
on the veins (rarely all over) beneath, rather thin; petiole usually with
stalked glands; stipules narrow, auricles straight. Fls 1–6, white, 3–5 cm.
diam.; pedicels 2–4 cm., with stalked glands, rarely smooth; buds short.
Sepals short (not 1 cm.), ovate, acuminate, the tip not expanded, often
purplish, soon caducous, lobes very few. Styles glabrous, equalling the
stamens. Fr. small, red, smooth, globose or sometimes ovoid or oblong.
Fl. 6–7. $2n=14*$. N.

Native. Woods, hedgebanks and scrub; ascending to 1250 ft. Rather
common in S. England and Wales, becoming very local in N. England, very
rare in Scotland, extending north to Stirling; throughout Ireland but rather
local, 73, H39. W. and S. Europe from Belgium, C. Germany (52° 38′ N.
in Hanover) and Hungary to N. Spain, S. France, Sicily and Greece.

*R. sempervirens L.

Lvs evergreen, coriaceous; lflets 2 pairs; stipules very narrow with divergent auricles. Our plant (var. *melvinii* Towndr.) is characterized by its narrowly elliptic lflets and glabrous styles (they are usually hairy in this sp.). $2n=14$. Naturalized in Worcester. Native of the Mediterranean region.

*R. multiflora Thunb., with pectinate stipules and many-fld infl., is rarely naturalized. $2n=14$. Japan, Korea.

Section 2. *Pimpinellifoliae* DC. Usually low shrubs with ± erect stems, and ± numerous straight prickles and stiff bristles. *Lflets* 3 *or more pairs*. Fls solitary; *bracts* 0 or small and scale-like. Sepals entire, erect and persistent on the fr. Styles free, short.

2. R. pimpinellifolia L. Burnet Rose.

R. spinosissima L., p.p.

Low, erect, deciduous shrub 10–40(–100) cm., *spreading by suckers and forming large patches; prickles numerous, straight, mixed with numerous stiff bristles and passing into them* (rarely stems almost unarmed). *Lflets 3–5 pairs*, small, 0·5–1·5(–2) cm., suborbicular or oval, obtuse, simply serrate, rarely doubly glandular-serrate, glabrous on both sides or sparingly pubescent beneath, rarely somewhat glandular; stipules narrow, auricles expanded and divergent. Fls cream-white, rarely pink, 2–4 cm. diam.; pedicels 1·5–2·5 cm., glandular-hispid or smooth. Styles woolly. *Fr.* 1–1·5 cm., subglobose, *purplish-black*. Fl. 5–7. $2n=28*$. N.

Native. Dunes, sandy heaths, limestone pavement, etc., especially near the sea, but ascending to 1700 ft. 96, H39, S. From Caithness and the Outer Hebrides southwards and throughout Ireland, but rather local, and absent from a number of counties. Europe from Iceland, S. Norway and C. Russia (Middle Dnieper and Transvolga regions) to N. Spain, Italy, Macedonia and the Caucasus; temperate Asia east to Manchuria and N.W. China, south to Asia Minor.

Hybrids between *R. pimpinellifolia* and the members of the section *Caninae* occur occasionally. They are usually recognizable by the presence of bristles and by the habit and lf-shape showing the influence of *R. pimpinellifolia*. The other parent is frequently impossible to determine exactly. The main groups are as follows:

R. pimpinellifolia × subsection *Eucaninae* (*R.* × *hibernica* Templeton etc.). Occasional.

R. pimpinellifolia × subsection *Villosae* (*R.* × *involuta* Sm. etc.). The most frequent group.

R. pimpinellifolia × subsection *Rubiginosae*. Rare.

Their cytology is incompletely known. *R.* × *sabinii* Woods ($2n=42$) behaves cytologically like the *Caninae*, having 14 bivalents and 14 univalents. It differs from them, however, in being highly sterile. It belongs to the *involuta* group.

Until other forms have been cytologically studied, it seems impossible to treat this group satisfactorily taxonomically.

Section 3. *Cinnamomeae* DC. Erect deciduous shrubs; prickles straight, bristles often present. Lflets 2–5 pairs. Fls in bracteate corymbs. Sepals usually entire, erect and usually persistent after fl. Styles free, short. Carpels on sides and bottom of receptacle.

*3. R. rugosa Thunb.

Deciduous shrub 1–2 m. *Stems densely prickly and bristly, tomentose.* Lflets 2–5 cm., 2–4 pairs, ± elliptic, dark green and rugose above, pubescent beneath. Upper stipules broad. *Fls* 6–8 *cm. diam.*, 1 or few, bright purplish-pink to white; pedicels bristly. *Fr. large* (2–2·5 cm.), red, crowned by the erect sepals. Fl. 6–7. 2*n*=14. N.

Introduced. Frequently grown in gardens and often used as a stock for other roses, sometimes occurring as an escape; naturalized in several places. Native of N. China and Japan; occasionally naturalized in C. Europe.

R. canina × *rugosa* (*R.* × *praegeri* W.-Dod) occurs spontaneously in Antrim.

Section *Carolinae* Crép. Differs from *Cinnamomeae* in: Sepals spreading and caducous after fl. Carpels confined to bottom of receptacle.

*R. virginiana Mill. (*R. lucida* Ehrh.)

Shrub to 2 m. Prickles straight or curved, slender, bristles confined to young shoots. Lflets 2–4 pairs, 2–6 cm., ± elliptic, coarsely serrate, dark green and shining above, glabrous or pubescent on the veins beneath. Fls bright pink, 5–6 cm. diam.; pedicels and receptacle glandular-hispid. Sepals lobed. Fr. red, 1–1·5 cm. diam. Sometimes grown in gardens and occurring as an escape. 2*n*=28. Native of eastern N. America.

Section 4. *Caninae* DC. Deciduous shrubs with erect or arching stems, and ± numerous, usually hooked, but sometimes straight prickles; bristles usually 0. Lflets 2–3(–4) pairs. Fls in bracteate corymbs. Outer sepals with pinnately arranged projecting lobes. Styles free or fused into a short column.

In the preceding sections reproduction is sexual and with normal pairing. The plants of this section, however, are all irregular polyploids conforming to the formula 2*n*=14+7*a*, or less frequently 28+7*a*, where *a*=2, 3 or 4 (?6), the 14 (or 28) chromosomes being composed of 7 (or 14) homologous pairs, the others having no partners. Reproduction appears often to be apomictic but sexual reproduction occurs ± frequently. Meiosis takes place in such a manner that the fertile egg has 7 (or 14)+7*a* chromosomes and the fertile pollen grain 7 (or 14) chromosomes only. The original number is thus maintained. Any new forms produced by cross-fertilization can continue apomictically and sexually.

The forms in this section are very numerous but can be grouped into a limited number of fairly well characterized species. Intermediate forms exist but are mostly rare. It is often merely a matter of taste to which of 2 spp. these forms are referred. Forms are therefore liable to be found which will scarcely be identifiable from the following short account. It should, however, be possible to refer one of these plants to its species in the vast majority of cases, provided that in doubtful plants consideration is given to the characters as a whole. A bush may fairly frequently deviate from that which is given as 'usual' in the description in one or two respects but if it does so in many it will probably be found to agree better with another sp.

Hybrids between spp. of this section have been recorded and undoubtedly occur. It is doubtful, however, how far these recent hybrids can be distinguished in a group, the spp. of which are all essentially hybrid in constitution. (In some cases it appears that the recent hybrids are more sterile.) It should be noted that the inverse hybrids between any pair of spp. are likely to be very different.

Subsection 1. *Stylosae*. (Crép.) Crép. Prickles stout, hooked. Lflets glabrous, or pubescent not tomentose or glandular. Pedicels usually glandular. *Styles at first fused into a rather short exserted column*, becoming free. Stigmas in a narrow head.

4. R. stylosa Desv.

R. systyla Bast.

Shrub 1–4 m.; stems arching; prickles hooked, some with very stout bases. Lflets 1·5–5 cm., 2–3 pairs, narrowly (rarely broadly) ovate to lanceolate, usually acuminate, simply (rarely doubly) serrate, usually pubescent beneath, at least on the veins, rarely also above, or glabrous on both sides, eglandular; *stipules and bracts rather narrow*. Fls 3–5 cm. diam., 1–8 or more, white or pale pink; pedicels long, usually glandular-hispid, rarely smooth. Sepals reflexed after fl., falling before fr. is ripe. *Stylar column glabrous, shorter than the stamens; stigmas in an ovoid head; disk conical, prominent*. Fr. c. 1–1·5 cm., ovoid, rarely globose, red, smooth. Fl. 6–7. $2n = 35^*$, 42. N. or M.

Native. Hedges, etc.; from Denbigh, Worcester, Leicester, and Suffolk (?Norfolk) southwards and from Limerick, Offaly and Wicklow southwards, local; Channel Is. 40, H 8, S. France, W. Germany (Rhineland), W. Switzerland, N.W. Italy (very rare), N.W. Spain (very rare).

Subsection 2. *Eucaninae* Crép. (*R. canina* agg.). Dog Rose. *Prickle scurved or hooked*, equal, usually stout. *Lflets glabrous or pubescent, eglandular or with a few glands on the main veins below*, rarely (only in *R. obtusifolia*) with numerous glands, the glands not strongly scented. Pedicels glabrous or less frequently glandular. Styles free.

5. R. canina L.

Shrub 1–3 m.; stems arching. *Prickles strongly curved or hooked*. Lflets 1·5–4 cm., 2–3 pairs, ovate, oval, obovate or elliptic, acute or acuminate, simply to doubly (and then often glandular-) serrate, *glabrous on both sides or* (less frequently) *pubescent* beneath (on the nerves or all over) or on both sides, *eglandular*, or rarely with a few, often deciduous glands on the main veins; upper stipules and bracts broad. Fls 1–4 or more, pink or white; *pedicels relatively long* (0·5–2 cm., usually more than 1 cm.), smooth or less often glandular-hispid. *Sepals falling before the fr. ripens and usually before it reddens, reflexed* or rarely spreading after fl.; *lobes narrow, entire or nearly so*. Petals c. 2–2·5 cm. *Styles glabrous or pilose, rarely villous* (more frequently so in the pubescent-lvd forms); *stigmas in a* (usually) *conical head, which is usually obviously narrower than the scarcely prominent disk*. Fr. usually c. 1·5–2 cm., globose, ovoid or ellipsoid, scarlet, smooth, very rarely glandular-hispid. The pubescent-lvd forms are often separated as *R. dumetorum* Thuill. but there seems little justification for this. An extremely variable sp.; Wolley-Dod recognizes over 60 varieties and forms (including those under *R. dumetorum*). Fl. 6–7. $2n = 35^*$. N. or M.

Native. Woods, hedges, scrub, etc., ascending to 1800 ft. 88, H40, S. Common in England, Wales and Ireland (in general much the commonest sp.), becoming rare in Scotland though reported from as far north as Orkney; Channel Is. Europe (except Iceland; to c. 62° N. in Norway); N. Africa; S.W. Asia (south to Palestine, north to Russian C. Asia); Madeira; naturalized in N. America.

6. R. dumalis Bechst.

R. coriifolia Fr.; *R. Afzeliana* auct.; *R. glauca* Vill. ex Lois., non Pourr.; *R. caesia* Sm.

Differs from *R. canina* as follows: Prickles usually smaller. Upper stipules and bracts usually broader, often red tinged. *Pedicels short*, 0·5–1·5 cm. (mostly less than 1 cm.) often concealed by the bracts. *Sepals persistent at least till the fr. reddens*, usually becoming ± erect after fl., but sometimes spreading or reflexed. *Styles villous*, not at all exserted; *stigmas in an almost flat, villous head, completely concealing the narrow disk*. Very variable though less so than *R. canina*; the hairy-lvd forms are often separated as *R. coriifolia* Fr. Fl. 6–7. Fr. earlier than *R. canina*. 2n=35*. N. or M.

Native. Woods, hedges, scrub, etc.; from Sussex and Gloucester northwards, very rare in S. England, becoming commoner in the north and largely replacing *R. canina* in the Scottish Highlands; Ireland (recorded from a few places in the north and west, but incompletely known). 62, H7. Europe from Iceland (very rare) and Scandinavia (c. 67° N.) to the Pyrenees, S. Alps and Greece; W. Asia (mountains).

7. R. obtusifolia Desv.

R. tomentella Léman

Differs from *R. canina* as follows: *Prickles usually more strongly hooked* with stout bases. *Lflets* 1·5–3·5 cm., ovate or oval, usually more rounded in outline and less acute or even obtuse, doubly glandular serrate or, less frequently, simply serrate, *always pubescent* at least on the veins beneath and usually so on both sides, *usually glandular on the main veins beneath* but often eglandular, rarely with glands on the whole lower surface. Fls usually white; pedicels 0·5–1·5 cm. *Sepals short; lobes broad, dentate or lobed*. Petals c. 1·5 cm. Styles pilose, rarely subglabrous. Fr. 1·0–1·8 cm., globose or ovoid, red. Fl. 6–7. 2n=35*. N. or M.

Native. Hedges, scrub, etc.; from Northumberland and Cheshire southwards, local; Ireland (distribution uncertain). 42, H3. Europe from S. Scandinavia to France, C. Italy, Roumania and Greece.

Subsection 3. *Villosae* (DC.) Crép. (*R. villosa* agg.). 'Downy Rose'. *Prickles straight or curved*, usually equal, rather slender. *Lflets* doubly glandular-serrate, *usually densely tomentose, frequently glandular below, the glands* often small, *always scentless or little scented*. Pedicels nearly always, and fr. often, glandular-bristly. Styles free.

8. R. tomentosa Sm.

Shrub 1–2 m.; *stems arching; young stems and lvs pale green; prickles* curved or nearly straight, *stouter* than in the other spp. of this subsection. Lflets

2–3 pairs, 1·5–4 cm., ovate, ovate-lanceolate or elliptic, rarely obovate, acuminate or acute, usually densely pubescent or tomentose on both sides but sometimes only thinly so or with the pubescence confined to the midrib, but then harsh to the touch, usually glandular all over beneath but sometimes eglandular; upper stipules not very broad, flat, their *auricles* short, triangular, *straight or diverging*. Fls c. 4 cm. diam., 1–4 or more, pink or white; *pedicels relatively long* (1–2 cm.), glandular-hispid. *Sepals falling before the fr. ripens* but frequently persisting until it reddens, erect, spreading or reflexed, relatively long, *markedly constricted at the attachment*, with projecting pinnately-arranged lobes. *Styles pilose or glabrous*, rarely villous. *Disk 4–6 times the diameter of the orifice*. Fr. c. 1–2 cm., ovoid, less frequently globose or ellipsoid, red, usually glandular-hispid but often smooth. Fl. 6–7. $2n = 35^*$. N.

Native. Woods, hedges, scrub, etc.; widespread and rather common in England, Wales and Ireland, becoming rare in Scotland but extending at least to Ross. 66, H40. Europe from S. Scandinavia and Finland to C. Portugal and N. Spain, C. Italy and Macedonia and S.W. Russia; Caucasus; Asia Minor.

9. R. sherardii Davies

R. omissa Déségl.

Intermediate in most respects between *R. tomentosa* and *R. villosa*. Habit intermediate, lower and more compact than in *R. tomentosa*, branches flexuous; *lvs and young stems somewhat glaucous*. Prickles and lvs as *R. tomentosa* but lflets usually more markedly doubly serrate. Fls often several, frequently deep pink; *pedicels* short (up to 1·5(–1·7) cm. but some of them, at least, usually less than 1 cm.). *Sepals usually shorter* than in either of the other spp., *slightly constricted at the attachment*, rounded on the back, lobed as in *R. tomentosa*, erect or ascending after fl., persistent until the fr. is ripe but finally falling. *Styles villous*, rarely pilose. *Disk not more than 3½ times the diameter of the orifice*. Fr. most frequently globose, obovoid or pyriform, less often ovoid. Fl. 6–7. $2n = 28^*, 35^*, 42^*$. N.

Native. Woods, hedges, scrub, etc.; throughout Great Britain, rather common in Scotland, becoming very rare in S. England; Ireland (only recorded from a few places in the north and west but incompletely known). 66, H7. Europe from S. Scandinavia to C. France, east to Poland and W. Switzerland.

10. R. villosa L.

R. mollis Sm.; *R. mollissima* Willd.

Erect shrub 40–100(–200) cm., somewhat suckering; *branches straight*; young stems and lvs glaucous; *prickles quite straight, slender, subulate, their bases little thickened*. Lflets (1·5–)2–4 cm., 2–3(–4) pairs, oblong, oval or elliptic, more rounded than in the other spp. of the subsection, subobtuse, tomentose on both sides, usually glandular beneath, but sometimes eglandular, markedly doubly glandular-serrate; upper stipules very broad, their *auricles falcate and incurved*. Fls 3–5 cm. diam., 1–3, usually deep pink or almost red; *pedicels short*, 0·5–1·0(–1·5) cm., glandular-hispid. *Sepals* 1·5–2·5 cm., *erect and persistent on the fr. till it falls or decays*, not constricted at the attachment, usually less lobed than in the other spp. of the subsection (lobes 1–3 pairs) and sometimes entire. *Styles villous*; stigmas in a broad flat head. Fr. 1–2 cm., red, usually globose, usually glandular-hispid, sometimes eglandular-hispid or smooth. Fl. 6–7. $2n = 28^*$. N.

Native. Woods, hedges, scrub, etc., ascending to over 2000 ft.; rather common in Scotland, extending south to Glamorgan, Hereford and Derby, occasionally naturalized farther south; scattered over Ireland but apparently local and mainly in the north. 65, H19. Europe from Scandinavia (69° 10′ N.) and Finland to the mountains of N.E. Spain, C. Italy, Albania, Bulgaria and the Caucasus; Asia Minor, Persia.

Subsection 4. *Rubiginosae* Crép. (*R. rubiginosa* agg.). Sweet-briar. Prickles usually hooked, rarely arched or straight, equal or unequal, sometimes mixed with a few stout bristles mainly on the fl. branches. *Lflets* always doubly glandular-serrate, glabrous or somewhat pubescent, never tomentose, ± *thickly clothed below with sweet-scented viscous brownish glands*. Pedicels glandular-hispid or less frequently glabrous. Styles free.

11. R. rubiginosa L.

R. eglanteria L., p.p.

Shrub 1–2 m.; *stems erect. Prickles* hooked, *usually unequal*, usually mixed with scattered stout bristles on the fl. branches below the infl. and sporadically elsewhere. *Lflets* 2–3(–4) pairs, 1–2(–2·5) cm., suborbicular to oval-elliptic, rarely ovate, *rounded at the base*, pubescent on the veins beneath and slightly so above, very glandular beneath. Fls 1–3, bright pink; *pedicels* rather short, c. 1 cm., *glandular-hispid. Sepals usually persistent at least till the fr. reddens, erect or spreading after fl.* Petals c. 1·5 cm. *Styles pilose*, short. Fr. sub-globose or ovoid, c. 1–1·5 cm., scarlet, smooth or glandular-hispid at the base, rarely bristly all over. Fl. 6–7. Differs from most spp. in having nectar secreted by the edge of the receptacle, but is apparently mainly visited for pollen. $2n = 35^*$. N.

Native. Scrub, rarely woods and hedges, mainly on calcareous soils; an early colonizer of chalk grassland; widespread in England and Wales and locally common, rarer in Scotland but reported to extend to Caithness; widespread in Ireland but rather local; Channel Is. 89, H29, S. Frequently planted and an introduction in many of its localities. Europe from S. Scandinavia and Esthonia to N. and E. Spain, Italy, Macedonia and the Caucasus; W. Asia (to N.W. India); naturalized in N. America.

12. R. micrantha Borrer ex Sm.

Shrub 1–2 m.; *stems arching. Prickles* hooked, ± *equal*; stout bristles rarely present. *Lflets* 1·5–3(–3·5) cm., 2–3 pairs, ovate, obovate, oval or elliptic, *rounded at base*, ± pubescent beneath, usually glabrous or nearly so above, glandular beneath. Fls 1–4, pink; *pedicels* rather long, 1–2 cm., *glandular-hispid*, very rarely smooth. *Sepals falling early, reflexed* to (rarely) suberect. Petals 1–1·5 cm. *Styles glabrous*, very rarely thinly pilose, somewhat exserted. Fr. c. 1–1·5 cm., ovoid, scarlet, smooth or glandular-hispid. Fl. 6–7. $2n = 35$, 42. N.

Native. Woods, scrub and hedges mainly on calcareous soils but more frequent off them than *R. rubiginosa*; widespread and rather common in

England and Wales; rare in Scotland and reported from only a few counties, north to Ross; in Ireland apparently confined to Kerry and Cork; Channel Is. 66, H4, S. Mediterranean region (not known from Crete, Sardinia, etc.) extending north to Poland, Germany and N. France.

13. R. elliptica Tausch

R. inodora auct.

Shrub 1–2 m.; *stems erect*. Prickles hooked. *Lflets* 2–3 pairs, obovate, *cuneate at base*, pubescent on both sides, glandular beneath. Fls solitary, white or pale pink; *pedicels smooth*, 6–12 mm. *Sepals persistent till the fr. is ripe, spreading or suberect after fl. Styles pilose or villous*, short. Fr. subglobose, red, smooth. Fl. 6–7. $2n=35$. N.

Native. Very rare; Somerset, Huntingdon, Warwick; Lough Derg. 3, H1. Mountains of C. Europe to N.W. Italy and the Carpathians; Macedonia, Greece; local everywhere.

14. R. agrestis Savi

R. sepium Thuill., non Lam.

Shrub 1–2 m.; *stems arching*. Prickles hooked, ± equal. *Lflets* 1–3(–5) cm., 2–3 pairs, narrowly obovate, oblong or elliptic, *cuneate at base*, glabrous or pubescent, glandular beneath. Fls 1–3 or more, white or pale pink; *pedicels* 1–2 cm., *smooth*. Sepals soon falling, reflexed, rarely suberect after fl. Petals c. 1·5 cm. *Styles glabrous or slightly pilose*, somewhat exserted. Fr. 1–1·5 cm., subglobose, ovoid or ellipsoid, red, smooth. Fl. 6–7. $2n=35$. N.

Native. Scrub, etc., mainly on calcareous soils; from Berwick southwards, local and rare; C. Ireland from E. Mayo and Westmeath to N. Tipperary and Kilkenny. 24, H7. Europe from Denmark (very rare), and Poland to C. Portugal and S. and E. Spain, Corsica, Italy and N. Greece; Crimea; N. Africa (mountains).

Subfamily 3. PRUNOIDEAE. Trees or shrubs. Lvs simple, stipules present. Receptacle flat or concave. Carpel 1, rarely 2 or 5, free from the receptacle; ovules 2 in each carpel, pendulous. Fr. a drupe.

17. PRUNUS L.

Fls hermaphrodite. *Petals and sepals* 5. Stamens usually 20. *Carpel* 1; style terminal. Fr. with stony endocarp, 1-seeded. Nectar secreted by the receptacle.

Nearly 200 spp. mainly north temperate, a few extending into tropical Asia and the Andes. A number of spp. are ± commonly planted.

Measurements of lvs and petioles are when mature; they are smaller at fl.

1	Fls in clusters or umbels or solitary.	*2*
	Fls in racemes.	*6*
2	Fls 1–3 from axillary buds; bracts 0 or small and few.	*3*
	Fls in few-fld umbels; infl. surrounded at base by the persistent bud scales, forming an involucre.	*5*
3	Twigs and lvs dull; plant usually ± hairy on twigs, lvs or pedicels; lvs nearly always broadest above the middle.	*4*

Twigs and upper surface of lvs somewhat glossy; plant glabrous except
  for petioles and lower part of midrib; lvs broadest above or below the
  middle.   **3. cerasifera**

4 Very thorny shrub; fls usually before the lvs, mostly solitary, rarely 2;
     petals 5–8 mm.; fr. 10–15 mm.   **1. spinosa**
   Unarmed or somewhat thorny shrubs or small trees; fls with the lvs, 1–3;
     petals 7 mm. or more; fr. usually 2 cm. or more.   **2. domestica**

5 Lvs light dull green, somewhat pubescent beneath; most of the infls without
     lf-like scales; petals obovate; usually a tree.   **4. avium**
   Lvs dark green, rather glossy, soon nearly glabrous beneath; all the infls
     with the inner scales lf-like; petals orbicular; usually a shrub.

                                                               5. cerasus

6 Lvs deciduous, rather thin, closely and sharply serrate; peduncles with
     1 or 2 lvs near base.   **6. padus**
   Lvs evergreen, thick and coriaceous, distantly serrate or subentire;
     peduncles lfless.   **7. laurocerasus**

Subgenus 1. PRUNUS.

Axillary buds solitary, terminal bud 0. Lvs rolled or folded in bud. Fls
solitary or in a few-fld axillary fascicle, appearing with or before the lvs.
Ovary with a groove down one side. Fr. usually with a whitish bloom
(pruinose). Our spp. belong to Section *Prunus* with rolled lvs, 1–3 stalked
fls and glabrous fr. *P. armeniaca* L. (Section *Armeniaca* (Lam.) Koch, lvs
rolled, fr. pubescent), the apricot, with broadly ovate lvs rounded or subcordate at base, subsessile fls and orange-yellow fr., is occasionally grown
for its fr.

1. P. spinosa L.   Blackthorn, Sloe.

Rigid, deciduous much-branched shrub 1–4 m., often suckering and forming
dense thickets. *Twigs shortly pubescent when young*, the pubescence usually
persisting for c. 1 year, *becoming blackish or dark brown or grey in the 1st winter*
(sometimes remaining green in shade), *dull, with numerous short lateral shoots
which become thorns*. Buds obovate, hairy. Lf-blades 2–4 cm., oblong-
obovate, elliptic-obovate or oblanceolate, cuneate at base, acute or obtuse,
crenate-serrate, ± pubescent, at least on the midrib beneath, dull above;
petiole 2–10 mm. *Fls usually appearing before the lvs, solitary (–2); pedicels
glabrous*, rarely sparingly pubescent. *Petals 5–8 mm.*, pure white, oblong-
obovate. *Fr. 10–15 mm.*, *globose*, blue-black, strongly pruinose, very astringent, flesh greenish, adherent to the stone; stone nearly globose, 7·5–10 ×
6–8 mm., little flattened, nearly smooth or slightly pitted. Fl. 3–5. Pollinated
by various insects, protogynous. $2n = 32$. N. or M.

Native. Scrub, woods and hedges on a great variety of soils (not acid peat),
not tolerant of dense shade; ascending to 1360 ft. in Yorks. From Sutherland
southwards and throughout Ireland. 108, H40, S. Europe (except Iceland,
Crete, etc.) to 60° N. in Scandinavia; Mediterranean region (to Persia); S.W.
Siberia (Upper Tobol region).

Var. *macrocarpa* Wallr. (?*P. fruticans* Weihe) is a rather taller, less thorny
shrub with larger lvs (–5 cm.) and fls, and larger 12–16(–20) mm. fr., the stone
more flattened and the fls often appearing with the lvs. It occurs sporadically

with the type and requires further study. It does not appear to have been examined cytologically.

*2. P. domestica L.

Deciduous shrub or small tree 2–6(–12) m., often suckering. *Twigs* usually ±pubescent when young, sometimes glabrous, but amount and persistence of pubescence very variable, *becoming grey or brown in the 1st winter*, rarely remaining green, *dull; thorns* 0 *or few*. Lf-blades 4–10 cm., obovate or elliptic, cuneate at base, usually acute, crenate-serrate, ±pubescent on both sides when young, becoming glabrous and dull above; petiole 5–20(–25) mm. *Fls appearing with the lvs*, 1–3 *together*; pedicels 0·5–2 cm., pubescent or glabrous. *Petals* 7–12(–15) *mm.*, white, obovate. *Fr.* globose, ovoid or oblong, 2–4(–8 in cultivated forms) *cm.*, blue-black, purple, red, green or yellow; flesh greenish; stone flattened, somewhat pitted. Fl. 4–5. Pollinated by various insects, protogynous or homogamous, some vars self-sterile, others self-fertile. $2n = 48$ (all 3 sspp.). M.

There is evidence to show that this sp. originated in the past from *P. spinosa* and *P. cerasifera* by hybridization and chromosome doubling.

Introduced. Hedges, etc. Widespread but usually in small quantity in England, Wales and Ireland; less so in Scotland, extending to Sutherland; Channel Is. Probably always ultimately descending from cultivated plants. Doubtfully wild anywhere, but the triploid hybrid from which it probably arose occurs in S.W. Asia. The following sspp. are rather ill-defined and intermediates occur. They are completely interfertile.

Ssp. insititia (L.) C. K. Schneid.   Bullace.
P. insititia L.

Usually a shrub, often somewhat thorny. *Twigs densely and conspicuously pubescent*, the pubescence persisting for 1 year or more. Lvs usually ±hairy above, more densely beneath. Pedicels conspicuously pubescent. Petals pure white. Fr. (1·5–)2–3(–4) cm., globose or shortly oval, usually blue-black and pruinose or purple. Stone bluntly angled, the flesh adherent to it.

More thoroughly naturalized and probably longer established than the other sspp. and often found more remote from houses, often considered native. 73, H34, S. The bullace does not appear to be much cultivated at the present day, but the damson is usually regarded as a cultivated form.

Ssp. italica (Borkh.) Hegi   Greengage.
Larger in all its parts than ssp. *insititia*, not thorny. Twigs finely hairy at least in the first year. Lvs obtuse. Petals usually pure white. Fr. 3–5 cm., globose or ovoid, usually green. Stone bluntly or sharply angled, the flesh adherent.

Apparently infrequent as an escape, but it has not usually been separated from ssp. *insititia*. Commonly grown for its fr.

Ssp. domestica   Plum.
Usually a small tree, not thorny. Twigs sparingly hairy in the first year, or glabrous. Lvs large, petiole long. Pedicels sparingly pubescent or glabrous. Petals tinged greenish (especially in bud). Fr. 4–8 cm., oblong-ovoid, of

varied colours. Stone much flattened, sharply angled, usually free from the flesh.

Widespread and rather frequent but mostly near houses. Very commonly grown for its fr. 72.

*3. P. cerasifera Ehrh.   Cherry-Plum.
P. divaricata Ledeb.

Deciduous shrub or small tree to 8 m., rarely thorny. *Twigs glabrous from the first, usually green* (often reddish on the exposed side) *in the 2nd year, rather glossy. Lf-blades* 3–7 cm., ovate, elliptic or obovate, rounded or broad cuneate at base, acute, unequally crenate-serrate with rather deep obtuse teeth, glabrous and *somewhat glossy above,* usually pubescent along the lower part of the midrib beneath, at least when young; petiole 5–10 mm. *Fls* appearing with the lvs, *usually solitary* (–2); pedicels 5–15 mm., glabrous. *Petals 7–11 mm.,* white, ovate. Fr. 2–2·5 cm., globose, yellow or reddish. Stone little flattened, ± orbicular, quite smooth on the faces. Fl. 3–4, earlier than *P. spinosa* and *P. domestica.* Fr. only produced in favourable years. $2n = 16$. M.

Introduced. Frequently planted in hedges in many parts of England, often far from houses but rarely naturalized. Russian C. Asia and Persia through the Caucasus to the N. Balkans (to Serbia). A form with pink fls and purple lvs (var. *atropurpurea* Jaeg. = var. *pissardii* (Carr.) L. H. Bailey) is very commonly grown in small gardens.

Subgenus 2. AMYGDALUS (L.) Focke
Axillary buds 3, the lateral ones fl.-buds, terminal bud present. Lvs folded in bud. Fls 1–3 in axillary fascicles. Fr. usually tomentose. Stone often deeply pitted. The two following are frequently cultivated.

*P. amygdalus Batsch   Almond.
Lvs 7–12 cm., ± lanceolate, broadest rather below the middle, serrulate. Fls pink, 3–5 cm. diam., appearing before the lvs. Fr. 3–6 cm., ellipsoid, compressed, green and dry when ripe, finally splitting. Native of W. Asia.

*P. persica (L.) Batsch   Peach.
Lvs 8–15 cm., oblanceolate-elliptic to narrowly oblong, broadest about or rather above the middle, serrate. Fls pink, 2·5–3·5 cm. diam., appearing before the lvs. Fr. 5–7 cm., subglobose, reddish on the sunny side, juicy, not splitting. The Nectarine is a glabrous-fruited form (var. *nectarina* (Ait.) Maxim.). Besides the forms cultivated for fr., forms with semi-double fls, ranging from white to red, are frequently grown. Native country uncertain, perhaps China, or derived in cultivation from the Chinese *P. davidiana* (Carr.) Franch.

Subgenus 3. CERASUS (Mill.) Focke

Lvs folded in bud. Terminal bud present. Fls solitary or in umbels or short racemes. Style (and sometimes ovary) grooved. Fr. usually glabrous, not pruinose.

Our spp. belong to Section *Cerasus* with solitary buds, teeth of lvs obtuse, bud scales persisting round the infl. as an involucre, fls in umbels and sepals reflexed.

Of the other spp. *P. serrulata* Lindl. (Section *Pseudocerasus* Koehne, sepals upright, teeth of lvs acute), Japanese Cherry, is commonly planted, usually in pink double-fld forms. Other spp. are less frequently grown.

4. P. avium (L.) L. Gean, Wild Cherry.

Deciduous *tree* 5–25 m., of rather open habit, suckering rather freely. Bark smooth, reddish-brown, peeling off in thin strips. Branches spreading or ascending. *Lf-blades* 6–15 cm., obovate-elliptic, acuminate, crenate-serrate, *light dull green* and glabrous above, sparingly but persistently appressed-*pubescent beneath*, rather thin, drooping; *petiole to* 5 *cm.* with 2 large glands near the apex. Fls 2–6, cup-shaped, in usually sessile umbels; *bud scales* greenish, persisting below the infl., those of most of the infls *not lf-like*, but some infls with 1 or 2 inner ones with small lf-like blades; pedicels 2–5 cm. *Receptacle constricted at apex.* Sepals oblong, reflexed, entire. *Petals* 8–15 mm., white, *obovate, gradually narrowed to the base*, emarginate or retuse. Fr. c. 1 cm., subglobose, bright or dark red, glabrous, sweet or bitter. Fl. 4–5. Pollinated by various insects, homogamous, all vars completely self-sterile. $2n = 16$. MM. or M.

Native. Woods and hedges on the better soils, rather common in England, Wales and Ireland, becoming rare in N. Scotland but extending to Caithness. 105, H33. Europe from Scandinavia (64° N.) and S.W. Russia southwards (only Sardinia of the Mediterranean Is.); N. Africa (mountains); W. Asia. The garden Sweet Cherries are referable to this sp. but sometimes have an extra chromosome ($2n = 17$), probably because of past hybridization with *P. cerasus.*

*5. P. cerasus L. Sour Cherry.

Differs from *P. avium* as follows: *Usually a shrub*, rarely a small tree to c. 7 m., of rather bushy habit, suckering very freely. Branches often somewhat drooping. *Lf-blades* 5–8 cm., relatively broader, *dark green* and somewhat shining *above, soon glabrous* or nearly so *beneath*, firm, spreading; *petiole* 1–3 *cm.*, the glands smaller or 0. Fls flat; umbels usually shorter-stalked; *inner bud scales* of all the infls *with lf-like blades*; pedicels 1–4 cm. *Receptacle not or scarcely constricted.* Sepals broader, sometimes crenate. *Petals* ± *orbicular, rounded at the base*, entire, emarginate or irregularly notched. Fr. bright red, acid. Fl. 4–5. Pollinated mainly by bees, protogynous (or homogamous?). $2n = 32$. M.

Introduced. Hedges, etc.; widespread in S. England, Wales and Ireland but local everywhere, rarer in N. England and E. Scotland (to Banff); apparently not reported from N. and W. Scotland. 58, H39, S. Origin uncertain. The garden Sour and Morello Cherries belong to this sp., and all our wild plants are, in all probability, descended from them.

Subgenus 4. PADUS (Mill.) Focke

Lvs folded in bud. Terminal bud present. Fls in long racemes. Style not grooved. Fr. glabrous, not pruinose.

Section *Padus.* Lvs deciduous. Peduncle lfy. Sepals deciduous.

6. P. padus L.   Bird-Cherry.

Deciduous tree 3–15 m. Bark brown, strong smelling, peeling. Twigs brown or grey. Lf-blades 5–10 cm., elliptic or obovate, acuminate, rounded or cordate at base, closely and sharply serrate, glabrous except sometimes for tufts of hairs in the axils of the primary veins beneath; petiole 1–2 cm., with a gland on each side of the apex. Fls 10–40 in long, loose, ascending to drooping, glabrous racemes, 7–15 cm.; peduncle with 1 or 2 lvs. Receptacle hairy within. Sepals short, obtuse, ascending, gland-fringed, deciduous. Petals 4–6(–10) mm., white, irregularly toothed. Fr. 6–8 mm., ovoid, black, astringent. Fl. 5. Pollinated by various insects, mainly flies, protogynous. $2n = 32$. M. or MM.

Native. Woods, etc., ascending to 2000 ft., widespread and rather common in Scotland from Caithness southwards, and in N. England and Wales, extending as a native tree to Gloucester, Derby and E. Yorks, occasionally planted in S. England; widespread in Ireland but rather local. 69, H 32. Europe from northernmost Scandinavia to the mountains of Spain and Portugal (absent from W. France), N.W. Italy, Croatia, Bulgaria and the Caucasus; Asia from W. Siberia to the Himalaya and N.E. Asia Minor; Morocco (Atlas, very rare).

Section *Calycopadus* Koehne. Lvs deciduous. Peduncle lfy. Calyx persistent.

*P. serotina Ehrh.

Deciduous tree to 30 m. Lf-blades 5–12 cm., oblong-ovate or oblong-lanceolate, cuneate or rounded at base, serrulate with appressed teeth, shining above. Receptacle glabrous within. Calyx persistent in fr. Sometimes planted among native vegetation. Native of eastern N. America.

Section *Laurocerasus* (Durham.) Hook. f. Lvs evergreen. Peduncle lfless. Calyx deciduous.

*7. P. laurocerasus L.   Cherry-Laurel.

Evergreen glabrous shrub or small tree 2–6 m. Lvs 5–18 cm., oblong-obovate, dark green and very glossy above, acuminate, cuneate or rounded at base, distantly serrate or subentire, thick and coriaceous; petiole 5–10 mm., green. Fls numerous, in ± erect racemes, 5–12 cm.; peduncle lfless. Petals white, c. 4 mm. Fr. c. 8 mm., ovoid, purple-black. Fl. 4–6. $2n = c. 176$. M.

Introduced. Commonly planted in gardens, woods, etc., frequently self-sown and ± naturalized. Native of S.E. Europe and S.W. Asia.

*P. lusitanica L.   Portugal Laurel.

Evergreen shrub. Lvs 6–12 cm., oblong-ovate, dark green, serrate; petioles 15–25 mm., deep red. Less frequently planted than *P. laurocerasus* and rarely naturalized. $2n = 64$. Native of Spain, Portugal, Madeira and the Canaries.

Subfamily 4. POMOIDEAE. Trees or shrubs with long and short shoots, the latter sometimes modified as thorns. Lvs simple or pinnate; stipules present, usually small and caducous. Receptacle deeply concave. Sepals 5, imbricate. Petals 5. Carpels 1–5, united to the receptacle at least in the lower half so that the ovary is inferior or ½-inferior. Ovules (1–)2(–20) in each carpel, axile or basal. Fr. a pome or drupe, usually crowned by the persistent calyx.

A very distinct subfamily, sometimes regarded as a separate family. Besides the morphological characters, the basic chromosome number (17) is constant and is not found elsewhere in the Rosaceae.

18. Cotoneaster Medic.

Shrubs, rarely small trees, *not thorny. Lvs entire*, shortly petioled. Fls solitary or few or in corymbs. Sepals short, ± triangular. Petals imbricate in bud. Stamens c. 20. *Carpels 2–5, free on the inner side, the wall stony in fr.; ovules* 2 *in each carpel*, similar; styles free. Fr. with mealy flesh, red, purple or black with 2–5 stones. Nectar secreted by the inner wall of the receptacle.

About 50 spp., north temperate Old World. A number are grown in gardens.

1   Petals erect, pink; lvs 10 mm. or more, broadest near the middle.   2
    Petals spreading, white; lvs 8 mm. or less, broadest near the apex.
                                            3. microphyllus

2   Lvs grey-tomentose beneath, rounded at base.   **1. integerrimus**
    Lvs green and sparsely appressed-pubescent beneath, broad cuneate at
    base.   **2. simonsii**

Section 1. *Cotoneaster.* Petals erect, usually pink.

1. C. integerrimus Medic.

C. vulgaris Lindl.

Bushy deciduous shrub 15–100(–200) cm. *Twigs tomentose when young, soon glabrous. Lvs* 1·5–4 *cm.*, ovate, oval or suborbicular, *rounded at base, obtuse or acute*, usually mucronate, green and glabrous above, persistently *grey-tomentose beneath*; petioles 2–4 mm. Fls 1–4 in short cymes. Petals pink, c. 3 mm. Styles 2(–5). Fr. c. 6 mm., subglobose, red. Fl. 4–6. Visited by various insects, especially wasps, protogynous or homogamous, self-pollination possible. Fr. 8. $2n=68$. N.

Native. Limestone rocks on Great Orme's Head (Caernarvon) in very small quantity. 1. Rocky places in Europe from Scandinavia and Finland to C. and E. Spain, Sicily and Macedonia; Crimea and the Caucasus; N. Asia Minor, N. Persia.

*2. C. simonsii Baker

Deciduous or half-evergreen shrub 1–4 m., stems suberect. *Twigs strigose-pubescent, the pubescence persisting for 2 or 3 years. Lvs* 1–3 *cm.*, ovate or broadly elliptic, broadly cuneate at base, acute, deep green and glabrous or slightly pubescent above, paler *green and sparsely strigose beneath*, somewhat coriaceous; petioles 2–4 mm. Fls 2–4 in short cymes. Sepals pubescent. Petals pink. Fr. c. 8 mm., obovoid, scarlet. Fl. 5–7. Fr. 10. N. or M.

Introduced. Commonly planted and naturalized in many places. 23. Native of the Khasia Hills (India).

*C. horizontalis Decaisne

Low shrub with horizontally spreading stems with the branches flattened into a horizontal plane. Lvs, fls, etc., much like those of *C. simonsii* but lvs smaller. Native of W. China; commonly cultivated and rarely naturalized.

Section 2. *Chaenopetalum* Koehne. Petals spreading, white.

*3. C. microphyllus Wall. ex Lindl.

Low evergreen shrub to 1 m., with rigid spreading or drooping branches. Twigs strigose-pubescent. *Lvs 5–8 mm., obovate-cuneiform,* cuneate at base, *obtuse to retuse,* dark green and glabrous above, glaucous beneath and appressed hairy, coriaceous; petiole 1–2 mm. Fls 1(–3), c. 1 cm. diam. Anthers purple. Styles 2. Fr. c. 6 mm., globose, crimson. Fl. 5–6. Fr. 9–10. $2n = 68$. N.

Introduced. Commonly grown in gardens (introduced 1824) and now naturalized in many places, especially on limestone near the sea. 36, H5, S. Native of the Himalaya.

*C. frigidus Wall. ex Lindl.

Deciduous tall shrub or small tree to 6 m. Lvs ± oblong, 6–12 cm., dull green, glabrous or nearly so when mature. Fls numerous, in corymbs. Fr. red.

Native of the Himalaya. Frequently planted and has occurred as an escape.

PYRACANTHA M. J. Roem.

Differs from *Cotoneaster* as follows: Thorny. Lvs usually toothed. Fls always in many-fld corymbs. Carpels always 5. Petals white, spreading. Several spp. are grown in gardens, the commonest being ***P. coccinea** M. J. Roem. Lvs 3–4 cm., lanceolate to oblanceolate, acute, glabrous or slightly pubescent beneath. Infl. pubescent. Fls c. 8 mm. diam. Fr. 5–6 mm., subglobose, scarlet.

Sometimes escaping, rarely ± naturalized. Native of S. Europe from Italy eastwards to W. Asia.

19. CRATAEGUS L.

Deciduous trees or shrubs, usually *with thorns. Lvs lobed or serrate.* Fls in corymbs (rarely solitary). Sepals short, ± triangular. Stamens 5–25. *Carpels 1–5, free at the apex, united* at least *at the base on the inner side, the wall stony in fr.; ovules* 2, *the upper sterile;* styles free. Fr. usually with mealy flesh, red, yellow or black. Nectar secreted by a ring in the receptacle; fls strong-smelling, visited by various Diptera, Hymenoptera and Coleoptera, proto-gynous.

Numerous spp. in the north temperate zone, about 90 in the Old World and a large number of highly critical N. American spp. They have mostly larger, less divided lvs and larger fls than our spp. and are sometimes confused with *Sorbus* spp. A number are cultivated and some have been found as escapes. The pink and double-fld forms so commonly grown are forms of our native spp.

Lobes of lvs of short shoots broader than long, usually rounded, sinuses not reaching half-way to midrib; styles mostly 2 (often 1 or 3 on some fls).
1. oxyacanthoides

Lobes of lvs of short shoots longer than broad, ± triangular, sinuses reaching more than half-way to midrib; style 1 (sometimes 2 on some fls).
2. monogyna

1. C. oxyacanthoides Thuill.   'Midland Hawthorn.'

Thorny much-branched shrub or small tree 2–10 m. Twigs glabrous or with a few long hairs when young. *Lf-blades* of short shoots 1·5–5 cm., obovate in outline, 3–5-lobed, glabrous except for scattered hairs on the main veins on both sides when young, *without axillary hair-tufts; lobes shallow (rarely reaching half-way to the midrib), usually rounded in outline, broader than long, serrate* except in the sinuses; main veins straight or curving upwards; those of the long shoots usually more deeply lobed, with conspicuous lf-like stipules. Fls rarely more than 10. Receptacle and pedicels glabrous (?rarely woolly). Sepals deltoid, about as broad as long. Petals 5–8 mm., white. Anthers pink or purple. *Styles mostly* 2 (often 1 or 3 on some fls). Fr. (6–)8–10(–13) mm., deep red, stones mostly 2. Fl. 5–6, about a week earlier than *C. monogyna.* $2n=34$. M.

Native. Woods, less frequently scrub or hedges, much less common than *C. monogyna* though locally more frequent in woods and more tolerant of deep shade; mostly on clay or loam. In England local and mainly in the east, extending west to Somerset, Carmarthen, Denbigh and Cumberland, rare in the north; very rare in Scotland and doubtfully native; scattered over Ireland. 43, H 18. Europe from S. Sweden (probably introduced in Norway), Finland and the Baltic States southwards, apparently not extending east of Poland.

2. C. monogyna Jacq.   Hawthorn.

Habit of *C. oxyacanthoides.* Twigs glabrous. *Lf-blades* of short shoots 1·5–3·5 cm., ovate or obovate in outline, 3–7-lobed, glabrous except for *patches of hairs in the axils of the lower veins beneath* or with occasional scattered hairs; *lobes deep (usually reaching more than half-way to the midrib), tapered to an acute or subobtuse apex, longer than broad, entire or sparingly serrate near their apices*; main veins curving downwards; those of the long shoots more deeply lobed, with conspicuous lf-like stipules. Fls often up to 16, sometimes more. Receptacle and pedicels glabrous to woolly. Sepals triangular, 1–2 times as long as broad. Petals 4–6 mm., white. Anthers pink or purple. *Style* 1 (or 2 in a few fls). Fr. (6–)8–10(–13) mm., deep red, calyx appressed or spreading; stone 1, rarely 2. Fl. 5–6. $2n=34$. M.

Native. Scrub, woods and hedges, ascending to 1800 ft.; the commonest scrub dominant on most types of soil, rare only on wet peat and poor acid sands; the shrub most commonly planted for hedges. 112, H 40, S. Throughout the British Is., very common in England, rare in N. Scotland. Europe (except Iceland); Mediterranean region to Afghanistan.

C. monogyna × oxyacanthoides. Hybrids, intermediate in various ways between the two spp., are usually to be found in quantity where *C. oxyacanthoides* occurs.

*C. orientalis Pall. ex Bieb.

Lvs pinnatifid with narrow lobes, grey-tomentose beneath. Styles 4–5. Fr. 1·5–2 cm. diam., orange-red. Self-sown in one or two places. Native of S.E. Europe and S.W. Asia.

20. Mespilus L.

Deciduous tree, sometimes thorny. Lvs entire or finely serrulate above. *Fls solitary*, large. *Sepals large, lf-like.* Stamens 30–40. *Carpels 5, almost completely united, the walls stony in fr.*; ovules 2, the upper sterile; styles free. Fr. brown, crowned by the persistent calyx, at first hard, finally soft (bletting). Nectar secreted by a ring in the receptacle, homogamous.

One sp.

*1. M. germanica L.   Medlar.

Pyrus germanica (L.) Hook. f.

Thorny shrub 2–3 m. (in cultivation a thornless tree to 6 m.). Twigs densely pubescent when young, glabrous and blackish in the second year. Lvs 5–12 cm., lanceolate or oblanceolate, cuneate at base, acute, entire or serrulate near the apex, appressed pubescent on both sides or subglabrous above; petiole c. 2 mm. Pedicels 5 mm. or less, densely pubescent like the receptacle and calyx. Sepals 10–16 mm., linear-lanceolate-triangular, somewhat lf-like. Petals c. 12 mm., suborbicular. Anthers red. Fr. 2–3 cm., subglobose, crowned by the calyx. Fl. 5–6. $2n = 34$. M.

Introduced. Grown for its fr. and naturalized (recorded 1597) in hedges in S. England, north to Middlesex, Oxford and Gloucester, and in Stafford and S. Yorks, rare; Channel Is. 18, S. Native of S.E. Europe and S.W. Asia (to Turkistan); probably only naturalized elsewhere in Europe but possibly native in the south.

21. Amelanchier Medic.

Deciduous shrubs or small trees, not thorny. Lvs simple, serrate. *Fls in racemes*, rarely solitary. Sepals ± triangular. Petals narrow. Stamens 10–20. Carpels (2–)5, free from the receptacle above and so ½-inferior, *walls cartilaginous in fr.; ovules* 2 in each carpel, *separated by a false septum*; styles united below or free. Fr. bluish- or purplish-black, usually sweet and juicy.

About 25 spp., north temperate, mainly N. America; others, in addition to the following, are sometimes grown.

*1. A. laevis Wieg.

A. canadensis auct.; *A. confusa* Hyl.

Shrub or small tree to 12 m. Lf-blades 3–7 cm., ovate-elliptic or ovate-oblong, acute, shortly acuminate, rounded or subcordate at base, finely and sharply serrate, glabrous or when very young tomentose beneath, purplish when unfolding. Infl. slender, spreading or drooping, glabrous or nearly so, many-fld. Petals 10–22 mm., linear-oblong, white. Top of ovary glabrous. Fr. globose, blackish-purple. Fl. 4–5. Fr. 6–7. M.

Introduced. Grown in gardens and sometimes becoming naturalized on light acid soils; common over several sq. miles in the Hurtwood (Surrey). Native of eastern N. America from Newfoundland to Wisconsin and Georgia.

22. SORBUS L.

Deciduous trees or shrubs, not thorny. Lvs pinnate or simple and lobed or toothed. *Fls in compound corymbs.* Sepals ± triangular. Petals white (pink in some foreign spp.). Stamens 15–25. *Carpels* 2–5, united at least to the middle, inferior or ½-inferior, *walls cartilaginous in fr.*; ovules 2, without false septum; styles free or joined below. Fr. variously coloured, with 1 or 2 seeds in each cell.

About 100 spp., north temperate regions. Several foreign spp. are some-times planted.

The British spp. of this genus consist of three widespread diploids, *S. aria*, *S. aucuparia* and *S. torminalis* and a number of polyploids, some of which are certainly apomictic. There are three main groups of polyploids: (A) resem-bling *S. aria*, (B) intermediate between (A) and *S. aucuparia* (*S. intermedia* agg.), and (C) those intermediate between (A) and *S. torminalis* (*S. latifolia* agg.). The groups (B) and (C) probably originated as hybrids of *S. aria* or other species of group (A) with *S. aucuparia* and *S. torminalis* respectively. In contrast to *S. aria* which is very variable, the polyploid spp. are ± constant.

The most useful characters are:

Lvs. Depth and character of toothing and lobing of lvs of short shoots. The depth of the lobes in the following account is measured perpendicular to the midrib. The main types of toothing are (*a*) ± symmetrical teeth pro-jecting perpendicular to the lf margin (e.g. *S. porrigentiformis*), and (*b*) with the outer margin longer than the inner and somewhat curved so that the tooth is directed more towards the lf-apex (e.g. *S. aria*).

Colour of upper and lower surfaces of lvs.

Fls. These have been little studied. The colour of the anthers is often constant in a sp., and the petal size is another useful character. Other char-acters will probably also be available when the fls have been studied further.

Fr. Colour: Usually a constant and important character. The colour called 'crimson' in the following account is perhaps scarcely a true crimson but has much more blue in it than that called 'scarlet' which is a true scarlet. Size and shape of fr. Number, size and distribution of the lenticels on the fr.

The following key is intended for use primarily on fruiting specimens which are the best for beginning the study of the genus.

1   Lvs pinnate with 4 or more pairs of lflets, the terminal lflet ± equalling
     the lateral ones (Fig. 45A).   **1. aucuparia**
     Lvs simple or with up to 3 pairs of free lflets at the base and with the
     terminal part several times as large.   2

2   Lvs or many of them with at least one free lflet.   3
     Lvs without free lflets.   4

3   Veins of lvs 7–9(–10) pairs; lvs (Fig. 45C) 5·5–8·5 cm., mostly 1·5–1·7 times
     as long as broad; fr. longer than broad; Arran.   **2. pseudofennica**
     Veins of lvs 10–12 pairs; lvs (Fig. 45B) 7–11 cm., 1·6–2·3 times as long
     as broad; fr. subglobose; occurring occasionally as a single tree.
                                                              aria × aucuparia

4   Lvs (Fig. 47B) green on both sides, subglabrous beneath, except when
     very young, deeply lobed; lobes acuminate; fr. brown.   **20. torminalis**
     Lvs persistently grey- or white-tomentose beneath.   5

5 Fr. red.   *6*
   Fr. orange or brown; lvs grey-tomentose beneath, with ±triangular
   lobes or at least doubly serrate with the primary teeth acuminate and
   very prominent.   **(latifolia agg.)** *19*

6 Lvs grey beneath, lobed, with the deepest lobes extending at least ⅓ of
   the way to the midrib.   **(intermedia agg.)** *7*
   Lvs white beneath, not lobed or shallowly crenately lobed with the lobes
   not reaching more than ⅛ of the way to the midrib (and then lvs very
   white below).   **(aria agg.)** *11*

7 Lobes of the lvs extending nearly half-way to the midrib or more.   *8*
   Lobes of the lvs not extending to more than ⅓ of the way to the midrib
   (sometimes to half-way on a few lvs).   *9*

8 Lvs (Fig. 45 D) ±elliptic, acute, with 7–8(–9) pairs of veins; fr. longer
   than broad; petals c. 4 mm.; Arran.   **3. arranensis**
   Lvs (Fig. 45 E) ±oblong, obtuse or subacute, with (8–)9–10 pairs of veins;
   fr. subglobose; petals c. 5 mm.; Brecon.   **4. leyana**

9 Lvs (Fig. 45 F) ±elliptic, 6–8 cm., mostly 1·8–2·2 times as long as broad;
   fr. 6–8 mm., subglobose, scarlet; petals c. 4 mm.; Brecon.
                                                              5. minima
   Lvs 7–12 cm., mostly not more than 1·8 times as long as broad; fr.
   7–15 mm.; petals c. 6 mm.   *10*

10 Lvs (Fig. 45 G) ±elliptic, many of them rounded at the base, yellowish-
   grey-tomentose beneath; fr. scarlet, much longer than broad; anthers
   cream.   **6. intermedia**
   Lvs (Fig. 45 H) ±obovate, nearly all cuneate at the base, whitish-grey-
   tomentose beneath; fr. crimson or crimson-scarlet, subglobose or
   broader than long; anthers pink or tinged pink.   **7. anglica**

11 Fr. longer than broad.   *12*
   Fr. subglobose or broader than long.   *15*

12 Lvs (Fig. 46 F) obovate, tapering gradually to a cuneate base from about
   the middle of the lf and entire in the basal ¼ or more, rounded above;
   veins 8–9(–10) pairs; S. Somerset and N. Devon.   **16. vexans**
   Lvs not tapering at the base for a long distance (or if so also tapering
   above) and entire at the base for ⅙ or less; veins rarely less than 10 pairs
   (except in *S. wilmottiana*).   *13*

13 Lvs with the terminal tooth and those terminating the main veins
   markedly prominent (except in *S. wilmottiana*), greenish-white-
   tomentose beneath, veins rarely more than 11 pairs; fr. with the
   lenticels mainly towards the base.   *14*
   Lvs (Fig. 45 J) with the terminal tooth and those terminating the main
   veins not or scarcely prominent, pure white-tomentose beneath; veins
   10–14 pairs; fr. ±scarlet with scattered lenticels; lvs dull yellow-green
   above.   **8. aria**

14 Fr. scarlet; lvs (Fig. 45 K) yellow or dark green above, mostly 1·5–1·7
   times as long as broad, teeth somewhat curved on the outer margin,
   basal margins of the lf not arched inwards; veins c. 11 pairs; Brecon.
                                                            9. leptophylla
   Fr. crimson; lvs bright green above, mostly 1·6–2·0 times as long as
   broad, teeth tending to be curved on the outer margin, basal margins
   of lf not arched inwards; veins usually 8–9 pairs; Avon Gorge.
                                                            10. wilmottiana

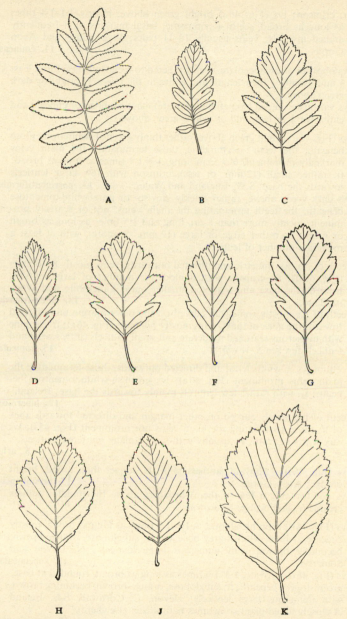

Fig. 45. Leaves of *Sorbus*. A, *S. aucuparia*; B, *S. aria × aucuparia*; C, *S. pseudo-fennica*; D, *S. arranensis*; E, *S. leyana*; F, *S. minima*; G, *S. intermedia*; H, *S. anglica*; J, *S. aria*; K, *S. leptophylla*. × ⅓.

Fr. crimson; lvs (Fig. 46A) bright green above, mostly 1·1–1·4 times as long as broad, teeth ±symmetric, basal margins of lf frequently arched inwards; veins usually 10–11 pairs; Wye Valley and Avon Gorge.   **11. eminens**

15   Lvs obovate or oblanceolate, evenly tapered to a cuneate base for at least ⅓ and usually ½ their length and entire in the lower ¼, bright or dark green above.   16
      Lvs variously shaped (sometimes obovate), base rounded or cuneate and entire for ⅕ or less, dull or yellow-green above.   18

16   Lvs (Fig. 46C) bright green above, rather thinly greenish-white-tomentose beneath, the teeth ±symmetric, those terminating the main veins markedly prominent, 6–9·5 cm. long, 1·3–1·7 times as long as broad; fr. rather small (12 mm. or less), crimson with few large lenticels towards the base; S.W. England and Wales.   **13. porrigentiformis**
      Lvs dark green above, rather thickly greyish- or pure white-tomentose beneath, the teeth terminating the main veins, not or scarcely prominent, mostly more than 8 cm. long and 1·5 times as long as broad or more; fr. rather large or large (12 mm. or more), with at least a moderate number of lenticels.   17

17   Teeth of lvs ±symmetric and directed outwards, veins mostly 8–10 pairs (Fig. 46D); fr. crimson with a moderate number of large lenticels towards the base and scattered smaller ones; Lancashire and Westmorland.   **14. lancastriensis**
      Teeth of lvs mostly somewhat curved on the outer margin and directed towards the apex of the lf, veins mostly 7–9 pairs (Fig. 46E); fr. carmine with numerous scattered moderate and small lenticels (if fr. scarlet and lenticels few see *S. vexans*).   **15. rupicola**

18   Teeth of lvs ±symmetrical and directed outwards, those terminating the main veins prominent (Fig. 46B); lvs greenish-white-tomentose beneath; fr. with rather few lenticels mainly towards the base; Ireland.
                                                                                                          12. hibernica
      Teeth of lvs mostly curved on outer margin and directed towards apex of lf, those terminating the main veins not prominent (Fig. 45J); lvs very white beneath; fr. usually with few or many scattered lenticels.
                                                                                                          8. aria

19   Lvs (Fig. 46J) ±obovate; anthers pink; fr. longer than broad, bright orange; Avon Gorge.   **17. bristoliensis**
      Lvs not (or only a few of them) broadest above the middle; anthers cream; fr. dull orange to brown.   20

20   Lvs (Fig. 46H) rhombic-elliptic, c. 1·6–1·9 times as long as broad, mostly cuneate at base, whitish-grey beneath; fr. subglobose, orange-brown, becoming brown, with numerous large lenticels; N. Devon and Somerset.   **18. subcuneata**
      Lvs (Fig. 46G) ±ovate, 1·3–1·6 times as long as broad, rounded at base, greenish-grey beneath; fr. subglobose, orange-brown, becoming brown, with numerous large lenticels; Devon, E. Cornwall, S.E. Ireland (a closely allied plant sometimes naturalized elsewhere).
                                                                                                          19. devoniensis

      Not as above; lower margin of lf often arched slightly inwards near base (Fig. 47A); fr. in most plants longer than broad, with few small lenticels; Wye Valley.   **aria × torminalis**

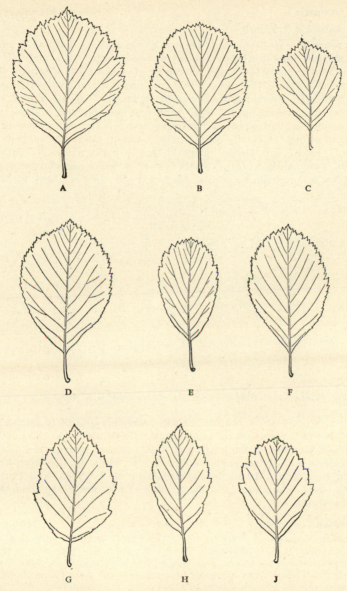

Fig. 46. Leaves of *Sorbus*. A, *S. eminens*; B, *S. hibernica*; C, *S. porrigentiformis*; D, *S. lancastriensis*; E, *S. rupicola*; F, *S. vexans*; G, *S. devoniensis*; H, *S. subcuneata*; J, *S. bristoliensis*. ×⅓.

1. S. aucuparia L.   Rowan, Mountain Ash.

Pyrus aucuparia (L.) Ehrh.

Slender tree to 15(–20) m. with narrow crown and ± ascending branches. Bark greyish, smooth. Twigs pubescent when young, then glabrous and greyish-brown. Buds 10–15 mm., ovoid or ovoid-conic, dark brown, somewhat pubescent. *Lvs* (Fig. 45 A) 10–25 cm., *pinnate; lflets* (4–)6–7(–9) *pairs, the terminal lflet ± equalling the lateral* (never larger), 3–6 cm., oblong, acute or subacute, ± rounded at the often somewhat unequal base, serrate, sometimes doubly so, dark green and glabrous above, subglaucous beneath and pubescent at first, especially on the midrib, usually becoming subglabrous; petiole 2–4 mm. Infl. dense, many-fld, woolly-pubescent in fl. Petals c. 3·5 mm., oval. Anthers cream. Styles 3–4. Fr. 6–9 mm., subglobose, scarlet, with a few inconspicuous lenticels. Fl. 5–6. Fr. 9. $2n = 34$. MM. or M.

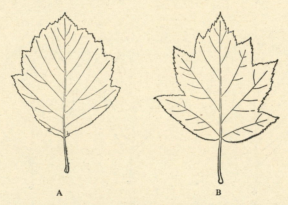

A   B

Fig. 47. Leaves of *Sorbus*. A, *S. aria × torminalis*; B, *S. torminalis*. × ⅓.

Native. Woods, scrub, mountain rocks, ascending to 3200 ft. (higher than any other tree), mainly on the lighter soils, rare or absent on clays and soft limestones, common in the north and west, rare and perhaps not native in some eastern and central English counties. 111, H40. Europe from Iceland and N. (not arctic) Russia to the mountains of C. Spain and Portugal, Corsica, Italy, Macedonia and the Caucasus; Morocco (high mountains); N. Asia Minor.

S. domestica L.   Service Tree.

Resembling *S. aucuparia* in its lvs but a larger tree. Branches spreading. Bark rough, scaling. Fls 16–18 mm. diam. Styles 5. Fr. c. 2·5 cm., apple- or pear-shaped, green or brown with numerous large lenticels and stone cells in the flesh. $2n = 34$. A single tree formerly grew (probably introduced) in the Wyre Forest (Worcester) but it is apparently not now known as a wild tree in this country.

Native of S. Europe, N. Africa, Asia Minor.

S. aria × aucuparia = *S. × thuringiaca* (Ilse) Fritsch

Lvs (Fig. 45 B) 7–11 *cm.*, ± oblong in outline, 1·6–2·3 *times as long as broad, mostly with* 1–3 *pairs of free lflets at the base* but on some trees many of the lvs are lobed nearly to the base but without free lflets, upper part with ± oblong lobes, serrate, obtuse, dull green and glabrous above, greenish-grey-tomentose beneath; *veins* (including free lflets) 10–12 *pairs*; petiole 1·5–3 cm. Infl. woolly-pubescent. Petals 4–6 mm. Anthers cream, more rarely pink. Styles 2–3. *Fr.* 8–10 cm., *subglobose*, scarlet with few inconspicuous lenticels. Occurring very rarely as a single tree with the parents. Sometimes also planted. A fertile hybrid showing marked segregation from seed but F_2 plants have not been found wild in this country.

2. S. pseudofennica E. F. Warb.

S. fennica auct. angl.

Small slender tree. *Lvs* (Fig. 45 c) 5·5–8·5 *cm.*, oblong or ovate-oblong in outline, (1·3–)1·5–1·7(–2·2) *times as long as broad, mostly with* 1(–2) *pairs of free lflets at the base* but some of the lvs without free lflets, lobed above, serrate with sharp rather variously directed teeth most marked at the apex of the lobes, obtuse or acute at the broad apex, at maturity dark yellowish-green and glabrous above, unevenly and rather thinly grey-tomentose beneath; *veins* 7–9(–10) *pairs*; petiole 8–20 mm. Petals c. 4 mm. Anthers cream. *Fr.* 7–10 mm., *longer than broad*, scarlet, with few inconspicuous lenticels. Fl. 5–6. Fr. 9. M.

Native. Steep granite stream-bank. 1. Glen Catacol (Arran). Endemic.

*S. hybrida L.

S. fennica Fr.

More robust than *S. pseudofennica*. Lvs larger, 7·5–10·5 cm., deep bluish-green above, more densely tomentose and with a whiter tomentum beneath. Fls much larger, petals c. 6 mm. Anthers pink (?). Fr. globose, crimson. $2n=68$.

Occasionally planted but much less frequently than *S. × thuringiaca* with which it is often confused. Native of Scandinavia and Finland.

(3–7). S. intermedia agg.

Lvs lobed in various ways to at least ⅓ of the way to the midrib, grey-tomentose beneath; veins 7–10(–11) pairs. Styles 2. Fr. red.

3. S. arranensis Hedl.

Pyrus arranensis (Hedl.) Druce

Small slender tree. *Lvs* (Fig. 45 D) 6–9 cm., *elliptic or rhombic-elliptic*, 1·5–2·1(–2·6) *times as long as broad, acute*, cuneate at base, *lobed* with oblong or oblong-lanceolate acute lobes, *the deepest lobes extending* (⅓–)½–¾ *of the way to the midrib*, and occasionally nearly to the base, sharply serrate, the teeth mainly on the upper part of the lower margin of the lobe and directed towards its apex and mostly with straight outer margin; at maturity dark yellowish-green, glabrous and somewhat glossy above, usually rather evenly grey-tomentose beneath, when young with a few scattered hairs above and rather dense whitish-grey tomentum beneath; *veins* 7–8(–9) *pairs*; petiole 9–18 mm. Infl. small and rather narrow, pubescent at first; receptacle densely tomentose. Sepals deltoid. Petals c. 4 mm. Anthers cream or pink. *Fr.* 8–10 mm., *ovoid*, scarlet with few inconspicuous lenticels. Fl. 5–6. Fr. 9. M.

Native. Steep granite stream banks. 1. Glen Easan Biorach and Glen Catacol, Arran. Endemic; Norwegian plants referred to this sp. differ.

4. S. leyana Wilmott

Shrub c. 2 m. *Lvs* (Fig. 45 E) 7–9 cm., *oblong, ovate-oblong or elliptic-oblong*, 1·3–1·7(–1·9) *times as long as broad, obtuse or subacute*, cuneate at base, lobed with oblong or oblong-lanceolate acute or obtuse lobes, *the deepest lobes extending at least to nearly half-way to the midrib*, except on occasional lvs, and sometimes nearly to the base, serrate, the teeth mainly on the upper part of the lower margin of the lobe, directed rather irregularly and mostly with curved outer margin, at maturity dull dark yellowish-green and glabrous above, rather unevenly grey-tomentose beneath; *veins* (8–)9–10 *pairs*; petiole 10–20 mm. Petals c. 5 mm. Anthers pinkish. *Fr.* c. 10 mm., *subglobose*, scarlet with few small lenticels. Fl. 5–6. Fr. 9. N.

Native. On carboniferous limestone crags. 1. Near Dan-y-Graig, Brecon. Endemic.

5. S. minima (A. Ley) Hedl.

Pyrus minima A. Ley

Shrub to c. 3 m. of more slender habit than allied spp. *Lvs* (Fig. 45 F) 6–8 cm., *elliptic or oblong-elliptic* (1·4–)1·8–2·2 *times as long as broad*, acute or sub-acute, cuneate or somewhat rounded at base, shallowly lobed with obtuse or acute lobes, *the deepest lobes extending* ⅕–⅓(–½) *of the way to the midrib*, serrate, the teeth mostly curved on the outer edge and directed towards the apex of the lobe; at maturity dull green and subglabrous above, ± evenly and rather thinly grey- (not whitish- or yellowish-) tomentose beneath, when young sparsely pubescent above; veins (7–)8–9(–10) pairs; petiole 12–17 mm. Infl. small, rather narrow and round-topped, thinly tomentose, densely so on the receptacle. Sepals triangular-lanceolate. *Petals c. 4 mm.* Anthers cream. *Fr. small, 6–8 mm., subglobose, scarlet*, with few small lenticels. Fl. 5–6. Fr. 9. 2*n*=51. M. or N.

Native. Carboniferous limestone crags. 1. Near Crickhowell, Brecon. Endemic.

*6. S. intermedia (Ehrh.) Pers.

Pyrus intermedia Ehrh.; *Sorbus scandica* (L.) Fr.; *S. suecica* (L.) Krok.

Tree to 10 m., with spreading branches and rather broad crown. Twigs stout. *Lvs* (Fig. 45 G) 7–12 cm., *elliptic or oblong-elliptic*, (1·5–)1·6–1·8(–1·9) *times as long as broad*, obtuse or acute, *rounded or broad cuneate at base* (*some lvs always rounded*), lobed with broad acute ascending lobes, *the deepest lobes extending* ¼–⅓(–½) *of the way to the midrib*, serrate, the teeth ascending and nearly all with curved outer margin; dark yellowish-green and glabrous above (very sparingly pubescent when young), rather evenly *yellowish-grey-tomentose beneath*, very densely when young; *veins* 7–9 *pairs*; petiole 10–15 mm. Lvs of long (especially sucker) shoots and sometimes also those of short shoots when shaded sometimes lobed nearly to the base or even with a free lflet. Infl. rather large, densely tomentose in fl. Sepals triangular-lanceolate. *Petals c. 6 mm. Anthers cream. Fr.* 12–15 *mm., oblong, much longer than broad, scarlet with few scattered small lenticels.* Fl. 5. Fr. 9. 2*n*=68. MM. or M.

Introduced. Rather commonly planted, sometimes bird-sown and freely naturalized in a few places. Native of S. Sweden (to c. 61° N.), Bornholm, Baltic states and N.E. Germany.

7. S. anglica Hedl.

S. Mougeotii Soy.-Will. & Godr. var. *anglica* (Hedl.) C. E. Salmon

Shrub c. 1–2 m. Twigs rather stout. *Lvs* (Fig. 45H) 7–11 *cm., obovate or rhombic-obovate,* 1·3–1·7(–1·8) *times as long as broad,* obtuse or acute, *nearly all cuneate at base,* lobed with acute ascending lobes, the *deepest lobes extending c.* $\frac{1}{6}$–$\frac{1}{4}$ *of the way to the midrib,* serrate, the teeth projecting and acuminate or with straight outer margin; dark green and glabrous above (very sparingly pubescent when young), rather evenly or somewhat unevenly *whitish-grey-tomentose beneath; veins* (7–)8–10(–11) *pairs*; petiole 12–20 mm. Infl. small to rather large, thinly tomentose in fl., receptacle densely tomentose. Sepals triangular-lanceolate. *Petals c. 6 mm. Anthers pink or tinged pink,* at least in bud. *Fr.* 7–12 *mm., subglobose, crimson or crimson-scarlet, with few to rather many small, or small and moderate, lenticels mainly near the base* (the fr. varies considerably in size, colour and number of lenticels in different localities but is constant in each). Fl. 5 (rather earlier than most spp.). Fr. 9. 2*n*=68*. M. or N.

Native. Crags and rocky woods, nearly always carboniferous limestone, very local. 10, H1. Devon, N. Somerset (Cheddar), Avon Gorge, Wye Valley, Brecon, Montgomery, Shropshire, Denbigh; Kerry (Killarney). Endemic; closely allied spp. occur in Norway and the Alps, Pyrenees, etc.

(8–16). S. aria agg.   White Beam.

Lvs variously toothed, rarely lobed and then with rounded lobes extending no tmore than $\frac{1}{5}$ of the way to the midrib, white-tomentose beneath; veins 7–14 pairs. Styles 2. Fr. red.

8. S. aria (L.) Crantz

Pyrus Aria (L.) Ehrh.

Tree to 15(–25) m. with wide dense crown, or a large shrub. Bark dark grey, shallowly fissured. Twigs tomentose when young with rather numerous small lenticels, becoming chestnut-brown, later dark grey. Buds up to 2 cm., ovoid, greenish, glabrous or tomentose, the scales ciliate. *Lvs* (Fig. 45J) very variable in size and shape, 5–12 cm., *ovate, elliptic or oval,* very rarely obovate, commonly 1·1–1·6 times as long as broad but up to 2·4 times and on some trees constantly more than twice as long as broad, obtuse or acute, rounded or cuneate at base, *the sides mostly gradually curved and the upper and lower portions of the lf rarely triangular,* doubly crenate-serrate or shallowly lobed, then the lobes broader than long and not reaching more than $\frac{1}{5}$ of the way to the midrib, *the apical tooth scarcely projecting beyond its neighbours* and those terminating the main veins broader based than in allied spp. and not projecting markedly beyond the longest of the others, the teeth all acute, broad-based, mostly somewhat curved on the outer margin which is longer than the upper, thus directed upwards; *lf entire for the basal ⅓ or less; dull yellow-green above*, subglabrous or tomentose when young, usually glabrescent but sometimes remaining thinly tomentose, *densely and evenly pure white-tomentose beneath; veins* (9–)10–14(–15) *pairs*, usually impressed above at maturity; petiole 7–20 mm. Infl. large and broad, tomentose in fl., remaining so or glabrescent in fr. Sepals deltoid. Petals c. 6 mm., oval. Anthers cream or, less frequently, pink. *Fr.* 8–15 mm., *usually longer than broad* but sometimes subglobose, *scarlet* but varying in shade, with usually numerous but sometimes few *scattered small lenticels* sometimes with a few moderate ones in addition but *not markedly more numerous towards the base*, often somewhat woolly at base and apex. Fl. 5–6. Fr. 9. $2n=34*$. MM. or M. A very variable sp., contrasting markedly in this respect with the other British spp. and, unlike them, varying considerably in the same locality. It is thus sometimes difficult to separate from them except by considering the characters as a whole. Any plant of the agg. not agreeing with the description of one of the other spp. is probably to be referred here.

Native. Woods and scrub usually on chalk or limestone, and on these soils common within its range, more local on sandstone hills. 26, H3. Kent and Herts (?native in Norfolk and Suffolk) to Dorset and the Wye Valley (and ?Worcester); Galway; elsewhere rather frequently planted and sometimes becoming naturalized. C. and S. Europe from France and C. Germany to the mountains of S. Spain, Corsica, Italy, Macedonia and Transylvania (but southern limit uncertain owing to confusion with allied spp.); ?N. Africa (mountains) (perhaps not this sp.).

9. S. leptophylla E. F. Warb.

Near *S. aria*. Shrub. *Lvs* (Fig. 45к) (8–)9–12(–14) *cm., usually obovate*, (1·3–)1·5–1·7(–1·9 or –2·5 on sterile shoots) *times as long as broad*, acute, cuneate at base; *usually straight-sided for about ⅓ of lf at base and apex so that the apical and basal portions of the lf are ± triangular*, doubly crenate-serrate, the teeth increasing in size upwards, the toothing sharper and coarser than is usual in *S. aria*, especially near the apex, and the secondary teeth fewer, *apical tooth projecting markedly* (c. 3 mm.) *beyond its neighbours*, those terminating the main veins also prominent, larger and longer than those of *S. aria*; *lf entire for basal ⅓ or less; above yellow- or dark-green*, loosely and sparsely tomentose when young, glabrescent; *beneath more thinly tomentose than in S. aria and greenish-white*; thin in texture; *veins* (9–)11(–13) *pairs*, not markedly impressed above. Sepals triangular-lanceolate. Petals c. 6 mm. Anthers cream or slightly pink-tinged. *Fr.* c. 20 mm., *scarlet, longer than broad, with rather few scattered moderate and small lenticels, the former mostly in the lower half*, patchily woolly at base and apex. Fl. 5. Fr. 9. $2n=68*$. M.

Native. Shady limestone crags. 1. In 2 localities in Brecon; probably also in Montgomery. ?Endemic.

10. S. wilmottiana E. F. Warb.

Shrub or small tree. Lvs 7–12 cm., *elliptic or obovate-elliptic*, 1·6–2·0(–2·7) *times as long as broad*, acute, cuneate at base, doubly serrate with rather sharp teeth, the teeth mostly directed somewhat upwards, the *apical tooth* and those terminating the main veins *scarcely prominent; lf entire for basal ⅓ or less;*

lvs bright green and somewhat glossy above, evenly greenish-white-tomentose beneath, veins 8–9(–10) pairs, not markedly impressed; petiole 1–2 cm. *Fr.* 10–13 mm., *longer than broad, crimson, with rather few large lenticels mostly towards the base*. Fr. 9. M.

Native. Rocky limestone, woodland and scrub. 2. Avon Gorge. ?Endemic.

11. S. eminens E. F. Warb.

Shrub or small tree. *Lvs* (Fig. 46A) of fertile shoots (5·5–)7–9(–12) cm., *ovate-orbicular, obovate-orbicular or subrhombic*, 1·1–1·4(–1·7) *times as long as broad*, obtuse or subobtuse, mostly broadly cuneate at base, *often with one or both margins arched inwards, the margins becoming rounded where the teeth begin at c. ⅓ of the distance or less*, upper part of lf rounded in outline or becoming broadly triangular near the apex, doubly serrate with rather shallow primary teeth, the *teeth terminating the main veins markedly longer than the others, acute or subacuminate, ±symmetric and directed outwards; lf bright green above*, with some tomentum at fl., soon glabrescent and glossy, evenly *greenish-white-tomentose beneath*; lvs of sterile short shoots often smaller and narrower, up to 1·8 times as long as broad and more obovate in outline; veins (9–)10–11(–12) pairs, not markedly impressed above at maturity; petiole 10–20 mm. Infl. c. 15–20-fld, loosely and patchily woolly in fl., receptacle densely woolly, glabrescent. Petals c. 5 mm. Anthers pink. *Fr.* c. 20 mm., *slightly longer than broad, crimson, with a moderate number of large and small lenticels mostly towards the base*, slightly woolly at the apex. Fl. 5–6. Fr. 9. $2n=68^*$. M. or N.

Native. Woods on carboniferous limestone. 4. Wye Valley and Avon Gorge. ?Endemic.

12. S. hibernica E. F. Warb.

Small tree of much more slender and open habit than *S. aria*. *Lvs* (Fig. 46B) (7–)8–10(–11) cm., *oval to obovate*, (1·1–)1·2–1·5(–1·8) *times as long as broad, rounded at apex*, broadly cuneate or more rarely rounded at the base, *the basal cuneate portion short, the margins becoming curved where the teeth begin at c. ⅓ of the distance to the apex*, doubly serrate, *the teeth triangular, acute, ±symmetrical*, directed outwards, becoming longer and narrower towards the apex, *those terminating the main veins larger and longer than the others* at least towards the apex of the lf; *lf thinly tomentose above* when young, glabrescent, *at maturity dull green, beneath* evenly tomentose and *whitish-green;* veins (8–)9–11(–12) pairs; petiole 10–20 mm. Petals c. 4–5 mm. Anthers tinged pinkish. *Fr.* c. 15 mm., *broader than long, pinkish-scarlet, with rather few moderate lenticels especially near the base and a few small ones*. Fl. 5. Fr. 9. M.

Native. Rocky open woods and scrub on carboniferous limestone. H15. Across C. Ireland from Wicklow and Dublin to Clare, Mayo and Sligo, rather local. Endemic.

13. S. porrigentiformis E. F. Warb.

S. porrigens Hedl., p.p.

Shrub, more rarely a small tree to c. 5 m. *Lvs* (Fig. 46C) (5–)6–9·5 cm., *obovate*, 1·3–1·7 *times as long as broad, rounded in the upper part to an obtuse or shortly acuminate apex, tapering uniformly to a cuneate base from near the middle of the lf* (occasionally more rounded on fertile shoots), doubly and rather finely serrate, *the teeth terminating the main veins markedly longer than the others,* acute or acuminate, ± *symmetric and directed outwards, entire or nearly so for at least* ($\frac{1}{4}$–)$\frac{1}{3}$ *of the distance from the base, lf bright green above,* slightly tomentose at fl., soon glabrescent and glossy, evenly and *rather thinly greenish-white-tomentose beneath; veins* (7–)8–10(–11) *pairs* not markedly impressed above at maturity; petiole 8–20 mm. Infl. ± woolly at fl., glabrescent. Sepals triangular-lanceolate. Petals c. 5–6 mm. Anthers pink or tinged pink. *Fr.* 8–12 *mm., subglobose or broader than long, crimson, with rather few large lenticels towards the base of the fr.* Fl. 5–6. Fr. 9. $2n = 51^*, 68^*$. N. or M.

Native. Crags and rocky woods on limestone, usually carboniferous. 9. S. Devon (Babbacombe), N. Somerset (Mendips), Wye Valley, Wales. ?Endemic.

14. S. lancastriensis E. F. Warb.

Shrub. *Lvs* (Fig. 46D) (6·5–)8–11(–12·5) *cm., obovate,* (1·4–)1·5–1·8(–2·0) *times as long as broad, rounded in the upper part to a usually obtuse apex, tapering to a cuneate base from about* $\frac{1}{3}$ *of the distance above the base* on many of the lvs but some lvs usually more rounded (though always cuneate at the attachment), unequally serrate, *the teeth terminating the main veins scarcely prominent* and the toothing coarser than in *S. porrigentiformis,* acute or subacuminate, ± *symmetric and directed outwards; lf entire or nearly so in the basal* $\frac{1}{4}$, *dark green* and glabrous *above* at maturity, usually evenly and *rather thickly greyish-white-tomentose beneath,* the tomentum occasionally becoming patchy at fr.; *veins* (7–)8–10 *pairs;* petiole 10–20 mm. Infl. woolly at fl. particularly on the receptacle, usually with some persistent wool particularly at the base and apex of the fr. Sepals triangular-ovate. *Fr.* 12–15 *mm., subglobose or broader than long, crimson, with a moderate number of large lenticels towards the base of the fr. and scattered smaller ones.* Fl. 5–6. Fr. 9. N. or M.

Native. Carboniferous limestone rocks. 2. Near Morecambe Bay (Lancashire and Westmorland). Endemic.

15. S. rupicola (Syme) Hedl.

Pyrus rupicola (Syme) Bab.; *S. salicifolia* (Hartm.) Hedl.

Shrub c. 2 m., rarely a small tree, of rather stiff habit. *Lvs* (Fig. 46E) (6–)8–14·5 *cm., obovate or oblanceolate,* (1·4–)1·6–2·1(–2·4) *times as long as broad* (shade lvs often much larger and broader than sun lvs), *rounded in the upper part to a usually obtuse apex, mostly tapering uniformly to a cuneate base from the middle of the lf or above, coarsely* and unequally *serrate, the teeth terminating the main veins not prominent, teeth* acute or subacuminate, *mostly somewhat curved on the outer margin and directed towards the apex of the lf; lf entire or nearly so in the basal* $\frac{1}{3}$–$\frac{1}{2}$, *dark green above,* slightly tomentose at first, soon glabrescent, *rather thickly white-tomentose beneath,* the tomentum often becoming patchy at fr.; *veins* (6–)7–9(–10) *pairs,* not markedly impressed above at maturity; petiole 8–20 mm., usually reddish at fr. Infl. woolly in fl., especially on the receptacle, glabrescent. Sepals triangular-ovate or triangular-lanceolate. Petals c. 7 mm. Anthers cream or slightly pinkish-tinged. *Fr.*

12–15 mm., *broader than long, carmine*, when ripening green on one side reddish on the other (like many apples), *with numerous scattered moderate and small lenticels*. Fl. 5–6 (rather later than most spp.). Fr. 9. $2n=68*$. N. or M.

Native. On crags and among rocks, usually limestone and nearly always basic, very local, ascending to 1500 ft. 22, H6. S. Devon, Wales, Pennines from Stafford and Derby to Durham and Cumberland, S.E. Scotland, Scottish Highlands; N. and W. Ireland. Scandinavia (to over 66° N. in W. Norway).

16. S. vexans E. F. Warb.

Small tree. *Lvs* (Fig. 46F) (7–)8–10(–11) cm., *obovate*, (1·4–)1·5–1·9(–2·0) times as long as broad, rounded in the upper part to an obtuse apex, *tapering uniformly to a cuneate base from about the middle of the lf*, coarsely and unequally, sometimes doubly serrate, the teeth terminating the main veins not or scarcely prominent, *teeth* acute, *mostly somewhat curved on the outer margin and directed towards the apex of the lf; lf entire or nearly so in the basal* ¼–⅓, *yellow-green above* and glabrous at maturity, *white-tomentose beneath; veins* 8–9(–10) *pairs*, not markedly impressed above at maturity; petiole 8–15 mm. Infl. somewhat woolly in fl., receptacle densely so, with some tomentum remaining on the pedicels in fr. Sepals deltoid. Petals c. 6 mm. Anthers cream. *Fr.* 12–15 mm., *longer than broad, scarlet, with few moderate lenticels towards the base of the fr. and a few small ones above.* Fl. 5. Fr. 9. M.

Native. Rocky woods near the coast. 2. Between Culbone (Somerset) and Lynmouth (Devon). Endemic.

(17–19). S. latifolia agg.

Lvs lobed with ± triangular, acute or acuminate lobes, or doubly serrate with the teeth terminating the main veins very prominent, straight, acute or acuminate, grey-tomentose beneath; veins 7–9(–10) pairs. Styles 2, sometimes joined below. Fr. orange or orange-brown, often becoming brown when fully ripe. (Description including *S. aria × torminalis*.)

17. S. bristoliensis Wilmott

Tree. *Lvs* (Fig. 46J) (6–)7–9(–10·5) cm., *obovate, oblong-obovate or rhombic-obovate*, 1·4–1·7(–2·0) times as long as broad, subacute, broadly cuneate or slightly rounded at base, lobed mainly above the middle with short ascending broadly triangular-acute lobes, the deepest lobes extending c. ⅛–⅙ of the way to the midrib, finely serrate, the teeth mainly curved on the outer margin and directed upwards; at maturity bright yellowish-green and somewhat glossy above, subglabrous except when very young, densely and evenly grey- (not whitish-)tomentose beneath; veins (7–)8–9(–10) pairs; petiole 12–20 mm. Infl. rather small, rather sparsely woolly in fl., except the densely woolly receptacle, glabrescent. Sepals triangular-lanceolate. Petals c. 6 mm. *Anthers pink. Fr.* 9–11 mm., intermingled with considerably smaller fr., *longer than broad, bright orange, with fairly many moderate and small lenticels mainly towards the base of the fr.* Fl. 5–6. Fr. 9. $2n=51*$ (apomictic). M.

Native. Rocky woods and scrub on carboniferous limestone. 2. In the Avon Gorge. Endemic.

18. S. subcuneata Wilmott

Small rather graceful tree. *Lvs* (Fig. 46 H) 7–10·5 cm., *rhombic-elliptic*, (1·5–)1·6–1·9(–2·5) *times as long as broad*, acute, *cuneate or somewhat rounded at the base but always tapered*, lobed in the upper ⅔ with ascending broadly triangular acute or subacuminate lobes, the deepest lobes extending (⅐–)⅙–¼(–⅓) of the way to the midrib, sharply serrate, the teeth with mainly straight outer margin but directed upwards, those terminating the main veins projecting markedly; at maturity bright green above (deep green in shade), subglabrous except when very young, densely and evenly *whitish-grey*-tomentose *beneath*, the tomentum rather creamy when young; veins (6–)8–9(–10) pairs; petiole 12–25 mm. Infl. large but narrow, rather sparsely woolly in fl., except the densely woolly receptacle, glabrescent. Sepals triangular-ovate. Petals c. 6 mm. *Anthers cream. Fr.* 10–13 mm., *subglobose*, few on each infl. or if more numerous then smaller, *brownish-orange becoming brown, with numerous lenticels which are large towards the base of the fr.* becoming smaller upwards. Fl. 5–6. Fr. 9. M.

Native. Open sessile oakwoods. 2. Minehead (S. Somerset) to Watersmeet (N. Devon). Endemic.

19. S. devoniensis E. F. Warb.   French Hales.

Tree with dense crown. *Lvs* (Fig. 46 G) 7–11(–12) cm., *ovate or oblong-ovate*, occasionally a few of them obovate, 1·3–1·6(–1·8) *times as long as broad*, acute or subacuminate, *rounded at base*, shallowly lobed (sometimes ¼ of the way to the midrib but usually much less) with very broadly triangular acute or acuminate lobes, the lowest often spreading, or lf scarcely lobed but doubly serrate, with very prominent straight acuminate teeth terminating the main veins in the upper part, the other teeth much smaller, straight-sided; lf at maturity deep green above, subglabrous almost from the first, rather unevenly *greenish-grey*-tomentose *beneath*; veins 7–9 pairs; petiole 10–30 mm. Infl. large, rather sparsely woolly in fl., except the densely woolly receptacle, glabrescent. Sepals triangular-lanceolate. Petals c. 7 mm. *Anthers cream.* Styles 2, united below. *Fr.* 10–15 mm., *subglobose, brownish-orange becoming brown, with numerous lenticels which are very large towards the base of the fr.,* becoming smaller upwards. Fl. 5–6. Fr. 9. MM. or M.

Native. Woods, etc. 3, H 3. Widespread in Devon where it is the commonest member of the genus; E. Cornwall; Kilkenny, Wexford, Carlow. Endemic.

An allied form differing in the lvs scarcely ever lobed, with broader teeth terminating the main veins, the surface raised between the veins at maturity and the veins (8–)9–10(–11) pairs is rather frequently planted and sometimes becomes naturalized. Its name and origin are uncertain at present. Other forms of this agg. are occasionally found bird-sown and ±naturalized.

S. aria × torminalis = S. × vagensis Wilmott

A variable hybrid. Tree. *Lvs* (Fig. 47 A) 6–12 cm., ovate to elliptic or rhombic-elliptic, (1·1–)1·2–1·7(–2·1) times as long as broad, acute or subacute, rounded or cuneate at the base, *on many plants with the margins arched slightly inwards towards the base*, variously lobed with ± triangular acute or acuminate lobes, from ⅐ to over ¼ of the way to the midrib, finely serrate with the teeth usually small and somewhat appressed; usually yellow-green above, closely yellowish- or greenish-grey-tomentose beneath, in some plants partially denuded at maturity; veins (6–)7–10(–12) pairs; petiole (15–)20–40 mm. *Fr.* 7–12 mm., *longer than broad or subglobose, brownish-orange to brown, usually with few small lenticels* but sometimes with a moderate number of large ones. $2n=34*$.

Not infrequent in woods on limestone in the Wye Valley with the parents; not known elsewhere in Britain (the parents rarely occur together).

20. S. torminalis (L.) Crantz   Wild Service Tree.

Pyrus torminalis (L.) Ehrh.

Tree to 25 m. with wide crown and spreading branches. Bark dark grey, shallowly fissured. Twigs woolly-pubescent when young, soon glabrous and dark brown. Buds greenish, glabrous. *Lvs* (Fig. 47 B) 7–10 cm., ± ovate in outline up to 1·3 times as long as broad but often broader than long, acuminate, rounded or cordate (rarely broadly cuneate) at base, *deeply lobed with acuminate lobes, the lowest pair of lobes much deeper than the others* (usually half-way to the midrib), and *spreading* often at right angles to the midrib, the others ascending, finely and often doubly serrate, the teeth small; *lf green on both sides*, somewhat pubescent beneath when young mainly near the base of the midrib, subglabrous at maturity; veins 4–6 pairs; petiole 15–40 mm. Infl. moderate, sparsely woolly in fl., glabrescent, the receptacle tomentose. Sepals deltoid. Petals c. 6 mm. Anthers cream. Styles 2, united to about half-way. *Fr.* 12–16 mm., considerably longer than broad, *brown*, with numerous large but not very conspicuous lenticels. Fl. 5–6. Fr. 9. $2n=34$. MM.

Native. Woods, usually on clay, sometimes on limestone, local, and usually with only a few trees in any locality. 53. From Westmorland and Lincoln southwards. Europe from France, Germany (not north-west), Denmark and S.W. Russia (Middle Dnieper region) to the mountains of C. Spain and Portugal, Sardinia, Sicily, N. Greece and the Crimea; Caucasus, Asia Minor, N. Syria; Algeria.

23. PYRUS L.

Deciduous trees or shrubs, sometimes thorny. Lvs simple. *Fls in short, umbel-like, simple corymbs.* Sepals ± triangular. Stamens 20–30. *Carpels 2–5, completely united with each other and with the receptacle, the walls cartilaginous in fr.*; ovules 2, without false septum; *styles free.* Fr. brownish or greenish, not deeply indented at base, *flesh with numerous grit-cells.* Nectar secreted by receptacle.

About 20 spp., north temperate Old World.

Petals 10–15 mm.; fr. 2·5 cm. or more, the calyx persistent; infl. rhachis short
  (rarely more than 1 cm.).   **1. communis**
Petals 8–10 mm.; fr. 12–18 mm., the calyx finally caducous; infl. rhachis 1 cm.
  or more; Devon and Cornwall.   **2. cordata**

1. P. communis L.   Pear.

Tree of broadly pyramidal outline, or a large shrub 5–15 m., sometimes thorny. Bark rather deeply fissured, scaly. Twigs and buds glabrous or slightly pubescent, yellowish-brown; short shoots numerous. Lf-blades 2·5–6 cm., orbicular-ovate, ovate or oval, rounded or subcordate at base, cuspidate, acuminate or acute, crenate-serrulate or subentire, woolly-tomentose when young, especially beneath, finally glabrous or sparsely tomentose beneath; petiole nearly as long as blade. *Infl.* a short *umbel-like* corymb but with the *rhachis* usually obvious but *rarely more than* 1 cm.; fls c. 5–9; pedicels 1·5–3 cm., woolly-tomentose to subglabrous. Sepals c. 5 mm. *Petals 10–15 mm.*, white, obovate or orbicular, shortly clawed. Anthers purple. *Fr.* 2–4 cm. (much larger in cultivated forms), pyriform or globose, rounded or tapered at base, brownish, with numerous lenticels, *calyx persistent.* Fl. 4–5, earlier than *Malus sylvestris.* Pollinated by various Diptera and Hymenoptera, protogynous, self-pollination finally possible, some varieties self-sterile, others self-fertile. $2n = 34$, 51, rarely 68 (all from garden vars). MM. or M.

?Introduced. Hedges, etc. Widespread in England and Wales but usually as isolated trees; rarely reported in Scotland (north to Perth) and Ireland, even as an escape; Channel Is. 57, S. The cultivated pears belong to this sp. Europe from France and Germany southwards (not S. Greece, Crete, Corsica or Sardinia); N. Asia Minor, N. Persia, Russian C. Asia.

2. P. cordata Desv.

Differs from *P. communis* as follows: Thorny shrub 3–4 m. Buds glabrous. Lf-blades 1–4 cm., ovate or rounded or occasionally subcordate at base, conspicuously crenate-serrulate, nearly glabrous from the first; petiole often longer than blade. *Infl. more slender, not or scarcely umbel-like* and with the *rhachis more elongate* (1–3 cm.). *Petals 8–10 mm.*, obovate. *Fr. 10–18 mm.*, globose or obovoid, not tapered into the pedicel, *calyx finally caducous.* Fl. 4–5. M.

?Native. Hedges near Plymouth (in Devon and Cornwall) and there very rare (plants from the Wye Valley reported as this sp. are referable to *P. communis*). 2. W. France; a plant said to be the same occurs in Macedonia, Thessaly and Persia.

24. MALUS Mill.

Deciduous trees or shrubs, occasionally thorny. Lvs simple, toothed or lobed. *Fls in short umbel-like corymbs.* Sepals ± triangular. Stamens 15–50. Carpels 3–5, completely united with each other and with the receptacle, *the walls cartilaginous in fr.*; ovules 2, without false septum; *styles united below.* Fr. green to red, usually deeply indented at base, *flesh without grit-cells* (or in a few foreign spp. with some). Nectar secreted by receptacle.

About 25 spp., north temperate regions. Several spp. and a number of hybrids are grown for ornament.

1. M. sylvestris Mill.   Crab Apple.

Pyrus Malus L.; *M. pumila* Mill.

Small tree with dense round crown, or shrub 2–10 m., ± pubescent or nearly glabrous. Bark grey-brown, irregularly fissured and scaly. Twigs reddish-brown with numerous short shoots. Lf-blades 3–4 cm., ovate or oval, broadly cuneate or rounded at base, crenate-serrate; petiole c. half as long. Infl. almost an umbel, fls c. 4–7; pedicels 1–3 cm. Sepals c. 3–7 mm., tomentose within. Petals 1·3–2·8 cm., obovate, white, usually ±suffused with pink. Anthers yellow. Fr. 2 cm. or more, subglobose with a depression at each end, yellowish-green often speckled or flushed with red; calyx persistent. Fl. 5. Pollinated by various Diptera and Hymenoptera, protogynous; most vars partially self-sterile, self-pollination sometimes possible. $2n = 34, 51$ (garden vars). M.

Native. Woods (especially of oak), hedges, and scrub; ascending to 1250 ft. Throughout England, Wales and Ireland, rather common; north to Ross but rare in C. and N. Scotland; Channel Is. 99, H40, S.

Native in many places but often descended from cultivated apples. Europe from Scandinavia (63° 50′ N.) and N. Russia to Spain and Portugal, Sicily and Greece; S.W. Asia.

Both the following sspp. are widespread but the details of their ranges are not worked out.

Ssp. sylvestris

M. acerba Mérat; *Pyrus Malus* var. *sylvestris* L.; *P. Malus* ssp. *acerba* (Mérat) Syme

Usually thorny. Twigs and buds glabrous or loosely hairy when young. *Lvs* acuminate or cuspidate, sparingly hairy on the veins on both sides when young, soon *glabrous. Pedicels, receptacle and outside of calyx glabrous* or nearly so. Petals 2 cm. or less. Fr. 2–2·5 cm. diam., sour.

Native, but probably often introduced. Nearly the whole range of the sp. (apparently not the southern Balkans nor S.W. Asia).

*Ssp. mitis (Wallr.) Mansf.

Pyrus Malus var. *mitis* Wallr.; *M. domestica* Borkh.

Rarely thorny. Twigs tomentose at first, later glabrous, buds hairy. *Lvs* obtuse to acuminate, ± *persistently pubescent beneath. Pedicels, receptacle and outside of calyx tomentose.* Fr. usually large, often sweet.

Introduced. Apparently always descended from cultivated trees. Most of the garden apples are to be referred here. Native of S.E. Europe and S.W. Asia.

The two following frequently cultivated genera differ from all the preceding by the numerous ovules in each cell.

CHAENOMELES Lindl. Styles united below; sepals deciduous; lvs toothed.

*C. speciosa (Sweet) Nakai; *C. lagenaria* (Makino.) Koidz.; *Cydonia japonica* auct.

Japonica.

Twigs smooth. Lf-blades 3–8 cm., ovate, sharply serrate. Fls few in clusters, 3·5–5 cm. diam., scarlet or, less often, pink or white. Fr. 3–7 cm., globose or ovoid, green becoming yellow. Commonly grown in gardens. Native of China.

CYDONIA Mill. Styles free; sepals persistent; lvs entire.

*C. oblonga Mill.   Quince.

Lf-blades 5–10 cm., ovate or oblong, villous below. Fls solitary, 4–5 cm. diam., white or pink. Fr. large, pyriform, villous, green becoming yellow when ripe. Cultivated for its fr., also used as a stock for pears; it has occurred as an escape. Native of C. Asia. The only sp.

53. PLATANACEAE

Deciduous trees with bark coming off in large flakes, giving a mottled appearance.
Buds within base of petiole. Lvs alternate, palmately lobed, stipulate. Fls monoecious
in unisexual infls consisting of 1–7 short-stalked dense globose heads arranged in
a raceme. Perianth inconspicuous. Female fl.: hypogynous; calyx a cup or of free
sepals; petals present or 0; staminodes 3–4; carpels 5–9, free, in 2–3 whorls, styles
long, slender, the stigma along the inner face; ovules 1–2. Male fl.: calyx cupular
or 0; petals present; stamens 3–5, alternating with petals; rudimentary carpels sometimes occur in some fls. Fr. a group of nutlets; endosperm scanty. Wind-pollinated.

One genus with about 6 spp., S.E. Europe, S.W. Asia, N. America.

*Platanus × hybrida Brot. (*P. occidentalis* L. × *orientalis* L.; *P. × acerifolia* (Ait.)
Willd.)   London Plane.

Tree to 35 m. Lvs 12–25 cm., 3–5-lobed; ± glabrous, truncate or cordate at base;
lobes ± triangular, entire or somewhat toothed, extending c. ⅓ of way to lf-base;
petiole 3–10 cm. Female heads (1–)2(–3) per infl.; calyx cupular, petals present,
strap-shaped. Male fls: sepals present, petals fleshy, 3-lobed. Fr.-heads c. 2·5 cm.
diam. with numerous hairs surrounding the frs.

Originated c. 1700, much planted, especially in towns, and occasionally self-sown
and ± naturalized.

54. CRASSULACEAE

Herbs or undershrubs, usually succulent. Lvs exstipulate, usually simple and
entire. Infl. usually cymose. Fls actinomorphic, hermaphrodite or rarely
dioecious, 3–32-merous, usually 5-merous, hypogynous or slightly perigynous.
Petals free or united. Stamens as many as petals and alternate with them or
twice as many; anthers introrse. Carpels as many as petals, free or united
at the base, each with a nectariferous scale at the base; ovules usually numerous
(rarely 1 or few), anatropous. Fr. a bunch of follicles or a capsule; carpels
dehiscing along upper edge; seeds very small; embryo straight; endosperm
scanty or 0.

Over 30 genera and 1250 spp., cosmopolitan, mainly in dry warm temperate regions. Spp. of several genera are grown under glass.

A very natural family usually easily recognized by the succulent lvs and the
free or nearly free carpels equal in number to the petals.

1   Petals free, or united only at base; lvs not peltate.   2
     Petals united into a cylindrical tube for more than half their length; lvs
       peltate.   4. UMBILICUS

2   Stamens twice as many as petals; fls ± conspicuous.   3
     Stamens as many as petals; small annuals with inconspicuous fls.
                                        3. CRASSULA

3   Fls 4–5-merous; plants with ± elongated lfy stems.   1. SEDUM
     Fls 8–18-merous; lvs (except on fl. stems) crowded into a basal rosette.
                                2. SEMPERVIVUM

1. SEDUM L.

Herbs or undershrubs of varied habit. Lvs usually alternate but sometimes
opposite or whorled, succulent. *Fls cymose, 3–10- (usually 5-)merous*, usually
hermaphrodite but occasionally unisexual. Sepals shortly united at base,

somewhat fleshy. *Petals free* (shortly united at base in some foreign spp.). *Stamens twice as many as petals* (rarely as many in a few foreign spp.), free or adnate to petals below. Scales small, entire or slightly toothed. Carpels free or shortly united at base; ovules many (rarely few or 1). Fr. a group of follicles.

About 500 spp., northern hemisphere, mainly warm temperate, a few in the mountains of E. Africa, Peru and Bolivia, 1 in Madagascar. Many are cultivated and several have occurred as escapes.

1	Lvs broad, flat, usually toothed.	2
	Lvs terete or, if flat above, linear, entire.	4
2	Stems erect, several from a central stock.	3
	Stems numerous, creeping and rooting; no central stock.	**3. spurium**
3	Stock very thick and fleshy, scaly; fls 4-merous, dioecious, greenish. *yellow.*	**1. rosea**
	Stock not fleshy, short, bearing carrot-like tubers, not scaly; fls 5-merous, hermaphrodite, reddish-purple. *spp fabia*	**2. telephium**
4	Petals pink; lvs glandular-pubescent, alternate.	**11. villosum** K.M Pl.33
	Petals yellow or white (sometimes tinged pink); lvs glabrous or, if Pl. Pl 40 glandular-pubescent, opposite.	5
5	Petals white.	6
	Petals yellow.	8
6	Lvs mostly opposite, usually pubescent.	**4. dasyphyllum**
	Lvs alternate, glabrous.	7
7	Lvs 3–5 mm., glaucous; infl. with 2 main branches.	**5. anglicum**
	Lvs 6–12 mm., green; infl. with several branches.	**6. album** K.M Pl.33 POL Pl.39
8	Lvs small, 7 mm. or less, ovoid or cylindric, obtuse.	9
	Lvs 8 mm. or more, linear, terete or flat above, acute or apiculate.	10
9	Lvs ovoid, imbricate.	**7. acre** POL. Pl.40
	Lvs linear-cylindric, spreading.	**8. sexangulare**
10	Lvs flat above, forming a congested head at the apex of the sterile shoots.	**9. forsteranum**
	Lvs terete; sterile shoots equally lfy throughout.	**10. reflexum**

Section 1. *Rhodiola* (L.) S. F. Gray. Perennial herbs. Stock fleshy, erect, crowned with lvs which are often reduced to scales in the axils of which the annual lfy shoots are borne. Lvs alternate, flat. Fls 4- or 5-merous, dioecious or hermaphrodite.

1. S. rosea (L.) Scop.   Rose-root, Midsummer-men.
S. Rhodiola DC.   *POL Pl. 40*

Glabrous glaucous perennial 15–30 cm. *Stock thick, fleshy,* branched, projecting above ground-level, marked with the scars of old stems, with a crown of brownish chaffy scales at apex; roots not tuberous. *Stems erect,* simple, usually several from each branch of the stock. Lvs 1–4 cm., alternate, numerous, dense, obovate or obovate-oblong, acute or sometimes obtuse, rounded at base, sessile, dentate near apex or subentire; decreasing in size downwards, the lowest brown and scale-like. Infl. terminal, compact. *Fls dioecious, 4-merous, greenish-yellow.* Male fl. c. 6 mm. diam.; sepals linear;

petals linear c. $1\frac{1}{2}$ times as long as sepals, both often tinged purple on back; stamens rather longer than petals, filaments yellow, anthers purple; scales emarginate, conspicuous; abortive carpels erect, shorter than petals. Female fl. with petals equalling sepals; stamens 0; scales as in male; carpels 4–6 mm., erect, greenish; style short, spreading, reflexed in fr. Fl. 5–8. Visited by small flies. $2n=22, 36$ (N. America). Hp.

Native. Crevices of mountain rocks, ascending to 3850 ft., and of sea cliffs in W. Scotland and Ireland; common in such habitats; N. and S. Wales and from Yorks, Lancashire and the Isle of Man northwards; all the Irish mountains (thus absent from the centre). 45, H17. Arctic Europe, Asia and N. America and in the mountains of these continents south to the Pyrenees, Italian Alps, Macedonia, Japan and New Mexico.

Section 2. *Telephium* S. F. Gray. Perennial herbs. Stock short, without lvs or scales, bearing thick tuberous roots. Stems annual from buds near the apex of the stock. Lvs flat. Fls 5-merous, hermaphrodite.

2. S. telephium L.   Orpine, Livelong.

Glabrous subglaucous perennial 20–60 cm., often tinged with red. Stock short, stout; *roots tuberous, carrot-like. Stems erect*, simple, clustered. Lvs 2–8 cm., alternate, numerous, oval-oblong to oblanceolate-oblong, sessile or shortly stalked, rounded or cuneate at base, obtuse to subacute, irregularly dentate except at base. Infl. terminal, composed of compact subglobose terminal and axillary cymes. Fls 9–12 mm. diam. Sepals green, lanceolate, acute. *Petals* spreading, *reddish-purple*, lanceolate, acute. Scales yellow, ligulate, emarginate. Carpels erect, purple; styles very short. Fl. 7–9. Visited by Diptera and Hymenoptera; protandrous. Hp.

Native. Woods and hedgebanks, ascending to 1500 ft.; spread over the greater part of the British Is. but rather local and becoming very rare in N. Scotland; certainly native in many places but often an escape. 95, H24. Europe from Scandinavia to the Pyrenees, Italian Alps and Thrace, temperate Asia, N. America. The habitats and distributions (either as native or introductions) of the two following ssp. in this country are not well known.

Ssp. **purpurascens** (Koch) Syme

S. purpurascens Koch; *S. purpureum* auct., ?(L.) Schult.

Lvs sessile; upper oval-oblong, rounded at base; lower obovate-oblong, cuneate at base. Carpels grooved on back. $2n=36$.

Europe and Asia (to Japan).

Ssp. **fabaria** Syme

S. fabaria Koch

Lvs shortly stalked or subsessile, all cuneate at base, obovate-elliptic to oblanceolate-oblong. Carpels not grooved. Europe.

Section 3. *Sedum*. Perennial evergreen herbs. Stock 0. Sterile stems numerous, creeping and rooting; fl.-stems mostly ascending, annual. Fls hermaphrodite, 5-merous.

*3. S. spurium M. Bieb.

S. stoloniferum auct.

Creeping perennial forming large mats with numerous lfy ascending puberulent branches; fl.-branches c. 15 cm., sterile ones shorter. *Lvs* 2–3 cm., opposite, *flat*, *obovate-cuneiform*, obtuse, *crenate-serrate* above the middle, ciliate, otherwise glabrous, tapered into a short petiole. Infl. a flat dense terminal cyme. Fls 9–12 mm. long, sessile or subsessile. *Petals pink*, less frequently white or crimson, *suberect*. Carpels erect. Fl. 7–8. $2n = 28$. Chh.

Introduced. Commonly planted and frequently escaping, probably increasing. Native of the Caucasus and Transcaucasia.

4. S. dasyphyllum L.   'Thick-leaved Stonecrop.'

Small tufted perennial 2–5 cm., glaucous, often tinged pink. Sterile branches short, ascending, with imbricate lvs; fl.-stems taller, with distant lvs. *Lvs opposite* (or some alternate on fl.-stems), 3–5 mm., *ovoid or obovoid*, slightly flattened above, entire, sessile, ± *glandular-pubescent*, rarely glabrous, readily caducous and then often rooting and forming fresh plants. Infl. 2–4-fld. Fls c. 6 mm. diam. *Petals white*, tinged pink on back, oblanceolate, spreading. Carpels erect. Fl. 6–7. Visited by small Diptera and Hymenoptera; protandrous. $2n = 28, 42, 56$. Chh.

?Introduced. Occasionally naturalized on old walls, mainly in S. England; possibly native on limestone rocks in Cork. 25, H3. S. Europe (not Portugal), extending north to C. France and S. Germany; N. Africa.

5. S. anglicum Huds.   'English Stonecrop.'

Glabrous glaucous evergreen perennial often tinged red, 2–5 cm., with numerous slender creeping and rooting stems forming mats, and ascending barren and fertile branches. *Lvs* 3–5 *mm.*, *alternate*, ovoid or shortly ellipsoid, terete, entire, spreading, clasping the stem and with a small spur at the base. *Infl. with* 2(–3) *main* branches each with 3–6 fls, and with a fl. at the fork. Fls c. 12 mm. diam. Pedicels short, stout, c. 1 mm. Sepals ovate, obtuse, free. *Petals white*, tinged pink on back, lanceolate, spreading. Scales red. Carpels ascending in fl., erect in fr., pale at first, becoming red after fl. Fl. 6–8(–9). Visited by various insects. $2n = 48, 144$. Chh.

Native. Rocks, less frequently dry grassland, ascending to 3500 ft.; also dunes and shingle, absent from strongly basic soils; common in the west. From the Hebrides southwards extending east to Perth, N.W. Yorks, Derby and Worcester; also in a few places near the south and east coasts; throughout Ireland, but rare in the centre. 66, H32, S. Scandinavia through W. and C. France to Spain and Portugal; Morocco; the southern forms differ from ours.

6. S. album L.   'White Stonecrop.'

Glabrous bright *green*, evergreen perennial, sometimes tinged red, 7–15 cm., with numerous creeping stems forming large mats, fl.-branches ascending. *Lvs* 6–12 *mm.*, *alternate*, obovoid or oblong, terete or somewhat flattened above, obtuse, entire, sessile but not clasping or spurred. *Infl. with several main terminal and axillary branches* forming a much-branched, *many-fld*, flat-

topped cyme 2–5 cm. across. Pedicels slender, 1–4 mm. Fls 6–9 mm. diam.
Petals white, sometimes tinged pink on back, ovate or lanceolate, spreading.
Scales yellow. Carpels erect, pale, becoming tinged with red after fl. Fl. 6–8.
Visited by various insects, protandrous. $2n=32, 64$. Chh.

?Introduced. Walls and rocks, possibly native in the Malverns, Mendips
and Devon, naturalized in a number of places elsewhere. 58, H27, S. Europe
(except Iceland); N. and W. Asia; N. Africa.

Ssp. **album**

Lvs oblong. Petals 3–4 mm. The common form.

*Ssp. **micranthum** (DC.) Syme

Lvs oblong-ovate to subglobose. Petals 2–3 mm. Rare and certainly introduced. Only in the southern part of the range of the sp. abroad.

*S. lydium Boiss.

Small, bright green. Lvs c. 6 mm., terete, linear. Infl. compact, flat. Petals white.
Commonly cultivated and reported as naturalized. Native of Asia Minor.

7. S. acre L. *yellow*   Wall-pepper.

K.M. Pl. 33
POL. Pl. 40

Glabrous green evergreen perennial 2–10 cm., with numerous creeping stems
forming mats, and ascending or erect sterile and fl.-branches; *taste hot and
acrid. Lvs 3–5 mm.*, alternate, *ascending, imbricate* (except sometimes on the
fl.-stems) *ovoid-trigonous, obtuse*, entire, sessile, spurred at base. Infl. with
2 or 3 main branches, each with 2–4 fls at the fork. Fls c. 12 mm. diam.
Sepals lanceolate, obtuse. *Petals bright yellow*, lanceolate, acute, spreading.
Scales whitish. Carpels ascending in fl., spreading in fr. Fl. 6–7. Pollinated
by Diptera and Hymenoptera; protandrous, perhaps sometimes selfed.
$2n=16$, c. 48. Chh.

Native. Dry grassland, dunes, shingle and walls, especially on basic soils;
ascending to 1500 ft. Throughout the British Is. except Shetland, common.
111, H40, S. All Europe; N. and W. Asia; N. Africa; naturalized in N.
America.

*8. S. sexangulare L.   'Insipid Stonecrop.'

yellow

S. mite Gilib.; *S. boloniense* Lois.

Differs from *S. acre* as follows: *Without biting taste. Lvs 3–6 mm., linear-
cylindric*, terete, *spreading, dense, but not imbricate*, obtuse. Fls c. 9 mm. diam.
Scales yellow. Carpels erect in fl., spreading in fr. Fl. 7–8. Chh.

Introduced. Naturalized on old walls in a few places in England and Wales.
14. Europe from Sweden and Finland to N. Spain, Italy and Macedonia.

9. S. forsteranum Sm. K.M Pl.33   'Rock Stonecrop.'

yellow

S. rupestre L., p.p.

Glabrous, glaucous or green, evergreen perennial 15–30 cm., with numerous
creeping stems forming mats, and short ascending sterile branches and long
fl. ones. *Sterile shoots* clothed with withered lvs below, *rosette-like above* with
the very crowded lvs. *Lvs 8–20 mm.*, alternate, ascending, *linear* or oblan-

ceolate-linear, *flat on upper surface*, *apiculate*, entire, sessile, spurred at base; on the fl. shoots more distant and broader, ascending. Infl. a many-fld umbel-like cyme with about five main branches, drooping in bud, becoming erect and flat-topped or convex in fl., concave in fr. Fls occasionally 6–8-merous, c. 12 mm. diam. Sepals triangular, nearly free. *Petals bright yellow*, oblong-linear, obtuse, spreading. Scales yellow. Carpels erect in fl. and fr. Fl. 6–7. Chh.

Native. Rocks and screes, very local, ascending to 1200 ft.; Devon, Dorset, Somerset, Gloucester, Hereford, Wales; often planted and sometimes naturalized elsewhere. c. 19 (as native). W. Europe from the Netherlands and W. Germany to Spain and Portugal; Morocco.

The two following sspp. are incompletely understood and appear to have much the same general range in this country.

Ssp. elegans (Lej.) E. F. Warb.

S. elegans Lej.; *S. rupestre* ssp. *elegans* (Lej.) Syme
Glaucous. Infl. flat-topped in fl. Robust or slender. Dry rocks.

Ssp. forsteranum ~~Rock Stonecrop~~

Usually green, sometimes glaucous. Infl. convex in fl. Slender. Damp rocks.

*10. S. reflexum L. *K.M. Pl. 33 POL Pl. 40*

Differs from *S. forsteranum* as follows: Usually more robust. *Sterile shoots equally lfy for a considerable distance* so that the tips are not rosette-like, dead lvs not persistent. *Lvs terete* or nearly so, sometimes reflexed on the fl.-stems. Fls rather larger. Petals linear-lanceolate, acute, bright or pale yellow. Fl. 6–8. Visited by various insects; weakly protandrous. $2n = 34, 68, c. 112$. Chh.

Introduced. Commonly cultivated; naturalized on old walls and rocks in many places. 67, H 21, S. Europe from Scandinavia and Finland to C. Spain, Sicily, Albania and Thrace. A variable sp. which might repay detailed cytological investigation.

11. S. villosum L. *POL. Pl. 40 K.M. Pl. 33 white* 'Hairy Stonecrop.'

Small, usually reddish, not glaucous perennial 5–15 cm., perhaps sometimes biennial, *glandular-pubescent*. Stem erect, often branched at the base, the branches both sterile and flowering; sterile ones short. *Lvs 6–12 mm., alternate*, linear-oblong, flat above, obtuse, entire, sessile, not spurred. Infl. lax, rather few-fld, with terminal and axillary branches; pedicels long, slender. Fls c. 6 mm. diam. Sepals lanceolate, obtuse. *Petals pink*, ovate, apiculate, spreading. Scales yellowish. Carpels erect. Fl. 6–7. Chh.

Native. Streamsides and wet stony ground, etc., in the mountains; ascending to 3600 ft. From Yorks and Lancashire to Inverness and Argyll, widespread within its range but rather local. 34. Europe from Iceland, Scandinavia (71° 06′ N.) and Finland to the mountains of Spain and Portugal, Sardinia, N. Italy, Serbia and Roumania; Morocco, Algeria (rare); Greenland.

*S. dendroideum Sessé & Moç. ex DC.: small shrub with lvs 2–5 cm., oblanceolate, thick; infl. a large panicle; petals 4–8 mm., yellow; naturalized in Jersey. Mexico.

2. SEMPERVIVUM L.

Herbs. *Lvs thick and fleshy, alternate, mostly densely crowded into a radical rosette,* which often persists for several years and reproduces by stolons, finally flowering and dying. Fl.-stems erect, bearing a branched infl. of scorpioid cymes often forming a panicle. *Fls 6–18-merous. Petals fused for a short distance at the base. Stamens twice as many as petals,* epipetalous. Scales toothed to fimbriate. Carpels free; ovules many. Fr. a group of follicles.

About 25 spp. (or possibly many more), Europe, Atlas and Asia Minor, mainly in the mountains. Some are grown on rock gardens.

*1. S. tectorum L.   Houseleek, Welcome home husband, however drunk you be. *show pt.*
Rosettes 5–14 cm. diam. Lvs 2·5–6 cm., obovate-lanceolate, mucronate, strongly tinged dull red above, ciliate, otherwise glabrous. Fl.-stems 30–60 cm., glandular-pubescent above. Infl. a many-fld panicle of scorpioid cymes. Fls 2–3 cm. diam., 8–18- (mostly c. 12-)merous. Petals dull red, lanceolate. Inner whorl of stamens mostly sterile, often replaced by carpels. Fl. 6–7. Visited by various insects; protandrous. $2n = 72*$. Chh.
Introduced. Has been planted for centuries on old walls, roofs, etc., but scarcely naturalized. The type plant (with partly aborted stamens), described above, extends over Europe but is probably nowhere wild. Other sspp. or allied spp. occur over the whole range of the genus.

3. CRASSULA L.

Herbs or shrubs of varying habit. *Lvs opposite,* usually entire. Fls in cymes or, more rarely, solitary and axillary, usually small, (3–)5(–9)-merous. Sepals free or nearly so. *Petals free or joined at base only. Stamens as many as petals,* alternate with them. Scales usually small. Carpels free or joined at base; ovules 1–many; styles short, apical; *stigmas apical,* small. Fr. a group of follicles.

About 300 spp., cosmopolitan but the great majority in S. Africa.

Stems very densely clothed with lvs and fls; lvs 1–2 mm., ovate-oblong,
    concave; fls mostly 3-merous, petals smaller than sepals.   **1. tillaea** *Keble n*
Lvs and fls distant; lvs 3–5 mm., linear; fls 4-merous, petals larger than sepals. */ Pl. 33*
                                                                       2. aquatica

1. C. tillaea L.-Garland

Tillaea muscosa L.

Small, often tufted, glabrous, usually reddish annual 1–5 cm. Stems decumbent or ascending, with short axillary branches very densely clothed above by the lvs and fls. *Lvs 1–2 mm., ovate-oblong,* thick, concave, connate in pairs, subacute. Fls solitary, sessile in the axils of nearly all the lvs and about the same length, whitish, 3(–4)-merous. Sepals ovate, mucronate. Petals smaller, lanceolate, acute. Scales 0. Carpels 2-seeded. Fl. 6–7. Self-pollinated. Th.

Native. Bare sandy or gravelly ground, very local; S. Devon, N. Somerset, Dorset, S. Wilts, Hants, Norfolk, Suffolk, Nottingham, Channel Is. 11, S. Mediterranean region and W. Europe to W. and C. France and N.W. Germany; Macaronesia.

2. C. aquatica (L.) Schönl.

Tillaea aquatica L.

Small, slender, glabrous annual 2–5 cm. Stems decumbent, simple or branched from the base, with few axillary branches, *the lvs and fls distant. Lvs 3–5 mm., linear*, somewhat fleshy, connate, subacute. Fls 1–2 mm., solitary, subsessile in the lf-axils, whitish, 4-merous. Sepals ± triangular, obtuse. Petals larger, ovate, acute. Seeds numerous. Fl. 6–7. Probably self-pollinated. $2n = 42$. Th.

?Native. Muddy margins of a pool near Adel Dam, Yorks, where it was first found in 1921; it had apparently disappeared in 1945. N. and C. Europe from Iceland and Finland to C. Germany and Lower Austria; N. Asia; N. America.

4. Umbilicus DC.

Perennial herbs with *tuberous stock. Lvs alternate*, stalked, *peltate or cordate*, fleshy. *Infl. a narrow terminal spike or raceme. Fls 5-merous. Petals fused for most of their length* into a cylindric or nearly campanulate tube. *Stamens* 10 (rarely 5). Scales small. Carpels free; styles short; ovules numerous. Fr. a group of follicles; seeds very small.

About 10 spp. in Mediterranean region extending to Persia, Abyssinia, Macaronesia and W. Europe.

1. U. rupestris (Salisb.) Dandy   Pennywort, Navelwort. K.M Pl. 33

U. pendulinus DC.; *Cotyledon Umbilicus-veneris* L., p.p.

Glabrous perennial herb 10–40 cm. Stock tuberous, subglobose. Stem usually solitary, simple or occasionally branched in the infl. Lvs mostly radical, blades 1·5–7 cm. diam., orbicular, peltate, depressed at the junction with the long petiole, crenate with very broad teeth, or sinuate; stem lvs few, becoming smaller upwards, with shorter stalks, the uppermost often not peltate. Infl. long, usually occupying more than half the length of the stem, many-fld, sometimes branched below; pedicels 2–4 mm.; upper bracts linear, entire, somewhat exceeding pedicels, lower often lf-like; fls drooping, sepals lanceolate, acute, less than half as long as corolla. Corolla 8–10 mm., whitish-green, tubular; lobes ovate, short. Stamens inserted on corolla-tube, included. Fl. 6–8(–9). Protandrous, but probably usually selfed, only thrips known as visitors. Hs.

Native. Crevices of rocks and walls, especially acid; ascending to 1800 ft. Rather common in W. England and Wales, rarer eastwards, extending to Kent, Berks, Leicester and W. Yorks; S.W. Scotland from Kirkcudbright to Argyll and the mid Inner Hebrides; throughout Ireland but local in the centre. 58, H 40, S. Mediterranean region and W. Europe to C. and W. France; Madeira, Azores.

55. SAXIFRAGACEAE

Herbs. Lvs usually alternate, exstipulate or with small sheathing stipules. Fls hermaphrodite, actinomorphic or more rarely somewhat zygomorphic, usually 5-, sometimes 4-merous, hypogynous to epigynous, often perigynous in varying degrees. Petals sometimes 0. Stamens equalling and alternate with

[handwritten notes: "only one plane of symmetry", "radially symmetrical more than one plane of symmetry"]

Stamens in 2 whorls [handwritten]

the petals or twice as many and obdiplostemonous. Carpels 2, very rarely 3–5, free to united, very frequently united at the base, free above and gradually tapering to the stigmas, free from or adnate to the receptacle; ovules numerous, placentation various; styles usually free. Fr. a capsule; seeds numerous with small straight embryo and copious endosperm. Nectar secreted by ovary.

About 35 genera and 500 spp., cosmopolitan but mainly temperate. A family often easily recognized by its perigynous fls and partly fused carpels. Distinguished from the *Rosaceae* by the capsular fr., the almost constantly 2 carpels and the usually fewer stamens. It is often regarded as including Parnassiaceae, Hydrangeaceae, Escalloniaceae and Grossulariaceae, q.v. for the distinctions.

Spp. and hybrids of *Bergenia, Astilbe, Heuchera*, etc., are commonly grown in gardens.

Fls 5-merous; petals present.   1. Saxifraga
Fls 4-merous; petals absent.   2. Chrysosplenium

1. Saxifraga L.

pentá [handwritten] Usually perennial herbs. Lvs often thick, simple. *Bracteoles present*. Fls 5-merous (very rarely 4-merous in foreign spp.). Receptacle flat or cup-shaped and adnate to the ovary. Sepals imbricate. *Petals present, narrow at the base. Stamens* 10 (very rarely 8). *Carpels* (usually) 2, ± *united below; placentation axile*; styles free, at first erect then spreading. Capsule opening along the inner edges of the carpels. Nectar secreted at the base of the ovary or by an epigynous disk.

Over 300 spp. north temperate, arctic and in the Andes. Many spp. and hybrids are grown in rock gardens, etc.

1 Lvs alternate; fls white or yellow.   2
   Lvs opposite; fls purple.   **16. oppositifolia**

2 Fls yellow.   3
   Fls white, sometimes spotted with yellow or purple.   4

3 Ovary superior; sepals reflexed in fr.; fls 1(–3); lower lvs with petiole as
      long or longer than blade; fl.-stems reddish-pilose above.   **3. hirculus**
   Ovary partly inferior; sepals spreading; fls several, rarely 1; lvs sessile;
      no red hairs.   **15. aizoides**

4 Ovary superior; sepals reflexed at or soon after fl.; fl.-stem naked.   5
   Ovary partly inferior; sepals erect to spreading.   8

5 Lvs nearly sessile, with distant teeth; stock not woody, short; filaments
      subulate; capsule inflated.   **2. stellaris**
   Lvs stalked, with contiguous teeth; stock somewhat woody; filaments
      clavate; capsule not inflated.   6

6 Lvs cuneate at base, glabrous; petiole 2 mm. broad or more; petals
      spotted with red.   7
   Lvs cordate at base, hairy on both sides; petiole scarcely 1 mm. broad;
      petals often without red spots.   **6. hirsuta**

7 Teeth of lvs obtuse, terminal tooth shorter than its neighbours; petiole
      densely ciliate throughout, usually shorter than blade; styles ascending
      in fr.   **4. umbrosa**

Teeth of lvs acute, terminal tooth as long or longer than its neighbours; petiole sparsely ciliate at base only, longer than blade; styles divaricate in fr. **5. spathularis**

8   Fl.-stem naked; lvs ±spathulate, coarsely and closely crenate; plant purple-tinged. **1. nivalis**
Fl.-stem lfy; lvs not as above; plant not purple-tinged. *9*

9   Plant without numerous procumbent or rosette-like lfy barren shoots. *10*
Plant with numerous procumbent or rosette-like lfy barren shoots, of mossy habit. *13*

10 Annual; lvs cuneate at base. **7. tridactylites**
Perennial; basal lvs cordate at base. *11*

11 Ovary c. ¾-inferior; basal lvs crenate-lobulate with 7 or more teeth; fls 2–12; lowland. **8. granulata**
Ovary less than ½-inferior; basal lvs palmately 3–5(–7)-lobed; fls 1–3; high alpine. *12*

12 Plant with numerous red bulbils in the axils of the bracts; petals 8 mm. or more; fl.-stems erect. **9. cernua**
Bracts without bulbils; petals 5 mm. or less; fl.-stems ascending. **10. rivularis**

13 Lobes of lvs ± oblong, the largest always more than 1 mm. broad, obtuse or subacute, sometimes mucronate, never aristate; sterile shoots erect or ascending, without axillary bulbils. *14*
Lobes of lvs ± linear, rarely more than 1 mm. broad, acute or acuminate, aristate; sterile shoots procumbent, usually with axillary fusiform lfy bulbils. **14. hypnoides**

14 Plant very compact; lobes of lvs very obtuse; glandular hairs short and dense. **11. cespitosa**
Plant rather loose; lobes of lvs subacute; glandular hairs short and dense. **12. hartii**
Plant compact to rather loose; lobes of lvs subacute; glandular hairs few or 0. **13. rosacea**

Section 1. *Micranthes* (Haw.) D. Don. Perennial herbs with basal rosette and few or *no stem lvs*. Lvs alternate, toothed, without pits or lime glands. Fls usually in compact racemes or panicles. Receptacle flat to cup-shaped. Sepals often reflexed after fl. *Petals ± white*. Ovary free or united to the receptacle for ⅓ or less. *Capsule ± inflated, dehiscing to the middle or below*. Seeds oblong.

1. S. nivalis L.   'Alpine Saxifrage.'

Perennial herb 3–15 mm., with short thick stock and 1 or few basal rosettes of lvs. *Lf-blades* 1–2 cm., orbicular- or obovate-spathulate, the petiole broad, about as long as blade, *coarsely* and closely *crenate-serrate*, obtuse, green above and glabrous at maturity, *± purple below* and *glandular-pubescent near the margins*, rather thick. Fl.-stems naked, glandular-pubescent, often purple. Fls 3–12 in a dense head-like infl., bracts linear; pedicels very short. *Sepals* triangular-ovate, often purplish, *erect or spreading*. Petals c. 3 mm., *greenish-white, not spotted*, reddish or purplish on the back, obovate. Anthers dull orange. *Ovary about ½-inferior*, adnate to the receptacle below; styles divergent; carpels often purple or reddish, in fr. 3–4 mm., ovoid. Fl. 7–8. Visited by flies, slightly protandrous to slightly protogynous. $2n = 60$. Hr.

Native. Wet rocks on mountains from 1200 to 4300 ft., very local and rare; Caernarvon, Lake District, Dumfries, Scottish Highlands from Dunbarton and Stirling to Sutherland, Skye; Sligo. 16, H1. Arctic Europe, Asia and eastern N. America (incl. Greenland); Sudeten Mountains.

2. S. stellaris L.   'Starry Saxifrage.'

Perennial herb with *short stock* and 1 to several basal rosettes of lvs often becoming elongated into lfy shoots, *not purple-tinged*. *Lvs* 0·5–3 cm., obovate-cuneiform, obovate-spathulate, or oblanceolate, *scarcely petioled*, acute or subacute, *remotely serrate or dentate, with scattered long hairs* especially above, rather thick. Fl.-stems naked, with scattered long hairs below and glandular hairs above. Infl. an open panicle up to 12- or more-fld, the branches cymose, but often simple and few-fld; bracts linear or linear-lanceolate; pedicels 4–10 mm., slender. *Sepals* ovate- or oblong-lanceolate, *reflexed*. Petals 4–5 mm., ovate-lanceolate, pure white with two yellow spots near the base. Anthers orange-red, filaments subulate. *Ovary superior*, free; styles somewhat divergent; fruiting carpels ovoid, c. 6 mm. Fl. 6–8. Visited mainly by flies; protandrous to protogynous. $2n = 28$. Hr. or Chh.

Native. Common on mountains by streams, in springs, on wet rock ledges and wet stony ground, ascending to the top of Ben Nevis. Wales; Yorkshire and Westmorland to Caithness and the Outer Hebrides; most of the Irish mountains. 45, H18. N. Europe from Iceland to arctic Russia and in the mountains of Europe, south to the Sierra Nevada, Corsica, N. Apennines and Macedonia; Greenland, Labrador.

Section 2. *Hirculus* (Haw.) Tausch. Perennial herbs. Stems usually lfy. Lvs alternate, entire or nearly so, without pits or lime glands. Fls solitary or in panicles. Receptacle flat or nearly so. Sepals spreading or reflexed. *Petals yellow*, often with 2 projections near the base. *Ovary free* or nearly so. Capsule not inflated, dehiscing only near the apex. Seeds fusiform.

3. S. hirculus L.   'Yellow Marsh Saxifrage.'

Perennial herb 10–20 cm., with short prostrate or ascending sterile shoots and erect lfy fl.-stems. Lower lf-blades 1–3 cm., lanceolate, oblanceolate, or oblong, obtuse, entire, narrowed below into a sheathing *petiole as long or longer than blade*; upper lvs shorter and narrower. *Fl.-stems reddish-pilose above*. *Fls* 1(–3), terminal. Sepals oblong, ciliate with reddish hairs, spreading in fl., reflexed in fr. Petals 10–13 mm., narrowly oblong or obovate, yellow, often with orange spots near the bi-tuberclate base. *Ovary superior*, free; styles divergent. Fr. c. 1 cm., oblong-ovoid. Fl. 8. Visited mainly by flies; protandrous. $2n = 32$. Hp.

Native. Wet grassy ground on moors, very local and rare; ascending to over 2000 ft. Cheshire (extinct); N. Yorks and Westmorland to Midlothian and Lanark; Perth, Banff, Caithness; Ireland from N. Tipperary and Leix to Mayo and Antrim. 10, H6. Arctic and subarctic Europe, Asia and America extending south to the Jura, S. Germany, Switzerland (very rare), Carpathians, Caucasus, Himalaya and Colorado.

Section 3. *Gymnopera* D. Don. Perennial herbs with a rather woody stock and short branches bearing rosettes of lvs. Lvs alternate, toothed, without pits or lime glands; *stem lvs* 0. Fls in panicles. Receptacle flat. *Sepals reflexed* after fl. *Petals white.* Filaments clavate. *Ovary superior*, free. Capsule not inflated, dehiscing only near the apex. Seeds ovate or elliptic.

*4. S. umbrosa L.

Perennial herb 8–40 cm. *Lf-blades* 1–6 cm., spreading at maturity, *oval to obovate-oblong, cuneate at base*, subtruncate at apex, crenate or crenate-serrate with 4–10 *obtuse teeth* on each side, *the terminal tooth shorter* and broader *than the lateral, glabrous*, dark green, thick, with a broad cartilaginous border; *petiole usually shorter than blade, flat, broad* (2–4 mm.), *densely long-ciliate.* Scape reddish, glandular-pubescent especially above. Sepals oblong, obtuse. *Petals* 3–4 mm., elliptic, white with a yellow spot at the base and a few small *crimson spots* above. Fr. 5–6 mm., oblong-ovoid, reddish; styles ascending. Fl. 6–7. Visited by Diptera and Hymenoptera. $2n=28$. Hr.

   Introduced. Heseldon Gill, Yorks, where it has been known since 1792. Native of W. and C. Pyrenees.

*S. spathularis × umbrosa   London Pride.

S. umbrosa L. var. *crenoserrata* Bab.

Differs from *S. umbrosa* in larger, less spreading, subacute lvs with up to 12, more angular teeth on each side. Scape taller, less glandular. Petals large with more numerous crimson spots. Fr. very rarely produced. Commonly grown in gardens, sometimes escaping and becoming naturalized. Of unknown origin; not known wild.

*S. hirsuta × umbrosa = *S.* × *geum* L. may occur occasionally as an escape.

5. S. spathularis Brot.   St Patrick's Cabbage.

S. umbrosa auct. brit. p.p.; *S. umbrosa* var. *serratifolia* D. Don

Differs from *S. umbrosa* as follows: Lvs ascending at maturity, suborbicular or spathulate to oval, obtuse or acute, dentate with 4–7 *acute triangular teeth* on each side, *terminal tooth as long or longer than lateral; petiole much longer than blade, sparsely ciliate towards the base.* Scape glandular-pilose throughout. Petals 4–5 mm., white with 1–3 yellow spots at base and numerous crimson spots above. Fr. oblong, greenish; *styles divaricate.* Fl. 6–8. Hr.

   Native. Among rocks, mainly acid, in sun or shade in mountain districts, very local but often abundant, ascending to over 3400 ft.; Kerry, Cork, Waterford, S. Tipperary, Galway, Mayo, Donegal, Wicklow. H12. N.W. Spain and mountains of N. Portugal.

S. hirsuta × spathularis = *S.* × *polita* (Haw.) Link

S. hirsuta auct. brit., non L.

Fertile and very variable, combining the characters of the parents in various ways. Common with the parents and often commoner than *S. hirsuta*; found also in Galway and Mayo where *S. hirsuta* is unknown.

6. S. hirsuta L.   'Kidney Saxifrage.'

S. Geum L., 1762, non 1753; incl. *S. lactiflora* Pugsl.

Perennial herb 12–30 cm. *Lf-blades* 1–4 cm., ascending, *orbicular or reniform, cordate at base*, obtuse, crenate or crenate-dentate with 6–13 obtuse or api-culate teeth on each side, *with long hairs on both surfaces*, bright green, not very thick, with a narrow cartilaginous border; *petiole 2–4 times as long as blade, not flattened, slender* (scarcely 1 mm. broad), pilose. Scape glandular-pubescent. Sepals oblong, obtuse. Petals c. 4 mm., elliptic, white with a yellow basal blotch, with or without crimson markings. Fr. 4–5 mm., oblong-ovoid, green; styles divergent. Fl. 5–7. $2n=28$. Hr.

Native. Shady rocks, and by mountain streams, in Kerry and W. Cork, locally common, ascending to 3000 ft.; sometimes naturalized in Great Britain. H3. Pyrenees and N. Spain.

Section *Cymbalaria* Griseb. Annual or biennial herbs; stems short, usually very lfy and much branched. Lvs rather thick, usually reniform, coarsely toothed, without pits or lime glands. Fls small, on long pedicels. Sepals often reflexed. Petals yellow. Ovary free or united to the receptacle for $\frac{1}{3}$ or less. Capsule globose or oblong, dehiscing only near the apex. Seeds globose.

*S. cymbalaria L.

Annual or biennial to 30 cm. Lvs 5–25 mm., reniform, crenate or dentate with 5–13 teeth. Sepals ± spreading. Petals c. 5–6 mm., yellow, 3–4 times as long as sepals. Ovary superior. $2n=18$.

Occasionally ± naturalized as a garden weed. Native of the Carpathians, Asia Minor, Caucasus, Algeria.

Section 4. *Tridactylites* (Haw.) Griseb. *Annual or biennial herbs*. Stems lfy. Lvs alternate, entire to palmately lobed, without pits or lime glands. Receptacle cup-shaped or campanulate. Sepals erect. *Petals white. Ovary fused to receptacle, almost completely inferior*. Seeds fusiform or oblong.

7. S. tridactylites L.   Rue-leaved Saxifrage.

Erect annual herb 2–15 cm., simple or with axillary branches, ± glandular hairy all over. *Basal lvs* ± congested, ± *spathulate*, with a broad petiole, blade 1 cm. or less, orbicular or obovate, palmately 3–5-lobed or -fid, or the smaller lvs entire; upper smaller, passing into entire bracts. Fls in simple or branched monochasial cymes or solitary and terminal. Pedicels slender, twice as long as fl. or more. Receptacle campanulate. Sepals ovate. Petals 2–3 mm., white, obovate-cuneiform. Fr. subglobose. Fl. 4–6. Probably usually self-pollinated. $2n=22$. Th.

Native. Dry open habitats in sandy grassland, walls, etc., mainly on basic soils, rather local, usually lowland but ascending to 2400 ft. Throughout England and Wales; E. Scotland north to Caithness; Inner and Outer Hebrides; almost throughout Ireland; Channel Is. 90, H35, S. Europe from Scandinavia (64° 13′ N.) southwards; Mediterranean region to W. Persia.

Section 5. *Saxifraga. Perennial herbs, often overwintering by axillary bulbils, without numerous sterile rosettes*. Lvs alternate, without pits or lime glands, the basal *reniform or ovate, with long petioles*, toothed or palmately lobed, stem lvs few, often smaller than the basal but always present. Receptacle cup-shaped. Sepals spreading or erect. *Petals* usually *white*. Ovary from about ⅓ to almost completely inferior, fused to receptacle below. Seeds oblong or fusiform.

8. S. granulata L.   'Meadow Saxifrage.'

Perennial herb 10–50 cm., *overwintering by bulbils produced in the axils of the basal lvs*. Basal lvs in a rosette, sometimes dead at fl., *blade* 0·5–3 cm., reniform, cordate at base, obtuse, *crenate-lobulate* with 7 or more teeth, with scattered flexuous long white hairs and glandular-ciliate; petiole several times as long as blade; cauline lvs few, decreasing in size upwards, more acutely toothed, cuneate at base, petioles short. Bulbils brown, ovoid, c. 5 mm. Fl.-stems solitary, simple, pilose below, glandular especially above. Infl. a loose, open, terminal cyme; *fls* 2–12; bracts small, linear; pedicels 4–20 mm., densely glandular. Receptacle densely glandular. Sepals ovate. *Petals* 10–17 *mm.*, white, obovate. *Ovary c. ¾-inferior*; styles long, ascending; stigmas conspicuous. Fr. 6–8 mm., ovoid or subglobose. Fl. 4–6. Visited by various insects; protandrous. $2n=46, 48*, 49*$ to 60 (very variable). Hs.

Native. Basic and neutral grassland ascending to 1500 ft., rather local and with an eastern tendency; usually on well-drained soils. From Moray and Renfrew southwards (not Pembroke or Cornwall), probably introduced in Devon; in Ireland native round Dublin, probably introduced elsewhere. 89, H4. Europe from Scandinavia (c. 70° N.) to C. Spain and Portugal, Corsica, Sicily and Montenegro, east to Poland; sspp. or allied spp. in S. Spain, Morocco, Sardinia and the S. Balkans.

9. S. cernua L.   'Drooping Saxifrage.'

Perennial herb 3–15 cm., overwintering by bulbils produced in the axils of the lvs. Basal lvs in a rosette, *blade* 5–18 mm., reniform, shallowly cordate at base, *palmately 3–5-lobed*, with ovate, acute lobes, subglabrous; petiole 2–3 times as long, slightly dilated below, not sheathing or toothed; cauline on shorter petioles, passing into the sessile, oblong or linear, entire, bulbiliferous bracts. Basal bulbils c. 5 mm., solitary, narrowly ovoid. *Fl. stems solitary, erect. Fls solitary and terminal or more frequently* 0, *being wholly or partially replaced by numerous small red bulbils* several together in the axils of the bracts. Sepals ovate, spreading. *Petals* 8–13 *mm.*, white, obovate-oblong. *Ovary less than ½-inferior*. Fr. unknown. Fl. 7. Visited by flies, but apparently never fruiting. $2n=50, 60, 64$, c. 66. Hs.

Native. Among basic rocks on mountains, from 3000 to nearly 4000 ft., very rare. Perth, Argyll, Inverness. 3. Arctic Europe, Asia and eastern N. America (including Greenland); Alps, Carpathians; mountains of C. Asia, Himalaya, Japan; Rocky Mountains.

10. S. rivularis L.   'Brook Saxifrage.'

Perennial herb 2–8 cm., sometimes producing rooting stolons. Basal lvs scarcely forming a rosette, *blade* 5–20 mm., reniform, shallowly cordate at base, *palmately 3–7-lobed* with ovate, obtuse or subacute lobes, subglabrous; petiole 3–10 times as long, dilated at base and sheathing, with a tooth on each side of the expanded part at the junction with the petiole; cauline lvs sometimes entire, petioles shorter. Basal bulbils small, concealed by lvs or 0. *Fl.-stems usually several, ascending*, not or scarcely longer than basal lvs, *not bulbiliferous. Fls* 1–3; pedicels glandular-pubescent above, variable but mostly long; bracts ovate or oblong, entire. Sepals ovate, ascending, subobtuse. *Petals* 3–5 *mm., white*, obovate or obovate-oblong. *Ovary less than ½-inferior.* Fr. ovoid or ovoid-oblong; styles spreading. Fl. 7–8. Visited by flies, slightly protogynous. $2n = 26$, 52, 56. Hs.

Native. Wet rocks on mountains from 3000 to nearly 4000 ft.; very rare and local. Perth, Aberdeen, Inverness, Ross, Argyll. 6. Arctic Europe, Asia and eastern N. America (including Greenland); Cascade, Rocky and White Mountains.

Section 6. *Sedoides* Gaudin. *Perennial herbs with numerous sterile lfy rosettes or elongated creeping lfy shoots. Lvs* alternate, without pits or lime-glands, *palmately* 3–5(–11)-*lobed or -fid with narrow segments, or narrow and entire.* Lvs on the fl.-stems few. Receptacle cup-shaped or campanulate. Sepals spreading or erect. *Petals* usually *white. Ovary* nearly completely *inferior*, fused to receptacle. Seeds oblong or fusiform.

For an account of the British species of this section see Webb in *Proc. Roy. Irish Acad.* vol. **53** (1950), Sect. B, pp. 207–40.

11. S. cespitosa L. 'Tufted Saxifrage.'

Compact perennial 4–10 cm., *forming dense cushions. Sterile shoots short*, erect, densely lfy, thickly clothed with dead lvs below the living ones, which are in a compact rosette. Axillary bulbils 0. *Lvs* to c. 1 cm., 3–5-lobed, narrowed at base into a short petiole, densely and uniformly *clothed with short glandular hairs*, lobes of 3-fid lvs spreading at about 50° or less; *lobes* oblong, *obtuse* more than 1 mm. broad, upper cauline lvs entire. Stem, pedicels and calyx shortly glandular. Fls 1–5, pedicels mostly shorter than calyx. Sepals oblong-ovate, obtuse. *Petals* c. 5 mm., obovate, *impure white*. Fl. 5–7. Visited by flies; usually protogynous. $2n = 80$. Chc.

Native. High mountain rocks from 2000 to 3500 ft., very local and rare. Caernarvon, Banff, Aberdeen, Inverness. 4. Arctic Europe, Asia and America.

12. S. hartii D. A. Webb

Perennial herb, *forming loose tufts.* Axillary bulbils 0. *Lvs* larger than in *S. cespitosa*, 5- or more-lobed, *densely clothed with short glandular hairs; lobes subacute. Petals* larger than in *S. cespitosa*, *pure white*. Protandrous. $2n = $ c. 48*. Chc or Chh.

Native. Sea-cliffs. Arranmore Island (Donegal). Endemic.

13. S. rosacea Moench

S. decipiens Ehrh. ex Pers.; including *S. hirta* Donn ex Sm., non Haw., *S. affinis* D. Don, *S. incurvifolia* D. Don, *S. Sternbergii* Willd., *S. Drucei* E. S. Marshall

Perennial herb 4–20 cm. More robust and less compact than *S. cespitosa* and forming looser tufts, but very variable. *Sterile shoots* short or long (to 10 cm.), *erect or ascending* at fl., densely or laxly lfy. Axillary bulbils 0.

Lf-blades 1–1·5 cm., 3–7-lobed, narrowed at base into a petiole about as long as blade, very variable in indumentum but often with long flexuous eglandular hairs; with *few (or* 0) *rather long glandular hairs*; lobes of 3-fid lvs spreading at c. 50° or less; *lobes* oblong or oblong-lanceolate, *subacute*, sometimes mucronate, or those of the lowest lvs obtuse, the largest more than 1 mm. broad; upper cauline lvs entire. Stem, pedicels, and calyx densely or sparingly glandular. Fls 1–8, pedicels short or long. Sepals triangular-ovate, subobtuse or acute. *Petals* 6–8 mm., obovate, *pure white*. A very variable plant. Fl. 6–8. Protandrous. $2n = $ c. 48*, c. 64*. Chc. or Chh.

Native. Damp rock ledges, stream-sides, etc., on mountains; ascending to 3500 ft. Caernarvon (extinct); Kerry, Tipperary, Limerick, Clare, Galway, Mayo, very local but often abundant. H7. Iceland, Faeroes; mountains of S. Central Germany and W. Czechoslovakia; Vosges.

14. S. hypnoides L.   Dovedale Moss.

Incl. *S. platypetala* Sm.; *S. sponhemica* auct. brit.

Perennial 5–20 cm., forming mats with flowering rosettes and *numerous axillary procumbent or decumbent sterile shoots* (up to 15 cm.), *usually bearing axillary fusiform lfy bulbils*. Lf-blades of flowering rosettes to c. 1 cm., 3–5(–9)-lobed, contracted at base into a petiole longer than the blade, sparsely ciliate on the petiole with long flexuous eglandular hairs, otherwise subglabrous, the lobes of 3-fid lvs spreading at c. 75°; *lobes* linear or linear-oblong, *acute or acuminate*, *aristate*, usually c. 1(–1·5) mm. broad. Lvs of barren shoots all, or at least those in the middle of the shoot, entire, rarely all 3-lobed, elliptic-linear, acute, aristate. Upper cauline lvs entire. Stem glabrous or sparingly glandular, pedicels and calyx more glandular. Fls 1–5, pedicels mostly slender, longer than fl. Sepals triangular or triangular-ovate, acute or acuminate, aristate. Petals 4–10 mm., oblong or obovate, pure white. Fl. 5–7. Protandrous. $2n = $ c. 48*, c. 64*. Chh.

Native. Rock-ledges and scree, especially basic, and stony basic grassland in hilly districts, rarely on dunes, ascending to 4000 ft., locally common; Somerset, Wales and the border counties, Derby and Lancashire to Caithness; local in the Irish mountains from Clare, Waterford and Wicklow northwards. 55, H11. Faeroes, Iceland; Norway (1 small area); Vosges (1 locality).

S. hypnoides × *rosacea* occurs occasionally.

Section 7. *Xanthizoon* Griseb. Perennial herb with numerous lfy stems. Lvs alternate, entire, with a pit at the apex, but not secreting lime. Receptacle cup-shaped. *Sepals spreading. Petals yellow* (rarely red). *Ovary* ¼- to ⅓-*inferior*, fused to receptacle. Seeds fusiform.

15. S. aizoides L.   'Yellow Mountain-Saxifrage.'

Perennial herb 5–20 cm. with numerous decumbent sterile, and ascending fl.-stems. *Lvs* 1–2 cm., numerous, dense on the sterile stems, less so towards the top of the fl. ones, oblong-linear, acute, *sessile*, usually remotely ciliate with stiff hairs, otherwise glabrous, rather thick. Fl.-stems simple or branched above, pubescent. *Fls* 1–10, *in a loose terminal cyme*; bracts like the lvs but rather smaller. Sepals triangular-ovate, obtuse, spreading. Petals yellow, often spotted with red, distant, 4–7 mm., obovate, oblanceolate, or oblong. Styles divergent. Fr. c. 7 mm., ovoid. Fl. 6–8(–9). Pollinated mainly by flies; protandrous. $2n = 26$. Chh.

Native. Stream sides and wet stony ground on mountains; ascending to 3850 ft., locally common. Caernarvon; Yorks, Westmorland and Isle of Man to Orkney (not S.E. Scotland nor Outer Hebrides); Sligo to Donegal, Antrim. 33, H6. Arctic Europe, Greenland, eastern N. America, W. Asia; high mountains of Europe to Pyrenees, Apennines, Montenegro and Roumania; N. Rocky Mountains, Vermont.

Section 8. *Porphyrion* Tausch. Perennial herbs with numerous lfy stems. *Lvs opposite*, small, entire, with 1 or more pits near the apex, *secreting lime*. Receptacle cup-shaped. Sepals erect or spreading. *Petals purple* or pink, rarely white. Ovary about $\frac{1}{2}$-inferior, fused to receptacle below.

16. S. oppositifolia L.   'Purple Saxifrage.'

Perennial herb with numerous stems with long prostrate and short erect branches forming loose mats or flat cushions. Lvs 2–6 mm., very dense, in 4 rows, oval, oblong or obovate, concave, the apex thickened and flattened, with a single lime-secreting pit, dark blue-green, setulose-ciliate, otherwise glabrous, sessile. Fls solitary, terminal on stems 1–2 cm., which are less densely lfy than the sterile ones. Sepals ovate, suberect. Petals 6–10 mm., rosy-purple, ovate or ovate-oblong. Fr. 3–6 mm., ovoid; styles spreading. Fl. 3–5 (sometimes a few fls again in 7). Visited mainly by Lepidoptera, probably often self-pollinated; protandrous to protogynous. $2n = 26^*$, 52. Chh.

Native. Damp rocks and stony ground on mountains, ascending to nearly 4000 ft., local but often abundant. Brecon, N. Wales, Yorks, Lake District, Dumfries, Peebles; Stirling northwards; Galway to Londonderry. 35, H7. Arctic Europe, Asia and America; high mountains of Europe to Sierra Nevada, Apennines and Macedonia (the more southern forms belong to distinct sspp.); mountains of C. Asia to Kashmir (a ssp.); Rocky Mountains, Vermont.

2. Chrysosplenium L.

Herbs. Lvs simple, stalked. Fls in dichotomous cymes, epigynous. Receptacle cup-shaped, adnate to the ovary. *Sepals* 4–5. *Petals* 0. Stamens 8–10. Carpels 2, united; *placentae parietal*; styles free, short. Capsule opening along the inner edges of the carpels. Nectar secreted by a disk round the styles.

About 60 spp., north temperate regions and temperate S. America, mainly in E. Asia.

Lvs opposite; basal lvs truncate or broad-cuneate at base.   **1. oppositifolium**
Cauline lvs alternate (usually only 1); basal lvs cordate at base.
                                                          2. alternifolium

1. C. oppositifolium L.   'Opposite-leaved Golden Saxifrage.'

Perennial herb 5–15 cm., with numerous *decumbent lfy rooting stems* forming large patches; fl.-stems ascending. *Lvs opposite*; blades of lower 1–2 cm., *orbicular, entire or shallowly sinuate-dentate,* very obtuse, *truncate or broad-cuneate at base,* often shortly decurrent, with scattered appressed hairs above; *petiole as long as or shorter than blade*; lvs of the fl.-stems smaller, on shorter petioles, usually glabrous, 1–3 pairs. Bracts similar but smaller and decreasing upwards, bright greenish-yellow. Infl. loose below, dense above. Calyx 3–4 mm. diam., coloured like the bracts; sepals mostly 4, usually 5 in some fls. Fl. 4–7. Visited by various small insects; protogynous, self-pollination possible. $2n=42$. Chh.

Native. Stream-sides, springs, wet rocks, wet ground in woods, usually in shade, ascending to 3400 ft.; throughout the British Is. but unrecorded from Cambridge, Huntingdon and Shetland; common except in S.E., E. and C. England. 109, H40, S. W. and C. Europe from Norway (64° 30′ N.), Denmark and N. and W. Germany to C. Spain and Portugal, N. Italy.

2. C. alternifolium L.   'Alternate-leaved Golden Saxifrage.'

Similar to *C. oppositifolium* but rather more robust and without creeping lfy stems but spreading by *lfless stolons*. Radical lvs 2–4, *blades* 1–2·5 cm., reniform, *crenate, cordate at base,* not decurrent; *petiole several times as long as blade*; lvs on the fl.-stems usually only 1, smaller, truncate or shallowly cordate at base, petiole shorter. Bracts like those of *C. oppositifolium* but rather larger and more deeply toothed. Calyx 5–6 mm. diam. Fl. 4–7. Pollinated by small insects; ± homogamous. $2n=48$. Hs.

Native. In similar places to *C. oppositifolium,* but local, ascending to 3250 ft.; Inverness and Argyll southwards but absent from the extreme west (Wigtown, W. Wales and Cornwall), from Ireland, and from all the smaller islands. 81. Europe from Scandinavia (c. 65° N.) to C. France, Apennines and Macedonia; N. and C. Asia to Caucasus and Himalaya.

The two following monotypic genera of rhizomatous herbs with shallowly lobed broadly ovate-cordate lvs, racemose fls with a large perigynous zone and parietal placentation are grown in gardens and occasionally naturalized. Western N. America.

*Tolmiea menziesii (Pursh) Torr. & Gray

Perigynous zone ±cylindric, greenish, tinged with purple. Sepals 5, unequal. Petals 4(–5), filiform, brown. Reproducing by bulbils on the lvs.

*Tellima grandiflora (Pursh) Dougl. ex Lindl.

Perigynous zone broadly campanulate or sometimes narrowed at mouth, glandular. Fls regular, 5-merous. Petals lanceolate in outline, laciniate, pale green becoming dark purple. Bulbils 0.

56. PARNASSIACEAE

Perennial herbs. Lvs simple, alternate, exstipulate; the radical numerous, those on the fl.-stems few. Fl. solitary, terminal, the stem from the axil of a radical lf. Fls actinomorphic, hermaphrodite, hypogynous or perigynous, 5-merous. A ring of 5, large, usually fringed, staminodes, each with nectaries on the upper surface, occurs inside and opposite the petals. Fertile stamens 5, alternating with the staminodes. Carpels (3–)4(–5) united into a superior or ½-inferior ovary, septate below, free above; placentae axile below, parietal above, with numerous anatropous ovules; style short, thick or 0; stigmas usually 4. Fr. a loculicidal capsule.

One genus.

An isolated type probably most closely allied to the Saxifragaceae, in which it is often included, but showing also resemblances to the Droseraceae and Hypericaceae. The staminodes are characteristic.

PARNASSIA L.

About 45 spp., north temperate regions.

1. P. palustris L.   Grass of Parnassus.

Glabrous perennial herb 10–30 cm., with short vertical stock, often somewhat tufted. Radical lvs ± numerous, blades 1–5 cm., ovate, cordate at base, sub-acute, entire, petiole longer than the blade. Fl.-stem erect, straight, with 1 sessile deeply cordate lf near the base. Petals 7–12 mm., oval or oblong, white, with conspicuous veins. Staminodes c. $\frac{1}{3}$ as long as petals, ± spathulate at the apex with 7–15 long setaceous processes tipped with yellowish glands. Ovary ovoid, superior. Fl. 7–10. Pollinated by various insects, protandrous. $2n = 18$. Hs.

Var. *condensata* Travis & Wheldon has very numerous rather leathery radical lvs on shorter petioles, shorter (2·5–15 cm.) fl.-stems and rather larger fls.

Native. Marshes and wet moors; ascending to 2600 ft. Widespread but rather local; Cardigan, Glamorgan, Hants (extinct in Dorset and Isle of Wight), Berks and Essex northwards (absent from Isle of Man); Ireland (absent from Cork, Kerry, Wexford and Down); 90, H33. Var. *condensata* in dune-slacks in Lancashire, Cheshire and N. Ireland (Antrim to Sligo) and perhaps elsewhere. Europe from S. Scandinavia to Spain, Italy and Greece; Morocco (Atlas); temperate Asia.

57. HYDRANGEACEAE

Shrubs or trees. Lvs opposite (very rarely alternate) simple, exstipulate. Fls herma-phrodite or some sterile, actinomorphic, 4–5(–10)-merous, perigynous or epigynous. Stamens twice as many as petals and obdiplostemonous, or more numerous. Carpels (2–)3–5(–10), united; styles free or united; ovules numerous (very rarely only 1), anatropous; placentation axile or parietal. Fr. a capsule, rarely a berry; seeds usually numerous; endosperm copious.

Over 15 genera and 200 spp., mainly north temperate, a few in tropical Asia (to New Guinea) and C. and S. America. Spp. and hybrids of *Hydrangea* and *Deutzia* are commonly grown in gardens. Differs from the Saxifragaceae in the woody habit, opposite lvs, tendency to more numerous carpels and stamens and to a more com-plete fusion of the ovary.

1. PHILADELPHUS L.

Fls 4–(6)-merous, all fertile. Petals contorted in bud. Stamens numerous. Ovary inferior, (3–)4(–5)-celled with numerous ovules on axile placentae.

Over 50 spp. in E. Asia and N. America, 1 in Europe. A number of spp. and hybrids are commonly cultivated.

*P. coronarius L.   Syringa, Mock Orange.
Deciduous shrub to 3 m., with brown peeling bark. Buds hidden in the base of the
petiole. Lvs 4–8 cm., ovate, remotely toothed, glabrous except for tufts of hairs in
the axils beneath, and sometimes pubescent on the veins. Fls 5–7, in terminal
racemes, white, 2·5–3·5 cm. diam., very fragrant. Calyx glabrous. Styles united to
about half-way, shorter than stamens. Fl. 6. $2n=26$.
Commonly planted, sometimes among native vegetation and possibly sometimes
naturalized. Native of S.E. Europe and Caucasus.

58. ESCALLONIACEAE

Shrubs or trees. Lvs alternate, rarely opposite or whorled, often with glandular
teeth, exstipulate or rarely with small stipules. Fls hermaphrodite, rarely
dioecious, actinomorphic (4–)5(–9)-merous, perigynous or epigynous, mostly
with a well-marked disk. Stamens as many as corolla-lobes. Carpels 2–6,
united, very rarely free, superior to inferior; ovules numerous (rarely 1 or 2).
Fr. a capsule or berry; seeds usually numerous; embryo straight; endosperm
copious.
Over 20 genera and 150 spp., southern hemisphere (rare in continental
Africa) and S.E. Asia, a very few in Mexico and eastern N. America. Differs
from Hydrangeaceae in the alternate, often glandular lvs and few stamens.

1. ESCALLONIA Mutis ex L. f.

Shrubs. Lvs alternate, usually evergreen, glandular-toothed. Fls in racemes
or panicles, 5-merous. Petals with erect claws and spreading limb, pink or
white. Stamens inserted under the edge of the disk. Ovary completely
inferior, of 2–3 carpels, placentation parietal; style 1; ovules numerous.
Fr. a septicidal capsule.
About 50 spp., S. America; some spp. and especially hybrids are frequently
grown in gardens.

*1. E. macrantha Hook. & Arn.

Evergreen shrub 2–4 m. Twigs pubescent and glandular, sticky. Lvs 2·5–
7·5 cm., oval or obovate, doubly serrate, dark green, shining and glabrous
above, dotted with resinous glands beneath, sessile. Infl. 5–10 cm. Fls c. 15 mm.
long and across, bright pinkish-red. Calyx campanulate, lobes narrow, acute,
glandular. Fl. 6–9. $2n=24$.
Commonly planted as a hedge near the sea in S.W. England and Ireland
and sometimes self-sown, as in Kerry. Native of the Island of Chiloe.

59. GROSSULARIACEAE

Shrubs. Lvs alternate, simple, often palmately lobed, stalked, exstipulate or
with small stipules adnate to the petiole, plicate or convolute in bud. Fls in
racemes, hermaphrodite or dioecious, actinomorphic, epigynous, with the
conspicuous and often coloured receptacle prolonged beyond the ovary,
4–5-merous. Sepals coloured like the receptacle. Petals usually shorter than
sepals. Stamens equal in number to and alternating with petals. Ovary

inferior, of 2 carpels with parietal placentae; styles 2, connate below; stigmas entire; ovules numerous. Fr. a berry, the calyx persistent at the apex; endosperm copious, fleshy; embryo rather small.

One genus.

Nearest *Escalloniaceae*, usually recognizable by the conspicuous receptacle and small petals.

1. RIBES L.

The only genus. About 150 spp., in the north temperate zone and on mountains in C. and S. America. Some spp., in addition to the following, are cultivated. Nectar secreted by an epigynous disk.

1 Unarmed; fls 5 or more in racemes.   2
  Branches with spines at the nodes; fls 1–2(–3).   **5. uva-crispa**

2 Fls hermaphrodite, usually in spreading or drooping racemes; bracts less
    than half as long as pedicel.   3
  Fls dioecious; racemes erect in fl. and fr.; bracts exceeding pedicels.
                                                              4. alpinum

3 Lvs and ovary eglandular, not strong smelling; fr. red (rarely white)
                                                         (**rubrum** agg.) 4
  Lvs and ovary with sessile glands, strong smelling; fr. black. **3. nigrum**

4 Receptacle obscurely pentagonal, with raised rim round the style, saucer-
    shaped; connective broad; lvs cordate at base with narrow sinus.
                                                            1. sylvestre
  Receptacle circular, without raised rim, cup-shaped; connective obsolete;
    lvs truncate at base, or cordate with wide sinus.   **2. spicatum**

Subgenus 1. RIBES.

Unarmed shrubs. Fls in many-fld racemes.

(1–2). R. rubrum agg.   Red Currant.

Deciduous shrub 1–2 m. Lf-blades 3–10 cm., rather broader than long, 3–5-lobed, pubescent at least when young, *not strong-smelling* or *glandular*; lobes triangular-ovate, dentate, with broad-ovate apiculate teeth; petiole usually with stalked glands, somewhat expanded and ciliate with long, often gland-tipped, hairs. *Fls 6–20 in ascending to drooping racemes* (sometimes erect in fl., very rarely in fr.); *bracts* ± ovate, somewhat curved, *less than half as long as pedicel*. *Fls hermaphrodite*, greenish, not pubescent nor glandular; petals cuneiform, very small; stamens short. *Fr.* 6–10 mm. diam., *red*, rarely brownish-white, globose. Fl. 4–5. Visited by Hymenoptera, homogamous.

1. R. sylvestre (Lam.) Mert. & Koch

R. rubrum var. *sativum* Rchb.; *R. vulgare* Lam.; *R. domesticum* Jancz.; *R. rubrum* auct. angl., L., p.p.

Lf-blades to 6 cm., deeply *cordate at base with narrow sinus*, glabrous at maturity or slightly pubescent beneath. Infl. drooping or spreading and curving downwards; rhachis and pedicels glabrous or slightly pubescent. Fls c. 5 mm. diam., pale green or slightly tinged with purple; *receptacle saucer-shaped, obscurely pentagonal with a raised pentagonal rim round the style*; sepals inserted on lobes of receptacle, broader than long, contracted at base; *stamens* slightly inclined inwards, inserted on lobes of receptacle, *with a broad connective* whose breadth equals that of the anther-cells. Young fr. not contracted above. $2n = 16$. N.

?Native. Woods and hedges, ascending to 1500 ft. From Caithness southwards (not recorded from the Scottish islands) widespread (except in N. Scotland); locally frequent and probably native by streams in woods etc., and in fen carr in many parts of England and Wales; probably introduced in Scotland (where many of the records may apply to *R. spicatum*); certainly so in Ireland. 91, H3. France, Belgium and perhaps elsewhere in W. Europe but of doubtful status everywhere and possibly of garden origin. The majority of cultivated red and white currants belong to this sp.

2. R. spicatum Robson

R. petraeum auct. (incl. Sm.), non Wulf.; *R. pubescens* (Hartm.) Hedl.

Lf-blades to 10 cm., *truncate at base to shallowly cordate with wide sinus*, usually persistently pubescent beneath but sometimes nearly glabrous. Infl. erect, ascending or spreading in fl., sometimes somewhat arched, but not drooping, except in fr.; rhachis and pedicels ± pubescent and with minute whitish glands. Fls c. 7 mm. diam., pale green, usually tinged with brownish-purple; *receptacle cup-shaped*, *perfectly circular*, *without raised rim*; sepals ± orbicular, contracted at base; stamens erect, inserted on margin of receptacle, with a *connective very narrow on the inner side* (though broad on the outer). Young fr. contracted above. $2n = 16$. N.

Native. Woods on limestone, ascending to 1400 ft. From Lancashire and Yorks to Caithness, local; an occasional escape elsewhere, but rarely grown for fruit. 16. N. Europe and Asia from Scandinavia to Manchuria, south to N. Germany and Poland.

R. spicatum × sylvestre.

Some cultivated red currants belong here and it might occur as an escape.

Hybrids of *R. sylvestre* with foreign spp. are also occasionally grown for fr.

3. R. nigrum L. Black Currant.

Deciduous shrub 1–2 m. Lf-blades 3–10 cm., rather broader than long, 3–5-lobed, cordate at base, glabrous above, pubescent on the veins beneath, at least when young, *with scattered sessile brownish glands*, especially beneath, *strong smelling*; lobes deltoid or ovate-deltoid, acute, irregularly and coarsely serrate-dentate; petiole pubescent, below conspicuously expanded and with sessile brownish glands. *Racemes lax, drooping*, 5–10-fld; rhachis and pedicels pubescent, often glandular; *bracts* ovate, straight, *less than half as long as pedicel*. *Fls hermaphrodite*, c. 8 mm., dull purplish-green; *ovary with sessile glands; receptacle* broadly campanulate, *pubescent outside*; sepals oblong, pubescent, recurved at apex; petals whitish, ovate, c. half as long as sepals; stamens about equalling petals. *Fr. black*, globose, 12–15 mm. diam. Fr. 6–7. Fl. 4–5. Visited by bees, etc., but visits few; often self-pollinated; homogamous. $2n = 16$. N.

Native. Woods and hedges from Sutherland southwards (not recorded from the Scottish Is.) widespread (except in N. Scotland); commonly cultivated for its fr. and often an escape but probably native by streams in woods, etc., and in fen carr, in many places; not native in Ireland. 89, H4. Europe from Scandinavia (69° 30′ N.) to E. France, Italian Alps and Bulgaria; N. and C. Asia to the Himalaya.

*R. sanguineum Pursh   Flowering Currant.
Lvs obtusely lobed, doubly serrate, pubescent beneath. Fls bright pink, 10–15 mm.; receptacle tubular. Fr. bluish-black, very pruinose. Commonly planted and sometimes naturalized.
Native of western N. America.

4. R. alpinum L.   'Mountain Currant.'
Deciduous shrub 1–3 m. Lf-blades 3–5 cm., as broad as long, deeply 3(–5)-lobed, truncate to subcordate at base, with scattered hairs; lobes dentate with acute teeth; petiole scarcely expanded below, with stalked and subsessile glands. *Fls dioecious, in erect racemes*, the male 20–30-fld, female 8–15-fld; rhachis and pedicels densely glandular with subsessile glands, otherwise glabrous; *bracts* lanceolate or linear, acute, *longer than pedicels*. Fls yellowish-green, 4–6 mm. diam., female smaller than male; receptacle nearly flat in male fl., cup-shaped in female; sepals ovate; petals very small, cuneiform; stamens short; style present in male fl. and abortive anthers in female. Fr. red, globose, 6–8 mm. diam., insipid. Fl. 4–5. Pollinated by Diptera and Hymenoptera. Fr. 7. $2n = 16$. N.
    Native. Cliffs and rocky woods on limestone; ascending to 1250 ft. Wales, Staffs, Lancashire, Yorks, Westmorland and Cumberland, very local; an occasional escape elsewhere but not now often cultivated. c. 15. Mountains of Europe from Scandinavia to N. Spain, C. Italy, Montenegro and Bulgaria; Morocco (Atlas).

Subgenus 2. GROSSULARIA (Mill.) A. Rich.
Twigs with spines at the nodes. Fls solitary or in 2–4-fld racemes.

5. R. uva-crispa L.   Gooseberry.
R. Grossularia L.
Deciduous much-branched shrub c. 1 m. Spines 1–3 at each node, 5–15 mm. Twigs glabrous, pubescent or bristly, with many short axillary shoots. Lf-blades 2–5 cm., as broad or rather broader than long, 3–5-lobed, broadly cuneate to subcordate at base, glabrous or pubescent; lobes ovate-rhombic, obtuse, deeply and obtusely dentate in the upper part. Fls 1–2(–3) on each peduncle, drooping; rhachis pubescent and glandular; bracts small, ovate. Fls c. 1 cm., hermaphrodite, greenish, purple tinged; ovary glandular-bristly or pubescent, rarely glabrous; receptacle and calyx hairy on both surfaces; receptacle broadly campanulate; sepals oblong, reflexed; petals small, whitish; stamens twice as long as petals. Fr. globose or ovoid, 10–20 mm. (more in cultivated forms), green, yellowish-green or reddish-purple, bristly, pubescent or smooth. Fl. 3–5. Visited mainly by Hymenoptera, protandrous. $2n = 16$. N.
    Native. Woods and hedges from Caithness southwards, widespread (rare in N.W. Scotland and not recorded from the Scottish Is.); commonly culti-

vated for its fr. and often an escape but probably native by streams in woods, etc., in many places; not native in Ireland. 101, H6, S. Europe from Scandinavia to the mountains of N. Spain, C. Italy and Macedonia; Morocco, Algeria (mountains, rare); Caucasus.

60. PITTOSPORACEAE

Trees, shrubs or woody climbers with resin in the cortex. Lvs alternate. Fls usually hermaphrodite, regular, hypogynous, 5-merous. Sepals and petals imbricate. Petals clawed. Stamens 5, alternate with petals. Carpels 2(–5), placentation parietal or rarely axile; ovules usually with one integument, numerous; style simple. Fr. a capsule or berry; endosperm copious.

Nine genera and 200 spp., warmer parts of Old World, especially Australia.

1. PITTOSPORUM Banks ex Soland.

Trees or shrubs. Ovules numerous on parietal placentas. Fr. a leathery capsule. Seeds not winged.

About 160 spp. with the range of the family.

*1. P. crassifolium Soland. ex Putterl.

Evergreen shrub or small tree to 10 m., of erect habit. Lvs 4–8 cm., narrowly obovate, obtuse, coriaceous, dark green and shining above, white- or buff-tomentose beneath, margins recurved. Fls 1–6 in terminal umbels, deep purplish-red. Fr. 1·5–3 cm., woody, tomentose, (2–)3(–4)-valved.

Planted in the warmer parts of the country and naturalized in the Scilly Is. New Zealand (N. Island).

61. DROSERACEAE

Perennial glandular insectivorous herbs, sometimes shrubby below. Lvs often circinate in bud. Fls actinomorphic, hermaphrodite, often in unbranched circinate cymes, 4–8-merous. Sepals and petals imbricate in bud, sometimes connate at base. Stamens 4–20, often 5; anthers 2-celled, extrorse. Ovary superior, 1-celled, placentation parietal or sub-basal; styles 3–5, rarely connate below. Fr. a loculicidal capsule with numerous small seeds.

Four genera with about 100 spp., cosmopolitan in acid sandy, stony and boggy places.

1. DROSERA L.

Slender insectivorous herbs. Lvs densely glandular and fringed with long glandular hairs ('tentacles'), in a rosette or alternate; stipules usually present. Fls in circinate cymes, rarely solitary, 4–6- or 8-merous. Stamens as many as the petals. Ovary 1-celled; styles 3–5, free or connate below. Fr. a many-seeded capsule; seeds minute, testa often inflated, forming a wing.

About 100 spp. in temperate and tropical regions, specially abundant in S. Africa and Australia.

The British spp. are usually cleistogamous. For accounts of the carnivorous habits of the genus see Darwin, *Insectivorous Plants* and Lloyd, *The Carnivorous Plants*.

1   Lvs not appreciably longer than broad, narrowed abruptly at base.

                                                         1. rotundifolia

    Lvs distinctly longer than broad, gradually narrowed at base.   2

2   Lvs narrowly obovate to linear-oblong; scape in fl. up to twice as long
      as lvs, apparently arising from the centre of the rosette.   **2. anglica**

    Lvs obovate; scape in fl. little longer than lvs, arising laterally below the
      rosette.   **3. intermedia**

1. D. rotundifolia L.   Sundew.

A slender reddish scapigerous perennial 6–25 cm. Stem short or rarely (in floating forms) sending out slender stolons. *Lvs* long-petioled, *spreading horizontally* and forming a rosette, *blade orbicular*, up to 1 cm. diam.; *petioles hairy. Scape* erect, simple or rarely branched above, 2–4 *times as long as the lvs*, appearing to arise from the centre of the rosette. Fls 5 mm., white, shortly pedicelled, usually 6-merous, in 2 rows. Capsule acute, equalling or exceeding the sepals; seeds reticulate, winged. Fl. 6–8. $2n=20$. Hr. or Hel.

     Native. In bogs and wet peaty places on heaths and moors, often amongst *Sphagnum*, occasionally forming a floating fringe to small ponds. 108, H40, S. Throughout the British Is. wherever suitable habitats exist. Europe, except the Mediterranean region; N. Asia; N. America.

2. D. anglica Huds.   'Great Sundew.'

A reddish scapigerous perennial 10–30 cm. Stems short. Lvs ± erect, *blade up to c. 3 cm., narrowly obovate or linear-oblong*, gradually attenuate into the long nearly or quite glabrous petiole. *Scape* erect, *up to twice as long as lvs*, appearing to arise from the centre of the rosette. Fls similar to those of *D. rotundifolia*, 5–8-merous. Capsule obovoid, exceeding the sepals; seeds reticulate, winged. Fl. 7–8. $2n=40$. Hr. or Hel.

     Native. Usually amongst *Sphagnum* in the wetter parts of bogs. 63, H35. Scattered throughout the wetter parts of the British Is. Europe, except the Mediterranean region; N. Asia.

D. × obovata Mert. & Koch, a hybrid between *D. rotundifolia* and *anglica*, occurs occasionally. It resembles *D. intermedia* in general appearance but has a straight scape 2–3 times as long as the lvs, arising from the centre of the rosette. It is sterile.

3. D. intermedia Hayne   'Long-leaved Sundew.'

D. longifolia L., p.p.

A reddish scapigerous perennial 5–10 cm. Stem short, slender. Lvs ± erect, *blade c. 1 cm., obovate*, gradually attenuate into a long glabrous petiole. *Scape curved or decumbent at base, then erect, little longer than the lvs, arising laterally from below the terminal rosette.* Fls 5–8-merous, similar to those of *D. rotundifolia*. Capsule pyriform, equalling or slightly exceeding the sepals; seeds granulate, not winged. Fl. 6–8. $2n=20$. Hr.

     Native. In damp peaty places on heaths and moors, but often in rather drier habitats than the two preceding spp., though occasionally in wet *Sphagnum* bogs especially in Ireland. 58, H18. Mainly in the west of the British Is., rather local but often abundant in suitable habitats. W. and C. Europe north to Norway, Sweden and Finland; Asia Minor.

62. SARRACENIACEAE

Insectivorous herbs with radical lvs forming tubular 'pitchers' with a small blade at the top. Fls hermaphrodite, solitary or in few-fld racemes, nodding. Sepals 4–5, free, imbricate, persistent, often coloured. Petals 5 or 0, free, imbricate. Stamens numerous, free. Ovary superior, 3–5-celled; style simple, often much enlarged and peltate at apex. Ovules numerous, axile, anatropous. Capsule loculicidal. Seeds small, endosperm fleshy.

Three genera and about 10 spp. in Atlantic N. America, California and Guiana.

1. SARRACENIA L.

Fls solitary, bracteolate. Sepals 5, spreading, coloured. Petals 5, connivent. Ovary 5-celled; style greatly dilated and umbrella-like above.

Seven spp. in Atlantic N. America, some widely cultivated and not infrequently naturalized. Hybrids are of frequent occurrence.

*1. S. purpurea L.   Pitcher-plant.

A stout low-growing perennial herb. Pitchers 10–15 cm. Scape, 15–40 cm. Fl. c. 5 cm. diam. Sepals and petals purple. Umbrella-like top of style c. 3 cm. diam. Fl. 6. Hel.

Introduced. Abundantly naturalized in bogs in C. Ireland, where it was planted in 1906. Roscommon and Westmeath. Native of Atlantic N. America; naturalized in Switzerland and probably elsewhere in Europe.

63. LYTHRACEAE

Herbs, shrubs, or, less frequently, trees. Lvs usually opposite or in whorls; stipules 0 or minute. Fls usually actinomorphic, hermaphrodite. Calyx tubular, teeth valvate, often with appendages alternating with them. Petals present or 0, inserted at or near the top of the calyx-tube, crumpled in bud. Stamens 4 or 8(–12), inserted below the petals; filaments usually inflexed in bud. Ovary superior, usually 2–6-celled; style simple; ovules numerous, placentation axile. Fr. usually a capsule, opening by a transverse slit, or by valves, or irregularly. Seeds small, endosperm 0; embryo straight.

More than 20 genera and 500 spp., throughout the world except for cold regions.

Stems erect or ascending, not rooting at nodes; lvs linear to ovate; fls 5 mm. or more; calyx-tube straight.   1. LYTHRUM
Stems prostrate, rooting freely at nodes; lvs obovate; fls c. 1 mm.; calyx-tube obconical.   2. PEPLIS

1. LYTHRUM L.

Herbs or small shrubs. Stems 4-angled, at least when young. Lvs quite entire, usually opposite. Fls axillary, solitary or in small cymes, purple or pink. Calyx-tube straight, teeth 4–6. Petals 4–6, rarely 0. Stamens 6–12, in 1 or 2 whorls, sometimes unequal in length. Ovary sessile, 2-celled; style filiform.

About 25 spp., cosmopolitan.

Perennial; fls 10–15 mm., magenta, arranged in axillary cymes.   **1. salicaria**
Annual; fls c. 5 mm., pale pinkish-purple, solitary.   **2. hyssopifolia**

1. L. salicaria L.   Purple Loosestrife.

A ± pubescent *perennial* 60–120 *cm.* Stem erect, simple or with relatively
short, slender branches. Lvs 4–7 cm., sessile, lanceolate to ovate, acute, shortly
pubescent to glabrescent, ± cordate at base, the lower opposite or in whorls
of 3, the upper sometimes alternate. Infl. up to 30 cm. or sometimes more,
spike-like, usually dense and many-fld. *Fls 10–15 mm., arranged in whorls* in
the axils of bracts. Bracteoles caducous. Calyx pubescent, tube c. 6 mm.,
cylindrical, ribbed; teeth 2–3 mm. *Petals* 8–10 mm., obovate, *purple*. Stamens
12. Capsule 3–4 mm., oblong-ovoid, enclosed in calyx. Fl. 6–8. $2n = 60^*$;
50*. Hel. or Hp.

Three forms of fls are found on different plants: 1. Style short, stamens long and
medium. 2. Style medium, stamens long and short. 3. Style long, stamens short
and medium. Three sizes of pollen grain are found, one in each of the three types
of stamen.

Native. In reed-swamp at the margins of lakes and slow-moving rivers,
and in fens and marshes, often forming large stands. 99, H40, S. England,
Wales and Ireland, locally abundant, less frequent in Scotland and absent
from most of the north. Europe, W. and N. Asia, N. Africa, N. America.

2. L. hyssopifolia L.   Grass Poly.

A glabrous *annual* 10–25 cm. Stem erect or decumbent, usually much-
branched. Lvs 1–1·5 cm., lower oblong-ovate, ± obtuse, often opposite, upper
linear or linear-lanceolate, subacute, alternate, all sessile and rounded or
narrowed at base. *Fls c. 5 mm., solitary* in the axils of the lvs, subsessile, not
heterostylous. Bracteoles persistent. Calyx-tube c. 4 mm., narrowly funnel-
shaped, teeth c. 0·75 mm. *Petals* c. 2 mm., *pale pinkish-lilac*. Stamens 6.
Capsule c. 4 mm., subcylindrical, enclosed in the calyx. Fl. 6–7. $2n = 20$. Th.

Native. On bare ground or among open vegetation in hollows where water
lies in winter, very local and uncertain in appearance from year to year.
15, S. Recorded from a number of counties, undoubtedly a casual in most,
but probably native in some southern and eastern districts. C. and S. Europe,
W. Asia, Africa, America, Australia.

2. PEPLIS L.

Small weak annual herbs. Lvs alternate or opposite, entire. Fls solitary,
small, most often 6-merous. *Calyx campanulate.* Petals fugacious. Stamens
inserted about the middle of the calyx-tube; filaments short, included. Ovary
2-locular; style short, stigma capitate.
Three spp. in north temperate regions.

1. P. portula L.   'Water Purslane.'

A glabrous *creeping* annual 4–25 cm. *Stems* 4-angled, branched, *rooting
freely at nodes. Lvs* 1(–2) cm., *obovate-spathulate*, opposite, rather thick;
stipules minute, gland-like. Fls c. 1 mm., subsessile, solitary in the axils of

almost all the lvs. Calyx-tube obconical; teeth 6, triangular, about as long as tube, appendages longer or shorter, setaceous. Petals 6 (5 or 0), fugacious. Stamens 6 or 12. Capsule c. 1·5 mm., subglobose. Fl. 6–10. $2n=10$. Th.

Native. At muddy margins of pools or puddles, in open communities, or, more often, on bare ground, locally common but always absent from calcareous soils. 105, H 34, S. Scattered throughout the British Is. from Orkney southwards, rare in Scotland. Europe, chiefly central and northern but not arctic, east to the Caucasus.

64. THYMELAEACEAE

Trees or shrubs, rarely herbs. Lvs usually alternate, entire, sessile or shortly stalked, exstipulate. Infl. a raceme, spike, head, umbel, or fls solitary. Fls regular, usually hermaphrodite, 4–5-merous, with an elongated, often coloured receptacle. Calyx often coloured like the receptacle. Petals scale-like or 0. Stamens usually twice as many as sepals, in 2 whorls, sometimes as many, rarely only 2. Ovary superior, 1-celled with a single apical pendulous ovule, rarely 2-celled; style simple or 0; stigma usually capitate. Fr. a nutlet, drupe or berry, very rarely a capsule. Seed with a straight embryo; endosperm present or 0.

About 40 genera and 500 spp., cosmopolitan (except the Arctic). Usually easily recognized by the elongated, often coloured, receptacle and single ovule. Placed by Bentham & Hooker among the apetalous families, but its true affinity seems to be with the Lythraceae which has a similar receptacle.

1. DAPHNE L.

Shrubs. Lvs usually alternate, shortly stalked. *Fls* terminal or axillary in short racemes or umbels, hermaphrodite. *Receptacle* brightly coloured or green, *cylindric or campanulate*, usually caducous, *without or with an inconspicuous scale-like or ring-like outgrowth at the base of the ovary*. Sepals 4, well-developed, coloured like the receptacle. *Petals* 0. *Stamens 8, inserted near the top of the receptacle*; filaments very short. Ovary 1-celled; *stigma large, capitate, sessile or nearly so*. Fr. a drupe, endosperm sparse.

About 50 spp. in Europe and Asia. Several spp. are cultivated.

Fls purple, appearing before the lvs, in 2–4-fld clusters in the axils of the
    fallen lvs of the previous year; lvs thin, light green; fr. red.   **1. mezereum**
Fls green, in short racemes in the axils of the persistent lvs; lvs coriaceous,
    dark green; fr. black.   **2. laureola**

1. D. mezereum L.   Mezereon.

Deciduous shrub 50–100 cm., with few ± erect branches. *Lvs* 3–10 cm., oblanceolate, acute, *light green, not coriaceous*, glabrous. *Fls* 8–12 mm. diam., *appearing before the lvs*, 2–4 *in a subsessile cluster in the axil of a fallen lf of the previous year, purple* rarely white, pubescent outside, fragrant. Sepals about as long as tube of receptacle. *Fr.* 8–12 mm., ovoid, *scarlet*. Fl. 2–4. Pollinated by Lepidoptera and long-tongued bees, probably often selfed. Fr. 8–9. $2n=18$. N.

Native. Woods on calcareous soils in England from Yorks and Westmorland southwards (not in Devon or Cornwall) very local and rare; has probably decreased considerably owing to transference to cottage gardens where it is often grown; a rare alien in Wales and Scotland. 31. Europe from Scandinavia (67° 15′ N.) to C. Spain, the Apennines and N. Greece; temperate Asia (east to the Altai Mountains).

2. D. laureola L.   Spurge Laurel.

Evergreen shrub 40–100 cm., with erect little-branched stems, the lvs often all near the top of the plant. *Lvs* 5–12 cm., obovate-lanceolate or oblanceolate, acute or subacute, *dark glossy green, coriaceous*, glabrous. *Fls* 8–12 mm., *in short axillary* 5–10-*fld racemes, green*, glabrous outside; peduncle c. 1·5 cm.; bracteoles obovate-lanceolate, obtuse, caducous; pedicels 1–3 mm. Sepals $\frac{1}{3}$–$\frac{1}{2}$ as long as tube of receptacle. Anthers yellow. *Fr.* c. 12 mm., ovoid, *black*. Fl. 2–4. Pollinated by Lepidoptera and humble bees. $2n = 18$. N.

Native. Woods, mainly on calcareous soils, widespread and rather common in England, extending north to Lancashire and S. Northumberland but seldom abundant; N. Wales, Pembroke, Glamorgan; Channel Is.; alien in Scotland. 61, S. W. and S. Europe from Belgium and W. Germany to Spain, Corsica and Macedonia; Asia Minor; N. Africa (rare); Azores.

D. laureola × *mezereum* is reported.

65. ELAEAGNACEAE

Trees or shrubs, densely covered with peltate or stellate silvery-brown scale-like hairs. Lvs entire, exstipulate. Fls solitary or in few-fld clusters or racemes, actinomorphic, hermaphrodite or variously unisexual, regular, with a well-developed receptacle. Sepals 2 or 4, valvate. Petals 0. Stamens equal in number to sepals and alternate with them, or twice as many. Ovary superior, 1-celled with a single basal anatropous ovule; style long; stigma simple. Fr. drupe-like consisting of the dry true fr. surrounded by the lower half of the receptacle which becomes fleshy; endosperm 0 or scanty; embryo straight.

Three genera and about 30 spp., north temperate and subtropical regions, extending to tropical E. Asia.

Species of *Elaeagnus* L., with 4 sepals and stamens, hermaphrodite or polygamous fls and broad lvs, are often cultivated.

Resembling Thymelaeaceae but easily distinguished by the characteristic indumentum and basal ovule.

1. HIPPOPHAË L.

Thorny deciduous shrubs or trees. *Lvs alternate*, narrow. Fls dioecious borne on the wood of the previous year, the female in short axillary racemes, the axis often subsequently developing into a thorn; the male in short spikes, the axis usually deciduous. Female fl. with conspicuous elongated receptacle, bearing 2 minute sepals at the apex, disk 0. Male fl. with short receptacle and *two* large *sepals*, disk small. Stamens 4. Style filiform; stigma cylindric.

Two spp., the second in the Himalaya.

1. H. rhamnoides L. Sea Buckthorn.

Much-branched shrub 1–3 m., suckering freely. Lvs 1–8 cm., linear-lanceolate, subsessile, covered with silvery scales on both sides or becoming subglabrous and dull green above. Fls before the lvs, greenish, very small. Fr. 6–8 mm., subglobose or ovoid, orange. Fl. 3–4. Wind pollinated. Fr. 9, often persisting through the winter. $2n = 24$. M. or N.

Native. Fixed dunes and occasionally on sea cliffs. From Yorkshire to Sussex, local but sometimes dominant; often planted on other parts of the coast. 8. Coasts of Atlantic and Baltic from Norway (67° 56′ N.) to the Channel; S. and E. Spain; river shingles in C. Europe in the Rhone Valley, Alps, etc.; Black Sea coast; temperate Asia to Kamchatka, Japan and the N.W. Himalaya.

66. ONAGRACEAE

Annual or more usually perennial herbs, and often marsh or water plants; sometimes woody as in the genus *Fuchsia*. Lvs spirally arranged, opposite and decussate, or whorled; simple, usually exstipulate. Fls solitary, axillary, or in racemes; usually hermaphrodite but sometimes dioecious in *Fuchsia*; usually actinomorphic but zygomorphic in *Chamaenerion*, *Lopezia*, *Zauschneria*, etc.; epigynous and usually also perigynous, there being an often brightly coloured perigynous tube (*hypanthium* or 'calyx-tube'); commonly 4-merous, sometimes 2-merous or 5-merous. Calyx of 2, 4 or 5 free *valvate sepals*; petals 2, 4 or 5, free, rarely 0; stamens 1, 2, 4, 5 (10) or, most commonly, 8 in two whorls; *ovary inferior*, (1–)2-, 4- or 5-celled with 1–many anatropous ovules in each cell; style single with a ± entire stigma. Nectaries are usually present and the fls are visited by insects, many being pollinated by night-flying moths. Some *Fuchsia* spp. are wind-pollinated, but most are pollinated by humming-birds; some are di- or tri-stylic. Fr. usually a loculicidal capsule, but a berry in *Fuchsia* and indehiscent with barbed bristles in *Circaea*; seeds non-endospermic, with a chalazal plume of hairs in *Epilobium*, *Chamaenerion* and *Zauschneria*.

About 470 spp. in 19 genera.

Commonly cultivated in this country are several species of the woody genus *Fuchsia* (q.v.); *Zauschneria californica* C. Presl, suffruticose, with brilliantly scarlet fuchsia-like fls; *Oenothera* spp. (q.v.); *Clarkia unguiculata* Lindl. (*C. elegans* Dougl.) and its hybrids, resembling *Epilobium* but with non-plumed seeds and differing from *Oenothera* in the basifixed, not versatile, anthers; *Godetia* spp., now included in *Clarkia* from which only the non-clawed petals distinguish them; and *Gaura* spp., tall herbs with spikes or racemes of usually white fls and indehiscent fr. lacking barbed bristles. (*Trapa*, with 3 spp. of free-floating aquatic annuals from C. Europe to Tropical Africa and E. Asia is best excluded from Onagraceae because of its ½-inferior ovary and different embryo-development. *T. natans*, 'Water Chestnut', has floating lvs with swollen spongy stalks and caltrop-like mature fr. with a hard wall and 4 spiny projections.

1	Shrubs with pendulous crimson and violet fls.	5. FUCHSIA
	Herbs.	2
2	Petals 0; stamens 4; a ±creeping aquatic herb (resembling *Peplis*) with solitary axillary fls only 3 mm. diam.	1. LUDWIGIA
	Petals 2 or 4.	3

3 Petals 2, white; stamens 2; fr. indehiscent, covered with barbed bristles.
                                                                        6. CIRCAEA
   Petals 4; stamens 4+4; fr. a capsule.   4
4 Fls yellow; seeds not plumed.   4. OENOTHERA
   Fls rose or white; seeds plumed.   5
5 Lvs all spirally arranged; fls held horizontally; petals slightly unequal and
     stamens and styles deflexing, so that fls are somewhat zygomorphic;
     'calyx-tube' 0.   3. CHAMAENERION
   At least the lower lvs in opposite pairs; fls ±erect and actinomorphic;
     'calyx-tube' short.   2. EPILOBIUM

1. LUDWIGIA L.

Chiefly perennial aquatic herbs, often creeping or free-floating. Lvs opposite
or spirally arranged, simple, entire or toothed. Fls solitary axillary, often
forming a terminal lfy raceme; usually 4-merous, not perigynous. Sepals 4,
free; petals 4, free, often minute, or 0; stamens 4, opposite the sepals; ovary
4-celled, with a short style and a capitate 4-furrowed stigma. Fr. a capsule
opening by 4 valves or by apical pores; seeds numerous, non-endospermic,
not plumed.

Twenty to twenty-five spp. in temperate and warmer regions, especially of
N. America.

1. L. palustris (L.) Ell.

Isnardia palustris L.

A perennial aquatic herb with slender glabrous reddish stems 5–30(–60) cm.,
prostrate and rooting below, ascending or floating above. Lvs 1·5–3(–5)×
0·5–2 cm., in *opposite and decussate pairs, ovate to broadly elliptical,* narrowed
abruptly into a short stalk, shortly acuminate, *blunt, entire,* glabrous; lvs of
wholly submerged shoots elliptical to oblanceolate. Fls 3 mm., solitary, axillary, subsessile, with 2 bracteoles. Sepals broadly ovate-acuminate, spreading,
green, often red-margined. *Petals 0.* Stamens shorter than the sepals. Capsule 2·5–3·5 mm., obovoid, 4-angled, crowned by the persistent spreading
calyx. Fl. 6. Self-pollinated. Hel.–Hyd.

Native. Shallow pools in acid fen, very local. Known only from the New
Forest (Hants) and Jersey, but formerly in Sussex. Europe (especially W.
and S.); W. Asia; N. Africa; Cape; N. and C. America.

2. EPILOBIUM L.

Herbs, or rarely suffruticose plants, with *at least the lower lvs in opposite and
decussate pairs* or in whorls of 3; upper lvs or bracts commonly spirally
arranged. *Fls* usually ± erect, solitary, axillary, or in terminal bracteate racemes
or spikes; actinomorphic, 4-merous, perigynous with a *very short 'calyx-
tube',* a break across which causes the shedding of the floral parts on withering.
Sepals free, borne on the rim of the short calyx-tube. *Petals* purple, rose,
or white, *equal,* usually 2-lobed. Stamens 4+4, the antepetalous longer.
Ovary 4-celled with numerous ovules; stigma club-shaped or 4-lobed. *Fr.
a long capsule* dehiscing loculicidally into 4 valves; *seeds* tubercled, each *with
a* chalazal *plume of long hairs,* and thus freely carried by wind.

About 160 spp. in temperate and arctic-alpine regions of both hemispheres.

1 Stem quite prostrate, slender, rooting at the nodes; lvs 2·5–8 mm., almost circular; fls solitary, axillary.　　**12. nerterioides** (see also **pedunculare**)

   Stem not wholly prostrate; fls in terminal and axillary racemes, rarely solitary and then terminal.   2

2 Stem with spreading hairs, both glandular and non-glandular; stigma 4-lobed.   3

   Stem glabrous or with ± appressed non-glandular hairs (spreading hairs, if present, glandular only); stigma either 4-lobed or entire and club-shaped.   4

   (Stem with hairs spreading for part of their length, then turning upwards. Hybrids: see p. 477.)

3 Lvs usually 6–12 cm., somewhat clasping and slightly decurrent; petals more than 1 cm., deep rose.   **1. hirsutum**

   Lvs usually 3–7 cm., neither clasping nor decurrent; petals less than 1 cm., pale rose.   **2. parviflorum**

4 Stigma 4-lobed.   5

   Stigma entire, club-shaped.   6

5 All lvs opposite, ovate-lanceolate, short-stalked, toothed throughout, including the rounded base.   **3. montanum**

   Only the lower lvs opposite; lvs lanceolate, stalked, toothed except for the entire cuneate base.   **4. lanceolatum**

6 Lvs usually with stalks c. 0·5 cm.   **5. roseum**

   Lvs sessile or short-stalked.   7

7 Stem with no ridges or raised lines decurrent from the lvs, though sometimes with 2 rows of crisped hairs; lvs narrowly lanceolate-elliptical; slender below-ground stolons, ending in bulbils, produced in summer.   **9. palustre**

   Stem with 2–4 ridges or raised lines decurrent from the lvs; lvs usually lanceolate to ovate or, if narrowly lanceolate, then not markedly narrowed at base.   8

8 Herbs with erect or ascending stems, usually 25–80 cm.; top of infl. and buds erect.   9

   Small alpine herbs with procumbent or ascending stems, 5–20(–30) cm.; top of infl. nodding in fl. and young fr.   11

9 Stem with numerous slender spreading glandular hairs as well as appressed crisped hairs; fls 4–6 mm. diam.   **6. adenocaulon**

   Spreading glandular hairs 0 or only on the 'calyx-tube'; fls 6–12 mm. diam.   10

10 Glandular hairs present on 'calyx-tube'; capsule 4–6 cm.; elongating above-ground stolons formed in summer.   **8. obscurum**

   Plant wholly without glandular hairs; capsule 7–10 cm.; almost sessile lf-rosettes formed at base of stem in autumn.   **7. tetragonum**

11 Stem very slender (1–2 mm. diam.); lvs 1–2 cm., yellowish-green, ± lanceolate; stolons above-ground, with small green lvs; fls 4–5 mm. diam.   **10. anagallidifolium**

   Stem 2–3 mm. diam.; lvs 1·5–4 cm., bluish-green, ± ovate; stolons below ground, with distant yellowish scales; fls 8–9 mm. diam.   **11. alsinifolium**

1. E. hirsutum L.   Great Hairy Willow-herb, Codlins and Cream.

A tall perennial herb producing in summer *white fleshy underground stolons.*
Stem 80–150 cm., erect, almost terete, branching above, ± densely glandular-

pubescent and *with numerous spreading hairs. Lvs* 6–12 × 1·5–2·5 cm., smaller above and on branches, mostly opposite, oblong-lanceolate, acute, *sessile, semi-amplexicaul and slightly decurrent*, ± hairy on both sides and especially on the veins, the margins ciliate and with distant, unequal, slender incurved teeth. Bracts alternate, resembling the upper lvs. *Fls* 15–23 mm. diam., *erect in bud*, in ± corymbose racemes terminating the main stem and branches. Sepals 7–9 mm., hooded and shortly apiculate. Petals 12–16 mm., deep purplish-rose, broadly obovate, distinctly but not deeply notched. Outer stamens twice as long as inner. *Stigma of 4 revolute lobes exceeding the longest stamens.* Capsule 5–8 cm., downy. Seeds c. 1 mm., oblong-obovoid, acute at the base, brownish-red, densely and acutely tubercled. Fl. 7–8. Protandrous and visited chiefly by bees and hover-flies; ± homogamous forms with stigmas about equalling the stamens have been reported. $2n=36$. H. Variable in hairiness from densely villous to subglabrous.

Native. Stream-banks, marshes, drier parts of fens, etc., to 1200 ft. in Derbyshire. 100, H40, S. Common throughout most of Great Britain and reaching Caithness, but absent in the extreme N.W.; not in Hebrides, Orkney or Shetland. Europe northwards to S. Sweden; temperate Asia; N., E. and S. Africa. Introduced in N. America.

2. E. parviflorum Schreb.   'Small-flowered Hairy Willow-herb.'

A fairly tall perennial herb producing in autumn basal lf-rosettes, often reddish, which are at first ± sessile but later terminate *short above-ground lfy stolons. Stem* 30–60(–90) cm., erect, ± terete, glandular-pubescent above and ± covered *with soft short spreading hairs* throughout; branching above. *Lvs* commonly 3–7 × 1–1·5 cm., but variable in size and usually smaller above and on branches; the lower ones opposite but those above the middle usually alternate; oblong-lanceolate, acute, rounded but *neither amplexicaul nor decurrent* at the sessile base, softly hairy on both sides, the margins shortly ciliate and ± denticulate with short and distant horny teeth. Bracts alternate, like the lvs but smaller. *Fls* 6–9 mm. diam., *erect in bud*, in ± corymbose terminal racemes. Sepals 4–6 mm., acute. Petals 6–9 mm., pale purplish-rose, obovate, deeply notched. *Stigma of 4 non-revolute spreading lobes, about equalling the stamens.* Capsule 3·5–6·5 cm., subglabrous to downy. Seeds 0·9–1 mm., narrowly obovoid, rounded at the base, brownish-red, densely and acutely tubercled. Fl. 7–8. Homogamous. Visited sparingly by hive-bees, etc., and often self-pollinated. $2n=36$. H.–Ch. Very variable in hairiness. The subglabrous form has been named *E. rivulare* Wahl. but does not deserve specific rank. Sometimes the lvs are almost all alternate.

Native. Stream-banks, marshes and fens, to 1200 ft. in Derby. Common; throughout the British Is. except Shetland. 110, H40, S. Europe northwards to S. Sweden; N. Africa; W. Asia to India.

3. E. montanum L.   'Broad-leaved Willow-herb.'

A perennial herb producing in late autumn short stolons which may be underground with fleshy pink and white scales, or above ground, very short and terminating in subsessile lf-rosettes. Stem (5–)20–60 cm., erect, slender, terete, often reddish, subglabrous, or with ± sparse short curved hairs. *Lvs* commonly 4–7 × 1·5–3 cm., mostly opposite or occasionally in whorls of 3, ovate-lanceolate to ovate, acute, rounded at the *short-stalked* base, the stalks to 6 mm., and with narrow slightly connate wings, sharply and irregularly toothed, subglabrous but usually hairy on the margins and veins. Bracts alternate, like the lvs but smaller. Fls 6–9 mm. diam., in terminal lfy racemes, ± *drooping in young bud*. Sepals 5–6·5 mm., ± acute, often reddish. Petals 8–10 mm., pale rose, longer than broad, deeply notched. *Stigma of 4 short non-revolute lobes*, exceeded by the longer stamens. Capsule 4–8 cm., downy with short curved hairs. Seeds 1–1·2 mm., reddish-brown, narrowly obovoid, blunt at the base, densely but shortly tubercled. Fl. 6–8. Homogamous. Sparingly visited by insects and commonly self-pollinated. $2n = 36^*$. H.–Ch.

Native. In woods on the more base-rich soils, hedgerows, walls, rocks, and as a weed in gardens; to 2600 ft. in Wales. 112, H40, S. Common throughout the British Is. Europe to Norway and Finland; W. Asia, Siberia and Japan.

4. E. lanceolatum Seb. & Mauri   'Spear-leaved Willow-herb.'

A fairly tall perennial herb producing in late autumn short above-ground stolons terminating in spreading lf-rosettes which appear subsessile. Stem 20–60(–90) cm., erect, simple or slightly branched, with 4 hardly raised lines; subglabrous below and downy with short crisped hairs above. *Lvs* commonly 2–5 × 0·8–1·6 cm., only the lower usually opposite, elliptical to elliptical-lanceolate, blunt, *cuneate* at the base, narrowing gradually into a *stalk 4–8 mm.*; hairy on the veins and margins, and with small ± equal rather distant marginal teeth except at the entire base. Bracts alternate, like the lvs but smaller. Fls 6–7 mm. diam., in a terminal raceme drooping in bud. Sepals c. 4 mm., lanceolate, ± acute, often reddish. Petals 6–8 mm., pale pink becoming deeper, shortly 2-lobed. *Stigma of 4 short spreading lobes*. Capsule 5–7 cm., downy. Seeds c. 1 mm., reddish-brown, narrowly oblong-obovoid, finely tubercled. Fl. 7–9. Probably self-pollinated. $2n = 36^*$. H.–Ch.

Native. Roadsides, railway-banks, walls, dry waste places, etc. 24, S. S. and S.W. England from Cornwall, Devon, Dorset, Somerset, Gloucester and Monmouth to Sussex, Surrey, and Kent; Glamorgan. W. and S. Europe from France and Belgium to the Balkans; N. Africa; Caucasus and the Near East.

5. E. roseum Schreb.   'Small-flowered Willow-herb.'

A rather slender perennial herb producing in late autumn lax subsessile lf-rosettes from very short stolons. Stem 25–60(–80) cm., erect, fragile, ± branched, usually with two distinct and two indistinct raised lines; glabrous below but with whitish crisped hairs and numerous spreading *glandular hairs* above. Lvs c. 3–8 × 1·5–3 cm., the lower opposite or all alternate, ovate-elliptical to lanceolate-elliptical, narrowed both to the acute apex and to the cuneate base and ± *long stalk*, 3–20 mm.; *glabrous* or hairy only on the veins, the margin finely and sharply toothed. Fls 4–6 mm. diam., their buds cuspidate, drooping. Sepals 3–3·5 mm., lanceolate, acute. Petals 4–5 mm., at first white then streaked with rose-pink, shortly 2-lobed. *Stigma entire*, about equalling the style. Capsule 4–7 cm., downy with crisped and glandular hairs. Seeds c. 1 mm., oblong-obovoid, ± blunt at the base, finely tubercled. Fl. 7–8. $2n = 36$. H.–Ch.

Native. Damp places, woods and copses, railway banks and cultivated ground; lowland. 66, H5, S. Throughout lowland Great Britain northwards to Perth and Angus. N.E. Ireland. Europe northwards to Norway and S. Sweden; Asia Minor.

Distinguished from other British spp. by the long-stalked lvs, from *E. montanum* and *E. lanceolatum* also by the entire stigma, from *E. adenocaulon* by the cuneate base of the lvs and the paler fls, and from the next 3 spp. by the crisped and glandular hairs on the upper part of the stem.

*6. E. adenocaulon Hausskn.

A tall perennial *non-stoloniferous* herb, usually reddish, with numerous ± erect branches above, abundant small fls, and sessile or subsessile basal rosettes, formed in late summer, having at first small fleshy rounded lvs but later developing normal lvs. Stem (30–)60–90(–150) cm., strictly erect, with 4 raised lines except at the base where there are only 2; ± glabrous below, ± densely clothed with *crisped hairs* and with *short spreading glandular hairs above*. Lvs 3–10 × 1·8–3 cm., all but the uppermost opposite, oblong-lanceolate, gradually tapering to an acute apex, suddenly narrowed at the rounded to subcordate base into a *short-stalk* (1·5–3 mm.); ± glabrous, with numerous small irregular forwardly directed teeth. Fls 4–6 mm. diam., erect in bud, numerous, in racemes terminating the main stem and branches. Petals 3–6 mm., pale pink edged with purplish-rose, divided almost half-way into 2 ± parallel lobes. *Stigma entire, less than half the length of the style.* Capsule 4–6·5 cm., spreading, covered with crisped and glandular hairs or becoming ± glabrous when ripe. Seed c. 1 mm., pale reddish-brown, acute at the base, when ripe with a usually paler translucent beak below the plume of hairs; densely tubercled. Fl. 6–8. Probably self-pollinated. $2n = 36^*$. H.–Ch.

Introduced. First record 1891. Damp woods, copses, stream-sides, railway-banks, gardens, waste places; spreading rapidly, especially in S.E. England. 31. From Hants to Kent and northwards to Wilts, Gloucester, Hereford and Worcester, and to Essex, Bedford, Huntingdon and Derby. Reported also from Coll (Inner Hebrides) and S. Uist (Outer Hebrides). Well established in Scandinavia, Poland, and the Baltic States. Native in N. America.

Very close to the N. American *E. glandulosum* Lehm., but a much less robust plant. Distinguishable from *E. roseum* by the shorter-stalked lvs with a rounded base, and the deeper-coloured smaller fls; and from *E. adnatum*, and *E. obscurum* by the numerous glandular hairs (few or 0 in those spp.), the crisped (not ± appressed) non-glandular hairs, and the stigma much shorter than the style instead of ± equalling it as in those spp. The pellucid beak of the seed is found elsewhere amongst British spp. only in *E. palustre*, *E. anagallidifolium* and *E. alsinifolium*.

7. E. tetragonum L.   'Square-stemmed Willow-Herb.'
E. adnatum Griseb.

A fairly stout narrow-leaved perennial herb producing in late autumn several lax lf-rosettes on very short stolons. *Stem* 25–60(–80) cm., erect, *firm*, tough, branched, with 4(–2) conspicuously raised lines or wings; ± glabrous below but downy above with silky ± appressed whitish hairs. *Lvs* 2–7·5 × 0·3–1 cm., the lower and usually the middle lvs opposite, the rest alternate, *strap-shaped to narrowly oblong-lanceolate*, blunt at the apex, narrowed to the ± *sessile* base and decurrent into the raised lines of the stem; *shining* and greasy-looking above, glabrous or slightly hairy on the veins, the margin strongly and irregularly denticulate. Bracts alternate, like the lvs but smaller. Fls 6–8 mm. diam.; buds erect, acute. Sepals c. 4 mm., lanceolate, acute; '*calyx-tube' without glandular hairs*. Petals 5–7 mm., pale lilac, shallowly notched. *Stigma entire, equalling the style.* Capsule 7–9(–11) cm., downy with short crisped hairs. Seeds 1–1·2 mm., reddish, densely but rather bluntly tubercled. Fl. 7–8. Homogamous, the anthers dehiscing before the fls open. Self-pollinated. H.–Ch.

The above description is of ssp. **tetragonum**, the widespread form. A much less frequent form, perhaps confined to the southern half of Great Britain, is often treated as a separate sp. (*E. lamyi* F. Schultz) but seems to merit no more than subspecific rank as ssp. **lamyi** (F. Schultz) Léveillé. It differs from ssp. *adnatum* in having at least the *upper lvs narrowed into a short stalk*. The lower lvs are somewhat shorter (2–5 cm.), less acute and less strongly toothed than in ssp. *adnatum* and the fls are often somewhat larger (10–12 mm. diam.), but these differences are by no means constant.

Native. Damp woodland clearings and hedgebanks, stream and ditch-sides etc., and cultivated ground; lowland. 69, S. Great Britain northwards to Inverness and Argyll, locally common in the south but rare in the north; casual and rare in Ireland. Europe northwards to S. Sweden; W. Asia.

Ssp. *lamyi* in S. England from Cornwall and Kent northwards to Brecon, Hereford, Worcester and Leicester, eastwards to Bedford and Huntingdon; Outer Hebrides. Europe northwards to S. Sweden; Asia Minor; Madeira.

E. obscurum is distinguished by the presence of glandular hairs on the 'calyx-tube', by its shorter capsules and by the elongated stolons which it produces in summer.

8. E. obscurum Schreb.

A tall perennial herb producing in late summer slender ± *elongated stolons* above- or below-ground, bearing ± distant pairs of small lvs which do not form distinct rosettes. Stem 30–60(–80) cm., erect from a curving base, with 4(–2) distinctly raised lines; glabrous below but somewhat downy above with ± appressed hairs. *Lvs* commonly 3–7 × 0·8–1·7 cm., the lower opposite, the upper usually alternate, lanceolate to ovate-lanceolate or the lower oblong-lanceolate, tapering to a usually blunt apex, rounded at the *sessile* base and suddenly contracted to become *decurrent* into the raised lines of the stem; *dull* above, glabrous or hairy only on the margins and veins, the margins with a few distant irregular small teeth. Bracts alternate, like the lvs but smaller. Fls 7–9 mm. diam.; buds erect, acute. Sepals 3–4 mm., lanceolate, acute; '*calyx-tube' with glandular hairs*. Petals 5–6 mm., deep rose, shortly 2-lobed. Stigma entire, equalling the style. Capsule 4–6 cm., downy with short crisped hairs. Seeds 1 mm., reddish, densely and acutely tubercled. Fl. 7–8. Homogamous and self-pollinated. H.–Ch.

Native. Marshes, stream- and ditch-banks, moist woods; to 2550 ft. in

Ireland. 111, H40, S. Locally common throughout the British Is. except Orkney and Shetland. Europe northwards to C. Norway; N. Africa; Madeira; Caucasus.

Recognizable by the glandular 'calyx-tube', the elongated summer stolons and the short capsules.

9. E. palustre L.   'Marsh Willow-herb.'

A perennial narrow-leaved plant producing in summer *filiform hypogeal stolons* bearing distant pairs of yellowish scale-lvs and terminating in autumn in a *bulbil-like bud* with fleshy scales. Stem 15–60 cm., erect from a curved base, simple or branched, *terete* and without raised lines but often with 2 rows of crisped hairs; subglabrous or downy with short crisped hairs. Lvs 2–7 × 0·4–1(–1·5) cm., mostly opposite, lanceolate to linear-lanceolate, narrowed to the blunt apex, *cuneate* at the base, ± sessile or the uppermost very shortly stalked; sometimes limp and somewhat drooping, subglabrous or with crisped hairs on the margins and veins, entire or very obscurely denticulate. Bracts alternate, like the lvs but smaller. Fls 4–6 mm. diam., held almost horizontally; buds blunt, initially erect, soon drooping so that the top of the raceme hangs over to one side. Sepals c. 4 mm., lanceolate, acute. Petals 5–7 mm., pale rose or lilac, rarely white, shortly notched. Stigma entire, shorter than the style. Capsule 5–8 cm., downy with short crisped hairs. Seeds 1·6–1·8 mm., ± fusiform but narrowed from above the middle to an acute base, rounded above and with a short beak formed by the projecting inner integument; pale reddish-brown, finely and acutely tubercled. Fl. 7–8. $2n = 36^*$. H.

Native. In marshes and acid fens, ditches, etc.; calcifuge; reaching 2500 ft. in Ireland. 112, H40, S. Locally common throughout the British Is. Europe northwards to Iceland and Lapland; Asia; N. America and Greenland.

10. E. anagallidifolium Lam.   'Alpine Willow-herb.'

E. alpinum L., p.p., et auct. plur.

A perennial alpine herb with a stem 4–10(–20) cm., slender (1–2 mm. diam.), ascending from a decumbent base, ± glabrous except for 2 lines of hairs down the 2 faint ridges; producing in summer numerous ± prostrate slender *epigeal stolons* at first forming small subsessile rosettes but soon elongating and then with *distant pairs of small green lvs*. Lvs usually 1–2 cm., often yellowish-green, mostly opposite, lanceolate or elliptical-lanceolate, gradually narrowing into a short stalk-like base, entire or distantly and faintly sinuate-toothed, ± glabrous. Fls c. 4–5 mm. diam., 1–3, the top of the raceme drooping in fl. and young fr. Sepals 2·5–4 mm., reddish. Petals 3·5–4·5 mm., rose-red. Ripe capsule 2·5–4 cm., reddish, on an erect stalk 2·5–5 cm. Seeds c. 1 mm., obovoid, not beaked, acute below, very faintly tuberclate. Fl. 7–8. Probably self-pollinated. $2n = 36$. H.

Native. By streams and springs on mountains, commonly with *Montia fontana* and *Philonotis fontana*; from 500 to 4000 ft. in Scotland. Great Britain northwards from N.W. Yorks and Durham; Inner and Outer Hebrides. 23. N. Europe and mountains of C. Europe; Asia; Greenland; N. America.

11. E. alsinifolium Vill. 'Chickweed Willow-herb.'

E. alpinum L., p.p.

A perennial alpine herb with a stem 5–20(–30) cm., rather slender (2–3 mm. diam.), ascending from a decumbent base, ± glabrous except for 2 rows of hairs down the 2 faint ridges and a few crisped hairs above; producing in summer slender *yellowish hypogeal stolons* with *distant pairs of yellowish scale-lvs*. Lvs usually 1·5–4 cm., somewhat bluish-green and shining above, mostly opposite, ovate to ovate-lanceolate, rounded at the base and narrowed into a short stalk, distantly sinuate-toothed, ± glabrous. Fls 8–9 mm. diam., usually 2–5, the top of the raceme drooping in fl. and young fr. Sepals c. 4–6 mm. Petals 7–9 mm., bluish-red. Ripe capsule 3–5 cm., subglabrous, on an erect stalk 2–3 cm. Seeds c. 1 mm., narrowly obovoid with an apical beak upon which the plume of hairs is borne, minutely tuberclate. Fl. 7–8. Probably self-pollinated. H.

Native. By streams and springs on mountains; from 400 ft. on Eigg to 3600 ft. in Perth. Caernarvon, and northwards from Durham, N.W. Yorks and Cumberland; Inner and Outer Hebrides; Shetlands. Ireland (one locality in Leitrim). 27, H1. N. and C. Europe.

*12. E. nerterioides A. Cunn.

E. pedunculare auct., non A. Cunn.

A *prostrate* perennial herb with slender creeping stems to 20 cm., *rooting at the nodes*. Lvs 2·5–8 mm., opposite, broadly ovate to suborbicular, very short-stalked, entire or faintly sinuate-toothed, ± glabrous. Fls 3–4 mm., solitary in the axils of the lvs, on stalks 0·5–4 cm. Sepals 2–3 mm., often reddish. Petals 3–4 mm., pink. Ripe capsule 2–4 cm., glabrous, on an erect stalk to 6 cm. Seeds c. 0·7 mm. Fl. 6–7. Probably self-pollinated. $2n = 36$. H.

Introduced. On moist stony ground, rocky beds and sides of streams, etc. Spreading rapidly especially in N. and N.W. England where it is already well established in the Lake District, N.W. Yorks and Westmorland, the N. Yorkshire Moors. N. Wales, Scotland, Ireland. Native in New Zealand.

*E. linnaeoides Hook. f. is naturalized in several places. Like *E. nerterioides* but lvs sharply denticulate and fr.-stalks 5–10 cm.

Hybrids arise freely within the genus *Epilobium*, a high proportion of all the possible combinations having been reported from the British Is. They are almost invariably sterile and can be recognized by their failure to set good seed as well as by being intermediate in morphology between the putative parents. Of special interest are the hybrids between species with spreading and appressed hairs respectively, in which the hairs spread for part of their length and then turn upwards.

3. CHAMAENERION Adans.

Perennial herbs resembling *Epilobium* but with *lvs all spirally arranged*, ± entire, and *fls held horizontally*. Perigynous tube very short; petals spreading, the upper 2 broader than the lower 2, so that the corolla is somewhat *zygomorphic*; stamens and style exserted and ultimately bending downwards.

About 8 spp. in temperate and arctic regions of the northern hemisphere.

1. C. angustifolium (L.) Scop.   Rosebay Willow-herb, Fireweed.

Epilobium angustifolium L.

A tall perennial herb with long horizontally spreading roots which give rise to erect lfy stems 30–120 cm., subterete, glabrous below, ± pubescent above. Lvs 5–15 cm., spirally arranged, numerous, ± ascending, narrowly oblong-lanceolate or oblong-elliptical, narrowed at each end, entire or with small and distant horny teeth, often ± waved at the margins, glaucous beneath with conspicuous veins, lateral veins numerous, joining into a continuous wavy intra-marginal vein. Fls 2–3 cm. diam., numerous, in a long rather dense spike-like bracteate raceme; stalks 1–1·5 cm., ascending, pubescent. Sepals 8–12 mm., acuminate, dark purple. Petals 10–16 mm., obovate, clawed, entire or slightly notched, the upper pair broader than the lower, rose-purple. Stigmas finally exceeding the anthers, at first spreading then recurved or revolute. Capsule 2·5–8 cm., ± 4-angled, pubescent, at first spreading, then ± erect. Seeds c. 2 mm., almost smooth, with a white plume. Fl. 7–9. Protandrous; visited by various insects for nectar secreted by the epigynous disk. $2n = 36$. H.

Variable. '*Epilobium angustifolium* var. *macrocarpum* Syme' was distinguished from '*var. brachycarpum* Syme' in fr.-length. The latter was probably a sterile cultivar, thus accounting for its short capsules. Efforts to relate these differences to supposed wild and naturalized forms in Britain have failed. Although certain British populations, especially in mountainous districts, are native the status of the common lowland plants is in doubt. There are, however, no constant morphological differences between these and the undoubtedly native populations.

Native. Rocky places, scree slopes, wood-margins and wood-clearings, disturbed ground, gardens, bombed sites, etc.; to 3200 ft. in Scotland. Throughout the British Is., but commonest in the south. A century ago the species was a local plant though scattered throughout the country, especially in rocky places and on scree. Its phenomenal spread in the last few decades may be related to the increasing areas of cleared woodland and waste land. 108, H 11.

Europe, Asia, N. America. Often cultivated as an ornamental garden plant.

*C. dodonaei (Vill.) Schur (*C. rosmarinifolium* Moench), with linear, indistinctly veined lvs, shorter lfy raceme, and unclawed petals, is sometimes grown in gardens and has been reported as a casual.

4. OENOTHERA L.   Evening Primrose.

Annual to perennial herbs, rarely suffruticose, with *spirally arranged* exstipulate *lvs* and usually large fls, solitary or in small groups in the axils of the upper lvs, forming a large lfy spike. *Fls deeply and narrowly perigynous*, 4-merous. Sepals usually strongly reflexed, often caducous; petals broad, overlapping, commonly yellow, rarely white or rose; stamens 4+4, equal or the inner shorter than the outer; stigma entire or ± 4-lobed. Fls usually opening in the evening when they become fragrant and are visited by moths, but some are self-pollinated or even cleistogamous. Fr. a capsule splitting loculicidally into 4 valves; *seeds not plumed*.

About 100 spp., some of them taxonomically difficult owing to the peculiar genetical situation arising from the formation of chromosome-rings. Originally American but a few have arisen by hybridization in Europe and elsewhere.

The following species are naturalized in this country, especially on dunes, but several others occur not infrequently as casuals.

1 Petals less than 30 mm.   *2*
      Petals more than 30 mm.   *3*

2 Stem bearing long hairs whose bases become enlarged and reddish; stem-lvs narrowly lanceolate, rather fleshy and thick, subglabrous; tips of sepals in bud standing apart down to their bases; petals 11–18 mm.
                                                                          4. parviflora
      Stem usually without red-based hairs (but these present in the related *O. rubricaulis*); stem-lvs ± broadly lanceolate, hairy; tips of sepals in bud closely connivent at their bases; petals 18–25 mm.
                                                                 1. biennis (and **rubricaulis**)

3 Stem with many red bulbous-based hairs; lvs broadly oblong-lanceolate, strongly crinkled, ± pubescent; petals 35–50(–60) mm. remaining yellow throughout.   **2. erythrosepala**
      Stem without red-based hairs; lvs linear-lanceolate, with waved margins; petals 30–40 mm. yellow at first, then turning wine-red.   **3. stricta**

*1. O. biennis L.   Evening Primrose.

A usually biennial herb with a large fleshy root-stock and erect robust lfy stems 50–100 cm., ± pubescent but *without hairs with red bulbous bases*. Basal *lvs narrowly oblanceolate*, stalked, hairy; stem-lvs lanceolate to ovate-lanceolate, subsessile, denticulate; midrib green at first, later turning reddish. *Infl. erect* throughout. Sepals green. Petals 18–25 mm., obcordate, broader than long, yellow throughout. Stigmas equalling the anthers. Capsule 30–35 mm., ± cylindrical, sessile, pubescent but with no red-based bulbous hairs. Fl. 6–9. $2n = 14$. Hs. (biennial).

Introduced. Dunes, roadsides, railway-banks, waste places; naturalized locally; formerly widespread as a casual but apparently decreasing. Great Britain northwards to Perth and Angus; Ireland; Channel Is. Native of N. America but widely established in Europe and New Zealand.

*O. rubricaulis Klebahn resembles *O. biennis* but has bluish-green lvs with reddish veins, the stems and capsules bear red-based hairs and the young infl. is purplish-red. Petals c. 24 mm., about as wide as long. Capsule at first red-striped. Perhaps established on dunes in S.W. England. Not known wild.

*O. grandiflora L-Hérit. (*O. suaveolens* Pers.; *O. biennis* ssp. *grandiflora* (L'Hérit.) Stomps) is a closely related but larger plant whose lvs have white midribs throughout and whose petals are 35–45 mm. The fls have a strong scent of orange-blossom. From *O. erythrosepala*, also large-fld, it may be distinguished by the absence of bulbous-based hairs, the usually non-crinkled lvs, and the thinly and shortly pubescent non-angular fl.-buds.

Introduced. A casual, escaping from gardens to waysides and waste places. Native of N. America.

*2. O. erythrosepala Borbás

O. Lamarkiana De Vries, non Ser.; *O. Vrieseana* H. Lév.

A usually biennial herb with erect robust lfy stems 50–100 cm., ± pubescent
with short hairs and ± densely *clothed in* longer *hairs with red bulbous bases.*
Basal *lvs broadly elliptic-oblanceolate* to ovate-lanceolate, stalked, strongly
crinkled and distantly toothed; stem lvs ovate-lanceolate, subsessile; midrib
usually white, sometimes red. Infl. erect throughout. Fls angular in bud, the
sepals red-striped and bearing both long and short hairs. *Petals* 40–50 mm.,
broader than long, *yellow throughout.* Stigmas exceeding the anthers, even in
bud. Capsule 25–40 mm., tapering upwards, pubescent and with red-based
bulbous hairs. Fl. 6–9. $2n = 14$. Hs. (biennial).

Introduced. Roadsides, railway banks, waste places, dunes; naturalized
locally and apparently spreading. Great Britain; Ireland; Channel Is. Long
believed to have arisen in Europe; probably not native in N. America.
Established locally in C. and N. Europe.

*3. O. stricta Ledeb. ex Link

O. odorata auct.

An annual or biennial herb with an erect simple or sparingly branched slender
stem 50–90 cm., pubescent and glandular but without bulbous-based hairs.
Lvs 3–10 cm., diminishing up the stem; basal *lvs linear-lanceolate,* narrowed
into a stalk-like base; stem lvs broader, subsessile; all with the margins waved
and ciliate and distantly but sharply toothed. Infl. erect throughout. Sepals
hairy, becoming reddish. *Petals* 30–40 mm., *yellow at first then turning red.*
Stigmas about equalling the anthers. *Capsule* 1·8–2·5 cm., *clavate,* with both
silky and glandular hairs. Fl. 6–9. $2n = 14$. Th.–Hs. (biennial).

Introduced. Dunes. Established in several localities northwards to Selkirk,
but especially in S.W. England and the Channel Is. Chile.

*4. O. parviflora L.   'Small-flowered Evening Primrose.'

A biennial to perennial herb with simple or rarely branching stem 10–80 cm.,
sparsely hairy, red below and usually with some spreading hairs with red
bulbous bases, at least above. Rosette-lvs spathulate; *stem-lvs narrowly
lanceolate, ascending, rather fleshy and thick, subglabrous beneath,* grading into
the *persistent lf-like bracts; veins reddish* and young lvs and bracts with
narrow reddish margins. Young infl. somewhat drooping for a short time.
Sepals hairy, at first green but becoming red-streaked, their *tips spreading
and free to the base.* Petals 11–18 mm., obovate, yellow. Ripe capsule
6–10 mm., thick at base. Fl. 6–9. $2n = 14$. Hs.

Introduced. Established on dunes in S.W. England. Eastern N. America;
long introduced into Europe.

The closely related ***O. ammophila** Focke, with neither veins nor margins
of lvs reddish and with the young infl. drooping for a considerable distance
behind the tip, has also been reported from S.W. England.

Other *Oenothera* spp. often recorded as casuals include *O. laciniata* Hill (*O. sinuata*
L.), N. America, a prostrate plant with coarsely toothed or pinnatifid lvs and small
yellow fls later turning reddish; *O. rosea* Ait., western N. America to Peru, a shrubby

plant, 15–40 cm., with ± lyrate lower and elliptical upper lvs, all stalked, small pink fls 1·5 cm. diam. and very characteristic 8-angled clavate fr.; *O. indecora* Cambess. (*O. argentinae* Lév. & Thell.), S. America, with narrow lanceolate toothed lvs, small rose fls and long narrow fr.; and *O. perennis* L. (*O. pumila* L.), E. N. America, 15–50 cm., with entire oblanceolate lvs, small yellow fls open during the day, and small clavate fr.

5. Fuchsia L.

Shrubs or small trees with usually opposite or whorled glabrous simple stalked lvs with deciduous stipules. *Fls* 1 to several in the axils of the lvs, usually stalked and *pendulous*; actinomorphic, epigynous and with a perigynous tube, hermaphrodite or dioecious. *Sepals* 4, *spreading, often red*; *petals* 4, free, *erect*, or 0; *stamens* usually 8, *exserted*; stigma capitate or 4-lobed. Nectar-secreting and pollinated by insects or small birds. *Fr. a* 4-*celled berry* with numerous seeds.

About 100 spp. in tropical and subtropical America and a few in New Zealand; 1 in Tahiti. Many spp. are cultivated but most are frost-tender.

*1. F. magellanica Lam. var. *macrostema* (Ruiz & Pav.) Munz

F. Riccartonii auct.; *F. gracilis* Lindl.

A small shrub to 3 m., with lvs 2·5–5·5 cm., short-stalked, ovate-oblong, acuminate, ± toothed, often purplish, slightly pubescent on the margin and underside of the midrib. Fls solitary, pendulous, their stalks c. 5 cm. Perigynous tube 5–8 mm., red. Sepals 12–15 mm., ovate-lanceolate, red. Petals 6–10 mm., obovate, violet. Stamens much exserted. Berry black, 4-angled. Fl. 6–9. Protandrous. $2n = 22, 44$. N. or M.

Introduced. Much cultivated, especially in the west as a hedge plant, and established in S.W. England and W. Ireland. S. Chile.

Variable, some forms probably of hybrid origin. 'F. gracilis Lindl.', with longer perigynous tube and narrower sepals, is a variant established in Ireland.

6. Circaea L.

Perennial *stoloniferous* herbs with stalked ovate lvs in opposite and decussate pairs and terminal bracteate *racemes* of small white or pinkish shortly perigynous 2-*merous fls*. Sepals 2, free, caducous; petals 2 (front and back), 2-lobed; stamens 2 (lateral); ovary inferior, 1–2-celled, with 1 ovule in each cell; style 1, with an entire, emarginate or 2-lobed stigma. *Fr. indehiscent*, 1–2-seeded, *covered with* ± *hooked bristles*.

Ten spp. in north temperate regions of Europe, Asia and N. America.

1 Infl.-axis not elongating until fls have dropped (open fls in a terminal
   cluster); petals 0·6–1·5 mm.; perigynous tube (hypanthium) c. ¼ as long
   as ovary; ovary 1-celled.   **3. alpina**
   Infl.-axis elongating before fls drop (open fls spaced along axis); petals
   2–4 mm.; perigynous tube (hypanthium) at least ½ as long as ovary;
   ovary with 2 equal or unequal cells.   2

2 Fls with a conspicuous dark disk protruding c. 0·3 mm. above the summit
   of the perigynous tube; lvs not or slightly cordate, sinuate-toothed or

remotely denticulate, gradually acuminate; ovary equally 2-celled; fr.
ripening.   **1. lutetiana**
   Disk inconspicuous or 0; lvs strongly cordate, dentate, shortly acuminate;
ovary unequally 2-celled; fr. rarely ripening.   **2. × intermedia**

1. C. lutetiana L.   Common Enchanter's Nightshade.

Rootstock slender. Stem usually 20–70 cm., erect or ascending, swollen at
the nodes, sparsely glandular-pubescent. *Lvs* usually 4–10 cm., ovate, *gradually
acuminate, rounded or slightly cordate at base* (but distinctly cordate in the
more glabrous f. *cordifolia* Lasch), *sinuate-toothed or remotely denticulate*,
thin, dull above, paler and shining beneath, subglabrous or with the margins
and undersides of the veins hairy; *stalk* 1–10 cm., *furrowed above, not winged,
pubescent all round. Infl.-axis elongating before the fls drop*, so that open fls
are well spaced. Bracteoles usually 0. Fl.-stalks glandular-pubescent, reflexed
in fr. Sepals glandular-pubescent. *Petals 2–4 mm.*, ± rounded at base, *notched
to at least half-way. Stigma deeply 2-lobed.* Fr. c. 3 mm. diam., obovoid,
equally 2-celled, densely covered with stiff hooked white bristles; *fr. ripening
freely*. Fl. 6–8. Visited by small Diptera. $2n = 22$. Grh.
   Native. Woods and shady places on a moist base-rich soil, to 1200 ft. in
Yorkshire; common. Throughout Great Britain; Inner Hebrides; Ireland.
99, H 40, S. Europe and eastwards to Iran and the Altai Mts; N. Africa.

2. C. × intermedia Ehrh.   'Intermediate Enchanter's Nightshade.'
(C. alpina × C. lutetiana)

Intermediate between the parents and frequently confused with *C. alpina*.
Rootstock slender. Stem 15–40 cm., sparsely glandular to glabrous. *Lvs*
usually 3–8 cm., ovate, *abruptly acuminate, distinctly cordate* at base, *strongly
but distantly toothed*, thin, ± translucent, subglabrous; *stalks furrowed above,
rarely winged, pubescent only above. Infl.-axis elongating before the fls drop.*
Bracteoles setaceous. Fl.-stalks sparsely glandular to glabrous, somewhat
reflexed in fr. Sepals sparsely glandular to glabrous. *Petals 2–4 mm.*, cuneate
at base, *notched to at least half-way. Stigma deeply 2-lobed.* Fr. 1·5–3 mm.,
obovoid, *2-celled but with 1 cell small and aborting*, densely covered with
softer bristles than in *C. lutetiana; very rarely setting viable seed.* Fl. 7–8.
$2n = 22$. Grh.
   Native. Shady rocky places; woods in mountainous districts. In W. and
N. Great Britain from Gloucester and S. Wales eastwards to Stafford, Derby,
W. Yorks and Berwick and northwards to Sutherland; Ireland. 52, H 7.
C. and N. Europe.

A hybrid of *C. alpina* and *C. lutetiana*, which only rarely sets fr. and has completely
sterile pollen, so that it can spread only vegetatively. Fertile plants should be
investigated carefully.

3. C. alpina L.   Alpine Enchanter's Nightshade.

Rootstock tuberous. Stem (5–)10–30 cm., erect or ascending, subglabrous.
Lvs 2–6(–8) cm., ovate, *acute or acuminate, cordate* at base, *strongly but distantly toothed*, thin, ± translucent, somewhat shining above, glabrous; *stalk
flat above, winged,* ± *glabrous. Infl.-axis not elongating until after the fls have
dropped*, so that open fls are in a terminal cluster. Bracteoles setaceous.

Fl.-stalks glabrous, little reflexed in fr. Sepals glabrous. *Petals* 0·6–1·5 *mm.*, narrowly cuneate at base, *shallowly notched. Stigma entire or notched. Fr.* 1–1·5 mm. diam., oblong to narrowly obovoid, 1-*celled*, covered with soft bristles which are shorter and less consistently hooked than in *C. lutetiana*; *fr. ripening freely* but tending to fall early. Fl. 7–8. 2*n*=22. Grh.

Native. Local in woods in mountainous districts and shaded rocky places. In western Great Britain from Glamorgan north to N.-W. Yorks, Westmorland and Cumberland; Argyll; Arran. 10. Other records are erroneous. Europe; Asia; N. America. Our plants belong to ssp. *alpina*.

67. HALORAGACEAE

Herbaceous plants, usually aquatic or subaquatic, often very large. Lvs spirally arranged, opposite and decussate or whorled, very variable in size and shape, exstipulate, sometimes with intravaginal scales (*Gunnera*). Fls inconspicuous, solitary axillary, in axillary dichasia or in terminal spikes, racemes or panicles; hermaphrodite or unisexual (monoecious or polygamous), actinomorphic, epigynous. Sepals 4, 2 or 0, free, small; petals free, as many as the sepals, but often much larger, commonly caducous, sometimes 0; stamens 4+4, 4, or 2, free, filaments short; ovary inferior, usually 4-celled with 1 pendulous anatropous ovule in each cell, but 1-celled with 1 ovule in *Gunnera*; styles 1–4, often very short, stigmas 1–4, feathery or coarsely papillose. Apparently anemophilous. Fr. a nut or drupe, sometimes separating into 1-seeded nutlets (*Myriophyllum*). Seed with much endosperm.

About 160 spp. in 7 genera, cosmopolitan but with a concentration in the southern hemisphere.

The relationships of the family are obscure, though the basic floral diagram, as seen in *Haloragis*, is that of the Onagraceae.

Water plants, submerged except for the infl.; lvs whorled, pinnate, with capillary segments. 1. MYRIOPHYLLUM
Large rhubarb-like marsh plants with enormous ±peltate exclusively radical lvs. 2. GUNNERA

1. MYRIOPHYLLUM L.

Perennial *aquatic* herbs, free-floating or with rhizomes in the substratum and the lfy shoots submerged apart from the infl. *Lvs in whorls of* 3–6, exstipulate, *pinnately divided into unbranched capillary segments*; aerial lvs and bracts sometimes simple, toothed or entire. Fls in lfy or bracteate terminal spikes, sessile in whorls in the axils of lvs or bracts; hermaphrodite, polygamous or monoecious, rarely dioecious, the upper fls commonly male, the lower female. Calyx inconspicuous, of 4 small lobes in the male fl., minute in the female fl.; corolla of 4 boat-shaped caducous petals in the male fl., minute or 0 in the female fl.; stamens usually 8, sometimes 4 or 6; ovary 4-celled; style very short or 0; stigmas 4, subsessile, oblong, recurved, persistent. Anemophilous. Fr. separating into 1-seeded nutlets, usually 4, sometimes fewer by abortion.

About 36 spp., cosmopolitan.

1 Lvs usually 5 in a whorl, commonly much exceeding the internodes; specialized winter-buds (turions) formed in autumn; bracts shortly pinnate or pectinate as long as the fls even near the tip of the spike. **1. verticillatum**

Lvs usually 3, 4 or 6 in a whorl, rarely 5, commonly equalling or shorter than the internodes; turions 0; all but the lowest bracts entire or toothed and usually shorter than the fls.   2

2   Lvs in whorls of 4 or 6, the uppermost simple, entire or toothed (like the bracts); stamens 4 (rare alien).   **4. heterophyllum**
     Lvs in whorls of 3 or 4 (rarely 5), even the uppermost pinnate or pectinate; stamens 8.   3

3   Lvs in whorls of 3; bracts broadly ovate to elliptic, serrate to subentire; fls all hermaphrodite; ripe nutlets longitudinally ridged and tubercled on the outer face (rare alien).   **5. verrucosum**
     Lvs mostly in whorls of 4; upper bracts narrowly ovate to lanceolate, entire; some fls unisexual; ripe nutlets smooth or rugose but not tubercled on the outer face.   4

4   Lvs with 6–18 capillary segments; spike short, at first drooping at the tip; lowest bracts lf-like; upper fls often alternate, not whorled; petals yellow with red streaks.   **3. alterniflorum**
     Lvs with 13–35 capillary segments; spike erect throughout; all bracts equalling or falling short of the fls; all fls whorled; petals dull red.
                                                                  2. spicatum

1. M. verticillatum L.   'Whorled Water-milfoil.'

Rhizome elongated, creeping in the muddy substratum. Lfy shoots 50–300 cm., branched. *Lvs* 2·5–4·5 cm., *usually 5 in a whorl*, rarely 4–6, commonly much exceeding the internodes, simply pinnate with 25–35 rather distant segments. Spike 7–25 cm., emergent. *Fls usually in whorls of 5 in the axils of shortly pinnate or pectinate bracts* of very variable length, from little shorter than the lvs to little longer than the fls, but never entire and never shorter than the fls even at the tip of the spike. A few hermaphrodite fls usually present between the male and female fls. Petals of female fls 0; of male fls 4, c. 2·5 mm., greenish-yellow, rarely reddish, caducous. Stamens 8. Fr. c. 2 mm., subglobular, 4-lobed, at length separating into 4 nutlets. Fl. 7–8. Perennation and vegetative reproduction by clavate turions, 0·5–5 cm., with closely appressed lvs. $2n=28$. Hyd.

Native. Ponds, lakes and slow streams of lowland districts, especially in base-rich water; not common. 53, H26. England and Wales; Ireland. Europe northwards to the Arctic Circle; N. Africa; Asia; N. and S. America.

2. M. spicatum L.   'Spiked Water-milfoil.'

A rhizomatous water-plant with branching lfy shoots 50–250 cm., naked below through decay of lvs. *Lvs* 1·5–3 cm., *usually 4 in a whorl*, rarely 3 or 5, about equalling the internodes, simply pinnate *with 13–35 segments*. Spike 5–15 cm., emergent, erect throughout even in bud. Fls usually in whorls of 4 in the axils of *bracts all but the lowest* of which are *entire and shorter than the fls*, the lowest usually pectinate and somewhat larger than the fls. About 4 basal whorls are of female fls with 4 very small petals, then 1 whorl of hermaphrodite fls, the upper whorls being of male fls with larger dull red caducous petals, c. 3 mm. Stamens 8. Fr. subglobular, 4-lobed. Turions 0. Fl. 6–7. $2n=28, 36$. Hyd.

Native. Lakes, ponds, ditches, etc., to 1562 ft. in the Lake District; locally common, especially in calcareous water. 108, H39, S. Throughout most of the British Is. Europe, Asia, N. Africa, N. America.

3. M. alterniflorum DC.   'Alternate-flowered Water-milfoil.'

A rhizomatous water-plant with slender branching lfy shoots 20–120 cm., not exceeding 2 mm. diam., naked below through decay of older lvs. *Lvs* 1–2·5 cm., *usually* 4 *in a whorl*, sometimes 3, about equalling the internodes, simply pinnate with 6–18 *segments* 6–20 mm. long. *Spike* 1–2(–3) cm., emergent, its *tip drooping in bud.* Basal whorl usually of 3 female fls, with rudimentary petals and stamens, in the axils of lf-like pinnate bracts; then other female fls, solitary or in groups of 2–4 in the axils of short pectinate bracts; next herma-phrodite fls, and in the upper half of the spike *male fls, usually c.* 6, *solitary or in opposite pairs in the axils of entire bracts shorter than the fls*; petals 2·5 mm., yellow with red streaks; stamens 8. *Fr.* 1·5–2 mm., *longer than wide*, separating into 4 nutlets. Turions 0. Fl. 5–8. $2n=14$. Hyd.

Var. *americanum* Pugsl. has lvs only 3–5 mm., with segments 2–4 mm.

Native. Lakes, streams, ditches, etc., to 2350 ft. in Wales and Scotland; locally common but especially in the west and north and in base-poor and peaty water. 100, H39, S. Throughout the British Is. Europe eastwards to C. Russia and northwards to S. Scandinavia and Finland; Iceland; Green-land; Azores. Var. *americanum* in base-rich water in Lough Neagh and Lough Beg (N.E. Ireland), and in eastern N. America.

*4. M. heterophyllum Michx.

Rhizomatous, with stoutish stems bearing lvs in whorls of 4 and 6. Sub-merged lvs 2–5 cm., with 15–20 capillary segments; *uppermost* (*emergent*) *lvs lanceolate to elliptic, simple, entire or toothed*; lvs pinnately cut in the transi-tional region. Spike 3–35 cm., emergent, with fls in whorls of 4 or 6, herma-phrodite or the lower female and upper male; bracts lanceolate, entire or toothed. Stamens 4. Fr. 1–1·5 mm., sub-globular, each nutlet beaked and with the outer face 2-ridged and minutely papillose. Fl. 6–9. Hyd.

Introduced. Established in a canal near Halifax (S.W. Yorks). 1. Eastern N. America.

*5. M. verrucosum Lindl.

Rhizomatous, with branching stems to 60 cm., 1–2 mm. diam., glabrous, rooting below: terrestrial forms growing on mud or gravel have decumbent stems only to 10 cm. *Lvs* up to 1 cm., *in whorls of* 3, simply pinnate, with 8–18 capillary segments; lvs on emersed plants often in opposite pairs or alternate. Spike to 18 cm. Fls 1–3 per node, solitary in the axils of *broadly ovate to elliptical bracts* which are usually in whorls of 3, occasionally opposite; *all fls hermaphrodite.* Petals 4, pinkish-green. Stamens 8. Fr. to 1·5 mm., 4-angled, each nutlet with a prominent median longitudinal ridge down the outer face which is also *minutely and bluntly tubercled.* Fl. 7–9. Hyd.

Introduced. In gravel-pits near Eaton Socon (Beds). 1. Australia.

2. GUNNERA L.

Perennial, often gigantic, herbs with creeping rhizomes and stalked ovate or orbicular lvs, exclusively radical, with numerous intravaginal scales. Infl. a large racemose panicle, sometimes very dense and exceeding 2×1 m. Fls monoecious or polygamous, the lower female, upper male, with or without intervening hermaphrodite fls. Sepals 2–3 or 0; petals 2, free, hooded, or 0, stamens 1–2; ovary inferior, 1-celled with 1 ovule; styles and stigmas 2. Fr. a 1-seeded drupe.

About 30 spp. in S. and C. America, S. and S.E. Africa, Madagascar, East Indies, New Guinea, Tasmania and New Zealand. Some of the spp. are cultivated as waterside plants of spectacular dimensions. Frequently cultivated and occasionally escaping are:

*1. G. tinctoria (Molina) Mirb. (*G. chilensis* Lam.; *G. scabra* Ruiz & Pav.) with short rhizome, round, cordate, *palmately-lobed lvs* to 2 m. diam. with hispid stalks, and a compound infl. to 1 m. Chile, Ecuador, Colombia.

*2. G. manicata Linden ex André (*G. brasiliensis* Schindler), a larger plant with a long-creeping rhizome, and orbicular, *peltate, pedately-lobed lvs* sometimes exceeding 2 m. diam., whose stalks bear *red spiny hairs*. S. Brazil.

68. HIPPURIDACEAE

An aquatic herb with whorled linear exstipulate lvs and solitary axillary fls, hermaphrodite, or unisexual, epigynous. Perianth a rim round the top of the ovary. Stamen 1, anterior, median. Ovary inferior, 1-celled; ovule solitary, pendulous, anatropous; integument single; micropyle closed, the pollen-tube passing laterally through funicle and integument; style 1, long and slender with stigmatic papillae throughout its length. Anemophilous. Fr. an achene.

One sp.

Sometimes included in Haloragaceae but differing in many features of vegetative and floral morphology and unlikely to be closely related.

1. HIPPURIS L.

The only genus, with the characters of the family.

1. H. vulgaris L. Mare's-tail.

A perennial usually aquatic herb with a stout creeping rhizome from which arise lfy shoots 25–75(–150) cm. if wholly or partly submerged but only 7–20 cm. in terrestrial forms. *Lvs* 1–7·5 cm. × 1–3·5 mm., 6–12 *in a whorl, linear, sessile, entire, glabrous, with a hard acute tip*; submerged shoots have longer, thinner, more flaccid and more translucent lvs than those of emerged shoots, and the internodes are longer and less rigid, so that the name mare's-tail appears very appropriate, especially to the luxuriant submerged shoots of flowing water. Fls small, greenish, in the axils only of the emerged lvs. Stamen with reddish anther, sometimes 0. Fr. ovoid, smooth, greenish. Fl. 6–7. $2n = 32$. Hyd.

Native. In lakes, ponds and slow streams, to 1800 ft. in Scotland; especially in base-rich water, local. 101, H40, S. Throughout the British Is., Europe; W. and N. Asia; N. Africa.

69. CALLITRICHACEAE

Annual or perennial herbs, usually aquatic or subaquatic, with *filiform stems* and *opposite, entire, linear to ovate, exstipulate lvs*. Monoecious. Fls axillary, either solitary or a male and female fl. in the same axil. Bracteoles 2, crescent-shaped, or 0; *perianth* 0; *stamen* 1, with a long slender filament and reniform anther; ovary syncarpous, 4-celled by secondary septation, 4-lobed; styles 2, long, free, papillose; ovules solitary in each cell, anatropous, pendulous; integument 1. Fr. 4-lobed, the lobes ± keeled or winged, separating at maturity into 4 *drupelets*; seeds with a fleshy endosperm.

One genus.

1. CALLITRICHE L.

Submerged lvs commonly linear, floating lvs narrowly spathulate to ovate, often forming a terminal rosette; lvs with or without stomata.

About 25 spp., cosmopolitan.

A taxonomically troublesome genus owing to the dependence of lf-shape upon whether the lvs are submerged or floating and upon the depth and rate of movement of the water. For this reason a key based on lf-shape is unreliable for plants from non-typical habitats. One based on fr. is more reliable but often cannot be used owing to the sterility of some species, especially when growing in deep water. The following key makes use of both vegetative and fr. characters and should be used only when healthy plants with fls and fr. are available.

1	Lvs all ± linear, submerged.	2
	Lvs not all ± linear; upper lvs ovate or spathulate, forming a floating rosette.	4
2	Lvs dark blue-green, mostly less than 10 × 1 mm., very slightly tapering to a truncate, feebly emarginate tip; stems reddish; fr. (rarely seen) c. 1 mm., its lobes not keeled.	6. truncata
	Lvs pale or yellowish green, mostly more than 1 cm. long, with a distinctly emarginate tip; lobes of fr. keeled or winged.	3
3	Lvs commonly 15–25 × 0·5–1 mm., ± parallel-sided then abruptly expanded at the emarginate tip and shaped like a bicycle-spanner; fr. 0·8–1·3 mm., its lobes keeled but not winged.	4. intermedia
	Lvs commonly 10–20 × 1–2 mm., widest at the base and distinctly tapering above to the narrow emarginate tip; fr. 2 mm., its lobes broadly winged.	5. hermaphroditica
4	Rosette-lvs ± rhomboidal; submerged lvs narrower, sometimes linear; fr. c. 1·5 × 1 mm., its lobes with rounded margins and with a very shallow groove between members of a parallel pair.	3. obtusangula
	Rosette-lvs broadly ovate to narrowly elliptical, not rhomboidal; submerged lvs linear or not; lobes of fr. keeled, with a distinct groove between members of a parallel pair.	5
5	Rosette-lvs broadly ovate to almost circular; submerged lvs often not very different in shape and never linear; fr. 1·6–2 mm., somewhat broader than long, conspicuously keeled, its lateral grooves deep.	1. stagnalis
	Rosette-lvs broadly to narrowly elliptical; submerged lvs usually narrowly elliptical or linear.	6

6 Rosette concave above; stamens and stigmas submerged; lower lvs linear, parallel-sided but with an expanded and deeply emarginate (bicycle-spanner) tip; fr. almost circular, c. 1·2 mm. diam. **4. intermedia**
  Rosette ± convex above; stamens and stigmas not submerged; lower lvs linear or narrowly elliptical, emarginate but not expanded at the tip; fr. somewhat longer than broad, c. 1·5 mm. diam. **2. platycarpa**

Section *Callitriche*: typically with narrow, often linear, submerged and broader spathulate to obovate floating lvs, all with stomata; usually not fruiting under water; subparallel lobes of the fr. joined at least in the basal half.

1. C. stagnalis Scop.

Annual to perennial. Stem to 60 cm. in water or 15 cm. when prostrate on mud. *Lowest lvs elliptical or spathulate* (never linear), slightly emarginate, sometimes persisting to flowering but usually lost in shallow water and on mud; *uppermost lvs forming a floating rosette*, each c. 10–20 mm., with a *broadly elliptical or almost circular blade narrowed rather abruptly into the stalk*; blade usually 5-veined, rounded or slightly emarginate at the tip. Fls in the axils of rosette-lvs only; bracts falcate, persistent. Stamen c. 2 mm. Stigmas 2–3 mm., soon becoming arcuate-recurved, only occasionally persistent in fr. *Fr.* 1·6–2 mm., *suborbicular, conspicuously keeled, with deep lateral grooves*; *seeds* broadly winged, *divergent*. Fl. 5–9. Male fls in axils of upper lvs and maturing before the female. Usually anemophilous but perhaps also hydrophilous. $2n = 10^*$. Th.–Hyd.
  Var. *serpyllifolia* Lönnr. is the name given to very small terrestrial forms with smaller and narrower thyme-like lvs.
  Native. Ponds, ditches, streams, wet mud; to c. 3000 ft. in Wales. Common throughout the British Is. 112, H40, S. Europe; Canary Is.; N. Africa.

2. C. platycarpa Kütz.

C. polymorpha auct., *C. verna* auct.
Perennial. Stem to 100 cm. in water, 1–15 cm. when prostrate on mud. *Lowest lvs linear to narrowly elliptical*, emarginate but *not expanded at the tip; uppermost lvs in a floating* ± *convex rosette*, each 10–20 mm., with a *broadly elliptical blade* ± *gradually narrowed into the stalk*; blade 3(–5)-veined, slightly emarginate at the tip. Fls in the axils of rosette-lvs; bracts falcate, persistent. Stamen c. 4 mm. Stigmas 3–4(–8) mm., erect, at least their bases persistent in fr. *Fr.* 1·5–1·8 × 1·3–1·5 mm., *suborbicular to shortly elliptical, distinctly keeled and with fairly deep lateral grooves*; seeds winged, *parallel*. Fl. 4–10. Anemophilous. $2n = 20^*$. Th.–Hyd.
  Native. Ponds, ditches, streams, etc. Probably common throughout lowland Great Britain but distribution imperfectly known; Ireland. Europe, northwards to S. Sweden and S. Finland.
  Most British material named *C. polymorpha* Lönnr. and *C. palustris* L. belongs here. There are no authentic records for either species.

3. C. obtusangula Le Gall

Stem 10–60 cm. in water, 2–12 cm. on mud. *Lowest lvs* 10–40 × 0·5–2 mm., *linear, deeply emarginate but not widened at the apex; middle* (submerged) *and* *floating lvs* 8–20 × 3–7 mm., *rhomboid-spathulate*, 3–7-veined, blunt or ± retuse; *floating lvs in a well-marked rosette*; linear lvs commonly not persisting; mud forms may have all lvs linear or lanceolate. Fls only in the axils of floating lvs; bracts falcate, persistent. Stamen c. 5 mm. Stigmas c. 4 mm., their erect bases persisting in fr. *Fr.* c. 1·5 × 1 mm., *with broadly rounded lobes and barely discernible grooves*; seeds unwinged. Fl. 5–9. Anemophilous. $2n=10$*. Hyd.

Native. Ponds, ditches, lakes; locally frequent in S. England, rare in the north. Great Britain northwards to Lancs and Yorks; Wigtown and W. Inverness; scattered throughout Ireland. 62, H16, S. France, Belgium, Holland, W. Germany, Corsica, Italy, Sardinia, Sicily, Greece; N. Africa.

Hardly distinguishable from *C. stagnalis* when sterile, but very different in its fr.

'Var. *lachii* Warren' appears to be a hybrid between *C. obtusangula* and *C. intermedia*, with rosette-lvs like those of *C. obtusangula* and submerged lvs like those of *C. intermedia*.

4. C. intermedia Hoffm.

C. hamulata Kütz. ex Koch

Stem 20–80 cm. in water, 10–25 cm. on mud. *Lowest lvs* 10–25(–40) × 0·5–1·3 mm., linear, usually *widened and deeply emarginate at the apex*, like a bicycle-spanner, 1-veined; upper submerged lvs 11–25 × 1·5–5 mm., narrowly oblanceolate-spathulate. Fls in the axils of both submerged and floating lvs; bracts falcate, soon falling. Stamen c. 1 mm. Stigmas 1·5–2 mm., soon re-flexing and falling or only *the appressed bases persisting*. *Fr.* c. 1·25 mm., *suborbicular, keeled, with fairly deep lateral grooves*; seeds winged. Fl. 4–9. Hydrophilous. $2n=38$*. Th.–Hyd.

In the absence of more precise cytogenetical and ecological information it seems best to recognize 2 sspp. differing chiefly in the fr.

Ssp. **hamulata** (Kütz. ex Koch) Clapham has sessile fr. 1·3 × 1·1 mm. and its linear lvs are widened and deeply emarginate at the apex.

Ssp. **pedunculata** (DC.) Clapham has sessile or stalked fr. c. 0·8 × 1 mm., the stalks 0–50 mm., those of the lowest fls usually longest, and its linear lvs are emarginate, truncate or blunt, not appreciably widened at the apex and when emarginate often forming unequal apical teeth which are shorter and less claw-like than in ssp. *hamulata*. Usually smaller, fruiting earlier in the summer and growing more commonly in pools which dry up in summer than ssp. *hamulata*. Treated as a distinct sp. by many continental botanists.

Native. Lakes, pools, ditches and slow streams; to 3250 ft. in Scotland. 99, H24, S. Throughout the British Is. Ssp. *pedunculata* in a few scattered localities, but distribution imperfectly known. Ssp. *hamulata* throughout Europe; ssp. *pedunculata* in W. and S. Europe, from Norway and Denmark to Spain and Portugal, Italy and the Balkans; Iceland; Faeroes; Morocco; Palestine; Caucasus.

Section *Pseudocallitriche* Hegelm.: lvs all similar, submerged, linear to linear-lanceolate, without stomata; fr. of 4 readily separating lobes.

5. C. hermaphroditica L.   'Autumnal Starwort.'

C. autumnalis L.

Stem 15–50 cm., branched, *yellowish*, entirely submerged. *Lvs* 8–18 × 1–2 mm., those *in middle of stem* longest, linear-lanceolate, *widest at the base* and *distinctly tapering above* to an emarginate apex, 1-veined, pale green, becoming olive to blackish on drying. *Bracts* 0. Filaments short, little exceeding the anthers. Styles commonly longer than the ovary, spreading or reflexed, soon falling. Fr. (1·5–)2 mm. diam., ± orbicular, of 4 readily separating *broadly and acutely winged lobes*. Fl. 5–9. Hydrophilous. Fruits more freely than any other British sp. $2n=6*$. Hyd.

Native. Lakes and streams; to 1250 ft. in Yorks. N. Britain from Anglesey, Stafford, Cheshire and Yorks northwards (i.e. north of c. 53° N.) to Caithness; Inner and Outer Hebrides; Orkney; Shetland. Throughout Ireland, but common in the north. 43, H20. Europe northwards to Finmark and N.E. Russia; Iceland.

6. C. truncata Guss.

Stem 8–20 cm., very slender, branched, entirely submerged, *commonly reddish*. *Lvs* 5–11 × 0·8–1·3 mm., linear, almost *parallel-sided*, imperceptibly tapering to a truncate and shallowly emarginate apex, 1-veined, dark green. Bracts 0. Stamens and styles as in *C. hermaphroditica*. Fr. c. 1 mm. diam., almost orbicular, of 4 readily separating *blunt unwinged lobes*; sessile or shortly stalked (1–3 mm.). Fl. 5–9. Hydrophilous. Rarely fruits in the British Is. Hyd.

British material all falls in var. *occidentalis* (Rouy) Druce, which differs from the type in being more robust and more lfy, fruiting poorly, and having subsessile frs (fr.-stalks in type 2–4(–9) mm.).

Native. Pools and ditches; local and only south of c. 53° N. Somerset, Sussex, Kent, Gloucester, Notts; Wexford (Ireland); Guernsey. 6, H1, S. S. and W. Europe from Crete, Greece and Dalmatia to Spain and Portugal and northwards to Brittany, Normandy and Belgium.

Distinguished from *C. hermaphroditica*, even when sterile, by the shorter, imperceptibly tapering lvs.

70. LORANTHACEAE

Mostly shrubs partially parasitic on trees. Lvs exstipulate, usually opposite or whorled, entire, sometimes reduced to scales. Fls actinomorphic, hermaphrodite or unisexual. Perianth of 2 whorls, the inner often brightly coloured, the outer sometimes suppressed and the inner then sepaloid; petals free or united. Stamens epipetalous. Rudimentary ovary often present in male fls, staminodes in female. Ovary inferior; ovules usually not differentiated from the placenta. Style simple or 0. Seed solitary, devoid of testa; embryo large, sometimes up to 3 in a seed; endosperm usually copious.

About 25 genera and 600 spp. in tropical and temperate regions.

1. VISCUM L.

Hemiparasites. Fls unisexual. Sepals much reduced, petals sepaloid, usually 4. Stamens sessile, opening by pores. Berries white, viscous.

About 20 spp. in the Old World.

1. V. album L. Mistletoe.

A somewhat woody evergreen, parasitic on the branches of trees. Stems up to c. 1 m., green, much-branched, branching apparently dichotomous. Lvs 5–8 cm., narrowly obovate and often somewhat falcate, obtuse, rather thick and leathery, yellowish-green, narrowed at base into a short petiole. Infl. a small compact cyme of 3–5 subsessile fls. Bracts united to the short pedicels. Fls usually dioecious; male fls: calyx 0; female fls: calyx small, indistinctly 4-toothed. Fr. c. 1 cm. diam. Fl. 2–4. Fr. 11–12. $2n = 20$.

Native. On the branches of a great variety of deciduous trees, most commonly apple, rarely on evergreens and very rarely on conifers, more abundant on calcareous soils. From Cornwall and Kent to Denbigh and mid-west Yorks, common in S. England and the W. Midlands, rather infrequent elsewhere and absent from Scotland and Ireland. 54. From S. Scandinavia southwards to N. Africa and east to C. Asia and Japan.

71. SANTALACEAE

Trees, shrubs or herbs, sometimes hemiparasitic. Lvs alternate or opposite, entire, exstipulate. Fls actinomorphic, hermaphrodite or unisexual. Perianth of one whorl, sepaloid or petaloid, often fleshy, lobes 3–6, valvate. Stamens the same number as the per. segs and opposite them; anthers 2-celled, opening lengthwise. Ovary inferior or ½-inferior, 1-celled; ovules 1–3, pendulous from a basal placenta. Fr. indehiscent, nut-like or drupaceous. Seed devoid of testa; embryo often oblique, straight, cotyledons usually terete; endosperm copious.

About 30 genera and 300 spp., in tropical and temperate regions.

1. THESIUM L.

Perennial hemiparasitic herbs having haustoria on their roots by means of which they attach themselves to the roots of other plants. Lvs narrow, alternate. Fls small, greenish, hermaphrodite, usually in small dichotomous cymes. Perianth funnel-shaped or campanulate, segments 5 (–4), persistent. Stigma capitate. Ovules 3.

About 240 spp., widely distributed.

1. T. humifusum DC. 'Bastard Toadflax.'
T. Linophyllon auct.

A slender glabrous yellowish-green perennial 5–45 cm. Rootstock woody. Stems spreading or prostrate, herbaceous or woody at base, angled, angles slightly rough. Lvs 5–15(–25) mm., linear, acute or obtuse, 1-nerved, lower lvs scale-like, distant. Infl. ± branched, terminal. Bracteoles 3, linear-lanceolate, serrulate, inserted at the base of the short stout pedicel. Fls

c. 3 mm. diam., yellowish. Per. segs triangular, acute. Fr. c. 3 mm., ovoid, ribbed, green, crowned by the persistent, inrolled per. segs. Fl. 6–8. Fr. 7–9. Ch.

Native. Parasitic on the roots of various herbs in chalk and limestone grassland, local. 24, S. N. Somerset to Kent and Bedford (except Middlesex), E. Gloucester, Shropshire, S. Lincs; Glamorgan. Belgium, France, Spain, but not in N. or C. Europe.

72. CORNACEAE

Trees or shrubs, rarely herbs. Lvs simple, rarely exstipulate. Fls small, usually in panicles, sometimes umbels or heads, regular, hermaphrodite, rarely dioecious, epigynous, 4(–5)-merous. Sepals small, sometimes obsolete. Petals usually valvate, rarely 0. Stamens equalling in number and alternate with petals. Disk usually cushion-like. Ovary inferior, 1–4- (usually 2-)celled; ovules solitary in each cell, pendulous from apex, anatropous; integument 1; style simple or divided, very rarely styles free. Fr. a drupe, rarely a berry; embryo straight; endosperm copious.

Fifteen genera and about 100 spp., mainly in north temperate regions and S.E. Asia, a few in Africa, S. America and New Zealand. *Aucuba japonica* Thunb. with opposite, evergreen, laurel-like lvs, small dull purple dioecious fls and red berries is commonly grown, usually as a form with small yellow spots on the lvs.

Shrub; fls in corymbose panicles without involucral bracts.
                                                      1. THELYCRANIA
Herb; fls in umbels with 4 large white petaloid involucral bracts.
                                              2. CHAMAEPERICLYMENUM

1. THELYCRANIA (Dum.) Fourr.

Trees or shrubs, usually deciduous. Lvs usually opposite, entire, exstipulate. *Fls in corymbose cymes*, hermaphrodite, 4-merous; *bracts* 0. Calyx-teeth small, but always present. *Petals valvate. Ovary 2-celled*; style filiform or columnar, stigma capitate or truncate. *Fr. a drupe; stone* 1, 2-celled. Nectar secreted by the disk.

About 45 spp., north temperate regions (to Himalaya and Guatemala); 2 in Peru and Bolivia. Some spp. are cultivated.

1. T. sanguinea (L.) Fourr.   Dogwood.

Cornus sanguinea L.

Deciduous shrub ¼–4 m. Twigs purplish-red at least on the sunny side. Lvs opposite, blade 4–8 cm., ovate or oval, cuspidate or acuminate, rounded at base, appressed pubescent on both sides, paler beneath, usually becoming purplish-red in autumn; main veins 3–4 pairs from below the middle of the lf, curving round towards the apex; petiole 8–15 mm., grooved above. Infl. many-fld, ± flat-topped, pubescent; peduncle 2·5–3·5 cm.; pedicels to 6 mm. Calyx-teeth very small. Petals 4–6 mm., oblong-lanceolate, obtuse, appressed-pubescent outside. Style clavate above. Fr. black, subglobose, 6–8 mm. diam. Fl. 6–7. Pollinated by various insects. Fr. 9. $2n=22$. M. or N.

Native. Woods and scrub on calcareous soils, occasionally locally dominant in chalk scrub. From Durham and Cumberland southwards, widespread and common on suitable soils; probably introduced farther north, extending to Dunbarton and Fife; rather local in Ireland and native only in the south midlands and north-west; Jersey. 68, H12, S. Europe from S. Scandinavia to C. Spain and Portugal, Sicily and Greece; S.W. Asia (very rare).

*T. sericea (L.) Dandy (*Cornus stolonifera* Michx.)

Deciduous shrub to 2·5 m. with numerous decumbent suckering branches. Twigs deep blood-red. Lf-blades 6–12 cm., ovate to oblong-lanceolate, acuminate, veins c. 5 pairs; petiole 1–2·5 cm. Style not dilated above, crowned by the discoid stigma. Fr. white, globose, c. 5 mm. diam., stone as broad or broader than high, rounded below.

Frequently planted and perhaps naturalized in a few places. Native of eastern N. America.

*T. alba (L.) Pojark. (*Cornus alba* L.)

Differs from *T. sericea* in the acute lf-blades 4–8 cm., in not or scarcely suckering, and in the stone being higher than broad, acute at both ends.

Planted, and reported as an escape. Native of N.E. Asia.

*Cornus mas L.   Cornelian Cherry.

Shrub or small tree to 8 m. Lvs 4–10 cm., ovate. Fls small, yellow, produced before the lvs expand, in umbels subtended by 4 yellowish bracts. Fr. ellipsoid, scarlet. Frequently planted, rarely naturalized. C. and S. Europe, W. Asia.

2. CHAMAEPERICLYMENUM Hill

Herbs with creeping rhizome and annual stems. Lvs opposite, entire, exstipulate. *Fls in umbel-like cymes, surrounded by* 4 *large white involucral bracts.* Otherwise as in *Thelycrania*.

Two spp., arctic and mountains of north temperate zone.

1. C. suecicum (L.) Aschers. & Graebn.   'Dwarf Cornel.'

Cornus suecica L.

Perennial herb 6–20 cm. Stems erect, often a few together, simple or with short axillary branches from the uppermost pair of lvs, glabrous or appressed pubescent. Lvs 1–3 cm., ovate or ovate-elliptic, subsessile, acute or very shortly acuminate, 3–5-veined from the base, green and appressed pubescent above, subglaucous and glabrous beneath. Infl. terminal, fl.-like, of 8–25 blackish-purple fls, surrounded by 4 white ovate involucral bracts, which are 5–8 mm. long; pedicels 1–2 mm., appressed-pubescent like the receptacle. Sepals small but obvious, deltoid. Petals 1–2 mm., ovate-triangular, acute. Fr. red, c. 5 mm., subglobose or ovoid. Fl. 7–8. $2n = 22$. Hp.

Native. Moors on mountains, usually under heather or bilberry; ascending to 3000 ft. Very local and rare in England; Lancashire, Yorks, Northumberland; Dumfries, Peebles; more frequent in the Scottish Highlands from Argyll and Angus to Shetland but local and mainly in the east (not recorded from the Inner or Outer Hebrides). 20. Arctic Europe, Asia and America, extending south to N. Germany and N. Japan.

73. ARALIACEAE

Trees, shrubs or woody climbers, occasionally herbs. Lvs usually alternate, simple or compound, usually stipulate. Fls small, in umbels, heads, racemes or spikes often forming compound infl., regular, hermaphrodite or unisexual, epigynous, usually 5-merous. Sepals small or obsolete. Petals free or united, 3–many, valvate or imbricate, sometimes falling off together as a cap. Stamens 3–many, usually equal in number and alternate with petals, occasionally more. Disk flat or swollen. Ovary inferior, cells usually as many as petals, sometimes fewer (2–4), very rarely 1 or many; ovules solitary in each cell, pendulous from apex, anatropous; styles free or united or 0. Fr. a drupe or berry; embryo small; endosperm copious.

About 60 genera and 700 spp., mainly tropical, some temperate. Resembling Cornaceae and not separable by any single character but with tendencies to have alternate and compound and lobed lvs, more compound infls and free styles. Some genera are grown in gardens, spp. of *Fatsia* Decne. & Planch. and *Aralia* L. being the most common.

1. HEDERA L.

Evergreen *woody climbers*, climbing by roots. *Lvs simple*, coriaceous, glabrous, alternate, petioled; stipules 0. Fls in terminal umbels, often arranged in a panicle; *pedicels not jointed*. Calyx with 5 small teeth. *Petals 5, valvate with broad base*, free. *Stamens 5. Ovary 5-celled*; styles joined into a column. Fr. berry-like; *endosperm ruminate*. Nectar secreted by the disk.

About 5 spp., north temperate Old World.

1. H. helix L.   Ivy.

Woody climber sometimes climbing to 30 m. or creeping along the ground and forming carpets, flowering only in sun at the top of whatever it is climbing on. Stems up to 25 cm. diam., densely clothed with adhesive roots. Young twigs stellate-pubescent. Lvs glabrous, dark green above, often with pale veins, sometimes tinged purple, paler beneath; blades of those of the creeping or climbing stems 4–10 cm., palmately 3–5-lobed with ± triangular, entire lobes; those of the fl. branches entire, ovate or rhombic. Fls in subglobose umbels, the umbels often arranged racemosely and forming a terminal panicle; peduncle, pedicels and receptacle stellate-tomentose. Calyx-teeth small, deltoid. Petals yellowish-green, 3–4 mm., triangular-ovate, somewhat hooded at apex. Fr. black, globose, 6–8 mm. Fl. 9–11. Pollinated by flies and wasps; homogamous or protandrous. $2n=48$. MM., M., N. or Ch.

Native. Climbing in woods, hedges or on rocks and walls or creeping in woods, on all but very acid, very dry or water-logged soils, very tolerant of shade; ascending to 2000 ft. Common throughout the British Is. 112, H40, S. Europe from Norway (60° 32′ N.) and Latvia southwards (absent from most of Russia); Asia Minor to Palestine and N. Persia.

74. HYDROCOTYLACEAE

Perennial herbs, small and often creeping. Lvs entire or 3–5-foliate. *Fls in one or more superposed whorls*, bracts small or 0. Sepals 5, small. Petals 5. Stamens 5, alternating with the petals, inflexed in bud. Ovary inferior, 2-celled; ovules pendulous, solitary in each cell; styles 2, filiform. Fr. strongly laterally compressed, commissure narrow; *inner layer of fr. wall woody, carpophore* 0. Vittae 0 or slender and sunk in the primary ridges of the fr. and absent from the furrows.

One genus with c. 70 spp. in temperate and tropical regions.

Resembles Araliaceae in the woody layer of the fr. wall and Umbelliferae in the herbaceous habit.

1. HYDROCOTYLE L.

The only genus.

1. H. vulgaris L.   Pennywort, White-rot.

A slender, *creeping* or sometimes floating perennial, rooting freely at the nodes. Petioles 1–25 cm., erect, sparsely hairy. *Lvs* 1–5 cm. diam., *peltate, orbicular, crenate*. Peduncles shorter than petioles. *Umbels* 2–3 *mm. diam. 2–5-fld, or sometimes peduncle with several successive whorls of fls*; bracts triangular. Fls c. 1 mm., pinkish-green, subsessile, usually self-pollinated. Fr. 2 mm. diam., carpels with 2 ridges on each face, covered with brownish resinous dots. Fl. 6–8. Fr. 7–10. $2n=$ c. 96. H.

Native. In bogs, fens and marshes, usually on acid soils. 112, H40, S. Generally distributed in suitable localities throughout the British Is., ascending to 1750 ft. Europe from southern Scandinavia to Portugal and Greece, eastwards to the Caspian; Algeria and Morocco; introduced in New Zealand.

75. UMBELLIFERAE

Annual or perennial herbs, rarely shrubs. Stems often furrowed, pith wide and soft or internodes hollow. Lvs alternate, exstipulate, usually much divided, petioles sheathing at base. Infl. usually a compound umbel, sometimes a simple umbel, rarely capitate or very reduced and cymose with solitary fls surrounded by a whorl of bracteoles; bracts and bracteoles usually present, whorled. Fls hermaphrodite or unisexual, usually with nectar, often strongly protandrous. Calyx-teeth usually small, 5, sometimes unequal, often 0. Petals 5, valvate or slightly imbricate, often notched with an inflexed or incurved point, sometimes very unequal, often white, sometimes pink or yellow, rarely blue. Stamens 5, alternating with the petals, inflexed in bud. Ovary inferior, 2 (rarely 1)-celled; ovules pendulous, solitary in each cell; styles 2, often with an enlarged base (*stylopodium*). Fr. dry, of 2 indehiscent dorsally or laterally compressed carpels separated by a narrow or broad commissure; carpels adnate to or suspended from a slender simple or divided axis (*carpophore*), usually separating when ripe, mostly prominently 5- or 9-ribbed, and generally with four resinous canals (*vittae*) between the primary ridges (rarely in them) and 2 on the commissural face. Pollinated by various insects, particularly small beetles and flies.

About 200 genera and 2700 spp., cosmopolitan but chiefly in the north temperate region.

Conspectus of Tribes and Genera

Subfamily 1. SANICULOIDEAE

Fr. with a soft parenchymatous inner layer of the wall containing crystals, outer layer with scales, spines or bristles, rarely smooth. Styles long, surrounded by the annular stylopodium. Vittae in the primary ridges or several or scattered, or inconspicuous or lacking.

Tribe 1. SANICULEAE. Both loculi of the ovary fertile. Fr. with a flat commissure. Vittae usually present.

1. **Sanicula.** Lvs palmately lobed. Bracts small. Fr. with hooked spines.
2. **Astrantia.** Lvs palmately lobed. Bracts very large. Fr. with wrinkled, toothed or plaited inflated ridges.
3. **Eryngium.** Lvs spiny. Fls densely capitate. Fr. with straight spines or scaly.

Subfamily 2. APIOIDEAE

Fr. with a soft parenchymatous inner layer of the wall, middle layer sometimes hard. Style from the apex of the stylopodium. Vittae in the furrows in young ovaries, later variously arranged.

Tribe 2. SCANDICEAE. Fls in the main umbels all hermaphrodite or irregularly male and female. Seed deeply grooved or hollow on the commissural face. Crystals present in the fr. wall.

Subtribe SCANDICINAE. Fr. cylindrical and beaked, smooth or with short spines.

4. **Chaerophyllum.** Fr. very shortly beaked, ridges broad and rounded.
5. **Anthriscus.** Fr. distinctly beaked, ridges confined to beak.
6. **Scandix.** Beak longer than fertile part of fr., ridges broad or slender, not extending into beak.
7. **Myrrhis.** Fr. very long, beak short, ridges almost winged.

Subtribe CAUCALINAE. Fr. ovoid to subspherical with bristles or spines overhanging the furrows.

8. **Torilis.** Ridges slender, ciliate, furrows thickly beset with spines or tubercles. Calyx-teeth small.
9. **Caucalis.** Primary ridges slender and inconspicuous, smooth or hispid; secondary ridges prominent, bearing 1–3 rows of spines.

Tribe 3. CORIANDREAE. Fr. broadly ovoid, nut-like, smooth, ridges visible only on drying; wall devoid of crystals.

10. **Coriandrum.** Secondary ridges broader than primary. Vittae solitary, obscure.

Tribe 4. SMYRNIEAE. Fr. broadly ovoid, rarely elongate; wall devoid of crystals. Carpels rounded on the back, commissure narrow, ridges prominent or inconspicuous.

11. **Smyrnium.** Fr. ovoid; carpels subterete or angular with 3 prominent sharp ridges; vittae numerous.
12. **Physospermum.** Fr. inflated, didymous; carpels smooth, ridges inconspicuous; vittae solitary in the furrows.

13. **Conium.** Fr. broadly ovoid to suborbicular; carpels with 5 prominent ridges; vittae 0.

Tribe 5. AMMIEAE. Primary ridges all equal. Seed semicircular in section, flat on the commissural face.

Subtribe CARINAE. Ridges inconspicuous, commissure narrow.

    (*a*) Petals entire, acute, tip sometimes shortly inflexed.

14. **Bupleurum.** Lvs simple. Fls yellow, hermaphrodite.
15. **Trinia.** Lvs compound. Plant usually dioecious. Fls white.
16. **Apium.** Lvs compound. Fls white, hermaphrodite.

    (*b*) Petals notched, tip inflexed.
        Petals slightly notched.

17. **Petroselinum.** Calyx-teeth 0. Ridges slender. Vittae slender, solitary in the furrows.
        Petals deeply notched.
18. **Sison.** Calyx-teeth 0. Ridges slender. Vittae very short, solitary in the furrows.
19. **Cicuta.** Calyx-teeth distinct. Ridges broad. Vittae long, solitary in the furrows.
20. **Ammi.** Bracts long, pinnatifid. Petals unequally 2-lobed. Calyx-teeth 0. Ridges slender. Vittae long, solitary in the furrows.
21. **Falcaria.** Bracts entire, usually large. Calyx-teeth distinct. Ridges blunt. Vittae slender, solitary in the furrows.
22. **Carum.** Root fibrous or fusiform. Calyx-teeth minute. Ridges slender. Vittae broad, solitary in the furrows.
23. **Bunium.** Root a solitary tuber. Calyx-teeth minute. Ridges slender. Vittae slender, solitary in the furrows.
24. **Conopodium.** Root a solitary tuber. Calyx-teeth 0. Ridges slender. Vittae several in each furrow.
25. **Pimpinella.** Root not tuberous. Bracts 0; bracteoles 0 or few. Calyx-teeth small or 0. Ridges slender. Vittae numerous in each furrow.
26. **Aegopodium.** Rhizomatous. Bracts and bracteoles few or 0. Calyx-teeth 0. Ridges slender. Vittae 0.
27. **Sium.** Bracts and bracteoles numerous. Calyx-teeth distinct. Ridges obtuse or thickened, lateral ones marginal. Vittae 3 or more in each furrow, superficial.
28. **Berula.** As *Sium* but lateral ridges not marginal and vittae deeply seated.

Subtribe SESELIINAE. Ridges very prominent, sometimes winged; commissure broad, the prominent marginal ridges making it appear still broader.

29. **Crithmum.** Lvs fleshy. Calyx-teeth 0. Ridges thick, acute. Vittae several in each furrow.
30. **Seseli.** Calyx-teeth distinct. Carpels dorsally compressed. Ridges thick, the marginal somewhat more prominent than the dorsal. Vittae solitary, rarely 2 or 3 in each furrow.
31. **Oenanthe.** Calyx-teeth distinct. Carpels ½-terete. Lateral ridges grooved or thickened, often forming a corky rim round the carpels. Vittae solitary in the furrows.
32. **Aethusa.** Calyx-teeth small or 0. Carpels dorsally compressed. Ridges very broad, lateral narrowly winged. Vittae solitary in the furrows.
33. **Foeniculum.** Fls yellow. Petals entire with an obtuse incurved point. Calyx-teeth 0. Carpels ½-terete. Ridges stout. Vittae solitary in the furrows.
34. **Silaum.** Fls yellowish. Petals broad and truncate at base, tip incurved. Calyx-teeth minute. Carpels ½-terete. Ridges slender, lateral winged. Vittae numerous, irregular.
35. **Meum.** Petals acute, narrowed to base, tip sometimes shortly inflexed. Calyx-teeth 0. Carpels ½-terete. Ridges slender, acute. Vittae 3–5 in each furrow.

36. **Selinum.** Petals notched, point inflexed. Calyx-teeth 0. Carpels ½-terete. Ridges winged, the lateral broadly so. Vittae solitary in each dorsal furrow.
37. **Ligusticum.** Petals entire, inflexed. Calyx-teeth small or 0. Ridges acute or winged. Vittae many in each furrow or obscure.

Tribe 6. PEUCEDANEAE. Marginal ridges much wider than the usually inconspicuous dorsal ridges, making a wing round the carpels. Seed narrow in section.

Subtribe ANGELICINAE. Marginal wings of opposite carpels not appressed.

38. **Angelica.** Petals with a short inflexed point. Calyx-teeth minute or 0. Carpels strongly dorsally compressed.

Subtribe FERULINAE. Fr. with a thin wing. Marginal wings of opposite carpels closely appressed.

39. **Peucedanum.** Petals with a long inflexed point. Calyx-teeth small or 0. Carpels strongly dorsally compressed. All ridges equidistant. Marginal wing thin. Vittae 1–3 in each furrow, as long as the fr.
40. **Pastinaca.** Petals entire, involute. Calyx-teeth minute. Carpels strongly dorsally compressed. Dorsal ridges equidistant, lateral distant from them. Marginal wing thin. Vittae solitary in the furrows, as long as the fr.

Subtribe TORDYLIINAE. Fr. with a thickened wing, wings of opposite carpels closely appressed.

41. **Heracleum.** At least the larger petals notched. Calyx-teeth small, unequal. Carpels strongly dorsally compressed. Vittae solitary in the furrows, club-shaped and shorter than the fr.
42. **Tordylium.** At least the larger petals notched. Calyx-teeth unequal or 0. Carpels strongly dorsally compressed. Vittae 1–3 in each furrow, linear, as long as the fr.

Tribe 7. DAUCEAE. Secondary ridges over the vittae similar to the primary ones or larger. Fr. covered with spines or papillae.

43. **Daucus.** Petals notched. Calyx-teeth small or 0. Carpels convex. Primary ridges filiform, secondary stouter and more prominent, all, or the secondary only with rows of spines. Vittae solitary.

Key to Genera

1	All lvs entire or palmately lobed; if pinnately lobed then margins spiny.	2
	Lower lvs at least 1-pinnate or -ternate; if pinnately lobed then margins not spiny.	5
2	Lvs spiny; fls in dense heads.	3. ERYNGIUM
	Lvs not spiny.	3
3	Lvs linear to ovate, entire.	14. BUPLEURUM
	Lvs palmately lobed.	4
4	Bracts small, inconspicuous.	1. SANICULA
	Bracts conspicuous, coloured, forming an involucre.	2. ASTRANTIA
5	Aerial lvs few or 0; aquatics with finely divided submerged lvs.	6
	Aerial lvs numerous and well-developed at fl.	7
6	Plant slender; rays 1–2(–3).	16. APIUM
	Plant stout; rays 5 or more.	31. OENANTHE
7	Lower lvs or lowest aerial lvs simply pinnate or pinnately lobed.	8
	Lower lvs 2–4-pinnate or ternate.	21

8	Plant ± hairy, sometimes minutely so.	9
	Plant quite glabrous.	13
9	Calyx-teeth conspicuous, about equalling petals.	10
	Calyx-teeth minute or 0.	11
10	Lvs glabrous above; inner fls long-pedicelled.	9. CAUCALIS
	Lvs hispid on both surfaces; all fls subsessile.	42. TORDYLIUM
11	Fls yellow.	40. PASTINACA
	Fls white or pinkish.	12
12	Bracteoles 0.	25. PIMPINELLA
	Bracteoles several, setaceous, reflexed.	41. HERACLEUM
13	Petioles fistular; partial umbels in fr. ± globose.	31. OENANTHE
	Petioles not fistular; partial umbels ± flat, or irregular.	14
14	Lf-segments many together, appearing as if whorled.	22. CARUM
	Segments of at least the lower lvs lanceolate to suborbicular.	15
15	At least some bracts ½ as long as the shorter rays.	16
	Bracts all much shorter than the rays, or 0.	18
16	Bracts subulate; stem solid.	17. PETROSELINUM
	Bracts broader, often divided; stem hollow.	17
17	Lf-segments 4–6(–8) pairs.	27. SIUM
	Lf-segments 7–10 pairs.	28. BERULA
18	Bracts 0 or umbels lf-opposed.	19
	Bracts present; umbels not lf-opposed.	18. SISON
19	At least some umbels subsessile and lf-opposed.	16. APIUM
	Umbels all distinctly peduncled, terminal.	20
20	Bracteoles 0.	25. PIMPINELLA
	Bracteoles present.	10. CORIANDRUM
21	Plant hairy, hairs often minute and sometimes confined to peduncles or rays or margins and nerves of upper surface of lvs.	22
	Plant entirely glabrous; lf margins sometimes serrulate with spinous-ciliate teeth.	36
22	Stems at or after fl. with a large hollow in the middle.	23
	Stems solid or nearly so, even after fl.	29
23	Only the rays and pedicels pubescent.	39. PEUCEDANUM
	Plant pubescent elsewhere.	24
24	Segments of lower lvs 3 cm. or more, not or slightly pinnatifid.	25
	Segments of lower lvs smaller, frequently and deeply pinnatifid.	27
25	Bracteoles 0.	25. PIMPINELLA
	Bracteoles several.	26
26	Lvs ternately divided; stems terete, nearly smooth.	38. ANGELICA
	Lvs pinnately divided; stems ridged.	41. HERACLEUM
27	Plant not aromatic.	28
	Plant strongly aromatic.	7. MYRRHIS
28	Rays 3 or more; bracteoles entire or 0.	5. ANTHRISCUS
	Rays 1(–2); some bracteoles 2–3-fid.	6. SCANDIX
29	Stems, at least in the younger parts, with short straight appressed, downward-pointing hairs, not purple-spotted.	8. TORILIS
	Stems purple-spotted or glabrous, or hairs short and crisped or longer and pointing in various directions, not appressed.	30

30   Stems and sheaths glabrous; lvs sparsely and minutely hairy on nerves
        and near margins on upper surface.   12. PHYSOSPERMUM
     Stems pubescent, at least near nodes.   31

31   Hairs minute, crisped.   32
     Hairs spreading or pointing in various directions, or stems purple-
        spotted.   33

32   Bracts 0; stock not fibrous.   25. PIMPINELLA
     Bracts present; stock crowned with fibres.   30. SESELI

33   Bracts 7–13, large, ternate or pinnatifid.   43. DAUCUS
     Bracts 0–5, small, simple.   34

34   Bracteoles large, green, some 2–3-fid.   6. SCANDIX
     Bracteoles all entire, often ±scarious.   35

35   Stem purple-spotted or entirely purple; rays (4–)8–25.
                                                 4. CHAEROPHYLLUM
     Stem not purple-spotted; rays 2–3(–5).   9. CAUCALIS

36   Fls yellow or yellowish-green.   37
     Fls white, pinkish-, purplish- or greenish-white.   42

37   Lf-segments linear, entire.   38
     Lf-segments not linear, but pinnatifid, lobed or toothed.   40

38   Lvs ternate, segments long-acuminate; pedicels long, filiform.
                                                   39. PEUCEDANUM
     Lvs pinnate, segments filiform or acute; pedicels not exceptionally long
        or slender.   39

39   Lf-segments fleshy; bracts and bracteoles numerous; bushy plant up to
        c. 30 cm.   29. CRITHMUM
     Lf-segments very narrow, not fleshy; bracts and bracteoles 0–1; erect
        plant up to c. 130 cm.   33. FOENICULUM

40   Segments of lower lvs crenate or serrate, obtuse.   11. SMYRNIUM
     Segments of lower lvs pinnatifid or lobed, acute.   41

41   Sheathing bases of petioles with broad hyaline margins; lf margins not
        serrulate.   17. PETROSELINUM
     Margins of petiole bases not or narrowly hyaline; lf margins minutely
        serrulate.   34. SILAUM

42   Stems ±glaucous, purple-spotted.   13. CONIUM
     Stems not purple-spotted.   43

43   Bracteoles 0 or 1.   44
     Bracteoles 2 or more.   46

44   Segments of lower lvs serrate but not pinnatifid; rhizomatous.
                                                   26. AEGOPODIUM
     Segments of lower lvs pinnatifid; rhizomes 0.   45

45   Radical lvs soon withering; petioles long and slender, largely under-
        ground; perennial, with tubers.   24. CONOPODIUM
     Radical lvs persistent; petiole short, stout and sheathing, arising above
        ground; biennial, with a fusiform root.   22. CARUM

46   Umbels with 1 or 2 stout rays; bracteoles spinous-ciliate, usually bifid
        or pinnatifid.   6. SCANDIX
     Umbels with (2–)3 or more rays; bracteoles not as above.   47

47   Lf-segments up to 30 cm., narrow, regularly and sharply serrate,
        ±falcate; bracts and bracteoles numerous, linear.   21. FALCARIA
     Not as above.   48

48   Bracts at least ½ as long as rays, some or all pinnatifid with linear
        segments.   20. AMMI
        Bracts shorter or 0, never pinnatifid, rarely 3-cleft.   *49*

49   Bracteoles 3–4 on outside of each partial umbel, conspicuous, green,
        downward-pointing.   32. AETHUSA
        Not as above.   *50*

50   Lower lvs soon withering; stem tapering downwards from the surface of
        the earth to junction with tuber.   *51*
        Not as above.   *52*

51   Stem hollow after fl.; styles suberect (common but almost absent from
        chalky soils).   24. CONOPODIUM
        Stem always solid; styles deflexed in fr. (rare, on chalk in S.E. England).
                                      23. BUNIUM

52   Lf-segments very narrow; stock crowned by abundant persistent fibres.   *53*
        At least the lower lvs with broad lf-segments; fibres 0 or very scanty.   *54*

53   Stem solid; plant glaucous; dioecious.   15. TRINIA
        Stem hollow; plant neither glaucous nor dioecious.   35. MEUM

54   Segments of lower lvs ovate, obovate or suborbicular, little lobed, at
        least some 2 cm. or more.   *55*
        Segments of lower lvs smaller and usually narrower, often deeply pin-
        natifid.   *57*

55   Lower lvs 3–4-pinnate, segments many.   31. OENANTHE
        Lower lvs 1–2-ternate, segments few.   *56*

56   Lf-segments not serrate in lower half; rays 8–12.   37. LIGUSTICUM
        Lf-segments serrate nearly to base; rays 20–25.   39. PEUCEDANUM

57   Lf-segments linear-lanceolate, sharply serrate but not lobed; bracts 0;
        bracteoles strap-shaped.   19. CICUTA
        Lf-segments broader and variously lobed or else quite entire.   *58*

58   Partial umbels often very dense; pedicels not longer than the oblong fr.;
        roots with several tubers, or else plant aquatic, often with much-
        divided submerged lvs.   31. OENANTHE
        Partial umbels not very dense, pedicels distinctly longer than the fr. or
        else fr. globose; roots not tuberous; submerged lvs 0.   *59*

59   Bracts 4 or more.   *60*
        Bracts fewer than 4.   *61*

60   Stems hollow, strongly ridged and ±angled.   39. PEUCEDANUM
        Stems solid or nearly so, striate, terete.   12. PHYSOSPERMUM

61   Lf-segments with entire margins; rays (2–)3–5(–10), smooth.
                                  10. CORIANDRUM
        Lf-segments minutely serrulate; rays 10–20, rough on the ridges.
                                   36. SELINUM

1. SANICULA L.

Erect perennial herbs. Rootstock short, creeping. *Lvs palmately lobed.*
Umbels small, irregularly compound; partial umbels subglobose; bracts few,
small, lfy; bracteoles few. Fls male and hermaphrodite. Calyx-teeth pungent,
longer than the deeply-notched inflexed petals. Fr. ovoid, clothed with

hooked bristles, commissure broad; carpels ½-terete, ridges inconspicuous; vittae 1 in each groove; styles filiform.

About 40 spp., cosmopolitan except for Australia.

1. S. europaea L.   Sanicle.

An erect glabrous perennial 20–60 cm. *Radical lvs* 2–6 cm., *3–5-lobed, lobes cuneate, coarsely and acutely serrate*; petiole 5–25 cm. Umbels of few (often 3) few-fld partial umbels, each 4–7 mm. diam.; bracts 3–5 mm., 2–5, simple or pinnatifid; bracteoles simple. Fls pink or white, outer male and shortly pedicelled, inner hermaphrodite and nearly sessile. Fr. c. 3 mm., animal dispersed. Fl. 5–9. Pollinated mainly by small flies and beetles; self-pollination also possible. $2n=16$. Hr.

Native. In woods; forming societies in chalk beechwoods and in oakwoods on loams. 110, H40. Throughout the British Is. except Orkney, Shetland and Channel Is. Wooded regions of Europe, Asia Minor, Syria, Caucasus, Persia and N. Africa, only in mountain woods (at some altitude) in the Mediterranean region; mountains of tropical Africa (to 2500 m. in the Cameroons); S. Africa; S., C. and E. Asia, East Indies.

2. ASTRANTIA L.

Perennial herbs. Rootstock short, creeping. Lvs palmately lobed or cut. Umbels simple in our sp.; bracts many, conspicuous, often coloured. Fls male and hermaphrodite, smaller later umbels often male only; pedicels of male fls longer than those of hermaphrodite. Calyx-teeth triangular, acuminate, longer than the notched inflexed petals. Fr. ovoid or oblong, commissure broad; carpels somewhat dorsally compressed with equal, plaited, wrinkled or toothed inflated primary ridges; vittae 1 in each groove; styles filiform.

About 9 spp. in Europe and W. Asia.

*1. A. major L.

An erect glabrous perennial 30–75 cm. Radical lvs 8–15 cm. diam., long-petioled, with 3–7 coarsely serrate lobes. Bracts 1–2 cm., lanceolate, acuminate, entire or slightly toothed at apex, whitish beneath, pale greenish-purple above, equalling or exceeding the umbel. Fls whitish or pinkish, in a convex umbel; pedicels filiform; male fls usually outnumbering hermaphrodite. Fr. 6–8 mm. Fl. 5–7. Pollinated by various insects, especially beetles. $2n=14$. Hs.

Introduced. Naturalized in meadows and at margins of woods in several localities; established for over 100 years near Stokesay Castle, Shropshire. 18. S.W. Spain through S. and C. Europe to the Caucasus.

3. ERYNGIUM L.

Rigid, often glaucous and spiny, perennial herbs. Lvs toothed, lobed or dissected. Fls bracteolate, sessile in dense heads surrounded by rigid lfy bracts. Calyx-teeth rigid, acute or pungent, longer than the petals. Petals narrow, deeply notched, tip inflexed. Fr. ovoid, subterete, commissure broad; carpels obscurely ridged; vittae slender.

About 220 spp. in temperate and subtropical regions, especially S. America; absent from S. Africa. The fls have nectar and are pollinated by various insects.

Plant glaucous; radical lvs suborbicular, 3-lobed; involucral bracts oblong-cuneate, spinous-serrate. **1. maritimum**
Plant pale green; radical lvs pinnate; involucral bracts linear-lanceolate, nearly or quite entire. **2. campestre**

1. E. maritimum L. Sea Holly.

An *intensely glaucous* glabrous branched perennial 30–60 cm. *Radical lvs* 5–12 cm. diam., stalked, *suborbicular*, *3-lobed; cauline* sessile, *palmate;* both spinous-toothed with a thickened cartilaginous margin. Heads 1·5–2·5 cm., becoming ovoid. Primary involucre of 3, partial of 5–7, *oblong-cuneate, spinous-serrate bracts; bracteoles linear, spinous,* 3-*fid*, often purplish-blue, *somewhat exceeding the fls. Fls c. 8 mm. diam.*, bluish. Fr. covered with hooked papillae. Fl. 7–8. 2n = 16. Hs.

Native. On sandy and shingly shores. 54, H19, S. Round the coasts of the British Is. north to Shetland. Atlantic coasts of Europe from Portugal to Shetland; S. Scandinavia; North Sea and Baltic coasts to the Gulf of Bothnia; Mediterranean and Black Sea.

2. E. campestre L.

A *pale green* glabrous branched perennial 30–60 cm. *Radical lvs* stalked, *pinnate; cauline* sessile, *subcordate* at the base and ± *clasping stem,* less spiny than in *E. maritimum.* Heads 1·0–1·5 cm., ovoid. *Bracts linear-lanceolate,* spinous, *nearly or quite entire; bracteoles subulate,* spinous, *entire,* 2–3 *times as long as the fls. Fls* 2–3 *mm. diam.*, purplish or white. Fl. 7–8. 2n = 14, 28. Hs.

Native. In dry grassy places. c. 9. In a few scattered localities in S. England, probably introduced in some, but certainly established at Plymouth before 1670; rare. Europe from the Mediterranean region to the N. German plain, east to C. Russia; Persia and Afghanistan; N. Africa; introduced in N. America.

4. CHAEROPHYLLUM L.

More or less hairy herbs. Lvs usually 2–3-pinnate. Umbels compound; bracts few or 0, bracteoles several. Fls white (in our spp.). Calyx-teeth subulate or 0; petals notched with an inflexed point. Fr. oblong or oblong-ovoid, scarcely beaked, laterally compressed, commissure narrow; carpels subterete, ridges broad and rounded; vittae solitary in the furrows.

About 36 spp. in north temperate regions.

Lvs dark green; fr. c. 5 mm.; style as long as stylopodium. **1. temulentum**
Lvs yellow-green, shortly ciliate at margins, slightly pubescent beneath; fr. c. 12 mm.; style twice as long as stylopodium. **2. aureum**

1. C. temulentum L. 'Rough Chervil.'

A rough erect biennial 30–100 cm. *Stem* solid, somewhat grooved, *clothed with short stiff hairs*, swollen below the nodes, purple-spotted or entirely purple. *Lvs* up to 20 cm., 2–3-pinnate, *pubescent on both surfaces*; segments ovate, lower shortly stalked, pinnately lobed, lobes coarsely serrate. *Umbels* 3–6 cm. diam., *rather irregular*, nodding in bud, the earliest terminating the main stem and subsequently overtopped by branches arising from axils of lvs immediately below it; rays (4–)8–10(–15), 0·5–4 cm., slender; bracts 0–2; *bracteoles* 5–8, usually shorter than pedicels, lanceolate to ovate, aristate, fringed, spreading in fl., *deflexed in fr.* Fls 2 mm. diam. *Fr.* 5–7 mm., oblong-ovoid, *narrowed upwards*, often purple. Fl. 6–7. $2n=22$. Hs.

Fls just after *Anthriscus sylvestris*; readily recognized by the rough purple-spotted stems.

Native. In hedge-banks and in grassy places. 103, H23, S. Common and generally distributed in Great Britain, except N. Scotland; Ireland, frequent in the east, almost absent from the west. Europe, except N. Scandinavia and N. Russia; only on mountains in Greece; Caucasus, Dahuria; western N. Africa; introduced in N. America and New Zealand.

*2. C. aureum L.

An erect somewhat hairy perennial somewhat resembling *Anthriscus sylvestris* in general appearance, but stouter. *Stem* solid, slightly grooved, ± rough, swollen below the nodes, ± purple-spotted. *Lvs* 3-pinnate, *slightly pubescent beneath, margins shortly ciliate*, segments lanceolate, acuminate, lobed, lobes serrate or subentire. *Umbels* 5–8 cm. diam., *nearly regular; rays* 15–25, 1·5–3 cm.; bracts 0–3, subulate, up to 2 cm.; *bracteoles* 5–8, *often equalling or exceeding pedicels*, lanceolate, aristate, fringed, *spreading in fl. and fr.* Fls 3–4 mm. diam. *Fr. c.* 12 *mm., oblong, contracted near the apex.* Fl. 7. $2n=22$. Hs.

Introduced. Naturalized in meadows at a few places in Scotland, particularly Callander, W. Perth. N. Spain, France, Italy, C. Europe, Danube basin, Balkan Peninsula, S. Russia, Caucasus and Persia.

5. ANTHRISCUS Pers.

Annual or biennial ± hairy herbs. Lvs usually 2–3-pinnate. Umbels compound; bracts 0 or rarely 1, bracteoles several. Fls white. Calyx-teeth minute or 0; petals notched, with an inflexed point. Fr. ovoid or oblong, beaked, commissure constricted; carpels sub- or ½-terete, ridges confined to the beak; vittae 0 or very slender and solitary in the furrows.

About 12 spp. in Europe, temperate Asia and N. Africa.

Stems glabrous; fls 2 mm. diam.; pedicels thicker than rays in fr.; fr. 3 mm., muricate. **1. caucalis**
Stems uniformly pubescent below; fls 3–4 mm. diam.; pedicels not thicker than rays; fr. 5 mm., smooth. **2. sylvestris**
Stems pubescent above nodes; fls 2 mm. diam.; pedicels not thicker than rays; fr. 10 mm., smooth. **3. cerefolium**

1. A. caucalis Bieb. 'Bur Chervil.'

A. scandicina Mansf.; *A. vulgaris* Pers., non Bernh.; *A. neglecta* Boiss. & Reut.; *Chaerophyllum Anthriscus* (L.) Crantz

A sparsely hairy, branched and spreading annual 25–50(–100) cm. *Stems* hollow, striate, *glabrous*, sympodial, often curved below and purplish towards the base, somewhat thickened below the nodes. Lvs up to 10 cm., usually smaller, 3-pinnate, glabrous above, with stiff scattered hairs beneath; segments up to 5 mm., ovate, pinnatifid, lobes obtuse. Umbels 2–4 cm. diam., shortly peduncled, lf-opposed; rays 3–6, 0·5–1·5 cm., glabrous, slender, often divaricate or recurved in fr.; bracts 0 or rarely 1; bracteoles several, about 2 mm., ovate, aristate, fringed. *Fls c. 2 mm. diam.*, subsessile, *pedicels* elongating and *becoming thicker than the rays in fr. Fr. 3 mm., ovoid,* clothed with hooked spines, *beak short, glabrous; pedicels with a ring of hairs at the top; stigmas subsessile.* Fl. 5–6. $2n = 18$. Th.

Our plant is var. *caucalis.*

Native. In hedgebanks, waste places and on sandy ground near the sea. 89, H26, S. Scattered throughout the British Is. except the extreme north, rather local. Nearly the whole of Europe from S. Sweden to S. Russia; Asia Minor, Caucasus, Siberia; western N. Africa; introduced in N. America and New Zealand.

2. A. sylvestris (L.) Hoffm.   Cow Parsley, Keck.

Chaerophyllum sylvestre L.

A ± downy, erect biennial often perennating by means of offsets, 60–100 cm. *Stems* hollow, furrowed, *downy below, glabrous above.* Lvs up to 30 cm., 2–3-pinnate, somewhat pubescent, at least beneath; segments commonly 15–25 mm., ovate, pinnatifid and coarsely serrate. Umbels 2–6 cm. diam., terminal, the earliest somewhat overtopped by lateral branches; *rays* 4–10, 1–4 cm., *glabrous;* bracts 0; *bracteoles* several, 2–5 mm., ovate, aristate, fringed, *spreading or deflexed,* often pink. *Fls* 3–4 *mm. diam.;* pedicels about equalling bracteoles, elongating but scarcely thickening in fr. *Fr.* 5 mm., oblong-ovoid, smooth, *very shortly beaked; styles slender, spreading.* Fl. 4–6. $2n = 16$. Hs.

By far the commonest of the early-flowering umbellifers in the southern half of England.

Native. By hedgerows, at edges of woods and in waste places. 112, H40, S. Generally distributed and often extremely abundant throughout the British Is. N. and C. Europe, mountains of S. Europe; Caucasus, Siberia, Dahuria; N. Africa, Abyssinia; introduced in N. America.

*3. A. cerefolium (L.) Hoffm.   Chervil.

Chaerophyllum sativum Lam.

An erect branched somewhat pubescent annual 30–50 cm. *Stems* hollow, striate, *pubescent above the nodes.* Lvs 3-pinnate; segments pinnatifid, somewhat pubescent beneath. Umbels 2–5 cm. diam., shortly peduncled or sessile, often lateral and lf-opposed, *peduncles pubescent; rays* 3–5, *pubescent;* bracts 0; *bracteoles* few, *linear,* ciliate. Fls c. 2 mm. diam., pedicels shorter than bracteoles, elongating but scarcely thickening in fr. *Fr.* 10 mm., smooth, oblong-ovoid, with a *rather long slender beak; styles rather stout, erect.* Fl. 5–6. $2n = 18$. Th.

Introduced. On hedgebanks and in waste places, escaped from cultivation and naturalized in some scattered localities. 37. Danube basin, S. and C.

Russia; Asia Minor and Caucasus, Armenia, Persia to the Urals; introduced in the rest of Europe, N. Africa, E. Asia, N. and S. America, and New Zealand.

6. SCANDIX L.

Annual herbs. Lvs 2–3-pinnate, segments narrow and short. Umbels simple or compound; bracts 1 or 0, bracteoles several, entire or variously lobed. Fls white. Calyx-teeth minute or 0; petals often unequal, point inflexed or 0. Fr. subcylindric with a very long beak, commissure constricted; carpels subterete, ridges broad or slender; vittae solitary in the furrows.

About 15 spp. in Europe and the Mediterranean region.

1. S. pecten-veneris L.   Shepherd's Needle.

An erect branched nearly glabrous annual 15–50 cm. Stems becoming hollow when old, striate, ± pubescent with short scattered hairs. Lvs 2–3-pinnate, oblong or narrowly deltoid; segments up to 5 mm., spathulate, margins denticulate or subentire. *Umbels simple or of 2 stout rays; bracteoles 5–10 mm., bifid or pinnatifid, sometimes entire, margins spinous-ciliate.* Fls 1 mm. diam., subsessile, pedicels elongating and becoming thickened in fr. *Fr. 30–70 mm., scabrid,* ridges broad; styles very short, erect. Fl. 4–7. $2n=16$. Th.

Native. A weed of arable land. 103, H38, S. Generally distributed throughout the British Is.; but rare and a casual in Wales and the north. Europe from S. Sweden to C. Russia and the Mediterranean region; eastwards to the western Himalaya; introduced in S. Africa, N. America, Chile, New Zealand, etc.

7. MYRRHIS Mill.

Finely pubescent perennial herbs. Lvs pinnate. Umbels compound; bracts few or 0, bracteoles several, membranous. Fls white, male or hermaphrodite. Calyx-teeth minute or 0; petals notched, with a very short inflexed point. Fr. linear-oblong, beaked, commissure broad; carpels with prominent ridges; vittae solitary or 0.

One sp. in Europe and (probably introduced) in Chile.

1. M. odorata (L.) Scop.   Sweet Cicely.

A stout erect sparingly pubescent perennial 60–100 cm. Stems hollow, somewhat grooved. Lvs up to c. 30 cm., 2–3-pinnate, pale beneath; segments oblong-ovate, pinnatisect, lobes coarsely serrate; petioles of stem lvs sheathing. Umbels 1–5 cm. diam., terminal; *rays 5–10, those of partial umbels bearing hermaphrodite fls 2–3 cm., stout; those of partial umbels bearing only male fls shorter, slender*; bracteoles c. 5, lanceolate, aristate. Fls 2–4 mm. diam., petals unequal. *Fr. 20–25 mm.,* strongly and sharply ridged, ridges often scabrid; styles slender, diverging. Fl. 5–6. $2n=22$. Hs.

The whole plant has a strong aromatic smell. The commonest spring-flowering umbellifer in many northern districts, especially near dwellings.

?Native. In grassy places, hedges and woods. 72, H16. Northwards from Glamorgan and Lincoln, rare and probably casual in the southern part of its range, common in northern England and southern Scotland, becoming rarer northwards; naturalized in Ireland and rare except in the north-east. Europe, mainly in mountainous districts; probably frequently an escape from cultivation though now well naturalized; Chile, probably introduced.

8. TORILIS Adans.

Annual herbs. Lvs 1–3-pinnate. Umbels compound; bracts several or 0, bracteoles several, subulate. Fls white or pinkish. Calyx-teeth small, triangular-lanceolate, persistent; petals with an inflexed point. Fr. ovoid, narrowed at the commissure; carpels with slender ciliate ridges, furrows thickly beset with tubercles or spines; vittae solitary in the furrows.

About 15 spp. in Europe, N. Asia and N. Africa.

Umbels long-peduncled, 5–12-rayed; bracts 4–12.	**1. japonica**
Umbels long-peduncled, (2–)3–5(–8)-rayed; bracts 0–1.	**2. arvensis**
Umbels nearly sessile, lf-opposed, very shortly 2–3-rayed; bracts 0.	**3. nodosa**

1. T. japonica (Houtt.) DC.   'Upright Hedge-parsley.'

T. Anthriscus (L.) C. C. Gmel.; *Caucalis Anthriscus* (L.) Huds.

An erect shortly appressed-hispid annual 5–125 cm. Stems solid, striate, hairs deflexed. Lvs 1–3-pinnate; *segments* 1–2 cm., *ovate* to lanceolate, pinnatifid, serrate. Umbels 1·5–4 cm. diam.; *rays* 5–12, hairs forward-pointing; *bracts* 4–6(–12), unequal, up to nearly as long as rays; bracteoles about equalling pedicels. Fls 2–3 mm. diam., pinkish- or purplish-white, outer petals larger than inner. *Fr. 3–4 mm., spines hooked; styles glabrous, recurved in fr.* Fl. 7–8. Germ. spring and autumn. 2n=12, 16. Th.

In many places the commonest roadside umbellifer flowering in July, just after *Chaerophyllum temulentum.*

Native. In hedges and grassy places. 109, H 40, S. Generally distributed throughout the British Is. except the extreme north of Scotland. Throughout Europe, except the north and a few Mediterranean islands; Caucasus, N. and E. Asia; N. Africa; introduced in N. America, S. Asia and Java.

2. T. arvensis (Huds.) Link   'Spreading Hedge-parsley.'

T. infesta (L.) Spreng.

An erect annual 10–40(–100) cm. Stems solid, terete below, ridged and ± angled above, sparsely deflexed-appressed-hispid or almost glabrous. Lvs 1–2-pinnate; *segments* 0·5–3 cm., *lanceolate*, pinnatifid or coarsely serrate, sparsely appressed-hispid on both surfaces. Umbels 1–2·5 cm. diam.; rays (2–)3–5(–8), hairs forward-pointing; *bracts* 0 *or* 1; bracteoles densely hispid, nearly equalling the few-fld partial umbels. Fls c. 2 mm. diam., white or pinkish, petals unequal. *Fr. 4–5 mm., spines curved but not hooked, thickened at tip; styles ± hairy, spreading in fr.* Fl. 7–9. Germ. autumn and spring. Th.

Native or introduced. In arable fields, scattered throughout England and Wales, probably often introduced with seed and now less frequent than formerly. 60. C. Europe (?native) to the Mediterranean region, eastwards to Persia and Turkistan; probably introduced elsewhere.

3. T. nodosa (L.) Gaertn.   'Knotted Hedge-parsley.'

An erect or more often prostrate annual 5–35 cm. Stems solid, striate, hairs sparse, spreading, deflexed. Lvs 1–2-pinnate; *segments ovate, deeply pinnatifid, lobes linear-lanceolate*. *Umbels* 0·5–1 cm., *subsessile, lf-opposed; rays 2–3, very short*; bracts 0; *bracteoles exceeding the subsessile fls*. Fls 1 mm. diam., pinkish, petals subequal. *Fr. 2–3 mm., outer carpels with straight, rarely hooked, spreading spines, inner tubercled*. Fl. 5–7. Germ. autumn and spring. 2*n* = 22. Th.

Native. On dry rather bare banks and arable fields. 82, H 28, S. Scattered throughout the British Is., locally common in the south. C. and S. Europe; W. Asia; N. Africa.

9. Caucalis L.

Annual herbs. Characters of *Torilis* but differing in the conspicuous herbaceous calyx-teeth and the fr. with inconspicuous, smooth or hispid primary ridges and prominent secondary ridges bearing 1–3 rows of spines.

About 8 spp. in north temperate regions.

Lvs 2–3-pinnate; fls c. 2 mm. diam.; bracts 0 (1–2).   **1. platycarpos**
Lvs 1-pinnate; fls c. 5 mm. diam.; bracts 3–5.   **2. latifolia**

*1. C. platycarpos L.   'Small Bur-parsley.'

C. Lappula Grande; *C. daucoides* L. 1767, non 1753.

An erect slightly pubescent annual 8–30 cm. *Stems solid, ± angular. Lvs 2–3-pinnate, nearly glabrous; segments* 0·5–1 cm., oblong, *pinnatifid*. Umbels 1–2 cm. diam.; *rays* 2–3(–5); *bracts* 0 *or* 1–2, *linear; bracteoles* 2–3, *linear, acute. Fls c. 2 mm. diam.*, white or pink; *inner of each partial umbel subsessile, hermaphrodite; outer long-pedicelled, male; petals subequal. Fr.* 8–10 mm. with 1 *row of spreading spines on the secondary ridges*. Fl. 6–7. 2*n* = 20. Th.

Introduced. A casual or ± naturalized in arable fields and waste places particularly on chalky soils; rather rare and less frequent than formerly. 48. Europe, from the Netherlands to C. Russia; Mediterranean region east to Persia.

*2. C. latifolia L.   'Great Bur-parsley.'

An erect hispid annual 15–60 cm. *Stems hollow, terete, striate. Lvs simply pinnate; segments* 0·5–2 cm., narrowly oblong, *coarsely serrate, nearly glabrous above, shortly hispid beneath*. Umbels 4–6 cm. diam.; *rays* 2–5; *bracts* 3–5, *broadly lanceolate; bracteoles* 3–5, *oblong, scarious. Fls* 4–5 mm. diam., pink; *inner of each partial umbel long pedicelled, male; outer subsessile, hermaphrodite; petals unequal. Fr.* 6–9 mm. with 2–3 *rows of spreading spines on the secondary ridges*. Fl. 7. 2*n* = ?32. Th.

Introduced. A casual in arable fields and waste places, rare. C. and S. Europe; Siberia and W. Asia; N. Africa.

10. Coriandrum L.

Slender glabrous annual herbs. Lvs 1–2-pinnate. Umbels compound or rarely simple; bracts 0, bracteoles few, lanceolate. Fls white or pink. Calyx-teeth acute, unequal; petals with an inflexed point. Fr. nearly spherical;

carpels ½-terete, primary ridges low, slender; secondary broader; vittae obscure, solitary beneath each secondary ridge.

Two sp. in the Mediterranean region.

*1. C. sativum L. Coriander.

An erect annual 20–70 cm. Stems solid, ridged. Lower lvs 1–2-pinnate or lobed; segments ovate, cuneate at base, pinnately lobed or toothed; upper lvs 2-pinnate, pinnatifid. Umbels 1–3 cm. diam.; rays 3–5(–10). Petals unequal, the larger 2–3 mm. *Fr.* 3–4 mm., *hard, red-brown, carpels adhering firmly to one another*; styles slender. Forms with 3 carpels and styles occur. Fl. 6. 2*n*=22. Th.

The whole plant smells strongly of bed-bugs.

Introduced. A casual in waste places, sometimes escaped from cultivation. Probably native of the eastern Mediterranean region; widespread as a weed.

11. SMYRNIUM L.

Erect glabrous biennial or perennial herbs. Radical lvs ternately compound, segments broad. Umbels compound; bracts and bracteoles small, few or 0. Fls yellow or greenish. Calyx-teeth minute or 0; petals with a short inflexed point. Fr. ovoid, laterally compressed, didymous, commissure narrow; carpels subterete or angular with 3 prominent sharp ridges; vittae numerous; styles short, recurved.

Seven spp. in Europe and N. Africa.

*1. S. olusatrum L. Alexanders.

A stout biennial 50–150 cm. Stem solid, becoming hollow when old, furrowed, branches in the upper part often opposite. *Lvs dark green and shiny, radical c. 30 cm., 3-ternate; segments rhomboid, obtusely serrate or lobed, stalked*; upper stem lvs often opposite, simply ternate, petiole base sheathing. Umbels axillary and terminal, subglobose; rays (3–)7–15, 1–5 cm.; bracts and bracteoles few. *Fls* c. 1·5 mm. diam., *yellow-green*, shortly pedicelled. *Fr.* c. 8 mm., *broadly ovoid, nearly black*. Fl. 4–6. 2*n*=22. Hs.

Formerly cultivated as a pot-herb; the young stems have somewhat the taste of celery.

Introduced. Extensively naturalized in hedges, waste places and on cliffs. 72, H40, S. Northwards to Dunbarton and Banff, especially near the sea, commoner in the south; throughout Ireland, but rather local except near the coast. S.W. Europe; Mediterranean region east to the Caucasus; Canary Is.

*S. perfoliatum L., a biennial with simple ovate or suborbicular denticulate or crenulate amplexicaul upper lvs, is occasionally naturalized. S. Europe to Asia Minor and the Caucasus.

12. PHYSOSPERMUM Cusson

Erect perennial herbs. Root fusiform. Lvs ternately compound with cuneate segments. Umbels compound; bracts few, narrow. Fls all hermaphrodite, white. Calyx-teeth shorter than the inflexed petals, sometimes 0. Fr. inflated, didymous, broader than long, commissure narrow; carpels smooth, ridges inconspicuous; vittae 1 in each groove.

About 5 spp. in Europe and W. Asia.

1. P. cornubiense (L.) DC.   Bladder-seed.

Danaa cornubiensis (L.) Burnat

A nearly glabrous perennial 30–75 cm. Stem ribbed. Lvs long-petioled, ternate; *segments* stalked, *cuneate, laciniate, puberulent on margins and nerves on both sides.* Umbels 1·5–5 cm. diam., long-peduncled; rays (6–)10–12(–15); bracts and bracteoles lanceolate, acute, entire. Fls c. 2 mm. diam., long-pedicelled; styles rather stout, spreading, recurved in fr. Fr. 4–5 mm., dark brown. Fl. 7–8. Hs.

Native. In bushy places. Cornwall and S. Devon; Buckinghamshire, perhaps introduced; very local. 3 or 4. Portugal and Spain; Corsica, W. and C. Italy; Balkans, Crimea, Asia Minor, Cyprus, Syria and the Caucasus.

13. Conium L.

Glabrous biennial herbs. Lvs pinnately compound. Umbels compound; bracts and bracteoles few, small; fls white, strongly protandrous. Calyx-teeth 0; petals obtuse or tip shortly inflexed. Fr. broadly ovoid to suborbicular, laterally compressed, commissure rather narrow; carpels with five prominent ridges; vittae 0; styles short, recurved.

Two spp. in north temperate regions and S. Africa. Very poisonous.

1. C. maculatum L.   Hemlock.

An erect branched foetid biennial up to 2 m. *Stems* furrowed, *smooth, purple-spotted, glaucous.* Lvs up to 30 cm., 2–3-pinnate, ovate to deltoid; segments 1–2 cm., oblong-lanceolate to deltoid, coarsely and deeply serrate. Umbels 2–5 cm. diam., terminal and axillary, shortly peduncled; rays 10–20, 2–3 cm.; bracts 2–5 mm., few, reflexed; bracteoles similar but smaller and only on one side of the partial umbel. Fls 2 mm. diam., petals with short inflexed tips. *Fr.* c. 3 mm., *suborbicular, ridges wavy-crenate.* Fl. 6–7. 2*n*=22. Hs.

Native. In damp places, open woods and near water. 108, H40, S. Throughout most of the British Is., less common in the north. Europe, Asia, N. Africa within the area bounded by Algeria, Norway, Finland, Altai, Baikal region, Persia, Abyssinia; also in the Canary Is.; chiefly on mountains in the eastern part of its range; introduced in eastern N. America, California, Mexico, West Indies, temperate S. America, and New Zealand.

14. Bupleurum L.

Annual or perennial glabrous herbs or rarely shrubs. Lvs simple, entire. Umbels compound, of few to many often unequal rays; bracts and bracteoles various. Fls yellow. Calyx-teeth 0; petals with inflexed points. Fr. laterally compressed, commissure broad; carpels usually with prominent ridges; vittae 1–5 in each groove, often disappearing as fr. ripens; styles short, reflexed.

About 100 spp. in Europe, Asia, Africa and N. America.

1 Shrub, ± evergreen.   **1. fruticosum**
    Herbs.   2

2 Lvs elliptic to suborbicular, upper perfoliate; bracts 0; bracteoles
    ± connate.   **2. rotundifolium**[1]
    At least the stem lvs linear-lanceolate to spathulate, not perfoliate; bracts
    present.   3

3 Perennial; stem hollow; bracteoles shorter than fls.   **5. falcatum**
    Annual; stem solid; bracteoles longer than fls.   4

4 Bracteoles ovate, concealing fls; fr. not granulate.   **3. baldense**[2]
    Bracteoles subulate, not concealing fls; fr. granulate.   **4. tenuissimum**

*1. B. fruticosum L.

A ± *evergreen* glaucous *shrub* up to 2·5 m. *Twigs slender, purplish when young.*
Lvs 5–8 cm., oblong-obovate to narrowly elliptic, mucronate. *Umbels*
7–10 *cm. diam.*, peduncles slender; bracts and bracteoles reflexed. Fl. 7–8.
$2n=14$. N.
  Introduced. Established in a few localities, especially in Worcestershire.
S. Europe, N. Africa and Syria.

2. B. rotundifolium L.   Hare's-ear, Thorow-wax.

An erect glaucous annual 15–30 cm. *Stem hollow. Lvs 2–5 cm., elliptic to*
suborbicular, apiculate, lower attenuate into petiole, *upper perfoliate. Umbels*
1–3 *cm. diam.*; rays 3–8, up to 1 cm.; *bracts 0; bracteoles ovate, margins green,*
connate at least at base, longer than the shortly pedicelled fls and suberect
in fr. Fls 1·5 mm. diam. Fr. 2–3 mm., blackish and somewhat pruinose, not
granulate, ovoid, ridges slender. Fl. 6–7. $2n=16$. Th.
  Native or introduced. In cornfields. From Devon and Kent to Wigtown
and N. Yorkshire, local and rarer in the north. 53. From the Mediterranean
to C. Europe, eastwards to Poland, S.W. Russia, the Caucasus and the trans-
Caspian plain; N. Africa; usually a weed of arable land; introduced in N.
America, Australia and New Zealand.

3. B. baldense Turra

B. aristatum auct.; *B. opacum* (Ces.) Lange

A slender erect annual 2–10(–25) cm. *Stem solid*, simple or divaricately
branched. *Lvs 1–1·5 cm., spathulate*, acute. *Umbels 5–10 mm. diam.*; rays
1–4, 2–5 mm.; *bracts few*, linear-lanceolate, acute, exceeding the rays;
bracteoles free, concealing fls, ovate, pungent, rigid, *margins whitish.* Fls
2 mm. diam., subsessile. Fr. 2–3 mm., ridges slender. Fl. 6–7. Th.
  Native. On dry banks, rocky slopes and grey dunes near the sea. 2, S.
S. Devon, E. Sussex and Channel Islands. S.W. Europe and Mediterranean
region east to Italy.

4. B. tenuissimum L.   'Smallest Hare's-ear.'

A slender erect or procumbent branched annual 15–50 cm. *Stem solid,*
flexuous, wiry. *Lvs 1–5 cm., linear-lanceolate to spathulate*, acute or acuminate,

[1] See also *B. subovatum* p. 512.   [2] See also *B. fontanesii*, p. 512.

rigid. *Umbels up to 5 mm. diam., axillary, sessile or shortly peduncled*; rays 1–3, unequal, often very short, partial umbels few-fld; *bracts* 2–5, 2–4 mm., *subulate*; bracteoles similar, exceeding fls. Fls 1 mm. *Fr.* 2 mm., black, *suborbicular, granulate*, ridges crenulate. Fl. 7–9. 2*n*=16. Th.

Native. In salt marshes and waste places. 30. From Cornwall and Kent to Lancashire and Durham, usually near the coast; local and rarer in the north. Europe from the Mediterranean to southern Scandinavia and east to the Caspian and Persia.

5. B. falcatum L.

An erect *perennial* 50–130 cm. *Stem hollow. Lvs* 3–8 cm., *lower petioled, narrowly obovate, upper linear-lanceolate, ½-amplexicaul.* Umbels 1–4 cm. diam., peduncled; rays 5–11, 5–20 mm., unequal; bracts 2–5, 2–7 mm., lanceolate-acuminate, unequal; *bracteoles* similar, 4–5, somewhat *shorter than the pedicelled fls.* Fls 1 mm. *Fr.* 3–4 mm., *oblong*, ridges prominent. Fl. 7–10. 2*n*=16, 28. H.

?Native. Waste places and hedgebanks. 2. Surrey (extinct), E. Essex. Scattered throughout Europe and Asia north to the subarctic regions and eastwards to Japan.

*B. fontanesii Guss. ex Caruel with lanceolate-acuminate, membranous, prominently nerved bracteoles exceeding the fls, and umbels c. 5 cm. diam., with 5–8 rays, and **B. lancifolium* Hornem. (*B. subovatum* auct.) which somewhat resembles *B. rotundifolium* L. but has narrow lvs and granulate fr., also occur as casuals. Both are native of the Mediterranean region.

15. TRINIA Hoffm.

Glabrous usually dioecious herbs. Lvs pinnately compound. Umbels compound; bracts and bracteoles 1–3, sometimes 0. Fls white. Calyx-teeth minute or 0; petals acute, point sometimes inflexed. Fr. broadly ovoid, laterally compressed or didymous, commissure narrow; carpels subterete with thick, often prominent ridges; styles short, recurved.

About 12 spp. in S. Europe and temperate Asia.

1. T. glauca (L.) Dum.   Honewort.

T. vulgaris DC.; *Apinella glauca* (L.) Caruel

An erect *glaucous* perennial 3–20 cm. *Root* stout, *with fibrous remains of petioles at top. Stem* solid, deeply grooved, *branched from base.* Lvs 3-pinnate; *segments linear or filiform*; petioles slender. *Male umbels* c. 1 cm. diam., *flattopped*, rays 4–7, c. 5 mm., equal; *female* c. 3 cm. diam., *rays irregular*, up to 4 cm. Bracts 0 or 1, 3-cleft; bracteoles 2–3, simple. Fls minute. Fr. c. 2 mm., broadly ovoid, ridges rather prominent, smooth and broad. Fl. 5–6. 2*n*=18*. Hs.

Native. In dry limestone grassland. 3. S. Devon, N. Somerset and Bristol. S.W. Europe and eastwards to Austria and S.W. Germany.

16. APIUM L.

Annual, biennial or perennial glabrous herbs. Lvs pinnate or ternate. Umbels often lf-opposed; bracts few or 0, bracteoles several or 0. Fls white. Calyx-teeth 0 or minute; petals entire, acute, point sometimes shortly inflexed. Fr. broadly ovoid or elliptic-oblong, laterally compressed, commissure narrow; carpels with 5 equal or rarely somewhat unequal ridges.

About 45 spp., cosmopolitan.

1 Lower lvs with deltoid to rhomboid segments, lower segments stalked; bracteoles 0. **1. graveolens**
  Lower lvs with sessile or capillary segments; bracteoles present. *2*

2 Lower lvs with deeply lobed segments, lobes usually capillary or linear; segments of upper lvs cuneate, often 3-lobed; fr. elliptic-oblong. **3. inundatum**
  Lvs with serrate, sometimes slightly lobed segments, never linear or capillary; fr. broadly ovoid. **2. nodiflorum**

1. A. graveolens L. Wild Celery.

An erect strong-smelling biennial 30–60 cm. Stem grooved. Radical lvs simply pinnate; *segments* 0·5–3 cm., *deltoid to rhomboid*, lobed and serrate, *lower stalked,* upper sessile; upper stem lvs ternate, narrowly rhomboid to lanceolate, subentire. Umbels terminal and axillary, shortly peduncled or sessile in the axil of a small ternate lf; rays unequal; *bracteoles* 0. Fls 0·5 mm., greenish-white. Fr. c. 1·5 mm., broadly ovoid. Fl. 6–8. $2n=22$. Hs.

Native. In damp places, by rivers and ditches, especially near the sea. 62, H25, S. Mainly in the maritime counties, rather local, rarer in the north and absent from the extreme north of Scotland; Ireland, around the whole coast, local. Europe from Denmark to S. Russia; W. Asia to the East Indies; N. and S. Africa; S. America; introduced in other temperate countries.

The cultivated celery (*var. **dulce** (Mill.) DC.) is sometimes found by roads; celeriac or turnip-rooted celery is another variety (var. *rapaceum* (Mill.) DC.).

2. A. nodiflorum (L.) Lag. Fool's Watercress.

Helosciadium nodiflorum (L.) Koch

A procumbent or ascending perennial 30–100 cm. Stems finely furrowed, slender or stout, flowering ones rooting at base only. Lvs simply pinnate, bright green, shiny; *segments* 1–3·5 cm., 4–6 pairs, *lanceolate to ovate,* serrate or crenate, often slightly lobed, *sessile.* Umbels lf-opposed, sessile or shortly peduncled; rays unequal, spreading or recurved; bracts usually 0; *bracteoles* c. 5, narrowly lanceolate, as long as the shortly pedicelled fls. Fls 0·5 mm., white. *Fr.* c. 2 mm., *ovoid, longer than broad,* ridges all equal. Fl. 7–8. $2n=22*$. Hel.

Sometimes mistaken for watercress, with which it often grows, and eaten, apparently without ill effects.

Native. In ditches and shallow ponds. 98, H40, S. Common and generally distributed except in north Scotland. C. and S. Europe; S.W. Asia to Persia; N. Africa, Abyssinia; introduced in N. America and Chile.

A. repens (Jacq.) Rchb. f. is probably not specifically distinct from the preceding sp. The stems root at every node, the lf-segments are usually broader and the fr. is broader than long with alternate large and small ridges. Recorded from Oxford, E. Yorks and Fife. W. Europe.

3. A. inundatum (L.) Rchb. f.

Helosciadium inundatum (L.) Koch

A straggling often submerged or floating perennial 10–50 cm. Stems slender, nearly smooth. Lvs pinnate; *segments of lower lvs deeply pinnatifid, lobes capillary in submerged lvs*, linear in floating or aerial lvs; *segments of upper lvs* 0·5–1 cm., *cuneate, often 3-lobed*, sessile. Umbels lf-opposed, peduncles 1–3·5 cm.; rays 2–3(–4), 0·5–1 cm., somewhat unequal; bracts 0; bracteoles 3–6, lanceolate, obtuse, unequal. Fls 1 mm., white, subsessile. *Fr.* 2 mm., *elliptic-oblong*. Fl. 6–8. $2n = 22$. Hyd.

Native. In lakes, ponds and ditches. 104, H39, S. Local but widely distributed throughout the British Is. Europe from Spain to S. Sweden, eastwards to C. Russia and Italy; Tunisia and Algeria; rare in the south and east.

A. × moorei (Syme) Druce resembles *A. inundatum* but is generally larger and has the segments of the lower lvs linear or strap-shaped; it is a sterile hybrid between that sp. and *A. nodiflorum*. Frequent in Ireland, rare elsewhere.

17. PETROSELINUM Hill

Annual or biennial herbs, root fusiform. Lvs pinnate, segments broad. Umbels compound; bracts and bracteoles several. Fls white or yellowish. Calyx-teeth 0; petals scarcely notched with a small inflexed point. Fr. ovoid or subspherical, laterally compressed, carpels with 5 slender ridges; vittae solitary in the furrows.

About 5 spp. in Europe.

Lvs 3-pinnate; fls yellowish.   **1. crispum**
Lvs 1-pinnate; fls white.   **2. segetum**

*1. P. crispum (Mill.) Nyman   Parsley.

P. sativum Hoffm.; *Carum Petroselinum* (L.) Benth.

A stout erect glabrous biennial 30–75 cm. Tap-root stout, fusiform. Stem terete, solid, striate, *branches ascending, strict*. Lvs *deltoid*, 3-*pinnate*, shiny; *segments* 1–2 cm., *cuneate*, lobed, often much crisped in cultivated forms, upper stem lvs often ternate. *Umbels* 2–5 cm. diam., *flat-topped*; rays 8–15, 1–2 cm.; bracts 2–3, erect, entire or 3-lobed, with a broad sheath-like lower part with hyaline margins; bracteoles c. 5, linear-oblong to ovate-cuspidate, often with hyaline margins. *Fls* 2 mm., *yellowish*. Fr. 2·5 mm., ovoid. Fl. 6–8. $2n = 22^*$. Hs.

Introduced. In grassy waste places, on walls and rocks, escaped from cultivation and naturalized. 47, H26, S. Scattered throughout the British Is. except northern Scotland. Probably native of S. Europe from Sardinia to Greece. Escaped from cultivation and ± naturalized in almost all temperate regions.

2. P. segetum (L.) Koch   'Corn Caraway.'

Carum segetum (L.) Benth. ex Hook. f.

A slender glabrous dark green ± glaucous biennial 30–100 cm. Tap-root rather slender, fusiform. *Stem* terete, striate, solid, *divaricately branched*, lower branches prostrate or nearly so. *Lvs linear-oblong, simply pinnate; segments* 0·5–1 cm., *ovate*, subsessile, serrate or sometimes lobed, margins thickened, teeth with forward-curving cartilaginous points. *Umbels* 1–5 cm. diam., *very irregular*; rays 2–5, 0·2–4 cm.; bracts and bracteoles 2–5, subulate; partial umbels of few (often 3–5) fls. *Fls* 1 mm., *white*, subsessile to long- (15 mm.) pedicelled. Fr. 3–4 mm., ovoid. Fl. 8–9. $2n=18$. Hs. Stem and fr. have a smell reminiscent of parsley.

Native. In hedgerows and grassy places. 51, S. Cornwall and Kent to Stafford and S. Yorkshire; Monmouth, Glamorgan, Pembroke, Denbigh. W. and S. Europe; Asia Minor; N. Africa; Canary Is.; introduced in C. Europe.

18. SISON L.

Biennial herbs. Roots fusiform. Lvs pinnate, segments broad. Umbels compound; bracts and bracteoles few, rarely 0. Fls white. Calyx-teeth 0; petals suborbicular-cordate, notched with an inflexed point. Fr. broadly ovoid or subglobose, laterally compressed; carpels with 5 slender ridges; vittae very short, in the upper half of the carpel only.

Two spp. in Europe and the Mediterranean region.

1. S. amomum L.   Stone Parsley.

An erect glabrous biennial 50–100 cm., *with a nauseous smell somewhat resembling that of nutmeg mixed with petrol*. Stems solid, finely striate, branches ascending. Lvs simply pinnate, lower 10–20 cm., long-petioled, segments 2–7 cm., sessile or subsessile, oblong-ovate, serrate and often lobed, margins thickened, teeth with forward-curving cartilaginous points; upper stem lvs usually ternate with spathulate or linear, lobed or toothed segments. Umbels 1–4 cm. diam., terminal and axillary, peduncles slender; rays 3–6, 0·5–3 cm., slender; bracts and bracteoles 2–4, rarely 0, subulate, bracts usually spreading or reflexed. Fls 1 mm., white or greenish-white. Fr. 3 mm., subglobose. Fl. 7–9. Hs.

Resembles *Petroselinum segetum* but differs in the smell, the bright green colour, the larger lvs, the finely divided upper stem lvs, the larger less irregular umbels with more fls in the partial umbels, and the broader fr.

Native. On hedgebanks and roadsides. 57. Throughout England and most of Wales north to Cheshire and S. Yorkshire. S. and S.W. Europe from France to the Mediterranean; Asia Minor; Algeria.

19. CICUTA L.

Perennial glabrous herbs. Lvs pinnate. Umbels compound, rays many; bracts few or 0, bracteoles many, narrow. Fls white. Calyx-teeth acute; petals with inflexed points. Fr. orbicular or broadly ovoid; carpels with

5 equal broad flattened ridges; vittae solitary in the furrows. Pollinated by insects.

About 6 spp. in north temperate regions. Highly poisonous.

1. C. virosa L.   Cowbane.

A stout erect perennial 30–130 cm. Stem somewhat ridged, hollow. Lvs up to 30 cm., deltoid, 2–3-pinnate; segments 2–10 cm., linear-lanceolate, acutely serrate, unequal at base; petiole long, stout, hollow. Umbels terminal, and lateral and lf-opposed, 7–13 cm. diam., flat-topped; rays many, 1–5 cm., partial umbels many-fld, dense; bracts 0; bracteoles many, strap-shaped, longer than the pedicels. Fls 3 mm., calyx-teeth ovate. Fr. c. 2 mm., broader than long. Fl. 7–8. $2n = 22$. Hel. or Hyd.

Native. In shallow water, ditches and marshes. 36, H16. Scattered throughout Great Britain, very local; north-east and north-central Ireland, extending to Meath and Galway. N. and C. Europe, rare or absent in the south; Asia eastwards to Japan.

20. AMMI L.

Branched annual herbs. Lvs pinnate or pinnatifid. Umbels compound; bracts several, long, pinnatifid; bracteoles numerous. Fls white. Calyx-teeth 0; petals obovate, irregularly and unequally 2-lobed with an inflexed notched point. Fr. oblong-ovoid, laterally compressed; carpels with slender ridges; vittae solitary in the furrows.

About 7 spp. from S. Europe to tropical Africa and Macaronesia.

*1. A. majus L.

A glaucous glabrous annual 15–100 cm. Lvs 1–2-pinnate or pinnatifid, segments 1–3 cm.; radical lvs 1–2-pinnate; segments oblong-ovate or spathulate, sometimes pinnatifid; stem lvs 2-pinnate or pinnatifid, segments narrowly lanceolate or linear-lanceolate; segments of all lvs serrate, teeth ending in fine cartilaginous points. Umbels terminal, 3–6 cm. diam.; rays 10–30, 1–4 cm.; bracts several, $\frac{1}{2}$–$\frac{2}{3}$ length of rays, mostly pinnatifid with linear segments; bracteoles subulate, about equalling the pedicels. Fls 3 mm. Fr. c. 2 mm., oblong or ovoid, ridges prominent, pale. Fl. 6–10. $2n = 22$. Th.

Introduced. A casual in waste places, rather rare. Mediterranean region from the Canary Is. to Persia; Abyssinia; introduced in most temperate regions.

21. FALCARIA Bernh.

Biennial or perennial usually glabrous herbs. Lvs ternately pinnate. Umbels compound; bracts and bracteoles numerous, usually large. Fls white. Calyx-teeth small but distinct; petals nearly equal, obcordate with an inflexed point. Fr. oblong-ovoid; carpels with 5 low, blunt ridges; vittae slender, solitary in the furrows.

Species 2–3 in C. and S. Europe and western Asia.

*1. F. vulgaris Bernh.

A glaucous much-branched perennial 30–50 cm. *Lvs ternate or 2-ternate, segments linear-lanceolate or linear, somewhat falcate, strongly sharply and*

regularly serrate, up to 30 *cm. in the lower lvs. Bracts and bracteoles linear.*
Fl. 7. $2n = 22$. H.

Introduced. Naturalized in East Anglia, S.E. England and Guernsey. Europe from S. Sweden south and eastwards; western Asia to the Altai.

22. Carum L.

Perennial or biennial herbs. Root fibrous or fusiform. Lvs pinnate, segments usually narrow. Umbels compound; bracts and bracteoles several, few or 0. Fls white or pinkish. Calyx-teeth minute; petals deeply notched with an inflexed point. Fr. ovoid or oblong, laterally compressed; carpels with 5 slender ridges.

About 25 spp. in Europe, N. Africa, temperate Asia and N. America.

Lvs 1-pinnate, segments palmately multifid, lobes capillary; bracts and
bracteoles several, reflexed.   **1. verticillatum**
Lvs 2-pinnate, segments pinnatifid, lobes linear-lanceolate or linear; bracts
and bracteoles 0 or 1, not reflexed.   **2. carvi**

1. C. verticillatum (L.) Koch   'Whorled Caraway.'

An erect glabrous perennial 30–60 cm. *Root of fusiform fibres thickened downwards. Stem little branched,* striate, *solid, nearly lfless. Lvs simply pinnate,* linear-oblong; *segments palmately multifid; lobes* up to 5 mm., *capillary, appearing as if whorled.* Umbels 2–5 cm. diam., flat-topped; rays 8–12, 2–5 cm.; *bracts and bracteoles several, lanceolate,* acuminate or aristate, short, *reflexed.* Fls c. 1 mm., white or pinkish. *Fr. c.* 2 *mm., ovoid, ridges prominent, acute.* Fl. 7–8. $2n = 20$. Hr.

Native. In damp grassy places, calcifuge and with one exception found only in the west of the country. 31, H6, S. E. Cornwall, Devon, Dorset, Surrey, Wales, Cumberland to Inverness; Outer Hebrides; Jersey; Kerry, W. Cork, Donegal, Antrim and Londonderry. From Germany west of the Rhine (very rare) through the Netherlands, Belgium, France to western Spain and Portugal.

2. C. carvi L.   Caraway.

An erect *much-branched* glabrous biennial 25–60 cm. *Tap-root fusiform. Stem* striate, *hollow, lfy. Lvs* all radical in the first year, 2-*pinnate,* narrowly triangular to linear-oblong; *segments deeply pinnatifid, lobes linear-lanceolate or linear.* Umbels 2–4 cm. diam., rather irregular; rays 5–10, 0·5–3 cm.; *bracts and bracteoles* 0 *or* 1, *setaceous.* Fls 2–3 mm., outer rather larger than inner, white. *Fr.* 3–4 *mm., oblong,* strong smelling when crushed, *ridges low, obtuse.* Fl. 6–7. $2n = 20$. Hs.

Seeds used for flavouring.

Perhaps native in some south-eastern counties and naturalized in waste places. 49, H26. Scattered throughout the British Is., rather rare. N. and C. Europe south to the Pyrenees, N. Italy and Serbia; Asia: throughout Siberia and south to N. Persia and the Himalaya; Morocco; introduced elsewhere.

23. BUNIUM L.

Perennial herbs. Root a solitary tuber. Lvs 2–3-pinnate, segments narrow. Umbels compound; bracts and bracteoles several. Fls white. Calyx-teeth minute; petals obovate, emarginate, point inflexed. Fr. oblong-ovoid, laterally compressed; carpels with 5 slender ridges. Seedlings with only one cotyledon.

About 30 spp. in Europe and temperate Asia.

1. B. bulbocastanum L.

A glabrous erect perennial 30–70 cm. *Tuber globose*, black, 1–2·5 cm. diam., edible. *Stems* terete, striate, *solid*. Lvs 10–15 cm., broadly triangular, 3-pinnate, mostly from the lower part of the stem, rarely a few radical, usually dead by flowering time; primary divisions stalked; segments 1–1·5 cm., spathulate to subulate with a blunt cartilaginous point. Umbels 3–8 cm. diam., flat-topped; rays 10–20, 1·5–3 cm.; bracts and bracteoles numerous, small, lanceolate to linear-lanceolate, acute. Fls 2 mm., white. *Fr.* 3 mm., *oblong-ovoid*, dark brown with paler inconspicuous ridges; *styles* short, *recurved*. Fl. 6–7. $2n = 22$. G.

Native. On the chalk in rough grassland and on banks, very local. 4. Hertford, Buckingham, Cambridge and Bedford. The Netherlands, W. Germany, Belgium, eastern France, W. Switzerland to Italy.

24. CONOPODIUM Koch

Glabrous or pubescent perennial herbs. Rootstock tuberous. Lvs ternately divided, segments narrow. Umbels compound; bracts and bracteoles few or 0, membranous. Fls white; lobes of the disk conical. Calyx-teeth 0; petals notched with an inflexed point, those of the outer fls often unequal. Fr. ovoid or oblong often shortly beaked, commissure constricted; carpels subterete, ridges slender, inconspicuous; vittae several in each furrow.

About 17 spp. in Europe, N. Africa and temperate Asia.

1. C. majus (Gouan) Loret Pignut, Earthnut.

C. denudatum Koch; *Bunium flexuosum* Stokes.

A slender erect glabrous perennial 30–50(–90) cm. Tuber 1–3·5 cm., dark brown, irregular. *Stem hollow* after fl., finely striate, flexuous, much attenuated near its junction with the tuber. Radical lvs 5–15 cm., soon withering, broadly deltoid, petioles slender, flexuous, mostly subterranean; primary divisions stalked, 2-pinnate, segments deeply pinnatifid, lobes linear-lanceolate; stem lvs smaller, petioles short and sheathing, lobes linear, the terminal much the longest. Umbels 3–7 cm. diam., terminal, nodding in bud; rays 6–12, 1–3 cm., slender; bracts and bracteoles 0–5, membranous, subulate, variable in length. Fls 1–3 mm., usually unisexual, disk lobes conical. *Fr.* 4 mm., *narrow-ovoid*, beaked; *styles* short, *erect*. Fl. 5–6. G.

The radical lvs wither about flowering time and the stem lvs by the time the fr. is ripe. Tubers edible raw or cooked. Resembles *Bunium bulbocastanum* but differs in the hollow stems, less numerous bracts and bracteoles, erect styles and the shape of the fr.

Native. In fields and woods. 112, H 40, S. Generally distributed throughout the British Is.; common except on chalk and in fens. Norway, France, Spain and Portugal to N. and W. Italy and Corsica.

25. PIMPINELLA L.

Perennial, rarely annual herbs. Lvs usually pinnate. Umbels compound; bracts 0, bracteoles 0 or few. Fls white or pinkish in our spp. Calyx-teeth small or 0; petals with a long inflexed point. Fr. ovoid or oblong, laterally compressed, commissure broad; carpels with 5 slender ridges; vittae numerous in each furrow; styles long or short. The fls are strongly protandrous and usually hermaphrodite, though sometimes unisexual. Pollinated by insects.

About 90 spp. mainly in the Mediterranean region, but a few spp. widespread in Europe and Asia.

Stem puberulent, subterete, tough.   **1. saxifraga**
Stem glabrous, strongly ridged or angled, brittle.   **2. major**

1. P. saxifraga L.   'Burnet Saxifrage.'

An erect usually puberulent rather slender perennial 30–100 cm. Rootstock rather slender, usually crowned with fibrous remains of petioles. *Stem subterete, slightly ridged, usually rough, tough.* Radical lvs and upper stem lvs usually simply pinnate, *lower stem lvs usually* 2-*pinnate*, very variable; segments 1–2·5 cm., ovate or deltoid to linear-lanceolate, obtusely or acutely serrate to pinnatifid, *sessile or subsessile,* those of the radical lvs broader and less divided than those of the stem lvs; uppermost stem lvs small or 0, petioles sheath-like, often purplish. Umbels terminal, 2–5 cm. diam., flat-topped; rays 10–20, 1–3 cm.; bracts and bracteoles 0. Fls 2 mm., white; *styles much shorter than petals.* Fr. 3 mm., broadly ovoid. Fl. 7–8. $2n=18, 36$. Hs.

Native. In dry grassy places. 107, H 37. Scattered throughout the British Is., rarer in the north; mainly calcicole. Most of Europe but absent from much of the south and from the arctic; Asia Minor; W. Siberia; ?E. Asia.

2. P. major (L.) Huds.   'Greater Burnet Saxifrage.'

P. magna L.

An erect stout nearly glabrous perennial 50–120 cm. Rootstock rather stout, usually devoid of fibrous remains of petioles. *Stem prominently ridged or angled, glabrous, brittle,* often magenta towards base. *Lvs all simply pinnate* (rarely bipinnate); *segments* 2–8 cm., those *of radical lvs* ovate, subcordate, *shortly stalked,* those of stem lvs narrower, sessile, all coarsely serrate and sometimes 3-lobed; uppermost stem lvs small, often ternate, rarely 0, petioles sheath-like, green. Umbels terminal, 3–6 cm. diam., flat-topped; rays 10–20, 1–3 cm.; bracts and bracteoles 0. Fls 3 mm., white or pinkish; *styles about as long as petals.* Fr. 4 mm., ovoid. Fl. 6–7. $2n=18$. Hs.

Native. In grassy places at margins of woods and in hedgebanks. 54, H 15. Scattered throughout the British Is. north to W. Perth and Clackmannan, local and rarer in the north, probably absent from Wales; in S.W. Ireland extending to Waterford, Leix and Mayo. Europe, except the extreme north, Portugal and S. Balkans; Caucasus.

26. Aegopodium L.

Glabrous rhizomatous perennial herbs. Lvs 1–2-ternate, segments broad. Umbels compound; bracts and bracteoles few or 0. Fls white. Calyx-teeth 0; petals somewhat unequal, point inflexed. Fr. ovoid, laterally compressed; carpels with 5 slender ridges; vittae 0.

Two spp. in Europe and Siberia.

1. A. podagraria L. Goutweed, Bishop's Weed, Ground Elder, Herb Gerard.

A stout erect glabrous perennial 40–100 cm. *Rhizomes far-creeping*, white when young. Stem hollow, grooved. Lvs 10–20 cm., deltoid; *segments 4–8 cm.*, sessile or shortly stalked, lanceolate to ovate, acuminate, often oblique at base, *irregularly serrate*; petioles much longer than blade, bluntly triquetrous. Umbels 2–6 cm. diam., terminal; rays 15–20, 1–4 cm.; bracts and bracteoles usually 0. Fls 1 mm. Fr. 4 mm., ovoid; styles slender, reflexed. Fl. 5–7. $2n=22, 44$. Hs.

?Introduced. Said to have been introduced and cultivated as a pot-herb, now very well naturalized in waste places near buildings and a persistent weed in gardens. 112, H40, S. Generally distributed throughout the British Is. Nearly the whole of Europe except parts of the south; Asia Minor; Caucasus, Siberia.

27. Sium L.

Glabrous herbs. Lvs pinnate, segments broad. Umbels compound, terminal or lateral; bracts and bracteoles many. Fls white. Calyx-teeth 5, small, acute; petals with a short inflexed point. Fr. ovoid, laterally compressed or constricted at the commissure; carpels with 5 obtuse or thickened, prominent ridges, the lateral ones marginal; vittae 3 or more in each furrow, superficial. Stylopodium depressed.

North temperate regions and tropical Africa.

1. S. latifolium L.   Water Parsnip.

A stout erect perennial up to 2 m. *Stem* hollow, *grooved*. Lvs c. 30 cm., long-petioled, simply pinnate; *segments* 2–15 cm., sessile, 4–6(–8) *pairs*, oblong-lanceolate to ovate, *all regularly serrate* or when submerged finely divided; petioles sheathing at base, fistular above. *Umbels* 6–10 cm. diam., *terminal, flat-topped; rays usually* 20 *or more*, 2–5 cm.; bracts and bracteoles variable, often large and lfy. Fls c. 4 mm. diam. *Fr.* 3 mm., ovoid, *longer than broad*. Fl. 7–8. $2n=20$. Hel.

Native. In fens and other wet places. 37, H12. Scattered throughout the British Is. northwards to southern Scotland, rare; Ireland, recently recorded only from the basins of Shannon and Erne. Most of Europe, except from Norway to N. Russia and Portugal, Greece and Turkey.

28. Berula Koch

Similar to *Sium* but differing in the ± orbicular, nearly didymous fr. with the lateral ridges not marginal; vittae deeply embedded; stylopodium shortly conical.

North temperate regions.

1. B. erecta (Huds.) Coville　　　　　'Narrow-leaved Water-parsnip.'

Sium erectum Huds.; *S. angustifolium* L.

An erect or decumbent stoloniferous perennial 30–100 cm. *Stem* hollow, *striate*. Lvs up to 30 cm., long-petioled, simply pinnate, dull, somewhat blue-green; *segments* 2–5 cm., sessile, (5–)7–10 *pairs*, oblong-lanceolate to ovate, sharply serrate or slightly lobed; *stem lvs small, usually very irregularly serrate*. *Umbels* 3–6 cm. diam., *lf-opposed, rather irregular; rays usually* 10–15, 1–3 cm.; bracts and bracteoles many, lf-like, often trifid. Fls 2 mm. *Fr.* 2 mm., almost orbicular, *broader than long.* Fl. 7–9. $2n=18$; 12. Hel.

Native. In ditches, canals, ponds, fens and marshes. 89, H 38. Throughout the British Is. north to Shetland but uncommon in Scotland and N. Ireland. Europe except the Artic and subarctic; W. and C. Asia; N. America.

The lvs of this plant are often confused with those of *Apium nodiflorum* (p. 513). They differ in colour and in the usually greater number of pairs of segments.

29. CRITHMUM L.

A fleshy glabrous herb, woody at the base. Lvs 1–2-ternate. Umbels compound; bracts and bracteoles numerous. Fls yellowish-green. Calyx-teeth 0; petals with a long inflexed point. Fr. ovoid-oblong, terete, commissure broad; carpels with thick, acute primary ridges; vittae several in each furrow.

One sp. on the Atlantic coasts of Europe from Britain southwards, the coasts of the Mediterranean, the Black Sea, and Macaronesia.

1. C. maritimum L.　　　　　　　　　　　　　　　Rock Samphire.

A branched perennial 15–30 cm. Stems solid, striate. Lvs deltoid, *segments* 1–4 cm., terete, *subulate or subfusiform, fleshy*, acute; petioles short, sheaths long, membranous, enfolding the stem. Umbels 3–6 cm. diam.; rays 8–20, rather stout; *bracts and bracteoles numerous*, lanceolate to linear-lanceolate, acute, ± membranous, spreading, at length reflexed. Fls c. 2 mm. diam. Fr. c. 6 mm., corky, olive-green to purplish. Fl. 6–8. $2n=20$, 22. Hp.

Lvs make a good pickle.

Native. On cliffs and rocks or more rarely shingle or sand by the sea. 33, H 20, S. Coasts of Lewis and Islay, and from Ayr to Cornwall and eastwards to Kent and Suffolk; coasts of Ireland, local in the north. Distribution of the genus.

30. SESELI L.

Biennial or perennial herbs. Lvs usually 2–3-pinnate. Umbels compound; bracts many, few or 0; bracteoles many, entire. Fls white. Calyx-teeth prominent (in our sp.); petals notched, with a long inflexed point. Fr. ovoid or oblong, subterete, commissure broad; carpels dorsally compressed, ridges prominent; vittae solitary, rarely 2 or 3, in the furrows.

About 60 spp. in Europe, N. Africa, temperate Asia and Australia.

1. S. libanotis (L.) Koch

An erect somewhat pubescent or nearly glabrous perennial 30–60 cm. *Rootstock* stout, *crowned with fibrous remains of petioles*. Stems solid, somewhat ridged. Lvs c. 10 cm., 2-pinnate; segments sessile, ovate, pinnatisect, lobes oblong, mucronate, ± ciliate. Umbels 3–6 cm. diam., terminal, convex in fl.; *peduncles* long, *pubescent at top*; rays 15–30, 1–3 cm., pubescent; *bracts* and bracteoles *many*, subulate, pubescent, spreading or deflexed. Fls 1–1·5 mm. diam., calyx-teeth subulate. Fr. 3 mm., ovoid, pubescent. Fl. 7–8. $2n=18$, 22. Hs.

Native. In rough grassy and bushy places on chalk hills. 4. Sussex, Hertford, Cambridge and Bedford, very local. Europe eastwards to C. Russia but absent from the north-west; Asia Minor and Persia; N. Africa. An allied sp. eastwards to China and Japan.

31. Oenanthe L.

Glabrous usually marsh or aquatic herbs. Lvs 1–3-pinnate, rarely some reduced to a fistular petiole. Umbels compound; bracts several, few or 0; bracteoles (in our spp.) many. Fls white. Calyx-teeth acute; petals notched, with a long inflexed point. Fr. ovoid, cylindrical or globose, commissure broad; carpels ½-terete, 2 lateral ridges grooved or thickened, sometimes obscure; vittae solitary in the furrows.

About 35 spp. in the north temperate regions of the Old World.

1	Segments of upper lvs spathulate to linear.	*2*
	Segments of upper lvs lanceolate to ovate.	*4*
2	Pinnate part of stem lvs shorter than the hollow petiole.	**1. fistulosa**
	Pinnate part of stem lvs longer than the solid or flattened petiole.	*3*
3	Roots with rounded tubers towards their ends; stem solid; rays 0·5–1 mm. diam. in fr.; bracteoles subulate; pedicels thickened at top.	**2. pimpinelloides**
	Roots fusiform or tuberous from base; stem hollow; rays 1–2 mm. diam. in fr.; bracteoles ovate-acuminate; a strong constriction at junction of fr. and pedicels.	**3. silaifolia**
	Roots cylindrical or somewhat fusiform; stem solid (rarely hollow); rays c. 0·25 mm. diam. in fr.; bracteoles oblong-lanceolate; pedicels neither thickened nor constricted at top.	**4. lachenalii**
4	Roots tuberous; umbels not lf-opposed, most with peduncles longer than the rays.	**5. crocata**
	Roots fibrous; at least some umbels lf-opposed with peduncles shorter than the rays.	*5*
5	Segments of submerged lvs (if present) capillary; fr. 3–4 mm., about twice as long as styles.	**6. aquatica**
	Segments of submerged lvs cuneate, cut at ends; fr. 5–6 mm., at least 3 times as long as styles.	**7. fluviatilis**

1. O. fistulosa L.   Water Dropwort.

An erect perennial 30–60 cm. *Roots with fusiform tubers. Stems* slender, *fistular*, often constricted at nodes, rooting at lower nodes. Lvs 1- (2-)pinnate; segments of lower lvs shortly stalked, ovate, lobed; of upper 0·5–2 cm., linear-lanceolate or subulate, entire, distant; *petioles* long, *fistular*. Umbels terminal, peduncled; *rays* 2–4, 1–2 cm., stout, spreading. *Partial umbels* c. 1 cm. diam., dense, flat-topped in fl., *spherical in fr.; bracts* 0; *bracteoles* many, *shorter than pedicels.* Fls c. 3 mm. diam., petals of outer unequal. Fr. 3–4 mm., angular; *pedicels not thickened* at top; *styles 4–5 mm., slender, diverging.* Fl. 7–9. $2n=22$. Hel. or Hs.

Native. In marshy places and shallow water. 77, H32, S. England and southern Scotland north to Dunbarton and Perth; Wales, local; Ireland, very rare in the west. Europe, from Gotland southward; Asia Minor; S.W. Asia; western N. Africa.

2. O. pimpinelloides L.

An erect branched perennial 30–100 cm. *Roots with rounded tubers towards their ends. Stem solid,* furrowed. Lvs 2-pinnate, segments of lower lvs c. 5 mm., stalked, ovate or lanceolate, lobed, of upper 1–3 cm., linear-lanceolate to linear, usually entire. Umbels 2–5 cm. diam., terminal; *rays* 6–15, 1–2 cm., *rather stout (0·5–1 mm. diam.) in fr.*; partial umbels dense, flat in fl. and fr.; *bracts several,* subulate, unequal; *bracteoles* many, *subulate, about equalling pedicels.* Fls 3–4 mm. diam., petals of outer unequal. Fr. c. 3 mm., cylindrical, strongly ribbed; *pedicels thickened at top; styles* 3 *mm., stout, erect.* Fl. 6–8. $2n=22$. Hs.

Native. In meadows and damp grassy places. 19, H1. Cornwall and Kent to Worcester and Essex; Ireland, recorded from Cork only. Europe from Belgium southward and east to Asia Minor.

3. O. silaifolia Bieb.

A perennial similar in general appearance to *O. pimpinelloides* but larger. *Roots usually thickened and fusiform from base. Stem hollow,* ribbed and striate. Lvs 2-pinnate, lower soon withering, *lobes* linear-lanceolate, *acute.* Umbels with 4–10 *rays, becoming very stout (1–2 mm. diam.) in fr.; bracts few or 0; bracteoles ovate, acuminate. Fr.* c. 4 mm., *subcylindrical, strongly constricted at the junction with the pedicel; styles* 1–2 *mm., rather slender,* spreading or suberect. Fl. 6. $2n=22$*. Hs.

Native. In damp rich meadows, usually near rivers. 21. Dorset and Kent to Worcester, Nottingham and Suffolk. France, Portugal, Italy, S.E. Europe; S.W. Asia; N. Africa.

4. O. lachenalii C. C. Gmel.   'Parsley Water Dropwort.'

A perennial similar in general appearance to *O. pimpinelloides. Roots cylindrical or somewhat fusiform. Stem solid or with a small cavity when old,* ribbed and striate. Lvs 2-pinnate, lower soon withering, *lobes* spathulate or linear, *obtuse or subacute.* Umbels of 5–15 *rays, slender (c. 0·25 mm. diam.) in fr.,* bracts few or several, subulate; *bracteoles oblong-lanceolate. Fr.* 2 *mm., ovoid; pedicels neither thickened nor constricted at top,* very short; styles c. 1 mm., rather slender, spreading or suberect. The partial umbels have a number of long-pedicelled, usually sterile fls round the outside. Fl. 6–9. $2n=22$. Hs.

Native. In brackish and freshwater marshes and fens. 79, H24, S. Scattered throughout the British Is. north to Inverness and Lewis chiefly near the coasts. S. Sweden and Germany to Portugal, eastwards to Switzerland and Macedonia; Caucasus and Caspian region; Algeria.

5. O. crocata L.   'Hemlock Water Dropwort.'

A stout erect branched perennial 50–150 cm. *Root tubers 2–3 cm. diam.*, sweetish-tasting but poisonous. Stems hollow, grooved. *Lvs 30 cm. or more, deltoid, 3–4-pinnate*; segments ovate or suborbicular, cuneate at base, 1–2-lobed, serrate, teeth obtuse or subacute with a minute apiculus, segments of stem lvs narrower; *petioles entirely sheathing*. Umbels 5–10 cm. diam., terminal; peduncles usually longer than rays; rays 12–40, 2–7 cm.; *bracts and bracteoles many, caducous, linear-lanceolate*. Fls c. 2 mm. diam., petals of outer unequal. *Fr.* 4–6 *mm.*, cylindrical; pedicels neither thickened nor constricted at top; *styles c.* 2 *mm., erect.* Fl. 6–7. $2n=22$. Grt.

Native. In wet places. 98, H37, S. Scattered through the British Is. north to Ross, but mainly in the south and west, and usually calcifuge. France, Spain, Portugal, Italy, Sardinia, Morocco.

6. O. aquatica (L.) Poir.   'Fine-leaved Water Dropwort.'

O. Phellandrium Lam.; *Phellandrium aquaticum* L.

A stout erect stoloniferous perennial 30–150 cm. *Root fibrous.* Stem hollow, striate, often very stout. Lvs 3-pinnate, all aerial or *lower submerged; submerged lvs with capillary segments*; aerial lvs with deeply lobed segments c. 5 mm., *lobes* lanceolate to ovate, *acute*. *Umbels* 2–5 cm. diam., terminal and *lf-opposed; peduncles usually shorter than rays*; rays 4–10, 0·5–2 cm.; bracts 0 or 1; bracteoles several, subulate. Fls c. 2 mm., *petals of outer subequal.* *Fr.* 3–4 *mm., ovoid or oblong-ovoid.* Individual plants are biennial, perennation being by means of offsets. Fl. 6–9. $2n=22$. Hyd.

Native. In slow-flowing or stagnant water, sometimes attaining great size in nearly dry fen ditches. 57, H22. From Somerset and Kent to Cumberland and E. Lothian; Ireland, frequent in the centre, local elsewhere. Europe, except the extreme north; W. Asia.

7. O. fluviatilis (Bab.) Coleman

Similar to *O. aquatica*, but stems submerged and ascending, then erect. Lvs 2-pinnate; *submerged lvs with cuneate segments cut at the ends* into longer or shorter narrow lobes; aerial lvs with rather shallowly lobed *segments up to* 1 *cm. or more, lobes* ovate *obtuse or subacute. Fr.* 5–6 *mm.* Fl. 7–9. Hyd.

Native. In streams and ponds. 35, H17. Somerset and Kent to Anglesey and Lincoln; Ireland: widespread but very local. Denmark and Germany, very rare.

32. AETHUSA L.

A glabrous annual herb. Lvs ternately 2-pinnate. Umbels compound; bracts 0 or 1, bracteoles 1–5, deflexed. Fls white. Calyx-teeth small or 0; petals notched, point inflexed. Fr. broadly ovoid; carpels dorsally compressed, ridges very broad, lateral narrowly winged; vittae solitary in the furrows.

One sp. in Europe; S. Caucasus; Algeria; introduced in N. America. Usually contains coniine and is consequently poisonous.

1. A. cynapium L.   Fool's Parsley.

A branched lfy annual 5–120 cm. Stems hollow, finely striate, somewhat glaucous. Lvs deltoid, segments ovate, pinnatifid. Umbels 2–6 cm. diam., terminal and lf-opposed; rays (4–)10–20, 1–2 cm.; bracts usually 0; *bracteoles usually 3–4 on the outer side of the partial umbels*, subulate, up to 1 cm. Fls 2 mm. diam., petals unequal. Fr. 3–4 mm. Fl. 7–8. $2n=20, 22$. Th.

Native. A weed of cultivated ground. 107, H39, S. Common and generally distributed throughout the British Is. but becoming rarer northwards and absent from the extreme north of Scotland. Distribution of the genus.

33. FOENICULUM Mill.

Tall, glabrous, biennial or perennial herbs. Lvs pinnately divided, segments narrow. Umbels compound; bracts and bracteoles 0 or few. Fls yellow. Calyx-teeth 0; petals with an obtuse incurved point. Fr. ovoid or oblong, subterete, commissure broad; carpels ½-terete, ridges stout; vittae solitary in the furrows.

Two or three spp. in Europe and the Mediterranean region.

1. F. vulgare Mill.   Fennel.

F. officinale All.

A stout erect rather glaucous perennial 60–130 cm. Stem solid, developing a small hollow when old, striate, polished. Lvs much divided; *segments 1–5 cm., not all in one plane, capillary* with cartilaginous points. Umbels 4–8 cm. diam., terminal and often lf-opposed; rays 10–30, 1–6 cm., glaucous; bracts and bracteoles usually 0. *Fls 1–2 mm., yellow.* Fr. 4–6 mm., ovoid. Fl. 7–10. $2n=22$. Hp.

The whole plant has a strong and characteristic smell. Lvs used for culinary purposes.

?Native. On sea cliffs and naturalized or as a casual in waste places inland. Coasts of England and Wales from Denbigh and Norfolk southwards; introduced further north; Ireland, mainly in the south and east. Native in the Mediterranean region and possibly elsewhere in Europe. Naturalized in most temperate countries.

34. SILAUM Mill.

Glabrous perennial herbs. Lvs 1–3-pinnate. Umbels compound; bracts few or 0, bracteoles several, small. Fls yellowish. Calyx-teeth minute; petals broad and truncate at base with an incurved tip. Fr. ovoid or oblong, commissure broad; carpels ½-terete, ridges slender, lateral winged; vittae numerous, irregular.

About 8 spp. in the north temperate regions of the Old World.

1. S. silaus (L.) Schinz & Thell.   'Pepper Saxifrage.'

Silaus flavescens Bernh.; *S. pratensis* Bess.

An erect branched perennial 30–100 cm. Tap-root cylindrical, stout and woody with a few fibrous remains of petioles at the top. Stem solid, striate. Lower lvs deltoid, 2–3-pinnate, *segments 1–1·5 cm.*, entire or pinnatisect, *exceedingly finely serrulate*; upper lvs few, small, 1–2-pinnate or reduced to petioles. Umbels 2–6 cm. diam., terminal and axillary, long peduncled; rays 5–10, 1–3 cm., rather unequal; bracteoles about equalling pedicels, linear-lanceolate with scarious margins. *Fls* c. 1·5 mm. diam., *yellowish*. Fr. 4–4·5 mm., oblong-ovoid. Fl. 6–8. $2n=22^*$. Hs.

Native. In meadows and on grassy banks. 71. From Devon and Kent to Islay and Fife, rather local and becoming rare northwards. S. Sweden, C. and S. Europe, rare in the Mediterranean region and absent from Portugal and most of the Balkan peninsula; western Siberia.

35. Meum Mill.

A glabrous very aromatic perennial herb. Lvs 3–4-pinnate, segments capillary. Umbels compound; bracts few or 0, bracteoles 3–8. Fls white or purplish. Calyx-teeth 0; petals acute, narrowed to base, sometimes with a short inflexed point. Fr. ovoid, commissure broad; carpels ½-terete, ridges slender, acute; vittae 3–5 in each furrow.

Two spp. in the mountains of western Europe.

1. M. athamanticum Jacq.   Spignel, Meu, Baldmoney.

A tufted branched perennial 20–60 cm. *Rootstock crowned by coarse fibrous remains of petioles.* Stems hollow, striate. Lvs mostly radical, *divisions capillary ± whorled*, ultimate ones c. 0·5 cm. Umbels 3–6 cm. diam.; rays 6–15; bracts 0 or 1–8, linear; bracteoles linear, shorter than pedicels. Fr. 6–10 mm. Fl. 6–7. $2n=22$. Hr.

Native. In grassy places in mountain districts. 31. Merioneth, Caernarvon, Lancashire, W. Yorks, Westmorland and Northumberland to Argyll and Aberdeen, local. From S. Norway to N. Spain, Germany and the N. Balkans. The closely related *M. nevadense* Boiss. occurs in the Sierra Nevada.

36. Selinum L.

Perennial herbs. Lvs pinnate. Umbels compound; bracts few or 0, bracteoles several. Fls white. Calyx-teeth 0; petals notched, point inflexed. Fr. oblong to broadly ovoid, dorsally compressed, commissure broad; carpels ½-terete, ridges winged, the lateral broadly so; vittae solitary in each dorsal furrow.

About 16 spp. in north temperate regions.

1. S. carvifolia (L.) L.

An erect glabrous perennial 30–100 cm. *Stem solid, ridged, ridges acutely angled or almost winged.* Lvs 2–3-pinnate, *segments* 3–10 mm., entire or pinnatifid, *minutely serrulate*, lanceolate to ovate, *mucronate or aristate*. Umbels 3–7 cm. diam., terminal, long-peduncled; *rays* 10–20, *slightly rough on the ridges*; bracteoles subulate, equalling or exceeding pedicels. Fls 2 mm. diam. Fr. 3–4 mm., ovoid. Fl. 7–10. $2n=22$. H.

Native. In fens and damp meadows. 3. Cambridge, N. Lincs and Nottingham. N. and C. Europe southwards to France, N. Italy and N. Balkans, east to the Urals, Altai and perhaps Lake Baikal; introduced in N. America.

37. Ligusticum L.

Glabrous perennial herbs. Lvs ternately pinnate. Umbels compound; bracts many, few or 0; bracteoles many. Fls all hermaphrodite, white, pink or yellow. Calyx-teeth small or 0, shorter than the entire, inflexed petals. Fr. oblong or ovoid, subterete or dorsally compressed, commissure broad; carpels with prominent, acute or winged primary ridges; vittae many, slender or obscure.

About 5 spp. in north temperate regions, Chile and New Zealand.

1. L. scoticum L. Lovage.

A glabrous, shiny, bright green perennial 15–90 cm. Rootstock stout. Stem terete, ribbed, little-branched, often magenta towards the base. Lvs 5–10 cm., 2–ternate, *segments* 3–5 cm., ovate-cuneate, sometimes lobed, *serrate in upper half.* Umbels 4–6 cm. diam., *rays* 8–12, 1·5–4 cm.; bracts 1–5, subulate, entire; bracteoles several, linear, entire, shorter than the pedicels. Fls c. 2 mm., greenish-white, sometimes tinged with pink. Calyx-teeth small; styles very short and stout. Fr. 4 mm., oblong, subterete, ridges acute. Fl. 7. $2n=22$. Hs. Lvs sometimes eaten as pot-herb.

Native. On rocky coasts. 29, H6. Northumberland and Kirkcudbright northwards, local; Ireland: Down to Donegal; Galway. N.W. Europe from Denmark to c. 71° N.

38. ANGELICA L.

Tall perennial herbs. Lvs 2–3-pinnate, segments broad and large. Umbels compound, many-rayed; bracts few or 0; bracteoles usually many. Fls greenish, white or pinkish. Calyx-teeth minute or 0; petals acuminate, incurved. Fr. ovoid, dorsally compressed, commissure broad; carpels flat, with 2 broad marginal wings and 3 dorsal ridges. Fls protandrous, sometimes unisexual, insect-pollinated.

About 70 spp. in the northern hemisphere and New Zealand.

1. A. sylvestris L. Wild Angelica.

A stout nearly glabrous perennial (7–)30–200 cm. or more. *Stem* hollow, *usually purplish and pruinose*, striate, pubescent towards base. Lvs 30–60 cm., 2–3-pinnate, deltoid, lower primary divisions of radical lvs long-stalked; segments 2–8 cm., obliquely oblong-ovate, acutely serrate; petioles laterally compressed, deeply channelled on upper side, dilated and sheathing at base; *upper lvs reduced to inflated sheathing petioles which ± enclose the fl. buds.* Umbels terminal and axillary, 3–15 cm. diam., peduncles puberulent; *rays* many, 1–5 cm., *puberulent*; bracts 0 or few, caducous, bracteoles few, setaceous, as long as the pedicels, persistent. Fls 2 mm., white or pink, petals suberect, incurved, calyx-teeth 0. Fr. 5 mm., wings scarious. Fl. 7–9. $2n=22$. Hs.

Native. In fens, damp meadows and woods. 112, H40, S. Generally distributed and common throughout the British Is. Europe, rare in the south; Asia Minor, Caucasus, Siberia from the Urals to Lake Baikal; introduced in Canada.

*A. archangelica L. (*Archangelica officinalis* Hoffm.), Angelica, is somewhat similar but has the *stems usually green*, the lf-segments somewhat decurrent, the *fls greenish-white or green* with small calyx-teeth, and the wings of the fr. corky. $2n=22$.

Formerly cultivated and now ± naturalized on river banks and in waste places, locally abundant. Greenland, Iceland, Scandinavia (to 71° 10′ N.), Denmark, E. Germany, N. Austria, Poland and C. Russia; naturalized in many other parts of Europe.

*Levisticum officinale Koch, a large herb differing from *Angelica* in its yellowish fls and in having all 10 ribs of the fr. narrowly winged, is occasionally naturalized.

39. PEUCEDANUM L.

Perennial, rarely biennial or annual herbs. Lvs pinnate or ternate. Umbels compound; bracts many, few or 0, bracteoles many. Fls white, yellow or pinkish. Calyx-teeth small or 0; petals with a long inflexed point. Fr. orbicular, ovoid or oblong, dorsally compressed, commissure very broad; carpels with marginal ridges forming a narrow or broad wing, dorsal ones filiform, all equidistant; vittae 1–3 in each furrow, as long as the fr.

About 170 spp. in Europe, Asia, Africa and America.

Stem solid; lf-segments 4–10 cm., narrow-linear, quite entire.   **1. officinale**
Stem hollow; lf-segments pinnatifid, lobes c. 0·5 cm., finely serrulate.
                                                            2. palustre
Stem hollow; lf-segments 2–10 cm., broad, lobed and coarsely serrate.
                                                          3. ostruthium

1. P. officinale L.   Hog's Fennel, Sulphur-weed.

An erect glabrous perennial 60–120 cm. Root very stout and woody. Root-stock crowned by persistent fibrous remains of petioles. *Stems solid*, striate. *Lvs 4–6 times ternately divided, segments 4–10 cm., sessile, narrow-linear, attenuate at both ends*, quite entire, spreading in all directions. Umbels 5–15 cm. diam.; rays 10–40; *bracts 0–3, setaceous*; bracteoles several, setaceous, shorter or longer than the *filiform pedicels. Fls* 2 mm. diam., *yellow. Fr.* 5–10 *mm.*, elliptic-ovoid. Fl. 7–9. 2n=c. 66. H.

Native. On banks near the sea, rare and local. 2. E. Kent and Essex. W. and S. Germany to N. Spain and Portugal, eastwards to C. Russia and the Balkans.

2. P. palustre (L.) Moench   Hog's Fennel, Milk Parsley.

A glabrous biennial 50–150 cm. Stems hollow, strongly ridged. *Lvs 2–4-pinnate, segments pinnatifid, lobes c. 0·5 cm., oblong-lanceolate, subacute or obtuse and mucronulate, margins exceedingly finely serrulate.* Umbels 3–8 cm. diam.; *rays 20–40, minutely pubescent or rough; bracts 4 or more, lanceolate to linear-lanceolate, deflexed; bracteoles lanceolate-acuminate,* equalling or shorter than the *minutely pubescent or rough pedicels.* Fls c. 2 mm., greenish-white. *Fr. 4–5 mm., ovoid, with a fine white webbing over the commissural vittae.* Fl. 7–9. 2n=22. Hs.

All parts of the plant yield a watery-milky juice when young.

Native. In fens and marshes. 11. Somerset, Sussex, Buckingham, Cambridge, Suffolk, Norfolk, Lincoln, S. Lancashire and S.W. Yorkshire. Most of Europe from C. Scandinavia to N. Spain and eastwards to the Altai and Urals.

*3. P. ostruthium (L.) Koch Master-wort.

An erect ± downy perennial 30–100 cm. Stem hollow, ridged. *Lvs 1–2-ternate, segments 4–10 cm., few, broad, lobed, serrate, usually downy beneath.* Umbels 5–10 cm. diam.; rays 20–50; *bracts* 0; bracteoles few, setaceous. Fls white or pinkish. *Fr.* 4–5 mm., *suborbicular*. Fl. 7–8. 2*n*=22.

Introduced. Formerly cultivated and now naturalized in moist meadows and on river banks. 29, H4. From Carmarthen, Stafford, Lancashire and mid-west Yorkshire northwards to Shetland; N.E. Ireland; local. S. Europe from the Pyrenees to the N. Balkans; introduced in many temperate countries.

40. Pastinaca L.

Herbs with pinnate lvs. Umbels compound; bracts and bracteoles 0 or 1–2, caducous. Fls yellow (in our sp.). Calyx-teeth small or 0; petals with a truncate involute point. Fr. broadly ovoid or orbicular, strongly dorsally compressed; carpels with a broad flattened rather narrowly winged margin, dorsal ridges filiform, distant from lateral; vittae 1 in each furrow, as long as the fr.

About 15 spp. in the north temperate regions of the Old World.

1. P. sativa L. Wild Parsnip.

Peucedanum sativum (L.) Benth. ex Hook. f.

An erect ± pubescent strong-smelling biennial 30–150 cm. Stems hollow, furrowed and ± angled. *Lvs simply pinnate*, segments ovate, lobed and serrate. Umbels 3–10 cm. diam.; rays 5–15. *Fls* c. 1·5 mm. diam., *yellow*. Fr. 5–8 mm., vittae conspicuous, tapering at the ends. Fl. 7–8. 2*n*=22. Hs.

Native. Roadsides and grassy waste places. 62, H27, S. Scattered throughout England and locally abundant, particularly on chalk and limestone in the south and east; Wales; in Scotland and Ireland only as an escape from cultivation. Europe, except the extreme north and Portugal, eastwards to the Caucasus and Altai; introduced in N. and S. America, Australia and New Zealand.

41. Heracleum L.

Annual or perennial herbs, sometimes of great size. Lvs 1–3-pinnate, segments broad. Umbels compound; bracts 0 or caducous, bracteoles several. Fls white or pinkish. Calyx-teeth small, unequal; petals often very unequal, at least the larger notched with an inflexed point. Fr. orbicular, obovate or oblong, strongly dorsally compressed, commissure very broad; carpels nearly flat, marginal ridges forming a broad wing, dorsal ones slender; vittae solitary in the furrows, conspicuous and swollen at their lower ends, shorter than the fr.

About 70 spp. in north temperate regions and mountains of tropical Africa.

1. H. sphondylium L. Cow Parsnip, Hogweed, Keck.

A stout erect hispid biennial 50–200 cm. Stems hollow, ridged, hairs deflexed. *Lvs* 15–60 cm., *simply pinnate*, hispid on both surfaces, segments 5–15 cm., variously lobed or pinnatisect, serrate, ovate to linear-lanceolate, lower stalked. Umbels 5–15 cm. diam., terminal and axillary; rays 7–20, 1–6 cm., stout; bracts few or 0, subulate; *bracteoles several, setaceous, reflexed*. Fls 5–10 mm. diam., *white or pinkish*, petals deeply notched, those of the outer fls very unequal. Fr. 7–8 mm., suborbicular, whitish; styles short, erect, becoming reflexed in ripe fr. Fl. 6–9. $2n=22*$. Hs.

Lvs edible, much esteemed by herbivorous animals.

Native. In grassy places, roadsides, by hedges and in woods. 112, H40, S. Common and generally distributed throughout the British Is., ascending to 3300 ft. in Argyll. Europe southwards from c. 61° N.; W. and N. Asia; western N. Africa; introduced in N. America.

Two forms occur which differ markedly in the shape of the lf-segments, the common form having ovate segments and the less frequent one, var. *angustifolium* Huds., linear-lanceolate segments.

*H. mantegazzianum Somm. & Lev. is easily recognized by its enormous size. Stem up to 3·5 m., hollow, up to 10 cm. diam., red-spotted, ridged. Lvs up to 1 m., pinnately divided. Umbels up to 50 cm. diam.; rays very numerous, c. 20 cm. Larger petals of outer fls up to 12 mm. Fr. c. 13 mm., ovate-elliptic; vittae very prominent, swollen and more than 1 mm. wide at their base; calyx-teeth prominent, triangular-lanceolate. The plant has a strong resinous aromatic smell somewhat resembling that of *Cistus*. Fl. 6–7.

Introduced. Naturalized in waste places, especially near rivers. Caucasus.

*H. persicum Desf. is sometimes seen in gardens and may escape. It is about as large as *H. mantegazzianum* but has oblong-elliptic fr. with vittae less than 1 mm. wide and obscure blunt calyx-teeth.

42. Tordylium L.

Pubescent annual herbs. Lvs simple or pinnate. Umbels compound; bracts and bracteoles several, few, or 0, subulate. Fls white or pinkish. Calyx-teeth subulate and unequal or 0; petals with an incurved point, unequal. Fr. orbicular or oblong, strongly dorsally compressed; carpels with the lateral ridges much thickened, the dorsal ones slender; vittae 1–3 in each furrow.

About 16 spp. in Europe, N. Africa and temperate Asia.

1. T. maximum L.

An erect branched annual 20–125 cm. Stems hollow at base, ridged, rough with stiff deflexed appressed hairs. Lvs simply pinnate; *segments of lower lvs broadly ovate or almost orbicular*, lobed and crenate; of upper lanceolate, obtusely serrate, appressed hispid on both surfaces. Umbels flat, long-peduncled; rays 5–8(–15), hispid with short stiff forward-pointing hairs; bracts and bracteoles several, hispid, the latter exceeding the subsessile fls. Fls white or pinkish; *calyx-teeth conspicuous*, about ½ as long as the petals. *Fr.* 5–8 mm., broadly oblong, hispid, *margins much thickened*, *whitish*, glabrous; vittae 1 in each furrow. Fl. 6–7. $2n=22$. Th.

?Introduced. Naturalized or perhaps native in a few places. Recorded from hedge- and river-banks in Essex, Middlesex, Oxford and Buckinghamshire, now seldom seen. S. Europe from Portugal and Spain to the Crimea; Asia Minor, Persia and the Caucasus.

43. DAUCUS L.

Annual or biennial hispid herbs. Lvs pinnately divided. Umbels compound; bracts and bracteoles numerous or 0, entire or cut. Fls white. Calyx-teeth small or 0; petals notched, point inflexed, often unequal. Fr. ovoid or oblong; carpels convex, primary ridges 5, filiform, secondary 4, stouter and more prominent than primary, all, or the secondary only, with rows of spines; vittae solitary under each secondary ridge.

About 60 spp. in Europe, N. Africa, W. Asia and America.

1. D. carota L.   Wild Carrot.

An erect ± hispid biennial 30–100 cm. Stem solid, striate or ridged. Lvs 3-pinnate; segments pinnatifid, lobes c. 5 mm., lanceolate to ovate and acuminate to obtuse. Umbels 3–7 cm. diam.; rays very numerous; bracts 7–13, about equalling pedicels, ternate or pinnatifid, conspicuous, margins broadly scarious. Fls white or the central one of the umbel often red or purple. Fr. 2·5–4 mm., oblong-ovoid to broadly ovoid, primary ridges ciliate, the secondary spiny. Fl. 6–8. $2n = 18^*$. Hs. (biennial).

Ssp. **carota**: Root tough, not fleshy. Lvs ovate in outline, umbels strongly concave in fr.

Ssp. **gummifer** Hook. f. (*D. gingidium* auct.): Lvs narrower in outline, umbels ± flat in fr.

*Ssp. **sativus** (Hoffm.) Hayek, with a thick and fleshy tap-root in the first year, is the cultivated carrot.

Native. Ssp. *carota* in fields and grassy places, particularly near the sea and on chalky soils; ssp. *gummifer* on cliffs and dunes, local and mainly on the south coast. 110, H40, S. Throughout the British Is., except E. Sutherland and Shetland. Scandinavia (from c. 65° N.) to N. Africa and the Canary Is., eastwards through Siberia to Kamchatka in the north and eastern India in the south; naturalized in most other temperate and many tropical countries.

76. CUCURBITACEAE

Herbs, rarely shrubs or small trees, and mostly annual tendril-climbers but also prostrate annuals, climbing perennials and spiny herbs and shrubs. All usually with very sappy often hispid stems. Tendrils spirally coiled, often branched, each arising at the side of a lf-axil. Lvs spirally arranged, palmately-veined and often palmately-lobed, exstipulate. Fls usually unisexual, monoecious or dioecious, actinomorphic, epigynous, and often also perigynous, solitary or in infl. of various kinds. Calyx with 5 narrow sepals; corolla of 5 free or basally united petals; androecium rarely of 5 free stamens each with 2 pollen-sacs, more usually of 2 pairs of ± completely united stamens with 1 free stamen; often the anthers are borne on a common connective-column, and frequently they are variously curved and twisted; ovary usually inferior, 1- or 3-celled; the commonly 3, parietal placentae sometimes meeting in the centre; ovules numerous, rarely few, anatropous; style 1, rarely 3; stigmas 3, commissural. Entomophilous. Fr. usually a berry or pepo. Seed non-endospermic with a straight embryo.

About 700 spp. in 90 genera mainly in tropical and subtropical regions.

Although commonly gamopetalous the Cucurbitaceae appear more closely related to the Begoniaceae and other polypetalous families than to any families of the Metachlamydeae.

A family of some economic importance since it includes *Cucumis melo*, the melon, and *C. sativus*, the cucumber, *Citrullus lanatus*, the water-melon, *Cucurbita pepo*, the various kinds of marrow, pumpkin and squash, as well as other edible fruits. The hard-shelled pepos of *Lagenaria* and *Cucurbita* yield calabashes and gourds, and the fibrous tissue of the fr.-wall of *Luffa aegyptiaca* is the loofah sponge.

1. BRYONIA L.

Perennial tendril-climbing dioecious herbs with palmately-lobed lvs and fls in axillary cymes or clusters. Corolla of 5 petals free or united below into a short tube. Androecium of 2 pairs of stamens united by their filaments and with 2 pollen-sacs, and 1 free stamen with 1 pollen-sac. Stigmas 3, each bifid. Fr. a small smooth globular berry.

Eight spp. in Europe, N. Africa and Canary Is.

1. B. dioica Jacq.   White or Red Bryony.

Stock erect, tuberous, massive, branched. Stem very long, branching especially from near the base, brittle, angled, hispid with swollen-based hairs, climbing by simple spirally coiled tendrils arising from the side of the lf-stalk. Lvs palmately (3–)5-lobed with the lobes sinuate-toothed, cordate at the base, with a curved stalk shorter than the blade. Fls in axillary cymes; those of the male plant stalked, corymbose, of 3–8 pale greenish fls 12–18 mm. diam., with triangular spreading sepals, oblong hairy distinctly net-veined petals 2–3 times as long as the sepals, and yellow anthers; those of the female plant ± sessile, umbellate, of 2–5 greenish fls 10–12 mm. diam., with sepals and petals as in the male fls but smaller, prominent bifid stigmas and a smooth broadly ellipsoidal ovary separated by a short constriction from the perianth. Berry 5–8 mm. diam., red when ripe, with 3–6 large compressed seeds yellowish with black mottling or vice versa. Fl. 5–9. Visited by various insects including many bees. $2n=20$. G.

Native. Hedgerows, scrub, copses; avoided by rabbits and common in warrens; locally common, especially on well-drained soils. 59. England and Wales northwards to N.W. Yorkshire and Northumberland; introduced locally in S. Scotland. Europe, especially central and south, absent from Scandinavia; W. Asia; N. Africa.

Ecballium elaterium A. Rich. (Squirting Cucumber) is naturalized in a few places on the S. coast. It is a hispid, somewhat glaucous, perennial, monoecious herb with prostrate succulent shoots, 20–60 cm., and fleshy triangular-cordate lvs, 7–10 cm., whitish-tomentose beneath, sinuate-toothed to obscurely lobed; tendrils 0. Fls c. 2·5 cm. diam., yellow; female fls usually solitary, axillary; male in racemes; stamens with free anthers. Fr. oblong, 3–5 cm., hispid, greenish, detaching itself explosively at maturity and squirting the seeds to some distance from the aperture at the proximal end. A weed throughout the Mediterranean region.

77. ARISTOLOCHIACEAE

Herbs or woody climbers. Lvs alternate, simple, entire or rarely lobed, stalked, exstipulate. Fls solitary or in racemes or clusters, stalked, hermaphrodite, actinomorphic to strongly zygomorphic, usually 3-merous. Perianth usually in 1 whorl, rarely in 2 whorls, petaloid, often brown or lurid purple, united below into a tube, regularly 3- or 6-lobed or with a unilateral entire or lobed limb. Stamens 6–many in 1 or 2 whorls, free or united with the stylar column; anthers extrorse. Ovary inferior, rarely ½-inferior, 4–6-celled, with numerous axile ovules; styles short, thick, 3–many, free or united into a column with a 3–many-lobed stigma. Fr. a capsule, opening variously, rarely indehiscent; embryo very small in a fleshy endosperm.

Seven genera and about 380 spp., tropical and temperate.

Fls regular.	1. Asarum
Fls strongly zygomorphic.	2. Aristolochia

1. Asarum L.

Herbs with creeping rhizome. Fls solitary, terminal. *Perianth regular in 1 whorl*, 3-lobed, persistent on the fr. Stamens 12 in 2 whorls, free or nearly so; filaments short. Ovary 6- (rarely 4-)celled with numerous (rarely 1) ovules in 2 rows in each cell; styles 6 (rarely 4) free or united into a column. *Fr. a subglobose capsule* opening irregularly; *seeds flat*.

About 60 spp., north temperate zone, mainly E. Asia.

1. A. europaeum L.   Asarabacca.

Perennial evergreen herb with thick creeping rhizome. Stems short, 2–5 cm., pubescent, with usually 2 lvs and 2 brown ovate scales (1–2 cm.). Lvs 2·5–10 cm., reniform, broader than long, deeply cordate at base, very obtuse, entire, dark green, pubescent on the veins above; petiole much longer than blade. Fl. solitary, terminal. Perianth brownish, c. 15 mm., pubescent outside; lobes ± deltoid, acuminate, about half as long as tube. Fl. 5–8. Pollinated by small flies or probably more often self-pollinated. $2n = $ c. 24, 40. Grh.

?Native. Rare. Perhaps native in woods, etc., in a few localities from Dorset and Bucks to Denbigh, Lancashire and Durham; sometimes found as an escape in Scotland, north to Lanark and Aberdeen; formerly much grown as a medicinal plant. 13. Europe from Belgium, Germany (not N.W.) and Finland to N. and E. Spain, C. Italy, Macedonia and the Crimea; W. Siberia.

2. Aristolochia L.

Herbs or woody climbers. *Perianth zygomorphic* with long tube, the *base swollen*, the upper part narrower, ± cylindric, straight or variously curved; limb equally 3-lobed or unilateral and entire or 2- or 3-lobed or 3–6-toothed, caducous. Stamens usually 6, rarely more or fewer, in 1 whorl, joined to the stylar column consisting of (5–)6(–12) styles. Ovary (4–)5–6-celled. *Fr. a septicidal capsule*.

About 300 spp., tropical and temperate. Several are grown in greenhouses for their fantastic fls.

*1. A. clematitis L.   Birthwort.

Glabrous foetid perennial herb 20–80 cm. Rhizome long, creeping. Stems erect, simple, numerous. Lvs 6–15 cm., broadly ovate-cordate, sinus open, obtuse, finely denticulate, with a petiole about half as long. Fls 4–8 in axillary fascicles; pedicels short. Perianth 2–3 cm., dull yellow, glabrous; tube somewhat curved, the basal swelling globose; limb entire oblong or ovate, about as long as tube. Capsule 2–2·5 cm., pyriform. Fl. 6–9. Protogynous, pollinated by small flies which are trapped in the basal swelling till the stamens mature. $2n = 14$. Grh.

Introduced. 12. Long cultivated as a medicinal plant and naturalized in a number of places mainly in E. England from Sussex to Yorks, extending west to Oxford, Hereford and Stafford; Stirling. C. and S. Europe from France, S. Austria and C. Russia (Volga-Kama region) to C. Spain and Portugal, S. Italy, Albania, Thrace and the Caucasus; N. Asia Minor.

*A. rotunda L.

Stock a subglobose tuber. Lf-sinus closed or nearly so; petiole short or 0. Fls solitary. Perianth yellowish with brownish limb, puberulent. Rarely naturalized. Mediterranean region.

78. EUPHORBIACEAE

Trees, shrubs or, as in all the British spp., herbs. Lvs usually alternate and stipulate, simple or compound. Infl. usually compound. Fls regular or nearly so, unisexual, usually hypogynous. Perianth of 1 whorl or 0, rarely of 2 whorls. Stamens 1–many. Ovary usually 3-celled, with 1 or 2 anatropous, pendulous ovules in each cell; raphe ventral. Fr. usually a capsule separating into three parts and leaving a persistent axis; rarely a drupe. Seeds with abundant endosperm and large embryo, usually carunculate.

Over 280 genera and 7000 spp., mainly tropical. The British genera are not closely related and give a very inadequate idea of this large and varied family of which the ovary and fr. and unisexual fls are the most constant features. Its relationships are obscure and it is quite likely to be polyphyletic.

Perianth present, 3-merous; fls in clusters, often spicate; without milky juice.
                                                                1. MERCURIALIS
Perianth 0; fls very small, one female and several male fls borne in a specialized
  4–5-lobed perianth-like involucre; juice milky.   2. EUPHORBIA

1. MERCURIALIS L.

Herbs with watery juice. Lvs opposite, stipules small. Male fls in clusters arranged on long pedunculate axillary spikes, per. segs 3, sepaloid, stamens 8–15. Female fls axillary, solitary or in clusters, pedunculate or subsessile, perianth as in the male, 2 or 3 sterile filaments present. Ovary of two 1-seeded cells; styles 2, free. Fr. dehiscing by 2 valves.

Eight spp. in Europe, temperate Asia and Mediterranean region. Windpollinated.

Hairy perennial with creeping rhizome; stems simple.   **1. perennis**
Glabrescent annual; stems branched.   **2. annua**

1. M. perennis L.   Dog's Mercury.

Hirsute perennial with long creeping rhizome. Stems erect, *simple*, 15–40 cm.
Lvs 3–8 cm., elliptic-ovate or elliptic-lanceolate, crenate-serrate, with a short
(3–10 mm.) petiole. Dioecious, rarely monoecious. Fls 4–5 mm. across;
female 1(–3) *on long peduncles.* Fr. hirsute, 6–8 mm. broad. Fl. 2–4. 2*n*=c. 64.
Hp.

Native. Woods on good soils and shady mountain rocks, ascending to
3400 ft. on Ben Lawers; frequently dominant in the field layer especially in
beechwoods on chalk. Common over most of Great Britain but absent from
Orkney, Shetland and the Isle of Man; rare in Ireland and usually introduced,
though possibly native in Clare; Jersey. 108, H13, S. Europe from Scandinavia (65° 26′ N.) to C. Spain and Portugal, Corsica, Sicily, Greece and the
Caucasus; Algeria (very rare); S.W. Asia.

2. M. annua L.   'Annual Mercury.'

Glabrescent annual. Stems erect, *branched*, 10–50 cm. Lvs 1·5–5 cm., ovate
to elliptic-lanceolate, crenate-serrate, with a petiole 2–15 mm. Dioecious,
occasionally (var. *ambigua* (L. fil.) Duby) monoecious. *Female fls subsessile*
in the lf axils. Fr. hispid, 3–4 mm. broad. Fl. 7–10. 2*n*=16. Th.

?Native. Waste places and as a garden weed, often only a casual, widespread but local in S. England, rare in Wales and N. England, extending to
Lancashire and Northumberland; Fife, Angus; S. Ireland, north to Dublin
and Clare; Channel Is. 58, H13, S. Europe (not Iceland, probably only
casual in Scandinavia, etc.), Mediterranean region, Azores.

2. EUPHORBIA L.   Spurge.

Herbs with milky juice. Lvs usually alternate and exstipulate. Monoecious.
Fls naked, in compound cymes, primary branches usually in an umbel, sometimes with axillary branches below the primary umbel, subsequent branching
usually dichotomous. Partial bracts conspicuous, usually differing from the
lvs, opposite. Fls very small, a number of male fls and a single female fl.
being grouped within a cup-shaped perianth-like involucre (cyathium) with
4 or 5 small teeth alternating with 4 or 5 conspicuous glands. Male fl. consisting of a single stamen on a jointed pedicel. Female fl. consisting of the
3-celled ovary on a pedicel which elongates in fr. Styles 3, stigmas often
bifid. Ovules 1 in each cell. Fr. a 3-valved capsule.

About 1600 spp. in tropical and temperate regions.

All the spp. appear to ripen fr. about a month after fl. The chromosome
numbers given by different authors (and given below) differ rather oddly and
cannot be considered satisfactory. All our spp. are probably pollinated by
flies.

The above description applies to the British spp. The foreign spp. include trees and
shrubs and many are cactus-like without or with very reduced caducous lvs. Some
of these are cultivated in greenhouses, as is *E. pulcherrima* Willd. (Poinsettia) with
conspicuous red bracts. A few spp. not differing greatly from ours are sometimes
grown out of doors and some of these and others have been found as escapes and
casuals.

1   Lvs and bracts not differing, very unequal on the two sides at base; stipules subulate; procumbent maritime plant.   **1. peplis**
    Lvs exstipulate, equal at the base; bracts often markedly different from lvs; plants ± erect.   **2**

2   Lvs opposite.   **2. lathyrus**
    Lvs alternate.   **3**

3   Involucral glands rounded on outer edge.   **4**
    Involucral glands with concave outer edge, the tips ± prolonged into horns.   **10**

4   Perennials with ± numerous stems; lvs oblong or lanceolate, cuneate at base.   **5**
    Annuals with single stems; lvs either obovate and very obtuse or oblong- or obovate-lanceolate and deeply cordate at base.   **8**

5   Capsule smooth or minutely tuberclate; lvs usually ± pilose on both sides (rarely glabrous above).   **6**
    Capsule strongly warted; lvs glabrous above, sometimes pilose beneath.   **7**

6   Bracts of the umbel oval, c. 2 cm., resembling the yellowish partial bracts; capsule glabrous or sparingly pilose.   **3. pilosa**
    Bracts of the umbel oblong-lanceolate c. 7–8 cm., resembling the lvs; partial bracts green or reddish tinged; capsule woolly-pilose.
    4. corallioides

7   Partial bracts subcordate or rounded at base, yellowish; glands yellowish, finally brown.   **5. hyberna**
    Partial bracts truncate at base, green; glands at first green, soon purple.
    6. dulcis

8   Lvs and bracts acute or subacute; lvs cordate at base; capsule warted.   **9**
    Lvs and bracts very obtuse; lvs tapered to base; capsule smooth.
    9. helioscopia

9   Capsule shallowly grooved, almost globose, its warts ± hemispherical; lower partial bracts similar in shape to upper, markedly different from the bracts of the umbel.   **7. platyphyllos**
    Capsule trigonous, its warts cylindrical or conical, as long as or longer than broad; lower partial bracts intermediate in shape between the upper and the bracts of the umbel.   **8. stricta**

10   Bracts all free; glabrous plants; stems annual.   **11**
    Bracts (except those of the umbel) connate in pairs; stems biennial, hairy.   **17. amygdaloides**

11   Annuals with single stems; lvs few.   **12**
    Biennials or perennials with ± numerous stems, some of them usually sterile, tufted or from a creeping rhizome; lvs many.   **13**

12   Lvs oval or obovate, stalked, green.   **10. peplus**
    Lvs linear, sessile, glaucous.   **11. exigua**

13   Lvs ovate to oblanceolate, all less than 2 cm., ± coriaceous; plant without creeping rhizome; umbel with 6 rays or fewer; maritime plants.   **14**
    Lvs linear-lanceolate or linear, mostly over 2 cm. (or if all less than 2 cm. then narrowly linear), thin; rhizome creeping; umbel with 6 or more rays; inland plants.   **15**

14   Lvs obovate or oblanceolate, slightly coriaceous, midrib prominent beneath; seeds pitted.   **12. portlandica**

Lvs ovate or oblong, very coriaceous, midrib obscure; seeds smooth.
                                                              13. paralias
15   Lvs of fl. stems 4 mm. broad or more.   *16*
       Lvs of fl. stems 2 mm. broad or less.   **16. cyparissias**

16   Lvs broadest below or slightly above the middle, tapering to the apex,
       little narrowed at the base; bracts of the umbel 12–35 mm.
                                                            14. uralensis
       Lvs broadest near the apex, less tapering above, gradually tapered to a
       narrow base; bracts of the umbel 5–15 mm.   **15. esula**

Section 1. *Anisophyllum* Dum. Lvs all opposite, stipulate. Cyathia in the
axils of the lvs and in the forks, solitary.

1. E. peplis L.   'Purple Spurge.'
Glaucous, often purplish annual *with* usually 4 *procumbent branches* 1–6 cm.
long *from the base*. Lvs 3–10 mm., ± oblong in outline, obtuse or retuse, *with
a large rounded auricle on one side at the base* (except sometimes the basal lvs)
entire, shortly petioled; *stipules* divided into subulate segments. Branches
branching dichotomously with cyathia in the forks and axils. Cyathia 1–2 mm.,
stalked; glands suborbicular, entire. Capsule 3–5 mm., trigonous, glabrous,
smooth. Seeds c. 3 mm., pale grey, smooth, not caruncled. Fl. 7–9. Th.
   Native. Shingle beaches, formerly found in fair quantity in a few places
in Devon, Cornwall and the Channel Is., now almost extinct and occurring
only sporadically (as it probably always did to some extent); isolated specimens have also been found in several other counties from Cardigan to Sussex
and in Waterford. 11, H1, S. Shores of Atlantic from France southwards
and of the Mediterranean.

Section 2. *Esula* Pers. Lvs exstipulate. Infl. cymose, the primary branches
umbellate. Seeds with caruncle (in the British spp.).

2. E. lathyrus L.   Caper Spurge.
Glabrous glaucous biennial 30–120 *cm.*, forming a short erect lfy stem the first
year, elongating and flowering in the second. *Lvs* 4–20 cm., *opposite* or subopposite in 4 rows, narrow-oblong, the lower smaller, obtuse or retuse, often
mucronulate, rounded at base, entire, sessile. Umbel 2–6-rayed, its bracts
triangular-lanceolate. Partial bracts 1·5–7 cm., triangular-ovate, acute, cordate
at base. Glands lunate, the horns blunt. Fr. 8–20 mm., trigonous, glabrous,
smooth. Seeds c. 5 mm., brown, reticulate. Fl. 6–7. 2n = 20. Not satisfactorily
referable to any standard life-form, appearing in winter as N., but biennial
and herbaceous.
   ?Native. Perhaps native in woods in a few places in England and Wales
from Somerset and Kent to Denbigh, etc., more common but local as a
garden weed and escape in waste places north to Lanark and Fife. 44, S.
Formerly cultivated for its fr. S. Europe from Spain, N. Italy and Greece
northwards to France and Germany but doubtfully native in C. Europe;
Morocco (rare); Azores; status often doubtful.

3. E. pilosa L.   'Hairy Spurge.'

Perennial 30–100 cm., *with stout rhizome and ± numerous stems* with axillary
sterile branches above. *Lvs* 4–10 cm., alternate, oblong or oblong-lanceolate,
obtuse to subacute, broad-cuneate at base, sessile, *serrulate near apex*, entire
near base, *softly pilose on both sides* or nearly glabrous above, dense and with
the uppermost less than its own length below the umbel. Umbel 4–6-rayed
often with axillary infl. branches below, its *bracts c. 2 cm. oval.* Partial bracts
8–12 mm., oval or suborbicular, obtuse, mucronulate, *rounded at base, yellowish*, glabrous or sparingly pilose. *Glands* transversely oval, *entire. Capsule*
4–8 mm., subglobose, scarcely grooved, *glabrous or sparingly pilose, smooth
or minutely tuberclate. Seeds brown*, smooth. Fl. 5–6. 2n = 18. Hp.

?Native. In a wood near Bath and neighbouring hedgebanks, where it has
been known since 1576, but is now probably extinct. W., C. and S. France,
N., C. and E. Spain, N. Italy, Algeria.

*4. E. corallioides L.   'Coral Spurge.'

Differs from *E. pilosa* as follows: Plant *not or scarcely rhizomatous, with few
stems and descending root.* Uppermost lf usually more than its own length
below the umbel. *Bracts of the umbel c. 7–8 cm., oblong-lanceolate*, similar to
the lvs. *Partial bracts green or tinged with red*, usually more pilose, the upper
oval or ovate, the lower intermediate between the upper and those of the
umbel. *Capsule densely pilose*, appearing woolly, minutely tuberclate. Fl.
6–7. Hp.

Introduced. Shady places in two stations in Sussex where it has been
known since 1808; casual at Oxford. Native of S. Italy and Sicily.

5. E. hyberna L.   'Irish Spurge.'

Perennial 30–60 cm., with thick rhizome and ± numerous erect simple stems.
Lvs 5–10 cm., alternate, oblong, elliptic-oblong or oblanceolate-oblong, obtuse or retuse, cuneate at base, sessile or subsessile, *entire, glabrous above,*
sparingly pilose beneath especially near midrib, or more rarely glabrous,
eventually turning red. Umbel 4–6-rayed, often with axillary infl. branches
below; its bracts 3–6 cm., elliptic-oblong. *Partial bracts* 8–30 mm., ovate or
ovate-elliptic, obtuse to subacute, not mucronulate, *subcordate or rounded at
base, yellowish. Glands* 5, *yellowish*, finally brown, *entire*, reniform. *Capsule*
5–6 mm., subglobose, grooved, glabrous, *with prominent cylindric warts.* Seeds
pale brown, smooth. Fl. 4–7. Hp.

Native. Woods, hedgebanks and rough pastures on lime-free soils; locally
common in S.W. Ireland from Kerry to Waterford and Limerick, ascending
to 1800 ft.; also in a few isolated localities in Galway, Mayo and Donegal
and in Cornwall, Devon and Somerset. 4, H11. W. and C. France, N. Spain
and Portugal, N.W. Italy.

*E. ceratocarpa Ten.

Woody at base. Lvs lanceolate, entire, glaucous, green. Bracts ovate or rhombic-
ovate, yellowish. *Capsule with horn-like warts on the angles.* Naturalized at Barry
Docks (Glamorgan). Native of S. Italy and Sicily.

*6. E. dulcis L.

Perennial 20–50 cm., more slender than the three preceding spp., with creeping nodular rhizome and ± numerous erect simple stems. Lvs 3–5 cm., alternate, oblanceolate-oblong, obtuse, tapered to base, subsessile, entire or finely serrulate near apex, glabrous or with a few scattered hairs beneath and occasionally above. Umbel usually 5-rayed, often with axillary infl. branches below, its bracts 2–3 cm., oblong-elliptic. *Upper bracts* 7–20 mm., ovate-deltoid, subacute, not mucronulate, denticulate, *truncate at base, green.* Glands green, *soon purple*, obovate-orbicular, *entire. Capsule* 2–3 mm., sub-globose grooved, glabrous (in Britain), *with prominent cylindric warts.* Seeds brown, smooth. Fl. 5–7. $2n = 12, 28$. Hp.

Introduced. Naturalized in shady places in a few localities, mainly in Scotland. 7. C. Europe from Belgium, C. Germany and Upper Dnieper region to N. Spain and Portugal, C. Italy and Macedonia.

7. E. platyphyllos L.   'Broad Spurge.'

Annual 15–80 cm., glabrous or pubescent, with single vertical root and simple stems. *Lvs* 1–4½ cm., alternate, obovate-lanceolate or oblong-lanceolate, acute, *deeply cordate or auricled at base*, sessile, serrulate except at base. Umbel usually 5-rayed, with axillary infl. branches below, the secondary forks often 3-rayed; its bracts 2–3 cm., elliptic-oblong. *Partial bracts* 5–15 mm., deltoid, acute or obtuse, mucronate, *the lowest not differing from the upper and markedly different from those of the umbel.* Glands suborbicular, *entire. Capsule* 2–3 mm., subglobose, *shallowly grooved, with hemispherical warts. Seeds c.* 2 mm., *olive-brown, smooth.* Fl. 6–10. $2n = 28, 36$. Th.

Native. In arable land and waste places, widespread but local in S. England from N. Somerset and Kent to Norfolk and Gloucester; rare in N. England (Lincoln, Nottingham, Yorks); Glamorgan. 35. C. and S. Europe from the Netherlands and Middle Dnieper region to N. Spain, Corsica, Sicily and Crete; W. Caucasus.

8. E. stricta L.   'Upright Spurge.'

Differs from *E. platyphyllos* as follows: Usually more slender, always glabrous. Lvs usually smaller. Umbel 2–5-rayed. *Partial bracts usually becoming relatively narrower downwards and passing into those of the umbel. Capsule* 2 mm. *or less, deeply grooved*, with prominent *cylindrical warts which are longer than broad. Seeds c.* 1½ mm., *red-brown.* Fl. 6–9. $2n = 28$. Th.

Native. Limestone woods in W. Gloucester and Monmouth. 2. C. Europe from N. France and Upper Dnieper region to the Pyrenees, N. Italy, Macedonia and the Caucasus; N. Persia, Aral–Caspia.

9. E. helioscopia L.   'Sun Spurge.'

Glabrous *annual* 10–50 cm., with slender vertical root and single stems, simple or with a few branches below. *Lvs* 1·5–3 cm., alternate, *obovate, very obtuse*, tapered from near apex to a narrow base, serrulate above. Umbel 5-rayed without axillary infl. branches below. Bracts all similar to lvs but less tapered below and often yellowish-tinged. *Glands transversely oval, entire*, green. *Capsule* 3–5 mm., subglobose, somewhat trigonous, *smooth. Seeds c.* 2 mm., brown, *reticulate.* Fl. 5–10. $2n = 42$. Th.

Native. Common in cultivated ground throughout the British Is., ascending to 1470 ft. 112, H40, S. Europe from Scandinavia (over 70° N.) southwards; Mediterranean region; C. Asia.

10. E. peplus L.   'Petty Spurge.'

Glabrous *green annual* 10–30 cm., with slender vertical root and simple or branched stems. *Lvs* 0·5–3 cm., alternate, *oval or obovate, obtuse,* shortly stalked, entire. Umbels 3-rayed. Bracts like the lvs but subsessile. *Glands lunate* with long slender horns. Capsule c. 2 mm., trigonous, each valve with 2 narrow wings on the back. Seeds pale grey, pitted. Fl. 4–11. $2n = 16$. Th.

Native. Very common in cultivated and waste ground throughout the British Is., ascending to 1350 ft. 110, H40, S. Europe (except Iceland); Mediterranean region, Azores; Siberia (to Lake Baikal).

11. E. exigua L.   'Dwarf Spurge.'

Glabrous, *glaucous annual* 5–30 cm., with slender vertical root and single simple or branched stem. *Lvs* 0·5–3 cm., alternate, *linear,* acute or more rarely obtuse, mucronate, sessile, entire. Umbel 3(–5)-rayed. Lower bracts triangular-lanceolate, upper triangular-ovate, all subcordate at base. *Glands lunate,* with long slender horns. Capsule c. 2 mm., trigonous, smooth or slightly rough on back. Seeds pale grey, tubercled. Fl. 6–10. $2n = 24, 28$. Th.

Native. Common in arable land in England, Wales and S. Scotland to Lanark and Angus, very rare further north and absent from Orkney, Shetland and the Hebrides; common in E. Ireland, rare in the west; Channel Is. 87, H34, S. Europe (except Iceland, Norway, Finland, etc.); Mediterranean region; Azores.

12. E. portlandica L.   'Portland Spurge.'

Glabrous *glaucous biennial or short-lived perennial* 5–40 cm., with ± *vertical root and several* usually simple *stems* from the base, all flowering or some sterile. *Lvs* 0·5–2 cm., alternate, dense, numerous, often caducous on fl. stems, *somewhat coriaceous, obovate to oblanceolate,* acute to apiculate, *mucronate, tapered to base,* sessile or subsessile, entire, *midrib prominent beneath.* Umbel 3–6-rayed, its bracts oval or obovate. Partial bracts 5–8 mm., triangular-rhombic, broader than long, mucronate. *Glands lunate,* with long horns. Capsule c. 3 mm., trigonous, granulate on the back of the valves. *Seeds* grey, *pitted.* Fl. 5–9. $2n = 16, 40$. ?Ch.

Native. Sea sands and young dunes on the S. and W. coasts from Sussex to Wigtown, very local; all round the Irish coast but rare in the north-west; Channel Is. 25, H17, S. W. coast of Europe from France to Portugal.

13. E. paralias L.   'Sea Spurge.'

Glabrous glaucous *perennial* 20–40 cm., with short woody stock, *vertical root and several* simple fertile and sterile *stems. Lvs* 0·5–2 cm., alternate, numerous, dense, often imbricate, *very thick and fleshy,* entire, somewhat concave, *ovate or oblong,* obtuse to subacute, *not mucronate, with broad sessile base; midrib obscure.* Umbel 3–6(–8)-rayed, its bracts ovate. *Glands lunate,* with short horns. Partial bracts 5–10 mm., orbicular-rhombic, thick and fleshy, mucronulate. Capsule c. 4 mm., trigonous, granulate. *Seeds* pale grey, *smooth,* with very small caruncle. Fl. 7–10. Fr. 8–11. $2n = 16$. ?Ch.

Native. Sea sands and mobile dunes from Wigtown and Norfolk southwards, rather local; all round the Irish coast but rare in the north and west; Channel Is. 32, H15, S. Atlantic coast from Belgium to Morocco; whole Mediterranean coast.

*14. E. uralensis Fisch. ex Link

E. virgata Waldst. & Kit., non Desf.

Glabrous *perennial* 30–80 cm., *with long creeping rhizome* and *numerous* erect fertile and sterile *stems*, often forming large patches. Fertile stems with long (7–20 cm.) axillary sterile or occasionally fl. branches above. *Lvs* alternate, *numerous, linear-lanceolate* or occasionally linear-oblanceolate, on the main stems 2–8 *cm.* × 4–6 *mm., broadest below or somewhat above the middle,* tapered to the acute, sometimes mucronate apex, *not or scarcely tapered to the base,* sessile, entire. Umbel 6–12-rayed with axillary infl. branches below; *its bracts* 12–35 *mm.,* lanceolate or linear-oblong. Infl. rather wide. Partial bracts 5–10 mm., ± deltoid, apiculate, yellowish. *Glands lunate* with rather long horns. Capsule 2–3 mm., trigonous, slightly roughened on back of valves. Seeds brown, smooth. Fl. 5–7. Hp.

Introduced. Naturalized in waste and grassy places in numerous localities in England and E. Scotland, north to Moray. 34, S. Native of E. Europe from Bohemia and N. Russia to Macedonia and the Caucasus; Siberia.

The taxonomy of this and the two following spp. is in need of revision. More than 3 spp. may occur and it is possible that our plant (or plants) are not *E. uralensis.*

15. E. esula L.

Differs from *E. uralensis* as follows: Smaller; sterile stems relatively more numerous. Axillary branches shorter, 2–10 cm. Lvs *linear-oblanceolate,* on the main stems 2–4·5 cm. × 4–7 mm., *broadest near the* subacute mucronate *apex, gradually tapered to a narrow base. Bracts of the umbel* 5–15 *mm.,* linear-oblong. Infl. narrower. Partial bracts smaller (to 7 mm.), green. Glands with rather short horns. Fl. 5–7. $2n = 64$. Hp.

Probably introduced. In a few places in woods and by streams in Scotland, where it is possibly native, and in England, but apparently very rare, many records applying to *E. uralensis.* Europe from Scandinavia and the Upper Dnieper region to E. and S. Spain, C. Italy and Macedonia.

16. E. cyparissias L.   'Cypress Spurge.'

Glabrous perennial 10–30 cm., *with long creeping rhizome* and numerous erect fertile and sterile stems, often forming large patches. Fertile stems usually with axillary branches above, often overtopping the infl., the stem then appearing bushy. *Lvs* alternate, very numerous, *linear,* on the main stems 1·5–3 *cm.* × 1–2 *mm.,* obtuse to subacute, sessile, entire. Umbel 9–15-rayed, often with axillary infl. branches below, its bracts ± oblong. Partial bracts 3–6 mm., deltoid or reniform, yellowish or becoming reddish. *Glands lunate* with very short horns. Capsule c. 3 mm., trigonous, slightly roughened. Seeds brown, smooth. Fl. 5–8. $2n = 20, 40$. Hp.

?Native. Possibly native in calcareous grassland and scrub in a few places from Kent and Wilts to Westmorland; more commonly occurring as a garden-escape or casual in waste places and as such extending to Sutherland. 46 (including introductions). Europe from Scandinavia (doubtful native) and the Upper Volga region to N. Spain, Italy and Macedonia.

17. E. amygdaloides L.   Wood Spurge.

Pubescent perennial 30–80 cm., with thick stock and tufted stems, sterile the first year, elongating and flowering the second, the lvs of the first year, persisting at flowering in a cluster at the top of the first year stem. Lvs of the first year 3–8 cm., oblanceolate, obtuse to subacute, gradually tapered at base into a short petiole, entire, dark green; of the fl.-stems oblong or obovate-oblong, not tapered at base, subsessile. Umbel with 5–10 rays with axillary infl. branches below, its bracts oval. *Partial bracts* 5–10 mm., reniform, *connate* in pairs for about half the width of their bases, yellowish. *Glands lunate* with converging horns. Capsule c. 4 mm., somewhat trigonous, granulate. Seeds grey, smooth. Fl. 3–5. $2n=18$. Not satisfactorily referable to any standard life-form, usually given as Hp., but appearing in winter as N.

Native. Damp woods, sometimes co-dominant in recent coppice; ascending to c. 1400 ft. Throughout England and Wales, rather common in the south, very local in the north; Channel Is.; in Ireland only in Cork and Donegal, probably introduced; not known in Scotland. 55, H3, S. C. and S. Europe from the Netherlands, C. Germany and the Upper Dnieper region to N. Spain and Portugal, Sicily, Macedonia and the Caucasus; Algeria (in the mountains).

79. POLYGONACEAE

Herbs, shrubs, climbers, or rarely trees. Lvs usually alternate and usually with sheathing stipules (ochreae). Fls hermaphrodite or unisexual. Per. segs 3–6, sepaloid or petaloid, free or connate, persistent, imbricate in bud. Stamens usually 6–9; anthers 2-celled, opening lengthwise. Ovary superior, syncarpous, unilocular, with a solitary, basal, orthotropous ovule. Fr. indehiscent, hard, trigonous or lenticular, usually enveloped in the perianth.

About 800 spp. distributed throughout the world, but mainly in the temperate regions.

1 Tiny annual; most lvs sub-opposite; stipules c. 1 mm.; fls c. 1·2 mm., nearly sessile in the lf-axils.   1. KOENIGIA
  Not as above.   2

2 Lvs reniform; per. segs 4.   4. OXYRIA
  Lvs not reniform; per. segs usually 5 or 6.   3

3 Infl. unbranched or nearly so; perianth petaloid, usually pink or white; per. segs not enlarging in fr., usually 5.   4
  Infl. usually with many long branches; per. segs 6, in 2 whorls, the inner enlarging in fr., or fls dioecious.   5. RUMEX

4 Panicle shortly branched; lvs triangular, hastate or cordate, fr. much longer than perianth; annual.   3. FAGOPYRUM
  Spike simple or little-branched; lvs various, but, if at all like the above, plant a stout perennial, 1–2 m.; fr. shorter or little longer than perianth.
                                                            2. POLYGONUM

1. KOENIGIA L.

Annual herb, most lvs sub-opposite. Fls clustered; bracts small. Per. segs 3, sepaloid. *Stamens 3, alternating with 3 gland-like staminodes.* Ovary trigonous; styles 2.

One sp., Arctic and subarctic, mountains of C. Asia, Tierra del Fuego.

1. K. islandica L.

1–6 cm. Stems reddish, flexuous; branches short. Lvs 3–5 mm.; blade broadly elliptic to suborbicular, rounded at apex, rather fleshy; petiole 0·5–1 mm., thick; stipules c. 1 mm., sheathing, hyaline. Fls in terminal and axillary clusters; pedicels 0·5 mm., jointed, pale green. Per. segs c. 1·2 mm., pale green, rounded, joined in lower ⅓. Stamens alternating with per. segs, inserted on tube; anthers reddish; staminodes yellow. Fl. 6–8. Th.

Native. On bare stony ground and among open vegetation, 1500–2360 ft. in Skye and Mull. Distribution of the genus.

2. POLYGONUM L.

Annual or perennial herbs, rarely shrubby. Stipules fused into a tube (ochrea). Perianth monochlamydeous, *segments 3–6, usually 5, spirally arranged, outer sometimes enlarging a little in fr. but never tubercled.* Stamens 4–8. Styles 2–3; *fr.* triquetrous or compressed and lenticular, ± *enclosed in the persistent perianth.* Insect- or self-pollinated, sometimes cleistogamous.

About 175 spp., cosmopolitan, but particularly in temperate regions.

The following hybrids, with characters ± intermediate between those of the parents, are recorded, but none seems to be particularly frequent:

P. hydropiper × minus; × mite; × nodosum; × persicaria. P. minus × mite; × persicaria. P. mite × persicaria.

1	Stems slender, twining; lvs cordate-sagittate.	2
	Stems not twining; lvs not cordate-sagittate.	3
	Stems slender, scrambling, 4-angled, furnished with weak deflexed prickles; lvs sagittate.	**18. sagittatum**
2	Pedicels in fr. 1–2 mm., jointed above middle; fr. not shining.	**16. convolvulus**
	Pedicels in fr. up to 8 mm., jointed at or below middle; fr. shining.	**17. dumetorum**
3	Infl. 1–6-fld, all axillary.	4
	Infl. usually many-fld, some or all terminal.	9
4	Fr. dull, striate, enclosed by or slightly longer than the persistent perianth. (**aviculare** agg.)	6
	Fr. smooth, distinctly shining, as long as or longer than the persistent perianth.	5
5	Ochreae with 4–6 unbranched veins, shorter than internodes; fr. 5–6 mm., exceeding the persistent perianth.	**5. raii**
	Ochreae with 8–12 branched veins, as long as upper internodes; fr. 4–4·5 mm., enclosed by or slightly projecting from the persistent perianth (very rare).	**6. maritimum**
6	Branch-lvs much smaller than stem-lvs; persistent perianth divided almost to base; fr. trigonous with 3 concave sides.	7

Branch- and stem-lvs ±equal; persistent perianth divided for half its
length; fr. with 2 sides convex, 1 concave.   **4. arenastrum**

7   Stem-lvs narrow, linear-lanceolate, 1–4 mm. broad; per. segs and fr.
    narrow, fr. exserted.   **3. rurivagum**
    Stem-lvs 5–18 mm. broad; per. segs overlapping; fr. broad.   8

8   Stem-lvs ovate-lanceolate, subsessile; petioles c. 2 mm., included in the
    ochreae; fr. 2·5–3·5 mm.   **1. aviculare**
    Stem-lvs obovate-spathulate; petioles up to 4–8 mm., projecting from
    ochreae; fr. 3·5–4·5 mm. (Shetland).   **2. boreale**

9   Plant rarely exceeding 50 cm., if more then stems soft, ±decumbent at
    base, and nodes often swollen.   10
    Stout erect herbs with somewhat woody stems 1–4 m. high.   18

10 Stems unbranched, never floating; rootstock stout.   11
    Stems usually branched, if nearly simple generally floating; root fibrous. 12

11 Radical lvs truncate or subcordate at base; petiole winged in upper half;
    infl. stout, without bulbils.   **8. bistorta**
    Radical lvs tapering at base; petiole not winged; infl. slender, lower part
    bearing bulbils.   **7. viviparum**

12 Lf-base cordate or rounded; stamens exserted.   **9. amphibium**
    Lvs narrowed to base; stamens included.   13

13 Plant with glands, sometimes sparse, on per. segs or peduncle.   14
    Infl. and peduncle entirely without glands.   16

14 Infl. slender, nodding; per. segs conspicuously glandular; peduncles
    eglandular.   **13. hydropiper**
    Infl. stout, erect; per. segs sparsely glandular; peduncles glandular. 15

15 Only the uppermost ochreae shortly fringed; peduncles sparsely glan-
    dular; fr. 2·5–3 mm.; fls greenish-white, rarely pink.   **11. lapathifolium**
    All ochreae shortly fringed; peduncles densely glandular; fr. 2 mm.
    or less; fls usually pink.   **12. nodosum**

16 Infl. dense, stout, obtuse; lvs often with a dark blotch.   **10. persicaria**
    Infl. slender, few-fld, acute; lvs never blotched.   17

17 Lvs usually 10–25 mm. broad; infl. slightly nodding; fr. 3–4 mm.
                                                            14. mite
    Lvs usually 5–8 mm. broad; infl. erect; fr. 2–2·5 mm.   **15. minus**

18 Lvs lanceolate, c. 3 times as long as broad.   **21. polystachyum**
    Lvs ovate, usually less than 1½ times as long as broad.   19

19 Lvs truncate at base, cuspidate.   **19. cuspidatum**
    Lvs weakly cordate at base, acute.   **20. sachalinense**

Section 1. *Polygonum*. Annual or perennial. Stem branched, ±prostrate.
Ochreae ± silvery or membranous. Infl. axillary, few-fld. Perianth ± petaloid.
Stamens 5–8, filaments of inner dilated at base. Fr. trigonous or subtrigonous.

P. aviculare agg. (spp. 1–4)   Knotgrass.

A glabrous annual 3–200 cm. Lvs up to 5 cm. × 18 mm., elliptic, spathulate,
lanceolate or linear, acute or subobtuse; *ochreae silvery*, lacerate; veins
indistinct. *Infl.* 1–6-fld, *axillary*. Stamens 5–8. *Fr.* 1·5–4·5 mm., trigonous,
sometimes with one side much narrower than the others, brown, usually
punctate or striate, *dull or slightly shining*, usually *enclosed within the persistent
perianth*. Fl. 7–10. Th.

Native. In waste places, arable land and on the sea shore. 112, H40, S. Common and generally distributed. Cosmopolitan except for the Arctic and Antarctic.

The aggregate includes the following four species which are all very variable according to habitat:

1. P. aviculare L.

P. heterophyllum Lindm.; *P. littorale* auct., p.p.

Plant robust, up to 2 m., erect or spreading, *heterophyllous when young. Lvs* 2·5–5 × 0·5–1·5 cm., *lanceolate to ovate-lanceolate,* subacute; those of main stems 2–3 times as long as those of flowering branches; *petioles up to 2 mm.,* included in the ochreae. Ochreae c. 5 mm. *Per. segs united at the base only,* margins pinkish or white. *Fr.* 2·5–3·5 × c. 1·5 mm., punctate, dull brown, *with 3 equal sides.* $2n = 60*$.

Native. Roadsides and waste places and on the coast. Very common and generally distributed. 112, H40, S. Europe, Asia; introduced into N. America, Australasia, S. America.

2. P. boreale (Lange) Small

Plant up to 100 cm., erect or suberect, *simple or sparingly branched,* markedly heterophyllous. *Stem-lvs oblong-obovate* to *spathulate,* the lower 3–5 cm. × 5–18 mm.; *petioles 4–8 mm., projecting from ochreae. Per. segs broad, inclined to be open;* margins pinkish. *Fr.* 3·5–4·5 × c. 2·5 mm., included, punctate, light brown, *with 3 broad sides.* $2n = 40*$.

Native. Known only from Shetland, where it seems largely to replace *P. aviculare.* Greenland, Iceland, Faeroes, N. Scandinavia; E. Canada.

3. P. rurivagum Jord. ex Bor.

Plant seldom more than 30 cm., very slender, flexuous, suberect, heterophyllous. *Lvs* 1·5–3·5 cm. × 1–3 mm., *linear-lanceolate to linear,* acute. Ochreae often c. 10 mm., brownish-red below. Infl. 1–2-fld, seldom more. *Per. segs narrow,* reddish. *Fr.* 2·5–3·5 × 1·5–2 mm., *exserted from the persistent perianth,* scarcely shining. $2n = 60*$.

Native. Usually in arable fields on chalky soil in the south of England, local. 5. Western Europe and Scandinavia.

4. P. arenastrum Bor. 'Small-leaved Knotgrass.'

P. aequale Lindm., *P. littorale* auct., p.p.; incl. *P. microspermum* Jord. ex Bor. and *P. calcatum* Lindm.

Plant usually 5–30 cm.; forming a dense prostrate mat, rarely erect. Lvs up to 20 × 5 mm., elliptic or elliptic-lanceolate, not heterophyllous. Infl. 2–3-fld. *Per. segs united for half their length,* greenish-white or pink. *Fr.* 1·5–2·5 mm., dull, but sometimes shining on the edges, *with 2 broad and one narrow sides,* or sometimes biconvex, brown to black. $2n = 40*$.

Native. Waste places, roadsides, common throughout the British Isles, though less so than *P. aviculare* and often in drier places than the latter species. 112, H40, S. Throughout Europe, introduced into N. America.

5. P. raii Bab.   'Ray's Knotgrass.'

A straggling, prostrate, glabrous, sometimes glaucous annual 10–100 cm.
Stems ± woody at base. Lvs 1–3·5 cm., elliptic-lanceolate to linear-lanceolate,
usually flat. *Ochreae with few unbranched veins.* Infl. 2–6-fld. Per. segs
c. 3 mm., with broad pink or white margins. *Fr.* 5–6 × 3–3·5 mm., *light brown,*
almost flat, *shining, exceeding the persistent perianth.* Fl. 7–10. $2n=40$*. Th.

Native. A decreasing Atlantic species on sandy shores or fine shingle above
high water spring-tides. S. and W. coasts to the Hebrides; formerly on the
east coast from Durham to Angus; ?Lincoln; all round the Irish coast.
54, H19, S. Coast of Europe from Norway to N. France; ?E. Canada.

6. P. maritimum L.   'Sea Knotgrass.'

A glabrous, glaucous, procumbent perennial 10–50 cm. Stems stout and
woody at base. Lvs 0·5–2·5 cm., elliptic-lanceolate, margins usually revolute.
Ochreae often *as long as the internodes, with 8–12 branched veins.* Infl. 1–4-fld.
Per. segs c. 2–2·5 mm., with broad pink or white margins. *Fr.* 4–4·5 × 2·5 mm.,
*dark reddish-brown, shining, as long as, or slightly exceeding, the persistent
perianth.* Fl. 7–10. $2n=20$*. Chh.

Native. On sand and fine shingle at, and just above high water spring-
tides; ?extinct in England, local in the Channel Is.; formerly in Cornwall,
Devon, N. Somerset and Hampshire. 6, S. W. coast of Europe from France
southwards; Mediterranean region; Macaronesia.

*****P. cognatum** Meisn., a ± prostrate perennial with the fr. perianth with a very thick
urceolate tube which is twice as long as the orbicular lobes, silvery stipules and
nerveless lvs, is naturalized in a few places. Asia Minor to Tibet.

Section 2. *Bistorta* (Mill.) DC. Perennial, rhizomatous. Stem unbranched,
erect. Ochreae truncate. Infl. terminal, spicate. Perianth petaloid. Stamens 8.
Fr. triquetrous.

7. P. viviparum L.

A slender erect *glabrous* perennial 6–30 cm. *Rhizome* rather stout, *not contorted. Lvs* 1·5–7 cm., *linear-lanceolate, tapering at base,* sometimes oval or
subrotund, lower lvs petioled, upper sessile; *petioles not winged;* ochreae
obliquely truncate, ± laciniate. *Infl.* terminal, *rather lax, slender* (4–8 mm.
diam.), *lower part with purple bulbils.* Fls 3–4 mm., white, in upper part of
infl. only, sometimes very few. Fl. 6–8. $2n=88$, c. 100, 110. ?Hp.

Native. In mountain grassland and on wet rocks. 35, H4. Mountain
districts of N. Wales, northern England and Scotland; Ireland: Kerry, Sligo,
Leitrim, Donegal, rare. In Scotland from sea-level in Sutherland to 4000 ft.
Arctic and northern Europe, Asia and America and on the higher mountains
further south in these continents.

8. P. bistorta L.   Snake-root, Easter-ledges, 'Bistort.'

An erect almost glabrous perennial 25–50 cm. *Rhizome* very stout and *contorted. Radical lvs* 5–15 cm., *broadly ovate,* obtuse, *truncate or subcordate at
base, puberulent on nerves beneath,* folded longitudinally in bud and showing
'creases' when mature, *petioles winged in upper part;* upper lvs triangular,
acuminate, petioles sheathing; ochreae obliquely truncate, ± laciniate. *Infl.*

terminal, *dense*, spicate, *stout* (10–15 *mm. diam.*). Fls 4–5 mm., pink, rarely white, numerous. Fl. 6–8. 2*n*=46. Hs.

The young lvs are eaten as Easter-ledge pudding in the Lake District.

Native. In meadows and grassy roadsides, commoner on siliceous soils, often forming large patches. 102, H 20. Scattered throughout the British Is., but rare and perhaps not native in the south-east or in Ireland. Northern and C. Europe (not Scandinavia), mountains of S. Europe; Asia Minor; C. Asia.

*P. amplexicaule** D. Don, a perennial with red fls and amplexicaul upper lvs, is naturalized in W. Ireland and locally elsewhere. Himalaya.

Section 3. *Persicaria* (Mill.) DC. Annual, rarely perennial. Stem branched, erect or decumbent. Ochreae truncate, often fringed. Infl. spicate. Perianth ± petaloid. Stamens 4–8, usually 8. Fr. compressed and lenticular or trigonous.

9. P. amphibium L.   'Amphibious Bistort.'

A glabrous or pubescent perennial commonly 30–75 cm. Aquatic and terrestrial forms differ considerably in vegetative features. Aquatic: glabrous; stems floating and rooting at nodes; *lvs* floating, 5–15 cm., ovate-oblong, subacute, *truncate or subcordate at base*; petioles 2–6 cm. Terrestrial: stems ascending or erect, rooting only at lower nodes, glabrous or slightly pubescent; *lvs* 5–12(–15) cm., appressed-hispid or rough, ciliate, oblong-lanceolate, obtuse, usually ± narrowed to the *rounded base*; ochreae hispid. Infl. 2–4 cm., usually terminal, many-fld, dense, lfless and peduncled. Fls pink or red. Per. segs eglandular; *stamens 5, exserted*; styles 2, united one-fourth the way up. Fr. c. 2 mm., orbicular, biconvex, brown, ± shining. Fl. 7–9. 2*n*=c. 66. Hyd. or Hp.

Native. In pools, canals and slow-flowing rivers; the terrestrial form on banks by water. 111, H 40, S. Generally distributed throughout the British Is. Europe, Asia, N. America, S. Africa.

10. P. persicaria L.   Red Shank, Willow weed, 'Persicaria'.

A branched erect or ascending nearly glabrous annual 25–75 cm. Stems reddish, swollen above the nodes. Lvs 5–10 cm., lanceolate, ciliate, sometimes woolly beneath, often black-blotched; *ochreae* truncate, *fringed. Infl. stout, obtuse, continuous* or somewhat interrupted, lfless or with a single lf at base. *Fls pink. Per. segs and peduncle eglandular*; styles 2 or 3, united below. *Fr.* c. 3 mm., *biconvex or bluntly trigonous* with concave faces, shining. Fl. 6–10. 2*n*=44. Th.

Native. In waste places, cultivated land and beside ponds. 112, H 40, S. Generally distributed throughout the British Is., common. Temperate regions of the northern hemisphere.

11. P. lapathifolium L.   'Pale Persicaria.'

A branched erect ± pubescent annual up to 100 cm. Stems usually greenish, swollen above nodes. Lvs 5–20 cm., lanceolate, ciliate, sometimes woolly beneath, ± hispid on the midrib, often black-blotched; petioles short, ± hispid; *ochreae* truncate, *only the uppermost very shortly fringed. Infl. stout, obtuse, continuous, some long-peduncled*, lfless. *Fls greenish-white*, rarely pink. *Per.*

segs and peduncle rather sparsely glandular; styles 2, separate nearly to base. *Fr.* 2·5–3 mm., *usually suborbicular, flattened, biconcave*, sometimes ± trigonous, shining. Fl. 6–10. 2n = 22. Th.

Native. In waste places, cultivated ground and beside ponds. 110, H 38, S. Throughout most of the British Is., common. Temperate regions of the northern hemisphere; S. Africa.

12. P. nodosum Pers.

P. maculatum (S. F. Gray) Dyer ex Bab.; *P. petecticale* (Stokes) Druce

Similar to *P. lapathifolium* but usually smaller and often ± decumbent. *Ochreae shortly fringed. Infl. rather lax, slender and ± interrupted*, though not as much so as in *P. hydropiper*, acute and shortly peduncled. *Per. segs usually pink, sparsely glandular; peduncles and undersides of lvs densely dotted with yellow glands. Fr.* 2 *mm. or slightly less*, flattened, suborbicular, shining. Fl. 7–9. 2n = 22. Th.

Native. In waste places and beside ponds and lakes. 57, H 12. Scattered through the British Is. north to Fife and Inverness. Temperate regions of the northern hemisphere; S. Africa. Perhaps only a ssp. of the preceding.

13. P. hydropiper L.   Water-pepper.

A nearly glabrous erect annual, 25–75 cm. Lvs usually 5–10 cm., lanceolate or narrowly lanceolate, subsessile, ciliate; *ochreae* truncate, *not or shortly fringed. Infl.* slender, *nodding*, acute, *interrupted and lfy in lower part. Fls greenish. Per. segs covered with yellow glandular dots; peduncles eglandular*; styles 2, free nearly to base. *Fr. c.* 3 *mm.*, ovoid, biconvex or subtrigonous, punctate, dark brown or black, *not shining*. Fl. 7–9. 2n = 20. Th. *Plant acrid and burning to the taste.*

Native. In damp places or in shallow water in ponds and ditches. 108, H 40, S. Generally distributed throughout the British Is., except the north of Scotland; common. Europe, except the arctic; N. Africa; temperate Asia; N. America.

14. P. mite Schrank

P. laxiflorum Weihe

Similar to *P. hydropiper* in general appearance but *ochreae conspicuously and coarsely fringed* and lvs rather abruptly narrowed at base. *Infl. nearly erect*, slender, interrupted but scarcely lfy, *eglandular. Fls pink*, rarely white. *Fr.* 3–4 *mm.*, broadly ovoid, biconvex, *shining*. Fl. 6–9. 2n = 40. Th. *Plant not acrid and burning*.

Native. In ditches and beside ponds and rivers. 46, H 8, S. Scattered throughout the southern part of Great Britain north to Stirling and N.E. Yorks; N. and W. Ireland; local. Europe north to S. Scandinavia; Asia Minor.

15. P. minus Huds.

A spreading branched decumbent or ascending annual 10–30(–40) cm. *Lvs usually* 2–5 × 0·5–0·8 *cm.*, narrowly lanceolate, ± obtuse, ciliate, subsessile; ochreae conspicuously and coarsely fringed. *Infl.* slender, *erect*, ± interrupted, eglandular. Fls pink or rarely white. *Fr.* 2–2·5 *mm.*, ovoid, biconvex, black and shining. Fl. 8–9. 2n = 40. Th.

Native. In wet marshy places and beside ponds and lakes. 66, H 26, S. Scattered through the British Is. north to W. Perth and Aberdeen; local. Europe north to C. Scandinavia and Finland; temperate Asia.

Section 4. *Tiniaria* Meisn. Annual or perennial. Stem usually twining. Ochreae truncate, margin entire. Lvs cordate at base. Infl. axillary. Perianth ± sepaloid, becoming keeled or winged in fr. Stamens 8. Fr. triquetrous.

16. P. convolvulus L.   Black Bindweed.

A somewhat mealy scrambling or climbing annual 30–120 cm. Stem angular, mealy on the angles. Lvs 2–6 cm., ovate, acuminate, cordate-sagittate at base, mealy beneath, nearly smooth above, longer than the petiole; ochreae obliquely truncate, ± laciniate. Infl. peduncled or subsessile, interrupted; *pedicels 1–2 mm., jointed above the middle.* Per. segs 5, *the 3 outer obtusely keeled or* (var. *subalatum* Lej. & Court.) *narrowly winged in fr.*, rough on back. *Fr.* 4 mm., *dull black*, minutely punctate. Fl. 7–10. 2*n*=40. Th.

Native. In waste places, arable land, and gardens. 111, H 40, S. Generally distributed throughout the British Is., except Shetland; common. Europe (except the Arctic), N. Africa, temperate Asia; introduced in N. America and S. Africa.

17. P. dumetorum L.

A somewhat mealy climbing annual up to nearly 2 m. Similar to *P. convolvulus*, but fruiting *pedicels up to* 8 *mm.*, capillary, *jointed at or below the middle* and deflexed. *Outer perianth segments broadly winged in fr.*, and strongly decurrent on the pedicels. *Fr.* 3 mm., black and *shining*. Fl. 7–9. 2*n*=20. Th.

Native. In thickets and hedges. 19. From N. Somerset and Worcester to Kent and Essex; Caernarvon; local. Europe north to S. Scandinavia; northern and western Asia.

Section 5. *Echinocaulon* Meisn. Annual herbs. Stems weak, 4-angled, furnished with deflexed prickles. Lvs cordate or sagittate at base. Ochreae truncate. Infl. subcapitate. Perianth petaloid. Stamens 5–8. Fr. lenticular or triquetrous.

*18. P. sagittatum L. Up to 1·5 m. *Stems, petioles and midribs with short, slender, deflexed prickles.* Lvs oblong-lanceolate, sagittate. Infl. subcapitate.
Native of N. America, naturalized in ditches in Kerry.

Section 6. *Pleuropterus* (Turcz.) Benth. Large, erect, perennial herbs. Infl. of terminal and axillary panicles. Outer per. segs keeled or winged in fr.

*19. P. cuspidatum Sieb. & Zucc. (*P. Sieboldii* hort.). *Lvs* broadly *ovate, cuspidate, truncate at base*, glabrous; ochreae much shorter than internodes. Infl. lax, some of the axillary ones much exceeding the petioles.
Native of Japan, sometimes cultivated and commonly naturalized, particularly in the south.

*20. P. sachalinense F. Schmidt. *Lvs ovate, acute, weakly cordate at base,* glabrous; ochreae much shorter than internodes. Infl. dense, the axillary ones shorter than or about equalling the petioles.

Native of Sakhalin, sometimes cultivated and occasionally naturalized.

*21. P. polystachyum Wall. ex Meisn. *Lvs lanceolate, acuminate, cordate to cuneate at base,* ± hispid beneath; ochreae of upper lvs nearly as long as or sometimes longer than internodes. Infl. much longer than petioles, divaricately branched.

Native of the mountains of Assam and Sikkim from c. 7000–12,000 ft., occasionally naturalized, commonly so in W. Ireland and S.W. England.

*P. campanulatum Hook. f., a perennial with pale pink fls and lvs whitish to buff tomentose beneath, is naturalized in S. England, W. Scotland and W. Ireland. Himalaya.

*P. baldschuanicum Regel, a woody climber with ovate-cordate to hastate lvs and white fls, is often cultivated and sometimes escapes. Bokhara.

3. Fagopyrum Mill.

Annual or perennial herbs similar to *Polygonum* in many respects. Stems erect. *Lvs cordate at base.* Perianth petaloid. Fls heterostylous. Stamens 8, styles 3. *Fr. much exceeding the perianth.*

About 6 spp. in Asia, introduced elsewhere and frequently cultivated.

*1. F. esculentum Moench   Buckwheat

F. sagittatum Gilib.; *Polygonum Fagopyrum* L.

An erect little-branched nearly glabrous annual 15–60 cm. Lvs cordate-sagittate, acuminate, as long as or longer than broad. Infl. a cymose panicle. Fls pink or white. Fr. c. 6 mm., 2–3 times as long as perianth, triquetrous, angles acute, entire. Fl. 7–8. $2n=16$. Th.

Introduced. Cultivated as a crop, particularly in the fens, and occurring as a casual in waste ground. Said to be native in C. Asia. Largely cultivated in some countries as a substitute for cereals, or for green fodder.

4. Oxyria Hill

Perennial herbs. Infl. branched, lfless. *Per. segs sepaloid, 2 + 2, inner enlarging in fr., not tubercled*; anthers versatile; stigmas 2. *Fr. lenticular, broadly winged.*

Two spp. in arctic regions and on mountains in the northern hemisphere.

1. O. digyna (L.) Hill   'Mountain Sorrel.'

Rheum digynum (L.) Wahlenb.

A glabrous often tufted perennial 5–30 cm. Rootstock stout. Stem erect. *Lvs* 1–3 cm., almost all radical, *reniform*, rounded or retuse, rarely subhastate, rather fleshy; petioles long. Panicle lfless. Pedicels slender, jointed about the middle and thickened towards the top. Outer per. segs spreading or reflexed, inner pressed to fr., spathulate or ± rhomboid. Fr. 3–4 mm., broadly winged. Fl. 7–8. $2n=14$. Hr.

Native. In damp rocky places on mountains, especially beside streams, locally common. 31, H8. Merioneth and Caernarvon; Westmorland, Cum-

berland; Dumfries, Arran and Angus northwards except for Kintyre and
Islay; Ireland, on the mountains of the west from Kerry to Donegal and in
the Galtees. Mountains of the N. temperate zone.

5. RUMEX L.

Annual, biennial or perennial herbs, usually with long stout roots, sometimes
rhizomatous. Lvs alternate, stipules tubular (ochreae). Fls hermaphrodite
or unisexual, arranged in whorls on simple or branched infls., anemophilous.
Per. segs 3+3, the outer always small and thin, *the inner* (fr. per. segs)
enlarging and usually becoming hard in fr. Fr. per. segs with or without swollen
corky spherical or ovoid tubercles on their midribs; tubercles developing as
the fr. ripens. Stamens 3+3; anthers basifixed. Fr. triquetrous with a woody
pericarp.

About 150 spp. in the temperate regions of the world.

Hybrids are of frequent occurrence and, although some (e.g. R. crispus ×
obtusifolius) are common, have not been included in the following key and
descriptions. The majority may be easily recognized by their fairly high degree
of sterility in both pollen and fr. They usually occur in close proximity to the
parents and are intermediate in character between them. Hybrid swarms do
not occur.

The following hybrids have been reported from Britain:

R. aquaticus × obtusifolius.
R. conglomeratus × crispus; × hydrolapathum; × maritimus; × obtusifolius; × pulcher.
R. crispus × longifolius; × obtusifolius (R. pratensis Mert. & Koch, R. acutus L.);
    × patientia; × pulcher; × sanguineus.
R. hydrolapathum × obtusifolius (R. maximus auct.).
R. longifolius × obtusifolius.
R. maritimus × obtusifolius.
R. obtusifolius × palustris; × patientia.
R. pulcher × rupestris; × sanguineus.

Really ripe fr. is essential for the accurate identification of many spp.

Much additional information is given by J. E. Lousley, *Rep. Bot. Soc. and Exch.
Club* (1938), pp. 118–57; (1941–2), pp. 547–85.

1	Foliage acid; lvs hastate; fls usually unisexual.	2
	Foliage not or scarcely acid; lvs not hastate; fls usually hermaphrodite.	4
2	Lvs about as long as broad, upper petioled. **5. scutatus**	
	Lvs several times as long as broad, if nearly as broad as long then upper subsessile and clasping stem.	3
3	Lobes of lvs spreading or forward-pointing, upper lvs not clasping stem. **acetosella** agg. (spp. 1–3)	
	Lobes of lvs ± downward-directed, upper lvs clasping stem. **4. acetosa**	
4	All fr. per. segs without tubercles.	5
	At least one fr. per. seg. with a distinct tubercle.	8
5	Plant slender; whorls up to 6-fld; fr. per. segs with 3–5 hooked teeth. **21. brownii**	
	Plant stout; whorls many-fld; fr. per. segs entire.	6
6	Pedicels of fr. with an almost imperceptible joint. **8. aquaticus**	
	Pedicels of fr. distinctly jointed.	7

7   Rhizomatous; lvs about as long as broad; fr. per. segs ovate, truncate
    at base.   **7. alpinus**
    Not rhizomatous; lvs distinctly longer than broad; fr. per. segs reniform.
                                                               9. longifolius

8   Fr. per. segs distinctly toothed (teeth more than 1 mm.).   9
    Fr. per. segs entire or denticulate (teeth not exceeding 1 mm.).   13

9   Branches making a wide angle with main stem, forming an entangled
    mass in fr.; lvs usually panduriform, lamina rarely exceeding 10 cm.
                                                               15. pulcher
    Branches usually making a narrow angle with main stem, not becoming
    entangled in fr.; lvs very rarely panduriform, lamina often exceeding
    10 cm.   10

10 Fr. per. segs 4–7 mm., reddish or brown.   11
    Fr. per. segs 2–3 mm., yellow or golden.   12

11 Fr. per. segs 5–6 × 3 mm., triangular, truncate at base, one (rarely all 3)
    with a prominent tubercle.   **14. obtusifolius**
    Fr. per. segs 4–5 × 4–5 mm., roundish with a short triangular apex, all 3
    with small tubercles.   **10. stenophyllus**
    (See also **11. cristatus**)

12 Anthers c. 1 mm.; outer per. segs with claw-shaped forward-curved
    apices; fr. per. segs with ligulate obtuse apices, teeth rigid, setaceous,
    shorter than segs; fr. 2–2·5 mm.   **19. palustris**
    Anthers c. 0·5 mm.; outer per. segs horizontally spreading or weakly
    reflexed; fr. per. segs with triangular, ± acute apices, teeth very fine and
    capillary, some longer than segs; fr. 1–1·5 mm.   **20. maritimus**

13 Lvs narrowly obovate, thick and coriaceous; rhizome far-creeping; infl.
    little-branched (dunes).   **22. frutescens**
    Not as above.   14

14 Primary infl. after flowering overtopped by secondary ones arising from
    lower axils; tubercles of fr. per. segs rugose.   **23. triangulivalvis**
    Not as above.   15

15 Infl. dense, whorls crowded, ± confluent; fr. per. segs (3·5–)4·5(–8) mm.;
    tubercles ½ as long as fr. per. segs or less.   16
    Infl. lax, whorls distant; fr. per. segs up to 3 mm., or if more then tubercles
    ¾–¾ length of fr. per. segs.   19

16 Fr. per. segs triangular, truncate or subcuneate at base. **6. hydrolapathum**
    Fr. per. segs broadly ovate or orbicular, cordate at base.   17

17 Lvs lanceolate, usually narrow, undulate and crisped; plant up to
    100 cm.; fr. per. segs 3–5 × 2–5 mm.   **13. crispus**
    Lvs ovate- or oblong-lanceolate, not crisped; plant 100–200 cm.; fr. per.
    segs 5–7 mm. broad.   18

18 Veins in middle of lf at an angle of 45–60° with midrib.   **12. patientia**
    Veins in middle of lf at an angle of 60–90° with midrib.   **11. cristatus**

19 Lvs glaucous; fr. per. segs c. 4 mm., all with prominent tubercles ¾–¾
    as long (S.W. England and S. Wales, by the sea).   **18. rupestris**
    (See also **15. pulcher**)
    Lvs not glaucous; fr. per. segs up to 3 mm.   20

20 Stems usually almost straight, branches making an angle of 20°(–40°);
    lowest whorls on branches subtended by lvs; tubercle (usually one)
    globular.   **16. sanguineus**

> Stems usually flexuous, branches making an angle of 30–90°; whorls usually subtended by lvs for ⅔ length of branches; tubercles (usually 3) oblong. **17. conglomeratus**

Subgenus 1. ACETOSELLA Raf.

Fls dioecious, rarely polygamous. Fr. per. segs of female fls not enlarged in fr. or, at most, twice as large as nut, all without tubercles. Lvs often hastate or sagittate.

R. acetosella agg. (spp. 1–3). Sheep's Sorrel.

An erect or decumbent perennial up to 30 cm., producing adventitious buds on horizontal roots. *Lvs* up to 4 cm., several times as long as broad, lanceolate to linear, often hastate with *spreading or forward-pointing lobes; upper lvs ± distinctly petioled, not clasping the stem; ochreae hyaline, lacerate.* Infl. up to c. 15 cm., lfless or nearly so. Fls dioecious. *Outer per. segs pressed to inner. Fr.* 0·8–1·5 *mm.* Fl. 5–8. Hs. or Grt. Lvs slightly acid and bitter.

Native. On heaths, in grassland and cultivated land, common on acid but infrequent on calcareous soils. 112, H40, S. Generally distributed throughout the British Is. Europe from Scandinavia southwards; temperate Asia; N. and S. Africa; Macaronesia; temperate America; Australia; Greenland; perhaps introduced in the southern hemisphere.

1. R. angiocarpus Murb.

Plant variable in size, stems erect; lower lvs 3–4 times as long as broad (excluding lobes), margins flat. Per. segs thin, adherent to fr. Ripe fr. up to c. 1 mm. and nearly as broad as long. $2n=14$.

S. England, apparently local. S. and C. Europe, N. Africa.

2. R. tenuifolius (Wallr.) Löve

Plant small, stems ± decumbent. Lower lvs 7–10 times as long as broad (excluding lobes); margins inrolled. Per. segs thick, not adherent to fr. Ripe fr. c. 1×0.7 mm. $2n=28$.

Native. Probably widespread but rather uncommon, occurring mainly on the poorest soils. C. and N. Europe, Arctic.

3. R. acetosella L.

Plant variable in size, stems erect. Lower lvs 3–4 times as long as broad (excluding lobes); margins flat. Per. segs thick, not adherent to fr. Ripe fr. c. 1.5×0.8 mm. $2n=42$.

Native. Widespread and common, avoiding the poorest soils and less strongly calcifuge than *R. tenuifolius*. C. and S.E. Europe to northern Scandinavia and Iceland.

Subgenus 2. ACETOSA (Mill.) Rech. f.

Fls dioecious or polygamous. Fr. per. segs of female fls many times longer than nut, without tubercles or with small recurved tubercles at the base. Lvs often hastate or sagittate.

4. R. acetosa L. Sorrel.

An erect nearly glabrous perennial up to 100 cm., though often less. *Lvs* up to 10 cm., usually less, 1½–several times as long as broad, oblong-lanceolate, obtuse or subacute, hastate, *lobes ± downward-directed; upper lvs subsessile, ± clasping stem; ochreae fringed.* Infl. up to 40 cm., lfless or nearly so. Fls dioecious. *Outer per. segs reflexed and appressed to pedicels after flowering; fr. per. segs 3–4 mm., orbicular-cordate. Fr.* 2–2·5 mm., shining. Fl. 5–6. $2n = 14$ (female), 15 (male); triploid and hexaploid forms also occur. Hs.

Our common plant is ssp. *acetosa.* The occurrence of other sspp. in Britain needs further investigation.

Lvs acid, sometimes used in salads and sauces.

Native. In grassland and in open places in woods. 112, H40, S. Generally distributed and common throughout the British Is. Europe, temperate Asia, N. America, Greenland.

*5. R. scutatus L.

A glabrous perennial up to 50 cm. *Lvs* c. 3–4 cm., *about as broad as long, broadly ovate or panduriform,* obtuse, hastate, lobes diverging; upper lvs petioled; *ochreae bifid, ± toothed but not fringed.* Infl. up to c. 15 cm., lfless or nearly so. Fls polygamous. Outer per. segs pressed to inner; fr. per. segs c. 5 mm., orbicular, cordate. Fl. 6–7. $2n = 20$. Hp. Lvs acid, sometimes used in salads and sauces.

Introduced. Naturalized on old walls and in pastures in a few scattered localities. C. and S. Europe; W. Asia, N. Africa.

Subgenus 3. RUMEX.

At least most of the fls hermaphrodite. Fr. per. segs many times longer than nut, with or without tubercles; tubercles never recurved. Lower lvs cuneate, rounded, or cordate at base, never hastate or sagittate.

Section Rumex. Annuals, biennials, or perennials. Primary infl. not normally overtopped by secondary infl. arising from the axils of the lower lvs.

6. R. hydrolapathum Huds.   'Great Water Dock.'

A stout erect perennial up to 200 cm. Lvs up to 110 cm., lanceolate to ovate, acute or acuminate, narrowed, cuneate or rarely somewhat cordate at base, margins flat or somewhat undulate. Infl. usually much-branched, somewhat lfy below, branches strict. Whorls ± crowded. *Fr. per. segs* 6–8 *mm., triangular, truncate or subcuneate at base, each with a prominent elongated tubercle,* margin usually with a few short teeth towards the base. Fr. 4 mm. Fl. 7–9. Germ. spring. $2n = 200$. Hyd. or Hel.

Native. In wet places and shallow water. 81, H36, S. Fairly generally distributed throughout the British Is., though rarer in Scotland. Western Europe from S. Scandinavia to Spain and eastwards to Italy and S. Russia.

*7. R. alpinus L.   Monk's Rhubarb.

A stout erect rhizomatous perennial 30–80(–150) cm. *Rhizome extensively creeping. Lvs* 20–40 cm., *about as broad as long, rounded,* cordate, margins ± undulate. Infl. interrupted below, little branched. Whorls confluent, lfy in the lower part. Pedicels 5–10 mm., slender, deflexed in fr. *Fr. per. segs c.* 6 *mm., ovate, truncate at base, margins entire, tubercles* 0. Fr. 3 mm. Fl. 7. $2n = 20$. Hs.

Introduced. Near buildings and beside streams and roads in hilly districts.
15. From Stafford and Derby northwards. Mountain regions of C. and S.
Europe, Asia Minor and Caucasus.

8. R. aquaticus L.

A tall stout perennial 80–200 cm. Lvs cordate-ovate or triangular, $1\frac{1}{2}$–$2\frac{1}{2}$
times as long as broad and broadest near the base; base cordate, rarely
truncate; apex rounded or ± acute. *Pedicels of fr. very slender with an almost
imperceptible joint near base. Fr. per. segs 5–8·5 mm., ovate-triangular, truncate
at base*, apex obtuse, margins entire, *tubercles* 0. Fr. c. 3·5 mm. Fl. 7–8.
$2n = c. 200$. Hyd. or Hel.

Native. In water at the margins of alder swamps. 2. Loch Lomond, Stirling
and Dunbarton. Europe except for Italy and southern Balkans, infrequent
in the south-west; N. Asia.

9. R. longifolius DC.

R. domesticus Hartm.; *R. aquaticus* auct., non L.

A stout erect perennial up to c. 120 cm. Lvs up to 80 cm., ovate to lanceolate,
obtuse or subacute, ± cordate at base, margins undulate and crisped. *Infl.
dense, compact, fusiform.* Whorls confluent, lfy at base. *Fr. per. segs 6 mm.,
thin, reniform*, entire, *tubercles* 0. Fr. 3–4 mm. Fl. 6–7. $2n = 60$. Hs.

Native. Besides rivers, in ditches and in damp grassy places. 48. Derby, and
from Yorks and N. Lancashire northwards. Western Europe from Scandinavia to the Pyrenees; Caucasus, C. Asia; N. America; Greenland.

*10. R. stenophyllus Ledeb.

A perennial 100–150 cm. Lvs oblong, often truncate at base, apex rather obtuse,
margin undulate; upper lvs lanceolate. Infl. dense, with ascending branches. *Fr. per.
segs roundish, 4–5 mm. broad, with a short triangular apex, all with small tubercles,
margin with 4–6 teeth less than 5 mm.* Hs. Lvs resemble those of *R. crispus*. Sometimes confused with *R. crispus × obtusifolius* but quite fertile.

Introduced. Apparently well established at Avonmouth Docks. Austria, Hungary,
S. Russia to C. Asia and Siberia. Introduced elsewhere in Europe.

*11. R. cristatus DC.

R. graecus Boiss. & Heldr.

A tall perennial similar to *R. patientia*. Veins in middle of lf at an angle of 60–90°
with midrib. *Fr. per. segs* roundish-cordate, 6–7 mm. broad, *with frequent irregular
acute teeth up to 1 mm.*, usually all tubercled, one tubercle 2–3 mm., the others
smaller or sometimes 0. Fl. 6–7. Hs.

Introduced. In waste places, apparently well established near Cardiff and London.
S. Balkans, W. Asia Minor, Cyprus.

*12. R. patientia L.

An erect perennial 100–200 cm. Lvs ovate- or oblong-lanceolate from a
truncate or subcuneate, rarely subcordate base, apex acute. Veins in middle
of lf at an angle of 45–60° with midrib. Whorls crowded. *Fr. per. segs broadly
ovate- or suborbicular-cordate, one with a large tubercle the others without or
with small tubercles; margins entire, crenulate or minutely denticulate.* Fl. 6–7.
$2n = 60$. Hs.

Introduced. In waste places. Well established in a few localities in S. England. S. Hungary, E. Balkans, S. Russia and Asia Minor. Naturalized elsewhere in Europe.

The two following sspp. occur in Britain:

Ssp. **patientia**. Stem usually purplish or reddish-brown. Fr. per. segs 5–6(–7) mm. broad, only one with a tubercle 1·5 mm.

Ssp. **orientalis** (Bernh.) Danser. Stems usually pale. Fr. per. segs 8–10 mm. broad, one with a tubercle 2–3 mm., the others sometimes with smaller tubercles.

13. R. crispus L.   'Curled Dock.'

Incl. *R. elongatus* Guss.

An erect perennial 50–100 cm. *Lvs* up to 30 cm., lanceolate or oblong-lanceolate, narrowed or ± rounded at base, tapering from about the middle to an obtuse point; *margins usually undulate and strongly crisped*, rarely (var. *uliginosus* Le Gall) flat or nearly so. Infl. usually nearly simple or little-branched, branches strict, usually subtended by linear-lanceolate much crisped lvs. Whorls close, distinct, or confluent towards the ends of the branches. *Fr. per. segs* (3·5–)4·5(–6) *mm., broadly ovate-cordate*, usually all three with tubercles, one larger than the other two or (particularly in sea-shore plants) all equal; margin entire or minutely denticulate. Fr. 2·5–3 mm. Fl. 6–10. 2*n*=60. Hs.

Very variable. The commonest British sp. and a serious agricultural weed.

Native. In grassy places, waste ground, cultivated land, dune-slacks, and shingle beaches. 112, H40, S. Generally distributed and common throughout the British Is. Europe, Azores, and most of Africa. Naturalized in most other parts of the world; rare in the north.

14. R. obtusifolius L.   'Broad-leaved Dock.'

An erect branched perennial 50–100 cm. Underside of lvs and veins usually hairy. Lower lvs up to c. 25 cm., ovate-oblong, cordate at base, apex obtuse, margins undulate; upper lvs ovate-lanceolate to lanceolate. Infl. branched, lfy in the lower part, branches rather spreading. Whorls distinct. *Fr. per. segs* 5–6 *mm., triangular, one* (rarely all three) *with a prominent tubercle*; margin with 3–5 long teeth. Fr. 3 mm. Fl. 6–10. Germ. spring. 2*n*=40. Hs.

The above description applies to ssp. *obtusifolius*, which appears to be the only ssp. native in this country. Sspp. *transiens* (Simonk.) Rech. f. and *sylvestris* (Wallr.) Rech. occur as rare aliens in the neighbourhood of London.

Native. On waste ground, in hedgerows, and at margins of fields, generally in disturbed ground. 112, H40, S. Common and generally distributed throughout the British Is., though less frequent towards the extreme north. Western Europe from N. Spain to southern Scandinavia and eastwards to C. Germany and Hungary; Azores (ssp. *agrestis*).

15. R. pulcher L.   'Fiddle Dock.'

A spreading much-branched perennial up to c. 50 cm. *Branches making a wide (–90°) angle with the main stem, rather slender and drooping in fl., forming an entangled mass in fr.* Lower lvs puberulent, panduriform or oblong from a cordate base, blade rarely exceeding 10 cm., apex obtuse; upper lvs lanceolate acute. Whorls distant. Pedicels jointed about the middle. *Fr. per. segs* oblong- or ovate-triangular, acute or obtuse, *all tubercled* (often unequally), *margin with c. 4 teeth not longer than ½ width of per. seg.*; tubercles verrucose. Fl. 6–7. $2n=20$. Hs.

The above description applies to ssp. *pulcher*, which is the only native form. Sspp. *anodontus* (Hausskn.) Rech. f. and *divaricatus* (L.) Murb. have occurred as rare aliens.

Native. In dry sunny habitats on sandy soils and, less commonly, on chalk and limestone. 35, H2 (perhaps more), S. England south of a line drawn from the Humber to the Severn; Wales: coastal districts of the south and north-west (doubtfully native); Ireland: S.E. coast, perhaps introduced. Western Europe from England southwards; Mediterranean region; Hungary, Balkans and Black Sea.

The following rather uncommon casuals can be confused with *R. pulcher*:

***R. dentatus** L. Annual (–biennial). Stem erect, branches usually ascending, straight. Lower lvs glabrous, cuneate to truncate (rarely weakly cordate) at base. Pedicels longer than fr. per. segs, jointed below the middle. Tubercles smooth. S. Europe, Africa, S. Asia.

***R. obovatus** Danser differs from the preceding in the obovate, thick, ± coriaceous lower lvs with rounded to subcuneate base, the more flexuous stems and the large verrucose tubercles. Argentina and Paraguay.

16. R. sanguineus L.   'Red-veined Dock.'

R. condylodes Bieb.; *R. nemorosus* Schrad. ex Willd.

An erect perennial up to c. 100 cm. *Stem usually almost straight, branches making an angle of about 20°(–45°) with it.* Lvs ovate-lanceolate from a rounded or subcordate base, upper narrowly lanceolate, all acute. *Infl.* much-branched, *branches with lowest whorls subtended by lvs.* Whorls rather distant. *Fr. per. segs c. 3 mm., oblong, obtuse, one with a globular tubercle c. 1·5 mm. diam., the others devoid of tubercles or with less developed ones*; margins entire. *Fr. 1·25–1·75 mm.* The common form is var. *viridis* Sibth., with green or occasionally rusty-red veins and usually thin lvs. Var. *sanguineus* is an uncommon plant which has lvs with purple veins and stems and panicle branches often suffused with purple. It was formerly widely cultivated for medicinal purposes and is now naturalized in several places. It is suggested that var. *sanguineus* arose as a mutant which bred true and has been spread by human agency. Fl. 6–8. Germ. throughout the year. $2n=20$. Hs.

Native. On waste ground, in grassy places, and in woods. 104, H40, S. Common in S. England, Wales and Ireland, less frequent in N. England and in Scotland. Europe north to southern Scandinavia; Caucasus; Asia Minor; C. Asia; N. Africa; introduced in N. and S. America.

17. R. conglomeratus Murr.   'Sharp Dock.'

R. glomeratus Schreb.; *R. acutus* sensu Sm.

An erect biennial or perennial up to c. 100 cm. *Stem usually distinctly flexuous, branches making an angle of* 30–90° *with it.* Lvs oblong or frequently panduriform from a rounded or subcordate base, upper lanceolate, all acute. Infl.

much branched. *Whorls* distant, *usually subtended by lvs for $\frac{2}{3}$ length of branches. Fr. per. segs* c. 3 mm., ovate to oblong, obtuse, *all with oblong tubercles* 1·25–1·75 *mm.*; margins entire. *Fr.* 1·75–2 *mm.* Fl. 7–8. Germ. 5–6. $2n = 20$. Hs.

Very variable and sometimes with difficulty distinguishable from *R. sanguineus.*

Native. In damp grassy places and less frequently in woods. 106, H40, S. Common and generally distributed throughout the lowland districts of England, Wales and Ireland; Scotland, rare. Europe north to southern Scandinavia; Asia Minor; N. Africa; Azores; introduced in N. America.

18. R. rupestris Le Gall   'Shore Dock.'

An erect branched perennial up to c. 70 cm. *Branches strict.* Lvs oblong, narrowed at base, obtuse, *glaucous, blade usually much longer than petiole. Whorls* distinct, *usually only the lowest on each branch subtended by lvs. Fr. per. segs* c. 4 *mm.*, oblong, obtuse, all with oblong tubercles 2·5–3 *mm.*; margins entire. *Fr.* 2·5 *mm.* Fl. 6–8. Germ. 4–6. Hs.

Native. On sea cliffs, rocky shores, and in dune-slacks. 6, S. Scilly, Cornwall, S. Devon, Dorset, Glamorgan, Pembroke. Western France (Normandy, Brittany, Vendée), N.W. Spain (Galicia).

19. R. palustris Sm.   'Marsh Dock.'
R. limosus auct. angl., non Thuill.

An erect branched annual, biennial or short-lived perennial up to c. 100 cm., usually much less. Plant yellowish in fr. Lvs oblong-lanceolate, tapering at base, apex subacute; upper lvs linear-lanceolate or linear, acute. Whorls distant below, crowded towards ends of branches. *Anthers* 0·9–1·3 *mm. Pedicels of fr. rather thick and rigid, usually not longer than fr. per. segs. Outer per. segs herbaceous, longer than $\frac{1}{2}$ diam. of fr. per. segs, apex claw-shaped and forward-curved. Fr. per. segs* 2–3 *mm., with ligulate obtuse apices, all tubercled, tubercles obtuse in front; margins with rigid setaceous teeth shorter than the segments. Fr.* 2–2·5 *mm.* Fl. 6–9. $2n = 40$. Th. or Hs.

Native. On bare muddy ground beside lakes and at margins of reservoirs, in dried-up ponds and, more rarely, in damp grassy places. 32. England, local becoming rarer in the north; apparently absent from Ireland. Europe from Jutland southwards; temperate Asia.

20. R. maritimus L.   'Golden Dock.'

Similar in general appearance to *R. palustris* but plant usually golden-yellow in fr. Whorls usually more crowded. *Anthers* 0·45–0·62 *mm. Pedicels of fr. very slender, mostly longer than the fr. per. segs. Outer per. segs thin, shorter than $\frac{1}{2}$ diam. of fr. per. segs, horizontally spreading or weakly reflexed. Fr. per. segs with triangular ± acute apices, tubercles acute in front; margins with very fine capillary teeth some of which are longer than the segments. Fr.* 1–1·5 *mm.* Fl. 6–9. $2n = 40$. Th. or Hs.

Native. In similar situations to *R. palustris.* 57, H5, S. England, local but rather commoner and extending further north than *R. palustris*; Ireland, very local. Europe to c. 60° N., Caucasus, N. and S. America (?introduced).

*21. R. brownii Campd.

A slender simple or little-branched perennial 25–60 cm. *Lvs* ovate or lanceolate, *usually narrowed abruptly a little above the cuneate to subcordate base*, apex acute or obtuse, margins finely crisped; *upper lvs* much narrower, acute and *sometimes hastate. Whorls* remote, *few-* (*up to* 6-) *fld. Fr. per. segs* 2·3–3·5 mm., deltoid, *apex acuminate, hooked; margins with* 3–5 *hooked teeth usually longer than* ½ *width of seg.; tubercles* 0. Fl. 9–11. Hs.

Introduced. On waste ground, persisting for several years; seeds introduced with wool. Recorded from Yorks, Beds, Selkirk and Roxburgh. Australia, Tasmania, Java, ?New Zealand.

Section *Axillares* Rech. f. Perennials. Primary infl. overtopped after flowering by a succession of secondary infl. arising from the axils of the lvs below the primary infl.

*22. R. frutescens Thou.

R. magellanicus auct. angl.; *R. cuneifolius* Campd.

A rhizomatous perennial c. 25 cm. *Rhizome c.* 1 *m. sending up shoots at intervals. Lvs thick and coriaceous, narrowly obovate,* ± *obtuse, often cuneate at base*, undulate, margins finely crisped. *Infl.* dense, *with few short simple branches*, secondary infl. sometimes produced from axils of lvs below primary infl. and ultimately overtopping it. *Fr. per. segs* 4–5 *mm., ovate-deltoid,* ± *acute, coriaceous, all tubercled; margins entire.* Fl. 7–8. ?Hs.

Introduced. Well naturalized in dune-slacks and near ports. Western England, S. Wales, S. Scotland. Native of Argentina, Uruguay, Chile, Bolivia, Peru; introduced in U.S.A. and western Europe.

*23. R. triangulivalvis (Danser) Rech. f.

R. salicifolius agg.

An erect perennial 30–50 cm. Stems flexuous, usually several to each plant, often decumbent at base. *Lvs linear-lanceolate*, gradually narrowed at both ends, *acute*, shortly petioled or subsessile, pale green, all cauline except in seedling stage. Infl. with simple, arcuate-ascending branches. The terminal infl. when in fr. overtopped by lateral infl. which arise in the axils of lvs below the primary infl. *Fr. per. segs* 3 *mm., olive, deltoid, each with a muricate tubercle; margins entire or finely denticulate.* Fl. 7–8. Germ. 4–5. ?Hp.

Introduced. On waste ground and rubbish-tips, especially near ports. Scattered throughout England, Wales and southern Scotland; Ireland only near Dublin. N. America; introduced in Europe.

80. URTICACEAE

Herbs, small shrubs or rarely soft-wooded trees, often with stinging hairs, usually with cystoliths in the epidermal cells. Lvs alternate or opposite, simple. Stipules usually present. Fls small, generally unisexual and cymose. Perianth 4–5-merous, often enlarged in fr. Male fls with 4–5 stamens inserted opposite the per. segs, rudimentary ovary usually present; stamens inflexed in bud, springing open and scattering the pollen at maturity; anthers 2-celled, opening lengthwise. Female fls often with small staminodes; ovary 1-celled, free or adnate to perianth; style simple; ovule solitary, erect. Fr. an achene (in our spp.) or drupe. Seeds usually with endosperm; embryo straight.

About 41 genera and 500 spp., generally distributed.

Some spp. yield valuable fibres.

1 Plant normally with stinging hairs; lvs toothed; stems ridged or 4-angled.
                                                                    3. Urtica

   Plant without stinging hairs; lvs entire; stems terete.   *2*

2 Stems spreading or decumbent, not rooting at nodes; most of the lvs 1 cm.
     or more, distinctly petioled; fls clustered.   1. Parietaria

   Stems creeping, rooting at nodes; lvs not exceeding 6 mm., subsessile;
   fls solitary.   2. Helxine

1. Parietaria L.   Pellitory.

Annual or perennial herbs. *Lvs alternate*, entire; stipules 0. *Infl.* cymose, axillary, cymes clustered, dichotomous, 3–*several-fld.* Fls hermaphrodite or unisexual, bracteolate, green. Perianth of female fls tubular, 4-toothed, of male and hermaphrodite fls 4-partite. Fr. enclosed in per. segs.

About 10 spp., mainly in temperate regions.

1. P. diffusa Mert. & Koch   Pellitory-of-the-Wall.

P. officinalis auct.; *P. ramiflora* auct.

A softly hairy perennial 30–100 cm. Stems terete, much-branched, spreading or decumbent, usually reddish. Lvs up to c. 7 cm., lanceolate to ovate, obtuse to acuminate; petiole shorter than blade, slender. Fls usually unisexual. Female fls terminal. Male fls lateral, surrounded by a calyx-like involucre of 1 bract and 2 bracteoles. Fl. 6–10. $2n=14$. Hp.

Native. In cracks in rocks and old walls, and in hedgebanks. 94, H39, S. Widely distributed but rather local in England, Wales and Ireland, rare in Scotland and absent from the north. W. and S. Europe.

2. Helxine Req.

A perennial herb. *Lvs alternate*, entire; stipules 0. *Fls* unisexual, green, *solitary*, axillary, surrounded by an involucre of 1 bract and 2 bracteoles. Perianth 4-lobed. Fr. enclosed in perianth and involucre.

One sp. in the Balearic Is., Corsica and Sardinia.

*1. H. soleirolii Req.   Mind-your-own-business, Mother of thousands.

A slender creeping puberulent herb forming dense evergreen mats. Stems 5–20 cm., very slender, rooting freely at nodes. Lvs 2–6 mm., suborbicular, subsessile. Female fls enclosed in the ± connate involucre; perianth tubular, narrowly and shortly 4-lobed. Male fls with a 4-lobed perianth. Fl. 5–10. Ch.

Introduced. Commonly planted in rock gardens and cool greenhouses and naturalized on walls and damp banks, particularly in S.W. England and S.W. Ireland.

3. Urtica L.   Nettle.

Annual or perennial *herbs usually with stinging hairs*. Stems ridged or 4-angled. *Lvs opposite*; stipules free. Infl. lateral, arising from an often suppressed lfy branch, usually spike-like with clustered cymes. Fls green, unisexual. Perianth 4-merous. Female fls with unequal per. segs, the larger enclosing the fr.

About 30 spp. in temperate regions.

Annual; plant monoecious; lower lvs shorter than their petioles.   **1. urens**
Perennial; plant dioecious; lower lvs longer than their petioles.   **2. dioica**

1. U. urens L.   Small Nettle.

An *annual* herb 10–60 cm. *Lvs* 1·5–4 cm., ovate or elliptic, obtuse to acuminate, incise-dentate, the *lower shorter than their petioles*. Infl. c. 1 cm.,
borne on usually well-developed lfy lateral branches. *Achene c.* 1·6 *mm.* Fl.
6–9. 2*n*=24, 26, 52. Th.

Native. In cultivated ground and waste places, particularly on light soils.
Not uncommon but rather local throughout the British Is. 111, H40, S.
North temperate regions.

2. U. dioica L.   Stinging Nettle.

A coarse hispid *perennial* 30–150 cm. Roots much-branched, very tough,
yellow. Stems creeping and rooting at the nodes, giving rise to erect shoots
in spring. *Lvs* 4–8 cm., ovate, acuminate, dentate, usually cordate at base,
the *lower longer than their petioles*. Infl. up to c. 10 cm., lateral branches
usually suppressed. *Achene c.* 1·2 *mm.* Fl. 6–8. 2*n*=48, 52. Hp.

Occasionally devoid of stinging hairs. The young shoots are eaten like
spinach.

Native. In hedgebanks, woods, grassy places, fens, and near buildings,
especially where the ground is covered with litter or rubble. 112, H40, S.
Abundant and generally distributed throughout the British Is.; ascending to
2750 ft. on Ben Lawers. Temperate regions of the world.

*****U. pilulifera** L., the Roman Nettle, is a monoecious annual or biennial with the
female fls in dense spherical heads up to 1 cm. diam. *Achene c.* 2·4 *mm.* It formerly
occurred as a rare alien but appears now to be extinct.

81. CANNABIACEAE

Herbs without latex. Lvs usually lobed, stipulate. Fls dioecious, axillary.
Male fls pedicelled; perianth 5-partite, segments imbricate; stamens 5, erect
in bud, anthers longer than their filaments; rudimentary ovary 0. Female fls
sessile; perianth entire, closely enfolding ovary; ovary sessile, unilocular;
style central, stigmas 2, caducous; ovule solitary, pendulous. Fr. an achene,
enveloped in the persistent perianth. Seed with fleshy endosperm; embryo
curved or spiral. Two genera, widely distributed. The infructescences of
Humulus are used in brewing; *Cannabis* yields a valuable fibre (hemp), and
a narcotic resin which is largely used by Asiatics as a stimulant.

1. HUMULUS L.

Climbing perennial herbs rough with deflexed hairs. Lvs opposite, palmately
lobed or entire, petioled. Fls pendulous. Infl. glandular, axillary or the upper
forming a lfy terminal panicle. Male infl. much-branched, fls shortly pedicelled. Female infl. a peduncled cone-like spike; bracts broad, membranous,
imbricate, persistent in fr.

Three spp., one probably native in Europe and W. Asia but now cultivated in all temperate countries, one in N. America, and the third in eastern Asia.

1. H. lupulus L.   Hop.

An herbaceous perennial climber 3–6 m. Stems climbing by twisting in a clockwise direction. Lvs (4–)10–15 cm., broadly ovate, ±cordate at base, usually deeply 3–5-lobed and coarsely dentate, lobes acuminate; petiole shorter than to about as long as blade. Male fls c. 5 mm. diam. Female 'cone' 15–20 mm., enlarging to 5 cm. in fr.; bracts c. 10 mm., ovate, acute, pale yellowish-green. Fls 2–3 in the axils of the bracts, each subtended by a bracteole. Fl. 7–8. $2n=20$. Hp.

A valuable constituent of the best beers.

Native. In hedges and thickets, extensively cultivated in certain districts. Widely distributed, but doubtless often an escape from cultivation—probably always so in Scotland. 100, H23, S. Europe, W. Asia.

*Cannabis sativa L., Hemp, an erect annual with usually alternate lvs which are palmately 5–7-lobed almost to the base, is occasionally cultivated and occurs rarely as a casual in waste places.

82. ULMACEAE

Trees without latex. Lvs alternate, simple, often asymmetrical at base; stipules caducous. Fls clustered, arising from the 1-year-old twigs, hermaphrodite or unisexual. Perianth herbaceous, shortly 4–8-lobed, lobes imbricate. Stamens the same number as the perianth lobes and opposite to them, erect in bud. Ovary of 2 connate carpels, 1–2-celled. Ovules solitary, pendulous. Fr. compressed, dry or slightly fleshy, often winged. Endosperm 0; embryo straight.

About 13 genera and 150 spp., mostly in north temperate regions.

1. ULMUS L.   Elm.

Large trees, usually with suckers. Lvs ± asymmetrical at base, serrate, hairy or glabrous above, hairy beneath, at least in the axils of the main veins ('axillary tufts'). Fls protandrous, appearing before the lvs. Perianth campanulate, persistent, usually 4–5-lobed. Anthers reddish. Ovary usually 1-locular. Fr. compressed, broadly winged, wing notched at top.

About 30 spp. in north temperate regions and on mountains of tropical Asia.

As in most trees, lvs of sucker shoots or those on rapidly growing branches and on young trees may differ greatly from those on slow-growing laterals on mature trees ('short shoots'). The latter show less variation within a sp. than the former and are therefore more useful in identification. The description of lvs in the following account refers only to the distal and subdistal lvs of short shoots of mature trees, and the description of serration to the teeth at $\frac{1}{3}$ the distance from the apex to base of such lvs. Lvs from suckers, lammas shoots etc. of spp. whose mature lvs are glabrous are hairy.

1 Lvs medium-sized to large (5–16 cm.), broadly ovate, elliptic or obovate, very rough above, petiole very short, less than 3 mm. **1. glabra**
Not as above. *2*

2 Lvs medium-sized to large (6–12 cm.), broad, smooth above.
(× **hollandica**) *3*
Lvs small to large (2–10 cm.), if broad less than 5 cm. or rough above, otherwise rough or smooth above. *4*

3 Main branches crooked, spreading upwards; strongly developed cork flanges on the epicormic shoots. × **hollandica** var. *hollandica*
Main branches straight, ascending; cork not conspicuous on the epicormic branches. × **hollandica** var. *vegeta*

4 Lvs broadly ovate to suborbicular, rough, dark green. **2. procera**
Lvs narrow to broad; if broad, smooth and light green. **3. carpinifolia**

1. U. glabra Huds. Wych Elm.

U. montana Stokes; *U. scabra* Mill.

A rounded tree up to c. 40 m., often dividing into 2 or 3 large branches near the base. Branches ascending and spreading and forming a closed canopy; suckers few or 0. Twigs stout, coarsely pubescent becoming smooth and ashy-grey by the third year; short shoots making an angle of about 90° with branches. Winter buds with rufous hairs on scales. *Lvs* (Fig. 48 A) *8–16 cm.*, suborbicular or broadly obovate to elliptic, long-cuspidate, scabrid above, coarsely to finely pubescent beneath; base unequal, with the *long side forming a rounded auricle which overlaps and often hides the short petiole*; serration sharp, secondary teeth 2–5 on the basal and often 1 on the apical side; lateral nerves 12–18 pairs. Fr. 15–20 mm., broadly obovate to elliptic; seed central; pedicel 2–3 mm. Fl. 2–3. Fr. 5–6. $2n=28$. MM.

Ssp. **glabra** has lvs suborbicular to broadly obovate with basal auricle well developed and is southern in its distribution.

Ssp. **montana** (Lindq.) Tutin has narrow lvs with the basal auricle not so well marked and is northern and western in its distribution.

Forma *cornuta* (David) Rehd., which has one or two prominent acuminate lobes on each shoulder of the leaf, is occasional and sometimes planted. Similar lobing of the leaves also occurs as a juvenile character.
Forma *pendula* (Loud.) Rehd. has branches horizontally spreading and pendulous and is the commonly planted weeping elm.
Forma *exoniensis* (C. Koch) Rehd. has erect branches forming a narrow columnar head with coarsely serrate, obovate ± puckered and twisted leaves and is occasionally planted.

Native. In woods, hedges and beside streams. Scattered throughout the British Is., but commoner in the west and north. Europe to 67° N.; N. and W. Asia; N. Africa (?introduced).
The following are assumed, on scanty evidence, to be *U. carpinifolia* × *glabra*:

U. × hollandica Mill. var. *hollandica* (Dutch Elm). A rounded tree up to 40 m. with the habit of *U. glabra* but usually a more open canopy. *Suckers* and epicormic shoots *numerous, with corky flanges* up to 2 cm. wide. Twigs ± as

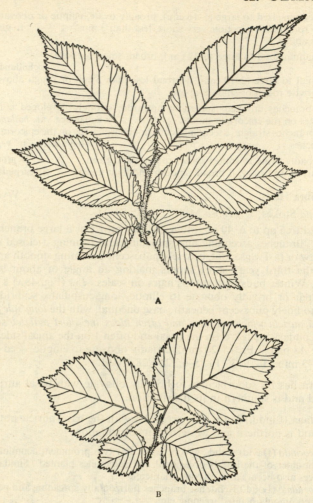

Fig. 48. Short shoots of *Ulmus*. A, *U. glabra* ssp. *montana*; B, *U. procera*. × ½.

in *U. glabra* but sparsely pubescent at first and becoming smooth and chestnut brown in the second year. *Lvs 6–12 cm.*, ovate, *acute*, minutely scabrid above, finally shining, glabrous; scabrid and glandular beneath, with axillary tufts ± prominent; basal auricle not overlapping and shorter than the petiole, serration coarse, broad based, rather blunt; lateral nerves 10–14 pairs.

U. × hollandica Mill. var. *vegeta* (Loud.) Rehd. (Huntingdon Elm). Tree up to 30 m. with short trunk and several ascending branches spreading fanwise to form a rounded head. *Suckers less numerous and not corky*. Twigs sparsely pubescent at first, dark brown in the second year. *Lvs 8–14 cm.*, ovate to

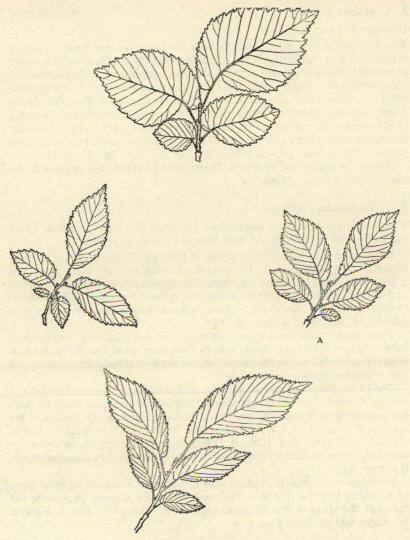

Fig. 49. Short shoots of *Ulmus carpinifolia*. A, var. *cornubiensis*.

ovate-lanceolate, *acuminate*, minutely scabrid above and on the nerves beneath, margin at the base of the short side often ending in the lateral nerve, serration rather sharp, lateral nerves 12–18 pairs.

U. × *elegantissima* Horwood, though sometimes assumed to be a form of *U. glabra* × *carpinifolia* var. *plotii*, may well be a curious form of *U. carpinifolia*.

2. U. procera Salisb.

English Elm.

U. campestris auct. angl.

An erect tree up to c. 30 m., usually with few but large and heavy branches in the lower part. Upper branches spreading, forming a crown with a number of rounded lobes and a closed canopy; suckers and epicormic shoots numerous, sometimes corky. *Twigs* rather stout, *persistently pubescent. Lvs* (Fig. 48 B) 4·5–9 *cm., suborbicular to ovate, acute, scabrid to glabrescent above* but always rough, *uniformly pubescent* with axillary tufts *beneath*; base unequal, rounded to subcuneate; lateral nerves 10–12 pairs; serration rather sharp, forwardly arched, secondary teeth 1–3. Fr. 10–17 mm., orbicular, rarely produced; seed centred $\frac{2}{3}$ from base; pedicel 1 mm. Fl. 2–3. Fr. 5–6. $2n=28$. MM.

Native. In hedges and by roads. Throughout England, less frequent in the north, often planted. ?Endemic.

3. U. carpinifolia Gleditsch

U. nitens Moench; incl. *U. angustifolia* (Weston) Weston; *U. stricta* (Ait.) Lindl., *U. coritana* Melville, *U.plotii* Druce, *U. diversifolia* Melville

A tree up to 20(–30) m., very variable in habit and lf shape. Branches subfastigiate (var. *cornubiensis*) to spreading; twigs slender, often long and ± pendulous. Lvs (2·5–)4–10(–11·5) cm., oblanceolate to suborbicular, smooth and shining or, at most, slightly rough above; base ± unequal; lateral nerves (5–)9–12(–14) pairs; serrations obtuse or sometimes acute. Fr. rarely produced. Fl. 2–3. Fr. 5–6. $2n=28$. MM.

There appear to be many clones differing from one another in lf-shape and habit. Various clones and clonal groups have been given specific rank but it is doubtful to what extent the discontinuities between these correspond to the discontinuities observable in the field.

One of the best-marked varieties is var. *cornubiensis* (Weston) Rehd. (*U. stricta*), Cornish Elm, with subfastigiate upper branches and ovate lvs with few secondary teeth and subequal bases. This is a characteristic feature of the landscape of parts of S.W. England and is commonly planted in towns. Another form (var. *plotii* (Druce) Tutin), with the leading shoot arched to one side, forms scattered populations which, however, vary considerably in lf morphology.

Probably introduced in thickets and hedgerows. Frequent in most parts of S. and E. England, the N.E. Midlands and in the Cornish peninsula; less frequent elsewhere in the Midlands and in S. Wales. C. and S. Europe, N. Africa and the Near East.

83. MORACEAE

Trees or shrubs, rarely herbs, with milky latex. Lvs usually alternate, simple, stipulate. Fls small, unisexual, in dense infls often with an expanded axis; per. segs usually 4; stamens usually equalling in number and opposite per. segs; carpels 2, 1 usually aborted; ovule 1, usually pendulous; styles usually 2. Fr. a drupe or achene; frequently a multiple fr. is formed from the infl. by the axis becoming fleshy. Endosperm present or 0.

About 55 genera and 1000 spp., mainly tropical.

1. Ficus L.

Trees or shrubs. Lvs convolute in bud; stipules intrapetiolar, at first forming a cup, soon falling and leaving a circular scar. Fls borne inside a fleshy hollow axis with a small opening, usually monoecious. Per. segs 2–6 in male, more in female. Stamens usually 1–2. Multiple fr. consisting of the fleshy axis enclosing the individual frs.

About 800 spp., mainly tropical.

*F. carica L. Fig.

Deciduous shrub or small tree to 10 m. Lvs 7–20 cm., palmately lobed into 3–5 obovate lobes, scabrid above, tomentose beneath. Infl. pear-shaped, in fr. 5–8 cm., green, brownish or purple. Pollinated by a gall-wasp. Fr. 7–10. MM.

Cultivated for its fr., occasionally self-sown and ± naturalized.

84. JUGLANDACEAE

Trees. Lvs alternate, pinnate, exstipulate. Fls monoecious, solitary in the axils of bracts. Male fls in lateral, usually drooping catkins; bracteoles 2, rarely 0; perianth 0 or small and 1–5-lobed; stamens 3–40, filaments short; rudimentary ovary sometimes present. Female fls solitary or in spikes, terminal; bracteoles and perianth as in the male. Ovary inferior, of 2 carpels, 1-celled; ovule 1, basal, orthotropous; style short, stigmas 2. Fr. a drupe or small nut; endosperm 0.

Six genera and over 50 spp., northern hemisphere, temperate and on mountains in the tropics and in S. America. Spp. of *Carya* Nutt. (Hickory, Pecan) and *Pterocarya* Kunth ('Wing-nut') are sometimes planted.

1. Juglans L.

Deciduous trees. Pith chambered. Buds with few scales. *Male catkins drooping*. Female fls in erect terminal few (sometimes 1)-fld spikes. Bracteoles and *perianth present* in both sexes; perianth 1–5-lobed in the male, 4-lobed in the female. Bracteoles united with the ovary, not persisting in fr. Stamens 8–40. *Stigmas simple*; carpels median. *Fr. a large indehiscent drupe*; stone (walnut) incompletely 2–4-celled; cotyledons ruminate.

About 15 spp., north temperate regions, Jamaica, S. America.

*1. J. regia L. Walnut.

Large tree to 30 m., with spreading crown; main branches tortuous. Bark grey, smooth for many years, finally fissured, not scaling. Buds c. 6 mm., broadly ovoid, blackish, glabrous or the terminal one greyish and tomentose. Twigs greenish or grey, glabrous; lf-scars Y-shaped. Lflets (2–)3–4(–6) pairs, 6–12(–15) cm., obovate or elliptic, acute or acuminate, pubescent when young becoming glabrous except in the axils of the veins beneath. Male catkins 5–15 cm. Fr. 4–5 cm., subglobose, green, gland-dotted, aromatic; stone ovoid, acute, somewhat wrinkled, 4-celled below, easily splitting into halves. Fl. 6. Wind-pollinated, some plants protandrous, others protogynous. $2n = 32$. MM.

Introduced. Planted for its fr., sometimes in wild situations and sometimes ± naturalized in S. England. Native of S.E. Europe and W. and C. Asia to China.

85. MYRICACEAE

Trees or shrubs. Lvs alternate, simple, dotted with resinous glands, aromatic, exstipulate. Fls monoecious or dioecious, solitary in the axils of bracts, forming catkins. Perianth 0. Male fl. usually without bracteoles; stamens 2–16, filaments free or united below. Female fl. with 2 or more small bracteoles; ovary 1-celled; ovule 1, basal, orthotropous; style short; stigmas 2, filiform. Fr. a drupe or nut; embryo straight; endosperm 0.

One genus.

1. Myrica L.

The only genus. About 60 spp., almost cosmopolitan.

1. M. gale L.   Bog Myrtle, Sweet Gale.

Deciduous shrub 60–250 cm., spreading by suckers. Twigs red-brown, with scattered yellowish glands, becoming dark, ascending at a narrow angle. Buds small, ovoid, obtuse, reddish-brown, with several scales. Lvs 2–6 cm., oblanceolate, cuneate at base, subsessile, obtuse or acute, serrate near the apex or subentire, grey-green and usually glabrous above at maturity, usually ± pubescent beneath, with conspicuous scattered shining yellowish sessile glands on both sides. Fls mostly dioecious but plants may change sex from year to year and monoecious and hermaphrodite plants also occur. Catkins lateral, on the last year's shoots, ascending, forming panicles at the ends of the shoots, appearing before the lvs; male 7–15(–30) mm., bracts glabrous, red-brown, bracteoles 0, stamens c. 4, anthers red; female 5–10 mm. in fr., bracts like the male, bracteoles 2, styles red. Fr. dry, compressed, gland-dotted, 2-winged by the adnate, accrescent bracteoles. Fl. 4–5, wind-pollinated. Fr. 8–9. $2n=48$. N.

Native. Bogs, wet heaths and fens ascending to 1800 ft., often abundant and sometimes locally dominant. Throughout the British Is. except Shetland and the Channel Is., but absent from a number of counties, especially in the Midlands, N.E. England and S.E. Scotland. 87, H39. W. Europe from Scandinavia (69° N.) through W. France to Portugal; and across N. Germany to N.W. Russia (Ladoga-Ilmen region); N. America, from Labrador and Alaska south to N. Carolina and Oregon; a ssp. or allied sp. in E. Asia.

*M. caroliniensis Mill.   Bayberry.

M. pensylvanica Lois.

Deciduous shrub to 3 m. Lvs 4–10 cm., obovate or oblong, obtuse or subacute, shallowly and obtusely dentate near the apex or subentire, pubescent on both sides, gland-dotted. Catkins on the old wood, appearing with but below the lvs. Fr. 3·5–4·5 mm., subglobose, coated with greyish white wax. $2n=16$.

Sometimes cultivated; naturalized in the New Forest (Hants). Native of eastern N. America from Newfoundland to N. Carolina.

86. BETULACEAE

Deciduous trees or shrubs. Buds scaly. Lvs alternate, simple; stipules caducous. Fls monoecious, the sexes in different infl. Male fls in drooping catkins, three together in the axil of each bract, bracteoles 2–4 to each group of fls; perianth present; stamens 2 or 4. Female fls in erect cylindric or ovoid catkins, 2–3 in the axil of each bract, bracteoles 2–4 to each group of fls; perianth 0; ovary 2-celled; ovules 1 in each cell, anatropous, pendulous; styles 2, free. Fr. a flattened nutlet, 1–3 on the surface of a scale formed from the accrescent fused bract and bracteoles, in a dense cylindric or cone-like fr. catkin; seed 1; endosperm 0; embryo straight, cotyledons flat. Wind-pollinated.

Two genera.

Stamens 2, bifid below the anthers; scales of the cylindric fr. catkin 3-lobed,
  falling with the fr.; fls with the lvs. 1. BETULA
Stamens 4, entire; scales of the cone-like fr. catkin 5-lobed, persistent;
  fls before the lvs. 2. ALNUS

1. BETULA L.

Trees or shrubs. Buds with several scales. Bracteoles 2 to each group of fls. *Stamens 2, bifid below the anthers*; perianth minute. *Female fls 3 to each bract. Fr. catkins cylindric, the scales 3-lobed, relatively thin, falling with the fr.* Fr. winged, styles persistent.

About 40 spp., north temperate regions.

1 Lvs ±ovate, subacute to acuminate, serrate or serrate-dentate; petiole
     7 mm. or more.   2
   Lvs orbicular, rounded at apex, crenate; petiole 3 mm. or less.   3. nana
2 Lvs acuminate, sharply doubly serrate with prominent primary teeth,
     glabrous; trunk black and fissured into ± rectangular bosses at the base,
     white above.   1. pendula
   Lvs acute or subacute, irregularly or evenly serrate or serrate-dentate
     without markedly projecting primary teeth; trunk ± smooth throughout.
                                                                        2. pubescens

1. B. pendula Roth   Silver Birch.

B. verrucosa Ehrh.; *B. alba* L. p.p.

Tree to 25 m. with single stem; *bark smooth and silvery-white above, peeling, ± suddenly changing near the base of the trunk to black and fissured into rectangular bosses. Branches ±pendulous. Twigs glabrous*, brown, somewhat shining, with ±conspicuous pale warts, especially well developed on vigorous or sucker shoots. Buds long, acute, not viscid. *Lf-blades* (2–)2·5–5(–7) cm. ovate-deltoid, *acuminate*, truncate or broadly cuneate at base, *sharply doubly serrate with the primary teeth very prominent and somewhat curved towards the apex of the lf, glabrous*; petiole 1–2 cm., glabrous. Male infl. 3–6 cm., ± drooping. Female infl. in fr. 1·5–3(–3·5) × c. 1 cm.; scales with short broad cuneate base, lateral lobes broad, spreading and curving downwards, middle lobe deltoid, obtuse; *wings of fr. 2–3 times as broad as achene, the upper edge surpassing the stigmas.* Fl. 4–5. Fr. 7–8. 2n = 28 (42). MM.

Native. Woods especially on the lighter soils, rare on chalk, colonizing heathland and often forming woods there as a successional stage to sessile oakwood, sometimes also forming pure woods in the Scottish Highlands, apparently more tolerant of dry conditions than *B. pubescens* and commoner in S. England, rare in N. Scotland, absent from Shetland. 110, H 28. Europe from Scandinavia and N. Russia to C. Spain, Corsica, Sicily (Etna), Macedonia and the Crimea; W. Siberia, N.E. and C. Asia Minor; Morocco (Rif).

B. pendula × pubescens occurs but is probably rare.

2. B. pubescens Ehrh.   Birch.

Tree to 20 m. with single stem, or shrub with several stems; *bark smooth, brown or grey, rarely white, not markedly different at the base of the trunk, sometimes with deep grooves but never broken up into rectangular bosses. Branches spreading or ascending* (sometimes pendulous in mountain forms). *Twigs ± pubescent* or glabrous, becoming dark brown or blackish, not or scarcely shining, with or without brown resinous warts (? sucker shoots always pubescent). Buds viscid or not. *Lf-blades* 1·5–5·5 cm., very variable in shape, ovate, orbicular-ovate or rhombic-ovate, *subacute or acute*, rounded or cuneate at base, *coarsely and sometimes irregularly serrate or serrate-dentate, the teeth not curved towards the apex of the lf*, usually pubescent at least on the veins beneath or in their axils, of a duller green than *B. pendula*; petiole 7–15(–20) mm. Male infl. 3–6 cm., drooping. Female infl. in fr. 1–4 × 0·5–1 cm.; scales with short or long cuneate base, lateral lobes rounded or nearly square, ± spreading or ascending, terminal lobe long or short, narrow, oblong- or triangular-lanceolate; *wings of fr.* 1–1½ *times as broad as achene, the upper edge not surpassing the stigmas, often not projecting beyond the achene*. Fl. 4–5. Fr. 7–8. 2*n* = 56. MM. or M.

Native. In similar places to *B. pendula* but more tolerant of wet and cold conditions, ascending to 2500 ft.; less common than *B. pendula* in S. England and there only abundant on wet soils, commoner in the north and west and forming pure woods above the limit of oakwood. 105, H 40. Europe from Iceland and N. Russia to N. Portugal and N.W. Spain, Italian Alps, Serbia, Roumania and the Caucasus; Siberia (to Yenisei region), N.E. Asia Minor.

A very variable sp. perhaps divisible into several sspp. The two following may be provisionally recognized but intermediates occur.

Ssp. pubescens

Tree with single stem, branches never drooping. Twigs usually conspicuously pubescent, warts few or 0. Buds not viscid. Lvs relatively large (c. 3–4 cm.).
    Mainly in southern and lowland areas.

Ssp. odorata (Bechst.) E. F. Warb.

B. odorata Bechst.; *B. alba* ssp. *odorata* (Bechst.) Dippel; *B. tortuosa* auct.

Often shrubby or with several stems. Twigs not or sparsely pubescent, covered when young with brown resinous warts as are the young lvs. Buds viscid. Plant with a resinous smell as the lvs unfold. Lvs relatively small.
    Northern and mountain areas; perhaps the only ssp. in the Scottish Highlands.

3. B. nana L. 'Dwarf Birch.'

Shrub to 1 m., the stems mostly procumbent or spreading widely, stiff. Twigs pubescent, not warty, dull dark brown. *Lf-blades* 5–15(–20) mm., *orbicular or obovate-orbicular, rounded at both ends, deeply crenate*, glabrous at maturity; *petiole up to 3 mm.* Male infl. c. 8 mm. Female infl. in fr. 5–10 mm.; *scales with* cuneate base and 3 ± *equal narrow erect lobes at the apex; wings of fr. very narrow.* Fl. 5. Fr. 7. 2n=28. N.

Native. Mountain moors from 800 to 2800 ft., very local; Northumberland; Peebles; Perth and Argyll to Sutherland (on the mainland). 14. Arctic and northern Europe and Asia from Iceland to W. Siberia south to N. Germany, Roumania and the Upper Dnieper region; Ardennes, Jura, mountains of Germany, Switzerland, Austria and Bohemia; Greenland.

B. nana × pubescens = B. × intermedia Thomas ex Gaud. occurs rarely.

2. ALNUS Mill.

Trees or shrubs. Bracteoles 4 to each group of fls. *Stamens 4, not bifid. Female fls 2 to each bract. Fr. catkins ovoid or ellipsoid, cone-like, the scales 5-lobed, thick and woody, long persistent after the fall of the fr.* Fr. usually winged.

About 30 spp., northern hemisphere, mainly temperate, and S. America.

Lvs truncate or retuse at apex, bright green on both sides, veins 4–7 pairs; bark dark brown. 1. glutinosa
Lvs acute or acuminate, pale or glaucous beneath, veins 10–15 pairs; bark pale grey. 2. incana

1. A. glutinosa (L.) Gaertn. Alder.
A. rotundifolia Stokes

Tree to 20(–40) m. with ± oblong crown. *Bark dark brown, fissured.* Twigs glabrous. Buds purplish, obtuse, stalked, with a large outer scale almost hiding the inner ones. *Lf-blades* 3–9 cm. suborbicular or broadly obovate, *truncate or retuse*, rarely rounded, at apex, cuneate at base, irregularly and often doubly serrate-dentate, *bright green and glabrous on both sides* except for tufts in the axils of the veins beneath, glutinous when young; *veins* 4–7 *pairs*; petiole 1–3 cm. Fls before the lvs. Male catkins 2–6 cm., 3–6 together at the ends of the twigs. Female catkins in fl. 1 cm. or less, in fr. 7–28 mm., ovoid or ellipsoid. Wing of fr. narrower than the achene. Fl. 2–3. 2n=28*, 56. MM.

Native. Wet places in woods and by lakes and streams, ascending to 1600 ft., often forming pure woods in succession to fen or marsh; common almost throughout the British Is. (only planted in Shetland). 111, H40, S. Europe from Scandinavia (65° 30′ N.) and N. Russia southwards; W. Asia from W. Siberia to N. and W. Asia Minor; N. Africa.

*2. A. incana (L.) Moench 'Grey Alder.'

Tree to 20 m., often shrubby. *Bark pale grey, smooth.* Twigs pubescent. *Lf-blades* 3–10 cm., ovate to elliptic, *acute or acuminate*, usually rounded at base, regularly doubly serrate, dull green above, *grey-green or glaucous* and

± pubescent *beneath*, not glutinous when young; *veins* 10–15 *pairs*; petiole 1–2·5 cm. Fls before the lvs. Male catkins 2–4 together. Female catkins in fr. c. 2 cm. Wings of fr. about as broad as achene. Fl. 2–3. $2n=28$. MM. or M.

Introduced. Sometimes planted, especially for shelter or on poor soils, mainly in Scotland and sometimes ± naturalized. N. Europe from Scandinavia (70° 30′ N.) to N. Russia south in the mountains to the French Alps, Apennines, Albania, Bulgaria and the Caucasus; W. Siberia; a ssp. in eastern N. America.

87. CORYLACEAE

Deciduous trees or shrubs. Buds scaly. Lvs alternate, simple; stipules caducous. Fls monoecious, the sexes in different infl. Male fls in drooping catkins, solitary in the axil of each bract; bracteoles 2, united to the bract, or 0; perianth 0; stamens 3–14. Female fls in pairs in the axil of each bract; bracteoles present; perianth present, small, irregularly lobed; ovary inferior, 2-celled; ovules 1 in each cell, anatropous, pendulous; styles 2, filiform, free or joined at base. Fr. a nut, surrounded or subtended by a lf-like involucre formed from the fusion of the accrescent bract and bracteoles; seed 1; endosperm 0; cotyledons large and fleshy, embryo straight.

Four genera and about 50 spp., north temperate regions and C. America.

Female fls numerous, in drooping catkins; fr. involucre 3-lobed, unilateral.
                                                                                  1. CARPINUS
Female fls few, in short erect bud-like spikes; fr. involucre irregularly lobed,
    surrounding the nut.   2. CORYLUS

1. CARPINUS L.

Trees, rarely shrubs. *Male fls without bracteoles. Stamens c.* 10, *bifid below the anthers; anthers with hair-tufts at the apex. Female fls in terminal drooping catkins; carpels median. Fr.* a small nut *subtended by a large, unilateral, 3-lobed, bract-like involucre.* Cotyledons folded.

Over 25 spp., north temperate regions and C. America.

1. C. betulus L.   Hornbeam.

Tree to 30 m. but usually less, with ovoid bushy crown, often coppiced as a shrub. Trunk fluted; bark smooth, grey; branches ascending at an angle of 20–30°. Buds 5–10 mm., narrowly oblong, pointed, pale brown, with numerous scales. Twigs brown, sparsely pubescent. Lvs 3–10(–12) cm., ovate, acute or acuminate, cordate or rounded at base, sharply and doubly serrate, folded along the veins in bud, glabrous except for long appressed hairs on the main veins (7–15 pairs) below; petiole 5–15 mm. Male catkins 2·5–5 cm., bracts ovate, greenish. Female catkins c. 2 cm. in fl., 5–14 cm. in fr.; involucre 2·5–4 cm., 3-lobed, the middle lobe much longer than the laterals, lobes entire or serrate. Fr. 5–10 mm., greenish, ovoid, compressed, strongly veined, crowned by the persistent perianth. Fl. 4–5. Wind-pollinated. $2n=64$. MM. or M.

Native. Woods and hedgerows, often dominant as a coppiced shrub in oakwoods on sandy or loamy clays in S.E. England; extending as a native to Sussex, Oxford and Cambridge with isolated occurrences in Somerset and Monmouth; planted elsewhere (north to Sutherland). Europe from S. Sweden (57° 11′ N.) to the French Pyrenees, Italy, Greece and the Caucasus; N. Asia Minor, Persia.

2. CORYLUS L.

Shrubs, rarely trees. *Male fls with 2 bracteoles.* Stamens c. 4, bifid below the anthers; anthers with hair-tufts at the apex. *Female fls few, in erect short bud-like spikes. Carpels transverse. Fr.* a large nut *surrounded or enclosed by a lobed involucre.* Cotyledons not folded.

About 15 spp., north temperate regions.

1. C. avellana L. Hazel, Cob-nut.

Shrub 1–6 m. with several stems, usually seen coppiced, rarely a small tree. Bark smooth, coppery-brown, peeling in thin papery strips. Buds c. 4 mm., ovoid, obtuse; scales several, ciliate. Twigs thickly clothed with reddish glandular hairs. Lvs 5–12 cm., suborbicular, cuspidate, cordate at base, sharply doubly serrate or lobulate, slightly pubescent above, more so beneath, becoming nearly glabrous; petiole 8–15 mm., glandular-hispid. Fls appearing before the lvs. Male catkins 1–4 together, 2–8 cm., bracts ovate, anthers bright yellow. Female spikes 5 mm. or less, styles red. Fr. in clusters of 1–4, 1·5–2 cm., globose or ovoid, brown, with hard woody shell, surrounded by an involucre about as long or rather longer than itself; involucre deeply divided into usually toothed lobes. Fl. 1–4. Wind pollinated, protandrous, homogamous or protogynous. Fr. 9–10. $2n=22$. M.

Native. Woods, scrub and hedges, on damp or dry basic and damp neutral or moderately acid soils; the common shrub-layer dominant (as coppice) of lowland oakwoods, sometimes also dominant in the shrub-layer of ashwoods or forming scrub on exposed limestone. Ascends to over 2000 ft. Common almost throughout the British Is. but absent from Shetland. 111, H40, S. Europe (not Iceland, to over 68° N. in Norway); Asia Minor.

*C. maxima Mill. Filbert.

A larger shrub (to 10 m.). Involucre tubular, about twice as long as fr., lobed at the apex. Sometimes planted for its nuts.

Native of S.E. Europe and W. Asia.

88. FAGACEAE

Trees or shrubs. Buds scaly. Lvs alternate, simple; stipules usually caducous. Fls monoecious, usually in different infl. Male fls in catkins or in many-fld tassel-like heads (or in *Nothofagus* solitary or in threes); perianth 4–6-lobed; stamens usually twice as many as perianth lobes; rudimentary ovary sometimes present. Female fls in groups of 1 or 3, arranged in spikes or at the base of the male infl.; each group surrounded at the base by an involucre bearing numerous scales; perianth 4–6-lobed; ovary 3- or 6-celled; ovules 2 in each cell, pendulous, anatropous; styles 3 or 6. Fr. a 1-seeded nut, in groups of 1–3, surrounded or enclosed by accrescent scaly or spiny 'cupule' formed from the involucre; endosperm 0, cotyledons fleshy.

Six genera and over 800 spp., absent from tropical and S. Africa, Australia (except S.E.), E. tropical S. America, etc.

1 Male fls in tassel-like heads; lvs nearly entire; buds fusiform; fr. completely enclosed by a woody 4-valved cupule. 1. FAGUS
   Male fls in catkins; lvs toothed or lobed, if entire then evergreen and tomentose beneath; buds ovoid. 2

2 Lvs serrate with aristate teeth, not lobed or evergreen; male catkins erect, female fls in lower part of male catkin; cupule prickly, completely enclosing the fr., splitting into 2–4 valves. 2. CASTANEA
   Lvs lobed, or if entire or serrate, evergreen and densely tomentose beneath; male catkins drooping, female fls in separate spikes; cupule cup-like enclosing only the lower half of the fr., not prickly or splitting into valves. 3. QUERCUS

1. FAGUS L.

Deciduous trees. Buds fusiform, acute. *Male fls in tassel-like heads on long peduncles*; perianth campanulate, 4–7-lobed; stamens 8–16. Female fls usually paired, surrounded below by the peduncled cupule bearing numerous long scales; styles 3. *Nuts triquetrous, 1 or 2, enclosed in a woody regularly 4-valved cupule* covered with projecting scales or prickles. Germination epigeal.
   Ten spp., north temperate regions.

1. F. sylvatica L. Beech.

Large tree to 30(–40) m. with broad dense crown. Bark grey, smooth. Twigs brownish-grey in second year. Buds 1–2 cm., fusiform, reddish-brown. Lvs 4–9 cm., ovate-elliptic, acute, broadly cuneate or rounded at base, obscurely sinuate, glabrous except for the long-ciliate margins and silky hairs on the veins and axils beneath; veins 5–9 pairs, prominent beneath; petiole 5–15 mm., hairy. Male fls numerous, peduncles 5–6 cm. Fr. 12–18 mm., brown; cupule c. 2·5 cm., brown, with spreading subulate scales. Fl. 4–5. Fr. 9–10. Wind-pollinated, protogynous. $2n = 24$. MM.
   Native. Woods, etc., native in S.E. England north and west to about Hertford, Gloucester, Brecon, Glamorgan, Somerset and Dorset; elsewhere (north to Caithness, throughout Ireland, Channel Is.) planted and often naturalized (regenerating freely in Aberdeen); the characteristic dominant of chalk and soft limestone in S.E. England, also frequently dominant on well-drained loams and sands. Europe from S. Scandinavia (60° 30′ N. in Norway) to the mountains of C. Spain, Corsica, Sicily and Greece, east to W. Russia (Upper Dnieper region); Crimea.

2. CASTANEA Mill.

Deciduous trees, rarely shrubs. Buds with 3–4 scales. *Fls in erect* terminal *catkins*, the female fls at the base, the rest male. Male fls with 6-lobed perianth and 10–20 stamens. Female fls usually 3 in each cupule; ovary 6-celled; styles 7–9, cylindric, stigma small. *Fr.* large, brown, 1–3, rarely more, *enclosed, except for the styles, in a symmetrical prickly cupule which dehisces rather irregularly by 2–4 valves.* Germination hypogeal.
   About 10 spp., north temperate regions.

*1. C. sativa Mill. Sweet Chestnut, Spanish Chestnut.
C. vesca Gaertn.

Large tree to 30 m. with wide crown. Bark dark brownish-grey, fissured, the longitudinal fissures often spirally curved. Twigs olive-brown, glabrous or slightly pubescent, with prominent lenticels. Buds c. 4–5 mm., yellowish-green tinged brownish, ovoid, obtuse. Lvs 10–25 cm., oblong-lanceolate, acute or acuminate, broad-cuneate to subcordate at base, coarsely and regularly serrate-dentate with aristate teeth, glabrous above, pubescent below when young, finally glabrous; petiole 0·5–3 cm. Catkins 12–20 cm., conspicuous because of the yellowish-white anthers; male fls numerous; female fls few. Fr. 2–3·5 cm., deep brown, shining; cupule green, densely covered with long branched spines. Fl. 7. Fr. 10. Pollen-fls, visited by various insects. $2n = 24$. MM.

Introduced. Commonly planted (often in pure stands) throughout the British Is.; extensively naturalized in S.E. England. S. Europe north to Austria, S.W. Germany and France; Algeria; Asia Minor, Caucasus, W. Persia.

3. QUERCUS L.

Deciduous or evergreen trees or sometimes shrubs. Buds with numerous scales. Male and female fls in separate infl. *Male in drooping catkins*, perianth 4–7-lobed; stamens 4–12 usually 6. Female fls in each involucre solitary or in spikes; ovary usually 3-celled; styles variously shaped, the stigma on the ± flat inner surface. *Fr.* (acorn) large, solitary, *surrounded below by a cup-like indehiscent symmetrical cupule.* Germination hypogeal. Wind-pollinated.

Over 500 spp., northern hemisphere, temperate, subtropical and mountains of the tropics, extending south to the E. Indies and the Andes of Colombia.

1 Lvs evergreen, dark green above, grey-tomentose beneath, entire or remotely serrate or dentate.   **2. ilex**
   Lvs deciduous, green and glabrous or pubescent beneath, lobed.   **2**

2 Cup-scales subulate, spreading; buds surrounded by long subulate persistent stipules; lvs rough above, lobes acute.   **1. cerris**
   Cup-scales ovate, appressed; buds without subulate stipules; lvs smooth above, lobes obtuse.   **3**

3 Petiole short (less than 1 cm.) or 0; lvs glabrous beneath, with small reflexed auricles at the base; peduncle 2–8 cm.   **3. robur**
   Petiole 1 cm. or more; lvs stellate-pubescent along either side of the midrib beneath, without reflexed auricles; peduncle short (1 cm.) or 0.
                                                                          4. petraea

Section 1. *Cerris* Dum. Styles long, linear, acute. Shell of acorn usually tomentose within, abortive ovules in the lower part; cup-scales usually long, linear, spreading or reflexed.

*1. Q. cerris L. Turkey Oak.

Deciduous tree to 35 m.; bark dark grey, fissured; branches ascending, ultimately forming a wide crown. Twigs pubescent at first, brown and subglabrous in the second year. *Buds* small, ovoid, pubescent, *with long* (to 25 mm.) *subulate stipules below.* Lvs 5–10(–18) cm., very variable in shape and lobing,

± oblong in outline, cuneate to rounded at base, usually acute, pinnately lobed to lyrate-pinnatifid, greyish pubescent on both sides when young, later *dull green and somewhat rough with very small scattered hairs above*, somewhat paler beneath and finely stellate-pubescent especially on the veins; *lobes* 7–8 pairs, unequal, ovate-triangular, *acute or subobtuse*, mucronulate, some of them often lobed; petiole 1–2·5 cm.; stipules long, subulate, persistent. Male catkins 5–8 cm., stamens usually 4. Female spikes 1–5-fld, styles 4. Fr. spike short (0–2 cm.). Fr. ripening the second year after flowering, 20–35(–50) mm., ± oblong, reddish-brown; *cup* covering c. ½ the fr., 10–15(–30) mm. diam., *clothed with long* (to 1 cm.) *spreading or somewhat reflexed subulate scales.* Fl. 5. Fr. 9. 2*n*=24. MM.

Introduced. Commonly planted and quite naturalized in many places, at least in southern England on acid soils. Native of Europe and S.W. Asia from E. and S. France (rare), S. Switzerland (Ticino), Austria, Moravia and Roumania southwards to Sicily, Crete, Syria and east throughout Asia Minor.

Q. × *hispanica* Lam. (*Q. cerris* × *Q. suber* L.) differing in the half-evergreen, smaller, less deeply lobed lvs and somewhat corky bark is rather frequently planted in parks, etc.

Section 2. *Quercus*. Stigmas subsessile or on short styles, thick. Shell of acorn glabrescent within, abortive ovules in lower part; cup scales usually short, appressed. Fr. ripening the first year after flowering.

*2. Q. ilex L.   Evergreen Oak, Holm Oak, Ilex.

Evergreen tree to 30 m. with broad dense crown; bark grey, scaly. Twigs densely tomentose when young, grey and subglabrous in the second year. Buds small, ovoid, tomentose, without subulate stipules below. *Lvs* 2–6(–9) cm., very variable in shape and toothing even on the same tree, ovate, oblong, elliptic or lanceolate, cuneate to subcordate at base, obtuse or acute, *entire or remotely serrate or dentate*, the teeth often mucronate or aristate, *dark green and glabrous above, densely grey-tomentose beneath*; petiole 5–10(–15) mm.; stipules linear, caducous; lvs of young trees or on sucker shoots often sinuately spine-toothed, ± holly-like, green and stellate-pubescent beneath. *Male catkins c. 3–5 cm. Female spikes* 1–4-*fld*, styles 3–4. Fr. spike short. Fr. 2–4 cm., ovoid or ellipsoid; *cup* usually covering less than ½ the fr., grey-tomentose, *clothed with small ovate appressed imbricate scales.* Fl. 5. Fr. 9. 2*n*=24. MM.

Introduced. Commonly planted, sometimes naturalized in S. England. Native of the Mediterranean region extending north in W. Europe to Brittany.

3. Q. robur L.   Common Oak, 'Pedunculate Oak.'

Q. pedunculata Ehrh. ex Hoffm.

Large *deciduous* tree to 30(–40) m. with broad crown; bark brownish-grey, fissured. Twigs glabrous, greyish-brown. Buds 2–5 mm., ovoid, obtuse, glabrous, without subulate stipules below. *Lvs* 5–12 cm., obovate-oblong, obtuse, rounded to cordate at base *with a reflexed auricle on either side* (at least on most of the lvs), pinnately lobed, *dull* green and glabrous *above*, paler beneath and sometimes with simple hairs when young, soon *glabrous*; lobes

3–5(–6) pairs, obtuse, rather unequal; stipules linear, caducous; *petiole to* 5 (rarely 10) *mm., often almost* 0. Male catkins 2–4 cm.; stamens (4–)6–8(–12). Female spikes 1–5-fld, pedunculate; styles 3, obvious. *Fr. peduncles 2–8 cm.* Fr. 1·5–4 cm., ellipsoid or oblong, brown. *Cup* 1·5–2 cm. diam., covering $\frac{1}{3}$–$\frac{1}{2}$ of the fr., *clothed with small ovate appressed imbricate scales.* Fl. 4–5. Fr. 9–10. $2n = 24$. MM.

Native. Woods, hedgerows, etc.; reaching nearly 1500 ft. in Derry and on Dartmoor, but rarely occurring over 1000 ft. From Sutherland southwards and throughout Ireland; the characteristic dominant tree of heavy and especially basic soils (clays and loams) and thus of most of S., E and C. England, sometimes also dominant or co-dominant with *Q. petraea* on the damper acid sands; in other parts of the British Is. mainly on alluvium; not thriving on acid peat or shallow limestone soils; often planted and perhaps not native in some northern and western areas. 102, H21, S. Europe from over 63° N. in Norway and the Ladoga-Ilmen and Dvina-Pechora areas of Russia south to the mountains of N. Portugal and C. Spain, Italy, Macedonia, Greece, Crimea and the Caucasus, east to the Urals.

Q. petraea × *robur* is rather frequent where the parents occur together and is thus especially frequent on certain sands. Fertile and consequently variable.

4. Q. petraea (Mattuschka) Liebl.   Durmast Oak, 'Sessile Oak.'

Q. sessiliflora Salisb.

Differs from *Q. robur* as follows: Usually branching higher; crown rather narrower. Buds often slightly larger (to 6 mm.) with ciliate scales. *Lvs* cuneate to cordate at base, *without reflexed auricles, somewhat shining above,* persistently *stellate-pubescent* with large hairs *along either side of the base of the midrib* and main veins *beneath,* sometimes with smaller stellate hairs on the surface also; lobes more regular, 4–6 pairs; *petiole 1–2·5 cm.* Stigmas subsessile. *Fr. peduncle 0–1 cm.* Fl. 4–5. Fr. 9–10. $2n = 24$. MM.

Native. Woods from Sutherland southwards (not Channel Is.) and throughout Ireland mainly on acid soils; the characteristic dominant tree of the siliceous soils of N. and W. England, Wales, Scotland and Ireland, sometimes dominant or co-dominant with *Q. robur* on acid soils in southern England; forming woods up to 1500 ft. in Cumberland. 99, H24. Europe from nearly 62° N. in Norway to the mountains of N.W. Spain and N.E. Portugal, Corsica, N. Apennines, Albania and N. Greece east to W. Poland and S.W. Russia (western Middle Dnieper and Black Sea regions and Crimea).

Section *Erythrobalanos* Spach. Styles long, spathulate. Shell of acorn tomentose within; abortive ovules apical; cup-scales appressed. Fr. usually ripening in the 2nd year after flowering.

***Q. borealis** Michx. f. var. *maxima* (Marsh.) Ashe (*Q. rubra* L. sec. Duroi, *Q. maxima* (Marsh.) Ashe)   'Red Oak.'

Deciduous tree to 25(–50) m. Lvs 12–22 cm., oblong, with 3–5 pairs of triangular acute bristle-pointed lobes reaching about half-way to mid-rib, glabrous except for brownish tufts of hairs in the axils of the main veins beneath, often turning dark red in autumn. Fr. 2·5–3 cm.; cup saucer-shaped, enclosing only base of acorn.

Often planted in parks, etc., for ornament and occasionally for forestry. Rarely fruiting. Native of N. America from Nova Scotia to Florida and Texas.

89. SALICACEAE

Deciduous trees or shrubs. Lvs alternate, simple, stipulate. Fls dioecious, in catkins, each fl. solitary in the axil of a bract (scale). Bracteoles 0. Perianth 0, but fls (of both sexes) usually with a cup-like disk or 1 or 2 small nectaries which possibly represent a perianth. Stamens 2–many, filaments long and slender, occasionally fused. Ovary of 2 carpels, 1-celled with 2 or 4 parietal placentae; ovules numerous (rarely 4), anatropous; styles 1 or rarely 2; stigmas 2, often bifid. Fr. a 2-valved capsule; seeds enveloped by long silky hairs arising from the funicle; endosperm 0; embryo straight.

Three genera, the third (*Chosenia*) with 1 sp. in N.E. Asia.

Catkin scales toothed or laciniate; stamens numerous; fls with a cup-like disk;
    buds with several outer scales.   1. POPULUS
Catkin scales entire; stamens 5 or fewer; fls with 1 or 2 nectaries, without
    disk; buds with 1 outer scale.   2. SALIX

1. POPULUS L.

Trees. Buds with several outer scales; terminal bud usually present. Fls appearing before the lvs, in pendulous catkins, *scales toothed or laciniate*; each fl. *with a cup-like disk; stamens 4–many*, anthers red or purple. Wind pollinated.

About 30 spp., north temperate regions. Some spp. and hybrids, besides the following, are sometimes grown and some modern hybrids are now being grown commercially and are likely to become commoner.

1 Bark smooth, grey (except at base of old trunks) with conspicuous rhomboid lenticels; lvs (of normal shoots) obtuse, rarely acute, coarsely toothed or lobed with up to c. 10 teeth; catkin scales ciliate with long hairs.   2

   Bark fissured (if smooth above, branches suberect and without conspicuous rhomboid lenticels), dark; lvs acuminate, crenate or serrate with numerous (c. 20 or more) teeth; catkin scales glabrous.   4

2 Lvs, at least at the ends of the long shoots, persistently tomentose beneath; catkin scales not divided to half-way; stigmas greenish; petiole scarcely compressed laterally.   3

   Lvs (except of sucker shoots) glabrous; catkin scales divided to more than half-way; stigmas purple; petiole strongly compressed laterally.
                                            3. **tremula**

3 Lvs of sucker shoots and of ends of long shoots deeply palmately 5-lobed; those of short shoots persistently tomentose; catkin scales dentate.
                                            1. **alba**

   Lvs of sucker shoots ±evenly toothed, not or shallowly lobed, those of short shoots becoming subglabrous; catkin scales laciniate.
                                       2. **canescens**

4 Lvs with narrow translucent border, not scented when unfolding, cuneate, truncate or (rarely) subcordate at base; petiole flattened laterally.   5

   Lvs without translucent border, strongly balsam-scented when unfolding, cordate at base; petiole terete.   6. **gileadensis**

5   Branches suberect forming a very narrow crown.   **4. nigra** var. *italica*
    Branches forming a wide crown.   6

6 Trunk and larger branches usually with swollen bosses; branches spreading
   and arching downwards; lvs without glands at the junction with the
   petiole.   **4. nigra**
   Trunk and larger branches without swollen bosses; branches ascending
   and curving upwards; some of the lvs with 1 or 2 glands near the
   junction with the petiole.   **5.** × **canadensis** var. *serotina*

Section 1. *Populus*. Bark smooth, grey, with conspicuous rhomboid lenticels, rough only at the base of old trunks. Buds not scented when unfolding. Lvs lobed or coarsely toothed, glabrous or tomentose; petiole laterally compressed or scarcely compressed. Catkin scales ciliate with long hairs. Stamens 5–12(–20).

*1. P. alba L.   White Poplar, Abele.

Tree to 25 m., freely suckering, with wide spreading crown. Buds and young twigs white, cottony-tomentose, not viscid. *Lf-blades of short shoots* 3–5 cm., ovate-orbicular, *broadest below the middle*, often broader than long, obtuse, rounded or subcordate at base, coarsely dentate or shallowly lobed with c. 4, obtuse, ± triangular teeth on each side, dark green above and hairy when young, *white-tomentose beneath, becoming grey and less tomentose; petiole* 1–3 cm., *not or scarcely compressed; summer lf-blades* (*especially of the suckers*) 4–10 cm., *palmately lobed*, the lobes c. 5-toothed, *densely and persistently white-tomentose beneath*. Male catkins 4–8 cm.; stamens 6–10. Female catkins 2–6 cm.; *stigmas linear, greenish-yellow. Scales irregularly crenate or dentate at apex with rounded or ± deltoid teeth*, long-ciliate. Fl. 3. $2n = 38$ (57). MM.

    Introduced. Frequently planted, especially in S. England, but not naturalized. Europe from S. and E. Germany, Italy and Sicily (doubtful native of Spain, Portugal, France and Switzerland) eastwards; W. Asia from W. Siberia to the Himalaya and Palestine; N. Africa.

2. P. canescens (Ait.) Sm.   Grey Poplar.

Tree to 35 m., resembling *P. alba* but usually larger. Buds thinly tomentose or glabrescent, not viscid. Twigs stouter than in *P. alba*, less tomentose. *Lf-blades of short shoots* 3–6 cm., ovate or orbicular-ovate, *broadest below the middle*, mostly longer than broad, obtuse or subcordate at base, coarsely dentate with c. 4–6 obtuse, ± triangular teeth on each side, *white- or grey-tomentose beneath when young, becoming subglabrous* in summer; petiole 1·5–4 cm., not or scarcely compressed; *summer lf-blades* (*especially of the suckers*) 5–10 cm., *ovate or triangular-ovate, coarsely* and often doubly *dentate, not lobed, persistently grey- or white-tomentose beneath*. Male catkins 5–10 cm.; stamens 8–15. Female catkins 2–10 cm., stouter than in *P. alba*; *stigmas oblong, greenish-yellow. Scales laciniate, divisions ± lanceolate, not reaching half-way to base*. Fl. 2–3. $2n = 38$ (57). MM.

   ?Native. Probably native in damp and wet woods in S., E. and C. England, west and north to Dorset, Shropshire and Derby; planted elsewhere; often forming groups of many trees by vegetative spread. 68, S (incl. introductions).

Europe from C. and E. France east to S. Russia (Middle Dnieper region to Caucasus), south to Italy and Macedonia.

P. canescens × *tremula* occurs occasionally. Lvs intermediate in shape, broadest near the middle as in *P. tremula*.

3. P. tremula L.   Aspen.

Tree to 20 m., suckering freely. Buds glabrous, somewhat viscid. Twigs rather slender, glabrous (except sometimes when very young). *Lf-blades* of normal shoots 2·5–6 cm., orbicular, *broadest about the middle*, often broader than long, obtuse, acute or rarely apiculate, rounded at base, coarsely crenate-dentate or serrate-dentate with c. 8–10 sometimes curved teeth on each side, silky pubescent when young, soon *glabrous*, or glabrous from the beginning; *petiole* 2·5–6 cm., *strongly compressed laterally* causing the lvs to tremble; lf-blades of sucker shoots up to 15 cm., ovate, often acuminate at apex and cordate at base, with more numerous shallower teeth, greyish pubescent beneath. Male catkins 4–8 cm.; stamens 5–12. Female catkins 3–8 cm.; *stigmas* broadened above, *purple. Scales deeply laciniate, divisions reaching more than half-way to base.* Fl. 2–3. 2*n* = 38 (57). MM.

Native. Woods, especially on the poorer soils where it is sometimes locally dominant, forming thickets by vegetative spread; ascending to 1650 ft. in Yorks. Throughout the British Is., common in the north and west, rather local in the south and east. 111, H36, S. Europe from Iceland, Scandinavia (70° 55′ N. in Norway) and N. (not arctic) Russia to S. Spain, Sicily, Greece and the Caucasus; temperate Asia to Japan, south to Asia Minor; Algeria.

Section 2. *Popularia* Dum. Bark furrowed. Buds viscid but not scented when unfolding. Lvs green on both sides, glabrous or pubescent when young, toothed, teeth numerous, with a narrow translucent border; petiole laterally compressed. Catkin scales laciniate, glabrous. Stamens 8–60. Stigmas greenish, stout.

4. P. nigra L.   Black Poplar.

Tree to 35 m., rarely suckering. Bark nearly black, with long deep fissures; *trunk and larger branches usually bearing large swollen bosses; branches spreading, arching downwards*, forming a wide crown. Twigs terete, glabrous or pubescent when young, yellowish, shining, becoming greyish. Buds long, reddish-brown curving outwards at apex. Lf-blades 5–10 cm., rhombic-ovate or deltoid-ovate, truncate or cuneate at base, crenate-serrate with numerous teeth; *petioles* 3–6 cm., *without glands at their apices*, glabrous or pubescent when young. Male catkins 3–6 cm.; stamens 8–20. Female catkins 6–7 cm., becoming 10–15 cm. Fl. 4. 2*n* = 38 (57). MM.

?Native. Possibly native in wet woods and stream-sides in E. and C. England from Essex and Lincoln to Gloucester and Shropshire; planted else-where, but much less commonly than *P.* × *canadensis*. 48 (incl. introductions). Mediterranean region north to Ladoga-Ilmen region of Russia, E. Germany and France but northern limit as a native tree very uncertain; temperate Asia, east to the Yenisei region and the Himalaya.

Var. *italica* Duroi (*P. italica* (Duroi) Moench), Lombardy Poplar, differs mainly in its narrow fusiform outline but the bark is paler, bosses 0, and the lvs smaller. It is nearly always male. Frequently planted in the lowlands on damp soils.

Of unknown origin but reported as wild in C. Asia.

*5. P. × canadensis Moench var. *serotina* (Hartig) Rehd. Black Italian Poplar.

P. deltoides Marsh. × *nigra* L.

Differs from *P. nigra* as follows: Tree to 40 m. *Trunk without bosses; branches ascending, curving upwards forming a wide fan-like crown.* Buds longer. Lf-blades 6–10 cm., deltoid, shortly acuminate, truncate or subcordate at base, margins ciliate with short hairs, more deeply crenate-serrate, *with 1–2 glands at the base* of at least some of the lvs of each twig. Male catkins 4–7 cm.; stamens 20–25. Female unknown. Fl. 4. $2n = 38$. MM.

The above is much the commonest form of *P. × canadensis* in Britain but others are occasionally planted; var. *marilandica* (*Poir.*) Rehd. is a female form with more spreading branches and cuneate-based lvs.

Introduced. Originated c. 1750, now commonly planted in many situations, mainly on damp soils throughout the British Is. and commoner than *P. nigra*. *P. deltoides* is a native of eastern N. America.

Section 3. *Tacamahaca* Spach. Bark furrowed. Buds very viscid, strongly balsam-scented when unfolding. Lvs usually very pale beneath, glabrous or slightly pubescent, toothed, teeth numerous, without a translucent border; petiole not flattened. Catkin scales laciniate, glabrous. Stamens c. 20–60. Stigmas stout.

*6. P. gileadensis Rouleau Balm of Gilead.

P. Tacamahacca auct.; *P. candicans* auct., non Ait.

Tree to 20(–30) m., often suckering. Branches spreading. Twigs stout, pubescent, brown. Buds large, very viscid. *Lf-blades* 6–16 cm., triangular-ovate, acuminate, cordate at base or the later lvs truncate, crenate-serrate, ciliate, densely *pubescent on the veins beneath*, glabrous or slightly pubescent and very pale on the surface; petioles 3–6 cm., pubescent. Male fls unknown. Female catkins 7–16 cm.; stigmas yellowish at first, becoming pink; capsule stalked. Fl. 2–3. $2n = 38$. MM.

Introduced. Sometimes planted by streams, ponds, edges of woods, etc., and spreading by suckers. Origin unknown but probably *P. deltoides* Marsh. × *tacamahacca* Mill. Cultivated before 1755.

2. SALIX L.

Trees or shrubs. Buds with 1 outer scale, terminal bud 0. Fls appearing before or after the lvs, in usually erect catkins, *scales entire; each fl. with 1 or 2 small nectaries*; stamens 2–5(–12), usually 2. Insect pollinated.

About 300 spp., mainly north temperate and arctic regions, some tropical and south temperate but absent from Australasia and the E. Indies.

Though there is now general agreement as to the number of spp., the determination of plants of this genus is often a matter of some difficulty. This arises from various causes: (*a*) the freedom with which they hybridize and from the planting of some of the hybrids away from their parents, reproduction by cuttings being very easy; (*b*) the fact that the sexes occur on different trees and the lvs are not fully developed at flowering time (in some spp. not expanded at all); (*c*) the variability of many of the spp. in such features as lf-shape. Plants in mature lf, or, for those spp. which flower with the lvs, in fl., are not, in the vast majority of instances, difficult to determine when once an acquaintance with the spp. has been obtained. For hybrids and unusual forms, however, it is often necessary to have fls and lvs from the same plant before a reasonably certain determination can be made, and the parentage of many hybrids is still in doubt. In the following account an attempt has been made to include descriptions of those hybrids likely to be met with away from their parents, and the more common ones have been included in the key except where the parents are closely allied (e.g. *S. alba* × *fragilis*). For more detailed accounts of the hybrids see Linton, *Journ. Bot. Suppl.*, 1913 and Moss, *Camb. Brit. Fl.* (1914), vol. II.

In order to facilitate identification of plants of this genus, three keys are included, but it must be emphasized that caution is needed in using them (male plants of those spp. flowering before the lvs expand are especially difficult). If more than one key can be used, either because both sexes are present or because both fls and lvs can be obtained, it is desirable to do so.

Numerous hybrids occur in the genus, some of them involving 3 spp. Only those which occur as osiers and a few of the most frequent of the others are included in the following account.

Key to Male Plants

1   Catkins terminal, appearing after the lvs; lvs orbicular or oval, rounded
      or retuse at apex; dwarf creeping alpine shrubs.   **2**
      Catkins lateral on the previous year's wood, appearing with or before
      the lvs; lvs not as above (except rarely in *S. repens* which fls before
      the lvs).   **3**

2   Lvs green on both sides, veins not impressed above.   **18. herbacea**
      Lvs glaucous beneath; veins strongly impressed above.   **19. reticulata**

3   Scales of catkins uniformly yellowish.   **4**
      Scales of catkins dark at apex.   **8**

4   Stamens 5 (rarely 4–12); buds and young lvs viscid.   **1. pentandra**
      Stamens 2–3; buds and young lvs not viscid.   **5**

5   Stamens 3; bark smooth, peeling off in patches.   **4. triandra**
      Stamens 2 (very rarely 3); bark rough, not peeling, occasionally flaking
      off when old.   **6**

6   Lvs appressed silky hairy on both surfaces; branches ascending at 50°
      or less forming a narrow crown, twigs not fragile.   **2. alba**
      Lvs glabrous or soon glabrescent; branches spreading at 60° or more,
      forming a broad crown, twigs fragile at the junctions.   **7**

7   Twigs olive, never reddish-tinged; bark not flaking.   **3. fragilis**
      Twigs reddish on exposed side; bark at length flaking off.   **decipiens**

8   Twigs dark purple, with a whitish waxy bloom.   **6. daphnoides**
     Twigs without a whitish waxy bloom.   *9*

9   Catkins clothed with yellow hairs; Scottish Highlands.   **15. lanata**
     Catkins with white or greyish hairs.   *10*

10 Stamens completely united and appearing as 1 stamen.   **5. purpurea**
     Stamens united below, free above.   **purpurea × viminalis**
     Stamens 2, free.   *11*

11 Scales brownish or reddish at apex; tall shrub (3 m. or more); lvs linear-
     lanceolate, soon glabrous; stamens occasionally 3.   **triandra × viminalis**
     Scales blackish at apex or, if brown, low shrub (60 m. or less) on moun-
     tains; stamens never 3.   *12*

12 Catkins appearing with the lvs on lfy stalks (often short); lvs glabrous
     to sparingly pubescent (if lvs conspicuously hairy see *S. nigricans* or
     S. repens); (not occurring south of Lancashire and Yorks).   *13*
     Catkins appearing before (rarely with) the lvs, subsessile with at most
     a few bracts at the base; lvs ±conspicuously hairy; (if filaments
     glabrous and twigs shining see *S. phylicifolia*).   *15*

13 Stamens purple; lvs equally green on both sides.   **17. myrsinites**
     Stamens yellow, sometimes tinged with red; lvs pale or glaucous beneath.   *14*

14 Scales brown at apex; low spreading shrub, 60 cm. or less; catkins usually
     less than 1·5 cm.   **16. arbuscula**
     Scales blackish at apex; ±erect shrub, 1–4 m.; catkins 1·5–2·5 cm.
          12. phylicifolia

15 Rhizome creeping below the ground; catkins 5–20 mm., slender.
          13. repens
     Without creeping rhizome; catkins 1 cm. or more.   *16*

16 Anthers reddish before dehiscence, filaments glabrous; lvs silvery silky-
     woolly on both sides; mountain shrub 30–150 cm.   **14. lapponum**
     Anthers not reddish before dehiscence (except sometimes in *S. cinerea*
     with filaments hairy near the base); lvs not silky above; usually 1 m.
     or more.   *17*

17 Twigs without raised striations under the bark (rarely with a few short
     ones), stout; filaments glabrous or nearly so.   *18*
     Twigs with raised striations under the bark of the 2-year old wood,
     more slender.   *19*

18 Twigs pubescent, long, straight, ±erect, flexible; lvs lanceolate or linear-
     lanceolate, 7 or more times as long as broad.   **7. viminalis**
     Twigs glabrous, long, straight, ±erect; lvs ovate-lanceolate or lanceolate,
     3–4½ times as long as broad.   **caprea × viminalis**
     Twigs glabrous, not markedly long, straight or erect; lvs oval to obovate;
     not more than twice as long as broad.   **8. caprea**

19 Lvs often present at fl., blackening on drying; catkins 1·5–2·5 cm.
     (Yorks and Lancashire northwards).   **11. nigricans**
     Lvs rarely present at fl., not blackening on drying.   *20*

20 Twigs pubescent or glabrous, not angled or divaricate; catkins 2–3 cm.
          9. cinerea
     Twigs glabrous, angled, usually divaricate; catkins 1–2 cm.   **10. aurita**
     (Hybrids of *S. cinerea* and *S. aurita* with *S. viminalis* are scarcely dis-
     tinguishable from *S. cinerea* and *S. aurita* without lvs. They have
     longer twigs and relatively narrower catkins.)

Key to Female Plants

1 Catkins terminal, appearing after the lvs; lvs orbicular or oval, rounded
   or retuse at apex; dwarf creeping alpine shrubs.   2
   Catkins lateral on the previous year's wood (sometimes on long lfy
   peduncles); lvs not as above (except rarely in *S. repens* which fls before
   the lvs).   3

2 Lvs green on both sides, veins not impressed above.   **18. herbacea**
   Lvs glaucous beneath, veins strongly impressed above.   **19. reticulata**

3 Scales of catkins uniformly yellowish.   4
   Scales of catkins dark at apex.   8

4 Bark peeling off in patches; scales of catkins persistent.   5
   Bark not peeling; scales of catkins soon falling.   6

5 Lvs glabrous, $2\frac{1}{2}$–$7\frac{1}{2}$ times as long as broad; pedicel finally 3–4 times as
   long as nectary.   **4. triandra**
   Lvs silky when young, $5\frac{1}{2}$ times as long as broad or more; pedicel only
   slightly longer than nectary.   **× lanceolata**

6 Buds and young lvs viscid; lvs 2–4 times as long as broad.   **1. pentandra**
   Buds and young lvs not viscid; lvs rarely less than 4 times as long as
   broad.   7

7 Lvs appressed silky on both surfaces; branches ascending at 50° or less,
   forming a narrow crown, twigs not fragile; pedicel not longer than
   nectary.   **2. alba**
   Lvs glabrous almost from the first; branches spreading at 60° or more,
   forming a wide crown, twigs fragile at the junctions; pedicel finally
   2–3 times as long as nectary.   **3. fragilis**

8 Scales reddish or brownish at apex; tall shrub; lvs linear-lanceolate,
   soon glabrous.   **triandra × viminalis**
   Scales blackish at apex or, if brown, low shrubs with relatively broad lvs.   9

9 Twigs dark purple with a whitish waxy bloom.   **6. daphnoides**
   Twigs without a whitish waxy bloom.   10

10 Catkins clothed with yellow hairs (finally fading to greyish); lvs ovate,
   ± silky-tomentose on both sides; Scottish Highlands.   **15. lanata**
   Catkins with white or greyish hairs; lvs not as above.   11

11 Capsule sessile or subsessile or with pedicel not longer than the nectary
   (if tall shrubs, capsule sessile or subsessile).   12
   Capsule with pedicel longer than the nectary (sometimes only as long or
   slightly shorter, then tall shrubs).   17

12 Catkins sessile or subsessile.   13
   Catkins on lfy peduncles; low shrubs on mountains.   16

13 Style very short or 0; lvs, or some of them, subopposite.   **5. purpurea**
   Style moderate or long; lvs all alternate.   14

14 Low shrub to 150 cm.; lvs silky on both sides, 2–4 times as long as broad
   (mountains).   **14. lapponum**
   Tall shrubs with long straight twigs; lvs not silky above, 5 times as long
   as broad or more.   15

15 Lvs silky only when young; style moderate, not $\frac{1}{2}$ as long as ovary.
                                                                  purpurea × viminalis
   Lvs persistently and densely silky beneath; style very long, more than $\frac{1}{2}$ as
   long as ovary even in fr. and as long in fl.   **7. viminalis**

16   Lvs pale or glaucous beneath; nectary quadrate. **16. arbuscula**
      Lvs equally green on both sides; nectary linear. **17. myrsinites**

17   Rhizome creeping below the ground; lvs appressed silvery-silky at least
      beneath, not more than 5 cm. long and usually much less. **13. repens**
      Rhizome not creeping; lvs not silky or rarely somewhat silky and then
      more than 5 cm. **18**

18   Pedicel twice as long as nectary or less; lvs (2½–)3 times as long as broad
      or more, tapered at both ends, hairy beneath, usually rather silkily so
      (hybrids of *S. viminalis*). **19**
      Pedicel three times as long as nectary or more (except sometimes in
      S. phylicifolia with glabrous lvs); lvs rarely more than 3 times as long
      as broad, not silky. **23**

19   Twigs without raised striations under the bark; pedicels much longer
      than nectary. **caprea × viminalis**
      Twigs with raised striations under the bark; pedicels as long or rather
      longer than nectary. **20**

20   Twigs soon glabrous, branches usually divaricate. **aurita × viminalis**
      Twigs persistently pubescent, branches long, straight, ±erect. **21**

21   Stipules small; twigs finely pubescent; lvs 5–12 cm.; style moderate.
                                            cinerea × viminalis
      Stipules conspicuous; twigs tomentose; lvs 7–18 cm.; style long. **22**

22   Tomentum of older twigs black; twigs very stout; catkins 3–7·5 cm.;
      stigma shorter than style. **calodendron**
      Twigs brown, less stout; catkins 2·5–5 cm.; stigma longer than style.
                                              × stipularis

23   Style long; northern plants. **24**
      Style very short or 0. **25**

24   Lvs and twigs glabrous; capsule usually pubescent; lvs not blackening.
                                          12. phylicifolia
      Lvs and usually twigs pubescent; capsule usually glabrous; lvs blackening
      when dried. **11. nigricans**

25   Twigs without (rarely with a few short) striations under the bark,
      glabrous, catkins finally 3–7 cm. **8. caprea**
      Twigs with conspicuous raised striations under the bark. **26**

26   Twigs glabrous or pubescent, not or scarcely angled, not divaricate;
      catkins finally 3–5·5 cm. **9. cinerea**
      Twigs glabrous, angled, usually divaricate; catkins finally 1–2·5 cm.
                                          10. aurita

Key to Plants with Mature Leaves

1   Small prostrate, creeping shrubs, with lvs rounded or retuse at apex. **2**
      Lvs not rounded at apex (or if so tall shrubs). **4**

2   Lvs densely appressed silky, at least beneath. **13. repens**
      Lvs glabrous or nearly so. **3**

3   Lvs green on both sides, veins not impressed above. **18. herbacea**
      Lvs glaucous beneath, veins strongly impressed above. **19. reticulata**

4   Lvs, or some of them, subopposite, obovate-oblong to oblanceolate-
      linear, glabrous, dull and slightly bluish-green. **5. purpurea**
      Lvs all alternate. **5**

5   Lvs glabrous or very nearly so.   **6**
    Lvs hairy at least beneath.   **16**

6   Trees or ± erect shrubs 1 m. or more, mainly lowland.   **7**
    Low shrubs 60 cm. or less, branches procumbent to ascending; Scottish
       mountains.   **14**

7   Lvs 2–8 cm., acute, not long-tapering to apex, glaucous beneath, $1\frac{1}{2}$–$2\frac{1}{2}$(–5)
    times as long as broad.   **12. phylicifolia**
    Lvs 5–15 cm., acuminate or gradually tapering to an acute apex (except
       sometimes in *S. purpurea* × *viminalis*).   **8**

8   Twigs purple with a whitish waxy bloom.   **6. daphnoides**
    Twigs without a whitish waxy bloom.   **9**

9   Bark peeling off in patches (if lvs more than 8 times as long as broad
    see *S.* × *lanceolata*).   **4. triandra**
    Bark not peeling.   **10**

10   Twigs very brittle at the junctions (if red see *S. decipiens*); lvs lanceolate,
    usually asymmetric at apex, glaucous or less frequently pale green
    beneath.   **3. fragilis**
    Twigs not brittle at the junctions; lvs not or scarcely asymmetric at apex,
    usually green beneath.   **11**

11   Lvs 2–4 times as long as broad, dark green and very glossy above, green
    beneath, subcoriaceous.   **1. pentandra**
    Lvs more than 4 times as long as broad, not especially dark, or very
    glossy, not subcoriaceous.   **12**

12   Twigs bright yellow.   **2. alba** var. *vitellina*
    Twigs not bright yellow.   **13**

13   Shrub; lvs dull.   **purpurea × viminalis**
    Shrub; lvs somewhat glossy above.   **triandra × viminalis**
    Tree with branches ascending at a narrow angle.   **2. alba** var. *caerulea*

14   Lvs equally green on both sides.   **17. myrsinites**
    Lvs glaucous or pale beneath.   **15**

15   Main branches usually ± decumbent, often diverging in a crown; lvs
    rarely more than 2 cm., often sparingly appressed-pubescent.
                                          16. arbuscula
    Main branches spreading or erect, not forming a crown; lvs more than
    2 cm., glabrous.   **12. phylicifolia**

16   Lvs lanceolate or linear-lanceolate, more than 5 times as long as broad;
    indumentum of appressed silky hairs; trees or shrubs 3 m. or more.   **17**
    Lvs 4 times as long as broad or less (rarely more in *S. repens*); if trees or
    tall shrubs, indumentum not silky.   **18**

17   Lvs moderately and ± equally silky on both sides; tree.   **2. alba**
    Lvs 10–25 cm. dark green above, densely silvery-silky beneath, $(4\frac{1}{2}$–)7–18
    times as long as broad (if lvs less than 6 times as long as broad and
    not very silky see hybrids under *27* below).   **7. viminalis**

18   Lvs with closely appressed silky hairs; shrubs 150 cm. or less, lvs 5 cm.
    or less.   **19**
    Hairs not closely appressed.   **20**

19   Rhizome creeping below the ground; lvs densely silky beneath, often
    silky above.   **13. repens**
    Without creeping rhizome; lvs glabrous above and with few hairs
    beneath; Scottish mountains.   **16. arbuscula**

20 Lvs clothed beneath and usually above with dense white matted silky-
      woolly hairs; low bushy shrubs (to 150 cm.); Scottish mountains.   21
      Indumentum not as above.   22

21 Lvs ovate or obovate, not more than twice as long as broad; branches
      stout; buds large.   **15. lanata**
      Lvs elliptic or oblong, 2–4 times as long as broad; branches not markedly
      stout; buds small.   **14. lapponum**

22 Lvs thinly pubescent beneath, frequently only on the veins, not or rarely
      glaucous, blackening when dried, hairs not rust-coloured.
                                                            11. nigricans
      Lvs not blackening when dried, rather densely tomentose or pubescent
      all over beneath, if thinly pubescent then glaucous and usually with
      some rust-coloured hairs.   23

23 Lvs less than 3 times as long as broad (rarely more in *S. cinerea*), apex
      obtuse or shortly acuminate or cuspidate, rarely acute.   24
      Lvs more than 3 times as long as broad, gradually tapered to an acute
      or acuminate apex.   27

24 Lvs rugose above, 2–3 cm., grey-tomentose beneath; usually low shrub
      (1–2 m.) with divaricate branches.   **10. aurita**
      Lvs not rugose, 3 cm. or more; shrubs or small trees 2 m. or more,
      branches not divaricate.   25

25 Lvs thickly and softly tomentose beneath without rust-coloured hairs.   26
      Lvs thinly pubescent beneath with at least some of the hairs rust-coloured.
                                              9. cinerea ssp. **atrocinerea**

26 Lvs oval or ovate-oblong, or sometimes obovate; twigs soon glabrous,
      without raised striations under the bark.   **8. caprea**
      Lvs obovate, oblanceolate or elliptic; twigs tomentose, with raised
      striations under the bark.   **9. cinerea** ssp. **cinerea**

27 Twigs not striate under the bark, glabrescent.   **caprea** × **viminalis**
      Twigs with raised striations under the bark.   28

28 Twigs soon glabrous, usually divaricate; lvs often twisted at apex,
      margins undulate.   **aurita** × **viminalis**
      Twigs persistently pubescent, ± erect; lvs not twisted at apex, margins
      not undulate (slightly so in *S.* × *stipularis*).   29

29 Stipules small, often caducous; lvs 5–12 cm., with some rust-coloured
      hairs beneath.   **cinerea** × **viminalis**
      Stipules large, at least on vigorous shoots; lvs 7–18 cm., without rust-
      coloured hairs.   30

30 Lvs oblong-lanceolate to obovate-lanceolate, greyish and softly pubescent
      beneath.   **calodendron**
      Lvs lanceolate, rather silvery and silky beneath.   × **stipularis**

Section 1. *Salix*. Trees or tall shrubs. Lvs ± lanceolate and tapered to
the apex. Catkins appearing with or after the lvs, on lfy peduncles, from
lateral buds of the preceding year. Scales yellowish, uniformly coloured.
Stamens 2 or more, free. Male fls with 2 nectaries, female fls with 1 or 2
nectaries.

1. S. pentandra L.   Bay Willow.

Shrub or small tree 2–7 m. Bark grey, fissured. *Twigs* glabrous, green to brown, shining as if varnished, *not fragile. Buds* ovoid, *viscid*, shining. *Lvs* 5–12 cm., *2–4 times as long as broad*, elliptic-ovate to elliptic-lanceolate or obovate-lanceolate, acuminate, scarcely asymmetric at apex, rounded or broadly cuneate at base, glandular serrate, *glabrous, dark green and glossy above, paler beneath, somewhat coriaceous* at maturity, *viscid and fragrant when young; petiole* c. 1 cm. or less, *with* 1–3 *pairs of glands* near the top; stipules ± ovate, usually small, soon caducous. Catkins cylindric, appearing rather later than the lvs, on lfy pubescent stalks, rhachis pubescent; male 2–6 × 1·5 cm., dense; female 2–5 × c. 1 cm., more slender. Scales oblong, pubescent near the base, glabrous above, caducous. *Stamens* (4–)5(–12); anthers golden yellow before dehiscence; filaments pilose towards the base. Female fl. with 2 nectaries; ovary narrowly ovoid, glabrous, 5–6 mm. in fr.; style short; stigmas 2-lobed; pedicels rather longer than nectaries (to twice as long in fr.). Fl. 5–6. Fr. 6–7. $2n = 76$. M.

Native. Stream-sides, marshes, fens and wet woods; ascending to 1500 ft. in Derby; native from about N. Wales, Derby and Yorks to Sutherland and in N. Ireland, rather common; planted elsewhere. 52, H34. Europe from N. Norway (70° 25′ N.) and N. Russia to the Pyrenees, Italian Alps, Macedonia and the Caucasus; Siberia; N.E. Asia Minor.

S. alba × pentandra = *S. × ehrhartiana* Sm.

Small tree. Twigs pubescent when very young, soon glabrous and polished, olive-brown. Lvs 5–10 cm., usually oblong-lanceolate, cuneate at base, finely serrate, pubescent at first, soon glabrous, ± glossy above, petiole with 2–4 glands. Scales ± resembling those of *S. pentandra*, catkins more slender. Male fl. with 3–6 stamens, most frequently 6. Ovary ovoid-conic; style short; pedicels about as long as nectaries or slightly longer. Fl. 5. Rare.

S. fragilis × pentandra = *S. × meyerana* Rostk. ex Willd.

S. cuspidata K. F. Schultz

Small tree or shrub to 12 m. Habit nearer *S. fragilis*. Lvs 5–13 cm., 3–5 times as long as broad, oblong-lanceolate, longer acuminate than *S. pentandra*, usually asymmetric at apex, cuneate at base, serrate, glabrous, shining above, not viscid; petiole 2–3 cm., with glands near top; stipules sometimes persisting. Catkins appearing with the lvs (male to 9 cm.). Scales usually thinly hairy to apex. Stamens 3–5, most frequently 4. Style short; pedicels longer than nectaries. Fl. 5. Rare.

2. S. alba L.   White Willow.

Tree 10–25 m., *branches ascending at* 30–50°, *forming a narrow crown*, appearing silvery-grey in lf, often pollarded. *Bark deeply fissured*, greyish, *not peeling. Twigs* silky when young, later glabrous and olive, *not fragile.* Buds oblong, usually pubescent for some time. *Lvs* 5–10 cm., (5–)5½–7½(–10) *times as long as broad*, lanceolate, acuminate, not or slightly asymmetric at apex, cuneate at base, finely serrate, *covered with white silky appressed hairs on both surfaces*; petiole c. 5–8 mm., not glandular; stipules small, lanceolate, usually caducous. Catkins appearing with the lvs on pubescent stalks with entire lvs, dense, cylindric; male 2·5–5 cm. × c. 6 mm.; female smaller in fl. reaching 6 cm. in fr. Scales ovate-lanceolate or oblong-lanceolate, straw-yellow, pubescent near the base and on the margins, caducous. *Stamens* 2; anthers yellow; filaments hairy towards the base. Female fl. with usually 1 nectary; capsule ovoid-conic, obtuse, glabrous; style short; stigmas rather thick, deeply 2-lobed; *pedicel at first* 0, *finally almost as long as the nectary*. Fl. 4–5. Fr. 6. $2n=76$. MM. Bark rich in salicylates and used for tanning.

Native. By streams and rivers, marshes, fens and wet woods on the richer soils, throughout the British Is., not recorded from Shetland; commonly planted along rivers, etc., but probably also native except in N. and W. Scotland. 107, H38, S. Europe from Norway (perhaps introduced but possibly native in one area, planted to 63° 52′ N.) and N. Russia southwards; N. Africa; W. Asia from W. Siberia to the Himalaya and Palestine.

Var. *coerulea* (Sm.) Sm. Cricket-bat Willow.

Branches ascending at a narrower angle. Lvs less hairy (often glabrescent) at maturity, of a bluer green above and somewhat glaucous beneath. $2n=76$. Quick growing. The best willow for cricket bats and frequently planted for that purpose, mainly in E. England.

Var. *vitellina* (L.) Stokes

Smaller tree (to 15 m.). *Twigs bright yellow or orange* for their first year, very conspicuous in winter. Lvs soon nearly glabrous above, finally thinly hairy and glaucous beneath. Scales narrower. Rather frequently planted as an osier.

S. alba × fragilis = S. × rubens Schrank

S. viridis auct.

Variable. Habit ± intermediate. Twigs somewhat fragile, but less so than in *S. fragilis*. Lvs silky when young, finally glabrous or glabrescent but darker and duller than in *S. fragilis*, serrations intermediate. Ovaries somewhat obtuse, not tapering, the pedicel about 1½ times as long as the nectaries.

Apparently widespread and not uncommon.

*S. babylonica L. Weeping Willow.

Tree to 20 m., *with long recurving branches drooping nearly to the ground*. Lvs 8–16 cm. × 8–15 mm., narrowly lanceolate or linear-lanceolate acuminate, serrulate, glabrous at maturity; petiole 3–5 mm. Male not known in Britain. Female catkins to 2 cm. × 3–4 mm., curved, sometimes with some male fls. Scales concolorous, glabrescent. Female fl. with 1 nectary; capsule glabrous, sessile; style short; stigmas emarginate. Fl. 5. Commonly planted by lakes and rivers mainly in S. England, but scarcely naturalized. Native country uncertain, probably China.

S. alba × babylonica = S. × sepulcralis Simonk. and *S. babylonica × fragilis = S. × blanda* Anderss. are occasionally planted. They have both much of the weeping habit of *S. babylonica*.

3. S. fragilis L. Crack Willow.

Tree 10–25 m., branches spreading widely at 60–90°, forming a broad crown, appearing green in lf, often pollarded. *Bark deeply fissured, greyish, not peeling. Twigs* sometimes slightly pubescent when young, soon glabrous and *olive, very fragile at the junctions.* Buds brown, ovate, glabrescent, ± viscid. *Lvs* 6–15 × 1·5–4 cm., (3½–)4½–9 times as long as broad, lanceolate, long acuminate, *usually asymmetric at apex*, cuneate at base, rather coarsely serrate, thinly silky when very young, soon *glabrous*, bright green above, *glaucous or less often paler green beneath*; petiole 1–3 cm., usually with 2 small glands at top; stipules ± lanceolate, usually soon caducous except on sucker shoots. Catkins appearing with the lvs, drooping, on pubescent stalks with entire lvs, rather dense, cylindric; male 2·5–6 cm., up to 1 cm. broad; female 3–7 cm. Scales quickly caducous, ± oblong, obtuse or truncate, straw-yellow, with long straight hairs. *Stamens* 2; anthers yellow; filaments hairy towards base. Female fl. usually with 2 nectaries; capsule ovoid-conic, tapered at apex, glabrous; style rather short but longer than stigmas; stigmas bifid, divaricate; *pedicel finally 2 or 3 times as long as nectaries.* Fl. 4 (earlier than *S. alba*). $2n = 76$, 114. MM.

Native. By streams and rivers, marshes, fens and wet woods, more tolerant of poor soils than *S. alba*; commonly planted but commoner as a native tree than *S. alba* in England, Wales and S. Scotland (to about Perth); always planted further north; throughout Ireland but local and probably introduced. 105, H23, S. Europe from S. Sweden (planted in Norway to 65° 5′ N.) and N. Russia southwards; W. Siberia to Persia.

S. decipiens Hoffm.

Differs from *S. fragilis* as follows: Smaller tree. *Bark flaking off when old. Twigs red on the exposed side when young*, later yellowish-grey and shining. Buds shorter, dark brown, often becoming blackish. Lvs somewhat smaller (to 9 cm.) subglaucous beneath. Stamens occasionally 3. Pedicel of the capsule shorter.

Variously regarded as a variety of *S. fragilis* and as a form of the hybrid *S. fragilis × triandra* or possibly *S. fragilis ×* some other sp. In view of this uncertainty it is here treated separately. Perth southwards, local, sometimes grown as an osier, probably always planted, nearly always male, the female rather doubtfully British; Londonderry. 39, H1.

S. fragilis × triandra = S. × speciosa Host

Small tree or large shrub. Twigs brown, shining. Lvs 7–14 cm., 3½–8 times as long as broad, oblong-lanceolate, often somewhat asymmetric at apex, cuneate at base; petiole 10–15 mm.; stipules usually small and caducous except on sucker shoots where they are large and persistent. Catkins 3–6 cm. Scales tardily caducous, oblong or obovate, hairy at base, ciliate or glabrous above. Stamens 2 or 3. Female fl. usually with 1 nectary; capsule somewhat tapered; style short; stigmas short; pedicel finally 3–4 times as long as nectaries. Fl. 4.

Apparently very rare but not easily identified. Recorded for only 3 or 4 English and 2 Irish counties. *S. decipiens* (see above) is very likely a form of this hybrid.

4. S. triandra L.   'Almond Willow.'

Shrub, rarely a small tree, 4–10 m. *Bark smooth, peeling off in patches.* Twigs glabrous, except at the very first, olive- or reddish-brown, rather fragile. Buds brown, ovoid, glabrous. *Lvs* very variable, 5–10 *cm.*, (2½–)3½–7½ *times as long as broad*, oblong-ovate or oblong-lanceolate, acute or shortly acuminate (not asymmetric), rounded at base, serrate, *glabrous*, dark green and somewhat shining above, glaucous or less often pale green beneath; petiole 6–15 mm., glabrous, with 2 or 3 small glands near top; stipules ± ovate, dentate, rather large (5–10 mm.), usually persistent, especially conspicuous on the sterile shoots. Catkins appearing with the lvs, erect on short lfy ascending stalks, cylindric; male 3–5 cm., slender; female rather denser and shorter. *Scales concolorous*, greenish-yellow, *persistent*; male obovate, hairy at base, glabrous above; female oblong, hairy at base, usually ciliate above. *Stamens* 3; anthers yellow; filaments hairy at base. Female fl. with 1 nectary; capsule ovoid-conic, obtuse, glabrous; style very short or 0; stigmas short, thick, emarginate, divaricate; *pedicel finally* 3–4 *times as long as the nectary*. Fl. 3–5 (often again 7–8). Fr. 6. $2n = 38, 44, 88$ (no taxonomic difference observed). M.

Native. Sides of rivers and ponds, marshes, etc., often planted as an osier; widespread and rather common in England, less so in Wales; very local in Scotland and only in the south-east, north to Perth and Kincardine, doubt-fully native; south Ireland, northwards to Clare and Kildare; Westmeath, Donegal, Armagh and Down; probably native in the south-east; Channel Is. 73, H 14, S. Europe from 63° 28′ N. in Norway and N. (not arctic) Russia southwards (absent from Mediterranean Is.); temperate Asia to Japan, N. Persia and N. and E. Asia Minor; Algeria.

S. × lanceolata Sm. (*S. triandra* × ?)

S. undulata Ehrh. ?

Large shrub. *Bark peeling.* Twigs olive, glabrescent (pubescent when very young). *Lvs* 7–12 cm., 5½–8½ *times as long as broad*, lanceolate, gradually narrowed to apex, serrulate, silky at first, soon glabrous, dull green above, paler beneath; petiole c. 1 cm.; stipules usually persistent, large on the sucker shoots. Female catkins (male unknown) appearing with the lvs on short lfy stalks, to 8 cm., dense. Scales concolorous, ovate or obovate, obtuse, hairy, especially near apex. Capsule glabrous; style short; stigmas 2-lobed; *pedicel rather longer than nectary*. Fl. 4–5. Fr. 5–6.

Variously referred to a form of *S. triandra* × *viminalis* (see below) and to *S. alba* × *triandra*.

Mainly in E. England and Midlands, extending north to Perth. Grown as an osier and usually (?always) planted.

S. triandra × **viminalis** = *S. × mollissima* Ehrh.

S. hippophaëfolia Thuill.

Shrub 3–5 m. Twigs olive, glabrescent. *Lvs* 5–12 *cm.*, 5½–9 *times as long as broad*, *linear-lanceolate*, acute or acuminate, obscurely serrulate, pubescent at first, soon *glabrous, dark green and somewhat shining above*, paler beneath; petiole to 1 cm.; stipules ± lanceolate, usually caducous. Catkins with the lvs, subsessile or shortly stalked, c. 2·5 cm., occasionally androgynous. *Scales brownish or reddish at apex*, villous especially at apex, ovate-oblong, obtuse. Male fls with 2(–3) stamens. Capsule finely pubescent (rarely glabrescent); style rather long; stigmas bifid; pedicel about as long as nectary. Fl. 4–5.

Midlands from Gloucester and Glamorgan to Derby and Nottingham; sometimes grown as an osier.

Section 2. *Vimen* Dum. From low shrubs to small trees. Lvs variously shaped. Catkins before or with the lvs on lfy peduncles or subsessile with a few bracts at the base, from lateral buds of the preceding year. Scales blackish or dark brown at the apex, markedly 2-coloured. Stamens 2, free or united. Male and female fls with 1 nectary.

5. S. purpurea L.   'Purple Willow.'

Shrub 1½–3 m., rather slender. Bark bitter. Twigs slender, straight, glabrous, shining, usually purplish at first, becoming olive or yellowish-grey. Buds oblong, acute, glabrous. *Lvs* 4–10 cm., 3·3–15 times as long as broad, *obovate-oblong to oblanceolate-linear*, acute or acuminate, gradually narrowed to the base, finely serrulate, *glabrous* (or pubescent for a very short time), *dull and slightly bluish-green above*, paler and often glaucous beneath, often turning black when dried, usually *subopposite* at least near the apex of the twigs; petiole 3–10 mm.; stipules small, very quickly caducous. *Catkins* appearing before the lvs, 2–3·5 cm., dense, cylindric, *subsessile* with a few small lvs at base, suberect to spreading, often arcuate. *Scales blackish at the tip*, reddish below, obovate, pilose, persistent. *Stamens* 2, *connate and appearing as if only* 1; anthers reddish or purplish before dehiscence; filaments villous near the base. Ovary ovoid, obtuse, tomentose, sessile; *style very short or* 0; stigmas thick, short, entire or slightly emarginate, often purplish. Fl. 3–4. Fr. 5. $2n = 38$. M.

Very variable in width of lf. Broad-leaved forms have been distinguished as var. *lambertiana* (Sm.) Koch and narrow-leaved ones as var. *helix* (L.) Koch.

Native. Fens, where it is sometimes locally dominant, marshes and by rivers and ponds from Caithness southwards, ascending to 1450 ft., sometimes planted as an osier; common in some districts, rare in others; throughout Ireland. 98, H36, S. C. and S. Europe from Belgium, N. Germany and the Ladoga-Ilmen region of Russia southwards (not Crete, etc.), (introduced in Scandinavia); W. Asia from S.W. Siberia to N. and E. Asia Minor; C. Asia to Japan; N. Africa.

S. purpurea × viminalis = S. × rubra Huds.

Shrub 1–3 m. Twigs pubescent when young, becoming glabrescent and yellowish. Buds ovate-oblong, puberulent. *Lvs* 7–15 *cm.*, 5–12 *times as long as broad*, obovate-oblong to linear-lanceolate, ± acuminate, ± denticulate, silky when young, *glabrous and dull above at maturity*, paler and glabrous to rarely slightly pubescent beneath, sometimes subglaucous; petiole 5–10 mm.; stipules small, quickly caducous. Catkins dense 2–3 cm., subsessile, with a few lvs at base; male ovoid-oblong; female cylindric. Scales dark at tips, obovate-oblong, villous. Male fls with 2 stamens, *filaments ± connate, often for half their length*; anthers reddish before dehiscence. Capsule ovoid-conic, tomentose, sessile; *style moderate*; stigmas entire, about equalling style or shorter. Fl. 4. $2n = 38, 57$.

Throughout England, frequently grown as an osier, but probably also occurring naturally; rare in Scotland (north to Perth) and Ireland, and always planted.

*6. S. daphnoides Vill.

Large shrub or tree 7–10 m. *Twigs covered with a bluish waxy bloom* for about 3 years, *purple. Lvs* 5–10 *cm.*, 2½–8 times as long as broad, *linear-lanceolate or oblong-ovate*, acute or acuminate, glandular-serrulate, soon glabrous, dark green and shining above, glaucous beneath; petiole 2–4 mm.; stipules large, semicordate, usually caducous. Catkins 3–4 cm., appearing before the lvs, subsessile, cylindric, dense. Scales blackish at apex, obovate, pilose. Stamens sometimes slightly fused at base; anthers yellow; filaments glabrous. Capsule ovoid-conic, glabrous, subsessile; style long, slender; stigmas entire to bifid. Fl. 2–3. Fr. 5–6. MM.

Introduced. Sometimes planted in wet places, scarcely naturalized.

Ssp. daphnoides

Lvs oblong-lanceolate or oblong-ovate, 2½–6 times as long as broad; stipules large, semicordate. Catkins cylindric. $2n=38$.

Commoner in Britain than ssp. *acutifolia.* Native of Europe from Scandinavia, E. France and N. Apennines to Roumania and W. Russia.

Ssp. acutifolia (Willd.) Dahl

S. acutifolia Willd.

Lvs linear-lanceolate; stipules lanceolate. Catkins ovoid-oblong. Capsules longer and narrower than in ssp. *daphnoides.* $2n=38$. Native of Finland, Russia, Siberia and C. Asia.

7. S. viminalis L. Common Osier.

Shrub 3–5(–10) m., *with long straight flexible branches. Twigs* densely pubescent, becoming glabrous later, *not striate under the bark.* Buds ovoid-oblong, acuminate, pubescent. *Lvs* 10–25 cm., (4½–)7–18 *times as long as broad, lanceolate or linear-lanceolate*, gradually narrowed to the apex, cuneate at base, *dark green and glabrous above, silvery silky-tomentose beneath*; margins undulate, revolute when young, entire or nearly so; petiole 4–12 mm.; stipules small, linear-lanceolate, usually caducous. *Catkins* appearing before the lvs (or female with the lvs), *subsessile*, with or without a few small lvs at base, dense; male 2·5–4 cm. ovoid-oblong; female cylindric, to 6 cm. in fr. Scales blackish at the apex, ovate-oblong, pilose. Nectaries long, linear. Stamens free; anthers yellow; filaments glabrous. Capsule ovoid-conic, silky tomentose, sessile or subsessile; style very long, more than ½ as long as ovary in fr., and as long in fl.; stigmas slender, entire, rarely bifid. Fl. 4–5. Fr. 5. $2n=38$. M.

Native. By streams and ponds, in marshes and fens, and commonly planted as an osier; ascending to 1350 ft. Throughout the British Is., common in lowland areas, rare in hilly ones. 110, H40, S. Europe from Scandinavia (status doubtful), the Netherlands and France through C. Europe to the Urals, south to Montenegro and Thrace (introduced in Spain, Italy, etc.); temperate Asia to Caucasus, Himalaya and Japan.

The five plants which follow are very difficult to separate from each other. All are intermediate between *S. viminalis* and the group *S. caprea*, *S. cinerea* and *S. aurita*. The differences between these 3 spp., *inter se*, being so much less than those between any of them and *S. viminalis*, they become obscured in the hybrids.

S. caprea × **viminalis** = *S.* × *sericans* Tausch ex A. Kerner

Shrub, rarely small tree, 3–5 m. *Branches long, straight, ± erect. Twigs* rather stout, tomentose when young, glabrescent, *not striate under the bark.* Buds ovoid, pubescent at first. *Lvs* 5–12 cm., 3–4½ *times as long as broad, ovate-lanceolate or lanceolate,* acute, margins flat to somewhat undulate, crenulate to entire, green and glabrescent above, softly tomentose beneath, the veins prominent; petiole 6–15 mm., tomentose; stipules semicordate or ovate-falcate, rather large, ± toothed. Catkins appearing before the lvs (or female with the lvs), subsessile or shortly stalked, rather large; male 2·5–4 cm., ovoid; female cylindric, finally 4–5 × 1 cm. Scales blackish at the apex, ± obovate, long silky-pilose. Capsule tomentose, large; style moderate; stigmas about as long as style; pedicel finally about twice as long as nectary. Fl. 3–4. Fr. 5.

In scattered localities in England, Scotland and Ireland.

S. cinerea × **viminalis** = *S.* × *smithiana* Willd.

S. geminata Forbes

Shrub 2–4 m. Branches long, ± erect. *Twigs* rather stout, *persistently pubescent, ± erect, striate under the bark.* Buds ovoid, pubescent. *Lvs* 5–12 *cm.,* lanceolate or oblanceolate, acute or acuminate, *not twisted at apex, margins flat,* ± crenate, dull green and pubescent above, grey silky pubescent *often with some rust-coloured hairs beneath, finally usually glabrescent*; petiole 1–1·5 cm., pubescent; *stipules* ovate-lanceolate, *often caducous, rather small,* subentire. Catkins appearing before the lvs, subsessile, moderate; male 2·5–3·7 cm., ovoid or oblong-ovoid; female finally c. 5 cm. Scales blackish at apex, obovate, long silky-pilose. Capsule conic, pubescent; *style moderate*; stigmas about as long as style; pedicels finally about as long as nectary. Fl. 4. Fr. 5.

In scattered localities in England, Wales, S. Scotland (to Perth) and Ireland.

S. aurita × **viminalis** = *S.* × *fruticosa* Doell

Shrub 1·5–3 m. *Branches usually divaricate. Twigs* rather slender, pubescent at first, *soon glabrous, striate under the bark.* Buds ovoid, pubescent at first. Lvs 4–12 cm., ± lanceolate, 3–6 times as long as broad, *acute and often twisted at apex, margins undulate,* ± crenate, somewhat rugose, somewhat pubescent above, later glabrous; tomentose or pubescent beneath, sometimes becoming glabrous; petiole 5–7 mm.; stipules ± lanceolate, caducous or not. Catkins appearing before lvs, subsessile or shortly stalked, rather small; male 1·5–3 cm., female 2·5–3·5 cm., finally 5 cm. × 8 mm. Scales blackish at apex, ± oblong, silky-pilose. Capsule conic, tomentose; style moderate or short; stigmas about as long; *pedicels finally longer than nectary.* Fl. 4. Fr. 5.

From Perth southwards, very local; very rare in Ireland.

*S. calodendron Wimmer

S. acuminata auct.; *S. dasyclados* auct.

Shrub or small tree 3–6 m. Branches ± erect. *Twigs* stout, *densely velvety tomentose*, the tomentum persisting and turning black in winter, *striate under the bark. Lvs* 7–18 *cm.*, 2½–5 times as long as broad, *oblong-lanceolate or obovate-lanceolate*, acute, crenate or crenate-serrate, dull green and somewhat pubescent above, *greyish-glaucous and softly pubescent beneath*; *stipules large, ± lanceolate*; petiole 1–1·5 cm. Catkins appearing with lvs, subsessile; male not known in Britain; *female* 3–7·5 *cm.* × c. 12 mm., curved. Scales oblong-obovate, silky-villous. Capsule to 6 mm. stout, ovoid-conic, tomentose; *style long*; stigmas stout, shorter than style, entire; pedicels about equalling nectary. Fl. 3–4. M.

Variously regarded as a species or as *S. caprea × viminalis* or one of its allies. If it is a hybrid, the parentage *S. caprea × cinerea × viminalis* is the most likely.

Introduced. By rivers, etc.; E. England and the Midlands, local; E. Scotland (very rare). Germany, Denmark, Sweden.

S. × stipularis Sm.

Shrub 2–4 m. Branches long, ± erect. *Twigs* densely *tomentose, brown in winter, striate under the bark. Lvs* 9–18 cm., *lanceolate*, c. 4–5 times as long as broad, attenuate to acute apex, margins slightly undulate, obscurely crenate or subentire, glabrescent above, *rather silvery-silky pubescent beneath; stipules large and conspicuous* (to 2·5 cm.) at least on the strong shoots, ± lanceolate, cordate on one side at base, toothed; petiole often shorter than stipules. *Catkins* before the lvs; male unknown; female 2·5–5 *cm.*, stout, cylindrical, *subsessile*. Scales ovate-oblong, densely long silky-pilose. Capsule tomentose; style long; stigmas longer than style, entire; pedicel shorter than nectary. Fl. 5.

Variously regarded as a form of *S. cinerea × viminalis* or *S. calodendron × viminalis*.

In scattered localities in England, Wales and Scotland.

8. S. caprea L.   Great Sallow, Goat Willow.

Shrub or small tree 3–10 m. Bark coarsely fissured; branches numerous, ascending. *Twigs rather stout*, pubescent at first, *glabrous* or nearly so by autumn, *without striations under the bark* of the 2-year-old wood. Buds ovoid-conic, trigonous, acute, pubescent at first, later glabrous. *Lvs* 5–10 *cm.*, 1·2–2 *times as long as broad, oval to ovate-oblong or obovate*, obtuse to subacute or shortly and often obliquely acuminate, subcordate to cuneate at base, margin somewhat undulate, crenate to entire, dark green and finally glabrous or glabrescent above (pubescent at first), *persistently softly and densely, almost woolly, grey-tomentose beneath* and strongly reticulate-veined; petiole c. 1–1·5 cm.; stipules semicordate, dentate. *Catkins* appearing before the lvs, dense, *subsessile*, with a few small bracts at base; male 2–3·5 × 1·5–2 cm. oblong-ovoid; *female finally lax and* 3–7 *cm.* Scales blackish at apex, obovate, densely clothed with long silky hairs. Stamens free; anthers yellow; filaments glabrous or slightly hairy near base. Capsule 6–8 mm., tomentose, ovoid-conic; *style short*; stigmas suberect, emarginate, more rarely bifid; *pedicel finally 4–6 times as long as nectary.* Fl. 3–4. Fr. 5. $2n = 38^*, 76^*$ (no taxonomic difference observed). M.

Native. Woods, scrub and hedges, etc., ascending to 2800 ft. in Scotland, common throughout the British Is. (apparently absent from Outer Hebrides). 111, H40, S. Europe from Scandinavia (70° 57′ N. in Norway) and N. (not arctic) Russia to the mountains of S. Spain, the Apennines and Greece; temperate Asia to Syria and Turkestan.

Ssp. caprea

Lvs usually oval to elliptic-oblong, broadest about the middle, rounded at the base, nearly glabrous above at maturity except on the veins; stipules usually present.

The common lowland form; it is uncertain how far it occurs in the Scottish Highlands. Range of the sp.

Ssp. sericea (Anderss.) Flod.

S. coaetanea (Hartm.) Flod.

Lvs obovate or obovate-elliptic, acuminate, broadest above the middle, cuneate at the base, thinly appressed cobwebby-silky above at maturity, more silvery beneath than ssp. *caprea*; stipules usually 0.

Scottish Highlands. Scandinavia, Finland; ?elsewhere.

9. S. cinerea L.   Common Sallow.

Shrub or more rarely small tree, 2–10 m. often with long straight suberect branches. *Twigs rather stout, usually persistently pubescent or tomentose* but sometimes glabrescent, not *divaricate, with raised striations on the 2-year old wood* if the bark is peeled off. Buds ovoid. *Lvs 2·5–7(–10) cm., (1·7–)2–4 times as long as broad,* obovate, oblanceolate or rarely elliptic, obtuse, acute, apiculate or shortly cuspidate at apex, usually cuneate at base, subentire to somewhat serrate or crenate-serrate, pubescent above at least when young, persistently pubescent or tomentose beneath; petiole 5–15 mm.; stipules usually persistent. *Catkins appearing before the lvs, dense, subsessile* with very few small bracts; *male 2–3 × 1–2 cm.,* ovoid, or oblong-ovoid; *female finally 3–5·5 cm.* Scales blackish at apex, ± obovate, obtuse, clothed with long silky hairs; anthers yellow, often tinged reddish when young; filaments pilose near base. Capsule ovoid-conic, tomentose; *style short*; stigmas rather stout, bifid, usually short, rarely longer; *pedicel finally 2–5 times as long* as nectary. Fl. 3–4. Fr. 5–6. M.

Ssp. cinerea

Shrub, usually of dense bushy habit with branches less erect than ssp. *atrocinerea. Twigs persistently velvety tomentose,* greyish-brown. Buds densely tomentose. *Lvs* tending to be less obovate and more toothed than in ssp. *atrocinerea,* silvery pubescent above when young, *beneath glaucous, grey-tomentose and soft to the touch at maturity, the hairs not rust-coloured and uniformly distributed on and between the veins*; stipules usually 2–8 mm. $2n = 76$.

Native. Mainly in E. England in fens, etc. and often dominant in carr, where it frequently replaces ssp. *atrocinerea*; occasionally in damp woods, etc. in other parts of England but apparently restricted to wet habitats and very local. Europe from Scandinavia and N. (not arctic) Russia to N. and E. France, Corsica, Italy, Greece and the Caucasus; W. Asia from W. Siberia to N. Asia Minor and N. Persia; Tunisia.

Ssp. **atrocinerea** (Brot.) Silva & Sobr.

S. atrocinerea Brot.

Shrub, rarely small tree, usually with long straight suberect branches. *Twigs shortly pubescent at first, brown, becoming reddish in winter and remaining thinly pubescent or becoming glabrous.* Buds thinly pubescent or glabrous. Lvs usually buff- or reddish-pubescent above when young; *beneath thinly and shortly pubescent mainly on the veins, at maturity glaucous and with some or all of the hairs rust-coloured,* not soft to the touch as in *S. caprea*, *S. aurita* and ssp. *cinerea*, ±prominently reticulately veined; stipules usually small, c. 3 mm. or less. $2n = 76$.

Native. Woods and heaths, by ponds and streams, in marshes and fens; ascending to 2000 ft.; common throughout the British Is. 112, H40, S. W. and C. France, Spain, Portugal, N.W. Morocco.

10. S. aurita L. 'Eared Sallow.'

Shrub 1–2(–3) m. with numerous spreading branches. Twigs rather slender, pubescent at first, soon *glabrous,* brown, *usually angular and divaricate, with raised striations under the bark.* Buds oval, subobtuse, glabrous or puberulent. *Lvs 2–3 cm., 1·6–2·7 times as long as broad,* obovate or oblong-obovate, shortly cuspidate, the cusp often oblique and decurved, ±cuneate at base, undulate, serrate or dentate to subentire, *rugose,* dull grey-green and pubescent or glabrescent above, *grey-tomentose beneath*; petiole short, to 8 mm.; stipules subcordate or reniform, large and conspicuous, persistent, dentate. *Catkins* appearing before the lvs, *subsessile* with a few bracts at base; *male 1–2 cm.,* ovoid; *female 1–2·5 cm.,* cylindric. *Scales blackish at apex,* oblong or obovate, pilose. Stamens free; anthers yellow, *filaments hairy near base.* Capsule narrowly ovoid-conic, tomentose; *style very short*; stigmas short, thick, emarginate; pedicel finally 3–4 times as long as nectary. Fl. 4. Fr. 5–6. $2n = 38^*, 76^*$ (no taxonomic difference observed). N.

Native. Damp woods, heaths, rocks by streams and on moors, etc., on light acid or slightly basic soils; ascending to 2600 ft. Common throughout the British Is. 111, H40, S. Europe from Scandinavia (66° 27′ N. in Norway) to Portugal (very rare) and N. and E. Spain, Corsica, N. Italy and Macedonia; in Russia only in the south (Trans-Volga, Black Sea, Crimea and Lower Don regions).

11. S. nigricans Sm. 'Dark-leaved Willow.'

S. Andersoniana Sm.

Shrub 1–4 m., or spreading and procumbent. *Twigs pubescent,* the hairs often conspicuous and white and rather coarse, *or glabrescent, dull,* blackish, brownish or olive-green, striate under the bark. Buds oval, pubescent at least when young. *Lvs 2–7(–10) cm., 1½–3(–4) times as long as broad,* very variable in shape, orbicular-ovate to lanceolate, elliptic or obovate, acute or shortly acuminate, ±rounded at base, uniformly serrate or crenate-serrate to apex, or subentire, glabrescent, deep green and *rather dull above, paler but not usually glaucous and ±pubescent beneath at least on the veins*, usually *turning blackish when dried*; stipules semicordate, usually well developed and rather large at least on the strong shoots, sometimes 0 on the weaker ones; petiole 6–20 mm., ±pubescent. *Catkins with or before the lvs*; *male* 1·5–2·5 *cm.*, subsessile, ovoid or oblong; *female* 1·5–3 *cm.* in fl., lengthening to 5–7 cm., rather lax, usually on short lfy tomentose stalks. *Scales* ovate, obovate or oblong, *blackish at apex*, pilose. Stamens free; filaments hairy near base (or sometimes glabrous?); anthers yellow. *Capsule* long, *glabrous* or (more rarely) pubescent; *style long*; stigmas emarginate or bifid, spreading, nearly as long as style; *pedicel 3 or 4 times as long as the nectary*. Fl. 4–5. Fr. 5–6. 2*n*=114. N.

Native. By lakes and streams and on damp rock-ledges; ascending to 3000 ft. From Lancashire and Yorks to Sutherland and the Outer Hebrides, rather local; Antrim, Londonderry. 33, H3. N. Europe from Scandinavia and N. Russia to Germany (not N.W.) and in the mountains south to the Vosges, French Alps, Apennines, Croatia, Bulgaria and S. Russia; Siberia (east to the Yenisei region).

S. nigricans × phylicifolia = *S.* × *tenuifolia* Sm.

S. tetrapla Walker

A common and very variable hybrid where the two parents occur together. It is possible that plants with glabrous ovaries otherwise resembling *S. phylicifolia* and plants with hairy ovaries or glabrous filaments otherwise resembling *S. nigricans* should be referred to it, besides the more obvious intermediates.

12. S. phylicifolia L.   'Tea-leaved Willow.'

Shrub 1–4 *m. Twigs glabrous* or slightly pubescent when young, *polished*, brown at maturity, striate under the bark. Buds narrow ovate, acute, yellowish, glabrous or puberulous. *Lvs* 2–8 *cm.*, 1·7–2·5(–5) *times as long as broad*, ovate to oblong-lanceolate, elliptic or obovate, ±*acute*, rounded or cuneate at base, serrate or crenate-serrate, usually subentire near base and apex, or subentire throughout, *glabrous and shining above*, somewhat glaucous and *glabrous beneath* except sometimes when very young, rather coriaceous, not blackening when dried; stipules small, usually quickly caducous; petiole 6–12 mm., pubescent. *Catkins appearing with*, or rarely before, *the lvs, subsessile or shortly stalked*, lfy at base; male 1·5–2·5 cm., ovoid; female 2·5–6 cm., rather lax in fr. *Scales blackish at apex*, oblong, obovate or oblong-lanceolate, pilose. Stamens free; filaments glabrous; anthers yellow. *Capsule* lanceolate-ovoid, pubescent, rarely glabrous; *style long*; stigmas bifid, large, usually shorter than style; *pedicel 2–4 times as long as nectary*. Fl. 4–5. Fr. 5–6. 2*n*=88, 114. N.

Native. By lakes and streams and among wet rocks; ascending to 2300 ft. From Lancashire and Yorks to Orkney and the Outer Hebrides, rather local; Mayo, Sligo, Leitrim, Donegal, Antrim, Londonderry. 41, H6. N. Europe from Iceland and the Faeroes to arctic Russia. Allied spp. or sspp. in the mountains of C. and S. Europe.

13. S. repens L.   'Creeping Willow.'

Shrub 30–150 *cm., with creeping rhizome.* Stems prostrate to erect, slender.
Twigs silky-pubescent when young with a few, fine striations under the bark.
Buds ovate, silky-pubescent at first, soon glabrous. *Lvs* 0·5–4·5 *cm.,* 1½–3½(–8)
times as long as broad, very variable in shape, oval, oblong, elliptic or lanceolate,
obtuse or acute, and straight or twisted or recurved at apex, rounded or
cuneate at base, entire to serrulate, sometimes somewhat revolute, *appressed
silvery-silky* on both sides when young, persistently so *beneath,* becoming
glabrous above or not, prominently reticulate-veined beneath, often blackening
when dried; petiole 2–5 mm.; stipules ± lanceolate, usually quickly caducous,
or 0. Catkins appearing before the lvs, subsessile or on short lfy stalks; *male*
5–20 *mm., slender,* ovoid to oblong; female 8–25 mm., globose to oblong.
Scales brownish at apex, ± obovate, hairy. Stamens free; filaments glabrous;
anthers yellow. Capsule ± conic, tomentose to glabrous; style short or
moderate; stigmas variable in length, entire or bifid; pedicels finally 3–4 times
as long as nectary. Fl. 4–5. N.

Throughout the British Is., but rather local and absent from a few counties.
106, H35, S. Europe from Scandinavia to N. Spain and Portugal, N. Italy
and the Dobruja; W. and C. Asia, Siberia.

Ssp. **repens**

Rhizome long, stems prostrate to ascending or (var. *fusca* (L.) Wimm. &
Grab.) rhizome short, stems erect. Lvs usually silky beneath only, those of
the sterile stems not markedly large, up to c. 25 × 10 mm. (rarely longer, then
narrow). Female catkins subsessile, with few lfy bracts even at maturity.
$2n = 38$.

Damp and wet heaths, usually lowland, but ascending to 2800 ft.; var.
fusca, fens.

Ssp. **argentea** (Sm.) G. & A. Camus

S. arenaria auct., ?L.

Rhizome long, branches ascending. Lvs of vigorous first-year sterile shoots
large, up to 45 × 25 mm., mostly rounded at base or even subcordate, those
of the fertile shoots considerably smaller, usually silky on both sides. Female
catkins at maturity on lfy stalks, those of both sexes tending to be larger than
in ssp. *repens.* $2n = 38$.

Dune-slacks, sometimes dominant, sometimes also on rocky heaths in N.
Scotland.

14. S. lapponum L.   'Downy Willow.'

Shrub 30–150 *cm.,* of compact habit, branching from base. Branches ascending,
pubescent at first, glabrous, dark brown and shining at maturity. Buds ovoid,
obtuse, pubescent or glabrous. *Lvs* 2–5 cm., 2–4 *times as long as broad, ovate,
elliptic or oblong,* acute or acuminate, rounded or cuneate at base, entire,
sometimes undulate, *grey-green or dark green with appressed silky hairs above,
grey and densely silky-tomentose beneath,* veins somewhat impressed above,
prominent beneath; petiole 3–7 mm.; stipules small, quickly caducous. Catkins
appearing before the lvs, subsessile, without or with a few small lvs at base;

male c. 2·5 cm., oval; female 2·5–5 cm., cylindric; rhachis woolly. Scales
± oblong, dark brown or blackish at apex, densely villous with long white
hairs. Stamens free; filaments glabrous; *anthers* yellow *tinged brownish or
reddish before fl.* Capsule ovoid-conic, whitish-tomentose, subsessile; style
long or moderate; stigmas usually rather shorter, entire or bifid; nectary
linear, reaching middle of capsule. Fl. 5–7. Fr. 7–8. $2n = 38$, 76 (no taxonomic difference observed). N.

Native. Wet rocks on mountains from 650–3500 ft.; Westmorland; Argyll,
Stirling and Angus to Sutherland, on the mainland, local. 16. N. Europe
from Scandinavia to arctic Russia and in the mountains south to Poland and
the Middle Dnieper region of Russia; Siberia.

15. S. lanata L.   'Woolly Willow.'

Shrub 50–100 cm., of bushy habit. Branches stout, woolly-pubescent when
young, dark brown somewhat shining when 1 year old. *Buds large*, woolly-
pubescent. *Lvs* 2·5–6·5 cm., $1\frac{1}{4}$–2 *times as long as broad, ovate or obovate*,
shortly and often obliquely acuminate, cordate to cuneate at base, entire,
sometimes undulate, *silky-tomentose above* when young, finally dull green and
glabrescent, *persistently woolly-silky-tomentose beneath*, subglaucous and prominently reticulate-veined; petiole to c. 1 cm., stout, woolly; stipules large,
4–12 mm., ± ovate. Catkins appearing with the lvs, subsessile or female
shortly stalked, without or with a few small lvs at base, beautifully golden-
yellow when young; male to 5 cm., ovoid-oblong or cylindric; female 5–8 cm.,
cylindric. *Scales* ± obovate, blackish at tip, *densely clothed with long golden-
yellow* (fading to greyish) *silky hairs*. Stamens free; filaments glabrous,
yellow; anthers yellow. Capsule subulate-conic, glabrous; style long; stigmas
linear, entire or bifid; pedicel finally about as long as the linear nectary.
Fl. 5–7. Fr. 7. $2n = 38$. N.

Native. Damp ledges of basic rocks on mountains from about 1800–
3000 ft.; Perth, Angus, Aberdeen, very local and rare. 3. Arctic and subarctic
Europe and Asia, south to the Altai and Dahuria; absent from C. Europe.

16. S. arbuscula L.

*Shrub 30–60 cm., with prostrate to ascending branches, often diverging in a
crown from the base.* Twigs glabrous (or slightly pubescent when very young)
becoming dark brown and shining, striate under bark. Buds ovoid, acute,
soon glabrous. *Lvs* 0·5–2(–4) *cm.,* $1\frac{1}{3}$–3 times as long as broad, ovate to
ovate-lanceolate or elliptic, acute to obtuse, cuneate or rounded at base,
glandular-serrate to subentire, appressed pubescent when young, *glabrous
and shining above, somewhat glaucous* and sometimes appressed pubescent
beneath at maturity; veins slightly prominent on both surfaces or beneath
only, sometimes blackening when dried; petiole 2–4 mm.; stipules small or 0.
Catkins appearing ± with the lvs, slender; *male* subsessile, *lfy at base, usually
less than* 1·5 *cm.; female on long lfy stalks, to* 3 *cm. in fr. Scales* ± obovate,
brownish at apex, clothed with crisped white hairs. Stamens free; filaments
glabrous; *anthers reddish-yellow before dehiscence.* Capsule ovoid-conic,
pubescent; styles long or moderate; stigmas small, bifid, the divisions sometimes filiform; *pedicel shorter than the linear nectary or* 0. Fl. 5–6. Fr. 6.
$2n = 38$. N.

Native. Damp ledges of basic rocks on mountains from c. 1500 to 2700 ft.; Perth, Argyll, very local but sometimes abundant. 3. N. Europe from Scandinavia to arctic and N. Russia (not Iceland); Siberia. Allied spp. or sspp. in the mountains of Europe.

17. S. myrsinites L.

Shrub 10–40 cm. high, with decumbent or ascending branches. Twigs pubescent at first, soon glabrous, becoming brown and shining, not striate under bark, the withered remains of the previous year's lvs often persistent. Buds ovoid, pubescent at first, later glabrous. Lvs 1·5–3(–5) cm., 1½–2½ times as long as broad, ovate or elliptic, acute or subacute, rounded or cuneate at base, glandular-serrate, *bright green and shining on both sides* at maturity, sometimes blackening when dried, veins prominent on both sides; petiole short, 3 mm. or less; stipules usually conspicuous on the strong shoots, ± ovate, c. 6 mm. (rarely to 12 mm.), often caducous on normal shoots. *Catkins* appearing with or slightly after the lvs *on lfy* pubescent *stalks; male* 1·5–2·5 *cm.*, oblong, on short stalks; female 2–5 cm., oblong-cylindric on long stalks. Scales ovate-oblong, brownish-purple, finally blackish at apex, clothed with long silky hairs. Stamens free; filaments glabrous or pubescent at base; *anthers purple.* Capsules usually purplish, narrowly ovoid-conic, thinly pubescent; style long, often reddish; stigmas usually large but shorter than style, purplish, ± bifid; *pedicel shorter than or equalling the quadrate nectary.* Fl. 5–6. Fr. 6–7. $2n = 38, 152, 190$ (no taxonomic difference observed between 38 and 190). N.

Native. Wet rocks, mainly basic, on mountains from 300 to 3800 ft.; Dumfries; Argyll, Perth and Angus to Orkney, rare and local. 13. N. Europe from Scandinavia to arctic and N. Russia; N. Asia. Allied spp. in the mountains of Europe.

Section 3. *Chamaetia* Dum. Prostrate creeping shrubs. Lvs orbicular or oval. Catkins on lfless peduncles from terminal buds of the preceding year. Scales not dark at apex, concolorous. Stamens 2, free. Fls with 1 or 2 nectaries.

18. S. herbacea L.   'Least Willow.'

Shrub with long creeping branched underground rhizome, aerial branches few (2–5)-lvd., usually very short (2–3 cm.), procumbent to erect, glabrous. Buds ovoid, glabrous. *Lvs* 6–20 mm., less than 1½ times as long as broad, and sometimes broader than long, *orbicular or broadly ovate or obovate, rounded or retuse,* cordate or rounded at base, crenate-serrate, *glabrous, bright green and shining, and with the veins rather prominent on both sides*; petioles c. 5 mm. or less; stipules 0 or small and quickly caducous. Catkins appearing after the lvs, terminal on short lfless pubescent peduncles, 5–15 mm., few (2–12)-fld. Scales ± obovate, yellowish-green sometimes reddish on margin, glabrous or thinly pubescent. Male fl. with 1 nectary; filaments glabrous; anthers yellow, sometimes red-tipped. Female fl. with 2 nectaries (or the outer obsolete); capsule ovoid-conic, obtuse, glabrous, often reddish; style short; stigmas large, bifid; pedicels shorter than the ±linear nectaries. Fl. 6–7. Fr. 7. $2n=38$. Ch.

Native. Mountain tops, where it is often abundant, and rock-ledges; occurring on almost all the higher mountains; usually at high altitudes (to 4300 ft.) but descending to 300 ft. in Sutherland. Wales; Yorks and Westmorland to Ayr and Peebles; Kintyre, Stirling and Angus northwards; Ireland (on all the higher mountains). 38, H17. Arctic Europe from Iceland to Russia, high mountains of C. Europe south to the Pyrenees, C. Apennines and Bulgaria; Greenland; arctic America south in the mountains to New Hampshire.

19. S. reticulata L.   'Reticulate Willow.'

Shrub with a short branched rhizome or creeping and rooting stem, branches ascending, short, few-lvd, pubescent when young, soon glabrous. Buds ovoid, pubescent at first. *Lvs* 1–3(–5) cm., less than twice as long as broad, orbicular or oval, rounded to retuse at apex, cordate to (rarely) cuneate at base, entire, coriaceous, clothed with long silky hairs on both sides when young, at maturity *dark green, glabrous and rugose with the impressed veins above, glaucous glabrous or sparingly hairy and prominently reticulate-veined beneath*; petiole long, 5–15 mm., reddish; stipules 0. Catkins appearing after the lvs, terminal on long (to 2·5 cm.) lfless pubescent peduncles; male 12–20 mm.; female 12–30 mm. Scales obovate, light brown, villous within. Male fls with 1 nectary; filaments pilose at base; anthers red or purple before dehiscence. Female fls with a deeply 2–4-partite nectary, embracing the base of the ovaries; capsule ovoid, sessile or subsessile, tomentose; style short. Fl. 6–7. $2n=38$. Ch.

Native. Basic rock ledges on mountains, from about 2000–3600 ft., very local; from Argyll and Angus to Sutherland. 7. Arctic and subarctic Europe and Asia (not Iceland) and in the high mountains south to Pyrenees, Italian Alps, Macedonia, Urals and Altai; arctic N. America from Labrador to Alaska, and south in the Rocky Mountains to Colorado.

METACHLAMYDEAE

90. ERICACEAE

Shrubs or rarely trees with simple leaves, stipules 0. Fls hermaphrodite, regular or rarely slightly zygomorphic, (3–)4–6(–7)-merous. Calyx often small, persistent. Corolla usually gamopetalous, rarely petals free, on the edge of a fleshy disk. Stamens free from corolla, usually twice as many as corolla-lobes and then obdiplostemonous, rarely more or fewer; anthers usually dehiscing by apical pores, often with awn-like appendages; pollen in tetrads. Carpels usually as many as the corolla-lobes, united to form a usually 4–5-celled superior or inferior ovary, placentation axile, ovules 1–many, anatropous; style simple, stigma capitate. Fr. a capsule, berry or rarely a drupe. Seeds small with copious endosperm and a usually small straight embryo.

About 80 genera and 1500 spp., arctic, temperate and mountains in the tropics (very few in Australia and only in the south-east).

A very natural family, often of characteristic heath-like aspect with corollas appearing waxy in texture. Apart from this, usually easily recognized among the Gamopetalae by the obdiplostemonous free stamens. The dehiscence of the anthers by pores is also nearly constant. Many species have endotrophic mycorrhiza. All the British spp., except *Arbutus unedo*, are calcifuge.

Spp. of a number of non-British genera are sometimes cultivated as ornamental plants.

1 Ovary superior.   *2*
   Ovary inferior.   14. VACCINIUM

2 Petals free; lvs rusty tomentose beneath.   1. LEDUM
   Petals ± united; lvs not rusty tomentose.   *3*

3 Corolla caducous after fl.; lvs alternate, rarely opposite and then not
      imbricate.   *4*
   Corolla persistent in fr.; lvs whorled or opposite and imbricate.   *13*

4 Lvs opposite; creeping mountain under-shrub.   3. LOISELEURIA
   Lvs alternate.   *5*

5 Corolla campanulate, slightly zygomorphic.   2. RHODODENDRON
   Corolla urceolate, regular.   *6*

6 Fls 4-merous in lax racemes; lvs white tomentose beneath. W. Ireland.
                                                                     5. DABOECIA
   Fls normally 5-merous; lvs at most glaucous beneath (and then fls in
      clusters).   *7*

7 Lvs linear or linear-elliptic, entire or serrulate, at most 5 mm. broad;
      fr. dry; fls in clusters.   *8*
   Lvs ovate, obovate or elliptic, if less than 5 mm. broad, clearly though
      remotely serrate; fr. fleshy.   *9*

8 Calyx and pedicels glandular; corolla ovoid, purple (Perth).
                                                                    4. PHYLLODOCE
   Calyx and pedicels glabrous; corolla subglobose, pink. Bogs.
                                                                    6. ANDROMEDA

9 Lvs rounded to cordate at base; fr. a capsule enclosed in the fleshy calyx;
      suckering shrub.   7. GAULTHERIA
   Lvs cuneate at base; fr. a drupe or berry; not suckering.   *10*

10 Tree or tall shrub; lvs 4 cm. or more; fr. warty (W. Ireland).
                                                                      9. ARBUTUS
    Shrubs to 2 m. but usually prostrate; lvs 2(–2·5) cm. or less; fr. smooth.   *11*

11 Erect shrub; lvs ovate or lanceolate, acute, aristate; anthers with short
       erect awns.   8. PERNETTYA
    Prostrate shrubs; lvs ± obovate, obtuse or subacute, not aristate; anthers
       with long reflexed awns.   *12*

12 Lvs persistent, coriaceous, entire; fr. red.   10. ARCTOSTAPHYLOS
    Lvs deciduous, thin, serrulate; fr. black.   11. ARCTOUS

13 Calyx longer than and coloured like the corolla; lvs opposite.
                                                                     12. CALLUNA
    Calyx much shorter than corolla; lvs whorled.   13. ERICA

Subfamily 1. RHODODENDROIDEAE. Ovary superior. Corolla caducous after fl. Fr. a septicidal capsule. Anthers without awns.

1. LEDUM L.

Evergreen shrubs. Buds scaly, conspicuous. Lvs alternate, entire, shortly petioled. *Fls on long slender pedicels in umbel-like terminal racemes, 5-merous.* Calyx-lobes short. *Petals free.* Stamens 5–10, anthers opening by pores. Capsule opening from base. Seeds narrow, flat with a wide border.

Three to six spp., cold north temperate zone.

1. L. palustre L.

Shrub to 1 m. Twigs rusty tomentose. Lvs 1–4·5 cm., linear to linear-oblong, dark green above, rusty-tomentose beneath but with the midrib ± visible, margins revolute. Pedicels glandular. Fls 1–1·5 cm. across. Petals cream, oval. Stamens (7–)10(–11). Capsule 4–5 mm., oblong. Fl. 6–7. $2n=52$. N.

Possibly native in bogs near Bridge of Allan (Stirling and Perth), a rare escape elsewhere. 2. N. Europe and Asia from Scandinavia to Sakhalin and N. Japan, south to C. Germany, Austria, C. Russia and the Altai.

*L. groenlandicum Oed. (*L. latifolium* Jacq.) differs in its elliptic or oblong lvs with tomentum hiding midrib beneath, the puberulous pedicels, stamens 5–8, capsule 5–6·5 mm. $2n=26$.

A rare escape. Native of N. America.

2. RHODODENDRON L.

Shrubs, rarely trees. Buds scaly, conspicuous. Lvs alternate, usually entire, shortly petioled. Fls usually in terminal racemes, usually 5-merous. Calyx usually small. *Corolla campanulate or funnel-shaped, slightly zygomorphic.* Stamens from as many to twice as many as corolla-lobes; anthers opening by pores. *Seeds small, numerous, flat, bordered.*

More than 600 spp., mainly E. Asia (to New Guinea and N. Australia), a few in W. Asia, Europe and N. America.

Many spp. and especially hybrids are ± commonly cultivated. *Azalea,* formerly regarded as a distinct genus, is now included.

*1. R. ponticum L.

Evergreen shrub to 3 m., nearly glabrous. Lvs 6–12 cm., elliptic to oblong, dark green above, paler beneath, acute, cuneate at base. Fls numerous. Corolla widely campanulate, c. 5 cm. across, dull purple spotted with brown. Stamens 10. Fl. 5–6. $2n=26$. M.

Introduced. Commonly cultivated and often planted in woods, etc., and has become thoroughly naturalized in many places on sandy and peaty soils, sometimes becoming locally dominant both as the shrub layer of woods and in the open. Native of C. and S. Portugal, S. Spain, Thrace, Asia Minor and Lebanon.

*R. luteum Sweet (*Azalea pontica* L.)

Deciduous shrub. Twigs and infl. glandular. Lvs ± oblong, hairy. Corolla funnel-shaped, yellow. Stamens 5.

Often cultivated and naturalized in one or two places. Native of Asia Minor and the Caucasus.

3. LOISELEURIA Desv.

Small creeping evergreen shrub. Buds small. *Lvs opposite*, entire, shortly petioled. Fls 1–5 in terminal clusters, 5-merous. *Corolla widely campanulate* with spreading lobes. *Stamens* 5; *anthers opening by slits. Ovary 2–3-celled.* Seeds numerous, small. Nectar secreted by ring at base of ovary.

One species.

1. L. procumbens (L.) Desv.   'Loiseleuria.'

Intricately branched, glabrous procumbent under-shrub. Lvs dense, 3–8 mm., oval or oblong, obtuse, coriaceous, convex above, margins revolute. Pedicels shorter than fl. Calyx-lobes ovate-lanceolate, reddish, about half as long as corolla. Corolla 4–5 mm., pink, paler inside. Capsule 3–4 mm., ovoid. Fl. 5–7. Pollinated by various insects, but probably more often selfed; weakly protogynous. $2n = 24$. Chw.

Native. Mountain tops and moors at high altitudes from 1300 ft. in Orkney to over 4000 ft. From Stirling and Dunbarton northwards. 19. N. Europe (to 71° 10′ in Norway) and Asia from Iceland and the Faeroes to N. Japan; mountains of C. Europe to Pyrenees, S. Alps and Croatia; N. America from Newfoundland and Alaska, south to the mountains of New Hampshire; Greenland.

KALMIA L.

Distinguished from the other genera of the family by the 10 pouches in the widely campanulate or saucer-shaped corolla which hold back the anthers till the corolla expands. Several spp. are cultivated.

*K. polifolia Wangenh. A straggling shrub to 70 cm.; lvs 2–3·5 cm., oblong, opposite, dark green above and glaucous beneath; fls 1–1·5 cm. diam., rose-purple.

Naturalized in a bog in Surrey. Native of N. America.

4. PHYLLODOCE Salisb.

Low heath-like evergreen shrubs. Lvs alternate, linear, usually serrulate, shortly petioled. *Fls* on slender pedicels, *in terminal clusters, 5-merous. Corolla urceolate* or campanulate. Stamens 10. Anthers opening by pores. Seeds oval with a narrow border. Nectar secreted by a ring at base of ovary.

About 7 spp., arctic regions and north temperate mountains.

1. P. caerulea (L.) Bab.

Menziesia caerulea (L.) Sw.; *Bryanthus caeruleus* (L.) Dippel

Low bushy shrub to 15 cm. Branches ascending. Lvs 5–9 mm., dense, linear, obtuse, serrulate. Fls 2–6, nodding. Pedicels and calyx reddish, glandular. Calyx-lobes nearly free, triangular, lanceolate. Corolla 7–8 mm., urceolate, ovoid, purple. Capsule ovoid, glandular. Fl. 6–7. $2n = 24$. Chw.

Native. Rocky moorland at c. 2400 ft. on the Sow of Atholl (Perth) in small quantity. 1. N. Europe and Asia from Iceland to Japan; very rare in Alps and Pyrenees; Greenland to Quebec and the mountains of New Hampshire.

5. DABOECIA D. Don

Low heath-like evergreen shrubs. Lvs alternate, entire, subsessile. *Fls in terminal racemes*, 4-*merous*. Calyx small. *Corolla urceolate*, lobes very short, recurved. Stamens 8; anthers opening by pores. Seeds numerous, small, tuberclate, not bordered.

Two spp., the other in the Azores.

1. D. cantabrica (Huds.) C. Koch   'St Dabeoc's Heath.'

D. polifolia D. Don; *Menziesia polyfolia* Juss.; *Boretta cantabrica* (Huds.) O. Kuntze

Straggling shrub to 50 cm. Twigs, pedicels and calyx glandular-hairy. Lvs 5–10 mm., elliptic or elliptic-linear, acute, revolute, dark green above with scattered glandular hairs, white-tomentose beneath. Racemes lax, 3–10-fld. Pedicels c. 5 mm., fls nodding. Corolla 8–12 mm., rose-purple, ovoid. Capsule oblong, glandular-hairy. Fl. 7–9. $2n=24$. Chw.

Native. Heaths and rocky ground in Mayo and W. Galway, ascending to 1900 ft., locally common. H3. Commonly cultivated in several colour forms and occasionally found as an escape in England. W. France, N., W. and C. Spain, N.W. Portugal.

Subfamily 2. ARBUTOIDEAE. Ovary superior. Corolla caducous after fl. Fr. a loculicidal capsule, berry or drupe. Seeds not bordered.

6. ANDROMEDA L.

Low evergreen shrubs. Lvs oblong or linear, alternate, entire, shortly petioled. Fls in terminal clusters, 5-merous. *Calyx small, not becoming fleshy.* Corolla urceolate, lobes short. Stamens 10. Anthers opening by pores, 2-awned. *Fr. a capsule.* Seeds oval, smooth. Nectar secreted by swellings at the base of the ovary.

Two spp., cold north temperate and Arctic.

1. A. polifolia L.   'Marsh Andromeda.'

Glabrous shrub to 30 cm. with creeping rhizome and scattered erect little-branched stems. Lvs 1·5–3·5 cm., linear or elliptic-linear, acute, revolute, dark green above, glaucous beneath. Fls 2–8, nodding, on pedicels 2–4 times as long as corolla. Corolla 5–7 mm., subglobose, pink. Capsule subglobose, glaucous. Fl. 5–9. Pollinated by humble-bees and butterflies, also selfed, homogamous. $2n=48$. Chw.

Native. Bogs, rarely wet heaths; ascending to 1750 ft. From Somerset and Norfolk to Perth and the S. Inner Hebrides, local and decreasing at least in the south; extinct in Hunts; in Ireland, rather common in the Central Plain, and extending as far as Antrim, E. Mayo, Limerick and Wicklow. 34, H27. N. and C. Europe from Scandinavia to Normandy, the Auvergne,

S. Alps and C. Russia, mainly in the mountains in the southern part of its range; N. Asia to Japan; Greenland; N. America south to New York and Idaho.

7. GAULTHERIA L.

Evergreen shrubs. Lvs usually alternate, shortly petioled. Fls 5-merous. *Calyx accrescent and fleshy in fr.* Corolla urceolate or campanulate. Stamens 10; awns short, erect, or 0. Fr. a capsule surrounded by the fleshy calyx and thus appearing berry-like. Seeds small, numerous.

More than 100 spp. America, E. Asia, Australia and New Zealand. Several others are cultivated.

*1. G. shallon Pursh

Shrub to 1 m., creeping underground and forming large dense patches. Stems ascending, glandular-hirsute when young. Lvs 5–12 cm., alternate, broadly ovate, acute, rounded to cordate at base, serrulate, glabrous. Fls in terminal glandular panicles. Calyx-lobes triangular. Corolla c. 1 cm., urceolate, ovoid, white tinged with pink. Fr. c. 1 cm., purplish-black, hairy. Fl. 5–6. Fr. 9–10. $2n=88$. N.

Introduced. Commonly planted for pheasants and naturalized on sand or peat in a number of places. Native of western N. America from Alaska to California.

*G. procumbens L.

Creeping underground. Aerial stems to 15 cm. Lvs 1·5–4 cm., obovate, crenate-serrate. Fls solitary. Fr. red. One patch in a pinewood in W. Inverness. Native of eastern N. America. Formerly the source of Oil of Wintergreen.

8. PERNETTYA Gaudich.

Evergreen shrubs. Lvs alternate. Fls 5-merous. Corolla urceolate. Stamens 10, *awns short, erect*, rarely 0. *Fr. a berry* with dry, rarely fleshy calyx at base.

About 25 spp. America from Mexico southwards, Tasmania and New Zealand.

*1. P. mucronata (L. f.) Gaudich. ex Spreng.

Branched suckering erect shrub to 2 m. forming a dense thicket. Lvs 1–2 cm., ovate to elliptic-lanceolate, acute and aristate at apex, remotely serrate with mucronate teeth. Fls solitary, axillary, white; pedicels c. 5 mm. Fr. c. 1 cm., crimson, purple, pink or white. Fl. 5–6. N.

Introduced. Commonly planted in gardens and naturalized in a number of places especially in Ireland. Native of Chile.

9. ARBUTUS L.

Evergreen trees or shrubs. Lvs alternate, petioled. Fls in terminal panicles, 5-merous. Calyx deeply lobed. Corolla urceolate. Stamens 10. Anthers with long reflexed awns, opening by pores. Ovules numerous in each cell. *Fr. a ± warty, globose berry.*

About 12 spp. in W., N. and C. America and in the Mediterranean region to W. Ireland and the Canary Is. Two spp., in addition to the following, are sometimes cultivated.

1. A. unedo L.   Strawberry Tree, Arbutus.

Erect shrub or tree to 12 m., glabrous or nearly so. Bark reddish-brown, thin, somewhat rough. Lvs 4–10 cm., elliptic-oblong or elliptic-obovate, acute, cuneate at base, serrate, dark green and shining above, paler beneath. Petiole c. 6 mm. Infl. c. 5 cm., many-fld. Calyx-lobes triangular. Corolla c. 7 mm., ovoid, or subglobose, creamy-white sometimes tinged pink. Fr. 1·5–2 cm., red, warty. Fl. 9–12. Fr. 9–12 (the following year). $2n = 26$. M.

Native. Rock crevices and between boulders, both sandstone and limestone, persisting in scrub and developing oakwood but not in mature oakwood. Kerry, W. Cork and Sligo (Lough Gill) but formerly more widespread; locally abundant. H4. Mediterranean region to S.W. France; Brittany.

10. ARCTOSTAPHYLOS Adans.

Evergreen shrubs. Lvs alternate, usually entire. Fls in terminal racemes or panicles, normally 5-merous. Calyx deeply lobed. Corolla urceolate. Stamens 10; anthers with long reflexed awns, opening by pores. Ovary 5–10-celled, each cell with 1 ovule. *Fr. a drupe with rather dry flesh.* Nectar secreted by a ring surrounding the ovary.

About 50 spp., confined to N.W. and C. America except for the following.

1. A. uva-ursi (L.) Spreng.   Bearberry.

Prostrate shrub with long rooting branches often forming mats. Twigs glabrous or nearly so. Lvs 1–2 cm., obovate or obovate-elliptic, usually obtuse (to subacute), cuneate at base, dark green above, paler beneath, conspicuously reticulately veined, entire. Fls 5–12 in short dense racemes; pedicels 3–4 mm. Corolla 4–6 mm., ± globose, white tinged with pink. Fr. 6–8 mm., red, globose, glossy. Fl. 5–7. Pollinated by humble-bees or selfed, weakly protogynous. Fr. 7–9. $2n = 52$. Chw.

Native. Moors, often covering rocks or banks; ascending to 3000 ft. Common in the Scottish Highlands, extending south to Westmorland, Derby and Yorks; N. and W. Ireland from Donegal to Antrim and Clare, local. 34, H8. Europe from Iceland to the mountains of N. Portugal, Spain, Italy and Macedonia; N. Asia to N. Japan; Caucasus, Himalaya; N. America from Labrador and Alaska to the mountains of Virginia, New Mexico and N. California.

11. ARCTOUS (A. Gray) Niedenzu

Deciduous prostrate shrubs. Lvs alternate, shortly petioled. Fls few, in terminal clusters, normally 5-merous. Calyx deeply lobed. Corolla urceolate. Stamens 10; anthers awned, opening by pores. Ovary 5-celled, each cell with 1 ovule. *Fr. a juicy drupe.*

Two spp. Cold north temperate and arctic regions.

1. A. alpinus (L.) Niedenzu 'Black Bearberry.'

Arctostaphylos alpinus (L.) Spreng.

Prostrate intricately branched shrub. Twigs glabrous. Lvs 1–2 (–2·5) cm., obovate, obtuse to subacute, cuneate at base, bright green, conspicuously reticulately veined, serrulate, usually ciliate. Infl. 2–4-fld. Corolla c. 4 mm., white. Fr. 6–10 mm., globose, black. Fl. 5–8. Fr. 8–10. $2n=26$. Chw.

Native. Mountain moors, ascending to 3000 ft. From Inverness to Shetland (not in Hebrides), very local. 9. N. Europe and Asia from Scandinavia to Japan; high mountains of C. Europe to Pyrenees, S. Alps and Montenegro; Greenland to the mountains of New Hampshire.

Subfamily 3. ERICOIDEAE. Ovary superior. Corolla persistent in fr. Fr. a capsule.

12. CALLUNA Salisb.

Evergreen shrub. Lvs opposite, very small. Fls 4-merous, axillary, forming a loose terminal raceme-like infl. *Calyx large, deeply lobed, of the same colour and texture as the corolla.* Corolla smaller, campanulate. Stamens 8, awned; anthers opening by pores. *Capsule septicidal,* few-seeded. Nectar secreted by 8 swellings between the filament bases.

One species.

1. C. vulgaris (L.) Hull Ling, Heather.

Diffuse evergreen shrub to 60 cm. (rarely to 1 m.), with numerous tortuous, decumbent or ascending branched stems, rooting at the base and bearing numerous axillary short shoots. Lvs 1–2 mm., linear, sessile with two short projections at base, margins strongly revolute making the lf trigonous, glabrous, pubescent or (var. *hirsuta* S. F. Gray) densely grey-tomentose, those of the main stems distant, those of the short shoots densely imbricate in 4 rows. Fls solitary, axillary, on the main axis and on the short shoots, with 4 ovate bracteoles forming a calyx-like involucre under each fl.; forming a raceme- or panicle-like infl. 3–15 cm. Calyx c. 4 mm., somewhat scarious; lobes ovate-oblong, pale purple. Capsule 2–2·5 mm., globose. Fl. 7–9. Pollinated by various insects and wind, weakly protandrous. Fr. 11. $2n=16$. N. or Chw.

Native. Heaths, moors, bogs and open woods on acid soils; ascending to 2400 ft. Common throughout British Is. 112, H40, S. Dominant over large areas on well-drained acid soils (heaths and moors), mainly on the east side of Great Britain and of Ireland, also becoming dominant in the field layer of open woods on similar soils; var. *hirsuta* abundant in some places near the sea (as on the Culbin Sands, Moray, where it is dominant over a large area to the exclusion of the type), scarce inland. Europe from Iceland and Finland to Spain, N. Italy, Greece and the Urals, common in the west, much decreasing in abundance eastwards; N.W. Morocco; Azores; eastern N. America (rare, probably introduced).

13. ERICA L.

Evergreen shrubs. Lvs whorled, small, entire, shortly petioled, revolute, dense. Infl. various. Fls 4-merous. *Calyx much shorter than corolla,* deeply lobed, not petaloid. Corolla urceolate, campanulate or cylindric; lobes short.

Stamens 8; anthers opening by pores. *Capsule loculicidal,* many-seeded. Nectar secreted by a ring round the base of the ovary.

About 500 spp., mainly S. Africa, a few in Europe and the Mediterranean region. The British spp. and hybrids are all ± commonly cultivated, often in forms varying in fl. colour, as are two or three other European spp. Some of the S. African spp. are grown in greenhouses and often sold in pots. White-fld forms of all our native spp., except *E. mackaiana,* occur.

1	Stamens included in corolla-tube.	2
	Stamens exserted, at least partly; lvs glabrous.	7
2	Lvs and sepals ciliate with long usually glandular hairs.	3
	Lvs and sepals glabrous.	5
3	Fls in terminal umbel-like clusters; anthers awned.	4
	Fls in racemes; anthers awnless (Dorset to Cornwall).	**3. ciliaris**
4	Lvs grey-pubescent above as well as ciliate, revolute nearly to the midrib, shorter and more distant below infl. (common).	**1. tetralix**
	Lvs dark green and glabrous above, somewhat revolute but leaving much of the white under surface exposed, not different below infl. (Galway and Donegal).	**2. mackaiana**
5	Corolla urceolate, pink or purple, rarely white (and then white in bud).	6
	Corolla narrowly campanulate, pink in bud, white when open (naturalized in Dorset and Cornwall).	**lusitanica**
6	Low shrub with short axillary lfy shoots; infl. long, raceme-like (common native).	**4. cinerea**
	Erect shrub without short shoots; fls in terminal umbel-like clusters (naturalized in Derry).	**terminalis**
7	Pedicels shorter than fls; corolla ± tubular; stamens about half exserted (W. Ireland).	**5. mediterranea**
	Pedicels longer than fls, corolla widely campanulate; stamens long exserted (Cornwall and Fermanagh).	**6. vagans**

1. E. tetralix L.   Cross-leaved Heath, Bog Heather.

Diffuse shrub to 60 cm., with numerous tortuous, ascending, branched stems, rooting at base, without short axillary shoots. Twigs pubescent, purplish, often glandular-hirsute. *Lvs 4 in a whorl,* 2–4 mm., linear, *glandular-ciliate, grey-pubescent above, margins revolute nearly to midrib hiding under surface, below the infl. more distant and usually shorter. Fls 4–12 in terminal, umbel-like clusters,* nodding in fl., becoming erect in fr. Pedicels c. 2 mm., pubescent, bracteoles about the middle. *Calyx-lobes* c. 2 mm., oblong-lanceolate, *pubescent on the surface and ciliate with long wavy glandular hairs. Corolla* 6–7 mm., *urceolate, ovoid, rose-pink,* sometimes somewhat paler beneath; lobes very short. *Anthers included, awned.* Capsule pubescent. Fl. 7–9. Pollinated by small insects or selfed. Fr. 10. $2n = 24$. N. or Chw.

Native. Bogs and wet heaths and moors, rarely on drier heaths; ascending to 2400 ft. Common in suitable habitats throughout the British Is., sometimes locally dominant. 112, H40, S. W. Europe from Scandinavia to C. Spain and Portugal (not in E. France or E. Spain), and across N. Germany (rare in the east) to the Baltic States and N.W. Poland.

2. E. mackaiana Bab.

E. Mackaii Hook.

Differs from *E. tetralix* as follows: Habit denser. Twigs almost hispid when young, soon glabrous. *Lvs oblong-lanceolate, dark green and glabrous above, margins somewhat revolute but leaving white under-surface exposed, not different below infl.* Pedicels with scattered long hairs. *Calyx-lobes* oblong-ovate, *glabrous except for the shorter straight glandular cilia on the upper half. Corolla deeper and brighter,* usually white beneath. Fl. 8–9. Seed not produced in Ireland. N. or Chw.

Native. In blanket bog in W. Galway and W. Donegal, very local. H2. N.W. Spain.

3. E. ciliaris L. Dorset Heath.

Diffuse shrub to 60 cm., with numerous ascending branched stems rooting at the base, with some axillary shoots. Twigs pubescent. *Lvs 3 in a whorl,* 1–3 mm., *ovate, glandular-ciliate, otherwise glabrous above,* margins somewhat revolute but leaving white under-surface exposed, distant below infl. *Fls axillary, forming unilateral terminal racemes,* 5–12 cm. Pedicels 1–2 mm., bracteolate about the middle. Calyx-lobes oblong-lanceolate, ciliate, slightly pubescent. *Corolla* 8–10 mm., *urceolate, somewhat curved above and inflated below,* mouth oblique, deep pink. *Anthers included, awnless.* Ovary glabrous. Fl. 6–9. N. or Chw.

Native. Heaths in Dorset, S. Devon and W. Cornwall, very local but sometimes abundant. 3. W. France, Spain (except east), Portugal, N.W. Morocco.

4. E. cinerea L. Bell-heather.

Diffuse shrub to 60 cm., with numerous ascending branched stems rooting at base, *with numerous short lfy axillary shoots,* often appearing as bunches of lvs. *Lvs 3 in a whorl,* 5–7 mm., linear, *glabrous,* dark green, margins strongly revolute. Fls in short terminal racemes and on the upper axillary shoots, the whole forming a terminal infl. 1–7 cm. Pedicels c. 3 mm., puberulous, bracteolate immediately below fl. Calyx-lobes lanceolate, glabrous, usually purple, keeled. *Corolla* 5–6 mm., *urceolate,* ovoid, usually crimson-purple; lobes very short, reflexed. *Anthers included, awned.* Ovary glabrous. Fl. 7–9. Pollinated by humble-bees or selfed. Fr. 10. $2n=24$. N. or Chw.

Native. Heaths and moors, usually dry; ascending to 2200 ft. Common throughout the British Is. 108, H40, S. W. Europe from Norway (62° 20′ N.) and the Faeroes to N. Spain and Portugal and N.W. Italy; Madeira.

*E. terminalis Salisb.

E. stricta Donn ex Willd.

Erect shrub 50–100 cm., with strict branches. Lvs 4 in a whorl, 4–6 mm., linear, glabrous. Fls in terminal 3–8-fld clusters. Corolla 6–7 mm., urceolate, ovoid, rose-pink. Stamens included.

Sometimes cultivated; naturalized on Magilligan dunes (Derry). Native of Corsica, Sardinia, S. Italy, S.W. Spain, N.W. Morocco.

*E. lusitanica Rudolphi

Erect shrub 1–2 m., with strict branches. Lvs 3–4 in a whorl, 5–6 mm., setaceous, glabrous. Fls axillary, forming long pyramidal panicles. Corolla c. 4 mm., narrowly campanulate, pink in bud, white when open. Stamens included.

Sometimes cultivated; naturalized in Dorset and on railway banks in Cornwall. Native of S.W. France, N.W. Spain, Portugal.

5. E. mediterranea L.   'Irish Heath.'

E. hibernica (Hook. & Arn.) Syme

Glabrous shrub 50–200 cm. Stems several, erect, with strict branches without short axillary shoots. Lvs 4 in a whorl, 5–8 mm., linear, dark green; margins strongly revolute. Fls axillary, nodding, forming a dense unilateral lfy raceme. *Pedicels much shorter than fls*, bracteolate about the middle. Calyx-lobes lanceolate. *Corolla* 5–7 mm., ± *tubular*, dull purplish-pink; lobes broad, obtuse, erect. *Anthers c. ½-exserted*, deep purple, awnless; the cells not separated. Fl. 3–5. N.

Native. Bogs, especially when relatively well-drained as at the edges of lakes and streams. W. Galway and W. Mayo, locally plentiful. H2. W. France (Gironde), N., W. and S. Spain, Portugal.

6. E. vagans L.   Cornish Heath.

Glabrous diffuse shrub 30–80 cm., with numerous ascending stems with erect branches, without short axillary shoots. Lvs 4–5 in a whorl, 7–10 mm., linear bright green, margins strongly revolute. Fls axillary forming a dense cylindrical lfy raceme 8–16 cm., often terminated by lvs. *Pedicels* 3–4 *times as long as fls*, bracteolate about or below the middle. Calyx-lobes triangular-ovate. *Corolla* 3–4 mm., *widely campanulate*, pale lilac; lobes deltoid, erect. *Anthers fully exserted*, deep purple, awnless; the cells separate except at the base. Fl. 7–8. $2n=24$. N.

Native. Heaths round the Lizard (Cornwall), where it is often dominant; probably also native in Fermanagh; commonly cultivated and an occasional escape elsewhere. 1, H1. W. and C. France, N. Spain.

The following hybrids occur: *E. ciliaris* × *tetralix* = *E.* × *watsonii* Benth. common with the parents; *E. mackaiana* × *tetralix* = *E.* × *stuartii* E. F. Linton, common with the parents; *E. tetralix* × *vagans* = *E.* × *williamsii* Druce, once found.

Subfamily 4. VACCINIOIDEAE. Ovary inferior. Corolla caducous after fl. Fr. a berry, the calyx-lobes persistent at the apex.

14. VACCINIUM L.

Shrubs. Lvs alternate, shortly petioled. Fls 4–5-merous, in racemes, or axillary and solitary, or clustered. Calyx-lobes short. Stamens 8 or 10. Anthers opening by pores. Nectar secreted by a swelling at the base of the style.

About 130 spp., arctic and north temperate regions and high mountains in the tropics. A number of spp. are sometimes cultivated.

1 Corolla campanulate or urceolate with small lobes; aerial stems neither creeping nor filiform.   2

  Corolla divided nearly to base, lobes reflexed; stems creeping and filiform.   4

2 Lvs persistent, dark green and glossy; fls in racemes; corolla campanulate.

                                                      1. vitis-idaea

  Lvs deciduous; fls 1–4, axillary; corolla urceolate.   *3*

3 Lvs serrulate, acute, bright green; twigs angled.   **2. myrtillus**

  Lvs entire, obtuse, blue-green; twigs terete.   **3. uliginosum**

4 Lvs ± ovate, acute, to 8 mm.; infl. terminal.   *5*

  Lvs oblong, obtuse, 6–18 mm.; infl. rhachis terminated by a lfy shoot.

                                                 6. macrocarpon

5 Pedicels puberulous; lvs ± oblong-ovate, equally wide for some distance
    at the base.   **4. oxycoccos**

  Pedicels glabrous; lvs ± triangular-ovate, widest near base.

                                             5. microcarpum

Subgenus 1. *Vaccinium*. Stems ± erect. Corolla lobes short. Ovary 4–5-celled.

Section 1. *Vitidaea* Dum. Fls in racemes. Corolla campanulate. Filaments pubescent; anthers awnless. Lvs persistent.

1. V. vitis-idaea L.   Cowberry, Red Whortleberry.

Evergreen shrub to 30 cm., with creeping rhizome and numerous ± erect, often arching, much-branched stems. Twigs terete, puberulent when young. *Lvs* 1–3 cm., obovate, obtuse or emarginate, *dark green and glossy above*, paler and *gland-dotted beneath*, glabrous, coriaceous, ± 2-ranked, margins somewhat revolute, obscurely crenulate. Fls c. 4 in short terminal drooping racemes. Calyx-lobes ovate-orbicular, reddish. *Corolla* c. 6 mm., *campanulate*, white, tinged with pink; lobes ± revolute, c. half as long as tube. *Fr.* 8–10 mm., *red*, globose, acid, edible. Fl. 6–8. Pollinated by bees but often selfed; homogamous. Fr. 8–10. $2n=24$. N. or Chw.

Native. Moors and woods on acid soils, occasionally locally dominant; ascending to over 3500 ft. Common in Scotland, extending south to Yorks, Leicester and through Wales to Somerset and Devon (perhaps extinct in S.W. England); Ireland except the S.W., rare in centre. 73, H20. N. Europe, and Asia from Iceland and the Faeroes to Japan and in the mountains of C. Europe to the Pyrenees, N. Apennines and Macedonia; a ssp. in N. America.

V. myrtillus × **vitis-idaea** = *V.* × *intermedium* Ruthe. Evergreen or nearly. Lvs elliptic, serrulate, with inconspicuous glands beneath, intermediate in colour. Fr. rare, purple.

  In some quantity with the parents on Cannock Chase (Staffs) and in Derby, rare elsewhere.

Section 2. *Vaccinium*. Fls 1–4, axillary. Corolla urceolate. Filaments glabrous; anthers awned. Lvs deciduous.

2. V. myrtillus L.   Bilberry, Blaeberry, Whortleberry, Huckleberry.

Glabrous *deciduous* shrub to 60 cm., with creeping rhizome and numerous erect stems and branches. *Twigs angled, green. Lvs* 1–3 cm., ovate, *acute, serrulate, bright green*, conspicuously reticulately veined. Fls 1(–2) in each axil. Calyx limb scarcely more than sinuate. Corolla 4–6 mm., globose, greenish-pink; lobes very short, reflexed. *Fr.* c. 8 mm., *black with a glaucous bloom*, globose, sweet, edible. Fl. 4–6. Pollinated mainly by bees, more rarely selfed; weakly protandrous. Fr. 7–9. 2*n* = 24. N. or Chw.

Native. Heaths, moors and woods on acid soils, more tolerant of exposure and of shade than *Calluna* and becoming dominant in a zone higher on the mountains and in the field layer in more shaded woods; ascending to over 4000 ft. Common throughout most of the British Is. but becoming local in England towards the south-east and absent from several counties in the east and E. Midlands. 102, H40. Europe from Iceland to the mountains of N. Portugal, C. Spain, Corsica, Italy, Macedonia, and the Caucasus; N. Asia.

3. V. uliginosum L.   'Bog Whortleberry.'

Deciduous shrub to 50 cm., with creeping rhizome and numerous bushy stems with ± spreading branches. *Twigs terete, brownish*, glabrous or puberulent. Lvs 1–2·5 cm., obovate or oval, *entire, obtuse, blue-green*, conspicuously reticulately veined, glabrous to somewhat pubescent. Fls 1–4 in each axil. Calyx-lobes short, broad, obtuse. Corolla c. 4 mm., subglobose or ovoid, pale pink, lobes very short, reflexed. *Fr.* c. 6 mm., *black with a glaucous bloom*, globose, sweet, edible in small quantity. Fl. 5–6. Pollinated by bees; weakly protandrous. Fr. 8–9. 2*n* = 48. N. or Chw.

Native. Bilberry moors, ascending to 3500 ft. From Durham and Cumberland to Shetland (absent from Hebrides), local but sometimes abundant. 23. N. Europe and Asia from Iceland to Japan and in the high mountains to the Sierra Nevada, N. Apennines, Albania and Bulgaria, Caucasus and Altai; Greenland; N. America, south to the mountains of New York.

Subgenus 2. *Oxycoccus* (Hill) A. Gray. Evergreen prostrate undershrubs with filiform stems. Fls 4-merous, on long slender pedicels. Corolla lobed nearly to base, lobes reflexed. Ovary 4-celled.

4. V. oxycoccos L.   Cranberry.

Oxycoccus palustris Pers.; *O. quadripetalus* Gilib.

Stems prostrate, rooting, filiform, usually widely separated from each other. Lvs 4–8 mm., distant, *oblong-ovate, equally broad for some distance from the base or broader towards the middle, acute*, dark green above, glaucous beneath, *strongly revolute. Fls 1–4, in a terminal raceme. Pedicels* 1·5–3 cm., *puberulous*, 2-bracteolate about or below the middle; bracteoles pubescent. Calyx-lobes ciliate. Corolla pink; lobes 5–6 mm. Filaments pubescent on the edges, glabrous or slightly pubescent outside. Fr. 6–8 mm., globose or pyriform, red- or brown-spotted, edible. Fl. 6–8. Fr. 8–10. 2*n* = 48. Ch.

Native. Bogs, more rarely wet heaths, local, but occurring through most of the British Is., absent from Channel Is., Cornwall, Pembroke, Isle of Man, Orkney, Shetland and Outer Hebrides and from a number of other counties especially in the Midlands. 76, H34. Europe from Scandinavia (70° 20′ N.) (not Iceland), to C. France, N. Italy and Roumania; N. Asia to Japan; Greenland; N. America to N. Carolina, Wisconsin and British Columbia.

5. V. microcarpum (Rupr.) Hook. f. 'Small Cranberry.'

Oxycoccus microcarpus Turcz. ex Rupr.

Differs from *V. oxycoccos* as follows: *Lvs* 3–5 mm., *triangular-ovate*. Fls 1–2. *Pedicels* and usually bracteoles and calyx *glabrous*. Corolla deeper pink. Filaments often pubescent outside. Fr. lemon-shaped or pyriform. Fl. 7. $2n=24$. Ch.

Native. Bogs in the Scottish Highlands from Perth to Inverness; ascending to 2200 ft. 6. Iceland and Scandinavia (71° 04′ N.), south to N. Germany, east to Sakhalin; Alps and Carpathians; Rocky Mountains to British Columbia and Alberta (distribution imperfectly known); absent from Greenland; extending further north in general than *V. oxycoccos.*

*6. V. macrocarpon Ait. 'Large Cranberry.'

Oxycoccus macrocarpos (Ait.) Pursh

Relatively robust. Stems prostrate, rooting, ascending at ends, slender but scarcely filiform. *Lvs* 6–18 *mm., oblong or elliptic-oblong, obtuse,* slightly glaucous beneath, *flat or slightly revolute.* Fls 1–10 in a raceme, *the rhachis continuing as a lfy shoot.* Pedicels 2-bracteolate near apex. Corolla pink; lobes 6–10 mm. Filaments much shorter than anthers. Fr. 10–20 mm., red, edible. Fl. 6–8. Fr. 9–11. $2n=24$. Ch.

Introduced. Sometimes grown for its fr. and naturalized in a few places. Native of N. America from Newfoundland to Saskatchewan, N. Carolina and Minnesota.

91. PYROLACEAE

Evergreen perennial herbs with creeping rhizome. Fls hermaphrodite, regular, 5-merous. Petals free. Stamens obdiplostemonous, free, anthers opening by pores, pollen usually in tetrads. Ovary incompletely 5-celled, with thick fleshy axile placentae with numerous small anatropous ovules. Style simple; stigma capitate. Fr. a loculicidal capsule. Seeds very small, numerous; endosperm copious; embryo undifferentiated.

Four genera and about 35 spp., north temperate and arctic regions. A small family distinguished from Ericaceae, with which it is often united, by the herbaceous habit, incompletely septate ovary and undifferentiated embryo.

All the spp. are partial saprophytes and tend to be associated with raw humus and are especially characteristic of pine woods.

1	Fls in terminal racemes.	*2*
	Fls solitary.	3. MONESES
2	Fls secund, greenish-white; lvs acute, petiole c. 1 cm.	2. ORTHILIA
	Fls not secund, pinkish or pure white; lvs obtuse, rarely subacute, petiole 2 cm. or more.	1. PYROLA

1. PYROLA L. Wintergreen.

Herbs with a slender creeping rhizome and short, often distant, aerial stems frequently reduced to a basal rosette of lvs. *Lvs alternate. Fls in racemes, not secund,* on scapes usually bearing a few scales. *Disk* 0. *Anthers with very short tubes bearing the pores. Pollen in tetrads. Valves of the capsule webbed at the edges.* Nectar secreted by base of petals.

About 25 spp., north temperate and arctic regions.

1	Style straight; fls ± globose in outline.	*2*
	Style strongly curved; corolla nearly flat.	**3. rotundifolia**
2	Style 1–2 mm., not thickened below the stigma.	**1. minor**
	Style c. 5 mm., thickened into a ring below the stigma.	**2. media**

1. P. minor L. 'Common Wintergreen.'

Stem very short or lvs all radical. Lvs 2·5–4 cm., ovate or oval, obtuse or subacute, crenulate, light green, petioles 2·5–3 cm., shorter than blade. Scape 10–30 cm. Racemes rather dense. *Fls* c. 6 mm., *± globose in outline.* Calyx-lobes deltoid, acute. *Petals pinkish. Style* 1–2 mm., *straight, included, shorter than stamens and ovary, without a ring below the stigma; stigma with 5 large spreading lobes.* Fl. 6–8. Pollinated by insects or selfed; homogamous. $2n=46$. Hr. or Chh.

Native. Woods, moors, damp rock-ledges and dunes; ascending to 3750 ft. in Perth. Rather local in Scotland, becoming more local and more confined to woods southwards but extending over most of England; absent from Channel Is., Devon and Cornwall, Isle of Man, Outer Hebrides, Orkney and Shetland; in Wales only in Glamorgan and Caernarvon; scattered over Ireland but very local (mainly north). 82, H12. N. Europe and Asia from Iceland to N. Japan, south to the mountains of C. Spain, C. Italy and Thessaly; N. America from Labrador to Alaska, New England and California.

2. P. media Sw. 'Intermediate Wintergreen.'

Lvs all radical, 3–5 cm., orbicular or ovate-orbicular, obtuse, obscurely crenulate, dark green; petiole 2·5–5·5 cm., about equalling blade or longer. Scape 15–30 cm. Racemes rather lax. *Fls* c. 10 mm., *± globose in outline.* Calyx-lobes triangular-ovate. *Petals white, tinged pink. Style* c. 5 mm., *straight, exserted, longer than stamens and ovary, expanded into a ring below the stigma which has 5, moderate, erect lobes.* Fl. 6–8. $2n=92$. Hr.

Native. Woods and moors; ascending to 1800 ft. Sussex; S. Wilts; Worcester and Shropshire to S. Lancs; Westmorland and Yorks to Shetland (not Orkney or Outer Hebrides); N. and W. Ireland to Down and Clare; very local, commonest in the Scottish pinewoods. 48, H11. Europe from Scandinavia (70° 41′ N.) to the mountains of E. France, N. Apennines, Serbia, Bulgaria and the Caucasus; Asia Minor.

3. P. rotundifolia L. 'Larger Wintergreen.'

Lvs all radical, 2·5–5 cm., orbicular or oval, obscurely crenulate, dark green and glossy; petiole 3–7 cm., longer than blade. Scape 10–40 cm. Racemes lax. Fls c. 12 mm. diam., ± flat. *Petals pure white. Style decurved then curving back, longer than the petals, stamens and ovary,* expanded into a ring below the stigma; stigma with five small erect lobes. Fl. 7–9. Pollinated by various insects or selfed; homogamous. $2n=46$. Hr.

Ssp. **rotundifolia.** Scales on scape 1–2. Pedicels 4–6 mm. Calyx-lobes usually triangular-lanceolate, acute, 2–3 times as long as broad. Style 7–8 mm.

Native. Bogs, fens, damp rock-ledges and woods; ascending to 2500 ft. Sussex and Worcester to Orkney, very local and with a distinct eastern tendency (absent from Wales except Glamorgan, N.W. England and the Hebrides and very rare in W. Scotland); Westmeath; Channel Is. 33, H1, S. Europe from Iceland to the mountains of C. Spain, N. Apennines, Serbia and Bulgaria; N. Asia to the R. Lena and the Altai Mountains; Asia Minor.

Ssp. **maritima** (Kenyon) E. F. Warb. Scales on scape 2–4. Pedicels 2–5 mm. Calyx-lobes ovate, subobtuse, scarcely twice as long as broad. Style 5–7 mm., usually more thickened at apex and more curved than in ssp. *rotundifolia*.
Native. Dune-slacks in Flint and Lancashire. 2. Atlantic coast from W. Germany to N.W. France.

2. ORTHILIA Raf.
(*Ramischia* Garcke)

Differs from *Pyrola* as follows: *Racemes secund. Disk consisting of 10 small glands. Anthers without tubes. Pollen grains free.*
Three or four spp., north temperate and arctic regions.

1. O. secunda (L.) House   'Serrated Wintergreen.'
Pyrola secunda L., *Ramischia secunda* (L.) Garcke
Stems short, 2–10 cm. *Lvs* 2–4 cm., *ovate, acute*, serrulate, light green, petioles c. 1 cm., shorter than blade. Scape 5–12 cm., with 1–5 scales. *Racemes secund*, dense. Fls c. 5 mm. Calyx-lobes ovate, obtuse. *Petals greenish-white*, ± erect. Style c. 5 mm., exserted, without a ring, but somewhat thickened below the stigma, which has five rather small spreading lobes. Fl. 7–8. Pollinated by insects or selfed. $2n = 38$. Chh?
Native. Woods and damp rock-ledges, ascending to 2400 ft. Sutherland and Skye to Yorks and Westmorland, local; Monmouth, Glamorgan, Brecon; Fermanagh, Antrim, Derry. 33, H3. N. Europe and Asia from Iceland to Japan and in the mountains south to the Pyrenees, Sicily and Thessaly; N. America from Newfoundland to Alaska, New Jersey and California.

3. MONESES Salisb.

Differs from *Pyrola* as follows: *Lvs opposite. Fls solitary. Disk obvious, 10-lobed. Anthers with relatively long tubes. Valves of capsule not webbed.* No nectar.
Two spp., north temperate regions.

1. M. uniflora (L.) A. Gray   'One-flowered Wintergreen.'
Pyrola uniflora L.
Stem 1–5 cm. Lvs 1–2·5 cm., orbicular, serrate, light green, decurrent down the petiole which is shorter than blade. Scape 5–15 cm., usually with 1 scale and a bract below the fl. Fls wide open, c. 15 mm. diam. Petals white. Pollen in tetrads. Style without a ring below the stigma; stigma with five large spreading lobes. Fl. 6–8. $2n = 26$. Pollinated by insects and ?selfed. H. or Chh.

Native. Pinewoods, etc.; E. Scotland from Perth and Kincardine to Orkney, only Argyll in the west (extinct in Dumfries), very local and rare. 11. N. Europe and Asia from Iceland to N. Japan, south to the mountains of N.E. Spain, Corsica, Italy and Macedonia; N. America from Newfoundland to Alaska, Pennsylvania and New Mexico.

92. MONOTROPACEAE

Differ from Pyrolaceae in being saprophytic herbs without chlorophyll, anthers opening by longitudinal slits and the free pollen grains. They are occasionally (but not in our spp.) gamopetalous and with the ovary 1-celled with parietal placentae.

Nine genera and 16 spp. in north temperate zone, mainly N. America extending to the mountains of Colombia and Malaya.

1. MONOTROPA L.

Fls in short racemes, 4–5-merous. Sepals free, large, oblong-spathulate. Petals free, saccate at base. Disk of 8–10 glands. Style columnar, stigma ± lobed.

About 3 spp., in north temperate zone to the mountains of Colombia.

Filaments, style and inside of petals with rather stiff hairs; petals 9–12 mm.
                                                                                        1. hypopitys
Filaments, style and inside of petals glabrous; petals 8–10 mm.   **2. hypophegea**

M. hypopitys agg.   Yellow Bird's-nest.

Whole plant uniformly yellowish or ivory-white, of waxy appearance, with simple stems 8–30 cm. Lvs 5–10 mm., scale-like, ovate-oblong, entire, ± erect, numerous, especially at the base of the stem. Infl. drooping in fl., erect in fr. Fls 10–15 mm., on short pedicels, 4–5-merous. Fl. 6–8. Pollinated by insects; homogamous. Grh.

Native. Woods, especially beech and pine, and on dunes among *Salix repens*, etc.; widespread but rather local in England from Somerset and Kent to Westmorland and Yorks; in Wales only in Glamorgan, Caernarvon, Anglesey and Flint; Aberdeen, E. Inverness, Moray; in Ireland local, mainly in the west but extending to Kildare, Dublin and Derry. Europe from Scandinavia to Portugal (very rare) and C. Spain, Italy and Greece; temperate Asia to Japan, the Himalaya and N. Syria; N. America (to Mexico).

The distribution and ecology of the two following are little known in this country and further observations of these and of the differential characters are needed. Both have a wide European range.

1. M. hypopitys L.

Hypopitys multiflora Scop.

Infl. dense, up to 11-fld. Upper part of stem, sepals and outside of petals pubescent or glabrous. Petals 9–12(–13) mm., somewhat spreading at their apices. *Inside of petals, filaments, style* and sometimes ovary *covered with rather stiff hairs.* Ovary and capsule of lateral fls often longer than broad. $2n = 48$.

Apparently less common than *M. hypophegea.*

2. M. hypophegea Wallr.

M. Hypopitys var. *glabra* Roth; *Hypopitys glabra* (Roth) DC.

Infl. looser and usually fewer (up to 6)-fld than in *M. hypopitys.* Upper part of stem, sepals and outside of petals glabrous. Petals 8–10 mm., straight. *Inside of petals, filaments, ovary and style glabrous.* Ovary and capsule subglobose. $2n = 16$.

93. EMPETRACEAE

Small evergreen heath-like shrubs. Lvs alternate, entire, margins strongly revolute, exstipulate. Fls axillary or in terminal heads, small, usually dioecious, sometimes monoecious or hermaphrodite, regular, hypogynous, 2–3-merous. Per. segs 4–6, in two ± similar whorls. Stamens half as many. Ovary 2–9-celled; ovules 1 in each cell, basal, anatropous; integument 1; raphe ventral; style short; stigmas as many as carpels. Fr. a drupe with 2–9 stones; endosperm copious, fleshy; embryo straight, cotyledons small.

Three genera and about 9 spp., north temperate and arctic regions, southern S. America and Tristan da Cunha.

1. EMPETRUM L.

Low shrubs. *Fls* 1–3, *axillary*, with scale-like bracts below. *Per. segs* 6. *Stamens* 3; anthers introrse. *Carpels* 6–9. Stigmas toothed. Fr. juicy.

About 10 closely allied spp., distribution of the family but absent from the warmer parts of the north temperate zone.

Fls dioecious; young stems reddish; margins of the lvs almost exactly parallel,
lvs 3–4 times as long as broad. **1. nigrum**
Fls hermaphrodite; young stems green; margins of lvs somewhat rounded;
lvs 2–3 times as long as broad. **2. hermaphroditum**

E. nigrum agg. Crowberry.

Low shrub 15–45 cm., with numerous procumbent and ascending stems. Lvs dense, 4–6 mm., oblong or linear-oblong, obtuse, shortly stalked, glandular on the margin when young, otherwise glabrous. Fls c. 1–2 mm. diam., pinkish or purplish. Fr. c. 5 mm., black, subglobose. Fl. 5–6.

1. E. nigrum L.

Stems relatively long and slender, prostrate and rooting round the edge of the tuft. *Young twigs reddish*, becoming red-brown. *Lvs* oblong or oblong-linear, *parallel-sided*, c. 3–4 *times as long as broad. Fls dioecious* (very rarely hermaphrodite). Wind-pollinated. $2n = 26*$. Chw. or N.

Native. Moors, mountain-tops and the drier parts of blanket bogs, extending to at least 2500 ft. (upper limit uncertain because of confusion with *E. hermaphroditum*), often abundant and sometimes locally subdominant.

Common in Scotland, N. England and Wales, rare in S.W. England (absent from Cornwall) and absent south-east of Dorset, Monmouth, Stafford, Leicester (extinct in Sussex and N. Lincoln); throughout Ireland but absent from a few counties. 77, H34. Europe from Iceland and Scandinavia (63° N.) to the Pyrenees, C. Apennines, Montenegro and Bulgaria; W. Siberia (sspp. or allied spp. in E. Asia); N. America.

2. E. hermaphroditum Hagerup

Stems not prostrate nor rooting round the edge of the tuft, so that the tufts are more rounded and taller than in *E. nigrum*. Internodes shorter. *Young twigs green, becoming brown. Lvs* oblong or oval-oblong, *margins somewhat rounded, c. 2–3 times as long as broad. Fls hermaphrodite* (the stamens on some of them often persist round the fr. for some time). $2n = 52$. Chw. or N.

Native. Mountain tops and moors mostly at high altitudes, but at sea level in Shetland, dominant in the Cairngorms and elsewhere in a zone at varying heights from 2000 to 3500 ft. Caernarvon; Lake District; Dumfries; Scottish Highlands from Perth to Shetland (distribution incompletely known). 9. N. Europe from Iceland and the Faeroes eastwards, Alps of France and Switzerland, Urals; arctic Siberia; W. Greenland (to 79° N.); Canada; (incomplete).

94. DIAPENSIACEAE

Perennial herbs or undershrubs. Fls hermaphrodite, regular. Calyx deeply 5-lobed. Disk 0. Corolla gamopetalous, deeply 5-lobed. Fertile stamens 5, alternating with the corolla-lobes, inserted on the corolla; staminodes 5 or 0. Ovary superior, 3-celled, deeply 3-lobed; ovules anatropous, on axile placentas; style simple with a 3-lobed stigma or divided above into 3. Fr. a loculicidal capsule. Seeds small; endosperm copious, fleshy; embryo small, cylindric.

Six genera and about 10 spp., north temperate and arctic regions.

1. DIAPENSIA L.

Cushion-like evergreen *undershrubs* with dense rosettes of leaves. Fls solitary ± stalked. Calyx leathery. *Anthers broadly ovate, diverging, opening length-wise.* Style slender, stigma small, 3-lobed.

Four spp., the following and 3 in the Himalaya and W. China.

1. D. lapponica L.

Cushion-like undershrub up to c. 5 cm. Lvs 5–10 mm., obovate-spathulate, entire, obtuse, tapered to the base, leathery, densely crowded. Peduncles 1–3 cm., usually with 1 bract near the middle and 2 bracteoles close below the fl. Sepals c. 5 mm. Petals up to 1 cm., white, obovate. Filaments broad and flat. Staminodes 0. Fl. 5–6. Protogynous. Visited by flies. $2n = 12$. Chw.

Native. On an exposed mountain at about 2500 ft. in W. Inverness, where it was first found in 1951. 1. Arctic regions and extending south to c. 60° in Norway to the Urals and to New York; replaced by a ssp. in N.E. Asia (S. to Japan) and N.W. America.

95. PLUMBAGINACEAE

Perennial or rarely annual herbs or shrubs with simple exstipulate usually spirally arranged lvs often confined to a basal rosette. Infl. usually cymose, the cymes sometimes closely aggregated into heads. Bracts scarious. Fls hermaphrodite, actinomorphic, hypogynous, 5-merous. Calyx tubular below, scarious and often pleated above, persistent. Petals free, slightly joined at the base, or with a long basal tube. Stamens 5, free or epipetalous, opposite the petals. Ovary superior, 1-celled, with 1 basal anatropous ovule. Styles 5, or 1 with 5 stigma lobes, opposite the sepals. Fr. dry with a thin papery wall, opening with a lid or irregularly, or remaining closed. Seed with a straight embryo in mealy endosperm.

About 10 genera and 260 spp., cosmopolitan. Chiefly plants of seashores and salt-steppes, but there are some arctic and alpine species. A very natural family distinguishable from the Primulaceae by the 5 styles or stigma-lobes and the single basal ovule, and usually recognizable by the scarious and persistent brightly coloured calyx. The British members belong to the tribe Staticeae, with complex infls, a spreading coloured scarious calyx, epipetalous stamens and styles free almost to the base. The tribe Plumbagineae includes the cultivated genera *Plumbago* and *Ceratostigma* and differs in the simple infl., only slightly scarious calyx, longer corolla-tube, hypogynous stamens, and styles free only in the upper part.

Fls in terminal panicles; styles glabrous throughout.   1. LIMONIUM
Fls in dense ± globular heads; styles hairy below.   2. ARMERIA

1. LIMONIUM Mill.

Perennial herbs, rarely annuals, with woody stocks and with lvs confined to a basal rosette. Fls shortly stalked in 1–5-fld cymose spikelets, each with three scale-like bracts, the *spikelets further aggregated into spikes of varying length and compactness which terminate the branches of the infl.* Calyx funnel-shaped, 5–10-ribbed at the base, expanding above into a scarious, usually coloured, 5-lobed limb, with or without smaller teeth between the lobes. Corolla with a very short tube. Stamens inserted at the base of the corolla. *Styles glabrous,* free quite or nearly to the base. Fr. opening transversely above or irregularly below. The fls secrete nectar and are visited by various insects. Some species are heterostylous and many have dimorphic pollen and stigmas. In such dimorphic species, which include *L. vulgare* and *L. bellidifolium,* plants with 'cob' stigmas have pollen of type A, while those with 'papillate' stigmas have type B pollen (Fig. 50). These species are self-incompatible, effective pollination being of cob stigmas by B pollen or of papillate stigmas by A pollen. Other species are monomorphic and self-compatible, all plants being A/papillate (e.g. *L. humile*), or, more rarely, all B/cob. There are finally some species which are monomorphic and apomictic. These have cob stigmas with either A pollen or empty anthers (*L. binervosum* agg.), or papillate stigmas with B pollen (*L. dodartii,* a non-British species related to *L. binervosum*), or there may be some populations with the one and some with the other combination (*L. auriculae-ursifolium*). These features are observable only with the microscope, but they are often valuable in confirming identification.

About 120 maritime and salt-steppe spp. all over the world, but especially numerous in W. Asia.

Several non-British species are grown in gardens, and their infls are sold for house decoration in winter on account of the persistent coloured calyx ('Statice').

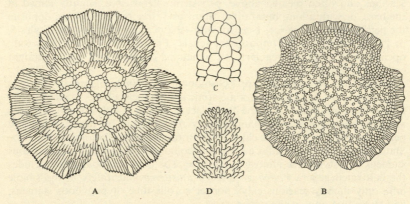

Fig. 50. Pollen and stigmas of dimorphic *Limonium*. A, B, the two pollen types; C, 'cob' stigma; D, 'papillate' stigma. A and B × c. 800, C and D × c. 200. (After H. G. Baker.)

1 Lvs pinnately veined; calyx-lobes with intermediate teeth.   2
   Lvs not pinnately veined; calyx-lobes without intermediate teeth.   3

2 Stem usually not branched below the middle; infl. ±corymbose, the
   spikelets crowded into short spreading spikes; outer bract rounded on
   the back; dimorphic (A/cob and B/papillate).   **1. vulgare**
   Stem usually branched below the middle; infl. not corymbose, the
   spikelets distant in long ±erect spikes; outer bract usually keeled;
   monomorphic (A/papillate).   **2. humile**

3 Infl. with very numerous slender zig-zag barren branches below; outermost bract entirely scarious.   **3. bellidifolium**
   Infl. with few or no barren branches below; outermost bract scarious only
   at the margin.   4

4 Lvs broadly obovate-spathulate, 5- or more veined, narrowed below into
   a 5–9-veined stalk.   **4. auriculae-ursifolium**
   Lvs narrowly obovate- or lanceolate-spathulate or oblong-linear, with a
   1–3-veined stalk (**binervosum** agg.).   5

5 Spikes sub-capitate, each of only 2 (1–3) spikelets of which the upper has
   2 or fewer bracts.   **8. paradoxum**
   Spikes elongated, of numerous spikelets each with 3 bracts; spikelets
   arranged in 2 rows on the upperside of the spike.   6

6 Spikes straight, usually ±erect; spikelets distant or at least far enough
   apart for their outer and intermediate bracts not to overlap those of the
   next spike in the same row.   **5. binervosum**

Spikes straight or recurved, spreading; spikelets so closely set that their outer and intermediate bracts may overlap those of the next spike in the same row.   7

7   Lvs narrowly obovate-spathulate with a winged 3-veined stalk; petals contiguous or imbricating.   **6. recurvum**
    Lvs oblanceolate or linear-oblong, narrowed gradually into a 1-veined stalk-like base; petals narrow, distant.   **7. transwallianum**

1. L. vulgare Mill.   Sea Lavender.
Statice Limonium L.

A perennial herb with deep tap-root and branched stout woody stock. *Lvs 4–12(–25) cm. strongly pinnate-veined,* variable in shape from broadly elliptic to oblong-lanceolate, acute or obtuse, usually mucronate, narrowing gradually into a long slender stalk. *Flowering stems* 8–30(–40) cm., erect, somewhat angular, *corymbosely branched usually well above the middle;* lowest branches sometimes barren; *spikes short, dense, spreading, ± recurved; spikelets closely set* in 2 rows on the upper side of the spike; length from base of spikelet to apex of innermost bract 3·5–5 mm.; *outer bract rounded on the back.* Calyx-teeth acute, entire or jagged, with small intermediate teeth. Corolla 8 mm. diam., blue-purple, the petals broad and rounded. Anthers yellow. Fl. 7–10. Protandrous. Visited for nectar by various bees, flies and beetles. Hetero-stylous, *dimorphic* and self-incompatible, A/cob plants having styles exceeding the stamens, and B/papillate plants styles equalling or falling short of the stamens. $2n = 36*$. Hr.

Native. Muddy salt-marshes; often an abundant or dominant sp. in inter-mediate zones. 42, S. Great Britain northwards to Fife and Dumfries. W. and S. Europe; N. Africa; N. America.

Plants with the flowering stem branching near or below the middle and with the infl. lax and less corymbose than in the type have been placed in f. *pyramidale* C. E. Salmon. There appear, however, to be all intermediates between this and the type, and the variation is probably determined environmentally, not through genetic differences; f. *pyramidale* is characteristic of the higher and drier zones of salt-marshes.

2. L. humile Mill.   'Lax-flowered Sea Lavender.'
Statice bahusiensis Fr.; *rariflora* Drejer

A perennial herb resembling *L. vulgare* but with oblong-lanceolate and obscurely-veined lvs. *Flowering stems branched from below the middle; infl. not corymbose; spikes long, lax, erect or somewhat incurved; spikelets distant;* length from base of spikelet to apex of innermost bract 5–7·5 mm.; *outer bract usually keeled.* Calyx-teeth acute, denticulate, with small intermediate teeth. Anthers short, reddish. Fl. 7–8. *Monomorphic* (A/papillate, with styles shorter than stamens) and self-incompatible. $2n = 36$. Hr.

Native. Muddy salt-marshes. 27, H23. Great Britain northwards to Dumfries and Northumberland; all round the Irish coast, but rare in the north. W. Europe from Brittany to Norway and Sweden.

Fertile hybrids between *L. humile* and *L. vulgare* (*L.* × *neumanii* C. E. Salmon), with intermediate characters, have been found where both parents are present.

3. L. bellidifolium (Gouan) Dum.   'Matted Sea Lavender.'

Statice reticulata auct. angl., non L.; *S. caspia* Willd.

A perennial herb with deep tap-root and much-branched woody stock. Lvs 1·5–4 cm., few in each rosette, obovate- or lanceolate-spathulate, blunt, narrowed into a slender stalk, dying before flowering is over. *Flowering stems* 7–30 cm., *scabrid, decumbent,* spreading in a circle, much branched from near the base, with *numerous repeatedly forked barren branches below; fertile spikes* only on the uppermost branches, *dense, spreading, recurved*; spikelets closely set in 2 rows on the upper side of the spike; bracts with very broad membranous margin, the outermost almost entirely scarious. Calyx-teeth ovate, cuspidate, denticulate, with no intermediate teeth. Corolla small, 5 mm. diam., pale lilac. Fl. 7–8. Dimorphic; self-incompatible. $2n = 18$. Hr.

Native. Drier parts of sandy salt-marshes. 3. Coasts of Norfolk and Suffolk, and formerly also of Lincoln. Shores of Mediterranean, Black Sea, Caspian Sea, and in E. Asia.

4. L. auriculae-ursifolium (Pourr.) Druce

Statice lychnidifolia Girard

A *robust* plant with *large broadly obovate-spathulate* usually apiculate *lvs, glaucous and often viscid, 5–9-veined*, narrowed into a broadly winged *5–9-veined stalk.* Flowering stems 10–45 cm., stout, branched from near the base, rarely with barren branches below; spikes stout, dense, spreading or nearly horizontal, recurved; spikelets crowded; *inner bract* more than twice as long as the outer, with a membranous *bright red margin.* Calyx-teeth shallow, blunt. Corolla large, violet-blue. Fl. 6–9. Monomorphic (A/cob or B/papillate in different populations) and apomictic. $2n = 25^*$. Hr.

Native. Maritime rocks. Jersey and Alderney. France, Spain, Portugal, N. Africa.

5. L. binervosum (G. E. Sm.) C. E. Salmon   'Rock Sea Lavender.'

Statice binervosa G. E. Sm.; *S. auriculaefolia* auct.; *S. occidentalis* Lloyd?

A perennial herb with ascending branched woody stock and erect glabrous flowering stems. Lvs 2–12·5 cm., numerous, very variable in shape from obovate-spathulate to narrowly oblanceolate, *3-veined below,* acute or blunt, apiculate or not; narrowing below into a winged *obscurely 3-veined stalk* about the same length as the blade. *Flowering stems* 5–30(–50) cm., slender, wavy, *branched from near the base,* with a *few barren branches below; spikes* usually *slender, ± straight and erect; spikelets* in 2 rows on the upperside of the spike, usually *not closely set,* so that their *outermost and intermediate bracts never overlap those of the next spikelet in the same row; innermost bract* about twice as long as the outermost, with a membranous *pink-tinged margin.* Calyx-teeth blunt, entire. Corolla c. 8 mm. diam., violet-blue, the petals imbricating, obovate, emarginate. Fr. narrow with a smooth reddish seed. Fl. 7–9. Monomorphic, but some plants are male-sterile; some races may be apomictic. $2n = 35^*, 36^*$. Hr.

Native. Maritime cliffs, rocks and stabilized shingle. 28, H 12, S. Great Britain northwards to Wigtown and Lincoln. France, Spain and Portugal.

A very variable species with many named varieties of which the most distinct is var. *intermedium* (Syme) Druce, with obovate-spathulate lvs, *no barren branches, spikes short, dense, somewhat spreading*, with the *spikelets closely set* but not so closely that their bracts imbricate with those of the next in the same row.

Certain broad-leaved forms which show some failure of chromosome pairing may be segregates from a cross with *L. bellidifolium.*

The following closely related local types, all monomorphic (A/cob) or male-sterile and all probably apomictic, have been given specific rank because of striking morphological features:

6. L. recurvum C. E. Salmon

Lvs narrowly obovate-spathulate, blunt, narrowed into a broadly winged 3-*veined stalk. Flowering stem* 5–20 cm., asperous, rigid, *branched only above the middle; barren branches* 0; *spikes suberect or spreading, ± recurved*, rather *short and very dense*, terminating the short branches; *spikelets closely set* so that their *outermost and intermediate bracts overlap those of neighbouring spikelets in the same row. Calyx-teeth shallow*, blunt or truncate. Corolla 6 mm. diam.; petals narrow, contiguous, emarginate. $2n=27*$. Fl. 7–9.

Native. Apparently endemic to the British Is., occurring only at Portland, Dorset.

7. L. transwallianum (Pugsl.) Pugsl.

Lvs 2·5–4·5 cm., *narrowly oblanceolate or linear-oblong*, blunt, minutely mucronate, *narrowing gradually into a* 1-*veined stalk-like base. Flowering stems* 8–15(–30) cm., smooth, *branched from near the base* in large plants; *usually no barren branches; spikes ± spreading or recurved, short and very dense, usually crowded at the ends of branches; spikelets closely set* so that their *outer and intermediate bracts often overlap those of the next spikelet in the same row. Calyx-teeth deep*, triangular, obtuse or subacute. Corolla 4 mm. diam., violet-blue; petals distant, narrow, oblong, emarginate. $2n=35*$ (Pembroke), $27*$ (Clare). Fl. 7–9.

Native. Apparently endemic in the British Is., occurring only on maritime cliffs in Pembroke. A population at Poulsallagh, Co. Clare, consists of similar but more robust plants differing in chromosome number.

8. L. paradoxum Pugsl.

Lvs 3–4·5 cm., oblanceolate or oblong-spathulate, blunt, mucronate or muticous, narrowed below to a winged 1-*veined stalk. Flowering stems* 5–15(–20) cm., slightly asperous, usually *branched from near the base*, with *short erect branches which do not reach the apex of the stem; barren branches* 0; fls in *subglobose heads* each consisting of 2 (rarely 1 or 3) irregular spikelets, of which the lower has 3 bracts while the upper usually has 1, but may have 2 or none. Calyx-teeth deep, triangular, obtuse. Corolla 4 mm. diam., clear violet with a pink mid-vein to each of the imbricating petals. $2n=33*$ (St David's). Fl. 7–9. Probably male-sterile and apomictic.

Native. Apparently endemic to the British Is., occurring only on basic igneous rocks at St David's Head, and perhaps also near Malin Head, Donegal (Ireland).

2. ARMERIA Willd.

Perennial herbs with branched woody stocks and *basal rosettes of long very narrow entire lvs.* Fls stalked in a *solitary terminal hemispherical head* consisting of a close aggregate of bracteate cymose spikelets with a *scarious involucre*; top of flowering stem enclosed by a downward prolongation of the connate bases of the outer involucral bracts, so forming a *tubular sheath* round the growing zone *just beneath the head.* Calyx with a funnel-shaped 5–10-ribbed tube expanding upwards into a spreading scarious pleated limb; *petals free except at the extreme base, persistent*; stamens inserted at the base of the petals, their filaments broadened below; *styles* free to near the base, *hairy below.* Some spp. are dimorphic in respect of the size of the stigmatic papillae and the sculpturing of the pollen grains. Fr. enclosed in the persistent corolla, with 5 radiating ribs at the top, dehiscing transversely above or irregularly below.

About 60 spp. of maritime, arctic and alpine habitats in temperate Europe, Asia, N. Africa, N. America and Chile.

Lvs linear, 1–3-veined; calyx-teeth acute or shortly awned.   **1. maritima**
Lvs linear-lanceolate, 3–5-veined; calyx-teeth with awns half their length.
                                                                      2. arenaria

1. A. maritima (Mill.) Willd.   Thrift, Sea Pink.
Statice Armeria L.; *S. maritima* Mill.
Rootstock erect, stout, woody, branched. *Lvs* 2–15 cm., *linear*, acute or blunt, 1- (*rarely* 3-)*veined*, somewhat fleshy, punctate, glabrous, ciliate or pubescent. *Scapes* 5–30 cm., erect, lfless, usually *shortly pubescent*, rarely glabrous. Involucral sheath 8–14 mm.; *outermost involucral bracts* ± green on the back, mucronate or not, *shorter than the head*; innermost wholly scarious, blunt; heads 1·5–2·5 cm. diam. Fl.-stalks ± equalling the *calyx-tube* which has 5 *hairy ribs* and may or may not be hairy also between the ribs; *calyx-teeth acute or with awns less than half their overall length.* Corolla 8 mm. diam., rose-pink or white. Fr. exceeding the calyx-tube. Fl. 4–10. Fls fragrant, slightly protandrous or homogamous; visited for nectar and pollen by various insects. Dimorphic (A/cob and B/papillate: see under *Limonium*, Fig. 50) and self-incompatible. $2n = 18$. Hr.–Ch.

Salt-marsh forms seem ecotypically different from those on hard substrata, whether maritime or inland. Forms with broadly linear, often 3-veined, lvs (var. *planifolia* Syme) occur in pure populations or together with narrow-leaved forms on mountains in Scotland and the Hebrides and also, less frequently, in maritime localities.

It does not seem possible to uphold the distinction between *A. maritima* (Mill.) Willd., with calyx-tube hairy only on the ribs, and *A. pubescens* Link, with hairs also between the ribs, since both types may be found on the same plant, in the same head, or even on the same calyx-tube.

The above description is of ssp. **maritima**, which is the widely distributed native ssp. in coastal habitats and inland. Ssp. **elongata** (Hoffm.) Bonnier, a more robust plant with *quite glabrous scapes* up to 50 cm. and often with acuminate outer involucral bracts, occurs inland in a few localities in one small area and may perhaps be native there.

Native. Ssp. *maritima* in coastal salt-marshes and pastures, on maritime rocks and cliffs and also to 4200 ft. on mountains inland. In suitable localities throughout the British Is. 82. H27, S. Ssp. *elongata* only on sandy soil inland near Ancaster (Lincs). ?2. *A. maritima* sensu lato is widely distributed in the northern hemisphere. Ssp. *maritima* has a restricted W. European range from Iceland and N. Norway to N. Spain. Ssp. *elongata* is still more restricted, ranging from S. Denmark, N.W. Germany, Holland and Bavaria eastwards to S. Sweden, S. Finland, western C. Russia and Poland.

2. A. arenaria (Pers.) Schult.   'Jersey Thrift.'

A. plantaginea Willd.

A more robust and more rigid plant than *A. maritima*, more densely tufted, with broader, flat, *linear-lanceolate*, acuminate, 3–5(–7)-*veined lvs. Scapes* 20–60 cm., erect, *glabrous*, rough. Involucral sheath 20–40 mm.; *outermost involucral bracts acuminate*, wholly scarious or with a long herbaceous point, ± *equalling or exceeding the head*; innermost with broad scarious margins. Heads c. 2 cm. diam. Calyx-tube with hairy ribs; *calyx-teeth* with *very long awns about half their overall length*. Fl.-stalks shorter than the calyx-tube. Corolla c. 1 cm. diam., deep rose. Fl. 6–9. Visited for nectar by various insects. Dimorphic. H.–Ch.

Native. On stable dunes in Jersey. C. and S. Europe.

A hybrid between *A. maritima* and *A. arenaria* has been described by Syme. It resembles *A. arenaria* but has the lvs 1-veined or indistinctly 3-veined and ciliate, and its flowering stems are densely and shortly pubescent. It flowers earlier than *A. arenaria*.

*****A. pseudarmeria** (Murr.) Lawrence (*A. latifolia* Willd.), with limp broadly ob-lanceolate to oblong-lanceolate 5–7-veined lvs up to 18 mm. wide and stout scapes bearing heads up to 50 mm. across with bright pink fls (sometimes deep rose or white), is often grown in gardens and has become established locally. Portugal.

96. PRIMULACEAE

Perennial or sometimes annual herbs, rarely under-shrubs. Lvs exstipulate. Fls actinomorphic, bracteate, 5 (4–9)-merous, often heterostylous. Corolla present (0 in *Glaux*), rotate, campanulate, or funnel-shaped. Stamens inserted in the corolla-tube and opposite its lobes, sometimes alternating with sta-minodes. Ovary superior ($\frac{1}{2}$-inferior in *Samolus*), 1-celled, with a free-central placenta; style simple. Ovules numerous. Fr. a capsule dehiscing by valves or transversely. Seeds endospermic, embryo small, straight.

About 30 genera and 550 spp., cosmopolitan but mainly in the northern hemisphere.

1	Lvs all radical.		2
	Cauline lvs present.		3
2	Corolla-lobes incurved or spreading; stock not a corm.	1. PRIMULA	
	Corolla-lobes strongly reflexed; stock a corm.	3. CYCLAMEN	
3	Water plant; lvs submerged, pinnate; fls lilac.	2. HOTTONIA	
	Land plants; lvs not pinnate; fls not lilac.		4

4   Fls yellow.   5
    Fls not yellow.   6

5   Fls not in axillary racemes.   4. LYSIMACHIA
    Fls in dense axillary racemes.   5. NAUMBURGIA

6   Fls white; stems erect.   7
    Stems prostrate to ascending; fls not white (except rarely in *Glaux*, and
        then apetalous).   8

7   Lvs mostly in 1 whorl; fls solitary or few; ovary superior.
                                                              6. TRIENTALIS
    Lvs not whorled; fls numerous in a lfy raceme; ovary ½-inferior.
                                                              9. SAMOLUS

8   Capsule dehiscing transversely; corolla present.   7. ANAGALLIS
    Capsule dehiscing by 5 valves; corolla 0; calyx petaloid.   8. GLAUX

1. PRIMULA L.

Perennial scapigerous herbs. Fls bracteolate, in umbels or whorls, rarely
apparently solitary, white, yellow, pink or purple. *Calyx 5-toothed.* Corolla
funnel- or salver-shaped, 5-lobed, lobes incurved or spreading. Stamens
included. Styles filiform; stigma capitate. *Capsule dehiscing by valves.* Fls
usually heterostylous and cross-pollinated by bees or Lepidoptera.

About 250 spp., mainly in temperate or mountainous districts of the
northern hemisphere, a few in temperate S. America.

1   Lvs mealy beneath; fls lilac or purple.   2
    Lvs not mealy beneath; fls yellow.   3

2   Lvs crenulate; fls lilac, heterostylous; corolla-lobes distant; calyx-teeth
    ± acute; capsule cylindrical, much exceeding calyx. (N. England and
        S. Scotland.)   **1. farinosa**
    Lvs not crenulate; fls purple, not heterostylous; corolla-lobes contiguous;
        calyx-teeth obtuse; capsule ovoid, slightly exceeding calyx. (N. Scotland
        and Orkneys.)   **2. scotica**

3   Scape distinct; pedicels finely pubescent; limb of corolla concave, rarely
        exceeding 20 mm. diam.   4
    Scape 0 or nearly so; pedicels with shaggy hairs; limb of corolla flat,
        usually more than 30 mm. diam.   **5. vulgaris**

4   Calyx uniformly pale green; corolla with folds in the throat; fr. ovoid,
        enclosed in the calyx.   **3. veris**
    Calyx with midribs conspicuously darker green than remainder; corolla
        without folds in the throat; fr. oblong-ovoid, exceeding calyx.   **4. elatior**

Section 1. *Aleuritia* Duby. Lvs mealy beneath, not wrinkled. Calyx-tube
terete. Fls lilac or purple in our spp.

1. P. farinosa L.   'Bird's-eye Primrose.'

Lvs 1–5 cm., obovate-spathulate, obtuse or subacute, *crenulate*, glabrous
above, with white or sulphur-coloured meal beneath. Scape up to 15 cm.,
mealy when young. *Pedicels* slender, c. 5 mm., *elongating in fr.* Fls c. 1 cm.
diam., *rosy lilac*, erect or spreading, *heterostylous. Calyx* 3–4 × 2 *mm.*, mealy,

teeth acute or subacute. Corolla-tube 5–6 *mm.*, throat yellow, contracted; lobes flat, *distant from each other*, cuneate, bifid, *segments linear-oblong*, obtuse. *Fr.* 5–7 mm., *cylindrical, much exceeding the calyx.* Fl. 5–6. $2n=18$*, 36. Hr.

Native. In damp grassy and peaty places on basic soils. 14. Derby, Lancashire and Yorks to Cumberland and Northumberland; Peebles, E. Lothian and Midlothian, locally abundant. Europe, except Greece and Ireland; N. Asia to the shores of the N. Pacific and south to the Altai and Tien Shan. ?Caucasus. Other records are incorrect.

2. P. scotica Hook.

Similar in general appearance to *P. farinosa. Lvs* obovate-spathulate, usually broader than those of *P. farinosa, quite entire.* Scape rarely exceeding 10 cm. *Pedicels* stouter and rather shorter, *not or scarcely elongating in fr. Fls purple, not heterostylous. Calyx* 4–6 × 3 *mm., teeth obtuse. Corolla-tube* 7–10 *mm.; lobes contiguous,* bifid, *segments broadly ovate or suborbicular. Fr.* 5–6 mm., *ovoid, slightly exceeding calyx.* Fl. 6–9. $2n=54$*. Hr.

Native. In damp pastures, locally abundant. 3. W. Sutherland, Caithness and Orkney. Endemic. The Norwegian plants referred to this sp. are usually *P. scandinavica* Bruun.

Section 2. *Primula.* Lvs not mealy beneath, pilose or pubescent, wrinkled. Calyx-tube 5-angled. Fls yellow and normally heterostylous in our spp.

3. P. veris L.   Cowslip, Paigle.

A ± *glandular-pubescent* perennial. Rhizome short, stout, ascending, covered with swollen persistent lf-bases and bearing many stout roots. *Lvs* 5–15(–20) cm., ovate-oblong, obtuse, crenulate or ± serrate, *finely pubescent* on both sides, abruptly contracted at base; petioles about as long as blade, winged. *Scape* 10–30 cm., *finely pubescent,* 1–30-fld. Pedicels 1 cm. or more. *Fls* 10–15 *mm. diam., deep yellow or buff* with *distinct orange spots* at base of lobes, nodding but *scarcely secund. Calyx* 12–15 × 6–8 *mm., finely pubescent, teeth* 2–3 *mm., ovate, obtuse and apiculate.* Corolla-tube c. 15 mm., mouth with folds; lobes strongly concave, notched. *Fr.* c. 10 mm., *ovoid, enclosed by the enlarged calyx;* pedicels erect in fr. Fl. 4–5. Fr. 8–9. $2n=22$*. Hr.

Native. In meadows and pastures on basic and especially calcareous soils, locally abundant. 101, H38, S. Generally distributed but rarer in the north and absent from a number of Scottish counties. Europe, except the extreme north; temperate Asia to the upper Amur.

P. veris L. × vulgaris Huds.   Common Oxlip.

Intermediate between the parents and variable owing to back-crossing. The commonest form (presumably the F_1 cross) differs from *P. veris* in having the lvs not abruptly contracted at base; larger, paler yellow fls with a less concave limb; and longer and more shaggy pubescence. From *P. vulgaris* it differs in having a distinct scape and shorter pedicels; smaller, deeper yellow fls with a more concave limb; and shorter less shaggy pubescence. It occurs, though not usually in great quantities, where the parents grow together.

4. P. elatior (L.) Hill   Oxlip, Paigle.

Rhizome similar to that of *P. veris*. *Lvs* 10–20 cm., similar in shape but usually longer than those of *P. veris*, irregularly serrate, *crisped-pubescent*, particularly beneath; petioles about as long as blade, winged. *Scape* 10–30 cm., *crisped-pubescent*, 1–20-fld. Pedicels c. 1 cm. *Fls 15–18 mm. diam., pale yellow with diffuse orange markings* in the throat, *secund*, somewhat nodding. *Calyx c.* 15 × 4–5 *mm., narrower at base, crisped-pubescent, teeth c.* 4 *mm., lanceolate, ± acuminate.* Corolla-tube c. 18 mm., *throat without folds*; lobes concave, shallowly notched. *Fr. oblong-ovoid, exceeding calyx*; pedicels erect in fr. Fl. 4–5. Fr. 7. $2n = 22^*$. Hr. The total absence of folds in the throat of the corolla, the secund infl., the lvs abruptly contracted into the petiole and the smaller, paler fls distinguish this sp. from *P. veris × vulgaris* which is sometimes confused with it.

Native. In woods on chalky boulder clay. 7. Essex, Herts, Suffolk, E. Norfolk, Cambridge, Bedford and Huntingdon, abundant in a small area from most of which *P. vulgaris* is absent. Europe from S. Sweden to the Alps and S. Russia. Other sspp. in S. Europe and W. Asia.

Hybrid swarms with *P. vulgaris* occur in woods where the two spp. grow together.

5. P. vulgaris Huds.   Primrose.
P. acaulis (L.) Hill

Rhizome similar to that of *P. veris*. *Lvs* 8–15(–20) cm., obovate-spathulate, obtuse, irregularly serrate, pubescent beneath, *glabrous above* except on the veins, *narrowed gradually at base*; petiole short or 0. *Scape* 0 *or very short. Pedicels* 5–10 *mm.*, with shaggy indumentum. *Fls c.* 30 *mm. diam.*, yellow (rarely pink), ascending. *Calyx* 15–17 × 4 *mm., nearly cylindrical*, with shaggy hairs, *teeth* 4–6 *mm., narrowly triangular, acuminate.* Corolla-tube c. 15 mm., *mouth contracted with thickened folds* and with greenish stellate markings; lobes flat, shallowly notched. *Fr.* ovoid, *shorter than calyx; pedicels recurved in fr.* Fl. 12–5. Seeds arillate, sticky when fresh. Fr. 3–8. $2n = 22^*$. Hr.

The roots are stated to be a strong and safe emetic. Cultivated forms with pink or white fls are sometimes found growing in hedges, and the pink form in woods in Wales, where it may be native.

Native. In woods and hedgebanks and, in the west, in open grassy places, common. Generally distributed throughout the British Is. but rarer in the north. 112, H40, S. Now much less common than formerly in some areas (e.g. the home counties and the Chilterns) owing to the depredations of 'flower-lovers'. Western Europe from C. Norway to S. Portugal, Italy, Balkans and Crimea; N. Africa and Asia Minor.

2. HOTTONIA L.

Floating herbs. Lvs submerged. Fls racemose, whorled, heterostylous, white or lilac. *Calyx 5-partite.* Corolla salver-shaped, throat thickened; lobes 5, fringed at base. Stamens 5, included. Style filiform, stigma capitate. *Capsule dehiscing by valves which adhere at the top.* Fls contain nectar and are pollinated by various insects.

Two spp. in Europe, Siberia, and N. America.

1. H. palustris L. Water Violet.

A pale green nearly glabrous perennial. Stems floating and rooting, nodes not inflated. Lvs up to c. 10 cm., apparently whorled, 1–2-pinnate, lobes linear. Scape up to 40 cm. or more, erect, subaerial. Fls 20–25 mm. diam., lilac with a yellow throat, 3–8 in a whorl, sometimes cleistogamous. Pedicels 1–2 cm., finely glandular-pubescent, ascending in fl., deflexed in fr.; bracteoles 5–10 mm., subulate. Calyx 5–10 mm., divided almost to base; teeth linear or linear-oblong, subacute, equalling corolla-tube. Fr. c. 5 mm., globose, 5-valved. Fl. 5–6. $2n=20$. Hyd.

Native. In ponds and ditches. 54, H2. Widely distributed in England and Wales but local and rather rare except in the east; Inverness; Down and Antrim, probably introduced. N. and C. Europe from S. Sweden, southwards to Italy and east to Siberia.

3. CYCLAMEN L.

Herbs with large *corms* from which the lvs arise. Fls nodding, solitary, on long erect lfless stems which are spirally coiled in fr. Corolla-tube short, throat thickened; *lobes* large, *reflexed*. Stamens inserted at base of corolla-tube, included; anthers cuspidate. Style short; stigma simple. Capsule dehiscing by valves which become reflexed.

About 15 spp. in S. Europe, N. Africa, and W. Asia, mainly in mountains. *C. persicum* Mill. is much cultivated in pots and varies greatly in colour, etc.

*1. C. hederifolium Ait. Sowbread.

C. europaeum auct.; *C. neapolitanum* Ten.

A perennial, glabrous except for the puberulent infl. Corm 2–10 cm. diam., rooting over the upper surface. Lvs 4–8 cm., ovate-cordate, ±strongly 5–9-angled, dark green with a whitish border above, often purplish beneath, appearing in autumn after the fls; petioles long. Fl. stalks 10–30 cm. Calyx-lobes ovate-lanceolate, equalling corolla-tube. Corolla pink or rarely white; tube 5-angled; lobes c. 2·5 cm. Fl. 8–9. $2n=34$.

Introduced. Naturalized in hedgebanks and woods in a number of localities in England and Wales, very rare. 11. Southern Europe.

4. LYSIMACHIA L.

Herbs with opposite or whorled, rarely alternate, quite entire, sometimes gland-dotted lvs. Fls usually 5-merous, axillary or in terminal panicles, solitary or in racemes, yellow (in our spp.). *Corolla rotate*, lobes spreading or connivent. Stamens included. *Staminodes usually* 0. *Capsule* subglobose, 5-*valved*, *many-seeded*. *Seeds* rugose, ± *margined*. The fls are visited mainly by bees, for the sake of their abundant pollen.

About 160 spp. in temperate and subtropical regions of both hemispheres.

1	Plant prostrate or procumbent.	2
	Plant erect.	3
2	Lvs acute; calyx-teeth subulate.	**1. nemorum**
	Lvs obtuse; calyx-teeth ovate.	**2. nummularia**

3 Plant bearing numerous elongate bulbils in axils of lvs; corolla yellow
      with dark dots or streaks; lvs narrow-lanceolate.   **6. terrestris**
      Plant without bulbils; corolla plain yellow; lvs lanceolate to ovate.   *4*

4 Corolla-lobes glandular-ciliate.   **5. punctata**
      Corolla-lobes not ciliate.   *5*

5 Lvs dotted with orange or black glands; pedicels 1 cm. or less; margins of
      calyx-teeth orange; corolla-lobes not glandular.   **3. vulgaris**
      Lvs not gland-dotted; pedicels 2–4 cm.; margins of calyx-teeth green;
      corolla-lobes densely glandular towards base.   **4. ciliata**

1. L. nemorum L.   'Yellow Pimpernel.'

A slender glabrous *procumbent* perennial up to 40 cm. *Lvs* 2–4 cm., *ovate,
acute*, rounded at base; petioles short. *Fls c.* 12 *mm. diam.*, axillary, solitary,
on *capillary pedicels* which equal or exceed the lvs. *Calyx-teeth* c. 5 mm.,
subulate. Corolla rotate, lobes spreading, not ciliate. Fr. c. 3 mm. diam.,
globose. Fl. 5–9. 2*n*=18. Chh.

Native. In woods and shady hedgebanks. 110, H40, S. Throughout almost
the whole of the British Is., though rare in the drier parts. W. and C. Europe
from Norway (62° 44′ N.) to Spain, eastwards to the C. Carpathians and the
Caucasus.

2. L. nummularia L.   Creeping Jenny.

A glabrous *creeping* perennial up to 60 cm. *Lvs* 1·5–3 cm., *broadly ovate or
suborbicular, obtuse*, rounded or almost truncate at base, *gland-dotted*;
petioles short. *Fls* 15–25 *mm. diam.*, axillary, solitary on rather *stout pedicels*,
usually shorter than lvs. *Calyx-teeth* 8–10 mm., *ovate* acuminate. *Corolla
subcampanulate, lobes gland-dotted, minutely fringed.* Fr. very rarely produced
in Britain. Fl. 6–8. 2*n*=36. Chh.

Native. In usually moist hedgebanks and grassy places, though this sp. is
less intolerant of drought and full sunlight than *L. nemorum.* 84, H25, S.
Scattered throughout the British Is., though rare and usually an escape from
cultivation in the north. From C. Sweden to S. Spain, Italy and N. Greece,
eastwards to the Caucasus.

3. L. vulgaris L.   'Yellow Loosestrife.'

An *erect* pubescent rhizomatous perennial 60–150 cm. *Lvs* 5–12 cm., opposite
or in whorls of 3–4, lanceolate to ovate-lanceolate, acute, subsessile, *dotted
with orange or black glands. Fls c.* 15 *mm. diam.*, in terminal panicles; *pedicels*
1 *cm. or less*, slender. *Calyx-teeth* triangular-lanceolate, ciliate, *margins
orange.* Corolla subcampanulate, lobes not ciliate. Filaments connate in
lower ⅓. Fr. globose. Fl. 7–8. 2*n*=28. Hel. or Hp.

Native. In fens and beside rivers and lakes, locally common. 91, H40.
Scattered throughout most of the British Is., becoming rarer in the north and
absent from N. Scotland. Europe and Asia except the extreme north and
south.

*4. L. ciliata L.

An almost glabrous *erect* rhizomatous perennial 30–100 cm. *Lvs* up to
c. 10 cm., ovate acuminate, rounded or subcordate at base, *finely ciliate,*

not gland-dotted; *petioles c. 2 cm., with a few long, ± crisped cartilaginous cilia.* Fls c. 25 mm. diam., axillary, *pedicels* 2–4 *cm.,* capillary. Calyx-teeth linear-lanceolate, acute, glabrous, margins green. Corolla rotate, *lobes* suborbicular, usually shortly cuspidate, *densely glandular towards the base.* Filaments free, alternating with small staminodes. Fl. 6–7. Hp.

Introduced. Naturalized near buildings in a few places in N. England and Scotland. N. America.

*5. L. punctata L.

A somewhat pubescent perennial 30–90 cm. *Lvs* up to 10 cm., opposite or whorled, ovate, subacute, *shortly petioled, margins ciliate.* Fls axillary, peduncles usually 0, pedicels 15–20 mm., shorter than the subtending lvs except at top of infl. *Calyx-teeth* 8–10 mm., oblong-lanceolate, acute or acuminate, *glandular-pubescent. Corolla* c. 35 mm. diam., *glandular-ciliate.* Fl. 7–10. $2n=30$. Hp.

Introduced. More or less naturalized in marshy fields and beside rivers in a few localities. From Austria to the Caucasus and Asia Minor.

*6. L. terrestris (L.) Britton, Stearns & Poggenb. (*L. stricta* Ait.)

An erect glabrous herb *usually bearing long bulbils in the axils of the narrow lanceolate lvs. Fls* (infrequently produced) *in a terminal raceme.* Calyx-teeth c. 2 mm., ovate, acute. Corolla c. 10 mm. diam., lobes lanceolate, yellow, streaked or dotted with purple. Hp.

Introduced. Well naturalized in damp places on the Lancashire shore of Windermere and perhaps elsewhere. N. America.

5. NAUMBURGIA Moench

Herb with many of the characters of *Lysimachia.* Fls in dense axillary racemes. *Stamens subexserted. Staminodes present,* small, alternating with the corolla lobes. *Seeds smooth, scarcely margined.*

One sp. in north temperate regions.

1. N. thyrsiflora (L.) Rchb.   'Tufted Loosestrife.'

Lysimachia thyrsiflora L.

An erect glabrous rhizomatous perennial, 30–60 cm. Lvs 5–10 cm., opposite, rarely whorled, oblong-lanceolate, obtuse, sessile, ½-amplexicaul, densely dotted with black glands; lower lvs small and ± scale-like. Infl. 5–10 cm., 2–3 in the axils of the lvs about the middle of the stem, many-fld, dense, bracteate. Fls c. 5 mm. diam., on slender pedicels 2–3 mm. Calyx-teeth narrow lanceolate, subacute. Corolla campanulate, lobes subobtuse, erect, with black glands. Stamens somewhat exserted. Fr. ovoid, gland-dotted, shorter than calyx. Fl. 6–7. $2n=$ c. 40. Hel.

Native. In wet marshes and shallow water by ditches and canals, rare. 18. Hants and Bucks and scattered throughout the northern part of the country from Lincoln to Angus. Temperate regions of Europe, Asia and N. America; in Europe from 69° 16′ N. to the Alps.

6. TRIENTALIS L.

Erect, unbranched, glabrous herbs with slender rhizomes. Lvs in one whorl of 5–6 at the top of the stem, with a few small alternate lvs below. *Fls* white, *solitary,* ebracteolate. Calyx 5–9-partite. Corolla rotate, 5–9-partite. Stamens 5–9. *Capsule* globose, 5-*valved, few-seeded.* Probably cross-pollinated by insects.

Two species in north temperate regions.

1. T. europaea L.   Chickweed Wintergreen.

A slender erect perennial 10–25 cm. Lvs 1–8 cm., obovate to obovate-lanceolate, stiff and shining, acute or obtuse, entire or finely serrulate in the upper part, base cuneate; petiole short or almost 0. Fls erect, few, usually 1, 15–18 mm. diam.; pedicel 2–7 cm., filiform. Calyx-teeth 4–5 mm., usually 7, linear-acuminate. Corolla-lobes usually 7, ovate, acute or apiculate. Fr. c. 6 mm., valves very deciduous. Fl. 6–7. $2n=$ c. 112; c. 160. Grh.

Native. In pine woods, among moss in grassy places, usually rooting in humus, locally common; ascending to 3500 ft. 44. E. Suffolk, S. Lancs and Yorks northwards, infrequent in the southern part of its range. N. Europe, N. Asia.

7. ANAGALLIS L.

Slender annual or perennial herbs. Lvs opposite, entire. Fls axillary, solitary, ebracteolate. *Corolla rotate or funnel-shaped,* 5-lobed. *Stamens inserted at base of corolla-tube;* filaments pubescent. *Capsule* globose, *dehiscing transversely.*

About 20 spp. in Europe, Asia, Africa, and America.

1   Plant 1–4(–7) cm.; corolla not more than 1 mm. diam., exceeded by calyx-
     teeth.   **3. minima**
     Plant 5–30 cm.; corolla at least 6 mm. diam., not exceeded by calyx-teeth.   *2*

2   Stems subterete, rooting at nodes; lvs broadly ovate, obtuse; corolla 2–3
     times as long as calyx.   **1. tenella**
     Stems quadrangular, not rooting at nodes; lvs lanceolate to ovate, acute;
     corolla less than twice as long as calyx.   **2. arvensis**

1. A. tenella (L.) L.   Bog Pimpernel.

A slender prostrate glabrous perennial 5–15 cm. *Stems subterete, rooting at nodes. Lvs c. 5 mm.,* ovate or suborbicular, *shortly petioled.* Fls on filiform pedicels much exceeding the lvs. Calyx-teeth linear-lanceolate, acuminate. *Corolla* up to 14 mm. diam., *funnel-shaped, 2–3 times longer than the calyx,* pink. Capsule c. 3 mm. diam. Fl. 6–8. $2n=22^*$. Chh.

Native. In damp peaty and grassy places and in bogs. 104, H40, S. Throughout most of the British Is., though rare in S.E. England and the Midlands, and absent from Radnor, much of southern Scotland and E. Sutherland. Extinct in Middlesex. Atlantic Europe from the Netherlands southwards, east to N. Italy; Crete, Crimea (very rare); N. Africa, Morocco to Algeria.

2. A. arvensis L. Scarlet Pimpernel, Shepherd's Weather-glass.

A procumbent or ascending glabrous annual or perennial 6–30 cm. *Stems quadrangular*, gland-dotted. *Lvs* 15–28 *mm.*, ovate to lanceolate, *sessile*, dotted with black glands beneath. Fls on slender pedicels. *Calyx-teeth* narrow-lanceolate, apiculate, *not much shorter than the corolla. Corolla rotate*, up to 14 mm. diam., lobes entire, crenulate, or denticulate. Capsule c. 5 mm. diam. Fl. 6–8. Th. or Chh. Throughout the greater part of the world with the exception of the tropics.

Ssp. **arvensis.** Fr. pedicels longer than the lvs. Fls usually red or pink, more rarely blue or lilac, up to 14 mm. diam. Corolla-lobes broadly obovate, over-lapping, margins densely fringed with 3-celled glandular hairs. Calyx-teeth not concealing corolla in bud. Capsule 5-veined. $2n=40*$.
Native. On cultivated land, by roadsides, and on dunes. 112, H40, S. Common and widely distributed throughout the British Is. but rare in the extreme north.

Ssp. **foemina** (Mill.) Schinz & Thell. Fr. pedicels shorter than or equalling lvs. Fls blue, up to 12 mm. diam. Corolla-lobes narrowly obovate, not over-lapping, margins very sparingly fringed with 4-celled glandular hairs. Calyx-teeth concealing corolla in bud. Capsule more than 5-veined.
Native. In arable fields in the south and west of England, rare; elsewhere as a casual.

3. A. minima (L.) E. H. L. Krause Chaffweed.

Centunculus minimus L.

A glabrous annual 2–7 cm. Lvs 3–5 mm., ovate, subsessile, obtuse or apiculate, entire. Calyx divided nearly to base, teeth lanceolate, acute, longer than corolla. Corolla less than 1 mm. diam., white or pink. Fr. c. 1·5 mm. diam., exceeded by calyx-teeth. Fl. 6–7. $2n=22$. Th.
Native. In damp sandy places in rather open communities on heaths and by the sea, often on somewhat disturbed ground, local. 76, H20, S. In suitable habitats throughout the British Is. north to Lewis. Europe; Asia east to Baikal; N. Africa; N. and S. America.

8. GLAUX L.

A small glabrous succulent herb. Stems creeping and rooting. Lvs quite entire, decussate. *Fls axillary, sessile. Calyx* 5-partite, *white or pink. Corolla* 0. Stamens 5, hypogynous, alternating with calyx-lobes. *Capsule* globose, *5-valved, few-seeded.* Self-pollinated and, perhaps, sometimes cross-pollinated by Diptera.
One species, coasts of the north temperate regions and saline districts inland.

1. G. maritima L. 'Sea Milkwort', 'Black Saltwort.'

A small procumbent or suberect perennial 10–30 cm. Lvs 4–12 mm., elliptic-oblong to obovate, obtuse or acute, subsessile. Fls 5 mm. diam. Calyx with obtuse, usually pink lobes with hyaline margins. Fr. c. 3 mm. Fl. 6–8. $2n=30$. Hp.

Native. In grassy salt-marshes, in crevices of rocks or at the foot of cliffs by the sea or estuaries, also in saline districts inland, locally common. 84, H27, S. Coasts of the British Is. North temperate regions.

9. Samolus L.

Herbs. Lvs alternate or mostly radical. *Fls white, bracteate.* Calyx 5-toothed. Corolla 5-lobed, subcampanulate. *Stamens 5, alternating with staminodes, filaments very short. Ovary ½-inferior. Capsule ovoid, 5-valved.* Usually self-pollinated.

About 10 spp., cosmopolitan but mainly in the southern hemisphere.

1. S. valerandi L.   Brookweed.

A glabrous perennial 5–45 cm. Stem simple or little-branched, lfy. Lvs 1–8 cm., obovate to spathulate, entire, obtuse. Infl. racemose, simple or branched. Pedicels 5–15 mm., with small lanceolate bracts adnate to about the middle, straight in fl., often geniculate in fr. Calyx campanulate, teeth triangular, acute. Corolla 2–3 mm. diam., lobes short, obtuse. Stamens included. Fr. 2–3 mm. diam. Fl. 6–8. $2n=$ c. 24, 36. Hs.

Native. In wet places, especially near the sea, locally common. 86, H40, S. Scattered throughout the British Is., though absent from a number of inland counties. Cosmopolitan, though usually near the sea.

97. BUDDLEJACEAE

Shrubs or trees, rarely herbs, with simple opposite (except *Buddleja alternifolia*) stipulate or exstipulate lvs; glandular hairs present. Fls hermaphrodite, actinomorphic, almost always 4-merous. Disk small or 0. Stamens alternate with corolla-lobes, inserted on tube. Ovary superior (very rarely ½-inferior), of two united carpels with usually numerous ovules on axile placentas. Style 1. Fr. a capsule or berry. Seeds with endosperm. Vascular bundles collateral.

About 10 genera and 130 spp., mainly tropical, extending to temperate America, China, S. Japan and S. Africa.

1. Buddleja L.

Trees or shrubs, rarely herbs, with interpetiolar stipules often reduced to a line. Indumentum of stellate and glandular hairs. Infl. cymose, often head-like or forming long terminal panicles. Calyx campanulate. Corolla with ± cylindrical usually straight tube and ± spreading limb. Stamens included or slightly exserted. Fr. a capsule with numerous very small seeds.

About 100 spp., tropical Asia to China, Japan and New Guinea, E. and S. Africa, tropical and warm temperate America.

Several other spp. are cultivated, *B. globosa* Lam. from Chile, with orange fls in globose heads, rather commonly.

*1. B. davidii Franch.

B. variabilis Hemsl.

Shrub 1–5 m. Twigs somewhat angled, pubescent, pithy. Lvs 10–25 cm., ovate-lanceolate or lanceolate, acuminate, serrate, dark green and glabrescent above, white-tomentose beneath. Fls in dense many-fld cymes forming a somewhat interrupted narrow terminal panicle, 10–30 cm. Rhachis, peduncles, pedicels and calyx ± tomentose. Bracts and bracteoles linear, inconspicuous. Corolla lilac or violet with orange ring at mouth; tube cylindric, c. 1 cm., somewhat pubescent outside; lobes 1–2 mm. Fl. 6–10. Commonly visited by butterflies. M.

Introduced. Commonly grown in gardens (introduced c. 1890), now naturalized in waste places, rarely in woods, in a number of places in S. England; increasing. Common on bombed sites. Native of China.

98. OLEACEAE

Trees or shrubs. Lvs usually opposite, exstipulate. Infl. usually cymose. Fls hermaphrodite or unisexual, actinomorphic, usually 4-merous. Calyx small. Corolla gamopetalous, rarely polypetalous or 0, lobes valvate or imbricate, not twisted in bud. Stamens 2, very rarely more, usually adnate to corolla-tube. Ovary superior, 2-celled, with usually 2 ovules in each cell. Seeds with or without endosperm, embryo straight.

About 21 genera and 400 spp., widely distributed but mainly in Asia.

1 Tree; lvs pinnate; bark grey; buds black; fr. winged. 1. Fraxinus
    Shrubs; lvs simple; bark and buds brown or greenish; fr. not winged. *2*

2 Lvs 5–12 cm., ovate; fls usually lilac; fr. a capsule. 2. Syringa
    Lvs 3–6 cm., lanceolate; fls white; fr. a berry. 3. Ligustrum

1. Fraxinus L.

Trees with simple or pinnate lvs. *Fls polygamous or dioecious.* Calyx 4-lobed or 0. Petals 4, connate at base or 0. Stamens 2. *Fr. a compressed 1–2-celled samara winged at tip*, cells 1-seeded.

About 50 spp. in Europe, N. Africa, temperate Asia and N. America.

1. F. excelsior L. Ash.

A deciduous tree 15–25 m. Bark grey, smooth, becoming fissured on old branches. Buds large (terminal ones 5–10 mm.), black. Lvs up to 30 cm., opposite, decussate, imparipinnate or rarely simple; lflets c. 7 cm., 7–13, lanceolate to ovate, apiculate to acuminate, serrate. Fls polygamous, in axillary panicles, appearing before the lvs, purplish, devoid of perianth. Fr. c. 3 cm., pale brown. Fl. 4–5. Wind-pollinated. Fr. 10–11. Germ. spring. $2n=46$. MM.

Native. Forming woods on calcareous soils in the wetter parts of the British Is.; in oakwoods, scrub and hedges; common, but less frequent on acid soils. 111, H40, S. Generally distributed throughout the British Is. except for Shetland, perhaps introduced in some northern counties. Europe from c. 64° N., N. Africa, W. Asia.

2. SYRINGA L.

Shrubs or small trees, usually deciduous. Lvs opposite, usually entire. Fls hermaphrodite, sweet-scented, in terminal or lateral panicles. *Fr. a coriaceous, 2-celled, loculicidal capsule with 2 winged seeds in each cell.*
   About 30 spp. in Asia and Europe.

*1. S. vulgaris L.   Lilac.
A shrub or small tree up to 7 m., suckering freely. Bark fibrous. Lvs 5–12 cm., ovate. Fls lilac or white, in pyramidal panicles, 10–20 cm. Corolla-tube c. 1 cm. Fr. 1–1·5 cm., acute, smooth. Fl. 5. Pollinated mainly by bees. $2n = 46$. N. or M.
   Introduced. Much planted in gardens and sometimes ± naturalized in hedges, thickets and shrubberies. S.E. Europe.

3. LIGUSTRUM L.

Shrubs or small trees. Lvs often evergreen, entire, opposite. Fls hermaphrodite, in terminal panicles, strong-scented. Calyx caducous. Corolla funnel-shaped. *Berry with oily flesh,* 2-celled, cells 1–2-seeded.
   About 50 spp., Asiatic and Indo-Malay, one in Europe.

1. L. vulgare L.   Common Privet.
A tardily deciduous shrub up to 5 m. Bark smooth. Branches slender, *young twigs puberulent.* Lvs 3–6 cm., *lanceolate,* obtuse to acute, shortly petioled. Panicle 3–6 cm., *puberulent.* Fls white, shortly pedicelled, 4–5 mm. diam.; *corolla-tube as long as limb; anthers exceeding tube, shorter than limb.* Fr. 6–8 mm., black, shining. Fl. 6–7. Pollinated by various insects. Fr. 9–10. $2n = 46$. N. or M.
   Native. Common in hedges and scrub, particularly on calcareous soils, north to Durham, naturalized elsewhere. 99, H 39, S (incl. introductions). Europe (not north), N. Africa.

L. vulgare is now nearly supplanted for hedging by ***L. ovalifolium** Hassk., a shrub with elliptic-oval to elliptic-oblong lvs which persist longer than those of *L. vulgare* except in dirty industrial districts. Young twigs and panicle branches glabrous. Corolla-tube 2–3 times as long as limb; anthers as long as limb. Fl. 7. A variety with golden-yellow lvs is commonly cultivated.
   Introduced. Planted, usually near houses. Japan.

99. APOCYNACEAE

Woody plants, often climbing, rarely herbs, with milky latex and internal phloem. Lvs usually opposite, rarely whorled or spiral, entire, usually exstipulate. Fls hermaphrodite, actinomorphic, hypogynous, 5-merous, solitary or in cymose infls. Corolla gamopetalous, contorted in bud. Stamens as many as the corolla-lobes and alternating with them, epipetalous, with short filaments; anthers convergent on the stylar head. Ovary of 2 carpels, free below but with a common style; ovules numerous on the ventral sutures of the carpels. Fr. various; seeds often winged or plumed, usually endospermic.

About 130 genera and 1100 spp., chiefly in tropical and subtropical regions.

Funtumia elastica yields the Lagos Silk Rubber and lianas of the genus *Landolphia* the Landolphia Rubber of commerce. Seeds of *Strophanthus* spp. (also lianas) yield the cardiac stimulant strophanthine. Various spp. with poisonous latex, and especially *Acokanthera venenata*, have been used for arrow poisons. *Nerium oleander*, the Oleander, a native of the Mediterranean region, is often seen in conservatories.

1. VINCA L.

Creeping shrubs or perennial herbs with evergreen lvs in opposite pairs. *Fls solitary in the axils of the lvs*. Corolla blue or white, salver-shaped, with 5 broad asymmetric lobes and an obconic tube fluted and hairy within; stamens 5, with short sharply kneed filaments and introrse anthers terminating in *broadly triangular hairy connective flaps* which meet over the stylar head; *styles united to a column* which is slender below but with an *enlarged head* tapering upwards and *surmounted by a plume of white hairs*; stigmatic surface as a band round the broadest part of the stylar head; ovary of 2 free carpels united only by their styles. Fr. of 2 follicles each with several long narrow seeds. There are two fleshy nectaries at the base of the gynoecium, alternating with the carpels, and the fls are insect-pollinated.

Five spp. in Europe and W. Asia.

Fls 25–30 mm. diam.; calyx-lobes glabrous.   **1. minor**
Fls 40–50 mm. diam.; calyx-lobes ciliate.   **2. major**

1. V. minor L.   Lesser Periwinkle.

A procumbent shrub with trailing stems 30–60 cm., *rooting at intervals*, and short erect flowering stems. *Lvs* 25–40 *mm., very shortly stalked, lanceolate-elliptic*, quite *glabrous*. *Flowering stems each with* 1 *axillary fl.*, rarely 2. Fl. 25–30 mm. diam. *Calyx-lobes lanceolate, glabrous*. Corolla blue-purple, mauve or white. Ripe follicles (rarely seen in Britain) 25 mm., divergent, each with 1–4 blackish seeds. Fl. 3–5. Pollinated by long-tongued bees and bee-flies. $2n = 32$; ?46. Chw.

Doubtfully native. Found locally in woods, copses and hedgebanks throughout Great Britain northwards to Caithness. 85. Widespread in Europe from Denmark southwards, and in W. Asia. In C. Europe occurs in ash and oak-hornbeam woods on the better soils.

*2. V. major L.   Greater Periwinkle.

A semi-procumbent shrub with trailing or somewhat ascending stems 30–100 *cm., rooting only at their tips*, and short erect flowering stems. *Lvs* opp. 20–70 *mm., with stalks c.* 1 *cm., ovate*, ± acute, somewhat *cordate* at the base, *ciliate*. Flowering stems erect, to 25 cm., each with 1–4 *axillary fls* at successive nodes. Fl. 40–50 mm. diam. *Calyx-lobes long, subulate, ciliate*. Corolla blue-purple, rarely white. Ripe follicles (rarely seen in Britain) 4–5 cm., each with 1–2 dark brown seeds. Fl. 4–6. Pollinated by long-tongued bees, especially *Anthophora pilipes*. $2n = 16$. Ch.

Introduced. Copses and hedgerows. S. England. Occasionally in Ireland. Native in C. and S. Europe and N. Africa.

*V. herbacea Waldst. & Kit., with herbaceous *non-rooting trailing shoots*, 15–60 cm., *dying back in winter*, thinnish ciliate *lvs*, all but the lowermost *narrowly elliptical to linear-lanceolate*, and long-stalked blue-violet (rarely pink or white) fls, 2·5–3 cm. diam., with ciliate calyx-teeth, is grown in gardens and occasionally established. S.E. Europe.

100. GENTIANACEAE

Herbs, usually glabrous. Lvs opposite, entire, exstipulate, usually sessile. Infl. usually a dichasial cyme. Fls hermaphrodite, regular, usually 4- or 5-merous. Calyx-lobes imbricate. Corolla persisting round the capsule, lobes contorted in bud. Stamens epipetalous, equalling in number and alternating with the corolla-lobes. Ovary superior, unilocular with 2 parietal placentas, each with numerous anatropous ovules, sometimes 2-celled by the intrusion of the placentas; style simple with a simple or bilobed stigma or two stigmas. Fr. usually a septicidal capsule; seeds small, numerous, with copious endosperm and small embryo.

About 65 genera and 800 spp., cosmopolitan, mainly temperate.

A natural family, unlikely to be confused with any other. The opposite lvs and unilocular ovary with two many-ovuled parietal placentae and actinomorphic fls are sufficient to distinguish the British members from the other Gamopetalae. Many species contain bitter principles. Vascular bundles bicollateral.

1   Fls 6–8-merous, yellow; cauline lvs connate in pairs.   4. BLACKSTONIA
    Fls 4–5-merous; lvs not connate.   2
2   Corolla pink or yellow, rarely white; style distinct, filiform, caducous.   3
    Corolla blue or purple, rarely white; style 0 or ovary gradually tapering
        into style; stigmas persistent.   5
3   Calyx-lobes deltoid, less than half as long as tube; corolla yellow; stigma
        peltate.   1. CICENDIA
    Calyx divided nearly to base, lobes linear; corolla pink; stigma bifid or
        stigmas 2.   4
4   Anthers ovate, not twisted; calyx-lobes flat; fls 4-merous. (Guernsey.)
                                                                                2. EXACULUM
    Anthers linear, twisting after fl.; calyx-lobes keeled; fls usually 5-merous.
        (Widespread.)   3. CENTAURIUM
5   Corolla with small lobes between the large ones, without a fringe at the
        throat, blue.   5. GENTIANA
    Corolla fringed at the throat, without small lobes, purple, rarely white.
                                                                                6. GENTIANELLA

1. CICENDIA Adans.

Small annuals. Fls 4-merous. *Calyx campanulate with short deltoid lobes.* Corolla with ovoid tube and short spreading lobes, yellow. *Anthers cordate, not twisted. Style filiform with peltate stigma, caducous.*

Two spp., the second in California and temperate S. America.

1. C. filiformis (L.) Delarbre

Microcala filiformis (L.) Hoffmanns. & Link

A slender annual 3–12 cm., simple or somewhat branched. Branches strict. Lvs 2–6 mm., linear, few. Pedicels 1–5 cm. Fls 3–5 mm. Capsule ovoid, c. 5 mm. Fl. 6–10. Fr. 8–10. Th.

Native. Damp sandy and peaty places not far from the sea, very local; Cornwall to Sussex; W. Norfolk; Pembroke, Caernarvon; Kerry, W. Cork and W. Mayo; Channel Is. 10, H3, S. W. and S. Europe from Denmark to Spain and Portugal, Sardinia, Sicily and Thrace; Asia Minor; N. Africa; Azores.

2. EXACULUM Caruel

Small annual. Fls 4-merous. *Calyx deeply divided into 4 flat linear lobes.* Corolla with cylindric tube and spreading lobes, pink (in Britain). *Anthers ovate, not twisted. Style filiform with bifid stigma, caducous.*

A single sp.

1. E. pusillum (Lam.) Caruel

Cicendia pusilla (Lam.) Griseb.

A slender annual 3–12 cm., with divaricate branches. Lvs linear, c. 6 mm. Pedicel slender. Capsule fusiform. Fl. 7–9. Th.

Native. Sandy commons in two spots in Guernsey. W., C. and S. France, Spain, Portugal, Corsica, Sardinia, Italy; N. Africa.

3. CENTAURIUM Hill   Centaury.
(*Erythraea* Borkh.)

Annual, rarely perennial, herbs. Fls usually 5- (rarely 4-)merous. Calyx deeply divided into keeled linear lobes. Corolla ± funnel-shaped, pink, rarely white. *Anthers linear or oblong-linear, twisting spirally after dehiscence.* Placentae intrusive, nearly meeting. *Style filiform with 2 stigmas, caducous.*

About 30 spp., cosmopolitan except tropical and S. Africa.

1 Erect annuals without decumbent sterile shoots; corolla-lobes 7 mm. or less.   2
   Perennial with decumbent sterile shoots with orbicular shortly petioled lvs; corolla-lobes 8–9 mm. (Pembroke and Cornwall.)   **6. portense**
2 Fls pedicelled, not clustered; corolla-lobes 3–4 mm.; plants without basal rosette.   3
   Fls sessile or subsessile, ±clustered; corolla-lobes 5–7 mm.; plants with basal rosette.   4
3 Branches spreading at a wide angle; infl. lax; internodes 2–4.   **1. pulchellum**
   Branches strict; infl. rather dense; internodes 5–9.   **2. tenuiflorum**
4 Lvs ligulate or linear-spathulate; basal 5 mm. broad or less.   **5. littorale**
   Lvs ovate, obovate or oblong; basal more than 5 mm. broad (rarely only 4 mm.).   5
5 Stamens inserted at top of corolla-tube; infl. various.   **3. erythraea**
   Stamens inserted at base of corolla-tube; infl. always very dense. Coastal.   **4. capitatum**

1. C. pulchellum (Sw.) Druce

Erythraea pulchella (Sw.) Fr.

A glabrous erect *annual, without a basal rosette,* varying in habit from very slender, 2–3 cm., unbranched and 1-fld, to 15 cm., much branched above with dichotomous *widely spreading branches. Cauline internodes* 2–4. Lvs 2–15 × 1–10 mm., ovate or ovate-lanceolate, acute, the upper usually longer than the lower. *Fls in a loose dichasial cyme on pedicels c. 2 mm.,* sometimes 4-merous in small plants. Corolla-tube usually exceeding calyx; *lobes 2–4 mm.* Stamens inserted at top of corolla-tube. Capsule about equalling calyx. Variable. Fl. 6–9. Pollinated by insects and self. Germ. spring. $2n$=c. 34*, 42. Th.

Native. Damp grassy places usually in rather open habitats. 53, H6, S. Common near the sea in S. and C. England, more local inland, rarer northwards and more confined to coast; Scotland: Dumfries, E. Lothian and S. Inner Hebrides; S. and E. coast of Ireland from Cork to Dublin, local, Leix. Europe (except Iceland); Mediterranean region; W. and C. Asia to the Punjab and Tien Shan; Madeira; naturalized in N. America.

2. C. tenuiflorum (Hoffmanns. & Link) Fritsch

Erythraea tenuiflora Hoffmanns. & Link

A glabrous erect *annual* 10–35 cm., *without a basal rosette, with strict branches above. Cauline internodes* 5–9. Lvs 10–25 × 8–12 mm., ovate or oval, obtuse or subacute, sessile, the upper usually longer than the lower. *Fls in a rather dense dichasial cyme on pedicels c. 2 mm.* Corolla-tube slightly exceeding calyx; *lobes 3–4 mm.* Stamens inserted at top of corolla-tube. Capsule about equalling calyx. Fl. 7–9. Germ. spring. Th.

Native. Damp grassy places near the sea in the Isle of Wight and Dorset; very rare. 2. Coasts of W. Europe from N. France southwards and of the Mediterranean.

C. latifolium (Sm.) Druce

Fls small with corolla-lobes 3–4 mm. as in the preceding spp., but sessile and in a dense infl. as in the following. Lvs broadly ovate. Formerly found on sand dunes in Lancashire but now extinct. Endemic.

3. C. erythraea Rafn   Common Centaury.

Erythraea Centaurium auct.; *Centaurium umbellatum* auct.; *C. minus* auct.

A glabrous *annual* 2–50 cm., *with a basal rosette of lvs* and usually solitary but sometimes several erect stems, branched above. *Basal lvs* 1–5 cm. × (4–)8–20 *mm., obovate or elliptic,* often somewhat spathulate, usually obtuse, prominently 3–7-veined; cauline shorter, sometimes narrower and acute, but *never parallel-sided. Fls sessile or subsessile,* often clustered, *forming a ± dense* corymb-like *cyme.* Corolla-tube longer than calyx, limb ± flat; *lobes 5–6 mm.* Stamens inserted at top of corolla-tube. Capsule exceeding calyx. Very variable; the maritime var. *subcapitatum* (Corb.) Gilmour is only certainly distinguishable from *C. capitatum* by the insertion of the stamens. Fl. 6–10. Pollinated by insects and self. Germ. autumn. $2n$=42. Th.

Native. Dry grassland, dunes, wood margins, etc.; common in England

and Ireland, less so in Scotland but extending to Ross and the Outer Hebrides. 102, H40, S. Europe from Sweden southwards; Mediterranean region; Azores; S.W. Asia to the Pamir-Alai region; naturalized in N. America.

4. C. capitatum (Willd.) Borbás

Erythraea capitata Willd.

A glabrous erect *annual* 2–5(–10) cm., *with a basal rosette of lvs.* Basal lvs c. 2 cm. × 8 mm., ovate, obtuse, prominently 3–5-veined; cauline usually smaller. *Fls sessile in a dense head-like cyme.* Corolla-tube about equalling calyx; *lobes 5–6 mm. Stamens inserted at base of corolla-tube.* Capsule exceeding calyx. Fl. 7–8. Germ. autumn. Th.

Native. Dry grassland, usually calcareous, near the sea, very local; Cornwall to Kent, Wales, Lancashire, Yorks, Northumberland, Guernsey. 18, S. Normandy, Denmark, Öland, Germany (very rare).

5. C. littorale (D. Turner) Gilmour

C. Turneri Druce; *C. vulgare* Rafn; *Erythraea compressa* Kunth; *E. littoralis* (D. Turner) Fr.; *E. Turneri* Wheldon & C. E. Salmon

An erect *annual* 2–25 cm., usually scaberulous, *with a basal rosette of lvs* and solitary or frequently several erect stems, branched above. *Basal lvs* 1–2 cm. × 3–5 mm., *linear-spathulate,* obtuse, indistinctly 3-veined; *cauline shorter, ligulate,* obtuse. *Fls sessile, clustered in ± dense corymb-like cymes,* relatively few. Corolla-tube not longer than calyx, limb concave; *lobes* 6–7 mm. Stamens inserted at top of corolla-tube. Capsule much exceeding calyx. Variable. Fl. 7–8. Germ. autumn. $2n = 38^*$, c. 56^*. Th.

Native. Dunes and sandy places near sea, local, almost confined to the coasts of Wales, N.W. and N. England, and Scotland, from South Wales to Northumberland, one isolated locality in Hants; Derry. 38, H1. Narrow-lvd forms of *C. erythraea* have frequently been erroneously recorded as this species. Coasts of W. Europe from Scandinavia to N. France and of the Baltic; inland in C. Europe from Austria to S. Russia (Upper Dnieper region).

6. C. portense (Brot.) Butcher   'Perennial Centaury.'

C. scilloides var. *portense* (Brot.) Druce; *Erythraea portensis* (Brot.) Hoffmanns. & Link

A glabrous perennial herb with numerous decumbent sterile stems and ascending fl.-stems. Lvs of sterile stems to 1 cm., *suborbicular,* narrowed into a short petiole; of the fl. stems oblong, sessile. Fls 1–6, pedicelled. *Corolla-lobes* 8–9 mm. Stamens inserted at top of corolla-tube. Capsule rather longer than calyx. Fl. 7–8. Chh.

Native. Grassy cliffs near Newport (Pembroke); W. Cornwall (status unknown). N.W. France, N.W. Spain, N.W. Portugal. The allied *C. scilloides* (L. f.) Sampaio, of which this is often regarded as a variety, is endemic in the Azores.

The hybrids *C. erythraea* × *littorale* (*C.* × *intermedium* (Wheldon) Druce) and *C. erythraea* × *pulchellum* occur rarely.

4. BLACKSTONIA Huds.

(*Chlora* Adens.)

Annuals. *Fls 6–8-merous*. Calyx deeply divided into linear lobes. *Corolla rotate with short tube*, yellow. Anthers oblong or linear, sometimes slightly twisted after fl. *Style filiform with 2 deeply bilobed stigmas, caducous.*

About 5 spp. in Europe and Mediterranean region.

1. B. perfoliata (L.) Huds.   Yellow-wort.

Chlora perfoliata (L.) L.

An erect glaucous annual 15–45 cm., with basal rosette of lvs. Stems simple or branched above. Basal lvs 1–2 cm., obovate, obtuse, free; cauline ovate-triangular, acute, each pair connate by nearly the whole base. Fls in a loose dichasial cyme. Corolla 10–15 mm. diam. Capsule oval. Fl. 6–10. Self-pollinated. $2n=44^*$. Th.

Native. Calcareous grassland and dunes, rather common in S. England, extending north to Durham and Lancashire; Kirkcudbright; S. Ireland to Meath and Sligo; Jersey. 65, H27, S. Europe from Holland and S.W. Germany southwards (only Crimea in Russia); S.W. Asia; Morocco.

5. GENTIANA L.

Perennial, rarely annual, herbs. Glabrous. Fls 5-merous. *Calyx* ± tubular; *teeth joined by a membrane* which forms the upper part of the tube. *Corolla*-tube of varied shape; limb spreading, not fringed, usually blue, *with small lobes between the 3-veined large ones. Anthers* not twisted, *not versatile. Ovary gradually tapering into the style or style* 0. *Stigmas 2, persistent on the capsule. Nectaries at the base of the ovary.*

Over 200 spp., northern hemisphere, mainly in the mountains, a few in the Andes.

1 Corolla sky-blue, tube obconic; lvs linear, 1·5–4 cm.   **1. pneumonanthe**
   Corolla brilliant deep blue, tube cylindric; lvs ovate or ovate-oblong,
   1·5 cm. or less.   *2*
2 Perennial with numerous rosettes of lvs; corolla c. 15 mm. across. (Teesdale and W. Ireland.)   **2. verna**
   Erect annual; corolla c. 8 mm. across. (Scotland.)   **3. nivalis**

1. G. pneumonanthe L.   'Marsh Gentian.'

A perennial 10–40 cm., with a few suberect simple stems. Lowest lvs scale-like; the rest 1·5–4 cm., *linear*, obtuse. Fls 1–7, terminal and axillary, forming a rather dense infl., pedicels to 2 cm., but usually much less. Calyx-tube c. 5 mm., obconic, not angled; lobes linear, acute, about as long. *Corolla sky-blue* with 5 green lines outside; *tube obconic*, 2·5–4 cm.; lobes broadly ovate, ascending. Capsule ellipsoid, stipitate. Fl. 8–9. Pollinated by humble-bees; protandrous. $2n=26$. Hp.

Native. Wet heaths from Dorset and Kent to Carmarthen, Anglesey, Yorks and Cumberland, very local, decreasing. 30. Europe from Scandinavia (59° 15′ N.) to N. Spain and Portugal, C. Italy, Macedonia and the Caucasus; Siberia.

2. G. verna L. 'Spring Gentian.

A perennial 2–6 *cm.*, with few or many underground stems from a short stock, each ending in a rosette of persistent lvs, *the rosettes forming a ± dense tuft or cushion. Rosette lvs* 5–15 *mm.*, *ovate or ovate-oblong*, obtuse to subacute; cauline few, smaller, elliptic or oblong. Fls solitary, terminal. Calyx-tube c. 1 cm., ± cylindric, strongly 5-angled, lobes triangular-lanceolate, much shorter than tube. *Corolla* 1·5–2·5 cm., *brilliant deep blue, tube subcylindric*; limb 1·5–2 cm. across, spreading; lobes ovate, obtuse. Stigma white. Capsule oblong, subsessile. Fl. 4–6. Pollinated by Lepidoptera. $2n=26, 28$. Chh. or Chc.

Native. Stony grassy places on limestone; Yorks, Durham, Westmorland, Cumberland, 1500–2400 ft.; Clare, Galway, Mayo, 0–1000 ft.; very local. 4, H5. Mountains of C. Europe to Sierra Nevada, Apennines, Albania and Bulgaria; Caucasus, N. and C. Asia (sspp.); Morocco (a ssp.).

3. G. nivalis L. 'Small Gentian.'

An erect slender annual 3–15 cm., simple or branched. *Lvs* 2–5 *mm.*, *ovate or obovate*, lower sometimes forming a rosette. Fls terminal on stem and branches. Calyx-tube c. 8 mm., ± cylindric, 5-angled, lobes triangular-lanceolate, shorter than tube. *Corolla* 1–1·5 cm., *brilliant deep blue, tube subcylindric*; limb c. 8 mm. across, spreading; lobes ovate. Capsule ellipsoid, subsessile. Fl. 7–9. Probably usually self-pollinated. $2n=14$. Th.

Native. Rock ledges in the mountains of Perth and Angus from 2400 to 3450 ft. 2. N. Europe from Iceland eastwards; mountains of C. Europe to Pyrenees, Apennines, Bosnia and Bulgaria; N. Asia Minor, Caucasus; arctic N. America; Greenland.

6. GENTIANELLA Moench

Annual, biennial or (in foreign spp.) perennial herbs. Glabrous. Fls 4- or 5-merous. *Calyx* ± tubular; *membrane* 0. Corolla-tube cylindric or obconic; limb spreading, fringed at the throat in the British spp., purple or whitish (blue in some foreign spp.); *lobes 5–9-veined, without small lobes between. Anthers* not twisted, *versatile. Ovary gradually tapering into the style or style* 0. *Stigmas 2, persistent on the capsule. Nectaries on the corolla.*

Over 120 spp., northern hemisphere, S. America, S.E. Australia, Tasmania, New Zealand.

Individual plants of *G. amarella* agg. may show unusual features and determinations should be based on small samples of a population.

1 Calyx-lobes 4, the two outer much larger than the inner and overlapping and enclosing them. **1. campestris**
   Calyx-lobes 4 or 5 (often on the same plant) equal or somewhat unequal, not (or rarely slightly) overlapping. 2

2 Corolla 25–35 mm., twice as long as calyx or more; internodes 9–15 (small annual plants with smaller corollas may occur mixed with the others). **2. germanica**
   Corolla (12–)13–20(–23) mm., less than twice as long as the calyx; internodes 0–9(–11). **(amarella** agg.) 3

3   Uppermost internode and terminal pedicel together forming less than
    half the total height of the plant (usually much less).   4
    Uppermost internode and terminal pedicel together forming more than
    half the total height of the plant.   8

4   Fls uniformly coloured, usually dull purple.   5
    Fls creamy-white within, purplish-red outside (Scotland).   7

5   Fl. (July–)August–October; internodes 4–11, uppermost ±equalling to
    much shorter than others.   6
    Fl. March–May(–July); internodes 3–5, uppermost rather longer than
    others (Cornwall).   **4. anglica** ssp. **cornubiensis**

6   Corolla usually 16–18 mm.; cauline lvs lanceolate or ovate-lanceolate
    (widespread in Great Britain).   **3. amarella** ssp. **amarella**
    Corolla usually 19–22 mm.; cauline lvs linear-lanceolate or lanceolate
    (Ireland).   **3. amarella** ssp. **hibernica**

7   Internodes 6–7, uppermost very short; corolla equalling or slightly longer
    than calyx.   **3. amarella** ssp. **septentrionalis**
    Internodes 2–5(–6), uppermost about equalling others; corolla about
    1½ times as long as calyx.   **3. amarella** ssp. **druceana**

8   Upper cauline lvs lanceolate; calyx-teeth ±appressed to corolla up to
    1·5 mm. broad (England; chalk grassland).   **4. anglica** ssp. **anglica**
    Upper cauline lvs ovate or ovate-lanceolate; calyx-teeth ±spreading, up
    to 3 mm. broad. (Wales; dune-slacks)   **5. uliginosa**

1. G. campestris (L.) Börner

Gentiana campestris L.; incl. *G. baltica* auct.

An erect annual or biennial 10–30 cm., simple or branched. Basal lvs 1–
2·5 cm., ovate, lanceolate or spathulate, ± obtuse; cauline lvs 2–3 cm., lingulate,
oblong, ovate-lanceolate or lanceolate, obtuse to acute. Fls 4-merous. *Calyx
divided nearly to base into 4 lobes; 2 outer lobes ovate to ovate-lanceolate, acute
to acuminate, overlapping and hiding most of the 2 lanceolate inner ones.*
Corolla 15–25(–30) mm., bluish-lilac, rarely white, tube equalling or longer
than calyx, lobes oblong. Fl. 7–10, visited by humble-bees and Lepidoptera,
sometimes selfed. Hs. or Th. $2n = 36$.

Native. Pastures and dunes, usually acid or neutral, ascending to 2600 ft.
Common in Scotland, N. England and Wales, very local in S. and E. England
and there absent from several counties; throughout Ireland but rare in the
centre. 101, H30. Europe from Iceland and N. Russia to E. Pyrenees, N.
Apennines and Bohemia.

There appear to be no satisfactory grounds for maintaining *G. baltica* as
a separate species. None of the characters attributed to it appear to be
constant or correlated.

2. G. germanica (Willd.) Börner

Gentiana germanica Willd.

Usually biennial (5–)7–35(–50) cm., producing in the first year a rosette of lvs
dying in autumn, but annual plants often occur sporadically among the
biennials (these are much smaller in all their parts than the biennials and are
excluded from the following description). Stems usually branched above the
middle; *internodes 9–15, all equal or the uppermost shorter*. Basal lvs spathulate,

obtuse, usually dead at fl. time; *cauline lvs* 1–2·5 cm., *ovate or ovate-lanceolate*, subacute or acute, subcordate at base and *tapering from a wide base to the apex*. Fls 5-merous. Calyx-teeth ± equal, ± spreading. *Corolla* 25–35 *mm*. (down to 15 mm. in annual plants), *bright bluish-purple*, twice as long as calyx or more, tube obconic. Fl. (8–)9–10. Pollinated by humble-bees, perhaps sometimes selfed. Hs.(–Th.).

Native. Chalk grassland and scrub in open habitats often among tall grasses. From Hants and Wilts to Middlesex and Bedford, very local. 8. Belgium and N.E. France to Silesia, Moravia and Switzerland.

3. G. amarella (L.) Börner   Felwort.

Gentiana amarella L.; *G. axillaris* (Schmidt) Rchb.

Biennial (3–)5–30(–50) cm., producing in the first year a rosette of lanceolate to lingulate lvs dying in autumn. *Internodes* (2–)4–9(–11), *all almost equal or the upper one shorter* (frequently much shorter). Basal lvs (of second year) obovate or spathulate; *cauline lvs* 1–2 cm., *ovate to linear-lanceolate*, ± acute. Fls 4- or 5-merous, often on same plant. *Calyx-teeth ± appressed to corolla. Corolla* (12–)14–22 *mm., twice as long as calyx or less*, narrower than in *G. germanica*. Fl. (late 7–)8–9(–10). Pollinated by humble-bees. Fr. 9–10. $2n = 36$. Hs.

Native. Basic pastures, usually among short grass, and dunes, rather common over most of the British Is. but absent from S.W. Scotland, N.E. Ireland, Isle of Man and Channel Is. 89, H 30. Iceland and Scandinavia to N. France, N. Hungary, the Caucasus and the Yenisei region.

Ssp. **amarella**. Internodes 4–9(–10), all ± equal or the upper one shorter. Cauline lvs lanceolate or ovate-lanceolate. Calyx-teeth subequal to markedly unequal. Corolla (14–)16–18(–20) mm., dull purple, more rarely greenish-white, 1·25–1·8 times as long as calyx.

Basic pastures and dunes, from Islay and Angus southwards. 68. Foreign distribution not yet worked out but certainly occurring on the nearer parts of the Continent.

Ssp. **hibernica** Pritchard. Internodes 7–11, uppermost always very short (c. 1 mm.). Cauline lvs linear-lanceolate or lanceolate. Calyx-teeth subequal. Corolla (17–)19–22 mm., dull purple, rarely white, pink or pale blue, 1·5–1·8 times as long as calyx.

Ireland. H 16 (but probably all Irish plants belong to this ssp.). Endemic.

Ssp. **septentrionalis** (Druce) Pritchard (*G. septentrionalis* (Druce) E. F. Warb.; *Gentiana septentrionalis* (Druce) Druce). Internodes 6–7, uppermost very short (frequently c. 1 mm.). Middle and upper cauline lvs ovate (or ovate-lanceolate), acute, wide at the base. Calyx-teeth almost always markedly unequal. Corolla (12–)14–16 mm., creamy-white within, purplish-red outside, equalling or slightly longer than calyx; lobes more erect than in spp. *amarella*.

Dunes and limestone pastures. 4. Outer Hebrides, W. Ross, W. Sutherland (from Tongue westwards), Shetland. ?Iceland.

Ssp. **druceana** Pritchard. Internodes 2–5(–6), uppermost about equalling others. Middle and upper cauline lvs ovate-lanceolate or lanceolate (–linear-lanceolate), not markedly wider at base. Calyx-teeth usually equal or sub-equal. Corolla (13–)15–17 mm., about 1½ times as long as calyx; colour and lobes as in ssp. *septentrionalis*.

Dune-slacks from N. Aberdeen to Sutherland (Bettyhill), Orkney; pastures on metamorphic limestone in Perth, Angus and Aberdeen. 11. Endemic.

G. amarella × *germanica* (*G.* × *pamplinii* (Druce) E. F. Warb.) and *G. amarella* × *uliginosa* occur with the parents, or with *G. amarella* alone in places where the other parent formerly occurred.

4. G. anglica (Pugsl.) E. F. Warb.

Gentiana anglica Pugsl.; *G. amarella* var. *praecox* Towns.; *G. lingulata* var. *praecox* (Towns.) Wettst.

Biennial 4–20 cm. *Internodes 2–5, the uppermost equalling or longer than the others* (rarely short and then the next long). Basal lvs (of second year) spathulate, obtuse. Lower stem-lvs linear or linear-lanceolate, upper often lanceolate. Fls 4- or 5-merous. *Calyx-teeth* ± *appressed*, not more than 1·5 mm. broad. Corolla 13–20 mm., dull purple, about 1½ times as long as calyx. Fl. (3–)4–6(–early 7). Hs.

Ssp. **anglica**. Usually branched from base. Internodes 2–3(–4); uppermost internode c. 1½ times as long as others (rarely short and then next long). Basal lvs narrow. Upper stem-lvs lanceolate, acute, lower narrower. Terminal pedicel forming about ½ total height of plant. Calyx-teeth unequal. Corolla 13–16 mm. Fl. (4–)5–6(–7).

Native. Chalk grassland from S.E. Devon and W. Sussex to N. Wilts and S. Lincoln, local; dunes in N. Devon, perhaps extinct. 15. Endemic.

Ssp. **cornubiensis** Pritchard. Branched from middle. Internodes 3–5, all ± equal. Basal lvs broad. Stem-lvs all linear or linear-lanceolate, obtuse or subobtuse (rarely uppermost subacute). Terminal pedicel somewhat longer than internodes but not attaining ⅓ total height of plant. Calyx-teeth sub-equal. Corolla (15–)17–20 mm. Fl. (3–)4(–7).

Native. Cliffs on the N. coast of W. Cornwall. 1. Endemic.

5. G. uliginosa (Willd.) Börner

Gentiana uliginosa Willd.

Annual or biennial 1–15 cm., the *biennials with long branches from the base giving a pyramidal habit*, the annuals (found mixed with the biennials) small and consisting of 1 or 2 fls from the basal rosette. *Internodes* 0–2(–3). Basal lvs of annuals (and first year rosettes of biennials) lanceolate, of second year rosettes obovate or spathulate. *Cauline lvs ovate or ovate-lanceolate*, acute or subacute, wide at the base. *Terminal pedicel very long, together with the terminal internode forming at least ½ total height of plant.* Fls 4–5-merous. *Calyx-teeth* ± *spreading, very unequal*, the largest up to 3 mm. broad. Corolla 10–20(–23) mm. (smaller in annuals than biennials), dull purple. Fl. (7–)8–11. Hs. or Th.

Native. Dune-slacks in Glamorgan, Carmarthen and Pembroke, very local. 3. N. France and Belgium to N. Russia.

101. MENYANTHACEAE

Differ from Gentianaceae as follows: Aquatic or bog plants. Lvs alternate, except sometimes on fl. stems. Corolla caducous, lobes valvate in bud. Vascular bundles collateral.

About 5 genera and 35 spp., cosmopolitan.

Often united with the Gentianaceae but a small natural group of distinct habit.

Lvs ternate; fls pink or white.	1. MENYANTHES
Lvs simple, orbicular; fls yellow.	2. NYMPHOIDES

1. MENYANTHES L.

Aquatic or bog plant. *Lvs ternate*, all alternate. *Fls in a raceme* on a lfless scape, dimorphic, 5-merous. Capsule opening by 2 valves, subglobose.

One species.

1. M. trifoliata L.   Buckbean, Bogbean.

A glabrous aquatic or bog plant with the lvs and fls raised above surface of water; rhizome creeping. Lflets obovate or elliptic, 3·5–7 cm., obtuse to sub-acute, entire; petioles 7–20 cm. with long sheathing base. Scape 12–30 cm., c. 10–20-fld; pedicels 5–10 mm., longer than the ovate bracts. Calyx-lobes ovate, somewhat recurved. Corolla pink outside, paler or white within, c. 15 mm. across, much fimbriate. Fl. 5–7. Pollinated by various insects, heterostylous. Fr. 8. $2n = 54$. Hel.

Native. Ponds, edges of lakes and in the wetter parts of bogs and fens, ascending to 3000 ft., sometimes locally dominant in shallow water. 111, H40, S. Rather common throughout the British Is. Europe from Iceland and Scandinavia to C. Spain and Portugal, Italy and Macedonia; N. and C. Asia; N. Morocco; Greenland; N. America.

2. NYMPHOIDES Hill

Aquatic. *Lvs simple*, orbicular, deeply cordate, those of the fl. stems opposite. *Fls* on long pedicels *in clusters* in the lf-axils, 5-merous. Capsule opening irregularly, ovoid, beaked.

About 20 spp., mainly tropics and subtropics.

1. N. peltata (S. G. Gmel.) O. Kuntze   'Fringed Waterlily.'

Limnanthemum nymphoides (L.) Hoffmanns. & Link; *L. peltatum* S. G. Gmel.

A glabrous aquatic with floating lvs and fls; rhizome creeping, its lvs alternate. Fl. stems long, floating, their lvs opposite. Lvs orbicular, 3–10 cm., deeply cordate at base, entire or sinuate, purplish below and purple spotted above, long-petioled. Fls in 2–5-fld axillary fascicles; pedicels 3–7 cm. Calyx 5-partite, lobes oblong-lanceolate, c. 1 cm. Corolla yellow, c. 3 cm. across, lobes fimbriate-ciliate. Fl. 7–8. Pollinated by various insects, heterostylous. $2n = 54$. Hyd.

Native. Ponds and slow rivers in E. and C. England from Sussex and Kent to Berks, E. Gloucester, Shropshire and S. Yorks, local; occasionally introduced elsewhere. 23. C. & S. Europe from S. Sweden, Denmark to N. Spain and Portugal, Sardinia, Italy and Thrace; N. and W. Asia to the Caucasus, Himalaya and Japan; naturalized in N. America (District of Columbia).

102. POLEMONIACEAE

Annual or perennial herbs and a few shrubs. Lvs alternate or opposite, exstipulate. Fls hermaphrodite, actinomorphic, hypogynous. Sepals 5. Corolla with basal tube and 5 free lobes above. Stamens 5, epipetalous, alternating with the corolla-lobes. Ovary superior, 3-celled with a single style and usually 3 stigmas. Fr. a capsule. Seeds endospermic.

About 12 genera with 270 spp., chiefly N. American.

1. POLEMONIUM L.

Usually perennial herbs with *alternate pinnate lvs* and showy fls. Corolla rotate; *stamens inserted all at the same height* in the corolla-tube, the *broadened downwardly-curved hairy bases of the filaments almost closing the throat of the corolla.*

25–30 spp., chiefly in America but some in Europe and Asia.

1. P. caeruleum L.　　　　　　　　　　　　　　　　　　Jacob's Ladder.

A perennial herb with short creeping rhizome and erect simple lfy stem 30–90 cm., hollow, angled, ± glandular-pubescent above. Lvs 10–40 cm., pinnate, with terminal and 6–12 pairs of lateral lflets; lower lvs with long slender winged stalks; upper lvs smaller, subsessile; lflets 2–4 cm., ovate-lanceolate or oblong, acuminate, entire, glabrous. Infl. corymbose, ± bractless. Fls 2–3 cm. diam., many, drooping. Calyx campanulate with acute teeth. Corolla blue or white, shortly tubular (2 mm.), with spreading ovate subacute lobes. Stamens exserted. Style ultimately exceeding the stamens, with 3 slender stigmas. Nectar secreted from a fleshy ring round the base of the ovary. Capsule erect, included in the calyx-tube. Seeds 4–6 in each cell, angular, rugose, shortly winged. Fl. 6–7. Fls protandrous, visited by hover-flies and humble-bees. The population at Malham Cove is gynodioecious. $2n = 18$. Hs.

Native. Locally on grassy slopes, screes and rock-ledges on limestone hills; to 1900 ft. in N. England. England from Stafford and Derby northwards to the Cheviots. 6. Widely introduced as a garden-escape. N. and C. Europe. Caucasus; Siberia; N. America.

Species of *Gilia*, with *equally inserted* stamens whose filaments are *not downwardly curved*, and of *Collomia*, with stamens *inserted at different heights* in the corolla-tube, are sometimes found as casuals. *Phlox* spp., *with opposite entire lvs*, are commonly grown in gardens.

103. BORAGINACEAE

Herbs, sometimes shrubs or trees, often hispid or scabrid, sometimes glabrous. Stems usually terete. Lvs alternate, very rarely opposite, exstipulate, entire, sometimes sinuate. Fls often in scorpioidal cymes, actinomorphic or sometimes zygomorphic. Calyx 5-toothed, sometimes deeply so. Corolla 5-lobed, rotate, funnel-shaped or campanulate, often pink in bud then bright blue in our spp.; throat often closed by scales or hairs. Stamens 5, inserted on the corolla and alternating with its lobes. Ovary superior, 2-celled or (in our spp.) 4-celled by false septa and deeply 4-lobed; style simple, terminal or from the middle of the 4 lobes (gynobasic). Fr. of 4 nutlets, rarely a drupe. Seeds usually without endosperm, embryo straight or curved.

About 85 genera with 1600 spp., cosmopolitan but specially abundant in the Mediterranean region and E. Asia.

1 Lvs all petioled, rounded or cordate at base.   2. OMPHALODES
   At least the upper lvs sessile, though sometimes narrowed towards the base.   2

2 Calyx-lobes toothed, enlarging considerably in fr. and forming a compressed 2-lipped covering round the fr.; plant procumbent; fls 3 mm. diam.   3. ASPERUGO
   Calyx not or only slightly enlarging in fr., not compressed and 2-lipped, lobes not toothed.   3

3 Nutlets covered with hooked or barbed bristles; calyx-teeth spreading nearly horizontally, not concealing fr.   1. CYNOGLOSSUM
   Nutlets without bristles; calyx ±concealing fr.   4

4 Some or all the stamens long exserted; plant hispid.   5
   Stamens all included or if slightly exserted then plant quite glabrous.   7

5 Stamens all exserted; corolla regular.   6
   Stamens not all exserted; corolla somewhat zygomorphic.   13. ECHIUM

6 Stamens glabrous; anthers 8–10 times as long as broad; connective prolonged; plant annual.   5. BORAGO
   Stamens hairy; anthers 2–3 times as long as broad; connective not prolonged; plant perennial.   6. TRACHYSTEMON

7 Plant glabrous, very glaucous; lvs punctate (sea-shores in the north).   12. MERTENSIA
   Plant ±pubescent or hispid, not glaucous; lvs not punctate.   8

8 Fls nodding.   4. SYMPHYTUM
   Fls erect.   9

9 Corolla with hairy or papillose oblong scales or folds in the throat.   10
   Corolla with glabrous rounded scales in the throat.   10. MYOSOTIS

10 Calyx divided for $\frac{1}{4}$–$\frac{1}{2}$ its length; fl. spring.   9. PULMONARIA
    Calyx divided almost to base; fl. summer.   11

11 Corolla with long hairy folds in the throat.   11. LITHOSPERMUM
    Corolla with conspicuous scales in the throat.   12

12 Perennial; lvs ovate; corolla-tube straight.   7. PENTAGLOTTIS
    Annual; lvs linear-lanceolate to oblong; corolla-tube curved.   8. ANCHUSA

1. CYNOGLOSSUM L.

Annual or biennial hispid or silky herbs. Fls in cymes, usually ebracteate. Calyx 5-partite. Corolla funnel-shaped, mouth closed with prominent scales, tube fairly long, lobes obtuse. Stamens included. *Nutlets 4, flat or convex, covered with hooked or barbed bristles, attached to the conical receptacle by a narrow outgrowth of the lower surface.* The fls have nectar and are pollinated mainly by bees; self-pollination also occurs.

About 60 spp. in temperate and subtropical regions, especially Asia.

Lvs grey with silky ±appressed hairs; fr. with a thickened border.
                                                                    1. officinale
Lvs green, sparsely hispid, upper-surface nearly glabrous; fr. without a
    thickened border.   **2. germanicum**

1. C. officinale L.   Hound's-tongue.

An erect *soft grey-pubescent* biennial 30–90 cm. Radical lvs up to c. 30 cm., lanceolate to ovate, usually acute, petioled; upper stem-lvs sessile, lanceolate, usually acute; *all with rather silky ±appressed hairs on both surfaces*, upper surface of old lvs sometimes rough with papillae. Cymes usually branched, lengthening to 10–25 cm. after flowering. *Pedicels c. 1 cm.*, stout, recurved in fr. Calyx-lobes oblong or ovate, obtuse, 5–7 mm. in fr. Corolla c. 1 cm. diam., dull red-purple, rarely whitish. *Nutlets 5–6 mm.*, flattened, ovate, *surrounded by a thickened border* and covered with short barbed *spines, all of about the same length.* Fl. 6–8. $2n=24$. Hs. (biennial). Plant smells of mice.

Native. In grassy places and borders of woods on rather dry soils, on sand, gravel, chalk or limestone, particularly near the sea. 77, H14, S. Widely distributed but local, north to Angus; S. and E. Ireland from Cork to Down. Europe, probably introduced in the extreme north and nearly absent from the Mediterranean region; Asia; N. America, probably introduced.

2. C. germanicum Jacq.   'Green Hound's-tongue.'
C. montanum auct.

Similar to *C. officinale*, but *green*, rough and usually more slender. *Lvs sparsely hispid with short spreading hairs beneath, nearly glabrous above. Pedicels c. 5 mm.* Calyx-lobes oblong, obtuse, up to c. 10 mm. in fr. Corolla 5 mm. diam. *Fr. without a thickened border but with marginal spines longer than the others.* Fl. 5–7. Hs.

Native. In woods and hedgebanks, rare and apparently diminishing. 19. Recorded from a number of counties in S. England and the Midlands, particularly around London. C. France to N. Spain, eastwards to the Carpathians, Caucasus and Asia Minor; introduced in eastern N. America.

2. OMPHALODES Mill.

Annual or perennial, glabrous or nearly glabrous herbs. Calyx deeply 5-toothed. Corolla rotate, blue or white, tube very short, lobes obtuse, *throat closed by 5 obtuse scales.* Stamens included. *Nutlets smooth, compressed, margins membranous, inrolled leaving a groove round the edge of the nutlet, margin ciliate or dentate; nutlet attached to receptacle by the inner margin.*

About 24 spp. in Europe, Asia, Algeria and Mexico.

*1. O. verna Moench   Blue-eyed Mary.

A shortly pubescent far-creeping stoloniferous perennial 10–30 cm. high. Lvs ovate, acuminate, rounded or (the lower) cordate at base; petioles c. 10 cm. Flowering stems erect or ascending, with few lvs. Cymes short, very lax, ebracteate. Fr. pedicels recurved. Calyx with appressed hairs. Corolla 10–15 mm. diam., bright blue, 1–2 times as long as calyx. Nutlets with ciliate margins. Fl. 3–5. $2n=42$. Hs.

Introduced. Frequently cultivated and sometimes naturalized in woods near dwellings. Lombardy to Hungary; naturalized in France and probably elsewhere.

3. ASPERUGO L.

A hispid procumbent annual. Cymes axillary, 1–3-fld. Fls small, blue, on short recurved pedicels. *Calyx* deeply 5-lobed, *lobes lf-like, enlarging and forming a compressed, 2-lipped covering round the fr.* Corolla funnel-shaped, throat closed by scales, lobes rounded. Stamens included. *Nutlets 4, laterally compressed, ovate, finely warty, attached to the convex receptacle by the margin.* Fls contain nectar but are usually self-pollinated.

One sp. in Europe and Asia.

*1. A. procumbens L.   Madwort·

A hispid herb 15–60 cm. Stems angled, with stiff downward directed hairs on the angles. Lvs 2–7 cm., ovate to oblong-elliptic, obtuse, lower petioled, upper sessile, subopposite or ± whorled. Fls c. 3 mm. diam., at first purplish becoming blue, tube and scales white. Fr. calyx-lobes c. 8 mm., deltoid, toothed, conspicuously veined. Fl. 5–7. $2n=48$. Th.

Introduced. In waste places, margins of arable fields and near ports, rare and usually casual but ± naturalized in a few localities. Europe from Scandinavia southwards; W. Asia; N. Africa.

4. SYMPHYTUM L.

Hispid perennial herbs. Radical lvs petioled, cauline usually sessile or decurrent. Fls in terminal forked scorpioidal cymes, nodding. Calyx campanulate or tubular, 5-toothed. *Corolla* funnel-shaped or subcylindrical, *shortly and broadly 5-lobed. Scales 5, linear or subulate, ciliate, connivent,* included, rarely exserted. Stamens included. *Nutlets 4, ovoid, smooth or granulate, base annular, toothed, teeth clasping the receptacle.*

About 25 spp. in Europe, Asia Minor, western Siberia and Persia.

1	Rhizome slender, far-creeping; fl. stems rarely more than 20 cm.	**6. grandiflorum**
	Rhizome short, stout; fl. stems taller.	2
2	Rhizome swollen and tuberous; roots fibrous; stems simple or with 1 or 2 short branches near the top; middle cauline lvs considerably larger than lower ones.	**5. tuberosum**
	Roots thick and tuberous; stems much branched; lower lvs largest.	3
3	Calyx-teeth at least as long as tube.	4
	Calyx-teeth c. ½ length of tube.	**4. orientale**
4	Upper lvs shortly petioled; calyx-teeth obtuse.	**2. asperum**
	Upper lvs sessile, often decurrent; calyx-teeth acute.	5

5 Cauline lvs strongly decurrent, stem broadly winged; fls yellowish-white,
      sometimes purplish or pinkish.   **1. officinale**
   Cauline lvs slightly decurrent, stem narrowly winged; fls blue or purplish-
      blue.   **3. × uplandicum**

1. S. officinale L.   Comfrey.

An erect hispid perennial 30–120 cm. Root thick, fleshy, fusiform, branched.
Stem clothed with long deflexed conical hairs, branched, *winged with decurrent*
lf-bases. Lvs sparsely pilose, rarely with tuberclate bristles; the lower 15–
25 cm., ovate-lanceolate, petioled; upper oblong-lanceolate, *broadly decurrent*
at base. Calyx 7–8 mm., teeth lanceolate-subulate, twice as long as tube.
Corolla 15–17 mm., whitish, yellowish-white, purplish or pink. Scales triangular-subulate, scarcely longer than the stamens. Nutlets shining, black.
Fl. 5–6. $2n=36$. Hs. Still used in country districts as a poultice.

   Native. In damp places, especially beside rivers and streams. 105, H19, S.
Generally distributed throughout Great Britain, though less common in the
north and probably not native there; common but probably nowhere native
in Ireland. C. Scandinavia to Spain and eastwards to western Siberia and
Turkey.

*2. S. asperum Lepech.   'Rough Comfrey.'

S. asperrimum Donn ex Sims

A scabrid perennial up to c. 150 cm. Root thick, branched. *Stems* much
branched, *covered with short stout hooked bristles*. Lvs ovate or elliptic,
scabrid or with tuberclate bristles; lower 15–19 cm., cordate or rounded at
base, petioled; *upper very shortly petioled*, cuneate at base. *Calyx 3–5 mm.*,
accrescent, covered with short stout bristles; *teeth* linear-oblong, *obtuse*, becoming ± triangular in fr., 1–2 times as long as tube. Corolla 9–14 mm., at
first pink, becoming clear blue. Scales lanceolate, about equalling stamens.
Nutlets granulate. Fl. 6–7. $2n=40$. Hs.

   Introduced. Formerly cultivated, now occasionally naturalized in waste
places. Caucasus and Persia.

3. S. × uplandicum Nyman (*S. asperum × officinale*)   'Blue Comfrey.'

Incl. *S. peregrinum* auct.

Similar in general appearance to the two preceding and showing various
combinations of their characters. Stems hispid to scabrid. Upper lvs usually
shortly and narrowly decurrent. Calyx sometimes accrescent, teeth acute or
subacute. Corolla blue or purplish-blue. Fl. 6–8. $2n=36*$. Hs. *S. pere-*
grinum auct., ?Ledeb. is a ± central form of this hybrid swarm.

   The commonest *Symphytum* of roadsides, hedgebanks, woods, etc., but
usually absent from the waterside habitats occupied by *S. officinale*. Distribution imperfectly known through confusion with *S. officinale*, which appears
to be much less common.

*4. S. orientale L.

A *softly pubescent* perennial up to c. 70 cm. Root fusiform, branched. Stems
sparsely puberulent and pilose, branched. Lvs softly pubescent, ovate or

oblong, subacute, base cordate, truncate or rounded; lower up to 14 cm., often less, petioled, petiole narrowly winged at top; upper subsessile. *Calyx* 7–9 mm., tubular, *teeth c.* ½ *length of tube*, ovate or oblong, obtuse. Corolla 15–17 mm., white. Scales broad-subulate, slightly exceeding the stamens. Nutlets tuberclate, dark brown. Fl. 4–5.

Introduced. Naturalized in hedgebanks and grassy places, not uncommon in some districts. Turkey.

5. S. tuberosum L. 'Tuberous Comfrey.'

A hispid perennial 20–50 cm. Root fibrous; *rhizome stout, tuberous. Stems* covered with reflexed bristles, *simple or with one or two short branches near the top.* Lvs puberulent and densely hispid; lower small, ovate or spathulate, narrowed at base, petioled; *middle cauline 10–14 cm., considerably larger than the lowest,* ovate-lanceolate or elliptic, shortly petioled; upper sessile. Calyx 7–8 mm., teeth lanceolate-acute, 3 times as long as tube. Corolla 12–16 mm., yellowish-white. Scales broadly triangular-subulate, acuminate, somewhat exceeding the stamens. Nutlets minutely tuberclate. Fl. 6–7. $2n=$ c. 72.

Native. In damp woods and hedgebanks. 61. Scattered throughout Great Britain, local but commoner in the north; an escape from cultivation in Ireland. Germany to Spain and eastwards to S.W. Russia and Turkey.

*6. S. grandiflorum DC.

Hispid perennial with *long, slender rhizomes.* Lower lvs elliptic to ovate, long-petioled; blade 5–10 cm.; cauline lvs smaller, uppermost sessile. *Calyx* 4–5 mm., accrescent, *divided nearly to base,* teeth linear-lanceolate, obtuse. Corolla 10–15 mm., yellowish-white. Style c. 20 mm., persistent. Fruiting calyx c. 10 mm. Hs. Fl. 4–5 and sporadically later.

Introduced. Not infrequently naturalized in hedges and woods in S. England and the Midlands. Sometimes mistaken for *S. tuberosum.* Native of the Caucasus.

5. BORAGO L.

Annual or perennial hispid herbs. Fls blue, bracteate, in lax forked cymes. Calyx 5-partite. Corolla rotate, with notched scales in the throat, lobes acute. Stamens inserted in throat of corolla; *filaments broad and flattened with a narrower obtuse prolongation parallel to the anthers; anthers oblong, mucronate, exserted, connivent. Nutlets 4, rugose, base concave; receptacle flat, fleshy.* The fls have nectar and are pollinated by bees.

Three or four spp. in the Mediterranean region, one extending into C. Europe.

*1. B. officinalis L. Borage.

A stout erect hispid annual 30–60 cm. Lvs 10–20 cm., ovate, obtuse or acute, lower petioled, upper sessile. Cymes axillary and terminal, few-fld; pedicels 2–4 cm.; bracts linear or lanceolate, lower lf-like. Calyx-teeth c. 1 cm., linear-lanceolate, very hispid. Corolla 2 cm. diam., bright blue. Anthers purple-black. Fl. 6–8. $2n=16$. Th.

Introduced. A garden-escape on waste ground near houses. C. Europe and the Mediterranean region; introduced in America.

6. Trachystemon D. Don

Similar to *Borago* but fls with a longer corolla-tube and anthers much shorter than filaments and without a prolongation of the connective.

Two spp. in the eastern Mediterranean region.

*1. T. orientalis (L.) G. Don

A hispid perennial herb with a stout mucilaginous rhizome up to c. 5 cm. diam., clothed with blackish, persistent lf-bases. Radical lvs long-stalked with an ovate, ± obtuse blade c. 30 cm. Fls similar to those of *Borago* but rather smaller. Stamens hairy. Fl. 4–5. Hs.

Introduced. Naturalized in damp woods in Devon, Kent, Yorks and probably elsewhere. Eastern Mediterranean region.

7. Pentaglottis Tausch

Perennial hispid herbs. Lvs ovate, net-veined. Fls actinomorphic, blue, in bracteate scorpioidal cymes. Calyx 5-toothed. *Corolla* rotate; *tube straight*, shorter than or as long as the obtuse lobes; *throat closed by scales*. Stamens included. *Nutlets concave at base*, with a small stalked attachment.

One sp. in Europe.

1. P. sempervirens (L.) Tausch   Alkanet.

Anchusa sempervirens L.; *Caryolopha sempervirens* (L.) Fisch. & Trautv.

A somewhat hispid perennial with the habit of a *Symphytum*, 30–100 cm. Lvs ovate, acute or acuminate, entire and scarcely undulate, lower petioled, up to 30 cm. Cymes very hispid, in long-peduncled, axillary, subcapitate pairs, each cyme subtended by a lf-like bract 1·5–3 cm. Fls subsessile. Calyx-teeth linear. Corolla c. 10 mm. diam., bright blue, tube shorter than lobes; throat with white scales. Nutlets reticulate. Fl. 5–6. $2n=22$. Hs.

?Introduced. Naturalized in hedgerows and at borders of woods near buildings, possibly native in a few places in S.W. England. 79, H12, S. Widely distributed but rather rare, except in S.W. England. Western Europe from Belgium to Spain and Portugal; Lombardy.

8. Anchusa L.

Annual or perennial, usually hispid, herbs. Fls in usually bracteate scorpioidal cymes. Calyx ± deeply divided into 5 lobes. Corolla-tube straight or curved, at least as long as the regular or oblique limb; throat closed by scales or hairs. Stamens included. *Nutlets* deeply concave at base, *not stalked*. About 30 spp. in Europe and Asia.

1. A. arvensis (L.) Bieb.   Bugloss.

Lycopsis arvensis L.

An erect very hispid annual or biennial 15–50 cm. Hairs with swollen bulbous bases. Lvs up to c. 15 cm., obovate-lanceolate to linear-oblong, obtuse or apiculate, margins undulate, distantly and irregularly toothed; lower lvs narrowed into a petiole, upper sessile, ½-amplexicaul. Cymes simple or forked, at first subcapitate, elongating after flowering. Fls subsessile, bracts lf-like. Calyx-teeth linear-lanceolate, enlarging a little in fr. Corolla 5–6 mm. diam., bright blue, tube abruptly curved; scales white. Nutlets 4 mm., reticulate. Fl. 6–9. $2n = 54$. Th. or Hs.

Native. On light sandy and chalky soils in arable fields, sandy heaths and near the sea; probably introduced in the north and perhaps elsewhere. 110, H 14, S. Widely distributed and locally common in Great Britain; Ireland, chiefly near the sea in the east. Throughout the greater part of Europe and Asia eastwards to Tibet; absent from the extreme north and south.

In addition to the above, 2 spp., *A. officinalis* L. and *A. azurea* Mill. (*A. italica* Retz.), with cymes forming a terminal panicle, occur occasionally in waste places. *A. officinalis* has ovate-lanceolate bracts and the calyx divided about to the middle; *A. azurea* has linear-lanceolate bracts and the calyx divided almost to the base.

9. PULMONARIA L.   Lung-wort.

Perennial herbs with creeping rhizome usually ending in sterile shoots. Flowering stems simple, cymes terminal. Fls usually heterostylous, purple or blue, often pink in bud. Calyx 5-angled at base, cylindrical or campanulate in fl., enlarging somewhat and becoming strongly campanulate in fr., lobes erect. *Corolla funnel-shaped, with 5 tufts of hair alternating with the stamens.* Stamens included. *Nutlets with a raised ring round the base; receptacle flat.* Fls have nectar and are pollinated chiefly by humble-bees.

About 10 spp. in Europe and W. Asia.

Radical lvs lanceolate, gradually narrowed at base; corolla 5–6 mm. diam.;
    nutlets compressed, crested.   **1. longifolia**
Radical lvs ovate, abruptly contracted at base; corolla c. 10 mm. diam.;
    nutlets ovoid, acute.   **2. officinalis**

1. P. longifolia (Bast.) Bor.

P. angustifolia auct., non L.

A pubescent perennial 20–40 cm. Stems not scaly at base. *Radical lvs in autumn reaching 40–60 cm., 8–10 times as long as broad, lanceolate, gradually attenuate at base,* spotted with white; cauline lvs lanceolate or ovate-lanceolate, sessile, ½-amplexicaul. Cymes short, scarcely elongating after flowering. *Calyx c. 10 mm., campanulate, teeth c. ½ length of calyx,* lanceolate, acute. *Corolla 5–6 mm. diam.,* pink then blue. *Nutlets strongly compressed, ovate, crested,* shining. Fl. 4–5. $2n = 14$. Hs.

Native. In woods and thickets on clay soils, very local. 3. Dorset, S. Hants and Isle of Wight. W. Europe from Sweden to France.

*2. P. officinalis L.

A pubescent perennial 10–30 cm. *Radical lvs c. 10 cm., 1½ times as long as broad, ovate, cuspidate or shortly acuminate, often cordate at base, abruptly narrowed into a winged petiole,* white-spotted or unspotted; cauline lvs ovate, sessile, ½-amplexicaul. Cymes short, scarcely elongating after flowering. *Calyx 6–7 mm., nearly cylindrical, teeth ⅓–¼ length of calyx,* triangular, sub-

obtuse. *Corolla c.* 10 *mm. diam.*, pink then blue. *Nutlets ovoid, acute.* Fl. 3–5.
$2n = 14$. Hs.

Introduced. Naturalized in woods and on hedgebanks in a number of
localities, mainly in southern England. 26. C. and northern Europe; Caucasus.

10. MYOSOTIS L.

Forget-me-not, Scorpion Grass.

Annual or perennial ± hairy herbs. Cymes terminal, scorpioidal, sometimes
bracteate at base. Calyx 5-toothed, sometimes divided nearly to base. Corolla
rotate, usually pink in bud, ultimately blue; *throat closed by 5 short notched
scales*; lobes emarginate or entire, flat or concave, contorted in bud. Stamens
included. Style short, stigma capitate. *Nutlets small, shining, lenticular or
subtrigonous, base narrow; receptacle small.* The fls contain nectar and may
be cross-pollinated by insects though self-pollination is also possible and
probably more frequent; it is known to result in good seed-production in
several spp.

About 40 spp. in the temperate regions of both hemispheres.

1   Hairs on calyx-tube appressed, rarely almost 0.   *2*
     At least some hairs on calyx-tube short, stiff, hooked or crisped, not
     appressed, calyx never subglabrous.   *6*

2   Corolla 4 mm. or more in diam., if less then lobes emarginate; plant
     perennial.   *3*
     Corolla not more than 4 mm. diam., lobes rounded and entire, not emar-
     ginate; plant annual or biennial.   *5*

3   At least most lvs 3–5 times as long as broad; fls bright blue.   *4*
     Lvs scarcely more than twice as long as broad; fls very pale blue.
     3. brevifolia

4   Fr. pedicels 1–2 times as long as calyx; calyx-teeth not more than ⅓ length
     of calyx; cymes ebracteate.   **1. scorpioides**
     Fr. pedicels 3–5 times as long as calyx; calyx-teeth ½ length of calyx;
     cymes bracteate below.   **2. secunda**

5   Calyx pubescent; corolla-lobes flat; style ½ length of calyx-tube (rarely
     more); nutlets truncate at base (wet places, common).   **4. caespitosa**
     Calyx almost glabrous; corolla-lobes concave; style equalling calyx-tube;
     nutlets rounded at base (dunes, Jersey).   **5. sicula**

6   Fr. pedicels as long as or longer than calyx; cymes in fr. shorter or not
     much longer than lfy part of stem.   *7*
     Fr. pedicels shorter than calyx; if as long as or slightly longer than calyx,
     then cymes in fr. much longer than lfy part of stem and nutlets truncate
     at base.   *9*

7   Corolla 4–10 mm. diam., lobes flat.   *8*
     Corolla usually smaller, lobes concave.   **8. arvensis**

8   Fr. pedicels about as long as calyx, lower slightly longer; nutlets black
     (alpine rocks, rare).   **6. alpestris**
     Fr. pedicels 1½–2 times length of calyx; nutlets dark brown (damp woods,
     locally abundant).   **7. sylvatica**

9 Corolla at first yellow or white; tube at length about twice as long as calyx; calyx-teeth ultimately nearly erect; nutlets rounded at base, dark brown; cymes in fr. not much longer than lfy part of stem.   **9. discolor**
   Corolla never yellow; tube shorter than calyx; calyx-teeth spreading; nutlets truncate at base, pale brown; cymes in fr. much longer than lfy part of stem.   **10. ramosissima**

1. M. scorpioides L.   'Water Forget-me-not.'

M. palustris (L.) Hill

A ± rhizomatous often stoloniferous perennial 15–45 cm. Stem decumbent or erect, angular, ± hairy. Lower lvs up to c. 7 cm., often less, oblong-lanceolate to obovate-lanceolate, usually obtuse, attenuate at base but scarcely petioled, subglabrous or with short appressed hairs; upper lvs narrower and often apiculate. *Cymes ebracteate. Fr. pedicels 1–2 times as long as calyx*, rarely longer, spreading or reflexed. Calyx campanulate, hairs appressed; *teeth* triangular, $\frac{1}{4}$–$\frac{1}{3}$ *length of calyx.* Corolla (3–)4–10 mm. diam., sky-blue, rarely white; lobes flat, emarginate. Style equalling calyx-tube to longer than calyx. *Nutlets c.* 1·5 × 1 *mm., narrowly ovoid*, obtuse, slightly bordered, not keeled, *black* and shining. Fl. 5–9. $2n = 64$. Hel. or Hs.

Native. In wet places by streams and ponds. 108, H40, S. Common and generally distributed throughout the British Is., to 1650 ft. in Perth. Europe north to c. 68°; Asia, south to N. India; N. Africa; N. America.

2. M. secunda A. Murr.   'Water Forget-me-not.'

M. repens auct.

An erect ± pubescent perennial 20–60 cm., perennating by means of stolons. Rhizome short, scarcely creeping. Stem erect, with spreading hairs below; *decumbent or prostrate stems arising from its base, the sterile ones rooting at the nodes.* Lower lvs ovate-spathulate, obtuse, sparsely hairy, ciliate below; upper lvs oblong-lanceolate, obtuse or subacute. Cymes lax, bracteate below. *Fr. pedicels 3–5 times as long as calyx*, reflexed. Calyx with appressed hairs, campanulate in fr.; *teeth* lanceolate, acute, $\frac{1}{2}$ *length of calyx.* Corolla 4–6 mm. diam., blue, lobes slightly emarginate. Style equalling or slightly exceeding calyx-tube. *Nutlets c.* 1 × 0·75 *mm., broadly ovoid*, obtuse, slightly bordered, not keeled, *dark brown*, shining. Fl. 5–8. Hel. or Hs.

Native. In wet, often peaty, places, usually avoiding calcareous soils. 102, H37, S. Fairly generally distributed and commoner in mountain districts than *M. palustris.* Ascends to 2640 ft. on Carnedd Llewelyn (Caernarvon). N. Europe.

3. M. brevifolia C. E. Salmon

An erect ± pubescent *dark bluish-green* perennial 12–20(–30) cm. *Erect stems producing roots and stolons from their lower nodes, the stolons bearing small lvs; hairs appressed*, slightly spreading-ascending towards base. *Lvs* short, broad, rounded, obtuse or emarginate, *scarcely more than twice as long as broad, only the lowest tapering at base.* Branches of infl. from below the middle of the stem, usually ebracteate. Fr. pedicels spreading or recurved, 1–2 times as long as calyx. Calyx narrowly campanulate, hairs appressed;

teeth oblong, rounded or obtuse, $\frac{1}{2}$–$\frac{2}{3}$ length of calyx. Corolla c. 5 mm. diam., pale blue; lobes entire or emarginate. Style slightly exceeding calyx-tube. Nutlets c. 1 mm., ovoid, obtuse, olive-brown. Fl. 6–8. Hel. or Hs.

Native. In wet places in mountainous districts. 9. Yorks, Durham, Westmorland, Cumberland, Dumfries and perhaps elsewhere in the north. ?Endemic.

4. M. caespitosa K. F. Schultz   'Water Forget-me-not.'

M. laxa Lehm. ssp. *caespitosa* (K. F. Schultz) Hyl.

An erect ± pubescent annual or biennial 20–40 cm. *Stems* simple or branched from the base, *terete*, faintly ribbed, *glabrous or with appressed hairs. Lvs* lanceolate, narrowed at base, obtuse, *with appressed hairs above and, at least the lower, subglabrous beneath.* Cymes usually bracteate at base. Fr. pedicels spreading, 2–3 times as long as calyx. Calyx campanulate in fr., hairs appressed; teeth triangular-ovate, subacute, $\frac{1}{2}$–$\frac{3}{4}$ length of calyx. *Corolla* 2–4 *mm. diam.*, sky-blue, rarely white; *lobes rounded. Style c.* $\frac{1}{2}$ *length of calyx-tube*, rarely equalling it. *Nutlets* c. 1·3 × 1 mm., broadly ovoid, *truncate at base*, obtuse, slightly bordered, not keeled, dark brown, shining. Fl. 5–8. $2n=$c. 80. Hel. or Hs.

Native. In marshes and beside streams and ponds. 112, H40, S. Common and generally distributed throughout the British Is. Europe; Asia north to Siberia and east to the Himalaya; N. Africa.

5. M. sicula Guss.

An erect or decumbent ± appressed-pubescent annual or perhaps biennial 5–10(–30) cm. *Stems* solitary or several, simple or with *divaricate flexuous branches, glabrous or nearly so below, with scattered hairs above. Lower lvs* 2·5–3 × 0·5–1 *cm.*, oblong-spathulate, obtuse, glabrous or subglabrous beneath, with scattered hairs above; upper lvs 1–2 × 0·3–0·4 cm., linear-oblong, obtuse, hairy on both surfaces. Cymes elongate, flexuous, rarely short, lower branches usually divaricate; bracts sometimes present. *Fr. pedicels* 1–3 times as long as calyx, spreading or reflexed, *thickened above. Calyx usually with a few appressed hairs at base*, oblong-campanulate; *teeth* oblong, obtuse, c. $\frac{1}{2}$ length of calyx, *subconnivent in fr. Corolla* 2$\frac{1}{2}$–3 *mm. diam.*, blue; *lobes concave*, entire. Style equalling calyx-tube. Nutlets 1–1·25 × 0·6–0·75 mm., narrowly ovoid, obtuse, slightly bordered, not keeled, pale or dark brown, shining. Fl. 4–6. Th. or ?Hs.

Native. In damp places on fixed dunes. Jersey, very local. France, Portugal, Spain, Italy, Corsica, Sardinia, Sicily, Albania, Greece, Bulgaria and Serbia.

6. M. alpestris Schmidt

A pubescent rhizomatous perennial 5–20 cm. Stems stiff, hairs spreading. Lvs oblong-lanceolate, acute or subacute, the lower long-petioled, the upper sessile, all with ± spreading hairs on both surfaces. *Cymes ebracteate, rather short. Fr. pedicels ascending, as long as the calyx or the lower slightly longer. Calyx* campanulate, *rather silvery*, hairs ± spreading, *with a few short stiff hooked ones on the tube*; teeth narrow-lanceolate, $\frac{1}{2}$–$\frac{3}{4}$ length of tube, erect or somewhat spreading in fr. Corolla 4–8 mm. diam., blue; lobes flat, rounded. Style longer than calyx-tube. *Nutlets* roundish-ovoid, keeled on one surface, *black*. Fl. 7–9. $2n=24$. Hs.

Native. On basic mountain rocks, rare and local. Mickle Fell, Westmorland; Ben Lawers, Perth, 2400–3900 ft.; ?Angus. Alpine and subalpine Europe, Asia and N. America.

7. M. sylvatica Hoffm.   'Wood Forget-me-not.'

An erect pubescent perennial 15–45 cm. Stems simple or branched, hairs spreading. Lower lvs ovate-spathulate, obtuse, forming a rosette; stem-lvs lanceolate to oblong, ± acute, sessile, all with ± spreading hairs on both surfaces. *Cymes* ebracteate, *lax, much elongated after flowering. Fr. pedicels spreading*, $1\frac{1}{2}$–2 *times length of calyx. Calyx* campanulate, *with short crisped or hooked hairs on the tube*; teeth oblong-lanceolate, acute, $\frac{2}{3}$–$\frac{3}{4}$ length of calyx, spreading in fr. *Corolla* 6–10 *mm. diam.*, rarely less, bright blue rarely white; *lobes flat*, rounded; *tube equalling or exceeding calyx. Style longer than calyxtube. Nutlets* c. 2×1 mm., ovoid, acute, bordered, keeled on one face, *dark brown*, shining. Fl. 5–6(–9). $2n=18, 32$. Hs.

Native. In damp woods, locally abundant. 61. Scattered throughout Great Britain, rare in the north. Europe to 58° N.; Asia to the Himalaya.

8. M. arvensis (L.) Hill   'Common Forget-me-not.'

An erect pubescent annual occasionally perennating, 15–30 cm. Stem branched, hairs spreading. Lower lvs roundish-ovate, stalked, obtuse, forming a rosette; stem lvs oblong-lanceolate, acute, sessile, all with ± spreading hairs on both surfaces. Cymes ebracteate, lax, elongated after flowering and then shorter or slightly longer than the lfy stem. Fr. pedicels spreading, up to twice length of calyx. Calyx campanulate, with abundant short crisped or hooked hairs on the tube; teeth narrowly triangular, $\frac{2}{3}$–$\frac{3}{4}$ length of calyx, spreading in fr. *Corolla up to* 5 *mm. diam., usually less*, bright blue; *lobes concave; tube shorter than calyx. Style shorter than calyx-tube.* Nutlets c. $1\cdot5 \times 0\cdot75$ mm., ovoid, acute, scarcely bordered, keeled on one face, dark brown, shining. Fl. 4–9. $2n=24$, c. 48, c. 54. Th.

Native. In cultivated ground, by roads, in woods and on dunes. 112, H40, S. Common and generally distributed throughout the British Is. Europe, Siberia, W. Asia.

9. M. discolor Pers.   'Yellow and blue Forget-me-not.'

M. versicolor Sm.

A slender erect pubescent annual 8–25 cm. Stem with spreading hairs below and appressed hairs above. Lvs oblong-lanceolate, lower obtuse, narrowed below, upper acute, sessile, all hairy on both surfaces. *Cymes* ebracteate, lax, *in fr. not much longer than lfy part of stem. Fr. pedicels* ascending, *shorter than calyx. Calyx* campanulate, *closed or nearly so in fr.*, tube covered with hooked hairs; *teeth* oblong-lanceolate, c. $\frac{1}{2}$ length of calyx, *ultimately nearly erect. Corolla* c. 2 *mm. diam., at first yellow* or *white*, usually becoming blue; lobes concave; *tube at length about twice as long as calyx.* Style equalling or exceeding calyx. *Nutlets* c. 1·25 × 0·5 mm., ovoid, obtuse, not keeled, scarcely bordered, *dark brown or almost black*, shining. Fl. 5–9. $2n=$ c. 60. Th.

Native. In grassy places, usually on fairly light soils and in open communities. 112, H 40, S. Generally distributed and locally common throughout the British Is. Europe from S. Scandinavia southward and east to N. Italy and the Carpathians; Azores.

10. M. ramosissima Rochel   'Early Forget-me-not.'

M. collina auct.; *M. hispida* Schlecht.

A slender erect or decumbent pubescent annual 2–25 cm. Stems with spreading hairs below and appressed hairs above. Lower lvs ovate-spathulate, obtuse, forming a rosette, upper oblong, sessile, all hairy on both surfaces. *Cymes* ebracteate, lax, *in fr. much longer than the lfy part of stem.* Fr. pedicels ascending, shorter than to slightly longer than the calyx. *Calyx* campanulate, *open in fr.*, tube covered with hooked hairs; teeth narrow-triangular, $\frac{1}{3}$–$\frac{1}{2}$ length of calyx. *Corolla* c. 2 mm. diam., *blue*, rarely white; lobes concave; *tube shorter than calyx. Nutlets* c. 0·75 × 0·5 mm., ovoid, obtuse, *truncate at base*, not keeled, narrowly bordered, *pale brown*, shining. Fl. 4–6. $2n=$48. Th.

Native. On dry shallow soils, locally common. 98, H 14, S. Fairly generally distributed in the drier parts of the British Is., absent from much of Wales, N.W. England, N. Scotland and W. Ireland. Europe from C. Norway southwards; S.W. Asia; N. Africa.

11. LITHOSPERMUM L.

Annual or perennial herbs or small shrubs, hispid or softly hairy. Fls subsessile (in our spp.) in bracteate cymes. Calyx 5-partite. Corolla rotate or funnel-shaped, *throat with hairy longitudinal folds or small scales.* Stamens included. *Nutlets very hard*, smooth or warty, *base truncate, receptacle flat.* Fls pollinated mainly by bees.

About 50 spp. in temperate regions.

1 Fls 12–15 mm. diam., reddish-purple, then blue; sterile stems creeping,
    flowering erect.   **1. purpurocaeruleum**
    Fls 3–4 mm. diam., white, very rarely blue; all stems erect.   *2*

2 Lvs with prominent lateral nerves; nutlets smooth, shining, white; plant
    perennial.   **2. officinale**
    Lateral nerves not apparent; nutlets warty, greyish-brown; plant annual.
                                                        3. arvense

1. L. purpurocaeruleum L.   'Blue Gromwell.'

A pubescent perennial with a *creeping woody stem from which spring long creeping sterile shoots* and shorter erect flowering ones up to 60 cm. Lvs up to c. 7 cm., narrow-lanceolate, acute, subsessile, rough and dark green above, light green beneath; lateral nerves not apparent. Cymes terminal, elongating after flowering. *Corolla twice as long as calyx*, 12–15 *mm. diam., at first reddish-purple, then bright blue.* Nutlets ovoid, obtuse, white, somewhat shining. Fl. 5–6. $2n=16$. Chh.

Native. In bushy places and at margins of woods on chalk and limestone.

8. S. Devon, N. Somerset, Kent, W. Suffolk, Monmouth, Glamorgan, and Denbigh, very local. Europe from Luxembourg and C. Germany to the Mediterranean, eastwards to the Black Sea, Asia Minor and Persia.

2. L. officinale L. Gromwell (Grummel).

An *erect* pubescent slightly rough perennial 30–80 cm., much branched above. Lvs up to c. 7 cm., lanceolate (rarely ovate), acute, sessile, *lateral nerves conspicuous*. Cymes terminal and axillary, elongating after flowering. *Corolla* not much exceeding calyx, 3–4 mm. diam., *yellowish- or greenish-white*. Nutlets ovoid, obtuse, white, shining. Fl. 6–7. $2n=28$. Hp.

Native. In hedges, bushy places and borders of woods mainly on basic soils. 85, H 35, S. Fairly generally distributed in England and Wales; rare in Scotland and not found north of Inverness; rare but widespread in Ireland. Europe except the extreme south-west and north, eastwards to the Caucasus, Persia and Baikal.

3. L. arvense L. 'Corn Gromwell', 'Bastard Alkanet.'

An erect pubescent somewhat rough *annual* 10–50(–90) cm. Stems usually simple or little branched. *Lvs* up to 3(–5) cm., the *lower obovate, obtuse, narrowed into a petiole, the upper linear-lanceolate or oblong-lanceolate*, acute or subacute, sessile, *lateral nerves not apparent*. Cymes terminal, short. *Corolla* not much exceeding calyx, 3–4 mm. diam., *white, rarely bluish, tube violet or rarely blue. Nutlets trigonous-conical, greyish-brown, warty*. Fl. 5–7. $2n=28$. Th.

Native. Mainly in arable fields. 91, H 18, S. Fairly common and generally distributed in England, rare and perhaps not native in the rest of the British Is. Europe, probably not native in the north; Asia to N.W. India.

12. MERTENSIA Roth

Perennial, often glabrous and glaucous herbs. Cymes terminal. Fls blue-purple or blue and pink, heterostylous. Calyx 5-lobed. Corolla-tube cylindric, lobes campanulate, throat without scales, sometimes with 5 folds. *Stamens* inserted towards the top of the corolla-tube, *slightly exserted*. Style filiform. *Nutlets rather fleshy*, smooth or rough, *narrowed towards the flat base; receptacle small, 2–4-lobed, flat*.

About 30 spp. in north temperate regions.

1. M. maritima (L.) S. F. Gray 'Northern Shore-wort.'

A decumbent glabrous glaucous rather fleshy perennial up to 60 cm. Stems purple-pruinose, lfy. Lvs 2–6 cm., in 2 rows, ovate or obovate, obtuse or apiculate, lower attenuate into a petiole, upper sessile, punctate on the upper surface. Cymes bracteate, often branched, rather short. Pedicels 5–10 mm., elongating somewhat and becoming ± recurved in fr. Calyx-lobes ovate. Corolla c. 6 mm. diam., pink then blue and pink; throat with 5 folds. Nutlets c. 6 mm., flattened, fleshy, outer coat becoming inflated and papery. Fl. 6–8. Probably self-pollinated. $2n=24$. Hp.

Native. On shingle beside the sea. 37, H 7. Norfolk, Caernarvon, Anglesey, N. Lancashire and Northumberland to Shetland, very rare in the south, local in the north; Ireland, E. and N. coasts from Wicklow to Donegal, rare and local; apparently decreasing everywhere. Atlantic coast of Europe from Jutland northwards, Iceland.

13. ECHIUM L.

Usually stout hispid or scabrid herbs, sometimes shrubby. Fls in spiked or panicled unilateral cymes. Calyx 5-partite. *Corolla funnel-shaped with a straight tube and open throat, lobes unequal. Stamens unequal, at least some exserted. Nutlets 4, attached to the flat receptacle by their flat nearly triangular bases.*

About 30 spp. in Europe, the Mediterranean region, Madeira, the Canaries and the Azores.

Lvs rough and harsh, the radical with no apparent lateral veins, the upper
    cauline rounded at base; fls blue; 4 stamens long-exserted.   **1. vulgare**
Lvs not particularly rough, the radical with prominent lateral veins, the
    upper cauline cordate at base; fls red, becoming purple-blue; 2 stamens
    long-exserted.   **2. lycopsis**

1. E. vulgare L.   Viper's Bugloss.

An erect *very rough* hispid biennial 30–90 cm. *Lvs* up to 15 cm.; radical petioled, *with a prominent midrib and no apparent lateral* veins; cauline sessile, lanceolate or oblong, rarely ovate, acute, *rounded at base. Fls 15–18 mm.*, subsessile. Cymes short, dense, elongating after flowering, arranged in a large terminal panicle. Calyx-teeth linear-lanceolate, shorter than corolla-tube. *Corolla* pinkish-purple in bud, *becoming bright blue*, rarely white. *Four stamens long-exserted*, the 5th included. Nutlets angular, rugose. Fl. 6–9. Fls protandrous and visited by a great variety of insects. $2n = 16, 32$. Hs.

Native. In grassy places on light dry soils, sea cliffs and dunes, locally common. Scattered throughout England and Wales; Scotland, rare and perhaps not native, absent from the north; Ireland, mainly coastal and native only in the east. 98, H16, S. Europe from C. Scandinavia to C. Spain and eastwards to the Urals and Asia Minor.

2. E. lycopsis L.   'Purple Viper's Bugloss.'

E. plantagineum L.

An erect biennial 20–60 cm. Somewhat similar in general appearance to *E. vulgare* but much softer. Radical lvs ovate, *lateral veins distinct*; cauline lvs oblong to lanceolate, *the upper cordate at base*; all with appressed hairs, *not very rough to the touch.* Cymes usually distinctly stalked, becoming very long (up to 25–30 cm.) after flowering. *Corolla up to 25–30 mm.* and nearly as broad, *red becoming purple-blue. Two stamens long-exserted*, the others included or only slightly exserted. Fl. 6–8. $2n = 16$. Hs.

Native. On cliffs and sandy ground near the sea. 2 or 3, S. Jersey, Cornwall (and ? S. Devon); occasionally elsewhere as a garden-escape. Mediterranean region, Caucasus, Canaries.

104. CONVOLVULACEAE

Herbs or shrubs, rarely trees; stems often climbing, juice usually milky. Lvs alternate, exstipulate. Fls actinomorphic, hermaphrodite, often large and showy, in terminal or axillary racemes or heads, sometimes solitary. Sepals usually free, 5. Corolla funnel-shaped or sometimes campanulate, usually shallowly 5-lobed or -angled. Stamens 5, inserted towards the base of the corolla-tube and alternating with the lobes. Ovary 1–4-celled, ovules solitary or in pairs in each cell; style terminal. Fr. a capsule, 2–4-valved or splitting transversely near the base, sometimes fleshy and indehiscent. Seeds with endosperm which surrounds the usually folded or bent embryo.

About 40 genera and 1000 spp., mainly tropical but also in the temperate regions.

1	Green plants with large lvs; fls more than 1 cm. diam., solitary or few together, peduncled.	*2*
	Reddish or yellowish parasites with minute scale-like lvs; fls less than 5 mm. diam., sessile or subsessile in dense axillary heads. 3. CUSCUTA	
2	Bracteoles small, distant from the calyx. 1. CONVOLVULUS	
	Bracteoles large, overlapping the calyx. 2. CALYSTEGIA	

*Dichondra repens J. R. & G. Forst., a small softly pubescent plant rooting at the nodes, with ± reniform entire lvs 4–7 × 3–7 mm., and small solitary greenish-white fls with silky calyces, is established in Cornwall.

1. CONVOLVULUS L.

Annual or perennial herbs. Stems often climbing, sometimes woody at base. Lvs alternate, large, green. Fls large, in 1–few-fld axillary or terminal corymbs; *bracteoles small, distant from the fl.* Calyx 5-lobed, divided nearly to base. Corolla funnel-shaped or campanulate, 5-angled, entire or sinuate-lobed. Stamens 5, inserted at bottom of corolla-tube, included; anthers dilated near base. Style filiform, *stigmas 2, linear, slender.* Capsule 2-celled, each cell 1–2-seeded.

About 180 spp., chiefly in temperate regions.

1. C. arvensis L.   Bindweed, Cornbine.

A scrambling or climbing perennial 20–75 cm. Rhizomes stout, often ± spirally twisted, penetrating the earth to depths of 2 m. or more. Stems slender, climbing by twisting in a counter-clockwise direction round the stems of other plants. Lvs 2–5 cm., oblong or ovate, hastate or sagittate, obtuse, often mucronate, ± pubescent when young; petiole shorter than blade. Fls c. 2 cm., solitary or 2–3 together; peduncles exceeding lvs; pedicels with 2 subulate bracteoles c. 2 mm., not overlapping the calyx. Calyx c. 4 mm. Corolla up to c. 3 cm. diam., white or pink, soon withering. Capsule c. 3 mm. diam. Fl. 6–9. The fls have nectar and are scented; they are visited by many different insects which may bring about cross-pollination, though self-pollination also occurs. $2n = 50$. Hp. or Grh.

An exceedingly persistent and noxious weed.

Native. In cultivated land, waste places, beside roads and railways, and in short turf especially near the sea; commonest on the lighter basic soils. 109, H40, S. Generally distributed and common throughout England and Wales, becoming rarer in Scotland and absent from Caithness, Orkney and Shetland; generally distributed in Ireland, but probably introduced in some localities. Throughout the temperate regions of both hemispheres.

2. CALYSTEGIA R.Br.

Characters of *Convolvulus* but *bracteoles large, arising immediately below the calyx and* ± *enclosing it, and stigmas broad.*

About 10 spp. in temperate and subtropical regions.

Stems 1–3 m., climbing; lvs ovate-cordate or -sagittate; bracteoles longer
    than calyx.   **1. sepium**
Stems 10–60 cm., not climbing; lvs reniform; bracteoles shorter than calyx.
                                                        2. soldanella

1. C. sepium (L.) R.Br.   Bellbine, 'Larger Bindweed'.

A *climbing* rhizomatous and stoloniferous glabrous or slightly pubescent perennial 1–3 m. Rhizomes far-creeping, seldom more than 30 cm. deep. Stolons penetrating the soil and rooting at their ends. Stems climbing by twisting in a counter-clockwise direction. *Lvs up to* 15 *cm., ovate-cordate to -sagittate*, obtuse and apiculate to acuminate, petiole usually shorter than blade. Fls solitary, 3·5–7 cm., white or pink; *bracteoles longer than calyx*, nearly flat to strongly inflated, margins overlapping or not. Calyx c. 1 cm. Corolla funnel-shaped, odourless, open by day, closed at night. Capsule 7–12 mm.; seeds 4–7 mm., dark brown. Fl. 7–9. Hp. Self-sterile. Widely distributed in temperate regions.

Ssp. **sepium**: ± glabrous. Corolla 35–40 × 30–35 mm., white, rarely pale pink. Bracteoles not or scarcely inflated, their margins not overlapping and therefore usually not completely hiding the calyx. Filaments c. 10 mm., 1–1·5 mm. wide at base, with a few short glandular hairs in the lower ¼. $2n=24$.
    Native. In hedges, bushy places, edges of woods and in fens; introduced in gardens. Common in southern England and throughout Ireland, becoming local northwards and usually introduced in Scotland. Europe (except the north), W. Asia, N. Africa, N. America.

*Ssp. **pulchra** (Brummitt & Heywood) Tutin: hairy, at least on the peduncles, upper part of petioles and lf sinuses. Corolla intermediate in size between the other 2 sspp., bright pink, usually with white bands inside. Bracteoles slightly inflated, distinctly longer than broad, with overlapping edges and ± hiding the calyx. Filaments intermediate in size between those of ssp. *sepium* and ssp. *sylvatica* but with longer glandular hairs than the latter.
    Introduced. Rather uncommonly naturalized in the neighbourhood of gardens. E. Siberia; similar, or possibly identical, plants in N. America.

*Ssp. **silvatica** (Kit.) Maire: ± glabrous. Corolla 55–60 × 60–75 mm., white, rarely with 5 pale pink stripes on the outside. Bracteoles strongly inflated, about as broad as long, with overlapping edges and completely concealing the calyx. Filaments c. 17 mm., 3·5 mm. wide at base, thickly glandular-hairy in lower $\frac{1}{3}$. Intermediates between this and ssp. *sepium* occur not infrequently and may well be hybrids.

Introduced. Widely naturalized in hedges and waste places, commonest in S.E. England and the Midlands. S.E. Europe, N. Africa, Caucasus.

2. C. soldanella (L.) R.Br.   'Sea Bindweed.'

A procumbent glabrous perennial 10–60 cm. Rhizomes slender, far-creeping. *Stems not climbing. Lf blade* 1–4 *cm., reniform; petiole usually longer than blade.* Fls c. 5 cm., solitary; *peduncles* longer than lvs, *sharply quadrangular; bracteoles* 10–15 mm., *broadly oblong, rounded at tip, shorter than calyx.* Corolla 2·5–4 cm. diam., funnel-shaped, pink or pale purple. *Capsule ovoid, acute,* incompletely 2-celled. Fl. 6–8. Pollinated mainly by humble-bees, but also sometimes self-pollinated. $2n = 22$. Hp.

Native. On sandy and shingly sea shores and dunes. 47, H14, S. Locally common in suitable habitats around the shores of the British Is., though rare in N. Scotland and N. and W. Ireland. Atlantic and Mediterranean coasts of Europe from Denmark southwards; N. Africa; Asia; N. and S. America; Australia and New Zealand.

3. CUSCUTA L.   Dodder.

Slender, *twining,* pink, yellow or white, annual *parasites,* attached to the host plant by suckers. *Lvs small and scale-like. Fls in lateral,* often bracteate, *heads or short spikes.* Calyx 5 (rarely 4)-lobed. Corolla urceolate or campanulate, 5- (rarely 4-)lobed; throat usually with a ring of 4–5 small dentate or laciniate petaloid scales below the insertion of the stamens. Stamens 4–5, inserted on the corolla-tube. Styles 2, free or connate; stigmas linear or capitate. Capsule 2-celled, splitting transversely near the base, cells 2-seeded; seeds angled, cotyledons 0.

About 100 spp. in tropical and temperate regions.

Styles shorter than ovary; stamens included; scales small, applied to the
    corolla-tube and not closing it.   **1. europaea**
Styles longer than ovary; stamens somewhat exserted; scales large, connivent,
    $\pm$ closing corolla-tube.   **2. epithymum**

1. C. europaea L.   'Large Dodder.'

A reddish parasite twisting in a counter-clockwise direction. *Stems up to c.* 1 *mm. diam.* Fls c. 2 mm., diam., in dense *bracteate* heads 10–15 mm. diam.; pedicels very short, fleshy. *Calyx-teeth obtuse. Corolla pinkish-white, lobes about* $\frac{1}{2}$ *as long as tube, spreading,* rather obtuse; *scales very small, applied to the corolla-tube and not closing it. Stamens included. Styles shorter than ovary;* stigmas linear. Fl. 8–9. $2n = 14$. Th.

Native. On *Urtica dioica, Humulus,* and rarely various other plants. 42. England, mainly in the south; Glamorgan; Scotland, very rare and probably not native. Rare and probably decreasing. Europe from C. Scandinavia; N. Africa; Asia to the Himalaya and Tibet; introduced in N. America.

*****C. epilinum** Weihe, which is similar to *C. europaea,* parasitizes flax (*Linum usitatissimum*) and has yellowish fls in ebracteate heads and incurved corolla-lobes c. $\frac{1}{2}$ length of tube; it is sometimes introduced with flax seed. $2n = 42$.

2. C. epithymum (L.) L.   'Common Dodder.'

Incl. *C. Trifolii* Bab.

Similar to *C. europaea* but much smaller. *Stems* reddish, *very slender* (c. 0·1 mm. diam.). Fls pinkish, scented, sessile in dense heads, 5-10 mm. diam. *Calyx open-campanulate, divided for ¾ of its length, lobes acute.* Corolla-lobes acuminate, somewhat spreading; *scales* connivent, *closing the corolla-tube, the spaces between their bases narrow and acute. Stamens exserted* but shorter than corolla. *Styles longer than ovary*; stigmas linear. Fl. 7–9. 2*n*= 14. Th. (often perennating).

Native. On *Ulex, Calluna* and various other plants. 77, H 14, S. Locally common in England and Wales; S. Scotland and Ireland, very local. Europe from southern Norway to N. Spain and eastwards to the Caucasus and Altai. Introduced in S. Africa.

105. SOLANACEAE

Lvs alternate, sometimes paired through adnation; stipules 0. Fls usually actinomorphic. Calyx (3–)5(–10)-lobed, usually persistent. Corolla usually 5-lobed. Stamens inserted on the corolla-tube, when isomerous alternating with corolla-lobes. Ovary 2-locular, cells sometimes divided by false septa; ovules numerous; placentation axile. Fr. a capsule or berry. Seeds with endosperm; embryo straight or bent.

About 75 genera and 2000 spp., widely distributed but chiefly tropical.

Closely related to Scrophulariaceae which, however, usually have zygomorphic fls. The two families are best distinguished by the presence of phloem on both sides of the xylem in the vascular bundles of the Solanaceae.

1 Stamens prominent; anthers much longer than filaments, opening by
       apical pores and forming a cone.   4. SOLANUM
     Anthers shorter than filaments, opening by slits, stamens often included,
       never forming a cone.   2
2 Shrubs, often spiny.   1. LYCIUM
     Herbs.   3
3 Plant viscid pubescent; fls in a scorpioidal cyme.   3. HYOSCYAMUS
     Plant not viscid; fls not in a scorpioidal cyme.   4
4 Fls up to 3 cm., drooping; lvs entire.   2. ATROPA
     Fls c. 7 cm., erect; lvs ± toothed.   5. DATURA

1. LYCIUM L.

Erect or pendulous spinous or unarmed *shrubs*. Lvs alternate or fascicled, entire. Calyx campanulate, (3–)5-lobed. Corolla funnel-shaped, 5-lobed. *Stamens inserted at mouth of corolla-tube, long-exserted, anthers short. Fr. a berry*, seated in the persistent calyx.

About 120 spp. in temperate and subtropical regions.

Usually spiny; lvs lanceolate, grey-green; corolla-lobes shorter than tube,
    rarely equalling it.   **1. halimifolium**
Usually unarmed; lvs ovate or rhomboid, bright green; corolla-lobes as long
    as or longer than tube.   **2. chinense**

*1. L. halimifolium Mill.   Duke of Argyll's Tea-plant.

A shrub with arching greyish-white often spinous stems up to 2·5 m. *Lvs* up to c. 6 cm., *grey-green, lanceolate*, obtuse or subacute, tapering gradually into the short petiole. Fls solitary or few, usually on very short axillary shoots often bearing a number of small lvs. Calyx 3–5 mm.; teeth as long as tube, acute. *Corolla* c. 1 cm., rose-purple turning pale brown; *lobes shorter than tube, rarely equalling it*. Fr. 1–2 cm., scarlet, ovoid. Fl. 6–9. Fr. 8–10. $2n = 24$. N.

Introduced. Naturalized in hedges and on walls and waste ground. Long cultivated and widely naturalized, but probably native of S.E. Europe and W. Asia.

*2. L. chinense Mill.

L. rhombifolium (Moench) Dippel

Similar to *L. halimifolium* but less spiny. Lvs ovate or rhomboid, bright green. Corolla-lobes equalling or exceeding tube. Fr. 1·5–2·5 cm. Fl. 6–9. Fr. 8–10. $2n = 24$. N.

Introduced. Naturalized in similar places to *L. halimifolium* but commoner near the sea. E. Asia.

2. ATROPA L.

Tall much-branched *herbs* with entire lvs. Fls solitary or few. Calyx 5-partite. Corolla campanulate, slightly zygomorphic. *Stamens inserted at base of corolla-tube*. Ovary 2-celled; stigma peltate. Fr. a many-seeded, 2-celled *berry* subtended by the spreading calyx.

Two spp., in Europe, W. Asia and N. Africa.

1. A. bella-donna L.   Dwale, Deadly Nightshade.

A glabrous or pubescent and glandular perennial up to 150 cm. Lvs up to 20 cm., alternate or in unequal pairs, ovate-acuminate, narrowed into the petiole. Fls 25–30 mm., peduncled, drooping, axillary or in the forks of the branches; peduncles 1–2 cm. Calyx somewhat accrescent. Corolla lurid violet or greenish, lobes obtuse. Anthers pale. Berry 15–20 mm. diam., black. Fl. 6–8. Fr. 8–10. $2n = 50, 72$. Hp.

A powerful narcotic and very poisonous. Contains the alkaloids atropine and hyoscyamine.

Native. In woods and thickets on calcareous soils; naturalized near old buildings and in hedges, rather rare. 44. England and Wales, on chalk and limestone from Westmorland southwards to Somerset and Kent, local. C. and S. Europe; W. Asia to Persia; N. Africa.

3. HYOSCYAMUS L.

Annual or biennial, often viscid, herbs. Fls axillary or in bracteate scorpioidal cymes. *Calyx urceolate*, 5-toothed. *Corolla somewhat zygomorphic*, campanulate or funnel-shaped; lobes obtuse. Stamens inserted at base of corolla-tube. Ovary 2-celled; stigma capitate. *Fr. a many-seeded capsule enclosed in the calyx and constricted in the middle, circumscissile near the top.*

About 10 spp., in Europe, Asia and N. Africa.

1. H. niger L.   Henbane.

A viscid pubescent strong-smelling annual or, more usually, biennial up to 80 cm. Hairs soft, glandular. Stem stout, rather woody at base. Lvs with a few large teeth or nearly entire; lower 15–20 cm., petioled, oblong-ovate; cauline smaller, amplexicaul, oblong. Fls c. 2 cm., subsessile in 2 rows in a scorpioidal cyme; bracts lf-like. Calyx-teeth triangular-acuminate, points becoming rigid in fr. Corolla 2–3 cm. diam., lurid yellow usually veined with purple. Anthers purple. Calyx-tube subglobose in fr., 15–20 mm. diam., strongly veined. Fl. 6–8. $2n = 34$. Hs.

Poisonous and narcotic. Contains the alkaloids hyoscyamine and scopolamine.

Native. In sandy places, specially near the sea, elsewhere usually on disturbed ground in farmyards, etc. 83, H32, S. Native in S. England and S.E. Ireland and widely scattered throughout the British Is. north to Sutherland as a casual, local. Europe; W. Asia; N. Africa.

*__Nicandra physalodes__ (L.) Gaertn., an annual with a strongly accrescent bladdery calyx which is cordate at base and with bright blue fls 3–4 cm., occurs as a casual. Native of Peru, widely naturalized in N. temperate regions.

*__Physalis alkekengi__ L., a somewhat pubescent perennial with dirty white fls and a calyx similar to that of *Nicandra* enclosing an orange or red berry the size of a cherry, is occasionally naturalized. C. and S. Europe.

4. SOLANUM L.

Herbs, shrubs or small trees. Lvs alternate or in pairs. Fls solitary or in cymes, white, purple or blue. Calyx 5–10-fid. *Corolla rotate*; lobes 5–10. *Stamens inserted on the throat of the corolla-tube; filaments short; anthers long, exserted, connivent or cohering in a cone, opening by apical pores.* Ovary 2–(3–4)-celled. Fr. a many-seeded berry.

About 1500 spp. mainly in the tropics.

S. tuberosum, the potato, is widely cultivated in temperate countries. The tomato, *Lycopersicum esculentum*, belongs to a closely related genus.

Woody perennial; fls c. 1 cm., purple (very rarely white); fr. ovoid, red.
1. dulcamara
Herbaceous annual; fls c. 0·5 cm., white; fr. globose, black.   **2. nigrum**

1. S. dulcamara L.   Bittersweet, Woody Nightshade.

A glabrous, pubescent or sometimes tomentose *scrambling woody perennial* 30–200 cm. Lvs up to c. 8 cm., ovate, entire, or with 1–4 deep lobes or stalked pinnae at base, *apex acuminate, base rounded, cordate or sometimes hastate*; petiole shorter than blade. Cymes lf-opposed, peduncled, branched, umbellate; pedicels erect in fl., recurved in fr. *Fls c. 1 cm.* Calyx with broad, shallow, rounded lobes. *Corolla purple* or very rarely white; *lobes 3–4 times as long as calyx*, at first spreading then revolute. *Anthers yellow, cohering in a cone. Fr.* c. 1 cm., *ovoid, red.* Fl. 6–9. $2n = 24$. N. or Chw.

A form which may be genetically distinct (var. *marinum* Bab.) with fleshy lvs and prostrate stems occurs on the south coast.

Native. In hedges, woods, waste ground and on shingle beaches, common. 100, H39, S. Throughout the British Is. except the extreme north of Scotland. Europe; Asia; N. Africa.

2. S. nigrum L. 'Black Nightshade.'

A glabrous or pubescent annual up to c. 60 cm. *Lvs* ovate or rhomboid, entire or sinuate-dentate, acute, *cuneate at base*; petiole shorter than the *decurrent blade*. Cymes extra-axillary, scarcely branched, umbellate; pedicels erect in fl., deflexed in fr. *Fls c. 0·5 cm.* Calyx scarcely accrescent, lobes obtuse. *Corolla white, lobes about twice as long as calyx*, at first spreading, then revolute. *Anthers* yellow, *not or scarcely cohering. Fr.* c. 8 mm., *globose, black.* Fl. 7–9. $2n = 72$. Th.

Native. In waste places and a weed in gardens. 68, H14, S. Throughout England, becoming rarer northwards; very local in Wales, S. Scotland and Ireland; absent from most of Scotland. Throughout most of the world as a weed of cultivation.

*S. sarrachoides Sendtn., resembling *S. nigrum* but with the berries green when ripe and closely surrounded by the strongly accrescent calyx in the lower part, occurs as a weed in E. Anglia and the Scilly Is. and as a casual on rubbish dumps. Native of Brazil.

*S. triflorum Nutt., easily recognized by the pinnatifid lvs and large (10 mm. or more diam.) berries, is naturalized in Norfolk. Native of western N. America.

*S. pseudocapsicum L., Jerusalem Cherry, is often cultivated and is sometimes found as a casual. Glabrous, ±shrubby (though usually treated as an annual); lvs oblanceolate, petioles short. Fls white, solitary or in small axillary clusters. Fr. red, cherry-like, c. 1·5 cm. diam. Probably S. American, though known from Madeira for c. 300 years.

5. DATURA L.

Herbs, shrubs, or small trees. Lvs alternate. Fls large, often terminal, pendulous or erect. *Calyx long, tubular, often 5-angled*, often splitting transversely after flowering, the upper part deciduous, the lower persistent and accrescent. *Corolla funnel-shaped, with a long cylindrical tube* and ± spreading, sometimes unequal lobes. *Stamens included. Fr. a capsule, dehiscing by 4 valves or irregularly, or a berry.*

About 20 spp. in warm temperate and tropical regions, mostly in C. America.

*1. D. stramonium L. Thorn-apple.

A stout erect dichotomously branched herb up to 1 m. Lvs up to c. 20 cm., ovate, sinuate-dentate or coarsely toothed, acute or acuminate, often unequal at base; petiole up to c. 7 cm. Fls 6–8 cm., erect, solitary; peduncle short. Calyx c. 4 cm., pale green, 5-angled, splitting transversely after flowering; teeth c. 0·5 cm., narrow-triangular. Corolla white or sometimes purple; lobes c. 1 cm., narrow, acuminate, ± erect. Fr. a many-seeded capsule 4–5 cm., ovoid, densely clothed with long sharp spines or rarely unarmed, dehiscing by 4 valves. Fl. 7–10. $2n = 24$. Th. Narcotic and very poisonous. Contains the alkaloids hyoscyamine, hyoscine and scopolamine.

Introduced. An uncommon casual or ± naturalized in waste places and cultivated ground. Throughout most of the temperate and subtropical parts of the northern hemisphere.

Salpichroa origanifolia (Lam.) Baillon is naturalized in the Channel Is. and along the south coast of England. Somewhat woody perennial. Lvs 1·5–2·5 × 1–2 cm., ovate-rhomboid. Fls less than 1 cm., solitary, nodding; corolla urceolate, white, with a ring of hairs above the insertion of the stamens within. E. temperate S. America.

106. SCROPHULARIACEAE

Herbs, sometimes hemiparasites, sometimes shrubs, rarely trees, with exstipulate lvs. Fls hermaphrodite, zygomorphic, hypogynous, basically 5-merous (except gynoecium). Calyx 5-lobed or sometimes 4-lobed, the top lobe being absent. Corolla gamopetalous, imbricate in bud, not plicate, very variable, from regularly 5- (rarely to 8-)lobed to strongly 2-lipped with the lobes obscure. Stamens 5, epipetalous, alternating with corolla-lobes or more frequently 4, the upper being absent or represented by a staminode; sometimes only 2; anthers introrse. Ovary 2-celled with numerous anatropous ovules on axile placentae, the septum transverse; style terminal, simple or bilobed. Fr. a capsule (rarely a berry, but not in British spp.). Seeds usually numerous, with fleshy endosperm and straight or slightly curved embryo. Vascular bundles collateral.

Over 200 genera and 2500 spp., cosmopolitan.

A family of very varied aspect, distinguishable from allied families by the zygomorphic fls and structure of the ovary. Some foreign Solanaceae can only be distinguished by the bicollateral bundles and either the oblique septum of the ovary or the plicate aestivation of the corolla, but the British spp. are all nearly actinomorphic (see also Acanthaceae).

Species of *Alonsoa, Calceolaria, Nemesia, Phygelius, Paulownia, Penstemon*, etc., are ± commonly grown in gardens for their fls.

1   Stamens 5; fls many in a large erect terminal raceme or panicle, yellow or white.   1. VERBASCUM
    Stamens 4 or 2.   2

2   Stamens 2; fls blue, more rarely white or pinkish.   13. VERONICA
    Stamens 4; fls never a true blue.   3

3   Corolla-tube saccate or spurred at base.   4
    Corolla-tube not saccate or spurred.   8

4   Lvs lanceolate to linear or oblong, cuneate or tapered at base, not or shortly stalked.   5
    Lvs ovate, orbicular, reniform or hastate, rounded to cordate at base, stalked.   7

5   Corolla saccate at base.   2. ANTIRRHINUM
    Corolla spurred.   6

6   Fls in a terminal raceme, the bracts much shorter than the lvs; mouth of corolla closed; capsule opening by valves.   3. LINARIA
    Fls axillary, the bracts scarcely differing from lvs; mouth of corolla slightly open; capsule opening by pores.   4. CHAENORHINUM

7   Lvs entire or hastate, pinnately veined; decumbent annuals; fls yellow
with upper lip purple.   5. KICKXIA
  Lvs lobed, palmately veined; creeping perennials; fls lilac with orange
spot on palate.   6. CYMBALARIA

8   Plants creeping and rooting at nodes, or lvs all radical; corolla rotate
with very short tube, 5 mm. diam. or less; fls solitary.   9
  Plants not creeping and rooting, lfy; fls in infls.; corolla with distinct
tube, strongly zygomorphic (except *Erinus*).   10

9   Lvs oblong to subulate, entire; anthers 1-celled.   9. LIMOSELLA
  Lvs reniform, dentate; anthers 2-celled.   10. SIBTHORPIA

10   Calyx 5-lobed, not inflated.   11
  Calyx 4-lobed or if 5-lobed strongly inflated after flowering and lobes
usually lf-like.   14

11   Lvs opposite.   12
  Lvs alternate.   13

12   Corolla with nearly globular tube and small lobes.   7. SCROPHULARIA
  Corolla with broad straight tube and large lobes.   8. MIMULUS

13   Corolla-tube narrow, scarcely longer than calyx, lobes 5, nearly equal;
low tufted plant.   11. ERINUS
  Corolla bell-shaped, tube several times as long as calyx, lobes small;
tall erect plant.   12. DIGITALIS

14   Calyx inflated, at least after flowering.   15
  Calyx not inflated.   16

15   Lvs alternate, pinnatisect; fls pink.   14. PEDICULARIS
  Lvs opposite, crenate or dentate; fls yellow or brown.
                                15. RHINANTHUS

16   Upper lip of corolla laterally compressed, mouth nearly closed; fls
± yellow.   16. MELAMPYRUM
  Upper lip of corolla not compressed, mouth open.   17

17   Upper lip of corolla 2-lobed, only slightly concave; corolla white or lilac
with violet lines and usually yellow spot on lower lip, rarely purple or
yellow.   17. EUPHRASIA
  Upper lip of corolla entire or nearly so, arched.   18

18   Fls small, 4–8 mm., in unilateral spikes, pink.   18. ODONTITES
  Fls larger, more than 1 cm., spikes not one-sided.   19

19   Seeds large, winged or ribbed; plant perennial; fls purple.   20. BARTSIA
  Seeds minute; plant annual; fls yellow.   19. PARENTUCELLIA

Subfamily 1. PSEUDOSOLANOIDEAE. Autotrophic. Lvs alternate. Fls only
slightly zygomorphic. Stamens usually 5. Nectaries 0 or on the corolla.

1. VERBASCUM L.   Mullein.

Herbs (rarely shrubs), usually biennial, with rosette of radical lvs and tall
erect stems with alternate lvs. *Fls in terminal racemes or panicles.* Calyx
deeply 5-lobed, nearly regular. *Corolla* usually *yellow, rotate with* 5 *nearly
equal lobes and very short tube.* Stamens 5, the filaments all or at least the
three upper hairy, the two lower longer than upper. Capsule septicidal. Seeds
small, numerous.

Over 250 spp., mainly in the Mediterranean region but extending to N. Europe and N. and C. Asia.

Besides those given below, several other spp. have been recorded as casuals.

1 Hairs on the filaments white.   2
   Hairs on the filaments purple.   4
2 Lower filaments glabrous or much less hairy than the upper, with adnate
     or obliquely inserted anthers; fls large (15–30 mm. diam.) in a dense
     spike-like raceme.   **1. thapsus**
   Filaments all equally hairy, the anthers transversely inserted; fls smaller
     (less than 20 mm.) in a laxer raceme or panicle.   3
3 Stem angled; lvs nearly glabrous above, mealy beneath.   **2. lychnitis**
   Stem terete; lvs mealy on both sides.   **3. pulverulentum**
4 Anthers all transversely inserted, equal; lvs and stems hairy.   **4. nigrum**
   Lower anthers obliquely inserted, larger than the upper, lvs and stems
     glabrous or nearly so.   5
5 Pedicels longer than the calyx; fls always solitary.   **5. blattaria**
   Pedicels shorter than the calyx; fls 1–5 together.   **6. virgatum**

1. V. thapsus L.   Aaron's Rod.

An erect biennial or rarely annual 30–200 cm., *densely clothed with soft whitish wool*. Lvs obovate-lanceolate to oblong, acute or subacute, crenate; the radical 15–45 cm., attenuate into a narrowly winged petiole; the cauline smaller, *base decurrent* nearly to the next lf below. *Fls in a dense terminal spike-like raceme* rarely with axillary racemes from the upper lvs. Bracts triangular-lanceolate, apex with a long acuminate point, usually rather longer than the fls. Pedicels very short (to 2 mm.) or almost 0. Sepals triangular-ovate, acuminate. Corolla yellow, 1·5–3 cm. diam., concave. *Three upper filaments clothed with whitish or yellowish hairs; two lower glabrous or sparingly hairy, their anthers obliquely inserted and partially adnate* to the filaments. Capsule ovoid, rather longer than calyx. Fl. 6–8. Pollinated by various insects or selfed, self-fertile. $2n=36$. Hs.

Native. Rather common on sunny banks and waste places usually on a dry soil. 102, H37, S. Throughout England, Wales, Ireland and S. Scotland; rare in N. Scotland and perhaps not native but extending to Orkney. Europe except Iceland, Faeroes, Crete, etc. (to 64° 12′ N. in Norway); Asia, south to Caucasus and Himalaya and east to W. China; naturalized in N. America.

*V. thapsiforme Schrad. differing in the larger (3–5 cm. diam.) quite flat corolla and the shorter filaments with completely decurrent anthers, is sometimes found as a casual, rarely naturalized. $2n=32$.
   Native of C. Europe.

*V. phlomoides L. differing from *V. thapsiforme* in its non-decurrent lvs, is another occasional casual, rarely naturalized. $2n=32$. Native of C. and S. Europe, W. Asia.

2. V. lychnitis L.   'White Mullein.'

An erect biennial 50–150 cm. *Stem angled, with a short powdery stellate pubescence. Lvs dark green above* and nearly glabrous, *with a dense white powdery stellate-pubescence beneath*, crenate; the radical 10–30 cm., oblong or lanceolate, narrowed into a short petiole; cauline smaller, ovate, acuminate; the upper sessile, often narrower. Fls in a narrow panicle, 2–7 in the axil of each bract. Bracts and bracteoles linear-lanceolate or linear, shorter than the fls. Pedicels varying from as long to twice as long as calyx. Sepals linear, acute, very woolly. Corolla white or yellow, 15–20 mm. diam. *Filaments all clothed with whitish hairs. Anthers equal, transversely inserted.* Capsule ovoid or pyramidal, longer than calyx. Fl. 7–8. Pollinated by various insects or selfed, homogamous. $2n=32$. Hs.

Native. Waste places and calcareous banks. 21. Very local from Devon and Kent to Merioneth, Flint and Norfolk; casual farther north. Europe from Belgium and N. Germany to C. Spain, C. Italy, Thrace and the Caucasus; W. Siberia (Upper Tobol region); Morocco; naturalized in N. America.

The common British form is white-fld (var. *album* (Mill.) Druce), the yellow-fld form being known only from Somerset.

3. V. pulverulentum Vill.   'Hoary Mullein.'

An erect biennial 50–120 cm., *thickly clothed with a mealy white wool which is easily rubbed off. Stem terete*, weakly striate. *Lvs mealy on both sides*, crenate; the radical 20–50 cm., broadly oblong, narrowed to a very short petiole; cauline, smaller, sessile, the upper cordate. Fls in a pyramidal panicle, 4–10 in the axil of each bract. Pedicels at first all short, some becoming longer later. Sepals lanceolate, acute. Corolla yellow, less than 2 cm. diam. *Filaments all clothed with whitish hairs. Anthers equal, transversely inserted.* Capsule ovoid, rather longer than calyx. Fl. 7–8. $2n=32$. Hs.

Native. Roadsides in Norfolk and Suffolk, very local; a rare casual elsewhere. W. and S. Europe from Holland and the Rhineland to C. Spain and thence to Switzerland, Portugal, Roumania and Greece.

*V. speciosum Schrad.

Whole plant thickly clothed with persistent white wool. Stem angled. Lvs entire; radical oblanceolate, obtuse, gradually tapered to base; cauline ovate, acute, cordate at base. Infl. a panicle. Corolla 2–2·5 cm. diam., yellow. Stamens as in *V. lychnitis* and *V. pulverulentum.*

Occasionally ± naturalized. Native of S.E. Europe, Asia Minor, Caucasus.

4. V. nigrum L.   'Dark Mullein.'

An erect biennial 50–120 cm. *Stem angled, stellate-pubescent. Lvs* dark green above, thinly pubescent, pale *beneath* and more *conspicuously stellate-pubescent*, crenate; the radical 10–30 cm., long-petioled, ovate to lanceolate, cordate at base; cauline smaller, broad-cuneate at base, the upper nearly sessile. *Fls* in a terminal raceme sometimes with axillary racemes from the upper lvs, 5–10 *in the axil of each bract*. Upper bracts linear, acute, lower somewhat lf-like. Pedicels of varying length. Sepals linear, acute. Corolla yellow, sometimes cream, with small purple spots at the base of each lobe, 12–22 mm. diam. *Filaments* all *clothed with purple hairs. Anthers equal, transversely inserted.* Capsule ovoid, truncate at the apex, rather longer than calyx. Fl. 6–10. Pollinated by various insects, self-sterile, homogamous. $2n=30$. Hs.

Native. Waysides and open habitats on banks, etc., usually on calcareous soil. 53, S. Rather common in S. England, extending north to Caernarvon and Nottingham; naturalized further north. Europe from Scandinavia (c. 65° N.) to N. and E. Spain, C. Italy and Macedonia and the Caucasus; Siberia (east to the Yenisei region).

*V. chaixii Vill.

Allied to *V. nigrum* but with lvs rounded at base, infl. branched, fls 2–5 in each axil. Rarely naturalized. Native from S. Europe to W. China.

*V. sinuatum L.

Radical lvs sinuate-pinnatifid, tapered to base. Infl. a panicle. Corolla c. 2 cm. diam., yellow. Stamens as in *V. nigrum*. A rather frequent casual. Native of the Mediterranean region.

5. V. blattaria L.   'Moth Mullein.'

An erect biennial to 1 m., *glabrous below, the infl. glandular.* Stem angled. *Lvs glabrous*; the radical 10–25 cm., oblong, gradually narrowed to the base, crenately or sinuately lobed, the lobes often toothed; cauline smaller, the upper triangular, cordate at base, sessile, with acute triangular teeth. *Fls in a terminal raceme occasionally with axillary racemes in addition, solitary in the axil of each bract.* Bracts lanceolate, acuminate, the lower somewhat lf-like. *Pedicels much longer than calyx.* Sepals linear-elliptic, subacute. Corolla yellow, rarely whitish, 2–3 cm. diam. *Filaments all clothed with purple hairs. Anthers of the three upper filaments small, of the lower very obliquely inserted and adnate to the filament.* Capsule subglobose, equalling or rather longer than calyx. Fl. 6–10. Pollinated by various insects or selfed, homogamous. $2n = 30, 32$. Hs.

?Introduced. Rather rare, in waste places. 40, H1. S. England extending north to Merioneth, Cheshire and Durham, rarely persisting in the same spot for more than a few years; near Cork City. Europe from the Netherlands and C. Spain eastwards; W. and C. Asia, from Upper Tobol region south to Palestine and Afghanistan; N. Africa; naturalized in N. America.

6. V. virgatum Stokes   'Twiggy Mullein.'

Differs from *V. blattaria* as follows: More glandular. *Fls 1–5 in the axil of each bract. Pedicels much shorter than calyx.* Fl. 6–8. $2n = 32$. Hs.

Native. Native only in Cornwall and S. Devon. 3 (39), (H3), S. Elsewhere with a similar status to *V. blattaria* but recorded as far north as Westmorland and Midlothian; near Cork City and Kenmare (Kerry). W., C. and S. France, Spain, Portugal; widely naturalized elsewhere.

*V. phoeniceum L., similar in general aspect to the last two species but with purple fls, solitary on long pedicels and anthers all equal, is a rare escape from cultivation. Native of E. Europe and W. Asia.

The following hybrids occur, none is common: *V. lychnitis* × *nigrum*, *V. lychnitis* × *thapsus*, *V. lychnitis* × *pulverulentum*, *V. nigrum* × *pulverulentum*, *V. nigrum* × *thapsus* *V. phlomoides* × *thapsus*, *V. pulverulentum* × *thapsus*.

Subfamily 2. ANTIRRHINOIDEAE. Autotrophic. Stamens 4 or 2. Nectaries at the base of the ovary, usually ring-shaped.

Tribe 1. ANTIRRHINEAE. At least the lower lvs opposite. Fls axillary or in terminal spikes or racemes. Corolla spurred or saccate at base, the two upper lobes outside the lateral ones in bud. Stamens 4.

2. ANTIRRHINUM L.

Herbs with lower lvs opposite, upper alternate. Fls solitary and axillary, or forming terminal racemes. Calyx deeply 5-lobed. *Corolla strongly* 2-*lipped, the lower lip* 3-*lobed with a projecting 'palate' closing the mouth* of the corolla; the upper 2-lobed. *Tube* broad, *gibbous or saccate at base. Capsule* with 2 unequal cells, *opening by* 3 *pores.*

About 40 spp., mainly in the Mediterranean region, a few extending farther north, and in western north America.

Annual; calyx-lobes linear; corolla 10–15 mm.   **1. orontium**
Perennial; calyx-lobes ovate; corolla 3–4 cm.   **2. majus**

1. A. orontium L.   Weasel's Snout, Calf's Snout.

Misopates orontium (L.) Raf.

An *erect annual* 20–50 cm., simple or branched, usually glandular-pubescent above. Lvs 3–5 cm., linear or narrowly elliptic, entire, attenuate at base. *Fls subsessile in axils of upper lvs* forming a lfy terminal raceme. *Calyx-lobes linear*, unequal, *equalling or longer than corolla. Corolla* 10–15 *mm.*, pinkish-purple. Capsule shorter than calyx. Fl. 7–10. Pollinated by bees or selfed, homogamous. 2*n*=16. Th.

Native. In cultivated ground. 54, H5, S. From Cumberland and Yorks southwards, local; Cork. Europe (except Iceland and Faeroes, casual in Norway) and the Mediterranean region, extending east to the Himalaya and south to the Canaries and Abyssinia.

*2. A. majus L.   Snapdragon.

A *perennial* 30–80 cm., often becoming woody at base but usually not of long duration, branched at the base and of rather bushy habit, glabrous below and glandular-pubescent above. Lvs 3–5 cm., lanceolate or linear-lanceolate, entire, attenuate at base. *Fls in a terminal raceme, in the axils of short, sessile, ovate bracts. Calyx-lobes ovate. Corolla* 3–4 *cm., several times as long as calyx*; in the wild continental plant reddish-purple, more rarely yellowish-white. Capsule longer than calyx. Fl. 7–9. Pollinated by humble-bees, self-fertile or self-sterile, homogamous. 2*n*=16. Hp.

Introduced. Naturalized, usually on old walls, in many parts of England and in Ireland. Mediterranean region; naturalized in C. Europe. Commonly cultivated in gardens in many colour and habit varieties.

ASARINA Mill.

Differs from *Antirrhinum* as follows: Lvs opposite, palmately veined. Capsule with two equal cells, opening by 2 pores. One sp.

*A. procumbens Mill. Viscid perennial with long procumbent stems. Lvs 2–7 cm., orbicular, crenately lobed. Fls c. 3 cm., pale yellow, solitary, axillary. Rarely naturalized on walls, etc. S. France, N. Spain.

3. Linaria Mill.

Herbs. *Lvs* all opposite or whorled or more commonly the upper alternate, *pinnately veined*. *Fls in terminal racemes*, the bracts small. Fls as in *Antirrhinum* but *corolla spurred*. *Capsule opening by* 4–10 *apical valves* of varying length.

About 150 spp., mainly Mediterranean, a few in C. Europe, temperate Asia, temperate America.

In addition to the following, a number of spp. are cultivated and several are from time to time found as garden-escapes or among ballast.

1	Plant glabrous or only infl. glandular; corolla (excluding spur) 8 mm. or more.	*2*
	Whole plant viscid-pubescent; corolla 4–6 mm.	**6. arenaria**
2	Corolla violet or whitish striped with purple.	*3*
	Corolla yellow.	*5*
3	Annual; fls few; spur nearly as long as rest of corolla, straight (Jersey).	**1. pelisseriana**
	Perennial; fls numerous; spur short or strongly curved.	*4*
4	Infl. dense; fls violet; spur c. ½ rest of corolla, curved.	**2. purpurea**
	Infl. lax; fls whitish striped with purple; spur c. ¼ rest of corolla, straight.	**3. repens**
5	Erect perennial; sepals ovate or lanceolate, acute.	**4. vulgaris**
	Decumbent annual; sepals linear, obtuse.	**5. supina**

1. L. pelisseriana (L.) Mill.

A glabrous *annual* 15–30 cm.; stems erect, usually simple, sometimes several from the base with short sterile branches at base. Lower lvs and those of sterile stems in whorls of 3, elliptic; upper lvs 1–3 cm., alternate, linear. *Fls few* (10 or less) in a raceme which is at first short and dense, later elongating. Bracts linear. Pedicels longer than or equalling bracts and calyx. Sepals linear-lanceolate, acute. *Corolla* 10–15 mm., *violet*, with a white palate; *spur slender, straight, nearly as long as corolla*. *Capsule* broader than long, 2-lobed, *shorter than calyx*. *Seeds flat, tuberclate on one face only, winged*. Fl. 5–7. Th.

Native. Heathy places in Jersey only. Rarely occurring as a casual elsewhere. S. Europe and W. Asia from W. and C. France (not Portugal or W. Spain), eastwards to Transcaucasia and Palestine; Algeria.

*2. L. purpurea (L.) Mill.   'Purple Toadflax.'

A glabrous glaucous *perennial* 30–90 cm.; stems branched above. Lvs linear or linear-lanceolate, to 4·5 cm. *Fls numerous* (15–40) in dense racemes terminal on the stem and branches. Pedicels shorter than the linear bracts. Sepals linear, acute. *Corolla* c. 8 mm., *violet*, rarely bright pink; *spur long, incurved, more than half as long as corolla*. *Capsule longer than calyx*. *Seeds angled, wingless, reticulate*. Fl. 6–8. Pollinated by bees, self-sterile. $2n=12$. Hp.

Introduced. Much cultivated in gardens and sometimes naturalized on old walls and in waste places in England and Ireland. C. and S. Italy, Sicily.

3. L. repens (L.) Mill. 'Pale Toadflax.'

A glabrous glaucous *perennial* 30–80 cm., with a creeping underground rhizome and numerous *erect stems*, branched above. Lvs 1–4 cm., linear, whorled below, alternate above. Fls 10–30 in rather long lax racemes terminal at the end of the stem and branches. Pedicels usually rather longer than calyx and linear bracts. Sepals linear-lanceolate, acute. *Corolla* 7–14 mm., *white or pale lilac, striped with violet veins*, palate with orange spot, *spur* short, *straight, about a quarter as long as corolla. Capsule* subglobose, *longer than calyx. Seeds angled, wrinkled, wingless.* Fl. 6–9. Pollinated by bees, self-sterile. $2n = 12$. Hp.

Native. Local in dry stony fields and waste places, usually calcareous. 43, H6, S. England and Wales extending north to Yorks and Lancashire; Cork, Carlow, Louth, Donegal, Armagh and Down; introduced in Scotland. Belgium, France, N.E. Spain, N.W. Italy. Naturalized in Scandinavia, Germany, Switzerland.

L. repens × vulgaris = L. × sepium Allman

Corolla 12–21 mm., usually yellowish, striped with violet. Lvs broader than in *L. repens*. Partially fertile and thus variable. Not uncommon where the parents occur together.

4. L. vulgaris Mill. Toadflax.

A glaucous *perennial* 30–80 cm., glabrous except for the sometimes glandular infl.; rhizome creeping; *stems* numerous, *erect*, branched above. Lvs 3–8 cm., linear-lanceolate, rarely lanceolate. *Fls numerous (c.* 20) *in a long dense raceme.* Pedicels usually rather longer than calyx and linear bracts. *Sepals ovate or lanceolate, acute. Corolla* 15–25 mm., *yellow* with orange palate; spur ± straight, nearly as long as corolla. *Capsule* ovoid, *more than twice as long as calyx. Seeds papillose, winged.* Fl. 7–10. Pollinated by large bees, self-sterile. $2n = 12$. Hp.

A 'peloric' form of this species, with a regular 5-spurred corolla, sometimes occurs.

Native. Common in grassy and cultivated fields, hedgebanks and waste places in England and S. Scotland to Ross; widespread in Ireland but less common and possibly introduced. 99, H27, S. Europe from c. 70° N. in Norway to the Pyrenees, Corsica, Italy and Greece; W. Asia, east to the Altai; naturalized in N. America.

*L. dalmatica (L.) Mill., similar to *L. vulgaris* but with broadly lanceolate subcordate lvs and wingless seeds, is rarely naturalized. S.W. Europe, etc.

5. L. supina (L.) Chazelles

A glaucous *annual* 5–20 cm., glabrous except for the glandular infl., branching at the base, *the branches decumbent*, the ends ascending. Lvs 1–3 cm., linear. *Fls few* (10 or less) *in a short dense raceme.* Pedicels shorter than calyx and linear bracts. *Sepals linear, obtuse. Corolla* 10–15 mm., *yellow* with orange palate; spur almost straight, nearly as long as corolla. *Capsule* subglobose, *not much longer than calyx. Seeds smooth, winged.* Fl. 6–9. Self-sterile. $2n = 12$. Th.

?Native. In sandy places near Par, Cornwall; naturalized in waste places round Plymouth (Devon) and elsewhere in Cornwall. France, Spain, Portugal, N.W. Italy; Morocco.

*6. L. arenaria DC.

A *viscid-pubescent annual* 5–15 cm. with a bushy habit. *Lvs* 3–10 mm., *lanceolate*, attenuate at base into a short petiole. Fls few in a short, ultimately lax, raceme. Pedicels shorter than calyx. Sepals linear-lanceolate. *Corolla* 4–6 mm., *yellow*, spur rather shorter than corolla, slender, often violet. *Capsule obovoid, about equalling calyx.* Seeds black, smooth, narrowly winged. Fl. 5–9. Th.

Introduced. Naturalized on dunes at Braunton Burrows (N. Devon). Native of W. France from Dunkirk to the Gironde.

The hybrids *L. purpurea × repens = L. × dominii* Druce, *L. repens × supina = L. × cornubiensis* Druce are reported.

4. Chaenorhinum (DC.) Rchb.

Differs from *Linaria* in the *axillary fls, the mouth of the corolla being somewhat open*, and the unequal cells of the *capsule which open by pores*.

About 20 spp., all except the following confined to the Mediterranean region.

1. C. minus (L.) Lange   'Small Toadflax.'
Linaria minor (L.) Desf.

An erect annual 8–25 cm., usually glandular-pubescent, rarely glabrous; branches ascending. Lvs 1–2·5 cm., alternate, linear-lanceolate to oblong, entire, obtuse, narrowed at base into a short petiole. Fls solitary in the axils of the rather smaller upper lvs. Pedicels much longer than fls. Sepals ± linear, obtuse. Corolla 6–8 mm., purple outside, paler within; spur short, obtuse. Capsule ovoid, shorter than calyx. Seeds ovoid, ridged longitudinally. Fl. 5–10. Probably self-pollinated. $2n = 14$. Th.

Native. In arable land and waste places, particularly common along railways. 53, H35. Common in S. England becoming rarer northwards, extending to Kincardine and Argyll, but only casual in the north of its range; widespread in Ireland but rare in the north. Europe (except Mediterranean islands, Russia, Iceland and Faeroes); W. Asia to the Punjab.

*C. origanifolium (L.) Fourr.

Biennial or perennial with ± numerous ascending stems. Lvs thick; lower oval, 1 cm. or less, close together; upper narrower. Fls bright violet, 8–15 mm. Rarely naturalized. S. France, Spain, Portugal.

5. Kickxia Dum.   Fluellen.

Annual herbs with *pinnately veined* stalked *lvs* which are alternate, except the few lowest. *Fls axillary*, the bracts like the lvs but rather smaller, otherwise as in *Linaria*. *Capsule opening by pores with deciduous lids*.

About 30 spp. in Europe, Africa and W. Asia to India.

Lvs rounded or cordate at base; pedicels villous.	**1. spuria**
Lvs hastate; pedicels glabrous.	**2. elatine**

1. K. spuria (L.) Dum.

Linaria spuria (L.) Mill.

A hairy and glandular annual, decumbent and branched from base, stems 20–50 cm. *Lvs* shortly petioled, *ovate or orbicular*, entire or the lower slightly toothed, *rounded or subcordate at base*, the lowest stem lvs to 6 cm., the upper bracts less than 1 cm. *Pedicels* 1–2(–2·5) cm., *villous*. Sepals ovate. Corolla 8–11 mm., yellow with deep purple upper lip; spur curved, about equalling the corolla. Capsule globose, glabrous, shorter than calyx. Seeds pitted. Fl. 7–10. Self-fertile, cleistogamous fls occur. Th.

Native. Rather local in arable land, usually cornfields, on light soils. S. England and Wales, extending north to Cardigan, Nottingham and Lincoln. 45, S. Europe (except Iceland and Faeroes, only Crimea in Russia; casual in Scandinavia); Mediterranean region; Macaronesia.

2. K. elatine (L.) Dum.

Linaria Elatine (L.) Mill.

Differs from *K. spuria* as follows: Usually more slender, less hairy and scarcely glandular. *Upper and middle lvs hastate. Pedicels glabrous.* Sepals narrower. Corolla 7–9 mm., rather smaller, upper lip paler purple; spur straight. Fl. 7–10. Probably self-sterile. $2n=18, 36$. Th.

Native. Rather local in the same habitat as *K. spuria* but extending north to Yorks, Cumberland and the Isle of Man; S. and W. Ireland, from Galway to Wexford. 60, H9, S. Europe, except Iceland and Faeroes; Mediterranean region; Macaronesia; naturalized in N. America.

6. CYMBALARIA Hill

Creeping perennial herbs with *palmately veined* stalked lvs which are alternate above, opposite below. *Fls axillary* (bracts not differentiated), otherwise as in *Linaria. Capsule opening by* 2 *lateral pores, each pore with* 3 *valves.*

About 10 spp. in the Mediterranean region.

*1. C. muralis Gaertn., Mey. & Scherb.　　　Ivy-leaved Toadflax.

Linaria Cymbalaria (L.) Mill.

A glabrous perennial with trailing or drooping, rooting, often purplish stems 10–80 cm. Lvs 2·5 cm., nearly all alternate, (3–)5(–7)-*lobed*, thick, sometimes purplish beneath; petiole longer than blade. Pedicels recurved, rather long (c. 2 cm.). Sepals linear-lanceolate, about half the length of corolla-tube. Corolla 8–10 mm., lilac (rarely white) with a whitish palate with yellow spot at the mouth and darker lines on the upper lip; spur curved, about ⅓ length of corolla. Capsule globose. Seeds ovoid with thick flexuous ridges. Fl. 5–9. Pollinated by bees, self-fertile. $2n=14$. Chh.

Introduced. First recorded in 1640, now common on old walls, rarely on rocks, almost throughout the British Is. 93, H40, S. S. Europe from Portugal to the Crimea and Crete. Naturalized in C. and N. Europe; N. Africa, etc.

*C. pallida (Ten.) Wettst.

Linaria pallida (Ten.) Guss.

Differs in being pubescent. Lvs mostly opposite, lobes more obtuse. Corolla larger, 10–15 mm.; spur relatively longer. $2n=14$. Naturalized on shingle at Bardsea (Lancashire). Native of S. Italy.

Tribe 2. CHELONEAE. Lvs usually opposite. Fls in cymes, often forming a terminal panicle. Corolla with well-developed tube, not saccate or spurred at base, the two upper lobes outside the lateral ones in bud. Stamens usually 4.

7. SCROPHULARIA L.

Herbs with square stems and opposite lvs. *Fls in cymes* in the axils of the upper lvs or of bracts and then forming a terminal panicle. Calyx 5-lobed. *Corolla* usually dingy in colour, *with a nearly globular tube and 5 small lobes,* the two upper united at the base. *Fertile stamens* 4, bent downwards, the fifth usually represented by a staminode inserted at the base of the upper lip of the corolla, sometimes 0. Stigma capitate. *Capsule septicidal.* Seeds small, ovoid, rugose.

About 120 spp. in the temperate northern hemisphere. All spp. proto-gynous, mostly pollinated by wasps.

1   Sepals obtuse, with a scarious border; fls brownish or purplish; staminode
     present.   2
     Sepals acute, without a scarious border; fls yellowish; staminode 0.
                                                                  5. vernalis
2   Lvs and stems glabrous or nearly so.   3
     Lvs (on both surfaces) and stems downy.   **4. scorodonia**
3   Stems 4-angled, not winged; sepals with narrow scarious border.
                                                                   1. nodosa
     Stems 4-winged; sepals with broad (0·5–1 mm.) scarious border.   4
4   Lvs crenate; staminode orbicular or reniform.   **2. aquatica**
     Lvs serrate; staminode 2-lobed with diverging lobes.   **3. umbrosa**

Section 1. *Scrophularia.* Staminode present, scale-like.

1. S. nodosa L.   Figwort.

A perennial 40–80 cm., *glabrous* except for the glandular infl. (rhachis and pedicels); rhizome short, swollen and nodular. *Stem sharply quadrangular but not winged. Lvs* 6–13 cm., ± ovate, *acute,* coarsely and unequally *serrate,* ± truncate at base but usually slightly and often unequally decurrent down the *petiole, which is not winged.* Infl. a panicle made up of cymes borne in the axils of the bracts. Lowest pair of bracts like the lvs; one or two more pairs often slightly lf-like; the upper small, linear, alternate (rarely all lf-like). Pedicels 2 or 3 times as long as the fl. *Calyx-lobes* ovate, obtuse, *with very narrow,* often scarcely visible, *scarious border.* Corolla to c. 1 cm., with greenish tube and reddish-brown upper lip (rarely wholly green). *Staminode* broader than long, *retuse.* Capsule 6–10 mm., ovoid, acuminate. Fl. 6–9. Pollinated by wasps, less frequently by bees. $2n=36$. Hp.

Native. Common in damp and wet woods and hedgebanks throughout the British Is., except Shetland; ascending to 1500 ft. in Yorks. 111, H40, S. Europe from 69° 48′ N. in Norway to Spain, N. Italy and Greece; temperate Asia, east to the Yenisei region.

2. S. aquatica L.   Water Betony.

A perennial 50–100 cm., *glabrous* except for the somewhat glandular infl. and the sometimes puberulent lvs. Rhizome not nodular. *Stem 4-winged.* Lvs 6–12 cm., ± ovate, *obtuse, crenate,* often with 1 or 2 small pinnae at the base, which is subcordate to broadly cuneate; *petiole winged.* Infl. a panicle made up of cymes in the axils of bracts. Upper bracts oblong, lower somewhat lf-like but all differing markedly from the lvs. Pedicels about equalling the fls. *Calyx-lobes* rounded *with broad* (0·5–1 *mm.*) *scarious border.* Corolla to c. 1 cm., brownish-purple above, greenish on the underside. *Staminode suborbicular* or rather broader than long, *entire.* Capsule 4–6 mm., subglobose, apiculate. Fl. 6–9. Pollinated by wasps, less frequently by other insects. $2n=80*$. Hp. or Hs.

Native. Common on the edges of ponds, streams, and in wet woods and meadows in the south of England extending north to Ayr and Midlothian but rare in Scotland; over the whole of Ireland, but rather local in the north. 79, H38, S. From Belgium and W. Germany to Morocco and Tunisia; Azores.

3. S. umbrosa Dum.

S. Ehrhartii Stevens; *S. alata* Gilib.

A *glabrous* perennial 40–100 cm. Differs from *S. aquatica* as follows: Stem more broadly winged. *Lvs serrate, acute or subobtuse,* never cordate at base. Cymes laxer. Bracts larger, lf-like. *Staminode with two divaricate lobes,* so that it is much broader than long. Fl. 7–9. $2n=$ c. 52. Hp.

Native. Rare, in damp shady places in England and S. Scotland, absent west of Wilts, Gloucester, and Denbigh and north of Lanark and Angus; Dublin, Kildare, Fermanagh, Derry. 34, H4. C. Europe from S. Sweden, Belgium and E. France eastwards, south in the Balkans to Montenegro and Thrace; Asia, east to Tibet and south to Palestine.

4. S. scorodonia L.   'Balm-leaved Figwort.'

A perennial 60–100 cm., *whole plant greyish pubescent. Stem quadrangular, not winged.* Lvs 4–10 cm., ovate, petiolate, cordate at base, obtuse to acute, *doubly dentate, with mucronate teeth, rugose; petiole not winged.* Infl. a panicle of lax few-fld cymes in the axils of ± lf-like bracts. Pedicels 2 or 3 times as long as the fls, arcuate, divaricate. *Sepals* rounded, *with a broad scarious border.* Corolla 8–11 mm., dull purple. *Staminode suborbicular, entire.* Capsule 6–8 mm., subglobose or ovoid, apiculate. Fl. 6–8. Hp.

Native. Hedgebanks, etc., in Cornwall, Devon and Jersey, locally frequent; naturalized in Glamorgan. W. France, Spain, Portugal, Madeira, Azores, N.W. Morocco.

*S. canina L. Differs from all the above in its pinnatifid lvs. Naturalized at Barry Docks (Glamorgan). $2n=26$. Native of C. Europe and the Mediterranean region.

Section 2. *Venilia* G. Don. Staminode 0.

*5. S. vernalis L. 'Yellow Figwort.'

A biennial or perennial 30–80 cm., *softly glandular-hairy*. Stem obscurely quadrangular. Lvs 4–15 cm., thin, broadly ovate, cordate, petioled, acute, deeply toothed. *Fls in cymes in the axils of the upper lvs.* Cymes long-peduncled, many-fld and compact, with lf-like bracts. Pedicels shorter than calyx. Sepals oblong-lanceolate, *subacute, without a scarious border. Corolla* 6–8 mm., *greenish-yellow, contracted at the mouth,* the lobes very small, nearly equal. *Staminode* 0. Stamens finally protruding. Capsule 8–10 mm., ovoid-conic. Fl. 4–6. Visited by bees. $2n=40$. Hp.

Introduced. In plantations and waste places, usually in shade. Very local but found in a number of places in Great Britain, mainly in the south-east but extending north to Aberdeen and west to Cornwall. 36. Native of C. and S. Europe in the mountains from the Vosges and the Volga-Kama region to the Pyrenees, Sicily, Serbia and the lower Don region; naturalized in the plains.

Tribe 3. GRATIOLEAE. Lvs opposite, at least below. Fls solitary or in spikes or racemes. Corolla not saccate or spurred at base, the two upper lobes outside the laterals in bud. Stamens 4 or 2.

8. MIMULUS L.

Herbs with opposite lvs. Fls on peduncles from the axils of the lvs or bracts. Calyx tubular, 5-angled and 5-toothed. *Corolla with a long tube, hairy in the throat, 2-lipped;* the upper lip 2-lobed; the lower longer, 3-lobed, the lobes all flat. Stamens 4; anthers 2-celled. Stigma with 2 flat lobes. *Capsule* included in the calyx, *loculicidal.* Seeds small, numerous.

About 114 spp., mostly in temperate America (mainly California), a few in the Old World in E. Asia, Australia, New Zealand and southern Africa.

1 Plant glabrous or pubescent only above; corolla 2·5–4·5 cm.   2
  Plant viscid-hairy all over; corolla 1–2 cm.   **3. moschatus**

2 Calyx and pedicels pubescent; pedicels mostly 1·5–3 cm.; corolla with small red spots only, markedly 2-lipped, throat nearly closed.
    1. guttatus

  Calyx and pedicels glabrous; pedicels mostly 3·5 cm. or more; corolla with large red blotches or variegated with purple, only slightly 2-lipped, throat wide open.   **2. luteus**

*1. M. guttatus DC. Monkey-flower.

M. luteus auct. angl. p.p.; *M. Langsdorffii* Donn ex Greene

A perennial (5–)20–50 cm., *glabrous below,* ±*pubescent* (often glandular-pubescent) *above* at least on the pedicels or calyx (in America sometimes entirely glabrous). Fl. stems ascending or decumbent. Lvs 1–7 cm., irregularly dentate; lower stalked, ovate or oblong; upper sessile, ovate or orbicular-ovate. Infl. on well-developed plants many-fld, the upper bracts much smaller than the lvs but the lower passing into the lvs, but sometimes few-fld and with the bracts more like the lvs. *Pedicels* (1–)1·5–3 (the lower sometimes up to 5·5) *cm*. Calyx becoming inflated in fr.; *teeth ± deltoid, the upper much longer than the others. Corolla* 2·5–4·5 *cm., yellow with small red spots in the throat and sometimes at the top of the lower lip, markedly* 2-*lipped with the lower lip much longer than the upper and with a prominent palate nearly closing the throat.* Capsule oblong, obtuse. Fl. 7–9. Pollinated by bees, homogamous. $2n = 28, 48*$. Hp., Hs. or Hel.

Introduced. First recorded in 1830, now rather common on the banks of streams, etc., through nearly the whole of the British Is. and appearing quite native. 100, H27, S. Western N. America (Alaska and Montana to N.W. Mexico). Naturalized also in Europe and in the eastern United States.

*2. M. luteus L.   Blood-drop Emlets.

Differs from *M. guttatus* as follows: *Glabrous all over* (except inside the calyx and corolla). Fl. stems usually decumbent. Lvs usually with fewer and more regular teeth. Fls always few (up to c. 10). *Pedicels* (3–)3·5–6(–10) *cm. Corolla yellow with small red spots in the throat and large red spots on the lobes or with the lobes ± variegated with pinkish-purple* (sometimes, but apparently not in Britain, coloured as *M. guttatus*), *lower lip little longer than the upper, throat open.* Fl. 6–9. Hp. or Hel.

In similar places to *M. guttatus* but much less common and mainly in Scotland. Native of Chile; naturalized also in C. Europe.

Hybrids involving the two preceding spp., *M. cupreus* L., and perhaps other spp., occasionally occur ± naturalized.

*3. M. moschatus Dougl. ex Lindl.   Musk.

A viscid-hairy perennial with decumbent stems 10–40 cm. Lvs 1–4 cm., all alike, shortly petioled, ovate or elliptic with small distant teeth. Pedicels equalling or shorter than lvs. *Calyx-teeth triangular-lanceolate, subequal. Corolla* 1–2 *cm.,* yellow, not blotched. Capsule acute. Fl. 7–8. Hp.

Introduced. Commonly cultivated in gardens and occasionally naturalized in wet places in England and E. Ireland. Native of N. America (British Columbia to Montana and California). Naturalized also in C. Europe and the eastern United States. Formerly much cultivated for the musky scent of all parts of the plant. All the plants in this country today, however, appear to be scentless.

9. LIMOSELLA L.

Annual herbs, creeping by runners, *the lvs all radical.* Fls small, axillary, solitary. Calyx 5-toothed. *Corolla rotate, nearly regular,* 4–5-*lobed with short tube.* Stamens 4, with 1-*celled anthers. Capsule septicidal,* septum incomplete.

Eleven spp., spread over the greater part of the globe.

Lvs of mature plants elliptic; calyx longer than corolla-tube.   **1. aquatica**
Lvs all subulate; calyx shorter than corolla-tube.   **2. subulata**

1. L. aquatica L.   'Mudwort.'

A glabrous annual, creeping by runners which are at first upright then become horizontal and produce fresh rosettes at the nodes. *Upper lvs* 5–15 mm., *elliptical, with petiole several times as long as blade,* dark green, the lower

lanceolate-spathulate, the lowest subulate. *Calyx longer than corolla-tube.*
Corolla 2–5 mm. diam., white or lavender sometimes splashed purple on the
back; *tube* 1·5 *mm., campanulate*; lobes triangular, acute, with few long hairs.
Style short, stigma medium-sized. Capsule round-oval to ellipsoid. Fl. 6–10.
$2n=40^*$. Th.

Native. In wet mud at the edges of pools or where water has stood. Local
in England and Wales and absent from a number of counties, particularly in
the west, ascending to 1500 ft. in Yorks; still less common in Scotland (four
counties, north to Dunbarton and Kincardine); in Ireland only known from
the limestone in Clare and S. Galway. 56, H2. Europe from Iceland and
Scandinavia to C. Spain and Portugal, N. Italy and Montenegro; Egypt;
arctic and temperate Asia to Japan and India; Greenland; N. America from
Labrador to North-West Territory and south in the mountains to Colorado
and California.

L. aquatica × subulata

Often perennating. Upper lvs lanceolate-spathulate, dark green, lower subulate.
Calyx about equalling corolla-tube. Corolla up to 4 mm. diam., white, sometimes
with purple marks on back; tube c. 2 mm., greenish-yellow, cylindric; lobes ligulate,
hairs moderate in size and number. Style medium, stigma large. Fr. not formed.
$2n=30^*$.

Morfa Pools, Glamorgan.

2. L. subulata Ives

L. aquatica var. *tenuifolia* auct.

Runners arching, often below the soil surface, occasionally perennating.
Lvs 1–2·5 cm., *all subulate*, light green. *Calyx c. ⅔ as long as corolla-tube.*
Corolla up to 4 mm. diam., white with orange tube; *tube c.* 3 *mm., contracted
below the insertion of the stamens*; lobes ovate-ligulate with numerous short
hairs. Style long; stigma small. Fr. nearly globose. Fl. 6–10. $2n=20^*$. Th.

Native. In similar habitats to *L. aquatica*, very rare. 3. Glamorgan
(3 stations), near the River Glaslyn in Caernarvonshire and Merioneth.
Eastern N. America from Labrador to Maryland.

Tribe 4. DIGITALEAE. Corolla not saccate or spurred, its lobes ± spreading,
the two upper inside the lateral ones in bud. Stamens 2–4(–8).

10. SIBTHORPIA L.

Creeping herbs with reniform radical and alternate lvs. Fls small, axillary,
solitary. Calyx 4–8-lobed. *Corolla* 5–8-fid, nearly regular, *rotate; tube very
short. Stamens equalling or one fewer than corolla lobes*, with 2-celled anthers.
Capsule loculicidal.

About 6 spp. in S. and W. Europe, N. and W. Africa and the Andes of
S. America.

1. S. europaea L.   'Cornish Moneywort.'

A creeping hairy perennial, rooting at the nodes. Lvs 0·5–2 cm. diam., in
clusters at the nodes, alternate from the upper part of the filiform stem, reniform, long-stalked, crenately 5–7-lobed. Calyx 4- or 5-lobed. Corolla 1–2 mm.

diam., 5-lobed, the two upper lobes yellowish, the three lower broader, pink. Stamens 4. Fl. 7–10. Chh.

Native. Very local in moist shady places. Sussex, Somerset to Cornwall, S. Wales (in Glamorgan, Carmarthen and Cardigan); Lewis, probably introduced; Dingle Peninsula (Kerry) ascending to 1700 ft.; Channel Islands. 9, H2, S. W. France, W. Spain, Portugal.

11. ERINUS L.

Low tufted perennial with alternate lvs. Fls in terminal bracteate racemes. Calyx deeply 5-lobed. *Corolla with a slender narrow cylindric tube* about equalling the calyx; lobes 5, spreading, nearly equal, deeply emarginate. *Stamens* 4. Capsule loculicidal.

Two spp., the following and a Moroccan endemic.

*1. E. alpinus L.

A tufted perennial 5–15 cm.; stems numerous, simple, hairy, ascending. Lvs c. 1·5 cm., obovate or spathulate, attenuate into a petiole, crenate or dentate, green, glabrous or hairy, the lower forming a rosette. Raceme corymbiform. Bracts entire. Calyx-lobes linear. Corolla purple. Capsule ovoid, shorter than calyx. Fl. 5–10. Pollinated probably mainly by Lepidoptera, self-pollination possible. Chh.

Introduced. Naturalized on walls and in rocky woods in a few places in England, Scotland and Ireland. Mountains of France, Switzerland, Austria, Italy, Spain, Sardinia; Algeria and Morocco.

12. DIGITALIS L.

Tall biennial or perennial herbs with alternate lvs, the lowest forming a rosette. Fls nodding, in terminal, unilateral, bracteate racemes. Calyx deeply 5-lobed. *Corolla with a long campanulate tube,* constricted near the base, shortly 5-lobed, the two upper forming an obscurely lobed upper lip, the lowest usually longer and more prominent. *Stamens* 4. Capsule septicidal with many small seeds.

About 20 spp. in Europe and the Mediterranean region to C. Asia.

Several spp. besides our own are cultivated and *D. lutea* L. with a yellow corolla to 2 cm. and *D. grandiflora* Mill. with yellow corolla 3–4 cm. have been recorded as escapes.

*D. lanata Ehrh.

Biennial or perennial. Lvs glabrous or ciliate, ±lanceolate. Infl. woolly. Corolla 20–25 mm., yellowish-white with purplish network; lowest lobe very long, whitish. Rarely naturalized. S.E. Europe.

1. D. purpurea L.   Foxglove.

An erect biennial, rarely perennial, 50–150 cm. Stem usually simple, greyish tomentose or the lower part glabrous. Lvs 15–30 cm., ovate to lanceolate, crenate, green and softly pubescent above, grey-tomentose beneath attenuate at the base into a winged petiole. Raceme 20–80-fld. Bracts lanceolate, sessile, entire, decreasing in size upwards. Pedicels tomentose, longer than calyx. Lower sepals ovate, upper lanceolate, all acute. Corolla 4–5 cm., 3 or 4 times as long as calyx, pinkish-purple with deeper purple spots on a white ground inside the lower part of the tube, rarely white and spotted or unspotted, shortly ciliate and with a few long hairs within. Stamens and filiform style included. Capsule ovoid, rather longer than calyx. Fl. 6–9. Pollinated by humble-bees, protandrous. $2n=56^{*}$. Hs.

Native. Common in open places in woods, etc., and on heaths and mountain rocks, ascending to 2900 ft. On acid soils throughout the British Is. (not Shetland), often becoming dominant in clearings and burnt areas in woods on light dry soils. 110, H40, S. W. Europe from C. Norway to Bohemia, Spain and Sardinia (not Switzerland or Italy); Morocco. The drug digitalin, obtained from this plant, is still extensively used for heart complaints.

13. Veronica L.

Annual or perennial herbs or low shrubs with opposite lvs. Fls blue, rarely white or pinkish, in axillary or terminal racemes or solitary in the axils of lvs similar to the stem-lvs but alternate. Calyx with 4 lobes, the upper being absent (or rarely 5-lobed, the upper much smaller). *Corolla with a very short tube and rotate 4-cleft limb*, the upper lobe the largest (representing 2 lobes of the 5-lobed genera), the lower the smallest. *Stamens 2.* Capsule ± laterally compressed. Seeds few.

About 200 spp. in temperate regions of both hemispheres. A number of spp. are cultivated in gardens.

1	Fls in axillary racemes.	2
	Fls in terminal racemes or solitary.	8
2	Plant glabrous or, if hairy, lvs linear-lanceolate; growing in wet places.	3
	Plant ± hairy, lvs ovate or oblong; growing in drier places.	6
3	Racemes alternate (i.e. from one only of a pair of lvs).	**4. scutellata**
	Racemes opposite.	4
4	Lvs petioled, obtuse.	**1. beccabunga**
	Lvs sessile, acute.	5
5	Corolla pale blue; pedicels ascending after flowering.	**2. anagallis-aquatica**
	Corolla pinkish; pedicels spreading after flowering.	**3. catenata**
6	Pedicels 2 mm. or less, shorter than bract and calyx; racemes rather dense, pyramidal.	**5. officinalis**
	Pedicels 4 mm. or more, as long as or longer than bract and calyx; raceme lax.	7
7	Stem hairy all round; capsule longer than calyx; lvs stalked, petiole 5–15 mm.	**6. montana**
	Stem hairy on two opposite sides only; capsule shorter than calyx; lvs sessile or petiole less than 5 mm.	**7. chamaedrys**
8	Fls in dense many-fld, terminal racemes; corolla-tube longer than broad.	**8. spicata**
	Fls in lax racemes, or few-fld head-like racemes, or solitary; corolla-tube very short, much broader than long.	9
9	Fls in racemes, the bracts often passing gradually into the lvs but the upper, at least, very different from them.	10

Fls solitary in the axils of lvs resembling the cauline lvs, though the upper sometimes rather smaller. 19

10   Lvs glabrous or finely puberulent, entire or obscurely crenulate; perennial (except *peregrina*). 11

Lvs conspicuously pubescent, often glandular, dentate or lobed (but only obscurely crenulate in *acinifolia*); annual. 15

11   Perennial; bracts (at least the upper) shorter or scarcely longer than fls. 12

Annual; bracts all much longer than fls. **13. peregrina**

12   Shrubby at base; corolla bright blue, c. 1 cm. diam. **9. fruticans**

Herbs; corolla white, pale or dull blue or pink, smaller. 13

13   Pedicels much longer than bracts; fls pink. **10. repens**

Pedicels shorter than bracts; fls not pink. 14

14   Pedicels longer than calyx; fr. broader than long, scarcely exceeding calyx, style about as long; fls white or pale blue. **12. serpyllifolia**

Pedicels shorter than calyx; fr. longer than broad, much exceeding calyx, style very short; fls dull blue. **11. alpina**

15   Pedicels much shorter than calyx. 16

Pedicels longer than calyx. 17

16   Lvs toothed. **14. arvensis**

Lvs pinnatifid, the lobes longer than the entire portion. **15. verna**

17   Lvs toothed or nearly entire; fr. longer than calyx. 18

Lvs digitately 3–7-lobed; fr. shorter than calyx. **18. triphyllos**

18   Lvs obscurely crenate; pedicels 2–3 times as long as calyx; fr. broader than long. **16. acinifolia**

Lvs conspicuously dentate; pedicels not twice as long as calyx; fr. longer than broad. **17. praecox**

19   Lvs with 5–7 large teeth or small lobes near base; sepals cordate at base. **19. hederifolia**

Lvs regularly crenate-serrate; sepals narrowed at base. 20

20   Decumbent annuals; lvs not reniform; pedicels not twice as long as lvs. 21

Creeping perennial often forming mats; lvs reniform; pedicels several times as long as lvs. **23. filiformis**

21   Lobes of fr. divergent; fls 8–12 mm. diam. **20. persica**

Lobes of fr. not divergent; fls 4–8 mm. diam. 22

22   Sepals ovate, acute or subacute; fr. with short crisped glandless hairs and a few longer glandular hairs; corolla usually uniformly blue. **21. polita**

Sepals oblong, obtuse or subobtuse; fr. without short crisped hairs; lower lobe of corolla usually white or pale. **22. agrestis**

Section 1. *Beccabunga* (Hill) Dum. Fls in opposite axillary racemes. Capsule loculicidal. Normally perennial.

1. V. beccabunga L.   Brooklime.

A glabrous perennial 20–60 cm.; stems creeping and rooting at base, then ascending, fleshy. *Lvs* 3–6 cm., rather thick and fleshy, *oval or oblong, obtuse,* base rounded, shallowly crenate-serrate, *shortly stalked. Racemes opposite,* rather lax, 10–30-fld. Bracts linear-lanceolate, ± equalling the slender pedicels. Calyx-lobes narrowly ovate, acute. Corolla 7–8 mm. across, blue. Capsule

± orbicular, retuse, shorter than calyx. Fl. 5–9. Pollinated by various Diptera and Hymenoptera, protogynous, often selfed. $2n = 18$. Hel. or Hp.

Native. In streams, ponds, marshes and wet places in meadows. Common throughout the British Is. 112, H40, S. Europe (except Iceland, Corsica, Crete, etc.), to 65° 08′ N. in Norway; N. Africa; temperate Asia extending to Japan and the Himalaya.

2. V. anagallis-aquatica L.　　　　　　　　　'Water Speedwell.'

A perennial or sometimes annual herb, *glabrous* except for the sometimes glandular infl. Stems 20–30 cm. high, shortly creeping and rooting at base, then ascending, fleshy, usually branched, green. *Lvs 5–12 cm., ovate-lanceolate or lanceolate, acute, semi-amplexicaul*, remotely serrulate. *Racemes opposite,* rather lax, 10–50-fld, ascending. *Bracts linear*, acute, *shorter than or equalling pedicels at flowering.* Calyx-lobes ovate-lanceolate, acute. *Corolla* c. 5–6 mm. across, *pale blue. Pedicels ascending after flowering.* Capsule ± orbicular, slightly emarginate, usually slightly longer than broad. Fl. 6–8. Visited by Diptera, easily self-pollinated. $2n = 36$. Hp. or Hel.

Native. In ponds, streams, wet meadows and wet mud, rather common. Probably throughout the British Is. All Europe; Asia to Japan and the Himalaya; N.E. and S. Africa; N. and S. America; New Zealand.

3. V. catenata Pennell

V. aquatica Bernh., non S. F. Gray; *V. comosa* auct.

Differs from the last as follows: Stem usually purplish-tinged. Racemes laxer, more spreading. *Bracts broader*, c. 1·5 mm., usually lanceolate, less acute, *longer than pedicels at flowering.* Calyx-lobes elliptic or oblong, widely spreading after flowering. *Corolla pink* with darker lines. *Pedicels spreading at right angles after flowering.* Capsule more definitely emarginate, usually broader than long. Fl. 6–8. $2n = 36$. Hp. or Hel.

Native. In similar places to the last, with which it often grows, perhaps rather commoner but the distribution of the two spp. is still imperfectly known. Europe from S. Sweden to Bulgaria and Albania; N. and E. Africa; N. America; temperate Asia.

V. anagallis-aquatica × catenata is most easily recognized by its complete sterility, the capsules never developing. Not uncommon where the parents occur together or sometimes alone.

Section 2. *Veronica.* Fls usually in alternate axillary racemes. Capsule septicidal. Perennial.

4. V. scutellata L.　　　　　　　　　　'Marsh Speedwell.'

A *glabrous perennial*, or sometimes glandular, or sometimes (var. *villosa* Schumach.) densely pubescent. Stems creeping below and then ascending, simple or slightly branched, 10–50 cm. high. *Lvs 2–4 cm., linear-lanceolate* or linear, acute, semi-amplexicaul, remotely denticulate, yellowish-green often tinged with purple, midrib impressed and conspicuous above. Fls in very lax, slender, few (up to 10)-fld *alternate racemes.* Bracts linear. Pedicels slender, more than twice as long as the bracts, spreading at right angles in fr. Sepals ovate. Corolla c. 6–7 mm. across, white or pale blue with purple lines. *Capsule flat*, broader than long, deeply emarginate, much longer than calyx. Fl. 6–8. $2n = 18$. Hp. or Hel.

Native. In ponds, bogs, wet meadows, etc., often on acid soils. 110, H 40, S. Rather common throughout the British Is.; var. *villosa* less common, often in rather drier habitats. Europe from Iceland and Scandinavia (70° N.) and arctic Russia to C. Spain and Portugal, C. Italy, Serbia and Bulgaria; N. Asia, east to Kamchatka.

5. V. officinalis L. 'Common Speedwell.'

A perennial herb with *stems* 10–40 cm., creeping and rooting and often forming large mats, ascending above, *hairy all round*. Lvs 2–3 cm., oblong to obovate-elliptic, crenate, subacute, cuneate at base, subsessile, hairy on both sides. *Racemes* from the axils of one or, more rarely, both of a pair of lvs, long-stalked, *rather dense and pyramidal*, 15–25-fld. *Bracts* linear, *about twice as long as pedicels* which are 1–2 mm. Calyx-lobes lanceolate. *Corolla* c. 6 mm. across, *lilac*. Filaments, anthers and style ± lilac. Capsule obovate or obcordate, longer than calyx. Fl. 5–8. Pollinated by various Diptera and Hymenoptera, homogamous, protandrous or protogynous. $2n = 18$ (rare), 36. Chh.

Native. In grassland, heaths and open woods, often on dry soils. Common throughout the British Is. 112, H 40, S. Europe from Iceland and Scandinavia (70° 48′ N.) and arctic Russia to C. Spain and Portugal, Sardinia, Sicily and N. Greece; Asia Minor and Caucasus; Azores; eastern N. America (probably introduced).

6. V. montana L. 'Wood Speedwell.'

A perennial herb; *stems* 20–40 cm., *hairy all round*, creeping and rooting, the ends of the branches ascending. Lvs 2–3 cm., ovate or orbicular-ovate, sub-obtuse, truncate or broad-cuneate at base, coarsely crenate-serrate, light green, hairy on both sides; *petiole 5–15 mm.* *Racemes* from the axil of one or both of a pair of lvs, long-stalked, *lax, 2–5-fld.* *Bracts* small, linear, *much shorter than pedicels* which are 4–7 mm. Calyx-lobes obovate. *Corolla* c. 7 mm. across, *lilac-blue*. Filaments and anthers paler. *Capsule* nearly flat, ± orbicular, retuse or emarginate, *longer than calyx*, glandular-ciliate. Fl. 4–7. Pollinated by various Diptera and Hymenoptera. $2n = 18^*$. Chh.

Native. In damp woods. Rather local throughout England, Wales and Ireland and extending north to Ross. 92, H 38. Europe from S. Sweden and the Volga-Kama region to C. Spain, Corsica, C. Italy, Macedonia and the Caucasus; mountains of Algeria and Tunisia.

7. V. chamaedrys L. 'Germander Speedwell.'

A perennial herb 20–40 cm.; *stems* prostrate and rooting at the nodes, ascending above, *with long white hairs in two lines on opposite sides*, glabrous between. Lvs 1–2·5 cm., triangular-ovate, *sessile or shortly stalked* (petiole to 5 mm.), subobtuse, subcordate to cordate at base, coarsely crenate-serrate, dull green, hairy, especially on margins and veins below. *Racemes* from the axils of one or, more rarely, both of a pair of lvs, long-stalked, lax, 10–20-fld. *Bracts* lanceolate, *about equalling or shorter than pedicel*. Pedicels 4–6 mm.

Calyx-lobes ± lanceolate, hairy. *Corolla* c. 1 cm. across, deep *bright blue* with white eye. Filaments and style blue, anthers pale. *Capsule obcordate, shorter than calyx*, broader than long, ciliate and pubescent. Fl. 3–7. Pollinated by various Diptera and Hymenoptera, homogamous. $2n = 32$. Chh.

Native. In grassland, woods, hedges, etc. Very common throughout the British Is. 112, H 40, S. Europe from Scandinavia (68° 30′ N.) to C. Spain and Portugal, N. Italy and Greece; N. and W. Asia; naturalized in N. America.

Section 3. *Pseudolysimachium* Koch. Fls in terminal racemes. Corolla-tube longer than broad. Capsule septicidal. Perennial.

8. V. spicata L.   'Spiked Speedwell.'

A pubescent perennial 8–60 cm., with shortly creeping, somewhat woody rhizome and erect flowering stems. Lowest lvs ovate or oval, petioled, passing gradually into the lanceolate or linear, sessile upper ones, all ± crenate or crenate-serrate or the upper entire. *Fls in a many-fld terminal spike-like raceme*. Pedicels very short. Bracts and calyx-teeth ± lanceolate. Corolla violet-blue with a rather long tube. Capsule ± orbicular, about as long as calyx-lobes, retuse or emarginate. Fl. 7–9. Pollinated by various insects. $2n = 34, 68*$. Ch.

Europe from S. Scandinavia to C. Spain, Italy and Greece; W. and C. Asia to China and Japan.

Ssp. **spicata**. Plant 8–30 cm. Lowest lvs 1·5–3 cm. × 8–12 mm., sparingly crenate mostly near the middle, usually broadest near the middle, gradually narrowed into a narrow petiole.

Native. In dry grassland on basic soils in the breckland of E. Anglia and even there rare. 3.

Ssp. **hybrida** (L.) E. F. Warb. Plant 15–60 cm., more robust. Lvs 2–4 × 1–2 cm., more deeply crenate or crenate-serrate nearly all round, usually broadest below the middle, abruptly narrowed into a broad petiole.

Native. On limestone rocks; Avon Gorge (Bristol), Wales, W. Yorks and Westmorland; very local. 12.

The two following rare garden-escapes are sometimes confused with *V. spicata* from which they can be distinguished as follows:
*V. spuria** L. Taller (40–80 cm.). Lvs sharply serrate. Pedicels about as long as bracts and calyx. Fr. glabrous. C. Europe, etc.
*V. longifolia** L. Lvs sharply serrate. Racemes dense, usually arranged in terminal panicles. Pedicels much shorter than bracts and calyx. Rarely naturalized. Europe, etc.

Section 4. *Berula* Dum. Fls in terminal racemes, the lower bracts often lf-like. Capsule septicidal. Seeds flat, convex or cup-shaped. Annual or perennial.

9. V. fruticans Jacq.   'Rock Speedwell.

V. saxatilis Scop.

A perennial 5–20 cm., *woody at base*, glabrous below, puberulent above, branches ascending, numerous. Lvs c. 1 cm., obovate or oblong, entire or slightly crenulate, cuneate at subsessile base, coriaceous. Raceme lax, up to 10-fld. Bracts narrow, not passing into lvs, shorter than pedicels. Sepals narrow, oblong, obtuse. *Corolla c. 1 cm. across, deep bright blue*, purplish in the middle. Capsule ovate or elliptic, longer than sepals, about equalling style, apex entire. Fl. 7–8. Visited by various insects but apparently often selfed, homogamous. $2n=16$. Chw.

Native. On alpine rocks from 1600 to 3000 ft. Perth, Angus, E. Aberdeen, W. Inverness; very local. 5. Arctic Europe from Iceland to Russia; alpine Europe south to the Pyrenees, Corsica, the Apennines and Bosnia; Greenland.

*10. V. repens Clarion ex DC.

A *perennial herb* 4–10 cm., glabrous below, glandular-puberulent above; stems slender, creeping and rooting, with short lateral flowering branches. Lvs ovate-orbicular, shortly stalked, entire or obscurely crenulate. *Racemes* terminal, *lax, 3–6-fld. Pedicels longer than bracts and calyx.* Calyx-lobes elliptic. *Corolla pink.* Capsule obovate, emarginate; style nearly 3 times as long. Fl. 4–5. $2n=14$. Chh?

Introduced. Sometimes found as an escape and ± naturalized in a few places in N. England and S. Scotland. Native of mountains of Corsica and S. Spain.

11. V. alpina L.   'Alpine Speedwell.'

A *perennial herb* 5–15 cm., glabrous below, glandular-pubescent above; stems shortly creeping at base with very few branches, the flowering ones ascending. Lvs ovate, entire or serrulate, cuneate at base, subsessile. *Racemes 4–12-fld, dense*, head-like. *Bracts* not passing into the lvs, *longer than the pedicels.* Calyx-lobes elliptic, subacute. *Corolla dull blue*, about twice as long as calyx. *Capsule* obovate, glabrous, slightly emarginate, *longer than calyx and much longer than the very short (1 mm.) style.* Fl. 7–8. Probably usually self-pollinated; insect visits few (flies). $2n=18$. Chh?

Native. Damp alpine rocks in Scotland from Dumfries to Aberdeen and Inverness, 1600–3700 ft., local. 10. Arctic Europe and N. America; high mountains of Europe south to the Sierra Nevada, C. Apennines and Macedonia; Siberia, Manchuria, Korea; mountains of N. America south to New England and Colorado.

12. V. serpyllifolia L.   'Thyme-leaved Speedwell.'

A *perennial herb* 10–30 cm.; stems creeping and rooting at nodes, the flowering usually ascending, puberulent. Lvs 1–2 cm., oval or oblong, entire or weakly crenulate, rounded at both ends, subsessile or shortly stalked, glabrous, light green. *Racemes* terminal, up to 30-fld, often long (10 cm. or more), *lax.* Upper *bracts* narrowly oblong, the lower larger and broader, passing into the lvs, *longer than the pedicels.* Calyx-lobes oblong. *Corolla white or pale blue* with darker lines. Filaments and style white, anthers slaty-violet. *Capsule* obcordate, ciliate, broader than long, *about equalling calyx and style.* Infl. sometimes glandular. Fl. 3–10. Visited by flies.

Europe from Iceland and Scandinavia (71° N.) to C. Spain and Portugal, Sicily, Albania and Thrace; mountains of N. Africa; Madeira and Azores; temperate Asia; N. and S. America.

Ssp. **serpyllifolia**. Shortly creeping, flowering branches ascending. Corolla c. 6 mm. across, white with slaty-violet lines. Infl. rarely glandular. $2n=14$.

Native. In grassland, heaths, waste places, etc., sometimes as a garden weed, often on rather moist ground. 112, H40, S. Common throughout the British Is.

Ssp. humifusa (Dicks.) Syme

V. tenella All.; *V. humifusa* Dicks.; *V. borealis* (Laest.) Hook. f.

Decumbent; stems rooting for most of their length. Lvs usually broader, often ± orbicular, nearly entire. Corolla blue, rather larger. Infl. and capsule glandular. Racemes few-fld. $2n=14*$.

Native. Damp places in mountainous districts, rather local. 22. Wales and Welsh border, N.W. England, Dumfries, Scottish Highlands. N. Europe, mountains of C. Europe.

*13. V. peregrina L.

A *glabrous annual* 5–25 cm.; stems erect, simple or with spreading branches. Lvs ovate or oblong, entire or weakly toothed, tapering at base to a short petiole. Racemes terminal, long and lax. *Upper bracts* lanceolate, entire, *much longer than the fls.* Pedicels very short. Calyx-lobes lanceolate, about 6 times as long as the pedicel. Corolla blue, shorter than calyx. *Style almost* 0. Capsule glabrous, scarcely emarginate, about as broad as long. Fl. 4–7. Self-pollinated. $2n=52$. Th.

Introduced. Thoroughly naturalized on cultivated ground in N.W. Ireland in Donegal, Derry and Tyrone; extending south to Galway, Roscommon and Monaghan, but less well established; also in a few places in England. Native of America; naturalized also in W. and C. Europe.

14. V. arvensis L.   'Wall Speedwell.'

An *erect annual* 5–25 cm., but very variable in size, simple or branched at base with ascending branches, pubescent and sometimes glandular. *Lvs* to 1·5 cm. but usually less, triangular-ovate, *coarsely crenate-serrate*, the lowest stalked, the upper sessile. Racemes long, lax, occupying the greater part of the length of the stem. *Upper bracts* 4–7 mm., lanceolate, entire, ciliate, *longer than the fls*, lower gradually passing into the lvs. *Pedicels very short* (less than 1 mm.). Calyx-lobes like the upper bracts but smaller. Corolla blue, shorter than calyx. Style c. 1 mm. or less. Capsule about as long as broad, obcordate, ciliate, shorter than calyx. *Seeds flat*. Fl. 3–10. Visited by small bees, probably often selfed. $2n=16$. Th.

Native. In cultivated ground and grassland and on heaths in ± open habitats, usually on dry soils. 112, H40, S. Common throughout the British Is. Europe (to c. 68° N. in Sweden); C. and W. Asia; N. Africa; Macaronesia; naturalized in N. America.

15. V. verna L.   'Spring Speedwell.'

An *erect annual* 3–15 cm. Differs from *V. arvensis* as follows: *Lvs pinnatifid* with 3–7 lobes. Raceme denser, always glandular. Capsule broader than long. Fl. 5–6. $2n=16$. Th.

Native. In open habitats in dry grassland in the breckland of Norfolk and Suffolk and local there. 4. Europe from Scandinavia (c. 65° N.) to C. and E. Spain, Corsica, N. Italy and Greece; W. Asia, east to the Altai; Morocco.

*16. V. acinifolia L.

An *annual* 5–15 cm., usually with erect or ascending branches from the base, ± *glandular-pubescent. Lvs* 1 cm. or less, oval, *obscurely crenate* or almost entire. Infl. lax. Upper *bracts* elliptic-oblong, entire, lower gradually passing into lvs, *equalling or shorter than pedicels. Calyx-lobes* ovate-oblong, ⅓ or ½ *as long as pedicel.* Corolla blue, rather longer than calyx. *Capsule broader than long,* 2-lobed with a deep sinus, slightly longer than calyx. *Seeds flat.* Fl. 4–6. 2n = 14. Th.

Introduced. Sometimes occurring as a casual in cultivated ground and perhaps naturalized in one or two places in England. Native of C. and S. Europe and Asia Minor.

17. V. praecox All.

An *erect annual* 5–20 cm., simple or with erect or ascending branches, glandular-pubescent. *Lvs* ovate, *deeply toothed,* shortly stalked. Racemes lax. *Upper bracts* elliptic-oblong, entire, lower passing into the lvs, *slightly shorter than the pedicels. Calyx-lobes* oblong, hairy, *slightly shorter than the pedicel.* Corolla blue, rather longer than calyx. Style c. 2 mm. Capsule obovate, longer than broad, emarginate, rather longer than calyx. *Seeds cup-shaped.* Fl. 3–6. 2n = 18. Th.

?Introduced. Known only from a few cultivated fields in W. Norfolk and Suffolk, where it was first found in 1933. 2. C. and S. Europe from Belgium and Gotland to E. Spain, Sicily and Crete; Asia Minor, Caucasus; N. Africa.

18. V. triphyllos L.   'Fingered Speedwell.'

An *annual* 5–20 cm., suberect with spreading or decumbent branches, glandular-pubescent. *Lvs digitately* 3–7-*lobed,* lobes spathulate or oblong, to c. 1 cm., the lower stalked, upper sessile. *Uppermost bracts* usually entire, *shorter than the slender pedicels,* passing gradually but rather soon into the lower lfy ones. Calyx-lobes spathulate, obtuse, shorter than pedicel. Corolla deep blue, shorter than calyx. Style 1 mm. Capsule about as long as broad, deeply lobed, shorter than calyx. *Seeds cup-shaped.* Fl. 4–6. Visited by small bees, often selfed. 2n = 14. Th.

Native. In sandy arable fields in Norfolk and Suffolk, and rare even there; Yorks (extinct); sometimes occurring as a casual elsewhere. 6. Europe from S. Sweden and the Volga-Don region to N. Portugal and C. Spain, C. Italy and Greece; Asia Minor, Caucasus; N. Africa.

Section 5. *Pocilla* Dum. Fls solitary in the axils of the upper lvs. Capsule septicidal. Seeds cup-shaped. Annual. The last two spp. of the preceding section are transitional to this one.

19. V. hederifolia L.   'Ivy Speedwell.'

An annual, branched at the base, with decumbent branches 10–60 cm.; stem hairy. *Lvs* to c. 1·5 cm., ± reniform, *with 2 or 3 large teeth or small lobes on each side near the base,* rather thick, light green, stalked, obtuse at apex, ± truncate at base, 3-veined, ciliate and with a few scattered short hairs below. Upper lvs somewhat smaller than lower. Pedicels usually rather shorter than lvs. *Sepals* ovate, strongly *cordate at base. Corolla* shorter than calyx, *pale lilac.* Capsule glabrous, scarcely compressed, scarcely emarginate, broader than long. Fl. (3–)4–5(–8). Visited by various insects, often selfed. $2n=56$. Th.

Native. Common in cultivated ground throughout the British Is. 108, H40, S. Europe (except Iceland, etc.); temperate Asia to Japan; N. Africa; Madeira; naturalized in N. America.

*20. V. persica Poir.   'Buxbaum's Speedwell.'

V. Buxbaumii Ten., non Schmidt; *V. Tournefortii* auct.

An annual, branched at the base, with decumbent branches 10–40 cm.; stem hairy. Lvs 1–3 cm., triangular-ovate, shortly stalked, coarsely crenate-serrate, light green, hairy on the veins below and minutely ciliate. *Pedicels longer than lvs* but not twice as long, decurved in fr. Calyx-lobes 5–6 mm. at flowering time, ovate, ciliate, accrescent and strongly divaricate in fr. *Corolla 8–12 mm. across, bright blue,* the lower lobe often paler or white. *Capsule* 2-lobed, *lobes* sharply keeled and *divergent,* so that the capsule is nearly twice as broad as long, ciliate. Fl. 1–12. Visited by various insects, often selfed. $2n=28$. Th.

Introduced. First recorded 1825. Now common in cultivated land throughout the British Is. and the commonest species of the genus in this habitat. 112, H40, S. C. and S. Europe north to C. Scandinavia, Mediterranean region; originally from W. Asia.

21. V. polita Fr.   'Grey Speedwell.'

V. didyma auct.

A *pubescent annual,* branched at base, with decumbent branches. *Lvs* 5–15 mm., *ovate,* shortly stalked, the lower broader than long, upper longer than broad, dull green, obtuse, ± truncate at base, coarsely and irregularly crenate-serrate. *Pedicels equalling or shorter than lvs,* decurved in fr. *Sepals ovate, acute or subacute,* accrescent, conspicuously veined. *Corolla 4–8 mm. across, usually uniformly bright blue,* rarely the lower lobe paler. *Capsule* rather broader than long, *lobes erect,* not keeled, *clothed with short, crisped, glandless hairs* and some longer glandular ones. Fl. 1–12. Usually self-pollinated. $2n=14*$. Th.

Native. In cultivated ground throughout the British Is., common in S. England, much less so elsewhere. 103, H40, S. Europe (not Iceland), rare in the north and perhaps only casual in Norway, etc.; temperate Asia; N. Africa.

22. V. agrestis L.   'Field Speedwell.'

Differs from *V. polita* as follows: Lvs all longer than broad, rather lighter green, more regularly crenate-serrate. *Sepals oblong or ovate-oblong, obtuse or subobtuse,* faintly veined. *Corolla* usually *pale blue with the lower lobe or three lobes white or very pale,* less frequently all white, or pink above. *Capsule* lobes obscurely keeled, with long glandular hairs, often with rather shorter glandless ones but *without short crisped hairs.* Fl. 1–12. Visited by Diptera and Hymenoptera, homogamous, often selfed. $2n = 28$. Th.

Native. In cultivated ground, probably throughout the British Is., common in the north, very local in S. England. 112?, H40, S. Europe (except Iceland and S. Balkans) but rare in the south, extending to 64° 48′ N. in Norway; mountains of N. Africa; Asia Minor.

*23. V. filiformis Sm.

A pubescent *perennial with numerous creeping stems,* often forming large patches. *Lvs* c. 5 mm., *reniform,* shortly stalked, crenate. *Pedicels filiform, several times as long as lvs.* Calyx-lobes oblong, obtuse. Corolla resembling that of *V. persica* but slightly purplish. Fr. suborbicular, very rare in Britain. Fl. 4–6. $2n = 14$. Chh.

Introduced. Grown in gardens and now extensively naturalized in grassy ground by streams, roads, on lawns, etc., throughout the British Is., especially in S. England; first recorded as an escape in 1927. Native of Asia Minor and the Caucasus.

*V. crista-galli Stev.

Annual resembling *V. persica* in habit. Lvs ovate 1–2 cm., the lower stalked. Sepals connate in pairs, each pair broader than long, divided near apex into two acuminate lobes. Corolla small, pale blue. Naturalized in Sussex. Native of Caucasus, etc.

HEBE Comm.

Differs from Veronica in its truly shrubby habit and dorsally compressed capsule.

About 100 spp., mainly in New Zealand, a few in Australia and the extreme south of South America. Numerous species and hybrids are cultivated (Shrubby Veronicas) and the two following are reported as naturalized:

*H. lewisii (Armstrong) Cockayne & Allan

Shrub 1–2 m. Twigs pubescent. Lvs 4–7 cm., ±oblong, margins pubescent. Infl. a simple raceme, pubescent. Fls c. 8 mm. diam., pale blue. Native of New Zealand (South Island).

*H. salicifolia (Forst. f.) Pennell

Nearly glabrous shrub 1–3 m. Lvs 5–15 cm., ±lanceolate. Infl. a simple raceme, 8–25 cm. Fls 3–5 mm. diam., pale lilac to nearly white. Native of New Zealand.

Subfamily 3. RHINANTHOIDEAE. Hemiparasitic. Nectaries at base of ovary, unilateral. Stamens (2–)4. Fls in spikes or racemes. Corolla 2-lipped, the upper lobes inside the laterals in bud. Capsule loculicidal.

14. PEDICULARIS L.

Herbs with alternate pinnatisect lvs. Fls in terminal lfy-bracted spikes or racemes. *Calyx tubular* in fl., soon *becoming inflated,* with 2–5 lf-like lobes. *Upper lip of corolla laterally compressed,* entire or with 2 or 4 small teeth near the end; lower lip 3-lobed. Stamens 4, included in the upper lip. Capsule compressed, with a few large seeds in the lower part. Nectar secreted by a swelling at the base of the ovary; pollinated by humble-bees.

About 250 spp. in north temperate regions and the Andes.

Annual; calyx pubescent outside; upper lip of corolla with 4 teeth.
                                                                    1. palustris
Perennial; calyx usually glabrous outside; upper lip of corolla with 2 teeth.
                                                                    2. sylvatica

1. P. palustris L.   Red-rattle.

Annual 8–60 *cm.* high, nearly glabrous. *Stem single,* branching from near the base to the middle. Lvs 2–4 cm., oblong in outline, pinnatisect, the lobes dentate. Bracts similar but smaller. Pedicels short. *Calyx pubescent,* at least near the upper edge, often reddish, with 2 broad, short, irregularly-cut lf-like lobes. Corolla 2–2·5 cm., purplish-pink; *upper lip with a tooth on each side at the tip and another lower down.* Capsule curved, longer than the calyx. Fl. 5–9. $2n=16*$. Th.

Native. Wet heaths and meadows, rather common throughout the British Is., ascending to 2800 ft. 112, H40, S. Europe from the Faeroes and Scandinavia to the Urals, south to C. France, N. Italy, Serbia, Bulgaria and the Caucasus.

2. P. sylvatica L.   Lousewort.

Nearly glabrous *perennial* 8–25 *cm., with many decumbent branches from the base,* and with thick tap-root. Lvs to 2 cm., pinnatisect, oblong or linear in outline, the divisions dentate. Racemes terminal, 3–10-fld. Bracts trisect, lf-like, passing into the lvs below. Pedicels short, stout. *Calyx* ± cylindric in fl., 5-angled with 4 small lf-like 2–3-lobed teeth, the 5th (upper) tooth small and linear, arising at a lower level, *glabrous* or (in Ireland and the Hebrides) pubescent *outside,* pubescent on the lobes within. Corolla 2–2·5 cm., pink; *upper lip with a tooth on each side near the tip only;* lower lip with three ± orbicular lobes. Capsule obliquely truncate, about equalling the calyx. Fl. 4–7. $2n=16$. Hp.

Native. Damp heaths, bogs and marshes, rather common throughout the British Is., ascending to 3000 ft. 112, H40, S. W. and C. Europe from Scandinavia to Spain, N. Italy, Poland and W. Russia (Upper Dniester region).

15. RHINANTHUS L.

Annual herbs with opposite toothed lvs. Fls in terminal lfy-bracted spikes. *Calyx flattened and inflated* in fl., accrescent, with 4 entire teeth. *Upper lip of corolla laterally compressed* with 2 teeth at the end; lower lip 3-lobed. Stamens 4, included in the upper lip. Capsule compressed, with a few large, usually winged seeds. Nectar secreted at the base of the ovary.

●About 25 spp. in north temperate regions.

The spp. of this genus are critical and difficult of determination. The following account must be regarded as only provisional, as the British forms are certainly not yet completely worked out.

'Intercalary lvs' are the lvs between the topmost branches and the lowest bracts. They are often transitional between the lower lvs and bracts in shape and toothing. Bracts in the following descriptions exclude the two lowest pairs which are often transitional to the lvs.

1 Teeth of upper lip of corolla long (c. 2 mm.), twice as long as broad;
   corolla-tube curved upwards. **1. serotinus**
   Teeth of upper lip of corolla short, not longer than broad; corolla-tube
   straight. 2

2 Calyx hairy only on the margin; usually lowland plants. **2. minor**
   Calyx hairy all over; mountain plants. 3

3 Stem unbranched or with short branches; lvs oblong or oblong-linear,
   ± parallel-sided; plant never purple-tinged. **3. borealis**
   Stem frequently branched; lvs linear-lanceolate, tapering; plant often
   purple-tinged. **× gardineri**

1. R. serotinus (Schönh.) Oborny ssp. **apterus** (Fr.) Hyl. 'Greater Yellow-Rattle.'

R. major Ehrh., non L.; incl. *R. serotinus* ssp. *vernalis* (Zinger) Hyl.

Plant ± robust 20–60 cm., stem black-spotted, simple or more frequently with short or long branches from about the middle. Stem-lvs 2·5–7 cm. × 4–15 mm., lanceolate or linear-lanceolate, often strongly crenate-serrate with prominent teeth but sometimes the teeth less prominent, scabrous; branch lvs usually smaller. Intercalary lvs 0–1(–2) pairs. Main inflorescence 8–18-fld. Bracts yellowish-green, ovate or rhombic-ovate, usually acuminate, at least the basal teeth usually deep and sharp. Calyx glabrous except for the pubescent margin. *Corolla* 15–20 mm., yellow with violet teeth, *tube curved upwards, mouth closed; teeth c. 2 mm., conical, twice as long as broad.* Seeds varying from broadly winged to wingless. Fl. 6–9. Pollinated by humble-bees, self-pollination not possible. $2n = 22^*$. Th.

Native. In cornfields, less commonly in meadows and on sandhills, in scattered localities in Scotland and N. England; rare in S. England and Wales; in Ireland known only from Limerick. 43, H1.

Ssp. *serotinus*, much branched below, lvs narrow tapering, intercalary lvs 2–5(–10) pairs, does not appear to occur in the British Is.

Europe from Scandinavia to C. France, Bosnia and Roumania; Siberia.

2. R. minor L. Yellow-Rattle.

Stem simple or branched, usually black-spotted. Lvs crenate-dentate, scabrous. Bracts triangular-ovate to triangular-lanceolate with triangular acute or shortly aristate teeth. *Calyx shortly hairy on margin, otherwise glabrous. Corolla c.* 12–15 mm.; yellow or brown with violet, rarely white teeth; *tube straight,* mouth somewhat open; *teeth rounded, short, not longer than broad.* Seeds always winged. Fl. 5–8, pollinated by humble-bees, selfed if not visited. $2n = 22^*$. Th.

Grassland, etc., common throughout the British Is. 112, H40, S. Europe from Iceland and Scandinavia to C. Spain, C. Italy, Macedonia and the Caucasus; W. Siberia; S. Greenland, Newfoundland.

Intermediates between the following sspp. occur, ssp. *calcareus* being the most distinct.

Ssp. **minor**. Stem 12–40 cm., usually with only *very short flowerless branches but often with longer suberect flowering branches from the middle and upper part of stem.* Internodes all (except the lowest) ± equal. *Lvs of main stems*

(1–)2–4(–5) cm. × (3–)5–7 mm., *usually oblong* (less often oblong-linear) and parallel-sided for the greater part of their length. *Intercalary lvs* 0(–1) *pairs* (rarely more). *Lowest fls usually from 6th–9th node.* Fls yellow with violet (rarely white) tooth. Fl. 5–7.

Pastures, especially on dry basic soils, probably throughout the British Is.; common in England, becoming rarer in Scotland and there often tending towards ssp. *stenophyllus*. Almost throughout the range of the sp.

Ssp. **stenophyllus** (Schur) O. Schwarz (*R. stenophyllus* (Schur) Druce). *Stem* (15–)25–50 *cm.*, usually with *long arcuate-ascending flowering branches* and shorter flowerless branches from the lower and middle part of the stem. Lower internodes usually short, equalling or shorter than lvs, upper much longer. *Lvs* of main stems 1·5–4·5 cm. × 2–5(–7) mm., *narrowly lanceolate or linear-lanceolate*, ± tapering from near the base. *Intercalary lvs* (0–)1–2(–4) *pairs. Lowest fls usually from* (8th–)10th–13th(–15th) *node.* Fls yellow with violet tooth, sometimes becoming brown. Fl. 7–8.

Damp grassland, fens, etc., common in Scotland and N. and W. England and Wales, rare in S.E. England; probably throughout Ireland. Europe from Scandinavia to the mountains of France, Italy, Croatia and Roumania.

Ssp. **monticola** (Sterneck) O. Schwarz (*R. monticola* (Sterneck) Druce; *R. spadiceus* Wilmott). *Plant often tinged purple. Stem* (5–)10–20(–25) *cm.*, *usually with short or moderate flowerless branches from near the base, sometimes with 1–3 longer flowering branches in addition.* Lower internodes usually very short, upper much longer. *Lvs* of main stems 1–2(–2·5) cm. × 2–4 mm., *linear-lanceolate*, ± tapering from near base, tending to be more erect than in ssp. *stenophyllus*. *Intercalary lvs usually* 1–2(–3) pairs. *Lowest fls usually from* (7th–)8th–11th(–12th) *node.* Fls dull yellow *becoming treacle-brown, or treacle-brown throughout*; tooth violet. Fl. 7–8.

Grassy places in mountain districts from Yorks and Dumfries to Shetland; Kerry, Derry. Alps.

Ssp. **calcareus** (Wilmott) E. F. Warb. (*R. calcareus* Wilmott). Stem 25–50 cm., usually with *long arcuate-ascending flowering branches from about the middle.* Lower internodes short, upper very long. *Lvs of main stems* 1–2·5 *cm.* × 1·5–3 *mm., linear*, ± spreading. *Intercalary lvs usually* (2–)3–6 *pairs. Lowest fls usually from 14th–19th node.* Fls yellow; tooth violet.

Chalk and limestone downs from Dorset and Sussex to Gloucester and Northants, local; Clare. ?Endemic.

3. R. borealis (Sterneck) Druce

Stem (5–)9–20(–28) cm., unspotted or spotted, *unbranched* or occasionally with short axillary flowerless (very rarely flowering) branches and then intercalary lvs 0. Stem-lvs 1–3 cm. × 3–7 mm., oblong or oblong-linear, ± parallel-sided. Lowest fls from 5th–7th(–8th) node. *Calyx hairy all over.* Corolla bright yellow; teeth violet or white. Bracts and corolla-shape as in *R. minor*. Fl. 7–8. Th.

Native. Grassy places on mountains, local, ascending to 3200 ft. Caernarvon; Dumfries to Shetland and the Outer Hebrides; Kerry. 18, H1. Alaska, Greenland, Iceland, ?Scandinavia.

R. borealis × **minor** = **R.** × **gardineri** Druce

R. Drummond-hayi auct.; incl. *R. Lintonii* Wilmott, *R. lochabrensis* Wilmott, *R. Vachellae* Wilmott

A series of hybrids between *R. borealis* and *R. minor* ssp. *monticola* and ssp. *stenophyllus*. Differs from the latter in the calyx hairy all over; from the former in the frequent purple tinge, the linear-lanceolate tapering lvs and in the frequent presence of longer flowerless or flowering branches and of intercalary lvs.

Grassy places on mountains often forming locally uniform hybrid swarms to the exclusion of its parents and sometimes occupying a zone between that of *R. minor* (lower down) and *R. borealis* (higher up) where neither parent occurs.

16. MELAMPYRUM L.   Cow-wheat.

Annual herbs with opposite mostly entire lvs. Fls in terminal lfy-bracted racemes or spikes, the bracts often toothed. *Calyx tubular*, 4-toothed. Corolla 2-lipped; *upper lip laterally compressed*; lower lip 3-lobed, shorter, with prominent palate nearly closing the mouth. Stamens 4, included in the upper lip. *Capsule* compressed, *with* 1–4 ovoid *seeds*. Nectary at the base of the ovary. All our spp. are pollinated by humble-bees but can be selfed if not visited.

About 25 spp., in Europe, temperate Asia and eastern N. America.

The spp. of this genus are very variable though *M. cristatum* and *M. arvense* show little variation in Britain.

1	Fls in ± dense spikes.	2
	Fls in pairs in the axils of lf-like bracts, forming a very lax, secund, interrupted raceme.	3
2	Spikes 4-sided, very dense; bracts cordate, recurved.	**1. cristatum**
	Spikes conical or cylindrical, laxer; bracts not cordate or recurved.	**2. arvense**
3	Corolla-tube much longer than the calyx; lower lip of corolla straight.	**3. pratense**
	Corolla-tube equalling or shorter than the calyx; lower lip of corolla deflexed.	**4. sylvaticum**

1. M. cristatum L.   'Crested Cow-wheat.'

Puberulous annual 20–50 cm., simple or with a few spreading branches. Lvs 5–10 cm., sessile, lanceolate. *Fls in a dense 4-sided spike. Bracts cordate, recurved, the base bright rosy-purple and finely pectinate* (teeth less than 2 mm.); *lower with long lf-like green entire points*, upper with short points. Calyx-tube with two lines of hairs; *teeth shorter than tube, unequal.* Corolla 12–16 mm., pale yellow variegated with purple, the palate deeper yellow, the tube longer than the calyx. Capsule normally 4-seeded. Fl. 6–9. $2n = 18^*$. Th.

Native. Very local, at the edges of woods. From Wilts, Hants and Essex to Nottingham and S. Lincoln. 14. N. and C. Europe from Norway and N. Russia to N. Spain, C. Italy and Thrace; W. Asia, east to the Yenisei region.

2. M. arvense L.   'Field Cow-wheat.'

Pubescent annual 20–60 cm., with spreading branches. Lvs 3–8 cm., sessile, lanceolate, the upper usually with a few, long teeth at the base. *Fls in a rather lax cylindrical or conical, scarcely 4-sided spike. Bracts lanceolate, ± erect, pink at first, pinnatifid, with long* (to 8 mm.) *slender teeth, without lf-like points.* Calyx-tube pubescent; *teeth longer than tube, nearly equal.* Corolla 20–24 mm., with a pink tube, yellow throat and deep pink lips, the tube about equalling the calyx. Capsule normally 2-seeded. Fl. 6–9. 2*n*=18. Th.

Native. Cornfields from Isle of Wight and Sussex to Bedford and Lincoln; very local and rare. 8. Europe from Sweden and the Volga-Kama region to N. Spain, Italy, Thrace and the Caucasus; W. Siberia (upper Tobol region).

3. M. pratense L.   'Common Cow-wheat.'

Very variable. Annual 8–60 cm., glabrous or somewhat hispid, with suberect to spreading branches. Lvs 1·5–10 cm., sessile or very shortly stalked, ovate-lanceolate to linear-lanceolate, entire. *Fls in the axils of each of distant opposite green lf-like bracts, both fls turned to the same side of the stem,* ± horizontal. Bracts varying from entire to pectinate, the lower always less toothed than the upper and often closely resembling the lvs. *Calyx with appressed, linear-setaceous lobes,* usually rather longer than the tube. Corolla 11–17 mm. to tip of upper lip, varying from deep yellow to whitish, sometimes variegated with red or purple, *the tube about twice as long as the calyx,* mouth closed or somewhat open; *lower lip directed forward.* Capsule normally 4-seeded. Fl. 5–10. 2*n*=18*. Th.

Native. In woods, heaths, etc.; ascending to over 3000 ft., common in many districts, local in others. Throughout the British Is. 108, H 39. Europe from Scandinavia and arctic Russia to C. Spain and Portugal, C. Italy, Serbia and Bulgaria. W. Asia, east to the Yenisei region.

A very variable sp. The characters which have been relied upon to distinguish varieties are the habit, whether glabrous or hispid, shape of lvs, toothing of the bracts, colour of the fls both when fresh and when fading and the relative length of the anther-hairs and anther-appendages.

4. M. sylvaticum L.   'Wood Cow-wheat.'

Annual 5–35 cm., glabrescent or pubescent, with a few suberect to spreading branches. Lvs linear-lanceolate, entire, sessile or shortly stalked. *Infl. as in M. pratense but fls suberect.* Bracts entire or the upper with a few small teeth. *Calyx with rather spreading linear-lanceolate lobes,* longer than the tube. *Corolla* 8–10 mm., deep brownish-yellow, the *tube about equalling the calyx,* mouth wide open, lower lip deflexed. Capsule normally 2-seeded. Fl. 6–8. 2*n*=18. Th.

Native. Mountain woods, ascending to 1300 ft. From Yorks to Inverness, very local; Antrim, Derry. 12, H2. Northern and mountain Europe from Iceland and arctic Russia to the Pyrenees, Alps and Macedonia.

17. EUPHRASIA L.

Annual herbs. Lvs opposite or the upper alternate, rather small. Fls sessile in the axils of the bracts (floral lvs) which usually differ somewhat from the cauline lvs, thus forming a terminal spike. Calyx campanulate, 4-toothed.

Upper lip of corolla slightly concave with 2 porrect or reflexed lobes, the lower 3-lobed with emarginate lobes. Stamens 4, the anther cells pointed at the base. Seeds small, numerous, oblong or fusiform, furrowed.

Over 130 spp., temperate northern hemisphere, temperate S. America, Australia and New Zealand. All the north temperate spp. belong to the section *Semicalcaratae*.

The determination of the spp. is not easy. They have, however, definite habitats and geographical distributions. The best time for determinations is when the fls are well out and some capsules have formed. Several plants should be used and depauperate or damaged plants neglected. Very dwarf, compact, unbranched forms of many spp. occur, especially in N. Scotland. These are very difficult to determine and no allowance has been made for them in the key.

The most important characters are those derived from the habit of the plant, the form, toothing and indumentum of the lvs, size and colour of the corolla and shape and size of the capsule. Unbranched individuals will usually be found in any colony, even of normally branched spp.

Certain characters, not included in the generic diagnosis are common to all the British spp. These have not been repeated in the detailed descriptions below. They are as follows:

Stem clothed with crisped whitish hairs. Lower cauline lvs with 1–2 teeth on each side, bracts usually with 5–6 teeth. Indumentum of the calyx like that of the lvs. Corolla (which may be white, of various shades of blue or purple, or rarely yellow) with a yellow blotch and purple lines on the lower lip.

The measurement of the corolla given is the length from the base of the tube to the apex of the upper lip. This does not in many cases give a satisfactory guide to the size of the corolla as it appears to the observer, which depends largely on the dimensions of the lower lip. Unfortunately, there are insufficient measurements of the lower lip available for it to be used in this account. Unless otherwise stated all the British spp. have a corolla-tube which does not elongate much during flowering.

Numerous varieties have been described. These have only been included in the following account if of importance ecologically, or if liable to be confused with other spp.

Hybrids and hybrid swarms seem to occur commonly, sterile hybrids being comparatively rare. The swarms sometimes occur where one of the parents is not present, though it can usually be found in the district. Some of these supposed hybrids may perhaps prove to be spp. but genetical work on the genus is needed. Hybrids have not been included in the key, but a list of them is given at the end of the genus.

For a full account of the British spp. see Pugsley, *Journ. Linn. Soc.* XLVIII (1930), 467, on which the following account is mainly based. The larger-fld spp. appear to be pollinated by bees or other insects, the small-fld ones usually selfed.

1   Bracts generally more than ½ as broad as long, with contiguous teeth; capsule ciliate with long straight hairs (*officinalis* agg.).   2
    Bracts generally less than ½ as broad as long, with more distant, narrow teeth; capsule glabrous or rarely ciliate with a few weak bristles.
                                                                    25. salisburgensis

2　Stalked glands present, at least on the bracts.　　　　　　　　　　24
　　Stalked glands 0 on all parts of the plant.　　　　　　　　　　　*3*

3　Lvs and bracts glabrous, or with short bristles confined to the margins
　　and to the veins beneath.　　　　　　　　　　　　　　　　　*4*
　　Lvs and bracts with long hairs, or with short bristles on the surface as
　　well as the margins and veins.　　　　　　　　　　　　　　　*15*

4　Cauline internodes all much longer than the lvs; plant simple or with
　　(usually few) branches from near the middle.　　　　　　　　　　*5*
　　Lower (or all) cauline internodes shorter or scarcely longer than the lvs;
　　plant branched from the base of the stem (rarely simple).　　　　*10*

5　Plant robust, 10–35 cm.; lvs 8–15 mm., oval or ovate, obtuse and with
　　obtuse teeth; fls rather large (6–8 mm.).　　　　　**17. borealis**
　　Plant slender, 4–15(–30) cm.; lvs 4–7(–9) mm.; fls small or rather small
　　(4–7(–8) mm.).　　　　　　　　　　　　　　　　　　　*6*

6　Bracts ± orbicular; capsule large and broad, emarginate; cauline lvs all
　　obtuse with obtuse teeth; stem simple or with few branches (Scotland).　*7*
　　Bracts narrower; capsule smaller, truncate or retuse; upper cauline lvs
　　with acute teeth; stem simple or with few or numerous branches.　*8*

7　Upper cauline internodes very long, those of the infl. short; corolla white
　　or with lilac upper lip, lower lip much longer than the upper (moun-
　　tains).　　　　　　　　　　　　　　　　　　　**4. frigida**
　　Cauline internodes all ± equal, not much longer than the lvs; corolla
　　usually violet, lower lip not much longer than upper (near the sea).
　　　　　　　　　　　　　　　　　　　　　　5. foulaensis

8　Lower lip of corolla far exceeding upper; capsule usually longer than its
　　bract.　　　　　　　　　　　　　　　　　　　　　　　*9*
　　Lower lip of corolla scarcely exceeding upper; capsule usually shorter
　　than its bract.　　　　　　　　　　　　　　　　**2. scottica**

9　Branches, lvs and bracts suberect; lvs usually strongly purple-tinged;
　　fls small.　　　　　　　　　　　　　　　　　**1. micrantha**
　　Branches and lvs spreading; bracts often recurved; lvs not or slightly
　　purple-tinged; fls moderate.　　　　　　　　　　**13. nemorosa**

10　Fls large (8–)9–11 mm. (chalk downs).　　　　**16. pseudokerneri**
　　Fls 8 mm. or less, small or moderate.　　　　　　　　　　　　*11*

11　Stem slender, decumbent or ascending, usually much branched below
　　but sometimes simple.　　　　　　　　　　　　**15. confusa**
　　Stem erect or nearly so, usually stout.　　　　　　　　　　　*12*

12　Stem with few short branches or simple, short (2–8(–12) cm.); capsule
　　emarginate (N. Scotland).　　　　　　　　　　**5. foulaensis**
　　Stem usually with ± numerous branches from the base; capsule rounded,
　　truncate or rarely retuse.　　　　　　　　　　　　　　　*13*

13　Plant stiff; lvs thick, ± fleshy; infl. very dense; capsule exceeding the
　　calyx-teeth.　　　　　　　　　　　**12. occidentalis** var. *calvescens*
　　Plant more slender; branches usually flexuous; lvs not fleshy; infl. laxer:
　　capsule rarely exceeding calyx-teeth.　　　　　　　　　　　*14*

14　Cauline lvs with obtuse shallow teeth; corolla 4–5 mm.; capsule narrow
　　(Rhum and W. Ross).　　　　　　　　　　**14. heslop-harrisonii**
　　Cauline lvs with acute or subacute teeth; corolla usually 5 mm. or more,
　　capsule not very narrow (common).　　　　　　　**13. nemorosa**

15 Lvs and bracts dull or grey-green, $\pm$ densely clothed on both sides with
    long bristles.   *16*
    Hairs short or few, or long hairs confined to the veins.   *20*

16 Stem slender, simple or with a few slender suberect branches from near
    the middle; lowest fls from 3rd to 5th node; cauline internodes longer
    than (rarely equalling) lvs; lower lip of corolla scarcely longer than
    upper (Rhum).   *17*
    Stem rather stout, branched at the base, or with short branches from the
    middle; lowest fls at c. 7th node or higher; cauline internodes, at least
    the lowest, shorter than or equalling lvs; lower lip of corolla usually
    much longer than upper.   *18*

17 Stem usually branched; lvs narrow, oblong or oblong-obovate; capsule
    4–5·5 mm. not very broad, obovate-oblong.   **3. rhumica**
    Stem always simple; lvs broad, oval or obovate-orbicular; capsule
    3·5–4 mm. very broad, obovate-orbicular.   **6. eurycarpa**

18 Lvs suborbicular, rounded at base; stem with few short branches from
    the middle.   **8. rotundifolia**
    Lvs oval, obovate or oblong, cuneate at base; stem with long branches
    from near the base.   *19*

19 Fl. spike very dense, with imbricate bracts; lvs large (rarely less than
    8 mm.); capsule 6–8 mm., calyx accrescent; hairs on lvs stout.
    9. marshallii
    Fl. spike rather lax, bracts not imbricate; lvs rather small (rarely more
    than 7 mm.); capsule 4·5–5·5 mm., calyx scarcely accrescent.   **10. curta**

20 Lvs with thick whitish bristles on margins and veins beneath, shortly
    pubescent near apex above (Lewis).   **7. campbelliae**
    Indumentum not as above.   *21*

21 Plant robust, stem 5–35 cm.; lvs large 6–12(–18) mm.; fls rather large,
    6–8 mm.   **18. brevipila** var. *subeglandulosa*
    Plant more slender, stem rarely more than 15 cm.; lvs small, 4–7(–10) mm.;
    fls small or moderate, 6 mm. or less (–8 mm. in *E. frigida*).   *22*

22 Cauline internodes very long, those of the infl. short; lvs without long
    hairs except sometimes on the veins beneath.   **4. frigida**
    Internodes, at least the lower, shorter than the lvs, or if longer, lvs with
    long hairs on surface.   *23*

23 Stem erect, lowest fls from about 7th node; capsule rather narrow,
    truncate or retuse.   **10. curta**
    Stem ascending, plant usually very dwarf (less than 2·5 cm.); lowest fls
    from about 4th node; capsule broad, emarginate (Welsh mountains).
    11. cambrica

24 Glandular hairs short, straight, often few.   *25*
    Glandular hairs long, flexuous, numerous.   *28*

25 Lowest fls from 2nd to 4th node; stem slender, often flexuous, simple or
    with 1–3 branches near the base; cauline internodes all much longer
    than lvs.   **4. frigida**
    Lowest fls from 6th to 8th node; branches usually more numerous, lower
    cauline internodes shorter than lvs (except in *E. brevipila*).   *26*

26 Stem slender, ascending or decumbent, usually with numerous slender
    flexuous branches from near the base; lvs small (3–7 mm.), narrow
    (rarely glandular).   **15. confusa**
    Stem robust, erect; lvs large, 6–13 mm. or more.   *27*

27　Plant compact, the internodes all shorter than the lvs; lvs thick and
　　　somewhat fleshy; bracts imbricate; fls small, 5–7 mm.　**12. occidentalis**
　　Plant tall, with at least the upper cauline internodes much longer than
　　　the lvs; lvs not fleshy; bracts not imbricate; fls rather large, 6–8(–9) mm.
　　　　　　　　　　　　　　　　　　　　　　　　　　　　18. brevipila

28　Cauline internodes shorter than the lvs; stem rather robust with many
　　　long branches from near the base.　　　　　　　　　　**22. anglica**
　　Cauline internodes (except sometimes the lower) longer than the lvs;
　　　branches usually few or 0; if several, then from near the middle of the
　　　stem.　　　　　　　　　　　　　　　　　　　　　　　　　　　29

29　Corolla small, 5–8 mm.; stem erect, robust, stiff, unbranched or with
　　　short suberect branches.　　　　　　　　　　　　　　**24. hirtella**
　　Corolla large, (6–)7–13 mm.; stem slender, unbranched or with long or
　　　slender branches; capsule 6·5 mm. or less.　　　　　　　　　　30
　　Corolla large, 8–9 mm.; stem robust often with several branches; capsule
　　　6–9 mm.　　　　　　　　　　　　　　　　　　**18. brevipila** vars

30　Corolla usually purple; internodes rarely twice as long as lvs (Devon
　　　and Cornwall).　　　　　　　　　　　　　　　　　**23. vigursii**
　　Corolla not purple; usually some internodes at least twice as long as lvs
　　　(absent from Devon and Cornwall).　　　　　　　　　　　　31

31　Corolla very large, tube elongating during flowering, finally 7–13 mm.　32
　　Corolla 8–9 mm., tube scarcely elongating during flowering, plant very
　　　small and slender.　　　　　　　　　　　　　　　　**21. rivularis**

32　Lower internodes shorter than lvs; upper cauline lvs with acute teeth;
　　　branches from near base of stem.　　　　　　　　**19. rostkoviana**
　　All cauline internodes longer than lvs; all cauline lvs with obtuse teeth;
　　　branches from near the middle of stem.　　　　　　**20. montana**

Subsection *Ciliatae* Jorg. (*E. officinalis* agg.).　　　　　　　Eyebright.

Bracts generally more than half as broad as long, acutely or more rarely
bluntly serrate with ± contiguous teeth. Capsule ciliate with long straight
hairs.

1. E. micrantha Rchb.

E. gracilis (Fr.) Drej.

Stem (2–)5–30 cm., *slender*, erect, strict, purplish, sometimes simple but *usually
producing ± slender suberect branches from about the middle. Lower internodes
much longer than lvs*, becoming shorter upwards. *Lvs* 4–6(–8) mm. × 1–3 mm.,
narrow, dark green or more frequently purplish, glabrous or with minute bristles
on the margins and veins beneath; lower oblong, obtuse with 1–2 blunt teeth
on each side, upper broader with 3–4 acute teeth; bracts about equalling lvs,
ovate, acute, with about 4 acute or acuminate teeth. Lowest fls at about
7th node. Calyx with short finely acuminate teeth, scarcely accrescent in fr.
Corolla small, 4–6(–7) mm., 3–5 mm. across lower lip, white, or more usually
lilac, violet or purple; upper lip with very small subentire or retuse lobes;
lower lip longer than upper with three narrow emarginate segments, the middle
the longest. *Capsule* 4–6 mm., *narrowly oblong*, truncate or more rarely retuse,
glabrous except the ciliate margins, usually *exceeding* the calyx-teeth and *its
subtending lf*, but often rather shorter. Fl. 7–9. $2n = 44^{*}$. Th.

　　Native. On moors and heaths on light, not very wet soils, occasionally in small bogs, throughout the British Is., common in the north and west, very local in the south and east. 63, H 20. Faeroes and Scandinavia to Bohemia and W. France.

2. E. scottica Wettst.

Differs from the last as follows: Stem flexuous, green or tinged with red, *simple* or with a few slender suberect branches mostly from below the middle. *Lvs light green* or occasionally purplish, larger on an average (5–9 × 3–5 mm.) and more cuneate at base. Flowering from 5th or 6th node. *Corolla* usually *white*, sometimes violet; *lower lip* with 3 subequal lobes, *scarcely exceeding the upper lip. Capsule* retuse or truncate, usually *not exceeding* the calyx or *its subtending lf.* Fl. 7–8. 2*n* = 44*. Th.

Native. Wet moors; widespread in the Scottish mountains, ascending to over 3000 ft., rarer in the mountains of N. England and N. and C. Wales. In Ireland from Wicklow, Louth and Tipperary to Galway, Mayo and Sligo. 29, H 12. Faeroes, Norway, Sweden.

3. E. rhumica Pugsl.

Stem 7–10 cm., *very slender*, suberect, flexuous, purplish *with very slender suberect branches from near the middle. Internodes all longer than the lvs* (except the uppermost), the lowest very long. *Lvs* 4–7 mm., *narrow*, dull green, *with many fine whitish bristles on both sides* and on the margins; cauline *oblong to oblong-obovate*, apex very obtuse, with narrow but obtuse teeth; bracts sometimes broader, with less obtuse or subacute teeth. Lowest fls at 4th or 5th node. Calyx with short, acute teeth, not accrescent. *Corolla small,* 4–5 mm., white, tinged blue above; upper lip obscurely lobed; *lower scarcely longer than upper*, with three narrow retuse lobes, the middle equalling or longer than the lateral. *Capsule* 4–5·5 mm., *obovate-oblong*, retuse or emarginate equalling or exceeding the calyx-teeth. Fl. 8. Th.

Native. Isle of Rhum (Inner Hebrides). Endemic.

4. E. frigida Pugsl.

E. latifolia auct., non L.

Stem 5–20 cm., erect usually flexuous, *slender*, green or reddish, *simple or with* 1–3 *slender erect basal branches. Cauline internodes very long, upper usually the longest, floral* (at least the upper) *very short.* Lvs 3–7 mm., *rather broad*, thick, light or dull green, nearly *glabrous*, but with minute bristles on the margins or on the upper surface also and with or without longer hairs on the veins beneath, sometimes with small short-stalked glands, ciliate or hairy on nerves beneath, sometimes minutely glandular; cauline oblong to broadly ovate with obtuse teeth; upper *bracts broader* and usually larger, orbicular, with more acute teeth. Fls beginning from 2nd to 4th node. Calyx with long teeth, accrescent in fr. *Corolla* small or moderate, 5–8 mm., 5–6 mm. across lower lip, *white* usually with lilac upper lip; upper lip rather narrow; *lower lip longer* than upper, with emarginate lobes, the median usually rather longer and narrower than lateral. *Capsule large*, 5–8 mm., elliptic or obovate, usually *deeply emarginate*, ciliate and ± pilose, equalling or exceeding calyx-teeth. Fl. 7–8. 2*n* = 44. Th.

Native. Damp alpine rock-ledges, grassland and mountain tops usually above 2000 ft. and ascending to over 3500 ft.; widespread in the Scottish Highlands; ?Ireland. 13. Arctic and subarctic from Labrador to W. Siberia (Ob region).

5. E. foulaensis Towns. ex Wettst.

Stem 2–8(–12) cm., erect, rather slender, reddish, simple or with a few short suberect branches. *Cauline internodes about as long as or rather longer than lvs,* floral shorter (or all short in compact forms). Lvs 4–7 mm., rather broad, thick, glabrous or with minute bristles on the margins and less often a few on the surfaces also; cauline lvs oval or broadly obovate, obtuse, with 1–3 obtuse teeth on each side; bracts broader, with obtuse or acute teeth. Fls beginning at from 3rd to 8th node. Calyx with short broad subacute teeth, accrescent in fr. Corolla small, 5–7 mm., usually violet, sometimes white; lower lip rather longer than upper with emarginate subequal lobes. *Capsule large*, 5–7 mm., elliptic, *emarginate*, ciliate and sometimes pilose, exceeding the calyx-teeth, often greatly. Fl. 7–8. 2*n*=44*. Th.

Native. Sea cliffs and coastal pastures from Moray to Shetland and Outer Hebrides. 7. Faeroes.

6. E. eurycarpa Pugsl.

Stem 2–6 *cm.*, suberect, *slender*, ± *flexuous*, purplish, *simple. Internodes equalling or exceeding the lvs* (except the uppermost). *Lvs broad, very small,* 2·5–5 mm., thick, dark or dull green, *with ± numerous whitish bristles on both surfaces* and on the margins; cauline *oval to roundish obovate* with very obtuse apex and obtuse teeth; bracts broader, with the teeth sometimes less obtuse. Fls beginning at the 3rd to 5th node. Calyx with broad teeth, accrescent in fr. *Corolla very small*, 3–4 mm., white with bluish upper lip; upper lip rounded with obscure lobes; *lower scarcely longer than upper*, with subequal retuse lobes. *Capsule very broad*, 3·5–4 mm. long and nearly as wide, roundish obovate, *deeply emarginate*, slightly exceeding calyx-teeth. Fl. 8. Th.

Native. Isle of Rhum (Inner Hebrides). Endemic.

7. E. campbelliae Pugsl.

Stem 5–12 cm., *erect and rather slender*, dusky purplish, *simple* or with 1–4 suberect branches from about the middle. *Internodes about equalling the lvs*, the floral lvs scarcely imbricate. *Lvs* to 9 mm., dark or greyish green, *shortly pubescent near the apex above and with rather thick whitish bristles on the margins and nerves beneath; cauline oblong to oblong-ovate or oval*, cuneate-based, very obtuse and with obtuse teeth; bracts broader, the upper subacute with acute teeth. Lowest fls from 5th to 7th node. Calyx with blackish nerves and short teeth, somewhat accrescent. Corolla medium, 6–8 mm., narrow in front view, white, usually with a purple upper lip; upper lip with narrow retuse lobes; lower longer than upper with narrow, apically dilated, emarginate lobes, the median the longest. *Capsule moderate*, c. 6 mm., oblong to elliptic-oblong, retuse or slightly emarginate, ciliate and sparingly pilose above, about equalling calyx-teeth. Fl. 7. Th.

Native. Moors near the sea. Isle of Lewis (Outer Hebrides). Endemic.

8. E. rotundifolia Pugsl.

Stem 4–10 cm., erect, strict, *robust*, purplish, *simple or with a very few short suberect branches from about the middle. Lower internodes short, about equalling lvs*; upper shorter so that bracts are densely imbricate. *Lvs* 4–7 mm., *thick, broad*, dull green, *densely hirsute*, especially beneath, *with stout-based white bristles* of varying length; *cauline oval to broadly ovate or orbicular*, rounded at base, crenate with 1–3 *rounded obtuse teeth* and a very obtuse apical lobe; bracts similar, but apex less obtuse. Lowest fls at about 7th node. Calyx with short teeth, accrescent in fr. Corolla small, c. 6 mm., villous externally, white; upper lip with short emarginate lobes; lower lip somewhat exceeding upper, with emarginate lobes, the median the longest and dilated at the apex. *Capsule large* and broad, 6–8 mm., oblong-elliptic, retuse, ciliate, otherwise glabrous, exceeding the short calyx-teeth. Fl. 7. Th.

Native. Sea-cliffs, Sutherland, Shetland and Outer Hebrides; very local. 3. Iceland.

9. E. marshallii Pugsl.

Stem 5–15 cm., erect, *robust*, purplish, *often much-branched with long erect-spreading branches from near the base*. Internodes as in last sp. *Lvs large*, 8–14 mm., rather thick, dull green, *densely hairy with waved strong whitish bristles; cauline oval or rhomboidal to obovate*, with long *cuneate base*, obtusely toothed; bracts ovate or suborbicular, scarcely cuneate, obtuse with subacute teeth. Lowest fls at from 7th to 10th node. Calyx with rather long teeth, accrescent in fr. Corolla moderate, 6–7 mm., villous externally, white, rarely lilac or purple; upper lip with short denticulate lobes; lower lip much exceeding upper, with emarginate lobes, the median the longest and narrowest. *Capsule large* and broad, 6–8 mm., oblong-elliptic, retuse, ciliate and pilose, equalling calyx-teeth. Fl. 7–8. $2n=44^*$. Th.

Native. Very local on grassy sea-cliffs; Sutherland to Shetland; Outer and Inner Hebrides; Mull of Galloway (Wigtown). 7. Endemic.

10. E. curta (Fr.) Wettst.

Stem 5–15(–35) cm., *erect, moderately robust or rather slender*, purplish or green, usually *with long slender ascending or suberect branches from near the base or from the middle. Lower internodes shorter than the lvs*, the upper relatively longer so that the fl.-spike is lax. *Lvs rather small*, 4–7(–10) mm., *greyish-green*, strongly veined beneath when dry, *densely hirsute on both sides with long waved whitish hairs*, or less frequently with short bristles; lower cauline lvs oblong-cuneate below, with obtuse teeth becoming broader upwards; bracts broader but not longer, ovate, rounded, acute with deep acute or shortly aristate teeth. Lowest fls at about 7th node. *Calyx* with rather short but finely acuminate teeth, *scarcely accrescent. Corolla small*, (3–)5–6 mm., 3–5 mm. across lower lip, white; upper lip with retuse, sometimes reflexed lobes; lower lip longer with subequal emarginate lobes. *Capsule small* or medium, 4·5–5·5 mm., oblong, rounded-truncate or slightly retuse, pilose and ciliate, about equalling calyx-teeth. Fl. 6–9. $2n=44^*$. Th.

Native. Grassy slopes on mountains, very local, and near the sea, local;

coasts from Isle of Wight, Devon and Northumberland to Shetland; Welsh and Scottish mountains; widespread on Irish coast. 26, H16. Iceland to W. Russia, N.E. Bohemia and Brittany; Quebec.

11. E. cambrica Pugsl.

Dwarf. *Stem* 1–2·5(–15) cm., *ascending, flexuous, slender*, usually *with slender flexuous branches below*, greenish. *Internodes usually shorter than lvs*. Lvs up to 5(–8) mm., *bright green, hirsute with scattered whitish hairs on both sides*; cauline oblong to obovate, ±cuneate below, rounded-obtuse with obtuse teeth; bracts broader, less obtuse with subacute teeth. Lowest fls at about 4th node. Calyx with acuminate teeth, accrescent in fr. *Corolla small*, 4–6·5 mm., whitish, sometimes the upper lip violet-tinted; tube longer than lips; upper lip with subentire lobes; lower lip scarcely longer than upper, with narrow retuse or emarginate lobes, the median the longest. *Capsule broad*, c. 5 mm., elliptical, emarginate, shortly pilose and ciliate, longer than calyx-teeth. Fl. 7. Th.

Native. Steep grassy slopes and rock-ledges on mountains of Caernarvon, where it is widespread, and of Brecon and possibly also in Westmorland. 2. Endemic.

12. E. occidentalis Wettst.

Stem 5–15(–20) cm., *erect, robust*, purplish, usually with *basal* ascending or suberect *stout branches*. *Internodes all shorter than lvs, so that the fl.-spike is dense. Lvs rather large*, 5–13 mm., dull green, *thick*, with ±revolute margins, *clothed with* short whitish hairs and *shortly stalked glands*, sometimes (var. *calvescens* Pugsl.) hairs and glands very few or 0; cauline lvs oblong-obovate to obovate, with obtuse apex and teeth, the upper often much larger than the lower; bracts broadly ovate, apex and teeth obtuse or acute, or those of the upper shortly aristate. Lowest fls at about 8th node. Calyx with triangular subulate teeth, somewhat accrescent in fr. *Corolla small*, 5–7 mm., white, upper lip with short retuse lobes; lower lip longer than upper with rather broad emarginate lobes, the median the longest. *Capsule rather large*, 6–8 mm., oblong or elliptic-oblong, truncate or slightly retuse, ciliate and usually pilose, normally exceeding calyx-teeth. Fl. 5–8. $2n=44^*$. Th.

Native. Local on grassy sea cliffs round the greater part of the British Is., commonest in the south-west; also rarely on limestone pastures inland in Somerset, Dorset and W. Gloucester. 28, H7, S. Brittany.

13. E. nemorosa (Pers.) Wallr.

Stem (10–)15–20(–40) cm., *erect, moderately robust*, usually purplish, with *numerous long, slender*, suberect or ascending *branches*. *Lower internodes shorter than lvs but upper internodes often much longer than the upper cauline and floral lvs. Lvs medium-sized*, 6–12 mm., *dark* or occasionally purplish *green*, veins ±prominent beneath when dry, *glabrous* or with minute marginal bristles, sometimes with scattered longer hairs on margins and veins beneath; *lower cauline* lvs oblong or oval, obtuse *with subacute teeth*; upper becoming larger, ovate and more acute; bracts shorter than upper cauline lvs, ovate, acute and with acute or shortly aristate teeth. Lowest fls from about 10th node. Calyx with acuminate teeth, scarcely accrescent. *Corolla small* or moderate, 5–7·5(–8) mm., white or with bluish upper lip; upper lip rounded, with small notched lobes; *lower lip longer than upper*, deflexed, with emarginate, rather narrow lobes, the median the longest. *Capsule small or medium*, 5–6 mm., *oblong*, rounded-truncate or retuse, ciliate and usually slightly pilose, about equalling calyx-teeth. Fl. 7–9. $2n=44^*$. Th.

Native. Woods, heaths, downs and pastures; common in England and Wales, more local but widespread in Scotland and Ireland. 84, H17. Belgium and N. France to Bohemia and S.E. Germany.

A variable sp., a number of varieties of which have been described.

Var. *transiens* Pugsl. is of more slender habit, with long flexuous branches and smaller, dull green lvs much shorter than the internodes. Bracts often arcuate-recurved. Corolla c. 6 mm., usually purple-tinted. It has been confused with *E. micrantha* and has been found in several widely separated localities.

The following varieties are distinct ecologically as well as morphologically. When better known they may very likely be better treated as sspp.

Var. *calcarea* Pugsl. More robust and shorter (5–20 cm. high), branching nearer base and flowering lower. Internodes mostly shorter than lvs. Lvs thick, rarely with aristate teeth. Calyx somewhat inflated or accrescent in fr., with broader teeth, usually exceeding capsule. Chalk downs and similar places in England and Wales.

Var. *collina* Pugsl. Usually shorter (5–30 cm.), sometimes with fewer, more spreading branches. Internodes, except the lowest, longer than lvs. Lvs thicker, less acutely toothed. Calyx rather inflated in fr. Corolla larger (6–7 mm.) usually with bluish upper lip and much larger lower lip. Capsule often exceeding calyx. Mainly hilly districts, in Wales, N. and W. England, Scotland and Ireland, largely replacing var. *nemorosa*.

Var. *sabulicola* Pugsl. Plant small with rather slender greenish stem, 5–10 cm. high, and several long erect branches. Lowest fls from 6th or 7th node. Corolla smaller (c. 4–5 mm.), lower lip scarcely exceeding upper. Sandy places near the sea in Scotland, rare.

14. E. heslop-harrisonii Pugsl.

Stem 8–14 mm., *suberect, slender, flexuous, with slender flexuous branches from near the base. Internodes shorter than or equalling the lvs. Lvs rather small* (7·5 mm.), dull green, *glabrous* or with minute marginal bristles; *cauline* oblong or obovate-oblong, rounded, obtuse, *with obtuse shallow teeth*; bracts broader, ovate with subacute or acute but very rarely aristate teeth. Calyx with short acute teeth, accrescent in fr. *Corolla very small*, 4–5 mm., white with pale blue upper lip; upper lip rounded, with short subtruncate lobes; *lower lip scarcely longer than upper*, with equal, subtruncate or retuse lobes. *Capsule narrow*, 5–6 mm., *oblong* and somewhat contracted below, rounded-truncate, about equalling the calyx-teeth. Fl. 8. Th.

Native. Isle of Rhum (Inner Hebrides) and in a salt-marsh in W. Ross. Endemic.

15. E. confusa Pugsl.

E. minima auct. angl., non Jacq. ex D.C.

Rather small. *Stem 2–10(–20) cm., ascending or decumbent, slender,* greenish, *usually with numerous slender flexuous branches from near the base. Lower internodes shorter than cauline lvs,* upper longer, about equalling floral lvs. *Lvs small,* 3–7 mm., *narrow,* green, with well-marked veins beneath; *glabrous* except for minute marginal bristles and rarely a few very short glandular hairs; cauline oblong to oblong-obovate, obtuse, cuneate below, with obtuse teeth, the upper the largest; *bracts alternate,* broader, with subacute apex and teeth. Lowest fls at about 7th node. Calyx with finely acuminate teeth, scarcely accrescent. *Corolla small* or moderate, 4·5–8(–9) mm., white, purple or yellow, upper lip with subentire or retuse lobes; lower lip longer than upper, with emarginate lobes, the median the longest. *Capsule rather small,* 4·5–5·5 mm., *broadly oblong,* emarginate or retuse, ciliate and slightly pilose, generally exceeding calyx-teeth. Fl. 7–9. $2n=44*$. Th.

Native. Grassy cliffs and moorlands on siliceous soils, sometimes also on carboniferous limestone; widespread but local in Great Britain; Wicklow. 55, H1. Faeroes.

16. E. pseudokerneri Pugsl.

E. Kerneri auct. angl., non Wettst.

Stem (5–)10–15(–25) cm., *erect, robust,* purplish, with ± *numerous ascending branches from near the base. Internodes all short,* shorter or slightly longer than lvs. *Lvs small* (to 8 mm.), *thick, dark or purplish green,* often paler beneath, glabrous or with minute marginal bristles; lower cauline lvs cuneate-oblong, obtuse, obtusely toothed; upper longer, broader, more acutely toothed; bracts ovate, acute or cuspidate with acuminate or aristate teeth. Lowest fls at 10th to 15th node. Calyx with acuminate teeth, not accrescent. *Corolla large* (8–)9–11 mm., 8–10 mm. across lower lip, white or with bluish upper lip, rarely all bluish; upper lip long, about equalling tube, with reflexed emarginate lobes; lower lip much longer than upper (c. 2 mm.), with broad deeply emarginate lobes, the median longest. *Capsule small,* c. 5 mm., oblong, truncate or slightly retuse, ciliate and slightly pilose, usually shorter than calyx-teeth. Fl. 7–9. $2n=44*$. Th.

Native. Calcareous downs from S. Lincoln and Kent to N. Somerset and Isle of Wight. 18. Endemic.

17. E. borealis Wettst.

Stem 10–35 cm., erect, robust, greenish or tinted with dull red, *with a few long suberect branches from near the middle of the stem. Cauline internodes much longer than lvs,* becoming shorter upwards so that the internodes are shorter than the upper bracts. *Lvs large and broad* (8–15 mm. long), *dark green,* thick, *glabrous* or with minute marginal bristles, more rarely with scattered hairs on the veins beneath; lower cauline lvs oval, shortly cuneate below, rounded, obtuse, and with obtuse teeth; *upper cauline lvs* ovate, rounded below, *with obtuse* (or the lower acute) *teeth;* bracts nearly as large, lower broadly ovate, rounded-cordate below, obtuse, teeth obtuse near apex, acute near base of lf; upper more rhomboidal, ±cuneate below with acute or acuminate teeth. Lowest fls at about 6th pair of lvs. Calyx with triangular subulate teeth, strongly accrescent in fr. *Corolla rather large*, 6–8(–10) mm., *white* or with bluish upper lip, rarely all bluish; upper lip broad, with emarginate lobes; lower lip much longer than upper, with broad spreading emarginate lobes, the median the longest. Capsule large, 6–8 mm., oblong with broad rounded or retuse apex, ciliate and slightly pilose, about equalling calyx-teeth. Fl. 7–8. Th.

Native. Rough pastures. Widespread but local in Scotland, becoming rarer southwards but extending to Sussex; very local in Ireland. 34, H9. Faeroes, Norway, Denmark.

18. E. brevipila Burnat & Gremli

Stem 5–35 cm., erect, *robust*, usually red-tinted, *with few or more numerous long, ascending or suberect branches from below or about the middle of the stem. Internodes, except sometimes at the base of the stem, much longer than the lvs*; lower internodes of infl. longer than the bracts, upper shorter. *Lvs large*, 6–12(–18) mm., *light green, clothed with a varying number of short glands and short bristles*, either distributed all over or confined to the margins and bases (sometimes of the bracts only), sometimes glands 0 (var. *subeglandulosa* Bucknall); lower cauline lvs oblong, obtuse, with obtuse teeth, often caducous; upper lvs oval, obtuse usually with subacute teeth; bracts ovate, subacute, with acute or aristate teeth. Lowest fls at about 6th to 8th node. Calyx with acuminate or aristate teeth, accrescent. *Corolla rather large*, 6–8(–11) mm., 7–8 mm. across lower lip, *lilac, or white with lilac upper lip*; upper lip broad with retuse or denticulate lobes; lower lip much longer than upper, with broad emarginate lobes, the median the longest. Capsule large, 6–9 mm., oblong with rounded or retuse apex, ciliate and sparsely pilose, equalling or exceeding calyx-teeth. Fl. 6–8. $2n=44*$. Th.

Native. Meadows, pastures, etc. Widespread and often common in Scotland, Ireland, Wales and N. England, very local in S. England. 66, H32. N. and Alpine Europe from Iceland, Scandinavia and W. Prussia eastwards and in the French, Swiss and Austrian Alps; Siberia east to the Yenisei region; Newfoundland and Quebec.

Two varieties with *long* glandular hairs occur. Both are rare.

Var. *notata* (Towns.) Pugsl. Lvs broader with more obtuse teeth and glandular hairs waved and relatively long. Corolla large, 8–9 mm., white with bluish-tinted upper lip. Capsule elliptic-oblong, broader. Scottish Highlands. $2n=44*$.

Var. *reayensis* Pugsl. Internodes short, upper floral lvs often imbricate. Lvs large, broader, bracts generally triangular-reniform, nearly truncate below, all grey-green with dense long and short glandular and eglandular hairs. Calyx-teeth broader. Corolla c. 8 mm., whitish with lilac or blue upper lip. Caithness. $2n=44*$.

19. E. rostkoviana Hayne

Stem 10–40 cm., suberect from decumbent base, *slender* and flexuous, green or slightly reddish, with *rather few slender flexuous branches from near the base of the stem. Lower internodes shorter than lvs, upper longer. Lvs* rather small, 6–10 mm., *bright green*, rather thin, often conspicuously veined, ± *densely clothed with long and short glandular hairs* on both sides; lower cauline oblong, cuneate below, rounded, obtuse, with obtuse teeth, upper broader and more acutely toothed; bracts smaller, ovate, cuspidate or subacute, with acute or aristate teeth. Lowest fls from 8th to 12th node. Calyx with long finely acuminate teeth, scarcely accrescent. *Corolla large*, 6–8 mm., with tube *elongating* during flowering, then 7–10 mm., white, upper lip often lilac tinted; upper lip with broad emarginate lobes, lower lip much longer than upper (to 10 mm.), with broad spreading emarginate lobes, the median much the longest. *Capsule medium or small*, 4·5–6·5 mm., elliptic or oblong-elliptic with emarginate or retuse apex, ciliate and sparingly pilose, equalling or shorter than calyx-teeth. Fl. 7–8. 2*n* = 22. Th.

Native. Moist meadows, very local; Wales and Border counties, N. England; Ireland (mainly in the west). 15, H6. Sweden to France, N. Italy and Russia (Volga-Don region).

20. E. montana Jord.

Stem 5–35 cm., erect, *rather slender*, green or slightly reddish, *unbranched* or sometimes with few slender branches from near the middle of the stem. Internodes much longer than the lvs (except the upper floral). *Lvs* moderate, 6–10(–14) mm., *bright green*, rather thin, ± *densely clothed with long and short glandular hairs* on both sides; lower cauline lvs oblong or elliptic, ± cuneate below, apex rounded-obtuse, with obtuse teeth; upper ovate, with obtuse teeth, rounded below; bracts as large roundish or ovate-triangular, rounded below, obtuse or more rarely acute, with obtuse, acute or rarely acuminate teeth, upper bracts narrower and more acutely toothed. Lowest fls at from 3rd to 6th node. Calyx with long triangular-subulate teeth, slightly accrescent. *Corolla very large*, 8–10 mm., *elongating* to 10–13 mm., white, often with lilac upper lip; upper lip with broad, retuse or denticulate lobes; lower lip much longer (to 10 mm.) with broad spreading emarginate lobes, the median the longest. *Capsule medium*, 5·5–6·5 mm., elliptic with broad retuse or emarginate apex, ciliate and sparingly pilose, equalling or shorter than calyx-teeth. Fl. 6–7. 2*n* = 22*. Th.

Native. Grassland in mountain districts; N. England; Brecon; Shropshire; Dumfries. 7. Sweden to France, N. Italy and Russia (Volga-Don region).

21. E. rivularis Pugsl.

Stem 5–15 cm., erect, *very slender*, ± purplish, *unbranched* or with few slender branches from near the middle. *Internodes* longest near the middle of stem, but all (except the uppermost) *longer than the lvs. Lvs very small* (rarely to 7 mm.), *dark green or purple-tinted; clothed*, often sparingly, *with long and short glandular hairs*; lower cauline lvs oval, ± cuneate below, apex rounded, obtuse, teeth obtuse; upper broader, rounded below; bracts broadly ovate, subacute with acute teeth. Lowest fls from 4th to 6th node. Calyx with rather short triangular-subacute teeth, not accrescent. *Corolla large*, 7–9 mm., white tinted with lilac, with purplish upper lip; upper lip with retuse lobes; lower lip longer than upper, with broad spreading retuse lobes, the median the longest. *Capsule small*, c. 5 *mm.*, elliptic with retuse apex, ciliate and sparingly pilose, about equalling calyx-teeth. Fl. 5–7. $2n = 22*$. Th.

Native. Mountain slopes in N. Wales and Lake District. 4. Endemic.

22. E. anglica Pugsl.

Stem 10–40 cm., ascending from a decumbent base, *flexuous but rather robust*, purplish, with few or many *long flexuous branches from near base of stem*, rarely unbranched. *Internodes shorter than cauline lvs*, about equalling bracts. *Lvs* moderate, 5–10 mm., broad, *greyish-green*, rather thick, ± *densely clothed with long and short glandular hairs*; cauline lvs oblong to oval (upper ovate), rounded-obtuse with obtuse teeth; bracts as large, broadly ovate or ovate-triangular, obtuse or acute with broad acute teeth. Lowest fls from about 6th node. Calyx with triangular-subulate teeth, scarcely accrescent. *Corolla large*, 5·5–9 mm., whitish or lilac-tinted with lilac upper lip; upper lip with broad emarginate lobes; lower lip with broad emarginate lobes, the median the longest. Capsule moderate, 5–7 mm., elliptic with emarginate, retuse or rarely subtruncate apex, ciliate and sparingly pilose, about equalling calyx-teeth. Fl. 5–9. $2n = 22*$. Th.

Native. Grassy places on wet or heavy soils and on heaths; widespread and rather common in S. England extending north to Lancashire and Nottinghamshire; Isle of Man; S.E. and C. Ireland. 36, H7. Endemic.

23. E. vigursii Davey

Stem 5–25 cm., ± *erect, usually flexuous, with flexuous branches from near the middle*, rarely unbranched. *Internodes* (except the upper floral) *longer than lvs*. *Lvs* 5–8 mm., greyish-green, ± *densely clothed with long and short glandular hairs* (rarely eglandular), ovate, with 3–4 teeth on each side, lower with obtuse, upper with acute teeth. Lowest fls from about 6th to 8th node. Calyx with narrow triangular teeth, not accrescent. *Corolla large*, (6–)7–8·5 mm., *purple*, rarely lilac or white with lilac upper lip; upper lip with emarginate lobes; lower lip with broad emarginate lobes the median the longest. Capsule 3–6 mm., elliptic with emarginate apex, ciliate, shorter or longer than calyx-teeth. Fl. 6–9. $2n = 22*$. Th.

Native. Heaths in Cornwall and S. Devon, local. 3. Endemic.

Appears to have originated by hybridization between *E. anglica* and *E. micrantha*, but has regained the diploid chromosome number and seems best treated as a sp.

24. E. hirtella Jord. ex Reut.

Stem 5–20 cm., erect, *strict*, rather *robust*, greenish or purple-tinted, *simple* or occasionally with 1–4 short suberect branches. *Lower internodes longer than lvs, upper shorter*. *Lvs* moderate (to 8 mm.), greyish-green or sometimes reddish-tinted beneath, often paler and plicate when dry, *densely clothed with long and short glandular hairs*; cauline oblong to oblong-lanceolate, obtuse, with subacute teeth; bracts generally larger, ovate, acute with acute or acuminate teeth. Lowest fls at from 5th to 8th node. Calyx with short subulate teeth, slightly accrescent. *Corolla rather small* or medium, 5–9 mm., white with bluish-tinted upper lip; upper lip with short denticulate or retuse lobes; lower lip longer with narrow retuse or emarginate lobes, the median longest. Capsule moderate, 5–6 mm., oblong-ovate, retuse or emarginate, ciliate and slightly pilose equalling or slightly exceeding the calyx-teeth. Fl. 6–9. $2n=22^*$. Th.

Native. Mountain pastures in N. and C. Wales; Leicestershire (Charnwood); Perth; Wexford. 5, H1. Spain and S. France to Himalaya and Mongolia.

Subsection *Angustifoliae* (Wettst.) Jorg. Bracts generally less than half as broad as long (sometimes much narrower) with distant narrow acute teeth. Capsule glabrous or ciliate with weak marginal hairs.

25. E. salisburgensis Funck

Stem 2–10 cm., erect, slender, with numerous spreading branches. Internodes about equalling lvs. *Lvs* small (up to 7 mm.), *narrow*, green or tinted brown or purple; glabrous or with minute bristles on margins and veins beneath; cauline linear-oblong, obtuse; *bracts* at least as large, *lanceolate, acuminate* with *distant* spreading acute to long-aristate *teeth*. Calyx with long finely-pointed teeth, somewhat accrescent in fr. Corolla small, 5–7 mm., white; upper lip with small subentire lobes; lower lip with retuse lobes, the median longest. *Capsule* moderate, 4–6 mm., oblong, retuse, *glabrous* or very rarely with a few weak marginal bristles. Fl. 7–8. $2n=44$. Th.

Native. Among limestone rocks and on dunes. Widespread in W. Ireland from Limerick to Donegal. H10. Mountains of Europe from Gotland and W. Russia (lower Dniester region) to Balkans, Italy, Corsica and S. Spain.

The following hybrids are recorded from Britain on reliable authority; others doubtless occur:

E. anglica × *micrantha*; *E. anglica* × *nemorosa* = *E.* × *glanduligera* Wettst.; *E. anglica* × *rostkoviana*. *E. borealis* × *marshallii*. *E. brevipila* × *confusa*; *E. brevipila* × *curta* = *E.* × *murbeckii* Wettst.; *E. brevipila* × *marshallii*; *E. brevipila* × *micrantha* = *E.* × *difformis* Towns.; *E. brevipila* × *nemorosa*—widespread and variable; *E. brevipila* × *rotundifolia*; *E. brevipila* × *scottica* = *E.* × *venusta* Towns. *E. confusa* × *frigida*. *E. confusa* × *micrantha*; *E. confusa* × *nemorosa*; *E. confusa* × *occidentalis*; *E. confusa* × *rostkoviana*. *E. curta* × *nemorosa*. *E. foulaensis* × *micrantha*; *E. foulaensis* × *occidentalis*. *E. frigida* × *micrantha*; *E. frigida* × *scottica*. *E. marshallii* × *micrantha*; *E. marshallii* × *rotundifolia*. *E. micrantha* × *nemorosa*; *E. micrantha* × *scottica* = *E.* × *electa* Towns. *E. nemorosa* var. *calcarea* × *pseudokerneri*. *E. occidentalis* × *pseudokerneri*.

18. ODONTITES Ludw.

Annual herbs with opposite lvs. Fls in a unilateral terminal spike-like raceme, in the axils of bracts resembling the lvs but smaller. Calyx 4-toothed, tubular or campanulate. Corolla 2-lipped; *the upper lip concave, entire or emarginate*; the lower 3-lobed with entire lobes. Stamens 4, the anther cells mucronate. *Seeds* small, rather few, oblong or fusiform, *furrowed*. *Hilum basal*.

About 40 spp., in Europe, W. Asia, N. Africa and S. America.

1. O. verna (Bell.) Dum.　　　　　　　　　　　　　　'Red Bartsia.'

Bartsia Odontites (L.) Huds.; *O. rubra* Gilib.

Erect branching pubescent annual, often purple-tinted, up to 50 cm. high.

Lvs sessile, lanceolate to linear-lanceolate, remotely toothed. Racemes terminal on the stem and upper branches. Bracts like the lvs but smaller. Calyx campanulate, 4-toothed. Corolla c. 9 mm., purplish-pink, tube about equalling calyx. Anthers coherent, slightly exserted. Style long, filiform. Capsule ± equalling the calyx. Pollinated by bees. Th.

Common in cultivated and uncultivated fields and waste places throughout the British Is., ascending to over 1300 ft. 112, H40, S. Europe (except Iceland, Greece, Crete, etc.); N. and W. Asia.

Ssp. **verna**. Plant 10–30 cm. high. *Branches coming off at an angle of less than 45°, ±straight. Lvs lanceolate,* rounded at base, distinctly toothed. *Bracts longer than fls.* Fl. 6–7. $2n = 40$.

Much less common in S. England than ssp. *serotina* but common in N. Scotland. More northern in Europe than ssp. *serotina* and absent from S. Spain, S. Italy, etc.; N. Asia.

Ssp. **serotina** (Wettst.) E. F. Warb. Plant 20–50 cm. high. *Branches spreading at a wide angle,* sometimes nearly at right angles, their tips often upcurved. *Lvs linear-lanceolate,* somewhat narrowed at the base, obscurely toothed. Intercalary lvs 2 or more pairs. *Bracts shorter than or equalling the fls.* Fl. 7–8. $2n = 20$.

Common in S. England, rare or absent in N. Scotland. Europe, W. Asia. Extending further south in Europe than ssp. *verna*.

19. PARENTUCELLIA Viv.

Annual herbs with opposite lvs. Fls in the axils of bracts similar to the lvs, forming terminal spike-like racemes. Calyx tubular, 4-toothed. Corolla 2-lipped; *the upper lip* entire or emarginate, *forming a hood*; lower with three entire lobes. Stamens 4, the anthers aristate at the base. Capsule lanceolate. *Seeds* many, minute, smooth, *with basal hilum*.

Two spp. in W. Europe and the Mediterranean region.

1. P. viscosa (L.) Caruel
'Yellow Bartsia.'

Bartsia viscosa L.; *Eufragia viscosa* (L.) Benth.

Erect viscid-hairy annual 10–50 cm., usually unbranched. Lvs 1·5–4 cm., lanceolate, dentate, acute or subacute, sessile. Calyx-teeth triangular-lanceolate, subacute, about as long as the tube. Corolla c. 2 cm., yellow, the lower lip much longer than the upper. Stamens included in the upper lip, anthers hairy. Capsule slightly longer than the calyx-tube. Fl. 6–10. $2n = 48*$. Th.

Native. Damp grassy places usually near the south and west coasts. 30, H9, S. From W. Kent to Argyll but only common in the south-west; has recently been found in Norfolk and in several inland counties but is probably impermanent in them; in Ireland also chiefly south-western, from Waterford to Kerry, also in Galway, Sligo, Donegal and Londonderry. Western Europe from Dept. of the Eure (France) southwards; all round the Mediterranean and near the Black and Caspian Seas; Azores, Canaries.

20. BARTSIA L.

Perennial herbs with opposite lvs. Infls and fls as in *Parentucellia*. Capsule broad. *Seeds few, large, with strong longitudinal ribs or wings and a lateral hilum.*

About 6 spp., in Europe, N. Africa; only the following more widespread.

1. B. alpina L.   'Alpine Bartsia.'

Pubescent perennial 10–20 cm., with short rhizome and erect simple stems. Lvs 1–2 cm., sessile, ovate, crenate-serrate, obtuse. Infl. short, few-fld, glandular; bracts purplish. Calyx-teeth ovate-lanceolate, obtuse, about as long as the tube. Corolla c. 2 cm., dull purple, the upper lip much longer than the lower. Anthers hairy. Capsule about twice as long as the calyx. Fl. 6–8, pollinated by humble-bees. $2n = 24$. Hp.

Native. Mountain meadows and rock-ledges on basic soils; from 800 to 3000 ft. Yorks, Durham, Westmorland, Perth and Argyll. 6. Arctic and subarctic regions from Labrador to European Russia and in the mountains of Europe, extending south to the Pyrenees, S. Alps and N. Balkans.

107. OROBANCHACEAE

Annual to perennial herbaceous root-parasites devoid of chlorophyll, usually with erect scaly aerial shoots bearing fls in dense terminal bracteate racemes or spikes. Fls hermaphrodite, zygomorphic, hypogynous. Calyx tubular below, 2–5-toothed or laterally 2-lipped above; corolla 2-lipped with a somewhat curved tube; stamens 4, in two pairs, their anthers connivent; ovary superior, 1-celled, with numerous ovules on 4 (2–6) parietal placentae; style single with a 2-lobed stigma. Fr. a capsule dehiscing into 2 incompletely separating valves; seeds very small and numerous with minute undifferentiated embryo in oily endosperm.

About 11 genera with 130 spp., chiefly in the warm temperate zone of the Old World.

Close to Scrophulariaceae but devoid of chlorophyll and with a 1-celled ovary.

Plants with a scaly rhizome; calyx equally 4-lobed; lower lip of corolla almost
    parallel to upper, not spreading.   1. LATHRAEA
Plants not rhizomatous; calyx laterally 2-lipped; lower lip of corolla divergent
    from upper, spreading.   2. OROBANCHE

1. LATHRAEA L.

Herbaceous *root-parasites with* branched creeping *rhizomes* covered *with* broad whitish fleshy *imbricating scales* and bearing rootlets which are swollen where they are attached to roots of the host plant. Fls borne singly in the axils of scales. *Calyx* campanulate, *equally* 4-*lobed* above; corolla 2-lipped, the upper lip strongly concave, entire, the lower smaller and 3-lobed; stamens subexserted; style exserted, curved downwards at its tip. Nectary at base of ovary, crescent-shaped. Capsule opening elastically from the apex into two valves. Plants blackening on drying.

Five spp., in temperate Europe and Asia.

Aerial shoot 8–30 cm.; fls white or dull purple, short-stalked.   **1. squamaria**
No aerial shoot; fls bright purple, long-stalked.   **2. clandestina**

1. L. squamaria L.   Toothwort.

A perennial herb with a *stout simple erect flowering shoot* 8–30 *cm.*, white or pale pink, slightly pubescent above and with a few whitish ovate scales below. Infl. a *one-sided raceme* with scaly bracts, at first drooping, but straightening later. Fls shortly stalked, each in the axil of a broadly ovate bract. *Calyx glandular-hairy*, tubular below, with 4 broadly triangular teeth above. *Corolla white ± tinged with dull purple, slightly longer than the calyx. Capsule ovoid* acuminate with *numerous seeds*. Fl. 4–5. Visited by humble-bees. $2n=42*$; 36. Grh.

Native. On roots of various woody plants, especially of *Corylus* and *Ulmus*, in moist woods and hedgerows on good soils, and locally common in some limestone areas; to 1000 ft. in N. England. 72, H30. Great Britain northwards to Perth and Inverness. Throughout Europe and W. Asia to the Himalaya.

*2. L. clandestina L.

A perennial herb with *no aerial shoot. Fls in corymbose clusters, long-stalked, arising* singly in the axils of fleshy rhizome-scales *at or just beneath the soil surface. Calyx glabrous*, campanulate, with 4 short triangular lobes above. *Corolla bright-purple, twice or more as long as the calyx. Capsule globose* with *4–5 seeds*. Fls 4–5. Visited by humble-bees. $2n=42$. Grh.

Introduced. Naturalized in a few localities in England on roots of *Populus* and *Salix* in damp shady places. Native in Spain, Italy, W. and C. France, and Belgium.

2. OROBANCHE L.

Annual to perennial herbaceous *root-parasites* ('broomrapes') *with underground tubers* attached to roots of the host plant. From the tubers arise erect scaly flowering shoots with terminal spikes of ± sessile fls. *Calyx* laterally 2-*lipped*, the lips entire, toothed, or with 2 slender connivent lobes, a fifth posterior tooth sometimes present; corolla with a curved tube, 2-lipped, the upper lip erect, ± 2-lobed, the lower spreading, 3-lobed; stamens included in the corolla-tube; style curved downwards near its tip; stigma-lobes 2, fleshy. Valves of capsule sometimes remaining attached at their tips.

About 100 spp., cosmopolitan.

1	Fls each with 1 bract and 2 bracteoles (apart from the calyx-lobes); valves of fr. free above.	2
	Fls each with 1 bract only (apart from the calyx-lobes); valves of fr. united above.	3
2	Fls 10–16 mm., yellowish-white tinged marginally with blue-purple; stem usually branched; chiefly on *Cannabis* and *Nicotiana*. **1. ramosa**	
	Fls 18–30 mm., dull blue-purple; stem usually simple, bluish; chiefly on *Achillea millefolium*. **2. purpurea**	
3	Stigma-lobes yellow, at least at first.	4
	Stigma-lobes purple, red or brown throughout.	7

4   Corolla 18–25 mm.   5
    Corolla 10–20 mm.   6

5   Upper lip of corolla almost entire; stamens inserted close to base of
      corolla-tube, their filaments glabrous below; parasitic on shrubby
      Papilionaceae, especially *Ulex* and *Sarothamnus*.   **3. rapum-genistae**
    Upper lip of corolla finely toothed; stamens inserted well above base of
      corolla-tube, their filaments hairy below; parasitic chiefly on *Centaurea*
      scabiosa.   **6. elatior**

6   Stem reddish or purple; parasitic on *Hedera*.   **10. hederae**
    Stem yellow.   **8. minor** var. *flava*

7   Stem and corolla both purplish-red (rarely paler); parasitic on *Thymus*
      and perhaps other Labiatae.   **4. alba**
    Stem and corolla variously coloured but not both purplish-red; not
      parasitic on *Thymus* or other Labiatae.   8

8   Corolla 20–30 mm.; parasitic on *Galium mollugo*; Kent and Argyll only.
                                                             5. caryophyllacea
    Corolla 10–20(–22) mm.   9

9   Stem purple; calyx usually shorter than corolla-tube; lower lip of corolla
      unequally 3-lobed, the reniform middle lobe the largest; usually
      coastal and parasitic chiefly on *Daucus*.   **11. maritima**
    Stem yellowish, tinged with red or purple; calyx usually about equalling
      corolla-tube; lower lip of corolla about equally 3-lobed.   10

10 Corolla 10–16(–18) mm., its back regularly curved throughout. **8. minor**
    Corolla 15–22 mm., its back strongly curved near the base, then almost
      straight (or curved also over the upper lip).   11

11 Corolla with pale glands; filaments densely hairy below; on *Picris* and
      Crepis spp. in S. England and S. Wales.   **9. picridis**
    Corolla with dark glands; filaments glabrous or sparsely hairy below;
      on *Cirsium* and *Carduus* in Yorks.   **7. reticulata**

Section 1. *Trionychon* Wallr. Stem simple or branched; fls each with 1 bract
and 2 bracteoles; capsule with the valves free above.

*1. O. ramosa L.   'Branched Broomrape.'

Phelypaea ramosa (L.) C. A. Mey.

Flowering *stem* 5–30 cm., *usually with a few branches near the base*, rarely
simple, yellowish-white or tinged with blue, slender, with very few scales near
the base, glandular-pubescent above. Fls in lax spikes; *bracts* shorter than
the calyx, *ovate-lanceolate*. Calyx hairy, tubular, with 4 lanceolate-acuminate
teeth somewhat shorter than the tube. *Corolla* 10–15 mm., twice as long as
the calyx, *yellowish-white, the lips usually tinged marginally with blue-purple*;
corolla-tube constricted just below the middle, curved in the upper half;
upper lip of corolla *with 2 rounded lobes*, lower with 3 blunt ciliate lobes.
Stamens inserted just below the constriction in the corolla; anthers ± glabrous.
Stigma-lobes white or slightly bluish. Fl. 7–9. Apparently self-pollinated.
$2n=24$. ?Th.

    Introduced in some eastern and southern counties from Norfolk to Devon.
Parasitic chiefly on hemp (*Cannabis*) and tobacco, but occurs on several other

plants in France, including potato and tomato. C. and S. Europe, Caucasus, N. Africa, and W. Asia; also as var. *interrupta* in S. Africa.

2. O. purpurea Jacq. 'Purple Broomrape.'

O. caerulea Vill.; *O. arenaria* auct.; *Phelypaea caerulea* (Vill.) C. A. Mey.

Flowering *stem* 15–45 cm., *simple*, rarely branched, *bluish*, fairly stout, with a few narrow scales towards the base, glandular-pubescent especially above. Fls in a lax spike; *bracts* somewhat shorter than calyx, *lanceolate*. Calyx hairy, tubular, with 4 lanceolate-acute teeth usually shorter than the tube. *Corolla* 18–30 mm., twice as long as the calyx, *dull bluish-purple* suffused with yellow at the base; corolla-tube constricted just below the middle, slightly curved in the upper half; *upper lip* of corolla *with 2 acute lobes*, lower with 3 subacute lobes. Stamens inserted just below the constriction in the corolla; anthers ± glabrous. Stigma-lobes white. Fl. 6–7. Apparently self-pollinated. G.

Native. Parasitic on *Achillea millefolium* and a few other Compositae. Common on walls in the Channel Is. A rare plant of S. England from Nottingham and Norfolk southwards; S. Wales. 9, S. C. and S. Europe, N. Africa, and W. Asia to the Himalaya.

O. arenaria Borkh. with fls 25–35 mm. long and hairy anthers, parasitic on *Artemisia campestris*, has been reported in error from Alderney. Its fls are pale purple or lavender with no yellow at the base. The Alderney plant is a form of *O. purpurea*.

Section 2. *Orobanche*. Stem simple; fls with 1 bract and no bracteoles; capsule with the valves coherent above.

3. O. rapum-genistae Thuill. 'Greater Broomrape.'

O. major auct.

Flowering *stem* 20–80 cm., simple, *yellowish*, stout, *with numerous scales* near the base, glandular-hairy throughout. Fls in a long compact spike; bracts lanceolate-acuminate, exceeding the fls. Calyx laterally 2-lipped, each lip usually equally bifid and nearly equalling the corolla-tube. Corolla 20–25 mm., yellowish tinged with purple, glandular-pubescent; corolla-tube campanulate, inflated at the base in front, its back curved throughout; *upper lip of corolla almost entire* with spreading margins, lower 3-lobed, the middle lobe much larger than the lateral, all ciliate; both lips waved and indistinctly denticulate. *Stamens inserted at the base of the corolla-tube; filaments glabrous* at base, glandular above. *Stigma-lobes distant, pale yellow.* Fl. 5–7. Visited by bees. Corolla often perforated at the base by *Bombus terrestris*. G.

Native. Parasitic on the roots of shrubby Papilionaceae, chiefly on *Ulex* and *Sarothamnus*, occasionally on *Genista tinctoria*, etc. Throughout England and Wales but only in S.W. Scotland and S. and S.E. Ireland. 69, H5, S. W. and C. Europe, Italy, Sicily, and N. Africa.

4. O. alba Steph. ex Willd. 'Red Broomrape.'

Incl. *O. rubra* Sm.

Flowering *stem* 8–25(–35) cm., simple, *purplish-red*, rather stout, *with numerous* reddish *scales* near the base, glandular-hairy throughout. Fls rather few in a lax spike; *bracts* ovate-lanceolate, acuminate, *shorter than the fls*. Calyx 2-lipped, the *lips entire*, subulate, about equalling corolla-tube. *Corolla* 15–20 mm., *dull purplish-red*, sparingly glandular-pubescent; corolla-tube campanulate, back strongly curved only at the basal and distal ends; *upper lip of corolla notched* with somewhat spreading margins, lower 3-lobed with the lobes almost equal; both lips crisped and distinctly denticulate. *Stamens inserted 1–2 mm. above base of corolla-tube*; filaments slightly hairy below, glandular above. *Stigma-lobes contiguous, reddish.* Fl. 6–8. Fls fragrant, visited by humble-bees. G.

Native. Parasitic on *Thymus* and perhaps other Labiatae on rocky slopes, screes, etc., chiefly on basic rocks. Cornwall, Glamorgan, W. Scotland from Wigtown to W. Ross, Inner and Outer Hebrides; also in Lincoln, Yorks and Fife; N. and W. Ireland; Sark. 16, H10, S. Europe from Gotland and Belgium southwards; W. Asia to Himalaya.

Most British specimens have deep reddish fls (*O. rubra* Sm.), but the form common on the Continent, with much paler fls, is sometimes found.

5. O. caryophyllacea Sm.   'Clove-scented Broomrape.'

O. vulgaris Poir.

Flowering *stem* 15–40 cm., simple, yellowish usually tinged with purple, *with numerous* yellow or purple-brown *scales* below, glandular-hairy. Fls rather few in a lax spike; *bracts* ovate-lanceolate, acuminate, *shorter than the fls. Lateral lips of calyx* ovate, ± equally bifid, *shorter than the corolla-tube.* Corolla 20–30 mm., yellowish or tinged with reddish-brown or purplish, densely glandular-pubescent; corolla-tube campanulate, back curved throughout; upper lip notched with the lobes at first erect; lower 3-lobed with the lobes nearly equal; all lobes crisped and denticulate, those of the lower lip almost fimbriate. *Stamens inserted 1–2 mm. above base of corolla-tube; filaments hairy below*, glandular above. *Stigma-lobes distant, purple.* Fl. 6–7. Fls said to smell of cloves; visited for nectar by humble-bees. G.

Native. On *Galium mollugo* in Kent, but also on other Rubiaceae on the Continent. Very local; Kent and Argyll. Europe from Denmark southwards, Algeria, Caucasus, Siberia.

6. O. elatior Sutton   'Tall Broomrape.'

O. major auct.; incl. *O. Ritro* Gren. & Godr.

Flowering stem 15–70 cm., simple, yellowish or reddish, stout, with numerous acuminate scales below, glandular-hairy throughout. Fls numerous, in a rather dense spike. Bracts lanceolate-acuminate, as long as the fls. *Lateral lips of calyx* somewhat unequally bifid, usually *shorter than corolla-tube.* Corolla 18–25 mm., pale yellow usually tinged with purple, glandular-pubescent, rather widely tubular, back curved throughout; upper lip ± 2-lobed usually spreading; lower lip 3-lobed, the lobes nearly equal; all *lobes crisped, denticulate, not ciliate. Stamens inserted 4–6 mm. above base of corolla-tube; filaments hairy below*, glandular throughout. *Stigma-lobes yellow.* Fl. 6–7. Nectar-secreting but scentless. G.

Native. Parasitic chiefly on *Centaurea scabiosa* on dry calcareous soils;

rare but widely distributed. 45, S. England and Wales northwards to Cumberland and N. Yorks. Europe northwards to Denmark and S. Sweden; Caucasus; Asia Minor to India.

7. O. reticulata Wallr. 'Thistle Broomrape.'

Flowering stem 15–50 cm., simple, yellowish to purplish, glandular-hairy, with a few ovate-lanceolate scales below. Fls numerous, in a spike compact above but lax below; bracts narrowly triangular, about equalling the fls. Lateral lips of calyx entire or unequally bifid, somewhat shorter than the corolla-tube. *Corolla* 15–22 mm., yellowish, ± purple-tinged marginally, *with sparse dark glands*; tube somewhat widened above the insertion of the stamens, its back strongly curved near the base and into the upper lip; upper lip notched, with spreading lobes; lower with 3 ± equal lobes, the central truncate or squarish; all lobes denticulate. Stamens inserted 3–4 mm. above the base of the corolla, their filaments glabrous or sparsely hairy below and glabrous or sparsely glandular above. *Stigma-lobes dark purple.* Fl. 6–8. Fls unscented. $2n=38$. G.

The description is of ssp. *pallidiflora* (Wimm. & Grab.) Hegi, which in its var. *procera* (Koch) Hegi includes all British specimens. Ssp. *reticulata* has the corolla yellowish only at the base (elsewhere bright reddish-purple with darker veins) and densely dark-glandular, and the stamen-filaments densely glandular above. It is found on a wider range of hosts than ssp. *pallidiflora*.

Native. On *Cirsium arvense*, *C. eriophorum* and other spp. of *Cirsium* and *Carduus*; very local. Reported only from several stations in Yorks. C. and S. Europe.

8. O. minor Sm. 'Lesser Broomrape.'

O. apiculata Wallr.; ?*O. major* L.

Flowering stem 10–50 cm., simple, usually yellowish tinged with red, ± glandular-hairy, with rather sparse ovate to lanceolate brownish scales. Fls fairly numerous, in a spike lax below; bracts ovate-acuminate, equalling or exceeding the fls. Lateral lips of calyx unequally bifid or sometimes entire, many-veined, about equalling the corolla-tube. Corolla 10–16 mm., yellowish ± tinged and veined with purple, sparsely glandular; tube not widened upwards, its *back regularly curved throughout*; upper lip notched, with the lobes suberect or forwardly directed; lower lip ± equally 3-lobed, the central lobe broadly reniform; all lobes crisped and denticulate. Stamens inserted 2–3 mm. above the base of the corolla-tube, their *filaments subglabrous throughout*. Stigma-lobes distant, purple. Fl. 6–9. $2n=38$. Th.–G.

Var. *lutea* Tourlet differs from the type in its yellow stem and fls and yellow stigma-lobes.

Var. *compositarum* Pugsl. has suberect fls with corolla 12–18 mm., narrower than in the type (3–4 mm. instead of 5 mm.), with paler violet tinting and fewer glandular hairs, and with filaments ± densely hairy below.

Native. Parasitic chiefly on *Trifolium* spp. and other herbaceous Papilionaceae; var. *compositarum* on *Crepis capillaris*, *Hypochoeris radicata* and other Compositae, etc. Throughout England and Wales and common on clover crops in the south but becoming rare in the west and north and reported only from Fife in Scotland. 57, H21, S. Var. *flava* in Channel Is and Oxford. S. and W. Europe northwards to Belgium and W. Germany; Asia Minor; Socotra; N. Africa; Madeira.

9. O. picridis F. W. Schultz ex Koch   'Picris Broomrape.'

Flowering stem 10–60 cm., simple, pale yellowish often tinged with purple, glandular-pubescent or nearly glabrous, with a few brownish acuminate scales below. Fls numerous, in a spike lax below; bracts narrowly ovate-acuminate, equalling or exceeding the fls. Lateral lips of calyx entire or unequally bifid, equalling or somewhat exceeding the corolla-tube. Corolla 15–20 mm., yellowish-white tinged and veined with pale purple, glandular-pubescent; tube narrowly campanulate, its back curved at the base then nearly straight; upper lip retuse to notched, folded in the middle, erect; lower lip ± equally 3-lobed, the middle lobe squarish; all lobes crisped and denticulate but not ciliate. *Stamens inserted 3–5 mm. above the base of the corolla-tube; filaments densely hairy below*, ± glabrous above. *Stigma-lobes* just touching, *purple*. Fl. 6–7. G.

Native. In grassland, on *Picris* and *Crepis* spp. and especially on *Picris hieracioides*; rare and local. 15. England from Devon and Kent northwards to Norfolk, Cambridge, Bucks and Worcester; Brecon, Pembroke. S. and C. Europe; N. Africa; Palestine.

10. O. hederae Duby   'Ivy Broomrape.'

Flowering stem 10–60 cm., simple, purplish, glandular-pubescent, with a few brownish acuminate scales below. *Fls rather few, in a long lax spike; bracts* ovate-lanceolate acuminate ± *serrate* near the tip, equalling or exceeding the fls. Lateral lips of calyx entire or unequally bifid, 1-veined, about equalling the corolla-tube. Corolla 12–20 mm., cream strongly veined with purple, sparsely glandular; tube inflated near the base then gradually narrowed upwards, its back straight except at the base; upper lip entire to notched, the lobes forwardly directed or spreading, lower 3-lobed with the middle lobe truncate; all lobes crisped and denticulate, not ciliate. *Stamens inserted 3–4 mm. above the base of the corolla-tube; filaments almost glabrous below*, glabrous or sparsely glandular above. *Stigma-lobes partially united*, yellow. Fl. 6–7. G.

Native. On *Hedera helix*. A local plant of S. England and Wales from Cornwall to Kent and northwards to Anglesey and Leicester; commonest in coastal districts; local in Ireland. 26, H29, S. W. and S. Europe northwards to Belgium; N. Africa; Asia Minor.

11. O. maritima Pugsl.   'Carrot Broomrape.'

O. amethystea auct. angl., non Thuill.

Flowering *stem* 10–50 cm., simple, *purple*, glandular-pubescent or nearly glabrous, with a few brownish-purple scales below. *Fls numerous, usually in a dense spike*; bracts purple, lanceolate-subulate from a broad base, about equalling the fls. *Lateral lips of calyx* unequally bifid or rarely entire, commonly *shorter than the corolla-tube*. Corolla 12–17 mm., dingy pale yellow veined with purple, sparsely glandular or glabrous; corolla-tube cylindrical, its back strongly curved near the base then ± straight; upper lip notched, with spreading lobes, lower unequally 3-lobed with the *reniform middle lobe the largest*; all lobes crisped and denticulate, not ciliate. *Stamens inserted 2–3 mm. above base of corolla-tube; filaments hairy below*, glabrous above. *Stigma-lobes* partially united, *purple*. Fl. 6–7. Th.–G.

Native. Coastal; parasitic on *Daucus carota* and very rarely on *Plantago coronopus* and *Ononis repens*. 10. A rare plant in Kent, Dorset, Devon, Cornwall; Channel Is. Recognized from Gibraltar but extra-British distribution otherwise unknown.

O. amethystea Thuill. grows on *Eryngium campestre*, has longer and narrow scales and bracts, longer and more acuminate calyx-segments, and a larger corolla, 15–23 mm., paler in colour. *Orobanche* on *Eryngium maritimum* in Isle of Wight and Channel Is. is *O. minor*.

108. LENTIBULARIACEAE

Insectivorous aquatic, bog or terrestrial herbs (or epiphytic), scapigerous. Lvs alternate or all radical. Fls solitary or in a raceme. Calyx 5-lobed or 2-lipped with the lobes obscure. Corolla gamopetalous, 2-lipped, spurred, imbricate, upper lip 2-lobed, lower 3-lobed or the lobes obscure. Stamens 2, inserted on base of corolla, 1- or 2-celled, introrse. Carpels 2, forming a unilocular ovary with numerous ovules on a free-central placenta. Stigma often sessile, 2-lobed with 1 lobe much reduced. Fr. a capsule opening irregularly or by 2 or 4 valves. Seeds small, numerous, without endosperm.

Five genera and about 250 spp., cosmopolitan.

A small family, clearly distinguished from the Scrophulariaceae by being insectivorous and by the placentation.

Lvs entire, in a basal rosette; insectivorous by sticky glands covering the
whole plant.   1. PINGUICULA
Lvs divided into filiform segments, alternate; insectivorous by special bladders
borne on the lvs.   2. UTRICULARIA

1. PINGUICULA L.

Perennial scapigerous herbs with lvs all forming a radical rosette, clothed all over (except corolla) with sticky glands which catch insects. Lvs entire, sessile, margins involute. Fls solitary on naked scapes. Calyx 5-lobed, the lobes unequal. Corolla 2-lipped, upper lip 2-, lower 3-lobed, spurred, open at mouth. Capsule opening by 2 valves.

About 30 spp., northern hemisphere and S. America.

1   Fls pale lilac or white, 10 mm. or less (excluding spur); spur 2–4 mm.   2
    Fls bright violet, 10 mm. or more; spur more than 4 mm.   3

2   Corolla pale lilac, 6–7 mm.; spur cylindric.   **1. lusitanica**
    Corolla white, 8–10 mm.; spur conic.   **2. alpina**

3   Corolla 10–15 mm., lobes of lower lip widely separated. Widespread.
       3. vulgaris
    Corolla 15–20 mm., lobes of lower lip contiguous or overlapping. S.W.
    Ireland.   **4. grandiflora**

1. P. lusitanica L.   'Pale Butterwort.'

Overwintering as a rosette. Lvs 1–2 cm., oblong, yellowish, tinged purple. Scapes 3–15 cm., very slender. Calyx-lobes suborbicular or ovate, obtuse. *Corolla 6–7 mm., pale lilac*, yellow in the throat, lobes of upper lip suborbicular; of the lower emarginate; *spur* 2–4 *mm.*, bent downwards, ± *cylindric*, obtuse. Capsule globose. Fl. 6–10. Self-pollinated. Hr.

Native. Bogs and wet heaths; ascending to 1600 ft.; local. Cornwall to Wilts and Hants; Pembroke; Isle of Man; W. Scotland from Kirkcudbright to Orkney and Outer Hebrides; throughout Ireland but rare in the centre. 33, H 34. W. France, W. Spain, Portugal, N.W. Morocco.

2. P. alpina L.

Roots long, thick. Lvs 1–2 cm., elliptic. Scapes 5–10 cm. *Corolla 8–10 mm., white* with yellow spot at mouth; *spur* 2–3 *mm.*, bent downwards, *conical*. Capsule ovoid. Fl. 5–8. 2*n* = 32. Hr.

Formerly occurred in Ross, now believed extinct. 1. N. Europe from Iceland to Finland and Baltic States; Pyrenees, Jura, Alps (to Croatia); Himalaya.

3. P. vulgaris L.   Common Butterwort.

Overwintering as a rootless bud. Lvs 2–8 cm., ovate-oblong, bright yellow-green. Scapes 5–15 cm. Calyx-lobes ovate or oblong, broad-based, obtuse or subacute, upper lip divided to middle or less. *Corolla violet*, 10–15 × c. 12 mm., *with a short broad white patch at the mouth; lobes of lower lip* deep, *much longer than broad*, flat, entire, *divergent; spur* 4–7 *mm.*, directed backwards or somewhat downwards, straight or somewhat curved, slender, *acute*. Capsule ovoid. Fl. 5–7. Pollinated by small bees. 2*n* = 64. Hr.

Native. Bogs, wet heaths and among wet rocks; ascending to 3200 ft. Common throughout the British Is. except S. England and S. Ireland where it is rare and absent from several counties. 100, H 38. Europe from Iceland to C. Spain and Portugal, Corsica, Italy and Macedonia (only in the mountains in the south); N. Asia; N. America south to New York and British Columbia; N. Morocco (Rif).

4. P. grandiflora Lam.

Overwintering as a rootless bud. Lvs 2–8 cm., ovate-oblong, bright yellow-green. Scapes 8–20 cm. Calyx-lobes ovate or oblong, upper lip divided nearly to base. *Corolla* 15–20 × 25–30 *mm., violet with a long cuneate white patch at the mouth, lobes of lower lip* shallow, *broader than long*, somewhat undulate, *contiguous or overlapping; spur* 10 *mm. or more* directed backwards, straight, rather stout, sometimes slightly bifid. Capsule subglobose. Fl. 5–6. 2*n* = 32. Hr.

Native. Bogs and wet rocks, locally common in Kerry and W. Cork, extending to E. Cork and Clare, ascending to 2800 ft.; naturalized in Cornwall. H 5. Jura, French Alps, Pyrenees and mountains of N. Spain.

P. grandiflora × *vulgaris* = *P.* × *scullyi* Druce occurs where the parents grow together.

2. Utricularia L.

Perennial rootless herbs overwintering by turions. Stems long, lfy. Lvs divided into filiform segments, bearing small bladders which trap animals. Fls in short racemes on lfless scapes. Calyx 2-lipped, divided nearly to base, lips entire or obscurely toothed. Corolla 2-lipped, spurred; lips entire, lower with a projecting palate, larger than upper. Capsule globose, opening irregularly.

More than 200 spp., cosmopolitan, mainly tropical.

1   Stems of one kind, bearing green floating lvs furnished with numerous bladders.   (**vulgaris** agg.)   *2*
    Stems of two kinds (*a*) bearing green floating lvs with or without bladders, and (*b*) colourless, bearing bladders on much reduced lvs, often in substratum.   *3*

2   Lower lip of corolla with deflexed margins; pedicels 6–17 mm.; fr. freely produced.   **1. vulgaris**
    Lower lip of corolla ± flat; pedicels 11–26 mm.; fr. very rare.   **2. neglecta**

3   Lf-segments denticulate, with bristles on the teeth; bladders confined or nearly confined to colourless stems.   **3. intermedia**
    Lf-segments entire, without bristles; bladders present on green lvs.
           4. minor

(1–2). U. vulgaris agg.   'Greater Bladderwort.'

Free floating. *Stems* 15–45 cm., *all alike, bearing green lvs with numerous bladders*. Lvs 2–2·5 cm., broadly ovate in outline, pinnately divided; segments denticulate, with small bristles on the teeth; bladders c. 3 mm. Scape 10–20 cm., 2–8-fld. *Corolla* 12–18 mm., *bright yellow*, spur conic. Upper lip ± ovate.

Native. Lakes, ponds and ditches usually in relatively deep water; ascending to 2000 ft. Throughout the British Is. (except Channel Is.) local; flowering sporadically and fls unknown from the north part of its range and, as the two following are not certainly determinable without fls, their complete distribution is unknown. 97, H40. Europe, temperate Asia, N. Africa, N. America.

1. U. vulgaris L.

Lf-segments with groups of bristles. *Pedicels 6–17 mm.*, rather stout. *Upper lip of corolla about as long as palate; lower lip with margins deflexed ± at right angles. Fr. freely produced* (if the plants fl.). Fl. 7–8. Pollinated by bees. $2n = 36$–40. Hyd.

Native. North to Dunbarton and Angus, rare in the west; Ireland. 36, H3. Europe from Faeroes and Scandinavia to Spain, Sicily and Macedonia; temperate Asia.

2. U. neglecta Lehm.

U. major auct.

Lf-segments with solitary, rarely grouped bristles. *Pedicels 11–26 mm.*, rather slender. *Upper lip of corolla about twice as long as the less projecting palate; lower lip ± flat, somewhat undulate. Fr. very rare.* Fl. 7–8. Pollinated by bees. $2n = 36$–40. Hyd.

Native. North to Perth, rare in the east; Ireland. 30, H1. Europe from Sweden to C. Portugal, W. and C. France, N. Italy and the Dobruja.

3. U. intermedia Hayne   'Intermediate Bladderwort.'

Incl. *U. ochroleuca* auct. angl. non, R.W. Hartm.

Stems slender, 10–25 cm.; *of two kinds* (*a*) bearing green lvs without or occasionally with very few bladders, and (*b*) colourless, often buried in substratum, bearing bladders on very much reduced lvs. Lvs distichous, 4–12 mm., orbicular in outline, palmately divided; *segments denticulate, with* 1–2 *small bristles on the teeth.* Bladders c. 3 mm. Fls very rarely produced, 2–4, on a scape 9–16 cm. *Corolla* c. 8–12 mm., *bright yellow*, marked with reddish-brown lines, spur 5–6·5 mm., conic. Fl. 7–9. Hyd.

Native. Lakes, pools and ditches, usually in shallow peaty water, very local; ascending to 3250 ft. (but rare at high altitudes); commonest in N. Scotland and Ireland. Dorset, Hants, Suffolk, Norfolk; Westmorland to Wigtown; Fife and Kintyre to Shetland; widespread in Ireland (rare in the east). 39, H23. Europe from Scandinavia to W. and C. France, N. Italy and Serbia; N. Asia; N. America from Newfoundland and British Columbia to New Jersey and California.

4. U. minor L.   'Lesser Bladderwort.'

Stems 7–25 cm., slender; *of two kinds* (*a*) bearing green lvs with a few bladders, and (*b*) colourless, often buried in substratum, bearing bladders on very much reduced lvs. Lvs 3–6 mm., orbicular in outline, palmately divided, *segments entire, without bristles.* Bladders c. 2 mm. Scapes 4–15 cm., 2–6-fld. *Corolla* 6–8 mm., *pale yellow*; spur very short, obtuse. Fl. 6–9. Hyd.

Native. Ponds and ditches, bog- and fen-pools; ascending to 2250 ft. Throughout the British Is. but local and absent from a number of counties. 87, H39, S. Europe from Iceland and Scandinavia to N.W. Spain, N. Italy, Albania and the Dobruja; Morocco (Atlas Mountains); N. Asia; N. America south to Pennsylvania and California.

109. ACANTHACEAE

Differs from the Scrophulariaceae chiefly in the indurated funicle, which often develops into a 'jaculator' for ejecting the seeds, and almost always in the absence of endosperm. In addition the lvs are always opposite, coloured bracts are often present, the corolla is strongly zygomorphic, the stamens often unequally 2-celled or only 1-celled and the style unequally bilobed.

About 220 genera and over 2000 spp., tropical and warm temperate regions.

1. ACANTHUS L.

Herbs (or shrubs) usually with a large radical rosette of lvs. Fls in large terminal bracteate spikes. Calyx 4-partite into large upper and lower lips and small lateral lobes. Corolla 1-lipped (upper lip 0), 3-lobed; tube very short. Stamens 4, shorter than the corolla, filaments thick, bent near the top; anthers 1-celled, connivent in pairs. Style filiform, stigma bifid. Jaculators well developed.

More than 20 spp. in the Mediterranean region and tropical Asia and Africa. Several, all similar in general appearance, are grown in gardens.

*1. A. mollis L. Bear's Breech.

Perennial herb 30–80 cm. Stems several, stout, simple. Radical lvs 25–60 cm., oblong-ovate, pinnatifid into oblong, lobed divisions, glabrescent, petiolate. Bracts ovate, spinous-dentate, tinged purple. Calyx glabrous. Corolla 3–5 cm., whitish with purple veins. Capsule glabrous, ovoid, with 2–4 seeds. Fl. 6–8. Pollinated by humble-bees; protandrous. $2n = 56$. Hs.

Introduced. Naturalized in waste places in W. Cornwall (including Scilly Is.), first recorded 1820. Native of S. Europe from Spain to Thrace.

110. VERBENACEAE

Herbs, shrubs, trees and woody climbers usually with opposite or whorled exstipulate lvs. Fls solitary, axillary or more usually in many-fld infl. of various kinds; hermaphrodite, ± zygomorphic, hypogynous. Calyx of 5 or rarely 4 sepals, sometimes 2-lipped, often enlarging in fr.; corolla gamopetalous, commonly 2-lipped; stamens antesepalous, epipetalous, rarely 5, usually 4, sometimes 2, the missing stamens sometimes represented by staminodes; *ovary* syncarpous, initially 1-celled but becoming 2-, 4- or 5-celled by ingrowth of the placentae, and *most commonly* 4- (8- *or* 10-)*celled through formation of false septa between the placentae* (as in Labiatae); *style terminal*, 1 or dividing above into 2, 4 or 5 stigmatic lobes; ovules anatropous with the micropyle directed downwards, usually 1 per cell. *Fr. usually a drupe* with 1, 2 or 4 stones each 4–1-celled with 1 seed per cell; sometimes a capsule (*Avicennia*) and rarely of four 1-seeded nutlets (*Verbena*). Seeds usually non-endospermic, with a straight embryo.

About 800 spp. in 80 genera, chiefly tropical and subtropical.

Teak (*Tectona grandis* L. f.), a large timber tree of Burma, Malaya, the East Indies, etc., is by far the most important member of the family commercially. *Clerodendrum* includes several small shrubs grown for their attractive fls whose corolla and persistent calyx have contrasting colours. *Lantana* spp. are pestilential weed-shrubs in many tropical and subtropical countries. *Avicennia* is a genus of mangroves with pneumatophores and viviparous fr., and *Verbena* includes many herbs grown for their brightly coloured fls.

1. VERBENA L.

Annual to perennial herbs or dwarf shrubs usually with opposite or whorled lvs and spikes, corymbs or panicles of smallish fls. Calyx tubular below, unequally 5-toothed above; corolla-tube straight or curved, downy within, the limb ± 2-lipped with 5 spreading lobes; stamens 4, in pairs, rarely 2 or 5, not exserted, ovary 4-celled with 1 ovule in each cell; style slender, somewhat 2-lobed above. Fr. of 4 nutlets separating at maturity.

About 200 spp. chiefly in America, many of them weeds of arable land and waste places in N. America.

The commonly cultivated *Verbena* is *V.* × *hybrida* Voss, probably of mixed hybrid origin with *V. peruviana* (L.) Druce as one of the parents.

1. V. officinalis L.   Vervain.

A perennial herb with a woody stock and several stiffly erect tough stems 30–60 cm., ± hispid, paniculately branched above. Lvs 2–7·5 cm., oblanceolate or rhomboidal in outline, pinnatifid, with acute or blunt, ovate to oblong lobes, the lowest sometimes much the largest; upper lvs narrower, less divided, sometimes ± entire; all dull green, hispid. Fls in slender terminal spikes, at first dense, then elongating to become lax in fr. Bracts c. 2 mm., ovate-acuminate, ciliate, reaching half-way up the calyx. Fls 4 mm. diam., subsessile. Calyx 2–3 mm., ribbed, shortly hairy. Corolla-tube almost twice as long as the calyx, limb pale lilac. Some fls have only 2 fertile anthers. Nutlets 4, reddish-brown, truncate, granulate on the inner face. Fl. 7–9. Homogamous; visited by small bees, hoverflies and butterflies, and automatically self-pollinated. $2n = 14$. H.

Native. Waysides and waste places; local. Throughout England and Wales, and in Fife. 72, H 26, S. Europe northwards to Denmark; N. Africa; W. Asia to Himalaya. Introduced in N. America.

111. LABIATAE

Herbs, less often shrubs, with ± quadrangular stems and opposite exstipulate simple lvs. Infl. basically of cymose type, borne in the axils of opposite bracts which vary from being like the cauline lvs to very different from them, usually much contracted so that the two opposite cymes form a whorl-like infl. (referred to in the following account as a whorl), the whorls themselves often close together at the apex of the stem and forming a spike-like or head-like infl. (referred to as spikes or heads in the following account). Fls sometimes solitary in the axils of each bract. Bracteoles usually small, sometimes 0. Fls hypogynous. Calyx usually 5-toothed, often 2-lipped with the upper lip 3-, the lower 2-toothed, rarely with the lips entire. Corolla with a ± well-developed tube, basically 5-lobed, but with the two upper lobes nearly always closely united to form a single lip, its double origin sometimes apparent only from the venation; lower 3 lobes often forming a ± distinct lower lip but never so closely united as the upper, the middle lobe usually larger than the 2 lateral; rarely (*Teucrium*) the 5 corolla lobes form a single lower lip; aestivation imbricate. Stamens 4 (the upper absent), didynamous, more rarely reduced to 2, epipetalous, alternating with the corolla-lobes; anthers introrse. Carpels 2, each with 2 ovules but ovary apparently equally 4-lobed because of a secondary division developing later. Ovules anatropous, usually erect and basal. Style simple below, branched into 2 above, usually gynobasic. Fr. of 4 nutlets; endosperm 0 or scanty. Nectar secreted at the base of the ovary.

About 170 genera and 3000 spp. in tropical and temperate regions, rare in the Arctic.

A very natural family, usually recognizable at sight. The 4 nutlets are characteristic, but are found also in the Boraginaceae; these usually have alternate lvs and actinomorphic fls and the infl. and appearance are very different. The quadrangular stems and opposite lvs are found also in certain Scrophulariaceae, e.g. *Scrophularia*, *Rhinanthus*, the latter especially somewhat resembling the Labiatae in appearance but with a very different ovary.

The Labiatae are frequently aromatic and have a characteristic smell (which may be either pleasant or unpleasant); many common pot-herbs belong to this family.

The fls are normally hermaphrodite but in several genera female fls occur either on the same or more commonly on different plants. The corolla of these fls is smaller than that of the hermaphrodite ones.

Commonly cultivated non-British genera are: *Lavandula* (Lavender), *Rosmarinus* (Rosemary), *Physostegia*, *Coleus* and *Monarda*.

1	Corolla with 4 nearly equal lobes, small.	2
	Corolla obviously 2-lipped.	3
	Corolla 1-lipped or the upper lip represented by 2 short teeth.	24
2	Stamens 4; lvs entire or toothed.	1. Mentha
	Stamens 2; lvs pinnately lobed.	2. Lycopus
3	Calyx 2-lipped, lips entire, the upper with a scale on the back.	
		24. Scutellaria
	Calyx ±equally 5- (rarely 10-)toothed or 2-lipped with toothed lips; scale absent.	4
4	Stamens included in the corolla-tube.	23. Marrubium
	Stamens exserted.	5
5	Stamens 2, each with a unilocular anther.	11. Salvia
	Stamens 4 with bilocular anthers.	6
6	Stamens, at least the longer pair, longer than the upper lip of the corolla, diverging.	7
	Anthers placed under the upper lip of the corolla, filaments converging or parallel.	9
7	Fls in a loose terminal panicle; bracteoles ovate, conspicuous.	
		3. Origanum
	Fls in a dense head or spike; bracteoles linear.	8
8	Calyx 2-lipped; fls purplish-pink. Common native.	4. Thymus
	Calyx nearly equally 5-toothed; fls violet-blue. Rare alien.	
		5. Hyssopus
9	Stamens curved, connivent; upper lip of corolla flat (or weakly concave in *Melissa*).	10
	Stamens straight; upper lip of corolla usually concave.	14
10	Lvs linear, entire. Rare alien on old walls.	6. Satureja
	Lvs ovate to elliptic, toothed (sometimes obscurely so).	11
11	Corolla-tube straight.	12
	Corolla-tube curved upwards.	10. Melissa
12	Calyx-tube straight; fls in opposite axillary peduncled cymes, pale.	
		7. Calamintha
	Calyx-tube curved; fls in axillary whorls without a common peduncle.	13
13	Whorls many-fld, dense; calyx-tube not gibbous at base; fls rosy-purple.	
		9. Clinopodium
	Whorls 3–8-fld; calyx-tube gibbous at base; fls violet.	8. Acinos
14	Outer pair of stamens longer than the inner (sometimes curving downwards after dehiscence and then appearing shorter).	15
	Outer pair of stamens shorter than the inner.	23

15 Calyx 2-lipped. *16*
    Calyx ± equally 5-toothed. *17*

16 Calyx broadly campanulate, open after flowering, lobes of lower lip
    obtuse; corolla 2·5 cm. or more. 12. MELITTIS
    Calyx closed after flowering, lobes of lower lip acute; corolla 1·5 cm.
    or less. 13. PRUNELLA

17 Lower lvs deeply lobed. 19. LEONURUS
    Lvs not lobed. *18*

18 Calyx funnel-shaped (i.e. with a spreading rim below the base of the
    teeth), teeth short and broad. 16. BALLOTA
    Calyx tubular or campanulate, teeth usually longer. *19*

19 Lateral lobes of lower lip of corolla short, obscure, with one or more
    small teeth. 18. LAMIUM
    Lateral lobes of lower lip of corolla well developed. *20*

20 Corolla deep yellow (see also *Phlomis*); stoloniferous perennial. In
    woods. 17. GALEOBDOLON
    Perennials with purple fls or annuals of cultivated ground with purple
    or pale yellow fls. *21*

21 Corolla with 2 bosses at base of lower lip; upper lip of corolla laterally
    compressed, helmet-shaped; annual. 20. GALEOPSIS
    Corolla without bosses; upper lip concave but not laterally compressed;
    annual or perennial. *22*

22 Perennials or biennials with numerous stem-lvs, or else annuals; corolla
    with a ring of hairs within. 14. STACHYS
    Perennial with lvs mostly in a basal rosette and few stem-lvs; corolla
    with scattered hairs within. 15. BETONICA

23 Fls many in each whorl, forming a terminal infl., white. 21. NEPETA
    Fls 2–6 in whorls in the axils of the lvs, violet. 22. GLECHOMA

24 Corolla with a single 5-lobed lip; tube glabrous inside. 25. TEUCRIUM
    Upper lip of corolla consisting of 2 short teeth; lower lip conspicuous,
    3-lobed; tube with a ring of hairs inside. 26. AJUGA

Subfamily 1. STACHYOIDEAE. Style gynobasic. Nutlets with small basal attachment. Seeds erect. Radicle short, straight. Calyx 5- (rarely 10-)toothed.

Tribe 1. SATUREJEAE. Calyx 10–15-nerved. Corolla of 4 nearly equal lobes or 2-lipped, the upper lip flat or nearly so. Stamens 4, the outer pair longer than the inner, diverging or curved and connivent under the upper lip, rarely 2. (Genera 1–10.)

1. MENTHA L.   Mint.

Perennial herbs with a characteristic pleasing smell and creeping rhizome. Fls small, purple, pink or white, in axillary whorls, often forming a terminal spike or head. Bracteoles small or 0. Calyx tubular or campanulate, 10–13-nerved, with 5 nearly equal teeth. Corolla-tube shorter than the calyx; lobes 4, nearly equal but the upper usually broader and often emarginate. Stamens 4, diverging, ± equal in length, usually exserted, but sometimes, especially in hybrids, included in the corolla; anther cells parallel. Nutlets ovoid, rounded at apex, smooth.

About 25 spp. in north and south temperate regions of the Old World.

The species are very variable and hybridize freely. Several are commonly cultivated for flavouring and often escape. Propagation is largely vegetative and the hybrids are thus often found away from their parents and the parentage of many of them is in doubt. Small-fld female plants of most spp. occur; hermaphrodite fls protandrous; pollinated by various insects.

The following account is confined to the spp. and hybrids, the numerous varieties being mostly omitted. Hybrids are included in the key.

1	Lvs less than 1 cm. broad, entire or obscurely toothed; throat of calyx hairy; calyx-teeth unequal, the two lower narrower and longer than the three upper.	2
	Lvs more than 1·5 cm. broad; throat of calyx naked; calyx-teeth equal.	3
2	Lvs ovate-orbicular, 3–5 mm.; stems filiform, creeping; whorls 2–6-fld; corolla-tube not gibbous.	**1. requienii**
	Lvs oval or oblong, 5–15 mm.; stems not filiform; whorls many-fld; corolla-tube somewhat gibbous below the mouth.	**2. pulegium**
3	Whorls all axillary, the axis terminated by lvs, or with very few fls in the axils of the uppermost pair.	4
	Whorls forming terminal spikes or heads.	7
4	Calyx campanulate.	5
	Calyx tubular.	6
5	Calyx-tube hairy, teeth scarcely longer than broad, ±deltoid though sometimes acuminate; scent not pungent; stamens usually exserted.	**3. arvensis**
	Calyx-tube glabrous, teeth c. twice as long as broad, subulate; scent pungent; stamens usually included.	×**gentilis**
6	Calyx-tube hairy; scent not pungent; stamens usually included.	×**verticillata**
	Calyx-tube glabrous; scent pungent; stamens usually exserted.	×**smithiana**
7	Fls in a head, often with axillary whorls below.	8
	Fls in a spike.	9
8	Calyx-tube hairy; stamens usually exserted; scent not of lemon.	**4. aquatica**
	Calyx-tube glabrous; stamens included; smelling of lemon.	×**piperita** var. *citrata*
9	Lvs stalked.	10
	Lvs sessile or subsessile (petioles less than 3 mm.).	12
10	Pedicels and calyx-tube usually glabrous; lvs lanceolate; stems and calyces often red; taste of peppermint.	×**piperita**
	Pedicels and calyx-tube hairy; lvs ovate or oblong; stems and calyces usually greenish; taste not of peppermint.	11
11	Uppermost lvs and bracts usually ovate; corolla pure lilac.	×**dumetorum**
	Uppermost lvs and bracts usually suborbicular; corolla reddish-lilac.	×**maximilianea**
12	Pedicels and calyx-tube glabrous; lvs glabrous or thinly hairy, ±con-colorous; smelling of spearmint.	13
	Pedicels and calyx-tube hairy; lvs ±densely hairy at least below; smell not of spearmint.	14

13 Lvs more than twice as long as broad, ±lanceolate, not rugose.

5. spicata

Lvs less than twice as long as broad, ovate or ovate-oblong, ±rugose.

× **cordifolia**

14 Lvs lanceolate or narrowly oblong, not or scarcely rugose.   **6. longifolia**

Lvs broadly oblong, ovate or suborbicular, ±rugose.   *15*

15 Lvs rarely more than 4 cm., rounded at apex or minutely cuspidate, crenate-serrate.   **7. rotundifolia**

Lvs 3–10 cm., subobtuse to acuminate, sharply serrate.   × **niliaca**

Section 1. *Audibertiae* Briq. Whorls axillary, 2–6-fld. Calyx turbinate-campanulate, weakly 2-lipped, the 2 lower teeth narrower than the 3 upper; throat hairy within. Corolla-tube not gibbous.

*1. M. requienii Benth.

Low perennial 3–12 cm., very strongly scented, glabrous or slightly hairy, the *stems* creeping and rooting at the nodes, *filiform*, forming mats, the fl. stems ascending. *Lvs 3–5 mm., ovate-orbicular*, entire, petiolate. Corolla pale lilac, little exserted. Fl. 6–8. $2n=18$. Chh.

Introduced. Grown in gardens for its scent and sometimes naturalized as by numerous rills at c. 1000 ft. on Slieve Gullion (Armagh). Native of Corsica and Sardinia.

Section 2. *Pulegium* (Mill.) DC. Whorls axillary, many-fld. Calyx tubular, weakly 2-lipped, the 2 lower teeth narrower than the 3 upper; throat hairy within. Corolla-tube gibbous on the lower side.

2. M. pulegium L.   Penny-royal.

Perennial 10–30 cm. *Stems* prostrate, less often erect (var. *erecta* Martyn), pubescent, often red, *relatively stout*. *Lvs 0·8–2 × 0·6–1 cm., oblong or oval*, rarely suborbicular, obtuse, ±cuneate at the base, minutely puberulous, gland-dotted, obscurely crenate-serrate with 1–6 teeth on each side, shortly petiolate. Whorls distant. Bracts like the lvs, but becoming smaller upwards, all longer than the fls. Pedicels and calyx shortly pubescent. Corolla hairy outside, glabrous within, lilac. Fl. 8–10. $2n=10, 20^*, 30, 40$. Hp.

Native. Wet places on sandy soil. Very local but widespread in S. England, becoming rarer northwards; in Scotland only known from Ayr and Berwick; in Ireland mainly in the S.W. and N.E., also Wexford and Dublin; Channel Is. 52, H13, S. C. and S. Europe, Mediterranean region, Macaronesia.

Section 3. *Mentha*. Calyx equally or nearly equally 5-toothed, not hairy in the throat. Corolla-tube not gibbous.

3. M. arvensis L.   'Corn Mint.'

Very variable perennial 10–60 cm., with erect or ascending, simple or branched, ±hairy stems; *smell not pungent*. Lvs (1·3–)2–6·5 × 1–2(–3·2) cm., often elliptic but varying from lanceolate to suborbicular, usually obtuse, cuneate or rounded at the base, shallowly crenate or serrate, petioled, ±hairy on both sides. *Fls in distant axillary compact whorls; the bracts like the lvs*, gradually decreasing in size upwards but *always much longer than the fls.* Pedicels glabrous or hairy. *Calyx campanulate, hairy all over, the teeth short, scarcely longer than broad,* often deltoid, obtuse to acuminate. Corolla lilac, hairy outside. *Stamens normally exserted.* Fl. 5–10. $2n=72*$.

Native. Common in arable fields, etc., in rides in woods and in damp places, ascending to 1200 ft.; throughout the British Is. 111, H40, S. Europe from Scandinavia to Spain and C. Italy, N. Asia to the Himalaya.

M. × verticillata L. (*M. aquatica × arvensis*)

M. sativa L.

Very variable. Perennial 30–90 cm., often more robust than *M. arvensis*, ± hairy; *smell not pungent.* Lvs 2–6·5(–8) × (1–)1·5–4 cm., usually ovate or oval but sometimes elliptic, oblong or lanceolate, acute or obtuse, cuneate to subcordate at base, serrate, petioled, ± hairy on both sides. *Fls in whorls,* all distant or the upper crowded; whorls less compact than in *M. arvensis* and often stalked; bracts like the lvs but often decreasing in size more than in *M. arvensis*, the upper sometimes shorter than the fls. *Pedicels and calyx hairy. Calyx tubular, teeth usually about twice as long as broad,* ± triangular, apex acuminate sometimes with a subulate point. Corolla lilac, hairy. *Stamens included,* very rarely exserted. $2n=42*, 84*, 120*, 132*$. Rather common throughout the British Is., often occurring where only one or neither parent is present. 104, H40, S.

M. × gentilis L. (*M. arvensis × spicata*)

Incl. *M. gracilis* Sole and *M. cardiaca* (S. F. Gray) Baker

Variable. Perennial 30–90 cm., *with pungent smell.* Stem erect, glabrous or hairy, sometimes red. *Lvs* (2–)3–6(–7·5) × 1–2·5 cm., *ovate-lanceolate, lanceolate, elliptic or oblong,* usually thinly hairy on both faces, sometimes subglabrous above and with the hairs restricted to the veins beneath, obtuse to shortly acuminate, cuneate or rarely rounded at the base, shortly petioled. *Whorls* usually *all separate*; the bracts ± like the lvs, gradually becoming smaller upwards in a variable degree. *Pedicels and calyx-tube* usually *glabrous,* but glandular. *Calyx campanulate,* the teeth about twice as long as broad, subulate, acuminate or with a setaceous point. Corolla lilac, c. 3 mm. *Stamens included.* $2n=54*, 60*, 84*, 96*, 108*, 120*$.

A very variable plant apparently including amphidiploids as well as primary hybrids and back-crosses. There is insufficient evidence at present to separate the various types morphologically.

Sides of ditches, waste ground, etc.; throughout Great Britain but rather local. Ireland. 74, H4.

M. × smithiana R. A. Graham (*M. aquatica × arvensis × spicata*)

M. rubra Sm., non Mill.

Stem 30–150 cm., erect, simple or branched above, usually red or purple, glabrous or more rarely thinly hairy; *smell sweetly pungent.* Lvs (2–)3–6 × 1·5–3·5 cm., *ovate,* rarely oval or oblong, usually glabrous or nearly so, sometimes thinly hairy on both faces, obtuse or acute, rounded or often ± cuneate at base, serrate, shortly petioled, often purple-veined. *Whorls separate; bracts* like the lvs, but gradually smaller, ovate or suborbicular, *the uppermost usually longer* but occasionally shorter *than the fls. Pedicels and calyx glabrous*, except for the shortly ciliate teeth, but glandular. *Calyx tubular*, the teeth about twice as long as broad. Corolla lilac, c. 5 mm. Stamens usually ± exserted. Fl. 8–10. $2n=120*$. Hp.

Probably originating as a hybrid between *M.* × *verticillata* and *M. spicata*, but with a double chromosome number.

Native? Sides of ditches, damp hedgebanks, waste ground, etc.; in scattered localities throughout the British Is., local. 76, H5.

M. × **muellerana** F. W. Schultz (*M. arvensis* × *rotundifolia*)

Like *M. arvensis* but much more robust, lvs broader, bracts suborbicular or broadly ovate, markedly decreasing in size upwards. Calyx-teeth lanceolate. Stamens included. $2n=60*$.

Formerly in 1 locality in S. Devon, now extinct.

4. M. aquatica L. 'Water Mint.'

Variable. Perennial 15–90 cm., strongly but not pungently scented. Stems ± erect, simple or branched, often reddish in exposure. Lvs 2–6(–9) × (1·2–)1·5–4 cm., usually ovate, ± hairy on both faces, or sometimes subglabrous, obtuse or acute, cuneate to subcordate at base, serrate or crenate-serrate, petiolate. *Infl. consisting of a terminal head c. 2 cm.* across composed of 1–3 whorls, and *usually 1–3 axillary whorls below.* Lower bracts lf-like, upper lanceolate or linear-lanceolate, hidden by the fls. *Pedicels and calyx hairy.* Corolla lilac. *Stamens normally exserted.* Fl. 7–10. $2n=96*$. Hp. or Hel.

Native. In swamps, marshes, fens and wet woods and by rivers and ponds, ascending to 1500 ft. 112, H40, S. Common throughout the British Is. Europe, S.W. Asia, N. and S. Africa, Madeira.

M. × **piperita** L. (*M. aquatica* × *spicata*) Peppermint.

Perennial 30–80 cm., with a pungent smell and taste (of peppermint). Stem erect, usually branched, reddish or purple, very thinly hairy. *Lvs* (2–)3·5–8·5 × 1·5–3·8 cm., usually lanceolate, less often ovate, acute, usually cuneate to (rarely) subcordate at base, sharply serrate, subglabrous to thinly hairy, *petiolate. Infl. a terminal oblong spike* (2–)3·5–6 cm., usually interrupted at the base; bracts lanceolate, about as long as the fls, the lowest 1–2 pairs somewhat lf-like. *Pedicels and calyx-tube usually glabrous,* ± glandular. Calyx tubular, often strongly tinged with red, teeth subulate, long-ciliate. *Corolla reddish-lilac. Stamens included.* $2n=66*, 72*$.

Var. *crispa* (L.) Koch. Lvs subsessile, deeply cut, curled and rugose. $2n=96$.

Var. *citrata* (Ehrh.) Briq. Resembling *M. aquatica* in infl. but with glabrous pedicels and calyx, glabrous or very thinly hairy lvs and included stamens and a characteristic lemony scent. $2n=84*, 120*$.

Sides of ditches, etc., and damp roadsides; throughout the British Is. but rather local. Cultivated as the source of peppermint. The varieties much rarer. 101, H34, S.

M. × dumetorum Schult (*M. aquatica × longifolia*)

M. palustris Sole, non Mill.; *M. pubescens* auct.

Stems erect, with ascending branches, densely hairy, not or little reddish-tinged. *Lvs 2–4 × 1·5–2·5 cm., ovate or the lower oblong*, obtuse to cuspidate, at least some of them usually subcordate at the base, usually less than twice as long as broad, green and hairy above, tomentose beneath, petiolate. Infl. a terminal oblong obtuse spike, rarely a head, interrupted below. *Upper lvs and lower bracts usually ovate, acute.* Upper bracts lanceolate or setaceous, shorter than the fls. *Pedicels and calyx hairy*, not or little reddish-tinged. *Corolla pure lilac.* Stamens included. Rare, mainly in S. and W. England.

M. × maximilianea F. W. Schultz (*M. aquatica × rotundifolia*)

Not tasting of peppermint. Stems erect or ascending, branched, densely hairy, not or little reddish-tinged. *Lvs 2·5–7 × 2–5 cm., broadly ovate* or rarely oblong, *mostly subcordate at base*, green and hairy above, somewhat greyish-tomentose beneath, petiolate or sessile. *Upper lvs and lower bracts suborbicular.* Infl. a terminal oblong spike, sometimes interrupted below, rarely a head with axillary whorls below. *Pedicels and calyx hairy, not or little reddish-tinged. Corolla reddish-lilac.* Stamens usually included. $2n = 60^*, 108^*$.

Very rare, only known from Devon, Cornwall and Jersey, plants from different localities differing considerably.

*5. M. spicata L.   Spearmint.

M. viridis auct.

Perennial 30–90 cm., with a pungent smell. *Stem* erect, usually branched, *glabrous. Lvs 4–9 × 1·3–3 cm., lanceolate* or oblong-lanceolate, acute or acuminate, sharply serrate, green and *glabrous* on both faces or with a few hairs below, *sessile or subsessile* (petiole not more than 3 mm.), not rugose. *Infl. a terminal cylindrical spike* 3–6 cm., whorls often becoming ± separate. Spikes often clustered at the top of the main axis. Bracts linear-setaceous, longer than fls, lowest pair sometimes lf-like. *Pedicels and calyx glabrous* (except the sometimes ciliate calyx-teeth). Corolla lilac, glabrous. Stamens exserted. Fl. 8–9. $2n = 36^*, 48^*, 84.$ Hp.

Introduced. The sp. most commonly cultivated as a pot-herb and naturalized in many places by roadsides and in waste places, usually damp. 73, H7, S. C. Europe.

M. × cordifolia Opiz (*M. rotundifolia × spicata*)

M. rubra Mill., p.p.

Stems 30–90 cm., stout, erect, thinly hairy or glabrous. *Lvs 2–7 × 1·5–4 cm., ovate or oblong*, cuspidate, cordate or subcordate at base, irregularly serrate, ± rugose, *glabrous except for a few hairs on the veins beneath, sessile or subsessile. Infl. a terminal* cylindrical *spike*, 2·5–5 cm., interrupted below, sometimes clustered. Bracts linear-lanceolate, shorter than the fls, lowest pair lanceolate, somewhat serrate. *Pedicels and base of calyx glabrous* but often glandular. Calyx-teeth shortly ciliate. Corolla lilac or pale pink. Stamens usually exserted; anthers reddish-purple. Waste places, local.

6. M. longifolia (L.) Huds. Horse-mint.

M. sylvestris L.

Perennial 60–90 cm., with creeping underground stems; smell pungent or not. Aerial *stem* erect, simple or branched above, ±*hairy*. *Lvs* 3–8 × 1–3 cm., *lanceolate* or oblong-lanceolate, acute or acuminate, serrate, green and ± hairy above, *grey and felted or tomentose beneath*, rounded or subcordate at base, *sessile*, not or scarcely rugose. *Infl. a terminal cylindrical spike*, 3–10 cm., becoming interrupted and slender after fl. Bracts subulate, longer than fls. *Pedicels and calyx hairy*. Corolla lilac, hairy outside, glabrous within. Stamens normally exserted. Fl. 8–9. $2n = 24, 36*, 48*$. Hp.

Doubtfully native. Scattered over the British Is. in damp roadsides and waste places but rather local. 81, H9, S. C. and S. Europe, Mediterranean region. W. Siberia.

M. × niliaca Juss. ex Jacq. (*M. longifolia × rotundifolia*)

Very variable. *Stem* stout, erect, 40–150 cm., usually branched, *densely hairy, at least above*. *Lvs* 3–10 × (1–)1·5–4(–6·5) cm., *ovate or oblong, cuspidate or acuminate, serrate*, green and ± hairy above, *usually greyish or whitish and felted or tomentose beneath* but sometimes green and less hairy, cordate or subcordate at base, sessile or less often subsessile, ± rugose. Fls in terminal cylindrical spikes, (2·5–)3–8 cm., often interrupted below, often clustering or forming a terminal panicle. Bracts subulate, usually shorter than fls. *Pedicels and calyx hairy*. Corolla lilac or pinkish-lilac, hairy outside. Stamens included or rarely exserted.

A very variable hybrid of which the two following forms are widespread and distinct.

Var. *villosa* (Huds.) J. Fraser (*M. nemorosa* auct.) resembling *M. longifolia* but with broader (1·5–4 cm.) oblong or ovate-oblong rugose lvs. $2n = 36*, 48$.

Var. *alopecuroides* (Hull) Briq. Very robust. *Lvs* oval or suborbicular 3–9 × 2–6·5 cm., woolly below, *sharply serrate*, cordate at base, *apex not rounded*. Sometimes confused with *M. rotundifolia*. $2n = 36*$.

Roadsides, etc., local but widespread (var. *alopecuroides* 53, H2, S).

M. scotica R. A. Graham perhaps belongs here. Like *M. longifolia* but lvs broader, more oblong, more shallowly serrate, grey above, densely very softly white-tomentose beneath. Spikes shorter. Stamens included. $2n = $c. $44*$. Perth, Angus and Moray.

7. M. rotundifolia (L.) Huds. Apple-scented mint.

Perennial 60–90 cm., strongly fragrant, stoloniferous. *Stem* erect, usually branched above the middle, *densely clothed with white hairs*. *Lvs* 2–4 × 1·5–3 cm., *oblong, ovate or suborbicular* on the same stem, *rounded or minutely cuspidate at apex, crenate-serrate*, the teeth sometimes cuspidate, pubescent above, *grey or white tomentose beneath*, subcordate at base, *sessile, rugose*. *Fls in dense spikes*, 3–5 cm., falcate when young, ± interrupted below, often forming a panicle. Bracts lanceolate-subulate, longer than the fls. *Pedicels and calyx hairy*. Corolla pinkish-lilac or whitish, hairy outside. Stamens usually exserted. Fl. 8–9. $2n = 24*$. Hp.

Native in S.W. England and Wales. Ditches, roadsides and waste places. Throughout England and Wales but local, still more so in Scotland and Ireland. 60, H25, S. C. and S. Europe, Mediterranean region, Azores.

2. LYCOPUS L.

Perennial odourless herbs with creeping rhizome. Fls in many-fld distant axillary whorls; bracts not differentiated from the lvs. Bracteoles small. Fls small. Calyx campanulate, 13-nerved with 5 equal teeth. *Corolla as in Mentha.* Stamens 2, diverging, longer than the corolla. Anther cells parallel. *Nutlets tetrahedral, apex truncate,* bordered.

About 7 spp. in temperate regions of the northern hemisphere and in Australia.

1. L. europaeus L.   Gipsy-wort.

Stems 30–100 cm., erect, somewhat hairy, usually with ascending branches. Lvs to 10 cm., ovate-lanceolate or elliptic, acute, shortly stalked, pinnately lobed, with numerous triangular, acute lobes, the lower often reaching to the midrib, but sometimes all shallow. Calyx-teeth lanceolate, hairy, spiny-pointed. Corolla white with a few small purple dots on the lower lip, c. 3 mm. long and across. Fl. 6–9. Pollinated by various insects; protandrous; small female fls occur. $2n = 22$. Hp. or Hel.

Native. On the banks of rivers and ditches and in marshes and fens. Common in England, Wales and Ireland, rarer in Scotland, but extending to Ross and the Outer Hebrides; Channel Is. 98, H40, S. Europe and the Mediterranean region; N. and C. Asia; naturalized in N. America.

3. ORIGANUM L.

Aromatic perennial. Fls in dense corymbose cymes, forming a terminal panicle. Bracteoles ovate, exceeding the calyx, imbricate. *Calyx* ovoid-campanulate, 13-nerved, *with 5 nearly equal teeth. Corolla 2-lipped. Stamens 4, straight, diverging, longer than the corolla* (at least the longer pair). Anther cells divergent. Nutlets ovoid, smooth.

About 6 spp., north temperate Old World, mainly Mediterranean.

1. O. vulgare L.   Marjoram.

Stems 30–80 cm., erect, branched above, and often with very short axillary sterile branches below, somewhat hairy. Lvs 1·5–4·5 cm., ovate, stalked, entire or obscurely toothed, with scattered appressed hairs on both sides. Panicle lax or dense but the cymes always dense. Bracts similar to the lvs but smaller, the upper nearly sessile, entire. Bracteoles conspicuous, ovate, purple, longer than the calyx. Calyx hairy within; teeth short. Corolla 6–8 mm., rose-purple, tube longer than the calyx. Fl. 7–9. Pollinated by various insects; protandrous; small female fls occur. $2n = 30^*, 32$. Chh.

Native. Dry pastures, hedge-banks and scrub, usually on a calcareous soil. Common in England and Wales; local in Scotland, extending to Caithness; in Ireland, common in the south, local in the north. 96, H38, S. Europe, N. and W. Asia.

4. Thymus L.   Thyme.

Low aromatic shrubs. Fls in few-fld whorls, forming a terminal spike or head. Bracts ± differentiated. Bracteoles minute. *Calyx 2-lipped*, the upper lip with three ± equal teeth, the lower 2-lobed. *Corolla, stamens* and fr. *as in Origanum.*

About 50 spp. in the north temperate Old World.

T. vulgaris L., 'Garden Thyme', an erect greyish shrublet, is the sp. usually cultivated as a pot-herb. Some other spp. are also cultivated for their aromatic lvs.

Plants with the fls only female occur rather commonly in the British spp.; these are normally smaller-fld than the hermaphrodite plants. Pollinated by various insects, protandrous.

1   Fl.-stem sharply angled, with long hairs on the angles only (2 faces shortly pubescent); plant tufted with the branches ascending; lateral veins of lvs not prominent beneath when dry; infl. elongated on normally developed plants.   **1. pulegioides**

Fl.-stem obscurely angled, hairy on at least two sides; plant with long creeping branches, forming a mat; lateral veins of leaves prominent beneath when dry; infl. usually capitate.   *2*

2   Fl.-stem hairy on two opposite sides only or more hairy on two opposite sides than on the other two; stomata 25–28 μ.   **2. drucei**

Fl.-stem equally hairy all round; stomata 22–24 μ.   **3. serpyllum**

1. T. pulegioides L.   'Larger Wild Thyme.'

T. Chamaedrys Fr.; *T. ovatus* Mill.; *T. glaber* Mill.

Tufted, *strongly aromatic* undershrub to 25 cm., *with ascending fl.-branches* and short creeping branches (more prostrate if heavily grazed). *Fl.-stems*, below the infl., *sharply 4-sided with long hairs only on the angles*, two opposite faces narrow and shortly pubescent, the other two broader and glabrous. *Lvs* 6–10(–17) mm. × 3–6 mm., ovate to elliptic, obtuse, shortly stalked, ciliate at base, otherwise glabrous, *slightly folded upwards about the midrib; lateral veins slender, not prominent beneath when dry*; stomata 21–23 μ, very few on upper surface. *Infl. elongated* and interrupted below when normally developed, but capitate on grazed plants. Calyx 3–4 mm., ± hairy, the teeth long-ciliate. Corolla rose-purple. Fl. 7–8. Fr. 9. 2n=28*. Chw.

Native. Dry grassland, usually calcareous, common in S.E. England, extending to Devon, Monmouth and Yorks; Wigtown, Angus; Cork, Cavan. 34, H2. Europe from S. Scandinavia (64° 30′ N.) and Russia to Spain, Serbia and Bulgaria.

2. T. drucei Ronn.

T. arcticus (E. Durand) Ronn., *T. britannicus* Ronn., *T. neglectus* Ronn., *T. zetlandicus* Ronn., *T. pseudolanuginosus* Ronn., *T. carniolicus* auct. brit., *T. pycnotrichus* auct., *T. serpyllum* auct. brit., p.p.

Mat-like, faintly (to moderately) *aromatic undershrub*, rarely more than 7 cm., *with long creeping branches*, the fl.-stems in rows on the branches. *Fl.-stems*, below infl., *4-angled with 2 opposite sides densely hairy with hairs of varying* *length, the other 2 sides less hairy or glabrous* (very rarely all equally hairy). *Lvs* 4–8(–11·5) mm. × 1·5–4 mm., usually borne horizontally (rarely upright), suborbicular to elliptic or oblanceolate, obtuse, shortly stalked, glabrous or hairy as well as ciliate, *flat, lateral veins prominent beneath when dry*; stomata 24·5–28 μ, numerous on both surfaces (though more so beneath). *Infl. capitate*, rarely somewhat elongated. Calyx 3·5–4 mm., ± hairy, the upper teeth long-ciliate. Corolla rose-purple. Very variable, several ecotypes occurring. Fl. 5–8, earlier than the other spp. when growing together. $2n = 50$–56^*. Chw.

Native. Dry grassland, heaths, dunes, screes and among rocks. Common throughout the British Is., ascending to 3700 ft. in Perth. 107, H38. Greenland, Iceland, Faeroes, W. Norway, France, N.W. Spain; closely related and perhaps conspecific plants occur in the Alps, Pyrenees, etc.

3. T. serpyllum L.

Similar to *T. drucei* but much less variable in Britain. Small. Faintly aromatic. *Fl.-stems* below infl. scarcely angled *with short white hairs evenly distributed all round.* Lvs 4–5(–7) mm. × 1·5 mm., borne upright, oblanceolate-spathulate, ciliate, usually otherwise glabrous; *stomata* 22–24 μ. Infl. capitate. Fl. 7–8. $2n = 24^*$. Chw.

Native. Open sandy heath and grassland in the Breckland of E. Anglia. 2. S. Sweden and N. Russia to C. France, Hungary and N. Roumania; divisible into several ssp. and with other ssp. in Siberia and China.

5. Hyssopus L.

Aromatic perennials, woody at the base. Lvs lanceolate or linear, entire. Fls in few-fld whorls, forming terminal unilateral spike-like infl. Bracteoles small, linear. *Calyx* tubular, 15-*nerved*, with 5 nearly equal teeth. *Corolla* 2-*lipped. Stamens* 4, *ascending and curved at the base, then suddenly diverging*, longer than the upper lip of the corolla; anther cells divergent. Nutlets ovoid-trigonous, smooth.

About 6 spp. in the Mediterranean region.

*1. H. officinalis L. Hyssop.

Somewhat woody, 20–60 cm. high, nearly glabrous, green. Branches erect, glabrous or puberulent. Lvs 1·5–2·5 cm., oblong-lanceolate to linear, subobtuse. Infl. rather dense and long. Bracts about as long as the fls. Calyx 4–5 mm., with ovate, shortly aristate teeth 2–3 mm. long. Corolla 10–12 mm., violet-blue. Fl. 7–9. Pollinated by bees; protandrous. $2n = 12$. N.

Introduced. Formerly much cultivated as a herb and still sometimes grown for ornament. Has become naturalized on old walls in a few places as at Beaulieu Abbey. S. Europe, W. Asia, Morocco.

6. Satureja L.

Aromatic herbs or low shrubs. *Fls in few-fld axillary whorls*, forming a terminal unilateral infl. Bracteoles small. *Calyx* 10-*nerved*, ± equally 5-toothed. *Corolla* 2-*lipped, tube straight*, naked within. *Stamens* 4, *shorter than the corolla, converging*. Style-branches nearly equal, subulate. Nutlets ovoid, smooth.

About 15 spp. in the Mediterranean region.

*1. S. montana L. Savory.

Undershrub 15–40 cm. high, with erect or ascending stems having stiff erect branches. Lvs 1–2 cm., lanceolate-linear, coriaceous, mucronate, ciliate, acute, longer than the internodes, entire. Fls in small shortly peduncled 2–4-fld cymes, forming a long terminal unilateral infl. Bracts like the lvs. Bracteoles shorter than the fls, mucronate. Calyx-teeth triangular-lanceolate, acuminate. Corolla white or pink, tube 6–7 mm., much longer than the calyx. Fl. 7–10. $2n=30$. N.

Introduced. Formerly much cultivated as a pot-herb, rarely naturalized on old walls, as at Beaulieu Abbey. S. Europe, Algeria.

7. Calamintha Mill.

Perennial herbs. *Fls in opposite axillary peduncled cymes*; bracts similar to cauline lvs but becoming much smaller above. *Calyx* tubular, 13-*nerved*, 5-toothed, tube straight, *hairy within. Corolla* 2-*lipped; tube straight*, naked within. *Stamens* 4, *shorter than corolla, curved and convergent.* Style-branches unequal, the upper subulate, the lower longer and broadened. Nutlets ovoid, smooth.

About 30 spp. in north temperate regions.

Forms with small female fls only occur. The measurements given apply to the hermaphrodite fls. Measurements of the lvs apply to the cauline lvs of the main stem.

1 Corolla more than 15 mm.; lvs 3·5–6 cm.; Isle of Wight. **1. sylvatica**
  Corolla not more than 15 mm.; lvs rarely more than 4 cm. *2*
2 Hairs in throat of calyx included; calyx-teeth long-ciliate; partial peduncles
    0 or very short; lvs on main axis mostly 2–4 cm. **2. ascendens**
  Hairs in throat of calyx protruding after fl.; calyx-teeth shortly and sparsely
    ciliate; partial peduncles obvious; lvs on main axis mostly 1–2 cm.
                                     3. nepeta

*C. grandiflora (L.) Moench with larger (3–4 cm.) fls and more deeply toothed lvs than our native spp. has occurred as a garden-escape. Native of S. Europe, S.W. Asia and Algeria.

### 1. C. sylvatica Bromf.					'Wood Calamint.'

C. intermedia auct.; *Satureja sylvatica* (Bromf.) Maly

Perennial with creeping rhizome. *Stems* erect, 30–60 cm., *little branched*, with long spreading hairs. *Lf-blades* 3·5–7 × 3–4·5 cm., ovate, obtuse or subobtuse, broad-cuneate or rounded at base and often somewhat decurrent down the petiole (5–15 mm. long), *serrate with* 7–10 *teeth on each side*, green, with long ± appressed hairs on both surfaces. Lower peduncles rather long (6–10 mm.), up to 9(–12)-fld, upper shorter, spreading after fl.; partial peduncles short or 0. Calyx 7–10 mm., with long hairs on the nerves and sessile shining glands between; *upper teeth spreading at c.* 90°; *lower teeth much longer, long-ciliate* (cilia longer than width of tooth), curved upwards; *throat with ring of hairs included. Corolla* 17–22 *mm.*, purplish-pink, much variegated with purple on lower lip. Fl. 8–9. Hp.

Native. Chalky banks in Isle of Wight. 1. France and W. Germany to C. Spain, Algeria and N. Syria.

### 2. C. ascendens Jord.					'Common Calamint.'

C. officinalis auct. angl. (?Moench); *Satureja ascendens* (Jord.) Maly; *C. baetica* auct. (incl. Pugsl.).

Perennial with shortly creeping rhizome. *Stems* erect, 30–60 cm., *little branched*, with long spreading hairs. *Lf-blades* (1·5–)2–4(–5) × 1·5–3·5 cm., ovate or orbicular-ovate, obtuse, truncate or broad-cuneate at base, sometimes somewhat decurrent, *obscurely crenate-serrate with 5–8 teeth on each side*, green, with long ± appressed hairs on both surfaces; petiole c. 1 cm. Peduncles all short or 0, or the lower longer (to 1 cm.), up to 9(–12)-fld, spreading after fl.; *partial peduncles 0 or very short.* Calyx 6–8 mm., with long hairs on the nerves and sessile shining glands between, often purple-tinted; *upper teeth spreading at about 45–90°; lower teeth much longer, long-ciliate,* curved upwards; *throat with ring of hairs included* (occasionally a few hairs protrude). *Corolla* 10–15 *mm.,* pale reddish-purple or lilac with darker spots on lower lip. Fl. 7–9. Hp.

Native. Dry banks, usually calcareous; from Yorks, Cumberland and Isle of Man southwards and from Louth and Donegal southwards; rather local. 66, H26, S. C. and S. Europe from France to the Caucasus; N. Africa.

3. C. nepeta (L.) Savi   'Lesser Calamint.'

Satureja Nepeta (L.) Scheele

Perennial, appearing more greyish than any of the preceding spp., with long creeping rhizome. *Stems* erect, 30–60 cm., *much branched*, of rather bushy appearance, grey with long soft spreading hairs. *Lf-blades* 1–2 cm., ovate, obtuse, usually broad-cuneate at base, *obscurely crenate or crenate-serrate with 5 or fewer teeth on each* side, greyish, above with short scurfy and few long hairs, beneath with more numerous long hairs; petiole usually 5 mm. or less. Peduncles short, up to 15-fld, ascending after fl.; *partial peduncles present.* Calyx 4–6 mm., with short hairs often obscuring the shining sessile glands; *upper teeth nearly straight* or somewhat spreading; *lower teeth* somewhat *longer, shortly and sparsely ciliate* (cilia less than width of tooth at its base), straight or nearly so; *throat with ring of hairs protruding.* Corolla 10–15 *mm.,* lilac, scarcely spotted. Fl. 7–9. Visited by various insects, mainly bees. Hp.

Native. Dry banks, usually calcareous; from Kent to Sussex, Gloucester, Pembroke, Caernarvon and Yorks; local. 30, S. S. Europe and Mediterranean region from W. and C. France to S. Russia; N. Africa, N. Syria and N. Persia.

8. ACINOS Mill.

Herbs. *Fls in c. 6-fld axillary whorls.* Bracts not differentiated. Bracteoles minute. *Calyx tubular, 13-nerved, tube curved, gibbous at base, hairy within. Corolla, stamens and style as in Calamintha.*

About 7 spp., in north temperate zone.

1. A. arvensis (Lam.) Dandy   Basil-thyme.

Calamintha Acinos (L.) Clairv.; *Acinos thymoides* Moench; *Satureja Acinos* (L.) Scheele

Usually annual, occasionally lasting for more than one year, branched from the base. Stems 10–20 cm., ascending, pilose. Lf-blades 0·5–1·5 cm., ovate to elliptic, petioled, subacute, cuneate at base, obscurely crenate, glabrescent.

Fls in 3–8-fld axillary whorls, forming a loose terminal infl. Calyx pilose; tube contracted in the middle; teeth subulate, the upper one broad-based. Corolla 7–10 mm., violet with white markings on the lower lip. Fl. 5–9, pollinated by bees. $2n=18$. Th.

Native. Arable fields, open habitats in grassland or rocks on dry usually calcareous soils, rather local and becoming increasingly so northwards but extending to Sutherland; Ireland, only in the south-east and centre extending to Dublin, Longford and Tipperary, very local. 80, H9. Europe from Scandinavia and N. Russia to C. Spain, Sicily, Albania and Thrace; Asia Minor, Caucasus.

9. CLINOPODIUM L.

Herbs. *Fls in remote many-fld terminal and axillary whorls.* Bracts not differentiated. Bracteoles subulate, conspicuous. *Calyx cylindric, 13-nerved; tube curved, not gibbous, glabrous or weakly hairy within. Corolla, stamens and style as in Calamintha.*

About 4 spp., in north temperate zone.

1. C. vulgare L.　　　　　　　　　　　　　　　　　　　　　　　Wild Basil.

Calamintha vulgaris (L.) Druce; *Calamintha Clinopodium* Benth.

Perennial, pilose, scent weak. Rhizome shortly creeping. Stems 30–80 cm., erect, simple or sparingly branched. Lf-blades 2–5 cm., ovate, petioled, sub-obtuse, rounded or broad-cuneate at base, shallowly and remotely crenate-serrate. Bracteoles numerous, long-ciliate. Calyx somewhat 2-lipped, the 3 upper teeth broad-based, the 2 lower longer. Corolla 15–20 mm., rose-purple. Fl. 7–9. Pollinated by bees and Lepidoptera; protandrous; small female fls occur. $2n=20$. Hp.

Native. Hedges, wood borders and scrub; less often in grassland on dry, usually calcareous soils; ascending to 1300 ft. Common in England, becoming more local northwards, extending to Inverness and Argyll; absent from Ireland except as a rare alien. 97, (H7), S. Europe, C. and W. Asia, Siberia, N. Africa, Azores, N. America.

10. MELISSA L.

Perennial herbs. Fls in axillary whorls. Bracteoles ovate or obovate. Calyx campanulate, 2-lipped, 13-nerved. *Corolla 2-lipped*, upper sometimes slightly concave; *tube curved upwards and dilated above the middle. Stamens 4, shorter than the corolla, curved, converging.* Style-branches equal. Nutlets obovoid, smooth.

Four spp., in Europe, W. Asia, N. Africa.

*1. M. officinalis L.　　　　　　　　　　　　　　　　　　　　　　Balm.

Sweet-scented perennial herb with short rhizome and erect, branched, some-what hairy stems 30–60 cm. high. Lf-blades 3–7 cm., ovate, with a petiole more than $\frac{1}{2}$ as long, deeply crenate or serrate, glabrescent, passing into the smaller, more sharply toothed bracts. Calyx with long spreading hairs; upper teeth broadly triangular, lower triangular-lanceolate, all with subulate points.

Corolla c. 12 mm., white or pinkish. Fl. 8–9, pollinated mainly by bees; protandrous to protogynous; small female fls occur. $2n = 32$. Hp.

Introduced. Cultivated for its sweet scent; a not uncommon garden-escape in S. England and naturalized in some places; rare in N. England and Scotland; very rare in Ireland. 50, H2. C. and S. Europe, W. Asia, N. Africa.

Tribe 2. SALVIEAE. Corolla 2-lipped, lobes usually very unequal. Upper lip often concave. Stamens 2, connective much elongated, with a single linear anther-cell.

11. SALVIA L.

Annual or perennial herbs or undershrubs. Fls in axillary whorls forming a ± interrupted terminal spike, the bracts differentiated. Bracteoles usually small. Calyx tubular or campanulate, 2-lipped, the upper with 3 teeth, lower with 2. Stamens 2, the filaments short, the connective much elongated, one anther cell aborted so that the stamen appears branched with an anther cell on the end of the longer branch. Nutlets ovoid-trigonous, smooth.

About 500 spp., in tropics and temperate zones.

A number of spp. are cultivated, including *S. officinalis* L. (Sage) from S. Europe, a greyish-leaved undershrub much cultivated as a pot-herb, and *S. splendens* Sellow ex Nees from Brazil with brilliant scarlet fls much used as a bedding plant. About 15 other spp. have been recorded as garden-escapes or casuals.

1. Upper lip of calyx with conspicuous, ± equal teeth; corolla-tube with a ring of hairs within.   **1. verticillata**
   Upper lip of calyx with very short teeth, the two lateral connivent over the middle one; corolla-tube without a ring of hairs.   2

2. Calyx pubescent and glandular but without long white hairs; corolla 10–25 mm., the smaller fls female only; upper lip glandular outside.   **2. pratensis**
   Calyx glandular and pilose with long white hairs; corolla 6–15 mm.; fls all hermaphrodite, the smaller cleistogamous; upper lip not glandular outside.   3

3. Radical lvs usually less than twice as long as broad; corolla with two white spots at base of the lower lip.   **3. horminoides**
   Radical lvs usually more than twice as long as broad; corolla without white spots. (Guernsey.)   **4. verbenaca**

Section 1. *Hemisphace* Benth. Calyx tubular, upper lip with 3 nearly equal teeth. Corolla-tube with a ring of hairs within; upper lip weakly convex. Whorls many-fld.

*1. S. verticillata L.

Perennial 30–80 cm., hairy, foetid. Lf-blades 5–15 cm., broadly ovate, obtuse, cordate at base, irregularly crenate-dentate, sometimes pinnatifid, petiolate. Bracts small, brown, reflexed. Pedicels 3–5 mm. Calyx c. 5·5 mm., tubular-campanulate, 12-nerved, *the upper lip with conspicuous teeth, the middle tooth slightly broader and shorter than the lateral.* Corolla 10–15 mm., violet; tube exserted, 2-lobed, contracted at the base into a short claw. Fl. 6–8. Pollinated by bees; protandrous; small female fls occur. $2n = 16$. Hs.

Introduced. In waste places, recorded from many localities, usually a casual but sometimes naturalized. Mountains of S. Europe from S. Alps and Carpathians to C. Spain, C. Italy and Greece and of S.W. Asia.

Section 2. *Plethiosphace* Benth. Calyx campanulate; upper lip with 3 short teeth, the middle one very small (less than 0·3 mm.), the two lateral connivent over it. Corolla-tube without a ring of hairs within; upper lip concave. Whorls c. 6-fld.

2. S. pratensis L.   'Meadow Clary.'

Perennial 30–100 cm., pubescent, glandular above, aromatic. Radical lf-blades 7–15 cm., ovate or oblong, obtuse, cordate at base, irregularly doubly crenate or occasionally lobed, rugose, with long petioles; cauline 2–3 pairs, smaller, the upper sessile. Bracts green, ovate, acuminate, entire, shorter than the calyces. Fls hermaphrodite or female, usually on separate plants, never cleistogamous. *Calyx* pubescent and glandular but *without long white hairs*, c. 6·5 mm. in the hermaphrodite fls. *Corolla of the hermaphrodite fls 15– 25 mm., violet-blue*; the tube exserted; *upper lip laterally compressed, forming a hood, falcate, glandular outside; style long-exserted.* Female fls much smaller, sometimes only 10 mm. Fl. 6–7. Pollinated by long-tongued bees; protandrous or homogamous. $2n=18$. Hs.

Native. On calcareous grassland from Kent and Wilts to Monmouth, Shropshire, S. Yorks and Lincoln, very local; also as a casual in waste places but even as such not recorded outside England; rare. 24. Europe from Scandinavia to C. Spain, C. Italy, Serbia, Bulgaria and the Crimea; Morocco.

*S. sylvestris L. differing in its more numerous, oblong-lanceolate stem-lvs, purple bracts and numerous whorls of smaller (10–14 mm.) fls is naturalized in one place at Barry Docks (Glamorgan), and occurs as a casual elsewhere. Native of C. and S. Europe, W. Asia.

3. S. horminoides Pourr.   Wild Clary.

S. Verbenaca auct. brit., p. max. p.

Perennial 30–80 cm., little-branched, pubescent, glandular above, slightly aromatic. *Radical lf-blades* 4–12 cm., oblong or ovate, obtuse, *usually less than twice as long as broad*, very variable in toothing, from crenate-serrate or sinuate to incised or pinnately lobed with the lobes crenate-serrate, rugose, petiolate. Stem-lvs usually 2–3 pairs, smaller, the upper sessile. Upper part of stem and calyces strongly tinted with dull blue-purple. Bracts as in *S. pratensis. Fls* very variable on the same or different plants, *most commonly small, cleistogamous and shorter than the calyx* but sometimes open and much larger, all hermaphrodite. *Calyx* c. 7 mm., pubescent, glandular and *pilose with long white hairs* which occur especially near the base and round the sinuses, with more prominent nerves and longer points than in the preceding sp. *Corolla up to 15 mm.* in the largest open-fld forms, violet-blue *with 2 white spots at the base of the lower lip*, and then with the upper lip somewhat compressed and falcate and the style slightly exserted; *in the cleistogamous fls as little as 6 mm.*, the lips nearly equal and connivent, the upper nearly straight and the style included. Fl. 5–8. $2n=64$. Hs.

Native. Local in dry pastures and roadsides. Widespread in S. England, less so in Wales and N. England, rare in Scotland, extending to Ayr and E. Ross; Ireland, mainly in the south-east, extending to Cork, Clare and Louth. 69, H10, S. S. Europe to C. and W. France; Algeria.

4. S. verbenaca L.

S. clandestina auct. angl.; *S. Marquandii* Druce

Differs from *S. horminoides* as follows: More slender. Radical *lvs* oblong, *usually more than twice as long as broad*, coarsely crenate-serrate. Upper part of stem and calyx usually less glandular and more pilose, much less deeply and strongly purple-tinted and of a more pinkish tinge. *Fls* 12–15 *mm.*, *usually open, rarely cleistogamous. Corolla* of a more lilac tint and *without white spots at the base of the lower lip.* Fl. 6–8. 2*n*=54, 64. Hs.

Native. Dunes at Vazon Bay, Guernsey. Mediterranean region to W. and C. France and Spain.

Tribe 3. STACHYEAE. Calyx 5–10-nerved. Corolla strongly 2-lipped; upper lip concave, often falcate or forming a hood. Stamens 4, the outer pair longer than the inner, filaments parallel, anthers ovate, connivent in pairs under the upper lip of the corolla.

12. MELITTIS L.

Perennial herb. Fls in 2–6-fld axillary whorls; bracts not differentiated. *Calyx open in fr., 2-lipped, upper lip with 2–3 small irregular teeth, the lower with 2 rounded lobes. Upper lip of corolla only slightly concave*; tube broad, naked within. Anther cells diverging. Nutlets ovoid, smooth.
One sp.

1. M. melissophyllum L.   Bastard Balm.

Strong smelling. Stems erect 20–50 cm., hirsute. Lf-blades 5–8 cm., ovate, petiolate, crenate. Corolla very large, 2·5–4 cm., tube much exceeding the calyx, pink, or white spotted with pink. Fl. 5–7. Pollinated by humble-bees and hawk-moths; protandrous. Hp.

Native. In woods and hedgebanks, very local. From Cornwall to Sussex, Gloucester, Pembroke, Merioneth and Denbigh. 15. C. and S. Europe.

13. PRUNELLA L.

Perennial herbs. Fls in few- (c. 6-)fld whorls forming a dense terminal oblong infl. Bracts orbicular, sessile, differing markedly from lvs. *Calyx strongly 2-lipped, closed in fr.; the upper truncate with 3 very short aristate teeth, the lower with 2 long teeth. Upper lip of corolla very concave*; tube straight, obconic. Anther cells diverging; filaments with a subulate appendage below the apex. Nutlets oblong, smooth.

Five spp., *P. vulgaris* almost cosmopolitan, the remainder in Europe and the Mediterranean region.

Upper lvs entire or shallowly toothed; fls normally violet; sinus between upper calyx-teeth gradually rounded.   **1. vulgaris**
Upper lvs lyrate or pinnatifid; fls normally cream; sinus between upper calyx-teeth parallel-sided.   **2. laciniata**

1. P. vulgaris L. Self-heal.

Sparingly pubescent perennial with short rhizome and ascending or erect stems, 5–30 cm. high. *Lf-blades* 2–5 cm., ovate, *entire or shallowly dentate*, petioled, cuneate or rounded at base. Bracts and calyx with long white hairs, usually purplish-tinged. Upper lip of calyx with 2 lateral teeth ill-developed, their points diverging, *sinus between them and the middle tooth rounded*, sometimes obsolete, teeth of lower lip shortly ciliate. *Corolla* 10–14 mm., *violet*, rarely pink or pure white. Appendages of longer stamens nearly straight. Fl. 6–9. Pollinated mainly by bees; protandrous or homogamous; small female fls occur. $2n=32$. Hs.

Native. Very common in grassland, clearings in woods and waste places, mainly on basic and neutral soils; ascending to 2500 ft. Throughout the British Is. 112, H40, S. Europe, temperate Asia, N. Africa, N. America, Australia.

2. P. laciniata (L.) L.

Differs from *P. vulgaris* as follows: *Lvs with more abundant and softer hairs*, somewhat whitened beneath, narrower, the lower oblong-lanceolate, often entire, *the upper very variable, from deeply pinnatifid to toothed with a single lobe on each side near the base*. Lateral teeth of upper lip of calyx better developed, their points nearly erect, *sinus between them and the middle tooth narrow, parallel-sided and well marked*; teeth of lower lip longer and long-ciliate. *Corolla larger*, c. 15 mm., *cream-white*, rarely pink or pale blue-violet. Appendages of longer stamens curved. Fl. 6–8. Pollinated by bees; protandrous; small female fls occur. $2n=32$. Hs.

Probably introduced. First recorded 1887. In dry calcareous grassland from Somerset and Gloucester to Kent and Cambridge, very local; appearing quite native; Alderney. 14, S. C. and S. Europe, W. Asia, N. Africa.

P. laciniata × vulgaris (*P. × intermedia* Link) is often found where the parents grow together.

14. STACHYS L.

Annual or perennial herbs *without well-marked basal rosette. Stem-lvs ± numerous. Calyx tubular or campanulate with 5 narrow equal teeth. Upper lip of corolla concave, but not helmet-shaped.* Corolla-tube usually (always in British spp.) with a ring of hairs within. *Outer stamens diverging laterally from corolla after fl. Anther-cells divaricate* (very rarely parallel). *Nutlets obovoid, rounded at apex.*

About 200 spp. cosmopolitan, except Australia and New Zealand.

1	Fls yellowish-white; annual.	**1. annua**	
	Fls purple.		2
2	Perennial or biennial; corolla large, 12 mm. or more.		3
	Annual; corolla small, 7 mm. or less.	**2. arvensis**	
3	Stem and lvs densely clothed with long white silky hairs.	**3. germanica**	
	Lvs green.		4
4	Bracteoles nearly as long as the calyx-tube (very rare).	**4. alpina**	
	Bracteoles very short, scarcely longer than the pedicel (common).		5

5   Lvs lanceolate, with short petioles or subsessile; fls dull purple.

                                                                         5. palustris

  Lvs ovate, with long petioles; fls claret-coloured.   **6. sylvatica**

*1. S. annua (L.) L.

Annual 10–30 cm., much-branched, glabrescent. *Lvs* 2–6 cm., *oblong*, obtuse, *cuneate at base*, shallowly crenate, the lower shortly petioled, the upper sub-sessile. Whorls 3–6-fld, in the axils of lanceolate acute bracts, passing into the lvs below. Bracteoles linear, very small. Calyx c. 8 mm., tubular-campanulate, hirsute, teeth narrowly triangular-lanceolate, mucronate. *Corolla* 11–13 mm., *yellowish-white*. Fl. 6–10. Visited by humble-bees. $2n=34$. Th.

    Introduced. A casual recorded from many localities in waste places and arable land but rarely naturalized. C. and S. Europe, W. Siberia, Orient.

*S. recta L., with yellowish-white fls like the preceding but differing in the perennial habit and broader calyx-teeth, is naturalized at Barry Docks (Glamorgan). Native of C. and S. Europe, Asia Minor and the Caucasus.

2. S. arvensis (L.) L.   'Field Woundwort.'

Annual 10–25 cm. with slender ascending stems usually branched at the base, hirsute. *Lf-blades* 1·5–3 cm., *ovate*, obtuse, *truncate* or cordate at base, crenate-serrate, petioled. Whorls 2–6-fld, in the axils of bracts resembling the lvs but becoming much smaller and subsessile above, forming a very lax spike, much interrupted below. Bracteoles linear, very small. Calyx 4–6 mm., tubular-campanulate, hirsute, teeth triangular-lanceolate or triangular-ovate, mucronulate. *Corolla 6–7 mm., pale purple*. Fl. 4–11. Self-pollinated, insect visits few. $2n=10$. Th.

    Native. Arable fields on non-calcareous soils; ascending to 1250 ft. Throughout the British Is., rather common in the west, local in the east. Norway to the Azores, Cape Verde, N. Africa, Crete and Palestine. Naturalized in America.

3. S. germanica L.   'Downy Woundwort.'

Perennial or biennial 30–80 cm. *Whole plant densely covered with long white silky hairs*, giving it a whitish appearance. Lf-blades 5–12 cm., ovate-oblong to lanceolate, cuneate to cordate at base, crenate, the lower with long petioles, the upper sessile or nearly so; venation reticulate, conspicuous. Whorls many-fld, forming a dense terminal spike interrupted below; bracts lanceolate, passing into the lvs. *Bracteoles linear, nearly as long as the calyx. Calyx* 9–11 mm., tubular, very silky, with triangular mucronate, somewhat unequal *teeth less than half as long as the tube. Corolla c. twice as long as the calyx, pale rose-purple*, hairy outside. Fl. 7–8. Pollinated by bees; protandrous; small female fls occur. $2n=30$. Hs.

    Native. Pastures and hedgebanks, very rare; now believed extinct except in Oxford but formerly also in Hants, Kent, Bucks, Northampton, Denbigh and Lincoln. 10. C. and S. Europe, N. Africa, Orient.

4. S. alpina L.

Perennial 40–100 cm., *green*, hirsute, glandular above. Lf-blades 4–16 cm., ovate, cordate at base, crenate-serrate; petioles 3–10 cm. Whorls many-fld, distant in the axils of subsessile ovate or lanceolate bracts, the lower crenate-serrate, lf-like, the uppermost smaller, entire. *Bracteoles entire, nearly as long as the calyx. Calyx* c. 8 mm., tubular, glandular-hairy; *teeth less than half as long as the tube*, triangular-ovate, mucronate, somewhat unequal. *Corolla 15–20 mm., dull reddish-purple*, hairy. Fl. 6–8. 2n=30. Hp.

Native. Very rare in open woods in Gloucester and Denbigh. 2. C. Europe.

5. S. palustris L.   'Marsh Woundwort.'

Perennial herb with long creeping rhizome producing small tubers at the apex in autumn, *green*, hairy, almost hispid, *odourless*. Stems 40–100 cm., simple or slightly branched, hollow. *Lvs* 5–12 cm., *oblong-lanceolate* or linear-lanceolate, acute, rounded or subcordate at base, crenate-serrate, *the lower very shortly petioled* (petiole 5 mm. or less), *the upper sessile*. Whorls c. 6-fld, forming a terminal spike, dense above, interrupted below. Lower bracts resembling the lvs but smaller, upper small, shorter than the fls, entire. *Bracteoles* linear, *scarcely reaching the base of the calyx. Calyx* c. 8 mm., tubular-campanulate, pilose, *eglandular* or sparingly glandular, *the teeth triangular-subulate, more than half as long as the tube. Corolla 12–15 mm., dull purple*, pubescent outside. Fl. 7–9. Pollinated mainly by bees; protandrous. 2n=c. 64, 102. Gt.

Native. Common by streams and ditches and in swamps and fens, sometimes also in arable land; ascending to 1500 ft. Throughout the British Is. 112, H40, S. Europe from Norway to Portugal, Spain, Corsica, Italy and Greece; temperate Asia (to Japan); N. America.

6. S. sylvatica L.   'Hedge Woundwort.'

Perennial herb with long creeping rhizome not producing tubers, *green*, almost hispid, *foetid* when bruised. Stems 30–100 cm., often branched, solid. *Lf-blades* 4–9 cm., *ovate*, acuminate, cordate at base, coarsely crenate-serrate, *all petioled* (petioles 1·5–7 cm.). Whorls c. 6-fld, forming an interrupted terminal spike. Bracts shortly petioled, the lower ovate-lanceolate, toothed, the upper lanceolate, entire. *Bracteoles* linear, *scarcely reaching the base of the calyx. Calyx* c. 7 mm., campanulate, hairy and *glandular, the teeth triangular-lanceolate, more than half as long as the tube. Corolla 13–15 mm., claret-coloured*, pubescent outside. Fl. 7–8, pollinated by bees; protandrous. 2n=66. Hp.

Native. Common in woods, hedgebanks and shady waste places on the richer soils; ascending to 1500 ft. Throughout the British Is. 111, H40, S. Norway to N.E. Portugal, C. Spain, Sicily, Albania, Thrace, Caucasus, Kashmir and the Altai.

S. palustris × *sylvatica* = *S.* × *ambigua* Sm. has oblong shortly-petioled lvs and is normally sterile. It is widespread and not uncommon with the parents.

15. BETONICA L.

Differs from *Stachys* as follows: *Perennial herbs with well-marked basal rosette and few stem-lvs.* Upper lip of corolla nearly flat; tube long, usually without a ring of hairs (but with scattered hairs) within. *Outer stamens not diverging from corolla after fl. Anther-cells nearly parallel.*

About 8 spp., Europe, W. and C. Asia, N. Africa.

1. B. officinalis L.   Betony.

Stachys Betonica Benth., *S. officinalis* (L.) Trev.

Sparingly hairy perennial 15–60 cm., with a short woody rhizome and erect stems, simple or somewhat branched below. Lf-blades 3–7 cm., oblong or ovate-oblong, cordate at base, obtuse, coarsely crenate; the radical numerous on very long (to 7 cm.) petioles; cauline 2–4 pairs, distant, several times their own length apart, petioles becoming shorter upwards, the uppermost sub-sessile. Infl. often interrupted below. Bracts ovate or lanceolate, entire; the lowest pair crenate-serrate, rather lf-like. Bracteoles lanceolate, aristate, about equalling the calyx. Calyx 7–9 mm., the teeth triangular-lanceolate, aristate. Corolla c. 15 mm., bright reddish-purple, the tube longer than the calyx. Fl. 6–9. Pollinated mainly by bees; protandrous or homogamous. $2n=16$. Hs.

Native. Open woods, hedgebanks, grassland and heaths, usually on the lighter soils; ascending to 1500 ft. Common in England and Wales, local in Scotland, where it extends to Perth and the Inner Hebrides, and in Ireland; Jersey. 83, H13, S. Europe from S. Sweden and N. Russia to C. Spain, Italy, Greece and the Caucasus, but not in the Mediterranean islands; Algeria.

16. BALLOTA L.

Perennial herbs. Fls in many-fld axillary whorls, bracts not differentiated. *Calyx funnel-shaped,* 10-nerved, *with 5 broadly ovate acuminate, mucronate ± equal teeth. Upper lip of corolla somewhat concave*; tube shorter than the calyx, with a ring of hairs within. Stamens 4, parallel, the outer pair longer; anther cells diverging. *Nutlets oblong, rounded at apex.*

About 25 spp., mainly Mediterranean, the following more widespread, and 1 in S. Africa.

1. B. nigra L. ssp. foetida Hayek   Black Horehound.

Hairy perennial with an unpleasant smell. Rhizome short, stout. Stems 40–100 cm., branched. Lf-blades 2–5 cm., ovate or orbicular, broad-cuneate to cordate at base, petioled, coarsely crenate. Whorls numerous, many-fld in the axils of bracts resembling the cauline lvs but smaller. Bracteoles subulate. Fls subsessile. Calyx c. 1 cm., teeth broadly ovate, suddenly acuminate, 1–2 mm. Corolla 12–18 mm., purple, hairy. Fl. 6–10. Pollinated mainly by bees; protandrous. $2n=22^*$. Hp.

Native. Common on roadsides and hedgebanks in England and Wales, local in Scotland and not recorded north of Renfrew and Moray; local in Ireland and not native. 76, H25, S. Scandinavia to Morocco, Palestine and N. Persia; Azores.

*Ssp. **nigra** (*B. nigra* var. *ruderalis* (Swartz) Koch) differs in the shorter calyx-tube and longer (2–4 mm.) lanceolate, gradually tapered calyx-teeth. It is rare and not native.

17. GALEOBDOLON Adans.

Stoloniferous perennial herbs. Fls in dense axillary whorls, the bracts like the lvs. *Calyx tubular-campanulate with 5 nearly equal mucronate teeth.* Corolla 2-lipped; *upper lip laterally compressed, helmet-shaped*; lower lip 3-lobed, *the middle lobe only slightly larger than the lateral; tube* longer than the calyx, straight, *dilated above*, with a ring of hairs within. *Anther cells divaricate, glabrous. Nutlets trigonous, truncate at apex.*

About 4 spp., in Europe and temperate Asia.

1. G. luteum Huds.   Yellow Archangel.
Lamium Galeobdolon (L.) L.
Perennial herb with long lfy stolons sometimes not produced till after fl., sparingly hairy. Stems 20–60 cm. Lf-blades 4–7 cm., ovate, acute or acuminate, truncate or rounded at base, stalked, irregularly crenate-serrate or serrate, those of the stolons shorter and broader than those of the fl.-stems. Calyx c. 10 mm. Corolla c. 2 cm., yellow with brownish markings. Fl. 5–6. Pollinated by bees; homogamous. $2n=18$. Chh.

Native. In woods, usually on the heavier soils, sometimes becoming locally dominant, especially after coppicing. Common in England and Wales; rare in Scotland, only in the south and doubtfully native; in Ireland only in the south-east in Wexford, Carlow, Wicklow and Dublin. 69, H4, S. Sweden and N. Russia to C. Spain, Italy, Albania, Bulgaria and Persia.

18. LAMIUM L.

Annual or perennial herbs without aerial stolons. Fls in dense axillary whorls, the bracts ± like the lvs. Calyx tubular or tubular-campanulate with 5 nearly equal mucronate teeth. Corolla 2-lipped, the upper lip laterally compressed forming a hood; lower lip 3-lobed, *the lateral lobes very small, each with a small tooth* (in all the other genera the lateral lobes are well developed); *tube dilated above. Anther cells divaricate, hairy. Nutlets trigonous, truncate at apex.*

About 40 spp., in Europe, temperate Asia, N. Africa.

1	Annuals; corolla not more than 15 mm.; tube not suddenly contracted near the base.	2
	Perennials; corolla 2 cm. or more; tube suddenly contracted near the base.	5
2	Bracts, at least the upper, sessile, differing somewhat from the lvs.	3
	Bracts stalked, resembling the lvs.	4
3	Calyx in fl. 7 mm. or less, tube densely clothed with somewhat spreading white hairs, teeth connivent in fr.	**1. amplexicaule**
	Calyx in fl. 8 mm. or more, tube with a moderate number of stiff appressed hairs, teeth spreading in fr.	**2. moluccellifolium**
4	Lvs irregularly incised-dentate; corolla without or with a faint ring of hairs towards the base.	**3. hybridum**

Lvs crenate-serrate; corolla with a conspicuous ring of hairs towards the
 base.   **4. purpureum**
5 Corolla white; tube with an oblique ring of hairs towards the base.
                                                                     5. album
  Corolla purple; tube with a transverse ring of hairs towards the base.
                                                                  6. maculatum

Section 1. *Lamiopsis* Dum. Annuals. Corolla-tube regularly cylindrical
below, enlarged above into a wide throat, with or without a ring of hairs.

1. L. amplexicaule L.   Henbit.

Finely pubescent annual 5–25 cm., with ascending branches from the base.
Lf-blades 1–2·5 cm., *orbicular* or ovate-orbicular, obtuse, truncate, rounded
or subcordate at base, *crenate-lobulate*, with a long petiole (3–5 cm. in the
lower lvs). *Bracts* similar to lvs, often larger and lobed, but *sessile* (rarely
lowest stalked), semi-amplexicaul, often broader than long. Whorls few,
rather distant. *Calyx* 5–6 *mm.*, tubular, densely *clothed with white,* ± *spreading
hairs*; the *teeth usually rather shorter than the tube, connivent in fr. Corolla*
when well developed c. 15 mm., pinkish-purple, long-exserted, or small, included and cleistogamous; the *tube glabrous within*. Fl. 4–8. Insect visitors
rare (bees), mainly self-pollinated. $2n = 18$. Th.

Native. Cultivated ground, usually on light dry soils through nearly the
whole of the British Is., rather common in Great Britain, local in Ireland.
105, H 26, S. All Europe, except the extreme north, extending to the Azores,
Canaries, N. Africa, Palestine, Persia and the Usuri region. Naturalized in
N. America.

2. L. moluccellifolium Fr.   'Intermediate Dead-nettle.'
L. intermedium Fr.

Differs from *L. amplexicaule* as follows: Usually more robust. Lvs somewhat
deltoid. Bracts not amplexicaul, the lowest shortly stalked. Infl. denser.
Calyx 8–12 *mm.* in fl., *with appressed hairs; the teeth longer than the tube,
spreading in fr.* Corolla-tube scarcely longer than calyx, with a faint ring of
hairs within. Fl. 5–9. $2n = 36$. Th.

Native. In cultivated ground. Widespread in Scotland, rare in England
and not known south of Huntingdon, Derby and Lancashire; in Ireland
mainly in the north and west but extending to Cork and Carlow. 42, H 25.
N. Europe from Iceland and Scandinavia to N. Germany.

3. L. hybridum Vill.   'Cut-leaved Dead-nettle.'
L. incisum Willd.; *L. dissectum* With.

Differs from *L. purpureum* as follows: More slender, less pubescent. Lvs often
smaller, *the upper truncate at the base and* ± *decurrent down the petiole,* all
irregularly incised-dentate. Corolla-tube less exserted *without or with a faint
ring of hairs* towards the base. Fl. 3–10. Visited by bees. $2n = 36$. Th.

Native. In cultivated ground, spread over the whole of the British Is. but
local. 92, H 28, S. W. and C. Europe from the Faeroes and Scandinavia to
Russia (Volga-Kama region), N.W. Italy and C. Spain; Morocco, Algeria
(rare).

4. L. purpureum L.   Red Dead-nettle.

Pubescent annual 10–45 cm., branched from the base, often somewhat purple-tinted. *Lf-blades* 1–5 cm., *ovate*, obtuse, *cordate at base*, *regularly crenate-serrate*, petiolate. *Bracts* similar, *rounded or truncate at base*, *petiolate*, or the upper subsessile. Infl. rather dense. Calyx 5–6 mm., tubular-campanulate, pubescent, the teeth about as long as the tube, spreading in fr. *Corolla* 10–15 mm., pinkish-purple, the *tube longer than the calyx*, *with a ring of hairs near the base*. Fl. 3–10. Pollinated by bees (rarely other insects) or selfed; homogamous. $2n=18$. Th.

Native. Very common in cultivated ground and waste places throughout the British Is.; ascending to 2000 ft. 112, H40, S. Europe, in the south only in the mountains and absent from several of the Mediterranean islands, to Palestine and the Yenisei region.

Section 2. *Lamium*. Perennials. Corolla-tube cylindrical at the base for a very short distance, then suddenly enlarged and gibbous, the upper part curved, with a ring of hairs within at the enlargement.

5. L. album L.   White Dead-nettle.

Hairy perennial 20–60 cm., with a creeping rhizome and erect stems. Lf-blades 3–7 cm., ovate, acuminate, cordate at base, coarsely and often doubly serrate or crenate-serrate, petiolate. Bracts similar. Whorls mostly distant. Calyx c. 10 mm., tubular-campanulate, the teeth slightly longer than the tube. *Corolla* c. 2 cm., *white*, the *tube with an oblique ring of hairs near base*, lateral lobes of lower lip with 2–3 small teeth, upper lip long-ciliate. Fl. 5–12. Pollinated by long-tongued Hymenoptera, mainly humble-bees; homogamous. $2n=18$. Hp.

Native. Hedgebanks, roadsides and waste places. Common in England, absent from the Scottish islands and very rare in Scotland north of the Caledonian Canal; in Ireland mainly in the east, local and not native. 100, H25, S. Scandinavia to C. Spain, Italy, Macedonia, Himalaya and Japan, only in the mountains in the southern part of its range.

*6. L. maculatum L.   Spotted Dead-nettle.

Differs from *L. album* as follows: Lf-blades 2–5 cm., usually acute, often with a large whitish blotch. Calyx-teeth relatively rather shorter. *Corolla pinkish-purple*, the *tube with a transverse ring of hairs within*, lateral lobes of lower lip with a single tooth, upper lip shortly ciliate. Fl. 5–10. Pollinated by humble-bees; homogamous. $2n=18$. Hp.

Introduced. Commonly cultivated in gardens in the form with a whitish blotch on the lf and sometimes found as an escape. The Netherlands and Germany to Portugal, C. Spain, Italy, Greece and Persia.

19. LEONURUS L.

Perennial herbs. Fls in dense-fld, distant, axillary whorls. *Calyx campanulate*, *5-nerved with 5 nearly equal spiny-pointed teeth*. *Corolla-tube* shorter than the calyx, *not widened at the throat*. Anther cells opposite, opening by a common slit. *Nutlets trigonous*, *truncate at apex*.

About 8 spp., in Europe and temperate Asia, 1 widespread in the tropics.

*1. L. cardiaca L.   Motherwort.

Pubescent perennial 60–120 cm. with stout rhizome and branched stems. Lf-blades 6–12 cm., cordate at base; the lower ovate-orbicular, palmately 5–7-lobed, the lobes irregularly dentate; the upper trifid, lanceolate, cuneate at base, passing into the lf-like bracts; all petiolate. Whorls numerous, distant, many-fld. Calyx 5-angled, teeth ovate-lanceolate, the 2 lower deflexed. Corolla c. 12 mm., white or pale pink with small purple spots. Fl. 7–9. Pollinated by bees. $2n=18*$. Hp.

Introduced. Waste places and hedge-banks, rare, scattered over the British Is. Europe from Scandinavia and W. Russia (upper Dnieper region) to N. Spain, Italy and Greece.

PHLOMIS L.

Small shrubs or large herbs with large fls, many in a whorl. Differs from all the British members of the tribe in the unequal style branches.

*P. fruticosa L.   Jerusalem Sage.

Whitish-tomentose shrub to 1 m. Lvs 3–6 cm., ovate or oblong, entire, shortly stalked. Bracts ovate or lanceolate. Corolla 2–3 cm., bright yellow. Often cultivated, and naturalized in Somerset and Devon. Mediterranean region.

*P. samia L.

Herb c. 1 m. Lvs 8–15 cm., ovate, crenate, green above, grey-tomentose beneath, lower long-stalked. Bracts subulate. Corolla c. 3 cm., purple. Rarely naturalized. S. Balkans and Asia Minor.

20. GALEOPSIS L.   Hemp-nettle.

Annual herbs. Fls in dense whorls, terminal and axillary, the bracts like the cauline lvs, but often smaller. *Calyx tubular or campanulate with 5 somewhat unequal spiny-pointed teeth.* Corolla 2-lipped; *upper laterally compressed, helmet-shaped; lower lip* 3-lobed *with two conical projections at the base* (not found in the other genera of the tribe); tube longer than the calyx, straight, dilated above, with a ring of hairs within. *Anther cells parallel, superposed, opening separately,* ciliate. Nutlets trigonous, rounded at the apex.

Nine spp., in Europe and temperate Asia.

1	Stem softly pubescent or glabrescent, not swollen at the nodes.	2
	Stem hispid, swollen at the nodes.	4
2	Lvs and calyx hairy but not silky; fls rose-purple.	3
	Lvs beneath and calyx velvet-silky; fls pale yellow.	**3. segetum**
3	Lvs linear-lanceolate to oblong-lanceolate with 1–4 teeth on each side.	
		1. angustifolia
	Lvs ovate or ovate-lanceolate with 3–8 teeth on each side.	**2. ladanum**
4	Corolla 13–20 mm., pink, purple or white, rarely pale yellow with violet spot, the tube nearly always scarcely exceeding the calyx. (**tetrahit** agg.)	5
	Corolla c. 30 mm., pale yellow usually with a violet spot on the lower lip, the tube much exceeding the calyx.	**6. speciosa**

5   Middle lobe of lower lip of the corolla entire, its network of dark markings
     restricted to the base, never reaching the margin.   **4. tetrahit**
     Middle lobe of lower lip of corolla deeply emarginate, its network of
     markings reaching the margin, or the whole lip dark.   **5. bifida**

Subgenus 1. LADANUM (Gilib.) Rchb.
Stem softly pubescent or glabrescent, not swollen at the nodes.

1. G. angustifolia Ehrh. ex Hoffm.   'Narrow-leaved Hemp-nettle.'
G. Ladanum auct., p.p.
Stem 10–80 cm., *pubescent or glabrescent, not swollen at the nodes.* Lvs
1·5–8 cm., *linear-lanceolate to oblong-lanceolate*, less than 1 cm. broad, acute,
attenuate at base into a short petiole, often abundantly appressed-hairy, *with
1–4 small serrations on each side.* Bracteoles often as long as the calyx. Calyx
tubular, ± hairy, often with appressed whitish hairs, usually eglandular or
with few glands. Corolla 1·5–2·5 cm., rosy-purple, the tube usually much
longer than the calyx. Fl. 7–10. Insect or self-pollinated. $2n=16$. Th.
    Native. Rather common in arable land in England extending to Dun-
barton and Moray. In Ireland only in the east, extending to Kilkenny and
Westmeath. 77, H6. Europe from S. Sweden to Spain, Italy and Bosnia.

*2. G. ladanum L.

Differs from *G. angustifolia* as follows: *Lvs* 1–3 cm. broad, not silky, *ovate-
oblong or ovate-lanceolate*, cuneate at base, serrate, with 3–7 *prominent teeth
on each side.* Bracteoles shorter. Calyx green, hirsute and glandular, usually
longer. Corolla-tube often scarcely exceeding calyx. Fl. 7–10. Insect or
self-pollinated. $2n=16$. Th.
    Introduced. Rare and usually impermanent in waste or cultivated ground.
Europe from Scandinavia to N. Spain, Italy and Macedonia. Russian Asia.

3. G. segetum Neck.   'Downy Hemp-nettle.'
G. ochroleuca Lam.; *G. dubia* Leers
Differs from *G. ladanum* as follows: *Lvs velvety silky, especially beneath*, with
more prominent veins. Bracteoles much smaller. Whorls often fewer-fld.
Calyx velvety silky and glandular. *Corolla* 2–3 cm., *pale yellow*, large,
c. 4 times as long as the calyx. Fl. 7–10. Insect pollinated (but not self-sterile).
$2n=16$. Th.
    Native. In arable land in a few places in England and Wales extending
north to Isle of Man, Cheshire and Durham; rare. 12. France, N. and E.
Spain, W. Germany, W. Switzerland, Denmark.

Subgenus 2. GALEOPSIS.
Stem hispid, swollen at the nodes.

(4–5). G. tetrahit agg.   'Common Hemp-nettle.'
Stem 10–100 cm., with ascending branches, hispid and with glandular hairs.
Lf-blades 2·5–10 cm., ovate-lanceolate or ovate, acuminate, cuneate at base,
crenate-serrate, somewhat hairy. Calyx ± hispid with rather prominent veins.

Corolla 13–20 mm., usually pink, purple, or white; tube scarcely longer than calyx (rarely much longer, because of the short calyx-teeth).

Native. Arable land, less often in woods, fens and wet heaths; ascending to 1470 ft. Common throughout the British Is. 112, H40, S.

The British distribution of the two following spp. is incompletely known but both are widespread and common.

4. G. tetrahit L.

Stems hispid, especially below the nodes, and with red-tipped glandular hairs in a group below the nodes. Teeth of lvs usually larger and fewer than in *G. bifida*. Corolla 15–20 mm., purple, pink, or white with darker markings; *middle lobe of lower lip of corolla* flat, *entire*, often nearly as broad as long, *network of dark markings restricted to base, never reaching margin.* Fl. 7–9. Usually self-pollinated. $2n=32$. Th.

Iceland, Faeroes to European Russia, south to C. Spain, Montenegro and Macedonia. Naturalized in N. America.

A plant indistinguishable from this sp. has been produced artificially by crossing *G. speciosa* and the continental *G. pubescens* Bess. and back-crossing a triploid raised from this hybrid with *G. pubescens* (see Müntzing, *Hereditas*, 1938).

5. G. bifida Boenn.

G. Tetrahit var. *bifida* (Boenn.) Lej. & Court.

Usually shorter and more slender than *G. tetrahit*. Stem more evenly hispid, *glandular hairs red- or yellow-tipped, distributed all over internodes*, though mainly in the upper half. Teeth of lvs usually smaller and more numerous than in *G. tetrahit*. Corolla smaller, c. 13–14 mm., purple, pink, white, or pale yellow, with dark, often violet markings on lower lip (sometimes coloured as in *G. speciosa*); *middle lobe of lower lip* convex, *deeply emarginate, network of dark markings reaching the margin or the whole lip dark without a network.* Fl. 7–9. Usually self-pollinated. $2n=32$. Th.

N. and C. Europe; N. Asia. Probably originated in a similar (but not identical) way to *G. tetrahit*.

The hybrid *G. bifida × tetrahit* occurs rather commonly on the Continent and probably also in this country.

6. G. speciosa Mill.   'Large-flowered Hemp-nettle.'

Differs from *G. tetrahit* agg. as follows: Usually more robust. Stem uniformly hispid, glandular hairs yellow-tipped, mainly on the upper half of the internodes. Veins of calyx less prominent. *Corolla 22–34 mm., pale yellow with lower lip mostly violet*, occasionally all yellow, *tube about twice as long as the calyx*, lower lip entire or slightly emarginate. Fl. 7–9. Insect pollinated (but not self-sterile). $2n=16$. Th.

Native. In arable land, often on black peaty soil; ascending to 1470 ft. Scattered over the whole of the British Is. but rather local. 99, H22. Norway to S. France, N. Italy, Albania, Bulgaria and Siberia (Yenisei region).

Tribe 4. NEPETEAE. Calyx 15-nerved. Corolla as Stachydeae, but upper lip sometimes flat. Stamens 4, outer pair shorter than inner, otherwise as in Stachydeae.

21. NEPETA L.

Perennial herbs. *Fls in many-fld axillary whorls forming a terminal infl.* Calyx tubular, 5-toothed. Upper lip of corolla flat; *tube rather suddenly curved and dilated at the middle,* glabrous within. *Anther cells divergent, opening by a common slit.* Nutlets obovoid.

About 150 spp., in Europe, Asia and N. Africa.

N. × *faassenii* Bergmans ex Stearn (*N. Mussinii* hort. p.p.) and *N. mussinii* Spreng. are extensively used as edging-plants.

1. N. cataria L.　　　　　　　　　　　　　　　　　　　　Cat-mint.

Strongly scented. Stems 40–100 cm., erect, branched, hoary pubescent. Lf-blades 3–7 cm., ovate, petioled, cordate at base, coarsely serrate, white-tomentose beneath; bracts similar but much smaller, mostly shorter than fls. Upper whorls crowded, lower more widely spaced. Calyx pubescent, teeth lanceolate-subulate, the upper the longest, the two lower the shortest; tube ovoid. Corolla c. 12 mm., white with small purple spots. Nutlets smooth. Fl. 7–9. Pollinated by bees; protandrous. $2n = 32, 36$. Hp.

Native. Hedgebanks and roadsides usually on calcareous soil from Westmorland and Northumberland southwards, rather local; Wigtown, Stirling; scattered over Ireland, except the south-west, but not native; Jersey. 66, H18, S. Europe from Scandinavia to Portugal, Spain, Corsica, Sicily and Greece. W. and C. Asia to Kashmir. Naturalized in N. America and S. Africa.

22. GLECHOMA L.

Perennial herbs. *Fls in few* (2–4)-*fld secund axillary whorls, the bracts not differing from the foliage lvs.* Calyx tubular, somewhat 2-lipped. Upper lip of *corolla* flat; *tube narrowly obconic, straight,* hairy within at base of lower lip. *Anther cells at right angles, each opening by a separate slit.* Nutlets obovoid, smooth.

About 6 spp., in Europe and Asia.

1. G. hederacea L.　　　　　　　　　　　　　　　　　　Ground Ivy.

Nepeta hederacea (L.) Trev.; *N. Glechoma* Benth.

Softly hairy or nearly glabrous perennial with stems creeping and rooting. Flowering branches 10–30 cm., ascending. Lf-blades 1–3 cm. diam., reniform to ovate-cordate, obtuse, crenate; petiole long. Corolla 15–20 mm., violet with purple spots on lower lip. Nutlets smooth. Fl. 3–5. Pollinated mainly by bees; protandrous; small female fls occur commonly. $2n = 18, 24, 36^*$. Hp.

Native. Woods, grassland and waste places, usually on the damper and heavier soils, sometimes becoming locally dominant in damp oakwoods, especially after coppicing; ascending to 1300 ft. Common throughout the British Is. except in N. Scotland where it is rare (absent from Orkney and the Outer Hebrides). 107, H40, S. Europe, W. and N. Asia to Japan.

Tribe 5. MARRUBIEAE. Corolla 2-lipped. Stamens 4, like the style included in the corolla-tube.

23. MARRUBIUM L.

Perennial herbs. Fls in many-fld axillary whorls, bracts not differentiated. Bracteoles linear. Calyx tubular, 10-nerved, with 5 or 10 teeth. Upper lip of corolla nearly flat; tube shorter than the calyx, naked or with a poorly developed ring of hairs within. Anther cells diverging. Nutlets ovoid, smooth.

About 35 spp., in Europe, temperate Asia and N. Africa.

1. M. vulgare L.   White Horehound.

White-tomentose perennial 30–60 cm., with short stout rhizome and branched stems. Lf-blades 1·5–4 cm., orbicular or ovate-orbicular, crenate, cordate or cuneate at base, lower long-, upper short-stalked, obtuse, rugose, green above. Whorls many-fld, broader than high. Calyx with 10 small hooked teeth. Corolla c. 1·5 cm., whitish. Fl. 6–11. Pollinated mainly by bees or selfed; homogamous or weakly protandrous; small female fls occur. $2n = 34^*$, 36. Hp.

Native. Local on downs, in waste places and by roadsides from Dunbarton and Moray and from Dublin and Galway southwards, perhaps only native near the S. coast of England. 75, H12, S. Europe, C. and W. Asia, N. Africa, Azores, Canary Is.

Subfamily 2. SCUTELLARIOIDEAE. Style gynobasic. Nutlets with small basal attachment. Seeds ± transverse. Radicle bent, lying along one of the cotyledons. Calyx 2-lipped with entire lips. Corolla 2-lipped, the upper lip helmet-shaped. Stamens 4, the outer pair longer than the inner.

24. SCUTELLARIA L.

Perennial herbs. Fls in axillary pairs, sometimes forming a terminal raceme. Calyx campanulate, 2-lipped, closed after flowering; the lips entire, the upper bearing a small scale on the back. Corolla 2-lipped, the lateral lobes small, ± free from the lips; tube naked within, dilated above, straight or curved below. Anthers of outer pair of stamens with 2 nearly parallel cells, of the inner with 1 cell. Nutlets subglobose, smooth or tuberclate.

About 180 spp., spread over the whole world, except S. Africa.

1	Fls blue-violet, 10–20 mm.	**2**
	Fls pale pinkish-purple, 6–10 mm.	**2. minor**
2	Lvs cordate at base, toothed; fls not forming a well-marked raceme; corolla-tube slightly curved; common.	**1. galericulata**
	Lvs hastate at base, otherwise entire; fls forming a short terminal raceme; corolla-tube strongly curved; Norfolk.	**3. hastifolia**

1. S. galericulata L.   Skull-cap.

Pubescent or glabrescent perennial 15–50 cm. high, with simple or branched stems from a slender creeping rhizome. *Lvs* 2–5 cm., ovate-lanceolate or oblong-lanceolate, obtuse or subacute, cordate at base, *remotely and shallowly crenate*; petiole short. Bracts like the lvs but gradually decreasing in size upwards; *fls not in a well-marked raceme*. Pedicels very short. Calyx glabrous or pubescent. *Corolla* 10–20 mm., *blue-violet*, several times as long as the calyx, the tube somewhat curved below. Fl. 6–9. Visited by various insects; homogamous; small female fls rare. $2n = $ c. 32. Hp.

Native. On the edges of streams and in fens and water meadows; ascending to 1200 ft. Throughout the British Is. (not Shetland, etc.), common. 108, H38, S. Europe from Scandinavia (69° 10′ N.) to C. Spain and Portugal, Italy and Greece, N. and W. Asia, Algeria, N. America.

2. S. minor Huds.   'Lesser Skull-cap.'

Smaller in all its parts than *S. galericulata*. Stems 10–15(–30) cm., usually more glabrous. Lvs 1–3 cm., entire except near the base, lower sometimes ovate. *Bracts* often truncate at the base, *quite entire. Corolla* 6–10 mm., *pale pinkish-purple with darker spots*, 2–4 times as long as the calyx, tube nearly straight. Fl. 7–10. Hp.

Native. On wet heaths, etc.; ascending to 1500 ft. Throughout the greater part of the British Is., local, but absent north-east of Stirling and the Hebrides and of Dublin and Sligo. 77, H17, S. S.W. Sweden, France, W. Germany, N.W. Spain, Portugal, Switzerland, N. Italy, Azores.

S. galericulata × minor = S. × hybrida Strail (*S. × nicholsonii* Taub.) sometimes occurs with the parents.

*3. S. hastifolia L.

A subglabrous ± erect perennial herb 20–40 cm.; rhizome, slender, creeping. *Lvs* 1–2·5 cm., ovate-lanceolate, truncate and *hastate* at the base, *otherwise entire*, acute to subobtuse, shortly stalked. Bracts smaller than the lvs, the uppermost c. 5 mm.; *fls forming a well-marked* short and rather dense *terminal raceme.* Pedicels very short. Calyx glandular-hairy. *Corolla blue-violet*, 15–25 mm.; *tube strongly curved* near the base. Fl. 6–9. $2n = $ c. 32. Hp.

Introduced (first found 1948). In semi-natural oakwood near Brandon, Norfolk. 1. Europe from Sweden and Finland to N.W. Spain, C. Italy, Macedonia and the Caucasus but rare in the west; Asia Minor.

*S. altissima L. (*S. Columnae* auct.). Lvs large, ovate, long-petioled, strongly crenate, hairy only on the veins beneath. Bracts small, lanceolate. Corolla 15–18 mm., upper lip bluish-purple, lower whitish. Naturalized in Somerset. Native of S.E. Europe and S.W. Asia.

Subfamily 3. AJUGOIDEAE. Style not gynobasic. Nutlets with large lateral-ventral attachment. Calyx 10-nerved, 5-toothed. Corolla 1-lipped or nearly so. Stamens 4, exserted, the outer pair longer than the inner.

25. TEUCRIUM L.

Herbs or undershrubs. Infl. various. Calyx tubular or campanulate, equally 5-toothed or the upper tooth larger. *Corolla of one 5-lobed lip* (the lower), the 4 upper lobes short, the middle lobe large; tube usually included, without a ring of hairs within. Nutlets obovoid, smooth or reticulate.

About 100 spp., cosmopolitan, mainly Mediterranean.

1 Fls in terminal racemes, the bracts very different from the lvs; upper tooth of calyx much larger than the others; corolla pale yellowish-green.

4. scorodonia

Fls in axillary whorls, sometimes forming a terminal spike, the bracts similar to the lvs though sometimes smaller; calyx-teeth equal or nearly so; corolla purple.   *2*

2 Lvs pinnatifid.   **3. botrys**
Lvs toothed.   *3*

3 Whorls forming a terminal spike, upper bracts shorter than the fls though somewhat gibbous (on walls).   **1. chamaedrys**
Whorls distant, all the bracts longer than the fls (in wet places).

2. scordium

Section 1. *Chamaedrys* (Mill.) Dum. Whorls 2–6-fld, forming terminal spikes. Calyx tubular-campanulate, teeth subequal, tube not saccate.

*1. T. chamaedrys L.   'Wall Germander.'
Perennial, almost woody at base, without a creeping rhizome, forming low tufts. Stems 10–30 cm., many, ascending, hairy, rooting at the base. *Lvs* 1–3 cm., ovate, obtuse, attenuate at base into a short petiole, *deeply crenate* or lobulate, dark green and shining above, ± hairy. *Bracts similar but smaller, the upper shorter than the fls*, subsessile. Infl. somewhat secund, short. Pedicels c. 2 mm. Calyx c. 6 mm. *Corolla pinkish-purple*; the tube c. 5 mm., lip c. 8 mm., the upper 4 lobes acute, ciliate, the middle broad, obovate-cuneate, crenulate. Fl. 7–9. Pollinated by bees, self-pollination possible; protandrous. $2n = 64$. Chw.
    Introduced. Grown in gardens, and sometimes naturalized on old walls in England and Wales, very rarely in Scotland and Ireland. 22, H3. C. and S. Europe, Orient, Morocco.

Section 2. *Scordium* (Mill.) Benth. Whorls 2–6-fld, axillary. Calyx tubular, teeth subequal, tube saccate at the base on the lower side.

2. T. scordium L.   'Water Germander.'
Softly hairy *perennial* with creeping rhizome or lfy stolons. Stems 10–60 cm., rooting at the base, ascending. *Lvs* 1–5 cm., oblong, sessile or subsessile, usually rounded at base, *coarsely serrate*, not shining above. *Bracts scarcely different*. Whorls distant, secund. Pedicels short. *Corolla* c. 12 mm., *purple*. Fl. 7–10. Pollinated by bees, self-pollination possible; protandrous. Hp.
    Native. Banks of rivers and ditches on calcareous soil, and dune-slacks. Rare in England and only recorded from Devon, Berks, Oxford, Norfolk, Cambridge and Lincoln and (extinct) Yorks; in Ireland widespread along the R. Shannon, but only known from two other stations, ranging from Limerick to Leitrim; Guernsey. 9, H8, S. Sweden to France, N. Italy, Serbia and the Dobruja; W. Siberia and the Aral-Caspian region.

3. T. botrys L.   'Cut-leaved Germander.'
Biennial or annual 10–30 cm., softly hairy. *Lf-blades* 1–2·5 cm., ± ovate in outline, petiolate, *pinnatifid*; the segments oblong, often lobed, obtuse, 1–2 mm. broad. *Bracts smaller but longer than the fls*, once-pinnatifid. Whorls all along

the branches, secund. Pedicels 3–4 mm. Calyx c. 7 mm., reticulately veined. *Corolla pinkish-purple*; the tube c. 6 mm.; lip deflexed, c. 8 mm. Fl. 7–9. Pollinated by bees, self-pollination possible. $2n = 10$. Th. or Hp.

Native. Chalky fallow fields and open habitats in chalk grassland. Wilts, Hants, Kent, Surrey, Gloucester; very rare. 6. C. and S. Europe, Algeria.

Section 3. *Scorodonia* (Hill) Benth. Fls in pairs in terminal racemes, the bracts very different from the lvs. Calyx campanulate, the upper tooth orbicular-ovate, much broader than the others, tube gibbous.

4. T. scorodonia L.   Wood Sage.

Pubescent perennial with creeping rhizome. Stems 15–30 cm., erect, branched. Lf-blades 3–7 cm., ovate, cordate at base, petiolate, subacute, crenate, rugose. *Bracts ovate-lanceolate, entire, shorter than the fls.* Pedicels short. Calyx c. 5 mm. *Corolla pale yellowish-green*; tube c. 8 mm.; lip 5–6 mm., deflexed. Fl. 7–9. Pollinated by bees; protandrous. $2n = 32, 34^*$. Hp.

Native. Woods, grassland, heaths and dunes usually on dry not strongly calcareous soils; ascending to 1800 ft. Common in Great Britain, local but widespread in Ireland. Channel Is. 110, H40, S. W. Europe from Norway (but not Sweden) to Portugal and C. Spain, N. Italy, Croatia and Germany.

26. AJUGA L.

Herbs with 2–many-fld whorls, sometimes forming a terminal infl. Bracts differentiated or not. Bracteoles small or 0. Calyx tubular-campanulate, nearly equally 5-toothed. *Corolla with a very short upper lip and conspicuous 3-lobed lower lip*; tube ± exserted, with a ring of hairs within. Nutlets obovoid, reticulate.

About 45 spp., in the Old World.

1	Lvs divided into 3 linear lobes; corolla yellow.	**1. chamaepitys**
	Lvs not divided; entire or toothed; corolla blue or violet.	2
2	With aerial stolons; stem usually hairy on 2 opposite sides.	**2. reptans**
	Without aerial stolons; stem usually hairy all round.	3
3	Upper bracts shorter than the fls; radical lvs withering before fl. time.	
		3. genevensis
	Upper bracts longer than the fls; radical lvs green at fl. time.	
		4. pyramidalis

Section 1. *Chamaepitys* (Hill) Benth. Whorls 2-fld, axillary, very rarely 4-fld. Corolla often yellow.

1. A. chamaepitys (L.) Schreb.   Ground-pine.

Hairy annual 5–20 cm., smelling of pine when crushed, branched below, branches ascending. *Lf-blades 2–4 cm., divided into 3 linear, obtuse lobes*, attenuate into a petiole; radical lvs withering early, entire or toothed. Bracts not differentiated. Whorls 2-fld, axillary, many, much shorter than the lvs. Calyx c. 8 mm., campanulate, hairy. *Corolla yellow*, lower lip red-spotted; tube included. Fl. 5–9. Visited by bees. $2n = 28^*$. Hs.

Native. Very local in chalky arable fields and open habitats in chalk grass-land in S.E. England, extending to Hants, Bedford and Cambridge. 9. C. and S. Europe, Orient, N. Africa.

Section 2. *Ajuga.* Whorls 6–many-fld, forming a terminal spike. Fls never yellow.

2. A. reptans L. Bugle.

Perennial 10–30 cm., with short rhizome and *long lfy and rooting stolons.* Stems simple, ± hairy, often on two opposite sides only. Radical *lvs* forming a rosette; blades 4–7 cm., *glabrescent*, obovate or oblong, *entire to obscurely crenate*, obtuse, attenuate at base into a long petiole; cauline shorter, sub-sessile, few. Upper bracts shorter than their fls, ± blue-tinted, ovate, entire. Calyx c. 5–6 mm., campanulate, teeth rather shorter than the tube. *Corolla blue*, rarely pink or white, tube exserted. Fl. 5–7. Pollinated by bees, self-pollination possible; usually homogamous. $2n = 32*$. Hs.

Native. Common in woods, usually damp, and in damp meadows and pastures; ascending to 2200 ft. Throughout the British Is. 112, H40, S. Europe from Scandinavia and N. Russia to C. Portugal, Sicily, Greece and the Caucasus. S.W. Asia, E. Algeria, Tunisia.

*3. A. genevensis L.

Perennial 1–30 cm., *creeping by underground stolons* and emitting fl. stems at intervals, *without aerial stolons.* Stems simple, with long white hairs on two opposite sides or all round. Radical *lf-blades* 5–12 cm., *hairy*, obovate, *crenate or dentate*, obtuse, attenuate into a *long petiole, withering before flowering time*; cauline oblong, scarcely shorter, shortly petioled. *Upper bracts shorter than their fls*, usually dentate, ± blue-tinted. Calyx 4–5 mm., campanulate, teeth about equalling the tube. *Corolla bright blue.* Fl. 5–7. Pollinated by bees, self-pollination possible; protandrous. $2n = 32$. Hp.

Introduced. Chalk pasture in Berks where it was first found in 1918; dunes near Hayle (Cornwall). Europe from S. Sweden to France, Italy and Mace-donia and the Caucasus, S.W. Asia.

4. A. pyramidalis L. 'Pyramidal Bugle.'

Perennial 10–30 cm., with *short rhizome, without stolons.* Stems simple, hairy all round. Radical *lvs* hairy or glabrescent, obovate, obtuse, obscurely crenate, attenuate at base into a *short petiole, persistent at fl. time*; cauline obovate-oblong, very shortly petioled or sessile, much shorter than the radical. *Bracts all much longer than fls*, purple-tinted; entire, deep violet or purple. Calyx c. 8 mm., teeth longer than the tube. *Corolla pale violet-blue.* Fl. 5–7. Pollinated by bees, self-pollination possible; protandrous or homogamous. $2n = 32$. Hs.

Native. Crevices of usually basic rocks; ascending to 1750 ft. Westmor-land, Dumfries; Inverness and Argyll to Orkney and the Hebrides; Clare, Galway. 15, H2. N. Europe, mountains of C. Europe (rare in the plains) to C. Spain, Italian Alps, Montenegro and Bulgaria.

A. pyramidalis × reptans (*A. × hampeana* Braun & Vatke) is found in Clare and Sutherland.

112. PLANTAGINACEAE

Annual or perennial, usually scapigerous herbs. Lvs usually all radical and spirally arranged, rarely cauline and spiral or opposite. Scapes axillary. Fls bracteate, small, usually in racemose heads or spikes, actinomorphic, hermaphrodite or rarely unisexual, usually 4-merous. Perianth green or ± scarious, lobes imbricate. Sepals persistent, fused at base, the lower sometimes ± united. Corolla gamopetalous, scarious. Stamens inserted on the corolla-tube, rarely hypogynous, filaments usually long, anthers large. Ovary superior, 1–4-celled; ovules 1–many in each cell, axile or basal. Fr. a circumscissile capsule or 1-seeded, hard and indehiscent. Seeds often mucilaginous when wet.

Three genera and about 200 spp., cosmopolitan, but mainly in the temperate regions.

Terrestrial; stolons 0; fls hermaphrodite, many together; fr. dehiscent.
                                                                    1. PLANTAGO
Aquatic; stolons present; fls unisexual, male solitary, female few; fr.
  indehiscent.   2. LITTORELLA

1. PLANTAGO L.   Plantain.

Terrestrial herbs. Fls 4-merous, in heads or cylindrical spikes, mostly hermaphrodite. Stamens inserted on the corolla-tube. Ovary 2–4-celled; ovules 2–many. Capsule circumscissile. Fls mostly protogynous and wind pollinated.

About 200 spp., in the temperate regions of both hemispheres, a few in the tropics.

1 Stems long; lvs opposite; lower bracts with lf-like tips.   **6. indica**
  Stems short; lvs spirally arranged, forming a radical rosette; bracts not
    normally lf-like.   2
2 Corolla-tube glabrous; lvs not linear or pinnatifid.   3
  Corolla-tube pubescent; lvs linear or pinnatifid.   5
3 Scape deeply furrowed; corolla-lobes with prominent brown midrib.
                                                                   3. lanceolata
  Scape not furrowed; corolla-lobes without a midrib.   4
4 Lvs abruptly contracted at base, petiole usually as long as blade; scape
    scarcely exceeding lvs; capsule 8–16-seeded.   **1. major**
  Lvs gradually narrowed at base, petiole much shorter than blade; scape
    much exceeding lvs; capsule c. 4-seeded.   **2. media**
5 Lvs 3–5(–7)-nerved, never pinnatifid; bracts obtuse or subacute; corolla-
    lobes with a brown midrib; capsule 2-celled, 2-seeded.   **4. maritima**
  Lvs 1-nerved, often pinnatifid; bracts often acuminate; corolla-lobes
    without a midrib; capsule 3–4-celled, 3–4-seeded.   **5. coronopus**

Subgenus 1. PLANTAGO.

Lvs spirally arranged, usually in a radical rosette.

1. P. major L.   'Great Plantain.'

A stout glabrous or pubescent perennial. *Lvs* 10–15(–30) cm., ovate or elliptic, entire or irregularly toothed, *abruptly narrowed into the petiole; petiole usually about as long as blade.* Infl. (1–)10–15(–50) cm.; *scape scarcely exceeding lvs*, not furrowed. Bracts acute, brownish with a green keel. Fls c. 3 mm., corolla yellowish-white, lobes triangular. Anthers at first lilac, later dirty yellow. *Fr.* c. 5 mm., 2-celled, 8–16-*seeded.* Fl. 5–9. Wind pollinated. $2n=12$. Hr. Variable.

Native. In farmyards, by roads, in cultivated ground and rarely in grassy places but always in open habitats. 112, H40, S. Generally distributed throughout the British Is. Europe; N. and C. Asia; naturalized throughout most of the world.

2. P. media L.   'Hoary Plantain.'

A finely pubescent perennial. *Lvs* 4–6(–30) cm., elliptic to ovate, weakly and irregularly toothed, 5–9-ribbed, gradually narrowed into a *very short petiole or subsessile.* Infl. 2–6(–8) cm.; *scape much exceeding lvs*, often 30 cm., not furrowed. Bracts acute, membranous at margins. Fls c. 2 mm., scented, corolla whitish, lobes lanceolate, obtuse. *Filaments purple*, anthers lilac or white. *Fr.* c. 4 mm., 2-celled, *usually* 4-*seeded.* Fl. 5–8. Insect-pollinated. $2n=24$. Hr.

Native. In grassy places on neutral and basic soils. Fairly generally distributed in S. England and the Midlands, becoming rarer northwards and often introduced in Scotland; introduced in Ireland. 90, H19. S. Europe; temperate Asia.

3. P. lanceolata L.   Ribwort.

A glabrous or somewhat pubescent perennial. Stem with long silky hairs. Lvs (2–)10–15(–30) cm., usually lanceolate or ovate-lanceolate, entire or weakly and distantly toothed, 3–5-nerved, gradually narrowed into a petiole, rarely subsessile; petiole usually about half as long as blade. Infl. 1–2(–5) cm.; *scape deeply furrowed*, much exceeding lvs, up to 45 cm. *Bracts ovate-acuminate*, points scarious. Fls c. 4 mm., corolla brownish, *lobes ovate, acute, with a prominent brown midrib* reaching the tip. Stamens white. Fr. c. 5 mm., 2-celled, 2-seeded. Fl. 4–8. Wind-pollinated. $2n=12$. Hr.

Native. In grassy places on neutral and basic soils. Generally distributed throughout the British Is. 112, H40, S. Europe north to Iceland; N. and C. Asia; introduced in most other temperate countries, less frequent in the tropics.

4. P. maritima L.   'Sea Plantain.'

A glabrous or rarely hairy perennial. Stem stout, woody, sometimes with long silky hairs. *Lvs* 5–15(–30) cm., *narrow-linear*, fleshy, entire or slightly toothed, *faintly* 3–5-*nerved*, rarely up to 15 mm. broad and 7-nerved. Infl. 2–6(–12) cm.; scape equalling or exceeding lvs, not furrowed. *Bracts ovate, obtuse or subacute*, appressed. Fls c. 3 mm.; 2 *lower sepals united, not winged; corolla* brownish, *lobes* ovate, acute, *with a rather broad indistinct brown midrib* reaching the tip. Stamens pale yellow. *Fr.* c. 4 mm., 2-*celled*, 2-*seeded; seeds slightly winged at one or both ends.* Fl. 6–8. Wind-pollinated. $2n=12$*. Hr.

One of the alpine forms is sometimes considered to be a distinct sp., *P. hudsoniana* Druce.

Native. In salt marshes, in short turf near the sea and beside streams on mountains. 86, H29, S. Round most of the coasts of the British Is. and on the Snowdon range, the Pennines and the higher mountains of Scotland; inland on limestone in W. Ireland. Western Europe from N. Scandinavia to C. Spain and eastwards to Hungary and Russia, rarely inland.

5. P. coronopus L.   Buck's-horn Plantain.

A pubescent or occasionally glabrous biennial or perhaps sometimes annual or perennial. Lvs usually 2–6 cm., very variable, narrow linear, nearly entire, toothed or (*most often*) *1–2-pinnatifid*, 1(–3)-*nerved*. Infl. 0·5–4 cm.; scape somewhat longer than lvs, often curved below, not furrowed. *Bracts ovate, often long-acuminate with spreading points*, sometimes obtuse and appressed. Fls c. 3 mm.; 2 *lower sepals united and winged; corolla* brownish, *lobes* ovate, acute or acuminate, *without a midrib*. Stamens pale yellow. *Fr.* c. 4 mm., 3–4-*celled*, 3–4-*seeded*. Fl. 5–7. Wind pollinated. $2n=10*$. Hr.

Native. In dry and ± open habitats on sandy and gravelly soils and in cracks in rocks, most common near the sea. 100, H27, S. Through most of Great Britain in suitable habitats but absent from some inland counties; strictly maritime in Ireland. Coasts of C. and S. Europe from S. Sweden, N. Africa, W. Asia, Azores; introduced in N. America, Australia, New Zealand.

Subgenus 2. PSYLLIUM (Mill.) Harms.

Lvs opposite; stem long, lfy, often branched.

*6. P. indica L.

P. Psyllium L., nom. ambig.; *P. ramosa* Aschers.; *P. arenaria* Waldst. & Kit.

An erect or spreading ± pubescent *usually much-branched* annual 15–30 cm. Lvs up to c. 10 cm., narrow-linear, entire or obscurely toothed, *the lower with short very lfy shoots in their axils*. Infl. c. 1 cm.; scape slender, exceeding the subtending lf. *Lower bracts ending in lf-like points*, upper rounded, margins hyaline. Fls c. 4 mm.; corolla brownish-white. Fl. 7–8. $2n=12$. Th.

Introduced. In disturbed ground and sometimes on dunes, widely distributed but infrequently as a casual, ± naturalized in a few localities. S. and C. Europe, S.W. Asia.

2. LITTORELLA Berg.

Perennial scapigerous *aquatic herbs. Fls monoecious. Male fls* 4-merous, *solitary on short scapes; stamens hypogynous*, rudimentary ovary small. Female fls 3–4-merous, solitary or few at base of male scape; *ovary 1-celled*; style long, rigid; ovule 1, rarely 2, erect, campylotropous. *Fr. indehiscent, hard*.

Two spp., one in Europe (excluding the Mediterranean region), and the Azores; the other in temperate S. America.

1. L. uniflora (L.) Aschers.   Shore-weed.

L. lacustris L.

An abundantly stoloniferous perennial often forming an extensive turf in shallow water. Stolons slender, far-creeping, rooting and producing rosettes of lvs at the nodes. Lvs 2–10(–25) cm., ½-cylindrical and linear-subulate or sometimes flattened and broader, sheathing at base, not septate. Scape shorter than lvs, rarely equalling them, slender, bracteate below the middle. Male fl. 5–6 mm.; stamens 1–2 cm. Female fls 4–5 mm., subsessile; style c. 1 cm. Fl. 6–8. Wind-pollinated. $2n=24*$. Hyd. Fls are produced only when the plant is exposed.

Native. In shallow water down to about 4 m. or just exposed on sandy and gravelly shores of non-calcareous lakes and ponds. In suitable habitats throughout the British Is., but commoner in the north. 97, H36, S. C. and N. Europe to c. 68° N. in Norway; Azores.

113. CAMPANULACEAE

Mostly herbs, nearly always with latex. Lvs usually alternate, simple, exstipulate. Fls hermaphrodite, actinomorphic, often showy. Calyx-tube adnate to ovary. Corolla gamopetalous, campanulate or sometimes with a very short tube; lobes valvate. Stamens as many as corolla-lobes and alternate with them, free, inserted towards base of corolla or on the disk. Ovary inferior or rarely superior, 2–10-celled; placentation usually axile. Ovules numerous, rarely few. Fr. a capsule or fleshy.

About 35 genera and 700 spp., throughout most of the world.

Certain genera (e.g. *Phyteuma* and *Jasione*) approach the Compositae in having numerous small fls arranged in compact bracteate heads, and in their pollination mechanism.

1 Fls solitary or in racemes or panicles; corolla-lobes usually shorter than tube, ovate or broadly triangular; style not or only shortly exserted.   2
    Fls capitate or spicate, very numerous; corolla-lobes much longer than tube, linear; style long-exserted.   4

2 Stem creeping, slender; lvs all similar, long-petioled.
                                          1. WAHLENBERGIA
    Stem erect or ascending; upper lvs much smaller or narrower than basal, sessile or nearly so.   3

3 Perennials or biennials; ovary and capsule ovoid or subglobose.
                                      2. CAMPANULA
    Annual; ovary and capsule subcylindrical.   3. LEGOUSIA

4 Plant glabrous or very nearly so; fl. buds curved; stigmas linear (on chalk).   4. PHYTEUMA
    Plant pubescent or hispid; fl. buds straight; stigmas short, stout (on lime-free soils).   5. JASIONE

1. WAHLENBERGIA Schrad. ex Roth

Annual or perennial herbs of varied habit. Fls 5-merous, usually blue and nodding. *Calyx-tube hemispherical or oblong-obconic.* Corolla campanulate or subrotate. Anthers free. Ovary 2–5-celled; stigmas 2–5, filiform. *Capsule dehiscing by* 2–5 *apical loculicidal valves*, alternating with the persistent calyx-teeth.

About 80 spp., mostly south temperate particularly in S. Africa, a few in tropical America and the temperate regions of the Old World.

1. W. hederacea (L.) Rchb.   'Ivy Campanula.'

Campanula hederacea L.

A slender glabrous creeping perennial. Stems up to c. 30 cm., little-branched, weak. Lvs all petioled, upper often subopposite; blade 5–10 mm., suborbicular, angled or obscurely lobed, ± cordate. Peduncles up to c. 4 cm., much longer than petioles, filiform, 1-fld. Fls ± nodding. Calyx 2–3 mm., teeth subulate, erect, much longer than tube. Corolla 6–10 mm., campanulate, pale blue; lobes c. ½ as long as tube, ovate ± acute. Capsule c. 3 mm., turbinate, erect. Fl. 7–8. H. or perhaps Chh.

Native. In damp acid peaty places on moors, heaths and by streams. Cornwall to W. Kent and S. Essex, Berks, and west of a line from Gloucester to Durham; west coast of Scotland from Kirkcudbright to Argyll; S. Ireland north to Dublin and Mayo; local everywhere and rare except from Cornwall to Somerset, Pembroke to Glamorgan, Caernarvon, Isle of Man, Wexford and S. Kerry. 49, H 10. W. Europe from Denmark to Spain and Portugal; Dalmatia.

2. CAMPANULA L.   'Bellflower.'

Herbs of varied habit, usually perennial, rarely annual or biennial. Fls 5-merous, blue or purplish, rarely white, usually in racemes or panicles, sometimes cymose. *Calyx-tube ovoid or subglobose,* lobes flat or folded at the sinus. Corolla rotate or campanulate. Anthers free. Ovary 3–5-celled; *style clavate, hairy opposite the anther cells,* stigmas 3–5, filiform. *Capsule* ovoid or turbinate, 3–5-celled, *dehiscing by lateral pores or valves.*

About 300 spp., in north temperate regions and on mountains in the tropics. Several are cultivated.

The fls are strongly protandrous, the pollen being shed in bud and deposited on the hairs of the style. As the fl. opens the stamens wither and the style presents the pollen to insects which come for the nectar, which is protected by the persistent triangular bases of the stamens. The stigmas eventually separate and finally curl right back so that, if cross-pollination fails, self-pollination occurs (cf. Compositae).

1	Hispid biennial; calyx with broad-cordate, reflexed appendages between the teeth; stigmas 5.	**9. medium**
	Calyx without appendages; stigmas 3.	2
2	Fls sessile.	**5. glomerata**
	Fls distinctly pedicelled.	3
3	Middle stem-lvs ovate (2–4 times as long as broad).	4
	Middle stem-lvs linear, linear-lanceolate or oblong.	6
4	Calyx-teeth spreading in fl.; fls nodding; plant with spreading roots producing numerous adventitious buds.	**3. rapunculoides**
	Calyx-teeth ± erect in fl.; fls erect or inclined; no root buds.	5
5	Stem obscurely and bluntly angled; radical lvs decurrent on petiole, rarely somewhat cordate, blade 10 cm. or more; stem-lvs sessile; corolla 40–55 mm.; plant softly hairy.	**1. latifolia**
	Stem sharply angled; radical lvs broadly ovate, deeply cordate, blade 10 cm. or more; stem-lvs petioled; corolla 25–35 mm.; plant ± hispid.	**2. trachelium**

6   Fls nodding; radical lvs orbicular, cordate; lower stem-lvs stalked.

                                          6. rotundifolia

  Fls erect or suberect; radical lvs not orbicular or cordate; lower stem-lvs
    sessile.   7

7   Plant glabrous; infl. 1–8-fld; corolla 25–35 mm.   **4. persicifolia**

  Plant pubescent or scabrid; infl. typically many-fld; corolla not more than
    20 mm.   8

8   Root slender, fibrous; radical lvs narrowed at base and decurrent on
    petiole; infl. with many long branches; bracteoles at middle of fl. stalks.

                                            7. patula

  Root swollen, fleshy; radical lvs abruptly contracted at base; infl. simple
    or nearly so; bracteoles at base of fl. stalks.   **8. rapunculus**

1. C. latifolia L.   'Large Campanula.'

A stout erect softly and sparsely pubescent perennial 50–120 cm. *Stem*
simple, *obscurely and bluntly angled.* Lvs (Fig. 51 A) ovate to ovate-oblong,

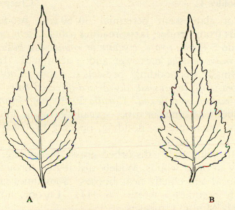

Fig. 51. Leaves of *Campanula*. A, *C. latifolia*; B, *C. trachelium.* × ⅓.

acuminate, rounded or narrowed at base, margins irregularly 1–2-serrate;
radical lvs petioled, blade 10–20 *cm.*, *usually decurrent on the petiole*, rarely
cordate at base; *upper lvs sessile.* Infl. a raceme, sometimes with short
branches below, lower bracts lf-like. Fls suberect or inclined; pedicels c. 2 cm.
Calyx-tube 5–7 mm., 5-ribbed, glabrous or nearly so; *teeth* 15–25 mm., narrowly triangular, acuminate, *erect. Corolla* 40–55 *mm.*, blue-purple or white;
lobes slightly shorter than tube, suberect, acute or acuminate, ciliate.
Stigmas 3. Capsule 12–15 mm., ovoid, nodding, opening by basal pores.
Fl. 7–8. $2n = 34$. Hs.

    Native. In woods and hedgebanks. Widely distributed but local in Great
Britain and much commoner in the north than the south. 82. Europe, to
c. 68° N. in Norway; W. Asia; Siberia.

2. C. trachelium L.   Bats-in-the-Belfry.

An erect ± *hispid* perennial 50–100 cm. *Stem simple, sharply angled. Radical lvs* long-petioled, blade *c.* 10 *cm., broadly ovate* or almost an equilateral triangle in outline, *deeply cordate* at base, tapering to a rounded or subacute apex, coarsely dentate, teeth rounded at apex and irregularly crenate-serrate; *stem-lvs* (Fig. 51 B) *shortly petioled*, smaller, ovate to ovate-oblong, ± acuminate, irregularly and coarsely 1–2-serrate, teeth obtuse. Infl. a lfy panicle with short branches bearing 1–4 fls. Bracteoles linear to linear-lanceolate. Fls suberect or inclined; pedicels up to c. 1 cm. Calyx-tube c. 3–5 mm., usually hispid; *teeth* c. 10 mm., triangular, acute, *erect. Corolla* 25–35 *mm.*, blue-purple; lobes shorter than tube, suberect, acute, ciliate or sparsely hispid. Stigmas 3. Capsule c. 7 mm., hemispherical, nodding, opening by basal pores. Fl. 7–9. $2n = 34$. Hs.

Native. In woods and hedge-banks, usually on clayey soils. Scattered throughout Great Britain, north to Fife, rather local; S.E. Ireland. 61, H5. Europe to S. Scandinavia; Siberia; N. Africa.

*3. C. rapunculoides L.   'Creeping Campanula.'

A puberulous or glabrescent perennial 30–60 cm., producing numerous adventitious buds from slender, far-spreading roots. *Stem subterete. Radical lvs* petioled, blade 5–8 *cm., ovate, cordate or rounded at base*, subacute, margins serrate; *stem-lvs sessile*, narrower and ± acuminate. Infl. a raceme or panicle, bracts small. Fls nodding; pedicels c. 5 mm. *Calyx-tube* 3–4 mm., *covered with short, stiff, appressed, downward-directed hairs; teeth* c. 8 mm., linear, acute, *spreading or reflexed. Corolla* 20–30 *mm.*, funnel-shaped, blue-purple; lobes about as long as tube, spreading, acute, ciliate. Stigmas 3. Capsule c. 7 mm., hemispheric, nodding, opening by basal pores. Fl. 7–9. $2n = 102$. Hs.

Introduced. In fields and ± disturbed grassy places (e.g. railway banks), and occasionally in open woods. Widely distributed throughout the British Is., but rather local and chiefly near houses. Sometimes grown in gardens where it speedily becomes a weed. 56, H11. Europe, north to c. 65° in Scandinavia; Asia Minor; Caucasus.

*C. lactiflora Bieb., a tall (up to 1·5 m.) perennial with large loose panicles and generally 3-fld peduncles, fls 25–35 mm. diam., and ovate hispid calyx-lobes, is established in S. Aberdeen. Caucasus, Armenia, Asia Minor and N. Persia.

*4. C. persicifolia L., *a glabrous perennial* with linear-lanceolate to subulate, finely and remotely crenate stem-lvs, is often cultivated and is naturalized in a few places. *Raceme few-* (1–8-)*fld.* Fls suberect. *Corolla* 25–35 *mm.* and about as broad as long, blue, rarely white; lobes broadly ovate, acute, about ½ as long as tube. Fl. 6–8. $2n = 16$. Hs.

Introduced. In open woods and on commons. S. Devon, Gloucester, Berks, well established; also recorded from several other English and Scottish counties. Europe, western and northern Asia.

5. C. glomerata L. 'Clustered Bellflower.'

An erect downy perennial 3–20(–60) cm. Radical lvs long-petioled, blade 2–4(–8) cm., ovate, obtuse, cordate or rounded at base, margin serrulate, teeth obtuse; stem-lvs sessile, ½-amplexicaul, or lower petioled, ovate to lanceolate, obtuse or subacute, rounded or narrowed at base, entire or serrulate. Infl. subcapitate, often with several ± distant fls or short few-fld branches below the terminal head. *Fls erect, sessile.* Calyx-tube c. 3 mm., obconical, 5-ribbed; teeth rather longer than tube, triangular. Corolla 15–20 mm., bright blue-purple, rarely white; lobes nearly as long as tube, suberect, eventually spreading, acute. Stigmas 3. Capsule c. 3 mm., erect, opening by apical pores. Fl. 5–9. 2*n*=30, 34, 68. Hs.

Native. In grassy places on calcareous soils, particularly in chalk grassland, less commonly on sea-cliffs or in woods. From Dorset and Kent to Cumberland and Kincardine, locally common. 54. Europe, temperate Asia.

6. C. rotundifolia L. Harebell; Bluebell (in Scotland).

A slender nearly or quite glabrous perennial 15–40(–60) cm., producing slender underground stolons. Stems decumbent at base. *Radical lvs* long-petioled, blade 5–15 mm., ovate or *suborbicular*, crenate, ± *cordate* at base; *lvs on decumbent part of stem petioled*, lanceolate, entire or serrate, acute; lvs on erect part of stem sessile, narrowly linear, entire, acute. Infl. a ± branched panicle, or reduced to a solitary terminal fl.; bracts small, linear. Buds erect, *fls nodding*, pedicels very slender. Calyx-tube c. 2 mm.; teeth c. 5 mm., setaceous, spreading. Corolla c. 15 mm., broadly campanulate, narrowed at base, blue, rarely white; lobes about ½ as long as tube, broadly ovate, sub-acute. Stigmas 3. Capsule c. 5 mm., subglobose, nodding, opening by basal pores. Fl. 7–9. 2*n*=34, 55–56, 68. Hs.

Native. In dry grassy places and on fixed dunes, often in poor shallow soils, locally common. Throughout the British Is., except for the Channel Is., Orkney and parts of C. and S. Ireland. 111, H28. North temperate regions to over 70° N. in Norway.

7. C. patula L.

A *scabrid* biennial or perennial 20–60 cm. *Root slender, fibrous.* Stems slender, angled. *Lower lvs* c. 4 cm., obovate-oblong, narrowed below into a short petiole and *decurrent* on it, apex obtuse, margins crenate; stem-lvs smaller and narrower, sessile, acute, margins serrate or nearly entire. *Infl.* cymose, *much-branched*, branches up to c. 20 cm., spreading. *Bracteoles inserted about the middle of the fl.-stalk.* Fls erect, stalks 2–5 cm., slender. Calyx-tube c. 4 mm., funnel-shaped; teeth c. 1 cm., triangular to setaceous. Corolla 15–20 mm., broadly campanulate, narrowed at base, purple; lobes about as long as tube, ovate, acute, spreading. Capsule c. 1 cm., obconical, erect, opening by pores at top; calyx-teeth erect in fr. Fl. 7–9. 2*n*=20*. Hs.

Native. In shady woods and hedge-banks. Scattered throughout England and Wales, from Dorset and Kent to Shropshire and Durham, local. 33. C. and N. Europe to Scandinavia but not Iceland.

*8. C. rapunculus L.   Rampion.

A biennial somewhat similar to *C. patula*, is naturalized in a number of places.
Root swollen, fleshy. Radical lvs abruptly contracted at base. Infl. simple
or with a few short branches in the lower part. Bracteoles at base of fl.-stalk.
Corolla-lobes about $\frac{1}{2}$ as long as tube. Fl. 7–8. $2n=20$. Hs.

Introduced. In fields and hedge-banks, usually on gravelly soils. Scattered
throughout England and S. Scotland, rare and local. 22, H1. Europe from
the Netherlands southwards.

*9. C. medium L.   Canterbury Bell.

A stout erect very hispid biennial, occurs sometimes as a casual. Fls suberect; pedicels short. Bracteoles as long as the calyx-teeth. *Calyx with large,
broad-cordate, reflexed appendages between the teeth.* Corolla 4–5 cm., inflated-
campanulate, dark violet-blue. *Stigmas* 5. Fl. 5–6. $2n=34$. Hs. Many
colour and other garden forms exist.

Introduced. Naturalized, particularly on railway banks in S.E. England
and the E. Midlands, and an impermanent casual elsewhere. N. and C. Italy,
naturalized elsewhere in Europe.

*C. alliariifolia Willd., a perennial with nodding creamy fls, appendages between
the calyx-teeth, and 3 stigmas, is naturalized in Cornwall. Native of Caucasus and
Asia Minor.

3. Legousia Durande

Characters of *Campanula* but ovary and capsule much elongated and the
corolla $\pm$ rotate.

About 10 spp., in north temperate regions and S. America.

1. L. hybrida (L.) Delarb.   Venus's Looking-glass.

Specularia hybrida (L.) A.DC.

An erect hispid annual 5–30 cm. Stems simple or $\pm$ branched. Lvs 1–3 cm.,
sessile, oblong or oblong-obovate, margin undulate. Fls erect, mostly in
terminal few-fld cymes. Calyx-teeth elliptic-lanceolate to linear-lanceolate,
acute or obtuse, c. $\frac{1}{2}$ as long as ovary. Corolla 8–15 mm. diam., scarcely $\frac{1}{2}$ as
long as calyx, reddish-purple or lilac. Capsule 15–30 mm., subcylindrical,
opening by valves just below the calyx-teeth. Fl. 5–8. $2n=20^*$. Th.

Native. In arable fields, locally common, becoming rarer in the north. 52.
Cornwall and Kent to Westmorland and Northumberland; absent from
Wales, Scotland and Ireland. C. and S. Europe, W. Asia, N. Africa.

4. Phyteuma L.

Perennial herbs. *Fls 5-merous, small, numerous, usually sessile in terminal
heads or dense spikes*, rarely shortly pedicelled and subumbellate. *Corolla
divided nearly to base, lobes linear*, at first cohering near the top to form a
tube, later spreading. *Anthers free.* Ovary 2–3-celled. *Stigmas 2–3, linear.
Capsule opening by lateral pores or small valves between the ribs.*

About 45 spp., in Europe (especially the Mediterranean region), and temperate Asia. Pollination mechanism similar to *Campanula* except that the

pollen is held in the tube formed by the corolla-lobes and pushed out by the elongating style.

Infl. capitate; fls violet.	**1. tenerum**
Infl. spicate; fls yellowish-white.	**2. spicatum**

1. P. tenerum R. Schulz   'Round-headed Rampion.'

P. orbiculare auct. angl., non L.

A glabrous or slightly hairy erect perennial 5–50 cm. Root with a deeply-buried fusiform enlargement. *Radical lvs* long-petioled, blade 2–4 cm., lanceolate to ovate-oblong, acute or obtuse, crenate, *narrowed, rounded or rarely subcordate* at base, lateral veins prominent when dry; stem-lvs much smaller, sessile, linear-lanceolate, few and scattered. *Infl.* 1–2 cm. diam., *depressed-globose.* Bracts narrowly triangular, much shorter than infl. *Corolla* c. 8 mm., *deep violet*, curved and cylindrical in bud; lobes narrow, eventually free nearly to base and spreading or reflexed. Style c. 1 cm.; stigmas usually 2. Fruiting head c. 2 cm., ovoid or cylindrical. Capsule c. 5 mm., ovoid, crowned by the short, stiff, triangular, erect calyx-teeth. Fl. 7–8. Hs.

Native. In chalk grassland, locally abundant. 10. Dorset and Wilts to Kent. C. Europe (not Scandinavia).

2. P. spicatum L.

A robust glabrous perennial 30–80 cm. Root swollen, fleshy, fusiform. *Radical lvs* long-petioled, blade 3–5 cm., ovate, obtuse, crenate or serrate, *deeply cordate at base*; lower stem-lvs petioled, as large as or even larger than radical, upper smaller, sessile, lanceolate to linear. *Infl.* 3–8 cm., *cylindrical.* Bracts subulate, about as long as diam. of infl. *Corolla* c. 1 cm., *yellowish-white,* curved and cylindrical in bud; lobes eventually separating nearly to base. Style c. 1 cm.; stigmas usually 2. Fruiting head 5–12 cm., cylindrical. Capsule similar to that of *P. tenerum,* but calyx-teeth subulate, spreading. Fl. 7–8. $2n = 36$. Hs.

Native. In woods and thickets, very local. E. Sussex. 1. C. and S. Europe, except the Mediterranean region.

5. JASIONE L.

Perennial, biennial, or perhaps sometimes annual herbs. *Fls small, numerous, sessile or subsessile in a terminal head.* Calyx-tube ovoid or turbinate, 5-partite. *Corolla* blue, rarely white, *5-partite nearly to base, lobes narrow,* spreading. *Anthers shortly connate at base.* Ovary 2-celled. *Stigmas 2, short, stout. Capsule dehiscing loculicidally by 2 short valves within the persistent calyx-teeth.*

About 5 spp., in Europe and the Mediterranean region. Pollination mechanism similar to *Phyteuma* except that the anthers cohere and form a tube while the corolla-lobes spread out soon after the fl. opens. (Cf. Compositae.)

1. J. montana L.   Sheep's-bit.

A pubescent spreading or ascending herb, usually biennial but probably sometimes perennial or annual. Stems 5–50 cm., usually decumbent at base, stout or slender, simple or branched, naked in the upper half. Lvs up to

c. 5 cm., linear-oblong to linear-lanceolate, obtuse or subacute, undulate or
crenate, ciliate, radical narrowed into a short petiole, cauline sessile. Infl.
5–35 mm. diam., depressed-globose. Bracts numerous, shorter than infl.,
triangular-acute to ovate-cuspidate. Corolla c. 5 mm., persistent, blue, rarely
white. Calyx-teeth subulate, about as long as the unopened corolla. Fl. 5–8.
$2n = 12, 14$. Hs.

Native. In grassy places on light sandy or stony lime-free soils, in rough
pastures, on heaths, cliffs and banks. 85, H31, S. Scattered throughout the
British Is., generally local, but abundant in some areas (e.g. Cornwall and
Shetland). Distribution of the genus.

114. LOBELIACEAE

Herbs or rarely small trees, often with latex. Lvs alternate, simple, exstipulate.
Fls hermaphrodite or rarely unisexual, zygomorphic. Calyx-tube adnate to
ovary. Corolla gamopetalous, 1–2-lipped, lobes valvate. Stamens 5, alternate
with the corolla-lobes and inserted on the corolla or free; anthers cohering
in a tube round the style. Ovary ± inferior, 2–3-celled. Style simple, surrounded by a ring of hairs; stigma capitate or 2-lobed. Ovules numerous,
placentation axile. Fr. fleshy or a capsule.

About 25 genera and 400 spp., mainly in the tropics and subtropics.

1. LOBELIA L.

Herbs or rarely shrubs. Fls solitary in the axils of lvs or arranged in terminal
bracteate racemes. Corolla-tube oblique and curved, split to base along the
back, 2-lipped, rarely with subequal connivent lobes. Stamens not epipetalous;
2 or all of the anthers bearded at the tip. Stigma shortly 2-lobed. Capsule
dehiscing loculicidally by 2 valves within the calyx-teeth.

About 250 spp., in all the warm and temperate parts of the world except
C. and E. Europe and W. Asia.

A number of spp. are cultivated for ornament, *L. erinus* L., the annual blue-fld sp.
used for bedding-out, being the commonest.

Terrestrial; stems lfy; lvs toothed; fls erect or spreading, blue.   **1. urens**
Aquatic; stems lfless except for a few small scales; lvs entire; fls nodding,
    pale lilac.   **2. dortmanna**

1. L. urens L.   'Acrid Lobelia.'
A nearly or quite glabrous erect perennial 20–60 cm. Juice very acrid. *Stems*
slender, solid, angular, *lfy*. Lvs up to c. 7 cm., obovate or obovate-oblong,
upper linear-oblong, all acute or obtuse, irregularly *serrate*. Raceme up to
c. 20 cm., lax. *Fls erect or spreading*, shortly pedicelled, bracts linear, ± equalling calyx. Calyx c. 8 mm., teeth rather longer than or about equalling the
narrow obconic tube, subulate, spreading. *Corolla* 10–15 mm., *blue* or purplish. Anthers shortly exserted. Capsule erect. Fl. 8–9. $2n = 14$. Hs.

Native. In rough pastures, on grassy heaths and at margins of woods, on
damp acid soils, very local but apparently increasing. 6. E. Cornwall to
S. Hants, E. Sussex, Hereford. France, Spain, Portugal, Madeira, Azores.

2. L. dortmanna L. 'Water Lobelia.'

A glabrous erect stoloniferous perennial 20–60 cm. *Stems* slender, terete, fistular, *lfless* except for a few small scales. *Lvs* 2–4(–8) cm., linear, obtuse, ± recurved, quite *entire*, submerged. Racemes up to c. 10 cm., very lax and few-fld, emersed. *Fls nodding*, pedicels up to c. 1 cm.; bracts oblong, obtuse, much shorter than pedicels. Calyx c. 5 mm., teeth shorter than tube, ovate, obtuse, erect. *Corolla* 15–20 mm., *pale lilac*. Anthers included. Capsule nodding. Fl. 7–8. $2n = 16$. Hyd.

Native. In stony lakes and tarns with acid water. 45, H22. Locally common in Wales, the Lake District and most of Scotland and Ireland. Western Europe from Brittany northwards to c. 68° in Scandinavia.

115. RUBIACEAE

Woody plants or herbs with opposite and decussate or whorled simple entire lvs whose stipules may stand between the lvs (interpetiolar) or between lf and stem (intrapetiolar). Fls usually small in terminal and axillary cymes, sometimes in heads; usually hermaphrodite, actinomorphic, epigynous, 4–5-merous. Sepals usually free, often very small or represented only by an annular ridge; corolla funnel-shaped or rotate with the petals joined below into a longer or shorter tube; stamens isomerous, epipetalous, alternating with the petals; style simple or bifid; ovary inferior, usually 2-celled, with 1–many anatropous ovules on an axile placenta in each cell. Fr. a capsule, berry or drupe, or dry and schizocarpic; seeds endospermic with a usually straight embryo.

About 5500 spp. and 450 genera, cosmopolitan but chiefly tropical.

The British representatives all belong to the tribe Galieae with 2 or more lf-like interpetiolar stipules so that there appear to be 4 or more lvs in a whorl, of which, however, only 2 have axillary buds. The fls are very small and the fr. is usually of 2 separating indehiscent 1-seeded parts.

1   Fls in small terminal heads with a basal involucre of lf-like bracts; corolla-tube at least twice as long as the free lobes.   2
    Fls not in involucrate heads; corolla-tube not twice as long as the free lobes, often much shorter.   3

2   Calyx of 4(–6) distinct teeth, persistent in fr.; corolla pinkish-lilac.
                                  1. SHERARDIA
    Calyx an inconspicuous ridge; corolla whitish or bright blue.
                                  2. ASPERULA

3   Corolla pinkish, with tube at least as long as the free lobes.   2. ASPERULA
    Corolla white, yellow or greenish, with tube usually shorter than the free lobes, often very short.   4

4   Corolla with 4(–3) free lobes; fr. dry.   3. GALIUM
    Corolla with 5 free lobes; fr. a berry.   4. RUBIA

1. SHERARDIA L.

Lvs (and lf-like stipules) 4–6 in a whorl. *Fls few in small terminal head-like clusters with a basal involucre of lf-like bracts. Calyx with 4–6 distinct sepals; corolla funnel-shaped* with 4 lobes; stamens 4; style bifid with capitate stigmas.

One sp.

1. S. arvensis L. Field Madder.

An annual herb with slender reddish roots and numerous prostrate or decumbent spreading stems 5–40 cm., simple or branched, ± glabrous, with 4 rough angles. Lower lvs 4 in a whorl, obovate-cuspidate, soon withering, upper 5–18 mm., 5–6 in a whorl, elliptical-acute, ± glabrous but the margins and underside of the midrib scabrid with forwardly directed prickles. Fls 3 mm. diam., subsessile, 4–8 in terminal heads. Involucre of 8–10 lanceolate lf-like bracts longer than the fls. *Sepals* 4–6, green, triangular-lanceolate, at first small but persistent and usually *enlarging in fr*. Corolla 4–5 mm., pale lilac, funnel-shaped, with a long slender tube about twice as long as the 4 lobes. Fr. 4 mm., crowned by the sepals, of 2 obovoid mericarps rough with short appressed bristles. Fl. 5–10. Said to be gynomonoecious or -dioecious in Continental Europe. Visited chiefly by flies. $2n = 22$. Th.

Native. Arable fields and waste places, up to 1200 ft. in Scotland. Common throughout the British Is. 112, H40, S. Probably originally native in the Mediterranean region but now established throughout Europe and in extra-tropical regions all over the world.

2. ASPERULA L.

Usually perennial herbs with lvs and lf-like stipules in whorls. Fls in heads or panicles, 4-merous. *Calyx an indistinct annular ridge* or of 4–5 small teeth, not persistent in fr.; *corolla funnel-shaped*, its tube longer than the 4(–3) lobes; styles ± connate; stigmas capitate. Fr. of two 1-seeded dry mericarps, with the testa of the seed adherent to the fr. wall.

About 90 spp. in Europe, Asia and Australia.

Upper lvs ovate-lanceolate; fls in involucrate heads, white tinged with
   yellowish-pink. **1. taurina**
Upper lvs linear; fls pink, not in involucrate heads. **2. cynanchica**

*1. A. taurina L. 'Pink Woodruff.'

A perennial herb with far-creeping reddish rhizome and an erect or ascending, usually simple, 4-angled stem, 15–60 cm., sparsely hairy with a hairy ring beneath each node. Lvs 4–5 × 1–2 cm., ovate-lanceolate, acute, sparsely pubescent and ciliate. *Fls* 5 mm. diam., andromonoecious, *in a* terminal *head-like corymb* with an *involucre of bracts which exceed the fls*. Corolla 10 mm., white tinged with yellowish-pink, with the slender tube almost 3 times as long as the 4 lobes. Stamens exserted, with violet anthers. *Fr.* 3 mm., brownish, *glabrous*, roughly punctate. Fl. 5–6. Protandrous. Visited by flies and bees. $2n = 22$. Grh.–Hp.

Introduced. Waste places. Established in a few localities in Leicestershire, Westmorland, etc. S. Europe from N. Spain to Italy, with a different ssp. in S.E. Europe.

2. A. cynanchica L. Squinancy Wort.

A perennial herb with a woody branching non-creeping stock producing numerous slender prostrate or ascending shoots 8–40 cm., much branched, 4-angled, glabrous. Lower lvs elliptical to obovate, 4 in a whorl, withering early; middle and upper *lvs* 6–25 mm., *linear*, 4(–6) in a whorl, often very unequal; all mucronate, glabrous, firm, recurved. Fls 3–4 mm. diam., in lax, long-stalked, few-fld, terminal and axillary corymbs. Bracts lanceolate, mucronate. *Corolla* c. 6 mm., funnel-shaped, the 4(–3), acute lobes almost as long as the tube, *white within, pale pink-lilac and rough on the outside.* Fr. 3 mm., densely tubercled and somewhat wrinkled. Fl. 6–7. Gynodioecious; homogamous. Vanilla-scented and visited by various small insects. $2n=44$. Hp.

Native. Dry calcareous pastures and calcareous dunes, to 1000 ft. in Ireland. 40, H8, S. Locally abundant in S. England and S. Wales and extending northwards to S.E. Yorks and Westmorland; S.W. Ireland to Waterford and Galway. C. and S.E. Europe; Caucasus.

*A. arvensis L., an annual herb with an erect 4-angled glabrous branching stem 8–30 cm., bearing persistent cotyledons and distant whorls of linear blunt ciliate lvs, 12–25 cm., 6–9 in a whorl; bright blue fls, 4 mm. diam., in involucrate sessile terminal heads, the ciliate bracts exceeding the fls; and brown glabrous ± smooth fr. 2 mm., has occurred as a casual in arable land and waste places in various localities in England. S. and C. Europe; N. Africa; Asia Minor.

*Phuopsis stylosa (Trin.) B. D. Jackson (*Crucianella stylosa* Trin.) is much grown in gardens and frequently escapes. It is a procumbent annual (rarely longer-lived) 15–25 cm., with lanceolate hispid lvs, 12–18 mm., 8–9 in a whorl, and small 5-merous deep pink fls in rounded terminal heads, 10–15 mm. diam.; styles long-exserted, 2-cleft; ovules 1 per cell. Persia.

3. GALIUM L.

Annual to perennial herbs with whorls of 4–10 lvs and lf-like stipules. Fls in terminal and axillary cymes, rarely solitary; small, hermaphrodite or polygamous. Calyx a minute annular ridge; corolla usually rotate with a very short tube and 4 (3–5) lobes, sometimes funnel-shaped with the tube almost equalling the lobes; stamens 4, exserted; styles 2, short, connate below; stigmas capitate. Nectar-secreting and insect pollinated. Fr. didymous, of two 1-seeded mericarps, glabrous, hairy or with hooked bristles, with the testa of the seed adherent to the pericarp.

About 300 spp., almost cosmopolitan.

1	Lvs 3-veined, in whorls of 4.	2
	Lvs 1-veined, or with strong midrib and weaker diverging laterals.	3
2	Fls yellow; fr. smooth, glabrous.	**1. cruciata**
	Fls white; fr. rough with hooked hairs.	**3. boreale**
3	Corolla funnel-shaped, white, with tube almost equalling the lobes.	**2. odoratum**
	Corolla rotate, yellow or white, with lobes spreading horizontally from the top of a very short tube.	4
4	Lvs blunt or acute but never mucronate.	5
	Lvs cuspidate or mucronate.	6
5	Lvs linear-lanceolate to broadly oblanceolate; submerged autumn lvs similar to lvs on aerial shoots; infl. pyramidal or oblong, its branches soon spreading.	**9. palustre**
	Lvs narrowly linear; submerged lvs very narrow, flaccid, up to 2 cm.; infl. obconical, its branches erect-ascending throughout.	**10. debile**

6 Stems markedly rough with recurved prickles on the angles.   7
   Stems smooth or slightly rough on the angles.   11

7 A perennial plant of wet peaty places; fls white.   **11. uliginosum**
   Annual plants; fls green, cream or reddish, not pure white.   8

8 Fls in axillary 3-fld cymes shorter than the subtending lvs; fr.-stalks
      strongly recurved.   **12. tricornutum**
   Fls in axillary cymes longer than the subtending lvs; fr.-stalks straight.   9

9 Fls 2 mm. diam.; fr. 4–6 mm., covered with hooked hairs having swollen
      bases.   **13. aparine**
   Fls not more than 1 mm. diam.; fr. 1–3 mm., smooth or with hairs lacking
      swollen bases.   10

10 Lvs with forwardly-directed marginal prickles; fls 0·5 mm. diam., reddish
      outside; fr. 1 mm., glabrous, granulate.   **15. parisiense** ssp. **anglicum**
   Lvs with backwardly-directed marginal prickles; fls about 1 mm. diam.,
      greenish-white; fr. 1·5–3 mm., usually with white hooked hairs, rarely
      glabrous.   **14. spurium**

11 Fls yellow.   12
   Fls white.   13

12 Fls golden yellow.   **5. verum**
   Fls very pale yellow.   × **pomeranicum** (*mollugo* × *verum*)

13 Robust decumbent to erect plants with fls in large terminal pyramidal
      panicles; corolla-lobes long-cuspidate; fr. finely wrinkled.   **4. mollugo**
   Slender ± decumbent plants with fls in small axillary and terminal
      corymbs; corolla-lobes acute; fr. with small acute tubercles or low
      dome-shaped papillae.   14

14 Lvs of flowering shoots obovate to oblanceolate, their margins with
      forwardly-directed prickles.   **6. saxatile**
   Lvs on flowering shoots oblanceolate to linear, their margins with few
      to many backwardly-directed prickles.   15

15 Fr. covered with acute tubercles.   **8. sterneri**
   Fr. covered with low dome-shaped papillae.   **7. pumilum**

1. G. cruciata (L.) Scop.   Crosswort, Mugwort.

Valantia Cruciata L.; *Cruciata chersonensis* (Willd.) Ehrendorf.

A perennial herb with slender creeping stock and slender 4-angled hairy
decumbent lfy stems 15–70 cm., much branched near the base. *Lvs* up to
2·5 cm., largest in the middle of the stem, *ovate-elliptical*, *3-veined, hairy on
both sides*, 4 *in a whorl, yellowish-green*. Fls 2–2·5 mm. diam. in axillary,
about 8-fld cymes usually shorter than the subtending lvs. *Corolla pale yellow*
with 4 acute lobes. *Fr.* 1·5 mm., ± globose, glabrous, smooth, ultimately
blackish, *on recurved stalks*. Fl. 5–6. Andromonoecious, with the terminal
fls hermaphrodite, ± protandrous, and the laterals male; honey-scented and
visited by bees and flies. 2*n* = 22. Hp.

Native. Open woodlands and scrub, hedges, waysides and pastures
especially on calcareous soils, to 1550 ft. in Wales. 95, H1. Throughout
Great Britain northwards to Moray and Inverness; Inner Hebrides; Ireland
(introduced in Down). Europe northwards to the Netherlands, Brandenburg
and S. Poland; Caucasus; Asia Minor; Siberia.

Sometimes excluded from the genus *Galium* because of the very short axillary cymes and the andromonoecious fls.

2. G. odoratum (L.) Scop. Sweet Woodruff.

Asperula odorata L.

A perennial herb, *hay-scented* when dried, with slender branched far-creeping rhizomes and *erect simple* 4-angled *stems* 15–45 cm., hairy beneath the nodes, otherwise glabrous. Lvs firm, in distant whorls, lanceolate or elliptical, ±cuspidate, glabrous but with forwardly-directed marginal prickles; lowest lvs small, 6 in a whorl; middle lvs 2·5–4 × 0·6–1·5 cm., 6–8(–9) in a whorl. Fls c. 6 mm., diam., short-stalked, in long-stalked terminal and 1–2(–4) lateral cymes, together forming a ±umbellate infl. Bracts small, linear-lanceolate. *Corolla* 4–6 mm. *4-lobed to c. half-way*, pure white, the lobes blunt, slightly recurved, downy within. Fr. 2–3 mm., rough with hooked black-tipped bristles. Fl. 5–6. Homogamous. Fragrant and visited chiefly by flies and bees. $2n = 44$. Grh.–Hp.

Native. Locally abundant in woods on damp calcareous or base-rich soils, up to 2100 ft. in Scotland. 109, H40. Throughout the British Is., except the Outer Hebrides and Orkney. N. and C. Europe and montane woods in Italy and the Balkans; N. Africa; Siberia.

There is difficulty in drawing a line between *Galium* and *Asperula*. *G. odoratum* has funnel-shaped fls but with a corolla-tube hardly longer than the lobes. The closely related *G. glaucum* L., commonly placed in *Asperula*, has an even shorter corolla-tube and hybridizes with *G. verum*. It seems desirable to transfer both spp. to *Galium*, although they lack the rotate corolla typical of the genus.

3. G. boreale L. 'Northern Bedstraw.'

A perennial herb with creeping stock and erect, rigid, 4-angled, glabrous or pubescent stems 20–45 cm., with erect-ascending branches. *Lvs* 1–4 cm., lanceolate or elliptical, *3-veined, rough on the margins and the underside of the midrib, 4 in a whorl, bright green*, turning black when dried. Fls 3 mm. diam., in a ±pyramidal terminal lfy panicle whose ascending branches exceed their bracts. *Corolla white* with 4 apiculate lobes. *Fr.* 2·5 mm., olive-brown, *densely hispid with hooked bristles*. Fl. 7–8. Hermaphrodite, slightly protandrous. Visited by various small insects. $2n = 44$. Hp.

Native. Rocky slopes and stream-sides, moraine, scree and shingle, stable dunes, etc., to 3480 ft. in Scotland. Locally common in Wales and N. Britain northwards from Lancashire and Yorks; Inner Hebrides; doubtfully native in Orkney and Shetland; W. and N. Ireland, extending to Waterford, Westmeath and Down. 46, H24. N. and C. Europe southwards to N. Italy and Thrace; Caucasus; Asia Minor.

4. G. mollugo L. 'Hedge Bedstraw.'

G. elatum Thuill., non J. F. Gmel; incl. *G. erectum* Huds. 1778, non 1762.

Perennial herb with stout stock and decumbent to erect, glabrous or pubescent, 4-angled stems 25–120 cm., *not blackening when dried*. Lvs 8–25 mm., 6–8 in a whorl, linear to obovate, mucronate to cuspidate, 1-veined, glabrous or pubescent, rough on the *margins with stout forwardly-directed ±appressed prickles*. Fls in rather lax cymes forming a lax terminal panicle. *Corolla* *white*, its lobes cuspidate. Fr. glabrous, rugulose, blackening when dry; fr.-stalks divaricate. Hermaphrodite, protandrous. Visited chiefly by small flies. $2n = 44^*$; 22, 55, 66. Hp.

Very variable. The two following subspecies may be recognized, but intermediate forms are frequent and further work is needed.

Ssp. mollugo 'Great Hedge Bedstraw.'

Fl. stems weak, *decumbent or ascending, diffusely branched*, swollen below the nodes. *Lvs up to 25 mm., obovate, oblanceolate* or rarely linear, cuspidate. *Fls* 3 mm. diam., in a *panicle with spreading branches*. Fr. 1 mm.; *basal infl.-branches divaricate in fr.*

Native. Hedge-banks, open woodland, scrub and grassy slopes on base-rich soils, to 1200 ft. in Scotland. 95, H21, S. Common in the south and reaching Sutherland and Orkney; scattered throughout Ireland. Europe, except arctic Scandinavia and Russia; Caucasus, N. Asia.

Ssp. erectum Syme

Fl. stems ± erect with erect branches and linear-lanceolate mucronate *lvs*. Fls 4 mm. diam., in a *narrow panicle with ascending branches*. Fr. 1·5–2 mm.; *basal infl.-branches ascending in fr*. Fl. 6–7 and 9.

Native. Pastures, dry grassy slopes, waste land, etc., especially on calcareous soil; to 800 ft. in England. Throughout lowland Great Britain; Orkney; Ireland (except N.W.); Channel Is. 66, H16, S. S. Europe northwards to Sweden (introduced); N. Africa; Caucasus; Near East.

5. G. verum L. Lady's Bedstraw.

A perennial stoloniferous herb with a slender creeping stock and erect to decumbent, glabrous or sparsely pubescent, bluntly 4-angled stems 15–100 cm., with numerous ascending branches; *blackening when dried*. Lvs $6-25 \times 0.5-2$ mm., *linear*, mucronate, 1-veined, dark green and rough above, pale and pubescent beneath, *with revolute margins*, 8–12 in a whorl. Fls 2–4 mm. diam., in a terminal lfy compound panicle. *Corolla bright yellow*, its 4 lobes apiculate. Fr. 1·5 mm., smooth, glabrous, ultimately black. Fl. 7–8. Hermaphrodite, protandrous. Coumarin-scented and visited by various small insects, especially flies. $2n = 22, 44$. Hp.

Native. In grassland on all but the most acid soils, hedge-banks, stable dunes, etc., to 2150 ft. in Scotland. 112, H40, S. Abundant throughout the British Is. All Europe except Russia, north to Iceland; W. Asia. The fls are used for coagulating milk and the stolons yield a red dye.

The hybrid between *G. mollugo* and *G. verum*, **G. × pomeranicum** Retz. (*G. ochroleucum* Wolf), has pale yellow fls and remains green when dried.

6. G. saxatile L. 'Heath Bedstraw.'

G. harcynicum Weigel; *G. hercynicum* auct.; *G. montanum* Huds.

A perennial mat-forming herb with numerous prostrate non-flowering branches and decumbent or ascending flowering shoots (5–)10–20(–30) cm., 4-angled, glabrous, smooth, much branched. *Plant turning black when dried*. Lvs (5–)7–10(–11) mm., 6–8 in a whorl, obovate on non-flowering and obovate to oblanceolate on flowering shoots, mucronate, with small straight *marginal prickles pointing obliquely forwards*. Fls 3 mm. diam., in *few-fld ascending cymes* longer than their subtending lvs but *shorter than the stem internodes* and forming a cylindrical panicle. Corolla pure white, its 4 lobes acute, not cuspidate. *Fr.* 1·75–2 mm., glabrous, *covered with high acute tubercles*. Fl. 6–8. Hermaphrodite, protandrous, self-incompatible. Visited by flies and other small insects. $2n=44*$. Hp.

Native. Heaths, moors, grasslands and woods on acid soils, to 4300 ft. in Scotland. 111, H 40, S. Common throughout the British Is. Strictly calcifuge. W. Europe from S. Scandinavia to N. Spain and eastwards to Bohemia; Newfoundland. An Atlantic species.

Plants intermediate between this species and *G. sterneri* occur on some Scottish mountains and may be hybrids, but they require further investigation.

7. G. pumilum Murr.

G. pumilum ssp. *vulgatum* (Gaud.) Schinz & Thell.

A perennial herb with ± prostrate non-flowering and a few decumbent stout flowering shoots (15–)20–30(–40) cm., little branched, glabrous or hairy. Lower stem-internodes long, so that lf-whorls are distant. *Plant drying pale green.* Upper lvs on flowering shoots (12–)14–18(–20) mm., 5–7 in a whorl, linear-lanceolate, mucronate, their margins thickened and generally with a *few small prickles, mostly directed backwards*. Fls 4 mm. diam., in open *flat-topped cymes* which are spreading and *longer than the stem-internodes*, forming a long open panicle. Corolla creamy white. *Fr.* 1·5 mm., glabrous, *covered with low dome-shaped papillae*. Fl. 6–7. Hermaphrodite, protandrous, self-incompatible. Visited by flies and other small insects. $2n=88*$. Hp.

Native; introduced in a few localities. In chalk and limestone grassland. 16. A number of small isolated populations, mostly in S.E. England but extending to Cornwall and Gloucester and northwards to Lincoln. Widespread and variable in W. and C. Europe from France eastwards to Poland and Austria. A central European species.

The large population at Cheddar, N. Somerset, differs from all the others in its more compact habit and infl., but other populations in S.W. England may connect it with typical British *G. pumilum.*

8. G. sterneri Ehrendorf. ssp. sterneri

G. sterneri ssp. *septentrionale* (Sterner) Hyl.; *G. pumilum* ssp. *septentrionale* Sterner ex Hyl.

A perennial herb overwintering as a mat (more prostrate and compact than *G. pumilum*) and later with some prostrate non-flowering and many erect or ascending flowering shoots (10–)15–25(–30) cm., glabrous or hairy, much branched at the base, giving a tangled growth. Lower stem-internodes short, so that lf-whorls are crowded. Plant turning dark green when dried. Lvs on flowering shoots (8–)10–14(–16) mm., 6–8 in a whorl, oblanceolate to linear, mucronate, their *margins with many curved backwardly-directed prickles*. Fls 3·25–4 mm. diam., in compact ascending *cymes longer than the stem-internodes* and forming a compact pyramidal panicle. Corolla creamy white. *Fr.*

1·25 mm., glabrous, *covered with acute tubercles*. Fl. 6–7. Hermaphrodite, protandrous, self-incompatible. Visited by flies and other small insects. $2n=44*$. Hp.

Native. On grassy calcareous slopes and on calcareous or basic igneous rocks; to 3000 ft. in Scotland. 28, H10. Abundant on Carboniferous Limestone in Brecon, Derbyshire, Yorkshire, Durham and W. Ireland; occasional on other basic outcrops from Caernarvon and Northumberland to Sutherland and Orkney, and in N.E. Ireland.

The plants from Caernarvon, from the Durness Limestone in Sutherland and from W. Ireland are diploid ($2n=22*$). They are more slender and smaller in all parts (flowering shoots 10–15(–20) cm.; lvs 7–9 mm.; fls 3·25 mm. diam.; fr. 1–1·15 mm.), the infl. is more open and the cymes smaller and more compact; but there is overlap with the tetraploid in all these features.

G. sterneri and its close allies are confined to N.W. Europe (incl. Iceland), ssp. *sterneri* being also in Denmark and S. Norway and ssp. *oelandicum* (Sterner & Hyl.) Hyl. ($2n=22$) on Öland.

9. G. palustre L.   'Marsh Bedstraw.'

A perennial herb with a slender creeping stock above or below the surface of the substratum and decumbent or ascending non-flowering and flowering shoots up to 120 cm., 4-angled, glabrous, smooth or more usually ± scabrid

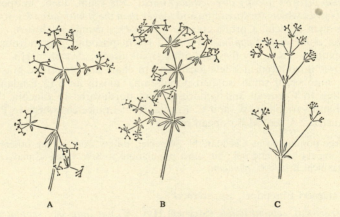

Fig. 52. Inflorescences of *Galium palustre* (A),
G. uliginosum (B), and *G. debile* (C). × ½.

on the angles, weak, in taller forms being supported by surrounding vegetation, *turning black when dried*. Lvs up to 3·5 cm., broadly oblanceolate, rarely ovate to narrowly elliptical-oblong, *blunt* or sometimes subacute but never mucronate, usually rough on the margins with small backwardly directed prickles, 1-veined, 4–6 *in a whorl*. Fls 3–4·5 mm. diam. in ± *spreading* axillary *cymes forming a* loose *pyramidal panicle* (Fig. 52A). Corolla white, its 4 lobes acute. Anthers red. Fr. 1·2–1·7 mm., glabrous, rugose, ultimately black.

Fl. 6–7. Hermaphrodite, protandrous. Visited chiefly by bees. $2n=24*$, $48*$, $96*$. H.–Hel.

Native. Marshes, fens, flushes, ditches, stream-banks, etc., to 2000 ft. in N. England. 112, H40, S. Common throughout the British Is. Europe to 70° 29′ N. in Scandinavia; Asia Minor.

A polymorphic species represented in the British Is. by at least three cytologically distinct subspecies:

(*a*) Ssp. **elongatum** (C. Presl) Lange. (incl. 'var. *lanceolatum* Uechtr.' auct. angl., p. max. p.): stems stout but weak, to 120 cm. high, diffusely branched, supported by surrounding vegetation, usually ± rough, rarely smooth on the angles; lvs usually 1·5–2 cm., oblanceolate, blunt, rough on the margins, rarely obovate or narrowly elliptical and rarely acute; *fls* 4·5 *mm. diam.* in large lax pyramidal panicles whose branches are erect-ascending in fl., spreading but not reflexed in fr.; *mericarps* 1·6 *mm. diam.*, coarsely rugose, their stalks divaricate. Common in reed-swamps, etc., and usually in standing water. $2n=96*$. Modal length of stomata of fully grown lvs 35–39 μ.

(*b*) Ssp. **palustre** (incl. 'var. *witheringii* Sm.', and 'var. *angustifolium* Druce' auct. angl., p. max. p.): stems slender, 15–30(–50) cm. high, ± erect, ± rough on the angles, rarely smooth; lvs usually 0·5–1 cm., linear-oblanceolate, blunt, rarely subacute, rough on the margins; *fls* 3 *mm. diam.* in lax oblong panicles whose branches are erect-ascending in fl. but ± reflexed in fr.; *mericarps* 1·2 *mm. diam.*, finely rugose, their stalks markedly divaricate. Common in marshy or peaty areas with standing water only in winter. Flowers 2–3 weeks earlier than ssp. *elongatum*. $2n=24*$. Modal length of stomata of fully grown lvs 25–30 μ.

(*c*) Ssp. **tetraploideum** Clapham: intermediate in stature and in size of parts between sspp. *elongatum* and *palustre*. Known certainly only from the margin of one pond in Devon, but probably widespread though less common than the two other sspp. $2n=48*$. Modal length of stomata of fully grown lvs 29–33 μ.

10. G. debile Desv.   'Slender Marsh Bedstraw.'
G. constrictum Chaub.

A perennial herb with slender 4-angled glabrous stems 15–40 cm., prostrate below then ascending, smooth or slightly rough on the angles, not blackening when dried. *Lvs* 0·5–1 cm., 4–6 in a whorl, *linear*, broadest beyond the middle, subacute, sometimes minutely apiculate but never mucronate, glabrous but the margins rough with forwardly directed prickles; *autumn lvs of submerged shoots* 1–2(–3) *cm., very narrow, flaccid.* Fls 2·5 mm. diam., in long-stalked erect-ascending axillary corymbs forming an *obconical panicle* whose branches do not spread in fr. (Fig. 52 c). Corolla pinkish-white, its 4 lobes acute. Fr. 1·2 mm., the mericarps 1 mm. across, glabrous, granulate; stalks not divaricate. Fl. 5–7. $2n=22$. Hel.–Hyd.

Native. Pond margins. Devon and the New Forest (Hants); Channel Is. 2, S. S.W. and S. Europe from France, Spain and Portugal to S. Balkans and Crete.

11. G. uliginosum L.   'Fen Bedstraw.'

A perennial herb with slender creeping stock and weak decumbent or ascending glabrous 4-angled stems 10–60 cm., very rough on the angles with downwardly directed prickles, *not blackening when dried*. Lvs 0·5–1(–1·5) cm., (4–)6–8 *in a whorl, linear-oblanceolate, mucronate*, 1-veined, glabrous, the margins rough with backwardly directed prickles. Fls 2·5–3 mm. diam., in small axillary corymbs forming a *narrow panicle* (Fig. 52 B). Corolla white, the 4 lobes acute. Anthers yellow. Fr. 1 mm., glabrous, rugulose, ultimately dark brown; stalks deflexed in fr. Fl. 7–8. Hermaphrodite, protandrous. Coumarin-scented and visited by small insects. $2n=22^*$, 44. Hel.

Resembles small types of *G. palustre* but is readily distinguishable by the 6–8 mucronate lvs in each whorl, and by remaining green when dried.

Native. Fens, to 1650 ft. in N. England. 100, H18. Locally frequent throughout Great Britain northwards to Ross; Ireland: northwards to Mayo and Meath, mostly near the coast. Europe from N. Spain, N. Italy and N. Balkans to 70° 27′ N. in Scandinavia.

12. G. tricornutum Dandy   'Rough Corn Bedstraw.'

G. tricorne Stokes, p.p.

An annual herb with decumbent or ascending-scrambling stems 10–40(–60) cm., glabrous, sharply 4-angled, the *angles very rough* with downwardly directed prickles. Lvs 2–3 cm., 6–8 in a whorl, linear-lanceolate, mucronate, 1-veined, glabrous, the margins with backwardly directed strong hooked prickles. Fls 1–1·5 mm. diam., in 1–3-fld stalked axillary cymes. Corolla cream-coloured, the 4 lobes acute. Fr. 3–4 mm., wider than the corolla, pale-coloured, often of a single 1-seeded ± spherical mericarp by abortion of the second, *granulate* with large papillae, not bristly; their *stalks strongly recurved* (Fig. 53 A). Fl. 6–9. Hermaphrodite, ± homogamous. Little visited by insects. $2n=44$. Th.

Doubtfully native. Cornfields, chiefly on calcareous soils. Local in England and Wales and native or long established northwards to Northumberland and Westmorland, but probably introduced recently further north to Moray. Probably originally native in the Mediterranean region but now established throughout Europe.

*G. valantia Weber (*G. saccharatum* All.), an annual herb resembling *G. tricornutum*, occurs as a casual. It is readily distinguishable by its less scabrid stems, lvs 5–6 in a whorl with forwardly directed marginal prickles, usually 3-fld axillary cymes with only the terminal fl. fertile, fr. 3 mm., covered with long acute conical papillae, and deflexed fr.-stalks.

13. G. aparine L.   Goosegrass, Cleavers, Hairif, Sticky Willie.

An annual herb with prostrate or, more usually, scrambling-ascending diffusely branched stems 15–120 cm., glabrous or hairy above the nodes, 4-angled, the angles very rough with downwardly directed prickles. Lvs 12–50 mm., 6–8 in a whorl, linear-oblanceolate or narrowly elliptical mucronate, 1-veined, glabrous or bristly above, the *margin with prickles backwardly directed* except those near the tip. Fls 2 mm. diam. in stalked 2–5-fld axillary cymes, the stalks bearing a *whorl of lf-like bracts*. Corolla whitish, with

4 acute lobes. *Fr.* 4–6 mm., wider than the corolla, olive or purplish, covered *with white hooked bristles with tuberclate bases*, their stalks divaricate (Fig. 53 B). Fl. 6–8. Hermaphrodite, protandrous. Sparingly visited by small insects. $2n = 22, 44, c. 66, c. 88$. Th.

Native. Hedges, waste places, drained fen peat, limestone scree, maritime shingle, etc., to 1200 ft. in Yorks. 112, H40, S. Abundant throughout the British Is. Europe to 69° N. in Scandinavia; N. and W. Asia. Widely introduced.

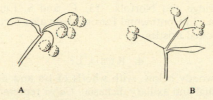

Fig. 53. Partial infructescences of *Galium tricornutum* (A) and *G. aparine* (B). × 1.

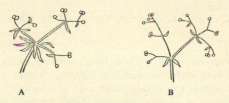

Fig. 54. Partial infructescences of *Galium spurium* (A) and *G. parisiense* (B). × 1.

14. G. spurium L.   'False Cleavers.'

Incl. *G. vaillantii* DC.

An annual herb resembling *G. aparine* but with somewhat narrower linear-lanceolate mucronate lvs (having marginal backward-directed prickles) and *greenish fls* 1 mm. *diam.* in axillary *cymes* of 3–9 fls *with* only 2(–3) *lf-like bracts*, not a whorl of 4–8 as usually in *G. aparine*. Fr. 1·5–3 mm., rugulose, blackish (Fig. 54 A). Fl. 7. $2n = 20$. Th. Var. *spurium* has glabrous fr. and is a rare casual. More frequently seen is:

(Var. *vaillantii*) DC. Gren. *Fr.* thickly covered *with white hooked hairs with non-tuberclate bases.*

Doubtfully native. Established in arable fields near Saffron Walden (Essex) and in a few other scattered localities in S. England and northwards to Warwick, Stafford and Leicester. Europe, except arctic Scandinavia and Russia; W. Asia; N. and C. Africa.

15. G. parisiense L. ssp. **anglicum** (Huds.) Clapham   'Wall Bedstraw.'

G. anglicum Huds.

An annual herb with decumbent or ascending, usually much-branched, weak, slender, glabrous stems 10–30 cm.; 4-angled, the angles rough with small downwardly directed prickles; blackening when dried. *Lvs* 3–12 mm., 5–7 in a whorl, linear-oblong, shortly mucronate, 1-veined, at first spreading then *reflexed*, glabrous, the margin with *forwardly directed prickles*. *Fls* c. 0·5 *mm. diam.*, in few-fld axillary stalked corymbs forming a long narrow panicle. Corolla greenish inside, reddish outside, its 4 lobes acute. Fr. 1 mm., wider than the corolla, glabrous, granulate, blackish (Fig. 54B). Fl. 6–7. $2n=22$, 44, 66. Th.

Native. Walls and sandy places. Rare and local in S.E. England from Kent to Suffolk, Cambridge and Norfolk. 13. C. and S. Europe; Canary Is.; N. Africa; W. Asia. In Continental Europe there also occur forms whose fr. is covered with hooked bristles.

4. RUBIA L.

Perennials, often woody below, with whorls of lvs and lf-like stipules. Fls yellowish, in terminal and axillary dichasia. *Calyx* represented by *an annular ridge*; *corolla rotate with* a very short tube and usually 5 *lobes*; stamens 5; styles 2, connate below; stigmas capitate. *Fr. succulent*, usually globose and 1-seeded and derived from only 1 cell of the ovary, rarely didymous with the parts not separating.

About 38 spp., chiefly in temperate regions of the Old and New Worlds.

1. R. peregrina L.   Wild Madder.

A perennial evergreen plant with a long creeping stock and trailing or scrambling-ascending glabrous stems 30–120 cm., woody and terete below, sharply 4-angled above, the angles rough with downwardly directed prickles. Lvs 1·5–6 cm., ovate- to elliptic-lanceolate, rigid, leathery, shining above, the cartilaginous margins and the underside of the midrib rough with curved prickles, 1-veined, 4–6 in a whorl. Fls 5 mm. diam. in terminal and axillary cymes forming a lfy panicle. Corolla pale yellowish-green, its 5 lobes long-cuspidate. Fr. 4–6 mm. diam., subglobose, black. Fl. 6–8. $2n=66$, ?132. H.–Ch.

Native. In hedges, thickets, scrub and on stony ground. Only in S. and S.W. England (Kent, Sussex, Hants, Dorset, Devon, Cornwall, Somerset, Gloucester, Hereford, Monmouth) and the coastal counties of Wales northwards to Caernarvon. 25, H17, S. W. and S. Europe; N. Africa.

*R. tinctorum L., Madder, was formerly grown in England and still occurs as a casual. It has longer and narrower lanceolate lvs with a conspicuous network of lateral veins on the underside, brighter yellow fls with the corolla-lobes acute but not cuspidate, and a reddish-brown berry.

Native in the Mediterranean region and Near East. Madder is obtained from the roots.

116. CAPRIFOLIACEAE

Shrubs, rarely herbs. Lvs opposite. Stipules 0 or small and adnate to petiole, very rarely conspicuous. Infl. of a cymose type. Fls hermaphrodite, usually 5-merous (except ovary). Calyx often small. Corolla gamopetalous, lobes imbricate, sometimes 2-lipped. Stamens inserted on corolla-tube, equalling in number and alternate with corolla-lobes, rarely 1 stamen suppressed;

anthers introrse (except *Sambucus*). Ovary inferior. Carpels 2–5 (sometimes two of them sterile) with 1–many axile ovules. Fr. a berry, drupe or achene, rarely a capsule. Seeds with fleshy endosperm and usually small straight embryo.

Thirteen genera and about 400 spp., cosmopolitan, but mainly north temperate regions.

Very near Rubiaceae and chiefly distinguished by the stipules. There is, however, no close resemblance between the British members of the two families. The woody habit, inferior ovary and stamens equalling in number the corolla-lobes are not found in combination in any other British member of the Metachlamydeae.

Besides members of the British genera, species and hybrids of *Weigela* are commonly cultivated. They have conspicuous large usually pink or white fls and differ from all the other genera except *Diervilla* in the capsular fr. *Abelia* and *Dipelta* are less frequently cultivated.

1 Lvs pinnate.   1. SAMBUCUS
   Lvs simple.   2
2 Fls in compound cymes.   2. VIBURNUM
   Fls in short spike-like racemes.   3. SYMPHORICARPOS
   Fls in pairs, heads or whorls.   3
3 Prostrate creeping undershrub; fls on long pedicels.   4. LINNAEA
   Erect or climbing shrubs; fls sessile.   5. LONICERA

1. SAMBUCUS L.

Deciduous shrubs or small trees, rarely herbs, with large pith. Buds with several pairs of scales. *Lvs pinnate.* Stipules present or not. Fls regular, in compound umbel-like or panicle-like cymes, usually 5-merous. Calyx-limb very small. Corolla rotate with short tube and flat spreading limb. Anthers extrorse. *Ovary 3–5-celled*, with 1 pendulous ovule in each cell. Style short, with 3–5 stigmas or branches. Fr. a drupe. Seeds compressed.

About 20 spp., temperate and subtropical regions (not C. and S. Africa).

1 Cymes flat-topped; fr. black.   2
   Cymes panicle-like; fr. red.   3. racemosa
2 Herb; stipules conspicuous.   1. ebulus
   Shrub; stipules 0 or very small.   2. nigra

1. S. ebulus L.   Danewort.

Perennial, foetid, glabrous *herb* 60–120 cm., with creeping rhizome and numerous stout erect grooved simple or little-branched stems. Lflets 5–15 cm., 7–13, oblong or oblong-lanceolate, acuminate, sharply serrate. *Stipules conspicuous*, ± ovate. *Infl. flat-topped*, 7–10 cm. diam., *with 3 primary rays*. Corolla white, sometimes pink-tinged outside. *Anthers purple*. *Fr.* globose, *black*. Fl. 7–8. Pollinated by insects. $2n = 36$. Hp.

?Native. Local by roadsides and in waste places, scattered over the greater part of the British Is., but very rare in N. Scotland though extending to Orkney. 88, H35, S. C. and S. Europe from the Netherlands and Germany southwards, Mediterranean region, W. Asia to Himalaya; Madeira (probably introduced).

2. S. nigra L.   Elder.

Shrub or more rarely a small tree to 10 m. often with straight vigorous erect shoots from the base; branches often arching. Bark brownish-grey, deeply furrowed, corky. Twigs stout, greyish, with prominent lenticels. Lflets 3–9 cm., (3–)5–7(–9), ovate, ovate-lanceolate or ovate-elliptic, acuminate, rarely (var. *rotundifolia* Endl.) orbicular or (var. *laciniata* L.) deeply dissected, sparingly hairy on veins beneath, serrate. *Stipules* 0 *or very small and subulate. Infl. flat-topped*, 10–20 cm. diam., *with 5 primary rays*. Corolla c. 5 mm. diam., cream-white. *Anthers cream. Fr.* 6–8 mm., globose, *black*, rarely greenish. Fl. 6–7. Pollinated by small flies, etc. Fr. 8–9. $2n = 36$. M.

Native. Woods, scrub, roadsides and waste places, especially characteristic of disturbed, base-rich and nitrogen-rich soils; very resistant to rabbits; comparatively infrequent in closed communities; ascending to over 1500 ft. Throughout the British Is., common except in N. Scotland. 112, H40, S. Europe from Scandinavia southwards; W. Asia; N. Africa; Azores.

*3. S. racemosa L.

Shrub to 4 m., glabrous. Lflets 4–8 cm., 5–7, ovate, ovate-lanceolate or elliptic, acuminate, serrate. *Stipules represented by large glands. Infl. a dense ovoid panicle*, 3–6 cm. Corolla cream-white. Anthers cream. *Fr.* c. 5 mm., globose, *scarlet*. Fl. 4–5. Pollinated by insects. Fr. 6–7. $2n = 36$. M.

Introduced. Commonly planted and sometimes naturalized, especially in Scotland. Native of Europe and W. Asia.

2. Viburnum L.

Shrubs or small trees. *Lvs simple*. Stipules 0 or small. *Fls regular, in compound umbel-like, rarely panicle-like cymes*, 5-merous (except ovary). Calyx-teeth very small. Corolla rotate, funnel-shaped or campanulate. *Ovary 1-celled* with 1 pendulous ovule. Stigmas 3, sessile. *Fr. a drupe*, stone usually compressed.

About 120 spp., north temperate zone extending to C. America and Java. A number of spp. are cultivated.

Fls all alike; lvs serrulate; buds naked.   **1. lantana**
Outer fls sterile and much larger than inner; lvs lobed; buds scaly.   **2. opulus**

1. V. lantana L.   Wayfaring Tree.

Deciduous shrub 2–6 m. *Twigs and naked buds* greyish, *scurfily stellate-pubescent. Lf-blades* 5–10 cm., oval, ovate or obovate, usually acute, cordate at base, *serrulate*, rugose above and sparingly stellate-pubescent, densely stellate-tomentose beneath; petiole 1–3 cm., exstipulate. Infl. 6–10 cm. diam., umbel-like, dense, on a short peduncle. *Fls all alike and fertile*. Corolla c. 6 mm., cream-white. Fr. c. 8 mm., oval, compressed, at first red, finally changing quickly to black. Fl. 5–6. Pollinated by insects or self-pollinated. Fr. 7–9. $2n = 18$. M.

Native. Scrub, woods and hedges on calcareous soils; common in S. England, becoming rarer northwards and in Wales; not usually considered native north of Yorks, but extending as an introduction to Lanark and Inverness. 51. C. and S. Europe from Belgium, C. Germany and S. Russia to C. Spain, Italy, Greece and the Caucasus; N. Asia Minor; Morocco and Algeria (very rare).

2. V. opulus L. Guelder Rose.

Deciduous shrub 2–4 m. *Twigs* greyish, *glabrous*, slightly angled. Buds scaly. *Lf-blades* 5–8 cm., *with* 3(–5) *acuminate, irregularly dentate lobes*, glabrous above, sparingly pubescent or glabrescent beneath, usually reddening in autumn; petiole 1–2·5 cm., with subulate stipules and discoid glands. Infl. 5–10 cm. diam., rather loose; peduncle 1–4 cm. *Inner fls fertile, c. 6 mm. diam.*; *outer* 15–20 *mm., sterile,* white. Fr. c. 8 mm., subglobose, red. Fl. 6–7. Pollinated by insects or selfed. Fr. 9–10. 2n=18. M.

Native. Woods, scrub and hedges especially on damp soils; rather common in England, Wales and Ireland, less common in Scotland but extending to Caithness. 105, H 40. Europe from Scandinavia (68° N.) and Finland to C. Spain, Italy and Thrace; N. and W. Asia; Algeria (very rare).

A form with all the fls sterile (var. *roseum* L., Snowball Tree) is commonly grown in gardens.

*V. tinus L. (Laurustinus) with evergreen entire lvs, fls pinkish outside, white within, and metallic blue fr. is frequently planted and occasionally self-sown. Native of the Mediterranean region.

3. SYMPHORICARPOS Duham.

Deciduous shrubs. Buds scaly. Lvs entire or lobed, exstipulate. *Fls regular, in terminal spike-like racemes or clusters,* 4–5-merous. Calyx-teeth small. Corolla tubular or campanulate. *Ovary with 2 fertile cells,* each with 1 ovule *and 2 sterile cells* with numerous ovules. Style slender, with capitate stigma. *Fr. a berry.*

About 15 spp., in N. America and 1 in China.

*1. S. rivularis Suksdorf Snowberry.

S. racemosus auct.; *S. albus* (L.) Blake var. *laevigatus* (Fern.) Blake

Shrub 1–3 m., spreading underground, with numerous rather slender erect stems. Twigs slender, yellowish-brown, ascending, glabrous. Lvs of the twigs 2–4 cm., oval or ovate, obtuse, cuneate or rounded at base, entire or a few sinuately lobed, dull green and glabrous or sparsely pilose below; petiole c. 5 mm.; lvs of the sucker shoots often conspicuously lobed. Fls 3–7, in terminal spike-like racemes, 2 cm. or less. Bracts and bracteoles ovate, small. Corolla 5–6 mm., campanulate, pink, hairy at the throat within. Fr. 1–1·5 cm., globose, white. Fl. 6–9. Pollinated by bees, wasps and syrphids. Fr. 9–11. 2n=c. 54. N.

Introduced. Commonly planted and ± naturalized in many places often spreading by suckers and forming large thickets, apparently rarely spreading by seed. Native of western N. America from Alaska and Alberta to California and Colorado.

4. LINNAEA L.

Evergreen creeping undershrub. Lvs small, exstipulate. *Fls in pairs on long peduncles, terminal on short lateral branches.* Calyx lobes 5, narrow-lanceolate, caducous. Corolla campanulate, 5-lobed. Stamens 4. *Ovary with 1 fertile cell* with 1 ovule *and 2 sterile cells* with numerous ovules. Style filiform, with capitate stigma. *Fr. an achene.*

One sp.

1. L. borealis L.   'Linnaea.'

Prostrate plant with slender pubescent stems, often forming large mats. Lf-blades 5–15 mm., broadly ovate to orbicular, subobtuse, crenate-dentate in the upper half, sparingly hairy, tapered at the base into a petiole 2–3 mm. Peduncles 3–7 cm., pubescent and glandular, with 2 small lanceolate membranous bracts at apex. Pedicels 1–2 cm., pubescent and glandular, with 2 small lanceolate bracteoles near the fl. Corolla pink, often beautifully marked, c. 8 mm., hairy within. Fr. c. 3 mm., densely glandular-hairy, but now very rare. Fl. 6–8. Pollinated by insects. $2n = 32$. Chw.

Native. Woods, especially pine, and in the shade of rocks; ascending to 2400 ft. From Yorks (but now extinct in England) to Sutherland, confined to the east, very local and rare, especially in the southern part of its range. 18. N. Europe and Asia (from 71° 10′ N. in Norway) to N. Germany; Alps; Carpathians, Caucasus; sspp. in N. America.

5. LONICERA L.

Deciduous, rarely evergreen, shrubs or woody climbers. Buds scaly. Lvs usually entire, exstipulate. Fls sessile, either in axillary pairs on long peduncles or in heads or whorls. Calyx with 5 small teeth. *Corolla* either (as in all the British spp.) *strongly 2-lipped* with 4-lobed upper and entire lower lip or with a nearly regular 5-lobed limb; tube long or short. Stamens 5. *Ovary 2-celled* with numerous axile ovules. Style slender, stigma capitate. *Fr. a few-seeded berry.*

About 180 spp., north temperate zone, extending to Mexico and Java. Many spp. are grown in gardens, including *L. nitida* Wils. from W. China, with evergreen lvs, c. 1 cm., now much used for hedges.

1 Upright shrub; fls in pairs.	**1. xylosteum**	
Woody climbers; fls in heads or whorls.		*2*
2 Lvs all free; bracts small.	**2. periclymenum**	
Upper lvs and bracts connate in pairs; bracts large.	**3. caprifolium**	

Section 1. *Xylosteon* (Mill.) DC. Upright shrubs. Fls in pairs. Corolla-tube short.

1. L. xylosteum L.   Fly Honeysuckle.

Deciduous *bushy shrub* 1–2 m. Twigs grey, somewhat pubescent. Lvs 3–6 cm., ovate to obovate, acute, broad-cuneate or rounded at base, greyish-green, ± pubescent, especially beneath; petiole 3–8 mm. *Fls in pairs*, sessile in the axils of subulate bracts about as long as ovaries, on pubescent peduncles 1–2 cm. Bracteoles ovate, about ½ as long as ovaries. Ovaries glandular.

Corolla c. 1 *cm.*, yellowish, often tinged reddish, pubescent outside. Fr. red, globose. Fl. 5–6. Pollinated by humble-bees. Fr. 8–9. $2n=18$. N.

Native. Woods and hedges in one or two places in Sussex; elsewhere probably introduced, occurring in a number of places in England and Wales and a very few in E. Scotland (to Ross) and Ireland. 39, H3. Europe from Scandinavia (c. 64° N.) to Spain, Sicily, Macedonia and the Caucasus (rare in W. France); N. and W. Asia.

Section 2. *Nintooa* (Spach) Maxim. Climbing shrubs. Fls in pairs. Corolla-tube long.

*L. japonica Thunb.
A half-evergreen twining shrub with white fls 3–4 cm., in peduncled axillary pairs and black fr. Naturalized in Devon. Native of E. Asia.

Section 3. *Lonicera.* Climbing shrubs. Fls in heads or whorls. Corolla-tube long.

2. L. periclymenum L. Honeysuckle.

Twining shrub reaching 6 m. but often low and trailing or scrambling, glabrous or somewhat pubescent. *Lvs* 3–7 cm., ovate, elliptic or oblong, dark green above, glaucous beneath, usually acute, lower shortly petioled, upper sub-sessile, smaller, *all free. Fls in terminal heads. Bracts small, not exceeding ovary. Corolla* 4–5 *cm.*, with long tube and spreading limb, cream-white within, turning darker after pollination, purplish or yellowish and glandular outside. Fr. red, globose. Fl. 6–9. Pollinated by hawk-moths, also humble-bees, etc. Fr. 8–9. $2n=18*, 36*$. M. or N.

Native. Woods, hedges, scrub and shady rocks; ascending to 2000 ft., common throughout the British Is. 112, H40, S. Europe from Scandinavia (62° 44′ N.) and Germany (not Finland or Poland) to Portugal and Spain, Italy and Greece; N. and C. Morocco.

*3. L. caprifolium L. 'Perfoliate Honeysuckle.'

Twining glaucous shrub. Lvs 4–10 cm., ovate or oblong, dark green above, glaucous beneath, obtuse, lower shortly petioled, 2 *or 3 pairs below infl. connate by the greater part of their bases. Fls in terminal heads* often with axillary whorls in addition. *Bracts like the upper lvs* but smaller. Corolla like the last sp., but not glandular. Fr. red, globose. Fl. 5–6. Pollinated by hawk-moths. Fr. 8–9. $2n=18$. M.

Introduced. Hedges, etc., in a number of places in S. and E. England and a few in N. England and S. Scotland to Perth. 33. C. and S. Europe from Bohemia and N.E. France to Italy and Greece, Crimea, Caucasus; Asia Minor.

LEYCESTERIA Wall.

Deciduous shrubs. Fls in whorls in the axils of lf-like bracts forming drooping terminal spikes. Fls regular, 5-merous. Corolla funnel-shaped. Ovary 5-celled. Six spp. in Himalaya and China.

*L. formosa Wall. Lvs ±ovate, acuminate, 5–18 cm. Bracts purplish. Corolla 1·5–2 cm., purplish. Fr. deep brownish-purple. Commonly cultivated and some-times escaping. Native of the Himalaya and S.W. China.

117. ADOXACEAE

Perennial rhizomatous herb. Lvs ternate, exstipulate. Fls c. 5 in terminal heads. Terminal fl. with 2-lobed calyx (or bracts) and 4-lobed corolla (or perianth); stamens 4, epipetalous, alternating with the corolla-lobes but divided to the base and appearing as 8 but each bearing ½-anther only; lateral fls with 3-lobed calyx (or bracts), 5-lobed corolla (or perianth) and 5 (apparently 10) stamens. Ovary 3–5-celled, ½-inferior, tapered above into the styles; ovule 1 in each cell, anatropous, pendulous from the inner angle. Fr. a drupe; endosperm copious; embryo small.

One genus and sp. of obscure relationships.

1. ADOXA L.

The only genus. One sp. Nectar secreted by a ring round the base of the stamens.

1. A. moschatellina L.   Moschatel, Townhall Clock, Five-faced Bishop.

Glabrous perennial herb 5–10 cm. Rhizome far-creeping with fleshy whitish scales at the apex. Radical lvs ternate, on long petioles, light green, dull above, rather glossy beneath; lflets 1–3 cm., long-stalked, ternate or trisect, the divisions often 2–3-lobed, lobes oval or oblong, obtuse, mucronate, Fl.-stems erect, unbranched; cauline lvs 2, opposite, 8–15 mm., shortly stalked, ternate or trisect, the terminal lflet often 3-lobed. Infl. c. 6 mm. diam.; calyx and corolla light green, like the lvs. Anthers yellow. Fr. green, rarely produced. Fl. 4–5. Visited by various small insects; homogamous or slightly protogynous, self-pollination possible. $2n = 36, 54, 72$. Grh.

Native. Woods, hedge-banks and mountain rocks; ascending to 3600 ft. From Sutherland southwards, widespread but rather local; absent from the Isle of Man and Channel Is.; Antrim only in Ireland. 105, H1. Europe from Scandinavia to the mountains of C. Spain, Italy, Montenegro and Bulgaria; Morocco (mountains of N.W.); N. and C. Asia to the Caucasus, Himalaya and Kamchatka; N. America.

118. VALERIANACEAE

Herbs, sometimes woody at the base, often with strong-smelling rhizomes. Lvs opposite or radical, entire or pinnatifid, exstipulate. Infl. cymose, often capitate. Fls generally small, hermaphrodite or unisexual, often somewhat zygomorphic. Calyx annular or variously toothed, often inrolled in fl. and forming a feathery pappus in fr. Corolla funnel-shaped, base equal, saccate, or spurred on one side; lobes 5 (3–4), imbricate, obtuse. Stamens 1–4 (1–3 in our spp.), inserted towards the base of the corolla-tube. Ovary 3-celled, one cell fertile, the two sterile cells usually small or almost 0; ovule solitary, pendulous. Fr. dry, indehiscent.

About 8 genera and 350 spp., generally distributed except for Australia.

Annual; corolla not spurred or saccate at base; stamens 3; calyx not forming
a feathery pappus in fr.; fls pale lilac.   1. VALERIANELLA

Perennial; corolla saccate at base; stamens 3; calyx forming a feathery pappus
in fr.; fls pale pink. 2. VALERIANA
Perennial, often rather woody at base; corolla spurred; stamen 1; calyx forming
a feathery pappus in fr.; fls red (less often white). 3. CENTRANTHUS

1. VALERIANELLA Mill.
Lamb's Lettuce, Corn Salad.

Small *annual* herbs with apparently dichotomous branching. Fls solitary in
the forks of the branches and in terminal bracteate cymose heads. *Calyx
neither inrolled in fl. nor forming a pappus in fr., but a toothed or funnel-
shaped rim, or sometimes almost 0. Corolla funnel-shaped, regular, not spurred
or saccate at base,* lobes 5. Stamens 3. *Fr. of one 1-seeded cell and two distant
but sometimes small sterile empty cells.* Self-pollination is probably the rule.
About 55 spp., in the temperate parts of the northern hemisphere.
Ripe fr. is essential for the determination of the spp.

1 Fr. compressed or nearly quadrangular in section; calyx in fr. very small
and inconspicuous. *2*
Fr. ovoid or oblong, flat on one face, convex on the other; calyx in fr.
distinct. *3*
2 Fr. compressed, suborbicular in side view, c. 2·5 × 2 mm., fertile cell corky
on back. **1. locusta**
Fr. oblong, nearly quadrangular in section, c. 2 × 0·75 mm., fertile cell not
corky on back. **2. carinata**
3 Cymes rather lax; calyx in fr. minutely toothed, c. ⅓ as broad as the ovoid
fr.; sterile cells together larger than the fertile. **3. rimosa**
Cymes very dense; calyx in fr. deeply 5–6-toothed, as broad as the oblong
fr., strongly veined, teeth subequal; fertile cell many times larger than
sterile ones. **4. eriocarpa**
Cymes rather lax; calyx in fr. ½ as broad as ovoid fr., scarcely veined, one
tooth much larger than others; fertile cell many times larger than sterile
ones. **5. dentata**

1. V. locusta (L.) Betcke Lamb's Lettuce, Corn Salad.
V. olitoria (L.) Poll.

A slender erect nearly glabrous annual 7–40 cm. Stems rather brittle, much-
branched, weakly angled, slightly pubescent below. Lvs 2–7 cm., entire or
sometimes dentate, lower spathulate, obtuse, upper oblong, obtuse or subacute. Cymes capitate, bracts shortly ciliate. Fls small, pale lilac. Calyx
indistinct, 1-toothed. *Fr.* (Fig. 55A) *c.* 2·5 × 2 *mm., compressed, suborbicular
in side view, fertile cell corky on the back,* sterile cells confluent. Fl. 4–6.
$2n = 14$. Th.
Native. On arable land, hedge-banks, rocky outcrops and dunes, usually
on dry soils; much the commonest British sp. Throughout nearly the whole
of the British Is., though rather local. 108, H 40, S. Europe, except the
extreme north; Madeira; N. Africa; W. Asia; introduced in N. America.

2. V. carinata Lois.

An annual very similar in general appearance to *V. locusta*. Calyx indistinctly
1-toothed. *Fr.* (Fig. 55 B) 2 × 0·75 *mm., oblong, nearly quadrangular in section,*

fertile cell not corky on the back, sterile cells nearly confluent. Fl. 4–6. $2n = 18$. Th.

Native. On arable land, banks, old walls and rocky outcrops. 38, H5, S. Local in England and Ireland, occurring mainly in the south and east. C. and S. Europe to the Caucasus; N. Africa.

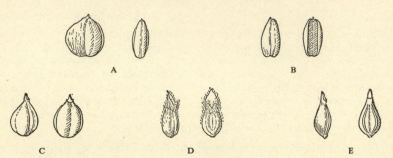

Fig. 55. Fruits of *Valerianella*. A, *V. locusta*; B, *V. carinata*; C, *V. rimosa*; D. *V. eriocarpa*; E, *V. dentata*. × 5.

3. V. rimosa Bast.

V. Auricula DC.

An annual similar in general appearance to *V. locusta*. Stems rather rough on the angles. Upper lvs and bracts usually linear. *Cymes rather lax. Calyx in fr. distinct, minutely toothed, c. $\frac{1}{3}$ as broad as the fr. Fr.* (Fig. 55c) *c. 2 mm., ovoid, the sterile cells together larger than the fertile one*, confluent. Fl. 7–8. $2n = 14$. Th.

Native. In cornfields, local. 47, H20. Scattered throughout England, Wales and Ireland, mainly in the south and east; Scotland, recorded from Fife only. C. and S. Europe from Denmark southward; N. Africa.

*4. V. eriocarpa Desv.

An annual similar in general appearance to *V. locusta*. Stems rather rough. *Cymes very dense. Calyx in fr. distinct, as broad as the fr., oblique, strongly net-veined, deeply 5–6-toothed, teeth subequal. Fr.* (Fig. 55D) *c. 1 mm., oblong, ± hispid, fertile cell many times larger than the distant sterile ones.* Fl. 6–7. $2n = 14$. Th.

Introduced. Naturalized on banks and old walls in a few places, mainly in southern England; Wales: Pembroke; Scotland: Midlothian. 17, S. Mediterranean region from Portugal to Greece; N. Africa.

5. V. dentata (L.) Poll.

Differs from *V. eriocarpa* in its laxer cymes, the calyx $\frac{1}{2}$ as broad as the ovoid glabrous fr., scarcely veined, with one tooth much longer than the others (Fig. 55E). Fl. 6–7. $2n = 14$. Th.

Native. Locally common in cornfields. 85, H33, S. Fairly generally distributed throughout the British Is., but infrequent in the west and north. C. and S. Europe; Macaronesia; N. Africa; W. Asia.

2. VALERIANA L.

Perennial herbs, mostly with a bitter taste and peculiar smell, particularly evident when dry and very attractive to cats. Lvs entire, pinnate or pinnatifid. Infl. cymose, terminal and usually subcapitate in fl. Fls hermaphrodite or unisexual, bracteolate. Calyx inrolled in fl., enlarging and forming a pappus in fr. *Corolla* funnel-shaped, *tube slightly saccate at base,* lobes 5, rarely 3 or 4, unequal. *Stamens* 3. *Ovary apparently of* 1 *carpel.* Fr. a unilocular, 1-seeded nut.

About 180 spp., in Europe, Asia, Africa and America.

All lvs imparipinnate; fls hermaphrodite; stolons 0 or very short.   **1. officinalis**
Lower lvs simple; upper simple or pinnate, cordate, deeply toothed; fls
  hermaphrodite; stolons 0.   **2. pyrenaica**
Lower lvs simple and quite entire, upper pinnatifid; fls dioecious; stolons long.
                                                                     3. dioica

1. V. officinalis L.   Valerian.

An erect nearly glabrous perennial 20–150 cm., *rarely producing short stolons.* Stem hairy below. *Lvs* up to c. 20 cm., *all imparipinnate,* lower long-petioled, upper nearly sessile; *lflets lanceolate, entire or distantly and irregularly toothed.* Fls 4–5 mm., hermaphrodite. Corolla c. 5 mm. diam., pale pink. Fr. 4 mm., ovate-oblong. Fl. 6–8. $2n=28*, 56*$. Hs. A variable sp. the extreme forms of which have been named *V. mikanii* Syme and *V. sambucifolia* Mik. The tetraploid and octoploid forms differ somewhat in, but appear not to be certainly distinguishable by, morphological features.

Native. In rough grassy and bushy places, usually on damp soils but also in dry habitats. The tetraploid apparently confined to the south and parts of the Midlands, on chalk, oolite and limestone; the octoploid throughout the British Is., in damp valleys in the south and both damp and dry habitats in the north. 111, H40. Europe except the extreme north and south; temperate Asia to Japan.

*2. V. pyrenaica L.   'Pyrenean Valerian.'

An erect dark green ± pubescent perennial c. 100 cm. Stolons 0. Stem pubescent at nodes. *Lvs* 8–20 cm. broad, *simple or the upper with a few small lateral lflets and a large terminal one, ovate or suborbicular, cordate, deeply and irregularly dentate,* lower obtuse, upper acuminate. Fls 5–6 mm., hermaphrodite. Corolla c. 5 mm. diam., pale pink. Fr. 5–6 mm., linear-oblong. Fl. 6–7. Hs.

Introduced. Naturalized in woods in a number of localities in the western parts of the British Is. Pyrenees and Hautes Corbières.

3. V. dioica L.   'Marsh Valerian.'

An erect nearly glabrous bright green perennial 15–30(–50) cm. *Stolons present, long.* Stem slightly pubescent at nodes. Basal lvs long-petioled, *blade* 2–3 cm., *elliptic to ovate, obtuse and quite entire; stem-lvs* sessile or subsessile, *pinnatifid.* Fls dioecious, male c. 5 mm. diam., female c. 2 mm. diam., pinkish. Fr. c. 3 mm., elliptic. Fl. 5–6. $2n=16$. Hs.

Native. In marshy meadows, fens and bogs. 83. Scattered throughout

most of Great Britain, local, and rarer in the south than in the north. From
S. Norway (1 locality) and S. Sweden to N. Spain, eastwards to N. Italy and
C. Russia.

3. CENTRANTHUS DC.

Perennial or annual glabrous herbs. Fls in terminal panicled cymes, red,
pink or white, bracteolate. Calyx inrolled in fl., enlarging and forming a
pappus in fr. *Corolla-tube with a spur or small conical projection at base,*
lobes 5. *Stamen 1. Ovary apparently of 1 carpel.* Fr. a 1-seeded nut.

About 12 spp., in Europe and the Mediterranean region.

*1. C. ruber (L.) DC.   Red Valerian.

An erect somewhat glaucous perennial 30–80 cm. Lvs c. 10 cm., ovate or
ovate-lanceolate, entire or the upper sinuate-dentate, lower narrowed at base
into a petiole, upper sessile. Fls c. 5 mm. diam., red or less commonly white,
scented; corolla-tube 8–10 mm., slender; spur as long to twice as long as
ovary. Stamen exserted. Fl. 6–8. The fls are protandrous and are cross-
pollinated by long-tongued insects, mainly Lepidoptera. $2n=14$. Hp.

Introduced. Frequently cultivated and well naturalized on old walls, cliffs,
dry banks and in waste places, locally abundant, particularly in S.W. England
and S.E. Ireland. 54, H12, S. C. and S. Europe, N. Africa and Asia Minor;
introduced in Madeira and doubtless elsewhere.

119. DIPSACACEAE

Herbaceous or rarely suffruticose plants with opposite or whorled exstipulate
lvs. Infl. usually a head (*capitulum*) with a calyx-like involucre of bracts;
rarely a spike of false whorls (*Morina*). Fls hermaphrodite, zygomorphic,
epigynous, each surrounded at its base by an epicalyx or 'involucel' of united
bracteoles, and often subtended by a receptacular bract. Calyx small, cup-
shaped or ± deeply cut into 4–5 segments or into numerous teeth or hairs;
corolla sympetalous with the tube often curved and the 4–5 lobes ± equal or
forming 2 lips; stamens 4 or 2, epipetalous, alternating with the corolla-lobes,
exserted, anthers free; ovary inferior, 1-celled; ovule solitary, anatropous,
pendulous; style slender with a simple or 2-lobed stigma. Insect-pollinated.
Fr. dry, indehiscent, 1-seeded, enclosed in the involucel and often surmounted
by the persistent calyx; endosperm present; embryo straight.

About 155 spp. in 9 genera, chiefly in the Mediterranean region and Near
East.

Readily distinguished from Compositae by the exserted stamens with free
anthers, and by the fr. enclosed in an involucel.

1   Stem spiny or prickly; involucral and receptacular bracts spine-tipped.
                                                              1. DIPSACUS
    Stem not spiny or prickly; involucral and receptacular bracts not spine-
        tipped.   2
2   Receptacle hemispherical, hairy; receptacular bracts 0; calyx of 8(–16)
        teeth or bristles.   2. KNAUTIA
    Receptacle elongated, not hairy; receptacular bracts present; calyx of
        5 teeth or bristles.   3

3 Stem-lvs pinnatifid or pinnate; marginal fls larger than the central; corolla 5-lobed; involucel ending above in a pleated scarious funnel-shaped cup.   3. SCABIOSA

   Stem-lvs entire or faintly toothed; fls equal; corolla 4-lobed; involucel ending above in 4 herbaceous teeth.   4. SUCCISA

1. DIPSACUS L.

Biennial or perennial herbs, often large, with *spiny or prickly stems*, opposite pairs of often *connate lvs*, and capitulate, usually conical or cylindrical infl. *Involucre of* 1–2 rows of linear to lanceolate, erect or spreading, *spine-tipped bracts* which usually considerably exceed the ± spine-tipped receptacular bracts. Involucel ± distinctly 4-angled, united with the ovary below and ending above in a very short ± 4-lobed cup. Calyx a 4-angled cup with a ciliate upper margin. Corolla with a long tube and 4 unequal lobes. Stigma oblique, entire. Fr. crowned by the persistent calyx.

About 12 spp., in Europe, W. Asia and N. Africa.

Basal lvs oblong, short-stalked; stem-lvs with connate bases; head conical, usually pale lilac.   **1. fullonum**
Basal lvs ovate, long-stalked; stem-lvs not connate; head spherical, white.   **2. pilosus**

1. D. fullonum L.   Teasel.

A biennial herb with a stout yellowish tap-root and erect glabrous angled stems 50–200 cm., prickly on the angles, branching above. *Basal lvs* in a rosette dying early in the 2nd season, *oblong to elliptical-oblanceolate*, shortly-stalked, blunt or acute, entire, glabrous but with scattered swollen-based prickles; *stem-lvs* narrowly lanceolate, *connate at the base* into a water-collecting cup, prickly only on the midrib beneath, entire, crenate or distantly toothed. *Heads* 3–8 cm., *conical*, blunt, *always erect*. Involucral bracts linear-lanceolate to subulate, spiny. Corolla rose-purple, rarely white. Fr. 5 mm. Fl. 7–8. Visited by bees and long-tongued flies. $2n = 18$. Hs. (biennial).

There are 2 sspp.:

Ssp. fullonum   Wild Teasel.
D. sylvestris Huds.

Heads 5–8 cm. *Involucral bracts curving upwards*, unequally long but the longest *equalling or exceeding the head. Receptacular bracts exceeding the fls*, ciliate, ending in a long *straight spine*. Corolla-tube 9–11 mm.

Native. Copses, stream-banks, roadsides, rough pasture, etc., especially on clay soils; locally common in the south. Great Britain northwards to Perth; Ireland, local. 77, H20, S. Europe northwards to S. Norway; Near East; N. Africa; Canary Is.

This is the wild plant whose receptacular bracts are too flexible for use in combing cloth.

*Ssp. sativus (L.) Thell   Fullers' Teasel.
D. sativus (L.) Scholler

Heads 3–8 cm. *Involucral bracts spreading ± horizontally*, not overtopping the head. *Receptacular bracts almost equalling the fls*, spinous-ciliate, ending in a *stiff recurved spine*. Corolla-tube 13 mm. The cultivated plant whose heads are used for raising the nap of certain kinds of cloth.

Introduced. An occasional escape from cultivation. Still grown as a crop in Somerset.

2. D. pilosus L.   'Small Teasel.'

Cephalaria pilosa (L.) Gren.

A biennial herb with an erect angled and furrowed stem 30–120 cm., with sparse weak prickles on the angles. *Basal lvs* in a rosette, *ovate*, acute or shortly acuminate, narrowed into a *long stalk*, crenate-toothed, hairy, sometimes prickly on the midrib beneath; *stem-lvs short-stalked*, ovate to narrowly elliptical, simple or the upper lvs usually with a basal pair of small, often unequal, ± free, elliptical lflets, entire to crenate-toothed, sparsely hairy. *Heads* 2–2·5 cm. diam., *spherical, at first drooping*, later erect, on long weakly prickly stalks. Involucral bracts narrowly triangular, spine-tipped, sparsely covered with long silky hairs, spreading or slightly reflexed, falling short of the fls. Receptacular bracts equalling the fls, obovate, abruptly contracted into a straight spiny point, ciliate with long silky hairs. Corolla 6–9 mm., whitish. Anthers dark violet. Fr. 5 mm., hairy. Fl. 8. Sparingly visited by small bees and flies. $2n = 18$. Hs. (biennial).

Native. Damp woods, hedge-banks, ditch-sides, etc., especially on chalk or limestone, to 840 ft. in Derby; local and infrequent. 54. England and Wales northwards to S. Lancs and Yorks. C. Europe from Spain, N. Italy, N. Balkans and S. Russia northwards to the Netherlands, Denmark, Sweden and Hungary. W. Asia; Japan.

Unlike other *Dipsacus* spp. in habit and sometimes included in *Cephalaria* which also has globular heads but lacks stem-prickles and spinous tips to the bracts. *C. gigantea* (Ledeb.) Bobrov (*C. elata* Schrad.) (Russia, Caucasus and W. Siberia), to 2 m. high, with creamy-white heads, is often cultivated and sometimes escapes.

2. KNAUTIA L.

Annual to perennial herbs with lvs in opposite pairs. Heads long-stalked, flat, with an involucre of numerous lanceolate herbaceous bracts. *Receptacle hemispherical, hairy; receptacular bracts* 0. Involucel ± 4-angled, ending above in a very short obscurely toothed cup. *Calyx* a short cup below, *prolonged upwards as* 8(–16) *usually suberect teeth or bristles*. Corolla ± unequally 4–5-lobed, those of the marginal fls often larger than the central. Stigma emarginate. Calyx deciduous in fr. Fr. dispersed by ants.

About 40 spp., in Europe, W. Asia and N. Africa.

1. K. arvensis (L.) Coult.   'Field Scabious.'

Scabiosa arvensis L.

A perennial herb with a branched ± erect stock and tap-root. Stem 25–100 cm., erect, ± terete, simple or branched, rough at least below with downwardly directed bristles. Basal lvs in an overwintering rosette from the base of the old flowering stem, oblanceolate, short-stalked, commonly simple, sometimes lyrate-pinnatifid, entire or crenate-toothed; stem-lvs usually deeply pinnatifid with an elliptical terminal segment and linear-oblong lateral segments; some of the uppermost usually ± entire; all dull green, ± hairy. Heads (1·5–)3–4 cm. diam., stalks hairy. Involucral bracts ovate-lanceolate, hairy, in 2 rows, shorter than the fls. Fls bluish-lilac, the *marginal larger than the central*. Calyx with 8 ciliate teeth. Corolla unequally 4-lobed. Fr. 5–6 mm., densely hairy. Fl. 7–9. Protandrous and visited by bees and butterflies; some plants have smaller heads of female fls (gynodioecious). $2n = 40$; 20. Hs.

Very variable, especially in the form of the lvs.

Native. Dry grassy fields, dry pastures, banks, etc., to 1200 ft. in Derby; less common in the north. 109, H40, S. Throughout the British Is. except the Outer Hebrides and Shetland. Europe northwards to 68° 50′ N.; Caucasus; W. Siberia.

3. SCABIOSA L.

Annual to perennial herbaceous or rarely suffruticose plants with lvs in opposite pairs. Heads long-stalked, convex, with an involucre of numerous herbaceous bracts in 1–3 rows. *Receptacle elongated*, with pale linear-lanceolate *non-spinous receptacular bracts*. *Involucel* cylindrical with 8 furrows, at least above, *ending* upwards *in a ± scarious pleated funnel-shaped cup*. *Calyx* cup-shaped below, usually *prolonged upwards as 5 ± spreading setaceous teeth*. Corolla with a short tube and 5 *unequal lobes*, the marginal fls usually larger than the central. Stigma ± 2-lobed. Fr. crowned by the persistent calyx, wind-dispersed.

About 80 spp. in Europe, N. Africa and W. Asia, chiefly in the Mediterranean region.

1. S. columbaria L. 'Small Scabious.'

A perennial herb with a long tap-root prolonged upwards into an erect branching stock. Flowering stems 15–70 cm., lfy, slender, terete, sparsely hairy, branching at or below the middle. Basal lvs ± long-stalked, obovate to oblanceolate, simple and crenate or ± lyrate-pinnatifid; stem-lvs successively shorter-stalked and more deeply pinnatifid with narrower segments, the uppermost with linear segments; all subglabrous to pubescent. Heads 1·5–3·5 cm. diam., their long slender stalks usually pubescent. Involucral bracts c. 10, in 1 row, linear-lanceolate, shorter than the fls. *Corolla* bluish-lilac, rarely pinkish or white; those *of outer fls much larger than of central fls*. Fruiting head ovoid. Fr. 3 mm., its involucel deeply 8-furrowed, hairy, with the scarious cup 1·5 mm.; calyx-teeth 3–6 mm., setaceous, blackish. Fl. 7–8. Homogamous to protandrous and visited by bees and Lepidoptera. $2n = 16$. Hs.

Native. Dry calcareous pastures, banks, etc., to 2000 ft. in N. England; locally common. 72, S. Great Britain northwards to Stirling and Angus. Europe to the Arctic Circle; W. Asia and Siberia; N. Africa.

*S. atropurpurea L. (S. Europe), an annual herb with branching stems 30–60 cm., lyrate-pinnatifid coarsely toothed basal lvs and pinnatifid stem-lvs, and long-stalked heads of dark purple, rose or white fls, is much grown in gardens and is naturalized near Folkestone and Newquay (Cornwall).

S. caucasica Bieb. (Caucasus) is the large blue scabious much grown in gardens and for cutting.

4. SUCCISA Haller

A perennial herb with lvs in opposite pairs. Heads long-stalked, hemispherical, with an involucre of numerous herbaceous bracts in 2–3 rows. *Receptacle elongated, not hairy,* with *non-spinous receptacular bracts. Involucel* 4-angled with 2 furrows in each face, *ending above in 4 erect triangular herbaceous lobes. Calyx* a short cup *prolonged upwards as* 5(–4) *setaceous teeth. Corolla almost equally 4-lobed,* those of the *marginal fls not appreciably larger than* those of the *central.* Calyx persistent in fr.

One sp.

1. S. pratensis Moench   Devil's-bit Scabious.

Scabiosa Succisa L.

Rootstock short, erect, *premorse,* with long stout roots. Stem 15–100 cm., erect or ascending, ± terete, subglabrous or appressed-hairy above. Basal lvs 5–30 cm., in a rosette, obovate-lanceolate to narrowly elliptical, blunt or acute, narrowed into a short stalk, firm, reticulate beneath, usually entire, sparsely hairy, rarely glabrous; stem-lvs few, resembling the basal but narrower, entire or distantly toothed, sometimes all ± bract-like. Gynodioecious. Heads 1·5–2·5 cm. diam., the female usually smaller than the hermaphrodite. Involucral bracts broadly lanceolate, pubescent, ciliate. Receptacular bracts elliptical, herbaceous, purple-tipped, much exceeding the calyx-teeth. Corolla 4–7 mm., usually smaller in female heads, mauve to dark blue-purple, rarely white or pink. Anthers 2–2·5 mm., red-purple and much exserted in hermaphrodite heads; smaller, whitish and not or hardly exserted in female heads. Fr. 5 mm., downy, surmounted by the involucel and the 5(–4) persistent reddish-black calyx-teeth which fall short of the lf-like bracts. Fl. 6–10. Protandrous; visited by butterflies and various bees. There is a variable degree of abortion of the anthers in 'female' heads. $2n=20$. H.

Native. Marshes, fens, meadows and pastures, damp woods; common. Throughout the British Is. 112, H40, S. Europe northwards to 68° 21′ N.; Caucasus; W. Siberia; N. Africa.

120. COMPOSITAE

Herbaceous or sometimes woody plants of very diverse habit; often with latex or oil-canals, rarely with both. Lvs exstipulate. Fls small (*florets*), aggregated into heads (*capitula*) simulating a single larger fl. and surrounded by a calyx-like *involucre* of one or more rows of bracts which vary greatly in number, arrangement, form and consistency. Receptacle of the head expanded and concave, flat, convex or elongated-conical, with or without *receptacular scales* each subtending a floret. Florets all similar (head *homogamous*) or central and marginal florets differing (head *heterogamous*), and then the central florets usually hermaphrodite or rarely male, the outer female or rarely neuter, but there are many variations. Calyx never typically herbaceous but represented by a *pappus* either of numerous simple or feathery hairs in one or more rows, or of a smaller number of membranous scales, or of teeth or bristles, or of a continuous membranous ring; sometimes 0. Corolla gamopetalous, variable in form but of three main types: (*a*) *tubular*, actinomorphic, the corolla-tube being surmounted by 5 ±short equal teeth; (*b*) tubular, 2-lipped; (*c*) *ligulate*, the corolla-tube being prolonged only along one side as a strap-shaped '*ligule*', usually 3- or 5-toothed at its tip. In homogamous heads the florets may all be of any of these types, but in heterogamous heads the central or disk-florets are usually tubular and actinomorphic and the marginal or ray-florets usually ligulate or, if tubular, distinctly larger and more conspicuous than the disk-florets. Stamens 5, epipetalous; their anthers, often sagittate or tailed below and often with terminal appendages, are usually united laterally so that they form a closed cylinder round the style; dehiscence introrse, into the interior of the cylinder. Ovary inferior, 1-celled, with 1 basal anatropous ovule having a single integument; style single below but branching above into 2 stigmatic arms of variable length and shape. Pollination entomophilous, rarely anemophilous (*Artemisia, Xanthium,* etc.); some genera have apomictic spp. (*Taraxacum, Hieracium*). Typically the hermaphrodite floret is protandrous, pollen being shed into the anther-tube either above the closed style-arms or upon 'collecting hairs' on their outer side or on the stylar shaft below them. Pollen is thus presented to a visiting insect during or after emergence of the style and before the style-arms diverge, so that cross-pollination is favoured. Self-pollination is often also possible when, at a later stage , the style-arms curl downwards and bring the stigmatic surfaces into contact with residual pollen on the collecting hairs, but self-incompatibility is common. Fr. an achene crowned by the pappus, sometimes with a slender 'beak' between them. Dispersal usually by wind, but achenes with spiny or barbed bristles (*Bidens*, etc.) or whole heads with hooked bristles or spines on the involucre (*Arctium, Xanthium,* ect.) may be carried by animals. Seeds non-endospermic, usually oily.

Over 900 genera and 14,000 spp., throughout the world.

The largest family of flowering plants, comprising examples of almost every ecological type and life-form: trees, shrubs, perennial and annual herbs; aquatic, alpine and desert plants; climbers, succulents and spiny shrubs. There is nevertheless a striking constancy in characters of the fl. and infl., though dioecism, anemophily or reduction of the head to a single fl. are encountered in a small minority of genera.

Many spp. are cultivated as vegetables, the most familiar being:

Lactuca sativa L. (lettuce), *Cichorium endivia* L. (endive), *Helianthus tuberosus* L. (Jerusalem artichoke), *Cynara cardunculus* L. (globe artichoke and cardoon), *Tragopogon porrifolius* L. (salsify), *Scorzonera hispanica* L. (scorzonera). *Helianthus annuus* L. (annual sunflower), *Madia sativa* Mol. and *Sigesbeckia orientalis* L. are grown for their oily seeds, *Cichorium intybus* L. yields chicory, *Carthamus tinctorius* L. (safflower) an orange or red dye, and *Taraxacum bicorne* Dahlst. (*T. kok-saghyz* Rodin) is being grown extensively for its rubber-yielding latex. Many genera, including *Achillea, Aster, Calendula, Callistephus, Centaurea, Cosmos, Dahlia, Echinops, Erigeron, Gaillardia, Helenium, Helianthus, Rudbeckia, Tagetes,* etc., are valuable garden plants.

Synopsis of Classification[1]

A. *Subfamily* CARDUOIDEAE: no latex present; corolla of disk-florets never ligulate.

Tribe 1. HELIANTHEAE: lvs usually opposite; involucral bracts not broadly scarious-margined; receptacular scales present; heads usually with tubular hermaphrodite or male disk-florets and ligulate female or neuter ray-florets, sometimes of tubular florets only and then commonly dioecious; anthers rounded to acute at the base, not tailed; style-arms flat, with hairs on the outside which rarely extend below the bifurcation, each arm with 2 marginal stigmatic strips; pappus various, never of hairs, commonly of 2–3 scales or awns.

All disk-florets subtended by receptacular scales.
 Ray-florets usually present, their corollas ligulate, rarely tubular; entomophilous.
  Both disk- and ray-florets setting fr.
   Corolla of ray-florets deciduous.
    Achenes of disk-florets 4–5-angled or laterally compressed.
     Ray-florets ligulate; receptacle flat or somewhat convex.
                                                  1. HELIANTHUS
     Ray-florets ligulate; receptacle markedly conical.   2. RUDBECKIA
     Ray-florets tubular; outer involucral bracts linear-spathulate, glandular.
                                                  (SIGESBECKIA)
     Ray-florets 0; outer involucral bracts ovate-lanceolate, not glandular.
                                                  (SPILANTHES)
    Achenes of disk-florets dorsiventrally compressed.
     Corolla hairy outside at the base.   (GUIZOTIA)
     Corolla glabrous outside.
      Disk-florets with pappus 0 or of 2–3 bristles with no downwardly directed barbs.
       Perennial herbs with root-tubers; outer involucral bracts ±lf-like; heads inclined or drooping.   (DAHLIA)
       Roots not tuberous; outer involucral bracts not lf-like; heads erect.
                                                  (COREOPSIS)
      Disk-florets with a pappus of 2–3 bristles with downwardly directed barbs.
       Ray-florets yellow, white or 0.   3. BIDENS
       Ray-florets red or purple.   (COSMOS)
      Disk-florets with a pappus of several fimbriate scarious scales.
                                                  4. GALINSOGA
   Corolla of ray-florets persistent in fr.
    Lvs entire, sessile.   (ZINNIA)
    Lvs entire, stalked.   (SANVITALIA)
    Lvs toothed.   (HELIOPSIS)
  Disk-florets not setting fr.   (SILPHIUM)
 Corolla of ray-florets rudimentary or 0; anemophilous.
  Male and female florets in the same head.   (IVA)
  Male and female florets in different heads.
   Involucral bracts of male head joined below; fruiting head 1-seeded with a ring of 4–8 spines.   5. AMBROSIA
   Involucral bracts of male head free; fruiting head 2-seeded with 2 distal beak-like appendages.   6. XANTHIUM

[1] Genera with names in brackets include species found as infrequent casuals. Brief descriptions of these, but not of the genera, will be found after the accounts of native members of their tribes.

Only the outermost row of disk-florets subtended by receptacular scales.

(MADIA)

Tribe 2. HELENIEAE: like Heliantheae but lvs often spirally arranged and receptacular scales 0; conspicuous oil-glands commonly present.

Conspicuous ligulate ray-florets present.
  Lvs opposite; plants strongly aromatic.   (TAGETES)
  Lvs spirally arranged.
    Receptacle naked; involucral bracts herbaceous, not coriaceous at the base.

(HELENIUM)

    Receptacle bristly; involucral bracts herbaceous or lf-like, coriaceous at
      the base.   (GAILLARDIA)
Ray-florets 0 or few and very shortly ligulate.   (SCHKUHRIA)

Tribe 3. SENECIONEAE: lvs spirally arranged; involucre usually of a single row of
equal bracts with or without a few shorter outer bracts; margins of bracts not
scarious; receptacular scales usually 0; heads with tubular hermaphrodite disk-
florets, sometimes not setting fr.; marginal female florets ligulate, tubular or 0;
anthers and style-arms as in Heliantheae; pappus of many simple hairs.

Disk-florets not setting fr.
  Heads all similar.   9. TUSSILAGO
  Heads subdioecious.   10. PETASITES
Disk-florets fertile.
  Marginal florets with obliquely truncate filiform tube.   11. HOMOGYNE
  Marginal florets ligulate or heads homogamous.
    Receptacle convex; involucral bracts in several equal rows. 8. DORONICUM
    Receptacle flat; involucral bracts in 1 row with or without smaller ones at
      the base.   7. SENECIO

Tribe 4. CALENDULEAE: lvs usually spirally arranged; receptacular scales 0; heads
with tubular hermaphrodite disk-florets which commonly do not set fr. and
usually with ligulate female ray-florets; anthers acute or shortly tailed at the
base; style-arms usually truncate at the ends, otherwise as in Heliantheae; achenes
large, often spiny and irregularly shaped; pappus 0.

Disk-florets not setting fr.   (CALENDULA)
Disk-florets setting fr.   (DIMORPHOTHECA)

Tribe 5. INULEAE: lvs spirally arranged, often with white woolly hairs; heads
usually with tubular hermaphrodite disk-florets and ligulate or tubular-filiform
female ray-florets, sometimes ±completely dioecious; anthers with basal tail-like
appendages; style-arms various; pappus usually of simple or feathery hairs.

Ray-florets tubular-filiform; receptacular scales present in 1 row within the
  marginal florets.   14. FILAGO
Like *Filago* but receptacular scales 0; disk-florets fertile. 15. GNAPHALIUM
Like *Gnaphalium* but disk-florets sterile.   16. ANAPHALIS
Dioecious; pappus-hairs of male heads clavate, like the antennae of a butterfly.

17. ANTENNARIA

Ray-florets ligulate; receptacular scales 0; pappus of simple hairs. 12. INULA
Like *Inula* but pappus with an inner row of hairs and an outer row of small
  scales.   13. PULICARIA

Tribe 6. ASTEREAE: lvs spirally arranged; involucre usually of numerous herbaceous
bracts in several imbricating rows, progressively shorter towards the outside;
heads heterogamous with the marginal florets ligulate or filiform with short or no
ligule, or homogamous with all florets tubular; anthers usually blunt at the base
with basally inserted filaments; style-arms broad and flat, densely hairy outside

towards their tips, each with 2 well-marked marginal stigmatic strips; pappus various or 0.

Pappus present.
  Pappus of 2–8 deciduous awn-like bristles; ray-florets yellow.

(GRINDELIA)

  Pappus of persistent hairs or bristles, sometimes with an outer row of scales.
    Shrubs; pappus of many unequal bristles; ray-florets whitish.   (OLEARIA)
    Herbs.
      Ray-florets yellow, ligulate; pappus-hairs equal.   18. SOLIDAGO
      Ray-florets usually blue, red or white, ligulate, or if filiform then with
        definite ligules at least equalling the pappus.
      Ray-florets in 1 row, not filiform; pappus of hairs only.   19. ASTER
      Ray-florets in 1 row, not filiform; like *Aster* but with lf-like involucral
        bracts and a row of concrescent scales outside the pappus-hairs.

(CALLISTEPHUS)

      Ray-florets in 2 or more rows, ligulate, narrow, or if filiform at least the
        outermost with long ligules; pappus of hairs.   20. ERIGERON
      Ray-florets filiform with or without tiny ligules (whitish in *C. canadensis*)

21. CONYZA

      Ray-florets 0; all florets tubular, yellow.   19. ASTER
Pappus 0.   22. BELLIS

Tribe 7. EUPATORIEAE: lvs often opposite; florets all tubular and hermaphrodite, never pure yellow; anthers blunt at the base with basally inserted filaments; style-arms hairy outside, each with 2 short marginal stigmatic strips.

Pappus of many simple hairs.   23. EUPATORIUM
Pappus of 5–20 scales.   (AGERATUM)

Tribe 8. ANTHEMIDEAE: lvs usually spirally arranged and commonly pinnatifid; involucral bracts usually with a broadly scarious margin; receptacular scales present or 0; heads with tubular, hermaphrodite, usually fertile, disk-florets; ray-florets ligulate, and female or neuter, sometimes 0; anthers blunt or rounded at the base, not tailed; style-arms of the disk-florets usually truncate at the end, otherwise as in Heliantheae; pappus rudimentary or 0.

Receptacular scales present.
  Ray-florets tubular or 0; corolla-tube with a basal appendage covering the
    top of the achene.
    Woolly maritime herb.   27. OTANTHUS
    Strongly aromatic dwarf shrubs.   28. SANTOLINA
  Ray-florets ligulate, rarely 0 (and then corolla-tube lacking basal appendage).
    Achenes not strongly compressed.
      Achenes slightly compressed, weakly 3-veined on the posterior face;
        corolla-tube of disk-florets saccate at base (or spurred), so that it covers
        the top of the achene.   25. CHAMAEMELUM
      Achenes not or scarcely compressed, distinctly ribbed all round; corolla-
        tube not saccate at base.   24. ANTHEMIS
    Achenes strongly compressed.
      Achenes unwinged; ligules short and broad.   26. ACHILLEA
      At least the marginal achenes strongly winged; ligules longish.

(ANACYCLUS)

Receptacular scales 0.
  Heads small, drooping, inconspicuous, anemophilous, usually in elongated
    racemose panicles (solitary or 2 in a very rare Scottish alpine); florets all
    tubular.   33. ARTEMISIA

Heads ±conspicuous, not drooping, entomophilous, not in elongated racemose panicles.
  Ray-florets with conspicuous ligules.
    Receptacle hemispherical to conical; lvs repeatedly pinnatisect with the ultimate segments linear.
      Achenes horizontally truncate above and with truly basal insertion-scar, 3-ribbed on the posterior face and with 2 distinct dark-coloured oil-glands on the anterior face.   29. TRIPLEUROSPERMUM
      Achenes obliquely truncate or rounded above and with basal-lateral insertion-scar, 3–5-ribbed on the posterior face; oil-glands 0.
                                          30. MATRICARIA
    Receptacle flat; lvs simple to pinnatifid or pinnate but the ultimate segments not linear.   31. CHRYSANTHEMUM
  Ray-florets very inconspicuously ligulate or tubular or 0.
    Florets 4-merous; heads solitary, hemispherical.   32. COTULA
    Florets 5-merous.
      Receptacle flat; heads in a dense flat corymb.   31. CHRYSANTHEMUM
      Receptacle conical; heads not in a dense corymb.   30. MATRICARIA

Tribe 9. ARCTOTIDEAE: lvs usually spirally arranged; involucral bracts in many rows, the outer usually shorter, commonly spiny and often scarious-margined; receptacle often bristly, but scales 0; heads with hermaphrodite tubular disk-florets and usually neuter ligulate ray-florets; anthers blunt to acuminate at the base, but not tailed; style commonly with very short arms and with a ring of hairs well below the bifurcation; pappus of scales, a membranous ring or 0.

Involucral bracts free; rays usually white or pink; lvs cottony.   (ARCTOTIS)
Involucral bracts united below; rays yellow or orange.   (GAZANIA)

Tribe 10. CYNAREAE: lvs spirally arranged, often spinous; involucral bracts in many rows, often spinous; receptacle often bristly but subtending scales 0; florets usually all tubular and all hermaphrodite or the outermost female or neuter; anthers usually tailed at the base; style with a ring of hairs or swelling below the bifurcation; pappus usually of many hairs or scales.

Heads 1-fld, aggregated into globular secondary heads.   (ECHINOPS)
Heads with several florets, not aggregated into secondary heads.
  Achene with a straight basal insertion.
    Achene covered with silky hairs, not bordered round its upper end.
                                        34. CARLINA
    Achene usually glabrous, bordered round its upper end.
      Receptacle bristly.
        Outer involucral bracts hooked at the tip.   35. ARCTIUM
        Outer involucral bracts not hooked.
          Stamen-filaments free.
            Receptacle not fleshy.
              Pappus of simple hairs.   36. CARDUUS
              Pappus of feathery hairs.   37. CIRSIUM
            Receptacle fleshy.   (CYNARA)
         Stamen-filaments united.   38. SILYBUM
      Receptacle not bristly.
        Lvs spiny.   39. ONOPORDUM
        Lvs not spiny.   40. SAUSSUREA
  Achene with an oblique lateral insertion.
    Heads surrounded at the base by spinous lf-like bracts.
      Pappus of 1 row of scales or bristles.   (CARTHAMUS)
      Pappus of 2 unequal rows of bristles.   (CNICUS)

Heads not surrounded by lf-like bracts.
Involucral bracts with a terminal scarious or spinous appendage.
                                                            41. CENTAUREA
Involucral bracts without terminal appendages.   42. SERRATULA

B. *Subfamily* CICHORIOIDEAE: latex present; all florets ligulate, the ligules
5-toothed at their tips.

Tribe 11. CICHORIEAE: lvs radical or spirally arranged; receptacular scales present
or 0; all florets hermaphrodite; anthers acute at the base, sometimes tailed;
style-arms long, flattened and stigmatic above, rounded and hairy beneath; pappus
various but commonly of many hairs.

Thistle-like plants.   (SCOLYMUS)
Not thistle-like.
  Long filiform bracts borne beneath the involucre.   (TOLPIS)
  Filiform extra-involucral bracts 0.
    Pappus 0.
      Florets blue, rarely pink or white.   43. CICHORIUM
      Florets yellow, orange or reddish.
        Stems lfy.   44. LAPSANA
        Lvs all radical; stems markedly swollen beneath the heads.
                                                             45. ARNOSERIS
    Pappus of hairs or scales present.
      Pappus-hairs feathery.
        Receptacular scales present.   46. HYPOCHOERIS
        Receptacular scales 0.
          Involucral bracts in 1 row.   49. TRAGOPOGON
          Involucral bracts in many rows.
            Lvs all radical or stems only with a few small bracts.
                                                             47. LEONTODON
            Stems lfy.
              Stems rough with hook-tipped bristles; pappus-hairs not inter-
              locking, readily deciduous.   48. PICRIS
              Stems not rough; pappus-hairs interlocking, not deciduous.
                                                             50. SCORZONERA
      Pappus-hairs simple or rough, not feathery.
        Achenes ± strongly compressed.
          Achenes beaked or at least strongly narrowed upwards.
            Involucral bracts in many unequal rows; pappus-hairs in 2 equal rows.
                                                             51. LACTUCA
            Involucral bracts and pappus-hairs both in 2 unequal rows, inner long,
            outer short.   52. MYCELIS
          Achenes neither beaked nor much narrowed upwards.
            Florets yellow; pappus hairs in 2 equal rows.   53. SONCHUS
            Florets blue; pappus-hairs in 2 unequal rows, inner long, outer short.
                                                             54. CICERBITA
        Achenes not strongly compressed.
          Achenes truncate above, neither narrowed nor beaked; pappus brown,
          brittle.   55. HIERACIUM
          Achenes narrowed above or beaked.
            Achenes beaked and conspicuously muricate below the beak.
                                                             57. TARAXACUM
            Achenes beaked or not, not muricate.   56. CREPIS

Key to Native and Naturalized Genera

1   Plant with milky latex; florets all ligulate (subfamily Liguliflorae).   58
     Plant without milky latex; at least the central florets tubular (subfamily Tubiflorae).   2

2   Heads with distinct ligulate ray-florets.   3
     Florets all tubular (or the outermost very slender and minutely ligulate).   28

3   Ray-florets yellow.   4
     Ray-florets not yellow.   16

4   Stem-lvs all or mostly in opposite pairs.   5
     Stem-lvs all or mostly alternate or lvs all radical.   6

5   Heads not exceeding 2 cm. diam.; pappus of 2–4 persistent rigid barbed bristles.   3. BIDENS
     Heads at least 5 cm. diam., pappus of 2 deciduous scale-like awns.   1. HELIANTHUS

6   Pappus of hairs, with or without an outer row of scales.   7
     Pappus not of hairs, or 0.   12

7   Head solitary, terminal on a scaly scape and appearing before the exclusively radical true lvs.   9. TUSSILAGO
     Head not terminal on a scaly scape.   8

8   Involucral bracts in 2 or more ± imbricating rows, progressively shorter towards the outside.   9
     Involucral bracts in 1 or 2 rows, all of equal length or with a few much shorter bracts at the base.   11

9   Heads 5–10(–20) mm. diam., usually in elongated racemes or racemose panicles.   18. SOLIDAGO
     Heads 10 mm. diam. or more, usually solitary or in cymose infls.   10

10   Pappus of an inner row of hairs and an outer row of scales.   13. PULICARIA
     Pappus of 1 row of hairs.   12. INULA

11   Heads 4 cm. or more in diam.; involucral bracts in 2 equal rows.   8. DORONICUM
     Heads not exceeding 3·5 cm. diam.; involucral bracts equal in 1 row or with a few much shorter basal bracts.   7. SENECIO

12   Heads tiny (c. 4 mm. diam.) in dense compound corymbs; ray-florets 4–6.   26. ACHILLEA
     Heads at least 2·5 cm. diam.   13

13   Heads at least 6 cm. diam.; receptacular scales present.   14
     Heads less than 6 cm. diam. (or if as large, then with no receptacular scales).   15

14   Lvs simple; receptacle not strongly conical.   1. HELIANTHUS
     Lower lvs commonly pinnatifid; receptacle strongly conical.   2. RUDBECKIA

15   Lvs green, deeply pinnatifid with toothed lobes, densely hairy beneath; receptacular scales present.   24. ANTHEMIS
     Lvs glaucous, toothed or pinnately lobed, glabrous; receptacular scales 0.   31. CHRYSANTHEMUM

16   Lvs opposite; heads less than 1 cm. diam.; ray-florets usually only 5, very small, white.   4. GALINSOGA
     Lvs alternate or all radical.   17

17   Pappus of hairs.   18
      Pappus not of hairs, or 0.   22

18   Shrub with coarsely toothed lvs and small heads with whitish ray-florets.
                                             Olearia macrodonta (p. 846)
      Herbs.   19

19   Ray-florets not extremely narrow, in 1 row.   20
      Ray-florets very narrow, usually in 2 or more rows.   21

20   Ray-florets blue, reddish or white; stem lfy.   19. ASTER
      Ray-florets yellowish-white; lvs mostly radical.   7. SENECIO

21   Heads solitary (–3) or in ±lax corymbs; ray-florets with evident and
      often spreading ligules.   20. ERIGERON
      Heads small, numerous, in elongated panicles; ray-florets with tiny erect
      whitish ligules.   21. CONYZA

22   Lvs all radical; heads solitary.   22. BELLIS
      Lvs not all radical.   23

23   Receptacular scales present.   24
      Receptacular scales 0.   26

24   Heads 4–6 mm. diam. or, if larger, then lvs simple, narrow, finely
      toothed; ray-florets with short broad ligules; achenes strongly com-
      pressed.   26. ACHILLEA
      Heads exceeding 1 cm. diam.; lvs finely divided into narrow segments;
      ray-florets with long narrow ligules; achenes not or little compressed.   25

25   Annual; tube of disk-florets flattened and somewhat winged below;
      achenes ribbed on both faces; truncate above.   24. ANTHEMIS
      Perennial; tube of disk-florets not flattened or winged; achenes ribbed
      only on 1 face, rounded above.   25. CHAMAEMELUM

26   Lvs finely divided into linear segments; involucral bracts equal.   27
      Lvs not divided into linear segments; outer involucral bracts shorter than
      inner.   31. CHRYSANTHEMUM

27   Heads with ray-florets soon reflexed; receptacle conical from the first,
      hollow; achene with 3–5 slender ribs on one face; oil-glands 0.
                                     30. MATRICARIA
      Heads with ray-florets spreading until near end of flowering; receptacle
      becoming conical in fr., solid; achenes with 3 broad corky ribs on
      one face and 2 conspicuous dark oil-glands on the other.
                             29. TRIPLEUROSPERMUM

28   Heads unisexual, greenish, made conspicuous only by the yellow anthers;
      female heads axillary, solitary or clustered, their involucres either
      densely covered with hooked spines or with a single ring of straight
      spines.   29
      Heads not both unisexual and spiny.   30

29   Fruiting heads covered densely with hooked spines and with 2 prominent
      terminal processes.   6. XANTHIUM
      Fruiting heads with a single ring of straight spines.   5. AMBROSIA

30   Heads small, inconspicuous, campanulate to ellipsoidal, usually numerous
      in long racemose panicles; lvs in native and naturalized spp. pin-
      natifid or pinnate (rarely heads 1(–3), and then a very dwarf Scottish
      alpine with silky-white subpalmate lvs).   33. ARTEMISIA
      Heads not as above, or, if so, then lvs not pinnatifid or pinnate.   31

31 Stem-lvs all or mostly in opposite pairs. *32*
    Stem-lvs all or mostly alternate, or all lvs radical. *33*

32 Florets pale pink; pappus of hairs.   23. EUPATORIUM
    Florets yellow; pappus of 2–4 stiff barbed awns.   3. BIDENS

33 Heads purplish or white in stout racemes appearing early in the year,
    before or with the large cordate radical lvs.   10. PETASITES
    Not as above. *34*

34 Florets red, purplish or blue, rarely white. *35*
    Florets not red, purplish or blue; commonly yellow. *44*

35 Radical lvs long-stalked, reniform, to 4 cm. across; stem with c. 2 distant
    ± scale-like lvs and a solitary pale violet head; a rare Scottish alpine.
                11. HOMOGYNE
    Not as above. *36*

36 Involucral bracts slender, spreading, hook-tipped, the fruiting head
    forming a ± globular adhesive bur containing several achenes.
                35. ARCTIUM
    Involucral bracts not slender, spreading and hook-tipped. *37*

37 Involucral bracts with a scarious or spinous terminal appendage, dif-
    fering in colour and texture from the basal part; pappus 0, or of scales,
    or of hairs not exceeding the length of the achenes.
                41. CENTAUREA
    Involucral bracts spinous or not, but not appendaged; pappus of hairs
    exceeding the length of the achenes. *38*

38 Pappus-hairs feathery, i.e. with slender lateral hairs readily visible (when
    dry) to the naked eye. *39*
    Pappus-hairs simple or toothed but not feathery to the naked eye. *41*

39 Lvs spinous at least at the margins; involucral bracts usually distinctly
    spine-tipped (or sometimes appressed and mucronate but not evidently
    spine-tipped).   37. CIRSIUM
    Neither lvs nor involucral bracts spinous. *40*

40 Heads usually 4–10 in a close corymb; lvs entire or distantly toothed;
    a rare alpine.   40. SAUSSUREA
    Heads usually solitary, rarely 2–3; lvs simple (rarely pinnate-lobed),
    fringed with fine prickles.   37. CIRSIUM

41 Neither lvs nor involucral bracts spinous.   42. SERRATULA
    Lvs and involucral bracts spinous; thistles. *42*

42 Lvs glabrous, conspicuously white-veined above; stem not spinous-
    winged.   38. SILYBUM
    Lvs not conspicuously white-veined above, decurrent down the stem in
    spinous wings. *43*

43 Lvs white with dense cottony hairs above and beneath; stem cottony-
    white; achenes 4-angled.   39. ONOPORDUM
    Lvs not white on both sides; achenes not 4-angled.   36. CARDUUS

44 Involucre spinous. *45*
    Involucre not spinous. *46*

45 Lvs spinous; inner involucral bracts yellow, scarious, spreading in fr.
                34. CARLINA
    Lvs not spinous; stem with wavy non-spinous wings; involucral bracts
    with a terminal spinous appendage.   41. CENTAUREA

46 Lvs simple, entire or toothed.   *47*
      Lvs pinnately lobed or divided.   *54*

47 Lvs glabrous.   19. ASTER
      Lvs not glabrous.   *48*

48 Pappus 0; a very rare white-woolly maritime plant.   27. OTANTHUS
      Pappus of hairs.   *49*

49 Lvs hairy, at least beneath, but not woolly or cottony beneath; involucral
        bracts neither woolly nor wholly scarious (though they may have
        scarious margins).   *50*
      Stem-lvs linear to broadly lanceolate or elliptical, cottony or woolly at
        least beneath; heads usually small and inconspicuous, but if as much
        as 1 cm. diam., then with scarious or woolly bracts.   *51*

50 Heads c. 1 cm. diam., in a corymb; lvs ovate to ovate-oblong, pubescent
        especially beneath.   12. INULA
      Heads 3–5 mm. diam., in a long panicle; minute whitish ray-florets
        present; lvs narrowly lanceolate to linear, hairy and ciliate.
                                                            21. CONYZA

51 Annual woolly herbs with usually erect stem-lvs and cymosely branching
        stems, the main stem and the successively overtopping branches each
        ending in a small roundish cluster of heads; outer involucral bracts
        herbaceous, ±woolly; receptacle with marginal scales.   14. FILAGO
      Perennial or rarely annual herbs; heads variously arranged, but if as in
        Filago then with spreading stem-lvs; all involucral bracts scarious;
        receptacular scales 0.   *52*

52 Perennial herbs, 5–25 cm., with basal rosettes of obovate-spathulate lvs
        and long lfy stolons; stem-lvs erect, 1–2 cm.; heads in a close terminal
        cluster; involucral bracts white or pink; dioecious.
                                                     17. ANTENNARIA
      Not as above.   *53*

53 Perennial herb, 25–100 cm., with short stolons; stem-lvs spreading,
        5–10 cm.; heads in a cymose panicle; involucral bracts shining white;
        subdioecious.   16. ANAPHALIS
      Perennial herbs with heads in spike-like racemes, or annuals with heads
        in roundish clusters; involucral bracts straw-coloured or brownish;
        not dioecious.   15. GNAPHALIUM

54 Aromatic shrub with whitish-tomentose pinnatifid lvs.
                                                            28. SANTOLINA
      Herbs.   *55*

55 Pappus of hairs.   7. SENECIO
      Pappus not of hairs.   *56*

56 Florets 4-merous; heads solitary, hemispherical.   32. COTULA
      Florets 5-merous.   *57*

57 Lvs repeatedly divided into linear segments; receptacle conical.
                                                            30. MATRICARIA
      Lvs not divided into linear segments; receptacle flat; heads numerous,
        in a flat-topped corymb.   31. CHRYSANTHEMUM

58 Pappus of scales or 0.   *59*
      Pappus, at least of the central achenes, of hairs.   *61*

59 Heads blue; pappus of scales.   43. CICHORIUM
      Heads yellow; pappus not of scales.   *60*

60   Stem lfy.   **44. LAPSANA**
       Lvs all radical.   **45. ARNOSERIS**

61   All pappus-hairs, or the inner of 2 rows, feathery, i.e. with slender lateral
       hairs visible (when dry) to the naked eye.   *62*
       Pappus-hairs simple or shortly toothed but not feathery to the naked eye.   *66*

62   Involucral bracts in 1 row; stem-lvs markedly sheathing; heads yellow or
       purple.   **49. TRAGOPOGON**
       Involucral bracts in more than 1 row; heads yellow.   *63*

63   Outer involucral bracts either narrow and lax or broadly cordate; stems
       rough with hooked bristles.   **48. PICRIS**
       Outer involucral bracts appressed, not broadly cordate.   *64*

64   Stem-lvs linear-lanceolate to linear, entire; lateral hairs of pappus inter-
       locking. Very rare marsh plant.   **50. SCORZONERA**
       Stem-lvs minute or 0.   *65*

65   Receptacular scales present.   **46. HYPOCHOERIS**
       Receptacular scales 0.   **47. LEONTODON**

66   Achenes strongly compressed.   *67*
       Achenes not strongly compressed.   *70*

67   Achenes beaked or at least markedly narrowed upwards.   *68*
       Achenes neither beaked nor markedly narrowed upwards.   *69*

68   Involucre of many unequal rows of bracts; pappus-hairs in 2 unequal
       rows.   **51. LACTUCA**
       Involucre of an inner row of equal long bracts and an outer row of much
       shorter bracts; pappus-hairs in 2 rows, the inner long, the outer short.
                                 52. MYCELIS

69   Heads yellow.   **53. SONCHUS**
       Heads blue or lilac.   **54. CICERBITA**

70   Lvs all radical; achenes strongly muricate at the base of the beak.
                             57. TARAXACUM
       Stem with at least 1–few small lvs; achenes not muricate.   *71*

71   Involucre commonly of many unequal rows of bracts; achene not or
       scarcely narrowed upwards; pappus-hairs brownish, brittle (but see
       also *Crepis paludosa*).   **55. HIERACIUM**
       Involucre usually of bracts of equal length except for some shorter ones
       near the base; achene usually distinctly narrowed upwards; pappus
       usually white and soft (but *C. paludosa* has the achenes not or scarcely
       narrowed above and a yellowish-white pappus: it differs from most
       Hieracium spp. in being quite glabrous).   **56. CREPIS**

Subfamily CARDUOIDEAE

Tribe 1. HELIANTHEAE

1. HELIANTHUS L.

Tall annual to perennial herbs with simple opposite or spirally arranged lvs
and large heads. Bracts herbaceous, often with lf-like tips, in 2 to several imbricating rows. *Receptacle* flat to conical, *with scales which partially enclose
the achenes at maturity*. Disk-florets hermaphrodite, tubular, yellow to brown;
ray-florets neuter, ligulate, yellow, deciduous. Achenes somewhat flattened,
slightly angled; pappus of 1–4 deciduous bristles or scales.

About 80 spp., chiefly in N. America.

The following 3 spp. are found as escapes from cultivation:

1 Annual; lvs spirally arranged, broadly ovate, the lower cordate; heads
    10–30 cm. diam. **1. annuus**
  Perennial; lvs mostly opposite, not cordate; heads not exceeding 10 cm. 2

2 Stolons tuberous; lvs ovate-acuminate with winged stalks; bracts
    lanceolate-acuminate; rarely flowering. **2. tuberosus**
  Stolons not tuberous; lvs broadly lanceolate, subsessile; bracts broadly
    elliptical; free-flowering. **3. rigidus**

*1. H. annuus L. Common Sunflower.

A large *annual* herb with a stout erect stem 75–300 cm., often unbranched.
Lvs spirally arranged, stalked, broadly ovate, the lower *cordate* at the base,
3-veined, sinuate-toothed, hispid with stiff appressed hairs above and beneath.
Heads 10–30 cm. diam., ± drooping. Involucral scales ovate-acute, ciliate.
Disk-florets brownish; ray-florets golden-yellow. Achenes 7–17 mm., obovoid-
compressed, pubescent, variable in colour but often white with black streaks;
pappus falling early. Fl. 8–10. Visited for nectar and pollen by many bees
and long-tongued flies. $2n = 34$. Th.

Introduced. Commonly cultivated for its oil-yielding achenes or as an
ornamental plant, and occasionally escaping. There are many cultivated races
differing in stature, diam. and colour of head, etc. Probably native of Mexico.

*2. H. tuberosus L. Jerusalem Artichoke.

A large perennial herb overwintering by stolons irregularly tuberized at their
tips. Stem 100–250 cm., usually branched above, scabrid below, ± pubescent
above. *Lvs* mostly opposite, ovate-acuminate, narrowed below into the
winged stalk, coarsely toothed, scabrid above and scabrid-pubescent beneath.
Heads 4–8 cm. diam., erect. Bracts lanceolate-acuminate, ciliate. All florets
yellow. Achenes with 1–4 ciliate awns. Fl. 9–11. Visited by many bees and
Diptera. $2n = 102$. Gr.

Introduced. Much cultivated for its inulin-containing tubers, called 'Jerusalem artichokes'. Frequently escaping from cultivation. Native throughout
much of Canada and U.S.A., and introduced into England in 1616. Flowers
only after a long hot summer.

*3. H. rigidus (Cass.) Desf. Perennial Sunflower.

H. scaberrimus Ell.

A large perennial herb 50–200 cm., with *non-tuberized stolons*, further distinguishable from *H. tuberosus* by the *broadly lanceolate, subsessile* ± *entire lvs,
very scabrid* above and beneath, faintly 3-veined; and by the broadly elliptical
bracts. Heads 6–10 cm. diam., long-stalked. Disk-florets becoming dark-
brown. Fl. 8–10. Visited by many bees and Diptera. $2n = 102$. Hp.

Introduced. A good deal cultivated for its bright yellow heads and often
escaping. Eastern U.S.A.

The cultivated form of *H. decapetalus* L. (*H. multiflorus* hort.) resembles *H. rigidus*
but has ovate to ovate-lanceolate *serrate* lvs with *narrowly winged stalks*, and *yellow
disk-florets*. It is frequently grown and sometimes escapes. Native in N. America.

2. RUDBECKIA L.

Annual or perennial herbs, with 1 or few large long-stalked heads. Involucre of many imbricating herbaceous bracts. *Receptacle strongly convex to elongated-conical, covered with rigid acute scales.* Disk-florets hermaphrodite, tubular, brown or purple; ray-florets neuter, ligulate, yellow, orange or red. Achenes prismatic, ± 4-angled; pappus a short cup or 0.

About 30 spp. in N. America.

Lvs variable, some pinnatifid or pinnate.   **1. laciniata**
Lvs all simple, ± entire.   **2. hirta**

*1. R. laciniata L.

A tall perennial herb with a branched creeping rhizome and erect glabrous stems 50–250 cm. *Lvs* spirally arranged, the lower simple to twice pinnatifid with acute lobes, the middle *divided into* 2–3 *± pinnatifid segments*, the uppermost ovate, simple; all entire or coarsely toothed, glabrous or with scattered hairs. Heads 7–12 cm. diam., long-stalked. Bracts ovate-oblong, with reflexed acute tips. Receptacle conical. Disk-florets brownish-black. Ray-florets golden-yellow, twice as long as the involucre. Receptacular scales equalling the achenes. Achenes 5 mm., ± 4-angled, glabrous; pappus a small 4-toothed cup. Fl. 7–10. Visited by various bees and hoverflies. $2n = 38, 76$. G.

Introduced. Grown in gardens, sometimes escaping and locally naturalized. Introduced into Europe early in the seventeenth century and now established in many parts of C. Europe.

*2. R. hirta L.

An annual to perennial herb with an erect hispid stem 5–60 cm. *Lvs* spirally arranged, oblong-lanceolate to lanceolate, the lowest narrowed into a stalk-like base, the middle and upper sessile; all hispid, *entire to remotely toothed*, distinctly 3-veined. Heads 6–8 cm., long-stalked. Disk-florets brownish-black; ray-florets bright yellow, much exceeding the involucre. Achenes 4-angled; pappus 0. Fl. 7–9. $2n = 38$. Th.–H.

Introduced. Grown in gardens and occasionally escaping but not naturalized. Established in Germany and elsewhere in C. Europe.

3. BIDENS L.

Annual to perennial herbs with opposite lvs. Heads solitary or in panicles or corymbs, heterogamous or homogamous; *bracts in* 2 *rows, the outer usually lf-like*; receptacle flat or convex, scaly. Ray-florets ligulate, neuter, or 0; disk-florets tubular, hermaphrodite. Achenes 4-angled, ± compressed, sometimes beaked; *pappus represented by* 2–4 *stiff barbed bristles*.

About 200 spp., cosmopolitan but mostly American; 5 in Europe.

1   Lvs simple, sessile; heads drooping; achenes usually with 3–4 bristles.
       1. cernua
  Most or all lvs deeply lobed or compound; heads ± erect.   2

2   Most lvs deeply cut into 3(–5) lobes; achenes greenish-brown, with downwardly-directed barbs on the angles and with few or no tubercles on the faces.   **2. tripartita**
  Most lvs pinnate with 3–5 separate lflets; achenes blackish, their angles not barbed but the faces strongly and densely tubercled.   **3. frondosa**

1. B. cernua L.   'Nodding Bur-Marigold.'

An annual herb with erect glabrous or sparsely hairy stems 8–60 cm., simple or branched above. *Lvs* 4–15 cm., *simple, lanceolate-acuminate, sessile*, the members of an opposite pair slightly connate, coarsely serrate, glabrous or slightly hairy, pale green. *Heads* 15–25 mm. diam. (without ray-florets), long-stalked, *drooping*, usually solitary at the ends of the stem and main branches. Outer bracts 5–8, lanceolate, lf-like, much longer than the inner, spreading; inner broadly ovate, dark-streaked. Ray-florets usually 0; when present (var. *radiata* DC.) c. 12 mm., spreading, yellow; disk-florets numerous, yellow; receptacular scales c. 8 mm., oblanceolate, scarious. *Achenes* 5–6 mm., straight-sided, broadening upwards, compressed, the outer with 3 and the inner with 4 barbed angles which are *prolonged upwards into* 4 (rarely 3) *barbed bristles*. Fl. 7–9. Visited sparingly by hive-bees and flies. $2n = 24$. Th.

Native. A locally common plant of ponds and stream-sides, and especially of places with standing water in winter but not during the growing season. 90, H40, S. Throughout the British Is. northwards to Moray and Argyll. Europe northwards to 60° 40′ N.; Caucasus; N. Asia. Introduced in N. America.

2. B. tripartita L.   'Tripartite Bur-Marigold.'

An annual herb with an erect usually much-branched glabrous or somewhat downy stem 15–60 cm. *Lvs* 5–15 cm., usually 3(–5)-*partite*, sometimes simple, narrowed below into a *short-winged stalk*; lflets lanceolate, acute, coarsely serrate, the terminal lflet sometimes broader and 3-lobed. *Heads* 15–25 mm. diam., solitary, *suberect*. Outer bracts 5–8, oblong, lf-like, spreading, inner broadly ovate, brownish. Receptacular scales broadly linear, equalling the achenes. Ray-florets usually 0; disk-florets yellow. *Achenes* 7–8 mm., obovoid, oblong, much compressed, glabrous, with barbed angles, 2 *of which are prolonged upwards into barbed bristles* with or without 1–2 shorter ones. Fl. 7–9. Visited by bees and hover-flies. $2n = 48$. Th.

Native. A locally common plant of ditches, pond and lake margins, stream-sides, etc. 90, H40, S. Throughout the British Is. northwards to Moray and Kintyre; Inner Hebrides. Throughout Europe to 63° 45′ N. and in W. Asia. Introduced in Australia.

*3. B. frondosa L.   Beggar-Ticks, Stick-tight.

An annual herb with erect glabrous stems 10–80 cm. or more. *Lvs* 5–15 cm., *pinnate*, with terminal lflet and 1–2 pairs of somewhat smaller lateral lflets, all lanceolate to ovate-lanceolate, sharply and irregularly toothed, stalked. Heads 15–20 mm. diam., solitary, erect, long-stalked. Outer involucral bracts 5–8, linear-lanceolate, green, lf-like; inner 6–12, oblong, blunt, brown or livid. Ray-florets usually 0; when present yellow; disk-florets orange-yellow; receptacular scales equalling the florets. *Achenes* (5–)7–10 mm., obovoid-compressed, *blackish*, their *faces tubercled* and their *angles upwardly ciliate* (but *not downwardly barbed*), 2 of them prolonged into barbed bristles. Fl. 7–10. Th.

Introduced. Established locally in wet places and waste ground. N., C. and S. America. Widely introduced.

4. GALINSOGA Ruiz & Pav.

Annual herbs with simple 3-veined lvs in opposite and decussate pairs and small heads in dichasial cymes. *Involucre of a few ovate bracts.* Heads with few, yellow, hermaphrodite, tubular, disk-florets and 4–8 white female ligulate, ray-florets. Achenes obovoid-prismatic; *pappus of several distinct scarious scales.* About 5 spp., S. America.

Stem ±glabrous; stalks of heads with short ascending hairs (less than 0·5 mm.) and some spreading glandular hairs; receptacular scales usually trifid; pappus scales not awned. **1. parviflora**

Stem hairy, its upper part and the stalks of heads ±densely clothed with spreading flexuous simple hairs (0·5 mm. or more) and glandular hairs; receptacular scales not trifid; at least the longest pappus scales narrowed into an awn. **2. ciliata**

*1. G. parviflora Cav. Gallant Soldier, Joey Hooker.

An annual herb with an erect much-branched ±*glabrous* stem 10–75 cm. Lvs opposite, stalked, ovate-acuminate, undulate and remotely toothed at the margin. Heads 3–5 mm. diam., in terminal and axillary dichasia. Involucre of 1–2 rows of ovate bracts, glandular or not. *Receptacle shortly conical, with trifid scales* finely toothed above. Ray-florets usually 5, the ligule c. 1 mm., about as broad as long, 3-lobed, somewhat exceeding the involucre. Central achenes c. 1 mm., ovoid, black, covered with short white bristles, *pappus-scales* 8–20, ±*equalling corolla and achene*, silvery, lanceolate, blunt or acute, fimbriate, *not awned*; marginal achenes flattened, 3-angled, ±curved, bristly only above, their pappus only about ¼ as long as the achene or abortive. Fl. 5–10. $2n=16*$. Th.

Introduced. Well established as a weed of arable land and waste places in London and elsewhere in S. England. Native in S. America and common in Peru, but now a cosmopolitan weed.

*2. G. ciliata (Raf.) Blake (*G. quadriradiata* auct.)

Closely resembles *G. parviflora* in general appearance but *stem* ±*hairy*, ±thickly clothed *with spreading flexuous simple hairs, 0·5 mm. or more,* as well as spreading glandular hairs above. Involucre hairy and glandular. *Receptacular scales not trifid,* finely toothed above. *Pappus scales* of central achenes *usually shorter than corolla and achene,* at least the longest narrowed above into *distinct awns*; those of the ray achenes similar but shorter. Achenes hispid. $2n=32*$, 36. Th.

Introduced. A casual recently recognized in S. England and S. Wales. Native in America from Mexico to Chile, and introduced in the United States and Europe.

5. AMBROSIA L.

Annual or perennial aromatic *monoecious* herbs with opposite or spirally arranged lvs, often lobed or dissected. Fls inconspicuous, greenish, in unisexual heads: *male heads in racemes or spikes with female heads* in the axils of lvs or bracts *at their base.* Male heads: involucre campanulate or turbinate, of joined bracts enclosing 5–20 fls with tubular corollas. Female heads: involucre obovoid or turbinate, cupule-like, enclosing a single fl. with neither corolla nor pappus; *fr. a beaked bur with a single row of prickles near the top.* About 20 spp., chiefly in America.

Lvs 1–3 times pinnatifid, some opposite, some alternate; fr. 4–5 mm., with
    subulate beak 1–2 mm.   **1. artemisiifolia**
Lvs entire or palmately 3(–5)-cleft, all in opposite pairs; fr. 6–12 mm., with
    conical beak 2–4 mm.   **2. trifida**

*1. A. artemisiifolia L.   Roman Wormwood, Ragweed.
Annual. Stem 20–90 cm., erect, bluntly 4-angled, densely covered with mostly appressed hairs. *Lvs mostly in opposite pairs, some alternate*; short-stalked, ovate in outline, *deeply pinnatifid*, dark green and appressed-hairy above, grey-felted beneath. Male heads 4–5 mm. diam., hemispherical, short-stalked, drooping, in a slender bractless terminal spike-like raceme; fls 10–15; anthers almost free. Female heads singly or in clusters of 2–4 in the axils of the upper lvs. *Fr. 4–5 mm. with a subulate beak* 1–2 mm. Fl. 8–10. Anemophilous. 2*n*=36. Th.
    Introduced. A casual, locally established. N. America (Ragweed), where it is an important cause of hay-fever.

*2. A. trifida L.   Great Ragweed.
A robust annual 50–300 cm., roughly hairy. *Lvs all in opposite pairs, entire or palmately* 3(–5)-*cleft*, the lobes ovate-lanceolate, toothed. Male heads in 1 or several racemes. *Fr.* 6–12 mm., *with a conical beak* 2–4 mm. Fl. 7–9. 2*n*=24. Th.
    Introduced. A casual. N. America.

6. XANTHIUM L.

Annual monoecious herbs with spirally arranged triangular or palmately lobed lvs, often with spines at the base of the lf-stalk. Heads solitary or in axillary clusters, the *male heads in the upper part of the plant and the female below.* Male heads: involucre subglobose, of 1 row of free bracts; receptacle cylindrical, with scarious scales; florets numerous, tubular; anthers connate, ovary rudimentary, style bifid. Female heads: *involucre ovoid, ending above in 2 horn-like processes, bracts* united and *covered with recurved hooks*; florets 2, tubular, filiform, completely enclosed by the involucre, styles protruding through lateral holes in the terminal processes; stamens 0. Anemophilous. Achenes compressed, remaining enclosed in pairs in the hardened involucres which are animal-dispersed. Five spp., weeds of cultivation in temperate and warmer regions throughout the world; probably native in S. America and spread in wool, etc.

The following spp. are frequently found as casuals in the British Is.

1 Plant with 1–2 trifid spines at the base of each lf-stalk.   **2. spinosum**
    Plant not spiny.   2
2 Fruiting involucre grey-green with straight terminal processes; plant
    ± softly hairy, hardly hispid, not aromatic.   **1. strumarium**
    Fruiting involucre brown, with incurved terminal processes; plant hispid
    and strongly aromatic.   **3. echinatum**

*1. X. strumarium L.   Cocklebur.
Plant not aromatic, grey-green. Stem 20–75 cm. Lvs long-stalked, ± triangular, cordate at the base, coarsely toothed to palmately lobed. Stem and lvs *somewhat*

rough with short weak hairs, the lvs rarely glandular. Male heads 5–6 mm. diam., spherical. *Fruiting heads* 14–18 mm., ellipsoidal, ± *straight-beaked,* pubescent and densely covered with hooked spines except near the apex. Fl. 7–10. 2n=36. Th.

Introduced. A casual, locally established. Probably native in America, but already well established in Europe in 1542 (Fuchs) and a ruderal in most warm temperate and subtropical countries.

*2. X. spinosum L.   Spiny Cocklebur.

Stem 15–70 cm., much branched, glabrous or with short hairs, and with *strong yellow trifid spines* (1–3 cm.) at the base of each lf-stalk. *Lvs narrowly rhomboidal, cuneate at the base* and short-stalked, usually with 3–5 narrow triangular lobes; dark green above, *white-felted beneath.* Male heads 4–5 mm. diam., spherical. Fruiting heads 8–12 mm., ellipsoidal, shortly straight-beaked, somewhat pubescent and fairly densely clothed throughout with hooked spines. Fl. 7–10. 2n=36. Th.

Introduced. A casual, locally established. Native in America. Introduced in C. and S. Europe, S. Africa, Australia and New Zealand.

*3. X. echinatum Murr.

Plant yellowish-green, *aromatic.* Stem 30–75 cm. Lvs stalked, broadly ovate in outline with three blunt primary lobes, each further lobed and denticulate. Stem and lvs *very rough with stiff conical hairs,* and the *lvs covered with ± sessile yellowish glands.* Male heads 6–8 mm. across, conical. *Fruiting heads* 18–25 mm., ovoid-ellipsoidal *with strongly incurved terminal processes,* densely covered throughout with hooked spines equalling the diam. of the head and ± straight up to the hooked tip. Fl. 7–10. Th.

Introduced. Casual. Native in America. Introduced in C. and S. Europe. The closely related *X. orientale* L. has the spines of the fruiting head shorter than its diam. and each curved from near the middle.

Several other members of the Heliantheae occur as casuals. Amongst the most frequently encountered are:

*Guizotia abyssinica (L. f.) Cass. An annual herb with erect stems 1–2 m., glabrous below, glandular-hairy above. Lvs opposite or the uppermost alternate, sessile, lanceolate, distantly toothed. Heads 2–3 cm. diam.; outer involucral bracts 5, ovate-lanceolate, herbaceous or lf-like, the inner scarious. Disk-florets tubular, herma-phrodite; ray-florets c. 8, shortly ligulate, female; all with yellow corollas hairy outside at the base. Achenes 3·5–5 mm., 3–4-angled, broadened upwards; pappus 0. E. Africa; cultivated there and in India, W. Indies, etc. for the oil from the achenes which is used as a food, as burning oil and in soap manufacture.

*Spilanthes oleracea L. An annual ± glabrous herb with branched ascending stems to 30 cm. Lvs opposite, stalked, broadly ovate-deltoid to ovate-cordate, sinuate-toothed. Heads c. 14 mm., solitary, axillary, long-stalked. Florets yellow to brownish, all tubular and hermaphrodite. Achenes ciliate; pappus of 2 bristles. Cultivated in the tropics and subtropics as a salad plant and formerly grown in Europe as an antiscorbutic and medicinal herb.

*Sanvitalia procumbens Lam. An annual ± procumbent herb with trailing hairy stems rising to c. 15 cm. above the ground. Lvs 2–4 cm., opposite, short-stalked, ovate-rhomboid, entire, conspicuously 3-veined, scabrid. Heads 2–3 cm. diam., solitary, outer involucral bracts herbaceous above. Disk-florets tubular, herma-phrodite, almost black; ray-florets ligulate, female, golden-yellow. Marginal achenes angled and winged and crowned with 1–3 stout bristles, those of the disk flattened and often awnless. Mexico. Often grown in gardens.

*Sigesbeckia orientalis L. An annual glandular-pubescent aromatic herb with an erect stem 25–90 cm., and opposite lanceolate to rhomboid lvs 3–10 cm. Heads small, in a lax panicle; outer involucral bracts linear-spathulate, spreading, glandular, the inner shorter, boat-shaped, they and the receptacular scales clasping the marginal and central achenes respectively. Disk-florets tubular, hermaphrodite; ray-florets shortly ligulate, female; all yellow. Pappus 0. Throughout the warmer parts of the world and grown in India as a medicinal herb.

*Madia capitata Nutt. An annual herb, smelling of bitumen, with erect glandular-viscid stem 30–90 cm.; lvs lanceolate, acute, entire, sessile and semi-amplexicaul, densely glandular, the lower opposite, upper alternate. Heads small, short-stalked; involucre of 1 row of ovate concave bracts which enclose the marginal achenes; receptacular scales in 1 row between the disk- and ray-florets. Disk-florets tubular, hermaphrodite; ray-florets shortly ligulate, deeply 3-lobed, female; all yellow. Achene c. 7 mm., broadening upwards, blackish; pappus 0. S. America. Cultivated in many warmer parts of the world as a pasture plant and for the edible oil expressed from the achenes. Western N. and S. America.

*M. glomerata Hook. has the heads clustered, with 5 or fewer ray-florets in each.

*Hemizonia pungens Torr. & Gray. An annual herb with a branched hairy stem c. 50 cm., and spinous subulate lvs c. 1 cm., with long white hairs. Heads small, sessile, with narrow spinous involucral bracts and deeply 3-lobed pale yellow ray-florets; disk-florets sterile. Achenes pyramidal; pappus 0. N. America ('Spike-weed').

*Iva xanthifolia Nutt. (*Cyclachaena xanthifolia* (Nutt.) Fresen.)

An annual anemophilous herb, related to *Ambrosia*, with erect branching stems 90–180 cm., glabrous below, sparsely hairy above. Lvs to 15 × 10 cm., ± opposite, broadly ovate-deltoid, stalked, coarsely and irregularly toothed, hairy beneath. Heads small, greenish, clustered in terminal and axillary bractless panicles. Florets monoecious, each head with 8–20 male florets surrounding 1–5 female florets, the latter with abortive corollas. Achenes glabrous, unwinged; pappus 0. N. America (Prairie Ragweed): an important cause of hay-fever.

Several garden plants which occasionally escape are also members of the Heliantheae: *Coreopsis tinctoria* Nutt. and other *Coreopsis* spp., *Heliopsis scabra* Dunal, *Cosmos bipinnatus* Cav., *Dahlia pinnata* Cav. and *D. merckii* Lehm., *Zinnia elegans* Jacq., etc.

Tribe 2. HELENIEAE

The related tribe Helenieae has no native members but *Helenium nudiflorum* Nutt., *H. autumnale* L., *Gaillardia pulchella* Foug., *G. aristata* Pursh, *Tagetes patula* L., *T. erecta* L. and *T. tenuifolia* Cav. (*T. signata* Bartl.) are familar garden plants. Occasionally found as casuals are:

*Tagetes micrantha Cav. Differs from the African and French marigolds (*T. erecta* and *T. patula*) in its filiform subulate entire lflets and smaller fls. Mexico. ***T. minuta** L. is a still smaller plant.

*Schkuhria pinnata (Lam.) Thell. An annual herb with slender stems 30–50 cm., and alternate pinnate lvs with a few distant linear punctate pinnae. Heads small, turbinate, long-stalked, in a lax panicle; involucral bracts few, punctate, with scarious margins. Disk-florets 3–9, tubular, hermaphrodite; ray-florets 0–2, very shortly ligulate, female; all yellow. Achenes 4-angled, broadening upwards, crowned by 8 scarious pappus scales, with laciniate margins, alternately awned and awnless. Mexico.

In var. *pinnata* the pappus scales are all awnless.

Tribe 3. SENECIONEAE

7. SENECIO L.

Annual or perennial herbs, shrubs or trees, some climbers and many succulents; lvs spirally arranged. Heads solitary or in corymbs, usually heterogamous; *bracts mostly in* 1 *row* with a few outer short ones; *receptacle naked*. Ray-florets ligulate, female, or 0; disk-florets tubular, hermaphrodite. Achenes cylindrical, ribbed; *pappus of simple hairs*, rarely 0.

About 2000 spp., cosmopolitan: many in S. Africa, the Mediterranean region, temperate Asia and America.

1	Plant with elongated twining stem; lvs ivy-shaped with 5–7 angles or deltoid lobes and cordate or hastate base; ray-florets 0.	**16. mikanioides**
	Plant not twining; lvs not angled or deltoid-lobed.	*2*
2	Lvs ± deeply pinnatifid or pinnatisect, or some linear and some pinnatifid.	*3*
	Lvs simple, entire or toothed, not linear.	*13*
3	Lvs 3–4 cm., linear, entire, toothed or with short spreading lobes, their margins revolute; plant ± glabrous.	**17. inaequidens**
	Lvs not linear with revolute margins.	*4*
4	Robust glabrous perennial, to 2 m.; lvs broad in outline, pinnatifid, pale beneath, the stalks of the lower with broadened clasping base; heads numerous, small, with only 3–4 ray-florets.	**18. tanguticus**
	Not as above.	*5*
5	Lvs densely white-felted beneath.	**14. cineraria**
	Lvs glabrous, pubescent or cottony beneath.	*6*
6	Heads with ligule of ray-florets short (5 mm. or less) and soon becoming revolute, or 0.	*7*
	Heads with ray-florets conspicuous and spreading.	*10*
7	Heads at first subsessile in dense clusters, later stalked; ray-florets usually 0 (but short in var. *radiatus*).	**8. vulgaris**
	Heads stalked from the first, in loose corymbs; ray-florets present.	*8*
8	Involucre conical; ligule of ray-florets very short and strongly revolute from the first.	*9*
	Involucre broadly cylindrical; ligule of ray-florets 4–5 mm., at first erect or spreading, then revolute; achene 3–3·5 mm.	**5. cambrensis**
9	Stem and lvs very viscid with glandular hairs; involucral bracts c. 20, the outermost 2–3 more than one-third as long as the rest; achenes glabrous.	**7. viscosus**
	Stem and lvs not or slightly glandular; involucral bracts c. 13, the outermost very short; achenes pubescent.	**6. sylvaticus**
10	Heads in dense flat-topped corymbs.	*11*
	Heads in irregular loose corymbs or cymes.	*12*

11 Lower stem-lvs with small narrow acute terminal lobe; small outer
 involucral bracts about half as long as the rest; all achenes hairy.

 3. erucifolius

 Lower stem-lvs with blunt terminal lobe; small outer involucral bracts
 less than half as long as the rest; achenes of ray glabrous, of disk
 hairy.   **1. jacobaea**

12 Weed of waste places, banks, walls, etc.; 5–13 small outer involucral
 bracts; all involucral bracts conspicuously black-tipped.   **4. squalidus**

 Biennial marsh plant; 2–5 small outer involucral bracts; bracts not black-
 tipped.   **2. aquaticus**

13 Ray-florets whitish.   **15. smithii**

 Ray-florets yellow.   *14*

14 Involucre with a few short outer bracts forming a kind of epicalyx at its
 base.   *15*

 Involucre with no short outer bracts.   *17*

15 Lvs cottony beneath; ray-florets 10–20.   **9. paludosus**

 Lvs glabrous beneath; ray-florets fewer than 10.   *16*

16 Stem-lvs thick, fleshy, decurrent; ray-florets 4–6.   **10. doria**

 Stem-lvs thin, not decurrent; ray-florets 6–8.   **11. fluviatilis**

17 Infl. a simple terminal corymb; ray-florets c. 13; achenes hairy.

 13. integrifolius

 Infl. compound, of a terminal and several lateral corymbs; ray-florets
 c. 21; achenes glabrous.   **12. palustris**

1. S. jacobaea L.   Ragwort.

A biennial to perennial *non-stoloniferous* herb with a short erect stock and
erect flowering stems 30–150 cm., furrowed, glabrous or cottony, branched
above the middle. Basal lvs in a rosette 7–15 cm. diam., usually dying before
flowering, stalked, lyrate-pinnatifid with a large ovate blunt terminal lobe
and 0–6 pairs of much smaller oblong lateral lobes, all sinuate-toothed or
pinnatifid; lower *stem-lvs* stalked, upper semi-amplexicaul, *pinnatifid to
bipinnatifid*, the *terminal lobe* not much longer than the laterals, *blunt*; all lvs
± glabrous or sparsely cottony beneath, crisped, firm, dark green. *Heads*
15–25 mm. diam. *in a large flat-topped dense compound corymb.* Bracts
oblong-lanceolate acute ± glabrous with a few subulate outer bracts less than
one-quarter as long. Ray-florets 12–15, rarely 0 (var. *floxulosus* DC.) bright
golden yellow like the disk-florets. *Achenes* 2 mm., c. 8-ribbed; those *of the
ray glabrous, of the disk hairy*; pappus twice as long as the achenes, readily
falling. Fl. 6–10. Visited by various bees and flies. $2n=40$. Hs.

 Native. A weed of waste land, waysides and neglected or overgrazed
pastures on all but the poorest soils; dunes. 112, H40, S. Abundant
throughout the British Is. Reaches 2200 ft. in Scotland. Europe to 62° 30′ N.
in Scandinavia; Caucasus; W. Asia; N. Africa. Introduced in N. America
and New Zealand.

 Poisonous to stock when eaten in quantity.

2. S. aquaticus Hill   'Marsh Ragwort.'

A usually biennial *non-stoloniferous* herb with a short ± erect premorse stock
and erect flowering stems 25–80 cm., glabrous or cottony above, often reddish, with *ascending branches* above. Basal lvs long-stalked, elliptical to

ovate, undivided, or lyrate-pinnatifid with a large ovate to ovate-oblong terminal lobe and 1–several pairs of much smaller oblong lateral lobes; lower *stem-lvs* stalked, ±lyrate-pinnatifid, middle and upper semi-amplexicaul, pinnatifid, *with the lateral lobes directed forwards*; all crenate to coarsely serrate, ± glabrous, firm, slightly waved, often purplish below. *Heads* 2·5–3 cm. diam. *in irregular lax corymbs.* Involucral bracts narrowly acuminate, green with white margins, with a few narrower and much shorter bracts at the base. Ray-florets 12–15, golden yellow. *Achenes* 2·5–3 mm., *all ±glabrous*; pappus about twice as long as the achene, readily falling. Fl. 7–8. Visited by flies. Hs. Very variable in the shape of the lvs and especially of the basal lvs.

Native. Marshes, wet meadows and ditches. 111, H40, S. Common throughout the British Is. Reaches 1500 ft. in England and Ireland. W. and C. Europe from N. Italy northwards to 62° 47′ N. in Scandinavia and eastwards to Posen and Lower Silesia.

S. erraticus Bertol., close to *S. aquaticus* but with divaricate branching, basal lvs always lyrate-pinnatifid and lateral lobes of stem-lvs spreading at right-angles to the lf-axis, is very doubtfully claimed for S. England and the Channel Is.

3. S. erucifolius L.   'Hoary Ragwort.'

A perennial *stoloniferous* herb with short creeping stock and erect sparsely cottony furrowed stems 30–120 cm., with ascending branches above. Basal lvs stalked, obovate-lanceolate, ± pinnately lobed; *stem-lvs* ovate-oblong in outline, deeply pinnatifid with a *small narrow acute terminal lobe* and long parallel *linear-oblong lateral lobes*, entire or with a few teeth; lower stem-lvs stalked, upper sessile, clasping the stem with their small entire basal lobes; all firm, somewhat revolute at the margins, *cottony, especially beneath.* Heads 15–20 mm. diam., in terminal and axillary corymbs. Bracts lanceolate-acute, cottony, the 4–6 *outer about half as long as the rest.* Ray-florets 12–14, bright yellow. *Achenes* 2 mm., *all with hairy ribs*; *pappus* about three times as long as the achenes, *persistent.* Fl. 7–8. Visited by flies and bees. 2n=40. Hs.

Native. Roadsides, field-borders, shingle-banks, grassy slopes, etc., chiefly on lowland calcareous and heavy soils. 71, H5, S. Locally common through most of England and Wales but rare in S. Scotland northwards to Fife; Ireland, only in the east. C. and S. Europe, northwards to Denmark, S. Sweden and Lithuania; W. Asia.

*4. S. squalidus L.   Oxford Ragwort.

An annual herb, rarely biennial or perennial, with ± glabrous stems 20–30 cm., decumbent at the base then erect, tough, almost woody below, flexuous, branched. Lower lvs narrowed into a winged stalk, upper semi-amplexicaul; all ± glabrous, usually deeply pinnatifid, the lobes oblong, entire or toothed, rather distant, the basal lobes of the upper lvs clasping the stem. Heads 16–20 mm. diam. in an irregular lax simple or compound corymb. *Involucre campanulate*, broadening below but remaining constricted above in older heads; inner *bracts* usually 21; outer 5–13, very short; *all conspicuously black-tipped.* Ray-florets usually 13, bright yellow, broad. Achenes

1·5–2 mm., brownish, hairy in the grooves; pappus long. Fl. 5–12. $2n=20*$. Th.–H. Very variable in the dissection of the lvs, which may be merely toothed.

Var. *leiocarpus* Druce has glabrous achenes.

Introduced. On old walls, waste ground, railway embankments, waysides, bombed sites, etc., throughout S. England to Lancs and Yorks, locally in S. Scotland and S. Ireland. 54, H6, S. Native in Sicily and S. Italy, and probably introduced from the Oxford Botanic Garden. First recorded on walls in Oxford in 1794, and now spreading very rapidly.

5. S. cambrensis Rosser

Intermediate between *S. squalidus* and *S. vulgaris* but more robust than either (to 50 cm. high). *Heads broadly cylindrical*, 6 mm. diam. (10 mm. with rays expanded). *Ray-florets* 8–15, bright yellow, short (c. 5 mm.) and broad, *soon becoming revolute*. *Achenes* 3–3·5 mm. Fl. 5–10. $2n=60*$. Hs.

Native. Roadside in 1 locality in Flint. 1. A naturally arisen allopolyploid from *S. squalidus* × *vulgaris* (which has low fertility and achenes less than 2·5 mm.).

6. S. sylvaticus L.   'Wood Groundsel.'

An annual herb with an erect slender furrowed stem 30–70 cm., ± cottony or pubescent, *not or somewhat glandular*; branches ascending. *Lvs yellow-green*, deeply and irregularly pinnatifid, the lobes unequal, ± cut, or toothed; lower lvs oblanceolate or obovate, narrowed into a short stalk, upper oblong, sessile or clasping the stem with enlarged auricle-like basal lobes; all cottony at first but *becoming ± glabrous*. Heads 7–9 × 5 mm. in a large flat-topped terminal corymb. Involucre conical, the inner bracts narrow, green, the *outer less than one-quarter their length*; all glandular-hairy. Ray-florets 8–14, bright yellow, very short and revolute. *Achenes* 2·5 mm., dark green, *stiffly hairy* on the ribs; pappus whitish, nearly twice as long as the achenes. Fl. 7–9. Visited by flies, etc. $2n=40$. Th.

Native. In open vegetation on sandy non-calcareous substrata. 110, H40, S. Locally common throughout the British Is. except the Outer Hebrides and Shetland. C. Europe from Spain, C. Italy and N. Balkans to 62° N. in Scandinavia; W. Asia.

7. S. viscosus L.   'Stinking Groundsel.'

An annual *foetid* herb with 1 or more erect *very viscid* glandular-hairy stems 10–60 cm., somewhat flexuous, usually with many spreading branches. *Lvs dark green*, glandular-pubescent and *very viscid*, deeply pinnatifid with nearly equal toothed or pinnatifid lobes; lower lvs obovate in outline, narrowed into a short stalk; upper oblong, sessile, not or slightly amplexicaul. Heads 10–12 × 8 mm., long-stalked, in a large irregular rounded compound corymb. Involucre ovoid-conical, densely glandular, the inner bracts green, linear, acute, *the outer almost half as long as the inner*. Ray-florets about 13, yellow, short and revolute. *Achenes* 3–4 mm., yellowish, strongly ribbed, *glabrous*; pappus white, very long. Fl. 7–9. Visited by flies and bees. $2n=40$. Th.

Probably native. Waste ground, railway banks and tracks, sea-shores, etc.

83, H4, S. Locally common and scattered throughout lowland Great Britain; increasing. Introduced in Ireland. Europe to 61° N. in Scandinavia and eastwards to Karelia; Asia Minor. Introduced in N. America.

8. S. vulgaris L. Groundsel.

An annual or overwintering herb with erect or ascending weak rather succulent stems 8–45 cm., glabrous or with non-glandular hairs, irregularly branched. Lvs glabrous or cottony, pinnatifid with distant, oblong, blunt, irregularly toothed lobes; lower lvs lanceolate or obovate in outline, narrowed into a short stalk, upper oblong, semi-amplexicaul, auricled. *Heads* 8–10 × 4 mm., at first *subsessile*, later stalked, in dense terminal and axillary corymbose clusters. Involucre ± cylindrical, usually glabrous, the inner *bracts* green, linear-acute, *the outer about ¼ as long*, black-tipped. Ray-florets usually 0, rarely (var. *radiatus* Koch) up to 11, shortly revolute, yellow. Achenes 1·5–2 mm., densely hairy on the ribs; pappus white, long. Fl. 1–12. Little visited by insects. $2n=40*$. Th. Very variable in the dissection of the lvs hairiness, etc.

Native. Cultivated ground and waste places. 112, H40, S. Abundant throughout the British Is., reaching 1750 ft. in Scotland. Europe to 69° N.; Asia; N. Africa. Widely introduced. Sometimes a troublesome arable weed.

9. S. paludosus L. 'Great Fen Ragwort.'

A perennial herb with shortly creeping stock and erect cottony stems 80–200 cm. *Lvs* 7–15 × 1·3–2 cm., narrowly elliptical, acute, sharply serrate; the basal lvs narrowed into a short broad stalk-like base, upper sessile, very slightly amplexicaul; all ± *cottony beneath*. Heads 3–4 cm. diam., in a usually simple terminal corymb. Involucre campanulate, the inner bracts lanceolate, the outer numerous, almost half as long as the inner, all glabrous or cottony. Ray-florets 10–16, spreading, bright yellow. Achenes 3 mm., all glabrous; pappus three times as long as the achene. Fl. 5–7. Visited by various insects. $2n=40$. Hel.

Formerly native. Fen ditches in Lincoln, Cambridge, Norfolk and Suffolk, now apparently extinct. C. Europe from Spain, N. Italy and the Balkans northwards to the Netherlands, S. Sweden and N. Russia; N. Asia.

*10. S. doria L.

A perennial herb with an oblique premorse stock and erect glabrous stems 40–150 cm., branched above. *Lvs* narrowly elliptical blunt or acutish, *entire or denticulate, glabrous, glaucous, coriaceous*; basal lvs narrowed into a short stalk, *stem-lvs* sessile, ± *decurrent*. Heads 1·5–2 cm. diam., numerous, in a large compound corymb. *Involucre* narrowly campanulate, ± *glabrous*, the bracts 6–8 mm., lanceolate, the 4 outer bracts about half the length of the inner. *Ray-florets* 4–6, bright golden-yellow, short, broad, spreading. Achenes 3 mm., ribbed, all glabrous; pappus at least three times as long as the achenes, whitish. Fl. 7–9. $2n=40$. Hel.

Introduced. Wet meadows and stream-sides. Established in a few scattered localities. S. and S.E. Europe; N. Africa.

*11. S. fluviatilis Wallr.   'Broad-leaved Ragwort.'

S. sarracenicus L., p.p.; *S. salicetorum* Godr.

A perennial herb with creeping stock and *long stolons*. Flowering stems 80–150 cm., erect, very lfy, glabrous below, slightly downy and sometimes glandular above, corymbosely branched above. *Lvs* 10–20 × 2–5 cm., elliptical acute, sessile, hardly amplexicaul, ± *glabrous*, the margins with small rather irregular forwardly directed and somewhat incurved cartilaginous teeth. Heads 3 cm. diam., numerous, in a large compound corymb. Involucre ovoid-campanulate, about as long as broad, the 12–15 inner bracts 6–8 mm., with about 5 outer *bracts* half as long, all *pubescent*. *Ray-florets* 6–8, bright yellow, spreading. Achenes 3–4 mm., ribbed, glabrous; pappus three times as long as the achenes. Fl. 7–9. Visited by bees and flies. Hel.

Introduced. Stream-sides, fens and fen-woods. 45, H14. Naturalized in many localities scattered throughout Great Britain northwards to Ross, and in Ireland. C. and S. Europe from Spain, N. Italy, N. Balkans and S. Russia northwards to the Netherlands and Esthonia; Siberia.

12. S. palustris (L.) Hook.   'Marsh Fleawort.'

Cineraria palustris (L.) L.; *S. congestus* (R.Br.) DC.

A biennial or sometimes perennial herb with a short, stout, erect stock and stout erect *woolly* flowering stems 30–100 cm., furrowed, hollow, very lfy, branched only above. Lvs 7–12 cm., broadly lanceolate-acute, entire or sinuate-toothed, pale yellowish-green, glandular-woolly; rosette lvs of the first season narrowed into a short stalk-like base, dying before flowering; stem-lvs sessile with a broad semi-amplexicaul base. Heads 2–3 cm. diam., short-stalked in small dense terminal and axillary corymbs. Involucral bracts all equal, about 10 mm., oblong-lanceolate acuminate, ± glandular-woolly. Ray-florets about 20, sulphur-yellow, spreading. Achenes glabrous, 10-ribbed. Fl. 6–7. $2n=48$. Hel.

Formerly native. Fen ditches. East Anglia, Lincoln, Sussex; but now apparently extinct. C. Europe from N. France, Denmark, S. Sweden and N. Russia to Alsace, Bohemia, Hungary and Ukraine; Siberia.

13. S. integrifolius (L.) Clairv.

S. campestris (Retz.) DC.; incl. *S. spathulifolius* auct. angl.

Perennial herb with short ± erect stock and erect stems, ± cottony, branched only in the infl. Basal lvs in a rosette, ovate to ovate-spathulate, entire to coarsely toothed, with a *winged stalk*; stem-lvs few, oblong-lanceolate to broadly lanceolate, sessile, ± amplexicaul. *Heads 1–12 in a simple sub-umbellate corymb. Involucral bracts all equal*, cottony below. Ray-florets about 13, bright yellow. Achenes 3–4 mm., ribbed, hairy. Fl. 6–7. $2n=48*$. Hs.

Two forms may be recognized in Britain, but their relationship to the numerous continental forms has not been elucidated:

Var. *integrifolius*: stems 7–30 cm.; *basal lvs* 3–5 cm., broadly ovate, *rounded at the base, entire or remotely denticulate*, narrowing into a winged stalk shorter than the blade; *stem-lvs few*, oblong-lanceolate, entire, *sometimes slightly amplexicaul*; all pubescent above and beneath, cottony when young; heads 1·5–2·5 cm. diam., 1–6, short-stalked; involucre 6–8 mm.

Native. Chalk and limestone grassland and calcareous banks. Local in England from Dorset and Kent northwards to Gloucester and N. Lincoln. C. Europe from the Alps northwards to E. Norway, S. Sweden and N. Russia.

Var. *maritimus* (Syme) Clapham (*S. spathulifolius* auct. angl. vix Turcz.): more robust, 25–60(–90) cm.; basal lvs ovate-spathulate, ± *coarsely toothed*, truncate or subcordate at base, *narrowed abruptly into a short broadly winged stalk; stem-lvs several, broadly lanceolate*, at least the lower *distinctly amplexicaul*; heads 2–2·5 cm. diam., 3–12, involucre 8–12 mm.; flowers about a fortnight earlier than var. *integrifolius*.

Native. On maritime cliffs near South Stack, Holyhead Island, and near Brough, Westmorland. Similar forms, but with longer lf-stalks, occur in N. France.

*14. S. cineraria DC.   'Cineraria.'
Cineraria maritima (L.) L.

A perennial plant woody below with erect branching white-woolly stems 30–60 cm. *Lvs* ovate or ovate-oblong, the lowest coarsely toothed or shallowly pinnate-lobed, the upper deeply pinnatifid or pinnate, short-stalked or ± sessile, with 4–6 oblong blunt often 3–5-lobed ± equal segments, densely *white-felted beneath, cottony but green above*. Heads 8–12 mm., in large flattish dense corymbs. Involucre white-woolly, the bracts oblong-lanceolate. Ray-florets 10–12, bright yellow. Achenes glabrous. Fl. 6–8. $2n = 40$. Ch.–N.

Introduced. Naturalized on maritime cliffs in S. and S.W. England, S. Wales (Glamorgan) and Ireland (Dublin); Jersey. W. Mediterranean region.

Hybrids of *S. jacobaea* with *S. aquaticus* and *S. cineraria*, and of *S. squalidus* with *S. vulgaris* and *S. viscosus* have been found in the British Is., and are intermediate between the parents and usually or always sterile. Other reported hybrids require confirmation.

*15. S. smithii DC.

A robust perennial 60–120 cm., with erect hollow stem branched only in the infl. Basal lvs with blade 5–25 cm., oblong, acute, cordate at base, long-stalked, irregularly toothed; stem-lvs acuminate, sessile, crenate-toothed; all loosely woolly at least beneath. Heads 6–12 or more in an irregular corymb. Involucre 10–12 mm., broadly campanulate. Ray-florets c. 15, yellowish-white. Achenes hairy, ribbed.

Introduced. Established in a few places on the coast in Scotland. Chile and Patagonia.

*16. S. mikanioides Otto ex Walp.   German Ivy.

A much-branched quite glabrous woody twiner reaching 6 m. Lvs 5(–7)-angled or deltoid-lobed, hastate or cordate at base, in shape resembling those of ivy, somewhat fleshy, stalked. Heads in terminal and axillary compound corymbs. Involucre of 8–9 narrow bracts. Ray-florets 0; disk-florets 8–10, yellow, much exceeding the involucre. Achenes with hispid ribs.

Introduced and rarely established. S. Africa.

*17. S. inaequidens DC.

Annual to biennial subglabrous herb with erect angled much-branched stems, 30–60 cm. Lvs 3–4 cm., *linear*, acute, *entire, toothed or pinnatifid* with 2–4 pairs of short spreading lobes, *margins revolute*, often clasping the stem with basal auricles. Heads in a lax corymb. Involucre glabrous, of c. 20 bracts with some shorter ones at the base. Ray-florets c. 10, yellow; disk-florets 60–70, yellow. Achenes puberulous.

Introduced. Occasionally established. S. Africa.

*18. S. tanguticus Maxim.

Robust perennial subglabrous herb with stems up to 3 m. *Lvs* 12–18 cm., *broadly ovate or deltoid* in outline, *pinnately divided* into narrow coarsely toothed segments, *dark green above, pale beneath*; lower lvs with stalks broadened into a ±clasping base. *Heads small, very numerous, in a terminal pyramidal panicle.* Ray-florets 3–4, yellow; disk-florets 3–4, with revolute corolla-lobes.

Introduced. Grown in gardens and sometimes establishing itself as an escape. W. China.

Amongst the several other *Senecio* spp. which occur as casuals, ***S. lautus** Sol. (temperate Australia) is the most frequently encountered and may become temporarily established. It is an erect glabrous perennial 10–70 cm., with fleshy lvs 1–6 cm., very variable in shape, from linear entire to deeply pinnatifid with linear lobes, the broader lvs usually amplexicaul. Heads loosely corymbose. Involucre 5–8 mm., its bracts brown-tipped and 2-ribbed when dry, with several small outer ones. Ray-florets 8–14, 5–10 mm. Achenes pubescent.

S. cruentus DC., 'Cineraria', is much grown in greenhouses for its red, violet, purple or blue heads.

8. DORONICUM L.

Perennial herbs with *locally tuberized stolons*, spirally arranged simple lvs and large long-stalked yellow heads. Bracts herbaceous in 2–3 almost equal rows. Receptacle convex, often hairy. Heads heterogamous; disk-florets hermaphrodite, tubular, yellow; ray-florets in 1 row, female, ligulate, yellow. Achenes ±cylindrical, ribbed; pappus of 1–2 rows of simple hairs, or 0 in marginal florets.

About 34 spp. in Europe, Asia and Africa and especially in the mountains of temperate Asia.

Basal lvs broadly ovate-cordate; uppermost lvs ovate, amplexicaul; heads
    usually several, 4–6 cm. diam. **1. pardalianches**
Basal lvs ovate-elliptical narrowed gradually into the stalk; uppermost lvs
    elliptical to lanceolate, sessile, somewhat decurrent; head usually solitary,
    5–8 cm. diam. **2. plantagineum**

*1. D. pardalianches L.   'Great Leopard's-bane.'

Herb perennating by numerous stout hypogeal stolons which are tuberized at their tips and give rise in spring to rosettes of long-stalked *broadly ovate-cordate* ± entire *lvs*, ciliate and hairy on both sides. Flowering rosettes produce erect lfy stems 30–90 cm., ±woolly with spreading hairs. Lower stem-lvs with a long stalk winged above and amplexicaul below; middle stem-lvs panduriform, amplexicaul; uppermost *ovate-amplexicaul*; all pale green, thin, hairy, with entire, crenate or sinuate-toothed margins. *Heads 4–6 cm. diam., usually several.* Involucre saucer-shaped, its bracts triangular-subulate, glandular-ciliate. Receptacle pubescent. Florets bright yellow. Achenes black, 10-ribbed; those of the disk pubescent and with a pappus; those of the ray glabrous, with no pappus. Fl. 5–7. Visited by flies, beetles and Lepidoptera. $2n = 60$. Hs.

Introduced. Woods, plantations; local. 61, H6. Scattered throughout Great Britain northwards to W. Ross and Moray. W. Europe, eastwards to Italy, W. Switzerland and W. Germany.

Formerly much cultivated as a medicinal drug.

*2. D. plantagineum L. Leopard's-bane.

Like *D. pardalianches* but stems more slender and less woolly, glandular-pubescent above. *Basal lvs ovate-elliptical narrowed gradually* into the long stalk; *uppermost lvs elliptical to lanceolate, sessile, slightly decurrent*; all with prominent curving lateral veins rather as in *Plantago major*. Heads 5–8 cm. diam., commonly *solitary*; receptacle glabrous. Fl. 6–7. $2n = c.\ 120$. Hs.

Introduced. Grown in gardens and rarely naturalized. 28. Scattered throughout Great Britain northwards to Banff. S.W. Europe (Portugal, Spain, S. France, Italy).

The common early-flowering *Doronicum* of gardens has somewhat cordate leaves and a pubescent receptacle. It is (?) always sterile. Its status is unknown.

9. TUSSILAGO L.

Heads solitary, heterogamous; involucral bracts numerous, mostly in 1 row; receptacle slightly concave, naked. Ray-florets ligulate, numerous (up to 300), in many rows, female; disk-florets few, male. Pappus of many rows of long simple hairs. One sp.

1. T. farfara L. Coltsfoot.

A perennial with long stoutish white scaly stolons, these and their short branches terminating in rosettes of lvs in whose axils arise the flowering shoots of the following season. Lvs 10–20(–30) cm. across, all radical, roundish-polygonal, very shallowly 5–12-lobed, the lobes acute and with small distant blackish teeth; ± deeply cordate at the base, with a stalk broadly furrowed above; at first white-felted above and beneath, later only beneath. *Heads* 15–35 mm. diam., *solitary, terminal on purplish scaly and woolly flowering shoots* 5–15 cm., *opening long before lvs appear*, erect in bud. Bracts numerous, linear, blunt, green or purplish, somewhat hairy, mostly in 1 row but with a few broader basal scales transitional to the scale lvs of the flowering stem. Florets bright pale yellow. After flowering the head first droops and then re-erects when the fr. is ripe, the stem meanwhile lengthening to c. 30 cm. Achenes 5–10 mm., glabrous, pale; pappus white, much longer than the achenes. Fl. 3–4. Visited chiefly by flies and bees. The fls close at night. $2n = 60$. Grh.–Hr.

Native. Abundant, especially on stiff soils, in arable fields (often a troublesome weed), waste places, banks, landslides, boulder-clay cliffs, etc.; also on dunes, screes and stream-side shingle and in seepage-fens on hillsides. 112, H40, S. Throughout the British Is. Reaches 3500 ft. on Ben Lawers. Europe northwards to 71° N. in Norway; W. and N. Asia; N. Africa. Introduced in N. America.

The dried lvs were formerly smoked as a remedy for asthma and coughs.

10. PETASITES Mill.

Perennial rhizomatous herbs with large radical lvs produced with or after the fls. ± Dioecious. Heads in spike-like racemes or panicles, small, heterogamous; bracts irregularly 2–3-rowed; receptacle flat, naked. 'Male' heads with 0–5 filiform tubular or shortly ligulate female marginal florets; 'female' heads with several rows of filiform female marginal florets and 1–5 sterile 'hermaphrodite' central florets. Achenes cylindrical, glabrous; pappus of long slender simple hairs, numerous in the female, few in the sterile florets.

Fifteen spp., in Europe, N. Asia and N. America.

1   Fls lilac, vanilla-scented, appearing at the same time as the lvs; corolla
     of marginal florets shortly but distinctly ligulate.   **3. fragrans**
     Fls not vanilla-scented, appearing before the lvs; corolla of marginal
     florets tubular, not at all ligulate.   *2*

2   Fls pale reddish-violet; lvs up to 60(–90) cm. diam., greyish beneath at
     maturity, with the lower part of each basal lobe bounded by a lateral vein.
                                                                    1. hybridus

     Fls whitish; lvs up to 30 cm. diam., white-woolly beneath at maturity,
     with no part of their basal lobes bounded by lateral veins.   **2. albus**

Section 1. *Petasites*. Corolla of the marginal (female) florets filiform, obliquely truncate, not ligulate.

1. P. hybridus (L.) Gaertn., Mey. & Scherb.   Butterbur.

Tussilago Petasites L. (male) and *T. hybrida* L. (female); *P. ovatus* Hill; *P. officinalis* Moench; *P. vulgaris* Desf.

Rhizome stout, ± horizontal, with branches up to 150 cm. *Lvs* 10–90 cm. across, mostly radical, long-stalked, roundish, deeply cordate, at first downy on both sides but later green above and *greyish beneath*; stalks stout, hollow, channelled above; blade with larger distant teeth and smaller intervening teeth; *lower part of each basal lobe bordered by a lateral vein*. Flowering stems 10–40 cm. (–80 cm. in fr.), appearing before the lvs, stout, purplish below, covered with greenish lanceolate scales often with a rudimentary blade. Heads 1–3 in the axils of linear acute bracts, pale reddish-violet; 'male' heads 7–12 mm., very short-stalked, with 0–3 female and 20–40 sterile 'hermaphrodite' florets; 'female' heads 3–6 mm., lengthening in fr., longer-stalked, with about 100 female and 1–3 sterile florets. Bracts narrow, blunt, glabrous, purplish. Achenes 2–3 mm., yellowish-brown, cylindrical; pappus whitish. Fl. 3–5. Visited chiefly by bees; only the sterile florets secrete nectar. $2n = 60$. G.

Native. In wet meadows and copses and by streams to 1500 ft. in Scotland. 110, H40, S. The male plant is locally common throughout the British Is. The female plant is not uncommon in Lancs, Yorks, Cheshire, Derby and Lincoln, but rare or absent elsewhere. Europe to 63° 26′ N. in Scandinavia; N. and W. Asia. Introduced in N. America.

*2. P. albus (L.) Gaertn.   'White Butterbur.'

Tussilago alba L.

Like *P. hybridus* but smaller. Lvs 15–30 cm. across, roundish, deeply cordate, long-stalked, glabrous above and *white-woolly beneath* when mature; margin with conspicuous large teeth (more prominent than in *P. hybridus*), the intervening spaces sharply denticulate; *basal lobes not bordered by lateral veins.* Flowering stems 10–30 cm. (–70 cm. in fr.), appearing before the lvs, becoming ± glabrous, covered with pale green lanceolate scales. Infl. not much longer than broad. Heads whitish, the male larger and more crowded. Bracts linear, ± acute, pale green, glandular. Achenes 2–3 mm.; pappus white. Fl. 3–5. Visited by various insects. $2n = 60$. G.

Introduced. Locally in waste places, roadsides, plantations and woods from Leicester, Warwick and Merioneth northwards to Aberdeen, Ireland (Down); C. and N. Europe from Spain, Italy and the N. Balkans to S. Scandinavia and C. Russia; Caucasus; W. Asia.

***P. japonicus** (Sieb. & Zucc.) F. Schmidt, which is naturalized in Devon, Bucks, Cheshire and Westmorland, is a native of Sakhalin. It is a robust herb whose orbicular sharply sinuate-toothed glabrescent lvs, reniform-cordate at the base, may attain 1 m. diam., with stalks up to 2 m. The flowering stems, appearing before the lvs, bear numerous oblong bracts and dense corymbs of whitish fl.-heads.

Section 2. *Nardosmia* (Cass.) Fiori. Corolla of the marginal female florets with a distinct but short ligule.

*3. **P. fragrans** (Vill.) C. Presl   Winter Heliotrope.

Rhizome far-creeping. *Lvs* 10–20 cm. across, long-stalked, roundish, deeply cordate, *equally serrate*, green on both sides, slightly pubescent beneath; basal lobes bordered for a short distance by lateral veins; *persistent through the winter* and until the next season's lvs appear. Flowering stems 10–25 cm. with a few scales often with rudimentary laminae. Infl. short, lax, of about 10 *pale lilac vanilla-scented heads*. Bracts narrow, acute. Male heads chiefly of tubular florets, *female of slender florets with a short broad ligule*. Fl. 1–3. Visited by flies and hive-bees. $2n = 52$. G.

Introduced. Streamsides, banks, and waste places. An escape from gardens, now naturalized in scattered localities throughout Great Britain and Ireland. 52, H 39. W. Mediterranean region.

11. HOMOGYNE Cass.

Perennial rhizomatous herbs with *reniform radical lvs* and scapes bearing a few small stem-lvs and terminating in a *solitary head* (rarely 2 or more). Involucre of a single row of bracts. *Heads heterogamous*: *central florets tubular*, hermaphrodite, fertile; *marginal with short slender obliquely truncate tube*, female, fertile. Receptacle naked. Achenes cylindrical, slightly ribbed; pappus of several rows of simple hairs.

Three spp. in the mountains of C. Europe and the Balkan Peninsula.

*1. **H. alpina** (L.) Cass.

Tussilago alpina L.

Rhizome slender, covered with woolly scales, emitting lfy stolons. Radical lvs to 4 cm. across, long-stalked, orbicular to reniform, deeply cordate at base, shallowly sinuate-toothed, dark green and glabrous above, paler and often purplish beneath, hairy on the lf-stalk and on the veins beneath. Scape 10–30 cm., cottony below and with crisped and glandular hairs above, bearing a few ovate-lanceolate scales of which the lowest may have a small blade and a sheathing base. Head 10–15 mm., usually solitary. Involucral bracts linear-lanceolate, cottony below, purplish at margins and tip. Florets somewhat exceeding the involucre, pale violet. Achenes 4–5 mm., with pure white pappus. Fl. 5–8.

Almost certainly introduced. Clova (where it was recently rediscovered); 1 locality in Outer Hebrides. 2. Mountains of C. Europe from Pyrenees to N. Balkans.

Tribe. 4. CALENDULEAE

There are no native members of the tribe Calenduleae, but several *Calendula* spp. occur as garden-escapes or casuals.

*C. officinalis L. (Pot Marigold) has the lower lvs oblong-ovate, the fruiting heads erect, and all or most of the achenes boat-shaped. This is the familiar garden plant, often escaping. ***C. arvensis** L. has all the lvs oblong-lanceolate and ±toothed, the fruiting heads drooping, the outer achenes narrow and not greatly curved, the middle boat-shaped and the innermost curved into a complete ring. A frequent casual. Both spp. are apparently native in C. Europe and the Mediterranean region.

Dimorphotheca sinuata DC. (*D. aurantiaca* hort.) is a related sp., often grown in gardens and differing from *Calendula* in that the disk-florets set fr.

Tribe 5. INULEAE

12. INULA L.

Annual to perennial herbs with spirally arranged usually simple lvs. Heads rather large, solitary or in corymbs or panicles, heterogamous; bracts herbaceous, imbricate, in many rows; receptacle flat or slightly convex, naked. Ray-florets in 1 row, ligulate, female, rarely almost tubular through reduction of the ligulate limb; disk-florets numerous, tubular, hermaphrodite; all florets yellow. Many spp. apomictic. Achenes ±cylindrical or angled; *pappus-hairs in 1 row.*

About 120 spp. in temperate and subtropical regions of the Old World.

1 Heads 6–8 cm. diam.; outer bracts broadly ovate; stem-lvs large, ovate.
                                                                    1. helenium
   Heads less than 5 cm. diam.; outer bracts narrow.   2

2 Plant viscid with glandular hairs; achenes not ribbed, narrowed above into a neck then expanding into a cupule on which the pappus-hairs are inserted.   3
   Plant not glandular-viscid; achenes ±distinctly ribbed, with neither neck nor cupule.   4

3 Stem-lvs lanceolate, amplexicaul; ray-florets much exceeding the involucre.
                                                                     6. viscosa
   Stem-lvs linear, not amplexicaul; ray-florets not exceeding the involucre.
                                                                   7. graveolens

4 Marginal florets inconspicuous, not ligulate; lvs pubescent, resembling those of *Digitalis*.   **4. conyza**
   Ligulate ray-florets present.   5

5 A maritime plant with ± linear succulent lvs often 3-toothed at the apex;
      heads up to 2·5 cm. diam.   **5. crithmoides**
   Plants not maritime; lvs lanceolate or broader, not succulent; heads
      usually more than 2·5 cm. diam.   6

6 Lvs elliptical, glabrous; outer bracts lanceolate, glabrous.   **2. salicina**
   Lvs narrowly lanceolate, softly hairy below; outer bracts linear, hairy.
                                                                  3. britannica

Section 1. *Inula.* Outer bracts broad, lf-like. Achenes 4-sided.

*1. I. helenium L.   Elecampane.

A large perennial herb with a branched tuberous stock and stout erect furrowed stems 60–150 cm., simple or corymbosely branched above, downy.
Basal lvs 25–40 cm., elliptical, narrowed into a long stalk; *stem-lvs* ± sessile,
ovate-cordate, acute, amplexicaul; all finely and irregularly toothed, ± glabrous
above, softly tomentose beneath. *Heads 6–8 cm. diam.*, solitary or 2–3 in a
corymb. *Outer bracts broadly ovate*, lf-like, softly hairy, with spreading tips;
inner oblong, scarious. Ray-florets ligulate, narrow, spreading, bright yellow
like the disk-florets. Achenes 4–5 mm., strongly 4-ribbed, rhomboidal in
section, glabrous; pappus longer than the achene, reddish. Fl. 7–8. Visited
by many bees and hoverflies. $2n=20$. H.
   Introduced. Widely scattered but uncommon in fields, waysides, waste
places, copses, etc., throughout Great Britain and Ireland, probably as an
escape from cultivation; Orkney. 73, H27, S. Probably native in C. Asia
but widely naturalized in Europe and W. Asia, and also in N. America and
Japan.

Formerly much grown for medicinal purposes, its rootstock (*Radix Helenii* or
Enulae) being used as a tonic and also as a diaphoretic, diuretic and expectorant.
The lvs are said to have been used as pot-herbs and the candied stock for sweetmeats.
The name *elecampane* is from 'Enula campana'.

Section 2. *Enula* Duby. Outer bracts narrow. Achenes with 5–12 ± distinct
ribs.

2. I. salicina L.   'Willow-leaved Inula.'

A perennial herb with white slender underground stolons and stiffly erect
brittle very lfy stems 25–50 cm., simple or corymbosely branched above,
± glabrous. *Lvs* 3–7 cm., the lowest oblanceolate narrowed to the base, the
upper elliptical sessile, cordate and half-amplexicaul at the base, acute; all
*firm, entire or remotely toothed, stiffly ciliate, glabrous or usually hairy beneath
the veins.* Heads 2·5–3 cm. diam., usually solitary or 2–5 in a corymb. Outer
bracts lanceolate, lf-like, with somewhat spreading tips, inner linear; all
glabrous with ciliate margins. Ray-florets ligulate, narrow, twice as long as
the involucre, golden yellow like the disk-florets. Achenes 1·5 mm., cylindrical, faintly ribbed, glabrous; pappus whitish. $2n=16$. Fl. 7–8. H.
   Native. Confined to the stony limestone shores of Lough Derg, Galway
and Tipperary (Ireland), growing with *Schoenus nigricans, Sesleria varia
Filipendula ulmaria*, etc. Europe to 62° 42′ N. in Scandinavia; W. Asia.
A gregarious fen plant, often with *Molinia, Carex panicea, Schoenus* spp., etc.

*3. I. britannica L.

A perennial herb with an oblique stock and an erect usually simple stem 20–60 cm., densely covered with appressed tuberclate hairs or ± glabrous. Lvs lanceolate to broadly oblong, entire or denticulate, blunt or subacute; the basal lvs narrowed into a short stalk, the upper sessile and rounded or subcordate at the base, amplexicaul; all sparsely hairy above, densely hairy to almost glabrous beneath. Heads 2–4(–5·5) cm. diam., usually solitary, rarely 2–3 in a corymb. *Bracts* in 2 rows, *linear, hardly imbricating, green, softly hairy* and ± glandular. Ray-florets ligulate, almost twice as long as the involucre, golden-yellow like the disk-florets. Achenes 1·3 mm., appressed-hairy, sometimes glandular. Fl. 7–8. $2n = 16$, (24), 32. H.

?Extinct. Established (1894) on the border of Cropston Reservoir, Leicester; probably introduced from C. Europe by waterfowl. Europe northwards to Denmark and S. Scandinavia; W. Asia. A plant of moist meadows, stream-sides, ditches, wet woods, etc.

4. I. conyza DC.   Ploughman's Spikenard.

Conyza squarrosa L.; *I. squarrosa* (L.) Bernh., non L.

A biennial to perennial herb with an oblique irregular thickened stock and erect or ascending stems 20–130 cm., simple or branched above, often reddish, softly pubescent. Basal lvs ovate-oblong, narrowed into a flattened stalk; stem-lvs elliptical to lanceolate, subsessile, acute; all irregularly denticulate, downy, especially beneath. Heads c. 1 cm. diam., numerous, in a terminal corymb. Outer *bracts lanceolate*, green, pubescent, with spreading or recurved tips; inner longer, narrower, ± scarious, ciliate, often purple. Marginal florets tubular or with a very short ligule, shorter than the inner bracts; all yellowish. Achenes c. 2 mm., dark brown, strongly ribbed, sparsely hairy; pappus-hairs reddish-white. Fl. 7–9. $2n = 32$. H.

Native. A calcicolous plant of dry or rocky slopes and cliffs and of open scrub-woodland, locally common in England and Wales northwards to Westmorland, and Northumberland. 65, S. C. and S.E. Europe northwards to Denmark; Near East; Algeria. The basal lvs are often mistaken for those of the Foxglove.

5. I. crithmoides L.   Golden Samphire.

A perennial glabrous *maritime* herb with a branched woody stock and ascending very *fleshy stems* 15–90 cm., branched above. *Lvs* 2·5–6 cm., *glabrous, fleshy, linear or oblanceolate*, narrowed below to the sessile base, those on the main stem often 3-*toothed at the apex*, otherwise entire; lateral lf-clusters in the axils of the upper lvs. Heads few, c. 2·5 cm. diam., terminating the main stem and branches in a corymb, the lvs below them small and bract-like; bracts linear-lanceolate, herbaceous, glabrous, not spreading, the inner with scarious margins. Ray-florets numerous, narrow, almost twice as long as the inner bracts, golden yellow; disk-florets orange-yellow. Achenes 2·5 mm., grey-brown, cylindrical, faintly ribbed, pubescent; pappus whitish. Fl. 7–8. $2n = 18$. H.–Ch.

Native. On salt-marshes, shingle banks and maritime cliffs and rocks on the south and west coasts of Great Britain from Essex round to Wigtown and

Kirkcudbright. S. and E. Ireland from Kerry to Dublin. 20, H5, S. Coasts of Europe and W. Asia.

Section 3. *Cupularia* (Gren. & Godr.) Benth. Achenes not ribbed, narrowed above into a neck and then expanded into a cupule upon which the pappus-hairs are inserted.

*6. I. viscosa (L.) Ait.

A perennial plant whose erect *glandular viscid stems* 40–100 cm., woody below, have a strong resinous odour. *Lvs lanceolate*, entire or distantly toothed, acute, glandular above and beneath; *stem-lvs* sessile, ± *amplexicaul.* Heads in long racemes, each with yellow ligulate *ray-florets twice as long as the involucre* whose outer bracts are herbaceous, inner scarious. Fl. 9–10. 2*n*=18. H.

Introduced. Naturalized in a few localities. Native in the Mediterranean region.

*7. I. graveolens (L.) Desf.

An annual *glandular-pubescent, viscid, strongly smelling* herb with much-branched very lfy shoots 20–50 cm. *Lvs ± linear*, entire or denticulate, sessile, *not amplexicaul.* Heads in a long racemose panicle, the shortly ligulate, often violet, ray-florets not exceeding the involucre of linear-lanceolate scarious bracts; disk-florets yellow. Fl. 8–10. Th.

Introduced. Established in a few localities. S. Europe, northwards to Paris; N. Africa.

13. PULICARIA Gaertn.

Annual to perennial herbs resembling *Inula* spp. but with a *2-rowed pappus*, the outer row of short scales, free or joined into a cup, the inner of hairs. Bracts in many rows, imbricating. Receptacle naked. Ray-florets often with much reduced ligule. Achenes cylindrical, ribbed.

About 45 spp. in Europe, Africa and W. Asia to the Altai Mountains and Tibet.

Stem-lvs distinctly cordate; ray-florets much exceeding the disk-florets; outer
    pappus-scales united in a denticulate cup.   **1. dysenterica**
Stem-lvs not or hardly cordate; ray-florets hardly exceeding the disk-florets;
    outer pappus-scales free to the base.   **2. vulgaris**

1. P. dysenterica (L.) Bernh.   Fleabane.

Inula dysenterica L.

A perennial stoloniferous herb with erect sparsely hairy stems 20–60 cm., corymbosely branched above. Basal lvs oblong, narrowed to the base; middle and upper *stem-lvs oblong-lanceolate to lanceolate, cordate at base, amplexicaul*; all 3–8 cm., entire or distantly sinuate-toothed, densely and softly hairy, especially beneath. Heads 1·5–3 cm. diam., c. 2–12 in a loose corymb; bracts linear, herbaceous with long fine scarious tips, covered with long hairs, glandular. *Ray-florets* numerous, ligulate, linear, *almost twice as long as the involucre and disk-florets*, golden yellow like the disk. Achenes

1·5 mm., hairy; outer pappus a small denticulate or crenate cup, inner of long hairs. Fl. 8–9. Visited by many insects, chiefly flies. $2n = 20$. H.

Native. A common plant of marshes, wet meadows, ditches, etc., throughout the British Is. northwards to Stirling and Kintyre; Inner Hebrides. 84, H40, S. Europe northwards to Denmark and C. Russia; N. Africa; Caucasus; Asia Minor.

2. P. vulgaris Gaertn.   'Small Fleabane.'

Inula Pulicaria L.

An annual herb with much-branched slightly glandular-pubescent stems 8–45 cm., the branches overtopping the main stem. Basal lvs oblanceolate narrowed into a stalk-like base; middle and upper *lvs* elliptical or lanceolate, rounded at the base, half-amplexicaul but *not or hardly cordate*; all 2·5–4 cm., undulate at the margin, entire or distantly sinuate-toothed, glandular-pubescent or ± glabrous. Heads c. 1 cm. diam., numerous, in a lax sub-corymbose panicle. Bracts linear, herbaceous with long fine scarious points, glandular-pubescent. *Ray-florets* in 1 row, shortly ligulate, ± erect, *hardly exceeding the involucre* and pale yellow disk-florets. Achenes 1·5 mm., hairy; outer pappus of irregular distinct narrow scales; inner of hairs little longer than the achenes. Fl. 8–9. $2n = 18$. Th.

Native. A rather rare plant of moist sandy places, pond-margins, etc., where water stands in winter but not during the growing season. 24. Decreasing. S. England and Wales north to Norfolk, Huntingdon, Leicester, Worcester, Montgomery and Merioneth. Europe north to Denmark and S. Sweden; Caucasus; W. Asia; N. and E. Africa.

14. FILAGO L.

Mostly annual herbs with the stems and spirally arranged lvs covered with woolly hairs. Heads heterogamous, small, in roundish terminal and axillary clusters; involucre 5-angled, of numerous imbricating bracts, the outer herbaceous ± woolly, the inner scarious; *receptacle conical, scaly round the margin.* Florets all tubular; outer female, with filiform corollas, in several rows, subtended by *receptacular scales like the inner involucral bracts*; central hermaphrodite, with broader corollas, often failing to set fr. Achenes not much compressed; pappus of inner florets of several rows of simple hairs, of the outer of 1 row or 0.

Twelve spp. in Europe, N. America and Argentina.

1	8–40 heads in each cluster; bracts cuspidate, not spreading in fr. 2
	Fewer than 8 heads in each cluster; bracts not cuspidate, spreading like a star in fr. 4
2	Lvs not apiculate, often undulate; 20–40 heads in each cluster; clusters not overtopped by their basal lvs. **1. germanica**
	Lvs apiculate, hardly undulate; 8–20 heads in each cluster; clusters overtopped by their 1–5 basal lvs. 3
3	Each cluster of heads with 3–5 lvs overtopping it and resembling an involucre; outer bracts with recurved points; plants white-woolly, branching from the base. **3. spathulata**

Clusters with 1–2 overtopping lvs; bracts with straight points; plant yellowish-woolly, branching above the middle. **2. apiculata**

4 Lvs linear-subulate, the uppermost much exceeding the clusters of heads. **4. gallica**

Lvs linear-lanceolate to lanceolate, the uppermost not exceeding the clusters of heads. 5

5 Bracts hairy below but scarious, glabrous and blunt at the apex; branches long, ± erect. **5. minima**

Involucral bracts woolly to the acute apex; branches short. **6. arvensis**

1. F. germanica (L.) L. Cudweed.

F. canescens Jord.

An annual herb with erect or ascending densely woolly stems 5–30(–45) cm., simple or branched at the base, main stem and branches with further branches immediately beneath the terminal clusters of heads. *Lvs* 1–2(–3) cm., erect, lanceolate, blunt or tapering to an acute apex, entire and usually *undulate*, covered with *white* woolly hairs. Heads 20–40 in ± sessile clusters c. 12 mm. diam. terminating main stem and branches and half sunk in white woolly hairs, each head c. 5 mm. *Bracts* linear, longitudinally folded, cuspidate, erect, in 5 rows; the outer bracts short, densely woolly, straight-pointed, the inner longer, scarious, yellowish, *with a yellow awn-like point*. Florets small, yellow, several rows of female surrounding the central hermaphrodite florets. Achenes 0·6 mm., somewhat compressed, papillose; pappus of inner achenes scabrid, longer than the achenes; of outer 0. Fl. 7–8. 2*n*=28. Th.

Native. A fairly common plant of heaths, dry pastures, fields and waysides, usually on acid sandy soils, throughout England, Wales, Ireland and S. Scotland and reaching E. Ross but rare in N. Scotland and not in the Hebrides, Orkney or Shetland. 97, H34, S. C. and S. Europe northwards to Denmark and Poland; Siberia and W. Asia; N. Africa; Canary Is. Introduced in N. America.

2. F. apiculata G.E.Sm. 'Red-tipped Cudweed.'

An *erect* annual herb resembling *F. germanica* but with broader oblong-oblanceolate, *apiculate*, hardly undulate *lvs*, the stems and lvs covered with *yellowish* woolly hairs. *Heads* 10–20 in a cluster, half sunk in woolly hairs and *overtopped by* 1–2 *lvs*. *Bracts* keeled, *with erect reddish points*. Fl. 7–8. Th.

Native. A rare plant of sandy fields in S. and C. England northwards to Worcester and S. Yorks. 22. Europe to S. Sweden, W. Asia, N. Africa.

3. F. spathulata C. Presl 'Spathulate Cudweed.'

An annual herb 6–30 cm., *much branched near the base*, the several *stems decumbent or ascending* with almost horizontally spreading branches above. *Lvs* 10–15 mm., obovate-spathulate to spathulate-lanceolate, entire, usually *apiculate*, hardly undulate. Stem and lvs covered with whitish woolly hairs. *Heads* c. 5 mm., 8–15 in sessile clusters, 6–12 mm. diam., ± *overtopped by their basal lvs*. *Involucre sharply 5-angled*, bracts keeled, cuspidate; the *outer* densely woolly *with* a somewhat *recurved yellowish point*; the inner longer, scarious, whitish, shining. Florets small, yellowish. Achenes 0·6 mm.; pappus of inner florets longer than the achenes, of outer florets 0. Fl. 7–8. $2n=28$. Th.

Native. A rare plant of sandy fields and waysides in Cornwall, Devon and W. Gloucester, and in S. and E. England from Dorset and Oxford to Bedford, Cambridge and Norfolk. Jersey. 27, S. S. and S.E. Europe; Near East; N. Africa; Canary Is.

*4. F. gallica L.   'Narrow Cudweed.'

An annual herb with erect slender much-branched stems 8–20 cm. *Lvs linear-subulate*, 8–20 × 1 mm., entire. Stem and lvs covered with greyish silky hairs. *Heads* 4 mm., 2–6, *in clusters much overtopped by the lvs at their base.* Outer bracts triangular-acute, woolly below, glabrous, scarious and yellowish at the tip; *inner folded, saccate below, each enclosing a marginal floret.* Achenes 0·5 mm., with transparent papillae, all with pappus-hairs in several rows. Fl. 7–9. Th.

Introduced. Naturalized in dry gravelly places in Surrey, Bucks, Herts, Essex and Suffolk, and near Edinburgh, but probably extinct in some of these localities. Netherlands, Belgium, S.W. Germany, Switzerland, S. Europe; Near East; N. Africa.

5. F. minima (Sm.) Pers.   'Slender Cudweed.'

An annual herb with slender erect or ascending stem 5–15(–30) cm., branched above the middle, the branches ± erect. Lvs 5–10 mm., linear-lanceolate, acute, entire; stem and lvs covered with greyish silky hairs. *Heads* c. 3 mm., 3–6, *in small clusters which exceed the lvs at their base.* Bracts blunt, lanceolate, woolly, *with glabrous, scarious and yellowish tips*, spreading like a star in fr. Achenes 0·5 mm., papillose; pappus of inner achenes many-rowed, deciduous, of outer 0. Fl. 6–9. $2n=28$. Th.

Native. A locally common calcifuge of sandy heaths and fields throughout Great Britain, but rare in the extreme north and not in the Hebrides, Orkney or Shetland; scattered throughout Ireland. 102, H19, S. Europe northwards to Scandinavia; Siberia.

*6. F. arvensis L.

An annual herb with an erect stem 10–40 cm., simple below but with *short lateral branches above.* Lvs lanceolate, acute. Stem and lvs white-woolly. Heads 4–5 mm., 2–7 in small clusters usually overtopped by the lvs at their base. *Bracts* linear, blunt, *with woolly hairs extending to the tip*, spreading like a star in fr. Achenes 0·6 mm.; pappus deciduous, that of the inner florets many-rowed, of the outer 1-rowed. Fl. 7–9. $2n=28$. Th.

Introduced. A casual, resembling *F. minima* but with a raceme-like arrangement of the short lateral branches and with the bracts woolly to the tip. S. and C. Europe, Scandinavia; W. Asia; Canary Is.

15. GNAPHALIUM L.

Annual to perennial herbs or dwarf shrubs with spirally arranged simple lvs usually covered with white woolly hairs. Heads small in terminal or axillary spikes, corymbs or clusters, heterogamous; *bracts* imbricate in several rows, *scarious*, about equalling the florets, *spreading in fr.*; *receptacle flat, naked.* Outer florets female, tubular, very slender, in 1 or more rows, with long slender style-arms. Achenes not or little compressed; pappus of 1 row of slender brittle hairs.

About 120 spp., chiefly in temperate regions and tropical mountains, some arctic-alpines.

1 Perennials with heads in spikes or racemes, sometimes solitary; bracts
    with a ± marginal dark band.   2
    Annuals with heads in dense clusters; bracts ± concolorous.   4

2 A densely tufted plant usually only 2–12 cm.; heads 1–7; outermost bracts
    more than half as long as the head.   **3. supinum**
    Erect not densely tufted plants usually more than 12 cm.; heads many,
    in long spikes; outermost bracts hardly as long as the head.   3

3 Lvs linear to linear-lanceolate, 1-veined, steadily diminishing in size
    upwards; heads in axillary clusters forming a long interrupted spike
    more than half the length of the stem.   **1. sylvaticum**
    Lvs lanceolate, 3-veined, not diminishing in size until well above the
    middle of the stem; spike short, compact above, sometimes interrupted
    below, about ¼ the length of the stem.   **2. norvegicum**

4 Lvs linear-lanceolate, acute; clusters of heads overtopped by lvs at their
    base; bracts brownish.   **4. uliginosum**
    Lvs oblong, lower blunt, upper subacute; clusters of heads not overtopped
    by lvs; bracts straw-coloured.   **5. luteoalbum**

Section 1. *Gamochaeta* (Wedd.) Benth. Pappus-hairs united in a basal ring and falling together.

1. G. sylvaticum L.   'Wood Cudweed.'

A perennial herb with a short oblique woody stock producing short lfy non-flowering shoots and erect simple lfy flowering shoots 8–60 cm., covered with whitish woolly hairs. Lvs of basal rosettes and lower stem-lvs 2–8 cm., lanceolate-acute, 1-veined, narrowed into a long stalk-like base; middle and upper *stem-lvs linear-lanceolate to linear, acute,* 1-*veined*, sessile, *diminishing steadily in size up the stem*; all glabrous above, woolly beneath. *Heads* c. 6 × 1·5–2 mm., solitary or in small clusters of 2–8 in the axils of the upper stem-lvs, *forming an interrupted lfy spike more than half the length of the stem.* Bracts with a central green stripe and a broad scarious margin, brown towards the apex, the outer bracts ± woolly below, the *inner about equalling the florets.* Florets pale brown, almost all female, with 3–4 central hermaphrodite florets. Achenes 1·5 mm., cylindrical, hispid; pappus reddish. Fl. 7–9. Little visited by insects. $2n = 14$, 58–60. H.

Native. Locally common in dry open woods, heaths and dry pastures on acid soils throughout the British Is. 108, H33. Europe, Caucasus, N. America.

2. G. norvegicum Gunn.   'Highland Cudweed.'

A perennial herb resembling *G. sylvaticum* in habit but only 8–30 cm. and with broader *lanceolate-acuminate* 3-*veined stem-lvs* which diminish abruptly in length only in the shorter more *compact spike*, continuous or interrupted below, occupying about ¼ *of the length of the stem.* Heads 6–7 mm., solitary or 2–3 in the axils of the upper lvs. Bracts with an olive central stripe and *dark brown* scarious margins, the *inner shorter than the florets.* Achenes 1·5 mm., cylindrical, hispid; pappus white. Fl. 8. 2*n* = 56. H.

Native. A rare plant of alpine rocks to 3600 ft. on mountains in Perth, Angus, Aberdeen, Inverness and E. Ross; also reported from Caithness. 5. Mountains of C. Europe, Balkans and Caucasus; arctic and subarctic Europe.

3. G. supinum L. 'Dwarf Cudweed.'

A *dwarf perennial tufted herb* with a slender creeping branched stock producing numerous short lfy non-flowering and simple erect or ascending flowering stems 2–12(–20) cm. Basal lvs linear-oblanceolate, stem-lvs linear acute; all 0·5–1·5(–2) cm., entire, woolly above and below. *Heads 1–7 in a short ± compact terminal spike* somewhat lengthening in fr., each campanulate, c. 6 mm. long and 8 mm. diam. Bracts in 3–4 rows, broadly elliptical with a woolly olive-coloured central stripe and broad brown scarious margins, the inner bracts almost equalling the florets. Female florets in 1 marginal row. Achenes 1·5 mm., spindle-shaped, compressed, shortly hairy. Fl. 7. 2*n* = 28*. Chh.

Native. An alpine plant of cliffs and moraines of Scottish mountains from Stirling to Sutherland, and in Skye. 16. Mountains of C. Europe and W. Asia; arctic Europe; Greenland; N. America.

Section 2. *Gnaphalium.* Pappus-hairs free to the base and falling separately.

4. G. uliginosum L. 'Marsh Cudweed.'

An annual herb with decumbent or ascending *stems* 4–20 cm., *much and diffusely branched near the base*, densely covered with woolly hairs. Lvs 1–5 cm. × 2–4 mm. narrowly oblong or spathulate, narrowed below, blunt or acute, entire, woolly on both sides. *Heads* 3–4 mm., in dense ovoid sessile terminal clusters of 3–10, *overtopped by the lvs at their base.* Bracts lanceolate, scarious, pale brown and woolly below, darker and glabrous towards the tip. Florets yellowish, mostly female with a few hermaphrodite in the centre. Achenes 0·5 mm., glabrous (or with hair-like papillae in var. *pseudopilulare* Scholz); pappus 1-rowed, deciduous. Fl. 7–8. Little visited by insects. 2*n* = 14. Th.

Native. Common throughout the British Is. in damp places in sandy fields, heaths, waysides, etc., on acid soils. 112, H40, S. Europe; W. Asia. Introduced in N. America.

5. G. luteoalbum L. 'Jersey Cudweed.'

An annual herb with erect stem 8–45 cm., usually with decumbent then *erect branches* from near the base, the main stem and branches simple below but *corymbosely branched above*, all very densely covered with white woolly hairs. Basal lvs broadly oblanceolate, usually blunt; stem-lvs oblong-amplexicaul, ± undulate, acute; all 1·5–3(–7) cm., woolly on both sides. *Heads* in dense terminal lfless clusters of 4–12, *not overtopped by basal lvs*; each head 4–5 mm., ovoid. *Bracts* elliptical, largely scarious, *shining, straw-coloured,* only the

outermost woolly below. Florets yellowish with red stigmas. Achenes
0·5 mm., brown, tubercled. Fl. 6–8. Visited by various flies and bees.
$2n = 14$. Th.

Native in the Channel Is.; probably introduced in the British Is. A rare
plant of sandy fields and waste places in Jersey and Guernsey and established
in a few localities in Hants, Norfolk, Suffolk and Cambridge where it may
be native; casual elsewhere. Europe northwards to Holstein, S. Sweden and
Lithuania and in warm temperate regions all over the world.

*G. undulatum L. (S. Africa), 20–80 cm., more robust than *G. luteoalbum*,
and with the *lvs decurrent, green and asperous above*, is established in the
Channel Is.

16. ANAPHALIS DC.

Perennial woolly herbs closely resembling *Antennaria* spp. but usually more
robust and less completely dioecious.

About 30 spp., chiefly in temperate and tropical Asia.

*1. A. margaritacea (L.) Benth.   Pearly Everlasting.

Gnaphalium margaritaceum L.; *Antennaria margaritacea* (L.) S. F. Gray

A perennial subdioecious herb with erect robust lfy flowering stems 30–
100 cm., woolly with white hairs. Lvs 6–10 × 1–1·5 cm., elliptical, acute,
± entire, woolly beneath, becoming ± glabrous above. Heads 9–12 mm.
diam., numerous in terminal corymbs. *Bracts* oblong, brown below, the
outer woolly, inner glabrous, all *with a shining white scarious rounded apex*.
Florets of the male plants all without ovaries; those of the female without
anthers except for a few central hermaphrodite florets; corolla yellowish.
Achenes spindle-shaped, papillose; pappus-hairs slender in fertile florets but
thickened above (as in *Antennaria*) in male florets. Fl. 8. $2n = 28$. H.

Introduced. A N. American plant long cultivated in gardens as an 'everlasting' and naturalized in moist meadows, by rivers, on wall-tops and in
sandy and waste places in many localities but especially in Monmouth,
Glamorgan and Merioneth, and in Aberdeen and Selkirk.

17. ANTENNARIA Gaertn.

Perennial herbs or dwarf shrubs with spirally arranged narrow simple entire
lvs, the stem and lvs covered with whitish woolly or silky hairs. *Heads
± dioecious, homogamous, aggregated into umbels or clusters, bracts* closely
imbricate, *scarious, white or coloured*, often perianth-like, *not spreading in fr.*;
heads of female plants with filiform tubular florets each with many-rowed
pappus-hairs; *of male plants* with broader hermaphrodite but usually sterile
tubular florets, each *with a pappus of a few hairs thickened above like the
antennae of a butterfly*. Achenes hardly compressed.

About 15 spp., arctic-alpines, chiefly of the northern hemisphere.

Close to *Gnaphalium* but dioecious or nearly so.

1. A. dioica (L.) Gaertn.   Cat's-foot.

Gnaphalium dioicum L.

A perennial with above-ground creeping woody stock producing lfy *stolons rooting at the nodes* and erect simple woolly flowering shoots 5–20 cm. *Lvs* 1–4 cm., mostly in rosettes at the ends of the stock and stolons, obovate-spathulate, blunt or apiculate; upper stem-lvs erect and appressed, lanceolate to linear, acute; all green and glabrous or sparsely hairy above, *white-woolly beneath*. Heads short-stalked, 2–8 in a close terminal umbel; those on female plants c. 12 mm. diam., on male (hermaphrodite) plants c. 6 mm. diam. Outer bracts woolly below, scarious and glabrous at the tip; of the male heads obovate-spathulate blunt, usually white, sometimes pink, spreading above like ray-florets; of the female heads linear-lanceolate ± acute, usually rose-pink, erect. Achenes 1 mm.; pappus white. Fl. 6–7. Visited by various insects but apomictic. 2*n*=28 (34). Chh.

In var. *hyperborea* (D. Don) DC. the lvs are broader and white with hairs on both sides. The status of this variety is not yet clear, but it may merit subspecific rank.

A. hibernica Br.-Bl., described as having long-stalked heads (stalks 0·5–2 cm.) with white, not pink, involucral bracts which are rounded or emarginate at the apex, is reported from limestone areas in Ireland. It seems insufficiently distinct to warrant specific rank.

Native. On heaths, dry pastures and dry mountain slopes throughout the British Is., but rare in the south; usually over limestone, basic igneous rock or base-rich boulder-clay, perhaps most commonly where the surface-soil is somewhat leached. 94, H39. Reaches 3000 ft. in Scotland. Var. *hyperborea* in the Inner and Outer Hebrides. N. and C. Europe (not the Mediterranean region), Siberia, W. Asia, N. America.

Tribe 6. ASTEREAE.

18. SOLIDAGO L.

Perennial herbs with ± sessile simple lvs. Heads small, yellow, usually in racemose panicles of scorpioid cymes; involucre ± cylindrical with many rows of imbricating ± lfy bracts; receptacle flat, naked, often pitted. Ray-florets in 1 row, female or neuter; disk-florets hermaphrodite. Style-arms with terminal papillose cones. *Achenes many-ribbed, not compressed; pappus-hairs* shortly ciliate, *in* 1(–2) *row*.

About 120 spp., chiefly American.

1. S. virgaurea L.   Golden-rod.

A perennial herb with a stout obliquely ascending stock and an erect simple or somewhat branched lfy stem 5–75 cm., terete, glabrous or pubescent. Basal lvs 2–10 cm., obovate or oblanceolate, narrowing to a short stalk-like base, usually toothed; stem-lvs elliptical or oblong-lanceolate, ± acute, entire or obscurely toothed. Heads 6–10 mm. in fl., short-stalked, in a panicle with straight erect branches, or in a raceme. Bracts greenish-yellow, linear, acute, glabrous or slightly downy, with scarious margins. Florets all yellow; ray-florets 6–12, spreading. Achenes c. 3 mm., brown, pubescent; pappus whitish.

Fl. 7–9. Much visited by various bees and flies, and automatically self-pollinated. $2n=18$. H.

Very polymorphic, with many named varieties differing in stature, pubescence, size, shape and serration of lvs, branching of infl. and size of individual heads. Mountain forms only 5–20 cm. with simple racemes of large heads have been named var. *cambrica* (Huds.) Sm. They seem to grade into the type but closer investigation may lead to the recognition of 2 or more sspp.

Native. A common plant of dry woods and grassland, rocks, cliffs and hedge-banks, dunes, etc., on acid or calcareous substrata, throughout the British Is., but rare in the south-east. 111, H40, S. Reaches 3550 ft. in Scotland. Europe, Asia, N. America.

The common Golden-rod of gardens is *S. canadensis L., a tall rhizomatous herb with stems 60–250 cm., pubescent throughout, and lanceolate 3-veined lvs roughly hairy on both sides or only beneath, toothed except sometimes at the base. Heads 5 mm. diam., golden-yellow, in dense one-sided recurved axillary partial infls to form a pyramidal panicle. Ray-florets barely exceeding the involucre and about equalling the disk-florets. Fl. 8–10. $2n=18$. Much cultivated and often escaping. Native of N. America.

*S. gigantea Ait. (*S. serotina* Ait.), resembling *S. canadensis* but with stems glabrous below and ray-florets distinctly exceeding the involucre and disk-florets, and *S. graminifolia (L.) Salisb. (*S. lanceolata* L.) with linear-lanceolate 3–5-veined entire lvs, glabrous except on the veins beneath, and erect corymbose panicles of golden yellow heads whose ray-florets do not exceed the disk-florets, also occur as garden-escapes or casuals. N. America.

19. ASTER L.

Perennial, rarely annual, herbs with spirally arranged *simple lvs.* Heads heterogamous or homogamous; *involucre of many rows of imbricating herbaceous or scarious bracts; receptacle flat, naked,* pitted, the pits with toothed membranous borders. *Ray-florets ligulate, blue, red or white,* 1-rowed, female or neuter, sometimes 0; disk-florets tubular, yellow, hermaphrodite. Style-arms of ray-florets linear, of disk-florets short with terminal papillose cones. *Achenes compressed, not ribbed; pappus of* 1, 2 *or several rows of shortly ciliate hairs.*

About 500 spp. in America, Asia, Africa and Europe.

Crinitaria Cass. (*Linosyris* Cass., non Ludw.) with ray-florets 0 and 2-rowed pappus, and *Galatella* Cass., with neuter ray-florets and many-rowed pappus, seem insufficiently distinct to warrant generic separation.

1   Glabrous maritime plant with markedly fleshy faintly 3-veined lvs; heads
      with or without ray-florets.   **1. tripolium**
      Lvs not markedly fleshy or plant not glabrous.   *2*

2   Lvs linear to linear-lanceolate; ray-florets 0; very rare plant of limestone
      rocks near the sea.   **2. linosyris**
      Lvs variously shaped; ray-florets bluish, reddish or white; Michaelmas
      Daisies.   *3*

3   Basal and lower stem-lvs cordate and with slender unwinged stalks.
                                                                    3. macrophyllus

        Lvs never both cordate and slender-stalked (sometimes ±cordate but
            then sessile or with broadly winged stalks).   4

4   Stem-lvs ± amplexicaul with a distinctly broadened sessile base or with
        winged stalks broadened and clasping at the base.   5
    Stem-lvs not broadened at the base, not or very slightly amplexicaul.   10

5   Stems ± hispid or villous-hispid throughout.   6
    Stems ± glabrous to pubescent, at least below.   7

6   Stem glandular and plant smelling like *Calendula*; lvs entire; rays
        reddish-purple, pink or whitish.   **4. novae-angliae**
    Stem not glandular; all or most lvs conspicuously toothed; rays lilac-blue
        to white.   **5. puniceus**

7   At least the outer bracts distinctly spreading.   8
    All bracts appressed.   9

8   Bracts in 2 rows; lvs 0·5–1·2 cm. wide.   **6. longifolius**
    Bracts in more than 2 rows; lvs 1–2 cm. wide.   **7. novi-belgii**

9   Lower stem-lvs ovate-lanceolate with broadly winged stalks; stems and
        stem-lvs pruinose.   **8. laevis**
    Lower stem-lvs not or shortly stalked; stem and stem-lvs not pruinose.
                                                                          9. × versicolor

10 Outer bracts less than half as long as inner; stem-lvs rounded at base
        and slightly amplexicaul; rays blue-purple or whitish.   **10. lanceolatus**
    Outer bracts at least half as long as inner; stem-lvs narrowed into a
        non-amplexicaul base; rays white at first, becoming blue-purple later.
                                                                          11. salignus

1. A. tripolium L.   'Sea Aster.'

A short-lived perennial *maritime* herb with a short swollen suberect rhizome
and stout erect stems 15–100 cm., glabrous. *Lvs* 7–12 cm., *fleshy, glabrous*,
faintly 3-veined, entire or obscurely toothed; basal lvs oblanceolate to obovate,
narrowing into a long stalk; stem-lvs narrowly oblong to linear. Heads
8–20 mm. diam., in corymbs, their stalks with 1–2 small bracts. Involucre of
not many appressed narrow blunt bracts, the outer scarious-tipped, the inner
longer, largely scarious. Ray-florets spreading, blue-purple or whitish, or 0
(var. *discoideus* Rchb. f.); disk-florets yellow. Achene 5–6 mm., brownish,
hairy; pappus 10 mm., brownish. Fl. 7–10. Visited by many flies and bees,
etc.; self-pollination possible. $2n=18$. Hel.
    Native. A common salt-marsh plant occurring also on maritime cliffs and
rocks, all round the coasts of the British Is., but almost confined to estuaries
in N. England and Scotland. 84, H28, S. Rarely found inland at saltworks.
Var. *discoideus* is frequent southwards from N. Lancs and N. Yorks and
occurs in two isolated localities in Scotland. Most European coasts but not
in Iceland and the Faeroes; N. Africa; Caspian Sea; Lake Baikal; inland
saline areas of Europe and C. Asia.

2. A. linosyris (L.) Bernh.   Goldilocks.

Chrysocoma Linosyris L.; *Linosyris vulgaris* DC.; *Crinitaria Linosyris* (L.) Less.
A perennial herb with a woody stock and erect slender glabrous stems
10–50 cm., wiry, very lfy. *Lvs* (1–)2–5 cm., *very numerous, linear*, acute, narrowing to the base, 1-veined, entire and glabrous but rough at the margins

and punctate above. Heads 12–18 mm. diam., in dense corymbs. Involucre lax, of many acute bracts; the outer linear, spreading at the tip, lf-like; the inner oblong, yellowish, scarious-margined. Florets exceeding the bracts, bright yellow. Achenes 5 mm., brown, pubescent; pappus about equalling the achene, reddish. Fl. 8–9. $2n=18$. Hp.

Native. A rare plant of limestone cliffs in S. Devon, N. Somerset, Glamorgan, Caernarvon and on the N. Lancashire coast. 5. C. Europe and Mediterranean region.

The following spp. are Michaelmas Daisies, mostly natives of N. America, which have become naturalized, especially by rivers and streams. Several others are grown in gardens and may occasionally escape.

*3. **A. macrophyllus** L., with glandular infl. axes, viscid involucres and white or purple-flushed rays, is established in Renfrew.

*4. **A. novae-angliae** L., very distinct in its roughly hairy glandular stem and usually reddish rays is commonly grown and has become established in several localities throughout Great Britain. $2n=10$.

*5. **A. puniceus** L. (*A. hispidus* Lam.), markedly hispid with the $\pm$ coarsely toothed lvs scabrid above and hispid on the midrib beneath, and with a smell of juniper berries when crushed, is established at Berwick, Dundee and Perth.

*6. **A. longifolius** Lam., with long narrowed lvs, a 2-rowed involucre, and rays white at first then becoming bluish, is established in Oxford, Merioneth and Perth.

*7. **A. novi-belgii** L., the most commonly cultivated and most widely established species, is divided by Thellung into three subspecies all of which are naturalized in the British Is. $2n=54$.
Ssp. **novi-belgii**. Infl. corymbose-paniculate, the lowest branches long and ascending; lvs lanceolate to oblong; heads 2·5–4 cm. diam.; bracts c. 1 mm. wide, unequal. Widely naturalized.
Ssp. **floribundus** (Willd.) Thell. Infl. closely corymbose-paniculate; heads as in ssp. *novi-belgii* but smaller and bracts only c. 0·5 mm. wide. Naturalized in Surrey, Essex, Herts, Oxford and Orkney.
Ssp. **laevigatus** (Lam.) Thell. Infl. racemose-paniculate, the lowest branches short, spreading and with 1 or a few heads; bracts not very unequal. Naturalized in Middlesex, Bucks and Glamorgan.

*8. **A. laevis** L., recognizable by its broad lower lvs with long winged stalks, its pruinose stem and stem-lvs and its rigid coriaceous whitish but green-tipped appressed bracts, is naturalized by Lough Neagh, Tyrone (Ireland). A smaller variety, var. *geyeri*, A. Gray has been recorded from Somerset.

*9. **A. × versicolor** Willd. appears to be a hybrid between *A. laevis* and *A. novi-belgii* and resembles the former in its broadish lvs and appressed bracts but its lower lvs are at most shortly stalked and its stem and stem-lvs are not pruinose; rays white at first, then becoming bluish. Much grown and established in Surrey, Hants and Leicester, etc.

*10. **A. lanceolatus** Willd. (*A. paniculatus* auct.; *A. Lamarckianus* auct., ?Nees) was probably the species to which the name Michaelmas Daisy was first applied in the mid-seventeenth century. A tall plant, to 2 m., with linear-lanceolate to lanceolate-acuminate lvs, the stem-lvs often slightly amplexicaul; infl. narrowly pyramidal; bracts c. 0·5 mm. wide, very unequal. Much grown in gardens and naturalized in Esthwaite Fen (N. Lancs), Surrey, Berks and Oxford.

*11. A. salignus Willd. differs from the preceding in that its stem-lvs are narrowed into a sessile non-amplexicaul base; its bracts, similarly narrow, are much less unequal, and its rays are white at first then becoming blue-purple. Naturalized on Wicken Fen (Cambs) and also in Surrey, Middlesex, Oxford, Cornwall, Roxburgh, Angus, etc.

The S. European *A. amellus* and *A. acris*, shorter plants which flower earlier than the American spp., are much grown in gardens. *A. amellus* has oblong lvs; *A. acris* linear gland-dotted lvs. The China Aster of gardens is *Callistephus chinensis* (L.) Nees.

20. ERIGERON L.

Perennial, rarely annual, herbs with spirally arranged simple lvs. Heads heterogamous; involucre of narrow, equal, scarcely imbricating herbaceous bracts; *ray-florets* female, usually *numerous and in 2 or more rows, with narrow ligules*; disk-florets hermaphrodite. Achenes compressed, usually pubescent and 2-veined; pappus usually of 1 row of simple hairs with much shorter ones intermixed, or with an outer row of small bristles or chaffy scales. In section *Trimorpha* (Cass.) Rehb. the female florets have filiform tubes but at least the outermost have definite ligules equalling or exceeding the pappus. (In *Conyza* ligules, if present, are inconspicuous, shorter than the filiform tubes and scarcely exceeding the pappus.)

Close to *Aster* but differing in the more numerous and narrower ligulate florets.

About 180 spp., chiefly American but with some in the Old World, including several arctic-alpines.

1	Heads usually several in a panicle or corymb.	*2*
	Heads solitary, terminal, rarely 2–3.	*3*
2	Annual or biennial; lvs all entire, softly hairy; ray-florets pale purple-blue.	
		1. acer
	Slender branching perennial; some of the lower lvs usually 3-lobed or 3(–5)-toothed at the apex; lvs sparsely hairy or ± glabrous; ray-florets white above, purplish beneath.	**3. mucronatus**
3	Heads 1·5–2 cm. diam.; rare alpine plant.	**2. borealis**
	Heads 3–4 cm. diam.; lowland alien.	**4. glaucus**

1. E. acer L.   'Blue Fleabane.'

An annual or biennial herb with an erect slender stem 8–40 cm., usually branched above, rough with long hairs, reddish. Basal lvs 3–7·5 cm., obovate-lanceolate, stalked; stem-*lvs* numerous, linear-lanceolate, semi-amplexicaul; all *entire*, hairy. Heads 12–18 mm. diam., 1–several in a corymbose panicle; bracts linear, glabrous or hairy, not glandular, red-tipped. *Ray-florets* pale purple, very slender, *erect, not much exceeding the yellow disk-florets*; outer disk-florets filiform, female; inner broader, hermaphrodite. Achenes 2–3 mm., yellowish, hairy; pappus much longer than the achenes, reddish-white. Fl. 7–8. $2n = 18^*$. Th.

Native. A locally common plant of dry grassland, dunes, banks and walls, especially on calcareous substrata, throughout England and Wales but in Scotland only in Angus and Banff; local in Ireland. 72, H17, S. Reaches 1400 ft. in Banff. Temperate regions of the northern hemisphere.

2. E. borealis (Vierh.) Simmons   'Boreal Fleabane.'

E. alpinus auct. angl., non L.

A perennial alpine herb with a short creeping woody stock and erect flowering stems 7–20 cm., usually unbranched, hairy. Lvs mostly in a basal rosette, 1·5–3 cm., narrowly oblanceolate, narrowing to a long winged stalk; stem-*lvs* few, linear-oblong, sessile, somewhat amplexicaul; all *very hairy, ciliate towards the base*, entire. Heads c. 18 mm. diam., usually solitary, rarely 2–3; *bracts* linear-acute, *hairy, not glandular. Ray-florets* numerous, *purple*, slender, *spreading*, much exceeding the yellow disk-florets; outer disk-florets very slender, female, inner broader, hermaphrodite. Achenes yellowish, downy; pappus about equalling the achene, reddish. Fl. 7–8. $2n=18$. Ch.

Native. A rare and local plant of alpine rock-ledges between 2400 ft. and 3500 ft. on Ben Lawers and other mountains in Perth, Angus and S. Aberdeen. 3. Scandinavia, Iceland, Greenland.

E. uniflorus L., a perennial alpine herb closely resembling *E. borealis* but rather less tall (5–15 cm.), with subglabrous basal lvs, fewer and smaller stem-lvs which are not or hardly ciliate, the heads always solitary with *whitish ray-florets which later turn purple-blue*, and *disk-florets all similar*, yellow, has been reported from rock-ledges on Rhum (Inner Hebrides) but needs confirmation. $2n=18$. Mountains of C. Europe; Scandinavia, Iceland, Greenland.

*3. E. mucronatus DC.

E. Karvinskianus DC. var. *mucronatus* (DC.) Aschers.

A perennial with branched slender lfy stems 10–25 cm., somewhat woody below, sparsely hairy. Lower lvs obovate-cuneate, often 3-lobed or coarsely 3(–5)-toothed at the apex, the teeth mucronate; upper lvs linear-lanceolate, ± entire; all sparsely hairy to nearly glabrous, often ciliate towards the base. Heads c. 1·5 cm. diam. in a lax corymb, bracts linear-acute, hairy. Ray-florets in 2 rows, white above, purple beneath. Achenes 1·5 mm., reddish-brown, shining, somewhat hairy; pappus whitish, longer than the achene. Fl. 7–8. Ch.

Introduced. Naturalized for over 80 years on old walls at St Peter Port, Guernsey and now well established in the Channel Is. and S.W. England. Native of Mexico.

*4. E. glaucus Ker-Gawl.   Beach Aster.

A perennial herb with a stout woody stock bearing a rosette of somewhat fleshy finely puberulent or glabrescent lvs, 2·5–10 cm., obovate-spathulate, blunt, entire or with a tooth on each side of the apex; other rosettes terminating prostrate woody offsets. Stems erect, 10–20(–30) cm., pubescent to villous, bearing a few small stem-lvs and terminating in a solitary head, 3–4 cm. diam., or less commonly branching and bearing 2 or more heads. Involucre shaggy and somewhat glandular; ray-florets broadish, lilac or violet; disk-florets yellow. Fl. 5–8.

Introduced. Established in a few places. Coast of Pacific N. America.

The lf-rosettes resemble those of *Limonium vulgare*.

21. CONYZA Less.

Perennial or annual herbs with spirally arranged simple lvs. *Heads with a few central hermaphrodite florets surrounded by numerous female florets with filiform tubular corollas; ligulate florets 0 or with inconspicuous ligules* shorter than the filiform corolla-tube and scarcely if at all exceeding the pappus. Otherwise like *Erigeron*.

It is difficult to draw a distinct line between *Erigeron* and *Conyza*: restriction of the latter to spp. with no ligulate florets separates obviously close relatives. *C. canadensis* does not seem closely related to any species of undoubted *Erigeron* and has the habit of many other N. American weeds falling clearly into *Conyza*.

*1. C. canadensis (L.) Cronq.　　　　　　　　　　Canadian Fleabane.

Erigeron canadensis L.

An annual herb with stiffly erect very lfy stems 8–100 cm., much branched, sparsely hairy or ± glabrous. Basal lvs obovate-lanceolate, stalked, entire or ± toothed, soon dying; *stem-lvs* 1–4 cm., numerous, *narrowly lanceolate or linear*, acute, entire or obscurely toothed, hairy and bristly-ciliate. Heads 2·5–5 mm. (3–5 mm. diam.), numerous, in long panicles; involucral bracts linear-attenuate, glabrous, with broad scarious margins. *Ray-florets whitish* to pale lavender, in several rows, narrow, erect, female; disk-florets pale yellow, hermaphrodite. Achenes 1·5 mm., pale yellow, downy; pappus yellowish, longer than the achene. Fl. 8–9. Visited by small insects and said not to be self-pollinated. $2n = 18$. Th.

Introduced. A local weed of waste ground, waysides, cultivated land on light soils, dunes, walls, etc., throughout England and Wales but rare in the north and in only a few localities in S. Scotland. 46. N. America, but widely naturalized.

*Olearia macrodonta Baker, a shrub or small tree with a strong musky scent, native in mountainous areas of New Zealand, is much grown in gardens and has established itself locally. Its branchlets are tomentose and the spirally arranged, ovate to narrowly oblong, acute, rounded-based, rigid lvs 5–10 × 2·5–4 cm., coarsely and sharply toothed on the waved margins, are glabrescent above and white-tomentose beneath. Heads 6–8 mm., numerous, in large compound corymbs; involucral bracts pubescent; ray-florets 3–5, whitish, their ligules short and narrow; disk-florets 4–7. Achenes pubescent, grooved; pappus dirty-white or reddish.

22. BELLIS L.

Annual to perennial herbs with spirally arranged lvs often confined to a basal rosette. Heads solitary; involucre of many lf-like bracts in (1–)2 rows; receptacle conical, pitted. Ray-florets in 1 row, white or pink, female; disk-florets hermaphrodite. Style arms short, thick, with terminal papillose cones. *Achenes obovate, compressed, bordered, not ribbed; pappus 0.*

Fifteen spp., in Europe and the Mediterranean region.

1. B. perennis L.　　　　　　　　　　　　　　　　Daisy.

A perennial herb with a short erect stock and stout fibrous roots. Lvs 2–4(–8) cm., confined to a basal rosette, obovate-spathulate, broad and rounded at the end, crenate-toothed, narrowed abruptly into a short broad stalk, sparsely hairy. Scapes 3–12(–20) cm., naked, hairy. Head 16–25 mm. diam.; bracts oblong, blunt, green or black-tipped, hairy. Ray-florets numerous, narrow, spreading, white or pink; disk bright yellow. Achenes 1·5–2 mm., pale, strongly compressed, distinctly bordered, ±downy. Fl. 3–10. Visited by many small insects. $2n=18*$. Hr.

Native. An abundant plant of short grassland throughout the British Is. 112, H40, S. Reaches 3000 ft. in Scotland. Europe, W. Asia.

To the Astereae belongs the frequent casual:

*Grindelia squarrosa (Pursh) Dunal. A perennial viscid herb with yellowish-green ovate-oblong amplexicaul toothed lvs. Heads hemispherical; involucral bracts narrow with cylindrical spreading tips, those of the outer bracts strongly recurved. Ray-florets numerous, yellow. Achenes with a pappus of 2–8 ±smooth deciduous bristles. N.W. America; naturalized in Australia. The sticky lvs are said to be used for healing wounds.

Tribe 7. EUPATORIEAE.

23. EUPATORIUM L.

Perennial herbs or shrubs with usually opposite lvs. *Heads few-fld white, pink or purplish,* in terminal corymbs or panicles. Bracts few, loosely imbricate in 2–3 rows; *receptacle naked, flat. Fls all tubular,* hermaphrodite, actinomorphic, 5-merous. Achenes 5-angled; *pappus-hairs in 1 row,* denticulate.

About 400 spp., chiefly American.

1. E. cannabinum L.   Hemp Agrimony.

A large perennial herb with a woody rootstock and erect downy striate shoots 30–120 cm., simple or with short branches. Basal lvs oblanceolate, stalked; stem-lvs subsessile, 3(–5)-partite with elliptical-acuminate toothed segments 5–10 cm.; branch lvs simple, ovate or lanceolate; all lvs opposite, shortly hairy and gland-dotted. Heads in dense terminal corymbs; each head with 5–6 reddish-mauve or whitish florets and c. 10 oblong purple-tipped involucral bracts, the inner c. 6 mm., narrow, ±scarious, the outer much shorter. Styles white, long. Achenes blackish, 5-angled, gland-dotted; pappus whitish. Fl. 7–9. Protandrous. Visited chiefly by Lepidoptera and by some flies and bees; automatic cross-pollination between different fls in the same head occurs. $2n=20$. H.–Hel.

Native. A common gregarious plant of marshes and fens, stream-banks and moist woods throughout most of the British Is. but less common in Scotland and not in the Outer Hebrides, Orkney or Shetland. 102, H40, S. Channel Is. Europe, W. and C. Asia, N. Africa.

Some species of the related genus *Ageratum*, differing from *Eupatorium* in having the pappus of free or basally united scarious scales, are often cultivated. Best known is *A. houstonianum* Mill. (*A. mexicanum* Sims), with broadly ovate ±cordate lvs and blue (sometimes pink or white) fls, much grown as an edging-plant and in window-boxes and occasionally escaping.

Tribe 8. ANTHEMIDEAE.

24. ANTHEMIS L.

Annual to perennial usually strongly scented herbs with spirally arranged 1–3 *times pinnately divided lvs* whose *ultimate segments* are *linear*. Heads solitary, usually heterogamous; bracts imbricate, blunt, usually scarious and often dark-coloured at the margins; *receptacle* flat to conical, *with narrow flat scarious scales* subtending some or all of the florets. Ray-florets ligulate, female or neuter, yellow or white, sometimes 0; disk-florets tubular, hermaphrodite, yellow, their corolla-tubes compressed and winged. *Achenes with at least* 10 *ribs*, not or little compressed in the anterior-posterior plane; *pappus* represented by a *small often oblique membranous rim, or* 0.

About 100 spp., chiefly in the Mediterranean region and Near East.

1 Ray-florets yellow. **1. tinctoria**
    Ray-florets white. 2

2 Plant glabrous or slightly hairy, foetid; receptacle-scales linear-acute; ray-florets usually without styles; achenes tubercled. **3. cotula**
    Plant usually pubescent or woolly, aromatic; receptacle-scales lanceolate-cuspidate; ray-florets with styles; achenes strongly ribbed, not tubercled. **2. arvensis**

Section 1. *Anthemaria* Dum. Receptacle hemispherical in fr., its scales keeled; achenes truncate above, somewhat compressed.

*1. A. tinctoria L. Yellow Chamomile.

A biennial to perennial herb with erect or ascending woolly stems 20–60 cm., usually branched. Lvs 4–7 cm., deeply pinnatisect with pinnately lobed or toothed segments united only by a narrow wing along the rhachis, ± glabrous above, white-woolly beneath. Heads 2·5–4 cm. diam., solitary, long-stalked. Bracts lanceolate, acute, scarious-tipped, ± woolly at the back, with brown ciliate margins. *Receptacle hemispherical*, its scales lanceolate, cuspidate, not exceeding the disk-florets. *Ray-florets* female, *golden-yellow* like the disk-florets, rarely 0 (var. *discoidea* Willd.). Achenes c. 2 mm., 4-angled, faintly ribbed on each face, glabrous; pappus represented by a membranous border. Fl. 7–8. Much visited by a great variety of insects. $2n=18$. H.

Introduced. A garden plant naturalized in Bucks and found as a casual in waste places, waysides, banks, etc., in many English and a few Scottish counties. 31, S. S. and C. Europe northwards to Scandinavia and Finland; W. Asia. Introduced in N. America. The fls yield a yellow dye.

Section 2. *Anthemis*. Receptacle long-conical in fr., its scales lanceolate or oblong, not keeled, blunt or cuspidate; achenes ± turbinate, not compressed, strongly ribbed.

2. A. arvensis L. Corn Chamomile.

An annual aromatic herb with decumbent or ascending pubescent stems 12–50 cm., much-branched below, the branches simple or irregularly branched above. Lvs 1·5–5 cm., 1–3 times pinnate, the *ultimate segments* short, *oblong*, acute, *hairy or even ± woolly beneath*, especially when young. Heads 20–

30 mm. diam., solitary, long-stalked. *Bracts* oblong, scarious-tipped, pale, *downy*. *Receptacle conical, with lanceolate-cuspidate scales* just exceeding the disk-florets. *Ray-florets* female (*with styles*), white, spreading; disk-florets yellow. *Achenes* 2–3 mm., whitish, ribbed all round, glabrous, *not tubercled*, rugose on top; border crenate. Fl. 6–7. Fragrant and much visited by bees and flies. $2n=18$. Th.

Var. *anglica* (Spreng.) Syme, with fleshy bristle-pointed lf-segments and a 'flat receptacle' was probably a maritime form or ecotype of the Durham coast, now apparently extinct.

Native. A locally common calcicolous plant of arable land and waste places throughout the British Is. to Caithness; Orkney; introduced in E. Ireland. 88, H4, S. Europe northwards to S. Norway, C. Sweden and Lake Onega; Asia Minor; N. Africa. Introduced elsewhere.

Section 3. *Maruta* (Cass.) Boiss. Receptacle long-conical in fr., its scales narrowly lanceolate or subulate, usually confined to the apical part of the receptacle; achenes not compressed, truncate above, strongly ribbed.

3. A. cotula L. Stinking Mayweed.

An annual *foetid* herb with erect *sparsely hairy* stems 20–60 cm., usually branched below and corymbosely branched above. *Lvs* 1·5–5 cm., 1–3 times pinnate, the ultimate segments narrowly linear, acute, ±*glabrous*. Heads 12–25 mm. diam., solitary, rather short-stalked. Bracts oblong, blunt, ± glabrous, with a green central stripe and broad scarious margins. *Receptacle long-conical* with *linear-subulate scales* only near the apex. *Ray-florets* usually neuter (*without styles*), white, at first spreading, later *reflexed*; disk-florets yellow. *Achenes* 2 mm., yellowish-white, 10-ribbed, *tubercled* on the back; membranous border crenate, inconspicuous. Fl. 7–9. $2n=18$. Th.

In var. *maritima* Bromf. the stems are prostrate and the lvs fleshy.

Native. A locally common weed of arable land and waste places especially in S. and C. England and on heavy soils, rarer in the north but reaching Perth and Dunbarton and introduced in the Outer Hebrides, Orkney and Shetland. 79, H37, S. S. Europe northwards to S. Norway; N. and W. Asia. Introduced elsewhere.

Several spp. of *Anthemis* native in S. and E. Europe occur as casuals.

25. CHAMAEMELUM Mill.

Like *Anthemis* but *disk-florets with corolla-tube saccate at base*, especially in the anterior-posterior plane, *so that it covers the top of the achene* (in *C. mixtum* prolonged into a posterior spur); and *achenes laterally compressed*, *unribbed* (but with 3 slender vascular bundles on the posterior face).

Three spp. in W. Europe and Mediterranean region.

1. C. nobile (L.) All. Chamomile.

Anthemis nobilis L.

A perennial *pleasantly scented* herb with a short much-branched creeping stock and decumbent or ascending branched hairy stems 10–30 cm. Lvs

1·5–5 cm., 2–3 times pinnate, the ultimate segments short, linear-subulate, sparsely hairy. Heads 18–25 mm. diam., solitary, long-stalked. Bracts oblong, downy, with broadly scarious and laciniate white margins. *Receptacle* conical, *scales oblong, concave, blunt,* often laciniate at the apex. *Ray-florets* female (*with styles*), broad, white, spreading, rarely 0; disk-florets yellow, with the *enlarged and persistent base of the corolla-tube enveloping the apex of the achene.* *Achene* 1–1·5 mm., obovoid, *narrowed and rounded above,* smooth except for 3 faint striae on the inner face; border inconspicuous. Fl. 6–7. $2n = 18$. H.

Native. A local plant of sandy commons and pastures and grassy road-sides throughout England, Wales and Ireland; doubtfully native in Scotland but casual there and in the Hebrides, Orkney and Shetland. 57, H21, S. W. Europe from Belgium southwards; N. Africa; Azores.

26. ACHILLEA L.

Usually perennial herbs with spirally arranged usually pinnatisect, rarely simple, lvs. *Heads in corymbs,* rarely solitary, heterogamous; bracts imbricate, many-rowed, scarious-margined; *receptacle* small, flat or slightly convex, *with narrow scarious scales.* Ray-florets female, ligulate, usually short and broad, white, yellow or reddish; disk-florets hermaphrodite, tubular, white or yellow. *Achenes strongly compressed, truncate above, not ribbed; pappus 0.*

About 100 spp., chiefly in temperate regions of the Old World; 1 in N. America.

Close to *Anthemis* but with the achenes strongly compressed and the smaller heads usually in corymbs.

1	Ray-florets yellow.	**3. tomentosa**
	Ray-florets white.	*2*
2	Lvs finely pinnatisect; heads 4–6 mm. diam., numerous, in a dense corymb.	**2. millefolium**
	Lvs simple, broadly linear, serrate; heads 12–18 mm. diam., few, in a lax corymb.	**1. ptarmica**

Section 1. *Ptarmica* (Mill.) Dum. Ray-florets 6–25, female, white, almost as long as the involucre; receptacle convex or hemispherical, not much elongated in fr.

1. A. ptarmica L.   Sneezewort.

A perennial herb with a creeping woody stock and erect angular stems 20–60 cm., simple or branched above, glabrous below but hairy above. *Lvs* 1·5–8 cm., *linear-lanceolate,* sessile, acute, ± glabrous, sharply *serrulate,* the serrations with a cartilaginous and denticulate margin. *Heads 12–18 mm. diam., not numerous, in a rather lax corymb.* Involucre hemispherical, its bracts lanceolate to oblong, blunt, green-centred, ± woolly, with reddish-brown scarious margins. Ray-florets 8–13, ovate, white, as long as the involucre; disk-florets greenish-white. Achenes 1·5 mm., pale grey. Fl. 7–8. Freely visited by bees and flies. $2n = 18$. H.

Native. Common throughout the British Is. in damp meadows and marshes, and by streams, reaching 2400 ft. in the Lake District. 112, H 38. Europe except the southern Mediterranean region; Asia Minor; Caucasus; Siberia. Introduced in N. America.

Double forms are much grown in gardens as Bachelors' Buttons. Formerly used as a salad plant. Infusions of the lvs and fls were also used medicinally.

Section 2. *Achillea*. Involucre ovoid; ray-florets 4–6, white, shorter than the involucre; receptacle long-conical in fr.

2. A. millefolium L. Yarrow, Milfoil.

A perennial strongly scented far-creeping stoloniferous herb with erect, fur-rowed, usually simple, ± woolly stems 8–45(–60) cm. *Lvs* 5–15 cm., lanceolate in outline, 2–3 *times pinnate*, the ultimate segments linear-subulate; basal lvs long, stalked; upper shorter, sessile, often with 2–3 small axillary lvs. *Heads 4–6 mm. diam., numerous, in dense terminal corymbs*. Involucre ovoid, bracts rigid, oblong, blunt, keeled, ± glabrous, with a broad brown or blackish scarious margin. Ray-florets usually 5, about half as long as the involucre and as broad as long, 3-toothed at the apex, white, rarely pink or reddish; disk-florets white or cream-coloured. Achenes c. 2 mm., shining greyish, some-what winged. Fl. 6–8. Much visited by a great variety of insects. 2n = 18, 36, 54, 88. Chh. Very variable in hairiness and in the colour of the bracts.

Native. A common plant of meadows and pastures, grassy banks, hedge-rows and waysides on all but the poorest soils throughout the British Is. 112, H 40, S. Europe to 71° 10′ N.; W. Asia. Introduced in N. America, Australia and New Zealand. Used medicinally for a great variety of purposes from early times, and still officinal in Austria and Switzerland as *Herba Millefolii* or *Flores Millefolii*. It contains the two alkaloids achillein and moschatin and an ethereal oil which is responsible for its aromatic scent.

*A. distans Waldst. & Kit. ex Willd. (*A. tanacetifolia* All., non Mill.). Native in the mountains of C. Europe from the Alps to the Carpathians, has been recorded from Derbyshire, presumably a garden-escape. It is a perennial far-creeping stolon-iferous herb resembling *A. millefolium* but taller (to 120 cm.). The basal *lvs* are *elliptical* in outline *with small toothed pinnae between the primary pinnae* which are deeply pinnatifid *with triangular toothed lobes*; stem-lvs lanceolate, once-pinnate with oblong toothed segments; all with *winged rhachis*. Heads c. 5 mm. diam., very numerous in a large dense compound corymb. Ray-florets 5, roundish, 3-lobed at the end, white to reddish, half as long as the involucre. Achenes 1·5 mm.

*A. nobilis L. Native of S. and C. Europe and W. Asia, is a not infrequent casual. It is recognizable by the ovate-lanceolate twice-pinnatisect lvs whose slightly winged rhachis has small teeth between the primary pinnae, the small heads (3 mm. diam.) and the yellowish-white ray-florets which are broader than long and only ⅓ to ¼ as long as the involucre.

*3. A. tomentosa L.

A perennial herb with very short stolons and erect or ascending *densely woolly stems* 8–30 cm. *Lvs linear-lanceolate* in outline, *twice-pinnatisect, with* crowded pinnatifid or entire *mucronate* ± *woolly segments*. Heads c. 3 mm. diam., numerous, in dense compound corymbs, with 4–6 *bright yellow ray-florets* half as long as the involucre, and yellow disk-florets.

Introduced. Established in a few localities in Scotland and N.E. Ireland. S. Europe northwards to C. France and Tirol; W. Asia.

27. OTANTHUS Hoffmanns. & Link

A maritime herb with corymbs of yellow heads. Involucre campanulate-hemispherical, of *numerous imbricating woolly bracts*. Receptacle shortly conical, with ovate-acuminate scales. *Florets all hermaphrodite and tubular*, the *tube prolonged downwards into 2 auricle-like spurs which almost enclose the ovary*. Achenes compressed, longitudinally ribbed; pappus 0.

One sp.

1. O. maritimus (L.) Hoffmanns. & Link   Cotton-weed.

Athanasia maritima (L.) L.; *Diotis maritima* (L.) Desf. ex Cass.

A perennial maritime herb with a long woody branching stock and *white woolly* ascending *stems* 15–30 cm., stout, branched only above. *Lvs* oblong to oblong-lanceolate, crenate, *white above and beneath with a dense cottony felt*. Heads 6–9 mm. diam., several in a dense corymb, short-stalked. Involucre of a few densely white-felted ovate bracts and numerous glabrous ones with woolly tips. Florets yellow, little exceeding the involucre. Achenes 4–5 mm., curved, smooth, with 5 thick longitudinal ribs, permanently covered by the decurrent spurs of the corolla-tube. Fl. 8–9. $2n = 18$. Ch.

Native. Sandy sea-shores and stable shingle; very rare and decreasing. Wexford, ?Waterford; Jersey. Formerly in Cornwall and ?Sussex. H2, S. Mediterranean region and W. Europe from Cherbourg to Palestine and Syria; N. Africa.

28. SANTOLINA L.

Suffruticose aromatic plants with pinnatifid lvs. Heads long-stalked, ± homogamous, the marginal florets tubular or very shortly ligulate; receptacular scales present. *Corolla-tube* compressed, winged, and *with a basal one-sided appendage enclosing the apex of the achene*. Achene ± 3–5-angled, glabrous; pappus 0.

About 8 spp., in S.W. Europe, eastwards to Dalmatia, and N. Africa.

*1. S. chamaecyparissus L.   Lavender Cotton.

A small evergreen strongly aromatic shrub with pubescent branching woody stems to 50 cm. Lvs only 2–3 mm. wide, linear in outline, whitish-tomentose, pinnatifid with 4 rows of ±crowded fleshy blunt lobes, 1–2 mm. Heads 10–15 mm. diam., solitary, terminal, globular; involucral bracts ± glabrous, somewhat scarious-tipped. Florets all tubular, yellow. Fl. 7–8. $2n = 18$. Ch.–N.

Introduced. A garden-escape, sometimes establishing itself. Mediterranean region from Spain to Dalmatia; N. Africa.

29. TRIPLEUROSPERMUM Schultz Bip.

Annual to perennial herbs with spirally arranged *repeatedly pinnatisect lvs*, the ultimate *segments narrowly linear*. Heads solitary, usually heterogamous; involucre of usually 2 rows of equal imbricating bracts with scarious margins;

receptacle hemispherical to ovoid-conical, solid, naked. Ray-florets ligulate, white, female, rarely 0; disk-florets tubular, yellow, hermaphrodite, the corolla-tube laterally compressed and winged below. *Achenes turbinate*, truncate above, *with truly basal scar of insertion* on the receptacle, strongly dorsiventral, with 3 *strong ribs on the posterior face and 2 conspicuous oil-glands high up on the anterior face*; pappus represented by a membranous rim or 0.

One sp. in northern hemisphere.

1. T. maritimum (L.) Koch   Scentless Mayweed.

Matricaria maritima L.; incl. *M. inodora* L.

Annual to perennial almost scentless herbs with erect, decumbent or prostrate, usually branching, glabrous stems 10–60 cm. Lvs oblong in outline, 2–3 times pinnate, glabrous, the primary pinnae ovate in outline and the ultimate linear segments variable in length, breadth and acuteness. Heads 1·5–4(–5) cm. diam., solitary, long-stalked. Involucre hemispherical, its bracts oblong, blunt, with a narrow brown scarious margin. Receptacle slightly convex, solid. Ray-florets 12–30, white, spreading; disk-florets yellow. *Achenes* 2–3 mm., *with 3 broad ribs on the inner face and 2 dark brown oil-glands at the top of the outer face*; all achenes crowned with a membranous border. Fl. 7–9. Visited by bees and flies. $2n = 18, 36$. Th.–H.

A very variable group whose British forms require closer study. They may be divided tentatively into 2 ssp., each with varieties.

Ssp. **maritimum**. A *perennial maritime herb* with a branching woody stock and usually *prostrate or decumbent diffusely branched stems* 10–30 cm. Lvs with the ultimate segments *short, blunt, ± cylindrical, fleshy*. Heads 3–4·5(–5) cm. diam., long-stalked. Ray-florets 20–30. Achenes 2–2·5 mm., c. $\frac{1}{2}$ as long again as broad, with 2 *much elongated oil-glands* at the top of the outer face. $2n = 18$.

Var. *phaeocephalum* (Rupr.) Hyl. has the scarious margins of the involucral bracts broader and almost black.

Ssp. **inodorum** (L.) Hyl. ex Vaarama. Usually an *annual ruderal herb* with ± *erect stems* 15–60 cm., simple or corymbosely branched above. Lvs with the ultimate *segments long, slender, acute or bristle-pointed*, not fleshy (except in var. *salinum*). Heads 1·5–3·5(–4) cm. diam. Ray-florets c. 12–22. Achenes 2–3 mm., up to twice as long as broad, with 2 *circular or slightly elongated oil-glands* at the top of the outer face. $2n = 18, 36$.

Var. *salinum* (Wallr.) Clapham is a biennial (rarely perennial) usually decumbent maritime plant with ± erect lvs whose crowded linear segments are short, fleshy, and shortly mucronate, but whose heads and achenes resemble those of ssp. *inodorum*.

Native. Ssp. *inodorum* is abundant throughout the British Is. as a weed of arable and waste land on all kinds of soils; var. *salinum* on sand, shingle, rocks, walls, etc., by the sea, chiefly in S. England and S. Wales. Ssp. *maritimum* is a locally common maritime plant of the drift-line at the foot dunes, on shingle beaches, maritime rocks and cliffs, walls, etc., in N. Wales, N. England, Scotland and Ireland; var. *phaeocephalum* is a rare plant of

N. Scotland and Shetland. 112, H40, S. N. and C. Europe from N. Spain, N. Italy and N. Balkans northwards to Iceland, Scandinavia and Finland; ssp. *inodorum* is not in Iceland and extends eastwards to the Caucasus and W. Asia.

30. MATRICARIA L.

Usually annual herbs with finely pinnatisect lvs, the ultimate segments narrowly linear. Closely resembling *Tripleurospermum* but heads sometimes homogamous (lacking ray-florets), *receptacle* often *hollow* and *achenes* strikingly different, these being *obovoid*, somewhat compressed laterally, obliquely truncate above and with the *basal insertion-scar oblique and distinctly lateral* (posterior), *rather inconspicuously 3–5-ribbed* on the posterior face and *lacking oil-glands* on the anterior face; pappus represented by a membranous rim or 0.

About 70 spp. in the Mediterranean region and W. Asia, some in S. Africa and Pacific N. America.

Matricaria and *Tripleurospermum* differ in embryo-sac development, achene anatomy and orientation of the cotyledons within the seed, in addition to the differences noted above.

Ray-florets 0; plant strongly aromatic. **2. matricarioides**
Ray-florets white, soon reflexed; plant pleasantly aromatic. **1. recutita**

1. M. recutita L. Wild Chamomile.

M. Chamomilla auct.

An annual *pleasantly aromatic* herb resembling *Tripleurospermum maritimum*, with erect glabrous stems 15–60 cm., usually much branched, and 2–3 times pinnate lvs whose ultimate segments are narrowly linear and bristle-pointed. Heads 12–22 mm. diam. Involucral bracts linear, blunt, yellowish-green with the *narrow scarious margins much the same colour* (not brown as in *T. maritimum*). Receptacle markedly conical from the first, hollow. *Ray-florets* c. 15, white, 6–9 × 2–3 mm., *reflexed soon after flowering begins*, sometimes 0; disk-florets yellow. *Achenes* 1–2 mm., pale grey, slender, obliquely truncate and *unbordered* above, with 4–5 ribs on the inner face, and *lacking oil-glands* on the outer face. Fl. 6–7. Freely visited by flies and some small bees. 2*n* = 18. Th.

Native. Locally abundant as a weed of sandy or loamy arable soil or waste places throughout England and Wales; probably introduced further north and in Ireland. 69, S. Europe (probably introduced in the north), W. Asia to India. Introduced in N. America and Australia. Used as a substitute for true Chamomile, *Chamaemelum nobile*, Oil of Chamomile being distilled from the flower heads.

*2. M. matricarioides (Less.) Porter Pineapple Weed, 'Rayless Mayweed.'

M. discoidea DC.; *M. suaveolens* (Pursh) Buchen., non L.

An annual *strongly aromatic* herb with erect glabrous stems 5–30(–40) cm., much branched above, the branches rigid. Lvs 2–3 times pinnate with the ultimate segments linear, bristle-pointed. *Heads* 5–8 mm. diam., solitary, *short-stalked.* Involucral *bracts* oblong, blunt, *with broad scarious margins.*

Receptacle conical, hollow. Ray-florets 0; disk-florets dull greenish-yellow. Achenes 1·5 mm., with 4 inconspicuous ribs on the inner face and an obscure rim at the apex. Fl. 6–7. Little visited by insects. $2n=18$. Th.

Robust plants from W. Ireland with larger heads (to 12 mm. diam.) and achenes having a large, often unilateral, toothed or lobed crown at the apex, have been named *M. occidentalis* Greene (California), but they seem to grade into typical *M. matricarioides*.

Introduced. An abundant and increasing weed of waysides and waste places, and especially of tracks, paths and trampled gateways, throughout the British Is. 90, H40, S. Probably native in N.E. Asia but established in N. America, throughout Europe, in Chile and in New Zealand.

*****M. disciformis** DC. (Asia Minor), like *M. matricarioides* but with long-stalked rayless heads, is sometimes found as a casual.

31. Chrysanthemum L.

Annual to perennial herbs or shrubs with spirally arranged toothed or variously dissected lvs. Heads solitary or in corymbs, heterogamous or homogamous; involucre hemispherical, of a few rows of imbricating bracts with scarious margins; *receptacle flat or slightly convex, naked*. Ray-florets, when present, ligulate, female, yellow or white; disk-florets tubular, hermaphrodite, their corolla-tubes compressed and winged below or ± cylindrical. Achenes of disk-florets cylindrical or obconical, often angled and sometimes compressed, ribbed all round; of ray-florets similar or strongly 2–3-angled and then often broadly winged; pappus 0 or represented by a complete or partial membranous rim.

About 150 spp. in the northern hemisphere and S. Africa.

Chrysanthemum has been variously split by taxonomists but satisfactory lines of demarcation are difficult to draw.

1   Ray-florets 0 (or with extremely short and inconspicuous ligules).

                                                                    5. vulgare
    Ray-florets clearly present.   2

2   Ray-florets yellow.   **1. segetum**
    Ray-florets white.   3

3   Lvs pinnatisect or pinnate, with pinnatifid segments; heads usually
      1·5–2 cm. diam.; plant very strongly aromatic.   **4. parthenium**
    Lvs at most pinnatifid, usually only crenate or dentate; heads 2·5–10 cm.
      diam.; plant not very strongly aromatic.   4

4   Basal and lower stem-lvs roundish to obovate-spathulate, contracted
      ± abruptly into the long stalk; heads 2·5–5 cm. diam.   **2. leucanthemum**
    Basal and lower stem-lvs elliptical-lanceolate to long-oblanceolate, nar-
      rowing gradually into the long stalk; heads 5–10 cm. diam.
                                                               3. maximum

Subgenus Chrysanthemum. Annuals. Heads heterogamous with all florets fertile: ray-florets female, disk-florets hermaphrodite. Achenes of two kinds, those of ray-florets with 2–3 often winged angles, of disk-florets cylindrical; all ribbed. Pappus a membranous rim or 0.

1. C. segetum L.   Corn Marigold.

An annual glabrous and glaucous herb with erect simple or branched stems 20–50 cm. Lvs 2–8 cm., *glabrous, glaucous, somewhat fleshy*; lower lvs oblong-cuneate, narrowed into a winged stalk, coarsely toothed or pinnatifid; upper oblong, sessile, half-amplexicaul, toothed or ±entire. *Heads* 3·5–6·5 cm. diam., solitary, *with a long stalk thickened upwards. Bracts* broadly ovate, glaucous, *with very broad pale brown scarious margins. Ray-florets* broadly linear, *golden yellow,* about equalling the involucre, rarely 0; disk-florets golden yellow. Achenes of the ray-florets with 2 narrow lateral wings, those of the disk-florets ±cylindrical, unwinged; all c. 2·5 mm., pale, strongly ribbed, with no pappus. Fl. 6–8. Freely visited by various insects, especially flies. $2n = 18, 36$. Th.

Probably introduced. A locally common weed of acid arable soils throughout the British Is. 111, H40, S. Often a troublesome weed on loamy and sandy soils, but much reduced in abundance by liming. Probably a native of the Mediterranean region and W. Asia, but well established throughout Europe to 70° N. in Norway. Introduced in N. and S. America and N. Africa.

The annual Summer Chrysanthemums of gardens are *C. carinatum* Schousb., with the ray-florets differently coloured in their outer and inner halves, the bracts keeled and the achenes winged; and *C. coronarium* L., with yellow or white ray-florets, non-keeled bracts and angled but not winged achenes, and their hybrids.

Subgenus LEUCANTHEMUM (Mill.) Hegi. Annuals or perennials. Heads usually heterogamous with female ray-florets and hermaphrodite disk-florets (rarely homogamous, all florets hermaphrodite); ray-florets fertile or sterile. Achenes all similar, or those of the ray-florets compressed; all c. 10-ribbed. Pappus present or 0.

2. C. leucanthemum L.   Marguerite, Moon-Daisy, Ox-eye Daisy.

Leucanthemum vulgare Lam.

A perennial herb with slender branched oblique woody stock producing non-flowering lf-rosettes and erect simple or branched flowering stems 20–70 cm., sparsely hairy or almost glabrous. Basal and lower stem-lvs roundish to obovate-spathulate, crenate to dentate, long-stalked; upper stem-lvs oblong, blunt, toothed to pinnatifid, sessile, half-amplexicaul; all dark green, very sparsely hairy. *Heads* 2·5–5 cm. diam., *solitary,* long-stalked. Bracts lanceolate to oblong, green, with narrow dark purplish scarious margins and tips. *Ray-florets* long, *white,* rarely 0; disk-florets yellow. Achenes 2–3 mm., cylindrical, pale grey, with 5–10 strong ±equal ribs; ray-achenes with a border, disk-achenes not bordered. Fl. 6–8. Freely visited by a great variety of bees, flies, beetles, butterflies and moths. $2n = 18*$. H.

Native. A common plant of grassland on all the better types of soil. Throughout the British Is., though less common in Scotland. 112, H40, S. Throughout Europe to Lapland; Siberia. Introduced in N. America and New Zealand.

*3. C. maximum Ramond   Shasta Daisy.

Perennial herb with numerous non-flowering lf-rosettes and erect simple or sparingly branched flowering stems 35–75 cm., resembling a very robust

C. leucanthemum. Basal and lower stem-lvs up to 25 cm., the elliptical-lanceolate to oblong-oblanceolate blunt blade narrowing gradually into a long stalk; remaining stem-lvs with short winged stalks or sessile; all coarsely and bluntly toothed. Heads 5–7·5(–10) cm. diam., long-stalked. Bracts pale, with brownish or greenish centres. Ray-florets with broad white ligules.

Introduced and often escaping from gardens; Pyrenees.

Subgenus TANACETUM (L.) Hegi. Perennials, rarely annuals. Heads homogamous with all florets hermaphrodite or heterogamous with female ray-florets and hermaphrodite disk-florets. Achenes all similar, obconical, with 5–10 not very prominent ribs and pappus represented by a membranous rim.

4. C. parthenium (L.) Bernh. Feverfew.

Matricaria Parthenium L.

A perennial *strongly aromatic* herb with a ± vertical rootstock and an erect somewhat downy stem 25–60 cm., corymbosely branched above. Lvs 2·5–8 cm., yellowish-green, downy to subglabrous; lower lvs long-stalked, ovate in outline, pinnate, the narrowly ovate pinnatifid lflets with toothed or lobed segments, upper lvs shorter stalked and less divided. *Heads* 12–22 mm. diam., long-stalked, *in ± lax corymbs.* Involucre hemispherical; bracts lanceolate to oblong, bluntly keeled, downy, with narrow pale scarious margins and laciniate tips. *Receptacle hemispherical. Ray-florets white,* short and broad; disk-florets yellow. Achenes 1·5 mm., 8–10-ribbed, all with a short membranous border. Fl. 7–8. Visited by bees and flies. $2n=18$. H.

Probably introduced. A frequent plant of walls, waste places, hedgerows, etc., throughout Great Britain, and in several localities in Ireland but not in the Outer Hebrides, Orkney and Shetland. 108. Formerly cultivated as a medicinal herb and used as a febrifuge, whence the common name Feverfew. Probably native in S.E. Europe, Asia Minor and the Caucasus, but now established throughout Europe and in N. and S. America.

5. C. vulgare (L.) Bernh. Tansy.

Tanacetum vulgare L.; *C. Tanacetum* Karsch, non Vis.

A perennial strong-smelling herb with a creeping stoloniferous stock and stiffly erect lfy stems 30–100 cm., tough, angled, usually reddish, subglabrous, corymbosely branched above. *Lvs* 15–25 cm., the lower stalked, oblong in outline, *pinnate,* with c. 12 pairs of deeply pinnatifid lflets, their segments lanceolate-acute, ± sharply toothed; the uppermost sessile, semi-amplexicaul, simply pinnate with long narrow acute sharply toothed pinnae; all dark green, *gland-dotted and strongly fragrant. Heads* 7–12 mm. diam., *discoid,* numerous, in a dense flat-topped ± compound corymb. Involucre hemispherical, pale green, glabrous; its outer bracts ovate-lanceolate, the inner narrower, all with scarious margins especially near the blunt tip. Florets golden yellow, the obliquely truncate marginal female florets sometimes 0. Achenes 1·5 mm., cylindrical-obovoid, greenish-white, 5-ribbed; pappus a short unevenly toothed membranous cup. Fl. 7–9. Freely visited by a great variety of small insects. $2n=18$. H.

Native. Roadsides, hedgerows, waste places, etc., to 1200 ft. in Scotland; common. 112, H40, S. Throughout the British Is. Europe. Caucasus; Armenia; Siberia. Formerly much cultivated as a medicinal and pot-herb.

The florist's Chrysanthemum is *C. morifolium* Ramat. (*C. sinense* Sabine), probably a hybrid of Chinese origin and now with a great variety of cultivated races. *C. coccineum* Willd. is the Common Pyrethrum of gardens, and *C. cinerariifolium* (Trev.) Vis., Dalmatian Pyrethrum, is grown on a commercial scale in many parts of the world for the insecticidal powder prepared from its dried fl.-heads.

Costmary, *C. balsamita* L., formerly grown for the mint-like smell of its lvs, is sometimes placed in a separate genus, *Balsamita* Desf.

32. Cotula L.

Annual to perennial marsh and water plants usually with pinnatifid lvs and small long-stalked rayless heads of yellow florets. Receptacular bracts 0. Central florets tubular, hermaphrodite; marginal florets usually with so short a ligule as to appear tubular or with no corolla. Achene compressed; pappus 0.

About 50 spp. chiefly in Africa, Australia and S. America.

*1. C. coronopifolia L.

An annual to perennial strongly aromatic glabrous herb with a procumbent or ascending branched stem 8–30 cm. *Lvs* 2–5 cm., alternate, sessile, oblong-lanceolate in outline, deeply toothed to irregularly pinnatifid with long narrow lobes, rarely entire, *broadening below into a whitish sheathing base*. *Heads* 6–10 mm. diam., *hemispherical* on stalks exceeding the lvs; involucre of 2 rows of bracts with scarious margins and rounded ends. Disk-florets with a white corolla-tube and 4 short yellow lobes; marginal florets in 1 row with a short style and ± abortive corolla. Achenes stalked, papillose on the inner face; those of the disk narrowly, of the marginal row broadly winged. Fl. 7–8. $2n = 20$. Th.–H.

Introduced. A casual, occasionally establishing itself. Probably native in S. Africa but now widely distributed in both hemispheres.

33. Artemisia L.

Suffruticose or herbaceous perennials, rarely shrubs or annual herbs, with spirally arranged pinnatisect or palmatisect lvs. *Heads small, numerous, in racemes or racemose panicles*, rarely few or only 1, homogamous or heterogamous; involucre cylindrical or globular, of many imbricating bracts with narrow scarious margins; *receptacle ± flat, naked*, glabrous or hairy. *Marginal florets tubular, female or* 0; disk-florets tubular, hermaphrodite. Achenes cylindrical or somewhat compressed, not strongly ribbed; pappus 0.

About 200 spp., chiefly in the steppes and prairies of E. Europe, Asia and N. America.

1 Plant only a few cm. high with a single head (rarely 2) c. 12 mm. diam.;
     lvs ± palmately cut; a rare Scottish alpine.   **4. norvegica**
   Plant usually more than 20 cm. high with numerous heads in a racemose
     panicle.   2

2 Ultimate segments of lvs broadly linear or lanceolate, at least 2 mm. wide.   3
   Ultimate segments of lvs narrowly linear or subulate, c. 1 mm. wide or less.   6

3 Lf-segments glabrous or nearly so above, whitish-tomentose beneath;
     heads ovoid to ellipsoid.   4
   Lf-segments whitish, with hairs on both sides; heads broadly campanulate.   5

4 Plant caespitose or with short underground shoots; stems usually
     glabrescent, with large central pith; lvs with only the larger veins
     translucent; infl.-branches strict, straightish; fl. 7–9.   **1. vulgaris**
   Plant with long rhizomes; stems persistently pubescent, with narrow
     central pith; lvs with even the smaller veins translucent; infl.-branches
     arcuate-divaricate; fl. 10–11.   **2. verlotorum**

5 Stems, lvs and heads densely white-felted; heads 5–7 mm. diam.; longer
     than wide.   **3. stellerana**
   Stems, lvs and heads covered with silky white hairs but not densely felted;
     heads 3–5 mm. diam., wider than long.   **5. absinthium**

6 Plant hardly scented, ± glabrous or sparsely silky.   **7. campestris**
   Plant strongly aromatic; lvs whitish-woolly beneath.   **6. maritima**

Section 1. *Artemisia*. Marginal florets female, central hermaphrodite, all fertile; receptacle glabrous.

1. A. vulgaris L.   Mugwort.

A perennial caespitose aromatic herb with a branching root-stock and erect, *glabrescent or sparsely pubescent*, reddish, grooved and angled *stems* 60–120 cm., with large *white central pith* and narrow green peripheral tissues. *Lvs* 5–8 × 2·5–5 cm., *dark green and* ± *glabrous above, whitish-tomentose beneath;* basal lvs short-stalked, ± lyrate-pinnatifid, auricled; stem-lvs ± sessile, amplexicaul, bipinnate, the uppermost simply pinnate; *ultimate segments shortish, lanceolate* to oblong, 3–6 mm. wide, toothed or entire; *only the main veins translucent.* Heads 3–4 mm., 2–2·5(–3·5) mm. diam., narrowly campanulate to ovoid, ± erect, numerous, in dense sparsely lfy racemose panicles, the *infl.-branches strict and nearly straight.* Involucral *bracts* lanceolate to oblong, ± *densely arachnoid-pubescent* with broad scarious margins. Florets usually reddish-brown, all fertile. Achenes c. 1 mm., glabrous. Fl. 7–9. Wind-pollinated. $2n = 16, 18$. Hp.

Variable in the degree of dissection of the lvs and branching of the infl.

Native. Common in waste places, waysides, hedgerows, etc., throughout the British Is. Most of the temperate regions of the northern hemisphere to 70° N. in Norway and 74° N. in Siberia.

*2. A. verlotorum Lamotte   'Verlot's Mugwort.'

A perennial herb resembling *A. vulgaris* but differing in the following ways: plant more strongly and pleasantly aromatic, not caespitose but with *long rhizomes; stem* usually *more densely and persistently pubescent* and with a *small central pith* and relatively broader green peripheral tissues; *ultimate lf-segments* conspicuously *elongated, linear-lanceolate to linear,* even the *smaller veins clearly translucent;* heads 3·5–5 mm., 2·5–3(–4) mm. diam., ellipsoid, the *infl.-branches arcuate-divaricate* and the whole infl. very lfy;

involucral *bracts* only *thinly arachnoid-pubescent*; achenes not as yet seen in Britain. Fl. 10–11. $2n = 54$. Hp.

Introduced. Established in Surrey, Middlesex, Herts and Kent. 4. S.W. China.

*3. A. stellerana Bess.   Old Woman, Dusty Miller, Beach Sagewort.

A. ludoviciana Hort.

A perennial non-aromatic herb with a creeping woody stock and *densely white-felted stems* 30–60 cm. *Lvs* pinnate or pinnatifid, the uppermost ± entire, the lobes broad, blunt, *densely white-felted above and beneath*. Heads 5–9 mm. diam., broadly campanulate, somewhat longer than broad, numerous, in a racemose panicle. Bracts oblong to ovate, densely felted. Receptacle glabrous. Marginal florets female, central hermaphrodite, all yellow, fertile. Fl. 7–9. Wind-pollinated. H.

Introduced. A common garden plant grown for its white foliage, occasionally escaping and established in a few localities in Cornwall, Devon, Hants, etc. and near Dublin. 5, H1, S. N.E. Asia and the Atlantic coast of N. America from Massachusetts to Delaware; Kamchatka.

4. A. norvegica Fr. var. *scotica* Hultén   'Norwegian Mugwort.'

A dwarf tufted aromatic perennial herb with flowering stems 3–6 cm., hairy. *Basal lvs* c. 2 cm., stalked, *subpalmate*, with deeply 3–5-toothed *cuneate lobes*; stem-lvs ± sessile; all *silky-white*. *Heads usually solitary* (–3), c. 12 mm. diam., nodding. Involucral bracts with green midrib and broad dark-brown scarious margins. Fl. 7–9. $2n = 18$. Hs.

Native. At 1800–2500 ft. in *Rhacomitrium* heath on two mountains in W. Ross. 1. Other vars. in Norway and the Ural Mts.

Section 2. *Absinthium* (Mill.) DC. Like *Abrotanum*, but receptacle hairy.

5. A. absinthium L.   Wormwood.

A perennial aromatic plant with non-flowering rosettes and erect silky-hairy grooved and angled stems 30–90 cm., ± woody below. Lvs 2·5–5(–10) cm., those of the barren rosettes and the lower stem-lvs tripinnate, middle-stem *lvs* bipinnate, and the uppermost simply pinnate or undivided; ultimate segments lanceolate or linear-oblong, c. 2–3 mm. wide, usually blunt, *punctate, silky-hairy on both sides*. *Heads* 3–4 mm. diam., *broadly campanulate to globular*, rather broader than long, *drooping, numerous*, in a much-branched racemose panicle. Involucre silky-hairy, the outer bracts linear, inner ovate, all blunt and broadly scarious-margined. *Receptacle with long hairs*. Marginal florets female, central hermaphrodite, all fertile, yellow. Achenes 1·5 mm., glabrous. Fl. 7–8. Wind-pollinated. $2n = 18$. H.–Ch.

Native. A not infrequent plant of waste places throughout the British Is. to Aberdeen and Ross; Orkney. 81, H20, S. Temperate Europe and Asia northwards to Lapland, Karelia and S. Siberia. Introduced in N. and S. America and New Zealand.

Section 3. *Seriphidium* Bess. All fls hermaphrodite and fertile; receptacle glabrous; rich in the vermifuge santonin.

6. A. maritima L.   'Sea Wormwood.'

A perennial strongly aromatic herb with a short usually branching vertical woody stock producing non-flowering rosettes and decumbent then erect flowering shoots 20–50 cm., usually downy, branched above. Lvs 2–5 cm., mostly bipinnate, the ultimate segments linear, c. 1 mm. wide, blunt; lower lvs stalked, auricled; upper *lvs* simply pinnate, sessile; uppermost pinnatifid or entire; all ± *woolly on both sides*, not punctate. Heads 1–2 mm. diam., ovoid, longer than wide, erect or drooping, numerous, in lfy racemose panicles with short branches. Bracts oblong, outer herbaceous, downy, inner with broad scarious margins. Receptacle glabrous. Florets yellowish or reddish, all hermaphrodite, but the central ones sometimes sterile. Achenes apparently very rarely produced. Fl. 8–9. Wind-pollinated. $2n=18$. H.–Ch.

All British plants belong to ssp. **maritima**, with few non-flowering rosettes, persistently woolly and scarcely woody stems, and usually a broad panicle. Forms with the panicle branches short, erect, crowded and the individual heads erect and crowded are placed in var. *subgallica* Rouy (*A. gallica* auct., non Willd.), but no clear distinction can be made.

Native. Locally common on the drier parts of salt-marshes and sea-walls. 53, H8. All round the coast of Great Britain from N. Aberdeen to Dunbarton, but absent in the extreme north; very local on the east and west coasts of Ireland. Var. *subgallica* is found scattered amongst the type. Ssp. *maritima* is confined to the coasts of N.W. Europe from W. France to Denmark and S. Sweden, but other sspp. occur in the W. Mediterranean littoral from Spain to Dalmatia, in W. Switzerland (on dry calcareous substrata), in the inland saline areas of Germany, Hungary, etc., and from the Black Sea coasts of S. Russia, and the Caucasus, Georgia and Armenia across C. Asia to Lake Baikal.

Section 4. *Dracunculus* Bess. Marginal florets female, fertile; central hermaphrodite but mostly sterile: receptacle glabrous.

7. A. campestris L.   'Field Southernwood.'

A perennial *scentless* plant with a branched creeping woody stock producing tufts of short non-flowering shoots and decumbent then ascending ± glabrous flowering shoots 20–60 cm., ± woody below, paniculately branched above. Basal and lower stem-lvs 2–3 times pinnate, stalked, auricled; the upper stem-lvs less divided and shorter stalked upwards; the uppermost linear, entire, sessile; ultimate segments long, linear, c. 1 mm. wide, mucronulate, at first silky-hairy on both sides, later glabrous. Heads 3–4 mm. diam., broadly ovoid, ± erect, numerous in a narrow elongated racemose panicle. Bracts ovate to oblong, green or reddish, glabrous, with scarious margins. Receptacle glabrous. Florets yellow or reddish; the marginal female, fertile; the central hermaphrodite but mostly sterile. Achenes glabrous. Fl. 8–9. Wind-pollinated. $2n=18$, 36. Ch.

The British plant belongs to ssp. **campestris**.

Native. A very local plant confined to the 'breckland' heaths of S.W. Norfolk, N.W. Suffolk and E. Cambridge, though casual elsewhere. Casual near Belfast. 4. The various sspp. are distributed throughout the northern hemisphere to 75° N. in Novaya Zemlya and to 8800 ft. in the Alps.

Several spp. of *Artemisia* occur as casuals or garden-escapes. Frequently encountered are *A. abrotanum* L. (section *Abrotanum*), Southernwood or Lad's Love, a strongly aromatic shrubby plant up to 1 m. high, with 1–3 times pinnate lvs (uppermost simple) whose filiform punctate segments are glabrous above and ± hairy beneath, and very small drooping heads; much grown in gardens: *A. biennis* Willd. (section *Abrotanum*) has simply pinnate lvs with long narrowly lanceolate pinnatifid or sharply toothed segments, glabrous on both sides, and erect heads on short ± erect branchlets, the whole infl. very narrow; a frequent casual from S.E. Europe, Asia and N. America: *A. pontica* L. (section *Abrotanum*), Roman Wormwood, shrubby, with fine-dissected lvs white-woolly beneath: *A. scoparia* Waldst. & Kit. (section *Dracunculus*), a biennial with usually glabrous filiform lf-segments and pendulous broadly ovoid heads not exceeding 2 mm. diam. E. and S.E. Europe; Asia.

Also belonging to the Anthemidae is the genus *Anacyclus* which closely resembles *Anthemis* in habit and in having receptacular scales but differs in the broadly membranous-winged achenes. Occasionally found as casuals are *A. radiatus* Lois., with yellow ray-florets, *A. clavatus* (Desf.) Pers., with white ray-florets, and *A. valentinus* L. with ray-florets 0 or so shortly ligulate as to be inconspicuous.

Tribe 9. ARCTOTIDEAE.

The tribe Arctotideae resembles Anthemidae in that the heads have ligulate ray-florets, but has the style of Cynareae, with a ring of hairs below the bifurcation. There are no native members, but *Gazania* spp. and *Arctotis stoechadifolium* are commonly grown in gardens.

Tribe 10. CYNAREAE.

Echinops sphaerocephalus L. (Globe Thistle), a native of C. and S. Europe and W. Asia, is much grown in gardens and occasionally establishes itself. It is a rigidly erect thistle-like perennial herb 50–200 cm., with ovate-oblong pinnatifid lvs, green, densely glandular and sparsely bristly above, whitish-tomentose beneath, the ± triangular lobes ending in a strong spine and having spinous-toothed margins. Heads 1-flowered, aggregated into globular secondary heads 4–6 cm. diam. General involucre of laciniate straw-coloured scales hidden by the downwardly-directed heads; individual involucre of outer bristle-like bracts only half as long as the broader fimbriate inner ones which are glandular on the back. Florets tubular, pale blue or whitish; anthers blue-grey. Achenes 7–8 mm., silky; pappus represented by a cupule of partially joined hairs.

34. CARLINA L.

Thistle-like herbaceous or suffruticose plants with ± pinnatifid *spinous lvs*. Heads homogamous, solitary or in corymbs; outer bracts lf-like, *inner longer, scarious, coloured, shining, spreading* in dry weather; receptacle flat, pitted, with stiff bristles bordering the pits. Florets all tubular, hermaphrodite; style-arms short and almost closed, forming a pubescent cone. Achenes cylindrical, covered with appressed forked hairs; *pappus of feathery hairs in 1 row, united at the base in groups of 2 to 4*, deciduous.

About 20 spp., in Europe and W. Asia; Canary Is.

1. C. vulgaris L.   Carline Thistle.

A biennial herb with a tap-root and stiffly erect flowering stems 10–60 cm., usually purplish, somewhat cottony, usually branched above. Rosette lvs of the first season dying before flowering, 7–13 cm., oblong-oblanceolate, acute, narrowing to the sessile base, cottony especially beneath; stem-lvs shorter and broader-based, semi-amplexicaul, ± glabrous; all with undulate and somewhat lobed margins carrying numerous short weak spines. Heads 2–4 cm. diam., 2–5 or more in a corymb, rarely solitary. Outer bracts broadly lanceolate, cottony, green or purple-tinged, spiny at the tips and margins, shorter than the numerous long linear *straw-yellow inner bracts which spread when dry and simulate ray-florets*. Achenes 2–4 mm., covered with rusty hairs; pappus two or three times as long as the achenes. Fl. 7–10. Visited chiefly by bees and hoverflies. $2n=20$. H. (biennial).

Native. Almost always in calcareous grassland. Locally common throughout lowland Great Britain northwards to Ross; reaches 1500 ft. in Westmorland. 87, H30, S. Europe northwards to 60° 44′ N. in S. Norway, and Karelia; Siberia; Caucasus; Asia Minor.

35. ARCTIUM L.

Tall biennial herbs with long stout tap-roots and large spirally-arranged ovate *non-spinous lvs*. Heads solitary or in racemes or corymbs, homogamous; involucre ovoid-conical to globose, *bracts* numerous, imbricating, subulate, with appressed bases and *long stiff spreading hooked tips*; receptacle flat, with subulate rigid scales. Florets all tubular, hermaphrodite, purple or white; anthers acuminate above and prolonged into filiform tails below; style-arms cuneate. Achenes obovoid-oblong, compressed, grooved; *pappus of scabrid hairs in several rows, free to the base*. Fr.-heads animal-dispersed.

About 5 closely related spp. in Europe and Asia.

Petioles of basal lvs solid; heads few, sub-corymbose; wider upper part of
  corolla distinctly shorter than the filiform basal part; achenes 5–7 mm.
                                           1. lappa
Petioles of basal lvs hollow; heads few to many, ± racemose; wider upper
  part of corolla about equalling the filiform basal part.   **2. minus**

1. A. lappa L.   'Great Burdock.'

A. majus Bernh.

Stems 90–130 cm., stout, furrowed, often reddish, ± woolly, with many spreading-ascending branches. Basal lvs with ovate-cordate, entire or distantly toothed usually blunt blades to 40 cm., green and sparsely cottony above, grey-cottony beneath; *petiole solid*, up to 30 cm., furrowed above. *Heads* 3–4 cm. overall diam., *long-stalked, few, in subcorymbose clusters* terminating the main stem and branches; globular in bud, hemispherical and *widely open above in fr*. Involucral bracts green, glabrous or slightly ciliate at the base (cottony in var. *subtomentosum* Lange), the tips of the inner bracts about equalling the florets. Florets reddish-purple, *the wider upper part of the corolla distinctly shorter than the filiform lower part*. Achenes 5–7 mm., pale, somewhat wrinkled above. Fl. 7–9. Visited by bees and Lepidoptera. $2n=32, 36$. H.

Native. Waste places and waysides, rarely in woods. 65, H7. Scattered through lowland Great Britain northwards to Argyll and Fife. Europe northwards to S. Norway and Karelia; N. Asia Minor. Introduced in N. America.

2. A. minus Bernh., sensu lato   Lesser Burdock.

Stems 60–130 cm., furrowed, often reddish, somewhat woolly, with many spreading or down-curving branches. Basal lvs as in *A. lappa* but narrower (ovate-oblong) and acute, nearly glabrous above, sparsely cottony but green beneath; *petiole hollow. Heads in racemes* or the top 2–4 subcorymbose. Florets red-purple, *the wide upper part of the corolla about equalling the filiform lower part*. Achenes brownish. Fl. 7–9. H.

Very variable. The following 3 sspp. have been recognized as native, but their delimitation is difficult and more work is needed.

Ssp. **minus**: Plant slender. Heads 1·5–2·2 cm. overall diam. (including spines), ovoid in bud and *contracted at the top in fr., short-stalked or subsessile, in racemes. Inner involucral bracts* 10–13 mm., longer than the receptacular scales (7–10 mm.) but *equalled or exceeded by the corollas*. Achenes 5–7 mm., brownish, often with black blotches. $2n=32, 36$.

Ssp. **pubens** (Bab.) J. Arènes (*A. pubens* Bab.): Like ssp. *minus* but *heads* 2–3 cm. overall diam., *open at the top in fr., with stalks to 15 cm.*

Ssp. **nemorosum** (Lej.) Syme (*A. nemorosum* Lej., *A. vulgare* sensu A. H. Evans): Plant robust. Heads 3–4 cm. overall diam., globular in bud and *open at the top in fr., short-stalked or subsessile*, in racemes but with *the top 2–4 in a subcorymbose aggregate. Inner involucral bracts* 14–15 mm., considerably exceeding the receptacular scales (9–10 mm.) and *equalling or exceeding the corollas*. Achenes 6–8 mm., dark brown. $2n=36$.

Native. Waste places, waysides and hedgebanks, scrub, woodland margins and clearings. Throughout the British Is.; ssp. *minus* to 1250 ft. in Yorks, sspp. *nemorosum* and *pubens* perhaps more restricted to lowland localities. 106, H38, S. Europe to c. 65° N. in Scandinavia; N. Africa; Caucasus.

*__A. tomentosum__ Mill., with very cottony heads in corymbs and with the reddish tips of the broad inner involucral bracts not hooked and falling short of the florets, is a Continental sp. not native in Britain but occasionally found as a casual. $2n=36$.

36. Carduus L.

Annual to perennial herbs (thistles) whose spirally arranged simple or pinnate lvs have ± spiny margins. Heads solitary or in clusters at the ends of the main stem and branches, homogamous; bracts imbricating, many-rowed, narrow, usually spine-tipped; receptacle deeply pitted, densely bristly. Florets all tubular, hermaphrodite; anthers tailed below; style-arms united below, forming a shortly 2-lobed column with a ring of hairs at the base. Achenes obovate-oblong, grooved, glabrous; *pappus* of many rows *of simple* but scabrid (not plumose) *hairs*, united below, and deciduous.

About 120 spp., in Europe, N. Africa, and Asia, with a few in the Canary Is. and tropical Africa.

1 Heads oblong-cylindrical, falling when the achenes are ripe; corolla
    equally 5-lobed.   2
    Heads ovoid or hemispherical, not falling when the achenes are ripe;
    corolla distinctly 2-lipped, with one entire and one 4-lobed lip.   3

2 Stem lfy to close beneath the fl.-heads, and with continuous broad spinous
    wings; lvs somewhat cottony beneath; heads 2–10(–20) in a cluster;
    inner bracts equalling or exceeding the florets.   **1. tenuiflorus**
    Stem naked close beneath the fl.-heads, with interrupted narrow spinous
    wings; lvs densely cottony beneath; heads solitary or 2–4 in a cluster;
    inner bracts falling short of the florets.   **2. pycnocephalus**

3 Heads 3–5 cm. diam., usually solitary, drooping; bracts lanceolate above
    then contracted abruptly into an oblong base; middle and outer strongly
    reflexed, inner erect.   **3. nutans**
    Heads 1–3 cm. diam., solitary or clustered, erect; bracts linear-subulate,
    straight or the outermost recurved at the tip, not contracted above the
    base.   **4. acanthoides**

1. C. tenuiflorus Curt.   'Slender Thistle.'

An annual or biennial herb with a stout tap-root and erect broadly and
continuously spinous-winged stems 15–120 cm., ±cottony, branched above,
the branches erect-ascending. Basal lvs oblanceolate in outline, blunt, narrowed to the base; stem-lvs decurrent, acute; all sinuate-pinnatifid, spinous-
margined, ±cottony beneath. *Heads* about 15 × 8 mm., *cylindrical*, sessile,
in dense terminal clusters of 3–10(–20), with the stem winged close beneath
them. Bracts ovate-lanceolate, acuminate, glabrous, broadly scarious-
margined, 1-veined, terminating in a ±outwardly curved spine; inner bracts
scarious, equalling or exceeding the florets. Florets pale purple-red, rarely
white. Achenes 3·5 mm., fawn, shining, with fine transverse wrinkles, prominently tubercled above; pappus long, white. Fl. 6–8. Visited by bees.
Th.–H.

Native. Waysides and waste places, especially near the sea. 77, H34, S.
Locally common throughout lowland Great Britain, except in parts of the
W. Midlands, northwards to the Clyde Is. and Moray; throughout Ireland.
W. Europe from the Netherlands to Spain and Portugal, Italy, Dalmatia, the
Balkans, Crimea.

*2. C. pycnocephalus L.

An annual or biennial herb closely resembling *C. tenuiflorus* but with *stem
narrowly and discontinuously spinous-winged, naked close beneath the fl.-heads;
lvs densely cottony beneath*; heads 2 × 1 cm., distinctly stalked, *solitary* or 2–4
in a cluster; bracts cottony, tips spiny, straight and erect, *inner bracts falling
short of the florets.* Fl. 6–8. Th.–H. (biennial).

Introduced. Waste places. Established at Plymouth and a rare casual elsewhere. S. Europe from Spain to the Balkans; Canary Is.; N. Africa; W. Asia.

3. C. nutans L.   'Musk Thistle.'

A usually biennial herb with an erect interruptedly spinous-winged *stem*
20–100 cm., *naked for some distance beneath the heads*, ±cottony throughout,

simple or with spreading ascending branches above. Basal lvs elliptical, narrowing into a stalk-like base, sinuate; stem-lvs oblong-lanceolate, decurrent, deeply pinnatifid with triangular 2–5-lobed spine-tipped segments; all with undulate and spinous margins, sparsely hairy on both sides, woolly on the veins beneath. *Heads* 3–5 *cm. diam.,* *solitary,* or 2–4 in a loose corymb, hemispherical, ± *drooping.* Involucre cottony, often purplish, the outer and middle *bracts* ± strongly reflexed, the inner erect; all spine-tipped, *lanceolate-acuminate, contracted abruptly into an oblong base.* Florets red-purple, distinctly 2-lipped. Achenes 3–4 mm., fawn, with fine transverse wrinkles; pappus long, whitish. Fl. 5–8. Slightly musky and visited by many bees, hover-flies and Lepidoptera. $2n=16$. H.

Native. Pastures, waysides, arable fields and waste places on calcareous soils, up to 1650 ft. in Yorks. 85, H7, S. Locally common throughout Great Britain northwards to Ross, but sparser in the north. Inner Hebrides; S. and W. Ireland. Europe northwards to S.E. Norway and Esthonia; Siberia; Caucasus; Asia Minor; N. Africa.

4. C. acanthoides L.   'Welted Thistle.'

A biennial herb with a slender tap-root and an erect cottony *stem* 30–120 cm., usually branched above, *usually naked just beneath the fl.-heads* but otherwise with a continuous narrow undulate spinous-margined wing. Basal lvs elliptical in outline, sinuate-pinnatifid, narrowed into a stalk-like base; stem-lvs lanceolate in outline, decurrent, deeply pinnatifid with 3-lobed ovate segments whose terminal lobe is longest; all dull green, cottony beneath, with weakly spinous margins. *Heads about* 20×10–20 *mm.,* erect, ± spherical, usually *in* dense *clusters of* 3–5, rarely solitary. Involucre roundish-ovoid, slightly cottony, the *bracts linear-subulate, not contracted above the base,* ending in a weak slender spine, the outermost somewhat spreading, green, the inner erect, purplish, shorter than the florets. Florets red-purple or white, 2-lipped. *Achenes* 3 mm., fawn, *with* fine transverse wrinkles and *a prominent terminal tubercle which is not 5-angled*; pappus long, whitish. Fl. 6–8. Visited by many bees, hover-flies and Lepidoptera. $2n=16$. H.

Native. Damp grass verges and stream-sides, hedgerows, waste places. 100, H19, S. A lowland plant common in the south but becoming rarer in the north though reaching Ross; Inner Hebrides; E. Ireland. Guernsey. Often with *Urtica dioica, Galium aparine* and *Poa trivialis.* Europe northwards to 69° 22′ N. in Scandinavia; Siberia; Caucasus. Introduced elsewhere.

The hybrid of *C. acanthoides* and *C. nutans,* intermediate between the parents, is frequently found.

*C. crispus L. closely resembles *C. acanthoides* L. and the two are perhaps best regarded as sspp. of a single species. It may be distinguished by the stronger spines on the margins of the stem-wings, and by the wings usually extending up to the heads, not falling short of them as is more usual in *C. acanthoides.* The lvs are green and ± glabrous beneath; when pinnatifid the terminal segment is smaller than the laterals, and the marginal spines are stiffer than in *C. acanthoides.* The heads are 3(–4) cm. diam. and usually solitary; the involucre is subspherical, and its linear-lanceolate bracts are recurved at the tip and end in a stout spine. The achenes are olive-green, with the terminal tubercle 5-angled. Fl. 6–10. $2n=22$. H.

Introduced and at most a rare casual. Further investigation is needed; many of the records are of forms of *C. acanthoides* with large solitary heads. Europe, especially S. and S.E.; Caucasus.

There has been doubt about the allocation of the two names *C. crispus* L. and *C. acanthoides* L.

37. CIRSIUM Mill.

Annual to perennial herbs (thistles) with spirally arranged simple or pinnately lobed or divided lvs, usually with prickly margins. Heads solitary or in corymbs or dense clusters, homogamous; the florets all tubular, hermaphrodite or female. Involucre, receptacle and florets as in *Carduus* but achenes with a *pappus of* many rows of *feathery hairs* united at the base.

About 120 spp. throughout the northern hemisphere.

1	Plant stemless, with 1–3 usually sessile heads at the centre of the lf-rosette.	**6. acaulon**
	Plant with elongated stem.	*2*
2	Lvs roughly hairy and prickly on the upper surface, and therefore dull, not shining.	*3*
	Lvs not prickly (though sometimes hairy) on the upper surface and often ± glossy.	*4*
3	Stem not winged; involucre very cottony.	**1. eriophorum**
	Stem interruptedly spinous-winged; involucre not or scarcely cottony.	**2. vulgare**
4	Stem continuously spinous-winged.	**3. palustre**
	Stem not winged, or, if so, then only for short distances.	*5*
5	Heads yellow, overtopped by large pale ovate bract-like uppermost lvs; most stem-lvs ovate-acuminate, sessile and clasping, hardly spinous on the margins.	**5. oleraceum**
	Heads red-purple or pale purple, rarely white.	*6*
6	Heads 1·5–2·5 cm.; florets with the slender basal tube of the corolla longer than the broad upper part which is 5-cleft almost to its base.	**4. arvense**
	Heads 2·5–5 cm.; florets with the slender basal tube of the corolla about equalling the broad upper part which is 5-cleft to about the middle.	*7*
7	Lvs densely white-felted beneath with softly prickly-ciliate margins; head 3·5–5 cm.	**7. heterophyllum**
	Lvs at most whitish-cottony beneath, not densely felted; head 2·5–3 cm.	*8*
8	Lvs deeply pinnatifid, green beneath; heads solitary or 2–4 in a cluster; roots swollen, spindle-shaped; rare plant of calcareous pastures.	**9. tuberosum**
	Lvs sinuate-toothed, sometimes ± lobed, whitish-cottony beneath; head usually solitary; roots not swollen; fens and bogs.	**8. dissectum**

1. C. eriophorum (L.) Scop. ssp. **britannicum** Petrak   'Woolly Thistle.'

Carduus eriophorus L.; *Cnicus eriophorus* (L.) Roth

A biennial herb with a thick tap-root and a stout erect *unwinged* furrowed *stem* 60–150 cm., corymbosely branched above, cottony, not prickly. Basal lvs up to 60 cm., ovate-oblong in outline, narrowing into a short stalk, deeply pinnatifid and strongly undulate, the narrowly lanceolate distant spine-tipped *segments* usually 2-*lobed with one lobe directed upwards and one downwards;*

stem-lvs similar but sessile, auricled, *semi-amplexicaul, not decurrent*; all prickly-hairy but green above, white-cottony beneath, spiny and ciliate on the margins. *Heads* 3–3·5 × 4–7 cm., usually *solitary* ± erect, usually with a few small lvs close below. *Involucre very cottony*, its bracts lanceolate-acuminate ending in a long narrow ± spreading reddish-ciliate point usually with a slight dilation just below its apex; the outermost spine-tipped. Florets pale red-purple. Anthers blue-purple. Achenes 6 mm., buff mottled with black, smooth, shining; pappus very long, shining white. Fl. 7–9. Visited by long-tongued bees and Lepidoptera to whom alone the nectar is accessible. $2n = 34$. Hs. (biennial).

Native. Grassland, open scrub and roadsides on calcareous soil, to 850 ft. in Yorks. 48. Local in England and Wales northwards to Westmorland and Durham, and introduced further north to Argyll and Aberdeen. Ssp. *britannicum* is endemic in Great Britain but other sspp. range through C. Europe from France, Belgium and the Netherlands to N. Balkans and Upper Volga.

2. C. vulgare (Savi) Ten. 'Spear Thistle.'

Carduus lanceolatus L.; *Cirsium lanceolatum* (L.) Scop., non Hill

A biennial herb with a long tap-root and erect *interruptedly spiny-winged* cottony furrowed *stems* 30–150 cm., branched above. Basal lvs 15–30 cm., obovate-lanceolate in outline narrowed into a short stalk-like base, ± deeply pinnatifid and undulate, the segments usually 2-lobed, the upper lobe toothed near the base, the lower entire, lobes and teeth tipped with long stout spines; *stem-lvs* similar but sessile and *decurrent, with a long narrow terminal segment; all prickly-hairy above*, rough or cottony beneath; earliest lvs not undulate and only slightly pinnatifid. Heads 3–5 × 2–4 cm., ovoid-oblong, solitary or 2–3 in a cluster, short-stalked. Involucre slightly cottony, its bracts green, lanceolate-acuminate with the long neither ciliate nor dilated point recurved and spine-tipped in the outer, erect and scarious in the inner bracts. Florets pale red-purple. Achenes 3·5 mm., yellow streaked with black; pappus long, white. Fl. 7–10. Visited chiefly by long-tongued bees, hover-flies and butterflies. $2n = 68$. Hs. (biennial).

Var. *hypoleucum* (D.C.) Clapham (*C. nemorale* Rchb.) is a form of uncertain taxonomic status. Lvs hardly undulate, densely cottony beneath; branches few, strict, or 0; heads few, ovoid-globose.

Native. Fields, waysides, gardens, waste places, to 2050 ft. in N. England. 112, H40, S. Common throughout the British Is. Europe to 67° 50′ N. in Scandinavia; W. Asia; N. Africa. Introduced in N. America and Chile.

3. C. palustre (L.) Scop. 'Marsh Thistle.'

Carduus palustris L.

A biennial herb with a short erect premorse stock and an erect, furrowed, narrowly but continuously spiny-winged, hairy and cottony stem 30–150 cm., usually with short ascending branches. Basal lvs narrowly oblanceolate in outline, stalked, pinnatifid, the lobes shallow with spinous margins; *stem-lvs* sessile, *long-decurrent*, deeply pinnatifid and undulate, the segments each with 2–3 spine-tipped and spiny-ciliate lobes; *all hairy above*, slightly cottony beneath; earliest rosette lvs flat, hardly pinnatifid, cottony above and beneath.

Heads 1·5–2 × 1–1·5 cm., short-stalked in crowded lfy clusters at the ends of main stem and branches. Involucre ovoid, slightly cottony, its bracts purplish, lanceolate, appressed, the outer mucronate, inner acuminate. Florets dark red-purple, rarely white. Achenes 3 mm., pale fawn, smooth; pappus 2–3 times as long as the achenes, dirty white. Fl. 7–9. Visited by many bees, flies and Lepidoptera, the nectar being more readily accessible than in *C. vulgare* and *C. eriophorum*; said to be gynodioecious in continental Europe. 2*n* = 34. Hs. (biennial).

Native. Marshes, moist grassland, hedgerows, woods, to 2500 ft. in Scotland. 112, H40, S. Abundant throughout the British Is. Europe to 67° 50′ N. in Scandinavia; W. Asia; N. Africa.

4. C. arvense (L.) Scop. Creeping Thistle.

Serratula arvensis L.; *Carduus arvensis* (L.) Hill.

A *perennial* ± *dioecious* herb initially with a slender tap-root producing *far-creeping whitish lateral roots which bear numerous adventitious* non-flowering and flowering *shoots*, the latter 30–90(–150) cm., erect, furrowed, *unwinged*, glabrous or cottony above, usually branched. Basal lvs not in a compact rosette, oblong-lanceolate in outline, narrowed to a short stalk-like base, usually ± pinnatifid and undulate with triangular toothed and spiny-ciliate lobes ending in strong spines; middle and upper lvs similar but sessile, semi-amplexicaul, more deeply pinnatifid, not or very slightly decurrent; all ± glabrous on both sides or cottony beneath. Heads 1·5–2·5 cm., short-stalked, solitary or in terminal clusters of 2–4, together forming an irregular corymb. Involucre purplish, glabrous or somewhat cottony, of male heads ± spherical, of female ovoid; *bracts* numerous, appressed, the outer short, ovate-mucronate with ± spreading spiny points, the inner longer lanceolate-acuminate with erect scarious tips. Florets dull pale purple or whitish, *the broad upper part of the corolla shorter than the slender basal tube and 5-cleft to its base.* Achenes 4 mm., dark brown, smooth; pappus very long, brownish. Fl. 7–9. The male heads have abortive ovaries but sometimes ripen a few fr.; in the female heads the anthers are abortive. Strongly honey-scented and visited freely by a great variety of insects. 2*n* = 34. Gr.

Very variable, especially in the lvs. Plants with broad, flat, hardly lobed and weakly spiny lvs, green and ± glabrous beneath, have been placed in var. *setosum* C. A. Mey. or var. *mite* Wimm. & Grab., but intermediates are found. Var. *incanum* (Fisch.) Ledeb. with similar but narrower lvs, densely white-cottony beneath, is a S. European form which may merit subspecific rank. It occurs in Britain as a casual.

Native. Fields, waysides and waste places to 2100 ft. in N. England. 112, H40, S. Abundant throughout the British Is. and a very troublesome weed of cultivated land because of its capacity for regenerating from fragments of root. Europe to 68° 50′ N. in Scandinavia; Asia; N. Africa. Introduced in N. America.

*5. C. oleraceum (L.) Scop.

Cnicus oleraceus L.

A perennial herb with an obliquely ascending stock and erect furrowed unwinged stems 50–120 cm., simple or rarely branched above, ± glabrous.

Basal lvs ovate to broadly elliptical in outline, narrowed to a stalk-like base, simple or ± deeply pinnatifid with triangular-acuminate lobes; *middle and upper lvs* usually *unlobed, ovate-acuminate, sessile and amplexicaul* with large rounded basal auricles; all green and flaccid, ± glabrous, sharply toothed, with ciliate but hardly spinous margins, not decurrent. Heads 2·5–4 cm., ovoid, erect, clustered on short cottony stalks, *exceeded by the yellowish ovate-acuminate ciliate bract-like uppermost lvs.* Involucre slightly cottony, its bracts linear-lanceolate, erect, but with ± spreading tips, the outer ending in a short spine. *Florets yellowish-white*, rarely reddish. Achenes 4 mm., pale grey, angled. Fl. 7–9. The usually hermaphrodite heads are freely visited by bees and butterflies. $2n = 34$. H.

Introduced. Established in a few localities in England and Scotland. In marshes, fens, flushes, stream-sides and wet woods in C. Europe from C. France, N. Italy and N. Balkans northwards to 61° 15′ N. in Norway and to C. Russia; Siberia.

6. C. acaulon (L.) Scop.   'Stemless Thistle.'

Carduus acaulos L.

Perennial with short rhizome, long stout roots, and a rosette of spiny lvs usually with a *few sessile heads* at its centre, there being only rarely (f. *caulescens* Rchb.) an elongated unwinged aerial stem to 30 cm., simple or branched. Lvs 10–15 × 2–3 cm., oblong-lanceolate in outline, stalked, ± deeply pinnatifid, strongly and stiffly undulate, the segments with 3–4 triangular teeth or lobes, stoutly spine-tipped and spiny-ciliate, ± glabrous above, hairy on the veins beneath. Heads 3–4 × 2·5–5 cm., ovoid, 1–3, usually sessile on the rosette. Involucre glabrous, purplish, the outer bracts ovate with a short spiny mucro, the inner oblong-lanceolate blunt; all appressed. Florets bright red-purple with the broad upper part of the corolla somewhat shorter than the slender basal tube, 5-cleft to about the middle. Achenes 3–4 mm., smooth; pappus very long, whitish. Fl. 7–9. Gynodioecious, the hermaphrodite heads somewhat larger. Visited by hover-flies, bees and Lepidoptera. $2n = 34$. Hr.

Native. In closely grazed pastures, especially on chalk or limestone, to 1275 ft. in Derby. 46, S. Locally common in England northwards to Yorks; Glamorgan, Montgomery and Denbigh. N. Spain, N. Italy and N.W. Balkans to S. Scandinavia and Esthonia; W. Asia.

7. C. heterophyllum (L.) Hill   'Melancholy Thistle.'

Carduus heterophyllus L.; ?*C. helenioides* L.

A perennial *stoloniferous* herb with obliquely ascending stock and erect usually simple grooved *cottony unwinged stems* 45–120 cm. Basal lvs 20–40 × 4–8 cm., elliptic-lanceolate, long-stalked, finely toothed; lowest *stem-lvs* narrowed to the base, sometimes ± pinnatifid with forwardly directed lobes, the remainder usually unlobed, lanceolate-acuminate, broad-based, *amplexicaul, with rounded auricles*, entire or ± toothed; all *flat and flaccid, green and glabrous above, white-felted beneath with softly prickly-ciliate margins*, not decurrent. Heads 3·5–5 × 3–5 cm., solitary or rarely 2–3 in a terminal cluster. Involucre broadly ovoid, its bracts ovate-lanceolate, the outer mucronate, innermost blunt, all appressed, glabrous or finely pubescent, purplish-tipped. Florets red-purple, rarely white, the broad upper part of the corolla longer than the slender basal tube, 5-cleft to about the middle. Achenes 4–5 mm., fawn, smooth; pappus very long, whitish. Fl. 7–8. Visited chiefly by bees. $2n=34$. H.

Native. Hilly pastures and stream-sides, upland scrub and open woodland, from 300 ft. in Yorks to 3200 ft. in Scotland. 55, H1. From Sutherland and Caithness southwards to Merioneth, Stafford and Derby; Inner and Outer Hebrides; Fermanagh. N. Europe from 71° 10′ N. in Scandinavia southwards to Schleswig and Pomerania, and in the mountains of C. Europe from the Pyrenees to Roumania; C. Russia; Siberia. Introduced in N. America.

8. C. dissectum (L.) Hill   'Meadow Thistle', 'Marsh Plume Thistle.'

Carduus dissectus L.; *C. pratensis* Huds.; *C. anglicus* Lam.

A perennial shortly stoloniferous herb with short obliquely ascending stock, *cylindrical roots* and an erect, usually simple, terete cottony unwinged *stem* 15–80 cm., usually *with few small bract-like lvs above the middle*. Basal lvs 12–25 × 1·5–3 cm., elliptic-lanceolate, long-stalked, sinuate-toothed or ± pinnatifid; stem-*lvs* usually only 3–5, like the basal but oblong-lanceolate and semi-amplexicaul with basal auricles; all green and hairy above, *whitish cottony beneath*, the *margins with soft prickles*, longest on the teeth or lobes; not decurrent. Heads 2·5–3 × 2–2·5 cm., usually solitary. Involucre ovoid, purplish, cottony, bracts lanceolate, appressed, the outer spine-tipped, the inner acuminate. Florets dark red-purple. Achenes 2·5 mm., pale fawn, smooth; pappus long, pure white. Fl. 6–8. Florets hermaphrodite. Hel.

Native. Fens and bog-margins, always on wet peat, to 1650 ft. in W. Ireland. 53, H40, S. Local in England and Wales northwards to Yorks. Throughout Ireland, common in the west. W. Europe from Spain to the Netherlands and N.W. Germany.

9. C. tuberosum (L.) All.   'Tuberous Thistle.'

Carduus tuberosus L.

A perennial *non-stoloniferous* herb with an obliquely ascending stock, *swollen fusiform roots* and an erect grooved cottony unwinged stem 20–60 cm., lfy chiefly below the middle, simple or with very long erect branches. Basal *lvs* long-stalked broadly elliptical in outline, usually *deeply pinnatifid*, the segments distant, each with 2–5 spreading oblong lobes; stem-lvs few, less deeply divided, ± sessile, semi-amplexicaul but with auricles small or 0; all *green on both sides*, slightly cottony beneath, undulate and spinous-ciliate, not decurrent. Heads 2·5–3 cm., usually solitary. Involucre subglobose, cottony below, its bracts appressed, oblong-lanceolate, the outer shortly mucronate. Florets dark red-purple. Achenes 3 mm., pale fawn, smooth; pappus long, white. G.

Native. On chalk downs and other calcareous pastures. 5. Very rare and local in England and S. Wales (Wilts, Cambridge, E. Gloucester and Glamorgan). W. and C. Europe eastwards to S. Sweden, Saxony, Bohemia, Tirol and N. Italy.

Closely related to *C. dissectum* but readily distinguishable by the tuberous roots and absence of stolons.

The following hybrids have been reported from Britain: *C. acaulon × arvense*,

C. acaulon × *dissectum*, *C. acaulon* × *tuberosum*, *C. acaulon* × *vulgare*, *C. arvense* × *palustre*, *C. arvense* × *vulgare*, *C. dissectum* × *palustre*, *C. heterophyllum* × *palustre*, *C. palustre* × *vulgare*.

38. Silybum Adans.

Annual or biennial thistles with *white-veined or otherwise variegated lvs.* Heads homogamous; involucre of many rows of spiny bracts; *receptacle hairy,* not pitted. Florets all tubular and hermaphrodite, red-purple; *stamen-filaments united at their base into a tube;* anthers with short terminal points; style-arms connate. Achenes obovoid compressed, crowned with a membranous border; pappus of many rows of rough hairs united below into a basal ring.

Two spp. in the Mediterranean region, C. Europe, and the Near East. Differs from *Carduus* in the connate stamen-filaments.

*1. S. marianum (L.) Gaertn.   Milk-Thistle.

Carduus Marianus L.

An annual to biennial herb with an erect grooved slightly cottony unwinged stem 40–120 cm., simple or branched above. *Lvs* oblong, sinuate-lobed or pinnatifid with strongly spinous margins, glabrous, *pale shining green variegated with white along the veins above;* basal lvs narrowed into a sessile base, stem-lvs amplexicaul with rounded spinous-ciliate auricles, hardly decurrent. Heads 4–5 × 1–2 cm., solitary, erect or ± drooping. Involucre ovoid, glabrous, the outer bracts with an ovate-oblong base surmounted by a triangular spinous ciliate lf-like appendage which, in all but the basal bracts, ends in a stout *yellowish spine, long and spreading or recurved in the middle bracts,* shorter and erect in the innermost. Florets red-purple. Achenes 6–7 mm., blackish, grey-flecked and with a yellow ring near the apex, transversely wrinkled; pappus long, pure white. Fl. 6–8. Visited by various bees. $2n = 34$. H. (biennial)

Introduced. Naturalized in waste places locally and a common casual throughout lowland Great Britain and Ireland, northwards to Aberdeen. 60, H25. S. Europe from Spain to S. Russia; N. Africa; Caucasus; Near East. Introduced in C. Europe to Denmark, and in N. and S. America and S. Australia.

39. Onopordum L.

Biennial thistles with sinuate or pinnatifid lvs. Heads homogamous; involucre of many rows of coriaceous spiny bracts; *receptacle naked, deeply pitted, the pits with toothed membranous borders.* Florets all tubular and hermaphrodite, usually red-purple: anthers with terminal subulate appendages and short basal tails; style-arms connate. Achenes obovoid, compressed or 4-angled; pappus of many rows of rough hairs united into a basal ring.

About 20 spp. in the Mediterranean region and Near East. Close to *Carduus,* but lacking receptacular scales.

1. O. acanthium L.   Scotch Thistle, Cotton Thistle.

A large biennial thistle with a stout tap-root and a stiffly erect *continuously and broadly spinous-winged white-woolly stem* 45–150 cm., shortly branched above. Lvs sessile, elliptic-oblong, with sinuate teeth or triangular lobes ending in strong spines, cottony above and beneath; stem lvs decurrent. Heads 3–5 × 3–5 cm., usually solitary. Involucre subglobose, cottony, bracts lanceolate-subulate, green, tipped with yellowish spines, spreading or reflexed. Florets pale purple, rarely white. Achenes 4–5 mm., grey-brown with darker mottling, transversely wrinkled; pappus pale reddish, up to twice as long as the achene, its hairs strongly toothed. Fl. 7–9. Visited chiefly by bees. $2n = 34$. H. (biennial).

Doubtfully native. Fields, roadsides and waste places. Scattered throughout Great Britain northwards to Ross, but rare in Scotland. 74, S. Europe northwards to S. Scandinavia and C. Russia; W. Asia. Introduced in N. America.

40. Saussurea DC.

Perennial herbs with spirally arranged *non-spinous lvs*. Heads solitary or corymbose, homogamous; involucre with imbricating non-spinous bracts in many rows; *receptacle* flat, *with dense chaffy bristles*. Florets all tubular, hermaphrodite; *anthers with* long acute terminal appendages and *basal feathery tails*; style-arms connate. Achenes cylindrical, 4-ribbed, glabrous, smooth or wrinkled; *pappus of an outer row of persistent rough hairs and an inner row of deciduous feathery hairs united below into a basal ring*.

About 130 spp., chiefly alpine herbs of C. and E. Asia, others in Europe and N. America, 1 in Australia.

Differs from *Serratula* in the long anther-tails and the inner row of feathery hairs of the pappus.

1. S. alpina (L.) DC.   'Alpine Saussurea.'
Serratula alpina L.

A perennial herb with a ± horizontal scaly stock producing *short stolons* ending in lf-rosettes and an erect lfy grooved somewhat cottony simple flowering stem, 7–45 cm. Basal lvs 10–18 cm., ovate to lanceolate, stalked; stem-lvs diminishing upwards, the lower short-stalked, the uppermost narrow, sessile; all sharply toothed or ± entire, becoming glabrous above, white-cottony beneath. Heads 1·5–2 cm., ± sessile in a small dense terminal corymb; involucre ovoid-cylindrical, the outer bracts ovate, concave, sparsely hairy, the inner ovate-lanceolate covered with long dense grey hairs, all blunt, purplish. Florets exceeding the bracts, white below purple above. Anthers dark purple. Achenes 4 mm., brown, with pale ribs; pappus very long, whitish. Fl. 8–9. Protandrous. Fragrant and visited by flies and bees. $2n = 54$. Chh.

Native. Alpine and maritime cliffs, from 150 to 3900 ft. in Scotland. 31, H9. N. Wales, Lake District, Scotland, Inner and Outer Hebrides, Orkney, Shetland. Higher mountains of Ireland. N. Europe and the mountains of C. Europe; reaching 75° N. in Siberia; N. America.

Cnicus L.

One sp. in the Mediterranean region.

*C. benedictus L. (Blessed Thistle.) An annual thistle-like herb with an erect *pubescent* stem 10–40 cm., branched above. Basal lvs stalked; stem-lvs sessile,

± amplexicaul, shortly decurrent; all oblong, toothed or pinnatifid with spreading or backwardly directed lobes, pubescent, spinous-ciliate, with the *veins white and prominent on the underside.* Heads up to 4 × 2 cm., solitary at the end of the main stem with or without others on branches overtopping the primary head. *Involucre* ovoid, *woolly, equalled or exceeded by a rosette of lvs just beneath the head*; bracts ovate-acuminate, the point spiny and with spreading lateral spines. Receptacle with shining filiform scales. *Florets yellow,* the marginal very small, neuter. Achenes yellow-brown, shining, ribbed; pappus longer than the achenes. Fl. 5–9. Th.

Introduced. A casual. Native of the Mediterranean region and Near East. Contains a bitter glucoside, cnicin, and was formerly much used as a tonic and as a cure for gout.

41. Centaurea L.

Herbaceous (rarely suffruticose) plants with spirally arranged usually non-spinous lvs. Heads homogamous or heterogamous; *bracts* numerous, imbricating, each *with a membranous or scarious terminal appendage* which is laciniate, ciliate, toothed, pectinate or spiny, rarely entire; receptacle ± flat, bristly. Florets tubular, all similar, hermaphrodite, or more often the *marginal florets larger and neuter*; anthers with a long terminal appendage, and with or without basal tails; style-arms connate below. Achenes compressed, smooth, with an *oblique attachment-scar*; pappus of *several rows of rough hairs free to the base,* the innermost shortest and sometimes scaly, or 0.

About 400 spp., chiefly in the Mediterranean region and Near East, some in N. Europe, N. Asia and S. America and 1 in N. America and Australia.

1	Bracts with a non-spiny terminal appendage which is at most minutely prickly.	2
	Bracts with a simple, palmate or pinnate distinctly spiny appendage.	6
2	Appendages of bracts decurrent, i.e. extending some way down the sides of the basal part of the bract.	*3*
	Appendages not or very slightly decurrent.	*4*
3	Florets red-purple; perennial.	**1. scabiosa**
	Florets blue; annual.	**2. cyanus**
4	Involucre 5–6 mm. diam.; appendage of bracts small, slightly decurrent; middle stem-lvs deeply pinnatifid with ± linear lobes.	**3. paniculata**
	Involucre at least 10 mm. diam.; appendages not decurrent; middle stem-lvs entire, sinuate-toothed or shallowly lobed, not deeply pinnatifid.	5
5	Appendage of bracts pale brown, scarious, the broad pale margins entire to laciniate but not pectinate; pappus 0.	**4. jacea**
	Appendages dark brown or blackish, those of outer bracts deeply pectinate; pappus present.	**5. nigra**
6	Appendages of bracts spreading or reflexed with short equal palmately arranged spines.	**6. aspera**
	Appendages with a long terminal and much shorter lateral spines.	7
7	Florets reddish-purple.	**7. calcitrapa**
	Florets yellow.	8
8	Spines of bracts palmate, the short lateral spines confined to the base of the long terminal spine; corolla not glandular.	**8. solstitialis**
	Spines pinnate, the short lateral spines extending half-way up the terminal spine; corolla glandular.	**9. melitensis**

Section 1. *Cyanus* (Mill.) Pers. Bracts appressed, with ciliate or pectinate non-spinous appendages which are decurrent along the sides of the basal part of the bract but do not reach its base.

1. C. scabiosa L. 'Greater Knapweed.'

A perennial herb with a stout woody oblique branching stock, enclosed above in fibrous scales, and erect grooved ± pubescent stems 30–90 cm., usually branched above the middle. *Basal lvs* 10–25 cm., oblanceolate in outline, stalked, entire, toothed or more usually ± *deeply pinnatifid* with entire or pinnatifid lobes; stem-lvs sessile, ± deeply pinnatifid; all firm, usually sparsely hispid above and beneath, often shining above. Heads 3–5 cm. diam., solitary on long ± glabrous stalks. Involucre ovoid-globose, bracts with *blackish-brown horseshoe-shaped pectinate decurrent appendages*, which do not completely conceal the green basal parts. Florets red-purple, with or rarely without a large neuter marginal row. Achenes 4–5 mm., greyish, pubescent; pappus stiff, whitish, about equalling the achene. Fl. 7–9. Freely visited by various bees and flies. $2n = 14^*, 20^*, 24$. Hs.

A very variable plant split by Continental taxonomists into several sspp.

Native. Dry grassland, hedgebanks, roadsides, cliffs, etc., especially on calcareous substrata, to 1050 ft. in Derby. 85, H21, S. Throughout lowland Great Britain to Sutherland, common in the south but rare in Scotland. Europe northwards to 67° 56′ in Scandinavia, Finland and Karelia; Caucasus; W. Asia.

2. C. cyanus L. Cornflower, Bluebottle.

An annual or overwintering herb with an erect wiry grooved cottony stem 20–90 cm., usually with many slender ascending branches. Lower lvs 10–20 cm., stalked, usually lyrate-pinnatifid with narrow distant lobes, rarely oblanceolate, distantly toothed or entire; *upper lvs smaller, sessile, linear-lanceolate*; all *greyish with cottony hairs*. Heads 1·5–3 cm. diam., solitary on main stem and branches, on long cottony stalks. Involucre ovoid, cottony, its *bracts* green below *with narrow decurrent appendages* cut half-way into long spreading narrowly triangular teeth, those of the outer bracts usually *silvery white*, of the middle bracts brown with white-edged teeth. *Florets of the marginal row large, bright blue,* of the centre red-purple. Anthers purple. Achenes 3 mm., silvery grey, finely pubescent; pappus reddish, shorter than the achene. Fl. 6–8. Freely visited, especially by flies and bees. $2n = 24$. Th.

Native, *fide* Godwin. Cornfields and waste places to 1250 ft. in Scotland. 100, H23, S. Formerly common throughout Great Britain but now rare owing to greater care in cleaning seed-grain. Probably native in most of Europe and the Near East, but widely introduced as a cornfield weed.

*3. C. paniculata L. 'Panicled Knapweed.'

A biennial herb with erect slender sharply angled cottony stems 20–70 cm., with panicled branches. Lvs 2·5–8 cm., pinnatifid to bipinnatifid with ± linear acute lobes, cottony above and beneath. Heads about $2 \times 1\cdot5$ cm., in a long panicle. Involucre ovoid-oblong, 5–6 mm. diam., bracts striate, with small pale brown shortly decurrent triangular pectinate-ciliate appendages. Florets purple, the marginal row larger and neuter. Achenes 2 mm., silvery-white, glabrous; pappus of very short scaly bristles. Fl. 7–8. H. (biennial).

Introduced. Established in Jersey and a rare casual elsewhere in the British Is. Native in the W. Mediterranean region from Spain to Italy.

Section 2. *Jacea* (Mill.) Pers. Involucral bracts appressed, with laciniate or pectinate non-spinous and non-decurrent appendages.

*4. C. jacea L.   'Brown-rayed Knapweed.'

A perennial herb with an oblique branching stock and erect or ascending grooved glabrous or cottony stems 15–60 cm., usually with a few long slender subcorymbose branches above. Basal and lower stem-lvs oblanceolate narrowing into a stalk-like base, entire, coarsely toothed or somewhat pinnatifid; upper stem-lvs lanceolate, sessile, entire or with 1–2 teeth towards the base; all roughly hairy. Heads 10–20 cm. diam., solitary and subsessile; peduncles not or only slightly thickened immediately beneath heads. Involucre ovoid-globose, bracts ovate to oblong, bases quite concealed by the border, with *rounded pale brown scarious non-decurrent appendages* whose margins are *irregularly but usually not deeply laciniate*, those of the innermost entire. Florets red-purple, the marginal row usually larger. Achenes 3 mm., greyish, shining; *pappus* 0. Fl. 8–9. Male and female as well as hermaphrodite heads have been reported from Continental Europe. Freely visited by bees, flies and Lepidoptera. $2n = 44$. H.

Very variable. The usual British form is var. *angustifolia* Rehb., described above, but forms with broader basal lvs occur as casuals.

Introduced. Grassland and waste places. 19, H1, S. Established but rare in S. England and casual elsewhere. Formerly near Belfast. Europe northwards to 63° 41′ N. in Scandinavia and Karelia; N. Africa; N. and W. Asia.

C. jungens (Gugl.) C. E. Britton is the 1st generation hybrid of *C. jacea* and *C. nigra*; and *C. pratensis* Thuill., *C. drucei* C. E. Britton and probably *C. surrejana* C. E. Britton are segregants from this cross. All have characters intermediate between those of the parent spp., and in particular have pale but deeply laciniate or pectinate scarious bract-appendages.

5. C. nigra L.   'Lesser Knapweed', Hardheads.

A perennial herb with a stout branching oblique stock and erect tough rigid stems 15–60(–90) cm., grooved, usually ± roughly hairy, branched above. Basal lvs stalked, entire, sinuate-toothed or somewhat pinnatifid; stem-lvs sessile, entire or with a few teeth towards the base; all softly or roughly hairy, often cottony beneath when young. Heads 2–4 cm. diam., subsessile, solitary. Involucre ovoid-globose, bracts with *brown or blackish triangular non-decurrent ± deeply pectinate appendages*. Florets red-purple, those of the marginal row sometimes larger and neuter. Achenes 3 mm., pale brown, ± pubescent; *pappus of short bristly hairs*. Fl. 6–9. Freely visited by a great variety of insects. $2n = 44$. H.

Native. Grassland, waysides, cliffs, etc., to 1900 ft. in Wales. 111, H40, S. Throughout the British Is. except Shetland.

There are 2 sspp. in the British Is.:

Ssp. **nigra** (*C. obscura* Jord.)

Plant hispid, with broadly lanceolate usually toothed or shallowly pinnatifid lvs. Stem stout, *conspicuously swollen beneath the heads.* Appendages of bracts blackish-brown, ± *completely concealing the pale basal parts of the bracts;* those of the outer bracts very broadly triangular, *teeth about equalling the undivided central portion.* Heads dark purple, rather rarely with enlarged marginal florets. This is the prevalent northern form and that of heavier and moister soils in the south.

Ssp. **nemoralis** (Jord.) Gugl. (*C. nemoralis* Jord.)

Plant pubescent, more branching than ssp. *nigra*, with narrowly lanceolate, entire or sinuate-toothed, rather softly hairy lvs, diminishing rapidly up the stem. Stem slender, *not much swollen beneath the heads.* Appendages of bracts brown or brownish-black, *not completely concealing the pale basal parts of the bracts;* those of the outer bracts about equilaterally triangular, *teeth longer than the undivided portion.* Heads usually pale purplish-red, not uncommonly with enlarged marginal florets. This is the prevalent southern form, especially common on calcareous soils and usually on lighter soils than ssp. *nigra.* Intermediates between these sspp. are often reported, but on the whole they are geographically and ecologically distinct.

The aggregate species is native in W. Europe from Portugal, Spain and Italy northwards to 64° 13′ N. in Norway, and eastwards to the Netherlands, W. Germany and Switzerland. Introduced in N. America and New Zealand.

C. microptilon Gren. appears to be a segregant from *C. jacea* × *nigra* and shows intermediate characters, with narrow acute dark-coloured bract-appendages.

Section 3. *Seridia* (Juss.) Pers. Bracts with horny non-decurrent appendages palmately divided into short ± equal spines.

6. C. aspera L.   'Rough Star Thistle.'

A perennial herb with a slender stock and ascending stems 20–90 cm., sparsely hairy below, cottony above, with many slender spreading branches. Basal lvs usually lyrate, narrowed to a stalk-like base; stem-lvs 2–4 cm., narrowly oblong, pinnatifid, sinuate-lobed, or more usually toothed or entire, sessile or ± amplexicaul; all sparsely hairy. Heads 2·5 cm. diam., solitary, subsessile. Involucre ovoid-globose, glabrous or slightly cottony, its *bracts yellowish, leathery, with spreading or reflexed reddish-brown appendages having 3–5 subequal palmately-arranged short spines* about 3 mm. Receptacular scales white. Florets pale red-purple, the marginal neuter but little larger than the central. Achenes 3–5 mm., greyish-white, pubescent; pappus white, shorter than achene. Fl. 7–9. $2n = 20, 22$. H.

Doubtfully native in Channel Is. Dunes and waste places. A rare plant of Jersey and Guernsey, naturalized in S. Wales (Glamorgan) and casual elsewhere. S. and W. France, Spain, Portugal, Corsica and Sardinia, Italy.

Section 4. *Calcitrapa* (Hill) Pers. Bracts with non-decurrent appendages ending in a long terminal spine and smaller lateral or basal spines.

7. C. calcitrapa L.   Star Thistle.

A biennial herb with a stout tap-root and branched erect stock producing erect or ascending grooved ± glabrous flowering stems 15–60 cm., with stiffly flexuose divaricate-ascending branches arising from just beneath the heads. Basal and lower stem-lvs to 8 cm., deeply pinnatifid with narrow distant entire or toothed lobes; upper stem-lvs sessile, irregularly toothed or entire; all sparsely hairy above and beneath, the lobes and teeth bristle-pointed. Heads 8–10 mm. diam., subsessile; branches successively over-topping the older heads so that these appear lateral. Involucre ovoid, glabrous, its bracts with the *appendage ending in a stout spreading spine* 2–2·5 cm., yellow and channelled above, *with shorter spines at its base.* Florets pale red-purple, *glandular*, the marginal no larger than the central. Achene 3–7 mm., ovoid, whitish with or without brown mottling, glabrous; pappus 0. Fl. 7–9. Visited by bees and flies. $2n = 20$. H. (biennial).

Probably introduced. Waysides and waste places on sandy and gravelly soils and on chalk (especially near the Sussex coast). 17. A rare plant of S. England from Cornwall to Northampton and Kent, and in S. Wales (Glamorgan); E. Anglia; a casual near ports and elsewhere over a wider area. S. and C. Europe north to the Netherlands, C. Germany and C. Russia; N. Africa and Canary Is.; W. Asia.

*8. C. solstitialis L.   St Barnaby's Thistle.

An annual or rarely biennial herb whose erect or ascending stiff-cottony stems 20–60 cm., have *broad and continuous wavy wings* as do the many slender ascending branches. Basal lvs deeply lyrate-pinnatifid with distant, narrow, toothed or entire lobes; middle and upper lvs lanceolate, ± entire, sessile and decurrent into the wings of the stem; all cottony above and beneath. Heads 12 mm. diam., solitary, stalked. Involucre ovoid-globose, cottony, rarely ± glabrous, all but the innermost bracts with a *palmately spinous appendage*, the *terminal spine spreading, yellow*, channelled above, short in the lowest bracts but 10–20 mm. in the middle ones. *Florets pale yellow, eglandular*, the marginal no larger than the central. Achenes 2·5–3·5 mm., obovoid, the central yellowish mottled with brown, and with a white pappus equalling the achene; the marginal dark brown with no pappus. Fl. 7–9. Visited chiefly by bees. $2n = 16$. Th.

Introduced. Cultivated land, especially in lucerne and sainfoin fields. Rare in S. and E. England where it may persist for many years; casual elsewhere. S. and S.E. Europe; W. Asia. Introduced in C. and N. Europe.

*9. C. melitensis L.   'Maltese Star Thistle.'

An annual herb resembling *C. solstitialis* but with involucral bract-appendages whose slender brownish 5–8 mm. terminal spine bears *pinnately arranged short lateral spines in its lower half*. Heads subsessile. Florets with yellow *glandular corolla*, the marginal no larger than the central. Achenes 2–2·3 mm. pale greenish-grey with whitish stripes; pappus shorter than the achene, whitish. Fl. 7–9. $2n = 22$. Th.

Introduced. Waste places and roadsides. A not infrequent casual. Mediterranean region eastwards to Greece and Tunis; Madeira; Canary Is. Widely introduced in C. and N. Europe, N. and S. America, S. Africa and Australia.

Several other *Centaurea* spp., especially natives of the Mediterranean region and Near East, occur as casuals.

42. SERRATULA L.

Perennial herbs with spirally arranged *non-spinous lvs*. Heads solitary or corymbose, homogamous, often gynodioecious or ± dioecious; involucre with imbricating acute but not spinous bracts in many rows; receptacle flat, with *dense chaffy scales*. Florets all tubular, similar or sometimes the central hermaphrodite and the marginal female; *anther-tails short or* 0; style-arms connate or free. Achenes oblong, slightly compressed, glabrous; *pappus of many rows of stiff, rough, deciduous, simple hairs, all free to the base*, the outermost shortest.

About 40 spp. in Europe, N. Africa and temperate Asia. Like *Saussurea* but with the anthers hardly tailed and the pappus-hairs non-feathery.

1. S. tinctoria L. Saw-wort.

A perennial glabrous subdioecious herb with a short stout ± erect stock and an erect, slender, wiry, grooved stem 30–90 cm., with a few ascending branches above. Lvs 12–25 cm., ovate-lanceolate to lanceolate in outline, glabrous or with a few scattered hairs, their *margins with fine bristle-tipped teeth*, very variable in degree of dissection from undivided to lyrate-pinnatifid or almost pinnate with narrow lateral lobes and a very long narrowly elliptical terminal lobe; basal and lower stem lvs stalked, upper sessile. Heads 1·5–2 cm., loosely corymbose, stalked, or few, crowded, subsessile ('var. *monticola* (Boreau) Syme'), the female heads larger than the male; involucre of female heads ovoid-cylindrical, the florets exceeding the bracts; of male heads oblong-cylindrical, the florets about equalling the bracts; outer bracts ovate-acute, downy on the margins, inner narrower, rough on the margins; all appressed, purplish where exposed. Florets reddish-purple, rarely white; the female with corolla swollen in the middle, white abortive anthers and spreading style-arms; the male with corolla not swollen, dark-blue anthers and appressed style-arms. Achenes 5 mm., fawn, slightly rough; pappus yellowish. Fl. 7–9. Visited by flies and bees. $2n = 22^*$. H.

Native. Wood margins, clearings and rides and open grassland on moist basic soils over limestone or chalk; to 1250 ft. in Wales. 70, H1. Local throughout England and Wales, and in Dumfries and Kirkcudbright. C. Europe from N. Spain, C. Italy and N. Balkans northwards to C. Scandinavia, Estonia and C. Russia; Siberia; Algeria.

Amongst other members of the Cynareae which occur as casuals the most frequently encountered are:

*Carthamus tinctorius L. An annual to biennial herb with an erect simple or little branched stem 10–60(–100) cm., furrowed, pale yellow, glabrous. Lower lvs ovate-oblong narrowed into a short stalk; upper lvs ovate-lanceolate, sessile, with a ± cordate amplexicaul base; all glabrous, finely and softly spinous-toothed. Heads 2–3 cm. diam., surrounded by a cluster of spreading lf-like bracts passing over into

± appressed involucral bracts with green spinous-toothed terminal appendages. Florets bright reddish-orange. Achenes obovoid, shining white; pappus of numerous narrow scales. Not known as a wild plant, but still cultivated for the red and saffron dyes from its flowers (Safflower, False saffron) and for the oil from its achenes.

*C. lanatus L. An annual herb 20–60 cm., with a terete, pale yellow, ± cottony stem and sessile to amplexicaul lanceolate coarsely sinuate-toothed or pinnatifid lvs with long stout marginal spines. Heads surrounded by narrow spreading spinous bracts; involucre of narrow appressed bracts with a spinous terminal appendage. Florets yellow. Achenes 4-angled, rugose, dark brown; pappus of linear scales.

    C., S. and S.E. Europe, Mediterranean region and Canary Is.

Subfamily CICHORIOIDEAE

Tribe 11. CICHORIEAE.

43. CICHORIUM L.

Usually perennial herbs with spirally arranged often runcinate basal lvs. Heads terminal and axillary; involucre cylindrical, its *bracts in 2 rows*, the inner longer; receptacle ± flat, usually naked. *Florets* usually *blue*; anthers without basal tails; style-arms hairy. *Achenes* obovoid, ± angled, flat-topped, crowned *with* 1–2 *rows of short blunt scales*. Nine spp. in the Mediterranean region with 1 reaching N. Europe and another in Abyssinia.

1. C. intybus L.   Chicory, Wild Succory.

A perennial herb with a long stout tap-root, a short vertical stock and stiffly erect, tough, grooved stems 30–120 cm., with stiff spreading-ascending branches, all roughly hairy or ± glabrous. Basal lvs short-stalked, oblanceolate in outline, runcinate-pinnatifid or toothed; lower stem-lvs similar but sessile and ± amplexicaul; upper stem-lvs lanceolate, entire or distantly toothed, sessile, clasping the stem with the pointed auricles of their broadened base; all ± glabrous or roughly hairy beneath, glandular-ciliate. *Heads* 2·5–4 cm. diam., solitary, terminal on a somewhat thickened stalk and in *subsessile clusters of* 2–3 *in the axils of upper lvs*. Outer bracts about 8, broadly lanceolate, spreading above; inner about 5, twice as long, narrower, erect; all green, herbaceous. *Florets large, bright blue*, rarely pink or white. Achenes 2–3 mm. irregularly angular, pale brown often with darker mottling, pappus of fimbriate scales $\frac{1}{10}$–$\frac{1}{8}$ as long as the achene. Fl. 7–10. The heads open in early morning and close soon after mid-day. Visited chiefly by bees and hover-flies. $2n=18$. H.

    Probably native. Roadsides and pastures to 900 ft. in Scotland. Locally common especially on calcareous soils in England and Wales, but rare and probably introduced in Scotland and Ireland. 72, H 32, S. Europe northwards to C. Scandinavia, Esthonia and C. Russia; W. Asia; N. Africa. Introduced in E. Asia, N. and S. America, S. Africa, Australia and New Zealand.

    The dried and ground roots yield the chicory of commerce. In some countries the roots are boiled and eaten as a vegetable. The plant is sometimes included in seed mixtures on shallow chalky soils in England, both because cattle readily eat its lvs and for the effect of its deep tap-root in breaking up the subsoil.

The cultivated endive, *C. endivia* L. ssp. *endivia*, closely resembles *C. intybus* but its lvs are glabrous, the basal lvs are less deeply lobed and often merely sinuate-toothed, the stalk of the terminal head is conspicuously thickened, the bracts are eglandular, and the larger achenes (2·5–3·5 mm.) have a pappus of scales $\frac{1}{6}$–$\frac{1}{2}$ as long as the achene. It is widely grown as a salad plant, often in varieties with strongly crisped lvs.

The wild ssp. *pumilum* (Jacq.) Hegi (Mediterranean region) is a low-growing herb with runcinate and hairy basal lvs but with the long pappus-scales of the cultivated form.

44. LAPSANA L.

Annual to perennial herbs with *lfy flowering stems* and small yellow heads in loose panicles or corymbs. Involucral bracts erect, in 1 row with a few very small basal scales; *receptacle flat, naked.* Anther-lobes without basal tails; style-arms slender, hairy. *Achenes somewhat compressed, about 20-ribbed, rounded above, glabrous,* the outermost longer than the central and distinctly curved; *pappus* 0.

About 20 spp. in temperate Europe and Asia.

Upper stem-lvs usually ovate-rhomboid to lanceolate, distantly toothed
  (or sometimes entire); heads 1·5–2 cm. diam.   **1. communis**
Upper stem-lvs broadly linear, subentire; heads 2·5–3 cm. diam.
                                             2. intermedia

1. L. communis L.   Nipplewort.

Lampsana communis auct.

Annual. Stem 20–90 cm., erect, usually with spreading hairs below, ± glabrous above, paniculately branched in its upper half, the branches ascending. Lower lvs long-stalked, lyrate-pinnatifid, with a very large ovate, acute or obtuse, often cordate, ± coarsely sinuate-toothed terminal lobe and a few small lateral lobes or merely a narrow wing below it; *upper lvs ovate-rhomboid to lanceolate, acute, short-stalked, distantly but sharply sinuate-toothed* or sometimes entire; all thin, usually hairy. *Heads* 1·5–2 *cm. diam.*, usually 15–20, on slender stalks in a ± corymbose panicle. *Involucre* 6–8 *mm.*, ovoid, glabrous or glandular, of 8–10 linear bracts suddenly contracted to a blunt apex, erect and strongly keeled in fr., with a few ovate-lanceolate basal bracts hardly 1 mm. long. Florets 8–15, pale yellow. *Achenes* 2·5–5 mm., pale brownish, *fully half as long as the involucre.* Fl. 7–9. The heads close in mid-afternoon and do not open in dull weather. Visited by small bees and flies. $2n = 12^*$. Th.

Native. Waysides, hedges, wood-margins, walls and waste places to 1450 ft. in England. 112, H40, S. Common throughout the British Is. Europe to 64° 30′ N. in Scandinavia; N. Africa; W. and C. Asia. Introduced in N. America. Formerly used as a salad plant.

*2. L. intermedia M. Bieb.

Annual to perennial herb closely resembling *L. communis* but with the *upper lvs broadly linear* (up to 1 cm. wide) and *almost entire*; *heads* 2·5–3 *cm. diam.* and usually 20–25 in number, with *involucre* 8–9 *mm. long*; *achenes less than half as long as the involucre*; $2n = 14^*$; Th.–H.

Introduced. On a chalky railway-embankment at Totternhoe (Beds). 1. Caucasus, Crimea, Thrace, Hungary.

45. ARNOSERIS Gaertn.

A small annual *scapigerous* herb with yellow heads whose numerous bracts are in 1 row, apart from a few tiny basal scales; receptacle flat, naked, pitted marginally. Anther-lobes without basal tails; style-arms short, blunt, hairy. *Achenes* broadly obovoid, strongly ribbed, *crowned by a very short membranous border.*

One sp.

Differs from *Lapsana* in habit and in the membranous border of the achenes.

1. A. minima (L.) Schweigg. & Koerte   Lamb's or Swine's Succory.

Hyoseris minima L.; *Arnoseris pusilla* Gaertn.

Lvs 5–10 cm., *all radical*, obovate-oblong, narrowed into a short stalk, distantly and rather coarsely toothed, ciliate, glabrous or ± hairy on both sides. *Scapes* 7–30 cm., many, ± glabrous, simple or sparingly branched above, the branches, in the axils of minute bracts, curving upwards and finally overtopping the main stem; all fistular, enlarging upwards and so *markedly clavate* beneath the heads. Heads 7–11 mm. diam., solitary, terminal. Involucre campanulate, ± glabrous, its 15–20 *bracts* narrowly triangular-acuminate with a *prominent pale keel, connivent in fr.* with the extreme tips slightly spreading; basal scales tiny, subulate. Florets yellow, half as long again as the bracts. Achenes 1·5 mm., with 5 strong and 5 weaker intermediate ribs, transversely wrinkled between the ribs; border very short, sometimes obscurely toothed above the ribs. Fl. 6–8. Sparingly visited by flies. $2n = 18^*$. Th.

Probably native. Arable fields on sandy soils. 30. A local plant of eastern Great Britain from Dorset and Kent northwards to Moray. C. Europe from Spain and Portugal, Corsica, N. Italy Hungary, Roumania and S. Russia northwards to S. Sweden, N. Poland and C. Russia. Introduced in N. America, Australia and New Zealand.

46. HYPOCHOERIS L.

Annual to perennial usually scapigerous herbs, commonly with the branches of the scape thickened distally beneath the heads. Involucre of many imbricating rows of lanceolate bracts. *Receptacle flat, with numerous lanceolate scales.* Florets yellow. Anthers shortly tailed below and with rounded terminal appendages. Achenes cylindrical, ribbed, at least the *inner beaked; pappus* usually of 1 *row of feathery hairs*, but sometimes also an outer row of simple hairs.

About 70 spp., 12 in Europe and the Mediterranean region, the remainder in S. America.

1 Annual; lvs ± glabrous; florets about equalling the involucre, their ligules
    only about twice as long as broad; heads opening only in full sun.
                                                           2. glabra
  Perennial; lvs hispid; florets exceeding the involucre, their ligules four
    times as long as broad; heads opening in dull weather.   *2*

2 Heads commonly solitary, sometimes 2–4; scape unthickened above or
thickened only immediately beneath the heads; lvs obovate-oblong,
often purple-spotted; pappus of 1 row of feathery hairs.   **3. maculata**
Heads usually several; scape thickened for some distance beneath the
heads; lvs ± broadly oblanceolate-oblong, not spotted; pappus hairs in
2 rows, the inner feathery, the outer simple.   **1. radicata**

Section 1. *Hypochoeris*
Pappus hairs in 2 rows, the inner feathery, the outer shorter, simple.

1. H. radicata L.   Cat's Ear.

A perennial scapigerous herb with a short ± erect branching premorse stock
and fleshy roots. *Lvs* 7–25 cm., in a basal rosette, ± broadly oblong-lanceolate,
narrowed gradually into the broad stalk-like base, sinuate-toothed to sinuate-
pinnatifid, usually *hispid with simple hairs*, dull green above and somewhat
glaucous beneath. *Scapes* 20–60 cm., usually several from each rosette, erect
or ascending, lfless or with 1–2 small lvs below, usually *forking, enlarged below
the heads and bearing numerous small scale-like bracts*. Heads 2·5–4 cm. diam.
Involucre 18–25 mm., cylindrical-campanulate; bracts lanceolate-acuminate,
dull green, bristly on the midrib, the outermost somewhat lax. Florets bright
yellow, exceeding the involucre, the ligules about 4 times as long as broad,
the outer ones greenish or grey-violet beneath. Achenes 4–7 mm. (excl. beak),
orange, strongly muricate, narrowed above into a beak, that of the central
achenes exceeding, of the outer equalling or falling short of, the achenes or
sometimes 0. Fl. 6–9. Visited freely by many kinds of insects, especially bees,
and not automatically self-pollinated. $2n = 8$. Hr.
   Native. Meadows and pastures, grassy dunes, waysides, etc., to 2000 ft.
in Ireland; common. 112, H40, S. Throughout the British Is. Europe northwards to 62° 47′ N. in Scandinavia and to N. Russia; Asia Minor; N. Africa.

2. H. glabra L.   'Smooth Cat's Ear.'

An *annual* herb with a basal rosette of oblanceolate *lvs* 1–20 cm., narrowed
gradually to the stalk-like base, sinuate-toothed to sinuate-pinnatifid, ± *glabrous*, pale green and sometimes reddish near the margin. Scapes 10–40 cm.,
usually several from each rosette, erect, ascending or decumbent, lfless or
with 1–2 small lvs, ± branched, the branches slightly enlarged beneath the
heads and bearing a few scale-like bracts. *Heads* rather small, *opening widely
only in full sunlight*. Involucre 12–15 mm., cylindrical; its bracts very unequal,
lanceolate-acuminate, with whitish margins and dark points. *Florets* bright
yellow, *about equalling the involucre*, their ligules only about twice as long as
broad. Achenes 4–5 mm. (excl. beak), reddish-brown, muricate; those of
central florets with long slender beak, of outer usually unbeaked (all unbeaked in var. *erostris* Coss. & Germ.). Fl. 6–10. Visited by bees, etc.
$2n = 10$. Th.
   Native. Grassy fields, derelict arable land, heaths, fixed dunes, etc., on
sandy soils; locally frequent. 57, H1, S. Great Britain northwards to
Moray and Inverness. In Ireland confined to Derry. Europe northwards to
S. Scandinavia and N. Poland; Asia Minor and Syria; N. Africa. A charac-

teristic plant of open communities on non-calcareous sand, on dunes with *Corynephorus canescens* and *Jasione montana*, and on arable land with *Scleranthus* spp., *Teesdalia nudicaulis*, *Arnoseris minima*, etc.

Section 2. *Achyrophorus* Duby

Hairs of pappus in 1 row, all feathery.

3. H. maculata L.   'Spotted Cat's Ear.'

A perennial herb with a stout cylindrical blackish stock and a basal rosette of obovate-oblong *lvs* 4–15(–30) cm., *usually spotted with dark purple* and with reddish midribs, narrowed to the stalk-like base, ± sinuate-toothed, hispid. *Scapes* 20–60 cm., 1 or a few from each rosette, erect, *simple* or less commonly with 1–3 branches, lfless or with 1–2 small lvs, not enlarged above or enlarged only immediately below the heads and with *scale-like bracts few or* 0. Heads 3–4·5 cm. diam., solitary or 2–4. Involucre 18–23 mm., campanulate, blackish-green; its outer bracts lanceolate, hispid, the middle and inner linear-lanceolate with woolly margins. Florets lemon-yellow, twice as long as the involucre. Achenes 5–7 mm. (excl. beak), slightly muricate, transversely ridged, all rather shortly beaked. Fl. 6–8. $2n=10$. Hr.

Native. Calcareous pastures and grassy cliffs; rare and decreasing. 11. E. England from Essex and W. Suffolk to Northants and Lincoln; Cornwall, Caernarvon, Lancs, Westmorland. Europe from the Pyrenees, S. France, N. Italy, N. Balkans and S. Russia northwards to 64° 7′ N. in Scandinavia, to Karelia and C. Russia. A 'pontic' plant in C. Europe, growing commonly with *Filipendula vulgaris, Peucedanum oreoselinum, Asperula cynanchica, Orchis ustulata*, etc.; usually calcicolous but sometimes found in *Calluna* heath.

47. Leontodon L.

Usually perennial *scapigerous* herbs with rosettes of entire or pinnatifid lvs, the lvs and scapes commonly with forked hairs. Involucral bracts in several imbricating rows. *Receptacle naked*, pitted, the pits often with toothed or ciliate margins. Anthers not tailed, with blunt terminal appendages. Achenes little compressed, narrowed above and sometimes beaked, strongly ribbed. *Pappus usually of* 2 *rows of hairs, the inner feathery, the outer simple* or sometimes 0; that of the outermost row of achenes sometimes represented by a small cup of scarious scales.

About 45 spp. in Europe, C. Asia, N. Africa and the Azores.

1 Lvs glabrous or with simple hairs; scape usually branched and bearing 2 or more heads; pappus of a single row of feathery hairs.
                                                                                    1. autumnalis
   Lvs usually with forked hairs; scape simple with a single terminal head; pappus of central achenes consisting of an inner row of feathery and an outer row of shorter scabrid simple hairs.   *2*

2 Scape usually densely hairy above; involucre exceeding 10 mm.; outer florets orange or reddish beneath, rarely pale grey-violet; achenes all with feathery hairs.   **2. hispidus**
   Scape sparsely hairy, especially below; involucre not exceeding 10 mm.; outer florets grey-violet beneath; outermost achenes surmounted by a cup of scarious scales.   **3. taraxacoides**

Section 1. *Oporinia* (D. Don) Koch

All achenes with a single row of feathery hairs, dilated at the base; lvs glabrous or with simple hairs.

1. L. autumnalis L. 'Autumnal Hawkbit.'

A perennial herb with an oblique usually branched premorse stock, each branch terminating in a rosette of ± oblanceolate *lvs* varying from distantly sinuate-toothed to deeply pinnatifid, and *glabrous or with simple hairs. Scapes* 5–60 cm., decumbent below then erect or ascending, usually *branched*, rarely simple, somewhat enlarged and hollow above and bearing *numerous scale-like bracts just beneath the heads*, glabrous or sparsely hairy, especially below. Head 12–35 mm. diam., erect in bud. Involucre c. 8 mm., ovoid-cylindrical, dark green, glabrous to woolly; its bracts linear-lanceolate, acute. *Fls* golden yellow, the outer *with reddish streaks beneath*. Achenes 3·5–5 mm., reddish-brown, slightly narrowed above, longitudinally ribbed and with numerous small transverse ridges; beak 0; pappus of a single row of feathery hairs. Fl. 6–10. Freely visited by a great variety of insects, and automatically self-pollinated. $2n=12^*$, 24. Hr.

Very variable. Var. *pratensis* (Link.) Koch, with the involucre thickly covered with usually blackish woolly hairs, is specially characteristic of mountainous districts, though not confined to them; but it is not clearly separable from the type. Its alpine forms have commonly a single head but in cultivation the scapes branch.

Native. Meadows, pastures, waysides, screes; to 3200 ft. in Scotland and Ireland. Abundant throughout the British Is. 112, H40, S. Throughout Europe except Greece; N. and W. Asia; N.W. Africa; Greenland. Introduced in N. America.

Section 2. *Leontodon*

All achenes with 2 rows of hairs, the inner feathery, the outer shorter, simple, scabrid; lvs usually with forked hairs.

2. L. hispidus L. 'Rough Hawkbit.'

A perennial herb with an erect or oblique usually branched premorse stock, each branch terminating in a rosette of ± oblanceolate *lvs* varying from distantly sinuate-toothed to runcinate-pinnatifid, narrowed into a long stalk-like base, *usually hispid with forked hairs*, rarely subglabrous. *Scapes* 10–60 cm., 1 or a few from each rosette, erect or ascending, *simple*, slightly enlarged above and with 0–2 small bracts beneath the *solitary head*, usually densely hairy at least above. Head 25–40 mm. diam., drooping in bud. Involucre 10–17 mm., ovoid, dark to blackish green, hispid to nearly glabrous; its bracts linear-lanceolate with the outermost lanceolate and somewhat spreading. Fls golden yellow, the outermost orange or reddish, rarely grey-violet, beneath. Achenes 5–8 mm., pale brown, fusiform, narrowing for the upper $\frac{2}{3}$ of their length but not beaked, with muricate longitudinal ribs and numerous distinct transverse ridges; pappus dirty white, of 2 rows of hairs, the inner feathery, the outer shorter, simple, scabrid. Fl. 6–9. Freely visited by a variety of insects, especially bees and flies. $2n=14^*$. Hr. Very variable, especially in hairiness. Almost or wholly glabrous plants corresponding with var. *glabratus* (Koch) Bischoff (*L. hastilis* L.) are occasionally found in this country.

Native. Meadows, pastures, grassy slopes, etc., especially on calcareous soils, to 2000 ft. in N. England. Throughout Great Britain and locally abundant; Ireland northwards to Meath and Mayo; not in Orkney and Shetland. 93, H25. Europe from N.E. Spain, S. Italy and Greece northwards to Norway, Sweden and Karelia; Asia Minor; Caucasus; N. Persia.

Section 3. *Thrincia* (Roth) Dum.

Pappus of marginal achenes represented by a cup of small scarious scales; that of the central achenes usually of 2 rows of hairs, the inner feathery with dilated bases, the outer shorter, simple, scabrid; lvs with simple or forked hairs.

3. L. taraxacoides (Vill.) Mérat   'Hairy Hawkbit.'

L. Leysseri Beck; *Crepis nudicaulis* auct.; *Thrincia hirta* Roth; *Leontodon hirtus* auct., non L.

A perennial, rarely biennial, herb with a short erect premorse stock and a basal rosette of narrowly oblanceolate *lvs* gradually narrowed into the long stalk-like base, remotely sinuate-toothed to runcinate-pinnatifid, ± glabrous or *with* ciliate margins and *rather dense forked hairs*. Scapes 2·5–30 cm., 1 to several from each rosette, ascending from a decumbent base, slender, bractless, hardly thickened beneath the *solitary head*, ± glabrous or sparsely hairy below with forked hairs. Head 12–20 mm. diam., drooping in bud. Involucre 7–9 mm., its bracts narrowly lanceolate, the inner equal, the outer shorter and imbricating; all glabrous or with bristly midribs and ciliate margins. *Fls* golden yellow, the *outermost grey-violet beneath*. Achenes c. 5 mm., those of the central florets chestnut, attenuate above and ± short-beaked, straight, with strongly muricate longitudinal ribs and a brownish-white pappus of feathery hairs; those of the outer row pale brown, ± cylindrical, curved, with fainter transversely wrinkled longitudinal ribs, and surmounted by a cup of small scarious scales. Fl. 6–9. Visited by many bees and syrphids, etc. $2n=8^*$, 10. Hr.

The description applies to *Thrincia hirta* Roth, *sens. strict.*, the S. European plant being *T. hispida* Roth, usually annual and with long-beaked central achenes. Recent authors distinguish the two as *L. nudicaulis* ssp. *taraxacoides* (Vill.) Schinz & Thell. (including the British plant) and ssp. *nudicaulis* respectively.

Native. Dry grassland, especially on base-rich soils; fixed dunes; reaching 1350 ft. in Ireland. Throughout the British Is. northwards to Aberdeen and Argyll. Inner and Outer Hebrides. 85, H40, S. Europe northwards to S. Scotland, Denmark, Gotland and C. Russia.

48. Picris L.

Annual to perennial *stiffly hairy* herbs with flowering *stems bearing* spirally arranged entire or sinuate-toothed *lvs*. Heads solitary or in corymbs; bracts imbricate in many rows, those of the innermost row longest, erect and equal, the outermost sometimes very broad and resembling an epicalyx; receptacle flat, naked, pitted. Florets yellow; anther-lobes with short basal tails; style-arms hairy. Achenes curved, ribbed with transverse wrinkles between the ribs, beaked or not; *pappus* of 2 rows, the inner always of feathery *deciduous* hairs, the outer similar or of rough simple hairs.

About 50 spp., in the Mediterranean region and in temperate Europe and Asia, with 4 spp. in Abyssinia.

1 Outer bracts 3–5, large, ovate-cordate, resembling an epicalyx; achenes long-beaked.   **1. echioides**
   Outer bracts small, narrow, ±spreading; achenes not or very shortly beaked.   2

2 Heads distinctly stalked; bracts obscurely keeled, densely covered with mostly simple bristles, with a few forked and hooked bristles.
                                                                          2. hieracioides
   Heads ±sessile, in dense terminal and axillary clusters; bracts distinctly keeled in fr., the keel almost spinous with whitish forked and hooked bristles.   **3. spinulosa**

Section 1. *Helmintia* (Juss.) O. Hoffm. Outer bracts large, ovate-cordate, resembling an epicalyx; achenes beaked, the upper part of the beak falling with the pappus; central achenes straight, marginal curved.

1. P. echioides L.   'Bristly Ox-Tongue.'
Helmintia echioides (L.) Gaertn.

An annual or biennial herb with a stout furrowed erect irregularly forked stem 30–90 cm., covered with short rigid hairs which are tuberclate at the base, trifid and minutely hooked at the apex. Basal and lower stem-lvs oblanceolate, narrowed into a stalk-like base; *middle and upper lvs* lanceolate to oblong, *sessile, ± cordate, amplexicaul* or shortly decurrent; all coarsely toothed or sinuate, bristly-ciliate, and *very rough with scattered bristles on white tuberclate bases.* Heads 2–2·5 cm. diam., somewhat crowded on short lateral stalks in an irregular corymb. *Outer bracts 3–5, lf-like, broadly cordate-acuminate,* bristly-ciliate and rough like the lvs, not quite equalling the lanceolate, awned, bristly inner bracts. Florets almost twice as long as the involucre, yellow, the outermost purplish beneath. Achenes 2·5–3·5 mm. (excluding the beak), the central red-brown, ± straight, glabrous, the marginal whitish, curved, downy on the ventral side; all transversely wrinkled, with a *slender beak about as long as the achene; pappus of pure white feathery hairs, falling with the end of the beak.* Fl. 6–10. Visited by hive-bees, but said to be apomictic. $2n = 10*$. Th.–H. (biennial).

Doubtfully native. Roadsides, hedge-banks, field margins and waste places, especially on stiff and calcareous soils. 69, H10, S. Locally common in lowland England and Wales and in extreme S.E. Scotland but perhaps recently introduced further north to Aberdeen. S. and E. Ireland. Native in the Mediterranean region, Canary Is. and S.W. Asia; perhaps introduced in C. Europe northwards to the Netherlands and Denmark. Introduced in N. America.

Section 2. *Picris.* Outer bracts short, narrow; achenes all similar, not or very shortly beaked.

2. P. hieracioides L.   'Hawkweed Ox-Tongue.'

A biennial to perennial herb with a stout furrowed erect stem 15–90 cm., rough with short forked and hooked bristles especially below, usually branched above, the branches spreading, corymbose. Basal and lower stem-lvs 10–20 cm., oblanceolate, narrowed into a stalk-like base; middle and upper *lvs* lanceolate, usually *broadened at the base and ± amplexicaul*, sometimes narrowed; all *± sinuate-toothed and undulate*, bristly-ciliate and bristly at least on the veins beneath. Heads 2–3·5 cm. diam., solitary, terminal on the main stem and branches, which are bracteate, roughly hairy, and somewhat thickened distally. Involucre ovoid, its inner bracts 8–15 mm., lanceolate, obscurely keeled, with bristles and short white hairs down a central strip, the margins ± glabrous; *outer bracts short, narrow*, usually spreading or recurved, with blackish mostly simple hairs. Florets bright yellow. Achenes 3–5 mm., fusiform, slightly curved, reddish-brown, with fine interrupted transverse wrinkles; *beak very short*; pappus of cream-coloured, deciduous, feathery hairs. Fl. 7–9. Freely visited by flies and bees, but said to be apomictic. $2n = 10$. H. Very variable, especially in hairiness and mode of branching.

Native. Grassland, especially on calcareous slopes, waysides, etc. 65, H5, S. Locally common in lowland England and Wales, and reaching S. Scotland in Wigtown and Roxburgh. Introduced on railway banks in Ireland. Europe, from N. Spain, Italy, N. Balkans and S. Russia northwards to Denmark, S. Sweden and Karelia; W. and C. Asia. Introduced in N. America, Australia and New Zealand.

*3. P. spinulosa auct., very closely related to *P. hieracioides* and often treated as a ssp., may be distinguished by its subsessile heads crowded in terminal and axillary clusters, and by the distinct and almost spinous keel of its bracts whose bristles are very stiff and mostly forked and hooked. Fl. 7–9. H. (biennial).

Introduced. A Mediterranean species established in a few villages in W. Kent.

49. TRAGOPOGON L.

Herbs with copious latex and *linear or linear-lanceolate entire long-pointed sheathing lvs* resembling those of leeks. Heads large, yellow or purple. *Involucre conical in bud, of 1 row of lanceolate-acuminate bracts* united at their base. Receptacle naked. Anthers shortly tailed below and with short terminal appendages. Achene fusiform, 5–10-ribbed; *beak long*, ending upwards in a hairy ring; *pappus*, at least of the central achenes, of 1 *row of hairs simple below and densely feathery at the tips* except for 5 which exceed the remainder and are simple throughout; marginal achenes sometimes with a pappus of stiff bristles.

About 45 spp., in Europe and W. Asia.

Fls yellow.   **1. pratensis**
Fls purple.   **2. porrifolius**

1. T. pratensis L.   Goat's-Beard, Jack-go-to-bed-at-noon.

An annual to perennial herb with a long brownish cylindrical tap-root surmounted by the remains of old lvs. Stem 30–70 cm., erect, simple or little branched above, glabrous or slightly woolly when young, somewhat glaucous.

Basal lvs linear-lanceolate, long-pointed, entire, glabrous, broadened and somewhat sheathing at the base, with conspicuous white veins; stem-lvs similar but more abruptly narrowed into the long acumen from a broad semi-amplexicaul base. *Heads* large, *yellow*, terminal on the main stem and its few branches which are slightly enlarged just beneath the heads. Involucre 2·5–3 cm., of 8 or more equal lanceolate-acuminate bracts, glabrous or with some woolly hairs at their base. *Florets falling short of, rarely equalling, the spreading involucre*. Achenes 10–22 mm., yellowish, those in the centre commonly smooth and the outer ones scaly-muricate on the ribs and sometimes also tuberclate between the ribs, but sometimes all ± smooth or all ± muricate; *beak about equalling the achene*; pappus *very large*, with the feathery hairs interwoven. Fl. 6–7. Visited by various insects and ultimately self-pollinated; the heads close round noon ('Jack-go-to-bed-at-noon'). $2n = 12*$. H.–G.

The species has been divided into 3 sspp. of which ssp. *minor* is the commonest in this country:

Ssp. **minor** (Mill.) Wahlenb., with *bright yellow florets* only *about half as long as the red-bordered involucral bracts*, anthers brownish above, achenes 10–12 mm., the outer scaly-muricate on the ridges, tuberclate between. Fls closing in dull weather. This has the most westerly continental distribution and may be the only ssp. native in the British Is.

Ssp. **pratensis** has the *pale yellow florets almost or quite equalling the pale-bordered involucral bracts*, the anthers are yellow below and dark violet at their tips, the achenes are 15–20 mm., the outer smooth or slightly scaly-muricate. Fls usually remaining open in dull weather. This is the main ssp. in C. Europe north of the Alps and the Danube. In Britain it is much less frequent than ssp. *minor*, if indeed it occurs at all as a native.

*Ssp. **orientalis** (L.) Vollmann has *golden yellow florets* which *equal or exceed the whitish-bordered involucral bracts*, the anthers are yellow with dark brown lines, and the *achenes* are large, *twice as long as the beak*, the marginal ones muricate with cartilaginous scales. It is more easterly in distribution and is known only as a casual in this country.

These three sspp. differ in several morphological features and in geographical range, but their ranges appear to overlap and individuals are not infrequently found which combine features of different sspp. It seems inadvisable, therefore, to treat them as distinct species.

Native. Meadows, pastures, dunes, roadsides, waste places, etc., to 1200 ft. in Derby; locally common. 94, H27, S. The British Is. northwards to Sutherland and Caithness. Europe from C. Spain, C. Italy, Macedonia and S. Russia northwards to Denmark, S.W. Norway, S. Sweden and Karelia. Caucasus; Armenia; Persia; Siberia.

*2. T. porrifolius L.   Salsify.

An annual or biennial glabrous and ± glaucous plant, with a branched irregularly cylindrical tap-root and an erect branching stem 40–120 cm.

Resembles *T. pratensis* but the lvs taper more gradually and the stem and *branches* are *conspicuously enlarged just beneath the heads*. Involucre 3–5 cm., usually of 8 bracts. *Fls purple*, varying from half as long to as long as the involucre. Achene c. 4 mm., faintly 10-ribbed, scaly, gradually narrowed upwards into a beak somewhat exceeding the achene. Fl. 6–8. Visited by various insects. $2n=12$. Th.–H. (biennial).

Introduced. Cultivated for its tap-roots and occasionally escaping. Native in the Mediterranean region but widely cultivated. Hybrids between *T. porrifolius* and *T. pratensis* are found in the British Is.

50. SCORZONERA L.

Herbs with copious latex and simple entire linear to ovate-lanceolate lvs and yellow or purple heads, closely resembling *Tragopogon* but with the involucre of *many rows of imbricating bracts*, and the *pappus of several rows of hairs*, all feathery or the outermost simple.

About 90 spp., in Europe and Asia, chiefly in the Mediterranean region and the Near East.

1. S. humilis L.   'Dwarf Scorzonera.'

A perennial herb with a black often branched cylindrical rootstock, scaly beneath the basal lvs of the current season. Stem 7–50 cm., erect or ascending, usually simple, woolly when young but becoming ± glabrous. Basal lvs 10–20 cm., narrowly lanceolate to elliptical, long-acuminate, narrowed into a long stalk-like half-sheathing base, at first woolly then becoming green and glabrous; stem-lvs narrower but more abruptly broadened below into the semi-amplexicaul base. Head 2·5–3 cm. diam., usually solitary. Involucre 2–2·5 cm., woolly below; its outer bracts ovate, inner oblong-lanceolate, all blunt. Fls pale yellow, twice as long as the involucre. Achenes 7–9 mm., with smooth longitudinal ribs; pappus dirty white. Fl. 5–7. Visited by various bees and other insects, and automatically self-pollinated. $2n=14$. H.

?Native. Marshy fields in Dorset and Warwickshire, very local. 2. Europe from Portugal, C. Spain, S. France, N. Italy, N. Balkans and S. Russia northwards to Denmark, S. Sweden, Karelia and C. Russia; Caucasus.

51. LACTUCA L.

Annual to perennial herbs with copious latex and panicles of smallish heads. Involucre cylindrical, of many imbricating rows of bracts, the outermost row sometimes resembling an epicalyx. Receptacle naked, pitted. Fls yellow or bluish. Anthers short-tailed below and with short terminal appendages. *Achenes strongly compressed, abruptly contracted into a* short or long *beak* which terminates in a small disk; *pappus of two equal rows of* soft white *simple hairs*.

About 70 spp. distributed through the greater part of the world, but chiefly in drier temperate and subtropical regions.

1 Margins of lvs and lf-lobes prickly-toothed.   2
    Margins of lvs or lf-lobes entire or nearly so, not prickly-toothed.   3

2 Stem-lvs usually held vertically (often all in one plane), the lower often
     pinnatifid with clasping base and narrow distant lateral lobes shaped
     like a curved bill-hook; bracts with spreading auricles; ripe achenes
     olive-grey, shortly bristly at apex.   **2. serriola**
   Stem-lvs not held vertically, obovate-oblong or pinnatifid with broad
     lobes; bracts with appressed rounded auricles; ripe achenes blackish,
     ± glabrous at the apex.   **3. virosa**
3 Stem-lvs simple, ovate to orbicular; infl. a dense corymbose panicle;
     florets often violet-streaked.   **1. sativa**
   Upper stem-lvs linear-lanceolate with sagittate base; basal lvs undivided
     to pinnatifid with narrow distant lobes; infl. a narrow spike-like panicle
     with short erect branches; florets often reddish beneath.   **4. saligna**

*1. L. sativa L.   Garden Lettuce.

An annual or biennial herb with a slender tap-root, a dense basal rosette and
a tall erect flowering stem 30–100 cm., whitish, glabrous. Rosette lvs entire
or runcinate-pinnatifid, very short-stalked; *stem-lvs ovate to orbicular*, cordate-
amplexicaul, sessile; all glabrous *with entire margins, smooth* on the main
veins beneath. Heads numerous, in a *dense corymbose panicle* with small
sagittate scale-like bracts. Involucre 10–15 mm., bracts ovate-lanceolate,
blunt, brownish-green with pale margins. Fls few, exceeding the involucre,
pale yellow, often violet-streaked. Achene 3–4 mm., narrowly obovate, 5–7-
ribbed on each face, often finely muricate above; beak white, equalling the
achene. Fl. 7–8. Visited by flies and automatically selfed. $2n = 18, 36 + 1$.
Th.–H. (biennial).

There are many cultivated races of which most are included in var. *capitata*
L., the cabbage lettuce. Var. *crispa* L., the cos lettuce, has long, erect, crisped
and ± lobed lvs.

Introduced. Long cultivated as a salad plant and frequently escaping on
waste ground. Origin unknown but probably from S.W. Asia or Siberia.

2. L. serriola L.   'Prickly Lettuce.'

L. scariola L.

An overwintering or biennial herb with a tap-root and stiffly erect lfy stem
30–150 cm., glabrous or somewhat prickly below, whitish or reddish. Lower
lvs narrowly obovate-oblong in outline, undivided to ± deeply pinnatifid with
sagittate amplexicaul base, an acuminate terminal lobe, and a few pairs of
distant, narrow, acute, lateral lobes spreading at right-angles to the winged
lf-axis but curving distally back towards the lf-base so that each is shaped
like a curved bill-hook; upper lvs less lobed and the uppermost ± simply
hastate or sagittate; all rigid, ± glaucous, glabrous but *spinous-ciliate on the
margins and prickly on the underside of the white main veins*. The *stem-lvs* of
plants fully exposed to the sun are all *held vertically* in the north–south plane
('compass plant'). Infl. an elongated *pyramidal panicle*; *bracts* sagittate *with
spreading auricles*. Heads 11–13 mm. diam., closely spaced along the distal
halves of the panicle branches. Involucre 8–12 mm., narrowly cylindrical;
its bracts lanceolate, glabrous, glaucous and often violet-tipped. Fls few,
pale yellow, often mauve-tinged, exceeding the involucre. *Achenes* 3 mm.,
elliptical, *olive-grey* when ripe, 5–7-ribbed on each face, narrowly bordered,

shortly bristly at the apex; beak white, equalling the achene. Fl. 7–9. Little visited by insects and automatically selfed. $2n = 18$. Th.–H. (biennial).

Probably native. In waste places and on walls, sometimes on ± stable dunes. S. England and Wales. 30. S. and C. Europe, but established northwards to C. England, the Netherlands, N.W. Germany, Denmark, Gotland and C. Russia; W. Asia to the Altai and Himalaya; N. Africa from the Canary Is. to Abyssinia. Introduced in N. America. In C. Europe associated with steppe species such as *Stipa pennata* and *Artemisia campestris*.

3. L. virosa L.

An annual or more usually biennial herb with a branched tap-root and an erect white or reddish lfy stem 60–200 cm., glabrous or prickly below. Lvs obovate-oblong in outline (broader than in *L. serriola*), undivided or ± deeply pinnatifid, the basal lvs narrowed into a stalk-like base, the *stem-lvs* sessile and *cordate-amplexicaul with appressed auricles*; all rigid, ± glaucous, glabrous but spinous-ciliate and prickly on the underside of the main veins. Infl. an elongated *pyramidal panicle*; *bracts* amplexicaul with *appressed ± rounded auricles*. Heads 10 mm. diam. Involucre 8–12 mm., cylindrical-ovoid, glabrous; its numerous imbricating bracts glaucous with a white margin and crimson tip. Fls pale greenish-yellow, exceeding the involucre. *Achenes* 3 mm., narrowly elliptical, with a narrow wing-like border (rather broader than in *L. serriola*), *blackish* when ripe, 5-ribbed on each face, ± *glabrous at the apex*; beak white, equalling the achene. Fl. 7–9. $2n = 18$. Th.–H. (biennial).

Probably native. In grassy places by roads, canals, etc., and on banks near the sea, local. Scattered throughout Great Britain north to E. Perth. 58. S. Europe northwards to C. Scotland, Belgium, Saxony and C. Russia. N. Africa; W. Asia. A sub-Mediterranean species, associated in C. Europe with such thermophilous plants as *Acer monspessulanum*, *Seseli libanotis* and *Aster linosyris*.

4. L. saligna L.   'Least Lettuce.'

An annual rarely biennial herb with an erect stem 30–100 cm., whitish, glabrous or bristly below, with long slender steeply ascending branches, the lowest often arising from near the base of the stem. Basal lvs oblong, entire, sinuate-pinnatifid, or sometimes runcinate-pinnatifid with narrow distant acute lateral lobes and a long slender terminal lobe, narrowed below into a stalk-like base, withered at flowering; *stem-lvs linear-lanceolate, entire*, with a sagittate, amplexicaul base, or pinnatifid with a few distant narrow lobes, commonly *held vertically* and all ± in one plane; all lvs glabrous and ± glaucous with a conspicuous broad white midrib and ± *entire margins*. Heads borne singly or in small clusters in the axils of sagittate bracts on the long branches of the *narrow strict panicle*. Involucre 15 mm., narrowly cylindrical; its bracts linear-lanceolate, blunt, greenish with a narrow white margin. Fls few, pale yellow, often reddish beneath, exceeding the involucre, becoming deep blue when dry. Achenes 3–4 mm., ± ribbed on each face, very narrowly bordered, finely muricate above; beak white, twice as long as the achene. Fl. 7–8. $2n = 18$. Th. (H. biennial).

Probably native. In similar places to *L. virosa*, but more especially near the sea and very local. E. Cornwall, E. Sussex to Norfolk, Cambridge and Huntingdon. 10. Mediterranean and Near East, and northwards to S. England, Belgium, the Netherlands, W. Germany, Bohemia, Silesia and S. Russia. Introduced in Australia.

52. MYCELIS Cass.

Annual to perennial herbs with small heads in ± spreading panicles. Involucre cylindrical, of 2 rows of bracts, the outer row short and resembling an epicalyx. Receptacle naked. Fls usually 3–5, yellow. *Achenes* somewhat flattened, abruptly *beaked*. *Pappus of an inner row of long simple hairs and an outer row of shorter hairs.* The genus therefore differs from *Lactuca* in pappus structure in the same way that *Cicerbita* differs from *Sonchus*.

About 30 spp. in Europe, Asia and Africa.

1. M. muralis (L.) Dum.   'Wall Lettuce.'

Prenanthes muralis L.; *Lactuca muralis* (L.) Gaertn.

A perennial herb with a short premorse stock and an erect stem 25–100 cm., glabrous, paniculately branched above. Lower lvs lyrate-pinnatifid with long winged stalks, the lobes rhomboidal or hastate, the terminal lobe much larger than the laterals and itself often hastately 3-lobed; the middle and upper lvs sessile and ± amplexicaul, becoming successively smaller and less divided; all lvs thin, glabrous, often reddish, their lobes triangular-toothed. Heads in a large open panicle. Involucre 7–10 mm., narrowly cylindrical, the inner bracts linear, the outer very small and spreading, all blunt and often reddish. Fls usually 5, yellow, slightly exceeding the involucre. Achene 3–4 mm. (including the short pale beak), fusiform, blackish. Fl. 7–9. $2n=18$. Hp.

Native. On walls and rocks and sometimes in woods, particularly beech-woods on chalk; usually on base-rich soils. Scattered throughout the British Is., except the Channel Is. and most of the Highlands of Scotland and neigh-bouring islands. 79, H16. Europe northwards to 64° 12′ N. in Scandinavia, the Åland Is. and Karelia. N.W. Africa; Asia Minor; Caucasus.

53. SONCHUS L.

Annual to perennial herbs with copious latex, sometimes suffruticose, with amplexicaul stem-lvs and racemes or panicles of large heads. Involucre of several rows of imbricating bracts. *Receptacle naked*, pitted. Anthers not tailed at the base but with a short blunt distal appendage. Achene flattened, somewhat narrowed above and below, truncate above, ± strongly ribbed; *beak* 0; *pappus* white, *of two equal rows of simple hairs*, the outermost thickened near the base.

About 70 spp., throughout the Old World but especially in the Mediter-ranean region and Africa.

1	Annual to biennial; achenes with 3 longitudinal ribs on each face.	2
	Perennial: achenes with 5 longitudinal ribs on each face.	3
2	Stem-lvs with rounded auricles; achene smooth.	**4. asper**
	Stem-lvs with pointed auricles; achene rugose.	**3. oleraceus**

3 Plant with creeping rhizome; stem-lvs with rounded auricles; glands on
    involucre usually yellow; achenes brown.   **2. arvensis**
  Plant with ±erect stock; stem-lvs with pointed auricles; glands on
    involucre usually blackish-green; achenes yellow.   **1. palustris**

1. S. palustris L.   'Marsh Sow-Thistle.'

A tall perennial herb with a *short ± erect tuberous stock* and a stout erect
stem 90–300 cm., 4-angled, hollow, with the large central cavity square in
cross-section, glabrous below but glandular-hairy above. Basal *lvs* lanceolate-
oblong in outline with a *deeply and acutely sagittate* sessile *base*, pinnatifid
with a few distant lanceolate lateral lobes and a larger lanceolate-acute
terminal lobe, all spinous-ciliate and with spine-tipped teeth; stem-lvs becoming less pinnatifid, the uppermost simply linear-lanceolate with a deeply
sagittate amplexicaul base, the *auricles long, narrow, acute*. Heads to 4 cm.
diam., in a dense corymbose panicle whose branches, like the stalks and
involucres of the heads, are densely *covered with blackish-green* (rarely yellow)
glandular hairs. Involucre 12–15 mm., ovoid-cylindrical; its outer bracts
ovate-acuminate, blunt. Fls pale yellow, exceeding the involucre. *Achene*
4 mm., slightly flattened, *yellowish*, 5-ribbed on each face. Fl. 7–9. $2n=18$.
Hp.

    Native. Marshes, fens, stream-sides; rare and decreasing. 10. S.E. England
from Kent to Norfolk and Hunts; formerly in Oxford and Bucks. C. Europe
from Spain, Corsica, N. Italy, Serbia and S. Russia northwards to S. England,
the Netherlands, Denmark, S. Sweden and C. Russia; Caucasus; Armenia.

Differs from *S. arvensis* in being non-stoloniferous and in the pointed (not rounded)
auricles of the amplexicaul lvs.

2. S. arvensis L.   'Field Milk-Thistle.'

A perennial herb with *creeping underground stolons* and erect or ascending
stems 60–150 cm., furrowed, hollow with the small central cavity elliptical
in cross-section, glabrous or at first cottony below, glandular-hairy above.
Basal lvs oblong or lanceolate in outline, narrowed into a winged stalk,
runcinate-pinnatifid with short triangular-oblong spinous-ciliate and spine-
toothed lobes; *stem-lvs* similar but less divided and sessile, the *cordate-
amplexicaul base* having *rounded appressed auricles*. Heads 4–5 cm. diam. in
a loose corymb whose branches like the *involucres* are usually *densely covered
with yellowish glandular hairs*. Involucre 13–20 mm., campanulate; bracts
oblong-lanceolate, blunt. Fls golden yellow. *Achene* 3–3·5 mm., *dark brown*,
narrowly ellipsoidal with 5 strong ribs on each face. Fl. 7–10. Visited freely
by many kinds of insects, especially bees. $2n=64$. H.

    Native. Stream-sides, drift-lines on salt- and brackish-marshes, banks,
arable land, etc., to 1250 ft. in Wales; common. 112, H40, S. Throughout
the British Is. Throughout Europe to 70° 33′ N. in Scandinavia; W. Asia.
Widely introduced in Asia, America, Africa and Australia.

3. S. oleraceus L.   Milk- or Sow-Thistle.

Annual or overwintering herb with a long slender pale tap-root and stout
erect glabrous stems 20–150 cm., ± 5-angled, hollow except at the nodes,

branched above. Lvs very variable; basal lvs usually ovate, stalked; lower stem *lvs runcinate-pinnatifid* with the terminal lobe usually wider than the uppermost pair of laterals and with a short winged stalk and enlarged *acute spreading auricles*; uppermost with a reduced blade and more broadly winged stalk; all glabrous (or cottony only when young), ±glaucous, *dull, never spinous*. Infl. an irregular cymose umbel, its branches sometimes glandular-hairy. Heads 2–2·5 cm. diam. Involucre 1–1·5 cm., glabrous (or cottony in young bud), rarely glandular-hairy, its outer bracts broadly lanceolate, shorter and more acute than the inner. Florets yellow, the outer purple-tinged below. *Achenes* 3 mm., *oblanceolate*, compressed, never winged, first yellow then brown, longitudinally 3-ribbed on each face, *transversely rugose*. Fl. 6–8. Visited by various insects, especially bees and hover-flies. $2n=32*$ (16). H.

Native. Cultivated soil, waysides, waste places, etc., throughout the British Is. 112, H40, S. Europe to 66° 13′ N. in Scandinavia; N. and W. Asia; N. Africa and the Canary Is. Widely introduced as a weed of cultivation.

4. S. asper (L.) Hill   'Spiny Milk- or Sow-Thistle.'

An annual or overwintering herb closely resembling *S. oleraceus* but differing in the form and appearance of the lvs, which are less often pinnatifid and then have the terminal lobe narrower than the uppermost pair of laterals, show a less clear distinction between blade and stalk, have *rounded appressed auricles*, are usually *dark glossy green* above, and are commonly *crisped and spinous-ciliate* at the margin. In var. *inermis* Bisch. the lvs are all simple, obovate or ovate-lanceolate, with flat and softly spinous-ciliate margins. Florets usually golden-yellow, and so a deeper colour than is usual in *S. oleraceus*. *Achenes* 2·5 mm., obovate, compressed, sometimes winged (Scotland), usually brown but variable in colour, longitudinally 3-ribbed on each face, otherwise *smooth*. Fl. 6–8. Visited by bees and hover-flies, etc. $2n=18$. Th.

Native. Cultivated soil, waste places, etc., to 1300 ft. in England; common. 112, H40, S. Throughout the British Is. Europe northwards to 64° 5′ N. in Scandinavia; N. and W. Asia; N. Africa. Widely introduced as a weed of cultivated land.

A sterile hybrid between *S. oleraceus* and *S. asper* has been reported from Great Britain, but appears to be rare.

54. CICERBITA Wallr.

Perennial herbs usually with panicles of large heads whose involucres consist of an inner row of larger and an outer row of shorter bracts. Receptacle naked. Fls all ligulate, blue or yellow. Anthers tailed below and with short terminal appendages. Achenes flattened, narrowed above and below; *beak* 0; *pappus of 2 rows of simple hairs, the outer shorter*.

Eighteen spp., of which 14 are native in the mountains of Europe, Asia and N. Africa, and 4 in America. Differs from *Sonchus* in the outer row of shorter pappus hairs.

Plant with erect branched stock; lower lvs with triangular terminal lobe and
a few pairs of small lateral lobes; infl. an elongated simple or compound
raceme; heads blue.   **1. alpina**
Plant with far-creeping rhizome; lower lvs with cordate terminal lobe and
usually a single pair of lateral lobes; infl. corymbose; heads lilac.
2. macrophylla

1. C. alpina (L.) Wallr.   'Blue Sow-thistle.'

Sonchus alpinus L.; *Mulgedium alpinum* (L.) Less.; *Lactuca alpina* (L.) A. Gray

A tall perennial herb with a ± erect cylindrical rootstock and a stout erect
furrowed stem 50–200 cm., simple or branched, bristly below and with dense
reddish glandular hairs above. Lowest lvs stalked, lyrate- or runcinate-
pinnatifid with a large broadly triangular-acuminate terminal lobe and a few
pairs of much smaller ± triangular denticulate lateral lobes, narrowed into
a winged stalk; the succeeding lvs becoming smaller and less divided, the
winged stalk broadened into a *cordate-amplexicaul* base, the uppermost
± lanceolate; all lvs glabrous, somewhat glaucous beneath. *Heads* c. 2 cm.
diam., *pale blue*, in a simple or compound raceme, their stalks, like the infl.
axis, densely glandular-hairy. Involucre 1–1·5 cm., purplish-green, with long
glandular hairs. Achenes 4·5–5 mm., linear-oblong, with 5 strong and several
weaker ribs; pappus 7 mm. Fl. 7–9. $2n=18$. H.

Native. Moist places on alpine rocks; very rare. 2. Angus and S. Aberdeen. Mountains of C. Europe from the Pyrenees to the Carpathians; and
in Scandinavia and Karelia, to 71° 7′ N. in Norway. Reaches over 7000 ft.
in Switzerland, where it is associated with *Peucedanum ostruthium, Senecio
nemorensis, Adenostyles alliariae, Achillea macrophylla*, etc.

*2. C. macrophylla (Willd.) Wallr.

A tall perennial herb with *far-creeping rhizome* and erect stems 60–150 cm.,
glabrous below, glandular-hispid above. Lower lvs large, ± lyrate, with large
cordate terminal lobe and usually a *single pair* of small lateral segments decurrent into a winged stalk with broadened cordate-amplexicaul base; upper lvs
smaller and ± sessile but with cordate-amplexicaul base, the uppermost
lanceolate to linear; all glandular-bristly, somewhat glaucous beneath. Infl.
a loose ± *corymbose panicle* with glandular-bristly branches. Heads c. 30 mm.
diam.; involucral bracts glandular-hairy; *florets lilac-coloured*. Achenes
5 mm., glabrous, narrowly winged, 3-ribbed on each face. Fl. 7–9.

Introduced. Established in Derbyshire and elsewhere. Caucasus.

*C. plumieri (L.) Kirschl., a smaller plant, 60–130 cm., *wholly glabrous*, non-creeping,
with lyrate-pinnatifid amplexicaul lvs, blue heads in a loose corymb and achenes
6·5 mm., is often cultivated and sometimes escapes. Mountains of C. Europe.

55. HIERACIUM L.   Hawkweed.

Perennial herbs, sometimes stoloniferous (subgenus *Pilosella*), with vertical
to horizontal stocks and stout fibrous roots. Lvs spirally arranged on the
flowering stems or some or all in a basal rosette. Infl. a cymose and often
corymbose panicle or a few-fld forking cyme or the heads solitary, terminal.

Involucral bracts erect or incurved in bud, imbricate in few to several irregular rows, the outermost shortest. Stem, lvs, infl. and involucre glabrous or clothed with simple hairs, toothed hairs, glandular hairs, sessile micro-glands and soft white stellate hairs ('floccose') in varying proportions. Receptacle pitted, the scarious margins of the pits variously toothed or fimbriate; *receptacular scales* 0. Florets usually yellow, glabrous or sometimes hairy at the tips ('ciliate-tipped') and less commonly also on the backs of the ligules. *Achenes* 1·5–5 mm., *cylindrical*, 10-ribbed, *truncate above*, neither appreciably narrowed upwards nor beaked; *pappus-hairs* in 1 or 2 rows, *simple, rigid*, brittle, whitish to tawny, usually *pale brownish*.

Perhaps 10,000–20,000 'species', chiefly in temperate, alpine and arctic regions of the northern hemisphere but some in S. America (subgenera *Stenotheca* and *Mandonia*) and some in S. Africa, Madagascar, S. India and Ceylon (subgenus *Stenotheca*).

The genus *Hieracium* is the most difficult taxonomically of those represented in the British flora. This arises largely from the peculiar mode of reproduction now found in the British forms, seeds usually being produced apomictically, that is, independently of fertilization. This results in the perpetuation of individual differences arising by mutation, by specialized modes of segregation and also, presumably, by occasional crossing. Each hawkweed and its offspring have thus become more or less closed units whose features can never or only at infrequent intervals be shared with other units. There is therefore no out-breeding unit comparable with those which constitute ordinary 'amphimictic' species, and the situation becomes very complex.

British students of hawkweeds are fortunate in having available the recently published *Prodromus of the British Hieracia* by H. W. Pugsley. This provides an introduction to the study of the genus, a conspectus of the classification of British forms and detailed descriptions of 260 forms treated as 'species', a number which could certainly be greatly increased. In a Flora not primarily designed for specialists it is impracticable and unnecessary to deal with this great number, and the aim has been to make it possible at least to place a British hawkweed in its appropriate section, the classification following closely that adopted by Pugsley. A key to sections, with text-figures of representative species, is followed by a general description covering the species included in each section in turn, with brief descriptive notes on one or more of the most widespread or abundant of them. It should usually be possible to reach the correct section and to decide whether a specimen falls into one of the named species; and it is hoped that users of the Flora will be encouraged to record, instead of the despondent '*Hieracium* sp.' an identification at least as close as '*Hieracium* section *Oreadea*'.

1   Pappus hairs ±equal, in 1 row; achenes 1·5–2 mm., their 10 longitudinal ridges each ending above in a projecting tooth; plants overwintering by rosettes and often stoloniferous (as in the commonest British sp.).
Subgenus PILOSELLA   16
Pappus hairs of various lengths, in 2 rows; achenes 2·5–4·5 mm., their 10 longitudinal ridges merging above into a swollen ring round the top; plants overwintering by rosettes or lateral buds, never stoloniferous.   Subgenus HIERACIUM   2

2   No rosette formed, or rosette-lvs withered at time of flowering, rarely a few (often moribund) rosette-lvs persisting; stem-lvs 8–many, rarely fewer in small plants.   3
Rosette-lvs present at flowering; stem-lvs 0–8(–10).   8

3   Middle stem-lvs somewhat constricted just above the broad amplexicaul
    base; all lvs strongly reticulate and somewhat glaucous beneath;
    involucre and peduncles floccose and densely glandular; ligules ciliate-
    tipped; styles dark; achenes tawny or pale yellowish-brown.
                                                          5. *Prenanthoidea*
        Lvs not as above; involucre not or slightly floccose; achenes purplish- or
        blackish-brown.   4

4   Middle stem-lvs not or hardly amplexicaul.   5
        Middle stem-lvs distinctly though not broadly amplexicaul; involucre not
        or slightly glandular.   7

5   Lvs numerous, ±crowded, at least below, all ±similar in shape, com-
    monly lanceolate to linear-lanceolate, narrowed to a sessile base, their
    margins recurved and scabrid; heads in a ±umbellate corymb; involu-
    cral bracts blunt, uniformly green, ±glabrous, all except the inner with
    recurved tips; styles yellow.   1. *Umbellata*
        Lower lvs stalked; middle and upper lvs shortly stalked or sessile,
        lanceolate to broadly ovate, their margins not recurved; involucral
        bracts very rarely with recurved tips, usually dark green, ±glabrous
        or somewhat but not densely hairy or glandular; styles usually dark
        or with short dark hairs, sometimes yellow.   6

6   No rosette-lvs; stem-lvs numerous, crowded at least below; upper stem-
    lvs short with broad rounded bases; involucral bracts olive- or
    blackish-green.   2. *Sabauda*
        Rosette-lvs withering early or a few persisting until flowering; stem-lvs
        usually not very numerous, rather widely spaced in the upper half or
        sometimes confined to the lower half; all lvs narrowing to the base,
        usually toothed in the middle.   4. *Tridentata*

7   No rosette-lvs; stem-lvs fairly numerous, all ±amplexicaul or the lowest
    merely sessile, paler and ±glaucous and reticulate beneath; involucre
    sparsely hairy and glandular; ligules glabrous-tipped.   3. *Foliosa*
        Rosette-lvs stalked, withering early or a few persisting; stem-lvs not
        numerous, rather widely spaced, often confined to the lower half of
        the stem, the lowest often stalked; panicle lax, somewhat lfy; involucral
        bracts moderately glandular and hairy; ligules glabrous- or ciliate-
        tipped.   6. *Alpestria*

8   Stem-lvs 1–6, large, usually yellowish-green, the upper amplexicaul with
    large rounded auricles; whole plant glandular-viscid.
                                                        13. *Amplexicaulia*
        Plant not viscid; stem-lvs if amplexicaul not yellowish-green.   9

9   Stem-lvs 1–7, ±amplexicaul; all lvs glaucous with long stout denticulate
    hairs on both sides; heads 4·5–5·5 mm. diam.; involucral bracts in-
    curved in bud; florets often pale yellow; styles dark; robust plants of
    N. Britain, the Hebrides and Ireland.   14. *Cerinthoidea*
        Stem-lvs 0, or if present not amplexicaul.   10

10 Rosette-lvs small, not glaucous, always with small glandular hairs
    especially on the margins and usually with shaggy stalks; stem-lvs 0–3,
    often bract-like; stem and infl. glandular and involucre glandular and
    often shaggy with long whitish black-based hairs, or densely floccose;
    heads often solitary, never many, rather large; ligules ciliate-tipped
    and sometimes hairy also on the outside; small alpine plants, 5–
    15(–25) cm.   16. *Alpina*

Taller alpine plants with few rosette-lvs, larger stem-lvs, 2–10 smaller heads, but otherwise similar. 15. *Subalpina*
Not as above; lvs not glandular nor heads with long shaggy hairs. 11

11 Lvs glaucous, firm, often purple-blotched or purplish beneath, usually bristly with stiff stout-based hairs especially on and near the margins; stem forking above; heads usually few, large; involucral bracts erect in bud with glandular and stout black-based hairs, not densely floccose; styles usually yellow; ligules usually glabrous, occasionally ciliate-tipped. 12. *Oreadea*
Not as above; lvs hairy, floccose or often ± glabrous above, hairy but hardly bristly on the margins and underside; stem forked or corymbosely branched above; heads few and large or numerous and small; involucral bracts incurved in bud, usually densely floccose at least at the margins, usually ± glandular, often with slender hairs, sometimes with stout black-based hairs but never shaggy as in *Alpina*. 12

12 Stem-lvs 0–1(–3), rosette-lvs several to many, ± broad, often with concave-sided spreading or reflexed basal teeth, the base of the blade usually cordate, truncate or rounded and thus clearly demarcated from the often shaggy stalk; styles commonly yellow or discoloured. 13
Stem-lvs 2–5(–10); rosette-lvs not numerous, often very few, the blade broadly to narrowly lanceolate, narrowing gradually into the stalk, with outwardly or forwardly directed (not reflexed) teeth; styles commonly discoloured or dark. 15

13 Lvs green or glaucous, usually ± glabrous above; stem-lvs 0–1, bract-like or narrowly lanceolate; stem ± glabrous or slightly hairy below, ± forked above, with few medium to large heads; styles usually yellowish or discoloured. 11. *Bifida*
Lvs grass- or yellowish-green, ± hairy above and more so beneath; stem-lvs (0–)1–2(–3) of which one near the middle of the stem is usually fairly large and stalked; stem usually with abundant long hairs below, glandular above; heads 2–10, in a corymbose panicle, their involucres ± densely glandular and floccose, with or without hairs; styles yellow to dark. 14

14 Clothing of involucre predominantly glandular, with or without hairs. 10. *Glandulosa*
Involucre usually densely hairy but rather sparsely glandular or eglandular. 9. *Sagittata*

15 Lvs glaucous or pale green; stem-lvs 2–5(–8); stem subglabrous or hairy below, ± floccose above, not or slightly glandular; heads few, loosely panicled (or the stem forked above), with stalks floccose but with sparse hairs and glands; involucre hairy and floccose but not or sparsely glandular. 8. *Caesia*
Lvs grass- or yellow-green; stem densely hairy below, floccose and glandular above; heads panicled, few to many, with stalks and involucres usually ± densely glandular, hairy and floccose. 7. *Vulgata*

16 Stem-lvs 0; head 1; rosette-lvs white- or grey-felted beneath; ligules reddish beneath. 17. *Pilosella*
Stem-lvs 0–several; heads 2–many. 17

17 Stem not exceeding 20(–25) cm., ascending, 0–1(–2)-leaved; rosette-lvs ± glaucous; heads 2–7. 18. *Auriculina*
Stem exceeding 20(–25) cm., erect, with 1 to several lvs; heads 4–many. 18

18 Rosette-lvs yellow- or grass-green, elliptic to lanceolate, flaccid; stem-lvs
    usually 1–2; heads often in an umbel; ligules yellow to deep red.
                                                                         19. *Pratensina*
    Rosette-lvs glaucous, lanceolate to linear, rigid; stem-lvs 1–3(–7); heads
    panicled; stolons sometimes 0; ligules yellow.   20. *Praealtina*

Subgenus 1. HIERACIUM

Never stoloniferous. Ligules never marked with red beneath. Achenes
(2·5–)3–4·5(–5) mm., the ribs confluent into an obscure ring at the top.
Pappus hairs sub-biseriate, the long hairs mixed with shorter rigid ones.

Section 1. *Umbellata* F. N. Williams. Fig. 56 A.

Stem 25–120 cm., erect, wiry, often ±woody below, subglabrous or hairy,
not glandular; rosette-lvs 0; *stem-lvs numerous* (up to 50), crowded at least
below, diminishing in size upwards, *all similar, usually linear to linear-
lanceolate, with revolute scabrid margins*, subentire or ±toothed on the sides,
narrowed to a sessile base, hairy and floccose beneath; usually subglabrous
above. Heads few to many in a *±umbellate panicle*, their stalks thickened
just beneath the head, ±scaly with bracteoles, and floccose but not hairy or
glandular. *Involucre* 9–11 mm., *glabrous* or very sparsely floccose or hairy
or with a few tiny glands; *bracts* uniformly green or blackish, lanceolate,
blunt, all but the inner with spreading or *recurved tips*. Ligules glabrous-
tipped. *Styles yellow*, rarely dark; achenes 3–4 mm., brownish-black.
    Roadsides, banks, open woods and copses, heaths, rocks; chiefly lowland.

The two widespread spp. of this section are:

H. umbellatum L.

Stem commonly 30–80 cm., slender, usually hairy below but sometimes
±glabrous throughout. *Lvs dark green*, paler beneath, *linear to narrowly
linear-lanceolate*, subentire or with 2–3 distant teeth on each side. Involucral
bracts usually *blackish-green*. Achenes 3–4 mm.
    Var. *coronopifolium* (Hornem.) Fr. has 2–4 distant long and curved teeth
on each side of the lvs.
    Native. 68, H11. Throughout the British Is., except Outer Hebrides,
Orkney and Shetland. Europe; N. Asia; N. America.

H. bichlorophyllum (Druce & Zahn) Pugsl.

Stem commonly 25–50 cm. *Lvs clear green*, paler beneath; the *lower ovate-
lanceolate to oblong*, subentire or faintly sinuate-toothed, the *middle oblong
to lanceolate*, usually subentire, the upper lanceolate. Involucral *bracts* usually
olive-green. Achenes 3–3·5 mm.
    Native. 15, H5. S.W. England and Wales. Ireland. Channel Is. ?Endemic.

Section 2. *Sabauda* F. N. Williams. Fig. 56 B.

H. boreale auct. angl.

Stem 30–120 cm., erect, robust, usually reddish hairy at least below, not
glandular. Rosette-lvs 0. *Stem-lvs numerous* (up to 50), *crowded at least below*,

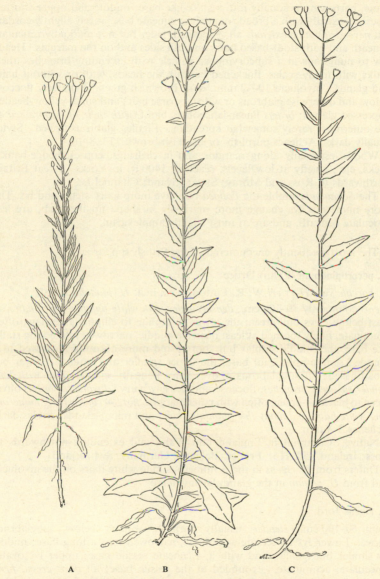

Fig. 56. *Hieracium.* A, *Umbellata* (*H. umbellatum*); B, *Sabauda*
(*H. perpropinquum*); C, *Foliosa* (*H. subcrocatum*). × ⅕.

decreasing rapidly in size upwards; lowest ovate-lanceolate to lanceolate, acute, narrowing gradually into a stalk-like base; middle and upper shorter, ± sessile, usually with a broad rounded or cuneate base, rarely slightly cordate but *never truly amplexicaul*; all ± toothed, paler but not markedly reticulate beneath and with stout-based hairs on both sides and on the margins. Heads few to numerous in a subcorymbose panicle with ascending branches, their stalks with many scales, thickened beneath the heads, with or without hairs and glands. Involucre 10–12 mm., dark brownish-green, ± sparsely floccose below and otherwise glabrous or with ± sparse hairs and small yellow-headed appressed glands; *bracts* linear-lanceolate, blunt, *with their tips appressed* or the outermost rarely somewhat spreading. Ligules glabrous-tipped. Styles usually dark. Achenes purplish- or blackish-brown. Fl. 8–10.

Woods, especially along margins and in clearings, copses, hedge-banks, rocks, etc.; chiefly at low levels, reaching 1400 ft. in Yorks. Great Britain northwards to Ross and Moray; Skye; Ireland; Channel Is.

The *Sabauda* resemble the *Foliosa* but have more hairy stems and lvs. The more numerous lvs change more markedly in shape up the stem, are less reticulate beneath, and are at most slightly amplexicaul.

The most frequently encountered of the British spp. are:

H. perpropinquum (Zahn) Druce

H. boreale var. *Hervieri* W. R. Linton, p.p.; incl. *H. bladonii* Pugsl.

Stems commonly 50–100 cm., *densely hairy with white bulbous-based hairs*, at least below. Lvs dull green, the lower lanceolate to elliptic-lanceolate, often acuminate, finely toothed at least near the middle, narrowed into a long stalk-like base, the middle similar but shorter and more abruptly narrowed to a rounded semi-amplexicaul base, the upper smaller and often narrower but broad-based; all subglabrous above, hairy beneath, ciliate. *Infl. with long slender suberect branches*, floccose and densely hairy. *Involucral bracts* olive-green to blackish-green, often with paler margins, *not floccose, with ± numerous long white hairs and many fine glandular hairs* and micro-glands. Styles dark. Achenes 3–4 mm.

Native. Common in England and Wales and extending northwards to Ross; Ireland. 50, H 5. France, Belgium, Germany, Switzerland.

Differs from *H. rigens* in the ± numerous long white hairs on the involucre and from *H. vagum* in the glandular involucre.

H. rigens Jord.

Stem 50–100 cm., *slender*, usually purplish, ± hairy below, subglabrous above. Lower lvs lanceolate-acuminate, narrowed into a long base; middle *lvs* similar but smaller and with a ± cuneate sessile base; upper lvs ovate-lanceolate, acuminate, ± rounded at the sessile base; all *dark green, firm*, sparsely hairy beneath, subglabrous above, slightly ciliate, finely and rather distantly toothed. *Infl. with long strict suberect ± floccose branches. Involucral bracts* olive- or blackish-green *with ± dense fine glandular hairs and few or no simple hairs.* Styles dark. Achenes c. 4 mm., purplish-black. Native. S. England and Wales. 12. France; C. Europe.

H. vagum Jord.

Stem commonly 50–80 cm., hairy below, becoming subglabrous above. *Lvs* deep green, *firm*, mostly *suberect*, the lower ovate to elliptic, narrowed into a long stalk-like base, the middle and upper ± ovate-elliptic with a rounded sessile base; all subglabrous above, hairy beneath, ± ciliate. *Infl.* often compact, *with short suberect branches*. Involucral *bracts blackish-green, concolorous*, very blunt, *subglabrous* or slightly floccose on the margins, *eglandular*. Styles dark. Achenes 3·5 mm.

Native. 38. Chiefly in C. and N. England and Wales, but extending from Brecon, Hereford and Hertford northwards to Moray. Spain, France, Switzerland.

Section 3. *Foliosa* Pugsl. Fig. 56 c.

Stem 30–120 cm., erect, often reddish, ± glabrous or hairy at least below. Rosette-lvs 0. Stem-*lvs* rather numerous (10–30), lowest oblanceolate, narrowed to a stalk-like base; middle and upper ovate-lanceolate to lanceolate, truncate or rounded, sessile and *semi-amplexicaul* at the base, *paler and ± reticulate beneath*. Heads in simple or compound corymbs, their stalks floccose and sometimes with a few hairs and glands. Involucre 8–12 mm., its scales usually blunt, dark or blackish-green with paler margins, not or sparsely floccose and hairy, ± glandular. Ligules glabrous-tipped. Styles yellow or dark. Achenes purplish or blackish-brown. Fl. 7–8.

Mountain districts from Wales, Stafford and Yorks northwards to Sutherland and Orkney; Inner Hebrides; N.E. Ireland. Mountains of C. and N. Europe; Caucasus; N. Asia; Asia Minor.

Probably derivatives of *Prenanthoidea* by hybridization with *Umbellata* and *Tridentata*, and distinguishable from the two latter by the semi-amplexicaul lvs and from the former by the lvs being less amplexicaul and the infl. and involucre much less glandular, and by the glabrous-tipped ligules and dark achenes.

The commonest British spp. are:

H. latobrigorum (Zahn) Roffey

H. auratum auct.

Stem commonly 30–80 cm., stout, reddish and ± hairy below, subglabrous to densely floccose above. *Lvs* dull or yellowish-green, ± conspicuously reticulate beneath; the lower oblanceolate to oblong, subentire or with small distant teeth, narrowed into a shortish winged stalk, the middle *elliptic-lanceolate*, ± distantly and shortly toothed, sessile with a shortly *cuneate to rounded base*, the upper ± broadly lanceolate, acuminate, sessile, with a rounded amplexicaul base; all subglabrous or floccose above, hairy and ± floccose beneath. Heads medium-sized, rounded below, 3–25 or more in a compact or diffuse corymb with densely floccose branches. Involucral *bracts dark olive-green*, blunt, hardly floccose, with *numerous unequal glands* and sometimes some long hairs. *Styles pure yellow*. Achenes 3–3·75 mm.

Native. 24, H 3. Great Britain from Stafford and Caernarvon northwards to Sutherland and Caithness; Inner Hebrides; Orkney; E. Ireland. C. Europe.

H. subcrocatum (E. F. Linton) Roffey (*H. neocorymbosum* Pugsl.) resembles *H. latobrigorum* but has *dark styles*.

Native. 13, H1. N. Britain from Caernarvon and Derby to Caithness; Outer Hebrides; Ireland (Wicklow). Endemic.

H. strictiforme (Zahn) Roffey

H. strictum auct. angl., non Fr.

Stem commonly 50–90 cm., reddish below, sparsely hairy, slightly floccose above. Lvs 15–25, deep green, paler and reticulate beneath; the lower ± broadly oblanceolate, subentire, narrowed into a ± short winged stalk, the middle ± *lingulate-lanceolate*, acute to acuminate, subentire or with small distant spreading teeth, with a *broad rounded sessile base*, the upper ovate-lanceolate with a broad rounded base; the lower hairy on both sides, ± ciliate, the remainder becoming glabrous. Infl. a corymbose panicle commonly with 6–12 large truncate-based heads, its branches floccose and sometimes sparsely hairy. Involucral *bracts dark olive-green*, blunt, scarcely floccose, with dense dark glandular hairs, some long black-based hairs and micro-glands. *Styles dark*. Achenes c. 3·5 mm.

Native. 18, H5. Montgomery; N. Britain from N.W. Yorks to Sutherland and Caithness; Ireland. Endemic.

H. reticulatum Lindeb. resembles *H. strictiforme* but has narrower *lvs* more strongly reticulate beneath and *with* longer *unequal ± cusped teeth*. *Styles yellow*.

Native. 18. Brecon; Scotland from Dumfries to Sutherland and Caithness. Scandinavia.

Section 4. *Tridentata* F. N. Williams. Fig. 57 A.

Stem 20–100 cm., erect, ± slender, rigid, often reddish below, subglabrous to ± densely hairy. Rosette lvs 0 or a few withering early and usually dead before flowering, sometimes persisting; stem lvs 3–15 or more, rather distant or lower lvs crowded, sometimes confined to the lower half of the stem, linear- to ovate-lanceolate, all narrowed to the base, at least the lowest stalked, all ± acuminate and usually with 3–5 teeth in the basal two-thirds of each side; all green, paler beneath, floccose and hairy at least on the veins beneath, ± glabrous above, the margins usually thickened and ciliate or scabrid. Heads few to many, panicled, their stalks floccose but not or scarcely hairy and glandular. Involucre 9–11(–15) mm., bracts incurved in bud, ± blunt, appressed, usually rather sparsely hairy, glandular, and floccose, sometimes densely hairy, dark green with paler margins. Ligules glabrous-tipped. Styles yellow to dark. Achenes 3·5–4 mm., blackish-brown.

Intermediate between *Umbellata* and *Sabauda* on the one hand and *Euvulgata* on the other, both in the number and position of the lvs and in the clothing of the involucral bracts.

Woods, copses, hedge-banks, rocky slopes, walls and waste places. Throughout the British Is. Europe. C. and N. Asia. N. America. From sea-level to the subalpine zone.

The spp. most commonly met with are:

H. trichocaulon (Dahlst.) Johans.

Stem commonly 40–60 cm., hairy especially below, ±floccose above. *Lvs*
6–15, often crowded below, all with a *few ± long and often somewhat curved*

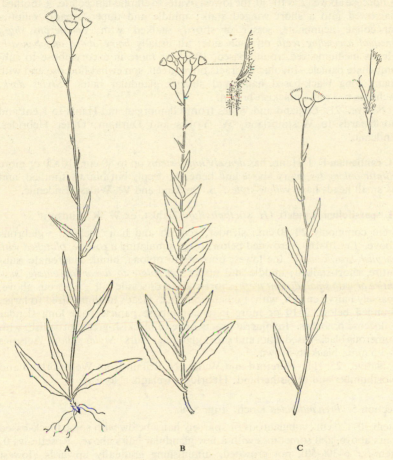

Fig. 57. *Hieracium.* A, *Tridentata* (*H. trichocaulon*); B, *Prenanthoidea*
(*H. prenanthoides*); C. *Alpestria* (*H. dewarii*). ×⅕.

teeth on each side; the lower elliptic-lanceolate, gradually narrowed into a
winged stalk, the middle and upper narrower, sessile; all ±*glabrous above,*
hairy beneath. Heads small, narrow, 5–20 in a corymbose panicle whose
branches overtop the terminal head. Involucral bracts olive-green with paler
tips and margins, narrow, with long black-based hairs, finer pale hairs, and
glandular hairs. Styles yellow to dark. Achenes c. 3 mm., blackish-brown.

Native. 27. England and Wales from Devon to Kent and northwards to Yorks.

H. eboracense Pugsl.

Stem commonly 50–100 cm., stout, often reddish, hairy especially below, ± floccose above. *Lvs* 10–30, the lowest ovate- to elliptic-lanceolate, ± toothed, narrowed into a short winged stalk; middle and upper broadly elliptic-lanceolate, acuminate, sessile or shortly stalked, with ± *numerous sharp unequal ascending teeth* on each side; all usually *hairy above and beneath.* Heads medium-sized, round-based, 10–30 or more in a corymbose to sub-umbellate panicle. Involucral bracts olive-green, somewhat floccose and with many long black-based hairs and shorter glandular hairs. *Styles dark.* Achenes c. 3 mm., blackish-brown.

Native. 21. England and Wales from Glamorgan and Hants to Kent and northwards to Westmorland, W. Yorks and Durham; Outer Hebrides. Endemic.

H. cantianum F. J. Hanb. has *densely hairy stems* up to 80 cm., 10–20 or more *elliptic-oblong lvs*, hairy above and beneath, deeply but bluntly toothed, and its small heads have *yellow styles.* S. England and W. Wales. Endemic.

H. sparsifolium Lindeb. (*H. stictophyllum* Dahlst. ex W. R. Linton)

Stem commonly 20–60 cm., slender, reddish and hairy below, ± glabrous above. *Lvs* 10–16, ± crowded below, often simulating a rosette, *blotched with purplish-brown above*; the lowest obovate to oblong, blunt, mucronate, sub-entire, short-stalked; middle and upper *lanceolate to linear-lanceolate, sub-entire or with small distant teeth*, short-stalked or sessile; all ± glabrous above, sparsely hairy beneath, with ± ciliate margins. Heads medium-sized to large, rounded below, 2–10 or more in a corymbose panicle with long slender ± floccose branches. Involucral bracts olive to blackish-green, broadish, with numerous black-based hairs and short glandular hairs. *Styles yellow.* Achenes 3–3·5 mm., blackish-brown.

Native. 25, H1. Hereford and Wales; N. Britain from Westmorland and Northumberland to Sutherland; Hebrides; Ireland. Norway.

Section 5. *Prenanthoidea* Koch. Fig. 57 B.

Stem 30–120 cm., subglabrous or sparsely hairy below with some black-based hairs above, and sometimes with a few glandular hairs above. Rosette-lvs 0; stem-*lvs* 6–30(–50), not crowded, diminishing gradually upwards; lowest oblanceolate, narrowed into a short broadly winged stalk; middle elliptic-oblong, usually ± *constricted above the wide base* (panduriform); upper ovate-lanceolate; all *amplexicaul*, subentire or distantly denticulate in the middle, green and ± glabrous above, ciliate at the margins, ± *glaucous, distinctly reticulate-veined* and usually somewhat hairy *beneath.* Heads small, numerous, in a ± compound subcorymbose panicle, their *stalks densely floccose and glandular. Involucre* 7–11 mm., dark brownish-green, floccose, *densely glandular*, rarely with a few black hairs; involucral bracts rather few, incurved in bud, in 2 sets, the outer short, ± acute, the inner long, blunt, with paler margins. Ligules ciliate-tipped. Styles dark. Achenes tawny to yellowish-brown. Fl. 8–9.

Subalpine Wales and N. Britain; Ireland. N. and C. Europe; N. Africa; N. Asia; S.W. Greenland.

A well-defined section characterized by the amplexicaul lvs which are somewhat glaucous and strikingly net-veined beneath, the densely glandular involucral bracts and peduncles, and the pale achenes.

Most British specimens are placed in:

H. prenanthoides Vill.   'Prenanth Hawkweed.'

Stem commonly 40–100 cm., erect, rather robust, reddish below. Lvs 12–30 or more, the lowest gradually narrowed into the winged stalk, often withered at flowering, the median often panduriform. Inner involucral bracts linear-oblong, sparsely floccose. Achenes 3·25–3·5 mm. $2n = 36$.

Native. Stream-sides and wooded steep-sided valleys in mountain districts. 11. Frequent in Wales and in N. Britain from Derby northwards to Sutherland. ?Ireland. French and Swiss Alps; Scandinavia.

Section 6. *Alpestria* F. N. Williams. Fig. 57c.

Stem 20–100 cm., erect, rather slender, usually with white stout-based hairs below, floccose and often also with black-based hairs above. Rosette-lvs few, usually withering before flowering, sometimes persistent; *stem-lvs* 2–10(–15), distant, the lowest narrowed to a ± stalk-like base, middle and upper *semi-amplexicaul*; all dark or grass-green, paler but *not clearly net-veined beneath*, subentire or distantly toothed, ciliate and often with stout-based hairs at least beneath. Heads 2–6(–12) in a corymbose panicle, their stalks floccose and often sparsely hairy and glandular. Involucre 8–12 mm., dark green, ± floccose and sparsely to fairly densely glandular and hairy, the hairs black at least below; involucral bracts lanceolate, the inner subacute. Ligules usually glabrous-tipped, sometimes ciliate-tipped. Styles yellow or dark. Achenes reddish-brown to black. Fl. 7–9.

Cliffs, rocks, hilly pastures, stream-sides, etc. Chiefly in Shetland with only 2 spp. extending southwards to C. Scotland. N.W. Europe; Carpathians.

A group of forms intermediate between *Prenanthoidea* and *Vulgata*, distinguishable from the former by the subpersistent rosette-lvs, the smaller number of less amplexicaul stem-lvs, hardly reticulate beneath the sparsely glandular infl., the usually glabrous-tipped ligules and the dark achenes; from the latter by the more numerous and semi-amplexicaul stem-lvs.

The commoner of the C. Scottish species is:

H. dewarii Syme

Stem hairy throughout, sparsely floccose above. Rosette-lvs few or 0, broadly elliptical with short hairy stalks, persistent; stem-lvs 6–15, lower narrowly elliptical, stalked, middle and upper ovate-lanceolate, sessile, semi-amplexicaul; all hairy above and beneath, ciliate, subentire or with small distant teeth. Infl. a corymbose panicle with rather long slender incurving branches, floccose and with a few glands and black-based hairs. Heads usually 3–15, medium sized. Involucral bracts floccose below, with short black hairs, longer white black-based hairs and many glandular hairs. Ligules somewhat ciliate-tipped. Styles dark. Achenes 3·5–4 mm., dark brown.

Native. Mountain glens; Fife, Perth, Argyll, Inverness. Apparently endemic.

Section 7. *Vulgata* F. N. Williams (Subsection *Euvulgata* Pugsl.). Fig. 58 A, B.

Stem 15–80 cm., usually slender and wiry, hairy below, ± glandular and floccose above. Rosette-lvs 2 to several, stalked, ± persistent until flowering; stem-*lvs* 2–8(–12), diminishing upwards, the lowest stalked, upper ± sessile; all *grass- or yellow-green*, lanceolate to elliptical, *narrowed at both ends* and passing gradually into the stalk-like base, usually with spreading or forwardly directed (not reflexed teeth), glabrous or sparsely hairy above, densely hairy on the margins and stalks and at least on the veins beneath. Heads few to many, small to medium-sized, panicled, their stalks like the *involucre* usually floccose and ± *densely glandular* but not or scarcely hairy. Involucre 8–12 mm., its bracts incurved in bud, dark or blackish-green usually with paler margins. Ligules glabrous-tipped. Styles usually dark. Achenes dark brown or black. Fl. 6–9.

Woods, hedgerows, banks, rocky slopes and cliffs throughout the British Is., but chiefly in S. England and Wales and becoming infrequent in Scotland and the Hebrides. Temperate and subarctic Europe and Asia. ?N. America.

A large group which includes some of the commonest *Hieracia* of lowland Britain. Distinguished from *Caesia* agg., by the non-glaucous lvs, and from *Sagittata*, *Glandulosa* and *Bifida* by the gradual narrowing of the lf-blade into the stalk and the non-reflexed basal teeth; but forming with these four a large group (*Vulgata* Pugsl.) characterized by the non-glandular rosette-lvs lacking marginal bristles, the incurved involucral bracts in bud, the usually mixed clothing (stellate down, stout black-based hairs, slender pale hairs, glandular hairs) of the infl., and the usually glabrous-tipped ligules.

The commonest British spp. are:

H. vulgatum Fr.   'Common Hawkweed.'

Stem 20–80 cm., hairy below. Rosette-lvs few, ovate-lanceolate, shortly but acutely toothed, ciliate, hairy, floccose and often violet beneath; stem-lvs 1–3(–5), elliptical-lanceolate, the lowest stalked, remainder subsessile; all subentire or commonly the stem-lvs with numerous acute forward-spreading teeth longest near the base, the stalks long-hairy. Infl. corymbose, of 1–20 heads, *branches* floccose and hairy but *not or sparsely glandular*. Involucre sparsely to moderately glandular and hairy, bracts broadly lanceolate, ± blunt, with floccose margins. Ligules glabrous-tipped. Styles yellowish at first, then darkening. Achenes 2·5–3 mm., blackish. Fl. 7–8. Very variable.

Native. Woods, banks, walls and rocks. 50, H 17. Common in N. Wales, N. England and Scotland, and ranging from Brecon, Montgomery and Bedford to Sutherland and Caithness; Hebrides; Ireland. Scandinavia; C. Europe.

H. maculatum Sm. 'Spotted Hawkweed.'

Stem 30–50 cm., reddish and very hairy below. Rosette-*lvs* few, oblong to elliptic-lanceolate, long-stalked, *usually with dark purple spots* or blotches (sometimes unspotted in shady habitats), variously and often deeply toothed especially towards the base; stem-lvs 2–5(–8), the lower stalked, upper sub-

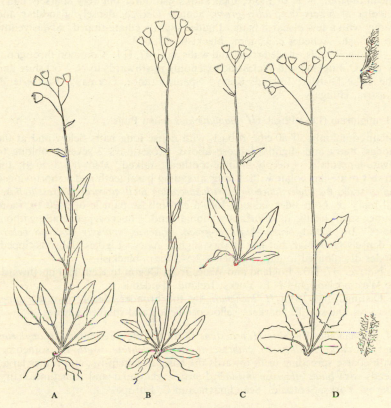

Fig. 58. *Hieracium.* A, B, *Vulgata* (A, *H. lachenalii*; B, *H. vulgatum*); c, *Caesia* (*H. cravoniense*); D, *Sagittata* (*H. oistophyllum*). × ⅕.

sessile, marked and toothed like the rosette-lvs; all ± stiffly hairy on both sides, ciliate, with villous stalks. Heads 6–20 in a lax panicle, their *stalks* floccose, *densely glandular* and often hairy. Involucre 9–13 mm., somewhat floccose and hairy, and densely glandular. Ligules usually glabrous-tipped. Styles dark. Achenes 2·5–3 mm., blackish. Fl. 6–8.

Doubtfully native. Walls, banks and rocks. 16, H2. Scattered throughout England and Wales from Devon to Kent and northwards to Westmorland and N.E. Yorks; Ireland. Germany; Austria.

H. lachenalii C. C. Gmel. (*H. sciaphilum* (Uechtr.) F. J. Hanb.)

Stem 60–90 cm., usually robust, densely hairy below. Rosette-lvs large, ovate to ovate-lanceolate, subglabrous above, with dense long hairs on the midrib beneath and on the stalk, strongly toothed and often laciniate or ± pinnatifid near the base; *stem-lvs* 4–7(–12), diminishing upwards, the lower ovate-lanceolate, stalked, uppermost *linear-lanceolate, subentire*, sessile. Heads medium-sized, 3–50 or more, their stalks glandular but very sparsely hairy. Involucral *bracts* deep *olive-green*, acute, *floccose*, densely glandular and rarely with a few blackish hairs. Ligules slightly ciliate-tipped. Styles yellow to darkish. Achenes c. 3·5 mm., blackish. Fl. 6–9.

Native. Woods, shaded rocks and walls, etc. 47, H1. Common throughout S. and C. England and Wales and extending northwards to N. Yorkshire and the Lake District; Ireland, Kerry. Spain, France, Germany, Switzerland, Austria, Hungary.

H. anglorum (Ley) Pugsl. (*H. diaphanoides* sensu Pugsl.)

Stem commonly 30–80 cm., robust, with dense long hairs below and at the nodes, hairy and slightly floccose above. Rosette-lvs 2–several, oblong to ovate-lanceolate, ± deeply sinuate-toothed, stalked; stem-*lvs* 3–5(–9), the lowest ovate-lanceolate with ± long spreading basal teeth and a short winged hairy stalk, the *upper lanceolate, with spreading teeth towards the sessile base*; all hairy on both sides, ciliate. Infl. a corymbose panicle of 3–20 or more rather small heads, its branches glandular and ± floccose, generally without hairs. Involucral *bracts blackish-green, ± acute, sparsely floccose below*, ± densely glandular, but not or very slightly hairy. Ligules glabrous-tipped. Styles discoloured to dark. Achenes 2·5–3 mm., blackish.

Native. 37, H3. England and Wales from Devon to Kent and northwards to Westmorland and N.E. Yorks; Ireland. Endemic.

Distinguished from *H. lachenalii* by the broader, deeply toothed upper stem-lvs, the smaller, darker ± efloccose heads and smaller achenes.

H. diaphanoides Lindeb. (*H. praesigne* (Zahn) Roffey) resembles *H. anglorum* but has *darker, dull green, glabrescent, sinuate-dentate rosette-lvs*, sometimes with coarse spreading teeth towards the base, 3–4 stem-lvs, and rather larger heads with *blunt efloccose involucral bracts*, glandular and sometimes hairy. Derby, Yorks, Scotland. Scandinavia and C. Europe.

Section 8. *Caesia* (Zahn) Clapham (subsection *Caesia* Zahn). Fig. 58 c.

Stem ± glabrous or hairy below, somewhat floccose but not or slightly glandular above. Rosette-lvs several, ovate-lanceolate to narrowly lanceolate, ± cuneate-based, stalked, toothed or ± pinnatifid, never truncate at the base nor with reflexed teeth; stem-lvs 1–3(–6), the lowest stalked, all more acute and more deeply toothed than the rosette-lvs; lvs commonly *glaucous* or pale green, sometimes purple-spotted or purple beneath, glabrous above, usually floccose and ± hairy beneath, ciliate and with hairy stalks. Infl. usually of few heads in a loose subcorymbose panicle, its branches ± floccose and hairy but not or sparsely glandular. *Involucre* 10–14 mm., usually ± densely floccose, moderately hairy but *not or scarcely glandular*; its bracts incurved in bud, lanceolate, blunt, with floccose margins. Ligules with glabrous or ciliate tips. Styles yellow, dingy yellow or dark. Fl. 6–8.

Chiefly on rocks. Mountain districts of W. and N. England, Wales and Scotland; Ireland. N. and C. Europe.

An aggregate of spp. resembling *Euvulgata* in the shape and position of the lvs but differing in their glaucous colour, in the usually smaller number of stem-lvs and the fewer and larger heads.

Closely related to *Bifida* to which it bears the same relation as do *Vulgata* to *Glandulosa* and *Sagittata*.

Most are rare and local, but the following are more widespread:

H. caledonicum F. J. Hanb. (incl. *H. rubicundum* F. J. Hanb. and *H. farrense* F. J. Hanb.)

Stem commonly 10–40 cm., robust, sparsely hairy. *Rosette-lvs* firm, deep *bluish-green*, sometimes purple-tinged, ovate to elliptic, blunt and mucronate, *subentire* or with small distant spreading teeth, *rounded below; stem-lvs* 1–2(–3), *the lower usually large, the upper bract-like*, ovate-lanceolate, toothed, shortly stalked; all ± hairy at least below, ciliate with ± bristly hairs. Infl. forking with 1–4 or more large heads with rounded or truncate base; branches floccose, glandular and hairy. Involucral bracts connivent in bud, dark green, blunt, often red-tipped, sparsely floccose and glandular, with many black-based hairs. Ligules ± glabrous tipped. Styles yellow to darkish. Achenes 4–4·5 mm., blackish.

Native. 17, H5. Great Britain from Hereford and S. Wales to Sutherland; Inner and Outer Hebrides; Orkney; Ireland. Endemic.

H. cravoniense (F. J. Hanb.) Roffey

Stem commonly 25–60 cm., slender, hairy especially below, ± floccose above. Rosette-lvs pale green, ovate to lanceolate, blunt, mucronate, irregularly and ± deeply sinuate-toothed towards the cuneate base; stalk short, winged; stem-lvs 2–4, ± toothed towards the base, the lowest ± stalked; all stiffly hairy on both sides, ciliate, with villous stalks. Infl. a ± forking corymb with 3–15 or more small heads, the terminal pair approximated; its branches densely floccose and long-hairy. Involucral bracts dark olive-green, ± blunt, floccose, densely long-hairy, not or sparsely glandular. Ligules often undeveloped so that the florets are tubular. Styles dark. Achenes 3–3·5 mm., reddish-black.

Native. 15. Local from Lancashire and Yorks northwards to Sutherland. Outer Hebrides. Endemic.

Section 9. *Sagittata* (Pugsl.) Clapham (subsection *Sagittata* Pugsl.). Fig. 58 D.

Stem 20–70 cm., rather robust, reddish and hairy below, floccose and somewhat hairy but not or sparsely glandular above. *Rosette-lvs many*, grass- or yellow-green, paler beneath; outer ovate to elliptical, blunt, denticulate, usually *truncate or subhastate at the base*; inner narrower with long spreading or reflexed teeth at the ± sagittate base; stem-lvs 0–1(–2), the lower, if more than 1, fairly large and stalked, ovate-lanceolate to lanceolate; sharply toothed; all lvs hairy beneath, on the stalks and on the margins. Heads

2–20 in a panicle with long lower and shorter incurved and subumbellate upper branches, their *stalks* floccose, ± *densely hairy* and not or somewhat glandular. Involucre 8–10 mm., ± densely hairy and not or somewhat glandular, its bracts usually incurved in bud, rather broad, blackish-green with paler floccose margins. Ligules glabrous or somewhat ciliate-tipped. Styles usually dark. Fl. 6–8.

Rocky woods and valleys, chiefly in the north. W. and N. England, Wales and Scotland; Hebrides; Orkney. N. Europe, southwards to Denmark and Poland. ?Labrador.

Distinguished from *Caesia* and *Vulgata* by the truncate to sagittate base of the lvs and stem lvs 0–1; and from *Glandulosa* by the predominantly hairy covering of the infl. and involucre, but there are intermediates difficult to place.

The British spp. are mostly uncommon or local. The following is the most widely distributed:

H. euprepes F. J. Hanb.

Stem commonly 20–50 cm., robust, ± densely hairy, floccose above. Rosette-lvs dull green, often reddish beneath; oblong to ovate-lanceolate, blunt and mucronate to acute, subentire to ± sinuate-toothed, cuneate or abruptly narrowed into short stalks; stem-lvs 1–2, the lower lanceolate, acuminate, ± finely toothed, ± stalked, the upper bract-like; all ± stiffly hairy on both sides, ciliate, with villous stalks. Infl. corymbose, of 2–16 heads, the terminal heads paired, branches floccose and hairy but not or sparsely glandular. Heads medium-sized, narrow. Involucral bracts few, connivent in bud, dark green, blunt, sparsely floccose, with dense black-based hairs but few or no glands. Ligules glabrous-tipped. Styles yellowish. Achenes c. 3·25 mm., blackish.

Native. 12, H2. Wales; Scotland northwards to Sutherland and Caithness; N.E. Ireland. Endemic.

Somewhat intermediate in lf-shape between *Sagittata* and *Caesia*.

Section 10. *Glandulosa* (Pugsl.) Clapham (subsection *Glandulosa* Pugsl.). Fig. 59 A.

Stem 20–60 cm., often hairy below, floccose and glandular especially above. Rosette-lvs several to many, variable in shape but usually broad, the blade cordate, truncate or rounded at the base (and so clearly demarcated from the stalk) and typically with spreading or reflexed basal teeth; stem-lvs 0–1(–3), the lowest near the middle of the stem, stalked, resembling the inner rosette-lvs, with sometimes an additional 1–2 smaller upper lvs; all *lvs grass- or yellow-green*, ± glabrous or hairy above, hairy beneath, ciliate and with densely hairy stalks. Infl. corymbose, forking or panicled, its branches floccose and densely glandular but not or scarcely hairy. *Involucre* usually 9–10 mm., ± ovoid, *densely glandular*, ± floccose, not or scarcely hairy; its bracts incurved in bud, usually blackish-green with paler ± floccose margins. Ligules usually glabrous-tipped. Styles usually dark. Achenes black. Fl. 6–8.

Woods, shaded rocks, stream-sides, banks, roadsides. Throughout Great Britain; Ireland. Throughout Europe. C. and N. Asia.

The species included here resemble *Sagittata* in the shape, position and

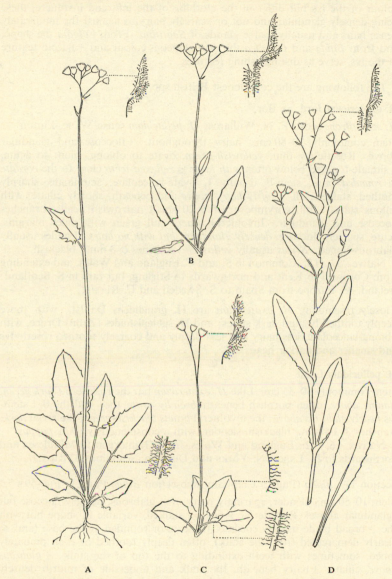

Fig. 59. *Hieracium.* A, *Glandulosa* (*H. exotericum*); B, *Bifida* (*H. sanguineum*); C, *Oreadea* (*H. hypochaeroides*); D, *Amplexicaulia* (*H. amplexicaule*). ×⅕.

colour of the lvs but differ in the clothing of the infl. and involucre, these being densely glandular and not or scarcely hairy as against the moderately dense hairs and usually sparse glands of *Sagittata*. From *Vulgata* the shape, and from *Bifida* and *Oreadea* the non-glaucous colour and ± flaccid texture of the lvs, serve as distinguishing features.

The following are the commonest British spp.:

H. exotericum Jord. ex Bor.

H. serratifrons sensu F. N. Williams; *H. pellucidum* sensu W. R. Linton
Stem commonly 30–80 cm., hairy throughout, ± floccose and glandular above. Rosette-*lvs* thin, *yellowish-green*, ovate to oblong, blunt to acute, ± sinuate-toothed below often *with large* ± *reflexed teeth* close to the *cordate to rounded base*; stem-lf usually 1, ovate-lanceolate, acuminate, sharply toothed, stalked; all ± *softly hairy above* and beneath, shortly ciliate, with villous stalks. Infl. corymbose, of 5–20 small narrowish heads, branches floccose and glandular. Involucral *bracts* dull green with paler margins, acute, sparsely floccose, *densely glandular*, *but with no hairs*. Ligules usually glabrous-tipped. *Styles* usually *yellowish*. Achenes c. 3 mm., blackish.
Native. 37, H5. Common in S. and C. England and Wales, and extending from Cornwall to Kent and northwards to Stirling, but rare in S. Scotland; Ireland. C. Europe from Spain to S. Sweden and C. Russia.

Closely resembling *H. exotericum* are **H. grandidens** Dahlst., with more deeply toothed lvs and dark styles; and **H. sublepistoides** (Zahn) Druce, with oblong-lanceolate, less hairy and less deeply and coarsely toothed rosette-lvs and smaller and darker heads.

H. pellucidum Laest.

Stem commonly 30–55 cm. Like *H. exotericum* but the rosette-*lvs dark green*, *shining above*, often purplish beneath, *broadly ovate*, they and the 0–1 stem-lvs *nearly glabrous above*; the involucral *bracts blackish-green*; the *styles dark*. Achenes 3–3·5 mm., blackish, slender, with a conspicuously *white pappus*.
Native. 16, H2. England and Wales, from Glamorgan to Middlesex and northwards to N. Lancs, N. Yorks and Durham, Scandinavia, Russia.

Section 11. *Bifida* (Pugsl.) Clapham (subsection *Bifida* Pugsl.). Fig. 59 B.

Stem 10–50 cm., slender, sparingly hairy or ± glabrous, densely floccose but eglandular above. Rosette-*lvs* 3–5(–10), stalked, variable in shape but with the ± broad blade usually cordate, truncate or rounded at the base (and so clearly demarcated from the stalk), often deeply toothed or even pinnately lobed, sometimes with teeth extending to the top of the stalk; ± *glabrous above*, ciliate, ± hairy beneath, the stalk and underside of midrib densely hairy; stem-lvs 0–1(–3), the lowest shortly stalked, floccose beneath; all commonly *glaucous*, often purplish beneath and sometimes purple-spotted. Infl. a forking or loosely paniculate corymb of 1–8 or more medium to large heads, usually rounded below, branches densely floccose and ± hairy and glandular. Involucre 9–11 mm., ± densely floccose but not or sparsely hairy and usually sparsely glandular, often only near the base; its bracts usually incurved in bud, blackish-green, the outer blunt, inner ± acute. Ligules glabrous-tipped or shortly ciliate. Styles yellow to dark. Achenes black. Fl. 6–8.

Rocks, banks and walls. Throughout the British Is. but common only in mountain districts. C. and N. Europe northwards from the Alps and N. Balkans, frequent in Scandinavia.

Closely related to *Caesia*, from which it differs in lf-shape and position and in the smaller number of heads; and connected by intermediates with *Oreadea* but differing typically in the less coriaceous and less bristly-margined lvs, in the more paniculate infl. and the bracts usually incurved in bud. From *Sagittata* and *Glandulosa* typical representatives are distinguished by the ± glabrous stem and upper surface of the lvs, the colour of the lvs and the clothing of the involucre.

The spp. are all uncommon and local. The following is locally frequent in N. England:

H. subcyaneum (W. R. Linton) Pugsl.

Stem commonly 20–50 cm., ± glabrous, somewhat floccose above. Rosette-lvs dull green, sometimes blotched with purple or purple-tinged beneath, ovate to elliptic-lanceolate, blunt and mucronate to acute, subentire to sinuate-toothed, often with a few large spreading teeth near the subtruncate base, ± hairy above and beneath, ciliate, and with villous stalks; stem-lf 0, occasionally 1, linear to lanceolate, stalked. Infl. a forking corymb of 1–5 or more large round-based heads, branches floccose and sparsely hairy but not or slightly glandular. *Involucral bracts erect in bud*, dark greyish-green, acute, floccose especially on the margins, with dense black-based hairs and sparse glands. Ligules glabrous or ciliate-tipped. Styles darkish. Achenes 4 mm., blackish.

Native. 2. Frequent on carboniferous limestone in Derby and W. Yorks. Endemic.

H. cymbifolium Purchas, placed by Pugsley in a separate subsection *Stellatifolia*, resembles *Bifida* but has lvs ± *floccose on both surfaces*. The infl. and involucre are also floccose but hairy and glandular as well. Styles yellow. Achenes 3·5 mm., reddish-black.

Native. 3. Carboniferous limestone of Derby, Stafford and W. Yorks. Endemic.

Section 12. *Oreadea* Zahn (incl. section *Suboreadea* Pugsl.). Fig. 59 c.

Stem 10–60 cm., ascending, slender, ± sparsely hairy and often reddish below, ± glandular and floccose above. *Rosette-lvs* up to 10, ovate to lanceolate, acute to acuminate, entire or toothed, usually *cuneate-based*, ± long-stalked, *glaucous*, usually thick and *firm*, commonly *bristly with stiff stout-based hairs on the margin*, often also *on the upper surface near the margin* or all over, hairy beneath (especially on the under-side of the midrib) and on the stalks; stem-lvs 0–3(–6), the lowest usually ± stalked, not amplexicaul, diminishing upwards. Infl. a forking corymb with (1–)2–12 large heads; branches floccose and ± glandular, but often not or sparsely hairy with black-based hairs.

Involucre 10–14 mm., ovoid; *bracts erect in bud*, ± floccose, usually densely glandular with long-stalked and tiny glands, and often also densely hairy with stout black-based hairs. Ligules glabrous or occasionally ciliate-tipped. Styles usually yellow. Achenes blackish. Fl. 5–8.

Rocks. Throughout mountain districts of the British Is. in Great Britain northwards from N. Devon and Somerset, S. Wales, Stafford, and Derby. C. and S. Europe, Norway, Sweden and Iceland.

The Section as defined here includes the *Suboreadea* of Pugsley, forms connecting the more typical *Oreadea* with *Bifida*, *Vulgata*, etc. The distinguishing features are the firm, glaucous, usually bristly-margined lvs, the large heads borne on forkings of the stem, the involucral bracts erect in bud, and the covering of the involucre, which includes micro-glands as well as stalked glands and black-based hairs.

The commonest British spp. include:

H. britannicum F. J. Hanb.

Incl. *H. stenolepis* auct. angl.

Stem commonly 20–45 cm., robust, sparsely hairy and floccose, slightly glandular above. *Rosette-lvs* ± glaucous, strongly veined, short-stalked, *broadly ovate to lanceolate-sagittate*, blunt and mucronate to acute, usually sharply toothed, the teeth very long and horizontal or reflexed near the abruptly narrowed or broad truncate base; *stem-lvs* 0 or 1, *small and bract-like*; all ± *glabrous above*, hairy beneath, ciliate, with villous or long-hairy stalks. Infl. a forking corymb with 2–6 or more largish heads; branches floccose, and fully glandular but sparsely hairy. Involucral *bracts* erect in bud, *grey-green*, long and narrow, acute, floccose on the margins, hairy and sparsely glandular. Ligules glabrous-tipped. Style yellowish or darkish. Achenes 3–3·5 mm., blackish. Variable in colour and shape of the rosette lvs, etc.

Native. 14, H3. Great Britain from N. Somerset and Brecon to Caernarvon and Derby and northwards to Fife; Inner Hebrides; Ireland. Endemic.

H. britannicum lacks the bristle-like marginal hairs of the true *Oreadea*.

H. hypochoeroides Gibson

H. Gibsonii Backh.

Stem commonly 20–40 cm., sparsely hairy, slightly glandular above. *Rosette-lvs* thick, *pale green, spotted with purplish-brown*, broadly ovate to elliptic-lanceolate, obscurely glandular-toothed or rarely coarsely toothed near the ± *subcordate base*; stem-lf 0 or 1, linear-lanceolate, hardly stalked; all ± glabrous above, hairy beneath, the margin *ciliate with somewhat bristly hairs*. Heads 2–4(–6), large, rounded below. Involucral *bracts* erect in bud, *dark green*, bluntish, floccose on the margins, ± densely hairy and glandular. Ligules glabrous-tipped. Styles yellowish. Achenes 3·5–4 mm., dark reddish-brown.

Native. 11, H2. W. and N. Britain from Gloucester and Brecon northwards to Sutherland; Ireland (Antrim and Cork). Endemic.

The beautifully spotted or marked subcordate lvs are distinctive.

H. argenteum Fr.

Stem commonly 15–40 cm., slender, ±glabrous or sparsely hairy below. *Rosette-lvs glaucous*, ± *narrowly lanceolate*, subacute, subentire to sinuate-toothed, narrowed gradually into short winged stalks; *stem-lvs* usually 2, linear-lanceolate, *spreading*, or small and bract-like; all *glabrous above*, glabrous or sparsely hairy beneath, ±ciliate, with long hairy stalks. Infl. forking, with 1–4(–14) medium-sized truncate-based heads; branches sparsely floccose, glandular and hairy. Involucral bracts erect in bud, dark green, blunt, sparsely floccose, with dark glandular hairs and a variable number of black-based hairs. Ligules usually glabrous-tipped. Styles yellow. Achenes 3·5 mm., blackish.

Native. Rocky river-banks, etc. in mountainous districts. 28, H5. Wales; N. Britain from S.W. Yorks to Sutherland and Caithness; Outer Hebrides; Orkney; Ireland. Scandinavia.

Recognizable by the narrow glaucous ±glabrous lvs and the markedly spreading stem-lvs.

H. schmidtii Tausch

Stem commonly 15–40 cm. *Rosette-lvs* glaucous, *roundish to ovate-lanceolate*, blunt and mucronate to acute, subentire or with sharp forwardly directed teeth in the lower half, *narrowed into long hairy stalks*; stem-lf usually 1, linear-lanceolate, sessile, sometimes bract-like; *all ±bristly on the margins and sparsely bristly on both sides*, or ±glabrous above. Infl. a corymb of 1–6 large round-based heads, its branches floccose, with numerous fine glandular hairs and some simple hairs. Involucral *bracts* erect in bud, *dark green*, acute, sparsely floccose, ±densely glandular and with many black-based hairs. Ligules glabrous- or ciliate-tipped. Styles yellow. Achenes 3–3·5 mm., black.

Native. 17, H3. Wales; N. Britain from N.W. Yorks and Durham northwards to Sutherland and Caithness; Outer Hebrides; Shetland; Ireland. Spain; France; Switzerland; Germany; Scandinavia.

H. lasiophyllum Koch

Stem commonly 10–20 cm., hairy and floccose throughout. *Rosette-lvs* firm, *very glaucous*, ovate to oblong, with *rounded mucronate tips*, *subentire* or with a few teeth near the *rounded base*; stem-lf usually 0, or 1, lanceolate, acute, short-stalked; *all long-bristly above*, ± *hairy and floccose beneath*, and *ciliate with long hairs* extending down the stalks. Infl. forking, of 1–6 medium-sized round-based heads; branches floccose, glandular and hairy. Involucral *bracts* erect in bud, *grey-green*, ±acuminate, floccose and with numerous glandular and black-based hairs. Ligules usually glabrous-tipped. Styles yellow. Achenes 3 mm., reddish-black.

Native. 20, H2. W. and N. Britain from Gloucester and S. Wales northwards to Perth and Aberdeen; N. Ireland. C. Europe.

*Section 13. *Amplexicaulia* Zahn Fig. 59 D.

Plant *glandular-viscid throughout*. Stem 10–60 cm., erect or ascending. Rosette-lvs numerous, ovate-spathulate to lanceolate, ±blunt, mucronate, narrowed below into a winged stalk-like base; *stem-lvs* 2–6(–12) large, the upper broadly ovate, usually *amplexicaul* with large rounded auricles; all lvs ± shortly toothed, densely covered on both sides with yellow-headed glands. Infl. corymbose, of 3–20 or more large heads; branches floccose and densely glandular. Involucre 12–16 mm., rounded or truncate below, ± floccose and densely glandular; bracts incurved in bud, lanceolate-acuminate. Ligules ciliate-tipped. Styles yellow or dark. Achenes 3–4 mm., brownish-black.

Old walls. S. Europe from Spain and S. France to Dalmatia; Alps; Jura.

Very distinct in being viscid throughout and in the large broad amplexicaul stem-lvs.

There are 3 introduced spp. in the British Is. *H. amplexicaule* L. (walls of the Oxford Botanic Garden) is devoid of simple non-glandular hairs and has yellow styles; *H. pulmonarioides* Vill. and *H. speluncarum* Arv.-Touv. have simple hairs at least on the base of the stem, lf margins, under-side of midrib and lf-stalk, and have dark styles. In *H. pulmonarioides* the lvs are ± glaucous, the upper stem-lvs linear-lanceolate; in *H. speluncarum* the lvs are yellowish-green, the upper stem-lvs being ovate and subcordate.

All 3 spp. are very local.

Section 14. *Cerinthoidea* Koch. Fig. 60 A.

Stem 10–60 cm., robust, hairy below, subglabrous above. Rosette-lvs ± persistent to flowering, obovate to elliptic-lanceolate, entire or toothed, with winged stalks, glabrous or hairy above, ciliate, hairy beneath, the stalk and under-side of the midrib *shaggy with long white denticulate hairs*; stem-lvs 1–7, ± *amplexicaul*; all lvs *glaucous*. Infl. of 1–8(–20) large heads (4·5–5·5 cm. diam.), terminating forkings of the stem; branches ± floccose with numerous black glandular and some white black-based hairs. Involucre 11–14 mm., broadly ovoid, sparsely floccose, and densely clothed with black-based hairs usually with some glandular hairs; bracts incurved in bud, in 5–6 imbricating rows. Ligules usually pale yellow, ± ciliate-tipped. Styles usually dark. Achenes black. Fl. 6–8.

Rocks. Throughout mountain districts of the British Is. northwards from Somerset and S. Wales, becoming much more frequent towards the north, and reaching the Outer Hebrides and Shetland. W. Europe from the Pyrenees to Switzerland and Iceland; not in Scandinavia.

The forms included here are chiefly found in S. France and Spain and are easily recognized by their robust habit, glaucous lvs with shaggy stalks, ± amplexicaul stem-lvs and very large pale yellow heads with dark styles.

The following are the two most widespread British spp.:

H. anglicum Fr.

Stem commonly 20–30 cm., robust, ± hairy. *Rosette-lvs* obovate to elliptical-ovate, ± acute, *narrowed* gradually *into long shaggy winged stalks*, entire to sharply toothed towards the base, hairy especially beneath, densely ciliate; *stem-lvs* 1 or more, *usually* 2, the lower ovate, narrowed to a semi-amplexicaul base, the upper linear-lanceolate, bract-like, sessile or slightly amplexicaul. Infl. of 1–4 large heads, rounded below; branches sparsely floccose but densely glandular and with ± numerous black-based hairs. Involucral bracts acuminate, dark green, usually greyish with pale-tipped black-based hairs, with few or no glandular hairs. Achenes 3·5–4 mm., reddish-black. Fl. 6–8.

Native. Rocks in mountain districts. 23, H16. Great Britain northwards from Westmorland, N. Yorks and Durham; Inner and Outer Hebrides; Orkney; Ireland. Endemic.

Variable in the shape and toothing of the lvs.

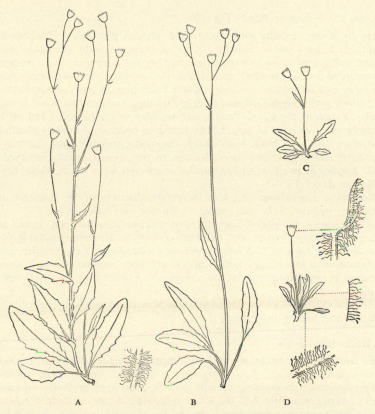

Fig. 60. *Hieracium.* A, *Cerinthoidea* (*H. iricum*); B, *Subalpina* (*H. gracilifolium*); C, D, *Alpina* (C, *H. hanburyi*; D, *H. holosericeum*). × ⅕.

H. iricum Fr.

Stem 20–60 cm., robust, hairy, ±floccose with black-based hairs above. *Rosette-lvs* ovate- or elliptic-lanceolate, acute, ±sinuate-toothed, hairy, *narrowed into* rather *short shaggy stalks; stem-lvs* 3–7, ovate to lanceolate, toothed, the lower narrowed into a stalk-like base, the upper sessile, all ±amplexicaul, sharply toothed, long-hairy especially beneath. Infl. of (1–)4–8(–16) large heads; branches floccose, densely glandular and somewhat hairy. Involucre truncate-based, bracts dark green, subacute, with dense black-based and ±numerous glandular hairs. Ligules bright yellow, ciliate-tipped. Achenes 3·5–4 mm., blackish. Fl. 6–8.

Native. 15, H12. Great Britain northwards from Cumberland and N. Yorks; Hebrides; Orkney; Ireland. Endemic.

A stouter plant than *H. anglicum*, with more numerous stem-lvs and deeper yellow ligules.

Section 15. *Subalpina* Pugsl. Fig. 60 B.

Stem 20–50 cm., usually with black-based hairs at least below, floccose and glandular at least below. Rosette-lvs not numerous, usually yellow- to grass-green, stalked, narrowly to broadly elliptical or the outermost more rounded, subentire to coarsely toothed; stem-lvs (0–)1–3 or more, the lowest resembling the inner rosette-lvs, the remainder usually bract-like; all hairy at least beneath and ciliate with long stout-based strongly toothed hairs and with *few to many stalked glands on the margins and sometimes also on the surfaces*. Infl. of 2–5 or more medium-sized to large heads, branches floccose, ± densely glandular and often also with black-based hairs. Involucre usually with ± numerous glandular and black-based hairs, floccose or not; bracts usually incurved in bud. Ligules often ciliate-tipped but not or sparsely hairy on the back. Styles yellow or dark. Fl. 7–9.

Subalpine and alpine zones in mountain districts of N. England and Scotland; Ireland. Subarctic and arctic Europe and Asia.

A heterogeneous assemblage of forms intermediate between *Alpina* and others, especially *Oreadea* and *Glandulosa*, but usually distinguishable from *Alpina* by the broader and less numerous rosette-lvs and the smaller, more numerous and less shaggy heads, and from other sections by the constant presence of marginal glands on the non-glaucous lvs, and by the ciliate-tipped ligules.

The *Subalpina* are almost confined to the Scottish mountains. The following are the most widely distributed:

H. lingulatum Backh.

Stem commonly 25–40 cm., slender, with spreading black-based hairs through-out, ± floccose above. *Rosette-lvs* few, pale green, mostly long, *narrowly oblong-lanceolate or lingulate*, usually *subentire*, narrowed gradually into short stalks; stem-lvs usually 2–3, the lower lanceolate, acute, ± toothed towards the sessile base, the upper linear, bract-like; all *with ± numerous stiff hairs and a few fine glandular hairs on both surfaces*, ciliate, with long-hairy stalks. Infl. of (1–)2–4 or more medium-sized to large heads; branches floccose, rather short, with many black-based hairs and a few fine glandular hairs. Involucral *bracts erect in bud, blackish-green*, broad, acute, slightly floccose below, with many black-based and a few glandular hairs. Ligules slightly ciliate-tipped. Styles yellow or dark. Achenes 3·5–4 mm., reddish-black.

Native. 11. Scottish Highlands from Perth and Angus to Sutherland. Endemic.

The lingulate lvs up to 20 cm. long distinguish *H. lingulatum* from all other spp.

H. senescens Backh. has ovate to lanceolate subentire to coarsely sinuate-toothed lvs and involucral bracts strongly floccose and hairy at the tips, especially in bud. Styles yellow. 7, H1. Scottish Highlands; Ireland, Down. Endemic.

H. gracilifolium (F. J. Hanb.) Pugsl. has the involucral bracts blackish-green, ± densely glandular but hardly floccose and not hairy. 5. Scottish Highlands. Scandinavia; C. Europe.

Section 16. *Alpina* F. N. Williams. Fig. 60 C, D.

Stem 5–15(–30) cm., erect or ascending, ± floccose especially above, with stout black-based and smaller glands, and with whitish black-based hairs often so numerous as to make the stem shaggy. Rosette-lvs numerous, usually small, narrowly spathulate to ovate-elliptical or obovate, narrowing below into a ± winged and shaggy stalk; outer lvs broader and often withering at time of flowering; stem-lvs 0–3, usually bract-like or the lowest resembling the inner rosette-lvs; all ± hairy but rarely floccose, with *fine yellowish glands at least on the margins*. Heads usually *large and solitary*, sometimes 2–4 terminating forkings of the stem. *Involucre* usually *shaggy* with whitish black-based hairs and usually with some black-based glands and smaller glands; bracts incurved in bud, linear to lanceolate, the outermost often lax. *Ligules ciliate-tipped and often also with white hairs on the back.* Styles yellow, brownish or very dark. Achenes large, black. Fl. 7–9.

Rocks on high mountains. Wales, N. England and Scotland. Arctic and subarctic Europe and Asia and mountains of C. Europe. Greenland. Labrador.

A group easily recognized by the small stature, glandular lvs, scapose stem usually with a single large head, and the long white-tipped black-based hairs which so commonly cover the stems, lvs and involucres.

H. holosericeum Backh.

Stem commonly 5–15 cm., *shaggy throughout with very long, spreading, silky, white-tipped, black-based hairs,* floccose and somewhat glandular-hairy above. *Rosette-lvs pale green,* the outer obovate-oblong, ± entire, rounded to retuse at the apex, narrowed into broadly winged stalks; the inner *linear-oblong, entire,* rounded to subacute at the apex, *gradually narrowed into ± long stalks;* stem-lvs 0–2(–3), narrow, entire, subacute, sessile; all shaggy above and beneath, on the margins and on the stalks, with long silky whitish hairs, and with small yellow glandular hairs on the blade. Head solitary. Involucral *bracts* few, incurved in bud, blackish-green, the outer lax, blunt, the inner narrow, ± acute, *all very densely covered with long silky white-tipped black-based hairs. Ligules strongly ciliate-tipped and hairy on the back. Styles yellow.* Achenes 3·5 mm., reddish-black.

Native. 14. High mountains. N. Wales; Lake District; Scotland from Galloway to Sutherland. Scandinavia; mountains of C. Europe.

A strikingly beautiful plant.

H. alpinum L. has *deep green obovate to obovate-lanceolate* subentire or denticulate bluntish *rosette-lvs* narrowed into *long slender stalks,* and 1–2 stem-lvs; all long-hairy and glandular above and beneath. Its heads are less silky than in *H. holosericeum.* Styles yellow. Achenes 3·5–4 mm., reddish-black.

Native. 4. Scottish Highlands (Aberdeen, Banff and Inverness). N. Europe and mountains of C. Europe; Greenland.

H. eximium Backh. has long (to 20 cm.) yellowish-green, densely hairy, oblong to linear-lanceolate *toothed* to subentire *rosette-lvs* and 1(–3) *large heads* with densely long-hairy involucre and *dark styles*.

Native. 9. Scottish Highlands from Perth and Angus to Sutherland. Norway; Germany.

H. hanburyi Pugsl. (*H. chrysanthum* Backh., non Ledeb.) has distantly sinuate-toothed to *deeply and irregularly incise-toothed* bright green long-stalked *rosette-lvs* and 1–4 *large heads* with blackish-green *velvety-hairy* and glandular *involucre* and *yellow styles*.

Native. 7. Scottish Highlands from Perth and Angus to Inverness. Endemic.

Subgenus 2. PILOSELLA (Hill) Tausch

Stoloniferous. Ligules yellow, red, or red beneath. Achenes 1·5–2(–2·5) mm., crenulate at apex owing to the ribs expanding into very short spreading projections. Pappus hairs in one series, with few shorter hairs mixed with the long ones.

Section 17. *Pilosella* Fr. Fig. 61 A.

Stoloniferous scapigerous herbs. Stem (scape) 5–30(–45) cm., ± floccose, hairy and glandular. Rosette-lvs linear-oblong to elliptical, ± entire, floccose or tomentose beneath, usually with scattered stiff hairs above; stem-lvs 0 or 1–2, linear, bract-like. Head solitary. Involucre usually floccose and ± hairy. Ligules yellow, the outer with reddish streaks on the under-side, glabrous-tipped. Styles yellow. Achenes 2–2·5 mm., purplish-black. Fl. 5–8.

Grassy pastures and heaths, banks, rocks, walls, etc. Throughout the British Is. except Shetland. Temperate and subarctic Europe (excluding Iceland) and Asia. N.W. Africa. The only native British spp. are:

H. pilosella L.   Mouse-ear Hawkweed.

Rhizome long, slender. *Stolons* numerous, *long*, epigeal, *bearing small distant lvs* and terminating in overwintering rosettes. Scape 5–30 cm., floccose especially above, with black-based glandular and simple hairs. Rosette-lvs linear- to obovate-oblong, ± blunt, narrowed below into a long stalk-like base, entire, floccose to tomentose beneath, with sparse long stiff white hairs on both sides, on the margins and on the stalk; stem-lvs 0, or 1(–3), small hairy and bract-like. Involucre 9–11 mm., floccose and with dense black-based glandular and simple hairs; its bracts up to 1·5 mm. wide. Ligules pale yellow, the outer reddish beneath. Styles yellow. Achenes 2 mm., purplish-black. Very variable in shape and hairiness of lvs and clothing of involucre.

Native. 111, H17. Throughout the British Is. except Shetland. Temperate and subarctic Europe and W. Asia.

H. peleteranum Mérat

Rhizome short and stout. *Stolons short*, curved, long-hairy, *bearing large crowded lvs*. Scape 10–20 cm., covered with long hairs, floccose and densely glandular above. Rosette-lvs deep green, lanceolate to elliptic, ± acute, rather thick, with dense long stiff whitish hairs above, tomentose beneath. Involucre 10–14 mm., rounded below, ± densely covered with white silky hairs, but not or scarcely floccose or glandular; its bracts up to 3 mm. wide. Ligules pale yellow, the outer red beneath. Styles yellow. Achenes 2·5 mm., purplish-black.

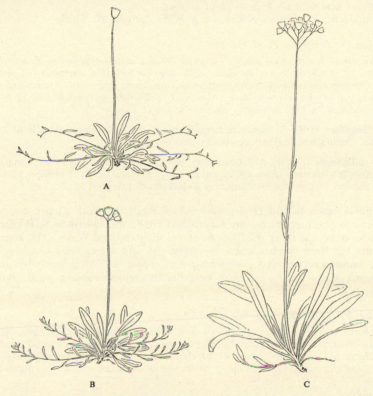

Fig. 61. *Hieracium*. A, *Pilosella* (*H. pilosella*); B, *Auriculina* (*H. lactucella*); C, *Pratensina* (*H. brunneocroceum*). × ⅕.

Native. Dry grassy banks. Local in Cornwall, Devon, Isle of Wight, Hereford, Montgomery, Merioneth, Stafford, Derby, Channel Is. W. Europe from Spain and Portugal to S. Scandinavia.

Section 18. *Auriculina* Zahn. Fig. 61 B.

Stoloniferous herbs. *Stem usually less than 20 cm.*, ± *sparsely hairy*, floccose and glandular. *Rosette-lvs* spathulate to linear-lanceolate, ± blunt, entire, *glaucous*, shining, ± *glabrous above*, not or sparsely floccose beneath; stem-lf usually 1, near the base. Heads (1–)2–5(–7) in a compact corymb; stalks short, floccose and glandular. Involucre 6–9 mm., ± densely floccose and glandular; bracts variable in width, blunt, dark- or blackish-green with pale margins. Ligules pale yellow on both sides. Styles yellow. Fl. 5–8.

*H. lactucella Wallr. (*H. Auricula* auct., non L.) is a naturalized alien with long prostrate lfy stolons, and is easily distinguished from native members of subgenus *Pilosella* by the small glaucous and ±glabrous rosette-lvs and the compact cluster of 1–5 small heads. Involucre 6–8 mm., blackish-green, hairy and glandular but not floccose. Achenes 1·5–1·75 mm., purplish-black.

Introduced. In a pasture at Keevil in S. Wilts. Europe; W. Siberia.

Section 19. *Pratensina* Zahn Fig. 61 c.

Stoloniferous herbs. *Stem usually exceeding* 20 *cm., densely long-hairy. Rosette-lvs* lanceolate to elliptic, *green, not rigid, stiffly hairy on both sides*; stem-lvs 1–4. Infl. a ±compact corymb of 2–30 smallish heads. Involucre long-hairy. Ligules yellow to deep red. Styles yellow to dark. Achenes 1·75–2·5 mm., purplish-black. Fl. 6–7.

Introduced, or 1 sp. doubtfully native. N. and C. Europe; W. Asia.

*H. flagellare Willd. is readily distinguishable by its subracemose panicle of 2–5 heads. Naturalized at Hanslope (Bucks) and near Edinburgh.

*H. colliniforme (Naegeli & Peter) Roffey has usually 15–30 heads in a ±compact subumbellate corymb, *yellow ligules* and *dark styles*. Naturalized, or possibly native, in a number of scattered localities from Worcester to Inverness.

H. aurantiacum L. and *H. brunneocroceum* Pugsl. are both garden escapes which have become extensively naturalized, the former chiefly in N. England and Scotland, the latter chiefly in S. and C. England and Wales. ***H. aurantiacum** has deep green *obovate to elliptic lvs* and *bright brick-red ligules*; ***H. brunneocroceum** has yellowish-green *oblong to oblanceolate lvs* and *brownish-orange ligules*, and easily becomes an aggressive garden weed. Both come from C. Europe.

Section 20. *Praealtina* Zahn

Tall herbs, stoloniferous or not. *Rosette-lvs linear to lanceolate, glaucous, rigid*, stem-lvs commonly 2–3. Infl. corymbose, ±compact, of 10–30 or more small heads. Involucre 6–8 mm., dark green, its bracts subacute to blunt, floccose, ±densely hairy and glandular. Ligules yellow. Styles yellow. Achenes 1·75–2 mm., purplish-black.

Europe.

There are 3 spp. naturalized in only one or two localities.

*H. praealtum Vill. is non-stoloniferous. Of the 2 stoloniferous spp. ***H. arvorum** (Naegeli & Peter) Pugsl. has bristly lvs and densely hairy involucres; ***H. spraguei** Pugsl. has ±glabrous lvs and less hairy involucres.

56. CREPIS L.

Usually herbaceous plants with spirally arranged often runcinate lvs and branching lfy flowering stems. Heads in panicles or corymbs, rarely solitary; bracts many, *equal, in* 1 *row*, apart from a few small basal bracts; receptacle flat, naked, pitted, the pits with toothed or hairy margins. Florets usually yellow, sometimes pink or white; anther-lobes without basal tails; style-arms slender, hairy. *Achenes* ±cylindrical, ribbed, *narrowed* above, with or without a beak; *pappus of simple hairs in many rows, usually white,* sometimes brown; marginal achenes sometimes without pappus.

About 200 spp. chiefly in temperate and subtropical Europe and Asia, some in N. and C. Africa and N. America, and 1 each in S. Africa and S. America.

Close to *Hieracium,* but in that genus the achenes are not or scarcely narrowed to the truncate apex and the pappus is brownish and brittle. *C. paludosa,* however, combines features of both genera, resembling *Crepis* in habit and in the distinct though slight narrowing of the achene towards its apex, but having the discoloured brittle pappus of *Hieracium.*

1 Perennials with blackish stocks and entire to coarsely toothed (but not pinnatifid) lvs; in shady and moist places of upland areas northwards from Glamorgan, Shropshire and Leicester, and in Ireland. 2

   Annuals or biennials with tap-roots and usually pinnatifid (sometimes merely toothed) lvs; weeds of arable lands and pastures, roadsides, banks, waste places, etc. 3

2 Lvs strongly toothed; achenes not or hardly narrowed upwards, about 10-ribbed; pappus yellowish-white, brittle. **8. paludosa**

   Lvs entire or shallowly and distantly sinuate-toothed; achenes distinctly narrowed upwards, 20-ribbed; pappus pure white, soft. **4. mollis**

3 Involucral bracts and upper parts of infl.-branches sparsely covered with long stiff almost spinous bristles. **3. setosa**

   Involucral bracts not covered with stiff bristles. 4

4 Heads 2·5–3·5 cm. diam.; stem-lvs usually not sagittate; achenes 13-ribbed. **5. biennis**

   Heads usually not exceeding 2·5 cm. diam.; at least some of the stem-lvs usually sagittate, amplexicaul; achenes 10-ribbed. 5

5 Heads drooping or inclined in bud; marginal achenes short-beaked and clasped by the inner involucral bracts, central achenes long-beaked. **1. foetida**

   Heads erect in bud; achenes all similar. 6

6 Outer involucral bracts ±appressed; styles yellow; heads commonly only 1–1·5 cm. diam. (but to 2 cm. in var. *glandulosa*). **6. capillaris**

   Outer involucral bracts spreading horizontally; styles brownish-green; heads usually 1·8–2·5 cm. diam. 7

7 Achenes long-beaked; fruiting pappus much exceeding the involucre. **2. vesicaria**

   Achenes narrowed above but not truly beaked; fruiting pappus hardly exceeding the involucre. **7. nicaeensis**

Section 1. *Barkhausia* (Moench.) Gaud. Achenes distinctly beaked, many-ribbed.

1. C. foetida L. 'Stinking Hawk's-beard.'

An annual or biennial *foetid* herb with erect main stem 20–60 cm., and usually several spreading-ascending corymbose branches from its base and lower half, all slightly furrowed, hairy. Lvs chiefly in a basal rosette, runcinate-pinnatifid, stalked; middle and upper stem-lvs few, small, lanceolate, cut or toothed below, sessile, semi-amplexicaul; all densely hairy on both sides. *Heads* 1·5–2 cm. diam., *drooping in bud,* solitary, terminal on long incurved bracteolate stalks somewhat thickened upwards. Involucre 10–13 mm., grey-downy and usually with some longer glandular hairs; *inner bracts* linear-lanceolate, *later hardening and enclosing the outer achenes*, finally spreading starwise; outer bracts short, narrow, ± spreading. Florets pale yellow, exceeding the bracts, the outermost purplish beneath. Styles yellow. Achenes 3–4 mm. (6–15 mm. including the beak); *marginal short-beaked, central long-beaked*; pappus 6–8 mm., white. Fl. 6–8. $2n=10$. Th.–H. (biennial).

Native. Waysides and waste places, especially on chalk and shingle. 12. A rare plant of S.E. England from Sussex and Kent to Cambridge and Norfolk; probably introduced in Worcester, Hereford, etc. W. and S. Europe from Spain and Portugal to Belgium, Germany, and Jugoslavia. The closely related *C. rhoeadifolia* M. Bieb. occurs in E. Europe and the Near East.

*2. C. vesicaria L. ssp. **taraxacifolia** (Thuill.) Thell.   'Beaked Hawk's-beard.'

C. taraxacifolia Thuill.

A biennial herb, sometimes annual or perennial, with 1 or more erect ± downy stems 15–80 cm., hispid and purplish below, branched above or from near the base. Basal lvs stalked, usually oblanceolate, blunt or acute, lyrate- or runcinate-pinnatifid with the lobes very variable in length and width, sometimes merely toothed; stem-lvs short-stalked or sessile, pinnatifid to ± entire, the middle ones amplexicaul; all finely pubescent on both sides. *Heads* 1·5–2·5 cm. diam., *erect in bud*, on slender, ± straight, non-bracteolate stalks in corymbs terminating the main stems and branches. Involucre 8–12 mm., cylindric-campanulate, tomentose and often glandular; *inner bracts* lanceolate blunt, hardening in fr. but *not enclosing the marginal achenes*; *outer bracts* narrow, *spreading*, scarious-margined. Florets yellow, the outermost brown-striped beneath, exceeding the involucre. Styles greenish-brown. Achenes 4–5 mm., pale brown, with 10 narrow ± rough ribs, *all gradually narrowed into a beak about as long as the achene*; pappus 4–6 mm., white, soft, exceeding the involucre. Fl. 5–7. $2n=8*$. Th.–H. (biennial).

Introduced. Waysides, walls, railway banks, waste places, especially on calcareous soils, to 500 ft. in England. 64, H 19, S. First recorded in 1713 and now locally common in England and Wales northwards to Yorks and Lancs and in C. Ireland; rapidly spreading. W. and S. Europe from Spain and Portugal to N.W. Germany and the Balkans; N.W. Africa. Part of a large complex of forms placed under *C. vesicaria* L.

*3. C. setosa Haller f.   'Bristly Hawk's-beard.'

An annual or biennial herb with an erect hispid stem 20–70 cm., branched from near the base or throughout, the branches long, spreading-ascending. Basal lvs stalked, oblanceolate, blunt or acute, runcinate-pinnatifid with lobes very variable in size, or merely toothed; stem-lvs lanceolate-acuminate, entire or toothed or cut near the base, amplexicaul with pointed auricles; all hispid. *Heads* 1–1·4 cm. diam., *erect in bud*, on slender erect deeply grooved hispid or prickly stalks in few-headed corymbs terminating the main stem and branches. *Involucre* 8–10 mm., cylindric-campanulate, *contracted above in fr.*, ± *prickly with yellow non-glandular bristles; inner bracts* lanceolate-acuminate, later hardening but *not enclosing the outer achenes*; outer bracts linear, keeled, up to half as long as the inner, spreading. Florets pale yellow, reddish beneath, exceeding the involucre. Styles blackish-green. *Achenes* 3·5–5 mm., pale brown, with 10 prominent ribs rough below, *all narrowed into a slender beak* up to half as long as the achene; *pappus* 2·5–5 mm., white, soft, *slightly exceeding the involucre*. Fl. 7–9. $2n = 8$. Th.–H. (biennial).

Introduced. A casual in arable fields, especially of clover. S. and S.E. Europe from S. France to the Crimea; W. Asia. Introduced northwards to Denmark and C. Russia.

Section 2. *Crepis*. Achenes not beaked; pappus white, silky.

4. C. mollis (Jacq.) Aschers.   'Soft Hawk's-beard.'
C. succisifolia (All.) Tausch; *C. hieracioides* Waldst. & Kit., non Lam.

A perennial herb with a short erect blackish premorse stock and an erect glabrous or hairy stem 30–60 cm., branched above. Basal lvs 5–10 cm., oblanceolate narrowed into a long winged stalk, blunt; middle and upper lvs oblong, sessile, *semi-amplexicaul with rounded auricles*; all sinuate-toothed or entire, glabrous or hairy. Heads 2–3 cm. diam., in a terminal corymb, erect in bud, on rather long, slightly incurved, slender stalks. *Involucre* 8–13 mm., subcylindric, ± sparsely *glandular-hairy*; inner bracts linear-lanceolate acute, outer very short, appressed. Florets yellow, almost twice as long as the involucre. Achenes 4·5 mm., yellow, with 20 smooth ribs; pappus 6 mm., pure white, soft. Fl. 7–8. Visited by bees. $2n = 12$. H.

Native. Stream-sides and woodlands in mountainous districts, very local; reaches 1200 ft. in N. England. 22. N. Wales and N. Britain from Yorks and N. Lancs to Banff. C. Europe from the Pyrenees, N. Italy and S. Russia northwards to Germany, Poland and C. Russia.

5. C. biennis L.   'Rough Hawk's-beard.'

A biennial herb with an erect stout grooved ± hispid stem 30–120 cm., often purplish below, corymbosely branched above. Basal lvs 15–30 cm., irregularly lyrate-pinnatifid, runcinate, stalked; *stem-lvs* mostly similar but smaller and sessile, *semi-amplexicaul but not or scarcely sagittate* at the base; all ± rough with scattered hairs. Heads 2–3·5 cm. diam., at first crowded, then with elongation of the stout hairy stalks forming corymbs terminating the main stem and branches. Involucre 10–13 mm., cylindric-campanulate, pubescent and often glandular; *inner bracts* linear-lanceolate, blunt, *downy within; outer* narrow, *spreading*, without scarious margins. Florets golden yellow, not reddish beneath, exceeding the involucre. Styles yellow. *Achenes* 7–12 mm., narrowed above but not beaked, reddish-brown, usually with 13 ±*smooth ribs* and numerous fine transverse wrinkles; *pappus equalling or slightly exceeding the involucre*, white, soft. Fl. 6–7. Visited freely, chiefly by bees and hover-flies, but probably often apomictic. $2n = 40$. H. (biennial).

Probably native. Pastures, waysides, clover and lucerne fields, waste places. Locally frequent in lowland Great Britain northwards to Aberdeen, often on calcareous soils; Orkney. Introduced in Ireland. 59, H23, S. C. Europe from C. Spain, N. Italy, C. Balkans and C. Russia northwards to Denmark, S. Scandinavia and Esthonia.

6. C. capillaris (L.) Wallr.   'Smooth Hawk's-beard.'

C. virens L.

A usually annual herb with one or more erect or ascending stems 20–90 cm., glabrous or ± hairy below, branched above, or with several spreading-ascending branches from the base upwards. Basal and lower stem-lvs 5–25 cm., very variable, usually oblanceolate or lanceolate, narrowed into a stalk-like base, lyrate- or runcinate-pinnatifid with distant often toothed lobes, or merely toothed; middle and upper *stem-lvs* lanceolate, acute, sessile, amplexicaul with a *sagittate base*; all glabrous or somewhat hairy on one or both sides. Heads 1–1·3(–2·5 in var. *glandulosa*) cm. diam., erect in bud, on slender glabrous or hairy stalks in lax terminal corymbs. Involucre 5–8 mm., cylindrical, contracted above in fr., usually shortly downy and often with black glandular bristles; *inner bracts* linear-lanceolate, acute, *glabrous within; outer bracts* ⅓ as long, *appressed.* Receptacle with shortly and sparsely ciliate pit-borders. *Florets* exceeding the involucre, bright yellow, the outermost often *reddish beneath.* Styles yellow. *Achenes* 1·5–2·5 mm., pale brown, with 10 ± *smooth ribs*; pappus equalling or slightly exceeding the involucre, snow-white, soft. Fl. 6–9. Freely visited by flies and bees. $2n=6$. Th.–H. (biennial).

Var. *glandulosa* Druce is a taller and more robust plant, usually hairy on the lower part of the stem. It has larger heads (about 20 mm. diam.), and the involucre is densely covered with blackish or blackish-green mostly glandular hairs, lacking the greyish tomentum characteristic of the type.

Native. Grassland, heaths, walls, waste places, etc., to 1460 ft. 112, H40, S. Common throughout the British Is. Var. *glandulosa* is also frequent and widespread. Most of Europe northwards to Denmark and S. Sweden; Canary Is. Introduced in N. America.

*7. C. nicaeensis Balb.   'French Hawk's-beard.'

A usually biennial herb with an erect strongly ribbed stem, hispid-pubescent at least below and often reddish below, branched above. Basal and lower stem-lvs oblanceolate, blunt, short-stalked, runcinate-pinnatifid; middle and upper *lvs* lanceolate, ± deeply toothed below, sessile, amplexicaul with a *sagittate base*; uppermost narrow, entire; all roughly hairy on both sides. Heads 2·5 cm. diam., erect in bud, on slender ± glandular-pubescent stalks in a terminal corymb. Involucre 8–10 mm., campanulate, ± glandular-pubescent and grey-tomentose; *inner bracts* lanceolate, blunt at the ciliate tip, *glabrous within*, strongly keeled in fr.; *outer bracts* about half as long, ± *spreading. Receptacle* with *elevated membranous fimbriate-ciliate pit-borders.* Florets yellow often red-tipped, exceeding the involucre. Styles greenish-brown. Achenes 3·5–4·5 mm., yellowish-brown, much narrowed above, with 10 smooth or ± rough ribs; pappus 4–5 mm., white, hardly exceeding the involucre. Fl. 6–7. $2n=8$. H. (biennial).

Introduced. Arable fields and leys. A casual introduced as a seed-impurity. S. Europe from S. France to N. Balkans; Caucasus. Introduced in C. Europe and N. America.

Section 3. *Catonia* Rchb. Perennial herbs; achenes not beaked, 10-ribbed; pappus of brownish-white stiff brittle hairs.

8. C. paludosa (L.) Moench   'Marsh Hawk's-beard.'

Hieracium paludosum L.; *Aracium paludosum* (L.) Monnier.

A perennial herb with an oblique blackish premorse stock and an erect striate glabrous stem 30–90 cm., often reddish below, branched only above. Basal and lower stem-lvs obovate-lanceolate narrowed into a short winged stalk; *middle lvs panduriform-acuminate, sessile, amplexicaul*; upper lvs ovate- to lanceolate-acuminate, amplexicaul with long-pointed auricles; all thin, sinuate- or runcinate-toothed, glabrous. Heads 15–25 mm. diam., few, on ± straight slender stalks in a terminal corymb. *Involucre* 8–12 mm., campanulate, *woolly and with many black glandular hairs*; inner bracts linear-lanceolate acute, blackish-green; outer bracts hardly half as long, ± appressed. Florets yellow. Styles greenish-black. *Achenes* 4–5 mm., brownish, cylindrical, *hardly narrowed at the ends*, 10-ribbed, smooth; pappus about as long as the achene, of *stiff brittle brownish hairs*. Fl. 7–9. Visited by bees and flies. $2n=12$. Hel.

Native. Stream-sides, wet copses, wet meadows and fens, to 3000 ft. in Scotland. 65, H37. Locally common in N. Britain to Caithness and extending southwards to Glamorgan and Worcester; throughout Ireland. Europe from Spain, C. France, N. Italy and N. Balkans northwards to 70° 30′ N. in Norway; Russia; W. Siberia.

C. rubra L. (Greece), a branching annual 15–45 cm., with glabrous runcinate-toothed lvs and red or whitish heads with a hispid involucre, is grown in gardens.

57. TARAXACUM Weber

Perennial herbs with simple or branched tap-roots upon which adventitious buds arise readily, and spirally arranged entire to pinnatifid commonly runcinate *lvs confined to a basal rosette. Scapes* 1 or more, *simple.* Heads solitary terminal; involucre cylindrical, the inner bracts erect, equal, in 2 rows, the outer shorter, erect, spreading or reflexed; *receptacle* flat, *naked*, pitted. Florets yellow; anther-lobes acuminate at the base but not tailed; style-arms slender, hairy. *Achenes* fusiform-cylindrical, ribbed, the ribs usually *muricate above*; ± *long-beaked; pappus* of many rows *of simple rough white hairs.* Heads freely visited by a great variety of insects. Many hundreds of distinct forms, mostly apomictic, have been described from the northern hemisphere, and there are 2 spp. in Australia, 1 of which also occurs in S. America.

The genus presents great difficulties to the taxonomist. There are a few regularly amphimictic species, some forms which are occasionally amphimictic and a vast number which seem exclusively apomictic, often with defective or quite abortive pollen. The apomictic types are polyploids which have presumably arisen from primary interspecific hybrids.

1   Inner bracts with an appendage on the outer side just beneath the apex,
      so that they appear bifid at the tip; achene abruptly contracted into a
      cylindrical cusp between its apex and the beak.   **4. laevigatum**
    Inner bracts not appendaged beneath the tip; achenes abruptly contracted
      into a cylindrical cusp or gradually narrowed into a conical cusp.   **2**

2   Outer bracts mostly narrowly lanceolate or linear, usually more than
      3 times as long as broad, and usually ± spreading or reflexed; achenes

3·5–4 mm. including the short conical cusp; beak 2·5–4 times as long
as the achene.   **1. officinale**
Outer bracts mostly broadly or narrowly ovate, usually less than 3 times
as long as broad, and commonly appressed; achene 4–6 mm. including
the cusp; beak 1·5–2·5(–3) times as long as the achene.   *3*

3 Outer bracts broadly ovate, appressed; achene contracted abruptly into
a long slender cusp; lvs narrow in outline.   **2. palustre**
Outer bracts narrowly ovate, appressed or spreading; achenes narrowed
gradually into a short conical cusp; lvs broad in outline, commonly not
deeply lobed, ± hispid above, midrib reddish.   **3. spectabile**

1. T. officinale Weber, sensu lato   Common Dandelion.

(Section *Vulgaria* Dahlst.)

This aggregate includes those forms in which the outer *bracts* are mostly
narrowly lanceolate or linear, more than three times as long as wide and
almost half as long as the inner, usually ± *spreading or reflexed*; inner bracts
not appendaged on the outer side just beneath the tip. Achene 3·5–4 mm.
(including cusp), brownish, greenish or grey, never purplish-red, moderately
to slightly muricate, with a short conical cusp and a beak 2·5–4 times as long
as the achene. The plants are usually robust with large variously toothed or
cut but commonly runcinate-pinnatifid lvs and heads 3–6 cm. diam. Fl. 3–10.
All British forms are apomictic. 2*n*=24 (triploid). Hr. Over 100 forms have
been recognized in the British Is.

Native. Pastures, meadows, lawns, waysides, waste places, etc., to 3850 ft.
in Scotland. Abundant throughout the British Is. Northern hemisphere.

2. T. palustre (Lyons) DC., sensu lato 'Narrow-leaved Marsh Dandelion.'

(Section *Palustria* Dahlst.)

Includes forms in which the outer *bracts* are *broadly ovate and firmly appressed*,
less than three times as long as wide; inner bracts *not appendaged* beneath the
tip. Achene 4–6 mm. (including cusp), variously coloured but often olive and
never purplish-red, ± strongly muricate, contracted abruptly into a *long
slender cusp* and a beak 1·5–2·5(–3) times as long as the achene. Plants fairly
robust with *narrowly oblanceolate lvs*, ± entire or sinuate-toothed in *T.
palustre* (Lyons) DC., very deeply and distantly pinnatifid in *T. balticum*
Dahlst. Heads 3–4 cm. diam. Fl. 4–6. All British forms are apomictic.
2*n*=32 (tetraploid). Hr.

Native. Marshes, fens and stream-sides, to 3000 ft. in Scotland. Locally
abundant and probably throughout the British Is. in its various forms.
Europe.

3. T. spectabile Dahlst., sensu lato   'Broad-leaved Marsh Dandelion.'

(Section *Spectabilia* Dahlst.)

Forms resembling *T. palustre* s.l. but with the outer *bracts narrowly ovate,
loosely appressed, spreading or somewhat reflexed* and achene narrowed
gradually upwards into a *short conical cusp. Lvs broadly oblanceolate to
obovate-lanceolate*, variously cut or lobed but commonly merely toothed,
± hispid above and commonly with the midrib reddish. Heads 3–5 cm. diam.
Fl. 4–8. All British forms are apomictic. 2*n*=32 (tetraploid), rarely 40. Hr.

Native. Marshes, stream-sides, moist or boggy mountain pastures and cliff ledges; to 3500 ft. in Scotland. Probably widely distributed in Great Britain, especially (but by no means exclusively) at high altitudes in the west and north. Ireland. Includes forms connecting *T. palustre* with *T. officinale*.

4. T. laevigatum (Willd.) DC., sensu lato   'Lesser Dandelion.'
(Sections *Erythrosperma* Dahlst. and *Obliqua* Dahlst.)

Forms differing from the foregoing in having a *short appendage on the outer side of the inner bracts, just beneath the apex,* so that the tip of the bract appears bifid; outer bracts narrowly ovate, appressed, spreading or somewhat recurved. *Achene* 3–4·5 mm. (including cusp), brownish, grey, or most commonly purplish-red, ± strongly muricate above, *narrowed abruptly into a slender cylindrical cusp* and a beak $\frac{1}{3}$–$2\frac{1}{2}$ times as long as the achene. Often small plants with narrow deeply cut lvs and pale yellow heads. Fl. 4–6. All British forms are apomictic. $2n = 24$ (triploid). Hr.

Native. Dry pastures on sandy or calcareous soils, heaths, waste ground, walls, etc. Locally abundant and probably throughout lowland Great Britain. Ireland. *T. lacistophyllum* Dahlst., with deeply cut lvs and purplish-red achenes, seems the most widespread form. Europe.

Of the several alien members of the Cichorieae occasionally found as casuals the most frequent are:

*Scolymus hispanicus** L. A biennial to perennial thistle-like herb with an erect branching stem 20–80 cm., pubescent, interruptedly spinous-winged. Lvs oblong-lanceolate in outline, white-veined, sinuate-pinnatifid with spinous lobes and teeth; stem-lvs decurrent. Heads axillary, subsessile, each surrounded by 3 long ascending lf-like spiny bracts; involucral bracts linear-lanceolate, acuminate. Florets yellow. Achene enclosed by the laterally winged receptacular scale; pappus of 2–3 deciduous bristles. Mediterranean region, Madeira, Canary Is. Cultivated for its salsify-like tap-root.

*S. maculatus** L. An annual herb with strongly spiny white-bordered lvs and no pappus. Mediterranean region.

*Tolpis barbata** (L.) Gaertn. An annual herb with an erect ± pubescent branching stem 20–50 cm. Lower lvs oblong-lanceolate short-stalked; upper lvs narrower, sessile; all toothed, ± pubescent. Heads small, in a cyme-like panicle; outer involucral bracts setaceous, spreading, exceeding the inner bracts and the florets. Florets all yellow or those in the centre reddish-brown. Marginal achenes with numerous pappus-bristles, those of the centre with 4–5 long hairs. Mediterranean region.

MONOCOTYLEDONES

121. ALISMATACEAE

Annual or perennial herbs, erect or sometimes with floating lvs, living in water or wet places. Lvs petioled, blades linear-lanceolate to ± rounded. Fls actinomorphic, usually hermaphrodite, often whorled, arranged in panicles or racemes. Perianth in 2 whorls, heterochlamydeous, outer whorl persistent. Stamens (3–)6 or more, free; anthers 2-celled. Ovary superior. Carpels free or rarely connate at base. Ovules 1 or several, basal or in the inner angle of the carpel. Fr. usually a head or whorl of achenes, rarely dehiscing at base. Seeds small, endosperm 0; embryo horseshoe-shaped.

Thirteen genera and about 80 spp. in temperate and tropical regions, mainly in the northern hemisphere.

An interesting family, showing some striking similarities with the Ranunculaceae and perhaps derived from it.

1   Fls solitary or umbellate, or in not more than 2 whorls; lvs not cordate
    at base.   2
    Fls in more than 2 whorls; if fewer, then fls unisexual or lvs cordate at
    base.   3

2   Stems and lvs not floating; lvs narrowly lanceolate, acute; fls not more
    than 1 cm. diam.   1. BALDELLIA
    Stems and lvs floating, lvs ovate, obtuse; fls 2·5–3 cm. diam.
                                                                  2. LURONIUM

3   Fls hermaphrodite, up to 1 cm. diam.; stamens 6; lvs never sagittate;
    plant not producing tubers (turions) in autumn.   4
    Fls unisexual, more than 1 cm. diam.; stamens numerous; lvs often
    sagittate; plant producing turions in autumn.   5. SAGITTARIA

4   Lvs narrowed or rounded at base; ripe carpels numerous, c. 1 mm., not
    beaked.   3. ALISMA
    Lvs cordate at base; ripe carpels 6–10, c. 13 mm., ending in a long beak
    and spreading stellately.   4. DAMASONIUM

1. BALDELLIA Parl.

A glabrous perennial scapigerous herb, sometimes stoloniferous. Lvs in a basal rosette. *Fls* long-pedicelled, *umbellate* or (rarely) in 2 simple whorls, sometimes solitary. Stamens 6. *Carpels numerous*, free, *forming a crowded head. Ripe carpels ovoid.*

One sp. in Europe and N. Africa.

1. B. ranunculoides (L.) Parl.   'Lesser Water-Plantain.'

Alisma ranunculoides L.; *Echinodorus ranunculoides* (L.) Engelm.

A variable plant, usually between 5 and 20 cm. Lvs long-petioled, blade 2–4 cm., linear to linear-lanceolate, acute, narrowed gradually into petiole. Stems usually erect or spreading, sometimes (var. *repens* Davies) decumbent and then rooting and producing tufts of lvs and (often solitary) fls at nodes. Fls up to 15 mm. diam., pale purplish, opening in succession. Pedicels unequal, up to c. 10 cm. Bracts small, scarious. Heads of achenes ± spherical.

Achenes c. 2 mm., ovoid, curved, strongly 3-ribbed on the back with a double rib at the ventral suture. Fl. 5–8. $2n = 16^*, 30^*$. Hel.

Native. In damp places beside streams, ponds and lakes and in fen ditches, locally common. 88, H40, S. Widely distributed throughout the British Is., but becoming rarer northwards and absent from the extreme north of Scotland. Europe to 60° N. in W. Norway; N. Africa.

2. LURONIUM Raf.

A slender herb. *Stems floating* and rooting at nodes. *Fls* 1(–5) *in the axils of the lvs*, long-pedicelled. Stamens 6. *Carpels* 10–12, crowded, free. *Ripe carpels oblong-ovoid.*

One sp. in western Europe, N. Germany, Poland and Bulgaria.

1. L. natans (L.) Raf.   'Floating Water-Plantain.'

Alisma natans L.; *Elisma natans* (L.) Buchen.

Stems 50 cm. or more, slender, floating but rooted at base. Lower lvs submerged, reduced to linear, flattened, translucent petioles up to c. 10 cm. × 2 mm.; other lvs floating, long-petioled, blade 1–2·5 cm., ovate or elliptic, rounded and obtuse at apex. Fls 12–15 mm. diam., white with a yellow spot in middle. Head of achenes hemispherical. Fl. 7–8. $2n = 38^*$. Hyd.

Native. In lakes, tarns and canals with acid water, very local but apparently increasing. 18. Glamorgan, Cardigan to Denbigh and Anglesey; Shropshire, Westmorland, Cumberland; Ayr; also here and there in canals in several other English counties, where it is of recent introduction. W. Europe to Denmark, S. Norway, Poland and Bulgaria.

3. ALISMA L.

Glabrous perennial scapigerous herbs with acrid juice. *Infl. much-branched*, branches whorled. *Carpels numerous, in one whorl*; style lateral. *Ripe carpels strongly compressed.*

Six spp. in north temperate regions and Australia.

1 Style spirally coiled; fr. broadest near apex; lvs usually linear.
                                                   3. gramineum
    Style ± straight; fr. ovate; lvs lanceolate to ovate.   *2*

2 Lvs ovate, rounded to subcordate at base; style arising below the middle
    of the fr.; anthers c. twice as long as broad.   **1. plantago-aquatica**
  Lvs lanceolate, gradually narrowed at base; style arising above the middle
    of the fr.; anthers about as long as broad.   **2. lanceolatum**

1. A. plantago-aquatica L.   Water-Plantain.

An erect glabrous perennial 20–100 cm. Stem stout, usually unbranched in lower half. *Lvs* long-petioled, blade 8–20 cm., *ovate*, subacute to acuminate, *rounded or subcordate at base*; the first lvs of land and water forms reduced to a ½-cylindrical petiole with small narrow blade; floating lvs sometimes occur in the water form. Infl. branches ± straight, usually ascending. Fls up to c. 1 cm., usually pale lilac, open from 1 to 7 p.m. *Outer per. segs oblong, inner rounded.* Stamens (in fl.) longer than carpels (excluding style); anthers about twice as long as broad. Fr. ovate; *style ± straight, long, arising below* *the middle of the fr.* Carpels c. 20, in a ± flat head. Fl. 6–8. $2n=12, 14^*$, 16, 24, 28. Hyd. or Hel.

Native. On muddy substrata beside slow-flowing rivers, ponds, ditches and canals, in damp ground or shallow water. 103, H40, S. Throughout nearly the whole of the British Is., rarer in the north and absent from the extreme north of Scotland, Orkney and Shetland. Distribution of the genus.

2. A. lanceolatum With.

Similar to *A. plantago-aquatica* in general appearance. *Lvs lanceolate, narrowed gradually into petiole.* Fls usually pink, open from 9 a.m. to 2 p.m. *Outer per. segs ovate, inner pointed.* Stamens (in fl.) somewhat longer than carpels (excluding style); anthers about as long as broad. Fr. ovate; *style ± straight, short, arising near the top of the fr.* $2n=26^*$, 28. Hyd. or Hel.

Native. In similar situations to the foregoing. Distribution imperfectly known, but apparently a much less frequent plant than *A. plantago-aquatica.* Recorded from S. and E. England north to Yorks and Oxford; Pembroke; Galway. Europe, N. Africa, W. Asia.

3. A. gramineum Lejeune ssp. gramineum

A. Plantago-aquatica L. var. *graminifolium* Wahl.

A perennial herb (5–)15–30(–60) cm. Stem stout usually ± curved, branched below the middle. *Lvs linear, ribbon-like,* the blade not or scarcely distinguishable from the petiole; later lvs sometimes with a short (15–40 mm.) linear-lanceolate or narrowly oblong blade. Floating lvs 0. Infl. branches ± recurved particularly in fr. Fls c. 6 mm. diam., open between 6 and 7.15 a.m.; petals caducous. Outer per. segs ovate, inner rounded. Stamens as long as carpels (excluding style); anthers broader than long. Fr. broadest near apex; *style coiled.* Fl. 6–8. $2n=14$. Hyd. or Hel.

?Native. Locally abundant in S. Lincs and in a shallow artificial pond near Droitwich where it was first recorded in 1920. 2. Perhaps elsewhere. C. Europe, Denmark and the Baltic States, Netherlands, France; ?elsewhere.

4. DAMASONIUM Mill.

Glabrous scapigerous annual herbs. Infl. of several, usually simple whorls. Inner per. segs entire. Stamens 6. *Carpels 6–10, in one whorl, connate at base;* style apical. Ovules 2–several in each carpel. *Ripe carpels* (1–)2–several-seeded, indehiscent or tardily dehiscent at base, *spreading stellately.*

Three spp., in Europe, N. Africa, W. Asia and Australia.

1. D. alisma Mill.   Thrumwort.

D. stellatum Thuill.; *Actinocarpus Damasonium* R.Br.

An erect herb 5–30(–60) cm. Lvs long-petioled, floating or sometimes submerged, blade 3–5 cm., ovate to oblong, obtuse, *cordate at base.* Fls c. 6 mm. diam., white. Inner per. segs caducous. *Ripe carpels c.* 13 *mm., tapering into a long beak,* usually 2-seeded. Fl. 6–8. Hyd.

Native. In gravelly ditches and ponds, very local and apparently decreasing. 15. Hants, Sussex, E. Kent, Surrey, Berks, Bucks, Middlesex, Hertford, S. Essex, Worcester, Shropshire, Leicester, S.E. Yorks. W. and S. Europe from France to Italy; S. Russia; N. Africa.

5. Sagittaria L.

Scapigerous herbs, usually perennial. Infl. of several simple or rarely slightly branched whorls. *Fls unisexual. Stamens numerous. Carpels numerous, spirally arranged* on a large receptacle, strongly *compressed.*

About 17 spp. throughout the temperate and tropical regions.

Lvs typically sagittate; inner per. segs with a dark violet patch at base.
1. sagittifolia
Lvs never sagittate; inner per. segs entirely white, or slightly yellowish at base.
2. rigida

1. S. sagittifolia L. Arrow-head.

An erect herb 30–90 cm., perennating by means of turions borne at the ends of slender runners. Turions c. 3 cm., ovoid or subcylindrical, bright blue with yellow spots. Submerged lvs linear, translucent; floating lvs lanceolate to ovate; *aerial lvs long-petioled, blade* 5–20 cm., *sagittate,* acute or obtuse, lateral lobes about as long as main portion of blade. Fls monoecious, 3–5 in a whorl, c. 2 cm. diam., female in lower part of infl., rather smaller than male. *Scape longer than lvs.* Pedicels of male fls c. 20 mm., of female fls c. 5 mm. *Inner per. segs white with a dark violet patch at base.* Mature head of carpels c. 15 mm. diam., hemispherical. Fl. 7–8. $2n=22$*. Hyd.

Native. In shallow water in ponds, canals and slow-flowing rivers on muddy substrata. 62, H23. Scattered throughout England, rather local and rarer in the north; Glamorgan, Pembroke, Denbigh, Flint and Anglesey; scattered throughout C. and N.E. Ireland, local; not native in Scotland. Europe, Asia, N. America.

*2. S. rigida Pursh

S. heterophylla Pursh, non Schreb.

May be distinguished from *S. sagittifolia* as follows: *Aerial lvs ovate or elliptic* (rarely sagittate in America). Plant monoecious or dioecious. *Scape shorter than lvs.* Pedicels of male fls c. 12 mm., female fls subsessile. *Inner per. segs entirely white or pale yellowish at base.*

Our plant has been described as *S. heterophylla* var. *iscana* Hiern.

Introduced. Naturalized in the river Exe, in and near Exeter; first observed in 1898. Native of N. America from Quebec to Tennessee and west to Minnesota.

122. BUTOMACEAE

Rhizomatous herbs, often with latex, living in water or wet places. Lvs various Fls solitary or in umbels. Perianth in 2 whorls. Stamens 6–9; filaments flattened; anthers basifixed, 2-celled, opening by lateral slits. Ovary superior. Carpels 6–9, free or cohering only at base. Ovules numerous, scattered over the inner surface of the carpel wall. Seeds small, endosperm 0; embryo straight.

Five genera in temperate and tropical regions.

1. Butomus L.

A scapigerous herb devoid of latex. Lvs linear, erect. Fls hermaphrodite, umbellate. Per. segs persistent. Stamens 6–9. Carpels 6–9, cohering at base. Embryo straight.

One sp. in Europe and temperate Asia.

1. B. umbellatus L.   Flowering Rush.

A glabrous rhizomatous perennial up to 150 cm. Lvs about as long as stems, in a basal rosette, triquetrous, twisted, acuminate, sheating at base. Stems terete. Umbel with an involucre of acuminate bracts. Pedicels up to c. 10 cm., unequal. Fls 2·5–3 cm. diam., opening in succession. Per. segs pink with darker veins, the outer somewhat smaller and narrower than the inner. Ripe carpels obovoid, crowned by the persistent style. Fl. 7–9. 2n = 26, 39. Hel. or Hyd.

Native. In ditches, ponds and canals, and at margins of rivers. England and Ireland, rather local; rare in Wales and doubtfully native in Scotland. 68, H20. Europe, temperate Asia; naturalized in N. America.

123. HYDROCHARITACEAE

Aquatic herbs, wholly or partially submerged, inhabiting fresh waters or the sea. Fls actinomorphic, usually dioecious, arranged in a spathe composed of one bract or two opposite free or connate bracts. Spathes sessile or peduncled, male fls usually numerous, female solitary. Per. segs usually in 2 series, 3 (rarely 2) in each series. Stamens (1–)3–numerous. Rudimentary ovary present in male fls. Staminodes sometimes present in female fls. Ovary inferior, unilocular, with 3–6 parietal placentae; ovules numerous.

About 16 genera and 70 spp., in the warmer parts of the world, a few in temperate regions.

1	Lvs petioled, orbicular-reniform, floating.	1. Hydrocharis
	Lvs sessile, submerged or partly so.	*2*
2	Lvs radical, much more than 2·5 cm.	*3*
	Lvs cauline, up to 2·5 cm.	3. Elodea
3	Lvs spinous-serrate, rigid, tapering from base.	2. Stratiotes
	Lvs denticulate at top, flaccid, ribbon-shaped.	4. Vallisneria

1. Hydrocharis L.

A floating herb. Lvs orbicular-reniform, petioled. Fls white, unisexual, male 2–3 in a spathe, female solitary, *both distinctly pedicelled. Sepals herbaceous, narrower and smaller than the petals.* Stamens 12, 3–6 outer usually sterile. Female fls with 6 staminoides; ovary 6-celled with 6 bifid styles. Fr. fleshy, indehiscent.

One sp. in Europe and Asia.

1. H. morsus-ranae L.   Frog-bit.

A floating stoloniferous herb with lvs in groups at the nodes. Roots in bunches at the nodes. Turions enclosed by 2 scale-lvs produced at ends of stolons in autumn. Lvs floating on the surface, blade c. 3 cm. diam.; stipules large, scarious. Fls c. 2 cm. diam., erect, aerial; petals broadly obovate, crumpled, with a yellow spot near base. Fr. rarely if ever produced in this country. Fl. 7–8. $2n = 28^*$. Hyd.

Native. In ponds and ditches, usually in calcareous districts, locally common. 52, H13, S. Scattered throughout England from Devon and Kent to S. Lancashire and Durham; Wales: Carmarthen, Glamorgan, Monmouth, Caernarvon and Flint; absent from Scotland; Ireland, mainly in the N. Centre extending to Clare, Dublin and Down. Distribution of the genus; generally local and probably diminishing.

A land-form occurs occasionally in dry seasons. The plant is probably monoecious but with male and female fls at different nodes and often separated by a long slender stem which usually breaks when removed from the water.

2. STRATIOTES L.

A submerged herb rising to the surface at flowering time. *Lvs sessile, tapering from the base.* Fls white, dioecious; *male* several in a spathe, *bracteolate, pedicelled; female* solitary, *sessile.* Stamens 12 fertile surrounded by numerous sterile. Fls otherwise much as in *Hydrocharis.*

One sp. in Europe and N.W. Asia.

1. S. aloides L.　　　　　　　　　　　　　　　　　　Water Soldier.

A stoloniferous herb with lvs in large rosettes, reproducing mainly by means of offsets. Lvs 15–50 cm., spinous-serrate, rigid, brittle, many-nerved, resembling those of an aloe. Fls 3–4 cm. diam., erect, aerial, petals suborbicular, thick. Ovary becoming deflexed after fl.; fr. never produced in this country. Fl. 6–8. $2n = 24$. Hyd.

Native. In Broads, ponds and ditches in calcareous districts, very local and probably diminishing. 21. Scattered throughout England, mainly in the east; Wales: Flint; Scotland: Angus (?introduced); Ireland, introduced in a few localities. Distribution of the genus. In the north the plants are entirely female, in the south predominantly or entirely male; both sexes occur in an intermediate area.

Only the female plant occurs normally in Britain, though plants with hermaphrodite fls have been recorded. The floating and submerging of the plant is said to be due to changes in the amount of calcium carbonate on the lvs.

3. ELODEA Michx.

Submerged herbs with sessile whorled lvs. Fls dioecious, solitary in tubular spathes which are sessile in the axils of the lvs. Petals narrower than sepals. *Stamens 3–9, filaments short or 0.* Female fls with a long slender axis. *Styles 3, lobed or notched,* free to base. Water-pollinated, the female fls reaching the surface by the elongation of the axis, the male usually breaking off and floating.

About 10 spp. in N. and S. America.

Plant dark green; lvs 3 (rarely 4) in a whorl, mostly 3–4 mm. wide, acute or
　　subobtuse; scales green, entire. (Common.)　　　　　　**1. canadensis**
Plant pale green; lvs 3–5 (often 4) in a whorl, mostly 1–1·5 mm. wide,
　　acuminate; scales brown, fringed (N. Lancs., W. Galway).　　　**2. nuttallii**

*1. E. canadensis Michx.   Canadian Pondweed.

Anacharis canadensis Planch.

A *dark green* pellucid submerged branched herb perennating by means of winter buds. Stems up to 3 m., usually much less, very brittle. *Lvs c.* 10 × 3–4 *mm.*; 3 (*rarely* 4) *in a whorl*, lower opposite, *oblong-lanceolate, acute or subobtuse*, translucent, *serrulate*. *Scales* on upper side of lf towards base minute, *entire, green*. Fls 5 mm. diam., floating, greenish-purple, axis filiform, up to 30 cm. Spathe 1–2 cm., bifid at tip. Male plant very rare in Britain. Fl. 5–10. 2*n*=24; 48. Hyd.

Introduced. Naturalized in slow-flowing fresh waters throughout most of the British Is. 87, H37, S. First introduced about 1836 in Ireland and 1842 in Britain, spread rapidly and attained great abundance so as to block many waterways, then diminished; at present widespread but seldom abundant. N. America; introduced in much of Europe to c. 66° N.

2. E. nuttallii (Planch.) St John

Hydrilla verticillata auct.; *H. lithuanica* Dandy, p.p.

A slender branched *pale green* herb perennating by means of winter buds. *Lvs c.* 10 × 2 *mm.*, 3–5 *in a whorl* or the lower opposite, *linear-lanceolate, acuminate*, translucent, *margins with small projecting teeth*. A pair of small *brownish fringed scales* occur on the upper surface of the lf towards the base. Fls not recorded from Britain. Hyd.

Native. In 1·5–2·6 m. of water on blue-green clayey mud with less than 15 % organic matter. 1, H1. Esthwaite Water, N. Lancashire; W. Galway. N. America.

The two following are locally abundant and may well spread. They are commonly grown in aquaria and have probably been introduced from this source.

*E. callitrichoides (Rich.) Casp. is naturalized in Middlesex. It may be recognized by the light green, narrowly lanceolate, acute or acuminate lvs up to 2·5 cm., mostly in rather distant whorls of 3. Native of S. America. 2*n*=16. Hyd.

An allied plant, ***Egeria densa** Planch. (*Elodea densa* (Planch.) Casp.), is naturalized in a canal near Manchester. Plant robust; lvs c. 2 cm., narrowly oblong and abruptly acute or apiculate, mostly in densely crowded whorls of 4. Fls c. 1 cm. diam., white. Native of Argentina. 2*n*=48. Hyd.

4. VALLISNERIA L.

Submerged perennial stoloniferous herbs. Fls unisexual, male many together in a tubular, 2-toothed, shortly peduncled spathe, female solitary in a *spathe borne on a long filiform peduncle. Per. of male fls single, of female double but petals rudimentary.* Stamens 1–3, usually 2. Stigmas 3, bifid. Capsule cylindrical. Water-pollinated, the male fls breaking off and floating to the surface when the spathe opens.

About 3 spp. in the warmer parts of the world.

*1. V. spiralis L.

Lvs radical, ribbon-shaped, obtuse and denticulate at top. Fls pinkish-white. Peduncle of female fls spirally twisted after fl. Fl. 6–10. $2n = 20$. Hyd.

Introduced. Naturalized in W. Gloucester, Worcester, S.W. Yorks, and S. Lancashire, often where water is heated by effluents from mills. Widely distributed in the warmer parts of the world, in Europe reaching as far north as C. France.

124. SCHEUCHZERIACEAE

Perennial marsh herb. Lvs linear, sheathing at base; ligule present. Infl. a few-fld, terminal raceme. Bracts present. Fls hermaphrodite. Per. segs 6, all sepaloid, persistent. Stamens 6, free. Ovary superior; carpels 3–6, shortly united towards the base, divaricate and free or nearly so in fr.; stigmas sessile; ovules 2 or few, basal, erect, anatropous. Fr. dehiscing on the curved, adaxial side.

One genus and 1 sp. in the colder parts of the northern hemisphere.

1. SCHEUCHZERIA L.

The only genus.

1. S. palustris L.

An erect perennial 10–20(–40) cm. Rhizome creeping, clothed with persistent lf-bases. Stems lfy. Lvs alternate, linear, slightly grooved, obtuse, with a conspicuous pore at the tip. Infl. 3–10-fld, very lax, overtopped by the lvs. Fls c. 4 mm. diam., yellowish-green; per. segs lanceolate, acute. Carpels usually 3. Fl. 6–8. $2n = 22$*. Hel.

Native. In very wet *Sphagnum* bogs, usually in pools. Perth, Argyll and Offaly, very rare; formerly in Shropshire and some north English counties. Colder parts of the northern hemisphere.

125. JUNCAGINACEAE

Annual or perennial scapigerous marsh or aquatic herbs. Lvs mostly radical, linear, sheathing at the base. Fls small, in spikes or racemes, hermaphrodite or unisexual, actinomorphic or slightly oblique, 2–3-merous. Bracts 0. Per. segs in 2 series, green or reddish. Stamens 6 or 4, filaments very short. Ovary superior; carpels 6 (sometimes 3 sterile) or 4, free or $\pm$ connate; style short or 0; ovules solitary, basal, anatropous. Fr. dehiscent or not. Fls protogynous, wind pollinated.

Four genera with few spp., widely distributed, particularly in the temperate and cold regions of both hemispheres.

1. TRIGLOCHIN L.

Rhizomatous herbs with fibrous roots and $\pm$ tuberous stems. Lvs erect, linear, $\frac{1}{2}$-cylindrical. Infl. a raceme. Fls 3-merous. Per. segs deciduous. Carpels all fertile or alternate ones sterile. Fr. dehiscing by the carpels separating from the central axis.

About 12 spp. in temperate regions.

Lvs ½-cylindrical, deeply furrowed on upper surface towards base; fr.
  8–10 × 1 mm., clavate, appressed to scape.   **1. palustris**
Lvs ½-cylindrical, not furrowed; fr. 3–4 × 2 mm., oblong-ovoid, not appressed
  to scape.   **2. maritima**

1. T. palustris L.   'Marsh Arrow-grass.'

A slender herb 15–50 cm. Rhizomes long, slender. Lvs ½-cylindrical, deeply
furrowed on upper surface towards the base. Raceme elongating after fl.
Per. segs purple-edged. *Fr. 8–10 × 1 mm., clavate, appressed to scape; carpels
3 sterile and 3 fertile, remaining attached to the triquetrous axis at the top* after
dehiscence. Fl. 6–8. 2n = 24. Hel. or Hr.

Native. In marshes, usually among tall grass. 112, H40, S. Throughout
the British Is. but very local in most southern and midland counties. Europe
to the Arctic, N. Africa, N. Asia, N. America.

2. T. maritima L.   'Sea Arrow-grass.'

A rather stout herb 15–50 cm. Rhizomes short, stout. Lvs subulate or
linear, up to c. 3 mm. wide, ½-cylindrical, not furrowed. Raceme scarcely
elongating after fl.; pedicels 1–2 mm., elongating after fl. *Fr. 3–4 × 2 mm.,
oblong-ovoid, not appressed to scape; carpels 6, all fertile, separating completely
at dehiscence.* Fl. 7–9. 2n = 48. Hel. or Hr.

Native. In salt marsh turf and grassy places on rocky shores. 80, H27, S.
In suitable habitats around the entire coast of the British Is. Europe to the
Arctic; N. Africa, W. and N. Asia, N. America.

126. APONOGETONACEAE

Aquatic herbs. Fls hermaphrodite or rarely unisexual, ebracteate. Per. segs
1–3, sometimes 0, sometimes petaloid, usually persistent. Stamens 6 or more,
free, persistent; pollen subglobose or ellipsoid. Ovary of 3–6, free, sessile
carpels. Ovules 2 or more, basal, anatropous. Seeds without endosperm and
with a straight embryo.

One genus and c. 25 spp. in the tropics of the Old World and S. Africa.

1. APONOGETON L. f.
The only genus.

*1. A. distachyos L. f.   'Cape Pondweed.'

A perennial herb with floating lvs and edible tuberous rootstock. Stems long,
green, spongy. Lvs c. 15 × 4 cm., oblong-elliptic, rounded at base and apex.
Infl. of two terminal, rigid, whitish spikes. Fls c. 10 on each spike, distichous,
white, fragrant; perianth of a single ovate segment 1–1·5 cm. Stamens 6–18.
Hyd.

Introduced. Frequently planted in ponds and sometimes ± naturalized.
S. Africa.

127. ZOSTERACEAE

Perennial submerged marine herbs with creeping or tuberous rhizomes. Stems flattened, slender. Lvs linear, sheathing at base, sheaths with stipule-like margins. Fls monoecious or dioecious, borne on one side of a flattened axis, ± enclosed in a lf-sheath; bracts 0. Perianth 0 or much reduced. Male fls of one, 1-celled, dorsifixed, sessile anther; pollen filiform. Female fls of an ovary with 2 stigmas; ovules solitary, pendulous, orthotropous. Fr. indehiscent or irregularly dehiscent.

Two genera and thirteen spp. in temperate seas throughout the world.

1. ZOSTERA L.

Rhizome monopodial, creeping and rooting at nodes, bearing alternate distichous lvs. *Squamulae intravaginales* occur on the stem just above the insertion of each lf. Flowering shoots annual, erect, sympodial, simple or branched, internodes long. Fls monoecious, male and female alternating in two rows, male sometimes with reduced perianth of bract-like structures (*retinaculae*). Pollination by water; reproduction mainly vegetative by the breaking up of the rhizome.

Eleven spp. in temperate seas of the world.

1 Flowering stems branched; sheaths entire; seeds ribbed.   2
  Flowering stems unbranched; sheaths split; seeds smooth.   **3. noltii**
2 Lvs usually 0·5–1 cm. broad; stigma twice as long as style; seed 3–3·5 mm.
                                                                      1. marina
  Lvs usually 0·2 cm. broad; style as long as stigma; seed 2·5–3 mm.
                                                                 2. angustifolia

1. Z. marina L.   Eel-grass, Grass-wrack.

A rhizomatous perennial with rather broad dark green grass-like lvs. Rhizome 2–5 mm. thick, internodes short; cortex with bundles of fibres in its outer layers. *Lvs* of sterile shoots 20–50(–100) cm., (2–)5–10 *mm. broad, rounded and mucronate at apex*; sheaths closed. Lvs of fertile shoots shorter and narrower, sometimes emarginate. Flowering stems up to 60 cm., much-branched. Infl. (4–)9–12(–14) cm., *membranous margins of sheath* 0·5–1·0 *mm. wide*, retinaculae 0, *stigma twice as long as style*. *Seed* 3·5 mm., ellipsoid, pale brown or bluish-grey, longitudinally ribbed. Fl. 6–9. Fr. 8–10. Germ. autumn. $2n=12^*$. Hyd.

Native. On fine gravel, sand or mud in the sea, from low-water spring tides down to 4 m., rarely in estuaries. Local but formerly covering large areas in suitable localities; decreased markedly in abundance about 1933. 63, H26, S. Coasts of the British Is., becoming rarer northwards. Europe from Norwegian Lapland (71° N.) to the Mediterranean; west Greenland (64° N.); Atlantic and Pacific coasts to N. America from N. Carolina to Hudson Bay and California to Unalaska Bay.

2. Z. angustifolia (Hornem.) Rchb.

Z. Hornemanniana Tutin; *Z. marina* L. var. *angustifolia* Hornem.

A slender rhizomatous perennial with narrow dark green lvs. Rhizome

1–2 mm. thick, rooting at the black slightly inflated nodes; cortex with bundles of fibres in its outer layers. Lvs of sterile shoots 15–30 cm., *2 mm. wide* in summer, in winter 5–12 cm., c. 1 mm. wide, obtuse and rounded at apex when young, *later emarginate*; sheaths closed. Lvs of fertile shoots 4–15 cm., 2–3 mm. wide. Flowering shoots 10–30 cm., compressed, c. 1 mm. wide, pale green or white, branched. Infl. 8–11 cm., *membranous margins of sheath* 1·5–2 *mm. broad*, retinaculae 0; *style and stigmas approximately equal in length. Seed* 2·5 *mm.*, ellipsoid, pale brown, longitudinally ribbed. Fl. 6–11. Fr. 7–12. Germ. autumn. $2n = 12^*$. Hyd.

Native. On mud flats in estuaries and in shallow water, from half-tide mark to low tide mark or rarely down to 4 m., in salinities between 25 and 42 gm. per litre. Distribution imperfectly known, but scattered round the coasts of the British Is. north to Orkney and probably not uncommon in suitable habitats. Europe, so far recorded only from Denmark and Sweden.

3. Z. noltii Hornem.

Z. nana Roth, p.p.

A slender shortly creeping rhizomatous perennial with narrow lvs. Rhizome 0·5–1 mm. thick; *cortex with bundles of fibres in its innermost layers. Lvs* of sterile shoots (4–)6–12(–20) cm., *up to c.* 1 *mm. wide*, emarginate; *sheaths open. Flowering stems unbranched* or occasionally with one or two branches from very near the base; peduncles 0·5–2 cm. Infl. 3–6 cm., *retinaculae present, sheath inflated. Seed* 2 mm., *smooth*, dark brown when ripe, sub-cylindrical. Fl. 6–10. Fr. 7–11. Germ. autumn. $2n = 12^*$. Hyd.

Native. On mud-banks in creeks and estuaries from half-tide mark to low-tide mark. Locally common in suitable habitats around the coasts of the British Is. north to Inverness, less frequent on the west coast. Europe from the Mediterranean to S.W. Norway and Sweden.

128. POTAMOGETONACEAE

Aquatic herbs, chiefly of fresh water, with alternate or opposite distichous usually 'stipulate' lvs. Fls in axillary or terminal bractless spikes, inconspicuous; hermaphrodite, actinomorphic, hypogynous. Per. segs 4; stamens 4, sessile on the claws of the per. segs; carpels (1–3) 4, free, each with 1 campylotropous ovule near the base of the ventral margin; stigma ± sessile. Fr. a small green or brownish drupe or achene, 4 or fewer from each fl.; seed non-endospermic; embryo with a massive hypocotylar 'foot'.

Two genera.

All lvs in opposite pairs or in whorls of 3; stipules 0, except for the involucral
    lvs subtending fl.-spikes.   2. GROENLANDIA
All or most lvs alternate and all stipulate.   1. POTAMOGETON

1. POTAMOGETON L.

Chiefly perennial and rhizomatous, and overwintering both by the rhizome and by specialized winter buds ('turions') which may be borne directly on the rhizome, on rhizome-stolons or on the lfy stems. In some spp. the

creeping stem rarely overwinters, and in others none is formed, perennation being only by the turions. Lvs all submerged, thin and translucent, or some floating lvs which are usually ±coriaceous and opaque; submerged lvs linear or with a ±broadened sessile or stalked blade; floating lvs usually narrowly to broadly elliptical-oblong. In most spp. the lf has in its axil a ±delicate membranous sheathing scale which may be free throughout ('stipule') or may be adnate to the lf-base in its lower part ('stipular sheath') and free above ('ligule'); in either case the basal part may be open and with overlapping (convolute) margins, or tubular. Spikes ovoid to cylindrical, dense, lax or interrupted; either submerged, with pollination by water, or emergent and wind-pollinated. Perianth of 4, free, rounded, shortly clawed, valvate segments, sometimes regarded as appendages of the connectives of the anthers.
About 90 spp. Cosmopolitan.

The genus is reputed to be taxonomically difficult because most species are very plastic in their vegetative morphology, varying greatly in the size and shape of their lvs at different stages of development and in different conditions of light intensity, mineral nutrient supply, speed of water-movement, etc. Spp. which can form floating lvs in sufficiently shallow water may fail to do so in deep water, and some form lvs only of the floating type when growing subterrestrially. Moreover many hybrids are known, some of them being fairly common plants. They are usually sterile but vegetative material must often be examined very closely before a confident judgment can be made. Descriptions are given only of the three most frequently encountered hybrids; others reported for Britain are listed, and it must be understood that they are intermediate in certain features between their putative parents.
Amongst the most important diagnostic features are the venation of the lvs, including the point at which lateral veins join the midrib and the angle of the join, the shape of the lf-apex, the denticulation or otherwise of the lf-margin, and whether the stipules of the young lvs are open throughout or tubular in their lower part. These features are best examined in fresh or soaked-out material by means of a strong hand lens or, better, a binocular dissecting-microscope.

1	Floating lvs present.	2
	Floating lvs 0.	10
2	Submerged lvs (phyllodes) all linear with no expanded blade; floating lvs with the margins decurrent for a short distance down the stalk, making a flexible joint. **1. natans**	
	Submerged lvs with a ±expanded translucent blade; floating lvs not jointed.	3
3	Stems compressed; submerged lvs all linear, sessile, with a well-marked band of air tissue bordering the midrib. **13. epihydrus**	
	Stems ±terete; submerged lvs not all linear.	4
4	All lvs thin, translucent, distinctly and finely net-veined; floating and most of the submerged lvs ±broadly elliptical, all short-stalked. **3. coloratus**	
	Floating lvs ±coriaceous, opaque, not translucent.	5
5	Submerged lvs mostly 10–16 × 3·5–5 cm., elliptical-lanceolate, transparent, very conspicuously but delicately net-veined; floating lvs of similar shape; all lvs long-stalked. **4. nodosus**	
	Submerged lvs translucent but not conspicuously net-veined unless held against the light; floating lvs usually broader than submerged lvs.	6

6　All lvs distinctly stalked; usually in very shallow water and then most
　　or all lvs commonly of the floating type, coriaceous and opaque,
　　broadly elliptical to elliptical-ovate.　　　　　**2. polygonifolius**
　　Some or all submerged lvs sessile.　　　　　　　　　　　　　　　7

7　At least the lower submerged lvs rounded and semi-amplexicaul at the
　　base; a sterile hybrid.　　　　　　　　　　　　　　**9. × nitens**
　　Submerged lvs narrowed to the base, sessile or short-stalked.　　8

8　Submerged lvs blunt, with quite entire margins, commonly 6–12 × 1–2 cm.,
　　often reddish; stipules 2–5 cm., broad, blunt,; peduncles not thickened
　　upwards.　　　　　　　　　　　　　　　　　　　**10. alpinus**
　　Submerged lvs acute, cuspidate or mucronate, with microscopically
　　denticulate margins; peduncles thickened upwards.　　　　　　9

9　Submerged lvs commonly 2·5–10 × 0·5–1 cm., acute or acuminate, sessile;
　　stipules 1–2 cm.; plant much branched at the base.　　**8. gramineus**
　　Submerged lvs commonly 8–12 × 2–3 cm., acuminate or cuspidate;
　　stipules 2–5 cm.; plant not much branched at the base.　　**6. × zizii**

10　Grass-leaved; lvs narrowly linear or filiform, not exceeding 6 mm. in
　　width, parallel-sided.　　　　　　　　　　　　　　　　　19
　　Lvs usually exceeding 6 mm. wide, or if narrower linear-lanceolate, not
　　parallel-sided.　　　　　　　　　　　　　　　　　　　11

11　At least the lower lvs ± amplexicaul.　　　　　　　　　　　12
　　Lvs not amplexicaul.　　　　　　　　　　　　　　　　　13

12　Lvs usually narrowly to broadly ovate, all cordate and amplexicaul;
　　stipules small, fugacious; fertile.　　　　　　　　**12. perfoliatus**
　　Lvs lanceolate, narrowed to a cuspidate tip; at least the lower lvs rounded
　　and semi-amplexicaul at the base; stipules usually ± persistent;
　　a sterile hybrid.　　　　　　　　　　　　　　　　**9. × nitens**

13　Stem compressed; lvs commonly 4–9 × 1–1·5 cm., linear-oblong, margins
　　distinctly serrate and often strongly undulate; beak equalling the rest
　　of the fr.　　　　　　　　　　　　　　　　　　　**22. crispus**
　　Stem ± terete; lf-margin not distinctly serrate to the naked eye; beak
　　shorter than the rest of the fr.　　　　　　　　　　　　　14

14　Lvs blunt; peduncles not or hardly thickened upwards.　　　　15
　　Lvs not blunt; peduncles distinctly thickened upwards.　　　　17

15　Lvs distinctly hooded at the tip and rounded at the base, commonly
　　10–18 × 2–4·5 cm., with microscopically entire margins.　**11. praelongus**
　　Lvs not hooded.　　　　　　　　　　　　　　　　　　16

16　Lvs rounded at base and apex, sessile, commonly 6–10 × 1·5–2 cm.,
　　margins microscopically denticulate at least when young; fr. abortive.
　　　　　　　　　　　　　　　　　　　　　　　7. × salicifolius
　　Lvs narrowed to the sessile base, commonly 6–12 × 1–2 cm.; margins
　　microscopically entire; fertile.　　　　　　　　　　**10. alpinus**

17　Plant much branched at the base; lvs commonly 2·5–10 × 0·3–1 cm.,
　　acute or acuminate, sessile; stipules 1–2 cm.　　　　**8. gramineus**
　　Plant not much branched at the base; lvs 8–20 × 2–5 cm.; stipules 2–6 cm.　18

18　Lvs commonly 8–12 × 2–3 cm., lanceolate to oblong, cuspidate or
　　rounded and mucronate.　　　　　　　　　　　　**6. × zizii**
　　Lvs commonly 12–20 × 3·5–5 cm., oblong-lanceolate, cuspidate or
　　acuminate.　　　　　　　　　　　　　　　　　　**5. lucens**

19 Lvs with 2 large air-filled longitudinal canals, one on each side of the
   midrib, occupying the greater part of their volume; stipules adnate
   below to the lf-base, forming a stipular sheath with a free ligule. 20
   Lvs not as above; stipules free from the lf-base throughout their length. 21

20 Stipular sheath open and convolute with a whitish margin; fr. 3–5 ×
   2–4 mm., with a short beak terminating the ventral margin.
                                                      24. pectinatus
   Stipular sheath tubular below when young; fr. 2–3 × 2 mm. with an
   extremely short almost central beak.   **23. filiformis**

21 Lvs with 3(–5) principal and many faint intermediate longitudinal strands;
   stems strongly compressed or even winged.   22
   Lvs usually with only 3–5 longitudinal veins or apparently 1-veined
   especially when very narrow, but never with many faint intermediate
   longitudinal strands; stems not or slightly compressed.   23

22 Lvs 10–20 cm. × 2–4 mm., usually rounded and cuspidate at the tip,
   sometimes acuminate; principal lateral veins usually distinctly joining
   the midrib below the tip of the lf; stipules blunt; fr. smooth, with a
   straight beak.   **20. compressus**
   Lvs 5–13 cm. × 2–3(–4) mm., finely acuminate; principal lateral veins
   usually not distinctly joining the midrib; stipules acute; fr. toothed
   near the base of the ventral margin and tubercled along the dorsal
   margin; beak recurved.   **21. acutifolius**

23 Lvs mostly 5-veined, the laterals closer to the margin and to each other
   than to the midrib; stipules tubular below.   **14. friesii**
   Lvs mostly 3-veined or apparently 1-veined, rarely 5-veined and then
   with equal spacing.   24

24 Lvs 2–3 mm. wide, 3(–5)-veined, the laterals joining the midrib at a wide
   angle close below the blunt mucronate tip; stipules open, convolute.
                                                      17. obtusifolius
   Lvs mostly less than 2 mm. wide.   25

25 Stipules tubular below.   26
   Stipules open, convolute.   27

26 Stipules tubular for at least ⅔ of their length; lf-tip gradually acute with
   a narrow blunt terminal cusp.   **16. pusillus**
   Stipules tubular only towards the base; lf narrowed to a sharply acuminate
   tip (Shetland and Outer Hebrides).   **15. rutilus**

27 Lvs rarely exceeding 1 mm. wide, subsetaceous, narrowing to the base
   and tapering to a long fine point; usually no air-filled lacunae bordering
   the midrib; fr. often toothed below and tubercled on the dorsal margin,
   usually 1 per fl.   **19. trichoides**
   Lvs usually exceeding 1 mm. wide, subacute to rounded and mucronate,
   with air-filled lacunae bordering the midrib; rarely narrower and
   acute; fr. smooth, usually 4 per fl.   **18. berchtoldii**

Subgenus 1. POTAMOGETON.

Lvs all submerged or some floating, alternate (only the involucral ones
opposite), variously shaped; stipules free from the lf throughout, or adnate
only at the very base; stigma with small papillae; fr. spike not or hardly
interrupted, its stalk and rhachis rigid; fr. drupaceous, with fleshy exocarp
and bony endocarp; wind-pollinated.

1. P. natans L.   'Broad-leaved Pondweed.'

Incl. *P. hibernicus* (Hagstr.) Druce

Rhizome extensively creeping. Lfy stems commonly to 100 cm. but reaching 500 cm. in deep water, ± terete, not or little branched. Submerged lvs (phyllodes) 15–30(–80) cm. × 1–3 mm., linear, channelled, with several longitudinal veins, rarely with a small blade. *Floating lvs* stalked, 15–60 cm. overall, with the blade 2·5–12·5 × 0·8–7 cm., elliptical to ovate-lanceolate, ± acute, rounded to subcordate at the base, *inrolled at the base* for a time after emergence, *coriaceous*, and with 2 coriaceous wings decurrent for a short distance down the stalk which therefore appears *jointed just below the blade*; longitudinal veins c. 20–25; transverse veins indistinctly visible against the light. Stipules 5–12(–18) cm., persistent, at length fibrous. Fr. spike 3–8 cm., cylindrical, dense; stalk 5–12 cm., axillary, stout, not enlarging upwards.

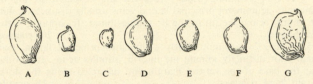

Fig. 62. Fruits of *Potamogeton*. A, *P. natans*; B. *P. polygonifolius*; C, *P. coloratus*; D, *P. lucens*; E, *P. gramineus*; F, *P. alpinus*; G, *P. praelongus*. × 3.

Fr. (Fig. 62A) 4–5 × 3 mm., olive green, obovoid, somewhat compressed; ventral margin convex, dorsal ± semi-circular and keeled when dry; beak short, straight. Fl. 5–9. Turions usually 0. 2*n* = 52. Hyd.

Native. Lakes, ponds, rivers, ditches, especially on a highly organic substratum, and usually in water less than 1 m. deep. 112, H40, S. Common throughout the British Is. Northern hemisphere.

Readily recognized by the jointed blade of the floating lf and the very long stipules.

P. × *fluitans* Roth (*P. lucens* × *natans*) and *P.* × *sparganifolius* Laest. ex Fr. (*P. gramineus* × *natans*) have been reported.

2. P. polygonifolius Pourr.   'Bog Pondweed.'

P. oblongus Viv.; *P. anglicus* Hagstr.

Rhizome extensively creeping. Lfy stems commonly to 20 cm., but to 50 cm. in deeper water, slender, terete, unbranched. *Lvs all with blade and stalk.* Submerged lvs with blade commonly 8–20 × 1–3 cm., ± narrowly lanceolate-elliptical but very variable in size and shape, membranous, translucent. *Floating lvs* with blade commonly 2–6 × 1–4 cm. and stalk one-half to twice as long, broadly elliptical to lanceolate, cuneate to cordate at the base, *not jointed below the blade*, subcoriaceous, not translucent; longitudinal veins c. 20; transverse veins plainly visible against the light. Stipules 2–4 cm., blunt. Fr. spike 1–4 cm., cylindrical, dense; stalk much exceeding the spike, slender, not widening upwards. Fr. (Fig. 62B) 2 × 1·5 mm., reddish, obovoid, slightly compressed; ventral margin convex, dorsal semi-circular and slightly keeled when dry; hardly beaked. Fl. 5–10. Turions little differentiated from lfy shoots. 2*n* = 26. Hyd.

Native. Ponds, bog-pools, ditches and small streams with acid and usually shallow water; to 2350 ft. in Wales. 112, H40, S. Common throughout the British Is. Europe; N.W. Africa; eastern N. America.

Very variable in size and shape of lvs with varying water depth. A state which has been called f. *cancellatus* Fryer has thin, translucent, finely reticulate-veined submerged lvs somewhat resembling those of *P. coloratus*.

3. P. coloratus Hornem. 'Fen Pondweed.'

P. plantagineus Du Croz ex Roem. & Schult.

Resembling *P. polygonifolius* in habit but *all lvs with stalk usually shorter than blade*, ± blunt, *thin, translucent, finely and distinctly reticulate-veined.* Submerged lvs 6–10(–18) cm., linear-lanceolate to narrowly elliptical, often subsessile. Floating lvs with blade 2–7(–10) × 1·5–5 cm., ovate-elliptical. Stipules 2–4 cm., blunt, ± persistent. Fr. spike 2·5–4 cm., cylindrical, dense; stalk 5–20 cm., slender, not widening upwards. Fr. (Fig. 62c) 1·7–2 × c. 1 mm., green, ovoid, compressed; ventral margin slightly convex, dorsal semicircular and slightly keeled when dry; beak very short, curved. Fl. 6–7. Turions little differentiated from lfy shoots. $2n=26$. Hyd.

Native. Shallow ponds and pools in fen peat, especially in calcareous water; 47, H31, S. Local throughout the British Is. northwards to Argyll; Hebrides. Ireland. Europe.

P. × *billupsii* Fryer (*P. coloratus* × *gramineus*) and *P.* × *lanceolatus* Sm. (*P. berchtoldii* × *coloratus*) have both been recorded from a few localities in the British Is.

4. P. nodosus Poir. 'Loddon Pondweed.'

P. petiolatus Wolfg.; *P. Drucei* Fryer

Rhizome creeping. Lfy stems commonly 20–30 cm. but reaching 2 m., terete, stout, simple. *Lvs all with a ± long-stalked blade. Submerged lvs* with blade 10–20 × 1·5–4 cm., broadly lanceolate, or elliptical-lanceolate, ± *equally narrowed* at each end, ± *cuspidate*, thin, translucent, *very finely and beautifully reticulate-veined.* Floating lvs with blade 6–15 × 2·5–6 cm., oblong-elliptical or ovate, shortly mucronate, coriaceous; transverse veins visible against the light. Stipules 7–10 cm., lanceolate. Fr. spike 2–6 cm., cylindrical, fairly dense but often with many abortive fr.; stalk long, stout, not thickening upwards. *Fr.* c. 3·5 × 2·5 mm., obovoid; ventral margin convex, dorsal almost semi-circular and *acutely keeled*; beak short but stout (c. 2 × 1·5 mm.). Fl. 8–9. Turions borne on slender stolons. Hyd.

Native. Gravelly shallows and deeper water or slow-flowing base-rich rivers. 7. Known only in the Avon (Somerset, Gloucester and Wilts), Stour (Dorset), Thames (Oxford, Bucks and Berks) and Loddon (Berks). Mediterranean region, Canary Is., Madeira, Azores; W. and C. Europe northwards to Poland and Germany.

Easily recognizable by the ± lanceolate beautifully net-veined submerged lvs, which are quite different from those of any other British sp.

5. P. lucens L.   'Shining Pondweed.'

Incl. *P. acuminatus* Schumach. and *P. longifolius* Gay ex Poir.

Rhizome creeping extensively. Lfy stems 0·5–2(–6) m., stout and tough. *Submerged lvs* commonly 10–20 × 2·5–6 cm., *oblong-lanceolate*, rarely ovate, *shortly stalked* with the blade decurrent on the stalk and so appearing sub-sessile, rounded and cuspidate or apiculate at the tip, or acuminate, margins minutely denticulate, thin, translucent, shining; longitudinal veins c. 11–13. *Floating lvs* 0. *Stipules* 3–8 cm., blunt, *prominently 2-keeled.* Fr. spike 5–6 cm., cylindrical, stout, dense; *stalk* 7–25 cm. or more, stout, *thickening upwards.* Fr. (Fig. 62D) c. 3·5 × 2·2 mm., olive-green, ovoid, swollen, hardly compressed; ventral margin almost straight, dorsal ± semicircular, hardly keeled; beak short. Fl. 6–9. Turions produced on the rhizome-system. $2n = 52$. Hyd.

Native. Lakes, ponds, canals and slow streams on base-rich inorganic substrata; commonly in calcareous water, when the lvs become chalk-encrusted; to 1250 ft. in Malham Tarn. 86, H33. Locally common through-out the British Is. except Cornwall and Devon, the Outer Hebrides and Shetland. Europe; W. Asia.

One of the largest-leaved British spp., known by the large oblong-lanceolate short-stalked submerged lvs with denticulate margins and the upwardly thickened stalk of the fr. spike. *P.* × *cadburyi* Dandy & Taylor (*P. crispus* × *lucens*) and *P.* × *nerriger* Wolfg. (*P. alpinus* × *lucens*) have been reported.

6. P. × **zizii** Koch ex Roth

P. angustifolius auct. mult.; *P. gramineus* × *lucens*

Rhizome creeping. Lfy stems 20–50(–150) cm., terete, stout, branching below. *Submerged lvs* ± sessile or very shortly stalked, their blades c. 3·5–15 × 0·7–2·5 cm. *oblanceolate to oblong-lanceolate, narrowed to the base* and either cuspidate or abruptly rounded and mucronate at the apex, margins *undulate* and irregularly denticulate throughout, thin, translucent. *Floating lvs* often produced, *oblong-lanceolate,* usually *cuspidate, distinctly stalked* but the stalk never exceeding the blade, coriaceous, with denticulate margins. Stipules 1·5–4 cm., broad, blunt, 2-keeled. Fr. spike 3–6 cm., cylindrical, with *mature achenes very variable in number, sometimes 0; stalk* commonly 7–12 cm., but very variable, stout, *thickened upwards.* Fr. c. 3 × 2 mm., ovoid, compressed; ventral margin almost straight, dorsal ± semicircular, strongly keeled and with lateral ridges; beak very short. Fl. 6–9. Turions at the ends of stolons borne on the rhizome. Hyd.

Native. Ponds, lakes, streams, etc.; local. 52, H24. Widely distributed throughout Great Britain from Surrey northwards to Ross; Ireland; Orkney. Europe.

Resembles *P. lucens* but distinguished by the presence (commonly) of floating lvs, the absence of a stalk in at least some of the submerged lvs, and by the smaller fr.

7. P. × **salicifolius** Wolfg.

P. lucens × *perfoliatus*; *P.* × *decipiens* Nolte ex Koch

Rhizome creeping. Lfy stems long, commonly to 1 m., rarely to 3 m., terete, stout, slightly branched above. *Submerged lvs* very variable; blades 3–20 ×

1·5–4 cm., *oblong, ± sessile and clasping at the base and ± rounded at the non-hooded apex,* margins *at first minutely denticulate* but sometimes becoming smooth later, thin, translucent. *Floating lvs* 0. Stipules c. 2·5–3 cm., blunt, slightly 2-keeled, persistent. *Spike* 2·5 cm., dense, *always sterile*; stalk to 8 cm., not or slightly thickened upwards. Fl. 6–9. Hyd.

Native. Ponds, canals, rivers, etc., with the parents. 29, H7. Local, but scattered throughout Great Britain from Dorset, Surrey and Kent northwards to Perth; Ireland from Clare to Dublin and Fermanagh; Inner Hebrides. Europe.

Resembles *P. lucens* but distinguished by its sessile and smaller lvs (commonly 6–10 × 1·5–2 cm.) with ± clasping base, its scarcely keeled stipules and its sterility.

8. P. gramineus L.   'Various-leaved Pondweed.'

P. heterophyllus Schreb.; incl. *P. lonchites* Tuckerm. and *P. graminifolius* H. & J. Groves

Rhizome creeping. Lfy stems to 1 m. or more, terete, slender, flexuous, with *very numerous short non-flowering branches near the base. Submerged lvs* 2·5–8(–18) × 0·3–1·2(–3) cm., usually *linear-lanceolate* or elliptical-oblong, *narrowed into a sessile base,* acuminate or cuspidate, with *minutely serrate* margins, thin, translucent; longitudinal veins 7–11. *Floating lvs few or* 0, *long-stalked*; blade 2·5–7 × 1–2·5 cm., usually broadly *elliptical-oblong,* ± rounded at the base, coriaceous, with transverse veins visible against the light. Stipules 2–5 cm., c. half as long as the lower internodes, broadly lanceolate, acute, not keeled. Fr. spike c. 2·5–5 cm., cylindrical, dense; *stalk* 5–8(–25) cm., stout, *thickened upwards.* Fr. (Fig. 62E) 2·5–3 × c. 2 mm., green, ovoid, ± compressed; ventral margin nearly straight, dorsal semicircular and slightly 3-keeled; beak short, ± straight. Fl. 6–9. Turions produced at the ends of stolons. $2n = 52$. Hyd.

Native. Lakes, ponds, canals, streams, etc., chiefly in acid water; to 2300 ft. in Scotland. 77, H36. Locally common throughout the British Is., though absent from Cornwall, Devon, and districts with calcareous water. Northern hemisphere.

Very variable, with numerous habitat forms to which many names have been given. Plants with small narrow submerged lvs not exceeding 5 cm. have been named var. *lacustris* (Fr.) Aschers. & Graebn., and others usually without floating lvs, with submerged lvs 5–15 cm., var. *fluvialis* (Fr.) Blytt.

P. × nericius Hagstr. (*P. alpinus × gramineus*) has been reported from Scotland.

9. P. × nitens Weber

P. gramineus × perfoliatus

Rhizome creeping. Lfy stems to 120 cm., terete, slender, *sparingly branched. Submerged lvs* 2–8(–15) × 0·5–1·5(–2) cm., *oblong-lanceolate, rounded, ± cordate* and usually *semi-amplexicaul* at the sessile base, ± acute or cuspidate, undulate and at first minutely denticulate at the margins (which may later become smooth), thin, translucent; longitudinal veins 7–15. *Floating lvs few or* 0, elliptical-oblong, long-stalked to sessile or semi-amplexicaul, coriaceous. Stipules 1–2 cm., lanceolate, acute, not keeled, persistent. *Spike* 1–2 cm., cylindrical, dense, *sterile*; stalk 2·5–10(–15) cm., variable in thickness, usually widest at or above the middle. Fl. 6–8. Turions produced at the ends of stolons from the rhizome. Hyd.

Native. Lakes, ponds, streams, etc., to 2000 ft.; rather local throughout the British Is. 43, H22. In Great Britain recorded from Devon and Surrey and from Cambridge, Northants and Anglesey northwards; scattered in Ireland Northern hemisphere.

A plant having much the habit of *P. gramineus* but lacking the dense basal branching and having the lvs rounded to semi-amplexicaul at the base; always sterile.

Very variable, and has been divided by Hagström into varieties differing in the form of the involucral lvs which subtend the spikes: in var. *subgramineus* (Raunk.) Hagstr. they are distinctly stalked; in var. *subperfoliatus* (Raunk.) Hagstr. sessile and rounded to semi-amplexicaul at the base, and in var. *subintermedius* Hagstr. short-stalked or sessile, sometimes with one rounded at the base.

10. P. alpinus Balb.   'Reddish Pondweed.'

P. rufescens Schrad.

Rhizome creeping. Lfy stems 15–200 cm., terete, slender, ± simple. *Submerged lvs* usually 6–15 × 1–2 cm., often reddish, *narrowly oblong-elliptical, narrowed to each end*, subsessile or shortly stalked, always *blunt, entire*, thin, translucent, with a conspicuously reticulate midrib; longitudinal veins 7–11. Floating lvs 3–8 × 0·8–2 cm., oblanceolate- to obovate-elliptical, narrowing to a short stalk, blunt, entire, subcoriaceous, with the transverse veins easily visible against the light; sometimes 0. *Stipules* 2–6 cm., shorter than the internodes, *ovate, blunt*, not keeled, robust. Fr. spike 2–4 cm., ± cylindrical, dense; stalk 5–18 cm., slender, not thickened upwards. Fr. (Fig. 62F) 3 × 2 mm., becoming pale reddish, ovoid-acuminate, somewhat compressed; ventral margin very convex, dorsal semicircular and sharply keeled, narrowed above subequally into the fairly long somewhat curved beak. Fl. 6–9. 2*n* = 52. Hyd.

Native. Lakes, ditches, streams, etc., especially in non-calcareous water and on substrata rich in organic matter; to 3350 ft. in Scotland. 88, H27, S. Throughout the British Is. except Cornwall, Orkney and Shetland, but local in S. and E. England. Northern hemisphere.

P. × griffithii A. Benn. (*P. alpinus × praelongus*), *P. × prussicus* Hagstr. (*P. alpinus × perfoliatus*) and *P. × olivaceus* Baagöe ex G. Fisch. (*P. venustus* A. Benn.; *P. alpinus × crispus*) have all been recorded from two or more localities.

11. P. praelongus Wulf.   'Long-stalked Pondweed.'

Rhizome creeping. Lfy stems 0·5–2(–6) m., terete, stout, somewhat branched above. *Submerged lvs* 6–18 × 2–4·5 cm., green, strap-shaped to oblong, usually 4–6 times as long as wide, *rounded at the sessile ± semi-amplexicaul base*, narrowing gradually to the *blunt and hooded apex*, slightly undulate, *entire*, thin, translucent, not shining; longitudinal veins c. 13–17, with 1 strong and 5–7 weaker veins on each side of the midrib. Floating lvs 0. Stipules 0·5–6 cm., often exceeding the internodes, blunt, thin, not ridged or keeled. Fr. spike 3–7 cm., cylindrical, dense; stalk 15–40 cm., fairly stout, not widening upwards. Fr. (Fig. 62G) 4–6 × 3–4 mm., green, asymmetrically obovoid, hardly compressed; ventral margin very slightly convex, dorsal ± semicircular and sharply keeled or winged; beak short, straight. Fl. 5–8. No specialized turions. $2n = 52$. Hyd.

Native. Lakes, rivers, canals and ditches, chiefly on substrata not very rich in organic matter; rather local. 63, H 17, S. To 3000 ft. in Scotland. In Great Britain northwards from Essex, Surrey, Berks and Radnor, but chiefly on the east side, though frequent in some of the English Lakes and in Ireland mainly near the north and west coasts. Northern hemisphere.

A very distinct species recognizable by its sessile, tapering, hooded lvs and its very long peduncles.

P. × *cognatus* Aschers. & Graebn. (*P. perfoliatus* × *praelongus*) and *P.* × *undulatus* Wolfg. (*P. crispus* × *praelongus*) have been reported from Britain, the latter also from Ireland.

12. P. perfoliatus L. 'Perfoliate Pondweed.'

Rhizome extensively creeping. Lfy stems 0·5–2(–3) m., terete, usually stout, branching above. *Submerged lvs* 2–6(–10) × 1·3–4(–6) cm., all *sessile and ± completely amplexicaul at the wide cordate base*, commonly ovate, but very variable in shape from lanceolate to orbicular, blunt or rarely mucronate, irregularly and microscopically denticulate, very thin, translucent; 5–7 strong longitudinal veins with fainter intermediate veins. *Floating lvs* 0. *Stipules* up to 1 cm., blunt, not keeled, very delicate and *soon disappearing* or persisting only where the lvs are opposite. Fr. spike 1–3 cm., cylindrical, stout, dense; stalk 3·5–10(–13) cm., stout, not thickened upwards. Fr. (Fig. 63 A) 3·5–4 × 2·5–3 mm., olive-green, hardly compressed; ventral side concave below but convex above, dorsal semicircular, obscurely keeled with faint lateral ridges. Fl. 6–9. Turions formed on stolons from the rhizomes. $2n = 52$. Hyd.

Native. Lakes, ponds, streams, canals, etc., especially on substrata of moderate but not very high organic content; to 2300 ft. in Scotland. 108, H 38, S. Common throughout the British Is. Northern hemisphere; Australia.

Recognizable by the wide-based cordate and amplexicaul lvs which are all submerged, and the fugacious stipules. The great variability in lf size and shape is largely if not wholly dependent on factors of the external environment. Narrow-leaved forms occur in low light intensity and on substrata of low calcium content, while very broad-leaved forms are characteristic of highly calcareous substrata. *P.* × *nitens* frequently has floating lvs and its submerged lvs are typically narrower than those of *P. perfoliatus*, and only semi-amplexicaul, with persistent stipules: the plant is, moreover, always sterile.

P. × *cooperi* (Fryer) Fryer (*P. crispus* × *perfoliatus*) has been reported from numerous scattered localities.

13. P. epihydrus Raf. 'Leafy Pondweed.'

P. Nuttallii Cham. & Schlecht.; *P. pensylvanicus* Willd. ex Cham. & Schlecht.

Rhizome creeping. *Stems* slender, *compressed*, mostly simple. *Submerged lvs* 8–20 cm. × 3–8 mm., *linear, sessile*, tapered to a blunt or subacute apex, thin, translucent; longitudinal veins (3–)5(–7); there are transverse veins but no fainter intermediate longitudinal veins; *midrib bordered by a well-marked band of lacunar tissue* which extends sideways towards or beyond the inner lateral veins. *Floating lvs oblong to elliptical*, blunt, tapered into the stalk, coriaceous, the blade $3.5-7 \times 1-2.5$ cm. *Stipules* up to 3·5 cm., *open*, broad, subtruncate or rounded. Spike 1–2·5 cm., cylindrical, dense and continuous; stalk 3–4 cm., slender, not thickened upwards. Fr. 2·5–3·5 mm., round-obovoid, 3-keeled, sides flat; beak short; *embryo subspiral* (*coiled more than 1 complete turn*). Fl. 6–8. Hyd.

Native. Lakes. 1. Known as a native plant only from the Outer Hebrides. Also introduced in S.W. Yorks and S.E. Lancs, in the R. Calder and canals in the neighbourhood of Halifax. Widely distributed in North America.

A very distinct species, known as an alien in S.W. Yorks for more than 40 years but only recently discovered as a native in the Outer Hebrides. The distribution of the species recalls that of *Eriocaulon septangulare*. The British plants are referable to var. *ramosus* (Peck) House.

A   B   C   D   E   F   G

Fig. 63. Fruits of *Potamogeton*. A, *P. perfoliatus*; B, *P. friesii*; C, *P. pusillus*; D, *P. obtusifolius*; E, *P. berchtoldii*; F, *P. trichoides*; G, *P. compressus*. × 3.

14. P. friesii Rupr.   'Flat-stalked Pondweed.'

P. compressus auct. mult.; *P. mucronatus* Schrad. ex Sond.

Rhizome 0. Stem 20–100 cm., arising from a turion, strongly compressed, slender, ± simple below, branching above, and producing *numerous very short lfy branches or fascicles of lvs* which later develop turions. Submerged lvs commonly 4–6·5 cm. × 2–3 mm., linear, sessile, subacute or blunt and abruptly mucronate, thin, pale green, translucent; *longitudinal veins usually* 5 (occasionally 3 or 7), the distance between the midrib and the nearest lateral veins being almost twice that between the laterals and between the farther lateral and the margin; there are transverse veins but no fainter intermediate longitudinal veins. Floating lvs 0. *Stipules* 0·7–1·5 cm., *tubular below* at first but soon splitting, fibrous-persistent, whitish. Fr. spike 0·7–1·5 cm., with 3–4 remote whorls of fr.; stalk 1·5–5 cm., flattened, somewhat thickened upwards. Fr. (Fig. 63 B) 2–3 × 1·5–2 mm., olive, asymmetrically obovoid, slightly compressed; ventral face strongly convex above, dorsal ± semicircular, bluntly 3-keeled; beak c. 5 mm., erect or recurved. Fl. 6–8. *Turions* terminal on short lateral branches, *prominently ribbed towards the base.* $2n = 26$. Hyd.

Native. Lakes, ponds, canals, etc., especially on a muddy substratum; lowland. 56, H7. Common in suitable lowland waters throughout Great Britain except Shetland, Devon and Cornwall; rare in Ireland. Northern hemisphere.

Distinguishable from *P. obtusifolius*, which sometimes has 5 veins with no faint intermediate, by the unequal spacing of the veins, the tubular stipules, the longer peduncles and the smaller fr.

P. × *lintonii* Fryer (*P. crispus* × *friesii*) is known from several localities. It resembles *P. crispus* in habit, and has irregularly and microscopically denticulate lvs whose stipules are tubular for a short distance. *P.* × *pseudofriesii* × Dandy & Taylor (*P. acutifolius* × *friesii*) has also been reported.

15. P. rutilus Wolfg. 'Shetland Pondweed.'

Rhizome 0. Stem 3–60(–100) cm., arising from a turion, compressed, very slender, much branched. *Submerged lvs* 3–6 cm. × 0·5–1 mm., very narrowly linear, sessile, *gradually tapering to a finely pointed apex*, bright green or reddish below, thin, translucent; longitudinal veins usually 3, sometimes 5 below, the laterals joining the midrib at some distance below the apex or vanishing without doing so; there are no faint intermediate longitudinal veins. Floating lvs 0. *Stipules* 1–2 cm., *tubular below*, acuminate, *strongly veined*, ± *fibrous-persistent*. Fr. spike 5–10 mm., shortly cylindrical, 6–8-fld; stalk c. 2–4 cm., slightly thickened upwards. Fr. 1–2 mm., brownish-red, semi-ovoid; ventral margin slightly convex, dorsal semicircular, not keeled; beak short, stout. Fl. 8. Turions formed at the tips of lateral branches. Hyd.

Native. Lakes. 4. Known from the Outer Hebrides and Shetland. Europe; W. Asia.

Distinguishable from *P. pusillus* by the stipules, which in *P. rutilus* are tubular for a shorter distance, firmer, more strongly veined and fibrous-persistent; and by the gradually tapered finely pointed lf-tip.

16. P. pusillus L.

P. panormitanus Biv.

Rhizome usually 0. Stem 20–100 cm., generally arising from a turion, slightly compressed, very slender, usually with many long branches from near the base. Submerged lvs 1–4(–7) cm. × 0·3–1(–3) mm., narrowly linear, sessile, narrowing gradually to a long but blunt tip, firm, translucent; longitudinal veins 3 (*rarely 5*), *the laterals joining the midrib, usually at a narrow angle and at a distance of* 2–3 lf-widths below the tip; midrib except of the uppermost lvs, usually *not bordered by pale bands of large-celled lacunae*, or sometimes by a single row on each side. Floating lvs 0. *Stipules* 0·6–1·7 cm., *tubular to above half-way*, splitting later, ± persistent, pale brown. Spikes 6–12 mm., terminating very short branches, cylindrical, interrupted, of c. 2–8 fls in 2–4 whorls; stalk 1·5–3 cm. or more, filiform, not thickened upwards. Fr (Fig. 63c) 2–2·5 × 1–1·5 mm., pale olive, obovoid, *smooth*; ventral face convex, dorsal more strongly convex, broadly and obscurely keeled when dry; beak c. 0·4 mm., almost centrally placed, ± straight. Fl. 6–9. *Turions* 10–15 × 0·5 mm., *narrowly fusiform, chiefly axillary* and produced first towards the base of the branches, later upwards. $2n=26$. Hyd.

Native. Lakes, ponds, canals, streams; especially in highly calcareous or even brackish waters; not uncommon. 72, H35. Throughout the British Is. Northern hemisphere; Africa.

Often confused with *P. berchtoldii* (*P. pusillus* auct., non L.) but distinguishable by the usual absence of lacunae along the midrib, the tubular stipules and the slender axillary turions. *P.* × *trinervius* G. Fisch. (*P. pusillus* × *trichoides*) has been reported.

17. P. obtusifolius Mert. & Koch 'Grassy Pondweed.'

P. gramineus auct. mult.

Rhizome 0. Stem 20–100 cm., slender, compressed (c. 2:1), frequently forked and with many short lateral lfy branches. *Submerged lvs* 3–9 cm. × 2–4 mm., linear, *narrowed to the sessile base*, rounded and shortly apiculate at the tip, dark green, thin, very translucent; longitudinal veins 3, sometimes 5, the laterals joining the midrib at a wide angle, sometimes greater than a right-angle, and usually close below the tip; faint intermediate longitudinal veins 0; midrib bordered by pale bands of elongated lacunae, especially towards the base. Floating lvs 0. *Stipules* 1·3–2 cm., *open, broad, blunt*, with many faint veins. Spike 0·6–1·3 cm., ovoid to stoutly cylindrical, dense and continuous, fruiting freely; stalk 0·8–2(–3·5) cm., slender, straight, not thickened upwards. Fr. (Fig. 63 D) 3–4 × c. 2 mm., brownish-olive, oblong-ovoid, slightly compressed; ventral face convex, dorsal semicircular, 3-keeled when dry; beak c. 0·6 mm., straight. Fl. 6–9. Turions 2–4 cm. × 3·5–7 mm., narrowly fan-shaped, terminal. $2n=26$. Hyd.

Native. Lakes, ponds, streams, canals, ditches, etc.; local. 74, H21. Throughout the British Is. from S. Devon and Kent northwards to Inverness and Argyll, but not in Cornwall or Somerset; Orkney. Northern hemisphere.

Distinguishable from *P. compressus* and *P. acutifolius* by the absence of faint intermediate longitudinal veins, and from *P. friesii* by having usually only 3 veins (or 5 evenly spaced), by the open stipules and the shorter peduncles.

P. sturrockii (A. Benn.) A. Benn., from Marlee Loch in E. Perth, has been regarded as a distinct sp., as a subsp. of *P. berchtoldii*, and as a hybrid of *P. obtusifolius* and *P. pusillus*. It more probably represents a slender state of *P. obtusifolius*, narrow-leaved forms of which may easily be confused with *P. berchtoldii*.

18. P. berchtoldii Fieb.   'Small Pondweed.'

P. pusillus auct. mult, non L.; incl. *P. tenuissimus* (Mert. & Koch) Rchb., *P. lacustris* (Pearsall & Pearsall f.) Druce and *P. Millardii* Heslop-Harrison

Rhizome usually 0. Stem 10–100 cm., usually arising from a turion, very slender, very little compressed (less than 2:1), almost simple to freely branched; internodes commonly 1–5 cm. Submerged lvs 2–5·5 cm. × 0·5–2 mm., linear, sessile, rounded to acute at the tip usually shortly mucronate, dark green, thin, translucent; *longitudinal veins always* 3, the laterals close to the margin, meeting the midrib almost at right angles and *at $\frac{1}{2}$–1 lf-width below the tip; no faint intermediate longitudinal veins*, but transverse veins often present; midrib bordered at least towards the base by pale bands of lacunae. Floating lvs 0. *Stipules* 3–10 mm. or more, *open*, convolute, *blunt*, faintly 6–8-veined, ± deciduous but persisting in the axils of the uppermost lvs. Fr. spike 2–8 mm., subglobose, continuous or slightly interrupted; stalk 0·5–3(–10) cm., filiform, hardly thickened upwards. Fr. (Fig. 63 E) 2–2·5 × c. 1·5 mm., dark olive, ± obovoid-acuminate, hardly compressed, ± tubercled below when dry; ventral margin convex, dorsal rounded, broadly and bluntly keeled when dry; beak short. Fl. 6–9. *Turions* 7–15 × 0·5–2·5 mm., *terminal*, fusiform, olive. $2n=26$. Hyd.

Native. Lakes, ponds, canals, streams, ditches, etc., in very calcareous to very acid waters. 106, H39. Common throughout the British Is. Northern hemisphere.

Names have been given to some of the numerous states of this sp. growing in different habitat conditions. Thus the name *P. tenuissimus* has been used for plants, usually of shallow ditches and pools, in which the lvs are very narrow, dark-coloured and with the midrib bordered by lacunae only towards the base. At the other extreme are plants found in deep, clear, nutrient-poor waters, as in the English Lakes, in which the lvs tend to be broad, very delicate and light-coloured, the apex rounded and minutely mucronate. Plants of this form were formerly confused with *P. sturrockii* and have also been treated as a distinct subspecies or species (*P. lacustris*). For differences from *P. pusillus* see under that species.

P. × *sudermanicus* Hagstr. (*P. acutifolius* × *berchtoldii*) is known from Dorset.

19. P. trichoides Cham. & Schlecht.   'Hair-like Pondweed.'

Rhizome filiform or 0. Stems 20–75 cm., filiform, terete or very slightly compressed, repeatedly and divaricately branched, most of the branches ultimately bearing spikes. *Submerged lvs* 2–4(–6·5) cm. × c. 0·5–1 mm., deep dull green, *narrowly linear to subsetaceous*, narrowed to the sessile or semi-amplexicaul base and tapering to a *long fine point*, spreading, ± rigid, translucent; *longitudinal veins* 3, the midrib thick and prominent but the *laterals faint* and often indistinct even under the microscope, *midrib* usually *not bordered by lacunae*. Floating lvs 0. *Stipules* 7–11(–20) mm., *open* and convolute, narrow, ± acute, somewhat rigid. Spikes 1–1·5 cm., 3–6-fld, ± ovoid, interrupted, usually with only 1–3 carpels *per fl.* and ripening only 1 *fr. per fl.*; stalk 5–10 cm., filiform, not thickened upwards, often curved above. Fr. (Fig. 63 F) c. 2·5 × 2 mm., ovoid, somewhat compressed, often with a tooth near the base of each side; ventral margin almost straight, often toothed near its base, dorsal strongly rounded, obscurely keeled and often ± tuberclate; beak short, straight. Fl. 6–9. 2*n* = 26. Hyd.

Native. Ponds, canals, ditches, etc. 34. Chiefly in S. and E. England from E. Cornwall and Gloucester to Sussex and Kent and then northwards to Lancs and Yorks; also in Anglesey, Dunbarton and Stirling. Europe; W. Asia; Africa.

Distinguished from *P. pusillus* and *P. rutilus* by the open stipules and from *P. berchtoldi* by the very narrow obscurely 3-veined finely pointed lvs with no lacunae bordering the thick midrib; and from all three by the single often tuberclate fr. from each fl.

P. × *bennettii* Fryer (*P. crispus* × *trichoides*) is known only from Stirling.

20. P. compressus L.   'Grass-wrack Pondweed.'

P. zosterifolius Schumach.

Rhizome terete or 0. *Stem* 0·5–2 m. × 3–6 mm., *strongly flattened and ± winged*, branched. *Submerged lvs* 10–20 cm. × 2–4 mm., linear, sessile, rounded and cuspidate or sometimes acuminate, thin, translucent; main longitudinal veins usually 5, *with many faint intermediate longitudinal strands* most of which approach the main laterals below the tip. Floating lvs 0. Stipules 2·5–3·5 cm., open, convolute, very blunt, 2-keeled, persistent. Spike 1–3 cm., cylindrical, rather dense, continuous; stalk 3–6 cm. or more, stout, compressed, not thickened upwards. *Fr.* (Fig. 63 G) 3–4·5 × 2–3 mm., ± obovoid, somewhat compressed; ventral margin convex, dorsal semicircular, smooth, bluntly 2-keeled, usually *not toothed*; beak very short, stout, almost centrally placed. Fl. 6–9. Turions with the inner lvs exceeding the outer by up to 10 mm. 2*n* = 26. Hyd.

Native. Lakes, slow streams, canals, ponds, ditches, etc., local. 30. C. and
E. England from Surrey and Essex northwards to Lancs and E. Yorks;
Montgomery; Angus. Europe, but rare in the Mediterranean region.

Readily distinguished from all other linear-leaved spp. except *P. acutifolius* by the
numerous longitudinal sclerenchyma strands. For differences from *P. acutifolius*
see under that species.

21. P. acutifolius Link   'Sharp-leaved Pondweed.'

P. cuspidatus Schrad.

Rhizome terete or 0. Stem 0·5–1 m. × 2–4 mm., strongly flattened, much
branched. Submerged lvs 5–13 × 2–3(–4) mm., linear, sessile, gradually acuminate or long-cuspidate, thin, translucent; main longitudinal veins 3, the

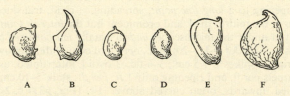

A   B   C   D   E   F

Fig. 64. Fruits of *Potamogeton*. A, *P. acutifolius*; B, *P. crispus*; C, *Groenlandia densa*;
D, *P. filiformis*; E, F, *P. pectinatus*. × 3.

laterals sometimes hardly stronger than the main faint intermediate longitudinal strands which usually have free ends just below the lf-tip. Floating lvs 0.
Stipules 1·5–2·5 cm., open, acute, many-veined, becoming fibrous and ± persistent. Spike 4–10 mm., ovoid to globose, 4–8-fld, dense; *stalk* 5–15(–35) mm.,
commonly *about equalling the spike*, slender, not thickened upwards. Fr.
(Fig. 64A) 3–4 × 2 mm., greenish-brown, half-obovoid, compressed; *ventral
margin* nearly straight, *toothed near the base and continuing upwards into the
beak, dorsal margin* curved into more than a semicircle, somewhat keeled
and ± *crenulate; beak* fairly long and *recurved*. Fl. 6–7. Turions with the
inner lvs usually not protruding beyond the outer. $2n=26$. Hyd.

Native. Ponds, streams, ditches, chiefly in calcareous water; rare. 16.
S. and E. England from Dorset to Kent and northwards to Gloucester,
Warwick, Northants, Lincoln and E. Yorks. Europe.

Closely resembles *P. compressus* in vegetative structure, but distinguished by the
usually narrower stem and lvs, the latter short and more gradually acuminate, with
many free-ending longitudinal strands and a better-marked band of air tissue along
the midrib, by the shorter and more acute stipules, and the more slender turions
with the inner lvs not protruding. The spike is shorter with a stalk about equalling it,
instead of about twice as long as in *P. compressus*, and the fr. differs considerably.

22. P. crispus L.   'Curled Pondweed.'

Incl. *P. serratus* auct. mult.

Rhizome creeping, perennating or arising from a turion. Stem 30–120 cm.,
slender, compressed, ± 4-angled with the broader sides furrowed when

mature, simple below, repeatedly forked above. *Submerged lvs* 3–9(–10·5) cm. × (5–)8–15 mm., *lanceolate to linear-lanceolate*, sessile, *rounded and blunt* or rarely acute and slightly mucronate, *serrate*, and often *strongly undulate* at the margin when mature, though commonly not so when young, often reddish, shining, translucent; longitudinal veins 3–5, the laterals close to the margin, with no faint intermediate longitudinal veins. Floating lvs 0. Stipules 10–20 mm., ± triangular, narrowed below, blunt, all but the uppermost evanescent, soon becoming torn and decayed. Fr. spike c. 1–2 cm., oblong-ovoid, rather lax; stalk 2·5–7(–10) cm., fairly slender, somewhat compressed, narrowed upwards and commonly curved. Fr. (Fig. 64B) 2–4 mm. (excluding beak), dark olive, ovoid-acuminate, compressed; ventral margin convex, dorsal strongly rounded and ± keeled; *beak about equalling rest of fr.*, tapering, ± *falcate*. Fl. 5–10. Turions 1–5 cm., with thick, horny, spiny-bordered, ± squarrose lvs. $2n = 52$. Hyd.

Native. Lakes, ponds, streams, canals, etc., to 1300 ft. in Scotland. 103, H40, S. Common throughout the British Is. except Shetland. Old World; introduced in N. America.

Readily recognized by its serrate and usually undulate lvs and by the highly characteristic fr. with its long curved beak. Young growths and other phases in which the lvs are flat and not crisped at the margins have often been named '*P. serratus*'.

Subgenus 2. COLEOGETON (Rchb.) Raunk.

Lvs all submerged, alternate (only the involucral ones opposite), narrowly linear, sheathing at the base, entire, with two wide longitudinal air-filled canals, one on each side of the midrib, occupying the greater part of their interior; stipules adnate below to the lf-base (forming the basal sheath), but free above as a ligule; spike few-fld, interrupted; stigma with large papillae; fr. drupaceous with fleshy exocarp and bony endocarp; water-pollinated.

23. P. filiformis Pers. 'Slender-leaved Pondweed.'

P. marinus auct.

Rhizome extensively creeping. Lfy stem 15–30(–45) cm., filiform, ± cylindrical, sparingly forked or branched at the base. *Submerged lvs* 5–20 cm. × 0·25–1 mm., *linear-setaceous, tapering, blunt* (*not mucronate*), yellowish-green, ± translucent; longitudinal veins 3, but *only midrib readily visible*, the laterals faint and actually in the margin. Floating lvs 0. *Stipular sheath* 0·5–3 cm., *tubular* below when young, its ligule 0·5–1·5 cm., acute, deciduous. Spike 4–12 cm., slender, interrupted, of 2–5 usually 2-fld whorls; stalk 5–25 cm., filiform not thickened upwards. Fr. (Fig. 64D) 2–3 × 2 mm., pale olive, ± obovoid, slightly compressed; *ventral margin convex*, dorsal semicircular, bluntly 3-keeled when dry; beak very short, almost central. Fl. 5–8. Perennation by tuberized winter buds borne on the rhizome. $2n = $ c. 78. Hyd.

Native. Lakes and streams, chiefly near the coast, and sometimes in brackish water. 21, H16. Anglesey, and Scotland from Berwick and Ayr northwards; Hebrides, Orkney, Shetland. N. Ireland. Northern hemisphere.

Distinguished from *P. pectinatus* by the tubular stipular sheath and the extremely short and ± central beak of the fr. and by the blunt (not mucronate) lvs.

P. × *suecicus* Richt. (*P. filiformis* × *pectinatus*) occurs in a number of localities in Scotland, and W. Yorks. It is intermediate in many features between the parent spp., and is sterile.

24. P. pectinatus L.   'Fennel-leaved Pondweed.'

Incl. *P. interruptus* Kit. (*P. flabellatus* Bab.)

Rhizome extensively creeping, usually arising from a tuberous winter bud. Lfy stem 0·4–2 m., very slender, ± cylindrical, usually much-branched. Submerged lvs 5–20 cm. × 0·25–2(–5) mm., setaceous to linear (stem lvs always broader than branch lvs, and lower lvs than upper), blunt and mucronate or tapering and acute, dark green, ± translucent; longitudinal veins 3–5, the laterals often indistinct in narrow lvs. Floating lvs 0. *Stipular sheath* 2–5 cm., *open and convolute with a whitish margin*; ligule long, ± blunt, deciduous. Spike 2–5 cm., of 4–8 usually 2-fld whorls, ± interrupted, especially below; stalk 3–10(–25) cm., filiform, not thickened upwards. Fr. (Fig. 64E, F) 3–5 × 2–4 mm., olive tinged with orange, obovoid to semi-obovoid, ± compressed; ventral margin straight or somewhat convex, dorsal semicircular, 1–3-keeled; beak continuing the ventral margin, short. Fl. 5–9. Usually perennates as tuberized lateral buds of the rhizome. 2*n*=78. Hyd.

Native. Ponds, rivers, canals, ditches, etc. 94, H29, S. Abundant in base-rich waters of the lowland zone throughout the British Is. but absent from mountainous districts of Wales and Scotland and from much of C. Ireland. Almost cosmopolitan.

Very variable in mode of branching, size and acuteness of the lvs, and size and shape of the fr. Robust plants much branched above so as to spread in a fan-like manner, with lower stem-lvs abruptly rounded to the subacute tip and reaching 3–5 mm. in width, upper stem-lvs much narrower, branch lvs filiform, acuminate, have been placed in var. *interruptus* (Kit.) Aschers. (*P. interruptus* Kit.; *P. flabellatus* Bab.). They have been described as having a much interrupted spike and large fr. differing from the type in having the ventral margin straight, not convex, and the dorsal gibbous above, with a strong central keel instead of the two distinct lateral keels of the type. There is, however, no close correlation between vegetative and fr. characters, nor between the various fr. characters, and the var. *interruptus* can be regarded as no more than a growth form.

2. GROENLANDIA Gay

Aquatic herbs with all lvs submerged and flowering-spike emergent. Like *Potamogeton* but *all lvs in opposite pairs* or rarely in whorls of 3, sessile and amplexicaul; *stipules* 0, except for the involucral lvs subtending spikes, where they are adnate to the lf-base and form 2 lateral auricles; spike few-fld; fr. an achene with thin (not bony or fleshy) pericarp. Wind-pollinated.

One sp. in Europe, W. Asia and N. Africa.

1. G. densa (L.) Fourr.   Opposite-leaved Pondweed.

Potamogeton densus L.; *P. serratus* L.

Rhizome creeping, much branched. Lfy stem 10–30 cm., ± cylindrical, forking above. *Submerged lvs* (0·5–)1·5–2·5(–4) × 0·5–1·5 cm., ovate-triangular to lanceolate, blunt or acute, with minutely serrate margins especially near the apex, commonly folded longitudinally and recurved, translucent; longitudinal veins 3–5, with sparse transverse connections. Floating lvs 0. *Spike usually of only 4 fls, ovoid in fr.*; stalk 5–15 mm., slender, shorter than the lvs, erect at first, then strongly recurved. Fr. (Fig. 64c) c. 3×2 mm., olive, orbicular-obovate, strongly compressed; ventral margin convex, dorsal semi-circular, sharply keeled; beak short, $\pm$ central, recurved. Fl. 5–9. $2n = 30$. Hyd.

Native. Clear swift streams, ditches, ponds. 66, H14. Locally abundant in lowland Great Britain northwards to Ross; Ireland, northwards to Dublin and Sligo. Europe; western N. Africa; W. Asia.

Variable in size and shape of the lvs. In some plants they are only $5-10 \times c.$ 3 mm.

129. RUPPIACEAE

Submerged aquatic herbs of salt or brackish pools, rarely in fresh waters. Lvs linear or setaceous, sheathing at the base. Infl. a short terminal raceme appearing subumbellate. Bracts 0. Fls hermaphrodite, small. Perianth 0. Stamens 2, opposite each other; filaments very short and broad. Ovary superior; carpels 4 or more, becoming long-stipitate in fr.; ovule solitary, pendulous, campylotropous. Fr. indehiscent.

One genus and about 3 spp. throughout temperate and subtropical regions and rarely in mountain lakes in the tropics.

1. Ruppia L.

The only genus.

Peduncle in fr. many times longer than pedicel.	**1. spiralis**
Peduncle in fr. about equalling pedicel.	**2. maritima**

1. R. spiralis L. ex Dum.

R. maritima auct. mult., non L.

A slender perennial 30 cm. or more. Stems filiform, much-branched. Lvs alternate or opposite, filiform (c. 1 mm. wide), dark green, sheaths dilated, brownish. *Peduncle 10 cm. or more, much elongated after flowering, several times as long as the pedicels of the carpels*, often spirally coiled. *Fr. ovoid, nearly symmetrical, beak slightly oblique.* Fl. 7–9. Hyd. $2n = 16$.

Native. In brackish ditches near the sea, rather rare. 27, H8. Scattered round the coasts of England, Wales and Ireland; Scotland; Wigtown, Orkney and Shetland. Distribution of the genus.

2. R. maritima L.

R. rostellata Koch

Similar to *R. spiralis* in general appearance. Lvs c. 0·5 mm. wide, light green, sheaths narrow. *Peduncle 0·5–5 cm., shorter than to twice as long as the pedicels of the carpels*, flexuous but not spirally twisted. *Fr. very asymmetrical, ventral side convex ending in a long beak, dorsal side strongly gibbous at base.* Fl. 7–9. Hyd.

Native. In brackish ditches and salt-marsh pools, local. 58, H21. Round most of the coasts of the British Is. north to Shetland, less frequent in the north. Europe (excluding Iceland), N. Africa, W. Asia, N. America.

130. ZANNICHELLIACEAE

Submerged perennial aquatic herbs of fresh, brackish or sea-water, with slender creeping rhizomes. Lvs alternate, opposite or ± whorled, linear, with sheathing bases and usually with a ligule at the junction of blade and sheath. Fls very small, solitary axillary or in cymes; unisexual (monoecious or dioecious), hypogynous. Perianth of 3 small scales or 0; stamens 3, 2 or 1, with 1–2-celled anthers and spherical or filiform pollen grains; carpels 1–9, free, each with 1 pendulous orthotropous ovule; style terminating in a capitate, peltate or spathulate stigma. Pollination hydrophilous. Fr. of 1–9 sessile or stalked achenes, each with 1 non-endospermic seed.

About 20 spp. in 6 genera, widely distributed, mainly in salt water.

1. ZANNICHELLIA L.

Perennial monoecious herbs of fresh or brackish water with slender simple or branched lfy stems. Lvs mostly opposite, linear, entire, with sheathing or free axillary stipules. Fls axillary in a hyaline deciduous cup-shaped spathe, 1 male and 2–5 female fls often in the same spathe. Male fl. of 1 stamen; pollen grains spherical. Female fl. of 1 carpel with a short or long style and a flattened peltate or lingulate stigma. Perianth 0. Fr. a stalked ± curved achene, entire or toothed along one or both margins.

Two spp., one cosmopolitan, the other in S. Africa.

1. Z. palustris L.   'Horned Pondweed.'

A submerged herb with a filiform rhizome and filiform much-branched lfy shoots commonly to 50 cm., rooting only near the base or creeping and rooting over much of their length. Lvs 1·5–5(–10) cm. × 0·4–2 mm., linear to capillary, tapering to a fine point, parallel-veined, translucent; stipule amplexicaul, tubular below, scarious, soon falling. Style stout below, tapering upwards; stigma peltate with a ± waved and crenate margin. Achenes 2–6, each 2–3 mm. excluding the persistent style (0·5–1·5 mm.), subsessile or with a short stalk up to 1 mm.; dorsal margin ± toothed. Fl. 5–8. $2n = 28$. Hyd.

Very variable. Var. *pedicellata* (Fr.) Wahlenb. & Rosen. (*Z. pedicellata* Fr., *Z. maritima* Nolte) has the 2–5 achenes distinctly stalked (1·5–2·5 mm.) and often on a common peduncle, the persistent style longer (2–2·5 mm.) and more slender, and the stigma lingulate instead of peltate. It is said to be a more slender plant with a preference for brackish water, and may need to be treated as a separate ssp. *Z. pedunculata* Rchb. and *Z. gibberosa* Rchb., the latter with both dorsal and ventral margins of the achenes muricate, appear to belong here.

Native. Rivers, streams, ditches and pools of fresh or brackish water; to 700 ft. in Ireland; locally common. 88, H27, S. Throughout the British Is. Cosmopolitan.

131. NAJADACEAE

Slender submerged fresh- or brackish-water herbs. Lvs subopposite or verticillate, linear, base sheathing; two small scales (*squamulae intravaginales*) within each sheath. Fls unisexual, small; plants monoecious or rarely dioecious. Male fl. enclosed in a spathe; perianth 2-lipped; stamen 1, anther sessile. Female fl. naked or with a very thin perianth; carpel one, 1-celled, stigmas 2–4; ovule solitary, basal, erect, anatropous. Fr. indehiscent.

One genus and about 35 spp. in temperate and tropical regions.

1. NAJAS L.

The only genus.

1 Lvs with many large spinous teeth (Norfolk Broads).   **3. marina**
  Lvs minutely denticulate or nearly entire.   2

2 Lvs very minutely denticulate or nearly entire, 2–3 together (lakes).
       1. flexilis
  Lvs minutely spinous-denticulate, mostly densely tufted on short lateral branches (Reddish Canal).   **2. graminea**

1. N. flexilis (Willd.) Rostk. & Schmidt

A slender brittle annual. Stems c. 30 cm., smooth. Lvs c. 20×0.75 mm., 2–3 together, translucent, apiculate, margin very minutely and remotely denticulate or nearly entire; sheaths ciliate. Fls 1–3 in the axils of the lvs. Fr. 3 mm., narrowly ovoid. Fl. 8–9. Hyd. $2n=24$.

Native. In lakes. 6, H7. N. Lancashire, Perth, Islay, Outer Hebrides, Kerry, Galway, Mayo and Donegal. Finland, Norway, Sweden and N. Germany; C. Russia; N. America.

*2. N. graminea Del.

Similar to *N. flexilis* but rather stouter. Lvs mostly densely tufted on short lateral branches, margins minutely spinous-denticulate; sheaths denticulate. Fr. 2 mm., narrowly ellipsoidal. Fl. 8–9. $2n=24$. Hyd.

Introduced. In water warmed by the outflow from a mill, Reddish Canal near Manchester. Probably introduced with Egyptian cotton. Tropics and subtropics of the Old World to Australia and New Caledonia.

3. N. marina L.

Stems with occasional teeth near the top, forked, rather stiff and brittle. Lvs strongly spinous-dentate and $\pm$ toothed on the back; sheaths entire, not ciliate. Fr. 6 mm., ellipsoid or ovoid. Fl. 7–8. Hyd. $2n=12*$.

Native. In slightly brackish water in a few of the Norfolk Broads. Cosmopolitan, except for the colder parts of the temperate regions; in Europe extending north to the Baltic.

132. ERIOCAULACEAE

Perennial, rarely annual, usually scapigerous herbs, sometimes woody at base. Lvs usually linear, crowded, mostly radical. Fls small, actinomorphic, unisexual, crowded in bracteate heads, male and female often mixed, or male in the middle, female surrounding them, very rarely male and female in separate heads. Perianth scarious or membranous, segments in 2 series, each 2–3-merous, outer free or ± connate, inner often tubular, rarely 0. Stamens as many as or double (rarely fewer than) the number of per. segs and opposite them; staminodes usually 0 in female fls. Ovary superior, 2–3-celled, style short, terminal, stigmas 2–3; ovules solitary and pendulous in each cell. Fr. membranous, loculicidal.

About 9 genera and 600 spp., distributed throughout most of the world except the continent of Europe, few in temperate regions, particularly abundant in S. America; frequently in swampy habitats.

1. ERIOCAULON L.

The only British genus. Male fls chiefly in the middle of the head. Stamens 6 or 4, anthers 2–3-celled. Inner per. segs free in both sexes.

Over 200 spp. with the distribution of the family.

1. E. septangulare With.   Pipe-wort.

A slender *scapigerous perennial herb with a creeping rootstock. Roots white, soft, jointed and worm-like. Stem very short, densely lfy.* Lvs 5–10 cm., *subulate, laterally compressed, translucent, septate.* Scape (5–)15–60(–150) cm., 6–8-*furrowed, twisted.* Heads 0·5–2 cm. diam.; *bracts obovate, obtuse, lead-coloured;* bracteoles black, obtuse. Fls 2-merous, outer per. segs lead-coloured, with a tuft of hairs at tip; inner pale with a black spot near top, ciliate. Fl. 7–9. $2n = 64^*$. Hyd. or Hel.

Native. In shallow water or bare wet ground, forming dense mats on peaty soil. 2, H7. Scotland: Skye and Coll; Ireland: from W. Cork to W. Donegal, locally abundant near the coast and ascending to 1000 ft. The same or a closely related sp. is widely distributed in N. America; our plant is probably endemic.

133. LILIACEAE

Herbs, rarely shrubby, sometimes climbing. Lvs alternate or in several whorls or all radical. Infl. usually racemose (an umbel in *Allium*). Fls usually hermaphrodite and actinomorphic, usually 3-merous (rarely 2-, 4- or 5-merous). Perianth usually petaloid, usually in two similar whorls. Stamens in two whorls, opposite per. segs, inserted on them or free. Ovary superior, usually 3-celled with axile placentation, rarely 1-celled with parietal placentation. Ovules usually numerous, in 2 rows on each placenta, anatropous. Fr. a capsule or berry. Seeds with copious fleshy or cartilaginous endosperm.

About 200 genera and 2500 spp., cosmopolitan.

A family of very diverse habit and appearance but of uniform floral structure. It might well be divided into several families but no satisfactory way of doing this has yet been devised, though the families recognized by Hutchinson (*Families of Flowering Plants*, II. Monocotyledons) appear to be natural groups. Of these Ruscaceae has been retained here in Liliaceae but Trilliaceae is separated; the others are not British. The Liliaceae probably represent a central type from which the other families of the order have been derived; for the distinctions, see these families.

1 Lvs normally developed; herbs with the aerial stems unbranched or
    branched only in infl.   *2*
  Lvs scale-like, the assimilating organs being lf-like or needle-like cladodes,
    borne in their axils; much-branched herbs or shrubs.   *17*

2 Fls racemose or solitary, rarely subumbellate and then bracts lf-like;
    plant not smelling of onion.   *3*
  Fls umbellate; infl. enclosed before flowering in a spathe, splitting at fl.
    into 1 or more, usually scarious, involucral bracts, or caducous; plant
    smelling of onion or garlic.   **18. ALLIUM**

3 Lvs alike on both surfaces, vertical, iris-like. Wet places.   *4*
  Lvs ± horizontal, the surfaces differing. Drier ground.   *5*

4 Styles 3, free; fls whitish; filaments glabrous.   **1. TOFIELDIA**
  Style simple; fls yellow; filaments woolly.   **2. NARTHECIUM**

5 Fl.-stem lfy or with lf-like bracts.   *6*
  Lvs all radical or lvs absent at flowering, or stem with a few scale-like lvs.   *11*

6 Per. segs 4, white; lvs cordate at base.   **6. MAIANTHEMUM**
  Per. segs 6; lvs not cordate.   *7*

7 Fls in the axils of the lvs, greenish-white.   **5. POLYGONATUM**
  Infl. terminal, or fls terminal and solitary, of other colours.   *8*

8 Per. segs recurved; anthers dorsifixed, versatile; infl. a raceme.
    9. LILIUM
  Per. segs not recurved; anthers basifixed; infl. not a raceme.   *9*

9 Fls solitary.   *10*
  Fls subumbellate, yellow.   **13. GAGEA**

10 Fls yellow, erect, 3–5 cm.; lvs broad.   **11. TULIPA**
  Fls white, erect, 1–2 cm.; lvs filiform. (Caernarvon.)   **12. LLOYDIA**
  Fls purple, rarely white, nodding, 3–4 cm.; lvs linear.   **10. FRITILLARIA**

11 Fls solitary, from the ground; plant lfless at fl.   **19. COLCHICUM**
  Fls in infl.; lvs present at fl.   *12*

12 Infl. a panicle; fls white (to c. 1 cm.). (If fls large and yellow or orange,
    see *Hemerocallis*.)   **3. SIMETHIS**
  Infl. a raceme.   *13*

13 Lvs ovate-lanceolate, petiolate; stock a rhizome.   **4. CONVALLARIA**
  Lvs linear, sessile; stock a bulb.   *14*

14 Perianth contracted at mouth, united for the greater part of its length;
    upper fls sterile.   **17. MUSCARI**
  Perianth not contracted at mouth; segments free or united only at base;
    fls all fertile.   *15*

15 Fls blue or purple, rarely pure white.   *16*
  Fls white, marked with green outside.   **14. ORNITHOGALUM**

16 Bracts 0 or 1 to each fl.; per. segs quite free, 8 mm. or less.   **15. SCILLA**
  Bracts 2 to each fl.; per. segs united at base, more than 1 cm.
    16. ENDYMION

17 Cladodes needle-like, not bearing the fls.   **7. ASPARAGUS**
  Cladodes broad, bearing the fls in the middle of one surface.   **8. RUSCUS**

Tribe 1. **NARTHECIEAE.** Stock a rhizome. Radical lvs numerous, vertical, distichous, the two surfaces alike. Stem-lvs few or 0. Fls in racemes. Per. segs free or nearly so, persistent in fr. Anthers introrse or extrorse. Fr. a capsule.

1. Tofieldia Huds.

Anthers ovate, *introrse*; filaments glabrous. Ovary trigonous, carpels free above. *Styles* 3, *free. Capsule septicidal.* Seeds oblong, numerous, very small.
About 14 spp., N. temperate and arctic regions and Andes.

1. T. pusilla (Michx.) Pers.   'Scottish Asphodel.'

T. palustris auct.; *T. borealis* (Wahlenb.) Wahlenb.

Glabrous. Rhizome short. Radical lvs 1·5–4 cm. × 1–2 mm., rigid, 3–5-veined; stem-lvs 0 or 1–2, much smaller. Fl.-stem 5–20 cm. Raceme short, dense, 5–15 mm. Bracts 3-lobed, scarious, shorter than or equalling pedicel. Pedicel 1 mm. or less. Per. segs greenish-white, obovate-oblong, c. 2 mm., obtuse. Fl. 6–8. Pollinated by insects and selfed. $2n = 30$. Hs.–Hel.

Native. By springs and streams on mountains from 700 to 3000 ft., local; N.W. Yorks and Durham; Arran and Stirling to Caithness. 18. Arctic and subarctic Europe from Iceland and the Faeroes eastwards to the Urals; Alps (from France to Croatia); E. Siberia; Arctic N. America and Canada.

2. Narthecium Huds.

Anthers linear, *extrorse*, versatile. Filaments woolly. Ovary trigonous, carpels completely united. *Style* 1, *simple. Capsule loculicidal.* Seeds numerous, elongated into a tail at each end.
Eight spp., north temperate regions.

1. N. ossifragum (L.) Huds.   Bog Asphodel.

Glabrous. Rhizome creeping; roots many, fibrous. Radical lvs 5–30 cm. × 2–5 mm., rigid, often curved, usually 5-veined; stem-lvs few, 4 cm. or less, sheathing. Fl.-stem 5–40 cm. Raceme 2–10 cm., rather dense, or lax below. Bracts lanceolate, entire, about equalling pedicel. Pedicel 5–10 mm., with a subulate bracteole above the middle. Per. segs 6–8 mm., yellow, linear-lanceolate, spreading in fl., erect in fr. Anthers orange. Stem, per. segs and ovary becoming uniformly deep orange after fl. Capsule narrow-ovoid, 6-grooved, mucronate, c. 12 mm. Fl. 7–9. Pollinated by insects. Fr. 9. $2n = 26$. Hs.–Hel.

Native. Bogs, wet heaths and moors and wet acid places on mountains, ascending to 3270 ft.; common over most of the British Is., becoming less common towards the south-east and absent from several eastern and midland counties; often abundant, and dominant in the hollows of the raised bog cycle, sometimes also locally dominant in blanket bogs. 98, H 40. Faeroes, Scandinavia (to 69° 42′ N.), the Netherlands, Belgium, N.W. Germany, W. and C. France, N. Spain and Portugal (in the mountains).

Tribe 2. Asphodeleae. Stock a short rhizome. Lvs usually all basal or stem-lvs few and reduced. Fls in racemes or panicles. Per. segs free or nearly so, sometimes persistent. Anthers introrse. Fr. a loculicidal capsule. Species of *Anthericum, Asphodeline, Asphodelus, Bulbinella* and *Eremurus* are more or less frequently cultivated.

3. SIMETHIS Kunth

Lvs all basal. Infl. a panicle. Per. segs free, spreading. *Filaments woolly*, somewhat thickened, *inserted into a pit* on the anther; *anthers versatile, dorsifixed.* Ovules 2 in each cell. Style simple, filiform.

One sp.

1. S. planifolia (L.) Gren. & Godr.

S. bicolor (Desf.) Kunth; *Pubilaria planifolia* (L.) Druce

Glabrous. Rhizome very short, erect, clothed with the brown fibrous remains of the lf-bases; roots tufted, fusiform, fleshy. Lvs linear, 15–45 cm. × c. 6 mm., glaucous. Scape about as long. Panicle lax. Bracts shorter than pedicels. Per. segs oblong, c. 8–10 mm., white within, purplish outside. Capsule c. 6 mm., subglobose, obscurely angled. Seeds black, shining. Fl. 5–7. Gr.

Native. Rough rocky furzy heath near the coast, over an area of 8 or 9 × 1–2 miles near Derrynane (Kerry); very rare among the pines near Bournemouth (Dorset), possibly introduced with *Pinus pinaster*; planted in Hants from Dorset stock and naturalized. 2, H1. W., C. and S. France, W. Spain, Portugal, Corsica, Sardinia, N.W. Italy, Morocco, Algeria, Tunisia.

The commonly cultivated genus *Kniphofia* (Red-Hot Poker) with numerous scarlet cylindrical fls belongs to the tribe *Kniphofieae* chiefly distinguished from the preceding by the cylindrical or campanulate perianth with united segments.

Tribe 3. HEMEROCALLIDEAE. Differs from Asphodeleae mainly in the funnel-shaped perianth united below and the fls usually slightly zygomorphic at least by the curving upwards of the stamens and style. Spp. of *Hosta* (*Funkia*) with ovate, petiolate lvs are commonly cultivated.

*Phormium tenax J.R. & G. Forst., New Zealand Flax, with fibrous distichous reddish-bordered lvs 1–3 m., and numerous reddish fls on a scape 1–5 m., is naturalized in W. England and W. Ireland. New Zealand.

HEMEROCALLIS L.   Day-Lily.

Rhizome very short, with ± fleshy roots. *Lvs long, linear. Infl. a panicle.* Stamens inserted on the perianth-tube, curved upwards; anthers oblong-linear. Ovary trigonous; ovules many in each cell; style long, simple, curved upwards.

About 15 spp., temperate Asia.

Perianth dull orange, 8–10 cm., segments undulate.   **fulva**
Perianth lemon yellow, 6–8 cm., segments flat.   **lilioasphodelus**

*H. fulva (L.) L.

Roots thick, not swollen. Lvs 1–2 cm. broad. Stems 50–100 cm., with a few scale-like lvs. Fls 8–10 cm., dull orange, scentless, per. segs elliptic-oblong, reticulately-veined, margins undulate. Sterile. Fl. 6–8. $2n = 33$.

Commonly grown in gardens and sometimes naturalized. A sterile triploid, not known wild. Allied diploids in China.

*H. lilioasphodelus L. (*H. flava* (L.) L.)

Roots with swollen tubers. Lvs 5–10 mm. broad. Stem 40–80 cm., naked or nearly so. Fls 6–8 cm., lemon-yellow, scented; per. segs oblong-lanceolate, without transverse veins, margins flat. Fl. 5–6. $2n = 22$.

Commonly grown in gardens and sometimes escaping. Native probably of E. Asia.

Tribe 4. CONVALLARIEAE. Stock a rhizome. Lvs clustered from the rhizome. Fls in a spike or raceme on a lfless scape from the axils of the lvs. Anthers introrse. Ovules few in each cell; style simple, long; stigma capitate. Fr. a berry.

4. CONVALLARIA L.

Rhizome long, creeping, much-branched, with clusters of roots from the nodes. Lvs in pairs on long petioles, that of the lower sheathing the upper so as to appear stem-like, with several scale-lvs at the base of the lf-pair. Scapes, from the axil of one of the scale-lvs. *Fls* nodding *in* unilateral *racemes. Perianth* globose-campanulate, *segments united* to about the middle. Stamens inserted on base of perianth, included. Ovary with 4–8 ovules in each cell.

Three spp., north temperate regions.

1. C. majalis L.   Lily-of-the-Valley.

Glabrous. Lvs 8–20 × 3–5 cm., ovate-lanceolate or elliptic, acute; petiole 5–12 cm. Infl. 6–12-fld. Bracts scarious, ovate-lanceolate, shorter than pedicels. Fls sweet-scented. Perianth white, c. 8 mm. Berry globose, red. Fl. 5–6. Pollinated by insects and selfed; weakly protandrous. $2n = 38$. Grh.

Native. Dry woods, mainly calcareous; local and with a distinct eastern tendency; widespread in England; in Wales only in the south-east and north; in Scotland more local, perhaps native as far north as Dunbarton and E. Inverness; not native in Ireland. 69. Commonly cultivated and sometimes escaping. Europe from Scandinavia (67° 17′ N.) and N. Russia (Ladoga-Ilmen region) to N. Spain, C. Italy, Greece and the Caucasus; N.E. Asia.

Two spp. of *Aspidistra* Ker-Gawl. belonging to the tribe *Aspidistreae*, chiefly differing from the preceding in the solitary or densely spicate fls and lobed stigmas, are commonly grown in pots in houses. They have solitary brownish fls at ground level and are said to be pollinated by snails.

Tribe 5. POLYGONATEAE. Stock a rhizome. Stem lfy, terminal on the rhizome. Fls axillary or in terminal racemes or panicles. Anthers introrse, basifixed. Style slender, ± 3-fid. Fr. a berry.

5. POLYGONATUM Mill.

Rhizome thick, long, creeping. Lvs all cauline, numerous. *Fls* 3-merous, nodding, *axillary, solitary, or in few-fld axillary racemes. Perianth* tubular-campanulate, *united into a tube* for the greater part of its length. Stamens included, inserted on perianth-tube. Anthers oblong, 2-lobed at base. Ovules 4–6 in each cell; style filiform, with a small, 3-lobed stigma. Plants glabrous or nearly so.

About 23 spp., north temperate regions.

1	Lvs in whorls.	**1. verticillatum**
	Lvs alternate.	*2*
2	Fls 1–2; perianth 18–22 mm., cylindric, not contracted in the middle; stem angled.	**2. odoratum**
	Fls 2–5; perianth 9–15 mm., contracted in the middle; stem terete.	**3. multiflorum**

1. P. verticillatum (L.) All. 'Whorled Solomon's Seal.'

Stem 30–80 cm., angled. *Lvs in whorls* of 3–6, 5–12 cm., linear-lanceolate, sessile. Fls 1–4 on a common peduncle. Perianth 6–8 mm., greenish-white, contracted in the middle. Filaments papillose. Fr. c. 6 mm., globose, red. Fl. 6–7. Pollinated by bees and selfed; homogamous. $2n=28$ (30, 60, 86–91). Grh.

Native. Mountain woods, very rare; Northumberland, Dumfries, Perth. 4. Europe from Scandinavia (69° 35′ N.) and Germany to the mountains of C. Spain, Apennines and Macedonia; Caucasus (not elsewhere in Russia); Asia Minor, Himalaya.

2. P. odoratum (Mill.) Druce 'Angular Solomon's Seal.'
P. anceps Moench; *P. officinale* All.

Stem 15–20 cm., *angled*, arching. *Lvs* 5–10 cm., *alternate*, subdistichous, ovate to elliptic-oblong, acute, sessile. *Fls scented*, 1 *or* 2 on a common peduncle. *Perianth* 18–22 *mm.*, *cylindric*, *not contracted in the middle*, greenish-white. *Filaments glabrous.* Fr. c. 6 mm., globose, blue-black. Fl. 6–7. Pollinated by humble-bees and selfed; homogamous. $2n=20$ (26, 28, 29, 30). Grh.

Native. Limestone woods, very local in N. and W. England and Wales from Devon, Hants, and Derby to Westmorland and Northumberland; S. Inner Hebrides. 22. Europe from Scandinavia (c. 66° N.), Finland and the Volga-Kama region, to the mountains of Spain and Portugal, C. Italy, Greece and the Caucasus; Morocco; Siberia, China, W. Himalaya.

3. P. multiflorum (L.) All. Solomon's Seal.

Stem 30–80 cm., *terete*, arching. *Lvs* 5–12 cm., *alternate*, subdistichous ovate to elliptic-oblong, acute, sessile. *Fls* 2–5 in axillary racemes. *Perianth* 9–15 *mm.*, *contracted in the middle*, greenish-white. *Filaments pubescent.* Fr. c. 8 mm., globose, blue-black. Fl. 5–6. Pollinated by humble-bees and selfed; homogamous. $2n=18$ (20, 24, 28, 29, 30). Grh.

Native. Local in woods over most of England and Wales; Jersey; reported as naturalized in Scotland. 51, S. Europe from S. Scandinavia, Finland and the Transvolga region to the mountains of N. Spain, Italy, Greece and N. Caucasus; temperate Asia to Japan (but not in Asiatic Russia).

The hybrid *P. multiflorum × odoratum=P. × hybridum* Brügg. is probably the commonest member of the genus in gardens. Fls 2–4 together, the size of those of *P. odoratum*, somewhat contracted in the middle. It is sometimes found ± naturalized and may occur wild.

6. MAIANTHEMUM Weber

Rhizome slender, creeping. Lvs 1–3. *Fls 2-merous*, in an erect *terminal raceme*. Per. segs free, ± spreading. Stamens inserted on the base of the per. segs, anthers ovoid. Ovules 2 in each cell. Style short, stigma obscurely lobed.

Three spp., north temperate regions.

1. M. bifolium (L.) Schmidt　　　　　　　　　　　　　　May Lily.

M. Convallaria Weber; *Unifolium bifolium* (L.) Greene; *Smilacina bifolia* (L.)
J. A. & J. H. Schult.

Glabrous except for the upper part of the stem which has stiff white hairs.
Stem 8–20 cm., flexuous. Lvs 3–6 cm., ovate, acute or shortly acuminate,
deeply cordate at base, the solitary radical lvs long-petioled, upper cauline
often sessile. Infl. 2–5 cm., 8–15-fld, rather dense. Perianth white, segments
1–2 mm. Fr. c. 6 mm., red. Fl. 5–6. Pollinated by insects and selfed; protogynous. $2n = 30, 32, 36, 38, 42$.

Native? Woods, very rare; N.E. Yorks, N. Lincoln, Middlesex, probably
native, at least in the first of these; recorded from several other counties but
probably only as an escape. 3. Europe from Scandinavia (69° 50′ N.) and
N. Russia to N., E. and C. France, N.E. Spain, N. Apennines, Bosnia and
S. Russia (Lower Don region); N. Asia to Kamchatka and Korea.

Tribe 6. ASPARAGEAE. Stock a rhizome. Stems erect or climbing, sometimes woody, much-branched. Lvs reduced to small scarious scales, bearing
in their axils 1 or more green assimilating cladodes. Fls solitary to racemose,
axillary, not borne on the cladodes. Pedicels articulated. Per. segs free or
nearly so. Stamens free. Anthers introrse, dorsifixed. Ovules few. Fr. a
berry.

7. ASPARAGUS L.

The only genus. About 100 spp., Old World.

1. A. officinalis L.　　　　　　　　　　　　　　　　　Asparagus.

Glabrous herb with creeping rhizome. Stems annual, erect or procumbent,
much-branched. Lvs scarious, whitish, thin; those of the main stem triangular-
lanceolate, to 5 mm.; those of the branches sagittate, ovate, less than 1 mm.
Cladodes in clusters (up to 10 in each), needle-like. Fls 1 or 2, axillary on
separate pedicels, dioecious by abortion, rarely hermaphrodite. Perianth
campanulate, 3–6 mm., male larger than female. Fr. red, globose. Fl. 6–9.
Pollinated by insects. Grh.

Ssp. **prostratus** (Dum.) E. F. Warb.

A. officinalis var. *maritimus* auct., non L.; *A. prostratus* Dum.

Stems ± procumbent, 10–30(–100) cm. Cladodes thick, rigid, glaucous,
(4–)5–10(–15) mm. Pedicels 3–6 mm. Per. segs of male fl. 7–8·5 mm., yellow,
tinged red at base, recurved at tip; of female fl. 4–4·5 mm., yellow to whitish-
green, slightly recurved. Fr. 5–7 mm., seeds 1–6, most frequently 2. Fl. 6–7.
$2n = 40*, ?80*$.

Native. Grassy sea-cliffs, very local and rare. Dorset, Cornwall, W.
Gloucester, Wales; maritime sands, Waterford, Wexford, Wicklow; Channel Is.;
decreasing. 9, H 3, S. West coast of Europe from N. Germany to S.W. France.

*Ssp. **officinalis**

Incl. *A. officinalis* var. *altilis* L.

Stems erect, 30–150 cm. Cladodes slender, often flexuous, green, 5–15(–20) mm.
Pedicels 6–10 mm. Per. segs of male fl. 5–6 mm., yellow, straight; of female fl.
c. 4 mm., yellow-green, straight. Fr. 7·5–8·5 mm., seeds 5–6. Fl. 7–9. $2n = 20*$.

Introduced. Commonly cultivated as a vegetable and naturalized in many localities in waste places and dunes, etc. Europe, etc., but native distribution obscure owing to frequent cultivation.

Tribe 7. RUSCEAE. Stock a rhizome. Stems woody, erect or climbing. Lvs reduced to small scarious scales, bearing in their axils a broad green assimilating flattened lf-like cladode. Fls on the surface or margin of the cladode or in free terminal racemes. Per. segs free or connate. Stamens 3 or 6; filaments connate into a column; anthers extrorse, sessile. Ovules 2 in each cell. Fr. a berry.

8. RUSCUS L.

Stems erect. Fls solitary or in clusters on the surface of the cladode. Fls dioecious. Inner per. segs smaller than outer. Stamens 3. Style short. Perianth persistent.

Three spp. in Mediterranean region, W. Europe, Madeira, Azores.

1. R. aculeatus L.   Butcher's Broom.

Glabrous. Rhizome creeping, thick, fibrous. Stems erect, 25–80 cm., stiff, much-branched, green, striate. Lvs ± triangular, less than 5 mm., scarious, thin, brownish. Cladodes 1–4 cm., ovate, entire, spine-pointed, thick, rigid, dark green, twisted at base. Fls 1–2 on the upper surface of the cladode in the axil of a small scarious bract. Perianth greenish, c. 3 mm. Female fl. with a cup representing the stamens. Fr. c. 1 cm., red, globose. Fl. 1–4. Fr. 10–5. Germ. summer. $2n = 40$. N.

Native. Dry woods and among rocks, widespread in S. England but rather local; extending north to Caernarvon, Leicester and Norfolk; commonly cultivated and sometimes escaping in N. England and Scotland. 41, S. Mediterranean region north to Transsylvania, S. and W. Switzerland and N. France; Azores.

Tribe 8. TULIPEAE. Stock a bulb. Stem lfy. Fls solitary or racemose, rarely subumbellate. Per. segs free, usually all alike. Anthers introrse. Fr. a loculicidal (rarely septicidal) capsule.

Species of *Erythronium* (Dog's Tooth Violet) are commonly grown in gardens.

9. LILIUM L.   Lily.

Bulb of numerous imbricate fleshy scales *without a tunic*. Lvs all cauline, numerous, alternate or whorled. Fls large, solitary or in a terminal raceme, erect to nodding. Perianth of various shapes; segments spreading or revolute, caducous, with a longitudinal groove-like nectary at the base. *Anthers dorsifixed, versatile*. Style long; stigma ± 3-lobed. Seeds numerous, flat.

About 90 spp., N. temperate regions. Many are grown in gardens.

Lvs whorled; fls purple.	**1. martagon**
Lvs alternate; fls yellow.	**2. pyrenaicum**

1. L. martagon L.   'Martagon Lily.'

Stem 50–100 cm., erect, scaberulous. Lvs 7–20 cm., *mostly in distant whorls* of 5–10, obovate-lanceolate, scaberulous on margins; upper alternate, smaller.

Fls c. 4 cm. across, alternate, nodding, 3–10 in a terminal raceme. *Perianth dull purple* with darker raised projections; segments oblong, clawed, strongly recurved. Anthers reddish-brown. Capsule obtusely 6-angled, obovoid. Fl. 8–9. Pollinated by lepidoptera and selfed; homogamous. $2n = 24$. Gb.

?Introduced. Commonly grown in gardens and quite naturalized in woods in a number of places from Somerset and Kent to Monmouth, Cumberland and Yorks, mainly in the south and possibly native in Surrey and Gloucester; Fife. Mountains of Europe from C. and E. France and C. Russia (Upper Volga region) to N. Spain and Portugal, C. Italy, Greece and the Caucasus; naturalized in Scandinavia; Siberia, N. Mongolia.

*2. L. pyrenaicum Gouan

Stem 40–80 cm., erect, glabrous. *Lvs alternate*, very dense, linear-lanceolate. Fls nodding, 1–8 in a terminal raceme. *Perianth yellow* with small black dots; segments oblong, strongly recurved. Anthers reddish. Capsule obtusely 6-angled, obovoid. Fl. 5–7. $2n = 24$. Gb.

Introduced. Grown in gardens, sometimes escaping and quite naturalized on hedgebanks in N. Devon, and elsewhere. Native of Pyrenees.

10. FRITILLARIA L.

Bulb small with few scales and thin white tunic. Fls solitary or paired, terminal or in whorls. *Perianth campanulate, nodding, caducous, with a large glistening nectary near the base of each segment. Anthers basifixed.* Style long, entire or with 3 stigmas. Seeds many, flat, often winged.

About 70 spp., north temperate regions. A number are cultivated, including *F. imperialis* L. (Crown Imperial) with numerous lvs partly in whorls and a single whorl of yellow or orange fls.

1. F. meleagris L.   Snake's Head, Fritillary.

Glabrous, somewhat glaucous. Stem erect, 20–50 cm. Lvs few (3–6), alternate, linear, 8–20 cm. (upper shorter) × 4–9 mm. Fls solitary, rarely paired. Perianth 3–5 cm., dull purple, chequered dark and pale, rarely cream-white, segments oblong. Stigmas linear. Capsule subglobose, trigonous. Fl. 4–5. Pollinated by humble-bees and selfed. $2n = 24$. Gb.

Native. Damp meadows and pastures from Somerset and Kent to Stafford and Lincoln, very local and in many places doubtfully native; commonly grown in gardens and sometimes escaping elsewhere. 28. Europe from Sweden and the Volga-Don region to France, Austria, Serbia and the Lower Volga region.

11. TULIPA L.

Bulb with numerous scales and brown tunic. Fls usually solitary, terminal. *Perianth ± campanulate or rotate, erect, segments without nectaries, caducous. Anthers basifixed.* Style usually 0, stigma 3-lobed. Seeds many, flat.

About 50 spp., C. and S. Europe, temperate Asia, N. Africa. The garden tulip (*T. gesnerana* L.) was introduced to Europe from Turkey where it was cultivated, in the sixteenth century. Its origin is unknown, no closely related certainly wild species having been found.

*1. T. sylvestris L. Wild Tulip.

Glabrous, slightly glaucous. Stem 30–60 cm., flexuous, with (most often) 3 lvs near the base. Lvs linear, 15–30 cm. × 6–12 mm. Fls drooping in bud, solitary or 2. Perianth yellow, greenish outside, 3–5 cm.; segments elliptic, acute, opening widely in the sun. Filaments villous at base. Fr. oblong-trigonous c. 3 cm.; rarely produced. Fl. 4–5. Pollinated by small insects and selfed; homogamous. $2n=48$. Gb.

Introduced. Naturalized in meadows and orchards, mainly in E. England and S.E. Scotland north to Fife, extending west to Somerset, Gloucester, Worcester and Lancashire. 35. France, Italy, Sardinia, Sicily; N. Balkans to Serbia and Bulgaria; naturalized in C. and N. Europe.

12. LLOYDIA Salisb. ex Rchb.

Bulb small, with a membranous tunic. Fls 1–2, terminal. *Perianth erect, segments persistent*, spreading, *with a small transverse nectary* at the base. Anthers basifixed. Style filiform, stigma obtuse. Seeds many, 3-angled.

About 12 spp. in north temperate and arctic regions.

1. L. serotina (L.) Rchb. 'Lloydia.'

Glabrous. Stem 5–15 cm., with 2–4 lvs. Lvs filiform, radical 15–25 cm., cauline much shorter, relatively broader. Perianth white with purplish veins, segments oblong, c. 1 cm. Fr. trigonous. Fl. 6. Pollinated by flies, etc.; protandrous. $2n=24$. Gb.

Native. Basic rock-ledges from 1500 to 2500 ft. in the Snowdon range (Caernarvon). 1. Arctic Russia, Alps, Carpathians, Bulgaria, Urals, Caucasus; Russian Asia (widespread); N. Japan, Himalaya, W. China; western N. America from Alaska to Oregon and New Mexico.

13. GAGEA Salisb.

Bulb tunicate. *Infl. a raceme or subumbellate, often subtended by lf-like bracts.* Stem with only the bracts or with 1–3 lvs in addition. Fls erect. Per. segs spreading, persistent, without nectaries. *Anthers basifixed.* Style trigonous, stigma obscurely lobed. Seeds many, subglobose.

About 100 spp., north temperate Old World, mainly Europe.

1. G. lutea (L.) Ker-Gawl. 'Yellow Star-of-Bethlehem.'

Nearly glabrous. Stem 8–25 cm., with a small lf-like bract 2–3 cm., subtending the infl. and another much larger lf or bract a short distance below, occasionally a third bract and 2 infl. are present. Radical lf single, (7–)15–45 cm. × 7–12 mm., suddenly contracted to a hooded acuminate apex, often curled. Fls 1–5, subumbellate. Pedicels somewhat unequal, 1·5–5 cm., glabrous or pubescent. Perianth yellow with a green band outside, segments 10–15 mm., narrow-oblong. Fl. 3–5. Pollinated by insects, weakly protogynous. $2n=72$. Gb.

Native. Damp woods and pastures, especially on basic soils. From Kent and Dorset to Inverness, extending west to Glamorgan, Cumberland, Ayr and Stirling. Very local, commonest in N. and C. England. 46. Europe from

Scandinavia (69° 03′ N.) and N. Russia to N. Spain, Corsica, Sicily, Macedonia and the Caucasus; temperate Asia to Kamchatka and Japan.

It may persist for years without flowering, but can be distinguished from *Endymion* by the 2 lateral ribs on the under surface of the lf.

Tribe 9. SCILLEAE. Stock a bulb with tunic. Lvs all radical. Fls racemose, rarely spicate. Per. segs all alike, free or connate. Anthers introrse, dorsifixed. Fr. a loculicidal capsule.

Species of *Camassia*, *Chionodoxa*, *Puschkinia*, *Hyacinthus* and *Galtonia* are ± commonly grown in gardens.

14. ORNITHOGALUM L.

Per. segs free, obscurely veined, persistent, usually white marked with green. Filaments flattened, hypogynous. Seeds ovoid or globose, numerous.

About 100 spp. in Europe, Africa and W. Asia.

1 Infl. corymbose, lower pedicels much longer than upper.   **1. umbellatum**
   Pedicels all ± equal.   2

2 Fls large, 2–3 cm., few (2–12); filaments with two teeth at apex.   **2. nutans**
   Fls small, 6–10 mm., many (more than 20); filaments without teeth.
                                                                            3. pyrenaicum

1. O. umbellatum L.   Star-of-Bethlehem.

Glabrous. Bulb c. 2·5 cm., with numerous bulbils. Lvs 15–30 cm. × c. 6 mm., linear, green, grooved, with a white stripe down the midrib. Scape 10–30 cm. *Infl. a corymbose raceme*, 5–15-fld; pedicels ascending, lower to 10 cm., upper shorter; fls erect. Bracts linear-lanceolate, acuminate, lower 2–3 cm., thin, whitish. Per. segs 1·5–2 cm., oblanceolate-oblong or linear-oblong, white with green band on the back. Stamens c. $\frac{1}{2}$ as long, filaments lanceolate, acuminate. Capsule 1–1·5 cm., obovoid, 6-angled. Fl. 4–6. Pollinated by insects and selfed. $2n = 18, 27, 36, 43, 45, 52, 54, 72$. Gb.

?Native. Probably native in E. England, elsewhere naturalized in grassy places, widespread in England but local, more local in Wales and Scotland extending north to Sutherland and S. Inner Hebrides; Howth (Dublin) not native. 58. Mediterranean region extending north to Sweden, Denmark and S.W. Russia (Middle Dnieper region) but very doubtfully native in the northern part of its range.

*2. O. nutans L.   'Drooping Star-of-Bethlehem.'

Glabrous. Bulb ovoid, c. 5 cm. Lvs 25–60 cm. × 6–10 mm., linear, with a broad whitish stripe, channelled. Scape 25–60 cm. *Infl. a 2–12-fld unilateral raceme*; pedicels 5–12 mm., curved, all ± equal, fls drooping. Bracts ovate-lanceolate, acuminate, thin, whitish, longer than pedicels. *Per. segs* 2–3 *cm.*, oblong-lanceolate, white with a green band covering most of the back. Stamens shorter than perianth; *filaments* broad, *deeply bidentate at the apex* with the anther in the sinus. Capsule ovoid, pendulous, 6-grooved. Fl. 4–5. Pollinated by insects and selfed; protandrous. $2n = 16, 30, 42$. Gb.

Introduced. ± Naturalized in grassy places in E., C. and N. England, usually in small quantity. C. and E. France (status doubtful) to Italy, Greece and S.W. Russia (Middle Dnieper region); Asia Minor; naturalized elsewhere in Europe.

3. O. pyrenaicum L.   Bath Asparagus.

Glabrous. Bulb c. 5 cm., ovoid. Lvs 30–60 cm. × 3–12 mm., linear, glaucous, withering early. Scape 50–100 cm. *Infl. a many-* (*more than* 20-)*fld* raceme; pedicels slender, spreading or ascending, all ± equal, 1–2 cm.; fls ± erect. Bracts lanceolate-acuminate, thin, whitish, shorter than pedicels. *Per. segs* 6–10 *mm.*, oblong-linear, greenish-white with deeper band. Stamens about ¾ as long as perianth; *filaments* lanceolate, *acuminate*. Capsule c. 8 mm., ovoid, 3-grooved. Fl. 6–7. $2n = 16, 32$. Gb.

Native. Woods and scrub, very local but often abundant, the young infls being sold in Bath market as 'Bath Asparagus'; Somerset, W. Gloucester, Wilts, Berks, Sussex, Norfolk, Bedford, a rare casual elsewhere. 10. Belgium, S. and W. Switzerland and Austria to the mountains of Spain and Portugal, Italy and Greece; Crimea; Asia Minor; Morocco (mountains).

15. SCILLA L.

Bulb persisting several years, scales not tubular. Bracts 0 or 1 to each fl. *Per. segs free*, spreading, *with a prominent midrib*, usually blue or purple. Filaments filiform or dilated at base, inserted on base of perianth. Seeds ovoid or subglobose, sometimes angled, black, few or many.

About 80 spp., Europe, Africa and temperate Asia. Several spp. are grown in gardens, notably *S. siberica* Andr. with about 3 nodding bright blue fls.

Bracts conspicuous; fl. Apr.–May.	**1. verna**
Bracts absent; fl. July–Sept.	**2. autumnalis**

1. S. verna Huds.   'Spring Squill.'

Glabrous. Bulb 1·5–3 cm. Lvs 3–6, linear, 3–20 cm. × 2–4 mm., produced before the fls. Scape 5–15 cm. Infl. 2–12-fld, dense, corymbose. Pedicels 3–12 mm., ascending. *Bracts solitary*, bluish, lanceolate, *usually longer than pedicels*. Per. segs 5–8 mm., violet-blue, ovate-lanceolate, acute, ± ascending. Filaments lanceolate, anthers violet-blue. Capsule c. 4 mm., subglobose, trigonous, the cells c. 4-seeded. Fl. 4–5. Pollinated by insects. Fr. 7. $2n = 22*$. Gb.

Native. Dry grassy places near the coast, very local; Cornwall and Devon, Wales, Isle of Man, Kirkcudbright to Ayr, Kintyre, Hebrides, Sutherland to Shetland, Moray to Aberdeen, Berwick and N. Northumberland; E. Ireland from Wexford to Derry. 30, H7. Faeroes, Norway, W. France, N. and C. Spain and Portugal.

2. S. autumnalis L.   'Autumnal Squill.'

Nearly glabrous. Bulb 1·5–3 cm. Lvs linear, 4–15 cm. × 1–2 mm., produced after fl. Scape 4–25 cm. Infl. 4–20-fld, dense becoming lax, not or slightly corymbose. Pedicels 2–20 mm., ± ascending. *Bracts* 0. Per. segs 4–6 mm., purple, oblong or oblong-lanceolate, obtuse or subacute, ± ascending. Filaments narrow, somewhat dilated below; anthers purple. Capsule c. 4 mm., ovoid-globose, obscurely trigonous, cells 2-seeded. Fl. 7–9. $2n=14, 28, ?44$. Gb.

Native. Dry short grassland usually near the sea, very local. Devon and Cornwall, Isle of Wight, Kent, Surrey, Essex, W. Gloucester, Glamorgan, Channel Is. 11, S. Mediterranean region and W. Europe to N. France.

16. ENDYMION Dum.

Differs from *Scilla* as follows: Bulb renewed annually; scales tubular. Bracts 2 to each fl. Perianth ± campanulate, segments united at base. Stamens, at least the outer, inserted about the middle of the perianth.

Three or four spp. in W. Europe and N.W. Africa.

Raceme drooping at tip; per. segs ± parallel below; anthers cream. Common.
1. non-scriptus
Raceme erect; per. segs not parallel below; anthers blue. Rare introduction.
2. hispanicus

1. E. non-scriptus (L.) Garcke   Bluebell, Wild Hyacinth.
Scilla non-scripta (L.) Hoffmanns. & Link; *S. nutans* Sm.
Glabrous. Bulb 2–3 cm., ovoid. Lvs linear, 20–45 cm. × 7–15 mm. Scape 20–50 cm. *Raceme drooping at tip*, 4–16-fld, *unilateral*, the fls erect in bud, nodding when fully open. Pedicels c. 5 mm., afterwards elongating to c. 3 cm., and becoming erect. Bracts paired, bluish, the lower linear-lanceolate, longer than the pedicel, the upper smaller. *Per. segs* 1·5–2 cm., violet-blue, rarely pink or white, ± *parallel* and erect, so that the lower part of the fl. appears cylindrical, the tips somewhat recurved. Filaments narrow, outer inserted about middle of perianth, inner lower. *Anthers cream.* Fr. c. 15 mm., ovoid; seeds several in each cell. Fl. 4–6. Pollinated by insects. Fr. 7. $2n=16$. Gb.

Native. Common in woods, hedgebanks, etc., rarely in pastures, throughout the British Is., except Orkney and Shetland; ascending to 2000 ft.; often dominant in woods on light acid soils. 110, H40, S. The Netherlands, Belgium, N., W. and C. France. Doubtfully native in N. and C. Spain and Portugal, N. Italy, and N.W. Germany.

*2. E. hispanicus (Mill.) Chouard
Scilla hispanica Mill.; *S. campanulata* Ait.
Differs from *E. nonscriptus* as follows: Lvs broader, 10–35 mm. *Racemes erect, not unilateral.* Pedicels ascending, not curving at fl. so that the fls are ± erect. Per. segs paler, ± spreading, so that the *perianth* is *campanulate*, tips not recurved, broader and somewhat shorter. Filaments all inserted about middle of perianth. *Anthers blue.* Fl. 5. Pollinated by insects. $2n=16$. Gb.

Introduced. Commonly grown in gardens, being usually supplied for the preceding, and naturalized in a few places. Native of Spain and Portugal, W. France, N. and C. Italy; N. Africa.

Hybrids occur in gardens and may be expected as escapes.

17. MUSCARI Mill.

Upper fls sterile. Perianth urceolate, the lobes very small. Stamens inserted on tube, included, filaments short. Ovary with 2 seeds in each cell. Style short. Capsule triquetrous. Seeds obovoid or globose, black.

About 40 spp., Europe, W. Asia, N. Africa. Several are commonly grown in gardens.

1. M. atlanticum Boiss. & Reut.   Grape-Hyacinth.

M. racemosum auct.

Glabrous. Lvs 3–5, 15–30 cm. × 1–3 mm., linear, semi-cylindrical, narrowly grooved above. Scape 10–25 cm. Fls many, in a dense terminal raceme, drooping at fl. Pedicels shorter than fl. Bracts very small. Perianth 3–5 mm., ovoid, mouth small, dark blue, the lobes white or pale, sterile fls smaller, bright blue, never opening. Capsule c. 4 mm., broader than long, emarginate. Fl. 4–5. Pollinated by insects and selfed; protogynous. $2n = 36, 45, 54$. Gb.

Native. Dry grassland in Norfolk, Suffolk and Cambridge, mainly in Breckland, probably also native in Oxford; often recorded as an escape elsewhere but sometimes in error for other species, as it is not now commonly grown in gardens. 5. Mediterranean region, north to Belgium, Germany and S. Russia (Middle Dnieper and Lower Don regions).

*M. comosum (L.) Mill.   Tassel Hyacinth.

Hyacinthus comosus L.

Lvs 6–15 mm. broad. Scape 20–50 cm. Infl. lax; fertile fls 7–8 mm., brown, obovoid-cylindric, spreading; sterile fls purple, on long ascending pedicels. Fl. 4–7. $2n = 18$. A rather frequent casual, naturalized in Glamorgan. Mediterranean region to France, C. Germany and S.W. Russia.

Tribe AGAPANTHEAE. Differs from *Allieae* in the rhizomatous stock. Species and hybrids of **Agapanthus**, robust plants with usually blue flowers with per. segs united below are frequently cultivated and *A. orientalis* F. M. Leighton is reported as naturalized. S. Africa.

Tribe 10. ALLIEAE. Stock a bulb or corm. Lvs all radical but often sheathing the stem for a considerable distance. Infl. an umbel enclosed before flowering in a spathe splitting at fl. and forming 1 or more involucral bracts. Per. segs all alike. Anthers introrse, dorsifixed. Fr. a loculicidal capsule.

This tribe is transferred by Hutchinson to the Amaryllidaceae on account of its infl.

Species of *Brodiaea* and allied genera are sometimes cultivated, and *Ipheion uniflorum (Grah.) Raf., with solitary bluish-white fls and with an onion-like smell, is naturalized in Jersey and elsewhere.

18. ALLIUM L.

Stock usually a bulb with tunic. *Plant smelling of onion or garlic. Perianth* campanulate or rotate; *segments free or nearly so*. Ovules usually 2, rarely more, in each cell. Seeds black. Bulbils often present between the fls, sometimes completely replacing them.

Over 500 spp., north temperate regions to Abyssinia and Mexico. Several spp. are grown as vegetables and others for their fls.

1 Lvs linear or cylindric, sessile.   2
  Lvs ± elliptic, stalked.   **12. ursinum**

2 Scape terete; perianth 12 mm. or less.   3
  Scape triquetrous; perianth 10–18 mm., white.   13

3 Infl. of bulbils only, without fls.   4
  Infl. with fls.   5

4 Spathe scarious, shorter than infl., usually 1-valved, often caducous.
  (Common.)   **5. vineale** var. *compactum*
  Spathe 2-valved with long lf-like points. (Rare.)   **6. oleraceum**

5 Inner filaments divided at apex into 3 long points, the middle one bearing
  the anther, outer entire.   6
  Filaments all entire or the inner with 2 small teeth at base.   10

6 Lvs flat, solid.   7
  Lvs cylindric or semicylindric, hollow.   9

7 Spathe 1-valved, caducous before fl.; plant robust with lvs 12–35 mm.
  broad. (Western.)   8
  Spathe 2-valved, present at fl.; plant rather slender with lvs 7–15 mm.
  broad. (Northern.)   **3. scorodoprasum**

8 Umbel subglobose, many-fld, without or with a few small bulbils.
                                                                       1. ampeloprasum
  Umbel irregular, few-fld with numerous large bulbils.   **2. babingtonii**

9 Inner stamens with lateral points shorter than anther; bulbils absent;
  perianth reddish-purple. (Bristol and Jersey.)   **4. sphaerocephalon**
  Inner stamens with lateral points longer than anther; bulbils nearly
  always present; perianth pink or greenish-white. (Widespread.)
                                                                       5. vineale

10 Spathe shorter than fls; per. segs 7–12 mm.   11
  Spathe with long lf-like points, much longer than fls; per. segs 5–7 mm.   12

11 Lvs cylindric, hollow.   **8. schoenoprasum**
  Lvs flat.   **9. roseum**

12 Anthers included.   **6. oleraceum**
  Anthers exserted.   **7. carinatum**

13 Lvs 2–5; infl. without bulbils.   **10. triquetrum**
  Lf 1; infl. with bulbils.   **11. paradoxum**

Section 1. *Allium*. Bulb subglobose or ovoid, rhizome 0. Lvs distichous, linear, their bases sheathing the lower ⅓ of the stem or more. Stem not inflated, terete. Spathe 1–2-valved, usually cast off early if longer than pedicels. Per. segs connivent. *Filaments dimorphic*; the outer slender, entire; the inner broad, divided above into three points, the middle one bearing the anther. Ovary with a small shelf-like projection above each nectary pit. Stigma entire. Seeds angular, without aril.

1. A. ampeloprasum L.   Wild Leek.

Bulb with offsets inside the tunic. *Lvs linear*, keeled, 15–60 cm. × 12–35 *mm.*, scabrid on margins and keel, glaucous. *Scape* cylindric, *stout*, 60–200 cm.

Spathe scarious, 1-*valved*, with a compressed beak, *caducous before fl.* Umbels *many-fld*, 7–10 cm. across, *compact, globose, without or* (in Guernsey) *with a few small bulbils.* Perianth campanulate, pale purple or whitish, c. 8 mm. *Stamens slightly exserted*; lateral points of inner filaments much longer than the antheriferous one which about equals undivided part of filament. Style exserted. Fl. 7–8. $2n=32*$. Gb.

?Native. Rocky and waste places near the coast, very rare. Cornwall, Steep Holm (Somerset), Flat Holm (Glamorgan), Pembroke, Guernsey; has been found as a casual elsewhere. 6, S. Whole Mediterranean region; Macaronesia.

The English form is without bulbils; the Guernsey plant with bulbils needs further study.

*A. porrum L. Leek.

Differs in the bulb usually without offsets, green herbaceous spathe with longer beak, filaments obviously exserted, the antheriferous tooth about half as long as the undivided part of filament. Style included. Infl. sometimes with bulbils, more often without. $2n=32$. Chh. Commonly cultivated as a vegetable. Origin not definitely known but probably derived from *A. ampeloprasum* in cultivation.

2. A. babingtonii Borrer

Differs from *A. ampeloprasum* as follows: *Umbel loose, irregular, few-fld with numerous large bulbils*, usually with some of the pedicels longer (5–10 cm.) and bearing secondary heads. Antheriferous point of filament rather shorter than undivided part. Fl. 7–8. $2n=48*$. Gb.

Native. Clefts of rocks and sandy places near the coast, Cornwall, Dorset; W. Ireland from Clare to Donegal, especially on the islands off the coast. 3, H6. Endemic.

3. A. scorodoprasum L. 'Sand Leek.'

Bulb with offsets. *Lvs linear*, flat, 15–20 cm. × 7–15 *mm.*, scabrid on the margins and keel. *Scape* 30–80 cm., cylindric, *rather slender*, sheathed by the lf bases in the lower half. *Spathe* 2-*valved*, shorter than umbel, with a very short beak. *Umbels with few fls and purple bulbils.* Perianth campanulate, reddish-purple; segments 5–8 mm., oblong-lanceolate, scabrid on keel. *Stamens included*; antheriferous point of inner filament much shorter than the lateral points and about half as long as undivided part. Fl. 5–8. $2n=16$, 24. Gb.

Native. Grassland and scrub on dry soils from N. Lincoln and Cheshire to Perth and Angus, very local; Kerry and Cork, probably introduced but naturalized; Wicklow (certainly introduced); casual elsewhere. 21, H6. Europe from Scandinavia, Finland and C. Russia (Upper Dnieper region) to S.E. France, C. Italy, Macedonia, and the Caucasus; Asia Minor, Syria.

*A. sativum L. Garlic.

Lvs flat not scabrid. Spathe 1-valved with long point, caducous. Fls whitish, mixed with bulbils. Stamens included; points of the inner filaments nearly equal. $2n=16$.

Native of C. Asia, sometimes cultivated for flavouring and occasionally escaping.

4. A. sphaerocephalon L.   'Round-headed Leek.'

Bulb with offsets within the tunic. *Lvs subcylindric*, grooved, *hollow*, 20–60 cm. × 1–2 mm. Scape 30–80 cm., cylindric, sheathed by the lf bases in the lower half. Spathe usually 2-valved, shorter than fls. *Umbel* dense, subglobose, 2–2·5 cm. across, many-fld, *without bulbils*; outer pedicels about as long as fls, inner longer. Perianth campanulate, reddish-purple; segments c. 5 mm., ovate, oblong, scabrid on back. Stamens exserted; *antheriferous point of inner filaments rather longer than lateral points*. Fl. 6–8. Pollinated by insects and selfed; protandrous. 2*n*= 16. Gb.

Native. Limestone rocks, St Vincent's Rocks, Bristol and at St Aubin's Bay, Jersey. 1, S. Mediterranean region north to Belgium, central W. Germany and C. Russia (Upper Dnieper region).

5. A. vineale L.   Crow Garlic.

Bulb with offsets. *Lvs subcylindric*, somewhat grooved, *hollow*, 20–60 cm. × c. 2 mm. Scape 30–80 cm. Spathe usually 1-valved and caducous, scarious, with a beak about as long as itself, not or scarcely exceeding fls. Umbels rather lax with (var. *vineale*) fls and bulbils or more commonly (var. *compactum* (Thuill.) Boreau) with bulbils and without fls, rarely with fls only (var. *capsuliferum* Koch); pedicels several times as long as fls. Perianth campanulate, pink or greenish-white; segments ± oblong, c. 5 mm., not scabrid on back. Stamens exserted; *antheriferous point of inner filaments about half as long as lateral points*. Fl. 6–7. Pollinated by insects; strongly protandrous. 2*n*= 32, 40. Gb.

Native. Fields and roadsides; ascending to 1500 ft. in Yorks. Rather common in England and Wales and a serious weed in parts of E. England and the S. Midlands; local in Scotland, extending north to Aberdeen and the S. Inner Hebrides; local in Ireland and mainly in the south and centre; Channel Is. 93, H15, S. Europe from Scandinavia and S.W. Russia to Spain, Sardinia, Sicily, Macedonia and the Crimea; N. Africa (rare); Caucasus, Lebanon.

Section 2. *Codonoprasum* (Rchb.) Endl. Differs from Sect. *Allium* in: Spathe 2-valved, persistent, the valves unequal, each from an ovate base drawn out into a long slender appendage much longer than the pedicel. Filaments all slender, entire. Ovary without conspicuous nectary pits or projections. Stigma entire. Seeds angular, aril 0.

6. A. oleraceum L.   'Field Garlic.'

Bulb ovoid, usually with offsets. Lvs semi-terete, grooved, usually hollow at least below, 15–30 cm. × 2–3 mm. or (var. *complanatum* Fr.) flat and to 4 mm. broad. Scape 25–80 cm., cylindric, sheathed by lf-bases in the lower half. Spathe 2-valved, with long lf-like points much longer than fls (often several times as long). Umbel loose, few-fld, or (var. *complanatum*) many-fld, with bulbils, rarely fls 0; pedicels unequal, much longer than fls. Perianth campanulate; *segments 5–7 mm.*, *oblong*, obtuse, pinkish, greenish or brownish. *Stamens included* in perianth. Fr. rarely or never produced. Fl. 7–8. 2*n*= 32. Gb.

Native. Dry grassy places, local from Devon and Kent to Wigtown and Moray, with a distinct eastern tendency; var. *complanatum* confined to the north; in Ireland very local and only in Wexford, Dublin and Antrim, probably introduced. 65, H3. Europe from Scandinavia and N. Russia (Ladoga-Ilmen region) to N. Spain, Corsica, C. Italy, Montenegro, Bulgaria and the Caucasus.

*7. A. carinatum L.

Differs from *A. oleraceum* as follows: Lvs always flat, somewhat grooved. Per. segs bright pink. *Stamens conspicuously exserted.* Fr. never produced. Fl. 8. 2*n* = 24 (triploid). Gb.

Introduced. Thoroughly naturalized in a number of places from Gloucester and Lincoln to Kirkcudbright and Angus and in N.E. Ireland. 12, H3. Scandinavia and Bornholm to E. France, Switzerland and Macedonia.

Section 3. *Schoenoprasum* Dum. Bulbs narrow, elongate, clustered on a short rhizome. Lvs distichous, linear, hollow, their bases sheathing the lower third or so of the stem. Stem not inflated, terete. Spathe 2–3 valved, persistent, shorter than the pedicels. Per. segs somewhat connivent. Filaments all slender, entire, connate at the base. Ovary with 3 distinct nectary pits but without projections. Stigma entire. Seeds angular, without aril.

8. A. schoenoprasum L. Chives.

A. sibiricum L.

Growing in tufts. Lvs cylindric, subglaucous, 10–25 cm. × 1–3 mm. Scape 15–40 cm., cylindric. Spathe usually 2-valved; valves ovate, shortly acuminate, scarious. Umbel subglobose, dense-fld, without bulbils; pedicels shorter than fl. *Per. segs* 7–12 *mm., spreading, pale purple or pink.* Stamens about half as long as perianth. Fl. 6–7. 2*n* = 16* (32 in a Siberian form). Gb.

Plants from Pembroke and Cornwall have been separated as a sp. under the name of *A. sibiricum* L. They appear to differ in their flexuous lvs, and may be subspecifically separable, but are not *A. sibiricum*, which is not itself specifically separable.

Native. Rocky pastures, usually on limestone, very local; Cornwall, Hereford, Brecon, Carmarthen, Radnor, Pembroke, Northumberland, Berwick, Westmorland; Lough Mask (Mayo). Sometimes cultivated for flavouring and occasionally escaping elsewhere. 12, H1. N. Europe and Asia from Scandinavia and arctic Russia to Japan, south in the mountains to N.W. Portugal, Corsica, C. Apennines, Greece, Asia Minor, Himalaya; N. America from Newfoundland and Alaska to New York and Washington.

Section 4. *Cepa* (Mill.) Prokhanov. Bulb subglobose or ovoid to elongate, attached to a rhizome in perennial spp. Lvs distichous, hollow, their bases sheathing the lower part of the stem. Stem hollow, inflated. Spathe about as long as the pedicels which are bracteolate at base. Per. segs spreading, ± equal. Filaments slender, entire or the inner bidentate at base, slightly connate. Ovary with 3 distinct nectary pits but without projections. Stigma entire. Seeds angular, without aril.

*A. cepa L.   Onion.

Whole plant robust. Bulb usually solitary. Lvs ± semicircular in section, slightly channelled. Scape inflated and fusiform below the middle. Infl. very many-fld. Fls greenish-white. Per. segs 4–5 mm. Inner filaments with a small tooth on each side. $2n = 16, 32$. Commonly grown as a vegetable. The Shallot, is a perennial form, not known wild but closely allied to *A. oschaninii* O. Fedsch. from C. Asia.

Section 5. *Phyllodolon* (Salisb.) Prokhanov. Bulbs narrowly ovoid to elongate, attached to a rhizome. Lvs and stem as in section *Cepa*. Spathe shorter than or as long as the pedicels which are without bracteoles. Per. segs connivent, unequal. Filaments slender, all entire. Ovary without conspicuous nectar pits or projections. Stigma entire. Seeds angular, without aril.

*A. fistulosum L.   Welsh Onion.

Lvs circular in section. Scape inflated near the middle. Infl. many-fld. Fls yellowish-white. Per. segs 6–8 mm. Sometimes grown as a vegetable. Not known wild but probably originating in E. Asia.

Section 6. *Molium* Koch. Bulb subglobose; rhizome 0. *Lvs spiral*, linear, flat, almost basal. Stem terete, not inflated. Spathe shorter than the pedicel. Per. segs spreading. Filaments all slender, entire, free. Ovary with nectary pits. Stigma entire. Seeds angular, without aril.

*9. A. roseum L.

Bulb with numerous offsets. *Lvs* 2–4, *linear*, 4–12 mm. broad, not keeled beneath. Scape 30–80 cm., cylindric, sheathed by lvs at base. Spathe 2–4-valved, scarious. Pedicels 2–3 times as long as fl. Perianth campanulate; *segments* 10–12 *mm.*, *pink*, becoming scarious. Stamens included. Fl. 6. Gb.
    Introduced. Naturalized in several places. Native of Mediterranean region; Azores.

Ssp. **roseum.** Umbel without bulbils. $2n = 32$. Rarely naturalized.

Ssp. **bulbiferum** (DC.) E. F. Warb. Umbel with fls and bulbils. $2n = 48$. More commonly naturalized.

Section 7. *Briseis* (Salisb.) Stearn. Bulb subglobose; rhizome 0. Lvs distichous or solitary, linear, so strongly keeled as to be almost 3-sided, almost basal. *Stem triquetrous*, not inflated. Spathe shorter than the pedicels. Per. segs connivent after fl. Filaments slender, entire, unequal, connate at the base with the per. segs. Ovary without deep nectary pits. *Stigma trifid.* Seeds angular, with aril.

*10. A. triquetrum L.   'Triquetrous Garlic.'

Bulb small, whitish. *Lvs* 2–5, linear, 12–20 cm. × 5–10 mm. Scape 20–50 cm. Spathe 2-valved, scarious; valves lanceolate. *Infl.* 3–15-fld, lax, *without bulbils*, somewhat unilateral; fls drooping; pedicels longer than fls. Perianth ± campanulate; *segments* 12–18 mm., oblong, acute, *white with green line*. Stamens shorter than perianth. Fl. 4–6. Seeds dispersed by ants. $2n = 18^*$. Gb.

Introduced. Thoroughly naturalized in hedge-banks and waste places in Channel Is., S.W. England, S. Wales, south Ireland and perhaps elsewhere, increasing. 11, H3, S. Native of W. Mediterranean region from S. Spain and Portugal and Morocco to W. Italy, Sicily and Tunisia.

*11. A. paradoxum (Bieb.) G. Don

Lf 1, brighter green than in *A. triquetrum*. *Infl.* 1–4-fld with *numerous bulbils*. Per. segs c. 10 mm., white. Fl. 4–5. Gb.

Introduced. Naturalized in a number of places. Native of the Caucasus, N. Persia and Mountain Turkmenia.

Section 8. *Ophioscorodon* (Wallr.) Endl. Bulb long, narrow; rhizome short. *Lvs* basal, *stalked; blade* lanceolate, *elliptic or ovate-elliptic*. Stem triquetrous, not inflated. Spathe 2-valved, shorter than the pedicels. Per. segs spreading. Filaments slender, entire, equal, free. Ovary deeply 3-lobed; nectary pits 0. Stigma entire. Seeds globose; aril 0.

12. A. ursinum L.   Ramsons.

Bulb narrow, solitary, consisting of a single petiole-base. *Lvs* 2(–3) *elliptic or ovate-elliptic*, 10–25 × 4–7 cm., acute, bright green; petiole 5–20 cm., twisted through 180°. Scape trigonous or semicylindrical and 2-angled, 10–45 cm., sheathed by petioles at base. Spathe scarious, valves ovate, acuminate, shorter than fls. Infl. 6–20-fld, flat-topped, without bulbils; pedicels longer than fls. Per. segs 8–10 mm., white, lanceolate, acute. Stamens shorter than perianth. Stigma obtuse. Fl. 4–6. Pollinated by insects and selfed; protandrous. $2n=14^*$. G.

Native. Damp woods and shady places, sometimes forming local societies; ascending to 1400 ft. Rather common throughout British Is., but absent from Orkney, Shetland and Channel Is. 109, H40. Europe from Scandinavia (c. 64° N. in Norway) and C. Russia (Upper Dnieper and Volga-Don areas) to C. Spain, Corsica, Sicily, Macedonia and the Caucasus; Asia Minor.

NOTHOSCORDUM Kunth

Stock a bulb with tunic. Plant without onion smell. Perianth usually campanulate; segments joined at base into a short tube. Ovules 4–12 in each cell. Seeds black.

About 15 spp., America.

*N. inodorum (Ait.) Nicholson

Allium fragrans Vent.

Bulb subglobose. Lvs basal, 25–30 cm. × 5–15 mm., linear. Scape 20–40 cm. Spathe 2-valved. Infl. many-fld, loose, fastigiate; fls scented. Per. segs 8–14 mm., dull white with greenish base and reddish midrib outside. Stamens included. $2n=16, 18, 24$.

Occurs on waste ground and as a garden weed in a number of places. Native of N. America; naturalized in many parts of the Old World.

Tribe 11. COLCHICEAE. Stock a corm. Lvs all radical. Fls 1–3, from the ground. Per. segs all alike, with long claws, connivent or connate into a tube. Anthers introrse, dorsifixed. Fr. a septicidal capsule.

19. Colchicum L.

Per. segs united below into a long tube. Styles 3, filiform, free from the base.
About 65 spp., Europe, W. and C. Asia, N. Africa. Some besides the following are sometimes grown in gardens.

1. C. autumnale L.   Meadow Saffron, Naked Ladies, Autumn Crocus.
Glabrous. Corm 3–5 cm., large, with brown outer scales. Plant lfless at fl., lvs produced in spring as the fr. ripens. Perianth pale purple, lobes oblong, 3–4·5 cm.; tube 5–20 cm. Scape elongating in fr. and appearing above the ground, sheathed by the lf-bases. Lvs 12–30 × 1·5–4 cm., oblong-lanceolate, bright glossy green. Fr. 3–5 cm., obovoid; seeds numerous. Fl. 8–10. Pollinated by insects and selfed, protogynous. Fr. 4–6. 2n=38, 42. Gt.
Native. Damp meadows and woods on basic and neutral soils, local but sometimes in quantity, from Devon and Kent to Durham and Cumberland; S.E. Ireland in Kilkenny, Wexford and Carlow; occasionally naturalized in Scotland. 55, H3. C. and S.E. Europe from the Netherlands and Denmark to N. Portugal, C. Spain, C. Italy and Macedonia.

134. TRILLIACEAE

Rhizomatous herbs with erect simple stems. Lvs opposite or in a single whorl near top of stem, reticulately veined. Fls solitary or umbellate, terminal, hermaphrodite, actinomorphic. Per. segs free, unequal, the outer often sepaloid, the inner petaloid. Stamens equalling in number and opposite to the per. segs; anthers basifixed. Ovary superior, 1-celled, with parietal placentation, or 3–6-celled with axile placentation. Ovules numerous. Fr. a berry or fleshy loculicidal capsule. Seeds with copious hard fleshy endosperm.
Four genera and about 25 spp.; north temperate regions.

A small natural group, closely allied to Liliaceae, differing mainly in leaf arrangement and perianth. Species of *Trillium*, similar to *Paris* but with 3 lvs and 3-merous fls, are frequently cultivated.

1. Paris L.

Rhizome creeping. Lvs 4 or more in a whorl. Fls solitary, 4–6-merous, outer per. segs sepaloid, inner petaloid, narrow. Filaments short, flat; anthers linear, connective elongated. Styles free. Fr. a fleshy loculicidal capsule.
About 6 spp., Europe and temperate Asia.

1. P. quadrifolia L.   Herb Paris.
Glabrous. Stems 15–40 cm. Lvs 6–12 cm., (3–)4(–8), obovate, shortly acuminate, cuneate at base, subsessile, 3–5-veined with reticulate veins between. Pedicels 2–8 cm. Fls 4(–6)-merous. Sepals 2·5–3·5 cm., green, lanceolate, acuminate. Petals subulate, nearly as long as sepals. Ovary 4–5-celled. Fr. berry-like, black, globose, finally dehiscent. Fl. 5–8. Pollinated by flies and selfed; strongly protogynous. 2n=20*. Grh.

Native. Damp woods on calcareous soils; ascending to 1175 ft. in West-morland. From Somerest and Kent to Caithness, local and with an eastern tendency, absent from W. Scotland north of Ayr and Lanark (except Inner Hebrides) and from Ireland and the Isle of Man. 77. Europe from Iceland and N. (not arctic) Russia to C. Spain, Corsica, Italy, Macedonia, and the Caucasus; Siberia.

135. PONTEDERIACEAE

Aquatic herbs. Lvs sheathing. Infl. a raceme or panicle, subtended by a bract resembling a lf-sheath. Fls hermaphrodite, somewhat zygomorphic, hypogynous, 3-merous. Perianth corolla-like, segs usually united into a tube. Stamens 6, 3 or 1, inserted on the perianth. Ovary 3-celled with axile placentas or with only one cell fertile with a single pendulous ovule; style long; stigma entire or shortly lobed. Fr. a capsule or nutlet; endosperm copious, mealy.

Six genera and about 30 spp., mainly tropical.

1. PONTEDERIA L.

Rhizome creeping. Fl.-stems erect with 1 lf (besides the bract). Infl. a spike-like raceme. *Perianth funnel-shaped, 2-lipped, with 3 lobes to each lip,* upper united c. $\frac{1}{2}$-way, lower free from each other. Stamens 6. *Ovary with 1 ovule.* Fr. a nutlet enclosed in the accrescent perianth.

About 8 spp., America.

*1. P. cordata L.   *Pickerelweed*

Stem erect, to 10 cm. Lvs 9–18 cm., ovate to lanceolate, cordate to broadly cuneate at base, entire. Infl. 5–15 cm. Perianth 12–17 mm., violet-blue, upper seg. with yellow markings.

Grown in gardens, rarely naturalized. Native of eastern N. America from Nova Scotia to Texas.

136. JUNCACEAE

Herbs, usually perennial, glabrous or sparsely hairy, frequently tufted or with creeping sympodial rhizomes. Lvs long and narrow, terete, channelled or grass-like, with sheathing base, sometimes reduced to scales. Fls few to many, usually in numerous crowded monochasial cymes, sometimes con-densed into heads (rarely solitary), regular, hermaphrodite, protogynous, wind-pollinated. Per. segs 6, in 2 whorls, equal or subequal, usually greenish or brownish. Stamens free, in 2 whorls of 3 or with the inner whorl missing. Pollen remaining in tetrads. Ovary syncarpous, 1- or ± completely 3-locular, forming a loculicidal capsule; stigmas 3, brush-like. Seeds 3 or many, often with appendages; dispersed by various agencies. Endosperm starchy.

Eight genera and about 350 spp., cosmopolitan, but chiefly in temperate or cold climates or at high altitudes in the tropics.

A natural family, resembling the Liliaceae in fl.-structure, but wind-pollinated and with a characteristic vegetative habit.

Lvs glabrous, various, seldom flat and grass-like; capsule many-seeded.
1. JUNCUS

Lvs sparsely hairy, at least when young, flat or channelled and grass-like; capsule 3-seeded.
2. LUZULA

1. JUNCUS L.

Glabrous perennial herbs to 1 m. or more, or dwarf annuals; erect, ± tufted, *often rhizomatous*. Stem *lvs with* usually *split sheathing base which is often produced above into auricles*; lamina channelled, compressed or ± terete, sometimes wanting. Infl. a cluster of terminal or apparently lateral cymes, sometimes condensed into a head, fls rarely few or solitary. *Capsule many-seeded*. Seeds sometimes with appendages; testa usually finely sculptured, often becoming mucilaginous.

About 225 spp., cosmopolitan.

1	Flowering stems bearing only brown sheaths below the infl., which is apparently lateral and exceeded by a bract continuing the stem.	*2*
	Lvs on flowering stems (when present) with green lamina; infl. obviously terminal, or if apparently lateral, then lvs with prickly points.	*7*
2	Stem very slender, 1 mm. or less in diam.; infl. at the middle of the apparent stem or lower. **12. filiformis**	
	Stem stouter; infl. above the middle of the apparent stem.	*3*
3	Far-creeping, growing in straight lines; infl. usually 5–20-fld. **13. balticus**	
	Densely tufted; infl. usually many-fld.	*4*
4	Pith interrupted; stem glaucous, with 12–16 prominent ridges. **9. inflexus**	
	Pith continuous, at least in the lower part of the stem; stem not glaucous, usually with more than 18 ridges or striae.	*5*
5	Stem with 18–45 ridges; capsule with no fertile seeds. **effusus × inflexus**	
	Stem usually with more than 40 ridges or striae; capsule fertile.	*6*
6	Stem when fresh strongly ridged especially just below the infl.; capsule shortly mucronate. **11. conglomeratus**	
	Stem when fresh striate, scarcely ridged; capsule not mucronate. **10. effusus**	
7	Annuals, rarely over 30 cm., readily uprooted (lowland).	*8*
	Perennials, usually over 30 cm., difficult to uproot (except for some small alpine plants).	*10*
8	Infl. a much-branched lfy panicle, occupying the greater part of the plant, rarely in terminal heads; seeds usually roundly ovoid. **8. bufonius**	
	Fls in 1 or a few terminal heads; seeds twice as long as broad.	*9*
9	Per. segs less than 4 mm., with fine, often recurved points; lf-sheath without auricles. **16. capitatus**	
	Per. segs 5 mm. or more, gradually tapering to an acute point; lf-sheath with pointed auricles. **24. mutabilis**	
10	Lvs and lowest bracts ending in stiff prickly points; infl. many-fld, apparently lateral.	*11*
	Lvs and lowest bract not prickly-pointed; infl. terminal.	*12*
11	Fls reddish-brown; capsule much exceeding the perianth. **15. acutus**	
	Fls straw-coloured; capsule not or barely exceeding the perianth. **14. maritimus**	
12	Fls solitary and terminal or in capitate clusters of 6 or fewer; seeds with appendages. (Alpine plants.)	*13*
	Fls neither solitary nor capitate, usually in panicles; seeds without appendages. (Lowland or alpine.)	*16*
13	Fl.-stems densely tufted; fls 1–3 together between axils of 2–3 long filiform bracts. **7. trifidus**	

Fl.-stems solitary or in small not dense tufts; the longest bract not or only shortly exceeding the fls.   *14*

14 Stoloniferous; outer per. segs acute.   **25. castaneus**
Not stoloniferous; outer per. segs obtuse or bluntly pointed.   *15*

15 Fls 2–3 in a head; lvs in section of 2 tubes.   **27. triglumis**
Fls 1–2 in a head; lvs in section of 1 tube.   **26. biglumis**

16 Lvs not septate, never setaceous, usually solid.   *17*
Lvs septate, sometimes setaceous, always hollow.   *22*

17 Lvs all radical.   **2. squarrosus**
Cauline lvs 1–4.   *18*

18 Stem 50–100 cm., with 1–4 cauline lvs; lowest bract much less than ½ as long as infl.   **1. subulatus**
Stem seldom exceeding 50 cm., with 1–2 cauline lvs; lowest bract at least ½ as long as infl.   *19*

19 Fls greenish or straw-coloured; per. segs very acute.   *20*
Fls dark brown; per. segs blunt.   *21*

20 Lf-sheaths prolonged above into elongate scarious auricles.   **3. tenuis**
Lf-sheaths with short brownish auricles.   **4. dudleyi**

21 Usually in salt-marshes; style as long as capsule; perianth nearly equalling fr.   **6. gerardii**
Usually not on saline soils; style shorter than capsule; perianth ½–⅓ fr.   **5. compressus**

22 Lvs setaceous, in section (use lens) of 2 tubes; plant less than 25 cm.   *23*
Lvs not setaceous, in section of one or of more than 2 tubes; plant usually more than 25 cm.   *24*

23 Lvs 8–12 cm.; stamens 6.   **23. kochii**
Lvs 3–10 cm.; stamens 3.   **22. bulbosus**

24 Lvs with longitudinal as well as transverse septa; perianth straw-coloured to light reddish-brown.   **17. subnodulosus**
Lvs without longitudinal septa; perianth dark brown or blackish, never straw-coloured.   *25*

25 Per. segs blunt, the outer mucronate; capsule obtuse, mucronate.   *26*
At least the outer per. segs acute; capsule acute or acuminate.   *27*

26 Some fls pedicelled; capsule barely equalling perianth.   **21. nodulosus** var. *marshallii*
All fls subsessile; capsule slightly exceeding perianth.   **20. alpinoarticulatus**

27 Lvs strongly laterally compressed; capsule fertile, abruptly acuminate.   **19. articulatus**
Lvs subterete, slightly compressed; capsule acute or, if acuminate, without fertile seeds.   *28*

28 Fls usually less than 7 in head; capsule sterile, either much shorter than perianth or longer and abruptly acuminate.   **acutiflorus × articulatus**
Fls 6–12 in head; capsule many-seeded, gradually tapered to a very acute point.   **18. acutiflorus**

*1. J. subulatus Forsk.

A tall perennial with horizontally creeping rhizome. *Stems* to 1 m., *bearing 1–4 lvs with subterete hollow lamina. Panicle* terminal, many-fld, *narrow, interrupted*, the *bracts* inconspicuous, *much less than* ½ *as long as the infl.*

Per. segs finely acuminate, straw-coloured, *slightly larger than the ovoid-trigonous, mucronate capsule.* Seeds without appendage.

Introduced? Recently found in a salt marsh at Berrow, N. Somerset, where it is well established. Mediterranean region from Portugal to Egypt and Syria.

2. J. squarrosus L.   'Heath Rush.'

A tough, wiry perennial, forming *dense low tufts. Lvs* 8–15 cm., *usually all radical,* from a very short upright stock, subulate, deeply channelled, rigid, *sharply reflexed above the sheathing base.* Fl. stems 15–50 cm., erect, very stiff. Lowest bract less than ½ the length of the infl., usually lf-like. Infl. lax, the branches ending in clusters of 2–3 fls. *Per. segs* 4–7 mm., *dark chestnut brown, lanceolate, bluntly pointed* or acute, slightly exceeding the obovoid mucronate capsule. Fl. 6–7. $2n=40$. Hr.

Native. On moors, bogs and moist heaths, confined to acid soils. Abundant throughout the British Is., except where there is a scarcity of suitable habitats. 109, H40. W., C. and N. Europe, from the mountains of S. Spain and the Alps to Scandinavia, east to the Dnieper; Morocco; Iceland; S. Greenland.

*3. J. tenuis Willd.

J. macer S. F. Gray

A rather weak perennial, very variable in stature. *Lvs* 10–25 cm., usually *all basal,* on a very short, usually upright stock, curved, flexible, narrowly linear, channelled, with a broad sheathing *base* which is *produced above into obtuse scarious auricles several times as long as wide.* Fl.-stems 15–35 cm., erect, slightly compressed. Fls in a terminal, usually loose *panicle, much exceeded by at least one of the very narrow lf-like bracts. Per. segs* 3–4 mm., greenish, becoming *straw-coloured, narrowly lanceolate, very acute.* Capsule ovoid, obtuse, mucronate, shorter than the perianth. *Seeds* becoming *very mucilaginous.* Fl. 6–9. Seeds dispersed on cart-wheels, etc. Germ. spring. $2n=30, 32$. Hs.

Naturalized (according to some, native in S.W. Ireland) and still extending its range in Britain (first recorded 1883), on roadsides, waste ground and by field paths. North to Inverness and the Outer Hebrides, absent from most of the Midlands and the east coast counties from the Thames to the Tees, locally abundant in parts of Wales, Scotland and Ireland. 62, H5, S. N. and S. America. Naturalized in many parts of Europe, Australia, etc.

*4. J. dudleyi Wiegand

Closely similar to *J. tenuis* but more stiffly erect and auricles of lf-sheaths firm, brown, not scarious, broader than long.

Naturalized. Well established in marshy ground near Crianlarich, Mid Perth; Rhum. N. America.

5. J. compressus Jacq.   'Round-fruited Rush.'

A small or medium-sized tufted perennial, *seldom forming extensive patches.* Rhizome horizontal, usually less than 5 cm., rarely far-creeping. *Lvs narrowly linear, dorsiventrally flattened. Fl.-stems* 10–30 cm., curved, not stiffly erect, compressed throughout their length, *bearing* 1–2 *lvs.* Panicle compound, terminal, subcymose, lax to compact, usually shorter than the lowest bract. Per. segs 1·5–2 mm., ovate, very obtuse, light brown. *Anthers slightly shorter than the filaments. Style shorter than the ovary. Capsule subglobose, obtuse,* c. 1½ *times as long as the perianth,* very shortly mucronate, very glossy. Fl. 6–7. Grh.

Native. In marshes, alluvial meadows and grassy places where the vegetation is kept low by mowing or grazing, chiefly on non-acid soils, north to Ross, uncommon and becoming rarer northwards. 61. Eurasia (except the Arctic); eastern N. America.

6. J. gerardii Lois. 'Mud Rush.'

Resembling *J. compressus* and differing in a combination of variable characters. Usually *taller, forming more extensive tufts or patches. Rhizome far-creeping.* Lvs 10–20 cm., dark green. Fl.-stems straight, stiffly erect, compressed below, triquetrous above. *Infl.* usually laxer and *with straighter, less spreading branches,* usually considerably exceeding the lowest bract. *Perianth dark brown to blackish. Anthers 3 times as long as the filaments. Style equalling or slightly longer than the ovary. Capsule* acuminate, *not or slightly exceeding the perianth.* Fl. 6–7. 2n=80. Grh.

Native. In salt-marshes, from just below high-water mark of spring-tides upwards, abundant and locally dominant; rare inland. Great Britain north to Shetland; Ireland. 94, H27, S. Coasts and inland salt areas of Eurasia; N. Africa; N. America.

7. J. trifidus L. 'Three-leaved Rush.'

A slender, densely tufted, grass-like perennial, often forming circular patches. Rhizome horizontal, usually less than 15 cm., thickly covered with dead branches and lf-sheaths. Stems rigid, erect, terete, finely striate. *Basal lvs mostly sheaths only,* prolonged above into *laciniate auricles,* some with filiform lamina to 8 cm. *Infl. of* 1 *terminal and* 0–3 *lateral, sessile or shortly stalked fls. Stem-lvs* (bracts) 2–3, *filiform,* 2–8 *cm.* Per. segs 2·5–3 mm., ovate-lanceolate, dark chestnut brown. Capsule ovoid, long beaked, longer than the perianth. Fl. 6–8. 2n=30. Hs.

Native. Detritus and rock-ledges on high mountain tops, where it is sometimes the most abundant plant over large areas. Scotland from Arran to the Outer Hebrides and Shetland. 21. High mountains of Europe from C. Spain to Caucasus and E. Siberia. Subarctic and Arctic to 71° 30′ N.; N. America. Two ssp. occur on the Continent of which only one (ssp. *trifidus*) is British.

8. J. bufonius L. Toad Rush.

A slender *annual of very variable dimensions. Stem* 3–25 cm., simple or much branched from the base, forked once or more above, erect or more or less prostrate, *filiform,* usually with 1 cauline and several radical lvs. Lvs 1–5 cm., *setaceous* from a sheathing base, deeply channelled. Infl. *a much-branched lfy panicle,* occupying the greater part of the plant. *Fls* on upper side of stem, *mostly sessile and solitary* (2–3 together in var. *fasciculatus* Koch), often closed till after pollination. *Per. segs* 2·5–6·0 mm., usually *lanceolate-acuminate,* finely pointed, *pale green with hyaline border,* usually longer than the *oblong blunt capsule.* Seeds somewhat variable but often roundly ovoid, 1½ times as long as broad. Fl. 5–9. Wind-pollinated or cleistogamous. Germ. all summer. Seeds mucilaginous, readily carried on cart-wheels, etc. 2n=c. 60; c. 120. Th.

Native. Paths, roadsides, arable land, mud by ponds, etc., abundant throughout the British Is. 112, H40, S. Cosmopolitan, but chiefly in the north and south temperate zones.

9. J. inflexus L.   Hard Rush.

J. glaucus Sibth.

Perennial in *large dense grey-green tufts.* Rhizomes horizontal, matted, with very short internodes. *Stem slender*, 25–60 cm. × 1–1·5 *mm.*, stiffly erect, *with 12–18 prominent ridges*, dull, glaucous, pith interrupted. *Lf-sheaths glossy*, dark brown or blackish. *Infl. apparently lateral*, of many ascending branches, lax. *Per. segs* 2·5–4 mm., *lanceolate* with subulate points, unequal. *Capsule dark chestnut-brown, glossy, ovoid-acuminate, mucronate, about equalling the perianth.* Fl. 6–8. Germ. spring. 2n = 40. Hs.

Native. Chiefly in damp pastures, preferring heavy basic or neutral soils. North to Caithness, abundant in most of England, Wales and Ireland, becoming local in Scotland and parts of Wales and Ireland. 96, H39, S. Europe north to S. Sweden and C. Russia, east to India and Mongolia; N. Africa; Macaronesian islands. A ssp. in S. Africa and E. Java. Introduced in New Zealand.

J. effusus × inflexus (*J. × diffusus* Hoppe) differs from *J. inflexus* in the green, not glaucous, scarcely grooved stems, with c. 18–45 coarse striae (cf. *J. effusus*); pith continuous or ± interrupted above. Infl. and perianth as in *J. inflexus.* Capsule much shorter than the perianth, with abortive ovules. Not uncommon with the parents, spreading vegetatively. *J. conglomeratus × inflexus* (*J. × ruhmeri* Aschers. & Graebn., ?*J. glaucus* Ehrh. var. *pseudodiffusus* Syme) has been doubtfully reported as British.

10. J. effusus L.   Soft Rush.

J. communis var. *effusus* (L.) E. Mey.

A *densely tufted*, stiffly erect perennial. *Stems* 30–150 cm. × 1·5–3 mm. (just below infl.), rather soft, *bright to yellowish green*, glossy and quite *smooth throughout their length when fresh*, *with* 40–90 *striae*; pith *continuous.* Lf-sheaths reddish to dark brown, not glossy. *Infl.* apparently lateral, *placed about ½ the distance from the top of the stem*, many-fld, lax or *condensed into a single rounded head* (var. *compactus* Hoppe), with ascending spreading and deflexed branches. *Per. segs* 2–2·5 mm., lanceolate, finely pointed. *Capsule* yellowish to chestnut-brown, *broadly ovoid, retuse, not mucronate.* Fl. 6–8. Occasionally cleistogamous. Germ. spring. 2n = 40. Hs.

Native. Very abundant and locally dominant in wet pastures, bogs, damp woods, etc., especially on acid soils. Throughout the British Is. 112, H40, S. North temperate zone, E. and S. Africa, Malaysia, Australia, New Zealand, etc.

11. J. conglomeratus L.

J. communis var. *conglomeratus* (L.) E. Mey.; *J. Leersii* Marsson

Closely resembling J. effusus and often confused with its var. *compactus*, but usually less robust and forming smaller tufts. *Stems* bright to greyish green, *not glossy, with numerous ridges which are especially prominent just below the infl.* Sheathing base of *bract widely expanded.* Infl. usually *condensed into a rounded head, more rarely of several stalked heads* (var. *subuliflorus* (Drej.)

Aschers. & Graebn. Capsule as in *J. effusus*, but with the *remains of the style on a small elevation in the hollowed top*. Fl. 5–7 (usually about a month earlier than *J. effusus*). Germ. spring. $2n = 40$. Hs.

Native. In similar habitats to *J. effusus*, but with a narrower ecological range and a more marked preference for acid soils. Owing to confusion with *J. effusus* var. *compactus* the distribution is not accurately known, but probably throughout the British Is.; locally abundant, but in some districts rare. Europe, extending east to Asia Minor and the Volga, north to Faeroes and 68° 55′ N. in Norway; N. Africa; Macaronesia; Newfoundland.

J. conglomeratus × *J. effusus*. Presumed hybrids, intermediate between the two species and highly sterile, have been found with both parents, but *J. conglomeratus* var. *subuliflorus* has sometimes been mistaken for them.

12. J. filiformis L.

A slender wiry perennial *in not very extensive tufts. Stems* 15–30 cm. × 0·75–1 mm., stiffly erect, very faintly ridged when fresh, *filiform*, bearing several *brownish lf-sheaths, of which the uppermost often has a short green lamina. Infl.* usually placed *half-way or lower down the apparent stem of 7 fls or fewer*, forming a ±compact head. Per. segs 2·5–3 mm., lanceolate, becoming straw-coloured. *Capsule almost spherical, very shortly mucronate*, not exceeding the perianth. Fl. 6–9. $2n = 40$; c. 80. Hs.

Native. On stony lake shores among *Molinia*, other *Junci*, etc. Very local; Lake District, Kinross, Kincardine, Stirling and Dunbarton. 6. Northern and arctic Eurasia, south to mountains of Portugal, northern Apennines and Caucasus; Iceland; N. America; Patagonia.

13. J. balticus Willd.

A far-creeping rhizomatous perennial, not forming large tufts. *Stem* 15–45 cm. × 1–2 mm., stiffly erect, dull green, *smooth when fresh; pith continuous. Infl.* ¼ or less from apex of apparent stem, appearing lateral, *about* 7–20-*fld, lax, with ascending branches.* Per. segs to 5 mm., brown, ovate-lanceolate, acute. *Capsule* dark chestnut-brown, glossy, *ovoid, abruptly mucronate*. Seeds oblong, without appendages. Fl. 6–8. Said rarely to reproduce by seed. Grh.

Native. In dune-slacks, rarely in other damp sandy places. East and north coast of Scotland from Fife to Sutherland and the Hebrides; Lancashire. Very rare inland. 17. Europe from Pyrenees to Faeroes, Scandinavia and N.W. Russia; Iceland; Japan; N. and S. America; New Caledonia. British populations belong to var. *balticus*.

14. J. maritimus Lam.   'Sea Rush.'

J. spinosus auct., vix Forsk.

An erect, densely tufted, *very tough* perennial. Stem 30–100 cm., light green, smooth when fresh, lfy only near the base; pith continuous. Lowest stem-lvs brown glazed sheaths, upper with terete green lamina, sharply pointed. *Infl.* appearing lateral, many-fld, *forming an interrupted irregularly compound panicle*, with ascending branches, *shorter than the sharply pointed bract*, rarely a dense head or diffuse and exceeding the bracts. *Per. segs* 3·0–4·5 mm., *straw-coloured, lanceolate*, outer acute, inner blunt. *Capsule ovoid-trigonous*, *mucronate, about equalling the perianth.* Seeds obliquely ovoid, *with a large appendage.* Fl. 7–8. 2*n*=40. Grh.

Native. On salt-marshes above high-water mark of spring tides, often dominant over large areas. Abundant on the coasts of the British Is., north to Inverness. 62, H26, S. Atlantic and Mediterranean coasts of Europe, extending into the Baltic and Black Sea and east to Sind; N. and S. Africa; N. and S. America; Australasia; inland in salt areas of S. Europe and C. Asia.

15. J. acutus L.   'Sharp Rush.'

A tall and *very robust* perennial, forming *dense prickly tussocks,* somewhat like *J. maritimus,* but with stouter and taller stems, 25–150 cm., bearing a few very sharply pointed foliage lvs, the *infl. compact, ± rounded, with more spreading branches,* much shorter than the bract. *Per. segs* 2·5–4 mm., *reddish-brown, becoming almost woody,* ovate-lanceolate, with broad scarious margin. *Capsule* turgid, *broadly ovoid, at least twice as long as the perianth.* Seeds as in *J. maritimus* but broader. Fl. 6. Hs.

Native. Local on sandy sea-shores and dune-slacks, less frequently on salt-marshes on the west, south and east coasts of England and Wales from Caernarvon to Norfolk; extinct in N.E. Yorks; south-east coast of Ireland from Cork to Dublin. 16, H6, S. Mediterranean region of Europe and N. Africa, extending up the Atlantic coast to N. France, east to Transcaucasia; Macaronesia; California; S. America; S. Africa. The British population belongs to ssp. *acutus.*

16. J. capitatus Weigel

A dwarf tufted annual 1–5 cm., becoming reddish in fr., with *setaceous, stiffly erect* stems branched only from the base. *Lvs* 0·5–4 cm. all radical, *setaceous from a short sheathing base without auricles,* ± channelled. Infl. of 3–8 sessile fls in a *single terminal head,* rarely with 1–2 lateral heads. Bracts 1–2, the longer much exceeding the infl. *Outer per. segs* 3–5 mm., greenish, becoming reddish-brown, *curved, ovate-lanceolate, with fine, ± recurved points*; inner shorter, mostly membranous. *Capsule broadly ovoid, truncate,* apiculate, *shorter than the perianth.* Seeds *narrowly obovoid,* about twice as long as wide. Fl. 5–6. Fls of two types, short-styled cleistogamic and long-styled wind pollinated. Th.

Native. On damp heaths, especially where water has stood during the winter and characteristically associated with *Radiola, Isoetes histrix,* etc. Rare; Cornwall, Anglesey; reported from Hebrides (Raasay, Barra, Rhum). 3, S. S. and W. Europe and sparingly through C. Europe to S. Sweden, Finland and N.W. Russia; east to lower Don; Africa; Newfoundland (?introduced), S. America and Australia.

17. J. subnodulosus Schrank   'Blunt-flowered Rush.'

J. obtusiflorus Ehrh. ex Hoffm.

A tall, erect, *rather soft* perennial, growing in extensive patches, *not densely tufted. Stems and lvs* similar, 50–120 cm., bright green, terete, *smooth,* hollow, *with longitudinal and transverse septa.* Sterile stems with 1 lf; fl.-stems with brown scale-lvs and 1–2 *foliage lvs,* each *with 35–60 transverse septa.* Infl.

repeatedly compound, of many heads each with 3–12 fls, the *branches* of the second order *diverging at an angle of more than* 90°, rarely congested. *Fls pale* straw-coloured, darkening to light reddish-brown. *Per. segs* 2·0–2·25 mm., *incurved, obtuse*. Stamens 6. *Capsule* light brown, *broadly ovoid*, trigonous, *acuminate, shortly beaked, slightly longer than the perianth*. Fl. 7–9. Germ. spring. 2n=40*. Hs. or Hel.

Native. Fens, marshes and dune-slacks with basic ground water, often on calcareous peat. Locally abundant and, in E. Anglia, dominant over large areas. North to S. Hebrides, Renfrew and E. Lothian; frequent in W. and C. Ireland, local elsewhere. 69, H32, S. W., C. and S. Europe, north to Faeroes, S. Sweden and Oesel, east to Vistula and Danube delta; Kurdistan; N. Africa; eastern N. America.

18. J. acutiflorus Ehrh. ex Hoffm.   'Sharp-flowered Rush.'

J. sylvaticus auct.

A tall, *stiffly erect* perennial with stout, far-creeping rhizome, not densely tufted. Stem 30–100 cm. with 2–4 deep green, *straight, subterete lvs*, each *with 18–25 conspicuous transverse septa*. *Infl.* richly branched, repeatedly compound, of *many small, shortly-stalked heads*, each of 6–12 fls; *branches of second order diverging at an acute angle*. Fls chestnut-brown. *Per. segs* 3·0–3·5 mm., lanceolate, acute, *tapering to awn-like points, the outer recurved at the tips*, the inner with narrow brownish margins. Stamens 6. *Capsule* chestnut-brown, *evenly tapered to an acute point*, longer than the perianth. Seeds about 12 per capsule. Fl. 7–9 (the latest of the common British *Junci*). Germ. spring. 2n=40*. Hs. or Hel.

Native. Wet meadows, moorlands and swampy woodlands, abundant throughout the British Is., especially on acid soils. 112, H40, S. W., C. and S. Europe, north to Denmark and east to Moscow, but rare north of S. Schleswig and from W. Prussia eastwards; eastern N. America.

19. J. articulatus L.   'Jointed Rush.'

J. lampocarpus Ehrh. ex Hoffm.

Very variable, especially in size and habit. Perennial, ascending, decumbent or prostrate, seldom stiffly erect, with slender rhizome, often subcaespitose. Stem to 80 cm., terete, bearing 2–7 *laterally compressed, usually curved deep green lvs, with 18–25 inconspicuous transverse septa*. Infl. repeatedly compound, often sparingly branched, with few to many stalked heads, each of 4–8 fls; *branches diverging at an acute angle*. *Fls* c. 3 mm., *dark chestnut-brown to almost black*. *Per. segs* lanceolate, acute, the inner with broad colourless margins. Stamens 6. *Capsule long-ovoid, contracted above to an acumen.* Seeds about 40 per capsule. Fl. 6–9. Germ. spring. 2n=80*. Hs. or Hel. Exists in several ecotypes.

Native. Wet ground, especially on acid soils, preferring meadows or moors which are grazed or mown. Abundant throughout the British Is. 112, H40, S. Europe and Asia (except the Arctic), eastern N. America, N. Africa. Introduced in S. Africa, Australia and New Zealand.

A large form with 80 chromosomes has been described by Timm & Clapham (*New Phytol.* **39**, 1–16) from near Oxford and may perhaps be best regarded as a distinct species. It differs chiefly in its large size, stiffly erect habit, large heads with more numerous (14–18) fls and more numerous seeds per capsule. Similar plants occur in other districts.

J. acutiflorus × *articulatus* (2*n* = 60*) is common in some districts. Slightly decumbent, stem to 50 cm. Lvs subterete, occasionally curved, with inconspicuous septa. Lf-sheaths often tinged with red. Fls large, 3–6 in head. Capsule without seeds.

20. J. alpinoarticulatus Chaix

J. alpinus Vill.

An erect perennial resembling small to medium-sized forms of *J. articulatus*. Rhizome shortly creeping. Stem 15–30 cm., smooth, usually terete, with *purplish* or brownish *scale-lvs*. Foliage lvs 2–3 with distinctly septate, subterete lamina. *Infl. with few* (rarely many) *branches arising from 2 points about 4 cm. apart, ascending or erect, straight. Fls subsessile*, in few or many heads of 3–6 fls. *Per. segs* 2·5–3·0 mm., *dark* brown or blackish, ± incurved; the *outer ovate, obtuse, mucronate; the inner obtuse not mucronate. Capsule ovoid, obtuse, apiculate or mucronate*, slightly exceeding the perianth. Fl. 7–9. 2*n* = 40. Hs.

Native. In gravelly stream beds and marshy places on mountains. Rare. Teesdale; Scotland. 9. Arctic and northern Eurasia, southwards to the mountains of southern Europe and the Caucasus; ?Greenland; N. America.

21. J. nodulosus Wahlenb. var. *marshallii* (Pugsl.) P. W. Richards

J. Marshallii Pugsl.; *J. alpinus* Vill. var. *Marshallii* (Pugsl.) Lindquist

Like *J. alpinoarticulatus* but *less stiffly erect* and more tufted, differing chiefly in the form of the infl. Panicles irregularly branched, with erect *often curved branches. Fls partly subsessile, partly on pedicels up to 6 cm. in many small imperfect and unequal umbels. Fls very small* (2 mm.), dark reddish-brown. Per. segs narrower than in *J. alpinoarticulatus*, obtuse, the outer mucronate. *Capsule dark reddish-brown, not blackish, obovoid*, about equalling the perianth. 2*n* = 80 (var. *nodulosus*).

Native. Very rare. Shore of Loch Ussie, near Conan, E. Ross (E. S. Marshall, 1892) and near Braemar, Aberdeenshire. Scandinavia; N. America.

22. J. bulbosus L.   'Bulbous Rush.'

J. supinus Moench

A *small grass-like perennial*, very variable in habit and stature, erect and *densely tufted*, procumbent and rooting at the nodes or floating and much branched. *Stems* slender, *slightly swollen at the base*, not forming a creeping rhizome. *Lvs* 3–10 cm., *setaceous or filiform* with sheathing base, with numerous *very indistinct septa. Infl.* simple or with few to many irregular spreading branches, often *with small shoots in the 2–6-fld heads* or proliferating without fls. Per. segs 2 mm., light reddish-brown, acute, mucronate. *Stamens* 3; *anthers elongated*, about equalling the filaments. *Capsule* 2·5–3 mm., yellowish-brown, *oblong, obtuse, not sharply trigonous.* Fl. 6–9. 2*n* = 40*. Hs., Hel. or Hyd.

Native. Moist heaths, bogs, cart-ruts and rides in woods, chiefly or exclusively on acid soils. Abundant throughout the British Is. 112, H40, S. Europe from Gibraltar to N. Norway, east to western Russia; N. Africa; Macaronesia. Doubtful in Newfoundland and Labrador.

23. J. kochii F. W. Schultz

Like *J. bulbosus* but *more robust* and upright and in smaller less dense tufts. Stem more swollen at base. Lvs 8–12 cm., stouter. Infl. branches more erect. Per. segs dark brown. *Stamens* 6, rarely 3 or 4; *anthers shortly oblong*, distinctly shorter than the filaments. *Capsule* c. 2 mm., brown, *obovoid, retuse, sharply trigonous.* Fl. 6–9. Hs.

Native. In similar habitats to *J. bulbosus*. Widespread in Britain, but distribution not accurately known. N.W. Europe from Portugal to southern Norway and Sweden, east to Harz Mountains.

24. J. mutabilis Lam.   'Dwarf Rush.'

J. pygmaeus Rich.

A *dwarf annual*, forming small spreading tufts, *often becoming suffused with purple*. Stems 2–8 cm., ascending or upright. Lvs mostly radical, narrowly subulate from *a sheathing base with 2 pointed auricles*, indistinctly septate. *Fls cylindrical or narrowly conical*, subsessile, in 1 or a few heads each of 1–5 fls; lowest bract much exceeding the infl. *Per. segs* 4·5–6 mm., *greenish or purplish, linear-lanceolate*, blunt, *with straight or incurved points. Capsule* 3–3·5 mm., pale, *obclavate. Seeds ovoid-pyriform*, about twice as long as broad. Fl. 5–6. Th.

Native. Damp hollows and cart-ruts on heaths. W. Cornwall (Lizard). 1. Reported from Hebrides (Raasay). Mediterranean region of Europe and N. Africa, extending up the Atlantic seaboard to the Kattegat; formerly in one locality in Sweden; east to Asia Minor.

25. J. castaneus Sm.   'Chestnut Rush.'

A medium-sized *stoloniferous* perennial, *not tufted*. Stem 8–30 cm., erect, smooth, terete, with 2–3 cauline lvs. *Lvs* 5–20 cm., soft, *subulate, ± channelled above*, from a *long sheathing base without auricles*, bluntly pointed. *Fls very large, dark chestnut-brown*, in 2–3 ± *approximated terminal capitate clusters*. Per. segs linear-lanceolate, outer acute, inner blunter. *Capsule* 6–8 mm., *elliptic-oblong, obtuse, mucronate, much longer than the perianth*. Seeds 2–3 mm., with long white appendages. Fl. 6–7. 2n=40. Hs. or Hel.

Native. Marshes by springs and streams in high mountains. Highlands of Scotland (chiefly Clova and Breadalbane). 11. Circumpolar, reaching 78° 30′ N. in Spitsbergen, extending south along the mountain ranges to the Alps, Urals, Altai, Shensi and Colorado; Newfoundland.

26. J. biglumis L.   'Two-flowered Rush.'

A small, neatly tufted perennial, with shortly creeping rhizome. *Stem* 5–12 cm., erect, *channelled along one side*. Lvs 3–6 cm., all radical, curved, subulate from a sheathing base *with very small auricles, in section of one tube. Infl.* usually *of 2 fls, one just below the other, usually exceeded by the lowest *bract. Per. segs* oblong, obtuse, *purplish-brown. Capsule* 3–4 mm., *turbinate, retuse,* much longer than the perianth. *Seeds* 1–1·25 mm., with a *short white appendage at each end.* Fl. 6–7. Hr.

Native. Wet stony places and rock-ledges on high mountains, often on micaceous soil, 2000–3000 ft. Scottish Highlands from Breadalbane to Argyll and Inner Hebrides. 5. Throughout the Arctic, reaching 83° 6′ N. in Greenland, extending south to Iceland, S. Norway, Altai, Kamchatka, Rocky Mountains to 40° N.

27. J. triglumis L.   'Three-flowered Rush.'

A small tufted perennial, forming *larger and taller tufts than J. biglumis. Stems* 5–10 cm., slender, stiffly erect, *terete.* Lvs 3–10 cm., all radical, curved, subulate from a *sheathing base with large auricles, in section of 2 tubes. Infl. a terminal cluster of 2–3 fls, almost at one level, usually exceeding the lowest bract. Per. segs* ovate-lanceolate, blunt, *light reddish-brown. Capsule* 5 mm., *ovate-oblong, obtuse, mucronate, shortly exceeding the perianth. Seeds* 1·75– 2 mm., *with a long white appendage* at each end. Fl. 6–7. $2n = 50$. Hr.

Native. Bogs and wet rock-ledges on high mountains, preferring non-calcareous rocks. Less alpine than *J. biglumis,* descending to 950 ft. in Perthshire and 200 ft. in Shetland; N. Wales, Teesdale, Lake District, Scotland. 25. Arctic and subarctic to 80° N., extending south to the Pyrenees, Alps, Caucasus, Himalaya, Colorado, Labrador.

2. Luzula DC.   Woodrush.

Tufted grass-like perennials, rarely more than 1 m. high, sometimes stoloniferous. *Lvs* mostly radical, *with closed sheathing base, without auricles, lamina flat or channelled, fringed with long colourless hairs.* Cymes few- or many-fld, sometimes condensed into a head. *Capsule* 1-*celled; seeds* 3, smooth and shiny, usually with a conspicuous appendage (aril).

About 80 spp., cosmopolitan, but chiefly in the cold and temperate regions of the northern hemisphere.

1	Fls borne singly in the infl., rarely in pairs.	*2*
	Fls in heads or clusters of 3 or more.	*3*
2	Infl. branches drooping to one side; capsule acuminate, not suddenly contracted above the middle.	**2. forsteri**
	Infl. branches spreading; capsule truncate, suddenly contracted above the middle.	**1. pilosa**
3	Perianth white or whitish, sometimes suffused with red.	**4. luzuloides**
	Perianth chestnut or yellowish-brown.	*4*
4	Robust; fl. stem often over 40 cm.; lvs 6–12 mm. broad.	**3. sylvatica**
	Fl. stem rarely to 40 cm.; lvs less than 6 mm. broad.	*5*
5	Infl. drooping, spike-like.	**5. spicata**
	Infl. of stalked clusters or if compact not drooping.	*6*
6	Dwarf alpine with channelled lvs; fls drooping in a subumbellate panicle of many clusters of 2–5 fls.	**6. arcuata**
	Lvs not channelled; fls in each cluster more numerous.	*7*

7   Stoloniferous; anthers 2–6 times as long as filaments.   **7. campestris**
    Not stoloniferous; anthers as long as or shorter than the filaments.   8

8   Fls 2·5–3 mm., chestnut-brown.   **8. multiflora**
    Fls 2 mm., pale yellowish-brown.   **9. pallescens**

1. L. pilosa (L.) Willd.   'Hairy Woodrush.'

L. vernalis (Reichard) DC.

A tufted perennial with short upright rootstock and slender stolons. *Radical lvs* 3–4 *mm. broad*, about ½ the length of the stem or longer, grass-like, sparsely hairy, with a small truncate swelling at the apex. Fl. stem 15–30 cm. Infl. a lax cyme, with unequal *spreading* capillary *branches, reflexed in fr. Fls* single, rarely in pairs, *dark chestnut-brown. Per. segs* 3–4 mm., ovate-lanceolate, acute, with broad hyaline border, *shorter than* or almost equalling the fr. *Capsule ovoid, very broad below, suddenly contracted above the middle to a truncate conical top. Seeds with a long hooked appendage.* Fl. 4–6 (sporadically later). $2n = 66, 72$. Hs.

Native. Woods, hedge-banks, etc., throughout the British Is. 112, H27. Europe, excepting the sclerophyll region of the Mediterranean; east to Caucasus and Siberia; N. America.

L. forsteri × pilosa (*L. × borreri* Bromf. ex Bab.) is not uncommon with the parents. Resembles *L. pilosa* but often taller. Lvs 2·5–5 mm. broad. Infl. branches fewer than in *L. pilosa,* spreading. Capsule much shorter than the perianth, always sterile.

2. L. forsteri (Sm.) DC.   'Forster's Woodrush.'

Tufted, like *L. pilosa*, but usually smaller and more slender. *Radical lvs* 1·5–3 *mm. broad. Infl. branches drooping to one side, remaining erect in fr. Fls reddish chestnut-brown. Per. segs* 3–4 mm., *lanceolate,* acute, narrower and more finely pointed than in *L. pilosa, usually longer than the ovoid acuminate mucronate capsule,* which is not suddenly contracted nor truncate. *Seeds with short straight appendage.* Fl. 4–6. $2n = 24$. Hs.

Native. Woods and hedge-banks in S. England and S. Wales, local, but abundant in some districts; north to Leicester. 32, S. S. and W. Europe, north to Belgium, east to Crimea, Syria and Persia; N. Africa; Macaronesia.

3. L. sylvatica (Huds.) Gaud.   'Greater Woodrush.'

L. maxima (Reichard) DC.

A tall robust perennial, forming bright green mats or tussocks, with short ascending rootstock and numerous stolons. *Radical lvs* 10–30 *cm. × 6–12 mm. or more*, spreading, glossy, broadly linear, gradually tapering to a very acute point, sparsely hairy. Fl. stems 30–80 cm., erect, with about 4 stem-lvs, the longest about 5 cm. *Fls chestnut-brown*, 3–4 together, *in a lax terminal cyme*, the branches spreading in fr. *Per. segs* 3–3·5 mm., lanceolate, acuminate, *about equalling the finely beaked, ovoid fr.* Seeds 1–2 mm., slightly shiny, tubercled at the tip. Var. *gracilis* Rostrup differs markedly in its much smaller dimensions. Lvs 3–7 cm. × 3–4 mm. Cyme little-branched, few-fld. Fl. 5–6. $2n = 12$. Hs.

Native. Woods (especially oak) on acid soil and peat and open moorlands, especially on rocky ground near streams. Throughout the British Is., but most abundant in the west and north. 110, H 40, S. Ascends to 3000 ft. or more. Var. *gracilis* in Shetland only. Europe, except the east, Caucasus, Asia Minor. Doubtfully native in S. America.

*4. L. luzuloides (Lam.) Dandy & Wilmott

L. albida (Hoffm.) DC.; *L. nemorosa* (Poll.) E. Mey., non Hornem.

A loosely tufted *medium-sized* perennial. Radical lvs grass-like, hairy, 3–6 mm. broad; *stem-lvs* 10–20 cm., *long and grass-like*. Fl. stem 30–60 cm., slender. *Infl. corymbose, lax*, shorter than the uppermost lf, with divaricate branches ending in *clusters of* 2–8 *fls. Fls dirty white*, sometimes tinged with pink or red. *Per. segs* 2·5–3 mm., lanceolate, acute, very unequal, *about as long as the ovoid beaked capsule*. Fl. 6–7. Hs.

Introduced. Naturalized in various parts of England, Wales and Scotland in woods and damp places by streams, chiefly on acid soils. Established in some localities for at least 70 years. C. and W. Europe to N. Germany, east to Moscow. Introduced with grass and other seed into Scandinavia, N. America, etc.

*L. nivea (L.) DC., a native of the mountains of W. and C. Europe, has been reported as planted or naturalized in various places. Differs from *L. luzuloides* in the snow-white fls, perianth twice as long as capsule (4·5–5·5 mm.), and the more compact infl. with fls in more numerous clusters.

5. L. spicata (L.) DC.   'Spiked Woodrush.'

A small *tufted alpine* perennial with short stolons. Rootstock clothed with the persistent lf-sheaths. *Radical lvs* 2–8 cm. × up to 2 mm., recurved, *somewhat channelled*, sparsely hairy. Fl. stems 2–20 cm., erect, very slender, ending in the *dense drooping spike-like infl.* of chestnut-brown fls. Per. segs equal, lanceolate, acute or subobtuse, with fine awn-like points, equalling or slightly longer than the broadly ellipsoid capsule. Seeds with a short appendage at the base. Fl. 6–7. 2*n*=24. Hs.

Native. Rocks, screes, alpine heaths, etc., on non-calcareous substrata, local. Lake District, Scottish Highlands, Outer Hebrides, Shetland. 21. Arctic and subarctic, extending south into the mountains of S. Europe, Corsica, Atlas Mts, Himalaya; N. America south to California and New England.

6. L. arcuata Sw.   'Curved Woodrush.'

A *dwarf alpine perennial in small neat tufts* with short stolons. Rhizome creeping, clothed with the persistent lf-sheaths. *Radical lvs* 2–5 cm. × less than 2 mm., narrowly linear, recurved, stiff, *deeply channelled, almost glabrous*. Fl. stems 3–8 cm. Infl. bent to one side when young, a *subumbellate panicle of* 2–5-*fld clusters*, the outer on recurved stalks. Per. segs equal, broadly lanceolate, acuminate, much longer than the broadly ovoid capsule. Seeds to 1·2 mm., oblong, with a very small basal appendage. Fl. 6–7. 2*n*=36. Hs.

Native. Open stony ground (*Juncetum trifidi*, etc.) on high mountains, chiefly over 3000 ft. Scottish Highlands from the Cairngorms to Ross. 6. Arctic and subarctic Europe, rarer in Asia and N. America; Greenland.

7. L. campestris (L.) DC. Sweep's Brush, 'Field Woodrush.'

L. campestris (L.) DC. ssp. *vulgaris* (Gaud.) Buchen.

A compact, usually *loosely tufted* perennial, with short stock and *shortly creeping stolons*, variable in its characters. Lvs usually 2–4 mm. broad, linear, grass-like, with a small truncate swelling at the apex, bright green, thinly clothed with long colourless hairs. *Fl. stem seldom more than* 15 *cm.*, bearing a loose sometimes drooping *panicle of* 1 *sessile and* 3–6 *stalked spherical obovate clusters of* 3–12 *fls.* Branches of infl. ±curved, reflexed in fr. Fls 3–4 mm., chestnut-brown. Per. segs lanceolate, the outer finely pointed, subequal, longer than capsule. *Anthers* 2–6 *times as long as the filaments.* Capsule 2·5–3 mm., obovoid, obtuse, apiculate. *Seeds nearly globose, with a white basal appendage up to* $\frac{1}{2}$ *their length.* Fl. 3–6. $2n=12; 24; 36$. Hs.

Native. Very common in grassy places throughout the British Is. 112, H40, S. Europe; N. Africa; N. America; Malaysia; New Zealand, etc.

8. L. multiflora (Retz.) Lej. 'Many-headed Woodrush.'

L. campestris (L.) DC. ssp. *multiflora* (Retz.) Buchen.

A *densely tufted perennial with few or no stolons, taller* than *L. campestris.* Lvs to 6 mm. broad, bright green, sparsely hairy. Fl. stem 20–40 cm., erect, wiry. *Infl.* somewhat umbellate *of up to* 10 *ovate or elongate* 8–16-*fld clusters on straight, slender erect branches* or (var. *congesta* (DC.) Lej.) subsessile *in a rounded or lobed head.* Fls 2·5–3 mm., chestnut-brown. Per. segs broadly lanceolate, longer than the capsule. *Anthers about as long as the filaments.* Capsule almost spherical, apiculate. *Seeds oblong, nearly twice as long as broad*, with a white basal appendage up to $\frac{1}{2}$ their length. Fl. 4–6 (later than *L. campestris*). $2n=36, 24$ (12). Hs.

Native. Heaths and moorlands, woods, chiefly on acid and peaty soils. Abundant throughout the British Is. 112, H40, S. Europe, N. Africa, Asia, N. America, New Zealand.

9. L. pallescens Sw.

Tufted, like *L. multiflora*, but *paler* in colour. Lvs light green, almost glabrous, to 4 mm. broad. Fl. stem 10–30 cm., slender. *Panicle umbellate, of* 5–10 *roundish-oblong clusters* of 8–15 fls, the central larger cluster subsessile, the rest on straight erect branches. *Fls small* (to 2 mm.), *pale yellowish-brown.* Per. segs broadly lanceolate, the outer longer than the capsule, the inner shorter than the outer and not so long-acuminate. *Anthers slightly shorter than the filaments. Capsule* 1·5 mm., *obovoid, with a blunt apiculus*, contracted towards the base. *Seeds* about *twice as long as broad* with a broad appendage less than $\frac{1}{2}$ the length of the seed. Fl. 5–6. $2n=12$. Hs.

Native. In open grassy places in fens, very local. Hunts. Introduced in Surrey. Central and northern Europe, east to Japan.

137. AMARYLLIDACEAE

Bulbous herbs. Lvs all radical. Fls solitary or umbellate, enclosed before flowering in a usually scarious spathe, hermaphrodite, usually actinomorphic, 3-merous. Perianth petaloid, in two whorls, sometimes with a corona. Stamens in 2 whorls, opposite per. segs, inserted on them or free; anthers introrse. Ovary inferior, 3-celled, with axile placentation. Ovules usually numerous, in 2 rows on each placenta, anatropous. Style simple; stigma capitate or 3-lobed. Fr. a capsule, rarely a berry. Seeds with copious fleshy endosperm.

About 60 genera and 500 spp., temperate and tropical, mainly warm temperate.

A family conventionally distinguished from the Liliaceae only by the inferior ovary. It has been divided by Hutchinson into several families based mainly on habit and infl. These are probably derived from the Liliaceae along independent lines. Hutchinson includes certain Liliaceae (i.e. Allieae and 2 allied tribes) in the Amaryllidaceae on account of their essentially similar infl. They appear to link the two families and are here retained in the Liliaceae. Apart from these, the above diagnosis applies to the Amaryllidaceae in Hutchinson's sense.

Numerous genera and spp. are cultivated, among which *Amaryllis*, *Nerine*, *Crinum* and *Hippeastrum* may be mentioned.

1	Perianth without a corona.	*2*
	Perianth with a trumpet-like or ring-like corona inside the 6 segments.	
		3. NARCISSUS
2	Per. segs all alike and equal.	1. LEUCOJUM
	Inner per. segs much smaller than outer.	2. GALANTHUS

1. LEUCOJUM L.

Fls solitary to several, nodding. *Perianth* campanulate, *segments all alike; no corona*; tube 0 or very short.

Nine spp., C. Europe and Mediterranean region.

Fls 1(–2); per. segs 20–25 mm.	**1. vernum**
Fls (2–)3–7; per. segs 14–18 mm.	**2. aestivum**

1. L. vernum L.   'Spring Snowflake.'

Bulb rather large, subglobose. Lvs 20–30 cm. × c. 1 cm., linear, bright green. Scape 15–20 cm. Spathe 3–4 cm. usually bilobed at apex, green in the middle with scarious borders. *Fls* 1(–2). *Per. segs* 20–25 *mm.*, white, tipped with green, obovate, bluntly acuminate. Capsule pyriform; seeds pale, appendaged. Fl. 2–4. Pollinated by bees and Lepidoptera; homogamous. Seeds distributed by ants. $2n=22$. Gb.

Possibly native. Very rare. Damp scrub and hedge-banks in 2 localities in Dorset and S. Somerset; quite often cultivated but very rarely recorded as an escape. 2. Hills of Europe from Belgium, N. and E. France and C. Germany to N. Spain, C. Italy and Serbia.

2. L. aestivum L. Loddon Lily, 'Summer Snowflake.'

Bulb large (c. 2·5 cm.) ovoid. Lvs 30–50 cm. × 10–15 mm., linear, bright green. Scape 30–60 cm. Spathe 4–5 cm., green at apex. *Fls* (2–)3–7. *Per. segs* 14–18 *mm.*, white, tipped with green, obovate, obtuse or very bluntly acuminate. Capsule pyriform; seeds black, not appendaged. Fl. 4–5. Pollinated by bees; homogamous. Seeds distributed by water. 2*n*=22. Gb.

Native. Wet meadows and willow thickets, very local; from Devon and Kent to Oxford and Suffolk and from Wexford and Cork to Antrim and Fermanagh; certainly native along the Thames and Shannon and probably elsewhere within the range given above. 14, H12. Commonly cultivated and sometimes found as an escape in other places. Europe from C. and S. France, S. Austria and S. and W. Hungary to C. Italy and Greece; Crimea, Caucasus; naturalized further north; N. Asia Minor, N. Persia.

2. GALANTHUS L.

Fls solitary, nodding. Outer per. segs somewhat spreading, separated; *inner whorl* campanulate, *much smaller; no corona*; tube 0 or very short.

About 8 spp., all except the following, confined to E. Mediterranean region.

1. G. nivalis L. Snowdrop.

Bulb ovoid, c. 1 cm. across. Lvs 10–25 cm. × c. 4 mm., linear, glaucous. Scape 15–25 cm. Spathe 2-fid, green in middle with broad scarious margins. Outer per. segs 14–17 mm., pure white, obovate-oblong, obtuse; inner about half as long, obovate, deeply emarginate, white with a green spot at the incision. Capsule ovoid. Fl. 1–3. Pollinated by bees; homogamous. Fr. 6. 2*n*=24. Gb.

Probably native. Local in damp woods and by streams from Cornwall and Kent to Dunbarton and Moray, probably native in some places in Wales and W. England but very commonly planted and usually only naturalized. 60, S. Europe from the Netherlands (?native), W. and C. France, S. Germany and the Carpathians to Pyrenees, Italy, Sicily, Macedonia; N. Syria, Asia Minor, Caucasus and S.E. Russia (Lower Don region); naturalized further north.

STERNBERGIA Waldst. & Kit.

Fls solitary, erect, yellow, with distinct tube and no corona. *S. lutea (L.) Ker-Gawl. ex Spreng. is often grown in gardens and is naturalized in Jersey.

Native of Mediterranean region.

3. NARCISSUS L.

Fls solitary or several, usually horizontal. Per. segs all alike, usually spreading; *corona trumpet-like or ring-like*, inserted between the per. segs and stamens; tube distinct.

About 35 spp. in Europe, W. Asia and N. Africa.

The Daffodils and Narcissi of our gardens are now mainly hybrids derived from the following and numerous other spp. A number of spp. are also cultivated. Several have been found as escapes and others, as well as garden hybrids, are likely to occur.

1 Corona about as long as perianth.   *2*
    Corona less than ½ as long as perianth.   *4*

2 Corona darker yellow than perianth, scarcely spreading at mouth;
    pedicels 3–10 mm., strongly deflexed.   **1. pseudonarcissus**
    Corona and perianth both deep yellow, corona with margin spreading at
    mouth; pedicels 10–35 mm., ± erect, usually somewhat curved but not
    deflexed.   *3*

3 Pedicels 10–15 mm.; perianth 3·5–4·5 cm., the segments not twisted.
                                                          2. obvallaris
    Pedicels 25–35 mm.; perianth 5–6 cm., the segments usually spirally
    twisted.   **3. hispanicus**

4 Fls solitary; corona with a red rim.   **4. majalis**
    Fls usually 2; corona concolorous or with pale rim.   **5. × biflorus**

1. N. pseudonarcissus L.   Wild Daffodil.

Bulb 2–3 cm., ovoid. Lvs 12–35 cm. × 5–12 mm., linear, erect, glaucous,
usually somewhat channelled. Scape 20–35 cm., ± compressed, 2-edged. Fls
solitary, drooping to nearly horizontal. *Pedicels 3–10 mm., strongly deflexed.*
Perianth 35–60 mm., *pale yellow*; tube 15–22 mm.; segments oblong-lanceolate
to ovate-lanceolate or elliptic, obtuse and mucronate to acuminate, ascending,
wavy or spirally twisted. *Corona as long as* or slightly shorter than *per. segs,
deep yellow, scarcely expanded or spreading at mouth*, irregularly lobulate, the
lobules numerous and toothed. Anthers without dark apical spot. Fr. 12–
25 mm., obovoid or subglobose, very obtuse, roundly trigonous to nearly
terete. Fl. 2–4. Pollinated by humble-bees and other insects; homogamous.
Fr. 6. Germ. winter. $2n=14$. Gb.

Native. Damp woods and grassland, throughout England and Wales and
in Jersey but rather local in most districts; naturalized in Scotland and Ireland.
69, S. Belgium, France, N. and C. Spain and Portugal, N. Italy, W. Germany
(to the Rhine), Switzerland (?native); naturalized in Scandinavia and further
east.

2. N. obvallaris Salisb.   'Tenby Daffodil.'

N. lobularis (Haw.) J. A. & J. H. Schult.

Bulb 2·5–3·5 cm., subglobose-ovoid. Lvs 20–30 cm. × 8–10 mm., linear, erect,
glaucous, flat, not twisted. Scape 20–30 cm., little compressed, 2-edged. Fl.
solitary, ascending or nearly horizontal. *Pedicels* 10–15 *mm., not deflexed,* slightly
curved. *Perianth* 35–45 mm., *deep yellow like the corona*; tube 12–15 mm.;
segments ovate, obtuse, mucronate, imbricate, spreading, *not twisted. Corona
rather longer than per. segs*, 25–30 mm. across, *dilated above with the margin
spreading or slightly reflexed, rather regularly 6-lobed*, the lobes rounded and
crenate or subentire. Anthers with a minute dark apical spot. Fr. 20–
25(–30) mm., oblong or oblong-obovoid, subtruncate, obscurely trigonous
with somewhat flattened sides, scarcely furrowed. Fl. 4. $2n=14$. Gb.

?Native. Pastures near Tenby, Pembroke. Formerly in Shropshire, and
reported from Carmarthen. Not known outside Britain but a variety occurs
(?native) in N. Italy. *N. obvallaris* may be of early garden origin or may yet
be found native in Europe.

*3. N. hispanicus Gouan

N. major Curt.

Bulb 4–5 cm., ovoid. Lvs 40–50 cm. × 10–12 mm., linear, erect, glaucous, flat, ± spirally twisted. Scape 40–60 cm., much compressed, 2-edged. Fl. solitary, suberect to nearly horizontal. *Pedicels 25–35 mm., erect,* curved above. *Perianth* 50–60 mm., *deep yellow like the corona.*; tube c. 18 mm.; *segments* oblong-lanceolate, subacute, slightly imbricate below only, spreading, *spirally twisted. Corona* c. 45 mm. across, *as long as per. segs, abruptly dilated above with the margin widely spreading, obscurely lobed* and deeply crenate-serrate. Anthers with minute black apical spot. Fr. 20–30 mm., oblong-ellipsoid, very obtuse, bluntly trigonous with shallow furrows. Fl. 3–4. $2n=21$ (14 in a var.). Gb.

Introduced. Formerly much grown in gardens and found ± naturalized in grass in a number of places, now rarely cultivated, having been replaced by larger-fld hybrids. Native of S.W. France, N. Spain, N. Portugal.

The common double daffodil of gardens, *N.* 'Telamonius plenus' is often regarded as a variety of this sp. It differs, however, in its lvs and no corresponding single-fld plant is known. It sometimes occurs as an escape or relic of cultivation.

*N. × incomparabilis Mill.

Lvs 8–15 mm., broad, glaucous. Fls solitary. *Perianth* 4–6 cm. across, *pale* yellow, *corona* 8–12 mm. deep yellow, cup-shaped, *about half as long as perianth.* Fl. 4. $2n=14$, 21. Gb.

A hybrid or series of hybrids between members of the section *Ajax* (to which the 3 preceding belong) and *N. poeticus* (agg.). Grown in gardens and sometimes found as an escape. Garden hybrids of similar origin are very numerous, the perianth varying in colour from white to deep yellow and the corona from very pale yellow to reddish orange.

*N. × infundibulum Poir.

N. odorus auct. angl.

Lvs 4–5 mm. broad, linear, semi-terete, green. Scape 30–40 cm., subcylindric. Fls 1–2(–3). Perianth c. 4–5 cm. across, bright yellow, tube longer than segments; segments separated, undulate, broadly elliptic. Corona c. $\frac{2}{3}$ as long as per. segs, funnel-shaped, deeper yellow with erect nearly truncate margin.

A hybrid between some member of section *Ajax* and *Narcissus jonquilla* L. or allied sp. Formerly naturalized in quantity near St Austell (Cornwall) but now much reduced. Origin unknown.

*4. N. majalis Curt.   Pheasant's Eye.

N. poeticus auct., p.p.

Bulb 2·5–3 cm., ovoid. Lvs 9–13 mm. broad, linear, erect, glaucous, grooved. Scape 40–50 cm., compressed, 2-edged. Fl. solitary. *Perianth* 5·5–7 cm. across, *white*, tube about equalling segments, green; segments imbricate below, margins reflexed, oblong or obovate-oblong, obtuse or subacute, ± spreading. *Corona* c. 3 *mm., cup-shaped*, yellow with a green base and a broad white zone below the *light red margin* which is denticulate-fimbriate. Stamens unequal, three outer anthers slightly exserted, three inner included in the perianth-tube and only just exceeding base of outer. Fr. c. 15 mm., triangular-obovoid, scarcely furrowed. Fl. 5. Pollinated by lepidoptera and selfed. Gb.

Introduced. ± Naturalized in a number of places. Formerly much cultivated, now little grown. Type native of S. France, the var. not known wild.

The above appears to be the only member of *N. poeticus* (agg.) to have become naturalized in Britain though *N. radiiflorus* Salisb. with greenish-white per. segs narrowed below, and nearly equal stamens (fl. 4) has been found as an escape. Other spp. and modern hybrids are now much more commonly grown. For a full account of the group see Pugsley, *Journ. Bot.* Suppl. (1915).

*5. N. × biflorus Curt.   Primrose Peerless.

Bulb 2·5–3·5 cm., subglobose. Lvs 25–45 cm. × 7–14 mm., linear, scarcely glaucous. Scape 30–60 cm., compressed, 2-edged. *Fls* (1–)2(–3). *Perianth cream or yellowish-white*, 3–5 cm. across; tube about as long as segments; segments broadly obovate, obtuse, mucronulate, imbricate. *Corona 3–5 mm., yellow, usually with crenate whitish margins.* Anthers aborted. Fl. 4–5. Fr. not produced. $2n = 17, 24$. Gb. Almost certainly a hybrid between *N. poeticus* (agg.) and *N. tazetta* L.

Naturalized in grassy places in a number of localities scattered over the whole of England and Wales and in S. Ireland to Dublin and Clare; Midlothian; Jersey; after *N. pseudonarcissus* the most frequent member of the genus. 40, H6, S. Naturalized also in S. Europe to W. France.

138. IRIDACEAE

Perennial herbs with rhizomes, corms or bulbs. Lvs not differentiated into blade and petiole, often ensiform, sheathing at base and equitant. Fls 3-merous, hermaphrodite, usually actinomorphic, sometimes zygomorphic, usually with 1 or 2 bracts forming a spathe at their base. Perianth in 2 series, petaloid, withering and persistent after flowering; segments usually connate at base into a longer or shorter tube; tube straight, or curved in zygomorphic fls. Stamens 3, free or partially connate. Ovary inferior, 3-celled with axile placentas or 1-celled with 3 parietal placentas; style 3-lobed, branches entire or divided, sometimes petaloid; stigmas terminal or on the underside of the branches; ovules numerous, rarely few or 1, anatropous. Capsule loculicidal, opening by valves, usually with a conspicuous circular scar at top marking point of attachment of the perianth.

About 70 genera and 1000 spp. generally distributed throughout the world. Many cultivated as ornamental plants.

1	Outer per. segs distinctly larger than inner, if subequal then outer recurved, inner incurved.	*2*
	All per. segs equal or subequal, never curving in opposite directions.	*3*
2	Rhizomatous plants; lvs flattened; ovary 3-celled.	2. Iris
	Plant with tubers; lvs tetragonal; ovary 1-celled.	3. Hermodactylus
3	Fls subsessile on the corm; tube long and slender; midrib of lf white, at least beneath.	4. Crocus
	Fls peduncled; tube short; midrib green on both surfaces.	*4*

4 Scapigerous herbs; fls terminal.   5
    Stems lfy; fls lateral on a secund spike-like infl.   6

5 Scapes flattened, winged; lvs ensiform; fls yellow or blue.
                                                1. Sisyrinchium
    Scapes terete; lvs almost setaceous; wiry; fls greenish outside, whitish
      or pale blue within.   5. Romulea

6 Fls deep orange; tube straight or nearly so.   6. Crocosmia
    Fls crimson-purple; tube distinctly curved.   7. Gladiolus

Tribe 1. Sisyrinchieae. Rhizomatous. Perianth-tube 0 or very short.
Style-branches undivided.

1. Sisyrinchium L.

Roots fibrous from a short rhizome. Lvs linear-ensiform. Scapes usually
flattened. Spathes 2–several-fld. Fls actinomorphic, blue or yellow, very
fugitive. Per. segs in 2 series, all similar; tube very short. Stamens inserted
in the throat of the perianth-tube. *Style-branches undivided.* Capsule exserted
from the spathe.

    About 75 spp. in N. and S. America, one in western Ireland.

Fls blue; capsule c. 5 mm.   **1. bermudiana**
Fls yellow becoming orange; capsule c. 12 mm.   **2. californicum**

1. S. bermudiana L.   Blue-eyed Grass.

S. angustifolium Mill. *S. graminoides* Bicknell

A glabrous erect perennial 15–45 cm. Lvs 7–15 cm., 1–3 mm. wide, ensiform.
Scape flattened and narrowly winged. Spathe (1–)2–4(–6)-fld. *Fls* c. 15 mm.
diam., *blue*; per. segs 7 mm., obovate, retuse, long-mucronate; filaments fused
almost to top. *Capsule c. 5 mm.*, globular-trigonous. Fl. 7. $2n = 64^*$. Hr.
    Native. In marshy meadows and lake shores, locally abundant. H10.
Western Ireland from Cork to Donegal, but apparently absent from Mayo;
naturalized in at least 14 counties in Great Britain and elsewhere in Europe.
The native plant is apparently endemic. A closely related sp. occurs in eastern
N. America.

*2. S. californicum (Ker-Gawl.) Ait. f.

Rather similar in general appearance to *S. bermudiana* but stouter. Lvs
3–6 mm. wide. Scape broadly winged. Spathe 3–5-fld. *Fls yellow becoming
orange*; per. segs oval; filaments free almost to base. *Capsule c. 12 mm.*,
ellipsoidal-trigonous. Fl. 6. $2n = 34$. Hr.
    Introduced. Naturalized in marshy meadows, Wexford; reputed to have
been introduced by wreckage. Native of California and Oregon.

Tribe 2. Irideae. Rhizomatous or with corms. Perianth-tube usually very
short, rarely longer than ovary. Style-branches divided and ± petaloid.

2. Iris L.

Perennial herbs with sympodial rhizomes or bulbs. Lvs ensiform, sometimes
very narrow, often distichous. Spathe usually with scarious margins. Fls
actinomorphic, large and showy. Per. segs in 2 series, outer ('falls') usually
deflexed and larger than inner ('standards') which are often erect, usually

with a well-marked limb and claw ('haft'); tube short. Stamens inserted at base of outer per. segs. *Style-branches* ('crest') *broad, petaloid, bifid at tip. Ovary 3-locular, placentation axile.* Capsule trigonous.

About 180 spp. throughout the temperate regions of the northern hemisphere. Numerous spp. are cultivated; the common garden irises are mostly hybrids of ± complex parentage.

1 Outer per. segs bearded, nearly equalling inner; tube longer than ovary.
   5. germanica
   Outer per. segs not bearded, larger than inner; tube not longer than ovary.   2

2 Fls yellow, or if purplish then dirty greenish- or yellowish-purple and plant growing in dry places.   3
   Fls violet or clear pinkish-purple; plant growing in wet places.   4

3 Scape angled on one side; pedicels 4–5 times as long as ovary; fls purplish-livid, rarely yellow; capsule clavate; seeds orange-red (dry places).
   3. foetidissima
   Scape terete, compressed; pedicels about as long as ovary; fls yellow; capsule elliptic, apiculate; seeds brown (wet places). **4. pseudacorus**

4 Claw of outer per. segs twice as long as limb; style-branches violet.
   1. spuria
   Claw of outer per. segs about equalling limb; style-branches very pale pinkish-purple.   **2. versicolor**

1. I. spuria L.

An erect glabrous perennial 20–60 cm. Rhizome c. 2 cm. diam. Lvs linear, shorter than the unbranched cylindrical scape. Spathes green with a narrow scarious margin, 1–3-fld. Pedicels short (c. 3 cm.). *Fls blue-violet and whitish,* not scented. *Outer per. segs* with limb almost orbicular, not bearded, abruptly contracted below, *claw twice as long as limb; inner* shorter than outer, narrow-obovate, *violet;* tube short. *Style-branches violet. Capsule narrowed into a long point at top with 2 ridges at each angle.* Fl. 6. 2*n*=22, 44. Hel.

Native or introduced. Known for more than 100 years beside fen ditches in Lincoln and perhaps native there; Dorset; rare. Denmark to Spain and C. Europe, S. Russia, Caucasus, N. Africa.

*2. I. versicolor L.

An erect glabrous perennial 50–100 cm. Rhizome c. 1 cm. diam. Lvs ensiform, shorter than the branched scape. Spathes herbaceous, 1–3-fld. Pedicels longer than ovary. *Fls rather pale pinkish-purple.* Outer per. segs with roundish limb gradually contracted below; *claw about as long as limb;* inner shorter than outer, oblanceolate; tube short. *Style-branches nearly white.* Fl. 6–7. Hel.

Introduced. Naturalized in reed-swamp in Ullswater (for more than 40 years), R. Calder (Yorks), Loch Tay (mid Perth), and probably elsewhere. N. America.

3. I. foetidissima L.   Gladdon, Stinking Iris.

A dark green glabrous perennial 30–80 cm., with a strong unpleasant smell when bruised. Lvs evergreen, ensiform, equalling or exceeding the unbranched *scape* which is *angled on one side*. Spathes with a narrow scarious margin, 2–3-fld. Pedicels 4–5 times as long as ovary. *Fls* c. 8 cm. diam., *purplish-livid*, rarely yellow. Outer per. segs obovate-lanceolate, not bearded; inner spathulate, yellowish, shorter than outer; tube very short. Style-branches spathulate, yellowish. *Capsule clavate; seeds orange-red*. Fl. 5–7. 2n=40. Ch.

Native. In hedge-banks, open woods and on sea cliffs, usually on calcareous soils. 54, S. S. England and Wales, fairly widely distributed in suitable habitats; naturalized in Scotland and Ireland. W. Europe from France southwards, east to Italy and Greece; N. Africa.

4. I. pseudacorus L.   Yellow Flag.

An erect glabrous rather glaucous perennial 40–150 cm. Rhizome often 3–4 cm. diam. Lvs 15–25 mm. broad, ensiform, about equalling the often branched *compressed terete scape*. Spathes with broadly scarious margins towards the top, 2–3-fld. *Pedicels about as long as the ovary*. Fls 8–10 cm. diam., *yellow*, varying from pale to almost orange. Outer per. segs variable in form, shortly clawed, often purple-veined with an orange spot near the base, not bearded; inner spathulate, smaller than outer; tube short. Style-branches yellow. *Capsule elliptic, apiculate; seeds brown*. Fl. 5–7. 2n=24, 32–34.

Native. In marshes, swampy woods, and in shallow water or wet ground at edges of rivers and ditches. 112, H40, S. In suitable habitats throughout the British Is. Europe to c. 68° N. in W. Norway; N. Africa; Caucasus and W. Asia.

*5. I. germanica L.

A stout glabrous perennial 30–100 cm. Rhizome often up to 5 cm. diam. Lvs up to c. 5 cm. broad, ensiform, shorter than the branched cylindrical scape. *Spathes scarious in the upper half*, 2–3-fld. Pedicels short. Fls c. 10 cm. diam., purple, scented. Outer per. segs ovate-oblong, dark purple, yellowish-white with brown-purple veins at base, *beard yellow; inner about equalling outer, incurved*, light purple; *tube longer than ovary*. Style-branches spathulate. Capsule ovoid. Fl. 5–6. H.

Introduced. Frequently cultivated and sometimes naturalized in waste places. Perhaps native in the Mediterranean region, but widely cultivated and naturalized in many places.

3. HERMODACTYLUS Mill.

Similar to *Iris* in many respects. Roots tuberous. *Lvs tetragonal in section. Fls solitary*. Outer per. segs much larger than inner. *Ovary 1-celled, placentation parietal.*

One sp. in the Mediterranean region.

*1. H. tuberosus (L.) Mill.   Snake's-head Iris.

Iris tuberosa L.

An erect glabrous perennial 20–40 cm. Rootstock short, producing fibrous roots and 2–4 oblong tuberous ones. Lvs linear, tetragonal, exceeding the slender scape. Spathes herbaceous, 1-fld. Outer per. segs not bearded, limb smoky purple; claw green, twice as long as limb. Inner per. segs greenish-yellow, thread-like, inconspicuous, incurved and shorter than style-branches. Anthers yellow. Capsule obovoid. Fl. 4. Grt. $2n=20$.

Introduced. Not infrequently cultivated. Naturalized in a few localities in Cornwall and Devon. Mediterranean region.

Tribe 3. CROCEAE. Corm. Lvs tufted. Fls solitary and subsessile or stalked on the corm. Perianth-tube distinct, often long.

4. CROCUS L.

Corm covered by the lf-bases of the previous year (*tunic*), which are often fibrous. *Foliage lvs tufted*, linear, channelled, surrounded by scarious sheaths, midrib white beneath. Fls actinomorphic, shortly peduncled, large and showy, enclosed in bud by 2 spathes, one at the base, the other at the top of the peduncle, or the basal spathe 0; peduncle elongating in fr. Per. segs in 2 series; *tube long and slender*. Ovary subterranean; style-branches cuneate or variously divided; stigmas terminal. Capsule on a long slender peduncle.

About 70 spp., in temperate regions of the Old World, particularly the Mediterranean.

Fls appearing in autumn, long before the lvs.   **1. nudiflorus**
Fls appearing in spring, with or shortly after the lvs.   **2. purpureus**

*1. C. nudiflorus Sm.   'Autumnal Crocus.'

Corm subglobose, *abundantly stoloniferous in spring; fibres of tunic parallel.* Lvs 3–5, linear, smooth, margins recurved. Fl. purple, appearing in autumn. Per. segs c. 5 cm.; *throat purple. Style-branches divided into numerous capillary lobes.* Fl. 9–10. Gt. $2n=c.\,46$.

Introduced. Naturalized in meadows and pastures, rare. Gloucester to Lancashire, Lincoln, Nottingham, Derby and S.W. Yorks. Native of S.W. France, Pyrenees, Spain.

*2. C. purpureus Weston   'Purple Crocus.'
C. vernus auct.; *C. albiflorus* auct..

Corm depressed-globose, not stoloniferous, *fibres of tunic coarsely reticulate.* Lvs 2–4, appearing with the fls. *Fls purple or white*, appearing in spring. Per. segs 3·5–5 cm.; *throat purple or white*. Style-branches shortly toothed, deep orange. Fl. 3–4. Gt. $2n=8,\,18,\,19$.

Introduced. Naturalized in meadows and pastures. Scattered throughout England, local; Wales: Carmarthen; Scotland: Ayr and Midlothian. Native of Italy and the Balkans.

In addition to the above, two spring-flowering spp. with nearly entire style-branches sometimes persist on light soils in and near derelict gardens: *C. flavus* Weston has a tunic of parallel fibres and 1–4 golden-yellow fls which appear with the lvs and *C. biflorus* Mill. has a tunic of rings and 1–3 fls, appearing after the lvs, tinged in shades of purple with a yellowish throat and 3 stripes of dark purple on the backs of the outer per. segs.

5. ROMULEA Maratti

Corm tunicated. Foliage lvs tufted, slender, linear. Scape simple or branched. *Fls* actinomorphic, solitary in the spathes, *long-peduncled.* Per. segs in two similar series; *tube short.* Style-branches linear, bifid. Capsule ovoid, 3-lobed.

About 70 spp., mainly in the Mediterranean region and S. Africa.

1. R. columnae Seb. & Mauri

R. parviflora Bubani; *Trichonema Columnae* (Seb. & Mauri) Rchb.

Corm small, ovoid, producing offsets freely. *Lvs* 3–6, 5–10 cm., *almost setaceous, wiry. Scape shorter than the lvs, recurved after fl.* Spathe of 2 sub-equal blunt bracts. Fls 1–3, 7–10 mm. *Per. segs* subacute, *greenish outside, purplish-white inside, yellow towards the base, veins purple.* Fl. 3–5. Gt. $2n = c.$ 60.

The British plant is var. *occidentalis* Béguinot, which is the form also found in western France.

Native. In short turf on sandy ground near the sea, very local. S. Devon and Channel Is.; apparently extinct in Cornwall. 1, S. W. Europe from northern France southwards; Mediterranean region.

Tribe 4. GLADIOLEAE. Corm. Stems lfy. Fls distinctly zygomorphic or limb oblique and tube ±curved. Tube widening gradually from base to mouth. Stamens usually bent over towards one side of the fl.

6. CROCOSMIA Planch.

Corm covered by reticulate fibrous tunic. Foliage lvs ensiform, distichous, radical and cauline. Infl. long, spike-like, secund. *Spathes membranous, emarginate.* Fls weakly zygomorphic, inclined or horizontal, showy. *Tube straight or nearly so*, usually shorter than the per. segs, widening gradually upwards. Capsule small, ovoid or oblong.

About 5 spp. in S. Africa.

*1. C. × crocosmiflora (Lemoine) N.E.Br.   Montbretia.

(*C. aurea* Planch. × *pottsii* (Baker) N.E.Br.)

An erect glabrous perennial 30–90 cm. Corms nearly 2 cm. diam., often several together in a row, freely stoloniferous. Stems slender, branched, equalling or exceeding the lvs. Lvs 5–20 mm. wide. Bracts small, oblong, reddish. Fls 2·5–5 cm. diam., deep orange suffused with red, funnel-shaped with rather spreading per. segs, longer or sometimes shorter than the tube. Style-branches short, simple. Fl. 7–8. Gt. $2n = 22, 33, 44$.

Raised at Nancy, France, by Victor Lemoine by crossing *C. pottsii* (female) with *C. aurea* (male); flowered for the first time in 1880.

Introduced. Commonly cultivated and very tolerant of shade; naturalized by sides of lakes, rivers and ditches in hedge-banks, on waste ground and in woods, particularly in the west; spreading by vegetative means and by seed. Garden origin; parents from S. Africa.

7. GLADIOLUS L.

Corm covered by fibrous tunic. Lvs ensiform, distichous, radical and cauline. Infl. long, spike-like, secund. *Spathes* usually *herbaceous*. Fls weakly zygomorphic, inclined or horizontal, showy, solitary in the spathes. Per. segs in 2 similar series; *tube short, curved*. Stamens bent towards the upper side of the perianth. Style-branches undivided. Capsule subglobose or oblong-trigonous.

About 120 spp. in Europe, W. Asia, Africa and the Mascarene Is.

1 Anthers c. ½ as long as their filaments; stigmas abruptly dilated towards
     their ends.   **1. illyricus**
     Anthers as long as or longer than their filaments; stigmas gradually
     widening from the base upwards.   2

2 The 3 upper per. segs nearly equal; seeds winged.   **2. byzantinus**
     The upper per. seg. distinctly larger than the lateral ones; seeds not
     winged.   **3. segetum**

1. G. illyricus Koch

An erect glabrous perennial 40–90 cm. Corm c. 1·5 cm. diam., producing numerous offsets. Lvs up to c. 30 × 1 cm., glaucous, acuminate. Scape simple, exceeding the lvs. Infl. 3–8-fld. Spathe of 2 subequal ± herbaceous often purple-tipped bracts. Per. segs c. 2·5–3 cm., obovate, long-clawed, crimson-purple; tube short, curved. Anthers shorter than their filaments. Style-branches linear at base, then abruptly dilated and ovate, not papillose on the margin throughout their length. Capsule obovoid with 3 acute angles; seeds narrowly winged. Fl. 6–8. 2*n*=60, 90. Gt.

Native. Among bracken on bushy heaths, very local. Dorset, S. Hampshire and Isle of Wight. Western France and Portugal to the Northern Balkans and Northern Austria; Asia Minor; Caucasus.

*2. G. byzantinus Mill.

Similar in general appearance to *G. illyricus* from which it differs in having the anthers as long as or longer than their filaments and in the larger (3 cm. or more) fls. The 3 upper per. segs nearly equal. Seeds compressed and winged.

Introduced. Sometimes cultivated; a garden weed or ±naturalized in waste ground near the sea chiefly along the S. coast. Mediterranean region.

*3. G. segetum Ker-Gawl.

Differs from *G. byzantinus*: upper per. seg. somewhat longer than and nearly twice as broad as the lateral. Seeds pyriform, not winged. 2*n*=120.

Introduced. Sometimes cultivated; a garden weed or ±naturalized in waste ground near the sea, chiefly along the S. coast. Mediterranean region.

139. DIOSCOREACEAE

Usually slender herbaceous or woody twiners with tuberous rhizomes or stocks or thick woody stem-tubers (above ground in *Testudinaria*). Lvs usually spirally arranged, often cordate, entire, lobed or palmately divided, with palmate main veins and a network of smaller veins. Fls small, in spikes, racemes or panicles; unisexual (usually dioecious), actinomorphic, epigynous.

Perianth campanulate, of 3 + 3 ± equal segments united below into a short tube; stamens in male fl. borne on the base of the perianth-tube, 3 + 3 or 3 with or without staminodes replacing the missing set; in the female fl. rudimentary or 0; ovary in the female fl. inferior, 3-celled, with 2 superposed anatropous ovules on axile placentae in each cell; styles 3, or 1 divided above into 3 stigmatic lobes; ovary in the male fl. rudimentary or 0. Fr. a 3-valved capsule or berry; seeds often flattened or winged, with horny endosperm and a small embryo.

170 spp. in 8–9 genera, widespread in tropical and warm temperate regions.

The large genus *Dioscorea* includes several spp. cultivated throughout the tropics for their starchy root-tubers or tuberous rhizomes (yams). The closely related *D. pyrenaica* Bub. & Bordère ex Gren. is endemic in the Pyrenees and is probably a Tertiary relict.

1. Tamus L.

Perennial dioecious herbs with large hypogeal stem-tubers, slender annual twining stems, and entire cordate lvs. Fls small, in axillary racemes or spikes. Perianth campanulate; stamens 6, rudimentary in the female fl.; style 1 with three 2-lobed stigmas. Fr. a berry, incompletely 3-celled, with few globose seeds.

Two spp. in the Mediterranean region, N. Africa, Caucasus and W. Asia.

1. T. communis L.   Black Bryony.

A tall herb with a large irregularly ovoid blackish tuber up to 20 (rarely to 60) cm. and 10–20 cm. below the soil surface. Stems 2–4 m., slender, angled, unbranched, glabrous, twining to the left. Lvs 3–10 × 2·5–10 cm., broadly ovate, deeply cordate at the base, finely acuminate, entire, dark shining green, with 3–9 curving main veins; stalk long, with 2 stipule-like emergences at its base. Fls yellowish-green in axillary racemes, the male stalked, long, erect or spreading, the female subsessile, recurved, shorter and fewer-fld than the male; bracts minute, subulate, scarious; male fls 5 mm., female fls 4 mm. diam. Perianth with 6 narrow, somewhat recurved, lobes. Stigmas recurved. Berry c. 12 mm., ovoid-ellipsoidal, apiculate, pale red, glabrous, with 1–6 pale yellow rugose seeds. Fl. 5–7. Visited for nectar by many insects, including small bees. $2n = 48$. G.

Native. Wood-margins, scrub, hedgerows, etc., on moist well-drained fertile soils; to 800 ft. in Derby; common in the south. 70, S. Throughout England and Wales northwards to S. Cumberland and S. Northumberland. In Ireland only near Lough Gill (Sligo and Leitrim), and probably introduced. S. and W. Europe northwards to England, Belgium, W. Germany, Austria, Hungary, Transsylvania and the northern shore of the Black Sea. N. Africa; Palestine and Syria; coastal regions of Asia Minor; Caucasus.

140. ORCHIDACEAE

Perennial herbs with rhizomes, vertical stocks or tuberous roots; terrestrial, epiphytic or sometimes saprophytic, and almost invariably mycorrhizic. Stems often swollen at the base ('pseudo-bulbs'), and, in epiphytic types, often with aerial roots. Lvs entire, spirally arranged or distichous, rarely opposite, often with a sheathing base and sometimes with dark spots or blotches; in saprophytic forms reduced to scales or membranous sheaths lacking chlorophyll. Infl. a spike, raceme or racemose panicle, commonly pendulous in epiphytic types. Fls zygomorphic, epigynous and usually hermaphrodite, often large and very striking in form and colour. Perianth of 6 segments in 2 whorls, usually all petaloid (though often green) or with the outer whorl sepaloid and the inner petaloid; the median (posterior) segment of the inner whorl (*labellum*) is commonly larger and different in shape from the remainder, and is usually on the lower side of the fl. and directed ± downwards owing to the inversion of the fl. in a pendulous infl. or to the twisting of the ovary or its stalk through 180° in types with an erect infl.; the labellum is often spurred at its base. The anthers and stigmas are borne on a special structure, the *column*; stamens 2 (Diandrae) or 1 (Monandrae), with ± sessile 2-celled anthers seated behind or upon the summit of the column, their pollen of single grains or more commonly of tetrads cohering in packets which are bound together with elastic threads, the contents of each cell forming 1–4 granular or waxy masses (*pollinia*) often narrowed at their apical or basal end into a sterile stalk-like *caudicle*; ovary inferior, 1-celled, with numerous minute ovules on 3 parietal placentae, or rarely 3-celled; style 0; stigmas 3 fertile (Diandrae) or 2 fertile and the other sterile and forming a ± beak-like process, the *rostellum*, between the anther and fertile stigmas (Monandrae); in some types the rostellum forms 1 or 2 viscid bodies (*viscidia*) to which the pollinia are attached, and it may be represented only by viscidia or may be quite lacking. Fr. a capsule opening by 3 or 6 longitudinal slits; seeds very numerous and very minute, with an undifferentiated embryo and no endosperm. Pollination usually by insects to whose bodies the pollinia become attached in such a way that they are placed on the stigmas of other fls; some types are automatically self-pollinated and others apomictic, although retaining the structural features of cross-pollinated types.

800 genera and 15,000–30,000 spp., and thus one of the largest families of Angiosperms. Largely tropical but spread over the whole world.

Synopsis of Classification

Subfamily CYPRIPEDIOIDEAE: fertile stamens 2; pollen granular, not united in pollinia; fertile stigmas 3.

Tribe CYPRIPEDIEAE: median stamen forming a large staminode; labellum large, deeply concave, with inrolled margins.   1. CYPRIPEDIUM

Subfamily ORCHIDOIDEAE: fertile stamen 1; pollen granular or waxy, united in pollinia; fertile stigmas 2, often confluent; median (sterile) stigma commonly extended as a beak-like rostellum.

A.   Caudicle 0 or at the apex of the pollinium, i.e. the end which is in contact with the rostellum; anther hinged to the summit or back of the column and often readily detachable.

Tribe NEOTTIEAE: pollen granular, soft; anther usually persistent; infl. always terminal.

Subtribe NEOTTIINAE: anther usually withering in situ; pollinia 2–4.

(*a*) Anther ±erect, exceeding the rostellum when present; labellum with a
concave basal hypochile separated by a constriction or fold from the
distal tongue-like epichile (Fig. 65).

Fl. erect; rostellum 0; labellum with an erect hypochile which clasps the
base of the column and a forwardly directed epichile; spur 0; pollen not
in tetrads.   2. CEPHALANTHERA

Fl. horizontal or pendulous; rostellum ±globular or 0; labellum with a
horizontal hypochile and downwardly directed epichile; spur 0; pollen
in tetrads.   3. EPIPACTIS

Saprophytic; fl. not inverted, with upwardly directed labellum and spur;
pollinia with caudicles attached to the rostellar viscidium.

4. EPIPOGIUM

(*b*) Anther erect, parallel to and equalling the conspicuous flattened rostellum; pollinia granular, not breaking into large angula rmasses; labellum
not differentiated into hypochile and epichile.

Per. segs all directed forwards; labellum embracing the column at its base;
fls in spirally twisted rows.   5. SPIRANTHES

Per. segs ±spreading; labellum not embracing the column; lvs 2, opposite,
borne some way up the stem.   6. LISTERA

Saprophytic; stem covered with numerous brown scales; per. segs connivent to form a helmet.   7. NEOTTIA

(*c*) Anthers erect or forwardly inclined, about equalling the rostellum;
pollinia breaking into several large angular masses; labellum various;
lvs often net-veined.

Spur 0; labellum with a saccate base and spout-like distal part which are
not clearly separated.   8. GOODYERA

Tribe EPIDENDREAE pollen waxy or bony; anther usually soon deciduous;
infl. terminal or lateral.

Subtribe LIPARIDINAE: infl. terminal, or axillary to the uppermost vs;
pollinia 4; caudicles 0.

Anther erect, persistent; column very short, straight; labellum directed
upwards.   9. HAMMARBYA

Anther inclined, deciduous; column curved, winged at the base; labellum
variously orientated.   10. LIPARIS

Saprophytic; anther inclined; column long, curved.   11. CORALLORHIZA

B. Caudicle at the base of the pollinium, attached to a viscidium;
anther erect, closely adnate to the summit of the column and never
deciduous; pollinia always granular.

Tribe ORCHIDEAE (the only tribe).

(*d*) Viscidia 2, naked, or with a delicate skin which is removed with
them; never enclosed in a pouch which remains behind after their
removal.

Viscidia covered by a delicate skin; caudicles very short; viscidia large.

12. HERMINIUM

Viscidia covered by a delicate skin; caudicles fairly long; viscidium hardly
wider than the caudicle.   13. COELOGLOSSUM

Viscidia naked, long and narrow, close together; caudicle attached at right
angles to the viscid surface; per. segs spreading; spur long, filiform.

14. GYMNADENIA

Like *Gymnadenia* but viscidia less elongated; per. segs connivent; spur
short, conical.   15. LEUCORCHIS

Viscidia naked, oval or circular, distant; caudicle attached laterally so as to be parallel to the viscid surface.   16. PLATANTHERA

(*e*) Viscidia 2, naked; stigmas free and often long.
Stigmas crescent-shaped, diverging upwards and leaving a broad flat plate between them; per. segs connivent into a helmet; spur short.
                                                                   17. NEOTINEA

(*f*) Viscidia 2, enclosed in 1 or 2 pouches which are left behind when the viscidia are removed.
Viscidia in 2 separate pouches; labellum insect-like, velvety; spur 0.
                                                                   18. OPHRYS
Viscidia in a single pouch.
  Viscidia 2 in a common pouch.
    Spur 0.   21. ACERAS
    Spur present.
      Labellum with a very long ribbon-shaped mid-lobe which is spirally curled in bud.   19. HIMANTOGLOSSUM
      Labellum not very long and ribbon-shaped.
        Root-tubers entire.   20. ORCHIS
        Root-tubers palmately lobed or forked.   22. DACTYLORCHIS
  Viscidium 1, strap-shaped; labellum with vertical 'guide-plates'.
                                                                23. ANACAMPTIS

Key to Genera

1   Plant lacking green lvs (saprophytic).   2
   Plant with green lvs (or with green bract-like scales on a green stem).   4

2   Labellum directed upwards, pink; spur fairly long, directed upwards; stem swollen above the base.   4. EPIPOGIUM
   Labellum directed downwards; spur 0 or very short and adnate to the ovary; stem not swollen.   3

3   Stem covered with numerous brownish scales; fls brown; per. segs connivent into an open hood.   7. NEOTTIA
   Stem with 2–4 long sheathing scales; fls yellowish-green with a whitish labellum; outer lateral per. segs curved downwards close to the labellum.   11. CORALLORHIZA

4   Fl. spurred.   5
   Fl. not spurred.   15

5   Labellum with a long ribbon-like mid-lobe 3–5 cm. × 2 mm., coiled in bud like a watch-spring; a tall plant with long spikes of unpleasantly smelling fls.   19. HIMANTOGLOSSUM
   Labellum not as above.   6

6   Fls greenish-white.   7
   Fls green, white, pink, red, etc., but not greenish-white.   8

7   Fls large, strongly fragrant, with spreading per. segs, entire strap-shaped labellum and long slender spur 15–30 mm.   16. PLATANTHERA
   Fls in a narrow cylindrical spike, very small, ± campanulate, faintly vanilla-scented, with connivent per. segs, a 3-lobed labellum and a short blunt conical spur.   15. LEUCORCHIS

8   Spur very short (c. 2 mm.); per. segs connivent into a hood.   9
   Spur exceeding 5 mm.; per. segs connivent or spreading.   11

9   Fls green, often red-tinged; labellum narrowly oblong with parallel distal lobes.   13. COELOGLOSSUM
    Fls not green; labellum with 3 or 5 ± spreading lobes.   *10*

10   Lvs unspotted; labellum 5-lobed; per. segs dark purplish-brown at first but becoming paler, so that the tip of the spike appears burnt or scorched; labellum white with dark spots.   20. *Orchis ustulata*
    Lvs sometimes with rows of small dots; labellum 3-lobed, the central lobe longer and forked; fls small, whitish or pink, in a dense spike 2·5–8 cm.; W. Ireland.   17. NEOTINEA

11   Spur c. 12 mm., filiform; labellum with 3 subequal lobes; 2 fertile stigmas borne on lateral lobes of the column; viscidia 1 or 2, elongated.   *12*
    Spur less than 12 mm., or, if as long, then not filiform; 2 fertile stigmas ± confluent on the front of the column; viscidia 2, globular.   *13*

12   Spike markedly conical; fls not sweet-scented; labellum with vertical plates decurrent on its base from the lateral lobes of the column; viscidium 1, strap-shaped.   23. ANACAMPTIS
    Spike ± cylindrical; fls strongly and sweetly scented; no vertical plates on labellum; viscidia 2, naked.   14. GYMNADENIA

13   All per. segs except the labellum connivent into a helmet over the column.   20. ORCHIS
    At least the outer lateral per. segs spreading or upturned.   *14*

14   Emerging fl.-spike enclosed by thin spathe-like lvs; bracts membranous; spur ± ascending.   20. ORCHIS
    Emerging fl.-spike not enclosed in spathe-like lvs; bracts lf-like; spur descending.   22. DACTYLORCHIS

15   Labellum concave, slipper-shaped; fls very large, solitary (–2); per. segs 6–9 cm.   1. CYPRIPEDIUM
    Labellum not strongly concave; fls normally more than 2; per. segs less than 6 cm.   *16*

16   Labellum with an insect-like ± velvety mid-lobe.   18. OPHRYS
    Labellum not insect-like and velvety.   *17*

17   Labellum on the upper side of the fl. and directed upwards in some or all of the small yellowish-green fls.   *18*
    Labellum on the lower side of the fl.   *19*

18   Lvs 0·5–1 cm., obovate, rounded and concave, the margin fringed with tiny green bulbils; labellum always on the upperside of the fl.; a small plant usually growing in *Sphagnum*.   9. HAMMARBYA
    Lvs 2·5–8 cm.; oblong-elliptical; labellum variously orientated; usually in fen peat.   10. LIPARIS

19   Fls small, whitish, in 1 or more spirally twisted rows; infl. narrowly cylindrical.   *20*
    Fls not whitish and in spirally twisted rows.   *21*

20   Stoloniferous; lvs ovate, stalked, conspicuously net-veined; labellum free from the column, with a saccate base and an entire narrow spout-like distal part.   8. GOODYERA
    Not stoloniferous; lvs not conspicuously net-veined; labellum ± adnate to and embracing the column at its base, its distal part entire, frilled, recurved.   5. SPIRANTHES

21   Labellum with a concave basal part (hypochile) separated by a constriction or fold from the tongue-like distal part (epichile) (Fig. 65).   *22*
    Labellum not as above; fls greenish.   *23*

22 Fls erect, ±sessile; hypochile embracing the column; ovary twisted.
                                        2. CEPHALANTHERA
    Fl. horizontal or pendulous, stalked; hypochile not embracing the
      column; ovary straight.   3. EPIPACTIS
23 Labellum shaped like a man, with slender lobes resembling arms and
      legs.   21. ACERAS
    Labellum not shaped like a man.   24
24 Lvs 2, ±opposite, borne some way up the stem; labellum forked distally,
      sometimes with a small tooth in the sinus.   6. LISTERA
    Lvs 2, radical, with 0–2 smaller stem-lvs; labellum 3-lobed, the central
      lobe longer than the laterals.   12. HERMINIUM

1. CYPRIPEDIUM L.

Herbs with creeping rhizomes, green lfy stems and 1–3 large fls. Per. segs
spreading. *Labellum large, concave, inflated, shoe- or slipper-shaped*; spur 0.
Column curving forwards so as partly to close the aperture of the labellum,
and terminating in a large shield-shaped staminode with 2 *lateral fertile
anthers* below and behind it; stigmas confluent, *discoid, stalked*, projecting
downwards below the fertile anthers and just behind the staminode; rostellum 0.
Pollen not in tetrads, granular. Ovary straight.

About 50 spp. in tropical and temperate regions of both Old and New
Worlds.

C. calceolus L.   Lady's Slipper.

Stem 15–45 cm., erect, somewhat pubescent, with basal sheathing scales.
Lvs 3–4, ovate-oblong, sheathing, acute or acuminate, slightly pubescent,
strongly furrowed above and ribbed beneath along the several parallel veins.
Fls 1(–2); bracts large, lf-like. Per. segs 6–9 cm., maroon; upper outer segment ovate-lanceolate; ±erect, outer lateral segments narrower and usually
±connate to form a single bifid downwardly directed segment below the
labellum; inner lateral segments narrowly lanceolate-acuminate, often somewhat twisted. Labellum rather shorter than the per. segs, obovoid, rounded
at the tip, pale yellow with faint darker veins, its interior and the column with
reddish spots. Ovary ±pubescent. Fl. 5–6. Cross-pollinated by insects which
crawl from the cavity of the labellum past the stigmatic disk and the fertile
anthers. $2n = 22$.

Native. Woods on limestone. 4. Extremely rare and nearly extinct, but
still persisting in Yorks or Durham where most of the old records were
made. Europe; N. Asia.

2. CEPHALANTHERA Rich. Helleborine

Herbs with shortly creeping rhizomes, lfy stems and a few large suberect
white or pink fls in lax spikes. Per. segs commonly connivent so that the
fl. never opens widely. Labellum with a constriction between the suberect
concave hypochile, which clasps the base of the column, and the forwardly
directed epichile with several interrupted longitudinal crests along its upperside and a recurved tip; basal bosses 0; spur 0. Column long, erect; stigma
large, oblong; *rostellum* 0. Anther hinged to the summit of the column;

pollinia 2, clavate, each ±completely divided into longitudinal halves; caudicles 0; pollen grains single, in powdery masses. Capsule erect.

Ten spp. in the north temperate zone.

1　Fls bright rose-pink; ovary pubescent.　　　　　　　　　**3. rubra**
　　Fls white; ovary ±glabrous.　　　　　　　　　　　　　　　　　　　2

2　Lvs ovate or ovate-lanceolate; bracts exceeding the ovary; outer per. segs
　　blunt.　　　　　　　　　　　　　　　　　　　　**1. damasonium**
　　Lvs lanceolate; bracts shorter than the ovary; outer per. segs acute.
　　　　　　　　　　　　　　　　　　　　　　　　2. longifolia

1. C. damasonium (Mill.) Druce　　　　　　　　'White Helleborine.'

C. latifolia Janchen; *C. grandiflora* S. F. Gray; *C. pallens* Rich.

Stem 15–50 cm., erect, rigid, angled, glabrous, with 2–3 brown membranous sheathing basal scales, the uppermost often green-tipped. Lvs 5–10 cm., the lowest short, ovate-lanceolate, the middle ovate-oblong, the upper lanceolate, passing into the bracts. *Fls* 3–12, the lowest distant and all but the uppermost *much exceeded by the lf-like lanceolate bracts, creamy-white*, scentless, except for a brief period closed and ±tubular. *Outer per. segs* oblong, *blunt*; inner lateral per. segs oblong-lanceolate, blunt, shorter than the outer; labellum shorter than the per. segs, with an orange-yellow blotch in the concavity of the hypochile and 3–5 interrupted orange-yellow, crested ridges running along the cordate, crenate epichile. *Ovary glabrous.* Fl. 5–6. Usually self-pollinated, the pollen falling on the stigma. $2n=32$; 36^*. Grh.

Native. Woods and shady places on calcareous soils; commonly under beech. 31. Local from Kent to Somerset, Wilts and Dorset northwards to E. Yorks and Cumberland. Europe northwards to S. Sweden and C. Russia; N. Africa; Caucasus; Asia Minor.

2. C. longifolia (L.) Fritsch　　　　　　　　'Long-leaved Helleborine.'

C. ensifolia (Schmidt) Rich.

Stem 20–60 cm., ±glabrous, slightly ridged above and with 2–4 whitish, sometimes green-tipped, loose-topped basal sheaths. Lvs c. 7–20 × 1·5 cm., lanceolate, longer and narrower than in *C. damasonium*, often folded, the uppermost linear and sometimes exceeding the spike. *Fls* 3–15, *exceeding all or all but the lowest bracts; pure white*, resembling those of *C. damasonium* but smaller. Outer per. segs lanceolate, *acute.* Ovary *glabrous.* Fl. 5–7. Probably cross-pollinated by small bees. $2n=32$. Grh.

Native. Woods and shady places especially on calcareous soils. 43, H10. Local and rare from Hants and Sussex to W. Ross and Inverness; Inner Hebrides; Ireland. Europe northwards to Scandinavia; N. Africa; W. Asia to Kashmir; Siberia; Japan.

3. C. rubra (L.) Rich.　　　　　　　　　　　　'Red Helleborine.'

Serapias rubra L.

Stem 20–50 cm., erect, striate and *glandular-hairy* above, with brownish sometimes green-tipped basal sheaths. Lvs few, the lowest oblong-lanceolate, the upper linear-lanceolate, all very acute. *Fls* 3–10 exceeding all or all but the lowest of the linear-lanceolate acute bracts, *bright rose-red*, scentless, opening fairly wide. Outer per. segs lanceolate acute, spreading, glandular-hairy outside; inner per. segs shorter and broader, connivent; labellum erect, white, its epichile lanceolate with red-violet margins and tip and 7–9 narrow crested ridges. *Ovary glandular-hairy* with violet ribs. Fl. 6–7. Cross-pollinated by small bees. Grh.

Native. Confined to a few beech-woods on oolite in Gloucestershire and on chalk in Bucks. 3. Europe northwards to Scandinavia; Near East.

3. Epipactis Sw. *Helleborine*

Herbs with horizontal or vertical, often very short, rhizomes, numerous fleshy roots, and rather inconspicuous spreading or pendulous fls in ± 1-sided racemes. Per. segs spreading or remaining closed, dull reddish or greenish. *Labellum usually in 2 parts separated by a narrow joint or a fold*, the basal part (*hypochile*) forming a *nectar-containing cup*, the distal part (*epichile*) a

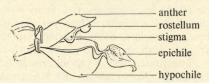

anther
rostellum
stigma
epichile
hypochile

Fig. 65. Labellum and column of *Epipactis helleborine*.

± cordate or triangular *downwardly-directed terminal lobe; spur* 0. Column short, with a shallow cup (*clinandrium*) at its apex; stigma prominent, ± transversely oblong. *Rostellum* placed centrally above the stigma, *large and globular, persistent or evanescent* (or 0 in some non-British spp.). Anther free, hinged at the back of the summit of the column, behind the stigma and rostellum; pollinia 2, tapering towards their apices near which they are attached to the rostellum, each ± divided longitudinally into halves; caudicles 0; pollen grains spherical, cohering in tetrads bound loosely together by weak elastic threads, so that the masses are very friable. Cross-pollinated by insects, or self-pollinated; perhaps sometimes apomictic. Ovary straight, looking like a twisted stalk.

Perhaps 20 spp. in the north temperate zone, but very imperfectly understood.

1 Fen plant with long creeping rhizome; fls brownish outside; hypochile with 2 erect lateral lobes; epichile connected by a narrow hinge.
                                                                   1. palustris
   Not fen plants; rhizome horizontal or vertical, short; fls greenish or purple outside; hypochile without lateral lobes; epichile connected by 1 or more folds.   2

2 Fls entirely dull reddish-purple; infl.-axis with dense whitish pubescence; epichile broader than long, with 3 large strongly rugose basal bosses; ovary pubescent.   **7. atrorubens**
   Fls not entirely reddish; infl.-axis glabrous to pubescent but not densely whitish-pubescent; basal bosses of labellum smooth or nearly so.   3

3 Lvs spirally arranged; rostellum prominent and persistent unless removed
   by insects with the pollinia.   *4*
   Lvs in 2 opposite ranks (distichous); rostellum 0 in mature fls; fls sometimes opening only partially or not at all.   *5*

4 Stems 1–3, rarely more; lvs broadly ovate to ovate-lanceolate, usually
   dark green; per. segs green to deep purplish; hypochile dark maroon or
   dark green inside.   **2. helleborine**
   Stems commonly 6–10 or more; lvs ovate-lanceolate to lanceolate, grey-
   green, often violet-tinged; per. segs pale whitish-green; hypochile
   ± mottled with violet inside.   **3. purpurata**

5 Infl.-axis glabrous or sparsely pubescent; fls hanging almost vertically
   downwards; hypochile greenish-white inside.   **6. phyllanthes**
   Infl.-axis markedly pubescent; fls spreading horizontally or hanging
   obliquely downwards; hypochile purplish inside.   *6*

6 Fls opening widely.   **4. leptochila**
   Fls rarely opening widely.   *7*

7 Fls remaining partially closed and therefore campanulate; epichile with
   a recurved tip; a rare plant of dunes in Lancs and Anglesey.
         5. dunensis
   Fls usually remaining closed; epichile flat; a rare plant of beechwoods in
   Gloucestershire.   **4. leptochila** var. *cleistogama*

1. E. palustris (L.) Crantz   'Marsh Helleborine.'

Serapias palustris (L.) Mill.

Rhizome long-creeping with stolon-like branches. Stem 15–45(–60) cm., erect, slender, wiry, pubescent above; often purplish below and with 1 or more sheathing basal scales. Lvs 4–8, the lower 5–15 cm., oblong-ovate to oblong-lanceolate, acute, concave above, often purple beneath; uppermost narrow, acuminate, bract-like; all ± erect and folded, with 3–5 prominent veins beneath. Raceme 7–15 cm., with 7–14 fls ± turned to one side; bracts lanceolate-acuminate, lowest about equalling the fls, the rest falling short. Fls drooping in bud and in fr., almost horizontal when open, scentless. *Outer per. segs* ovate-lanceolate, *brownish or purplish-green and hairy outside*, paler inside; inner per. segs somewhat shorter and narrower, whitish with purple veins at the base, glabrous. Labellum c. 12 mm.; *hypochile shallow with an erect triangular lobe at each side*, white with rose veins and many bright yellow raised spots down the middle; *epichile joined by a narrow hinge, white with red veins*, broadly ovate with *waved and frilled upturned sides*, and with a basal 4-sided furrowed boss, 3–4-toothed on its forward edge. Rostellum persistent. Ovary narrowly pear-shaped, puberulent. Fl. 6–8. Cross-pollinated by hive-bees and other insects. $2n=40$. G.

Var. *ochroleuca* Barla has yellowish fls with a white lip.

Native. Fens and dune-slacks. 74, H32. Locally frequent throughout England, Wales and Ireland and northwards to Perth; Inner Hebrides, Europe (except Arctic); temperate Asia; N. Africa.

2. E. helleborine (L.) Crantz   'Broad Helleborine.'

Serapias Helleborine var. *latifolia* L.; *E. latifolia* (L.) All.

Rhizome usually very short. Roots numerous, arising in a cluster from the base of the stem. Stems 1–3, rarely more, 25–80 cm., erect, solid, whitish with short hairs above, often violet-tinged below and with 2 or more basal sheathing scales. Largest *lvs* 5–17 × 2·5–10 cm., *spirally arranged*, somewhat crowded near the middle of the stem, usually *broadly ovate-elliptical or almost orbicular*, acute or shortly acuminate, with a few smaller but relatively broader lvs below, and with smaller and narrower lanceolate-acuminate lvs above, grading into the bracts; all dull green with c. 5 prominent veins beneath which are rough like the margins. Raceme 7–30 cm., of 15–50 greenish to dull purple fls, ± turned to one side, drooping, opening ± widely, scentless; bracts lanceolate-acuminate, the lowest equalling the fls, the upper the ovary. Outer lateral per. segs c. 1 cm., ovate to ovate-lanceolate, green to dull purple. Labellum shorter than outer per. segs; hypochile green outside and dark brown or green inside; *epichile* cordate to triangular, *broader than long*, purplish, rose or greenish-white, with a reflexed acute tip and 2(–3) *smooth or slightly rugose basal bosses. Rostellum* large, whitish, *persistent.* Ovary glabrous or with a few scattered hairs, pendulous in fr. Fl. 7–10. Said to be cross-pollinated by wasps. $2n = 36^*, 40$. G.

Very variable in breadth of lvs, colour of fls, shape of labellum, rugosity of basal bosses, and presence or absence of a third (central) boss.

Native. Woods, wood-margins, hedge-banks, etc.; local. 93, H36. Throughout the British Is. to Sutherland; Inner Hebrides. Europe; N. Africa; temperate Asia eastwards to Japan. Introduced in N. America.

3. E. purpurata Sm. 'Violet Helleborine.'

E. sessilifolia Peterm.; *E. violacea* (Dur. Duq.) Bor.

Rhizome vertical, often branched above, bearing fleshy roots at successive thickened nodes. *Stems* 1–*many*, 20–70 cm., erect, solid, shortly hairy above, violet-tinged below, with 2–3 brown sheathing basal scales. Largest *lvs* 6–10 cm., near the middle of the stem, *spirally arranged, ovate-lanceolate to lanceolate*, acute to shortly acuminate; lower lvs smaller but relatively broader; uppermost narrow and grading into the bracts; all *grey-green, often violet-tinged*, especially beneath; main veins 3–7; margins rough. Raceme 15–25 cm., dense, 1-sided; fls numerous, pale greenish-white, slightly fragrant; *bracts* narrow, acuminate, often violet-tinged, *equalling or exceeding the fls.* Outer per. segs 10–12 mm., lanceolate, bluntish, green outside, whitish inside; inner shorter, ovate-lanceolate, whitish and sometimes rose-tinged. *Labellum* slightly *shorter than the outer per. segs; hypochile* greenish outside, usually *mottled with violet within; epichile at least as long as broad*, triangular-cordate, dull white, with an acute reflexed tip and 2–3 ± confluent *smoothly plicate basal bosses*, the central one elongated and faintly violet-tinged. Rostellum whitish, persistent. Ovary rough with short hairs, held ± horizontally in fr. Fl. 8–9. Cross-pollinated by wasps. G.

Distinguishable from *E. helleborine* by the usually clustered stems, the narrower often violet-flushed lvs, longer bracts, paler fls, and longer and narrower epichile.

Native. Woods, especially of beech, on calcareous soils. Southern and Midland counties of England but extending northwards to Yorks and Cumberland; doubtfully in Scotland. France, Germany, Austria, Switzerland, Russia.

The 3 following species are characterized by having the rostellum rudimentary or lacking. The British forms whose fls are reduced in this way are taxonomically very difficult. There appear to be many local races differing in details of floral structure from their nearest relatives, a situation which may be presumed to arise from their being predominantly or exclusively self-pollinated. Some of the local forms have fls which are partially or completely closed throughout the flowering period.

4. E. leptochila (Godf.) Godf.   'Narrow-lipped Helleborine.'

Rhizome ± vertical, branched above, bearing numerous fleshy roots at successive nodes. Stems 1–several, 20–70 cm., ± pubescent, especially above, usually violet-tinged below; basal scales 2–3, violet, later becoming brown. Largest *lvs* 5–10 × 2–5 cm., broadly ovate to broadly lanceolate, acute, the lowest smaller, the uppermost smaller and narrower, grading into the bracts; all *arranged in 2 opposite ranks* (distichously), dull yellow-green or dark green, ± undulate at the margin, with c. 7 prominent veins beneath. Raceme long; fls many, large, yellow-green, scentless, all ± turned to one side; *lowest bracts much exceeding and the others about equalling the spreading widely open fls*. Outer per. segs 12–15 mm., lanceolate-acuminate, pale green; inner c. 10 mm., ovate-lanceolate, acute, whitish-green. Labellum with circular *hypochile* pale green outside and *mottled with red within; epichile longer than broad, somewhat concave*, yellow-green with a white border, narrowly cordate-acuminate with the long acute tip usually spreading, not reflexed, and with 2 smooth or somewhat rugose white or rose-tinged basal bosses separated by a deep channel. *Rostellum evanescent, leaving a small brownish mark.* Ovary covered with rough tubercles and sparse often blackish hairs. Fl. 6–8. Automatically self-pollinated by the pollinia falling on the stigma; perhaps rarely crossed by wasps in the earliest stages of opening. 2n = 36. G.

In var. *cleistogama* (C. Thomas) D. P. Young the lanceolate lvs are rather thick and somewhat leathery and the *green fls*, which are horizontal or obliquely hanging, usually *remain closed* throughout the flowering period but may open on weaker plants. The *epichile* is *green and flat*, with prominent white basal bosses; and pollination is said to take place not by intact pollinia falling on the stigma but by the separation of smaller masses of pollen (as in *E. dunensis*).

Native. Shady woods and dunes; local. S. England from Kent and Bedford westwards to Devon, S. Wales and Shropshire. Var. *cleistogama* only under beech on the steep western slope of the Cotswolds, near Wotton-under-Edge (Gloucestershire).

Extra-British distribution incompletely known, but recorded from Denmark.

5. E. dunensis (T. & T. A. Steph.) Godf.   'Dune Helleborine.'

E. leptochila var. *dunensis* (T. & T. A. Steph.) T. & T. A. Steph.

Rhizome short, slender, ascending, or more commonly reduced to a small irregular deeply buried mass; roots few, slender, wiry. Stems 1(–3), 20–40(–60) cm., slender, green and shortly pubescent above, violet-tinged and glabrous below, with 2–3 loose sheathing basal scales. *Lvs in 2 opposite ranks*, the largest near the middle of the stem, oblong-lanceolate acute; lowest broadly ovate, blunt; uppermost bract-like; all commonly yellowish-green, rigid and plicate, with undulate margins and c. 9 main veins. Raceme 5–12(–17) cm.; fls 7–20(–25), small, pale yellowish-green; *bracts* linear-lanceolate, *lowest longer than the spreading campanulate, incompletely opening fls.* Outer per. segs c. 7 mm., ovate, blunt, yellowish-green; inner smaller. Labellum with a ± circular *hypochile, mottled with red inside; epichile as broad as long*, whitish with a green or rosy tinge, *broadly triangular with a recurved tip*, and 2(–3) almost smooth basal bosses. *Rostellum* globular, if present, but *usually* 0. Ovary subglabrous to glabrous. Fl. 6–7. Usually self-pollinated by pollen falling over the edge of the stigma.

Native. In somewhat peaty but not very moist hollows in dunes, chiefly amongst *Salix repens* but sometimes beneath planted pines, and then taller and darker green. Known only from Lancashire and Anglesey.

Distinguishable from *E. leptochila* by the incompletely open fls and the broadly triangular epichile; from *E. phyllanthes* by the spreading, not drooping, fls.

6. E. phyllanthes G. E. Sm.

E. vectensis (T. & T. A. Steph.) Brooke & Rose

Rhizome short, horizontal or ascending. Stems solitary or rarely 2–3, (8–)20–45(–65) cm., stout, glabrous or very sparsely pubescent. Lvs 3·5–6(–7) cm., few (3–6), ± *in 2 ranks*, orbicular to lanceolate, the uppermost narrow and grading into bracts, glabrous or slightly pubescent above, not prominently ribbed, often undulate at the margin. Raceme of up to 35 fls; lower bracts much exceeding the fls. *Fls hanging ± vertically downwards, pale yellowish-green* or slightly violet-tinged. Labellum with the *hypochile whitish inside*, epichile greenish-white, usually longer than broad: sometimes no clear differentiation of hypochile and epichile. *Rostellum* 0. Ovary ± glabrous. Fl. 7–9. Self-pollinated or perhaps apomictic.

Native. A difficult aggregate of local forms differing in details of floral structure. D. P. Young (*Watsonia*, II, 1952) recognizes the following varieties between which intermediates sometimes occur:

Var. *phyllanthes*: labellum completely undifferentiated, ovate or lanceolate, resembling the other per. segs; fls sometimes not opening. Woods in S. England, often with vars *degenera* and *vectensis*.

Var. *vectensis* (T. & T. A. Steph.) D. P. Young: labellum embracing the stigma; hypochile small, hemispherical, entirely green; epichile jointed to the hypochile, cordate-acuminate, often whitish or pinkish, usually with 2 lateral bosses; fls sometimes not opening. Woods in S. England northwards to Cambridge, Warwick and Hereford, and in S. Wales.

Var. *degenera* D. P. Young: labellum imperfectly differentiated, no constriction dividing the shallow hypochile from the epichile; otherwise as in var. *vectensis*, but fls not usually opening widely. Woods in S. England northwards to Suffolk and Gloucester.

Var. *pendula* D. P. Young: labellum perfectly differentiated, large; hypochile usually held away from the column; epichile jointed to the hypochile, greenish, long, narrow, reflexed, rugose at the base or with 2 bosses; fls usually widely open. Woods, plantations and dunes in N.W. England and N. Wales.

E. cambrensis C. Thomas is a related form from Kenfig, Glamorgan.

7. E. atrorubens (Hoffm.) Schult.   'Dark-red Helleborine.'

E. atropurpurea Raf.; *E. rubiginosa* (Crantz) Koch

Rhizome ±horizontal, short, with many long slender roots. *Stem* 15–30(–60) cm., usually solitary, erect, *densely pubescent*, especially above, *violet below*, with 1–3 loosely sheathing scales. *Lvs* c. 5–10 *in 2 opposite ranks*, the largest 5–7·5(–10) × 2·5–4·5 cm., elliptical, acute, not or hardly sheathing, with a few smaller but broader sheathing lvs below, and others above, narrower, not sheathing, grading into bracts; all keeled, folded, many-veined, rough above and beneath. Raceme spike-like; *fls* 8–18, *red-purple*, faintly fragrant, short-stalked (3 mm.); bracts lanceolate, acute, only the lowest equalling or sometimes exceeding the fls. Outer per. segs c. 8 × 4 mm., ovate-acute, the laterals asymmetrical; inner c. 7 × 5 mm., blunt; *all* spreading-incurved, *reddish-violet outside*, the outer somewhat greenish especially within. Labellum shorter than the outer per. segs; hypochile green with a red margin and red-spotted inside; *epichile* 3 × 6 mm., deep reddish-violet with a small, acute, reflexed tip and 3 *brighter red, strongly rugose, confluent basal bosses*. *Rostellum an obvious whitish oval ledge. Ovary* 6–7 mm., pyriform, 6-ribbed, reddish-violet, *pubescent*; ripe capsule drooping. Fl. 6–7. Cross-pollinated by bees and wasps. $2n = 40$.

Native. Limestone rocks and screes, in woods or in the open; to 1000 ft. in W. Ireland. 17, H 4. Local and rather rare throughout Great Britain from Brecon and Derby northwards to Sutherland; Inner Hebrides. Ireland: Clare and Galway only. Europe northwards to Scandinavia; Caucasus; N. Persia.

4. EPIPOGIUM R.Br.

Lfless and rootless *saprophytic herbs with coralloid rhizomes.* Fls in racemes; drooping, spurred, with the *labellum uppermost* and the *spur directed vertically upwards.* Column pointing downwards with the horseshoe-shaped stigma on the overhanging base of its upper side. Rostellum large, cordate, at the apex of the column. Anther helmet-like, sessile in the concave summit of the column; pollinia 2, pear-shaped, with caudicles attached to their base; pollen in packets of tetrads. Ovary not twisted.

Two spp. in N. and C. Europe, N. Asia and Himalaya.

1. E. aphyllum Sw.   Ghost Orchid.

E. Gmelinii Rich.

Rhizome whitish with many very short bluntly 2–3-lobed fleshy branches and 1–2 long (5–7 cm.) filiform whitish stolons bearing at intervals buds which give rise to new rhizomes. Stem 10–20 cm., erect, ± translucent, white tinged with pink and with numerous short reddish streaks, weak, turgid, much swollen below then tapering suddenly to its very weak attachment to the rhizome. Lvs represented only by 2–3 brownish basal sheathing scales and 1–2 long close-fitting usually dark-edged sheaths higher up the stem. Fls 1–2–(4) distant, pendulous on slender stalks, the per. segs directed downwards, the labellum and spur upwards. Bracts oblong, blunt, membranous, translucent, equalling the fl. stalk. Outer per. segs linear, yellowish or reddish;

inner per. segs lanceolate, blunt, yellowish with a few short violet lines; all ± equally long, curving downwards and with upturned edges. Labellum with 2 short rounded basal lateral lobes and a large concave cordate terminal lobe, white with violet spots and irregularly tubercled crests which leave a deep channel down the middle. Spur about 8 × 4 mm., rounded at the tip, white tinged with yellow or reddish outside and with lines of violet spots within. The labellum is bent backwards near its middle and almost touches the spur. Capsule ± globular, drooping, opening by short slits which reach neither base nor apex. Fl. 6–8. Cross-pollinated by humble-bees and other insects. G. Saprophyte.

Native. In deep shade of oak or beech. 3. Found in Britain, on a few occasions, in Hereford, Shropshire, Oxford and Bucks. C. and N. Europe; Caucasus; Siberia; Himalaya.

5. SPIRANTHES Rich.

Small herbs with erect stocks, 2–6 ± tuberous roots, lfy stems and *small fls in spirally twisted spike-like racemes.* Outer and inner per. segs similar, the back outer segment cohering with the 2 inner to form the upper half of a 2-lipped trumpet-like tube round the column. *Labellum* frilled and ± recurved distally, *furrowed below and embracing the base of the column* to form the lower part of the perianth-tube, its edges being overlapped by the inner per. segs; spur 0; nectar secreted by 2 bosses at the base of the labellum. Column horizontal, with the circular stigma on its underside, facing the labellum. Rostellum narrow, projecting beyond the stigma and consisting of a narrowly elliptical viscidium supported between two long narrow teeth which are left behind like the prongs of a fork when the viscidium is removed. Anther hinged to the back of the column, resting on the rostellum: pollinia 2, each of 2 plates of coherent pollen tetrads, attached to the viscidium near their summits; caudicles 0.

Fifty spp. in the north temperate zone and S. America.

1 Flowering stem bearing only bract-like sheathing appressed scales, the true lvs of the current season being in a lateral basal rosette.

                                                                 1. spiralis

  Flowering stems bearing true lvs.   *2*

2 Fls in a single spirally twisted row.   **2. aestivalis**

  Fls in 3 spirally twisted rows.   **3. romanzoffiana**

1. S. spiralis (L.) Chevall.   'Autumn Lady's Tresses.'

Ophrys spiralis L.; *S. autumnalis* Rich.

Roots 2–3(–5), 1·5–2(–4) cm., ovoid or ovoid-oblong, radish-shaped, pale brown, hairy. Stem 7–20 cm., erect, terete, glandular-hairy especially above, with several pale green, lanceolate-acuminate, *appressed bract-like scales*, and sometimes with the withered remains of last season's rosette lvs still visible at its base. *Lvs* (4–5) of current season appearing with or after the fls *in a lateral basal rosette* which will flower next season; each c. 2·5 cm., ovate, acute, stiff, bluish-green, glabrous. Spike 3–12 cm., of 7–20 small (4–5 mm.), white, day-scented *fls in a single ± spirally twisted row.* Bracts lanceolate-cuspidate, concave and hooded, sheathing the ovary and about equalling it.

Outer per. segs oblong-lanceolate, blunt, translucent, white, ± ciliate, slightly glandular outside. Inner per. segs narrowly oblong, blunt. Labellum broadened, rounded and recurved distally, pale green with a broad white irregularly crenate or fringed margin. *Stigma ciliate below.* Ovary short, bent outwards at its apex, green, usually not twisted. Capsule 6 mm., obovoid. Fl. 8–9. Pollinated by humble-bees. $2n = 30$. Grt.

Native. Hilly pastures, downs, moist meadows and grassy coastal dunes, usually on a calcareous substratum. 63, H21, S. Throughout England and Wales northwards to Westmorland and N.E. Yorks, but common only in the west; S. Ireland, to Dublin and Mayo. Europe northwards to Denmark and C. Russia; N. Africa; Asia Minor.

2. S. aestivalis (Poir.) Rich.   'Summer Lady's Tresses.'

Roots 2–6, 5–8 cm., fleshy, fusiform-cylindrical. Stem 10–40 cm., erect, somewhat glandular-hairy above. *Lvs* 5–12 cm. × 5–9 mm., borne ± erect *on the flowering shoot of the current season*, linear-lanceolate, blunt, the lower narrowing downwards into a sheathing base, the uppermost very short, bract-like, appressed; all bright green, glossy on both sides, glabrous. Spike 5–8 cm., of 6–18 small, night-scented *fls in a single twisted row. Bracts* as in *S. spiralis*, but *exceeding the ovary. Fls* as in *S. spiralis* but somewhat larger, *pure white* and with the *stigma not ciliate below.* Ovary 9 mm., ± glandular-hairy. Capsule oblong. Fl. 7–8. Probably pollinated by moths. Grt.

Native. Marshy ground with sedges and rushes. 1, S. Known in Great Britain only in the New Forest (Hants), but perhaps now extinct. C. and S. Europe northwards to Belgium and Germany. N. Africa; Asia Minor.

3. S. romanzoffiana Cham.   'Drooping Lady's Tresses.'

Roots 2–6, long, fleshy, fusiform-cylindrical. Stem 12–25 cm., erect, bluntly 3-angled, sparsely pubescent above. Lower *lvs* 5–10(–15) cm. × 5–10 mm., *borne erect on the flowering shoot of the current season*, linear-oblanceolate, acute, glabrous; uppermost short, bract-like, acuminate, with a loosely sheathing base. Spike 2·5–5(–8) cm., slightly twisted, stout, dense, glandular-pubescent, with many large (8–11 mm.), white, hawthorn-scented *fls in 3 spirally curved rows.* Bracts lanceolate, acute or acuminate, concave, sheathing the ovary, the lowest about 3 cm. long, exceeding the fls, the upper exceeding the ovary. Ovary 10 mm., cylindrical, subsessile, glandular-hairy, not twisted, turned to one side and bent at the tip so that the fl. is held horizontally. Outer lateral per. segs 12 mm., lanceolate-acuminate, white, greenish below, glandular-hairy outside, united below and sheathing the base of the labellum. Inner per. segs linear, blunt. Labellum white with green veins, tongue-shaped, enlarging distally to the rounded frilled denticulate apex. Stigma crescent-shaped. Fl. 7–8. Cross-pollinated by insects. Grt.

Native. Moist meadows and pastures liable to flood in winter; drained peat; upland bogs. Cork, Kerry, W. Galway, around and downstream from Lough Neagh; S. Devon, Colonsay and Coll, W. Inverness. 3, H8. Not elsewhere in Europe. N. America; Kamchatka.

6. LISTERA R.Br.

Herbs with short rhizomes bearing numerous slender roots, erect stems usually with a *pair of ± opposite green lvs*, and spike-like racemes of inconspicuous greenish or reddish fls. Per. segs spreading or loosely connivent. Labellum long and narrow, forking distally into 2 narrow segments and sometimes with 2 lateral lobes near the base; spur 0; nectar secreted by the central furrow of the labellum. Column short, erect; stigma transversely elongated, prominent; rostellum broad, flat, blade-like, arching over the stigma and expelling its viscid contents explosively in a terminal drop when touched. Anther hinged to the back of the column; pollinia 2, club-shaped, each ± divided longitudinally into halves; caudicles 0; pollen friable in loosely bound tetrads.

About 10 spp. in north temperate and subarctic zones.

Plant 20–60 cm.; lvs large (5–20 cm.), broadly ovate-elliptical; spike long; fls many, green.   **1. ovata**
Plant 6–20 cm.; lvs small (1–2·5 cm.), ovate-deltoid; spike short; fls few, reddish-green.   **2. cordata**

1. L. ovata (L.) R.Br.   Twayblade.

Ophrys ovata L.

Rhizome horizontal, deep, with very numerous rather fleshy but slender roots. Stem 20–60 cm., rather stout, glabrous below, pubescent above; basal sheaths 2(–3), membranous. *Lvs* 5–20 cm., in a subopposite pair rather below the middle of the stem, *broadly ovate-elliptical*, sessile, with 3–5 prominent ribs; upper part of stem with 1–2 tiny triangular bract-like lvs. Raceme 7–25 cm., rather lax; fls numerous, short-stalked, yellowish-green, bracts minute, shorter than the fl.-stalks. Per. segs subconnivent, the outer laterals not quite contiguous with the others; outer segments ovate, bluntish, green, inner narrower, yellow-green. *Labellum* 10–15 *mm.*, *yellow-green*, directed forwards at the cuneate furrowed base, then turning abruptly and almost vertically downwards; lateral lobes 0 or rarely represented by 2 small erect teeth near the base, central lobe oblong, broadening slightly downwards and deeply forked almost to half-way into 2 narrowly strap-shaped parallel segments sometimes with an intervening tooth. Capsule ± globular on an ascending stalk. Fl. 6–7. Cross-pollinated by small insects which crawl up the nectar-secreting centre furrow of the labellum. $2n = 42*$; 32, 34*, 36, 38. G.

Var. *platyglossa* Peterm. has the labellum 6–8 mm. wide, with its distal segments slightly divergent.

Native. Moist woods and pastures on base-rich soils, and on dunes; common and locally abundant. 111, H40, S. Throughout the British Is., except Shetland. Var. *platyglossa* on dunes in S. Wales. Europe northwards to Iceland; Caucasus; Siberia.

2. L. cordata (L.) R.Br.   'Lesser Twayblade.'

Ophrys cordata L.

Rhizome creeping, very slender, with a few whitish filiform roots. Stem 6–20 cm., erect, slender, glabrous below, angled and slightly pubescent above;

basal scales 1–2, closely sheathing. *Lvs* 1–2(–2·5) cm., in a subopposite pair (rarely 3–4) rather below the middle of the stem, *ovate-deltoid* with rounded horny tip and basal corners and a *broadly cuneate* to subcordate sessile *base*; shining above, paler beneath, somewhat translucent. Raceme 1·5–6 cm., lax, of 6–12 fls, the 3 lowest sometimes in a whorl; bracts minute (1 mm.) triangular, greenish; fl.-stalks 1–2 mm. Outer per. segs green, the upper one broad, like a little hood, the 2 lateral narrower and forward-spreading like the inner segments, which are green outside but reddish inside. *Labellum* 3·5–4 *mm.*, *reddish*, twice as long as the per. segs, very narrow, with 2 linear-oblong lateral lobes close to the base, the central lobe forking about half-way into 2 widely diverging linear tapering segments. Capsule ± globular, strongly 6-ribbed, the ribs often reddish. Fl. 7–9. Cross-pollinated by minute flies and Hymenoptera. $2n=42*$; 38. G.

Native. Mountain woods, especially of Scots pine, and peaty moors, especially under heather; often amongst sphagnum; to 2700 ft. in Scotland. 65, H 26. Very rare in Devon, Somerset, Dorset and Hants; more frequent northwards from N. Wales, Shropshire and Derby, and locally common in Scotland to Sutherland and Caithness; Hebrides; Orkney; Shetland. Ireland, but rare in the south. Europe from the Pyrenees and Apennines to Iceland, Scandinavia and C. Russia; Transcaucasus; N. Asia; N. America.

7. NEOTTIA Ludw.

Perennial *saprophytic herbs* with short creeping rhizomes concealed in *a mass of short thick fleshy blunt roots* ('bird's nest'), *stems* densely *covered with brownish scales*, and spike-like racemes of pale brownish fls. Per. segs connivent into an open hood. Labellum saccate at the base, with 2 distal lobes; spur 0. Column long, slender, erect; stigma large, prominent; rostellum broad, flat, blade-like, arching over the stigma, expelling its viscid contents explosively when touched. Anther hinged to the back of the column, directed forwards; pollinia 2, ± cylindrical, slender, each divided longitudinally into halves; pollen tetrads loosely united by a few weak threads, so that the pollen masses are very friable.

Three spp. in temperate Europe and Asia.

1. N. nidus-avis (L.) Rich.   Bird's-nest Orchid.

Ophrys Nidus-avis L.

Roots cylindrical, pale fawn. Stem 20–45 cm., erect, robust, brownish, somewhat glandular above, clothed below with numerous brownish scarious sheathing scales. Green lvs 0. Raceme 5–20 cm., rather lax below; bracts falling short of the ovary, lanceolate-acuminate, scarious. Per. segs ovate-oblong, the inner slightly shorter than the outer. Labellum c. 12 mm., twice as long as the per. segs, darker brown, obliquely hanging, with 2 small lateral teeth near the base, and 2 ± oblong blunt distal lobes extending almost half-way and diverging, straight or outwardly curved. Capsule c. 12 mm., erect. Fl. 6–7. Cross-pollinated by small crawling insects which touch the sensitive rostellum, or self-pollinated. $2n=36*$.

Native. Shady woods, especially of beech and especially on humus-rich calcareous soils. 92, H 30. Throughout Great Britain northwards to Banff and Inverness; Ireland. Europe, northwards to Scandinavia; Caucasus; Siberia.

8. GOODYERA R.Br.

Small herbs with creeping rhizomes, *ovate stalked lvs* and small fls in *twisted one-sided spike-like racemes. Labellum in two parts*, a basal pocket-like hypochile and a distal narrow spout-like epichile; spur 0. Column horizontal, short, with the roundish stigma on its lower side, facing the labellum. Rostellum projecting beyond the stigma and consisting of two short curved horns enclosing the ± circular viscidium. Anther hinged to the back of the column, resting on the rostellum; pollinia 2, partially divided lengthways, attached to the rostellum just beneath their summits, usually without caudicles: pollen in tetrads cohering in packets.

Forty spp. in the north temperate zone, tropical Asia, Madagascar and New Caledonia.

1. G. repens (L.) R.Br.   'Creeping Lady's Tresses.'
Satyrium repens L.

Rhizome creeping and giving rise to a few short roots and pale slender stolons which end in lf-rosettes. Stem 10–25 cm., stiffly erect, terete below but angled above, glandular-hairy especially above, with 1 whitish sometimes green-tipped, sheathing, basal scale. Basal lvs 1·5–2·5 cm., in a lax rosette persisting through the winter, ovate to ovate-lanceolate, firm, narrowed into a winged *stalk-like base*, dark green often mottled with lighter green, with 5 main veins and a *conspicuous network of secondary veins*; uppermost ± bract-like, appressed to the stem. Fls cream-white, fragrant. Bracts 10–15 mm., exceeding the ovary, linear-lanceolate, acute, ciliate, green outside, glossy white within. Outer per. segs ovate, blunt, glandular-hairy, white or greenish, the lateral pair slightly spreading. Inner per. segs narrowly lanceolate, blunt, glabrous, whitish, connivent with the back outer per. seg. Labellum shorter than the outer per. segs, the hypochile deeply pouched, epichile narrow, furrowed, recurved. Ovary 9 mm., glandular-hairy, subsessile. Fl. 7–8. Cross-pollinated by humble-bees. $2n=30$. Grh.

Native. Locally in pine-woods, rarely under birch or on moist fixed dunes. 28. From Cumberland, E. Yorks (?extinct) and Northumberland to Sutherland and Orkney; Norfolk and Suffolk. C. and N. Europe from Pyrenees and Balkans to N. Scandinavia; Russia. Asia Minor; Afghanistan; Himalaya; Siberia; Japan; N. America.

9. HAMMARBYA O. Kuntze

A small green herb with a *pseudo-bulb* covered by pale sheathing scales and connected with the daughter pseudo-bulb above it by a short vertical stolon. Lvs few, short, broad. Raceme short, of several minute yellowish-green fls which are turned through 360° so that the *labellum* is *directed upwards*. Per. segs spreading. Labellum short, entire; *spur* 0. Column very short, stigma in a deep fold on its front; rostellum a membrane above the stigma and in front of the anther, surmounted by a small viscid mass. Anther hinged to the top of the column behind the rostellum; pollinia 2, each of 2 thin flat plates of waxy pollen, broad below and tapering upwards, standing in a clinandrium formed by 2 lateral membranous lobes of the column; pollen in tetrads which cohere firmly in the 4 plates; caudicles 0.

One sp. in central and north temperate Europe, N. Asia, and N. America.

1. H. paludosa (L.) O. Kuntze   Bog Orchid.

Ophrys paludosa L.; *Malaxis paludosa* (L.) Sw.

Old pseudo-bulb buried in moss or peat, ovoid, ± angled above, tapering below into a root; daughter bulb 1–2 cm. higher, enclosed by lvs. Stem 3–12 cm., slender, 3–5-angled above, glabrous. *Lvs* 0·5–1 cm., 3–5, of which the lowest 1–2 may have no blades, small, short, concave, *broadly rounded at the apex* and broadly sheathing at the base, 3–7-veined, pale green, usually *bearing a marginal fringe of tiny bulbils*. Raceme 1·5–5 cm., spike-like, becoming lax; its bracts lanceolate-acute, just exceeding the fl.-stalk. Outer lateral per. segs ovate-lanceolate, erect; outer median segment pointing downwards, rather longer and broader. Inner lateral segments linear-lanceolate, spreading with down-turned tips. Labellum lanceolate, acute, shorter than the outer per. segs, erect, with its base clasping the column. All floral parts greenish-yellow. Ovary turbinate, straight, its stalk twisted. Fl. 7–9. Cross-pollinated by small insects. $2n = 28$.

Native. Usually in wet sphagnum; to 1600 ft. in Scotland. 54, H12. Throughout the British Is., except Channel Is. and Shetland, but rare and local in S. and C. England. C. and N. Europe from France, Switzerland and C. Russia northwards to Scandinavia and Finland; Siberia; N. America.

10. LIPARIS Rich.

A large genus whose European representatives are small herbs with 2 *ellipsoidal pseudo-bulbs* (parent and daughter) side by side, an angled stem usually with 2 green lvs, and a raceme of small yellowish-green fls. Per. segs narrow, spreading. Labellum pointing upwards or downwards or in any intermediate direction, broad, entire; *spur* 0. Column long, slender; stigma transversely oblong, depressed, flanked by lateral wings of the column; rostellum minute with 2 evanescent viscidia. Anther on the top of the column, deciduous; pollinia 2, each of 2 flat plates of waxy pollen and each attached to one of the viscidia; caudicles 0.

About 100 spp., some epiphytic, widely distributed in temperate and tropical regions.

1. L. loeselii (L.) Rich.   'Fen Orchid.'

Ophrys Loeselii L.

Parent pseudo-bulb clothed in a pale network of old lf-bases, and giving rise to a short horizontal stolon on which the daughter stock is borne. Stem 6–20 cm., erect, glabrous, strongly 3(–5)-angled above and with 2–3 sheathing basal scales enclosing the developing pseudo-bulb. *Lvs* 2·5–8 cm., in *a sub-opposite pair*, oblong-elliptical, acute, keeled, many-veined, shining and greasy-looking, with long sheaths. Raceme 2–10 cm., rather lax, of 1–10(–15) fls; lower bracts lanceolate, often equalling or exceeding the fls, upper ones (or all) minute, falling short of the ovary. Outer *per. segs* linear-lanceolate, spreading; inner shorter and narrower; all *yellow-green. Labellum commonly directed ± upwards*, almost equalling the outer per. segs, oblong-obovate, furrowed, undulate or ± crenate, darker green. Column slightly inflexed. Capsule fusiform, almost straight and ± erect on an ascending twisted stalk. Fl. 7. Probably cross-pollinated by insects. $2n = 32$.

Var. *ovata* Riddelsd. is ex Godf. the dune-form, with broader ovate-elliptical blunt lvs.

Native. In wet fen-peat and dune-slacks. 8. Norfolk, Suffolk, Cambridge, Hunts; var. *ovata* Riddelsd. in Glamorgan and Carmarthen. Rare and decreasing. Europe from France, N. Italy and Austria to Germany, Denmark, S. Scandinavia, C. and S. Russia; N. America.

11. CORALLORHIZA Chatel.

Brown saprophytic herbs with coral-like much-branched fleshy rhizomes, stems with sheathing scales but *no green lvs*, and a few small fls in spike-like racemes. Per. segs ± spreading or the outer median and inner lateral subconnivent. *Labellum* short, *directed downwards*, ± 3-lobed, with the lateral lobes very small or 0; spur short, ± adnate to the ovary, or 0. Column long, erect; stigma discoid or triangular; rostellum small, globular. Anther terminal on the column, lid-like, deciduous; pollinia 4, subglobular; pollen waxy or powdery.

About 12 spp., in Europe (1 sp.), temperate Asia, N. America and Mexico.

1. C. trifida Chatel.   Coral-root.

Ophrys Corallorhiza L.; *C. innata* R.Br.

Rhizome cream-coloured or pale yellowish, with short rounded branches. Stem 7–25 cm., erect, slender, yellowish-green, glabrous, with 2–4 long, brown-veined, sheathing scales, often reaching half-way up the stem. Raceme lax, of 4–12 rather inconspicuous fls; its bracts very small, much shorter than the ovary, triangular, membranous. Outer lateral per. segs curved downwards close to the labellum, strap-shaped, with incurved margins; outer median per. segs ovate-lanceolate, ± connivent with the narrower inner segments; all yellowish- or olive-green, often with reddish borders or spots. Labellum c. 5 mm., about equalling the per. segs, oblong, 3-lobed with 2 very small rounded or tooth-like lateral lobes near its base; lateral lobes sometimes almost equalling the central, or 0; the whole whitish with crimson lines or spots, and with 2 broad longitudinal ridges near the base. Ovary c. 7 mm., straight, with a short twisted stalk. Fl. 5–8. Probably cross-pollinated by insects. $2n = 42$.

Native. Damp peaty or mossy woods, especially of birch, pine or alder, and moist dune-slacks; rare. 19. From Northumberland and Westmorland northwards to Inverness and Ross, especially in upland woods and on E. Scottish dunes. Europe from S. France and Dalmatia northwards; Siberia; N. America.

12. HERMINIUM Schaeff.

Small herbs with 1 fully developed entire root-tuber at flowering; *daughter-tubers at the tips of slender stolons.* Lvs commonly 2. Spike slender, dense; fls small, green, ±campanulate. Per. segs incurved, connivent. Labellum 3-lobed; spur 0. Column very short: stigma 2-lobed; rostellum represented by 2 *large viscidia* each covered by a delicate skin but not enclosed in a pouch. Anther adnate to the top of the column; pollinia 2, ovoid, diverging downwards and each attached basally by a *very short caudicle* to a viscidium which almost equals it in size; pollen in tetrads bound together by elastic threads.

About 4 spp. in Europe and temperate Asia.

Distinguishable by the spurless fls with 2 ±exposed viscidia but no rostellar projection.

1. H. monorchis (L.) R.Br.   Musk Orchid.

Ophrys Monorchis L.

Root-tubers small, globose; daughter-tubers at the tips of stolons up to 10 cm.; roots slender, few. Stem 7–15(–30) cm., erect, slender, glabrous, angled above, with 1–2 closely sheathing scales below. Lvs 2(–4), 2–7 cm., oblong or elliptical-oblong, ±acute, keeled; often with 1–2(–3) much smaller bract-like lvs higher up the stem. Spike 1–5(–10) cm., slender, cylindrical, often 1-sided, ±dense, with small drooping ±campanulate greenish fragrant fls; bracts about equalling the ovary or the lowest equalling the fls, lanceolate, green. Outer per. segs connivent, the upper ovate-oblong, the laterals lanceolate; inner per. segs longer but narrower, ±laterally lobed; all pale greenish-yellow. Labellum c. 4 mm., greenish, its lateral lobes short and widely divergent, the central lobe oblong, blunt; base of labellum saccate. Pollinia white. Capsule oblong, twisted, ±erect with the apex curved outwards. Fl. 6–7. Pollinated by minute flies, beetles and Hymenoptera, the viscidia becoming attached to their legs. $2n=40$.

Native. Chalk downs and limestone pastures; rare and local. 24. S. and E. England from Dorset to Kent and northwards to Monmouth, Gloucester, Oxford, Buckingham, Bedford and Norfolk. Europe; Caucasus; temperate Asia to Yunnan and E. Mongolia.

13. COELOGLOSSUM Hartm.

Small herbs with palmate root tubers, ovate or oblong lvs, and small green and brown fls whose outer and inner lateral per. segs are connivent into a hood. Labellum narrowly oblong; spur short, saccate, with free nectar. Column short, erect; stigma central, reniform, depressed; rostellum of 2 widely separated protuberances, one on each side of the upper edge of the stigma. Anthers adnate to the top of the column; pollinia club-shaped, converging above, but narrowing downwards into widely separated caudicles each attached basally to one of the 2 *oblong viscidia* which are covered by a delicate skin but only *partially enclosed in small pouches*; pollen in packets of tetrads bound by elastic threads.

Two spp. in the north temperate and arctic zones.

1. C. viride (L.) Hartm.   Frog Orchid.

Satyrium viride L.; *Habenaria viridis* (L.) R.Br.

Tubers 2, ovoid, usually palmately lobed with 2–4 tapering segments. Stem 6–25(–35) cm., angled and often reddish above, with 1–2 brown sheathing basal scales. Lower lvs 1·5–5 × 1–3 cm., broadly oblong, blunt, the lowest sometimes almost orbicular; upper lvs smaller, lanceolate, acute; all unspotted. Spike 1·5–6(–10) cm., cylindrical, rather lax-fld; bracts green, the lowest about equalling the fls. Fls greenish, inconspicuous, slightly scented. Outer per. segs c. 5 × 3·5 mm., brownish or greenish-purple, ± connivent into a hood; inner per. segs green, linear, ± hidden by the outer. Labellum 3·5–6 × 1·5–3 mm., oblong, straight and parallel-sided, hanging almost vertically, 3-lobed near its tip, the outer lobes narrowly oblong and parallel, the central usually much shorter, rounded or tooth-like; colour of labellum variable, uniformly green or edged with chocolate-brown or brownish in the distal half. Spur c. 2 mm., ovoid-conical, translucent greenish-white, containing free nectar. Fl. 6–8. Cross-pollinated by insects. $2n=40$.

Native. Pastures and grassy-hillsides, especially on calcareous soils; occasionally on dunes and rock-ledges; to 3300 ft. in Scotland. 108, H40. Throughout the British Is., though more frequent in the north. Europe (except Mediterranean zone); W. Asia; N. America.

14. Gymnadenia R.Br.

Herbs with usually palmately lobed root-tubers, lfy stems and dense spikes of small fragrant fls. Outer lateral per. segs spreading; outer median and inner lateral per. segs connivent into a hood. Labellum shortly 3-lobed, directed downwards; spur long and slender, with free nectar. Column short, erect, with the 2 *stigmas on oval lateral lobes*; rostellum elongated, projecting between the viscidia. Anther wholly adnate to the column; pollinia 2, convergent and narrowed below into caudicles each of which is attached basally to a *long linear viscidium* about equalling the caudicle; the viscidia lie close together and are naked, i.e. *not enclosed in a pouch*; pollen in packets of tetrads bound together by elastic threads.

Ten to twenty spp. in Europe and N. Asia.

1. G. conopsea (L.) R.Br.   Fragrant Orchid.

Orchis conopsea L.; *Habenaria conopsea* (L.) Benth., non Rchb. f.

Tubers compressed, with 3–6 tapering blunt segments. Stem 15–40(–60) cm., erect, glabrous, sometimes purplish above, with 2–3, close, brown basal sheaths. Lower lvs 3–5, c. 6–15 × 0·5–3 cm., ± narrowly oblong-lanceolate, keeled and folded, slightly hooded, blunt or subacute; upper lvs c. 2–3, smaller, lanceolate-acuminate, bract-like, appressed to the stem; all unspotted, with minutely toothed margins. Spike (3–)6–10(–12) cm., ± cylindrical, rather dense-fld; bracts green or violet-flushed, about equalling the fls. *Fls* small, reddish-lilac, rarely white or magenta, *very fragrant*. Outer lateral per. segs spreading horizontally or downwardly curved, their margins revolute. Labellum c. 3·5 × 4 mm., its 3 lobes subequal and rounded. *Spur* 11–13 mm., *very slender*, somewhat curved below, acute, *almost twice as long as the ovary*.

Fl. 6–8. Pollinated by moths to whose proboscis the viscidia become attached. $2n = 20*$; 40, 80.

Var. *densiflora* (Wahlenb.) Lindl. has the spike 8–16 cm., pyramidal at first, and the fls bright rose-red or magenta with the smell of cloves. Var. *insulicola* H.-Harr. is a smaller plant with dull reddish-purple fls having an unpleasant rubber-like smell.

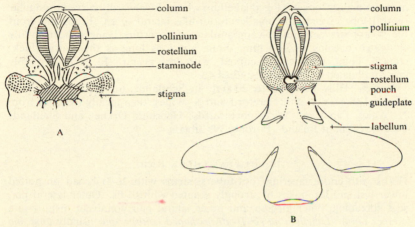

Fig. 66. Columns of A, *Gymnadenia conopsea*; B, *Anacamptis pyramidalis*.

Native. Base-rich grassland, especially on chalk or limestone, fens, and marshes; to 2100 ft. in Scotland. 107, H38. Locally abundant throughout the British Is. Var. *densiflora* in Anglesey and Isle of Wight; var. *insulicola* on Fuday (Outer Hebrides). Europe, N. and W. Asia.

G. odoratissima (L.) Rich., differing from *G. conopsea* in its narrower lvs, smaller fls, horizontally spreading (not downwardly curved) outer lateral per. segs, narrower labellum somewhat longer than broad, and shorter spur about equalling the ovary, has been reported more than once but seems extinct, if it ever occurred.

15. LEUCORCHIS E. Mey.

Herbs with root-tubers so deeply palmate that the several unequal ± cylindrical or tapering segments appear separate. Close to *Gymnadenia* but with the outer lateral *per. segs connivent* with the outer median and the inner per. segs *into a hood*, and the labellum ± connivent with the hood so that the *fl.* becomes almost *campanulate*. Spur short. Viscidia elliptical, less elongated and further apart than in *Gymnadenia*, but naked as in that genus.

One sp.

1. L. albida (L.) E. Mey.   'Small White Orchid.'

Satyrium albidum L.; *Habenaria albida* (L.) R.Br.; *Gymnadenia albida* (L.) Rich.

Stem 10–30 cm., solid, stiffly erect, glabrous, with 2–3 brownish or whitish sheathing basal scales. Lower lvs c. 4, 2·5–8 × 1–1·7 cm., ± oblong-oblanceolate, the lowest blunt, all keeled, firm, glossy above, unspotted but with rows of translucent dashes between the veins; upper lvs 1–2, narrow, acute, bract-like. Spike 3–6 cm., narrowly cylindrical, dense, often slightly curved; bracts lanceolate, green, about equalling the ovary. *Fls* 2–2·5 mm., *half-drooping*, turned to one side, *greenish-white, faintly vanilla-scented*. Outer and inner lateral per. segs connivent into a flattish short broad hood over the slightly longer 3-lobed labellum whose downwardly curved triangular central lobe exceeds the smaller tooth-like lateral lobes. From the front the fl. appears compressed-campanulate, its aperture being a horizontally elongated opening c. 3×1 mm. Spur not $\frac{1}{2}$ as long as the ovary, thick, conical, blunt, downwardly curved, with free nectar. Fl. 6–7. Probably cross-pollinated by tiny insects. $2n = 42$.

Native. Hilly pastures, to 2000 ft. in Scotland. 61, H25. Rare in S. England (Sussex, Kent), Hereford and S. Wales, then locally frequent from N. Wales, Lancs and Yorks northwards; Hebrides, Orkney and Shetland; local throughout Ireland. Europe, W. Siberia.

16. Platanthera Rich.

Herbs with entire tapering root-tubers, stems with 2(–3) broad unspotted lower lvs, and lax spikes of strongly scented whitish fls. Outer lateral per. segs spreading, outer median and inner lateral pair connivent in an ovate $\pm$ erect hood. *Labellum narrowly strap-shaped, entire; spur* usually *long and slender*, with free nectar. Column rather short; stigma transversely elongated, $\pm$ oblong, depressed; rostellum represented only by the laterally placed viscidia. Anthers adnate to the top of the column; pollinia 2, narrowed downwards into slender caudicles, each attached laterally, close to its base, either directly to one of the 2 *naked viscidia*, or indirectly through a short connecting stalk; pollen in packets of tetrads $\pm$ firmly tied by elastic threads.

Eighty spp. in the north temperate and tropical zones.

Spike $\pm$ pyramidal, greenish; fl. 18–23 mm. across; labellum 10–16 mm.; pollinia 3–4 mm., divergent downwards; viscidia c. 4 mm. apart, circular.
1. chlorantha
Spike $\pm$ cylindrical, whitish; fl. 11–18 mm. across; labellum 6–10 mm.; pollinia c. 2 mm., parallel; viscidia c. 1 mm. apart, oval.   **2. bifolia**

1. P. chlorantha (Cust.) Rchb.   Greater Butterfly Orchid.

Orchis chlorantha Cust.; *Habenaria chlorantha* (Cust.) Bab.; non Spreng., *H. virescens* Druce, non Spreng; *H. chloroleuca* Ridl.

Root tubers 2, ovoid, attenuate below. Stem 20–40(–60) cm., glabrous, $\pm$ angled above, with 1–3 brown basal sheathing scales. Lower lvs usually 2, 5–15(–20) × 1–5(–7) cm., elliptical or elliptical-oblanceolate, blunt; upper lvs 1–5, grading into bracts; all unspotted. *Spike* 5–20 cm., $\pm$*pyramidal*, lax, *greenish*; bracts ovate-lanceolate, acuminate, blunt, green, variable in length but usually about equalling the ovary. Fls 18–23 mm. across, heavily fragrant, especially at night. Outer lateral per. segs c. 10–11 × 5 mm., ovate, spreading; outer median per. seg. broadly triangular-cordate, blunt; inner lateral per. segs narrowly lanceolate, erect; all greenish-white. Labellum 10–16 × 2–2·5 mm., slightly tapering downwards, rounded at the tip, green below,

greenish-white distally. Spur 19–28 × c. 1 mm., curved downwards and forwards, sometimes almost into a semicircle. *Pollinia* 3–4 mm. in overall length,
sloping forwards and outwards and so *diverging* from c. 2 mm. apart above
to c. 4 mm. between the large *circular viscidia*. Fl. 5–7. Cross-pollinated by
moths to whose heads the pollinia become attached. $2n = 42$.

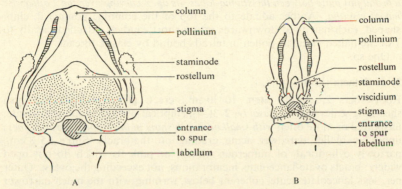

Fig. 67. Columns of A, *Platanthera chlorantha*; B, *P. bifolia*.

Native. Woods and grassy slopes, especially on base-rich and calcareous
soils, to 1500 ft. in N. England; local. 104, H40. Throughout the British Is.
except Orkney and Shetland. Europe; Caucasus; Siberia.

2. P. bifolia (L.) Rich.   Lesser Butterfly Orchid.

Orchis bifolia L.; *O. montana* Schmidt; *Habenaria bifolia* (L.) R.Br.
Closely resembling *P. chlorantha* but somewhat smaller in all its parts. Stem
15–30(–45) cm., with 2–3 brown or whitish sheathing basal scales. Lower lvs
usually 2, 3–9(–15) × 0·7–3(–4) cm., elliptical, short; upper lvs 1–5, smaller,
grading into the bracts; all unspotted. *Spike* 2·5–20 cm., usually lax, ± *cylindrical, whitish*. Fls 11–18 mm. across, rather less strongly night-scented than
in *P. chlorantha*. Outer lateral per. segs lanceolate. Labellum 6–10 mm.
Spur 15–20 mm., slender, almost horizontal. *Pollinia* c. 2 mm. in overall
length, *vertical, parallel, their small oval viscidia* c. 1 mm. apart. Fl. 5–7.
Pollinated by moths, to whose proboscis the pollinia become attached.
$2n = 42$.

Distinguishable from *P. chlorantha* at a distance by the narrower, almost cylindrical
spike of smaller, less distant and whiter fls. The fls are somewhat more sweetly
scented, lacking the slight pungency of those of *P. chlorantha*.

Native. Grassy hillsides and open woods on base-rich and especially on
calcareous soils; reaching 1260 ft. in N. England. 103, H39. Not infrequent
throughout the British Is. except Orkney and Shetland, but usually less
common than *P. chlorantha*, with which it often grows. Europe; Caucasus;
N. Asia; N. Africa.

17. NEOTINEA Rchb. f.

A small herb with entire root-rubers, usually spotted lvs and a short dense 1-sided spike of pale fls. Per. segs all connivent. Labellum directed forwards, 3-lobed; spur very short. Column very short and small; stigmas 2, large, ± crescentic, joined below, borne on lateral wings of the column; *rostellum a broad flat plate between the stigma-lobes, its apex curving over the 2 otherwise naked viscidia.* Anther adnate to the top of the column; pollinia 2, club-shaped, each narrowed downwards into a short caudicle attached basally to 1 of the naked viscidia; pollen in tetrads bound by elastic threads.

One sp.

1. N. intacta (Link) Rchb. f.

Root-tubers 2, ovoid. Stem 10–30 cm., erect, glabrous, with brownish sheathing scales below. *Basal lvs* 2–3(–4), elliptical-oblong, blunt, mucronate, unspotted or more often *with small brownish spots in interrupted parallel lines*; upper lvs smaller, narrower, acute, grading into the bracts. Spike 2·5–8 cm., narrowly cylindrical, of numerous whitish or pink fls which do not open widely; bracts ovate-lanceolate, membranous, not exceeding the ovary. Outer per. segs lanceolate-acute, coherent below, forming with the much narrower inner segments a long almost closed helmet over the column; outer segments sometimes purple-spotted. Labellum small, equalling or somewhat exceeding the outer per. segs, directed forwards or obliquely downwards, often with 2–3 purplish blotches at the furrowed base, 3-lobed with linear acute lateral lobes and a longer and broader central lobe, truncate or notched at the apex, sometimes with a tooth in the notch. Spur c. 2 mm., conical, blunt. Pollinia pale green. Capsule spindle-shaped, glabrous, twisted, ± erect. Fl. 4–6. Cross- or self-pollinated.

Native. Rocky and sandy calcareous pastures in W. Ireland; lowland. H 8. Galway, Clare, Mayo, Roscommon and Offaly. Atlantic and Mediterranean Europe from France and Portugal to Greece; Madeira; Canaries; N. Africa; Cyprus; Asia Minor.

18. OPHRYS L.

Herbs with entire ovoid to subglobose root-tubers, lfy stems and lax-fld spikes. *Per. segs spreading*, the inner lateral segments usually smaller than the outer. *Labellum large, entire or 3-lobed, often convex, velvety, usually dark-coloured and conspicuously marked*; spur 0; nectaries 0, but 2 shining eye-like projections commonly stand one on each side of the base of the labellum. Column long, erect; stigma single, large, central, depressed; rostellum represented by 2 *separated pouches* above the stigma. Anther adnate to the top of the column; pollinia 2, narrowed downwards into long caudicles which are attached basally to *separate ± globose viscidia* enclosed in the distinct rostellar pouches; pollen in packets of tetrads ± firmly united by elastic threads. Ovary not twisted.

About 30 spp. in Europe, W. Asia and N. Africa.

A very distinct genus easily recognizable by the large spurless labellum, often resembling an insect, and the 2 separate rostellar pouches (Fig. 68 B).

1 Outer per. segs green, herbaceous.   **2**
    Outer per. segs rose-pink or whitish, petaloid.   **3**

2 Inner lateral per. segs ±filiform, purplish-brown, velvety; labellum 3-lobed
    near the middle; the central lobe deeply bifid, purplish-brown with a
    glabrous slate-blue transverse blotch.   **4. insectifera**
    Inner lateral per. segs strap-shaped, greenish, undulate; labellum sub-
    entire or with small swollen basal lobes, broad, purplish-brown with
    yellowish glabrous markings.   **3. sphegodes**

3 Labellum strongly convex, terminating in two short truncate lobes and
    an intervening long narrow tooth turned up between and behind them,
    so as to be invisible from above.   **1. apifera**
    Labellum ±flat, terminating in a variously shaped, often cordate, ap-
    pendage which lies flat or turns up in front, so as to be clearly visible
    from above.   **2. fuciflora**

1. O. apifera Huds.   Bee Orchid.

Root-tubers subglobose. Stem 15–45(–60) cm., glabrous. Lvs 3–8 × 0·5–2 cm.,
elliptical-oblong, subacute, diminishing rapidly up the stem and grading into
the bracts; all unspotted. Spike 3–12 cm., of 2–5(–10) large, rather distant fls;

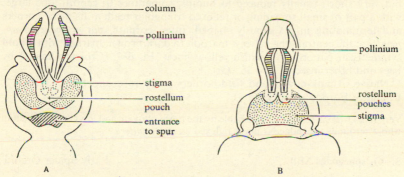

Fig. 68. Columns of *Orchis mascula* (A); *Ophrys fuciflora* (B).

bracts large and lf-like, commonly exceeding the fls. Outer per. segs 12–
15 mm., elliptical-oblong, rose-pink or whitish, greenish on the back; inner
lateral per. segs lanceolate, shorter than the outer, pinkish-green, ±downy
on the inner side. *Labellum* 12–15 mm., *resembling a humble-bee*, strongly
convex, semi-globose, 3-lobed, the lateral lobes small, hairy, ±rounded, with
basal protuberances and a blunt forwardly and upwardly directed apex;
central lobe much larger, curved downwards distally and terminating in
2 truncate lateral segments and a *long narrowly triangular central tooth which
is curled up behind* so as to be invisible from above; the central lobe velvety,
purplish- to dark brown, marked with glabrous yellow spots distally and a
greenish-yellow horseshoe or H basally; all lobes ±bordered with yellow.
Fls 6–7. The pollinia fall forwards and downwards on to the stigma while
remaining attached by their long slender caudicles, so that the fls are habitually
self-pollinated. $2n = 36$. Grt.

Var. *chlorantha* Godf. has the outer per. segs white and the labellum greenish-yellow and very narrow. Var. *trollii* (Hegetschw.) Druce (*O. Trollii* Hegetschw.) is a curious form in which the labellum is pale yellow with irregular brown markings, narrower than in the type, with the lateral lobes acute and ± reflexed and the terminal tooth long, acute and ± spreading.

Native. Pastures, field-borders, banks and copses on chalk or limestone, especially on recently disturbed soils; also on base-rich clays and calcareous dunes. Local throughout England, Wales, Ireland, and northwards to Lanark and Renfrew; Channel Is. 61, H34, S. C. and S. Europe; N. Africa.

2. O. fuciflora (Crantz) Moench   Late Spider Orchid.

O. arachnites (L.) Reichard

Root-tubers subglobose. Stem 10–35 cm., glabrous. Lvs 4–10 × 0·5–2·5 cm., elliptical-oblong, subacute. Spike lax, of 2–7 fls whose bracts exceed the ovaries. Outer per. segs 10–12 mm., ovate-oblong, spreading, rose-pink; inner lateral per. segs triangular, blunt, downy above, pinkish or with a green central stripe. *Labellum* 10–12 × 11–15 mm., *broader than long*, obovate-suborbicular or obscurely angled, *slightly convex*, 3-lobed; basal lobes broad but very short, usually reduced to humps, sometimes 0; central lobe large, with a pair of small spreading or reflexed triangular teeth at its distal corners and terminating in a variously shaped, often cordate, *flat or incurved appendage*; the central lobe velvety, maroon to dark brown, marked with a bold symmetrical pattern in greenish-yellow lines. Fl. 6–7. Cross-pollinated by bees and other insects. $2n = 36$. Grt.

Native. Chalk-downs and field-borders; very rare and local. Kent. 2. C. and S. Europe; Near East.

Very like *O. apifera* but distinguishable by the more obovate and less convex labellum whose terminal appendage is never reflexed as in *O. apifera*.

3. O. sphegodes Mill.   Early Spider Orchid.

O. aranifera Huds.

Root-tubers ovoid to subglobose. Stem 10–35 cm., glabrous. Lvs 4–10 × 0·5–1·5 cm., elliptic-oblong, the lower spreading. Spike lax, of 2–8 fls; lower bracts exceeding their fls. *Outer per. segs* 10–12 mm., ovate-oblong, *pale yellowish-green; inner per. segs* somewhat shorter, ± narrowly oblong, blunt, greenish-yellow, usually *undulate*, glabrous, rough or downy. *Labellum* suborbicular to ovate-oblong, strongly convex, subentire or 3-lobed at the base, the lateral lobes small, with basal protuberances, the central lobe entire or emarginate, usually *with no terminal appendage*; central lobe *velvety*, dull purplish-brown, later turning yellowish, with various *glabrous bluish markings*, often H- or horseshoe-shaped. Fl. 4–6. Visited sparingly by bees and setting seed only if visited. $2n = 36$. Grt.

There are two named varieties with intermediates. In var. *sphegodes* the inner lateral per. segs are ± glabrous and the labellum is usually lobed at the base. In var. *fucifera* Druce the inner lateral per. segs are rough to downy and the labellum is entire or only slightly 3-lobed.

Native. Grassy slopes, banks and field-borders on chalk or limestone; local. S. and E. England from Dorset and Wilts to Kent and northwards to Cambridge and Suffolk; Denbigh; Jersey. 10, S. C. and S. Europe; N. Africa.

4. O. insectifera L.   Fly Orchid.

O. muscifera Huds.

Root-tubers ovoid to subglobose. Stem 15–60 cm., ± glabrous. Lvs 4–12·5 cm., few, oblong to elliptical, subacute. Spike long, of 4–12 distant fls; the lower bracts much exceeding their fls. Outer per. segs 6–9 mm., oblong, yellowish-green; *inner lateral per. segs filiform, purplish-brown*, velvety. *Labellum* c. 12 mm., obovate-oblong, not convex longitudinally, 3-*lobed near the middle*; lateral lobes without basal protuberances, narrowly triangular, spreading; middle lobe obovate-oblong, ± *deeply bifid*; the whole purplish-brown, downy, with *a broad glabrous shining bluish transverse patch* or blotch just beyond the insertion of the lateral lobes. Fl. 5–7. Rather sparingly visited by bees and flies and setting seed only if visited. $2n = 36$. Grt.

Native. Woods, copses, field-borders, spoil-slopes, banks and grassy hillsides on chalk or limestone and in fens; not infrequent. Locally throughout lowland England from S. Devon, Somerset and Dorset to Kent and northwards to Westmorland, N. Yorks and Durham; Glamorgan, Denbigh, Anglesey; Perth; C. Ireland. 48, H 10. Europe northwards to Norway.

Hybrids of *O. sphegodes* with *O. apifera*, *O. fuciflora* and *O. insectifera* have been reported from Great Britain.

19. HIMANTOGLOSSUM Spreng.

Herbs with entire tubers, tall stout lfy stems, and long cylindrical spikes of large fls. All per. segs except the labellum connivent into a hood. Labellum 3-lobed, the lateral lobes short, the *central lobe very long and narrow, spirally coiled in bud like a watch-spring; spur very short*. Column rather short, erect; stigma large, single; rostellum beak-like, projecting above the stigma. Anther adnate to the top of the column; pollinia 2, narrowed below into caudicles both of which are attached to *a single ± 4-angled viscidium enclosed in a pouch* in the rostellum; pollen in packets of tetrads united by elastic threads.

Two spp. in C. Europe and the Mediterranean region.

1. H. hircinum (L.) Spreng.   Lizard Orchid.

Satyrium hircinum L.; *Orchis hircina* (L.) Crantz; *Loroglossum hircinum* (L.) Rich.

Root-tubers ovoid or ± globose. Stem 20–40(–90) cm., stout, glabrous, faintly mottled with purple. Lower lvs 4–6, 6–15 × 3–5 cm., elliptical-oblong, ± blunt; upper narrower, acute, clasping the stem; all unspotted. Spike 10–25(–50) cm., rather lax; bracts linear-lanceolate, acute, 3–5-veined, membranous, the lowest equalling the fls, the remainder exceeding the ovary. Fls numerous, large, greenish, untidy-looking, smelling strongly of goats. Outer per. segs cohering at the base, ovate, blunt, pale greenish often streaked and spotted with purple; inner per. segs linear, adhering laterally to the outer. Labellum 3–5 cm. × 2 mm., whitish with purple spots and furry with papillae near the cuneate base, pale brownish-green and glabrous distally; lateral lobes 5–10 mm., narrow, acute, curly; central lobe narrowly strap-shaped, c. 2 mm. wide, ± undulate and somewhat spirally coiled, truncate and notched or 2–4 toothed at the tip. Spur c. 4 mm., conical, blunt, slightly curved downwards. Fl. 5–7. Cross-pollinated by Hymenoptera. $2n = 24$.

Native. Wood-margins, by woodland paths, amongst bushes, on field-borders, and in grassland, chiefly on chalk or limestone; rare and local but apparently increasing. 26, S. S. and E. England from N. Devon, Gloucester, Somerset and Dorset to Kent and N.E. Yorks, but most frequently recorded from Kent and Sussex. C. and S. Europe from Holland, Belgium, France and Spain to N. Balkans and Greece; Sardinia, Sicily, N. Africa.

20. ORCHIS L.

Herbs with *rounded unforked root-tubers* and sheathing lvs, the basal *lvs in a rosette*. Fls in a spike which is usually enclosed by thin spathe-like lvs during emergence. *Bracts membranous. All per. segs except the labellum connivent into a helmet over the column, or only the outer lateral segs spreading or turned upwards.* Labellum usually 3-lobed, directed downwards, with descending or ascending spur. Column erect; stigma ± 2-lobed, roofing the entrance to the spur; *rostellum overhanging the stigma as a single pouch enclosing* 2 *separate ± globular viscidia.* Anther adnate to the top of the column; *pollinia 2, each narrowed below to a caudicle and attached by a basal disk to one of the viscidia*; pollen in packets of tetrads united by elastic threads. Capsule erect.

About 60 spp. in Europe, temperate Asia, N. Africa, Canary Is.

1 All per. segs except the labellum connivent to form a 'helmet' over the
     column; spur descending.   *2*
   Outer lateral per. segs erect or spreading; spur ± ascending.   *6*

2 Labellum shaped somewhat like a man, the central lobe much longer than
     the ± slender lateral lobes (arms) and forking distally into 2 branches
     (legs) often with a short tooth in the sinus between them; outer per. segs
     usually ± coherent.   *3*
   Labellum not shaped like a man; outer per. segs green-veined, not
     coherent.   **5. morio**

3 Helmet of young fls dark reddish-brown (maroon); outer per. segs free
     to the base; labellum white, c. 6 mm.; spur about ¼ the length of the
     ovary; bracts long.   **4. ustulata**
   Helmet not dark reddish-brown; outer per. segs coherent below; labellum
     at least 10 mm.; spur about ½ the length of the ovary; bracts minute.   *4*

4 Helmet heavily blotched with dark reddish-purple, almost black in un-
     opened fls; labellum usually tinged with pink or pale purple and with
     darker spots, its distal branches broadly rhomboidal.   **1. purpurea**
   Helmet greyish, flushed, veined and spotted with pale purple or rose; distal
     branches of labellum linear or oblong, not broadly rhomboidal.   *5*

5 Lobes of labellum all very narrow; helmet whitish, ± flushed or marked
     with rose.   **3. simia**

Distal branches of labellum oblong, shorter than and 3–4 times as wide as the basal lobes, red-purple or rose; helmet ± flushed or marked with pale purple.   **2. militaris**

6   Bracts 1-veined or the lowest sometimes 3-veined; spur equalling or exceeding the ovary; labellum 3-lobed, the central lobe largest; throughout British Is.   **7. mascula**

Bracts 3-veined; spur shorter than the ovary; labellum with the central lobe shorter or 0; only in Channel Is.   **6. laxiflora**

Section 1. *Orchis*. Labellum shaped like a man, its central lobe longer than the lateral lobes and forked distally, commonly with a short tooth in the sinus; per. segs connivent to form a helmet over the column, the outer usually ± coherent below; bracts usually short, membranous, 1-veined; hay-scented when drying.

1. O. purpurea Huds.   Lady Orchid.

Tubers ovoid. Stem 20–40(–90) cm., stout. Lvs 7–15 × 3–5 cm., 3–5(–7), ovate-oblong, blunt, the uppermost lanceolate; all unspotted, glabrous and very shining above, paler beneath. Spike 5–10(–15) cm., dense, almost black in bud; bracts very small. *Helmet* 10–12 mm., ovate, *dark reddish-purple*, becoming paler. Outer per. segs ovate, hooded, coherent below, pale greenish with purple mottling within; inner lateral per. segs linear, acute. Labellum c. 18 mm., hanging, whitish flushed with violet or rose, especially at the edges, and freely dotted with tufts of reddish-purple papillae; basal lobes linear, curved, dotted with red-purple; *central lobe broad*, widening downwards and *forking into 2 broadly rhomboidal segments* 4–5 times as broad as the lateral lobes, crenulate at their rounded or truncate apices and usually with a small tooth between them. Spur nearly half as long as the ovary, cylindrical, curved downwards and forwards, enlarged and truncate or notched at the tip. Fl. 5. $2n = 42$. Grt.

Native. Copses and open woods, rarely in open grassland, on chalk or limestone. 2, S. Rare and now apparently confined to Kent but formerly in Essex, Surrey and Sussex. Europe from Denmark southwards; Caucasus; Asia Minor.

2. O. militaris L.   Soldier Orchid.

Tubers ovoid to subglobular. Stem 20–45 cm. Lvs 3–10 × 0·5–2·5 cm., elliptical-oblong, the uppermost narrowly lanceolate, blunt, concave or folded, unspotted. Spike 3–8 cm., dense, blunt; bracts very small. *Helmet* c. 12 mm., ovate-lanceolate, *ash-grey* ± *flushed with rose or violet*. Outer per. segs acuminate, coherent below, veined and spotted with red-violet within; inner linear, acute. Labellum whitish in the middle, dotted with tufts of violet papillae and flushed with violet towards the edges; basal lobes linear, rose or violet; *central lobe narrowly oblong, forking into 2 widely divergent oblong segments* 2–4 times as broad as the basal lobes, entire or denticulate at their apices, with a short narrow acute tooth between them. Spur about half as long as the ovary, cylindrical, curved downwards and forwards, blunt. Fl. 5–6. $2n = 42$. Grt.

Native. Grassy hills, banks, borders of fields and edges of woods on chalk, rare. 2. Bucks, W. Suffolk; formerly in the Thames Valley, Oxford, Berks, Herts, Middlesex, Surrey and ?Kent. C. Europe from Gotland to Spain and Portugal; C. Italy, N. Balkans and S. Russia; Siberia; Caucasus; Asia Minor.

3. O. simia Lam.   Monkey Orchid.

O. tephrosanthos Vill.

Tubers ovoid or subglobular. Stem 15–30 cm., more slender than in *O. militaris*, which it closely resembles. Lvs as in *O. militaris* but somewhat smaller. Helmet whitish faintly streaked with rose, or pale violet, with minute violet dots and 2–3 raised white veins. Labellum white, rose, or deep crimson with a paler base; *basal lobes* linear, *crimson*; *central lobe* narrow, up to three times the width of the basal lobes, dotted with small tufts of violet papillae, *forking into 2 linear* crimson *segments*, as narrow as the basal lobes, rounded or truncate at their apices, with a conspicuous tooth between them. Spur as in *O. militaris*. Fl. 5–6. 2*n*=42. Grt.

Three forms have been described from Britain. That from Kent has a whitish hood and a very narrow deep crimson central lobe of the labellum; one from Oxford differs in its broader, white red-dotted central labellum-lobe; while the other (var. *macra* Lindl.) has a violet helmet, pink crimson-dotted central labellum lobe, and purplish (not crimson) basal lobes and distal segments.

Native. Grassy hills, bushy places and field borders on chalk, very rare. 3. Apparently confined to Oxford (Goring Gap) and Kent, but formerly in Berks. C. and S. Europe from Belgium to Spain, Italy, N. Balkans and Crimea; Caucasus; N. Africa; Near East.

4. O. ustulata L.   Dark-winged Orchid, Burnt Orchid.

Tubers ovoid. Stem 8–20(–30) cm., stout. Lvs 2·5–10 × 0·5–1 cm. elliptical-oblong to lanceolate, acute. *Spike* 2–5 cm., dense, small-fld, at first conical, *dark maroon* when the fls are unopened, *becoming paler and then whitish*; bracts at least half as long as the ovary. Helmet c. 5 mm., roundish, dark brownish-purple (maroon), ultimately white. Outer per. segs free to the base, ovate, subacute; inner narrower. Labellum c. 5–6 mm., white dotted with red-purple; basal lobes narrowly oblong, entire or notched at the apex; central lobe oblong, forking distally into 2 oblong segments about equalling the basal lobes, usually crenulate at their apices and commonly with a short tooth between them. Spur about ¼ as long as the ovary, cylindrical, curved downwards and forwards, blunt at the tip. Fl. 5–6. 2*n*=42. Grt.

Native. Chalk downs and limestone pastures; local. 44. Widely distributed throughout England from Devon to Kent and northwards to Cumberland and Northumberland; reaching 800 ft. in Bucks. Europe, from Gotland southwards; Caucasus; Siberia.

Section 2. *Morio* Dum. Labellum 3-lobed, not shaped like a man, the central lobe truncate, emarginate or 2-lobed; outer per. segs not coherent; bracts about equalling the ovary, membranous, 1-veined; not hay-scented on drying.

5. O. morio L.   Green-winged Orchid.

Tubers ± globular. Stem 10–40 cm. Lvs 3–9 × 0·5–1·5 cm., elliptical-oblong
to lanceolate, the lower ones spreading or recurved, the uppermost ± appressed, blunt or acute, unspotted. Spike 2·5–8 cm., rather lax; bracts
coloured. Per. segs connivent in a *helmet, purple with conspicuous green veins*,
rarely flesh-coloured or white. Outer per. segs ovate, blunt; inner narrow,
blunt. Labellum hardly exceeding the helmet, broader than long, 3-lobed;
the lateral lobes reddish-purple, ± folded back, broadly cuneate, crenulate;
central lobe pale reddish-purple dotted with deep red-purple, notched and
crenulate, about equalling the lateral lobes. Spur almost equalling the ovary,
cylindrical, horizontal or ascending, almost straight, enlarged and blunt at
the apex. Fl. 5–6. $2n=36$. Grt.

Native. Meadows and pastures, especially on calcareous soils; locally
abundant. 70, H22, S. Throughout England and Wales, but becoming rare
in the north. Europe southwards from Gotland and S. Sweden; Caucasus;
Asia Minor; Siberia.

6. O. laxiflora Lam.   'Jersey Orchid.'

Tubers ± globular. Stem 20–50(–90) cm., robust. Lvs 7–18 × 0·8–2 cm.,
lanceolate or linear-lanceolate, acute, keeled and ± folded, the uppermost
bract-like. *Spike 7–22 cm., lax, cylindrical, dark purple*; bracts somewhat
exceeding the ovary, ± membranous, coloured. Outer per. segs ovate, blunt,
dark crimson-purple, the lateral pair held erect and back to back, the upper
concave, curving upwards; inner shorter, narrower, blunt, forming a hood.
Labellum 10–15 mm., about as broad as long, dark crimson-purple, whitish
at the base, with 2 large, rounded, crenate or toothed, ± folded-back lateral
lobes; central lobe much shorter than the laterals and truncate, or 0. Spur
up to $\frac{2}{3}$ as long as the ovary, cylindrical, straight or slightly curving upwards,
horizontal or ascending, enlarged and truncate or notched at the tip. Fl. 5–6.
$2n=42$. Grt.

Native. Wet meadows. Frequent in the Channel Is. Europe from Belgium
southwards; N. Africa; W. Asia.

7. O. mascula (L.) L.   Early Purple Orchid, Blue Butcher.

Tubers ovoid to subglobular. Stem 15–60 cm., ± stout. Lvs 5–20 × 0·5–3 cm.,
broadly to narrowly oblanceolate-oblong, bluntish, usually with rounded
black-purple spots. Spike 4–15 cm., rather lax; bracts about equalling the
ovary, coloured, membranous, 1-veined. Outer per. segs ovate, ± acute, the
lateral pair at first spreading, later folded back, the upper connivent with the
narrowly ovate subacute inner per. segs; all purplish-crimson. Labellum
8–12 mm., about as broad as long, purplish-crimson, paler at the base and
dotted with darker purple, 3-lobed; lateral lobes crenate, ± folded back;
central lobe slightly larger, crenate with a ± distinct central notch and sometimes with a lateral notch on each side. *Spur at least equalling the ovary, stout*,
cylindrical, straight or curved upwards, *horizontal or ascending*, blunt or
truncate. Fl. 4–6. $2n=42$. Grt.

Native. Woods, copses and open pastures, chiefly on base-rich soils;
common. 110, H39, S. Throughout the British Is., reaching 2900 ft. in
Scotland. Europe, N. Africa, N. and W. Asia.

21. ACERAS R.Br.

A small *Orchis*-like plant with entire root-tubers, lfy stems, and long narrow spikes of greenish fls. All per. segs except the labellum connivent to form a hood over the column. *Labellum shaped like a man*, with long slender lateral lobes (the arms) near its base, and a longer narrow central lobe which forks distally into 2 slender segments (legs) rather shorter than the lateral lobes; *spur* 0, nectar being secreted from 2 tiny depressions at the base of the labellum. Column very short; stigma transversely elongated, forming the roofs and walls of a cavity at the base of the column; rostellum inconspicuous. Anther adnate to the top of the column; pollinia 2, narrowed downwards into caudicles whose bases are attached to the 2 *contiguous* ± *globular viscidia enclosed in a common pouch*; the viscidia are often coherent, and the pollinia are usually removed together, but occasionally they may be withdrawn separately; pollen in packets of tetrads united by elastic threads.

One sp., in Europe, W. Asia and N. Africa.

Distinguished only by the absence of a spur from *Orchis* section *Orchis* (*O. simia, militaris, purpurea*, etc.), sharing with them the shape of the labellum and the development of a hay-scent (coumarin) on drying, and hybridizing with them.

1. A. anthropophorum (L.) Ait. f.   Man Orchid.

Root-tubers ovoid or ± globose. Stem 20–40(–60) cm., glabrous, slightly ridged above, with sheathing scales at the base. Lower lvs several, crowded, 6–12 × 1·5–2·5 cm., oblong to oblong-lanceolate, ± acute, keeled, glossy on both sides; upper lvs smaller, erect, clasping the stem, grading into bracts; all unspotted. Spike up to half the length of the stem, narrowly cylindrical, many-fld, becoming lax; bracts membranous, shorter than the ovary. Fls greenish-yellow, often edged or tinged with reddish-brown. Outer per. segs forming a semi-globose hood c. 6–7 mm., ovate-lanceolate, blunt; inner per. segs somewhat shorter and much narrower. Labellum c. 12 mm., hanging almost vertically, greenish-yellow often tinged with maroon, rarely pure yellow; lateral lobes and distal segments of central lobe linear; sometimes with a small tooth in the sinus between the distal segments. Fl. 6–7. Cross-pollinated by small insects. $2n=42$. G.

Native. Field-borders, grassy slopes, and rarely in scrub and open woods, on chalk; also in old chalk-pits. 18. Local from Hants and Isle of Wight to Kent and northwards to Lincoln, Derby and Oxford; extinct in N. Somerset. Europe northwards to Belgium and Germany; Cyprus; N. Africa.

22. DACTYLORCHIS (Klinge) Vermeul.

Like *Orchis* but with ± *palmately lobed or divided root-tubers*, basal lvs not in a definite rosette at time of flowering, emerging spike not enclosed in spathe-like lvs, *bracts always lf-like, per. segs*, except labellum, erect or spreading, *not connivent into a helmet over the column; spur always* ± *descending*, usually less than 10 mm. in length.

About 20 spp. in Europe, temperate Asia, N. Africa, Canary Is.

1   Stem solid; (1–)2–6 or more small bract-like transitional lvs between true
lvs and bracts; lowest bracts narrow (3 mm. or less), usually not
exceeding the fls.   (**maculata** agg.)   2

Stem usually hollow; transitional lvs 0–1; lowest bracts broad (more than
3 mm.), usually exceeding the fls.   3

2   Lowest expanded lf broadly elliptic or obovate, usually blunt; lvs often
heavily marked with ±transversely elongated dark spots, but spots
sometimes faint or 0; transitional lvs 2–7; labellum with 3 subequal
lobes, the middle lobe usually longer than the laterals (Fig. 69 A).
                                                         3. fuchsii

All lvs lanceolate, acute, lightly marked with ±circular spots, but spots
sometimes very faint or 0; transitional lvs 1–3; labellum broad, its
middle lobe much smaller and usually shorter than the rounded lateral
lobes (Fig. 69 B).
                          2. maculata ssp. **ericetorum** (and **fuchsii** ssp. **rhumensis**)

3   A majority of plants with lvs spotted on both sides, the spots often
elongated parallel to the lf-margin; fls lilac-purple, the sides of the
labellum usually reflexed.   **1. incarnata** ssp. **cruenta**

Lvs not spotted on both sides.   4

4   Stem-cavity wide; lvs yellow-green, erect, keeled, tapering from the base
to a narrowly hooded tip, unspotted; sides of labellum usually strongly
reflexed.   **1. incarnata**

Stem-cavity moderate or narrow; lvs dark green or somewhat bluish,
often spreading, not or only slightly and broadly hooded, spotted or
not; sides of labellum reflexed or not.   (**majalis** agg.)   5

5   Plants usually with 3–5 narrow lvs; longest lf averaging less than 1·5 cm.
wide, rarely exceeding 2 cm.; lvs unspotted or with transversely
elongated spots; fls reddish-purple.   **7. traunsteineri**

Plants usually with 5–8 lvs, the longest averaging more than 2 cm. wide
and rarely less than 1·5 cm.   6

6   Labellum ovate or broadly diamond-shaped, subentire, usually less than
9 mm. wide (Fig. 69 E); fls usually rich deep purple; lvs usually short,
spreading, unspotted or with very small dots especially towards the tip.
                                                         5. purpurella

Labellum not ovate or diamond-shaped, usually more than 9 mm. wide.   7

7   Lvs 10–20(–30) cm., usually unspotted, rarely ring-spotted; fls rich reddish-
lilac; labellum subentire or with a short rounded middle lobe, often
concave, with darker dots or lines usually confined to a central elliptical
area (Fig. 69 D).   **4. praetermissa**

Lvs 4–10(–15) cm., often heavily spotted or blotched, sometimes ring-
spotted or unspotted; fls pale to dark purple; labellum usually distinctly
3-lobed, not concave, with irregular dark lines or dots usually extending
close to the somewhat angular and often notched lateral margins. (Fig.
69 c).   **6. majalis** ssp. **occidentalis** (and forms of **purpurella**)

1. D. incarnata (L.) Vermeul.   Meadow Orchid.

Orchis incarnata L. sec. Vermeul.; *O. latifolia* L. sec. Pugsl.; *O. strictifolia*
Opiz

Stem 15–50(–80) cm., *with a wide cavity. Lvs* 10–20(–30) × 1–2(–3) cm., *erect
or erect-spreading*, ±keeled, the *blade broadest at the base* and tapering
gradually to a *markedly and narrowly hooded tip, yellow-green*; lf-sheaths
loose-fitting. Transitional lvs between true lvs and bracts rarely more than 1,

C T & W

often 0. *Spike oblong*, dense-fld. *Lowest bracts more than 3 mm. wide and much exceeding the fls.* Outer per. segs erect at the base then spreading outwards at a small angle with the vertical. *Sides of labellum strongly reflexed* soon after the fl. opens, so that it appears very narrow from the front. Spur 6–8·5 mm., conical. Fl. 5–7. $2n=40$*. Grt.

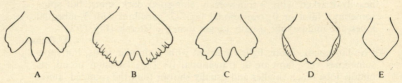

Fig. 69. Labella of *Dactylorchis* spp. A, *D. fuchsii*; B, *D. maculata* ssp. *ericetorum*; C, *D. majalis* ssp. *occidentalis*; D, *D. praetermissa*; E, *D. purpurella*.

Very variable. J. Heslop-Harrison has recognized the following British sspp. which are on the whole ecologically distinct, though mixed populations and intergrading forms occur not infrequently:

1   A majority of plants with lvs spotted on both sides, the spots often elongated parallel to the lf-margin; fls lilac-purple: highly calcareous peat in W. Ireland.   ssp. **cruenta** (O. F. Muell.) Vermeul.
Most or all plants with unspotted lvs.   *2*

2   Fls creamy or yellowish, lacking red or purple pigments; labellum unmarked, distinctly 3-lobed, the lateral lobes commonly ±notched: calcareous fens in E. Anglia and S. Wales.
                         ssp. **ochroleuca** (Boll) H.-Harr. f.
Fls with some red or purple pigment; labellum rarely with notched lateral lobes, often ±unlobed.   *3*

3   Fls flesh-pink.   *4*
Fls purple, magenta or red.   *5*

4   Plants usually not exceeding 40 cm. and with only 4–5 lvs; labellum less than 8 mm. wide, usually marked with lines: widespread.
                                   ssp. **incarnata**
Plants often much exceeding 40 cm. and with 6 or more lvs; labellum usually more than 9 mm. wide and marked with dots: fens in Norfolk and W. Ireland.   ssp. **gemmana** (Pugsl.) H.-Harr. f.

5   Fls magenta or purple: usually on mildly acid peat.
                   ssp. **pulchella** (Druce) H.-Harr. f. (incl.
                         O. latifolia var. *cambrica* Pugsl.)
Fls ruby- or crimson-red; plants commonly short (5–20 cm.): dune-hollows and dune-slacks.   ssp. **coccinea** (Pugsl.) H.-Harr. f.

Native. Ssp. *incarnata*, the commonest form, in wet meadows and marshes, less commonly on fen-peat; widely distributed in suitable habitats throughout Great Britain and Ireland. Ssp. *coccinea* occurs round the coasts of Wales, N. Britain and the Inner and Outer Hebrides. General distribution not certainly known but probably throughout Europe and the Near East and reaching 70° N. in Norway.

D. maculata agg.   Spotted Orchid.

Stem usually *solid*, 15–60 cm. Lvs usually spotted. *Transitional lvs* between the true lvs and bracts *several*, (1–)2–6 or more, erect and appressed to the stem. *Lower bracts less than 3 mm. wide, usually shorter than or equalling the fls*, sometimes exceeding them in large plants. *Infl. distinctly pointed above* until all the fls have opened. Colour of *fls* usually ranging from *bright rose-pink to white*, rarely darker; outer lateral per. segs ±horizontally spreading; labellum ± distinctly 3-lobed; spur ±cylindrical, straight. Fl. 6–8; usually the latest flowering of the palmate orchids. Visited by humble-bees, Syrphids and other Diptera and by beetles. Haploid or triploid embryos occasionally develop.

The aggregate is represented in Britain by two distinct spp. and by other named forms whose status is less clear.

2. D. maculata (L.) Vermeul. ssp. **ericetorum** (E. F. Linton) Vermeul. Moorland Spotted Orchid.

Orchis ericetorum (E. F. Linton) E. S. Marshall; ?*O. elodes* Griseb.

Stem 15–50 cm., slender. *Lvs* keeled and folded upwards, *all narrowly oblong-lanceolate to linear-lanceolate, ± acute, ± lightly marked with dark ± circular spots*, or unspotted. Transitional lvs usually 1–3. Labellum with central lobe much smaller and usually shorter than the broadly rounded, entire or notched lateral lobes; ± lightly marked with numerous small reddish dots and short lines on a pale pink or whitish background, the usual effect being of a pattern in dots rather than in lines (but sometimes with a bold and symmetrical line-pattern). *Spur 5–8 mm., very slender*, less than 1 mm. diam. in the middle. Fl. 6–8. $2n=80$. Grt.

Very variable, especially in size, shape and marking of the labellum, but rarely difficult to distinguish from *D. fuchsii*. A very pale form in Co. Clare with slightly glaucous unspotted lvs, very fragrant fls and buff anthers corresponds with *D. fuchsii* ssp. *okellyi* and like it is connected by intermediates with the typical form.

Native. The spotted orchid of moist acid peaty substrata throughout the British Is., to 3000 ft. in Scotland, and absent only from districts lacking suitable habitats. 85, H37, S. Ssp. *ericetorum* appears to occur in N. France, Belgium, Holland and Germany and at high altitudes in C. Scandinavia. The lowland form of S. Sweden is ssp. *maculata*, with wider lvs and larger spurs.

3. D. fuchsii (Druce) Vermeul.   Common Spotted Orchid.

Orchis fuchsii Druce; *O. maculata* var. *meyeri* Rchb. f.

Stem (5–)15–50(–70) cm. Lvs keeled but not or only slightly folded; *lowest lvs usually broadly elliptic to obovate-oblong, blunt*, the remainder oblong-lanceolate, subacute, becoming successively narrower up the stem; all marked with ± transversely elongated dark blotches, or occasionally unmarked. *Labellum distinctly and often deeply and narrowly divided into 3 subequal lobes, the middle lobe triangular and usually somewhat longer than the ± rhomboidal laterals*; usually clearly marked with a symmetrical pattern of ± continuous reddish lines on a paler pink or whitish background. Spur 5·5–8·5 mm.,

somewhat tapering downwards, 1·3–1·9 mm. diam. in the middle. Fl. 6–8.
2*n*=40*. Grt.

British forms may be grouped into 4 sspp. of which ssp. *fuchsii* is the most
widespread, the other 3 being much restricted in range.

Ssp. **fuchsii** Fls white to rose-pink; labellum 8–12 mm. wide, its lateral lobes
rhomboidal; spur 5·5–6·5 × 1·3–1·6 mm.

Ssp. **hebridensis** (Wilmott) H.-Harr. f. (incl. var. *cornubiensis* Pugsl.). Fls
pale rose to bright reddish-purple, mostly rosy magenta; *labellum* 10–15 mm.
wide, its *lateral lobes rounded-rhomboidal, crenate*; spur 7·0–8·5 × 1·6–1·9 mm.
Usually a stouter but shorter plant than ssp. *fuchsii*, 10–20(–35) cm., and
often with arcuate-recurved lvs. The labellum is broader and its middle lobe
often more rounded than in ssp. *fuchsii*. Outer and Inner Hebrides, Sutherland, W. Ireland, Cornwall. Some populations in the Inner Hebrides, W.
Ireland and Devon are intermediate between ssp. *hebridensis* and ssp. *fuchsii*.

Ssp. **okellyi** (Druce) Vermeul. Most plants with *unmarked fragrant white fls
and unspotted slightly glaucous lvs*; labellum 8–11 mm. wide; spur 5·5–6·5 ×
1·5 mm. On limestone in W. Ireland (Co. Clare) and N.W. Scotland (near
Inchnadamph). Some populations in W. Ireland grade into ssp. *fuchsii*.

Ssp. **rhoumensis** (H.-Harr. f.) Clapham, known only from Rhum, has lvs,
labellum and spur resembling those of *D. maculata* ssp. *ericetorum* but its
chromosome number, 2*n*=40*, is that of *D. fuchsii*.

Native. The spotted orchid of base-rich fens, marshes, damp meadows and
grassy slopes, and of woods on calcareous or other base-rich soils; to 1200 ft.
in Scotland. Ssp. *fuchsii* is common throughout S. and E. England and
reaches Caithness, but is local in Wales, N.W. England and Scotland;
common in Ireland; 85, H 33. The remaining sspp. are all of restricted western
distribution. Probably throughout most of Europe and Siberia except for
the Faeroes, Iceland and the extreme north. Triploid hybrids between
D. fuchsii and *D. maculata* ssp. *ericetorum* occur but are not common.

D. majalis agg.

Stem with a narrow cavity or almost solid. *Lvs broadest at a point above the
± closely sheathing base, not or slightly hooded* at the subacute tip. *Transitional
lvs rarely more than* 1, *often* 0. *Lowest bracts* 4–6 mm. *wide, exceeding the fls.*
Fls usually reddish or purple. Outer lateral per. segs ± erect. Labellum
entire or 3-lobed, often broader than long. Fl. 5–8. 2*n*=80*. Grt.

The British forms of this aggregate fall into 4 main groups which may be treated
as species in view of the correlations between morphology on the one hand and
geographical and ecological distribution on the other, but some populations are
intermediate or mixed.

4. D. praetermissa (Druce) Vermeul.   'Fen Orchid.'
Orchis praetermissa Druce

Stem 15–60 cm. or more. *Lvs* usually 6 or more, 10–20(–30) cm., the longest
averaging over 2 cm. and rarely less than 1·5 cm. in width; oblong-lanceolate,
broadest somewhat above the base and tapering gradually to a slightly hooded

tip; *unspotted*, rarely ring-spotted. Spike rather dense, with most of the fls exceeded by their bracts. *Fls* a fairly rich *reddish-lilac*, redder and paler than in *D. purpurella* or *D. incarnata* ssp. *pulchella*. *Labellum* 9–14 mm. wide, usually *wider than long, subentire or shortly* 3-*lobed*, usually ±concave through the upward curving of its sides, the lateral lobes becoming reflexed only at a late stage of flowering; marked in a central elliptic area with dots and fine lines. *Spur* 6–11 mm., sometimes curved, *conical, often very stout*, narrowing from 2–3·5 mm. diam. at its base to c. 1·5 mm. near the blunt ±oblique tip. Fl. 6–8. $2n = 80^*$. Grt.

Certain populations in S. and E. England contain individuals with heavily ring-spotted lvs but otherwise resembling typical *D. praetermissa*. These were the basis of *Orchis pardalina* Pugsl. They may be distinguished from the common hybrid *D. fuchsii* × *D. praetermissa* by their larger stem-cavity, smaller number of lvs, more acute lowest lf, less deeply lobed labellum and thicker spur, and by their chromosome number ($2n = 80^*$, in contrast with $2n = 60^*$ in the hybrid). Their labella are commonly more heavily marked than in *D. praetermissa* with unspotted lvs.

In localities where *D. praetermissa* and *D. fuchsii* both occur the hybrid between them may be found and may outnumber one or both parents. It is usually more robust than either parent, with heavily ring-spotted lvs and deeply marked labella, and is a strikingly conspicuous and beautiful plant. The hybrid *D. maculata* ssp. *ericetorum* × *praetermissa* is not infrequent, and locally there is evidence of hybridization between *D. praetermissa* and *D. traunsteineri*.

Native. Chiefly on wet calcareous or base-rich (fen) peat. Apparently confined to S. and E. England and Wales northwards to S. Lancs, Stafford and S. Yorks, where it is the common marsh orchid; Channel Is.; not in Scotland or Ireland. Distribution outside the British Is. not clearly known, but probably confined to parts of N.W. Europe from W. Germany to N. France.

5. D. purpurella (T. & T. A. Steph.) Vermeul.   'Northern Fen Orchid.'

Orchis purpurella T. & T. A. Steph.

Stem (5–)10–25(–40) cm., rigid. *Lvs* usually 5–8 in number, 5–10(–12·5) cm. long, the longest averaging c. 2 cm. and rarely less than 1·5 cm. in width, ±stiffly spreading, broadly lanceolate or oblong-lanceolate narrowing to the base and often slightly hooded at the subacute tip; dark green or grey-green, *unspotted or with rather few very small dark dots especially near the tip*, rarely more heavily spotted; lower internodes often very short. Spike short, broad, dense; bracts usually purplish. *Fls rich deep purple*, sometimes magenta or rose. *Labellum* 7–9 mm. wide, flat or slightly concave, *broadly diamond-shaped*, subentire or shortly 3-lobed; rather indistinctly marked with irregular dark reddish lines and spots. Spur 5–7·5 mm., stout, 2–3·5 mm. diam. at the base. Fl. 6–7. $2n = 80^*$. Grt.

In many Scottish and W. Irish populations the labellum is broader, more rounded and more distinctly 3-lobed and there are individuals with heavily spotted lvs, so that there is an approach to *D. majalis* ssp. *occidentalis*. Other populations, especially in fens in Scotland, resemble *D. praetermissa* in stature and in the absence of leaf marking (*Orchis praetermissa* var. *pulchella* Druce).

Native. Marshes, fens and damp pastures, especially on base-rich substrata. 47, H7. The northern counterpart of *D. praetermissa*, extending from Carmarthen, Derby and S.E. Yorks to the Hebrides and Shetland; in Ireland commonest in the north but extending southwards to S. Galway and E. Cork. General distribution unknown but occurs in S. Norway, S. Sweden and N. Denmark and perhaps also in N. Germany.

Hybrids of *D. purpurella* with *D. incarnata*, *D. praetermissa*, *D. majalis* ssp. *occidentalis*, *D. fuchsii*, *D. maculata* ssp. *ericetorum*, *Coeloglossum viride* and *Gymnadenia conopsea* occur not infrequently. *Orchis francis-drucei* Wilmott, with unspotted lvs, rich reddish-purple fls and a long tongue-like middle lobe of the labellum, is probably *D. purpurella* × *D. maculata* ssp. *ericetorum*.

6. D. majalis (Rchb.) Vermeul. ssp. **occidentalis** (Pugsl.) H.-Harr. f.   Irish Marsh Orchid.

Orchis occidentalis (Pugsl.) Wilmott; incl. *O. kerryensis* Wilmott

Stem 8–20(–30) cm. *Lvs* usually 6–8 in number, 4–15 cm., the longest averaging over 2 cm. and rarely less than 1·5 cm. in width; crowded, *spreading and mostly arcuate-recurved, rather broadly lanceolate*, at least the upper *distinctly narrowed into the base*, not or slightly hooded at the subacute tip; dark green or slightly glaucous, varying from unspotted to very heavily blotched or ring-spotted. Spike 3–5(–10) cm., dense-fld. *Fls dark purple to pinkish-mauve*. Outer lateral per. segs ± widely spreading. *Labellum* 9–12 mm. wide, *distinctly 3-lobed, with broad rounded* or bluntly angled often notched or incised *lateral lobes* and a *much smaller and narrower usually blunt central lobe*; marked with irregular dark lines and dots or only with dots. Spur 7–9 mm., almost cylindrical, 1–2 mm. diam. at the base, often strongly curved towards the tip. Fl. 5–6. $2n = 80^*$. Grt.

The British forms differ from continental *D. majalis* in being of shorter stature, in the short spike of deeply coloured fls and in the short central lobe of the labellum. There is a good deal of variability, especially in lf-marking and fl.-colour. Populations with a high proportion of plants having unspotted lvs, pale fls and rounded labella marked with dots rather than lines occur commonly in Cork and Kerry (W. Ireland) and were named *Orchis kerryensis* by Wilmott, but there is no clear morphological or geographical distinction. Populations intergrading with *D. purpurella* occur in W. Ireland, W. Scotland and the Hebrides.

Native. Marshes, fens, wet meadows, dune-hollows; the extreme western counterpart of *D. praetermissa*. In W. Ireland from Mayo southwards to Kerry and W. Cork; W. Sutherland; Outer and Inner Hebrides. Other sspp. of *D. majalis* occur widely in W. Europe and through C. Russia into Siberia and the Caucasus.

7. D. traunsteineri (Saut.) Vermeul.   Narrow-leaved Marsh Orchid.

Orchis traunsteineri Saut.; *O. traunsteinerioides* (Pugsl.) Pugsl.

Stem 15–45 cm., very slightly hollow, *slender*. *Lvs usually* 3–5, up to 15 cm., the longest averaging less than 1·5 cm. and rarely more than 2 cm. in width, *narrowly oblong-lanceolate or linear-lanceolate*, somewhat spreading to arcuate-recurved, slightly hooded at the blunt or subacute tip, *marked with small usually transversely elongated dark bars* or with rather faint ring-spots or unmarked. *Spike* variable but *commonly lax and few-fld*; bracts purplish. *Fls reddish-purple*, nearer the colour of *D. praetermissa* than of *D. purpurella*, though usually darker. Outer lateral per. segs slightly diverging from the vertical. *Labellum* 8–12 mm. wide, *narrow at the base and ± deltoid in outline*, flat or with the sides somewhat reflexed, 3-lobed, the bluntly triangular central lobe usually exceeding the lateral lobes; variously marked with fine dark purple lines and dots. Spur 8–10 mm., almost cylindrical, c. 2 mm. diam. at the base, straight. Fl. 5–7. $2n=80*$. Grt.

Native. Calcareous or very base-rich fens, marshes and dune-hollows. Distribution not yet fully known. In several scattered localities in Ireland from Wicklow and Antrim westwards to Clare, in N. Wales and in S. and E. England; perhaps also in N. England and S. Scotland. In some localities there is evidence of hybridization with *D. incarnata* and *D. fuchsii* and, especially in E. England, with *D. praetermissa*. Probably widespread in C., N. and E. Europe but taxonomic problems are not elucidated.

23. ANACAMPTIS Rich.

An *Orchis*-like herb with entire root-tubers, lfy stems and conical spikes of small fls. Outer lateral per. segs spreading. *Labellum* deeply 3-lobed *with 2 obliquely erect guide-plates* decurrent on its base from the lateral lobes of the column; spur long, slender, with no free nectar. Column short; *stigmas* 2, *on the rounded lateral lobes of the column* which are continued downwards as 'guide plates'; rostellum a centrally-placed pouch between the bases of the stigmas and partially closing the entrance to the spur. Anther adnate to the top of the column; pollinia 2, narrowed downwards into caudicles which are attached by their bases to a single transversely elongated narrowly saddle-shaped viscidium enclosed in the rostellum-pouch; pollen in packets of tetrads united by elastic threads.

One sp. in Europe and N. Africa.

Resembles *Gymnadenia* in the position of the 2 stigmas on lateral lobes of the column, and *Orchis* in the protection of the viscidium in a pouch, but differs from both in the attachment of both pollinia to a single viscidium, and in the 'guide-plates' at the base of the labellum (Fig. 68 B).

1. A. pyramidalis (L.) Rich.   'Pyramidal Orchid.'

Orchis pyramidalis L.

Root-tubers ovoid or subglobose. Stem 20–50(–75) cm., glabrous, slightly angled above, with 2–3 brown sheathing scales below. Lower lvs 8–15 cm. narrowly oblong-lanceolate, acute, keeled; upper lvs smaller, acuminate, grading into bracts; all unspotted. *Spike* 2–5 cm., *at first markedly conical*, dense-fld, with a foxy smell; bracts linear-lanceolate, slightly exceeding the ovary. Outer lateral per. segs broadly lanceolate, curved, spreading; outer median and inner lateral segments connivent into a hood; all deep rosy-purple, becoming paler. Labellum c. 6 mm., pale rose, broadly cuneate, with

3 subequal, oblong, truncate or rounded, ± entire lobes; 'guide-plates' convergent towards the base of the column. *Spur* c. 12 mm., *filiform*, acute, *equalling or exceeding the ovary*. Fl. 6–8. Pollinated by day- and night-flying Lepidoptera, the viscidium coiling tightly round their proboscides after withdrawal. $2n=40$. Grt.

Native. Grassland on chalk or limestone; calcareous dunes; to 800 ft. in England. 77, H40, S. Locally frequent in suitable habitats throughout Great Britain northwards to Fife; Colonsay (Inner Hebrides); Barra and Fuday (Outer Hebrides). Throughout Ireland. Europe northwards to S. Scandinavia and C. Russia; W. Asia; N. Africa.

141. ARACEAE

Herbs, often with tuberous or elongated rhizomes, rarely woody and climbing. Raphides often present. Fls small, usually crowded, arranged on a spadix which is generally enclosed in a spathe, either hermaphrodite or unisexual. Unisexual fls usually monoecious with the males on the upper part of the spadix, the females below, rarely dioecious. Perianth present in hermaphrodite fls, segments 4–6, free or connate into a truncate cup, usually 0 in unisexual fls. Stamens 2–4–8, anthers opening by pores or slits, free or united. Staminodes sometimes present. Ovary superior or sunk in the spadix, 1–many-celled. Fr. 1–many-seeded, fleshy or coriaceous.

About 115 genera and over 1000 spp., widely distributed throughout the world but by far the greatest number in the tropics.

1	Lvs without a clear division into lamina and petiole.	*2*
	Lvs with a well-marked lamina and petiole.	*3*
2	Lvs ensiform; spadix apparently lateral and spathe 0. 1. ACORUS	
	Lvs ovate-oblong; spadix terminal, spathe pale yellow, falling after fl.	
	LYSICHITON	
3	Lvs cordate; spathe flat, white inside (wet places). CALLA	
	Lvs hastate or sagittate; spathe pale green, closely wrapped round the lower part of the spadix (woods and hedgebanks). 2. ARUM	

1. ACORUS L.

Rhizomatous herbs with *linear lvs without petioles*. Scape flattened; *spathe* 0. *Spadix lateral*, terete, entirely covered with fls. *Fls all hermaphrodite*. *Per. segs* 6, free, membranous. Stamens 6. Ovary 2–3-celled, stigma sessile. Ovules orthotropous.

Two spp. in Europe, temperate Asia and America.

*1. A. calamus L.   Sweet Flag.

A stout glabrous rhizomatous aromatic perennial up to c. 1 m. Lvs 1–2 cm. wide, crowded, distichous, ensiform, acuminate, smelling of tangerines when bruised, midrib thick, margins waved. Scape reddish at base, ending in a long lfy point above the spadix. Spadix c. 8 cm., making an angle of c. 45°

with the scape, tapering upwards, obtuse at tip. Fls yellowish, tightly packed and completely covering the spadix. Fr. unknown in Britain. Fl. 5–7. $2n = 24, 36, 48$. Hel. or Hyd.

Introduced. In shallow water at margins of ponds, rivers and canals, local. 40. Native of S. Asia, C. and western N. America. Introduced into Europe by 1557 and recorded as naturalized in England by 1660.

*Lysichiton americanus** Hultén & St John, a rhizomatous marsh plant with sessile ovate-oblong lvs 30–70 cm. Spathes 10–15 cm., yellow, breaking off after fl. Fls all hermaphrodite. Fr. a green berry with 2 seeds. Fl. 4. Grown in gardens and sometimes naturalized. Western N. America.

*Calla palustris** L. A glabrous creeping rhizomatous perennial 15–30 cm. Rhizome stout, green, jointed and scaly. Lvs broadly cordate, cuspidate, entire; petiole with a long sheath. Spathe persistent, flat, not enclosing the spadix, oval, cuspidate, white within. Spadix stout, entirely covered with fls. Fr. a berry, red when ripe. Fl. 6–7. $2n = 36; 63, 69, 70; 72$. Hel.

Introduced. Naturalized in swamps and wet woods near ponds; first planted in 1861. Surrey. C. and N. Europe, N. Asia, N. America.

2. ARUM L.

Perennial herbs with tuberous rootstocks. Lvs net-veined, petioles sheathing at base. Spadix terminal, terete, *the upper part without fls; spathe convolute, margins not connate. Fls all unisexual*, female below, the upper sterile; male above, the upper sterile. Per. segs 0. Ovary 1-celled, stigmas sessile; ovules orthotropous.

About 20 spp. in Europe and especially the Mediterranean region.

Lvs appearing in spring, midrib dark green; terminal portion of spadix dull purple, rarely yellow; spathe twice as long as spadix (common).

                                                                    1. maculatum

Lvs well developed by December, midrib pale yellow-green; terminal portion of spadix always yellow; spathe three times as long as spadix (S. coast, very local).

                                                                    2. italicum

1. A. maculatum L.   Lords-and-Ladies, Cuckoo-pint.

An erect glabrous perennial 30–50 cm. Tuber c. 2 cm., a fresh one produced from the base of the stem each year. *Lvs appearing in spring*, long-petioled; blade 7–20 cm., often blackish-spotted, *triangular-hastate*, lobes acute or obtuse, *midrib dark green. Spathe 15–25 cm., erect*, pale yellow-green, edged and sometimes spotted with purple. *Spadix 7–12 cm., upper part dull purple*, rarely yellow. *Fr. c. 0·5 cm., scarlet*, fleshy; *fr. spike 3–5 cm.*, bursting the persistent base of the spathe. Fl. 4–5. Protogynous and cross-pollinated by small flies, particularly midges (for mechanism see Church, *Types of Floral Mechanism*, pp. 70–4). Fr. 7–8. $2n = 56^*, 84^*$. Gt.

Native. In woods and shady hedge-banks, especially on base-rich substrata, occasionally becoming a persistent weed in gardens; very shade-tolerant. Generally distributed throughout England, Wales and Ireland, less frequent in Scotland and perhaps not native in the extreme north. 91, H40, S. Europe northwards to S. Sweden; N. Africa.

2. A. italicum Mill.

Incl. *A. neglectum* (Towns.) Ridl.

An erect glabrous perennial 20–30 cm. Tuber c. 5 cm. Lvs well developed in December, long-petioled, not spotted, midrib pale yellow-green or white; winter lvs 15–30 cm., triangular-hastate; autumn lvs smaller. Spathe up to c. 40 cm., pale green, tip deflexed. Spadix c. ⅓ length of spathe, upper part yellow or orange-yellow. Fr. c. 1 cm., scarlet, fleshy; fr. spike c. 14 cm. Fl. 4–5, but rather later than *A. maculatum*. Fr. 8–9. $2n = 83^*, 84^*$. Gt.

The native plant with pale green lvs with yellow-green midrib has been separated as *A. neglectum* (Towns.) Ridl. The S. European plant, which is said to have dark green lvs and white midribs and to differ in other minor ways, is naturalized in a few localities.

Native. In light shade, often under brambles and in stony ground, usually within a mile of the sea. S. coast of England from Cornwall to W. Sussex, ?Glamorgan; Channel Is., very local. 8, S. Europe from Brittany southwards; N. Africa.

A. italicum × maculatum, ± intermediate between the parents but with spotted lvs, occurs occasionally.

142. LEMNACEAE

Small floating aquatic herbs. Roots simple or 0. Fls monoecious, naked, or at first enclosed in a sheath; perianth 0. Male fls consisting of 1–2 stamens with 1–2-celled anthers. Female fls consisting of a solitary, sessile, 1-celled ovary; ovules 1–7.

Two genera with about 22 spp., cosmopolitan in fresh waters.

Thallus ± flattened, with roots.   1. LEMNA
Thallus subglobose, rootless.   2. WOLFFIA

1. LEMNA L.   Duckweed.

Small aquatic herbs frequently forming a green carpet on the surface of stagnant water. Infl. minute, borne in a pocket in the margin of the thallus and consisting of 1 female and 2 male fls enclosed in a sheath. Anthers bilocular. Raphides occur in all spp., particularly abundantly in *L. minor* and *L. trisulca*. The thallus has been variously interpreted as a modified stem, a lf or partly lf and partly stem (see Arber, A. (1920), *Water Plants*, p. 73).

About 10 spp. distributed throughout the world.

1 Plants floating on the surface; thallus not stalked.   2
   Plants floating submerged; old thalli distinctly stalked.   **2. trisulca**

2 Several roots to each thallus.   **1. polyrhiza**
   One root to each thallus.   3

3 Thallus nearly flat on both sides.   **3. minor**
   Thallus convex, typically swollen below.   **4. gibba**

1. L. polyrhiza L. 'Great Duckweed.'

Spirodela polyrhiza (L.) Schleid.

Thalli 5–8(–10) mm. diam., flat and shiny, often *purplish below*, ovate or almost orbicular, *each with several roots* up to 3 cm.; thalli produce towards the end of summer purplish-brown reniform turions 2–4 mm. in diam., which become detached and often sink, rising to the surface in spring. Fl. 7 (recorded only from Somerset). $2n=40*$. Hyd.

Native. In still waters in ditches and ponds, local. England, except the extreme north and south-west; Wales: Glamorgan, Pembroke, Denbigh, Flint, Caernarvon and Anglesey; south Scotland, rare; Ireland: mainly in the east, but extending to Limerick, Galway and Fermanagh. 60, H17, S. Europe, north to 63°; Madeira; Africa; Asia; America; Australia; rather uncommon throughout most of its range.

2. L. trisulca L. 'Ivy Duckweed.'

Thalli submerged, translucent, tapering at the base into a stalk when mature, several thalli attached to each other by their stalks. Two young thalli arise on opposite sides of, at right angles to, and in the same plane as each old one. Thallus (5–)7–12(–15) mm., ± acute and serrulate at apex, narrowing abruptly to the stalk. Roots 1 to each thallus often ± hooked. Fertile thalli floating, smaller than sterile, pale green and often lacking the characteristic branching, ovate, producing stomata on the upper surface. Seed c. 1 mm., 12–15-ribbed. Fl. 5–7. $2n=44*$. Hyd.

Native. In ponds and ditches. England, except E. Cornwall and N. Devon; Wales; Scotland, north to Banff; Ireland except in the extreme south-west. 80, H36, S. Europe north to 68° 25′; N. Africa; Asia; N. America; Australia.

3. L. minor L. Duckweed, Duck's-meat.

Thalli floating, opaque, obovate or suborbicular, entire, subapiculate at point of attachment to parent thallus, 1·5–4·0 *mm. diam., nearly flat on both sides*, each with a single root up to 15 cm. Fls not uncommon, usually in shallow ditches fully exposed to the sun. Seed c. 0·6 mm. Fl. 6–7. Germ. 2. $2n=40*$. Hyd. By far the commonest sp.

Native. In still waters. Generally distributed throughout the British Is. except Sutherland and Shetland. 109, H40, S. Cosmopolitan except for the polar regions and tropics.

4. L. gibba L. 'Gibbous Duckweed.'

Thallus floating, convex above with reticulate markings just visible to the naked eye over most of the upper surface, typically strongly swollen beneath. Summer *thallus* 3–5 *mm.*, ovate, usually asymmetrical and rounded at base. Roots 1 to each thallus, up to 6 cm. Winter thalli dark green, rooted, not gibbous below, formed chiefly after flowering. Fls less frequently produced than in *L. minor.* $2n=64*$. Hyd.

Native. In still waters, local. England to N.E. Yorks and Lancashire, absent from extreme south-west; Wales: Monmouth, Glamorgan, Carmarthen, Caernarvon, Denbigh, Flint and Anglesey; Scotland: Ayr, Lanark, Midlothian, W. Lothian, Fife, Stirling and Caithness; Ireland: Wexford,

Waterford and Limerick, Dublin and Antrim to Fermanagh and Londonderry. 61, H16, S. Europe to 60° N., local in C. Europe, becoming commoner in the Mediterranean region; India, ascending to c. 10,000 ft. in Himalaya; extratropical Africa; N. and S. America, absent from tropics; Australia and New Zealand.

2. WOLFFIA Hork. ex Schleid.

Minute rootless floating aquatic herbs. Infl. borne in a hollow in the upper surface of the ovoid thallus, sheath 0; anthers unilocular.

About 12 spp., distributed throughout the world.

1. W. arrhiza (L.) Hork. ex Wimm.

Lemma arrhiza L.

Thallus 0·5–1·0 *mm.*, *ovoid to ellipsoid* or occasionally nearly globular, producing daughter thalli by budding from one end. Fl. unknown in Britain. No special resting thalli are produced, but the ordinary ones sink in winter. The smallest British flowering plant. $2n = c.$ 50*. Hyd.

Native. In still waters, rare. England: Somerset, S. Wilts, Sussex, Kent, Surrey, S. Essex and Middlesex; Wales: Glamorgan. 11. Europe north to 54°, local; Africa; Asia; America; Australia.

143. SPARGANIACEAE

Rhizomatous perennial aquatic herbs. Stems simple or branched, lfy. Lvs elongate-linear, sheathing at base, erect or floating. *Fls* unisexual, *crowded in separate globose heads* (capitula), the female towards the base in each infl. *Perianth of 3–6 membranous spathulate scales.* Male fls with 3 or more stamens, the filaments sometimes partially united; female with a 1-celled *sessile ovary. Fr.* dry, *indehiscent*, narrowed below, exocarp spongy. Wind-pollinated.

One genus with about 15 spp., generally distributed, but absent from Africa and S. America.

1. SPARGANIUM L.　　　　　　Bur-reed.

The only genus.

1 Infl. branched, male capitula on lateral branches; per. segs black-tipped; seed with 6–10 longitudinal ridges. **1. erectum**
　Infl. unbranched, all male capitula on the main axis; per. segs not black-tipped; seed smooth. *2*

2 Cauline lvs keeled at base, triangular in section; male capitula usually more than 3, distant. **2. emersum**
　Cauline lvs not keeled at base, flat in section; male capitula usually 1–2, crowded. *3*

3 Lf-like bract of lowest female capitulum 10–60 cm., at least twice as long as the whole infl.; male capitula usually 2. **3. angustifolium**
　Lf-like bract of lowest female capitulum 1–5(–8) cm., barely exceeding the infl.; male capitulum usually solitary. **4. minimum**

Subgenus 1. SPARGANIUM. Per. segs rather thick and firm, black-tipped. Seeds longitudinally ridged.

1. S. erectum L. Bur-reed.

An erect glabrous perennial (30–)50–150(–200) cm. Lvs usually all erect, triangular in section, 10–15 mm. wide. *Infl. branched* (rarely simple) with the *male capitula* borne above the female *on the branches*. Fl. 6–8. Hyd. or Hel.

Native. On mud or in shallow water in ponds, ditches and slow-flowing rivers and on ungrazed marshland. Common and widely distributed throughout the British Is., though not recorded from Cardigan and Shetland. 110, H40, S. N. temperate regions.

Four vars., differing in fr. shape and size and, to some extent, in distribution, occur in Britain. They may be distinguished as follows:

1 Fr. with a distinct shoulder, the upper part dark brown or black. 2
  Fr. spherical to ellipsoidal, without a distinct shoulder, uniformly light brown and shiny. 3
2 Fr. (5–)6–8(–10) mm. (excluding style) and (3–)4–6(–7) mm. wide at shoulder; upper part flattened. **var. erectum**
  Fr. 6–7(–8) mm. (excluding style) and 2·5–4·5 mm. wide at shoulder; upper part domed and wrinkled below style. **var. microcarpum**
3 Fr. 7–9×2–3·5 mm., ellipsoidal. **var. neglectum**
  Fr. 5–8×4–7 mm., ± spherical. **var. oocarpum**

Var. **erectum** (*S. ramosum* Huds.; *S. polyedrum* (Aschers. & Graebn.) Juzepczuk; ?*S. eurycarpum* Engelm.). Ovary usually bilocular. $2n = 30$. England south of the Wash. S. Sweden and Finland to the Mediterranean, east to C. Siberia.

Var. **microcarpum** (Neuman) C. Cook (*S. neglectum* ssp. *microcarpum* (Neuman) Aschers. & Graebn.). Ovary usually unilocular. Throughout the British Is. From the Arctic Circle to N. Africa and east to Siberia.

Var. **neglectum** (Beeby) Fiori & Paoletti. (*S. neglectum* Beeby). Ovary unilocular. $2n = 30$. In England common south of the Wash, but extending north to Westmorland. S. Sweden to N. Africa and east to the Caucasus.

Var. **oocarpum** (Čelak) C. Cook. Ovary unilocular, rarely bilocular; fertility low and perhaps var. *erectum × neglectum*. England south of the Wash. Distribution imperfectly known, but extending to N. Africa and Turkey.

Subgenus 2. XANTHOSPARGANIUM Holmberg. Per. segs thin, uniformly pale. Seeds smooth.

2. S. emersum Rehm. 'Unbranched Bur-reed.'

S. simplex Huds., p.p.

An erect or floating perennial 20–60 cm. Lvs 3–12 mm. wide, erect ones triangular in section; floating lvs usually present, keeled, sheathing at base but not inflated. *Infl. simple, with 3–10 remote male capitula on the main axis; anthers 6–8 times as long as broad*; female capitula 3–6, the lower often peduncled. Fr. 4–5 × 2–2·5 mm., ellipsoidal, often constricted in the middle. Fl. 6–7. 2*n* = 30. Hyd. or Hel.

Native. In shallow water in rivers, ponds and lakes, absent from very acid waters and from unsilted lakes. Throughout the British Is. north to Shetland with the exception of a few Scottish counties. 107, H 40, S. Throughout north temperate regions from the Arctic Circle to 40° N.

3. S. angustifolium Michx.   'Floating Bur-reed.'

S. affine Schnizl.

Perennial with long slender floating (rarely erect) stems. Lvs not keeled, sheathing and inflated at base. Infl. simple, with 1–2(–3) *male capitula* on the main axis; *anthers* 3–4 *times as long as broad*; female capitula 2–4, the lower peduncled. *Fr. c.* 8 *mm.*, ellipsoidal. Fl. 8–9. 2*n* = 30. Hyd.

Native. In peaty lakes, mainly in mountainous districts. Fairly widely distributed in the north and west of Great Britain, absent from much of the south and east and from most of C. Ireland. 53, H 25. Iceland to the Pyrenees and Alps, east to Japan; N. America.

4. S. minimum Wallr.   'Small Bur-reed.'

Stems 6–80 cm., usually floating. Lvs usually 2–6 mm. wide, flat, scarcely inflated at base. Infl. simple, *with* 1 (rarely 2 confluent) *male capitulum*; female capitula (1–)2–3, usually sessile. *Fr.* 3·5–4·5 *mm.*, obovoid. Fl. 6–7 2*n* = 30. Hyd.

Native. In lakes, pools and ditches on acid or alkaline substrata with a high proportion of organic matter. Throughout the British Is. in suitable habitats, north to Shetland. 75, H 37. Scattered throughout Europe and N. America, except for the Arctic.

144. TYPHACEAE

Stout rhizomatous herbs growing in shallow water. Stems erect, simple. Lvs distichous, coriaceous, or thick and spongy, elongate-linear, slightly spirally twisted, sheathing at base and mostly radical. *Fls* unisexual, very numerous, *densely crowded on a terminal spadix*, male above, female below. *Fls surrounded by slender jointed hairs or spathulate scales*, sometimes interpreted as per. segs. Male fls with 2–5 often monadelphous stamens; pollen often in tetrads; female with a 1-celled, *stipitate ovary*. *Fr.* dry, *at length splitting*. Wind-pollinated.

One genus with 9 spp., throughout the world from the Arctic Circle to 30° S.

1. TYPHA L.   Reedmace, Bulrush.

The only genus.

Lvs (7–)10–18(–22) mm. wide; male and female parts of spadix usually
  contiguous; female fls ebracteolate.   **1. latifolia**
Lvs c. 4 mm. wide; male and female parts of spadix usually distant; female fls
  bracteolate   **2. angustifolia**

1. T. latifolia L. 'Great Reedmace', Cat's-tail.

A robust perennial 1·5–2·5 m. *Lvs* (7–)10–18(–22) *mm. wide*, linear, over-topping the infl. *Male and female parts of infl. usually contiguous. Female fls ebracteolate.* Fr. cylindrical, tapering at base into a slender stalk. Fl. 6–7. Seeds shed 2–3. $2n = 30$. Hyd.

Native. In reed-swamps, often dominant, especially on inorganic substrata or where there is silting and rapid decay of organic matter, in lakes, ponds, canals and slow-flowing rivers. Generally distributed in suitable habitats throughout the British Is., though less frequent in the north. 103, H 40, S. From the Arctic Circle to 30° S., except for C. and S. Africa, S. Asia, Australia and Polynesia.

2. T. angustifolia L. 'Lesser Reedmace.'

A robust perennial 1–3 m. *Lvs* (3–)4–5(–10) *mm. wide*, convex on the back. *Male and female parts of infl.* about equal in length, *remote* (1–9 cm. apart), very rarely contiguous. *Bracteoles of female fls shorter than the stigmas.* Fl. 6–7. $2n = 30$. Hyd.

Native. In reed-swamps on ± organic soils in lakes, ponds, canals and slow-flowing rivers. Locally common, but much less generally distributed than *T. latifolia*; local and rare in Ireland. 71, H 15, S. Distribution of the genus. Our plant (ssp. *angustifolia*) has a wide distribution but is absent from America south of Louisiana and California and from Africa.

T. minima Hoppe has been reported from Britain from time to time since 1640, but none of these records has been substantiated and it is unlikely from its known distribution that the plant would occur in this country.

145. CYPERACEAE

Usually perennial, often rhizomatous herbs, most often growing in wet places. Stems usually solid, often trigonous. Lvs usually linear, some or all often reduced to sheaths. Ligule sometimes present. Fls hermaphrodite or uni-sexual, arising in the axil of a bracteole (glume) and arranged in 1–many-fld spikelets. Spikelets solitary, terminal, or grouped in branched or spike-like infl. often subtended by bracts. Female fl. sometimes enclosed by a modified glume (bracteole, perigynium) fused round it. Perianth of 1–many bristles or scales or, more often, 0. Stamens (1–)2–3(–6); anthers basifixed. Style simple; stigmas 3 or 2, linear, papillose. Ovary unilocular; ovule solitary, erect. Fr. indehiscent, globular or trigonous in plants with 3 stigmas, biconvex in plants with 2 stigmas. Seed erect, embryo small, endosperm abundant. Chromosomes often very small, little longer than broad. Fls wind-pollinated.

About 100 genera and 3000 spp., in all parts of the world.

The genera and spp. of Cyperaceae show a considerable diversity in both their vegetative and reproductive parts. Reduction appears to have occurred independently in many directions and any linear arrangement of genera is consequently bound to be artificial. The arrangement adopted here is based on the assumption that the following features are more likely to be primitive than the other variants which occur: lfy stems; many-fld spikelets; few or no sterile glumes; spiral arrangement of glumes; hermaphrodite fls; presence of perianth; 3 stamens; 3 stigmas; trigonous nut not enclosed by a perigynium or closely enfolding glume.

1   Fls all unisexual; male and female in separate spikes or separate parts
        of the same spike; female fls enclosed in perigynia or a closely enfolding
        inner glume.   *2*
        Fls predominantly hermaphrodite; female fls without perigynia or closely
        enfolding inner glume.   *3*

2   Female fls enclosed in perigynia, which often end in a beak (many spp.,
        several common).   15. CAREX
        Female fls without perigynia but closely enfolded by an inner glume
        (1 sp., very local).   14. KOBRESIA

3   Stem hollow; margins and keel of lf very rough, readily cutting the skin.
                                                                                    13. CLADIUM
        Stem solid; margins and keel of lf not rough enough to cut the skin.   *4*

4   Bristles long and silky, more than 6.   1. ERIOPHORUM
        Bristles shorter than glumes, rarely longer and then only 6.   *5*

5   Infl. of a solitary terminal spikelet; bract not forming an apparent
        prolongation of stem.   *6*
        Infl. of 2 or more spikelets, or (rarely) spikelet solitary but with a small
        stem-like bract beside it.   *8*

6   Water plant with elongate slender branched lfy stems.   9. ELEOGITON
        Bog or marsh plants; tufted or with creeping rhizome; stems nearly or
        quite lfless.   *7*

7   Uppermost sheath on flowering stems with a short blade.
                                                                                    2. TRICHOPHORUM
        Uppermost sheath on flowering stems without a blade, usually con-
        spicuous, rarely very thin and delicate.   3. ELEOCHARIS

8   Infl. with several flat or keeled lf-like bracts close together at base.   *9*
        Bracts not flat and lf-like or else solitary.   *10*

9   Spikelets terete, ovoid; glumes not distichous.   4. SCIRPUS
        Spikelets flattened, linear or oblong; glumes distichous (rare).
                                                                                    10. CYPERUS

10 Spikelets distichously arranged in a compressed oblong head.
                                                                                    6. BLYSMUS
        Spikelets spirally arranged, or solitary, infl. often apparently lateral,
        branched, or if a terminal head, then not compressed.   *11*

11 Infl. a compact blackish head, encircled at base by lowest bract; spikelets
        flattened; glumes distichous.   11. SCHOENUS
        Infl. reddish-brown or greenish; spikelets terete; glumes spirally arranged.   *12*

12 Stems lfy only at base or sometimes quite lfless; bracts stem-like; infl.
        apparently lateral.   *13*
        Stems lfy above base; bracts lf-like; infl. terminal.
                                                                                    12. RHYNCHOSPORA

13 Plant slender, seldom exceeding 15 cm.; infl. of 1–3 spikelets.
                                                                                    8. ISOLEPIS
        Plant stout, seldom less than 50 cm.; infl. of numerous spikelets.   *14*

14 Spikelets 5 mm. or more, few together, reddish-brown.
                                                    7. SCHOENOPLECTUS
    Spikelets 2–3 mm., crowded into dense globular heads, greenish (very
        local).   5. HOLOSCHOENUS

1. ERIOPHORUM L.   Cotton-grass.

Perennial rhizomatous herbs, either tufted or with a far-creeping rhizome.
Stems lfy. Lvs remaining green through the winter, but usually nearly or
quite dead at flowering time. Spikes many-fld, solitary or forming an umbellate infl. Fls hermaphrodite. Glumes spirally arranged, silver-slaty, membranous, lowest sterile. *Perianth of numerous bristles which elongate and
become cottony after flowering.* Stamens 3. Stigmas 3; style-base not swollen.
Nut compressed-trigonous.
    About 12 spp., mainly in arctic and north temperate regions.

1 Spikes several, ± nodding.   *2*
    Spike solitary, erect.   **4. vaginatum**
2 Plant tufted; lvs flat, uppermost without a ligule; bristles papillose at tip.
                                                        3. latifolium
    Plant not tufted; lvs channelled or involute, uppermost with a short
        ligule; bristles smooth at tip.   *3*
3 Stem subterete at top; peduncles smooth; glumes lanceolate, acuminate,
        1-nerved.   **1. angustifolium**
    Stem triquetrous at top; peduncles rough with short, forward-directed
        hairs; glumes ovate, subacute, usually several-nerved.   **2. gracile**

1. E. angustifolium Honck.   'Common Cotton-grass.'

E. polystachion L., p.p.

An *extensively creeping* rhizomatous perennial 20–60 cm. *Stem subterete*
(when fresh), smooth. *Lvs 3–6 mm. wide, channelled, narrowed into a long
triquetrous point. Uppermost lf with a ± inflated or funnel-shaped sheath and
a short ligule.* Spikes (1–)3–7, ± nodding. *Peduncles* simple, *smooth,* unequal.
Bracts several, shortly sheathing. *Glumes* c. 7 mm., 1-*nerved, lanceolate, acuminate,* brownish below, slate-coloured above, *margins broadly hyaline.*
Bristles up to c. 4 cm., *smooth at tip* (microscope). Nut 3 mm., obovoid-
trigonous, blackish-brown. Fl. 5–6. Fr. 6–7. $2n=58*$. Hel.
    Native. In wet bogs, shallow bog pools and acid fens. 112, H40, S.
Recorded from every vice-county in the British Is., but now very rare or even
extinct in some, owing to drainage; locally abundant, particularly in the north
and west. Europe (except the southern Mediterranean region), arctic regions,
Siberia, N. America.

2. E. gracile Roth

An extensively creeping rhizomatous perennial similar to *E. angustifolium* but
much more slender. Stems triquetrous (when fresh). Lvs 1–2 mm. wide,
short, obtuse. Spikes 3–5. Peduncles rough with short forward-directed
hairs. Glumes c. 5 mm., usually several-nerved, ovate, subacute, margins not
hyaline. Nut c. 2·5 mm., yellowish-brown. Fl. 6. Fr. 7–8. $2n=60, 76$. Hel.

Native. In wet acid bogs, rare and local. 6. S. Somerset, Dorset, Hants, Surrey, Northants; extinct in N.W. Yorks. C. and N. Europe, very local in the south and absent from the Mediterranean region and Hungarian plain.

3. E. latifolium Hoppe   'Broad-leaved Cotton-grass.'
E. paniculatum Druce

A *tufted* rhizomatous perennial 20–60 cm., similar in general appearance to *E. angustifolium*. Stems triquetrous (when fresh). Lvs 3–8 mm. wide, flat except for the short triquetrous point. *Uppermost lf without a ligule*, its sheath close-fitting and cylindrical. Spikes 2–12. *Peduncles* sometimes branched, *rough* with short, forward-directed hairs. *Glumes* 4–5 mm., 1-nerved, lanceolate, acuminate, blackish *with very narrowly hyaline margins. Bristles papillose at tip* (microscope). Nut 3–3·5 mm., narrowly obovoid-trigonous, reddish-brown. Fl. 5–6. Fr. 6–7. $2n=54$; 72. Hel.

Native. In wet places on base-rich soils. 78, H 24. Scattered throughout the British Is., but local and much less common than *E. angustifolium*. Throughout most of Europe, Asia Minor, Caucasus, Siberia, N. America.

4. E. vaginatum L.   Cotton-grass, Hare's-tail.

A *tussock-forming* rhizomatous perennial, 30–50 cm. Stems smooth, terete below, trigonous above. Lvs ± setaceous, up to 1 mm. wide, triquetrous. Stem-lvs 2–3, *blade* short, that *of uppermost lf almost* 0; *sheaths strongly inflated but narrowed at mouth. Spike solitary*, terminal, c. 2 cm. at flowering, ovoid or almost globular, *broad and rounded at base. Bracts* 0. Glumes c. 7 mm., 1-nerved, ovate-lanceolate, acuminate, silvery below, slaty-black above, translucent, middle ones 2 mm. broad. *Bristles* c. 2 cm., smooth, *pure white. Anthers* 2·5–3 *mm.* Nut 2–3 mm., rather broadly obovoid-trigonous, yellowish-brown. Fl. 4–5. Fr. 5–6. $2n=58$. Hel.

Native. In damp peaty places, especially on blanket bogs, locally abundant. 94, H 39. Throughout most of the British Is., but decreasing in the south and extinct in a number of counties. North temperate zone.

E. brachyantherum Trautv. & Mey., similar to *E. vaginatum* but with the upper stem-lf with a distinct blade, and yellowish bristles, has been reported from W. Ross, apparently in error.

2. TRICHOPHORUM Pers.

Perennial herbs. Stem terete or rarely trigonous. Transverse section of stem with air canals bounded by very thick-walled cells. *Lower sheaths lfless, only the uppermost with a short blade. Spikelet solitary, terminal*, the lowest glume generally fertile, though usually larger than the others. Fls hermaphrodite. Perianth bristles present, not more than 6, shorter or rarely longer than the glumes. Stamens 3. Stigmas 3.

Plant densely tufted; stems terete; bristles shorter than glumes.
                                                              2. cespitosum
Plant shortly creeping; stems trigonous; bristles longer than glumes (probably
    extinct).   **1. alpinum**

1. T. alpinum (L.) Pers.

Eriophorum alpinum L.; *Scirpus Hudsonianus* (Michx.) Fernald

A *shortly creeping* perennial 10–30 cm. *Stems* slender, *trigonous, rough.* Lvs setaceous, keeled. Spikelet 5–7 mm., 8–12-fld. Glumes obtuse, yellowish-brown, midrib green. *Bristles up to 2 cm., white, crumpled.* Nut 1 mm., obovoid-trigonous. Fl. 4–5. Fr. 6. $2n = 58$. Hel.

Closely resembles *T. cespitosum* when in fl., but may always be distinguished by the distinctly creeping rhizome.

Native. Formerly in a bog in Angus, but now apparently extinct through drainage. C. and N. Europe, Siberia, N. America.

2. T. cespitosum (L.) Hartman   Deer-grass.

Scirpus caespitosus L.

A *densely tufted* perennial 5–35 cm. *Stems* slender, *terete, smooth.* Lower sheaths lfless, light brown, shiny. Spikelet 3–6 mm., 3–6-fld. Glumes subacute, the two lower larger than the rest. *Bristles somewhat longer than fr. but shorter than glumes, brownish.* Nut c. 2 mm., ovoid, trigonous. Fl. 5–6. Fr. 7–8. Hs. or Hel.

Ssp. **cespitosum**

Basal sheaths shining; uppermost sheath (Fig. 70 A) fitting tightly round the stem (at least in fresh material), the opening c. 1 mm., hyaline margin narrow. Glumes brown with a yellowish-brown midrib, the lowest ending in a short, stout green point. $2n = 104$.

Ssp. **germanicum** (Palla) Hegi

T. germanicum Palla; *Scirpus germanicus* (Palla) Lindm.

Basal sheaths scarcely shining; uppermost sheath (Fig. 70 B) fitting loosely round the stem, the opening 2–3 mm., with broad hyaline margin. Glumes brown with a green midrib, the lowest ending in a stout, green, often almost lf-like, point which usually equals or exceeds the spikelet.

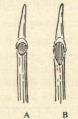

Fig. 70. Uppermost sheaths of *Trichophorum cespitosum.* A, ssp. *cespitosum*; B, ssp. *germanicum.* × 2·5.

Native. In damp acid peaty places, particularly blanket bogs and heaths, locally dominant. 104, H 40. The distribution of the sspp. is not known in detail, but ssp. *germanicum* is much the commoner; ssp. *cespitosum* is rare and its distribution is imperfectly known. The sp. is scattered throughout much of the British Is., but absent from base-rich soils. W. and N. Europe, local in C. Europe and rare in the south; Himalaya; N. America; Greenland.

3. ELEOCHARIS R.Br.

Perennial herbs. Stems terete or rarely 4-angular. Transverse section of stem with numerous approximately equal air canals without vascular bundles at the intersections of the strips of tissue separating the canals. *At least the upper sheaths entirely lfless.* Spikelet solitary, *terminal*, the lowest glume usually sterile, or at least different in shape from the others. Fls hermaphrodite. Perianth bristles present (rarely 0), shorter than or not much exceeding nut.

Stamens 3 or 2. Stigmas 3 or 2. Style usually with a swollen persistent base. The measurements of nut-length given in the descriptions exclude the style-base.

1 Lowest glume at least ½ as long as spikelet.   2
  Lowest glume much less than ½ as long as spikelet.   4
2 Upper sheath conspicuous, brownish; spikelet brown; style-base enlarged,
  persistent.   3
  Upper sheath very delicate and inconspicuous; spikelet greenish; style-base
  not enlarged.   **1. parvula**
3 Stems 4-angled; glumes 2 mm., obtuse, lowest sterile (wet sandy places,
  sometimes submerged).   **2. acicularis**
  Stems terete; glumes 5 mm., acuminate, lowest fertile (damp peaty places).
                                                                3. quinqueflora
4 Plant densely tufted; upper sheath obliquely truncate; stigmas 3; nut
  triquetrous.   **4. multicaulis**
  Plant not densely tufted; upper sheath almost transversely truncate;
  stigmas 2; nut biconvex.   5
5 Lowest glume not more than ½ encircling base of spikelet.   6
  Lowest glume ±completely encircling base of spikelet.   **7. uniglumis**
6 Bristles 4 (very rarely 0), style-base broader, obviously constricted at
  junction with nut.   **5. palustris**
  Bristles (4–)5(–6), style-base long and narrow, hardly constricted at
  junction with nut.   **6. austriaca**

1. E. parvula (Roem. & Schult.) Link ex Bluff., Nees & Schau.

Scirpus parvulus Roem. & Schult.; *S. nanus* Spreng., non Poir.

A slender glabrous perennial 2–8 cm. *Runners capillary, whitish*, forming small whitish tubers at their tips. Stems tufted, grooved, setaceous, soft; uppermost sheath very thin, brownish, lfless. Lvs setaceous, channelled, translucent and septate, about equalling stems. *Spikelet 2–3 mm.*, 3–5-fld. *Lowest glume sterile, more than ½ as long as spikelet*, encircling it at base. *Glumes* c. 2 mm., ovate, obtuse, *greenish-hyaline*. Bristles longer than fr. Stamens 3. Stigmas 3. Nut c. 1 mm., triquetrous, yellowish. Fl. 8–9. $2n=10*$. Hel.

Native. In wet muddy places near the sea, very local. 4, H3. S. Devon, Dorset; Merioneth, Caernarvon; N. Kerry, Wicklow, Londonderry. Coasts of the Atlantic and Mediterranean from C. Scandinavia southwards; N. and S. Africa, Japan, America.

2. E. acicularis (L.) Roem. & Schult.   'Slender Spike-rush.'

A slender rhizomatous perennial 2–10(–20) cm. *Rhizome creeping, brown. Stems few together, setaceous, 4-angled*, rather stiff; *sheaths brown, lfless. Spikelet 3–4 mm.*, 4–11-fld. *Lowest glume sterile, ½ as long as spikelet. Glumes* c. 2 mm., ovate, *obtuse, reddish-brown*. Bristles deciduous. Stamens 3. Stigmas 3. Nut c. 1 mm., finely longitudinally ribbed; swollen style-base separated from top of nut by deep groove. Fl. 8–10. $2n=56*$; 20; 30–38, 50–58. Hel.

Native. In wet sandy and muddy places at margins of lakes and pools, sometimes submerged and then usually sterile. 79, H25, S. Scattered throughout the British Is. north to Shetland, rather local. Europe, except the extreme north and south; N. Asia; Australia; N. and S. America

3. E. quinqueflora (F. X. Hartmann) Schwarz   'Few-flowered Spike-rush.'
Scirpus pauciflorus Lightf.; *E. pauciflora* (Lightf.) Link

A somewhat tufted glabrous perennial 5–30 cm. *Rhizome short, stout, producing slender runners.* Stems slender (usually less than 1 mm. diam.); sheaths lfless, pale reddish-brown, *upper rather obliquely truncate and obtuse.* Spikelet 5–7 mm., 3–7-fld. *Lowest glume fertile*, obtuse, *more than ½ as long as spikelet*, encircling it at base. *Glumes* 5 mm., ovate, *acuminate*, reddish-brown with broad hyaline margins. Bristles about as long as fr. Stamens 3. Stigmas 3. Nut 1·8–2·2 mm., trigonous, finely longitudinally ribbed, grey when dry, black when fresh; swollen style-base elongate, trigonous, slightly contracted at junction with nut. Fl. 6–7. Fr. 7–8. $2n=20^*$; 80, c. 100. Hel.

Somewhat resembles *Trichophorum cespitosum*, but is readily distinguished by the completely lfless sheaths.

Native. In damp peaty places on moors and in fens, always(?) with a fairly good supply of bases, local. 104, H35, S. Scattered throughout the British Is., rather commoner in the north than south. Europe, north to c. 71° in Scandinavia; temperate Asia; N. America.

4. E. multicaulis (Sm.) Sm.   'Many-stemmed Spike-rush.'

A densely tufted glabrous perennial 15–30 cm. Stems rather slender (c. 1–1·5 mm. diam.); sheaths all lfless, pale reddish or brownish, often straw-coloured, ±hyaline, *the upper very obliquely truncate and acute.* Spikelet c. 10 mm., many-fld, often proliferating vegetatively (viviparous). *Lowest glume sterile, less than ¼ as long as spikelet*, encircling it at base. Glumes c. 5 mm., ovate-oblong, obtuse, reddish-brown, ±hyaline, midrib green. Bristles shorter than fr. Stamens 3. *Stigmas* 3. Nut 1·2–1·5 mm., *triquetrous*, smooth, yellowish; swollen style-base forming a triquetrous cap to nut. Fl. 7–8. Fr. 8–10. $2n=20^*$. Hel.

Native. In wet, usually peaty places, particularly in acid bogs and on wet sandy heaths. 102, H31, S. Scattered throughout the British Is. but mainly in the south and west, locally common. S. and W. Europe, north to S. Scandinavia and east to Russia; western N. Africa; Azores.

5. E. palustris (L.) Roem. & Schult.   'Common Spike-rush.'
E. eupalustris Lindberg f.

A glabrous perennial 10–60 cm. *Rhizome far-creeping, producing single stems in the first season, then many small tufts.* Stems stout or slender (1–4 mm. diam.), reddish at base; sheaths all lfless, yellowish-brown, *the upper nearly transversely truncate.* Spikelet 5–20 mm., many-fld. Two *lowest glumes* sterile, much shorter than spikelet, *not more than ½ encircling base of spikelet.* Glumes ovate, margins hyaline. Bristles shorter or longer than fr., sometimes 0. Stamens 3. *Stigmas* 2. *Nut biconvex*, finely punctate or nearly smooth, yellow to deep brown; swollen style-base constricted at base. Fl. 5–7. Fr. 6–8. Hel.

Native. In marshes, ditches and at margins of ponds. 112, H40, S. Generally distributed and common throughout the British Is. Nearly cosmopolitan.

Ssp. **palustris**. Nut (1·1–)1·2–1·4(–1·5) mm. Glumes from middle of spike 2·75–3·5 mm., rather pale, often light brown and usually more easily detached than in ssp. *vulgaris*. Spike usually more densely-fld than in ssp. *vulgaris*. Stomatal length (middle of stem) (50–)55–60(–65)μ. $2n=16*$.

Ssp. **vulgaris** S. M. Walters. Nut (1·3–)1·45–1·8(–2·0) mm. Glumes from middle of spike 3·75–4·5 mm., variable in colour but usually ± dark brown. Stomatal length (middle of stem) (65–)70–80(–85)μ. $2n=38*$.

Common throughout the British Is. Widespread in Europe but rare or perhaps absent from the south and south-east.

Distribution imperfectly known, but apparently confined to south and east England and W. Midlands from Worcester, Wilts, and Sussex to Kent and Norfolk. Widespread in Europe but apparently absent from N. Scandinavia and extending farther south and east than ssp. *palustris*.

6. E. austriaca Hayek

Like *E. palustris* but stems weaker with fewer more widely-spaced vascular bundles, and with very little reddish coloration at base; spikelet short, dense, often conical; glumes small, readily caducous; ripe nut 1·2–1·4 mm., yellowish, with very narrow style-base, barely constricted at junction with nut; bristles (4–)5(–6), usually much exceeding nut, but very fragile.

Probably native. Marshy ox-bow of river Wharfe below Buckden, Yorks. 1. Alps, Carpathians, Caucasus, parts of European Russia to Urals, and parts of Siberia.

7. E. uniglumis (Link) Schult.

Incl. *E. Watsonii* Bab.

Differs from *E. palustris* as follows: Stems slender, usually shiny; lower sheaths reddish-tinged. *Lowest glume ± encircling base of spikelet*; glumes usually darker brown with narrower hyaline margins. Bristles frequently 0. Nut 1·4–2·2 mm., usually rather coarsely punctate-striate. $2n=16, 32; 40*$; $46*$; c. $68*$, $92*$. Hel.

A much less variable plant than *E. palustris*, often confused with that sp. and with *E. multicaulis*. From the former it may be distinguished by the lowest glume ± encircling the base of the spikelet, and from the latter by the obtuse sheath, 2 stigmas and biconvex nut.

Native. In marshes with rather open vegetation, not in ponds. 56, H14, S. Frequent near the coast; inland known definitely from Oxford, Berks, Cambridge and Yorks. Europe, W. Asia, N. Africa.

4. SCIRPUS L.

Stout perennial herbs with lfy stems. *Lvs with flat or keeled, well-developed blades. Infl. terminal, mostly much-branched.* Bracts several, lf-like. Perianth of 1–6 bristles, rough with backward-directed short hairs. Stigmas 3.

Lvs keeled; infl. dense; bracts much exceeding infl.; spikelets 10–20 mm., red-brown.   **1. maritimus**

Lvs flat; infl. lax; bracts about equalling infl.; spikelets 3–4 mm., green or greenish-brown.   **2. sylvaticus**

1. S. maritimus L. 'Sea Club-rush.'

A stout glabrous perennial, 30–100 cm. Rhizome producing short runners, tuberous at tip. Stems triquetrous, rough towards top, lfy. *Lvs* up to c. 10 mm. wide, *keeled*, margins rough. *Infl.* c. 5 cm., *dense*, corymbose. *Bracts* lf-like to setaceous, the larger *much longer than infl. Spikelets* 10–20 *mm.*, rather few, ovoid, *red-brown*, sessile or in groups of 2–5 at the ends of branches. *Glumes* c. 7 *mm.*, ovate, *apex bifid, awned from the sinus.* Bristles shorter than nut, brown. Stamens 3, filaments flattened. *Nut* 3 *mm., broadly obovate* from a cuneate base, *plano-convex, brown, shiny.* Fl. 7–8. Fr. 8–9. $2n = 86^*$; c. 104; 110. Hel.

Native. In shallow water at the muddy margins of tidal rivers and in ditches and ponds near the sea, locally abundant, rarely inland. 86, H27, S. Around the coasts of the British Is. north to E. Ross and Lewis. Cosmopolitan, except for the Arctic.

2. S. sylvaticus L. 'Wood Club-rush.'

A stout glabrous perennial 30–100 cm. Rhizome creeping. Stems trigonous, smooth, lfy. *Lvs* up to 20 mm. wide, *flat*, margins rough. *Infl.* up to 15 cm., *lax, spreading. Bracts* lf-like or the shorter ones setaceous, the larger *shorter than or about equalling infl. Spikelets* 3–4 *mm.*, very numerous, ovoid, *green or greenish-brown*, solitary or in small dense clusters at ends of branches. *Glumes* 1·5 *mm.*, ovate, *apex entire*, obtuse or subacute. Bristles equalling or longer than nut, brown, rough. Stamens 3, filaments flattened. *Nut* 1 *mm., ovoid or suborbicular, compressed-trigonous, yellowish, not shiny.* Fl. 6–7. Fr. 7–8. $2n = 64$; 62. Hel.

Native. In marshes, wet places in woods and beside streams, local. 84, H17. Scattered throughout the British Is. north to Banff and Argyll. Europe, except the extreme north and south; Caucasus; Siberia; N. America.

5. HOLOSCHOENUS Link

Tall perennial herbs. Infl. apparently lateral, the stem being continued by a $\frac{1}{2}$-terete bract. *Spikes several, globular, peduncled, each of numerous spirally arranged spikelets.* Perianth bristles usually 0. Stamens 3. Stigmas 3.

1. H. vulgaris Link

Scirpus Holoschoenus L.

A densely tufted perennial 50–100 cm. Stems terete, 2–4 mm. diam., smooth. Sheaths mostly without blades, the upper with $\frac{1}{2}$-terete rigid blades shorter than stem, margins rough. Infl. $\pm$ umbellate, subtended by (1–)2 long green bracts, one forming an apparent continuation of the stem. Heads 5–10 mm. diam., some peduncled, others sessile. Peduncles stout, flattened, margins rough. Glumes c. 2·5 mm., obovate, mucronate or almost 3-lobed, ciliate. Nut 1 mm., ovoid-trigonous, crowned by the persistent style-base. Fl. 8–9.

Native. On damp sandy flats by the sea, very local. N. Devon, N. Somerset, Glamorgan (extinct in Isle of Wight). Atlantic Europe from S.W. England to Portugal, Mediterranean region, S.E. Europe; Siberia; Canaries.

6. BLYSMUS Panz.

Perennial herbs. Stems subterete, lfy. *Infl. a terminal spike of several distichous spikelets.* Fls hermaphrodite, spirally arranged. Perianth of 3–6 bristles. Stamens 3. Stigmas 2; style persistent.

Lvs flat, keeled, rough; bracts (except lowest) many-ribbed, shorter than
    spikelet; glumes 3 mm., acute, reddish-brown.   **1. compressus**
Lvs involute, ± rush-like, smooth; bracts (except lowest) 1–3-ribbed, equalling
    spikelet; glumes 5 mm., obtuse, blackish-brown.   **2. rufus**

1. B. compressus (L.) Panz. ex Link   'Broad Blysmus.'

Scirpus planifolius Grimm; *S. Caricis* Retz.; *S. compressus* (L.) Pers., non Moench

A glabrous perennial with a far-creeping rhizome. Stems 10–35 cm., smooth. *Lvs* 1–2 mm. wide, rather shorter than stems, tapering from the base, *flat, keeled, margins rough.* Spikelets 5–7 mm., reddish-brown, 10–12 (rarely fewer). Infl. c. 2 cm. Lowest bract green, longer or shorter than infl., *other bracts several-ribbed, shorter than spikelet. Glumes* 3 mm., ovate, *acute, reddish-brown* with a pale midrib and narrow hyaline margin. *Bristles* brown, rough, *longer than nut, persistent.* Nut c. 2 mm., suborbicular, plano-convex, shortly stipitate, blackish. Fl. 6–7. Fr. 8–9. $2n=44$. Hel.

Native. In marshy places, usually in rather open communities, locally abundant. Scattered throughout England and S. Scotland to Midlothian and the Outer Hebrides; Caernarvon and Flint. 66. Europe, to c. 61° N.; temperate Asia.

2. B. rufus (Huds.) Link   'Narrow Blysmus.'

Scirpus rufus (Huds.) Schrad.

Similar to *B. compressus* but differing as follows: Lvs involute, ± rush-like, scarcely tapering, smooth. Spikelets dark brown, 5–8 in a spike. Bracts (except the lowest) 1–3-ribbed, equalling spikelet. Glumes 5 mm., ovate, obtuse, blackish-brown. Bristles white, short, caducous. Nut 4 mm., ovate, plano-convex, light brown. Fl. 6–7. Fr. 8–9. $2n=40$. Hel.

Native. Among short grass in salt-marshes, locally abundant. Scattered round the coasts of the British Is. from Suffolk and Kerry north to Shetland but very local in the south. 54, H 12. Denmark, N. Germany, Scandinavia to nearly 70° N. in W. Norway, the Baltic, the White Sea, Siberia.

7. SCHOENOPLECTUS (Rchb.) Palla

Mostly stout perennial herbs. *Stems* triquetrous or terete, *usually nearly or quite lfless.* Transverse section of stem with numerous approximately equal air canals; vascular bundles occur at the intersections of the strips of tissue separating the canals. *Lower bract making what appears to be a continuation of the stem beyond the infl.* Infl. apparently lateral, sessile, capitate or with short branches. Perianth of 6 (rarely fewer or 0) bristles rough with short downward-directed hairs. Stamens 3. Stigmas 3 or 2.

1 Stems triquetrous.   *2*
     Stems terete.   *3*

2 Upper sheath usually with a short blade; glumes with obtuse lateral lobes;
   bristles about equalling nut (very local).   **1. triquetrus**
   Two or three uppermost sheaths with blades up to 30 cm.; glumes with
   acute lateral lobes; bristles much shorter than nut (Jersey).
                                                             2. americanus

3 Stems not glaucous; glumes smooth (awn often papillose); stigmas
   usually 3.   **3. lacustris**
   Stems glaucous; glumes papillose on back at least near midrib; stigmas 2.
                                                             4. tabernaemontani

1. S. triquetrus (L.) Palla   'Triangular Scirpus.'

Scirpus triquetrus L.

A stout glabrous perennial, 50–150 cm. Rhizome creeping. *Stems triquetrous,
the upper sheath usually with a short blade*. Infl. a dense head, usually with
some lateral branches up to c. 4 cm. *Lower bract up to twice as long as mature
infl.* Spikelets 5–8 mm., ovoid, reddish-brown. *Glumes* 4 mm., broadly ovate,
fringed, brownish-hyaline, shallowly emarginate *with rounded lateral lobes*,
mucronate; midrib green. *Bristles about equalling nut*. Stigmas 2. Nut
2·5 mm. compressed, reddish-brown, shiny. Fl. 8–9. $2n=40$. Hel.

Native. Muddy banks of tidal rivers, very local. E. Cornwall, S. Devon;
Sussex, Kent, Surrey, Middlesex, probably extinct; Ireland; R. Shannon.
7, H2. S. and C. Europe (local in the north of its range); W. Asia; N. and
S. Africa; N. America.

2. S. americanus (Pers.) Volkart   'Sharp Scirpus.'

Scirpus americanus Pers.; *S. pungens* Vahl

A rather slender glabrous perennial 30–60 cm. *Stems triquetrous. Lvs up to
30 cm.*, 2–3, linear. Infl. a dense head of 2–6 sessile spikelets. *Lower bract*
up to 15 cm., *much exceeding infl.* Glumes 4 mm., broadly ovate, emarginate
with acute lateral lobes, mucronate. *Bristles much shorter than nut*, often 0.
Stigmas 2. Nut 2·5 mm., compressed, brownish, shiny. Fl. 6–7. Fr. 8–9.
$2n=$ c. 80. Hel.

Native. Margins of ponds near the sea, very local. Jersey, ?Lancashire.
S.W. Europe (to Venice, Belgium, the Netherlands and N. Germany);
America; Australia.

3. S. lacustris (L.) Palla   Bulrush.

Scirpus lacustris L.

A stout glabrous perennial 1–3 m., rarely less. Rhizome creeping, often producing tufts of submerged lvs. *Stems* up to 1·5 cm. diam., *terete, green*, the
upper sheath often with a short blade. Infl. a dense head or, more often,
with several branches up to c. 7 mm. *Lower bract usually shorter than mature
infl.* Spikelets 5–8 mm., ovoid, reddish-brown. *Glumes* 3–4 mm., broadly
ovate, often fringed, emarginate *with* rounded lateral lobes and a short, *often
papillose awn, otherwise smooth. Anthers bearded at the tip. Stigmas usually 3.*
Nut 2·5 mm., bluntly trigonous, grey-brown. Fl. 6–7. Fr. 8–9. $2n=38, 40,
42$. Hyd.

Native. In rivers, lakes and ponds, usually where there is abundant silt.
Generally distributed throughout the British Is., though apparently less

common in Wales than elsewhere. 104, H40. Europe (to Lapland), Asia, Africa, N. and C. America, Polynesia, Australia.

S. × carinatus (Sm.) Palla (*S. lacustris × S. triquetrus*) occurs occasionally with the parents and persists in Kent and Sussex, where *S. triquetrus* can no longer be found. It may be recognized by the stems being trigonous towards the top and terete below.

4. S. tabernaemontani (C. C. Gmel.) Palla   'Glaucous Bulrush.'
Scirpus Tabernaemontani C. C. Gmel.

Similar in general appearance to *S. lacustris*. Stems 50–150 cm., rather slender, glaucous. Glumes densely beset with small dark brown papillae on the back, specially near the midrib. Anthers not or scarcely bearded. Stigmas 2. Nut 2 mm., biconvex. Fl. 6–7. Fr. 8–9. $2n = 42$. Hel. or Hyd.

Native. In streams, ditches, pools and bogs, often in peaty places and specially near the sea. Scattered throughout the British Is., but less frequent and generally distributed than *S. lacustris*. 79, H26, S. Europe, except N. Russia, the Arctic and Portugal; temperate Asia.

8. ISOLEPIS R.Br.

Slender herbs. Stems terete. Lvs few, filiform, channelled. *Infl. apparently lateral*, a subterete green bract appearing as a prolongation of the stem. *Spikelets small*, 1–3 *together*. Fls spirally arranged, hermaphrodite. Perianth bristles 0. Stamens 2. Stigmas 3; style-base persistent.

Bract distinctly longer than infl.; nut longitudinally ribbed, shiny.   **1. setacea**
Bract shorter than or only slightly exceeding infl.; nut smooth, not shiny.
                                                                      2. cernua

1. I. setacea (L.) R.Br.   'Bristle Scirpus.'
Scirpus setaceus L.

A slender tufted perhaps annual herb with filiform stems 3–15(–30) cm. Lvs 1–2, usually shorter than stems. *Bract much exceeding infl.* Spikelets usually 2–3, up to 5 mm., rarely more, ovoid. Glumes 1·25 mm., ovate, mucronate, purple-brown, midrib green, margins hyaline. *Nut* 0·75 mm. *trigonous-obovoid, dark brown, ± shiny, longitudinally ribbed.* Fl. 5–7. Fr. 6–9. $2n = 26*$; 28. Hel. or Th.

Native. In damp places, sometimes among taller herbage in marshy meadows, more often in bare sandy or gravelly places and beside lakes. 111, H39, S. Not uncommon and recorded from every vice-county in the British Is., except Huntingdon and Kildare. Europe, except the extreme north and south; N. and S. Africa; N. and W. Asia; Australia.

2. I. cernua (Vahl) Roem. & Schult.   'Nodding Scirpus.'
Scirpus cernuus Vahl; *S. filiformis* Savi, non Burm. f.; *S. pygmaeus* (Vahl) A. Gray, non Lam.; *S. Savii* Seb. & Mauri

Similar to *I. setacea* but differs as follows: Bract shorter or slightly longer than infl. Spikelets usually solitary. Glumes greenish, often with a pale brown or reddish-brown (not purplish) spot on either side of the midrib. Nut broadly trigonous-obovoid or suborbicular, reddish-brown, smooth, not shiny. Fl. 6–8.

Native. In wet places, especially in ± bare sandy or peaty habitats near the sea, local. 34, H26, S. Cornwall to Hants and N. Somerset, E. Suffolk, E. and W. Norfolk, S. Lancashire, Isle of Man; coastal counties of Wales (?except Cardigan and Denbigh); Wigtown to W. Inverness including the adjacent islands, E. Ross and Lewis; in all the maritime counties of Ireland, but rarely far from tidal water. Western and southern Europe from France southwards; N. Africa.

9. ELEOGITON Link

Similar to *Isolepis*, but stem elongated, lfy, usually devoid of lfless sheaths, and spike always solitary, terminal and not over-topped by a stem-like bract.

1. E. fluitans (L.) Link   'Floating Scirpus.'

Scirpus fluitans L.

A slender, floating lfy perennial 15–40 cm. Stems compressed, branched. Lvs up to c. 5 cm., less than 1 mm. wide. Spike 2–3 mm., narrow-ovoid, 3–5-fld, solitary, terminal. Peduncles up to c. 5 cm., smooth, terete below, triquetrous above. Glumes c. 2 mm., ovate, subacute, pale greenish-hyaline. Bristles 0. Stamens 3. Stigmas 3. Nut 1·5 mm., obovate, compressed-trigonous, pale, crowned by the persistent style-base. Fl. 6–9. Fr. 7–10. $2n = 60$. Hyd.

Native. In ditches, streams and ponds, particularly those with peaty water. Widely distributed but local. 100, H36, S. Atlantic western Europe from S. Sweden and Denmark southwards; Italy; Africa; Asia; Australia.

10. CYPERUS L.

Perennial rhizomatous or rarely annual herbs. Stems usually lfy, sometimes winged. *Infl. umbellate or capitate. Bracts lf-like.* Spikelets many-fld. *Fls distichous*, hermaphrodite, rarely some male or lowest sterile. Perianth 0. Stamens 3, sometimes reduced to 2 or 1; filaments not elongating after flowering. Stigmas 3, sometimes reduced to 2. Nut trigonous or, in spp. with 2 stigmas, lenticular.

About 500 spp. in all the warmer parts of the world, rare in colder regions. *C. papyrus* L. is the source of the papyrus of the ancient Egyptians. The swollen tuberous rhizomes of several spp. are used as food.

Perennial 50–100 cm.; lvs 4–7 mm. wide, about equalling stems; infl. a
   compound umbel.   **1. longus**
Annual 5–20 cm.; lvs 1–3 mm. wide, shorter than stems; infl. subcapitate or
   a small dense umbel.   **2. fuscus**

1. C. longus L.   Galingale.

An erect glabrous *perennial 50–100 cm.* Rhizome shortly creeping, sympodial, not or scarcely swollen. Stems trigonous, smooth. *Lvs 4–7 mm. wide, about equalling stems. Infl. a compound umbel.* The larger bracts much longer than infl. *Primary branches* very unequal, the longer *c.* 10 *cm.*, each surrounded at base by a cylindrical, loose-fitting membranous sheath obliquely truncate at mouth. Bracteoles small, setaceous. Secondary branches surrounded at base by sheaths similar to those of primary branches. Spikelets c. 1 cm., linear or linear-lanceolate, compressed, distichous, crowded. Lower glumes sterile, c. 1 mm., ovate, ± truncate, hyaline. *Fertile glumes 2·5 mm.*, distichous, *closely imbricate*, lanceolate to ovate, *obtuse, keeled, reddish-brown*, keel green, *margins* hyaline towards base and *strongly decurrent forming a delicate wing to the rhachilla.* Stamens 3. Stigmas 3. Nut c. 1 mm., compressed-trigonous, reddish-brown. Fl. 8–9. Hel.

Native. In marshy places beside ponds and in ditches, very local. 17, S. Cornwall, N. Devon, N. Somerset, Dorset, S. Wilts, Hants, Isle of Wight, Berks, Oxford, Bucks, Surrey, Kent, Pembroke, Flint; Channel Is. Mediterranean region and in a few scattered localities north to the Lake of Geneva and Bodensee.

2. C. fuscus L.   'Brown Cyperus.'

A glabrous *annual* 5–20 *cm*. Stems triquetrous, soft, smooth. *Lvs 1–3 mm. wide, usually shorter than stems*, soon withering. *Infl. subcapitate or a small dense umbel with few short branches up to c.* 1 *cm*. Bracts 2–4 mm. wide, much exceeding infl. Branches with very loose membranous sheaths split almost to base. Spikelets up to c. 5 mm., oblong, compressed. *Glumes* usually all fertile, *spreading, scarcely imbricate, c.* 1 *mm.*, ovate, *acute, scarcely keeled, dark brown* with broad reddish-brown middle portion, *margins not hyaline nor decurrent.* Stamens usually 2. Stigmas 3. Nut 0·75 mm., compressed-trigonous, yellowish. Fl. 7–9. Fr. 8–10. $2n = $ c. 72. Th.

Native. In damp places, especially on bare ground left by the drying up of ponds and ditches, very local. 6. N. Somerset, Dorset, S. Hants, Surrey, Berks, Bucks; extinct in Middlesex. C. and S. Europe from Denmark and Gotland southwards; Asia; N. Africa; Madeira.

11. Schoenus L.

Perennial herbs. Stems terete. Upper sheaths with longer or shorter blades. *Spikelets 1–4-fld, in compressed, bracteate, terminal heads.* Bract of lowest spikelet encircling base of whole infl. *Glumes distichous, several lower sterile.* Fls hermaphrodite. Perianth of 1–6 bristles. Stamens 3. Stigmas 3; style somewhat thickened at base, usually deciduous. Nut trigonous or nearly globular.

About 70 spp., mainly in Australia and New Zealand, a few in Europe, Asia and America.

Lvs at least ½ as long as stems; lf-like point of lowest bract 2–5 times as long
    as infl.   **1. nigricans**
Lvs not more than ⅓ as long as stems; lf-like point of lowest bract shorter than
    or scarcely exceeding infl.   **2. ferrugineus**

1. S. nigricans L.   Bog-rush.

A densely tufted glabrous perennial 15–75 cm. Stems smooth, tough, wiry, lfy only at base. *Lvs at least ½ as long as stem*, wiry, subterete, margins involute. Lower sheaths dark reddish-brown or almost black, tough, shiny. Infl. 1–1·5 cm., dense, ovoid, blackish. *Lowest bract with a lf-like point 2–5 times as long as infl. Spikelets 5–8 mm.*, flattened, 5 *or more in an infl.* Glumes distichous, keeled, slightly rough on keel. Bristles 3–5, shorter than nut. Nut 1·5 mm., white, shiny, ± globular. Fl. 5–6. Fr. 7–8. $2n = 44^*$. Hs.

Native. In damp, usually peaty, base-rich places especially near the sea, sometimes in salt-marshes; on blanket bog in W. Ireland; locally abundant. 82, H 40, S. Widely distributed throughout the British Is. from Cornwall to Shetland. S. and C. Europe to Denmark, S.E. Norway (?extinct), Gotland and the Baltic States; S. Africa; America.

2. S. ferrugineus L.

Similar to *S. nigricans* but smaller and more slender. Stems 10–40 cm. Lvs never more than ⅓ as long as stem, usually much less. Infl. up to c. 1 cm., rather lax and narrow. Lowest bract with a lf-like point shorter or very little longer than infl. Spikelets 1–3 in an infl. Glumes quite smooth on keel. Bristles 6, longer than nut. Nut trigonous. Fl. 7. $2n=76$. Hs.

Native. At the wet peaty margin of Loch Tummel, mid Perth, in base-rich flushes, very local (?extinct). 1. C. Europe, north to Scandinavia, France, northern Balkans, Russia.

12. RHYNCHOSPORA Vahl

Lfy perennial herbs. Infl. of a compact terminal head or short spike with or without 1 or 2 ± distant, long-peduncled lateral heads. Bracts lf-like, the lower sheathing. *Spikelets 1–2-fld.* Fls hermaphrodite or the upper sometimes unisexual. *Glumes imbricate, several sterile. Perianth of 5–13 bristles.* Stamens 3 or 2. Stigmas usually 2, rarely 3; style-base enlarged, persistent, forming a beak to the nut. Nut biconvex or rarely trigonous.

About 150 spp., cosmopolitan but mainly tropical.

Plant without a creeping rhizome; spikelets whitish or pale brown; bracts not
    or little longer than terminal head; bristles 9–13.   **1. alba**
Plant with a creeping rhizome; spikelets dark reddish-brown; bracts 2–4 times
    as long as terminal head; bristles 5–6.   **2. fusca**

1. R. alba (L.) Vahl   'White Beak-sedge.'

A slender glabrous ± tufted perennial 10–50 cm. Stems terete or trigonous above, lfy. Lvs shorter than or about equalling stems, narrow, channelled, margins rough. *Lower sheaths* lfless, *often bearing bulbils in their axils. Bracts not or slightly exceeding terminal head. Spikelets* 4–5 mm., *whitish*, becoming pale reddish-brown, usually 2-fld. Terminal cluster of spikelets as broad as or broader than long. Glumes usually 4–5, 1 or 2 lowest sterile, 2–3 mm., the next fertile, the next sterile and the upper fertile, 4–5 mm. *Bristles 9–13*, shorter than or equalling fr. (including beak), rough with minute downward-directed hairs. Nut 1·5–2 mm., obovate, biconvex or trigonous, beaked. Fl. 7–8. Fr. 8–9. $2n=26$; 42. Hel.

Native. In wet, usually peaty places on acid soils, local. 73, H 37. Scattered throughout the British Is. from Cornwall to Shetland. Europe, except the Arctic and Mediterranean region; Siberia.

2. R. fusca (L.) Ait. f.   'Brown Beak-sedge.'

A slender glabrous perennial 10–30 cm., similar to *R. alba* but differing as follows: Rhizome far-creeping. Lvs usually much shorter than stems. Lower sheaths mostly with short blades, not bearing bulbils in their axils. Bracts 2–4 times as long as terminal heads. Spikelets dark reddish-brown. Terminal cluster of spikelets usually longer than broad. Bristles 5–6, longer than fr., rough with minute upward-directed hairs. Beak of fr. minutely pubescent. Fl. 5–6. Fr. 8–9. $2n=32$. Hs.

Native. On damp peaty soils on heaths and at margins of bogs, rare and local. 12, H22. S. Devon, N. Somerset, Dorset, S. Wilts, S. Hants, Surrey, Shropshire; Glamorgan, Cardigan; Kirkcudbright, W. Inverness, Argyll; W. Ireland from W. Cork to Donegal and east to Tipperary, Kildare and Cavan. N. and W. Europe (not the Arctic), Italy, S. and C. Russia.

13. CLADIUM P.Br.

Perennial rhizomatous herbs. Stems terete or nearly so, usually lfy. *Spikelets terete, few-fld.* Fls hermaphrodite, or the upper or lower sometimes male. *Glumes* imbricate, *lower sterile. Perianth* 0. Stamens 2–3. Stigmas 2–3; *style-base swollen*, deciduous. Nut trigonous or nearly globular.

About 30 spp. in tropical and temperate regions.

For an account of the autecology of our sp. see Conway, V. M., *New Phytol.* XXXV–XXXVII (1936–8).

1. C. mariscus (L.) Pohl   Sedge.

C. jamaicense Crantz; *Mariscus Mariscus* (L.) Borbás

A stout harsh perennial 70–300 cm. Rhizome creeping. Stems hollow, terete or bluntly trigonous, 1–4 cm. diam. at base (including sheaths). Lvs up to 2 cm. wide, evergreen, growing from base and dying away at top, tough, grey-green, keeled, serrate on margins and keel, ending in a long triquetrous point. Sheaths yellowish-brown, not shiny, very tough. Infl. much-branched, each branch terminated by a dense head of 3–10 spikelets, 5–10 mm. diam. Spikelets 3–4 mm., 1–3-fld, reddish-brown. Glumes smooth, rounded on back, imbricate, lower 2–3 small and sterile, the remainder fertile. Uppermost fl. sometimes male. Stamens usually 2. Stigmas usually 3; style-base much enlarged, deciduous. Nut 3 mm., ovoid, acuminate, dark brown, shiny. Fl. 7–8. Fr. 8–9. $2n=36$. Hel.

Native. Forming dense pure stands in reed-swamp and in fens, usually on neutral or alkaline soils, locally abundant. 54, H33, S. Thinly scattered over the British Is. north to Argyll and the Hebrides; frequent only in Norfolk and in W. and C. Ireland. Widely distributed in the warmer and damper parts of both hemispheres.

14. KOBRESIA Willd.

Perennial herbs. Lvs mostly basal. *Spikelets 1-fld, unisexual,* arranged in spikes, the male fls at top, the female below. *Female spikelets with 2 glumes, the inner enfolding but not fused round the flower.* Perianth 0. Stamens 3. Stigmas 3; style-base not enlarged. Nut trigonous.

About 15 spp. in north temperate regions of the Old World.

1. K. simpliciuscula (Wahlenb.) Mackenzie

K. caricina Willd.; *K. bipartita* auct.

A tufted glabrous perennial 5–20 cm. Stems trigonous, ± rough, stiff, erect. Lvs plicate, shorter than stems. Infl. up to 2 cm., of 4–10 crowded spikes.

Spikes male at top, female below. Stigmas 3. Nut 3 mm., trigonous, fusiform, acuminate. Fl. 6–7. Fr. 7–8. $2n=72$. Hs.

Native. On moors and damp banks in mountain regions, very local. 6. N.W. Yorks, Durham, Westmorland; Argyll, Perth. Pyrenees, Alps, Scandinavia (66° N.); Caucasus, Altai; Greenland; N. America.

15. CAREX L.

Perennial rhizomatous herbs. Stems solid, usually lfy, often triangular in section. Lvs usually linear, ± keeled or involute, less often flat, sometimes lanceolate with a distinct petiole; lf-base usually sheathing; sheaths entire; ligule present at junction of lf and sheath, 0 in spp. with petioled lvs. Infl. various, from a much-branched panicle to a simple spike. *Fls unisexual, borne in* 1-*fld spikelets*, each subtended by a glume. Male fls with 2–3 stamens; perianth 0. *Female fls surrounded by a globular, trigonous or compressed sac* (*perigynium*) usually crowned by a longer or shorter beak from which the stigmas project. Ovary trigonous and stigmas 3, or biconvex and stigmas 2. Fr. a trigonous or biconvex nut enclosed within the perigynium. Axis of spikelet occasionally prolonged beyond base of ovary (e.g. *C. microglochin*).

The male and female fls are variously arranged in the infl. Our spp. are, with one exception, monoecious. In the majority of spp. the terminal spike and sometimes some of the upper lateral spikes are male and the rest female. The female spikes frequently have a few male fls at the top and the male spikes less frequently a few female fls at the base. The other spp. have male and female fls in the same spike, the male fls being either at the top or base of the spike.

Probably about 1000 spp. throughout the world.

In the following descriptions the fr. includes the nut and the perigynium surrounding it, and measurements of length include the beak. The following arrangement of spp. has been adopted, putting those first which show the greatest number of characters which there is reason to believe are primitive. Our Carices fall into at least 34 sections; it is probable that the majority of these groups are not very closely related and consequently the sequence followed is often arbitrary.

Subgenus 1. CAREX.

Spikes several, unbranched, often peduncled, differing markedly from each other and normally unisexual. Peduncles subtended by sterile, often sheathing, bracts. Stigmas 3 or 2. (Sections 1–21.)

Section 1. *Elatae* Kükenth. Spikes 4–8(–many), upper 1–3 male (rarely male at base, female at top), others female (sometimes female at base, male at top), long-cylindrical, dense-fld, lower distant, long-peduncled. Bracts lf-like, sheathing. Female glumes brownish, lanceolate to ovate, acuminate or emarginate with ciliate apex. Perigynia elliptic to suborbicular and ± inflated, glabrous, many-ribbed; beak long, bifid. Nut trigonous. Stigmas 3. (Sp. 1.)

Section 2. *Spirostachyae* Kükenth. Spikes 3–5, terminal male, lateral female, oblong-cylindrical, dense-fld, distant, lower peduncled. Bracts lf-like, long-sheathing, erect. Female glumes purplish, or greenish- or reddish-brown, ovate, mucronate or acute. Perigynia ovoid or elliptic-trigonous, ± inflated, glabrous, many-ribbed, narrowed into a rather long, straight, bifid beak. Nut trigonous. Stigmas 3. (Spp. 2–5.)

Section 3. *Flavae* Christ. Spikes 2–6, terminal male (rarely female in middle), lateral female, ovate, dense-fld, upper usually contiguous; peduncles short or 0. Bracts lf-like, not or shortly sheathing, spreading or deflexed. Female glumes brownish or yellowish, lanceolate to suborbicular, acute or mucronate. Perigynia ovoid- to obovoid-trigonous, glabrous, ±ribbed, narrowed into a long curved, or rather suddenly contracted into a shorter, bifid beak. Nut trigonous. Stigmas 3. (Spp. 6–11.)

Section 4. *Hymenochlaonae* Kükenth. Spikes 4–many, terminal male, lateral female, cylindrical, ±dense and many-fld, or terminal or all female at base, male at top; peduncles long, filiform. Bracts lf-like, sheathing. Female glumes lanceolate to ovate, acute or acuminate. Perigynia elliptic- or obovoid-trigonous, glabrous, narrowed into a long slender bifid beak. Nut trigonous. Stigmas 3. (Sp. 12.)

Section 5. *Capillares* Mackenzie. Stems filiform. Spikes 3–4, terminal male, lateral female, slender, lax, few-fld; peduncles long, capillary. Bracts shortly lf-like. long-sheathing. Female glumes hyaline with brownish margins, obtuse, sometimes mucronate, caducous. Perigynia oblong-ovoid or subinflated-trigonous, glabrous; beak rather short, entire. Nut trigonous. Stigmas 3. (Sp. 13.)

Section 6. *Rhomboidales* Kükenth. Stems often lateral. Spikes 3–6, terminal male, lateral female, cylindrical or oblong-ovate, dense- or few-fld, remote, peduncled. Bracts sheathing, shortly lf-like. Female glumes brown with hyaline margins and green midrib. Perigynia rhomboidal or inflated-trigonous, pale brown or greenish, glabrous, many-ribbed, narrowed at base; beak long, subcylindrical, notched or deeply bifid. Nut trigonous. Style-base thickened into a pyramidal or annular cap; stigmas 3. (Sp. 14.)

Section 7. *Pseudocypereae* Kükenth. Stems tall, sharply angled, lfy nearly to top. Spikes 3–9, upper 1–3 male, rest female, cylindrical or oblong-cylindrical, dense-fld, upper contiguous, subsessile, lower 1–2 remote, long-peduncled, nodding. Bracts lf-like, nearly always without sheaths. Female glumes oblong-lanceolate or narrowly obovate, aristate. Perigynia spreading or deflexed, glabrous, many-ribbed, ±asymmetrical, ovoid or elliptic, ±inflated, green or yellowish; beak rather long, deeply notched. Nut trigonous. Stigmas 3. (Sp. 15.)

Section 8. *Vesicariae* Christ. Stems tall, lfy nearly to top. Spikes (2–)3–6(–11), upper 1–3(–7) male, rest female, oblong to ovoid, at least the lower peduncled. Bracts lf-like, usually sheathing. Female glumes usually acute, often dark brown or purplish. Perigynia spreading or deflexed, ovoid, ±inflated, glabrous, few-ribbed, usually brownish-green; beak fairly long, bifid. Nut trigonous or biconvex, style-base ±thickened; stigmas 3, rarely 2. (Spp. 16–19.)

Section 9. *Paludosae* Christ. Spikes 3–8(–many), upper 1–4 male, rest female, cylindrical or oblong, dense-fld, erect. Bracts lf-like, ±sheathing or not. Female glumes ovate or oblong, aristate or acuminate, often dark brown. Perigynia spreading, oblong-lanceolate to ovate, ±inflated-trigonous, glabrous, many-ribbed, narrowed at base; beak rather short, bifid. Nut trigonous. Stigmas 3. (Spp. 20–1.)

Section 10. *Maximae* Kükenth. Spikes 4–8, terminal male, lateral female, dense-fld, cylindrical. Lower bracts lf-like, exceeding the stem, sheathing or not. Female glumes brownish, often acuminate or mucronate. Perigynia ovate or elliptic, trigonous or ±inflated, glabrous; beak short, emarginate or slightly notched. Nut trigonous. Stigmas 3. (Sp. 22.)

Section 11. *Gracillimae* Mackenzie. Stems slender. Spikes 3–8, terminal male (sometimes male at base, female at top), lateral female, narrow-cylindrical, usually rather lax. Bracts lf-like, sheathing. Female glumes small, ovate or obovate, acute or mucronate. Perigynia oblong-trigonous, glabrous, exceeding the glumes; beak short, truncate or emarginate. Nut trigonous. Stigmas 3. (Sp. 23.)

Section 12. *Pallescentes* Christ. Stems slender, often pubescent. Lf-sheaths often pubescent. Spikes 2–4, mostly contiguous, lateral female, terminal male, oblong or ovate, dense-fld. Bracts shortly lf-like, not sheathing. Female glumes pale brownish, acuminate or mucronate. Perigynia ovoid to pyriform, glabrous or pubescent; beak 0 or very short. Nut trigonous. Style stout; stigmas 3. (Spp. 24–5.)

Section 13. *Paniceae* Christ. Spikes 3–6, terminal male, lateral female, oblong-cylindrical, lax and rather few-fld, erect, ± peduncled. Bracts lf-like, sheathing. Female glumes reddish-brown with a pale midrib, acute or mucronate. Perigynia obovoid or ovoid, ± inflated-trigonous, very minutely papillose, glabrous, exceeding the glumes; beak short, curved, entire or rarely notched. Nut trigonous. Stigmas 3. (Spp. 26–7.)

Section 14. *Limosae* Christ. Spikes 2–5, terminal male, lateral female, oblong or ovate, rather few-fld, nodding, peduncles long and slender. Bracts shortly lf-like, mostly very shortly sheathing. Female glumes brown or purplish. Perigynia ovate or elliptic, very minutely papillose, glabrous; beak very short, entire or emarginate. Nut trigonous. Stigmas 3. (Spp. 28–30.)

Section 15. *Glaucae* Christ. Spikes several, cylindrical, dense-fld, upper (1–)2–6 male, the rest female. Bracts lf-like, ± sheathing. Female glumes purplish or black, mucronate or aristate. Perigynia ovate or oblong, papillose or puberulent in the upper part; beak short, entire or emarginate. Nut trigonous. Stigmas 3. (Sp. 31.)

Section 16. *Carex*. Spikes 3–8, upper 1–4(–7) male, lower female, oblong or cylindrical, dense-fld, erect. Bracts lf-like, sheathing. Female glumes ovate or lanceolate, mucronate or aristate. Perigynia suberect or ± spreading, ovoid, ± inflated-trigonous, pubescent, many-ribbed, narrowed at base; beak long or rather short, bifid. Nut trigonous. Stigmas 3. (Spp. 32–3.)

Section 17. *Montanae* Christ. Stems slender, glabrous. Spikes 3–5, contiguous, sessile or subsessile, terminal male, lateral female, ovate or subglobose, rather dense-fld. Bracts glumaceous, or with a setaceous blade, rarely sheathing. Female glumes brown or rather pale, acute or often mucronate. Perigynia obovoid or ovoid, inflated-trigonous, pubescent, stipitate; beak short, conical, emarginate or slightly notched. Nut trigonous. Base of style ± swollen; stigmas 3. (Spp. 34–7.)

Section 18. *Digitatae* Christ. Stems slender, glabrous, often lateral. Spikes 3–6, terminal male, lateral female, lax and few-fld, included or exserted. Bracts glumaceous or shortly lf-like, sheathing. Female glumes brown or reddish, acute or mucronate. Perigynia pale green, obovoid-trigonous, pubescent, very rarely glabrous, stipitate; beak short, curved, subentire. Nut trigonous. Stigmas 3. (Spp. 38–40.)

Section 19. *Atratae* Christ. Spikes usually ovoid, contiguous, dense-fld, sessile or shortly peduncled, the terminal male at base female at top, clavate, longer than the lateral ones which are female or rarely male at base. Bracts shortly lf-like, enfolding the stem at base but not sheathing. Female glumes blackish or dark purple, acute. Perigynia glabrous and punctate or minutely papillose, elliptic or ovate; beak short, truncate or notched. Nut trigonous. Stigmas 3. (Spp. 41–3.)

Section 20. *Frigidae* Kükenth. Stems erect, nodding at the top. Spikes 3–6, terminal male or male at base, female at top, lateral female, ovoid or oblong, dense-fld, often pendulous, long-peduncled. Bracts shortly lf-like, sheathing. Glumes brown or purplish-black .Perigynia ellipsoid or trigonous, glabrous, without nerves; beak long, ± notched. Nut trigonous. Stigmas 3. (Sp. 44.)

Section 21. *Acutae* Christ. Spikes dense-fld, cylindrical, few to numerous, the upper 1–3 male, the rest female. Bracts lf-like, usually with dark auricles ± enfolding the stem, but not sheathing. Female glumes blackish or dark purple. Perigynia ovate or elliptic, plano-convex or biconvex, glabrous, conspicuously margined; beak short, truncate, nearly entire or rarely notched. Nut biconvex. Stigmas 2. (Spp. 45–50.)

Subgenus 2. VIGNEA (Beauv.) Kükenth.

Spikes sessile, all ± similar, usually bearing male and female fls, rarely the upper or lower ones unisexual. Branches of the 2nd and 3rd order nearly always without bracts. Stigmas 2, rarely 3. (Sections 22–30.)

Section 22. *Paniculatae* Christ. Spikes numerous, male at top, ovate, forming a ± contracted panicle, or rarely a spike-like infl. Bracts 0. Perigynia ± squarrose, broadly ovate, ± inflated, reddish-brown, glabrous, ± ribbed, base rounded and shortly stipitate; beak long, bifid. Nut plano-convex. Stigmas 2. (Spp. 51–3.)

Section 23. *Vulpinae* Christ. Spikes numerous, male at top, forming a ± branched but contracted panicle. Bracts small. Perigynia squarrose, ovate or lanceolate, plano-convex, rounded or subcordate at base, ± stipitate; beak long, slender. Nut biconvex. Style-base ± thickened, stigmas 2. (Spp. 54–5.)

Section 24. *Ammoglochin* Dum. Spikes several, unisexual or bisexual, forming an oblong head ± interrupted at base. Bracts small. Perigynia suberect, ovate-lanceolate, plano-convex, narrowed at base, glabrous, strongly ribbed, margin ± winged in upper part; beak bifid. Nut biconvex. Stigmas 2. (Spp. 56–7.)

Section 25. *Divisae* Kükenth. Spikes several, ovate, male at top (rarely dioecious), forming an oblong or ovate rather dense ebracteate head. Perigynia suberect, ovate, plano-convex, glabrous, shining, strongly ribbed; beak rather short, oblique or notched. Nut biconvex. Stigmas 2. (Spp. 58–9.)

Section 26. *Incurvae* Kükenth. Spikes few, male at top, crowded in a dense ovoid head. Glumes ovate, acute or mucronate, margins broadly hyaline. Perigynia spreading, ovoid, inflated, glabrous, shortly stipitate; beak short. Nut biconvex. Stigmas 2. (Sp. 60.)

Section 27. *Muehlenbergianae* Kükenth. Spikes rather few, male at top, ovate, ± distant, in a nearly simple infl. Bracts small. Perigynia squarrose or suberect, ovate, plano-convex, greenish, glabrous, not ribbed, narrowed at base; beak broad, serrulate. Nut biconvex. Stigmas 2. (Spp. 61–4.)

Section 28. *Stellulatae* Christ. Spikes several, male at base or rarely unisexual, rather dense-fld, lower remote, often bracteate. Perigynia suberect or squarrose, ovate or lanceolate, plano-convex, glabrous; beak long, notched or subentire, often serrulate. Nut biconvex. Stigmas 2. (Spp. 65–7.)

Section 29. *Canescentes* Christ. Spikes several, male at base or the lateral female only, ovate or subglobose, dense-fld, crowded into a head or the lower ± remote. Bracts small. Perigynia suberect, ovate, plano-convex, densely punctulate, narrowed at base, glabrous, ± obscurely ribbed; beak scarcely notched. Nut biconvex. Stigmas 2. (Spp. 68–9.)

Section 30. *Ovales* Christ. Spikes numerous, male at base, dense-fld, crowded into an ovoid head. Bracts small. Glumes brown. Perigynia appressed or slightly spreading, plano-convex, almost winged, glabrous, distinctly ribbed; beak bifid. Nut biconvex. Stigmas 2. (Sp. 70.)

Subgenus 3. PRIMOCAREX Kükenth.

Spike solitary terminal, monoecious or dioecious, rhachilla often present. Stigmas 2 or 3. (Sections 31–4.)

Section 31. *Petraeae* Kükenth. Spike male at top, female part few-fld. Female glumes persistent. Perigynia erect (rarely spreading), obovoid or ovoid, ± trigonous, glabrous or puberulent; beak short, truncate. Nut trigonous. Stigmas 3. (Sp. 71.)

Section 32. *Leucoglochin.* Dum. Similar to *Pulicares* but perigynia narrow-lanceolate, subtrigonous. Nut trigonous. Stigmas 3. (Spp. 72–3.)

Section 33. *Pulicares* Tutin. Spike male at top, lax and few-fld. Female glumes caducous. Perigynia narrowly ellipsoid, biconvex, stipitate, eventually reflexed, glabrous. Nut biconvex. Stigmas 2. (Sp. 74.)

Section 34. *Dioicae* Pax. Spike unisexual, ebracteate, male linear, female oblong-linear or oblong-ovate. Female glumes persistent. Perigyn iaeventually divaricate, glabrous; beak notched. Nut biconvex. Stigmas 2. (Sp. 75.)

List of Species

1. laevigata	28. limosa	53. diandra
2. distans	29. paupercula	54. obtrubae
3. punctata	30. rariflora	55. vulpina
4. hostiana	31. flacca	vulpinoidea
5. binervis	32. hirta	56. disticha
6. flava	33. lasiocarpa	57. arenaria
7. lepidocarpa	34. pilulifera	58. divisa
8. demissa	35. ericetorum	59. chordorrhiza
9. serotina	36. caryophyllea	60. maritima
10. scandinavica	37. montana	61. divulsa
11. extensa	38. humilis	62. polyphylla
12. sylvatica	39. digitata	63. spicata
13. capillaris	40. ornithopoda	64. muricata
14. depauperata	glacialis	65. elongata
15. pseudocyperus	41. buxbaumii	66. echinata
16. rostrata	42. atrata	67. remota
17. vesicaria	43. norvegica	68. curta
18. stenolepis	44. atrofusca	69. lachenalii
19. saxatilis	45. elata	70. ovalis
20. riparia	46. acuta	crawfordii
21. acutiformis	47. aquatilis	71. rupestris
22. pendula	48. recta	72. microglochin
23. strigosa	49. nigra	73. pauciflora
24. pallescens	50. bigelowii	capitata
25. filiformis	bicolor	74. pulicaris
26. panicea	51. paniculata	75. dioica
27. vaginata	52. appropinquata	davalliana

Key to Species

1   Spikes more than 1, lateral ones sessile or stalked.   *2*
    Spikes solitary, terminal (see also **59. chordorrhiza**).   *79*

2   Spikes dissimilar in appearance, the terminal one or more male (sometimes ±concealed among the female), the lateral wholly or mainly
female.   *3*
    Spikes all similar in appearance, one or more with both male and female
fls, rarely ±unisexual (**56. disticha, 57. arenaria**) but still similar in
appearance and forming a ±lobed head.   *56*

3   Fr. pubescent, or at least papillose towards the top.   *4*
    Fr. glabrous.   *14*

4   Fl. stems lateral, lfless with a few sheaths surrounding their bases;
female spikes c. 2 mm. diam., overtopping the male spike.   *5*

Fl. stems terminal from the centre of the rosette, lfy at the base; female
spikes 3–6 mm. diam., if narrower then concealed within a sheathing
bract.   6

5   Lowest female spike somewhat remote from the others; fr. about as long
     as the glume.   39. digitata
     Female spikes all close together; fr. distinctly longer than glume.
                                                                 40. ornithopoda

6   Lvs longer than stems; female spikes very slender, ±concealed by
     sheathing bracts.   38. humilis
     Lvs usually shorter than stems; female spikes 3–6 mm. diam., not at all
     concealed by sheathing bracts.   7

7   Sheaths usually pubescent, at least at mouth; lowest female spike from
     ¼ the way down the stem or lower, peduncle shortly exserted; fr.
     6–7 mm.   32. hirta
     Sheaths glabrous (rarely pubescent in *C. filiformis*); lowest female spike
     from upper ¼ of stem, if lower then peduncle long-exserted; fr. 4 mm.
     or less.   8

8   Lvs distinctly glaucous beneath; fr. papillose at top.   31. flacca
     Lvs not or scarcely glaucous; fr. distinctly pubescent.   9

9   Female spikes ±distant, not clustered round base of male.   10
     Most of the female spikes clustered round base of male.   11

10   Lvs setaceous, channelled; stem bluntly angled; glumes 4 mm., lanceolate,
     acuminate.   33. lasiocarpa
     Lvs 1·5–2 mm. wide, nearly flat; stems sharply angled; glumes 2 mm.,
     broadly ovate, mucronate.   25. filiformis

11   Lower bract sheathing.   36. caryophyllea
     Lower bract not sheathing.   12

12   Lowest bract not green and lf-like; glumes dark purple-brown, obtuse
     or mucronate.   13
     Lowest bract green and lf-like; glumes light brown, acuminate.
                                                                     34. pilulifera

13   Stems smooth; glumes obtuse, finely ciliate at top.   35. ericetorum
     Stems rough at top; glumes mucronate, not ciliate.   37. montana

14   Most or all of the sessile or short thick-stalked female spikes clustered
     round the base of the sessile or subsessile male spikes.   15
     Female spikes remote, or if clustered ±nodding and usually with long
     slender stalks; male spike usually distinctly stalked.   20

15   Lvs involute or channelled; stems scarcely angled (salt-marshes and dune
     slacks).   11. extensa
     Lvs flat; stems distinctly angled (not in brackish marshes).   16

16   Lower sheaths and lvs somewhat hairy beneath; lowest bract crimped at
     base.   24. pallescens
     Sheaths and lvs glabrous; lowest bract not crimped at base.   17

17   Fr. yellowish or golden; stigmas 3; nut trigonous (lowland).   18
     Fr. blackish; stigmas 2; nut biconvex (mountains).   50. bigelowii

18   Fr. 5–7 mm., arcuate-deflexed when ripe, gradually tapering into a beak
     2–3 mm. (very local).   6. flava
     Fr. 2–3 mm., straight, abruptly contracted into a beak 0·5 mm. or less
     (local).   19

19 Fr. ± inflated; nut filling ¾ of the perigynium; lvs longer than stem; male spike sessile. **9. serotina**

Fr. not inflated; nut completely filling perigynium; lvs shorter than stem; male spike usually peduncled. **10. scandinavica**

20 Female spikes 7–16 cm. × 5–7 mm., pendulous; peduncles included. **22. pendula**

Female spikes shorter or stouter, if pendulous then peduncles exserted. *21*

21 Most or all female spikes at about the same level, ± nodding; peduncles capillary; male spike ± surrounded or even overtopped by female spikes. *22*

Female spikes ± remote, scarcely reaching beyond the base of the terminal male spike. *25*

22 Female spikes ovoid; glumes blackish (very rare). **44. atrofusca**

Female spikes cylindrical or linear; glumes not blackish. *23*

23 Female spikes 5 mm. or more in diam., not overtopping the terminal male spike (lowland). *24*

Female spikes up to 3 mm. diam., overtopping the terminal male spike (base-rich mountains, very local). **13. capillaris**

24 Lower sheaths and lvs hairy beneath; fr. rounded at top; beak almost 0 (woods and grassy places). **24. pallescens**

Lvs and sheaths glabrous; fr. gradually tapering into a beak c. 2 mm. (ditches, etc.). **15. pseudocyperus**

25 Bracts not sheathing though sometimes partly enfolding stem. *26*

At least the lowest bract with an entire cylindrical sheath. *40*

26 Stigmas 3; nut trigonous; ripe fr. usually inflated or trigonous. *27*

Stigmas 2; nut biconvex; ripe fr. usually flattened. *35*

27 Female spikes ± nodding; peduncles slender, lowest usually at least as long as spike; beak of fr. very short, entire. *28*

Female spikes erect or suberect; peduncles 0 or lowest usually shorter than spike, stout; beak of fr. distinct, ± notched. *30*

28 Stem rough; lvs channelled or involute. **28. limosa**

Stem smooth; lvs flat. *29*

29 Glumes lanceolate, acuminate or aristate, much narrower than fr. **29. paupercula**

Glumes obovate, almost truncate, apiculate, wider than fr. **30. rariflora**

30 Plant not exceeding 50 cm.; female spikes ovoid, c. 3 times as long as broad; lvs c. 3 mm. wide (mountains). *31*

Plant 60–160 cm., female spikes ± cylindrical, c. 5–10 times as long as broad; lvs 6–15 mm. wide (lowland waters). *32*

31 Ligule 3–4 mm., acute; fr. ribbed, often sterile. **18. stenolepis**

Ligule c. 1 mm., rounded or almost truncate; fr. smooth. **19. saxatilis**

32 Fr. distinctly longer than glume. *33*

Fr. not or very little longer than glume. *34*

33 Stem bluntly angled; lvs glaucous; fr. abruptly narrowed into the beak. **16. rostrata**

Stem sharply angled; lvs dark green; fr. gradually narrowed into the beak. **17. vesicaria**

34 Male glumes obtuse; fr. c. 4 mm., ± flattened. **21. acutiformis**

Male glumes acuminate or with an excurrent midrib; fr. c. 8 mm., inflated. **20. riparia**

35   Lower spikes long-stalked (1–3 cm.); glumes in lower part of spike with a
     long excurrent midrib; lvs 3–5 mm. wide.   **48. recta**
     Lower spikes ± sessile, rarely long-stalked and then the lvs not more
     than 3 mm. wide; glumes sometimes acuminate but never with a long
     excurrent midrib.   36

36   Plant small (c. 10 cm.); lvs 4–5 mm. wide, recurved; stem very sharply
     angled (mountains, rarely below 2000 ft.).   **50. bigelowii**
     Plant larger or lvs narrower; stem usually not very sharply angled
     (mostly below 2000 ft.).   37

37   Lowest bract much shorter than infl.; margins of lf sheaths connected
     by filaments; densely tufted.   **45. elata**
     Lowest bract nearly as long to longer than infl.; margins of sheaths not
     connected by filaments; usually creeping.   38

38   Plant usually 20–40 cm.; if taller stem not more than 2 mm. diam.; culm
     lvs 1·5–3·0 mm. wide.   **49. nigra**
     Plant usually 50–100 cm.; stem 3 mm. or more in diam.; culm lvs
     3·0–6·0 mm. wide.   39

39   One bract (rarely more) as long as or longer than infl.; lower sheaths
     with short lvs; male spikes 2(–3); stem tough.   **46. acuta**
     Two or three bracts as long as or longer than infl.; lower sheaths lfless;
     male spikes 3–4; stem brittle.   **47. aquatilis**

40   Female spikes up to 6-fld; fr. 8 mm. (very rare).   **14. depauperata**
     Female spikes usually more than 6-fld; fr. smaller.   41

41   Lower bracts with close cylindrical sheaths; if loose, plant glaucous;
     beak deeply notched or very short and entire.   42
     Lower bracts with loose narrowly funnel-shaped sheaths; plant green or
     yellowish-green; beak distinct, obliquely truncate and shallowly
     notched (higher Scottish mountains).   **27. vaginata**

42   Beak not more than 0·5 mm., entire or nearly so; lvs glaucous or female
     spikes erect, c. 2 mm. diam.   43
     Beak usually more than 0·5 mm., distinctly notched; lvs rarely glaucous;
     female spikes at least 5 mm. diam. or else nodding.   45

43   Lvs glaucous; female spikes 4–6 mm. diam., the lowest rarely more than
     ⅓ way down the stem (common).   44
     Lvs green; female spikes c. 2 mm. diam., the lowest usually ½ way down
     the stem (rare).   **23. strigosa**

44   Male spike 1; female spikes lax; fr. 3·5–4 mm., smooth.   **26. panicea**
     Male spikes (1–)2 or more; female spikes dense; fr. 2–2·5 mm., minutely
     papillose, at least at top.   **31. flacca**

45   Male spikes 2 or more, one of the lateral at least ½ the length of the
     terminal.   46
     Male spike 1, rarely 2 and then the lateral much smaller than the terminal.   48

46   Spikes dark brown; lower lvs (8–)10–20 mm. wide.   **20. riparia**
     Spikes yellowish- or brownish-green; lower lvs c. 4(–7) mm. wide.   47

47   Stems bluntly angled; lvs glaucous; fr. abruptly narrowed into the beak.
                 16. rostrata
     Stems sharply angled; lvs dark green; fr. gradually narrowed into the
     beak.   **17. vesicaria**

48   Female spikes ovoid (c. twice as long as broad).   49
     Female spikes cylindrical (c. 4–5 times as long as broad).   51

49 Fr. arcuate-deflexed (on base-rich soils).   **7. lepidocarpa**
     Fr. not curved, the lowest sometimes deflexed (on acid soils).   *50*

50 Stems curved; lvs flat, 3·5–5 mm. wide; fr. ± inflated.   **8. demissa**
     Stems straight; lvs inrolled, up to 2 mm. wide; fr. not inflated.
                                                             10. scandinavica

51 Female spikes c. 3 mm. diam., overlapping.   **12. sylvatica** *52*
     Female spikes c. 5 mm. diam., not overlapping.

52 Lvs 5–10 mm. wide; female spikes mostly 3 cm. or more.   **1. laevigata**
     Lvs 2–4(–5) mm. wide; female spikes usually less than 3 cm.   *53*

53 Glumes with scarious margins; female spikes usually c. 1 cm., ± ovoid,
        with rather few fruits.   **4. hostiana**
     Margins of glumes not scarious; female spikes commonly 1·5–3 cm.,
        cylindrical, with numerous fruits.   *54*

54 Female spikes commonly 2–3 cm., dark reddish-brown; fr. with 2 pro-
        minent almost marginal nerves (mostly on heaths and mountains).
                                                              5. binervis
     Female spikes commonly 1·5–2 cm., greenish- or pale reddish-brown;
        fr. with several equal distinct nerves, or else smooth and shiny (mostly
        near the sea).   *55*

55 Fr. with several equal distinct nerves; beak rough.   **2. distans**
     Fr. smooth, shiny, faintly nerved when dry; beak smooth.   **3. punctata**

56 Fruiting spikes few (up to 5) never numerous small and squarrose nor
        forming a roundish or ovoid ± lobed head; glumes usually black or
        dark purplish with pale midrib (rare; high mountains and bogs,
        mostly in Scotland).   *57*
     Fruiting spikes usually numerous, either small and squarrose or forming
        a roundish or ovoid ± lobed head; glumes never black or dark purplish. *60*

57 Spikes clustered at top of stem, the lowest sometimes ± remote.   *58*
     Spikes not clustered; glumes pale.   **41. buxbaumii**

58 Spikes sessile or subsessile, not nodding.   **43. norvegica**
     Spikes ± peduncled, some or all nodding.   *59*

59 Bract subtending lowest spike not or scarcely sheathing (–5 mm.), free
        part usually 2 cm. or more; terminal spike usually female at top, male
        below.   **42. atrata**
     Bract subtending lowest spike with a long sheath (usually 10 mm. or
        more), free part usually less than 1 cm.; terminal spike commonly
        entirely male.   **44. atrofusca**

60 Terminal spike entirely male; rhizome far-creeping (sandy places by the
        sea, rarely inland).   **57. arenaria**
     Terminal spike partly, rarely entirely, female.   *61*

61 Male fls (usually easily recognized by the white filaments or narrower
        glumes) at the top of some or all the spikes.   *62*
     Male fls at base of some or all the spikes.   *74*

62 Rhizome far-creeping; stems not tufted.   *63*
     Rhizome short; stems in tufts.   *66*

63 Spikes forming a small scarcely lobed head; bracts glume-like.   *64*
     Spikes forming a ± lobed head; lower bracts lf-like or, if glume-like,
        with a distinct bristle-point.   *65*

64 Lvs curved, as long as or longer than the curved stems (sandy shores in
        the north).   **60. maritima**

Lvs straight, much shorter than the straight stems (bogs, W. Sutherland).
                                                           59. chordorrhiza

65   Stems slender, wiry; infl. up to c. 2 cm., fr. not winged.   **58. divisa**
     Stems rather stout, not wiry; infl. usually more than 2 cm.; fr. narrowly
       winged.   **56. disticha**

66   Spikes dark brown or reddish-brown.   67
     Spikes greenish.   70

67   Spikes sessile forming an oblong ± lobed head.   68
     At least the lowest spike stalked; infl. either narrow and ± interrupted
       or a ± spreading panicle.   69

68   Lvs 1–2 mm. wide; stems slender; fr. broadly ovate or suborbicular;
       not tufted.   **53. diandra**
     Lvs 5–12 mm. wide; stems stout; fr. ovate; tufted.   **55. vulpina**

69   Lvs 3–7 mm. wide; fr. scarcely nerved.   **51. paniculata**
     Lvs 1–2 mm. wide; fr. strongly nerved.   **52. appropinquata**

70   Lvs 4–10 mm. wide; fr. strongly ribbed.   **54. otrubae**
     Lvs 2–4 mm. wide; fr. smooth or feebly ribbed.   (*muricata* agg.) 71

71   Infl. usually more than 4 cm., lax; lowest spike up to 4 times its own
       length from next, sometimes stalked.   72
     Infl. rarely as much as 4 cm., rather dense; lowest spike not more than
       its own length from next, sessile.   73

72   Infl. usually with 1–2 short branches; fr. veined at base, nearly sym-
       metrical.   **61. divulsa**
     Infl. unbranched; fr. without apparent veins, asymmetrical. **62. polyphylla**

73   Fr. 4–5 mm., corky at base, greenish when ripe.   **63. spicata**
     Fr. 3–4 mm., distinctly veined at base, brown when ripe. **64. muricata**

74   At least the lower bracts lf-like.   **67. remota**
     Bracts never lf-like.   75

75   Spikes small, with up to c. 10 spreading fruits.   **66. echinata**
     Spikes larger, with more than 10 erect fruits.   76

76   Spikes more than 4, ± remote, forming an oblong head in fr.   77
     Spikes 3–4 (–9), contiguous, forming a ± lobed ovoid head in fr. 78

77   Spikes pale greenish-white; fr. 2–3 mm.; beak rough.   **68. curta**
     Spikes brown; fr. 3·5–4 mm.; beak smooth.   **65. elongata**

78   Plant 20–90 cm.; fr. 4–5 mm.; beak rough (common).   **70. ovalis**
     Plant up to 20 cm.; fr. 3 mm.; beak smooth (rare).   **69. lachenalii**

79   Spike unisexual or very nearly so; plant usually dioecious.   **75. dioica**
     Spike male at top, the lower half or more female.   80

80   Glumes persistent; fr. not deflexed when ripe; lvs c. 2 mm. wide, usually
       much curled.   **71. rupestris**
     Glumes caducous; fr. deflexed when ripe; lvs c. 1 mm. wide, straight or
       slightly curled at tips.   81

81   Stigmas 2; fr. dark brown, lanceolate (locally common).   **74. pulicaris**
     Stigmas 3; fr. yellowish, linear-lanceolate (mountain districts, uncommon). 82

82   Spike with 2–4 fruits; fr. gradually narrowed at base; persistent style
       protruding from top of fr.; glumes c. 4 mm. (Sphagnum bogs, local).
                                                         73. pauciflora
     Spike with 4–12 fruits; fr. abruptly narrowed at base; stout bristle from
       base of nut protruding along with stigmas from top of fr.; glumes
       c. 2 mm. (base-rich mountains, very rare).   **72. microglochin**

The following hybrids have, among others, been reported. They appear to be either rare or overlooked, and can, in general, be recognized by their sterility and by being ± intermediate between the parents.

C. acuta × elata; C. acuta × acutiformis; C. acuta × nigra. C. acutiformis × rostrata. C. aquatilis × bigelowii; C. aquatilis × elata; C. aquatilis × nigra; C. aquatilis × recta. C. bigelowii × nigra. C. binervis × demissa; C. binervis × hostiana; C. binervis × punctata. C. curta × echinata; C. curta × lachenalii (C. × helvola Blytt). C. demissa × hostiana; C. demissa × lepidocarpa. C. dioica × echinata. C. distans × extensa; C. distans × hostiana. C. divulsa × otrubae. C. elata × nigra. C. hirta × vesicaria. C. hostiana × lepidocarpa; C. hostiana × serotina. C. lasiocarpa × riparia. C. nigra × recta. C. otrubae × paniculata; C. otrubae × remota (C. axillaris Good., non L.). C. otrubae × spicata. C. paniculata × remota (C. × boenninghausiana Weihe). C. pseudocyperus × rostrata. C. remota × spicata. C. riparia × vesicaria. C. rostrata × vesicaria.

1. C. laevigata Sm.   'Smooth Sedge.'
C. helodes Link

A rather stout tufted glabrous perennial 30–100 cm. Stems erect, trigonous, slightly rough above. *Lvs 5–10 mm. wide*, shorter than stems, shallowly keeled, bright green, smooth; ligule c. 1 cm., ovate or triangular, ± obtuse.

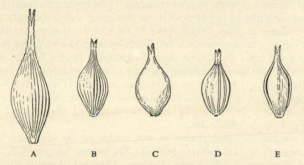

Fig. 71. Fruits of *Carex*. A, *C. laevigata*; B, *C. distans*; C, *C. punctata*; D, *C. hostiana*; E, *C. binervis*. × 5.

Lower sheaths lfless, brown, not fibrous. Male spikes 1(–2), 2–6 cm., slender, tapering below. Male glumes c. 5 mm. oblong-lanceolate, subacute or acute, brownish-hyaline. Female spikes 2–4, remote, 20–60 × 5–10 mm., lower pendulous, upper suberect. Peduncles filiform, smooth or slightly and remotely toothed. Bracts lf-like, all longer than their spikes but not exceeding infl., long-sheathing. Female *glumes* 3–3·5 mm., ovate, *acuminate*, brownish with a green midrib. *Fr.* (Fig. 71 A) 5–6 *mm.*, ovoid or suborbicular, rather inflated, green; *beak c.* 1·5 *mm.*, bifid, smooth or with a few slender teeth. Nut c. 2 mm., orbicular, trigonous, faces slightly concave, shortly stipitate. Fl. 6. Fr. 7–8. $2n = 72^*$. Hs.

Native. In marshes and damp woods, usually on acid but base-rich soils, local. 86, H33, S. Scattered throughout the British Is. north to Inverness. Western Germany, Belgium, France, Spain and Portugal, Corsica.

2. C. distans L.   'Distant Sedge.'

A densely tufted glabrous perennial 15–45 cm. Stems trigonous, rigid, smooth. Lvs 2–5 mm. wide, usually much shorter than stems, nearly flat, grey-green; ligule 2–3 mm., triangular. Lower sheaths brown, scarcely fibrous. Male spike 1(–2), 15–30 × 2–3 mm., nearly cylindrical. Male glumes c. 4 mm., ovate, obtuse or subacute, brownish-hyaline. Female spikes 2–3, 10–20 × 4–5 mm., distant, erect. Peduncles slender, ± included. Bracts lf-like, shorter than infl., long-sheathing. Female *glumes* ovate, mucronate, brown or greenish-brown with a pale midrib, *conspicuous in fr. Fr.* (Fig. 71 B) 3·5–4 mm., ellipsoid, *trigonous, equally many-ribbed*, not glossy, ± *erect*; beak bifid, rough. Nut 2·5 mm., elliptic, trigonous. Fl. 5–6. Fr. 6–7. $2n=74*$. Hs.

Native. In marshes and in cracks of wet rocks, mainly near the coast. 84, H 26, S. Scattered throughout the British Is. from Cornwall to Shetland, rather local. Europe to c. 60° N. in Scandinavia, western Asia, N. Africa.

3. C. punctata Gaud.   'Dotted Sedge.'

A glabrous rather tufted perennial, somewhat resembling *C. distans*, but usually shorter and more stiffly erect with lvs longer in proportion to the stems. Stems trigonous, smooth. Lvs 2–5 mm. wide, nearly flat, pale green, all ± basal; ligule c. 1 mm., almost truncate. Lower sheaths mostly with lvs, dark brown; sheaths of uppermost lf on flowering stems truncate, with two small appendages at sides. Male spike 1, 10–25 × 1–1·5 mm., stalked. Male glumes c. 3 mm., oblong-obovate, brownish-hyaline. Female spikes 2–4, distant, 5–25 × 4–5 mm. Peduncles long, included. Bracts lf-like, longer or shorter than infl., all long-sheathing. Female *glumes* broadly ovate, mucronate, pale reddish-brown, ± hyaline, *inconspicuous in fr. Fr.* (Fig. 71c) 2·5–3·5 mm., ovoid, *turgid, glossy*, pellucid-punctulate, *pale greenish, spreading or deflexed*; beak bifid, smooth. Nut c. 2 mm., obovoid-trigonous, shining and punctate. Fl. 6–7. Fr. 7–8. $2n=68*$. Hs.

Native. In marshes and cracks in wet rocks by the sea, very local. 13, H7, S. Cornwall to Hampshire; Suffolk; W. Wales, Isle of Man, Kirkcudbright and Wigtown; Cork, Kerry, Galway, Donegal. Western Europe from S.W. Norway southwards, Mediterranean region, rarely inland.

4. C. hostiana DC.   'Tawny Sedge.'

C. fulva auct.; *C. Hornschuchiana* Hoppe

A shortly creeping glabrous perennial 15–50 cm. Stems erect, trigonous, nearly smooth. Lvs 2–5 mm. wide, much shorter than stems, nearly flat, margins ± rough; ligule very short, truncate. Lower sheaths blackish, rather fibrous in decay. Male spike 1(–2), 10–20 × 1·5–2 mm., cylindrical, tapering at both ends. Male glumes 3–4 mm., ovate, obtuse or subacute, brown, margins broadly hyaline at top. Female spikes (1–)2–3, very distant, 8–12(–20) × 6–9 mm., ovoid or shortly cylindrical, erect, often male at top. Lower peduncles long, smooth, often long-exserted, upper short, ± included. Lower bracts lf-like, long-sheathing, longer than spikes; uppermost glumaceous with a setaceous point, shortly sheathing, shorter than spike. *Female glumes* 3–4 mm., broadly ovate, acute, dark brown *with a broad hyaline margin* and pale midrib. *Fr.* (Fig. 71 D) 5 mm., ovoid, *many-ribbed, rather inflated*; beak

c. 1 mm., slender, ± flattened, margins serrulate. Nut c. 2 mm., obovoid, trigonous. Fl. 6. Fr. 7. $2n = 56^*$. Hs.

Native. In fens and base-rich flushes, not uncommon. 105, H38. Throughout most of the British Is., but very local in the south and east. Europe, except the north-east and the southern Mediterranean region; W. Asia; N. America, probably introduced.

5. C. binervis Sm.   'Ribbed Sedge.'

A ± creeping glabrous perennial 30–60 cm. Stems erect, trigonous, nearly smooth. Lvs 2–5 mm. wide, shorter than stems, ± keeled, smooth; ligule very short, truncate. Lower sheaths brown, not fibrous. Male spike 1, 20–30 × 3–4 mm., narrowly clavate, sometimes with a few female fls at base. Male glumes 4 mm., oblong-obovate, obtuse, purplish-hyaline with a pale midrib. Female spikes 2–3, very distant, 15–35 × 5–7 mm., cylindrical, ± erect. Lower peduncles long, smooth, often long-exserted; upper short, often quite included. Lower bracts lf-like, long-sheathing, longer than spikes; uppermost glumaceous with a setaceous point, shortly sheathing, shorter than spike. Female glumes c. 3 mm., ovate, mucronate, dark purple-brown with a greenish midrib. *Fr.* (Fig. 71 E) c. 4·5 mm., ovoid, *subtrigonous with broad sharp lateral angles*, pale green or purplish, *with two prominent dark green submarginal nerves*; beak c. 0·5 mm., broad, flattened, bifid, slightly rough at edges. Nut c. 2 mm., ellipsoid or obovoid, trigonous. Fl. 6. Fr. 7. $2n = 74^*$. Hs.

Native. On heaths and moors and in rough pastures on acid soils. 110, H40, S. Throughout the British Is., common in suitable habitats. W. Europe from Spain to W. Norway and the Faeroes.

C. sadleri E. F. Linton (*C. frigida* auct. brit., non All.) needs further investigation. It appears to differ from *C. binervis* only in having an elliptic-lanceolate fr. with less well-marked submarginal nerves and longer beak, and in having an oblong-elliptic bluntly trigonous nut. It has been found on a few Scottish mountains. 8.

C. flava agg. (spp. 6–9).   'Yellow Sedge.'

A ± tufted perennial 5–95 cm. Stems smooth, trigonous or subterete. Lvs 2–7 mm. wide, channelled. Lower sheaths ± fibrous. Male spike 1, occasionally with female fls in the middle, sessile to long-peduncled. Male glumes 3–4 mm., linear-lanceolate to ovate, oblong or narrowly obovate, acute or obtuse, keeled or not, brownish. Female spikes 1–5, 5–15 × 4–12 mm., contiguous, remote or the lowest very distant, sessile or peduncled. Bracts lf-like, equalling or much exceeding infl., spreading or deflexed, lowest shortly sheathing. Female glumes 2–3·5 mm., lanceolate to ovate, acute or subacute, brownish or yellowish, ± hyaline. Fr. 2–7 mm., arcuate-deflexed or straight, obovoid to narrowly obovoid-trigonous, gradually narrowed or abruptly contracted into a beak; beak 0·5–3 mm., serrulate, ciliate or smooth. Nut 1–2 mm., narrowly obovoid-trigonous to almost round.

Native. In damp places throughout the British Is. 112, H40, S. Europe, Siberia and N. America.

6. C. flava L.

A ± tufted glabrous perennial 20–50(–95) cm. Stems rather stout, trigonous, smooth. Lvs 4–7 mm. wide, shorter than stems, transversely septate between veins, channelled, bright green, nearly smooth; ligule c. 5 mm., ovate, acute, shorter and rounded on flowering stems. Lower sheaths lfless, green or pale brown, becoming fibrous. *Terminal spike* 10–20 mm., wholly male or *frequently female in the middle*, trigonous, *scarcely peduncled*. Male glumes c. 4 mm., linear-lanceolate, keeled, brownish-hyaline, keel green, serrulate at top. *Female spikes* 3–4, *upper contiguous, sessile*, lower sometimes distant, shortly peduncled, 10–15 × 10–12 mm., peduncles slightly rough. *Bracts* lf-like *much longer than infl.*, *deflexed*, lower shortly sheathing. Female glumes 3·5 mm., lanceolate, acute, brownish-hyaline with a green midrib. *Fr.* (Fig. 72 A) 6–7 *mm.*, narrowly obovoid-trigonous, all but the upper *arcuate-deflexed when ripe*, greenish-yellow to golden; *beak* 2–2·5 mm., flattened, notched, edges serrulate or ciliate. Nut 1·5 mm., obovoid-trigonous. Fl. 6. Fr. 7. $2n = 60$*. Hs.

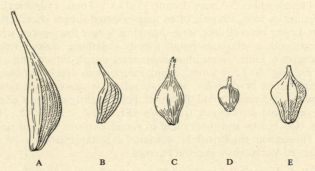

Fig. 72   Fruits of *Carex*. A, *C. flava*; B, *C. lepidocarpa*;
C, *C. demissa*; D, *C. serotina*; E, *C. extensa*. × 5.

Native. On damp peaty soils overlying carboniferous limestone, rare and very local. 2. N. Lancashire, Mid-west Yorks. Europe, north to Lapland; absent from the Mediterranean region; N. America.

7. C. lepidocarpa Tausch

A tufted glabrous perennial 20–60(–75) cm., rather similar to *C. flava* but generally smaller in all its parts and differing as follows: Stems usually slender. Lvs 2–3 mm. wide; ligule c. 1 mm., truncate or nearly so. *Terminal spike usually long-peduncled, wholly male*. Male glumes 3–3·5 mm., obovate or ovate-lanceolate, brownish-hyaline, midrib green or brown, not distinctly keeled. *Female spikes* 1–2(–4), 8–12 × 5–7 mm., *somewhat distant*, rarely contiguous, sessile or lower shortly peduncled. *Bracts* lf-like, *equalling or sometimes exceeding infl.* Female glumes 2 mm., ovate, acute, brown becoming yellow, ± hyaline. *Fr.* (Fig. 72 B) 3·5–4·5 *mm.*, obovoid-trigonous, *narrowed gradually into beak*, all but the upper *arcuate-deflexed when ripe; beak* 1·5 mm. Nut 1·5 mm., broadly obovoid-trigonous or almost round. Fl. 5–6. Fr. 7–8. $2n = 68$*. Hs.

Native. In wet places, often peaty and calcareous, probably always on base-rich soils. 38, H 8. Common in suitable habitats throughout the British ls., but apparently less widely distributed than *C. demissa*. Europe from Scandinavia (not arctic) southwards; Mediterranean region (very rare); eastern Canada.

8. C. demissa Hornem.

C. tumidicarpa Anderss.

A shortly creeping ± tufted perennial 5–30 cm. Stems subterete, smooth, ± spreading. Lvs 2–5 mm. wide, dark green, channelled, smooth; ligule very short, triangular. Lower sheaths grey-brown, rather fibrous. *Male spike* 1, 15–20 × 2 mm., acute, *distinctly peduncled*. Male glumes c. 4 mm., narrowly obovate, obtuse, brownish-hyaline. *Female spikes* 2–3, 7–10 × 6–8 mm., *upper two ± distant or sometimes contiguous, lowest often very remote*. Peduncles 0 or short. *Bracts* lf-like, one or more *about equalling infl.*, lowest ± sheathing. Female glumes c. 3 mm., ovate, acute or subacute, brown with a green midrib. *Fr.* (Fig. 72c) *c.* 3·5 *mm., lowermost sometimes deflexed*, obovoid *narrowed abruptly into the beak*, yellow-green; *beak c.* 1 *mm.*, notched. Nut similar to that of *C. lepidocarpa*. Fl. 6. Fr. 7–8. 2*n*=70*. Hs.

Native. In damp grassy and boggy places, stony margins of lakes, etc., on moderately acid soils. Common throughout much of the British Is. 74, H 14, S. Europe from Scandinavia south to N. Spain and Portugal; Greenland; eastern Canada.

9. C. serotina Mérat

C. Oederi auct., non Retz.

A glabrous tufted perennial 5–15(–30) cm. Stems trigonous, smooth. Lvs 2–3 mm. wide, usually much longer than stems, channelled, dark green; ligule 1–3 mm., triangular or rounded. Lower sheaths often becoming fibrous. *Male spike* 1, 5–10 × 2 mm., *sessile*, sometimes female at base. Male glumes 3–4 mm., ovate or oblong, acute or obtuse, brown with a pale or greenish midrib. *Female spikes* 3–5, 5–10 × 4–5 mm., *contiguous* and sessile *or the lowest sometimes remote* and peduncled. Peduncles 0 or included. Lower bracts lf-like, exceeding infl., sheathing. Female glumes 2–3 mm., ovate, subacute, keeled, yellow-brown, midrib green. *Fr.* (Fig. 72D) 2–3 *mm., not curved or deflexed*, inflated, *abruptly contracted into the short* (0·5 *mm. or less) beak*. Nut not filling the perigynium. Fl. 6–8. Fr. 7–9. 2*n*=70*. Hs.

Native. In damp places on base-rich soils, especially near the coast, rather local. 31, H 5, S. Western Europe from Iceland southwards; Siberia; Persia; N. Africa; Macaronesia.

10. C. scandinavica E. W. Davies

C. pulchella (Lönnr.) Lindm., non Berggr.

Similar to *C. serotina*. Lvs inrolled, narrow (0·5–2 mm. wide), usually shorter than stems. Male spikes usually shortly stalked; female spikes 2–3(–5), often rather distant. Female glumes dark brown with pale or green midrib. Fr. 1·5–2·5 mm., grey-green, not inflated. Nut completely filling the perigynium. Fl. 5–7. Fr. 6–8. 2*n*=70*. Hs.

Native. In damp habitats among rather open vegetation on lake shores, etc., in N. Scotland. 7. N. Russia, Finland, Scandinavia.

11. C. extensa Good.　　　　　　　　　　　　　'Long-bracted Sedge.'

A glabrous rather rigid perennial 20–40 cm. Stems trigonous or subterete, smooth, slender or rarely stout, sometimes curved. *Lvs* 2(–4) mm. wide or less, *channelled*, all ± basal; ligule c. 1 mm., ovate or truncate. Lower sheaths lfless, persistent, blackish. Male spike 1, 10–15 × 1–2 mm., sessile. Male glumes 3–3·5 mm., obovate, brownish-hyaline. *Female spikes 2–4, contiguous, or the lower somewhat distant*, 8–15 × 4–6 *mm.* Peduncles of lower spikes short, included. *Bracts lf-like, many times as long as spikes*, narrow, rigid, *spreading or deflexed*. Female glumes c. 2 mm., broadly ovate or suborbicular, mucronate, straw-coloured with brownish patches. *Fr.* (Fig. 72 E) 3(–4) mm.,

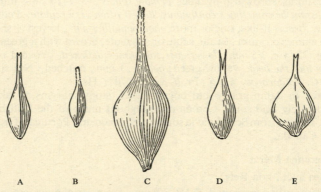

Fig. 73. Fruits of *Carex*. A, *C. sylvatica*; B, *C. capillaris*; C, *C. depauperata*; D, *C. pseudocyperus*; E, *C. rostrata*. × 5.

ovoid-trigonous, weakly veined, *greenish* or light brown; beak c. 0·5 mm., notched, smooth. Nut c. 2(–3) mm., broadly ovoid-trigonous. Fl. 6–7. Fr. 7–8. 2*n* = 60*. Hs.

Native. In grassy salt-marshes and on damp maritime cliffs and rocks. 66, H 24, S. Around the coasts of the British Is. north to Orkney, locally common. Europe to c. 60° N., W. Asia, N. Africa, N. America.

12. C. sylvatica Huds.　　　　　　　　　　　　　'Wood Sedge.'

A tufted glabrous perennial 15–60 cm. Stems slender, spreading or ± nodding, trigonous, smooth. Lvs 3–6 mm. wide, shorter than stems, slightly keeled, smooth, dark green; ligule c. 2 mm., triangular, subacute. Lower sheaths brown, fibrous. Male spike 1, 15–30 × 1–2 mm., tapering downwards. Male glumes 4–5 mm., narrowly obovate, obtuse or subacute, brownish-hyaline. *Female spikes 3–4, distant, ± nodding*, 20–50 × 3–4 *mm., rather lax-fld.* Peduncles filiform, rough, lowest often long-exserted. Lower bracts lf-like, sheathing, shorter than infl., upper setaceous. Female glumes c. 3 mm., ovate, acute, brown with hyaline margins and green midrib. *Fr.* (Fig. 73 A) 4–5 *mm.*, ellipsoid- or obovoid-trigonous, *green; beak* 1–1·5 *mm., bifid, smooth.* Nut 2–3 mm., ellipsoid- or obovoid-trigonous. Fl. 5–7. Fr. 7–9. 2*n* = 58*. Hs.

Native. On clayey soils in woods, and in grassland apparently as a woodland relict. 104, H 40. Generally distributed and common throughout the

British Is., except for the north of Scotland. Europe to S. Scandinavia (absent from S. Balkans); N. Africa; temperate Asia; N. America.

13. C. capillaris L. 'Hair Sedge.'

A tufted glabrous perennial 10–20 cm. Stems trigonous, smooth, rigid. Lvs 1–2·5 mm. wide, short, flat, recurved, all ± basal; ligule very short, truncate. Lower sheaths mostly with lvs, ± fibrous, dark brown. *Male spike* 1, 5–10 × 1 mm., *usually overtopped by female spikes*, peduncled. Male glumes 3·5 mm., oblong, obtuse, brownish-hyaline. *Female spikes* 2–3, 5–10 × 1 *mm., few-fld, all apparently arising from one bract. Peduncles slender, arcuate,* scabrid. Bracts lf-like, long-sheathing, lowest exceeding infl. *Female glumes* c. 2 mm., broadly ovate, mucronate, brownish-hyaline, *caducous.* Fr. (Fig. 73 b) c. 3 mm., dark brown, narrowly ovoid, tapering into the short slender smooth nearly entire beak. Nut 2 mm., obovoid-trigonous. Fl. 7. $2n = 54$. Hs.

Native. In moist turf, on hummocks in flushes and on rock-ledges of mountains on base-rich soils; very local. 21. Mountain districts of N. England and Scotland from Caernarvon and Mid-west Yorks to Caithness. Pyrenees, Alps, Carpathians, Caucasus, northern Europe; Arctic, Urals, Altai.

14. C. depauperata Curt. ex With.

A shortly creeping glabrous perennial 30–60 cm., forming large loose tufts. Stems trigonous, smooth, rather weak and slender. Lvs 2–3 mm. wide, shorter than stems, nearly flat, rough above; ligule 2–3 mm., ovate, acute. Lower sheaths lfless, fibrous, purplish. Male spike 1, 18–30 × 2–3 mm. Male glumes 5–6 mm., narrowly obovate, obtuse, brown with broad hyaline margins and pale midrib. *Female spikes* 2–4, 10–15 × 7–8 mm., 2–4(–6)-*fld*, erect, rather distant. Peduncles mostly longer than spikes, partly included, rough. Bracts lf-like, all longer than spikes and lower often exceeding infl., long-sheathing. Female glumes 5–6 mm., lanceolate, acuminate or aristate, brown with broad hyaline margins and green midrib. *Fr.* (Fig. 73 c) 8 *mm., rhomboidal, very bluntly trigonous*, narrowed at base, rather abruptly contracted into beak, *pale brown or greenish*, with many equal ribs; *beak* 3 *mm.*, straight, trigonous-cylindrical, angles rough, bifid, broadly hyaline in the notch. Nut 4 mm., broadly obovoid-trigonous. Fl. 5. Fr. 6. $2n = 44^*$. Hs.

A very distinct sp., readily recognized by the few-fld spikes and large fr.

Native. In dry calcareous woods and hedgebanks, very rare. 2 or 3. N. Somerset and Surrey, perhaps in Dorset; formerly in Kent and Anglesey. Europe from Belgium and S.W. Germany to the Balkans; Caucasus; Kamchatka.

15. C. pseudocyperus L. 'Cyperus Sedge.'

A glabrous tufted perennial 40–90 cm. Stems stout, sharply triquetrous, angles very rough. Lvs 5–12 mm. wide, longer than stem, bright green, margins rough; ligule 10–15 mm., triangular, acute. Lower sheaths dark brown, not fibrous. Male spike 1, 3–6 cm., sometimes female at top. Male glumes c. 6 mm., narrowly obovate with a long acumen, light brown with a paler midrib. *Female spikes* 3–5, *upper clustered, lowest ± remote*, 3–5 × 1 cm., *nodding*. Peduncles slender, rough. Bracts lf-like, lowest shortly sheathing, all but the uppermost exceeding infl. Female glumes 3·5–4·5 mm., similar in shape to those of the male fls, broadly hyaline with a narrow green area surrounding the slender whitish midrib; acumen serrulate. *Fr.* (Fig. 73 D) 5–6 mm., broader than glumes, *ovoid, tapering into the beak*, ± *asymmetrical*, green, many-ribbed, *shining, spreading or deflexed*; beak c. 2 mm., deeply notched, usually smooth. Nut 1·5 mm., ovoid-trigonous. Fl. 5–6. Fr. 7–8. $2n=66$. Hel.

Native. By slow-flowing rivers, in ditches, ponds and stagnant water in woods; local. 62, H 26, S. Scattered throughout England, Wales and Ireland; Scotland: recorded from Islay and Moray. Europe north to 62°, absent from the extreme south; Algeria; temperate Asia to N. Japan; Atlantic N. America; New Zealand (S. Island).

16. C. rostrata Stokes   'Beaked Sedge', 'Bottle Sedge'.
C. ampullacea Good.; *C. inflata* auct., non Huds.

A shortly creeping glabrous *rather glaucous* perennial 30–60 cm. or sometimes more. *Stems* erect, smooth and ½*-terete below*, rather rough and trigonous above. Lvs 3–7 mm. wide, longer than stems, inrolled or bluntly keeled, very glaucous above when young, rough; *ligule 2–3 mm., roundish-truncate*. Lower sheaths soon decaying, not fibrous. Male spikes 2–4, 20–70 × 1·5–2·5 mm., lower with a setaceous bract. Male glumes 5 mm., narrowly obovate, acute or subacute, light brown with a pale midrib. Female spikes 2–4, remote, 3–8 × 1 cm., suberect, upper often male at top. Peduncles short, smooth. Bracts lf-like, lower shortly sheathing, equalling or exceeding infl. *Female glumes 5 mm.*, oblong-lanceolate, *acute*, brown with a pale midrib. Fr. (Fig. 73 E) 5–6 mm., wider than glumes, ovoid, rather inflated, yellow-green; beak c. 1·5 mm., bifid, smooth. Nut 1·75 mm., obovoid-trigonous. Fl. 6–7. Fr. 7–8. $2n=c.\ 60, 76, 82$. Hel.

Native. In wet peaty places with a constantly high-water level, local. 110, H 40. Generally distributed throughout the British Is., but not recorded from Middlesex, Huntingdon and the Channel Is. Europe to c. 71° N., W. Asia to Altai; N. America.

17. C. vesicaria L.   'Bladder Sedge.'

A shortly creeping glabrous *dark green* perennial 30–60 cm. *Stems* erect, *triquetrous*, smooth below, *serrulate on the angles above*. Lvs 4–6 mm. wide, longer than stems, keeled, margins serrulate; *ligule 5–8 mm., ovate, acute*. Lower sheaths lfless, acute, developing a fibrous network with age, often reddish or purplish. Male spikes 2–3, 10–30 × 1–2 mm., lower with a setaceous bract. Male glumes c. 6 mm., oblanceolate, subacute, dark brown with a pale midrib. Female spikes 2–3, remote, upper erect, lower ± nodding, 15–35 × 7–10 mm. Peduncles filiform, smooth, lower up to 4 cm. *Bracts* lf-like, lower *exceeding infl., sheathing* or not. *Female glumes c. 3 mm.*, lanceolate, *acuminate*, dark brown with a pale midrib and hyaline acumen. Fr. (Fig. 74 A) 4–5 mm., much wider and longer than the glumes, ovoid, inflated, yellowish- or brownish-green, shiny; beak c. 1 mm., bifid, smooth. Nut c. 2 mm., obovoid-triquetrous. Fl. 6. Fr. 7. $2n=82$. Hel.

Native. In damp places liable to flooding, usually on humus-rich soils. 94, H 38, S. Scattered through the British Is. north to Orkney, local. Europe, N. Africa, temperate Asia, N. America.

18. C. stenolepis Less.

C. Grahamii Boott

A glabrous perennial 30–50 cm., somewhat resembling *C. vesicaria* but smaller. Angles of stems blunt and smooth below, acute and rough above. Lvs c. 3 mm. wide, scarcely keeled; *ligule* 3–4 *mm.*, ovate, *acute*. Male spikes 1–2, 10–35 × 1–2 mm., acute. Male glumes c. 5 mm., oblong-lanceolate, brown, hyaline at top. Female spikes 2–3, somewhat distant, 10–20 × 5–7 mm., lower 1–2 peduncled. *Lower bracts* lf-like, *shorter than infl.*, *not sheathing.* Female glumes c. 3 mm., ovate, obtuse or acute, dark purple-black with pale midrib and hyaline tip. *Fr.* (Fig. 74 B) 4–4·5 *mm.*, ovoid, inflated, *strongly ribbed*; beak 0·75 mm., bifid. Stigmas 2–3. Nut c. 2 mm., elliptic. Fl. 7. Hs.

Probably a hybrid between *C. saxatilis* and *C. hostiana* or *C. binervis.*

Native. In bogs on a few mountains in Argyll, mid Perth and Angus; local and rare. Norway.

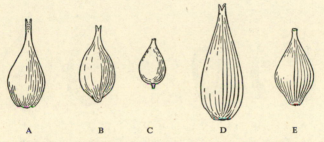

Fig. 74. Fruits of *Carex*. A, *C. vesicaria*; B, *C. stenolepis*; C, *C. saxatilis*; D, *C. riparia*; E, *C. acutiformis.* × 5.

19. C. saxatilis L.   'Russet Sedge.'

C. pulla Good.

A glabrous perennial 15–30 cm. Stems trigonous, smooth below, rough above, ± decumbent at base. Lvs 2–3 mm. wide, concave; *ligule c.* 1 *mm.*, broadly ovate or *almost truncate*. Lower sheaths lfless, persistent. Male spike 1(–2), 10–15 × 3 mm., fusiform. Male glumes 3–4 mm., ovate, obtuse, dark purple or almost black. Female spikes 1–2(–3), ± contiguous, 10–15 × 4–5 mm., erect, lower peduncled, upper nearly or quite sessile. *Lower bract* lf-like, *about as long as infl.*, not sheathing. Female glumes c. 3 mm., ovate, acute, dark purplish-brown, shorter than fr. *Fr.* (Fig. 74 C) *c.* 3·5 *mm.*, *smooth*, inflated, dark in upper half, pale below; beak c. 0·5 mm., notched. Stigmas 2, rarely 3. Nut 2 mm., subglobose. Fl. 7. $2n = 80$. Hs.

Native. In moderately base-rich flushes at 2000–3550 ft. on the higher Scottish mountains. 14. Local, from Dumfries to Caithness. Mainly in the mountains in arctic and subarctic Europe, Asia, America and Greenland.

20. C. riparia Curt.   'Great Pond-sedge.'

A large tufted glabrous perennial 1–1·6 m. Stems sharply triquetrous, rough. Lvs 6–20 *mm. wide*, longer than stem, sharply keeled, *glaucous*; ligule 10–15 mm., ovate, obtuse. *Male spikes* several, *often 5–6. Male glumes* c. 8 mm., oblong, *acuminate* or with an excurrent midrib, dark brown with pale margins and midrib. Female spikes 1–5, distant, upper suberect, lower nodding, (3–)6–9 × 1–1·5 cm., sometimes male at top, acute. Upper peduncles short or almost 0, lower rather long. Bracts lf-like, lowest shortly sheathing, over-topping stem. Female glumes c. 7 mm., oblong-lanceolate to oblong-ovate, brown; midrib strongly excurrent, pale. *Fr.* (Fig. 74 D) *c. 8 mm.*, ovoid, strongly convex on the back, weakly so on the front; beak c. 1·5 mm., bifid, smooth. Nut 2·5–3 mm., ovoid- or obovoid-trigonous, stipitate. Fl. 5–6. Fr. 7–8. 2*n* = 72. Hel. or Hs.

Native. By slow-flowing rivers, in ditches and ponds, more rarely on drier ground. 93, H 30, S. Common and generally distributed in S. England and the Midlands, rarer in the north; unrecorded from Radnor and Merioneth; local in Scotland and absent from many of the islands; scattered throughout Ireland. Europe, except the north and parts of the south, N. Africa, Caucasus, W. Asia.

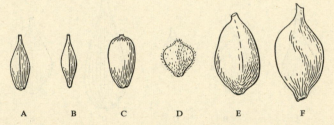

Fig. 75. Fruits of *Carex.* A, *C. pendula*; B, *C. strigosa*; C, *C. pallescens*; D, *C. filiformis*; E, *C. panicea*; F, *C. vaginata.* × 5.

21. C. acutiformis Ehrh.   'Lesser Pond-sedge.'

C. paludosa Good.

A shortly creeping glabrous somewhat glaucous perennial 60–150 cm. Stems sharply triquetrous, smooth below, rough above. *Lvs 7–10 mm. wide*, those of sterile shoots equalling or exceeding flowering stems, ± keeled, margins smooth below, serrulate above; ligule 1–3 cm., lanceolate, acute, margins often purplish. Lower sheaths lfless, brown, not fibrous. *Male spikes* 2–3, contiguous, 10–40 × 3–4 mm. *Male glumes* 5–6 mm., oblong-lanceolate, *obtuse or subacute*, dark brown with a slightly paler midrib. Female spikes 3–4, remote, 20–40 × 7–8 mm., erect, often male at top. Peduncles 0 or short, smooth. Bracts lf-like, lowest broad, exceeding infl., not or shortly sheathing. Female glumes 4–5 mm., oblong-lanceolate with a long, often serrulate acumen, purple-brown with a pale midrib. *Fr.* (Fig. 74 E) *c. 4 mm.*, elliptic, flattened, shortly stipitate, many-nerved, pale green; beak c. 0·2 mm., emarginate or notched. Nut 2 mm., obovoid-trigonous. Fl. 6–7. Fr. 7–8. Hel.

Native. Beside slow-flowing rivers, canals and ponds on clayey or peaty base-rich soils. 98, H 29. Scattered throughout the British Is. north to Caithness, locally abundant. Europe, except the north; N. and S. Africa; temperate Asia; N. America.

22. C. pendula Huds. 'Pendulous Sedge.'

A stout tufted glabrous perennial 60–150 cm. Stems triquetrous, smooth. *Lvs* 15–20 *mm. wide*, shorter than stems, ± keeled, yellow-green above, somewhat glaucous beneath, margins rough; ligule 3–6 cm., lanceolate, acute. Lower sheaths soon decaying. Male spike 1, 6–10 cm. × 5–7 mm. Male glumes c. 7 mm., lanceolate, acuminate, brownish-hyaline. *Female spikes* 4–5, *distant*, *pendulous*, 7–16 *cm.* × 5–7 mm., tapering somewhat towards base, dense-fld, often male at top. Lower peduncles included, upper 0. Lower bracts lf-like, about equalling infl., long-sheathing; upper glumaceous with setaceous points, scarcely sheathing. Female glumes 2–2·5 mm., ovate, acute or acuminate, reddish-brown. Fr. (Fig. 75 A) c. 3 mm., ellipsoid- or ovoid-trigonous, green-brown; *beak c.* 0·2 *mm.*, *slightly notched*. Nut 2 mm., obovoid-trigonous, shortly stipitate. Fl. 5–6. Fr. 6–7. $2n=58, 60$. Hs.

One of our most handsome spp., readily recognized by its large size, broad lvs, and long pendulous spikes.

Native. In damp woods and on shady stream banks, usually on clayey soils. 87, H 33, S. Scattered throughout the British Is. north to Moray and Mull, local. Europe from Denmark southwards; W. Asia; N. Africa.

23. C. strigosa Huds.

A shortly creeping glabrous perennial 35–70 cm. Stems trigonous, smooth, spreading and often nodding. *Lvs* 6–10 *mm. wide*, much shorter than stems, ± keeled, bright green, soft; ligule c. 10 mm., triangular, acute. Lowest sheaths lfless, brown, ± fibrous. Male spike 1, 30–40 × 2 mm., stalked. Male glumes 5–6 mm., narrowly obovate, brownish with a green midrib. *Female spikes* (3–)5–6, *distant*, the lowest near base of stem, (25–)40–80 × 1·5–2 mm., lax-fld, ± *erect*, *never pendulous*. *Peduncles* slender, *smooth*, nearly or quite included. Lower bracts lf-like, longer than spikes, long-sheathing; upper setaceous, shortly sheathing. Female glumes c. 2·5 mm., ovate or ovate-lanceolate, acute, green becoming brown. Fr. (Fig. 75 B) c. 3·5 mm., oblong or narrowly lanceolate, trigonous, rather suddenly narrowed at both ends, often ± curved, green; *beak* c. 0·2 *mm.*, emarginate. Nut c. 1·5 mm., oblong, trigonous. Fl. 5–6. Fr. 8–9. $2n=66*$. Hs.

Native. In somewhat open places in damp woods, especially beside streams on base-rich soils; local. 44, H 23. S. England, E. Anglia, Midlands, Cheshire, Lancashire, Yorks, Westmorland, Cumberland; Glamorgan and Denbigh; scattered throughout Ireland. W. and S.W. Europe from Denmark southward; apparently always local.

24. C. pallescens L. 'Pale Sedge.'

A tufted perennial 20–50 cm. *Stems* sharply triquetrous, somewhat rough and *shortly hairy*, especially on the angles. *Lvs* up to 5 mm. wide, weakly keeled, ± *hairy beneath*, particularly on the ribs and margins; ligule c. 5 mm., ovate, obtuse, margins usually brown and ± ciliate. *Sheaths* ± *hairy*, the lowest brown, lfless, acute. *Male spike* 1, c. 8 mm., ± *concealed by female spikes*. Male glumes c. 4 mm., obovate-oblong, subacute or mucronate, pale brown with a slender dark midrib. *Female spikes* 2–3, *contiguous* or the lower remote, 5–20 × 5–6 mm., sometimes with a few male fls at top, suberect or lower ± nodding. Peduncles slender, the lowest up to c. 2 cm. Lowest bract lf-like, crimped at the base, often overtopping infl., upper small, shorter than spike, often setaceous. Female glumes 3–3·5 mm., ovate, acuminate, hyaline with a broad green excurrent midrib. *Fr.* (Fig. 75 c) c. 4 mm., *ovoid-oblong, convex on both sides, bright green,* ± distinctly veined; *beak very short* or almost 0. Nut 2 mm., ellipsoid-trigonous. Fl. 5–6. 2*n* = 64*; 66. Hs.

Native. In damp woods, less frequently in damp grassy places. 104, H 32. Scattered throughout the British Is. north to Caithness; rather local but not uncommon. Europe, to 70° N. in Norway, absent from parts of the south; temperate Asia; N. America.

25. C. filiformis L.   'Downy Sedge.'

C. tomentosa auct.

A shortly creeping nearly glabrous perennial 20–45 cm. Stems triquetrous, rough towards top, slender, rigid, erect. *Lvs* 1·5–2 mm. wide, shorter than stems, *nearly flat,* rough, rather glaucous; ligule c. 2 mm., ovate. Lower sheaths lfless, purplish, persistent. Male spike 1, 12–25 × 1·5–2 mm., shortly stalked. Male glumes c. 5 mm., obovate-oblong, obtuse, apiculate, brownish-hyaline. *Female spikes* 1–2, 8–14 × 4–5 mm., *dense-fld, rather distant.* Peduncles very short. Lower bracts lf-like, equalling or exceeding spike, not sheathing. Female glumes 2 mm., broadly ovate, mucronate, purplish-brown, midrib green. *Fr.* (Fig. 75 d) 2 *mm.,* broadly pear-shaped, trigonous, *shortly tomentose; beak very short,* slightly emarginate. Nut 1·5 mm., pyriform, trigonous. Fl. 5–6. Fr. 6–7. 2*n* = 48*. Hs.

Native. In damp meadows and pastures, very local. 6. Wilts, Sussex, Surrey, Middlesex, Oxford, Gloucester. C. and S. Europe, except the Mediterranean region, to 60° N. in Sweden; Asia Minor; Caucasus.

26. C. panicea L.   Carnation-grass.

A shortly creeping glabrous *glaucous* perennial 10–40(–60) cm. Stems erect, trigonous, smooth. Lvs 2–5 mm. wide, shorter than stems, nearly flat, glaucous, slightly rough; ligule of old lvs much broader than long, truncate. Lower sheaths dull brown, not fibrous. *Male spike* 1, 10–20 × 1–2 mm. Male glumes c. 3 mm., ovate, subacute, brown with a pale midrib and hyaline margins. *Female spikes* 1–2(–3), distant, erect, 10–15(–25) × 4–6 mm., *rather few-fld* (up to c. 20), sometimes male at top. Peduncles rigid, shortly exserted or, more often, quite included. *Lower bract* lf-like, shorter or somewhat longer than spike, *closely sheathing.* Female glumes c. 2 mm., broadly ovate, acute, clasping base of fr., brown with a pale midrib and hyaline margins. *Fr.* (Fig. 75 e) 3–4 *mm., ovoid, inflated, asymmetrical, smooth,* olive-green, brownish or purplish; *beak short* (up to 0·5 mm.), *entire.* Nut 2 mm., broadly obovoid-trigonous. Fl. 5–6. Fr. 6–7. 2*n* = 32*. Hs.

Resembles *C. flacca,* with which it often grows, in many respects, but may be distinguished by the solitary male spike, larger swollen fr., fewer-fld spikes, and the truncate, not rounded, ligule.

Native. In wet grassy places and in fens. 112, H 40, S. Common and generally distributed throughout the British Is.; ascends to 4000 ft. in Scotland. Europe, except the south, temperate Asia; introduced in N. America.

27. C. vaginata Tausch

C. sparsiflora (Wahlenb.) Liljeb. ex Steud.

A shortly creeping glabrous *green or yellow-green* perennial 10–25(–40) cm. Stems trigonous, smooth, often curved. *Lvs* 3–5 mm. wide, shorter than stems, those *of fertile shoots very short, not glaucous,* ± keeled; ligule very short, truncate. Lower sheaths lfless, brownish, scarcely fibrous. Male spike 1, 10–15 × 4–5 mm., clavate. Male glumes 4–4·5 mm., obovate or oblong-obovate, obtuse, chestnut-brown, midrib green. Female spikes 1–2, 6–12 × 4–5 mm., distant, few-fld, but rather dense, erect. Peduncles smooth, ± included. Bracts lf-like, broad, abruptly contracted at apex, shorter or not much longer than spike, long-sheathing; *sheaths loose, funnel-shaped.* Female glumes 2·5–3 mm., broadly ovate, obtuse or mucronate, chestnut-brown, midrib pale. *Fr.* (Fig. 75 F) 3·5–4 mm., ovoid-trigonous, *not inflated, usually rather curved,* nerveless or with 2 prominent lateral nerves; *beak* c. 1 mm., *obliquely truncate, slightly notched,* smooth. Nut c. 2 mm., ellipsoid or ovoid, trigonous. Fl. 7. Fr. 8–9. 2n = 32. Hs.

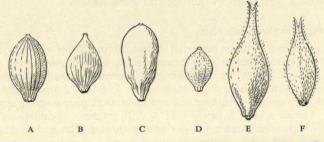

Fig. 76. Fruits of *Carex.* A, *C. limosa*; B, *C. paupercula*; C, *C. rariflora*; D, *C. flacca*; E, *C. hirta*; F, *C. lasiocarpa*. × 5.

Native. In wet grassy places, flushes and on rock-ledges on Scottish mountains. 16. On the higher mountains from Dumfries to Caithness, local. N. Europe, mountains of C. Europe to the Alps and Pyrenees; Siberia; N. Japan.

28. C. limosa L.   'Mud Sedge.'

A creeping glabrous perennial up to c. 30 cm. *Stems* triquetrous, *rough,* slender, rigid. *Lvs* 1–1·5 *mm. wide,* shorter than stems, *channelled, the greater part of the margins serrulate*; ligule 4–5 mm., acute. Lower sheaths lfless, reddish, not fibrous. Male spike 1, 10–20 × 2 mm., erect. *Male glumes* 4 mm., lanceolate, *acute or subacute,* reddish-brown. *Female spikes* 1–2(–3), 7–12 × 5–7 mm., nodding, *up to 20-fld.* Peduncles 0·5–2 cm., very slender, smooth. *Bracts with a lf-like point shorter than infl. or rarely equalling it,* shortly sheathing. *Female glumes* 3·5–4 mm., *ovate, acute,* brown or purplish with a green midrib. *Fr.* (Fig. 76 A) 3·5 *mm.,* elliptic, biconvex, *slightly broader than glumes,* strongly ribbed, pale brownish- or bluish-green; beak very short, entire. Nut 2–2·5 mm., obovoid-trigonous, shiny. Fl. 5–6. Fr. 7–8. 2n = 56, 64. Hel.

Native. In very wet bogs, rooting in peat detritus in shallow water, locally common. 44, H 25. Scattered throughout the British Is. from Dorset and Hampshire northwards, more frequent in the north and west. Europe north to c. 71°, only on mountains in the south; N. Asia; N. America.

29. C. paupercula Michx.

C. irrigua (Wahlenb.) Sm. ex Hoppe; *C. magellanica* auct., vix Lam.

A shortly creeping glabrous perennial up to 30 cm., resembling *C. limosa* but differing as follows: *Stems smooth. Lvs 2–3 mm. wide, flat, the greater part of the margins smooth. Female spikes 2–3, 5–12 × 4–8 mm., up to 10-fld. Lower bracts exceeding infl. Female glumes 5–5·5 mm., lanceolate, acuminate or aristate,* caducous, brown or purplish. *Fr.* (Fig. 76 B) 3 mm., ovate or sub-orbicular, biconvex, *much broader than glumes*, ribs not very conspicuous. Fl. 5–6. Fr. 6–7. 2n = 58. Hel.

Native. In very wet bogs, rare. 18, H 1. Denbigh; Durham, Northumberland, Cumberland; scattered through Scotland north to Lewis; Antrim. C. and E. Alps, Carpathians, N. Europe and arctic regions.

30. C. rariflora (Wahlenb.) Sm.

A shortly creeping glabrous perennial up to c. 20 cm. *Stems trigonous, smooth. Lvs 1–2 mm. wide, shorter than stems, flat, the greater part of the margins smooth;* ligule c. 3 mm., acute, upper part of sheaths broadly membranous. Lower sheaths soon decaying. Male spike 1, 6–10 × 2 mm., erect. Male glumes 4 mm., ovate, ± truncate and apiculate, dark purple with narrow pale midrib. Female spikes 2, rarely 1 or 3, c. 10 × 4 mm., nodding, lax, up to c. 10-fld. Peduncles slender. Lower bracts shorter than spike and usually shorter than peduncles. *Female glumes 3–4 mm., obovate, almost truncate,* reddish-purple, midrib narrow, pale. *Fr.* (Fig. 76 c) c. 3 mm., ellipsoid, *narrower and shorter than glumes,* strongly ribbed; beak very short, entire. Nut 2 mm., ellipsoid, trigonous. Fl. 6. Fr. 7. 2n = 50; 54. Hel.

Native. In small bogs and wet peaty places on the higher Scottish mountains, very local. 6. Perth, Angus, Aberdeen, Banff and Inverness. Arctic, Scandinavia, America, Kamchatka.

31. C. flacca Schreb.   Carnation-grass.

C. glauca Scop.; *C. diversicolor* Crantz, p.p.

A shortly creeping glabrous *glaucous* perennial 10–40 cm. Stems erect, bluntly trigonous or subterete, smooth. Lvs 2–4 mm. wide, shorter than stems, slightly keeled, becoming flat, pale green above, glaucous beneath, rough; ligule c. 2 mm., usually rounded. Lower sheaths not fibrous, reddish. *Male spikes* 2–3, very seldom 1, 10–30 × 1–2 mm. Male glumes 3–4 mm., narrowly obovate, rounded or subacute at apex, purple-brown, midrib broad, whitish, margins narrowly hyaline. *Female spikes* 2(–3), ± remote, erect or nodding, 15–40 × 4–6 mm., *dense-fld*, or rather lax towards base, often male at top. Peduncles slender, nearly smooth, variable in length. Bracts lf-like, shorter than infl., ± sheathing. Female glumes 1·5–2 mm., oblong-ovate, abruptly contracted to the acute or mucronate apex, purplish-brown or black, margins hyaline, midrib often pale. *Fr.* (Fig. 76D) 2–2·5 *mm.*, ellipsoid to obovoid, rather asymmetrical, *minutely papillose*, yellow-green, reddish or almost black; beak very short, entire. Nut c. 1·5 mm., obovoid-trigonous, faces ± concave. Fl. 5–6. Fr. 7–8. $2n=76^*$. Hs.

Native. In dry calcareous grassland, damp clayey woods, marshes, fens and bogs; common. 112, H40, S. Generally distributed throughout the British Is. Europe to c. 67° N. in Norway, N. Africa, Siberia; introduced in N. America, West Indies and New Zealand.

32. C. hirta L. Hammer Sedge.

A shortly creeping somewhat hairy perennial (15–)30–60 cm. Stems triquetrous, glabrous, shiny. *Lvs* 2–4 *mm.* wide, shorter than stems, channelled, *hairy on both surfaces; sheaths hairy*, *often densely so*; ligule 1–2 mm., rounded. Lower sheaths mostly with a short blade, reddish, nearly glabrous. Male spikes 2–3, 10–20 × 3–4 mm., fusiform. Male glumes c. 5 mm., obovate, mucronate, ± hairy, reddish-brown; midrib broad, pale; margins and tip hyaline. *Female spikes* 2–3(–5), 10–30 × 5–7 mm., *erect*, *distant*, *the lowest often near base of stem*, peduncles mostly included, glabrous. Bracts similar to lvs, all much longer than spikes, upper about equalling infl., all sheathing. Female glumes 6–8 mm., ovate or oblong, tapering rather abruptly into a long ciliate awn, pale greenish-hyaline. *Fr.* (Fig. 76E) 6–7 *mm.*, ovoid, many-ribbed, *pubescent*, greenish; *beak c.* 2 *mm.*, *bifid*, rough inside and outside notch. Nut 2–3 mm., obovoid-trigonous, often stipitate. Fl. 5–6. Fr. 6–7. $2n=112^*$. Hs.

Native. In rough grassy places, woods, damp meadows and damp sandy hollows, common. 103, H40, S. Throughout the British Is., though unrecorded from Cardigan, Montgomery, Isle of Man, and absent from the extreme north of Scotland, Orkney and Shetland. Europe, except the north; N. Africa; temperate Asia.

33. C. lasiocarpa Ehrh. 'Slender Sedge.'
C. filiformis auct. angl.

A rhizomatous nearly *glabrous* perennial 45–120 cm. Stems trigonous, slender, rigid, smooth or slightly rough above. *Lvs c.* 1 *mm.* wide, shorter than stem, *channelled*, stiff, grey-green; ligule 2–3 mm., ovate, obtuse, dark purplish at tip. Lower sheaths many, lfless, dark purplish-brown, margins sometimes becoming filamentous in decay. Male spikes 1–3, 30–70 × 2–3 mm. Male glumes 4–5 mm., lanceolate, acute, purplish-brown, midrib pale. Female spikes 1–3, 10–30 × 4–6 mm., ± remote, erect. Peduncles short or 0. Bracts lf-like, lower often exceeding infl., very shortly sheathing. Female glumes c. 4 mm., lanceolate, acuminate, chestnut-brown, midrib pale. *Fr.* (Fig. 76F) *c.* 4 *mm.*, ovoid, subtrigonous, *densely greyish-tomentose; beak c.* 0·5 *mm.*, deeply bifid. Nut c. 3 mm., ovoid-trigonous, stipitate. Fl. 6–7. Fr. 7–8. Hel.

Native. In reed-swamp in shallow water and in wet acid or alkaline peaty places, locally common; ascends to 1950 ft. 55, H26. Scattered through the British Is. from S. Devon to W. Sutherland. C. and N. Europe, rare in the south; N. Asia; N. America.

34. C. pilulifera L.   'Pill-headed Sedge.'

A nearly glabrous densely tufted rather rigid perennial 10–30 cm. Stems sharply triquetrous, slightly rough above, ± incurved. Lvs 2 mm. wide, shorter than the stems, recurved, yellow-green, nearly flat, rough especially near the top; ligule c. 1 mm., truncate or broadly triangular. Lower sheaths fibrous when old. Male spike 1, slender. Male glumes c. 4 mm., lanceolate, acute, brown with a pale or green midrib. *Female spikes* 2–4, *contiguous* or the lowest somewhat distant, 5–6(–8) × 4–6 mm., erect, *few-fld*. Peduncles 0. *Bracts* short or the lowest somewhat exceeding infl., *narrowly lf-like, green*, not sheathing. Female glumes 3–3·5 mm., broadly ovate, acuminate, brown with a green midrib. *Fr.* (Fig. 77 A) c. 2·5 mm., *almost globose, puberulent*, ribbed, grey-green, stipitate; stalk variable in length, stout; beak 0·5–0·75 mm., obliquely truncate becoming slightly notched, dark. Nut 1·5 mm., broadly obovoid, rounded-trigonous. Fl. 5–6. $2n=18*$. Hs.

Var. *longebracteata* Lange has slender, few (3–5)-fld female spikes, the lower 1 or 2 distinctly peduncled; lowest bract lf-like, about as long as infl.

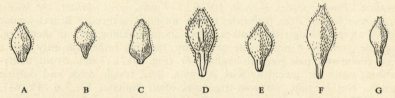

| | | | | | | |
| A | B | C | D | E | F | G |

Fig. 77. Fruits of *Carex*. A, *C. pilulifera*; B, *C. ericetorum*; C, *C. caryophyllea*; D, *C. montana*; E, *C. humilis*; F, *C. digitata*; G, *C. ornithopoda*. × 5.

Native. In grassy or heathy places, sometimes in open woods, on sandy or peaty acid soils; locally common. 109, H40, S. Throughout the British Is.; ascends to 3800 ft. on Cairntoul, Aberdeen. Europe, north to c. 68° in Norway, absent from parts of the south; N. Asia.

35. C. ericetorum Poll.

A nearly glabrous tufted perennial 5–15 cm. *Stems* slender, *smooth*, bluntly trigonous. *Lvs up to* 4 *mm. wide*, shorter than the stems, nearly flat; ligule very short, truncate. Lower sheaths brown, fibrous. Male spike 1, 6–12 × 2 mm., clavate. Male glumes 2·5–3 mm., obovate-oblong, obtuse, purplish-brown with a hyaline border at top. *Female spikes* 1–2(–3), contiguous, 5–8 × 3–4 mm., *dense-fld*. Peduncles 0. *Bracts glumaceous*, small, the lower with a short setaceous point. *Female glumes* 2·5 mm., broadly obovate-oblong, dark purplish-brown with a slender pale midrib; *apex rounded*, hyaline, *very finely ciliate. Fr.* (Fig. 77 B) c. 2·5 mm., obovoid-trigonous, pubescent, green below, *dark brown above*; beak c. 0·3 mm., rather stout, somewhat notched. Nut 1·5 mm., broadly obovoid-trigonous. Fl. 4–5. Fr. 6–7. $2n=30*$. Hs.

Native. In dry grassland on chalk and limestone, very local. 11. W. Suffolk, W. Norfolk, Cambridge, Lincoln, Derby, Yorks, Lancs, Durham, Westmorland. Scattered throughout Europe, local and absent from large areas, to c. 68° N. in Finland; Caucasus, Siberia.

36. C. caryophyllea Latour. 'Spring Sedge.'

C. verna Chaix, non Lam.; *C. praecox* auct., non Schreb.

A nearly glabrous creeping perennial 5–15(–30) cm. Stems triquetrous, slender. Lvs c. 2 mm. wide, shorter than stems, nearly flat, rough; ligule 1–2 mm., ovate, obtuse. Lower sheaths dark brown or blackish, fibrous. Male spike 1, rather long and stout (10–15 × 2–3 mm.), obtuse, occasionally with a few female fls at base. Male glumes c. 5 mm., oblong, subacute or mucronate, thin, brown above, hyaline below. Female spikes 1–3, contiguous or rarely the lowest distant, 5–12 × 3–4 mm., rarely male at top, erect. Peduncles 0 or rarely the lowest up to c. 1 cm. *Bracts* setaceous, *shortly sheathing*. Female glumes c. 2·5 mm., broadly ovate, obtuse and mucronate or ± acute, brown and shining, midrib green. *Fr.* (Fig. 77 c) *c.* 2·5 *mm.*, obovoid or ellipsoid, narrowed below, trigonous, pubescent, *olive-green*; beak very short, conical, slightly notched. Nut 2 mm., broadly obovoid-trigonous, stipitate, crowned by swollen style-base. Fl. 4–5. Fr. 6–7. $2n = 64*, 66*, 68*; 62$. Hs.

Native. In dry grassland, particularly calcareous, locally common. 106, H 40, S. Throughout the British Is. north to Orkney. Europe, except the Arctic; temperate Asia, N. America.

37. C. montana L. 'Mountain Sedge.'

A nearly glabrous shortly creeping and tufted perennial 10–30 cm. *Stems* slender, not rigid, *rough at top*. Lvs c. 2 mm. wide, shorter or longer than stems, sparsely pubescent beneath becoming glabrous when old; ligule c. 1 mm., ovate, ± acute. Lower sheaths dark reddish-brown, fibrous. Male spike 1, subclavate, acute. Male glumes c. 5 mm., oblong, obtuse or sub-acute, purplish-hyaline with a pale midrib often excurrent in a dark mucro. Female spikes (1–)2(–3), contiguous, 6–9 × 4–6 mm., erect, few-fld. Peduncles 0. *Bracts* glumaceous, the lower sometimes with a setaceous point, *not sheathing*. *Female glumes* 3–3·5 mm., broadly ovate or obovate, obtuse or retuse, *mucronate*, dark purple-brown with a pale midrib. *Fr.* (Fig. 77 D) 3–3·5 *mm.*, exceeding glumes, ellipsoid or obovoid, trigonous, pubescent, *pale brown or blackish* on the exposed face, stipitate; stalk variable, stout; beak c. 0·2 mm., dark, notched. Nut 2·5 mm., ovoid-trigonous, stipitate. Fl. 5. $2n = 38*$. Hs.

Native. In grassy and heathy places and in open woods, chiefly on lime-stone, locally abundant. 20. From Cornwall to Kent, north to Shropshire and Bucks, Derby; Glamorgan, Brecon and Denbigh; ascends to 800 ft. in the Mendips (Somerset). Europe, to c. 61° N. in Sweden, absent from parts of the Mediterranean region.

38. C. humilis Leyss. 'Dwarf Sedge.'

C. clandestina Good.

A small nearly glabrous tufted perennial. Stems 2–5(–15) cm., slender, flexuous, terminal. Lvs 1–1·5 mm. wide, longer than and ± concealing the stems, at first flat, later channelled, rough, ligule very short, truncate. Lower sheaths dark red-brown, fibrous. Male spike 1, c. 1 cm., tapering at both ends. Male glumes c. 6 mm., oblong, obtuse or subacute, purplish-brown with very broad hyaline margins. *Female spikes* 2–4, distant, 4–10 × 1–2 *mm.*, erect, *very lax*, 2–4-*fld. Peduncles short. Bracts glumaceous, sheathing,* ± *enclosing the spikes.* Female glumes 3 mm., ovate or suborbicular, ± acute, purplish-hyaline, margins incurved. Fr. (Fig. 77E) 2·5 mm., broader than glumes, obovoid-trigonous, puberulent; beak very short, entire or slightly notched. Nut 2 mm., obovoid-trigonous. Fl. 3–5. Fr. 5–7. 2*n* = 36*; 78. Hs.

Native. In short turf on chalk and limestone, locally abundant. 7. N. Somerset, W. Gloucester, Hereford, Wiltshire, Dorset, S. Hampshire. C. and S. Europe, Caucasus, Siberia.

39. C. digitata L.   'Fingered Sedge.'

A nearly glabrous tufted perennial. *Stems* 5–15(–25) cm., slender, flexuous, bluntly trigonous, *lateral, with a number of lfless sheaths enfolding the base.* Lvs 2–4 mm. wide, about as long as the stems, flat, yellowish-green, shortly and sparsely pubescent beneath, or almost glabrous; ligule 1–2 mm., ± triangular, obtuse or acute. Lower sheaths blackish, fibrous. *Male spike* 1, slender, *few-fld, overtopped by the upper* 1–2 *female spikes and therefore appearing lateral.* Male glumes 5 mm., oblong, obtuse or emarginate, whitish above, purplish-brown below, margins incurved. *Female spikes* 1–3, *rather distant,* 10–15 × 2 mm., very lax, 5–8-fld. Peduncles slender, rather short. Bracts glumaceous, sheathing, ± enclosing the peduncles and sometimes the base of the spike. Female *glumes* 3–4 mm., *enclosing the fr., broadly obovate,* ± *emarginate,* mucronulate, pale brown and shiny, midrib green. Fr. (Fig. 77F) 3–4 mm., obovoid-trigonous, puberulent, brown and rather shiny; beak c. 0·4 mm., nearly entire. Nut c. 2·5 mm., obovoid, stipitate, trigonous, faces nearly flat. Fl. 4–5. Fr. 6–7. 2*n* = 48*, 50*; 52. Hs.

Native. In rather open woods on chalk or limestone, usually in cracks of rock outcrops, or in cracks in ± wooded limestone pavement; very local and rather rare. 15. N. Somerset, N. Wilts, Gloucester, Monmouth, Hereford, Worcester; Nottingham and Derby to N. Yorks and Westmorland. Europe to c. 68° N. in Norway, absent from parts of the south; temperate Asia.

40. C. ornithopoda Willd.   'Bird's-foot Sedge.'

A tufted perennial resembling *C. digitata* but usually smaller. Lvs glabrous, margins recurved-serrulate. Female spikes 5–10 mm., contiguous, very lax, seldom more than 5-fld. Female *glumes* 2–2·5 *mm.*, folded round base of fr., *oblong, angled,* tapering abruptly to an *acute or subacute apex,* pale brown, margins hyaline above. *Fr.* (Fig. 77G) 3–4 mm., *much exceeding glumes.* Nut c. 2·5 mm., elliptic, stipitate, trigonous, faces concave. Fl. 5. Fr. 6–7. 2*n* = c. 46. Hs.

Native. In dry grassland on limestone slopes or in cracks in limestone pavement, rarely (?never) in shade; very local. 4. Derby, Yorks, Westmorland, ?Cumberland. C. and S. Europe north to c. 67° in Norway; Asia Minor; Urals.

C. glacialis Mackenzie, somewhat resembling *C. digitata* but with the lower bract lf-like and the fr. glabrous, has been reported from the island of Rhum, Inner Hebrides.

41. C. buxbaumii Wahlenb.

C. canescens L., nom ambig.; *C. fusca*; *C. polygama* Schkuhr auct., no nJ.F. Gmel.

A shortly creeping glabrous perennial 30–60 cm. Stems triquetrous, smooth, stiff. Lvs c. 2 mm. wide, shorter than stems, nearly flat, rather glaucous; ligule c. 3 mm., lanceolate, acute. Lower sheaths lfless, reddish; lf-sheaths reddish, margins often filamentous. *Spikes* 3–5, 7–15 × 5 mm., *erect, ± remote*, lower shortly stalked, upper sessile. Bracts lf-like, lowest sometimes exceeding infl. Female glumes 3·5–4 mm., ovate or oblong-ovate, dark purplish-brown with a pale *midrib excurrent in an arista*. Fr. (Fig. 78 A) c. 4 mm., ovate, much broader than glumes, becoming trigonous when ripe, indistinctly nerved, pale greenish; beak very short, notched. Nut 2·5 mm., elliptic, trigonous. Fl. 6–7. Fr. 7–8. $2n=$ c. 74. Hel.

Native. In a wet spongy bog. Inverness. 2. Europe, local; Algeria; N. Asia; N. America.

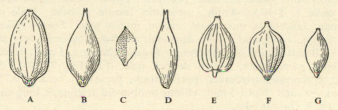

Fig. 78. Fruits of *Carex*. A, *C. buxbaumii*; B, *C. atrata*; C, *C. norvegica*; D, *C. atrofusca*; E, *C. elata*; F, *C. acuta*; G, *C. aquatilis*. × 5.

42. C. atrata L.   'Black Sedge.'

A glabrous perennial 30–50 cm. Stems triquetrous, smooth, often ± nodding at top. Lvs c. 5 mm. wide, keeled, rather glaucous; ligule 1–2 mm., ovate or ± truncate. Lower sheaths lfless, persistent, dark brown. *Spikes* 3–5, *contiguous, lower eventually nodding*, 10–20 × 3–5 mm., terminal one male at base only. Peduncles of lower spikes 1–2 cm. *Lowest bract often exceeding infl.*, others glumaceous. *Female glumes* c. 4 mm., ovate, *acute*, purple-black with pale midrib. Fr. (Fig. 78 B) 4·5 mm., narrowly elliptic or obovate, compressed, ± trigonous, minutely punctate; beak c. 0·2 mm., notched. Nut c. 2 mm., ovoid-trigonous. Fl. 6–7. $2n=54^*$; 56. Hs.

Native. On wet rock-ledges on mountains at 2400–3700 ft., local. 14. Caernarvon, Lake District and the higher Scottish mountains from Dumfries to Moray and Inverness. Europe from the Arctic to the Balkans and Pyrenees, on the higher mountains in the south; northern Asia and on mountains south to the Caucasus, Turkistan and Baikal; N. America: Rocky Mountains south to Utah.

43. C. norvegica Retz.

C. Halleri Gunn., p.p.; *C. Vahlii* Schkuhr; *C. alpina* Lijeb., non Schrank.

A glabrous perennial 15–30 cm., forming small tufts. Stems triquetrous, rough above, rigid. Lvs 1·5–3 mm. wide, shorter than stems, nearly flat; ligule c. 1 mm., rounded. Lower sheaths mostly with lvs, reddish, not fibrous.

Spikes 1–4, *contiguous, erect*, 5–8 × 4–5 mm., the lowest shortly stalked, the others sessile. *Lower bract short, lf-like.* Female glumes 1·5 mm., ovate, acute, purplish-black with pale midrib. Fr. (Fig. 78 c) 2 mm., obovoid-trigonous, minutely papillose, especially at the top; beak c. 0·25 mm., emarginate. Nut 1·5 mm., obovoid-trigonous. Fl. 6–7. Fr. 7–8. $2n = 54$, 56. Hs. or Hel.

Native. Among short grass on wet rock-ledges and by mountain streams, 2000–3200 ft. 3. Mid Perth, Angus and S. Aberdeen. Alps, mountains of Scandinavia to c. 71° N., Iceland.

44. C. atrofusca Schkuhr

C. ustulata Wahlenb.

A shortly creeping glabrous perennial 10–35 cm. Stems trigonous, smooth, often nodding at top. Lvs 3–5 mm. wide, much shorter than stems, nearly flat, rather soft; ligule 2–3 mm., rounded. Lower sheaths soon decaying. Male spike 1, 5–10 × 3–4 mm., sometimes female at base. Male glumes 3–3·5 mm., lanceolate, mucronate, brown, midrib pale. *Female spikes* 2–4, c. 10 × 7 mm., *contiguous, ovoid*, dense-fld, *nodding. Peduncles very slender, smooth.* Bracts setaceous, sheathing. *Female glumes* c. 3 mm., lanceolate, acuminate, *purple-black*, midrib slender, pale. Fr. (Fig. 78 D) 4·5–5 mm., elliptic, compressed-trigonous, purple-black, nerveless; beak c. 0·75 mm., notched, ± rough. Nut 1·5 mm., elliptic or obovoid, trigonous, long-stipitate. Fl. 7. Fr. 9. $2n = 36$. Hs.

Native. In micaceous mountain bogs, very rare. 2. Mid Perth and Inner Hebrides. Pyrenees, Alps, Scandinavia, Greenland, N. America.

45. C. elata All.   'Tufted Sedge.'

C. reticulosa Peterm.; *C. stricta* Good., non Lam.; *C. Hudsonii* A. Benn.

A tufted glabrous perennial up to c. 90 cm. Stems sharply triquetrous, slightly rough. Lvs (3–)4–6 mm. wide, sharply keeled, rather glaucous; ligule 5–10 mm., ovate, acute. *Lowest sheaths lfless*, reddish, acuminate, *edges becoming filamentous* in decay. Male spikes 1–2(–3). Male glumes c. 5 mm., narrowly obovate, obtuse, dark purplish-brown with a pale midrib and narrow hyaline border at top. Female spikes usually 2, rather distant, 15–40 × 5–7 mm., often male at top, erect; glumes and fr. arranged in conspicuous longitudinal rows. Peduncles 0 or very short. *Bracts setaceous, lower seldom more than half as long as infl. Female glumes* c. 3 mm., ovate, subacute, *nerve ceasing below or at apex*, purple-brown with paler midrib and *hyaline margins*. Fr. (Fig. 78 E) c. 3 mm., broadly ovate or suborbicular, flattened; beak c. 0·2 mm., rather stout, entire, smooth. Nut 2 mm., obovate, biconvex. Fl. 5–6. $2n = 74^*$; 78; 80. Hs. or Hel.

Native. By fen ditches and in wet places beside rivers and lakes; locally common. 56, H 37. Scattered throughout the British Is. north to S. Aberdeen and the Outer Hebrides. Europe except the Arctic and parts of the Mediterranean region; Algeria; Caucasus.

46. C. acuta L. 'Tufted Sedge.'

C. gracilis Curt.

A tufted or shortly creeping glabrous perennial 60–100 cm. *Stems sharply triquetrous*, somewhat rough. Lvs c. 5 mm. wide, broadly and bluntly keeled; *ligule* 2–3 *mm.*, broadly triangular or ± truncate, margins often brownish. *Lowest sheaths bearing short lvs*, not webbed. Male spikes 1–3. Male glumes 3–4 mm., oblong or obovate, obtuse or ± lacerate, purplish with pale midrib, black at tip. Female spikes 2–4, rather distant, 3–7 cm. × 4–5 mm., often male at top, nodding in fl., erect in fr. Peduncles short or upper 0. *Bracts lf-like, the lowest as long as or longer than infl.* Female glumes 2·5–3 mm., ovate, acute, sometimes obtuse, *nerve usually shortly excurrent*, purple-black with pale midrib in lower half. Fr. (Fig. 78 F) c. 2·5 mm., obovate or almost elliptic, flattened, rarely (var. *sphaerocarpa* Uecht.) 2 mm., orbicular; beak c. 0·2 mm., slender, entire, smooth. Nut 2 mm., obovate or suborbicular, biconvex. Fl. 5–6. 2*n*=c. 74; 84. Hs. or Hel.

Native. Beside water and in wet grassy places. 79, H 20. Scattered throughout the British Is. north to Caithness; common in many districts. Europe, W. and N. Asia, N. Africa, N. America.

47. C. aquatilis Wahlenb.

A shortly creeping, rather rigid glabrous perennial 30–100 cm. *Stems bluntly trigonous*, smooth, *rather brittle*. Lvs 3–5 mm. wide, concave, dark green and shiny beneath, light green above; *ligule c.* 10 *mm.*, ovate, acute. *Lower sheaths lfless*, persistent, often reddish. Male spikes 2–4, 15–40 × 1–2·5 mm. Male glumes 3–4 mm., obovate, obtuse, brown with a pale midrib and hyaline margin at top. *Female spikes* (2–)3–4(–5), rather distant, the lowest sometimes very distant, 2–6 cm. × 3–4 mm., *tapering downwards from about the middle*, sometimes male at top. Peduncles of lower spikes short, of upper 0. *Bracts all lf-like, that of lowest spike longer than infl.* (except when lowest spike is very distant), *of next spike equalling or exceeding infl.* Female glumes c. 2·5 mm., lanceolate to ovate, acute or subacute, dark purple-brown with a pale midrib and hyaline tip. Fr. (Fig. 78 G) c. 2·75 mm., broader than glumes, elliptic or obovate-elliptic, somewhat biconvex; beak c. 0·2 mm., entire, smooth. Nut 1·75 mm., obovate, biconvex. Fl. 7. 2*n*=76; 84. Hel.

Native. Margins of lakes and streams in mountainous districts; local. 31, H 15. Cardigan, Pembroke and Merioneth; Northumberland, Westmorland, Cumberland; not infrequent in Scotland; scattered throughout Ireland south to Kerry. Norway, Sweden, Finland, N. Russia, Siberia.

48. C. recta Boott

C. kattegatensis Fr. ex Krecz.; *C. salina* auct., non Wahlenb.

A shortly creeping glabrous perennial 30–60 cm. Stems trigonous, smooth. *Lvs 3–5 mm. wide*, shorter than or about equalling stems, weakly keeled, serrulate; ligule 2–3 mm., triangular, acute. Lower sheaths lfless, blackish-brown, not fibrous. Male spikes 1–3, 10–40 × 3–4 mm., lateral often peduncled. Male glumes 4–5 mm., oblong or lanceolate, obtuse or erose, brown with a pale midrib. Female spikes 2–4, 30–70 × 4–6 mm., erect, or spreading in fr. *Peduncles* 1–3 *cm.*, rough. Bracts lf-like, equalling or exceeding infl., not sheathing. *Female glumes* 4–5 mm., ovate or lanceolate, acute or obtuse *with a long excurrent midrib*, dark brown or purple, midrib pale. Fr. (Fig. 79 A) 2·5–3 mm., obovate or suborbicular, flattened, nerved; beak c. 0·2 mm., entire. Nut c. 2 mm., round or obovate, shortly stipitate, rather asymmetrical. Fl. 7. Fr. 8–9. $2n=84$. Hel.

Native. Estuaries of rivers in N. Scotland, on sand-banks with other sedges, very local. E. Inverness, Caithness. Norway, Sweden, Finland, Faeroes, Iceland; Atlantic coast of N. America from Labrador to Massachusetts; apparently always very local.

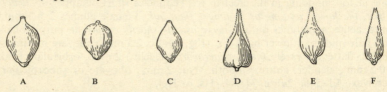

Fig. 79. Fruits of *Carex*. A, *C. recta*; B, *C. nigra*; C, *C. bigelowii*; D, *C. paniculata*; E, *C. appropinquata*; F, *C. diandra*. × 5.

49. C. nigra (L.) Reichard   'Common Sedge.'

C. angustifolia Sm.; *C. Goodenovii* Gay; *C. vulgaris* Fr.; *C. cespitosa* auct.; *C. fusca* All.; incl. *C. eboracensis* Nelmes

A creeping or occasionally densely tufted rhizomatous perennial 7–70 cm. Stems slender, not more than 2 mm. diam., triquetrous, smooth below, rough above. *Lvs 2–3 mm. wide*, longer or shorter than stems, margins often revolute when dry; ligule 1–3 mm., ovate, rounded or ± acute. Lower sheaths usually with lvs and soon decaying into a blackish fibrous mass. *Male spikes* 1–2, *lower usually much smaller than upper*. Male glumes 3–5 mm., oblong or narrowly obovate, obtuse or subacute, purplish with a pale midrib or rarely brown. Female spikes (1–)2–3(–4), ± contiguous or sometimes rather distant, 1–2 cm. × 4–5 mm., sometimes male at top. *Peduncles usually very short* or 0. Lowest bract lf-like, shorter or somewhat longer than infl., upper much shorter. *Female glumes* 3–3·5 mm., lanceolate to obovate-oblong, obtuse to acute or acuminate, *nerve ceasing below apex*, black, sometimes with a pale midrib and narrow hyaline border at top, rarely brown. Fr. (Fig. 79 B) 2·5–3 mm., broader than glumes, ovate to suborbicular, plano-convex, green or purplish; beak very short, entire. Nut 2 mm., elliptic, biconvex. Fl. 5–7. Fr. 6–8. $2n=74$; 84^*. Hs. or Hel.

A very variable and probably composite sp. which needs further investigation.

Native. In wet grassy places and beside water, on acid or sometimes basic soils. 112, H40, S. Widely distributed throughout the British Is.; common in suitable habitats; ascends to 3606 ft. on Lochnagar. Europe, W. Asia, N. America, Australia.

Very variable. *C. juncella* auct. angl., non T.M.Fr. (*C. subcaespitosa* (Kükenth.) Wiinstedt) is a densely tufted growth-form with very short ascending rhizomes and appears to be controlled by habitat conditions. *C. juncella* (Fr.) T.M.Fr. is mainly an arctic plant and has not been found in the British Is. It is completely non-rhizomatous.

50. C. bigelowii Torr., ex Schwein.   'Stiff Sedge.'

C. rigida Good., non Shrank.; ?*C. concolor* R.Br.; *C. hyperborea* Drej.

A shortly *creeping* rigid glabrous perennial 5–25 cm. Stems sharply tri-quetrous, usually rough towards the top. *Lvs* 2–4(–7) *mm. wide*, shorter than stem, *recurved*, keeled, shortly acuminate, margins revolute; ligule 1–2 mm., ± triangular, acute. Lower sheaths lfless, reddish, persistent. *Male spike* 1, rarely more. Male glumes 3–4 mm., obovate, purplish or almost black with pale base and midrib. Female spikes 2–3, contiguous or rather distant, 5–15 × 3–5 mm., erect. Peduncles 0 or very short. Lower bract lf-like, shorter than infl., other bracts small, all with large brownish auricles. Female glumes 2·5 mm., broadly ovate, obtuse, purple or blackish with a pale, sometimes inconspicuous midrib. Fr. (Fig. 79 c) c. 2·5 mm., little broader than glumes, ovate, obovate or suborbicular, weakly biconvex; beak very short, entire, smooth. Nut 2 mm., ovate, biconvex. Fl. 6–7. $2n = 70$. Hs.

Native. In damp stony places, nearly always above 2000 ft. 39, H 18. Mountain regions of N. Wales, N. England and Scotland; Ireland, chiefly in the south and west; from 100 ft. in Shetland to 4350 ft. on Ben Nevis. Alps (rare), mountains of C. Europe, mountains of Scandinavia to c. 71° N., Iceland, arctic Russia; N. America.

C. bicolor All., a small plant with usually 2–3 contiguous, peduncled and nodding, ovoid spikes, the terminal male in the lower half, the others entirely female, with dark purplish-brown glumes and pale grey-green suborbicular fr., has been reported from the island of Rhum, Inner Hebrides.

51. C. paniculata L.   'Panicled Sedge.'

A large *densely tufted* glabrous perennial often *building big tussocks* up to c. 1 m. diam. Stems 60–150 cm., triquetrous, rough, dark green, spreading. *Lvs* 3–7 *mm. wide*, shorter than stems, incurled or semi-cylindrical, stiff, *dark green*, margins serrulate; ligule 2–5 mm., rounded or nearly truncate. Lower sheaths dark brown, not fibrous. Infl. branched or rarely nearly simple, spikes numerous, few-fld, male at top. Bracts glumaceous with setaceous points. *Female glumes* c. 3 mm., *triangular-ovate*, sometimes mucronate, brownish-hyaline, embracing the fr. *Fr.* (Fig. 79 D) c. 3 mm., ovoid-trigonous, strongly corky at base, lateral angles acute, serrulate above, dorsal rounded, dark brown *with many faint veins at base; beak* c. 0·5 mm., broad, *bifid to base*, serrulate or ciliate on margins. Nut 1·5 mm., ovate, biconvex, almost truncate at top. Fl. 5–6. Fr. 7. $2n = 60, 62, 64$. Hs. or Hel.

Native. In wet, often shady places on peaty base-rich soils. 107, H 40, S. Scattered throughout the British Is. north to Orkney. Europe to c. 62° N., W. Asia, N. America.

52. C. appropinquata Schumach.

C. paradoxa Willd., non J.F. Gmel.

A *densely tufted* glabrous perennial 30–80 cm., rather similar to *C. paniculata* but smaller. Stems triquetrous, rough, *faces flat*. *Lvs* 1–2 *mm. wide*, shorter than or equalling stems, ± flat, rough, *yellow-green*; ligule c. 2 mm., rounded. Lower sheaths lfless, fibrous, blackish-brown. Infl. 4–6 cm., of 4–6 sessile or nearly sessile spikes. Bracts setaceous, shorter than spikes. *Female glumes* c. 3 mm., *ovate, acuminate,* reddish-brown. *Fr.* (Fig. 79 E) c. 2·5 mm., broadly ovate or suborbicular, *rather suddenly contracted into the beak,* plano-convex, *distinctly 3–7-nerved in lower half, some of the nerves reaching base of beak; beak* c. 1 mm., *notched,* margins serrulate. Nut c. 1·75 mm., broadly ovate, plano-convex, shortly stipitate. Fl. 5–6. Fr. 6–7. $2n=64$. Hs.

Native. In fens and damp places, on base-rich peaty soils; local. 12, H2. Middlesex, Herts, Bucks, Norfolk, Suffolk, Cambridge, Derby, S. Yorks; Peebles, Mull; Clare, Westmeath; ascends to 1250 ft. at Malham, Yorks. N. and C. Europe to c. 68° N., Balkans, S.W. Russia.

53. C. diandra Schrank

C. teretiuscula Good.

A *shortly creeping* glabrous perennial 25–40 cm. *Stems slender,* triquetrous, nearly smooth, *faces somewhat convex. Lvs 1–2 mm. wide,* shorter than or equalling stems, ± flat, smooth below, serrulate at tips, *grey-green*; ligule very short, truncate. Lower sheaths lfless, ± fibrous, brown. Infl. 1–4 cm., of several sessile spikes. Bracts glumaceous, the lowest sometimes with a setaceous point. *Female glumes* c. 3 mm., *broadly ovate, acute or mucronate,* brownish-hyaline. *Fr.* (Fig. 79 F) c. 3 mm., broadly ovate or suborbicular, narrowed into beak, plano-convex, *distinctly 2–5-nerved in lower half; beak* c. 1·5 mm., broad, notched, *split at back,* the halves overlapping, margins serrulate. Nut c. 2 mm., turbinate, plano-convex, stipitate. Fl. 5–6. Fr. 6–7. $2n=60$. Hs.

Native. In damp meadows and peaty places beside pools, local. 76, H34. Scattered throughout the British Is. from S. Devon to Orkney. Scattered in Europe to c. 71° N. (except most of the south), N. Asia, N. America.

54. C. otrubae Podp. 'False Fox-sedge.'

C. vulpina auct. occid., non L.

A stout tufted glabrous perennial up to 1 m. *Stems sharply* triquetrous but *not winged,* smooth below, rough above, spreading, *faces nearly flat. Lvs 4–10 mm. wide,* shorter than stems, *bright green becoming grey-green* when dry, channelled, margins rough; *ligule 10–15 mm., longer than broad, not overlapping lf-margins,* ovate, ± acute. *Hyaline front of sheaths neither gland-dotted nor wrinkled.* Lower sheaths brownish, soon decaying. *Infl.* compound, *branches sessile,* spikes numerous, up to 1 cm., male at top, *yellowish-green or light brown.* At least some bracts long and conspicuous. Female glumes 4–5 mm., ovate, acuminate, hyaline, with brownish margins and green midrib. *Fr.* (Fig. 80 A) 5–6 mm., ovate, plano-convex, *greenish, becoming dark brown, distinctly ribbed, smooth, not readily dropping at maturity,* tapering gradually into the bifid beak, margins serrulate in upper $\frac{1}{3}$. Nut 2·5 mm., oblong-ovoid, strongly compressed, shortly stipitate. Fl. 6–7. Fr. 7–9. $2n=60$. Hs.

Native. On clayey soils, usually in damp grassy places, more rarely in drier places by roads and in hedgebanks. 99, H40, S. Common and generally distributed throughout most of the British Is., absent from N. Scotland, and not recorded from Brecon, Radnor, and parts of C. Ireland. Europe, Mediterranean region, Caucasus, Asia Minor, N. Persia, C. Asia.

55. C. vulpina L. 'Fox Sedge.'

Similar in general appearance to *C. otrubae*, but differing as follows: Stems robust, very sharply angled or almost winged, faces ±concave. Lvs bright dark green, even when dry; ligule 2–5 mm., broader than long, overlapping lf-margins, truncate-deltoid. Hyaline front of lf-sheaths gland-dotted or transversely wrinkled. Infl. a warm, slightly reddish, brown. Bracts short and inconspicuous with prominent dark auricles. Fr. (Fig. 80 B) brown, ribs less conspicuous than in *C. otrubae* and usually 0 on the flat face, minutely papillose, readily dropping at maturity. Fl. 5–6. Fr. 6–7. $2n = 68$. Hs.

Native. In damp places near rivers and in ditches, local. 10. Sussex, Kent, Gloucester, Oxford, Berks, Surrey, Isle of Wight and S.W. Yorks. Europe, Caucasus, C. Asia, Siberia.

*C. vulpinoidea Michx. (Section Multiflorae Kükeath.), a N. American plant which occasionally occurs as an alien may be recognized as follows: stem 30–90 cm. Infl. spike-like, narrow, much-branched, branches short. Bracts setaceous, glumes acuminate or aristate, so that infl. has a 'bristly' appearance. Fr. 2 mm., ovate, plano-convex; beak short, bifid.

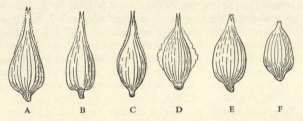

Fig. 80. Fruits of *Carex*. A, *C. otrubae*; B, *C. vulpina*; C, *C. disticha*; D, *C. arenaria*; E, *C. divisa*; F, *C. chordorrhiza*. × 5.

56. C. disticha Huds. 'Brown Sedge.'

An extensively creeping glabrous perennial 20–80 cm. *Stems* triquetrous, rough, *rather stout*. Lvs 2–4 mm. wide, about equalling stems, nearly flat; ligule 3–5 mm., ovate, obtuse. Lower sheaths lfless, not fibrous. Spikes numerous, contiguous, sessile, forming a ± dense infl. up to c. 5 cm. *Terminal spike female* or with a few male fls at top, *intermediate male, lower female*. Lower bracts glumaceous or sometimes narrowly lf-like and then exceeding infl. Female glumes c. 4 mm., ovate, acute, brownish-hyaline. *Fr.* (Fig. 80 c) 4–5 mm., ovate, plano-convex, distinctly many-ribbed, *narrowly winged and serrate in upper half*; beak c. 1 mm., bifid. Nut c. 1·75 mm., oval, biconvex, narrowed at top and bottom. Fl. 6–7. Fr. 7–8. $2n = 62^*$. Hs.

Native. In damp grassy places, fens, marshes and wet meadows. 95, H40. Throughout the British Is. north to Caithness, not uncommon but rather local. C. and N. Europe (except the Arctic), rare in S. Europe; Siberia.

57. C. arenaria L. 'Sand Sedge.'

An *extensively creeping* glabrous perennial 10–40 cm. Stems triquetrous, rough, often curved. Lvs 1·5–3·5 mm. wide, shorter than or equalling stems, nearly flat; ligule 3–5 mm., triangular, acute. Sheaths on horizontal rhizome soon becoming fibrous; lower sheaths on erect shoots lfless, brown, seldom fibrous. Spikes 5–12, contiguous, sessile, forming a ± dense infl. up to 4 cm. *Terminal spikes male, middle ones sometimes male at top, lower female.* Lower bracts glumaceous with setaceous points. Female glumes 5–6 mm., ovate, acute or acuminate, brownish-hyaline. *Fr.* (Fig. 80 D) 4–5 mm., ovate, plano-convex, distinctly many-ribbed, *broadly winged and serrate in upper half*; beak c. 1 mm., bifid. Nut c. 1·75 mm., compressed-cylindrical, abruptly contracted at top and bottom. Fl. 6–7. Fr. 7–8. $2n = 58$, 64. Hs.

Native. In sandy places usually near the sea, especially on fixed dunes, common. 71, H21, S. In all the maritime counties and inland in Hants, Surrey, Suffolk, W. Norfolk and S.E. Yorks. Coasts of Europe (except the Arctic), Black Sea, Siberia, N. America.

58. C. divisa Huds.   'Divided Sedge.'

A *creeping* glabrous perennial (15–)30–60(–80) cm. *Stems* triquetrous, ± rough, at least at top, *slender and wiry.* Lvs 1·5–3 mm. wide, shorter or sometimes longer than stems, nearly flat or ± involute; ligule 2–3 mm., ovate, obtuse. Lower sheaths lfless, soon decaying. *Spikes 3–7, contiguous or the lower 1–2 somewhat remote, the terminal male at top.* Infl. 1–2 cm. Lower bract lf-like or setaceous, shorter than to several times longer than infl. Female glumes 3·5–4 mm., ovate, shortly aristate, brownish-hyaline. *Fr.* (Fig. 80 E) 3·5–4 mm., broadly ovate or oval, plano-convex, *many-veined; beak* 0·5–0·75 mm., *parallel-sided, bifid, smooth, not winged.* Nut 2 mm., suborbicular. Fl. 5–6. Fr. 7–8. Hs.

Native. In grassy places and beside ditches near the sea or by estuaries, locally abundant but rarely inland. 34, H2, S. From Cornwall around the coasts to Kent and W. Gloucester; Middlesex and Essex to N.E. Yorks; Northumberland; Glamorgan, Carmarthen, Merioneth, Denbigh and Flint; Renfrew and Angus; inland in Surrey and Bedford. Atlantic coast of Europe from Belgium, Mediterranean region, Crimea, Himalaya, S. Africa; rarely inland.

59. C. chordorrhiza L. f.

A glabrous extensively creeping perennial 20–40 cm. *Rhizome obliquely ascending*, giving off lateral sterile shoots and usually terminating in a flowering stem. Stems stout, trigonous, smooth. *Lvs* 1–2 mm. wide, those *on the flowering stems few and much shorter than the stems, straight*, flat or ± involute; ligule c. 1 mm., rounded. Lower sheaths several, rather distant, lfless, not fibrous. *Spikes 2–4, male at top, sessile, contiguous, forming a small almost capitate infl.* c. 1 cm. Bracts glumaceous. Female glumes c. 4 mm., broadly ovate, acute or acuminate, brownish-hyaline. *Fr.* (Fig. 80 F) c. 5 mm., ovoid, *slightly compressed*, shortly stipitate, distinctly veined; *beak* c. 0·5 mm., ± *entire.* Nut 2 mm., cylindrical, tapered below, truncate at apex. Fl. 6–7. Fr. 7–8. Hel.

Native. In spongy bogs; very rare. W. Sutherland. Scattered throughout C. and N. Europe to Siberia; Iceland.

60. C. maritima Gunn. 'Curved Sedge.'

C. incurva Lightf.

An extensively creeping glabrous perennial. *Stems* 1–8 cm., *curved*, terete; smooth. *Lvs* c. 2 mm. wide, *curved, equalling or exceeding stems, involute*, serrulate, obtuse; ligule short, truncate. Lower sheaths brown or blackish, ± fibrous. *Spikes contiguous, forming an ovoid infl.*, 5–15 × 5–10 mm., male at top, male fls almost hidden. Bracts 0. Female glumes 3–4 mm., ovate, acute or obtuse and mucronate, brown with a pale midrib and broad hyaline margins. Fr. (Fig. 81 A) 4–4·5 mm., ovoid, rounded and shortly stipitate at base, brown or almost black; beak c. 0·5 mm., usually smooth. Nut 2 mm., elliptic, biconvex. Stigmas persistent. Fl. 6. Fr. 7. $2n = 60$. Hs.

Native. In damp hollows on fixed dunes, local. 18. Northumberland; east and north coasts of Scotland from the Firth of Forth northwards, Outer Hebrides, Orkney and Shetland. Alps, Jutland, W. and N. Scandinavia, Arctic, higher mountains of Asia and N. America.

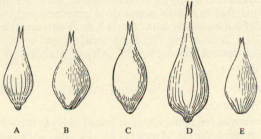

Fig. 81. Fruits of *Carex*. A, *C. maritima*; B, *C. divulsa*; C, *C. polyphylla*; D, *C. spicata*; E, *C. muricata*. × 5.

C. muricata agg. (spp. 61–64).

A glabrous tufted perennial 30–75 cm. Stems triquetrous, rough. Lvs 2–4 mm. wide, usually shorter than stems, channelled. Lower sheaths lfless, brown, becoming fibrous in decay. Infl. spike-like, rarely with 1–2 short branches at base, spikes ± crowded, sessile. Bracts setaceous. Glumes 3–4 mm., ovate, acuminate, pale brown with a green midrib. Fr. ovate, plano-convex, narrowed into a deeply-notched beak. Nut biconvex, shortly stipitate and almost truncate at top. Fl. 6–7. Fr. 7–8. Hs.

61. C. divulsa Stokes 'Grey Sedge.'

Infl. 5–10 cm., sometimes with 1 or 2 short branches at base, upper spikes crowded, lower distant. Fr. (Fig. 81 B) (3·7–)4–5 mm., not ribbed, gradually narrowed at the base and at the top into a long (c. 1 mm.) beak. Nut c. 1·8 mm. $2n = 58*$.

Native. In rough pastures, open woods and hedgebanks, local. 62, H 23, S. Scattered throughout England and Wales north to Westmorland and S. Northumberland, commoner in the west; Ireland: throughout most of the southern half but rarer in the west. Europe to c. 59° N.; N. Africa; temperate Asia; introduced in N. America.

62. C. polyphylla Kar. & Kir.

C. Leersii F. W. Schultz, non Willd.

Similar to *C. divulsa* but infl. never branched; fr. (Fig. 81 c) 3·7–4·8 mm., ribbed and rounded at base and usually asymmetrical; beak c. 1 mm.; nut 2·0–2·4 mm. $2n=58$.

Native. In dry calcareous grassland. 25, S. Somerset and Kent to Stafford and N.W. Yorks, local. Most of Europe, Mediterranean region, east to the Himalaya and N.W. Mongolia.

63. C. spicata Huds.   'Spiked Sedge.'

C. contigua Hoppe; *C. muricata* auct. angl., p.p.

Ligule longer than broad, rounded at apex. Infl. 2–4 cm., spikes contiguous or the lowest not more than its own length distant from the next, sessile. Bracts and lf-bases often tinged with purplish-red. Fr. (Fig. 81 d) 4·2–5·3 mm., greenish when ripe, corky and often furrowed when dry but not veined at base, gradually narrowed into a long (c. 1·5 mm.) beak; nut c. 2·2 mm. $2n=$ c. 58.

Native. In marshes or besides ponds or in grassland and on hedgebanks, on basic or acid but base-rich soils. 94, H7, S. Fairly generally distributed throughout Great Britain, but becoming local in the north; very local in Ireland. Europe (except the Arctic); N. Africa, Madeira, W. Asia; introduced in N. America.

64. C. muricata L.   'Prickly Sedge.'

C. pairaei F. W. Schultz

Similar to *C. contigua* but ligule about as broad as long, subacute, bracts and lf-bases never purplish-red; fr. (Fig. 81 e) 2·9–3·9 mm., dark brown or almost black when ripe, not corky or furrowed but clearly veined at base, some of the veins often extending halfway up the fr. rather abruptly narrowed into a short (c. 0·7 mm.) beak. Nut 1·7 mm.

Native. Dry grassy places usually on calcareous soils. 50, H8, S. Scattered throughout England and Wales, local; less frequent in Scotland and Ireland and absent from the north.

65. C. elongata L.   'Elongated Sedge.'

A tufted glabrous perennial 30–80 cm. Stems triquetrous, rough. Lvs 2–4 mm. wide, usually about equalling stems, nearly flat; ligule 2–3 mm., acute. Lower sheaths lfless, not fibrous. *Spikes* 5–15, *c.* 1 *cm.*, male at base, or terminal one rarely male at top, sessile, *brown, forming a* ± *lax infl.* 3–7 cm. Bracts glumaceous. Female glumes 2 mm., obovate, acute or obtuse, brownish-hyaline, midrib green. *Fr.* (Fig. 82 a) 3·5–4 *mm.*, lanceolate, planoconvex, often curved, many-ribbed, rounded at base, tapering upwards into the *smooth entire beak.* Nut 2 mm., compressed-cylindrical, narrowed at base, tapering slightly upwards and then truncate. Fl. 5. Fr. 6. $2n=$ c. 56. Hs. or Hel.

Native. In marshes and damp woods; local and rather rare. 23, H2, Dorset, N. Hants, Berks, Sussex, Surrey, E. Kent, N. Essex to Lincoln. Warwick, Shropshire, Cheshire to Mid-west Yorks, Westmorland, Cumberland; Dumfries, Kirkcudbright; Fermangh. C. and N. Europe (except the Arctic), rare in S. Europe; Siberia.

66. C. echinata Murr. 'Star Sedge.'

C. stellulata Good.

A tufted glabrous perennial 10–40 cm. Stems trigonous, rather slender, spreading, smooth. Lvs c. 2 mm. wide, rather shorter than stems, channelled; ligule c. 1 mm., ± triangular, apex rounded, margins incurved. Lower sheaths soon decaying. *Spikes* 3–4, somewhat remote, *c. 5 mm.*, sessile, *fr. spreading stellately.* Bracts small, glumaceous. Female glumes c. 2 mm., broadly ovate, acute, embracing the lower part of the fr., brown with a green midrib and broadly hyaline margins. Fr. (Fig. 82 B) 4 mm., yellow-brown, ovate, plano-convex, tapering gradually into the broad *notched serrate beak.* Nut c. 2 mm., obovoid, somewhat compressed. Fl. 5–6. Fr. 6–7. $2n = 56, 58$. Hs. or Hel.

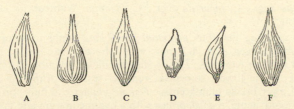

Fig. 82. Fruits of *Carex.* A, *C. elongata*; B, *C. echinata*; C, *C. remota*; D, *C. curta*; E, *C. lachenalii*; F, *C. ovalis.* × 5.

Native. In damp meadows and boggy places, on acid, humus-rich soils. 112, H40, S. Throughout the British Is., local in the south and east, common in the north and west. Europe, except the south, Azores; W. Asia, Siberia; N. America; Australia.

67. C. remota L. 'Remote Sedge.'

A densely tufted glabrous perennial 30–60 cm. Stems bluntly trigonous, smooth below, with 2 serrulate angles above, spreading. Lvs 2 mm. wide, nearly equalling stems, bright green, channelled; ligule 1–2 mm., triangular, acute or obtuse. Lower sheaths straw-coloured, not fibrous. *Spikes* usually 4–7, *remote*, male at base, upper mostly male, lower mostly female; lower 7–10 × 3–4 mm., upper smaller. Peduncles 0. *Lower bracts lf-like, exceeding infl.*, upper glumaceous. Female glumes c. 2·5 mm., lanceolate to ovate, acute, hyaline, midrib green below, brown above. Fr. (Fig. 82 C) c. 3 mm., ovate, plano-convex, greenish; *beak short, broad, serrulate, notched.* Nut 2 mm., ovate, biconvex. Fl. 6. Fr. 7. $2n = 62^*$. Hs.

Native. In damp shady places. 104, H40, S. Throughout most of the British Is., common in the south, rarer in Scotland and absent north of Ross. C. and N. Europe (absent from the Arctic), rare in S. Europe; N. Africa; temperate Asia; N. America.

68. C. curta Good. 'White Sedge.'

C. canescens auct., non L.

A tufted glabrous perennial 25–50 cm. Stems triquetrous, rough above. Lvs 2–3 mm. wide, usually about equalling stems, nearly flat, pale green; ligule

2–3 mm., lanceolate, acute. Lower sheaths lfless, brown, soon decaying. Infl. 3–4 cm., of 4–8 ± remote spikes 5–8 mm. *Bracts small, glumaceous*, the lower with a short setaceous point. *Female glumes* 2 mm., obovate to suborbicular, cuspidate, *hyaline with a green nerve. Fr.* (Fig. 82D) 2–3 *mm.*, ovate, plano-convex, *pale yellow-green with distinct yellow ribs; beak c.* 0·5 *mm., emarginate, rough.* Nut 1·5–2 mm., elliptic or obovate, biconvex. Fl. 7–8. Fr. 8–9. 2*n* = 56. Hs. or Hel.

Native. In bogs, acid fens and marshes, but never on limy soils, locally common. 91, H25. Scattered throughout the British Is. but commoner in the north. Europe, except the south; temperate Asia; N. and S. America.

69. C. lachenalii Schkuhr

C. tripartita auct.; *C. bipartita* auct.; *C. lagopina* Wahlenb.; *C. leporina* L., nom. ambig.

A shortly creeping glabrous *perennial* 10–20 *cm.* Stems trigonous, strongly ridged, often curved. Lvs 1–2 mm. wide, shorter than stems, nearly flat; ligule 2–4 mm., ± truncate. Lower sheaths lfless, not fibrous. *Spikes* 2–4, *contiguous* or the lowest somewhat remote, sessile, forming a dense infl., 1–2 cm. Lower bracts glumaceous. *Female glumes* c. 2 *mm.*, broadly ovate or ± rhomboid, acute, *reddish-brown* with a hyaline margin and green midrib. *Fr.* (Fig. 82E) *c.* 3 *mm.*, ovate, narrowed at each end, plano-convex, distinctly ribbed; *beak* c. 0·5 mm., *smooth*, split down the back, the halves overlapping. Nut c. 2 mm., oval, biconvex. Fl. 6–7. Fr. 7–8. 2*n* = 58, 64. Hs. or Hel.

Native. In flushes and on wet rock-ledges, rare. 6. Angus, S. Aberdeen, Banff, Argyll and Inverness. Pyrenees, Alps, Scandinavia north to Iceland and Lapland, arctic Russia, N. Asia, N. America.

70. C. ovalis Good.   'Oval Sedge.'

C. leporina auct., non L.

A densely tufted glabrous *perennial* 20–90 *cm.* Stems trigonous, smooth below, rough at top, stiff, often ± curved, lfy only near base. Lvs 2–3 mm. wide, shorter than stems, ± flat, bright green, margins rough; ligule c. 1 mm., broadly triangular. *Spikes* (1–)3–9, c. 10 × 5 mm., male at base, *contiguous*, sessile. Lower bracts often setaceous, exceeding spike, upper glumaceous, not more than half as long as spike. *Female glumes* 3–4 *mm.*, lanceolate, acute, *brownish-hyaline. Fr.* (Fig. 82F) 4–5 *mm.*, erect, elliptic-ovate, plano-convex, *almost winged*, distinctly nerved, light brown, margins rough near top; *beak* c. 1 mm., bifid, *rough*, green. Nut c. 2 mm., obovate- or elliptic-oblong, plano-convex, shortly stipitate, ± shiny. Fl. 6. Fr. 7–8. 2*n* = 64*; 66, 68. Hs.

Native. In rough grassy places on acid soils, locally common, especially along tracks. 112, H40, S. Throughout the British Is., but rather local in the south and east. Europe, except the Arctic and extreme south; Algeria; N. America.

*C. crawfordii Fernald, a N. American sp. somewhat resembling *C. ovalis* in general appearance, is naturalized in a few places. It may be distinguished as follows: lvs usually about equalling stems; spikes usually 7–15; glumes 2·5–3 mm.; fr. 4–5 mm., lanceolate.

71. C. rupestris All.

A shortly creeping tufted glabrous perennial 5–15 cm. Stems triquetrous, smooth. *Lvs c.* 2 *mm. wide*, about equalling stems, folded or involute, smooth, *usually much curled*; ligule 1–2 mm., ovate, obtuse. Lower sheaths lfless, brown, not fibrous. Spike 10–15 × 2 mm., few-fld, male at top. Bracts glumaceous. *Female glumes* c. 2·5 mm., broadly ovate, obtuse, sometimes mucronate, dark brown, *persistent. Fr.* (Fig. 83 A) *c.* 3 *mm.*, obovoid-trigonous, brown, *erect*; beak very short. Nut broadly elliptic, triquetrous. Fl. 6–7. Fr. 7–8. 2*n* = 50. Hs.

Native. On narrow ledges and in cracks of base-rich rocks in C. and N.W. Scotland, rare and local. 7. Perth, Angus, Aberdeen, Ross, Sutherland and Caithness. Arctic Europe, Alps, Pyrenees, Carpathians, Corsica; Urals, Siberia; Greenland.

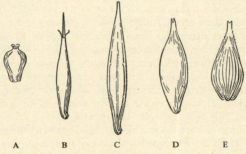

Fig. 83. Fruits of *Carex*. A, *C. rupestris*; B, *C. microglochin*; C, *C. pauciflora*; D, *C. pulicaris*; E, *C. dioica*. × 5.

72. C. microglochin Wahlenb.

A shortly creeping glabrous perennial c. 10 cm. Stems trigonous, smooth, stiff, straight. Lvs c. 1 mm. wide, usually shorter than stems, nearly flat; ligule very short truncate. Lower sheaths soon decaying. *Spike* 5–10 mm., *with* 4–12 *fr. which are deflexed when ripe.* Male glumes c. 3 mm., lanceolate, acute or obtuse, reddish-brown with pale midrib. *Female glumes c.* 2 *mm.*, *caducous*, ovate, obtuse, reddish-brown, midrib pale, tip hyaline. *Fr.* (Fig. 83 B) *c.* 6 *mm.*, *narrowly conical, abruptly contracted at base, yellowish*; beak c. 1 mm., stout, obliquely truncate. *A stout bristle arises at the base of the nut and protrudes from the top of the beak* together with the stigmas. Nut c. 1·5 mm. Fl. 7–8. Fr. 8–9. 2*n* = c. 56. Hel.

Native. In small micaceous bogs, 2500–2700 ft., very rare. Mid Perth. Mountains of C. Europe from the Alps eastward, N. Europe, Greenland, N. and C. Russia, Caucasus, Altai, Himalaya, Tibet.

73. C. pauciflora Lightf.　　　　　　　　　　　　'Few-flowered Sedge.'

A shortly creeping glabrous perennial c. 10 cm., somewhat similar to *C. micro-glochin*. Rhizome slender, branched, stems distant. Lvs of fertile shoots ± flat, of vegetative shoots setaceous. *Spike* c. 5 mm., *with* 2–4 *or rarely more fr.* which are deflexed when ripe. Male glumes c. 5 mm., lanceolate, ± acute, reddish-brown with hyaline margins. *Female glumes c.* 4 *mm.*, similar to, but rather broader than male, caducous. *Fr.* (Fig. 83 c) 5–6 *mm.*, narrow, *tapered at both ends*, yellowish. Nut 2 mm., oblong. The persistent style protrudes from the top of the fr. and should not be confused with the bristle in *C. microglochin*. Fl. 5–6. Fr. 6–7. Hel.

Native. On wet moors and in bogs. 33, H1. Caernarvon; N.E. Yorks and Cumberland to Caithness; Antrim; rather local. C. and N. Europe, northern Asia Minor, N. America.

C. capitata L., densely tufted, with a solitary terminal, dense conical spike, male at the top, female below, and with a compressed-ovoid fr. c. 3·5 mm., has been reported from South Uist, Outer Hebrides.

74. C. pulicaris L.　　　　　　　　　　　　　　　　Flea-sedge.

A shortly creeping glabrous perennial 10–30 cm., forming dense patches. Stems terete, slender, rigid, smooth. Lvs c. 1 mm. wide, nearly equalling stems, channelled, dark green, rather rigid; ligule very short, ovate, rounded. Lower sheaths soon decaying, upper auricled. *Spike* 10–25 mm., *top half male*, 1 mm. or less wide, acuminate; *female half lax*, 3–10-fld. Male glumes c. 5 mm., oblong, obtuse or subacute, brown. *Female glumes caducous. Fr.* (Fig. 83 D) 4–6 mm., lanceolate, flattened, dark brown, *shiny*, shortly and stoutly pedicelled, at length deflexed; *beak almost* 0. Nut c. 2·5 mm., compressed-cylindrical, almost truncate at top, narrowed to a short thick stalk at base. Fl. 5–6. Fr. 6–7. $2n=60$. Hs. or Hel.

Native. In damp calcareous grassland, fens, base-rich flushes, etc.; locally common. 112, H40, S. Recorded from every vice-county in the British Is., but uncommon in the south and east and now lost through drainage in many places. Scattered throughout Europe, except the Mediterranean region, Caucasus, Siberia.

75. C. dioica L.　　　　　　　　　　　　　　　　'Dioecious Sedge.'

A shortly creeping *usually dioecious* glabrous perennial 10–15 cm. Stems terete, smooth, rigid, erect. Lvs 0·5–1 mm. wide, shorter than stems, channelled, dark green, rather rigid; ligule very short, truncate. Lower sheaths dark brown, not fibrous. Male spike 10–15 × 2–3 mm., subclavate, acute, rarely female at base. Glumes 3–4 mm., ovate to oblong, acute or obtuse, brownish-hyaline. Female spike 10–15 × 5–7 mm., rather dense, 20–30-fld. Bracts 0. *Glumes* 2·5–3 mm., *persistent*, ovate, acute, brown with a paler patch surrounding the dark midrib, margins hyaline. *Fr.* (Fig. 83 E) c. 3·5 mm., spreading horizontally or slightly deflexed when ripe, compressed-ovoid, *tapering to the broad serrulate notched blackish beak*, greenish-brown *with numerous dark brown nerves*. Nut c. 1·3 mm., suborbicular, compressed. Fl. 5. Fr. 7. $2n=52*$. Hs.

Native. In fens, base-rich flushes, etc. 90, H22. Scattered throughout the British Is.; local, diminishing through drainage in the south. Ascends to 3250 ft. on Ben Lawers. Europe, Siberia, N. America.

C. davalliana Sm.　　　　　　　　　　　　　　　　'Davall's Sedge.'

A densely tufted dioecious perennial 15–25 cm. Lvs setaceous, rough. Female spike c. 15–20-fld. Glumes dark brown, except for the margins, persistent. Fr. c. 4–5 mm., dark brown, somewhat curved and deflexed when ripe. $2n=46$. Hs.

Formerly grew on Lansdown, near Bath, but now long lost through drainage. C. Europe, to N. France, the Altai and Asia Minor.

146. GRAMINEAE

Annual or perennial herbs, sometimes woody. Lvs with silicified cells (silica cells) on the lower epidermis in 1–2 rows beneath the vascular bundles. Stems usually branched at the base and, in perennials, of two sorts, flowering stems or *culms* and sterile or vegetative shoots; in annuals all the stems bear fls. Culms cylindrical or sometimes flattened, solid at the nodes and usually hollow in the internodes. Lvs solitary at the nodes and consisting of a *sheath*,

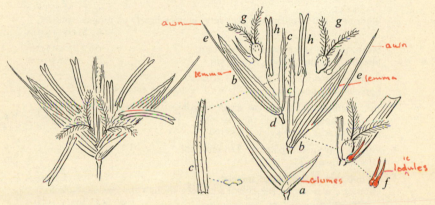

Fig. 84. A grass spikelet. *a*, glumes; *b*, lemma; *c*, palea; *d*, rhachilla; *e*, awn; *f*, lodicules; *g*, ovary and stigmas; *h*, stamen. × 5.

ligule and *blade*; sheath encircling the stem, margins overlapping or sometimes connate; ligule a small flap of tissue at the junction of sheath and blade, sometimes replaced by a ring of hairs, rarely 0; blade usually long and narrow, rarely broad, passing gradually into the sheath or rarely with a petiole-like constriction, sometimes with thickened projections (*auricles*) at each side of the base. Fls usually hermaphrodite, sometimes unisexual, consisting of 3 stamens (rarely 2 or 1), a 1-celled ovary with a solitary anatropous ovule and, usually, 2 styles with generally plumose stigmas and two small delicate scales (*lodicules*), which are occasionally absent. These are enclosed between 2 bracts, the lower (*lemma*)[1] membranous to coriaceous and the upper (*palea*)[2] generally thin and delicate, sometimes 0, the whole forming the floret. Florets 1 to many, distichous, ± imbricate and sessile on a short slender axis (*rhachilla*) usually with 2 bracts at the base (*glumes*), the whole forming

[1] =flowering glume, lower pale, or valve.
[2] =pale, upper pale, or valvule.

a *spikelet*. One or more of the florets sometimes reduced to an empty lemma and palea or only a lemma, which may be very small. Spikelets pedicelled in panicles or racemes, or sessile in spikes; panicles sometimes spike-like with short branches not readily apparent, bearing crowded spikelets. Fr. usually a caryopsis. A very large family with c. 600 genera and 8000–10,000 spp., generally distributed.

1	Ligule a ring of hairs.	*2*
	Ligule membranous or 0.	*6*
2	Infl. a panicle with distinctly pedicelled spikelets.	*3*
	Infl. of 2–several spikes; spikelets sessile or nearly so.	*5*
3	Plant large, reed-like; sheaths loose; florets with silky hairs at base.	
	2. PHRAGMITES	
	Not as above.	*4*
4	Stem-base swollen; spikelets numerous; glumes much shorter than spikelet.	3. MOLINIA
	Stem-base not swollen; spikelets few; glumes nearly as long as spikelet.	
	4. SIEGLINGIA	
5	Stout, tufted; spikelets at least 10 mm.	56. SPARTINA
	Slender, creeping; spikelets c. 2 mm.	57. CYNODON
6	Spikelets obviously of 2 or more florets.	*7*
	Spikelet of 1 floret, occasionally with vestiges of other florets.	*47*
7	Infl. a simple spike with the sessile or subsessile spikelets inserted singly or in opposite pairs on the rhachis.	*8*
	Infl. a panicle, or spikelets clustered.	*15*
8	Glume 1 in lateral spikelets, 2 in terminal.	*9*
	Glumes always 2.	*10*
9	Glume 1 in all the lateral spikelets.	9. LOLIUM
	Glumes 2 in some of the lateral spikelets, though the one next the axis often small.	8. × FESTULOLIUM
10	Tall perennials; plant tufted or with a far-creeping rhizome.	*11*
	Small annuals; plant neither tufted nor rhizomatous.	*14*
11	Spikelets edgeways on to rhachis or else terete.	*12*
	Spikelets broadside on to the rhachis, strongly flattened laterally.	*13*
12	Lemmas awned.	24. BRACHYPODIUM
	Lemmas awnless.	8. × FESTULOLIUM
13	Spikelets 2–3 cm., in opposite pairs; largest lvs c. 1 cm. wide.	
	26. ELYMUS	
	Spikelets (excluding awns) c. 1 cm., alternate; largest lvs c. 5 mm. wide.	
	25. AGROPYRON	
14	Glumes subequal; plant stiff and wiry.	13. CATAPODIUM
	Glumes unequal; plant not wiry.	
	10. VULPIA (see also 11. NARDURUS)	
15	Glumes largely hyaline, greenish-golden; uppermost lf usually less than 5 mm.; margins of lemmas of 2 lower florets fringed with hairs; plant smelling strongly of coumarin when crushed (rare).	
	51. HIEROCHLOË	
	Not as above.	*16*

16 Lemma awnless or with a straight awn from the tip, or from a notch at
        the tip.   *17*
       Lemma of at least some florets with an awn sticking out from the back
        below the tip; awn usually geniculate.   *40*

17 Spikelets silvery; upper glume about equalling 1st floret; tufted peren-
        nials with narrow dense ±parallel-sided panicles.   29. KOELERIA
       Not as above.   *18*

18 Spikelets crowded in dense one-sided masses towards the ends of the
        panicle branches; coarse tufted plant with strongly compressed
        vegetative shoots.   15. DACTYLIS
       Not as above.   *19*

19 Infl. a ±spreading panicle; at least some branches easily seen.   *20*
       Infl. spike-like; spikelets crowded in groups on short branches.   *36*

20 Spikelets not normally viviparous.   *21*
       Spikelets viviparous (mountain districts).   *35*

21 Lemmas awned or at least with a bristle-like point.   *22*
       Lemmas unawned.   *26*

22 Spikelets ovoid, subterete.   22. BROMUS
       Spikelets lanceolate, widening upwards or ±parallel-sided, ±strongly
        compressed.   *23*

23 Lower glume 1-nerved.   *24*
       Lower glume 3-nerved.   23. CERATOCHLOA

24 Spikelets up to 15 mm., lanceolate, somewhat compressed.
                                                                                7. FESTUCA
       Spikelets rarely less than 20 mm., parallel-sided or widening upwards,
        strongly compressed.   *25*

25 Annual; spikelets widening upwards.   21. ANISANTHA
       Perennial; spikelets ±parallel-sided.   20. ZERNA

26 Nerves of the lemma ±parallel, prominent and extending nearly or
        quite to the apex (±aquatic).   *27*
       Nerves of the lemma converging towards the tip or else inconspicuous
        or disappearing well below the apex.   *28*

27 Spikelets 5 mm. or more; lemma 7–9-nerved.   5. GLYCERIA
       Spikelets c. 2 mm.; lemma 3-nerved.   6. CATABROSA

28 Glumes nearly as long as spikelet; spikelets rather few; margins of
        sheaths connate.   18. MELICA
       Glumes much shorter than spikelet; spikelets usually numerous; margins
        of sheaths free.   *29*

29 Spikelets broadly conical or ovate, solitary on long, very slender pedicels;
        lemma cordate at base.   17. BRIZA
       Not as above.   *30*

30 Lemmas linear-lanceolate, acuminate.   7. *Festuca altissima*
       Lemmas broader, acute or obtuse.   *31*

31 Spikelets 10 mm. or more.   *32*
       Spikelets rarely more than 5 mm.   *34*

32 Lower lvs and sheaths quite glabrous.   *33*
       At least some of the lower lvs or sheaths with long scattered hairs.
                                                                      23. CERATOCHLOA

33 Spikelets subterete; florets 1 mm. apart on the rhachilla.
                                                      7. FESTUCA
      Spikelets strongly flattened; florets 2 mm. apart on the rhachilla.
                                                      20. *Zerna inermis*

34 Spikelets flattened; lemmas keeled; lvs flat when fresh (not salt-marshes).
                                                      14. POA
      Spikelets subterete; lemmas rounded on back; lvs usually rolled up
      (salt- and brackish-marshes, rarely elsewhere).   12. PUCCINELLIA

35 Lvs setaceous.   7. *Festuca vivipara*
      Lvs flat.   14. POA

36 Spikelets of 2 kinds; sterile lemmas stiff and distichous, fertile lemmas
      awned.   16. CYNOSURUS
      Spikelets all similar.   *37*

37 Lemmas ending in 5 small points; infl. bluish and shiny (local).
                                                      19. SESLERIA
      Not as above.   *38*

38 Lemmas awned.   10. VULPIA (see also 11. NARDURUS)
      Lemmas unawned.   *39*

39 Glumes subequal; wiry annuals.   13. CATAPODIUM
      Glumes very unequal; soft perennial.   8. × FESTULOLIUM

40 Spikelets (excluding awns) 9 mm. or more.   *41*
      Spikelets (excluding awns) 5 mm. or less.   *43*
      Spikelets viviparous; awn from just below the tip of the lemma.
                                                      35. *Deschampsia alpina*

41 Upper glume 2 cm. or more; spikelets eventually pendulous; annual.
                                                      31. AVENA
      Upper glume rarely more than 1·5 cm.; spikelets erect; perennial.   *42*

42 Lvs long, soft, 4–6 mm. wide; lower floret usually male, awned; upper
      floret female or hermaphrodite, usually unawned.
                                                      33. ARRHENATHERUM
      Lvs short, often rather stiff, c. 2 mm. wide; 2–4 lower florets all herma-
      phrodite and awned.   32. HELICTOTRICHON

43 Spikelets shiny.   *44*
      Spikelets whitish or pinkish, not shiny.   *46*

44 Annual; 5–20(–30) cm.; lvs mostly dead at fl. or soon after; sterile
      shoots 0.   36. AIRA
      Perennial; normally taller; lvs green at fl. and fr.; sterile shoots present.   *45*

45 Tip of lemma ± truncate and torn; spikelets silvery or purplish.
                                                      35. DESCHAMPSIA
      Tip of lemma acute, ending in 2 bristle points; spikelets usually yellowish.
                                                      30. TRISETUM

46 Awn tapering to a sharp point; lvs flat, not glaucous; plant ± hairy.
                                                      34. HOLCUS
      Awn thickened towards the tip; lvs setaceous, very glaucous; plant
      glabrous (very local).   37. CORYNEPHORUS

47 Spikelets inserted in 2 rows on the rhachis in groups of 2 or 3 at each node.   *48*
      Spikelets solitary or not inserted in 2 rows on the rhachis.   *49*

48 Central spikelet hermaphrodite, lateral male or sterile.   27. HORDEUM
      Central spikelet male or hermaphrodite, lateral hermaphrodite.
                                                      28. HORDELYMUS

49 Spikelets with 1–several conspicuous bristles from their pedicels.
                                                                60. SETARIA
      No bristles on pedicels.   50

50 Spikelets sunk in the rhachis of the solitary spike; small annuals;
         (clayey ground by the sea).   54. PARAPHOLIS
      Not as above.   51

51 Infl. ovoid, softly and densely hairy; annual.   47. LAGURUS
      Not as above.   52

52 Infl. of 1 or more spikes, the spikelets sessile or nearly so and arranged
         in 1 or 2 rows.   53
      Infl. not as above.   56

53 Spike solitary.   54
      Spikes more than 1.   55

54 Lemma awnless; small annual.   45. MIBORA
      Lemma awned; wiry perennial.   55. NARDUS

55 Spikes arranged in a raceme; spikelets usually awned.
                                                             58. ECHINOCHLOA
      Spikes digitate or nearly so; spikelets not awned.   59. DIGITARIA

56 Spikelets clustered on short branches and forming a dense cylindrical or
         ovoid spike-like infl.   57
      Spikelets ± spread out along the easily apparent branches of the panicle.   66

57 Tall sand-binding grasses; spikelets more than 10 mm.   58
      Not as above; spikelets less than 10 mm.   59

58 Infl. whitish, obtuse.   38. AMMOPHILA
      Infl. purplish-brown, acute.   39. ×AMMOCALAMAGROSTIS

59 Glumes very unequal, the lower much shorter than the spikelet; plant
         smelling of coumarin when crushed.   52. ANTHOXANTHUM
      Glumes subequal, both about as long as or longer than the lemma.   60

60 Glumes flattened, whitish with a green line, keel winged.   53. PHALARIS
      Not as above.   61

61 Glumes awned.   62
      Glumes unawned.   63

62 Awns conspicuous, 5 mm. or more; glumes deciduous.
                                                             43. POLYPOGON
      Awns c. 1 mm.; glumes persistent.   44. ×AGROPOGON

63 Glumes narrow, acuminate, swollen and rounded at the base.
                                                             46. GASTRIDIUM
      Glumes not swollen and rounded at the base.   64

64 Infl. very dense and regular, cylindrical or ovoid; glumes ciliate or silky. 65
      Infl. rather lax and irregular; glumes rough on their surfaces, falling
         with the fr.   43. POLYPOGON

65 Glumes tapering and curving inwards at the tip; awn of lemma exserted
         or spikelets silky; palea 0.   49. ALOPECURUS
      Glumes abruptly contracted to the outward-curving tip, or, if as above,
         then lemma awnless and spikelets not silky; palea present.
                                                             48. PHLEUM

66 Ligule long; infl. purplish; tall reed-like grass in damp places.
                                                             53. PHALARIS
      Not as above.   67

67 Glumes 0; spikelets strongly flattened (wet places, rare).   1. LEERSIA
    Glumes 2.   68

68 Lower glume small, ± encircling the base of spikelet; upper glume large,
      often awned and resembling the lemma.   58. ECHINOCHLOA
    Glumes ± equal.   69

69 Glumes rough on their surfaces, falling with the fr.   43. POLYPOGON
    Glumes smooth, persistent.   70

70 Glumes ovate, obtuse or subacute; lemma shiny, becoming very hard
      in fr.   50. MILIUM
    Glumes lanceolate, acuminate; lemma tough but not hard in fr.   71

71 Lemma awned; awn c. 5 mm.; annuals.   42. APERA
    Lemma awned or not; awn not more than 2 mm.; perennials.   72

72 Spikelets with silky hairs at base of floret.   40. CALAMAGROSTIS
    Spikelets without silky hairs at base of floret.   41. AGROSTIS

Tribe 1. ORYZEAE. Perennial or annual herbs. Lvs with transversely dumb-bell-shaped silica cells and slender 2-celled hairs; green tissue uniformly distributed between the vascular bundles; 1st foliage lf of seedling broad and horizontal. Ligule glabrous. Spikelets of 1 floret, compressed, apparently falling entire. Glumes very minute or 0. Lemma awnless or with a straight terminal awn, 3–9-nerved. Palea 3–9-nerved. Stamens 6, 3, 2 or 1. Fr. with a linear hilum and compound starch grains. Chromosomes small; basic number 12.

1. LEERSIA Sw.

Perennial herbs. Spikelets in a spreading panicle, strongly compressed, of 1 floret, apparently falling entire. Glumes 0. Lemma tough, 3-nerved. Palea tough, 3-nerved, equalling lemma. Lodicules 2; stamens 3, rarely 2. Ovary glabrous; styles very short. Fr. strongly compressed.

About 10 spp. in tropical and warmer temperate regions.

1. L. oryzoides (L.) Sw.   Cut-grass.

Oryza oryzoides (L.) Brand

An erect perennial, 30–60 cm. Rootstock creeping. Culms smooth, nodes shortly bearded. Lvs flat, scabrid, 4–6 mm. wide, acuminate, abruptly contracted at base. Sheaths smooth, more or less inflated, upper usually enclosing the panicle. Ligule c. 1 mm., truncate. Panicle 5–15 cm., branched; branches slender, smooth and flexuous. Spikelets 5–6 mm. Lemma half-ovate, apiculate, usually ciliate on the keel, margins thickened and ± fused with those of the palea. Palea subulate, usually ciliate on the keel. Anthers 0·5 mm., slender. Fl. 8–10. Chasmogamous and with larger anthers or cleistogamous and with panicle partially or wholly enclosed in uppermost lf-sheath, depending on climatic conditions. $2n = 48$. Hel.

Native. In wet meadows and beside rivers and ditches. 4. Surrey, Sussex, Hants, and Dorset. S. and C. Europe, north to S. Sweden and Finland; temperate Asia; N. America.

Tribe 2. ARUNDINEAE. Perennial herbs, usually tall and stout. Lvs with dumbbell-shaped silica cells; 2-celled hairs present. Ligule represented by a ring of hairs. Infl. a spreading panicle. Spikelets of 2–10 florets. Glumes membranous. Lemma 1–5-nerved, acuminate, awnless or awned from the tip, enveloped in long hairs arising from the rhachilla or from the back of the lemma. Lodicules 2. Stamens 3 or 2. Ovary glabrous, without an appendage. Fr. with compound starch grains. Chromosomes small, basic number 12.

2. PHRAGMITES Adans.

A stout, extensively creeping reed. Ligule replaced by a ring of hairs. Panicle large, lax, nodding. Spikelets, slender, acuminate, subterete. Glumes shorter than 1st floret, 3-nerved, upper twice as long as lower. Lemma 3-nerved, twice as long as upper glume. Palea not more than one-third the length of lemma. All the florets except the lowest with a tuft of long, silky hairs at the base. Stamens 1–3 in lowest floret, 3 in others. Lodicules, 2, oblong, obtuse. Ovary glabrous; styles short.

Three spp., cosmopolitan.

1. P. communis Trin.   Reed.

Arundo phragmites L.

A stout, erect, rhizomatous reed 2–3 m. Rhizome stout, extensively creeping. Lvs flat, 10–20 mm. wide, smooth, glaucous below, tapering to long slender points, deciduous in winter. Sheaths smooth, loose so that all the lvs point one way in the wind; auricles prominent. Ligule replaced by a ring of hairs. Panicle 15–30 cm., lax, nodding, soft, dull purple; branches smooth, usually with scattered groups of a few long silky hairs. Spikelets 10–15 mm., of (1–)3–6 florets; rhachilla hairy. Glumes lanceolate, acuminate. Lemma linear-lanceolate, acuminate. Palea ciliate in upper part. Silky hairs about as long as the lemma. The lvs break off at the junction with the sheaths in autumn; the dead culms and panicles stand throughout the winter. Fl. 8–9. $2n = 36$; 48; c. 96. Hel. or Hyd.

Native. In swamps and shallow water, but absent from extremely poor and acid habitats. 112, H40, S. In suitable habitats throughout the whole British Is. Cosmopolitan, except for a few tropical regions, e.g. the Amazon basin.

*Cortaderia selloana (J. A. & J. H. Schult.) Aschers. & Graebn., Pampas Grass, a very large densely tufted perennial with silvery panicles produced in late summer, is often cultivated and sometimes persists as a throw-out from gardens.

Tribe 3. DANTHONIEAE. Annual or perennial herbs. Lvs with deeply constricted dumbbell-shaped silica cells; 2-celled hairs 0; green tissue uniformly distributed between the vascular bundles. First foliage lf of seedling narrow and vertical. Ligule represented by a ring of hairs. Spikelets with 2–many florets, slightly laterally compressed or nearly terete, often shining. Glumes persistent. Lemma usually 3(–5)-lobed at apex, middle lobe often produced into an awn. Ovary glabrous, without an appendage. Fr. with a linear hilum c. $\frac{1}{2}$ length of grain; starch grains compound. Chromosomes small; basic number 6 or 9.

3. MOLINIA Schrank

Glabrous perennial herbs. Ligule represented by a ring of hairs. Panicle strict, branches long, slender. Spikelets subterete, tapering to a long point. Glumes subequal, membranous, about $\frac{1}{2}$ length of 1st floret, 1–3-nerved. *Lemma* cartilaginous, *3-nerved*. Palea equalling lemma, tough, obtuse. Lodicules 2, 1-toothed. Stamens 3. Ovary glabrous; styles terminal, very short. 1–2 spp.

1. M. caerulea (L.) Moench   'Purple moor-grass.'

An erect, wiry perennial 30–130 cm., often forming large tussocks. Rootstock ± creeping, roots very stout. Lvs flat, tapering from near the base, sparsely pilose, completely deciduous in winter. Culms smooth, base swollen and bulbous. Ligule 0, mouth of sheath ciliate. Panicle erect, 3–30 cm., green or purplish; pedicels short, finely ciliate. Spikelets 6–9 mm., tapering to a long point; florets 2–4. Glumes lanceolate, acute, 1-nerved or upper 3-nerved. Lemma tapering to an obtuse apex, bluntly 3-keeled. Anthers large, violet-brown. Fl. 6–8. $2n=18, 36, 90$. Hs.

Native. In damp or wet places in fens and heaths and on mountains. 112, H40, S. Throughout the British Is., locally abundant. Europe, except the extreme south; N.E. Asia Minor; Caucasus; Siberia; N. America.

4. SIEGLINGIA Bernh.

A perennial herb. Spikelets subterete, of 4–5-florets. *Glumes about equalling the spikelet*, membranous, subequal, bluntly keeled, 3-nerved, lateral nerves about $\frac{1}{2}$ as long as glume. *Lemma* coriaceous, *rounded on back*, 7-nerved, *apex with 3 short obtuse points*. Palea coriaceous, shorter than lemma. Rhachilla produced beyond the uppermost perfect floret and bearing a sterile rudiment. Lodicules 2, ovate, obtuse. Stamens 3. Ovary glabrous, stipitate; *styles* short, spreading, *inserted far apart*. One sp. in Europe.

1. S. decumbens (L.) Bernh.   'Heath Grass.'
Triodia decumbens (L.) Beauv.

A tufted, ± decumbent perennial 10–40 cm. Lvs flat, glabrous. Sheaths glabrous or slightly pubescent, with a ring of hairs at the mouth. Panicle 2–6 cm., of few (up to 10) spikelets. Spikelets 7–10 mm. Glumes lanceolate, obtuse. Lemma ovate, silky on the margin in the lower half, with a tuft of spreading silky hairs at the base, apex 3-toothed. Palea ovate, obtuse, ciliate. Fls cleistogamous or rarely chasmogamous. White compressed subterranean spikelets are borne singly in the axils of the basal sheaths. Fl. 7. $2n=36*$, $(124*)$. Hs.

Native. In acid grassland; locally on damp base-rich soils. 112. H40, S. Generally distributed throughout the British Is. in less peaty habitats than *Nardus*. Europe, except the extreme north; northern Asia Minor; Algeria; Madeira; only on mountains in the south.

Tribe 4. GLYCERIEAE. Perennial ± aquatic herbs. Lf-sheaths with connate margins. Lvs with elongate silica cells with several deep constrictions and stout unicellular papillae; 2-celled hairs 0; green tissue uniformly distributed between the vascular bundles; 1st foliage lf of seedling narrow and erect. Ligule glabrous. Spikelets with numerous florets, ± subterete. Glumes small. Lemma awnless, prominently (5–)7(–9)-nerved; nerves parallel. Lodicules connate, fleshy and swollen towards the top, truncate. Ovary glabrous, without an appendage. Fr. with linear hilum and compound starch grains. Chromosomes small; basic number 5.

5. GLYCERIA R.Br.

Perennial, *aquatic*, glabrous herbs. Rootstock creeping. Leaves flat; sheaths connate. Panicle simple or compound. *Spikelets subterete, of many florets. Glumes* hyaline, unequal, shorter than the 1st floret, 1-*nerved. Lemma membranous, prominently 7–9-nerved*, tip hyaline, nerves parallel, not quite reaching tip. Palea tough, 2-nerved, bifid, nearly or quite equalling lemma. Lodicules 2, connate, truncate, fleshy. Stamens 3. Ovary glabrous; styles terminal, short.
   About 40 spp. in temperate regions.

1	Apex of lemma distinctly toothed or lobed.	2
	Apex of lemma entire or nearly so.	4
2	Anthers 1 mm. or more; teeth of lemma obtuse, rather obscure.	3
	Anthers 0·5 mm.; teeth of lemma acute, prominent.	**3. declinata**
3	Anthers 1 mm.; lemma 3·5–4·5 mm.	**2. plicata**
	Anthers 1·2–1·5 mm.; lemma 5–5·5 mm.	**× pedicellata**
4	Spikelets more than 10 mm.	5
	Spikelets 10 mm. or less.	**4. maxima**
5	Anthers 2 mm.; lemma 6–7 mm.; sheaths smooth.	**1. fluitans**
	Anthers 1·5 mm. or less; lemma 5–5·5 mm.; sheaths minutely scabrid.	
		× pedicellata

1. G. fluitans (L.) R.Br.   Flote-grass.

An ascending or erect, glabrous perennial 25–90 cm. Stems creeping and rooting at the nodes, floating and then ascending or erect. Lvs flat, ± rough, rather abruptly contracted, acute. *Sheaths smooth.* Ligule up to 15 mm., torn. Panicle 10–50 cm., simple or little branched. Spikelets 15–30 mm., linear. Glumes oblong, lower acute, upper obtuse. *Lemma 6–7 mm.*, oblong, scabrid, *tip hyaline, entire, acute or subobtuse. Palea usually exceeding lemma*, lanceolate, with 2 short bristle points, shortly ciliate on margins towards top. *Anthers 1·5–2 mm.* Fl. 5–8. $2n=40^*$. Hel. or Hyd.
   Native. In stagnant or slow-flowing shallow water. 112, H40, S. Generally distributed throughout the British Is. Europe (except the Arctic), temperate Asia, Caucasus, Morocco, N. America.

2. G. plicata Fr.

Similar to *G. fluitans* but *sheaths minutely scabrid* on the ribs, lvs often rough, and tapering more gradually. Panicle compound, much branched, spreading, rarely nearly simple. Glumes obtuse. *Lemma 3·5–4·5 mm., rounded-truncate or obscurely lobed at apex*, strongly nerved. *Palea ovate glabrous, nearly equalling lemma. Anthers 1 mm.* Fl. 5–6. $2n=40^*$. Hel. or Hyd.

Native. In streams and ditches. 92, H 38, S. Generally distributed throughout the British Is. Europe, to c. 60° N. in Scandinavia, absent from N. Russia; W. Asia; N. Africa; N. America.

G. × pedicellata Towns. (*G. fluitans* (L.) R.Br. × *plicata* Fr.)

Ribs of sheaths minutely scabrid. Panicle ± branched. Lemma 5–5·5 mm., ± toothed, strongly nerved. Palea ovate, minutely ciliate on margins towards top, equalling lemma. Anthers 1·2–1·5 mm., indehiscent. Sterile. Fl. 6–7. Not uncommon beside slow-flowing rivers. Through England, also in Scotland and Ireland. Spreads vegetatively and is frequently abundant in the absence of the parents.

3. G. declinata Bréb.

Similar to *G. fluitans* but generally a smaller plant, (10–)15–30(–60) cm. Margins of sheaths hyaline (occasionally so in other spp.), ribs minutely scabrid. *Lvs* rough, *abruptly contracted, mucronate*. Panicle 5–30 cm., simple or branched. Spikelets 10–20 mm. Glumes ovate, obtuse. *Lemma 3·5–4·5 mm.*, ovate, scabrid, strongly nerved, *tip with 3–5 distinct, acute teeth*; nerves running out into hyaline tip and nearly reaching apex. Palea equalling or exceeding lemma, ovate, glabrous, bifid with often divergent subulate points. *Anthers c. 0·5 mm., ovate*. Fl. 6–9. 2*n*=20*. Hel.

Native. In swamps and muddy margins of ponds. 61, S. Fairly widely distributed though less common in most localities than *G. fluitans* and *G. plicata*. W. Europe (Sweden to S. Spain).

4. G. maxima (Hartm.) Holmberg   Reed-grass.

G. aquatica (L.) Wahlberg, non J. & C. Presl

A stout, erect perennial 60–200 cm. Lvs flat or ± folded, up to c. 2 cm. wide, smooth, rather abruptly contracted, acute; margins thickened, serrate. *Sheaths* rough, *with a reddish-brown band at the junction with the lf.* Ligule c. 5 mm., obtuse. Panicle 15–30 cm., compound, spreading. *Spikelets 5–8 mm., narrowly ovate*. Glumes keeled, subacute. *Lemma c. 3 mm.*, ovate, *entire*, obtuse, scabrid, narrowly hyaline at margins and tip, 9-nerved. Palea about equalling lemma, ovate, obtuse, very shortly bifid. Anthers 1 mm. Fl. 7–8. 2*n*=60*. Hel. or Hyd.

Native. In rivers, canals and large ponds, usually in deeper water than the other spp. 91, H 20. Scattered throughout the British Is. and abundant beside most lowland rivers; becoming rare in Scotland and absent from the north; Ireland, chiefly in the southern part of the central plain. Europe, except N. Scandinavia, N. Russia, Iberian peninsula and Greece; temperate Asia; Canada (probably introduced).

6. Catabrosa Beauv.

A glabrous perennial herb. Sheaths connate. Panicle lax, branches in half whorls of 3–5, successive whorls alternating. Spikelets subterete, usually of 2 florets. Glumes thin, much shorter than the spikelet, unequal; lower apparently nerveless, upper prominently 3-nerved. Lemma firm with a truncate or erose hyaline tip, prominently 3-nerved, nerves ± parallel. Palea as long as lemma. Lodicules 2, short, obovate, emarginate, connate below. Stamens 3. Ovary glabrous; styles terminal, very short.

One or two spp. in north temperate regions.

1. C. aquatica (L.) Beauv.   'Water Whorl-grass.'
An ascending rather soft perennial 5–70 cm. Stem creeping and often floating below, rooting at the nodes. Lvs smooth, soft, obtuse, often broad (up to 10 mm.), lower often floating. Ligule c. 5 mm., ovate, obtuse. Panicle up to 25 cm., lax, oblong, erect; branches smooth. Spikelets 3–5 mm., of (1–)2(–3) rather distant florets. Glumes hyaline or purplish, ovate, lower about $\frac{1}{2}$ length of upper. Lemma 2·5–3·5 mm., ovate-oblong, truncate with erose hyaline tip. Palea elliptic, glabrous or sparsely ciliate on nerves. Fl. 6–8. $2n=20^*$. Hel.

Native. In shallow streams and ditches, at the muddy margins of ponds and in wet sandy places near the sea. 105, H40, S. Widely scattered throughout the British Is., but local or rare in most districts; apparently less abundant than formerly. Europe, except S. Spain and Portugal; N. and W. Asia; Algeria; N. America (?a different ssp.).

Tribe 5. **Festuceae.** Annual or perennial herbs. Lvs with oblong or elliptic silica cells; 2-celled hairs 0; green tissue uniformly distributed between the vascular bundles; 1st foliage lf of seedling narrow, erect. Ligule glabrous. Spikelets of 2–many florets, laterally compressed or subterete, usually in panicles, rarely in distichous spike-like racemes. Glumes 2, rarely 1, usually shorter than the 1st floret on the same side. Lemma (3–)5–many-nerved, nerves convergent, often herbaceous to membranous, awnless, or awned from the entire or 2-lobed tip; awn not geniculate. Lodicules 2, thin, lobed. Stamens 3, rarely 2 or 1. Ovary glabrous, without an appendage. Fr. with linear to punctiform hilum and compound starch grains. Chromosomes large, basic number 7, rarely 5 or 9.

7. Festuca L.

Perennial herbs. Spikelets subterete, with 3 or more florets. Glumes membranous, lower 1-, upper 1–3-nerved, shorter than spikelet, unequal or subequal, acute. Lemma membranous or chartaceous, 5–7-nerved, usually mucronate or awned from the tip. Palea thin, about equalling the lemma, 2-nerved and shortly bifid. Lodicules 2, acute. Stamens 3. Ovary glabrous or rarely pubescent at the top; *styles* short, *terminal*.

About 100 spp. in temperate regions and on the higher mountains of the tropics. The taxonomy of spp. 5–11 is difficult and only the more distinct taxa are described here; a detailed account of the group is given by Howarth, W. O. (1924, 1925), *Journ. Linn. Soc. Bot.* XLVI, XLVII. Hybrids in the genus are discussed by Jenkin, T. J. (1933), *Journ. Genet.* XXVIII, 205.

1	Lvs of sterile shoots flat or sometimes folded but not setaceous.	2
	Lvs of sterile shoots setaceous.	9
2	Sheaths pubescent.	3
	Sheaths glabrous.	4

3 Lvs up to 4 mm. wide, usually much less; hairs on sheaths spreading,
    not deflexed.   **Zerna erecta** (p. 1147)
   Lvs 5 mm. wide, usually more; hairs on sheaths deflexed.
                                  Zerna ramosa (p. 1147)

4 Spikelets downy.   **7. juncifolia**
   Spikelets glabrous.   *5*

5 Culm lvs much wider than those of sterile shoots.
                               Zerna erecta (p. 1147)
   Culm lvs about the same width as those of sterile shoots.   *6*

6 Spikelets 6–7 mm.; ligule of uppermost lf c. 3 mm.   **4. altissima**
   Spikelets 8–20 mm.; ligule of uppermost lf very short.   *7*

7 Awn at least as long as lemma; spikelets green.   **3. gigantea**
   Awn 0 or shorter than lemma; at least some spikelets with a purplish
    tinge.   *8*

8 Basal sheaths thin, brownish; auricles glabrous; panicle branches solitary
    or one of each pair bearing a solitary spikelet.   **1. pratensis**
   Basal sheaths tough, white and persistent; auricles minutely and some-
    times sparsely ciliate; both panicle branches at each node usually with
    several spikelets.   **2. arundinacea**

9 Culm lvs flat.   *10*
   Culm lvs setaceous or convolute.   *11*

10 Not stoloniferous; panicle nodding; spikelets pale green, shiny; top of
    ovary pubescent.   **5. heterophylla**
   Usually stoloniferous; panicle erect; spikelets usually purplish; top of
    ovary glabrous.   **6. rubra**

11 Sheaths closed nearly to top when young; spikelets usually more than
    7 mm.   *12*
   Sheaths split more than half-way when young; spikelets usually less than
    7 mm.   (8–11) **ovina** agg.

12 Lvs obtuse or subacute; panicle usually reddish or purplish; anthers
    2–3 mm.   **6. rubra**
   Lvs acute; panicle greenish; anthers 5 mm.   **7. juncifolia**

1. F. pratensis Huds.   'Meadow Fescue.'

F. elatior L., p.p.

An erect glabrous perennial 40–80 cm. Stem-base usually clothed in decaying
remains of *dark brown sheaths*. *Lvs seldom exceeding* 4 *mm. wide*, smooth,
margins serrate. Sheaths smooth; auricles glabrous, inconspicuous. Panicle
8–15 cm., slender, nodding, secund; branches short, stout, triangular in
section and toothed at the edges, 1–2 together, *one of each pair usually bearing
only a single spikelet*; rarely spike-like with solitary subsessile distichous
spikelets; *spikelets* 10–15(–20) mm., *linear to lanceolate*, of 4–6(–10) rather
distant florets. Glumes unequal, linear-lanceolate, margins hyaline, lower 1-,
upper 3-nerved. Lemma lanceolate, hyaline at the apex, smooth, obscurely
nerved, middle nerve ending below the tip or rarely shortly excurrent. Palea
finely serrate on the nerves. Anthers 3 mm. Ovary glabrous. Fl. 6. $2n=14*$.
Hs. or Chh.

Native. In meadows and grassy places. 107, H35, S. Throughout the British Is., rare in northern Scotland. Europe north to c. 70°; temperate Asia. Introduced in N. America.

2. F. arundinacea Schreb.   'Tall Fescue.'

A stout tufted glabrous perennial 60–200 cm. Stem-base usually clothed in *tough whitish, sometimes scale-like sheaths*. Lvs flat or ±involute, *up to 10 mm. wide*, smooth, margins serrate. Sheaths smooth; auricles ciliate, prominent. Panicle (12–)20–30(–40) cm., usually effuse, subsecund, nodding; *branches* stout, flattened or triangular in section, edges serrate, 2 together *both with numerous spikelets*. *Spikelets* 10–15 mm., *lanceolate to ovate*, of (4–)5–6(–8) usually closely imbricate florets. Glumes unequal, margins hyaline; lower subulate acute, 1-nerved; upper lanceolate or oblong, acute or obtuse, 3-nerved. Lemma lanceolate, obscurely nerved, middle nerve often shortly excurrent. Palea finely serrate on the nerves. Anthers 4 mm. Ovary glabrous. Fl. 6–8. $2n=42*$. Hs.

Larger in all its parts than *F. pratensis* with which it hybridizes (Jenkin, T. J. *loc. cit.*).

Native. In grassy places. 100, H40, S. Generally distributed in the British Is. Europe north to 62°; western Siberia; N. Africa.

3. F. gigantea (L.) Vill.   'Tall Brome.'

Bromus giganteus L.

A stout erect glabrous perennial 50–150 cm. *Lvs up to 18 mm. wide, scabrid from above downwards*. Sheaths often scabrid from below upwards; *auricles prominent*, reddish. Panicle 12–40 cm., very lax, nodding; branches stout, long, bare below, scabrid, 1–2 together. Spikelets 10–15 mm., of 3–8 florets, lanceolate. Glumes unequal, lanceolate, acuminate, margins broadly hyaline; lower 1-, upper 3-nerved. Lemma lanceolate, scaberulous, *awn long, slender*. Palea glabrous. Anthers 2 mm. Ovary glabrous. Fl. 6–7. $2n=42$. Hs.

Native. In woods and shady hedge-banks. 104, H40. Throughout the British Is. except the extreme north of Scotland. Europe, except the Arctic, rarely south of the Alps; Asia.

4. F. altissima All.

F. silvatica Vill., non Huds.

An erect glabrous tufted perennial 80–120 cm. *Stem-base clothed with scale-like sheaths*. Lvs up to 10 mm. wide, smooth or slightly scabrid, margins finely serrate. Sheaths scabrid; auricles distinct. *Uppermost ligule up to 5 mm.*, torn. Panicle 10–16 cm., spreading or ±contracted in fr., subsecund, erect; branches slender, smooth, 1–2 together. *Spikelets* 5–8 mm., of 3–5 florets, *ovate*. *Glumes subequal*, subulate, 1-*nerved*. Lemma lanceolate, acuminate, scabrid, awnless. Palea scabrid. Anthers 2·75–3 mm. *Ovary pubescent at the top*. Fl. 6–7. $2n=14, 42$. Hs.

Native. In rocky woods mainly in the north, usually beside streams. 40, H21. Kent, Sussex; Gloucester to Stafford and Glamorgan to Hereford; Cardigan, Merioneth, Derby; N. Lancashire and the West Riding to

Cumberland and Durham, very local; more generally distributed in Scotland and Ireland though local and often rare. Scattered throughout Europe north to c. 65°.

5. F. heterophylla Lam.

An erect *tufted* perennial 60–120 cm. Culms slender, smooth. Radical lvs long, setaceous, 0·25–0·5 mm. wide, margins serrate; *cauline lvs flat*, 2–4 mm. wide, *with short spreading hairs* on the nerves above. Sheaths smooth. Ligule c. 1 mm., torn. *Panicle* 7–18 cm., slender, *nodding*, ± spreading. *Spikelets* 7–12 mm., of 3–4 florets, *shiny*; fls rather distant. Glumes unequal, subulate, margins hyaline, lower 1-, upper 3-nerved. Lemma lanceolate, thin, scaberulous, obscurely nerved, tapering into a slender awn. Palea glabrous. Anthers 2–3 mm. *Ovary pubescent at the top*. Fl. 6–8. $2n = 28, 42$. Hs.

Native, or more probably introduced. Naturalized in dry shady places. 12. Scattered and local in southern England; in a few localities elsewhere. Europe.

6. F. rubra L.   'Creeping Fescue.'

A ± erect perennial 10–70 cm. *Usually stoloniferous, some branches not arising from a scale and bearing several transitional lvs, creeping and then ascending* (*extravaginal*); *others arising from a scale without transitional lvs, ascending* (*intravaginal*). Lvs of sterile shoots setaceous, or ± convolute, sometimes rather stiff, the *culm lvs* ± *flat*, 0·5–3 mm. wide, glabrous or some-times pubescent, *obtuse or subacute*. *Sheaths* smooth, often ± puberulous, *closed almost to the top when young*, but readily tearing when older. Ligule short; auricles small. Panicle 3–15 cm., erect and ± spreading. *Spikelets* 7–14 *mm.*, of 4–8 florets, *often reddish or purplish*. Glumes unequal, acuminate, margins hyaline; lower 1-nerved, subulate; upper 3-nerved, linear-lanceolate. Lemma firm, lanceolate, obscurely nerved, shortly awned, glabrous or hairy. Palea finely ciliate on nerves. *Anthers* 2–3 *mm.* Ovary glabrous. Fl. 5–7. $2n = 14, 28*, 42*, 56, 70$. Hs.

Very variable. Var. *arenaria* Fr., far-creeping with stiff lvs and large pubescent spikelets, occurs on dunes; var. *fallax* (Thaill.) Hack., with stolons short or 0, is found locally; and many other forms occur. What appear to be hybrids with *Vulpia membranacea* and *V. myuros* are found occasionally with the parents.

Native. In grassy places, in salt-marshes, on dunes, in meadows and on mountains. 112, H40, S. Generally distributed and common throughout the British Is. Europe, north to Spitzbergen; N. Africa; temperate Asia; N. America. On mountains in the southern part of its range.

7. F. juncifolia St.-Amans

An extensively creeping, rather stout perennial 25–90 cm. *All the branches extravaginal. Lvs* 1–3 mm. wide, flat or convolute, *rigid, acute or pungent. Sheaths* smooth, glabrous, *closed when young*. Ligule short; auricles small. Panicle 10–15 cm., lanceolate to ovate. *Spikelets* 10–18 *mm.*, of 5–12 florets, *downy, greenish*. *Glumes* subequal, narrowly lanceolate, acuminate, ± *pubescent*; lower 1-, upper 3-nerved. Lemma chartaceous, lanceolate, at least the middle and marginal nerves ± prominent, pubescent particularly at the margin and tip, shortly awned or mucronate. Palea ciliate on the nerves in the upper half. *Anthers* 5 *mm.* Ovary glabrous. Fl. 7–8. Hs.

Native. On dunes. 9. S. Devon, Dorset, Suffolk, Norfolk, Lincoln, Durham, Fife and Angus. W. Europe.

F. ovina agg. (spp. 8–11).   Sheep's Fescue.

A tufted perennial 10–50 cm. *All branches intravaginal. Lvs up to* 1 *mm. wide, all setaceous,* subacute. *Sheaths* smooth, glabrous or puberulent, *split more than half way to base.* Ligule short, obtuse; auricles rounded. *Spikelets* 3–7(–10) *mm.*; lemma awned or mucronate. 112, H40, S.

1	Infl. viviparous.	**9. vivipara**	
	Infl. not viviparous.		*2*
2	Lf less than 0·7 mm. diam.	**8. ovina**	
	Lf more than 0·7 mm. diam.		*3*
3	Lvs glaucescent, rarely green, not pruinose, rhachis usually scabrid.		
		10. longifolia	
	Lvs, sheaths and nodes at least in part pruinose, rhachis smooth at least at the base.	**11. glauca**	

8. F. ovina L.

Lvs less than 0·7 mm. diam., not strongly pruinose nor usually glaucous. Panicle 1·5–10 cm., rather narrow, spreading at anthesis. Spikelets 3–7 mm., of 3–5 florets, greenish or purplish. Lemma mucronate or awned. Anthers 1·5–3 mm. Hs.

Ssp. **ovina.** Lvs usually short in proportion to the culms. Spikelets 4–7 mm., of 3–5 florets. Lemma shortly awned. Anthers 2–3 mm. Fl. 5–6. $2n=14$.

Ssp. **tenuifolia** (Sibth.) Peterm. Lvs very narrow, usually long in proportion to the slender culms. Spikelets 3–6 mm., of 3–4 florets. Lemma mucronate, not awned, often scaberulous above. Anthers 1·5–2 mm. Fl. 6–7. $2n=14$.

Native. Both sspp. are common and generally distributed in grassy places.

9. F. vivipara (L.) Sm.

Similar to *F. ovina* ssp. *tenuifolia* in vegetative characters. Spikelets always proliferating vegetatively ('viviparous'). Fl. 6–8. $2n=21, 28, 42$. Hs.

Native. Common on mountains and down to sea-level in N. and W. Scotland. Lake District, Wales, Scotland. Northern Europe.

10. F. longifolia Thuill.

F. trachyphylla (Hack.) Krajina, non Hack.; *F. duriuscula* auct., non L.

Lvs 0·7–1 mm. diam. or more. Panicle variable; rhachis usually scabrid. Spikelets 6–10 mm. Lemma awned. Fl. 6–7. $2n=42$. Hs. Railway banks and roadsides, etc. Probably introduced.

11. F. glauca Lam.

Similar to *F. longifolia* but the lvs pruinose at least at the base and the sheaths pruinose at least at the top. Panicle compact; rhachis smooth, at least below. Spikelets 3·5–5 mm. Lemmas aristate or mucronate, ± pruinose. Fl. 6–7. $2n = 14$. Hs. Very rare and probably introduced.

8. × FESTULOLIUM Aschers. & Graebn.

Hybrids between *Festuca pratensis* Huds. or *F. arundinacea* Schreb. and *Lolium perenne* L. combining some of the characters of both genera.

1. × **F. loliaceum** (Huds.) P. Fourn

F. pratensis × L. perenne

Spikelets nearly sessile in a simple or more rarely somewhat branched raceme, resembling those of *Festuca pratensis* but more compressed. Glume next to the axis usually present though much reduced in the lateral spikelets and occasionally 0 in some.

Native. In meadows, locally frequent with the parents and the commonest of these hybrids.

In addition the following hybrids occur more rarely: *F. arundinacea × L. perenne* ssp. *multiflorum*, *F. arundinacea × L. perenne* ssp. *perenne*.

9. LOLIUM L.

Glabrous herbs. *Infl.* normally *a simple spike* with the *spikelets edgeways on to the axis. Spikelets* of many florets, compressed, the *lateral ones with the lower glume suppressed*, the terminal one with two glumes. Glumes longer or shorter than the spikelet, membranous or chartaceous, 5–9-nerved, awnless. Lemma membranous, sometimes tumid in fr., awnless or awned, 5-nerved. Palea hyaline and tough, about equalling the lemma, 2-nerved and usually shortly ciliate on the nerves. Lodicules 2, lanceolate to ovate, acute or acuminate. Stamens 3; anthers stout, c. 3 mm. Ovary sparsely hairy at the top; styles distant, very short. *L. perenne* ssp. *multiflorum* and *L. remotum* are always cultivated or relics of cultivation, at least in Britain.

About 8 spp., in Europe, N. Africa and Asia; introduced elsewhere.

1 Glume equalling or exceeding the spikelet. **2. temulentum**
   Glume shorter than the spikelet. *2*
2 Perennial; lvs seldom exceeding 3 mm. wide; lemma awnless.
                                 1. perenne ssp. **perenne**
   Annual or biennial; lvs up to 10 mm. wide; lemma awned or plant annual
    and vegetative shoots 0. *3*
3 Lemma awned, wider than palea, not tumid in fr.
                             1. perenne ssp. **multiflorum**
   Lemma usually awnless, narrower than palea, tumid in fr. **remotum**

1. L. perenne L. ssp. **perenne**   Rye-grass, Ray-grass.

A wiry *perennial* 25–50 cm. Stems smooth, slender, often bent below. *Lvs* smooth, *up to 3 mm. wide, folded when young.* Sheaths smooth, basal persistent and often fibrous. Ligule c. 1 mm., truncate. Infl. 8–15 cm. Spikelets

5–15 mm., of (3–)8–11 florets. Glumes of terminal spikelet subequal; upper glume of lateral spikelets usually equalling 1st floret, sometimes longer, but always shorter than spikelet, linear-lanceolate or subulate, obtuse, acute or acuminate. Lemma linear-lanceolate, acute or subacute, scabrid below, margins hyaline in upper part. Palea almost equalling lemma, scabrid towards the base. Fl. 5–8. $2n=14^*$. Hs. Very variable; forms with the infl. branched, the axis shortened and the spikelets consequently crowded, or with viviparous spikelets, occur. A hybrid with *Festuca pratensis* is not infrequent (see p. 1130).

Native. In waste places, and sown for fodder. 112, H40, S. Generally distributed throughout the British Is. Europe, except the Arctic; temperate Asia; N. Africa; introduced in N. America and Australia.

*Ssp. multiflorum (Lam.) Husnot   Italian Rye-grass.
L. multiflorum Lam.

Annual or biennial. Lvs inrolled when young. Infl. up to 20 cm. Spikelets 10–20 mm., of 8–14 florets. Lemma broader and rather softer than in ssp. *perenne*, usually awned. Fl. 5–9. $2n=14$. Th. or Hs. The boundary between the sspp. is often obscured by the occurrence of hybrids, some of which are commonly sown in leys.

Introduced. Widely cultivated and well naturalized in temperate regions. Perhaps native in S. Europe.

*L. remotum Schrank (*L. linicola* A.Br.), a slender annual with no vegetative shoots and the lemma narrower than the palea and tumid in fr., occurs occasionally in cultivated fields. $2n=14^*$.

*2. L. temulentum L.   Darnel.

A stout erect annual 30–100 cm. *Stems ± scabrid below the infl.* Lvs usually scabrid, up to 10 mm. wide. Sheaths smooth or slightly scabrid. Ligule up to 3 mm., truncate. Infl. 10–25 cm. Spikelets 15–25 mm., of 5–8-florets. *Glumes equalling or exceeding the spikelet,* linear-lanceolate, obtuse or sub-acute, 7–9-nerved. Lemma lanceolate, bifid at apex and awned from the sinus or entire and awnless, tumid in fr. Palea slightly exceeding lemma. Fl. 6–8. $2n=14^*$. Th.

Introduced. Casual; in cultivated fields and on rubbish dumps. Scattered throughout the British Is., rare and inconstant. Europe except the north; N. Africa; temperate Asia; introduced in N. and S. America, S. Africa, and Australia.

10. VULPIA C. C. Gmel.

Glabrous *annuals with short convolute leaves* and slender, *little-branched secund panicles.* Spikelets of 4–6-florets, shortly stalked or almost sessile. *Glumes unequal; lower sometimes 0,* usually present but small. Lemma tough, convolute, nerves usually obscure; awn serrate, longer than the lemma which tapers gradually into it. Palea about equalling the lemma, thin, bifid at the apex, 2-nerved, finely ciliate in the upper part. Stamens 1(–3). Fls cleisto-gamous.

About 25 spp. in temperate regions, particularly the Mediterranean region and the Pacific coast of N. and S. America.

1 Upper sheath strongly inflated; lower glume 0 or very small; lemma with
1 prominent nerve.   **1. membranacea**
Upper sheath not or slightly inflated; lower glume at least $\frac{1}{6}$ length of
upper; lemma obscurely nerved.   *2*

2 Culm distinctly ridged; upper glume lanceolate with strong nerves.
                                                                           2. bromoides
Culm smooth or nearly so; upper glume setaceous or subulate, obscurely
nerved.   *3*

3 Panicle nodding; upper sheath not inflated; upper glume not more than
three times as long as lower.   **3. myuros**
Panicle erect; upper sheath somewhat inflated; upper glume 3–6 times as
long as lower.   **4. ambigua**

1. V. membranacea (L.) Dum.

V. uniglumis (Ait.) Dum.; *Festuca uniglumis* Ait.

A slender annual, usually somewhat decumbent at the base, 10–60 cm. *Sheaths inflated*, upper usually some distance below the panicle at flowering. Culms ribbed, ribs broad and shining. Panicle 2–12 cm., erect, dense, simple or rarely slightly branched at the base. Spikelets 10–15 mm. *Lower glume 0 or very small*; *upper* linear-lanceolate, *nearly equalling the first floret, broadly membranous, obscurely 3-nerved*, serrulate on the keel, acuminate or with a short (up to 5 mm.) awn. *Lemma subulate, chartaceous, with 1 prominent nerve, compressed and keeled in the upper part*, margins convolute; awn 2–3 times as long as lemma. Stamens usually 3. Fl. 6. $2n=14, 42^*$. Th.

Native. On dunes, usually in hollows, very local. 26, H5, S. From Devon and Kent to Lancashire and Norfolk; east coast of Ireland from Wexford Harbour to the Boyne. Western Europe and the Mediterranean region.

2. V. bromoides (L.) S. F. Gray   'Barren Fescue.'

Festuca sciuroides Roth

A slender annual, usually somewhat decumbent at the base, 10–25(–50) cm. Sheaths not inflated, upper usually some distance below the panicle at flowering. Culms ribbed, ribs broad and shining. Panicle 2–10 cm., erect, rather lax, usually compound. Spikelets 6–10 mm. Glumes green with a narrow hyaline margin; lower $\frac{1}{3}-\frac{1}{2}$ as long as upper, subulate; *upper lanceolate*, acuminate, shorter than the first floret, *with 3 prominent nerves*. Lemma chartaceous, obscurely nerved, terete, margins convolute; awn twice as long as lemma. Stamens usually 1. Fl. 5–7. $2n=14$. Th.

Native. In dry places on heaths, sandy and rocky places, walls and waste ground. 112, H40, S. Throughout the British Is., locally common. Europe from N. Jutland to the Mediterranean; Asia Minor; N. and S. Africa; N. America.

3. V. myuros (L.) C. C. Gmel.   'Rat's-tail Fescue.'

Festuca myuros L.

An erect annual 10–60 cm. Sheaths not inflated, upper usually reaching or enclosing the lower part of the panicle at flowering. Culms obscurely ribbed. *Panicle 4–20 cm., ± nodding*, lax, interrupted, compound. Spikelets 7–10 mm. *Glumes* setaceous, or upper subulate, nerves obscure, margins hyaline, *lower*

$\frac{1}{2}$ *as long as upper.* Lemma tough, terete, obscurely nerved, ±scabrid, sometimes ciliate on the margin towards the top; awn 2–3 times as long as lemma. Stamens usually 1. Fl. 5–7. $2n=14, 42$. Th.

Native. In sandy places and on walls. 63, H25, S. England, Wales, Ireland, mainly in the south, local; Scotland: probably introduced. C. and S. Europe; Asia to the Himalaya; N. Africa; Macaronesia; probably introduced in S. Africa, N. and S. America and Australia.

V. megalura (Nutt.) Rydb., differing from *V. myuros* mainly in the ciliate lemma, is established in a number of localities. N. America and western S. America.

4. V. ambigua (Le Gall) More

Festuca ambigua Le Gall

An annual, ±decumbent at the base, 10–30 cm. Sheaths often purplish, upper ±inflated, usually reaching or enclosing the lower part of the panicle at flowering. *Culms almost smooth.* Panicle 5–10 cm., erect, strict, often purplish, usually compound, pedicels very short, rather swollen. Spikelets 4–8 mm. *Glumes* setaceous, *upper* 3–6 *times as long as lower.* Lemma subulate, terete, smooth, obscurely nerved; awn 2–3 times as long as lemma. Stamens usually 1. Fl. 5–6. $2n=28*$. Th.

Native. On sandy heaths and by the sea in southern and eastern England. 12, S. S.W. Europe, Mediterranean region.

V. ciliata Link, resembling *V. ambigua* but readily distinguished by the long-ciliate lemmas, occurs, usually as a casual, in S. England. S. Europe. N. Africa, W. Asia.

11. NARDURUS (Bluff, Nees & Schau.) Rchb.

Annuals similar to *Vulpia* in general appearance. Infl. unilateral, simple or with few short branches. Spikelets of 3–9 florets, shortly pedicelled; pedicels thick, stiff, short, appressed to rhachis. Glumes subequal or unequal, lower always present. Lemma tough, obscurely nerved, awned or not. Stamens 3. Fls chasmogamous. Perhaps 6 spp., W. Europe to China.

1. N. maritimus (L.) Murb.

A rather stiff tufted annual 4–30 cm. Stems slender, geniculate, spreading, minutely hairy below the nodes. Lower lvs folded and setaceous, upper often flat and obtuse. Sheaths ±glabrous. Infl. 4–10 cm., slender, rather stiff. Spikelets c. 4 mm. (excluding awns). Glumes unequal, lower c. $\frac{1}{2}$ as long as upper. Lemma c. 2·5–3 mm., awned; awn short or up to 4 mm. Fl. 5. Fr. 6. Germ. autumn. $2n=14*$. Th.

Probably introduced. In ±open habitats in calcareous grassland and near the sea, local and perhaps overlooked. S. Devon and Kent to Oxford and Lincs. 7. Europe from Belgium and Switzerland southwards, eastwards to China.

12. PUCCINELLIA Parl.

Tufted or creeping perennial or annual herbs of variable habit. Panicle compound, of many subterete spikelets. Glumes unequal, shorter than the 1st floret, membranous with hyaline margins, 1–3-nerved, bluntly keeled.

Lemma membranous with hyaline apex, oblong or ovate, obtuse, 5-*nerved*, nerves stopping short of apex or the middle one shortly excurrent. Palea ovate or lanceolate, shortly bifid, 2-nerved, about equalling the lemma. Lodicules 2, membranous, lanceolate, acute. Stamens 3. Ovary glabrous; styles 0.

About 25 spp. in temperate regions; in muddy places near the sea, rarely inland. Hybrids have been recorded between several of the spp.

1	Lemma 3–4 mm.	2
	Lemma less than 2·5 mm.	3
2	Stoloniferous; panicle lax, not rigid; anthers 2 mm.	**1. maritima**
	No stolons; panicle compact, very rigid; anthers 1 mm. or less.	
		4. rupestris
3	Panicle branches 4–6 together, usually all long and bare below; midrib of lemma not reaching tip.	**2. distans**
	Panicle branches 2–4 together, at least the very short ones with spikelets to base; midrib of lemma reaching tip.	**3. fasciculata**

1. P. maritima (Huds.) Parl.   'Sea Poa.'

Glyceria maritima (Huds.) Wahlberg

A ± tufted stoloniferous perennial 10–80 cm. *Lvs* smooth, *narrow*, flat or convolute, somewhat obtuse. Ligule c. 1 mm., ovate, obtuse. *Panicle* 2–25 cm., narrow; *branches strict or somewhat spreading*, 2–3 *at each node*. Spikelets 7–12 mm., usually of 6–10 florets. Glumes broadly ovate. Lemma 3–4 mm., oblong, obtuse, or apiculate, slightly silky towards the base. Palea equalling or slightly exceeding the lemma, ovate, ciliate on nerves. Anthers 2 mm. Fl. 6–7. 2*n* = 14*, 49*, 63*, 77*; 56; 60; 70. Hs.

Native. In salt-marshes and muddy estuaries. 82, H25, S. Generally distributed around the coasts of the British Is. and common or dominant in suitable habitats. Western coasts of Europe; Sakhalin; N. America; Greenland. Probably often apomictic.

2. P. distans (Jacq.) Parl.   'Reflexed Poa.'

Glyceria distans (Jacq.) Wahlenb.

A tufted perennial 15–60 cm. *Stolons usually* 0. Lvs smooth or minutely scabrid, narrow, flat, acute. Ligule c. 1 mm., ovate, obtuse. *Panicle* 4–15 cm., lanceolate to triangular; *branches* strict before flowering, *spreading horizontally at flowering and ultimately deflexed*, 4–6 *together*, usually all long and devoid of spikelets below. *Spikelets* 4–6 mm., usually of 3–6 florets, ± *uniformly spaced*. Lower glume subacute, upper obtuse. *Lemma c.* 2 *mm.*, broadly ovate, obtuse or subacute, *midrib not reaching tip*. Palea equalling lemma, lanceolate to ovate, glabrous or finely and shortly ciliate on nerves in upper part. *Anthers* 0·5–0·75 *mm.* Fl. 6–7. 2*n* = 28; 14, 42. Hs.

Native. In salt-marshes, muddy estuaries and occasionally on sandy ground inland. 74, H13, S. Generally distributed around the coasts of the British Is. except for most of W. Ireland, common in suitable habitats. Europe, Siberia.

P. pseudodistans (Crép.) Jans. & Wacht. is ± intermediate between *P. distans* and *P. fasciculata*. Lvs scarcely revolute. Panicle symmetrical (not unilateral), branches not deflexed. Spikelets subsessile, extending to base of shorter panicle branches. Midrib of lemma usually excurrent. Anthers 0·4–0·5 mm.

Native. In salt-marshes. 2. E. and W. Kent. Western Mediterranean, W. France, Netherlands.

Sterile plants intermediate between *P. distans* and *P. fasciculata* and probably hybrids between the two, occur occasionally where these spp. grow together.

3. P. fasciculata (Torr.) Bicknell

Glyceria Borreri (Bab.) Bab.; *Poa fasciculata* Torr.

A tufted perennial 5–50 cm. Stolons 0. Lvs minutely scabrid, narrow, flat, hooded and apiculate. Ligule very short. *Panicle* 3–16 cm., lanceolate, unilateral; *branches spreading, 2–4 together, at least the 1 or 2 very short ones covered with spikelets to the base. Spikelets* 4–7 mm., usually of 4–8 florets, *crowded in groups.* Glumes ovate, obtuse or subacute. *Lemma* 1·5–2 mm., broadly ovate, obtuse, *with a very small rigid apiculus formed by the excurrent midrib.* Palea ovate, finely ciliate on the margins, equalling lemma. *Anthers* 0·75–1 *mm.* Fl. 6. $2n = 28*$. Hs.

Native. In muddy places near the sea; very local. 19, H3. Coasts of S. England; Scotland: Angus; Ireland: Waterford, Wexford and Dublin. Coasts of western Europe.

4. P. rupestris (With.) Fernald & Weath.   'Procumbent Poa.'

Glyceria rupestris (With.) E. S. Marshall; *G. procumbens* (Curt.) Dum.; *Sclerochloa procumbens* (Curt.) Beauv.

A procumbent, rarely ascending or erect annual or biennial 4–40 cm. *Stolons 0. Lvs rough*, flat, short and rather broad, hooded and apiculate. Upper sheaths ± inflated. Ligule c. 1 mm., obtuse. *Panicle* 2–8 cm., oblong to ovate, *stiff*; branches spreading, 2–3 together, short, rigid. Spikelets 5–7 mm., of 3–4 florets, crowded on the branches. Glumes ovate, subacute, both strongly nerved. Lemma 3–4 mm., ovate, obtuse with a small rigid apiculus formed by the excurrent midrib, distinctly 5-nerved. Palea equalling lemma, lanceolate, shortly ciliate on the nerves. *Anthers* 0·75–1 *mm.* Fl. 5–7. Th. or Hs.

Native. On clayey and stony sea-shores and beside brackish ditches. 28, S. Coasts of England and Wales, local. Coasts of western Europe, Syria.

13. CATAPODIUM Link

Glabrous annual herbs. Panicle rigid, simple or branched, secund. Spikelets somewhat compressed. Glumes chartaceous, subequal, 1–3-nerved. Lemma coriaceous, 5-nerved with 3 prominent and 2 faint nerves, rounded on the back. Palea shorter than or about equalling lemma, thin; lodicules lobed. Four or five spp. in the Mediterranean region extending to western Europe and Afghanistan.

Infl. usually somewhat branched and at least some of the spikelets distinctly stalked; rhachis slender, angled, not shiny; lemma 2–2·5 mm.   **1. rigidum**
Infl. usually unbranched and spikelets sessile, sometimes shortly stalked; rhachis stout, flattened, shiny; lemma 2·5–3·8 mm.   **2. marinum**

1. C. rigidum (L.) C. E. Hubbard   'Hard Poa.'

Poa rigida L.; *Festuca rigida* (L.) Rasp, non Roth; *Sclerochloa rigida* (L.) Link; *Scleropoa rigida* (L.) Griseb.; *Demazeria rigida* (L.) Tutin

A rigid, glabrous annual 5–20 cm. Lvs narrow, flat or convolute. Sheaths with broad hyaline margins towards the top. Ligule 1–3 mm., ovate. Culms smooth, often somewhat geniculate. Panicle 2–8 cm., strict, extremely rigid; *pedicels shorter than the spikelets*. Spikelets 2–4 *mm.*, of 3–6 florets, along one side of the branches; florets somewhat spreading. *Glumes acuminate,* strongly nerved, margins hyaline. Lemma ovate, obtuse, obscurely nerved. Fl. 5–6. Germ. autumn. 2*n* = 14. Th.

Native. On dry rocks, walls, banks, and sometimes on arable land; mainly in calcareous districts. 75, H 39, S. Scattered throughout the British Is. but rarer in the north, locally common in the south. W. Europe from Belgium southwards; Mediterranean region; Macaronesia.

2. C. marinum (L.) C. E. Hubbard   'Darnel Poa.'

Festuca marina L.; *F. rottboellioides* Kunth; *Poa loliacea* Huds.; *Catapodium loliaceum* Link; *Demazeria loliacea* Nym.; *D. marina* (L.) Druce

A stout, erect to decumbent or almost prostrate annual 1–13 cm. Lvs narrow, flat or convolute. Sheaths ribbed. Ligule rather variable in length, ovate. Culms broadly ribbed, ribs often shiny. Panicle 0·5–4·5 cm.; *pedicels very short* or almost 0, rarely longer, rhachis flattened. Spikelets 4–8 *mm.*, of 3–10 florets, distichous, all directed to one side; florets imbricate. Glumes lanceolate, margins hyaline; *lower acute*, keeled, narrower than upper; *upper obtuse*, rounded on back. Lemma ovate, obtuse or mucronulate, nerves distinct, margins hyaline. Fl. 6–7. 2*n* = 14. Th.

Native. By the sea on sand, shingle and rocks. 54, H 21, S. Coasts of the British Is., local. Mediterranean region and along the west coast of Europe to Britain.

14. Poa L.

Glabrous annual or perennial herbs with compound panicles. Spikelets compressed, of (1–)2–5(–7) florets. Glumes ± unequal, soft, keeled, lower 1–3-nerved, upper 3-nerved, awnless. *Lemma* keeled, 5–7-nerved, *herbaceous, often with a long tuft of cottony hairs at base, tip usually hyaline, awnless*. Palea bifid, 2-nerved, nerves often pubescent or shortly ciliate. Lodicules 2, ovate and swollen below. Stamens 3; anthers usually several times longer than broad. *Ovary glabrous*; styles short, terminal.

About 200 spp. in cold and temperate regions and on mountains in the tropics.

1 Lvs at least 5 mm. wide; panicle 10 cm. or more; plant large, densely tufted.   **13. chaixii**
    Lvs normally up to c. 3 mm. wide; plant smaller (except *P. palustris*).   2

2 Plant with distinct creeping rhizomes.   3
    Plant without rhizomes, tufted.   4

3 Stems and sheaths strongly compressed.   **9. compressa**
    Stems and sheaths ± terete.   **10. pratensis**

4 Stock stout, clothed with persistent ±fibrous and shining sheaths;
    spikelets often viviparous.   *5*
  Stock slender, not as above; spikelets never viviparous.   *8*

5 Stems bulbous at base, tapering above; fl. 4–5; plant lfless June–Oct.
    (sandy places or dry shallow soils usually by the sea).   **3. bulbosa**
  Stems cylindrical, not bulbous; fl. 7–8; plant green all summer (damp
    ledges and screes on higher mountains).   *6*

6 Lvs (or some of them) usually at least 3 mm. wide, tapering rather
    abruptly at apex; lower glume ⅔ length of upper; palea and lemma
    subequal.   **4. alpina**
  Lvs usually not exceeding 2 mm. wide, tapering gradually at apex;
    glumes subequal; palea ¾ length of lemma (very rare).   *7*

7 Spikelets viviparous; uppermost lf shorter than its sheath.
                                      5. × jemtlandica
  Spikelets not viviparous; uppermost lf about equalling its sheath.
                                        6. flexuosa

8 Lower panicle branches 1–2(–3) together.   *9*
  Lower panicle branches (3–)4–6 together.   *12*

9 Panicle ovate to triangular; some lvs usually transversely wrinkled;
    plant soft, annual though sometimes perennating.   *10*
  Panicle narrowly lanceolate to almost linear; lvs never transversely
    wrinkled; plant usually rather stiff, always perennial.   *11*

10 Lower panicle branches spreading or reflexed after flowering; florets
    closely imbricate (common and widespread).   **1. annua**
  Panicle branches not reflexed after flowering; florets distant, scarcely
    overlapping (Channel Is., Cornwall).   **2. infirma**

11 Ligule of uppermost lf 1 mm. or more.   **8. glauca**
  Ligule of uppermost lf not more than 0·5 mm.   **7. nemoralis**

12 Ligule of uppermost lf very short; panicle usually nodding.
                                        7. nemoralis
  Ligule of uppermost lf more than 1 mm.; panicle usually erect.   *13*

13 Ligule acute; lemma distinctly 5-nerved.   **11. trivialis**
  Ligule truncate; lemma very obscurely nerved.   **12. palustris**

1. P. annua L.   'Annual Poa.'

An erect or decumbent tufted annual or short-lived perennial 5–30 cm.
Stems sometimes creeping and rooting at the nodes. *Lvs* flat, slightly keeled,
smooth, abruptly contracted at the tip, *often transversely wrinkled.* Sheaths
smooth, compressed, the uppermost longer than its lf. Ligule 2–3 mm. Culms
smooth, with broad flat shining ridges. *Panicle* 1–8 cm., spreading or compact, ± *triangular; branches* smooth, 1–2(–4) together, *spreading or deflexed
after flowering.* Spikelets 3–5 mm. lanceolate, of 3–5 florets; *florets closely
imbricate.* Glumes herbaceous, unequal, boat-shaped, acute, margins hyaline,
lower 1-, upper 3-nerved. Lemma ovate, 5-nerved, glabrous or pubescent on
the nerves towards the base, margins and tip broadly hyaline, nerves not
reaching apex. Palea nearly equalling lemma, shortly ciliate on the nerves.
Unopened anthers 0·75 mm., 2–4 times as long as broad. Fl. 1–12. $2n = 28*$.
Th. or Hs.

Native. In waste places, gardens, cultivated land, grassland, on mountains and by water. 112, H40, S. Throughout the British Is., common; ascends to 3980 ft. on Ben Lawers. Throughout nearly the whole world, but in the tropics mainly on mountains.

2. P. infirma Kunth

P. exilis (Tomm.) Murb.

Similar to *P. annua* L. from which it differs in the small oval or oblong *panicle* with *branches not reflexed after flowering*, the long-linear *spikelets* of 5–6 florets, *with distant or slightly overlapping florets*, and *unopened anthers* 0·5 mm., *about as long as broad.* 2n=14*. Th.

Native. In sandy places near the sea. Cornwall, Channel and Scilly Is. 2, S. S.W. Europe and Mediterranean region.

3. P. bulbosa L.   'Bulbous Poa.'

An erect, tufted perennial 10–30 cm. *Stems swollen and ± bulbous at the base,* never creeping. Lvs usually narrow, flat, tapering to a long point. Basal sheaths light brown or purplish, persistent, uppermost much longer than its lf. *Ligule long (up to 5 mm.),* acute. Culms usually ± geniculate below, smooth, obscurely ribbed, ribs not shining. *Panicle compact, ovate,* 2–6 cm.; branches ± scabrid, 4–5 together. Spikelets sometimes viviparous; normal ones 3–5 mm., ovate, often purplish or glaucous, of 3–4 florets. Glumes subequal, ovate, acute or acuminate, serrate on margins and keel, 3-nerved, ± hyaline. Lemma ± hyaline, ovate, acute, 5-nerved with long cottony hairs on nerves towards the base. Palea about ¾ the length of the lemma, shortly ciliate on the nerves. Fl. 4–5. 2n=28*; 45*, 35 (viviparous). Hs. The lvs and stems soon wither after fl. and the 'bulbs' lie loose on or near the surface till autumn.

Native. On sandy shores and on limestone near the coast, inland in Oxford. 13, S. In southern England and Glamorgan; very local. Europe, mainly in the south and west, north to S. Scandinavia; temperate Asia; N. Africa.

4. P. alpina L.   'Alpine Poa.'

An erect, tufted perennial 10–40 cm. *Rootstock stout, clothed with the persistent, fibrous remains of basal lvs and sheaths. Lvs usually short, stiff, broad (up to 4 mm.), abruptly contracted at the apex,* mucronate, folded at the tip or throughout, often distichous, ± glaucous. Sheaths smooth, uppermost usually far distant from the panicle, longer than its lf. *Ligule long (c. 3 mm.),* ovate, ± laciniate. Culms smooth, terete, usually with broad shining ribs. Panicle 2–5 cm., erect, ± lax, ovoid or oblong; branches 1–2 together, nearly smooth. Spikelets viviparous in this country except in 2 localities; normal ones 4–7 mm., ovate, of 2–4 florets. Glumes thin, ovate, acute, *lower ⅔ the length of the upper,* scabrid on keel, obscurely 3-nerved, middle nerve reaching tip, broadly hyaline and often purplish. Lemma thin, ovate, ± acute, obscurely 5-nerved, ± pubescent on keel and 2 lateral nerves, margins and apex hyaline. Palea nearly as long as, to slightly longer than, the lemma, pubescent or shortly ciliate on the nerves. Fl. 7–8. $2n=28*$, $35*$ (viviparous); 42; 32–4; 38. Hs.

Native. In gullies and on rocks on mountains from 1000 to 4000 ft. 19, H2. Merioneth and Caernarvon; N.W. Yorks, and Lake District; Scottish Highlands; Ireland: Kerry and Sligo only. Arctic Europe, Asia and America; Alps, Caucasus. A diploid form ($2n=14$) occurs in the Caucasus, and a tetraploid ($2n=28$) in C. Europe and, rarely, in Britain; these apparently never have viviparous spikelets.

5. P. × jemtlandica (Almq.) Richt. (*P. alpina × flexuosa*)

A perennial resembling *P. alpina* in general appearance, but usually smaller and more slender. Rootstock similar to that of *P. flexuosa*. Lvs 1–2 mm. wide, folded, tapering gradually to the apex and not hooded, uppermost shorter than its sheath. Panicle slender, purplish, viviparous, spreading in fl., ± nodding; rhachis and branches not flexuous. Glumes subequal. Palea about ¾ length of lemma. Fl. 7–8. Hs.

Native. On damp stony slopes and rock-ledges, very rare. 2. Ben Nevis and Lochnagar. Scandinavia and Iceland.

6. P. flexuosa Sm.   'Wavy Poa.'

P. laxa ssp. *flexuosa* (Sm.) Hyl

A tufted perennial 6–20 cm. *Rootstock rather slender, clothed with the fibrous, persistent remains of lvs and sheaths.* Stem shortly creeping and rooting at the nodes, then ascending or erect. *Lvs thin*, narrow, *tapering gradually to the* ± involute, *often obliquely mucronate apex.* Sheaths smooth, compressed, uppermost usually shorter than its lf and often reaching nearly to the base of the panicle. Ligule c. 2 mm., acute, ± laciniate. *Culms slightly compressed*, smooth, ribs not shiny. Panicle 2–4 cm., erect, lax, ovate; branches 1–2 together, flexuous at least at the base. *Spikelets* 3–5 mm., *never viviparous*, ovate, of 2–3 florets. *Glumes* ovate-lanceolate, acuminate or mucronate, *subequal*, smooth, obscurely nerved, middle nerves sometimes excurrent, margins and apex hyaline, sometimes purplish. Lemma thin, ovate-lanceolate, obtuse, obscurely nerved, ± pubescent on keel and 2 lateral nerves, margins and apex hyaline. Palea ¾ the length of lemma, shortly ciliate on the nerves. Fl. 7–8. $2n=42*$. Hs.

Native. On scree on the higher Scottish mountains, very rare. 3. Ben Nevis, Lochnagar, Cairn Toul, Cairngorm. Arctic Europe, ?elsewhere.

7. P. nemoralis L.   'Wood Poa.'

An erect, slender perennial 30–90 cm. Stem shortly creeping at the base. *Lvs* narrow, smooth, *flaccid*. Uppermost sheath shorter than its lf. *Ligule very short or* 0. Culms slender, terete, smooth. *Panicle* 5–10 cm., *usually nodding*, lax, narrow; branches long, slender, scabrid, 2–5 together. Spikelets 2–4 mm., ovate, of 1–5 florets. Glumes unequal, lanceolate, acuminate, obscurely nerved, broadly hyaline, middle nerve reaching the tip, ± pubescent on the keel. Palea rather shorter than the lemma, nerves glabrous. Fl. 6–7. $2n=28–38$, $42*$, 43, 47–49; $56*$; c. 70. Hs.

Native. In shady places. 107, H26. Throughout most of the British Is., absent from the north of Scotland and local in Ireland, mainly in the east. Europe, only on mountains in the south and absent from Portugal; temperate Asia; N. America.

P. Parnellii Bab. appears to be a form of this sp.

8. P. glauca Vahl

P. caesia Sm.; *P. Balfourii* Parnell

An erect, *usually intensely glaucous, stiff*, ± tufted perennial 10–40 cm. Stem creeping and often decumbent at the base. Lvs ± involute, rough or smooth, tapering. Uppermost sheath rather longer than or about equalling its lf and far below the panicle. *Ligule c. 2 mm.*, blunt. Culms fairly stout, smooth. *Panicle* 2–10 cm., lax or dense, *erect*, narrow; branches scabrid, strict or spreading at flowering, 1–2(–3) together. Spikelets 4–7 mm., ovate, of 3–4 florets. *Glumes* subequal, *broadly lanceolate, or upper ovate*, acute or acuminate, scabrid on the keel, 3-nerved, broadly hyaline and often purplish. *Lemma ovate, ± acute*, obscurely 5-nerved, margins and apex hyaline, middle nerve not reaching tip, keel and 2 lateral nerves pubescent. Palea rather shorter than lemma, nerves shortly pubescent. Fl. 7–8. $2n=42^*$, 47, 49, 50, 56^*, 60, 63, 70^*; 65; 70–72 (some probably refer to mountain forms of *P. nemoralis*). Hs.

Native. Damp rock-ledges. 14. On the higher mountains of England, Scotland and Wales, from about 2000 to 3000 ft. Northern Europe and America; C. European mountains; Altai.

9. P. compressa L.   'Flattened Poa.'

A stiff, ± glaucous, erect or ascending perennial 20–40 cm. *Stem compressed*, creeping and stoloniferous, usually geniculate. Lvs flat, ± scabrid. *Sheaths* smooth, *compressed*, uppermost about equalling its lf. Ligule 1 mm. or less, truncate. *Culms strongly compressed*, smooth. *Panicle* 2–7 cm., *narrow*, compact or rarely spreading; *branches* short, strict, scabrid, 1–3 together. Spikelets 3–6 mm., lanceolate, of 2–7 florets. Glumes subequal, tough, lanceolate, acute or mucronate, scabrid on keel, 3-nerved, margins and tips hyaline. *Lemma oblong, obtuse*, keel minutely scabrid with a tuft of cottony hairs at the base, *obscurely nerved*, tip hyaline. Palea about as long as lemma or slightly exceeding it, nerves minutely scabrid. Fl. 6–8. $2n=35$, 42, 49; 56. Hs.

Native. On dry banks and walls. 82, H19, S. Throughout most of the British Is. except the north of Scotland and Ireland, where it is rare and doubtfully native. Europe except the Arctic and parts of the south; Asia Minor; Caucasus; N. America.

10. P. pratensis L.   Meadow-grass.

A tufted, erect, rather stiff perennial 15–80 cm. Stem creeping and stoloniferous. Lvs flat or ± involute, smooth or sometimes rough, very variable in length and breadth, usually abruptly contracted at the tip. Sheaths smooth, somewhat compressed, uppermost longer than its lf. *Ligule usually 1 mm. or less*, truncate. Culms smooth, terete. Panicle 2–10(–25) cm., spreading, ovate or oblong; branches nearly smooth, 2–5 together. Spikelets 4–6 mm., ovate, of 3–5 florets. Glumes membranous, unequal or subequal, lanceolate, subacute to acuminate, scabrid on the keel, margins sometimes hyaline. *Lemma lanceolate, acute or acuminate, distinctly 5-nerved, with an abundance of long cottony hairs on keel and marginal nerves*, margins hyaline in upper part. Palea and lemma equal or nearly so, nerves of palea pubescent. Fl. 5–7. $2n = 28$; 36–147. Hs.

Native. In meadows and grassy places and on dunes. 112, H40, S. Common and generally distributed throughout the British Is. Europe, but only on mountains in the south; temperate Asia; N. Africa; N. America.

A highly variable plant with a great range of chromosome numbers. A number of groups of forms can be recognized, though there is a great diversity in the combinations of characters which occur. In certain races apomixis is common and such genotypes consequently tend to remain distinct. The following key may help to distinguish the main groups which occur in this country.

1   Panicle with numerous spikelets; lower panicle branches usually 4–5
        together; glumes unequal, lower not acuminate.
     Panicle often with rather few spikelets; lower panicle branches often in
        pairs; glumes about equal, lower acuminate or subacuminate.
                                              ssp. **irrigata** (Lindm.) Lindb. f.

2   Panicle ovate or broadly pyramidal; lvs flat or channelled, 2–3 mm. broad,
        pure green; culms ascending, not densely tufted. Generally common
        and very variable.   ssp. **pratensis**
     Panicle oblong or narrowly pyramidal; lvs of sterile shoots filiform or
        folded, up to c. 1 mm. broad, pale green; culms usually erect, often
        many together and tufted. Rather local.
                                              ssp. **angustifolia** (L.) Gaud.

11. P. trivialis L.

A tufted yellow-green perennial 20–60 cm., with rather weak culms and flaccid lvs. Stem shortly creeping, not stoloniferous. Lvs flat, tapering gradually. Sheaths rough or smooth, uppermost longer than its lf. *Ligule long (up to 8 mm.), acute.* Culms smooth, terete. Panicle 5–10 cm., lax, broadly ovate to oblong; branches scabrid, 3–5 together. Spikelets 2–4 mm., ovate, of 2–4 florets. Glumes thin, lanceolate, acuminate, scabrid on keel, lower ¾ length of upper. Lemma lanceolate, acuminate, distinctly 5-nerved, keel silky with a mass of cottony hairs at the base, *marginal nerves glabrous*. Palea about equalling lemma. Fl. 6. $2n = 14^*$. Hs. or Chh.

Native. In meadows and waste places. 112, H40, S. Throughout the British Is., common. Europe, temperate Asia, N. Africa, Macaronesia.

12. P. palustris L.

P. serotina Ehrh. ex Hoffm.

A tall, erect, tufted perennial 50–100 cm. Stems shortly creeping at the base. *Lvs flat, rather narrow, tapering*, ± scabrid. *Sheaths smooth*, uppermost longer than its lf. Ligule 2–3 mm., truncate. Culms smooth, terete. Panicle 10–20 cm., effuse, oblong or ovate; branches scabrid, 4–6 together. Spikelets

3–5 mm., ovate, of 3–5 florets. Glumes thin, subequal, lanceolate, acuminate, obscurely nerved, keel smooth, margins broadly hyaline. Lemma lanceolate, ± acute, obscurely nerved, *keel silky with a mass of cottony hairs at the base, marginal nerves silky*, tip hyaline and often brownish. Palea and lemma subequal. Fl. 6–7. 2*n* = 28, 29, 30, 42. Hs.

Native. In a few fens in E. Anglia, very local; naturalized by rivers and in waste places in a number of scattered localities. 3 (25, H 3). Europe, temperate Asia, N. America.

*13. P. chaixii Vill.

A tall, stout, *densely tufted* perennial 50–100 cm. Stems shortly creeping at the base. *Lvs broad, hooded and apiculate, edges and midribs rough. Sheaths rough*, uppermost longer than its lf. Ligule very short, truncate. *Culms smooth, 2-edged*. Panicle 10–18 cm., narrow; branches smooth, short (5–7 cm.), 4–6 together. Spikelets 4–6 mm., ovate, of 3–5 florets. Glumes unequal, lanceolate, acute, keel scabrid at the top. Lemma lanceolate, acute, 5-nerved, serrate on the keel, *nerves all glabrous*. Palea rather shorter than the lemma. Fl. 6. 2*n* = 14. Hs.

Introduced. Naturalized in woods. 30. Scattered through the British Is., local. W. and C. Europe north to S. Scandinavia; mountains of Asia Minor; Caucasus.

15. Dactylis L.

Perennials. Vegetative shoots strongly compressed. Panicle compound, lower branches usually long. *Spikelets* compressed, shortly pedicelled and *crowded in dense masses at the ends of the branches*. Glumes subequal, 3-nerved, keeled. *Lemma* 5-nerved, keeled, *shortly awned*. Palea equalling lemma, acuminate, nerves finely ciliate. Lodicules 2, linear, acute, about ½ length of palea. Ovary glabrous; styles very short, terminal.

About 3 spp., in Europe, N. Africa and temperate Asia.

Lvs usually stiff and glaucous; keel of lemma ciliate.   **1. glomerata**
Lvs usually soft and green; keel of lemma glabrous.   **2. polygama**

1. D. glomerata L.   Cock's-foot.

A coarse, tufted, glabrous, usually glaucous, erect or decumbent perennial up to 1 m. Lvs flat, rough, ± keeled. Sheaths of lvs of vegetative shoots strongly plicate and flattened, of others keeled, rough or nearly smooth. Ligule 2–10 mm., acute, torn. Culms smooth. Panicle 3–15 cm., erect, lower branches usually long, distant, horizontal or reflexed in fl., erect in fr., upper very short. Spikelets 5–7 mm., secund, crowded at the ends of the branches, green and violet. Glumes lanceolate, mucronate, finely ciliate on keel. Lemma lanceolate, ciliate on keel, awn short, subterminal. Palea lanceolate. Fl. 5–8. 2*n* = 28*. Hs.

Native. In meadows, waste places, by roads and on downs. 112, H 40, S. Generally distributed and common throughout the British Is. Europe to c. 70° N. in Norway; temperate Asia; N. Africa; introduced in other temperate lands.

The infl. is sometimes reduced to a dense ovoid head with the spikelets subsessile on the rhachis (var. *collina* Schlechtd. (*congesta* Gren. & Godr.)).

*2. D. polygama Horvat.

D. Aschersoniana Graebn.

Similar to *D. glomerata* but differing in the softer, green or slightly glaucous lvs, the laxer panicles with smaller clusters of spikelets and the lemmas with glabrous keels and shorter points. Fl. 6–8. $2n=14$. Hs.

Introduced. Naturalized in woodland in a few places in S.E. England. C. Europe to W. Russia.

16. CYNOSURUS L. Dog's-tail Grass.

Erect, glabrous, annual or perennial herbs. *Panicle spike-like. Spikelets* nearly sessile, *dimorphic; upper spikelet on each branch fertile*, of few florets; *lower spikelets sterile, with rigid, distichous lemmas.* Fertile spikelets: glumes thin, subequal, acute, keeled. Lemmas terete, coriaceous, awned. Palea 2-nerved, shortly bifid at the tip, about as long as the lemma. Lodicules 2, short, acute. Stamens 3. Ovary glabrous; styles short, terminal.

About 5 spp. in temperate regions of the Old World.

Perennial; sheaths not inflated; panicle narrowly oblong. **1. cristatus**
Annual; sheaths inflated; panicle ovoid, squarrose. **2. echinatus**

1. C. cristatus L. 'Crested Dog's-tail.'

A wiry, erect, tufted perennial 15–75 cm. Lvs flat, smooth, c. 2 mm. wide. *Sheaths not inflated,* smooth, upper falling far short of the panicle. *Panicle* erect, dense and spike-like, *narrowly oblong.* Spikelets 3–4 mm., secund, compressed, dimorphic, sterile pectinate. *Sterile spikelets: glumes and lemmas complanate,* lower glume setaceous, upper linear, slightly pubescent towards the top; lemmas subulate, ciliate on the keel. Fertile spikelets: glumes subequal, linear-lanceolate, acuminate, strongly keeled, with broad, thin, hyaline margins; *lemma* terete, finely pubescent, lanceolate, coriaceous, obscurely nerved, *shortly awned.* Palea nearly as long as the lemma, lanceolate. Fl. 6–8. $2n=14$. Hs.

Native. In grassland on acid and basic soils. 112, H40, S. Common and generally distributed throughout the British Is. Europe, except the extreme north; Caucasus; northern Asia Minor; Azores.

*2. C. echinatus L.

An erect annual 10–50 cm. Lvs flat, smooth, up to 5 mm. wide. *Sheaths inflated,* smooth, upper falling short of the panicle. *Panicle* erect, *dense, ovoid, squarrose, shining.* Spikelets c. 6 mm., compressed, dimorphic, fertile of 1 floret, sterile 8–10 mm. with many lemmas. Sterile spikelets: glumes setaceous; *lower lemmas* subulate *with a long point,* keeled, ciliate, *upper ovate* with shorter points. Fertile spikelets: glumes subequal, narrowly lanceolate, acuminate, hyaline, subterete, keeled; lemma terete, broadly lanceolate, pubescent in the upper half, coriaceous, obscurely 5-nerved, *awn* subterminal, *longer than the lemma.* Palea as long as the lemma, thin, with 2 strong glabrous nerves. Fl. 6–7. $2n = 14$. Th.

Introduced. A casual; in waste places and on sandy shores. ?Native in Channel Is. Not infrequent in southern England, becoming rarer north-wards and absent from Ireland. 52, S. Southern Europe and Mediterranean region.

17. BRIZA L.   Quaking Grass.

Erect, glabrous, annual or perennial herbs. Panicle ± branched, pedicels slender. *Spikelets ovoid or broadly triangular,* compressed, *often pendulous. Glumes broad, obtuse, awnless,* ± obscurely nerved and rounded on the back, nerves not reaching the apex. *Lemma* ovate, obtuse, *cordate at the base, usually saccate,* obscurely nerved, nerves not reaching the apex, *awnless.* Palea shorter than the lemma or nearly equalling it, ovate, obtuse, concave, nerves shortly ciliate. Lodicules 2, lanceolate. Stamens 3. Ovary glabrous; styles short, terminal.

About 10 spp., in Europe, temperate Africa, Asia and S. America.

1	Perennial; ligule truncate, short (less than 1 mm.).	**1. media**
	Annual; ligule long (2–5 mm.).	*2*
2	Spikelets 2 mm., numerous.	**2. minor**
	Spikelets 10 mm. or more, few (3–8).	**3. maxima**

1. B. media L.   Quaking Grass, Doddering Dillies.

An erect, somewhat tufted *perennial* 20–50 cm. *Stock shortly creeping.* Lvs flat, narrow (c. 2 mm.), acute. *Ligule short (less than 1 mm.), truncate. Culms solitary.* Panicle 5–8 cm., effuse, compound; pedicels very slender, smooth, slightly thickened below the spikelets, and up to twice as long as them. *Spikelets* 4–5 × 4–6 *mm.,* ovoid, obtuse, usually purplish. Glumes unequal, making less than a right-angle with the pedicel, shorter than the lowest floret, ovate, obtuse, boat-shaped with a distinct midrib and fainter lateral nerves, margins white and shining. Lemma broadly saccate, obtuse, strongly cordate at the base, bluntly keeled in the lower half, the coriaceous shining back surrounded by a thinner purplish area with a broad thin hyaline margin. *Palea almost as long as the lemma,* thin, ovate, obtuse. Lodicules linear-lanceolate, acute. Fl. 6–7. $2n = 14$. Hs.

Native. In meadows and grassy places. 104, H 40, S. Occurring in varied habitats, ranging from wet and acid to dry and calcareous. Generally dis-tributed throughout the British Is. except N. Scotland. Europe, except the Arctic and parts of the south; temperate Asia.

2. B. minor L.

A slender, erect annual 10–50 cm. Lvs flat, acute. *Ligule long (c. 3 mm.), lanceolate, acute. Culms tufted.* Panicle 5–15 cm., diffuse, compound; pedicels very slender, smooth, slightly thickened below the spikelets and 2–5 times as long as them. *Spikelets* 2–4 × 4–5 *mm.,* numerous, broadly triangular, obtuse, green. *Glumes subequal,* spreading at right angles to the pedicels, *equalling or exceeding the* 1st *floret,* broadly saccate, obtuse, obscurely nerved, margins broadly hyaline. Lemma with an indurated shining area on the back, cordate at the base but otherwise similar to the glumes. *Palea about ½ the length of the lemma*, thin, ovate. Lodicules linear-lanceolate, acute. Fl. 7. 2*n*=10. Th.

Native. In dry arable fields in the south, rare and very local. 7, S. Cornwall to N. Somerset and Hampshire. S. Europe and Mediterranean region.

*3. B. maxima L.

An erect or ascending annual 25–50 cm. Lvs flat, broad (up to 8 mm.), tapering to a long fine point. *Ligule long (up to 5 mm.), obtuse and ± toothed or torn. Panicle of few (3–8) large ovoid spikelets*, simple or slightly branched below; pedicels very slender, smooth or slightly serrate in the lower part, swollen immediately below the spikelet, at least the lower longer than the spikelets. *Spikelets* 10–20 × 5–15 *mm. Glumes* making less than a right angle with the pedicels, shorter than the 1st floret, *hyaline*, shining, broadly ovate, obtuse, with 2 rounded keels; lower 5-nerved; *upper* 9-nerved, *twice as long as the lower. Lemma hyaline*, shining, broadly ovate, obtuse, 9-nerved. *Palea less than ½ as long as the lemma*, hyaline, tough, broadly ovate from a narrow base. Lodicules lanceolate, obtuse. Fl. 6–7. 2*n*=14. Th.

Introduced. Naturalized in waste places; Channel Is., Cornwall. Often cultivated for ornament. Mediterranean region.

18. MELICA L.

Slender perennial herbs with the margins of the sheaths connate. Infl. a panicle or raceme. *Spikelets* of 2–4 florets, *terete, upper floret sterile and club-shaped*, rhachilla glabrous. Glumes subequal, thin, awnless, 3–5(–7)-nerved, nearly as long as the spikelet. *Lemma* coriaceous, obtuse, *rounded on the back*, 7–9-nerved, *awnless*. Palea tough, 2-nerved, nerves pubescent. Lodicules very short, truncate, free or connate. Stamens 3. Ovary glabrous; styles short, spreading.

About 80 spp. in temperate regions and on mountains in the tropics.

Panicle broad, spreading; sheaths pubescent; glumes acute. **1. uniflora**
Panicle linear, secund; sheaths glabrous; glumes obtuse. **2. nutans**

1. M. uniflora Retz. Wood Melick.

An erect or ascending perennial 30–70 cm. *Lvs* flat, with long, scattered hairs above, *rough beneath. Sheaths pubescent*, with a ligule-like projection at the mouth, opposite the blade. *Panicle* 10–20 cm., lax, *spreading, ovate*, compound, lowest branch usually far distant from others. *Spikelets* 4–5 mm., *erect*, oblong. *Florets* 2, only the lower fertile. Glumes unequal, purplish-brown, lanceolate; *lower acute, 3-nerved; upper mucronate, 5-nerved*. Lemma oblong, obtuse, obscurely nerved, 2 marginal nerves on each side close together, margins narrowly hyaline in upper part. Palea equalling lemma. Fl. 5–6. 2*n*=18. Hp.

Native. In shady hedge-banks and woods. 99, H40. Scattered throughout the British Is. and locally dominant. Europe, to S. Scandinavia, absent from Portugal, S. Spain and most of Russia; Asia Minor; Caucasus; Algeria.

2. M. nutans L.   'Mountain Melick.'

An erect or ascending perennial 20–40 cm. *Lvs* flat, glabrescent, or sparsely hairy above, *smooth beneath*. *Sheaths glabrous*, somewhat rough, *lower with reduced lvs*. *Panicle* 5–15 cm., *nearly simple, linear, secund*, drooping. *Spikelets* 6–7 mm., *drooping*, ovate. *Florets* 3–4. *Glumes* subequal, purplish-brown and broadly hyaline, oblong, *obtuse, both 5-nerved*. Lemma ovate with an obtuse, hyaline apex, nerves prominent, equally spaced. Palea shorter than lemma. Fl. 5–6. $2n = 18$. Hp.

Native. In limestone woods and cracks in limestone pavement. 56. In suitable habitats throughout Great Britain, local and rather rare. Europe to c. 70° N., absent from the extreme south; Caucasus.

19. SESLERIA Scop.

Perennials. *Panicle ovoid and spike-like. Bracts sheathing base of lower panicle branches.* Spikelets shortly stalked and somewhat compressed, of few florets. Glumes subequal, keeled, longer than lemmas. Lemma boat-shaped, keeled, 5-nerved, 3 central nerves close together, *at least* 3 *of the nerves shortly excurrent*. Palea equalling lemma. Lodicules 2. Stamens 3. Ovary pubescent at top; *styles connate below, terminal.*

About 10 spp. in Europe and W. Asia.

1. S. albicans Kit. ex Schult.   'Blue Sesleria.'

S. caerulea (L.) Ard. ssp. *calcarea* (Čelak.) Hegi

An erect, wiry, tufted perennial 15–40 cm. Rootstock shortly creeping. Lvs flat, keeled, glaucous, smooth but scabrid on the margin, apex mucronate; uppermost lf usually very short and far below the infl. Sheaths keeled. Ligule very short. Panicle 1–2 cm., ovoid, blue-grey and glistening, with a small scale at its base. Spikelets 5–8 mm. Glumes hyaline, lanceolate, lower acuminate, upper mucronate. Lemma boat-shaped, pubescent and ciliate on the margins, slaty-blue and purplish towards the top, apex with 3–5 small points. Palea slaty-blue or purplish towards the top, ciliate on the nerves. Fl. 4–5. $2n = 28$. Hs.

Native. Abundant on calcareous hills and pastures in northern England and western Ireland, and rare on micaceous schists in Scotland; ?extinct in Brecon. 13, H14. C. and S.W. Europe, Iceland.

S. caerulea (L.) Ard. in fens, etc. in Scandinavia and E. Europe.

Tribe 6. BRACHYPODIEAE. Annual or perennial herbs. Lvs with oblong or dumbbell-shaped silica cells. First foliage lf of seedlings narrow, erect. Spikelets of several florets, laterally compressed, usually in panicles, sometimes in distichous racemes. Glumes 2, usually shorter than the 1st floret on the same side. Lemma 5-, 7-, 9- or 13-nerved, often awned from the sinus or tip. Lodicules 2, entire, thin. Ovary with a hairy terminal appendage, the styles inserted laterally; outermost cell-layer of nucellus becoming very thick-walled in fr. Fr. with linear hilum as long as the grain; starch grains simple. Chromosomes large or small; basic number 5, 7 or 9.

20. ZERNA Panz.

Perennial herbs. Panicle ± compound. *Spikelets* subterete to compressed, *narrower or only slightly broader at top than in the middle*, rhachilla jointed between the usually numerous florets. Glumes unequal, awnless, lower 1-nerved, upper 3-nerved. Lemma ± keeled on the back, usually 7-nerved; *awn* straight or slightly bent, *shorter than lemma or frequently* 0. Palea thin, entire, 2-nerved, glabrous or pubescent on the nerves. Lodicules entire. Ovary with a hairy appendage at the top, styles inserted below the top, usually c. ⅓ of the way down. Fr. linear-oblong, grooved, ± enfolded by the lemma, adhering to the palea.

About 25 spp. in temperate regions.

1   Lvs seldom more than 2 mm. wide, the upper broader than the lower; panicle with short, erect branches.   **1. erecta**
    Lvs seldom less than 5 mm. wide, the upper narrower than the lower; panicle with long ± nodding branches.   *2*

2   Lower panicle branches in pairs, each with 5–9 spikelets; upper sheath hairy.   **2. ramosa**
    Lower panicle branches (2–)3–5 together, the shortest with one spikelet; upper sheath ± glabrous.   **3. benekenii**

1. Z. erecta (Huds.) S. F. Gray   'Upright Brome.'

Bromus erectus Huds.

An erect perennial 60–100 cm. *Lower lvs convolute, upper flat, broader than lower*, glabrous or ± ciliate. Sheaths glabrous or with spreading hairs. Culms glabrous. *Panicle* 10–15 cm., *erect*, usually reddish or purplish, nearly simple; branches short, slightly scabrid. Spikelets 20–35 mm., narrowly oblong, slightly compressed. Glumes linear-lanceolate, subequal, acuminate, keeled and slightly serrate on the keel. *Lemma* 8–10 mm., *usually glabrous*, 7-nerved, margin broadly hyaline in upper ⅓, c. 3 times as long as awn. Palea slightly shorter than lemma, keels glabrous. Anthers 5–6 mm., orange. Fl. 6–7. $2n = 42, 56^*$. Hs.

Native. On dry grassy banks and downs, preferring chalk or limestone. 62, H 8, S. England and Wales, common and locally dominant in the south, rather local and rare in the north; doubtfully native in Scotland; Ireland: native in the centre, naturalized in a few other localities. Europe north to c. 63° in Norway; N. Africa; W. Asia; Canary Is.

2. Z. ramosa (Huds.) Lindm.   'Hairy Brome.'

Bromus asper Murr.; *B. ramosus* Huds.

An erect perennial commonly 100–140(–190) cm. *Lvs flat, lower broader than upper, sparsely hairy. All sheaths clothed with long downward-pointing hairs.* Culms puberulent. Panicle 15–30 cm. broad, often dark purplish or glaucous, compound; *branches in pairs, the lowest pair ± equal in length and each with 5–9 spikelets*, nodding, at least the lower with a *ciliate scale* at their base. Spikelets 20–30 mm., linear, compressed. Glumes unequal, often purplish, with broad hyaline margins; lower linear-lanceolate, acuminate; upper lanceolate, mucronate or shortly awned, nerves making prominent scabrid ridges towards the base. *Lemma* 10–13 mm., lanceolate, *broadest above the middle and rather abruptly narrowed* into an awn c. ½ its length, 7-nerved, *the 3 prominent nerves pilose in their lower half.* Palea half the length of the lemma, hyaline with 2 finely pubescent green nerves. Anthers 2·5–3 mm. Fl. 7–8. $2n=14, 42^*$. Hp.

Native. In hedges and woods. 107, H40, S. Throughout the British Is., except for the Isle of Man and parts of N. Scotland. Europe to S. Scandinavia; N. Africa; temperate Asia.

3. Z. benekenii (Lange) Lindm.

Bromus benekenii (Lange) Trimen

Similar to *Z. ramosa* in general appearance. Smaller (60–90 cm.). Uppermost sheath glabrous or nearly so. Panicle rather narrow and dense, contracted and only nodding in the upper part at maturity. Lowest branches (2–)3–5 together, unequal, the shortest with 1 spikelet. Scale at base of lowest panicle branches glabrous. Glumes usually green with a scarious margin. Lemma broadest at or below the middle, gradually acuminate. Fl. 7–8. $2n=28$. Hp.

Native. In hedges and woods. Monmouth and W. Kent to Shropshire and N.W. Yorks; Mid Perth; local but probably often overlooked and less common than *Z. ramosa*. C., S. and E. Europe, absent from Spain and southern Balkans.

*Z. inermis (Leyss.) Lindm.

Bromus inermis Leyss.

An erect perennial up to c. 150 cm. Panicle 15–25 cm., spreading, compound. Spikelets 15–25 mm., lanceolate, compressed. Glumes unequal, *at least the upper with an obtuse hyaline tip. Lemma* c. 10 mm., *awnless, tip obtuse, hyaline.* Palea and lemma subequal. Anthers 4–5 mm. Fl. 6. $2n=56$. H.

Introduced. A rare alien, naturalized in a few places. N. and C. Europe, south to Spain and N. Italy; temperate Asia to China; introduced in N. America.

21. ANISANTHA C. Koch

Differs from *Zerna* as follows: Annuals. Spikelets becoming distinctly broader towards the top. Awn longer than lemma. Keels of palea pectinate-ciliate.

About 15 spp.

1 Longer panicle branches with 4 or more spikelets.   **5. tectorum**
   Panicle branches with 1–2 spikelets.   *2*

2 Lemma less than 20 mm.; plant slender.   *3*
   Lemma more than 20 mm.; plant usually stout.   *4*

3 At least most panicle branches as long as or longer than the spikelets.
                                                           1. sterilis
   All or nearly all panicle branches shorter than the spikelets.
                                                   2. madritensis

4 Panicle lax, spreading; glumes with hyaline margins; lemma 25–30 mm.;
stamens 3 or 2.   **3. diandra**
Panicle dense, erect; glumes mostly hyaline, except for the nerves; lemma
22–25 mm.; stamens 2.   **4. rigida**

1. A. sterilis (L.) Nevski   'Barren Brome.'

Bromus sterilis L.

An untidy annual with erect or ± decumbent stems 30–100 cm. Lvs soft,
flat and downy, lower soon withering. Lower sheaths with short, often
downward-directed hairs. *Culms glabrous. Panicle 10–15 cm., drooping,*
simple or slightly branched; *branches scabrid, usually much longer than the
spikelets.* Spikelets 20–25 mm., strongly compressed. *Glumes glabrous,* somewhat unequal, margins broadly hyaline; lower almost linear, acuminate,
scabrid on keel; upper lanceolate, smooth. Lemma 14–18 mm., linear-
lanceolate, rounded and scabrid on the back, *nerves equally spaced,* margins
hyaline, lobes 1–2 mm.; awn 1½ to twice as long as lemma, serrate. Palea
hyaline, almost equalling lemma. Anthers 1 mm. Fl. 5–7. Fr. animal dispersed; germ. autumn. $2n = 14$. Th.
    Native. In waste places, by roads and as garden weed. 104, H35, S.
Widely distributed and common throughout most of Great Britain, except
N. Scotland; Ireland mainly on limestone, rare in the north-west. W. Europe
north to 65°. Mediterranean region east to Persia; introduced in N. America.

2. A. madritensis (L.) Nevski   'Compact Brome.'

Bromus madritensis L.

An erect or ± decumbent annual 10–40 cm. Lvs flat, puberulent, lower soon
withering. Sheaths puberulent. *Culms glabrous. Panicle 4–10 cm., erect,*
simple or slightly branched; *branches ciliate, short.* Spikelets 20–30 mm.,
strongly compressed. Glumes unequal, pubescent or glabrous, margins incurved, narrowly hyaline; lower subulate; upper linear-lanceolate, shorter
than 1st floret. Lemma 12–15 mm., linear-lanceolate, glabrous or pubescent,
2 lateral nerves on each side close together, margin narrowly hyaline, lobes
1–2 mm.; awn equalling or somewhat exceeding lemma. Palea almost equalling lemma, delicately pectinate-ciliate. Stamens 2. Anthers 0·75–1 mm. Fl.
6–7. Chasmogamous. $2n = 28, 42$. Th.
    Native. In dry rather open habitats on limestone and sand. 14, H3, S.
S. and W. England and Wales, very local. W. Europe to c. 51° N., Mediterranean region east to Persia and Arabia; Macaronesia; introduced in N. and
S. America, S. Africa and Australia.

3. A. diandra (Roth) Tutin   'Great Brome.'

A. Gussonii (Parl.) Nevski; *Bromus diandrus* Roth; *B. maximus* auct. angl.,
p.p.; *B. Gussonii* Parl.

An erect, usually stout, ± pubescent annual 30–60 cm. Lvs flat, sparsely
pubescent, with long hairs on both surfaces. Sheaths pubescent at least on
the margins. *Culms shortly pubescent at least near the top. Panicle* up to
c. 15 cm., erect, *at length nodding, spreading,* ± compound; *branches* 2–4 at
each node, ciliate, many *longer than the spikelets.* Spikelets 25–40 mm., of

4–8 florets, compressed. Glumes unequal, lanceolate-acuminate, glabrous, *margins hyaline*, upper equalling or nearly equalling the 1st floret. *Lemma* 25–30 *mm.*, lanceolate, scabrid or shortly hairy, nerves equally spaced, 3 middle ones usually prominent, margins hyaline, lobes 4–7 mm.; awn twice as long as lemma. Palea lanceolate, acuminate, nerves shortly and stoutly pectinate-ciliate. *Stamens* 3 *or* 2. Anthers 1·25–4 mm. Chasmogamous or cleistogamous. Fl. 5–6. $2n = 56$. Th.

It is suggested that this sp. arose by hybridization between *A. sterilis* and *A. rigida.*

Native on sandy shores in the Channel Is.; as a casual in waste places in S. England. S. and W. Europe; N. Africa.

4. A. rigida (Roth) Hyl.

Bromus rigidus Roth; *B. maximus* Desf.

A stout erect annual rather smaller than but similar in general appearance to *A. diandra.* Lvs rather densely hairy with short hairs above. *Panicle* c. 20 cm., *stiff, dense*, erect; *branches strict, shorter than the spikelets.* Spikelets 25–30 mm., of 4–5 florets. *Glumes* unequal, narrowly lanceolate, acuminate, *almost entirely hyaline*, except for the nerves. Lemma 22–25 mm., broadly hyaline, nearly smooth. *Stamens* 2. Fl. 5–6. Cleistogamous. $2n = 42$. Th.

Native. On sandy shores. Channel Is., occasionally as a casual elsewhere. S. and W. Europe; N. Africa.

*5. A. tectorum (L.) Nevski

A somewhat tufted annual 10–90 cm. Lvs flat, hairy. Panicle 5–15 cm., secund, drooping; *branches* ± pubescent, slender, the longer *bearing* 4 *or more spikelets.* Spikelets 10–15 mm., compressed, narrow. Glumes unequal, pubescent, margins and tips hyaline; lower subulate, upper lanceolate. Lemma c. 10 mm., with scattered short hairs and a hyaline margin; awn slender, equalling or exceeding lemma. Palea c. ¾ the length of lemma. Fl. 5–7. $2n = 14$. Th.

Introduced. Naturalized near Thetford (Norfolk); a casual elsewhere in waste places. Europe, except the extreme north.

22. Bromus L.

Annual grasses with simple or compound panicles. *Spikelets* somewhat compressed, *narrowing considerably towards the top*, ± ovoid; rhachilla jointed between the usually numerous florets. Glumes unequal, lower 3–5-nerved, upper 5–7-nerved. Lemma ± keeled on the back, up to 13-nerved; *awn* straight or ± divaricate, from *shorter to little longer than lemma.* Palea 2-nerved, pectinate-ciliate on nerves, entire or rarely bifid. Lodicules entire. Ovary with a hairy appendage at top, styles inserted laterally, below the top. Fr. linear-oblong, grooved, ± enfolded by lemma and adhering to palea.

About 50 spp. in temperate regions.

1 Caryopsis rolled and the lemma rolled round it, so that the whole is subterete; awns divaricate, particularly after flowering; spikelet disarticulating tardily.   **8. secalinus**

Caryopsis very thin, the lemma not completely enfolding it; awns rarely divaricate; spikelet disarticulating soon after flowering.   *2*

2   Panicle short, erect; at least some pedicels much shorter than their spikelets; nerves on back of lemma prominent; ligule hairy.   *3*

Panicle long, narrow, often nodding; most pedicels longer than their spikelets; nerves on back of lemma inconspicuous; ligule glabrous.   *7*

3   Many spikelets subsessile, often in groups of 3 at the end of a branch; palea split to base.   **7. interruptus**

Spikelets distinctly stalked, not in groups of 3; palea entire.   *4*

4   Lemma 5·5–6·5 mm., strongly angled, with broad hyaline margins.
   4. lepidus

Lemma 6·5–9 mm., rounded or weakly angled, margins narrowly hyaline.
   (mollis agg.)   *5*

5   Lemma 8–9 mm.; awn 5–8 mm., ±straight; spikelets usually pubescent.
   1. mollis

Lemma not exceeding 7·5 mm.; spikelets usually glabrous, if densely hairy then ovate-oblong in outline and awn 3–4 mm. and curving outwards in fr.   *6*

6   Spikelets densely hairy, ovate-oblong in outline; lemma 6·5 × 3·7–4 mm., broadly ovate; awn 3–4 mm. curving outwards in fr.   **2. ferronii**

Spikelets usually glabrous, lanceolate in outline; lemma 6·5–7·5 × 2·3 mm., ovate; awn 5–6·5 mm., ±straight.   **3. thominii**

7   Panicle erect, narrow; lemma c. 7 mm., rounded.   **5. racemosus**

Panicle usually nodding, broader; lemma c. 9 mm., bluntly angled.
   6. commutatus

B. mollis agg. (spp. 1–3).   Lop-grass.

Annual or biennial 5–80 cm. Culms erect or decumbent, stout or slender, pubescent or glabrous. Lvs flat, soft, ±pubescent; *ligule* short, truncate, *hairy. Panicle* 5–10 cm., *erect, usually dense,* rarely reduced to a single spikelet. *Pedicels short, at least some much shorter than their spikelets. Spikelets* 10–20 mm., compressed, lanceolate, glabrous or pubescent, *not in groups of* 3 *at ends of branches.* Glumes unequal, lower lanceolate, upper ovate, both acute. *Lemma* 6·5–9 mm., ovate, *weakly angled, nerves prominent*; awn up to 10 mm. Palea about equalling lemma. Anthers 0·2–2 mm. in normal (cleistogamous) fls, c. 3 mm. in chasmogamous fls.

Native. In meadows, waste places, and on dunes, shingle banks and cliffs, common throughout the British Is., but much less common in the north than south. 112, H40, S.

1. B. mollis L.

Culms usually stout. Panicle narrow, rather lax. Spikelets 15–20 mm., of 5–7(–11) florets, usually pubescent, rarely glabrous (var. *leiostachys* Hartm.), lanceolate in outline. Upper glume 5–7-nerved. *Lemma* 8–9 × 3–4·5 *mm.,* ovate, margins rounded. *Awn* 5–8 *mm.,* ±*straight.* Anthers 0·2–1·2 mm. Fl. 5–7. Fr. 6–8. Germ. autumn. $2n = 28$. Th.

Common. Apparently throughout the British Is. Nearly the whole of Europe; W. Asia; Macaronesia; introduced in N. and S. America.

2. B. ferronii Mabille

B. velutinus Schrad. var. *minor* Hook.; *B. molliformis* auct. brit., non Lloyd; *B. Lloydianus* auct. brit., non Nym.

Culms usually stout. Panicle ovate, dense, usually nearly simple. Spikelets (10–)13–20 mm. with 7–11(–13) florets, always pubescent, ovate-oblong in outline. Upper glume 7–9-nerved. *Lemma* 6·5–7·5 × 3·7–4 *mm.*, *broadly ovate*, margins bluntly angled. *Awn* 3–4 *mm.*, *curving outwards in fr.* Anthers 0·5–1 mm., stout. Fl. 5–7 (starting c. 2 weeks earlier than *B. mollis* and *B. thominii*). Fr. 6–8. Germ. autumn. Th.

On cliffs by the sea, very local. 10, S. W. Cornwall to E. Kent, Anglesey, Isle of Man and Channel Is. W. France, probably elsewhere.

3. B. thominii Hard.

B. hordeaceus L. sec. Holmberg

Culms usually slender. Panicle narrow, rather lax. Spikelets 10–15 mm., with 5–7(–9) florets, nearly always glabrous, lanceolate in outline. Upper glume 5–7-nerved. *Lemma* 6·5–7·5 × 2·5–3 *mm.*, *ovate*, margins bluntly angled. *Awn* 5–6·5 *mm.*, ± *straight*. Anthers 1–1·5 mm., slender. Fl. 5–7. Fr. 6–8. Germ. autumn. $2n = 28$. H. (biennial), rarely Th.

Rather less common than *B. mollis*. W. Europe from France to Scandinavia, probably elsewhere.

4. B. lepidus Holmberg

B. britannicus I. A. Williams

An erect biennial 15–70 cm. Culms often somewhat decumbent below and rooting at the lower nodes. Lvs flat, hairy; *ligule hairy*; sheaths glabrous or pubescent. Culms puberulent between the ridges. *Panicles* 3–10 cm., *erect*, linear-lanceolate to broadly lanceolate, rather lax. Pedicels mostly much shorter than their spikelets. Spikelets 10–15 mm., of (5–)7–9(–11) florets, glabrous (rarely hairy, var. *micromollis* (Krosche) C. E. Hubbard), compressed; *florets at first imbricate, later spreading*. Glumes unequal, ovate, acute or mucronate, margins broadly hyaline. *Lemma* 5–7-*nerved, sharply angled* about $\frac{2}{3}$ of the way from the base, upper $\frac{1}{2}$ with a broad hyaline shining margin; awn 4·5–7 mm., ± straight. Palea shorter than grain; grain about equalling lemma. Anthers 1–2 mm. Fl. 6–8. $2n = 28*$.

?Native. In waste places, less commonly in grassland. Fairly widely scattered throughout Great Britain, common in S. England and probably elsewhere. 51, H1, S. Scandinavia, Germany, Netherlands, Belgium, France, Austria.

5. B. racemosus L.   'Smooth Brome.'

An erect annual 20–90 cm. Culms rather slender, puberulent. Lvs flat, soft, ± hairy; *ligule glabrous*; sheaths ± hairy. *Panicle* 3–10 cm., *erect, narrow, usually simple*; pedicels up to 3 cm. Spikelets c. 15 mm., of 5–7 florets, lanceolate to linear-lanceolate, glabrous or nearly so; *florets* closely imbricate, 1–1·5 *mm. apart on the rhachilla*. Glumes unequal, lower lanceolate, upper ovate, both acute. *Lemma c.* 7 *mm.*, ovate, *nerves obscure, margins* hyaline, rounded; awn about as long as lemma. Palea about equalling lemma. Anthers 2–2·5 mm. Fl. 6. $2n=28*$. Th.

Native. In meadows and grassy places, sometimes on arable land. 74, H14, S. Scattered throughout the British Is., but rather uncommon. Nearly the whole of Europe to S. Scandinavia.

6. B. commutatus Schrad.   'Meadow Brome.'

B. pratensis Ehrh. ex Hoffm., non Lam.

An erect annual 30–90 cm. Culms usually rather stout, puberulent. Lvs flat, soft. *Ligule glabrous*; sheaths ± hairy. *Panicle* 7–20 cm., *usually nodding, broader than in B. racemosus*, usually simple; pedicels up to 7 cm. Spikelets 15–20 mm., of 5–8 florets, ovate-lanceolate, glabrous or nearly so (rarely pubescent, var. *pubens* Wats.); *florets* 1·5–2 mm. *apart on rhachilla.* Glumes unequal, lower lanceolate, upper ovate, both acute. *Lemma* c. 9 mm., broadly ovate, *nerves obscure*, margins broadly hyaline, bluntly angled; awn about as long as lemma. Palea often distinctly shorter than lemma (6–8 mm.). Anthers 1–1·5 mm. Fl. 6. $2n=14, 28, 56$. Th.

Native. In meadows, grassy places and on arable land. 97, H22, S. Scattered throughout the British Is. and commoner than *B. racemosus.* Nearly the whole of Europe; N. Africa; introduced in S. Africa.

7. B. interruptus (Hack.) Druce

An erect or somewhat decumbent annual or biennial 25–50 cm. Lvs flat, ± hairy, broad (up to 5 mm.); *ligule hairy*; sheaths ± hairy. Culms puberulent between the ridges. Panicle 3–5 cm., erect, usually interrupted, compound, *branches very short. Spikelets* c. 10 mm., hairy, often *subsessile in groups of 3 at ends of branches*, ovate, compressed; fls closely imbricate. Glumes somewhat unequal, acute or mucronate, margins hyaline; lower lanceolate, upper ovate. *Lemma* 7–8 mm., broadly ovate, *angled, nerves prominent*; awn about equalling lemma. *Palea bifid to base.* Anthers 0·75–1 mm. Fl. 6–7. $2n=28*$. Th. or H.

? Native. In arable fields, often associated with *Onobrychis viciifolia.* 28. Scattered throughout England, rare, and mainly in the south. Netherlands (? elsewhere).

*B. arvensis L.

An erect annual 30–80 cm. Lvs flat, hairy. *Panicle* 7–25 cm., *spreading*, most of *the branches much longer than their spikelets.* Spikelets 10–20 mm., compressed, glabrous and shiny; florets imbricate. *Lemma bluntly angled*, margins broadly hyaline in the upper ½; *tip bifid, acute*; awn about as long as lemma. Palea usually equalling lemma. *Fls chasmogamous*, rarely cleistogamous. *Anthers 3–4 mm.*, rarely only 2·5 mm. Fl. 6–7. $2n=14$. Th.

Introduced. A rare casual in waste places or hayfields, occasionally ± naturalized. Europe, Siberia, W. Asia.

*B. japonicus Thunb.

B. patulus Mert. & Koch

An erect annual 30–60 cm. *Panicle* 10–15 cm., *secund and nodding* particularly after flowering; branches mostly longer than their spikelets. Spikelets c. 20 mm., of 7–10 florets, glabrous. *Lemma* 7–8 mm., *angled*, ±enfolding fr.; *tip obtuse*; awns somewhat divaricate in fr. *Palea distinctly shorter than lemma.* Fls cleistogamous; anthers c. 1 mm. Fl. 7. 2*n*=14. Th.

Introduced. A rare casual in waste places. C. and E. Europe, W. Asia.

*8. B. secalinus L.   'Rye-Brome.'

An erect annual 30–60 cm. Lvs flat, ±pubescent. Sheaths pubescent or glabrous. Culms glabrous. *Panicle* 5–20 cm., *secund, nodding*, lax. Spikelets 10–15 mm., compressed, glabrous or (var. *hirtus* (F. Schultz) Aschers. & Graebn.) densely pubescent; florets at first imbricate, later spreading. Glumes unequal, margins hyaline; lower lanceolate, upper ovate. *Lemma* 7–9 mm., ovate, *curled round the rolled caryopsis and so strongly convex in fr.*; margins uniformly rounded; nerves obscure; awn rather shorter than lemma. Palea almost equalling lemma. Anthers 0·75–1 mm. Fl. 6–7. 2*n*=14, 28. Th.

Introduced. A casual in arable land, usually among winter wheat. 90, H15, S. Scattered throughout the British Is., rare in the north; E. Ireland. Europe; N. Africa; W. Asia; introduced in N. America.

23. CERATOCHLOA Beauv.

Perennial, rarely annual grasses. Panicle compound. *Spikelets lanceolate, strongly compressed, narrowed towards the top*, rhachilla jointed between the numerous florets. Glumes somewhat unequal, little shorter than lemmas and similar to them, lower 3–5-nerved, upper 5–7-nerved. Lemma strongly keeled; *awn 0 or shorter than lemma*. Palea thinly ciliate or shaggy on the nerves. Lodicules entire. Ovary with a small hairy appendage at top; styles inserted laterally. Fr. linear-oblong, channelled.

About 15 spp. in temperate N. and S. America and the Andes.

*1. C. carinata (Hook. & Arn.) Tutin

Bromus carinatus Hook. & Arn.

An erect perennial up to c. 80 cm. Culms stout, glabrous. *Lvs up to 20 mm. wide*, glabrous, *tough*, tapering to a long point. Sheaths pubescent at least at the mouth. Ligule c. 2 mm., truncate and torn. Panicle 15–30 cm., ±erect, branches long, spreading. *Spikelets 25–45 mm.*, of (4–)8–12 florets, *linear-lanceolate. Glumes* lanceolate, *acute.* Lemma 15–17 mm., lanceolate, scabrid and puberulent towards base, margins hyaline; awn 5–7 mm., terminal. Palea somewhat shorter than lemma. Fls cleistogamous or chasmogamous, rather distant (2–3 mm. apart on the rhachilla). Fl. 6–8. 2*n*=56. Hp.

Introduced. Well naturalized in some localities, particularly along the Thames near Kew and Oxford. Western N. America.

*C. unioloides (Willd.) Beauv.

Bromus unioloides Kunth; *B. catharticus* auct., ?Vahl

An erect or ascending perennial 20–70 cm. Upper sheaths glabrous, lower pubescent. Panicle 5–30 cm., ±spreading or lower branches deflexed. *Spikelets 20–30 mm., lanceolate. Glumes hooded at the tips, strongly keeled.* Lemma 15–18 mm., 13-nerved, uniformly rounded, mucronate or very shortly awned. Palea c. ½ as long as lemma. Fl. 6–9. Usually cleistogamous. $2n=28, 42$. H.

Introduced. A casual in waste places. S. America; introduced in all temperate regions.

24. BRACHYPODIUM Beauv.

Tufted perennial herbs with spike-like infl. *Spikelets subsessile, distichous, terete*, linear-lanceolate, *inserted edgeways on to the axis*, of many florets. Glumes chartaceous, shorter than the 1st floret, unequal, prominently 5–7-nerved. Lemma chartaceous, acuminate or awned from the tip, 7-nerved. Palea nearly as long as the lemma, tough, 2-nerved, ciliate, blunt or emarginate. Lodicules 2. Stamens 3. Ovary pubescent at the top; styles distant.

About 10 spp. in Europe, temperate Asia, N. and S. Africa.

Lvs soft, yellow-green; awn equalling or exceeding lemma.   **1. sylvaticum**
Lvs rigid, green or glaucous; awn much shorter than lemma.   **2. pinnatum**

1. B. sylvaticum (Huds.) Beauv.   'Slender False-brome.'

A tufted ± erect and *pubescent* perennial 30–90 cm. *Lvs* flat, broad (up to 13 mm.), *soft, ± drooping, yellow green*, ± scabrid, sparsely pubescent and ciliate. Sheaths pubescent. Ligule c. 2 mm., laciniate. Infl. 6–15 cm. Spikelets 12–25 mm., nearly straight, of 7–12 florets. Glumes lanceolate, acuminate, sparsely pilose and ciliate. Lemma linear-lanceolate, acute, pilose; *awn equalling or exceeding lemma*. Palea oblong, emarginate or rounded at tip. Rhachilla pubescent. Lodicules broad at the base, linear-lanceolate and fimbriate above. *Anthers 8–10 times as long as broad*. Fl. 7. $2n=18$. Hs.

Native. In woods and hedges, sometimes in grassland, then often relict from woodland. 111, H40, S. Throughout the British Is., except Shetland. Europe, N. Africa, Macaronesia, W. Asia to N.W. Himalaya.

2. B. pinnatum (L.) Beauv.   'Heath False-brome.'

A tufted erect, *glabrescent* somewhat rhizomatous perennial 30–60 cm. *Lvs* ± involute, rarely as much as 5 mm. wide, often *stiff and erect*, ± scabrid, nearly glabrous. Sheaths glabrous or pubescent. *Ligule very short*. Infl. 6–17 cm. Spikelets 20–35 mm., usually curved away from rhachis, of 8–16 florets. Glumes linear-lanceolate, acute, glabrous, lower 5-, upper 7-nerved. *Lemma* linear-lanceolate, *acuminate or usually shortly awned*, ± pilose at tip. Palea oblong, emarginate or rounded at apex. Rhachilla glabrous. Lodicules narrowly oblong, pubescent. Anthers stouter than in *B. sylvaticum*, 4–5 *times as long as broad*. Fl. 7. $2n=28$. Hs. or Chh.

Native. In grassland on chalk and limestone, locally dominant. 46, H16. From Northumberland to Cornwall and Kent; only Cambridgeshire in E. Anglia; Caernarvon; Mid Perth; scattered in Ireland. Europe to c. 62° N., N. Africa, Siberia.

Tribe 7. HORDEAE. Annual or perennial herbs. Lvs with oblong or elliptic silica cells; 2-celled hairs 0; green tissue uniformly distributed between the vascular bundles; 1st foliage lf of seedling narrow and erect. Ligule glabrous. Spikelets all alike or some male or sterile, of 1–many florets, solitary or in groups of 2–3, sessile or subsessile on alternate sides of the solitary spike, broadside on to the axis. Glumes large, persistent, placed laterally to the axis. Lemma chartaceous to coriaceous, usually 5–9-nerved, awnless or awned from the tip; awn not geniculate. Lodicules 2. Stamens 3. Styles free, very short. Ovary hairy at tip. Fr. with long linear hilum; starch grains simple. Chromosomes large, basic number 7.

25. AGROPYRON Gaertn.

Tough, usually rhizomatous, perennial herbs. *Infl. a spike of many distichous spikelets. Spikelets of 2–many florets, solitary at the nodes of the rhachis and broadside on to it. Glumes* somewhat unequal to subequal, membranous to chartaceous, 3–9-nerved, *shorter than spikelet.* Lemma chartaceous, 4–6-nerved, awned or awnless. Palea 2-, rarely 3-nerved. Lodicules 2, acute. Stamens 3. Ovary pubescent; styles distant, very short.

About 60 spp. in temperate regions.

1	Plant densely tufted; glumes persistent when fr. is shed.	2
	Plant with long rhizomes; glumes falling with the fr.	3
2	Awn usually longer than lemma; culms puberulous near some of the nodes (widespread).	**1. caninum**
	Awn much shorter than lemma; culms glabrous (Highlands, rare).	**2. donianum**
3	Ribs of lvs slender, numerous; lvs usually flat when dry, with scattered long hairs on upper surface.	**3. repens**
	Ribs of lvs prominent, broad ±concealing upper surface of lf; lvs convolute when dry, hairs 0 or very short.	4
4	Ribs of lvs with numerous short spreading hairs on upper surface.	**6. junceiforme**
	Ribs of lvs glabrous but sometimes scabrid.	5
5	Ribs of lvs scabrid; most spikelets overlapping by at least half their length.	**5. pungens**
	Ribs of lvs smooth; spikelets small, distant.	**4. maritimum**

A. caninum × *donianum* occurs with the parents. *A. junceiforme* × *repens* and *A. pungens* × *repens* are local and have lvs intermediate between those of the parents, and sterile pollen.

1. A. caninum (L.) Beauv.　　　　　　　　　　　'Bearded Couch-grass.'

An erect, *bright green perennial* 30–100 cm. *Rhizomes* 0. Culms smooth, *some of the nodes or culm near the nodes finely pubescent*, sometimes rather sparsely so. Lvs flat, up to c. 13 mm. wide, scabrid and glabrous below, scaberulous, with scattered hairs above, ribs very slender. Sheaths smooth. Spike 10–20 cm., often slender, ±flexuous and nodding, rhachis ciliate. Spikelets 10–17 mm., of 2–5 florets, lanceolate. Glumes membranous, unequal, glabrous, lanceolate, prominently nerved; lower 3-nerved, acuminate; upper 5-nerved, shortly aristate, not more than ⅔ length of spikelet. Lemma lanceolate, awned; awn slender, flexuous, scabrid, usually exceeding the lemma. Palea emarginate, 2-nerved, ciliate on nerves, slightly shorter than lemma. *Anthers 2–3·5 mm.* Fl. 7. $2n = 28*$. Hp.

Native. In hedges and woods. 101, H23. Scattered throughout the British Is. to Orkney though absent from most of N. Scotland; locally common. Europe, rare in the south; temperate Asia; N. America.

2. A. donianum F. B. White   'Don's Twitch.'

A tufted perennial. Lvs flat with slender ribs. Spike 10–13 cm., erect. Spikelets c. 20 mm., of 2–6 florets. Glumes 4–6-nerved, usually shortly awned, serrate on the nerves, margin and tip. Lemma lanceolate, margins hyaline towards the apex, 4–6-nerved, shortly awned. *Palea bluntly pointed with a distinct middle nerve towards the tip*; lateral nerves shortly excurrent in 2 teeth which fall short of the apex, densely ciliate. Fl. 8. $2n=28*$. Hp.

Native. On base-rich mountains. 2. Mid Perth and Sutherland, very local. Iceland (var. *stefanssonii* (Meld.) Tutin) and N. America.

3. A. repens (L.) Beauv.   Couch-grass, Scutch, Twitch.

An erect, *dull-green or ± glaucous perennial* 30–100 cm. *Rhizomes abundant and far-creeping. Culms and nodes glabrous.* Lvs similar to but usually narrower than those of *A. caninum*. Sheaths smooth, glabrous, or the lower pubescent. Spike 5–15(–20) cm., usually stiff and erect, rhachis pubescent or margins serrate. Spikelets 10–15(–20) mm., of 3–6 florets. Glumes subequal, $\frac{2}{3}–\frac{3}{4}$ length of spikelet, lanceolate, acute, chartaceous, margins hyaline, lower 3-, upper 5-nerved. Lemma lanceolate, apex variable, obtuse and apiculate, acute, or awned. Palea emarginate, slightly shorter than lemma. *Anthers 4–6 mm.* Very variable. Fl. 6–9. $2n=42*$. Hp.

Native. In fields and waste places. A noxious and persistent weed. 112, H40, S. Throughout the British Is., common. Europe, N. Africa, Siberia, N. America.

4. A. maritimum (Koch & Ziz) Jansen & Wachter

Like *A. repens* but glaucous with glabrous lvs, convolute when dry, with stout smooth ribs; spikelets small and few-fld, remote.

Native. Dunes on the S. and E. coasts of England. North Sea coasts, Siberia; ? Mediterranean.

5. A. pungens (Pers.) Roem. & Schult.   'Sea Couch-grass.'

A rather tufted, often glaucous, glabrous perennial 30–90 cm. Rhizomes far-creeping. Culms smooth, nodes glabrous. *Flowering and sterile shoots erect. Lvs ± convolute*, rather stiff, glabrous, *ribs broad and prominent, minutely scabrid.* Sheaths smooth, basal usually rather thin, brownish. *Spike* 5–12 cm., stout, stiff, erect, *resembling an ear of wheat; rhachis* flattened, *tough, not disarticulating at the nodes, margins serrate. Spikelets* 10–18 mm., of 3–8 florets, most of them overlapping each other by at least half their length. Glumes subequal, $\frac{1}{2}–\frac{2}{3}$ length of spikelet, lanceolate, acute ± keeled, 5–7-nerved, sometimes serrate on keel. Lemma lanceolate, obtuse, and apiculate, shortly awned, middle nerve rather prominent. Palea emarginate, equalling lemma, finely pubescent on nerves. Anthers 6–7 mm. Fl. 7–9. $2n=42*$. Hp.

Native. Locally dominant on dunes and in salt-marshes. 37, H8, S. Coasts and estuaries from Cornwall and Kent to Glamorgan and N.E. Yorkshire; Cheshire to Westmorland; Dublin to E. Cork, Limerick and S.E. Galway. Scottish records doubtful. W. and S. Europe (not Scandinavia), N.E. America. Var. *pycnanthum* (Godr.) Druce, a distinct-looking form with short broad spikes, larger densely imbricate spikelets (c. 20 mm.) and obtuse glumes, is found in the Channel Is.

6. A. junceiforme (A. & D. Löve) A. & D. Löve   'Sand couch-grass.'
A. junceum auct., non Beauv.

A glabrous, glaucous perennial 25–50 cm. Rhizomes abundant and far-creeping. Culms stout, smooth, *nodes frequently pruinose. Flowering stems ± erect, sterile shoots decumbent.* Lvs convolute, flat when damp, stiff, smooth; *ribs* broad and prominent above, nearly concealing the intervening lamina, *densely but shortly pubescent with spreading hairs.* Sheaths smooth, basal whitish and persistent, upper often purplish. Spike 5–15(–18) cm., stout, stiff, erect; *rhachis smooth, flattened, fragile and disarticulating readily at the nodes* when mature. Glumes somewhat unequal, tough, lanceolate, obtuse, 7–9-nerved, nerves not reaching tip. Lemma lanceolate, obtuse and apiculate, emarginate and mucronulate or subacute, middle nerve reaching tip, pubescent above towards the top. Palea emarginate, $\frac{2}{3}$–$\frac{3}{4}$ length of lemma, rarely more, pubescent on nerves. Anthers 5 mm. Fl. 6–8. $2n=28*$. Hp.

Native. On young dunes. 68, H19, S. Sandy coasts of the British Is. north to Shetland. Western and northern Europe to c. 63° N.

A. junceiforme × pungens (*A. acutum* auct. brit.) is a common hybrid. Sterile shoots decumbent and ascending. Flowering stems erect. Lvs convolute or some flat, ribs broad with short, spreading points. Spike 10–20 cm., erect, slender to stout; rhachis slightly serrate, sometimes fragile. Spikelets up to 20 mm., rather distant and spreading. Anthers 5–6 mm., sterile and rarely exserted. Fl. 6–9. $2n=35*$.

Native. With the parents, often forming an intermediate zone between *A. junceiforme* on the young dunes and *A. pungens* on the older ones.

Spp. of *Triticum* (Wheat), particularly **T. aestivum* L. (*T. vulgare* Host; $2n=42$), and **T. turgidum* L. ($2n=28$), are widely cultivated in temperate regions throughout the world. *Triticum* resembles *Agropyron* but has 1–2 hermaphrodite florets at the base of the spikelets and a number of male or sterile florets above, and the nerves of the lemma are not convergent. The spp. are annual or rarely biennial. The cultivated forms do not become naturalized and seldom persist for more than a year.

*Secale cereale L., Rye, is another closely related plant which is extensively cultivated. It has 2 hermaphrodite florets in the spikelet and the nerves of the lemma converge to form a long awn.

26. ELYMUS L.

Tall stout perennials. *Spikelets 2 or 3 together* broadside on to the rhachis, of 3–4 florets, upper floret sterile. Rhachilla pubescent and prolonged beyond the uppermost floret. *Glumes* equal, coriaceous, 5-nerved, *about equalling the spikelet and often placed side by side in front of it.* Lemma coriaceous, 5-nerved, rounded on the back, and bluntly keeled towards the apex. Palea tough, equalling lemma. Lodicules 2, entire. Stamens 3. Ovary hirsute; stigmas sessile.

About 45 spp. in temperate regions of the northern hemisphere.

1. E. arenarius L. Lyme-grass.

A stout, erect, glaucous perennial 1–2 m. Stem creeping and rooting freely below. Lvs broad, shortly hairy on ribs, rigid, pungent. Sheaths glabrous, smooth, ridged. Ligule very short. Spike 15–30 cm., stout. Spikelets c. 20 mm., in pairs. Glumes lanceolate, acuminate, ± pubescent towards the tip. Lemma pubescent, lanceolate, acute or obtuse. Palea linear-lanceolate, obtuse or shortly bifid, 2-keeled, serrate on the keels, pubescent. Fl. 7–8. $2n = 56$. Hp.

Native. On dunes, often with *Ammophila arenaria*. 55, H6, S. Around the coasts of the British Is., local; rare in Ireland. C. and N. Europe, to c. 71° N.; Siberia; N. America.

27. HORDEUM L.

Annual or perennial herbs. Ligule very short. Infl. a compressed or subterete spike. *Spikelets in alternate distichous groups of three*, subsessile and broadside on to the rhachis, *of 1 floret; the central one in each group* hermaphrodite, lateral male or sterile; all hermaphrodite in some cultivated races; rhachilla produced beyond the floret. Glumes equal, narrow, 1-nerved, awned, placed side by side in front of the florets. Lemma about equalling the glumes, tough, 5-nerved, rounded or dorsiventrally compressed, awned, awn terminal. Palea as long as the lemma, 2-nerved, ± 2-keeled. Lodicules 2, oblong or lanceolate, narrowed below, delicately fimbriate. Ovary subsessile.

About 25 spp. in temperate regions.

1	Sheath of uppermost lf not inflated.	**1. secalinum**
	Sheath of uppermost lf inflated.	2
2	Glumes of central spikelet ciliate.	**2. murinum**
	Glumes of central spikelet scabrid.	**3. marinum**

1. H. secalinum Schreb. 'Meadow Barley.'

H. pratense Huds.; *H. nodosum* auct.

A slender erect perennial 30–60 cm. Lvs narrow (less than 5 mm. wide), flat, or the lower ± involute, somewhat rough. Sheaths not inflated, lower ± pubescent, upper glabrous. Culms and nodes glabrous. Spike 2·5–5 cm., compressed. Spikelets c. 8 mm. *Glumes setaceous, serrate. Lemma lanceolate* subterete, obscurely nerved; *awn 2–3 times as long as lemma.* Palea subterete, with 2 rounded keels, nerves obscure. Lemma of sterile spikelets stalked, subulate, ending in a short point. Anthers many times as long as broad. Fl. 6–7. $2n = 14, 28^*$. Hp.

A hybrid with *Agropyron repens* occurs rarely.

Native. In meadows. 66, H12, S. England and Wales, locally abundant in the south becoming rarer in the north; Scotland, northwards to W. Perth, local; Ireland, south and east from Cork to Antrim, very local and mainly on coast. S. and W. Europe north to Denmark and S. Sweden; Caucasus; Asia Minor; N. and S. Africa; N. America.

2. H. murinum L. 'Wall Barley.'

A stout annual 20–60 cm., ± decumbent below. Lvs flat, pilose on both surfaces. Upper sheath glabrous and inflated, lower sometimes pubescent.

Culms and nodes glabrous. Spike 4–10 cm., compressed. Spikelets 8–12 mm. *Glumes of hermaphrodite spikelets subulate, ciliate; outer glumes of lateral spikelets setaceous, serrate; inner subulate, sparsely ciliate*; all awned. *Lemma ovate*, flattened and rounded on the back, obscurely nerved; awn 2–4 times as long as lemma. Palea lanceolate, obtuse or acute, distinctly nerved. Lemma of lateral fls lanceolate. Anthers about twice as long as broad. Fl. 6–7. $2n=14$. Th.

Native. In waste places, especially near the sea. 86, H12, S. Great Britain north to Caithness, but absent from much of northern Scotland; Ireland, doubtfully native though not uncommon in the east. C. and S. Europe, north to S. Sweden; N. Africa; W. Asia; N. America.

3. H. marinum Huds. 'Squirrel-tail Grass.'

H. maritimum Stokes

A glaucous annual 15–35 cm., ± decumbent and geniculate below. Lvs flat, stiff, short, scabrid. Sheaths inflated, upper glabrous, lower pubescent. Culms and nodes glabrous. *Spike 2–4 cm., oblong, subterete.* Spikelets 6–8 mm. *Glumes of central spikelet setaceous; inner glumes of lateral spikelets half ovate, outer setaceous; all scabrid*, not ciliate. *Lemma lanceolate*, flattened, obscurely nerved, *about equalling its awn.* Fl. 6. $2n=14, 28$. Th.

Native. In grassy places near the sea. 31, S. England and Wales, very local; Scotland: Wigtown and Kincardine. W. and S. Europe.

*****H. hystrix** Roth (*H. gussonianum* Parl.), resembling *H. marinum* but with the infl. readily breaking up at maturity and the glumes of the lateral spikelets nearly similar in shape, occurs as a casual. S. and S.E. Europe.

*****H. jubatum** L., a perennial with inflated sheaths and very long, silky, nearly smooth, spreading awns and small (c. 5 mm.) spikelets, occurs occasionally as a casual. N. America.

*****H. distichon** L. and *****H. vulgare** L., Barley, occur as relics of cultivation but do not persist or become naturalized.

28. HORDELYMUS (Jessen) Harz

A perennial herb with the characters of *Hordeum* except that the *lateral spikelets of each group are hermaphrodite* and the *central one hermaphrodite or sometimes male.* One species in Europe and Asia Minor.

1. H. europaeus (L.) Harz 'Wood Barley.'

Hordeum europaeum (L.) All.; *H. sylvaticum* Huds.; *Elymus europaeus* L.

A stout, ± tufted, erect perennial 40–120 cm. Lvs broad (up to c. 10 mm.), flat, smooth above, rough beneath. Sheaths pubescent, upper sometimes glabrous, not inflated. Culms glabrous, nodes pubescent. Spike 5–10 cm., compressed. Spikelets 10–15 mm. Glumes stiff, subulate, serrate on the margins, terminated by a bristle-like point. Lemma lanceolate, flattened on the back, margins inrolled, scabrid or shortly pilose; awn 2–3 times as long as lemma, serrate. Palea lanceolate, acute, prominently 2-nerved, nerves scabrid. Anthers many times longer than broad. Fl. 6–7. $2n=28$. Hs.

Native. In woods and shady places. 32, H1. England, local and usually on chalk or limestone; Scotland: Berwick; Ireland: Antrim. Europe, north to S. Sweden; Asia Minor; Caucasus.

Tribe 8. AVENEAE. Annual or perennial herbs. Lvs with oblong or elliptic silica cells and green tissue uniformly distributed between the vascular bundles; 2-celled hairs 0; 1st foliage lf of seedling narrow and erect. Ligule glabrous. Infl. a compound panicle. Spikelets slightly compressed to terete, of 2–5 florets, often shining; florets hermaphrodite or rarely the upper or lower male. Rhachilla usually disarticulating above the glumes, usually produced beyond the uppermost floret. Glumes tough, 1–3- (rarely 7–9-)nerved, at least the upper as long as the spikelet and ± enclosing the florets, rarely shorter, but then usually equalling the lowest floret (*Koeleria*). Lemma tough or coriaceous, usually 5–7-nerved, generally with a dorsal, often geniculate, awn. Palea 2-nerved, often tough. Lodicules 2. Stamens 3. Ovary without an appendage, glabrous, styles free. Fr. with a linear to punctiform hilum; starch grains compound. Chromosomes large, basic number 7.

29. KOELERIA Pers.

Perennial herbs. Panicle narrow, ± shining. Spikelets compressed, of 2–3(–5) florets. *Glumes* unequal, *firm, upper about equalling the* 1st *floret*, keeled, acute or aristate, lower 1-, upper 3-nerved. *Lemma* tough, *keeled*, obscurely 5-nerved, subobtuse to aristate. Palea hyaline, about as long as the lemma, 2-nerved, bifid at the apex and shortly ciliate on the nerves. Rhachilla pubescent, prolonged beyond the uppermost perfect floret, and bearing 1(–2) rudimentary florets. Lodicules connate, somewhat falcate and fimbriate. Stamens 3; anthers 3–6 times as long as broad. Ovary glabrous; *styles* short and rather stout, *terminal*.

About 30 spp. in temperate regions of both hemispheres.

Stem conspicuously swollen and reticulate at base.   **2. vallesiana**
Stem not as above.   **1. cristata**

1. K. cristata (L.) Pers.   'Crested Hair-grass.'

K. gracilis Pers.; incl. *K. albescens* auct.

A rather stiff, ± pubescent, often glaucous, erect perennial 10–40 cm. Stems sometimes creeping. Lvs usually narrow, convolute, ± hirsute. Sheaths pubescent, *basal brown,* entire or ± torn, not fibrous. *Culms pubescent or nearly glabrous.* Panicle 2–6·5 cm. narrowly oblong, ± lobed. Spikelets 3–5 mm., crowded, purplish, green or whitish. Glumes glabrous or puberulent, acuminate, mucronate or shortly aristate, broadly hyaline, scabrid on the keel, lower linear-lanceolate, ¾ the length of the upper, upper lanceolate to ovate. Lemma lanceolate, acuminate or shortly aristate. Fl. 6–7. 2n = 28, 30*. Hp.

Native. Sandy places and chalk and limestone pastures. 100, H30, S. Fairly generally distributed throughout the British Is. Northern hemisphere, between about 37° and 60° N. There is considerable variation in size and hairiness and two extreme forms have been regarded as sspp. They frequently grow together and appear to interbreed freely, as all combinations of characters are usually found within a population. It seems therefore that the division cannot be maintained.

2. K. vallesiana (Honck.) Bertol.

A densely tufted, glabrous or slightly pubescent perennial 10–40 cm. *Stems woody and conspicuously bulbous-thickened at the base, clothed with the fibrous-reticulate remains of the sheaths.* Lvs convolute, setaceous, those of the culms often flat, glabrous, short, rigid and glaucous. Sheaths glabrous or slightly pubescent, stiff. Panicle 1·5–6 cm., compact, scarcely lobed, usually broader than in the other spp., sometimes ovate. Spikelets 4–5 mm., shining green or brown, subsessile. Glumes ± hyaline, lanceolate-acuminate or shortly aristate, sometimes subobtuse, glabrous or hirsute, ciliate on the keel. Fl. 6–8. $2n=42$*. Ch.

Native. On rocky limestone slopes. 1. Brean Down, Uphill, Worle Hill, Purn Hill and Crook Peak, N. Somerset. France, Switzerland, south to N. Italy, Spain, Tunisia, and Algeria.

*Gaudinia fragilis (L.) Beauv., a softly hairy annual with a narrow fragile unbranched spike, with numerous appressed, 4–10-fld, distichous spikelets the lemmas of which have divaricate dorsal awns, occurs as a casual. S. Europe, N. Africa, W. Asia.

30. Trisetum Pers.

Perennials. Panicle compound, spikelets shining, compressed, of 2–4(–6) florets. Glumes unequal, hyaline, firm, keeled, 1–3-nerved. *Lemma* strongly keeled, 5-nerved, apex with 2 *bristle points; awn from above the middle,* geniculate. Palea slightly shorter than lemma, hyaline, nerves excurrent as 2 bristle points. Rhachilla silky, prolonged beyond the uppermost floret. Lodicules 2, lanceolate. Stamens 3. Ovary glabrous or pubescent at tip.

About 65 spp. in temperate regions and on mountains in the tropics.

1. T. flavescens (L.) Beauv.   'Yellow Oat.'

Avena flavescens L.

An erect stoloniferous perennial 20–50 cm. Lvs flat, rough or smooth beneath, pubescent above. Lower sheaths pubescent, upper glabrous. Culms smooth. Ligule very short, truncate. Panicle oblong or ovate, branches variable in length, ± rough. Spikelets 3–6 mm., shining, usually yellowish. Glumes ± scabrid on the keel; lower linear-lanceolate, acuminate, 1-nerved, about ⅔ length of upper; upper broadly lanceolate, acuminate, 3-nerved, almost equalling spikelet. Lemma lanceolate, scabrid on keel, margins broadly hyaline; awn scabrid and long-exserted. Anthers about 5 times as long as broad. The size of the spikelet, glumes and lemma is unusually variable. Fl. 5–6. $2n=24, 28$. Hp.

Native. In meadows and grassy places, especially on dry calcareous soils. 104, H39, S. Generally distributed throughout the British Is., but rarer in the north and absent from much of N. Scotland. Europe to S. Scandinavia; Algeria, temperate Asia; N. America.

31. AVENA L.

Stout *annual* herbs. Panicle spreading or secund. *Spikelets* large, terete, of 2–3 florets, *eventually pendulous*. Glumes membranous, subequal, lower 7-nerved, upper 9-nerved. Lemma coriaceous, 7-nerved, with a stout, scabrid, dorsal, geniculate awn from about the middle. Awn much exceeding the spikelet. Palea tough, rather shorter than the lemma, ciliate on the nerves. Rhachilla ± silky, and produced beyond the uppermost floret. Lodicules 2, lanceolate, acute. Stamens 3. Ovary pubescent; styles short, distant.

About 10 spp. in temperate regions.

1 Lemma shortly bifid or entire and hyaline at apex.   2
  Lemma with 2 long, scabrid bristles at apex.   **3. strigosa**

2 Rhachilla articulated between florets; second floret with callus-scar at
  base.   **1. fatua**
  Rhachilla not articulated between florets; second floret without scar at
  base.   **2. ludoviciana**

*1. A. fatua L.   Wild Oat.

A stout erect annual 30–90 cm. Lvs flat, slightly scabrid. Sheaths smooth. Ligule short (1–2 mm.), torn. Culms smooth, nodes glabrous or hairy. Panicle 15–20 cm., spreading, branches scabrid. *Spikelets* 20–25 *mm.*, with long spreading awns. Glumes exceeding the florets, lanceolate, acuminate. Lemma ovate-lanceolate, hyaline, the third (when present) awned. *Palea* subobtuse, *entire*. Rhachilla and lower ⅓ of lemma often clothed with silky, usually fulvous hairs, sometimes glabrous; rhachilla articulated between the florets, the second with a callus scar at base. Fl. 7–9. $2n=42$. Th.

Introduced. Naturalized in arable fields. 79, H5, S. Generally distributed and often common in suitable habitats; absent from Wales; very rare in Ireland. Europe except Greece; N. Scandinavia and N. Russia; N. Africa; Canaries; Asia.

*2. A. ludoviciana Durieu   Wild Oat.

A stout annual similar to *A. fatua* in general appearance. Lvs very scabrid on both surfaces. Spikelets 25–30 mm. Lemma ovate, lower ⅔ clothed with long silky hairs, the third (when present) awnless. Rhachilla not articulated between the florets, the second without a callus scar at base. Fl. 6–7. $2n=42$. Th.

Introduced. A weed of arable land, usually on heavy soils in S. England, locally abundant. Europe; Asia.

*3. A. strigosa Schreb.   Black Oat.

Similar to *A. fatua* but differing in the secund panicle; the glumes about equalling the fls; the lemma ending in 2 long, straight or flexuous (not geniculate) scabrid bristles; the shortly bifid palea; and the sparsely pilose rhachilla and lemmas. Fl. 7–9. $2n=14$. Th.

Introduced. Casual in cornfields. Rather rare and local. Europe.

*A. sativa L., the cultivated oat, resembles *A. fatua* but may be easily distinguished by the absence of the fulvous hairs. It is a relic of cultivation and does not usually persist.

32. Helictotrichon Besser

Erect, stout perennials. Panicle ± compound. *Spikelets* large, *erect*, of 2–3(–6) florets, subterete. Glumes somewhat unequal, nearly as long as the spikelet, membranous, bluntly keeled, lower 1–3-nerved, upper 5-nerved, lateral nerves short. Lemma coriaceous below, thin and hyaline above, rounded on the back, 5-nerved; awn dorsal, geniculate, stout, scabrid, much exceeding the spikelet. Palea thin, hyaline, shortly bifid at the apex. Rhachilla silky, prolongation slender, $\frac{2}{3}$ as long as the uppermost floret, with a terminal rudiment. Lodicules 2. Stamens 3. Ovary pubescent at the top; styles terminal, short.

About 50 spp. in temperate regions of both hemispheres.

Sheaths glabrous; lvs stiff, glaucous. **1. pratense**
At least the lower sheaths pubescent; lvs soft, green. **2. pubescens**

1. H. pratense (L.) Pilger　　　　'Meadow Oat.'

Avena pratensis L.

An erect, glabrous perennial 30–60 cm. *Lvs ± channelled, glaucous, stiff*, subobtuse, basal spreading. *Sheaths glabrous*, basal strict. Ligule 3–5 mm., acute. *Panicle 6–12 cm., strict, narrow*, nearly simple, *lower branches* 1–2 *together*. Spikelets 12–20(–25) mm. Glumes lanceolate, long-acuminate, both 3-nerved, margins and apex hyaline. *Lemma* lanceolate, *bi-aristate at the tip*, awn from about the middle. Palea shorter than lemma, lanceolate, ciliate. Rhachilla silky, particularly at the joints. Lodicules 2 mm., lanceolate, acuminate. Anthers 10–12 times as long as broad. Fl. 6. $2n=42^*$; 14, 28. Hp.

Native. In short turf on chalk and limestone, rarely on other formations. 83. In suitable habitats throughout Great Britain. C. and N. Europe, Apennines; Siberia.

2. H. pubescens (Huds.) Pilger　　　　'Hairy Oat.'

Avena pubescens Huds.

An erect perennial 30–70 cm. *Lvs flat, soft, ± pubescent*, obtuse. *Sheaths pubescent*. Ligule 1–4 mm., ± obtuse. *Panicle 6–14 cm., ± spreading*, simple or compound, *lower branches* 5 *together*. Spikelets 10–15 mm. Glumes hyaline, lanceolate, acuminate, serrate on keel, lower 1-nerved, upper 3-nerved. *Lemma* linear-lanceolate, subterete, ± scabrid towards the top *with 4 points at the apex*; awn from the middle. Palea nearly equalling lemma, glabrous. Lodicules short, obliquely truncate. Anthers stouter than in *H. pratense*, about 5 times as long as broad. Fl. 6–7. $2n=14^*$. Hp.

Native. On basic soils, usually in longer, rougher turf than *H. pratense*; occasionally on other formations. 105, H39, S. Generally distributed throughout the British Is. and locally abundant. C. and N. Europe, Iceland and arctic Russia, Balkans; Siberia.

33. Arrhenatherum Beauv.

Erect, rather stout, perennial herbs. Panicle lax, nodding. Spikelets slightly compressed, of 2 florets, *lower floret male, upper female or hermaphrodite*, sometimes both hermaphrodite. Glumes hyaline, unequal, keeled, lower

1-nerved, $\frac{2}{3}$ the length of the upper; upper as long as the florets, 3-nerved, lateral nerves about half as long as the glume. Lemma membranous, 7-nerved; *lower floret with a long geniculate awn* from near the apex. Palea hyaline, 2-nerved, nerve sciliate. *Rhachilla silky* with a slender prolongation beyond the upper floret. Lodicules 2, lanceolate. Stamens 3. Ovary pubescent; styles short.

About 6 spp. in Europe, N. Africa, and W. Asia.

1. A. elatius (L.) Beauv. ex J. & C. Presl   'Oat-grass.'

A. avenaceum Beauv.

An erect perennial 60–120 cm. Stems swollen or not at base; nodes glabrous or pubescent. Lvs flat, scabrid. Sheaths smooth. Ligule very short. Panicle 10–20 cm., lax, nodding, rather narrow. Spikelets 7–10 mm. Glumes lanceolate, acuminate, scabrid on keel. Lemma lanceolate, acute, broadly hyaline, nerves scabrid. A variable plant. Fl. 6–7(–11). $2n=28^*$. Hp.

Native. In rough grassy places, on scree and shingle. 112, H 40, S. Common and generally distributed throughout the British Is. Europe, except the Arctic and only on mountains in the south; N. Africa; W. Asia; introduced in N. America and Australia. *A. tuberosum* (Gilib.) Schultz is the form with strongly swollen tuberous stem-bases. It interbreeds freely with the other form, and every intermediate is found, so it cannot be maintained as a distinct sp.

34. HOLCUS L.

More or less pubescent perennial herbs. Panicle compound, rather close. Spikelets compressed, of 2 florets. Glumes membranous, subequal, longer than the spikelets, strongly keeled, lower 1-, upper 3-nerved. Lemma coriaceous, shining, obscurely 5-nerved, strongly keeled. *Lower floret hermaphrodite, awnless; upper male, with a dorsal awn* from just below the tip. Palea thin, slightly shorter than the lemma. *Rhachilla* shortly prolonged beyond the second floret, *glabrous*. Lodicules 2, lanceolate, acuminate. Stamens 3. Ovary glabrous; stigmas sessile, terminal.

About 8 spp. in Europe, temperate Asia, N. and S. Africa.

Nodes not bearded; awn included.   **1. lanatus**
Nodes bearded; awn exserted.   **2. mollis**

1. H. lanatus L.   Yorkshire Fog.

A soft, pubescent, tufted perennial 20–60 cm. Lvs flat, acute, pubescent on both surfaces. *Sheaths pubescent*, upper inflated. Ligule c. 1 mm., truncate. *Culms and nodes puberulous or glabrescent*. Panicle 4–10 cm., ovate, ± lobed, often pink; branches short, pubescent. Spikelets 3–4 mm., crowded. Glumes pubescent and ciliate on the keel; lower lanceolate, upper ovate, nerves prominent, both mucronate. *Lemma* coriaceous, smooth *with a few silky hairs at the base*, shining, keeled, lanceolate, ± acute; *awn* short, smooth, hooked, *included within the glumes*. Fl. 6–9. $2n=14^*$. Hp.

Native. In waste places, fields and woods. 112, H 40, S. Generally distributed and often abundant throughout the British Is. Europe, except the Arctic; temperate Asia; introduced in N. America.

2. H. mollis L.   'Creeping Soft-grass.'

A rather stiff, sparsely pubescent ± creeping perennial 20–60 cm. Lvs flat, acute, rough. *Sheaths glabrous or lower ± pubescent*, upper somewhat inflated. Ligule c. 1 mm., truncate. *Culms glabrous*, rather slender, *nodes with a tuft of downward-directed hairs*. Panicle 4–10 cm., rather lax, usually whitish; branches ciliate. Spikelets 4–5 mm. Glumes glabrescent, spinose-ciliate on the keels; lower lanceolate; upper ovate, obscurely nerved; both acuminate. *Lemma* coriaceous, smooth but *conspicuously silky at the base*, shining, keel scabrid; *awn* geniculate, *exserted*. Plants with short, stiff lvs and small (c. 2 cm.) panicles of few spikelets are found. Fl. 6–7. $2n = 28^*, 35^*, 42^*, 49^*$. Hp.

Native. On moderately acid soils; usually in woods. 111, H40, S. Generally distributed throughout the British Is., though less frequent than *H. lanatus*, and rare in calcareous districts. Europe, except the Arctic.

35. DESCHAMPSIA Beauv.

Tufted, glabrous, *perennial* herbs. Panicle compound, ± spreading. Spikelets of 2 florets, somewhat compressed, shining. Glumes subequal, about as long as the spikelet, 1-nerved, keeled, ± hyaline. *Lemma* subterete, *obscurely 5-nerved, truncate and jagged at the apex;* awn dorsal, straight, or geniculate; both florets hermaphrodite, equalling the lemma, hyaline. *Rhachilla* silky, *prolonged beyond* the *upper floret*. Lodicules 2, linear-lanceolate, stamens 3. Ovary glabrous; styles distinct.

About 35 spp. in temperate regions.

1   Lvs flat; awns not or scarcely exceeding glumes.   *2*
       Lvs setaceous; awn of lower floret distinctly longer than glume.   *3*

2   Awn from near base of lemma; infl. not (or rarely) viviparous.
                                                   1. cespitosa
       Awn from above the middle of lemma; infl. usually viviparous.
                                                   2. alpina

3   Upper sheaths rough; ligule c. 1 mm., truncate; florets close together.
                                                   3. flexuosa
       Upper sheaths smooth; ligule 4–5 mm., acute; florets separated by at least
       ⅓ the length of the lower lemma.   **4. setacea**

1. D. cespitosa (L.) Beauv.   'Tufted hair-grass.'
Aira cespitosa L.

A stout densely tufted perennial 50–200 *cm.* Rootstock stout. Lvs flat, scabrid above, smooth beneath. Sheaths smooth. Ligule 4–8 mm., obtuse. Panicle 15–50 cm., lax; branches long, slender, ± scabrid. Spikelets 3–4 mm., silvery or purplish. Glumes subequal, firm, hyaline, lanceolate; lower obtuse, upper acute. *Lemma* hyaline, lanceolate, *truncate and jagged at the apex; awn from near the base*, straight, about equalling the lemma. Palea truncate and jagged at the apex, serrate on the nerves. Fl. 6–8. $2n = 26, 28$. Hs.

Native. In damp meadows and woods, usually on badly drained clayey soils. 112, H40. Generally distributed and common throughout the British

Is. Europe (only on mountains in the south); W. and N. Asia, Himalaya, Abyssinia; Cameroons; N. America; Tasmania; New Zealand.

Var. *parviflora* (Thuill.) C. E. Hubbard has the spikelets 2–3 mm. long and is common in woodland, especially in S. England on heavy soils.

2. D. alpina (L.) Roem. & Schult.

Aira alpina L.

A tufted perennial 10–35 cm., lvs rather short and stiff. Infl. usually (?always) viviparous. Glumes acuminate; lemma usually with long, acuminate teeth at the apex; awn arising above the middle of the lemma. Fl. 7–8. Apomictic. $2n = 49^*$; 56^*; 39, 41. Hs. Doubtfully distinct from *D. cespitosa*.

Native on the higher mountains. 17, H3. Wales: Snowdon; Scotland: on mountains between 3000 and 4100 ft.; Ireland: Kerry and W. Mayo. Scandinavia, Iceland, N. Russia.

3. D. flexuosa (L.) Trin.   'Wavy hair-grass.'

Aira flexuosa L.

A tufted, *rather slender perennial 25–40 cm.* Lvs setaceous, scabrid on the margins. *Upper sheaths rough. Ligule c.* 1 *mm., truncate.* Panicle 5–10 cm., lax, branches long, flexuous, slightly rough. Spikelets 4–5 mm., silvery or purplish. Glumes rather unequal, ovate, acuminate, hyaline. Lemma lanceolate, tapering to a slightly jagged apex, hyaline in the upper half; *awn* geniculate, *arising near the base of the lemma* and exceeding it. Palea lanceolate, acute, nerves prominent, serrate. Rhachilla short, *florets close together*. Var. *montana* Huds. is smaller (10–30 cm.) with shorter lvs and a small (5 cm. or less) compact panicle with few rather large spikelets. Fl. 6–7. $2n = 28$. Hs.

Native. On acid heaths and moors and in open acid woods; var. *montana* in alpine pastures. 111, H38, S. Generally distributed throughout the British Is. in suitable habitats. Europe, N. Asia Minor, Caucasus, Japan, N. America; only on mountains in the south.

4. D. setacea (Huds.) Hack.

Aira uliginosa Weihe & Boenn.

A tufted slender perennial 25–60 cm. Resembles *D. flexuosa*, but has smooth sheaths, the cauline lvs broader than the basal, a long linear-lanceolate, acute ligule, c. 4–5 mm., a lemma with two long and two short points at the truncate apex, and a long rhachilla separating the florets by at least $\frac{1}{3}$ the length of the lower lemma. Fl. 6–7. $2n = 14^*$. Hs.

Native. In wet turfy bogs and at edges of pools. 29, H1. Scattered in suitable habitats throughout Great Britain, but very local; Ireland: W. Galway. W. Europe to S.W. Norway and S. Sweden.

36. AIRA L.

Slender glabrous, *annual* herbs with short, narrow lvs. Panicle compound. Spikelets somewhat compressed, of 2 florets. Glumes equal, as long as, or longer than the spikelet and enclosing it, firm, 1-nerved, slightly keeled. *Lemma* shorter than the glumes, firm, subterete, *3-nerved, apex with* 2 *setaceous points*, base with a tuft of hairs. Awn dorsal, geniculate, arising below the middle of the lemma. Palea thin, shorter than the lemma, bifid. *Rhachilla not produced beyond the 2nd floret.* Stamens 3; anthers scarcely longer than broad. Ovary glabrous; stigmas subterminal, sessile.

About 10 spp. mainly in Europe and temperate Asia.

Sheaths smooth; panicle compact; lemma smooth.   **1. praecox**
Sheaths scabrid; panicle spreading; lemma scabrid towards top.
                                                                    2. caryophyllea

1. A. praecox L.   'Early Hair-grass.'

An annual 5–12 cm. Lvs obtuse, sheaths smooth, ligule 2–3 mm., acute. *Panicle compact, oblong, 0·5–2 cm.; branches little longer than spikelets.* Spikelets c. 2 mm., crowded. *Glumes* as long as the spikelets ovate, acute, *smooth. Lemma* lanceolate, *smooth*; awn arising ⅓ the distance from base to tip. Fl. 4–5. 2*n*=14. Th.

Native. On dry rocky slopes, heaths and in dry fields especially on a sandy soil. 112, H40, S. Generally distributed throughout the British Is. in suitable habitats. S., C. and W. Europe; introduced in N. America.

2. A. caryophyllea L.   'Silvery Hair-grass.'

An annual 10–30 cm. Lvs ± oblique at tip, sheaths scabrid, ligule up to 5 mm., acute. *Panicle 2–8 cm., effuse, broadly ovate; branches very much longer than spikelets.* Spikelets 2·5–3 mm. *Glumes* longer than florets, ovate-lanceolate, acute, *serrate on keel. Lemma scabrid towards the top*; awn arising about ¼ the distance from base to apex. Fl. 5. 2*n*=14. Th.

Native. In dry gravelly and sandy places, shallow soils round rocks, and on walls. 112, H40, S. Generally distributed throughout the British Is. in suitable habitats, but usually less common than *A. praecox.* C. and S. Europe, Caucasus, Macaronesia, Abyssinia, Cameroons, S. Africa; only on mountains in the south.

*****A. multiculmis** Dum. resembles *A. caryophyllea* but is usually larger (up to 60 cm.) with numerous culms; panicle branches ascending, spikelets crowded at their ends, smaller (2–2·25 mm.) and awn longer. 2*n*=28. A casual. S. Europe, N. Africa.

37. CORYNEPHORUS Beauv.

Spikelets of 2 florets, compressed, in a compound panicle. Glumes, subequal, membranous, exceeding the florets, 1-nerved. Lemma firm, hyaline, rounded on the back, 1-nerved; *awn from near the base, clavate, bearded and geniculate about the middle.* Palea thin, shorter than lemma. Rhachilla silky, not produced beyond the 2nd floret. Lodicules 2, connate below, linear-acuminate and spreading above. Stamens 3. Ovary glabrous; styles short.

Three spp. in Europe and W. Asia.

1. C. canescens (L.) Beauv.

Aira canescens L.; *Weingaertneria canescens* (L.) Bernh.

A tufted, very glaucous perennial 10–30 cm. Lvs setaceous, rigid, scabrid, pungent. Sheaths smooth, often purplish, inflated. Ligule 2–4 mm., acute.

Culms slender, often geniculate. Panicle 1·5–6 cm., lanceolate but spreading at flowering. Spikelets 3–4 mm., purple and white. Glumes lanceolate, acuminate, slightly scabrid on keel, broadly hyaline. Lemma lanceolate, subobtuse; lower half of awn stout, chestnut-brown, upper half slender, white or purplish, awn bearded and geniculate at junction of 2 halves, exceeding lemma but shorter than glumes. Palea obtuse. Anthers orange or purple, c. 4 times as long as broad. Fl. 6–7. $2n=14$. Hs.

Native. In sandy places behind the seaward ridge of dunes. 8, S. Coasts of Norfolk, Suffolk and Channel Is.; perhaps not native in Glamorgan, S. Lancashire, Moray and W. Inverness. S. and C. Europe, north to southern Scandinavia.

Tribe 9. AGROSTEAE. Annual or perennial herbs. Lvs with oblong or oval silica cells and green tissue uniformly distributed between the vascular bundles; 2-celled hairs 0; 1st foliage lf of seedling narrow and erect. Ligule glabrous. Infl. an effuse or compact panicle of many, rarely few, usually small spikelets. Spikelets of 1 floret, subterete to strongly laterally compressed. Glumes equalling or more often exceeding the lemma, subequal, tough, 1–3-nerved. Lemma membranous, rarely hardening in fr., (3–)5(–7)-nerved. Lemma with an awn from the back. Palea thin, 2-nerved or nerveless, sometimes minute or 0. Rhachilla usually disarticulating above the glumes, sometimes below, sometimes produced as a small bristle. Lodicules 2, rarely 0. Stamens 3. Ovary glabrous, without an appendage; styles free. Fr. with linear to punctiform hilum; starch grains compound. Chromosomes large, basic number 7.

38. AMMOPHILA Host

Glabrous erect perennials, with convolute lvs. *Panicle dense, ± cylindrical. Spikelets large*, of one floret, compressed. Glumes longer than the lemma, membranous, keeled; lower 1-, upper 3-nerved. Lemma membranous, 5–7-nerved, the tip bifid, with a very short subterminal awn. Palea firm, nearly equalling the lemma. *Rhachilla silky, produced*. Lodicules 2, long-acuminate. Stamens 3. Ovary glabrous; styles short.

Two spp. around the coasts of Europe and N. Africa, and a third on the coasts of N. America.

1. A. arenaria (L.) Link　　　　　　　　　　　　　Marram Grass.

A. arundinacea Host; *Psamma arenaria* (L.) Roem. & Schult.

A stout, erect, perennial 60–120 cm. Stem extensively creeping and rooting at the nodes, binding the sand. Lvs convolute, terete, rigid, pungent, polished without and glaucous within. Sheaths smooth, persistent, upper somewhat inflated. Ligule very long, often exceeding 10 mm., acuminate. *Panicle* 7–15 cm., dense, stout, spike-like, cylindrical, *obtuse, whitish*. Spikelets 12–14 mm. Glumes unequal, whitish, linear-lanceolate, acuminate, keel serrate in upper part, margins hyaline. *Lemma 3 times as long as the hairs at its base*, scabrid, compressed, half-lanceolate, 5–7-nerved, nerves distinct, tip hyaline with 2 short bristle points, awn stout, not exceeding the bristles, keel and awn shortly ciliate. Palea compressed, lanceolate, acute, keeled, 2-nerved, nerves very close together, shortly ciliate on keel. Anthers c. 10 times as long as broad. Fl. 7–8. $2n=28$, 56. Hp.

Native. Abundant and often dominant on dunes. 75, H20, S. Round the coasts of the British Is. Coasts of W. Europe, except the Arctic.

39. × AMMOCALAMAGROSTIS P. Fourn.

A hybrid between *Ammophila* and *Calamagrostis*, combining some of the characters of both but more closely resembling *Ammophila* in general appearance.

1. × **A. baltica** (Schrad.) P. Fourn. (*Ammophila arenaria* × *Calamagrostis epigejos*)

Ammophila baltica (Schrad.) Dum.

Similar in habit and vegetative characters to *Ammophila arenaria*, but taller and stouter. Panicle ± lobed and interrupted, purplish, acute, with longer branches than in *Ammophila*. Lemma ± rounded on the back, 5-nerved, nerves, particularly the middle one on each side, obscure; awn slender, just exceeding the 2 bristle points. *Lemma 1–2 times as long as the hairs at its base.* Anthers shorter than in *Ammophila arenaria*, c. 8 times as long as broad. Fl. 7–8. $2n=28$, 42. Hp.

Native. On dunes. 5. Norfolk, Suffolk, Northumberland and W. Sutherland, locally abundant. W. Europe, from Scandinavia to France.

40. CALAMAGROSTIS Adans.

Erect, *perennial* herbs, often found in damp places. Panicles compound, rather narrow. Spikelets subterete or ± compressed, of 1 floret. Glumes longer than floret, chartaceous, subequal, lower 1-, upper 3-nerved. *Floret with a profusion of silky hairs at its base.* Lemma thin, ± hyaline, 3–5-nerved, bifid, *awned from the back or sinus*; awn slender, short. Palea shorter than lemma, thin, hyaline, bifid. Rhachilla not produced beyond the floret (*Epigejos* Koch), or with a slender, silky prolongation (*Deyeuxia* Beauv.). Lodicules 2, lanceolate. Stamens 3. Ovary glabrous; styles short. Probably a highly heterogeneous genus.

About 120 spp. in temperate regions.

1	Hairs longer than floret.	*2*
	Hairs shorter than floret.	*3*
2	Leaves scabrid but not hairy above.	**1. epigejos**
	Leaves hairy above.	**2. canescens**
3	Spikelet 3–4 mm.; glumes acute.	**3. stricta**
	Spikelet 4·5–5 mm.; glumes acuminate.	**4. scotica**

1. C. epigejos (L.) Roth   'Bushgrass.'

A stout, erect, glabrous perennial 60–200 cm. *Lvs* flat, *scabrid, long*, point long and slender. Sheaths smooth. *Ligule long* (*up to 12 mm.*), acute, torn. Culms scabrid just below the panicle. Panicle 15–30 cm., ± spreading, purplish-brown; branches and *pedicels scabrid*. Spikelets 5–7 mm., subterete. Glumes 2–3 times as long as floret, subulate, keeled in the upper part, serrate on keel, lower rather broader than upper. *Hairs at base of floret about twice as long as the floret. Lemma* hyaline, lanceolate, acuminate, bifid, *5-nerved; awn from near the base*, slender, *exceeding the lemma*. Palea ⅔ the length of the lemma. Fl. 7–8. 2n=28*; 42, 56. Hp.

Native. In damp woods, ditches and fens. 81, H4, S. Widely distributed in England; local in Wales and Scotland, north to Sutherland; W. and N. Ireland, very rare. Europe to 70° N.; temperate Asia.

2. C. canescens (Weber) Roth   'Purple Smallreed.'

C. lanceolata Roth

A rather slender, erect perennial 60–120 cm. *Lvs* ± convolute, long, *pubescent above*, slightly scabrid beneath. Sheaths and *culms smooth*. Ligule 1–2 mm., obtuse and torn. Panicle 5–15 cm., ± spreading, light brown or purplish; branches scabrid, *pedicels smooth*. Spikelets 4–5 mm., subterete. Glumes 2–3 times as long as floret, subulate, keeled in the upper part, keel smooth. *Hairs at base of floret little longer than the floret. Lemma* hyaline, lanceolate, *notched, with a very short slender awn from the sinus, 3-nerved*. Palea ½ as long as lemma. Fl. 6–7. 2n=28. Hp.

Native. In damp shady places and fens. 42. England, local but fairly widespread; Scotland: Kirkcudbright, and Aberdeen. C. and N. Europe. Absent from N. Scandinavia and N. Russia; Siberia.

3. C. stricta (Timm) Koeler   'Narrow Smallreed.'

C. neglecta auct.

A slender erect glabrous perennial 30–90 cm. Lvs flat or convolute, usually rather short, glabrous or hairy above, smooth, margins serrate. Sheaths smooth. Ligule 1–2 mm., obtuse. Culms scabrid just below the panicle. Panicle 10–20 cm., narrow. *Spikelets 3–4 mm.*, pale purplish and green becoming light brown when old. *Glumes lanceolate, acute* or sometimes acuminate, exceeding the lemma. Lemma hyaline towards the top, truncate and bifid at the apex; *awn from about the middle, equalling the lemma*. Hairs at the base of the lemma ⅔ length of the lemma. Rhachilla with a slender, silky prolongation. Var. *hookeri* Syme has a dense reddish-brown panicle usually less than 10 cm. and ± lobed, the uppermost ligule rather longer than broad and subacute, and the hairs ¾ length of lemma. Fl. 6–7. 2n=28. Hp.

Native. In bogs and marshes, rare. 9, H4. W. Norfolk, Cheshire, Yorks, Ayrshire, mid Perth and Caithness; extinct in Angus. Var. *hookeri*: W. Norfolk and Lough Neagh. N. Europe, scattered in C. Europe; N. and E. Asia; N. America.

4. C. scotica (Druce) Druce

C. strigosa auct. angl., non Hartm.

Similar to *C. neglecta*, but panicle up to 10 cm., rather dense. Spikelets 4·5–5·5 mm., rather darker than those of *C. neglecta*. Glumes narrowly lanceolate, long-acuminate, somewhat exceeding the lemma. Lemma awned from the base. Fl. 7.

Native. In bogs, very rare. Caithness. ?Endemic.

41. Agrostis L.

Tufted or creeping *perennials*. *Spikelets small*, of 1 floret, in effuse or dense panicles. Glumes equal or subequal, usually 1-nerved, membranous. Lemma shorter than or rarely equalling the glumes, ovate, truncate, or obtuse, 3–5-nerved, lateral nerves usually excurrent; *awn dorsal or* 0. Palea shorter than lemma, sometimes minute or 0, hyaline, 2-nerved or nerveless. Rhachilla sometimes produced, shortly bearded or glabrous, disarticulating above the glumes. Lodicules 2, lanceolate. Stamens 3. Ovary glabrous; styles short.

About 100 spp., mostly in temperate regions.

For a full account of the British spp. of the genus, their varieties and hybrids, as well as the aliens, see Philipson, W. R., *Journ. Linn. Soc. Bot.* LI (1937–38), 73–151.

1	Ligules of culm lvs acute; palea less than ¼ length of lemma.	2
	Ligules of culm lvs obtuse; palea more than ½ length of lemma.	3
2	Radical lvs with one ventral groove; glumes rough.	**1. setacea**
	Radical lvs with at least four ventral grooves; glumes smooth.	**2. canina**
3	Panicle closed in fr.; rhizomes 0; stolons present.	**5. stolonifera**
	Panicle open in fr.; rhizomes with scale-lvs present.	4
4	Ligule of sterile shoots shorter than broad.	**3. tenuis**
	Ligule of sterile shoots longer than broad.	**4. gigantea**

1. A. setacea Curt.   'Bristle Agrostis.'

A *densely tufted* perennial 20–60 cm. *Lvs setaceous*, scabrid, *stiff and glaucous*. Sheaths slightly scabrid. Ligule up to 4 mm., acute, ± torn. Culms scabrid, especially above. Panicle 3–10 cm., narrow, ± spreading at flowering. Spikelets 3–4 mm. *Glumes* lanceolate, acute, *finely scabrid*, keel serrate towards the top. Lemma c. ⅔ length of glumes, ovate, truncate, 5-nerved; awn from near the base, usually geniculate and exceeding the glumes. Palea minute. Rhachilla shortly bearded. Anthers 1·5–2·0 mm. Fl. 6–7. $2n=14^*$. Hs.

Native. On dry sandy and peaty heaths and downs in S. and S.W. England. 13. Spain, Portugal and W. France.

2. A. canina L.   'Brown Bent-grass.'

A tufted and shortly rhizomatous or stoloniferous perennial 10–60(–80) cm. *Lvs flat*, ± scabrid, tapering from the base to a long fine point. Sheaths smooth. Ligule up to 3 mm., acute, ± torn. Culms usually smooth, sometimes scabrid above. Panicle effuse, ovoid or pyramidal at flowering, contracted in fr. Spikelets 1·5–4 mm. *Glumes smooth*, lanceolate, acute, keel of lower glume serrate towards top, of upper smooth or nearly so, margins hyaline and shining. Lemma ⅔ length of glumes, ovate, truncate, finely scabrid towards top, 5-nerved; awn from about the middle, geniculate and twice as long as lemma, or shorter and straight, or rarely 0. *Palea minute*. Rhachilla shortly bearded. Anthers 1–1·5 mm. Fl. 6–7. Hp.

Ssp. **canina**. Culms usually decumbent, stolons present, often rooting at the nodes; rhizomes 0. Panicle little contracted after flowering. Usually in damp places and sometimes floating in ditches. $2n=14^*$.

Ssp. **montana** (Hartm.) Hartm. Tufted. Rhizomes present, stolons 0. Panicle usually strongly contracted and spike-like after flowering. On heaths and usually dry acid grassland. $2n = 28^*$.

Native. In acid grassland. 110, H39, S. Generally distributed in the British Is. Europe, and Asia from the Caucasus and Himalaya northwards. Introduced in N. America.

Hybrids with *A. tenuis* and *A. stolonifera* occur occasionally.

3. A. tenuis Sibth.   'Common Bent-grass.'

A. vulgaris With.

A tufted or rhizomatous perennial (2–)20–50(–100) cm. Lvs flat, slightly scabrid. Sheaths smooth. *Ligule* c. 1 mm., truncate or rounded, tha t *of sterile shoots shorter than broad.* Culms smooth, rarely rough below the panicle. *Panicle* 1–20 cm., effuse, ovoid, cylindrical or pyramidal, *spreading in fl. and fr.*, rarely spike-like. Spikelets 2–3·5 mm. Glumes lanceolate, acute, keel of lower ± toothed above. Lemma $\frac{2}{3}$ length of glumes, ovate, rounded, finely scabrid towards the base, 3–5-nerved; awn 0 or very short, rarely long, from near the top. *Palea $\frac{1}{2}$–$\frac{2}{3}$ length of lemma*, bifid. Rhachilla shortly bearded. Anthers 1–1·5 mm. Fl. 6–8. $2n = 28$. Hp.

Native. Usually in acid grassland. 112, H40, S. Generally distributed in the British Is. Europe, northern Asia and N. America. Introduced in Australia, New Zealand and Tasmania.

The hybrid with *A. stolonifera* is common.

4. A. gigantea Roth   'Common Bent-grass.'

A. nigra With.

A *rhizomatous* perennial 40–80(–120) cm. Lvs flat, often scabrid, strongly furrowed above. Sheaths smooth or scabrid. *Ligule* 1·5 mm. or more, rounded; that *of sterile shoots longer than broad.* Culms smooth, often procumbent and rooting at the lower nodes. *Panicle* up to 25 cm., effuse, pyramidal or cylindrical, *spreading in fl. and fr.* Spikelets 2–3 mm. Glumes lanceolate, acute, both toothed on the keel above. Lemma $\frac{2}{3}$ length of glumes, ovate, truncate, slightly scabrid towards the base, 3–5-nerved; awn 0 or short, rarely long and geniculate. Palea $\frac{1}{2}$–$\frac{2}{3}$ length of lemma, bifid. Anthers 1–1·5 mm. Fl. 6–8. $2n = 42^*$. Hp.

Native. In grassy places. A common weed of arable land, especially on light soils, also in damp woods. 47 (incomplete). Generally distributed in lowland districts in the British Is. Europe (except the north), C. and S. Russia, China, Japan, N. America. Introduced in Australia and New Zealand.

5. A. stolonifera L.   Fiorin.

A. alba auct.; *A. palustris* Huds.

A *stoloniferous, not rhizomatous perennial* 10–140 cm. Lvs flat, scabrid. Sheaths smooth or slightly scabrid. Ligule up to 5 mm., rounded, often torn. Culms smooth, frequently procumbent and rooting at the nodes. *Panicle* 1–30 cm., pyramidal or cylindrical, *spreading at flowering, contracted in fr.* Spikelets 1·75–3 mm. Glumes lanceolate, acute, keels ± serrate. Lemma $\frac{2}{3}$ length of glumes, ovate, truncate, slightly scabrid towards the base, 5-nerved; awn 0, or a short mucro, rarely long and geniculate. Palea $\frac{1}{2}$–$\frac{2}{3}$ length of lemma, bifid. Rhachilla shortly bearded. Anthers 1–1·5 mm. Fl. 7–8. 2n = 28, 35, 42. Hp.

Native. In grassy and waste places. 112, H40, S. Generally distributed in lowland districts in the British Is. Europe, C. Asia to Siberia, Japan and N. America. Introduced in Australia, New Zealand, S. Africa and the Falkland Is.

42. APERA Adans.

Annuals. Panicle compound, branches slender and scabrid, in alternating half-whorls. Spikelets small, shining, subterete, of 1 floret. Glumes membranous, unequal, acute, keeled, lower 1-nerved, upper 3-nerved, longer than the floret. *Lemma* chartaceous, terete, shortly bifid *with a long awn from the sinus.* Nerves obscure. Palea equalling lemma, hyaline, entire. Rhachilla shortly produced, glabrous. Lodicules 2, ovate-lanceolate, acute. Stamens 3. Ovary glabrous; styles 2, short, rather distant.

Three spp. in Europe and Asia.

Ligule acute, torn; panicle spreading, usually purplish; anthers linear.
1. spica-venti

Ligule truncate, toothed; panicle narrow, green; anthers ovate.   **2. interrupta**

1. A. spica-venti (L.) Beauv.   'Silky Apera.'

Agrostis spica-venti L.

A ± erect annual 20–70 cm. *Lvs flat, scabrid, long.* Ligule up to c. 7 mm., acute, torn. *Panicle* 10–25 cm., *effuse*, lanceolate, often purplish; *branches long.* Spikelets 2–3 mm. Glumes lanceolate, acuminate, keeled in fl., terete in fr., serrate on keel, margins hyaline, lower ¾ length of upper. Lemma scabrid towards the top, shortly bifid; awn from the sinus, slender, scabrid, ± flexuous, about 4 times as long as the lemma. *Anthers c. 4 times as long as broad.* Fl. 6–7. 2n = 14. Th.

Native. In dry sandy arable fields in E. Anglia; elsewhere as a casual. Europe, rare in the south; Siberia.

2. A. interrupta (L.) Beauv.

A ± erect, rather tufted annual 15–60 cm. *Lvs ± convolute, smooth, narrow, short.* Ligule up to c. 5 mm., truncate, toothed. *Panicle* 3–15(–20) cm., *narrow*, ± *interrupted*, green; *branches short*, strict. Spikelets 1·5–2 mm., very similar to those of *A. spica-venti* but glumes more broadly hyaline, awn rather shorter (c. 3 times as long as lemma), and *anthers* 1½–2 *times as long as broad.* Fl. 6–7. 2n = 14*. Th.

?Introduced. In dry sandy arable fields in E. Anglia and elsewhere, often only as a casual. S. Europe from S.W. France eastwards, north to Öland and Gotland; absent from the Balkans.

43. POLYPOGON Desf.

Glabrous annual or perennial herbs. *Panicle ± spike-like, dense. Spikelets falling entire when ripe*, small, of 1 floret, compressed. *Glumes* membranous, equal, *much exceeding the lemma*, concave, obtuse and ± *bifid*, 3-nerved, usually *awned from the sinus.* Lemma firm and silvery, obscurely nerved, notched or toothed and usually shortly awned from near the top. Palea hyaline, narrow, 2-nerved. *Rhachilla* produced, *disarticulating below the glumes*. Lodicules 2, small. Stamens 1–3. Ovary glabrous; styles 2, short.

About 10 spp. in warmer temperate regions.

Annual; glumes and lemma awned; glumes pilose.   **1. monspeliensis**
Perennial; glumes and lemma awnless; glumes scabrid.   **2. semiverticillatus**

1. P. monspeliensis (L.) Desf.   'Annual Beardgrass.'

A glabrous, ± tufted *annual* 5–80 cm. Lvs ± scabrid, flat, rather broad, acuminate. Sheaths smooth, upper somewhat inflated. Ligule up to c. 8 mm., obtuse. Panicle 2–8 cm., spike-like, or somewhat lobed, oblong to ovate, dense, yellowish and silky. Spikelets 1–2 mm. *Glumes* diverging at apex, 2–4 times as long as lemma, *pilose*, narrowly oblong, concave, shortly bifid; *awn* slender, scabrid, ± flexuous, 2–3 *times as long as glumes*. Lemma oblong, truncate, toothed, shining, very shortly awned. Palea equalling lemma. Fl. 6–7. $2n=28$. Th.

Native. In damp pastures near the sea in southern England and the Channel Is. c. 14, S. As a casual elsewhere. S.W. Europe, Mediterranean region, Azores, Abyssinia, S. Africa.

*2. P. semiverticillatus (Forsk.) Hyl.   'Beardless Beardgrass.'

Agrostis semiverticillata (Forsk.) C. Christ.; *A. verticillata* Vill.

A tufted or stoloniferous *perennial* 10–100 cm. Lvs flat, short, scabrid. Sheaths smooth. Ligule up to 5 mm., truncate and toothed. Culms smooth, often rooting at the lower nodes. Panicle up to 15 cm., pyramidal, lobed, dense. Spikelets 2–2·5 mm., readily deciduous. *Glumes* lanceolate, acute, *scabrid* at least near the keel, keels serrate; *awn* 0. Lemma c. ½ length of the glumes, ovate, truncate, smooth, 5-nerved; awn 0. Palea equalling lemma. Fl. 6–7. $2n=28$. Hp.

Introduced. Naturalized, in waste places, Channel Is. As a casual elsewhere in England. S. Europe, S.W. Asia, N. India, N. Africa, Canary Is. and Madeira. Introduced in W. France, N. and S. America, S. Africa and S. Australia.

44. × AGROPOGON P. Fourn.

Hybrid between *Agrostis* and *Polypogon* with characters ± intermediate between these genera.

× A. littoralis (Sm.) C. E. Hubbard (*Agrostis stolonifera* × *Polypogon monspeliensis*). 'Perennial Beardgrass.'

Polypogon littoralis Sm.; *P. lutosus* auct.

A glabrous tufted *perennial* 8–50 cm. Lvs ± scabrid, flat. Sheaths smooth, upper somewhat inflated. Ligule up to c. 5 mm., obtuse. Panicle 2–8 cm., narrow, dense, usually somewhat lobed, greenish or purplish, not silky. Spikelets 2–3 mm. Glumes not deciduous, diverging at the tips, slightly hairy, shiny, narrowly oblong, tapering and scarcely notched; awns equalling or shorter than the glumes. Lemma ⅔ the length of the glumes, lanceolate, notched; *awn shorter than in Polypogon monspeliensis*, sometimes shortly exserted. Palea nearly as long as lemma. Fl. 7–8. $2n = 28*$. Hp.

Native. In salt marshes. 9. Rare and local from Dorset to Norfolk; Gloucester and Glamorgan. W. Europe.

45. MIBORA Adans.

A *small annual. Infl. spike-like, linear*, secund, of up to 10 *subsessile, distichous spikelets.* Spikelets of 1 floret, compressed. Glumes membranous, equal, exceeding the lemma, concave, not keeled, 1-nerved, upper next the rhachis. Lemma 5-nerved. Palea equalling lemma. Rhachilla disarticulating above the glumes. Lodicules 2. Stamens 3. Ovary glabrous; styles long.

One sp. in western Europe.

1. M. minima (L.) Desv.

M. verna Beauv.; *Chamagrostis minima* Borkh.

A small, tufted, glabrous annual 2–7 cm. Lvs flat or convolute, obtuse. Culms filiform, naked. Infl. 0·5–1 cm. Glumes truncate, erose and hyaline at the tip. Lemma ovate, obtuse, woolly, margins fringed. Palea narrow, woolly. Fl. 4–5 (sometimes again 8–9). $2n = 14*$. Th.

Native. In wet sandy places near the sea. 1, S. Channel Is. and Anglesey; well naturalized in Hampshire. W. Europe, Greece, Algeria.

46. GASTRIDIUM Beauv.

Annuals. *Panicle ± spike-like*, dense and shining, pedicels enlarged and flattened at the top. *Spikelets* of 1 floret, *swollen and nit-like below.* Glumes coriaceous, subequal, ventricose below, compressed and keeled above, 1-nerved. *Lemma* tough, hyaline, $\frac{1}{3}$–$\frac{1}{4}$ *length of lower glume*, truncate, 3-nerved, rounded on back; awn geniculate, sometimes 0. Palea thin, hyaline, bifid, equalling lemma. Lodicules 2, ovate, acute. Stamens 3. Ovary glabrous; styles short.

Two spp. in the Mediterranean region and western Europe.

1. G. ventricosum (Gouan) Schinz & Thell.   'Nitgrass.'

G. lendigerum (L.) Desv.; *G. australe* Beauv.

A glabrous tufted annual 10–35 cm. Stems erect or ascending, slender, smooth and shining. Lvs flat, scabrid. Sheaths smooth, ± inflated. Ligule up to c. 3 mm., blunt. Panicle 2–10 cm., lanceolate, sometimes lobed; branches scabrid. Spikelets 2–3 mm., acuminate, swollen and nit-like below. Glumes lanceolate, acuminate, ventricose part very polished, keel serrate. Lemma ovate, apex 4-toothed, awn from just below the top, exceeding the glumes or sometimes 0. Anthers 2–3 times as long as broad. Fl. 6–8. $2n = 14*$. Th.

Native. In sandy places and on carboniferous limestone, usually near the sea. 25, S. Sometimes as a casual in arable fields. S. England and Glamorgan. S.W. Europe, Mediterranean region, Azores.

47. LAGURUS L.

A downy annual. *Panicle ovoid and spike-like, exceedingly soft and woolly.* Spikelets of 1 floret, strongly compressed, subsessile. *Glumes 1-nerved, pectinate-ciliate,* subequal. Lemma membranous, nearly equalling glumes, 5-nerved, rounded on the back; awn geniculate. Palea equalling lemma, thin. Lodicules 2. Stamens 3. Ovary glabrous; styles short.

One sp. in S.W. Europe and the Mediterranean region.

1. L. ovatus L.   Hare's-tail.

A downy, tufted, erect annual 7–30 cm. Culms slender, pubescent. Lvs flat, short, broad, downy, upper often triangular. Sheaths much inflated, woolly. Ligule c. 1 mm., truncate. Panicle 1–2 cm., dense, woolly, greyish. Spikelets c. 5 mm. Glumes very narrow, pectinate-ciliate. Lemma lanceolate, bi-aristate, nearly glabrous; awn from below the tip, about twice as long as lemma. Fl. 6–8. $2n = 14$. Th.

Native. In sandy places in the Channel Is. Naturalized in a few places in southern England. W. France, Mediterranean region, Madeira, Canary Is.

48. PHLEUM L.

Glabrous annuals or perennials. *Panicle cylindrical or ovoid, dense and spike-like.* Spikelets of 1 floret, strongly compressed. *Glumes* membranous, longer than lemma, *strongly keeled* and *pectinate-ciliate on keel,* 3-nerved, nerves close together, *shortly aristate,* margins overlapping or diverging in the upper half. *Lemma thin, truncate or obtuse,* 3–5-nerved. Palea equalling lemma, thin, 2-nerved. Lodicules 2. Stamens 3. Ovary glabrous; styles long, slender.

About 10 spp. in temperate regions.

For cytology see Gregor, J. W. and Sansome, F. W., *Journ. Genet.* XXII (1930), 373.

1	Perennial; glumes truncate or obliquely truncate.	*2*
	Annual; glumes tapering gradually.	**5. arenarium**
2	Panicle ovoid or stout-cylindrical; awn c. 3 mm.	**3. alpinum**
	Panicle cylindrical, rarely narrow-ovoid, but if so awn not exceeding 1 mm.	*3*
3	Lvs tapering gradually; glumes truncate; panicle branches smooth.	*4*
	Lvs abruptly contracted, scabrid; glumes obliquely truncate; panicle branches ciliate.	**4. phleoides**
4	Plant up to c. 50 cm.; panicle (0·8–)1–6(–8) cm. × 3–6 mm.; spikelets 2–3 mm. (excluding awn).	**1. bertolonii**
	Plant usually exceeding 50 cm.; panicle 6–15(–30) cm. × 6–10 mm.; spikelets 3–4 mm. (excluding awn).	**2. pratense**

1. P. bertolonii DC.   Cat's-tail.

P. nodosum auct.

A slender perennial 10–50 *cm.* Stems ascending, often tuberous at base. Lvs 2–3 mm. wide, tapering gradually. Sheaths smooth, lower brownish or blackish, ± fibrous and rather loose, upper ± inflated. Ligule c. 2 mm., blunt, torn. *Panicle* 0·5–8 *cm.,* mostly 1–6 cm. × 3–6 mm., dense, cylindrical, blunt;

branches smooth. Spikelets 2–3 mm. (excluding awn). Glumes oblong, truncate, shortly awned, keel densely pectinate-ciliate to base, margin of lower glume woolly, of upper glabrous, broadly hyaline; *awn* 0·4–1 *mm.* Lemma ½ as long as glumes, glabrous, obliquely truncate and toothed, 5-nerved. Fl. 7. $2n = 14^*$. Hp.

Native. In pastures and short rough grassland, common. Throughout the British Is. Europe, Algeria, N. Asia; introduced in N. America.

2. P. pratense L.   Timothy.

A stout erect *perennial* 50–100 *cm.,* similar in general to *P. bertolonii* but larger. Stem sometimes tuberous at base. Lvs 4–8 mm. wide, very rough. *Panicle up to* 30 *cm.,* mostly 6–15 cm. × 6–10 mm. Spikelets 3–4 mm. (excluding awn). *Awn* 1–2·5 *mm.* Fl. 7. $2n = 42^*$. Hp.

Native. In meadows, often sown for hay grass, common. Throughout the British Is., but rare in the north. Widely cultivated.

3. P. alpinum L.

P. commutatum Gaud.

A rather stout perennial 15–50 cm. Lvs smooth. Sheaths smooth, *basal dark brown or blackish, upper strongly inflated.* Ligule short, truncate. Panicle 1·5–5 cm., stout, ovoid or broadly cylindrical, obtuse, often purplish; branches slightly downy. Spikelets c. 3·5 mm. Glumes oblong, truncate, *awn longer than in any other British sp.* (c. 3 mm.), glabrous or nearly so, *never long-ciliate,* keel long pectinate-ciliate to base, margins scarcely hyaline, that of the lower glume woolly above. *Lemma* ½ as long as glumes, glabrous, broadly ovate, *truncate,* toothed, 5-nerved. Anthers 4–5 times as long as broad. Fl. 7–8. $2n = 28$. Hp.

Native. In damp places on the higher mountains. 19. Westmorland, Cumberland, Dunbarton, Perth, Angus, Aberdeen, Banff, Inverness and E. Ross. Arctic regions and higher mountains of the northern hemisphere, south to the Jura, Alps, Apennines, Caucasus, Urals, Altai and Himalaya, east to Formosa; N. America; Andes of S. America.

4. P. phleoides (L.) Karst.   'Böhmer's Cat's-tail.'

P. Boehmeri Wibel

An erect perennial 10–40 cm. *Lvs abruptly contracted to a truncate point,* scabrid, *margins whitish.* Sheaths smooth, lower brown or purplish, upper slightly inflated. Ligule 1 mm. or less, truncate. *Panicle* 1·5–9 cm., narrow, cylindrical, ± interrupted, tapering and often subacute; *branches ciliate.* Spikelets c. 3 mm. Glumes narrowly oblong, obliquely truncate, aristate, glabrous, keel sparsely ciliate-pectinate or scabrid in upper ½–⅔, *hyaline margins very silvery.* Lemma ¾ length of the glumes, narrow-elliptic, *obtuse, with scattered appressed hairs,* 3-nerved. Palea lanceolate, acute, silvery. Anthers 3 times as long as broad. Fl. 6–7. $2n = 14, 28$. Hp.

Native. In dry sandy and chalky pastures in S.E. England. 7. Europe to S.E. Scandinavia and the Baltic States, absent from the extreme south, N. Africa, Siberia, Turkistan.

5. P. arenarium L.   'Sand Cat's-tail.'

An *annual* 3–15(–30) cm. Lvs short, smooth. Sheaths smooth, *lower whitish, upper inflated, fusiform.* Ligule up to c. 4 mm., acute. Panicle 0·5–3 cm., narrowly ovoid to cylindrical, blunt; branches smooth. Spikelets c. 3 mm. *Glumes* lanceolate, shortly aristate, pectinate-ciliate on the keel and shortly ciliate on the margins and *diverging in the upper part*, broadly hyaline. *Lemma* $\frac{1}{3}-\frac{2}{3}$ as long as glumes, *pubescent*, broadly oblong, truncate. Palea pubescent, narrow, obtuse. Anthers 1–1½ times as long as broad. Fl. 5–6. $2n=14$. Th.

Native. On dunes and in sandy fields. 51, H17, S. Around the coasts of the British Is., locally common. W. Europe and the Mediterranean region, north to S. Sweden.

<h3 style="text-align:center">49. ALOPECURUS L.   Fox-tail.</h3>

Nearly glabrous annual or perennial herbs. *Spikelets* strongly compressed, crowded in narrow spike-like panicles, of 1 floret, *readily deciduous in fr.; rhachilla jointed below the glumes. Glumes* tough, subequal, equalling or slightly exceeding the floret, *often connate below the middle*, 3-nerved. Lemma hyaline, 3-nerved, awned from the back, rarely awnless, margins often connate below. *Palea* 0. Lodicules 0. Stamens 3. Ovary glabrous; styles usually connate below; stigmatic papillae short.

About 50 spp. in temperate regions.

1	Awn scarcely exceeding lemma or 0.	2
	Awn at least twice as long as lemma.	3
2	Panicle narrow-cylindrical, not conspicuously hairy.	**4. aequalis**
	Panicle ovoid or broadly cylindrical, conspicuously hairy.	**6. alpinus**
3	Annual; glumes connate for $\frac{1}{3}-\frac{1}{2}$ their length.	**1. myosuroides**
	Perennial; glumes free nearly to base.	4
4	Stem swollen and bulbous at base (salt-marshes).	**5. bulbosus**
	Stem not bulbous at base (meadows and wet places).	5
5	Glumes acute.	6
	Glumes obtuse.	7
6	Lemma acute.	**2. pratensis**
	Lemma obtuse.	× **hybridus**
7	Spikelet 2–3 mm.; lemma truncate.	**3. geniculatus**
	Spikelet 3·5–4·5 mm.; lemma obtuse.	× **hybridus**

1. A. myosuroides Huds.   Black Twitch.

A. agrestis L.

A tufted *annual* 20–70 cm. Stems decumbent at base. Lvs scabrid above, smooth beneath. Upper sheaths somewhat inflated. Ligule up to c. 5 mm., obtuse. *Panicle* 4–12 cm., narrow, *pointed.* Spikelets 4–7 mm. *Glumes connate for* $\frac{1}{3}-\frac{1}{2}$ *their length, shortly ciliate on nerves and keel* and at their base, keel narrowly winged, oblong to lanceolate, acute, narrowly hyaline in upper part. Lemma acute, margins connate for $\frac{1}{3}$ their length; awn from near the base, geniculate, twice as long as lemma. Anthers 4 times as long as broad. Fl. 6–7. $2n=14$. Th.

Native. A weed in arable fields and waste places. c. 45. Scattered throughout Great Britain, abundant in the south-east; a casual in the north; a rare casual in Ireland. Europe from Sweden southwards, Mediterranean region, W. Asia; introduced in N. America and New Zealand.

2. A. pratensis L.   'Meadow Foxtail.'

A rather stout erect perennial 30–90 cm. Stems ± geniculate at the lowest node. Lvs ± scabrid. Upper sheaths inflated. Panicle 3–6(–11) cm., tapering somewhat but obtuse. Spikelets 4–5 mm. *Glumes* ± hyaline, *long-ciliate on keel*, shortly ciliate or glabrescent on lateral nerves, lanceolate, acute or acuminate, free nearly to base. *Lemma acute*, margins connate for $\frac{1}{4}$ their length; *awn* from near the base, geniculate, *twice the length of lemma*. Anthers 5–6 times as long as broad. Fl. 4–6. Protogynous. $2n=28$. Hp.

Native. In damp meadows, pastures and grassy places. 112, H40, S. Generally distributed and common in most of the British Is., except Ireland, where it is local. Europe, except the Mediterranean region, Caucasus, N. Asia.

A. × hybridus Wimm. = *A. geniculatus × pratensis* has the stems ± creeping at base and rooting at nodes; the spikelets 3·5–4·5 mm.; the glumes subacute, slaty-grey at tips with hyaline margins; the lemma obtuse; and the anthers about 4 times as long as broad. Fl. 5–6. With the parents, in wet places.

3. A. geniculatus L.   'Marsh Foxtail.'

A perennial 15–40 cm. Stems ± creeping, rooting at the nodes, geniculate and ascending. Lvs scabrid above, smooth beneath. Upper sheaths distinctly inflated. Ligule up to c. 5 mm., obtuse. Panicle 2–4 cm., blunt. *Spikelets 2–3 mm.* Glumes free nearly to base, with silky hairs on nerves, lanceolate, obtuse, *broadly hyaline at the top. Lemma* truncate, *margins free; awn from near the base, geniculate, twice as long as lemma. Anthers* 4 *times as long as broad, ultimately violet-tinged.* Tips of glumes and lemmas slaty-grey. Fl. 6–7. $2n=28$. Hp.

Native. In wet meadows and at edges of ponds and ditches. 112, H39, S. Generally distributed throughout the British Is., frequent. Europe, temperate Asia, N. America.

4. A. aequalis Sobol.   'Orange Foxtail.'

A. fulvus Sm.

A ± tufted biennial or annual, perhaps sometimes perennial, 10–25 cm. Stems decumbent at base. Lvs scabrid above, smooth beneath. Sheaths glaucous, upper strongly inflated, close below the panicle. Ligule up to c. 4 mm., ± truncate and torn. Panicle 1·5–3 cm., narrow, cylindrical, obtuse. *Spikelets* 1–1·5 *mm.*, ovate. Glumes free nearly to base, shortly silky, long-ciliate on keel, lanceolate, obtuse, *margins hyaline to base. Lemma* elliptic, obtuse, *margins connate to middle; awn from just below the middle, straight, slightly exceeding lemma. Anthers* $1\frac{1}{2}$ *times as long as broad, at first white, later orange.* Glumes grey-green, sometimes with a very narrow slaty band at tips. Fl. 5–7. $2n=14$. ?Th. or Hp.

Native. In similar situations to *A. geniculatus* but far less common. 39. England, from Wilts and Kent to S. Lancashire and Yorks, local. Europe, local in the south; N. Asia.

5. A. bulbosus Gouan   'Tuberous Foxtail.'

A slender, tufted perennial 25–50 cm. *Stems swollen and bulbous at base*, erect or ± geniculate. *Lvs smooth.* Upper sheaths somewhat inflated, far below the panicle. Ligule c. 3 mm., obtuse. Panicle 1–3 cm., cylindrical, obtuse. Spikelets 2·5–3 mm. *Glumes* free nearly to base, shortly ciliate on keel and lateral nerves, lanceolate, *acute. Lemma* oblong, truncate, *margins free; awn from near the base, geniculate, c.* 3 *times as long as lemma*, pale below, dark above. Anthers 2–3 times as long as broad. Glumes slaty-grey at tip. Fl. 6. $2n=14^*$. Hp.

Native. In grassy salt-marshes. 27. Coasts of England and Wales from E. Cornwall and E. Sussex to Pembroke and Lincoln, Cheshire, S. Lancashire and N.E. Yorks; Sutherland and Caithness; local. Coasts of W. Europe from France to Italy and Algeria.

6. A. alpinus Sm.   'Alpine Foxtail.'

An erect perennial 15–40 cm. Stems shortly creeping then erect. Lvs smooth, short and broad. Sheaths strongly inflated. Ligule c. 1 mm., obtuse. *Panicle* 1–3 cm., *ovoid or broadly cylindrical. Spikelets* c. 3 mm., ovate, obtuse, *silky. Glumes* half-ovate, acute, *clothed with long silky hairs*, free nearly to base. Lemma membranous, ovate, obtuse, *margins fringed in upper part and connate below the middle; awn from the middle or above, short, sometimes* 0. Anthers c. 4 times as long as broad. Fl. 7–8. $2n=112$–130. H.

Native. In cold, moderately base-rich springs and rills on the higher Scottish mountains from Dumfries to E. Ross; N. England; very local. 10. Arctic Europe (not Scandinavia) north to Spitzbergen and Novaya Zemlya.

50. MILIUM L.

Annual or perennial glabrous herbs. *Spikelets of* 1 *floret*, subterete, rhachilla shortly produced. Glumes subequal, membranous, exceeding the floret, 3-nerved. *Lemma* coriaceous, *becoming extremely hard and shining in fr.*, margins folded round the palea, 5-nerved, awnless. Palea similar to lemma but smaller and 2-nerved. Lodicules 2, acute, toothed on one side. Stamens 3. Ovary glabrous; styles distinct, short, stout.

About 6 spp. in Europe, Asia and N. America.

Perennial 50–120 cm.; sheaths smooth; panicle spreading.   **1. effusum**
Annual 2·5–15 cm.; sheaths scabrid; panicle contracted.   **2. scabrum**

1. M. effusum L.   'Wood Millet.'

An erect, tufted perennial 50–120 cm. Culms smooth. Lvs flat, 5–10 mm. wide, thin, scabrid, acute. Sheaths smooth. Ligule c. 5 mm., subacute or torn. *Panicle* 10–25 cm., very lax and slender, *ovate or pyramidal; branches* capillary, *spreading or deflexed*, up to 6 together. Spikelets 2–3 mm. Glumes ovate, acute, scaberulous, green. Lemma coriaceous and shining, ovate, subacute. Fl. 6. $2n=28^*$. Hp.

Native. In damp, shady woods. 103, H25, S. Throughout the British Is., local and perhaps less frequent than formerly. Europe, except the Mediterranean region; Siberia, Himalaya; N. America.

2. M. scabrum Rich.

An *annual*, usually ± prostrate, 2·5–15 cm. Culms scabrid. Lower lvs 10–20 × 2 mm., uppermost 3–5 mm. *Sheaths scabrid*, somewhat inflated, often purplish. Ligule 2–3 mm., acute. *Panicle* 15–35 mm., *very narrow*, even at flowering; *branches* short, slender, flexuous, *erect*, 2–4 together. Spikelets 1·75–2 mm., green, sometimes purplish-tinged. Glumes ovate, acute, scaberulous, green with hyaline margins. Lemma coriaceous and shining, ovate, obtuse. Fl. 4. $2n=8^*$. Th.

Native. In nearly closed turf on fixed dunes, very local. Guernsey. W. Europe from the Netherlands southwards, Mediterranean region, W. Asia.

*Oryzopsis miliacea (L.) Benth. & Hook. ex Aschers. & Schweinf., differing from *Milium* in the longer more numerous whorled panicle branches and the narrower acute spikelets with awned lemmas, is sometimes cultivated and ± naturalized in a few places. Mediterranean region.

Tribe 10. PHALARIDEAE. Annual or perennial herbs. Lvs with oblong or dumbbell-shaped silica cells and green tissue uniformly distributed between the vascular bundles; 2-celled hairs 0; 1st foliage lf of seedling narrow and erect. Ligule glabrous. Infl. a panicle, sometimes compact and spike-like. Spikelets laterally compressed, of 3 florets, the lower 2 male or sterile, the uppermost hermaphrodite. Rhachilla disarticulating above the glumes and not between the florets, not produced beyond the uppermost floret. Glumes membranous or chartaceous, 1–3-nerved, equal or unequal, at least the upper as long as the spikelet. Lower 2 lemmas longer or shorter than the third, sometimes one or both reduced to minute scales, awnless or awned from the back; upper lemma awnless, 5–7-nerved. Lodicules 2 or 0. Stamens 2–3. Ovary glabrous; styles distinct or connate. Fr. with an elliptic or punctiform hilum; starch grains compound. Chromosomes large, basic number 5, 6 or 7.

51. HIEROCHLOË R.Br.

Perennial. Panicle ovate or pyramidal, branches 1–2 at each node. *Spikelets* compressed, shining, brownish, *of 3 florets, uppermost floret hermaphrodite, 2 lower male.* Glumes membranous, subequal and equalling the spikelet, keeled, 3-nerved, lateral nerves short. Lemma tough, 5-nerved, keeled, awned or not. Palea of male florets 2-nerved; of hermaphrodite floret 1-nerved. Lodicules 2, lanceolate, acuminate. Stamens of male florets 3, of hermaphrodite floret 2. Ovary glabrous; styles distinct, long.

About 20 spp. in temperate regions.

1. H. odorata (L.) Beauv.   Holy-grass.

H. borealis (Schrad.) Roem. & Schult.; *Savastana odorata* (L.) Scribner

A tufted, glabrous perennial, 25–50 cm. Lvs flat, acute, upper very short. Sheaths smooth, persistent, upper far distant from panicle. Ligule c. 2 mm., ovate, obtuse. Culms minutely scabrid. Panicle 5–8 cm., unilateral. Spikelets 4–5 mm., ovate. Glumes ovate, acute. Lemma of male florets ovate, acute, hyaline towards the top, very scabrid or shortly hispid, margins fringed; of hermaphrodite floret similar but narrower, glabrous and shining below, hispid above. Lodicules c. 2 mm. Fl. 4–5. $2n=28^*$; 42, 56. H.

Native. On wet banks in Caithness, Kirkcudbright and Renfrew; Ireland: Lough Neagh. Formerly in Angus, but now extinct. 3, H1. C. and N. Europe, Russia, Bulgaria, N. Asia, N. America.

52. ANTHOXANTHUM L.

Annual or perennial herbs, smelling of coumarin. Panicle compact, ovoid or oblong. *Spikelets* lanceolate, acute, compressed, *of 2 sterile florets and one uppermost hermaphrodite one.* Glumes thin, very unequal, lower 1-nerved, upper 3-nerved, shortly aristate, longer than the floret and enfolding it. *Sterile lemmas* thin, obtuse, bifid or toothed at apex, 3-nerved, *awned from the back.* Fertile lemma thin but firm, half as long as sterile one, almost orbicular, 5–7-nerved, awnless. Palea shorter than lemma, lanceolate, 1-nerved. Lodicules 0. Stamens 2. Ovary glabrous; styles long, distinct.

About 6 spp. in north temperate regions of the Old World.

Perennial; glumes pubescent; awn rarely exserted.   **1. odoratum**
Annual; glumes glabrous; awn always long-exserted.   **2. puelii**

1. A. odoratum L.   'Sweet Vernal-grass.'

A tufted perennial 20–50 cm. Lvs flat, short, sparsely hairy, acuminate. Sheaths smooth, glabrous or pubescent. Ligule up to c. 4 mm., uppermost often acute, lower ones truncate. Panicle (2–)4–6(–7) cm., compact, oblong, sometimes lobed below. Spikelets 7–9 mm. *Glumes* hyaline, keeled, *pubescent and minutely punctate;* lower ovate, acute, 1-nerved, $\frac{1}{2}$ length of upper; upper ovate-lanceolate, mucronate, exceeding the floret and enfolding it, 3-nerved. *Sterile lemmas curved, oblong, obtuse, bifid,* with brown silky hairs in lower half; *awn of lower nearly equalling or rarely somewhat exceeding upper glume,* of upper short. Fertile lemma glabrous, almost orbicular, half as long as sterile, awnless, 5–7-nerved. Palea shorter than lemma, lanceolate. Anthers 4 mm. Fl. 4–6. Protogynous. $2n=20^*$. Hp.

Smells strongly of coumarin, which gives the characteristic odour to new-mown hay.

Native. In pastures and meadows and on heaths and moors; equally common on acid or basic soils. 112, H40, S. Generally distributed and common throughout the British Is. Europe, Azores, western N. Africa, Asia Minor, Caucasus, N. Asia; only on mountains in the south; introduced in N. America, Australia and Tasmania.

2. A. puelii Lecoq & Lamotte

A. aristatum auct.

A slender, much-branched annual 10–20 cm. Lvs flat, short, glabrous or hairy, scabrid, acuminate. Sheaths smooth, somewhat inflated. Ligule up to c. 2 mm. Panicle 1–3 cm., rather lax, ovoid. Spikelets 6–7 mm. *Glumes glabrous,* minutely punctate, green near the keel and broadly hyaline, ovate; lower acuminate; upper shortly aristate, slightly serrate on keel in upper half. *Sterile lemmas straight, tapering to a blunt 3–4-toothed apex,* sparsely silky on back; *awn of lower always long-exserted.* Fertile lemma and palea similar to *A. odoratum.* Anthers 2·5–3 mm. Fl. 6–10. $2n=10$. Th. *Smell faint.*

Native or introduced. In sandy fields. 33. Scattered throughout England, local and chiefly in the south and east; rare in Wales and Scotland. Mediterranean region and northwards to W. France.

53. PHALARIS L.

Glabrous annual or perennial herbs. Panicle dense, ovoid or cylindrical, ± lobed and sometimes with branches spreading at anthesis. *Spikelets strongly compressed, with 1–2-rudimentary lower florets and a terminal hermaphrodite one,* or sometimes all florets sterile or male. *Glumes keeled,* ± *winged from the keel,* chartaceous, 3–7-nerved, subequal and exceeding the lemma, often straw-coloured with a green band on the keel. Lemma coriaceous, 5-nerved, keeled, ± enclosing the palea. Palea similar to lemma but slightly shorter and 3-nerved with hyaline margins. Lodicules 2, lanceolate, acuminate. Stamens 3. Ovary glabrous; styles long, connate.

About 20 spp., mainly in warm temperate regions.

1 Plant tall and reed-like, growing in damp places or in water; panicle distinctly lobed; branches spreading at flowering. **1. arundinacea**
 Plant smaller, annual, not reed-like, growing in dry places, usually on waste ground; panicle ovoid or cylindrical. *2*

2 Uppermost lf-sheath immediately below or enfolding the base of the panicle; at least some glumes long-acuminate. **4. paradoxa**
 Uppermost lf-sheath well below the base of the panicle; glumes acute. *3*

3 Spikelets nearly sessile; sterile lemmas lanceolate, equal, c. half as long as fertile. **2. canariensis**
 Pedicels at least one-fourth as long as spikelet; sterile lemmas very unequal, larger subulate, c. one-third as long as fertile. **3. minor**

1. P. arundinacea L. Reed-grass.

Digraphis arundinacea (L.) Trin.

A stout erect reed 60–120 cm. *Rhizomes far-creeping.* Lvs flat, smooth, c. 10 mm. wide, acuminate. Sheaths smooth. Ligule up to 16 mm. *Panicle* 10–15 cm., narrowly oblong, *lobed,* purplish; *branches spreading at flowering.* Spikelets c. 5 mm. Glumes lanceolate, acuminate, keel slightly rough. Lemma broadly lanceolate, acute, margins fringed towards the top. The lemma and palea are sometimes rather silky. Fl. 6–7. $2n = 28$; 27–31, 35; 42. Hel.

The dead lvs persist throughout the winter (compare *Phragmites*).

Native. In wet places. 112, H40, S. Generally distributed throughout the British Is. in suitable habitats. Europe, except the Mediterranean region; W., N. and E. Asia; N. America; S. Africa.

*2. P. canariensis L. Canary Grass.

An annual 20–60 cm. Lvs flat, scabrid, up to 12 mm. wide, acuminate. Sheaths smooth, upper strongly inflated. Ligule up to c. 5 mm., obtuse. Panicle 1·5–4 cm., dense, ovoid. Spikelets 5–8 mm., strongly compressed, ovate, acute. *Glumes half-ovate,* acute or apiculate, glabrous or hispid, 3-nerved, and *strongly winged in upper half.* Lemma coriaceous, shining, silky, ovate, acute, nearly enclosing the palea, 5-nerved. Palea similar to lemma but slightly shorter with hyaline margins. The 2 *lower florets repre-*

sented by lanceolate, coriaceous, 1-nerved *lemmas half as long as the fertile one*. Anthers 3 mm. Fl. 6–7. $2n = 12*$. Th.

Introduced. A casual, in waste places. Widely distributed in the south but not common. N. Africa and the Canary Is., widely naturalized elsewhere.

3. P. minor Retz.

A rather slender annual 10–30 cm., similar to *P. canariensis*. Lvs slightly scabrid, rather narrow. Upper sheaths somewhat inflated. Panicle 1·5–2·5 cm., ovoid or almost cylindrical. Spikelets 4–5 mm. *Glumes* lanceolate, acute, *with a rather narrow wing usually toothed near the top*. Lemma ovate, silky, acute, completely enclosing palea, obscurely nerved. *Sterile lemmas 2, 1 minute and scale-like, the other subulate, shortly silky*, $\frac{1}{3}$ the length of the fertile one. Anthers 1·5 mm. Fl. 6–7. Th.

Introduced and naturalized or perhaps native sandy places in the Channel Is.; a casual elsewhere. Mediterranean region and W. Asia.

*4. P. paradoxa L.

A ± decumbent annual 20–50 cm., similar to *P. canariensis*. Lvs scabrid, narrow. Lower sheaths somewhat scabrid, upper smooth, strongly inflated. Panicle 3–4 cm., cylindrical. Spikelets 5–7 mm., all in lower part, and many in upper part, of panicle usually sterile or male. *Glumes* variable; *those of male and sterile florets lanceolate*, acute, 3-nerved, *wing 2–3-toothed; of hermaphrodite florets* lanceolate, long-acuminate, *7-nerved, wing projecting from back as a single acute lobe. Lemma of hermaphrodite floret glabrous or slightly silky on keel*, ovate, acute, obscurely nerved, half enclosing palea. *Sterile lemmas 2 minute silky scales*. Anthers 1 mm. Fl. 6–7. $2n = 14$. Th.

Introduced. A casual, in waste places. Mediterranean region.

Tribe 11. MONERMEAE. Annual herbs. Lvs with dumbbell-shaped silica cells and green tissue uniformly distributed between the vascular bundles; 2-celled hairs 0. 1st foliage lf of seedling slender and erect. Ligule glabrous. Infl. a slender cylindrical spike, often disarticulating at maturity. Spikelets of 1 floret, sessile, solitary, alternate or rarely opposite, closely appressed to or sunk in hollows in the rhachis. Glumes ± equal, coriaceous, longer than lemma. Lemma hyaline, awnless, 1–3-nerved. Palea hyaline. Lodicules 2, entire. Stamens 3. Ovary with a lobed appendage at the apex; styles free. Fr. with a small narrowly oblong to elliptic hilum; starch-grains compound. Chromosomes large, basic number 7 or 13.

54. PARAPHOLIS C. E. Hubbard

Slender glabrous annuals. Lvs short, narrow. Spikelets of 1 floret, placed broadside on to the rhachis and embedded in its concavities. Glumes equal, placed side by side in front of the spikelet. Lemma and palea hyaline, 1-nerved. Stamens 3. Ovary glabrous.

Four spp. in western Europe and the Mediterranean region.

Plant usually more than 15 cm.; spikes straight or nearly so; uppermost
sheath not inflated. **1. strigosa**
Plant seldom exceeding 10 cm.; at least some spikes distinctly curved; uppermost sheath ± inflated. **2. incurva**

1. P. strigosa (Dum.) C. E. Hubbard   'Sea Hard-grass.'

Lepturus strigosus Dum.; *Lepturus filiformis* auct.; *Pholiurus filiformis* auct.

A slender annual 15–40 *cm*. Stems usually decumbent at base, geniculate and ascending, rarely erect, freely branched. *Uppermost sheaths not inflated*, distant from the base of the spike at maturity. Culms obscurely ridged, shining. Ligule very short. Spikelets 4–6 mm., rounded on the back. Glumes linear-lanceolate, ± asymmetrical, acuminate, 3–5-nerved, margins hyaline. Lemma and palea very delicate, lanceolate, acute, equal or subequal. Lodicules lanceolate, acute. Anthers 2–4 mm. Fl. 6–8. $2n=14^*$. Th.

Native. In salt-marsh turf and waste places by the sea. 55, H 20, S. Widely distributed but local round the coasts of the British Is., north to W. Lothian and Mull. Western Europe.

2. P. incurva (L.) C. E. Hubbard

Lepturus incurvus (L.) Druce; *Pholiurus incurvus* (L.) Schinz & Thell.

Similar in many respects to *P. strigosa*, but much smaller, commonly 5–8 *cm*. *Uppermost sheaths ± inflated*, reaching or enclosing the base of the spike at maturity. Culms distinctly ridged, not conspicuously shining. Spikes mostly distinctly curved. Anthers 0·5–1 mm. Fl. 6–7. $2n=$ c. 32, 36, 38. Th.

Native. In bare places, usually on clayey or muddy shingle, among taller vegetation by the sea. 10. S. England from Norfolk to Kent and Dorset, Bristol Channel. Western Europe and the Mediterranean region.

Tribe 12. NARDEAE. A densely tufted perennial herb. Lvs setaceous; silica cells mixed, some transversely dumbbell-shaped, others rounded; slender 2-celled hairs present; green tissue uniformly distributed between the vascular bundles. First foliage lf of seedling narrow and erect. Ligule glabrous. Infl. a unilateral spike. Spikelets of 1 floret, distichous, sessile in the notches of the concave rhachis. Lower glume very small, upper usually 0. Lemma coriaceous, 3-nerved, 2–3-keeled, wrapped round the palea; awn straight. Palea thin, hyaline, equalling lemma, 2-keeled. Lodicules 0. Stamens 3. Ovary glabrous, long and narrow, triangular in section narrowing into the single stout style; stigma papillose. Fr. with a linear hilum $\frac{1}{3}$ length of grain; starch-grains compound. Chromosomes large, basic number 13.

55. NARDUS L.

The only genus. One species.

1. N. stricta L.   Mat-grass.

A wiry, tufted perennial 10–30 cm. *Lvs setaceous, very hard*, scabrid, *erect when young, later spreading at right angles* to the sheaths, *whitish and persistent when dead*. Sheaths strict, smooth, persistent, whitish, the lower ones lfless. Ligule very short. Culms slender, smooth. Spike 2–10 cm. Spikelets 5–8 mm., narrow, acute. Lower glume very small, upper usually 0. Lemma subulate, with a short terminal awn, lateral nerves and awn serrate. Fl. 6–8. Apomictic. $2n=26$. Hs.

Native. Abundant on the poorer siliceous and peaty soils, covering great areas on moors and mountains. Rejected by sheep on account of the harsh foliage and therefore especially abundant in overgrazed areas. 111, H 39, S.

Recorded from every county in the British Is. except Huntingdon and West-meath but only abundant on moors and mountains. Europe, N. Asia, Caucasus, Asia Minor, Greenland; only on mountains in the southern part of its range and always calcifuge.

Tribe 13. SPARTINEAE. Perennial herbs inhabiting salt-marshes and mud-flats. Lvs with rounded silica cells and globular 2-celled hairs sunk in pits in the epidermis; 1st foliage lf of seedling narrow and erect. Ligule ciliate, hairs unicellular. Infl. of racemosely arranged spikes or rarely a solitary spike. Spikelets of 1 floret, distichous, disarticulating below the glumes. Glumes 3-nerved. Lemma (3–)5(–9)-nerved. Lodicules 0. Stamens 3. Ovary glabrous, styles connate below. Fr. with an ovate to elliptic hilum; embryo nearly as long as grain; starch grains compound. Chromosomes small, basic number 7.

56. SPARTINA Schreb.

Stout, erect, glabrous perennial herbs with tough, yellow-green lvs and yellowish or golden infl. Rootstock creeping. *Infl. of a number of erect spikes arranged in a raceme. Spikelets of 1 floret*, large, *distichous* and alternate on the flattened rhachis and pressed close to it, flattened on the side next the axis. Glumes membranous, keeled, unequal, 3-nerved, the upper exceeding the lemma. Lemma similar to upper glume, 3–5-nerved. Palea thin, hyaline, completely enfolding the stamens and ovary, 2-nerved, exceeding the lemma. Lodicules 0. Stamens 3. Ovary glabrous; styles very long, connate below. On tidal mud flats. The spp. are all protogynous.

About 14 spp., mainly in salt-marshes in temperate regions.

For cytology and the origin of *S. townsendii* see Huskins, C. L., *Genetica*, XII (1931), 531.

1 Spikelets 14 mm. or less; anthers 5 mm.   2
   Spikelets 15 mm. or more; anthers 10 mm., or 0.   **2. × townsendii**
2 Lvs falling short of infl.; rhachis scarcely extending beyond spikelets.
                                                                      1. maritima
   Lvs equalling or exceeding the infl.; rhachis prolonged beyond spikelets.
                                                                      3. alterniflora

1. S. maritima (Curt.) Fernald   Cord-grass.

S. stricta (Ait.) Roth

A stout, erect perennial 10–50 cm. Rootstock creeping. Lvs ± erect, rather soft, often purplish, c. 4 mm. wide, falling short of the infl., tapering to a rather stout point. Sheaths smooth. Ligule very short, truncate. *Infl.* 6–12 cm., *of 2–3 erect spikes* arranged in a raceme. Spikelets 10–14 mm., golden or yellow-brown, distichous and alternate on the flattened *rhachis* which is *scarcely prolonged beyond the spikelets*. Glumes keeled, pubescent; lower narrowly lanceolate to almost subulate, shorter than the lemma, acuminate; upper longer than the palea, nearly enfolding the floret, subacute, margins hyaline. Lemma similar to upper glume but shorter. Palea hyaline, longer than lemma, acute or subacute, glabrous. Anthers 5 mm. Fl. 8–9. $2n = 56$*. Hel.

Native. On tidal mud-flats. 18. S. Devon to Lincoln; Glamorgan and ?Cardigan; local. S.W. Europe.

2. S. × townsendii H. & J. Groves   Cord-grass.

Similar to *S. maritima* but stouter, 50–130 cm. Lvs up to c. 8 mm. wide, spreading ± horizontally, very stiff, falling short of the infl., tapering from the base to a rather long, slender point. *Infl.* 10–25 cm., *of (2–)4–5 spikes. Spikelets 15–20 mm.*, rhachis extending beyond the spikelets as a flexuous bristle. Glumes, lemma and palea larger but otherwise similar to those of *S. maritima*. A male-sterile hybrid (*S. alterniflora × maritima*), which arose in Southampton Water. Fertile plants derived from it (2*n*=126*) are also widespread and have anthers c. 10 mm. Fl. 6–8. Hel.

Native. Abundant on tidal mud-flats; frequently planted as a mud binder. 11. S. England, planted in many other districts. Introduced and planted in many temperate parts of the world. W. Europe, N. and S. America, New Zealand, Australia, etc.

3. *S. alterniflora Lois.

Similar to *S. maritima* but 60–100 cm. Lvs flat, up to c. 6 mm., wide tapering to a very long point and *equalling or exceeding the infl. Infl.* 10–20 cm., *of 6–8 spikes.* Spikelets 10–14 mm., rhachis extending beyond them as a flexuous bristle. *Glumes and lemma almost glabrous.* Anthers 5 mm. Fl. 8–9. 2*n*=56*; 70. Hel.

Introduced. Naturalized on mud-flats in Southampton Water. 1. Native of N. America.

Tribe 14. CHLORIDEAE. Annual or perennial herbs. Lvs with transversely dumbbell-shaped silica cells and swollen, ± club-shaped 2-celled hairs; green tissue not uniformly distributed between the vascular bundles; 1st foliage lf of seedling broad and spreading horizontally. Ligule ciliate, hairs multicellular. Infl. of solitary, digitate or racemosely arranged spikes or spike-like racemes. Spikelets laterally compressed, of 1 floret, in 2 rows along one side of the rhachis, disarticulating above the glumes. Glumes 1–3-nerved. Lemma 3-nerved. Lodicules 2. Stamens 3. Ovary glabrous, styles free. Fr. with a punctiform hilum; embryo short; starch grains compound. Chromosomes small, basic number 9 or 10.

57. CYNODON Rich.

Perennials. *Infl. digitate*, of usually 4–5 *unilateral spikes.* Spikelets of 1 floret, compressed. Glumes equal, shorter than or equalling the floret, membranous, 1-nerved. Lemma coriaceous, strongly compressed, obscurely 3-nerved, folded round the palea. Rhachilla glabrous, produced beyond the floret, disarticulating above the glumes. Lodicules 2, short, truncate. Stamens 3. Ovary glabrous; styles long.

About 7 spp. in the warmer regions of the world, mainly in S. Africa and Australia.

1. C. dactylon (L.) Pers.   Bermuda-grass.
Capriola dactylon (L.) O. Kuntze
An extensively creeping perennial 10–20 cm. Stems rooting at the nodes, culms ascending or erect. Lvs short, stiff, scabrid, acuminate, sparsely pilose.

Sheaths smooth, persistent, bearded at the mouth. Ligule very short. Infl. 2–6 cm., of (3–)4–5(–6) purplish spikes all arising at the same point. Spikelets c. 2 mm. Glumes spreading, subulate. Lemma broadly ovate, acute. Palea narrow. Fl. 8–9. $2n=36*$. Hp.

Native. On sandy shores in Dorset, Devon, Cornwall, N. Somerset, Isle of Wight, ?Surrey; Channel Is. 6, S. Elsewhere as a casual. Tropical and warmer temperate regions of the world.

Tribe 15. Paniceae. Annual or perennial herbs. Lvs with deeply dumb-bell-shaped silica cells and green tissue arranged in a ring round the vascular bundles; 2-celled hairs present; 1st foliage lf of seedling broad and spreading horizontally. Ligule glabrous. Infl. usually a spreading panicle. Spikelets usually dorsally compressed, often plano-convex, falling entire and readily deciduous in fr., of 2 florets, the lower sterile and represented by a lemma with or without a palea, the upper hermaphrodite. Glumes usually unequal, lower often minute, upper membranous, 3–5-nerved. Sterile lemma similar to upper glume, as long as the fertile one, 5-nerved. Fertile lemma coriaceous, very obscurely nerved, margins folded round the palea. Palea similar to fertile lemma. Lodicules 2. Stamens 3. Ovary glabrous, without an appendage. Fr. with a punctiform hilum; starch grains simple. Chromosomes small, basic number 5, 6, 9 or rarely 17.

Mainly tropical or subtropical, 1 sp. perhaps native in Britain, the rest aliens.

Two spp. of **Panicum** L., stout annuals whose panicles have numerous long branches, occur locally as casuals.

*P. mileaceum L., up to 1 m. with ascending panicle branches, often nodding at the tip, and mostly short-pedicelled spikelets 4–5 mm., is found on rubbish-dumps, etc. Widely cultivated.

*P. capillare L. var. *occidentale* Rydb. usually rather smaller, with widely spreading panicle branches and mostly long-pedicelled spikelets 2·5–3 mm., sometimes occurs abundantly as a weed in carrot fields in E. Anglia. N. America.

58. Echinochloa Beauv.

Panicle branched, rhachis triquetrous. Spikelets plano-convex, ± awned, shortly pedicelled *in clusters* along one side of the branches. Glumes very unequal, membranous, 3-nerved. *Sterile lemma* membranous, 5-nerved, ± *awned*. Fertile lemma coriaceous. Palea similar. Lodicules 2, fleshy. Stamens 3.

About 20 spp., in tropical and warm temperate regions.

*1. E. crus-galli (L.) Beauv.   Cockspur.

Panicum crus-galli L.

An erect glabrous annual 30–120 cm. Lvs smooth, margins thickened and finely serrate. Sheaths smooth. Ligule 0. Panicle 6–20 cm., rhachis flattened, ciliate, pedicels pubescent. Spikelets 3–4 mm., ovate. Lower glume pubescent, embracing the base of the spikelet and about half its length; upper as long as the lemmas, ± awned, hispid above. Sterile lemma hispid above; its palea shorter and hyaline. Fertile lemma white or straw-coloured, shiny, ovate, acuminate. Palea similar. Fl. 8–9. $2n = 36$, 54. Th.

Introduced. ?Naturalized in cultivated ground and waste places, local. Warmer regions of the earth.

59. DIGITARIA P. C. Fabr.

Annual or perennial herbs. *Infl. of shortly racemose or subdigitate spikes. Spikelets* small, compressed, falling entire, *arranged in 2 rows along one side of the flattened rhachis, usually in pairs,* one on a longer pedicel than the other, each of 2 florets, lower floret sterile and represented by a lemma, upper hermaphrodite. Glumes thin, membranous, very unequal, lower scale-like and minute or suppressed, upper ovate or lanceolate, 3-nerved. Sterile lemma membranous, as long as the fertile lemma, distinctly 5-nerved. Fertile lemma coriaceous, very obscurely 3-nerved, shiny and minutely ribbed, enfolding the palea. Palea similar to fertile lemma. Lodicules 2, fleshy, truncate. Stamens 3. Ovary glabrous; stigmas shorter than styles. Fr. compressed.

About 100 spp., in the warmer regions of the earth.

Lvs and sheaths all glabrous; upper glume as long as lemmas. **1. ischaemum**
At least lower lvs and sheaths hairy; upper glume up to half length of lemmas.
      2. sanguinalis

1. D. ischaemum (Schreb.) Muhl.   Red Millet.

D. humifusa Pers.; *Panicum glabrum* Gaud.

A procumbent tufted annual 10–25 cm. Culms slender, branched. *Lvs glabrous,* short. Sheaths smooth. Ligule short, truncate. Infl. 4–5 cm., of usually 3, sometimes more, shortly racemose or subdigitate spreading spikes. *Spikelets* 1·5–2 mm., plano-convex, ovate, acute. *Lower glume minute or 0, upper as long as lemmas,* prominently 3-nerved, ovate, acute, *downy.* Sterile lemmas similar to upper glume, 5-nerved. *Fertile lemma* coriaceous, *dark brown, shiny,* margins hyaline. Fl. 8. $2n = 36$. Th.

Native? In sandy fields. Possibly native in a few localities in S. England, introduced elsewhere; rare. Warm and warm-temperate regions of the earth.

*2. D. sanguinalis (L.) Scop.   Crab-grass.

Panicum sanguinale L.

An erect or ascending annual 25–50 cm. Culms ± branched, nodes often bearded. *Lvs* flat, at least the lower *pubescent. Lower sheaths pubescent* with many of the hairs arising from tubercles, upper glabrous, or pubescent. Ligule short, truncate. Infl. of 2–many, usually 4, spreading spikes 5–15 cm., all arising at the same point or in 2 distinct whorls. *Spikelets* 2·5–3 mm., strongly compressed, lanceolate, acute. *Lower glume scale-like, upper linear-lanceolate, up to ½ length of lemmas,* 3-nerved, *glabrous,* margins fringed. Sterile lemma lanceolate, acute, 5-nerved, with very short hairs on the nerves. *Fertile lemma green.* Fl. 8–9. $2n = 36$. Th.

Introduced. A casual; in waste places, near ports and in arable fields, rare. Warm and warm-temperate regions of the earth.

60. Setaria Beauv.   'Bristle-grass.'

Annual or perennial herbs. Lvs flat, sheaths smooth, ligule 0. *Panicle spike-like and cylindrical. Spikelets with 1–several long bristles from their pedicels,* plano-convex, of 2 florets, lower floret sterile, upper hermaphrodite. Glumes very unequal, membranous, lower 3-, upper 5-nerved. Sterile lemma membranous, 5-nerved. Fertile lemma coriaceous, minutely or coarsely rugose, margins folded round palea. Palea similar to fertile lemma. Lodicules, stamens and ovary as in *Digitaria*.

About 100 spp. in the warmer regions of the earth.

1	Fertile lemma conspicuously rugose.	**3. lutescens**
	Fertile lemma minutely rugose.	2
2	Teeth of bristles forward-pointing.	**1. viridis**
	Teeth of bristles backward-pointing.	**2. verticillata**

*1. S. viridis (L.) Beauv.   'Green Bristle-grass.'

Panicum viride L.

An erect, ± tufted annual 15–40 cm. Culms scabrid below panicle. Lvs scabrid, margins thickened. Sheaths smooth, margins ± ciliate. Panicle 3–5 cm. Spikelets c. 1·5 mm., ovate, obtuse, shortly pedicelled and subtended by a *cluster of* long, flexuous *bristles*. Bristles 5–8 mm., *teeth forward-pointing*. Lower glume $\frac{1}{3}$ length of spikelet, triangular, acute; upper equalling lemmas, ovate, acute. Sterile lemma ovate, obtuse. *Fertile lemma* coriaceous, ovate, obtuse, *minutely rugose*. Fl. 7–8. $2n = 18$. Th.

Introduced. A casual, in waste places and near ports, uncommon. Europe, except the north; Siberia, E. Asia; N. Africa.

*S. italica (L.) Beauv., Millet, differing from *S. viridis* in its large, lobed or interrupted panicle and in the persistent glumes and sterile lemma, occurs as a casual. Widely cultivated.

*2. S. verticillata (L.) Beauv.

Panicum verticillatum L.

Similar in general appearance to *S. viridis*. Panicle usually narrower. *Bristles solitary or in pairs, teeth backward-pointing. Fertile lemma minutely rugose.* Fl. 7–8. $2n = 18, 36$. Th.

Introduced. A casual, in waste places. Warmer regions of the earth.

*3. S. lutescens (Weigel) Hubbard

S. glauca auct.

Similar in general appearance to *S. viridis*. Stems less tufted, often solitary. Sheaths bearded at the mouth. Lvs with occasional, long scattered hairs. Sheaths glabrous. Panicle up to c. 9 cm. Bristles yellowish, tufted, teeth forward-pointing. Spikelets 2–3 mm. Upper glume shorter than lemmas. *Fertile lemma with conspicuous transverse ridges.* Fl. 7–8. $2n = 36, 72$. Th.

Introduced. A casual, in waste places. Warmer regions of the earth.

BIBLIOGRAPHY

The following is a list of the principal works referred to in the text, with the addition of a few other floras which have been freely consulted.

ASCHERSON, P. & GRAEBNER, P. *Synopsis der mitteleuropäischen Flora.* 8 vols. Leipzig: W. Engelmann, later Bornträger, 1896–1939.

BABINGTON, C. C. *Manual of British Botany* (10th ed., edited by A. J. WILMOTT). London: Gurney & Jackson, 1922.

BAILEY, L. H. *Manual of Cultivated Plants* (2nd ed.). New York: The Macmillan Company; London: Macmillan & Co., Ltd., 1949.

BENTHAM, G. *Handbook of the British Flora* (5th ed., revised by Sir J. D. HOOKER; 7th ed., revised by A. B. RENDLE). Ashford, Kent: L. Reeve & Co., Ltd., 1930.

BUTCHER, R. W. & STRUDWICK, FLORENCE E. *Further Illustrations of British Plants.* Ashford, Kent: L. Reeve & Co., Ltd., 1930.

CLAPHAM, A. R. *Check-List of British Vascular Plants.* (Reprinted from the *Journal of Ecology*, 33, no. 2 (1946), 308–47). London: Cambridge University Press, 1946.

COSTE, L'ABBÉ H. *Flore Descriptive et Illustrée de la France.* 3 vols. Paris: Libraire des Sciences et des Arts, 1900–6.

CRANE, M. B. & LAWRENCE, W. J. C. *The Genetics of Garden Plants* (3rd ed.). London: Macmillan & Co., Ltd., 1947.

DALIMORE, W. & JACKSON, A. B. *Handbook of Coniferae* (3rd ed.). London: Edward Arnold, 1948.

DANDY, J. E. *List of British Vascular Plants.* London: British Museum, 1958.

DARLINGTON, C. D. & WYLIE, A. P. *Chromosome Atlas of Flowering Plants.* London: George Allen & Unwin Ltd., 1955.

DRUCE, G. C. *British Plant List* (2nd ed.). Arbroath: T. Buncle & Co., Ltd., 1928.

DRUCE, G. C. *The Comital Flora of the British Isles.* Arbroath: T. Buncle & Co., Ltd., 1932.

FITCH, W. H., SMITH, W. G. and others. *Illustrations of the British Flora* (5th ed.). Ashford, Kent: L. Reeve & Co., Ltd., 1924.

Genetica. The Hague: M. Nijhoff, 1919– .

GILBERT-CARTER, H. *British Trees and Shrubs.* Oxford: Clarendon Press, 1936.

HEGI, G. *Illustrierte Flora von Mittel-Europa.* 7 vols. München: J. F. Lehmann, 1908–31. 2nd ed. 1957– .

Hereditas. Lund: Mendelska Sällskapat, 1920– .

HOOKER, J. D. *The Student's Flora of the British Islands* (3rd ed.). London: Macmillan & Co., 1884.

HYDE, H. A. & WADE, A. E. *Welsh Ferns* (2nd ed.). Cardiff: National Museum of Wales, 1940.

HYDE, H. A. & WADE, A. E. *Welsh Flowering Plants.* Cardiff: National Museum of Wales, 1934.

HYLANDER, N. *Förteckning över Skandinaviens växter* (Utgiven av Lunds Botaniska Förening). 1. *Kärlväxter* (3rd ed.). Lund: C. W. K. Gleerup, 1941.

HYLANDER, N. *Tillägg och rättelser till Förteckning över Skandinaviens växter.* 1. *Kärlväxter* (Lund, 1941). Smärre Uppsatser och Meddelanden. Uppsala, 1942.

HYLANDER, N. *Ytterligare tillägg och rättelser till Förteckning över Skandinaviens växter.* 1. *Kärlväxter* (1941). Reprinted from *Botaniska Notiser*, 1945. Lund: 1945.

HYLANDER, N. *Nomenklatorische und systematische Studien über nordische Gefäss-pflanzen.* Uppsala Universitets Årsskrift 1945: 7. Uppsala: A. B. Lundequista Bokhandeln; Leipzig: Otto Harrassowitz, 1945.

HYLANDER, N. *Nordisk Kärlväxtflora.* Stockholm: Almqvist & Wiksell, 1953– .

Journal of Botany, British and Foreign. London: Taylor & Francis, 1863–1942. *(Journ. Bot.)*

Journal of Ecology. Cambridge University Press, 1913–55; Blackwell (Oxford), 1956– .

Journal of Genetics. Cambridge University Press, 1910–57; Statistical Publishing Society, Calcutta, 1958– .

Journal of the Linnean Society of London (Botany). London: Longmans, Green & Co., Ltd., 1856– .

Journal of the Royal Horticultural Society. London: Royal Horticultural Society, 1866– . *(Journ. R.H.S.)*

KNUTH, P. *Handbook of Flower Pollination.* Translated from the German by J. R. Ainsworth Davis. 3 vols. Oxford: Clarendon Press, 1906–9.

KOMAROV, V. L. (editor). *Flora U.R.S.S.* Leningrad: Academy of Sciences of the U.S.S.R., 1934– .

LID, JOHANNES. *Norsk Flora* (2nd ed.). Oslo: Det Norske Samlaget, 1952.

LINDMAN, C. A. M. *Svensk Fanerogamflora* (2nd ed.). Stockholm: Norstedt & Söners, 1926.

LÖVE, ÁSKELL & DORIS. *Chromosome Numbers of Northern Plant Species.* University Institute of Applied Sciences, Department of Agriculture, *Reports*, Series B, no. 3. Reykjavík, 1948.

MAUDE, PAMELA F. *The Merton Catalogue: A List of the Chromosome Numerals of Species of British Flowering Plants.* Reprinted from *The New Phytologist*, 38, no. 1 (1939). London: Cambridge University Press, 1939.

MOSS, C. E. and others. *The Cambridge British Flora* (incomplete). Vol. I not issued; vol. II, 1914; vol. III, 1920. Cambridge University Press. *(Camb. Brit. Fl.)*

PRAEGER, R. LL. *The Botanist in Ireland.* Dublin: Hodges, Figgis & Co., 1934.

PUGSLEY, H. W. *A Prodromus of the British Hieracia.* *(Journal of the Linnean Society of London (Botany)*, vol. 54.) London: Longmans, Green & Co., Ltd., 1948.

REHDER, ALFRED. *Manual of Cultivated Trees and Shrubs* (2nd ed.). New York: The Macmillan Company, 1940.

ROUY, G. & FOUCAUD, J. *Flore de France.* 14 vols. Asnières, Rochefort and Paris, 1893–1913.

SCHINZ, H. & KELLER, R. *Flora der Schweiz.* Zürich: Albert Raustein, 1923.

SOWERBY, J. & SMITH, Sir J. E. *English Botany* (3rd ed.), by J. T. BOSWELL-SYME, 1863–72. Supplement to 3rd ed. by N. E. BROWN, 1892–3. London: Robert Hardwicke. (E.B. and E.B.S.)

TANSLEY, A. G. *The British Islands and their Vegetation.* Cambridge University Press, 1939.

The Botanical Magazine; or, *Flower-Garden Displayed* (by WILLIAM CURTIS). London: Stephen Couchman, 1787– .

The Botanical Society and Exchange Club of the British Isles. Reports. Arbroath: T. Buncle & Co., Ltd., 1880–1948. *(Rep.* B.E.C.)

WATSON, W. C. R. *Handbook of the Rubi of Great Britain and Ireland.* Cambridge University Press, 1958.

Watsonia: Journal of the Botanical Society of the British Isles. Arbroath: T. Buncle & Co., Ltd., 1949–53; Salisbury Press, 1953– .

Welsh Plant Breeding Station. Bulletins. Aberystwyth: Cambrian News (Aberystwyth) Ltd., 1920– .

AUTHORSHIP OF FAMILIES

The author responsible for the account of each family is given in the following list:

Acanthaceae	E.F.W.		Euphorbiaceae	E.F.W.
Aceraceae	E.F.W.		Fagaceae	E.F.W.
Adoxaceae	E.F.W.		Frankeniaceae	T.G.T.
Aizoaceae	T.G.T.		Fumariaceae	E.F.W.
Alismataceae	T.G.T.		Gentianaceae	E.F.W.
Amaranthaceae	T.G.T.		Geraniaceae	E.F.W.
Amaryllidaceae	E.F.W.		Gramineae	T.G.T.
Apocynaceae	A.R.C.		Grossulariaceae	E.F.W.
Aponogetonaceae	T.G.T.		Haloragaceae	A.R.C.
Aquifoliaceae	E.F.W.		Hippocastanaceae	E.F.W.
Araceae	T.G.T.		Hippuridaceae	A.R.C.
Araliaceae	E.F.W.		Hydrangeaceae	E.F.W.
Aristolochiaceae	E.F.W.		Hydrocharitaceae	T.G.T.
Azollaceae	E.F.W.		Hydrocotylaceae	T.G.T.
Balsaminaceae	E.F.W.		Hymenophyllaceae	E.F.W.
Berberidaceae	E.F.W.		Hypericaceae	T.G.T.
Betulaceae	E.F.W.		Iridaceae	T.G.T.
Boraginaceae	T.G.T.		Isoetaceae	T.G.T.
Buddlejaceae	E.F.W.		Juglandaceae	E.F.W.
Butomaceae	T.G.T.		Juncaceae	P.W. Richards
Buxaceae	E.F.W.		Juncaginaceae	T.G.T.
Callitrichaceae	A.R.C.		Labiatae	E.F.W.
Campanulaceae	T.G.T.		Lemnaceae	T.G.T.
Cannabiaceae	T.G.T.		Lentibulariaceae	E.F.W.
Caprifoliaceae	E.F.W.		Liliaceae	E.F.W.
Caryophyllaceae	A.R.C.		Linaceae	E.F.W.
Celastraceae	E.F.W.		Lobeliaceae	T.G.T.
Ceratophyllaceae	T.G.T.		Loranthaceae	T.G.T.
Chenopodiaceae	T.G.T.		Lycopodiaceae	E.F.W.
Cistaceae	E.F.W.		Lythraceae	T.G.T.
Compositae	A.R.C.		Malvaceae	A.R.C.
Convolvulaceae	T.G.T.		Marsiliaceae	E.F.W.
Cornaceae	E.F.W.		Menyanthaceae	E.F.W.
Corylaceae	E.F.W.		Monotropaceae	E.F.W.
Crassulaceae	E.F.W.		Moraceae	E.F.W.
Cruciferae	A.R.C.		Myricaceae	E.F.W.
Cucurbitaceae	A.R.C.		Najadaceae	T.G.T.
Cupressaceae	E.F.W.		Nymphaeaceae	A.R.C.
Cyperaceae	T.G.T.		Oleaceae	T.G.T.
Diapensiaceae	E.F.W.		Onagraceae	A.R.C.
Dioscoreaceae	A.R.C.		Ophioglossaceae	E.F.W.
Dipsacaceae	A.R.C.		Orchidaceae	A.R.C.
Droseraceae	T.G.T.		Orobanchaceae	A.R.C.
Elaeagnaceae	E.F.W.		Osmundaceae	E.F.W.
Elatinaceae	T.G.T.		Oxalidaceae	E.F.W.
Empetraceae	E.F.W.		Paeoniaceae	A.R.C.
Equisetaceae	E.F.W.		Papaveraceae	A.R.C.
Ericaceae	E.F.W.		Papilionaceae	T.G.T.
Eriocaulaceae	T.G.T.		Parnassiaceae	E.F.W.
Escalloniaceae	E.F.W.		Phytolaccaceae	E.F.W.

Pinaceae	E.F.W.	Saxifragaceae	E.F.W.
Pittosporaceae	E.F.W.	Scheuchzeriaceae	T.G.T.
Plantaginaceae	T.G.T.	Scrophulariaceae	E.F.W.
Platanaceae	E.F.W.	Selaginellaceae	E.F.W.
Plumbaginaceae	A.R.C.	Simaroubaceae	E.F.W.
Polemoniaceae	A.R.C.	Solanaceae	T.G.T.
Polygalaceae	T.G.T.	Sparganiaceae	T.G.T.
Polygonaceae	T.G.T.	Staphyleaceae	E.F.W.
Polypodiaceae	E.F.W.	Tamaricaceae	E.F.W.
Pontederiaceae	E.F.W.	Taxaceae	E.F.W.
Portulacaceae	A.R.C.	Thymelaeaceae	E.F.W.
Potamogetonaceae	A.R.C.	Tiliaceae	A.R.C.
Primulaceae	T.G.T.	Trilliaceae	E.F.W.
Pyrolaceae	E.F.W.	Typhaceae	T.G.T.
Ranunculaceae	A.R.C.	Ulmaceae	T.G.T.
Resedaceae	A.R.C.	Umbelliferae	T.G.T.
Rhamnaceae	T.G.T.	Urticaceae	T.G.T.
Rosaceae	E.F.W.	Valerianaceae	T.G.T.
Rubiaceae	A.R.C.	Verbenaceae	A.R.C.
Ruppiaceae	T.G.T.	Violaceae	E.F.W.
Salicaceae	E.F.W.	Vitaceae	T.G.T.
Santalaceae	T.G.T.	Zannichelliaceae	A.R.C.
Sarraceniaceae	T.G.T.	Zosteraceae	T.G.T.

NOTE ON LIFE FORMS

The Danish botanist Raunkiaer has classified plants, according to the position of the resting buds or persistent stem apices in relation to soil level, into a number of Life Forms, which have in turn been subdivided so as to convey more information about the plants included in the different groups. Life Forms are a convenient method of indicating how a plant passes the unfavourable season, and are also of interest because they show a correlation with climate.

The primary classes are:

1. Phanerophytes—woody plants with buds more than 25 cm. above soil level.
2. Chamaephytes—woody or herbaceous plants with buds above the soil surface but below 25 cm.
3. Hemicryptophytes—herbs (very rarely woody plants) with buds at soil level.
4. Geophytes—herbs with buds below the soil surface.
5. Helophytes—marsh plants.
6. Hydrophytes—water plants.
7. Therophytes—plants which pass the unfavourable season as seeds.

The subdivisions and abbreviations used in this book are as follows:

Phanerophytes

MM. Mega- and mesophanerophytes, from 8 m. upwards.
M. Microphanerophytes, 2–8 m.
N. Nanophanerophytes, 25 cm.–2 m.

Chamaephytes (Ch.)

Chw. Woody chamaephytes.
Chh. Herbaceous chamaephytes.
Chc. Cushion plants.

Hemicryptophytes (H.)

Hp. Protohemicryptophytes, with uniformly lfy stems, but the basal lvs usually smaller than the rest.
Hs. Semi-rosette hemicryptophytes, with lfy stems but the lower lvs larger than the upper ones and the basal internodes shortened.
Hr. Rosette hemicryptophytes, with lfless flowering stems and a basal rosette of lvs.

Geophytes (G.)

Gb. Geophytes with bulbs.
Gr. Geophytes with buds on roots.
Grh. Geophytes with rhizomes.
Grt. Geophytes with root tubers.
Gt. Geophytes with stem tubers or corms.

Helophytes (Hel.)

Not subdivided.

Hydrophytes (Hyd.)

Not subdivided.

Therophytes (Th.)

Not subdivided.

GLOSSARY
PRECEDED BY THREE EXPLANATORY FIGURES

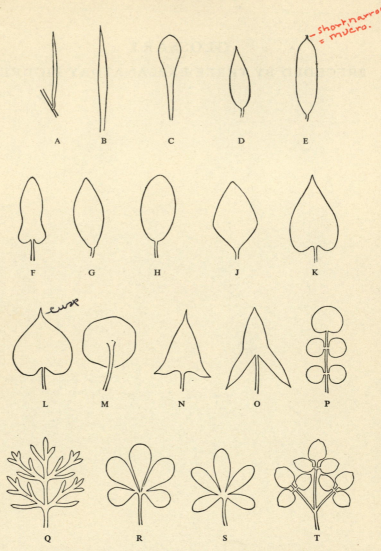

— short, narrow pt. = mucro.

— cusp

Fig. 85. A, linear; B, ensiform; C, spathulate; D, lanceolate, apex acute; E, oblong, apex mucronate; F, panduriform; G, elliptic; H, oval; J, rhomboid, base cuneate; K, ovate, base cordate; L, suborbicular, apex cuspidate; M, peltate; N, triangular, hastate, apex acuminate; O, sagittate; P, simply pinnate, segments orbicular; Q, bi-pinnatifid; R, pedate, leaflets obovate; S, palmate; T, biternate.

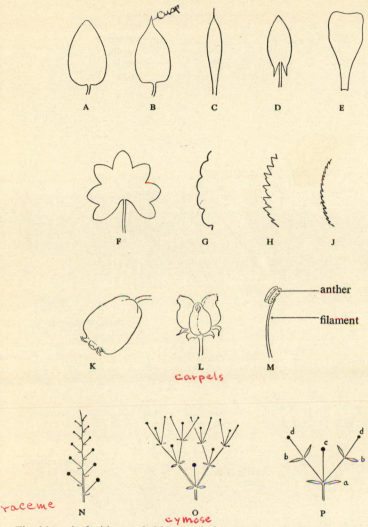

Fig. 86. A, leaf with rounded base and obtuse apex; B, leaf with truncate base and cuspidate apex; C, aristate apex; D, sagittate leaf with subacute apex; E, leaf with retuse apex; F, lobed margin; G, crenate margin; H, dentate margin; J, serrate margin; K, urceolate corolla; L, carpels; M, stamen; N, racemose inflorescence; O, cymose inflorescence; P, bracts and bracteoles: *a*, bracts of flower *d* and bracteoles of flower *c*; *b*, bracteoles of flower *d*.

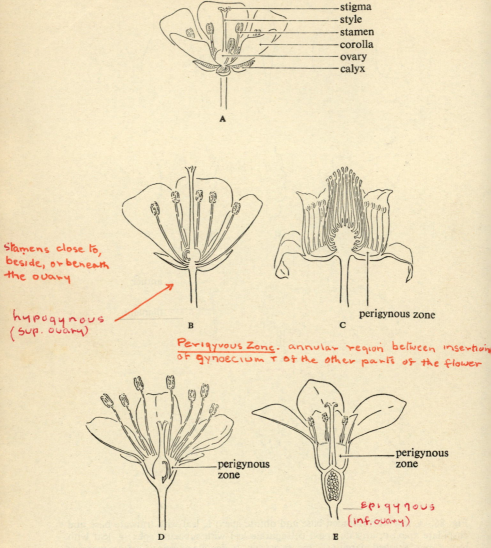

stigma
style
stamen
corolla
ovary
calyx

A

Stamens close to, beside, or beneath the ovary

hypogynous (sup. ovary)

B

perigynous zone

C

Perigynous Zone. annular region between insertion of gynoecium & of the other parts of the flower

perigynous zone

D

perigynous zone

Epigynous (inf. ovary)

E

Fig. 87. A, parts of the flower; B, hypogynous flower (ovary superior); C, D, perigynous flowers (ovary superior); E, epigynous flower (ovary inferior).

GLOSSARY

accrescent Becoming larger after flowering (usually applied to the calyx).

achene A small dry indehiscent single-seeded fr.

acicle See p. 372.

actinomorphic Radially symmetrical; having more than one plane of symmetry.

acuminate Fig. 85 N. **acumen**, the point of such a lf.

acute Fig. 85 D.

adnate Joined to another organ of a different kind.

alien Believed on good evidence to have been introduced by man and now ± naturalized.

allopolyploid A polyploid derived by hybridization between two different spp. with doubling of the chromosome number.

alternate Strictly, arranged in 2 rows but not opposite; commonly used (as often here) also to include spiral arrangement.

amphimixis Reproducing by seed resulting from a normal sexual fusion; adj. *amphimictic*.

amplexicaul Clasping the stem.

anastomosing Joining up to form loops.

anatropous (ovule) Bent over against the stalk.

andromonoecious Having male and hermaphrodite fls on the same plant.

anemophilous Wind-pollinated.

angustiseptate See pp. 112, 118.

annual Completing its life-cycle within 12 months from germination (cf. **biennial**).

annular Ring-shaped.

annulus Special thick-walled cells forming part of the opening mechanism of a fern sporangium, often forming a ring.

antesepalous Inserted opposite the insertion of the sepals.

anther The part of the stamen containing the pollen grains (Fig. 86 M).

antheridium The structure containing the male sexual cells.

apiculate With a small broad point at the apex.

apocarpous (ovary) Having the carpels free from one another.

apomixis Reproducing by seed not formed from a sexual fusion; adj. *apomictic*.

appressed Pressed close to another organ but not united with it.

arachnoid Appearing as if covered with cobwebs.

archegonium The structure containing the female sexual cell in many land plants.

arcuate Curved so as to form about ¼ of a circle or more.

aristate Awned (Fig. 86 C).

ascending Sloping or curving upwards.

asperous Rough to the touch.

attenuate Gradually tapering.

auricles Small ear-like projections at base of lf (especially in grasses).

autopolyploid A polyploid derived from one diploid sp. by multiplication of its chromosome sets.

autotrophic Neither parasitic nor saprophytic.

awn A stiff bristle-like projection from the tip or back of the lemma in grasses, or from a fr. (usually the indurated style, e.g. *Erodium*), or, less frequently, the tip of a lf (Fig. 86 C).

axile See placentation.

axillary Arising in the axil of a lf or bract.

base-rich   Soils containing a relatively large amount of free basic ions, e.g. calcium, magnesium, etc.

basic number   See **chromosomes**.

basifixed (of anthers)   Joined by the base to the filament and not capable of independent movement.

berry   A fleshy fr., usually several-seeded, without a stony layer surrounding the seeds.

biennial   Completing its life-cycle within two years (but not within one year), not flowering in the first.

bifid   Split deeply in two.

bog ('moss')   A community on wet very acid peat.

bract   Fig. 86 P.

bracteole   Fig. 86 P.

bulb   An underground organ consisting of a short stem bearing a number of swollen fleshy lf-bases or scale lvs, with or without a tunic, the whole enclosing the next year's bud.

bulbil   A small bulb or tuber arising in the axil of a lf or in an infl. on the aerial part of a plant.

caducous   Falling off at an early stage.

caespitose   Tufted.

calcicole   More frequently found upon or confined to soils containing free calcium carbonate.

calcifuge   Not normally found on soils containing free calcium carbonate.

calyx   The sepals as a whole (Fig. 87 A).

campanulate   Bell-shaped.

campylotropous (ovule)   Bent so that the stalk appears to be attached to the side midway between the micropyle and chalaza.

capillary   Hair-like.

capitate   Head-like.

capsule   A dry dehiscent fr. composed of more than one carpel.

carpel   One of the units of which the gynoecium is composed. In a septate ovary the number of divisions usually corresponds to the number of carpels (Fig. 86 L).

carpophore   See p. 495.

cartilaginous   Resembling cartilage in consistency.

caryopsis   A fr. (achene) with ovary wall and seed-coat united (Gramineae).

casual   An introduced plant which has not become established though it occurs in places where it is not cultivated.

cauline (of lvs)   Borne on the aerial part of the stem, especially the upper part, but not subtending a fl. or infl.

chartaceous   Of papery texture.

chasmogamous   Of fls which open normally (opposite of cleistogamous).

chlorophyll   The green colouring matter of lvs, etc.

chromosomes   Small deeply staining bodies, found in all nuclei, which determine most or all of the inheritable characters of organisms. Two similar sets of these are normally present in all vegetative cells, the number (*diploid number*, $2n$) usually being constant for a given sp. The sexual reproductive cells normally contain half this number (*haploid number*, n). Closely related spp. often have the same number, or a multiple of a common *basic number*.

ciliate   With regularly arranged hairs projecting from the margin.

cilium, cilia   A small whip-like structure by means of which some sexual reproductive cells swim.

circumscissile   Dehiscing transversely, the top of the capsule coming off as a lid.

cladode   A green ± lf-like lateral shoot.

clavate   Club-shaped.

claw   The narrow lower part of some petals.

cleistogamous Of fls which never open and are self-pollinated (opposite of chasmogamous).

commissure (in Umbelliferae) The faces by which the two carpels are joined together.

compound Of an infl., with the axis branched; of a lf, made up of several distinct lflets.

compressed Flattened.

concolorous Of approximately the same colour throughout.

connate Organs of the same kind growing together and becoming joined, though distinct in origin.

connivent Of two or more organs with their bases wide apart but their apices approaching one another.

contiguous Touching each other at the edges.

contorted (of perianth lobes in bud) With each lobe overlapping the next with the same edge and appearing twisted.

converging, convergent Of two or more organs with their apices closer together than their bases.

convolute Rolled together, coiled.

cordate Fig. 85 K.

coriaceous Of a leathery texture.

corm A short, usually erect and tunicated, swollen underground stem of one year's duration, that of the next year arising at the top of the old one and close to it; cf. **tuber**.

corolla The petals as a whole (Fig. 87 A).

corymb A raceme with the pedicels becoming shorter towards the top, so that all the fls are at approximately the same level; adj. *corymbose*.

corymbose cyme A flat-topped cyme, thus resembling a corymb in appearance though not in development.

cotyledon The first lf or lvs of a plant, already present in the seed and usually differing in shape from the later lvs. Cotyledons may remain within the testa or may be raised above the ground and become green during germination.

crenate Fig. 86 G; dimin. *crenulate*.

crisped Curled.

cultivated ground (or land) Includes arable land, gardens, allotments, etc.

cuneate Fig. 85 J.

cuneiform Wedge-shaped with the thin end at the base.

cuspidate Fig. 85 L, 86 B.

cymose Of an infl., usually obconical in outline, whose growing points are each in turn terminated by a fl., so that the continued growth of the infl. depends on the production of new lateral growing points. A consequence of this mode of growth is that the oldest branches or fls are normally at the apex (Fig. 86 O).

deciduous Losing its lvs in autumn; dropping off.

decumbent (of stems) Lying on the ground and tending to rise at the end.

decurrent Having the base prolonged down the axis, as in lvs where the blade is continued downwards as a wing on petiole or stem.

decussate (of lvs) Opposite, but successive pairs orientated at right angles to each other.

deflexed Bent sharply downwards.

dehiscent Opening to shed its seeds or spores.

deltoid Shaped like the Greek letter Δ.

dentate Fig. 86 H; dimin. *denticulate*.

diadelphous See p. 327.

dichasium A cyme in which the branches are opposite and approximately equal; adj. *dichasial*.

didymous Formed of two similar parts attached to each other by a small portion of their surface.

digitate   See **palmate**.

dioecious   Having the sexes on different plants.

diploid   A plant having 2 sets of chromosomes in its nuclei; similarly tetraploid, etc., plants having 4, etc., sets of chromosomes in their nuclei. See also **chromosomes**.

disk   The fleshy, sometimes nectar-secreting, portion of the receptacle, surrounding or surmounting the ovary.

distichous   Arranged in two diametrically opposite rows.

divaricate   Diverging at a wide angle.

diverging, divergent   Of two or more organs with their apices wider apart than their bases.

dominant   The chief constituents of a particular plant community, e.g. oaks in an oakwood or heather on a moor.

dorsifixed (of anthers)   Attached by the back.

drupe   A ± fleshy fr. with one or more seeds each surrounded by a stony layer, e.g. sloe, ivy.

ebracteate   Devoid of bracts.

ectotrophic   See **mycorrhiza**.

effuse   Spreading widely.

eglandular   Without glands.

ellipsoid   Of a solid object elliptic in longitudinal section.

elliptic   Fig. 85 G.

emarginate   Shallowly notched at the apex.

endemic   Native only in one country or other small area. If used without qualification in this book means confined to the British Is.

endosperm   Nutritive tissue surrounding the embryo in the seed of a flowering plant, formed after fertilization.

endotrophic   See **mycorrhiza**.

ensiform   Fig. 85 B.

entire   Not toothed or cut.

entomophilous   Insect-pollinated.

epicalyx   A calyx-like structure outside but close to the true calyx.

epigeal   Above ground; epigeal germination—when cotyledons are raised above the ground.

epigynous   Of the whole fl., or the perianth and stamens of a fl., with an inferior ovary (Fig. 87 E).

epipetalous   Inserted upon the corolla.

equitant   Of distichous lvs folded longitudinally and overlapping in their lower parts.

erose   Appearing as if gnawed.

escape   A plant growing outside a garden, but not well naturalized, derived from cultivated specimens either by vegetative spread or by seed.

exserted   Protruding.

exstipulate   Without stipules.

extravaginal   Of branches which break through the sheaths of the subtending leaves.

extrorse (of anthers)   Opening towards the outside of the fl.

falcate   Sickle-shaped.

fen   A community on alkaline, neutral, or slightly acid wet peat.

fertile   Producing seed capable of germination; or (of anthers) containing viable pollen.

filament   The stalk of the anther, the two together forming the stamen (Fig. 86 M).

filiform   Thread-like.

fimbriate   With the margin divided into a fringe.

fistular   Hollow and cylindrical; tube-like.

flexuous   Wavy (of a stem or other axis).

floccose See p. 897.

flush Wet ground, often on hillsides, where water flows but not in a definite channel.

follicle A dry dehiscent fr. formed of one carpel, dehiscing along one side.

fruit (fr.) The ripe seeds and structure surrounding them, whether fleshy or dry; strictly the ovary and seeds, but often used to include other associated parts such as the fleshy receptacle, as in rose and strawberry.

fugacious Withering or falling off very rapidly.

funicle The stalk of the ovule.

fusiform Spindle-shaped.

gamopetalous Having the petals joined into a tube, at least at the base.

geniculate Bent abruptly to make a 'knee'.

gibbous Of a solid object any part of which projects as a rounded swelling.

glabrescent Becoming glabrous.

glabrous Without hairs.

gland A small globular or oblong vesicle containing oil, resin or other liquid, sunk in, on the surface of, or protruding from any part of a plant. When furnished with a slender stalk usually known as a glandular hair.

glandular Furnished with glands.

glaucous Bluish.

glumaceous Resembling a glume.

glume See p. 1115.

gynobasic A style which, because of the infolding of the ovary wall, appears to be inserted at the base of the ovary.

gynodioecious Having female and hermaphrodite fls on separate plants.

gynoecium The female part of the fl., made up of one or more ovaries with their styles and stigmas.

gynomonoecious Having female and hermaphrodite fls on the same plant.

hastate Fig. 85 N.

heath A lowland community dominated by heaths or ling, usually on sandy soils with a shallow layer of peat.

hemiparasite See under parasite.

herb Any vascular plant which is not woody.

herbaceous (of a plant organ) Soft, green, having the texture of lvs.

hermaphrodite Containing both functional stamens and ovary.

heterochlamydeous Having the per. segs in two distinct series which differ from one another.

heterosporous Having spores of two distinct sizes.

heterostylous The length of the style in relation to the other parts of the fl. differing in the fls of different plants.

hexaploid See under diploid.

hilum The scar left on the seed by the stalk of the ovule.

hirsute Clothed with long, not very stiff, hairs.

hispid Coarsely and stiffly hairy, as many Boraginaceae.

homochlamydeous, homoiochlamydeous Having all the per. segs similar.

homogamous The anthers and stigmas maturing simultaneously.

homosporous Having all spores of approximately the same size.

hyaline Thin and translucent.

hybrid A plant originating by the fertilization of one sp. by another.

hybrid swarm A series of plants originating by hybridization between two (or more) spp. and subsequently recrossing with the parents and between themselves, so that a continuous series of forms arises.

hypogeal Below ground; hypogeal germination—when the cotyledons remain below the ground.

hypogynous   Of fls in which the stamens are inserted close to and beside or beneath the base of the ovary (Fig. 87 B).

imbricate, imbricating   Of organs with their edges overlapping, when in bud, like the tiles on a roof.

imparipinnate   Of a pinnate lf with a terminal unpaired lflet.

impressed   Sunk below the surface.

incurved   Bent gradually inwards.

indehiscent   Not opening to release its seeds or spores.

indumentum   The hairy covering as a whole.

indurated   Hardened and toughened.

indusium   A piece of tissue ±covering or enclosing a sporangium or group of sporangia.

inferior (ovary)   With perianth inserted round the top, the ovary being apparently sunk in and fused with the receptacle (Fig. 87 E).

inflexed   Bent inwards.

inflorescence (infl.)   Flowering branch, or portion of the stem above the last stem lvs, including its branches, bracts and fls.

intercalary lf   See p. 698.

internode   The length of stem between two adjacent nodes.

interpetiolar   Between the petioles.

interrupted   Not continuous.

intrapetiolar   Between the petiole and stem.

intravaginal   Within the sheath.

introduced   Not native; known to have been, or strongly suspected of having been, brought into the British Is. accidentally or intentionally by man within historic times.

introrse (of anthers)   Opening towards the middle of the fl.

involucel   See p. 796.

involucral   Forming an involucre.

involucre   Bracts forming a ±calyx-like structure round or just below the base of a usually condensed infl. (e.g. *Anthyllis*, Compositae); adj. *involucrate*.

involute   With the margins rolled upwards.

isomerous   The number of parts in two or more different floral whorls being the same, e.g. 5 petals, 5 stamens, and 5 carpels.

jaculator   See p. 728.

keel   A sharp edge resembling the keel of a boat; the lower petal or petals when shaped like the keel of a boat (*Fumaria*, Papilionaceae).

lacerate   Deeply and irregularly divided and appearing as if torn.

laciniate   Deeply and irregularly divided into narrow segments.

lamina   The blade of a lf or petal; a thin flat piece of tissue.

lanceolate   Fig. 85 D.

latex   Milky juice.

latiseptate   See pp. 113, 118.

lax   Loose; not dense.

lemma   See p. 1115.

lenticular   Convex on both faces and ±circular in outline.

ligulate   Strap-shaped.

ligule   A small flap of tissue or a scale borne on the surface of a lf or per. seg. near its base; see also p. 801 (Compositae).

limb   The flattened expanded part of a calyx or corolla the base of which is tubular.

linear   Fig. 85 A.

lingulate   Tongue-shaped.

lip   A group of per. segs ± united and sharply divided from the remaining per. segs.

lobed (of lvs)   Divided, but not into separate lflets (Fig. 86 F).

loculicidal   Splitting down the middle of each cell of the ovary.

lodicule   See p. 1115.

lyrate   Shaped ±like a lyre.   *macula = spot*

marsh   A community on wet or periodically wet but not peaty soils.

meadow   A grassy field cut for hay.

measurements   See Abbreviations (p. xlviii).

megaspore   A large spore giving rise to a prothallus bearing archegonia.

membranous   Thin, dry and flexible, not green.

mericarp   A 1-seeded portion split off from a syncarpous ovary at maturity.

-merous   E.g. in 5-merous (=pentamerous); having the parts in fives.

microspore   A small spore giving rise to a prothallus bearing antheridia.

monadelphous (of stamens)   United into a single bundle by the fusion of the filaments.

monochasium   A cyme in which the branches are spirally arranged or alternate or one is more strongly developed than the other; adj. *monochasial.*

monochlamydeous   Having only one series of per. segs.

monoecious   Having unisexual fls, but both sexes on the same plant.

monopodial   Of a stem in which growth is continued from year to year by the same apical growing point, cf. racemose.

moor   Upland communities, often dominated by heather, on dry or damp but not wet peat.

mucronate   Provided with a short narrow point (*mucro*), Fig. 85 E; dimin. *mucronulate.*

mull soil   A fertile woodland soil with no raw humus layer.

muricate   Rough with short firm projections.

muticous   Without an awn or mucro.

mycorrhiza   An association of roots with a fungus which may form a layer outside the root (ectotrophic) or within the outer tissues (endotrophic).

naked   Devoid of hair or scales.

native   Not known to have been introduced by human agency.

nerve   A strand of strengthening and conducting tissue running through a lf or modified lf.

node   A point on the stem where one or more lvs arise.

nodule   A small ±spherical swelling; adj. *nodular.*

nucellus   The tissue between embryo and integument in an ovule.

ob- (in combinations, e.g. obovate)   Inverted; an obovate lf is broadest above the middle, an ovate one below the middle (Fig. 85 R).

obdiplostemonous   The stamens in 2 whorls, the outer opposite the petals, the inner opposite the sepals.

oblong   Fig. 85 E.

obtuse   Blunt (Fig. 86 A).

ochrea   See p. 542.

octoploid   See under **diploid**.

opposite   Of two organs arising at the same level on opposite sides of the stem.

orbicular   Rounded, with length and breadth about the same (Fig. 85 P).

orthotropous (ovule)   Straight and with the axis of the ovule in the same line as that of the funicle.

oval   Fig. 85 H.

ovary   That part of the gynoecium enclosing the ovules (Fig. 87 A).

ovate   Fig. 85 K.

ovoid   Of a solid object which is ovate in longitudinal section; egg-shaped.

ovule   A structure containing the egg and developing into the seed after fertilization.

palate   See p. 677.

palea   See p. 1115.

palmate (of a lf)   Consisting of more than 3 lflets arising from the same point (Fig. 85s).

panduriform   Fig. 85f.

panicle   Strictly a branched racemose infl., though often applied to any branched infl.

papillae   Small elongated projections; adj. *papillose*.

parasite   A plant which derives its food wholly or partially (hemiparasite) from other living plants to which it is attached.

paripinnate   Of a pinnate lf with no terminal lflet.

partial infl.   Any distinct portion of a branched infl.

pasture   Grassy field grazed during summer.

pectinate   Lobed, with the lobes resembling and arranged like the teeth of a comb.

pedate   Fig. 85r.

pedicel   The stalk of a single fl.

peduncle   The stalk of an infl. or partial infl.

peltate   Of a flat organ with its stalk inserted on the under surface, not at the edge (Fig. 85m).

pentaploid   See under **diploid**.

perennating   Surviving the winter after flowering.

perennial   Living for more than 2 years and usually flowering each year.

perianth   The floral lvs as a whole, including sepals and petals if both are present.

perianth segment (per. seg.)   The separate lvs of which the perianth is made up, especially when petals and sepals cannot be distinguished.

perigynous   Of fls in which there is an annular region, flat or concave, between the base of the gynoecium and the insertion of the other floral parts (Fig. 87c, d).

perigynous zone   The annular region between the insertion of the gynoecium and of the other floral parts in perigynous or epigynous fls (Fig. 87c–e).

perisperm   The nutritive tissue derived from the nucellus in some seeds.

perispore   A membrane surrounding a spore.

petal   A member of the inner series of per. segs, if differing from the outer series, and especially if brightly coloured.

petaloid   Brightly coloured and resembling petals.

petiole   The stalk of a lf.

phyllode   A green, flattened petiole resembling a leaf.

pilose   Hairy with rather long soft hairs.

pinnate   A lf composed of more than 3 lflets arranged in two rows along a common stalk or rhachis (Fig. 85p); bipinnate (2-pinnate), a lf in which the primary divisions are themselves pinnate. Similarly, 3-pinnate, etc.

pinnatifid   Pinnately cut, but not into separate portions, the lobes connected by lamina as well as midrib or stalk (Fig. 85q).

pinnatisect   Like pinnatifid but with some of the lower divisions reaching very nearly or quite to the midrib.

placenta   The part of the ovary to which the ovules are attached.

placentation   The position of the placentae in the ovary. The chief types of placentation are: *apical*, at the apex of the ovary; *axile*, in the angles formed by the meeting of the septa in the middle of the ovary; *basal*, at the base of the ovary; *free-central*, on a column or projection arising from the base in the middle of the ovary, not connected with the wall by septa; *parietal*, on the wall of the ovary or on an intrusion from it; *superficial*, when the ovules are scattered uniformly all over the inner surface of the wall of the ovary.

pollen   The microspores of a flowering plant or Conifer.

pollinia   Regularly shaped masses of pollen formed by a large number of pollen grains cohering.

polygamous Having male, female and hermaphrodite fls on the same or different plants.

polyploid A plant having a chromosome number which is a multiple, greater than two, of the basic number of its group (see under **chromosomes**).

pome A fr. in which the seeds are surrounded by a tough but not woody or stony layer, derived from the inner part of the fr. wall, and the whole fused with the deeply cup-shaped fleshy receptacle (e.g. apple).

porrect Directed outwards and forwards.

premorse Ending abruptly and appearing as if bitten off at the lower end.

prickle A sharp relatively stout outgrowth from the outer layers. Prickles (unlike thorns) are usually irregularly arranged.

pricklet See p. 371.

procumbent Lying loosely along the surface of the ground.

prostrate Lying rather closely along the surface of the ground.

protandrous Stamens maturing before the ovary.

prothallus A small plant formed by the germination of a spore and bearing antheridia or archegonia or both.

protogynous Ovary maturing before the stamens.

pruinose Having a whitish 'bloom'; appearing as if covered with hoar frost.

pubescent Shortly and softly hairy; dimin. *puberulent, puberulous.*

punctate Dotted or shallowly pitted, often with glands.

punctiform Small and ±circular, resembling a dot.

pungent Sharply and stiffly pointed so as to prick.

raceme An unbranched racemose infl. in which the fls are borne on pedicels.

racemose Of an infl., usually conical in outline, whose growing points commonly continue to add to the infl. and in which there is usually no terminal fl. A consequence of this mode of growth is that the youngest and smallest branches or fls are normally nearest the apex (Fig. 86N).

radical (of lvs) Arising from the base of the stem or from a rhizome.

raphe The united portions of the funicle and outer integument in an anatropous ovule.

ray The stalk of a partial umbel.

ray-floret See p. 801.

receptacle That flat, concave, or convex upper part of the stem from which the parts of the fl. arise; often used to include the perigynous zone.

recurved Bent backwards in a curve.

regular **Actinomorphic** (q.v.).

reniform Kidney shaped.

replum See p. 110.

resilient Springing sharply back when bent out of position.

reticulate Marked with a network, usually of veins.

retuse Obtuse or truncate and slightly indented (Fig. 86E).

revolute Rolled downwards.

rhachilla See p. 1115.

rhachis The axis of a pinnate lf or an infl.

rhizome An underground stem lasting more than one growing season; adj. *rhizomatous.*

rhomboid Having ± the shape of a diamond in a pack of playing cards (Fig. 85J).

rotate (of a corolla) With the petals or lobes spreading out at right angles to the axis, like a wheel.

rounded (of lf-base) Fig. 86A.

rugose Wrinkled; dimin. *rugulose.*

ruminate Looking as though chewed.

runcinate Pinnately lobed with the lobes directed backwards, towards the base of the lf.

runner A special form of stolon consisting of an aerial branch rooting at the end and forming a new plant which eventually becomes detached from the parent.

saccate Pouched.

sagittate Figs. 85 O, 86 D.

salt-marsh The series of communities growing on intertidal mud or sandy mud in sheltered places on coasts and in estuaries.

samara A dry indehiscent fr. part of the wall of which forms a flattened wing.

saprophyte A plant which derives its food wholly or partially (partial saprophyte) from dead organic matter.

scabrid Rough to the touch; dimin. *scaberulous.*

scape The flowering stem of a plant all the foliage lvs of which are radical; adj. *scapigerous.*

scarious Thin, not green, rather stiff and dry.

schizocarp A syncarpous ovary which splits up into separate 1-seeded portions (mericarps) when mature; adj. *schizocarpic.*

scrub (incl. thicket) Any community dominated by shrubs.

secund All directed towards one side.

seed A reproductive unit formed from a fertilized ovule.

sepal A member of the outer series of per. segs, especially when green and ± lf-like.

sepaloid Resembling sepals.

septicidal Dehiscing along the septa of the ovary.

septum A partition; adj. *septate.*

serrate Toothed like a saw (Fig. 86 J); dimin. *serrulate.*

sessile Without a stalk.

setaceous Shaped like a bristle, but not necessarily rigid.

shrub A woody plant branching abundantly from the base and not reaching a very large size.

silicula See p. 110.

siliqua See p. 110.

simple Not compound.

sinuate Having a wavy outline.

sinus The depression between two lobes or teeth.

sorus A circumscribed group of sporangia.

spathulate Paddle-shaped (Fig. 85 C).

spermatozoid A male reproductive cell capable of moving by means of cilia.

spike A simple racemose infl. with sessile fls or spikelets; adj. *spicate.*

spikelet See p. 1116.

spine A stiff straight sharp-pointed structure.

sporangiophore A structure, not lf-like, bearing sporangia.

sporangium A structure containing spores; plur. *sporangia.*

spore A small asexual reproductive body, usually unicellular and always without tissue differentiation.

sporophyll A lf-like structure, or one regarded as homologous with a lf, bearing sporangia.

spur A hollow usually ± conical slender projection from the base of a per. seg., often a petal, or of a corolla; adj. *spurred.*

stamen One of the male reproductive organs of the plant (Figs. 86 M, 87 A).

staminode An infertile, often reduced, stamen.

standard See p. 327.

stellate Star-shaped.

sterile Not producing seed capable of germination; or (of stamens) viable pollen.

stigma The receptive surface of the gynoecium to which the pollen grains adhere (Fig. 87 A).

stipel A structure similar to a stipule but at the base of the lflets of a compound lf.

stipitate Having a short stalk or stalk-like base.

stipule A scale-like or lf-like appendage usually at the base of the petiole, sometimes adnate to it.

stolon A creeping stem of short duration produced by a plant which has a central rosette or erect stem; when used without qualification is above ground; adj. *stoloniferous*.

stoma A pore in the epidermis which can be closed by changes in shape of the surrounding cells; plur. *stomata*.

stomium The part of the sporangium wall (in the ferns) which ruptures during dehiscence.

striate Marked with long narrow depressions or ridges.

strict Growing upwards at a small angle to the vertical.

strigose With stiff appressed hairs.

strophiole A small hard appendage outside the testa of a seed.

style The part of the gynoecium connecting the ovary with the stigma (Fig. 87 A).

stylopodium See p. 495.

sub- (in combinations, e.g. subcordate) Not quite, nearly, e.g. subacute (Fig. 86 D), suborbicular (Fig. 85 L).

subulate Awl-shaped, narrow, pointed and ± flattened.

sucker A shoot arising adventitiously from a root of a tree or shrub often at some distance from the main stem.

suffruticose Adjective from *suffrutex*, a dwarf shrub or undershrub.

superior (ovary) With perianth inserted round the base, the ovary being free (Fig. 87 B–D).

suture The line of junction of two carpels.

sympodial Of a stem in which the growing point either terminates in an infl. or dies each year, growth being continued by a new lateral growing point, cf. **cymose**.

syncarpous (ovary) Having the carpels united to one another.

tendril A climbing organ formed from the whole or a part of a stem, lf or petiole. Most frequently the terminal portion of a pinnate lf, as in many Papilionaceae.

terete Not ridged, grooved or angled.

terminal Borne at the end of a stem and limiting its growth.

ternate lf A compound lf divided into 3 ± equal parts, which may themselves be similarly divided (2- or 3-ternate); Fig. 85 T.

testa The skin or outer coat of a seed.

tetrad A group of 4 spores cohering in a tetrahedral shape or as a flat plate and originating from a single spore mother-cell.

tetraploid See under **diploid**.

thallus The plant body when not differentiated into stem, lf, etc.

thorn A woody sharp-pointed structure formed from a modified branch.

tomentum A dense covering of short cottony hairs; adj. *tomentose*.

tree A woody plant with normally a single main stem (trunk) bearing lateral branches and often attaining a considerable size.

triangular Having ± the shape of a triangle. Fig. 85 N.

trifid Split into three but not to the base.

trigonous Of a solid body triangular in section but obtusely angled.

triploid See under **diploid**.

triquetrous Of a solid body triangular in section and acutely angled.

trisect Cut into 3 almost separate parts.

truncate Fig. 86 B.

tube The fused part of a corolla or calyx, or a hollow, cylindrical, empty prolongation of an anther.

tuber　A swollen portion of a stem or root of one year's duration, those of successive years not arising directly from the old ones nor bearing any constant relation to them; cf. corm.

tuberclate　With small blunt projections, warty.

tubercle　A ±spherical or ovoid swelling; see p. 550.

tunic　A dry, usually brown and ±papery covering round a bulb or corm; adj. *tunicated.*

turbinate　Top-shaped.

turion　A detachable winter-bud by means of which many water plants perennate.

umbel　An infl. in which the pedicels all arise from the top of the main stem. Also used of compound umbels in which the peduncles also arise from the same point. An umbrella-shaped infl.

unarmed　Devoid of thorns, spines or prickles.

undulate　Wavy in a plane at right angles to the surface.

unilocular　Having a single cavity; similarly bilocular, etc., having 2, etc., cavities.

urceolate (corolla)　±globular to subcylindrical but strongly contracted at the mouth (Fig. 86k).

valvate　Of per. segs with their edges in contact but not overlapping in bud.

vein　See nerve.

versatile　With the filament attached near the middle of the anther so as to allow of movement.

villous　Shaggy.

viscid　Sticky.

vitta　See p. 495.

viviparous　With the fls proliferating vegetatively and not forming seed.

waste place　Uncultivated ±open habitat much influenced by man. [NOTE: not used in Hooker's sense of almost any uncultivated place.]

whorl　More than two organs of the same kind arising at the same level; see also p. 730; adj. *whorled.*

wing　The lateral petals in the fls of Papilionaceae and Fumariaceae.

zygomorphic　Having only one plane of symmetry.

INDEX